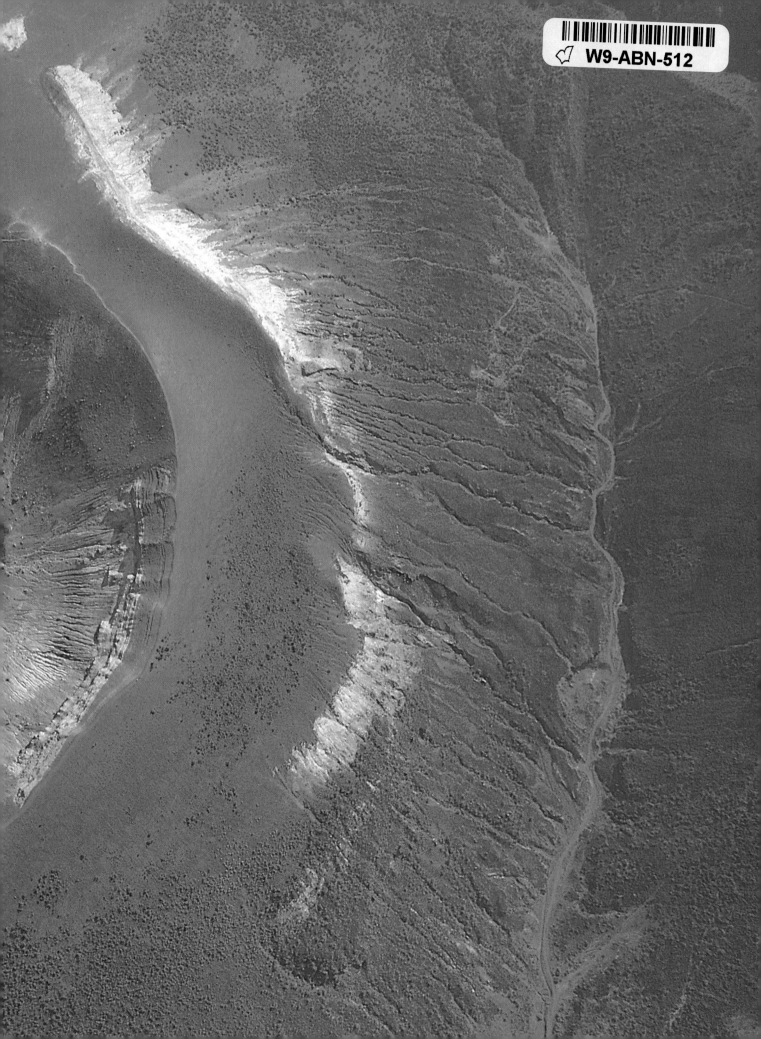

Guide to Places of the World

Inside front cover: Crater of Mount Vulcanello, Lipari Islands, Italy.
Inside back cover: Cultivation terraces, Judaea, Israel.
Preceding page: Waterfront scene, Calcutta, India.
Above: Masked tribal dancers, Papua New Guinea.

READER'S DIGEST

Guide to Places of the World

Published by The Reader's Digest Association Limited
London · New York · Sydney · Cape Town · Montreal

Contributors

Guide to Places of the World

was edited and designed by
The Reader's Digest Association Limited
London

Editor: John Palmer
Art Editor: Bob Hook

The publishers wish to express their gratitude
to the following who, as contributors or
consultants, assisted in the preparation of
Guide to Places of the World.

Consultant editor

Alan B. Mountjoy, MC, MA, FRGS
Reader in Geography
Royal Holloway and Bedford New College
University of London

Major contributors

Guy Arnold, MA

Kathleen Baker, BSc, PhD, AKC
Lecturer in Geography
School of Oriental and African Studies
University of London

David Burtenshaw, MA
Principal Lecturer in Geography
Portsmouth Polytechnic

Sylvia Chant, PhD
Lecturer in Geography
Liverpool University

Graham P. Chapman, MA, PhD
Lecturer in South Asian Geography
University of Cambridge

Hugh Clout, BA, MPhil, PhD, D de l'Université
Reader in Geography
University College London

John Davis, BA, PhD
Head of Department of Geography
Birbeck College, University of London

Harvey Demaine, PhD
Senior Lecturer in Regional Development
Planning, Asian Institute of Technology,
Bangkok, Thailand

F.C. Evans, MA
Senior Master
King's College School, Wimbledon

Alan Gilbert, BSc, PhD
Reader in Geography
University College London

F.E. Ian Hamilton, BSc, PhD
Senior Lecturer in Economics and Social
Studies of Eastern Europe, London School of
Economics and Political Science

Adrian Jansen, BA, MA, Diplome d'Etudes
Approfundis Sciences Politiques
Regional Director South, British Council

Russell King, BA, MSc, PhD
Reader in Geography
University of Leicester

B.J. Knapp, BSc, PhD
Head of Geography Department
Leighton Park School, Easley, Berkshire

David Livingstone, BA, DipEd, PhD
Research Officer, Department of Geography
The Queen's University of Belfast

Keith Lye, BA, FRGS

David McDowall, MA, MLitt

W.R. Mead, BSc, PhD
Emeritus Professor of Geography
University College London

Roy Millward, BA
Late Reader in Historical Geography
University of Leicester

Anthony O'Connor, BA, PhD
Reader in Geography
University College London

John H. Paterson, MA
Professor of Geography
University of Leicester

Deborah Potts, BSc
Lecturer in Geography
School of Oriental and African Studies
University of London

Stephen A. Royle, MA, PhD, FRGS
Lecturer in Geography
The Queen's University of Belfast

John Sargent, BA, FRGS
Reader in Geography
School of Oriental and African Studies
University of London

P.T.H. Unwin, MA, PhD
Lecturer in Historical Geography
Royal Holloway and Bedford New College
University of London

Lawrie Wright, BA, PhD
Lecturer in Geography
Queen Mary College, University of London

Consultants

Robert P. Beckinsale, MA, DPhil
Life Fellow of University College London
(Retired)

Ian Black, BA, PhD
Senior Lecturer in History
University of New South Wales

John Carter

D.J. Dwyer, BA, PhD
Professor of Geography
University of Keele

Valerie Fifer, BA, PhD
Head of Geography Department
Goldsmiths' College
University of London

T.W. Freeman, MA
Emeritus Professor of Geography
University of Manchester

Ivan Jolliffe, BSc, MSc, PhD, CompICE
Lecturer in Geography
Royal Holloway and Bedford New College
University of London

Hamish Keith, OBE

Richard I. Lawless, BA, PhD, FRAI
Senior Lecturer and Assistant Director
Centre for Middle Eastern and Islamic Studies
University of Durham

Robert B. Potter, BSc, PhD
Lecturer in Geography
Royal Holloway and Bedford New College
University of London

Craig J. Reynolds, BA, PhD
Senior Lecturer in History
University of Sydney

A.G. Terry, BSc
Second Master
Latymer Upper School, Hammersmith

Carl Thayer, BA, MA, PhD
Senior Lecturer in Politics
Australian Defence Force Academy

E.M. Yates, PhD, MSc
Reader in Geography
King's College, University of London

Top left: Marketplace, Quercy, France.
Top right: Poncho seller, Ecuador.
Above: Tuaregs and their flocks, Niger.

Illustrator

Gary Hincks

The publishers also wish to thank:

Africa Institute of South Africa, Pretoria,
South Africa

British Museum, Public Relations Office

Commonwealth Institute

The Institute of Petroleum, London

Royal Geographical Society

United Nations Information Centre, London

Numerous Embassies and High Commissions

The permanent Committee on Geographical
Names for British Official Use

SA Library, South Africa

Entries in this book are alphabetised word-for-word, with hyphenated entries read as one word. For example, entries appear in this order:

New Zealand
Newark
Newcastle *Australia*
Newcastle *South Africa*
Newcastle upon Tyne
Newcastle-under-Lyme

Features on nations and major dependent territories appear in approximately their correct alphabetical position, with a cross-reference inserted at the correct point in the main A to Z sequence, except where this would in any case appear on the same two-page spread as the feature. Cross-references to other entries are indicated by SMALL CAPITALS.

Above: Red Square, Moscow, Russia. Top right: Golden Gate Bridge, San Francisco, USA.

Contents

A - Z gazetteer

All the nations of the World, their major cities, towns and geographical features, described and mapped in nearly 8000 separate entries

The nations of the world and their dependencies

A to Z guide to the nations of the world, places of
interest and geographical terms in common use

aa Hawaiian term (pronounced ah-ah) for LAVA flow that has solidified into jagged, block-like masses. In English, it is called 'block lava'.

Aachen (Aix-la-Chapelle) *Germany* Westernmost German city located near the junction of the borders of Belgium and the Netherlands, about 70 km (43 miles) west of Bonn. It was founded by the Romans as Aquae Granni. It was the capital of Charlemagne (AD 742 to 814), the Frankish king who conquered most of Europe and who was crowned first Holy Roman Emperor by Pope Leo III in Rome in 800. Charlemagne's marble throne and a gold bust containing his relics are in the cathedral, built around the chapel that gives the city its French name. Thirty emperors were crowned here, from 1349 to 1531.

Aachen is a university city and spa, and its springs, the hottest in central Europe, have been enjoyed since Roman times. It is renowned for its engineering and food processing – especially chocolate and biscuit – industries, and for being a venue for international horse trials. Each year the city presents a Charlemagne award for an outstanding contribution to European unity.

Population 241 900
Map Germany – Bc

Aargau *Switzerland* Northern canton on the Swiss Plateau between Basle and Zürich. Its fertile valleys, cut by tributaries of the River Aare, support dairy, cattle and fruit farming. Nearby mineral springs are a tourist attraction. Products manufactured in the canton range from electrical and metal goods to machinery, textiles and furniture. A cement industry is based on plentiful lime resources. The inhabitants of the canton, which covers 1404 km² (542 sq miles), are mainly German-speaking. The principal town and capital is Aarau (population 15 900).

Population 572 000
Map Switzerland – Ba

Aba *Nigeria* Industrial town and capital of Abia State east of the Niger River. It has grown rapidly as state and private firms escape the overcrowded industrial centres, particularly Lagos. Aba lies near the oil fields of the Niger delta and Port Harcourt, and its products include cement, textiles, plastics, shoes, foodstuffs and chemicals.

Population 210 700
Map Nigeria – Bb

Abadan *Iran* Oil-refining port on an island in the Shatt al Arab waterway, about 50 km (30 miles) north of The Gulf. Besides a huge oil refinery, it also has a petrochemical complex which produces plastics and detergents. Abadan was badly damaged in the Iran-Iraq War (1980-8). Previously it had been the world's largest exporter of petroleum products.

Population 308 000
Map Iran – Aa

Abéché *Chad* Capital of the south-eastern prefecture of Ouaddaï, about 650 km (405 miles) east and slightly north of the capital, N'Djamena.

Abéché lies in a semi-desert region. It is a trading centre for nomad caravans, and has several markets and mosques serving the Islamic community. Gum arabic is produced locally.

Population 95 800
Map Chad – Bb

Abeokuta *Nigeria* Industrial town and capital of Ogun State, about 75 km (45 miles) north of Lagos. It was founded in 1830 as a buffer against invasion by Dahomeyans from what is now Benin. Its inhabitants adopted Western-style education after European missionaries arrived in 1847. It has many traditional craftsmen as well as large-scale industries such as quarrying, the manufacture of plastics, cement making, food processing and brewing. Most inhabitants, however, are cultivators of cocoa and palm oil.

Population 350 000
Map Nigeria – Ab

Abercorn *Zambia* See MBALA

Aberdare Mountains *Kenya* Mountain range north of Nairobi rising to 3994 m (13 104 ft) at Mount Lesatima, and dropping steeply in the west into the Great Rift Valley. Aberdare National Park occupies 788 km² (304 sq miles) of its lower slopes. Britain's Queen Elizabeth II was visiting the Treetops Hotel in the park when she acceded to the throne in February 1952.

Map Kenya – Cb

Aberdeen *Hong Kong* Small industrial town on the south-western coast of HONG KONG ISLAND. It is separated from the small island of Ap Lei Chau by a narrow stretch of water. Developed as a fishing village after the arrival of the British in Hong Kong in the mid 19th century, Aberdeen is famous for its concentration of boat people – Chinese who live on junks and sampans in the harbour – and for its floating restaurants. Some of these restaurants are like multistorey palaces, ablaze with lights after dark.

Population 92 550
Map Hong Kong – Cb

Aberdeen *United Kingdom* City on the east coast of Scotland, in Grampian region. The second largest fishing port in Scotland, it is known as 'the Granite City' because much of it is built of locally quarried stone. It has a university, founded before 1500, and three cathedrals, the oldest of which, St Machar's, dates back to the 14th century. It thrived in the late 1970s and 1980s as the headquarters and transport centre of the North Sea oil industry.

Population 211 080
Map United Kingdom – Db

Abertawe *United Kingdom* See SWANSEA

Abha *Saudi Arabia* Capital of Asir province, 575 km (357 miles) south of Jeddah. Set 2200 m (7218 ft) above sea level in the Asir Mountains, Abha, with its temperate climate with light summer rains and winter mists, is a popular

holiday resort. The old Arab town – one of the most picturesque in Saudi Arabia – has acquired a number of modern buildings, including hospitals and government offices.

Population 155 410
Map Saudi Arabia – Bc

Abidjan *Ivory Coast* The capital city before Yamoussoukro, the birthplace of former president Félix Houphouet-Boigny, became the new seat of government. Abidjan's economy and strategic importance were boosted in 1950 when the VRIDI CANAL was cut through a huge sandbar, which had blocked access to the open sea. The canal has turned Abidjan into a large, sheltered deep-water port, stimulating the whole economy.

Situated amid forests and lagoons, Abidjan has become a cosmopolitan city of many faces. Particularly impressive are the modern skyscrapers of its commercial centre, the Plateau; its administrative centre with superb views of Banco Bay; its exclusively African suburb of Adjamé; its elegant, lagoon-side suburb of Cocody Bay with its hotels and embassies; and Treichville, a crowded market and entertainment centre. Because of its reputation as one of Africa's most beautiful cities, Abidjan is a popular venue for conferences and international fairs. An industrial zone and port are located in the south, in the old suburb of Marcory and the new suburb of Koumassi. Adjamé has a busy food and cloth market.

Just north of Abidjan lies the Banco National Park, a protected wildlife region covering some 300 km² (116 sq miles).

Population 2 766 000
Map Ivory Coast – Bb

ablation Loss of snow and ice from the surface of an ice sheet or glacier, chiefly by melting and evaporation.

Åbo *Finland* See TURKU

Abomey *Benin* Trading and administrative town, about 110 km (68 miles) north of Cotonou. Founded in 1658 by the Fon tribe as the capital of their kingdom of Dahomey, Abomey is noted for its intriguing historic buildings and relics, and its craftsmanship.

Population 45 000
Map Benin – Ab

Aboukir *Egypt* See ABU QIR

abrasion Wearing away of part of the earth's surface by an abrasive material, such as sand, carried along by wind, water or moving ice.

Abruzzi *Italy* Highland Apennine region covering 10 794 km² (4167 sq miles) facing the Adriatic Sea in central Italy. Previously off the tourist track because of its remoteness, it has become increasingly popular since motorway links have made the area accessible from Rome. The uplands are favoured by tourists seeking skiing and hiking opportunites, while the coastal

resorts such as PESCARA, Francavilla al Mare and Ortona are more popular with summer visitors. Abruzzi embraces the GRAN SASSO and Maiella massifs, the highest of the Apennine Mountains, with Mount Corno the highest point at 2914 m (9560 ft). The southern part of the region is a superb national park and a favourite hiking and recreation centre for holidaying Italians. It is also the home of rare wolves and bears.
Population 1 249 390
Map Italy – Dc

Abruzzo National Park *Italy* A 400 km² (154 sq mile) park in the Apennine Mountains, rising to 2247 m (7372 ft) at Mount Petroso. It was founded in 1921 to protect rare animals such as the endangered Marsican bear, of which only about 100 survive, the Abruzzo chamois, the Apennine wolf and the wild cat.
Map Italy – Dd

Abu Dhabi *United Arab Emirates* Largest and richest of the seven emirates, and the prime mover in the formation of the UAE in 1971. Situated on the south-west coast of The Gulf, Abu Dhabi covers 67 350 km² (26 000 sq miles). In 1994 its oil reserves were estimated at 92 million barrels. It also has huge gas reserves. The main industry is petroleum – the emirate's oil exports account for 80 per cent of the oil exported by the UAE. Abu Dhabi also controls a large gas-processing plant on Das Island in The Gulf. Other industries include fishing and the harvesting and marketing of pearls, but they are of little significance in comparison with the revenue derived from oil.
Abu Dhabi City, the capital of the emirate and the UAE, has a colourful past, including a time as a pirates' hideaway. Since the 1960s it has grown dramatically into one of The Gulf's most modern cities – with wide boulevards, tall office blocks and high-rise apartments. Little remains of old Abu Dhabi City – except for the Old Palace, the city's oldest building (built over a well). The Abu Dhabi of the 1990s has three modern palaces, a modern Great Mosque and an airport that is a hub of international air traffic.
Population (emirate) 798 000; (city) 244 000
Map United Arab Emirates – Ba

Abu Qir (Aboukir; Abukir) *Egypt* Mediterranean bay 20 km (12 miles) north-east of the port of Alexandria. In August 1798, British admiral Horatio Nelson sailed into the bay and, in what subsequently became known as the Battle of the Nile, destroyed the French fleet of Napoleon Bonaparte, effectively ending Napoleon's plans to conquer the Middle East. In 1801, British general Sir Ralph Abercromby landed a force at the village of Abu Qir in the bay and drove the last of the French out of Egypt.
Map Egypt – Bb

Abu Simbel *Egypt* Site of two huge temples hollowed out and extending 63 m (207 ft) into a sandstone cliff beside the River Nile, near the border with Sudan. The temples were built for Pharaoh Rameses II, who ruled for 66 years during the 13th century BC. Inside one temple are four 9 m (30 ft) high standing figures of the pharaoh as the earthly embodiment of Osiris, the god of the afterlife. At the entrance are four seated figures of Rameses II, each soaring 20 m (66 ft) high. In 1964-8, the complete temples and

statuary were cut from the rock face and raised to a higher site, above the flood level of Lake Nasser created by the ASWAN HIGH DAM.
Map Egypt – Bd

Abuja *Nigeria* City which replaced Lagos as the federal capital and seat of government in 1991. It lies in the Federal Capital Territory some 525 km (325 miles) north-east of Lagos and was chosen because of its central position and because it was not associated with any one ethnic group. Initial plans for the city included the construction of a presidential palace, a large mosque, the National Assembly building and an international conference centre.
Population 400 000
Map Nigeria – Bb

Abukir *Egypt* See ABU QIR

Abyan *Yemen* District along the coast to the north-east of Aden. One of the country's richest agricultural regions, it has been developed as a cotton-growing area, with the aid of irrigation and land reclamation projects. The Abyan Dam is making further agricultural projects possible. Extensive limestone deposits in the district are used to make cement.
Map Yemen – Ab

Abydos *Egypt* Ruins of an ancient city and royal burial place on the west bank of the River Nile, about 450 km (280 miles) south of Cairo. Abydos was a cult centre of Osiris, the Egyptian god of the afterlife. Between 1600 and 1200 BC, pharaohs, who presented themselves as the earthly embodiment of the god, had temples and tombs built here. The most impressive temple was built by Setekhy I, father of Rameses II.
Map Egypt – Bc

abyssal Term applied to ocean depths of between 3000 m and 6000 m (9842 ft and 19 685 ft). The term 'Abyssal' also applies to the animal life of those depths.

abyssal plain Large, fairly level area, between 4000 m and 6000 m (13 123 ft and 19 685 ft) below the surface of the oceans, which makes up most of the ocean floor.

Acadia National Park *USA* Rugged area of forest and hills on Mount Desert Island, Isle au Haut and the coast of Maine, about 580 km (360 miles) north-east of New York City. It was established in 1919 as Lafayette National Park – the first United States national park to be proclaimed east of the Mississippi River – and covers 156 km² (60 sq miles).
Map United States – Mb

Acapulco *Mexico* Seaport and world-famous beach resort on the Pacific coast in the state of Guerrero. Acapulco enjoys sunshine throughout the year and temperatures of around 27°C (80°F). It has wide, sandy beaches that stretch for 16 km (10 miles) along its coastline. Luxury hotels are set against a backdrop of mountain slopes and evergreen tropical vegetation, including coconut groves.
From 1565 to 1815 galleons sailed annually from Acapulco to Manila in the Philippines, exporting silver, textiles and cocoa in exchange for porcelain, iron, silks and spices. From here

the goods were transported overland to Veracruz on the Gulf of Mexico and then by ship to Europe. Today, the city exports cotton, fruit, hides and tobacco.
In 1985 an earthquake measuring 8.2 on the RICHTER SCALE struck off the coast north-west of Acapulco, damaging hundreds of buildings and killing thousands of people in Mexico City, 290 km (180 miles) to the north. Acapulco itself was relatively unscathed.
Population 592 200
Map Mexico – Cc

Accra *Ghana* The country's capital, Accra is situated on the coast on a dry plain more suited to livestock than to crops. The site has been home to the Ga people since at least the 16th century. Their fishing, trading and farming economy was flourishing when the British, Dutch and Danes arrived in the 17th century to establish trading posts and forts. The increase in trade resulted in the transfer of the seat of colonial administration to Accra from Cape Coast, 120 km (75 miles) to the south-west, in 1877. A railway line into the interior was built soon afterwards.
The city's old core is centred on the harbour, where the landmark of Château Christiansborg (now known as Osu Castle), built in 1662 by Swedes, now serves as the seat of government. The business and shopping district lies to the east. Beyond, the elegant former dwellings of colonials, now occupied by the wealthier Ghanaians, are set in spacious grounds – a stark contrast to the cramped homes of the poorer residential areas.
Among Accra's many monuments to independence and unity are State House, a vast modern conference centre, and Independence or Black Star Square, where all of Ghana's main ceremonies are held. The city has a university, a museum and good hotels and restaurants.
Although luxury goods are sometimes in short supply, Accra's colourful local markets are always well stocked with cotton batik, baskets, wood carvings, and hand-made gold and silver jewellery.
Population (Greater Accra) 1 500 000
Map Ghana – Ab

acid rock An igneous rock containing 10 per cent or more free quartz. Silica (silicon dioxide or quartz) was formerly thought of as an acidic oxide, and an acidic rock was defined as one containing more than 66 per cent silica. The name has been retained.

Aconcagua *Argentina* Mountain in the ANDES about 1040 km (645 miles) west of Buenos Aires. It is the highest peak in the Americas, rising to 6959 m (22 831 ft). The western slopes are in Chile, but the summit itself lies wholly in neighbouring Argentina.
Map Argentina – Bb

Aconcagua *Chile* Region in the central valley, also known as Valparaíso. It has some of the country's most fertile soil and produces much of Chile's fruit, wine, hemp and tobacco. The principal towns are VALPARAÍSO, the regional capital, Quillota, Los Andes and San Felipe.
Population 1 373 900
Map Chile – Ac

Açores *Portugal* See AZORES

11

Afghanistan

FOUGHT OVER FOR THOUSANDS OF YEARS, PROSPECTS FOR PEACE IN THIS WILD MOUNTAINOUS LAND ARE ALWAYS ELUSIVE

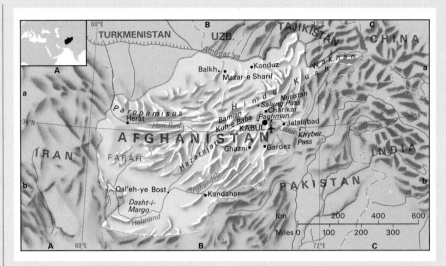

Afghanistan, ribbed and girdled by majestic mountains, is the hinge of Asia. On it pivots the gateway to India through the KHYBER PASS, and the back door to the ex-Soviet republics of Tajikistan, Uzbekistan, Kyrgyzstan and Kazakhstan – and so it has been traversed and fought over by invading armies since earliest times. Darius (about 522-486 BC) of Persia, Alexander the Great (356-323 BC), the Arabs, and the Tatars under Genghis Khan (1162-1227) and Tamerlane (1336-1405) all occupied the country. From 1839 to 1897, the British fought to maintain their hold on the Khyber Pass. Then, on Christmas Day 1979, the Soviets came.

They came to shore up a weak Marxist government; they stayed to endure the guerrilla warfare of fierce tribesmen used to handling guns since they were children, used to blood feuds and tribal battle. Only two things brought the tribes together: they were Muslims, and it was their country. Despite their modern weapons, there were large areas the Soviets could never control. After nine years they departed, but the sound of gunfire was still heard in the capital, KABUL,

Afghanistan's area of some 647 500 km² (250 000 sq miles), is about the size of Texas, and normally has a population of about 16 million. Following the Soviet invasion, some 4 million people escaped across the borders, mainly into Pakistan and Iran. Another 2 million crowded into the towns, where they were safe from Soviet bombing if not from guerrilla guns. No one – not even the Soviets – ever ventured far out of town at night, for armed tribesmen lurked not far away.

Afghanistan is a land of wild contours, hidden green valleys and baked mud villages. Little of the country is below 1200 m (about 4000 ft) and much of it remains unexplored. The mountains of the HINDU KUSH form a giant barrier across the centre of the land. Its jagged peaks rise to 7690 m (25 229 ft) in the east, forming a lunar landscape whose barren heights and windswept plains have always fascinated travellers. To the north lies the most productive farming area – along the banks of the AMUDAR'YA which forms a 960 km (595 mile) border with Uzbekistan and Tajikistan. Much of the south and south-west is desert.

The climate is one of extremes. Summer temperatures can soar to 49°C (120°F) in the south and winter temperatures may fall to –26°C (–15°F) in the mountains. The weather may also vary by as much as 30°C (86°F) in one day, and fierce winds bring terrible dust storms. What rain there is falls between March and May, but for most of the year there are clear, brilliant skies.

Afghanistan is one of the world's poorest countries. Only about 30 per cent of its people are literate and the average life expectancy is 44 years. Only about 12 per cent of the land is suitable for agriculture, yet 70 per cent of the workers are peasant farmers. Two million are nomads, following their flocks of fat-tailed sheep and goats – although their numbers are thought to have dropped drastically as a result of 13 years of war.

The Pathans (Pashtuns) are the largest ethnic group, accounting for half the population. With other groups – Tajiks, Hazaras, Turkmen, Uzbeks, Nuristanis, Baluchis, Persians and Kirghiz – they make a list which sounds like the roll call of invading armies that have passed through the land.

Tribal rivalries have always made the country difficult to govern – and continue to do so. Each group maintains its own individualism, ready to quarrel with its neighbours and unwilling to submit to any central government, at least not without constant protest. Only in recent years have attempts been made to move the country from its state of feudal tribalism. A five-year plan launched in 1956 to improve roads and develop mines and industry was followed by another which opened up hitherto inaccessible parts of the country.

In this development Afghanistan was careful not to favour East or West. A good example is the road, nearly 1600 km (1000 miles) long, that loops south from HERAT to KANDAHAR and then north to Kabul. It was built partly by the Americans and partly by the Soviets, then it was maintained by the Chinese.

Modern Afghanistan dates from the middle of the 18th century when Ahmad Shah (1747-73), a tribal leader, established a united state covering most of present-day Afghanistan. For most of the 19th century the country was subjected to big-power pressures from tsarist Russia in the north and the British in India to the east. Suspicions that its rulers were favouring the Russians led to the first two Afghan Wars with Britain (1839-42 and 1878-9). British forces were defeated both times, but returned in 1880 to enforce a treaty that gave them control of the Khyber Pass. By playing Britain and Russia off against each other,

Afghanistan managed to remain independent, though under British influence.

In 1919 Amanullah Khan (who styled himself king after 1926) invaded India when his demand for complete independence was not met, and after inconclusive fighting (the third Afghan War) this was granted in 1921. Afghanistan was neutral in both World Wars, but relations with Pakistan have been strained since the partition of India and Pakistan's independence; the Afghans wanted the Pathans of the North-West Frontier Province to be given the chance to join Afghanistan.

THE SOVIET INVASION

In 1973 King Muhammad Zahir was overthrown in a coup, the monarchy ended after nearly 50 years and Afghanistan was declared a republic. Five years later President Muhammad Daoud was murdered in a second coup, which led to a Marxist government under Muhammad Taraki. In September 1979, Taraki was ousted in yet another coup and replaced by Hafizullah Amin. But, bitterly opposed by the people of the countryside, the government found it could maintain itself only by calling on the USSR for help. The Soviets were happy to respond, realising an ambition to control their own 'back door' that goes back to tsarist days. In December 1979 they invaded, deposed Amin and installed Babrak Karmal in his place. He was replaced in May 1986 as leader of the Democratic People's Party by Dr Sayid Mohammed Najibullah, General Secretary of the People's Democratic Party of Afghanistan.

The rivalry between clans and their readiness to engage in fights or blood feuds now found a new outlet. The country was split between Marxist supporters, backed by the Soviet army, and resistance fighters providing fierce if uncoordinated opposition. Sometimes they fought among themselves. The Soviet troops found they were facing a brutal and bitter war against tribesmen renowned for their guerrilla tactics.

The Soviets relied mainly on bombing and ground sweeps, only to find that the guerrillas faded away when large numbers of soldiers appeared. They intensified the bombing to include crop-growing areas as well as villages; fields, stocks and herds were blitzed so that sowing times were missed and in some areas there was a serious danger of famine. Of the 22 000 farming villages that existed before the war, 12 000 were destroyed or abandoned. But anti-aircraft missiles supplied by the West made inroads into Soviet command of the air.

The war had been costing the USSR the equivalent of US$1500 million a year, yet the Soviets could only control the towns. They began seeking a way to a settlement. In May 1988, the Geneva Afghan Treaty took effect – 140 000 troops, the last of them departing on February 15, 1989, were flown back to the USSR.

On April 28, 1992, the Islamic State of Afghanistan was proclaimed by the victorious *mujaheddin* ('holy warrior') groups. In December of the same year the *mujaheddin* elected Burhanuddin Rabbani, of the *Jamiat-i-Islami* (Islamic Society) as president of Afghanistan and, in an effort to forestall factional fighting, Gulbuddin Hekmatyar of the *Hezb-i-Islami* (the Islamic Party) as prime minister, both for a two-year term to end in September 1994. But the coalition fell apart within weeks, the prime minister and president fighting each other for overall power. From then on, rockets slammed into Kabul almost daily. Little children scampered after government-controlled food trucks to collect whatever grains might fall to the ground from the vehicles; gun-toting teenagers manned checkpoints at the gates of the city. Approximately 4 million refugees remained in Pakistan, Iran, Tajikistan and Uzbekistan.

AGRICULTURE AND INDUSTRY

In good times, Afghanistan is able to produce sufficient food to sustain its people, although the balance of nature is always precarious. Wheat, fruit, cotton, vegetables, sugar, delicious melons, plums, quinces, apricots and figs ripen in the warm autumn days, and the peaches and grapes of Kandahar are famous. Many medicinal herbs, little known in the West, are found. The country has three rare domesticated animals: the yak, the two-humped Bactrian camel and the karakul lamb, whose soft, curly fleece is used for the treasured Persian lamb furs. Afghan rugs and sheepskin coats are also much valued.

Manufacturing is little developed, but includes textiles, cement, leather goods, glassware, bicycles and food products. There is a wide range of minerals, including coal, iron ore and copper, but most are so inaccessible that it is uneconomic to exploit them. One of the few exceptions is the natural gas found around Sheberghan and Sar-e Pol in the far north, which is exported to the Commonwealth of Independent States (CIS) countries. Some of the finest lapis lazuli, a blue semiprecious stone, comes from Badakhshan and the Panjshir Valley in the north-east.

For most Afghans, life is hard and simple. Some live in mountain valleys so high that they are cut off by snow for months. Sometimes whole communities will move from one village to another in search of better land. The herdsman with a rifle slung across his shoulder may look rugged and romantic in his turban and robes as he tends his flocks, but many people live in poverty, and when drought hits the land they suffer desperate hunger.

Afghanistan is an Islamic state, and education – where it exists – is often in the hands of the *mullahs* (religious leaders). Families are patriarchal, and large, but up to half the children die before they reach the age of five. Society remains male-dominated. Except in the towns, most women appear veiled in public – many may not be allowed out by their husbands at all; and in some parts a wife can be bought unofficially for as little as £3. Women are excluded from politics.

Rice or *nan* (a pancake-shaped bread) and mutton, followed by fruit, is the main food, and green tea, served with much sugar, the favourite drink. The teahouse is the place to

east the spectacular Khyber Pass, narrow and steep-sided, runs through the Safed Koh Mountains to link Peshawar, in Pakistan, with Kabul.

In the west, Herat lies on the old trade route from Persia to India. It is noted for its citadel, markets and decorated *gharries* (horse-drawn cabs). Its Great Mosque was founded in the 12th century, and it has exquisite minarets. North-west of Kabul, the small town of BAMIAN was once a Buddhist stronghold: its deserted Valley of the Gods, with huge stone-carved Buddhist figures, is awe-inspiring. The SALANG TUNNEL, at nearly 3400 m (11 150 ft), leads through the mighty Hindu Kush to the northern plains – and the world beyond, from which Afghanistan was once again totally isolated in 1994 because of relentless fighting among its holy warriors.

▲ **BEHIND THE VEIL In this Islamic country, women are mostly still veiled; here, a collection of veils hangs out to dry after being dyed blue.**

meet friends and gossip, or entertain strangers. Afghans – even the poorest – are very hospitable and strangers may be embarrassed to be given a needy family's choicest food. The bazaars and mosques are the real centres of activity; few prices are fixed and in the bazaars bargaining is automatic. (With the civil war breaking out anew, prices escalated; by the end of January 1994, they had trebled, putting most basic commodities beyond the reach of ordinary people.)

The great attraction for the visitor lies in the wild nature of the land: few countries have more rugged mountain scenery, higher passes or more stunning views. Kabul has a fascination of its own: it has some modern hotels and housing, but the interest lies in the old quarter's narrow, crooked streets and bazaars, and in seeing the races who mix there.

Legacies of former conquerors include the tomb and gardens of Babur, who captured Kabul in 1504 and made it the capital of the Mongol empire, and the mausoleum of Nadir Shah of Persia, who took Kabul in 1738. To the

AFGHANISTAN AT A GLANCE	
Official name Islamic State of Afghanistan	
Area 647 500 km² (250 000 sq miles)	
Population 16 494 150	
Per km² 25 (**Per sq mile** 66)	
Capital Kabul	
Government Islamic state (transitional government); legislature had not yet been approved by early 1995	
Currency 1 Afghani = 100 puls	
Languages Pushtu, Dari (Persian)	
Religion Muslim	
Climate Continental, with little rainfall in the mountains. Average temperature in Kabul ranges from –3°C (26.6°F) in January to 25°C (77°F) in July	
Land use Cultivation 12%, pastures 46%, forest and woodlands 3%, other 39%	
Main primary products Sheep, goats, fruit, wheat, cotton, nuts, vegetables; natural gas, oil, iron, copper, coal, salt	
Major industries Agriculture, fur and leather products, fertilisers, carpets, textiles, cement, glassware, bicycles, food processing	
Major exports Fruit, nuts, natural gas, carpets, wool, skins and hides, cotton	
Annual income per head (US$) 260	
Population growth (per thous/yr) 25	
Life expectancy (yrs) Male 45 **Female** 44	

Acre ('Akko) *Israel* Port on the Bay of Haifa, 13 km (8 miles) north of the city of Haifa. It has been a trading port since Phoenician times.

During the days of the New Testament it was known as Ptolemais. Acre changed hands frequently during the Crusades, but withstood a siege by French forces during Napoleon's Syrian campaign of 1799.

The remains of an underground 13th-century crusader town, including the huge crypt of the Knights of St John, who occupied the town from 1191 to 1291, are still visible today.

Population 37 700
Map Israel – Ba

Ad Dakhla (Villa Cisneros) *Western Sahara* Principal port on the Atlantic coast of the Sahara, lying on Río de Oro Bay, it was founded by Spanish colonists in 1884. The bay provides an open anchorage for shipping and the port has fuel-oil storage facilities for bunkering.

Population 5570
Map Western Sahara – Aa

Ad Dawhah *Qatar* See DOHA

Adamawa Massif *Cameroon* See ADOUMAOUA MASSIF

Adam's Peak (Samanala; Sri Padastanaya) *Sri Lanka* Mountain 72 km (45 miles) east of Colombo. It rises to 2243 m (7359 ft) and is sacred to the island's Buddhists, Hindus, Christians and Muslims. The summit has a mysterious depression shaped like a giant human footprint 1.6 m (5 ft 3 in) long. Buddhists believe it was made by the founder of their religion, Prince Gautama Siddhartha (about 563-483 BC); Hindus attribute it to their god, Shiva; Muslims contend that Adam stood on this spot on one foot for 1000 years as a penance for his expulsion from the Garden of Eden; Roman Catholics credit it to St Thomas, who preached in the region.

Map Sri Lanka – Bb

Adana *Turkey* City about 400 km (250 miles) south-east of the capital, Ankara. It was founded by the Hittites around 1400 BC and conquered by Alexander the Great in 334 BC. The 2nd to 3rd-century stone bridge across the Seyhan River, which runs through the city, was built by the Roman emperor Hadrian. Other ancient buildings include 14th and 16th-century mosques and a 15th-century covered bazaar. In the past, a major trade route led through the city; today, as the site of one of Turkey's largest airports, Adana has retained its position as an important transport hub.

The surrounding lowland province of the same name produces cotton, textiles, cement, agricultural machinery and vegetable oils. It covers 17 253 km² (6661 sq miles).

Population (city) 1 429 670; (province) 1 934 910
Map Turkey – Bb

Addis Ababa *Ethiopia* National capital, standing at the centre of the country at an altitude of more than 2440 m (8000 ft). The city, which is situated in the traditional homeland of the Amhara people, developed gradually during the 19th century, and became the capital in 1896. Among its attractions are an eight-sided Coptic cathedral, built in 1896, Africa Hall, built for the inauguration of the Organisation of African Unity (OAU) in 1963, the Mercato Market, one of the largest on the continent, the old imperial palace, the parliamentary buildings and a university (1961).

Population 2 419 000
Map Ethiopia – Ab

Adelaide *Australia* State capital of South Australia. Situated on Gulf St Vincent in the south-east of the state, it is the fifth largest city on the Australian continent. Founded in 1837 by Colonel William Light, modern Adelaide has remained faithful to its original design, with the broad streets of its city centre being surrounded by a ring of spacious parks. Historic buildings include Government House and Parliament House.

The city is an important railway junction and has a port from which a large part of Australia's exports – including wheat, wool and frozen meat – are shipped. Among its wide range of industries are motor-body assembly plants, railway workshops and the manufacture of electrical goods, household appliances and textiles. Adelaide has three universities.

Population 1 049 000
Map Australia – Fe

Adelsberg *Slovenia* See POSTOJNA

Aden *Yemen* The country's main port and a trade centre since ancient times, Aden lies 160 km (100 miles) east of the Bab al Mandab Strait. From the 7th to the 16th centuries it was held by Muslim Arabs. In 1538, it fell to the Ottoman Turks. British control started in 1802 when a treaty between Britain and Turkey to check Aden-based piracy was signed. In 1839, Aden was annexed by Britain, under whose control it became a coaling station on the sea route to India. Its importance as a trade centre grew further with the opening of the Suez Canal in 1869. In 1937 it became a crown colony. During the Second World War it was a fortified British naval base. In 1960 its inhabitants began a fight for independence, forcing the British to leave in 1968.

An oil refinery, which was nationalised when the British left, is still operating – but at a fraction of its capacity. Other industries include the manufacture of cement blocks, tiles, bricks, metal goods and salt, and food processing. A former Royal Air Force base at Khormaksar – 11 km (7 miles) north of the city – has been converted into an international airport. The present city is built in three sections around two peninsulas over which loom two extinct volcanoes: Crater, the old quarter, was built in the crater of one of the two extinct volcanoes; the business district is in at-Tawahi and the harbour area in Ma'alah.

In 1994, during the final stages of the civil war between the former South and North Yemen – now united again – Aden became the sought after prize for the northern forces. Before the city fell, it was encircled and continuously bombed.

Population 264 300
Map Yemen – Ab

Aden, Gulf of *Middle East* Major shipping lane linking the RED SEA (via the narrow strait of Bab al Mandab) to the ARABIAN SEA. The gulf is 880 km (547 miles) long and 480 km (298 miles) wide at its mouth. Geologically young, the gulf is getting wider as Arabia drifts north-eastwards away from Africa. The gulf's chief ports are ADEN (Yemen), DJIBOUTI and BERBERA (Somalia).

Map Yemen – Ab

Adige *Italy* Second longest river in the country, flowing 410 km (255 miles) from its source in the Alps near where the Swiss, Austrian and Italian borders meet, to the Adriatic Sea.

Map Italy – Cb

Adirondack Mountains *USA* Range in northeast New York State rising to 1629 m (5344 ft) at Mount Marcy. More than 200 lakes and hundreds of streams make it a popular tourist area. Minerals found in the range include iron ore, titanium, vanadium and talc.

Map United States – Lb

Adoumaoua (Adamawa) Massif *Cameroon* Plateau partly covered with volcanic rocks in central Cameroon. Between 150 and 300 km (95 and 185 miles) wide, it divides Cameroon into two, and is the source of most of the country's rivers. The valleys are densely wooded, and the grazing land between is used by more than 70 000 families, mainly of the Islamic Fulani group, with their herds of zebu cattle. The massif contains reserves of bauxite (aluminium ore).

Map Cameroon – Bb

Adrar *Algeria* Oasis town deep in the Sahara Desert, about 540 km (335 miles) south of Ain Sefra. It is a major staging post at the beginning of one of the most barren of all Saharan routes, the Route du Tanezrouft, which leads south to Gao in Mali.

Population 7500
Map Algeria – Ab

Adriatic Sea Arm of the Mediterranean, lying between the Balkan and Italian peninsulas and covering 160 000 km² (41 400 sq miles). The sea is 770 km (478 miles) long, with an average width of 160 km (99 miles). The Strait of OTRANTO, 76 km (47 miles) wide, separates Italy and Albania in the south. The Adriatic coast is popular with holidaymakers.

Map Italy – Ec

Aegean Sea Arm of the MEDITERRANEAN lying between Greece, Turkey and Crete with an area of 179 000 km² (69 110 sq miles). The narrow DARDANELLES in the north-east – only 1.2 km (0.7 miles) wide in places – is the gateway to the BLACK SEA. It separates the Aegean from the Sea of Marmara and the Bosporus. The Aegean is studded with islands, many of which have associations with Greek legend and history. They consist of three groups: the CYCLADES, Sporades and DODECANESE. These hilly peaks of a sunken plateau cover a land area of 9122 km² (3521 sq miles) and are fringed with whitewashed towns and dotted with the ruins of Byzantine monasteries. Seafaring, tourism, and the cultivation of vines, olives and fruits are the main activities. Earthquakes and volcanoes, notably at SANTORINI, are reminders that this is an unstable part of the earth's surface.

Map Greece – Db

Aegina (Aíyina) *Greece* Island resort in the Saronic Gulf 30 km (19 miles) south-west of Piraeus. Its pine-forested and vine and olive-growing interior drops down to fine bathing beaches. The hilltop Temple of Aphaia, dating from the 6th century BC, is a classic example of Doric architecture. Sculptures that once stood in it are now on display in a Munich museum. Aegina is famed for its pistachio nuts.

Population 11 200
Map Greece – Cc

aeolian Aeolian deposits are fine rock particles (such as sand in deserts) that have been carried by the wind and siltlike dust known as LOESS.

aeration zone See WATER TABLE

afforestation Conversion of open land, particularly heathland and moorland, into forests, usually by the planting fast-maturing coniferous trees. In Britain in particular, the timber produced from these forests constitutes a useful crop on marginal land.

Afghanistan See p 12

Africa Covering more than 30 million km² (11 million sq miles), Africa is, after Asia, the world's second largest continent. It is nearly bisected by the equator and both northern and southern Africa extend to the mid-30s of latitude, so that the continent's climatic zones mirror one another, from the tropical forests of the equatorial region through semi-tropical savannahs, dry steppelands and desert to the Mediterranean climates of the extremities.

With very high birthrates and rising life expectancy, it is expected that more than 850 million people will inhabit the continent by the year 2000 (up from the more than 700 million in the early 1990s). Still, the continent's average population density of 23 per km² (60 per sq mile) is low compared with a continent such as Asia, which has more than three times Africa's population density. Most of Africa's inhabitants are

poor, and disease, famine and drought are preva-lent. The continent also faces another threat from the spread of the HIV virus, which causes AIDS. By the mid-1990s, millions of Africans had already been affected.

Though Africa formed part of the Ancient World and was first charted, in rough outline, in the 15th century, most of its interior was not known to Europeans until much later – in the 19th century. European colonisation of Africa began much later than it did on other continents, but when European colonisers finally withdrew, granting independence after about 100 years of dominance and exploitation, in most cases they left behind them countries with borders drawn straight through traditional lands, their popula-tions insufficiently educated, and with vast social discrepancies between different ethnic groups – a recipe for civil war and economic hardship.

One reason for centuries of isolation was the continent's geographic layout: most of Africa consists of high plateaus that descend steeply to generally narrow coastal regions. Waterfalls and rapids flowing off the plateaus made exploration difficult, since the only way into the continent for explorers was by sailing upriver. Four of the world's greatest rivers are in Africa: the NILE, ZAIRE (Congo), NIGER and ZAMBEZI. But despite their size they did not permit much exploration of the continent's interior, for their courses, too, are studded with waterfalls and rapids.

There are few natural harbours and the coast-line is short in relation to the landmass; 14 out of the 47 mainland African nations (excluding Western Sahara) are landlocked.

Africa has some majestic mountains, often made more so by their isolation in vast plains. In the north, the Atlas range stretches 2250 km (1400 miles) from the Atlantic to Tunisia and reaches up to 4165 m (13 665 ft) at its highest point. In East Africa, the Ethiopian Highlands form a great massif of intersecting ranges, whose loftiest peak is Ras Dashen (4620 m, 15 157 ft). Farther south along the East African Rift Valley are the Ruwenzori Mountains, permanently snow-covered and climbing to 5109 m (16 762 ft), Mount Kenya (5199 m, 17 057 ft), Mount Meru (4565 m, 14 977 ft), and the mighty Kilimanjaro, Africa's highest mountain at 5895 m (19 340 ft). In South Africa, the Drakensberg range towers to nearly 3500 m (11 483 ft).

The East African Rift Valley, which is part of the GREAT RIFT VALLEY system, is the world's most spectacular geological depression. It stretches from Mozambique in south-east Africa, and divides in two north of Lake Malawi. One branch runs north along Lake Tanganyika, Lake Kivu, Lake Edward and Lake Albert (Lake Mobutu Sese Seko), while the other, east-ern branch bisects Kenya and Ethiopia, becomes the trench of the Red Sea and continues into south-west Asia. Scattered along the Rift Valley are some of the world's largest lakes: Lake Malawi (23 300 km², 8994 sq miles), the 676 km (420 mile) long Lake Tanganyika, and the sealike Lake Victoria (69 000 km², 26 635 sq miles).

Much of central Africa is occupied by a great expanse of tropical rain forest, second in size only to that of the Amazon basin. It covers large parts of the coastlands of West Africa from The Gambia to Cameroon, and continues into Gabon, Congo and half of Zaire.

Nearly one-third of the continent is desert. The SAHARA alone, which is the planet's largest arid area, covers 25 per cent of Africa – some 9.1 mil-lion km² (about 3.5 million sq miles) – while to the south are the Namib – the world's oldest desert – and the red-duned Kalahari. Because of climatic changes and overgrazing, the Sahara is advancing some of its borders at an estimated 5 km (3 miles) a year.

Africa is the hottest of the continents. Two-thirds of it has a tropical or subtropical climate. The forests receive heavy rains that leach the soil, while other parts receive less than 250 mm (10 in) of rain a year. Devastating droughts are frequent. All through the 1980s and again in the early 1990s, great belts of land across the north and south of the continent suffered the most savage droughts in history.

Area 30 334 592 km² (11 712 252 sq miles)

Population 713 000 000

Afsluitdijk *Netherlands* Dam, 30 km (18 miles) long, lying 62 km (39 miles) north of Amsterdam. It was completed in 1932 to seal off an arm of the North Sea known as the Zuiderzee, and create IJSSELMEER, a freshwater lake.

Map Netherlands – Ba

aftershock Vibrations of the earth's crust after the main waves of an earthquake have ceased. The shocks are caused by minor adjustments in ruptured rocks, and may go on for weeks.

Agadez *Niger* Oasis town in the centre of the mountainous region of Aïr, 380 km (235 miles) east of the border with Mali. Situated some 520 m (1705 ft) above sea level, Agadez has been a regular stopping place for Saharan travellers, including the nomadic Tuareg herdsmen, since the 16th century. A small but growing number of tourists visit the town to see the work of its leather and silver craftsmen, its beautiful Sudanic architecture, its clean streets made from sand, and its magnificent mud mosque.

Population 50 200

Map Niger – Ab

Agadir *Morocco* Port on the Atlantic coast 120 km (75 miles) south of Essaouira, founded by the Portuguese in the early 16th century. It was the scene of a sabre-rattling international incident in 1911 when Germany sent the gunboat *Panther* here during its struggle with France for control of north-west Africa. In 1960 the town was virtually destroyed by an earthquake that killed about 15 000 people.

The new town, built just south of the old, is a popular resort with 8 km (5 miles) of beaches. It is overlooked by the ruins of a 16th-century fortress. Half of the more than 500 000 tonnes of fish caught in Moroccan waters every year are landed here.

Population 779 000

Map Morocco – Ba

Agaña *Guam* The territory's capital, on the west coast. The main port, Apra Harbor, is 10 km (6 miles) south-west.

Population 900

agate A fine-grained QUARTZ marked with vari-coloured bands or with irregular clouding.

age Division of an epoch in the geological timescale.

agglomerate Mass of volcanic rock consisting mainly of fragments larger than 2 mm (about 0.08 in) in diameter which have been ejected from an explosive volcano.

Aggtelek *Hungary* Village near the Slovakian border in the Borsod limestone region, 40 km (25 miles) north of Miskolc. It is noted for its spec-tacular caves – part of a system that extends under the border to Domica in Slovakia.

Population 600

Map Hungary – Ba

Agno *Philippines* River 205 km (128 miles) long in northern Luzon. It rises in the mountains of the Cordillera Central, flows southwards across fertile plains (which it feeds with silt

during annual floods) and empties into Lingayen Gulf, an arm of the South China Sea. It is a rich source of freshwater fish, and its upper reaches have been dammed for hydroelectric power at Binga and Ambuklao.
Map Philippines – Bb

Agra *India* Mogul city and commercial centre on the Jumna River, about 200 km (124 miles) south and slightly east of Delhi. During the 16th and

▼ **LASTING MEMORIAL** The Mogul emperor Shah Jahan (1592-1666) built the Taj Mahal mausoleum at Agra for his favourite wife. He used craftsmen from many lands, including Italy and Turkey.

17th centuries, it was the capital of the Mogul empire. It contains what is said to be the world's most beautiful building – the Taj Mahal, a mausoleum built in 1631-53 in white marble to inter and commemorate Mumtaz-Mahal, the wife of Shah Jahan. Originally the mausoleum was inlaid with both precious and semiprecious stones, but the precious stones were stolen during the troubled period of the 18th century. Shah Jahan was deposed by his son, Aurangzeb, and locked in the Red Fort within sight of the Taj Mahal. Stone inlay work is still a speciality of Agra, which also produces carpets and glassware. In the early 1990s, pollution in Agra was threatening the Taj Mahal.
Population 956 000
Map India – Cb

Agram *Croatia* See ZAGREB

Agrigento *Italy* One of the great cities of classical antiquity, founded as Akragas in 581 BC on the south coast of Sicily. It prospered especially under the rule of Theron during the early 5th century BC, when it had a population of 200 000 – four times its present size – and was the home of the Greek philosopher Empedocles, who is said to have died by falling into the crater of Mount Etna to prove his immortality. The city was colonised by the Romans and destroyed by the Carthaginians.

The majority of its surviving monuments, mostly Greek, are to be found in the Valley of Temples in a superb setting of olive and almond orchards. The present town has a medieval

cathedral, founded in the 11th century, a 13th-century convent and an archaeological museum.

In contrast to the prosperity of 2500 years ago, when the surrounding area produced an abundance of wine, olive oil, wheat, livestock and sulphur for export, the economy of Agrigento province had become one of Italy's poorest by the 1990s.

Population 54 600
Map Italy – Df

Aguascalientes *Mexico* State in the north-west, on Mexico's central plateau. It is mountainous in the west and east, and flat (and therefore suitable for agriculture) in the centre. Its name means 'hot waters', after the hot mineral springs of its capital, which has the same name. Silver mining prompted the development of the town during colonial times, but it is now a centre in an area producing grapes, other fruits and wine.

Population (state) 719 650; (city) 506 380
Map Mexico – Bb

Agulhas, Cape *South Africa* Africa's southernmost point (34°50´S, 20°E) and village, 170 km (105 miles) south-east of Cape Town.

Population 290
Map South Africa – Bc

Ahaggar *Algeria* See HOGGAR

Ahmadabad (Ahmedabad) *India* Textile city in Gujarat state, about 551 km (342 miles) north of Bombay. In 1918 it was the scene of a violent anti-British rebellion. In 1930 the Indian nationalist leader Mahatma Mohandas Gandhi founded an *ashram* (home for the destitute and place of learning) in the city, from where he began his campaign of civil disobedience against British rule in India. Ahmadabad has many fine Hindu,

Muslim and Jain buildings and monuments, including the 15th-century Great Mosque, the mid-19th century Hathi Singh Jain Temple, and the tomb of Ahmad Shah, the Muslim ruler of Gujarat who founded the city in 1411.

Population 4 788 820
Map India – Bc

Ahmadi *Kuwait* Set in an oasis 35 km (21 miles) south of Kuwait City, Ahmadi was developed in the 1950s following the discovery of an oil field nearby. Until then, it was a maze of small roads with single-storey houses, green gardens and trees. Today it is a bustling oil town and commercial centre.

Population 25 000
Map Kuwait – Bb

Ahmedabad *India* See AHMADABAD

Ahvaz (Ahwaz) *Iran* Inland port about 150 km (95 miles) north of The Gulf, at the highest navigable point on the Karun River. The town stands on the site of a city dedicated to Ahura-Mazda, the Zoroastrian god of goodness and light, by the kings of the Sassanid dynasty (3rd to 7th century AD).

Population 725 000
Map Iran – Aa

Ahvenanmaa (Åland) *Finland* Province in the central Baltic at the entrance to the Gulf of Bothnia, covering an area of 1552 km² (599 sq miles) and consisting of 6554 islands, of which more than 100 are inhabited. It has a rich store of relics from the Stone, Bronze and Iron ages. Maarianhamina, the capital and principal port, has a medieval church. The village of Bomarsund has an 11th-century castle, Kastelholm, and an immense fortification

system, built by the Russians in the 1830s when Finland was a Russian grand duchy. It was destroyed by the British and French in 1854 during the Crimean War.

The people are Swedish-speaking, but despite a referendum in 1917, when a majority voted to secede to Sweden, the League of Nations – the forerunner of the United Nations – confirmed Finnish sovereignty in 1921. The islands were, however, granted a large measure of self-rule in 1920, and have their own flag, their own stamps, and their own representatives on the Nordic Council – the annual assembly of parliamentary representatives from Denmark, Finland, Iceland, Norway and Sweden. Their main sources of revenue are fishing, farming and tourism – and tax-free sales aboard ferries sailing between Finland and Sweden.

Population 25 000
Map Finland – Bc

Aigoual, Massif de l' *France* Tree-covered group of mountains in southern France, southeast of the Massif Central, containing the second highest peak – Mont Aigoual (1567 m, 5141 ft) – in the Cévennes Mountains. It is one of the wettest and foggiest places in France, despite being only 70 km (45 miles) from the Mediterranean.

Map France – Ed

Ain Salah *Algeria* Junction post, 694 km (431 miles) north of Tamanrasset, on the Route du Hoggar – the most frequently used north-

▼ **CAMEL TRAIN** A Tuareg caravan makes its way through the Aïr Mountains, which rise abruptly from the sands of the Sahara in northern Niger.

south route in the Sahara. The road branches here, with one track leading north of El Golea, the other west to Adrar. Ain Salah is a popular stopping point for trans-Sahara travellers.

Population 9300
Map Algeria – Bb

Aïr *Niger* Mountainous area in the Sahara Desert in north-central Niger. It extends for about 400 km (250 miles) from north to south and about 240 km (150 miles) from east to west. Its highest peak is Mount Gréboun at 1944 m (6378 ft). The chief occupation of the mostly nomadic inhabitants is cattle rearing. Ancient cave paintings in the mountains indicate that the region has had human occupants for at least 5000 years. The pictures portray the area as having had a much milder, wetter climate then than todays.

Map Niger – Ab

air mass Large body of air with only small horizontal variations of temperature, pressure and moisture content, usually covering hundreds of square kilometres and bounded by FRONTS. Air masses are usually classified according to the regions in which they originate, such as polar (cold) or tropical (warm), and according to whether they are maritime (moist) or continental (dry). These terms are combined, for example in 'polar maritime' (cold and moist) and in 'tropical continental' (warm and dry).

airstream Moving current of air at various altitudes, generally originating in an air mass. Drawn-out wisps of cirrus cloud show the presence of such currents at about 6 km (4 miles) above the earth's surface.

Aisén *Chile* Least developed and most sparsely populated of the country's regions, covering 107 153 km² (41 372 sq miles) in the extreme south. It has few natural resources and its economy revolves around sheep and cattle. Its capital, COIHAIQUE, is 67 km (42 miles) from Puerto Aisén, the main port.

Population 82 070
Map Chile – Ad

Aitutaki *Cook Islands* Second most populous island in the group, covering 18 km² (7 sq miles) and lying 225 km (140 miles) north of Rarotonga. Its hilly terrain rises to 120 m (400 ft), and it has a reef enclosing a large lagoon. It produces most of the islands' banana crop, which is exported, mainly to New Zealand.

Population 2400
Map Pacific Ocean – Fc

Aix-en-Provence *France* First Roman settlement in Gaul (2 BC) and now a thriving university city and tourist centre, 30 km (19 miles) north of Marseilles in Provence. By the 12th century it had become the capital of the counts of Anjou, who ruled Provence. It maintained this status until 1482, when Provence was united with France. The city kept its own parliament until 1789. An ancient cathedral and many fine 17th and 18th-century mansions survive, some of which have been converted into museums.

The city, the birthplace of artist Paul Cézanne (1839-1906), hosts an international music festival each year.

Population 126 900
Map France – Fe

Aix-la-Chapelle *Germany* See AACHEN

Aix-les-Bains *France* See BOURGET, LAC DU

Ajaccio *France* Main port and capital of Corse-du-Sud department on the west coast of Corsica. It was founded by Genoese colonists in 1492 and became French in 1768, when Genoa sold the island to France. The French emperor Napoleon was born here in 1769, and many members of his family are buried in the Palais Fensch, which also has a fine collection of Italian paintings.

Population 55 300
Map France – Hf

Ajanta *India* Small hill town about 400 km (249 miles) north-east of Bombay. A group of ancient Buddhist cave temples lie some 30 km (19 miles) to the north. The 29 caves, carved from solid rock, date from 200 BC to AD 650 and are up to 20 m² (215 sq ft) in size and 11 m (36 ft) high. The halls and sleeping quarters are richly decorated with carvings and frescoes.

Map India – Bc

Ajman *United Arab Emirates* The smallest of the seven emirates at only 250 km² (97 sq miles). It is bordered on three sides by the emirate of SHARJAH. The main port and capital, Ajman, lies 11 km (7 miles) north-east of Sharjah. Its docks service all the emirates, and a shipyard builds and repairs small vessels of up to 5000 tonnes. Traditional industries include fishing and the building of dhows (single-masted Arab boats).

Population (emirate) 76 000; (town) 48 000
Map United Arab Emirates – Ba

Ajmer *India* Industrial city almost 400 km (248 miles) south-west of Delhi. It is noted for its mosques, forts and artificial lake, the Ana Sagar, dating from 1135, which is a birdwatcher's paradise, being home to the fish eagle and other rare species. In the Aravalli range to the west is the sacred Lake Pushkar in which, according to legend, the Hindu god Brahma was born. Each October and November, hundreds of thousands of Hindus gather beside the lake to honour Brahma and to attend the ox and camel fairs. Ajmer is rich in modern temples – the original temples were destroyed by Muslim rulers in the 17th century. The origins of the city date back to 1100, and of the ancient buildings that survive, the 16th-century palace of the Mogul Akbar is one of the best known. Ajmer is a producer of goods such as salt, cotton cloth, footwear, oils and soaps. Many people are employed in the large railway workshops.

Population 1 723 080
Map India – Bb

Ajuda *Benin* See OUIDAH

Akademgorodok *Russia* See NOVOSIBIRSK

Akhelóös *Greece* Second largest river of Greece after the Aliákmon. It rises in the PÍNDHOS mountains and flows 220 km (136 miles) to the Ionian Sea. It is harnessed to produce hydroelectric power at the Akhelóös Dam, built in the late 1950s.

Map Greece – Bb

Akita *Japan* City near the north-west coast of HONSHU Island, about 440 km (275 miles) north of Tokyo, the national capital. It was founded in AD 734 as a garrison town. It is now a busy centre of industry and commerce, processing foods and timber.

Population 302 000
Map Japan – Dc

Akjoujt *Mauritania* See NOUAKCHOTT

'Akko *Israel* See ACRE

Akmola (Akmolinsk; Tselinograd) *Kazakhstan* City 190 km (120 miles) north-west of Karaganda, formerly called Akmolinsk and then Tselinograd. Originally a market town for cattle in semi-arid steppe country, it became a key base in the 1950s and 1960s for a Russian project to open up the so-called 'Virgin Lands' of Siberia for settlers and agriculture. Its industries include meat packaging, farm machinery and fertilisers.

Population 281 000
Map Kazakhstan – Da

Akmolinsk *Kazakhstan* See AKMOLA

Akosombo *Ghana* Dam built across the Volta River gorge in 1966 to provide electricity. It is 75 m (246 ft) high, and behind it lies Lake VOLTA. The small town of Akosombo nearby has a yacht club and a hotel with a particularly fine view of the dam and the lake.

Map Ghana – Bb

Akron *USA* City in Ohio about 45 km (28 miles) south of Cleveland, on the Little Cuyahoga River, which flows into Lake Erie. It is a leading manufacturer of rubber. About 15 km (9 miles) to the north-east is Kent State University, where four students were shot dead by National Guardsmen in May 1970 during a campus protest against American military involvement in Cambodia.

Population (city) 223 020; (metropolitan area) 657 580
Map United States – Jb

Akrotiri *Cyprus* Bay at the southern extremity of the island. Its cliffs are the nesting place of the rare Eleonora falcon and the location of the monastery of St Nicholas of the Cats, founded in AD 325. A British military base is situated on the shore of a salt lake to the west of the bay.

Map Cyprus – Ab

Aksai Chin *India* Cold, high desert plains covering some 37 550 km² (14 500 sq miles) north of the Himalayas. Although India claims them as part of the state of JAMMU AND KASHMIR, they have been occupied since the early 1950s by the Chinese, who built a road across them. Subsequent protests sparked the Indo-Chinese War of 1962, in which India was defeated.

Map India – Ca

Aksum *Ethiopia* See AXUM

Akure *Nigeria* Capital of Ondo State and a cocoa-marketing centre about 210 km (130 miles) north-east of Lagos. It is also a tourist centre for the seven Olumirin waterfalls in the Erinoke-Ijesha gorge of the Osse River just to the east of the town.

Population 114 400
Map Nigeria – Bb

Akureyri *Iceland* Port at the head of the 60 km (37 mile) long Eyjafjördur inlet. It is the third largest town in Iceland after Reykjavík and Kópavogur. Its industries include fish processing, ship repairing and the manufacturing of farm products, soaps, paints and textiles. The town is also a winter sports centre and the point of departure for tourist excursions to MYVATN Lake.

Population 14 000
Map Iceland – Ca

Akyab *Myanmar (Burma)* See SITTWE

Al Asnam (Ech-Cheliff; Orleansville) *Algeria* An important administrative and communications centre in the west, halfway along the Cheliff Valley between Algiers and Oran. Established in 1843 as a French military base, it is now the centre of a large agricultural region. It was struck by earthquakes in 1954 and 1980.

Population 125 000
Map Algeria – Ba

Al Ayn *United Arab Emirates* Town and oasis about 160 km (100 miles) west of the city of Abu Dhabi. The town, built around seven oases and noted for its greenery, is expanding rapidly. A chief source of revenue is a large cement plant capable of producing up to 750 000 tonnes a year. Al Ayn is the site of the palace of the crown prince of the United Arab Emirates (UAE), as well as of three historic forts: Jahili Fort, built in 1898 and now the headquarters of the UAE Defence Force, Al Ayn Fort, built in 1910, and Muwaiji Fort. The vice-president of the UAE is also based in the town.

The series of luxury hotels that have been built here bears testimony to the rapid growth of the local tourism industry. A museum in the town has exhibits from nearby archaeological sites, including household utensils and spearheads, some of which are 5000 years old. The Bronze Age site at nearby Hili contains reconstructed burial mounds and huge stone slabs arranged in a circle – much like Stonehenge (although only one stone is standing).

Population 103 000
Map United Arab Emirates – Ba

Al Basrah *Iraq* See BASRA

Al Fawr *Iraq* See FAO

Al Fahayhil *Kuwait* See FAHAHEEL

Al Fujayrah *United Arab Emirates* With an area of 1150 km² (444 sq miles), it is the only emirate on the coast of the Gulf of Oman (the other six are on The Gulf). In 1976 a road built through the HAJAR MOUNTAINS connected Al Fujayrah with the rest of the United Arab Emirates (UAE). The town of Al Fujayrah, with its modern airport, is overlooked by a 200-year-old fort, which has been converted into a museum.

Population (emirate) 63 000; (town) 760
Map United Arab Emirates – Ca

Al Furat *Middle East* See EUPHRATES

Al Hasa *Saudi Arabia* This subdivision of Eastern Province (known as the Emirate of Al Hasa) on the east coast has some of the world's largest oil fields. It is also the site of the world's biggest oasis, the Al Hasa oasis, which contains 50 villages, where every minute, thousands of litres of water flow from wells and irrigate more than 16 000 hectares (nearly 40 000 acres) of surrounding land. Major irrigation and drainage schemes are being employed to grow rice, dates, wheat, sesame seeds, citrus fruits, peaches, figs and vegetables. The main oasis town is Hofuf. Millions of trees have been planted to control the shifting sands.

Population 600 000
Map Saudi Arabia – Bb

Al Hillah *Iraq* Regional capital on the Euphrates River, about 100 km (60 miles) south of the national capital, Baghdad. It manufactures textiles and leather goods, trades in cereals and dates, and has a huge casino on the riverfront. The Hindya Barrage, 32 km (20 miles) to the north, diverts most of the Euphrates's waters to a channel farther west for irrigation.

Population 215 000
Map Iraq – Cb

Al Jazirah *Sudan* See EL GEZIRA

Al Kazimiyah *Iraq* See KADHIMAIN

Al Khums *Libya* See HOMS

Al Manamah *Bahrain* Capital and port at the north-east tip of Bahrain Island. It is the administrative area for the oil industry, an important banking centre for The Gulf region and a centre for boatbuilding and fishing. A city has existed here since the 15th century, but the oldest surviving buildings date back to the 18th century. Modern Al Manamah is a lively mixture of East and West, with a bustling Arab bazaar and some fine modern buildings.

Population 137 000
Map Bahrain – Ba

Al Mawsil *Iraq* See MOSUL

Al Qurnah *Iraq* City near the legendary site of the Garden of Eden, about 400 km (250 miles) south-east of Baghdad.

The marshland area, the Amarah marsh, to the north of the city is the home (and has been for more than 5000 years) of the Marsh Arabs, or Maadan. Until recently, the region teemed with wildlife, particularly birds. It was also the site of more than 1000 artificial islands constructed of mud and papyrus reeds by the Marsh Arabs, each island supporting huts made of reeds and reed matting. But by the end of 1993, an estimated 40 per cent of the marshland had been drained as part of a campaign of deliberate persecution by the Iraqi government. Many of the Marsh Arabs fled to Iran.

Map Iraq – Cc

Alabama *USA* Southern state that became part of the United States in 1819 after periods of French, English and Spanish rule.

A Confederate stronghold in the Civil War (1861-5), it later became a centre of resistance to black civil rights. Although it has always been associated with the cultivation of cotton, Alabama has also become an important centre for the farming of chickens, cattle, soya beans, peanuts and maize. Another important industry is coal mining, which, in turn, fuels the iron and steel industry in the city of Birmingham.

Oil and natural gas are also produced. Montgomery is the capital of the state, which covers 133 915 km² (51 692 sq miles).

Population 4 040 600
Map United States – Id

Alamo, The *USA* See SAN ANTONIO

Alamut *Iran* See QAZVIN

Åland *Finland* See AHVENANMAA

Alaska *USA* State covering the north-west tip of North America and separated from the rest of mainland United States by Canada. Alaska is the largest state of the Union, with an area covering 1 518 800 km² (586 400 sq miles) – 80 per cent of which is owned by the federal government. It includes the Aleutian Islands, the most westerly of which – Attu – is just 85 km (53 miles) across the Bering Strait from Russia. Formerly known as Russian America, Alaska was bought from tsarist Russia in 1867 for US$7.2 million, and became America's 49th state in 1959. The discovery of gold in the late 19th century and oil in the 20th century prompted population booms as tens of thousands of prospectors headed for the desolate wastes of the area. Oil accounts for about 85 per cent of the state's revenue. While Alaskans have not paid any state income tax since 1980, the state faced a budget deficit in the 1990s as a result of falling oil prices.

Although 80 per cent of the terrain is permanently frozen and only 0.3 per cent is farmed, the land remains an important source of revenue. Timber processing and the mining of natural gas, coal, tin, iron, silver and lead are important industries. Many Alaskans also make a living out of fishing.

▼ **RIVER OF ICE A glacier wends its way through the St Elias mountains to the Pacific Ocean in Alaska's Glacier Bay National Park.**

The southern mountains of the Alaska range – a growing tourist attraction – contain North America's highest peak, Mount McKinley (6194 m, 20 320 ft), in DENALI NATIONAL PARK. Alaska has 202 396 km² (78 127 sq miles) of national parks, and an additional 311 690 km² (120 316 sq miles) is classified as a wildlife refuge. Part of the Alaskan coastline was severely polluted in March 1989 when the oil tanker *Exxon Valdez* ran aground in Prince William Sound, causing the largest oil spill in United States history and an international outcry against pollution. The state capital is JUNEAU.
Population 550 000

Alaska, Gulf of Lying in the north-eastern PACIFIC OCEAN and covering more than 5200 km² (2000 sq miles), the Gulf of Alaska extends between the Alexander Archipelago and the Alaska Peninsula.

Beneath it the Pacific plate is slipping under the North American plate, causing volcanic eruptions and earthquakes. In 1964 an earthquake measuring 8.4 on the RICHTER SCALE caused a tsunami, which shattered the ports of ANCHORAGE, Seward and Valdez, all in the southern part of Alaska. The earthquake and the giant wave that followed killed 178 people.
Map Alaska – Cc

Alassio *Italy* Elegant resort on the Riviera di Ponente (western Riviera) about 85 km (52 miles) south-west of Genoa in the province of Liguria. Its mild climate draws many people to its excellent beach. The old part of the town has many shady alleys packed with interesting shops and lively bars.
Population 12 700
Map Italy – Bb

Álava *Spain* One of the three provinces of northern Spain making up the Spanish part of the BASQUE REGION, which is called Vascongadas in the Basque language.
Population 260 600
Map Spain – Da

Alba Iulia *Romania* Town and capital of Alba county, on the River Mures in south-west TRANSYLVANIA. Alba Iulia was the capital of the Roman colony of Dacia, and of Transylvania in the 16th and 17th centuries.

In 1918 the unification of Transylvania and Romania was proclaimed in the hall of the casino, which is now the Museum of the Union. Four years later, the town's Orthodox cathedral was the scene of the coronation of Ferdinand I, who was crowned King of All Romania after the country had doubled in size with the acquisition, after the First World War, of Transylvania, the Banat (an agricultural region, formerly in southern Hungary) and Bessarabia. The town has an early 18th-century citadel built by the Holy Roman Emperor Charles VI (1685-1740).
Population 50 900
Map Romania – Aa

Albania

AFTER CENTURIES OF TOTAL ISOLATION, ALBANIA IS CAUTIOUSLY OPENING ITS DOORS TO THE OUTSIDE WORLD

The people of Albania have always been physically isolated in their mountainous country, but after the Second World War they were virtually cut off from all contact with the outside world. Foreign news and books were restricted, foreign travel was forbidden and, although tourists were allowed into the country, Albanians were not encouraged to talk to them.

The country became a Stalinist-communist state in 1946, run until 1985 by Enver Hoxha (pronounced Hodja) and thereafter by Ramiz Alia. In 1991, a new constitution was passed and a unicameral People's Assembly, with 140 members, was established. Members are directly elected for a four-year term, and in turn elect the president of the republic as well as the Council of Ministers.

The land rises to 2694 m (8838 ft) in the north ALBANIAN ALPS, with broad grassy valleys running southwards into Greece and marshy plains on the coast of the Adriatic Sea. Its population is 90 per cent Albanian – who call themselves Shqiperie, which means 'sons of eagles'. There are also small groups of Greeks (8 per cent), and Serbs, Macedonians and Romanians. The Albanian language is of Indo-European origin and divides into two dialects – Gheg in the north and Tosk in the south.

After 400 years of Turkish rule, Albania became independent in 1912. Italy twice invaded the country, during the First World War and again under Mussolini in 1939-43. German troops followed the Italians and remained until 1945; the country was retaken by native communist guerrillas led by Hoxha and aided by Britain and the United States. This led to the establishment of the new state in January 1946, with Hoxha as prime minister.

CHINESE INFLUENCE

All relations with Western nations were broken off after the war. The USSR became Albania's chief ally, market and source of investment until the Soviet leader, Josef Stalin, died in 1953 and the new leader, Nikita Khrushchev, denounced Stalinism in 1961. Hoxha, for whom Stalin was a hero, thought the USSR's new communism was ideologically impure and switched his allegiance to China. Almost overnight, Chinese technicians replaced Soviets on engineering and industrial projects in Albania. But, as was the case in China, agricultural development took precedence over industrial development in Albania.

This Chinese style of communism, with decentralised administration encouraging rural development, suited Albania better than centrally controlled, Soviet-style industrialisation. However, in the 1970s disillusionment set in with China too.

Although factories were built and railways developed during the Stalinist era, it was agriculture which provided Albania's main success story, with results achieved mainly by hard work. For years, the country managed self-sufficiency in food, despite having very backward agricultural techniques. Two-thirds of the people lived in villages, and still do, and most of them worked on collective farms. Hours were long, and shirkers could be sent to jail. All children of school age had to spend a month of every year working in the fields.

END OF AN ERA

Under Hoxha's successor, Ramiz Alia, the country gradually resumed diplomatic relations with Yugoslavia, Greece and the West. The communist era ended only in 1992, however, when the Democratic Party won a clear majority. The new government faced the immediate task of removing both the economic and the social problems of 45 years of communist rule.

The country was on the brink of economic collapse and its people were facing bitter hunger. Food depots were plundered in the early 1990s and tens of thousands of Albanians fled to neighbouring countries such as Italy and Greece. Farmers had been unable to work the fields for lack of seed and machinery. In order

to reduce the mass emigration, the European Community (EC) sent in food aid.

Today, Albania is one of the poorest countries in Europe, despite generous EC aid. During 1991-2, the annual per capita income was a mere US$214 and unemployment soared to 70 per cent. Through large-scale privatisation and by encouraging foreign investment, the Albanian government is hoping to save the country from total collapse.

After decades of isolation, Albania has resumed ties with Europe and also with other Muslim nations, notably Turkey. Centuries of Turkish rule have left their mark on Albanian culture, though the Muslim religion was suppressed under the communists. About 60 per cent of the population were Muslim at the time of the communist takeover – but in 1967, the country was officially declared atheist. Mosques and churches were converted to museums and schools, warehouses and stables. Women were liberated from the veil and given equal pay and jobs in government.

Another inheritance from the communist days is the high literacy rate among the people – the communists had made it their goal to wipe out illiteracy, and by 1992 they had almost succeeded.

ALBANIA AT A GLANCE	
Official name Republic of Albania	
Area 28 748 km² (11 100 sq miles)	
Population 3 333 840	
Per km² 116 (**Per sq mile** 300)	
Capital Tiranë	
Government Republic	
Currency 1 lek = 100 quintars	
Language Albanian	
Religion Muslim (70%), Greek Orthodox (20%), Roman Catholic (10%)	
Climate Mediterranean on the coastal lowland, moderate to continental in the mountains, with rains in the winter months. Average temperature in Tiranë ranges from 7°C (45°F) in January to 25°C (77°F) in July	
Land use Cultivation 25%, meadows and pastures 15%, forest and woodlands 38%, other 22%	
Main primary products Cereals, potatoes, grapes, olives, tobacco, cotton, timber; petroleum and natural gas, coal, lignite, chromium, copper, nickel	
Major industries Agriculture, petroleum refining, mining, food processing, cement, fertilisers, textiles, tobacco processing	
Main exports Non-ferrous metal ores, petroleum and petroleum products, food, clothing, tobacco and tobacco products	
Annual income per head (US$) 710	
Population growth (per thous/yr) 12	
Life expectancy (yrs) Male 70 **Female** 76	

Albacete *Spain* Market town and provincial capital about 225 km (140 miles) south-east of Madrid, manufacturing chemicals, soap, furniture, processed foods, and knives and scissors. Its museum has a selection of 3rd-century Roman dolls with jointed limbs, found locally. The surrounding province of the same name is an important sheep-farming area.
Population (town) 130 020; (province) 342 680
Map Spain – Ec

Albania See above

Albanian Alps (Bjeshkët e Némuna; Prokletije) *Albania/Yugoslavia (Serbia and Montenegro)* Steep-sided mountains lying between the River Drin and Serbia and Montenegro, and dropping precipitously to the Shkodër lowlands. The forested slopes, deeply cut by short rivers, reach their highest point at Jezercë (2694 m, 8838 ft).
Map Albania – Ba

Albano, Lake *Italy* Lake covering 6 km² (2.3 sq miles) about 20 km (12 miles) south-east of Rome. It was formed by the fusion of two ancient volcanic craters and is fed by underground springs. Its artificial outlet was made by the Romans in the late 4th century. The lake is popular for local and international watersports events. It is overlooked by the town of Albano Laziale and the Pope's summer retreat, Castel Gandolfo.
Map Italy – Dd

Albany *Australia* Port and oldest town in Western Australia, 390 km (242 miles) south-east of Perth. Originally called Frederickstown, it was founded as a British penal colony in 1826 in order to forestall a possible French settlement on the west coast. It became a whaling port and a coaling station for ships trading between Australia and Europe. Its chief industries today are fish and meat canning.
Population 18 830
Map Australia – Be

Albany *USA* Capital of New York State and the oldest city in the country still operating under its original charter, granted in 1686. It was founded in 1624 by Dutch immigrants and became a distribution centre for the north-east United States after the Erie Canal opened in 1825. It stands on the Hudson River, 210 km (130 miles) north of New York City.
Population (city) 101 080; (metropolitan area) 861 420
Map United States – Lb

Alberobello *Italy* Town in Apulia, 55 km (35 miles) south-east of Bari. It has been classified a national monument because of its *trulli*, ancient conical stone houses of uncertain origin. They were built without mortar and have whitewashed walls and grey-tiled roofs topped with crosses and other symbols.
Population 10 660
Map Italy – Fd

Albert Canal *Belgium* Canal linking the Meuse River at Liège to Antwerp about 130 km (80 miles) away.
Map Belgium – Ba

Albert Edward Nyanza *Uganda/Zaire* See EDWARD, LAKE

Albert, Lake (Albert Nyanza; Lake Mobuto Sese Seko) *Uganda/Zaire* Lake covering some 5334 km² (2059 sq miles) in the western arm of the East African section of the Great Rift Valley. It is fed by the Semliki River, which drains Lake Edward, and by the Victoria Nile, which flows from Lake Kyoga, Uganda; it is drained by the Albert Nile which, in Sudan, is called the Bahr el Jebel.

A part of the Nile system, the lake was recorded in 1864 by the British explorers Sir Samuel Baker and his wife who named it Albert Nyanza after the prince consort, husband of Queen Victoria. It is also known as Lake Mobuto Sese Seko, after Zaire's president who took office in 1965. The lake is inhabited by crocodiles and hippopotamuses, while around its shores are many other animal species, including elephants and rhinoceroses.
Map Zaire – Ca

Albert National Park *Zaire* See VIRUNGA NATIONAL PARK

Albert Nyanza *Uganda/Zaire* See ALBERT, LAKE

Alberta *Canada* Most westerly of the prairie provinces, covering 661 190 km² (255 285 sq miles) and with a landscape ranging from prairies to lofty mountain peaks. Agriculture includes grain crops and beef and dairy cattle. The province has extensive forestry and large-scale output of oil, natural gas and coal. It also has unexploited oil reserves in the north. Western Alberta is a tourist area, with large areas of wooded country fringing the Rocky Mountains, including Banff and Jasper National Park. The main cities are EDMONTON, the capital, and CALGARY, which hosted the 1988 Winter Olympic Games.
Population 2 545 550
Map Canada – Dc

Albertville *Zaire* See KALÉMIÉ

Albi *France* Capital of the Tarn department, on the Tarn River 70 km (44 miles) north-east of Toulouse. It is an attractive town, with a 13th to 15th-century red-brick cathedral whose interior is adorned with designs by various Italian artists. The artist Henri de Toulouse-Lautrec (1864-1901) was born in Albi and some of his paintings are displayed in the bishop's palace.
Population 46 600
Map France – Ee

albite A widely distributed feldspar (a compound of sodium, aluminium and silica) found mainly in rocks such as granite.

Ålborg *Denmark* Fourth largest of Denmark's cities and third largest port, sitting astride the narrows of LIMFJORDEN, in northern JUTLAND, the fjord which runs from the North Sea to the Kattegat. Historic features on the waterfront include the Gothic-style Cathedral of St Budolfi (Botolph) and Jens Bang's house, one of the finest Renaissance homes in Denmark. Other places of interest include Lindholm Høje (a Viking burial site dating from about 1100), the North Jutland Museum of Art, Tivoli Korolinelund, the 15th-century Helligåndsklosteret containing early 16th-century frescoes, and the castle Ålborghus. Ålborg also has an airport, a university, several technical schools and the world-famous art gallery, Nordjyllands Kunstmuseum. East of the city are Denmark's largest cement plants. Ålborg has also acquired an international reputation for its akvavit (Danish schnapps).
Population 155 700
Map Denmark – Ba

Albufeira *Portugal* Small fishing town on the south coast of the Algarve, about 30 km (20 miles) west of Faro. It is a tourist resort, with a bustling daily market and a vibrant nightlife. A tunnel through the cliffs leads to a sandy beach below a Moorish castle dating from the 8th to the 13th centuries.
Population 12 000
Map Portugal – Bd

Albuquerque *USA* Largest city in New Mexico. Located on the banks of the upper Rio Grande River, it was originally a Spanish settlement but the old town has been swallowed up by the new town. It specialises in high-tech industries such as lasers, data processing and solar energy. It is the seat of the University of New Mexico.
Population (city) 385 000; (metropolitan area) 589 130
Map United States – Ec

Albury-Wodonga *Australia* Industrial twin cities straddling the Murray River and the state border, Albury being in New South Wales and Wodonga in Victoria. Albury-Wodonga was designated an inland growth centre in the 1970s. In addition to providing services for the surrounding farming country, the twin cities contain numerous industries, including textiles, furniture, brickworks and engineering. The cities lie on the major transport route between Sydney and Melbourne.
Population 63 610
Map Australia – Hf

Alcalá de Henares *Spain* Picturesque town 29 km (18 miles) north-east of Madrid. The birthplace of writer Miguel de Cervantes (1547-1616), author of *Don Quixote*, the town was devastated during the Spanish Civil War.
Population 159 360
Map Spain – Db

Aldabra *Seychelles* The largest coral atoll in the Indian Ocean, north of Madagascar. It covers 145 km² (56 sq miles), including the lagoon. Channels divide the ring into four low islands that stand about 30 m (100 ft) above the sea. They have remained isolated from any landmass throughout their existence and as a result are populated by some unique animals. The atoll, a World Heritage site managed by the Seychelles Island Foundation, is the home of some 150 000 giant land tortoises and the world's rarest bird – the Aldabran brush warbler, of which only about a dozen specimens survive. Cosmoledo Atoll, which includes Astove Island, 100 km (62 miles) to the east, forms part of the Aldabra group.
Map Indian Ocean – Bc

Alderney *English Channel* See CHANNEL ISLANDS

Aleksandropol *Armenia* See GYUMRI

Aleksandrorsk *Ukraine* See ZAPORIZHZHYA

Alençon *France* Capital of the Orne department in the fertile countryside of the Sarthe Valley, 48 km (30 miles) north-east of Le Mans. It was once an important lace-making centre, producing *point d'Alençon* work. A castle and the town's Church of Notre Dame date from the 15th century and the 15th to the 18th centuries respectively.
Population 30 000
Map France – Db

Alentejo *Portugal* Southern province to the east of the capital, Lisbon, and immediately north of the Algarve, covering about 30 per cent of the country (26 158 km², 10 097 sq miles). The area has a continental climate with almost dry, cool winters and dry, hot summers. The main rivers are the Tagus, Sado and Guadiana. Alentejo was the scene of many battles in the Middle Ages, its inhabitants fighting first against the Moors, then against Spanish invaders. Its mostly low, rolling landscape is dotted with cork oaks and wheat fields. The main source of income is agriculture – wheat, fruit, olives and cork oaks are cultivated and cattle and pigs reared. Following Portugal's political revolution in 1974, the province's large estates were turned into co-operatives, but many were later returned to private ownership. To the east, the hills of the Serra de São Mamede rise to 1025 m (3362 ft).
Population 563 600
Map Portugal – Bd

Aleppo (Halab) *Syria* Industrial city and capital of Aleppo province, 120 km (75 miles) north-east of Latakia. Aleppo dates from at least 2000 BC and, together with Damascus, is said to be the oldest continuously inhabited city in the world. The city is situated in the centre of a fertile region and at the crossroads of old trade routes, such as the Silk Road. As a result it was often under attack, changing hands many times over the centuries. The Arabs ruled it in the 7th century and the Ottomans from 1516 to 1918. Its citadel, ringed with ramparts, is situated in the centre of the city. Nearby, the modern bus station opens onto a large market, where the covered main souk (bazaar) is more than 800 m (2625 ft) long.
Population 1 355 000
Map Syria – Ba

Aletsch Glacier *Switzerland* A group of three glaciers, the largest of which is the Great Aletsch Glacier. It originates at the Konkordiaplatz on the southern slope of the Jungfrau massif, where three FIRN fields converge to become the largest glacier in the Alps. The Great Aletsch Glacier is 24 km (15 miles) long and covers 117.6 km² (45 sq miles); at Konkordia, the ice is about 900 m (2953 ft) thick. The glacier terminates below the timberline at 1550 m (5085 ft) near Brig, not far from the Rhône Valley.
Map Switzerland – Ba

Aleutian Islands See ALASKA

Aleutian Trench Lying in the North PACIFIC OCEAN, the Pacific plate is slipping beneath the North American plate along this trench. This causes the descending plate to melt, producing molten magma which fuels the volcanoes in the Aleutian island chain lying north of the trench. There are about 150 islands in the chain and 80 volcanoes, 35 of which are active. The greatest known depth of the trench is 7443 m (24 419 ft).
Map Alaska – Ac-Bc

Alexandra *South Africa* Once a black squatter settlement north-east of Johannesburg, Alexandra is now all but surrounded by formerly whites-only suburbs. A so-called 'black spot' earmarked for elimination, Alexandra received few amenities during the apartheid period and was allowed to decay into a slum.
Population 124 590
Map South Africa – Cb

Alexandretta *Turkey* See ISKENDERUN

Alexandria *Egypt* Largest port and second largest city of Egypt after the capital, Cairo. It lies at the western edge of the Nile delta, about 185 km (115 miles) north-west of Cairo. It is a modern industrial city with a deep-water dock handling more than 75 per cent of Egypt's imports and exports, with a flourishing tourist trade beside its Mediterranean beaches.
The city was founded by the Greek conqueror Alexander the Great in 332 BC (after whom it is named) as the capital of his empire. For more than 600 years it was one of the major cities of ancient times. The first, 180 m (591 ft) tall lighthouse, built of white marble in about 270 BC, stood on the island of Pharos in the bay and was one of the Seven Wonders of the Ancient World. The city's library of Greek manuscripts, founded in 300 BC when Alexandria already had a population of

more than half a million, made it the intellectual heart of the Graeco-Roman civilisation. After the death of the Egyptian queen, Cleopatra, in 30 BC, Alexandria became the seat of government of Rome's Middle Eastern empire. Decline set in after AD 300, when Constantinople (Istanbul) became the capital of the Byzantine Roman empire. In 642 Alexandria fell into Arab hands and further deteriorated during Turkish rule. Later silting up of the Nile delta led to the collapse of Alexandria as a port.
The city's regeneration began only in the 19th century, after Muhammad Ali, the governor who gained Egypt's independence from the Turkish Ottoman empire, had built the Mahmudiya Canal. The canal, which was opened in 1847, linked the city to the river and once again to the sea. In 1882 Alexandria was bombarded and then occupied by the British fleet.
Alexandria achieved literary fame as the setting for the famous Alexandria Quartet of novels written by British novelist Lawrence Durrell.
The city has tanneries, shoe-making factories, vehicle-assembly plants, an oil refinery and chemical plants, and handles the bulk of Egypt's cotton trade.
Population 4 360 000
Map Egypt – Bb

Alexandropol *Armenia* See GYUMRI

Alföld, Great (Nagyalföld) *Hungary* The monotonously flat 'Great Plain' east of the Danube River, covering 51 800 km² (19 995 sq miles) – nearly half of Hungary. The Tisza River meanders across the plain from north to south, cutting it roughly in half. The soils are mostly fertile black earths, and the prairielike *puszta* (grasslands) that once covered most of the plain was the traditional home of shepherds and of the colourful *csikós* (horse riders).
However, over the past 100 years, there have been dramatic changes. Most of the *puszta*, apart from the area known as HORTOBÁGY, has been divided into vast farms or market gardens for the growing of wheat, maize, potatoes and fruit. It produces sugar beet, flax, hemp and tobacco, which are Hungary's chief commercial crops. Villages, often surrounded by orchards, tend to be large. Many of the ancient towns, such as Békéscsaba, Cegléd and Debrecen, are 14th-century market towns. One of them, Szeged, has been inhabited since the 8th century.
Roads and railways radiate across the Great Alföld from the capital, Budapest, but travel in other directions is often slow.
Map Hungary – Ab-Bb

Alföld, Little (Kisalföld) *Hungary* Triangular plain covering the north-west of the country and bounded in the south-east by the Bakony Mountains. It occupies an area of 7000 km² (4320 sq miles). It is similar to the Great Alföld, and is a major producer of maize, sugar beet and wheat. GYÖR is the regional capital.
Map Hungary – Ab

Algarve *Portugal* Province of 4960 km² (1915 sq miles), stretching along the south Atlantic coast. It was the last part of Portugal to be wrested from the Moors, in 1253. The Algarve is now the country's most popular holiday area. In the west there are sandy beaches backed by cliffs; in the east, sandbanks protect the shore. Colourful fishing

villages dot the coast, while almonds, figs and citrus trees grow inland. FARO is the main town.
Population 344 900
Map Portugal – Bd

Algeciras *Spain* Port and winter resort on the country's southern tip, with a large export trade in cork from nearby forests. The Moors landed close to Algeciras when they began their invasion of the Iberian Peninsula in 711. The town was settled by Spanish refugees from Gibraltar after the rock was captured by the English in 1704.
Boat trips to TANGIER, CEUTA and GIBRALTAR leave from the harbour.
Population 101 230
Map Spain – Cd

Algeria See p 26

Alghero *Italy* Tourist resort in north-west Sardinia, traditionally popular with English visitors. Alghero's culture still reflects its colonisation by Spain in the 14th century. The people speak Catalan and there are many 14th-century Gothic buildings, including the cathedral and the bastions that partially encircle the old town.
Population 38 800
Map Italy – Bd

Algiers (El Djazair; Alger) *Algeria* The country's largest city, chief port and capital, lying at the centre of Algeria's coastline. Founded on four islands (now joined to the mainland) by the Phoenicians about 1200 BC, it became a thriving port in Roman times, only to fall into disuse with the fall of the Roman empire. It was refounded by Muslim Arabs in the 10th century. From the 16th to 18th centuries, the Barbary pirates, who preyed on vessels in the Mediterranean, made it the centre of their activities. In 1830 France invaded Algeria, which was declared French territory in 1848. Under the French, Algiers became a modern commercial port, increasing its population from a few thousand in 1830 to over 1 million by 1970. Wine, citrus fruits and iron ore are important exports, and the city has chemical, light engineering and consumer goods industries.
The *kasbah* (Arab city), with its citadel and old houses that have remained in the same families for generations, is a maze of twisting, picturesque streets. The Museum of Popular Arts stands at the foot of the *kasbah*. The modern town centres on the Admiralty Building. To the east is the fishing port and market. Above the city and overlooking the bay is the former summer palace of the governor, the Bardo Museum of Ethnography, which is housed in an 18th-century Turkish villa, and the Stephane Gsell Museum, which contains Roman antiquities, Islamic art and archaeological remains. There are several other museums and fine mosques.
Population (city) 2 168 000; (metropolitan area) 3 700 000
Map Algeria – Ba

Algonquin *Canada* Provincial park covering 7600 km² (2934 sq miles) and containing 2500 lakes in a beautiful wilderness in south-east ONTARIO. Most of the park is accessible only by canoe or on foot.
Map Canada – Hd

Aliákmon *Greece* Longest river of Greece, with a length of 297 km (184 miles). From a source in

▲ **PEACEFUL SHORE** Atlantic waves break in a cove between tall cliffs near the ancient port of Lagos in the Algarve, Portugal's busiest tourist area.

the PÍNDHOS mountains, it carves narrow gorges through the highlands of MACEDONIA, and is dammed to generate hydroelectric energy at the Aliákmon Barrage (built in 1973) before emptying into the Aegean Sea.
Map Greece – Ba

Alicante *Spain* Mediterranean port about 125 km (78 miles) south of the port of Valencia. Alicante exports wine, citrus fruits, vegetables, olive oil and almonds. Its industries include metalworking, oil refining, and the production of textiles, chemicals, tobacco products and marzipan. Tourism has become an important income earner and ferries sail from Alicante to the BALEARIC and CANARY ISLANDS.

The adjacent province of the same name is mainly barren, but irrigated areas produce wine and citrus fruits.
Population (town) 265 470; (province) 1 292 560
Map Spain – Ec

Alice Springs *Australia* Town in the Northern Territory, standing in the MACDONNELL RANGES in the great central desert, almost in the middle of the continent. It was founded as a staging point for the overland telegraph line in the 1870s and is now a centre for cattle, minerals (gas, oil, gold, copper and zinc) and tourism. It is situated on the Stuart Highway, which crosses the country from Darwin in the north to Port Augusta in the south. Alice Springs plays host each year to the Henley-on-Todd boating regatta, which is modelled on the annual event at Henley-on-Thames in England. The difference is that the Alice races take place when the Todd River is dry; competitors run along the dusty riverbed holding bottomless boats around them. The town is also a large base for the Flying Doctor Service and the School of the Air – radio-linked classes for children on remote outback properties without access to any schools.
Population 20 450
Map Australia – Ec

Aligarh *India* University town about 135 km (84 miles) south-east of Delhi. It contains the Muslim University, established in 1920 from the Anglo-Oriental college founded in 1875 by the Muslim reformer and educationist Saiyad Ahmad Khan, who modelled the buildings on the colleges of Oxford and Cambridge in England. It has a metal works and factories making carpets, locks and cotton.
Population 3 296 760
Map India – Cb

Ali-Sabieh *Djibouti* Settlement on the rail link between the port of Djibouti and the Ethiopian capital, Addis Ababa. In the early 1990s it was overwhelmed by tens of thousands of refugees fleeing drought and civil war in Ethiopia.
Map Ethiopia – Ba

Al-Kharijah *Egypt* See EL KHARGA

Al-Kharj *Saudi Arabia* Resort region 90 km (56 miles) south of the capital, Riyadh, through which pass the main road and railway to Dammam. It is a playground for the capital, with palm groves, gardens and other summer picnic spots. It also has a number of farms, including the Saudi Agriculture and Dairy Company which is said to be the largest dairy project in the Middle East. Al-Kharj is one of the country's major wheat-producing areas, relying heavily on extensive overhead sprinkler irrigation.

The main towns are As Salamiyah, Al Yamamah and Sulaymaniyah.
Map Saudi Arabia – Bb

Alkmaar *Netherlands* Town and industrial centre 33 km (20 miles) north and slightly west of Amsterdam. It is a tourist centre, with many of its 14th to 17th-century buildings and canals intact. The weekly cheese market is still held from the end of April to mid-September, when dealers, dressed in 17th-century costume, move the round, yellow cheeses on sledges. The town's products include machinery, chocolate, organs and paper.
Population 92 420
Map Netherlands – Ba

Allada *Benin* Market town about 50 km (30 miles) north-west of Cotonou. It was the capital of an old kingdom decimated by slave traders in the 18th and 19th centuries.
Map Benin – Bb

Algeria

FOLLOWING A SAVAGE WAR OF INDEPENDENCE WITH FRANCE, ALGERIA FINDS THE TRANSITION TO DEMOCRACY DIFFICULT

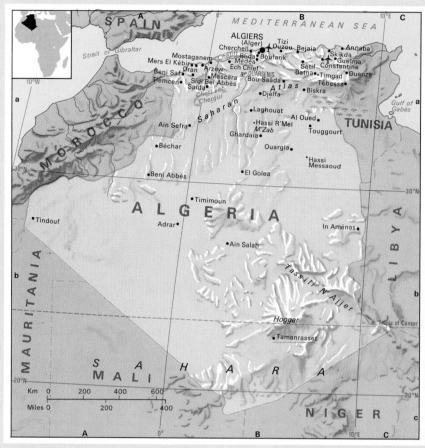

Providing a gateway to the SAHARA, the world's largest desert, Algeria is remembered as the home in times past of the elite French Foreign Legion. Immortalised in P.C. Wren's novel *Beau Geste*, the Legion fought Berber tribesmen under a blazing sun in the arid desert wastes. But today the desert has a new image.

Over four-fifths of Algeria is covered by the Sahara, and it was beneath its unproductive surface that enormous oil and gas fields were discovered in the 1950s. Today, Algeria is the third largest oil producer in Africa – only Nigeria and Libya produce more. Reserves of crude oil are estimated at 9.2 billion barrels, and Algeria also possesses the fourth largest reserves of natural gas in the world – a staggering 3090 thousand million m^3 (109 122 thousand million cu ft).

THE FACE OF THE LAND

Algeria is a huge country – the second largest in Africa (only Sudan is bigger) and the 11th largest in the world. It has two main geographical divisions – the ATLAS MOUNTAINS in the north and the Sahara to the south. The Atlas mountains – several parallel ranges running east-west with valleys between them – consist of three main areas: the Tell Atlas near the coast, the high plateaus, and the Saharan Atlas.

The Tell Atlas form a belt of narrow ranges, plateaus and massifs, between which are fertile lowlands, including the valley of the Cheliff, Algeria's longest river, the alluvial Mitidja plain around the capital ALGIERS, and the ANNABA plain – all of which form Algeria's chief farming regions. The Tell Atlas contain most of Algeria's people and all its larger towns – ORAN, CONSTANTINE, AL ASNAM (Ech-Cheliff, Orleansville), BLIDA, Annaba (Bône), as well as Algiers.

To the south are the high semiarid plateaus containing the Chott ech CHERGUI and the Chott al Hodna, large desert depressions with salt marshes. Farther south lie the Saharan Atlas mountains, which extend east of the Chott al Hodna as the Aures massif, rising to 2326 m (7631 ft). A line of oases, among which BISKRA is the most heavily populated, lies along the southern flanks of the Saharan Atlas.

The Sahara itself stretches for nearly 5000 km (more than 3000 miles) from the Atlantic to the Red Sea, and dominates 11 African countries. In the Algerian Sahara, great dunes of sand called *ergs* rise like waves in the north and east; the rest is covered with gravel or rocks. In the south, the HOGGAR massif reaches 2918 m (9573 ft) at Mount Tahat.

The climate on the coast is similar to that of the European countries that fringe the Mediterranean to the north – hot, dry summers and cool, moist winters. Summer temperatures can climb to 32°C (90°F) when the hot, dry, dusty wind known as the sirocco blows from the Sahara. Most of the rain falls between October and March. In the mountains, the weather is cooler and affected by altitude; some of the highest peaks are snow-covered in winter. Little or no rain falls in the Sahara; temperatures may reach 49°C (120°F) during the day but fall to 10°C (50°F) at night.

ALGERIA'S MANY INVADERS

The earliest known inhabitants were the Berbers, who became established across North Africa from about 3000 BC. About the 3rd century BC, Numidia, as the Berber kingdom was called, came under attack from the Carthaginians and later the Romans. Roman rule ended when Vandals from Spain swept across the country in AD 431. In the following century, the area was partly conquered by the Byzantines. Successive waves of Arabs swept over the country in the 7th century and introduced the Islamic religion and the Arabic language. Islam soon became dominant but the Berber language survived mainly in the mountains in the east of the country. From 1519, the Ottoman Turks held the coastlands, but their control was, at best, weak.

In 1830, on the pretext that the coastline was a haven for pirates – it was – French troops invaded the country, and in 1831, the French Foreign Legion was formed. Decades of fighting against the Arabs and Berber tribesmen followed. Most of the country was subdued by the early 1870s, though it was not until 1902 that France achieved control of the whole territory. In the period that followed, the French treated northern Algeria as an extension of France rather than as a colony, and settlers *(colons)* flocked to the country.

After the Second World War, a nationalist movement which had emerged during the 1920s and 1930s began to grow in strength. One major group, the FLN (the National Liberation Front), became dominant and in 1954 launched a bitter struggle for independence. By 1956 it controlled most of the countryside and was making terrorist attacks on the towns. The French fought back ruthlessly with 500 000 troops to protect the settlers.

In 1958, the French army in Algeria, dissatisfied with the conduct of the war, rebelled and helped General Charles de Gaulle to return to power. Both the army and the *colons* believed that de Gaulle would keep Algeria French. However, despite last-ditch stands, including counter-terrorism by the OAS (a secret organisation of army officers and *colons*), de Gaulle realised that Algeria could not be controlled by force indefinitely, and in 1962 independence was granted. The antagonism felt for the French by the Algerians resulted in more than a million settlers, most of them French, fleeing to France.

The first president of the new republic was Ahmed Ben Bella, an erratic left-wing radical. In 1965 he was deposed by his defence minister, the austere Houari Boumedienne. Foreign petroleum companies were nationalised and in

▲ **DESERT HEIGHTS** Glowing pinkish-brown in the sunlight, the Hoggar massif, in the south of the Algerian Sahara, reaches a peak of over 2900 m (more than 9500 ft) – the highest point in the country.

1976 Algeria, already a one-party socialist state, adopted a national charter which reaffirmed the goal of building a socialist society. President Boumedienne died in 1978 and was succeeded by Colonel Bendjeddid Chadli, who relaxed Algeria's rigid socialist policies. His government established a multiparty system in September 1989, and as of 31 December 1990, more than 30 legal parties existed.

In December 1991 the first free parliamentary elections in Algeria brought a 55 per cent victory to the Fundamentalist Islamic Salvation Front (FIS). The government cancelled the second round of elections in January 1992, nullified the first, and banned the FIS. President Chadli resigned and the military stepped in. Thousands of Islamic fundamentalists were arrested by the newly formed security council and detained in makeshift desert prisons. Meanwhile, the National Assembly had been suspended and a High Council of States headed by Mohamed Boudiaf ruled the country. Boudiaf was assassinated in June 1992 and succeeded by Ali Kafi.

This marked the beginning of a period of bloody confrontation – mainly between the Armed Islamic Movement (AIM) and the military. By late 1993, a political stalemate had been reached. Meanwhile, the public had lost all confidence in the government's ability to keep the peace. General Lamine Zeroual was appointed president in January 1994, following the dissolution of the High Council. Zeroual, at first willing to negotiate with the fundamentalists, dropped his dialogue policies with the rapid escalation of violence. In early 1994 the fundamentalists also began targeting foreign residents, who fled the country in their thousands – notably to France.

The situation was further compounded by two newly formed anti-Islamic organisations which joined in the fighting and claimed responsibility for numerous acts of terrorism. The country soon reached a political stalemate, with the exiled FIS demanding that all political prisoners be freed, and the Armed Islamic Movement calling for the punishment of all those who oppose Islam, 'especially journalists and foreigners'. Meanwhile, antifundamentalists threatened retaliation by counter-killing.

OIL-BASED ECONOMY

Algeria has many natural resources. Its crude oil refining capacity is the largest in Africa. Oil and gas account for 98 per cent of all exports and provide more than 65 per cent of revenue. Oil revenues and loans from abroad have financed major industrial developments, which include iron and steel plants as well as a substantial petrochemical industry. Iron ore is mined near Beni Saf, Zaccar Timezrit and near the eastern border at Ouenza and Boa Khadra. There is a steel complex at Annaba, gas liquefaction plants at ARZEW and SKIKDA, and several oil refineries and petrochemical plants. Algeria is one of the few African countries to have its own car-manufacturing industry and the country makes one-third of its commercial vehicles. Other industries include food processing, building materials, chemicals and textiles.

But over the years, corruption and mismanagement had virtually bankrupted the country, and the violence began scaring off investors. By 1993, Algeria had a foreign debt of US$27 billion – and in April 1994, the government took the decision to devalue the Algerian dinar by 28.6 per cent. The country faces a serious economic crisis – and though it survived the oil recession of the late 1980s and early 1990s, and the consequent fall in crude oil prices, better than many other oil-producing countries, and embarked on economic reform under Prime Minister Sid Ahmed Ghozali, most steps towards privatisation and a market economy were halted when Ghozali left politics in 1992.

A large proportion of Algerians (23 per cent) are still farmers, although arable land accounts for only 3 per cent of the country's total area. The rich coastland produces wheat, vines and olives, as well as early fruit and vegetables for export to the European market. Dates are also an important agricultural export – Algeria is the sixth largest producer in the world. In the mountains, where about 1 million people, mostly Berbers, live, more than 13 million sheep, as well as cattle and goats, graze on the plateau grasslands.

The rapid growth of industry has led to large numbers of people migrating from the countryside to urban areas. By the beginning of the 1980s, more than 60 per cent of the population lived in towns. Large-scale unemployment – as well as hopelessness – have turned many young Algerians militant.

ALGERIA AT A GLANCE

Official name People's Democratic Republic of Algeria
Area 2 381 741 km² (919 590 sq miles)
Population 29 000 000
Per km² 12 (**Per sq mile** 32)
Capital Algiers
Government Unitary republic under military-controlled transitional government
Currency 1 Algerian dinar = 100 centimes
Languages Arabic (official), Berber and French (commercial language)
Religions Muslim (Sunni) 99%
Climate Mediterranean on the coast, hot and dry in the south. Average temperature in Algiers ranges from 12°C (54°F) in January to 24°C (75°F) in July. Average midday temperatures in the south are 35-40°C (95-104°F)
Land use Grazing 13%, cultivation 3%, forest and woodlands 2%, other 82%
Main primary products Cereals, grapes, olives, citrus fruits, dates, vegetables, livestock, timber, fish; oil, natural gas, iron ore, zinc, phosphates, copper, lead, coal, salt
Major industries Oil and natural gas production and refining, petrochemicals, mining, cement, iron and steel, fertilisers, transport equipment, machinery, agriculture, wine production, food processing
Main exports Crude oil, petroleum products, natural gas, wine, fruit, vegetables, dates
Annual income per head (US$) 1830
Population growth (per thous/yr) 29
Life expectancy (yrs) Male 66 **Female** 68

Allahabad *India* City at the confluence of two holy rivers, the GANGES and the YAMUNA, midway between Delhi and Calcutta. Each year during the Hindu month of Magha (January/February), hundreds of thousands of Hindus visit the city to bathe in the holy waters and so purify their souls. In 1954 pilgrims stampeded to enter the waters and about 350 were crushed to death.

The Aryans, a prehistoric people from Central Asia who invaded India around 1500 BC, founded a city on the site called Prayag – 'the Place of Sacrifice' – which is still revered. The Buddhist emperor Asoka built a stone edict pillar there in 242 BC which still stands today, and the Mogul emperor Akbar built the fort there in 1575. In the 19th century, Allahabad became a British provincial capital, and was the scene of fierce fighting in the 1857 Indian Mutiny. The Indian National Congress (later to become the ruling Congress Party of India) was founded there in 1885.

There is a Great Mosque and a university, founded in 1887. Industries include flour milling and textiles, and a flourishing trade in sugar.
Population 4 909 920
Map India – Cc

Allegheny Mountains *USA* Part of the Appalachian system stretching 800 km (500 miles) from south-west Virginia to northern Pennsylvania and forming the most north-westerly main range of the system. They vary in height from about 600 m to 1481 m (2000 ft to 4860 ft), which they reach at Spruce Knob in West Virginia. They are rich in timber and coal and also contain iron ore, petroleum and natural gas.
Map United States – Jc

Allgäu *Germany* Alpine region of south-west Bavaria between Lake Constance, on the Swiss border, and the town of Füssen, about 75 km (45 miles) to the east. It is a skiing area, and around the town of Kempten lies Germany's main cheese-producing area.
Map Germany – De

Allier *France* Navigable river, 410 km (255 miles) long. It rises in the southern highlands of the massif central and flows north through a series of basins and the fertile Limagne plain to join the Loire near Nevers.
Map France – Ec

alluvial cones and fans Fan-shaped accumulations of alluvium deposited by mountain streams at the mouths of ravines onto adjacent plains. Fans have gentle slopes, of up to 6 degrees, while cones are steeper, up to 15 degrees. They are often found in arid and semi-arid areas.

alluvial plain See FLOOD PLAIN

alluvial terrace See RIVER TERRACE

alluvium Eroded particles of rock, usually sand and silt, transported by a river and deposited on its bed, its flood plain or in its delta or estuary. Much of the world's most fertile land consists of alluvium.

Alma-Ata (Almaty) *Kazakhstan* Capital city, 75 km (46 miles) north of Lake Issyk-Kul' near the Chinese border. A commercial and industrial centre in an agricultural and fruit-growing region, it trades in wheat, sugar beet, apples and grapes. Its industries include machinery, railway equipment, tanning, saw-milling, spinning, textiles, tobacco and food processing.

Alma-Ata was founded by the Russians as a fort in 1854 and grew rapidly after 1926 with the building of the Turkestan-Siberian Railway. In 1887 and 1911 it was virtually destroyed by earthquakes, but was rebuilt each time . It has an academy of science and several colleges and research institutes and is the country's cultural centre.
Population 1 151 300
Map Kazakhstan – Db

Almadies Point *Senegal* See VERDE, CAPE

Almaty *Kazakhstan* See ALMA-ATA

Almería *Spain* Ancient port 275 km (170 miles) east and slightly north of Gibraltar. It was used as a natural harbour by the Romans and was a base in the Middle Ages for Moorish pirates until the Moors were expelled from Spain in 1492. A mighty fortress from those days still remains. Modern Almería processes salt and exports grapes, fruit, iron and lead from the arid mountainous province of the same name. An agricultural centre, including hothouses and irrigation, has been built and the vegetables and fruit which ripen early in the season are exported to central European countries.
Population (town) 155 120; (province) 455 500
Map Spain – Dd

Almourol *Portugal* Romantically located castle on an island in the Tagus River, 105 km (65 miles) north-east of the capital, Lisbon. It was built in 1171 by the Knights Templar, the Crusaders who helped to drive the Moors from Portugal in the 13th century. The island was given to the knights as a reward by King Alfonso I (ruled 1139-85).
Map Portugal – Bc

Al-Oued (Al-Wad) *Algeria* Oasis town in the north-east corner of the country near the border with Tunisia. It is a major source of dates.
Population 52 000
Map Algeria – Ba

alp A shoulder high on a mountain side, especially the gentle, grassy slope above a U-shaped glaciated valley, often used as summer pasture. The term originally applied to such high slopes in the ALPS, but is now also applied elsewhere.

Alps *Southern and central Europe* A magnificent complex of mountains, lakes and glaciers, the Alps stretch for almost 1200 km (746 miles) in a great crescent between 130 and 220 km (81 to 137 miles) wide across southern Europe. They start near the Ligurian coast behind Monaco and end in the Hungarian lowlands. In between they take in south-eastern France, Liechtenstein, northern Italy, Switzerland, south-west Germany, and Austria.

The scenic attractions, winter sports and mountaineering activities they provide have lured generations of tourists and sportspeople from all parts of the world. Chamonix, Interlaken, Grindelwald, St Moritz, Zermatt and many other resorts are internationally famous. Some stunning lakes – including Geneva, Como, Lucerne and Garda – are strewn like jewels among the dazzling peaks. Highest peak of all is Mont Blanc (France), reaching 4807 m (15 771 ft) near Geneva; more daunting challenges can be found on the Eiger (Switzerland), 3970 m (13 025 ft) in the Bernese Oberland, and the mighty Matterhorn (Switzerland-Italy), towering 4478 m (14 692 ft) near Zermatt.

Nine major passes, six long tunnels (one under Mont Blanc itself) and numerous gaps ensure that the Alps do not form a total barrier between north and south. Cattle, sheep and goats graze the summer pastures below the snowline; vineyards are cultivated on the warmer, lower slopes. Many of the region's rivers and torrents have been harnessed to generate hydroelectric power for towns and villages.

The complex folds, ridges and pinnacles of the Alps were formed by earth movements about 40 to 60 million years ago, and further shaped by glaciation around 2 million years ago. The Alps still rise by approximately 1 mm (0.039 in) a year, but this is undone in about the same proportion by erosion. Rocky debris scoured from the land by glacier movement created moraines, or dams, to form long, deep lakes. Glaciers – the largest of which is the 24 km (15 mile) long Great Aletsch Glacier in the Bernese Alps – are the sources of many streams and rivers and cover a total area of 2909 km² (1123 sq miles).
Map Europe – Dd

Alsace *France* North-eastern region and former province, bordered by the Rhine and the Vosges Mountains. With LORRAINE, it was occupied by Germany from 1870 to 1919 and from 1939 to 1944. Its architecture and traditions reflect both French and German influences.

The fertile plain of Alsace is part of the Rhine Rift Valley. Famous white wines are produced from Riesling, Gewürztraminer and Sylvaner grapes in villages on the eastern slopes of the Vosges. The Grand Canal d'Alsace, running parallel to the Rhine, is navigable; and 10 power stations on the Rhine generate hydro-electric power for the region, which is a major industrial area. The main cities are STRASBOURG, MULHOUSE and COLMAR. Alsatian dogs – also known as German Shepherds – got their name from Alsace, where they were first bred and used as sheepdogs.
Population 1 624 000
Map France – Gb

Altai (Altay) *Central Asia* Group of mountain ranges where the borders of China, Kazakhstan, Russia and western Mongolia meet. It stretches more than 1600 km (1000 miles) south-eastwards from southern Siberia into the Gobi Desert. There are many remote summits more than 3000 m (9840 ft) high, including Mount Belukha (4506 m, 14 783 ft) in Siberia, and Taban Bogdo Ula (Youyi Feng; Mount Kuytun), which reaches 4356 m (14 290 ft) where the four borders meet.
Map Asia – Gd

Altamira *Spain* See SANTANDER

Altay Kray *Russia* Administrative territory *(kray)* of the Russian Federation in the basin of the upper Ob' River bordering on China and Mongolia. Deposits of gold, silver, lead, zinc and copper are mined in the south, in the ALTAI Mountains. The north is cultivated for wheat, maize, sugar beet and oilseeds. BARNAUL is the main city.
Population 2 728 000
Map Russia – Jc

Altiplano *Bolivia* Windswept, treeless plateau with an average altitude of 3670 m (12 040 ft). It lies between Bolivia's two main mountain ranges: the Western and Eastern Cordilleras of the Andes. It covers about 9 per cent (102 300 km², 39 000 sq miles) of the country and contains half its people. It is mainly arid, except around Lake TITICACA in the north. The typical animal found there is the llama, which is a source of meat, wool and leather, and is also used as a high-altitude pack animal. The potato, which is native to the region, is the staple diet of most of the Amerindian population.
Map Bolivia – Bb

altocumulus A middle-altitude cloud formation of bands of dense, fleecy balls or rolls, occurring 3-6 km (2-4 miles) above the earth. It usually indicates changeable weather.

altostratus A middle-altitude cloud formation of thick grey, or bluish sheets or layers, occurring 3-6 km (2-4 miles) above the earth. It usually indicates the approach of a warm front, and the arrival of rain within a few hours.

Altun Shan *China* Mountain range with peaks of more than 4000 m (13 000 ft) forming the northern edge of the Tibetan Plateau.
Map China – Cd

Aluvihara *Sri Lanka* See MATALE

Amacuro Delta *Venezuela* See ORINOCO

Amager *Denmark* Flat and fertile island in the Öresund (Danish Sound), immediately to the south of COPENHAGEN. Much of it is occupied by commercial and urban development which is an extension of the capital. Kastrup, Denmark's principal airport, built in 1925, lies here, only 10 km (6 miles) from the centre of Copenhagen. The Dutch were invited in 1521 to settle and reclaim low-lying Amager from the sea. The island has thriving horticulture and market gardening businesses as well as specialised handicrafts such as embroidery, weaving and ceramics. Dragør, the ferry port to Sweden, retains in its architecture something of the island's earlier picturesqueness
Population 149 500
Map Denmark – Cb

Amalfi *Italy* Historic town on the Bay of Salerno, about 60 km (38 miles) south-east of Naples, reached by a coastal road that is considered to be the most beautiful in Italy. Amalfi was one of the earliest Italian maritime republics, flourishing as early as the 9th century. Today it is an attractive tourist resort hemmed in by rocky hills and cliffs, which have prevented modern growth. The town's jewel is its cathedral, founded in the 9th century, remodelled in a striking Arab-Norman design in 1206 and modified again in the 18th century. The crypt holds the tomb of Andrew, brother of St Peter.
Population 5590
Map Italy – Ed

Amarillo *USA* City in north-west Texas, about 110 km (70 miles) east of the New Mexico border. It is a centre for the surrounding grain and live-stock-producing area, and for the area's coal and gas production. It has one of the world's biggest helium plants.
Population (city) 157 620; (metropolitan area) 187 550
Map United States – Fc

Amasya *Turkey* Market town about 300 km (185 miles) east of Ankara, and the capital of a province of the same name. The town was the capital of the kingdom of Pontus, which flourished from the 4th to the 1st centuries BC, and the tombs of the Pontic kings, including that of Mithradates the Great, can be seen hewn in rocky hills overlooking the town. The surrounding province produces apples, wheat, onions, tobacco, opium, wool, hemp, sugar beet and small amounts of lead, gold and silver.
Population (town) 53 200; (province) 358 900
Map Turkey – Ba

Amazon *South America* Second longest river in the world, after the River Nile. It is about 6440 km (4000 miles) long from its source in the Peruvian Andes to its delta in the northernmost corner of Brazil, and drains an area almost the size of Australia. For nearly half its length it flows through the world's largest rain forest – a jungle covering some 6.5 million km² (2.5 million sq miles), an area larger than the whole of Western Europe. In recent years, vast swathes of the forest have been destroyed as developers have moved in search of timber and minerals, and to clear the land for agriculture. As the forest has died, so have many of the Amerindians who lived here. Once they numbered more than a million; now only tens of thousands survive. Although attempts are being made to limit further destruction of the forest and to protect the remaining Amerindians, the destruction goes on.

Numerous huge rivers swell the Amazon's volume: the Negro, Branco and Japurá enter it from the north, and the Juruá, Purus, Madeira, Tapajós, Xingu and Tocantins from the south. From the Peruvian border the main river is known as the Solimões until it joins the Negro – itself 18 km (11 miles) wide – near the city of Manaus. Because of their speed, the two great rivers flow side by side for roughly 6 km (4 miles) – the acid

▼ **IMPENETRABLE FOREST A tributary snakes its way to the mighty Amazon – the easiest route for travellers in the vast rain forest of northern Brazil.**

Negro stained black like tea by the rotting leaves in the swamps of its early course through Colombia, contrasted with the yellow-brown Solimões – before the waters finally merge.

The huge river is navigable by ships of up to 6000 tonnes throughout its length in Brazil and beyond the border to the Peruvian jungle port of Iquitos, 3700 km (2300 miles) from the sea. For much of this course it is difficult to make out the banks, from the ships sailing down the middle of the river, because they are so far apart. The delta is made up of several channels, of which the main one is about 50 km (30 miles) wide. The volume of silt transported by the river is so great that it stains the sea for more than 200 km (124 miles) offshore.
Map Brazil – Cb

Ambato *Ecuador* City 140 km (85 miles) south of Quito, at the foot of the inactive volcano, Chimborazo, which is 6310 m (20 700 ft) high. Ambato was largely rebuilt after a severe earthquake destroyed much of the city and killed as many as 6000 people in 1949. Rugs and textiles are manufactured here.
Population 124 170
Map Ecuador – Bb

amber Translucent, yellowish fossilised resin from coniferous trees, found chiefly along the shores of the Baltic Sea. It is used for making ornaments and jewellery.

Ambon (Amboina) *Indonesia* Island of 813 km² (314 sq miles) in the Maluku (Molucca) group, south of the Philippines. Since the 16th century, it has been famous as the centre of the European spice trade, and cloves and nutmeg are still two of its main products. The Ambonese are one of Indonesia's few Christian communities. They resisted incorporation into the state of Indonesia in 1950 and many fled to the Netherlands, where they proclaimed the short-lived Republic of the South Moluccas.
Population 73 000
Map Indonesia – Gc

Amboseli *Kenya* Lake on the Tanzanian border, south of Nairobi. It is a dry bed of soda most of the year. The name has been given to a 380 km² (147 sq mile) national park east of the lake, and to a 3000 km² (1160 sq mile) game reserve, which is the home of Masai pastoralists and a safari centre for tourists.
Map Kenya – Cb

Ambre Mountains National Park *Madagascar* Humid forest park covering 18 200 hectares (about 44 970 acres) in the north, which is noted for lemurs and orchids. The volcanic hills rise to 1475 m (4839 ft).
Map Madagascar – Aa

Amer (Amber) *India* A series of hilltop palaces of the Maharajah of Jaipur, dating from about 1500 onwards, north of the city of Jaipur. It has ingenious means to store rainwater in a hollowed-out mountain top which has been roofed over. Amer has its own cannon foundry. A massive cannon said to be capable of hurling a ball 40 km (25 miles) stands untested here.
Map India – Bb

America Some old-style textbooks still have it that America is a single continent. Most modern geographers divide the landmass into two, NORTH and SOUTH AMERICA, with CENTRAL AMERICA usually separately identified.

America, Central Nearly a thousand years before Europeans came to what was to them the New World, the Mayas, or Mayans, had evolved a sophisticated culture in Central America, with

advanced art, architecture, engineering, mathematics and astronomy. Their still impressive buildings survive in the peninsula of Yucatán, and other parts of southern Mexico, and in much of Guatemala and western Honduras. In Mexico they built their splendid city-states of Chichén Itza and Uxmal, supported by a well-developed economy based on agriculture, commerce and crafts. But by 1519, when the Spanish arrived, the Mayans were already in decline.

Central America is sometimes defined as the land once occupied by the Mayas, but is more conveniently defined as the area between the northern border of Mexico and the Panamanian-Colombian border in the south. It therefore includes the nations of Mexico, Guatemala, Belize, Honduras, El Salvador, Nicaragua, Costa Rica and Panama, a largely uneasy grouping of buffer states along the isthmus between North and South America.

A chain of volcanic mountains along the western side of Central America connects the coastal range of California to the South American Andes. Many of the volcanoes are still active, among them Tajumulco in Guatemala (4210 m, 13 810 ft), the highest point in Central America. The whole area is subject to earthquakes. One particularly calamitous shock hit Mexico City in September 1985, killing about 10 000 people and destroying hundreds of buildings. Earthquakes caused serious damage to Nicaragua's capital, Managua, and to Guatemala City, in 1917, 1918, 1931 and 1972, and to Guatemala City again in 1976.

The climate of the area is tropical, although the mountains and plateaus are cooled by the altitude. North-east trade winds bring rain from the Gulf of Mexico, which falls on northern slopes. Around July the rainfall is particularly heavy, San Juan del Norte (Nicaragua) receiving a summer total of 3632 mm (143 in). Much of Central America's natural vegetation is dense rain forest. On the western, Pacific side, rainfalls are significantly lower and much of the vegetation is savannah. Cotton, coffee, sugar, beef, hardwoods and fishing have been mainstays since colonial times, and the gold and silver mines that enticed Spanish settlers are still worked.

Area (including Mexico) 2 481 364 km²
(957 840 sq miles)
Population 120 million

America, North Stretching more than 6500 km (more than 4000 miles) north to south, from the Arctic Ocean to the Gulf of Mexico, and more than 8000 km (5000 miles) from the western extremity of Alaska to Newfoundland, North America is the world's third largest continent. Though topographically and geologically it embraces all the land between Greenland (Kalaallit Nunaat) in the north-east and Panama in the south, most people refer to North America as the Anglo-American cultural region – in other words, to Canada and the United States, but not to Greenland and, further south, Mexico.

Despite intensive development, and overcrowding in some places, North America is relatively empty, with virgin wildernesses of tundra, mountain and desert.

The Atlantic coast in the east has a narrow coastal plain, narrowest and most densely populated in the north, where it is also most indented. Behind this, from Newfoundland to Alabama, lie chains of mountains: in the south, the Blue Ridge, the Alleghenies and Cumberlands form

parts of the Appalachian system, while farther north lie the Catskills and the New England ranges. Beneath the western edge of the Appalachians lie rich coal deposits.

North of the east coast mountains, and occupying a vast area of eastern Canada and the north-eastern United States, is the Canadian Shield, a rugged 4.8 million km² (1.9 million sq mile) plateau of low, worn hills. The southern part of the Shield is clothed in immense coniferous forests; to the north is the tundra, where the subsoil is permanently frozen, permitting only lichens and mosses to grow.

West of the eastern mountains lie the great central lowlands of North America, extending from the Mackenzie River in Arctic Canada to the subtropical Mississippi basin in the southern United States. The eastern part of the lowlands divides into a dairy-farming belt in the north, around the Great Lakes, a corn (maize) belt in the centre, and a cotton-growing area in the south. Farther west, and straddling the Canada-United States border, is a great area of spring and winter wheat, grown on what was once tall grass prairie. It also contains a wealth of oil and natural gas. Farther west again, stretching from Alberta in the north to Texas in the south, is semi-arid cattle country, which it is unwise to plough or plant except where irrigation water is available.

Far to the south, the 6019 km (3741 mile) long Mississippi-Missouri river system empties into the Gulf of Mexico, a beach-lined bay of the Atlantic sheltered by the long, low arm of Florida. Below the waters of the Gulf of Mexico lies a rich reservoir of oil, one of many such deposits which make North America a leading producer of crude oil. The south-western United States is arid, and driest in the deserts of Nevada and southern California. Across the region runs the monstrous slash or scar of the Grand Canyon of the Colorado River, 347 km (216 miles) long, and up to 29 km (18 miles) wide.

North America is dominated by the Rocky Mountains, the second longest mountain range in the world. The Rockies form the continental divide, separating drainage to the west and to the east.

To the west are the desert basins and then, guarding the approaches to the Pacific coast, the Sierra Nevada and Cascade ranges. These and the Rockies are part of a chain of mountains running down the entire length of North, Central and South America, from Alaska – where they reach their highest peak at Mount McKinley (6194 m, 20 320 ft) – to Tierra del Fuego at the tip of the Americas. The mountainous areas of British Columbia, Washington State and Oregon are famous for timber production, while central and southern California grow huge quantities of soft fruits, citrus, grapes, cotton and rice for consumption in the United States and for export.

North America has more than its share of nature's dramas. In 1980, the Mount St Helens volcano in Washington State spectacularly blew off its top. The Gulf of Mexico usually presents the eastern seaboard with at least one savage hurricane a year, and in San Francisco pessimists expect to be swallowed up at any moment in the San Andreas Fault – the line where two continental plates meet, creating an area prone to earthquakes.

The climate, too, has a tendency towards the dramatic, and is extremely varied. Snag, in Canada's Yukon, boasts North America's record

low temperature of –62°C (–80°F), and Death Valley in California its record high of 56.6°C (134°F). The High Plains have dust storms in summer and blizzards in winter. New York, like Washington, DC, can be steamy and humid in August, but it has a January wind honed to razor sharpness. Montreal and Buffalo record huge snowfalls, and Chicago's midwinter temperatures plunge in its mid-continental position, helped by a vicious wind whipping off the frozen Lake Michigan. Ellesmere Island, reaching out towards northern Greenland, has polar conditions, while Florida is subtropical and swampy.

Area (excluding Mexico and Greenland)
19 343 000 km² (7 466 610 sq miles)
Population 277 million

America, South The continent – or arguably, subcontinent – is usually considered to include all the countries below the southern border of Panama. An alternative term, Latin America, is generally held to include not only South America but also all the nations of CENTRAL AMERICA – an area where most of the countries have ties of Spanish or Portuguese language and culture. Because it is so isolated, Europeans did not learn of South America's existence until late into the 15th century – but when they did, they hastened to bring it under their control, and today the region's culture still has a strong resemblance to Spanish and Portuguese culture.

Unquestionably, South America's most striking feature is the great mountain chain of the Andes which runs from Venezuela in the north along the Pacific coast to Tierra del Fuego in the south, a distance of more than 7500 km (4660 miles). This cloud and snow-capped range spreads out to 600 km (373 miles) wide in Bolivia. Several of South America's greatest rivers, the Orinoco and the Amazon among them, rise in the Andes.

In the north and east of the continent are the Guiana and Brazilian highlands, a mass of block-like mountains and plateaus, some of which end in abrupt scarps. In Brazil lies the vast Amazon basin, laced by the mighty river, together with its many tributaries.

Rainfall varies greatly throughout the continent. The Amazon's rainfall is the heaviest (an annual 1500-2500 mm, 60-100 in), deposited by trade winds from the Atlantic. By the time they reach Peru and Chile, however, these winds have dropped their moisture, and there are parts of the Atacama Desert in Chile that have had no rain at all within living memory. Even the Peruvian capital, Lima, has no more than 40 mm (1.5 in) of rain every year.

Savannah – tropical grassland – lies to the north and south of the Amazon rain forest, which is now in danger of being chopped down faster than it can be replanted. Little effort is being made to reforest large tracts of eroded land. Farther south, the rainfall is heavier, the grassland more lush, and the soil more fertile. This area includes the *campo limpo* of the extreme south of Brazil, the purple grassland of Uruguay, and the *pampa* of Argentina. A large part of it is farmed for cereals, and it incorporates some of the finest grazing land anywhere in the world.

Area 17 832 000 km² (6 883 347 sq miles)
Population 302 million
Map See p 32

American Samoa See SAMOA, AMERICAN

Amersfoort *Netherlands* Industrial city 42 km (26 miles) south-east of Amsterdam. The medieval city centre was once moated and walled, but now only the gates remain. It is dominated by the 95 m (312 ft) Tower of Our Lady, built in the 15th century. The city's manufactures include metal goods, chemicals and foodstuffs.

Population 106 920

Map Netherlands – Ba

amethyst Purple or violet variety of transparent quartz, used as a gemstone. The purple colour comes from manganese in the quartz.

Amiens *France* Capital of the Somme department and an industrial and administrative city on the Somme River, 148 km (92 miles) north of Paris. In 1802, during the Napoleonic Wars, it was the scene of the signing of a peace treaty between Britain, France and their allies. However, the peace lasted barely a year – the war ended only with Napoleon's defeat at the Battle of Waterloo in 1815.

During the Second World War, the capture of Amiens by the Germans in 1940 led to the German blockade of the Allies in northern France and the evacuation of British, French and Belgian troops from Dunkirk.

The Gothic cathedral of Notre Dame is the largest and one of the most beautiful in France. Its industries include textiles, clothing, tyres and chemicals. A university was opened here in 1964.
Population 131 400
Map France – Eb

Amindivi Islands *India* See LAKSHADWEEP

Amman *Jordan* Capital and administrative and business centre of the country, situated 38 km (23 miles) north-east of the Dead Sea. The area around Amman has been the focus of human settlement since about 4000 BC. It was occupied during Biblical times and was conquered by Ptolemy II Philadelphus (ruled 285-246 BC) who renamed it Philadelphia. It continued to be occupied during Roman times. In AD 635 it was taken by the Arabs and by 1300 the city had disappeared. Popular lore has it that Circassian settlers from Russia arrived in the area in 1878; certainly they had built a village here by the 1880s. It developed a reputation as a commercial centre that was much enhanced by the completion in 1905 of the northern section of the HEJAZ railway which passed only 5 km (3 miles) east of the village. The status of Amman was settled in 1921 when the British established a protected emirate called Transjordan.

The Emir Abdullah, after trekking round the new emirate, established himself in Amman and made it his capital. The city, which is built on seven hills, began to expand rapidly after Jordanian independence in 1946.

Amman caters for an increasing flow of business, banking and diplomatic visitors, as well as many Palestinian refugees. It is the hub of the country's road and rail communications and has an international airport as well as the

▼ AMRITSAR The city's famous Golden Temple, its upper walls lined with beaten gold, is the spiritual centre of India's Sikh religion.

Queen Aliya International Airport at Zizya 30 km (19 miles) to the south. Despite Amman's modernity, it is easy to get lost in the city's many narrow, twisting streets and alleys. To cater for its growing cosmopolitan community, there are a number of special schools, such as the American Community School, as well as a range of foreign institutes – including the British Council and the French, American, Spanish, Turkish and Russian cultural centres. Amman is also the seat of the University of Jordan.
Population 1 625 000
Map Jordan – Ab

Amritsar *India* Manufacturing city and religious centre of the Sikhs, 446 km (277 miles) north-west of Delhi. It was founded in 1577 by the fourth guru (spiritual leader) of the Sikhs, Ram Das – and contains the Golden Temple, the Sikhs' most sacred *gurdwara* (temple). The shrine, a large complex of buildings built around a sacred rectangular lake, was occupied by armed Sikh extremists in 1983. The following year, Indian government troops stormed the temple and drove out the occupants in a bloody encounter.

An earlier violent episode in the city's history was the notorious massacre in 1919, when troops under the British general Reginald Dyer fired into a crowd of unarmed people protesting against anti-sedition laws. About 400 of the protesters were killed and many others injured.

The city is renowned for its crafts, particularly fine woollen cloths and carpets.
Population 2 501 730
Map India – Bb

Amsterdam *Netherlands* The country's capital, commercial centre, second port and largest city. It is built on piles, in sand and mud, where the Amstel River flows into IJsselmeer, once a branch of the sea, and is named after a protective dam built in about 1270.

Access for ocean-going ships is via the Noordzeekanaal, opened in 1876, and from the Rhine and Maas (Meuse) rivers by canals.

The old city, with its multitude of little bridges over encircling canals, dates mostly from 1650 to 1720, when it prospered as the capital of a world-wide commercial empire. Although its citizens suffered greatly during the German occupation from 1940 to 1945, the city itself escaped virtually unscathed, and attracts many tourists. Its diamond-cutting trade serves an international market, and other industries include shipbuilding, engineering, textiles and publishing.

Among the places of interest are the Rijksmuseum, with the world's greatest collection of Dutch paintings; Rembrandt House, where the painter Rembrandt van Rijn (1606-69) lived; the 17th-century royal palace; the 17th-century University of Amsterdam; and the house of Anne Frank, whose diary tells how the Jewish girl and her family hid there at the time of the German occupation during the Second World War.
Population (city) 719 860; (Greater Amsterdam) 1 091 340
Map Netherlands – Ba

Amsterdam Island *Indian Ocean* About 30 people in all work at a hospital, the Martin-de-Viviès Research Station and administrative offices on this near-barren volcanic island in the French Southern and Antarctic Territories. Only 54 km² (21 sq miles) in area, with a peak rising 911 m (2988 ft) near the centre, Amsterdam Island supports colonies of sea lions and herds of wild cattle. With its neighbouring island of St Paul, 100 km (62 miles) south, it lies far in the southern Indian Ocean, about 3700 km (2300 miles) south-east of Madagascar. Portuguese sailors are thought to have discovered it in 1522, but Anton van Diemen, a Dutchman, was first to visit the island in 1633, naming it after the Dutch city of Amsterdam. France annexed it in 1843. Crayfish are caught commercially off the island.
Map Indian Ocean – Cd

Amudar'ya *Central Asia* River in central Asia, 2540 km (1578 miles) long, measured over its greatest length. It rises in the Pamir Mountains and flows generally west, forming much of the border between Tajikistan, Uzbekistan and Afghanistan. It then turns north-west and drains into the Aral Sea by way of a 160 km (100 miles) long delta. Its middle course provides water for the Karakumskiy Canal and irrigates parts of the Turkmen and Uzbek deserts.

The river, known to the Greeks, Romans and Persians as the Oxus, is navigable for 1450 km (about 900 miles) from its mouth.
Map Afghanistan – Ba; Turkmenistan – Bb

Amundsen Sea Situated between Cape Dart and Thurston Island in ANTARCTICA, the sea is part of the South PACIFIC OCEAN. The Norwegian explorer Nils Larsen, who explored the area in 1929, named it after his compatriot Roald Amundsen (1872-1928), the first man to reach the South Pole.
Map Antarctica – Fc

Amundsen-Scott Station *Antarctica* American scientific base and landing strip at the geographic SOUTH POLE. It was founded in 1957.
Map Antarctica – d

Amur (Heilong Jiang) *Russia/China* River, 4510 km (2800 miles) long, formed on the Russo-Chinese border by the union of the Shilka and

Andorra

SMALL SKI RESORTS AND DUTY-FREE SHOPS BRING IN REVENUE TO THIS COUNTRY LYING IN A POCKET OF THE PYRENEES

High in the eastern Pyrenees, between France and Spain, lies the tiny state of Andorra, governed according to a system that dates from feudal times. According to tradition, the state was granted independence by Charlemagne, the first Holy Roman Emperor, in exchange for help against the invading Moors. It became a principality in 1278 – in fact, a co-principality, since it's shared by the French Comte de Foix and the Spanish Bishop of (Seo de) Urgel. The legacy of this arrangement is government by French and Spanish delegations (the co-princes are the President of France and, still, the Bishop of Urgel); there's also the obligation to pay feudal dues every second year to France (about US$2), and to Spain (about US$8), every other year, while to the bishop is due six hams, six cheeses and 12 hens every other year. A local General Council of the Valleys consisting of 28 members, which has formed the legislature since a new constitution was approved in March 1993, is elected by universal suffrage (18 years of age and older). Until Andorra became a parliamentary co-principality, only Andorrans of 21 years or older who were second and third-generation Andorrans, and all first-generation Andorrans of foreign parentage aged 28 years or older, could vote. Control was effectively in the hands of the heads of established families. Political parties are now constitutionally legal, while the old family loyalties still remain a strong influence.

The land consists of high valleys carved out by glaciers and drained by tributaries of the Balira, which itself flows into the Segre River in Spain. Andorra's mountain peaks reach heights between about 1800 and 2900 m (about 5900 and 9505 ft) and the scenery is spectacular. Cold winters and mild, sunny summers add to the co-principality's attractions.

About 13 million tourists come here every year to ski in the winter, walk in the mountains in the summer, and to take advantage of the duty-free goods on sale. Indeed, tourism and the duty-free trade are Andorra's principal sources of income – along with revenues from the sale of postage stamps and from advertising on its radio station, Radio Andorra, which broadcasts throughout Europe.

Because of substantial immigration in the 1960s and 1970s, only about one-third of the population of 61 960 are now native-born Andorrans. The official and most used language is Catalan – it resembles Provençal and is also spoken in north-eastern Spain and the adjoining area of France – but French and Spanish are also common.

PEOPLE OF THE VALLEYS

The majority of the people are Roman Catholic. They live in six valleys, including the capital, the market town of ANDORRA LA VELLA. Some who do not cater to tourists for a living raise sheep and cattle on the high pastures – which are common land – and spend summers in temporary homes on the mountainsides. Some grow cereals, vegetables, potatoes or tobacco in small, walled or terraced fields on the steep slopes. However, most food is imported. Smuggling into France and Spain is a thriving, though illegal, industry.

Many Andorrans feel that the powerful traditional family groupings and feudal system of government have held them back from the advances made by the rest of Europe. Women were not given the right to vote here until 1970 and political parties were not legally recognised until the first written constitution was approved in 1993.

ANDORRA AT A GLANCE	
Map France – De	
Official name Principality of Andorra	
Area 450 km² (174 sq miles)	
Population 61 960	
Per km² 138 **(Per square mile** 356)	
Capital Andorra la Vella	
Government Parliamentary co-principality	
Currency Spanish peseta, French franc	
Languages Catalan (official), French, Spanish	
Religion Christian (99% Roman Catholic)	
Climate Temperate, but modified by altitude. Average temperature in Les Escaldes ranges from 2°C (36°F) in January to 19°C (66°F) in July	
Land use Cultivation 2%, meadows and pastures 56%, forest and woodlands 22%, other 20%	
Main primary products Sheep, cattle, tobacco, potatoes, cereals (barley, rye), vegetables	
Major industries Tourism, duty-free trading	
Main exports Postage stamps	
Annual income per head (US$) 9834	
Population growth (per thous/yr) 33	
Life expectancy (yrs) Male 75 **Female** 81	

Argun rivers. The Amur flows south-east along the border, then north-east through Russia, entering the Sea of Okhotsk near Nikolayevsk.
Map China – Ja; Russia – Mc, Oc

amygdule Cavity in lava, caused by an escape of gas or steam, and filled by a mineral, such as quartz. An unfilled cavity is called a SCORIA.

An Bhlarna *Ireland* See BLARNEY

An Bhóinn *Ireland* See BOYNE, RIVER

An Cabhán *Ireland* See CAVAN

An Clár *Ireland* See CLARE

An Clochán *Ireland* See CLIFDEN

An Iarmhí *Ireland* See WESTMEATH

An Life *Ireland* See LIFFEY

An Mhi *Ireland* See MEATH

An Najaf *Iraq* Regional capital 160 km (100 miles) south of Baghdad. Shiite Muslims come on pilgrimage to the shrine of Ali Ibn Abi Talib, son-in-law of Muhammad, the founder of Islam.
Population 243 000
Map Iraq – Cc

An Tsionainn *Ireland* See SHANNON, RIVER

anabatic wind Local wind blowing up a valley during the day to replace the mountain air that rises in convection currents as the sun heats the slopes. Also called up-valley wind.

Anadyr' *Russia* Town, formerly Novomariinsk, in the Russian Far East on the estuary of the Anadyr' River. Its industries include coal mining and fish canning. The Anadyr' River, 1116 km (694 miles) long, rises in the mountains south of the Chukot range and flows north-east to the Gulf of Anadyr' in the Bering Sea.
Map Russia – Sb

Anatolia *(Anadolu) Turkey* Name used for the whole of Asiatic Turkey and also for the central wheat-growing plateau which makes up about 25 per cent of Turkey. It rises to between 800 and 2000 m (2625 and 6562 ft), and is surrounded on all sides by high mountains, such as Mt Ararat (5165 m, 16 945 ft) in the east. The plateau is crisscrossed by ravines and dotted with volcanic peaks. It was a battleground for numerous kingdoms of Asia Minor, from the Hittites (2000 to 1180 BC) to the Seljuk Turks who conquered it in the 11th century AD. The area embraces the site of ÇATAL HÜYÜKH, where one of the world's oldest cities was built 8500 years ago.
Map Turkey – Bb

Ancash *Peru* Western department lying between La Libertad and Lima. The dominant feature is the Santa River, which cascades from its source 4000 m (13 125 ft) up in the mountains at the lagoon of Conocha to the sea at the port of Chimbote. The famous archaeological site of CHAVÍN DE HUANTAR lies 64 km (40 miles) from the provincial capital, Huaráz. There are glacial lakes at Llaca and Llanganuco and, in the lower Santa Valley, the remains of the 'Great Wall of Peru' which dates from the time of the Chimú people who flourished between the 13th and the 15th centuries.
Population 983 200
Map Peru – Bb

Anchorage *USA* Largest city in Alaska and home to almost half the people living in Alaska. It lies at the head of Cook Inlet on the south Pacific coast, about 465 km (290 miles) from the Canadian border. Anchorage is the administrative and commercial heart of Alaska and the focus of the state's oil, coal and natural gas industries. The city was severely damaged by an earthquake in 1964. It is a fast-growing tourist centre.
Population 226 300
Map Alaska – Cb

Ancona *Italy* Seaport 180 km (110 miles) east of Florence on the 'elbow bend' of the Adriatic coast. Its name derives from the Greek word for elbow, *ankon*. It was founded in the 4th century BC by Greek colonists and became an important Roman and medieval port. In the Middle Ages it was an independent republic. Rail and road links in the 19th century led to new industrial expansion – and fishing is a major occupation. It is linked by ferry to Croatia and Greece.

The finely preserved Arch of Trajan, built in AD 115, stands on the quayside among railway sidings. Ancona's 11th to 13th-century cathedral church is an interesting fusion of Byzantine, Romanesque and Gothic styles built on the plan of a Greek cross. The medieval quarter which still shows some scars from an earthquake in 1972 has many churches and palaces, including Santa Maria della Piazza – 13th century, but of 5th to 6th-century origin – and the 15th-century Loggia dei Mercanti. The Mole Vanvitelliana is an 18th-century fort built in the harbour area in the shape of a pentagon.

Population 101 180
Map Italy – Dc

Andalucía *Spain* Mountainous region in the south of the country covering 87 268 km² (33 694 sq miles). Settled by the Phoenicians about 1000 BC, it was later ruled by the Carthaginians, the Romans, and finally (until 1492) the Moors. Two main mountain ranges run through the region – the Morena range, to the north, reaches 1797 m (5896 ft). The SIERRA NEVADA is the southern range and reaches over 3400 m (11 155 ft). Nestled against its northern flank is the city of GRENADA, well-known for its Moorish architecture. To the south lies Spain's Costa del Sol, a stretch of coastline which is overcrowded with tourists each summer. Andalucía is rich in minerals – including lead, copper and cinnabar – and grows olives, grapes, oranges and lemons. Its cities include SEVILLE, the regional capital, MÁLAGA and CÓRDOBA.

Population 6 903 000
Map Spain – Cd

Andaman and Nicobar Islands *India* Two groups of islands in the eastern Bay of Bengal covering 8249 km² (3185 sq miles) opposite the coast of Myanmar (Burma). They were given a unified administration by the British in 1872 and have been used as a penal colony from 1857 to the present day. Much of the area consists of evergreen forest and the economy is based largely upon timber and copra. Some of the original population still live here, catching fish and hunting. The capital of both groups is Port Blair on South Andaman Island.

Population 278 000
Map India – Ee

Andaman Sea An arm of the INDIAN OCEAN which covers an area of 777 000 km² (297 572 sq miles). It is bordered in the west by the Andaman and Nicobar Islands, and in the east by Myanmar (Burma) and Thailand. The Strait of MALACCA in the south, a key shipping lane, leads to the South CHINA SEA. The sea's chief port is YANGON (formerly Rangoon).

Map Indian Ocean – Db

Andes (Cordillera de los Andes) *South America* Mountain chain that extends like a spine for some 7500 km (4660 miles) along the entire west coast of South America from the Caribbean to Tierra del Fuego. For most of its length it con-

sists of two or three parallel ranges, separated by incisions or, in the central regions of South America, vast plateaus. Created by the collision of the Nazca plate of the eastern Pacific with the South American continent (SEE PLATE TECTONICS), it contains numerous volcanoes and is subject to violent earthquakes. The chain's two highest peaks are in Argentina: ACONCAGUA (6959 m, 22 831 ft) and OJOS DEL SALADO (6880 m, 22 572 ft). Many of the continent's major rivers, including the AMAZON and ORINOCO, rise here.

Map America, South – Bc-Bd

andesite Fine-grained volcanic rock named after its presence in the volcanoes of the Andes in South America. As a LAVA it is less fluid than BASALT and creates spectacular volcanoes.

Andhra Pradesh *India* State in the south-east covering 276 754 km² (106 830 sq miles) and bordering on the Bay of Bengal. In 1956 it took over much of the princely state of HYDERABAD, whose main city is now the state's capital. Its main products are cotton, sugar cane and rice. The low-lying coastal districts are liable to cyclone damage. The principal language is Telugu.

Population 66 304 900
Map India – Cd

Andorra See p 34

Andorra la Vella *Andorra* The capital, situated in the country's principal valley, though this lies 1059 m (3475 ft) above sea level. Surrounded by breathtaking mountain scenery, it is also a shoppers' paradise of duty-free goods, and a constant stream of cars makes parking difficult in the narrow streets.

It is practically joined to the larger town of Les Escaldes, a spa with hot sulphur springs.

Population 19 000
Map France – De

Andropov *Russia* See RYBINSK

Andros *Bahamas* Largest of the islands, covering 5957 km² (2300 sq miles) and lying about 56 km (35 miles) south-west of Nassau. Much of the interior is densely forested, and the western shore is fringed by mangrove swamps. The main occupations are farming and tourism.

Scuba diving amid spectacular underwater scenery is popular. A special feature of Andros is the 118 Blue Holes – deep circular holes, possibly collapsed caves, which rise up through the coral inland and on the coast.

Population 8160
Map Bahamas – Bb

Angara *Russia* Major tributary – 1826 km (1135 miles) long – of the YENISEY River. It flows from the south-west corner of Lake Baikal, past Irkutsk, to join the Yenisey at Strelka. It is navigable for most of its course and has several large hydroelectric power stations.

Map Russia – Lc

Angel Falls (Salto Ángel) *Venezuela* World's highest waterfall 979 m (3212 ft) high, named after American airman James Angel, who discovered them in 1937. The water plunges from the lip of the Auyán-Tepuí Plateau in the Guiana Highlands – bouncing only once, off a ledge.

Map Venezuela – Bb

Angers *France* Capital of Maine-et-Loire department, in the former province of Anjou, on the Maine River, 80 km (50 miles) north-east of Nantes. It is a commercial and industrial centre whose industries include food processing, wine, textiles, machinery and electronics.

There are a number of medieval buildings, including a vast 13th-century château which is now a museum of tapestries.

Population 146 200
Map France – Cc

Angkor *Cambodia* Ruins of the rich Khmer civilisation that dominated north-western Cambodia and eastern Thailand for about 500 years until the early 15th century. Its extensive network of temples and man-made lakes and irrigation canals, hidden by centuries of jungle undergrowth, was discovered in 1860. More than 600 Hindu temples, some as big as cathedrals, all with elaborately carved and decorated towers, were eventually uncovered and restored.

The two largest complexes were at Angkor Wat and Angkor Thom. At the heart of each complex is a pyramidal temple mountain with five ornate, soaring towers. The rectangular lakes – reservoirs also known as *barais* – and the network of canals functioned mainly to re-create on earth an image of the Hindu-Buddhist universe. Some images of the Buddha are thought to be portraits of Khmer kings.

The Angkor complex appears to have been abandoned by the Khmers after the Thais attacked Cambodia. A campaign to limit the damage done by fast-growing foliage and water erosion was halted when Pol Pot came to power in 1975 and was only resumed after the transitional government was established in 1993. Archaeologists fear, however, that the monuments may have been badly damaged.

Map Cambodia – Ab

Anglesey *United Kingdom* Island of 715 km² (276 sq miles) off the coast of north-west Wales, separated from the mainland by the Menai Strait – which narrows to only 200 m (650 ft) and is spanned by two bridges. Anglesey forms part of the county of Gwynedd. It was the sacred stronghold of the Celtic priests, the Druids, when their order was destroyed by the Romans in AD 61, and it remains a focus for Celtic and Welsh traditions and culture. It has excellent holiday beaches but is often regarded by outsiders simply as a stepping stone to the ferry port of Holyhead. Apart from the traditional agricultural industry, newly developed industries include a nuclear power station and an aluminium smelter.

Population 69 400
Map United Kingdom – Cd

Anglo-Normandes, Iles *United Kingdom* See CHANNEL ISLANDS

Angola See p 36

Angoulême *France* A historical fortified city and capital of Charente department, on a limestone hill overlooking the Charente River 105 km (65 miles) north-east of Bordeaux. It was an episcopal see in AD 379, and has many ancient houses and a 12th-century cathedral which was restored in the 19th century.

Population 42 700
Map France – Dd

Angola

PROSPERITY FROM DIAMONDS, OIL AND OTHER MINERALS DENIED BY A HORRIFIC CIVIL WAR

1 CUANZA NORTE
2 CUANZA SUL
3 BENGUELA
4 HUAMBO

Endowed with great mineral wealth and tremendous agricultural potential, Angola should be able to provide a prosperous living for roughly 11.5 million people in a territory covering 1 246 700 km² (481 350 sq miles). Instead, by 1994, more than two decades of civil war had caused extreme hardship and poverty.

The conflict started as a war of independence in this vast land of equatorial forests and savannah-covered plateaus. Three separate groups rebelled against the Portuguese colonial government in the 1960s and 1970s. The rebels combined to form a government after independence on November 11, 1975, but they were soon fighting again – this time among themselves.

The battle for power was won by the Popular Movement for the Liberation of Angola (MPLA), which has formed the government since 1976. But their main rivals, the National Union for the Total Independence of Angola (UNITA), launched a fierce guerrilla war, gaining control of about one-third of the country in the south and east.

Inevitably, the war brought in bigger powers on both sides. The Marxist MPLA was helped by an estimated 40 000 Cuban soldiers and Soviet arms and equipment, as well as Soviet and East German soldiers stationed in Angola as 'military advisers'. Pro-Western UNITA had support from the United States – this ended in May 1993 – and military assistance from South Africa, which sent troops into southern Angola to support UNITA while fighting one of its own enemies, the South West African People's Organisation (SWAPO), based in Angola. The South African troops were withdrawn in August 1988, as part of a complicated arrangement in which the Angolan government offered to send its Cuban troops home and control SWAPO, while pressing for Namibian independence. Meanwhile, UNITA has fought on under the leadership of Jonas Savimbi. In 1992, the United Nations Security Council imposed sanctions on UNITA, but UNITA already had enormous arms stockpiles, and sanctions were not effective.

With the extent of territory held by UNITA continually waxing and waning, the military activity almost totally disrupted peasant and commercial farming. At the height of the conflict, some villages were no longer able even to grow the crops needed to live on, such as maize, cassava, beans, millet and sorghum; fields were left studded with landmines, and raiding troops took all the food they could lay their hands on. This resulted in many thousands of people needing food aid to survive. Food aid also had to be flown in to government-held cities besieged by UNITA's forces. Many villagers fled to the towns; by 1994, for instance, the capital, LUANDA, had grown from 800 000 people at independence to 1.7 million.

In May 1991, the rivals signed a cease-fire agreement. But peace was short-lived, and in October 1992 fighting broke out anew. A second round of peace talks began in November 1993, and although the MPLA government made significant concessions, these were not enough to satisfy UNITA. In 1994 the United Nations brokered another cease-fire agreement between the warring parties. However, observers predicted that the attainment of real peace would take a long time.

COLONIAL HERITAGE

The Portuguese first landed on the coast in 1483 and began to settle in the territory in the late 1500s. They established the capital city of Luanda on the western coastal plain. At first the Portuguese had a reasonably amicable relationship with the local African tribes. They converted the Bakongo king, Afonso, to Catholicism, and another tribesman was consecrated as a bishop.

However, by the late 16th century the Portuguese were raiding the tribes for slaves to ship to South America. The Bakongo were the main victims; their numbers were greatly depleted and their kingdom was destroyed. It is estimated that three million slaves were taken from the 16th to the early 19th centuries. It was not until the 19th century that Portugal began to develop the interior of the country. Roads and railways were built, especially the strategic BENGUELA railway which runs for 1350 km (840 miles) from LOBITO, on the Atlantic coast, to the border with Zaire in the north-east. Diamond mines were opened in the Lunda region in the north-east during the 1920s; oil production began near Luanda in the 1950s and off the coast of CABINDA in the 1960s. But by then nationalist movements were growing in this land where 300 000 Portuguese settlers were ruling over 2.7 million black Africans.

The complexity of Angola's politics is partly a result of the diversity of its population, with many ethnic groups having different allegiances. Rebellion began in the north, among the Bakongo. A Bakongo nationalist movement, the National Front for the Liberation of Angola (FNLA), under Holden Roberto, led uprisings in 1961 and was one of the most important independence movements at that stage. In the late 1970s the FNLA threw in its lot with the ruling MPLA party.

The MPLA drew its main support from the central Angolan groups, including the Mbundu, Mbaka and Ndongo. Being nearest to the colonial capital, Luanda, they had had the most contact with Europeans and were among the best educated Angolans. Many people of mixed descent live in Angola's districts and towns. Those in Luanda, in particular, tended to support the MPLA and some took leading roles in the liberation struggle.

Altogether, there are 10 ethnic groups in Angola. The largest is the Ovimbundu, who make up 37 per cent of the population. They mostly live around Huambo on the Planalto, the central plateau, which has the best climate and soils and consequently has always attracted human settlement.

Traditionally an agricultural society, the Ovimbundu also controlled trade routes from the Atlantic across the central savannah and they were secure and prosperous before the Europeans arrived. Under colonial rule, they became involved in commercial farming and many of them were employed on running the Benguela railway.

At first the Ovimbundu showed little interest in becoming independent. However, they grew increasingly bitter as the Portuguese began forcing them into villages in an effort to control them more effectively. They also lost much land to the Europeans in the 1960s and 1970s. By the early 1990s, UNITA depended largely on their support.

The Lunda and Tshokwe occupy north-eastern Angola. In the past, they were hunter-gatherers who later also captured people from other tribes to sell as slaves. They are now mostly small farmers.

In the south, a group of related peoples, the Nganguela, have retained more of their traditions than other ethnic groups. The pastoralist Herero inhabit the dry south-west, and the Ambo (Ovambo), the dry steppe in the south-east, growing sorghum and millet, and herding their cattle.

The civil war not only disrupted the lives of these people, but also put a stop to farming – especially in the once rich agricultural region of the south and east – and caused unbearable hardship and hunger. Whole towns have been devastated.

Multiparty elections were held in September 1992, and although both the MPLA and UNITA fought them, when the MPLA won the elections, UNITA's Savimbi refused to accept the results and continued the war.

Estimates put deaths at 500 000 between 1992 and mid 1994; another 3 million people were displaced. At one stage, in November 1992, it was estimated that 2000 people were dying daily, as a direct or indirect result of war. Towns were depopulated. Cacusso, for instance, with a prewar population of 100 000, was reduced to 20 000 in 1994.

▲ **THE SPOILS OF WAR** Two decades of civil war have affected every aspect of Angolan society; here, the wreck of a jet fighter is put to peaceful use in a school playground in Luanda.

Commercial agriculture and industry had already suffered at independence when almost all the Portuguese who ran the concerns left the country. The new government nationalised the diamond and oil industries – although the multinational Gulf Oil company remains a major partner.

Many smaller businesses had to be taken over because the owners had disappeared. However, there was a shortage of people to manage them, particularly since education for the African population had always been limited. The civil war only added to these difficulties.

Angola's economy does have one success story, however, in the oil industry. It is by far the most important aspect of the economy; in fact, it is the only sector that has managed to survive through the war years. It generates hundreds of millions of dollars and 93 per cent of export revenue. Production has been increasing

steadily from offshore fields which are mainly north of the ZAIRE (Congo) River mouth, in Cabinda Province. Vast mineral wealth remains unexploited, including reserves of manganese (in the Cassala Quitungo district between Luanda and N'dalatando), copper (near NAMIBE, Ngunza Kabolo and in the Cambambe Dam area), iron ore and phosphates.

Angola has a high capacity for producing hydroelectric power, and the Cambambe Dam on the CUANZA River already provides the city of Luanda with its electricity. On the CUNENE River, the Ruacana Dam was built jointly by the South Africans and the Portuguese before independence, with the intention of using it for both irrigation and hydroelectric power.

Angola's physical environment provides great prospects for economic development – provided that the cease-fire signed in 1994 can be made to work.

ANGOLA AT A GLANCE

Official name Republic of Angola

Area 1 246 700 km² (481 350 sq miles)

Population 11 500 000

Per km² 9 (**Per sq mile** 24)

Capital Luanda

Government Unitary republic in transition

Currency 1 new kwanza = 100 kwei

Languages Portuguese (official), and various African languages

Religions Christian (53%), indigenous beliefs (47%)

Climate Tropical, with summer rains. Semi-arid to arid in the south, more humid to the north. Average temperature in Luanda is 24°C (75°F) throughout the year

Land use Forest and woodlands 43%, grazing 23%, cultivation 2%, other 32%

Main primary products Maize, cassava, bananas, palm oil, coffee, cotton, timber, sisal, fish; oil, diamonds, iron ore, phosphates, copper, manganese, feldspar, uranium

Major industries Agricultural products, mining and related products, forestry, fishing, chemicals, tobacco, food processing, cement, construction materials

Main exports Crude oil and petroleum products (93%), coffee, diamonds, sisal, fish and fish products, cotton, timber

Annual income per head (US$) 620

Population growth (per thous/yr) 28

Life expectancy (yrs) Male 43 **Female** 47

Anguilla *Leeward Islands* Coral and limestone island 91 km² (35 sq miles) in area, some 250 km (155 miles) east of Puerto Rico. It was named in 1493 by the Italian explorer Christopher Columbus, who called it *anguila*, the Spanish word for 'eel' – probably with reference to its long, narrow shape.

First colonised by the British in 1690, the island has since 1980 been a self-governing dependency of the United Kingdom. British troops intervened in 1969 to restore legal government after Anguilla broke away from domination by the St Kitts' administration in 1967. The people are mostly of African descent. Farming is poor, due to thin soil and scant rainfall. Fishing is a thriving industry – the catch being exported to neighbouring islands. Salt is produced by evaporating sea water, but a large part of the island's income is made up of money sent home by some 20 000 Anguillans living in Britain and the United States. Tourism has great potential as there are fine beaches and clear seas. In 1991, for instance, more than 93 000 tourists visited the island though many of them were day visitors. The capital is The Valley.

Population 8960
Map Caribbean – Cb

Angus *United Kingdom* See TAYSIDE

Anhui *China* Province of 130 000 km² (50 000 sq miles) in east-central China, astride the Chang Jiang and Huai rivers. The poorly drained northern part of the province in the North China Plain is prone to flooding. Wheat and rice are grown in the Chang Jiang Valley.

The southern part of the province is renowned for its mountain scenery, particularly the HUANG SHAN range. Although Anhui is first and foremost an agricultural province, it has reserves of coal, iron ore and copper. Cities include HEFEI, the provincial capital, and Bengbu, a centre for the engineering industry.

Population 58 340 000
Map China – He

Anjou *France* Former western province of France that straddles the lower Loire Valley. In the 12th century it was part of the Plantagenet domains incorporating England, Normandy and Aquitaine. It was claimed by Louis XI of France in 1481. ANGERS is the region's main market for local fruit, vegetables and wine – including the renowned Anjou Rosé.

Map France – Cc

Anjouan *Comoros* See NZWANI

Ankara *Turkey* The country's capital since 1923, lying at the centre of the Anatolian Plateau 350 km (220 miles) east and slightly south of Istanbul. It was founded in the 8th century by the Phrygians, who came from Thrace about 1200 BC and occupied central Turkey.

The city is Turkey's second largest after Istanbul, and is dominated by the hilltop mausoleum of Kemal Ataturk (1881-1938), the nationalist leader who founded modern Turkey after the collapse of Ottoman power in the First World War. Several buildings remain from Roman times, when the city was the provincial capital of Galatia. The city's archaeological museum contains the world's biggest collection of Hittite artwork.

Ankara's main industries are cement, textiles, leather goods and mohair – made from the wool of angora goats, from which the city probably gets its former name (Angora).

Population 2 559 470
Map Turkey – Bb

Anloga *Ghana* Coastal town on Keta Lagoon about 120 km (75 miles) east of the capital, Accra. It is growing in importance at the expense of its neighbour, Keta (population 12 000), which lies about 150 km (95 miles) east of Accra, on a narrow sandbar east of the lagoon.

The land on which Keta stands is being eroded steadily by the sea, and part of the town is already under water. Anloga is now taking over Keta's traditional market for cotton, cloth, fish, onions and salt – produced by evaporation from the neighbouring lagoon.

Population 15 000
Map Ghana – Bb

Annaba (Bône) *Algeria* Town and seaport on the Mediterranean coast near the border with Tunisia, it is Algeria's most important port after Algiers and Oran. It was built by Arabs in the 7th century on the site of Aphrodisium, the Ancient Roman port of Hippo, which was the bishopric of St Augustine in the 5th century.

Iron ore, phosphates, wine and cork are the chief exports. The town manufactures chemicals and is the site of Algeria's only iron and steel plant. It is an administrative and transport centre and has an airport. The Roman site of Hippo Regius, a rich city of Roman Africa until AD 300, lies 2 km (about 1 mile) to the south-west and many of the archaeological finds are displayed in the museum at Annaba.

Population 310 000
Map Algeria – Ba

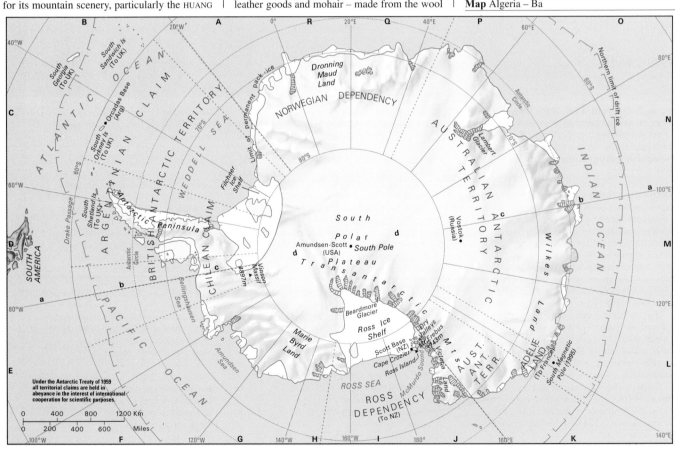

Annam *Vietnam* Name of the central MIEN TRUNG region in French colonial times.

Annamite Chain *South-east Asia* See MIEN TRUNG

Annapolis *USA* State capital of Maryland, standing on Chesapeake Bay. It was settled by the Puritans in about 1649 and became the Maryland colony's capital in 1694. In 1786, representatives of the Union met in the city for the Annapolis Convention, a precursor of the Constitutional Convention held the following year in Philadelphia. The United States Naval Academy was founded here in 1845.
Population 33 200
Map United States – Kc

Annapurna *Nepal* Mountain peak in the Annapurna Mountain Range in the Great Himalayas 175 km (110 miles) north-west of Kathmandu. At 8091 m (26 545 ft) it is one of the top 10 highest mountains in the world. First to reach the summit was a French expedition in 1950. In the west, the huge, precipitous south face of the main peak, Annapurna I, seems to tower above the town of Pokhara, though it is some 40 km (25 miles) distant. West, across the gorge of the Kali Gandak River, is the even higher peak of Dhaulagiri, reaching 8172 m (26 810 ft).
Map Nepal – Aa

Annecy *France* Capital of Haute-Savoie department, at the northern end of Lake Annecy, 100 km (60 miles) east of Lyons. It manufactures linen, yarn, paper, and precision instruments. A massive château, a 16th-century cathedral, old houses in narrow streets, and mountain scenery nearby have made the town popular with tourists.
Population 49 700
Map France – Gd

Anshan *China* Industrial city of Liaoning province about 180 km (110 miles) from the North Korean border. Iron and steel manufacturing was begun here during the Japanese occupation of the north-eastern region of Dongbei (Manchuria) between 1931 and 1932. Today Anshan produces one-fifth of China's crude steel output.
Population 1 450 000
Map China – Ic

Antakya (Antioch) *Turkey* City and provincial capital bordering Syria, about 500 km (300 miles) south of the capital, Ankara. Founded in 300 BC, it had a population of about 500 000 around the 2nd century BC, and was said to have had canalisation and lighted streets. It gained a reputation as a city of luxury and pleasure and became known as the 'queen of the east'. In the 1st century AD it was visited by St Peter, and his grotto is just outside the city.

The city became part of Syria in 1919, but was restored to Turkey 20 years later. Little of its former splendour remains following a series of earthquakes, but the Hatay Museum houses one of the world's finest collections of Roman mosaics.

The surrounding province of Hatay covers an area of 5403 km² (2086 sq miles) and produces cotton and grain; some iron is mined.
Population (city) 124 000; (province) 1 109 750
Map Turkey – Bb

Antalya *Turkey* Port and holiday resort on the Mediterranean Sea about 400 km (250 miles) south of the capital, Ankara. It was founded in about 150 BC by Attalus II of Bergama, one of the many princes striving to control the country. It has a 2nd-century decorated marble gate built to commemorate a visit by the Roman emperor, Hadrian. A project began in the 1980s to restore the old walled town, which had deteriorated badly. Antalya's main industries are grain, timber, canning, flour milling and tourism. The neighbouring lowland province of the same name covers an area of 20 591 km² (7950 sq miles) and produces grain and timber.
Population (city) 258 100; (province) 892 200
Map Turkey – Bb

Antananarivo *Madagascar* Capital of the republic founded in the 17th century and formerly called Tananarive. It stands in the centre of the island on a hill overlooking the Ikopa River floodplains, a fertile rice-growing area. It is the country's industrial heart, producing tobacco, processed food, textiles and leather goods. A university was founded here in 1961. Many public buildings date from the French occupation after 1895. Tourist attractions include an old royal palace and the colourful Zoma, a market which on Fridays expands to fill most of the city centre.
Population 802 000
Map Madagascar – Aa

Antarctic Circle Latitude 66°32′ south, along which, at the southern summer solstice (about 21 December), the sun does not set, and at the winter solstice (about 21 June) the sun does not rise all day.

Antarctic Ocean The 'Antarctic Ocean' or 'Southern Ocean' are names sometimes given to the waters around ANTARCTICA. But most geographers regard these waters as southern extensions of the ATLANTIC, INDIAN and PACIFIC oceans. The region is the only place in the world where a boat can sail in a straight line – east-west – indefinitely without hitting land.

Antarctic Peninsula *Antarctica* A largely ice-covered and mountainous, 1300 km (808 mile) long extension of Antarctica towards South America. The jagged coast is flanked by ice shelves and numerous offshore islands. The Antarctic Peninsula is disputed territory claimed by the United Kingdom, Argentina and Chile. It is sometimes called the Graham Land Peninsula.
Map Antarctica – Db

Antarctica A barren, empty and bitterly inhospitable area covering 13 975 000 km² (5 394 503 sq miles), Antarctica was the scene, in 1912, of one of the most dramatic contests in the history of world exploration – the race to reach the SOUTH POLE. That year, two rival expeditions – a British team led by Captain Robert F. Scott and a Norwegian team led by Roald Amundsen – reached the pole within 34 days of each other. With greater experience and superior technique – using dog sledges rather than relying on human muscle – Amundsen won the race. But the tragic deaths of the five Britons after discovering their defeat won them first place for heroism.

The culminating feat of Antarctic exploration was the first crossing of the continent, via the south pole, by the multinational Commonwealth

Trans-Antarctic Expedition in 1957-8. Today, most Antarctic researchers work for months or years at a time at permanent bases.

They work in remarkable international harmony despite the competing national interests that characterised early Antarctic history. This is due to the successful operation of the Antarctic Treaty, signed in 1959 by 12 nations which had some scientific interest in Antarctica.

This arrangement silenced the territorial claims of seven nations – Argentina, Australia, Chile, France, New Zealand, Norway and the United Kingdom – which each wanted a wedge-shaped piece of the Antarctica pie, in some cases overlapping. The treaty has kept the peace and many other nations have since signed it, bringing the number of signatories to 26 countries with voting rights and 13 adherents, though several countries have criticised its restricted membership. The possibility of exploiting Antarctic resources – from coal and iron ore to rich fisheries – is one reason for this criticism. In 1991 all the parties to the Antarctic Treaty imposed a ban on mineral exploitation for 50 years.

Antarctica has the harshest climate in the world. Temperatures of −89.6°C (−129.3°F) have been recorded in the dark winter, and winds of 145 km/h (90 mph) can blow for 24 hours at a stretch. In a whiteout, when driving snow hides everything, it is possible to be lost only a few metres from a camp hut. Even on calm and sunny days, when people can get sunburned, the temperatures hardly rise above freezing point.

The continent is covered by 90 per cent of the world's snow and ice. Ice forming on the SOUTH POLAR PLATEAU is slowly squeezed under its own accumulating weight outwards towards the edges of the landmass. The surface appears as a featureless ice cap, but sometimes great glaciers are created, such as the LAMBERT GLACIER (the world's biggest) and the BEARDMORE GLACIER, route of many journeys to the South Pole. The quantity and power of the moving ice is so great that in several places it is pushed beyond the landmass onto the sea. There it floats as an ice shelf as in the ROSS and WEDDELL seas, terminating in towering ice cliffs. From these cliffs, huge icebergs break off and are carried away into the ANTARCTIC OCEAN.

In some places, ice-free peaks protrude through the great white crust that covers Antarctica; the highest peak is the Vinson massif (4897 m, 16 066 ft). The Antarctic Peninsula, which stretches towards South America, is free of snow in some areas, giving geologists access to rocks which have yielded important fossil and mineral finds. It has been discovered that much of Greater Antarctica – the part of the main landmass east of the Greenwich Meridian – is a high continental shelf, above sea level under the ice. Once part of Gondwana, it has a granite and gneiss base (as do South Africa and Australia) covered with a thin layer of sandstone. To the west, Lesser Antarctica is a continuation of the South American ANDES, and would become islands if the ice vanished.
Map See opposite page

Antequera *Spain* Textile town 112 km (70 miles) north-east of Gibraltar. Nearby are the Menga Galleries, the site of mysterious prehistoric standing stones.
Population 38 770
Map Spain – Cd

Antigua and Barbuda

WHERE NELSON'S FLEET ONCE RULED THE WAVES, ANTIGUA NOW ATTRACTS TOURISTS TO ITS PALM-FRINGED BEACHES

Set on the eastern flank of the eastern Caribbean's LEEWARD ISLANDS, Antigua and Barbuda occupies a strategic position that has long been recognised by the major powers. The tiny state in fact comprises three separate islands – the two in its title plus the uninhabited Redonda.

In the 17th and 18th centuries, Antigua was one of Britain's vital naval bases in the West Indies, and its strategic importance has continued in this century. For instance, the United States built the island's large Coolidge Airport during the Second World War to defend the Caribbean and the Panama Canal.

Antigua's original inhabitants were Arawak people, none of whom survive today – they were killed or taken away as slaves by Carib tribesmen. Columbus, the first European to arrive here, named Antigua in 1493; in 1632 it was colonised by the British. They brought slaves from Africa whose descendants account for most of the present population of 66 200 people, although there are also tiny minorities of British, Portuguese, Lebanese and Asian stock. In 1981, Antigua and Barbuda became an independent member of the Commonwealth with the British monarch as its head of state.

General elections in March 1994 brought victory to the Antigua Labour Party (ALP) for the fifth consecutive term.

The staple of Antigua's economy today is tourism, accounting for 13 per cent of gross domestic product. Even for the Caribbean, the island is exceptionally well endowed with beaches, all of them palm-fringed with sparkling white or pink sand – but it is subject to hurricanes and tropical storms between July and October.

The island started to exploit tourism in the 1960s. Two gambling casinos attract wealthy tourists, mostly from North America.

Before tourism, Antigua's economy was almost exclusively based on sugar, which provided three-quarters of its export earnings. In the 1960s, however, the industry was faced with low world sugar prices, and private sugar plantations threatened to close down. The state stepped in and bought up the land in 1967 to keep the industry going. But four years later another government closed the industry down. Subsequent governments have encouraged farmers to grow grains and other food crops, including some cotton for export. They have also encouraged industrial development, and today local industries produce chiefly petroleum products (from imported raw materials), textiles and clothing, and electronic parts.

The low, heavily wooded island of Barbuda has a population of 1460 people, most of whom engage in small-scale farming and fishing, especially for lobsters for export. Tourists come for the beaches and the coral reefs surrounding Barbuda, where a number of spectacular wrecks of sailing ships can be seen. On the island itself the wildlife includes wild pigs and English fallow deer. They were introduced by the Codrington family who, from 1685 to 1870, leased Barbuda from the British Crown for the rent of 'one fat pig' a year.

Most of Antigua and Barbuda's places of historic and architectural interest are on the larger island, Antigua. ST JOHN'S, the capital, has some fine 18th-century administrative buildings and a splendid Baroque cathedral, built in the 1840s. Sugar mills – many of them now converted into houses – still dot the island's landscape and in the south-east of the island are ENGLISH HARBOUR and Nelson's Dockyard, named after Admiral Horatio Nelson, the British naval hero whose ships used the base. It has many restored 18th-century naval stores and houses.

ANTIGUA AND BARBUDA AT A GLANCE	
Map Caribbean – Cb	
Area 440 km² (170 sq miles)	
Population 66 200	
Per km² 150 (**Per sq mile** 389)	
Capital St John's	
Government Parliamentary monarchy	
Currency 1 East Caribbean dollar = 100 cents	
Language English (official); Creole	
Religion Christian (mainly Anglican)	
Climate Tropical, with a rainy season from May to November. Average temperature at St John's is 28°C (82°F) throughout the year	
Land use Cultivation 18%, meadows and pastures 7%, forest and woodlands 16%, other 59%	
Main primary products Cotton, fruit, vegetables, sugar cane, fish	
Major industries Tourism, petroleum products, cotton production and textiles, food processing, clothing manufacture, fishing	
Main exports Petroleum products, cotton, textiles, clothing, machinery, electronics, fruits, lobsters	
Annual income per head (US$) 4870	
Population growth (per thous/yr) 5	
Life expectancy (yrs) Male 71 **Female** 75	

Antibes *France* Seaport and luxury resort town on the Côte d'Azur, 16 km (10 miles) south-west of Nice. It trades in dried fruit, olives, tobacco, perfumes and wine. The Roman remains of an ancient settlement are in the vicinity. The museum contains works by Pablo Picasso.

Population 70 000
Map France – Ge

anticline A FOLD with strata sloping downwards on both sides from a common crest.

anticyclone Large high pressure system in the atmosphere with pressure diminishing outwards from the centre. Little vertical air movement makes for still weather conditions and poor visibility. In Europe, summer anticyclones give long, sunny but hazy days. In winter, anticyclones bring very cold weather. Autumnal fogs are usually associated with anticyclones.

Antigua *Antigua and Barbuda* The largest of the three islands making up this tiny Caribbean state – the other two being Barbuda and Redonda. Antigua, covering about 280 km² (108 sq miles) and rising to 405 m (1330 ft) at Boggy Peak, is a beach-lined, palm-fringed island which attracts thousands of tourists each year. Residents are proud that their island has 365 beaches, 'one for every day of the year'; watersports include sailing, deep-sea fishing, snorkelling and water-skiing. Until tourism became the primary source of income, Antigua's main income was from sugar, and sugar mills still dot the island's landscape. The capital and deep-sea port, ST JOHN'S, has several museums, a duty-free shopping complex and a casino. Antigua's carnival is held from the end of July to the first Tuesday in August each year; the chief event is J'ouvert – the morning when the main streets are filled with dancing people and a procession of steel and brass bands. An annual jazz festival is held in May.

Population 64 000
Map Caribbean – Cb

Antigua *Guatemala* City 20 km (12 miles) west of the capital, Guatemala City. It became the country's second capital in 1543 after Ciudad Vieja, 6 km (4 miles) away between the Agua (Water) and Fuego (Fire) volcanoes, was destroyed by an earthquake. In 1773, Antigua was also destroyed by an earthquake, and Guatemala City was founded as capital instead. Antigua was damaged again by an earthquake in 1976. It was one of the first planned colonial cities in the Americas, and some fine buildings survive.

Population 27 000
Map Guatemala – Ab

Anti-Lebanon (Jebel esh Sharqi) *Lebanon* Mountain range rising to 2740 m (about 9000 ft) and running north-south along the border with Syria. The range forms the east of the BEQA'A Valley, reaching the south at Mount HERMON.

Map Lebanon – Bb

Antioch *Turkey* See ANTAKYA

Antioquia *Colombia* The country's most populous state covering 63 612 km² (24 560 sq miles) in the north-west. It is centred on Medellin, Colombia's third largest city. Coal and gold are mined, and coffee and bananas grown.

Population 4 467 910
Map Colombia – Bb

Antivari *Yugoslavia* See BAR

Antofagasta *Chile* Northern region encompassing much of the ATACAMA DESERT. Most of its towns depend on water piped from the foothills of the mountains. Nitrates and copper are the basis of the economy. The open-cast copper mine at Chuquicamata 220 km (135 miles) north-east of the city of Antofagasta, the region's capital, is the largest in the world.

Population (region) 407 410; (city) 221 000
Map Chile – Ab

Antrim *United Kingdom* Town and county in Northern Ireland. The town stands near the north-east corner of Lough Neagh. There is a 9th-century round tower north of the town and a fine early 18th-century courthouse.

The county of Antrim (excluding Belfast and Lisburn) covers 2831 km² (1093 sq miles). It stretches north to the Atlantic Ocean, where its magnificent cliff scenery includes the GIANT'S CAUSEWAY. The chief town is Ballymena (population 28 110).

Population (town) 20 880; (county) 575 700
Map United Kingdom – Bc

Antwerp (Antwerpen; Anvers) *Belgium* Principal port and capital of the northern province of Antwerp, 45 km (28 miles) north of the national capital, Brussels, and the country's second largest city after the capital. It lies on the Schelde River estuary 88 km (55 miles) from the sea. One of the wealthiest trade centres in the world in the Middle Ages, Antwerp was heavily bombed during the Second World War, and now has to compete for trade with the Dutch ports of Rotterdam and Amsterdam. Its main industry is diamond cutting, but other major industrial plants produce chemicals, metals, and cars. The narrow streets of the old town contain a 14th-century cathedral, a 16th-century castle (now a museum) and a 16th-century town hall.

Population (city) 470 350; (metropolitan area) 1 597 300
Map Belgium – Ba

Anuradhapura *Sri Lanka* Ancient city about 160 km (100 miles) north-east of Colombo. The Sinhalese founded the city in about 500 BC and made it their royal capital until the Tamil in-vasions of the 11th century. By the 13th century it had almost disappeared. Its ruins then lay beneath the jungle until the British uncovered them in the 19th century. There is a wealth of fine carvings, temples, reservoirs and the dome-shaped Buddhist shrines known as *stupas*. One of the largest *stupas*, Jetawanarama Dagoba, built of bricks, is 76 m (249 ft) high. The city's sacred Bo tree, more than 2000 years old, is said to have been grown from a cutting taken from the tree in India under which the founder of Buddhism, Prince Siddartha Gautama, found enlightenment in the 6th century BC.

Population 36 200
Map Sri Lanka – Ba

Anvers *Belgium* See ANTWERP

anvil cloud A wedge-shaped cloud formed by the flattening of the top of a large convective cloud as it reaches the base of the stratosphere (see THE DIFFERENT LAYERS OF THE ATMOSPHERE), spread-ing towards high-level winds. Anvil clouds are normally storm clouds, bringing heavy showers.

Anyang *China* City of northern Henan Province about 470 km (300 miles) south-west of Beijing (Peking) in the WEI Valley. From 1300 to 1066 BC Anyang was the capital of the Shang civilisation of Ancient China.

Population 500 000
Map China – Hd

Aomori *Japan* Capital of HONSHU ISLAND's northernmost district. It is an industrial city and ferry port on the island's north coast and was extensively damaged during the Second World War. The city has been rebuilt and is home to the Nebuta festival in August, when papier-mâché dummies of men and animals are paraded through the streets.

Population 288 000
Map Japan – Db

Aóös Potamós *Greece/Albania* See VIJOSE

Aosta *Italy* Iron and steel town in the French-speaking autonomous Valle d'Aosta 80 km (50 miles) north of Turin. It is a tourist stage on the roads through the Grand St Bernard Pass into Switzerland and through the Little St Bernard Pass and the Mont Blanc tunnel into France.

The modern city faithfully preserves the Roman street pattern. Its Roman monuments include the theatre, the old city wall with 20 turrets, the Praetoria Gate, the Arch of Augustus and remains of the Forum. In the centre of the town is the large Piazza Chanoux, fronted by the 19th-century town hall.

Population (town) 35 900; (region) 115 400
Map Italy – Ab

apatite Naturally occurring form of calcium phosphate. It is the main source of phosphates for agricultural fertiliser.

Apatity *Russia* Mining town beside Lake Imandra, 120 km (75 miles) north of the Arctic Circle in the Kola Peninsula. Nearby is the world's largest deposit of apatite – a phosphate mineral used to manufacture fertilisers.

Population 74 100
Map Russia – Eb

Apennines *Italy* Mountains forming the back-bone of Italy, giving the peninsula its distinctive bootlike shape. They start behind the Genoese riviera and wind 1350 km (839 miles) to the tip of the Italian toe. The highest peak is Mount Corno (2914 m, 9560 ft), north-east of Rome.
Map Italy – Cb

Aphrodisias (Geyre) *Turkey* Ruined Roman city about 500 km (300 miles) south-west of Ankara. As the capital of the province of Caria during Byzantine times, it was named after Aphrodite, the Greek goddess of love. However, it was aban-doned after an earthquake in the 14th century, and remained virtually unknown until the 1960s.
Map Turkey – Ab

Apia *Western Samoa* Capital and main port on the north coast of Upolu Island. The main indus-tries are timber milling, handicrafts and the pro-cessing of copra and cacao for export. In March 1889, during a power struggle for the control of Samoa, four German and American warships were wrecked in Apia harbour during a fierce hurricane, two more were beached and damaged, and only one British ship survived; 144 people died. Robert Louis Stevenson, author of *Treasure Island*, was buried on the slopes of Mount Vaea, south of Apia, in 1894.

Population 36 000
Map (Upolu Island) Pacific Ocean – Ec

Apo, Mount *Philippines* See DAVAO

Apollonia *Libya* Ancient town of Cyrenaica, 30 km (19 miles) east of Al Bayda. It was the port for the Greek city of Cyrene and the birthplace in about 276 BC of the astronomer Eratosthenes, who first measured the size of the earth.
Map Libya – Ca

Appalachian Mountains (Appalachians) *North America* Eastern ranges 200-300 km (124-186 miles) wide and stretching 2570 km (1600 miles) south-west from the St Lawrence River in Canada to the state of Alabama in the United States. They rise to 2037 m (6684 ft) at Mount Mitchell in North Carolina, and are rich in timber and, in places, coal. The attractive ranges are dotted with national parks.
Map United States – Jc

Appenzell *Switzerland* Canton divided into the 'half-cantons' of Appenzell Innerrhoden, which is mainly Catholic, and Appenzell Ausserrhoden, which is Protestant. This was the result of the Wars of Religion in the late 16th century. It lies in scenic, mountainous country south of Lake Constance. The industrial town of Herisau (population 15 900) is the capital of Appenzell Ausserrhoden. The resort of Appenzell (5400) is the capital of Appenzell Innerrhoden. The canton covers 415 km² (160 sq miles).

Population (Innerrhoden) 14 800; (Ausserrhoden) 54 200
Map Switzerland – Ba

Appleton layer The F layer. See THE DIFFERENT LAYERS OF THE ATMOSPHERE

Apulia *Italy* See PUGLIA

Apurímac *Peru* Southern department in the Andes. It is one of the country's remotest and poorest regions, the economy depending largely on a declining sugar cane industry. Its capital, Abancay, is an administrative centre.

Apurímac is also the name of a river, 918 km (570 miles) long, rising in Lake Vilafro in the Andes. It flows north to unite with the Urubamba River at Atalaya and form the Ucayali River, an upper tributary of the Amazon. For short stretches in its lower course it is called the Ené and Tambo. Cofee is grown in the Ené Valley.

Population (department) 371 700
Map Peru – Bb

Apuseni Mountains *Romania* Mountain range in western TRANSYLVANIA, rising to a height of 1848 m (6063 ft). Known also as the Bihor Moun-tains, the range contains deep gorges, wooded valleys, and caves cut into the limestone rocks .
Map Romania – Aa

Aqaba *Jordan* The country's only port, situated 100 km (62 miles) south of Ma'an. Set at the head of the Gulf of Aqaba on the Red Sea, it handled large volumes of transit trade for Iraq during the Iraq–Iran War (1980-8). It is also the main export base for Jordan's phosphate industry. There has been some expansion on three fronts: to Aqaba's port, to its industrial zone, and to its growing resort area. The tropical climate provides ideal holiday weather for nine months of the year and the beautiful underwater scenery and clear waters for most of the year attract many divers.

In ancient times, Aqaba (then called Elath) was on the Arabian–Egyptian caravan route, and nearby archaeological remains show that it was an important settlement in 1000 BC. After spells under the Roman and Byzantine empires, Aqaba was taken in AD 639 by Omar, who made it Muslim. It flourished as a port for pilgrims to Mecca, and for a brief period in the 12th century was occupied by the Crusaders. However, it lost much of its importance with the opening of the Suez Canal in 1869.

Population 46 000

Map Jordan – Ab

Aqaba, Gulf of An arm of the RED SEA, 160 km (98 miles) long and 19-27 km (12-16 miles) wide. It is part of the GREAT RIFT VALLEY system and separates Egypt's Sinai Peninsula from north-western Saudi Arabia. In the north are the Israeli port of ELAT and Jordan's only port, AQABA. The gulf was an important waterway in Biblical and Roman times. Egypt's blockade of the gulf in 1967 was a major cause of the Six Day War, when Israel occupied Sinai.

Map Egypt – Cb

aquifer Water-bearing rock or rock formation.

Aquitaine *France* Lowland agricultural region – formerly a duchy and kingdom with slightly different borders – in the south-west, roughly triangular in shape and crossed by the Garonne River. It covers 41 407 km² (15 987 sq miles). BORDEAUX – a wine trade centre – and, for a period, TOULOUSE were its capitals, and are still major centres. The marriage in 1152 of Eleanor of Aquitaine to Henri Plantagenet, who became Henry II of England in 1154, heralded 300 years of English rule that ended with the Hundred Years' War (1337-1453).

Population 2 795 800

Map France – Cd

Ar Raqqah *Syria* Town on the east bank of the Euphrates River, 120 km (75 miles) north-west of Dayr az Zawr. Since the completion, in 1978, of the Euphrates Dam – 60 km (37 miles) downriver – the town has received a new lease of life. Irrigation has made possible farming and cattle rearing, among other agricultural developments.

Ar Raqqah contains the remains of a 9th-century palace built by the Abbassids – an early Arabic dynasty that took its name from its founder, Al-Abbas (AD 566-652), the uncle of the prophet Muhammad.

Population 218 000

Map Syria – Bb

Ar Rusayris *Sudan* See ER ROSEIRES

Arabian Gulf See GULF, THE

Arabian Sea Part of the INDIAN OCEAN, which is still dotted with Arab sailing vessels or dhows. It covers 3 860 000 km² (1 490 000 sq miles) and has two arms: the oil-rich GULF and the Gulf of ADEN. Leading ports on the sea include ADEN (Yemen), BOMBAY and COCHIN (both India), and KARACHI (Pakistan).

Map Indian Ocean – Cb

Arafura Sea An extension of the south-western PACIFIC OCEAN lying between Australia and New Guinea. In the west it merges into the TIMOR SEA, while the TORRES STRAIT in the east leads to the CORAL SEA. The Arafura Sea is 1290 km (700 miles) long and 560 km (350 miles) wide.

Map Indonesia – Hd

Aragón *Spain* Thinly populated region in the north-east of the country covering 47 669 km² (18 405 sq miles). It became an independent kingdom in 1035 and from here the Iberian Peninsula was recaptured from the Moors.

It was united with CATALONIA in the 12th century and with CASTILE in the 15th century. Within its borders are the modern provinces of HUESCA, TERUEL and ZARAGOZA.

Population 1 188 820

Map Spain – Eb

Arakan *Myanmar (Burma)* Coastal region covering 36 778 km² (14 200 sq miles) on the Bay of Bengal. It includes several offshore islands, of which the largest are Ramree and Cheduba. Arakan is cut off from the rest of the country by the heavily forested Arakan Hills, which rise over 1700 m (5600 ft). SITTWE (formerly Akyab) is the state capital.

Population 2 046 000

Map Myanmar – Bb

Aral Sea (Aral'skoye Morye) *Kazhakstan/ Uzbekistan* Once the fourth biggest inland sea, the Aral Sea has nearly halved in size within 10 years. It has shrunk from 65 500 km² (25 300 sq miles) to 37 000 km² (14 280 sq miles) as a result of the damming and diversion for irrigation of the rivers which feed it – mainly the Amudar'ya and Syrdar'ya.

The drying up of vast areas has had catastrophic results for those who lived on its shores and relied upon the sea for fishing and irrigation. What used to be fishing villages are now far inland, cut off from their former livelihood.

Map Kazakhstan – Cb; Uzbekistan – Aa

Aran Islands (Oileáin Arann) *Ireland* Three limestone islands in Galway Bay, 8 km (5 miles) from the mainland. The islands – Inishmore, Inishmaan and Inisheer – have been occupied since prehistoric times by fishermen and subsistence farmers, and fishermen still put to sea in canvas and tar currachs. Seaweed and sand have been laid on the bare limestone to make fields on which to grow potatoes; other crops grow behind high walls. The remains of a powerful pre-Christian fortress, Dun Aengus, stand on cliffs 90 m (295 ft) high on Inishmore. Tourism flourishes in this Irish-speaking area, and more than 200 000 visitors arrive every year by ferry or light aircraft.

Population 1320

Map Ireland – Bb

Ararat (Büjük Agri Dagi) *Turkey* Volcanic mountain massif in the east Anatolian highlands close to the Armenian and Iranian borders, some 1050 km (650 miles) east of Ankara. It is the Biblical landing place of Noah's Ark. The two main peaks are Great Ararat which, at 5165 m (16 945 ft), is Turkey's highest mountain, and Little Ararat, at 3907 m (12 818 ft).

Map Turkey – Cb

Arauca *Colombia/Venezuela* Tributary of the ORINOCO River, rising in Colombia and with a length of about 1000 km (620 miles), of which 790 km (490 miles) are navigable. It forms part of the Venezuela-Colombia border.

Map Venezuela – Bb

Araucanía *Chile* Region in the south, and one of the most picturesque lake districts in the world.

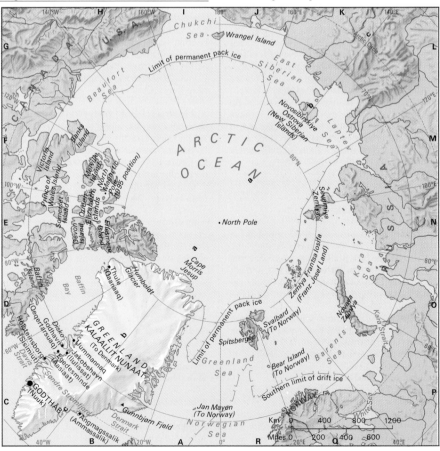

Perhaps the best known of the lakes is Villarrica and its resort Pucón, both overshadowed by the active Villarrica volcano (2840 m, 9317 ft). Araucanía is pioneer country – a land of medium-sized farms and a growing timber industry. The capital is TEMUCO.

Population 774 960

Map Chile – Ac

Arbe *Croatia* See RAB

Archaeozoic era The older Precambrian era, when the first life forms evolved. The name comes from the Greek and means 'primeval life'. The rocks of this period are at least 1750 million years old. See GEOLOGICAL TIMESCALE.

Archangel (Arkhangel'sk) *Russia* Major port on the North Dvina delta 52 km (32 miles) from the White Sea. It is ice-free for six months of the year. Its major export is timber, and its industries include shipbuilding, and pulp and paper making. The town takes its name from a nearby 16th-century monastery, named after the Archangel Michael.

Population 420 000

Map Russia – Fb

archipelago Group of islands clustered closely together.

Arcot *India* Agricultural town about 100 km (60 miles) west of the city of Madras. It is the former capital of the Nawabs, native rulers, of Carnatic – an independent region in south-east India bordering on the Bay of Bengal. In the 18th century it was the scene of bitter fighting between British troops led by soldier and statesman Robert Clive and the occupying French.

Population 114 880

Map India – Ce

Arctic Only the Inuit ('Eskimos') have learned to make their homes in the white wilderness of the high Arctic – the islands and the frozen sea north of the ARCTIC CIRCLE. And only a handful of people from any nation have reached the NORTH POLE at its heart, though thousands fly near it regularly on scheduled airline flights.

Apart from a number of Americans, Canadians and Russians in floating research stations, nobody lives on the ice pack. But people have for centuries made their homes along the Arctic coasts and in the tundra regions inland. Some of these polar peoples, such as the Inuit of North America and Greenland (Kalaallit Nunaat), and the Aleuts of the Aleutian Islands off Alaska, traditionally live by hunting. Others, including the nomadic Saame (Lapps) of northern Scandinavia and the Chukchis of north-eastern SIBERIA, keep herds of reindeer, from which they obtain meat, milk, fur, leather and (in the form of dried dung) fuel.

The polar Inuit, the world's most northerly people, live in a land of permanent ice in THULE, north-west Greenland, less than 1600 km (1000 miles) from the pole. Farther north still is the world's most northerly land: a tiny island off northern Greenland called Oodaq.

Despite the ferocious climate of the Arctic, some outsiders have moved in over the years. The motive for the early explorers' doing so was largely the hunt for a NORTHWEST PASSAGE – a route around North America which would nearly halve the sea distance from Europe to the Far East. In 1903-6 the Norwegian explorer Roald Amundsen made the first voyage through the Passage, but it was not until September 1969 that a giant ice-breaking oil tanker, the *SS Manhattan*, became the first ship to batter its way straight through the pack ice over the route.

Air, road and rail links have so far been more successful in opening up the lands of the Arctic to exploitation. In northern Alaska and Canada, the impetus has come largely from the discovery of Arctic oil fields, particularly on the North Slope of Alaska in 1968. The links have also made it easier for tourists to visit the far north.

On the other side of the Arctic, some 3000 Russian and Norwegian miners on SPITSBERGEN produce between them about 800 000 tonnes of coal a year from the bleak islands' huge reserves. And drilling ships from both nations are hunting for oil and gas in the chilly waters of the surrounding Barents and Norwegian seas.

Map See opposite page

Arctic Circle An imaginary circle around the earth at latitude 66°32´ north, with the North Pole at its centre. It represents the southernmost latitude at which, at the northern summer solstice (about June 21), the sun does not set, and at the winter solstice (about December 21), the sun does not rise all day.

arctic climates Those climates in which no month has a higher average temperature than 10°C (50°F). They are divided into tundra climates, where in summer temperatures there is some growth of vegetation, and polar climates, where no growth is possible. Similar climates occur at high altitudes as well as high latitudes such as the Arctic and Antarctic.

Arctic Ocean The world's smallest ocean, surrounded by northern Europe, Asia and North America and covering an area of 14 056 000 km² (5 425 770 sq miles). The North Pole is near its centre, and pack ice, 2-3 m (7-10 ft) thick, covers much of the ocean throughout the year.

Many rivers flow into the Arctic, whose greatest known depth is 5570 m (18 274 ft), and it has a lower salt content than the other oceans. The strongest currents flowing out of the Arctic are the East Greenland and Labrador currents, which carry icebergs from Greenland (Kalaallit Nunaat) into the North ATLANTIC OCEAN.

Balancing these outflowing cold currents is the warm water of the North Atlantic Drift which flows into the Arctic between Iceland and Scotland. Some water also flows through the BERING STRAIT, between Alaska and Russia. The Bering Strait is a key summer trade route for Russian merchant ships.

Ardabil (Ardebil) *Iran* Carpet-making market town about 70 km (43 miles) west of the Caspian Sea. Founded in the 5th century, it was destroyed by the Mongols in 1220, then in the late 13th century became a place of pilgrimage devoted to the Sunni mystic, Safi al-Din, who was born here in 1253. In 1828 the town was occupied by the Russians, who carried off his magnificent library, once the greatest in Persia. The mausoleum of Shah Ishmail (1486-1524), the founder of Shiite Islam, is in the town.

Population 311 000

Map Iran – Aa

Ardéche *France* River, 120 km (75 miles) long. Rising in the Cévennes Mountains, the river runs for 30 km (18 miles) of its course through deep limestone gorges as it descends to join the Rhône River. The most spectacular feature of these gorges is the 34 m (112 ft) high Pont d'Arc, a natural limestone bridge which arches some 59 m (194 ft) across the river.

The department (5523 km², 2132 sq miles), culminating in the peak of Mount Mézenc (1723 m, 5653 ft), lies west of the Rhône. The region is sparsely populated; agriculture includes vines and fruit and there are textile and engineering industries.

Population (department) 283 900

Map France – Fd

Ardennes (Ardenne) *Western Europe* Forested plateau in south-east Belgium, Luxembourg and north-east France. The plateau rises in eastern Belgium to form the Hautes Fagnes (Hohes Venn) range, which is largely a nature reserve. Botrange is the highest peak at 694 m (2277 ft). The sparsely populated Ardennes with its severe winters supports forestry, limestone quarrying and tourism. It was the scene of heavy fighting during both World Wars, particularly when German forces tried to turn back the Allied advance into Europe in late 1944.

Population 350 000

Map Belgium – Ba

Ardennes *France* See CHAMPAGNE-ARDENNES

Arecibo *Puerto Rico* Seaport on the north coast, about 70 km (43 miles) west of SAN JUAN. The main industry is rum distilling.

Population 93 390

Map Caribbean – Bb

Arenal *Costa Rica* Volcano 170 km (106 miles) north-west of the capital, San José. About 100 people died in the area in 1968 when the 1633 m (5358 ft) high volcano erupted for the first time in 500 years.

Map Costa Rica – Ba

Arendal *Norway* Old established port, chartered in 1723 on the Skagerrak coast of south Norway 65 km (40 miles) north-east of Kristiansand. It specialises in the timber trade and shipbuilding and a ferry to Denmark departs from it.

Population 38 400

Map Norway – Cd

Arequipa *Peru* Southern department, lying in a fertile plain fringed by volcanic cones. The provincial capital, also Arequipa, lying at the foot of the dormant volcano El Misti (5822 m, 19 100 ft) was an Inca town, refounded by the Spanish in 1540. It is sometimes known as the Ciudad Blanca ('White City') because most of the buildings are of white volcanic stone. A fine example is the Convent of Santa Catalina, built in 1580. The city is the second largest in Peru, the seat of a university and an industrial and commercial centre.

Population (department) 965 000; (city) 634 500

Map Peru – Bb

arête Sharp, narrow mountain ridge or spur, specifically a ridge between two adjacent CIRQUES, such as the Striding and Swirral Edges on Helvellyn in the English Lake District.

Argentina

THE SECOND LARGEST SOUTH AMERICAN COUNTRY HAS BEEN STRUGGLING FOR POLITICAL AND ECONOMIC STABILITY

Most Argentines think of themselves as different from other South Americans. They consider themselves more sophisticated and better educated. Indeed, European influence is very strong, and the capital city of BUENOS AIRES, where one-third of the population lives, is like a European city. Most of the people are descended from European immigrants, and there is a substantial British community in Buenos Aires – a legacy of the days when the British arrived to build the railways.

Once, Argentina was more prosperous than the other South American countries, but it has failed to fulfil its promise – and the rest of the continent has been catching up.

In 1914, Argentina was the source of much of the world's meat and cereals, the focus of a major migration of people from southern Europe, and seemed destined to join the ranks of the developed nations. Argentina was never to achieve this status, but it was left with a sense of its own importance which has not lessened now that it is only a middle-ranking nation in terms of power and wealth.

Argentina, the eighth largest country in the world (five times the size of France) and the second largest in South America (after Brazil), has a population of more than 33 million and a capital which is one of the world's great cities. The first Latin-American Nobel Prize went to an Argentinian, Carlos Saavedra (a Peace Prize in 1936).

Argentina rose to prosperity from unprepossessing beginnings. The Spanish conquerors who arrived here in the 16th century found no gold or silver, and little to interest them in the area that is now Argentina. Until 1713, Argentina was administered from LIMA, the capital of PERU, under the government of Spain. The first step towards independence was with the founding of the Viceroyalty of Río de la Plata (1776). After uprisings against the Spanish colonialists the United Provinces of Río de la Plata declared their independence in 1816. There followed a civil war (1816-53) – at the end of which Argentina as we know it today was constituted.

WEALTH FROM THE PLAINS

The main source of wealth for Argentina was the vast fertile plains of the PAMPAS, which occupies the central part of the country. The major exports were at first wool and mutton from these grasslands, and then beef and hides. Argentina also became one of the world's greatest sources of wheat and maize.

The great period of economic growth occurred between about 1853 and 1930. Britain poured money into the country, the railways were built and a tide of immigrants arrived, mostly Spaniards and Italians fleeing from poverty in Europe. The newcomers soon outnumbered the so-called 'Indian' tribes (Amerindians) and the *mestizos* – mixed-blood descendants of the Amerindians and early Spanish settlers.

The towns grew, especially Buenos Aires as the seat of national government and the port for the increasing exports and imports. The boom was marred by the First World War in Europe, and was virtually ended by the worldwide depression of the 1930s.

From 1914 onwards, Argentina started to establish manufacturing industries: iron and steel, motorcars, clothes and electrical goods. By 1950 the agricultural society had turned into an industrial one, and most Argentines earned their living in the cities. But the manufacturing sector was not able to compensate for a decline in agriculture as Argentina's share of world food exports shrank dramatically. Since the 1930s, Argentina's economy has been the source of grave social and political problems: the country fell from a position of seventh richest country in the world to 77th position in the 1960s. Some blame this decline on the populist leader Juan Perón, whose economic policy after the Second World War favoured urban workers to the detriment of agriculture. Others blame British and United States corporations for draining profits from the country.

POLITICAL TURBULENCE

Whatever its causes, economic trouble has always accentuated the political turbulence of Argentina. Governments were short-lived until the era of Colonel Perón and his wife Eva, from 1946 to 1955. Perón was a nationalistic, authoritarian president who endeared himself to

industrial workers by raising their wages repeatedly even though productivity was falling. His neglect of agriculture also helped speed the process towards economic decline. Perón's position was weakened by the death, in 1952, of his wife – a popular figure with the working masses – and a quarrel with the Roman Catholic Church, and in 1955 he was deposed by a military and civilian revolution.

A series of military regimes followed, until in 1973 Perón made a short-lived and controversial comeback. He was succeeded after his death in 1974 by his second wife, Maria, who was herself ousted in 1976. The followers of Perón are still one of Argentina's main political parties, although Maria resigned from leadership of the party in 1984.

Another succession of military governments was ended by the disastrous Falklands venture in 1982. The army invaded the British Falkland Islands, which Argentina calls the Malvinas and claims as its territory. British forces reinvaded during a 10-week war and established a garrison on the islands.

The conflict cost Argentina 750 lives and led to the ousting of a military junta headed by General Leopoldo Galtieri, and a return to civilian rule after a democratic election in December 1983. The new president, Raúl Alfonsín, ordered an inquiry into atrocities committed by previous military regimes. The inquiry revealed that 9000 people had been killed, a further 9000 imprisoned and tortured, and 2 million had fled the country.

The collapse of the economy led to an election victory, in 1989, for Carlos Saúl Menem, who tackled economic problems, for instance by reintroducing privatisation. Between 1989 and 1991, inflation dropped from 58 per cent to 2.2 per cent; between 1991 and 1993, the economy grew by an annual 8 per cent. This was not without a price – the rich became noticeably richer, the poor poorer. By the first quarter of 1994, more than 2 million Argentinians (18 per cent of the workforce) were unemployed, and the government was in arrears in paying what little social support it was still providing. Constituent Assembly elections in April 1994, while bringing victory to Menem and the Peronists, failed to bring them the desired majority vote.

LAND OF THE GAUCHOS

The geography of Argentina is on a grand scale. To the west, the massive chain of the ANDES forms the border with Chile and Bolivia. It is the longest chain of mountains in the world, containing South America's highest mountain, ACONCAGUA.

To the south stretches the huge windswept expanse of PATAGONIA and the breathtaking mountains and lakes around BARILOCHE. It is the geography and culture of the central plains, the pampas, that are best known abroad – a vast area of flat land that accounts for most of the population outside the cities.

The pampas are the home of Argentina's cowboy heroes, the *gauchos*, who ride the treeless plains on horses that never need to jump a fence. Miles of grass are punctuated by pylons and windpumps for underground water. A few trees may shelter the villages of the pampas,

and eucalyptuses often line the approaches to the cattle ranches where the *gauchos* live and work. The *gauchos*, mostly *mestizos*, wear dress which is romantic to foreigners: baggy knee-length trousers called *bombachas*, with a striped *chiripá* scarf wrapped around the waist and between the legs.

Away from the pampas are several other Argentinas with different lifestyles. The northwest resembles the high Bolivian plateau or *altiplano*. The population here is still mostly Amerindian and *mestizo*, and the economy agricultural. The land is lush with subtropical vegetation, and for this reason the area is known as the Garden of Argentina. The colonial towns with their Spanish churches and

market gardens, as well as some 5 million olive trees. An oil refinery nearby brought growth to the city of Mendoza. Three days' train journey south-west of Buenos Aires, the region around Bariloche rivals Switzerland with the grandeur of its blue lakes and snow-capped Andes, its waterfalls, glaciers and forests.

South of this alpine region the plateaus of Patagonia stretch their desolate way to TIERRA DEL FUEGO, through areas where colonies of elephant seals bask on the pebble beaches that line the Atlantic coast.

In 1865, settlers from Wales came to farm sheep in Patagonia, and today two languages are still spoken in some remote parts of the southern Andes: Spanish and Welsh.

▲ **MAN AND HIS DOGS** Sheep rearing has been the main occupation on the barren tablelands of Patagonia since Welsh settlers arrived there in the middle of the 19th century.

cathedrals recall the days when this region thrived before the rise of Buenos Aires and the shift of activity southwards.

The north-east is sheep-rearing and cotton country: the GRAN CHACO Plains. Farther east, between the PARANÁ and the URUGUAY rivers, is the country which grows bitter maté tea – the *gauchos'* favourite drink. And farther east still, at the point where Argentina, Brazil and Paraguay meet, lies one of South America's greatest natural wonders.

In the middle of the jungle, where the orange-red earth seems rusted by spray, the IGUAÇU FALLS thunder 82 m (269 ft) over a giant horseshoe 4 km (2.48 miles) wide – a plunging wall of water, higher than Niagara and several times as wide.

Lying in the foothills of the Andes in the west is the wine-growing region around MENDOZA. This fertile area, watered by the Mendoza River, is known as the Garden of the Andes. Though Chile today exports five times as much wine as Argentina, the Mendoza reds – Cabernet, Merlot and Malbec – are still well known and well loved worldwide.

Besides vineyards, the region has extensive

ARGENTINA AT A GLANCE	
Official name Argentine Republic	
Area 2 766 889 km² (1 068 296 sq miles)	
Population 33 533 256	
Per km² 12 (**Per sq mile** 31)	
Capital Buenos Aires	
Government Federal republic	
Currency 1 peso = 100 centavos	
Language Spanish (official), Welsh and others	
Religions Christian (94% Roman Catholic), Jewish and Muslim minorities	
Climate Tropical and subtropical in the north; subarctic in the south. Average temperature in Buenos Aires ranges from 9°C (48°F) in July to 23°C (73°F) in January	
Land use Grazing 52%, forest and woodland 22%, cultivation 13%, other 13%	
Main primary products Wheat, maize, potatoes, sorghum, sugar, grapes, apples, citrus fruit, soya beans, sunflower seed, cotton, cattle, sheep, fish, timber; coal, oil, natural gas, uranium, iron ore, copper, lead, zinc	
Major industries Steel, food processing, textiles, chemicals, machinery, motor vehicles, petroleum refining, mining, fishing, forestry, wood pulp and paper, wine	
Main exports Wheat, fodder (especially soya beans), meat, maize, wine, leather, wool, metals, machinery	
Annual income per head (US$) 6050	
Population growth (per thous/yr) 11	
Life expectancy (yrs) Male 68 **Female** 75	

A DAY IN THE LIFE OF A NAVAHO

The Grand Canyon in the American state of Arizona is striped pink and orange in the late afternoon sun as Rose makes her last sale of the day. Her stall, on a road approaching the canyon, offers a reasonable living out of selling Navaho crafts to the tourists who come to visit what is arguably the most magnificent natural wonder in North America.

Spread over the table in front of Rose is an assortment of silver and turquoise jewellery. The designs are traditional and modern Navaho. These artefacts are fitting symbols of the blend between native American traditions and the outside world.

Rose is dressed in a sweatshirt and pants; her little boy, Chuck, in T-shirt and jeans, fiddles with his personal stereo. Rose grew up on a reservation, among a people who had not yet adjusted to a life imposed on them by outsiders. The reservation land to which the Navaho's plateau had been reduced could not support the seminomadic existence they once knew – an existence of herding livestock.

In some ways, life is easier now. Native American rights to self-determination and to own their land have been recognised in Congress. Many today find work outside the reservations, but the migratory trend away from reservations has begun to slow. Quite a few Navaho are returning to their reservations, where jobs are becoming more plentiful.

Rose's husband, Jack, has a job as a driver on a large cattle ranch. They also have a small income from their share in the lease of mineral rights to outside companies that work on pieces of Navaho land. Industrial developers can only lease this land, they will never be allowed to buy it.

Rose and Jack built a neat, wood-lapped bungalow on the Navaho reservation where they grew up. It is important to protect this arid land from overexploitation and overgrazing: the Navaho believe that the earth is a living being – if you take something out of the ground, it is like digging into your own skin. Their house, though, has no piped water or mains sanitation. They make do with an earth closet and with trucking in water.

At home, Rose spends the evening assembling more jewellery; as a treat, she sometimes tells her son – in Navaho – stories about the famous Navaho chiefs, and about the spirits of the mountains. Chuck hears them at school, too; he goes to school with native American teachers and classmates. But mostly his lessons are the same as those given to other students in public schools. Rose hopes he will go to college one day but not lose his Navaho identity.

With Chuck in bed, Rose and Jack discuss how the Navaho nation might best use their land in the modern world. They have already won many battles for the preservation of their rights and territory.

Arezzo *Italy* City and province in Tuscany, 60 km (35 miles) south-east of Florence. The centre of the medieval city contains the church of Santa Maria, a rare example of unmodified Tuscan Romanesque, whilst the Gothic church of San Francesco contains frescoes by Piero della Francesca (1420-92). The house of the poet Petrarch, born in Arezzo in 1304, stands near the cathedral. The Archaeological Museum is built on the remains of a Roman amphitheatre. The province is agricultural, peppered with hilltop villages and small market towns. Shoes, clothing and furniture are the main industries.
Population (city) 90 580; (province) 312 940
Map Italy – Cc

Argentina See p 44

Árgos *Greece* Small market town built on the ruins of a 7th to 6th-century BC city on the Argolis Plain 10 km (6 miles) north-west of the port of Nauplia in the north-east Peloponnese. The market place and theatre of the Roman settlement have been excavated, but two citadels are partly buried by the modern town. Legend has it that Jason set sail from the port with the Argonauts to find the Golden Fleece.
Population 20 700
Map Greece – Cc

Argyllshire *United Kingdom* See HIGHLAND; STRATHCLYDE

Argyrokastron *Albania* See GJIROKASTER

Århus *Denmark* Once a Viking trading post, now the country's second largest city and port after Copenhagen, and the capital of Jutland. Its botanical garden has an open-air museum comprising an old town called Den gamle By, consisting of an interesting collection of re-erected 17th- and 18th-century buildings. Århus also has a fine cathedral, Denmark's first modern university, founded in 1934, a music academy and a concert hall. The museum of prehistory houses the 2000-year-old 'Grauballe Man' whose remains were found in the 1950s preserved in the bog of Grauballe Fen, 40 km (25 miles) west of Århus. Industries include clothing factories, processed foods and paper manufacturing.
Population 264 100
Map Denmark – Ba

Arialkhan *Bangladesh* Major branch of the Padma River (see JAMUNA).
Map Bangladesh – Cc

Arica *Chile* Coastal town near the Peruvian border. It is connected by rail to the Bolivian capital, La Paz, and by a modern highway to Tambo Quemado. It handles half of Bolivia's foreign trade. Its mild climate, almost constant sunshine and golden beaches have made the town a popular holiday resort. The Cathedral of St Marcos in the town's Plaza de Armas is built on an iron frame designed by the French engineer Gustave Eiffel (1832-1923).
Population 195 000
Map Chile – Aa

Arizona *USA* State covering 294 400 km² (113 642 sq miles), and the last state to become a part of the continental United States, in 1912. It lies in the south-west, and has some of the most famous scenery in the world, including the GRAND CANYON. The south is drained by the Salt and Gila rivers, which provide hydroelectricity and irrigation. Rainfall is low and there are extensive areas of desert and scrub. The chief farm products are cattle, dairy goods, cotton and vegetables. Arizona is the nation's principal producer of copper and also produces silver, gold, molybdenum and uranium.

Early in the 16th century, the Apache people moved into what is now Arizona and New Mexico from their original home in western Canada. About 25 per cent of the state is 'reserve' area, including Monument Valley straddling the border with Utah, which, with its enormous flat-topped rock formations, has been the setting for many famous Western movies. The capital of the state is PHOENIX.
Population 3 665 230
Map United States – Dd

Arkansas *USA* Central southern state lying immediately west of the Mississippi River. The broad plain of the Mississippi covers the east, from which the land rises to the forested Ozark Plateau and Ouachita Mountains. The state covers 137 539 km² (53 104 sq miles) and produces oil, gas, manganese, some diamonds and bauxite (aluminium ore). The main farm products are chickens, soya beans, rice and cotton. The capital is LITTLE ROCK.

The Arkansas River rises in the Rocky Mountains of Colorado and flows about 2335 km (1450 miles) east, then south-east through Kansas and Oklahoma into Arkansas, where it joins the Mississippi River.
Population (state) 2 350 000
Map (state) United States – Hc; (river) United States – Fc-Hc

Arkhangel'sk *Russia* See ARCHANGEL

Arlanda *Sweden* See STOCKHOLM

Arlberg Pass *Austria* Mountain pass, rising to 1793 m (5882 ft) in western Austria, which links Vorarlberg with the Tyrol. The Arlberg massif forms part of the watershed between the Rhine and the Danube basins. A 14 km (8.6 mile) long railway tunnel, opened in 1884, runs beneath the pass between Stuben in Vorarlberg and St Anton in the Tyrol.
Map Austria – Bb

Arles *France* Chief town in the south of the Roman province of Gaul, situated in south-east France, 95 km (59 miles) north-west of the city of Marseilles. It is now a centre for agricultural marketing and tourism. A vast Roman arena and necropolis can still be seen today and there are museums depicting Roman and Provençal life in the past. The Dutch painter Vincent van Gogh (1853-90) spent the last and most productive years of his life in the town – completing, in one frenzied 15-month period there, more than 200 pictures.
Population 52 600
Map France – Fe

Arlit-Akokane *Niger* Remote region about 275 km (170 miles) north-west of the oasis of Agadez. It is named after two towns: Arlit and Akokane. Uranium has been mined since 1971.
Map Niger – Ab

Armenia

THE COUNTRY THAT FINALLY GAINED INDEPENDENCE AND WENT TO WAR WITH AZERBAIJAN

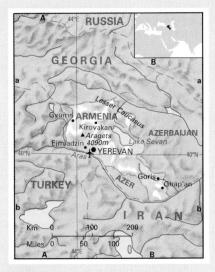

Situated strategically between the Mediterranean region and Central Asia, Armenia has been fought over by all the powerful nations that have been its neighbours. Its history is punctuated with conflicts – notably with Kurds, Turks and Persians. Originally Armenia included a much larger area, centred on Mount ARARAT, the traditional resting place of Noah's Ark. But the country was partitioned in AD 387, and is now divided between Turkey, Iran and Armenia.

After the Islamisation of the Near East, the Christian Armenians were ostracised by their Muslim neighbours and were persecuted well into the 20th century by Kurds and Turks, deported in their hundreds of thousands, and brutally murdered. Armenians today, like the Jewish people, are scattered all over the globe. At least as many live in other parts of the former Soviet Union, the Near East, France and the United States as live in Armenia.

Despite the years of persecution, deportation, massacres and occupation, Armenian culture has remained intact. More than 90 per cent of the country's population is Armenian, thus forming a cultural majority. The Armenian Church, in existence since the 2nd century, has also played a significant role in preserving culture and national unity. In the 4th century, Armenia became the first country in the world to make Christianity the state religion; and the Church remains a distinct sect of Orthodox Christianity.

Ninety per cent of the country lies at an altitude of over 1000 m (3281 ft); more than 50 per cent of its population is concentrated in the remaining lower-lying land. Here they live in crowded apartment blocks in YEREVAN, the capital, and on the fertile plains. By contrast,

the Lesser Caucasus in the north of the country, the plateau of Ararat in the west, the Armenian highlands, and the volcanic ranges in the central region are sparsely populated.

Compared with the barren highlands of neighbouring Anatolia and Iran, Armenia is well endowed with raw materials. The volcanic soil is fertile; and, though rainfall in the lower lying regions is no more than 200-500 mm (about 8-20 in) a year, there are numerous sizable mountain streams which provide water for hydroelectric energy and crop irrigation. Citrus fruits and cotton are grown and dairy farming is pursued. Grapes harvested around Yerevan are used for the local wine-making and brandy-distilling industry.

Copper, bauxite, zinc, molybdenum and some gold are among the mineral resources, yet mining is not well developed.

The country has a long tradition of producing crafts – the Armenian people were once renowned, way beyond their country's borders, for their weaving, carpet-making and gold-smithing skills. Today, Armenia has taken a leading position among members of the Commonwealth of Independent States (CIS) in the fields of mechanical engineering and electronics. Most industries are in Yerevan, but there are some industrial concerns in GYUMRI and Kirovakan, the two other large centres.

In 1988, Yerevan suffered a devastating earthquake which sent factories and apartment blocks tumbling down, burying people under rubble and killing about 25 000. The city's nuclear power plant, which had supplied about 40 per cent of the country's power needs, was destroyed, as was almost 10 per cent of Armenia's industrial capacity. The country has still not fully recovered from the disaster.

CONFLICT WITH AZERBAIJAN

Armenia's main exports are industrial goods, but it is largely dependent on the other ex-Soviet states for fuels and raw materials. Most of them are carried by rail through neighbouring Azerbaijan – but a dispute with Azerbaijan over Nagorno-Karabakh, an Armenian-populated enclave within the borders of Azerbaijan, saw this supply line being cut. War between the two states broke out shortly after Armenian

army troops had occupied Nagorno-Karabakh in February 1992 .The conflict has resulted in the deaths of thousands of inhabitants of both countries. Armenia has also suffered great economic loss as a result of sanctions imposed by its Islamic neighbours, Iran, Turkey and Azerbaijan.

Extensive land reform has helped reduce Armenian dependence on food imports. Nevertheless, when rebellion broke out in western Georgia and supplies were cut off from that country, Armenians faced serious shortages.

The history of Armenia goes back thousands of years. The Hayk, as the Armenians call themselves, have occupied the plateau between the Black Sea and the Caspian Sea, Caucasus and Taurus since the 7th century BC, and possibly even earlier. During their long and turbulent history they have had a few brief moments of sovereignty, when they could defend their country against foreign powers seeking to control it. When the country declared its independence from the Soviet Union in 1990, Armenians looked forward to a peaceful future – but prosperity remains elusive.

ARMENIA AT A GLANCE	
Official name Republic of Armenia	
Area 29 800 km² (11 506 sq miles)	
Population 3 481 210	
Per km² 117 (**Per sq mile** 303)	
Capital Yerevan	
Government Republic	
Currency 1 Russian rouble = 100 kopeks (transitional)	
Languages Armenian (official), Russian	
Religions Armenian Orthodox (94%); Muslim minority	
Climate Continental climate with cold winters and hot, dry summers. Average temperature in Yerevan ranges from –4°C (25°F) in January to 25°C (77°F) in July	
Land use Cultivation 29%, grazing 15%, other 56%	
Main primary products Copper, molybdenum, bauxite, zinc, some gold; cotton, citrus fruits	
Major industries Machinery, electronics, vehicles, textiles; food processing	
Main exports Electrotechnical industrial products, machinery, industrial tools, precision engineering, aluminium products	
Annual income per head (US$) 4710	
Population growth (per thous/yr) 12	
Life expectancy (yrs) Male 68 **Female** 75	

▼ **FERTILE VALLEYS Deep, fertile valleys cut through the volcanic mountain ranges of Armenia.**

Armagh *United Kingdom* City and county in Northern Ireland. The city lies 16 km (10 miles) from the Irish border. It has two cathedrals and is the ecclesiastical capital of the whole of Ireland. There is evidence of settlement from as early as 600 BC and the Irish patron saint, Patrick, is said to have founded the see here in the 5th century. The city has the province's oldest observatory and the only planetarium, as well as museums. Navan Fort, 3 km (2 miles) to the west, was probably a stronghold of the northern Celts until the 5th century AD.

The county of Armagh, with an area of 1254 km² (484 sq miles), is Northern Ireland's smallest. It has rich agricultural land and fruit growing is important. Lurgan and Portadown, both part of the new town of Craigavon, have textile industries.

Population (city) 14 270; (county) 152 780
Map United Kingdom – Bc

Armagnac *France* Plateau covering approximately 6000 km² (2300 sq miles) in Gers department, sloping from the Pyrenees towards the Garonne River, crossed by valleys with steep eastern sides. The region's main products are cereals, fodder, red and white wines – and a renowned brandy, named after the plateau, which is distilled around the town of Condom.
Map France – De

Armenia See p 47

Armero *Colombia* Town in the west-central department of Tolima, 80 km (50 miles) north-west of Bogotá. It lies to the east of the volcanic peak of Nevado del Ruiz. In November 1985 the volcano erupted, causing a huge mud avalanche that buried Armero, killing about half of its population of 21 000 and destroying most of its houses and buildings.
Map Colombia – Ba

Armidale *Australia* University town and agricultural centre in the NEW ENGLAND region of New South Wales, about 370 km (230 miles) north of Sydney. The town trades in high-quality wool, fruit and timber.
Population 21 610
Map Australia – Ie

Armorican Relating to Armorica, the part of north-west France now known as BRITTANY, or its people or language. In geology, it refers to the time about 200 to 250 million years ago when mountains were formed here and elsewhere in the world. It is also called HERCYNIAN.

Arnhem *Netherlands* Provincial capital of Gelderland, 82 km (51 miles) south-east of Amsterdam and situated on a northern branch of the Rhine, the Lek. It was the scene of a battle in September 1944, when British and Polish paratroops were dropped to capture its bridge and secure a path for an Allied advance in the northern Netherlands. The airborne troops were surprised by German armoured forces and defeated.

Now rebuilt, the town is a tourist centre with the Dutch national open-air museum. It produces rayon, and industries include electrical goods and shipbuilding.
Population (city) 133 270; (Greater Arnhem) 308 040
Map Netherlands – Ba

Arnhem Land *Australia* Area in the north of the Northern Territory, bordering the Gulf of Carpentaria. It is named after the Dutch ship *Arnhem* which carried the expedition that discovered the area in 1623. It is an Aboriginal trust land covering an area of 90 955 km² (35 110 sq miles). Following the discovery of minerals, such as manganese, bauxite and uranium, mining agreements were made with the Aboriginal people. The KAKADU NATIONAL PARK is in the west.
Map Australia – Ea

Arno *Italy* Principal river of Tuscany, 245 km (152 miles) long. Its upper reaches run through Apennine trenches parallel to the sea, but from Pontassieve, 20 km (12 miles) upstream of Florence, it flows more directly to the coast through Florence, Empoli, Pontedera, Cascina and Pisa, close to which it enters the sea. It has a tendency to flood – it did so in the 12th, 14th and 16th centuries, when bridges in Florence were swept away. But the worst flood occurred in November 1966, when water in the city's streets rose to 6 m (20 ft) and many art treasures were destroyed or severely damaged.
Map Italy – Cc

Arras *France* Capital of Pas-de-Calais department, on the navigable Scarpe River, 40 km (25 miles) south-west of Lille. Its industries include engineering, brewing, sugar refining, and the manufacture of vegetable oils and agricultural equipment. The old town was devastated during the First World War, but its town hall and other monuments have been restored to evoke its Flemish past – it was part of Flanders under the Spanish Habsburgs in the 16th and 17th centuries. The word 'arras' for a wall-hanging or screen made from tapestry comes from the phrase *drap d'Arras* (cloth of Arras) – the town was a famous tapestry centre during the Middle Ages.
Population 42 000
Map France – Ea

artesian structure Structural formation in sedimentary rocks, usually folded, that holds water in an AQUIFER below a layer of impermeable rock. Owing to the curvature of the aquifer, the water at the edge of the structure is higher than that at the centre. The water at the centre, therefore, is under a head of pressure from the height of the water table at the edge of the aquifer. If a well is sunk into the aquifer, the pressure forces the water to the surface.

Arthur's Pass *New Zealand* Mountain pass through the Southern Alps of South Island, linking the west coast by road and rail with Christchurch. It is approached from the west through the Otira gorge.
Map New Zealand – Ce

Artois *France* Territory in the north acquired by France from Spain in 1659, occupying a chalk plateau between Flanders and Picardy. It was the scene of three First World War battles between September 1914 and October 1915. Many military cemeteries and monuments may be visited there. Isolated farms and villages are set in rolling farmland producing cereals, sugar beet and fodder for dairy cattle. The area has many springs and artesian wells – so-called because they were first discovered here. ARRAS is the main city.
Map France – Ea

Aruba *Netherlands Antilles and Aruba* Island of 193 km² (75 sq miles) in the Caribbean, west of Curaçao and 30 km (19 miles) off the north coast of Venezuela. The island is hilly, but its highest point, Jamanota, is a mere 189 m (620 ft). The southern coast is lined with coral reefs. The capital city is ORANJESTAD.

Aruba was first visited by Europeans in 1499, and became a possession of the Netherlands in the 17th century. From 1845 it formed part of the NETHERLANDS ANTILLES, but on January 1, 1986 it adopted separate status within the Kingdom of the Netherlands. It has its own flag and uses the Aruba guilder instead of the Netherlands Antilles guilder as currency. However, Aruba continues to co-operate with the other Netherlands Antilles islands, particularly in the promotion of tourism to the area. Since Aruba's oil refinery was closed down, tourism has become the main industry of the island. Aloes are the only important cash crop – their extracts are used in the cosmetics and pharmaceutical industries.
Population 71 250
Map Caribbean – Ac

Arunachal Pradesh *India* Union territory in the north-east, covering 83 578 km² (32 262 sq miles) and bordering on Tibet. It consists mainly of mountains and dense jungle. The tribal inhabitants live off the soil and have little contact with the outside world. The capital is Itanagar.
Population 858 390
Map India – Eb

Arusha *Tanzania* Regional capital in the north-east, about 100 km (60 miles) south of the Kenyan border. Arusha is a tourist and conference centre at the foot of Mount Meru (4565 m, 14 977 ft), in a flourishing coffee-growing region. It has a natural history museum which was renovated in 1987 and contains displays on the origins of man. There is an international airport.
Population 55 300
Map Tanzania – Ba

Arzew *Algeria* Seaport 35 km (22 miles) east of Oran and one of the landing places of the American army in November 1942. It is a terminus for oil and gas pipelines from the Saharan fields, and is of key industrial importance because of its natural gas liquefaction and petrochemical processing plants.
Population 21 000
Map Algeria – Aa

Ascension Island *South Atlantic* Strategically vital volcanic speck in the ocean – one of the British St Helena Dependencies – that became an important air-sea staging post in the British recapture of the FALKLAND ISLANDS in 1982. Only 88 km² (34 sq miles) in area, it lies 1126 km (700 miles) north-west of St Helena. The United States maintains a satellite tracking station here. Vegetation is sparse, but islanders grow fruit and vegetables, besides fishing and raising livestock. Wildlife includes goats, rabbits, partridges, sea turtles and sooty terns.
Population (civilian) 1120
Map Atlantic Ocean – Ee

Aschaffenburg *Germany* Port on the River Main 35 km (22 miles) south-east of Frankfurt. The castle palace of Johannisburg, built in 1605-15, was for more than two centuries the seat of the local rulers, the electors of Mainz. The town's industries today include paper, textiles and machine tools.

Population 63 600

Map Germany – Cd

Ascoli Piceno *Italy* Town and tourist resort about 160 km (100 miles) north-east of Rome. It is rich in Roman remains such as the Solesta Bridge, the Gemina Gate, the theatre and, above all, the grid street layout. The cathedral, with an 11th-century crypt and a Renaissance façade, the town hall and the baptistry stand in the Piazza Arringo, once the Roman forum.

Population 52 370

Map Italy – Dc

Aseb (Assab) *Eritrea* Red Sea port 60 km (35 miles) north of the border with Djibouti. It has an oil refinery.

Population 25 000

Map Ethiopia – Ba

Ashdod *Israel* Port and industrial centre on the coast, 30 km (19 miles) south of Tel Aviv. Ashdod is the site of an ancient Philistine harbour, fort, and temple to the sea god, Dagon. It was also the site of the world's longest siege. According to the Greek historian Herodotus, it held out for 29 years in the 7th century BC before surrendering to Egyptian invaders.

Population 68 900

Map Israel – Ab

Ashgabat (Ashkhabad) *Turkmenistan* Capital, and commercial and cultural centre, of Turkmenistan, lying in a fertile oasis east of the Caspian Sea and 40 km (25 miles) from the Iranian border. It is one of Central Asia's industrial centres – including food processing, and the manufacture of silk, textiles, carpets, leatherware, skins and hides. The city is often shaken by earthquakes, and was nearly destroyed in 1948.

Population 402 000

Map Turkmenistan – Ab

Ashqelon *Israel* Port and resort about 50 km (30 miles) south of Tel Aviv. Originally a Philistine city-kingdom, it was captured by the Romans and then, in the 12th century, by the Crusaders. Today it is the Mediterranean terminal of an oil pipeline from the Red Sea. There are archaeological remains from many eras here – including excavations from the Bronze Age to the Roman period, the remains of the Crusader city wall, and ruins of two Crusader churches and one Byzantine church. Also of interest are an ancient synagogue and the old Arab quarter of Migdal.

Population 54 700

Map Israel – Ab

Asia In many respects, this is the ancient continent; the fabled continent, too, since, to Europeans for centuries it seemed to lie just at the edge of belief, between reality and myth. Many wonders and tall tales came out of it. Marco Polo, the 14th-century traveller in Asia and a fairly accurate reporter, said on his deathbed: 'I did not

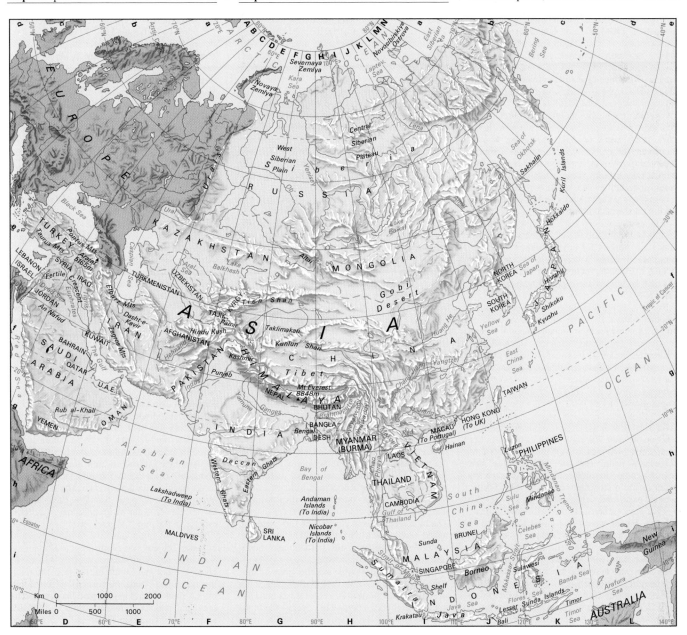

tell the half of what I saw' – for fear, apparently, of raising howls of derision.

Even in these prosaic days, the facts are breathtaking enough. The largest of the continents, Asia incorporates about one-third of the world's landmass and something like 60 per cent of its people. It embraces the permafrost of Siberia, the Malaysian jungles, the Arabian deserts and the symmetrical volcanoes of Japan. There is Everest, soaring to 8848 m (29 028 ft), and the earth's lowest surface point, the Dead Sea, 396 m (1229 ft) below sea level.

Asia takes in two-thirds of Russia and the ex-Soviet republics, and all of China. Babylon, Assyria, Persia, Arabia, the Levant, India, Mesopotamia, China and the Indus Valley were the nurseries of civilisation, and out of them emerged the great religions – Buddhism, Hinduism, Islam and Christianity.

If south-central Russia and ex-Soviet countries such as Turkmenistan, Tajikistan, Kazakhstan and Kyrgyzstan, which constitute the Central Asian region, are excluded, Asia can be divided into four regions, each with its own definite characteristics. There is EAST ASIA (the Far East), which includes China, Japan and also north-east Russia, SOUTH ASIA, whose chief feature is the Indian subcontinent, SOUTH-EAST ASIA containing the Pacific archipelagoes and Indo-China, and the MIDDLE EAST (West Asia), which borders on the Mediterranean and the Red Sea.

Despite vast jungles, near pristine stretches of coastline and enormous tracts of agricultural land, many parts of Asia are facing serious environmental problems, and it contains the most crowded, noisy, dirty and poisoned cities in the world. For instance, Delhi and Xian have the worst air pollution in the world, with 294 days out of the year for Delhi, and 273 for Xian, experiencing poor air conditions.

Life in an Asian city is a fierce struggle for survival where jobs are scarce, food is expensive, educational levels for urban children are dropping, and running water and electricity are a luxury. In Calcutta, for instance, only just over 55 per cent of houses are supplied with water and electricity.

In the countryside, large-scale deforestation has taken place. Between 1980 and 1990 alone, countries noted for their immense natural forests of tropical hardwood, such as Thailand, the Philippines and Malaysia, were chopping down this valuable resource at a rate of 2-3 per cent of the total area every year.

Area 44 385 000 km² (17 133 097) sq miles)
Population 3 184 000 000

Asia, East (Far East) Despite modern communications, the term 'Far East', with its hints of utter remoteness, remains rooted in European consciousness. Until the late 19th century, the region's comprehension of the West was sketchy. Most of the representatives of the West it met seemed to be bewildering barbarians.

East Asia embraces China, Mongolia, Korea, Taiwan, Japan, Macau and Hong Kong, and is about the same size as Europe. It stretches roughly 4000 km (2485 miles) from north to south and from west to east. Strictly speaking, East Asia also includes the Russian region in northern East Asia, between the Bering Strait and the mouth of the Ussuri River. Culturally, however, this region does not form part of East Asia; it is often referred to as 'the Russian Far East'.

The Philippines forms a sort of transition zone between East Asia and South-east Asia. East Asia's eastern boundary is probably the thousands of tiny islands that lie off Taiwan and Japan; the western border is the Pamir Knot, the gigantic crinkle of mountains that holds up the heavens over Central Asia. The plateaus of Tibet are encircled by two great mountain chains – the Himalayas and the Kunlun, which extends onward into China.

The climate of the Far East is as extreme as its landscape. The chief feature is the monsoon, which brings the region most of its rain. From June to September each year, for example, it brings 1220 mm (48 in) of rain to Tokyo. Parts of western China and Mongolia, on the other hand, receive less than 20 mm (0.75 in) in the whole year. Such conditions have created the Gobi and the Taklamakan deserts, which can support only wandering herdsmen.

Some narrow coastal strips, plains and valleys in Taiwan, Korea and China, however, well watered by the monsoon, grow prodigious quantities of rice, while colder northern China grows wheat and other cereals. In the highlands there is little cultivation and in some areas, especially in eastern China, the formerly densely wooded foothills have been eroded by over-intensive agriculture.

Since Buddhism forbids the slaughter of animals, cattle rearing plays an insignificant role in most of the region. In contrast to most other parts of the world, East Asian agriculture consists almost entirely of crop and garden cultivation.

East Asia contains some of the world's most thinly populated areas as well as some of the most crowded. Mongolia has 1.5 inhabitants per km² (about 4 people per sq mile), while Macau's density is 29 865.6 people per km² (79 641.6 per sq mile). Some of the world's largest cities are in the Far East, most obviously in China and particularly Japan, 76 per cent of whose people are now urban dwellers.

Area 11 761 902 km² (4 540 223 sq miles)
Population 1 354 000 000

Asia, South The region is a great kite-shaped wedge jutting into the Indian Ocean. It is some 3000 km (1800 miles) long, and about the same again at its broadest. It reaches from Pakistan in the west, through India and the Himalayan states of Nepal and Bhutan to Bangladesh in the east and to the island-nation of Sri Lanka to the south. These are among the most densely populated regions in the world – a teeming diversity of peoples with 18 principal languages, innumerable dialects and many religions.

Geographically, the area is no less diverse, but is generally divided into three major parts: the northern mountains, the Indus-Ganges Plain, and the southern peninsula, to which is added Sri Lanka. The Himalayas are young mountains, which are still being shoved and lifted by CONTINENTAL DRIFT, and consequently display more than 30 peaks that rise to over 8600 m (28 200 ft). The high tops are a wilderness of snow, ice and naked rock, but lower down there are alpine pastures, coniferous woods, bamboo and rhododendrons, and tropical forests.

South of the foothills are the great plains, stretching from Assam to the Punjab, then south to Pakistan and the near-deserts of Rajasthan. The northern plain is the land of the rivers – the Indus and the Ganges, the Brahmaputra, the Yamuna

and their many tributaries that all converge to run into the Bay of Bengal. The north and west are mainly fed by irrigation schemes, but the centre is watered by the summer monsoons. Here is the strong heart of India – the most fruitful agricultural lands, the most productive industries and the vibrant cities.

The third part, the Deccan, is a country of rolling hills and wide river valleys, bordered to the west by the spectacular peaks of the Western Ghats and to the east by the lower, interrupted range of the Eastern Ghats. The black volcanic soil and the red loam of the Deccan produce, with a boost from irrigation canals, principally beans, millet and cotton. To the south of the Deccan lie the scenic splendours of Sri Lanka. Geologically a splinter of India, with plains around its central mountains, it sits on India's continental shelf.

The average population density of the subcontinent is 257.7 per km² (667.7 per sq mile). Some areas are relatively empty – the northern foothills, for example, where there are fewer than 10 people per km² (26 per sq mile). Throughout the subcontinent, the majority of people live in rural areas – in Bangladesh, 90 per cent of them do so. But population growth also sends many thousands of people to the cities, most of which are desperately overcrowded.

Area 4 481 090 km² (1 729 750 sq miles)
Population 1 155 000 000

Asia, South-east The term South-east Asia was coined in about 1900 as a convenient means of defining a trading area; long before that, however, both the Chinese and the Japanese used a similar name, meaning 'the southern seas', to describe the region.

The area is both complex and diverse, reaching eastwards from the borders of India and embracing Myanmar (formerly Burma), Thailand, Laos, Cambodia, Malaysia, Singapore, Brunei, Indonesia, Vietnam and, on the region's edge, and considered part of either South-east Asia or East Asia, the Philippines. At the region's heart is a tangle of seas and straits – the Strait of Malacca between Malaysia and Sumatra, the Java and South China seas at the centre, the Makassar Strait (or Ujung Pandang Strait) between Borneo and Sulawesi, and the Banda, Sulu and Celebes seas to the east. Each of the mainland nations has a major river valley, the Ayeyarwady (Irrawaddy) of Myanmar, Thailand's Chao Phraya, the Mekong that runs through Laos, Cambodia and southern Vietnam, and the Red River in northern Vietnam. Their upper courses cut through forested mountains and end in great plains, which are subject to frequent flooding.

Suitably for such a dramatic collection of peninsulas and islands, South-east Asia is bordered by oceanic trenches of staggering depth – the Mindanao Trench which plunges down to about 10 670 m (35 000 ft) and the rather shallower chasm of the Java Trench. The wide mountains that run through Sumatra, Java and the Lesser Sunda Islands are volcanic – 70 have erupted in the past 200 years.

South-east Asia has two distinct types of climate. The equator neatly bisects Borneo and Sumatra, and there and on nearby islands the climate is hot and humid all the year round. To the north and south, however, it is influenced by the monsoons. They bring heavy rain to the mainland nations above the equator from May to October, while the islands to the south get their rain from November to April. The climatic patterns are reflected in the vegetation. The equatorial area, up to a height of 1000 m (about 3300 ft), is covered by the richest rain forest – though now much depleted – in the world; above are temperate forests and their more open dwarf trees and shrubs. Below, at the coasts, lie mangrove swamps. The monsoon areas with their wet and dry seasons have forests of teak and other deciduous trees, pine forests higher up, and wooded savannah.

The people of South-east Asia are as diverse as their world. This is hardly surprising. The islands of Indonesia alone stretch for more than 5000 km (3000 miles) over the ocean and harbour 300 ethnic groups who speak 250 languages and several hundreds of dialects. There is, too, an astonishing array of faiths. Buddhism is all-pervading in Myanmar (Burma) and Thailand. Islam is the chief religion in Malaysia, Brunei and Indonesia, while the Philippines are largely Roman Catholic, Bali is Hindu, and tribal gods are worshipped by the people of the hills.

Farming, for subsistence or cash, is the background to the region's way of life, except in the city state of Singapore, one of the world's great trading centres. More than 12 million families in South-east Asia live by subsistence and shifting cultivation, while the main cash crops are palm oil, rice, sugar, coffee and rubber. Maize, tea, cotton, copra and spices are also profitable exports; so too are hardwoods, though years of over-exploitation have significantly diminished this resource.

Malaysia and Indonesia supply nearly half the world's tin, and also possess tungsten and nickel. Vietnam has coal, the Philippines large reserves of copper and Myanmar and Thailand produce a major share of the world's sapphires and rubies. Indonesia's oil and natural gas are beginning to benefit that nation's development plans, while in neighbouring Brunei they are the mainstay of the country's wealth.

Area	4 479 918 km² (1 729 297 sq miles)
Population	453 000 000
Map	see p 49

Asir *Saudi Arabia* Mountainous coastal region in the south-west of the country, covering 104 000 km² (40 130 sq miles). The Asir range, which has rich copper and nickel deposits, runs for some 370 km (230 miles) along the coast; its average height is 1830-2134 m (6000-7000 ft). Though the passes are rugged and difficult to negotiate, this once almost inaccessible region is now linked by road to Mecca and Medina. At its foot lies the narrow coastal belt of TIHAMAH.

The region has Saudi Arabia's highest rainfall – 370 mm (15 in) a year – and there is much fertile agricultural land. Most of the country's wheat, maize and sorghum come from there. The farming potential has been increased by the Jizza Dam. Large sections of the mountains and the coastal plain have been included in the Asir National Park.

| Map | Saudi Arabia – Bc |

Asmara (Asmera) *Eritrea* Capital of Eritrea, about 700 km (435 miles) north of Addis Ababa, and 100 km (60 miles) from the Red Sea coast. It was the Italian colonial headquarters from 1889 to 1941. The United States built Africa's largest military communications centre there in the 1950s. Two years before Eritrea's independence

was finally decided in the 1993 referendums, the provisional government of the People's Liberation Front had its seat in Asmara. During the 1970s and 1980s Asmara was severely hit by drought and famine.

| Population | 430 000 |
| Map | Ethiopia – Aa |

Aso *Japan* National park on KYUSHU Island, 60 km (37 miles) south-west of the resort of Beppu. It covers 731 km² (282 sq miles) and contains the world's largest volcanic crater on Mount Aso. The crater contains five peaks and measures a total of 24 km (15 miles) north to south, and 18 km (11 miles) east to west. Inside the crater are towns, complete with infrastructures, and a total population of about 70 000 people.

| Map | Japan – Bd |

Assab *Eritrea* See ASEB

Assad, Lake *Syria* Lake, south-east of Aleppo, created by the building of the Euphrates Dam (1968-78). It is 80 km (50 miles) long and will eventually irrigate approximately 640 000 hectares (1 581 400 acres). The dam supplies 70 per cent of the country's hydroelectric power.

| Map | Syria – Ba |

Assam *India* North-eastern state of India constituted in 1950. It covers 99 680 km² (38 476 sq miles). The Assamese are descended from Tibeto-Burmans, and have a distinct language and culture, which includes the oldest prose literature in India. Immigration, much of it illegal, from Bengal caused resentment and led to violent clashes in the early 1980s. But after agreement on the status of illegal immigrants was reached between anti-immigration parties and the central government of India in 1985, the situation stabilised. Assam has great scenic beauty and huge economic potential. In the late 1980s, its oilfields furnished about half of India's crude oil, and three refineries produced petrochemicals. There are silk mills and cotton mills. Assam grows much of the world's tea; other major crops include rice, jute and timber. The chief town is GAUHATI.

| Population | 22 295 000 |
| Map | India – Eb |

Assen *Netherlands* Market town about 130 km (80 miles) north-east of Amsterdam. It was founded in the 13th century around a nunnery, which now serves as public buildings. Nearby are 4000-year-old megaliths (huge standing stones like those at Stonehenge in England), the 'Giants' Caves' mentioned by the Roman historian Tacitus (about AD 55-117), and a popular grand prix motorcycle circuit.

| Population | 51 710 |
| Map | Netherlands – Ca |

Assisi *Italy* Hillside town about 130 km (80 miles) north of Rome, immortalised by St Francis, patron saint of animals and founder of the largest monastic order of the Roman Catholic Church, the Franciscans. He was born here in 1182 and his life is recorded in a series of frescoes painted by Giotto in 1296 in the upper part of the massive Basilica of St Francis.

The 12th-century convent of St Damian stands 2 km (1.25 miles) to the south. It was built on the site where St Francis experienced the vision of the Crucifixion.

In an oak forest 4 km (2.5 miles) east is the Eremo delle Carceri, the saint's retreat, marked by a simple church surrounded by a 15th-century convent and hermits' grottoes. The ornate church of Santa Maria degli Angeli is built over the saint's death chapel, the Porziuncola, in a valley 5 km (3 miles) below the town.

| Population | 24 400 |
| Map | Italy – Dc |

Assyria *Middle East* Empire of the Ancient World whose capital was NINEVEH. It was founded some time before 3000 BC and controlled much of western Asia. Despite its armed might it was later conquered by the Babylonians and then the Romans. It stretched along the eastern bank of the middle TIGRIS River, in what is now part of modern Iraq.

asthenosphere The part of the mantle in the earth's interior that lies below the lithosphere. It is of low rigidity, thought to be caused by heating and partial melting. See CONVECTION CURRENTS.

Asti *Italy* City about 40 km (25 miles) south-east of Turin. It has mostly 18th and 19th-century buildings, but there are several medieval monuments, including a 14th-century cathedral, the 13th to 15th-century church of San Secondo and the 12th-century Baptistry of San Pietro. The Piedmont province of Asti is noted for its vines, and produces the sparkling *Asti Spumante* wine.

| Population | (city) 72 380; (province) 207 110 |
| Map | Italy – Bb |

Astrakhan *Russia* Port on the Volga River delta near the Caspian Sea, 375 km (235 miles) south-east of Volgograd. It trades in timber, fruit, cotton, grain, cereals and rice. Its industries include fishing (especially for sturgeon, prized for its caviar), cotton spinning, shipbuilding and the processing of karakul lamb's wool – used to make the fur named after the town. Astrakhan was founded by the Tatars in the 13th century. It has a 16th-century fortress built by Ivan the Terrible, who captured the city in 1556.

| Population | 512 000 |
| Map | Russia – Fd |

Asturias *Spain* Region in the north of the country, on the Bay of Biscay. It covers 10 565 km² (4080 sq miles) of mostly mountainous land. Largely agricultural, it also has deposits of lead, iron and copper. Its capital is OVIEDO.

| Population | 1 093 940 |
| Map | Spain – Ba |

Asunción *Paraguay* National capital and the only large city in the country. Standing on the east bank of the Paraguay River, it was founded as a fort in 1537 by Spanish colonists. Work on the fort started on the Feast of the Assumption (Asunción in Spanish), August 15, which is where the city's name comes from.

Little of the present city is older than the 19th century. During Paraguay's six-year war with Brazil, Argentina and Uruguay (1865-1870), the city was almost destroyed. It was rebuilt on a colonial grid pattern with tree-lined avenues. The government palace and buildings around the Plaza de los Héroes are modelled on Parisian public buildings. There are a number of open-air restaurants where the country's distinctive harp music can be heard. The main industries are

▲ **RUINS OF PASSION Plays were staged in the theatre dedicated to Dionysus, the god of wine and ecstasy, for the great spring festival in ancient Athens.**

cotton textile manufacturing and processing of agricultural products.

Population 800 000
Map Paraguay – Bb

Aswan *Egypt* City on the east bank of the River Nile, about 700 km (435 miles) south of Cairo. It lies near the quarries from which the Ancient Egyptians extracted granite for the obelisks and statues at Luxor and Karnak, which lie about 200 km (125 miles) to the north.

Aswan is now a resort. It has steel and textile industries, and some granite is still quarried. It stands by one of the most verdant stretches of the Nile, where cataracts tumble among rocky islands. One, Plants Island, is a forest of rare and introduced plants. On Elphantine Island is the Nilometer, the gauge by which, early in June each year, the Ancient Egyptians measured the rate of rise as the river began its annual flood.

Population 215 000
Map Egypt – Cc

Aswan High Dam (Sadd-el-Ali Dam) *Egypt* One of the world's largest dams, 13 km (8 miles) south of the city of Aswan. It holds back the flood waters of the Nile, controlling the flooding of

hundreds of kilometres of the Nile Valley to the north. Its water irrigates large stretches of land and supplies hydroelectric energy which is desperately needed for Egypt's industries. However, it is responsible for large tracts of land downstream becoming infertile through salination. The dam was completed in 1971; its wall is 3600 m (11 812 ft) long and 114 m (375 ft) high, and it holds back Lake Nasser, which covers about 5180 km^2 (2000 sq miles) of Egypt and the Sudan.

Map Egypt – Cd

Asyut *Egypt* Regional capital and university town on the left bank of the Nile, 380 km (186 miles) south of Cairo. Asyut is a modern city with wide, tree-lined avenues and the largest university and college complex of Middle and Upper Egypt, with a renowned agricultural college. Asyut is also a centre of traditional handcrafts, including pottery and hand-dyed cottons, sold in its lively souks (markets). A bridge dam controls the flow of the Ibrahîmich Canal and irrigates the land on the Nile's left bank. There is an airport and a railway link to Cairo. The tourist centre of Luxor lies 340 km (211 miles) to the north-west.

Population 313 000
Map Egypt – Bc

Atacama *Chile* Northern region including the southern part of the Atacama Desert. Its prosperity was originally founded on the silver mines at Chañarcillo, but today relies on both minerals

and farming. Copper produced at Potrerillos and El Salvador is exported from the ports of Barquito and Chañaral. Iron ore from El Algarrobo, in the Huasco Valley, is exported from Huasco and Los Lozes. The other major port, Caldera, serves the capital, COPIAPÓ, 80 km (50 miles) away. Copiapó and Caldera were linked by the Southern Hemisphere's first railway in 1852.

Population 230 780
Map Chile – Ab

Atacama Desert *Chile* Wasteland covering some 132 000 km^2 (51 000 sq miles) in the north of the country. It contains some of the driest spots on earth – parts of the desert went without rain for 400 years from 1570 to 1971. But it is rich in nitrates, which are exported for fertilisers, and in iodine, used in medical drugs.

Map Chile – Ba-Ab

Atakora (Atacora) Highlands *West Africa* Uplands stretching across central Togo and continuing into north-west Benin and south-east Burkina. Their peaks form the highest points in Togo (850 m, 2789 ft) and Benin (835 m, 2739 ft). The highlands are laced with spectacular gorges, waterfalls, lakes and rivers.

Map Benin – Aa

Atbara *Sudan* Industrial town about 350 km (220 miles) north and slightly east of the capital, Khartoum. It lies on the junction of two major road and railway lines to Khartoum (from Wadi

THE DIFFERENT LAYERS OF THE ATMOSPHERE

Earth's atmosphere is a multilayered mixture of gases bound to the planet by gravity. The protection it affords against harmful rays from the sun, debris from outer space and cosmic rays (the source of which is unknown) makes life on earth possible. Solar radiation and cosmic rays are absorbed or scattered harmlessly; meteors and METEORITES mostly burn up in the frictional heat generated by their entry into the atmosphere. Without this protection, the planet would also be subjected to extremes of temperature, for the atmosphere dissipates the sun's heat by day and insulates the earth, preventing too much loss of heat, during the night.

There are five major layers in the atmosphere. In ascending order from the earth, they are the troposphere, stratosphere, mesosphere, thermosphere and exosphere. Certain zones of nuclear, chemical and electrical activity occur within them, and some of the boundaries between layers in the atmosphere vary with latitude and season. Other layers occurring within the five major layers include the ionosphere and the magnetosphere.

Gases in the atmosphere include nitrogen (78 per cent), oxygen (21 per cent), argon (0.9 per cent) and carbon dioxide (0.03 per cent), there being also minute traces of helium, hydrogen, krypton, methane, neon, OZONE and xenon. Generally they are mixed in these proportions up to about 60-80 km (40-50 miles) high, but there is a concentration of ozone in the stratosphere at 25-27 km (15-17 miles) in low latitudes, falling in altitude in high latitudes. The lower atmosphere contains varying amounts of water vapour, which is virtually absent above about 8-18 km (5-11 miles).

Above the mesosphere, in the lower thermosphere, the atmosphere consists mainly of oxygen and nitrogen, but beyond 200 km (124 miles) oxygen predominates. In the exosphere, oxygen, hydrogen and helium constitute the very tenuous atmosphere. Ionised particles become more numerous and above 2000 km (1250 miles) only electrons are present. This is the magnetosphere, with an increased incidence of electrons at 3000 and 15 000 km (1850 and 9300 miles) – the Van Allen radiation belts. It is this part of the atmosphere that is influenced by the earth's magnetic field.

The atmosphere envelope weighs about 5000 million tonnes, but three-quarters is concentrated below the height of Mount Everest (8846 m, 29 022 ft). This weight of air exerts a constant pressure of about 1 tonne on the human body at sea level. However, air pressure varies. It increases as cold air sinks downwards and becomes more dense, and falls as warm air expands and rises. These changes in pressure affect the earth's weather systems.

The density of the atmosphere decreases rapidly with height – at 16 km (10 miles) above sea level, the density is only one-tenth of that at sea level.

Within the ionosphere there are several distinct layers of electrically charged particles called ions. Radio waves, which travel in straight lines, can be transmitted around the earth's curvature by bouncing them off ionised layers. The highest – and the most highly ionised – is the F2 layer, just above 300 km (186 miles). Below it is the F1 layer, at about 150 km (92 miles). This is the lowest level at which satellites can orbit, because the atmosphere below is too dense for free flight. At 90-160 km (56-100 miles) is the E layer, also known as the Heavyside-Kennelly layer, which is quite strongly ionised in daylight, but with the ions dissipating at sunset. In the mesosphere, which occurs at around 50-90 km (30-56 miles), is the D layer, where ionisation is weak and radio waves are poorly reflected.

EXOSPHERE Starts about 500 km (approx. 310 miles) above earth and extends via the magneto sphere into space at around 2000 km (1250 miles). In the very rarified atmosphere above about 2100 km (1500 miles) hydrogen particles predominate. Down to about 965 km (600 miles) helium and hydrogen are equal, while below this level there is some oxygen. Heavy solar radiation and cosmic rays penetrate to the ionosphere below.

THERMOSPHERE Extends from 80 km (50 miles) up to the exosphere. Gas molecules (mainly oxygen and nitrogen) within it are broken down by intense solar radiation, producing charged particles called ions. Glowing lights – aurorae – in earth's high latitudes result from these electrical disturbances due to the impact on the oxygen and nitrogen molecules of electrons and protons from the sun. Layers of ions at 100 – 300 km (about 60 – 190 miles) reflect radio waves.

MESOSPHERE A 30 km (20 mile) deep layer between the ionosphere and stratosphere. Temperature here falls sharply from 10°C (50°F) at 50 km (30 miles) to −80°C (−112°F) at 80 km (50 miles).

STRATOSPHERE Contains a band of ozone – oxygen with three, rather than two, atoms – which is vital to life on earth. It absorbs and filters out most of the solar ultraviolet radiation which is deadly to living organisms. Here, above the tropopause, temperature rises from −55°C (−67°F) at around 15 km (10 miles) to 10°C (50°F) at 50 km (30 miles). Meteors burn up in the frictional heat generated by their passage through a denser atmosphere. Meteoric dust forms noctilucent clouds. Here, too, cosmic rays are absorbed and scattered.

TROPOSPHERE The bottom layer of the atmosphere, and the most dense. It rises to an average of 15 km (10 miles) above earth, and all normal clouds and weather patterns are formed within it. Temperature falls about 2°C per 305 m (4°F per 1000 ft), stabilising at −55°C (−67°F) at the top, or tropopause, where high winds flow horizontally – the jet streams used to speed airliners.

Halfa and Port Sudan). Its industries include railway engineering and cement works. The government railways headquarters are in Atbara.

The town stands at the confluence of the Nile and the Atbara rivers – also called the Black Nile – which flows 1120 km (695 miles) north-west from Ethiopia. An Anglo-Egyptian army defeated followers of the Sudanese religious leader, the Mahdi, at Atbara in 1898.
Population 73 100
Map Sudan – Bb

Athabasca *Canada* River which rises in the Rocky Mountains and flows north 1231 km (765 miles) to Lake Athabasca, which drains into the Slave River. Important oil-sand deposits flank the river in ALBERTA.
Map Canada – Dc

Athens (Athínai) *Greece* Capital and largest city of Greece on the Attic Plain beside the Saronic Gulf. It is a sprawling, bustling, noisy city of concrete high-rise blocks and car-choked streets – around the classical majesty of the Acropolis. Ancient Athens, built on the rock of the Acropolis 156 m (512 ft) high, was a powerful Hellenic city-state in the 8th century BC, and for almost 1000 years was the classical centre of Western civilisation. Until the Roman general Sulla sacked the city in 86 BC, it was supreme in the arts, philosophy, science, literature and drama, unlike its main rival, militaristic Sparta.

Most of the magnificent architectural remains on the Acropolis were built during the Golden Age of the 5th century BC, under the political leadership of Pericles. These include the Parthenon (447-438 BC), Propylaia (437- 432 BC), the Temple of Athena Nike (427-424 BC), the Erechtheion (395 BC), the Theatre of Dionysus (6th to 5th century BC), and the Temples of Theseion (449 BC) and Olympeion (550-510 BC). Many philosophers made their home here, Socrates, Plato and Aristotle being best known.

Athens went into further decline after it was sacked by Germanic warrior tribes in AD 267, and it became a provincial Byzantine town. The Turks, who occupied it in 1458, were not driven out until 1833, and the following year Greece's new king, Otto, made it the capital.

Most of the present city is modern. Athens is now the main banking, shopping and communications centre of Greece, and a major industrial centre with its port at PIRAEUS.
Population (city) 885 700; (metropolitan area) 3 027 300
Map Greece – Cc

Athens *USA* City 100 km (60 miles) east and slightly north of Atlanta in north-eastern Georgia. It is the seat of the University of Georgia, the oldest state university in the country, whose charter dates back to 1785.
Population (city) 45 730; (metropolitan area) 126 260
Map United States – Jd

Atherton Tableland *Australia* Plateau region in northern Queensland to the west of Cairns, covering about 32 000 km² (12 350 sq miles) at the northern end of the Great Dividing Range. It is mostly about 600-900 m (2000-3000 ft) high and produces beef, peanuts, tobacco, maize and dairy products. Its rain forests are a tourist attraction.
Map Australia – Gb

Athlone (Baile Átha Luain) *Ireland* Market town in Westmeath at an ancient crossing of the River Shannon, 113 km (70 miles) west of Dublin, almost at the exact centre of Ireland. It is a fishing and cruising centre, with a 13th-century castle and ruined Franciscan abbey. It is named after Luain, who ran the riverside inn – Átha Luain means 'Luain's ford'.
Population 8170
Map Ireland – Cb

Áthos (Áyion Óros) *Greece* Self-ruling community of more than 20 monasteries on and around Mount Áthos, 2033 m (6670 ft) high, on the most eastern promontory of the CHALCIDICE Peninsula, many of them situated on almost inaccessible hilltops. Founded in AD 963, Áthos has an all-male population living in Byzantine monasteries and observing the Julian calendar, which is 13 days behind the Western European calendar. No women or 'beardless' boys are allowed, and the number of foreign male visitors is restricted to 20 new arrivals a day, each permitted to stay only four days. Its archives contain Byzantine art and manuscripts.
Population 1700
Map Greece – Da

Atitlán, Lake *Guatemala* Mountain lake 80 km (50 miles) west of the capital, Guatemala City, and 1562 m (5125 ft) above sea level. The lake is surrounded by volcanic cones soaring to 3500 m (11 500 ft). Atitlán, described by the English writer Aldous Huxley (1894-1963) as the world's most beautiful lake, is 18 km (11 miles) long, up to 10 km (6 miles) wide and more than 300 m (990 ft) deep. Hot springs around the lake and underwater thermal currents constantly change the shades of blue and green in the water. The biggest of 12 settlements on its shores is Panajachel, which has a Sunday market. A boat-ride across the lake is the village of Santiago. The village of Sololá, at an altitude of 2113 m (6933 ft), has a spectacular view of the lake.
Map Guatemala – Ab

Atlanta *USA* Capital and largest city of Georgia, in the north-west of the state. It was founded as a small rail town in 1837 and grew rapidly as a transport and cotton-manufacturing centre. It was a chief arsenal of the Confederacy during the Civil War (1861-5), and was burnt to the ground by General William Sherman's Union troops in 1864. Today the city is the industrial and financial centre of the south-eastern states and a major

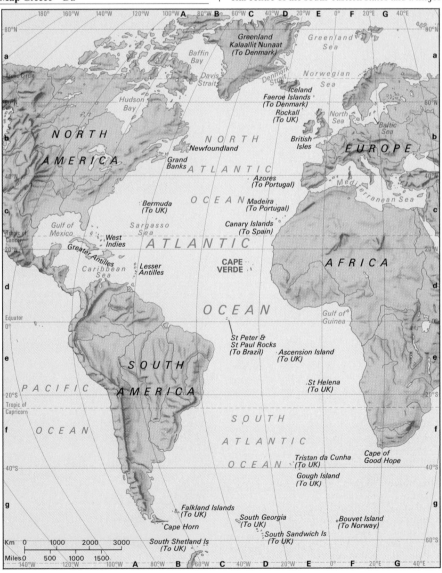

convention centre. It has one of the country's largest and most modern airports and was chosen as the site of the 1996 Olympic Games. Martin Luther King Jr (1929–68) was born there, as was Margaret Mitchell (1900–49), author of *Gone with the Wind*.

Population (city) 394 020; (metropolitan area) 2 959 950

Map United States – Jd

Atlantic Ocean The world's second largest ocean, after the Pacific. It is named after the Atlas Mountains in North Africa, which marked the western boundary between the known and the unknown world for the Ancient Greeks. It includes the Gulf of MEXICO and the CARIBBEAN, MEDITERRANEAN, NORTH and BALTIC seas. It covers more than 20 per cent of the earth's surface and contains about one-quarter of the earth's total marine water.

The Atlantic was formed within the past 175 million years as continents which previously formed one landmass drifted apart. It covers an area of 82 217 000 km² (31 736 663 sq miles) and is growing wider, by 10-20 mm (0.4-0.8 in) a year, and spreading outwards from the S-shaped ridge – the Central Atlantic Ridge – that runs through the centre of the ocean. A deep rift valley, 24-48 km (15-30 miles) wide, runs north-south through the centre of the ridge. As the continental plates of the sea floor move apart, new rocks are formed in this valley. Parts of the Central Atlantic Ridge reach the surface as volcanic islands, including ICELAND, the AZORES, ASCENSION, ST HELENA and TRISTAN DA CUNHA.

The Atlantic also contains abyssal plains and shallow continental shelves. Unflooded parts of the continental shelves form islands such as Newfoundland and the British Isles.

The average depth of the Atlantic is 3293 m (10 803 ft); its greatest depth, in the Puerto Rico Trench, is 9219 m (30 246 ft). In the South Atlantic the waters circulate in an anticlockwise direction (the South Equatorial Current, the Brazil Current and the Benguela Current). In the North Atlantic, the circulation is clockwise (the North Equatorial Current, the Gulf Stream and the Canary Current). Within these circulatory systems (or gyres) are areas of comparatively calm water, such as the SARGASSO SEA. The North Atlantic Drift, an extension of the Gulf Stream, conveys warm water into the ARCTIC OCEAN, while the East Greenland and Labrador currents bring cold Arctic water into the Atlantic.

The Atlantic provides about one-third of the world's fish and shellfish catch – though environmentalists and fishermen are expressing deep concern over the rapid depletion of fish stocks. The leading fishing grounds are over the North Atlantic's continental shelves, notably the Grand Banks off Newfoundland and the Dogger Bank in the North Sea. The Atlantic's continental shelves also contain oil reserves, notably in the Gulf of Mexico and the North Sea.

Map See opposite page

Atlas Mountains *North Africa* Mountain system extending north-eastwards about 2250 km (1400 miles) from the Atlantic coast of Morocco to northern Tunisia. It consists of several roughly parallel ranges. In Morocco these include, from south to north, the Anti-Atlas, High Atlas and Middle Atlas, and in Algeria the Saharan Atlas (which includes the Amour Mountains and

extends to the Aurès massif in the east) and, farther north, the Tell, or Maritime, Atlas. The highest peak in North Africa – Djebel Toubkal (4165 m, 13 665 ft) – stands in the High Atlas of Morocco.

Map Africa – Ca

atmosphere See THE DIFFERENT LAYERS OF THE ATMOSPHERE, p 53

atmospheric pressure Pressure exerted by the weight of the atmosphere on the earth, measured by a barometer and expressed in millibars (1000 millibars is equivalent to approximately 750.1 mm, or 29.53 in, of mercury). The average pressure at sea level is 1013.25 millibars, equivalent to the weight of a column of 760 mm (29.92 in) of mercury, or a pressure of about 1033 grams per cm² (14.66 lb per sq in). Pressure varies on the earth's surface with latitude, temperature and altitude, and these differences in pressure are responsible for winds blowing from areas of high to areas of low pressure, and thus for major climatic differences.

atoll Ringlike coral reef that almost or entirely encloses a lagoon.

Attock *Pakistan* Town and fort about 80 km (50 miles) west of Islamabad, at a strategic crossing point of the Indus River where the flow is constrained by a series of gorges. The clifftop fort was built by the Mogul emperor Akbar the Great, who ruled from 1556 to 1605. The British built a rail and road bridge across the river in the 19th century. The bridge has survived floods which have raised the water level by as much as 30 m (100 ft). A second bridge over the river offers excellent views of the surrounding countryside. There are several oil wells in the area and the town has a refinery.

Population 62 000

Map Pakistan – Da

Auckland *New Zealand* The country's largest city, chief port and former capital. Situated on a narrow isthmus between Manukau and Waitemata harbours in the north of North Island, it has spread over many extinct volcanoes. The city was founded in 1840 by British settlers on a Maori site and has the largest Polynesian population – some 205 000, about half Maori and half Polynesian Pacific islanders – of any city in the world. The Auckland Museum houses one of the finest collections of Polynesian, and especially Maori, artefacts in the world.

Maoris sold the site to the British in 1840 for clothes, food, tobacco and about £30 sterling. Today the port exports dairy produce, wool, hides and timber, and imports machinery, oil and fertilisers. Auckland's industries include the manufacturing of textiles and chemicals, as well as food processing, engineering, metalworking, and vehicle assembly.

Population 945 520

Map New Zealand – Eb

Audenarde *Belgium* See OUDENAARDE

augite Dark green to black PYROXENE mineral.

Augsburg *Germany* Bavarian city about 56 km (35 miles) north-west of Munich, where in 1555 the right of Europe's Protestants to freedom of

worship was recognised at the religious treaty, the Peace of Augsburg. Romans founded the city in 15 BC and named it after the emperor Augustus.

It grew rich by trading between Italy and the north, and in the 16th century was the banking capital of northern Europe. One banking family, the Welsers, were granted virtual sovereignty over Venezuela after the emperor Charles V could not repay his debts to them; another, the Fuggers, financed the Holy Roman Emperors for centuries. It was the Fuggers, too, who founded the Augsburg *Fuggerei* in 1515, the first social housing project in the world. In 1994, occupants of the project were still paying an annual rent of less than US$1.

The painters Albrecht Dürer (1471-1528) and Hans Holbein the Younger (about 1497-1543) were born in the city, and works by them hang in the Schaezler Palace. Rudolf Diesel (1858-1913), who invented the engine named after him, was also born there, as were the aircraft designer Willy Messerschmitt (1898-1978) and the playwright Bertolt Brecht (1898-1956).

Augsburg's cathedral, which dates from the 9th to the 14th centuries, claims to have the oldest stained-glass windows in Europe, fitted in the 11th century. The city has been a main textile-producing centre for almost 500 years, but now engineering industries have become relatively more important to the wealth of Augsburg.

Population 256 900

Map Germany – Dd

Augusta *Italy* Fortified peninsula town on the east coast of Sicily. It was founded by the Holy Roman Emperor Frederick II of Swabia in 1232. Today it is one of the Mediterranean's biggest oil-refining and petrochemicals centres.

Population 33 920

Map Italy – Ef

Augusta *USA* City in the state of Georgia on the Savannah River, which forms the border with South Carolina. It was founded as a military base in 1736 and is situated on one of the world's greatest kaolin belts. Today it is a market and textile-manufacturing centre and river port.

Population (city) 44 640; (metropolitan area) 415 180

Map United States – Jd

Augustów (Avgustov) *Poland* Boating and sailing centre amid the forested Suwalki lake district in the north-east, 35 km (22 miles) west of the border with Lithuania and Belarus. The Augustów Canal crosses the marsh and forest wilderness on the border. It was built in 1824-39, during the Russian occupation of the area, to link the Vistula and Neman river systems, bypassing what was then East Prussia.

Population 25 600

Map Poland – Eb

Aurangabad *India* Textile city about 400 km (249 miles) east and slightly north of Bombay. Originally known as Khadke, it was founded in 1610 and many of its buildings date from that period. The Mogul emperor Aurangzeb is buried on the outskirts of the city. It is also noted for its Buddhist cave temples.

One of India's fastest growing cities, it has a university and several colleges.

Population 2 209 050

Map India – Bd

Australia

A VAST AND SPARSELY PEOPLED ISLAND CONTINENT WHERE A YOUNG, COMPETITIVE SOCIETY FINDS AFFLUENCE AND FREEDOM

In the 1960s, Australia was dubbed 'the Lucky Country' – though not without some irony, the author implying that its prosperity owed more to luck than good judgment. That luck has not yet run out, although by the early 1990s Australia was facing a slightly more difficult period, with relatively high unemployment, a drop in overseas demand for its export commodities and a weakened currency. But despite these problems, it is still a fortunate country by almost any criterion, with most of its nearly 18 million people enjoying a life of affluence and freedom matched in few other parts of the world.

The good fortune stems from abundant mineral resources and an agricultural sector that, in most years, produces a large surplus of food to export. The wealth created by these primary industries is spread among a relatively small population and supports an easy-going lifestyle that contains many echoes of European and North American (particularly Californian) ways but has also developed a distinctive flavour of its own. Most recently, a decidedly Asian flavour has started to emerge.

However, the affluence and freedom have largely bypassed one group of Australians: the Aborigines who inhabited the country for at least 50 000 years before the first Europeans settled there in 1788, and whose rich and varied culture has the longest continuous history in the world.

THE GREAT SOUTH LAND

Whether you consider Australia – the 'South Land' – to be the world's largest island or its smallest continent, it is undoubtedly a land of superlatives. Despite its great size and reliance on rural production, it is one of the most highly urbanised countries in the world, with more than 70 per cent of the population living in the eight biggest cities. All but one of these – the federal capital, CANBERRA – are situated on the coast. Altogether, 86 per cent of Australians live in urban areas – an unusually high percentage by world standards.

It is difficult for most Europeans or even North Americans to appreciate the size and emptiness of Australia. PERTH, the capital of WESTERN AUSTRALIA, probably qualifies as the most isolated city in the world. It is more than 2100 km (about 1305 miles) from its closest neighbour, ADELAIDE, capital of SOUTH AUSTRALIA – farther than from London to Moscow or New York to Denver.

On the route north from Adelaide to DARWIN at the top end of the NORTHERN TERRITORY – a distance of some 3500 km (2200 miles) – the only town is ALICE SPRINGS, with 20 500 people. Large parts of Australia's huge 'red centre' – particularly in the west – are an inhospitable wasteland of desert and mallee (scrub), with little or no permanent population. Even in the semi-arid but more productive rough grazing country farther east and north, neighbours may live 100 km (60 miles) or more apart and measure their stock density in terms of hectares or acres per head of cattle rather than vice versa. In these isolated settlements, the Flying Doctor brings medical services, the School of the Air teaches by radio, and stockmen muster cattle by helicopter.

Superlatives apply to the climate, too. In the centre – and even in parts of the temperate south-east, inland from the GREAT DIVIDING RANGE – droughts are commonplace, and years can pass without a drop of rain. Other years bring huge downpours that can cause extensive flooding. In the north and north-east, such flooding is an annual occurrence during the summer 'wet', any time between October and May; then, roads may be impassable and tropical cyclones (hurricanes) may batter coastal settlements – like Cyclone Tracy, which devastated Darwin on Christmas Day, 1974. Only in relatively restricted areas, notably along the south-east coast, is rainfall at all reliable.

Australia's image as a sunburnt land contains a lot of truth, and as one consequence QUEENSLAND ('the Sunshine State') holds the unenviable record of having the world's highest incidence of skin cancer. Cloncurry in north-central Queensland holds Australia's temperature record: 53.1°C (127.6°F). Yet there are also ski resorts in the SNOWY MOUNTAINS (part of the Great Dividing Range), less than 400 km (250 miles) from both SYDNEY and MELBOURNE, the two biggest cities.

It is in this south-eastern part of the country, generally well watered and (except in the mountains) temperate, that most Australians live – in a broad arc from BRISBANE (the Queensland capital) in the north, through Sydney (the capital of NEW SOUTH WALES), Canberra and Melbourne (the capital of VICTORIA), to Adelaide. The only other major communities are Perth, HOBART in the southern island state of TASMANIA, and on parts of the Queensland coast from Brisbane through ROCKHAMPTON and TOWNSVILLE to CAIRNS.

Geologically, Australia is an ancient, stable land long worn down by wind and water, so that today a mere 13 per cent of its area is above 500 m (1640 ft). The main mountain range – the Great Dividing Range which runs down the entire east coast – peaks at Mount KOSCIUSKO (2228 m, 7310 ft) near the border of New South Wales and Victoria. This is the highest mountain on the island continent. There are numerous smaller ranges, such as the HAMERSLEY RANGE in the PILBARA of Western Australia, the MACDONNELL RANGES of the Northern Territory, and the Musgrave and FLINDERS ranges of South Australia, but they reach only 1000-1500 m (3300-5000 ft). On a smaller scale, there are spectacular gorges in most of these ranges whose stark beauty is often highlighted by the red and yellow rocks.

The Australian coastline – estimated to extend nearly 37 000 km (22 991 miles), much of it superb sandy beaches – has its own spectacular features. Greatest of all is the GREAT BARRIER REEF, a vast complex of coral reefs extending 2000 km (1250 miles) up the coast of Queensland almost to Papua New Guinea. The biggest of its kind in the world, in some places it is 300 km (190 miles) from the shore and more than 500 m (1650 ft) thick.

Although most of the continent consists of extensive plains and low, undulating downland – a somewhat monotonous landscape – there are also spectacular forests, especially along the east coast, in Tasmania and in the far southwest. Areas with heavy rainfall – both temperate and tropical – have dense rain forests with their closed canopy of foliage and climbing plants that scramble up the tall tree trunks to the light. In drier areas there are more open sclerophyll forests, whose trees have leathery leaves to conserve moisture.

Among the best known of these are the acacias (wattles) and the numerous species of eucalypts – about 900 have been identified. Many eucalypt leaves (which vary from blue-grey to dark green in colour) give off droplets of aromatic oil, causing a bluish haze in such areas as the blue mountains west of Sydney. Many species also have thick, hard bark able to withstand the bush fires that are a perennial hazard in most areas. One such fire in New South Wales turned into a major disaster in January 1994, threatening Sydney itself. About 607 200 hectares (1.5 million acres) of land were devastated and nearly 200 houses burned down in the city's suburbs.

Some native Australian plants positively benefit from fire; the seed heads of some banksia trees, for example, open and scatter the seeds only after a fire. Others have had to adapt to the dry conditions in the interior– for instance the baobab, whose swollen trunk stores water, and numerous other plants, whose seeds may remain dormant for years until rare rains enable them to germinate, grow and carpet the ground with flowers, all within a few days. There are even fish whose eggs are believed to remain dormant until it rains.

UNIQUE ANIMAL LIFE

Australia's long isolation from any other landmass has influenced evolution and given it unique zoological marvels. Some 200 million years ago, Australia was part of the great southern continent of GONDWANA, with Antarctica, South America and Africa. Small, primitive, warm-blooded animals lived there, and continued to do so and to evolve when Gondwana broke up, isolating Australia. They were the ancestors of today's egg-laying monotremes and pouched marsupials; almost everywhere else the more advanced placental mammals, which bear well-developed live young, ousted them.

Marsupial species are as diverse as the tree-climbing possums and cuddly looking koala, the ground-burrowing wombat and the heavy-haunched kangaroos and smaller wallabies – as well as marsupial 'mice' and 'cats', which superficially resemble their (unrelated) placental counterparts.

Among the monotremes, which hatch from leathery eggs, are the echidna or spiny anteater – well described by this latter name – and the duck-billed platypus. When British naturalists

▲ **SAILS IN THE SUN A lone surf-boarder skims past the sail-like roofs of the Sydney Opera House, a stone's throw from where the first European settlers landed – the outdoor life beside the symbol of Australia's new cultural awareness.**

first saw this animal with its flat bill, small clawed and webbed feet and furry body, they thought it was a taxidermist's joke.

Australia's bird life is also distinctive. There are two species of large flightless bird – counterparts of the African ostrich – the emu and the cassowary. The lovely lyrebirds are renowned mimics whose males display their lyre-shaped tail plumage during mating displays. But most characteristic of all are the numerous species of Australian parrot, ranging from budgerigars to large cockatoos.

Although Australia lacks large land predators like the lion, bear and wolf – marsupial carnivores are small and inefficient hunters – dangerous creatures abound. There are large crocodiles in the north, and venomous snakes and spiders (including some of the most dangerous of both) in all areas.

Sharks appear off the entire coast, but the major metropolitan beaches are protected by nets and beach patrols, and fatal shark attacks are rare. Tropical waters are most dangerous, being home to creatures such as the box jelly-fish and the stonefish, both of which are able to inflict fatal stings.

Feral animals have had a strong impact on the Australian environment and native fauna. Wild pigs, donkeys, cats and dogs all thrive in the wild and have few natural enemies. Rabbits introduced by a landowner in 1859 were a severe blight – competing with stock for grass – until myxomatosis and feral cats and foxes controlled their numbers. Dingoes, Australia's 'native' dogs, introduced thousands of years ago, are regarded as pests by sheep farmers since one dingo can kill 100 sheep in a night. The depredations of dingoes and hybrid dogs (for they interbreed) led to the building of the world's longest fence (8500 km, 5300 miles) in South Australia and Queensland.

Australia has the world's only wild camels, the descendants of animals that played an important part in the exploration and development of the inland areas in the 19th century. Some are today exported to the Middle East for breeding. A more unwelcome import is the poisonous cane toad, introduced to control a sugar cane beetle pest in the 1930s and now teeming over areas of north-eastern Australia.

CONVICTS AND FREE SETTLERS

The concentration of people in south-eastern Australia began after a Pacific voyage by the British navigator Captain James Cook. He was not the first European to 'discover' Australia, however. The 17th-century Dutch seamen Willem Janszoon and Abel Tasman explored the coasts of the GULF OF CARPENTARIA and Van Diemen's Land (now Tasmania) respectively.

The Englishman William Dampier surveyed the west Australian coast in 1699, but all these explorers wrote off New Holland (as Australia was then called) as of little value for trade or colonisation.

Cook had the good fortune to hit upon a much more hospitable part of the coast. In 1770, returning to England from a scientific expedition to Tahiti and after charting New Zealand, he made a landfall at an inlet on the east coast that he named BOTANY BAY because of its profuse and unusual plant life. He went on to chart almost the entire east coast (and was nearly wrecked on the Great Barrier Reef), claiming it for Britain and naming it New South Wales. Soon after, Britain lost its American colonies and, faced with overcrowded prisons, decided to start a penal colony at Botany Bay. The first fleet of 11 convict ships arrived in January 1788, found the land around Botany Bay too marshy, and landed instead a little farther north at a spot they named Sydney Cove, on the superb natural harbour of Port Jackson – which Cook had failed to explore.

Life for the first colonists was extremely tough, and the settlement barely survived. But it did, and over the next 50 years other penal settlements were founded on NORFOLK ISLAND and near present-day Hobart and Brisbane; free settlers began to arrive and by the early 1830s they outnumbered the convicts. The cities of Perth, Adelaide and Melbourne were founded.

These communities looked back to Britain for trade and aid, developing port facilities

which became the gateways into Australia and the main means of communication between themselves. They also became the focus of overland communications – by road and, from later in the 19th century, rail – which linked them with the small agricultural and mining settlements that spread over the country. Inevitably, manufacturing industries largely concentrated in these future state capitals and many immigrants ventured no farther. This is why Sydney, Melbourne, Brisbane, Adelaide, Perth and Hobart are so much bigger than the second cities of their states and why Australia has few large provincial centres. There are exceptions, notably Canberra (created as the national capital in the 20th century), NEWCAS-TLE and WOLLONGONG in New South Wales (important industrial centres built on coal-fields), and GEELONG in Victoria. A large part of Australia's wealth is created beyond the urban

frontier and outside the notice of many Australians. One of the worst droughts of this century, which ended in 1983 after temporarily crippling much of the farming sector, occurred without affecting the daily lives of 90 per cent of the population.

A high proportion of agricultural output is exported – for example, two-fifths of the annual beef total of 1.7 million tonnes and 95 per cent of the annual wool clip of around 800 000 tonnes. Coal is now the main export, but Australia still has 137 million sheep, producing 30 per cent of the world's wool.

The huge sheep and cattle stations (farms) extend over large parts of the interior, though in many areas the grazing is not rich enough to support a high density of stock. But since many sheep stations exceed 500 000 hectares (1.2 million acres) and the largest cattle station is over 3 million hectares (7.4 million acres,

almost as big as Belgium), enormous numbers can be raised even at only one sheep per 3 hectares (7.4 acres) or one head of cattle per 50 hectares (125 acres).

These are figures for the drier areas. Where the rainfall is higher and the grass more abundant, stock raising is much more intensive – especially in the south-east and the extreme south-west. Generally speaking, cattle are better able than sheep to adapt to hot conditions, and nearly half the country's 21 million beef cattle are reared in Queensland and the Northern Territory. The national herd used to be much bigger, but numbers had declined by one-third by the end of the 1982-3 drought. Some 2.8 million dairy cattle graze mainly in eastern New South Wales and Victoria.

Cereal growing in Australia is dominated by wheat, most of it produced to the west of the Great Dividing Range, in a belt from southern Queensland to Victoria, and in Western Australia. Annual output is about 10-17 million tonnes, most of which is exported, making Australia the world's fourth largest supplier of wheat. Other cereal crops include barley, sorghum, oats and about 930 000 tonnes of rice a year, grown in the Riverina area of New South Wales.

Because of the great range of climates, crops are equally diverse. The warmer areas of New South Wales and Queensland produce cotton, sugar cane, pineapples, bananas, peanuts, avocados and Australia's native macadamia nuts. Fruits flourish – especially peaches, apricots, grapes and citrus fruits along the MURRAY and Murrumbidgee rivers, and apples in Tasmania. All states produce wine, both ordinary and finer vintages – the older established wine areas are South Australia's BAROSSA VALLEY (settled largely by German immigrants in the last century) and the HUNTER VALLEY near Newcastle. In recent years the industry has boomed, with many new vineyards being planted. Timber is an important product, especially in Tasmania.

HUGE MINERAL WEALTH

In 1946, agriculture accounted for more than 75 per cent of Australia's export earnings – today, agricultural exports make up about 26 per cent of the country's total export earnings. Farm output has been overtaken by greatly increased exports of minerals. Massive, easily worked coal deposits provide 13 per cent of exports, sold mainly to Japan. Hydroelectric power (mainly from the Snowy Mountains and Tasmania) and limited reserves of oil and natural gas combine with coal to make Australia self-sufficient in energy. The country is also a leading producer of iron ore, copper, zinc, manganese, nickel, tungsten, uranium, tin and bauxite (supplying 40 per cent of the world's production). It is the world's biggest producer of lead ore, and of diamonds (35 per cent of world production). As an added bonus, there are silver, platinum and gold, and such strategically important minerals as zircon, titanium and cobalt.

A number of big towns – notably BROKEN HILL and MOUNT ISA – owe their existence to mining. Along 160 km (100 miles) of coast at Weipa on Cape York Peninsula, the world's largest bauxite deposit is worked. Iron-ore mining has been the basis of enormous developments since the 1960s, especially in the MIDDLEBACK RANGE of South Australia and the KIMBERLEY and Pilbara regions of Western Australia. Although the worldwide recession in the steel industry of the early 1990s affected production, Australia remains the major supplier to Japan and South Korea, supplying millions of tonnes to China as well. Uranium – of which Australia has one-third of the world's proven reserves – is also facing a difficult market, environmental pressures at home combining with reduced European and North American nuclear power programmes.

The mineral that put Australia on the map was gold. The first major strike was in New South Wales in 1851, but this was soon followed by bigger finds at BALLARAT and BENDIGO in Victoria. People flocked to the goldfields from around the world, Victoria's population quadrupling between 1851 and 1855 and the capital, Melbourne, growing from a small town to a thriving city. Forty years later, gold was discovered at KALGOOR-LIE in Western Australia, where mining still goes on. In fact, the increase in gold prices in recent years has resulted in some old mines being revived, and Australia today produces 13 per cent of the world's gold.

THE FORGOTTEN AUSTRALIANS

Much of Australia's recently found mineral wealth lies beneath land that is far from centres of European settlement – indeed, in places that few white people have visited and most Australians know only from their television screens. This is the true outback, a term that does not refer to a specific place but to any area isolated from civilisation. It is the ancestral home of many of Australia's largely forgotten and least privileged people, the Aborigines. The conflict between their claims to their own land and the mining companies' desire to exploit it has given rise to one of the most controversial issues in contemporary Australian politics: land rights.

Although Captain Cook declared the country empty, there were in 1770 between 500 000 and 1 million Aborigines and, in far northern Cape York, TORRES STRAIT Islanders, a Melanesian group, already inhabiting the island. The Aborigines had been in continuous occupation for at least 50 000 years, as shown by burial sites and other archaeological evidence. Indeed, rock engravings deep inside a cave on the Nullarbor Plain in South Australia have been dated to about 20 000 years ago – and are among the oldest rock art in the world.

When Britain set up its Australian colonies, all land became crown 'property' without compensation being made to the Aborigines. The European settlers also brought diseases such as measles, whooping cough and smallpox that devastated the indigenous population. A greater tragedy was the colonists' persecution of the Aborigines. Many were murdered, under the excuse that they posed a threat to the colony and were in any case less than human; others were exploited as forced labour. Confiscation of land was justified on the basis that the Aborigines did not 'use' the land efficiently. As many as 250 000 died between 1788 and the 1920s as a direct result of European colonisation; among the victims was almost the entire population of Tasmanian Aborigines, a distinct ethnic group.

By 1930, only about 70 000 Aborigines remained, but their population is now more than 265 000. They still suffer discrimination and poverty, however, with an average family income two-thirds that of white Australian families, an infant mortality rate double, an unemployment rate nearly four times higher, and a life expectancy 15-17 years shorter. Government payments account for almost three-quarters of the economy of the Aboriginal society as a whole.

Until the 1940s about half the Aborigines of the tropical and desert regions followed their old lifestyle of hunting and gathering, activities feasible in these barren lands only at very low population densities. Except in New South Wales, Tasmania and Victoria, many of them lived on reserve lands segregated from white Australians. Economic and population pressures led many Aborigines to migrate to towns and cities, and try to adjust to urban life. However, such 'detribalised' people lack a cohesive social structure, and many succumb to alcoholism and petty crime. Only about one-third of Aboriginal people still live in rural areas but over the last 20 years small groups of people, particularly in the Northern Territory, have been moving from the larger settlements to set up outstation communities in the bush. There they lead a traditional lifestyle, though often with modern aids such as tap water and telephones.

Whether the land rights campaign will repair the fabric of much of Aboriginal society remains to be seen. By the early 1990s, almost two-fifths of the Northern Territory (including AYERS ROCK), and one-fifth of South Australia was Aboriginal land. For years a political dispute raged as to whether Aboriginal landowners should be able to veto mining developments on their territory. Pro-land rights campaigners pointed to the Aborigines' deep – almost mystical – relationship to the land that has supported them for millennia. In January 1994 the Native Title Act came into force, granting Aborigines land rights, though in the face of serious opposition from mining interests and politicians from Western Australia, where land claims have been most numerous.

In the mainstream of Australian life there have been great changes in the population mix since the end of the Second World War. In 1945, out of the 7.5 million population (less than half the present figure) over 90 per cent were of British or Irish origin. Since then a large influx of Italian, Greek, Dutch, Polish, Turkish, German, Maltese and many other immigrants has begun to turn Australia into a multicultural society. This has been particularly true since the mid-1960s, when the so-called 'White Australia' immigration policy was abandoned. Now there are significant communities of Lebanese, Vietnamese and Cambodian settlers – many of them former refugees – as well as Cypriots, Egyptians, Pacific islanders, New Zealanders, Asians, and many others.

The ties with Britain, once so strong, are now much weaker, although British and Irish stock still predominate. Once the emphasis was on integrating the 'new Australians' into the Australian way of life. In the last 20 years, 'old' Australians have increasingly appreciated how the non-English-speaking immigrants can enrich Australian life. Where once steak and chips was standard restaurant fare, there is now cooking as diverse as Japanese, Cambodian, Fijian, Mexican, Caribbean, African and Lebanese, as well as every type of European cuisine. In literature and the arts, from folk festivals to newspapers and broadcasting, multiculturalism is official policy. Altogether, some 100 languages are spoken in today's Australia, and many of them can be read in daily or weekly newspapers or heard on radio stations or the television channel of the Special Broadcasting Service. More than

▲ RED CENTRE Ayers Rock – sacred Uluru to the Aborigines – glows in the sunset, adding to its dreamlike quality. Like the nearby Olgas, it is the tip of a sandstone hill half-submerged in a semi-arid plain.

2 million people speak a language other than English at home.

Over the years, as the population mix grew, so did some racial tension – largely due, many observers believe, to competition for jobs. Immigration was reduced as unemployment rose in the 1980s and 1990s, and many recent immigrants have fared less well than the previous generation. Naturally more prominent in a still predominantly white society, Asian immigrants have in the past been the victims of prejudice previously shown to Italians and Greeks. Today, however, Asian immigrants outstrip new arrivals from Europe: in 1993, 90 per cent of Australia's business migrants were Asians, and a growing number of Australians are learning to speak an Asian language and looking for trading links in the Australasian-Pacific realm. By 1993 Japan was Australia's largest trading partner, taking A$15.6 billion worth of exports that year.

STATE AND FEDERAL GOVERNMENT

Australia as a nation is less than a century old. The six states had been self-governing since 1850 (1890 in the case of Western Australia) when they formed the Commonwealth (federation) of Australia in 1901. The constitution divides power between the federal parliament (consisting of the Senate, with equal representation for each state, and the House of Representatives, with membership in proportion to the states' population) and the states, each one having its own parliament. Queen Elizabeth II is head of state and is

represented by the governor-general and in each state by a governor. The AUSTRALIAN CAPITAL TERRITORY (in which Canberra stands) was formed in 1911 and the Northern Territory was also placed under direct federal government control; the latter has had limited self-goverment since 1978; the Australian Capital Territory has been self-governing since 1988.

Power in Canberra has been held for varying periods by the Labor Party and by a coalition of the Liberal Party (conservative) and the National (formerly 'Country') Party, which is strongest in rural areas. A constitutional crisis arose in 1975 when the governor-general, Sir John Kerr, dismissed Gough Whitlam's Labor government after the Senate threatened to block budget legislation. Many Australians felt that Kerr exceeded his powers, and the incident increased support for those who want Australia to become a republic. The March 1993 elections brought victory for the Australian Labor Party (ALP) for the fifth consecutive term, with Prime Minister Paul Keating expressing the hope that by the year 2001 – the centenary of the federation of the Australian states – Australia will be a republic.

State and federal governments have often been at loggerheads, especially where different parties are in control. Recent areas of conflict have included welfare and education, which is partly funded from Canberra but run by the states. Schooling is compulsory to the age of 15 or 16, depending on the state.

THE OUTDOOR LIFE

Despite its recent influx of Asians, Australia is culturally and materially still part of the Western world. The average family regards a car as a necessity – there are more cars per head than in any major country except the United States – and owns a range of modern gadgetry and electronic goods (increasingly imported

from Japan). More than four-fifths of householders own or are buying their own home, mostly in huge sprawling suburbs.

Australians love the outdoor life, and millions spend sunny Sundays at the beach or having a barbecue in a suburban park. The fine climate provides opportunities for most sports – from surfing to skiing, football to fishing – giving an outlet for Australians' highly developed competitive urges, whether as participants or spectators.

There is enthusiastic support for cricket, rugby football, Australian rules football (using an oval pitch and ball), soccer, tennis and horseracing. The best known Australian horse-race is the Melbourne Cup, held each year on the first Tuesday in November. Today the focus of national betting, it was started in 1861 and now makes the entire country come to a virtual standstill on that day.

Nor is the artistic world ignored, despite Australia's old image overseas as a cultural wilderness. Most of the major cities have multifunction cultural centres, with theatres, art galleries and concert halls. Home-grown orchestras, opera, theatre and ballet companies and a thriving film industry offer Australian performers increasing opportunities, but remnants of the 'cultural cringe' – a kind of national inferiority complex regarding the arts – still linger on despite a more aggressive promotion of Australia's national identity.

AUSTRALIA AT A GLANCE	
Official name Commonwealth of Australia	
Area 7 682 300 km² (2 965 452 sq miles)	
Population 17 827 204	
Per km² 2.3 (**Per sq mile** 6)	
Capital Canberra	
Government Federal parliamentary monarchy	
Currency 1 Australian dollar = 100 cents	
Languages English, Aboriginal languages; numerous others among immigrant groups	
Religion Mainly Christian	
Climate Ranges from tropical monsoon to cool temperate; large areas are subtropical or warm temperate, and much of the centre is desert or semi-arid. Average temperature in Sydney ranges from 12°C (54°F) in July to 22°C (72°F) in January	
Land use Cultivation 18%, grazing 24%, forest and woodlands 39%, other 19%	
Main primary products Sheep, cattle, pigs, wheat, barley, oats, sorghum, rice, sugar cane, fruit, vegetables, tobacco, timber, fish; coal, iron ore, bauxite, copper, lead, zinc, diamonds, nickel, uranium, tungsten, tin, manganese, gold, opals, oil and gas	
Major industries Agriculture, mining, iron and steel, aluminium refining and smelting, vehicles, machinery, food processing, wool and hides, forestry, fishing, oil and gas production and refining, chemicals, cement, light engineering	
Main exports Coal, wool, wheat, iron ore, diamonds, alumina and aluminium, meat (mainly beef), minerals, petroleum products, dairy products, manufactured goods	
Annual income per head (US$) 17 080	
Population growth (per thous/yr) 16	
Life expectancy (yrs) Male 74 **Female** 80	

aureole Zone around an igneous rock intrusion which has been altered by the heat and chemicals generated during the intrusion of the hot molten rock (magma). It can be traced around the margins of a BATHOLITH.

Aurès Massif *Algeria* See ATLAS MOUNTAINS

aurora Brilliant sheets of coloured light appearing in the skies at high latitudes. They are caused by ultraviolet radiation and electrically charged particles from the sun interacting with the earth's atmosphere.

The radiation and particles are drawn by the earth's magnetic field towards the north and south poles, where they produce brilliant green, blue, white or red flashes by ionising gases in the earth's atmosphere, mainly at altitudes of some 95-145 km (60-90 miles). In the northern hemisphere the display is called the aurora borealis, or northern lights; in the southern hemisphere it is called the aurora australis.

Auschwitz (Oswiecim) *Poland* Chemical-producing town 54 km (33 miles) west of the city of Cracow. It was the site of the largest Nazi concentration camp, which operated from June 14, 1940 to January 27, 1945. Auschwitz was in fact composed of three main camps – Oswiecim, Brzezinka (Birkenau) and Monowice (Dory) – with 39 smaller camps nearby.

More than 4 million people of 39 nationalities were shot, gassed, starved or tortured to death in Auschwitz, and its crematoria burnt up to 12 000 bodies daily. Today it is preserved as the National Museum of Martyrology, which, with the world's largest burial ground at Brzezinka, is a place of pilgrimage.
Population 45 300
Map Poland – Cc

Austerlitz *Czech Republic* See SLAVKOV

Austin *USA* Capital of Texas, in the central southern part of the state, on the Colorado River. It is the commercial centre of a large farming and ranching region, and has the main campus of the University of Texas, an important centre for research, and St Edward's University.
Population (city) 465 600; (metropolitan area) 846 230
Map United States – Gd

Austral Islands *French Polynesia* See TUBUAI ISLANDS

Australasia Term used for Australia, New Zealand and Papua New Guinea, as well as the adjacent islands of the south Pacific, all of which are also part of OCEANIA.

Australia See p 56

Australian Alps *Australia* See SNOWY MOUNTAINS

Australian Capital Territory *Australia* Region around the capital, CANBERRA, bought from the state of New South Wales in 1911 by the Commonwealth of Australia government following the federation of the colonies in 1901. It has been a self-governing territory since 1988 and is located about 250 km (150 miles) south-west of Sydney. It covers 2400 km² (926 sq miles),

including a 70 km² (27 sq mile) naval base at JERVIS BAY on the New South Wales coast. The main rural activity is sheep farming.
Population 292 700
Map Australia – Hf

Austria See p 62

Auvergne *France* Largely rural region of some 26 000 km² (10 000 sq miles) in the Massif Central, where winters are harsh. It consists of glacially eroded uplands, volcanic peaks and rocky lowlands. The highest peak is the Puy de Sancy (1885 m, 6185 ft). Farming and tourism are the main occupations in the thinly populated mountains, with rich farming and modern industries such as rubber and electrical goods in the low-lying Limagne Plain.

The region contains many spas, including La Bourboule, Le Mont-Dore, and VICHY. CLERMONT-FERRAND is the main city.
Population 1 321 200
Map France – Ed

Auyuittuq *Canada* National park covering 21 471 km² (8288 sq miles) that was established in 1972 on the Cumberland Peninsula of eastern Baffin Island. It contains spectacular fjords as well as barren tundra and glaciers.
Map Canada – Ib

Avarua *Cook Islands* See RAROTONGA

Avebury *United Kingdom* Prehistoric monument in the English county of Wiltshire, about 120 km (75 miles) west of London. It is the largest stone circle in Europe, 410 m (1350 ft) across, with some 154 standing boulders; another 100 pairs lead off in an avenue to the south to a smaller circle, now destroyed. The circle was built in about 2500 BC, probably as a religious centre, and now encloses the village of Avebury. There are enormous earthworks, including burial mounds, nearby.
Map United Kingdom – Ee

Aveiro *Portugal* Quiet town on a lagoon, about 60 km (35 miles) south of the city of Oporto. It was a prosperous cod-fishing port until a violent storm silted up the harbour in 1575. The fishing industry revived in the 19th century, when a canal was built to the sea. Salt pans and rice cultivation are other mainstays of the town's economy, and there is a steelworks at Ovar, about 30 km (20 miles) to the north.
Population 29 200
Map Portugal – Bb

Avgustov *Poland* See AUGUSTÓW

Avignon *France* Capital of Vaucluse department, near the confluence of the Rhône and Durance rivers, 85 km (53 miles) north-west of Marseilles. It was the seat of seven popes during the 14th century and of the Avignon popes during the Western Schism (1378-1417). The Palace of the Popes still stands.

Also of interest are the city's well-preserved ramparts, built by Popes Innocent VI and Urban V between 1355 and 1365. Part of the famous bridge, immortalised in the folk song, also survives. The English philosopher John Stuart Mill (1806-73) is buried in the Avenue Cemetery.

Today, Avignon is one of the liveliest cities

in France, to which it was annexed in 1790. It flourishes on commerce and tourism, and holds a drama festival each July.
Population 89 440
Map France – Fe

Ávila *Spain* Town about 95 km (60 miles) west of Madrid. It was the birthplace and home of St Teresa (1515-82). After an ecstatic vision she founded the Discalced (barefoot) Order of Carmelites and many of the town's religious buildings are associated with her. Ávila started as a Roman settlement called Avela and fell to the Moors in about 714. It still has its 2.5 km (1.5 miles) of walls with 88 round towers. Its surrounding province, of the same name, is the highest in Spain, mostly lying more than 1000 m (3280 ft) above sea level.
Population (town) 45 980; (province) 174 380
Map Spain – Cb

Avon *United Kingdom* County covering 1338 km² (517 sq miles) of western England and named after the river which flows through it. It was formed in 1974 from parts of Gloucestershire and Somerset to give administrative unity to the towns of BRISTOL and BATH, formerly split between the two counties. It includes the Mendip Hills and the southern Cotswolds.

In the west, it stretches to the shore of the Bristol Channel, where there is a string of seaside resorts and commuter towns for Bristol, such as Weston-super-Mare.
Population 964 900
Map United Kingdom – De

Awash *Ethiopia* River rising 80 km (50 miles) west of the capital, Addis Ababa, and flowing about 600 km (375 miles) into Lake Abbe on the border with Djibouti. It provides hydroelectric power for the capital and vital irrigation on the arid eastern plains.
Map Ethiopia – Bb

Axios (Vardar) *Yugoslavia (Serbia and Montenegro)/Macedonia/Greece* Balkan river that rises in Serbia and Montenegro and flows 388 km (241 miles) through Macedonia – where it is called the Vardar – to the Thermaic Gulf near Salonica. It has long been a trade and invasion route. Its waters were diverted in the 1950s and 1960s to irrigate the large plain of Macedonia.
Map Macedonia – Bb; Greece – Ca

Axum (Aksum) *Ethiopia* One of Africa's most important historic towns, in Tigray province about 500 km (310 miles) north-east of the capital, Addis Ababa. It is thought to have been the royal city of the legendary Queen of Sheba, who visited the King of Israel, Solomon, in the 10th century BC, and to have been the city where the Jewish Ark of the Covenant was once kept. The ark was believed to have been taken from the temple in Jerusalem by Emperor Menelik I, the son of Solomon and the Queen of Sheba. Axum was certainly a royal capital from about 500 BC, and a centre of Christianity in AD 300.

The town has many granite obelisks up to 20 m (65 ft) high. The obelisks are carved with pictures of multistorey houses of a type still found across the Red Sea in Yemen, from where the early kings came.
Population 20 000
Map Ethiopia – Aa

Austria

THE NEUTRAL ALPINE REPUBLIC THAT WAS ONCE THE HUB OF A CENTRAL EUROPEAN EMPIRE, AND BECAME A MEMBER OF THE EUROPEAN UNION IN 1995

The wall of mountains that runs across the centre of Austria dominates the nation's economy as well as its landscape. The scenic land draws about 15 million visitors a year, making tourism a major industry. Tourists come in the summer to wander through the forests, pastures and mountains around glittering lakes and hundreds of tumbling rivers. And they come in the winter to ski the slopes above more than 50 ski resorts such as St Anton, KITZBÜHEL and INNSBRUCK, one of the venues for the Winter Olympics. But the yearly surge of tourism, though a boost to the economy, brings with it problems, such as traffic congestion and environmental degradation.

For the Austrians themselves – a conservative and largely Roman Catholic people – the mountains are an irreplaceable national asset, treasured and protected from garish modernity. Farmers, for instance, are subsidised by the government to encourage them to maintain the scenic appeal of their land. And many villages lay down stringent bylaws to make sure that new buildings fit in with the local style of the traditional wood-faced, broad-eaved chalet. In addition, some 3000 km² (nearly 1200 sq miles) of the ALPS have been set aside as protected areas.

The mountains are also a source of energy for factories and towns: around two-thirds of the country's power is generated from hydro-electric dams fed by streams from the Alpine snows. The mountains are also the source of Austria's best-known industry – tourism. The beauty of the Hohe Tauern and the neighbouring Niedere Tauern, which run east to west through the country, are accessible via magnificent north-south mountain passes such as the Grossglockner, which reaches an altitude of 2503 m (8213 ft) above sea level.

SMALL IS BEAUTIFUL

Industrially, Austria subscribes to the 'small is beautiful' philosophy of the 20th-century, German-born economist Ernst Schumacher. Most firms employ a very small staff; the average size of a factory workforce is just 61. The farms, too, tend to be small. Despite a spate of farm mergers in recent years, there are still about 124 000 farms.

The formula seems to work. Dairy products, beef and lamb from the hill farms supply the cities and contribute to exports as well. Truckloads of wheat and other crops flow in from the eastern lowlands. Wine, predominantly white, pours in from the vineyards, which are mostly in LOWER AUSTRIA (Niederösterreich) and BURGENLAND. However, Austria's national debt has risen, with the country importing goods to the value of US$7.2 billion per year more than it is exporting. In 1990, Austria had a foreign debt of US$11.8 billion.

One of the big difficulties for a small country in a world of big businesses is how to compete effectively when you have no really large home market. Austria is, after all, smaller than the US state of Maine. The Austrian solution has been to increase specialisation and the quality of its manufactured products.

The iron and steel industry, for instance, could hardly compete with the big producers of Germany, and so the answer of the state-run steel industry has been to develop more efficient ways of making steel.

As a result, the steel towns of LINZ and Donawitz have given their name to one of the most up-to-date methods of steel-making in the world: the Linz-Donawitz process.

Nearly half of Austria's annual production of 3.1 million tonnes of pig iron and 4.6 million tonnes of crude steel are exported. Other successful export industries are optical instruments, which are assembled near SALZBURG, electronics and chemicals, as well as paper which is manufactured near GRAZ from Austria's abundant supplies of timber. Forests cover more than 39 per cent of Austria – a greater proportion than in any other European country except Sweden and Finland.

Today there are four times as many Austrians employed in industry as there are working in more traditional lines of business. Many people outside the country still think of the Austrian economy as being largely carried by traditional hill farming; but revenue from agriculture is declining rapidly, and without the additional income coming from tourism many farmers would not be able to survive. Farmers in the Alpine foothills are somewhat better off, especially in and around the capital, Vienna, where wheat, fruit and grapes are grown.

INFLUENCE OF THE CHURCH

With about 85 per cent of the population being Roman Catholic, the Church exerts a powerful influence. Crucifixes and statues of the Virgin are conspicuous outdoor features, especially in the mountain villages. The rural areas also contain the best examples of the traditional farmhouses, which vary from one region to another. Examples of each type can be seen together in the Österreichisches Freilichtmuseum (Austrian open-air museum) near Graz.

The country's emphasis on regionalism – typical of a mountain culture where each valley developed largely in isolation from its neighbours – is reflected in Austria's political structure. The country is a federation of nine *Bundesländer*, or states, each with its own government. Each *Bundesland* is represented in the

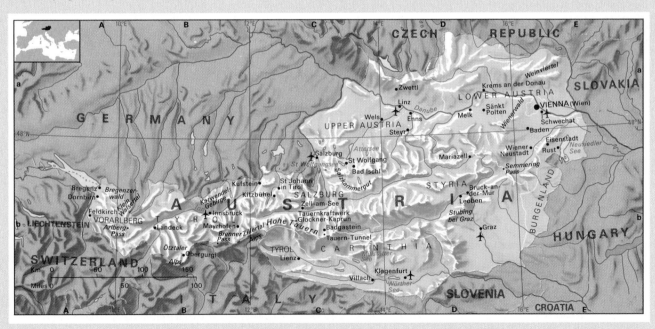

Federal Council, which is responsible for national and international affairs. National government has been in the hands of a coalition consisting of the Austrian Social Democratic Party and the Austrian People's Party. The government is heavily involved in providing public housing and in the profitable administration of the large proportion of nationalised industries, ranging from steel and chemicals to transport and banking.

Although socially and economically Austria clearly leans toward the West, politically it is firmly neutral. It is a member, however, of various international economic unions and of the Council of Europe and the United Nations. In January 1995, Austria became a member of the European Union (EU).

Neutrality, like the sparkling elegance of Vienna, is a product of Austria's turbulent history. Settled originally by Celts around 500 BC, the land that is now Austria was fought over by Romans, Vandals, Visigoths, Huns, Hungarian Magyars and Germanic tribes. Then, in 1246, a remarkable Swiss-Alsatian family came to power – the Habsburgs. They made Vienna their capital and built around it an empire that by 1530 included Hungary, the western part of Czechoslovakia (now Czech Republic), northern Italy, the Netherlands and the whole of Spain. They expanded through marriage between royal houses rather than war. The head of the family in each generation was usually elected Holy Roman Emperor, and Habsburg power endured for more than 600 years.

SHOT THAT STARTED A WAR

Conflict and bitterness between the various nationalities of the sprawling empire came to a head on June 28, 1914, when a pistol shot echoed round the world from SARAJEVO, now in Bosnia-Herzegovina. The shot, fired by a Serbian nationalist, killed the heir to the Habsburg throne, Archduke Franz Ferdinand. Austro-Hungarian forces invaded Serbia. Other European nations took sides in the conflict, and so started the First World War. Habsburg power did not survive the upheaval. By the treaty of Saint Germain-en-Laye in 1919, seven-eighths of the Habsburgs' Austro-Hungarian empire was parcelled up among its neighbours. One-eighth, the German-speaking rump, became the Republic of Austria.

The republic was a disaster. Economically, it teetered on the brink of bankruptcy, and chronic unemployment led to political unrest. Faction fighting between socialists and conservatives led to outright civil war in 1934 and a brief right-wing dictatorship under Engelbert Dollfuss. Nazi sympathisers assassinated him after only a few months, however, and in 1938 Adolf Hitler – himself an Austrian by birth – sent in German troops to annex the country.

Liberated by the Allies in 1945, Austria was allowed to set up its own government again in 1955, after guaranteeing that the country would be strictly neutral – an undertaking which remains in force today.

MAGNIFICENT CAPITAL CITY

Vienna, straddling the great waterway of the River DANUBE, is full of echoes of the past. Largely and magnificently rebuilt by the

▲ SKI PARADISE Colourful 16th-century houses line the north bank of the Inn at Innsbruck below the Nordkette, one of the Tyrol's many skiing areas. The city, popular with winter and summer visitors, hosted the Winter Olympics in 1964 and 1976.

Habsburg emperors in the late 18th and early 19th centuries, the city is laid out on an imperial scale quite disproportionate to Austria's present size and wealth. More than 6.5 million people visit the city each year – way over twice as many as live here.

Many of the tourists are from Eastern Europe, particularly Hungary, Austria's former imperial partner. Among the attractions are the Hofburg and the Schönbrunn palaces, the cavernous Stephansdom, the great church of Karlskirche, and the lavish opera, home of the Viennese waltz made famous by two composers, a father and a son both named Johann Strauss – and both forever remembered for the richness of their spirited music.

This city of music inspired other composers: Mozart, Beethoven, Brahms, Schubert, Haydn and Mahler all lived here for long periods.

Vienna has less highbrow attractions, too: comfortable pavement cafés and coffee houses, with the latest newspaper always available, and jovial, friendly, small winebars, often housed in farmhouses near the city, which specialise in selling *Heuriger*, the new wine.

The other great musical city of Austria is Salzburg, situated on the Salzach River near the border with Germany. It was the birthplace in 1756 of the composer Wolfgang Amadeus Mozart. His house still stands in the old town – a mecca for lovers of his memorable music. The centre of the city is a wonderland of narrow streets lined with intriguing shops selling,

among other things, the traditional male garb of leather *Lederhosen* and distinctive alpine hat, often with a feather stuck jauntily in the hatband, and, for women, the full-skirted, tight-bodiced dress known as the dirndl.

Music still resounds in Saltzburg, particularly in August when an annual music festival is held in the city.

AUSTRIA AT A GLANCE	
Official name Republic of Austria	
Area 83 853 km² (32 375 sq miles)	
Population 7 915 145	
Per km² 94 (**Per sq mile** 244)	
Capital Vienna	
Government Federal republic	
Currency 1 schilling = 100 groschen	
Languages German (99%), Magyar, Slovene	
Religion Christian (Roman Catholic 85%)	
Climate Temperate continental; average temperature in Vienna ranges from –1°C (30°F) in January to 20°C (68°F) in July	
Land use Cultivation 18%, meadows and pastures 24%, forest and woodlands 39%, other 19%	
Main primary products Cattle, sheep, wheat, maize, potatoes, hay and fodder, barley, sugar beet, vines, temperate fruits, timber; oil and natural gas, lignite, coal, iron ore, lead, copper, magnesite	
Major industries Agriculture, iron and steel, machinery, forestry and wood products, chemicals, textiles, oil and gas production and refining, wine, beer, food processing	
Main exports Machinery, iron and steel, metal goods, electrical appliances, textiles, chemical products, timber, paper, wine, meat and dairy produce	
Annual income per head (US$) 19 240	
Population growth (per thous/year) 6	
Life expectancy (yrs) Male 73 **Female** 80	

Azerbaijan

DREAMS OF A BETTER LIFE HAVE BEEN SHATTERED BY WAR AND A STRUGGLING ECONOMY IN THIS PREDOMINANTLY MUSLIM EX-SOVIET REPUBLIC

Shortly after declaring their independence in the early 1990s, the two former Soviet republics, Azerbaijan and Armenia, blazed into world headlines with a bloody war – despite having much in common.

Like the Armenians, the Azerbaijanis have fought invaders – Persians, Arabs, Tatars – through much of their history. What remains as Azerbaijan today is only a minute part of the historic region of Azerbaijan, which was divided between Persia (Iran) and Russia in the 19th century. Like the Armenians, there are many Azerbaijanis today living far from home, in countries less besieged and troubled.

The conflict between Azerbaijan and Armenia was carried over from previous power struggles between the Ottoman empire and Russia: in delimiting their spheres of interest the Ottomans and Russians ignored the fact that they were creating ethnic enclaves.

The main area of conflict revolved around Nagornyy-Karabakh in Azeri territory, an enclave with a largely Armenian-speaking population who sought reunification with Armenia. Following Armenia's advance on Nagornyy-Karabakh in early 1992, the conflict spread to the second area of conflict, the Autonomous Republic of Nakhichevan on the Turkish border, and threatened to involve Turkey on the Azeri side. Nakhichevan is separated from the rest of Azerbaijan by Armenian territory, and more than 90 per cent of its population are Azeri-speaking.

Conflict lasted on all through 1992-4, despite various attempts to arrange a cease-fire

between the two Transcaucasian countries. The fact that the Armenians are Christian and the Azerbaijanis Muslim made reconciliation extremely difficult.

In 1994 it was estimated that the conflict over Nagornyy-Karabakh was costing Azerbaijan as much as 25 per cent of its financial resources – a massive amount for one of the poorest ex-Soviet states.

Less developed industrially than the other Transcaucasian states, Azerbaijan has found the transition from a planned economy to a market economy to be fraught with difficulties. Unemployment and underemployment are high and the standard of living is low even by ex-Soviet standards.

A MINERAL-RICH LANDSCAPE

Azerbaijan's landscape can be divided into three main regions. The eastern tip of the Great Caucasus, rising to peaks of 4500 m (14 764 ft), shapes it in the north; it is arid to semi-arid and prone to drought. The land then tapers towards the coast, where the Abşeron Peninsula juts out into the great inland Caspian Sea. About 40 per cent of the country is lowland, and almost 10 per cent of Azerbaijan lies below sea level.

Further south and west, the land rises again towards the Lesser Caucasus, reaching to heights of more than 3500 m (11 483 ft). The range is so rich in minerals – iron ore, copper, cobalt, manganese – that it has been called 'the Azeri Urals'.

Between the lowland and the highland lies a vast central region consisting of the flood

▲ OIL RICHES The oilfields of Baku have been known since ancient times, and though supplies are now nearly exhausted, new offshore fields in the Caspian Sea have yet to be exploited.

plains of the Kura and Araks rivers. Large parts of it (14 010 km², 5408 sq miles) are under irrigation and it is a major cotton-producing area. Other crops are grains, rice, grapes, fruit, tea and tobacco. Cattle, pigs, sheep and goats are also reared. Agriculture employs 32 per cent of Azerbaijan's workforce.

The gently undulating Abşeron Peninsula, which contains the country's capital, BAKU, is the economic hub. Situated on the west coast of the Caspian Sea, it is the export harbour and the country's oil trading centre. From Baku, a pipeline runs to the port of Batumi on the Black Sea, carrying 'black gold' for export.

Baku's rich oil field has been tapped since antiquity; in about 1900, its oil, known for its high quality and low sulphur content, supplied about half of world demand. The deposits around Baku are now almost exhausted, but the city still has major oil-refining plants, and oil-field equipment, chemicals, shipbuilding, cement and textile industries.

Two offshore oil fields have recently been discovered in the Caspian Sea, and Western and Russian oil companies have been contracted to help develop them. However, very little other foreign investment has been made in the country since independence in 1991 because of the armed conflict in Nagornyy-Karabakh.

Oil has brought a relative degree of wealth to the region in the past, and limited investment since independence – but it has also brought major pollution. At the beginning of the 20th century, and under Soviet rule, environmental considerations hardly counted, and so the industrial district on the outskirts of Baku has been dubbed 'the Black Town'. Environmentalists refer to it as 'the most ecologically devastated area in the world'. Azerbaijani fishermen, too, suffer from the polluted Caspian Sea which provides ever-dwindling catches.

AZERBAIJAN AT A GLANCE	
Official name Azerbaijani Republic	
Area 86 600 km² (33429 sq miles)	
Population 7 573 440	
Per km² 87.5 (**Per sq mile** about 227)	
Capital Baku	
Government Republic	
Currency 1 manat = 10 Russian roubles (transitional)	
Languages Azeri (official), Russian	
Religions Muslim (87%), Eastern Orthodox (11%)	
Climate Continental, with mild winters and hot summers, and little rainfall at low altitude. Average temperature in Baku ranges from 4°C (39°F) in January to 26°C (79°F) in July	
Land use Cultivation 18%, grazing 25%, other 57%	
Mineral resources Oil, natural gas, iron ore, copper, cobalt, manganese, molybdenum	
Main primary products Cotton; oil, natural gas, iron ore	
Major industries Oil and oil products, iron and steel, cement, petrochemicals, textiles	
Main exports Oil, natural gas, chemicals; cotton, textiles	
Annual income per head (US$) 870	
Population growth (per thous/yr) 15	
Life expectancy (yrs) Male 67 **Female** 75	

Ayacucho *Peru* Southern department in the Andes Mountains which was the scene, on December 9, 1824, of the Battle of Quinua, in which Antonio José de Sucre defeated the Spaniards and won independence for Peru from Spain. Its capital, also Ayacucho, founded in 1539 by the Spaniards, still has numerous Spanish and colonial buildings. It produces textiles, wine and pottery. The site of Huari near Quinua has impressive pre-Columbian circular buildings, probably dating from the 8th century.
Population (department) 566 400; (city) 101 600
Map Peru – Bb

Ayers Rock *Australia* One of the largest monoliths in the world, lying in the Northern Territory 430 km (267 miles) by road to the south-west of Alice Springs. It rises 348 m (1142 ft) above the surrounding plain and is more than 8 km (5 miles) around the base.

The rock, the summit of a vast buried sandstone hill, is an Aboriginal sacred site, and caves in the rock contain Aboriginal paintings. It is named after Henry Ayers, a 19th-century premier of South Australia. In 1985, ownership of the rock was restored to the local Aborigines, whose name for it is Uluru.

More than 270 000 people visit the red rock each year. It is at its most spectacular at sunset, when it glows like a burning coal on the flat desert. Uluru National Park, covering an area of 1325 km² (512 sq miles), encompasses Ayers Rock and the nearby OLGAS.
Map Australia – Ed

Ayeyarwady (Irrawaddy) *Myanmar (Burma)* River running the length of Myanmar. One of the world's great rivers, the Ayeyarwady (formerly the Irrawaddy) is about 2090 km (1298 miles) long and drains the 430 000 km² (165 980 sq miles) of Myanmar's heartland. It rises in two major headstreams, the Mali Hka and the Nmai Hka, in the northern hills, and flows southwards through a series of narrow gorges before spreading into the lowlands near Mandalay.

The delta, covering 40 000 km² (15 440 sq miles), is Myanmar's main rice-growing zone; in places it is so wide that it contains large seasonal islands, which emerge as the river level drops during the dry season. They are used to grow crops – primarily rice – before the floods of the wet season return. The Ayeyarwady division, covering 35 139 km² (13 567 sq miles), administers the delta area.
Population (division) 4 991 000
Map Myanmar – Bc-Bb

Áyion Óros *Greece* See ÁTHOS

Áyios, Nikólaos *Greece* Town on the Mirabello Gulf on the northern coast of Crete. Fishing boats fill its little harbour and tourists swim and windsurf from the nearby sandy beaches at Eloúnda to the north and Lerapetra to the south. At Kritsá, 11 km (7 miles) to the south, is a 13th-century frescoed church and the 7th-century BC city of Lato.
Population 8100
Map Greece – Dd

Ayr *United Kingdom* Port, holiday resort and former county of south-west Scotland, in Strathclyde region 50 km (30 miles) south-west of Glasgow. The poet Robert Burns was born in the Ayr suburb of Alloway in 1759; his cottage can be visited.
Population 48 490
Map United Kingdom – Cc

Ayutthaya *Thailand* Town 80 km (50 miles) north of Bangkok, and the royal capital from 1350 to 1767, when it was sacked by the Burmese. Ayutthaya lies at the confluence of the Chao Phraya, Lopburi and Pa Sak rivers, which are joined to form a moat around the town.

As a royal capital, Ayutthaya became a busy international port, and several European trading posts were set up outside its walls in the 17th century. A late 17th-century Dutch visitor recorded that the gilding on the roofs of the temples and pagodas reflected the light so strongly that it hurt the eyes. Most of this magnificence was destroyed, together with manuscripts recording Thailand's early history, during the Burmese invasion. The pretext for the war between the two hostile kingdoms was the refusal of the Thai king to let his rival have one of his white elephants, which were treasured by royalty for their rarity and as symbols of good luck.

Some treasures, including some of the temples with their gold and silver ornamental contents, did survive the invasion, and most can be seen at the Chan Kasen Palace Museum. Together with the old city walls, they make Ayutthaya a tourist centre today. The town has Thailand's largest 'floating market' and there is also a market for rice grown in the surrounding province of the same name.
Population (town) 113 300; (province) 644 100
Map Thailand – Bc

Azad Kashmir *Pakistan* Disputed area in the former princely state of JAMMU AND KASHMIR which has been governed by Pakistan since a United Nations cease-fire between India and Pakistan in 1949. Azad means 'free'. The Karakoram Highway, built with Chinese help, passes through the region to connect Pakistan with the Xinjiang region of China.

The area is dominated by the towering peaks of the Himalayas, including K2 – at 8611 m (28 250 ft), the world's second highest mountain after Everest.
Map Pakistan – Da

Azarbaijan *Iran* Mountainous region in north-west Iran, separated from independent Azerbaijan by the Araks River. Its highest point is the volcano of Sabalan, east of the city of Tabriz, at 4811 m (15 783 ft).

The region is divided into two provinces: Azarbaijan Gharbi (capital Orumiyeh), bordering Turkey and Iraq, and Azarbaijan Sharqi (capital Tabriz), bounded by the Caspian Sea. The region's main farm products are sheep, wheat, cotton and tobacco. The population includes Armenians in the north and Kurds in the south. The principal language, Azeri, is derived from

Turkish and is also spoken across the border in the independent republic of Azerbaijan.
Population 4 613 000
Map Iran – Aa

Azerbaijan See p 64

Azores (Ilhas dos Açores) *Portugal* Block of volcanic islands in the North Atlantic, 1290 km (800 miles) off the west coast of Portugal. It consists of three groups: Flores (148 km², 57 sq miles) and Corvo (18 km², 7 sq miles) in the west, Faial (165 km², 64 sq miles), Pico (430 km², 166 sq miles), São Jorge (104 km², 40 sq miles), Graciosa (46 km², 18 sq miles) and Terceira (545 km², 210 sq miles) in the centre, and São Miguel (770 km², 297 sq miles) and Santa Maria (74 km², 29 sq miles) in the east.

The islands were colonised by the Portuguese in the mid 18th century. In the past, their economy was based on whaling; today revenue comes from agriculture, fishing and tourism.

Portugal's highest mountain, Pico Alto (2351 m, 7712 ft), is on the island of Pico. The capital is Ponta Delgada on São Miguel.

The Azores are still shaken by earthquakes, the last of which occurred in 1957-8.
Population 254 200
Map Atlantic Ocean – Dc

Azov *Russia* Small town and fishing port near the mouth of the Don River where it drains into the shallow Sea of Azov. The town, near the site of the ancient Greek colony of Tanais, was fortified by the Genoese in the 13th century and made a trading port for oriental goods. The Mongol ruler Tamerlane (about 1336-1405) sacked it in 1395, and in 1739 it became part of Russia. Its industries today include canning, textiles and farm machinery. The town's boats fish the Sea of Azov for herring, grey mullet, anchovy and sturgeon – the source of caviar.
Population 76 000
Map Russia – Ed

Azov, Sea of *Russia/Ukraine* Shallow sea covering 36 260 km² (14 000 sq miles) which is connected to the BLACK SEA by the narrow Strait of Kerch'. Its leading ports include KERCH', ROSTOV-NA DONU, TAGANROG and ZHDANOV. Although the Don and several smaller rivers feed into it, its average depth is only 10 m (33 ft) and at its deepest point it is only 15 m (49 ft) deep. Fishing is important in the south, but the evil-smelling marshes at the western end have earned it the name of Sivash ('Putrid Lake').
Map Russia – Ed; Ukraine – Cb

Azuay *Ecuador* Province of the southern Sierra containing the Cuenca basin, one of the country's most fertile areas. The provincial capital is CUENCA.
Population 506 090
Map Ecuador – Bb

azurite Azure-blue, glassy mineral composed of copper carbonate. It is a source of copper and is used as a gemstone.

Baalbek *Lebanon* The capital city of Beqa'a Province, situated in the Beqa'a Valley, 64 km (40 miles) north-east of Beirut. Baalbek was a city of some size and importance in ancient times. It was a centre of worship of the Semitic sun god Baal, which led the later Greek colonisers to call the city Heliopolis – 'City of the Sun'.

Baalbek has some of the most impressive Roman ruins in the world. The Great Temple dedicated to the god Jupiter (the supreme Roman deity) has six huge columns, 27 m (88 ft) high and 7 m (22 ft) round, still standing. Originally, it was surrounded by 54 such columns. There is also a temple to Bacchus, the god of wine, of which about half remains, and a temple to Venus, the goddess of love. The site contains the world's largest cut masonry stone, weighing 1500 tonnes and 18.2 m (60 ft) long.

Population 18 000
Map Lebanon – Ba

Babia Góra (Babia Hora) *Slovakia/Poland* See BESKIDY

Babylon *Iraq* Ruined city dating from before the 18th century BC, 90 km (56 miles) south of the capital, Baghdad. It is the site of the Hanging Gardens, one of the Seven Wonders of the Ancient World, and was rebuilt in the 6th century BC by Nebuchadnezzar II. The remains of the king's palaces are also visible, as are the foundations of the 90 m (300 ft) high ziggurat, or stepped pyramid, built during his reign and thought by some scholars to be the inspiration for the Biblical story of the Tower of Babel. The Macedonian conqueror Alexander the Great died here in 323 BC.

Map Iraq – Cb

backing Anticlockwise change of direction of a wind, for example, from south-west through south to south-east. A change in the opposite direction is called VEERING.

Bacolod *Philippines* See NEGROS

Bad Hofgastein *Austria* See BADGASTEIN

Bad Ischl *Austria* Spa town at the point where the Ischl flows into the Traun River, some 45 km (28 miles) south-east of Salzburg. It was the summer residence of the emperor Franz Josef (1830-1916), whose court was attended by such eminent composers as Johann Strauss and Franz Lehár. Lehár's house is open to the public. Each year, the town holds an operetta week.

Population 13 900
Map Austria – Cb

Badacsony *Hungary* Extinct volcano rising to 464 m (1522 ft) on the north shore of Lake Balaton in western Hungary. Striking columns of black basalt rock, up to 61 m (200 ft) high and fantastic lava formations adorn its flat top. The rich volcanic soils of the area produce fine dessert grapes and Badacsonyi white wines. The small lakeside resort village of Szigliget nearby is tucked beside a sandy beach. A Baroque 18th-century castle looms on a hilltop 183 m (600 ft) above the lake.

Map Hungary – Ab

Badajoz *Spain* Town 325 km (about 200 miles) south-west of Madrid on the Portuguese border. Its industries include brewing, distilling and food processing. Badajoz has many Roman remains, including two aqueducts. It was the capital of a Moorish kingdom in the 11th century, and was bloodily stormed and pillaged by the British under Sir Arthur Wellesley, later Duke of Wellington, in 1812, when they seized it from the French in the Peninsular War.

Badajoz Province is Spain's largest at 21 657 km² (8361 sq miles).

Population (town) 122 260; (province) 650 390
Map Spain – Bc

Baden *Austria* Spa town in Lower Austria, situated 26 km (16 miles) south-west of Vienna on the edge of the Vienna Wood (*Wiener Wald*). Its hot sulphur springs have been used for medicinal purposes since Roman times. The German composer Ludwig van Beethoven lived in the town in the early 19th century and there is a museum devoted to his life and works. After the Second World War, Baden was the headquarters of the Soviet zone of occupation, from 1945 to 1955.

Population 24 000
Map Austria – Ea

Baden-Baden *Germany* Spa town on the Black Forest slopes of the Rhine Valley, 70 km (43 miles) west of Stuttgart. The Romans built the first baths here above its hot springs; the town's modern bathhouse dates from 1821.

In the 19th century, Baden-Baden was known as the summer capital of Europe, because of the number of royalty visiting it. Its notorious casino was closed in 1872. Here, the Russian novelist Feodor Dostoevsky (1821-81) is said to have gambled many a wild night away during the time he was writing *The Idiot*.

Population 51 500
Map Germany – Cd

Baden-Württemberg *Germany* Industrial and farming state (35 751 km², 13 800 sq miles) in the south-west corner of the country, flanked by France and Switzerland. Its capital is STUTTGART. It contains the BLACK FOREST and the Swabian JURA as well as well-known centres such as Heidelberg and Mannheim. Its industries include machinery, chemicals, motorcars, watches and textiles, much of which is exported.

Population 9 888 000
Map Germany – Cd

Badgastein *Austria* Small spa town and ski resort on the northern edge of the Hohe Tauern Mountains, 76 km (47 miles) south of Salzburg. About 8 km (5 miles) to the north is the hot springs spa of Bad Hofgastein which is connected to Badgastein by means of a thermal pipe.

Population 5700
Map Austria – Cb

badlands Area of barren land characterised by roughly eroded ridges, peaks and plateaus. The name is derived from the Badlands of the western United States – in Nebraska, South Dakota and Wyoming – where the large stretches of heavy clay have been dramatically eroded.

Badlands National Monument *USA* Area of severely eroded land covering 985 km² (380 sq miles) in south-western South Dakota. Its steep gullies between flat-topped hills are etched out of brightly coloured, layered sandstones. The park contains many prehistoric animal remains.

Map United States – Fb

Baegdu Son *China/North Korea* See PAEKTU SAN

Bafatá *Guinea-Bissau* Commercial centre for the interior, at the highest navigable point on the Gêba River, 110 km (68 miles) from Bissau. It is in the middle of an area producing groundnuts, cattle, cotton and tobacco.

Population (region) 116 000
Map Guinea – Ba

▼ BADEN-WÜRTTEMBERG The fairytale spires of Hohenzollern Castle situated in the foothills of the Swabian Jura. The castle was the seat of the Hohenzollern family.

Baffin Bay Part of the ATLANTIC OCEAN, separating Baffin Island from Greenland (Kalaallit Nunaat), and named after the English navigator William Baffin (1584-1622), who first explored it in 1616. The cold Labrador Current, carrying dangerous icebergs, flows through the bay, which is ice free for only a short time in summer. Baffin Bay is 1130 km (about 700 miles) long, and 110-640 km (68-400 miles) wide.
Map Canada – Ia

Baffin Island *Canada* Fifth largest island in the world at 507 451 km² (195 927 sq miles), in the Canadian Arctic archipelago. It is inhabited mainly by Inuit (Eskimos), most of whom live in settlements scattered along the coast. Trapping and fishing are the chief occupations.
Map Canada – Ha

Bafing *West Africa* See SENEGAL River

Bafoussam *Cameroon* Industrial town and provincial capital 220 km (137 miles) north-west of the capital, Yaoundé. The town stands in an area of subsistence farming, but some coffee is grown as a cash crop and processed in Bafoussam for export. The Bamendjing Dam, with its restful lake, is a tourist attraction.
Population 99 400
Map Cameroon – Bb

Bagamoyo *Tanzania* Port 60 km (37 miles) north-west of Dar es Salaam and opposite the island of Zanzibar. In the 19th century it was a centre of the Arab slave and ivory trades, and a base for European explorers; now, it is overshadowed by Dar es Salaam.
Population 25 000
Map Tanzania – Ba

Bagan *Myanmar (Burma)* See PAGAN

Baghdad *Iraq* National capital, straddling the banks of the Tigris River. About 20 per cent of all Iraqis live in Baghdad, Iraq's political, economic and cultural metropolis. Founded in about AD 762, Baghdad became the city of the legendary tales of *The Thousand and One Nights*, the enduring stories which include the adventures of Aladdin, Sinbad the Sailor, and Ali Baba and the Forty Thieves.

Although it was known as 'The City of Peace', Baghdad has had a violent history. It was sacked in 1258 by the Mongols, conquered by the Tatar warlord Tamerlane in 1401, captured by the Ottoman sultan, Suleiman the Magnificent in 1534, taken again by the Turks in 1638, and captured by the British in the First World War.

In 1921, Baghdad became the capital of Iraq, which gained independence in 1932. Its industries today include distilling spirits for tourists (most Iraqis are Muslims and do not drink alcohol), tanning, tobacco and food processing, and the manufacture of clothing and cement.

The city's places of interest include colourful street bazaars, the 13th-century Mustansiriya School (Muslim University), remains of the city's ancient walls and gates, and the 14th-century Khan Murjin, the only completely roofed *khan* (inn) in Iraq.

Baghdad suffered considerable damage from enemy bombing during the 1991 Gulf War.
Population 3 845 000
Map Iraq – Cb

Bago (Pegu) *Myanmar (Burma)* Town and capital of the division of the same name (formerly Pegu), 70 km (43 miles) north-east of Yangon (Rangoon). The town was the capital of the Mon kingdom of Burma (as the country was called until 1989) at various times between 1250 and 1757, but little remains of its former glory. However, the Shwethalyaung, a reclining statue of the Buddha more than 55 m (180 ft) long and 15 m (49 ft) high, has been extensively restored.

The division, which covers 39 404 km² (15 214 sq miles), is largely agricultural. The main crops are rice and sugar cane.
Population (town) 254 800; (division) 3 800 000
Map Myanmar – Cc

Baguio *Philippines* Summer capital and resort on Luzon, in the southern Cordillera Central 210 km (130 miles) north of MANILA. It is set in pine forests and has many official buildings, hotels, restaurants and parks. Temperate fruit and vegetables grow well in the cool climate. At Banaue, 75 km (47 miles) to the north-east, extensive rice terraces are some 2000 years old.
Population 183 000
Map Philippines – Bb

Bahamas See p 68

Bahariya Oasis (Behariya) *Egypt* One of the great oases of the Western Desert, around the little town of Bawiti, 280 km (about 175 miles) south-west of Cairo. The land is watered from artesian wells and crops of dates are produced. Manganese and iron deposits occur nearby.
Map Egypt – Bb

Bahawalpur *Pakistan* Town and former princely state on the Sutlej River, 345 km (about 215 miles) south-west of Lahore. Most of the area is in the Thar Desert, which extends into India, but cotton is grown along the river.
Population 178 000
Map Pakistan – Cb

Bahia *Brazil* See SALVADOR

Bahia Blanca *Argentina* Although a port city, Bahia Blanca is actually built slightly inland with five ports on the Río Naposta. It is 550 km (about 340 miles) south-west of Buenos Aires. The port's main trade is in grain and fruit, which are shipped from the southern pampas.
Population 271 000
Map Argentina – Cb

Bahr el Ghazal *Sudan* Province covering some 134 700 km² (about 52 000 sq miles) in the southwest, on the border with the Central African Republic. The provincial capital is Wau. The province is mainly a cattle-raising area, but millet, oil seeds and bananas are grown, as well as rice at a government-sponsored project at Uwaye. The province contains the Southern National Park, in which is found a rich variety of species from elephants to rare butterflies.

The Bahr el Ghazal River (240 km, 150 miles long), which runs through the province, mostly disperses into the Sudd Swampland. Only in years of heavy rainfall does it flow on to join the White Nile at Lake No, about 700 km (435 miles) south and slightly west of the capital, Khartoum.
Population 1 500 000
Map Sudan – Ac

Bahrain See p 69

Bahrain Island *Bahrain* Largest of the 33 islands which make up the independent state of Bahrain. It lies in The Gulf, 25 km (16 miles) off the Al Hasa coast of Saudi Arabia, to which it is linked by a 25 km (16 mile) long causeway. Oil was found here in 1931. Although the island is mainly an arid limestone plateau, it is irrigated by wells, making agriculture possible. Its highest point is Jabal ad Dukhan, which rises 137 m (449 ft) above sea level.
Population 216 820
Map Bahrain – Ba

Baia Mare *Romania* City and capital of the north-west region of MARAMURES, about 40 km (25 miles) south of the Ukrainian border. It is set near a lake in a wooded section of the Carpathians.

It has a 15th-century belfry, an 18th-century cathedral, and an artists' quarter with a school of painting which has been renowned in Romania since the 19th century. Gold, silver, copper, lead and zinc ores are mined nearby and there are also zinc and lead smelting plants.
Population 152 100
Map Romania – Aa

Baikal (Baykal) *Russia* Freshwater lake covering 31 500 km² (12 150 sq miles) in south-east Siberia, near the Mongolian border. Fed by 336 rivers and drained by only one, the Angara, it is, in volume, the world's largest body of fresh water. It holds 23 000 000 m³ (812 720 850 cu ft) – as much as all five of North America's Great Lakes combined. It is also the world's deepest lake, reaching a depth of 1620 m (5315 ft).
Map Russia – Lc

Baile Átha Cliath *Ireland* See DUBLIN

Baile Átha Luain *Ireland* See ATHLONE

Baile Átha Troim *Ireland* See TRIM

Bairiki *Kiribati* See TARAWA

Baja (Lower) California *Mexico* Rugged, mountainous peninsula of desert and semidesert extending southwards to the Pacific Ocean from the United States border, mostly separated from the rest of Mexico by the Gulf of California. It is about 1300 km (more than 800 miles) long and on average 80 km (50 miles) wide, and rises to a height of 3078 m (10 100 ft) in the Sierra San Pedro Martir. Baja California is divided into the states of Baja California Norte (north of 28°N) and Baja California Sur (south of it).

Agriculture is confined to an irrigated area in the north and the valleys, where cotton, maize and wheat are grown. Since 1973, a highway running the length of Baja California from Tijuana in the north to Cabo San Lucas in the south has encouraged development of a thriving tourist industry along miles of sandy coastline. Other industries include fishing, as well as gold, silver, copper and iron mining. Baja California Sur contains Vizcaino Desert and the salt flats of GUERRERO NEGRO. La Paz became its capital in 1830.
Population (north) 1 458 000; (south) 290 000
Map Mexico – Ab

Bakarganj *Bangladesh* See BARISAL

Bahamas

PLAYGROUND OF THE NORTH AMERICANS, THIS CARIBBEAN ARCHIPELAGO HAS DEVELOPED TOURISM AND OFFSHORE BANKING

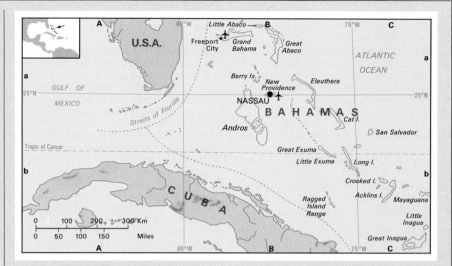

Nearly 500 years ago, the Genoan navigator Christopher Columbus made his first landfall in the New World on San Salvador (also called Watling Island) in the Bahamas. Precisely where on the island he landed is not known; however, San Salvador has no fewer than four monuments claiming to mark the exact spot where the great discoverer stepped ashore.

San Salvador is just one of the 700 islands and 2000 coral cays – or islets – that make up the Bahamas. Stretching in a 1200 km (about 750 miles) arc from the western edge of Haiti to within 80 km (50 miles) of south-eastern Florida, they range in size from ANDROS Island which measures 5957 km² (2299 sq miles) to tiny uninhabited dots in the ocean. The two most populated islands are NEW PROVIDENCE, where the capital, NASSAU, lies, and GRAND BAHAMA. New Providence, the more densely populated of the two, with 171 540 people, is slightly smaller than Grand Bahama, with 41 040 inhabitants. They account for over 75 per cent of the Bahamas' population.

DESCENDANTS OF SLAVES

More than 80 per cent of the Bahamas' population of 268 730 people are the descendants of slaves brought from Africa, largely in the 18th century. The original inhabitants of the islands were Arawak Indians, but they were all taken away by the Spanish to work as slaves in HISPANIOLA. The Spanish themselves never settled in the Bahamas; the first European settlers were a curious combination of Puritans, escaping religious persecution in 17th-century England, and pirates, taking advantage of the thousands of hidden, well-protected anchorages. They were later joined by Loyalists escaping from America after the American Revolution (1775-83) and still later by Southerners escaping from the American Civil War (1861-5). Except for a brief period during the 18th century, the Bahamas were occupied by the British from the 17th century until 1973, when they became an independent member of the Commonwealth.

The islands have few natural resources and for centuries most Bahamians were fishermen as well as small-scale farmers. The only industry of importance was sponge fishing in the coral reefs that surround the islands. But since the 1950s there has been a dramatic change in the country's economy. Taking advantage of their closeness to North America, a hot, sunny, though humid climate, white sand beaches, and waters kept warm by the Gulf Stream, the Bahamians and businessmen from outside the country began developing tourism on a vast scale. Now, more than 3.5 million tourists,

most of them from the United States, visit the Bahamas each year and the tourist industry has become the mainstay of the country's economy, employing 40 per cent of the workforce.

One of the most remarkable tourism projects has been the FREEPORT CITY-Lucaya development near the western end of Grand Bahama. Until the 1950s, this was a barren, unpromising area mostly covered with pine forest. But then an American financier named Wallace Groves came to an agreement with the government to develop the area for tourism and industry. Freeport City-Lucaya is now the largest single resort complex in the West Indies, with hotels, casinos, villas, fine beaches, and its International Bazaar, comprising 4 hectares (10 acres) of shops and restaurants.

A deep-water harbour has a number of industrial enterprises set round it, including a cement works, a pharmaceutical plant, a major oil refinery and oil transshipment terminals.

INAGUA, the southernmost inhabited island, has one of the world's largest complexes for producing salt by solar evaporation. A plant on New Providence processes frozen lobster tails for export, and the government is encouraging agriculture on a number of islands, especially Andros, to help make the country more self-sufficient in food. Thanks to Japanese funds, a huge citrus juice processing plant has been constructed, and 8090 hectares (20 000 acres) have been planted with citrus trees.

At the same time, Bahamian laws, which exempt individuals and corporations from income and inheritance taxes, have attracted more than 390 banks and trust companies to the country. Nassau is a Eurodollar trading centre that compares well with London.

The main centres in the Bahamas, New Providence and Grand Bahama, have been developed for tourism and business, but life on many of the other islands, known as 'the Family Islands' or 'Out Islands', has been little affected by the changes of recent years.

Off ELEUTHERA, for example, to the east of New Providence, is the little island of Spanish Wells. Many of its inhabitants are descendants of the first Puritan settlers and have managed to retain their separate identity for 300 years. Craftsmen on the Abaco group of islands,

between Eleuthera and Grand Bahama, are still building boats as their ancestors have done for years; and George Town, in the EXUMA group and almost on the Tropic of Cancer, still keeps the character of a tiny island capital.

Other islands include the tiny Bimini group, closest to Florida, where the American writer Ernest Hemingway lived for a while and actor Sidney Poitier spent much of his childhood. Cat Island, on the eastern edge of the Bahamas, has their highest point, Mount Alvernia – only 63 m (207 ft) above sea level – and Andros, west of New Providence, has more than 110 Blue Holes, which are found throughout the Bahamas but are particularly common on Andros. Blue Holes are large, steep-sided inland tunnels where the ocean rises through the coral foundations of the island. Even more impressive, though, is the Andros Barrier Reef off the island's eastern coast. It is 200 km (125 miles) long and is second in size only to Australia's Great Barrier Reef.

BAHAMAS AT A GLANCE	
Official name The Commonwealth of the Bahamas	
Area 13 940 km² (5381 sq miles)	
Population 268 730	
Per km² 19.27 (**Per sq mile** 49.94)	
Capital Nassau	
Government Parliamentary monarchy	
Currency 1 Bahamian dollar = 100 cents	
Language English, Creole	
Religion Christian (Baptist 32%, Anglican 20%, Roman Catholic 19%, Methodist 6%)	
Climate Tropical, with a rainy season from June to November. Average temperature at Nassau is 24°C (76°F) throughout the year	
Land use Cultivation 1%, forest and woodlands 32%, other 67%	
Main primary products Timber, fish and shellfish	
Major industries Tourism, offshore banking; oil refining, fishing, forestry, pharmaceuticals, cement	
Main exports Shellfish, rum, salt, timber, cement, pharmaceuticals	
Annual income per head (US$) 12 020	
Population growth (per thous/yr) 16	
Life expectancy (yrs) Male 68 **Female** 76	

Bahrain

LOOKING TO THE FUTURE, BAHRAIN IS SEEKING TO ATTRACT NEW INDUSTRIES TO OFFSET ITS DEPENDENCE ON OIL REVENUES

Known historically by the people of The Gulf as 'the island of a million palms' because of its abundant freshwater springs, Bahrain was the first Arab country to strike oil, in 1931. Today it has ambitions to make itself 'the Singapore of The Gulf' and diversify its economy before the oil runs out. Already it has an economically attractive free port and industrial zones – Mina Sulman and North SITRA – where no customs duties are charged on transshipped goods or on goods manufactured within the free area. And the government has begun to develop the non-oil manufacturing industrial sector. There is an aluminium smelting and processing plant – the largest industrial complex in The Gulf not connected with the oil industry – and the world's largest smelting plant outside the Commonwealth of Independent States (CIS); there is also a petrochemical industry, a major shipbuilding and repair yard handling supertankers, and a satellite communications centre.

Bahrain has established itself as a regional centre for offshore banking, and several prominent international banks now operate here. Following the Gulf War, many investors withdrew their capital, but by 1993 banking had regained its former strength. Other negative consequences – especially environmental – of the war will take years to overcome.

The emirate comprises 33 low-lying islands, of which Bahrain Island is the largest (562 km², 216 sq miles) and is gaining land each year through reclamation from the sea. It is connected to its neighbouring island of Muharraq by a 2.4 km (1.5 mile) causeway, while on its east coast another causeway links it to Sitra Island. Apart from a fertile strip in the north, Bahrain Island consists of limestone rock covered by varying depths of sand which is too poor and saline to support anything other than tough desert plants. Desertification is a major problem; and Bahrain is far from self-sufficient in food production. What little agriculture there is, is heavily subsidised. Drainage schemes have been started to reduce the salinity of the soil and a green belt has been created in the northern region, enriched with imported fertile soil, manures and chemical fertilisers, but altogether only 10 km², (4 sq miles) are under irrigation. Here, figs, pomegranites, bananas and a variety of vegetables are cultivated. Dates are grown around oases. Although Bahrain has underground springs (mostly in the north), they are being depleted rapidly and much of the country's water is obtained from desalination plants.

The country is oppressively hot between May and September, with temperatures usually rising to more than 40°C (104°F) during the day accompanied by high humidity. But the weather cools down quickly from October, after which December to March is the coolest period (21°C, 70°F).

About 80 per cent of the population lives in cities on the islands of Bahrain and Muharraq; about one-third are immigrants, among them Omanis, Indians, Pakistanis and Iranians, attracted by tax-free incomes.

The Bahrainis themselves are Arab Muslims. All are ruled by the emir, who governs with advice from his ministers. Political parties are illegal; however, in response to international pressure to democratise the system of government, plans were announced in 1992 to appoint a new consultative National Assembly – the first since 1975, when the previous elected National Assembly was suspended. The new, 30-member Council sat for the first time in 1993.

Some 80 per cent of foreign income comes from oil and oil products. Much of the revenue is spent on building schools, colleges and hospitals, and other public services. Education and health care are free, and much of the housing is subsidised. Pensions and sickness and unemployment benefits are provided. The country has highly developed communication and transport facilities.

Aluminium products bring in vast sums in export earnings. Tourism is another growth area, with the country's stretch of coastline being ideal for scuba diving and dhow boating expeditions. Government-initiated projects include an 18-hole golf course – as well as a desalination plant to water its greens.

Bahrain International Airport has been expanded, as has the Gulf Air airline company based in Bahrain. Apart from its new industries, Bahrain's traditional occupations still survive – though on a much smaller scale than in previous years. These include pearl fishing – Bahrain was once famed for the quality of its pearls – dhow building, basket and cloth weaving, and pottery making.

Like the other small states of The Gulf, Bahrain maintains close relations with Saudi Arabia, which it sees as the best guarantor of stability and protection. Indeed, a physical link has been established between the two countries with the opening of a US$900 million, 25 km (16 mile) causeway.

BAHRAIN AT A GLANCE	
Official name State of Bahrain	
Area 706.5 km² (273 sq miles)	
Population 538 090	
Per km² 761.6 (**Per sq mile** 1971)	
Capital Al Manamah	
Government Emirate	
Currency 1 Bahrain dinar = 1000 fils	
Languages Arabic (official); English widely spoken	
Religions Muslim (Shiite 51%, Sunni 34%), Christian (7%), Hindu (5%)	
Climate Very hot summers with high humidity but very little rainfall. Average daytime temperature ranges from 20°C (68°F) in January to 38°C (100°F) in August	
Land use Cultivation 4%, grazing 6%, other 90%	
Main primary products Oil and natural gas; dates, vegetables, livestock, fish	
Major industries Crude oil production and refining, aluminium smelting, shipbuilding and repairs, petrochemicals, banking, tourism	
Main exports Crude oil and petroleum products, natural gas, aluminium products	
Annual income per head (US$) 7150	
Population growth (per thous/yr) 30	
Life expectancy (yrs) Male 71 **Female** 76	

Bakhtaran (Kermanshah) *Iran* Trading city (formerly Kermanshah) lying about 400 km (248 miles) south-west of the capital, Tehran, on the ancient caravan route to Baghdad. The centre of a grain-producing area, Bakhtaran is home to many Kurdish people. It has an oil refinery.
Population 560 520
Map Iran – Aa

Bakony (Bakony Forest) *Hungary* Range of forested mountains in the north-west of the country between Lake Balaton and the Little Alföld, rising to 704 m (2310 ft) at Körishegy Peak. Its pretty valleys and ruined castles, its lakeside resorts and its extinct volcanoes make it a popular holiday area during the summer months. It has bauxite and manganese deposits.
Map Hungary – Ab

Baku (Bakı) *Azerbaijan* Port and national capital on the west coast of the Caspian Sea. It is the centre of a major oil field, and its industries include oil refining and shipbuilding; it also produces oil-field equipment, chemicals, textiles and cement. A pipeline runs to the port of Batumi on the Black Sea, carrying oil for export. Baku was founded in the 11th century, and was a major silk and saffron trading centre during the Middle Ages. It has palaces and mosques which date from the 14th and 15th centuries and was part of Persia (Iran) between the 16th and the 19th centuries. Baku is the largest city in the Transcaucasian region. Pollution is a major problem.
Population (city) 1 150 000; **(metropolitan area)** 1 757 000
Map Azerbaijan – Ba

Bakwanga *Zaire* See MBUJI-MAYI

Balaklava *Ukraine* See SEVASTOPOL

Balaton, Lake *Hungary* Central Europe's largest lake and a major holiday area covering 595 km² (230 sq miles). It is 110 km (68 miles) south-west of the capital, Budapest. The beautiful north shore, 80 km (50 miles) long, is lined with volcanic, forested and vine-clad hills, sandy coves and picturesque villages. By contrast, the south shore is flat and has a shallow beach packed with camp sites, modern hotels, private holiday homes and villas. Winter activities include ice-skating, ice-sailing and ice-fishing.
Map Hungary – Ab

Balatonfüred (Füred) *Hungary* Oldest and most elegant of the health resorts on the north shore of Lake Balaton, 110 km (68 miles) south-west of the capital, Budapest. Its rambling streets climb vine-clad hillsides, and its 11 medicinal springs are used for treating heart and nervous disorders. One of the springs bubbles through a fountain built in 1800 in Spa Square (Gyógy Tér), the town's centre. In summer, colourful equestrian pageants and jousts are held in the grounds of 15th-century Zichy Castle in the old town of Nagyvázsony, 16 km (10 miles) to the west.
Population 13 900
Map Hungary – Ab

Balboa *Panama* See CANAL AREA

Bâle *Switzerland* See BASLE

Balearic Islands *Spain* Island group off the east coast and one of Europe's prime holiday areas, with sweeping beaches, dramatic mountain scenery and a mild climate. The group consists of Majorca (Mallorca), Minorca (Menorca), Ibiza (Iviza), Formentera, Cabrera and many islets, making up the province of Baleares, whose capital is PALMA, in Majorca. Majorca also has the highest peak of the archipelago – Puig Mayor, at 1445 m (4731 ft). Besides tourism, the biggest income earner, there are fishing, pottery and metalware industries, and vines, cereals, olives and almonds are grown.
Population 709 140
Map Spain – Fc

Bali *Indonesia* Volcanic island of 5591 km² (2159 sq miles) off the east coast of Java and westernmost of the Lesser Sunda group of islands. It is one of the world's great tourist haunts because of the preservation of the Hindu civilisation of ancient Indonesia, with its rich traditions in art, architecture and music. Its highest point is Agung (3142 m, 10 310 ft) which, according to Bali myth, is the seat of the gods. The Balinese also worship the volcanoes which rise above the coastal belt, and which erupt at regular intervals. Rice, maize, tropical plants and fruit are grown in the fertile soil. Kuta Beach is the main tourist resort and DENPASAR the capital.
Population 2 782 000
Map Indonesia – Ed

Balikpapan *Indonesia* Port and city of East Kalimantan Province on Borneo. It is the supply centre for local oil fields.
Population 280 700
Map Indonesia – Ec

Balkan Mountains (Stara Planina) *Bulgaria* Range of mountains extending for more than 320 km (200 miles) across central Bulgaria from the Serbian border to the Black Sea. They are crossed by 20 passes, including SHIPKA PASS and the ISKUR Gorge. The highest point is Botev Peak (2367 m, 7766 ft).
Map Bulgaria – Ab

Balkans *Europe* Mountainous peninsula in the south-east between the Adriatic and Ionian seas in the west and the Black and Aegean seas in the east. It encompasses ALBANIA, BULGARIA, GREECE, ROMANIA, SLOVENIA, CROATIA, BOSNIA-HERZEGOVINA, YUGOSLAVIA (Serbia and Montenegro), MACEDONIA and European TURKEY.

Balkh *Afghanistan* Ancient ruined city 20 km (12 miles) west of Mazar-e Sharif in the north of the country. In pre-Christian times it was an important staging post on the caravan route between China and Rome; in the 7th century AD it became a cultural centre of Islam, but was destroyed by the Mongol ruler, Genghis Khan, in the early 13th century.
It was rebuilt by his successor, Tamerlane (about 1335-1405), and Tamerlane's son, but by the 18th century it had become a ghost town – although it still had a large bazaar where caravan trains stopped.
Map Afghanistan – Ba

Balkhash *Kazakhstan* Freshwater lake near the Chinese border, covering 17 000-22 000 km² (6500-8500 sq miles), depending on the season. It is frozen for five months of the year. Trout are caught in the lake; under Soviet rule, it provided a quarter of all the trout caught in the former USSR. The town of the same name on the lake's northern shore is a centre for processing copper ore that is mined nearby.
Population (town) 81 000
Map Kazakhstan – Db

Ballarat *Australia* City in Victoria, 112 km (70 miles) west of Melbourne. It was founded in 1851 at the start of the Victorian gold rush, and this era of the city's life has been recreated at Sovereign Hill Historical Park, which is a major tourist attraction. A monument at the site of the Eureka Stockade marks the place where local gold miners temporarily defied government tax collectors in 1854. This rebellion – and the flag the miners raised – became symbols of resistance to British rule in Australia. Ballarat, the third largest urban area in Victoria, is a major market and a commercial centre.
Population 64 980
Map Australia – Gf

Ballymena *United Kingdom* Town in Antrim county in Northern Ireland. It lies 16 km (10 miles) north of Lough Neagh in a fertile farming region.
The town, which hosts a large annual agricultural show has linen, wool fibre and engineering industries. Ballymena's population is largely of Scottish origin from 17th-century settlement, and is still largely Presbyterian.
Population 55 920
Map United Kingdom – Bc

Balsas *Mexico* Major river, 700 km (435 miles) long, rising as the Atoyac in the SIERRA MADRE ORIENTAL in the state of Tlaxcala. It flows south and then west, forming the boundary between the states of GUERRERO and MICHOACÁN, to Petacalco Bay on the Pacific. In 1966 the El Infiernillo Dam, used for hydro-electric power and irrigation, was built near its mouth.
Map Mexico – Cc

▼ BALI DANCERS The Barong (Balinese lion-dog) fights the Rangda (widow-witch) in this dance symbolising the eternal conflict between good and evil.

Baltic Sea Lying almost landlocked in northern Europe with a surface area of 385 000 km² (148 614 sq miles), it is linked to the NORTH SEA by narrow straits between Denmark and the Scandinavian Peninsula. The Baltic was once a freshwater lake, formed when glaciers melted at the end of the last Ice Age, about 10 000-17 000 years ago. Because so many rivers flow into the sea, it still has a low salt content, especially in the east. Its greatest depth is 459 m (1506 ft).

There are hardly any tides, which makes navigation easy – though storms can be a danger. On the other hand, much of it freezes in winter and icebreakers are needed to keep the shipping lanes open. Viking longships once looted the Baltic coastlands (between the 7th and the 10th centuries); in the late Middle Ages, the Hanseatic League, a confederation of German merchant cities, dominated Baltic trade.

Branches of the Baltic include the Gulf of BOTHNIA, the Gulf of FINLAND, the Gulf of Riga and the Gulf of GDAŃSK.
Map Europe – Ec

Baltimore *USA* Largest city of Maryland, standing on an arm of Chesapeake Bay, 60 km (37 miles) north-east of the national capital, Washington DC. It was founded by the Provincial Assembly in 1729. Its excellent harbour made it a centre for shipping tobacco and grain, and it has remained a major port and shipbuilding centre. It is also an important producer of aerospace equipment. In 1830 the country's first railway passenger station opened here.

Baltimore's industries include steel and chemicals. The city contains the country's oldest Roman Catholic cathedral, is the seat of Johns Hopkins University, founded in 1876, and has many fine houses and public buildings which escaped a devastating fire that destroyed more than 1000 of the city's buildings in 1904.
Population (city) 736 000; (metropolitan area) 2 382 200; (conurbation) 6 727 050
Map United States – Kc

Baluchistan *Pakistan* Province of the southwest, covering 347 190 km² (134 050 sq miles). The nomadic peoples who live here, mainly Muslims originally from Iran, have spread into both Iran and Afghanistan, paying little attention to the international borders drawn up under British rule. The province – whose capital is QUETTA – is mountainous, with peaks of up to 3400 m (11 155 ft), rocky, hot and arid. Goats and sheep are the mainstay of the economy, but vegetables are grown in irrigated valleys. Oil and gas have been discovered here. The fiercely independent tribes have often taken up arms against the government.
Population 14 870 000
Map Pakistan – Bb

Bamako *Mali* National capital, standing on the Niger River. It became Mali's first city in 1908 after a railway link was built from the town of Kayes near the Senegalese border in 1904. The line was extended to the Senegalese capital, Dakar, in 1923.

Bamako's city centre is attractive and spacious. The roads are wide, tree-lined and usually made from sand. Elegant buildings from the French colonial era, now mainly used as offices, stand alongside them. There are also many fine traditional mud-brick buildings, such as the old market. The Great Mosque, to which thousands throng each Friday, is modern, built with Arab money, but it recreates the traditional style.

The administrative quarter and presidential palace stand on top of the escarpment of the Manding Highlands north of the city and offer commanding views over the the city and its surroundings. Housing for the majority of Malians is cramped, built mainly of mud brick and with few facilities, in marked contrast to European buildings.

Bamako is the main market for cattle and kola nuts, which are used in making soft drinks.
Population 661 000
Map Mali – Bb

Bamberg *Germany* Port on the Regnitz River about 50 km (about 30 miles) north of Nuremberg. It is connected to a canal joining the Rhine, Main and Danube rivers. The four spires of the Romanesque cathedral, completed in the 13th century, soar over the town. Pope Clement II is buried in the choir.

The Bible of Alcuin (735-804), the Northumbrian theologian who became an adviser to Emperor Charlemagne, is in the city library.
Population 70 200
Map Germany – Dd

Bambouk Mountains *Mali* Mountain range in the far west. The mountains were once a major source of gold and were largely responsible for West Africa being named 'the Land of Gold' by medieval Arab historians.
Map Mali – Ab

Bamenda *Cameroon* Provincial capital in the west, about 270 km (165 miles) north-west of the capital, Yauondé. It stands on the high, cool Bamenda grasslands and is a market point for tea, coffee and bananas. It also has craft industries producing traditional masks, figurines, pottery, stone carvings and musical instruments of wood. The nearby Bamenda Falls are a source of hydroelectric power.
Population 138 000
Map Cameroon – Bb

Bamian *Afghanistan* Valley, known as 'the Valley of the Gods', in the Hindu Kush Mountains. Situated 112 km (70 miles) north-west of Kabul, and 2583 m (8474 ft) above sea level, it is the site of the small town of Bamian, with its hotel, several teashops and a bazaar. Two giant Buddhas are carved in the cliffs on the north side of the valley: one is 53 m (175 ft) high, the other 35 m (115 ft). There are many cliffside caves, where Buddhist monks once lived.
Map Afghanistan – Ba

Bamingui Bangoran National Park *Central African Republic* Savannah-covered park of 40 000 km² (15 440 sq miles) in the north. Its wildlife includes antelopes, buffaloes, lions, leopards and rhinoceroses.
Map Central African Republic – Ba

Ban Chiang *Thailand* Silk-weaving village 480 km (about 300 miles) north and slightly east of Bangkok. It caused an archaeological sensation in 1974, when Thai and American scholars found bronze artefacts believed to date from 3600 BC. Until then, archaeologists had almost universally believed that metalworking began in Mesopotamia in the Middle East several hundred years later. A museum displaying the archaeological finds has been established in the village.

Banaba (Ocean Island) *Kiribati* Small 5 km² (2 sq miles) raised atoll about 450 km (280 miles) south-west of Tarawa. Between 1900 and 1979 its rich deposits of phosphate rock were mined by a British company, leaving little productive land. The Banaban people were resettled on Rabi Island, Fiji, after the Second World War, but have been claiming substantial compensation from the British government.
Population 280
Map Pacific Ocean – Dc

Banana *Zaire* Seaport on the Atlantic coast, near the outlet of the Zaire River. It is linked by rail to Boma and Kinshasa. Nearby is Muanda which has an oil refinery and an offshore oil-loading terminal for crude oil extracted from oil wells in the ocean and the river's estuary. Banana was the first colonial trading post in present-day Zaire.
Map Zaire – Ab

Bananal (Ilha do Bananal) *Brazil* River island 480 km (about 300 miles) north-west of the capital, Brasília. At 18 130 km² (7000 sq miles) it is slightly smaller than Wales and is the world's largest river island. In fact, it is so large that it has rivers of its own, some of them more than 300 km (186 miles) long. It splits the Araguaia River for about 500 km (310 miles) of its course. Bananal gets its name from the giant wild banana trees which have developed here.
Map Brazil – Cc

Banco National Park *Ivory Coast* See ABIDJAN

Banda Sea Part of the PACIFIC OCEAN in southeastern Indonesia with an area of 738 150 km² (285 020 sq miles). Its greatest depth is more than 6400 m (20 990 ft). Plate movements in the area cause frequent undersea earthquakes.
Map Indonesia – Gd

Bandama *Ivory Coast* One of the country's three major rivers – the other two being the SASSANDRA and the KOMOE. With its two main tributaries, the Bandama Blanc and the Bandama Rouge, the Bandama drains some 50 per cent of the country. It rises as the Bandama Blanc in the highlands near Korhogo and flows south for 1050 km (652 miles) to enter the Atlantic Ocean at Grand Lahou. The Bandama Rouge rises in the north-west highlands and meets the Bandama Blanc about 33 km (20 miles) west of Bouaké. A hydroelectric plant at KOSSOU, just north of the confluence, provides power to Bouaké and the central part of the country. The river irrigates large areas of the rural centre.
Map Ivory Coast – Bb

Bandar Abbas *Iran* Iran's busiest port, situated on the Strait of Hormuz, opposite Oman. It was established in 1623 by Shah Abbas I to replace the island port of Hormoz, which was captured by the Portuguese in about 1614. During the 17th century Bandar Abbas was Persia's main port in trading with India. Today it is the focal point of the trade routes of south Iran, and its industries include cotton milling, fishing and fish canning.
Population 250 000
Map Iran – Bb

Bandar Seri Begawan *Brunei* National capital of the sultanate, set on the Brunei River in the midst of an alluvial coastal plain. Once known as Brunei Town, it is dominated by the vast Omar Ali Saiffudin Mosque, the largest in South-East Asia, while the enormous new royal palace is nearby. This and the modern commercial district of the town contrast markedly with the traditional riverbank homes on stilts of Kampong Ayer.
Population 58 000
Map Brunei – Ab

Bandiagara Scarp *Mali* Rockface south of the Niger River, with an almost sheer drop of 244 m (800 ft).
Map Mali – Bb

Bandipur *India* Wildlife sanctuary 80 km (50 miles) south of Mysore. It is noted for its tigers, elephants and leopards. In the 19th and early 20th centuries, it was a game reserve for the Maharajah of Mysore.
Map India – Be

Bandundu *Zaire* Capital of Bandundu region in the south-west and a busy commercial centre. Formerly called Banningville, it is 145 km (90 miles) from the Congo border on the Kwilu River just above its junction with the Kasai River. Bandundu region covers 295 658 km² (114 124 sq miles). In the north it is low-lying and densely forested and includes swamps and Lake Mai-Ndombe. The land rises to the savannah-covered south. The mostly Bantu-speaking peoples include the Teke in the north, Kwango in the centre, and Lunda in the south.
Population (town) 120 000; (region) 4 644 760
Map Zaire – Ab

Bandung *Indonesia* Provincial capital of West Java. Its main industry is textiles. It has a number of technical research institutions. In 1955 the Indian prime minister, Pandit Nehru, Premier Chou-Enlai of China, and Indonesia's President Achmed Sukarno founded the non-aligned movement at a conference in the city.
Population 1 462 700
Map Indonesia – Cd

Banff *Canada* The country's oldest national park, established in 1885 in south-west Alberta. It covers 6640 km² (2564 sq miles) and contains spectacular ROCKY MOUNTAINS scenery. Lake Louise is a major attraction. The town of Banff is the main tourist centre.
Map Canada – Dc

Banffshire *United Kingdom* See GRAMPIAN

Banfora *Burkina* South-western market town on the Comoé River at the foot of the Banfora Escarpment. It lies on the railway running from Burkina's capital, Ouagadougou, to Abidjan in Ivory Coast. The town is a collecting point for groundnuts and has a sugar mill. Wild honey is abundant. The area is rich in iron ore.
Map Burkina – Aa

Banfora Escarpment *Burkina* White sandstone escarpment 170 m (560 ft) high, which forms the southern edge of the Sikasso Plateau in south-west Burkina. It boasts spectacular waterfalls and stunning views from the top.
Map Burkina – Aa

Bangalore *India* Rapidly expanding industrial city in south-central India, 334 km (207 miles) west of the port of Madras. It was founded in the 16th century and today has booming aircraft and electronics industries, including a large number of computer companies, hence its nickname 'the Silicon Valley of India'.

In 1964 two flourishing educational centres were opened here – Bangalore University and the University of Agricultural Sciences. Bangalore is also the site of the All-India Institutes of Science and of Management.

Because of its altitude, Bangalore used to be a popular retirement centre for the British; Victorian-style mansions and the parliamentary buildings are reminders of those days gone by.
Population (city) 2 650 700; (metropolitan area) 6 000 000
Map India – Be

Bangkok (Krung Thep) *Thailand* The kingdom's capital, a wild and noisy city whose streets seem to be clogged with traffic almost around the clock and part of whose centre is garish with neon-lit strip clubs. The city sprawls beside the Chao Phraya River, some 30 km (19 miles) from the Gulf of Thailand. The old city, founded as a royal capital in 1782, is a place of *klongs*, or canals, and *wats* – temples or monasteries, of which there are some 300. This 'Venice of the East', as it was once known, is surrounded by a chaotic and mushrooming metropolis where nearly 6 million people live. Despite growing industries, the city has been unable to provide adequate productive employment for the many migrants who have arrived over the past 25 years. Many of the poorer people thus find work in marginal service industries, some of which have given Bangkok notoriety as the so-called 'sex capital of the world'.

The city is rich in Thai culture. The Grand Palace, begun in 1782 as the home of the Thai royal family, houses the Wat Phra Keo Temple containing the 460 mm (18 in) high Emerald Buddha. The statue is fashioned from translucent jasper, though when or where it was made is a mystery. The temple of Wat Po contains the Reclining Buddha, 50 m (164 ft) long, and the temple of Wat Trimitr contains a statue known as the Golden Buddha; the figure is said to contain more than 5 tonnes of gold.

Nearby is the Rose Garden, where a range of Thai culture – including traditional dancing and Thai boxing – is exhibited.

Bangkok means 'Village of the Wild Plum'; it is the name of part of the Thon Buri side of the river. Bangkok is not, strictly, the capital's correct name, although it is universally used by foreigners. Thais themselves call the city Krung Thep – meaning 'the City of Angels'.
Population (metropolitan Bangkok) 5 900 000
Map Thailand – Bc

Bangladesh See p 74

Bangor *United Kingdom* Town in Northern Ireland, about 20 km (12 miles) north-east of the capital, Belfast. Bangor's abbey is built on the site of a 6th-century monastery which, during the Dark Ages, sent missionaries to help to keep Christianity alive in Europe. There are some engineering factories, but Bangor is largely a residential centre for Belfast. Its harbour is an important yachting centre, and the town and the

district have abundant sporting facilities, mostly golf courses. Its popularity as a seaside resort has declined in recent years.
Population 52 440
Map United Kingdom – Cc

Bangui *Central African Republic* National capital and port on the Ubangi River, from which the city gets its name. It handles much of the country's cotton and timber exports. The modern city centre has a colourful market and a university which was founded in 1970.
Population 725 000
Map Central African Republic – Ab

Bangweulu *Zambia* Lake covering 9800 km² (3784 sq miles) 500 km (310 miles) north and slightly east of the capital, Lusaka. In 1873 the Scottish missionary David Livingstone travelled through the swampland south-east of the lake on his last journey of exploration.
Map Zambia – Bb

Bani Suwayf *Egypt* See BENI SUEF

Baniyas *Israeli-occupied Syria* Village 65 km (40 miles) south-west of Damascus in a region occupied by Israel since 1967. It is the site of the ancient Greek city of Paneas, which was renamed Caesarea Philippi by the Romans. It is also the site of a temple built by Herod the Great, and a shrine to the Greek god Pan. Near the Kibbutz Dan, to the south-west of the village, are the thermal springs which feed the headstreams of the River Jordan.
Map Syria – Ab

Baniyas *Syria* Port and terminal for the Iraq pipeline from Kirkūk, situated 144 km (89 miles) south-west of Aleppo. Spread around a small bay, Baniyas is overshadowed by a refinery and huge oil storage tanks. On the surrounding hills are the white domes of tombs, and on the highest hill to the east is the huge Marqab Fortress. Built of black basalt with 14 towers, this was an important military post for the Crusaders. In 1140 it was refortified by the Knights Hospitallers, or Knights of St John, an order founded in the 11th century to care for poor or sick pilgrims in Jerusalem. It could accommodate more than 1000 people as well as its defenders, and could hold enough provisions to withstand a five-year siege.

Baniyas was known as Leucas under the Greeks and as Balanea under Roman rule. Since ancient times it has been noted for its gardens. Tall cypresses protect the town and its surrounding orchards from the sea winds. In the Middle Ages the port exported wool, but today the harbour is silted up and used mainly by fishermen.
Population 11 000
Map Syria – Ab

Banja Luka *Bosnia-Herzegovina* City 148 km (92 miles) north-west of Sarajevo. A road and rail junction on the Zagreb-Sarajevo route, it lies where the Vrbas Gorge through the Dinaric Alps opens onto the Sava Plain. Banja Luka, a spa since Roman times, has Roman Catholic and Orthodox cathedrals, but all its mosques, many of them built by the Turks who fortified the town in the late 16th century, were destroyed in the civil war that began in 1992.
Population 195 000
Map Bosnia-Herzegovina – Ba

Banjarmasin *Indonesia* Provincial capital and fishing port of South Kalimantan Province on Borneo. It is the centre of a farming region which produces coconuts, rubber and rice.

Population 381 300
Map Indonesia – Dc

Banjul *The Gambia* Capital and commercial centre of The Gambia, standing on the low sandbanks or islands at the mouth of the Gambia River. British traders built a fort here in 1816 to control the slave trade. Merchants and missionaries followed, and the town that grew was named Bathurst after the Earl of Bathurst (1762-1834), then Secretary of State for the Colonies. It was renamed Banjul, meaning 'bamboo', at independence in 1965.

The town has outgrown the islands and spread out onto the mainland. There has been a vast improvement in living standards since the 1930s when flooding and overcrowding resulted in outbreaks of dysentery, yellow fever and malaria. Now there are wide streets in the town centre, which has been redeveloped to alleviate some of the overcrowding. However, in general, housing is still poor, with much of it built from corrugated iron. Water comes from standpipes and electricity supplies are minimal.

Despite the poverty, Banjul is being developed as a tourist resort, and is becoming popular with Europeans wanting a winter tan or a chance to explore a largely undeveloped part of Africa. A number of good hotels have been built alongside the miles of sandy beach and warm seas near the city. The rains arrive in June and last until October, with August being the wettest month.

Population 200 000
Map The Gambia – Ab

banner cloud Type of cloud that forms in clear skies on the lee side of a mountain as air rising to pass over the peak cools.

Banningville *Zaire* See BANDUNDU

Banská Bystrica *Slovakia* Regional capital 165 km (103 miles) north-east of Bratislava. In the Middle Ages it was a prosperous copper and silver-mining town. It has fine Gothic churches and a Gothic town hall. The town was the centre of the Slovak uprising against the Nazis (29 August to 27 October 1944). It is noted for its metal, textile and woodworking industries.

Population 73 300
Map Slovakia – Bb

Bantry (Beanntraí) *Ireland* Market town, tourist centre and fishing port in Bantry Bay, 72 km (45 miles) south-west of the city of Cork. The French sailed into Bantry Bay and made two unsuccessful attempts at landing – in 1689 to support James II and, during the 1798 rebellion, with Wolfe Tone, the exiled Irish nationalist. Offshore, Whiddy Island was developed as an oil terminal. Bantry House is home of the Earls of Bantry.

Population 2780
Map Ireland – Bc

Baoshan *China* Industrial town 8 km (5 miles) north of Shanghai. It is the site of China's most modern iron and steel works.

Population 8000
Map China – Ie

Baqubah *Iraq* Market town in the oil and fruit-producing province of Diyala. It lies 45 km (28 miles) north-east of the capital, Baghdad.

Population 114 520
Map Iraq – Cb

bar Ridge of mud, sand or shingle extending across a river mouth, harbour or bay, and lying roughly parallel to the coast.

Bar (Antivari) *Yugoslavia (Serbia and Montenegro)* Montenegrin fishing port 48 km (30 miles) from the Albanian border. Largely rebuilt after an earthquake in 1979, it is now a ferry port for Bari (Italy) and Corfu, and terminus of the spectacular Belgrade-Bar Railway, which threads through tunnels, along ravines and over gorges across the most rugged part of the Dinaric Alps.

Population 3600
Map Yugoslavia – Bc

Barania Gora *Slovakia/Poland* See BESKIDY

Barbados See p 76

Barbuda *Antigua and Barbuda* Island covering 160 km² (62 sq miles), lying about 45 km (28 miles) north of Antigua. It has wide beaches and a coral reef. The capital and the island's only settlement, Codrington, takes its name from an English family who owned the island between 1685 and 1870 and who are said to have experimented in slave breeding. The islanders grow crops – mainly maize and peas – and rear livestock for local consumption.

Population 1200
Map Caribbean – Cb

Barcelona *Spain* Seaport and industrial city on the northern Mediterranean coast. It is the capital of Catalonia and Spain's second largest city after Madrid. Barcelona is noted for its fascinating central avenue, the Ramblas, or Rambla de Barcelona, with its bookstalls, flowerstalls and caged birds for sale. Its industries include textiles, machinery, oil refining, chemicals, plastics, leather goods and tourism. It exports textiles, machinery, wine, olive oil and cork. Barcelona hosted the 1992 Olympic Games, as a result of which the city now has impressive sports facilities.

In the old city are Roman remains and 14th-century Gothic churches and palaces. The most striking features of the newer districts are the flamboyant buildings of the Spanish architect Antonio Gaudi (1852-1926), who echoes the shapes of natural phenomena, such as trees, in his designs. His most outstanding building in Barcelona is the Sagrada Familia (Sacred Family) Cathedral, begun in 1883 and still uncompleted.

Barcelona Province is the most densely populated and the most highly industrialised in Spain. It was a Republican bastion in the Spanish Civil War (1936-9).

Population (city) 1 643 540; (province) 4 654 410
Map Spain – Gb

Barcelos *Portugal* Market town on the Cavado River, 40 km (25 miles) north of the city of Oporto. The remains of a 15th-century palace overlook the river. The town produces painted pottery, notably the decorated pottery cockerels which are a national emblem.

Population 10 800
Map Portugal – Bb

barchan Crescent-shaped sand dune which is concave on the side sheltered from the prevailing wind and can exceed 30 m (about 100 ft) in height. It forms when the wind blows predominantly from one direction. The dune advances as the sand is blown from the windward side of the dune to fall down the dune face.

Barclay, Mount *Liberia* The country's first rubber plantation, established on a hill of the same name near Monrovia, the capital, by a British company in 1907. It is now part of the Firestone plantation centred on HARBEL.

Map Liberia – Aa

Barcoo *Australia* See COOPER CREEK

Barents Sea Part of the Arctic Ocean covering the continental shelf north of Norway and western Russia. It is 1290 km (about 800 miles) long and 1050 km (about 650 miles) wide. Its average depth is 229 m (751 ft) and its greatest depth is 465 m (1526 ft). The North Atlantic Drift keeps its southern waters relatively ice free, but dense fogs are common. Named after the Dutch explorer Willem Barents (1550-97), the sea has oil reserves and abundant fish.

Map Russia – Ea

Bari *Italy* Adriatic seaport 240 km (about 150 miles) east of Naples. Its industries include oil refining and engineering. Bari first grew to importance as the Byzantine capital of southern Italy. Later it was conquered by the Normans who built a great square castle, the cathedral and a church which is the prototype of all the Norman churches in the region.

The Fascist dictator Benito Mussolini (1883-1945) made Bari his headquarters for imperial ventures in the eastern Mediterranean.

Population 341 230
Map Italy – Fd

Bari Doab *Pakistan* Fertile region between the Sutlej and Ravi rivers in the Punjab. It was once semi-arid wasteland, but the British built the Upper Bari Doab Canal in 1857 to irrigate the area, and its resulting prosperity stimulated the growth of the city of Lahore.

Map Pakistan – Db

Bariloche (San Carlos de Bariloche) *Argentina* Holiday resort at the centre of Argentina's lake district, 1350 km (about 840 miles) south-west of Buenos Aires. Set beside the lake of Nahuel Huapí (area 531 km², 205 sq miles) and studded with Alpine-style chalets, it is a ski resort in winter. In summer, holidaymakers come here to explore the surrounding national park – 7850 km² (3030 sq miles) of forests and mountain meadows rich in red deer and waterbirds.

Population 60 000
Map Argentina – Bc

Barind *Bangladesh* Region in the Ganges Delta between the Ganges and Jamuna rivers. It is one of two parts of the delta (the other is Madhupur, north of Dhaka) which is 50 m (164 ft) higher than the rest. Heavy clay soils make agriculture more difficult than elsewhere in the delta, but the extra height means that the areas are not afflicted with floods. Part of Madhupur is still covered by forest.

Map Bangladesh – Bb

Bangladesh

EXPLOSIVE POPULATION GROWTH, NATURAL CATASTROPHES AND A DECLINING ECONOMY ARE THE HEAVY BURDEN CARRIED BY FORMER EAST PAKISTAN

Once a land of fabulous wealth, Bangladesh was for centuries known as Golden Bengal (Sonar Bangla). But rapacious rulers have squeezed it too hard, and it is now one of the poorest countries in the world.

Bangladesh – from 1947 to 1971 it was known as East Pakistan – is almost unbelievably flat. Apart from hills covered in bamboo forests in the south-east, the country is virtually one huge delta formed by the GANGES, BRAHMAPUTRA and Meghna rivers. It is so flat that from the air it is sometimes difficult to see where the land ends and the sea begins. The country is subjected to devastating floods; cyclones (hurricanes) that sweep in from the Bay of Bengal regularly exact huge death tolls. In 1970, for instance, a combined cyclone and tidal wave killed an estimated 300 000 people, and in 1991 another killed 140 000. In an effort to diminish the effets of such tragedies, the government introduced an early warning weather monitoring system, which has enabled it to issue alerts that have given hundreds of thousands of villagers a chance to move to higher ground before torrential rains begin.

Large and small rivers, often unstable and shifting in their courses, snake over the landscape. Whole villages may be washed away overnight as the Brahmaputra shifts its banks. Villagers have to be ready for rivers to suddenly rise 6 m (20 ft) or more. Most villages are built on mud platforms to keep them above water, and the individual houses are raised still higher on plinths.

The rainfall, sun and silt make the land productive: it is often possible to grow two or three crops a year. Irrigation could help to grow more crops in some areas during the long dry season from November to March, but there is a shortage of power to drive irrigation pumps.

LIVING OFF THE LAND

Bangladesh became world-famous for cotton in the 18th century; the special cotton cloth, muslin, was named after the Muslims of DHAKA who wore it. Agriculture concentrates on growing jute – the country shares the world jute market with India (the only other producer).

Processing jute is a highly specialised operation; only the outside fibres can be used for the manufacture of silk and cloths. The woody stems are used for fuel and for fencing, and the young green leaves as a vegetable. Some fibres from the bark are stripped by a process called retting (prolonged soaking in water), and go to make rope and sacks, and backing for carpets.

Jute is grown by small farmers who also grow rice as a subsistence crop to feed their families. The rice crops were heavily taxed by the Mogul emperors (1526-1857) and by the British when they ruled the territory as part of India (1857-1947).

Bangladesh is mainly an Islamic country, and its Muslim women are rarely seen away from the home. The rice fields, for instance, are strictly the responsibility of the men and boys. Around the homesteads there are many trees and bushes, shading small plots. Here the women tend fruits and gourds, and grow spices such as turmeric, garlic and chilli.

Many of the homesteads have an enduringly picturesque quality, with cows quietly chewing in the shade of a shelter made of bamboo overgrown with pumpkin vines. However, everything growing here has a purpose. The leaves of trees are used for fodder, the timber for building and for fuel and fruits such as mangoes for food. They also provide juices to preserve fish nets from rotting, or sap from which palm sugar is made. The bamboo plant is used for building posts and rafters, or for making plough tackle and mallet handles; it is also split and woven to make walls, fish nets and water scoops.

Cow dung plays an important part in the everyday lives of the rural population. It is a crucial element in the preparation of the threshing floor. A normal floor of mud would crack all over and steal the grains, so the top is skimmed with a slurry of mud and dung, which dries smooth and hard without cracking. Cow dung is also used as manure and for fuel.

Bangladeshi men do backbreaking work in the fields, clearing weeds, which are then used as cattle fodder. The leftovers of the rice plant once provided most of the fodder, but modern rice varieties have shorter straws and therefore do not yield as much.

A COUNTRY DIVIDED

Linguistically, culturally and geographically, Bangladesh and the Indian state of West Bengal are a single region. In former times they were either a single kingdom or a single viceroyalty of the Indian Gupta. But under the Islamic empires of India, the two Bengals became divided by religion, the majority in the east becoming Muslims, while the majority in the west remained Hindu.

The British ruled Bengal as a united province, apart from a short period at the beginning of the 20th century. But the state was divided when India and Pakistan gained independence in 1947. The eastern, mainly Muslim, half became the eastern wing of Pakistan.

The marriage of West and East Pakistan was short-lived, however, and ended in violence. At independence in 1947 East Pakistan's jute was the major export, but East Pakistan received little of the revenue it raised. The Pakistani government was dominated by West Pakistanis, particularly Punjabis, during years of military dictatorships. One of the greatest provocations to East Pakistan was a campaign by West Pakistan to impose Urdu as the national language at the expense of Bengali.

A movement for self-rule in East Pakistan was led by Sheik Mujibur Rahman. In the 1970

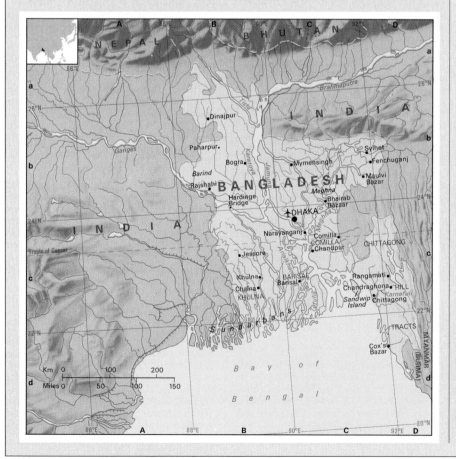

▲ RIVER OF LIFE Slow-moving boats carry grain and other vital foods on the Ganges River near the capital, Dhaka. The boats are built to a design that has not changed for hundreds of years.

elections which introduced civilian government to Pakistan, his Awami League won the greatest number of seats, all of them in East Pakistan. The second largest party was led by Zulfikar Ali Bhutto, whose Pakistan People's Party dominated the western part of the country.

The two failed to negotiate a constitution and in 1971 a rebellion broke out in East Pakistan. The West Pakistani army embarked on a ruthless campaign to put down the insurrection, and more than a million people were killed. Millions more fled across the border to India to escape the slaughter. India entered the conflict and defeated the West Pakistani army of occupation. In 1971 Sheik Mujibur Rahman became the leader of the new independent state of Bangladesh. He was assassinated in 1975, and most governments since then have been military dictatorships. Elections in 1991 showed a majority in favour of a return to a parliamentary democracy. The constitution was amended in September 1991, restoring a unicameral parliament.

Bangladesh is still an overwhelmingly rural country – only 20 per cent of the population lives in towns and 74 per cent is illiterate. Furthermore, it has a high population growth, with almost half of the country's population under the age of 15. This means that the state has an additional 3.2 million people to feed each year, a task that is near impossible without foreign aid. As it is, about 55 per cent of the population is undernourished.

Another major problem is the inadequate transportation network. Few villages are served by roads or rail, and more than 60 per cent of the country's internal trade is carried by boat. Very few mineral resources have been found; nevertheless, there is some modern industry, including a steel rolling mill and an oil refinery at CHITTAGONG, and a textile plant at Dhaka. A massive find of natural gas beneath the delta does, however, provide energy and raw material for the manufacture of fertiliser. Some cement is also produced.

The health standards of the rural areas are low. The principal scourges are malnutrition and stomach diseases, the most feared of which is cholera.

There is poverty in the towns as well as in the hinterland. The capital, Dhaka (formerly Dacca), which has grown to a city of 5 300 000 since independence in 1971, throngs with poor people trying to make something to live on. Many of them ply their trade as rickshaw pullers. Others are street hawkers, selling wares that would be rubbish in most cities: empty bottles, stripped-down parts from discarded cigarette lighters, and piles of old reading glasses. Those who have nothing to sell or are crippled will beg. Many people emigrate.

For all this, Bangladesh is renowned for its culture, and for producing men and women of great intellect.

An urban educated elite holds power in Bangladesh, even though rural villages send representatives to district development boards. The townspeople run a bureaucracy which is supposed to be helpful, but is often too elaborate to be useful. Development programmes have given some help in the form of funding the sinking of wells for irrigation and trials of new varieties of rice and wheat.

Bangladesh's future depends greatly on its relationships with its neighbours, particularly India and Nepal. The three countries need to agree on a policy for the Brahmaputra-Ganges basins to control floods, provide electricity and improve water supplies. The future also depends upon agricultural improvements for the millions of sharecroppers and small farmers, and a reduction in the birthrate.

BANGLADESH AT A GLANCE	
Official name People's Republic of Bangladesh	
Area 144 000 km² (55 598 sq miles)	
Population 122 254 850 **Per km²** 849 **(Per sq mile** 2190)	
Capital Dhaka	
Government People's republic	
Currency 1 taka = 100 poisha	
Language Bengali, or Bangla (official), English (commercial)	
Religions Muslim (83%), Hindu (16%), Christian, Bhuddist, other (1%)	
Climate Tropical; monsoon from June to September; dry season November to April. Average temperature from 29°C (84°F) in September to 19°C (66°F) in January	
Land use Cultivation 69%, meadows and pastures 4%, forest 16%, other 11%	
Main primary products Rice, jute, sugar cane, tea, wheat, tobacco, cattle, timber, fish; natural gas, salt, white clay, glass sand	
Major industries Jute spinning, textiles, sugar refining, tea, fertiliser, paper and leather processing, fishing	
Main exports Jute and jute products, leather, tea, fish, garments	
Annual income per head (US$) 220	
Population growth (per thous/yr) 23	
Life expectancy (yrs) Male 55 **Female** 54	

Barbados

AFTER INDEPENDENCE, BARBADOS, THE FIRST BRITISH COLONY IN THE CARIBBEAN, REMAINS LOYAL TO BRITAIN

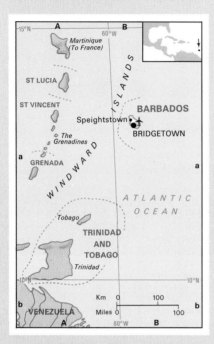

The Barbadian capital of BRIDGETOWN has a Trafalgar Square, just like London; and, rising in the centre of the square, is a statue of Nelson placed there in 1813. Barbados has an unmistakably English air and is sometimes called 'Little England', or 'Bimshire'('Bim' is slang for 'Barbados'). Churches that could have come straight from the English shires are scattered over its countryside and, just like the English, Barbadians ('Bajans') have an instinctive respect for tradition. Bridgetown's harbour police, for example, still wear sailor suits and straw hats.

Barbados is the easternmost island of the West Indies, lying well outside the arc of the Lesser Antilles. The original inhabitants were Arawak Indians who populated most of the West Indies. As elsewhere, they were captured in the 16th century and sold in the colonies by Spanish slave traders. In 1677, the first English settlers arrived. They founded Holetown on the island's western coast and, 17 years later, the colony's House of Assembly met for the first time. The island remained a British colony until 1966, when it became independent within the Commonwealth.

Most of Barbados, which is 34 km (21 miles) long and 23 km (14 miles) wide, is relatively flat. Its highest point, Mount Hillaby, near the centre, reaches only 340 m (1116 ft). The cooling North-east Trade Winds are felt in most places, and temperatures rarely rise above 30°C (86°F) or fall below 22°C (72°F). There are only two seasons: a dry season (December-May), and a wet season (June-November).

Barbados is densely populated with, on average, about 594 people per km² (1538 per sq mile) – compared with 239 people per km² (619 per sq mile) in the United Kingdom. As a result of its dense population, large numbers of Barbadians have had to emigrate.

For those who remain in Barbados, however, there is a stable and relatively high standard of living. Although much of the economy is still controlled by a tiny white minority, including the descendants of the British planters, political power is now in the hands of the black majority. The social services are well run, the standard of education is high, and the country has a 99 per cent literacy rate.

The Barbadian economy was built almost exclusively on the production of sugar and its by-products, rum and molasses. The industry has declined considerably in importance in recent years, but sugar is still the principal export. Steps taken by the government to remedy the ailing industry include restructuring, buying new machinery and moves towards privatisation. To reduce food imports, the Barbadian government has also encouraged farmers to grow food for local consumption.

The mainstay of the economy is tourism. Indeed, Barbados has attracted visitors for more than 200 years. In 1968, tourism overtook sugar as the main source of foreign currency and by the mid 1990s more than 400 000 tourists a year were enjoying the island. Barbados is almost entirely surrounded by spectacular pink and white sand beaches and coral reefs. The principal tourist area is the Platinum Coast, a sheltered strip on the western shore north of Bridgetown.

At the same time that tourism began to develop, the government started to promote industry. There are in excess of 400 factories, which employ 25 per cent of the work force and which are concentrated in nine industrial parks. The island has some petroleum and natural gas deposits, but certainly not enough to meet its fuel needs.

Despite industrialisation, unemployment is high. In 1993, it was estimated that 25.5 per cent of the work force was unemployed, prompting the government to announce some unpopular austerity measures, including reduced wages for the public sector, a revised tax system, and the proposed implementation of value-added tax (VAT).

▲ NATIONAL PASSION Youths take a break during sugar-cane cutting. In Barbados, as in the rest of the English-speaking Caribbean, devotees of cricket abound. Sir Garfield (Garry) Sobers, captain of the West Indies team from 1965 to 1974, was born here.

BARBADOS AT A GLANCE
Area 430 km²(166 sq miles)
Population 255 340
Per km² 593.8 (**Per sq mile** 1538)
Capital Bridgetown
Government Parliamentary monarchy
Currency 1 Barbados dollar = 100 cents
Language English (official)
Religion Christian (Protestant 67%, Catholic 4%); other 12%, none 17%
Climate Tropical maritime, with a rainy season in June-November. Average yearly temperature in Bridgetown is 25°C (77°F)
Land use Cultivation 77%, meadows and pastures 9%, other 14%
Main primary products Sugar cane, fish, shellfish; some petroleum and natural gas
Major industries Tourism, agriculture, textiles, chemicals, electrical parts, machinery
Main exports Electrical parts, sugar, molasses, chemicals, textiles, clothing, rum
Annual income per head (US$) 6350
Population growth (per thous/yr) 2
Life expectancy (yrs) Male 71 **Female** 76

Barisal (Bakarganj) *Bangladesh* Densely settled district covering 7299 km² (2818 sq miles) close to the Bay of Bengal.

Barisal port and town is the chief settlement of Barisal District, near the Tetulia mouth of the Ganges River. It exports fish and produces rice. The town is noted for a natural phenomenon known as the 'Barisal guns' which sounds like cannon-fire and is thought to be caused by earth movements deep underground.

Population (district) 4 667 000; (town) 142 100
Map Bangladesh – Cc

Barletta *Italy* Town on the Adriatic coast, 50 km (about 30 miles) north-west of Bari. It was an important port in Byzantine and Norman times. The Holy Roman Emperor Frederick II (1194-1250) built the imposing castle, which was added to by the Spanish in the 16th century. Outside the church of San Sepolcro is a bronze statue of a 4th-century Roman emperor known as Il Colosso, which is more than 5 m (16 ft) high.

Population 86 220
Map Italy – Fd

Barnaul *Russia* Russian industrial city and capital of the Altay Kray territory. It lies in the Altai foothills on the Ob' River, 175 km (109 miles) south of Novosibirsk. Its products include machinery, diesel engines, chemicals, textiles, processed foods, and timber.

Population 607 000
Map Russia – Jc

Barnsley *United Kingdom* Town in South Yorkshire, 18 km (11 miles) north of Sheffield. Originally a market town, it was for a century a leading centre of England's coal industry. However, all the mines close to the town have been shut – as have most of the railway lines that served them. In the early 1990s, major Japanese and European companies relocated to Barnsley.

Population 77 620
Map United Kingdom – Ed

Baroda *India* See VADODARA

Baroghil Pass *Afghanistan/Pakistan* Strategic pass in the Hindu Kush, connecting the northern tip of Pakistan and Azad Kashmir to the Wakhan region of Afghanistan.
Map Pakistan – Da

Barossa Valley *Australia* Area in South Australia 50 km (about 30 miles) north of Adelaide. Established by German immigrants in the 19th century, it is the principal non-irrigated wine-producing area of Australia, producing about a quarter of the country's wine.
Map Australia – Fe

Barotseland *Zambia* See WESTERN PROVINCE

Barquisimeto *Venezuela* Old city that has rapidly expanded into a modern industrial centre, 270 km (168 miles) west and slightly south of the capital, Caracas, on the Panamanian Highway. Its main industries are meat packing, cement manufacture and sugar refining.

Population (metropolitan area) 787 360
Map Venezuela – Ba

barrage Artificial obstruction in a watercourse, used especially to store water for irrigation and to prevent flooding. Similar structures are usually called dams if hydroelectricity is also generated.

Barranquilla *Colombia* Port on the Caribbean coast 18 km (11 miles) from the mouth of the Magdalena River. It was founded in 1629 and remained a small port until the 20th century, when the mouth of the river was deepened to make a seaport suitable for ocean-going vessels. Rapid growth in manufacturing industries such as textiles, chemicals and processed foods has helped to make it one of Colombia's largest cities and one of its major ports. Barranquilla has five universities, and each year in the week before Lent it holds a wild carnival, when it calls itself *la Ciudad Loca* – 'the Crazy City'.

Population 1 018 760
Map Colombia – Ba

barrier beach Long, narrow bar of sand built up parallel to a coastline by wave action and separated from the coastline by a lagoon. Also called barrier islands, barrier beaches are common along the eastern coast of the United States from New Jersey to Florida.

barrier reef Long, narrow CORAL REEF close to a coastline, but separated from it by a lagoon too deep or not clear enough for coral to grow.

Barrow Island *Australia* Island with an area of approximately 250 km² (97 sq miles) off the north-west coast of Western Australia, 1200 km (745 miles) north of Perth. Isolated from feral animals, it is a living museum of native wildlife coexisting with the workers on one of Australia's main oil fields.
Map Australia – Bc

Barumini *Italy* Village in the rolling wheatlands of central Sardinia, 60 km (37 miles) north of Cagliari. It is the site of Nuraghe Su Nuraxi, the most important Sardinian megalithic structure of Bronze Age round towers and dwellings. The central tower, made between the 13th and 9th centuries BC from uncemented volcanic blocks, is surrounded by 50 dwellings built in about the 7th century BC.

Population 1530
Map Italy – Be

basalt Fine-grained, dense, dark volcanic rock of BASIC ROCK composition and often having a glossy appearance. In its molten state it is highly fluid, and usually forms lava flows erupting from volcanic fissures. When it cools rapidly it may form hexagonal columns, as at the GIANT'S CAUSEWAY.

basaltic layer Crustal layer which forms the floors of the ocean basins. It is also the name sometimes given to the denser and lower of the two layers of the continental crust. The upper, less dense, layer is called the sial or granitic layer.

Basel *Switzerland* See BASLE

Bashkiria (Bashkir Republic) *Russia* Oil-rich autonomous republic of the Russian Federation covering 143 600 km² (55 430 sq miles) in the south-western foothills of the Ural Mountains. It is one of the richest regions in Russia – there is extensive mining of oil, natural gas, iron, copper and nickel ores. Its main industries are oil, coal, steel, chemicals, timber, paper and electrical appliances. Sugar beet, potatoes and cereals are grown in the area.

The republic gained its sovereignty in October 1990. Its capital is UFA. The indigenous people are mainly Russians, Tatars and Bashkirs (formerly nomadic Muslims).

Population 3 984 000
Map Russia – Gc

basic rock Dark-coloured igneous rock containing no free quartz. It usually has a low silica content (up to 52 per cent), as opposed to an ACID ROCK.

Basilan *Philippines* Island, covering about 1280 km² (495 sq miles), and city (formerly called Isabel) off the southern tip of the Zamboanga Peninsula of Mindanao. Geographically part of the Sulu Archipelago, it is administered from Mindanao; its people are mainly Muslim. Basilan produces timber, fish, rubber, palm oil and coconuts.

Population (city) 179 000
Map Philippines – Bd

Basilicata (Lucania) *Italy* Remote region of mountains and desolate hill country at the bottom of the Italian 'boot'. The region covers 9992 km² (3850 sq miles) and contains the provinces of POTENZA, with the regional capital of Potenza, and MATERA. It is agriculturally poor, except for the irrigated coastal plain.

Population 605 940
Map Italy – Ed

basin Large depression on the earth's surface which may be structural or caused by erosion. The term is also applied to an area drained by a river and its tributaries.

Baska *Croatia* See KRK

Basle (Bâle; Basel) *Switzerland* Mainly German-speaking city on the northern border with Germany and France. It is a river port at the highest point of navigation on the Rhine, and of all Swiss cities is second in size and economic importance only to Zürich. A fortified town since Roman times, Basle is now a world centre of the pharmaceutical industry. Other important products include food, metal goods and textiles. Together with two small communes, the city constitutes the half-canton Basel-Stadt, one of the richest in Switzerland. It is surrounded by half-canton Basel-Landschaft, a cherry-growing area where the main industries are chemical and biotechnological products, machines, clothing, and precision mechanics. Exceptionally rich in museums and art galleries, Basle also has fairgrounds at which the annual Swiss Trade Fair and many other fairs are held. It has a Gothic minster and Switzerland's oldest university, founded in 1460. Basel-Stadt covers 37 km² (14 sq miles) and Basel-Landschaft 518 km² (200 sq miles).

Population (city) 175 500; (half-canton Basel-Stadt) 196 600; (half-canton Basel-Landschaft) 248 500
Map Switzerland – Aa

Basque Region (País Vasco) *France/Spain* Area of south-western France and northern Spain, centred on the western end of the Pyrenees, whose native people have a unique language

unrelated to any other European tongue and a strong tradition of independence. It corresponds roughly to the French department of Pyrénées-Atlantiques with a population of about 100 000, and the Spanish provinces of Álava, Guipúzcoa and Vizcaya, with about 500 000 people. The main cities are BAYONNE in France and BILBAO in Spain. An underground Basque separatist movement in the Spanish Basque region – its militant wing calls itself ETA – gained popular support in Franco's days. Its aim is a Basque state independent of both France and Spain. The region was given a fair degree of autonomy in 1979. However, Basque separatists also carried out a number of shootings and bombings, prompting the government to clamp down on its activities.
Map France – Ce; Spain – Da

Basra (Al Basrah) *Iraq* The country's chief port before the two Gulf Wars, when its exports included dates, cereals, wool and oil. It lies on the Shatt al Arab waterway, 100 km (62 miles) from The Gulf, and 550 km (340 miles) south-east of the capital, Baghdad. An oil refinery was built here in 1974. Founded in 638 by the Caliph Omar I, it was the legendary starting point for the voyages of Sinbad the Sailor.
Population 700 000
Map Iraq – Cc

Bass Strait Situated between mainland Australia and Tasmania, the stormy strait contains offshore oil deposits. It was named after the British naval surgeon George Bass, who explored it in 1798-9. It is 290 km (180 miles) long and 130-240 km (about 80-150 miles) wide.
Map Australia – Hf

Bassar (Bassari) *Togo* Main town of the Bassari people, 350 km (about 220 miles) north of the capital, Lomé. It is a market where cattle, millet and sorghum from the north are exchanged for palm products, rice, root crops, kola nuts and coconuts from the south. Nearby are the forest reserves of Fazao and Malfakassa, whose inhabitants include warthogs, monkeys, antelopes, buffaloes, cheetahs and leopards.
Population 19 000
Map Togo – Bb

Basse Casamance *Senegal* National park covering 5000 hectares (12 360 acres), 60 km (37 miles) west of the river port of Ziguinchor in the south-west of the country. It extends to within 12 km (7 miles) of the sea. The park specialises in preserving threatened species and contains warthogs, monkeys, baboons and a wide range of birds – including guinea fowls, pheasants, snipes, ducks and bustards.
Map Senegal – Ab

Basse Terre *Guadeloupe* Seaport and capital of Guadeloupe, situated on the south-west coast of Basse Terre Island. It stands at the foot of the Soufrière Volcano, which erupted in 1797 and 1836 and was last mildly active in 1956. From the Soufrière, a chain of mountain peaks extends some 30 km (19 miles) up the spectacular western side of the island.
Basse Terre is the administrative centre of the French-ruled department of Guadeloupe; its old town dates from the 1640s.
Population (town) 14 000 ; (island) 141 000
Map Caribbean – Cb

▲THE WORLD OF ISLAM A bearded imam in the Iraqi city of Basra represents the powerful presence of Islam.

Bassein *Myanmar (Burma)* See PATHEIN

Basseterre *St Kitts and Nevis* Seaport and capital, lying on the south-west coast of St Kitts. Founded in 1627, the town was rebuilt after being destroyed by a fire in 1867.
Brimstone Hill, a fortress built on top of a 215 m (700 ft) cliff, lies 13 km (8 miles) to the west. One of the major historical buildings in the West Indies, it took a century to build, and was abandoned in 1852.
The chief industry of Basseterre is sugar refining and sugar is the main export
Population 12 610
Map Caribbean – Cb

Bastia *France* Principal commercial and industrial city of Corsica, 105 km (65 miles) north-east of Ajaccio. The old town is a maze of narrow streets. Fishing boats and yachts fill the old port, while ferries and commercial ships use the new harbour. Exports include wine and fish.
Population 31 800
Map France – He

Bastogne *Belgium* Town in the Ardennes region 69 km (43 miles) south of the city of Liège. It was the focus of the Second World War Battle of the Bulge in 1944, when the refusal of American troops to surrender it thwarted a German attempt to break through Allied lines.
Population 11 000
Map Belgium – Ba

Bas-Zaire *Zaire* Smallest and westernmost of Zaire's eight administrative regions, covering 53 920 km² (20 813 sq miles). It includes the towns of MATADI (its capital), Boma, Mbanza-Ngungu, Muanda and Banana. Behind the narrow coastlands, the land rises to the forested African plateau. The region contains the massive Inga hydroelectric scheme on the Zaire River.
Population 2 158 600
Map Zaire – Ab

Bata *Equatorial Guinea* Port and chief town of mainland Equatorial Guinea. It has no harbour, but there is offshore anchorage.
Population 30 700
Map Equatorial Guinea – Ab

Bataan *Philippines* Peninsula and province of Luzon 50 km (about 30 miles) long and 25 km (about 15 miles) wide, forming the western boundary of Manila Bay; off the southern tip is CORREGIDOR. Partly mountainous and densely wooded, Bataan was the site of fierce battles as American and Filipino troops resisted Japanese forces in 1942 during the Second World War. After its fall on April 9, 1942, prisoners of war were subjected to the infamous Death March to a prison camp near Cabanatuan, 130 km (about 80 miles) to the north-east. Today Bataan is being developed as an industrial and free trade zone.
A controversial project to build a nuclear power station in Bataan was halted by the government of Corazón Aquino in 1986 because the site was near a seismic fault. However, in 1992, the Filipino government decided to complete the project.
Population 263 300
Map Philippines – Bc

Batangas *Philippines* Province and city in south-western Luzon, 96 km (60 miles) south of Manila, facing the island of Mindoro. It is mountainous, and in the north has an active volcano – the Taal Volcano – rising from a crater lake, Lake Taal. Once important for growing coffee, the province contains a major steel works and an oil refinery.
Population (province) 1 032 000; (city) 185 000
Map Philippines – Bc

Batanghari *Indonesia* Largest river of Sumatra, 980 km (600 miles) long, most of it navigable.
Map Indonesia – Bc

Batéké Plateau *Congo* A vast plateau in the south-east of the country, stretching some 200 km (125 miles) northwards from Brazzaville. The region, deeply dissected by river gorges and valleys, is generally unproductive. Covered largely by savannah, with strips of forest along its rivers, the plateau contains impressive amphitheatres, formed by rockfalls eroded from steep hillsides. The Teké people occupy the area.
Map Congo – Ab

Bath *United Kingdom* Beautiful city in south-west England on the River Avon, 20 km (12 miles) south-east of Bristol. The suburbs cling to steep hills overlooking the city centre. Bath owes its fame to the Romans, who exploited its hot mineral springs and called it Aquae Sulis, and to 18th-century developers who made it a fashionable watering place by building a fine stone city of squares and crescents. The dandy Richard 'Beau' Nash (1674-1762) brought the elite of society to Bath, where he set new standards of dress.

Despite fierce opposition from conservationists, some of this Georgian city has been demolished to make way for new houses and offices because its narrow valley restricts expansion. But much remains: the Roman baths have been excavated. The Pump Room, built in 1790-5, stands across a square from the 15th to 16th-century abbey – and there is a costume museum in the Assembly Rooms (built 1769-71). Bath's modern university lies outside the city.
Population 84 100
Map United Kingdom – De

batholiths Large, irregularly shaped masses of intrusive igneous rock, usually granite. They extend to unknown depths and only their uppermost parts are exposed on the earth's surface by erosion. A batholith of relatively small size is sometimes called a stock.

Bathurst *Australia* City on the Macquarie River in New South Wales, about 160 km (100 miles) west of Sydney. It is the commercial centre for a sheep, wheat, beef and fruit-producing area, and has good fishing. It was founded in 1815, only two years after the first European crossing of the GREAT DIVIDING RANGE, and is the country's oldest inland town.
Population 24 700
Map Australia – He

Batinah *Oman* Narrow coastal plain facing the Gulf of Oman, 20 km (12 miles) wide and 350 km (217 miles) long. Protected by the HAJAR MOUNTAINS, it is the most fertile part of the country, with a good supply of underground fresh water from oasis springs. Date palms, limes,

oranges and breadfruit are among the most common crops. Because of its fertility, and also because of its good fishing, the coast has attracted many invaders. Traces of ancient civilizations, some going back to the 3rd millennium BC, can be found throughout the area.
Map Oman – Aa

Baton Rouge *USA* State capital of Louisiana, on the Mississippi River 316 km (196 miles) from its mouth, Baton Rouge is a busy deep-water port, easily reached by ocean-going vessels. It is the seat of Louisiana State and Southern universities, and a busy trade and manufacturing centre.
Population (city) 220 000; (metropolitan area) 528 300
Map United States – Hd

Battambang *Cambodia* Western province, bordering Thailand. It is Cambodia's main rice-growing region, and lies beside the shores of the lake of Tonle Sap. Historically, it is a disputed area between Thailand and Cambodia. The chief town, also called Battambang, is on the railway line between the Thai capital, Bangkok, and the Cambodian capital, Phnom Penh.
Population (province) 551 860; (city) 40 000
Map Cambodia – Ab

Batu Caves *Malaysia* Series of spectacular caves in the foothills of the Main Range, 11 km (7 miles) north of the capital, Kuala Lumpur. The caves, Malaysia's most popular tourist attraction, are reached by climbing 272 steps cut into the hillside. They open out, through brightly lit caverns of colourful stalagmites, into an underground temple dedicated to the Hindu god Subrahmanya. The main cave is Cathedral Cave. Beyond it lies a cavern called the Dark Cave, which quarrying has made unsafe.

In January or February each year, the caves are the site of the festival of Thaipuram, when Hindu mystics pierce their flesh with metal skewers, apparently without feeling pain.
Map Malaysia – Bb

Bat'umi (Batumi) *Georgia* Seaport, tourist resort and naval base in Georgia, on the east coast of the Black Sea, 20 km (12 miles) from the Turkish border. It is connected by rail and oil pipeline with T'bilisi and Baku on the Caspian Sea. Its industries include oil refining, chemicals and furniture manufacture.
Population 136 000
Map Georgia – Bb

Bauchi *Nigeria* Capital of the state of the same name, Bauchi lies some 820 km (510 miles) north-east of Lagos. It grew rapidly after the rail link to Port Harcourt was completed in 1914 and when demand for tin arose during the First World War. The state of Bauchi covers 64 605 km² (24 938 sq miles) and produces tobacco and tin.
Population (city) 60 800; (state) 4 294 410
Map Nigeria – Ba

Bautzen *Germany* Historic town midway between Dresden and the Polish border. It is the centre of the Slavonic people called Sorbs, who still wear their colourful traditional costumes. Industries include railway vehicle construction, and the manufacture of steel and electrical goods.
Population 49 100
Map Germany – Fc

bauxite Principal ore of aluminium. It is composed mainly of aluminium hydroxide. It forms as a result of leaching of the soil in tropical conditions when the clay minerals are decomposed and the silica is removed in solution, leaving an insoluble, alumina-rich soil. The name derives from Baux-en-Provence, in France, which had a tropical climate 60 million years ago when its deposits were formed.

Bavaria (Bayern) *Germany* The country's largest state, covering 70 554 km² (27 234 sq miles) in the south-eastern corner bordering Austria and the Czech Republic. It is a region of mountains (including the Bavarian Alps and the Alpine foothills), lakes, forests, farms, medieval towns and Baroque churches.

Industry clings to the two major cities, NUREMBERG in the north and MUNICH, Bavaria's historic capital, in the south. Toys, textiles, motor cars, paper, chemicals, porcelain, glass and musical instruments are manufactured here.

Bavaria was ruled independently by the Wittelsbach family from 1180 until 1918 – first as a duchy, then, from 1806, as a kingdom. It is a staunchly Catholic province, and has Catholic universities at Munich and WÜRZBURG.
Population 11 576 000
Map Germany – Dd

Bay For physical features whose names begin 'Bay', see main part of the name.

Bay, Laguna de *Philippines* Shallow crescent-shaped lake in central LUZON just south-west of Manila. It is the largest lake in the Philippines, covering 890 km² (344 sq miles), and is surrounded by fertile low-lying plains which are often flooded in the wet season. The plains on the southern side of the lake form the province of Laguna, whose fertile soil produces coconuts, rice, sugar cane and maize. The International Rice Research Institute at Los Baños, on the southern shore, has developed high-yielding, fast-maturing 'miracle' rice varieties.
Map Philippines – Bc

bay (bar) Ridge of mud, sand or shingle extending across a bay, usually linking two headlands.

Bayamon *Puerto Rico* Second city after San Juan and, by tradition, the first European settlement on the island. Spaniards settled here in 1508. It stands on the fertile plain, 10 km (6 miles) west of San Juan. The area produces pineapples, citrus fruit, tobacco, sugar and coffee.
Population 220 260
Map Caribbean – Bb

Bayern *Germany* See BAVARIA

Bayeux *France* Market town on the Aure River, 27 km (17 miles) west of Caen in Lower Normandy. It was the first town to be liberated during the Second World War, in June 1944. The 800-year-old Bayeux Tapestry, a strip of embroidered linen 70 m (230 ft) long and 500 mm (20 in) deep, recording the Norman conquest of England in 1066, is on display in the town's cultural centre.
Population 15 300
Map France – Cb

Baykal *Russia* See BAIKAL

Baykonyr (Baikonur) *Kazakhstan* See KAZAKH-STAN

Bayonne *France* Capital of the French Basque region, on the Adour River just inland from the Gulf of Gascony. Its ancient cathedral and low arcaded streets contrast with its busy port and modern industries. Bayonets were first made in Bayonne in the 17th century, and take their name from the town.
Population 129 730
Map France – Ce

bayou Marshy creek or backwater along a river or coast, chiefly in the southern United States.

Bayreuth *Germany* Town in northern Bavaria, 65 km (40 miles) north-east of Nuremberg, where the composer Richard Wagner (1813-83) built the *Festspielhaus*, or festival theatre. A summer season of his work is held here in July/August each year. Wagner and his wife Cosima, daughter of the composer Franz Liszt, are buried in the garden of their villa in the town.
  Binoculars and swimsuits are made in Bayreuth, which also produces almost half of Germany's cigarettes.
Population 72 000
Map Germany – Dd

Bcharre (Basharri; Besharre; Bisharri) *Lebanon* Small town 32 km (20 miles) south-east of Tripoli. It stands in the Lebanon Mountains, 1828 m (5997 ft) above sea level. Bcharre is the birthplace of the poet and mystic Khalil Gibran (1883-1931), author of *The Prophet*.
  The town is close to the Grove of Giant Cedars – one of the country's few remaining stands of cedar trees, which is classified as a national monument, though it is feared that the trees have been attacked by a disease and may not survive. It was of Lebanon cedar that Solomon (about 1015-977 BC), King of Israel, built the Great Temple in Jerusalem.
Map Lebanon – Ba

Beanntrai *Ireland* See BANTRY

Bear Island (Bjørnøya) *Norway* Uninhabited island (except for a few people employed at the meteorological station) covering 178 km² (69 sq miles) in the Arctic Ocean. It lies 386 km (240 miles) north of Norway and has been recognised internationally as Norwegian territory since 1925.
Map Arctic – Rb

Beardmore Glacier *Antarctica* One of the world's largest known valley glaciers, 200 km (125 miles) long and an average of 15 km (9 miles) wide. It descends at a rate of about 1 m (3 ft) per day from the South Polar Plateau to the Ross Ice Shelf, between the Queen Maud and Queen Alexandra ranges in the vast Transantarctic Mountains.
Map Antarctica – d

Beaufort scale Scale ranging from 0 to 12, devised in 1805 by the British Admiral Sir Francis Beaufort, and revised in 1926, to register the various velocities of wind, according to their effect at sea. Thus at scale 5, fresh breeze, 'white horses' appeared on the surface of the water. Today, the scale relates to exact wind speeds –

THE BEAUFORT WIND SCALE

Scale No.	Description	Speed (km/h)	(mph)	Observed effects
0	Calm	0	0	Smoke rises vertically
1	Light air	1-5	1-3	Wind direction shown by smoke drift, but not by vane
2	Light breeze	6-11	4-7	Wind felt on face; leaves rustle, vane moves
3	Gentle breeze	12-19	8-12	Leaves and small twigs in motion; flags are extended
4	Moderate breeze	20-29	13-18	Raises dust; small branches move
5	Fresh breeze	30-39	19-24	Small trees sway; small crests show on waves on lakes
6	Strong breeze	40-50	25-31	Large branches in motion; wind whistles in telephone wires
7	Near gale	51-61	32-38	Whole trees in motion
8	Gale	62-74	39-46	Breaks twigs off trees
9	Strong gale	75-87	47-54	Slight structural damage to houses
10	Storm/full gale	88-101	55-63	Trees uprooted; considerable structural damage
11	Violent storm	102-119	64-73	Widespread damage
12	Hurricane	Above 119	Above 73	Devastation

0 represents calm with little or no wind; 1 is 'light air', with winds of 1-5 km/h (1-3 mph), while 12 represents a hurricane with winds above 119 km/h (73 mph). See chart above.

Beaufort Sea A mostly ice-covered part of the ARCTIC OCEAN lying between northern Alaska and Banks Island (Canada). It was named after Admiral Sir Francis Beaufort (1774-1857), originator of the Beaufort scale of wind speeds. The Beaufort Sea has no islands and is partly unexplored, but large oil deposits have been found off the coasts of Alaska and north-western Canada. However, severe weather conditions make extracting and transporting the oil difficult. Depths of more than 3660 m (12 000 ft) have been recorded in the north.
Map Canada – Aa

Beaujolais *France* Region on the north-east edge of the Massif Central, on the Saône River between Mâcon and Lyons. Beaujolais is noted for its wines. The Beaujolais Mountains, making up much of the region, rise to 1012 m (3320 ft).
Map France – Fc

Beaune *France* Town in the Côte d'Or department, 35 km (22 miles) south of Dijon. It is the centre of the Burgundy wine-producing district. The town has 15th-century ramparts. A charity hospital – the Hôtel-Dieu or Hospice de Beaune – has been open since 1443 and is supported by a wine auction held every November.
Population 21 100
Map France – Fc

Beauvais *France* Capital of Oise department, on the Thérain River, 65 km (40 miles) north-west of Paris. It manufactures blankets, carpets, chemicals and tractors; other industries include food processing and mechanical engineering. It has a 13th-century cathedral, with massive buttresses, which survived German bombing in 1940, when most of the old town was destroyed.
Population 58 000
Map France – Eb

Béchar *Algeria* Oasis town with a picturesque fort, in the north-west Algerian Sahara, 467 km (290 miles) south-west of Oran and close to the Moroccan border. Béchar is an administrative centre at the beginning of the Route du Tanezrouft, one of the main trans-Saharan routes to the south, and also the start of the western trans-Saharan route to TINDOUF and Mauritania. The town is served by air and by rail; the railway ends just beyond Béchar at Kenadsa.
Population 65 100
Map Algeria – Aa

bed Layer or stratum of rock, usually sedimentary rock, divided from the layers above and below by bedding planes.

Bedford *United Kingdom* Town and county of south-central England. The county town stands on the Great Ouse River, 70 km (43 miles) north of London. The religious writer John Bunyan (1628-88) wrote *The Pilgrim's Progress* in Bedford in 1675 during a six-month spell in jail for his non-conformist teachings.
  The county covers 1235 km² (477 sq miles) and is largely agricultural, but there are extensive brickfields and engineering industries. The town of LUTON produces motor vehicles, electrical goods and scientific instruments. Woburn Abbey, an 18th-century stately home, is one of Britain's major tourist attractions.
Population (town) 77 020; (county) 532 400
Map United Kingdom – Ed

bedrock Solid unweathered rock that lies beneath all the layers of broken and weathered rock (regolith) and soil on the earth's surface.

Beechworth *Australia* Town in Victoria, 240 km (about 150 miles) north-east of Melbourne, in the foothills of the Victorian Alps. It was the centre of the Ovens gold-mining district in the 1850s and has several fine buildings dating from then, including a powder magazine. Wangaratta (population 15 980), 32 km (20 miles) farther west, is now the main centre of the area which produces

wool, wheat, tobacco and grapes. The outlaw Ned Kelly (1855-80) and his gang operated in the district in the 1870s.
Population 3140
Map Australia – Hf

Beersheba (Be'er Sheva) *Israel* Modern capital of the Negev (the southern part of Israel), 72 km (45 miles) south-west of Jerusalem. According to tradition, it is the spot where Abraham pitched his tent 3800 years ago. There are also archaeological remains and a Bedouin market. Ben Gurion University is at Beersheba. The city's industries include glass making and pottery.
Population 128 400
Map Israel – Ab

Behariya *Egypt* See BAHARIYA OASIS

Behistun *Iran* See BISOTUN

Beijing (Peking; Peiping) *China* The country's capital is, in part, a paradox: an imperial city carefully preserved and tended by an anti-imperial communist government. Lying on the extreme north-western edge of the North China Plain, it is both the ancient seat of government and a modern industrial and trading city.

Beijing did not become a seat of government until the 10th century AD when the Khitans, a semi-nomadic Mongol people, made it their headquarters. In the 13th century, the first emperor of the Yuan dynasty, Kublai Khan (1215-94), grandson of the great Mongol warlord, Genghis Khan, established his capital on a site to the north of the walls of the present-day city.

In 1421, Beijing became the imperial capital of the Ming dynasty (1368-1644). The third Ming emperor, Yong Le (ruled 1403-24), laid out a spacious walled city which, like many ancient Chinese cities, had its walls and streets oriented towards the cardinal points of the compass. Within the walls was the Imperial City and within that a further enclosure known as the Forbidden City – the emperor's court which ordinary people entered on pain of death. It is now open to tourists. Beijing has remained the capital of China since then, except for the period 1928-49, during which the Nationalist government was transferred to NANJING.

Today, Beijing spreads across 16 807 km² (6490 sq miles) in northern China, some 140 km (87 miles) from the shores of the Bo Hai (Gulf of Chihli). Its broad, straight streets are thronged with bicycles, buses and pedestrians. Most government buildings stand in Zhongnanhai Park. There are more than 50 institutes of higher education, including Beijing and Qinghua universities, the latter a centre for scientific research. Industry, based in the outer suburbs, ranges from engineering and iron and steel to refining.

Above all, Beijing is a city of great historical and cultural importance. In the middle is Tiananmen Square, which covers 40 hectares (100 acres) and is one of the largest public squares in the world. During political demonstrations in June 1989, more than 1000 students, workers and innocent bystanders were killed here by government troops. The demonstrations and massacre were televised internationally and provoked an outcry worldwide. On the square's western side is the Great Hall of the People, built in 1959, where the National People's Congress meets. To the south stands the massive Memorial

▲ **STAIRWAY TO HEAVEN** Visitors climb the long stairway leading to the Temple of Heaven in Beijing. Laid out in three terraces, the temple attracts the local faithful as well as tourists from abroad.

Hall of Chairman Mao, built in 1977 to house the remains of communist China's founding father.

Beyond Meridian Gate, north of the square, lie the palaces of the Forbidden City, the largest group of classical Chinese buildings in existence. They include the Hall of Supreme Harmony, formerly the main throne room of the Chinese emperors, and the Palace of Heavenly Purity, where emperors presided over affairs of state.

To the north is Jing Shan, sometimes called Prospect Hill, the site of the Pavilion of a Thousand Springs, which commands a magnificent view – air pollution permitting – of the Forbidden City's golden roofs.

Beijing has many parks, some of them laid out as private gardens by emperors. The largest is the Temple of Heaven Park. Among its 15th-century temples stands the Qiniandian, or Hall of Prayer for Good Harvest, a 41 m (134 ft 6 in) high, cone-shaped building with triple eaves and a beautiful deep blue roof.

Beijing is in an area rich in history. The GREAT WALL OF CHINA snakes along the mountains 80 km (50 miles) to the north. The road to the wall passes the Valley of the Ming Tombs, which contains the imposing mausoleums of 13 Ming emperors. The Summer Palace of the dowager empress, Ci Xi, an autocrat who governed China as regent from 1861 to her death in 1908, stands in the city's north-western suburbs. It is a late 19th-century complex of ornate halls, pavilions and pagodas set among artificial lakes.
Population (city) 7 000 000; (municipality) 10 870 000
Map China – Hc

Beira *Mozambique* Port in the south of the country, about 190 km (118 miles) south-west of the

Zambezi River mouth. Founded in 1891 to the north of the ancient Arab trading post at Sofala, it developed as an outlet for the landlocked countries in the interior, and is now linked by rail and oil pipeline to Zimbabwe and by rail to Malawi. Beira exports minerals and cash crops, including cotton and sugar.
Population 113 800
Map Mozambique – Ab

Beira *Portugal* Region in central Portugal which is divided into three provinces. The province of Beira Alta (10 525 km², 4063 sq miles) lies southeast of the city of Oporto and produces Dão wine. To the south, separated from it by the mountains of Serra da Estrêla, is the province of Beira Baixa (6675 km², 2576 sq miles), which includes Covilhào, a town producing woollen textiles. Castles, guarding the frontier with Spain, are dotted along the eastern border of both provinces.

To the west of Beira Baixa is the coastal province of Beira Litoral (6755 km², 2607 sq miles), which lies due south of the city of Oporto. Rice is grown in the low-lying areas near the coast, and olives and wheat are grown inland.
Population (Beira Alta) 629 200; (Beira Baixa) 234 200; (Beira Litoral) 1 059 300
Map Portugal – Bb, Cb-Cc

Beirut *Lebanon* Capital and principal port lying astride the small Beirut River in the shelter of the Lebanon Mountains. The city controls the main road and railway routes along the coast, as well as the links heading east via the Baider Pass through the mountains to Syria and Damascus.

Beirut was founded by the Phoenicians around the 14th century BC, and eventually it became important as a Greek and Roman trading centre. It was also noted as a place of learning – the ruins of a Roman law school can still be seen. Today, the tradition of learning continues in Beirut's four universities: the Lebanese University, the Arab Beirut University, the Université Saint-Joseph and the American University of Beirut. During the Crusades, the city fell to the Muslim leader Saladin, then to the Ottoman Turks in the 16th century. In 1918, it was captured by the French and it became the capital of the new state of Lebanon in 1920.

Called 'the Paris of the East', Beirut prospered as the chief financial and trade centre of the Middle East until the start of years of factional strife. After 1975, civil war and the Israeli invasion of Lebanon began taking their toll – much of the city was badly damaged as continued sporadic outbursts of fighting drove away business and tourism. Beirut's business life came to a standstill, and the city split in two – West Beirut, the Muslim section, and East Beirut, the Christian section, were divided by the so-called 'Green Line'. Until the 1989 Taif Peace Accord, many shops remained closed and, for a while, even the universities virtually discontinued their courses. With the elections that took place in August and September 1992, and the signing of the Israeli-Palestinian agreement in May 1994, Beirut's economic recovery began.

Population 1 500 000
Map Lebanon – Ab

Beit Eddine (Bayt Ad Din) *Lebanon* Settlement and former capital when Lebanon was under the rule of the Ottoman Turks. It lies in the Lebanon Mountains, 48 km (30 miles) south-east of Beirut and 792 m (2600 ft) above sea level. Beit Eddine contains the well-preserved palace of the Turkish emir, Beshir Chehab, Prince of Lebanon from 1804 until his death in 1840. One wing of the palace is now a folklore museum, and it is used as a summer residence for Lebanon's president.

Map Lebanon – Ab

Beit She'arim *Israel* Archaeological site 15 km (about 10 miles) west of Nazareth. The site's remains include a 2nd-century synagogue and 3rd-century catacombs. In the 2nd century AD, when the area was under Roman rule, it was the seat of the Sanhedrin, Israel's High Court of Judges, following the destruction of the Temple in Jerusalem in AD 70.

Map Israel – Ba

Beitbridge *Zimbabwe* Town on the border with South Africa. The Limpopo River marking the border is crossed by a road and rail bridge built in 1929 and named after Alfred Beit, who paid for it. It is the main route for freight between Zimbabwe and South Africa.

Population 1900
Map Zimbabwe – Bb

Bejaia (Bougie) *Algeria* Minor seaport in Carthaginian and Roman times, lying on the Mediterranean coast, 152 km (94 miles) north-west of Constantine, and surrounded on either side by a stretch of fine beaches. In the 5th century, it became the fortified capital of Genseric the Vandal for a brief period before falling into decline. Refounded by the Berbers in the 11th century, it became a major port and cultural centre in north-west Africa before declining into a notorious stronghold for pirates.

Today, Bejaia is a pipeline terminal, handling oil from the great HASSI MESSAOUD field. The port also exports citrus fruits, cereals, olive oil, iron ore and phosphates.

Population 150 000
Map Algeria – Ba

Békéscsaba *Hungary* Chief town of the south-eastern county of Békés, 178 km (111 miles) south-east of Budapest. It is typical of the Great ALFÖLD towns, in which single-storey, farmlike houses predominate. It is also a centre of Slav culture, where descendants of 18th-century Slovak settlers paint their houses pastel blue for distinction. Machinery and tool-making industries have been added to traditional textile making and food processing.

Population 67 910
Map Hungary – Bb

Belait *Brunei* River, 160 km (about 100 miles) long, in western Brunei. It gives its name to the country's largest administrative district (area 2727 km², 1053 sq miles) through which it flows. The coastal area of Belait district, in the south-west, is the traditional oil-producing area; the interior, composed mainly of low forested hills, is hardly developed.

Population (district) 56 000
Map Brunei – Ab

Belarus See p 83

Belau See PALAU

Belém *Brazil* River port and trade centre on the north coast, 145 km (90 miles) from the mouth of the Amazon. A religious festival called the Cirio is held at the cathedral each October. It honours a statue of the Virgin Mary said to have been removed several times from its original site – only to return miraculously on each occasion. The city's principal products are rubber, nuts, tropical hardwoods, fish and vegetable oils.

Population 1 235 600
Map Brazil – Db

Belfast *United Kingdom* Northern Ireland's capital, parts of which have been divided into bitterly opposed sectarian groups. The city stands at the mouth of the River Lagan. Its name – *Béal Feirsde* in Irish, meaning 'Approach to the Sandbank' – first appeared in the 7th century, but it was not until 1613 that the town received its first charter. By the beginning of the 19th century the population had grown to some 20 000, due largely to the cotton industry, which was later replaced by linen manufacture. Shipbuilding developed during the 19th century when the River Lagan was improved for navigation and reclaimed land was used for vast shipyards, notably those of Harland & Wolff. Engineering is now a major industry.

The sectarian troubles that have dogged the city date back to the coming of Scottish and English settlers during the 17th century. They introduced a new religious outlook into Catholic Ireland, and the rise of Belfast as an industrial centre in the 19th century drew in both Protestants and Catholics, sparking explosive rivalry. The most militant opposition to British rule in Belfast and other towns in Northern Ireland came from the Irish Republican Army (IRA), a nationalist organization which was founded in 1919.

The trademark of the IRA campaign of terror, which was intensified from 1968 onwards, was a series of bombings and assassinations in both Northern Ireland and mainland Britain. The IRA's violent rivalry with Unionist paramilitary groups also pushed up casualties.

In August 1994, after more than 3400 had been killed, the IRA announced a ceasefire. Peace

hopes were given a further boost when, shortly afterwards, Unionist paramilitary groups followed suit.

The city hall, completed in 1906, stands at the heart of Belfast. It is an ornate building whose interior has much Greek and Italian marble and a fine banqueting hall. Other gems include Queen's University, built in 1849 as a Tudor-style college, the Ulster Museum and Art Gallery in the Botanic Gardens, and Custom House on Donegal Quay, built of golden-coloured stone in Italian style, as well as the Presbyterian General Assembly's headquarters, and the government buildings built in 1928 at Stormont to the east. The Grand Opera House, built in 1895, has a splendid rococo interior. The fine modern Romanesque Church of Ireland cathedral was completed in 1981, after taking 80 years to build. There has been much redevelopment in Belfast in recent years, both of housing and in the commercial centre.

Population (urban area) 475 970
Map United Kingdom – Cc

Belfort *France* Town and region on the Savoureuse River, 40 km (25 miles) south-west of Mulhouse. The region remained French after 1870, when neighbouring Alsace was occupied by Germany. The town lies in a gap between the Vosges and Jura mountains, on an important trade route which links France to Germany and Switzerland. Its industries include electrical and mechanical engineering and textile manufacture.

Population (town) 76 200; (region) 132 000
Map France – Gc

Belgium See p 84

Belgrade (Beograd) *Yugoslavia (Serbia and Montenegro)* Capital of Serbia and Montenegro, founded by Celts on a strategic site where bluffs rise 60-70 m (200-230 ft) above the junction of the Sava River with the Danube. The city, whose name means 'white citadel', was developed, in turn, by Celts, Romans, Slavs, Byzantines, Turks, Austrians and Serbs to guard the Morava-Vardar-Axios route to the Aegean Sea against invaders from the north. The huge Kalemegdan Fortress, now enclosing a park with pavilions, a military museum and keep, stands on the original site.

Belgrade became capital of the Serbian kingdom in 1882 and later capital of Yugoslavia, until the collapse of the Yugoslav confederation in 1991. A new city centre of government buildings was developed to the south between the First and Second World Wars; after 1945, Belgrade expanded as a transport, international trade and industrial centre. An oil refinery was built, along with a new airport and factories producing machinery, electrical goods, chemicals, farm equipment, foods and consumer durables.

Little remains of the old city because during the Second World War Belgrade was devastated by German troops who captured it in April 1941. Yugoslav and Soviet forces liberated the city in October 1944. A 19th-century neo-Baroque cathedral with fine icons and tombs has survived; the Barjak Mosque is the only one of more than 100 mosques remaining. The Church of Sv Marko (St Mark) stands beside Tasmajdan Park. Belgrade's university dates from the 1860s; the national museum is one of Europe's finest.

Population 1 533 900
Map Yugoslavia – Cb

Belarus

A COUNTRY WITH A TURBULENT PAST IS BEGINNING TO COMPETE ON THE WORLD MARKET

Belarus (formerly Belorussia or 'White Russia') is part of a region that is wedged between the Baltic Sea and the Black Sea – and like its neighbours, present-day Estonia, Latvia, Lithuania, Russia, Poland and the Ukraine, can look back on a turbulent past. For centuries, the borders of Belarus kept shifting; MINSK, its capital, about 260 km (160 miles) from the Polish border, was founded in 1067 as Menesk, became part of Lithuania in 1326, was later ceded to Poland, then to Russia in 1793. By the 1930s, 40 per cent of its people were Jewish, but most of them were killed after invading German troops sent them to concentration camps during the Second World War.

Belarus gained independence in August 1991. Officially called the Republic of Belarus, it became a founder member of the Commonwealth of Independent States (CIS).

Belarus's other name – 'White Russia' – has nothing to do with colour: it indicates a point on the compass. The name was first recorded in the 14th century by the East Slavs, who referred to compass points as colours. For them, the colour *byeli* (white) meant north – and therefore the Slav people who settled to the north of the *rus* (empire) of Kiyev were called *byelorus* – the people of 'the White Empire'.

The issue of their origin became of interest only in the second half of the 19th century. Before then, the Belarussians' country had been split up between Poland and Lithuania. When Poland itself was divided up between 1772-95, the Russian tsars took control of Belarus.

Later, under the Soviets, the local culture was promoted. In 1919, the Belorussian Soviet Socialist Republic was founded and three years later it was incorporated in the Soviet Union. The Belorussians were granted their own government – for the first time in their history.

LAND OF LAKES AND RIVERS

Belarus is largely lowland and sparsely populated. It forms a bridge between Central and Eastern Europe. Climatically, too, it lies in a transitional zone, between the maritime climate of the Baltic countries and the continental climate of central Russia and the Ukraine. Its winters are milder than those farther east and it receives most of its rain in the summer.

Winters do bring abundant snowfalls, though; and melting winter snows cause widespread flooding. Belarus has more than 10 000 lakes and many rivers. The far north of the country is literally a sea of lakes. The White Russian Mountains, culminating at 346 m (1135 ft), are the only elevation.

Belarus has some timber and peat, potash and rock salt, plus small deposits of petroleum and natural gas. But it has no important natural resources apart from land and water. And yet it is one of the wealthiest members of the CIS and takes a leading position among all the ex-Soviet republics as an agricultural producer.

Agricultural workers have concentrated on a few crops that are successful, turning their country into a substantial exporter of meat, dairy products, flour and potatoes.

Its diverse industries have put Belarus in the ranks of the most advanced and economically stable among the CIS members. Industries started growing around the railway junctions and along the railway lines in the mid 19th century. Today, the industrial sector employs about 70 per cent of the country's workforce. Among its products are textiles, machinery and machine tools, and motor vehicles, especially tractors and trucks. Electrical goods and other consumer goods are manufactured, mostly for the home market. Petrochemicals, synthetic fibres and fertilisers are made for export. Under Soviet rule, Belarus produced nearly 30 per cent of the entire USSR's output of synthetic fibres.

Though Belarus is landlocked, the main traffic arteries linking Warsaw with Moscow, and the Baltic republics with the Black Sea, run through it, bringing additional income. The country is critically dependent upon the other ex-Soviet countries for imports of fuels and other raw materials. Most of its oil comes from Russia via the Druzhba Pipeline. In 1992, Belarus's gross domestic product (GDP) fell by about 13 per cent due to its dependence on the ailing Russian economy for raw materials. Yet in that year, although inflation rose to an estimated 1200 per cent, less than 1 per cent of its population was unemployed.

Belarus exports most of its products to the other CIS members. There was very little economic restructuring after independence, but a three-year programme to privatise 50 per cent of all state enterprises by 1996 was approved in June 1993. In April 1994, the Belarussian government signed an agreement with Russia to use the Russian rouble, thus giving up much of its economic independence.

In 1994, a new constitution introduced a presidential system of government. National presidential elections were held in June and July that year.

In contrast to other CIS states, nationalism has not been a major issue for the population, which is 66 per cent Belarussian; many prefer speaking Russian to using their own language. Besides, having been part of Poland-Lithuania and part of Russia, and having had, until the Second World War and the Holocaust, a vast number of Jewish people living in their cities, the Belarussians are well accustomed to foreigners and foreign cultures.

BELARUS AT A GLANCE	
Official name Republic of Belarus	
Area 207 600 km² (80 135.9 sq miles)	
Population 10 370 270	
Per km² 50 (**Per sq mile** 129.4)	
Capital Minsk	
Government Presidential republic	
Currency 1 rouble = 100 copeks	
Languages Byelorussian (66%), Russian (32%), other (2%)	
Religions Mainly Christian (Russian Orthodox 70%)	
Climate Moderate continental, with abundant summer rainfalls. Average temperature in Minsk ranges from −7°C (19°F) in January to 18°C (64°F) in July	
Land use Cultivation 29%, meadows and pastures 15%, other 56%	
Main primary products Peat, potash, salt, oil, natural gas, timber; dairy products, chickens, potatoes, grains	
Major industries Chemicals, motor vehicles, machinery; food processing	
Main exports Chemicals, fertilisers, synthetic fibres, motor vehicles, machinery; dairy products, meat, flour, potatoes	
Annual income per head (US$) 3120	
Population growth (per thous/yr) 3.4	
Life expectancy (yrs) Male 66 **Female** 76	

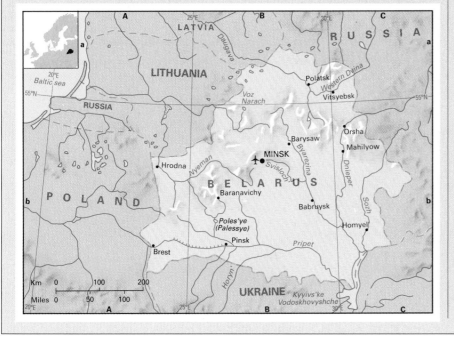

Belgium

EUROPE'S BATTLEGROUND HAS SURVIVED AGAINST THE ODDS, AND PROSPERS AS THE CENTRE OF EUROPEAN UNITY

Among the countries of Europe, Belgium is a surprise. Not on account of its scenery, which is rarely spectacular, nor for all its people, who are pleasantly average West Europeans with an above-average taste for beer and fried potatoes; not even for its small size, since, at about 30 500 km² (11 600 sq miles), it is ten times as big as Luxembourg to the south.

The surprise stems simply from its existence – that there should be a Kingdom of Belgium at all. It was born of a revolution within the larger Kingdom of the United Netherlands. At that time it had to import a king from Germany (Leopold I, 1831-65), because it had no royalty of its own. From the start, it brought together two quite dissimilar groups, with different languages, and has lived through more than a century of tension and rivalry between them. Divided permanently now into two regions, with a special status for the capital, BRUSSELS, Belgium has, perhaps surprisingly, survived.

UNHAPPY HISTORY OF CONFLICT

The Low Countries, comprising Belgium, the Netherlands and Luxembourg, have a more chequered history than almost any other European area of similar size. They have been held by Burgundians, Germans, Spaniards, Austrians and French. They have been fought over by Protestants and Catholics, revolutionaries and imperialists: Belgium, in particular, has an unhappy history as the ground where other nations staged their battles.

The last three conflicts in this 'cockpit' of Europe were the Waterloo campaign of 1815, when British and Prussians fought the French; the muddy trench warfare of the First World War, when the British and French fought with the Belgians against the Germans; and the Battle of the Bulge in the Second World War, when the Allies fought the Germans.

The territories and boundaries of Belgium took a long time to congeal into their modern form. The present borders date only from the 1919 Treaty of Versailles, and pay little regard to people. Leaving aside 65 000 German speakers brought in by boundary changes round EUPEN and MALMEDY in 1919, Belgium's population of 10 million divides into two: Flemish speakers (57.8%) and French speakers (32%), who are, confusingly, known as Walloons, although hardly any of them speak that language. The Flemish outnumber the Walloons by about three to two, and no other fact about modern Belgium is so important.

The long-standing geographical divide between Flemish and French runs from east to west roughly through Brussels. In the north, Flemish is spoken; it is a dialect of Dutch, and the area shares the cultural and religious traits of the southern Netherlands. South of the line, French is spoken and the cultural ties are with France.

HOLDING THE NATION TOGETHER

Nothing would appear simpler than to divide the country along the language line, and apportion the halves to the Netherlands and France. But for more than 160 years the Belgians have been opposing that division, and this despite the fact that, since the Second World War, holding the nation together has been an uphill struggle.

In 1830, when Belgium became independent, French was the language of government, society and scholarship, and Flemish considered fit only for peasants. The economic powerhouse of Belgium was in the south – on the coalfields that ran from MONS, through CHARLEROI to LIÈGE, and this was one of the most important industrialised areas in Europe.

Time brought changes – the decline of the southern coalfield, the search by newer industry for more space to develop, the impact of different birthrates. The birthrate in French Belgium fell; that of the Flemings was, and is, one of the highest in Europe. Furthermore, a Flemish cultural movement grew up to preserve and spread the language, and its success was marked by a law of 1932 making the Belgian government officially bilingual.

This achieved, the Flemish majority was set to consolidate its position. After years of turbulence, a law of 1970 effectively divided the country into four: the Flemish region in the north, and the Walloon region in the south, the former German areas, where German is still spoken, in the east, and Brussels, which remains officially bilingual, with every notice and street sign printed in both Flemish and French. Today there are ten Belgian provinces, five Flemish provinces in the north, and five French in the south – and in each group only one language is official.

To consolidate their progress, Flemish Belgians had to counter the former bias towards French in science, literature and higher education: otherwise, Flemish speakers would still be debarred from reaching the highest levels in academic and industrial life. So it was further enacted that every order given in a plant or office in a Flemish province must be in Flemish: everything from invoices to notices about not smoking. At the same time the universities, except that of Brussels, became single-language. Belgium's oldest university, LOUVAIN – now Leuven and Flemish – was split in two, library and all; the second half was established at Louvain-la-Neuve, south of the language line, to operate in French.

Today, Belgium has, in effect, four governments: a national government with separate ministries of education and separate cultural budgets, a Flemish regional government, a Walloon regional government, and a regional government for the capital, Brussels.

NATURAL DIVIDING LINE

There is a natural dividing line across Belgium as well as a language line, but the two do not coincide. The natural divide is the trenchlike valley of the MEUSE River. South of the river lie the uplands – among them the ARDENNES – rising by gentle slopes to almost 700 m (2300 ft): a region of sparse settlement, harsh winters, forests and tourism, where many old farm buildings have been turned into second homes and holiday centres. North of the river, where the rim of the valley rises to about 200 m (660 ft), there begins a slope

▲ GOTHIC GEM The Grand Place at the heart of Brussels, with its elegant pavement cafés, comes to life after dark. The richly carved Town Hall (left), built between 1402 and 1454, is topped by a statue of the archangel Michael, the city's patron saint.

which continues unbroken for 150 km (95 miles) to the sea. The slope is covered alternately by fertile loams and barren sands or gravels; the coastlands are flat and grassy. This is the Belgium of dense population and ancient cities, of tiny farms alongside modern factories, and of bicycles and canals.

In between the two lies the Meuse Valley, the industrial heart of Belgium from the time its Iron Age metalworkers first began making swords until the closure in the 1960s and 1970s of its hopelessly uneconomic coal mines. Steel output at Liège and Charleroi reached nearly 13 million tonnes in 1970. The valley floor became cluttered with industry and its paraphernalia. By 1983, output had fallen to 10.1 million tonnes a year; today, the mines have all been closed and Belgium's coal comes from the KEMPENLAND, farther north. Some of the industry has gone, too, and, although unemployment is high and many people have moved away, the Meuse Valley remains a place apart from the lands to the north or south.

The cities of northern Belgium were among Europe's earliest boom towns, drawing their wealth from trade and textiles. In spite of repeated devastation by war, they possess a legacy of marvellous old streets, squares and churches. Some of them, like ANTWERP and GHENT, have survived to become the metropolitan centres of today. Others, of which the largest is BRUGES, have central areas little changed since medieval times.

Belgian farming is focused on livestock production, and most of the field crops are for feeding animals. There are areas that specialise in orchards, wheat and sugar beet, and the glasshouses around Brussels are famous for their flowers and vegetables. But many farmers have left the land for urban occupations and many others are part-timers; only 3 per cent of the workforce is employed in agriculture, and farm labourers are hard to find.

Belgian industry has a long and proud tradition, particularly in view of its small home markets. It survived the Second World War almost intact and then set itself, in the words of one Belgian, to produce 'calves with five feet' – manufactured goods in which ingenuity counted for much more than mass production. In the old wool cities of the north there are new textile industries producing a range of goods from carpets to clothing. In the relatively empty spaces of the Kempenland there are steel, chemicals and motorcar industries. In the Meuse Valley the traditional skills in engineering and the manufacture of arms live on. The great port of Antwerp processes imports from the tropical world, builds and repairs ships, and cuts diamonds.

Brussels is not only a city of about a million people. It is also, in a sense, the capital of Western Europe. The unity which Brussels is unable to impose upon the Belgians is found, paradoxically, in the larger setting of the European Economic Community, now the European Union (EU). This owed its origins to an agreement of exiled wartime governments of Belgium, the Netherlands and Luxembourg to form an economic union, Benelux. The concept expanded to embrace France, Italy and West Germany (today, the reunited Germany) in the 1950s; Britain, Denmark and Ireland in 1973; Greece in 1981; Spain and Portugal in

1986; and, since January 1995, Sweden, Finland, and Austria. The city of Brussels has flourished as the administrative headquarters of Western Europe's supranational government. It has also become the headquarters of the North Atlantic Treaty Organization (NATO).

Today, Brussels is no longer just a place to change trains on the way to more glamorous locations. It is a centre of business, information and policy-making in its own right, and with a beautiful city centre to go with it!

BELGIUM AT A GLANCE	
Official name Kingdom of Belgium	
Area 30 519 km² (11 780.7 sq miles)	
Population 10 100 630	
Per km² 331 (**Per sq mile** 857.4)	
Capital Brussels	
Government Parliamentary monarchy	
Currency 1 Belgian franc = 100 centimes	
Languages Flemish (57.8%), French (32%), bilingual (9.5%), German (0.7%)	
Religion Christian (90% Roman Catholic)	
Climate Temperate; average temperature in Brussels ranges from 2°C (35.6°F) in January to 17.5°C (63.5°F) in July	
Land use Forest and woodland 21%, arable land 24%, meadows and pastures 20%, permanent crops 1%, other 34%	
Main primary products Cereals, sugar beet, potatoes, livestock, timber; coal	
Major industries Iron and steel, light and heavy engineering, textiles, petroleum refining, chemicals, cement, glass, food processing, diamond cutting, coal mining	
Main exports Machinery, motor vehicles, iron and steel, chemical and pharmaceutical products, textiles, cut diamonds, vegetables	
Annual income per head (US$) 15 540	
Population growth (per thous/yr) 2.3	
Life expectancy (yrs) Male 70 **Female** 77	

Belize See p 87

Belize City *Belize* The country's commercial centre, and its former capital beside the Caribbean Sea – but after serious hurricane damage in 1961 the seat of government was moved inland to Belmopan in 1970.

Belize City is the main port for Belize and parts of Mexico, exporting sugar, timber, fruit and maize. The town consists largely of wooden buildings raised on stilts above the mangrove swamp on both banks of Haulover Creek.

Belize is situated at the mouth of the Belize River. Inland along the river are the remains of Maya settlements dating from before the 10th century AD, of which Altun Ha, 50 km (about 30 miles) north of Belize City, is the largest.

Population 49 700
Map Belize – Bb

Belle-Ile *France* Island, 16 km (10 miles) long, 12 km (8 miles) south of the Quiberon Peninsula in Brittany. Its mild climate and varied landscape of cliffs, coves, sandy bays and tiny fishing harbours make it a popular destination for artists and tourists.

Population 4200
Map France – Bc

Bellingshausen Sea Lying off Antarctica in the South PACIFIC OCEAN between Thurston and Alexander islands. The Bellingshausen Sea was explored by a Russian expedition led by Captain von Bellingshausen in 1821.

Map Antarctica – Ec

Belluno *Italy* Town and province 80 km (50 miles) north of Venice. The town stands on a bluff overlooking the Piave River, which forms the central corridor of this mountainous province frequently shaken by earthquakes.

The province stretches north into the Alps and the Dolomites, with many peaks exceeding 3000 m (10 000 ft). The area is dotted with ski resorts such as CORTINA D'AMPEZZO. The city's Venetian-style architecture includes the porticoed Palazzo dei Rettori.

Population (town) 35 420; (province) 211 390
Map Italy – Da

Belmopan *Belize* Newly developed capital, 75 km (47 miles) south-west of Belize City and 80 km (50 miles) inland. The colony's government moved here in 1970 after Belize City, the former capital, was damaged by a hurricane in 1961. The town, which contains a number of modern public offices built in the style of Mayan architecture, is inhabited mainly by civil servants.

Population 4 500
Map Belize – Bb

Belo Horizonte *Brazil* The country's third largest city after São Paulo and Rio de Janeiro. It lies 720 km (447 miles) south-east of the national capital, Brasília. It is the capital of the state of Minas Gerais – a region rich in gold and gemstones. The city's manufactures include motorcars, steel, textiles, shoes, cement and electric trains.

Population 2 103 400
Map Brazil – Dc

Belorussia See BELARUS

Belostock *Poland* See BIALYSTOK

Belvoir, Vale of *United Kingdom* Lowland area some 20 by 10 km (12 by 6 miles) in extent in central England, straddling the borders of Leicestershire and Nottinghamshire. It is overlooked by medieval Belvoir (pronounced 'Beaver') Castle, which was extensively restored in the 19th century. The vale is famous for Stilton cheese.

Map United Kingdom – Ed

ben Scottish term for 'peak', used as a prefix to many Scottish mountain names, for example in 'Ben Nevis' or 'Ben More'.

Ben Nevis *United Kingdom* See GRAMPIANS

Benares *India* See VARANASI

Bendigo *Australia* City in Victoria, 130 km (about 80 miles) north-west of Melbourne. It was founded in 1851 during the Victorian gold rush and soon became one of the richest towns in the country. The Central Deborah Mine, which closed in 1954, has been restored as a museum. Bendigo's other industries include iron foundries and potteries. Originally called Sandhurst, the town was named Bendigo in 1891, after a local shepherd who took his nickname from the English pugilist Abednego William Thompson.

Population 57 430
Map Australia – Gf

Benevento *Italy* Ancient city 50 km (30 miles) north-east of Naples. It has a splendid Roman theatre, which could seat 20 000, the Arch of Trajan, from the 1st century AD, and the Basilica of Santa Sofia, founded in 700 AD.

Population 62 680
Map Italy – Ed

Bengal See BANGLADESH; WEST BENGAL

Bengal, Bay of Arm, 1600 km (about 990 miles) wide, of the INDIAN OCEAN, bounded by Sri Lanka, India, Bangladesh, Myanmar and the Andaman and Nicobar islands. It is known for its violent tropical cyclones, which have capsized large vessels and caused terrible floods in the delta of the GANGES and BRAHMAPUTRA rivers.

Map India – Dd

Benghazi *Libya* Major port and chief town of CYRENAICA at the eastern end of the Gulf of Sirte, situated on a tongue of land separated from the mainland by a salt lagoon (now silted up) and a stretch of marsh. It has one of the country's two international airports, is a road and rail centre and is the seat of the University of Garyounis. Benghazi served as one of Libya's two capitals during the monarchy. However, after the 1969 revolution, Tripoli was declared the sole capital.

Benghazi's past can be traced back to the 5th century BC, when Greek colonists established a settlement, Euhesperides, on the outskirts of the town. In AD 643, the Arabs conquered the town. Modern Benghazi stands on the site of the Graeco-Roman town of Berenice. A great deal of building was carried out by the Italians during their colonial occupation from 1912 to 1943. The city changed hands several times and was badly damaged during the North African campaigns of the Second World War. Many of the Italian and earlier Turkish buildings were destroyed, though a few remain in the old town.

Industries include tanneries, a shoe factory, leatherworks and textile factories; other products are carpets, olive oil, pasta and cement.

Population 591 000
Map Libya – Ca

Benguela *Angola* Port 410 km (255 miles) south of the capital, Luanda, and the capital of the province of the same name. Founded in 1617, it is one of Angola's oldest Portuguese settlements and was the centre of the slave trade with Brazil.

It has sugar-milling and fish-drying plants, and a rail link to the port of Lobito, 29 km (18 miles) away. Lobito is at one end of a 1415 km (880 mile) long railway network linking the inland republics of Zaire and Zambia to the coast. The railway has, in the past, been vital to Angola for the taxes it can charge on cargoes carried along it; it has also been vital to Zaire and Zambia because it is one of their main export trade routes. Since 1975, however, the railway has been out of commission as a result of sabotage by UNITA guerrillas.

Population (city) 155 000; (province) 584 000
Map Angola – Ab

Benguela Current *South Atlantic* Cold ocean current flowing northwards along the west coast of southern Africa, and named after the Angolan port of Benguela. See OCEAN CURRENTS.

Beni Abbès *Algeria* Oasis town 160 km (100 miles) south of Béchar, set among date palm groves at the beginning of the Route du Tanezrouft, the westerly trans-Saharan route to the south. Huge sand dunes surround the town, which has become a major centre for desert studies, including dune control and reclamation.

Population 4300
Map Algeria – Aa

Beni Suef (Bani Suwayf) *Egypt* Trading town beside the Nile, 100 km (62 miles) south of Cairo. Cotton and other products from nearby EL FAIYUM are processed here.

Population 150 000
Map Egypt – Bb

Benidorm *Spain* Holiday resort which sprang up after the Second World War around a Mediterranean fishing hamlet 100 km (62 miles) south of the port of Valencia. It has become one of the most popular Spanish resorts on the Mediterranean coast; during summer months its population swells to almost 1 000 000. The Sierra de Aitana towers 1558 m (5111 ft) above the town.

Population 42 440
Map Spain – Ec

Benin See p 88

Benin, Bight of Bay in the Gulf of GUINEA, once known as the Slave Coast. It extends westward from the NIGER RIVER mouth along the coasts of NIGERIA, BENIN, TOGO and GHANA.

Map Nigeria – Ab

Benin City *Nigeria* Capital of Edo State and centre of the Bini (or Edo) people, 240 km (150 miles) east of Lagos. It was founded in the 10th century and was the capital of the kingdom of Benin, whose territory in the 15th and 16th centuries stretched westwards to include what is now Benin, and eastwards to the Niger River. The city was visited by the Portuguese in 1485 and was

Belize

THE CENTRAL AMERICAN STATE THAT
OFFERS TOURISTS AN EXCITING
MIXTURE OF OLD AND NEW

British troops have been stationed in Belize ever since it gained independence from Britain in 1981. Self-governing since 1964, the country was previously British Honduras, and had been a British colony since 1862. The military presence was intended to protect Belize from the territorial claims of neighbouring Guatemala, which date back to 1921. But in 1991, Guatemala officially recognised its neighbour's sovereignty, though still disputing the Guatemalan-Belizean border.

In 1993, the British government reduced its garrison – at the time that Guatemala was going through an internal crisis. In the same year, general elections in Belize brought victory to the opposition United Democratic Party (UDP). The new government under Manuel

Esquivel suspended the territorial accord signed with Guatemala, claiming that the latter had received too many maritime concessions in exchange for formal recognition of his country's sovereignty. The accord had, for instance, given Guatemala unimpeded access to the Caribbean Sea.

The coast of Belize is approached through 550 km² (210 sq miles) of coral cays and reefs. The coastal belt and the northern half of the country are low-lying and swampy. Behind the coast the land rises through a series of ridges, some scrub-covered, others densely forested. Southern Belize consists mainly of a broad plateau backed by the Maya Mountains, a succession of serrated ranges with peaks of up to 1100 m (3600 ft).

The capital, BELMOPAN, in the mountainous and jungle-covered interior, was established in 1970 after the former coastal capital, BELIZE CITY, was damaged by a hurricane in 1961.

The climate is warm and humid, with cooling sea breezes. Rainfall is heavy, especially towards the south, but there is a relatively dry period from February to May. Hurricanes can occur in late summer. Dense tropical forest covers the wetter plains and highlands, with scrub-type savannah in the drier places. The forest provides valuable hardwoods such as mahogany, lignum vitae and cedars.

The country was settled in the 17th century by Spaniards and then by pirates and British woodcutters from Jamaica. Later, black slaves were imported to work on the sugar plantations. Some indigenous groups, descended from the Mayas, survived in the interior.

Today, 81 per cent of the people are black, Creole (a mixture of black and white races), or *mestizo* (a mixture of white and Amerindian). There are minorities of Amerindians and, mostly in the towns, Europeans.

Just under one-third of the workforce is employed on the land, growing the main export crop of sugar cane, as well as citrus fruits

(particularly grapefruit) and bananas. Others work in forestry or fishing.

Since the early 1990s, the former British colony has been developing its own industry, mainly clothing. This has been in part the cause of the recent economic upswing. Agriculture still accounts for 75 per cent of export earnings, though a small but increasingly important source of income and work is provided by tourists, drawn to Belize's splendid Caribbean beaches and its ruined temples of the Mayan civilisation (AD 300 to AD 900) buried deep in the tropical forest

BELIZE AT A GLANCE
Area 22 960 km² (8863 sq miles)
Population 203 960
Per km² 8.9 (**Per sq mile** 23)
Capital Belmopan
Government Parliamentary monarchy
Currency 1 Belize dollar = 100 cents
Languages English (official), Spanish, Creole, Maya
Religions Christian (Roman Catholic 62%, Protestant 30%)
Climate Tropical, with the main rainy season from May to October. Average temperature in Belize City is 26°C (78.8°F) throughout the year
Land use Cultivation 2%, meadows and pastures 2%, forest and woodlands 44%, other 52%
Main primary products Sugar cane, citrus fruits, maize, rice, bananas, timber, coconuts, fish
Major industries Agriculture, sugar refining, citrus concentrates, rum, forestry, fishing and aquaculture, clothing; tourism
Main exports Sugar, citrus fruits, timber, bananas, cultured shrimps, molasses, clothing
Annual income per head (US$) 2210
Population growth (per thous/yr) 24
Life expectancy (yrs) Male 66 **Female** 69

soon exporting slaves, leopard skins, elephant tusks, pepper, and coral beads in return for such goods as guns and gunpowder, used by the Beninese for slave raiding.

Today, Benin City is a market for palm oil and rubber. The museum houses magnificent 15th and 16th-century Yoruba bronzes and carved ivories.
Population 300 000
Map Nigeria – Bb

Benioff zone Zone of earthquake activity that marks the path of a descending oceanic plate as it moves deeper into the earth's interior.

Benmore *New Zealand* See WAITAKI

Benue (Bénoué) *Cameroon/Nigeria* River some 1390 km (about 865 miles) long. It rises on the Adoumaoua massif of central Cameroon and flows north, then west to the Niger River in Nigeria.

A project based on the Lagdo (or Lagdom) Dam near the town of Garoua in Cameroon irrigates 80 000 hectares (200 000 acres) for the cultivation of cereals, including rice and sugar cane, and provides hydroelectric power. The

Nigerian part of the river is navigable for about ten weeks each year, and also provides irrigation water and power for Nigerian farmers.
Map Cameroon – Bb

Benxi *China* Industrial city of Liaoning Province, 175 km (about 110 miles) from the North Korean border. It is a major producer of iron and steel and lies close to a coal-mining area.
Population 1 200 000
Map China – Ic

Beograd *Yugoslavia* See BELGRADE

Beppu *Japan* Resort and spa on the north-east coast of KYUSHU Island. Beppu attracts millions of visitors each year to its more than 3000 hot springs and mud pools, which have names such as 'Lake of Blood' and 'Waterspout of Hell'.
Population 130 000
Map Japan – Bd

Beqa'a (Biqa) *Lebanon* Fertile valley running down the centre of the country between the Lebanon Mountains and the Anti-Lebanon Mountains. It is 16 km (10 miles) wide and 160 km (100 miles) long. Lebanon's two most

important rivers rise in Beqa'a: in the north, the ORONTES flows into Syria, and in the south, the LITANI flows first south and then west. The waters of the rivers allow cotton and other crops, including citrus fruits, to be grown. The valley also receives some 380 mm (15 in) of rain a year and is the country's main cereal-growing area.

In Roman times, Beqa'a was one of the granary regions supplying the empire; it is also noted for its wine. During the Israeli occupation of Lebanon, the valley was the scene of fighting.
Map Lebanon – Ab

Berbera *Somalia* The country's second largest port after the capital, Mogadishu, standing on the north coast on the Gulf of Aden. It ships hides, gum and animals. The population of the town, which was founded before the 13th century, was swollen in the 1980s by refugees from the drought-stricken interior.
Population 70 000
Map Somalia – Aa

berg A single hill or mountain, or sometimes a range of mountains, as in the Drakensberg, Swartberg or Cederberg ranges of mountains in South Africa. See also INSELBERG and BERGWIND.

Benin

SLAVING AND VOODOO SACRIFICES ARE PART OF A SINISTER HISTORY, BUT BEAUTIFUL BEACHES AND A WEALTH OF WILDLIFE ARE ATTRACTING TOURISTS

Shaped like an ice-cream cone standing on the southern coast of West Africa, Benin is one of the region's smallest but most diverse nations. Its short southern coast has idyllic beaches of white sand, backed by coconut groves and lapped by surf from the Bight of Benin.

Behind the coast lie tranquil lagoons such as Nokoué and PORTO-NOVO, bordered by picturesque villages where square houses are built high on stilts beside the water. Rows of elegant dugout canoes rest by the water's edge – for here, people live mostly by fishing, ferrying and, more recently, from an increasing number of tourists

North of the lagoons are the fertile Terre de Barre lands where much of the tropical rain forest has been cleared for cultivation. These in turn give way to a deeply cut plateau covered with vast plantations of oil palms, coffee, cocoa and tobacco. In the north-west, from NATITINGOU to Pendjari Park, the ATACORA Mountains are grassy, savannah-covered plateaus up to about 800 m (2600 ft) high, and separated by steep forested valleys.

Most of the people live in the equatorial south, where the Fon and Yoruba peoples predominate. Many are subsistence farmers, growing yams, cassava, sweet potatoes, maize, rice, groundnuts and vegetables.

With at least nine rainy months a year, giving an annual rainfall of 1500 mm (59 in) in the south and south-west, there is little fear of crop failure. In the grassy savannahs north of the Atacora, the rains last 'only' five to six months. Here, farmers in scattered villages plant millet, sweet potatoes and sometimes other vegetables, with a few cash crops such as cotton and groundnuts.

In north and central Benin, the principal ethnic groups are the Bariba – great horsemen – and the Somba, who build square, fortress-like houses. The nomadic Fulani range through the northern territory.

Benin, which was called Dahomey until 1975, was once a principal source of slaves for the slave traders. It was also notorious for the human sacrifices demanded by its traditional animist religions – particularly the powerful Voodoo cult. Rival African kingdoms flourished in the area in the 18th and 19th centuries, and for a long time resisted colonisation. However, the country finally succumbed and was occupied by the French in 1868. It became a territory of French West Africa in 1904.

Independence from France in 1960 led to lengthy regional power struggles, but after several coups d'état, political stability came in 1972 with a takeover by Mattheu Kerekou, who established a one-party communist state. Faced with economic collapse, Kerekou renounced the Marxist orientation of his government in December 1989, and lifted restrictions on opposition parties. In March 1991 Kerekou was defeated in multiparty elections – the first time in mainland Africa's history that a head of state was voted out of office. It took more than a year for the newly elected president, Nicéphore Soglo, to gain the support of a majority in the National Assembly as each of the many parties represented there controlled only a minority of seats.

The country remains poor, underdeveloped, and overly dependent on agriculture, which employs two-thirds of the workforce, accounts for 35 per cent of gross national product (GNP), and earns a major share of export earnings. Most of the country's trade passes through the port of COTONOU. Efforts to diversify the economy in the past suffered from a lack of foreign investment, though foreign aid supported the construction of a large hydro-electric plant at Nangbeto which lies on the border with Togo.

Tourism is a major growth area, and although there are few hotels and other facilities, except in coastal towns, Benin has two of West Africa's most beautiful wildlife parks – the Pendjari, and the 'W' shared with Niger and Burkina. Both are home to hippopotamuses, crocodiles, elephants, lions, cheetahs, buffaloes and antelopes.

BENIN AT A GLANCE	
Official name Republic of Benin	
Area 112 620 k² (43 472.5 sq miles)	
Population 5 500 000	
Per km² 48.8 (**Per sq mile** 126.5)	
Capital Porto-Novo (official), Cotonou (economic and administrative)	
Government Republic	
Currency 1 CFA franc = 100 centimes	
Languages French (official); numerous African languages (such as Ewe, Fon, and Yoruba)	
Religions Animist (70%), Christian (15%), Muslim (15%)	
Climate Tropical, with two brief dry periods, in December/January and in August. Average temperature in Cotonou ranges from 26°C (78.8°F) in August to 29°C (84.2°F) in March	
Land use Forest 33%, cultivation 16%, grazing 4%, other 47%	
Main primary products Cassava, yams, maize, sorghum, coffee, cotton, palm products, groundnuts; oil	
Major industries Agriculture, palm oil and palm-kernel oil processing, textiles, cotton ginning, beverages	
Main exports Crude oil, coffee, cocoa, palm kernels and palm oil, cotton	
Annual income per head (US$) 410	
Population growth (per thous/yr) 30	
Life expectancy (yrs) Male 49 **Female** 53	

Bergama (Pergamum) *Turkey* Town near the Aegean coast, about 250 km (155 miles) south-west of Istanbul. Capital of the Hellenistic empire in the 3rd and 2nd century BC, it became part of the Roman empire in 133 BC. In the 8th century AD it was destroyed by Arabs, but excavations begun at the end of the 19th century have uncovered and partly restored many classical remains. Among them are shrines dedicated to Greek deities, a terraced gymnasium and a theatre which seats nearly 15 000 people.
Population 56 610
Map Turkey – Ab

Bergamo *Italy* City 50 km (about 30 miles) north-east of Milan, lying in the foothills of the Alps. Bergamo is divided into the new lower town, devoted to commerce and industry, and a smaller upper town, Bergamo Alta, which is enclosed within Venetian walls and holds historic monuments. These include the Communal and Gombino towers, the medieval Ragione Palace, Santa Maria Maggiore Church, and the startling Renaissance Colleoni Chapel, built in red and white marble. Works by painters of the Bergamese school are on display at the Carrara Academy, just below the old town. There is a theatre named after the opera composer Gaetano Donizetti (1797-1848), who began and ended his life in Bergamo.
Population 115 660
Map Italy – Bb

Bergen *Belgium* See MONS

Bergen *Norway* Second largest city in Norway after Oslo and the largest until the 19th century. Bergen, on the coast of south-west Norway, was founded by King Olaf III in about 1070. It is a busy port and tourist centre.

Bergen was a centre of the Hanseatic League – a mercantile association of mostly north German ports – until 1560. The old town focuses on the medieval harbour, near the former homes of Hanseatic traders and the restored Hall of King Håkon IV (1204-63). It has a 13th-century cathedral and some fine medieval churches, museums and galleries. There is also an open-air museum with old restored houses on display.

▲ HANSEATIC HARBOUR The medieval town of Bergen, with its 12th-century Mariakirche and houses of Hanseatic merchants, nestles beside the harbour with the modern port beyond.

The Norwegian composer Edvard Grieg (1843-1907) was buried at Troldhaugen, near Bergen, and the town has an annual music festival. Bergen's theatre, built in 1850, gained widespread recognition under such directors as the Norwegian playwright Henrik Ibsen (1828-1906), author of *Hedda Gabler*.

The city's main industries are shipbuilding, the production of steel, electrical equipment, office machinery and textiles, oil refining and fish processing. Bergen is also a centre for the North Sea oil industry.

The *Bergensbanen* (Bergen railway), a scenic rail route over the mountains to Oslo, is popular with visitors to the region.

Population 218 100
Map Norway – Bc

Bergen op Zoom *Netherlands* Town 108 km (67 miles) south and slightly west of Amsterdam, and close to the Belgian border. Formerly a fortified seaport, Bergen op Zoom has been barred from the sea by a dyke which is part of the Delta Plan for land reclamation, but has canal links to the Rhine, Maas and Schelde rivers. Much of the medieval centre survives. Industries include sugar refining and iron production.

Population 47 550
Map Netherlands – Bb

bergschrund Crevasse at the head of a glacier separating the moving ice from the ice adhering to the valley walls.

bergwind Hot, dry wind that blows from the plateau in South Africa down to the coast, warming as it descends.

Bering Sea Sea in the North PACIFIC OCEAN between Alaska and Siberia and bounded in the south by the ALEUTIAN and Komandorskiye islands. It covers 2 269 000 km² (956 700 sq miles); its greatest known depth is 5121 m (16 801 ft). It is named after the Dane Vitus Bering (1681-1741) who proved in 1728 that Asia and North America were separate continents. He was the first European to sight Alaska.
Map Alaska – Ac

Bering Strait Separating Asia from North America, the Bering Strait links the BERING and CHUKCHI seas. During the Pleistocene Ice Age (which began about 1.75 million years ago and ended about 10 000 years ago) the sea level fell, creating a land bridge between the continents. Crossing this bridge, perhaps 25 000 years ago, the ancestors of the native Americans began their settlement of North America.

The strait, which is 88 km (55 miles) wide and reaches a maximum depth of 52 m (170 ft), was discovered by Vitus Bering in 1728. In the middle, only 4 km (2.5 miles) apart, are the United States Little Diomede and the Russian Big Diomede islands.
Map Alaska – Ab

Berkshire *United Kingdom* County of southern England covering 1256 km² (485 sq miles), mostly to the south of the River Thames. The west has areas of fine open downland and chalk hills. It has three racecourses – Ascot, Newbury and Windsor – and a horse training centre at Lambourn. The east is lightly industrialised around the county town, READING, and the industrial estate of Slough, and has acquired a reputation as England's 'Silicon Valley'. The county is officially known as Royal Berkshire because it includes WINDSOR with its royal castle.
Population 753 000
Map United Kingdom – Ee

Berlin *Germany* Capital of Germany and, from 1871 to 1945, capital of the German Reich. After the Second World War, when Germany was divided into East and West Germany, Berlin was split up into four sectors with the French sector in the north, the British in the centre, the American in the south, and the Soviet in the east. Each of these countries kept a military presence in their sector. West Berlin remained a part of the West under the YALTA agreement of 1945, but it was an enclave in East German territory.

From June 1948 until October 1949, communist officials in the East shut all roads and rail links between West Germany and West Berlin, but the blockade was thwarted by a massive Allied airlift.

In 1961, the East Germans built a wall dividing the West from the Russian sector of the city, in a bid to stop a steady drain of East German people to the more affluent West. The adjacent mined belt, cleared of buildings and overseen by watchtowers, brought concentration-camp landscapes to the capital.

The Berlin Wall was torn down after almost 30 years, when each of the Second World War Allies relinquished their right to their city sector. Berlin was officially named the capital of the reunited, new Federal Republic of Germany on October 3, 1990. In June 1991, the *Bundestag* (German parliament) voted to shift the seat of government from Bonn back to Berlin.

Berlin was founded when Slavs and, later, Teutons settled at what was then a marshy crossing of the River Spree. From the early 15th century, Hohenzollern princes ruled the Brandenburg Province from Berlin, until it became the capital of united Germany in 1871, by which time Berlin had a population of 800 000. The *Tiergarten*, a 255 hectare (630 acre) former royal park, lies in the centre of the city alongside the Brandenburg Gate, a partly restored triumphal arch erected in the 18th century, and the old *Reichstag*, the former parliament, which was burnt down in 1933, just after the Nazi leader Adolf Hitler had become Chancellor. The *Kurfürstendamm*, which was once described by the American novelist Thomas Wolfe (1900-38) as the longest coffee shop in Europe, is the shopping and entertainment centre of the city.

Wholesale destruction during the Second World War and the split into East and West did little to reduce Berlin's role as a commercial and cultural metropolis. Many of the historic buildings on the *Unter den Linden* have been restored, including the opera house, national gallery, the Pergamon Museum (containing treasures from classical Greece, such as the altar of Zeus) and Humboldt University.

Culturally, Berlin, home of the Berlin Opera and the Berlin Philharmoniker, and many art galleries and museums of renown, is one of the leading cities in the world.

Traditionally, most of the population are employed in the public and municipal services; the city's manufactures include machinery, chemicals, electrical goods, textiles, furniture and food.

Population 3 438 000
Map Germany – Eb

Bermuda *Atlantic Ocean* Group of 150 small islands, of which only about 20 are inhabited, in the western Atlantic. Bermuda is situated 920 km (about 570 miles) east of Cape Hatteras in North Carolina, the nearest point on the North American mainland. The largest island, Great Bermuda, is 21 km (13 miles) long and is linked to the other main islands by bridges and causeways; their total land area is 53 km² (21 sq miles).

The hilly limestone islands are the caps of ancient volcanoes rising from the seabed. They were explored by a Spaniard, Juan Bermúdez, in 1503, but were uninhabited until 1609, when a party of British colonists led by Sir George Somers was shipwrecked here. In 1684 the British crown took over the settlement, and today Bermuda is a self-governing colony with one of the oldest parliaments in the Commonwealth, dating from 1620.

More than half the inhabitants are descendants of African slaves, introduced to work the land in the years before the abolition of slavery in 1834.

Bermuda's mild Gulf Stream climate, together with its proximity to the United States, is the basis of a flourishing tourist industry, though tourism declined by 16 per cent from 1987 to 1992, largely as a result of the recession in the United States. Many foreign banks and financial organisations operate to take advantage of the group's lenient tax laws for individuals and, to compensate for the drop in income from the tourist industry, the Bermudan government has been promoting these financial services. A substantial income also comes from the US naval and air bases on the islands, leased to the United States for 99 years in 1941. The capital, HAMILTON, stands beside a deep inland harbour on the main island of Great Bermuda.

Population 60 690
Map Atlantic Ocean – Bc

Berne (Bern) *Switzerland* Capital city lying on both sides of the River Aare, with the old town being located on a narrow bend of the river. Whereas Zürich is the business and industrial centre of Switzerland, Berne is the seat of the national government, and the city has a university, as well as a number of national museums and libraries.

The old town of Berne, now a UNESCO World Heritage area, was founded in the 12th century, but most of its buildings date from the 18th century. It is almost unspoilt, with its 15th-century minster, its clock tower, arcades and fountains, and a bear pit near the river where the bears cavort for the tourists. This animal is Berne's heraldic symbol, and the city gets its name from the German *Bär*, meaning 'bear'.

The surrounding canton of Berne covers an area of 5961 km² (2301 sq miles) and includes a little of every type of landscape found in the country – the ALPS, the JURA and the plateau.

The canton's main crops are maize, potatoes and sugar beet, along with dairy produce. Its main industries are metalworking, machinery, clothing, watchmaking and tourism, especially in the BERNESE OBERLAND. Most of its inhabitants are German-speaking and Protestant.

Population (city) 130 000; (canton) 938 200
Map Switzerland – Aa

Bernese Oberland *Switzerland* Mountain area south of the national capital, Berne. It stretches southwards to the Rhône Valley; in the west, it passes into the Alps of Waadt and in the east it is bordered by the mountains of Uri. Its heart is the massif dominated by the Jungfrau (4158 m, 13 642 ft), Eiger (3970 m, 13 024 ft), and the Finsteraarhorn (4274 m, 14 022 ft).

The area is one of the world's great tourist regions, with lakeside resorts such as Interlaken and mountain resorts such as Adelboden, Grindelwald and Gstaad.

Map Switzerland – Aa

Berre, Étang de *France* Shallow lake, 20 km (12 miles) long and 5-13 km (3-8 miles) wide, linked to the Gulf of Fos midway between Marseilles and Arles. Deep navigation channels reach its northern shores from the Mediterranean. Industries along the lake's shores include oil refining, petrochemicals, engineering and production of steel.

Map France – Fe

Berry *France* Former province dating from the 8th century. The region corresponds to today's Cher and Indre departments and occupies some 7000 km² (2700 sq miles) on a low plateau just north of the Massif Central. It is drained by the Cher, Creuse and Indre rivers. It contains numerous large farms which produce wheat and fodder. BOURGES is the main town.

Map France – Ec

Berwickshire *United Kingdom* See BORDERS

Berwick-upon-Tweed *United Kingdom* Northernmost English town, situated at the mouth of the River Tweed in Northumberland, 5 km (3 miles) south of the Scottish border. It changed hands 13 times in the turbulent border history of the two countries.

It is one of Britain's few remaining walled towns, the largely intact walls having been rebuilt last in the 1760s. The town is best known for the Tweed's salmon and for the Royal Border Bridge, a 28-arch structure built 38 m (125 ft) above the river by railway and structural engineer Robert Stephenson in 1850.

Population 12 900
Map United Kingdom – Ec

beryl Mineral, essentially aluminium beryllium silicate, usually bluish-green or green and occurring in hexagonal prisms. It is used as a gemstone (emerald) and is also the chief source of the metal beryllium.

Besançon *France* Fortress town originally settled in the Iron Age, on the flanks of the Jura Mountains, 75 km (47 miles) east of Dijon. Its massive citadel was built by Sébastien de Vauban, a 17th-century military engineer, to defend the eastern flanks of France from the Habsburg empire.

The capital of the department of Franche-Comté, Besançon also has Roman ruins, including an amphitheatre, forts, an aqueduct and a triumphal arch. Industries include textiles and watchmaking.

Population 120 800
Map France – Gc

Beskidy (Beskids; Beskydy) *Slovakia/Poland* Series of forested ranges which run for 300 km (186 miles) along the border of Slovakia and Poland and form the outer arc of the Carpathian Mountains. The highest peak is Babia Góra (Babia Hora), which rises to 1725 m (5659 ft) on the border between the two countries.

The Pieniny, a forested chain of limestone crags, lies on the border farther east and rises to 982 m (3222 ft) at Trzy Korony ('the Three Crowns'). It is cut by the DUNAJEC Gorge and includes a national park which spans the border and is home to a diversity of animal species, including bears and wolves.

Tourism has developed since the 1960s with the building of a 137 km (85 mile) circular road, the 'Beskidy Loop', that links it to the town of Lesko in Poland.

Map Slovakia – Bb-Cb; Poland – Dd

Bessarabia *Moldova* See MOLDOVA

Bet Guvrin *Israel* Ancient city in the foothills of Judaea, about 40 km (25 miles) south-west of Jerusalem. It has Roman, Byzantine and Crusader remains. At the Biblical town of Maresha nearby are 2nd-century burial caves.

Map Israel – Ab

Bethlehem *West Bank* Formerly part of Jordan which was occupied by the Israelis in the 1967 war, Bethlehem is in the West Bank region. In 1994, in the terms of an agreement between the Israeli government and Yasser Arafat's Palestinian Liberation Organisation (PLO) it was agreed that this area (including Bethlehem) would eventually fall under PLO rule. Situated 8 km (5 miles) south-west of Jerusalem, Bethlehem lies on a hill amid green, fertile country, and looks out to the Dead Sea and beyond. Bethlehem is regarded by Christians as the birthplace of Jesus Christ and the town has become an important pilgrimage and tourist centre

From this small town in Judaea, King David started his conquest of the northern kingdom of Israel, making Jerusalem the capital and religious centre of the new state. Later, Bethlehem fell into obscurity, ignored by all but those who looked to the town for the coming of the Messiah. The traditional place of the Nativity was desecrated by the Roman emperor Hadrian, who established a grove sacred to the god Adonis here. In AD 315 the Byzantine emperor Constantine, a Christian, replaced the heathen grove and built the Church of the Nativity in its place. The church was rebuilt in the 6th century and is now shared by monks of Latin, Greek and Armenian orders. The site of the manger in which Jesus was born is claimed to be the grotto beneath the church.

Population 30 000
Map Jordan – Ab

Béthune *France* Market and industrial town 27 km (17 miles) north-west of Arras. The old town was destroyed during the First World War but has been rebuilt. Béthune's industries include chemicals, plastics and tyres.

Population 259 700
Map France – Ea

Beuthen *Poland* See BYTOM

Beyrouth *Lebanon* See BEIRUT

Béziers *France* Leading centre of the Languedoc wine trade – and most occupations here are linked to the wine trade. Béziers lies on a hill 60 km (37 miles) south-west of Montpellier. It was once surrounded by defensive walls, which were destroyed in 1632. The town is dominated by the massive Gothic church of Saint Nazaire.

Population 83 200
Map France – Ee

Bhadgaon (Bhaktapur) *Nepal* Town 16 km (10 miles) east of Kathmandu. Many of the brick and timber buildings are medieval, ornately decorated with gods and animals. It has Buddhist temples and monasteries and a museum of Nepalese paintings. The town is noted for its pottery.

Population 40 100
Map Nepal – Ba

Bhadra *India* See TUNGABHADRA

Bhairab Bazaar *Bangladesh* Town at a crossing point of the Meghna River. The King George VI Bridge, which carries the Dhaka-Chittagong Railway, is one of only two major river bridges in Bangladesh, the other being the Hardinge.

Population 63 700
Map Bangladesh – Cb

Bhakra Dam *India* Dam on the Sutlej River in the Himalayan foothills, 205 km (127 miles) east of the city of Amritsar. It was completed in 1963 and is 226 m (742 ft) high and 518 m (1700 ft) long. It is part of a project to provide water for the INDIRA GANDHI Canal, one of the world's longest irrigation canals, which flows into the THAR Desert.

Map India – Bb

Bhaktapur *Nepal* See BHADGAON

Bhamo *Myanmar (Burma)* Town on the Ayeyarwady (Irrawaddy) River in north-eastern Myanmar, 30 km (19 miles) from the Chinese border. It is the upriver terminus for steamers on the Ayeyarwady and an important staging post for trade with China along the Burma Road, built between the two countries in the 1930s.

Population 65 000
Map Myanmar – Cb

Bhilai *India* Steel town in central India, about 260 km (162 miles) east of the city of Nagpur. Its vast steel works was built with Russian help in the 1950s.

Population 1 447 870
Map India – Cc

Bhlarna *Ireland* See BLARNEY

Bhopal *India* Industrial city in central India, 345 km (214 miles) north-west of the city of Nagpur. It produces heavy electrical equipment. Until 1956 it was the capital of the princely state of Bhopal, and today the nawab's lakeside palace is a tourist attraction. In December 1984, Bhopal was the scene of the world's worst chemical disaster at the Union Carbide plant, when some 3000 people were killed and many others permanently disabled by a gas leakage.

Population 1 350 300
Map India – Bc

Bhubaneshwar *India* Pilgrimage city 480 km (298 miles) south-west of Calcutta. It has some 500 Hindu temples, including the Great Temple dating from about AD 650. This contains a granite statue of the god Shiva, which is bathed daily with water, milk and *bhang* (a marijuana derivative). Originally some 7000 Hindu temples must have stood on the shores of the sacred lake. Bhubaneshwar is the capital of Orissa State.

Population 411 500
Map India – Dc

Bhumiphol Dam *Thailand* The country's first dam, built in 1964 across the Ping River. The dam provides water for irrigation, and also for a hydro-electric station with an output of 550 megawatts.

Map Thailand – Ab

Bhutan See p 92

Biafra *Nigeria* See SOUTH-EAST REGION

Biafra, Bight of Bay in the Gulf of Guinea extending eastwards from the NIGER River mouth along the coasts of Nigeria, Cameroon and Equatorial Guinea.

Map Nigeria – Bc

Biala Podlaska *Poland* Town situated 145 km (90 miles) east of the capital, Warsaw, and about 40 km (25 miles) from the Belarussian border. It has a number of fine Gothic and Baroque buildings dating from the 16th and 17th centuries, when it thrived on the wealth of the powerful Radziwill family, and on the production of carpets, cloth and glass.

Population 44 400
Map Poland – Eb

Bialowieza (Bjelowesch) *Poland/Belarus* Forest astride the Polish-Belarussian border. It was made a national park in 1921, 580 km² (220 sq miles) of it in Poland and 700 km² (260 sq miles) in Belarus. European bison, elks, wild tarpan ponies, deer, wild boar and lynx wander between oak, fir, pine, spruce and magnificent hornbeam trees up to 55 m (180 ft) tall.

Map Poland – Eb

Bialystok (Belostock) *Poland* Textile city in north-east Poland 55 km (about 35 miles) from the Belarussian border. It lay in the old Polish province of Podlasie and is the site of the 'Podlasian Versailles', an elegant Renaissance palace of 1728-58, which was built by the Branicki family.

Population 268 000
Map Poland – Eb

Bianco, Monte *France/Italy* See BLANC, MONT

Biarritz *France* Fishing port, coastal resort and spa town 8 km (5 miles) south-west of Bayonne. It became fashionable after Napoleon III, nephew of Napoleon Bonaparte, visited it in the 1850s.

Population 26 700
Map France – Ce

Bicol *Philippines* Winding peninsula forming the south-eastern part of LUZON. It is mountainous and has several active volcanoes. Its isolation and poor internal communications resulted in slow economic development until recently, but it has long produced copra and Manila hemp and has important copper, lead and zinc ore deposits.

The main centres are Legaspi, capital of Albay province in the south-east, and Naga, capital of Camarines Sur province in the north-east. Larap, situated in Camarines Norte province, has the Philippines' most important iron ore works. Tiwi, 45 km (28 miles) north of Legaspi, has thermal springs and a geothermal power plant. The southern part of the peninsula is drained by the 120 km (75 mile) long Bicol River.

Map Philippines – Bc

Bidar (Vidarba) *India* Fortified town about 500 km (310 miles) east and slightly south of Bombay. It was the capital of the Bidar kingdom in the 16th century. Bidar is noted for its traditional craft of inlaying iron with gold and silver to make decorative ware.

Population 1 251 060
Map India – Bd

Bié *Angola* See KUITO

Bielawa *Poland* See DZIERŻONIÓW

Bielefeld *Germany* Industrial city 90 km (56 miles) south-west of Hanover. It produces machinery and glass.

Population 319 000
Map Germany – Cb

Bhutan

*THE DRAGON KING RULES A REMOTE
MOUNTAIN LAND WHICH FEW
OUTSIDERS ARE ALLOWED TO SEE*

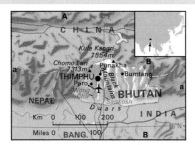

By world standards, the people of Bhutan
are among the poorest on earth. But
there is no starvation in their remote
Himalayan country; there is no unemployment, no begging and virtually no crime.
They are wary of wealth, because of the
damage it would cause to their cultural traditions. Although there is money in Bhutan –
mainly silver coins, with banknotes only
recently introduced – most trade is conducted
by barter.

Bhutan, known as the Land of the Dragon,
rises from the jungle-covered foothills which
overlook the BRAHMAPUTRA River to the white
southern slopes of the HIMALAYAS. Heights

range from 2000 to 7554 m (6500-24 783 ft).
The country has existed since the 10th century,
but apart from conflicts in defence of its territory, its contacts with the outside world have
been few and cautious. Bhutan has, in many
ways, stayed locked in the Middle Ages – a
real-life version of the fictional Shangri-la.

In a nation of 700 000 people, there are 1300
fortress monasteries, or *dzongs*, which dominate the countryside as castles once did in
Europe. Most of the people are Buddhist –
although pagan beliefs also influence their religion. The brightest sons of families are sent,
sometimes when they are only three years old,
to live in the *dzongs* and become monks. At
colourful religious festivals, masked dancers
enact legends in which Buddhist heroes van-
quish the forces of evil.

Monks ruled the country until 1907, when
they were replaced by an absolute monarch, the
Dragon King. The present ruler, Jigme Singye
Wangchuk, follows the regal customs established long ago, dispensing justice and even
marital advice to all supplicants. The country
has been compared to a Buddhist paradise.

A strictly limited 3000 tourists are allowed
in each year, and mountaineers are given access
to a new Himalayan climb every two years, as
long as they carry their own equipment and do
not use the Bhutanese as porters.

The few tourists and the sale of postage
stamps supply most of the foreign money that
comes in (an amount so small that in 1983 the
purchase of a second 17-seater aircraft for the
new national airline exhausted the country's
meagre foreign exchange reserves).

About 95 per cent of the Bhutanese workforce are farmers. Yaks on the high pastures
provide milk, cheese, meat and hair (used to
make ropes from which many local bridges are

constructed). Lower down, potatoes, wheat and
barley are grown and, lower still, rice is produced. But only 9 per cent of the country is
cultivated, and 70 per cent remains forested.
The demand for new farmland is low, partly
because many sons become monks. Little
forest has been cleared, and the Bhutanese hills
have not been exposed to soil erosion as elsewhere on the southern Himalayan slopes.

BHUTAN AT A GLANCE	
Official name Kingdom of Bhutan	
Area 47 000 km² (18 000 sq miles)	
Population 700 000	
Per km² 15 (**Per sq mile** 38.9)	
Capital Thimphu	
Government Constitutional monarchy	
Currency 1 ngultrum = 100 chetrum	
Languages Dzongkha (official); various Tibetan and Nepalese dialects	
Religions Buddhist (75%), Hindu	
Climate Climate moderated by high altitudes. High rainfalls. Average temperature in Thimphu ranges from 17°C (63°F) in July to 4°C (39°F) in January	
Land use Cultivation 2%, meadows and pastures 5%, forest and woodlands 70%, other 23%	
Main primary products Rice, millet, wheat, barley, maize, potatoes, oranges, apples, cardamom, timber, yaks; dolomite, coal, gypsum, calcium carbide	
Major industries Agriculture, forestry, handicrafts, mining	
Main exports Timber, cardamom, rice, barley, wool, and other agricultural products; postage stamps; gypsum	
Annual income per head (US$) 180	
Population growth (per thous/yr) 23	
Life expectancy (yrs) Male 51 **Female** 50	

Bielsko-Bialá (Bielitz) *Poland* Industrial city in
southern Poland, at the foot of the Carpathian
Mountains, 42 km (26 miles) south of the city of
Katowice. The twin towns of Bielsko and Bialá,
formally united since 1950, are separated by the
Bialá River.

The city, which specialises in making high-
quality woollen fabrics, also produces textile
machinery, electrical goods and engines for
Polski Fiat motorcars – with technical help from
the Italian auto giant, Fiat.
Population 180 000
Map Poland – Cd

Big Bend National Park *USA* Area covering
2866 km² (1106 sq miles) of Texas in a large bend
of the Rio Grande River on the Mexican border.
Desert, canyons and mountains predominate here.
Map United States – Fe

bight Indentation in the sea coast, similar to a bay
but with a wider, gentler curve.

Bihac *Bosnia-Herzegovina* See GRMEC PLANINA

Bihar *India* Densely populated north-eastern
state covering 173 877 km² (67 117 sq miles).
Although it is rich in minerals and accounts for
nearly 40 per cent of India's total mineral production, it is the country's poorest state and most
of its inhabitants are illiterate.

Some 80 per cent of the people are employed
on the land, growing rice, tobacco, sugar cane and
jute. Most of the population lives in crowded rural
settlements on the Ganges Plain. The main towns
are Bihar Sharif with its holy graves and mosques
– a place of pilgrimage for Muslims – PATNA, the
capital, and Gaya.
Population 86 338 900
Map India – Dc

Bijapur (Vijayapura) *India* Muslim city 386 km
(240 miles) south-east of Bombay. It has many
fine mosques, forts and mausoleums, one of
which, the Gol Gumbaz, is noted for its domes.
The central dome, at 38 m (125 ft) diameter, is
larger than St Paul's in London (33 m, 108 ft), but
smaller than St Peter's in Rome (42 m, 138 ft).
Population 2 914 670
Map India – Bd

Bikini *Marshall Islands* Atoll 750 km
(nearly 470 miles) north-west of Majuro where
the United States tested 23 nuclear weapons
(including the first hydrogen bomb) between
1946 and 1963. The people were resettled on
another island, but sued the United States government for compensation. In 1983 they were
promised more than US$150 million to be paid
over 15 years. The restoration of the atoll to its
former state could cost US$42 million.
Map Pacific Ocean – Db

Bilbao *Spain* North-eastern port and industrial
city 11 km (7 miles) south of the Bay of Biscay,
to which it is linked by the canalised River
Nervión. It exports iron ore, lead and wine and
is Spain's leading banking centre and stock
exchange. Its industries include iron and steel
production, engineering, shipbuilding and chemicals. Bilbao is the capital of the Basque province
of Vizcaya.
Population 369 840
Map Spain – Da

bill Long, narrow promontory or small peninsula.

billabong Australian term for a stagnant pool
or backwater extending from the main stream
of a river.

Billund *Denmark* See VEJLE

Binga Mountain *Mozambique* See CHIMOIO

Bingerville *Ivory Coast* Colonial-style town
overlooking the Ebrié Lagoon, 16 km (10 miles)
from Abidjan. It was the capital of Ivory Coast
from 1900 to 1934. Named after the first French
governor-general (from 1893), the naval officer
and explorer Louis-Gustave Binger, it has botanic
gardens and an agricultural research centre.
Population 20 000
Map Ivory Coast – Bb

Bintan *Indonesia* Island of the Riau Archipelago south of Singapore. It covers 1075 km² (415 sq miles) and has rich deposits of bauxite.
Map Indonesia – Bb

Bintulu *Malaysia* Coastal town in Sarawak, 350 km (217 miles) north-east of Kuching. Its petrochemical industry processes natural gas from the offshore Luconia fields for export, and also produces artificial fertiliser.
Population 59 000
Map Malaysia – Db

Bío-Bío *Chile* With a length of 380 km (235 miles), the longest river in the country. It rises in the ANDES of central Chile, near the Argentine border, and flows north-west to enter the Pacific Ocean near CONCEPCIÓN. It forms a natural divide between central and southern Chile. Its upper tributaries are harnessed to provide hydroelectric power.
Bío-Bío is also the name of a central region of Chile that is heavily industrialised around Arauco and Concepción, with coal mines at Lota, Lebu and Coronel and a steel plant at Huachipato.
In the foothills of the Andes, near Los Angeles, is an area of outstanding natural beauty. This is typified by the lake of Laja, its lovely river and falls, and the towering volcanoes of Antuco, Tolhuaca, and Copahué, which rises to 2969 m (9741 ft).
Population (region) 1 729 920
Map Chile – Ac

Bioko *Equatorial Guinea* Tropical volcanic island of 2017 km² (780 sq miles) in the Gulf of Guinea. Formerly called Fernando Póo, it was occupied by Portugal, Spain and Britain before independence in 1968. The national capital, MALABO, stands on the island, which produces cocoa, bananas, coffee, palm products, sugar and timber.
Population 57 000
Map Equatorial Guinea – Aa

Biorra *Ireland* See BIRR

biotite Dark green to black mica found in igneous and metamorphic rock.

Biratnagar (Morang) *Nepal* Industrial city in the far south-east of Nepal, near the border with BIHAR in India, about 120 km (75 miles) south-west of DARJILING. Like other border towns, Biratnagar trades heavily – and mostly illegally – with India, untaxed goods disappearing along forest and country trails.
A hot and humid place, the city is not on the tourist itinerary. It has jute and chemical factories and a sugar mill.
Population 90 000
Map Nepal – Ba

Birim Valley *Ghana* The country's largest diamond field, yielding industrial-grade stones. It lies in the Birim River valley, 80 km (50 miles) west of the southern part of Lake Volta. The field, discovered in 1919, yields 90 per cent of Ghana's total production.
Map Ghana – Ab

Birkenau *Poland* See AUSCHWITZ

Birkenhead *United Kingdom* See WIRRAL

Birmingham *United Kingdom* Second largest city of the UK after London, lying in west-central England. It is often called 'the heart of England', and is the main city of the West Midlands conurbation that extends north-west to Wolverhampton. Before it was overwhelmed by industrialisation in the 19th century, Birmingham was little more than a market town. With coal fields to the east and west, and canals linking mines to factories and factories to seaports, the area became known as 'the Black Country'.
One of Britain's earliest railways linked Birmingham with London, and the Scottish engineer James Watt produced the world's first rotary steam engines – the power source of the Industrial Revolution – in Birmingham in the 1780s.
The city still has a wide range of industries, including jewellery, guns, brassware, buttons, and motor vehicles. Chocolate manufacturing created a whole suburb, Bournville in the 1880s.
Birmingham's industrial importance led to the building of grandiose public offices in the 19th century, including the town hall, museum and art gallery. There is an 18th-century Anglican cathedral; its Roman Catholic counterpart dates from 1841. Some of Britain's earliest town planning and urban legislation originated in Birmingham.
Despite the decline of many of its older industries, the city remains one of Britain's leading commercial and industrial centres. It is home to the National Exhibition Centre (next to its airport), has an international convention centre which was opened in 1991 and three universities.
Population 1 006 500
Map United Kingdom – Ed

Birmingham *USA* Largest city in Alabama. It lies in the centre of the state in a valley of the Appalachian foothills. It was founded in 1871 and grew rapidly when local iron ore and coal were used on a large scale as a result of the expansion of the railroads. Steel is still produced here and has been joined by several other industries, including textiles and chemicals. The city was the scene of black civil rights protests in the 1960s.
Population (city) 265 970; (metropolitan area) 840 140
Map United States – Id

Birr (Biorra) *Ireland* Georgian market town, 113 km (70 miles) west and slightly south of Dublin. It was formerly known as Parsonstown, after the Parsons family who became Earls of Rosse, several of whom were noted astronomers. The 17th-century castle on the western edge of the town contains the tube of a telescope which, when built in 1845, was the world's largest. The castle has 200-year-old box hedges which, at 9 m (30 ft), are said to be the highest in the world.
Population 3280
Map Ireland – Cb

Biscay, Bay of Inlet of the ATLANTIC called the Gulf of Gascony by the French. It is noted for its high tides, sudden, severe storms caused by westerly and south-westerly gales, and strong currents. Lord Byron wrote of the 'rude' winds in 'Biscay's sleepless bay'. The bay stretches between the Isle d'Ouessant in BRITTANY in north-western France, and Cape Ortegal in north-western Spain. Its greatest width is 530 km (342 miles), its greatest depth 366 m (1200 ft).
Map Europe – Cd

Bishkek (Frunze) *Kyrgyzstan* Capital and industrial city between Lake Balkhash and the Chinese border. It stands at the foot of the snow-capped Kirghiz range, which rises more than 4400 m (14 450 ft) high. Originally called Pishpek, it was renamed Frunze from 1926 to 1991 in honour of its most distinguished native son, General Mikhail Frunze (1885-1925), one of the founders of the Red Army. Bishkek has tree-lined boulevards and spacious parks. Its industries include food processing and the manufacture of textiles, agricultural machinery and metal and electrical goods.
Population 616 000
Map Kyrgyzstan – Ba

Bisho *South Africa* Modern city built as capital of the former Republic of Ciskei (1981-94), which was recognised only by South Africa. Bisho lies 50 km (30 miles) west of East London in the Eastern Cape and adjoins King William's Town (founded 1835). In September 1992, Bisho made world headlines when Ciskei soldiers shot dead 29 people and wounded 200 during a protest march against the territory's military ruler, Brigadier Oupa Gqozo. The incident led to nationwide violence.
Population 6350
Map South Africa – Cc

Biskra *Algeria* Oasis town and communications and administrative centre, 190 km (118 miles) south-west of Constantine, just south of the Aurès Mountains. Biskra is situated at the start of the main eastern Algerian route to the far south and marks the start of the trans-Saharan route which cuts through the Agaggar Massif to the Niger River. Since Roman times, when it was the military base of Vescera, Biskra has been a fashionable resort because of its pleasant climate and hot springs. The Hamma Salahine spa at the entrance to the town is famous for its sodium sulphate waters, the hottest in the world after those of Iceland. The remains of the Roman town as well as a number of ancient dolmens (prehistoric stone tombs) are nearby. Huge palm groves, consisting of about 150 000 palm trees along the course of the oasis, are a tourist attraction.
Population 90 500
Map Algeria – Ba

Biskupin *Poland* See POZNAŃ

Bismarck *USA* State capital of North Dakota, standing on the Missouri River 273 km (170 miles) south of the Canadian border. It is an agricultural processing and servicing centre and has grown into a business and finance centre as well.
Population (city) 49 260; (metropolitan area) 83 830
Map United States – Fa

Bismarck Archipelago *Papua New Guinea* Group of mainly volcanic islands off the north-east coast of New Guinea. It includes NEW BRITAIN, NEW IRELAND and the Admiralty Islands, whose main island is Manus. The economy relies largely on fishing, but extensive coconut and cocoa plantations are also an important source of revenue. The group of islands was named after German statesman Otto von Bismarck and was a German colony from 1884 until the end of the First World War, when it was handed over to Australia as a Trust Territory. The inhabitants are mainly Melanesian, but there is a substantial Chinese and European population as well. The main town is RABUAL in East New Britain.

Population 402 650
Map Papua New Guinea – Ba

Bismarck Range *Papua New Guinea* Mountain range in the north-east of New Guinea Island. Its highest point, and the highest point in the country, is Mount Wilhelm at 4509 m (14 793 ft).
Map Papua New Guinea – Ba

Bismarck Sea Arm of the PACIFIC which was the scene of a victory by United States aircraft over a Japanese naval force in 1943. Bounded by the BISMARCK ARCHIPELAGO and New Guinea, it is 800 km (nearly 500 miles) wide.
Map Papua New Guinea – Ba

Bisotun (Behistun) *Iran* Sugar-refining town, 400 km (about 250 miles) south-west of the capital, Tehran. Outside the town, carved into a cliff face at a height of 60 m (200 ft), is a bas-relief more than 2000 years old depicting the triumph of the Persian king Darius I (ruled 522-486 BC) over his enemies. The inscription, repeated in three languages (Persian, Akkadian and Elamite), enabled 19th-century scholars to decipher Babylonian cuneiform script. Two Parthian bas-reliefs, dating from the 2nd century BC and the 1st century AD, are at the foot of the cliff.
Map Iran – Aa

Bissau *Guinea-Bissau* The country's capital, main commercial city and port, on the right bank of the Gêba River estuary. Founded by the Portuguese as a trading post and fort in 1692, it became the administrative capital of Portuguese Guinea in 1941. With good port facilities, it grew rapidly after 1945. Although MADINA DO BOE was named as the new national capital in 1980, Bissau continued its economic dominance; most of the country's few manufacturing industries – the processing of farm products and timber – are run from here. The city also handles 85 per cent of Guinea-Bissau's overseas trade.

An attractive town, Bissau has retained some of its colonial splendour, with wide, tree-lined streets and a seaside promenade. In spite of its poverty, peeling paintwork and crumbling masonry, it has a charming atmosphere and several points of interest, including Fort Sao José (built in 1693), a museum (with a good collection of early African carvings and artefacts) and a massive Roman Catholic cathedral. There are magnificent beaches nearby.
Population 900 000
Map Guinea – Aa

Bitola (Bitolj; Monastir) *Macedonia* The country's southernmost city situated 110 km (68 miles) south of Skopje and 15 km (9 miles) from the Greek border, on the edge of the fertile Pelagonia Valley. A major Turkish headquarters (Monastir) dating from 1820, it later became Macedonia's largest town. Kemal Ataturk (1881-1938), the founder of modern Turkey, was trained at its military academy. But its progress was blighted by the First World War – and later by the Balkan and Greek civil wars.

Its recovery was built on improved farming methods and drainage systems; today, the Pelagonia Valley produces wheat, maize, sugar beet, cotton, rice, tobacco and poppies for Bitola's processing plants. Other factories in the city manufacture carpets, clothing, leather, beer, metal goods and household appliances.

Bitola also serves as the headquarters for the tourist centres of Lake Prespa, Lake Ohrid and the national park around Mount Perister, which rises to 2600 m (8530 ft) west of the city. The park has magnificent beech and black pine forests, rivers stocked with trout and animals such as bears, boar and deer. The ruins of Heraclea (Lyncestis), a Roman town with outstanding mosaics, lie just south of the city.
Population 122 200
Map Macedonia – Ab

Bitter Lakes *Egypt* Two lakes – the Great and Little Bitter lakes – lying between Isma'iliya and Suez and forming part of the Suez Canal. Before the canal was cut between 1859-69, the lakes were marshy depressions 8-12 m (26-39 ft) below sea level.
Map Egypt – Cb

bitumen Group of solid or liquid hydrocarbons which occur naturally as asphalt (pitch), tar and crude oil, as in Trinidad's PITCH LAKE.

Biu Plateau *Nigeria* Granite plateau dotted with extinct volcanoes and rising to about 700 m (2300 ft) in the north-east of the country. Subsistence farmers in the area grow millet and vegetables and rear cattle.
Map Nigeria – Ca

Biwa *Japan* The country's largest lake. Situated on HONSHU Island, just north-east of KYOTO, it covers 675 km² (260 sq miles). The reflection of Hikone Castle in Biwa's shimmering waters is a famous attraction. The shores of the lake are dotted with temples, castles, historical sites and villages.
Map Japan – Cc

Bizerte (Bizert; Binzert) *Tunisia* Port and heavily fortified naval base situated on Tunisia's north coast. Lying on a channel leading from Lake Bizerta, it is the site of one of the earliest Phoenician settlements in 800-700 BC. It then became a Roman colony (Hippo Zaritus), was conquered by the Arabs in the 7th century, by Spain in 1535 and then, in 1881, by the French, who built a major naval base here. During the Second World War, Bizerte was occupied by the Germans until they were driven out by the Americans in 1943.

Said to be one of the safest harbours on the North African coast, Bizerte handles iron ore exports. Its chief industry is oil refining; cement and steel are also produced and exported. Among places of special interest are the old port, a huge 16th-century fort and the 17th-century Great Mosque.

The beaches attract many holidaymakers.
Population 400 000
Map Tunisia – Ad

Bjeshkët e Némuna *Albania/Yugoslavia* See ALBANIAN ALPS

Björneborg *Finland* See PORI

Black Country *United Kingdom* English Midlands industrial area centred roughly on the city of BIRMINGHAM. The area was named Black Country because of its concentration of coal mines and factories. It stretches from about Castle Bromwich in the east to WOLVERHAMPTON in the north-west, and encompasses most of the West Midlands. Today, clean air legislation and a decline of mining and traditional small metal-working industries have lightened the landscape and decreased the legacy of dereliction.
Map United Kingdom – Ed

Black Forest (Schwarzwald) *Germany* Pine-clad, low mountainous region in the south-west corner of the country, extending between the Swiss border and the city of Karlsruhe, 160 km (100 miles) to the north. Its highest point is Feldberg, at 1493 m (4898 ft).

▼ **LIGHT AND SHADE** Sunny pastures lie along the Wiese Valley in the Black Forest – so named after the dark conifers that clothe its heights.

The Roman name for the forest was Silva Nigra. Crisscrossed by paths, it is a hive of spa towns, camp sites, youth hostels and holiday homes. Scattered farming is carried out in the valleys – around distinctive Black Forest buildings, with eaves stretching almost to the ground. The main towns on its fringes are Freiburg im Breisgau, PFORZHEIM and BADEN-BADEN.
Map Germany – Be

Black Hills *USA* Range of hills on the border between the states of South Dakota and Wyoming, rising to 2207 m (7241 ft) at Harney Peak. The hills were sacred to the native Americans, but were overrun by white settlers in the 1870s when gold was discovered here.
Map United States – Fb

black ice (glaze ice) A thin, glassy coating of ice on the ground or on exposed objects, caused by freezing weather after a partial thaw, or by the freezing of rain or fog droplets on impact with a cold or freezing surface: also called 'glazed frost', 'silver frost' and 'silver thaw'.

Black Mountains *Bhutan* Mountain range, rising to almost 5000 m (16 400 ft) and forming a spur running south from the main Himalayas, effectively cutting Bhutan in two. The eastern half is largely inhabited by Indians, mostly from Assam; Tibetans form the majority in the western half. A road crosses the Pere-La Pass between Wangdu-Phodrang and Tongsa. At the northern extremity of the spur – and within the main Himalayan range – is Kula Kangri, the highest peak in Bhutan (7554 m, 24 783 ft).
Map Bhutan – Ba

Black Mountains *United Kingdom* Range on the border of England and southern Wales, to the north-east of the Usk River valley. It is good pony-trekking country, rising to 811 m (2660 ft). The Black Mountain is an upland of similar height, 40 km (25 miles) farther west, beyond the BRECON BEACONS.
Map United Kingdom – De

Black River (Song Da) *China/Vietnam* River rising in the Chinese province of Yunnan and flowing generally south-east for some 800 km (nearly 500 miles) to join the Red River, 50 km (30 miles) west of Hanoi.
Map Vietnam – Aa

Black Sea Lying between Europe and Asia, the tideless Black Sea, a landlocked arm of the MEDITERRANEAN, is the base for a Russian fleet. Because large rivers such as the Danube and Dnieper flow into it, the water in the top 90 m (295 ft) has a low salt content. Underlying this layer is salty, heavier water brought in by an underwater current from the Mediterranean. There is little mixing of the two layers. Whereas the salty water lacks oxygen and contains little life, the top layer is inhabited by many kinds of fish. The Romans called the sea Pontus Euxinus ('the Friendly Sea'), but severe storms and fogs can make it dangerous in winter. The sea covers 461 000 km² (175 960 sq miles), and its greatest known depth is 2245 m (7365 ft).
Map Europe – Gd

Black Volta *Burkina* The main headstream of the VOLTA River in western Burkina It flows north-

east, then east and south, forming part of the border between Burkina and Ghana, then between Ivory Coast and Ghana, before it swings eastwards to Lake Volta, some 645 km (400 miles) from its source. One of Burkina's most successful self-help projects is located at the Sorou (or Bagué), its northern tributary: here, a barrage built across the river by local labour stores water for irrigation.
Map Burkina – Aa

Blackburn *United Kingdom* Town in the county of Lancashire, 32 km (20 miles) north-west of Manchester. It dates back to Saxon times, and had a church as long ago as 596. In the 18th and 19th centuries, it prospered as a cotton town, but many of the mills have now closed. A museum of textile machinery catalogues the role played by the textile industry in the development of the town. The Blackburn of the 1990s is an important centre for the brewing industry; other industries include electronics, precision engineering and carpet-making. There is a large covered market in the centre of the town.
Population 110 250
Map United Kingdom – Dd

Blackpool *United Kingdom* Largest holiday resort in Britain. It lies on the coast of Lancashire, 65 km (40 miles) north-west of Manchester. Its beaches are backed by the greatest concentration of entertainment facilities in Britain, including a 157 m (515 ft) high tower, a half-sized copy of the Eiffel Tower, a promenade 11 km (7 miles) long, and a roller coaster.
Population 149 800
Map United Kingdom – Dd

Blagoevgrad *Bulgaria* City on the Struma River 70 km (44 miles) south of Sofia. Industries include ceramics, textiles, wood and tobacco. The centre of the tobacco-growing Struma Valley, it provides easy access to the Rila Mountains. Nearby, too, are the small town of Melnik – once

a feudal fortress and now an artists' retreat – and the ruins of the Roman spa of Sandanski, said to be the birthplace of Spartacus, the rebel gladiator. Formerly known as Gorna Dzhumaya, the city was renamed in 1950 after Dimitar Blagoev (1856-1924), founder of the Bulgarian Marxist movement.
Population 68 000
Map Bulgaria – Ab

Blagoveshchensk *Russia* City and capital of the far eastern region of Amur, on the Amur River near the Chinese border.
Its industries include food processing, flour milling, saw-milling, engineering, and the manufacture of footwear and furniture.
Population 208 000
Map Russia – Nc

Blanc, Cape *Mauritania/Western Sahara* Narrow peninsula, about 50 km (30 miles) long. The frontier between Mauritania and Western Sahara runs down the centre.
Map Mauritania – Ba

Blanc, Mont (Monte Bianco) *France/Italy* Highest mountain in Western Europe, at 4807 m (15 771 ft). It stands on the border between south-west France and Italy at the head of the Valle d'Aosta, 105 km (65 miles) north-west of the city of Turin.
Its lower slopes are a popular playground for skiers, with resorts such as COURMAYEUR and Entrèves in Italy, and CHAMONIX in France. A road tunnel runs 12 km (7.5 miles) through the mountain between Courmayeur and Chamonix.
Map Italy – Ab

▼ **PERPETUAL SNOW AND ICE The Mont Blanc massif has the highest peaks in the Alps. Its Bossons Glacier (right) slowly descends towards the French resort of Chamonix.**

Blantyre *Malawi* Capital of Southern Province and the country's largest city, standing in the Shire Highlands 1100 m (3600 ft) above sea level. When Scottish missionaries arrived here to spread the Gospel in 1876, they found a deserted village. But they stayed and started a mission station which they named after the Scottish birthplace of the explorer and missionary David Livingstone (1813-73).

In 1956, Limbe, 8 km (5 miles) away on the Beira-Lake Malawi rail route, became part of Blantyre, which now serves a major farming region and has food processing, textiles, and cement manufacturing industries.

Population 500 000
Map Malawi – Ac

Blarney (An Bhlarna) *Ireland* Town 8 km (5 miles) north-west of the city of Cork. Visitors to Blarney Castle who kiss the Blarney Stone, built into battlements on the castle's 15th-century keep, are said to acquire eloquence. The legend arose from a 16th-century Lord of Blarney, Cormac McCarthy, who used his verbosity to frustrate English demands for land reforms. Legend has it that Queen Elizabeth I remarked about his loquacity 'It's all Blarney; what he says, he never means.'

Though the eloquence may not be guaranteed, the views here are stunning.

Population 2040
Map Ireland – Bc

Blato *Croatia* See KORČULA

Bled, Lake (Blejsko Jezero) *Slovenia* Glacial lake in the Julian Alps in Slovenia, 48 km (30 miles) north-west of Ljubljana. The skiing centre of Bled lies at 475 m (1558 ft) above sea level on the north shore. It has a crimson-spired church on an islet and a Baroque castle clinging to a lakeside crag. The lake, popular with yachting enthusiasts, is a regular venue for regattas.

Map Slovenia – Ba

Bledow Desert *Poland* See KATOWICE

Blenheim (Blindheim) *Germany* Village on the Danube River, 90 km (55 miles) north-west of Munich. Here, in 1704, a British army led by John Churchill, first Duke of Marlborough, routed French and Bavarian troops during the War of the Spanish Succession.

Population 1600
Map Germany – Dd

Blida (El Boulaida) *Algeria* Administrative and trading centre on the Algiers-Oran rail route. It lies 48 km (30 miles) south-west of Algiers on the Mitidja Plain, in the middle of a citrus-growing area. It also has some light industry and manufactures flour, olive oil and soap.

Population 195 000
Map Algeria – Ba

Blindheim *Germany* See BLENHEIM

blizzard Violent windstorm accompanied by intense cold and driving, powdery snow or ice crystals.

block faulting Fracturing of part of the earth's crust into separate blocks by faulting.

block lava See AA

block mountain Also called a HORST, it is a mountain mass standing up prominently either because it has been elevated by earth movements between FAULTS or because the surrounding area

▲ **DREAMY BLUES Australia's Blue Mountains take their name from the blue haze of oil droplets given off by their eucalyptus trees. It clothes the landscape and can be seen for miles.**

has sunk between faults. The Harz Mountains and Black Forest in Germany, and the Vosges Mountains in France are examples of block mountains.

blocking high An ANTICYCLONE that remains relatively stationary and so blocks the passage of approaching DEPRESSIONS, or CYCLONES.

Bloemfontein *South Africa* Capital of Free State Province and judicial capital of South Africa. Founded in 1846, it is known to many Africans as Mangaung (Place of the Cheetah). Historic buildings include the council chambers of the old Orange Free State Republic, which lasted from 1854 to 1902 and was incorporated into the former Union of South Africa (1910-61).

Today, Bloemfontein is the home of the University of the Orange Free State. It is a commercial and light industrial centre, and a headquarters for agricultural administration.

Population (city) 126 870; (district) 300 150
Map South Africa – Cb

Blois *France* Capital of Loir-et-Cher department, on the Loire River, 55 km (34 miles) south-west of Orléans. It manufactures aircraft, precision instruments and footwear. From the 15th to the 17th century, Blois was a favourite haunt of French kings, who stayed in the town's great 13th to 15th-century château – one of the finest in the Loire Valley.

Population 61 100
Map France – Dc

Blood River *South Africa* Name given to the Ncome River, a short tributary of the Buffalo River in KwaZulu-Natal, after about 3000 Zulus died on its banks in an assault on a Boer settler ('voortrekker') wagon laager on 16 December 1838. On the Boer side, only three men were injured. The day was commemorated as a South African public holiday known first as Dingane's Day (after the Zulu chieftain, Dingane) and, later, as the Day of the Covenant to honour, as the Afrikaners put it, a 'compact' between the Boers and God. (In 1994 this public holiday was renamed the Day of Reconciliation.) A full-scale bronze replica of the wagon circle marks the site.
Map South Africa – Db

blowhole Almost vertical vent that reaches from the roof of a coastal cave to the cliff top. Incoming tides compress the air in the cave and blow it out through the vent. Water may also be blown out.

Blue Mountains *Australia* Part of the GREAT DIVIDING RANGE which stretches across the eastern side of the Australian continent from the Cape York Peninsula in the north to Melbourne in the south. The Blue Mountains are situated in New South Wales, 65 km (40 miles) inland from Sydney. Rising to 1360 m (4462 ft), the region is a plateau intersected by sheer-sided deep valleys. It takes its name at least partly from the blue haze given off by the eucalyptus trees, which can be seen from afar. The rugged terrain, numerous waterfalls and striking views make the mountains a popular tourist and residential area.

The range was first crossed by European colonists in 1813; until then, it represented a barrier to westward expansion of the colony. The Blue Mountains National Park, covering 2457 km² (948 sq miles), was established in 1959.
Map Australia – Ie

Blue Mountains *Jamaica* Thickly wooded mountain range, about 50 km (30 miles) long, in the eastern end of the country. Several peaks rise from a 1500 m (4900 ft) ridge, the highest being Blue Mountain Peak (2256 m, 7402 ft). High quality coffee is grown on the lower slopes.
Map Jamaica – Ba

blue mud Ocean-floor mud found on the continental slopes. It is washed down from land, and gets its dark slate-blue colour from tiny particles of iron sulphide.

Bluefields *Nicaragua* Caribbean port 320 km (199 miles) east of Managua. It exports tropical hardwoods such as mahogany, rosewood and black walnut. Named after the Dutch pirate, Abraham Blaauweld, the town grew up around a number of 18th-century British settlements. Its mainly Creole inhabitants, descendants from African and European settlers from Jamaica, speak English. In October 1988, Hurricane Joan destroyed nearly all of Bluefields, but a reconstruction programme began soon afterwards.
Population 20 000
Map Nicaragua – Ba

bluff Steep headland, promontory, river bank or cliff, often formed by a river cutting into the valley side on the outside of a MEANDER.

Bluff *New Zealand* Industrial town on the southern tip of South Island. It has the country's only aluminium smelter, which uses bauxite imported from Australia and is run by hydroelectricity generated in the Fiordland Mountains.
Population 2390
Map New Zealand – Bg

Bo *Sierra Leone* Southern regional capital, about 180 km (nearly 110 miles) south-east of the national capital, Freetown. Lying at the edge of the Sewa River swamps, it is a market town for local crops, including ginger and oil-palm kernels. It produces soap and palm oil.
Population 150 000
Map Sierra Leone – Ab

Bo Hai (Po Hai) Arm of the YELLOW SEA, 480 km (nearly 300 miles) long and 290 km (180 miles) wide, off the east coast of China, formerly called the Gulf of Chihli. It contains oil fields.
Map China – Hc

Bobo Dioulasso *Burkina* The country's second largest town after the capital, Ouagadougou, and former capital from 1934 to 1954, when the country was called Upper Volta. It lies some 310 km (193 miles) south-west of the present-day capital, and is a trade and industrial centre with groundnut-crushing mills, soap works, cotton ginneries and textile factories. Bobo Dioulasso was the terminus of the railway from Abidjan in Ivory Coast. It has a mosque, botanical gardens, lively markets and many restaurants and bars.
Population 228 700
Map Burkina – Aa

bocage Type of farmland divided into small fields by hedges and trees, or dry-stone walls. The term is mostly used to refer to such farmland in north-west France.

Bochum *Germany* Industrial city of the Ruhr, 20 km (12 miles) west of Dortmund. It produces coal, steel, motorcars and household goods.
Population 396 500
Map Germany – Bc

Bodensee *Germany* See CONSTANCE

Bodh Gaya *India* One of the four most important shrines of Buddhism – though the place is not visited to the same extent as Muslim Mecca or Hindu Varanasi. Here, in about 500 BC, the Lord Buddha achieved enlightenment while sitting in contemplation under the Bodhi tree.

The various world sects of Buddhism maintain temples here, so that it is possible to hear and see the chanting and services of Tibetan, Chinese, Japanese, Thai and Sri Lankan monks in close proximity. The town is small and there are no industries to disturb the peace of the temples.
Population 15 700
Map India – Dc

Bodø *Norway* Seaport, administrative centre and 'midnight sun' tourist resort on the Saltfjord, 80 km (50 miles) north of the Arctic Circle, and 180 km (about 110 miles) south-west of the port of Narvik. The city, which was badly damaged during the Second World War, is the capital of Nordland county. The port's main cargoes are copper ore and marble, exported from mines at Sulitjelma and quarries at Fauskeidet, Salten.
Population 37 900
Map Norway – Db

Bodrum *Turkey* Market town about 580 km (360 miles) south-west of the capital, Ankara. Formerly known as Halicarnassus, it was the site of the mausoleum of Halicarnassus, built in the 4th century BC and one of the Seven Wonders of the Ancient World. The tomb was commissioned by the widow of Mausolus – a local king whose name is the origin of 'mausoleum', the word for a grandiose tomb; it was destroyed by an earthquake some time before the 15th century. Stones from the ruins were subsequently used to build the fort of St Peter on an island in Bodrum Bay. The Greek historian Herodotus (484-425 BC) was born here.
Population 13 100
Map Turkey – Ab

bog Area of permanently waterlogged ground with a surface layer of decaying vegetation, particularly sphagnum moss, which forms highly acid peat.

Bog of Allen (Móin Aluine) *Ireland* A large peat bog covering about 950 km² (370 sq miles) in the eastern counties of Kildare, Offaly, Laois and Westmeath. Its deep, easily worked deposits are used in peat power stations including one at Allenwood, west of Dublin.
Map Ireland – Cb

Boğazkale (Boğazköy) *Turkey* Anatolian mountain village 145 km (90 miles) east of Ankara. It is on the site of the ancient city of Hattusas, capital of the Hittite empire from 1400 to 1200 BC. Among the Hittite remains are huge double-walled fortifications and an underground tunnel, Yen Kapi, which reveals signs of an earlier Iron Age period of occupation. Many of the objects found at the site are in a museum in the 'Village of the Pass'.
Map Turkey – Ba

Bogor (Buitenzorg) *Indonesia* City of West Java, 60 km (37 miles) south of the national capital, Jakarta. It was a Dutch colonial hill station and has a botanical garden with more than 2700 species of tropical plants which has become famous for research into cash crops of the region.
Population 250 000
Map Indonesia – Cd

Bogotá *Colombia* The country's capital and largest city, and capital of Cundinamarca department on a fertile plateau 2800 m (9200 ft) up in the eastern Andes. In 1538, the Spanish conquistador Gonzalo Jiménez de Quesada, in search of El Dorado, the legendary city of gold, marched onto the plateau, which was the home of the Chibcha people. After conquering the Amerindians, he founded the city, calling it Sante Fé de Bogotá – Sante Fé after his birthplace in Spain and Bogotá from Bacatá, which was the Amerindian name for the region.

From a population of 100 000 in 1905, Bogotá has grown into an expansive modern city whose industries include textiles, motor vehicles, engineering and chemicals. It also has a reputation for lawlessness.

The city centre is a mixture of Spanish colonial architecture and soaring skyscrapers laid out in a grid pattern. The house of the Venezuelan-born Simón Bolívar (1783-1830), who liberated Colombia from Spanish rule, stands just outside the city centre. The gold museum contains more than 20 000 gold artefacts created by the Chibcha. The museum also contains one of the world's largest emeralds, mined in Boyacá to the north.

The spectacular Tequendama Falls are 30 km (19 miles) south-west of Bogotá but unfortunately have become polluted with sewage. Here the River Bogotá, a tributary of the Magdalena, rushes through a gorge 18 m (60 ft) wide and plunges 156 m (515 ft) off the edge of the plateau.
Population 4 921 000
Map Colombia – Bb

Bogra *Bangladesh* Chief town of Bogra district, which covers 3888 km² (1501 sq miles) in the centre of the country. Parts of the district alongside the Brahmaputra River are prone to both flooding and erosion; whole villages have lost their lands. The Brahmaputra Right Bank Project, begun in 1963 to try to overcome this perennial problem, built 217 km (135 miles) of embankment, but has met with only partial success.
Population (district) 2 728 000; (town) 68 200
Map Bangladesh – Bb

Bohemia (Čechy) *Czech Republic* Mountainous, westernmost region of the country. It is the heartland of Czech culture and includes the capital, PRAGUE. Bohemia's forests and fertile Vltava and Elbe river basins produce timber, flax and hops, and contain deposits of coal, graphite, iron and uranium. The region has many fine castles, old towns and spas, including KARLOVY VARY (Carlsbad) and MARIÁNSKE LÁZNĚ (Marienbad). Industrial cities include HRADEC KRALÓVÉ, PLEZŇ, ÚSTÍ NAD LABEM and KLADNO
Population 6 329 600
Map Czech Republic – Ab

Bohemian Paradise *Czech Republic* See JICIN

Bohemian Switzerland *Czech Republic* See ÚSTÍ NAD LABEM

Bohinj, Lake *Slovenia* See JULIAN ALPS

Bohol *Philippines* One of the VISAYAN ISLANDS covering 3862 km² (1491 sq miles) to the east of Cebu; also the name of a province including some nearby minor islands, and of the sea between Bohol and Mindanao. The island is hilly, the so-called Chocolate Hills (which are brown in summer) being a tourist attraction. The main crops are coconuts and rice.
Population (province) 759 370
Map Philippines – Bd

Boirinn *Ireland* See BURREN, THE

Boise *USA* State capital of Idaho. It lies on the Boise River in the south-west of the state, 60 km (37 miles) east of the Oregon border. The city was a centre of the mining industry, but the economy has shifted to agriculture and tourism. It is famous for its hot springs.
Population (city) 125 700; (metropolitan area) 189 300
Map United States – Cb

Bokaro Steel City *India* City 260 km (161 miles) north-west of Calcutta. Its Russian-built steel plant is reputed to be the largest and most important in South Asia.
Population 416 000
Map India – Dc

Boksburg *South Africa* Mining, commercial and industrial town 22 km (14 miles) east of Johannesburg, which was established in 1887. Its East Rand Proprietory Mines (ERPM) is the largest gold mine in the world, with an area of around 4900 hectares (12 000 acres). Nearby are Boksburg Lake, Cinderella Dam and many parks. The determination of some of its white residents to maintain racially separate facilities attracted widespread attention in the early 1990s.
Population 195 910
Map South Africa – Cb

Bol *Chad* POLDER on the eastern shore of Lake Chad. It is a fertile agricultural area where millet, sorghum, vegetables, wheat and cotton are grown. A large percentage of the produce is smuggled across the lake to Nigeria, where higher prices can be obtained.
Map Chad – Ab

Bol *Croatia* See BRAČ

Bolama *Guinea-Bissau* A former colonial capital, which has fallen into decay. It is situated on the beautiful island of the same name, south of the Gêba River mouth. It serves as a port for local produce, including palm oil, coconuts and fruit.
Population 30 000
Map Guinea – Aa

Bolgatanga *Ghana* Capital of the Upper Region, 30 km (19 miles) from the border with Burkina, north-east of the national capital, Accra. It began as a trading centre and grew rapidly after 1937 when roads were improved to take motor transport. It markets millet, sorghum, groundnuts, vegetables, cattle, and animal products.
Population 32 500
Map Ghana – Aa

Bolívar *Venezuela* Cattle-ranching state in the south-east. It covers 239 250 km² (92 374 sq miles) – an area about the size of Britain. It has rich deposits of iron ore and bauxite, and reserves of oil. The provincial capital and largest city is CIUDAD BOLÍVAR.
Population 900 310
Map Venezuela – Bb

Bolivia See p 99

Bologna *Italy* City about 80 km (50 miles) north of Florence. In pre-Roman times, Bologna was the site of the Etruscan town of Felsina. The Roman colony founded here in the 2nd century BC was called Bononia, from which the city's name derives. By the 11th century, Bologna had emerged as one of northern Italy's first independent city-states; today it is the capital of the EMILIA ROMAGNA region.

Central Bologna preserves its Roman-style grid layout. It is surrounded by boulevards that mark the line of the medieval walls. Its chief landmarks are the twin 'leaning towers', survivors of more than 200 fortresses that once stood here. The centre also contains the church of San Petronio, one of northern Italy's largest Gothic buildings. The nearby church of San Domenico contains the tomb of St Dominic, founder of the Dominican monastic order, who died in Bologna in 1221.

Bologna's university was founded in the 11th century. The poets Dante (1265-1321) and

Petrarch (1304-74) were among its early scholars. Bologna is an agricultural, engineering and electrical equipment centre. It is noted for its pasta, which accounts for its nickname of *La Grassa*, the 'Fat City', and for its international children's book fair, held here each spring.
Population 404 320
Map Italy – Cb

Bolovens Plateau *Laos* Fertile plateau in the south, rising from the Mekong River plains east of Pakse to a height of 1570 m (5150 ft). The plateau has the highest yearly rainfall in Laos (4064 mm, 160 in). Coffee, cotton and tobacco are grown here.
Map Laos – Bc

Bolsena *Italy* Volcanic crater lake 90 km (56 miles) north-west of Rome. Its waters are rich in fish, especially eels. The lake covers 114 km² (44 sq miles) and reaches a depth of 151 m (495 ft). The village of Bolsena stands on the same site as the Etruscan village of Volsinii, most of which was destroyed in the 3rd century BC.
Population 4030
Map Italy – Cc

bolson Basin of inland drainage in an arid or semi-arid region, particularly among the high plateaus of the south-west United States, often containing a salt lake and ALLUVIAL FANS.

Bolton *United Kingdom* Town 16 km (10 miles) north-west of Manchester, formerly in the county of Lancashire but now part of Greater Manchester. It played an important part in the Industrial Revolution. It was the home of Samuel Crompton (1753-1827), who in 1779 invented the spinning mule – one of the machines that revolutionised the cotton industry. Crompton's home, Hall-i'-th'-Wood, is a 15th-century half-timbered building that serves as a folk museum. Bolton also has a fine art gallery and craft centre. Textiles are no longer dominant; other industries such as engineering have superseded it.
Population (metropolitan area) 269 000
Map United Kingdom – Dd

Bolu *Turkey* Anatolian town and provincial capital, lying 135 km (84 miles) north-west of the capital, Ankara. Founded in the 2nd century BC, it was built in an earthquake-prone zone and was destroyed by an earthquake in 1668. Today, the rebuilt town produces leather and timber.

The surrounding province of the same name, which covers 11 481 km² (4433 sq miles), includes the forested Bolu Mountains. The main products are cereals, flax, tobacco and opium.
Population (town) 52 100; (province) 536 870
Map Turkey – Ba

Bolzano (Bozen) *Italy* Alpine city 55 km (34 miles) south of the Austrian border and the BRENNER PASS, and centre of the autonomous province of Bolzano/Bozen. The arcaded Via dei Portici is the market centre of the town, which is a meeting place of Austrian and Italian culture, language, food and architecture. This can be seen in its museums, libraries and cultural institutions, in its music and folklore, its Gothic cathedral and many churches, and its farms and villages. The city produces steel, aluminium and vehicle parts.
Population (city) 100 700; (province) 441 700
Map Italy – Ca

Bolivia

PARADOXICALLY, EXTENSIVE MINERAL WEALTH HAS BROUGHT ONLY GRIEF AND POVERTY TO BOLIVIA'S MASSES

Poverty exists in many South American countries, but nowhere is it worse than in this landlocked Andean nation. Despite extensive mineral resources, the country and its people remain among the poorest in Latin America. The cause has been two wars and political instability since independence in 1825 which has resulted in 190 governments in 150 years.

The Amerindian peasants, who form more than 60 per cent of the population, live in thatched mud huts and raise just enough food – mainly maize and potatoes – to feed their families. Few have any education, and most are illiterate. Their main source of income is from working in the tin mines or on the gas fields – Bolivia is the fifth largest tin producer in the world; in 1990 its main export was zinc, followed closely by tin, silver and gold.

In the 16th century the Spanish conquistadores conquered the country, robbed the locals of their land, and forced them to work as slaves in the silver mines. Independence from Spain was achieved in 1825, but disputes among the liberators brought little improvement for the people or the country. Bolivia lost more than half its territory to more powerful neighbours – including its valuable nitrate deposits and the Pacific port of ANTOFAGASTA to Chile in the War of the Pacific (1879-83). More than 100 years later, in 1992, neighbouring Peru signed a treaty with Bolivia, allowing the latter a trade-free zone on the Pacific coast. However, an even larger area – most of the eastern lowland Chaco region – taken by Paraguay in the Chaco War of 1932-5 seems irretrievable.

The only political change that brought any real benefit to the Amerindian population was the revolution of 1952. It led to land reforms, some measure of emancipation for the native population and the nationalisation of the tin industry. However, this failed to transform either the industry or the economy. In 1991 tax and investment laws were liberalised to encourage foreign investment and revitalise the mining industry. In March 1994, the state companies running Bolivia's mines, as well as its railways, electricity and telecommunication, were privatised, with shares, in the form of contributions to pension schemes, given to all Bolivians aged 21 and over

Most of the poverty is concentrated in the west of the country, where half the population lives on the *altiplano*, the 3700 m (12 000 ft) high plateau between two chains of the Andes. Here the Aymaras scratch a living. Though the

Sugar cane, tobacco, cocoa and coffee grow readily in the lower *yungas*, and it is here, too, that coca leaves, from which cocaine is derived, have become an important – though illegal – crop. Though tin once comprised 33 per cent of Bolivia's exports, a vast illicit economy has now evolved, centred on coca growing. It is estimated that income from cocaine exports is at least as high as that from tin, natural gas and petroleum, silver and zinc combined. Although the Bolivian government has embarked upon a vast United States-backed anti-drug programme to eradicate coca leaf production, total eradication would threaten the livelihood of about 60 000 farmers. Trade in cocaine has become a major influence, not only on Bolivia's economic life, but on its politics too.

▲ **ON THE ALTIPLANO Aymara Indians bring their sheep to market on the cold high plateau. Barren soils and sparse rain make life hard.**

altiplano is the most densely populated part of Bolivia, it is still a vast emptiness, most of it untouched by human hand. Vegetation is sparse – just coarse grasses and a few shrubs. Condors sweep through the heights, and vicuñas, llamas and alpacas graze the land.

LA PAZ, the highest capital city in the world, lies on the *altiplano* to the south-east of Lake TITICACA – the world's highest navigable lake. To the east lie the lowlands that comprise 70 per cent of Bolivia's land but contain only 20 per cent of its population. Beyond the foothills of the Andes, the northern tropical forests yield timber, brazil nuts and tropical fruits. The rest of the lowlands, which stretch south to the Chaco, are wooded savannah and vast natural pasture lands, grazed by herds of cattle.

Between the *altiplano* and the lowlands, fertile valleys – the *yungas* – cut through the eastern chain of the Andes (the Cordillera Real) in the north. In the south lies the Puna Highland area, suitable only for grazing. The high *yunga* valleys are subtropical and yield hardwoods such as cedar, mahogany, walnut, cinchona (the source of quinine) and dyewoods. Farmers produce citrus and other fruits.

BOLIVIA AT A GLANCE	
Official name Republic of Bolivia	
Area 1 098 581 km² (424 163 sq miles)	
Population 7 500 000 **Per km²** 6.8 **(Per sq mile** 17.68)	
Capital Sucre (judicial); La Paz (administrative)	
Government Parliamentary republic	
Currency 1 Boliviano = 100 centavos	
Languages Spanish, Quechua, Aymara	
Religion Christian (92% Roman Catholic), other 8%	
Climate Tropical; cooler at altitude. Average temperature in La Paz is 10°C (50°F) throughout the year; in the north-eastern lowland, it is 26°C (78.8°F)	
Land use Forest and woodlands 52%, grazing 25%, cultivation 3%, other 20%	
Main primary products Coca leaves, potatoes, maize, sugar cane, rice, cassava, coffee, llamas, alpacas; tin, oil and natural gas, copper, lead, zinc, antimony, bismuth, tungsten, silver, gold, sulphur, iron	
Major industries Mining and smelting, oil and gas production, textiles, handcrafts, food processing, cement	
Main exports Zinc, tin, gold, silver, natural gas, petroleum; cocaine (illegal)	
Annual income per head (US$) 680	
Population growth (per thous/yr) 28	
Life expectancy (yrs) Male 51 **Female** 55	

Boma *Zaire* Port on the north bank of the Zaire River estuary, 80 km (50 miles) from the Atlantic coast. It was a bustling slave centre from the 16th to the 18th century. Today, it exports agricultural produce, including coffee, cotton, rubber and timber. Boma was the capital of the Congo Free State (controlled by King Léopold II of Belgium) between 1885 and 1908, and the capital of the Belgian Congo colony until 1929.

Population 32 000
Map Zaire – Ab

Bombay *India* City on the Arabian Sea, capital of Maharashtra State and former capital city of the now defunct Bombay State. It is India's most important commercial and industrial city, having replaced Calcutta, although Calcutta is still larger. It is also the centre of the Indian Parsi community, the Zoroastrians from Persia.

In the early 16th century the Portuguese ran a small trading colony on Bombay Island, a hot, humid, low and swampy island with a superb natural harbour, which was part of the dowry brought by Catherine of Braganza when she married Charles II of England in 1661. The British then leased it to the East India Company, which moved its headquarters here in 1708. The island is now a peninsula. Its hinterland is hemmed in by the majestic cliffs of the WESTERN GHATS, towering to 1524 m (5000 ft) just inland.

Bombay grew in prominence in the mid-19th century when the first roads and then railways were built across the Western Ghats, and trade in cotton became possible. The city grew in importance after the Suez Canal was opened in 1869,

▼ OFF TO WORK People hurrying to offices and shops in central Bombay throng a narrow street where a bewildering array of shop signs festoons the balconies of the crowded tenements.

when it became the gateway to India. An arch at the waterfront, known as 'the Gateway of India', commemorates the arrival of King (Emperor) George V and Queen (Empress) Mary in 1911.

Today, Bombay has a flourishing offshore oil industry, is the headquarters of a number of banks and has the country's largest stock exchange. Its port handles 40 per cent of India's cargo; its range of industries includes most consumer goods – motorcars, textiles and a flourishing film industry called Bollywood. It is the only city in India with significant daily commuting to the office district by train.

Hindu cave temples in the port at Elephanta Island (Gharapuri) are well worth visiting, but the city is noted more for its Victorian buildings than for its Hindu or Muslim architecture. Courthouses, schools, railway stations and museums are all Victorian adaptations of European and Oriental styles.

Population 9 909 500
Map India – Bd

Bomu (Mbomou; Mbomu) *Equatorial Africa* River 805 km (500 miles) long, forming part of the frontier between Zaire and the Central African Republic. It rises near the Sudanese border and joins the Uele at Yakoma to form the Ubangi. The first European to discover Bomu was the Greek explorer Potagos, in 1877.

Map Zaire – Ba

Bon, Cape (Ras el Tib) *Tunisia* Extreme point of the 80 km (50 mile) long Maouin peninsula that protrudes from the north-east corner of Tunisia into the Mediterranean Sea and shelters the harbour of Tunis. The eastern end of the ATLAS Mountains finally fades away in a series of low hills along the peninsula and ends in the rocks of El Haouria.

The region is noted for its links with the ancient Phoenicians, who founded nearby

CARTHAGE; most of the small towns on the peninsula have the ruins of ancient Phoenician counterparts alongside them.

In the 14th century the area attracted Arab refugees from ANDALUCÍA, fleeing from Spanish attempts to Christianise them. Cape Bon's mild climate has made it a vast fruit and flower garden dotted with vineyards, citrus groves and market gardens, developed during the period of Italian colonisation. Its excellent beach resorts are popular with tourists.

Map Tunisia – Ba

Bonaire *Netherlands Antilles* Arid island of 288 km² (111 sq miles) in the Caribbean, east of Curaçao and off the north coast of Venezuela. Its chief industries are textiles and clothing, salt produced by evaporating seawater, and tourism. Christopher Columbus was the first European to set foot on it. The Dutch claimed it in the 17th century. The capital is Kralendijk on the southwest coast.

Population 11 140
Map Caribbean – Bc

Bondi Beach *Australia* World-famous ocean beach in the eastern suburbs of Sydney, New South Wales, 8 km (5 miles) from the city centre. It is a classic surfing beach and attracts thousands of city dwellers. Australia's first official surf life-saving club was formed at Bondi in 1906.

Map Australia – Ie

Bondoukou *Ivory Coast* Market town on the eastern border with Ghana, 340 km (211 miles) north-east of Abidjan. One of the country's oldest towns, it was founded in AD 1466 by traders on the caravan route to the Niger River in the north. The French introduced cocoa as an important crop here in 1914.

Population 22 000
Map Ivory Coast – Bb

A DAY IN THE LIFE OF
A BOMBAY
DABBAWALLAH

Most office workers in Bombay never eat out or go home for lunch. Instead, their lunch arrives from home. Alok Chaudhry is one of the 2300 members of the Union of Tiffinbox Suppliers who form a network of relay teams that collect 100 000 lunch boxes at mid-morning from Bombay's suburbs and deliver them to offices in the city centre by lunchtime. The home-cooked lunches of curry, rice, vegetables and *chapatti* arrive still warm in aluminium containers stored in tin boxes, or *dabbas*.

Downtown Bombay is a booming, bustling place of crowded streets choked with taxis, honking scooters and scurrying pedestrians. Alok's job is to ride the trains from suburbs to city, handing out the food.

It is for a team mate in the lunch box relay race that Alok is waiting, by the clock at Malad Railway Station, 32 km (20 miles) from Bombay, at 11 am. Fifty other white-hatted *dabbawallahs*, or lunch-carriers, wait with him.

Reflecting on his role as a link between wives at home and working husbands, Alok thinks wistfully of his own wife, Meenakshi, in their village near the town of Pune, four hours away from his rented tenement room in Malad. He sees her and their three children only three or four times a year, but at least he has a job and can send his family 15 rupees a month out of the 90 he earns.

Alok's reverie does not last long. At 11:02, a bicycle hung with *dabbas* wheels into the station, ridden by Alok's friend, Rajneesh. The *dabbawallahs'* rendezvous are always punctual. Suburban homes expect the first in the chain of lunch-carriers at 10 am every day. These *dabbawallahs* arrive on foot and on time. Rajneesh, second in the chain, collects his lunch-boxes from them at the first meeting point at 10:20.

Rajneesh is already unhooking *dabbas* from his bicycle as Alok greets him. All the *dabbas* are marked with signs – dots, crosses, strokes and circles – the address that guides them through the stages of their journey. Written addresses would be useless; many of the *dabbawallahs*, like 62 per cent of India's entire population, cannot read. Alok is among these, despite a few childhood years at his village school. He selects the 39 *dabbas* marked with a red dot and packs them onto a wooden rack, then he boards the crowded train.

At each stop, other *dabbawallahs* get on and off the train and Alok checks his tins for symbols which mean they must be handed over to continue their journey. At midday the train draws into Churchgate Station. The last of Alok's *dabbas* are collected and he can stop to eat his own lunch – *chapatti* and vegetables wrapped in newspaper.

But his work is not over. The whole process of collection and delivery is reversed with the empty tins, so that the right *dabba* is returned to the right home by 4 pm.

bone bed Layer of sedimentary rock containing fossil bones, teeth and scales of vertebrates, especially fishes. It may be the result of a rapid catastrophe, such as an underwater earthquake, that killed all life simultaneously.

Bong Mountains *Liberia* Range of mountains in western Liberia which rise to 645 m (2146 ft) some 80 km (50 miles) north-east of the capital, Monrovia. The range contains substantial reserves of iron ore. The mining settlement of Bong Town (population 12 000) has grown since an iron ore mine was opened in the hills in 1965.
Map Liberia – Ba

Bonin Islands (Ogasawara-shoto) *Japan* Remote volcanic island chain stretching 800 km (500 miles) due south of Tokyo. Most people here make a living from fishing and growing fruit and vegetables, including bananas and mangoes.
Population 2300
Map Pacific Ocean – Ca

Bonn *Germany* Capital of West Germany from 1949 to 1991. It is situated on the bank of the Rhine, 30 km (18 miles) south-east of Cologne. Before 1945, Bonn was a dignified university town and the site of the two Baroque palaces of the archbishop-electors of Cologne. The composer Ludwig van Beethoven (1770-1827) was born here and his birthplace in Bonngasse is now a museum.

The destruction of Berlin during the Second World War and the subsequent division of Germany led to the choice of Bonn as the new capital in 1949. The West German parliament settled into a new *Bundeshaus*, formerly a teachers' training college. Diplomats poured into the Rhine valley, overflowing into nearby Bad Godesberg, an ancient spa town. In June 1991, however, Berlin became the capital and seat of parliament of the united Germany and Bonn lost its status as the national capital, as well as many of its inhabitants.
Population 292 200
Map Germany – Bc

Bonny *Nigeria* Seaport village on the east side of the Niger River delta. It has a sandy beach bordered by mangrove swamps. Bonny was a slave-trading centre in the 18th century; in 1961 an oil terminal was opened here.
Map Nigeria – Bc

Bonsa Valley *Ghana* The country's second largest diamond field after the one in Birim Valley. It lies 220 km (135 miles) west of the capital, Accra. Mining is frequently carried out by individuals or small family concerns. The diamond-bearing gravels are dug from shallow pits, washed and sorted by hand.
Map Ghana – Ab

Bophuthatswana *South Africa* Region and 'homeland' of the Tswana people which was made the 'independent' Republic of Bophuthatswana in 1977. Its independence was, however, recognised only by South Africa and in 1994, after that country's first democratic elections, the region was reincorporated into South Africa. The capital Mmabatho became the provincial capital of North-West Province in 1994, reverting to its old name of Mafikeng.
Map South Africa – Bb

bora Cold, violent north or north-east wind that blows along the Dalmatian coast of Croatia and in northern Italy. It occurs mainly in winter when atmospheric pressure is high over Central Europe and the Balkans and relatively low over the Mediterranean.

Borås *Sweden* Manufacturing town 69 km (43 miles) east of Gothenburg. Founded in 1622, it is the centre of the country's textile industry.
Population 102 800
Map Sweden – Bd

Bordeaux *France* Seaport, cultural centre and capital of the Gironde department, on the Gironde River estuary, 95 km (59 miles) from its mouth. The surrounding area – the Bordelais – is noted for its wines, known as Bordeaux or claret.

Bordeaux is an important wine-exporting trade centre. It has several medieval churches, an 11th to 15th-century cathedral, and a large university which was founded in 1441. The city produces chemicals and processed foods; other industries include shipbuilding, marine engineering and oil refining.
Population (city) 213 000; (metropolitan) 685 000
Map France – Cd

Bordelais *France* See BORDEAUX

Borders *United Kingdom* Region of southern Scotland covering 4713 km² (1819 sq miles) on the boundary with England. For centuries it was the scene of bloody clashes between the Scots and the English until their kingdoms were united in 1707, to cover the old counties of Berwick, Roxburgh, Selkirk and Peebles, as well as part of Midlothian.

This hill country is famous for sheep and for its cashmere and tweed products, for the ruined abbeys at Melrose, Jedburgh and Dryburgh, and for its association with the poet and novelist Sir Walter Scott (1771-1832), who set much of his work in the region. He is buried in Dryburgh Abbey.

Today the area is a major tourist and recreation area, with its fine scenery and clean air.
Population 102 650
Map United Kingdom – Dc

Bordighera *Italy* Riviera resort 10 km (6 miles) from the French border. Tourists flock here each year to enjoy the Mediterranean coastline. Scenic attractions include panoramic corniche roads and luxury villas with tumbling gardens of tangled vegetation. The English nonsense poet Edward Lear (1812-88) lived here in his last years. Date palms are cultivated around Bordighera, which is a flower-farming area.
Population 11 560
Map Italy – Ac

bore High wave, like a wall of water, which travels upstream in the tidal reaches of certain rivers. It is caused by the surge of a flood tide upstream in a narrowing estuary or by colliding tidal currents. As it surges upstream, the bore gradually dies out. There is a bore on the River Severn in England where, during spring tides, it is often 1 m (more than 3 ft) high and may reach heights of more than 2.5 m (nearly 9 ft). On some occasions it reaches beyond Gloucester, 33 km (20 miles) upstream. Bores are sometimes called 'eagres'.

boreal forest Collective name given to the northern forests of Canada and Eurasia. Boreal forests consist mainly of conifers, but also include hardy deciduous species, such as birch.

Borecka Forest *Poland* See MASURIA

Borgå *Finland* See PORVOO

Borkou-Ennedi-Tibesti (BET) *Chad* The country's largest prefecture, covering 600 350 km² (231 175 sq miles) in the north – almost half of the country's total area. Its capital is Faya-Largeau. Although it formed part of the French colony of Chad, its Muslim people always resisted French rule. When it was ceded to the N'Djamena government, it remained a centre of rebellion, and in the civil war in the 1980s was controlled by Libyan-backed anti-government forces. Most of its people are nomadic herdsmen who keep camels and goats, but there is a semi-sedentary population growing millet, wheat, vegetables and dates around oases.
Population 119 200
Map Chad – Ab

Borkum *Germany* See FRIESIAN ISLANDS

Borneo *South East Asia* One of the world's largest islands, bordering the South China Sea. With a total area of 746 951 km² (288 331 sq miles), Borneo is divided between a northern and a southern section, which are separated by a mountainous area stretching north-east to the massif of Mount Kinabalu. The northern section, formerly under British control, is today divided between the Malaysian states of SARAWAK and SABAH and the independent sultanate of BRUNEI. The southern section is the larger of the two. Formerly Dutch, it is now ruled by INDONESIA and known as KALIMANTAN.

Much of the coastal part of Borneo is low-lying, swampy land which is fairly sparsely populated. It has only recently been developed. There are deposits of gold, diamonds and coal as well as petroleum, but of these, only the petroleum deposits are substantial.
Map Indonesia – Db

Bornholm *Denmark* Island covering 588 km² (227 sq miles) in the south-west Baltic Sea and consisting mostly of rugged granite rocks containing kaolin, from which Copenhagen's famous pottery is made. It has fine examples of round, fortified medieval churches and contains the ruins of the impressive late medieval castle of Hammershus. The little island of Christiansø, off the north-east coast, was also fortified principally because of recurrent conflict with Sweden. Bornholm's principal town and ferry port is RØNNE; its fishing port is Neksø. In the Middle Ages, Bornholm was an important trading centre. Today, agriculture, fishing and fish processing and tourism are its chief sources of income.

The island was invaded by Germany during the Second World War and liberated by the Russians, who occupied it until April 1946.
Population 46 600
Map Denmark – Cb

Borno *Nigeria* See MAIDUGURI

Borobudur *Indonesia* Magnificent Buddhist temple 40 km (25 miles) north-west of the city of Yogyakarta in central Java. It was built between AD 750 and 850. The shrine, which consists of a series of terraces with carvings depicting Buddhist legends, was restored with United Nations help in the 1960s and 1970s.
Map Indonesia – Dd

Borovets *Bulgaria* Winter sports resort, on the northern slopes of the Rila Mountains. It has good pine-forested walking country, and is a departure point for Musala (2925 m, 9596 ft), the highest peak in the RHODOPE MOUNTAINS. It is Bulgaria's oldest mountain resort, and has several palaces and villas (now holiday homes) reflecting its former glory as the haunt of royalty and aristocracy before the Second World War.
Map Bulgaria – Ab

Borsod *Hungary* Attractive limestone upland, largely covered by deciduous forest, north of the city of Miskolc. The spectacular cave system at AGGTELEK lies near the Slovakian border. Sárospatak, the region's picturesque capital 70 km (43 miles) north-east of Miskolc, has a population of 15 200. Iron ore mined at the town of Rudabánya provides the raw material for the iron and steel works in Miskolc and Ózd.
Map Hungary – Ba

Börzsöny *Hungary* See MÁTRA

Bosnia-Herzegovina See p 104

Bosporus (Bosphorus) *Turkey* Strait 29 km (18 miles) long and up to 4 km (2.5 miles) wide linking the Black Sea and the Sea of Marmara and separating Europe from Asia; formerly known as the Bosphorus. ISTANBUL is at its southern end. Cars can drive across the Bosporus on a suspension bridge opened in 1973, and an even newer bridge opened in 1988.
Map Turkey – Aa

boss Small BATHOLITH with a roughly circular cross-section. Examples are Shap in Cumbria, England, and the northern part of Arran, Scotland.

Boston *USA* Port and capital of Massachusetts, 300 km (186 miles) north-east of New York City, and the largest city in New England. English colonists first settled here in 1630 – 10 years after the pilgrims established the nearby PLYMOUTH Colony. In 1632, it became capital of Massachusetts Bay Colony. The city was the scene of the so-called Boston Tea Party in 1773, when a party of angry citizens, disguised as native Americans, boarded tea ships and dumped their cargo overboard in a gesture of defiance against economic dominance, including a tax on tea, from Britain – an incident which finally sparked the American War of Independence.

Boston today is the centre of a conurbation that spreads north into southern New Hampshire. Its long-established clothing, footwear and textile industries have declined and have been replaced by a wide range of newer industries, including electronics.

The city is an important financial, business and educational centre. It is the birthplace of writer and poet Edgar Allan Poe (1809-49) and home of the world-famous Boston Symphony Orchestra. Nearby Cambridge is the seat of Harvard University (the oldest university in the United States, founded in 1636) and the Massachusetts Institute of Technology.
Population (city) 574 300; (metropolitan area) 3 227 710; (conurbation) 5 455 400
Map United States – Lb

Bosumtwi, Lake *Ghana* Mysterious circular lake 21 km (13 miles) south-east of Kumasi. It is 10 km (6 miles) across and more than 72 m (236 ft) deep. No one knows how it was formed, but a meteorite or volcanic explosion followed by subsidence are two possible explanations. There is much folklore and superstition associated with the lake – for instance, fishermen are forbidden to use boats on it. Instead, they sit astride planks.
Map Ghana – Ab

Botany Bay *Australia* Coastal inlet lying in the southern suburbs of Sydney, discovered by the English explorer Captain James Cook in April 1770. It was so named by Cook because of the rich variety of plant life growing here. The British originally planned to use the bay as a penal colony in 1788. But conditions were unsatisfactory, forcing them to move the site to Port Jackson (Sydney Harbour).
Map Australia – Ie

Bothnia, Gulf of Northernmost arm of the BALTIC SEA, lying between Sweden and Finland and covering 126 000 sq km² (48 637 sq miles). Ever since the end of the last Ice Age about 10 000 years ago its shoreline has been slowly rising – a consequence of the removal of the tremendous weight of the ice sheets. The gulf is covered by ice for up to five months every year. It has a low salt content as a result of the low evaporation rate during the short summer months and the many rivers which run into it.
Map Sweden – Cc

Botoşani *Romania* Capital of the province of the same name in northern MOLDAVIA, 40 km (25 miles) west of the Moldavian border. It was founded in the 14th century, and a medieval church remains from its early days. Today the trading centre of a fertile agricultural area, the town manufactures textiles, clothing and processed foods.
Population (town) 79 000
Map Romania – Ba

Botswana See p 103

Bottrop *Germany* City of the industrial Ruhr near the border with the Netherlands. It produces coal, textiles and steel.
Population 118 900
Map Germany – Bc

Bou Saâda *Algeria* Oasis town in the Ouled Nail mountain region, 250 km (155 miles) south-east of Algiers on the road to Biskra. With its old streets, citadel, its Ouled Attik and El Nekla mosques, and its many craft shops, it has become a major tourist centre.
Population 50 000
Map Algeria – Ba

Bouaké *Ivory Coast* Second largest city in the country, lying about 300 km (about 186 miles) north-west of Abidjan. The French established it as a military post in 1899 on the site of a former slave market. By the 1970s it had also developed

Botswana

THE LAND ALONG LIVINGSTONE'S 'MISSIONARY ROAD' IS A LAND OF UNTOUCHED NATURE AND ONE OF THE MOST AWE-INSPIRING SWAMPLANDS IN THE WORLD

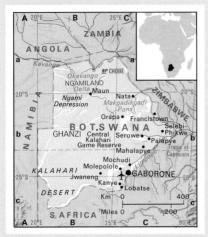

The explorer David Livingstone was one of several gospel-spreading Britons who trekked into the southern African interior in the mid 19th century. The track they trod became known as the Missionary Road, and it ran through the tribal lands of the people known as the Batswana. The friendly relations which they established led to the Batswana seeking British protection when the Boer republics to the south and west began taking a predatory interest in the territory. In 1885 the British set up a protectorate over the whole of a region that was then called Bechuanaland and is now Botswana.

The country maintained its ties with Britain until it became independent in 1966 under the presidency of a well-loved chief, Sir Seretse Khama, who died in 1980. As a young man, Khama studied in Britain in 1948, and married an English woman, Ruth Williams, who remained in Botswana after his death.

The early British administrators found themselves responsible for a large, landlocked country, sparsely populated and seemingly with little economic potential. Much of the western and south-western areas formed part of the KALAHARI Desert, a sea of red sands, with some sparse covering of grass and bush.

The eastern border zone contiguous with South Africa is the most populated area, because the land here is fertile and has enough rain to raise livestock and grow crops.

To the north, the landscape changes dramatically into the huge OKAVANGO DELTA of the Kavango River, fed by waters that rise in neighbouring Angola, and by heavier rainfall. This marshy area is a haven for wildlife, the home of elephants, lions, leopards, buffaloes, waterbuck, zebras and crocodiles. The delta forms an intricate pattern of channels, with palm-covered islands. In good rainy seasons it drains into the vast Makgadikgadi Pans, and when these fill with water spectacular parades of flamingos and pelicans are known to occur.

The people of Botswana are mainly farmers. By tradition their social status and wealth is judged by the size of their cattle herds. The population has always been small – it is now about 1.4 million – and although some white settlers established huge commercial ranches, the colonial history of the country largely followed a path of benign neglect.

After independence in 1966, Botswana's fortunes were changed dramatically by the exploitation of mineral resources. Diamonds have become the major foreign currency earner (between 1983 and 1993, diamond sales accounted for 70-80 per cent of all Botswanan exports); copper-nickel is also exported. The traditional meat exports were maintained and coal mining added to a temporary economic boom, aided by foreign investment. Recent years have brought a decline, as world markets became depressed, and several dry years in the 1980s and a severe drought in 1991-3 severely reduced grain crops. The world demand for diamonds, likewise, is depressed, and the government has begun to diversify the economy by increasing tourism, banking and manufacturing.

There are eight major Batswana tribes; the largest is the Ngwato, or Bamangwato, which makes up 38 per cent of the population. Other groups are the Kwena, Ngwaketse, Tawana, Kgatla, Lete, Rolong and Tlokwa. Water and good pasture are crucial factors in their daily lives and settlement patterns, although the inhospitable Kalahari is still inhabited by a few thousand hunter-gatherer Bushmen, or San.

An unusual feature of Batswana settlement patterns is the vast villages such as SEROWE and MOCHUDI, each with populations of around 30 000. In the past it was customary for households to have three different homes: one in the village, one on the family's agricultural land up to 45 km (28 miles) distant, and a third at far-off cattle posts, where young men spent most of the year tending herds.

The village centre is the *kgotla*, the meeting place where the chief and headmen used to order the times of sowing and harvesting. Poor

households could borrow cattle from the chief, which they tended on his behalf in return for the first calf. Nowadays the power of the chiefs is waning because people are beginning to leave the villages to live permanently on their land – partly to escape the pressure on resources of a growing population, but perhaps also to avoid the village headman's traditional right to call on them for labour.

The fickle rainfall also makes it necessary for families to earn money from activities other than farming – for example, brewing, trading or building. However, more than 70 per cent of the population is still rural; and most of the land in Botswana is tribal, which means that it is allocated by the chief and controlled to prevent overgrazing.

Botswana's main towns are GABORONE, expanded as a capital city after independence, the old colonial FRANCISTOWN, Lobatse, which has a large abattoir, and the various mining towns of Selebi-Phikwe, ORAPA, Letlhakane and Morupule. Maun is the centre of a lucrative safari industry to the Okavango Delta.

Botswana has a multiparty democracy with a president as head of state, a National Assembly, and a Cabinet. There is also a House of Chiefs, but its 15 members have no right of veto. In the past, the Botswanan government made notable protests against the South African government's policy of apartheid – despite its dependence on South Africa for export routes and most imports, as well as for employment for about 40 000 migrant Botswanan workers.

BOTSWANA AT A GLANCE		
Official name Republic of Botswana		
Area 581 730 km² (224 554 sq miles)		
Population 1 400 000 **Per km²** 0.41 **(per sq mile** 0.16)		
Capital Gaborone		
Government Parliamentary republic		
Currency 1 pula = 100 thebe		
Languages English and Setswana (both official); other local languages		
Religions Christian (40-50%), remainder tribal		
Climate Subtropical and dry; the dry season is from May until October. Average annual temperature in Maun is 22°C (72°F), average temperature for the country ranges from 26°C (78.8°F) in November to 15°C (59°F) in July		
Land use Grazing 77%, cultivation 2.4%, desert and marshes 18%, other 2.6%		
Main primary products Cattle, sheep, maize, sorghum; diamonds, nickel, copper, gold, manganese, coal, soda ash		
Major industries Cattle rearing, mining, meat processing, tourism		
Main exports Diamonds, beef, nickel, copper		
Annual income per head (US$) 2790		
Population growth (per thous/yr) 34		
Life expectancy (yrs) Male 60 **Female** 70		

into a commercial centre. After completion of the Abidjan-Ouagadougou railway line between Ivory Coast and Burkina, it became the focus of transport in the interior and an industrial centre.

The city is a busy market for local produce, and cotton, sisal, rice and tobacco are processed here. It is a collecting point for cocoa and coffee crops which are sent to Abidjan for export. The country's oldest textile mill (1922) is also to be

found here. There are many mosques as well as a Benedictine monastery, the city having long been a focal point for Catholic and Protestant missions.

Population 333 000

Map Ivory Coast – Bb

Bouba Ndjida National Park *Cameroon* Wildlife reserve on the border with Chad covering

about 2200 km² (850 sq miles). It is rich in such wildlife as giraffes, buffaloes, lions, leopards, cheetahs, rhinoceroses, elephants and many varieties of antelope, including the large and magnificent Derby eland. There is a camp on the banks of the Mayo Lidi River, where guides are available to take visitors on tours of the national park.

Map Cameroon – Bb

Bosnia-Herzegovina

CIVIL WAR HAS TURNED THIS COUNTRY INTO A WASTELAND AND DISPLACED HUNDREDS OF THOUSANDS OF PEOPLE

Peace is in short supply in Bosnia-Herzegovina, a country with a long tradition of conflict that was patched over during the years of communist rule, but broke out almost immediately after the old Yugoslavia – of which it formed a part – finally crumbled just one decade after the death of Yugoslavia's powerful leader and wartime guerrilla fighter against the Nazi occupation, Marshal Tito.

Made up of larger Bosnia, which covers some 42 010 km² (16 216 sq miles) in the north, and the smaller Herzegovina, covering 9119 km² (3520 sq miles) in the south, the country once formed the nucleus of Yugoslavia.

It declared its sovereignty within Yugoslavia in October 1991 and its independence, following a referendum, in March 1992. Soon after independence, however, the young republic became caught up in an escalating spiral of ethnic civil strife.

Bosnia-Herzegovina was the poorest of the republics, after Macedonia, in the old Yugoslav federation, and was the most central of the country's six republics, covering an area of 51 129 km² (19 736 sq miles). It is bordered in the north and west by Croatia, in the east by Serbia and in the south by Montenegro. Its tiny Adriatic coastline is no more than 20 km (12.4 miles) long.

It is mainly mountainous and deeply cut by rivers flowing north to join the Sava River or south to join the Neretva. In Bosnia, the Dinaric Alps give way northwards to rolling farmlands, which have become famous for plums and plum brandy (*slivovica*, or *rakija*). Herzegovina is limestone plateau country, pitted by large depressions which are fertile after flooding during the winter months.

In both Bosnia and Herzegovina the sirocco wind brings rain from the south-west, while the bora wind sweeps down from the north-east. However, the weather in Bosnia is generally mild for most of the year, although apt to be bitterly cold in winter; and in Herzegovina the summers tend to be extremely hot.

Although arable land covers about 50 per cent of the country's area, the fertile regions are in the north, where the soil is well suited to cereals, especially corn (maize). Herzegovina and the more sheltered areas of Bosnia produce soybeans, rapeseed, olives, grapes, figs, pomegranates, melons, mulberries, oranges, lemons, rice and tobacco.

Bosnia also possesses rich reserves of minerals including coal, iron, lead, copper, mercury, bauxite, manganese and silver; and the southern half of Herzegovina has asphalt and lignite mines.

Under Ottoman rule from 1463 until the Treaty of Berlin in 1878, when it came under the control of the Austro-Hungarian Empire, Bosnia-Herzegovina remained an impoverished region, mainly dependent on agriculture. Since the 19th century, however, mining and manufacturing have made a larger contributut& to the economy than agriculture, and prosperity was growing before war broke out in 1992.

After the 1992 declaration of independence, continued bitter ethnic warfare throughout the country caused production to plummet. In the war-torn centres to which refugees flocked, such as the capital, SARAJEVO, Gorazde or Tuzla, there were shortages of food, medicines and even water, and, everywhere, factories lay buried under rubble. Unemployment and inflation soared, and the toll in human misery and suffering escalated.

Many of the causes of the civil war can be found in the region's turbulent past. Bosnia was an independent kingdom in the 14th century, but soon lost land to the advancing Turks. Bosnia finally fell to the Turks in 1464. Herzegovina followed in 1482 and the Turks remained in control until 1878. The Turkish influence was enormous.

Today, many of the inhabitants are Muslim and, until the civil war broke out, there was a wealth of Islamic art and architecture, including mosques, covered bazaars, baths, fountains and bridges in towns such as Sarajevo, BANJA LUKA, MOSTAR, Travnik and Višegrad. But much of this came under Serbian fire after 1992 and has been destroyed.

The bloodiest fighting in 1993 took place between Croats and Muslims in western Herzegovina, but Muslims and Croats signed a peace accord in March 1994, in which they agreed to form a Muslim-Croat federation.

FROM THE HABSBURGS TO THE COLD WAR

Sarajevo, where an assassin's bullet sparked the First World War in 1914, was the capital of a Bosnia-Herzegovina which then formed part of the Habsburg Austro-Hungarian Empire. The assassin was a Bosnian-Serb nationalist, and his victim was Archduke Franz Ferdinand, heir to the Austro-Hungarian throne. Austria-Hungary declared war on Serbia, and other European powers joined the conflict.

Yugoslavia was created as the Kingdom of Serbs, Croats and Slovenes after the First World War. The name was changed to Yugoslavia in 1929. But the union was never really a happy one.

In Bosnia-Herzegovina tensions came violently to the surface when the Axis powers annexed the country to the Nazi puppet Independent State of Croatia in 1941. In the ensuing civil war Muslims, Serbs and Croats fought one another fiercely. The Croat Ustace regime persecuted the Serb population, sometimes aided by Muslims, and reprisals were carried out first by the royalist Serb Chetniks, led by Draza Mihajlovic, and later by Marshall Tito's communist partisans.

The fact that Yugoslavia lasted as long as it did after the Second World War owed a great deal to Marshal Tito's communist regime which emerged from it. Tito was confirmed as leader of Yugoslavia in 1945.

Tito led a sweeping programme to develop agriculture, industry and tourism and to open the country to the outside world.

Tito's refusal to obey the Moscow line alarmed Stalin, who ordered that the Yugoslavs were to be expelled from the Cominform, the international organisation of communist parties, in 1948.

From 1948 to 1955, the USSR and East European communist governments blockaded Yugoslavia. They hoped that by cutting off vital supplies such as Soviet and Czech machinery and Polish coal they would wreck the economy and bring about the collapse of the Tito regime. But the blockade served only to unite the Yugoslavs as never before.

A LAND OF MANY NATIONS

Tito once stated, 'I am a leader of one country which has two alphabets, three languages, four religions and five nationalities living in six republics surrounded by seven neighbours, a country in which live eight national minorities.'

Tito, in fact, understated the number of ethnic minorities. Strains inside the federation in the decade after his death led to Yugoslavia breaking up into five separate states, marked by conflicts within and between them.

In Bosnia-Herzegovina, Muslims made up 44 per cent of the population, Serbs accounted for 31 per cent and Croats for 17 per cent.

The Serbs – supported by neighbouring Serbia – were opposed to breaking away from the Yugoslav federation and boycotted the referendum on independence. In March 1992, war broke out in Bosnia.

Because all the ethnic groups were integrated, with Serbs, Croats and Muslims living as neighbours with one another, there was no possibility of simply carving up the country along ethnic lines – one region allocated to each of the three ethnic groups. Serbs began, instead, a systematic 'ethnic cleansing' of Muslims in the cities: by the end of 1994, they had driven the Muslim majority from many regions. The conflict had forced about half of the pre-war population of 4.4 million to flee their traditional homes.

The Serbs were well armed; the Muslims were much more poorly armed and hoping for UN support. But months of slaughter progressed with the West unable to agree or to find sufficient support for a peace-making strategy.

Sarajevo became the focus of international attention in 1992-4, when it was besieged by Bosnian Serb troops, despite the fact that it had been declared a 'safe' area by the United Nations Security Council and supplied with food and medical supplies. However, apart from sending in a force to guard supplies and keep some areas 'safe', the West did little to stop an almost daily bombardment from the Bosnian Serb-held hills above the city.

A vicious mortar bomb attack in February 1994 broke the impasse and a UN-NATO ultimatum – aided by Russian troops – eased the siege, although sporadic mortar and sniper fire still rained down on the city. During the siege, the city's population decreased by more than 150 000 – many elderly people, mothers and children were evacuated, almost 10 000 were killed, and 56 000 were injured. Six out of every 10 houses were either destroyed or badly damaged. Similar fates befell Banja Luka, Mostar, Gorazde and Tuzla.

The UN and the European Union (EU) tried to mediate. At one stage a peace proposal was accepted by the mainly Muslim government. Under this, Bosnia-Herzegovina would be divided, with 51 per cent going to the Croats and Muslims and 49 per cent to the Serbs. But the Serbs, whose forces had gained control of more than 70 per cent of Bosnia, rejected the plan and the haggling continued.

BOSNIA-HERZEGOVINA AT A GLANCE

Official name Republic of Bosnia and Herzegovina
Area 51 129 km² (19 736 sq miles)
Population 4 618 804 **Per km²** 90.3 **(Per sq mile** 234)
Capital Sarajevo
Government Republic
Currency 1 Croatian dinar = 1 old Yugoslav dinar = 100 paras
Languages Serbo-Croat
Religions Christian 50% (Orthodox 31%, Catholic 15%, Protestant 4%), Muslim 40%, other 10%
Climate Moderate continental, with an even spread of rainfall, except in the south, which has a transitional Mediterranean-continental climate. Average temperature in Sarajevo ranges from 19.5°C (67°F) in July to −1.4°C (29.5°F) in January
Before the war:
Land use Cultivation 22%, meadows and pastures 25%, forest 36%, other 17%
Main primary products Iron ore, coal, copper, chromium, lead, salt, zinc, bauxite, manganese; grapes, fruit, wheat, livestock
Major industries Mining, steel, textiles, oil refining, furniture, armaments, tobacco
Main exports Iron ore and other metal ores, metal products, machinery, textiles
Annual income per head (US$) 675
Population growth (per thous/yr) 7
Life expectancy (yrs) Male 72 **Female** 78

Bougainville *Papua New Guinea* Easternmost major island of the country, lying some 1000 km (620 miles) north-east of Port Moresby. Geographically it is part of the Solomon Islands, but politically it is not.

The island was discovered in the 18th century by the French navigator and explorer Antoine de Bougainville. It is densely wooded and mountainous, its highest point being the active volcano Mount Balbi (2743 m, 9000 ft). A huge open-cast copper mine at Panguna, inland from the capital, Arawa, is partially Australian-owned.

In 1987, a civil war began in Bougainville, triggered by landowners seeking compensation for damage from mining. Halting all mining activity, the Bougainville Revolutionary Army then demanded independence for their island. In 1994, the government of Papua New Guinea signed a ceasefire with the rebels.
Population (with outlying islands) 145 000
Map Papua New Guinea – Ca

Bouillon *Belgium* Town on the border with France, about 110 km (68 miles) south of Liège. Set in the gorge of the Semois River, it is dominated by a magnificent castle once owned by the Crusader Godfrey of Bouillon (1061-1100).
Population 5800
Map Belgium – Bb

Boukra (Bu Craa) *Western Sahara* Area containing one of the richest deposits of phosphates in the world, situated south-east of LAÂYOUNE in north Western Sahara. Reserves are estimated at 1000 million tonnes. Production, begun by the Spanish, was temporarily halted by the war between Morocco and the Polisario guerrillas of the Western Saharan independence movement, but resumed in 1982.
Map Western Sahara – Aa

boulder A large individual fragment of rock, greater than 250 mm (about 10 in) in diameter. It is larger than a COBBLE.

boulder clay Mass of unsorted material consisting of a mixture of clay loaded with sand, gravel and boulders, carried by a glacier and deposited when the ice melts. It is also known as 'till'.

Boulogne-sur-Mer *France* Largest French fishing port, in the north of the country 32 km (20 miles) south-west of Calais. Ferries and hovercraft provide good links to Britain.
Population 98 900
Map France – Da

Bourbonnais *France* Historic province of central France, 7340 km² (2833 sq miles) in area and crossed by the Allier River. It is situated just north of the Massif Central. It has been in existence since the 10th century and was the original seat of the royal family of Bourbon. Woodlands, small fields and hedged pastures where Charolais beef cattle are raised, make up its countryside. The chief towns are Moulins, Vichy and Montluçon.
Population 370 000
Map France – Ec

Bourg-en-Bresse *France* Chief town of Ain department, 60 km (37 miles) north-east of

Lyons. It has a busy livestock market and a 16th-century Gothic church, now a museum. Its industries include furniture and iron works.
Population 43 700
Map France – Fc

Bourges *France* Chief town of Cher department. It stands at the confluence of the Yèvre and Auron rivers, 65 km (40 miles) west of Nevers. It has a large cathedral, ancient town houses and museums. Nearby are metalworks and other industries producing aircraft, missiles, textiles and tyres.
Population 78 800
Map France – Ec

Bourget, Lac du *France* Largest lake in France, covering 45 km² (17 sq miles) and set in a steep-sided valley in the northern Alps. The spa town of Aix-les-Bains is on its eastern shore.
Map France – Fd

Bourke *Australia* Town in New South Wales, on the Darling River, 625 km (388 miles) north-west of Sydney. It is known as 'the Gateway to the Outback' and was first established as a fortified storage depot in 1835, to guard against attacks by Aborigines. It is the centre of one of the world's most productive wool areas.
Population 3000
Map Australia – He

bourne An intermittent or seasonal stream, especially in the chalk regions of southern England, which may flow in winter when the water table rises above the level of the valley floor.

Bournemouth *United Kingdom* Seaside town in Dorset county 150 km (93 miles) south-west of London. In 1850 it had fewer than 1000 inhabitants but it has grown into one of Britain's largest and best equipped resorts. Its conference centre is a favourite for political and trade union conventions. The port of Poole lies to the west on a large, shallow harbour; it dates from the 13th century and today mainly provides anchorage for leisure craft.
Population 158 800
Map United Kingdom – Ee

Boyacá *Colombia* Highland state covering 23 189 km² (8951 sq miles) in the eastern range of the Andes. The capital is TUNJA. Wheat and olives are grown in the warmer valleys towards the Magdalena River in the west. The Western world's largest emerald mine is at Muzo, some 80 km (50 miles) west of Tunja. Colombia's major iron and steel works is at Paz de Río, in the north-east.
Population 1 274 390
Map Colombia – Bb

Boyne (An Bhóinn) *Ireland* River which rises in the Bog of Allen 95 km (60 miles) west of Dublin and flows 115 km (71 miles) north-east to the Irish Sea at Drogheda. It is noted for its salmon. On 1 July 1690, in the river valley 6 km (4 miles) west of Drogheda, the Protestant William of Orange defeated the Catholic James II in the Battle of the Boyne, ensuring the Protestant succession to the British throne. West of the battlefield is the prehistoric passage grave of Newgrange.
Map Ireland – Cb

Boyoma Falls *Zaire* Seven cataracts above the river port of Kisangani, and the point where the Lualaba River becomes the Zaire River. There is a fall of 61 m (180 ft) over a distance of 90 km (56 miles). Formerly called the Stanley Falls, the Boyoma Falls were explored in 1877 by the British-American adventurer Sir Henry Morton Stanley, who established that the Lualaba River was the headstream of the Zaire River. On reaching the cataracts, his men dismantled his boat and carried it and all their supplies through thick forest and over difficult terrain, fighting off attacks by local inhabitants. Today a railway connects Kisangani with Ubundu above the cataracts.
Map Zaire – Ba

Bozen *Italy* See BOLZANO

Brabant *Belgium* Central province covering 3358 km² (1297 sq miles) around the capital, Brussels. A fertile lowland supporting cattle farming and market gardening, it is the country's most densely populated area and a major industrial centre with iron and steel, chemicals, paper, food processing and textile industries.

The historic Duchy of Brabant was formed in 1190. It later expanded into the South Netherlands – but the Habsburgs took it in the 15th century. Its prosperity in the Middle Ages was founded on wool and other textiles. In 1830, a revolt by South Brabant against Dutch rule, led to Belgian independence.
Population 2 252 610
Map Belgium – Ba

Brăc (Brazza) *Croatia* The largest Dalmatian island, covering 394 km² (152 sq miles), lying off the resort of Split. Its best known export is milk-white marble, used in Diocletian's Palace at Split, the White House in Washington, and the high altar of Liverpool's Catholic cathedral. The island's terraced slopes produce olives, figs, almonds and grapes. Fishing, too, is an important industry. However, labour for both farming and fishing is scarce as young people now migrate to mainland cities. Supetar is the chief town.

The island has some excellent beaches, such as at Bol, but tourism is underdeveloped because of a lack of freshwater, which has to be shipped from the mainland. In the early 1990s, civil war stopped all tourism in the region.
Population 13 700
Map Croatia – Cc

Bracciano *Italy* Volcanic crater lake about 30 km (20 miles) north of Rome. It covers 57 km² (22 sq miles) and is fed by streams and underground springs. It reaches a maximum depth of 165 m (541 ft).
Map Italy – Dc

brackish Term used to describe water that is slightly salty, but with a salt content lower than that of sea water.

Bradford *United Kingdom* City in West Yorkshire county, 50 km (31 miles) north-east of Manchester. It has been the headquarters of the English wool trade since the 19th century – and its prosperous past is reflected in its grandiose Italian-Gothic town hall. Moorside, one of its many defunct mills, has been turned into an industrial museum. Bradford has a university and a 15th-century cathedral.
Population 295 050
Map United Kingdom – Ed

Bragança *Portugal* Medieval town 170 km (106 miles) east of the city of Oporto. The old walls and castle overlook the modern town. The House of Braganza ruled Portugal from 1640 to 1910, the title Duke of Braganza being given traditionally to the heir to the Portuguese throne.
Population 13 900
Map Portugal – Cb

Brahestad *Finland* See RAAHE

Brahmaputra (Yarlung Zangbo) *China/India/Bangladesh* River, 2900 km (1798 miles) long, rising on the north (Chinese) side of the Tibetan Himalayas, where it is called the Yarlung Zangbo, and flowing east to the Dihang Gorges, 5500 m (1800 ft) deep, where it cuts south across the mountains. From here, it turns west to join the Ganges in Bangladesh, where its main channel is known as the Jamuna. Its discharge is the greatest of any Indian river, but extremely variable, depending on the season. At Gauhati in Assam, where it is already 1.5 km (1 mile) wide, it will rise by 5 m (16 ft) during the monsoon rains. The river constantly changes its course as it moves huge quantities of silt towards the Bangladesh delta, often released by landslips caused by earthquakes. In Tibet there is a navigable section of 640 km (397 miles) at an altitude of 3650 m (12 000 ft).
Map India – Eb; China – De

Brăila *Romania* Inland port and capital on the banks of the River Danube, 140 km (87 miles) west of the Black Sea port of Sulina. Brăila has been an important port since the late 14th century. Its manufactures include food, textiles,

▼ **CROWN OF THORNS** The roof of the Cathedral of Brasília was designed to represent Christ's crown; the vast, circular nave is below ground level.

rolling stock, shipbuilding and metalwork. The centre of the city is dominated by the spire of a 19th-century church.
Population 238 500
Map Romania – Ba

Brandenburg *Germany* Largest of the former East German states with an area of 29 107 km² (11 236 sq miles), Brandenburg State includes Berlin. It is also the name of a port and industrial town on the Havel River 60 km (37 miles) west of Berlin, which produces tractors, steel, machines, clothes, leather, and river boats.
Population (state) 2 574 900; (city) 90 400
Map Germany – Cb

Brandon *Canada* Manitoba's second largest city situated on the Assiniboine River, 200 km (124 miles) west of Winnipeg.
Founded by the Canadian Pacific Railway in 1881, it is the centre of a rich farming area; its industries include oil refining, and production of chemicals and farm implements. Brandon has a university and a residential school for native Canadians.
Population 38 580
Map Canada – Fd

Brasília *Brazil* Provincial and national capital, 970 km (623 miles) north-west of RIO DE JANEIRO. Founded in 1960, this strikingly modern city was built on the largely unpopulated open plateau known as the Cerrados. Its basic shape is that of an aeroplane, with wings formed by high-rise residential blocks and the fuselage by the Avenue of Ministries, culminating, at the nose, in one of the most impressive buildings around the Square of the Three Powers: the presidential palace (the Planalto), supreme court and Congress.
The 'Three Powers' in the square's name refer to the three arms of government: the executive, judiciary, and legislature. Brasília's main buildings were designed by the Brazilian architect Oscar Niemeyer (1907-   ).
Only light industry is permitted in the city. Its airport has frequent connections to all parts of the country, but Brazil's principal international airport remains in Rio de Janeiro.
Population (city) 1 841 000
Map Brazil – Dc

Braşov *Romania* Major industrial city 140 km (87 miles) north-west of Bucharest. The country's largest city after Bucharest, Braşov manufactures tractors, lorries, machinery, tools, cement, glass, paper and chemicals.
The city is set in a steep valley in TRANSYLVANIA and is overlooked by Mount Tâmpa, 957 m (3140 ft) high. It was founded in 1211 by the order of Teutonic knights – its Orthodox church dates from 1392 – and thrived on its trade with the Orient.
Braşov has large numbers of Hungarian and German people.
Population 352 300
Map Romania – Ba

Bratislava (Pressburg; Pozsony) *Slovakia* Capital of Slovakia and the country's largest city. It lies on the River Danube near the Hungarian and Austrian borders and was part of Hungary from 907 to 1918, when the federation of Czechoslovakia was founded. It was then given its present name after Bretislav, its 5th-century Slav founder. Following the Turkish invasions, Bratislava was the capital of Hungary from 1541 to 1784 and the seat of its Diet (parliament) until 1848. Many Hungarian kings were crowned in the city's Gothic cathedral of St Martin. The city's Komensky University dates from 1467.
Outside the city centre lie post-1950 suburbs, with oil refineries and chemicals, rubber, electrical goods and engineering industries. The village of Stupara, with a Roman camp nearby, lies 16 km (10 miles) to the north.
Population 444 500
Map Slovakia – Ab

Bratsk *Russia* Industrial town in south-east Siberia on the Angara River, 460 km (286 miles) north-west of Irkutsk. Its industries, based on one of Russia's largest hydroelectric power stations (capacity 4500 megawats), produce cement, timber, paper and aluminium.
Population 258 000
Map Russia – Lc

Braunschweig *Germany* See BRUNSWICK

Brazil See p 108

Brazza *Croatia* See BRAČ

Brazzaville *Congo* National capital set in luxuriant tropical forest on the Malebo (Stanley) Pool, a broad section of the ZAIRE River, opposite the Zairean capital, Kinshasa. Founded in 1880 by the French explorer and coloniser Pierre Savorgnan de Brazza, Brazzaville became the capital of French Equatorial Africa in 1910 and expanded rapidly after the Second World War, when it was the headquarters of the Free French Forces in Africa. In 1944, the city was the site of a major conference, set up at General Charles de Gaulle's initiative, at which the postwar future of France's overseas colonies was discussed.
Brazzaville is a major river port at the lowest navigable point on the Zaire-UBANGI river systems, and is linked to the port of POINTE-NOIRE by the Congo-Océan Railway. It has become a transshipment point; nearly half of the goods handled are for other central African countries. Industries include railway repair works, shipyards and factories producing consumer goods.
Brazzaville has an international airport at Maya-Maya. The University of Marien-Ngouabi was founded in 1972 as the Congo's national university. Near the city are scenic rapids on the lower part of the Zaire River, the BATÉKÉ PLATEAU and the village of Linzolo, site of Congo's first Christian church.
Population 630 000
Map Congo – Bb

breccia Angular fragments of different rocks cemented together. See CONGLOMERATE

Brecknock *United Kingdom* See POWYS

Brecon Beacons *United Kingdom* Twin sandstone peaks in the south of Powys county, forming the highest massif of south Wales. The higher of the two, Pen y Fan, reaches 886 m (2907 ft). The name 'Beacons' comes from their ancient use for displaying signal flares.
The peaks lie in the 1344 km² (519 sq mile) Brecon Beacons National Park, a paradise for hikers, pony trekkers and naturalists. Its hillsides are largely grazed by sheep but there are many wild plants and animals – particularly birds.
Map United Kingdom – De

Breconshire *United Kingdom* See POWYS

Breda *Netherlands* City 88 km (55 miles) south of Amsterdam. It is a route centre, and in former times was heavily fortified, withstanding several sieges in the 16th to 18th centuries. In 1660, the exiled Charles II of England issued his Declaration of Breda, offering amnesty to all those who had opposed him and his father, Charles I (though excluding those specifically excepted by parliament). The medieval castle now houses the Dutch military academy; the old part of town with its many canals is a tourist attraction. The city manufactures metal goods, chocolate, beer and artificial silk.
Population (city) 128 190; (Greater Breda) 164 510
Map Netherlands – Bb

breeze Wind between Force 2 (light breeze; 6-11 km/h, 4-7 mph) and Force 6 (strong breeze; 40-50 km/h, 25-31 mph) on the BEAUFORT SCALE.

Bregenz *Austria* Capital of Vorarlberg State and a port and yacht harbour. It lies at the eastern end of Lake Constance (the Bodensee), 125 km (78 miles) west of Innsbruck. Bregenz began as a Celtic settlement and was later an important Roman camp. It has a museum of Celtic and Roman remains. In the Middle Ages it was ruled by the Counts of Bregenz; at the beginning of the 16th century it was purchased by Austria. The town's industries include food, electrical goods and textiles.
Nearby is the dense and beautiful Bregenz Forest (*Bregenzerwald*), which lies between Lake Constance and the 1679 m (5508 ft) high Hochtannberg Pass. The highest peaks in the area are Hoher Ifen (2232 m, 7322 ft) and Kanisfluh (2047 m, 6715 ft).
Population 27 300
Map Austria – Ab

Breidamerkurjökull *Iceland* Glacier on the south-east coast, covering an area of 250 km² (97 sq miles). It forms part of Europe's largest glacier, the huge VATNAJÖKULL.
Map Iceland – Cb

Brela *Croatia* See MAKARSKA

Bremen *Germany* Port and moated city beside the River Weser, 95 km (59 miles) south-west of Hamburg. It is the capital of the state of Bremen (404 km², 156 sq miles), established in 1947. The city's wealth grew from 1358, when it became one of the leading members of the Hanseatic League, a powerful group of Baltic trading cities. Medieval ramparts, now turned to gardens, encircle the cathedral, town hall and market place of the old city. A 9 m (30 ft) high statue, dating from 1400, of Roland, nephew of the Holy Roman Emperor Charlemagne, stands at the town hall.
Today, Bremen is the country's second biggest port after Hamburg. Its industries include shipbuilding, food processing, oil refining and the manufacture of textiles, electronics and motor vehicles.
Population (city) 551 200; (state) 683 100
Map Germany – Cb

Brazil

COFFEE BUILT THIS ENORMOUS COUNTRY WHICH IS THE WORLD'S BIGGEST DEBTOR – BUT WHICH CONTAINS FABULOUS RICHES

There is fabulous RIO DE JANEIRO, its carnival, its Copacabana and Ipanema beaches, and its Sugarloaf Mountain. There are the limitless, primordial rain forests of the AMAZON basin, a new futuristic capital city in the middle of nowhere, empty deserts, lush savannahs, wild highlands. There is an awful lot of coffee, and almost everything else, in Brazil – including 158 million people, half of them under 25 years of age.

Riches it possesses, beyond imagination – but little money. Brazil is the world's biggest debtor, owing US$123 billion to foreign banks at the last count. Inflation officially ran at 1174 per cent in 1992 (the unofficial figure was closer to 2000 per cent), or at an average of 731 per cent between 1985 and 1992. The economic reform programme which was started in 1991 had little impact, and Brazil's Finance Minister, Henrique Cardoso, who had briefly raised hopes of a degree of financial recovery, resigned from his post in order to run for the October 1994 presidential elections – which he won.

Surviving, struggling and somehow managing to hope again – despite all – Brazil has a fun-loving, football-crazy, car-mad population of staggering ethnic complexity and every known complexion. All the soccer stadiums together hold 4 million fans; their best-known player, Pelé, is still a world figure, though long retired. Brazil has won the soccer World Cup four times and has bred three motor-racing world champions – Emerson Fittipaldi, Nelson Piquet and Ayrton Senna. Brazilians are among the world's greatest consumers of alcohol – when oil became scarce and expensive in the early 1970s, they began building cars that run on industrial alcohol made from their vast sugar cane resources.

Distances stretch in this enormous land. It runs 4320 km (2683 miles) from north to south and a little more from east to west. It has 7400 km (4595 miles) of coastline and more than 43 000 km (26 700 miles) of navigable waterways. It is, in fact, the fifth largest country in the world, sprawling across almost half of the South American continent , with an area of 8 511 965 km² (3 286 488 sq miles).

Brazil owes its great size to the fact that the Portuguese, and not the Spanish, 'discovered' it. After the subcontinent was divided between Spain and Portugal in 1494, Spain split its share into provinces, which later became independent countries. Portugal, more interested in its Eastern discoveries, was hesitant to colonise Brazil until the 16th century. Later, it imposed direct colonial rule, so that, despite internal rivalries, Brazil emerged as a single, Portuguese-speaking nation.

The Portuguese colonists mingled with native Indians they found there, not to mention African slaves they imported when the Indians proved reluctant workers. Around 4 million slaves were shipped to the region between the 16th century and 1888, when slavery was abolished.Massive immigration from Europe, the Middle East and the Far East added to the racial mix between 1820 and 1939. Whole German colonies were founded in the south; the Japanese grew pepper around the Amazon estuary and started market gardening outside the great city of SÃO PAULO. The aftermath of the Second World War brought more immigrants. Today around 50 per cent of the people are white, 38 per cent of mixed origin, 6 per cent black, and the remainder Asians or native Amerindian.

There is a yawning gulf separating the few rich from the many, many millions of very poor. But, rich or poor, they are seen at their extrovert, ebullient and talented best when they celebrate at carnival time. The pre-Lenten festival is celebrated in many Roman Catholic countries, but no place does it quite like Rio.

Rio de Janeiro, a city and metropolitan area of over 9 million people, lies in a magnificent setting, stretching 20 km (12 miles) along a narrow strip of land between green mountains and blue sea. Visitors sailing into the beautiful harbour are easily convinced that 'God made the world in six days; the seventh he devoted to Rio', as the locals proudly maintain. As though in affirmation, a gigantic statue of Christ looks down on the city.

Rio is the former capital and the major city of the east coast, though not the biggest overall. It lies about midway between SALVADOR in the north and PORTO ALEGRE in the south. Along this entire seaboard the land rises sharply in a great escarpment, leaving a coastal strip averaging only 100 km (60 miles) in width – occupying a mere 7.7 per cent of Brazil's area, yet supporting 30 per cent of its population. History as well as terrain are responsible for this: the early Portuguese settlers landed here and immediately began cultivating the rich alluvial soil.

Originally, much of the region was tropical forest, but most of that has been cleared over the centuries and the area is now Brazil's political and economic heartland. It forms one of the country's five great natural regions, and within it are the states of Minas Gerais, Espírito Santo, São Paulo and Rio de Janeiro, the two last containing the great cities of the same names. The coastal escarpment averages about 790 m (2600 ft) above sea level, while the uplands of the interior rise to about 2130 m (7000 ft).

THE COST OF DESTROYING FORESTS AND DISPLACING PEOPLE

Largest of Brazil's regions is the 5 300 000 km² (2 050 000 sq mile) Amazon River basin, which covers about one-third of the country, in the north and west. This vast lowland area is covered by rain forest and drained by the 6440 km (4000 mile) river and its tributaries, such as the Xingu and Negro. Not for nothing is the forest known as the *inferno verde* ('Green Hell'), with a year-round temperature of about 27°C (81°F) accompanied by exhaustingly high humidity and an annual rainfall of 3000 mm (118 in). However, the Trans-Amazonian Highway and other new roads and railways have opened it up for development.

Huge areas of the forest have been cleared for crop growing and ranching, as Brazil seeks to improve its economy by encouraging farmers from the desperately poor north-eastern region to move into the cleared lands. But, meanwhile, big business concerns have displaced small farmers, and incorporated the land into vast ranches, so the immigrants from the north-east have exchanged one poverty trap for another. More than 8 per cent of the Brazilian rain forest (404 000 km², 155 948 sq miles) has already been destroyed.

Besides the catastrophic environmental implications, these schemes are proving disastrous from an agricultural point of view as well. The luxuriant rain forest appears to be supported by fertile soils, but this is not so. The forest lives on its own dead remains, and removal of the forest yields farmland which becomes barren within two to three years.

At the end of 1990, the government approved a plan to replant 1 million hectares (2.47 million acres) of forest over 10 years; at a cost of US$3 billion. But in 1992 – the year Rio hosted the UN Earth Summit on the environment – the Brazilian president dismissed his Environment Minister, José Lutzemburger, for publicly criticising development in the Amazon basin. A year later, a seven-year plan was announced by the government to open up the region to chemical and pharmaceutical companies, since the rain forests contain 'many resources as yet untapped' – and unknown. Meanwhile, large parts of the Amazon basin have been contaminated with mercury – used by gold miners to separate gold particles from the river beds. It was estimated in 1994 that 2000 tons of mercury had already been thrown into the Amazon River.

The Amerindians are the group most affected by development; they have suffered most at the hands of the gold miners. Displaced native peoples, some still living a Stone Age existence, are being absorbed into the mainstream of society – a sometimes painful process which has caused disquiet among those opposed to the disturbance of the Amerindians' traditional cultures and ways of life, and to changing the ecology of the forest itself. Indeed, there have been times, from colonial days until the present, when the process has seemed more like elimination than assimilation. There were more than a million 'Indians' in Brazil when the Portuguese arrived. Many were killed or enslaved; many more succumbed to imported European diseases, such as influenza, measles and smallpox, to which they had no natural resistance. Even into this century they were still being enslaved to gather wild rubber, and killed when

▶ **GOLD RUSH Miners swarm over a human ant hill in the Serra Pelada range in Para state. Many came from Brazil's drought-stricken north-east when the huge gold ore deposit was discovered in the 1970s.**

no longer useful. The Indian National Foundation (FUNAI) has done much to mitigate their lot, but there is again strong evidence of maltreatment, and even murder, in the Amazon clearances, and it is feared that in the whole country only about 250 000 Amerindians remain. Indeed, in 1993, more than 70 Yanomamis were killed by illegal gold and tin miners prospecting and mining in their reserve. At the same time, mining companies and local politicians began intensifying their lobbying to abolish a decree granting the 94 000 km² (36 285 sq mile) reserve to the Yanomamis, and the granting of similar rights to other ethnic groups in the region.

The impoverished north-eastern region includes large stretches of semi-desert, covered by a thorny scrub called *caatinga;* here, years can pass with barely a drop of rain, but enough falls on the coastal area to encourage both evergreen and deciduous trees.

The country's fourth region is the southern plateau, incorporating the states of PARANÁ, Santa Catarina and Rio Grande Do Sul, on the border with Uruguay. Coniferous and broadleaved trees cloak the high country to the north; prairie grasslands similar to the *pampas* of Argentina covering the south. This is *gaucho* country, a fertile temperate land with cattle ranches in the west and huge coffee plantations in the north and north-east.

Finally, there is the west-central region, where savannah grasslands scattered with trees roll across another vast plateau. It contains three stock-raising states: MATO GROSSO, Mato Grosso Do Sul, and Goiás. And it was in the unpopulated wilderness of Goiás that President Juscelino Kubitschek chose to build his dream city, BRASÍLIA, which succeeded Rio as the new capital in 1960.

Despite its surrounding shanties, Brasília remains a modest-sized city of some 1.8 million – far behind São Paulo, the largest city in Brazil. São Paulo is one of the fastest growing cities in the world, and one of the most powerful financial centres. Its population of about 18 million is increasing at an estimated rate of over 150 000 a year.

A BETTER LIFE

But the tough, cosmopolitan, go-getting 'Paulistanos', as the citizens of São Paulo are called, love it all – the bustle, the opportunities for carving out a better life, the cool, temperate climate and easy access to splendid seaside resorts and the port of SANTOS. This, for South America, is where the action is – far removed from the small missionary settlement founded here by Jesuits in 1554, though a replica of their original church has been built. The region all around proved an agricultural gold mine. Its rich, red soil has proved perfect for large-scale coffee growing. A century ago São Paulo really started to expand, as wealth created by coffee was ploughed into new industries. Vast numbers of immigrants arrived from Europe, bringing their own skills and enterprise. Now this city alone accounts for some 40 per cent of Brazil's industrial production. The 247 898 km² (about 96 500 sq mile) state of São Paulo, with a population of around 33 million, produces 65 per cent of the nation's industrial output, 62 per cent of its sugar, 50 per cent of its cotton, 33 per cent of its coffee and 90 per cent of its motor vehicles.

It is perhaps fitting that São Paulo was the city where Brazilians first claimed independence from Portugal in 1822. Their whole history, before and since, has been extraordinary. A sea captain named Pedro Alvares Cabral 'found' the Brazilian coast in 1500 and claimed the country for Portugal, but it was 49 years before the Portuguese got around to appointing a governor-general. In 1572, the colony was divided into two, with Salvador – capital of what is now the state of Bahia – as the northern capital and Rio the southern. African slaves worked huge sugar plantation-estates for colonists along the coastal strip. Other settlers pressed inland to the São Paulo and Minas Gerais regions, looking for gold, precious stones and more slaves. Dutch attacks on Salvador and RECIFE were beaten off in the mid-17th century, but in general – apart from an unsuccessful revolution in Minas Gerais in 1789 – colonial life changed little until it was changed forever, in 1808, by Napoleon.

In that year the Portuguese royal family fled to Brazil ahead of Napoleon's invading armies – on Britain's advice and escorted by British ships. They landed in Rio, which became the seat of the Portuguese empire. King João VI remained in Brazil until 1821 (when Napoleon died on ST HELENA Island), then returned to Lisbon, leaving his son, Pedro, as regent. The next year, Pedro proclaimed Brazil as independent, with himself as Emperor Dom Pedro I.

His reign did not, however, continue as well as it began. Pedro I lost what is now Uruguay and abdicated in 1831, returning to Portugal to subdue his brother, Dom Miguel, who had usurped the throne. His five-year-old son, another Pedro, became ruler under a regent. At the tender age of 15, Pedro took over as Dom Pedro II, and immediately embarked upon liberal reforms. Communications and schools were improved, corruption was attacked. Immigrants poured in by the thousands. He won a war against a Paraguayan dictator, and was moved to declare that he would rather lose his crown than let slavery continue.

This declaration did, indeed, lose him his crown. His daughter, Princess Isabel, acting as regent during his temporary absence, freed the slaves in 1888. Plantation owners turned against him; so did the military. He was, like his father, forced to abdicate, and a republic was declared in 1889.

AFTER BRAZIL'S EMPEROR

Brazil continued to prosper after Pedro's departure – until 1929-30, when the coffee market went the way of most others in the Great Depression. Dr Getúlio Vargas seized power by organising the working classes, and ruled as dictator until he was deposed in 1945. He was legally re-elected president in 1950, but committed suicide four years later after members of his bodyguard were implicated in a bid to kill a political opponent. Kubitschek, founder of Brasília, took over in 1955. Inflation beat him and his successors, and the military seized power in 1964. General João Figueiredo, last of a line of military presidents, took office in 1979, and was committed to a slow, step-by-step return to democracy. New parties were formed and finally, in January 1985, the 686-member congressional electoral college voted in 74-year-old Tancredo Neves, the opposition candidate. It was a landslide victory by 300 votes over his military-backed rival, Salim Maluf.

In his victory speech, Neves said that, despite national frustration over the military-backed administration's refusal to allow public elections, '... we have made the electoral college the very instrument of the government's defeat!' He continued, amid wild applause from congress delegates: 'I have come to propose economic, social and political change...' – and change did come about, although Neves never lived to be sworn in. He died in April 1985, having suffered from a benign tumour and also a long-standing and widespread abdominal infection. It would seem that Neves put off proper treatment until after the election: to his admirers, he was a martyr for democracy.

Vice-President José Sarney took over and democracy marched on. In May, congress voted unanimously in favour of choosing the next president by direct elections. The new constitution came into effect in 1988. It allowed new parties to be formed without restrictions – meaning that hitherto banned communist parties would be permitted. Amerindian rights were addressed (the constitution pledged to return all their lands to them by 1993), and massive social programmes were announced, such as US$2500 million aid for the poor, which provided subsidised food for the 13 million worst off, school meals for 20 million children (about 700 000 of Brazil's children are street children), 250 000 new homes and 500 000 homes to be connected to mains services.

Hopes soared once more in the nation that had the courage to join the Allies against Germany in both the First and the Second World War (Brazilian troops fought in Italy in 1944-5). For, despite massive debts, the crushing poverty of millions and the glaring social inequalities, the country seemed to be on the move again. However, by 1992-3, many of the hopes had been dashed: 1993 saw the indictment of President Fernando Collor de Mello on charges of corruption and criminal association. Popular resentment at corruption in public life became widespread and caused mass demonstrations throughout the country. Violent crime (including kidnapping and murder) also began to increase – leading to the the murder of street children in Rio de Janeiro in 1992 and a similar incident in a Rio shanty town a year later.

MASSIVE SCHEMES

Brazil's government has been facing many problems. However, economically, the country has much to offer. Although it has already yielded up fabulous riches, they are but a small fraction of its almost limitless resources.

Its first export was a red wood called 'paubrasil' used for dyeing in the 16th century – and from which the country takes its name. Next, the Portuguese introduced sugar cane from Madeira; in the following century, gold, silver and precious gems of many kinds were found in Minas Gerais. However, it was not

until late in the 19th century that Brazil found its greatest asset, coffee, although the plant had been introduced as early as 1727. Brazil is still the world's biggest coffee producer. Rubber grew wild in the Amazon basin – between 1880 and 1920, the burgeoning tyre industries created an enormous demand until Malaya broke Brazil's monopoly with rubber grown from seeds smuggled out of Brazil by an Englishman.

The Brazilian government has started concentrating on creating a basis for more industrial and agricultural expansion. Great new electrification schemes are coming to fruition to provide power. In the north-east, a US$4600 million dam at Tucuruí came on stream in

January 1985; it took eight years to build, and by 1989 was churning out 8000 megawatts – then the fourth largest hydroelectric complex ever built. Among areas to benefit is Carajás, site of Brazil's biggest concentration of mineral resources – including one of the world's largest iron ore reserves and large deposits of bauxite, gold, nickel, copper and manganese.

A new railway passes from Carajás through land suitable for livestock and agriculture to the Atlantic port of São Luís. However, Tucuruí almost pales into insignificance alongside the ITAIPÚ Scheme on the Paraná River, at the border with Paraguay. Completed in 1988, it ranks among the world's largest hydroelectric plants, with a capacity of 12 600

megawatts. And work has started on 10 more hydroelectric schemes.

Money has been poured into mechanising agriculture – particularly in the central and southern areas, where beef production is expanding.

The vast undertakings in the Amazon basin are bringing new life to the extraordinary city of MANAUS. At the turn of the century, during the height of the rubber boom, the city was known as 'the Paris of the Tropics'. Here, glittering audiences in the huge, ornate opera house would hurl diamonds onto the stage to show their appreciation when great stars of their time came to perform. For even getting there was no mean feat – deep in the jungles of what is now the state of Amazonas, Manaus could be reached only up the Negro River, by way of the Amazon itself – a 1450 km (900 mile) journey from the Atlantic coast. Until recently the only alternative way was by air, but now a new highway connects the destination to Brazil's main road system.

Ocean-going ships can navigate the Amazon and Negro rivers to dock in this free port, which handles the products of the whole upper Amazon region – rubber (still), rosewood and other timbers, beef and hides, jute

▼ CARNIVAL IN RIO Five frantic days of parades, balls, street dancing and general revelry precede Lent in Rio de Janeiro. The throbbing beat of samba music specially composed for the festival fills the air.

and, lately, oil. Tourism is expanding; visitors can stroll through fine botanic and zoological gardens and a natural jungle park. The population numbers around 1 million and temperatures average 27°C (81°F). The city overlooks the Negro River and is laced with numerous creeks spanned by little bridges.

Manaus has also become the destination for cruise ships sailing up the long Amazon, taking wealthy Europeans and Americans deep into Brazil's green heartland in search of the new eco-tourism. The city has benefited from the current surge of interest among Western conservation groups in the great rain forests of Brazil and the important role they play in the ecology of the entire planet.

But perhaps the most significant development for modern Manaus is the discovery of oil in the area. Because, whatever else Brazil may have, it does not have enough oil – only about

one-fifth of its needs – hence the development of alcohol as an alternative fuel for the country's motor cars.

The country is also short of coal and natural gas, and currently generates only one-third of its total energy needs. But oil has always been a major problem: despite recent finds around Manaus, most of Brazil's supplies come from the Middle East, and price increases sparked by Arab-Israeli conflicts in the early 1970s led to a slowing down in the development of heavy industries, emphasis being switched to agriculture and light industry.

The great hydroelectricity schemes are designed to increase self-sufficiency, and the country is now opting more and more for nuclear power, with several nuclear power stations in operation since 1983. Brazil has large uranium and thorium reserves. The search for more oil and coal sources has been intensified.

Despite a large-scale population drift from countryside to towns and cities, about 20 per cent of the labour force still work in farming, forestry and fishing. The last available surveys reveal that one-fifth of all Brazilians are illiterate, and 2 million children aged 7 to 14 do not even have a school to attend. Even when they have one, only 60 per cent stay long enough to become literate.

Brazil is, according to the Vatican, the largest Catholic nation in the world. However, great numbers of Brazilians wear both a cross and a voodoo charm, and worship African gods as well as the Christian god.

BRAZIL AT A GLANCE		
Official name Federative Republic of Brazil		
Area 8 511 965 km² (3 286 488 sq miles)		
Population 158 000 000 **Per km²** 18.6 **(Per sq mile** 48)		
Capital Brasília		
Government Federal republic		
Currency 1 real = 100 centavos		
Language Portuguese		
Religions Christian (88% Roman Catholic, 6% Protestant), voodoo		
Climate Mainly tropical and subtropical, though cooler on the southern coast and on higher lands. Average temperature in Rio de Janeiro ranges from 20°C (68°F) in July to 26°C (78.8°F) in February		
Land use Forest and woodlands 67%, grazing 19%, cultivation 8%, other 6%		
Main primary products Cereals, cassava, soya beans, sugar, oranges, cocoa, coffee, rice, cotton, tobacco, bananas, rubber, timber, fish; iron ore, bauxite, manganese, crude oil and natural gas, coal, chromium, nickel, tin, zinc, gold, silver, diamonds, phosphates, salt, quartz crystal, beryllium, graphite, titanium, tungsten, asbestos		
Major industries Agriculture, mining, iron and steel, oil and mineral refining, chemicals, wood pulp and paper, motor vehicles, machinery, food processing, consumer goods, textiles, rubber processing, fertilisers		
Main exports Sugar, coffee, soya beans, beef, metal ores and non-ferrous metals, hardwoods, machinery, chemical products		
Annual income per head (US$) 2540		
Population growth (per thous/yr) 19		
Life expectancy (yrs) Male 62 **Female** 68		

Bremerhaven *Germany* North Sea port at the mouth of the River Weser, 55 km (34 miles) north of Bremen, from which it takes its name. The harbour, which can handle ships of all sizes, is also Europe's biggest fishing port, with huge weekday auctions of the catch, and the centre of Germany's marine and polar research.
Population 130 400
Map Germany – Cb

Brenner Pass *Austria/Italy* Lowest of the main trans-Alpine routes – 1371 m (4498 ft) high – which links Innsbruck, in Austria, with Bolzano, Italy. A railway runs through the mountain and a road and motorway pass over it.
The Brenner Pass became the border between Austria and Italy in 1919.
Map Austria – Bb; Italy – Ca

Brescia *Italy* City 40 km (25 miles) east of Milan. It has two fine cathedrals, one dating from the 9th century and the other from the 17th. It also has a 12th-century palace. It manufactures silks and linens, firearms and metal goods.
Population 200 720
Map Italy – Cb

Breslau *Poland* See WROCLAW

Bresse *France* Fertile agricultural area east of the Saône Valley, composed of pastures, lakes and woods. Dairy and poultry farming and cheese making are the main occupations.
Map France – Fc

Brest (Brzesc) *Belarus* Port on the Bug River where this forms the border with Poland. Founded by Poland in the 11th century as Brest-Litovsk, it was annexed by Russia in 1795. It was the scene of the signing of a Russo-German peace treaty in March 1918, under which the Russians ceded much territory. From 1919 to 1939 it reverted to Poland, but after being occupied by the Germans for most of the Second World War, it once again became part of Russia (by then, the USSR). Today it is an important industrial centre and one of the oldest cities in Belarus. Its industries include saw-milling, cotton spinning, food processing, and engineering. It trades in timber, grain and cattle.
Population 260 000
Map Belarus – Ab

Brest *France* Largest city of Finistère department, near France's westernmost tip. Its shipyards, naval schools and an arsenal were founded in the 17th century. It was badly damaged by bombing during the Second World War. Rebuilt Brest today is an important naval base, seat of a university, and an industrial centre with shipyards, engineering and chemical factories. Its 15th to 16th-century fortress, containing the naval museum, and some 17th-century fortifications, remain.
Population 205 600
Map France – Ab

Brezhnev *Russia* See NABEREZHNYYE CHELNY

brickfielder Hot, dry, dusty wind blowing in south-eastern Australia, especially in summer. It blows from the north, ahead of a depression or trough of low pressure, which brings temperatures that soar to more than 38°C (100°F) for days.

Bridgeport *USA* Connecticut's chief industrial city on the coast opposite Long Island. Its manufactures range from electronics and electrical goods to helicopters, hot-air balloons and stage sets for Broadway.
Population (city) 141 690; (metropolitan area) 443 040
Map United States – Lb

Bridgetown *Barbados* Capital, commercial centre and chief port, with a deep-water harbour, founded by the British in 1628 on the south-west coast of the island. A port of call for cruise ships, it exports the island's sugar crop, rum and molasses, as well as serving as a transshipment point for other places in the region.
Population 5930
Map Barbados – Ba

Brie *France* Agricultural area east of Paris, between the Marne and Seine rivers. It produces the cheese named after the area.
Map France – Eb

brigalow scrub Type of scrub, consisting mainly of a low-growing species of acacia known as 'brigalow', which covers much of central Queensland in Australia.

Brighton *United Kingdom* English coastal city 75 km (47 miles) south of London, in East Sussex county. It was originally a small fishing port called Brighthelmston which grew in the 18th century with the then novel habit of sea bathing. The prince regent, later King George IV (1762-1830), patronised the town and commissioned the magnificent Indian Mogul fantasy, the Royal Pavilion, finally completed in 1815. The modern University of Sussex is nearby; the Brighton Polytechnic became Brighton University in 1992.
Population 154 200
Map United Kingdom – Ee

Brijuni (Brioni) *Croatia* See PULA

Brindisi *Italy* Adriatic port on the heel of Italy, with sea links to Greece. The old town, on a peninsula between two creeks, has a castle built by the Holy Roman Emperor Frederick II in 1227, and an 18th-century Baroque cathedral rich in 12th-century mosaics. Modern suburbs have spread since the city became a petrochemicals centre in 1960.
Population 91 780
Map Italy – Fd

Brisbane *Australia* Third largest city and state capital of Queensland. Brisbane lies on the Brisbane River, 20 km (12 miles) from its mouth. On its lower course, the Brisbane River is navigable for ocean-going ships, which makes Brisbane a leading port in Australia. It was founded in 1824 as a penal colony; it became a town in 1834 and the capital on Queensland's foundation in 1859. In 1864, it was largely destroyed by fire. Among the few surviving original buildings is the Old Windmill, built by convicts in 1829 to a faulty design so that it became a treadmill instead.
Brisbane's factories produce textiles, agricultural machinery and sugar; other industries include oil refining and shipbuilding. The city has three universities.
Population 1 301 700
Map Australia – Id

Bristol *United Kingdom* City in the county of Avon, 170 km (105 miles) west of London. Bristol was once a principal port of western England. However, the River Avon, on which it stands, is too narrow to handle large ocean-going ships, and Avonmouth – downriver on the Bristol Channel – has become its main outlet.

Although Bristol was severely damaged in the Second World War, it is particularly rich in old churches. Its cathedral dates back to the 12th century; its Church of St Mary Redcliffe was built between the 13th and the 15th centuries. The suburb of Clifton has a suspension bridge 80 m (260 ft) high over the river gorge; it was started by the engineer Isambard Kingdom Brunel (1806-59) and completed after his death. From the river, the gorge's 18th and 19th-century houses have a Mediterranean look. There is an aircraft factory in the suburb of Filton, and large new industrial sites have been established to the north. Bristol University was founded in 1909.
Population 397 000
Map United Kingdom – De

Bristol Channel *United Kingdom* Inlet, 137 km (85 miles) long and 8 to 69 km (5 to 42 miles) wide, of the ATLANTIC OCEAN, between South Wales and south-western England. Ocean tides in the channel often rise by 12 m (40 ft). Periodically a BORE (high wave) sweeps up the River SEVERN, reversing the flow.
Map United Kingdom – De

British Columbia *Canada* Mountainous province on the Pacific coast covering 947 800 km² (365 974 sq miles). Most of its population lives in the valleys, especially in the lower FRASER RIVER region around VANCOUVER. Furs, fish and gold attracted the first settlers. Modern mines produce copper, gold, silver and coal. Hydroelectric power has led to the development of electrometallurgical industries. Limited agriculture is concentrated on fruit, vegetables and dairy produce, and there is a salmon-fishing industry.

Tourists are attracted to the province by its scenic beauty and the mild climate of the southwest. Vancouver is the largest city and VICTORIA the capital.
Population 3 282 060
Map Canada – Cc

British Indian Ocean Territory *Indian Ocean* Territory comprising the five mid-ocean coral atolls forming the Chagos Archipelago. They total only 52 km² (20 sq miles) in area and lie 1899 km (1180 miles) north-west of Mauritius. They were bought from Mauritius in 1965 for £3 million. In the 1970s, the US navy built a large base on the biggest island, Diego Garcia, under a British-American joint defence agreement. The base has given the island strategic importance as it is the only nuclear base in the INDIAN OCEAN.

The archipelago, which has no permanent population, is famous for the diversity of its species of fish, seabirds, coral and molluscs.
Map Indian Ocean – Cc

British Virgin Islands *Caribbean* See VIRGIN ISLANDS

Brittany *France* Historic peninsula region of north-western France, with 3500 km (about 2175 miles) of jagged coastline, formerly known as Armorica. In the 5th and 6th centuries it was occupied by Bretons, a Celtic people who fled the Anglo-Saxon invasion of Britain, and from whom it derives its name. The peninsula became part of France in 1491.

In western areas, Breton is still spoken and many picturesque folk traditions survive, including local festivals of music and folk dancing, and the wearing of headdresses. The main occupations are farming and fishing. Industries include motorcar manufacturing and electronics. RENNES and BREST are the main cities.
Population 2 708 000
Map France – Bb

Brno (Brünn) *Czech Republic* Central industrial city producing textiles, machinery and chemicals. It hosts an international trade fair every autumn at the time of the wine harvest. The city, which lies 190 km (118 miles) south-east of Prague, was once the capital of Moravia. Today, it is the administrative and cultural capital of South Moravia. Its old town includes the 13th-century hilltop Spilberk Castle, a university, and an opera house named after the Czech composer Leos Janácek (1854-1928), who wrote his 'Sinfonietta' for the city.
Population 392 600
Map Czech Republic – Cb

Brod *Croatia* See SLAVONSKI BROD

Broken Hill *Australia* Mining city in New South Wales, 920 km (571 miles) west of Sydney but only 420 km (260 miles) from Adelaide, capital of South Australia. It is known as 'the Silver City' and has some of the richest silver, lead and zinc deposits in the world.
Population 23 260
Map Australia – Ge

Broken Hill *Zambia* See KABWE

Bromberg *Poland* See BYDGOSZCZ

Broome *Australia* Port on the north-west coast of Western Australia, 1650 km (1025 miles) north and slightly east of Perth. It was founded in 1883 as a port for the Kimberleys. At about that time, pearl grounds were discovered offshore, which were soon to supply 80 per cent of the world's mother-of-pearl. Cultured pearls and, recently, tourism have prevented Broome from becoming a ghost town since plastic buttons killed off the mother-of-pearl industry. At the foot of the surrounding sandstone cliffs, the fossilised tracks of dinosaurs are still visible.
Population 8900
Map Australia – Cb

brown coal Brown fibrous deposit, intermediate between peat and black coal. See LIGNITE

Bruce, Mount *New Zealand* Bird reserve on North Island, about 100 km (60 miles) north-east of the capital, Wellington. Some of the country's rarest birds, such as the flightless takahe and the kakapo are bred there.
Map New Zealand – Ed

Bruges (Brugge) *Belgium* Historic lace-making city about 90 km (55 miles) north-west of the capital, BRUSSELS. Formerly the capital of FLANDERS, now the capital of West Flanders Province, Bruges prospered from the 13th to the 15th century as a centre of the wool trade and member of the Hanseatic League, but fell into disuse when silt blocked its waterway outlets to the North Sea. A 13 km (8 mile) long deep-water canal, opened in 1903 and linking Bruges to the port of ZEE-BRUGGE, led to the city's economic recovery. Besides lace, its main industries today are chemicals, electronics, metal and textile manufacturing and shipbuilding.

Because of its long economic slump, Bruges has kept much of its medieval character. Tourists are attracted to its beautiful buildings and the quiet canals of the old town.
Population 117 100
Map Belgium – Aa

Brugh Na Boinne *Ireland* Collection of prehistoric barrows in the Boyne Valley, 11 km (7 miles) south-west of DROGHEDA, the chief of which is Newgrange. A passage grave – 85 m (280 ft) in diameter and 13 m (43 ft) high – it dates from about 2500 BC and is made of 180 000 tonnes of

BRIDGING THE GAP BY ROAD AND RAIL

The world's first bridge with a main span of more than 1.6 km (1 mile) will be in Japan. The Akashi-Kaikyo Suspension Bridge, over the Akashi Strait, will form one link in the highway joining the main island of HONSHU to SHIKOKU Island. Completion is due in 1998. The longest multi-span bridge – the Second Lake Pontchartrain Causeway in Louisiana, in the United States – is 38.4 km (23.9 miles) long.

SOME OF THE WORLD'S LONGEST BRIDGE SPANS

Name	Location	Span (m/ft)	Open
Akashi-Kaikyo (suspension)	Akashi Strait, Japan	1990/6259	1998
Tsing Ma (suspension)	Hong Kong	1377/4518	1997
Humber Estuary (suspension)	Hull, UK	1410/4626	1981
Verrazano Narrows (suspension)	New York, USA	1298/4260	1964
Golden Gate (suspension)	San Francisco, USA	1280/4200	1937
Bosporus II (suspension)	Istanbul, Turkey	1090/3576	1988
Tagus (suspension)	Lisbon, Portugal	1013/3323	1966
Quebec Railway (cantilever)	Quebec, Canada	549/1800	1917
Forth Rail (cantilever)	Firth of Forth, UK	518/1700	1889
New River Gorge (steel arch)	West Virginia, USA	518/1700	1977
Bayonne (steel arch)	New York, USA	504/1652	1931
Sydney Harbour (steel arch)	Sydney, Australia	503/1650	1932

stone. The corbelled central chamber, with stones inscribed with spiral patterns, is reached through a passage 20 m (65 ft) long. Other, similar, passage graves have been discovered at nearby Knowth and Dowth.
Map Ireland – Cb

Brunei See box right

Brünn *Czech Republic* See BRNO

Brunswick (Braunschweig) *Germany* Ancient capital and trade centre of Saxony, founded in AD 861, 55 km (35 miles) east of Hanover. Much of the city was destroyed in the Second World War, but a 12th-century Romanic-Gothic dome, the 14th-century town hall (*Altstadtrathaus*) and the 16th-century drapers' hall (*Gewandhaus*) remain in the midst of the town's modern buildings. Brunswick produces motorcars and machinery, electronics, and processed foods. It is a publishing centre and a university town.
Population 258 800
Map Germany – Db

Brussels (Brussel; Bruxelles) *Belgium* National capital in the centre of the country and, since 1958, the administrative headquarters of the European Economic Community (EEC) which was later renamed the European Community (EC) and more recently the European Union (EU). The city, developed after AD 580 on a marshy island in the River Senne, takes its name from the Flemish words *brock*, meaning 'marsh', and *sali*, meaning 'building'. Most of its surviving buildings date back only to the 15th century. In the walled Lower City, with its narrow streets, the superb central square is still intact. The city hall, built between 1402 and 1480, dominates the square with its 96 m (315 ft) high spire.
More recent buildings of note are the Palais des Beaux Arts (which houses the national art gallery), the royal library and palace, and the Palais de la Nation parliament buildings, all mainly constructed during the reign of King Leopold II (1865-1909).
The city where every sign – by law – is in both Flemish and French produces mainly textiles, chemicals, electrical equipment, machinery, rubber goods and lace. Brussels sprouts – first developed from cabbages in the region in the 13th century – are named after the city.
Population (Greater Brussels) 951 000
Map Belgium – Ba

Bryansk *Russia* City on the Desna River, 340 km (210 miles) south-west of Moscow. It has a locomotive plant, ironworks, flour mills and sawmills, distilleries and glassworks.
Population 459 000
Map Russia – Ec

Bryce Canyon National Park *USA* Beautiful area of 144 km² (56 sq miles) on the edge of the Paunsaugunt Plateau in southern Utah. The park has strikingly colourful rocks, shaped into bizarre spires and pillars by erosion.
Map United States – Dc

Bu Craa *Western Sahara* See BOUKRA

Bubiyan Island *Kuwait* Large island at the head of The Gulf. Bubiyan is a source of conflict with neighbouring Iraq, which has long made claims

Brunei

THE 'OIL SHEIKDOM' OF THE FAR EAST, WHERE THE WEALTHY STILL CHOOSE TO LIVE IN PRIMITIVE HOUSES ON STILTS

Oil made Brunei so rich that it has been dubbed the 'Shellfare State' after the oil company that has dominated its economy. Before the 1985-6 oil slump, this Muslim country on the north-west coast of Borneo had an annual income of about US$14 000 per head – one of the highest in the world. Today its citizens earn about two-thirds of that, but still enjoy unheard-of privileges: there is no income tax, for example;. education and health services are free, and cheap housing loans are available.
Brunei is ruled by Sir Hassanal Bolkiah, an autocratic prime minister-cum-sultan who practises polygamy. An election in 1962 was won by the Brunei People's Party, but no parliament was called and a rebellion was put down by British Gurkha troops. This party has been banned since then, and the sultan and his brothers continue to rule by hereditary rights stretching back centuries.
The country is divided in two by a tongue of land – part of the Malaysian state of SARAWAK. Broad, tidal swamplands cover the coastal plains of both parts. Inland, much of Brunei is hilly and covered with tropical forest: rainfall is heavy and varies from 2500 mm (98 in) to 5000 mm (197 in).
In the 16th century the sultans of Brunei controlled all of northern Borneo, but their influence declined and in 1847 they signed a treaty with Britain to gain the British navy's support in fighting pirates. British influence has been evident since then. Brunei became a self-governing British protectorate in 1959 and independent on 1 January 1984. Britain's Prince Charles attended independence celebrations, for which 5000 foreign workers built the sultan a new US$300 million palace.

Before Brunei struck oil, the economy depended on rubber, but the rubber plantations are now neglected. Oil production began in the 1920s, and oil and natural gas still account for almost all exports. Until oil income slumped, Brunei spent less than half its annual revenue and today it is one of the few countries with no external debt. In fact, investments totalling several billion US dollars have been made overseas by residents of the country.
There is a drive to make the country self-supporting in food – at present it is only 80% self-sufficient; meanwhile, the sultan has bought a cattle ranch in Australia, larger than Brunei itself, to supply meat.
About 60 per cent of the people are Malay and 25 per cent Chinese. The Chinese, usually denied citizenship, run nearly all the shops. In the forest, the Iban people live in long houses that accommodate 100 people, and sometimes have human skulls, reminders of head-hunting days, hanging from the roofs.
Many Malays who run expensive cars and motor launches still choose to live in *kampongs* (water villages) such as Kampong Ayer, in primitive houses on stilts. On the busy waterways among the houses, boats ferry children to school, housewives to market and men to their places of work.
BANDAR SERI BEGAWAN, the capital, has 46 230 inhabitants; a feature of the town is a huge golden-domed mosque.
In this strict Muslim state, alcohol is served only in hotels for Europeans, mixed bathing is forbidden on public beaches and there is little night life, even in the capital.

BRUNEI AT A GLANCE		
Official name Negara Brunei Darrussalam		
Area 5770 km² (2227 sq miles)		
Population 276 980 **Per km²** 48		
(Per sq mile 124)		
Capital Bandar Seri Begawan		
Government Hereditary sultanate		
Currency 1 Brunei dollar = 100 cents		
Languages Malay (official); Chinese, English		
Religions Muslim (63%), Buddhist (14%), Christian (10%)		
Climate Tropical, very humid and wet. Average temperature is 28°C (84°F) throughout the year		
Land use Cultivation 2%, meadows and pastures 1%, forest and woodlands 79%, other 18%		
Main primary products Crude oil, natural gas, rice, cassava, bananas, cattle, fish, timber		
Main exports Crude oil, natural gas, refined petroleum products, rubber, spices		
Annual income per head (US$) 14 000		
Population growth (per thous/yr) 27		
Life expectancy (yrs) Male 69 **Female** 73		

to it. In 1977, a joint committee of the two countries was set up to deal with the border claims, none of which was resolved. However, after the 1991 Gulf War, Iraq accepted United Nations Resolution 687, thus ending its claims to Bubiyan.
Map Kuwait – Bb

Buçaco *Portugal* Ancient forest, 30 km (20 miles) north of the university city of Coimbra. It was the

site of a Carmelite monastery from the 6th century until 1834, when the monks were expelled following the abolition of all Portuguese religious orders. It is also the site of the British victory against the French military forces in 1810 as part of the campaign to oust Napoleon's forces from the Iberian Peninsula.
Today, the forest's 105 hectares (260 acres) contain some 700 species of plants.
Map Portugal – Bb

Bucaramanga *Colombia* Capital of Santander State, 310 km (193 miles) north-east of BOGOTÁ. It is known as the city of parks because of its green and spacious centre. The city manufactures cigars, cigarettes and straw hats.
Population (city) 349 400; (metropolitan) 700 000
Map Colombia – Bb

Bucegi Mountains *Romania* Mountain range at the south-eastern end of the Carpathians. They rise to 2505 m (8218 ft) and are renowned for the fantastic shapes of their eroded summits. The slopes are popular for skiing.
Map Romania – Aa

Bucharest (Bucureşti) *Romania* Known as 'the Paris of Eastern Europe', Bucharest has been the capital of the Romanian state since its formation in 1862. It lies on a tributary of the River Argeş and is a green and pleasant city of spacious parks, broad, tree-lined boulevards, terrace cafés and restaurants, and handsome 19th and 20th-century public buildings.

The main thoroughfare of the Parisian area is the elegant Calea Victoriei (Victory Road). It was originally built in 1702 and is bounded by the Operetta Theatre, the Stavropoleos Church with its beautifully decorated walls and carved doors, and the country's National History Museum with its magnificent displays of gold models, jewellery, precious stones, plates, goblets and weapons – some dating from the 4th century BC.

Also worth visiting are the early 18th-century Cretulescu Church and the former Royal Palace, now the National Art Gallery.

Nearby is the Park of Culture and Rest, spread over 210 hectares (470 acres) and containing a superb folk museum exhibiting some 70 farmsteads, watermills, windmills and rural homes from all over the country.

Bucharest is said to have been founded in the 15th century by a shepherd named Bucur. During the Turkish invasion which began in 1541, it was repeatedly besieged. In 1659, it became the permanent capital of the Ottoman province of WALACHIA, taking over from the previous capital, the nearby city of TIRGOVISTE. Bucharest's old town contains the ruins of the 15th-century Princely Palace and the fine 16th-century Curtea Veche (old church).

Since 1945, the city has expanded rapidly. Its modern outer suburbs bristle with high-rise apartment blocks. Under a controversial project, buildings in a large area of the town centre, including several 18th-century historic churches, were demolished to make room for a new administrative complex.

Bucharest is the country's most important economic centre and by far the largest Romanian city: its population is seven times that of BRAŞOV, the country's second largest city. Its industries include power stations, engineering and chemical plants, as well as food processing, textiles and furniture.

About 120 km (74 miles) to the north lies grim and imposing Bran Castle. It is associated with the 15th-century ruler Vlad Tepes – known as Vlad the Impaler because of his practice of impaling unwelcome visitors on stakes – upon whom the fictional Count Dracula is said to have been based. Today, his castle is a museum.
Population 2 325 000
Map Romania – Bb

Buchenwald *Germany* Town about 80 km (50 miles) south-west of Leipzig. Between 1933 and 1945, it was the site of a Nazi concentration camp where some 560 000 people died – many the victims of inhuman scientific experiments.
Population 6000
Map Germany – Dc

Buckingham *United Kingdom* Town and county of south-central England, lying to the east of Oxfordshire.

The town, sited in a bend of the River Ouse, was devastated by fire in 1725 and never fully recovered its importance. The new county town is Aylesbury, 60 km (37 miles) north-west of London. Another town of some importance is Milton Keynes, the home of the Open University, which gives courses by radio, television and correspondence, while Buckingham itself has Britain's only private university. The Chiltern

▼ ORTHODOX WORSHIPPERS Sunday churchgoers flock to Bucharest's Church of St John on Bulevardul 1848, named after Europe's 'Year of Revolutions'.

Hills and attractive villages have made the region popular with visitors and commuters from London.
Population (town) 6550; (county) 639 100
Map United Kingdom – Ed

Budapest *Hungary* The country's capital straddling the River Danube in the north of the country, some 60 km (37 miles) from the Slovakian border. The great city, which houses one-fifth of the country's population, dominates Hungarian cultural, educational, political and economic life, producing 22 per cent of the country's manufactured goods.

It has two distinct parts – Buda on the west side of the river, and Pest (pronounced Pesht) on the east. The two merged to become one city in 1872. Buda dates back to Roman times, when the Romans built a military camp here. In the 15th century, King Matthias Corvinus (1458-90) fortified the later medieval town and made it Hungary's capital. Castle Hill, at Buda's heart, has a mixture of Romanesque, Gothic and Baroque buildings, painstakingly restored after the near total destruction of 1944-5, when the occupying Germans held out against Soviet forces for seven weeks. Among its narrow streets

and steep alleys are the former royal palace, now a superb complex of museums, galleries and a library, the colourful Matthias (or Coronation) Church, the Fishermen's Bastion, built on Castle Hill over a former fishmarket and fishermen's village, and numerous medieval churches and fine town houses.

Gellért Hill to the south is topped by the Liberation Monument 32 m (105 ft) high, which commemorates battles of the Second World War and offers panoramic views of the city.

Pest, on the plain to the east, originally a small fortified Roman settlement in the 1st century AD and later a busy trade centre, expanded rapidly after it merged with Buda. It has concentric tree-lined boulevards crossed by avenues such as Rákóczi Út, one of the main shopping streets, radiating from the inner city of Pest. The inner city has remnants of its old walls. Within it stands a 12th-century parish church with a Moslem prayer niche, used as a mosque during the Turkish occupation (1541-1686), and the Vigadó concert hall, where the composers Liszt, Brahms, Bartók and Kodály performed.

Pest also has many modern riverside hotels. The enormous domed parliament building dominates the river frontage to the north, with ministries and law courts nearby. Pest's so-called 'Little Underground' Railway, begun in 1895, is Europe's second oldest after London's. Two later underground lines were built between 1970-84.

There are several elegant bridges – completely restored following wartime destruction – linking Buda with Pest. The oldest, the Chain Bridge, was designed and built by the British engineers William and Adam Clarke between 1839 and 1849. Margaret Bridge links both Buda and Pest with the Margaret Island resort.

Extensive suburbs, whose industries include engineering, electrical engineering, food processing, oil refining, and the manufacture of chemicals and pharmaceuticals, paper and textiles, ring outer Pest. They contrast with the Buda side of the city, where villas, second homes and resorts for summer walking and winter skiing sprawl in the Buda Hills, and also with the charming southern suburb of Budafok, where wine cellars have been dug into the slopes beside the river.

Population 2 015 960
Map Hungary – Ab

Budva *Yugoslavia (Serbia and Montenegro)* Fishing village and resort on the Montenegran riviera, about 40 km (25 miles) south-west of Podgorica. It is one of the oldest settlements along the Adriatic coast – its first records date back to the 4th century BC. Budva was ruled in turn by Greece, Rome, Byzantium, Venice and Austria. The walled old town has a slender *campanile* (bell tower) and red-roofed buildings, many of which date from the 15th century. The new town, built after an earthquake in 1979, has hotels among pines and cypresses behind sandy beaches and the warm Adriatic waters. Lovcen mountain to the north rises to 1749 m (5735 ft).

About 5 km (3 miles) south-east across the bay and joined to the mainland by a causeway is Sveti Stefan (St Stephen), once a fortified fishing village with a chapel huddled on a rock, but now a hotel complex with about 50 houses.

Population 11 700
Map Yugoslavia – Bc

Budweiss *Czech Republic* See CESKÉ BUDEJOVICE

Buea *Cameroon* Town and tourist resort near the coast, on the southern slopes of Mount Cameroon. It was the capital of the German protectorate of Cameroon (Kamerun) and, after 1919, of British Cameroon. The beautiful town has marvellous views over the Tiko-Missellelé plains and much German architecture.

Population 29 940
Map Cameroon – Ab

Buenaventura *Colombia* The country's largest Pacific port, about 330 km (205 miles) west and slightly south of Bogotá. It is the main port for exporting coffee, which is grown in the Andes. It also handles imports for much of the country.

Population 190 000
Map Colombia – Bb

Buenos Aires *Argentina* Capital city, political, economic and cultural centre as well as port on the Río de la Plata (River Plate). It is often called 'the Paris of Latin America', because its traditional architecture is similar to that of late 19th-century France.

Founded by Spanish settlers in 1536, it started expanding only towards the end of the 19th century. The city has long since outgrown the boundaries of Buenos Aires district (200 km², 77 sq miles) and covers an area close to 4000 km² (1544 sq miles), including its various ethnic residential

▲ **BUENOS AIRES** The capital of Argentina on the Rio de la Plata is a city of wide, modern thoroughfares, such as the Avenue of the 9th of July.

areas and suburbs, such as the Italian Boca quarter, its slums, and its industrial areas. The city centre is the square known as the Plaza de Mayo, around the president's residence, Casa Rosada – 'the Pink House'.

Buenos Aires is the country's most important industrial centre and transport hub and is likely to carry on growing and eating away parts of the fertile pampas. The products of the pampas, such as meat, skins, oilseeds and cereals, are processed here, but many other products are manufactured in its factories, including motor cars, machinery, chemicals, textiles and clothes.

The city's name is a shortened form of its original name: Ciudad de la Santísima Trinidad y Puerto de Nuestra Señora la Virgen María de los Buenos Aires (City of the Most Holy Trinity, and Port of Our Lady the Virgin Mary of Good Winds). The name was chosen because the city was founded on Trinity Sunday and because in Spain Mary was a patron saint of sailors, who needed favourable winds.

Population (city) 2 922 800; (metropolitan area) 11 382 000
Map Argentina – Db

Buffalo *USA* Port and industrial city at the eastern end of Lake Erie in New York State. The ERIE CANAL links it to the Hudson river and New York city, some 465 km (290 miles) to the south-east. Its industries include grain milling, feed manufacture, chemicals factories and steel works.

The spectacular NIAGARA FALLS, on the border between the Unites States and Canada, are 25 km (16 miles) to the north.
Population 328 120; (metropolitan area) 1 189 300
Map United States – Kb

Bug *Poland/Ukraine/Belarus* East European river which rises near the Ukrainian city of L'viv, and flows north to form 200 km (124 miles) of the Polish-Belarus border. It then turns west to the Zegrzynek recreational lake just north of Warsaw, where it joins the Narew and Vistula rivers. The Bug, which is 772 km (about 480 miles) long, is sometimes known as the Western Bug to distinguish it from the Južnyj Bug, a river flowing farther south in the Ukraine.
Map Poland – Eb

Bugac *Hungary* See KISKÚNSÁG

Buganda *Uganda* Former kingdom of the Baganda, or Ganda people, on the northern shore of Lake Victoria. It formed the nucleus of what became the British Protectorate of Uganda in the 1890s, and was given a considerable degree of autonomy in the constitution under which Uganda became independent in 1962. Indeed, the *kabaka* (king) of Buganda was the country's first president. However, when he was deposed in 1966, the monarchy was abolished, and the kingdom divided into a cluster of districts.
Map Uganda – Bb

Büjük Agri Dagi *Turkey* See ARARAT

Bujumbura (Usumbura) *Burundi* National capital and port near the northern tip of Lake Tanganyika. Founded by the Germans in 1899, it now exports coffee, cotton and tea.
Population 241 000
Map Tanzania – Aa

Bukavu *Zaire* Chief town of Kivu region, formerly called Costermansville. It stands at the southern end of Lake Kivu, beside the Rwandan border. Refugee camps were established in the area in 1994 to house some of the more than one million Hutus fleeing the civil war in Rwanda.

About 30 km (19 miles) to the north of Bukavu is the Kahuzi-Biega National Park, a sanctuary of the rare mountain gorilla.
Population 418 000
Map Zaire – Bb

Bukhara *Uzbekistan* City about 225 km (140 miles) west of Samarkand. One of the oldest central Asian cities, Bukhara was an Islamic centre from the 8th to the early 13th century, when it was destroyed by the Mongol chieftain Genghis Khan (about 1162-1227). However, it was rebuilt and remains a place of mosques and minarets. From 1868 to 1990, it was under Russian rule, first under the czars and then under the Soviets. Bukhara is a commercial centre with silk, wool and cotton industries; its carpets are renowned throughout the world.
Population 225 000
Map Uzbekistan – Bb

Bukidnon *Philippines* Province and plateau area in northern MINDANAO. Long isolated by lack of access and dense forests, its population is now expanding rapidly. It has good grazing land and produces maize, sugar cane and pineapples.
Population 532 820
Map Philippines – Cd

Bukit Timah *Singapore* Central district of Singapore Island, named after the island's highest point, 176 m (577 ft) above sea level. It has a nature reserve with jungle trails winding through tropical forest. Bukit Timah Road, which passes through the district, links Singapore City with the north of the island and the causeway to Malaysia.
Population 124 120
Map Singapore – Ab

Bükk *Hungary* Group of wooded limestone hills rising to 959 m (3146 ft) west of Miskolc. The rounded hills, deeply cut by trout rivers, descend steeply to the city of Eger and the Great ALFÖLD Plain to the south. The hills are hollow with caves where stone tools and bones dating from the Neolithic period have been found. Szilvásvárad, a small resort 28 km (17 miles) north of Eger, is noted for its waterfalls and for its stud and riding school of Lipizzaner horses.
Map Hungary – Ba

Bukoba *Tanzania* Regional capital on the western shore of Lake Victoria, about 40 km (25 miles) south of the Ugandan border. It is the centre of a coffee-growing region.
Population 30 000
Map Tanzania – Ba

Bulacan *Philippines* Province of central LUZON, immediately north of MANILA. Mainly a rice-growing area, it also produces cement. The provincial capital, Malolos, was chosen as the capital of the short-lived Philippine Republic during the revolution of 1896-9; its constitution was drawn up here, in the Barasoain Church, in 1898.
Population 1 050 000
Map Philippines – Bc

Bulawayo *Zimbabwe* The country's second largest city after the capital, Harare, and chief town of the province of Matabeleland. It is also the largest industrial town in the country. Founded on its present site in 1894, it now produces textiles, tyres, building materials, processed foods, agricultural machinery, furniture and radios.

Nearby are deposits of gold, asbestos, chromium ore and tin. Tourists have been drawn to the city for its climate, and to the Khami Ruins, the remains of a 13th to 15th-century civilisation, 21 km (13 miles) to the west. A large nature reserve is close by.
Population 621 000
Map Zimbabwe – Bb

Bulembu *Swaziland* Town, formerly known as Havelock, near the border with the South African province of Eastern Transvaal, 40 km (25 miles) north of Mbabane. It is the site of one of the world's largest asbestos mines, which was opened in 1939. The asbestos is exported by means of a spectacular 20 km (12 mile) aerial cableway over the mountains to the railhead at Barberton, a town in South Africa. Publicity exposing the mineral as a health hazard has

dimished world demand for asbestos, but mining continues, though on a smaller scale.
Population 4900
Map Swaziland – Aa

Bulgaria See p 119

Bumtang *Bhutan* The largest and oldest monastic settlement of central Bhutan. It is 3000 m (9850 ft) up in the mountains and consists of two monasteries and a school. It is also the home of some members of Bhutan's royal family.
Population 1000
Map Bhutan – Ba

Bundala *Sri Lanka* See WIRAWILA TISSA

Bundi *India* City about 540 km (335 miles) south-west of Delhi. The Indian-born English author Rudyard Kipling (1865-1936) wrote about its magnificent Rajput Palace.
Population 768 150
Map India – Bc

Bungoma *Kenya* Main town of a densely populated district in western Kenya, bordering Uganda. It lies about 90 km (56 miles) north-east of Lake Victoria and is a trading centre on a transport route across the border.
Population 35 000
Map Kenya – Ba

Bur Safaga (Safajah) *Egypt* Red Sea port about 120 km (75 miles) south of the entrance to the Gulf of Suez.
Map Egypt – Cc

Buraydah (Buraidah) *Saudi Arabia* Oasis town situated 320 km (about 200 miles) north-west of Riyadh. Buraydah stands on a wide plateau and with its fertile land and plentiful water is an important wheat and date-producing area. It is situated at an important road junction which has encouraged growth in the industrial sector. There are also some attractive old Arab buildings.

The inhabitants are more puritanical and reserved than in the rest of the country, and visitors are advised to dress modestly and not to take photographs.
Population 184 000
Map Saudi Arabia – Bb

Burgan *Kuwait* The country's largest oil field, 19 km (12 miles) north-west of Mina al-Ahmadi – to which it is connected by pipeline. Oil was found here in 1938. Because the oil rises under its own pressure, production costs are among the lowest in the world. Reserves are estimated at between 40 000 million and 72 000 million barrels.
Map Kuwait – Ab

Burgas *Bulgaria* Major seaport on the Black Sea coast, about 60 km (37 miles) north of the Turkish border. Founded in the 18th century as a fishing village on the site of the medieval village of Pyrgos, it developed as a port and industrial town after Bulgarian liberation from the Turks in 1878, and expanded greatly after the communists took over the country. Today, it has food processing, engineering and oil-refining industries, and international trade funnels to and from the Bulgarian interior.
Population 205 000
Map Bulgaria – Cb

Bulgaria

THIS BALKAN STATE – ONCE THE SOVIET UNION'S MOST LOYAL COMPANION – HAS TURNED TO DEMOCRACY

In less than 50 years, Bulgaria has been transformed from a country of poor peasants into a relatively advanced and prosperous state. The town has replaced the village as the centre of Bulgarian life. The urban population has increased from 1 in 5 in 1946 to more than 3 in 5 in 1994. SOFIA, with a population of more than 1.1 million, is six times larger than it was 40 years ago. New towns such as DIMITROVGRAD and Pernik house the miners who process coal and metal ores. The seven or eight-hour working day usually starts early and ends in mid-afternoon, leaving time for a long dinner and a siesta before the towns spring to life for the evening.

Centring on the Maritsa River valley, Bulgaria straddles the north-west approaches of the strategic land bridge linking Europe and Asia Minor. It has a 354 km (220 mile) coast on the Black Sea which helps to attract millions of tourists each year, and a climate ranging from Mediterranean in the south-facing valleys to extreme continental in the mountains and the north. It is a fertile land with substantial mineral and water resources, natural beauty and a charm that has survived the dramatic changes of the last 50 years.

BULGARS, TURKS AND SLAVS

The Bulgars are a Turkic-Tatar race who conquered a Slav civilisation between AD 500 and 700. The Bulgars became a ruling aristocracy, but absorbed the Slav language and culture. In 1018, the territory was overrun by the Byzantine Empire, but the Bulgars returned to power in 1186. Then, in 1396, the country fell to the Turks, who imposed serfdom and severe taxation on the local inhabitants. A Bulgarian uprising in 1876 was brutally crushed. International outrage, and the Russian defeat of the Turks in 1878, laid the foundation for an independent Bulgarian state and formed the roots of Bulgarian-Russian friendship. The last formal links with Turkey were severed after the First Balkan War of 1912, when combined Bulgar, Greek, Serb and Montenegrin forces ousted the Turks.

Bulgaria was an uneasy ally of Germany in the Second World War. When Russian forces invaded in 1944 the communists seized power and in 1946, the country became the independent People's Republic of Bulgaria. A substantial Turkish minority – about 8.5 per cent of the population – remains to this day.

The relatively good life was created by a revolution in farming. Food output quadrupled between the 1950s and 1970s, after 13 million parcels of peasant land were made into 3453 collective farms. These, in turn, were fused into 792 large units in 1959, averaging 7000 hectares (17 300 acres) each.

Workers earned wages for their work on a farm or in a factory, and many families fed themselves from small plots exempted from the collective farming programme, growing fruit and vegetables, and raising poultry and pigs.

The collective farms increased output by using more machinery, fertilisers and irrigation, and by specialising by area: irrigating rice along the DANUBE in the north, wheat, maize, sugar beet, vines and fruit on the north-sloping Danube plain, sunflowers around Razgrad in the north-east, roses for perfume in the KAZANLUK basin, and cereals, tobacco, cotton and fruit in the broad Maritsa valley.

Mulberry trees are grown to feed silkworms around Svilengrad near the frontier with Greece and Turkey, and tobacco crops flourish in the southern valleys. Livestock are reared in the BALKAN and RHODOPE mountains.

While smallholdings were forged into collective farms, some 160 larger collectives were developed into agro-industrial complexes. They combined modern, crop-growing estates with food processing plants covering, for example, the extraction of sugar from sugar beet, cotton ginning (separating the seeds and seed hulls from the cotton fibres), and canning, freezing and packing plants.

During the communist era, these advanced agricultural methods yielded Bulgaria large earnings from exports. At the end of the communist era plans were announced to return the land to its original owners. Privatisation in agriculture got off to a quick and successful start. But the industrial sector struggled as the dissolution of the Eastern Bloc led to a marked drop in exports and a consequent drop in industial production despite desperate attempts to develop new markets. Unemployment and inflation soared; in 1991, inflation was 500 per cent. Western investors were reluctant to move – even though Bulgaria had extensive mineral resources. However, it is dependent on imports of petroleum and natural gas, and most of this comes from the Commonwealth of Independent States (CIS).

For decades, the country was led by Todor Zhivkov, an orthodox communist and loyal Soviet. But in November 1989 Zhivkov retired and when the first free elections since 1946 were held in 1990, the former Bulgarian Communist Party under a new banner was voted in. However, in October 1991 parliamentary elections brought in the anti-communist opposition party with a clear majority.

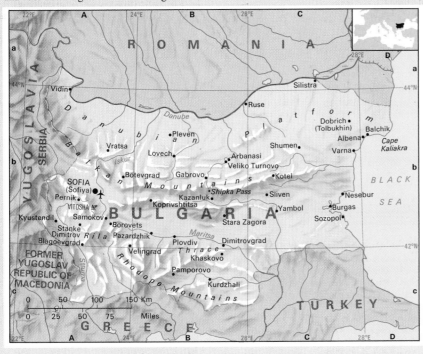

BULGARIA AT A GLANCE	
Official name Republic of Bulgaria	
Area 110 910 km² (42 822 sq miles)	
Population 8 831 168 **Per km²** 80 **(Per sq mile** 206)	
Capital Sofia	
Government Republic	
Currency 1 lev = 100 stotinki	
Languages Bulgarian (official), Turkish, Gypsy languages, Macedonian	
Religions Eastern Orthodox (85%), Muslim (13%), Roman Catholic and Jewish	
Climate Continental climate in the mountains and north; Mediterranean in the south-facing valleys. Average temperatures in Sofia range from -2°C (28.4°F) in January to 18°C (64.4°F) in July	
Land use Cultivation 37%, meadows and pastures 18%, forest and woodlands 35%, other 10%	
Main primary products Wheat, barley, maize, grapes, sunflower seeds, apples, tobacco, cattle, sheep, timber; coal, lignite, bauxite, lead, copper, zinc	
Main exports Tobacco, fruits, vegetables, wine, tinned foodstuffs, machinery, vehicles, metal goods, ores, clothing	
Annual income per head (US$) 1330	
Population growth (per thous/yr) Declining	
Life expectancy (yrs) Male 70 **Female** 76	

Burgenland *Austria* Eastern state noted for its agriculture and wine growing. Until 1921, it was part of Hungary; then it was transferred to Austria, following a referendum. However, Burgenland has retained much of its distinctive Hungarian character. Its capital is EISENSTADT.
Population 273 500
Map Austria – Eb

Burgos *Spain* Industrial city north of Madrid. It stands on the old pilgrimage road from France via the Pyrenees to Santiago de Compostela. Its products range from flour to tyres.

Burgos was the birthplace of El Cid (Rodrigo Diaz, Sire of Vivar, about 1043), a Spanish national hero who led the Christian campaign against the Moors. His remains are in the city's twin-spired 13th-century cathedral, one of Europe's finest Gothic buildings.

The surrounding Burgos Province is largely forested.
Population (town) 160 280; (province) 352 770
Map Spain – Da

Burgundy (Bourgogne) *France* Region of hills and valleys in east-central France covering 31 582 km² (12 194 sq miles) on both sides of the Saône river. It has more than 5000 vineyards producing the wines named after the region. For most of the 15th century, it formed part of a larger area ruled by the Dukes of Burgundy. Much of the region was annexed to France in 1482, other parts were held by Spain. DIJON is the main city.
Population 1 596 000
Map France – Ec

Burkina See p 121

Burma See MYANMAR

Burma Road *Myanmar (Burma)/China* A 1130 km (700 mile) mountain highway between Lashio, in Myanmar (formerly Burma), and Kunming, in the Yunnan Province of China. The Burma Road was built by the Chinese after the Japanese invaded their country and was designed to carry in military supplies from Burma. An amazing feat of engineering, it was started in 1937 and completed a year later.

It became increasingly important during the Second World War, when another strategic highway, the Ledo Road, was built in 1942-4 from Ledo, in northern India, to join the Burma Road through Myitkyina. By the early 1990s, the roads were no longer important, and were in a poor and neglected state.

burn A Scottish term for a small brook or stream.

Burren, The (Boirinn) *Ireland* Barren, windswept limestone region of about 500 km² (195 sq miles) on the west coast, on the south side of Galway Bay. The Burren rises in terraces to Slieve Elva, 345 m (1132 ft) high. Drainage is by underground streams, with fine cave systems. In sheltered areas there is rich flora and good grazing for sheep.

The region clearly had a large prehistoric population, for there are about 700 stone forts and dolmens – Stone Age burial chambers – including the impressive Cahermacnaghten Fort near Lisdoonvarna. The Burren is a national park centred on Kilfenora.
Map Ireland – Bb

Bursa *Turkey* Winter ski resort at the foot of Ulu Daği (2543 m, 1580 ft) and capital of the province of the same name, 100 km (62 miles) south of Istanbul. Known as Prusa in ancient times, it has a rich collection of mosques and tombs, and with its thermal springs is said to be among the most beautiful Turkish cities. Its main products are tobacco, textiles, carpets, silks, satin, tapestries and wine. Farmers in the surrounding province, which covers an area of 11 053 km² (4268 sq miles), mostly grow fruit, grain, grapes, cotton and tobacco.
Population (city) 775 400; (province) 1 603 140
Map Turkey – Aa

Buru *Indonesia* Island covering 8800 km² (3400 sq miles) of the Maluku group south of the Philippines. It is used as a penal colony for political prisoners.
Map Indonesia – Gc

Burutu *Nigeria* One of the ports in the Niger river delta port complex. It was created by the Niger River Transport Company in the 19th century, and has two ocean docks where goods are transferred to and from river boats. Burutu exports groundnuts and cotton brought down the Benue river from Cameroon and Chad. It is a base for offshore oil drilling.
Map Nigeria – Bb

Burundi See p 124

Bury St Edmunds *United Kingdom* Cathedral town, in the borough of St Edmundsbury, in the eastern county of Suffolk, 100 km (62 miles) north-east of London. It is named after the Anglo-Saxon King Edmund, who was killed by the Danes and reburied here in about AD 903.

An abbey was founded in 1020, but only its massive gatehouses remain substantially unaltered. A 15th-century church within the abbey precinct was raised to cathedral status in 1914. The bishopric is shared with Ipswich, 40 km (25 miles) to the south-east. The town has many fine old buildings, including the town hall, designed in the 1770s as a theatre by the architect Robert Adam.
Population (town) 31 180; (borough) 92 700
Map United Kingdom – Fd

Buryat Republic *Russia* Largely agricultural and forested autonomous republic of the Russian Federation. It lies in south-east Siberia, between Mongolia and Lake Baikal, and covers 351 300 km² (135 600 sq miles).

Although some crops are grown, cattle and sheep farming are the main occupations of the people, many of whom are Buddhists or Shamanists. The capital of the Buryat Republic is ULAN-UDE.
Population 1 056 000
Map Russia – Lc

bush Wild, uncultivated land, particularly a region covered with scrub or woodland, such as the vast uninhabited areas of Australia, New Zealand, southern Africa and parts of the United States.

Bushire (Bushehr) *Iran* Port on The Gulf, almost due south of the capital, Tehran. It exports wool, rugs and cotton. Enormous gas fields – Kangan (onshore) and Pars (offshore) – lie about 110 km (68 miles) to the south-east of Bushire As yet, they are undeveloped.
Population 133 000
Map Iran – Bb

Bushmanland *Namibia* See OTJOZONDJUPA

bushveld, bushvelt Area of open, uncultivated country in tropical and subtropical southern Africa with flora consisting mainly of tall grasses, low trees and thorn scrub.

Busia *Kenya* Town on the Ugandan border, about 45 km (28 miles) north of Lake Victoria. There is some agriculture around the town, but its main business is cross-border trade.
Population 35 000
Map Kenya – Ba

Butare *Rwanda* The country's second largest town after the capital, Kigali, 20 km (12 miles) from the border with Burundi. It is the seat of Rwanda's only university. Before the civil war in 1994, the population was 30 000.
Map Rwanda – Ab

Bute *United Kingdom* See STRATHCLYDE

Butha Buthe *Lesotho* Administrative and market town and the country's northernmost district. It covers 1767 km² (682 sq miles) of highlands where King Moshoeshoe I founded the Basotho nation in the early 19th century.
Population (town) 2500; (district) 92 200
Map South Africa – Cb

butte Small, prominent, flat-topped hill or isolated column of rock found in arid regions and rising abruptly from the surrounding area.

Butte *USA* Mining city in the Rocky Mountains in south-west Montana, 110 km (68 miles) from the Idaho border. It once had the reputation of being 'the richest hill on earth' and was one of the country's leading producers of copper, zinc, lead, silver and gold. It now produces cadmium and arsenic and has become one of the biggest inland ports in the United States.
Population 34 500
Map United States – Da

Butterworth *Malaysia* Port on the north-west coast of peninsular Malaysia facing Pinang Island. Ferries and the Pinang Bridge – opened in 1985 and with a total length of 13.5 km (8.4 miles) – link the port to George Town, Pinang's chief city. Butterworth has an oil refinery, tin smelters and an industrial estate at Mak Mandin. Nearby, the Royal Australian Air Force shares a base with Malaysian airmen.
Population 78 000
Map Malaysia – Ba

Buxar *India* City about 200 km (124 miles) east of Allahabad. Set on the southern bank of the Ganges River, it is of great importance to Hindus because of its connections with the early life of the Hindu god Rama. In 1764, it was the scene of a fierce battle in which the British defeated the Nawab Wazir of Oudh, a major semi-independent province of the Mogul empire, to secure their position in India.
Population 916 890
Map India – Dc

Burkina

OVERPOPULATION, OVERCULTIVATION AND SEVERE DROUGHT HAVE TURNED THIS ARID FRINGE OF THE SAHARA, FORMERLY UPPER VOLTA, INTO AN INFERTILE DUST BOWL

Nature seems to have loaded the dice against Burkina The good intentions of the leaders of this landlocked state in West Africa have not overcome the problems of its terrain and climate. Troubles have ranged from unemployment to famine, much of which can be blamed ultimately on the country's harsh environment.

A history of human folly has not helped, and the revolutionary government, which in 1984 changed the country's name from Upper Volta to Burkina (translating approximately as 'Country of Upright Men'), had difficulties tackling that which uprightness alone could not resolve.

Even the new name has caused confusion, many people referring to the nation as Burkina Faso. Shortly after the name change, however, the Director of the National Radio, Ouatamon Lamien, declared that the name of the country was Burkina – and that the word 'Faso' could be likened to the word republic, so that the use of 'Burkina Faso' together was similar to calling the country 'The Republic of Burkina'.

There is less argument about the state of Burkina's economy. The country's problems are partly a legacy of French colonial rule, which lasted from 1895 until independence in 1960. During this period, the French introduced a system of forced labour on the plantations in the neighbouring Ivory Coast, which robbed Upper Volta villages of their menfolk. A shortage of work in the country today means that many of the young men still go across the borders to Ghana and Ivory Coast to find jobs.

Since Burkina attained independence, military and civilian governments have followed one another in quick succession. The coup in August 1983 resulted in a mixed military and civilian government led by Captain Thomas Sankara; the new administration began work full of optimism for the future, embodied in the

change of name and new symbols of national identity – a new flag, a new national anthem and an amended constitution. The government promised to double the country's workforce by emancipating its women. But soon it banned all political parties and dissolved the National Assembly.

A further coup followed in October 1987 after Sankara called for the setting up of a single political party. This was too much for the Front Populaire headed by Captain Blase Compaore. A military commando unit assassinated Sankara and 13 of his colleagues, paving the way for the Front Populaire to take power. A year later Compaore introduced economic reforms encouraging the private sector and welcoming foreign investment

Little has changed in actual terms. In May 1992 the poverty-stricken country attempted a first, hesitant step towards a return to democracy. It held elections – but with a limited turnout. The poverty of the land remains a problem, and social conditions – particularly health, housing and water supplies – desperately need improvement. Moreover, 82 per cent of the people are illiterate, and, following the devaluation of the CFA franc in early 1994, inflation soared to more than 50 per cent.

Much of Burkina – both the north and west of the country – lies in the SAHEL, the arid (and ecologically vulnerable) fringe of the Sahara. Here, dusty, grey plains of infertile soil have been further impoverished by overgrazing and overcultivation.

THE VANISHING TREES

The land's deterioration has been rapid. Elders in the capital city of OUAGADOUGOU can remember a time when the central plateau of Mossi was covered with trees; now there are very few trees in a landscape that has become increasingly arid. In the extreme north the rains have failed year after year – six years running before 1973, followed by another drought in the mid-1980s and early 1990s. Food shortages have led to famine.

The rivers of the central plateau are the Black, Red and White Volta – headwaters of the main VOLTA River. They flow sluggishly (in years of drought, they sometimes dry up completely) through land which is swampy where it is not dry, towards the border with Ghana. Downstream, they flow more reliably through the wetter south where the landscape, though still flat, is greener with wooded savannah, against which the red soils look deceptively fertile. But the river valleys, which one might expect to be the most populated areas, are deserted: these are the breeding grounds of onchocerciasis (river blindness), bilharzia and malaria.

The country has considerable potential for tourism. The wildlife has had an easier time in Burkina than the country's 10 million people. The southern savannah is inhabited by elephants, hippopotamuses and many other creatures of the wild.

There is also some spectacular scenery in the south-west, where plateaus rise to a height of 150 m (500 ft) above the plains, and where waterfalls on the BANFORA Escarpment more than justify the long climb.

Some 80 per cent of the people live by farming. Food crops include sorghum, beans and maize, and cotton and livestock make up more than 70 per cent of the country's exports. Investors in Burkina's agriculture have often insisted on the growing of such cash crops, and this has aggravated the shortage of food for the local people – besides damaging the soil.

Unemployment and poverty have led to lawlessness; theft and muggings are frequent on the trains of Burkina's main railway line, which leads from Ouagadougou through Ivory Coast to the sea. Besides this danger, the railway is uncomfortable, crowded and not always reliable. Bush taxis, which drive from town to town, are a far preferable way of travelling long distances.

Most of Burkina's towns are concentrated along the railway line. Ouagadougou is not only the capital of the country but also the centre of the Mossi country – the homeland of the people who make up half of Burkina's population and who had kingdoms here as long ago as the 15th century. But all the peoples of Burkina are to be seen represented in Ouagadougou's bustling markets.

Tuareg traders from the north in flowing robes and elaborate head-dresses; Mossi, Bobo and Dioula farmers in bright cotton boubous or Western clothes; and tall, fine-featured Fulani or Peul. Non-Africans include mainly French and Americans; and there are Lebanese, too, who own and run many of the plentiful restaurants and bars.

About 80 per cent of the Burkinabé, as the people of Burkina are called, live in rural areas. In the villages, the conical huts of each family group are enclosed by a fence that sets them apart from other compounds. For most of these people, even in good years survival through the dry season from November to April is precarious, and is almost wholly dependent on the success of the previous harvest or, in recent years, on foreign aid.

BURKINA AT A GLANCE	
Area 274 200 km² (105 844 sq miles)	
Population 10 000 000 **Per km²** 36.5 **(Per sq mile** 94.5)	
Capital Ouagadougou	
Government Presidential republic	
Currency 1 CFA franc = 100 centimes	
Languages French (official), various indigenous languages	
Religions Indigenous beliefs predominate; Muslim (about 25%), Christian (10%)	
Climate Tropical; rains fall between May and November. Average temperature in Ouagadougou ranges from 25°C (77°F) in January to 33°C (91.4°F) in April	
Land use Cultivation 10%, grazing 37%, forest and woodlands 26%, other 27%	
Main primary products Sorghum, millet, maize, rice, livestock, groundnuts, shea nuts, sesame, cotton; manganese, gold, copper, bauxite, phosphates	
Major industries Agriculture, processed foods, textiles, mining, tyre manufacture	
Main exports Cotton, gold, livestock	
Annual income per head (US$) 290	
Population growth (per thous/yr) 28	
Life expectancy (yrs) Male 46 **Female** 50	

Buzău *Romania* Industrial city on the Buzău River, 100 km (62 miles) north-east of Bucharest. Set in a region of rich oil and gas fields, its industries produce chemicals, textiles, alcohol and processed foods.

Population 147 630
Map Romania – Ba

Bydgoszcz (Bromberg) *Poland* Central city, 235 km (about 145 miles) north-west of the capital, Warsaw.

It straddles the Brda River. 'Bydgoszcz Venice', a medieval complex of riverside granaries, mills and a royal mint, survives, along with Gothic churches and parts of the old town's fortifications near its confluence with the Vistula River. The city's factories produce furniture, paper, shoes, bicycles, refrigerators, machinery and books.

The first of the Nazi mass executions of Polish civilians began in Bydgoszcz market square on 'Bloody Sunday' in October 1939 when some 20 000 men, women and children were murdered by German troops.

Population 381 500
Map Poland – Bb

Byelorussia See BELARUS

Bystrá, Mount *Slovakia* See TATRA MOUNTAINS

Bytom (Beuthen) *Poland* Industrial city in the south, 13 km (8 miles) north-west of Katowice. It was founded in 1055 and, despite rapid industrialisation, with coal mining, steel making and metalworking in the 19th century, the medieval street plan and a restored church survive.

The town mines calamine, a zinc ore used in the manufacture of soothing skin lotions and creams. Other metals mined are iron ore, lead and silver. Today, Bytom has become the centre of the region's heavy industry.

Population 231 200
Map Poland – Cc

Byzantium *Turkey* See ISTANBUL

Burundi

A MOUNTAINOUS BUT CROWDED LAND THAT HAS BEEN DEVASTATED BY ETHNIC STRIFE

War, famine, disease, misery, massacres and waves of refugees have characterised the history of Burundi. As in neighbouring Rwanda, Burundi has suffered from the terrible problems of a border that cuts across ethnic enmities, leaving opposing groups of people unable to come to terms with living together in peace and harmony.

In 1972, in the first of many terrible encounters, tribal warfare resulting from centuries-old enmities killed perhaps 120 000 people in a tragic climax to 10 troubled years of independence – and an estimated total of 3 per cent of Burundi's population died.

This was, however, only the first of many such tragedies. Tensions came to a head in 1988 and again in 1991-2 in a new wave of massacres rooted in the traditional supremacy of the minority Tutsi tribe – a tall people originating from Ethiopia and Uganda who are vastly outnumbered by the shorter Hutu. Hutu make up 85 per cent of the population, the Tutsi only 14 per cent. The pygmy Twa, the original inhabitants, now constitute about 1 per cent of the population.

The farming Hutu displaced the Twa nearly 1000 years ago. The Tutsi arrived in the 16th century and established a feudal system under their *mwami* (king); the Hutu were reduced to serfdom, tending cattle on behalf of the Tutsi. In 1890 Burundi became part of German East Africa; from 1919 it was made part of the Belgian-administered territory of Ruanda-Urundi. Belgium's policy of ruling through the Tutsi aristocracy consolidated the traditional position of the Tutsi elite.

Burundi gained independence in 1962. Independent Burundi had several governments in four years while Tutsi and Hutu vied for power. A failed Hutu coup in 1965 led to Tutsi retaliation and the execution of most Hutu politicians and many others. The next year, after the king's son ousted his father, a military coup established a republic under Michel Micombero, another Tutsi. In 1972 he took dictatorial powers upon himself and there was an immediate Hutu uprising in which about 1000 Tutsi were killed.

True to Burundi's tragic history, the Tutsi retaliated brutally and systematically.

Some efforts have since been made to heal the wounds left by the war; Tutsi landlords have ceded some land to Hutu tenants. But the Tutsi continued to run the government and the army; when another military coup toppled Micombero in 1976, his successor was Jean-Baptiste Bagaza – yet another Tutsi. The first multiparty elections since 1965 were held in July 1993. Melchior Ndadaye took power, the first Hutu president in 30 years of independence, and embarked on an ambitious programme to end what he called 'Burundi's ethnic disease'.

Ndadaye's government was just 100 days old when the president was brutally murdered. This sparked another wave of violence in which about 150 000 people were killed and more than half a million people fled into neighbouring Zaire, Tanzania – and into Rwanda, which has been wracked by the same tribal warfare.

The violence continued in Burundi in April 1994 when an aircraft carrying Rwanda's president, Juvenal Habyarimani, and Burundi's new president, Cyprien Ntaryamira, returning from peace talks in Tanzania, was shot down. Ntaryamira, like Ndadaye, was a Hutu.

The October 1993 assassination followed by the death of Habyarimani and Ntaryamira led to an orgy of violence. The number of refugees fleeing the country shot up, with more than than 1 500 000 Burundians rushing headlong into disease-ridden refugee camps, many in neighbouring Zaire, where thousands died in front of a horrified world television audience. It was, as one journalist described, a scene from Hell.

Stability of a sort returned later in 1994 after an interim president was apointed to rule the country in conjunction with the prime minister and his cabinet. Many of the refugees began returning home – spurred on by the utter hopelessness of conditions in the camps

DENSE POPULATION AND INTENSE CULTIVATION

Burundi is a small, densely populated country that is intensively cultivated despite its mountainous terrain. West Burundi extends into the GREAT RIFT VALLEY, through which the Ruzizi River flows into Lake TANGANYIKA. Beyond a ridge to the east lie high grasslands. Much of the country is above 1500 m (4900 ft) and most of it slopes steeply. Soils are not rich, but there is enough rain to grow crops in most areas and the people manage to feed themselves from tiny plots.

The main subsistence crops are bananas, sweet potatoes, peas, lentils and beans in the mountains, and cassava near the lake and river banks. Some land holdings are big enough to grow coffee as a cash crop, and this is the source of 90 per cent of Burundi's very small export earnings. There are also small crops of tea and cotton. Cattle, goats and sheep are grazed on the grasslands. In 1991, the government began a programme of economic reform to diversify its agricultural exports and attract foreign investment. However, until the country is at peace, investors are likely to stay away.

One of the sources of tribal tensions may be population density and growth. The estimated average population density of more than 228 per km² (about 591 per sq mile) is extremely high for a country in which 90 per cent of the people live off the land – and the population is growing fast. The capital, BUJUMBURA, is the only large town. Add the problems of isolation – 2000km (1240 miles) to the Atlantic Ocean, 1400 km (870 miles) to the Indian Ocean – and difficult overland trade routes, and Burundi appears likely to be heavily dependent on foreign aid for many years.

BURUNDI AT A GLANCE	
Map Tanzania-Aa	
Official name Republic of Burundi	
Area 27 830 km² (10 746 sq miles)	
Population 6 350 000 **Per km²** 228 **(Per sq mile** 591)	
Capital Bujumbura	
Government Republic	
Currency 1 Burundi franc = 100 centimes	
Languages Kirundi and French (both official)	
Religions Christian (78%), tribal and Muslim	
Climate Equatorial; average temperature in Bujumbura is 24°C (75°F) through the year	
Land use Agriculture 52%, grazing 45%, forest and woodlands 3%	
Main primary products Cassava, sweet potatoes, maize, bananas, beans, coffee, tea, coconuts, cotton; nickel, uranium, vanadium, cobalt, copper, platinum (all of these not yet exploited), gold, tungsten	
Major industries Agriculture, forestry, fishing, beverages	
Main exports Coffee, tea, cotton	
Annual income per head (US$) 210	
Population growth (per thous/yr) 31	
Life expectancy (yrs) Male 39 **Female** 43	

Caacupé *Paraguay* Capital of La Cordillera department, about 50 km (30 miles) east of the national capital, Asunción. Founded in 1770 by the Spanish, it has become a popular tourist and pilgrimage centre, particularly in December on the Feast of the Blue Virgin of the Miracles. A modern basilica is dedicated to her.

The town produces oranges, tobacco, sugar and tiles.
Population 10 000
Map Paraguay – Bb

caatinga Thorn forest growing in the semi-arid region of north-east Brazil. It forms a virtually impenetrable deciduous jungle from which rise occasional taller drought-resistant trees such as acacias, wax-palms and giant cacti.

For three-quarters of the year it is a grey, leafless tangle, but after the short and unreliable rains, which come in December, it bursts into brilliant colour for two or three months, and then reverts to its dormant state.

Cabañas *El Salvador* Mountainous, coffee-producing department north-east of the capital, San Salvador, on the border with Honduras. Its capital is Sensuntepeque (population 10 000).
Population 136 290

Cabinda *Angola* Coastal province cut off from the rest of the country by the River Congo and a corridor, 30 km (19 miles) wide, belonging to neighbouring Zaire. The provincial capital, also called Cabinda, lies 390 km (242 miles) north of Luanda.

The province produces 75 per cent of Angola's oil following the discovery of offshore fields in 1966. Other main products are natural gas, hardwoods, cocoa, coffee, and palm oil.
Population (province) 163 000; (town) 13 500
Map Angola – Aa

Cabora Bassa Dam *Mozambique* See CAHORA BASSA DAM

Cacahuamilpa *Mexico* Largest natural caves in central Mexico, hidden in a valley near Taxco, 150 km (90 miles) south-west of Mexico City. They are carved by running water, and are part of a network of underground river channels that crisscross a limestone formation. Among the maze of passageways and chambers is a partly explored gallery 1380 m (4528 ft) long, 100 m (325 ft) wide and 70 m (230 ft) high, formed by slowly dripping water.
Map Mexico – Cc

Cáceres *Spain* Picturesque market town and provincial capital, 240 km (150 miles) south-west of Madrid founded in 74 BC by the Romans. It manufactures cork, leather goods, textiles and fertilisers. Its province, of the same name, produces wool, cereals and *embutidos* – red sausages – for which it is famous.
Population (town) 74 590; (province) 414 460
Map Spain – Bc

Cacheu *Guinea-Bissau* River in northern Guinea-Bissau with a length of about 200 km (125 miles). Cacheu is also the name of a region, and of a town on the left bank of the Cacheu estuary. Founded by the Portuguese in 1630, the town was the capital of Portuguese Guinea from the 17th century until 1879. It declined after this,

especially during the guerrilla war of independence (1964-74), but remains a regional capital, market town and fishing port.
Population (region) 150 000
Map Guinea – Aa

Cádiz *Spain* Seaport on the South Atlantic coast, just north-west of the Strait of Gibraltar.

The town of Cádiz is one of Europe's oldest, founded in 1100 BC as a trading post by the Phoenicians from the Middle East. It flourished under the Romans, who called it Gades, but declined after the fall of Rome in AD 476. It experienced a revival again after the discovery of the New World and became the centre of Spanish trade with the Americas – despite English attacks between the 16th and the 19th century. In 1587, Sir Francis Drake defeated a Spanish fleet here and in 1597, 10 years later, the Earl of Essex partly destroyed the town.

The town exports sherry, salt, olive oil and cork grown and produced in its adjacent province, also called Cádiz.
Population (town) 154 350; (province) 1 078 400
Map Spain – Bd

Caen *France* Capital of Calvados department in Normandy, 110 km (68 miles) north-west of Rouen. Parts of the castle built by William of Normandy – who led the Norman conquest of England in the 11th century – are still standing. The city was rebuilt after the Second World War following its devastation during the Allied invasion in 1944. It has a university as well as steel-making, electrical and mechanical industries.
Population 187 600
Map France – Cb

Caerdydd *United Kingdom* See CARDIFF

Caernarfon *United Kingdom* See GWYNEDD

Caesarea (Qesaria) *Israel* Seaport and resort 35 km (22 miles) south of Haifa, founded by Herod the Great in the 1st century BC. St Paul baptised the Roman centurion Cornelius in Caesarea and was imprisoned here in AD 57-59. The city's Roman remains include the harbour and aqueduct and a reconstructed theatre. The walls and moat of a crusader town on the site still survive, though the town was largely destroyed by Muslim invaders in 1265.
Population 37 000
Map Israel – Aa

Cagayan *Philippines* Largest river in LUZON, 350 km (220 miles) long, rising in the southern Sierra Madre and flowing northwards through a broad valley bordered in the west by the Cordillera Central, to the Babuyan Channel near Aparri. It is also called the Rio Grande de Cagayan to distinguish it from the Cagayan River of Mindanao. In all, with its tributaries, the Chico and Magat, the river drains an area of some 10 000 km² (3900 sq miles). Rice, tobacco and other crops grow in the Cagayan Valley, but little irrigation is practised. Cagayan Province comprises the lower part of the valley.
Map Philippines – Bb

Cagayan de Oro *Philippines* City on the north coast of MINDANAO, at the mouth of the Cagayan River. It is capital of the region and of Misamis Oriental Province. Largely an administrative,

transport and trading centre, Cagayan de Oro also has an excellent harbour.
Population 340 000
Map Philippines – Bd

Cagliari *Italy* City on the south coast of Sardinia, and capital of the island. It stands on a series of treeless low hills and is a place of stone and wind. It was founded by the Phoenicians and has a Roman amphitheatre, hewn out of solid rock. To the west, a canal port serves a large industrial estate which specialises in petrochemicals. The city's salt pans are the biggest in Italy.
Population 203 250
Map Italy – Be

Caguas *Puerto Rico* Commercial and industrial city lying in a fertile agricultural region, 25 km (15 miles) south of San Juan. It produces tobacco and sugar, and manufactures leather goods, cigars and cigarettes.
Population 123 450
Map Caribbean – Bb

Cahora Bassa Dam *Mozambique* Hydro-electric dam on the Zambezi River, 210 km (130 miles) upstream from the town of Tete. Begun in 1969, it was commercially operational by mid-l977. By the mid-1980s, South Africa took 98 per cent of the output. However, during the Mozambican civil war, guerrillas sabotaged the lines, with the result that no electricity could be exported. In 1993, a contract was signed to upgrade the facility so that exports to South Africa could be resumed.

The Portuguese government is financially responsible for the dam until the end of the 20th century, by which time the capital investment will have been paid off and Mozambique will take it over in its entirety. With an installed generating capacity of 2100 MW, Cahora Basso is surpassed in Africa only by the Inga (Zaire) and Aswan (Egypt) dams.

Cahora Bassa Lake, which is some 240 km (150 miles) long, extends from the dam almost to the Zambian border. It covers about 2660 km² (1027 sq miles).
Map Mozambique – Ab

Cainozoic era See CENOZOIC ERA

Cairngorm Mountains *United Kingdom* Range within the Grampian Mountains on the border of the Scottish Highland and Grampian regions. The highest peak is Ben Macdui, at 1309 m (4296 ft). The area has been developed for winter sports centred on the village of Aviemore. A form of yellow-brown quartz, known as cairngorm and used in jewellery, is found in the mountains.
Map United Kingdom – Db

Cairns *Australia* Tourist resort and port on the north-east coast of Queensland, about 1400 km (875 miles) north-west of Brisbane.

Founded in 1873 as a government customs collection point, it is a commercial centre for the surrounding sugar cane country, and has a bulk storage terminal for sugar exports. An access point for the northern section of the Great Barrier Reef Marine Park, the city is a well-known big-game fishing base. Rain forests to the north are a declared World Heritage area.
Population 64 460
Map Australia – Hb

Cairo (El Qahira) *Egypt* National capital and Africa's largest city, Cairo stands at the head of the Nile delta, about 160 km (100 miles) south of the Mediterranean Sea. It is a noisy, traffic-jammed, cosmopolitan city, a jostling crossroads of East and West. A forest of minarets proclaims its Islamic affinity, but beneath them swarms a mixed population of Arabs, Turks, Coptic Christians, Jews, black Africans and Europeans.

It is a city of two hearts. Old Cairo in the south is a walled enclave of old stone houses crowding together over narrow, crooked streets, some still unpaved, of packed bazaars as old as the city, and of the buzz of prayers chanted daily in its more than 500 mosques. Above it, on Moqattam Hill, dividing the old city from the new, stands the Citadel, a fortress built in AD 1177 by the sultan Saladin.

To the north and west is new Cairo, a working city of broad avenues and modern blocks – offices, banks, hotels, government ministries and an opera house that staged the first performance of Verdi's *Aida* in celebration of the opening of the Suez Canal. Here, too, is the famous Egyptian Museum where there are superb collections of Egyptian treasures and antiquities.

Cairo was originally a riverside military camp set up in AD 641. In 969, the Fatimid Muslim rulers of Libya conquered Egypt and began building the walled city, which they called al-Kahira, meaning 'the Triumphant One'. Saladin extended the city in the 12th century, when it became the capital of the Mameluke sultans until the Ottoman Turks took over Egypt in 1517. Egypt, and

▼ **MUSLIM METROPOLIS. The minarets of Sultan Hasan Mosque (left) pierce modern Cairo's skyline at sunset. The mosque was built in about 1361 when the city already had nearly half a million people.**

Cairo, regained independence under Muhammad, or Mehmet, Ali (1769-1849). Appointed governor by the Turks in 1806, Muhammad Ali ruled for 43 years, winning the country autonomy and creating modern Egypt.

At EL GIZA, 10 km (6 miles) south of the city on the west bank of the Nile, are the Pharaohs' burial pyramids, including the great pyramid of Cheops and the Sphinx.

Cairo's Al-Azhar University, founded in 972, is one of the oldest in the world.

Population (city) 10 361 000; (metropolitan area) 16 000 000	
Map Egypt – Bb	

Caiseal *Ireland* See CASHEL

Caithness *United Kingdom* See HIGHLAND

Cajamarca *Peru* Northern department centred around the wide valley of a tributary of the Río Marañón. The surrounding hills support a thriving dairy industry.

The capital, also Cajamarca, is where the Spanish conqueror Francisco Pizarro captured and killed the ruling Inca, Atahualpa, in 1533. The only Inca buildings to have survived are the Ransom Chamber and the Inca Bath. The cathedral and the churches of nearby Belén and San Francisco are notable examples of colonial architecture.

Population (department) 1 270 600; (city) 92 600	
Map Peru – Ba	

Calabar *Nigeria* Coastal town, capital of Cross River State and former capital of Southern Nigeria. Calabar lies 640 km (398 miles) southeast of Lagos, on the mouth of the Calabar River, which supplies the town with abundant catches of fish. Indeed, much of the population lives by fishing and subsistence farming. The surrounding land is still largely covered by rain forest.

Originally settled by the Efik and Efut peoples, Calabar became a slave-trading centre soon after the Portuguese arrived in the late 15th century. Portuguese traders were followed by German, Dutch, French and English missionaries, who introduced Christianity and Western education to the local population. Calabar remained a booming centre even after the abolition of slavery, when trade shifted to palm oil.

The city served as an important naval base during the Biafran War of 1967-70. Today, Calabar retains much of its colonial character. Each year in October, the city takes on a different face during the traditional *sekiapu* (masquerade) festival, when its streets are filled with masked drummers and dancers.

Population 126 000	
Map Nigeria – Bc	

Calabria *Italy* Region making up the toe of the Italian boot. It covers 15 080 km² (5820 sq miles) and takes in the provinces of COSENZA, Catanzaro and REGGIO DI CALABRIA. It is the poorest of Italy's regions, and its poverty has induced emigration to the more prosperous north.

Calabria has hardly any modern industry; agriculture, chiefly olive growing, is hampered by the mountainous terrain. Tourism is becoming increasingly important.

Population 2 037 690	
Map Italy – Fe	

Calais *France* Channel port on the extreme north coast, handling 4 million passengers and 1 million motor cars each year travelling to and from Britain. Submerged in Roman times, it was established in the 7th century when a channel was cut from a lagoon through the dunes to the sea. In 1347, it was conquered by Edward III of England after a year-long siege. It was regained by France only in 1558. It was largely rebuilt after its destruction during the Second World War. Chief industries now are clothing, lace making and food processing.

The French sculptor Auguste Rodin's monument *The Burghers of Calais* commemorates six citizens who offered their lives to spare the town following its capture by Edward III. They, and the town, were saved by Edward's queen, Philippa, who interceded on their behalf. The monument stands in front of the town hall.

Population 75 800	
Map France – Da	

Calbayog *Philippines* See SAMAR

calcite Crystalline form of natural calcium carbonate, the basic constituent of limestone, marble and chalk.

Calcutta *India* Port, capital of West Bengal, and India's largest city. It is situated in the north-east, 145 km (90 miles) from the Bay of Bengal and the mouth of the Hooghly River. It was founded in 1690 by the East India Company as a trading post on the river bank. The settlement prospered until it was attacked by the Nawab of Bengal in 1756. During the conflict, 126 Europeans, out of a total of 146, suffocated while being held captive in a confined room – the infamous 'Black Hole of Calcutta'. British troops under the command of Robert Clive set out from Madras to recapture the town, and by 1757 the British had re-established control of Bengal. In 1758, they

A DAY IN THE LIFE OF A RURAL MIGRANT IN CAIRO

Ahmed sits on his bunk bed and chews the end of his pencil. 'I bought a radio today,' he writes in the long curves of Arabic script, 'and I'm making friends with the three men who share my room.' He is not used to writing letters, and today's letter is especially difficult.

This is the first time he has slept away from his village in the Nile delta in all his 21 years. This morning he stepped off the train at Cairo's central railway station and into a new life. Lured from his village by the hope of work in the city, he ignored the anxious discouragement of his village friends. Now, as he writes to his family at the home he so recently left, he does not dare admit his growing fear: that a job may not be as easy to find as he had hoped. There are hundreds of thousands of people like Ahmed, all seeking a better life than the rural poverty trap of the Nile delta.

It was the images of city life on the village's new communal television that had filled Ahmed's head with dreams. The tall buildings of Cairo held out the promise that there was money there for the making. City dwellers in their smart Western clothes seemed to proclaim their prosperity by their very appearance. There must be jobs with good wages in the city.

In the village the piece of land Ahmed shared with his brothers was too small to offer them all a good living. So he took his modest savings, put his spare ankle-length *galabieh* (outer garment) in a bag, made his farewells and, clutching the address of a contact who could provide lodgings, he took the train that runs close to the River Nile, Egypt's lifeblood since the days of the pharoahs, northwards to Cairo.

The scale and the bustle of the city surprised him, but he found his way from the station to the contact's address – a small, dark dwelling in a narrow back street. He knocked on the door, asked for a place to stay and was shown to his room, where three other people were talking. He discovered that they all came from different villages in the country, and was impressed to see that all three had smart radios; they must have found a way to make money.

Talking to them, however, had revealed that none had a full-time job. Casual work sometimes came their way, or they would get goods from a wholesaler to sell on the street. But they had no trading licences, and there was always the risk of being picked up by the police. Ahmed wondered if he might find a job in a repair shop. But he was a farmer – and what skills could a farmer offer a mechanic?

But there was time yet to sort out these worries. And after all, a few small jobs might be enough. He might be able to save up the deposit on an airline ticket to Saudi Arabia. And everyone knows that a construction worker in Saudi Arabia can earn in a year what it would take a peasant farmer 20 years to earn in Egypt.

built Fort William, an impregnable stronghold. They kept the area around the fort clear, so that the Maidan, where today's racecourse and cricket fields are sited, was saved from being built over.

In 1774 Calcutta became the capital of British India; in 1887, in terms of a proclamation by the Empress of India, Queen Victoria, it became the imperial capital. The city's wealth grew rapidly as a result of the trade in cottons, silks, indigo, opium (to China) and, later, jute and tea. With the coming of jute processing, the town became an industrial centre. Although the imperial capital was moved to Delhi in 1911, Calcutta continued to prosper.

The downturn came shortly after India's independence in 1947. There were two factors that adversely affected the city: firstly, the arrival of millions of refugees from, and the loss of the natural trading hinterland of, East Bengal (now Bangladesh); secondly, the Hooghly River began to silt up as a result of the discharge of the Ganges shifting east over a number of years, and could no longer provide enough water for the city or adequate navigation. To counter this problem, India built the Farakka Barrage, which was completed in 1971.

Calcutta still produces and exports jute, as well as rice, tea, textiles, chemicals and paper. But the city has never been able to regain its premier position among Indian cities. To compound its problems, it has had to cope with rapid urbanisation and a population growth equalled by few cities in the world.

Population (city) 4 388 300; (metropolitan area) 10 916 300

Map India – Dc

Caldas *Colombia* Coffee-growing state covering 7888 km² (3045 sq miles) in the centre-west. It produces 30 per cent of Colombia's coffee, the country's chief export. Its capital is MANIZALES.

Population 909 850

Map Colombia – Bb

caldera Large, basin-shaped crater formed by the collapse of a volcanic cone, or by a volcanic explosion that has removed the top of a cone. The largest known caldera is ASO in Japan.

Caledon (Mohokare) *Lesotho/South Africa* River rising in the Drakensberg of northern Lesotho. It flows 480 km (300 miles) south-west to the Orange River in South Africa and forms much of Lesotho's western border with South Africa. Its fertile valley, now a maize-growing region, was once inhabited by San (Bushmen), many of whose cave paintings survive.

Map South Africa – Cc

Caledonian Term used for a geological episode of mountain building during the Palaeozoic era. See GEOLOGICAL TIMESCALE.

Calgary *Canada* Alberta's second largest city, at the junction of the Bow and Elbow Rivers. The ROCKY MOUNTAINS can be seen 80 km (50 miles) away and provide the city with a majestic background.

Established as a police post in 1875, Calgary has grown to be the hub of Alberta's oil business and a centre for agriculture, industry and commerce. It is surrounded by excellent arable and grazing land. Nearby, too, are rich sources of oil and natural gas. Industries include oil refining,

flour milling, timber processing and the manufacture of cement and bricks.

The city, which hosted the Winter Olympics in 1988, has a well-known university. Its Calgary Stampede is a renowned annual rodeo.

Population 710 680

Map Canada – Dc

Cali *Colombia* Capital of the rich agricultural department of Valle del Cauca. It straddles the Cali River, about 300 km (185 miles) south-west of Bogotá. Founded by the Spanish in 1536, Cali became the trade centre and colonial capital of a cattle-rearing and cotton and sugar-growing region. It experienced an economic boom after it was linked to the national railway network in the early 20th century. Since the 1950s, it has mushroomed into an industrial centre, manufacturing paper, textiles, pharmaceuticals and processed foods. More recently Cali has achieved notoriety as one of the major centres of Colombia's illegal cocaine trade.

Population 1 624 200

Map Colombia – Bb

Calicut (Kozhikode) *India* Port in the southwest, 493 km (306 miles) from the tip of India. It gave its name to calico, the coarse cotton cloth that has been made here since the 17th century. It also manufactures ropes and mats, and exports tea, coffee, coconuts and spices. The Portuguese navigator, Vasco da Gama, made his first Indian landing at Calicut in 1498.

Population 801 000

Map India – Be

California *USA* Most populous of all the states. It lies on the Pacific coast, and in 1850 became the first in the west to reach statehood, only two years after it was ceded to the United States by Mexico following the Mexican-American War (1846-8). The almost simultaneous discovery of gold in the SIERRA NEVADA sparked a huge rush to the area by money-hungry immigrants – the 'Forty-niners. California has the highest point in the mainland United States – 4418 m (14 494 ft) at Mount WHITNEY in the Sierra Nevada, and the lowest on the continent at Death Valley, 86 m (282 ft) below sea level. It contains the Central Valley, a 600 km (375 mile) strip of fertile farmland east of the COAST RANGES. Much of California is prone to earthquakes, especially along the San Andreas Fault.

California has a Mediterranean climate, with long, hot, dry summers and mild, moist winters, and irrigation is frequently essential for crops. Although only 10 per cent of the land is cultivated, it is the USA's leading agricultural state, producing oranges, grapes, lettuces, cotton, rice and dairy products. Californian wines produced in the Napa and Sonoma valleys in the north are well known.

California is a leading oil-producing state and is known for its electronics and computer industries based in the area south-east of SAN FRANCISCO known as Silicon Valley. Tourism is also important, with visitors flocking to the state's national parks, its famous coastal resorts, DISNEYLAND and the cities of San Francisco and LOS ANGELES, the film capital of the world.

California covers 411 015 km² (158 693 sq miles). SACRAMENTO is the state capital.

Population 29 760 000

Map United States – Bc

California, Gulf of Arm of the PACIFIC OCEAN in Mexico which was first named 'the Sea of Cortés', after the Spanish conquistador Hernán Cortés (1485-1547). It was later called *Mar Bermejo* ('the Vermilion Sea'). The gulf has oyster beds, pearl and sponge fisheries and big-game fish such as sharks and rays. Whales visit its quiet bays between January and March each year to give birth to their young. Storms can develop quickly and sailing can be hazardous. The gulf has an area of 161 870 km² (62 500 sq miles), and an average depth of 810 m (2660 ft).
Map Mexico – Ab

Callao *Peru* Main Peruvian port, handling most of the country's exports. It is situated near the capital, Lima, and has now grown to meet the ever-expanding capital. Today, it is part of Greater Lima.

During Peru's war of independence, Royalists under General Rodil made their last stand at Calloa's Real Felipe Fortress before surrendering in 1826. The city has a large oil refinery.
Population (city) 575 000; (Greater Lima) 6 404 500
Map Peru – Bb

Callejón de Huaylas *Peru* Fertile valley of the Río Santa in Ancash department. It lies between the snowless Black Cordillera (Cordillera Negra) and the snow-covered range of the Andes known locally as the White Cordillera (Cordillera Blanca), whose highest peak is Huascarán at 6768 m (22 205 ft). The valley has experienced many devastating avalanches that have taken their toll of its picturesque towns and villages.
Map Peru – Ba

calm Force 0 on the BEAUFORT SCALE, when there is virtually no horizontal movement of the air.

Caloocan *Philippines* See MANILA

Calvados *France* Largely agricultural department in Normandy. The D-Day landings by Allied troops took place on its beaches in 1944. The region is noted for its dairy farming and cider, and for having an apple brandy named after it. The chief town is CAEN.
Population 628 300
Map France – Cb

Cam Ranh Bay *Vietnam* Natural harbour, 20 km (12 miles) long and 16 km (10 miles) wide, on the central coast, about 290 km (180 miles) north-east of Ho Chi Minh City. It was developed as a base by the United States during the years of the Vietnam War (1965-75).
Map Vietnam – Bc

Camargue *France* An area of about 560 km² (216 sq miles) made up of marshes, lagoons and farmland within the Rhône delta. Rice, wheat and animal fodder are grown in the north. To the south lie the Etang de Vaccarès, a small artificial lake and a regional park, famed for its wild black bulls, white horses and flamingos. Stes-Maries-de-la-Mer is the main town.
Map France – Fe

Camarines *Philippines* See BICOL

Ca Mau *Vietnam* Swampy, lowland peninsula forming the southern tip of the country. River sediments swept by coastal currents make for an ever-changing shoreline which, in some parts, is estimated to be advancing by up to 75 m (about 250 ft) each year. Boatmen fish the streams and the muddy flatlands, which are fringed by mangrove forests and produce prolific rice crops.
Map Vietnam – Ad

Cambodia See p 127

Cambrian First of the six periods of the Palaeozoic era. See GEOLOGICAL TIMESCALE

Cambrian Mountains *United Kingdom* Range forming the backbone of Wales, and extending 140 km (85 miles) between SNOWDONIA in the north and the BRECON BEACONS in the south. The rugged plateau has a number of deep lakes and is cut by many river valleys. In ancient times, it was a formidable barrier to invasion from England. The highest peak is Aran Fawddwy, at 905 m (2970 ft).
Map United Kingdom – Dd

Cambridge *United Kingdom* City and county of eastern England with a world-famous university dating back to the founding of the first college, Peterhouse, in 1281. The earliest students came from Oxford to escape riots there. But unlike Oxford, whose colleges are scattered and sometimes hard to find, the glory of Cambridge is focused on a single sweep of architectural wonders along the east bank of the River Cam.

The city, which lies 80 km (50 miles) north and slightly east of London, is also an administrative centre, a market town, and a processor of farm produce. More recently, it has become a centre for computer and other high-technology industries, with new, small computer companies situated in the science and business parks to the north of the city. Many of the industries draw on the skills of the university, which has long excelled in science and mathematics.

The county, which is mostly agricultural and covers 3409 km² (1316 sq miles), includes the former county of Huntingdon, much of the southern FENS, the cathedral city of ELY and the industrial centre of PETERBOROUGH.
Population (city) 108 000; (county) 668 700
Map United Kingdom – Fd

Cameron Highlands *Malaysia* Hill station in west peninsular Malaysia, about 50 km (30 miles) south-east of the city of Ipoh. Its cool climate has made it popular among European expatriates. Vegetables and tea are grown.
Population 21 500
Map Malaysia – Bb

Cameroon See p 128

Cameroon, Mount *Cameroon* The country's highest volcano, rising to 4070 m (13 353 ft) near the northern end of the coast. It is still active, and its eruptions have made the surrounding plains extremely fertile. Crops include palm oil, rubber, tea, bananas and pepper.
Map Cameroon – Ab

Campania *Italy* Region covering 13 595 km² (5250 sq miles) centred on NAPLES. Campania has two distinct faces. The first is the densely settled coast with its fertile plains around the towns of CASERTA and SALERNO. It includes the thickly populated lower slopes of the active volcano VESUVIUS, and the AMALFI and SORRENTO coasts.

The other face is the remote and poor mountainous interior, an area that was severely damaged by an earthquake in 1980. Whereas the economy of the interior is rural, the coastal towns, especially Naples, are renowned for a variety of industries revolving around the production of steel and motorcars, ship-building, electrical appliances and food processing.
Population 5 589 590
Map Italy – Ed

Campeche *Mexico* Picturesque port and capital of Campeche State on the western coast of the YUCATÁN peninsula. Here, in 1517, the Spaniards first set foot on Mexican soil. The export of logwood – a rare source of dye – found growing in nearby forests soon made Campeche an exceptionally prosperous city. By the end of the 17th century, the thriving port had to be fortified against constant raids by French, Dutch and English pirates; its two main gates and seven of its original eight fortresses still remain.

Today, Campeche gives the impression that it has seen better times. The local economy relies heavily on fishing, textiles, wood and crafts.
Population (port) 99 000; (state) 529 000
Map Mexico – Cc

Campi Flegrei *Italy* See PHLEGREAN FIELDS

Campina Grande *Brazil* Market town 160 km (100 miles) inland from Recife. It is the gateway to a large inland semi-arid zone of north-eastern Brazil known as the *sertão*. The town's main products are leather, textiles and vegetable oils; there are also light engineering works.
Population 311 000
Map Brazil – Eb

Campinas *Brazil* Fast-growing modern town 75 km (47 miles) north-west of the city of São Paulo. More than 20 multinational companies are based here, many specialising in electronics. Other products include paper, textiles, machinery, cotton, maize and sugar.
Population 835 100
Map Brazil – Dd

Campine *Belgium* See KEMPENLAND

campos Type of savannah found in central Brazil, south of the equatorial forests of the Amazon basin, where it covers an area of some 2400 by 2750 km (about 1500 by 1700 miles). It consists of grass and low-growing trees.

Campos *Brazil* Industrial town on the coastal belt, 230 km (144 miles) north-east of Rio de Janeiro. It services Brazil's principal oil fields in the offshore Campos basin. Apart from oil, its main products are marble, cement, and sugar alcohol. Areas of rain forest near the town are the home of the rare golden lion monkey.
Population 366 700
Map Brazil – Dd

Can Tho *Vietnam* River port on the Mekong delta, 140 km (85 miles) south-west of Ho Chi Minh City. A waterway network links it to the delta's rice-growing districts.
Population 284 310
Map Vietnam – Bc

Cambodia

*DRAGGED INTO A NEIGHBOUR'S WAR –
AND RULED BY TERROR – THIS GREAT
EMPIRE OF ANCIENT TIMES IS NOW
ONE OF THE POOREST COUNTRIES IN
THE WORLD*

Darkness fell on Cambodia on 17 April 1975, when Pol Pot led his army of Khmer Rouge communists into the capital, PHNOM PENH, and began a reign of terror. He claimed to be a political saviour, creating a unique economy based on an agricultural society cut off from world capitalism. But his social experiment brought death and famine to a land which was once the rice bowl of Indo-China.

Under the Pol Pot regime, entire towns were evacuated and their populations, young and old, healthy and sick, were ordered to work on the land. The drive for a self-reliant peasant society received Chinese support, but was carried to brutal extremes. Enormous retraining centres were built and intellectuals were systematically eliminated. The population of Phnom Penh sank from 2 million to 23 000 and, even today, the population of towns and cities continues to fluctuate. Perhaps a million or more people were executed and over half a million died from disease and starvation. A further 600 000 people fled the country, exiled to Vietnam or to refugee camps in Thailand.

THE ALL-IMPORTANT MEKONG

Cambodia is dominated by the MEKONG River. It floods in the rainy season from May to October, providing fine rice-growing land. In its centre is TONLE SAP ('the Great Lake'), which quadruples in size in the rains and is one of the world's great inland fisheries.

Most of the people live in the central lowlands, which are surrounded by forested hills. Rice and maize are the main crops; there are rubber plantations north of Phnom Penh.

Temperatures average 27-32°C (81-90°F) and rainfall varies from 1300 mm (51 in) a year in the plains, to 5000 mm (196 in) in parts of the south-west.

Cambodia, known to its own people as Kampuchea, was overrun in the 6th century AD by the Khmer tribe from the north. The Khmer, whose empire flourished until about 1450, created the great Buddhist temple of ANGKOR WAT. Although the country was dominated alternately by its neighbours Annam (now Vietnam) and Thailand from the mid-15th century, Khmers still make up 90 per cent of the population. In 1863, Cambodia became a French protectorate, and during the Second World War the Japanese occupied most of it. Cambodia's King Norodom Sihanouk proclaimed independence from France in 1944, but the French returned after the war and Cambodia did not finally become independent until 1954.

Sihanouk abdicated and became prime minister; from 1960, he was head of state. He tried to steer his country on a neutral course during the Vietnam War, but when the North Vietnamese began using Cambodian territory to supply their forces in South Vietnam, Cambodia became embroiled in the conflict. In 1970, General Lon Nol, supported by the United States and 50 000 South Vietnamese troops, deposed Sihanouk. But the war continued and the population suffered. Over a seven-month period in 1973 the United States, in an effort to cut Viet Cong supply lines, dropped more bombs over Cambodia than they did on Japan during the entire Second World War. This finally provoked the Khmer Rouge uprising led by Pol Pot, which, in turn, led to the fall of Phnom Penh, in 1975.

THE UNITED NATIONS BRINGS SHORT-LIVED PEACE

Pol Pot's reign ended in January 1979 when Vietnamese troops occupied the capital. They set up a puppet government which began to revitalise the economy. But the civil war, which had already resulted in more than a million deaths, continued to ravage the country.

A UN peace plan, launched in 1991, was followed by elections in May 1993 – and the establishment of a transitional government in August of that year. The Cambodians proclaimed a kingdom, reinstated Sihanouk, and formed their new government.

Cambodia's slow return to economic recovery began with the promulgation of a new constitution. However, the country faced myriad problems: how to improve crop production, which had been disrupted for years, first during the civil war against the Khmer Rouge and then by the Pol Pot regime, was a major headache.

Furthermore, 25 years of civil war had crippled the rubber industry through shortages of

power and raw materials – reducing it to 40-50 per cent of pre-war capacity. Few shops remained, and most trading took place on the streets, with goods smuggled in through Thailand and sold for hoarded gold. The cities, cut off from the countryside by potholed and cratered roads, depended largely on overseas aid. Indeed, in its 1994 budget, almost half of Cambodia's revenue came from foreign aid.

Political stability (within the coalition government) led to some optimism about rapid economic recovery. However, Khmer Rouge guerrillas, who did not pariticipate in the UN-supervised elections, started a new assault nearly one year after the UN peace-keeping forces had withdrawn from the country. Khmer Rouge numbers were small – between 7000 and 10 000 – and their political power almost non-existent, but, as political analysts pointed out, they had the arms, and backing, to turn the new economic and political promise into another quagmire.

CAMBODIA AT A GLANCE		
Official name State of Cambodia		
Area 181 040 km² (69 899 sq miles)		
Population 9 898 900 **Per km²** 55 **(Per sq mile)** 141)		
Capital Phnom Penh		
Government Constitutional monarchy		
Currency 1 riel = 100 sen		
Languages Khmer (official), French		
Religions Buddhist (95%), other (5%)		
Climate Tropical; monsoon rains from May to October. Average temperature in Phnom Penh is 27°C (81°F) throughout the year		
Land use Cultivation 17%, meadows and pastures 3%, forest and woodlands 76%, other 4%		
Main primary products Rice, maize, bananas, rubber, livestock, tobacco, jute, timber, fish		
Major industries Agriculture, fishing, forestry		
Main exports Rubber, vegetables		
Annual income per head (US$) 200		
Population growth (per thous/yr) 21		
Life expectancy (yrs) Male 48 **Female** 51		

▶**RAVAGES OF TIME AND WAR Forest slowly engulfs the magnificent Khmer ruins of Angkor Wat; international conservation efforts came to a full stop during 25 years of civil war and the damage may now be irreparable.**

Cameroon

A DIVERSITY OF LANDSCAPES, PEOPLES AND LANGUAGES – BUT ONE OF THE MORE STABLE COUNTRIES IN AFRICA

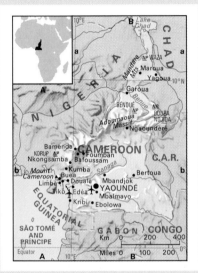

Centuries-old tribal customs and modern consumer society attitudes meet face to face in the African state of Cameroon. In rural areas, there are memorials and offerings of food to dead ancestors at almost every turn, and fetish figures of wood, clay or metal. In the towns, such as the capital, YAOUNDÉ, choice food is imported to satisfy sophisticated urban palates – even though Cameroon could be self-sufficient in food.

Diversity is perhaps the country's most outstanding quality. Starting in the north, it ranges from low-lying lands on the southern shores of Lake CHAD, through the semi-desert SAHEL, to dramatic mountain peaks (the Kapsiki Mountains north of Garoua). Then come grassy savannahs, rolling uplands (the ADOUMAOUA MASSIF), steaming tropical forests (including the KORUP forest, which is one of the richest remaining ecosystems in equatorial Africa), and coffee, cocoa, rubber, banana and hardwood plantations. Finally, there are rain-drenched volcanoes – the highest active one is Mount CAMEROON (4070 m, 13 353 ft). In 1986, a volcanic lake near Bamenda emitted a gas that killed 1700 villagers; in 1991, it threatened to give off clouds of deadly gas again. Coastal Cameroon has palm-fringed beaches at Kribi and LIMBE, and is hot with rain all year. The interior is drier.

The people of Cameroon are also a diverse mixture – the country has some 200 ethnic groups, including the Fulani in the north, the Kirdi whose villages teeter on the clifftops of the Kapsiki mountains, the Bamileke in the west, the Douala on the coast, and pygmies in remote parts of the jungle. The largest culture group, the Pangwe, is composed of Bantu-language peoples in the forested south. Some of these groups are quite small, but each speaks its own language.

The Portuguese were the first Europeans to arrive here, in the 15th century. They were followed by the Dutch, Spanish and British who came here in search of slaves until the 19th century; then, in 1884, the country became a German protectorate. After Germany's defeat in the First World War, Cameroon was handed over to the French and British; League of Nations mandates put four-fifths of the country under the control of the French in the east, while Britain was handed the western strip. The two parts were run separately as French Cameroon and British Cameroon until 1960. The eastern state then became independent as the Cameroon Republic, and was joined in 1961 by the southern section of the British territory to form the Cameroon federation. The north opted to join Nigeria, which accounts for the country's strange triangular shape. Full union of the two separately administered parts of the federal republic took place in 1972. The country was officially renamed Republic of Cameroon in 1984.

In 1966, the largest parties in the French and English-speaking parts of Cameroon merged and the country became a one-party state under the leadership of Ahmadou Ahidjo, the first president. After 16 years of stable but authoritarian rule, Ahidjo retired and was succeeded by the prime minister, Paul Biya. A period of increasingly oppressive rule followed – until the restrictions on opposition parties were finally lifted in December 1990. In multiparty legislative elections, held in March 1992, the ruling party gained a narrow victory over a divided opposition. Paul Biya himself won the presidential election, in the same year, on a minority vote.

LIFE IN THE VILLAGE

Most Cameroonians live in villages. In the south they grow maize and vegetables, and sell the products of the tropical trees – rubber and hardwood timber. In the dry northern savannah and Sahel, drought and hunger are not unknown. Here, a single, hard-won harvest of cereals depends on the short five or six-month rainy season, as do cash crops of groundnuts and cotton.

The country has one of the most stable economies on the African continent, and one of the highest incomes per head. A poor country at independence, Cameroon today has prospered to a middle-income economy. This is due mainly to discoveries of offshore oil, leading to comparatively rigorous economic growth during the 1970s and 1980s. In 1986, a fall in world market prices brought the economic boom to a stop. The devaluation of the CFA franc in 1994 was another setback.

However, new agricultural reforms and increased investment in the country again contributed to steady growth. Mineral resources – oil, gas and bauxite – still account for most of the country's prosperity, with oil bringing in 51 per cent of export revenue. Other main exports are coffee, beans, cocoa, aluminium products and timber.

CAMEROON AT A GLANCE	
Official name Republic of Cameroon	
Area 475 440 km² (183 525 sq miles)	
Population 13 500 000 **Per km²** 284 **(Per sq mile** 75.6)	
Capital Yaoundé	
Government Republic	
Currency 1 CFA franc = 100 centimes	
Languages French and English (official), African languages	
Religions Christian (52%), Muslim (22%), indigenous beliefs (26%)	
Climate Tropical; humid in the south, drier further north. Average monthly temperature in Douala is 26°C (78.8°F) throughout the year	
Land use Forest and woodlands 54%, grazing 18%, cultivation 15%, other 13%	
Main primary products Maize, millet, yams, sorghum, cassava, plantains, cocoa, coffee, sweet potatoes, groundnuts, cotton, palm oil, livestock, timber, rubber; oil and natural gas, bauxite	
Major industries Agriculture, mining, aluminium smelting, food processing, forestry, textiles	
Main exports Oil, coffee, cocoa, timber, cotton, aluminium, rubber	
Annual income per head (US$) 820	
Population growth (per thous/year) 28	
Life expectancy (yrs) Male 56 **Female** 56	

Cana (Kafr Kanna) *Israel* Village 7 km (4.3 miles) north-east of Nazareth, where Christ performed his first miracle, turning water into wine. Cana was also the home of Nathaniel, traditionally identified with St Bartholomew the Apostle.

The village has a 7th-century Franciscan and an 8th-century Greek church. St Nathaniel's Chapel stands on the traditional site of Nathaniel's house.

Map Israel – Ba

Canada See p 130

Canadian Shield *Canada* Rocky plateau stretching from the Great Lakes to the Arctic Ocean. It covers 5 million km² (1.9 million sq miles), more than half of Canada, and includes virtually all of Labrador, Quebec and Ontario, parts of Manitoba and most of the Northwest Territories.

It is rich in minerals and its rivers, lakes and waterfalls have been harnessed for hydroelectric power, which is distributed to cities and towns in the area. The soil is generally poor, which has led to the development of forestry as the region's major land use.

Map Canada – Eb

Canal Area *Panama* Strip of land, 8 km (5 miles) wide, on either side of the 82 km (51 mile) long PANAMA CANAL in Central America. It covers 1676 km² (647 sq miles) and includes the cities of Cristóbal and Balboa. It was called the Canal Zone from 1903 to 1979 while it was United States territory. Under a new treaty, which went into effect in October 1979, Panama assumed control of areas not needed for the defence and running of the canal. The entire area reverts to Panama in the year 2000.

Population 31 600

Map Panama – Ba

Cañar *Ecuador* West-central province in the Andes, just north of Azuay. Much of it is cold, dry and windswept, but around the provincial capital of Azogues (population 14 500), the land is farmed intensively. Azogues is a centre of the straw hat industry. About 25 km (15 miles) to the north lies the small town of Cañar, north of which is the Inca ruin of Ingapirca, the most important in Ecuador.
Population 189 350
Map Ecuador – Bb

Canary Islands (Islas Canarias) *Spain* Group of seven large and four smaller islands and many islets, 95 km (60 miles) off the north-west coast of Africa. Many of the islands, which are mountainous, are popular tourist resorts, particularly in winter. Generally, they have a mild climate, but are sometimes subject to drought and tornadoes. Irrigated farms grow grain, bananas, oranges, lemons, vegetables and vines. Industries include fishing and canning.
Known in ancient times as 'the Fortunate Islands', the Canary Islands were rediscovered and claimed by the Portuguese in 1341, but ceded to Spain in 1479.
The islands of Lanzarote, Gran Canaria and Fuerteventura form the province of Las Palmas; Tenerife, La Palma, Gomera and Hierro form the Santa Cruz de Tenerife Province. Between them, the islands cover an area of 7273 km² (2808 sq miles).
The main towns are Las Palmas (on Gran Canaria) and Santa Cruz (on Tenerife), both of which are duty-free ports. The group's highest point, and also the highest point in Spanish territory, is Pico de Teide, a volcanic cone on Tenerife, which rises to 3718 m (12 198 ft).
Population 1 601 810
Map Morocco – Ab

Canaveral, Cape *USA* Sandspit on the east coast of Florida, 300 km (185 miles) north of Miami. It is the site of the John F. Kennedy Space Center, and was for some years (1964-74) called Cape Kennedy. All the United States' manned space missions have been launched from here.
Map United States – Je

Canberra *Australia* The national capital and administrative centre, situated in the AUSTRALIAN CAPITAL TERRITORY, 250 km (155 miles) south-west of Sydney. The city was planned between 1911 and 1913 by Walter Burley Griffin, an American architect; the Australian parliament moved here from Melbourne in 1927. Since 1958, when foreign diplomats and federal government departments moved here from Sydney and Melbourne, the city's population has increased eightfold. The origin of the city's name is uncertain; it may come from an Aboriginal word *canberry*, meaning 'meeting-place'.
Population 302 500
Map Australia – Hf

Cancún *Mexico* Tiny, 22 km (14 mile) long island in the Caribbean, linked to mainland Mexico by a narrow causeway. Cancún is hot and sunny, and has fine beaches and clear seas – all of which have helped to make it a successful tourist resort. The splendid Mayan ruins of Tulum and Xel-Ha are nearby.
Population 70 000
Map Mexico – Db

Cango Caves *South Africa* Extensive limestone caves 26 km (16 miles) north of the town of Oudtshoorn, and 355 km (220 miles) east of Cape Town. They were once occupied by Stone Age people, but fell into disuse until rediscovered in 1780. The caves, known for their magnificent flowstone formations, attract as many as 200 000 visitors annually.
Their full extent is unknown, but passages have been explored for a distance of more than 3 km (2 miles). During the main tourist season, concerts are sometimes held in one of the enormous outer caverns.
Map South Africa – Bc

Çankiri *Turkey* Town lying about 100 km (60 miles) north-east of Ankara. It has a huge 16th-century mosque built by Sinan, who is regarded as one of Islam's greatest architects. The surrounding province, also called Çankırı, covers 8666 km² (3346 sq miles). Its chief crops are fruits and grains, and asbestos is one of its main minerals.
Population (town) 40 900; (province) 2 279 130
Map Turkey – Ba

Cannes *France* Seaport and fashionable resort town on the Côte d'Azur, 28 km (18 miles) south-west of Nice. It has casinos, palm-fringed boulevards and a yachting marina, but is probably best known for its film festival held each year in April/May. It manufactures soap, oils and perfumes, and exports fruit and anchovies.
Population 72 800
Map France – Ge

Cantabria *Spain* Province on the Bay of Biscay around the port of Santander, whose backbone is formed by the CANTABRIAN MOUNTAINS. Its main industries are coal and iron mining, as well as some zinc and lead extraction, fishing, and the rearing of cattle and sheep.
Population 527 330
Map Spain – Ca

Cantabrian Mountains (Cordillera Cantabrica) *Spain* Northern range extending about 480 km (300 miles) west from the Pyrenees. It is rich in minerals, especially coal and iron. Its highest peak is Peña Cerredo, 2648 m (8685 ft).
Map Spain – Ba-Ca

Cantal, Massif du *France* Volcanic upland in the Massif Central, about 80 km (50 miles) south and slightly west of CLERMONT-FERRAND. Its chief industries are the rearing of livestock, cheese making and tourism. The Plomb du Cantal, 1855 m (6086 ft), is the area's highest peak.
Map France – Ed

Canterbury *United Kingdom* English cathedral city in the county of Kent, 90 km (55 miles) south-east of London. It was the birthplace of the established Church in England, with its first archbishop appointed in AD 601, and is today the seat of the Primate of All England and the world headquarters of Anglicanism.
Pilgrims journeyed there for centuries to the shrine of Archbishop Thomas Becket (1118-70), who was murdered in his own cathedral following his opposition to King Henry II's attempts to control the clergy. These pilgrimages were the inspiration for Geoffrey Chaucer's 14th-century poetic masterpiece, *The Canterbury Tales*.

The city has retained much of its medieval character, including its narrow streets, quaint houses and a large section of its city wall. Canterbury's cathedral was begun in about 1070, and completed only in 1400. It stands in a wide close, surrounded by old monastic buildings and the King's School.
In the countryside outside the city is the modern University of Kent.
Population 39 740
Map United Kingdom – Fe

Canterbury Bight Inlet, 185 km (115 miles) wide, of the South PACIFIC OCEAN on the east coast of SOUTH ISLAND, New Zealand. The chief port is TIMARU.
Map New Zealand – Cf

Canterbury Plains *New Zealand* Grasslands on South Island stretching more than 150 km (90 miles) along the east coast, from the Kaikoura ranges in the north to the Waitaki River in the south. Since the 1840s, the plains, which are also the country's granary, have been grazed by sheep and cattle. The chief town is CHRISTCHURCH.
Map New Zealand – Cf

Canton *China* See GUANGZHOU

canyon Steep-sided narrow gorge, usually incised by a river flowing through it. Most canyons are found in arid or semi-arid areas, the source of the river being elsewhere in a well-watered area. The most famous canyon is the GRAND CANYON in the United States.

Canyon de Chelly National Monument *USA* Home of the Navaho, covering 335 km² (129 sq miles) in north-eastern Arizona. The monument contains many Indian remains dating back as far as AD 350, including cave dwellings in the base of sheer red sandstone canyon cliffs.
Map United States – Ec

Canyonlands National Park *USA* Spectacular area astride the Colorado River in south-eastern Utah. It is dotted with high flat-topped mountains called MESAS, deep canyons and natural arches. The park covers 1365 km² (527 sq miles).
Map United States – Ec

cape Promontory or headland projecting into the sea. For individual capes listed in this book, see the main part of the name.

Cape Breton Island *Canada* Part of NOVA SCOTIA, covering 10 311 km² (3980 sq miles) and linked to the mainland by a causeway over the Strait of Canso. Much of Cape Breton is ringed by the spectacular Cabot Trail road. The Cape Breton Highlands National Park in the north, covering 950 km² (367 sq miles), has spectacular forest and cliff landscapes.
Map Canada – Id

Cape Coast *Ghana* Coastal town 120 km (75 miles) west of Accra. It was the British colonial capital until 1876, and is now the capital of Central Region. The castle built here by the British is used today as government offices, while a fort has been converted into a lighthouse. Citrus farming and fishing are the basis of the economy.
Population 72 100
Map Ghana – Ab

Canada

BEARS AND WOLVES ROAM THE WILDERNESS AREAS OF THE WORLD'S SECOND LARGEST COUNTRY – A LAND OF FROZEN WASTES, FERTILE PRAIRIES AND ABUNDANT WATERS

A land of climatic and geographical extremes and with a great diversity of peoples, Canada is the second largest country in the world after Russia. The Trans-Canada Highway spans nearly 8000 km (5000 miles) from NEWFOUNDLAND to BRITISH COLUMBIA, and it takes four days and five nights to cross the country by train.

The average daytime temperature over most of Canada exceeds 24°C (75°F) in July. But in winter, nearly all the country is snow-covered and has temperatures below freezing point. Winter temperatures can dip to -62°C (-80°F) in the ARCTIC north, yet summer temperatures of 27°C (81°F) and more are commonplace in towns and cities to the south. In fact, the southernmost part of Canada is at the same latitude as Rome and northern California. Most of Canada's 27.3 million people live within a narrow strip along the United States border and almost two out of every three people live in ONTARIO or QUEBEC.

HUGE TERRITORY, FEW PEOPLE

With nearly four-fifths of this enormous land uninhabited, nature lovers can readily explore vast stretches of unspoilt countryside or pitch camp where the only neighbours may be black or grizzly bears. Moose, caribou, elk and wolves also roam the wilderness, and the inland waters abound with salmon, trout, bass and pike. Canada has more lakes than any other country.

The world's most famous police force, the Royal Canadian Mounted Police – the Mounties – stand as a symbol of Canada in their red ceremonial tunics. Today they are a 5600-strong federal force, but in 1873 when they were formed to check frontier lawlessness, there were only 300 of them.

The Mounties were involved in one of the most dramatic episodes in Canadian history, the KLONDIKE gold rush. Gold was discovered in Bonanza Creek, a tributary of the Klondike River in the YUKON TERRITORY, in 1896. When word of the strike reached the outside world it sent 100 000 men, and a few women, stampeding to the wild north-west of Canada. Over six years the Klondike yielded more than US$100 million in gold – worth many times that today. Many people became rich, some went mad with hardship and gold fever, while others died or were killed in a reckless society of gold robbers, gambling saloons and rowdy dance halls, policed by a couple of Mounties.

Canada is a sovereign state within the British Commonwealth. About 7.6 million Canadians, or 28 per cent of the population, are of British origin and about 6.3 million, or

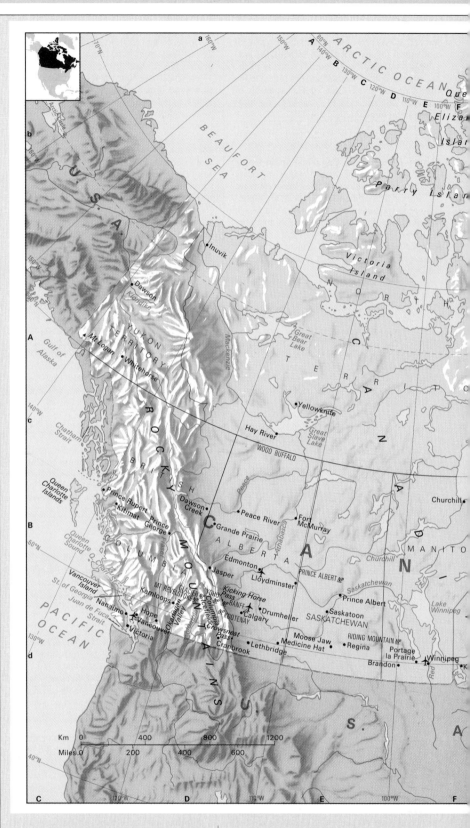

23 per cent, are of French descent; the 1991 census listed 470 615 Aboriginal Canadians and 49 255 Inuit (formerly referred to as 'Eskimos'), while more than 1 million people claimed at least partial aboriginal status. In addition, Canada is host to a wide range of immigrant nationalities. The large majority of French-speaking people live in the province of Quebec. Strong cultural and commercial influences are apparent across the 4200 km

GASPÉ Peninsula, NEW BRUNSWICK, NOVA SCOTIA, PRINCE EDWARD ISLAND and New-foundland. They are part of an old mountain range that extends northwards from the United States but seldom rises to more than 1000 m (3300 ft). Prince Edward Island in the Gulf of St Lawrence is an exception to the rugged terrain and of all the eastern provinces is best suited for agriculture.

The CANADIAN SHIELD, a plateau consisting of old hard rocks and extending over more than 4.65 million km² (1.8 million sq miles), forms the bedrock of half of Canada. The plateau is an area of forests, lakes and rocks which sweeps westwards from LABRADOR through most of Quebec and Ontario, embraces the northern parts of the prairie provinces of MANITOBA and SASKATCHEWAN, and extends into the NORTHWEST TERRITORIES.

The provinces of ALBERTA, Manitoba and Saskatchewan, bordered by the United States to the south and the ROCKY MOUNTAINS to the west, once the heart of waving prairies, today contain Canada's largest farming area. Forests cover their northern parts.

Much of the west is more than 2000 m (6600 ft) high; the country's highest peak, Mount LOGAN in the Yukon, reaches 5959 m (19 550 ft). A series of mountain ranges called the Western Cordillera runs down the west side of the country, covering most of British Columbia, the Yukon and parts of Alberta and the Northwest Territories. The Rocky Mountains form the eastern edge, towering to more than 3000 m (10 000 ft), and providing tourists with some of the most beautiful scenery to be found.

The west coast provides spectacular scenery with fjords, mountains and islands including VANCOUVER ISLAND and the QUEEN CHARLOTTE ISLANDS. The city of VANCOUVER stands in the main lowland area of the west coast, the lower FRASER valley.

FROZEN MINERAL WEALTH

The far north of Canada is inhabited mainly by small communities of Inuit. It is wild and inhospitable, with hardly any rainfall, little tree growth and large areas where the land is permanently frozen. But it holds great mineral wealth, including oil and gas. The Inuit live in the MACKENZIE delta, on the Arctic islands, and on the mainland coast of Labrador. Many of the native Canadians reside in 2250 reserves covering 2.6 million hectares (6.4 million acres). They are widely spread across the country and speak some 50 different languages belonging to 11 distinct linguistic families.

The first steps towards the establishment of a new territory for the Inuit were taken in 1976. In 1987 and 1990, boundary and land claim settlements were negotiated, and in 1992, the recommendations, accepted by the federal government, were presented in a plebiscite to the inhabitants of the Northwest Territories, and accepted by 54 per cent of voters. As a result, the Nunavut Act of 1993 was passed, according to which, a new territory by the name of Nunavut ('Our Land' in Inuktitut, the Inu language) is to be created in 1999, which will include the central and north-eastern region of the Northwest Territories.

The St Lawrence has long been a major

(2600 mile) border with the United States.

Most of the Canadian people and industries are in the GREAT LAKES region and the ST LAWRENCE lowlands. This heartland starts as a narrow strip near Quebec city, broadens out into southern Quebec and extends westwards across Ontario. The St Lawrence River flows through this corridor from the Great Lakes to the Atlantic Ocean. In the east, the APPALACHIAN highlands extend into Quebec's

trade route, especially since the building of the St Lawrence Seaway in 1954-9 enabled ocean-going vessels of up to 28 000 tonnes to sail into the Great Lakes. The major rivers in the west are the Fraser and Thompson, which run into the Pacific but are of limited transport value. Most other rivers flow to virtually uninhabited coasts of HUDSON BAY and the Arctic Ocean, and are frozen for much of the year.

Nevertheless, the huge cross-country net-work of rivers and lakes played an important role in opening up the country to explorers, trappers and fur traders. These adventurers were able, with only short hauls overland, to paddle their canoes from the St Lawrence River westward to central Alberta, north to the Mackenzie delta, and back again.

The first European settlers were the French. In 1605, they founded a community that is now Annapolis Royal on the coast of Nova Scotia; and in 1608, they sailed up the St Lawrence River to the site which is now the city of Quebec. British settlement started two years later, on Newfoundland. By the 1670s, French explorers, missionaries, traders and trappers had penetrated west to the MISSISSIPPI head-waters and Manitoba. The British set up the Hudson's Bay Company to trade in pelts and furs, and the scene was set for rivalry between the two powers.

BATTLES OF THE EARLY SETTLERS

Open rivalries began at the end of the 17th and the beginning of the 18th century. The French and British destroyed each other's settlements, with each side recruiting native Canadians as allies. The British navy raided French coastal forts, and French troops attacked British garrisons and trading posts. The issue was finally settled in the Seven Years' War of 1756-63. In 1759, a British force led by General James Wolfe routed the French in the Battle of the Plains of Abraham and took Quebec. In 1760, the French sur-rendered MONTREAL and the French colonies were ceded in 1763.

Despite Britain's military victory, the coun-try continued to support separate British and French cultures, languages and institutions. French settlers by far outnumbered the British in Quebec and the surrounding area, so it was decided to recognise the Roman Catholic reli-gion and retain French civil law and language (Quebec Act, 1774).

In 1982, the constitution of Canada, known as the British North America Act, was repatri-ated from Britain. A set of reforms and amend-ments, together with a *Charter of Rights and Freedoms*, was added to the constitution and agreed upon by all provinces except Quebec. In 1987, another package of amendments was agreed upon in principle at a constitutional conference at Meech Lake, in Quebec. It was ratified by Quebec but defeated by Manitoba and Newfoundland. Yet another attempt at constitutional reforms, known as the Canada Round, gave rise to the Charlottetown Accord of 28 August, 1992. This last attempt, how-ever, was defeated in a referendum in October 1992. The present government has no plans to reopen constitutional discussions.

After the American War of Independence (1775-83), more than 35 000 Loyalist refugees moved north from the new republic to New Brunswick, Nova Scotia, Quebec and Ontario, increasing the population fourfold in many areas. From 1815 onwards the British govern-ment encouraged settlers, and over the next 40 years more than a million immigrants landed in HALIFAX, SAINT JOHN'S and Quebec.

MAKING MODERN CANADA

After the American Civil War (1861-5), fears were expressed that the newly strength-ened United States might attempt to annex Canadian territory. As a defence, the provinces of New Brunswick, Nova Scotia, Ontario and Quebec signed the British North America Act, creating the autonomous Dominion of Canada in 1867. At the time, much of the west was still in the hands of the Hudson's Bay Company, but in 1869 the western territory was bought by the government, and the size of Canada was tripled overnight.

Two new provinces, Manitoba (1870) and British Columbia (1871), joined the expand-ing nation, the latter partly because it was promised a rail link to Ontario. The first rail line from the heartland to the Pacific (the Canadian Pacific Railway) was opened in 1885; the rail route eastwards from Montreal to Halifax had already been opened in 1876.

Prince Edward Island joined the Dominion in 1873; in 1905, Saskatchewan and Alberta threw in their lot with the provinces which had already joined the Dominion. The develop-ment of railways and the purchase of the Hudson's Bay Company lands opened the PRAIRIES to settlement and agriculture, bring-ing more European immigration.

Between 1900 and 1930 the population doubled. Many of the newcomers came from central and southern Europe. Most of them settled in the expanding cities and helped to give Canada its rich ethnic diversity.

During the First World War, Canada sent more than 600 000 men out of a total popu-lation of less than 8 million to fight in Europe. Canadians again hurried to the Allied side in September 1939, at the beginning of the Second World War. They also fought in the Korean War and provided peace-keeping forces for Cyprus, Suez and elsewhere.

Although Canada gained independence from Britain in 1931, it has kept many of its former political institutions that have British origins. It is a constitutional monarchy under Queen Elizabeth II; the crown's powers are administered by a governor-general (now always a Canadian).

There is a federal parliament in OTTAWA which has two houses – of which the House of Commons plays the dominant part. It has 295 members who are elected for a maxi-mum term of five years. The upper house, the Senate, has 104 members who are appointed by the governor-general.

The prime minister appoints a cabinet of ministers who are responsible to the House of Commons. The two houses have similar powers, but all major legislation is introduced and first debated in the House of Commons. The Senate has the power to veto any legisla-tion, but its members rarely choose to do so.

Each of the 10 provinces has its own elected parliament with a single house and a lieu-tenant-governor who represents the crown, and an elected executive council and legis-lature led by a premier. The Yukon and Northwest Territories have only limited self-government.

In any case, provincial laws are binding only within the province concerned, whereas federal legislation applies nationwide. At times there are conflicts between federal and provincial governments. For instance, a dis-pute in 1982 involved the right to develop oil and other mineral resources off the coast of Newfoundland. The provincial governments of Alberta and Saskatchewan have also been at odds with the federal government over the price of oil produced from their territories.

AGRICULTURE RESOURCES

More than 80 per cent of Canada's farmland is in the prairies that stretch from Alberta to Manitoba, and two-thirds of farm income is earned by prairie farmers. Despite the impor-tance of agriculture, only about one Canadian

in 33 is a farmer. The average farm is about 225 hectares (555 acres). Wheat and other grain crops cover three-quarters of the arable land. Canada regularly produces 6 per cent of the world's wheat and exports more than 50 per cent of its wheat production, which accounts for 16 per cent of the world's wheat exports. In the prairie provinces, tall narrow grain elevators are distinctive landmarks. Most elevators stand beside railways, and most grain is sent to market by rail.

In addition to grain, large prairie areas are devoted to raising cattle. In some other places, high-quality farmland is being used for industrial and urban developments, and this is causing concern, particularly in areas such as the Niagara peninsula in southern Ontario, where much fruit is grown. Another large fruit and wine-growing area is the Okanagan valley in British Columbia.

Canada's major natural renewable resource is the forests that extend across the country in a belt 500 to 2000 km (300 to 1250 miles) wide, and cover almost 40 per cent of the total land area. Most woodland is coniferous, con-

sisting of fir, pine and spruce. Canada is the world's largest exporter of wood pulp and paper – massive mills turn the wood by-product into newsprint for newspapers around the world – and one job in every 12 depends on wood. Nearly all the forests are owned by the provincial or federal governments.

FISHING INDUSTRY

Fish helped to attract the first Europeans to Canada, and fishing remains important. Until recently, about 30 000 commercial fishermen operated from the east coast in the Atlantic fishing grounds and about 15 000 workers sailed out from the west coast, where salmon is the chief catch. However, because of serious overfishing, the Canadian government in 1993-4 introduced a ban prohibiting fishing of certain species, among them cod, for an indefinite period of time but quite possibly till the year 2000.

Altogether 27 000 fishermen on the Atlantic coast were laid off as a direct result of the regulation. Fish exports, mainly to the United States, consequently began to drop.

▲ WINTER'S HOARY GRIP A herd of caribou crosses the frozen tundra near Repulse Bay in the Northwest Territories of Canada. Sparse reindeer moss – a type of lichen – is their only winter food.

Minerals range from asbestos to zinc; the most valuable are oil, natural gas, coal and iron ore. Alberta produces 90 per cent of the country's oil output and nearly half the coal total of 70 million tonnes a year. This mineral-rich province also has natural gas and large potential resources of oil from bituminous tar sands on the ATHABASCA river. The second most important coal-producing province is British Columbia, with about 40 per cent of national production. Canada is a net exporter of coal and natural gas, but imports some oil.

Iron ore is found on the north-western side of Lake SUPERIOR and on the Quebec-Labrador border. The world's second largest source of nickel is around SUDBURY, Ontario, and 220 km (135 miles) to the north, the town of Timmins is the main gold-mining centre in Canada.

Asbestos is produced in Quebec, and copper in Ontario and British Columbia. Canada is one of the world's leading producers of zinc and titanium, and also mines uranium, silver and molybdenum.

Electricity has usually been generated from water power, and 60 per cent still comes from hydroelectric stations; one of the world's largest hydroelectric power plants is at the Churchill Falls in Labrador. But potential sites for new dams tend to be in northern regions, far away from the users, so Canada has developed nuclear technology and has seven major nuclear power stations which produce 16.4 per cent of the country's electricity.

INDUSTRY AND TOURISM

Nearly one-fifth of Canadian workers are in manufacturing. Industry is highly mechanised and includes petroleum products, car manufacture, food and metal processing. Ontario contains most of the car factories, is the main steel producer and has important aircraft, electronics and electrical machinery sectors. Quebec produces textiles, paper and wood products, clothing, chemicals and machinery.

Inevitably, most international trade is with the United States, and more than one-third of Canada's manufacturing industry is owned by American firms. In 1961, United States President John F. Kennedy told the Canadian parliament: 'Geography has made us neighbours, history has made us friends, econom-

▼ **MAJESTIC GRANDEUR A valley cuts through the Rocky Mountains of Alberta in Banff National Park, Canada's first, established in 1885. Noted for hot springs, ice fields and glacial lakes, it is a game sanctuary and summer and winter sports area.**

ics has made us partners and necessity has made us allies.'

Over the past few years, Canada and the United States have continually been looking for ways to expand their neighbourly trade dealings: the North American Free Trade Agreement (NAFTA) between Canada, the USA and Mexico was signed in October 1992 and implemented on 1 January, 1994. It is designed to combine the capital, expertise and technology of Canada and the United States with the resources and low-cost labour of Mexico. However, it took many years before the Canadian electorate could be convinced that the benefits of such an agreement would outweigh the drawbacks; the Mexicans, on the other hand, welcomed the foreign investment – although their less-privileged population group rose up in protest against it the day it was to be implemented.

Trade between the USA and Canada has been going on and growing for years. The Americans are largely responsible for a recent surge in tourism, which has taken the industry into Canada's top eight export earners, employing about one person in 10. In 1993, a total of 36 100 461 tourists visited Canada, 32 622 746 of whom came from the USA.

Vast open spaces draw the hunter, fishing enthusiast, canoeist and camper, mostly from the United States, but many visitors are from Britain, Continental Europe and Japan. The need to safeguard areas of beauty has long been recognised, and the first national park was set up at BANFF in 1885.

NIAGARA FALLS and the Rocky Mountains are famous around the world, and tourists are also attracted to the historic city of Quebec, the old part of Montreal, modern TORONTO with its CN Tower – the world's tallest free-standing building – and the fishing ports and villages of the eastern Maritime Provinces.

Travel is mostly by car or aircraft. Railways carry most freight, but passenger trains are few outside the Montreal-Toronto-WINDSOR area. Exceptions are the popular routes through the Rockies, especially the Jasper-Vancouver line.

CANADA AT A GLANCE
Official name The Dominion of Canada
Area 9 970 610 km² (3 848 765 sq miles)
Population 27 296 860 **Per km²** 2.7 (**Per sq mile** 7)
Capital Ottawa
Government Federal parliamentary monarchy
Currency 1 Canadian dollar = 100 cents
Languages English (67%), French (15%), bilingual (16%)
Religion Christian (46% Roman Catholic, 30% Protestant)
Climate Continental; arctic in the north; maritime near the coast (especially in British Columbia). The average temperature in Ottawa ranges from -10.8°C (12.6°F) in January to 21°C (69.8°F) in July
Land use Forest and woodlands 35%, cultivation 5%, grazing 3%, other 57%
Main primary products Cereals, fruit and vegetables, livestock, rapeseed, tobacco, linseed, timber, fish; oil and natural gas, coal, copper, zinc, molybdenum, iron, lead, asbestos, silver, nickel, gold, salt, uranium, potassium
Major industries Agriculture, forestry, paper and other timber products, food processing, iron and steel, engineering, mining, transport equipment, chemicals, fertilisers, oil and gas refining, cement
Main exports Motor vehicles, machinery, paper, timber and wood pulp, wheat, oil, metal ores
Annual income per head (US$) 17 563
Population growth (per thous/yr) 10
Life expectancy (yrs) Male 74 **Female** 81

Cape Doctor In Cape Town, South Africa, a humorous name for the strong south-easterly wind that blows chiefly in summer, so called because it prevents stagnation of the air over Table Mountain and blows all the city's polluted air out to sea.

Cape Maclear *Malawi* Tourist resort on the southern shore of Lake MALAWI. Offshore is Thumbi Island, a nature reserve.
Map Malawi – Ab

Cape Peninsula *South Africa* Peninsula joined to the African mainland by the low-lying, sandy Cape Flats. At its northern tip lies Cape Town; to the south are suburbs, resorts, fishing villages, a naval base at Simon's Town and, at its southern tip, the Cape of Good Hope Nature Reserve which includes Cape Point, 56 km (35 miles) from Cape Town.
Map South Africa – Ac

Cape Province *South Africa* See WESTERN CAPE PROVINCE, EASTERN CAPE PROVINCE and NORTHERN CAPE PROVINCE

Cape Town *South Africa* Legislative capital and the country's second seaport after Durban, Cape Town was founded in 1652 as a supply depot for Dutch ships. It lies at the foot of the magnificent Table Mountain (1086 m, 3563 ft) and is the home of Groote Schuur Hospital, where the world's first human heart transplant was performed in 1967. Reputedly one of the world's most beautiful cities, Cape Town is also a major tourist resort surrounded by beaches, mountains and winelands. Places of interest are, close to the modern city centre, the Castle of Good Hope (1666-1810) – the country's oldest building – the 19th-century Parliament Buildings, several examples of gabled Cape Dutch architecture, and the Rhodes Memorial, which offers panoramic views over the city's suburbs. Farther afield lie Cape Town's Botanic Gardens at Kirstenbosch, and there are many scenic drives.

African towns and squatter settlements include Guguletu, Crossroads, Khayelitsha and Langa, where scenes of violent unrest were rife during the last years of apartheid, the official government policy of racial segregation from the late 1940s to the early 1990s.
Population 2 159 000
Map South Africa – Ac

Cape Verde See p 136

Capernaum (Kefar Nahum) *Israel* Ruins of an ancient town on the north shore of the Sea of Galilee (Lake Tiberias), where Jesus spent most of the period of his ministry and chose his first disciples (Andrew, Peter, James and John). There is a 2nd-century synagogue on a spot where Jesus is said to have preached. Nearby, St Peter's house has been excavated.

Rising to 91 m (368 ft) above Capernaum is the hill known as the Mount of the Beatitudes, where according to tradition, Christ preached his Sermon on the Mount.
Map Israel – Ba

Cap-Haitien (Le Cap) *Haiti* Seaport, and second largest city after Port-au-Prince, lying on the island's north coast. Founded in 1670 by the French, it was razed in 1802 by French troops.

Henri Christophe, the self-proclaimed King of North Haiti, rebuilt much of it, but an earthquake in 1842 caused widespread destruction. Today its modernised harbour handles exports of coffee, sugar cane, bananas, sisal and cocoa.

The ruined Sans Souci palace, built by Christophe in 1813 and wrecked by the 1842 earthquake, lies 16 km (10 miles) to the south. Another of his constructions, the Citadelle, lies a further 11 km (7 miles) to the south. It is a massive fortress rising from the sheer cliff face, with 200 cannons ranged around its 4 m (13 ft) thick walls. Begun in 1804, it took 13 years to complete and 20 000 slaves died building it.
Population 75 000
Map Caribbean – Ab

Capitol Reef National Park *USA* Area of 979 km² (378 sq miles) in south-central Utah, astride a tributary of the Colorado River. It incorporates 32 km (20 mile) long sandstone cliffs which rise abruptly from the desert and resemble the Capitol in Washington DC.
Map United States – Dc

Capo d'Istria *Slovenia* See KOPER

Cappadocia *Turkey* Vast volcanic plateau lying between Ankara and Kayseri, and distinguished by its many humpbacked rock formations. People have cut homes into its soft volcanic rock since at least 400 BC. Many of its cave houses are still in use, especially around the villages of Aucilar and Ürgüp. Cappadocia's Göreme valley contains numerous early Christian chapels and a Byzantine monastic complex, the inside of which is covered with 10th-century frescoes.
Map Turkey – Bb

Capri *Italy* Beautiful holiday island in the Bay of Naples, covering 10.4 km² (4 sq miles). It is garlanded with shrubs, vineyards and exotic gardens. Its main towns are Capri and Anacapri.

The island is renowned for its marine caves, particularly the Blue Grotto. For many years Capri was the home of the English entertainer Gracie Fields (1898-1979), whose villa is now a shrine to her memory. It was also a favourite haunt of the Roman emperor Tiberius.
Population 7050
Map Italy – Ed

Caprivi Region *Namibia* Strip of land, about 450 km (280 miles) long and 30-100 km (18-60 miles) wide, extending eastwards from the country's northern region. It lies between Angola and Zambia to the north and Botswana to the south and covers 19 352 km² (7470 sq miles). Formerly known as the Caprivi Strip, it was acquired from Britain in 1893 by an agreement with the German Chancellor Count Leo von Caprivi. Katima Mulilo on the Zambezi is the regional capital.
Map Namibia – Ba

Carabobo *Venezuela* State in the north, containing the cities of Valencia, its capital, and Puerto Cabello. In 1821, Carabobo was the scene of the battle which assured Venezuela's independence from Spain. There is a large monument to the victory under Simón Bolívar at Campo de Carabobo, south of Valencia. The state's principal products are sugar, citrus fruits and coffee.
Population 1 453 200
Map Venezuela – Ba

A DAY IN THE LIFE OF A CAPE TOWN UNIVERSITY LECTURER

Few people can wake up to a view as spectacular as the one Ray Wilson sees from the window of his bedroom every morning. His restored Victorian house in the inner Cape Town suburb of Tamboerskloof ('Drummer's Ravine') looks directly up to the majestic flat-topped Table Mountain. His home is in the so-called City Bowl area – a natural bowl formed by Table Mountain and the adjoining mountains of Devil's Peak and Lion's Head. Ray joins his wife Harriet and their two small sons, Sean and Anton, at the breakfast table. Harriet serves scrambled eggs to the children, while Ray swallows a cup of coffee.

After breakfast, Ray drives Harriet and the children to a private nursery school on his way to work. Harriet then travels the 6.5 km (4 miles) into town by bus, to the Social Services Office, headquarters of her job as a social worker. Blacks and whites can now travel on the same buses in Cape Town; between the 1950s and the 1980s, buses were segregated – one of the aspects of 'petty apartheid' that younger South Africans grew up with.

Ray, meanwhile, drives to the University of Cape Town (UCT), where he lectures in African History. This morning, Ray has a seminar for second-year students on the 19th-century frontier wars between the British and the Xhosa in the Eastern Cape. Inevitably, discussions on colonisation turn to the topic of exploitation by the colonisers – and the role of multinationals in Africa – and become heated.

Many of Ray's students, and some of his fellow lecturers, are black and still live in the townships that grew up far from the city centre during the apartheid years. They have to commute 30 to 40 km (about 19 to 25 miles) and more every day to get to the university. Ray and Harriet have invited some of his students, blacks and whites, to their house for dinner but, despite the abolition of apartheid, there is still little social mixing between the races. In the past, whites and blacks were not even permitted to eat together at the same restaurant, as some restaurants were designated for whites only, and some for 'non-whites'.

Ray spends the afternoon in his office, putting the finishing touches to a series of lectures. His faculty has recently restructured its courses to give African History an African – and less Eurocentric – slant and, although Ray has always tried to integrate both viewpoints in his lectures, the new policy means added work. But by 5 pm he has finished and, before going home for a quiet and relaxing evening, he drives down to the beach at nearby Camps Bay for a run on the warm sand and a quick dip in the cold Atlantic surf. Behind him, the sun illuminates the impressive backdrop of the Twelve Apostles, a chain of mountains that forms the backbone of the Cape Peninsula – one of the world's most beautiful city settings.

Cenozoic era The most recent geological era in which mammals and flowering plants first flourished. See GEOLOGICAL TIMESCALE

Central African Republic See p 143

Central Kalahari Game Reserve *Botswana* Part of the KALAHARI desert, a wilderness of sandy savannah with scattered thorn trees and inhabited by a few Bushmen, or San, hunting groups. The 51 800 km² (20 000 sq mile) park has no roads and can be entered only with special permission. Wildlife includes gemsbok, hartebeest and springbok. Joined to the park in the south is the Khutse Game Reserve, covering about 2590 km² (1000 sq miles). Visitors to either must have four-wheel drive vehicles and carry their own supplies, including water.
Map Botswana – Bb

Central Region *United Kingdom* Local government area of Scotland, created in 1975 from the former county of Clackmannan and parts of Stirlingshire and Perthshire. It covers 2635 km² (1017 sq miles). The region reaches into the Scottish Highlands, culminating at the peaks of Ben More (1174 m, 3851 ft) and Ben Lui (1130 m, 3707 ft). The principal towns are STIRLING, Falkirk, and the industrial centre of Grangemouth on the River Forth, with its oil-refining and chemical plants.
Both shores of the Forth have been extensively mined for coal, and some of Scotland's earliest iron works were situated here, but the field has now been largely worked out.
Population 267 960
Map United Kingdom – Cb

Centre *France* Farmland region to the south of Paris, astride the Loire River. It comprises Beauce, part of the LOIRE valley, and BERRY. The main towns are Orléans and Tours.
Population 2 371 000
Map France – Dc

Cephalonia (Kefallinia) *Greece* Largest of the IONIAN ISLANDS, covering 782 km² (302 sq miles). Its hilly countryside lined with fir trees and vine and olive groves, and its ragged coast of little bays attracts many tourists, as does the island's stalactite cave with its coloured underground lake, Melisani. White wine is produced on the island and perfumes made from its flowers.
Population 27 600
Map Greece – Bb

Ceram *Indonesia* See SERAM

Ceram Sea See SERAM SEA

Cernauti *Ukraine* See CHERNIVTSI

Cërrik *Albania* Principal oil-refining centre of Albania since 1950. It is linked to DURRËS port by a railway started by the occupying Italians during the Second World War and completed after the war with Russian aid.
Population 12 500
Map Albania – Bb

Cerro de Pasco *Peru* Town 4360 m (14 304 ft) up in the Andes and therefore one of the highest towns on earth. It lies 160 km (100 miles) north-east of Lima. Cerro de Pasco was founded in 1771

as a silver town, but now copper, zinc, bismuth and tungsten have made it the country's biggest mining centre overall. A new town, San Juan de Pampa, 2 km (1.25 miles) away, was founded in 1965, to cope with the growth.
Population 71 500
Map Peru – Bb

cerussite Naturally occurring lead carbonate. The mineral, often found in association with GALENA, is an important source of lead.

Cerveteri *Italy* Town 30 km (20 miles) north-west of Rome. Traces of Etruscan walls can be seen in the medieval castle. To the north lies an Etruscan necropolis, with many impressive monumental tombs dating from the 7th to the 1st century BC.
Population 12 100
Map Italy – Dd

Cervino *Italy/Switzerland* See MATTERHORN

Ceské Budejovice (Budweis) *Czech Republic* South Bohemian city and industrial centre on the Vltava River, 125 km (78 miles) south of Prague. The city is famous for its Budvar beer, produced here since the Middle Ages, which is known as Budweiser in its American version.
Hluboka Castle, built in Tudor style in the 19th century and modelled on England's Windsor Castle, is situated just north of Ceské Budejovice. To the north-west lies the village of Husinec, the birthplace of Jan Hus, or John Huss (about 1369-1415), the Czech hero and religious reformer who was burnt as a heretic.
Population 174 400
Map Czech Republic – Bb

Cesky Tesin *Poland* See CIESZYN

Cetinje *Yugoslavia (Serbia and Montenegro)* Major tourist centre and, from 1918 to 1946, the capital of the Yugoslav republic of Montenegro. More than 600 m (2000 ft) above sea level, it lies 18 km (11 miles) north-west of Lake Shkodër. The fortified monastery, home of the town's prince-bishops from 1515 to 1851, has fine collections of icons and books, including one of the first printed in a Slav language (1493).
The Biljarda, a palace built in 1836 for the poet Bishop Petar Njegos II, is now a museum housing fine ethnographic collections and the Museum of National Liberation; a second palace, once the home of Nicholas I, Montenegro's last princely ruler, is now home to the state museum.
Population 20 300
Map Yugoslavia – Bc

Ceuta *Spanish enclave in Morocco* Duty-free seaport covering 19.5 km² (7.5 sq miles) situated on the northernmost peninsula of Morocco, facing Europe. Administered as an integral, semi-autonomous part of Spain, it is directly represented in the Spanish government. The Mediterranean lies on two sides of the town, with its Jebel Musa (Mount Hacho) providing a spectacular view of the STRAIT OF GIBRALTAR; together with GIBRALTAR itself, this 842 m (2762 ft) mountain was known to the ancient Greeks as one of the Pillars of Hercules guarding the entrance to the Mediterranean.
Population 73 210
Map Morocco – Ba; Spain – Ce

Cévennes *France* Rocky mountainous area cut by deep gorges, along the south-east fringe of the Massif Central. Its highest peaks are Mont Lozère (1702 m, 5584 ft) and Mont Aigoual (1567 m, 5141 ft). Wide stretches of the area's hills are given over to sheep grazing and forestry. Sunny slopes are, however, terraced for vines, fruit and chestnuts. The national park of Cévennes covers 2360 km² (911 sq miles) of farmland and rough country in the area.
Map France – Ed

Ceylon See SRI LANKA

Chad See p 144

Chad, Lake *Equatorial Africa* Shallow lake in north-west central Africa, shared by Cameroon, Chad, Niger and Nigeria. Its maximum depth is about 7 m (23 ft), and much of its surface is choked with floating vegetation. Numerous rivers drain into it, but it has no outlet and so, during the wet season from May to October it expands, to cover about 26 000 km² (10 040 sq miles), making it the ninth largest freshwater lake in the world. However, evaporation reduces it to half that size by April, and sometimes it shrinks to 10 000 km² (3860 sq miles).
Geological evidence shows that during the Ice Ages the lake was as big as 310 000 km² (120 000 sq miles) – almost the size of the Caspian Sea – and even in historical times it extended as far as the Bodélé depression in Chad.
The area around Lake Chad has always been sparsely populated because of its arid soils and because it was once a no-man's land between rival states. However, the lake provides drinking water for animals, including herds of cattle. Rice is traditionally planted on flooded land, as the lake recedes. Groundnuts, sorghum and maize are also grown. The lake is a valuable source of fish, salt and potash.
Map Chad – Ab

Chaillu Mountains *Congo/Gabon* Range in south-eastern Gabon and south-western Congo. Named after the French-born American explorer Paul du Chaillu, who first explored the area in 1855-65, they include Mount Iboundji, Gabon's highest peak, at 1575 m (5167 ft).
The mountains form a watershed between the N'GOUNIE and OGOOUE river systems, which unite at LAMBARÉNÉ.
Map Congo – Ab

Chainat *Thailand* Market town and province 175 km (110 miles) north of Bangkok. The province is a fertile rice-growing area.
Population (town) 67 700; (province) 335 400
Map Thailand – Bb

Chalcidice (Khalkidhiki) *Greece* Wooded peninsula in MACEDONIA, which has three small, fingerlike peninsulas named Kassándra, Sithoniá and ÁTHOS reaching into the Aegean Sea south of SALONICA. Long, sandy beaches and attractive camping sites have made Chalcidice a popular tourist resort.
On the Gulf of Kassándra, 8 km (5 miles) west of Gerakini, lies the ancient city of Olynthos, which was destroyed by Philip II of Macedon in 348 BC and has been unoccupied ever since.
Population 79 000
Map Greece – Ca

Central African Republic

THE COUNTRY THAT WAS ONCE RULED BY A MEGALOMANIAC 'EMPEROR' REMAINS UNDEVELOPED AND POVERTY-STRICKEN

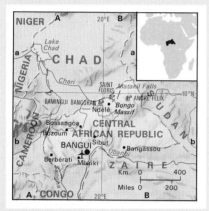

Until 1977, the remote, landlocked Central African Republic in the very middle of Africa was rarely mentioned in the world news. But in that year it shot into the headlines with the coronation on 4 December of its vainglorious leader, Jean-Bedel Bokassa, who had proclaimed himself emperor (and the country an empire) a year earlier. The bizarre ceremony was estimated to have cost US$22-30 million at a time when most of his people lived on an average income of around US$5 a week.

Formerly known as Ubangi Shari, a territory of French Equatorial Africa, the country became independent as the Central African Republic in 1960. Although it has rich mineral resources – diamonds alone account for 25 per cent of exports – 84 per cent of the people lead frugal lives as farmers, cultivating small plots of cassava, maize, millet, sorghum, sweet potatoes and yams. Only 3.2 per cent of the land is cultivated. Livestock rearing is done on a small-scale because of the prevalence of the disease-carrying tsetse fly.

Much of the country is rolling grassland and bush on a high plateau 600-900 m (about 2000-3000 ft) above sea level, with mountainous areas on the western border with Cameroon and the eastern border with Sudan. In the drier north, the plateau slopes down to the CHARI (or Shari) River across the northern border with Chad. In the south, it slopes towards the UBANGI (Oubangui) River, which forms much of the southern border with Zaire and flows into the ZAIRE (Congo) River. Most of the people live in the west and the hot, humid south and south-west – many of them in the dense south-western rain forest, where it rains nearly every day from March to November. The forest is the source of another major export – hardwood timber.

The largely uninhabited eastern half of the country contains two of the three national parks as well as various game reserves – but here, too, poaching has drastically reduced the wildlife population.

There are no railways in the republic. Much of the transport is by boat – there are some 7000 km (4350 miles) of inland rivers – particularly via the Ubangi, which is navigable for most of the year except for a few months at the height of the November-March dry season.

BANGUI, the capital and chief port, stands on the Ubangi, and from it goods are shipped some 1200 km (740 miles) to Brazzaville in the Congo Republic, then sent 500 km (310 miles) by rail to the Congo port of POINTE-NOIRE on the Atlantic coast.

Apart from the Babinga (pygmies) of the rain forest, there are eight main indigenous population groups, the two largest being the Banda and Baya, who make up half the population. Seven of the groups speak Sudanese languages, but the national language is Sango, the language of the Ubangians. Although they form a minority of the population, Ubangians have dominated the government and civil service since independence.

Former emperor Jean-Bedel Bokassa, a Ubangian; ruled for 14 years, in which time he reduced the country to bankruptcy and its elephant population by three-quarters. Indeed, much of his personal fortune was made from ivory and diamonds. Bokassa's extravagance, combined with accusations of brutality, led to his downfall. Riots occurred in 1979 when schoolchildren protested at having to buy school uniforms costing about US$20. Some 250 children were detained by the police and about 100 murdered.

Bokassa was deposed in a French-supported coup while visiting Libya in September 1979, and condemned to death in his absence. He returned to the Central African Republic in 1986, when he was tried, and again condemned to death – but his death sentence was later converted to life imprisonment. Late in 1993, he was released but stripped of rank and electoral rights.

The present government has instituted a number of reforms in its efforts to stamp out corruption and foster economic recovery. In 1992, the constitution was revised to allow a multiparty democracy. The 1993 elections were won – although not with an absolute majority – by President Ange-Felix Patasse and his Central African People's Liberation Party. Patasse reacted to the result by forming a coalition cabinet.

Despite long-standing government attempts to eradicate old customs and tribal affinities, traditional values remain strong outside the city and the Westernised elite. Only about one-third of the people can read and write, and more than half follow the old tribal religions. France remains the republic's chief trading partner and the main source of aid and investment. But while foreign interests – in the exploitation of diamonds and in plans to develop vast uranium deposits discovered near Bakouma, 480 km (300 miles) east of Bangui – influence the economy, very little of the wealth has reached the population at large.

CENTRAL AFRICAN REPUBLIC AT A GLANCE	
Area 622 980 km² (240 477 sq miles)	
Population 3 400 000 **Per km²** 5.4 **(Per sq mile** 14)	
Capital Bangui	
Government Republic	
Currency 1 CFA franc = 100 centimes	
Languages French (official), Sango, and other African languages	
Religions Indigenous beliefs (60%), Christian (35%), Muslim (5%)	
Climate Tropical, and very humid, but drier in the north. Average temperature in Bangui is 26°C (78.8°F) throughout the year	
Land use Forest and woodlands 64%, grazing 5%, cultivation 3%, other 28%	
Main primary products Cassava, groundnuts, bananas, plantains, sweet potatoes, maize, millet, coffee, cotton, timber; diamonds	
Major industries Agriculture, forestry, mining	
Main exports Diamonds, coffee, hardwoods, cotton	
Annual income per head (US$) 410	
Population growth (per thous/yr) 25	
Life expectancy (yrs) Male 47 **Female** 47	

---

chalk Soft, whitish limestone consisting almost entirely of calcium carbonate. Chalk is derived chiefly from the calcified remains of small marine organisms.

Challenger Deep See MARIANAS TRENCH

Châlons-sur-Marne *France* Capital of Marne department, 150 km (93 miles) east of Paris on the plain of Champagne, on a bend of the River Leysse. Chief town of the ancient Catalaunic people, it was occupied and fortified by the Romans. In AD 451, on the plains to the south, the Romans defeated the Barbarian invader Attila the Hun at the Battle of Châlons. The city today produces beer and champagne, processed foods, and manufactures electronics and textiles.
Population 51 500
Map France – Fb

Chambéry *France* Capital of Savoie department, 45 km (28 miles) north of Grenoble. It produces aluminium and cement. The old town, with its arcaded streets and imposing houses, is dominated by a huge 14th to 15th-century château, once the home of the dukes of Savoy and now open to the public.
Population 54 000
Map France – Fd

Chambord *France* Village 45 km (28 miles) south-west of Orléans, famous for its château, the largest in the Loire valley. The château, which is still unfinished, was begun in 1519 as a royal hunting lodge and has 440 rooms and 50 staircases.
Map France – Dc

Chad

NATIONAL SURVIVAL IS THE KEY ISSUE IN CHAD, ONE OF THE WORLD'S POOREST NATIONS, WHICH HAS BEEN FURTHER WEAKENED BY CIVIL WAR

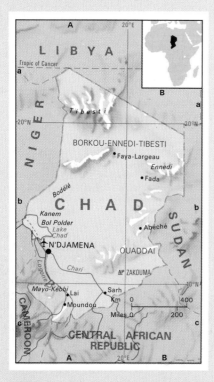

At the root of Chad's tragedy in modern times has been its geographical location. Landlocked in the centre of northern Africa and spanning the semi-arid SAHEL, the country stretches north, for some 1800 km (1100 miles), from the edge of the equatorial forests to the middle of the SAHARA. It is one of the world's most underdeveloped countries – lacking a proper infrastructure as well as resources. The country has suffered only slightly less than Ethiopia in the African drought that began in the 1970s and continued into the 1990s, apart from a spell of good rains in 1991, which accounted for an economic growth of more than 8 per cent that year.

With a journey of more than 1600 km (1000 miles) from the nearest sea-ports – in some cases through unfriendly neighbouring countries – plus woefully inadequate transport facilities, it has been extremely difficult to ferry in relief aid. This tragedy has been overlaid, as in Ethiopia, by internal political conflict. Chad stands at the junction of Arabic and black African cultures, and has a multiplicity of ethnic groups which, with entirely different ways of life, religions and allegiances, have long been political rivals.

Chad is more than twice the size of France, its former colonial ruler, and lies mostly in a vast basin draining (when there is rain) into Lake CHAD, on the western border. The most dramatic landforms are the TIBESTI Mountains in the far north, whose twisted peaks rise from the desert sands to more than 3000 m (10 000 ft); the extinct volcano of Emi Koussi, at 3415 m (11 204 ft), forms the highest peak.

The southern part of the country is the least poverty-stricken although it is the most densely populated. Relatively well watered, its wooded savannah has always been Chad's main arable region, but even here, the rains failed in 1984. In normal times it is farmed for cotton (the country's main cash crop), millet, sorghum, groundnuts, rice and vegetables, mainly by the predominant Sara. The older among these people are distinguished by the bold patterns of scars on their faces, a custom that is now dying out. Also in the south and south-west are groups who fish the rivers and Lake Chad for a living. These southerners mainly follow tribal religions or Christianity, while the north is strongly Muslim.

In physical features, the northern peoples resemble the Mediterranean Arabs, though some are very dark-skinned. They are largely nomadic or semi-nomadic graziers (grazing, that is, permitting) of cattle and sheep in the Sahel and mainly of camels in the far north. The drought has driven them farther south or to the outskirts of towns or oases, wherever they can find water.

The Islamic influence can be seen throughout Chad in the craftsmanship of its superb carpets, handwoven and embroidered cottons and other goods. Much more ancient art found on rocks, notably in the Tibesti region, shows that some 3000 years ago, the climate was wet enough for dense vegetation, and for hippos and rhinos to live where there is now desert.

Much later, from 1000 to 400 years ago, the Sao people from the Nile Valley settled near Lake Chad and left remains of distinctive pottery. In the 16th and 17th centuries, three major Muslim kingdoms were established in the Sahel region and became traders in slaves captured in the south. The most durable of these, the Wadai, was conquered by the French only in 1911. French forces had entered Chad in the 1890s and were welcomed as protectors by the southern tribes. Declared a separate French colony in 1920, Chad was granted independence in 1960. The northern prefecture of Borkou-Ennedi-Tibesti – always a centre of resistance to colonial rule – remained under French military control until 1965, when it joined the rest of Chad. Since then there has been continuous instability, fuelled by internal political and cultural differences, and in 1980 this developed into all-out civil war. Libya backed one faction, Egypt and Sudan backed the other, and French peace-keeping troops also became involved. In 1973, Libya occupied territory it claimed in northern Chad, including the Aozou Strip, only to be forced back by French-backed government forces. But the Libyans continued to claim and occupy the narrow Aozou Strip which runs parallel with the border between Chad and Libya. (There are indications of rich mineral deposits located in this piece of land covering 100 000 km² (38 601 sq miles).)

WITHIN SIGHT OF PEACE?

In 1994, the International Court of Justice based at The Hague ruled that the Aozou Strip belongs to Chad. In May that year, Libyan troops were evacuated from the area.

Internal politics have also moved closer to resolution. At the end of 1990, after a rebel army led by Idriss Deby seized control of the country, restrictions on political parties were lifted. In 1993 Deby formed a multiparty interim government committed to democracy. However, the country remained highly unstable.

CHAD AT A GLANCE	
Official name Republic of Chad	
Area 1 284 000 km² (about 496 000 sq miles)	
Population 6 400 000 **Per km²** 5 **(Per sq mile** 12.9)	
Capital N'Djamena	
Government Republic	
Currency 1 CFA franc = 100 centimes	
Languages French, Arabic, African languages	
Religions Muslim (about 50%); Christian, and indigenous beliefs	
Climate Desert climate; only the south receives regular summer rains. Average temperature in N'Djamena ranges from 24°C (75.2°F) in December to 32°C (89.6°F) in June	
Land use Grazing 36%, forest and woodlands 11%, cultivation 2%, other 51%	
Main primary products Cotton, cattle, millet, sorghum, rice, cassava, fish; natron (sodium carbonate)	
Major industries Agriculture, textiles, food processing, mining, fishing	
Main exports Cotton, cattle, meat	
Annual income per head (US$) 220	
Population growth (per thous/yr) 25	
Life expectancy (yrs) Male 47 **Female** 47	

Chamonix *France* Winter sports resort in the Arve valley, 57 km (35 miles) east of Annecy. It is the nearest town to Mont Blanc (4807 m, 15 771 ft), the highest Alpine peak. Six major glaciers and more than 20 smaller ones flow towards its valley. The first Winter Olympics were held here in 1924.
Population 9 700
Map France – Gd

Champagne-Ardennes *France* Region to the east and north-east of Paris, crossed by the Aisne, Aube, Marne and Meuse rivers. It is divided into Champagne sèche (dry) and Champagne humide (humid). The sparkling white wine which takes its name from the Champagne district was first produced here in the 18th century. Other manufactures include cereals, dairy goods and textiles. RHEIMS is the largest city. The wooded hills in the north of the region extend across the Belgian and Luxembourg borders to form the ARDENNES.
Population 1 348 000
Map France – Fb

Champassak *Laos* Southern, rice-producing province straddling the MEKONG River.
Population 403 000
Map Laos – Bc

Champion *Brunei* One of three offshore oil fields in the South China Sea, not far from an older, onshore production area, SERIA. Together with the South-west Ampa and Fairley fields, Champion now accounts for almost all the country's oil production. All three fields lie between 13 and 32 km (8-20 miles) north-west of Seria.

Chan Chan *Peru* Capital of the pre-Inca Chimú Empire, 5 km (3 miles) north of Trujillo. The adobe ruins cover some 20 km² (8 sq miles). There are nine great citadels, 300 m (985 ft) apart, each surrounded by a 10 m (33 ft) wall. Each compound contains temples, houses, workshops and store rooms, and most have a large well. One of the compounds has been reconstructed. Relics, including clothes and pottery, are exhibited in the museums of the country's capital, Lima.
Map Peru – Ba

Chandannagar (Chandernagore) *India* Town on the Hooghly River, on the north-west fringe of Calcutta. It was founded as a trading post in 1688 by the French East India company. In 1951, while still a French colony, it voted to become part of newly independent India.
Population 101 900
Map India – Dc

Chandernagore *India* See CHANDANNAGAR

Chandigarh *India* City 249 km (155 miles) north of Delhi. It was the capital of the states of Punjab and Haryana, although it belonged to neither. This came about after the British province of Punjab was divided between India and Pakistan in 1947. Pakistan retained the capital of Lahore, and India built the modern city of Chandigarh, designed by the controversial Swiss-born French architect Le Corbusier (1887-1965). However, Indian Punjab was Hindu in the east and Sikh in the west, and it finally split into the western state of Punjab, and what are now the states of Haryana and Himachal Pradesh. Although both Punjab and Haryana claimed Chandigarh, it was given to neither. Instead, it was designated a Union territory, a property of the federal government, until 1986, when it was to be ceded to Sikh Punjab. However, Haryana was not satisfied with the compensation offered and the scheduled transfer was put off.
Population 575 000
Map India – Bb

Chang Jiang (Yangtze) *China* Fourth longest river in the world, after the Amazon, the Nile and the Mississippi. It rises on the Xizang Gaoyhan (Tibetan Plateau) and flows 5980 km (3716 miles) to the East China Sea at Shanghai. The Chang Jiang drains the central part of China and is an important source of irrigation and hydro-electric power. It is navigable for ocean-going ships for 1100 km (680 miles) of its course to Wuhan, and for smaller craft for 2253 km (1400 miles), to Chongqing. The river carries between 60 and 70 per cent of the country's inland waterways trade. Flooding is frequent, but rarely disastrous because of a network of natural regulating lakes on its reaches, either side of Wuhan.
Map China – Ed; He

Changchun *China* Capital city and industrial centre of Jilin Province, about 900 km (560 miles) north-east of Beijing (Peking). The centre

of the motor-vehicle industry, it also produces locomotives and railway rolling stock. During the 1930s, while capital of the Japanese puppet state of Manchukuo, it was known as Xinjing.
Population 1 616 000
Map China – Ic

Chang-hua *Taiwan* Old city near the central west coast in the centre of a wealthy agricultural region that produces rice, asparagus and pineapples. A 20 m (66 ft) high statue of Buddha stands on a hill overlooking the city.
Population 1 272 000
Map Taiwan – Bb

Changi *Singapore* District at the eastern tip of Singapore Island. It is the site of Singapore's international airport, completed in 1981, partly on land reclaimed from the sea. Changi Jail, a notorious prison for Allied prisoners of the Japanese during the Second World War, is now a small museum.
Map Singapore – Ab

Changsha *China* Capital of HUNAN Province, about 550 km (340 miles) north of GUANGZHOU. It is a centre for traditional handcrafts such as ceramics, embroidery and bamboo carving, and for food processing, textiles, machine tools and precision engineering. The late Chinese communist leader Mao Ze-dong (1893-1976) was a student and teacher here from 1913 to 1923.
Population 1 300 000
Map China – Ge

Channel Islands *English Channel* Self-governing group of islands under British sovereignty, lying just off the north-west coast of France, where they are called Les Iles Anglo-Normandes. The islands were part of the Duchy of Normandy in the 10th century, and were linked with England by the Norman Conquest of 1066.

The largest island, Jersey (117 km², 45 sq miles), is little more than 20 km (12 miles) from France's Cotentin peninsula. Most place names are French, and many of the islanders speak French. The second largest island in the group is Guernsey (78 km², 30 sq miles), which is more English by inclination. Both Guernsey and Jersey are large enough for road networks and ports – St Helier on Jersey, St Peter Port on Guernsey. Alderney (8 km², 3 sq miles) has few roads and

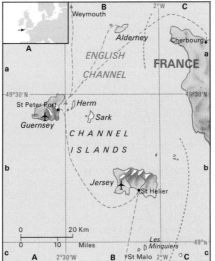

one village, St Anne, of granite-cobbled streets. There are no motorcars (only horse and tractor-drawn carriages) on Sark, which covers 5.5 km² (2.1 sq miles), or on tiny Herm.

The mild climate draws many tourists, and the low taxes attract wealthy British businessmen, but immigration is tightly controlled. The islands produce spring flowers and vegetables, especially tomatoes, for the mainland.

Jersey is expanding rapidly as a financial centre; this sector already contributes 50 per cent of the island's income, compared with 4 per cent earned from agriculture.
Population (Jersey) 84 080; (Guernsey) 58 870; (Alderney) 2130; (Sark) 600
Map Ab; Bb

Chanthaburi *Thailand* Market town and gem-mining centre 200 km (125 miles) south-east of Bangkok. The surrounding province of the same name produces tropical fruits.
Population (town) 93 250; (province) 381 950
Map Thailand – Bc

Chao Phraya *Thailand* River, about 1000 km (620 miles) long, running from the northern hills southwards through the capital and into the Bight of Bangkok. Its flood plains form the country's main rice-growing area.
Map Thailand – Bc

Chaouèn (Xauen) *Morocco* Small, attractive town perched on a rocky prominence in the RIF MOUNTAINS, 60 km (37 miles) south of Tétouan. Its name Chaouèn means 'the Town of Fountains' and is a reference to its many mineral water springs. Today the centre of the Riff Berbers, Chaouèn dates back to the 15th century, but only the city centre remains from that era. With its many mosques and sanctuaries, it is a Muslim holy city much frequented by pilgrims.
Population 24 000
Map Morocco – Ba

Chapala, Lake (Laguna de Chapala) *Mexico* The largest lake in Mexico, on the high plateau of Jalisco State, near GUADALAJARA. It covers 2460 km² (950 sq miles) but is now shrinking because water is increasingly being drawn off for irrigation. Tourist attractions include fishing, sailing and shooting.
Map Mexico – Bb

chaparral Dry scrub vegetation consisting of evergreen scrub oaks, vines and sparse grasses, found especially in the south-western United States and Mexico. It is similar to MAQUIS.

Chari (Shari) *Equatorial Africa* River rising in the Central African Republic and flowing some 1000 km (625 miles) north-west to enter Lake Chad. Chad's only permanent river, though navigable only in the wet season, the Chari is much valued for irrigation.
Map Central African Republic – Aa; Chad – Ab

Charleroi *Belgium* Industrial city some 50 km (30 miles) south of the capital, Brussels. It was given this name in 1666 by the Spanish governor-general of the Low Countries in honour of his king, Charles II. The main industries are steel, electrical machinery and glassware.
Population 229 000
Map Belgium – Ba

Charleston *USA* Chief port of South Carolina and one of the country's oldest cities, with the first settlers arriving here in 1670. The American Civil War broke out in its harbour when the Confederate army attacked Fort Sumter in 1861.

A number of old forts and 18th-century buildings remain, and there are many plantation homes in the area. The city is reputed to have given its name to the popular 1920s dance.

Population (city) 80 400; (metropolitan area) 506 880

Map United States – Kd

Charleston *USA* State capital of West Virginia. It lies in the west of the state, 50 km (30 miles) from the Ohio border, and was settled around the site at Fort Lee where the frontiersman Daniel Boone lived from 1788 to 1795. The city is the centre of a coal, oil, natural gas and chemical-producing and glass-manufacturing region.

Population (city) 57 290; (metropolitan area) 250 500

Map United States – Jc

Charlestown *St Kitts and Nevis* Chief town of Nevis, on the south-west coast. It is linked to Basseterre on St Kitts by ferry. South of the town, in Fig Tree Village, is St John's Church where Horatio Nelson, the British admiral, married Frances Nisbet on 11 March 1787.

The town's hot springs are its main tourist attraction.

Population 2090

Map Caribbean – Cb

Charleville-Mézières *France* Capital of Ardennes department on the Meuse River, only 12 km (7 miles) from the Belgian border. Charleville was founded in the early 17th century on the opposite bank of the river from the medieval walled town of Mézières. In 1966, the two towns and some smaller communes merged into a single city. The main industry is the manufacture of metal goods.

Population 59 500

Map France – Fb

▲ **HIGH GOTHIC Chartres cathedral, most of it built in less than 30 years, towers over the city and dominates the surrounding plain. Its stained glass is probably the finest in the world.**

Charlotte Amalie *US Virgin Islands* Capital of the territory, standing beside a large and protected harbour on the island of St Thomas.

Population 12 000

Map Caribbean – Cb

Charlottetown *Canada* Capital of PRINCE EDWARD ISLAND, situated on its south coast, overlooking Hillsborough Bay. It is a commercial centre built around fishing, tourism and industries producing knitwear, dairy products, canned goods and timber.

Population 15 400

Map Canada – Id

Chartres *France* Capital of the Eure-et-Loir department, in the plain of Beauce, 90 km (60 miles) south-west of Paris. Construction on its vast cathedral, the sixth on the site, started in 1194; it was consecrated in 1260. It is famous for the size and the beauty of its stained-glass windows – in particular for the luminous colour known as 'Chartres blue' which is used in them.

The cathedral's most precious relic is a veil said to have been worn by the Virgin Mary, and donated in AD 876 by Charles the Bald, King of the West Franks and grandson of Charlemagne.

Population 80 340

Map France – Db

Chatham *United Kingdom* See MEDWAY TOWNS

Chatham Islands *New Zealand* Island group in the Pacific, about 850 km (530 miles) east of Christchurch. It is made up of Chatham Island (964 km², 372 sq miles), Pitt Island (62 km², 24 sq miles), and several rocky islets. The chief town is Waitangi, on Chatham Island.

Population 760

Map Pacific Ocean – Ed

Chattanooga *USA* Industrial city and railway centre in the south-eastern corner of Tennessee, immortalised in the popular song, *Chattanooga Choo-Choo*. It straddles a bend in the Tennessee River near the Georgia border. The commercial bottling of Coca Cola soft drinks originated here.

Near the city is Lookout Mountain (654 m, 2146 ft), which provides spectacular views and where, in 1863, the Union army won a battle which proved decisive during the American Civil War.

Population (city) 152 470; (metropolitan district) 424 350

Map United States – Ic

Chaumont *France* Capital of Haute-Marne department, 250 km (155 miles) south-east of Paris. The town manufactures gloves, hosiery and textiles. In 1814, Britain, Austria, Russia and Prussia signed the Treaty of Chaumont in the town, binding themselves to pursue the war against the French emperor Napoleon. French president General Charles de Gaulle (1890-1970) died at Colombey les Deux Églises, 20 km (12 miles) to the north-west. His memorial is a huge Cross of Lorraine on a nearby hillside.

Population 27 600

Map France – Fb

Chaux-de-Fonds, La *Switzerland* Town about 50 km (30 miles) west of the capital, Berne. It is the centre of the Swiss watchmaking industry, and houses a watch and clock museum and a watchmakers' school. It was devastated by fire in 1795 and was rebuilt in a strictly rectangular pattern. It was the birthplace of the French architect Le Corbusier (1887-1965).

Population 36 840

Map Switzerland – Aa

Chaves *Portugal* City and spa about 120 km (75 miles) north-east of the city of Oporto and 10 km (6 miles) south of the border with Spain. It was fortified by the Romans and was known as Aquae Flaviae (the Flavian Springs) in Roman times. Roman remains include an 18-arch bridge and two inscribed columns. The city has several fine churches and a castle of the dukes of Braganza. Chaves is an agricultural and textile centre, and manufactures silks and linens.

Population 11 900

Map Portugal – Cb

Chavín de Huantar *Peru* Ruined fortress temple 3200 m (10 000 ft) up in the White Cordillera of the Andes, the centre of the Chavín culture which flourished from 800 to 200 BC. The Chavín people of South America worshipped a feline god that is sometimes depicted, mingled with characteristics of other animals, in stone carvings around the pyramids, plazas and temples. One of them, the Temple of Lanzón, contains a blackish, dagger-shaped stone monolith dating from about 800 BC.

Map Peru – Ba

Cheb (Eger) *Czech Republic* Bohemian bicycle-making town, with a medieval town centre, 5 km (3 miles) from the German (Bavarian) border and due west of Prague.

The water at Frantiskovy Lázne (Franzensbad) spa, just to the north, has the highest concentration in the world of Glauber's salt, or sodium sulphate.

Population 31 300

Map Czech Republic – Aa

Chechnya Republic *Russia* Autonomous republic of the Russian Federation, formerly Chechen-Ingush, covering 19 300 km² (7450 sq miles) on the northern slopes of the Caucasus Mountains, east of the Black Sea. Its industries, which are based on the Groznyy oil field, include engineering, chemicals and food canning. The capital is GROZNYY. In 1994 Russian troops moved into Chechnya to halt the republic's attempts to become independent, but met fierce resistance which continued into 1995.

Population 1 307 000
Map Russia – Fd

Cheju Do (Jejudo; Quelpart) *South Korea* Semi-tropical island covering 1828 km² (706 sq miles), making it the largest Korean island. It lies 90 km (56 miles) off Korea's south coast and is dominated by Halla San, at 1950 m (6398 ft) the highest mountain in South Korea. Centuries of isolation have resulted in the islanders developing their own cultural traditions and dress.

Population 463 000
Map Korea – Cf

Chelm (Chelm Lubelski; Kholm; Cholm) *Poland* Rapidly growing industrial city in the south-east, 25 km (16 miles) from the Ukrainian border. It was probably founded before the 13th century, and soon became a cathedral city and flourishing market for the rich farmlands around.

After years of Austrian and Russian rule, Chelm was returned to Poland after the Second World War. The Polish Committee of National Liberation, the embryo communist government of postwar Poland, was established in the city in July 1944.

Population 64 800
Map Poland – Ec

Chelmno (Kulm; Culm) *Poland* Medieval city on a promontory above the Vistula River, about 220 km (135 miles) north-west of the capital, Warsaw. The city was a member of the Hanseatic League and many buildings survive from those days of prosperity. Its encircling 13th-century ramparts with 17 bastions, its red-brick Gothic churches, and its Renaissance town hall and market square can still be seen today.

Population 21 400
Map Poland – Cb

Chelmno-nad-Nerem *Poland* See KONIN

Chelmsko Slaski *Poland* See WROCLAW

Chelyabinsk *Russia* Industrial city and major trans-port hub on the eastern fringe of the Ural Mountains, some 200 km (125 miles) south of YEKATERINBURG. After western Russian industries were relocated further east during the Second World War, Chelyabinsk, which was already connected by railway, grew rapidly. Its main products are iron and steel, motor vehicles, machinery and chemicals.

Population 1 148 000
Map Russia – Hc

chemical weathering Processes which cause the break-up of rocks through chemical reactions such as solution, oxidation and hydration.

Chemnitz *Germany* City 70 km (43 miles) south-east of Leipzig, named after the river that runs through it. In 1953, it was renamed Karl-Marx-Stadt in honour of the philosopher who had died 70 years previously, but in 1990, it reverted to its old name of Chemnitz.

Chemnitz is largely an industrial centre. Textiles have been produced here since the 14th century; in the 19th century, the city's machine construction industry was established. Germany's first machine tools as well as its first steam locomotive were made here. After being severely damaged during the Second World War, most of the city was restored and its textile and engineering industries re-established; it later branched out into chemicals and electronics.

Some of Chemnitz's historic buildings, dating back to the 12th century, now rub shoulders uneasily with modern monolithic architecture. For instance, the medieval town hall, with its vaulted *Ratsherrenstube* hall dating back to 1557, is flanked by the massive town hall which was completed in 1911.

Population 294 000
Map Germany – Ec

Chengde *China* Town in Hebei Province, 270 km (165 miles) north-east of Beijing. It is the site of the summer home of the Manchu emperors, the dynasty which ruled China from 1644 until 1912, when it was overthrown by republican Nationalists. The town has several palaces, pavilions and 18th-century temples, some of them in the Tibetan style.

Population 337 000
Map China – Hc

Chengdu (Chengtu) *China* Capital of Sichuan Province, in one of the richest agricultural regions in China. Its products include silk, electronics and timber. Chengdu zoo is famous for its giant pandas.

Population 2 810 000
Map China – Fe

Chenonceaux *France* Village in the Indre-et-Loire department, famous for its elegant Renaissance château which stands astride the Cher River. The château was begun in 1515 by Thomas Bohler, the French finance minister, and later enlarged by Diane de Poitiers – the mistress of the French King Henry II – and Catherine de Médicis, the king's wife. Every year, more than 800 000 visitors view the château's magnificent paintings and colourful tapestries and walk in the formal gardens.

Map France – Dc

Chenstokhov *Poland* See CZĘSTOCHOWA

Cheonam *South Korea* See CH'ONAN

Cherbourg *France* Channel port and naval dockyard on the north coast of the Cotentin peninsula. It became an important naval base and transAtlantic liner port after 1860, but now deals mainly with cargo boats and cross-Channel ferries. The town has been largely rebuilt since 1945, and has become a major industrial centre.

Population 89 200
Map France – Cb

Cherchell *Algeria* Coastal town and fishing port, it lies 95 km (59 miles) west of Algiers. It is built on the site of a Carthaginian settlement, which later became a Roman town called Caesarea in 25 BC. It has some fine Roman remains, including a theatre and an arena.

Population 17 000
Map Algeria – Ba

Chergui, Chott ech *Algeria* Marshy salt lake, 150 km (93 miles) long, situated south-west of Algiers in the High Plains between the Tell Atlas and the Saharan Atlas. A large depression, it provides rough pastureland for mountain pastoralists.

Map Algeria – Ba

Chernigov *Ukraine* See CHERNIHIV

Chernihiv (Chernigov) *Ukraine* Town and port on the Desna River, 130 km (80 miles) north-east of Kiyev. It is an agricultural centre and produces textiles, knitwear, footwear, chemicals, musical instruments and electrical goods. It has a 12th-century Byzantine cathedral.

Population 301 000
Map Ukraine – Ca

Chernikovsk *Russia* See UFA

Chernivtsi (Chernovtsy; Chernovitsi; Cernauti) *Ukraine* City near the Romanian border and 225 km (140 miles) south-east of L'viv. Its industries include saw-milling, engineering, food processing, and the production of textiles and rubber products.

The city has seen numerous name changes. Formerly an Austrian city named Czernowitz, it became part of Romania in 1919, when it was renamed Cernauti; it went to the USSR in 1940, and was called first Chernovitsy, then Chernovtsy. Now Ukrainian, it has been renamed Chernivsti.

Population 257 000
Map Ukraine – Bb

Chernobyl' *Ukraine* See KIYEV

Chernovtsy *Ukraine* See CHERNIVTSI

Cherrapunji *India* Until 1992, Cherrapunji was reputed to be the wettest inhabited place on earth. The village lies on the south side of the Khasi Hills in the north-eastern state of Meghalaya. Cherrapungi receives the full impact of the monsoon rains and has an average annual recorded rainfall of 11 314 mm (445 in).

The wettest place on earth is actually a station 16 km (10 miles) west of Cherrapunji called MAWSYNRAM, with an annual rainfall of 11 873 mm (467 in).

Map India – Ec

chert A crystalline form of silica in which the crystals are so small that they can be seen only under a powerful microscope. It usually occurs in bands in sedimentary rocks or as layers of pebbles. FLINT is one variety of chert.

Chesapeake Bay The largest inlet, 314 km (195 miles) long and between 5 and 48 km (between 3 and 30 miles) wide, on the east coast of the United States. It is rich in historical associations, for English settlers, who entered the bay in 1607, founded JAMESTOWN here. According to Captain John Smith, who explored and charted the bay in 1608, 'heaven and earth never agreed better to frame a place for man's habitation'.

Ports include BALTIMORE in the state of Maryland, and NORFOLK and Portsmouth in Virginia. Chesapeake Bay is popular with yachtsmen and is famed for its oysters, crabs and other seafood.

The Bay Bridge and Tunnel was built in the 1950s to span the mouth of the bay and increased industry and development on Maryland's eastern shore.

Map United States – Kc

Cheshire *United Kingdom* County covering 2322 km² (897 sq miles) of north-west England, stretching from the Welsh border and the Dee estuary across the River Mersey to the industrial centre of Warrington in the north and to the Pennine hills in the east.

Its undulating farmland is ideal for dairy herds – Cheshire cheese is produced in this region – and there is an unusual wealth of half-timbered houses. CHESTER, the county town, Runcorn, Ellesmere Port, CREWE and Macclesfield are its industrial centres.

Rock salt deposits have been worked for centuries at Nantwich and Northwich. Altrincham and STOCKPORT, formerly in Cheshire, were transferred to Greater Manchester in 1974.

Population 966 100
Map United Kingdom – Dd

Chester *United Kingdom* County town of the north-west English county of Cheshire, 55 km (35 miles) south-west of Manchester. It has been described as the best preserved walled city in Britain, and attracts many tourists.

Chester was founded by the Romans as Deva in the 1st century AD. The present walls, started in the 10th century, were built mainly on the foundations of the Roman fortifications.

The city also has one of the country's finest Roman amphitheatres, a castle, a red sandstone cathedral dating back to the 12th century, many fine timbered houses and unusual two-storey medieval shopping galleries called The Rows. During the 13th and 14th centuries, the city was a key regional port, but its importance declined largely because of silting in the River Dee.

Population (district) 118 000
Map United Kingdom – Dd

Cheviot Hills *United Kingdom* Range extending 60 km (about 37 miles) along the border of Scotland with the English county of Northumberland. It reaches 816 m (2676 ft) at The Cheviot. Most of the hills were open moorland until afforestation began after the First World War, creating among others Kielder Forest – one of the largest planted forests in Europe – at the south-western end of the range.

Map United Kingdom – Dc

Cheyenne *USA* State capital of Wyoming, in the south-east of the state, near the Colorado border. Named after the Cheyenne people, it was founded in 1867 and, as a rail junction, played a key role in opening up the west. The town history is commemorated in an annual Frontier Days festival. The town serves as a market for the sheep and cattle ranching in the area.

Population (town) 50 000; (metropolitan area) 73 140
Map United States – Fb

Chiang Mai *Thailand* See CHIENG-MAI

Chiang Rai *Thailand* Province and market town in the far north, 60 km (35 miles) from the borders of Myanmar (Burma) and Laos. Farmers in the province grow mainly rice, but much of the town's prosperity comes from cross-border trade and from smuggling links with the heroin-producing GOLDEN TRIANGLE region.

Population (province) 980 700; (town) 194 850
Map Thailand – Ab

Chianti *Italy* Hilly wine-making district in Tuscany, 40 km (25 miles) south of Florence. It is famous for its full-bodied red wines.

Map Italy – Cc

Chiapas *Mexico* State near the Guatemalan border in the extreme south of Mexico, situated on the Gulf of Tehuantepec. Now one of the poorest Mexican states, it was once at the core of the great civilisation of the Mayas which lasted from the 4th to the 10th century AD. It is largely inhabited by Amerindians – and several groups such as the Chamulas, Zinacantecans and Huistecos still speak their ancient languages and dress and live according to ancient customs, undisturbed by any outside influences.

In colonial times Chiapas formed part of Guatemala but its inhabitants always fought fiercely for their independence. On 1 January 1994, the Chiapans began another bitter uprising in protest at being exploited, and wilfully kept poor, as 'second-class' Mexican citizens. The uprising, which soon spread to other Mexican states, took on considerable political significance since it began on the day that the North American Free Trade Agreement (NAFTA) with Canada and the United States was to take effect – and close to national elections.

Chiapas is a mainly agricultural state, though agriculture is backward. It exports tropical products such as hardwoods, fruit, coffee, cotton, cocoa, chicle gum (from which chewing gum is made) and rubber. Its capital is Tuxtla Gutiérrez.

Population 3 204 000
Map Mexico – Cc

Chiba *Japan* Industrial city on HONSHU island, 35 km (22 miles) east of Tokyo. Large tracts of land have been reclaimed from Tokyo Bay for this fast developing city, which specialises in steel making and oil refining.

Population 829 000
Map Japan – Dc

Chicago *USA* The country's third largest city after New York and Los Angeles. It is situated in Illinois on the south-west shore of Lake Michigan, and is the country's transport hub. Railways radiate in every direction; it is a major Great Lakes port and is connected to the Mississippi River by a ship canal and the Illinois River; and its airport ranks amongst the busiest in the world.

Chicago is a popular venue for international conferences and trade fairs. Sears Tower, in the city, is the world's tallest building, at 443 m (1454 ft). It also has the world's largest building – Merchandise Mart. Chicago's stockyards and grain elevators handle huge quantities of foodstuffs. And its industries include iron, chemicals, steel, and electronics.

The city calls itself the Midwest's cultural centre, and indeed, the Art Institute of Chicago and its Museum of Science and Industry are among the most visited museums in the world. Chicago is also known as 'the Windy City'. During Prohibition in the 1920s, it became infamous for its alcohol bootleggers, such as the gangsters Al Capone and Bugsy Malone.

Population (city) 2 784 000; (metropolitan area) 7 410 860; (conurbation) 8 239 820
Map United States – Ib

Chichén Itzá *Mexico* Important archaeological site in Yucatán combining the building genius of the Mayas and the Toltecs. Its most remarkable feature is a four-sided pyramid called El Castillo. Light falling on its carefully constructed terraces at the spring and autumn equinoxes casts a shadow image of the feathered Toltec snake god Quetzalcóatl winding down its steps. Among its other notable buildings are the Caracol – an ancient Mayan observatory, and the Toltec Temple of the Warriors – a vast complex of columns in the form of serpents.

Map Mexico – Db

Chichester *United Kingdom* Cathedral city in southern England and county town of West Sussex, 90 km (55 miles) south-west of London. It was founded by the Romans as Noviomagus in the 1st century AD and still retains its Roman plan, with an almost complete ring of later city walls. Just to the west are the remains of Fishbourne Palace, a major Roman relic. Chichester's mainly 12th and 13th-century cathedral has a spire reaching 84 m (275 ft). There are many 18th-century buildings in the area, including Goodwood House, and a fine modern Festival Theatre in the city.

Population 27 240
Map United Kingdom – Ee

Chichicastenango *Guatemala* Market town and religious centre 90 km (56 miles) north-west of the capital, Guatemala City, and 2071 m (6795 ft) up in the Quiché Mountains. Each Thursday and Sunday thousands of Amerindians from the surrounding Maya villages, clad in magnificent costumes, descend on the town. They come for the markets and for the religious ceremonies – part Christian, part pre-Christian – held outside the 16th-century Santo Tomás Church.

Population 56 000
Map Guatemala – Ab

Chiclayo *Peru* Capital of Lambayeque department, 640 km (400 miles) north of Lima. It was originally an Amerindian stronghold, and then a Franciscan friary. Now it is a bustling modern city at the centre of a conurbation which includes the towns of Lambayeque and Ferreñafe.

Population 426 300
Map Peru – Ba

Chiemsee (Chiem) *Germany* Largest lake in Bavaria, covering 82 km² (32 sq miles), 65 km (40 miles) south-east of Munich. On Herrenchiemsee, one of three islands in the lake, is a palace, built in 1878-85 by Ludwig II of Bavaria and modelled on Versailles, near Paris. Every Saturday during the summer, concerts are held in the ballroom.

Map Germany – Ee

Chieng-Mai *Thailand* City in the north-west, 120 km (75 miles) from the border with Myanmar (Burma). It was founded in 1292, and still has some of its medieval walls. It is a tourist centre, being close to the traditional homes of the Hmong, Yao and Karen peoples.

The surrounding province of the same name grows garlic, tobacco, strawberries and other fruit on irrigated farms.
Population (city) 101 700; (province) 1 361 320
Map Thailand – Ab

Chihuahua *Mexico* The country's largest state sharing its northern border with the US states of New Mexico and Texas. It is known for a breed of small dog, with a round head and protruding eyes, which is named after the state.

Chihuahua is rich in minerals, especially zinc, lead and silver. Cattle ranching is also carried out here. Its capital city, also Chihuahua, is 1524 m (5000 ft) above sea level and ever since colonial times has grown wealthy from silver mining and cattle ranching. Today, mining has become secondary, and commerce and trade have become the main contributors to its wealth. The city was the home of Pancho Villa (1877-1923), the revolutionary leader. Father Miguel Hildago, leader of the Mexican Independence movement, was executed here by the Spaniards in 1811.
Population (state) 2 440 000; (city) 530 000
Map Mexico – Bb

▼ **CHICAGO SKYLINE** Skyscrapers, including the 100-storey John Hancock Center, hug the shore of Lake Michigan. The modern city arose from the ashes of the 1871 fire.

Chile See p 150

Chillán *Chile* Town 360 km (225 miles) south of Santiago. Situated in the fertile Central Valley, it is an important agricultural centre.

Although it was largely destroyed by earthquakes in 1835 and 1939, and damaged again in 1960, it still retains a few of its Spanish colonial buildings, including the birthplace of the revolutionary Bernardo O'Higgins. He was the son of the Irish-born Viceroy of Chile and Peru, Ambrosio O'Higgins, and liberated Chile from Spanish rule in 1817.

It has a modern cathedral and a market that specialises in the local handcrafts. Processing industries include shoe factories, flour mills and lumberyards.
Population 162 000
Map Chile – Ac

Chiltern Hills *United Kingdom* Ridge of chalk hills in southern England, which form an arc around the north and west of London and rise to more than 260 m (850 ft) near Wendover. London's suburbs have now reached into its valleys and have been restricted from further growth only by rural protection laws. The hills run between Luton in Bedfordshire and the River Thames west of Reading.
Map United Kingdom – Ee

Chi-lung *Taiwan* See KEELUNG

Chimborazo *Ecuador* Central Ecuadorian province lying in the Andes. It includes the impressive snow-capped inactive volcano of the same name, which at 6310 m (20 700 ft) is the highest point in Ecuador. Because Chimborazo lies near the equator, and the equator lies further from the centre of the earth than the poles, it also claims to be the world's highest summit (measured from the centre of the earth). However, measured from sea level, MOUNT EVEREST, at 8846 m (29 022 ft), is still the world's highest mountain.The summit of Chimborazo was first conquered in 1880 by the British mountaineer Edward Whymper.

The provincial capital is RIOBAMBA, which lies at a height of 2700 m (8860 ft) on a flat plain fringed by high, extinct volcanoes.
Population 364 680
Map Ecuador – Bb

Chimbote *Peru* The country's largest fishing port, at the mouth of the Río Santa, 400 km (250 miles) north of Lima. Peru's first iron and steel plant was opened here in 1958. Ore for the plant comes from southern Peru and the hydroelectricity to power it, from Cañon del Pato in the Santa Valley. A new port has been built especially to handle the steel produced here.

Pollution from heavy industry, coupled with a strong fishy smell emitted by fish-meal plants have become problems.
Population 297 000
Map Peru – Ba

chimney Term applied to a steep vertical cleft in a rock face, of great use to climbers. Sometimes also applied to the vent of a volcano.

Chile

BOUNDED BY THE MIGHTY ANDES AND THE RESTLESS PACIFIC OCEAN, CHILE'S DIVERSE SCENERY IS AMONG THE MOST BEAUTIFUL IN SOUTH AMERICA

Like the backbone of a huge animal, Chile runs its long, thin course down the South American continent's broad back – the Pacific coast – for 4200 km (2610 miles), never more than 400 km (about 250 miles) wide, and usually less than 200 km (about 125 miles) across. It has every kind of climate, from deserts to steaming forests, to icy wastes. The most hospitable area is in the centre of the country around the capital, SANTIAGO, especially the coastal lowlands and the great valley which lies between the double chain of the Andes. In the north lies the ATACAMA DESERT, a hostile expanse of sand and pebbles covering 132 000 km² (51 000 sq miles). The south, by contrast, is one of the stormiest places, an immense landscape of mountains, forests, volcanoes and desolate islands – a rain-drenched wilderness of fire, sea and ice.

Chile's cherished poet and Nobel Prizewinner Gabriela Mistral (1889-1957) described her country as a 'synthesis of the whole planet' – a comment on its astonishing variety. Chile has five distinct regions from north to south – the Norte Grande, the Norte Chico, the Central Valley, the Lake District and the Far South. But, despite its regional variety, the people of this land all have one thing in common – they are never very far from the sea.

NORTE GRANDE

The Norte Grande might have a barren surface, but it contains treasures worth fighting for – and Chile won the territory from Bolivia in the War of the Pacific (1879-83). The Atacama Desert is a rich source of nitrates – the raw material for fertilisers and explosives. Until other sources were found and cheaper synthetic nitrates became available in the early 20th century, Chile's national coffers were filled with the revenue from nitrate exports. While the wealth rolled in, boom towns flourished in the desert. Iquique and Pisagua in the far north were grand enough to have their own opera houses. Although the nitrate industry collapsed, mining did not. Copper, which the desert yields in abundance, brought in new riches, and now accounts for 45.5 per cent of the country's exports. Chile is the world's largest copper ore producer: it is estimated that the country contains about 20 per cent of the world's total copper deposits. Today, Chuquicamata is the largest open-cast copper mine in the world and supports a town of over 30 000 people. The copper is shipped out through the port of ANTOFAGASTA.

The teeming sea along the coast is rich in tuna and sardines, which are processed by numerous fish-canning factories. The sea also brings to the coastal towns a climate that is more moderate, though still hot, than the extreme conditions inland. The Peruvian Current sweeps cool water northwards along the coast, preventing the formation of rain clouds – the cause of the arid conditions inland. In the foothills of the Andes, the height cools the weather and people grow oranges, lemons and mandarins.

NORTE CHICO

The scrubland of the Norte Chico stretches 470 km (290 miles) south from Copiapo to Illapel as the desert slowly gives way to the fertile valleys of central Chile. Some copper and iron is mined here, but towards the south the valley dwellers grow fruit, olives, red peppers and barley, as well as alfalfa with which to fatten cattle.

Norte Chico has an Atacama and Calchaquí population of about 4000, who live in small stone villages where they cultivate maize, beans and tobacco. Some of them are skilled in crafts and make fine silver and copper ornaments.

CENTRAL VALLEY

Some 60 per cent of Chileans live in the 800 km (500 mile) stretch of land which runs from the valleys of ACONCAGUA Province to the BÍO-BÍO River in the province of CONCEPCIÓN. The Central Valley is Chile's heartland, with a mild Mediterranean climate like that of southern California. The landscape is filled with orchards and rich pastureland, fields of cereals, and Chile's main vineyards. Chilean wines are gaining importance as an export.

At Lota and Schwager, just south of Concepción, coal mines extend under the seabed for several kilometres. Inland lies the vast copper mine of EL TENIENTE, one of the largest in the world, where a community of 12 000 miners and their families live on the steep mountain slopes. Another vast copper mining project, at La Escondida, began production in 1990 and, similar to El Teniente, was predicted to produce 320 000 tonnes of copper per year. However, it is entirely owned by foreign companies, so that the Chileans' benefits from the project are few.

The main industrial cities also lie in this area. They include Concepción and the port of VALPARAÍSO – recently badly damaged by one of Chile's frequent earthquakes. The capital, Santiago, lies 97 km (60 miles) from the sea, in the foothills of the Andes.

Although Chile's Central Valley is fertile and its climate perfect for growing crops, there is still a great deal of hardship in the country around Santiago. Poverty has brought people flocking from the land into the prosperous industrial cities, causing an escalating housing problem. Since President Patricio Aylwin took office in 1990, Chile's economy has grown, with new highrise buildings mushrooming all over Santiago and other centres. The new government spends 15 per cent of the gross national product (GNP) on social programmes, and subsidises nine out of every 10 new dwellings. In 1993 alone, more than 100 000 houses and apartments were built.

The penury of farm workers remains a problem, however, and it is directly related to Chile's traditionally unequal system of land tenure. When the Spanish conquistadores occupied the region in the 16th century, they shared among themselves the land they had seized. Ever since, the farmland has been divided into vast estates owned by the very rich, but farmed by the poor.

During the 1960s, the Christian Democratic president, Eduardo Frei, introduced laws to split up the larger estates and establish a minimum wage structure for farm workers. But the pace of change under Frei did not satisfy the radical element in Chile. He had made more promises to the poor than he could fulfil, and

Most of the remaining Amerindians live in this region. Driven out of the areas farther north in the late 19th century, they survived to keep a community. They still speak their own language – Araucanian – instead of Spanish.

FAR SOUTH

South of Puerto Montt lies the fifth region – a great mountainous, forested wilderness, which is largely uninhabited. The scenery is spectacular, especially when seen by boat from among the fjords and rocky islands.

Most of the population eke out a living from this rainy and desolate land of forests and valleys by rearing cattle and sheep. However, oil and gas were discovered in the southern-most province of Magallanes after the Second World War and these now supply about half of Chile's fuel needs. The capital of the province is PUNTA ARENAS, the world's southernmost city, and an important fuelling station on the Magellan Strait before the opening of the Panama Canal. The indigenous population, the Fuegian Amerindians, who live in or near TIERRA DEL FUEGO, are mostly nomadic.

The largest of the region's islands, Chilo, 50 km (30 miles) south-east of Puerto Montt, is about the size of Corsica. Its community of fishermen and small farmers have to contend with the region's fogs and storms, and live mostly on a diet of potatoes, which grow wild and in abundance.

Chile's territory does not end at Tierra del Fuego. There are islands in the Pacific and Antarctic that are also possessions of Chile, among which perhaps the most interesting are the JUAN FERNANDEZ Archipelago, about 650 km (404 miles) west of Chile, and EASTER ISLAND with its strange stone faces, 3780 km (2350 miles) out in the Pacific.

CHILE AT A GLANCE	
Official name Republic of Chile	
Area 756 950 km² (292 191 sq miles)	
Population 13 739 760 **Per km²** 18 **(Per sq mile** 47)	
Capital Santiago	
Government Republic	
Currency 1 peso = 1000 centavos	
Language Spanish (92%), Amerindian languages	
Religion Christian (90% Roman Catholic); Jewish minority	
Climate Varies from dry desert in north to subarctic in south. Average temperature in Santiago ranges from 8°C (46.4°F) in July to 20°C (68°F) in January	
Land use Grazing 16%, forest and woodland 21%, cultivation 7%, other 56%	
Main primary products Wheat, maize, rice, sugar beet, potatoes, beans, timber, fruit, grapes, fish; copper, coal, iron, nitrates, molybdenum, zinc, manganese, lead, oil, gold, silver	
Major industries Steel, cellulose and wood pulp, cement, food processing, ceramics, glass, forestry, fishing	
Main exports Copper and other metal ores, fruit, vegetables, wine, wool, paper, fishmeal	
Annual income per head (US$) 2730	
Population growth (per thous/yr) 17	
Life expectancy (yrs) Male 71 **Female** 77	

▲ VALLEY OF THE MOON Gaunt and flanked by shadowy mountain peaks, Moon Valley lies in arid isolation in Chile's northerly Atacama Desert – one of the driest regions in the world.

in 1970 their votes brought in Salvador Allende to lead the world's first democratically elected Marxist government.

However, only one-third of the electorate voted for Allende, and a sizeable majority was opposed to his socialist reforms. The United States, ideologically hostile to the Allende administration, gave financial support to the big economic groups that opposed it. Using the state of the economy as an excuse, the military took over the government in 1973. Allende was killed in the coup.

A reign of terror, under Augusto Pinochet, began, which cost the lives of thousands of left-wing supporters and caused perhaps 1 million people to flee the country. A drastic reversal of all socialist policies followed – including the handing back of land to its original owners, and the privatisation of more industries than Allende had nationalised.

Harsh deflationary measures led to a brief economic revival. However, in the world recession of the late 1970s Chile slumped into inflation, accompanied by high unemployment and falling living standards.

Since the end of President Augusto Pinochet's military dictatorship, Chile has turned into one of the most stable of South American governments. The democratisation process has had a positive effect on the country's economy. Inflation has been stemmed and Chile has been able to pay off a significant portion of its national debt.

LAKE DISTRICT

The shock waves of the past political dramas were perhaps less strongly felt in the more sparsely populated areas of the south of Chile. Southwards from the River Bío-Bío to the town of PUERTO MONTT – a distance of about 600 km (370 miles) – lies the ravishingly beautiful Lake District, a region of mountains, lakes, rivers, waterfalls and now mostly extinct volcanoes. Its most beautiful lakes are Llanquihue and Todos Los Santos, both overlooked by the snow-capped OSORNO Volcano.

Chimoio *Mozambique* Market town and capital of the landlocked central province of Manica. Formerly called Vila Pery, the town lies on the railway linking the port of Beira – some 190 km (120 miles) to the south-east – with Harare, neighbouring Zimbabwe's capital.

Manica Province (population 666 800) covers 61 661 km² (23 801 sq miles). It rises to highlands on the Zimbabwe border, where Binga Mountain, Mozambique's highest peak, reaches 2436 m (7992 ft). Most of its people are subsistence farmers and cattle rearers.

Population 5000
Map Mozambique – Ab

China See p 154

china clay Greyish-white clay produced by chemical alteration of feldspar in granite, also known as kaolin. It is used in porcelain, white earthenware, tiles, and as a coating for paper.

China Sea Part of the PACIFIC OCEAN, consisting of two areas separated by the Formosa Strait. The East China Sea ('Tung Hai' in Chinese) lies between the north coast of China and Japan's KYUSHU and RYUKYU ISLANDS. The South China Sea ('Nan Hai') lies between southern China, the South-east Asian mainland countries of Vietnam, Cambodia and Thailand, and the region's islands including Borneo and the Philippines.

Once infested by pirates, the China Sea is now a major trading region, with such great ports as SHANGHAI, HONG KONG, MANILA, BANGKOK and SINGAPORE on its shores.

The East China Sea covers 1 248 000 km² (481 850 sq miles), and the South China Sea 2 318 000 km² (894 980 sq miles).

Map (East China Sea) Asia – Kf; (South China Sea) Asia – Jg

Chindwinn *Myanmar* Tributary of the Ayeyarwady (Irrawaddy) River. It rises in the far north of the country and flows south, parallel to the Ayeyarwady. The Chindwinn is navigable by small boats for more than half its 1130 km (700 mile) course. It joins the Ayeyarwady near the towns of Myingyan and Pakokku.

Map Myanmar – Bb

chine A fissure in a cliff composed of soft earthy material.

Chingola *Zambia* Commercial town about 300 km (185 miles) north-west of the capital, Lusaka. It was founded in 1943 to service the Nchanga copper mine, the site of some of Zambia's richest ores.

Population 201 000
Map Zambia – Bb

Chinhae (Jinhae) *South Korea* City on the south coast, 310 km (192 miles) south-east of the capital, Seoul. Its natural harbour has been a naval base since 1905.

Population 120 000
Map Korea – De

Chinhoyi *Zimbabwe* Market town about 100 km (60 miles) north-west of the capital, Harare. The Chinhoyi Caves, 8 km (5 miles) to the north-west, contain the deep-blue Sleeping Lake, lying 50 m (165 ft) underground. Sunlight filtering through their entrance illuminates the caves and lake,

whose African name, meaning 'pool of the fallen', refers to an incident in about 1830 when Angonni people migrating north murdered the local people and hurled the bodies into the caves.

Population 24 000
Map Zimbabwe – Ca

Chin-men Tao (Quemoy) *Taiwan* Group of two main islands and 12 islets in the Taiwan Strait, just off the south-east coast of the Chinese province of Fujian, close to city of Xiamen (Amoy). Like MA-TSU TAO, they are heavily fortified and defended by Chinese Nationalist forces, and were shelled from the mainland in 1954 and 1958. The non-military islanders are mainly farmers and fishermen.

Population 9600
Map Taiwan – Ab

Chinnampo *North Korea* See NAMP'O

chinook A warm, dry wind, similar to a FÖHN, that descends from the eastern slopes of the Rocky Mountains in North America, causing a rapid rise in temperature and, in winter, the melting of snow. In Calgary, Alberta, temperatures of 14°C (57°F) have been recorded during a chinook in January, when the normal average is -10°C (14°F).

Chios (Khíos) *Greece* Island in the Aegean Sea, 8 km (5 miles) from the Turkish coast, and the traditional home of the poet Homer. Chios flourished in the 5th century BC. Later, it was captured, first by Venetians (1204), then by Genoese (1261), and finally by the Turks (1566).

The massacre of the islanders by the Turks in 1822, during the Greek War of Independence, is re-created in a masterpiece by the French artist Eugène Delacroix, *The Massacre of Chios*, which hangs in the Louvre in Paris. Chios, which became part of modern Greece in 1912, has an 11th-century convent, Néa Moní, with fine mosaics.

Other attractions include Piryí, a medieval fortress town on the island, with decorated houses and streets spanned by arches. Mastic, a tree resin used as chewing gum and in varnish, is produced on the island.

Population 52 690
Map Greece – Db

Chipata *Zambia* Town near the border with Malawi, 520 km (325 miles) north-east of the capital, Lusaka. Formerly called Fort Jameson, it was founded in the 19th century as a military post to suppress the slave trade – and has since become the centre of a tobacco-growing region.

Population 15 400
Map Zambia – Cb

Chirripó Grande *Costa Rica* The country's highest mountain, situated in a national park, 50 km (30 miles) south-east of the capital, San José. It rises to 3820 m (12 533 ft).

Map Costa Rica – Bb

Chişnău (Kishinev) *Moldova* Capital lying about 160 km (100 miles) north-west of the Black Sea port of Odesa. Founded in 1466, it was part of Turkey-Romania until 1812 when it passed into Russian hands. From 1919 to 1940 and from 1941 to 1944 it was part of Romania. Devastated in 1944, it has now been rebuilt. A 19th-century

Orthodox cathedral lies at the heart of the city, which is noted for its food processing, wine, tobacco and textile industries.

Population 665 000
Map Moldova – Aa

Chitipa (Fort Hill) *Malawi* Town in Northern Province, 9 km (6 miles) from the Zambian border. It is the centre of a rural development project launched in 1977 with assistance from the World Bank to increase crop production and improve health and social services.

Population 1400
Map Malawi – Aa

Chittagong *Bangladesh* Chief city of the Chittagong division and the country's second largest city and main port, on the Karnafuli River, 20 km (12 miles) from its mouth. It exports jute and tea. Industries include oil refining, jute milling, engineering and textile, plywood and tobacco manufacture.

Originally part of an ancient Hindu kingdom, Chittagong passed to the Mogul empire of India in the 16th century. In 1760 it came under the control of the British, who linked it by rail to Dhaka, now the capital.

Population (city) 2 030 000; (division) 7 122 000
Map Bangladesh – Cc

Chittagong Hill Tracts *Bangladesh* Hilly rain forest district east of Chittagong, near the borders with India and Myanmar. It covers 8679 km² (3351 sq miles) and is inhabited mostly by minorities, including the Mro, Magh, Chakma, Tipra and Tenchungya peoples. In contrast with the mostly Muslim Bengalis, who dominate Bangladesh, these tribal groups are largely Buddhist or followers of traditional animist religions. Most have no formal education and depend for their survival on slash-and-burn cultivation and what they can gather in the forest. A significant number of inhabitants were displaced during the late 1950s by the building of the Karnafuli reservoir, which was completed in 1961. The area is politically unstable as a result of the demands by some of the groups, mainly the Chakma, for autonomy. In November 1992, the government and the separatist Chakma movement reached an agreement, but the refusal of Chakma refugees to return from India made its implementation difficult.

Population 580 000
Map Bangladesh – Cc

Chitwan (Royal Chitwan Park) *Nepal* The country's first wildlife park, established in 1973 to stop the encroachment of farmland. It covers 932 km² (360 sq miles) of the Tarai rain forest, near the Indian border. Among the protected animals here are tigers and rhinoceroses.

Map Nepal – Ba

Chobe National Park *Botswana* An 11 160 km² (4308 sq mile) park opened in 1968 in the extreme north-east of the country. It is home to an enormous variety of animals, including antelopes, elephants, hippopotamuses, lions, leopards and cheetahs. The Chobe River, a tributary of the ZAMBEZI River, runs through the park, which is best seen in the dry season – from April to November – when animals and birds congregate on the banks of the river.

Map Botswana – Ba

▲ **RIVER TAXIS** Cabbies carry both passengers and goods on the Karnafuli River in Chittagong. Like most of Bangladesh, the port lies in a vast delta washed down from hills far inland.

Cholon *Vietnam* Component town of HO CHI MINH CITY, known as District 5, and separated from the city proper by a narrow stream. It was established by Chinese migrants in 1778, and is the city's commercial and industrial quarter.
Map Vietnam – Bc

Cholula *Mexico* Town built on a vast Toltec archaeological site dating from 500 BC. It lies 8 km (5 miles) west of Puebla in south-east central Mexico. Seven successive Amerindian cultures have left their mark on the city, and seven different epochs have gone into building the largest pyramid in the world which rises 54 m (177 ft) above Cholula and has a base covering 17 hectares (42 acres).

The pyramid was built of earth, clay, brick and stone as a temple to the Toltec god Quetzalcóatl, the feathered serpent. Human sacrifices took place on the summit, which is now crowned with the Chapel of Los Remedios built by the Spanish in colonial times.
Population 160 000
Map Mexico – Cc

Ch'onan (Cheonam) *South Korea* Railway town 85 km (53 miles) south of the capital, Seoul. It is on the main line linking Seoul with the port of Pusan.

The city is a market place for locally grown fruit and vegetables.
Population 211 000
Map Korea – Cd

Ch'ongju (Jeonju) *South Korea* Central city 100 km (60 miles) south-east of the capital, Seoul. Cigarettes, silk and woollen yarn are amongst the products manufactured.
Population 517 100
Map Korea – Cd

Chongqing (Chungking) *China* Largest city of Sichuan Province, and a centre for steel and heavy engineering industries on the Chang Jiang River. It was the headquarters of Chiang Kai-shek's Nationalist government during the Second World War.
Population 2 218 000
Map China – Fe

Chonju *South Korea* One of the oldest cities in the Korean peninsula. Chonju lies in the fertile south-west, 200 km (125 miles) south of the capital, Seoul. It produces silk, paper, umbrellas and bamboo goods.
Population 517 000
Map Korea – Ce

Chonnam *South Korea* See KWANGJU

Chontales *Nicaragua* Cattle-raising department in the central highlands east of the capital, Managua. The departmental capital is Juigalpa.
Population 99 000
Map Nicaragua – Aa

Chorzów *Poland* Manufacturing and coal-mining city in the south, 8 km (5 miles) north-west of the city of Katowice. The Kosciuszko Steelworks, formerly the Prussian royal foundry (Königshütte), established in 1798-1802, dominates the 19th-century city centre.
Population 132 000
Map Poland – Cc

Chott al Hodna *Algeria* See ATLAS MOUNTAINS

Chott el Jerid *Tunisia* Huge salt lake, covering 4920 km² (1900 sq miles) and lying some 16 m (52 ft) below sea level in the west of the country, midway between Tunisia's coast in the north and its southern border. According to Greek legend, the goddess Athena was born here.

Freshwater irrigation schemes have been established in the region to help eliminate the salt in the ground.
Map Tunisia – Aa

Christchurch *New Zealand* University town and largest city on South Island. It lies on the edge of the Canterbury Plains, at the centre of one of the country's most productive wheat, grain and sheep-rearing regions. Its industries include plants for food processing and canning, woollen mills, railway workshops, and factories manufacturing carpets, fertilisers, footwear and furniture.

Christchurch was founded in 1850 as a Church of England settlement, and some of its finest buildings date from that period, such as the Provincial Government Buildings (1859-65) and the neo-Gothic-style cathedral (1864-1904). Often called 'the Garden City' of New Zealand, it has about 400 hectares (1000 acres) of public gardens and parkland, including the Queen Elizabeth II Park created for the 1974 Commonwealth Games.
Population 307 180
Map New Zealand – De

Christiania *Norway* See OSLO

Christiansand *Norway* See KRISTIANSAND

Christmas Island *Indian Ocean* Island 400 km (250 miles) south of the western end of Java, which covers 135 km² (52 sq miles). It was discovered by Europeans in 1643, annexed by Great Britain in 1888 and incorporated in the former Straits Settlements in 1889. During the Second World War, it was occupied by Japan (1942-5); in 1958, it passed into Australian control. Its inhabitants today are mostly Chinese and Malays, many of whom work its phosphate mine.

Between 60 and 70 per cent of Christmas Island consists of a national park. Tourism is now being promoted – in 1993, a huge hotel and casino complex were opened.
Population 1280
Map Indian Ocean – Ec

Christmas Island (Kiritimati) *Kiribati* Atoll in the Line Islands, more than 3200 km (2000 miles) east of TARAWA, Kiribati's most important island and administrative centre. Christmas Island covers 432 km² (167 sq miles) and is the largest atoll in the world. British navigator Captain James Cook landed here on Christmas Eve 1777. Its two main settlements are called London and Paris.

In 1956-62, the island was the base for British and US nuclear tests. Experts say contamination has dispersed and by the early 1990s efforts were being made to develop the tourist industry. The island also has copra and shrimp industries.
Population 2540
Map Pacific Ocean – Fb

Chubu Sangaku (Japanese Alps) *Japan* Mountainous national park in the centre of HONSHU Island, 200 km (125 miles) north-west of the capital, Tokyo. It covers 1698 km² (656 sq miles) and contains two of the highest peaks in Japan, Mount Hotaka (3190 m, 10 466 ft) and Mount Yari (3180 m, 10 434 ft). The mountains, also known as the Japan Alps, are similar in appearance to the European Alps.
Map Japan – Cc

Chulung Pass *India/Pakistan* Mountain pass over 4000 m (13 000 ft) high in the Karakoram range, on the 1949 ceasefire line between Indian and Pakistani Kashmir.
Map Pakistan – Da

China

AFTER CENTURIES OF ISOLATION, THE MOST POPULOUS NATION ON EARTH HAS INVITED THE WEST – AND THE EAST – OVER THE WALL TO HELP GET ITS ECONOMY ON THE MOVE

Few people believed the 13th-century explorer and merchant Marco Polo when he returned to Venice from China and claimed that he had discovered a land even richer and more sophisticated than his own. Yet he was speaking the truth. China's civilisation is one of the oldest in the world – in about 1700 BC the Shang dynasty was already ruling a society that worked bronze, knew the art of writing and supported a number of sizeable towns. By the 6th century BC, the Chinese regarded themselves as a special people inhabiting the 'Middle Kingdom' at the centre of the universe. Their emperor was the 'Son of the Heavens'.

For centuries the GREAT WALL, which writhes 3460 km (2150 miles) from the coast north-east of BEIJING to the deserts of Inner Mongolia, kept the surrounding 'barbarians' at bay. Even now, when the wall is a historic monument and tourist attraction, this sense of proud isolation survives, fostering many modern misconceptions.

LAND OF CONTRASTS

For a start, China is not the nation of blue boiler-suited conformity portrayed by the communist propaganda of the Cultural Revolution. The third largest country in the world after Russia and Canada, it is a land of contrasting landscapes, peoples and lifestyles. Even its cooking – recognised as among the most sophisticated in the world – is far more varied than most Westerners realise, with at least four major regional styles. Although officially atheist, China has large numbers of Buddhists, Muslims and Christians, as well as followers of two indigenous faiths, Confucianism and Taoism. And its population of more than 1000 million, which accounts for over one-fifth of the human race, embraces 55 officially designated minorities, such as Mongolians, Tibetans, Kazakhs, Uygurs, Zhuangs, Yis and Miaos – besides the Han Chinese, who account for 93 per cent of the population.

Apart from the languages spoken by its minorities, Chinese itself has many spoken dialects, which nevertheless share the same written ideograms and therefore can be understood by all who read Chinese.

The country can be divided into distinct western and eastern halves, and then further into four contrasting quadrants.

Western China accounts for roughly half of the nation's territory, yet due to the inhospitable terrain, it is home to a mere 5 per cent of the population. In the north, the deserts and steppes of XINJIANG (Sinkiang) stretch into neighbouring Mongolia and roll into the Central Asian republics. Nomadic peoples, such as the Uygurs, Kazakhs and Tajiks, herd cattle, sheep and horses, or settle beside oases along the fringes of the TAKLIMAKAN desert and, as followers of Islam, worship in mosques – a way of life that has been uninterrupted for centuries.

In the south-west, the ice-capped peaks of XIZANG (Tibet) – many of them more than 6000 m (20 000 ft) high – form one of the most remote wildernesses in the world, sparsely inhabited by a people who practise their own version of Buddhism – Lamaism.

Eastern China divides into two very different regions either side of a line running eastwards from the QIN LING Mountains, near the city of XIAN, to the sea. To the north, much of the land is comparatively inhospitable, with low rainfall and temperatures that rise to above 30°C (86°F) in summer and sink to well below freezing point in winter. Nowhere are these extremes more uncomfortable than in Beijing, the capital; it swelters during the summer months, is raked by winds from the icy wastes of Mongolia and Siberia in winter, and is carpeted by a yellow dust blown from the GOBI DESERT during the spring.

It is in the plains of northern China, on either side of the HUANG HE (Yellow River), that the country's population is most concentrated. In an area about the same size as France live more than 250 million people – more than four times France's population. Throughout the north there is very little rain, and the summer growing season is too short to permit widespread rice cultivation. Staple crops consist therefore of wheat, millet, maize and *gaoliang*, a Chinese variety of sorghum. Peanuts and cotton are also grown.

Whereas the north is generally brown, dusty and crisscrossed by cart tracks, the land to the south of the line is green and well watered, studded with lakes and traversed by an intricate pattern of rivers, creeks and canals. This is the China that has established itself in the minds of most Westerners – a land of paddy fields and terraced hillsides, water buffaloes and farmers in broad-brimmed hats. The fertile, mountain-ringed province of SICHUAN (Szechwan) alone supports an estimated 110 million people, while the mountains of neighbouring YUNNAN are a botanist's paradise, the home of many unique plants.

Rice is the staple crop of the south, but the happy combination of climate and landscape also favours cultivation of a wide range of subsidiary crops, including tea, oilseed plants, a range of vegetables, sugar cane and citrus fruits. In contrast to the north, which produces only one crop a year or, at best, three crops every two years, parts of the south produce two, three or even four main crops a year.

The well-watered lowlands in the valley of the CHANG JIANG (Yangtse River) and the warm and humid plains which surround the delta of the ZHU-JIANG (Pearl River) contain two of China's greatest cities and ports, SHANGHAI and GUANGZHOU respectively. But rather than the cities, it is the countryside, with its teeming villages and carefully tilled fields, that epitomises this part of China. Even today,

▲ INSPIRATION The weird hills of Guilin in south-western China look down on the town of Yangshuo, isolated by a moat of flooded paddy fields. Tourists can now cruise the Li Jiang River through this fantastic landscape which has inspired Chinese artists for generations.

farmers and their families make up more than 65 per cent of the entire population.

China's history over the past 100 years has been dramatic. After thousands of years of proud isolation under a series of dynastic emperors (including a century and a half of Mongol rule instituted by Kublai Khan), the country was desperately slow to react to the threat posed by the Western imperial powers in the 19th century. In 1838, the emperor

attempted to eliminate the lucrative traffic in opium enjoyed by European traders – with unfortunate results. China was humiliated by the British in the ensuing Opium War of 1840-2 and forced to cede HONG KONG. This was the beginning of a long, disastrous period for the nation. Not only did China fall increasingly under the control of the West and later Japan, to which it had to cede the island of Taiwan following a brief war, but it was also bitterly split from within.

THE TURBULENT YEARS

In 1911-12, a failed attempt by the Nationalist Party leader Sun Yat-sen to unify the country under a republic left China without an effective central government; for the next 37 years, Chinese Communists, Nationalists and local warlords fought a bitter and destructive struggle for power. On the verge of defeat in 1934, the Communist Red Army battled through the encircling Nationalist troops and embarked on a 9500 km (5900 mile) fighting retreat, known as the 'Long March'. Some 100 000 people (85 000 troops and supporting administrative staff) set out, but only 8000 survived the year-long ordeal.

Internal confusion was compounded when Japan, in pursuit of imperial ambitions, invaded the country in 1937, crippling cities and communications and causing massive loss of life. The defeat and withdrawal of the Japanese at the end of the Second World War left the same two rival groups – Nationalists and Communists – contending for power in China again. Chiang Kai-shek and the Chinese Nationalist Party, or Kuomintang, relied on the support of a middle class dominated by landlords, generals, industrialists and bankers. The Chinese Communist Party, led by Mao Ze-dong (Mao Tse-t'ung), appealed to the peasants who made up, and still make up, the great majority of the Chinese population.

When civil war between the Communists and the Nationalists broke out in 1947, the outcome was inevitable and cost between 11 and 12 million lives. Sweeping all before them, the Communist armies occupied the entire mainland. Chiang and his followers fled to the island of TAIWAN where he founded Nationalist China. On 1 October 1949, Mao Ze-dong proclaimed the People's Republic of China.

From 1949 to 1976, China was dominated by the wayward genius of Mao. Under his charismatic leadership, it turned first to the Soviet Union for advice on building a new socialist state – a recipe that called for the nationalisation of industry and the collectivisation of agriculture. To the West, this extension of the Communist Party's rule over all aspects of Chinese life seemed an ominous development; the Chinese takeover of Tibet in 1950, followed by its military intervention in the war in Korea and later border clashes with India, did nothing to dispel the fear that Mao's regime was a threat to world peace.

The alliance between China and the Soviet Union proved to be short-lived. Mao abandoned the Russian approach to modernisation and in 1958 launched the Great Leap Forward – a once-for-all attempt to mobilise the population and resources of China towards rapid economic growth. Peasants were organised into self-sufficient people's communes and pressured into meeting production targets laid down by the party.

Mao's Great Leap was a disaster. The havoc it wrought was partly repaired by forsaking some of his original ideals: incentives were reintroduced and the communes which were the essence of his peasant paradise were less rigidly enforced. However, this pragmatism in economic policy did not extend to the realm of politics. Mao was concerned that a self-perpetuating bureaucratic and intellectual elite would emerge in China, as he believed had happened in the Soviet Union. For this reason he launched the Great Proletarian Cultural Revolution in 1966 – a gigantic upheaval that was meant to transform China into a truly classless society.

Teachers and technicians, artists and engineers, and in particular moderate party and state officials were denounced as 'class enemies', sacked from their jobs and sent into exile in the countryside to work in the fields. Gangs of young thugs and students were mobilised into the makeshift Red Guards.

As they roamed from city to city, they dispensed a summary justice to those whom they believed to be the 'bourgeois elements' of the counter-revolution. One of their aims was to create as much chaos as possible; by 1968, however, Mao had had enough. He ordered the army to restore order, and it was now the turn of millions of Red Guards to be sent off for re-education by hard labour. The Cultural Revolution continued, though at a less violent level, for eight more years, until Mao's death.

Today the Cultural Revolution is referred to by the government as the 'decade of destruction', and the blame is put on the 'Gang of Four' – a group of extreme radicals led by Mao's wife, Jiang Qing – rather than on Mao himself. During the centenary celebrations of Mao's birth, in December 1993, Mao was hailed as the 'great patriot who brought to an end China's semi-colonial and semi-feudal society'; his part in the Cultural Revolution was referred to as nothing more than 'the mistakes of a great revolutionary and a great Marxist'.

Surprisingly, the Cultural Revolution had little effect on the conduct of Chinese foreign policy – which, if anything, became more diplomatic in tone. Hostility towards the Soviet Union remained acute and, despite opposition to American policy in South-east Asia, the Chinese leadership began to make friendly overtures towards the United States. In 1972, President Nixon visited Beijing and the door for *détente* was open. China has enjoyed 'most favoured nation' trading status with the United States since February 1980. Relations with Japan and the countries of Western Europe, too, have steadily improved, and Chinese trade with capitalist countries has grown steadily. As one modern Chinese saying goes: 'Our minds are on the left, but our pockets are on the right.'

In ancient times, earthquakes were seen as heralding the end of a dynasty, and the disastrous TANGSHAN earthquake, which preceded Mao's death in 1976, did just that. Power passed to a group of moderates led by Deng Xiaoping, vice-chairman. of the Communist Party. The new leadership jettisoned the ideological obsessions that had racked China's domestic policy for more than 20 years and, in a dramatic about-turn, embarked on the wholesale modernisation of the country.

For the first time since 1949, the West and Japan were invited to supply the expertise, capital and technology needed for economic growth. At the same time, foreign tourists were once again welcomed (not least as a source of foreign exchange) and material incentives were introduced to raise production within the country. As a result, China's economy has been transformed since the late 1970s.

AN ECONOMY ON THE MOVE

Agricultural output has soared; foreign trade has taken off; and attempts have been made to improve industrial organisation. China is hoping to achieve economically advanced nation status by the year 2049, the hundredth anniversary of its revolution.

Education is one of the keys to this advance. Today, nearly every Chinese child attends school and more than 2 million students attend the country's 700-plus universities and institutes of higher education. Even so, illiteracy is still widespread, but mainly among the old.

In the long term, China has sufficient fuel and mineral resources to support the growth of an industrial economy, but the colossal investment needed still holds back the exploitation of these fuel and mineral resources. Many of China's coal mines lack modern machinery; oil fields in the future will be expensive to develop, and many of the most valuable mineral deposits are to be found in the icy wastes of the Tibetan Plateau.

In the past, China's greatest single resource has undoubtedly been its people, but the rate of population growth has become a great cause for concern. In 1949, the population stood at 542 million; by 1969 it had risen to 807 million and was growing at over 2 per cent a year – enough to make it multiply by more than seven times over the next 100 years. So in 1979, the government set a quota of one child per couple, enforced with penalties such as reduced food rations and virtually compulsory abortion. The Chinese government thus aimed to keep the population below 1200 million until the year 2000, but in 1994, China's population already exceeded 1200 million. (Strangely enough, China exempted its ethnic minorities from its stringent birth control

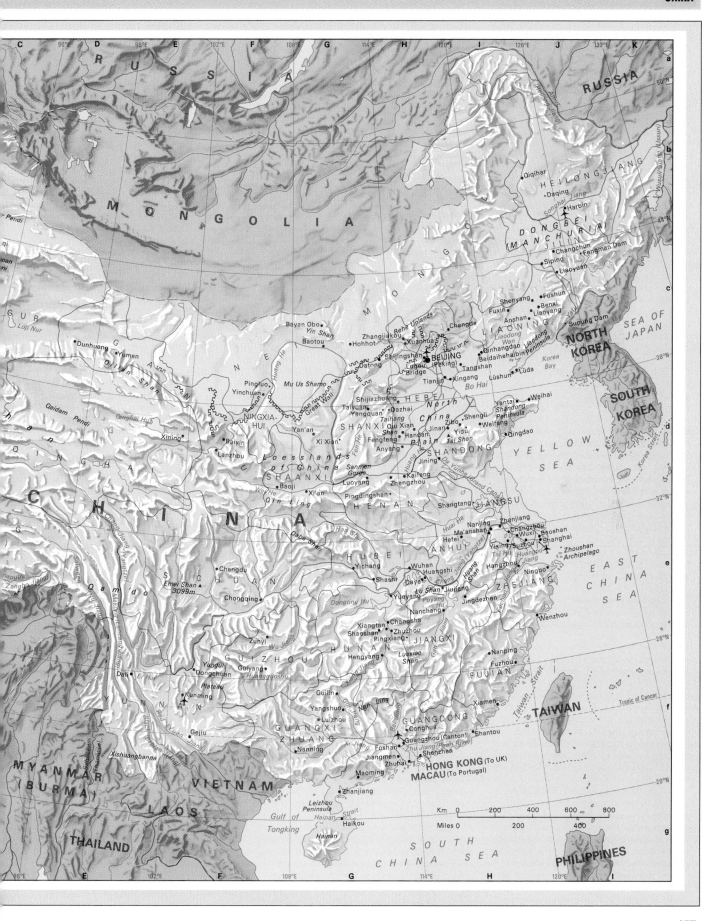

RUSSIA

MONGOLIA

Pendi

Lop Nur

GUR

Qi

ani

C H I N A

Dunhuang • Yumen
Qilian Shan

Qaidam Pendi
Qinghai Hu

Xining •
Baiyin •
Lanzhou •

QINGHAI

G

Qamdo

S
Emei Shan
3099m ▲

Dali • Er Hai

Plateau

Kunming

Xishuangbanna

MYANMAR
(BURMA)

THAILAND

LAOS

VIETNAM

Bayan Obo •
Yin Shan
Baotou •

Hohhot •
Datong •

Pingluo •
Yinchuan •

NINGXIA-
HUI

Yan'an •
Xi Xian •

Loesslands
of China

SHAANXI

Baoji •
Xi'an •
Wei He

Qin Ling

Daba Shan

S I C H U A N

Chengdu •

Chongqing •

Dongting Hu

Zunyi •

Wu Jiang

G U I Z H O U

Guiyang •
Huangguoshu

Dongchuan •

Y U N N A N

Gejiu •

Guilin •
Yangshuo •
Luizhou •

GUANGXI-
ZHUANG

Nanning •

Leizhou
Peninsula

Zhanjiang •

Gulf of
Tongking

Hainan
Strait

Haikou •

Hainan

SOUTH
CHINA SEA

Zhangjiakou •
Rehe Uplands
Xuanhua •

Great Wall

Mu Us Shamo

Huang He

Taiyuan •
Yangquan •

Shijiazhuang •
Dazhai •
Taihang

SHANXI Qiu Liang

Shen •
Fengfeng •
Anyang •

Sanmen
Gorge

Luoyang •

Pingdingshan •

HENAN

Han Shui

H U B E I

Yichang •

Shashi •

Daye •

Yueyang •

Lu Shan

Poyang
Hu

Nanchang •

Xiangtan • Changsha •
Shaoshan • Zhuzhou •
Pingxiang •
Hengyang •

Luoxiao
Shan

HUNAN

JIANGXI

Nan Ling

Chengde •
Xiangzhe •

Beijing
Lugou
Bridge

Skijingshan •
BEIJING
• Peking
Tianjin •

Xingang •

North
China
Plain

Jinan •
Zibo •
Handan •

Huang
He

Kaifeng •
Zhengzhou •

Shangtang •

Da Yunhe Grand Canal

Hai He

Nanjing •
Ma'anshan •
Hefei •

ANHUI

Yixing •

Hangzhou •

Chang Jiang/Yangtze

Wuhan •
Huangshi •

Huang
Shan

JIANGSU

Zhenjiang •
Changzhou •
Wuxi •

Suzhou •

Shanghai •
Baoshan •

Tai Hu

Ningbo •

ZHEJIANG

Jingdezhen •

Wenzhou •

Nanping •

Fuzhou •

FUJIAN

Xiamen •

Conghua •
Guangzhou (Canton) •
Foshan •
Jiangmen •
Zhuhai •

GUANGDONG

Shantou •

Shenzhen •

Zhu Jiang (Pearl River)

Xi Jiang

Bei Jiang

Maoming •

HONG KONG (To UK)
MACAU (To Portugal)

TAIWAN

Tropic of Cancer

Taiwan Strait

EAST
CHINA
SEA

M O N G O L I A

Qiqihar •
Daqing •

HEILONGJIANG

Songhai Jiang

Harbin •

DONGBEI
(MANCHURIA)

Changchun •
Fengman Dam

JILIN

Siping •
Liaoyuan •

Shenyang •
Fuxin •
Anshan •

Benxi •
Fushun •
Liaoyang •

LIAONING

Qinhuangdao •
Beidaihe Haibin •
Liaodong
Peninsula

Supung Dam

Liaodong
Wan

Lüshun • Luda •

Bo Hai

Korea
Bay

Yantai •
Shandong
Peninsula

Weihai •

Weifang •

Shengli •
Yidu •
Tai Shan ▲

SHANDONG

Jining •

Qingdao •

YELLOW
SEA

RUSSIA

Wussu Jiang (Ussuri)

Amur

Heilong Jiang

SEA OF
JAPAN

NORTH
KOREA

SOUTH
KOREA

Korea Strait

| Km 0 | | 200 | 400 | | 600 | 800 |
| Miles 0 | | | 200 | | 400 | |

PHILIPPINES

policy, possibly because they usually live in the remote, sparsely populated areas.)

One unforeseen effect of the birth control policy was the emergence of a generation of overindulged and overfed single children – known as 'little emperors'. However, China's big cities have managed to curb growth – many urban couples are now forgoing children in order to raise their living standards. In 1993, deaths in Shanghai outnumbered births, for the first time since 1949.

Thanks to a massive health care programme, life expectancy has more than doubled from the 1949 average of 32 years. Health care is in the hands of about a million doctors who practise Western-style medicine and about 300 000 practitioners of traditional healing methods. Acupuncture is widely practised by both groups. The treatment involves inserting needles up to 230 mm (9 in) in length into the patient's body. Specific locations are punctured to produce specific effects. The technique is used to anaesthetise and to treat many conditions including stomach ulcers, rheumatism and anxiety; it has proved so successful that it is gaining popularity in the West.

ON THE ROAD TO REFORM

In the short term, living standards in the country are likely to rise as much due to the government's agricultural reforms as to population control. In fact, for a wide section of the population the living standards have already risen appreciably. Since 1979 the regimentation of farm production through communes has gradually been replaced by the 'responsibility system', which has transformed China's peasants into tenant farmers. Either on their own or in small working groups, farmers have to meet a production quota but can sell any surplus on the open market.

Rural markets do a roaring trade in pigs, chickens, fruit and vegetables, all of which were in short supply during the days of Chairman Mao. Some of the profits are then used to set up small village factories and workshops producing, among many other things, umbrellas, fabrics and farm tools.

This 'responsibility system' has also been applied to industry. The result has been an improvement in the quality of manufactures and a wider range of products. China now produces more steel than any country in Western Europe and is the world's leading producer of cotton textiles.

The consumer goods industries which languished from neglect under the puritan Mao, are growing too. Over a million colour television sets are manufactured each year, and Chinese-made washing machines, electronic watches and tape recorders are to be seen in big city stores, though at prices unaffordable to many Chinese.

One problem that will be solved only by massive investment is the transport system. Most freight is hauled by steam trains, but the railways, even when supplemented by water transport on a network of rivers and canals, are simply unable to shift the growing traffic generated by China's economic expansion.

Motorcars are rare. Most are used by government officials or as taxis. Until recently, few motorcars were manufactured in China and even fewer imported. But a joint venture with the German Volkswagen company in Shanghai is said to be producing nearly 20 000 vehicles a year. The bicycle is the most widespread form of personal transport – there are estimated to be more than 100 million bicycles in the country – though mopeds are coming into wider use.

Despite Deng Xiaoping's liberal reforms, modern China remains a Marxist state with a strong central government that controls the economic, political and even personal activities of more than a thousand million people. In the early 1990s, a social revolution began to gather momentum as elements of Western culture – from Coca Cola to Mozart, once banned as 'decadent' – were allowed to infiltrate and take their place alongside traditional art forms such as Peking opera.

On the other hand, freedom of expression and freedom of action, though considerably greater than under Mao, remain limited. To change jobs, get married or simply travel to another city, an individual must ask for permission from the local *danwei* or work unit. The minutest details of everyday life are under constant scrutiny from the street committees which keep watch on every block of flats. Anyone foolish enough to break the rules is likely to be punished – usually by ostracism or other social sanctions.

And although the general standard of living has certainly improved since the late 1970s, modernisation has tended to accentuate the regional differences that exist in China.

Southern China, with its fertile soil and kind climate, has benefited most from the agricultural reforms and from the 'open-door' policy towards the West. Southerners with relatives in Hong Kong are assured of a steady supply of cash and consumer goods. Just across the border from Hong Kong in SHEN-ZHEN, one of several 'special economic zones' established with the help of foreign investment, workers take home tax-free wages that are the envy of Chinese labourers in other parts of the country. Shanghai and several commercial districts of Guangdong Province are thriving. In the northern and western provinces, however, every day is still a back-breaking struggle against a harsh, unyielding soil.

On balance, most of China's changes have had happier results. The country's population is better fed, better housed and probably more content than at any time since the revolution. In recent years, China's relations with capitalist countries have steadily improved. The ousting in 1987 of Hu Yaobang, who was tipped as a future leader, and checks in the process of liberalisation, indicated that 'old guard' Communists were still a powerful force in China, but when Deng Xiaoping retired later in that year, the choice of his eventual successor, Zhao Ziyang, seemed to confirm that reform would continue.

Despite his retirement, Deng continued to pull strings behind the scenes. When Zhao Ziyang became head of government in 1987 he continued economic reforms along the lines of Deng's philosophy of 'to get rich is glorious'. He even tended towards increasing democratisation. But then the Beijing mass demonstrations of 1989 ended in bloody suppression and Zhao Ziyang's political career came to an abrupt end – like that of all his compatriots who had advocated liberalisation. Demonstrators and their supporters were persecuted or otherwise harassed. The economic entrenchment was intensified and there were attempts to reverse some of the reforms. By mid-1990, however, it was recognised that the austerity measures could not be continued. By the end of 1991 it was apparent that the economy as a whole had recovered. Vice-premier Zhu Rongji continued the market-style economic reforms begun by Deng.

Although the country continues to open its doors to the West, in the early 1990s, when it became clear that protectionism from emerging Western trade blocs, such as the North American Free Trade Agreement (NAFTA), was a real threat, China started shifting its trade interests from the West to the East. Consequently, Japan has become a major trading partner. Also, about 12 000 Taiwanese companies have already invested in China (in 1992, Sino-Taiwanese trade accounted for 10 per cent of Taiwan's total trade earnings).

Hong Kong, too, has moved many of its labour-intensive industries to the mainland, providing an estimated 3 million mainland Chinese with employment. In 1992 it surpassed even the United States as China's biggest foreign investor – despite many of its fears related to the year 1997 when it is to be reunited with China.

CHINA AT A GLANCE
Official name People's Republic of China
Area 9 596 960 km² (3 704 532 sq miles)
Population 1 177 584 540 **Per km²** 122.7 **(Per sq mile** 317.8)
Capital Beijing
Government People's republic
Currency 1 yuan = 10 jiao = 100 fen
Languages Chinese (official); languages of the ethnic minorities
Religions Almost 70% of the population are officially atheist; Confucian (20%), Buddhist (5%), Taoist (2%), Muslim (2%), Christian (2%)
Climate Continental with extremes of temperature; temperate or subtropical in the south-east and central south; dry in the north and north-east. Average temperature in Beijing ranges from 26°C (78.8°F) in July to –5°C (23°F) in January
Land use Grazing 31%, forest and woodland 14%, cultivation 10%, other 45%
Main primary products Wheat, rice, cotton, sugar cane/beet, soya beans, jute and hemp, livestock, tobacco, tea, timber, fish; oil and natural gas, coal, iron ore, bauxite, copper, zinc, tin, lead, gold, nickel, manganese
Major industries Iron and steel, cement, fertilisers, mining, agriculture, machinery, machine tools, vehicles, textiles
Main exports Metal ores, crude oil, textiles, silk, tea, soya, skins and hides, machinery
Annual income per head (US$) 370
Population growth (per thous/yr) 14
Life expectancy (yrs) Male 66 **Female** 68

Chungking *China* See CHONGQING

Chung-yang Shan-mo *Taiwan* Principal range of Taiwan's central mountains, the Chung-yang Shan-mo is an area of spectacular mountain scenery. It has several peaks of more than 3000 m (10 000 ft), with Yu Shan being the highest (3997 m, 13 113 ft).
Map Taiwan – Bc

Chuquicamata *Chile* Mining town in the Andes, on an arid plateau nearly 3000 m (9850 ft) above sea level in Antofagasta region. It contains the world's largest open-cast copper mine, which employs 11 000 workers. The ore is processed on site before being transported to the port of Antofagasta for export.
Population 30 000
Map Chile – Bb

Churchill *Canada* Seaport and railway town on Hudson Bay. It stands at the mouth of the Churchill River in Manitoba. Churchill has the best harbour on Hudson Bay although it is ice free for just three months of the year.
   The Churchill River, which is 1600 km (993 miles) long, rises near the Saskatchewan-Alberta border and flows eastwards to drain into Hudson Bay at Churchill.
Population 1370
Map Canada (town) – Fc; (river) – Ec

Chuvashskaya *Russia* Autonomous republic of the Russian Federation. It is situated in the Volga Valley, 600 km (375 miles) east of Moscow, and covers 18 300 km² (7050 sq miles). Its main occupations are farming, forestry and timber processing, engineering, and the manufacture of chemicals and textiles. The capital is the university city of Cheboksary.
Population 1 346 000

Ciarraí *Ireland* See KERRY

Cicmany *Slovakia* Mountain village 137 km (85 miles) north-east of Bratislava. It is preserved as a working folk museum: its people still wear traditional, richly embroidered costumes and decorate the outsides of their wooden houses with carvings and paintings.
Map Slovakia – Bb

Ciechanów *Poland* Old market town in central Poland, 99 km (62 miles) north of the capital, Warsaw. Construction on its massive, rectangular castle with two round towers started in the 11th century. In 1650 it was ruined by invading Swedes. By the early 1990s, work at restoring it had begun.
Population 37 600
Map Poland – Db

Cienfuegos *Cuba* Seaport and capital of Cienfuegos Province, standing on Cienfuegos Bay, 225 km (140 miles) south-east of Havana. Industries include sugar refining, coffee processing, and cigar and soap making. The port handles sugar, tobacco, coffee and molasses.
Population (city) 124 000; (province) 356 700
Map Cuba – Aa

Cieszyn (Teschen) *Poland* Polish part of the city of Teschen, or Tesin, straddling the Olza River, which now forms part of the Polish-Czech border.

The Czech part of the city is known as Český Těšín The city, which was the old cultural capital of Upper Silesia, lies 64 km (40 miles) south-west of Katowice. Cieszyn's old town, centred on Castle Hill, has a 14th-century tower – the remnant of a prince's castle – and an 11th-century Romanesque chapel, which is one of the oldest stone buildings in Poland.
   Both Cieszyn and Český Těšín manufacture timber products, metal goods, chemicals, and electrical goods.
Population 36 600
Map Poland – Cd

Cilipi *Croatia* See DUBROVNIK

Cill Airne *Ireland* See KILLARNEY

Cill Chainnigh *Ireland* See KILKENNY

Cill Dara *Ireland* See KILDARE

Cill Mhantáin *Ireland* See WICKLOW

Cîmpulung Moldovenesc *Romania* A timber-processing and furniture-making centre in the north of the country, 50 km (30 miles) south of the border with Ukraine.
   It contains the remarkable Museum of Wooden Spoons, which has some 5400 individually carved spoons from different regions of Romania and various countries of the world – including France and the United States.
Population 20 000
Map Romania – Ba

Cincinnati *USA* City on the Ohio River, in the south-west of the state of Ohio, close to the borders with Kentucky and Indiana. It is an important industrial centre of the region and a major world producer of soap and playing cards. Its university, founded in 1819, is one of the oldest west of the Appalachian Mountains.
Population (city) 364 000; (metropolitan area) 1 744 100
Map United States – Jc

cinder cone Conical hill of fragmentary material around a volcanic vent.

cinnabar Heavy reddish mineral form of mercuric sulphide, the principal ore of mercury.

Cinque Ports *United Kingdom* Group of towns on the English Channel coast of Kent and Sussex that were granted special privileges by King Edward the Confessor in the 11th century in return for defending the coast. There were originally five (hence the name, derived from the French word *cinq*) – DOVER, Sandwich, Romney, HASTINGS and Hythe – but they were later joined by Winchelsea and Rye.

Circeo National Park *Italy* Park covering 83 km² (32 sq miles) near the coast, 75 km (45 miles) south-east of Rome. It was founded in 1934 to protect what remained of the classic Mediterranean vegetation of the Pontine Marshes when the rest were drained and settled in the 1930s under the supervision of Mussolini.
   Its forest is a unique area of untouched native vegetation; its lakes contain a great variety of birds and marsh vegetation.
Map Italy – Dd

Circum-Pacific Belt Narrow zone around the edge of the Pacific Ocean coinciding with the outer boundary of the adjoining Pacific, Nazca and Cocos plates of the earth's LITHOSPHERE. Because of the intense and frequent volcanic activity in the area it is also known as 'the Ring (or Circle) of Fire'.

cirque A steep-walled semi-circular rock basin occurring above or at the upper end of some mountain valleys. It results from erosion by ice, and in areas still under glaciation contains compacted snow and ice which is often the source of a glacier.
   A cirque may sometimes contain a small lake, or tarn. It is also called a coire, corrie, cwm (in Wales) or kar.

cirrocumulus A high cloud formation of fleecy, small white balls arranged in ranks 8 km (5 miles) or more above the earth. It usually indicates unsettled weather to come and is also known as 'mackerel sky'.

cirrostratus A high cloud formation like a fine whitish veil 8 km (5 miles) or more above the earth. If it grades into ALTOSTRATUS, it may herald the approach of a warm front, alternatively the onset of rain.

cirrus High clouds at altitudes of 8 km (5 miles), seen as wisps of white in a blue sky. Cirrus clouds often herald a DEPRESSION.

Ciskei *South Africa* In Eastern Cape Province, home of the Mfengu and of the Rharabe section of the Xhosa people. Declared an 'independent' republic in 1981, it was recognised only by South Africa, with which it was again amalgamated after the ending of apartheid in 1994.
   The area, which has fine coastal and mountain scenery, produces cereals, vegetables, pineapples and timber. Towns include Alice, home of the University of Fort Hare, and the Ciskei's former capital BISHO.
Map South Africa – Cc

Citlaltépetl (Pico de Orizaba) *Mexico* Snow-capped volcanic peak, 5747 m (18 855 ft) high, in the Sierra Madre Oriental Mountains in Veracruz state. It is Mexico's highest point and was first scaled in 1873.
   It is sacred to the native peoples of Mexico, who believe it contains the spirit of the feathered snake god, Quetzalcóatl.
Map Mexico – Cc

Citta Vecchia *Malta* See MDINA

Ciudad Bolívar *Venezuela* City on the south bank of the Orinoco River, 640 km (398 miles) south-east of the capital, Caracas. Previously called Angostura, it was renamed after the Venezuelan hero and liberator Simón Bolívar, who captured the town in 1817.
   The cocktail ingredient Angostura bitters originated here.
Population 285 980
Map Venezuela – Bb

Ciudad del Este *Paraguay* Tourist town formerly named after General Alfredo Stroessner who ruled Paraguay from 1954 to 1989. It is on the Brazilian border 320 km (200 miles) east of the capital, Asunción, and has prospered as a low-tax shopping centre and casino resort for foreign tourists and as the base for the nearby ITAIPÚ hydroelectric scheme.

Population 83 000
Map Paraguay – Bb

Ciudad Guayana *Venezuela* Industrial city, founded in 1961. It stands on the confluence of the Orinoco River and the Río Caroní, some 500 km (310 miles) south-east of the capital, Caracas. The country's principal steel and aluminium plants are based here.

Population 542 710
Map Venezuela – Bb

Ciudad Real *Spain* Market town about 160 km (100 miles) south of Madrid, trading in cereals, olive oil and wine. Its industries include flour milling, brandy distilling and textile manufacturing. The surrounding province of the same name is mostly a farming region, but there is also some mining of mercury and lead.

Population (town) 57 030; (province) 475 440
Map Spain – Dc

Clackmannanshire *United Kingdom* See CENTRAL REGION

Clare (An Clár) *Ireland* County on the west coast, bounded by the Shannon River and its fertile valley to the east and south, and including the barren limestone BURREN in the north. The county has played an important role in Irish history. First, Daniel O'Connell (1775-1847) was elected member of parliament for Clare in 1828, an act which led to Catholic emancipation in Ireland. Then, Eamon de Valera (1882-1975), the Irish statesman who fought against Britain during the Easter Rebellion in 1916, was elected member of parliament for East Clare in 1917, and later became prime minister of the Irish Free State and president of Ireland. The county town of Ennis (Innis) has a population of 13 730.

Population 90 920
Map Ireland – Bb

clastic rock Rock composed of fragments of other rock that have been broken off, transported and deposited elsewhere. Clastic rocks include SEDIMENTARY ROCKS, AGGLOMERATES, BRECCIAS, TUFF and CONGLOMERATES.

clay Fine-grained sedimentary rock with particles less than $1/256$ mm (0.00015 in) across. The term is also used for any sediment which is plastic when wet.

Clermont-Ferrand *France* Capital of the Puy-de-Dôme department, 135 km (84 miles) west of Lyons. The towns of Clermont and Montferrand were united in 1731. Main industries include tyre manufacturing, food processing and tourism.

Population 262 180
Map France – Ed

Cleveland *United Kingdom* County covering 583 km² (225 sq miles) of north-east England, named after the Cleveland Hills on its southern border. It was carved out of the counties of Durham and Yorkshire in 1974 to unite the urban and industrial areas along the River Tees. It contains the town of STOCKTON-ON-TEES, an old port and market town which was transformed after

1825 by the opening of the world's first steam railway, set up here by the locomotive pioneer George Stephenson. The iron and steel town of MIDDLESBROUGH on the south side of the Tees was founded in 1830. A vast petrochemical complex – one of the largest in Europe – is situated just outside Middlesbrough. Redcar, on the coast, has a modern steel works and there is a chemical works at Hartlepool.

Population 559 700
Map United Kingdom – Ec

Cleveland *USA* Port and industrial city in OHIO, on the south shore of Lake Erie. It was founded in 1796, and handles coal and iron ore. Its varied industries include steel motorcar parts, machine tools and oil refining. It has three universities.

Population (city) 506 000; (metropolitan area) 2 202 070
Map United States – Jb

Clifden (An Clochán) *Ireland* Main town of Connemara, located on the west coast, 66 km (41 miles) north-west of Galway. It is a tourist and fishing centre, and manufactures tweed. The remains of Marconi's first trans-Atlantic wireless station – destroyed during the Irish Civil War of the 1920s – are nearby. At Derrygimlagh Bog, south of the town, there is a memorial to John Alcock and Arthur Brown, British aviators who landed here in 1919 after making the first non-stop flight across the Atlantic.

Population 810
Map Ireland – Ab

climate The average weather conditions of a region over a period of 30 to 35 years, particularly of temperature and rainfall, including their seasonal variation. The type of climate depends on latitude, altitude, continental or maritime location, configuration of the land, soils and vegetation. Climatologists also study long-term trends, including those of the past.

clints In a limestone area, irregularly shaped blocks making up a flat, exposed 'pavement', and separated by deep grooves called GRIKES, are called clints. See HOW CAVES ARE FORMED, p 141.

Clonmacnoise (Cluain Mhic Nóis) *Ireland* Religious site on the River Shannon, 120 km (75 miles) west of Dublin. Founded in AD 548 by St Ciaran, it became a monastery and a seat of learning despite Viking, Anglo-Norman and English attacks. Abandoned in 1552 after the English finally sacked it, the site has a cathedral, seven churches, two round towers, three high crosses, and a castle dating from the 11th century.

Map Ireland – Cb

Clontarf *Ireland* Northern suburb of Dublin, lying on Dublin Bay. Here, on Good Friday 1014, the Irish defeated the Danes, breaking Viking power in Ireland. The Irish king, Brian Boru, then an elderly man, was slain at the moment of victory. His son and grandson also died in the battle.

Population 30 860
Map Ireland – Cb

cloudburst Sudden, concentrated downpour of rain, often associated with a thunderstorm.

clouds Clusters of very small droplets of water or, at high altitudes, ice crystals suspended in the

air. They are named according to their height, thickness and shape.

clough A steep-sided, narrow valley.

Cluain Mhic Nóis *Ireland* See CLONMACNOISE

Cluj-Napoca (Klausenburg) *Romania* Industrial city in north-west TRANSYLVANIA, 130 km (80 miles) south of the border with Ukraine. In Roman times, a town called Napoca stood on the same site. In the 15th century, under Hungarian rule, Cluj-Napoca became a cultural, educational and economic centre. Even after 1921, when it was ceded to Romania, it continued to cling to its Hungarian past. There is still a strong Hungarian atmosphere in the city, particularly in its architecture. Its Baroque, 18th-century Hungarian palace is now an art museum.

Cluj-Napoca has the country's largest Roman Catholic church, St Mikhail's. Built in the 14th to 15th century, the church is a Gothic building which had several parts added subsequently. Most of its sculptures and paintings were destroyed during the Reformation period. Cluj-Napoca is also the country's second university town, after Bucharest, the capital. It has a well-stocked botanic garden. Its industries include metalworking and chemical manufacturing.

Population 319 000
Map Romania – Aa

Cluny *France* Town 75 km (47 miles) north-west of Lyons. Its Benedictine abbey, founded in 910 by William I, the Duke of Aquitaine, became a major intellectual and artistic centre in medieval Christendom. Much of the vast church was demolished in 1798. More than 100 000 people visit the remnants each year. The town has a 13th-century church and some old houses, including the one where the painter Pierre Paul Prud'hon (1758-1823) was born.

Population 4700
Map France – Fc

cluse French term for a steep-sided valley cutting across a limestone ridge, especially in the Jura Mountains and the Alpine foothills at Savoy.

Clutha *New Zealand* South Island river rising in Lake Wanaka and flowing some 320 km (200 miles) into the Pacific Ocean near Balclutha, in the island's south-east. During the 1860s, some alluvial gold was found in the river; today, it is harnessed for hydroelectric power.

Map New Zealand – Bf

Clwyd *United Kingdom* County of north-east Wales, created in the 1970s from the old county of Flint and parts of Denbigh and Merioneth. It covers 2425 km² (936 sq miles) and stretches east to the suburbs of the English city of Chester and the banks of the River Dee, and west to include the popular seaside resorts of Rhyl and Colwyn Bay. The region was once a coal-mining centre – but today, there is only one working coal mine left, at Point of Ayr. Following the discovery of new oil and gas fields in Liverpool Bay, plans were passed in the early 1990s to develop these. There is fertile farmland in the Dee valley, but most of the area is now given over to sheep farming. The county town is Mold.

Population 413 500
Map United Kingdom – Dd

Clyde *United Kingdom* River in south-west Scotland, whose entire course is within Strathclyde region. It rises in the Lowther Hills and flows 170 km (105 miles) north and north-west across the Lanarkshire coal field to Glasgow, where its estuary, the Firth of Clyde, stretches more than 100 km (60 miles) to the sea. It is navigable to Glasgow and was one of the world's great shipbuilding centres.
Map United Kingdom – Cc

Clydebank *United Kingdom* See GLASGOW

Cnoc *Ireland* See KNOCK

coal Natural dark brown to black combustible deposit formed from the remains of swamp vegetation which has been converted to a solid state by the pressure of rocks deposited on top. As the vegetation compacted, water and gases were driven off, leaving mainly carbon. This conversion takes place in different stages, represented by peat, lignite, coal and anthracite.

Coanza *Angola* See CUANZA

coast Zone of land next to the sea, including the land behind the cliffs or their equivalent, as well as the beach and the shore. See EROSION.

Coast Ranges *USA* Mountains made up of 12 separate ranges stretching 4023 km (about 2500 miles) along the Pacific coast of North America. They extend from south-east Alaska to Baja in California and range in height from 610 to 6100 m (2000 to 20 000 ft).
Map United States – Bc

Coatbridge *United Kingdom* See STRATHCLYDE

cobble Rounded or semi-rounded rock fragment, larger than a pebble but smaller than a boulder, with a diameter of between 64 and 256 mm (about 2.5 to 10 in).

Cóbh *Ireland* Town on Great Island, the largest island in Cork Harbour on the south coast of Ireland, with road and rail connections to the mainland. Victims drowned in the sinking of the *Lusitania*, a Cunard passenger liner torpedoed by a German submarine off Old Head of Kinsale in 1915, are buried at Clonmel, 4 km (2 miles) from the town.
A monument stands beside the harbour, which was once a regular port of call for trans-Atlantic liners. Like Queenstown, Cóbh was once a major transshipment point for Irish immigrants to North America. Offshore is Spike Island, whose fort served as a prison from Victorian times.
Population 6230
Map Ireland – Bc

Coblenz *Germany* See KOBLENZ

Coburg *Germany* Bavarian town on the Itz River, 90 km (55 miles) north of Nuremberg. It was the seat of the Dukes of Saxe-Coburg-Gotha, and their ducal palace in the town dates from 1549. One of the family, Prince Albert, married Queen Victoria of England in 1840. Coburg has metal, glass and ceramics industries, and also produces toys.
Population 44 200
Map Germany – Dc

Cochabamba *Bolivia* City and department 230 km (144 miles) south-east of the capital, La Paz. The department lies at 2600 m (8530 ft) on the slopes of the Eastern Cordillera, bordering on the tropical lowlands. Its fertile valleys produce grain, fruit and beef for the towns of the *Altiplano* high plateau.
The city was founded by the Spanish in 1574, and its central square, with galleries on all four sides, is typically Spanish. There are a number of museums, including one that specialises in pre-Columbian relics from the cities of Incallacta and TIAHUANACO.
Cochabamba is the seat of an archbishopric and a university. It is a commercial and manufacturing centre, with a motorcar assembly plant, cement, textile, shoe, and furniture manufacturing, and fruit canning and milk processing industries. Its oil refinery supplies most of Bolivia's petrol and kerosene needs.
Population (city) 377 260; (department) 1 073 500
Map Bolivia – Bb

Cochin *India* Port on the Arabian Sea, 260 km (162 miles) from the country's southern tip. It was first visited by Portuguese explorers in 1500, who started the earliest European settlement here.
The Portuguese navigator Vasco da Gama, who discovered the sea route to India, died at Cochin on Christmas Eve, 1524. In the 17th century, the Dutch captured the town and fortress. The port, a now naval base, exports among other things, copra, rubber, tea and nuts.
Population 1 139 540
Map India – Be

Cochin China *Vietnam* Name of the MEKONG DELTA region in French colonial times.

Cockburn Town *Turks and Caicos Islands* Capital of the Caribbean Turks and Caicos island group, on the Island of Grand Turk. It is the main financial and business centre.
Population 3200
Map Caribbean – Aa

Cockpit Country *Jamaica* Deeply eroded limestone region in the north-west of the island, to the south-east of Montego Bay. Among the limestone features of caverns, underground streams and potholes are steep-sided hollows, or 'cockpits', some of which are more than 150 m (nearly 500 ft) deep.
The thickly wooded, jagged slopes of the area became a hiding place for escaped slaves during the mid-1600s. When the English captured the island from the Spanish in 1655, fugitive slaves waged a successful guerrilla war against the new conquerors, which resulted in a cease-fire on the ex-slaves' terms.
Today, the area is sparsely populated by the descendants of the slaves, called Maroons, who live in the Cockpit free of taxation and other government interference and have their rights guaranteed by treaty.
Population 5000
Map Jamaica – Ba

Cocos (Keeling) Islands *Indian Ocean* Group of 27 small coral islands arranged in two atolls, some 1100 km (680 miles) south-west of Java – scene of the sinking, in 1914, of the German cruiser *Emden* by the Australian cruiser *Sydney*.

Covering 14 km² (5.5 sq miles) of land, the islands were explored in 1609 by William Keeling, a mariner in the employ of the East India Company. They were claimed by Britain in 1857 and were put under Australian control in 1955. However, from 1831 to 1978, the entire group was owned by a family named Clunies-Ross. In 1991, Cocos Islands abandoned their outdated legal system and replaced it with Western Australia's.
The islanders, mostly Malays who work in the coconut plantations, live on the two main islands – Home Island, which consists mainly of plantations, and West Island, which is the administrative centre and has the airport. The biggest settlement is Bantam; the only export is copra.
Population 590
Map Indian Ocean – Dc

Cod, Cape *USA* Area of sand dunes and salt marshes on the coast of Massachusetts, forming a narrow curving peninsula that largely encloses Cape Cod Bay, to the south-east of Boston.
It was here that the pilgrims first landed in 1620 before setting up the PLYMOUTH Colony. Much of it is preserved as the Cape Cod National Seashore. The area is a tourist attraction and popular with summer holidaymakers.
Map United States – Mb

Cognac *France* Historic medieval town on the south bank of the Charente River. It once belonged to the English king, Richard the Lionheart. It is famous for its brandy, which has been distilled here for 400 years. Three-quarters of all its cognac is exported.
Population 32 400
Map France – Cd

Coihaique *Chile* Capital of Aisén region in the south. It was founded in 1931 in a grassy valley on the east flank of the Andes Mountains. It is a growing tourist centre and is used as a base for hiking and skiing in the nearby mountains.
Population 35 700
Map Chile – Ad

Coimbatore *India* Cotton textile and engineering city 300 km (186 miles) from the southern tip of India. It also produces tea and coffee.
Population 1 136 000
Map India – Be

Coimbra *Portugal* University city 175 km (110 miles) north-east of the capital, Lisbon, on a hill overlooking the Mondego River.
It has a fortress-like cathedral, which was built in the 12th century. Its university – the oldest in Portugal – was founded in Lisbon in 1290, moved to Coimbra in 1308, and became established here permanently in 1537. The older part of Coimbra is a maze of steep, narrow, cobbled streets.
The city's proximity to Lisbon has turned it into a centre for trade and industry.
Population 75 000
Map Portugal – Bb

coire See CIRQUE

col Pass across a ridge or range of mountains providing a routeway from one side to the other.

cold climates Those climates with a long cold season – having more than 6 months with temperatures below 6°C (43°F).

cold continental climate Climate experienced inland at high latitudes, namely at about 60°N and further north, as in Siberia, where winters are long and severe. Rainfall is light and most of the precipitation comes in the form of snow in winter. The climate corresponds to the BOREAL FOREST region of the northern hemisphere.

cold desert Term applied to tundra and polar regions where vegetation is non-existent or restricted by low temperatures.

cold front Surface of separation between a cold air mass and a warm air mass which the colder air is undercutting and uplifting. As the front moves, colder air replaces the warm air at ground level. The pushing up of the warm air produces CUMULUS and CUMULONIMBUS clouds along the line of the front, often accompanied by heavy showers of rain or even thunderstorms.

cold wave Sudden burst of cold weather brought on by a cold AIR MASS following a DEPRESSION.

Colditz *Germany* Town 40 km (25 miles) south-east of Leipzig. Between 1939 and 1945, its forbidding castle was used by the Nazis as a prison for Allied officers – but the inmates disproved its 'escape-proof' reputation several times. The town also has a 16th-century town hall. It produces machinery and china.

Population 2500	
Map Germany – Ec	

Coleraine *United Kingdom* Town in Northern Ireland, 76 km (47 miles) north-west of the capital, Belfast, and located at the lowest crossing point of the River Bann. It is a port for goods traffic, a lively market centre and a university town with some engineering, textile and agricultural industries.

Mountsandel Fort, 2 km (a little over 1 mile) south, was a stronghold, first of the Celts and then of the Normans. Stone Age remains in the area date back to about 7000 BC.

Population 20 720	
Map United Kingdom – Bc	

Colima *Mexico* Small state on the Pacific coast in south-west Mexico which produces citrus fruit, cotton, sugar cane, tobacco, rice, and fish. Its main port, Manzanillo, is a modern, bustling tourist resort, with fine beaches.

The town of Colima is the capital of Colima State, and lies 64 km (40 miles) inland from Manzanillo. It is a sleepy colonial town with graceful arcades, and markets selling cattle and agricultural produce.

Population (state) 425 000; (town) 150 000	
Map Mexico – Bc	

Colmar *France* Capital of Haut-Rhin department. It stands on the Lauch River, 38 km (24 miles) north of Mulhouse, and like Mulhouse, produces textiles. The old town has narrow streets with attractive timbered houses. Colmar holds a wine festival each year in the late summer.

Population 84 000	
Map France – Gb	

Cologne (Köln) *Germany* City, inland port and road and rail junction on the banks of the Rhine, 25 km (15 miles) north-west of Bonn. The Romans established a walled colony here in

A DAY IN THE LIFE OF A GERMAN SCHOOLTEACHER

The alarm clock wakes Klaus Fütterer at 6:30 each morning in his three-bedroomed, semi-detached house in Marienburg, a prosperous suburb on the outskirts of Cologne in Germany. There is a luxury villa around the corner but Klaus's terraced house is more modest.

Klaus, 39, tiptoes downstairs to prepare his breakfast while his wife Sibylle, 34, who runs a translation agency, is still asleep. Klaus hopes their only child, 18-month-old Nils, is also asleep.

A schoolday in Germany begins early. The first lesson is at 7:55 am in the Freirherr von Stein Secondary School in Bonn, 19 km (12 miles) away down the motorway – the A-3 Autobahn. Klaus is a teacher at the Secondary School, his subjects being English, geography and religion.

To get there, he has a choice of transport: the bus, the family's ageing Volkswagen camper van, or an Uno which the Fütterers bought last year. It takes about 17 minutes on the motorway.

The school has some 400 boys and girls aged 10 to 16. Klaus's first stop is the staffroom. The atmosphere among the 22 teachers is fairly formal by European standards. Some of the older teachers still shake hands on arrival and address each other by their surnames, but the younger generation are on first-name terms.

Klaus prefers the more formal approach however. 'I like the old-fashioned formality,' he says. 'It reflects a serious approach to teaching.'

On two mornings of the working week, Klaus's first lesson is English – which he teaches to mixed classes of all ages. As a senior teacher, he is also in charge of Class 8B, consisting of 23 pupils aged between 13 and 15.

When Klaus began his teaching career, some of his classes had 30-35 students each, but the declining birthrate in Germany has reduced their size.

Still, with 23 pupils, Klaus has his hands full keeping them quiet and occupied *and* interested. The babble of teenage chatter dies away as Klaus enters and says crisply in English, 'Good morning'. The class replies in unison, 'Good morning, sir'. They appear disciplined and well-mannered.

But gone are the days when 'discipline' was no problem – and though Klaus only teaches 17 hours a week and spends most afternoons at home relaxing or playing table tennis with his colleague, Dieter, he finds his job quite stressful.

Both Klaus and his wife earn enough money to meet their monthly commitments for their car and mortgage and still go on holiday twice a year – usually to Mallorca or one of the Aegean islands.

His school and family are the centre of Klaus's interests. 'I'm content with my lot,' he says, sipping lemonade on his back lawn, 'and I certainly would not describe myself as upwardly mobile.'

AD 50 – and the city takes its name from the Latin word for 'colony', *colonia*. Remains of the Roman governor's residence have been found beneath the city hall.

On the south side of the cathedral is an excellent Romano-Germanic museum with many Roman relics, while the city's Wallraf-Richartz museum and art gallery contains one of Germany's finest collections of paintings.

Cologne's cathedral, which took 600 years to build, was completed only in 1880 and survived the bombing of the Second World War. It is one of Europe's most imposing Gothic buildings, with twin towers topped by lacy spires rising to 157 m (515 ft).

The city's university, founded in 1388, is one of the oldest in Europe. Konrad Adenauer (1876-1967), West Germany's first Chancellor, was a former mayor of the city.

Population 953 600	
Map Germany – Bc	

Colombia See p 164

Colombo *Sri Lanka* The country's capital and largest city. It lies on the west coast and is the major port, handling 90 per cent of Sri Lanka's overseas trade. It was founded by Arab traders and captured by the Portuguese in 1517. The Dutch ousted the Portuguese in 1658, and built a fort in the central district now called Fort. The British took over in 1796. Colombo expanded rapidly when railways were built to new tea plantations inland. A new harbour was added in the 19th century.

The city has many fine avenues, parks, old colonial buildings and Hindu and Buddhist temples, as well as newer skyscrapers. Its has three universities and one of the finest zoological gardens in Asia. It has excellent bazaars and gemstone markets. However, there are slums too, and, on the swampy ground around the city, squatter settlements. In the early 1990s work was started on a new seat of government at the town of Kotte just east of Colombo.

Population 615 000	
Map Sri Lanka – Ab	

Colón *Panama* The country's second largest city after the capital, Panama City. Colón lies at the Caribbean end of the Panama Canal, 80 km (50 miles) north-west of Panama City. Its name is the Spanish version of 'Columbus' and its port and twin town is Cristóbal – Spanish for 'Christopher' – so that the two places take their names from the great Genoese explorer.

Colón's free trade zone is a showpiece of the Panamanian economy, but away from the business centre with its imposing civic buildings, there are extensive shanty towns.

Population 140 910	
Map Panama – Ba	

Colón Archipelago *Pacific Ocean* See GALAPAGOS ISLANDS

Colonial National Historical Park *USA* See JAMESTOWN; WILLIAMSBURG; YORKTOWN

Colorado *USA/Mexico* River rising in the Rocky Mountains north-west of Denver, in the US state of Colorado, and flowing 2330 km (1450 miles) south-west to the Gulf of California in Mexico. It flows through deep gorges, notably the GRAND

▲ **DOWNTOWN COLOMBO Grand offices built during British rule house government firms and the city's many banking, insurance and brokerage houses.**

CANYON, and is dammed in several places, one of the most dramatic being the Hoover Dam.
Map United States – Dd-Ec

Colorado *USA* Western state covering an area of 268 373 km² (103 595 sq miles) between Wyoming to the north and New Mexico to the south. Eastern Colorado forms part of the GREAT PLAINS, and of the COLORADO PLATEAU. Its plains have an average height of 1500 m (4900 ft) above sea level. The Rocky Mountains in the west have more than 1000 peaks rising above 3000 m (10 000 ft), with the highest being Mount Elbert (4399 m, 14 431 ft).
Cattle, sheep, wheat and maize are the state's main farm products, and oil, gas and coal its chief minerals, with some deposits of silver, tin, vanadium, uranium, lead and molybdenum. Many high-technology industries have been established in the state. The headquarters of the North American Aerospace Defense Command are situated near COLORADO SPRINGS. Colorado has attracted many immigrants, making it a state with one of the country's fastest growing populations. Tourists come for its fine scenery and skiing, notably at Aspen, one of the largest ski resorts in the United States. DENVER is the state capital.
Population 3 294 400
Map United States – Ec

Colorado Plateau *USA* Area of spectacular scenery in Colorado, Utah, New Mexico and Arizona. It lies 1520-3962 m (5000-13 000 ft) above sea level and is composed of sedimentary and volcanic rocks. The plateau is deeply cut by the canyons of the Colorado River and its tributaries, including the Grand Canyon.
Map United States – Dc

Colorado Springs *USA* Resort and spa city in Colorado, some 100 km (60 miles) south of the state capital, Denver, at the foot of Pikes Peak

(4300 m, 14 110 ft). It has undergone extensive economic and population growth since the 1970s and now has a wide variety of businesses. The United States Air Force Academy is nearby.
Population (city) 281 140; (metropolitan area) 397 010
Map United States – Fc

Columbia *Canada/USA* River, 1953 km (1214 miles) long, which rises in the Rocky Mountains in eastern British Columbia, Canada. From here it flows north, then turns south through the states of Washington and Oregon in the United States, and empties into the Pacific Ocean at Portland. It has been harnessed for irrigation and hydroelectricity and is a tourist attraction.
Map Canada – Dd; United States – Ba

Columbia *USA* State capital and largest city of South Carolina, 170 km (105 miles) north-west of Charleston. Founded in 1786, it was burnt down in 1865 during the American Civil War. Today, it is a cotton and textile-manufacturing and market centre and is the seat of the University of South Carolina. Buildings of interest include President Woodrow Wilson's boyhood home.
Population (city) 98 050; (metropolitan area) 453 300
Map United States – Jd

Columbia, District of *USA* See WASHINGTON DC

Columbia Icefield *Canada* Largest permanent ice field outside the Arctic and Antarctic, covering 300 km² (116 sq miles). It lies in the ROCKY MOUNTAINS and can be reached from the Icefields Parkway, near the boundary of the Banff and Jasper national parks. Tourist excursions are made by snow tractor.
Map Canada – Dc

Columbus *USA* State capital of Ohio. It lies in the centre of the state, 160 km (100 miles) south of Lake Erie. The city has metalworking and printing industries and three universities.
Population (city) 632 900; (metropolitan area) 1 345 450
Map United States – Jb

columnar structure Natural structure of columns, such as the GIANT'S CAUSEWAY in Northern Ireland, formed when a pattern of regular joints develops in an igneous rock as it cools and contracts.

Comayagua *Honduras* Town situated 65 km (40 miles) north-west of Tegucigalpa, which replaced it as the national capital in 1880. Founded in 1537, Comayagua has a wealth of Spanish architecture, including a 17th-century cathedral containing a 12th-century Moorish clock. It is the centre of a cattle-raising area.
Population 30 600
Map Honduras – Ab

combe See COOMBE

Comilla *Bangladesh* District covering 6599 km² (2548 sq miles) in the south-east near the Indian border consisting chiefly of an alluvial plain cut by rivers.
Low, forest-clad hills rise in the east, where some tea and cotton are grown. Rice, jute, oilseeds and vegetables are the main crops on the plain. Its chief town, also Comilla, trades in rice, jute, oilseeds, cane and bamboo, hides and skins, and tobacco. Its industries include engineering and metalworks, a match factory, a jute mill and a soap factory.
The town is noted for its many tanks, or reservoirs, nearby. There are more than 400; one, dating from the 15th century, measures 1.6 km (1 mile) in circumference.
Population (district) 8 925 000; (town) 210 000
Map Bangladesh – Cc

Communism Peak (Pik Kommunizma) *Tajikistan* The country's highest peak – and the highest peak in the Commonwealth of Independent States (CIS) – rising to 7495 m (24 590 ft) in the Pamir Mountains near the Afghan border. Until 1962 it was called Stalin Peak, but it was renamed after the regime of the Soviet leader Joseph Stalin was posthumously denounced by the then Soviet authorities. Communism Peak, which is heavily glaciated, was first conquered by a Soviet team in 1933.
Map Tajikistan – Bb

Como *Italy* Resort town at the south-western tip of Lake Como, 35 km (22 miles) north of Milan. It is one of many noted lakeside resorts.
Population 85 960
Map Italy – Bb

Como, Lake *Italy* Alpine lake about 35 km (22 miles) north of Milan. Covering 146 km² (56 sq miles) the lake is shaped like an inverted Y, with the towns of Lecco and COMO on its southern prongs. Bellagio, a beautiful little resort town, stands on a promontory at the fork of the Y. The lake reaches a depth of 410 m (1345 ft).
Map Italy – Ba-Bb

Comodoro Rivadavia *Argentina* See PATAGONIA

Comoros See p 166

Comorin, Cape *India* Southernmost tip of India, where the Arabian Sea meets the Bay of Bengal. The cape is named after its temple of the Hindu virgin goddess Kumari.
Map India – Be

Colombia

ONCE HOME TO THE LEGEND OF EL DORADO, COLOMBIA HAS BECOME NOTORIOUS FOR DRUGS AND VIOLENCE

Crime and violence go hand in hand with the illegal drugs trade and cast a long shadow over Colombia. The perils involved with living here affect the daily lives of the Colombians; and even the tourists who venture into this crime-riddled land may find themselves in trouble, especially in the cities where pickpocketing and mugging are all too frequent occurrences.

Despite the crime and violence, Colombia is a fascinating and beautiful country, with a rich mixture of cultures. Traditions survive, from the Amerindians who preceded colonisation, from the Spanish colonists and from the African slaves brought by the Europeans; the influence of these cultures is apparent in the country's food, music, art, architecture and literature. Recently, modern industry has been added to this mixture as Colombia's recipe for future prosperity.

Thirty-five thousand relics of Colombia's past prosperity are on show at the gold museum in the capital city of BOGOTÁ. Before the Spanish came to Colombia in the 16th century, the area was inhabited by Amerindian peoples who left an impressive legacy. The Chibchas in particular, who lived in the area around Bogotá, were skilled craftsmen in gold. Their jewellery and ceremonial objects were encrusted with emeralds, for which Colombia is renowned; indeed, the jewels are still mined at Muzo near Chiquinquira – one of the world's largest unpolished emeralds, now at Bogotá's gold museum, comes from the region – and today's Amerindians still make fine gold and silver artefacts.

The Chibchas gave rise to the legend of *El Dorado*, meaning 'the gilded one' in Spanish. Chibcha chiefs underwent a ritual in which they were smeared with resin and then rolled in gold dust; clad in nothing but this precious powder they dived into the sacred Lake Guatavita and were ceremonially cleansed. Rumours of this ceremony reached the gold-hungry Spaniards, who found not only the precious metal but also a rich, fertile land and beautiful scenery.

FROM SNOW-CAPPED MOUNTAINS TO RAIN FOREST

Colombia has almost every kind of climate and country, with the snow-covered peaks of the ANDES Mountains in the west, grassland in the centre, and tropical jungle to the east and south. There is even desert, between the Andes and the Pacific coast.

The Andes take the form of a three-pronged fork, with three ranges – the Cordilleras Occidental, Central and Oriental – running north and largely disappearing towards the Caribbean. Between the forks are the fertile valleys of the Cauca River, with rich volcanic soils, and the MAGDALENA River. The two rivers join before flowing into the Caribbean at BARRANQUILLA, Colombia's chief port. Large estates in this area account for many of Colombia's agricultural exports. Sugar is grown in the Cauca Valley, while cotton and cattle come from the Magdalena Valley and the coastal plain. By contrast, coffee, Colombia's main export crop (apart, unofficially, from drugs), is grown on small farms in the mountains. Small farms also raise most of the home produce, including rice, maize, barley, fruit and potatoes.

Half of the country lies east of the Andes. Much of this land takes the form of hot plains covered with tropical grassland. Farther east, in the Amazon basin, tropical forest takes over. To the west of the mountains, the Pacific coastal plain rises to the Serranía de Baudó. There are no seasons here; climate depends on altitude, so that much of the hot Pacific coast is desert, the slopes of the Andes offer eternal spring, and the summits are always wintry and frozen. The north along the Caribbean coast has summer all year round.

This wide range of climate supports an extraordinary range of plants and wildlife. There are many species of orchids, one of which is Colombia's national flower. It is said, too, that Colombia has more species of birds than any other country in the world – some 1500, including North American migrants.

But it is the illegal drugs trade that has brought undesirable fame to Colombia. Marijuana is widely grown in the northern coastal region, and in the central region cocaine is produced from Colombian coca plant leaves. Large amounts of coca leaves are also imported illegally from Bolivia and Peru for processing and exporting as cocaine. Obviously, the drug 'mafias' do not provide figures for their trade, but it is believed that drugs are the main export. The United States Narcotics Bureau has been working with the Colombian government in an attempt to stamp out the trade by destroying crops of marijuana, dumping hoards of processed cocaine in the Caribbean and investigating the drug syndicates; in 1992, the government proclaimed a

state of emergency in a bid to clamp down on the drug trade. On assuming the presidency in June 1994, the new president, Ernesto Samper Pizzano, committed himself to the anti-drug war – 'all the way'.

SHARING THE POWER

Bogotá is the seat of one of the most unusual governments in the world. The democratically elected president appoints a cabinet of which half is from his own party, and half is from the opposition. All public appointments are similarly divided; half the government employees, from the highest civil servants to the road sweepers, belong to the Conservative Party, and half belong to the Liberals. This system results from a long history of civil conflict. Between 1899 and 1901 the two parties fought the War of a Thousand Days, in which 100 000 people were killed. Then again, between 1948 and 1953, they clashed in a civil war known as 'the Violence' in which about 200 000 Colombians died. To end this strife, the two parties agreed to their unique system of power sharing. The agreement to govern in this way ended formally in 1974, although the Liberals,

▼ **HIGH-LEVEL FARMING Amerindian farmers grow beans, barley, wheat and potatoes on the uneven fields lying on the high, arid plateau around the city of Pasto in south-west Colombia. Pasto is the centre for the agricultural area, which rises to 2560 m (8400 ft) above sea level.**

at present the ruling party, have again included the opposition in their cabinet.

Today a new enemy faces this united government. Besides the drug lords, backed by radical right-wing groups, who are engaged in private warfare with the government, guerrillas are fighting for social reforms, and especially for a fairer distribution of farmland. Large estates hold most of the land – notably the sugar plantations and the cotton and cattle farms. In contrast, many peasant farmers have so little land that they can hardly grow enough to live on. As a result, people are drifting away from the countryside and 70 per cent of the people now live in towns.

The main cities of Bogotá, MEDELLIN, CALI and Barranquilla are busy, sophisticated centres with impressive architecture and modern roads. But they all suffer from unemployment and housing shortages.

In part, the social problems are due to a high birthrate, although family planning programmes are slowly beginning to work. They have lowered the average number of children a woman expects to have in the course of her lifetime from seven in the 1960s to six in the 1970s, to three in 1993.

The country's future depends on its ability to generate more exports. Coffee is still the most important export – and every car that enters Colombia is sprayed against diseases that might damage or destroy the coffee crop. In an attempt to decrease dependence on coffee as the sole agricultural export, the government has diversified agricultural production. Bananas are now a major export.

Colombia also supplies 85 per cent of the world's emeralds, is the ninth largest producer of gold and exports salt. Oil was discovered recently near the border with Venezuela and provides another growing export, along with natural gas. Colombia is South America's leading coal exporter, from the giant deposits found near the Caribbean town of Riohacha. The coal has greatly helped Colombians develop their own steel industry.

Many of Colombia's manufactures, including iron and steel, motorcars, refrigerators and other domestic appliances, are in the hands of international corporations, though the state is involved in some important industries, such as the textile industry. The country's wealth is still under-exploited, not least its potential for tourism. Among the many interesting places is the fortified city of CARTAGENA on the Caribbean coast.

Colombia is a land of festivals. Medellín, the orchid capital, has its own flower festival; Caribbean Baranquilla goes wild with parades and parties at carnival time. The south has a particular reputation for fun, and the first week of the year is one long fiesta. 'Black Day', when practical jokers dip their hands in grease and smear it on each other's faces, is on 5 January. The next day is 'White Day' – when a great deal of flour or talcum powder is thrown at passers-by.

The influence of Spain is clear from another popular pastime – bullfighting. The bullfighting highlight is the second week in January, in the mountain town of MANIZALES.

The plains – *llanos* – east of the Andes are rich in wildlife, including jaguars, pumas, deer, boars and tapirs. But visitors have to beware of snakes, alligators and piranha fish. Piranhas can strip a deer – or human being – to a skeleton in a few minutes.

COLOMBIA AT A GLANCE		
Official name Republic of Colombia		
Area 1 138 910 km² (439 632 sq miles)		
Population 34 942 770 **Per km²** 30.7 **(Per sq mile** 79.4)		
Capital Bogotá		
Government Presidential republic		
Currency 1 Colombian peso = 100 centavos		
Language Spanish; Amerindian languages and dialects		
Religion Christian (98% Roman Catholic)		
Climate Tropical on the coast, temperate on the plateau. Average temperature in Bogotá (altitude 2610 m, 8563 ft) is 14°C (57°F) throughout the year		
Land use Forest and woodlands 49%, pasture 29%, cultivation 6%, other 16%		
Main primary products Coffee, potatoes, rice, cassava, sugar, bananas, maize, cattle, timber; oil and natural gas, emeralds, gold, platinum, silver, iron ore, coal, copper		
Major industries Agriculture, mining, iron and steel production, food processing, cement, oil refining, paper, textiles		
Main exports Coffee, oil, and oil products, emeralds, gold, mining products, bananas, flowers, cotton, clothing		
Annual income per head (US$) 1280		
Population growth (per thous/yr) 31		
Life expectancy (yrs) Male 69 **Female** 75		

Comoros

INDEPENDENT TROPICAL ISLANDS WHICH ARE DEPENDENT ON THEIR EARNINGS FROM AROMATIC FLOWERS AND FRUITS

The volcanic Comoran Islands lie in the Indian Ocean, between Mozambique and the northern tip of Madagascar. In their turn, Malay, Malagasy, Persian, Arab and African peoples came to the islands, resulting today in a population of mixed blood, with Islamic culture and religion being dominant. France had administered, but largely neglected, Comoros since the middle of the 19th century.

Comoros comprises four main islands. Three of them changed their names following independence: NJAZIDJA (formerly Grande Comore), MWALI (Mohéli), and NZWANI (Anjouan). MAYOTTE, to the east, remains a French dependency.

When the three islands became independent in 1975 they became the Federal Islamic Republic of the Comoros. Since independence, political life has been in turmoil, with a series of coups, attempted coups and the assassination of two Comoran presidents, President Ali Soilih in 1978 and President Abdallah Abderrahman in 1989. The introduction of a multiparty system in recent years does not seem to have improved matters and the republic's politics remain highly unstable.

The Comoran islands were originally mostly forested, but in recent years, large parts of the rain forest were cleared for cultivation. Most of the land belongs to foreign plantation owners and small, traditional power groups.

The principal cash crops are vanilla (extracted from the vanilla bean), perfume oil (from the flowers of the ylang-ylang tree) and cloves (produced from the flowering clove tree). Approximately 80 per cent of the island's workforce are employed in agriculture and forestry.

The economy of the islands is precarious, based on subsistence agriculture and increasingly dependent on foreign aid. Unemployment is high (about 16 per cent), and with poor standards of education – more than 50 per cent of the population are illiterate – this is not likely to change soon. The population is growing fast, and many emigrants seek work in Madagascar, East Africa and France. However, tourism has become a growing industry.

COMOROS AT A GLANCE	
Map Madagascar – Aa	
Official name Federal Islamic Republic of the Comoros	
Area 1862 km² (719 sq miles)	
Population 564 000 **Per km²** 302.9 **(Per sq mile** 784.4)	
Capital Moroni	
Government Federal republic	
Currency 1 Comoran franc = 100 centimes	
Languages French and Arabic (both official); Comoran	
Religions Muslim (99%), Christian	
Climate Tropical, with heavy monsoon rains from November to April; average temperature in Moroni is 25°C (77°F) through the year	
Land use Cultivation 43%, forest 16%, grazing 7%, other 34%	
Main primary products Vanilla, perfume oils, cloves, coconuts, rice, maize, sweet potatoes, cassava, timber	
Major industries Agriculture, fishing, forestry, food processing, perfume oil extraction, tourism	
Main exports Vanilla, cloves, perfume oils, copra, coconut fibre, sisal, spices, ylang-ylang	
Annual income per head (US$) 510	
Population growth (per thous/yr) 34	
Life expectancy (yrs) Male 56 **Female** 56	

at Huachipato, and industries producing wool and cotton textiles, glass and cement. Nearby Talcahuano offers excellent port facilities and is a fishing centre.

Concepción was founded in 1550 on the Bío-Bío River, and despite severe earthquakes – most recently in 1939 and 1960 – it has expanded to become one of Chile's largest cities, with a modern university.

It has an attractive central square – the Plaza de Armas – and a colonial fort; however, as a result of earthquakes, few other colonial buildings remain. San Pedro, across the river, has lagoons and forests and is a tourist resort.

Population (province) 762 800; (city) 306 500
Map Chile – Ac

Concord *USA* State capital of New Hampshire. It lies in the south-central part of the state, 100 km (60 miles) north-west of the city of Boston. It has electronics, leather and printing industries. Its surrounding forests are dotted with lakes and are popular with tourists. It is famous for its granite, and quarries nearby provided stone for the Library of Congress building in Washington DC.

Population 36 000
Map United States – Lb

condensation Process by which a substance changes from the vapour state to the liquid state. For instance, clouds are formed by the condensation of water vapour into water droplets in the atmosphere. Cooling reduces the capacity of air to hold water vapour and sufficient cooling will cause SATURATION followed by condensation. Occasionally condensation does not take place and the air is then described as supersaturated.

condensation trail (vapour trail; contrail) Long cloudlike trail that forms behind an aircraft, particularly a jet, resulting from the condensation of water vapour produced by the combustion of its fuel. Trails sometimes form at the wing tips of high-flying aircraft passing through a region of supersaturated air.

conglomerate Rock composed of rounded rock fragments larger than 2 mm (¹⁄₁₆ in) in diameter cemented together by calcium carbonate, silica, iron oxide, clay or other fine-grained material. It is commonly known as puddingstone.

Congo See p 168

Congo River *Equatorial Africa* Old name for the ZAIRE River. The old name is still used in the Republic of the Congo.
Map Congo – Bb

coniferous forest Forest consisting mainly of evergreen cone-bearing trees with needle-shaped leaves and softwood timber. See also TAIGA and BOREAL FOREST.

Connaught (Cuige Chonnacht; Connacht) *Ireland* Former kingdom and one of Ireland's four ancient provinces – the other three being Leinster, Munster and Ulster. Connaught lies in the west of the country between the bays of Donegal and Galway. 'To Hell or Connaught,' was Oliver Cromwell's cry as he drove off landowners here to make space for his land grants to soldiers and supporters who had helped

Compiègne *France* Town surrounded by woodland on the Oise River, 72 km (45 miles) northeast of Paris. Joan of Arc was imprisoned in its 12th-century tower before her execution at the stake in 1431. In the 18th century, Compiègne's ancient château was enlarged into a royal palace, and became the favourite residence of Emperor Napoleon III. The armistice ending the First World War, and that between France and Germany in 1940, was signed just outside the town. It is an important tourist centre.

Population 43 300
Map France – Eb

Compostela *Spain* See SANTIAGO DE COMPOSTELA

Conakry *Guinea* The country's capital. It stands on the Atlantic coast, partly on Tumbo Island and partly on the adjacent mainland, the two being linked by causeway. Originally a Susu settlement, the city in its present form was laid out by the French in 1889. The administrative and commercial centre is on the island and has broad, straight streets and tree-lined avenues, while the main residential areas – where grand colonial villas contrast sharply with the crowded housing conditions of most local people – lie on the mainland. Here also is a large market place and the National Assembly building.

Conakry has a deep natural harbour sheltered by the offshore Iles des Los. Its trade was limited until the early 1950s, when export in iron ore began. As the terminus of Guinea's main railway, it now handles most of the country's exports of bananas, citrus fruits, pineapples, kola nuts, groundnuts and coffee.

Population 920 000
Map Guinea – Bb

Conamara *Ireland* See CONNEMARA

Concepción *Chile* Province and city in Bío-Bío region, 516 km (320 miles) south of Santiago. It is one of the country's most heavily industrialised areas, with coal fields, an oil refinery, a steel plant

finance his conquest of Ireland following the rebellion of 1641-42.

Because of its untouched nature the province is a popular tourist attraction.

Population 423 030

Map Ireland – Bb

Connecticut *USA* New England state on the Atlantic coast opposite Long Island. It was first settled by Europeans in 1635-36 and was one of the 13 original states and the first US state with a written constitution.

Connecticut covers 12 621 km² (4872 sq miles) and is an undulating, wooded area divided by the Connecticut River. Agriculture is based on dairy produce, fruit, vegetables and tobacco. Industries include aerospace technology, computers, submarines and machinery. The state capital is HART-FORD; the south-western part of the state is largely a commuter settlement of people working in New York City.

Population 3 287 100

Map United States – Lb

Connemara (Conamara) *Ireland* Wild, sparsely populated region of lakes, mountains and streams on the west coast, north of Galway Bay. It is an ice-scoured, rock-strewn country, mostly covered with peat bog.

Connemara is dominated by The Twelve Pins – a group of hills which rise to 730 m (2395 ft) – and by the Maumturk Mountains, which rise to 668 m (2193 ft). Known for its austerely grand scenery and tweeds, there is also an annual summer show and fair at CLIFDEN which trades in the famous tough Connemara ponies.

Population 20 190

Map Ireland – Bb

Conrad discontinuity The discontinuity within the earth's crust between the granitic and lower layers at which the velocity of earthquake waves changes abruptly.

Constance (Konstanz) *Germany* Tourist resort on the north shore of Lake Constance, a few kilometres from the Swiss border. The city played an important part in the history of the Church by ending the schism in the Roman Catholic Church which, from 1378, had led to the existence of two rival papacies – one in Avignon and one in Rome.

The same Council of Constance which was responsible in 1417 for restoring papal unity, also condemned the reformer Jan Hus (also known as John Huss), who attacked the practice of buying church offices. Hus was burnt at the stake in 1415 and his ashes scattered on the lake.

The city was the birthplace of airship inventor Count Ferdinand von Zeppelin (1838-1917); a memorial to Zeppelin stands by the lakeside quay. Offshore lies the island of Mainau, famous for its profusion of flowers.

Population 74 500

Map Germany – Ce

Constance, Lake (Bodensee) *Germany/ Switzerland/Austria* Lake covering 540 km² (208 sq miles), with depths up to 252 m (827 ft), which straddles the borders of Germany, Austria and Switzerland.

The Rhine, which rises in the Swiss Alps nearby, enters the lake in the south, traverses it and drains it in the north-west to continue along its course through Germany and the Netherlands.

A DAY IN THE LIFE OF A CHILEAN WINEGROWER

Antonio Perez Sanchez is a winegrower just outside the town of Concepción in southern central Chile. From his estate, nestling between the Andes Mountains and the Pacific Ocean, come some of Chile's more than 1000 varieties of wine, many of which are now exported.

The 162 hectare (400 acre) estate gives Antonio more than enough to do. Much of his time is spent touring the estate on horseback checking the vines, testing the grapes, overseeing the maintenance of the land and generally directing the work of the 20 or so workers he employs.

The employees live in small houses on the estate and are paid partly in money and partly in wine which is used for their own consumption.

Occasionally, Antonio gets into his ageing Ford and drives the 48 km (30 miles) to Concepción, Chile's third largest industrial city, to arrange some business affairs or to see his younger brother, Juan Alberto, a lawyer who takes care of the legal side of the estate for Antonio.

Slowly, the grapes ripen on the vines and the highlight of the year approaches – the harvest at the end of March or beginning of April. For a while, there is no rest for anyone. The estate's regular workers are supplemented by casual labourers.

When the harvesting is done Antonio throws a big feast which lasts three or four days. He provides wine, and spreads of chicken, fish and beef dishes, stuffed eggplant, tomatoes and peppers, chillied beans, cheese and fruit. The guests sing, someone inevitably brings out a guitar, and usually there is some dancing in the open air.

But at the end of a normal working day, Antonio returns straight home where his wife, Maria Isabel, will be preparing dinner, a lavish, tasty and leisurely meal of Chilean cooking.

Home is a large whitewashed building of two storeys made of sun-baked bricks in the Spanish colonial style, with a verandah round the outside and with nearby outhouses and stables. This is not only the hub of activity on the estate but also the centre of a close-knit family life which Antonio and Maria share with their three youngest children – two boys and a girl – and Antonio's parents.

Over dinner the children chatter about their day at the private highschool in Concepción, to which one of Antonio's employees drives them every morning.

After the children have gone off to watch television or do their homework, Antonio and his father sit quietly together and discuss estate business.

They might also share the latest news from Antonio's two eldest sons, José Antonio and Felipe Alberto. Both boys are living in the capital, Santiago, where José is studying at university and Felipe, who has always wanted to be a soldier, is at the military training college.

On its north shore lies the German resort town of CONSTANCE.

Map Germany – Ce

Constanta *Romania* Major port, resort and industrial city on the Black Sea, some 200 km (124 miles) east of Bucharest. It was founded in the 6th century BC by the Greeks, who called it Tomis. At the end of the 3rd century AD, it was destroyed by the Goths; in the 4th century, the new town on the site was named after the Roman emperor Constantine the Great. From 1413 to 1978, Constanta was under Turkish rule.

Tourists to Constanta can see many reminders of the city's colourful past. In AD 8 the Roman poet Ovid was banished here by the Emperor Augustus for an undisclosed indiscretion. Ovid spent the last years of his life in exile in the area, writing poems to Rome – collected as *Tristia*, or 'Sadness' – pleading in vain to be forgiven.

Today, Constanta has a square named after the poet, containing a bronze statue of him. Nearby is the national history and archaeological museum, whose exhibits include superb Greek statuettes, as well as artefacts from the Stone Age. There are also remains of Roman walls, shops, warehouses and baths, and a magnificent Roman mosaic measuring 2000 m² (21 528 sq ft).

Population 312 500

Map Romania – Bb

Constantine (Qacentina) *Algeria* Third largest city of Algeria, centre of communication and capital of the eastern part of the country, renowned for its spectacular setting. It stands on a rocky plateau surrounded on three sides by the Rhumel River whose gorges are a natural wonder. It was rebuilt in AD 312 by the Roman emperor Constantine on the site of Cirta, the ancient capital of the Princes of Numidia, which was destroyed in AD 311. Following the fall of the Roman Empire, it was occupied by the Vandals and then the Byzantines before it was taken by the Arabs in 710. Roman remains include arcades, part of an aqueduct and a bridge. The city has an old Arab quarter which has preserved its character and colour. When the French moved into Algeria in 1830, Constantine put up fierce resistance but finally fell in 1837. Constantine is well known for its leatherwork, and is a centre of the grain trade. It is an industrial centre, where woollen textiles, machine tools, motor vehicles and tractors are manufactured.

Population 600 000

Map Algeria – Ba

Constantinople *Turkey* See ISTANBUL

constructive plate margins Boundary zone between plates which are moving apart. See PLATE TECTONICS .

Contadora, Isla de *Panama* One of the smallest of the Pearl Islands, an archipelago of 227 Pacific islands, largely uninhabited, lying 75 km (47 miles) off Panama City. The island was once famous for its huge pearls found just offshore. However, at the beginning of the 20th century, the oysters died out because of overfishing.

Isla de Contadora now attracts deep-sea fishermen and, in recent years, world fishing records for red snapper, marlin and sailfish have all been achieved here.

Map Panama – Ba

Congo

OIL DISCOVERIES BOLSTERED THE AILING ECONOMY OF A CHRONICALLY UNSTABLE COUNTRY

Straddling the Equator in west-central Africa, the Republic of the Congo has suffered acute political instability since achieving independence from France in 1960. A left-wing group seized power in 1963, and in 1968, after a military coup, the country became a people's republic based on the Soviet model. The 1960s and 1970s were marked by numerous coups, countercoups and political assassinations.

After 1979, when the Congolese approved their fourth national constitution since independence, Congo entered a politically calmer and economically more active phase. The Communist era came to an end with multiparty parliamentary elections held in June 1992 – after 24 years of military rule.

But democracy did not bring an end to the country's problems. The party supporting the newly elected president, Pascal Lissouba, failed to win an overall majority of seats in the National Assembly where it was defeated in a no-confidence debate. New elections were held in May 1993. Lissouba's party won again, this time with a narrow overall majority, but further instability followed as dissatisfied opposition supporters continued to clash with the security forces.

Though the country was the first in Africa to declare itself a Communist state, its pursuit of Marxism-Leninism did not prevent it from turning to the capitalist West, particularly France, in search of finance to help develop local industry. In 1983, Congo's foreign debt totalled US$1370 million, and by 1991, US$4.1 billion.

First occupied by France in 1880, Congo was the poorest country in French Equatorial Africa and two-thirds of its people still eke out a meagre living by subsistence farming. Even

so, less than 1 per cent of the land is cultivated – and every year the percentage drops further. Much of the soil is poor because nutrients are washed out by heavy rains. Rainfall averages 1780 mm (70 in) a year in the north and about 1200 mm (47 in) in the south. Most Congolese farms are small, worked by women, and devoted almost entirely to food crops. Ninety per cent of the country's food is imported.

Behind the coastal plain are forested ridges, plateaus with savannah vegetation, mountains and some fertile valleys, notably the Niari basin in the south-west, which is linked to the capital, BRAZZAVILLE, and the country's main port, POINTE-NOIRE, by railway. The north is a vast plain drained by the Sangha, Likouala and other tributaries of the UBANGI and ZAIRE (locally referred to as the Congo) rivers. Swamps and flooded forest cover much of this sparsely populated region.

The country's oil industry, which had declined in the late 1960s, was brought to life again with the discovery of vast offshore oil deposits in 1969. By 1984, oil accounted for more than 90 per cent of export revenues and Congo's economy was on a finer footing again until the oil price drop of 1985-86. In 1990, oil accounted for 72 per cent of exports but prices were still low, and consequently, economic growth has been slow.

Timber, including the valuable hardwood limba and softwood okoumé, is the second most important product after oil. Diamonds (from Zaire), sugar and coffee are also exported. Cash crops, such as coffee and cocoa, are grown mainly on large plantations, state-owned land and collective farms.

Tourism is on a small scale, though Congo has a wide range of wildlife which includes elephants, rhinoceroses, giraffes, cheetahs and, in the forests, gorillas, wild boars and bongos – a type of antelope. Although creeping deforestation and poaching are endangering the wildlife – especially the Congo gorillas – some wildlife is protected in the Lefini and Divenie game reserves and in the Odzala National Park.

CONGO'S PEOPLE

The Congolese are divided into several ethnic groups. Besides about 12 000 Binga (pygmies), 8500 Europeans and a few northerners who speak Sudanese languages, most people speak Bantu languages. The largest Bantu cluster is the Kongo group, which accounts for about 45 per cent of the population. Their language, Kikongo, is Congo's second official language. Most Kongo are farmers, living between Brazzaville and the sea. The Téké dialect cluster includes the peoples of the plateaus east and north of Brazzaville, where they farm, hunt and fish. These make up 20 per cent of the population.

Other large groups are the Sangha cluster in the north (15 per cent of the population), and the Ubangi (or M'bochi) cluster in central Congo (12 per cent).

▶ **THREE MEN IN A BOAT The setting sun shows a golden way home across the Congo River for three traders.**

Traditional society, with its belief in folk medicine and indigenous religions, was weakened in the colonial era and is largely rejected by the governing elite. This consists of army officers, officials and politicians, most of whom are young, educated abroad and French-speaking, as also is the small urban middle class. Christianity has its strongest roots in the urban areas.

CONGO AT A GLANCE	
Official name Republic of the Congo	
Area 342 000 km² (132 000 sq miles)	
Population 2 700 000 **Per km²** 7.9 **(Per sq mile** 20.45)	
Capital Brazzaville	
Government Republic	
Currency 1 CFA franc = 100 centimes	
Languages French and Kikongo (both official); local languages, including Lingala, and other Bantu languages	
Religions Christian (Roman Catholic 33%), indigenous beliefs (over 50%), Muslim (1%)	
Climate Tropical, with a dry season from June to September. Average temperature in Brazzaville is about 26°C (78.8°F) throughout the year	
Land use Forest and woodlands 62%, grazing 29%, cultivation 2%, other 7%	
Main primary products Cassava, sweet potatoes, groundnuts, pineapples, bananas, plantains, sugar cane, coffee, cocoa, timber; crude oil, natural gas, lead, zinc, potash	
Major industries Crude oil production and refining, mining, forestry, food processing, textiles, cement, chemicals	
Main exports Crude oil, timber and timber products, coffee, cocoa, sugar, diamonds (from Zaire)	
Annual income per head (US$) 1030	
Population growth (per thous/yr) 32	
Life expectancy (yrs) Male 51 **Female** 51	

THE STRUCTURE OF THE CONTINENTS

The outer layer of the solid crust on which we live is composed of 92 per cent igneous and metamorphic rock (much of which is granite), and 8 per cent sedimentary rocks. This rides on top of a denser layer whose composition is not altogether certain, but from a very few outcrops that occur here and there in the world, it appears to be a metamorphosed form of the crust that underlies the oceans. The foundations of continents, comprising both layers, may be nearly 60 km (40 miles) thick, but those of the oceans, consisting only of a denser layer of basalt, may be no more than 5 km (3 miles) thick.

Of the two, the continental structure is by far the more ancient and complex. Broadly, it is made up of three principal geological features – shields, continental platforms, and mountain belts.

Shields, such as the Canadian Shield and much of the Australian interior, are areas of low-lying, ancient and much eroded rock. Platforms, the plains that make up most of the remainder of continental interiors, are areas of shield overlaid by up to several thousand metres depth of sediments. Mountain belts are formed by the collision of two plates and the ensuing folding and volcanic activity. In the case of young mountain belts, like the Andes, volcanic and earthquake activity continues still, while older fold mountains, such as the Western Highlands of Scotland, are now completely stable.

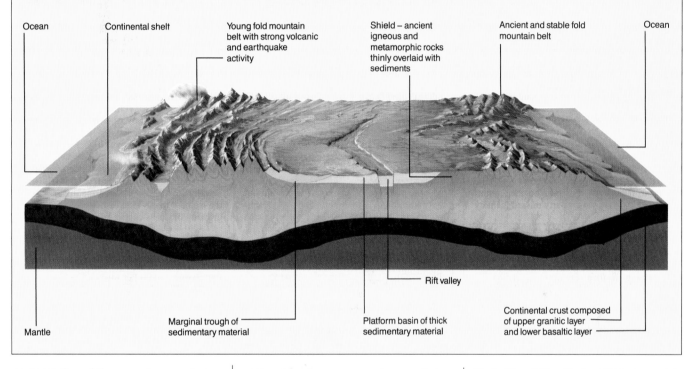

Ocean | Continental shelf | Young fold mountain belt with strong volcanic and earthquake activity | Shield – ancient igneous and metamorphic rocks thinly overlaid with sediments | Ancient and stable fold mountain belt | Ocean

Mantle

Marginal trough of sedimentary material

Rift valley

Platform basin of thick sedimentary material

Continental crust composed of upper granitic layer and lower basaltic layer

---

continent One of the seven larger, unbroken land masses of varying sizes, into which the earth's surface is divided. The seven are: AFRICA, NORTH AMERICA, SOUTH AMERICA, ANTARTICA, ASIA, AUSTRALIA and EUROPE.

continental air mass An AIR MASS originating over a continent.

continental divide An extensive stretch of high ground which separates rivers flowing to opposite sides of a continent. In North America, the Great Divide, formed by peaks of the Rocky Mountains, separates rivers flowing west to the Pacific and east and south to the Atlantic.

continental drift See p 171

continental shelf Seabed at the edge of a continent extending from the shoreline to the first steepening of the CONTINENTAL SLOPE. Where there is no obvious beginning of the continental slope, the outer edge of the continental shelf is set at the 200 m (about 660 ft) depth line.

continental slope Steep underwater slope between the CONTINENTAL SHELF and the ocean floor, sloping at between 3 degrees and 6 degrees to the horizontal.

contour Line drawn on a map to connect places at the same selected height above mean sea level. See ORDNANCE DATUM.

convection currents Circulating currents in the air, water or the mantle of the earth caused by local heating. Heated material expands, and rises, allowing cooler material to flow in below to take its place.

convection rain Rainfall resulting when warm moist air from a heated land surface expands and rises, and then cools, causing condensation into CUMULUS clouds, within which some of the droplets grow large enough to fall as rain.

Coober Pedy *Australia* Opal-mining town in the Stuart range in central South Australia, 750 km (470 miles) north-west of Adelaide. The town yields about half of the world's opals. The first stones were found in 1911.

Many of the inhabitants of Coober Pedy live in underground homes carved out of the hillsides to insulate them from the fierce desert sun, where temperatures can reach 52° C (125° F). The name comes from the Aboriginal language and means 'hole in the ground'.

Population 2490
Map Australia – Ed

Cook, Mount *New Zealand* Highest mountain in the country, rising to 3754 m (12 316 ft) in the Southern Alps of South Island, about 200 km (125 miles) west of the city of Christchurch. The Maori name for it is Aorangi – meaning 'Cloud-Piercer'. The triple-peaked mountain lost some 11 m (36 ft) of its height in December 1991 when an avalanche tore down its east face.

Mount Cook is a popular venue for mountaineering and skiing.
Map New Zealand – Ce

Cook Islands See p 170

Cook Strait Narrow strait between New Zealand's NORTH ISLAND and SOUTH ISLAND, noted for its strong currents. A mere 26 km (16 miles) wide at its narrowest point, it was charted by the British explorer Captain James Cook in 1769.
Map New Zealand – Ed

cool temperate climate Mid-latitude (about 50°N and 40°S) climate with a marked seasonal temperature rhythm and a winter that imposes a resting period on plants.

There are two main types. The first – cool temperate maritime, with rainfall throughout the year – is found on the western margins of continents, for example, in the British Isles. The second –

Copiapó *Chile* Capital of Atacama region, located south of the Atacama Desert on the Copiapó River. When silver was discovered in 1832 at nearby Chañarcillo, Copiapó became one of the country's most prosperous cities. The first railway in the southern hemisphere was completed in 1852 to link its copper mines with the port of Caldera 80 km (50 miles) to the west.

Population 75 790
Map Chile – Ab

Copperbelt *Zambia* The country's smallest province, covering 31 328 km² (12 096 sq miles) about 225 km (140 miles) north of Lusaka. It has known copper reserves of more than 497 million tonnes around the towns of Ndola, Kitwe-Nkana, Chingola and Chililabombwe. The area also produces cobalt, lead and zinc.

Population 1 579 540
Map Zambia – Bb

coppice Small wood, often of oak or hazel trees grown from sprouts rather than seed, which are periodically cut back almost to ground level, thus encouraging growth from the stumps to provide wood for fencing and fuel.

Coquimbo *Chile* Region in central Chile, between the Atacama Desert and the Chilean heartland around Santiago and Valparíso. Its economy revolves around copper and manganese mining, agriculture, sheep and cattle. The chief towns are the capital, LA SERENA, its port, Coquimbo, and Ovalle.

The region is noted for the the La Silla observatory near La Serena, and for its shrine of the Virgen del Rosario de Andacollo, a place of pilgrimage about 16 km (10 miles) south of the town of Coquimbo.

Population (region) 502 460; (town) 117 000
Map Chile – Ac

coral Lime-secreting marine polyp found living in colonies in tropical seas, characterised by calcareous skeletons massed in various shapes and often forming reefs or islands.

The term is also used for the hard, rocklike structure which is the final accumulation of their skeletons.

coral reef Marine ridge or mound built up along a shoreline by CORAL polyps. Fringing reefs border the land or are separated from it by a

▲ **DEEP BLUE SEA** The Coral Sea laps Australia's Great Barrier Reef, which is about 2000 km (1250 miles) long and strewn with islets and atolls. Ocean shipping reaches port via channels through the reef.

shallow lagoon. Barrier reefs (such as Australia's GREAT BARRIER REEF) are separated from the land by a deep lagoon. See also ATOLL.

Coral Sea Name given to an arm of the PACIFIC OCEAN, but like many other bodies of sea water it has rather indeterminate boundaries. Most geographers agree, however, that it washes the shores of north-eastern Australia (fringed by the GREAT BARRIER REEF), Papua New Guinea, Vanuatu and New Caledonia.

From 4 to 8 May 1942, during the Second World War, the Coral Sea was the scene of a fierce naval battle between the Japanese navy and United States carrier-borne aircraft. The US victory probably saved PORT MORESBY, New Guinea, and possibly Australia from being invaded by the Japanese army.

Map Pacific Ocean – Cc

Corcaigh *Ireland* See CORK

cordillera Series of broadly parallel mountain ranges, especially the main mountain system of a large land mass. In South America, the term applies to a small mountain range.

Cordillera Cantabrica *Spain* See CANTABRIAN MOUNTAINS

Cordillera Central *Philippines* Major mountain range of northern LUZON, lying between the Cagayan Valley and the South China Sea. The range extends about 320 km (200 miles) north to south and is up to 88 km (55 miles) wide. It consists of three parallel ranges: the Malayan, Central and Polis ranges. Mount Pulog, the highest peak (2928 m, 9606 ft), is in the Polis range.

The mountains form a massive barrier, making access to many areas difficult, only small groups of hill people, known as the Igorots, live here; they are ethnically distinct from the Malay Filipinos. BAGUIO, summer capital of the Philippines, lies in the southern foothills of the mountains.
Map Philippines – Bb

Córdoba *Argentina* The country's second largest city after Buenos Aires, about 660 km (410 miles) north-west of the capital, standing at the foot of the Sierra Chica. Archbishopric and seat of the oldest Argentinian university, Córdoba is the country's second cultural centre, after Buenos Aires, as well as a main railway junction and industrial centre.

The city's factories produce motorcars, textiles and processed foods. Farmers in the surrounding province of the same name mostly grow wheat, maize and soya beans, and rear livestock.
Population (city) 1 167 000; (province) 2 408 000
Map Argentina – Cb

Córdoba (Cordova) *Spain* City and provincial capital about 115 km (70 miles) north-east of the southern city of Seville. Standing on the Guadalquivir River, it was ruled by the Moors from the 8th to the 11th century, during which time it had a population of between 300 000 and 500 000. Its cathedral was the main mosque of Western Islam until the Christians captured the city in 1236. Much of its Moorish atmosphere has remained.

Modern Córdoba's industries include engineering, brewing, distilling and tourism. Its surrounding province, also called Córdoba, is mainly agricultural, but coal, copper, lead and silver are also mined.
Population (city) 302 150; (province) 754 450
Map Spain – Cd

core Central part of the earth's interior lying below its MANTLE. See JOURNEY TO THE CENTRE OF THE EARTH, p 207.

Corfu (Kérkira) *Greece* Most northerly of the IONIAN ISLANDS, and their capital. The island was the principal Venetian arsenal in Greece from AD 1386 to 1797, and Venetian fortifications and two citadels with magnificent views dominate the harbour town of Kérkira. Italian is still spoken, as well as Greek.

Corfu is linked to the Greek mainland and the Italian port of BRINDISI by ferry – many tourists on their way to mainland Greece step ashore here

and get no further. With a beach-lined coast, and offshore specks of islets, a variety of good hotels, hostels and camping sites, a marina and a lush interior, Corfu is popular with tourists. It produces red wine, fruit and olives.
Population 105 040
Map Greece – Ab

Corinth (Kórinthos) *Greece* Modern town in the north-east Peloponnese 3 km (2 miles) from the Corinth Canal, and 5 km (3 miles) east of the ancient city which was destroyed by an earthquake in 1858. From its hilltop site, the old town dominated the 6 km (4 mile) wide isthmus and only land route onto the Peloponnese. Archaeologists have uncovered a forum and a theatre, a temple of Apollo and the Sacred Fountain of Peirene.

Modern Corinth is now a port and centre of a thriving trade in currants. The resort of Loutraki to the north is famed for its spring water. The Corinth Canal, built in 1882-93, cuts through the isthmus, linking the Gulf of Corinth with PIRAEUS. It can take ships of up to 10 000 tonnes.
Population 23 000
Map Greece – Cc

Cork (Corcaigh) *Ireland* County covering 7459 km² (2880 sq miles), and the Republic of Ireland's second largest city – after the capital, Dublin. The county has broad river valleys and a flourishing dairy industry based on cooperative creameries and factories.

The county town of Cork stands on the River Lee, about 220 km (140 miles) south-west of Dublin – in fact, the old city and heart of Cork stands on an island in the Lee, originally a Viking stronghold invaded by the Anglo-Normans in AD 1172. Historically, Cork has an unenviable reputation of backing losers, notably Perkin Warbeck, rebel pretender to the English throne in 1492 (as a result of whose defeat the city's mayor lost his head), the Royalists in the English Civil War against Oliver Cromwell, and James II, who was defeated by William of Orange at the Boyne in 1690. Another mayor was a victim of the fight for Irish independence: Terence McSwiney died on hunger strike in Brixton Jail, London, in 1920.

Cork has a Protestant and a Catholic cathedral, a fine customs house, a modern city hall and the National University of Ireland. The city produces beer, spirits, chemicals, motorcars and tractors, and clothes and hosiery.

Cork Harbour, to the south of the city, is a centre for oil exploration off the Irish coast.
Population (county) 410 370; (city) 127 250
Map Ireland – (county) Bb-Bc; (city) Bc

Cornwall *United Kingdom* County covering 3546 km² (1369 sq miles) occupying the south-west tip of England. It belongs to the Celtic fringe of western Europe and has many physical and cultural similarities to Brittany, in France, and to Wales. It has its own language, and true Cornishmen regard crossing the River Tamar into Devon as 'going up to England'.

Tin and copper mined in Cornwall were traded with the Phoenicians of the Mediterranean as long ago as the 6th century BC, and continued to be exploited under the Romans. Today, virtually all the mines are closed, but china clay is quarried around St Austell. The economy relies mainly on tourists attracted by the beaches and spectacular coastal scenery, including LIZARD

POINT and LAND'S END. Thanks to its mild climate and subtropical vegetation, the south coast is known as the Cornish Riviera, while the Isles of SCILLY have some of the mildest winters in the British Isles. The county town is Truro.
Population 474 100
Map United Kingdom – Ce

Coromandel Coast *India* Coast in the southeast. Its main ports are Madras and Pondicherry; however, there are no good natural harbours. The coast's low, sandy shoreline is constantly lashed by heavy seas, particularly from October to April – the season of the monsoon.
Map India – Ce

Coromandel Peninsula *New Zealand* Mountainous area of North Island stretching from the town of Waihi to Cape Colville, 65 km (40 miles) east of Auckland. European traders were first drawn here by the magnificent forests, especially the huge kauri trees, but in the 1890s, they came for gold as well.

The chief occupations today revolve around timber, sheep and cattle, although tourism is becoming increasingly important. Along the coast around Christsmas, the red flowers of the pohutukawa trees are a magnificent sight.
Map New Zealand – Eb

corona A series of coloured circles surrounding the sun or moon, ranging from blue (inside) to red (outside), and caused by the diffraction of light by water droplets in middle altitude clouds.

Coronet Peak *New Zealand* Mountain on South Island rising to 1646 m (5400 ft) above Queenstown, some 160 km (100 miles) north-west of the port of Dunedin. It is one of the country's leading ski resorts.
Map New Zealand – Bf

Corpus Christi *USA* Industrial port on the coast of Texas, about 200 km (125 miles) north of the Mexican border. An oil refining and petrochemicals centre, it also processes fish. However, the city's main employer is the naval air station. Nearly 170 km (105 miles) of beaches on offshore sand spits have turned Corpus Christi into a booming tourist town.
Population (city) 257 000; (metropolitan area) 349 000
Map United States – Ge

Corregidor *Philippines* Island covering 5 km² (2 sq miles) at the entrance to Manila Bay, off the southern tip of the BATAAN Peninsula. The Spanish fortified it in the 18th century to guard Manila, and United States forces built extensive fortifications and tunnels in the rock after 1900.

Bataan fell to the Japanese army in April 1942, though American and Filipino troops on Corregidor continued to defend it against overwhelming numbers until 6 May 1942.
Map Philippines – Bc

corrie See CIRQUE

Corsica (Corse) *France* The country's largest island, covering 8680 km² (3350 sq miles). It lies in the Mediterranean, 160 km (100 miles) from the south-east coast of mainland France. It was ruled by the Genoese for 400 years until 1768, when it was sold to France.

Costa Rica

A HAPPY ODDITY IN CENTRAL AMERICA: A STABLE DEMOCRACY THAT HAS DISBANDED ITS ARMY TO PRESERVE THE PEACE

In a region of political instability and military rule, Costa Rica became renowned in the early 1990s for its stable democracy. All men and women over 18 years are obliged to vote, and their voting actually determines which party takes power. Civilian government has lasted since 1902. The last civil war occurred in 1948. It lasted only two months, and afterwards the army was disbanded, so military coups are now impossible. The country has had its own constitution based on democratic principles since 1949 – and the 1949 constitution forbids the establishment of armed forces. School attendance has been compulsory since about 1890, and since 1949, primary and secondary education has been free. Only 7 per cent of the adult population are illiterate.

Forming a bridge between the Pacific Ocean and the Caribbean, Costa Rica is a buffer between less tranquil countries: Panama to the south and Nicaragua to the north. Costa Rica – 'the Rich Coast' – has a tradition of neutrality which its population welcomes.with

enthusiasm The only cloud on its horizon is the risk of neighbouring conflicts spilling over its borders, but in 1987, Costa Rica signed a peace accord with four of its neighbours – El Salvador, Nicaragua, Guatemala and Honduras – which brought peace to the region at a critical time and won the Costa Rican president of the time, Oscar Arias Sanchez, the Nobel Peace Prize.

In 1991, Costa Rica became a member of the General Agreement on Tariffs and Trade (GATT); and in 1993, it signed a free trade agreement with other Central American countries, as well as forming a regional anti-drug commission. The 1994 presidential election brought victory to José María Figueres of the National Liberation Party.

BUILT ON VOLCANIC SOIL

Much of the country consists of volcanic mountain chains, running north-west to south-east, that reach their highest point at CHIRRIPÓ GRANDE (3820 m, 12 533 ft).

More than half of its population of 3.1 million people live in the Valle Central, a high-land basin which was the first area settled by the Spanish in the 16th century. The rich volcanic soils of the upland areas are good for the cultivation of coffee and the slopes provide lush pastures for cattle.

Lowland swamps cover most of the two coasts; the Pacific coast is drier and cooler than that of the Caribbean. The north-west region consists partly of savannah and partly of lowland forest. Increasingly, people are settling along the river and mountain valleys. About one-third of the country remains forested.

Costa Rica differs from its neighbours in that almost 90 per cent of the people are white and most of the rest *mestizo* (of mixed European and Amerindian ancestry). When Europeans arrived in the 16th century they introduced diseases, including measles, to which the local inhabitants had no resistance. Thousands died and they have never recovered their numbers.

Costa Rica was the first country in Central America to grow coffee and bananas commercially; they are are still its major agricultural exports. An enlightened 19th-century

government offered free land to anyone willing to grow coffee for export. This policy made the country and its people prosperous, and avoided the Central American pattern of a few rich landowners and a large class of poverty-stricken peasants.

Growth brought benefits to almost everyone, imports increased – until the recession of the 1980s which led to high inflation and depressed prices for the country's coffee and banana exports. An austerity programme, introduced in 1989 to boost the economy, did not produce the desired results, but by 1993, there was again a growth rate of 6 per cent.

There is potential for further prosperity from large deposits of bauxite, sulphur and iron ore – at present, only silver and gold are mined. The mountainous terrain provides attractive scenery for a growing tourist industry, as well as hydroelectric power which has made the country almost self-sufficient in electricity.

COSTA RICA AT A GLANCE
Area 51 100 km² (19 725 sq miles)
Population 3 264 780 **Per km²** 63.9
(Per sq mile 165.5)
Capital San José
Government Republic
Currency 1 colón = 100 céntimos
Language Spanish (98%)
Religion Roman Catholic (92%)
Climate Tropical; temperate in highlands. Average annual temperature in San José is about 20°C (68°F)
Land use Grazing 45%, forest and woodlands 34%, cultivation 13%, other 8%
Mineral resources Bauxite, iron ore, gold, silver, sulphur
Main primary products Coffee, bananas, maize, rice, cattle, oranges, sugar cane, cocoa; gold, silver, sulphur, bauxite, iron ore
Major industries Agriculture, aluminium smelting, fishing, chemicals, fertilisers, textiles, forestry, food processing
Main exports Coffee, bananas, beef, sugar, cocoa, textiles, timber
Annual income per head (US$) 2000
Population growth (per thous/yr) 24
Life expectancy (yrs) Male 75 **Female** 79

The interior, rising to 2710 m (8890 ft) at Monte Cinto, is covered by maquis scrub and woodland. A regional park of some 200 km² (77 sq miles) lies on the west coast, and includes the Golfe de Porto, a beautiful coastline of red rocks. The eastern plain has rich farmland, and fishing villages and there are small resorts along the coast. The main towns are AJACCIO, birthplace of the French Emperor Napoleon (1769-1821), and BASTIA.

Population 240 000
Map France – He

Cortina d'Ampezzo *Italy* Ski resort in the DOLOMITES, 120 km (75 miles) north of Venice. It is 1211 m (3973 ft) above sea level and is surrounded by mountain peaks reaching heights of more than 3200 m (10 500 ft). The 1956 Winter Olympics were held here.

Population 8100
Map Italy – Da

Corubal *Guinea/Guinea-Bissau* One of the main rivers of Guinea-Bissau, flowing about 400 km (250 miles) on a winding course from the Fouta Djallon of western Guinea across southern Guinea-Bissau to the estuary of the Gêba River. The Corubal was formerly called the Rio Grande, and in Guinea it is sometimes called the Koumba. It is a busy transport artery.

Map Guinea – Ba

corundum Extremely hard mineral (aluminium oxide) which sometimes contains iron, magnesia or silica, and occurs in various gemstone varieties such as ruby and sapphire, and in a common grey, brown or blue form that is used chiefly as an abrasive.

Corunna (La Coruña) *Spain* Seaport on the country's north-west corner specialising in sardine fishing and canning, but also manufacturing cigars, cotton goods and glassware.

The Spanish Armada called here during its disastrous voyage to invade England in 1588; a year later the town and harbour came under British attack and were destroyed. In 1809, Corunna was the scene of a British victory over French troops.

The town's Tower of Hercules, 58 m (190 ft) high, is the only Roman lighthouse in the world still in use. The farming and fishing province of Corunna falls in Spain's heaviest rainfall area, receiving an average of 800 mm (32 in) a year.

Population (town) 246 950; (province) 1 096 970
Map Spain – Aa

Cosenza *Italy* Calabrian city at the junction of the Busento and Crati rivers, 230 km (145 miles) south-east of Naples. It dates back to pre-Roman times, and flourished as a centre of learning until the 15th century. It has a Norman-Swabian castle and a recently restored Gothic cathedral. In AD 410, the Visigoth King Alaric I died and was buried in the bed of the Busento River after his followers had temporarily diverted its waters. Cosenza's industries include furniture and textiles.

Population 106 100
Map Italy – Fe

Costa Brava *Spain* Spain's principal tourist area of cliffs and bays – its name means 'Wild Coast' – stretching along the northern Mediterranean shore from the Spanish-French border to just north-east of Barcelona. Most of it is built up, with hotels, restaurants and flats reaching right down to its beaches. But there are pieces of coastline that can still be seen in their original beauty of red-brown headlands, cliffs and pine forests.
Map Spain – Gb

Costa Rica See p 174

Costa Smeralda *Italy* The 'Emerald Coast', so named after the colour of the sea, stretches for about 50 km (30 miles) along the coast of north-east Sardinia. It is a centre for yachting, water sports and the luxury tourist trade. Its main resorts are Baja Sardinia, Porto Cervo and Cala di Volpe.
Map Italy – Bd

Costermansville *Zaire* See BUKAVU

Cotabato *Philippines* Region and province in south-western MINDANAO comprising the swampy lowlands of the Mindanao River and its tributaries, and the surrounding forested uplands. Inaccessible and malaria-infested until the late 1930s, it has since been developed, and development has led to a rapid population increase. Most newcomers to the area have been Christian Filipinos from the VISAYAN ISLANDS, who now outnumber the original Muslim Moros people.
Population (province) 472 300
Map Philippines – Bd

Côte d'Azur *France* The heart of the French Riviera, one of Europe's prime tourist areas. Sprawled along the sunny Mediterranean, it contains rocky headlands, bays, beaches and yacht marinas. Its main resorts are CANNES and NICE.
Map France – Ge

Cotonou *Benin* Business capital of Benin and deep-water port west of the Couta Canal, linking Lake NOKOUE with the open sea. Well served by local and international transport links, its flourishing industries, based in a suburb east of the lake's outlet, produce palm oil, cement, soap, furniture, beer, and soft drinks; motorcars are also assembled. The city, founded by the French in the early 20th century, has a wealth of hotels, restaurants, boutiques and colourful markets. Although it also has beautiful beaches, undercurrents can make swimming dangerous.
Population 488 000
Map Benin – Bb

Cotopaxi *Ecuador* West-central province, lying in the Andes to the south of the capital city, Quito. It is dominated by Mount Cotopaxi which, at 5897 m (19 347 ft), is the world's highest active volcano. It last erupted in 1877, setting off avalanches of mud that reached the provincial capital, Latacunga, 30 km (19 miles) to the south and destroyed large areas of farmland.
Population 276 320
Map Ecuador – Bb

Cotswold Hills *United Kingdom* Highest section of a range of limestone hills which stretch across England from the county of Avon in the southwest to beyond the Humber estuary in the northeast. The hills rise to 329 m (1080 ft) at Cleeve Cloud near Cheltenham in Gloucestershire. Their eastern slopes feed the headwaters of the River Thames. The open, plateaulike summits, steep, wooded valleys and stone-built villages are a major tourist attraction; 60 parishes have been designated conservation areas. Cirencester is the main town.
Map United Kingdom – De

couloir Deep mountainside gorge or gully, especially in the Alps.

country rock The existing rock into which igneous rocks have intruded.

Courmayeur *Italy* Oldest of the country's alpine and ski resorts. It stands in the French-speaking Valle d'Aosta at the foot of Mont BLANC, (60 miles) north-west of Turin. It was first frequented in the early 19th century by the Piedmontese nobility seeking escape from the summer heat.
Population 2470
Map Italy – Ab

Courtrai (Kortrijk) *Belgium* Textile town near the French border, 80 km (50 miles) west of the capital, Brussels. Originally a medieval cloth-making centre using Flanders wool, it turned to linen made from local flax in the 17th century. Synthetics are manufactured, along with lace.
Population 77 300
Map Belgium – Aa

Coventry *United Kingdom* Motorcar-manufacturing city in the West Midlands of England, 30 km (20 miles) south-east of Birmingham. Founded by the Saxons in about the 7th century, it was immortalised by the 11th-century legend of Lady Godiva, who rode naked through the streets to persuade her husband, Earl Leofric, to ease the tax burden on the citizens.

During the Civil War of the 1640s, Coventry was a Parliamentary stronghold. Royalists captured in the Midlands were imprisoned in the 14th-century Church of St John in the city's Bablake district. They were 'sent to Coventry' in the words of an account written in 1647 – a phrase that came to mean 'shunned by society'.

German bomber aircraft devastated the city in 1940-1. After the war, the architect Sir Basil Spence designed a magnificent new cathedral and the artist Graham Sutherland provided a stunning tapestry of Christ 28 m (75 ft) high for the altar.

Tourism, pedestrian shopping precincts, and two universities have provided new life for the old city.
Population 305 600
Map United Kingdom – Ed

Cox's Bazar *Bangladesh* Small fishing and cigar-producing town and coastal resort near the border with Myanmar. Its beach – largely undeveloped – stretches for 90 km (56 miles) and is one of the longest in the world. Small offshore islands near the town were badly hit by a cyclone in 1985.
Population 29 800
Map Bangladesh – Cd

Cozumel *Mexico* Caribbean island just off the north-east coast of Yucatán. In Mayan legend, it was the home of the god of fertility; today, it is a tourist resort. Just offshore is Palancar Reef – the second largest coral reef in the world.
Map Mexico – Db

Crac des Chevaliers (Qal'at al Hisn) *Syria* Magnificent ruined Crusader castle. Set 650 m (2132 ft) above sea level 65 km (40 miles) west of Homs, Crac des Chevaliers controlled the gap between the Nusayriyah Mountains to the north and the Lebanese Mountains which linked inland Syria with the coast. For two centuries it played a major role in the Crusader-Muslim Wars. Much of it was destroyed, but intensive restoration work has taken place since the 1930s, and the vast evocative remains of ramparts, towers, halls and fortifications still grip the imagination.
Map Syria – Bb

▼ **KNIGHT'S CASTLE** The Knights Hospitallers of St John built Crac des Chevaliers (1142-1271) to hold a garrison of 2000. Its two concentric, towered walls have defied 12 sieges.

Cracow (Kraków) *Poland* The country's main tourist city, and third largest city, after Warsaw and Lodz. It lies in the south, 251 km (156 miles) south-west of the capital. Cracow grew where the Vistula River narrows between limestone hills, at the crossroads of major trade routes. It was Poland's capital from the 14th to the 16 century. Its university, the country's oldest, dates from 1364. Nicolaus Copernicus (1473-1543), the Polish astronomer famous as the proponent of the theory that the earth orbits the sun, which forms the basis of modern astronomy, was a student here from 1491 to 1494.

According to legend, Krak, a cobbler, founded Cracow and became its first prince after killing a virgin-devouring dragon living in Wawel Cave. Above, on Wawel Hill commanding the Vistula Narrows, stand a royal castle and cathedral erected by King Casimir the Great (1330-70) on the remains of a 10th-century fort and church.

The castle, rebuilt in the 16th century around an arcaded courtyard, has 71 restored rooms, and houses the world's only surviving 16th-century Arras tapestries.

The cathedral, former see of Cardinal Karol Wojtyla (who became Pope John Paul II), contains the tombs of Polish kings and those of St Stanislaus, a former bishop and the country's patron saint, and the nationalist Tadeusz Kosciuszko (1766-1817), who led an abortive uprising against occupying Russians in 1793.

Cracow's central market square, in the old part of the city north of Wawel Hill, is dominated by the 16th-century Cloth Hall, the town hall tower (1383), and the Church of the Virgin Mary, from whose tower, a bugle call is sounded every hour (and at noon on Polish radio). The call is always broken off in mid-blast – a custom commemorating a bugler whose throat was pierced by a Tatar arrow in 1241 while he was trying to alert the city to a Tatar attack.

Cracow's main industries today include amongst others steel (at nearby NOWA HUTA), chemicals, printing and ceramics.

Population 750 000
Map Poland – Cc

crag and tail Glacial land form, consisting of a steep-faced rocky outcrop on one side (the crag) with a gentle slope on the other where there are deposits of moraine (the tail).

Craigavon *United Kingdom* See LURGAN

Craiova *Romania* Rapidly expanding industrial city in the south-west of the country, 55 km (34 miles) north of the border with Bulgaria. Craiova produces agricultural machinery, motorcars, railway equipment, petrochemicals, textiles, food and electrical goods.

A magnificent green-roofed, 19th-century palace which is now a museum of the arts stands in the city's main square. Several old churches and mansions can also be seen.

Population 297 600
Map Romania – Ab

crater Bowl-shaped depression, either at the mouth of a volcano, or caused by the collision of a meteorite with the earth, for example or with another planet or moon.

crater lake Lake which has formed in the crater of an extinct volcano.

Crater Lake National Park *USA* Area of 641 km² (247.5 sq miles) on the crest of the Cascade range in south-west Oregon, 90 km (55 miles) north of the Californian border. The focal point of the park is a stunning blue CRATER LAKE, 589 m (1932 ft) deep and about 10 km (6 miles) across, and encircled by walls of multi-coloured lavas, 650 m (2130 ft) high.
Map United States – Bb

Craters of the Moon National Monument *USA* Volcanic area covering 217 km² (84 sq miles) in south-central Idaho, 145 km (90 miles) north of the Utah border. Its lava flows, fissures, cinder cones and craters bear a marked resemblance to the surface of the moon.
Map United States – Db

craton Large section of the earth's crust that has remained stable and immobile for millions of years. It is also called a shield or kraton.

Cremona *Italy* City in the Po valley, 75 km (45 miles) south-east of Milan. It is famous for its violins: the Stradivari, Amati and Guarneri families worked here between the 16th and 18th centuries. Violins are still made in the city.

Cremona's best known landmark is its magnificent 13th-century cathedral – Gothic inside with a Renaissance façade – which has a 111 m (365 ft) high bell tower, the Terrazzo.

Population 75 500
Map Italy – Cb

Cretaceous period Follows the Jurassic Period and precedes the Tertiary Period of the Mesozoic era in the earth's time scale. This was the period during which the chalk was formed. See GEOLOGICAL TIMESCALE.

Crete (Kriti) *Greece* The largest Greek island, covering 8366 km² (3229 sq miles) and lying in the south Aegean Sea. Crete was the heart of the rich Minoan civilisation of 3000-1100 BC. Its huge palace at KNOSSOS, near the main town and capital of HERAKLION (Iráklion), has been excavated and restored.

The island flourished under Venetian occupation from 1210 to 1669, producing some of Greece's finest medieval painting, poetry and drama. Later, after being impoverished under Turkish rule (1669-1896); it was united with Greece in 1912-13. In May 1941, fierce fighting broke out in northern Crete, when in the first massive paratroop operation of the Second World War, German forces seized Maleme airfield and went on to defeat 40 000 British, Commonwealth and Greek troops.

The island rises east to west to a limestone backbone, peaking at Mount Ida at 2456 m (8058 ft) in the White Mountains. The scenery varies from wild gorges to fertile plains where herbs and wild orchids abound. Olives, grapes, citrus fruits and wheat are grown. Besides tourism, industries include weaving and embroidery.

Population 536 980
Map Greece – Dd

crevasse Deep vertical crack or fissure in a glacier. See GLACIATION.

Crewe *United Kingdom* English industrial town and railway centre in Cheshire, 45 km (28 miles) south of Manchester. The railway works were opened in 1840, and turned a small market town into a flourishing engineering centre. Today, Crewe has become one of the most rapidly developing business and industrial centres in north-western England, with high-technology industries concentrated in the nearly 35 hectares (87 acre) Crewe Business Park.

Population 59 350
Map United Kingdom – Dd

Crimea (Krym) *Ukraine* Peninsula, resort area and autonomous republic in southern Ukraine, separating the Sea of Azov from the Black Sea and covering 25 500 km² (9843 sq miles). It consists mainly of dry but fertile steppe land. The Crimean Mountains stretch for 150 km (95 miles) along its southern coast, providing shelter for the resort region known until 1991 as the 'Soviet Riviera', and now as the 'Crimean Riviera', which is centred on YALTA. The range's highest peak is Roman Kosh, which reaches 1545 m (5069 ft). The industrial city of SIMFEROPOL' is the capital.

The Crimea, which was settled in the 6th century BC by Greek traders, became part of Russia in 1783. In 1853-6 the Crimean War between Russia and the forces of England, France and Turkey took place here. The ill-fated Charge of the Light Brigade took place near SEVASTOPOL, today one of the area's main towns. Many of the wounded were tended by the English nurse Florence Nightingale (1820-1910), who took 38 nurses to the Crimea and set up hospitals at Balakliya (formerly Balaklava) and Scutari (now USKUDAR in Turkey).

During the Second World War the Crimea was occupied by German troops in 1941-44. It was made a region of the Ukraine in 1954.

The northern parts are under intensive agriculture and wheat, maize and sunflowers are the main crops. In the south are vinyards.

Population 2 400 000
Map Ukraine – Cb

Cristóbal *Panama* See COLÓN and CANAL AREA

Crna Gora See YUGOSLAVIA

Crnojevica *Yugoslavia (Serbia and Montenegro)* See SHKODËR, LAKE

Croagh Patrick *Ireland* Holy mountain just south of Clew Bay on the west coast. It is the site of an annual pilgrimage believed to be the oldest in the Western world. On the last Sunday in July each year thousands of pilgrims, many of them barefooted, walk to the summit (765 m, 2510 ft), where Mass is celebrated in a small, modern chapel. Ireland's patron saint, St Patrick, is said to have prayed and fasted on the summit for 40 days and nights during the Lent of AD 441.
Map Ireland – Bb

Croatia See p 178

Crocodile *Southern Africa* See LIMPOPO

Crocodilopolis *Egypt* See EL FAIYUM

Cronstadt *Russia* See KRONSHTADT

Crossroads *South Africa* African residential area in transition from a squatter community. It lies close to Cape Town's international airport

and the national highway leading east to Port Elizabeth and Durban. In the 1980s, a local authority, established in terms of legislation passed by the white South African government, was largely rejected by inhabitants, and the area was wracked by frequent bloody clashes between young, radical and older, conservative members of the community.
Population 150 000
Map South Africa – Ac

Crotóne *Italy* Coastal town in southern Italy. It was famous in antiquity as the ancient Greek colony of Kroton, founded in 710 BC. The 6th-century BC mystic and philosopher Pythagoras lived and taught here. It was the site of one of the ancient world's great schools of medicine, and from 588 BC it became renowned for its successes in the Olympic games. Milo, the town's wrestler, was Olympic champion six times. Today, the town's main industry is its zinc works.
Population 55 630
Map Italy – Fe

Crowsnest Pass *Canada* Pass rising to 1357 m (4450 ft) through the Rocky Mountains in south-eastern British Columbia. In 1897, the Canadian Pacific Railway company received government funds to build a railway through the pass.
Map Canada – Dd

Crozet Islands *Indian Ocean* Archipelago consisting of five islands and 15 islets set in the Indian Ocean, about midway between Antarctica and Madagascar. It covers a land area of 300 km² (116 sq miles). Its main islands are called Apostles, Pigs and Penguins in the western part, and Possession and Eastern in the eastern part of the group. They are in the French Southern and Antarctic Territories, and France maintains a scientific and meteorological station on rocky Possession Island, which rises to 934 m (3064 ft). Rugged and volcanic in origin, the islands' steep cliffs have seen many a shipwreck.

The Crozet Islands were discovered in 1772 by Captain Nicolas-Thomas Marion-Dufresne and his ship's mate, Crozier, who annexed them for the French King Louis XV.
Map Indian Ocean – Be

Crozier, Cape *Antarctica* Eastern tip of Ross Island, abutted by the Ross Ice Shelf. The pressure of the ice against land squeezes the ice into vast ridges. The cape is the home of large colonies of emperor and Adélie penguins.
Map Antarctica – Jc

Cruphádraig *Ireland* See CROAGH PATRICK

crust Solid exterior portion of the earth above the Mohorovičić Discontinuity. See JOURNEY TO THE CENTRE OF THE EARTH, p 207.

Crystal Mountains *West Africa* Low range in north-west Gabon and south-east Equatorial Guinea. They form part of the edge of the African plateau, where rivers plunge down through rapids and waterfalls. The highest point is Mount Koudaké (980 m, 2920 ft).
Map Gabon – Ba

crystalline rocks Rocks in which the constituent minerals form a mass of crystals which have solidified from a molten state or which have been produced by metamorphism. Hence, igneous and metamorphic rocks are crystalline.

Cuando Cubango (Kwando) *Angola* Province whose capital, Menongue, lies about 800 km (500 miles) south-east of Luanda. From 1975, the year Angola became independent, until the early 1990s, the province was controlled by rebel UNITA guerrillas. Although the inhabitants here continued to grow subsistence crops such as maize, millet, beans and groundnuts, fierce fighting in the civil war devastated agriculture and caused great hardship for the population.
Map Angola – Bb

Cuango *Angola/Zaire* See KWANGO

Cuanza (Coanza; Kuanza; Kwanza) *Angola* River which flows for about 970 km (600 miles) from the central plateau to the Atlantic, 56 km (35 miles) south of the capital, Luanda. It is navigable for about 190 km (120 miles) upstream and in the 19th century was a main route to the interior for explorers.

The two provinces on either bank of the river are also called Cuanza – Cuanza Norte in the north, and Cuanza Sul in the south. The capital of Cuanza Norte, N'dalatando, is about 190 km (120 miles) south-east of Luanda. The province produces beans, coffee, cotton and tobacco.

Cuanza Sul's capital, SUMBE, lies 270 km (170 miles) south of Luanda. This province produces coffee, cotton and palm oil.
Population (Cuanza Norte) 378 000; (Cuanza Sul) 651 000
Map Angola – Aa; Ab

Cuba See p 180

Cubango *Southern Africa* See KAVANGO

Cuchilla de Haedo *Uruguay* Range of hills running for 200 km (125 miles) from the Brazilian border across the north and west of the country between the valleys of the Negro and Uruguay rivers. At its highest, it reaches 700 m (2300 ft). A second and lower range of hills, the Cuchilla Grande, lies east of the Negro River.
Map Uruguay – Ab

Cúcuta *Colombia* Capital of Norte de Santander department. It lies on the Venezuelan border, about 410 km (255 miles) north-east of Bogotá.

It was from Cúcuta in 1813 that the Venezuelan-born revolutionary leader Simón Bolívar (1783-1830) launched his troops into the battle for Colombia's independence from Spanish rule. The town was destroyed by earthquake in 1875, but was later rebuilt in its former style. Today, it is the centre of a cattle-rearing and tobacco and coffee-growing region.
Population 516 000
Map Colombia – Bb

Cuelap *Peru* Ruins of a walled fortress city discovered in 1843 near Tingo in the northern department of Amazonas. It is thought to have been the last outpost of a light-skinned race, known as the Chachas or Sachupoyans, who retreated from invading Incas in about 1460.
Map Peru – Ba

Cuenca *Ecuador* Capital of Azuay province, founded by the Spanish in 1557 on the banks of the Río Torrebamba, about 330 km (205 miles) south of Quito. The city's new and old colonial cathedrals stand in the Parque Calderón, the central square. Nearby are also many fine colonial churches and Inca ruins.

The city manufactures panama hats, fine gold jewellery, textiles and leather, and is a major centre for Ecuadorian handcrafts.
Population 195 000
Map Ecuador – Bb

Cuenca *Spain* Picturesque town 139 km (86 miles) south-east of the capital, Madrid. It is situated on the confluence of the rivers Júcar and Huecar and is noted for its 'hanging' riverside houses. The town's cathedral dates back to the 12th century.

About 30 km (18 miles) to the north-east is the 'Enchanted City', an area of limestone rocks eroded into strange, fascinating shapes. The surrounding province of Cuenca is sparsely populated, producing mainly wheat and wines.
Population (town) 42 820; (province) 205 200
Map Spain – Db

Cuernavaca *Mexico* Old and exotic resort town of tropical flowers and luxurious homes, known as 'the City of Eternal Spring' in reference to its warm, sunny climate. It lies 80 km (50 miles) south of MEXICO CITY, amid scenic mountains, and is the capital of the state of Morelos. The town has a 16th-century palace, an impressive Baroque cathedral and botanical gardens that were founded in the 18th century.

The CACAHUAMILPA caves, the largest (though not the deepest) system in Mexico, lie 70 km (44 miles) to the south-west.
Population 557 000
Map Mexico – Cc

cuesta Spanish term that has been widely adopted for an upland area with a gentle slope on one side, known as the dip slope, and a much steeper one on the other side, known as the scarp face or escarpment. This common relief feature is usually found where there is a tilted outcrop of resistant sedimentary rocks.

The dip slope corresponds with the tilt of the rocks, while the scarp face is cut by erosion across them. The North Downs and the Chilterns in England are examples of cuestas.

Cuige Laighean *Ireland* See LEINSTER

Cuige Mumhan *Ireland* See MUNSTER

Cuige Ulaidh *Ireland/United Kingdom* See ULSTER

Culebra *Puerto Rico* Small island, 26 km² (10 sq miles) in area, lying off the east coast. It is noted for its crescent-shaped beaches and the underwater life on its reefs. It has a deep and sheltered harbour.
Population 1540
Map Caribbean – Bb

Culiacán (Culiacán-Rosales) *Mexico* Capital of Sinaloa state in north-west Mexico. It is a mining city and an agricultural centre with an elaborate irrigation system which helps to provide the rest of Mexico with most of its winter vegetables.
Population 602 110
Map Mexico – Bb

Croatia

AN EX-YUGOSLAV STATE WITH A SPECTACULAR COASTLINE – WRACKED BY CIVIL WAR IN THE EARLY 1990s

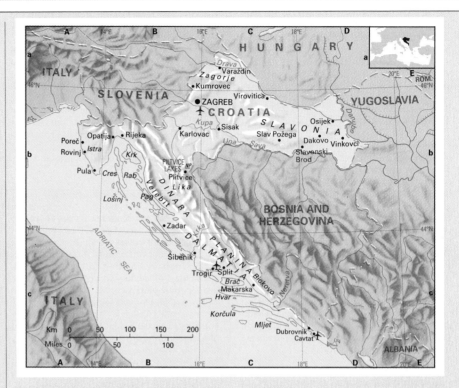

In the 1970s, when Marshal Tito was still firmly in charge, it might have seemed that the Serb and Croat nations were inseparably interwoven. This assumption was to be proved tragically untrue. Croatia, which lies between the Adriatic Sea and the Hungarian plain, declared itself independent in June 1991, and was recognised by the European Community in January 1992. Hitherto, the Croatians had experienced periods of lasting peace; with the coming of independence, however, war erupted.

First, the Serb-dominated Yugoslav federal government resisted the country's independence, and fierce fighting broke out. A cease-fire was agreed in January 1992, and Croatia proceeded to organise itself as an independent state. By then, however, the Serbs held a large portion of Croatian territory and soon began ridding the Serb-Croat mixed areas of Croats by means of 'ethnic cleansing', driving thousands of people from their homes, and proclaiming an independent 'Republic of Serbian Krajina' (RSK) – a state occupying about one-third of Croatia's territory. The 'ethnic cleansing' was marked by acts of extreme barbarism.

When the conflict worsened, the United Nations sent a Protection Force to the region. But fighting continued until March 1994, when the Croatian government and the government of the self-declared RSK signed a cease-fire agreement. However, the Croatian government stated categorically that it would not grant independence to the RSK – nor would it permit the RSK to reunite with Serbia.

This caused UN peacekeeping forces to maintain a presence in the region, mainly to control the buffer zone between Croatian and nationalist Serbian troops. The Croats gave the UN notice to leave by the end of March, 1995, provoking fears of a fresh round of warfare.

There were Croats, too, who had nationalistic aspirations when the federal republic of Yugoslavia disintegrated. In 1992, the Croat minority in BOSNIA-HERZEGOVINA declared their wish to be reunited with the Croats living in Croatia. They joined the other population groups – the Muslims and the Serbs – in Bosnia-Herzegovina in what quickly developed into one of the most brutal civil wars recorded in human history. In February 1994, they signed a cease-fire with the Bosnian Muslims and the following month agreed to form a federation with them, but tensions persisted.

EUROPEAN INFLUENCE

Historically, the region inhabited by the Croats reached south of the Sava flood plain to Slavonia, and included the coastal plains and south-western Herzegovina, and even parts of Montenegro. Croats have lived on the Balkan Peninsula since the 7th century AD, and were Christianised by Roman Catholics based at Aquileia. Croatia was united with Hungary by a personal union of thrones in 1091; later, Habsburgs ruled the territory.

The Venetians established colonies on the Adriatic Sea; from the east came Turks to try to incorporate their land in the Ottoman Empire. Croatia has strong political links with Western Europe, and religious ties with Rome.

Culturally, however, Croats like to think of themselves as Central Europeans instead of belonging to the Balkans. Many of their cities can be compared to Austrian or Hungarian cities; the Adriatic coastal towns, on the other hand, with their mild, moist winters and hot sunny summers, have a typically Mediterranean flavour and have always been favourite tourist places.

The Croats (along with the Slovenes and Macedonians) preserved their distinctive language, often under severe foreign pressure. Serbo-Croat, recognised as the main language in old Yugoslavia, is two separate tongues having many common elements but differing in alphabet, pronunciation and everyday words. Despite nearly 40 years of compulsory education, younger Croats and Serbs still have difficulty in reading each other's writing. The Roman Catholic Croats (and Slovenes) use latin script while the Orthodox Christian Serbs (and Montenegrins and Macedonians) use the Cyrillic alphabet.

It was perhaps due to their cultural independence that from the start, many Croats were unhappy with the kingdom of Serbs, Croats and Slovenes when it was created at the end of the First World War, in December, 1918. The Croats knew some degree of independence during the Second World War, when, supported by the invading Germans and Italians, they became 'independent' for a while, setting up a fascist state. Tito's Communist regime started as a partisan uprising against German invaders in the Second World War.

Tito led guerrilla forces who fought against the Germans and against a rival group of partisans, the *Chetniks*. By the time the country was freed in 1944, Tito had become a popular hero and he proclaimed himself head of an independent Yugoslavia.

During this period, Croatia became one of six socialist republics in the Yugoslav federation. Tito led a sweeping programme to develop agriculture, industry and tourism and to open Yugoslavia to the outside world. Croatia, with its coastline of 4012 km (2493 miles), benefited enormously.

Tourists soon started flocking to the sunny, island-strewn coast of DALMATIA and to the Istrian Peninsula, bringing with them valuable foreign currency. Chief ports here are PULA, RIJEKA, SPLIT, ZADAR, SIBENIK.

Predominantly rural until 1945, Croatia also developed its industrial sector. Industries based on timber, oil and gas, hydroelectricity, bauxite, limestone and iron ore are concentrated in the capital ZAGREB, OSIJEK, SLOVANSKI BROD, VARAZDIN and Sisak.

EFFECTS OF CIVIL WAR

With its 1994 land area, Croatia is slightly smaller than West Virginia. It can be divided into three main natural regions.

Lower Croatia consists of fertile, low-lying flood plains along the Sava and Drava rivers fringed by forested mountain ranges – an area of rolling farmland, orchards, forested plains and hills, including the beautiful Papuk reaching 953 m (3127 ft) west of the city of Osijek. Except for Osijek and SLOVANSKI BROD, the towns are small and little more than agricultural market places.

South and west of Lower Croatia lies Upper Croatia, a windswept region of high rainfall, consisting of rocky, largely infertile highland. Dalmatia, the third region, which stretches along the sunny Adriatic coast has a typically Mediterranean climate, ideal for the cultivation of figs, olives and lemons. Until the outbreak of the civil war, the region attracted hundreds of thousands of tourists each year.

Before the war, Croatia was the second most prosperous and industrialised of the Yugoslav republics and the most frequently visited. But both tourism and industry began to suffer heavily as a result of the brutal, bitter fighting. In the past, the region had little difficulty in producing a surplus of food. But, between 1991-1994, exports stopped. In 1994, the industrial sector was struggling to pick up the pieces of a shattered economy. For instance, hydroelectric supplies and transport and communications links were seriously disrupted in 1992, with consequent damage to the economy, particularly in Dalmatia.

The tourists boycott was a particularly bitter blow to the economy. At the height of the conflict, hotels along the coast housed refugees instead of holidaymakers. In 1993, it was estimated that there were some 500 000 refugees residing in Croatia – mainly Croats from Bosnia-Herzegovina and from Krajina. Solving the refugee problem, in particular, housing them, took the largest chunk out of the national budget after defence.

An economic stabilisation programme, launched in the second half of 1993, managed to decrease consumer inflation slightly (50 per cent per month in 1992).

CROATIA AT A GLANCE
Official name Republic of Croatia
Area 56 538 km² (21 824 sq miles)
Population 4 694 400 **Per km²** 83
(Per sq mile) 215)
Capital Zagreb
Government Republic
Currency 1 Croatian dinar = 100 paras
Languages Serbo-Croat
Religions Christian (Roman Catholic 77%, Orthodox 11%), Muslim
Climate Moderate continental climate in the north, with rainfall throughout the year; Mediterranean climate in the south, with dry summers. Average temperature in Zagreb ranges from 0°C (32°F) in January to 22°C (71.6°F) in July
Before the war:
Land use Forest and woodlands 15%, cultivation 52%, meadows and pastures 18%, other 15%
Main primary products Oil, bauxite, coal, iron ore; wine, fruits, cereals, olives, livestock
Major industries Mining, machinery, chemicals and plastics, paper and wood products, shipbuilding, oil refining and food processing
Main exports Machinery and transport equipment, plastics and chemicals, ships, wine, olive oil, fuels
Annual income per head (US$) 676-2695
Population growth (per thous/yr) 1
Life expectancy (yrs) Male 70 **Female** 77

Culloden *United Kingdom* Moor 8 km (5 miles) east of Inverness in the Scottish Highlands. A decisive battle was fought here in 1746, when the English army under the Duke of Cumberland defeated the forces of Bonnie Prince Charlie, the 'Young Pretender', forcing him to flee. This put an end to the Stuart family's attempt to regain the British crown.
Map United Kingdom – Cb

Culm *Poland* See CHELMNO

Cumaná *Venezuela* Capital of the state of Sucre, about 320 km (200 miles) east of the national capital, Caracas. Founded by Spaniards in 1520, it is the oldest Hispanic city in South America – although few old buildings remain as a result of earthquake damage. From here, plans were made for the conquest of the rest of the continent.

Straddling both banks of the Manzanares River, Cumaná is a port exporting, among other things, tobacco, cacao (from which cocoa is made), cotton textiles and coffee. The city also has huge fish processing plants. A major cotton mill is located near the city, and it has an important sardine-canning industry.
Population 269 590
Map Venezuela – Ba

Cumberland *United Kingdom* See CUMBRIA

Cumberland Gap *USA* Pass through the Cumberland Plateau of the Appalachians, near the junction of Virginia, Kentucky and Tennessee. Discovered in 1750, it was used by the pioneer Daniel Boone and became the main route for migration to the west until it was abandoned in the 1840s. The pass, 500 m (1640 ft) above sea level, lies in a national historical park covering 82 km² (32 sq miles).
Map United States – Jc

Cumbria *United Kingdom* County of north-west England, covering 6809 km² (2629 sq miles) on the Scottish border. It is made up of the former counties of Cumberland, Westmorland and the part of Lancashire known as Furness, and includes Britain's LAKE DISTRICT. The old Cumberland coalfield on the coast was the basis for industry in the towns of Whitehaven and Workington and the shipbuilding centre of Barrow-in-Furness where submarines are now built. The county town is CARLISLE. Hill farming and Lake District tourism are the main livelihoods inland.
Population 489 200
Map United Kingdom – Dc

Cumbrian Mountains *United Kingdom* See LAKE DISTRICT

cumulonimbus A cloud that builds up to a great height, rising in towering, turbulent clusters from a flat base, often spreading out at the top into an anvil-shaped head. Cumulonimbus clouds grow out of CUMULUS clouds in unstable air, bringing heavy showers and thunderstorms.

cumulus Vertical cloud formation of fluffy, cauliflower-shaped clusters rising from a flat base, often seen during the middle and latter part of the day and usually associated with sunny spells. They are distinguished from CUMULO-NIMBUS clouds by their rounded tops.

Cundinamarca *Colombia* Central department that includes the country's capital, Bogotá. It is made up of a variety of climatic areas, from the tropical to the very cold, and consequently a wide range of crops are grown, including coffee, maize, bananas and potatoes.
Population 6 579 750
Map Colombia – Bb

Cunene (Kunene) *Angola/Namibia* River rising in west-central Angola and flowing some 970 km (603 miles) south-west, then west along the Angolan-Namibian border to the Atlantic Ocean. Hydroelectricity is produced at the Matala Dam in Angola, and at the 107 m (351 ft) high Ruacana Falls on the border.
Map Namibia – Aa

Curaçao *Netherlands Antilles* Largest, most populous and most highly developed of the Netherlands Antilles, lying off the northern coast of Venezuela between Aruba and Bonaire. It covers 444 km² (171 sq miles) – most of which is limestone – and there is little agriculture because of inferior soils and lack of water.

The chief industries are oil refining, electronical engineering and tourism. Phosphates are mined for export. A deepwater harbour, the Schottegat, is located in the south-west beside WILLEMSTAD, the capital of the island and of the Netherlands Antilles. An orange-flavoured liqueur, Curaçao, is named after the island, where it was first made. Dried orange peel is used as a base for the famous drink.
Population 173 400
Map Caribbean – Bc

Curepipe *Mauritius* Town, health resort and sugar production centre set in the middle of the island, 16 km (10 miles) south of Port Louis.
Population 70 000
Map (Mauritius) Indian Ocean – Bd

Curitifa *Brazil* Regional capital 330 km (206 miles) south-west of the city of São Paulo. Founded in 1693, it is today one of the country's most prosperous and rapidly developing industrial cities and is becoming internationally known for its advanced approach to environmental problems. Its products include tea, tobacco, furniture and metals.
Population 1 248 000
Map Brazil – Dd

current See OCEAN CURRENTS

Curtea de Argeş *Romania* Town on the River Argeş, 150 km (93 miles) north-west of Bucharest. It became the capital of the former principality of Walachia in the 14th century and contains the ruins of the princely palace. The town's Byzantine Church of St Nicolae Donnesce is the oldest feudal monument in southern Romania, dating from the 14th century.
Population 32 600
Map Romania – Aa

Curzola *Croatia* See KORČULA

Cuscatlán *El Salvador* Department to the east of San Salvador. It produces cotton, sugar cane and livestock. Its capital is Cojutepeque (population 43 560).
Population 167 290

Cuba

LARGEST OF THE WEST INDIES ISLAND STATES, CUBA HAS EMBARKED ON ECONOMIC REFORMS – THOUGH RELUCTANTLY

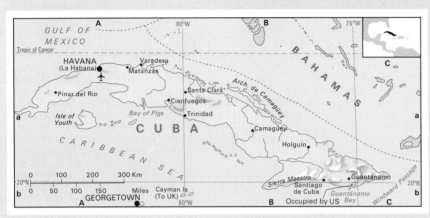

In 1962, Cuba became the fuse that nearly sparked off a Third World War. Its Marxist leader, Fidel Castro, had accepted plans to establish Russian missile bases on the island, a mere 145 km (90 miles) from the United States coast of Florida.

The United States navy mounted a blockade to stop Russian ships taking missiles to Cuba, and the American and Russian leaders, John F. Kennedy and Nikita Khrushchev, glowered at each other across their nuclear weapons. As the world watched with bated breath, Khrushchev backed down. For years to follow, Cuba continued to loom large in Communist intrigues, exporting revolution and aid to Marxists in developing countries as far apart as Angola, Ethiopia and Nicaragua.

COLONIALISTS AND REVOLUTIONARIES

When Christopher Columbus landed on Cuba in 1492, he found it inhabited by people whose ancestors had come from South America several centuries before. There were about 100 000 natives on the island when the Spanish built their first city at Baracoa in 1512; by 1570, they had all disappeared – the victims of genocide, forced labour and diseases. Then, in 1762, the British conquered HAVANA but a year later they returned it to Spain in exchange for Florida.

Apart from this brief British interlude, Spain retained its hold on Cuba until defeated by the United States in the Spanish-American War of 1898. United States troops occupied Cuba until 1902, when an independence treaty was implemented, guaranteeing the United States continued rights to keep military bases on the island.

American interests continued to dominate Cuban economic life as well – and the island soon became a popular resort for Americans attracted by its sunny beaches, and its exotic night clubs and casinos. In fact, over the years, Cuba's dependence on the United States increased.

Meanwhile, the island state was ruled by a series of dictators. The last, before Castro, was Fulgencio Batista, whose corrupt and intolerant regime lasted on and off from 1933 until 1959 when a group of revolutionaries led by Fidel Castro, a law student, and his lieutenant, Che Guevara, ousted him after a two-year guerrilla campaign. Mass arrests and executions followed, and thousands of Cuban exiles poured into the United States. Cuba turned to the USSR for economic aid and Castro proclaimed a Communist state.

In 1961, American CIA-trained Cuban exiles landed in the BAY OF PIGS in an unsuccessful attempt to launch a rebellion; the following year saw the Soviet missile crisis. During this time relationships between Cuba and the United States hit an all-time low and during the 1960s and 1970s Cuba became a satellite of the Soviets. More recently, Cuba has attempted to re-establish links with the US.

Cuba is as big as all the other 3700 Caribbean islands put together. The island is 1250 km (780 miles) long and 30 to 190 km (20-120 miles) wide, and consists mostly of extensive plains; forests still cover 25 per cent of the total land area.

The climate is warm, although northwestern Cuba has winter spells below 10°C (50°F). Many hurricanes pass close by but few have caused serious damage. Most of the island receives more than 1140 mm (45 in) of rain a year, and most of its soil is fertile.

Sugar provides 80 per cent of export earnings – Cuba is the world's third largest sugar cane producer. In 1837, Latin America's first railway was built to link Havana, the capital, with the cane fields.

Before the revolution, nearly all the cane was grown by small tenant farmers, whose land was owned by the sugar companies. In 1959, all land over 30 caballerías (13.4 hectares, 33 acres) was nationalised and turned into state-run farms, and in 1963, private land holdings were further reduced to a maximum of 5 caballerías (2.23 hectares, 5.5 acres).

Since 1971, steps have been taken to diversify agriculture. Tobacco, once the second agricultural export, has been overtaken by

▲ HAVANA'S PAST Colonial buildings line the Santos Suares in Havana's Old City; many influences have converged to make this one of the most beautiful cities in the Caribbean.

citrus fruits – but in 1992, Cuba still exported 100 million cigars. To reduce dependence on imported food, rice growing and the rearing of beef and dairy cattle have been increased. Modern fishing fleets in 1990 caught more than 20 times the amount of fish harvested in 1948 and are still increasing their catches.

land, 90 per cent of essential services and 40 per cent of the sugar industry. There was widespread corruption and unemployment, as well as a vast population of poor people; Castro's Communists tackled corruption and unemployment vigorously, regulating wages and taking control of production. About 10 per cent of Cuba's trade was with Communist countries, with the USSR buying sugar and nickel at well over the world price, and supplying cut-price oil and cheap loans to the islands. Soviet aid ran into millions of dollars – about US$5 billion per year at its peak.

But subsidies from the Soviets came to an end in 1990, and though Cubans today have free education (including tertiary education) and health services, they lack basic amenities such as soap, toilet paper, and bread. Furthermore, the ongoing United States embargo makes it difficult for Cubans to obtain machinery, oil – and even medicines and food. The lowest sugar cane crop in a decade was harvested in 1992, with the low yields put down to a shortage of fertiliser, pesticides and equipment. In 1993, output dropped even further – to 50 per cent; shortages of oil and harvesting machinery meant that 70 per cent of harvesting had to be done manually.

The economic catastrophe has forced the Cuban government to launch some reforms, such as lifting the 30-year ban on foreign currency, and legalising self-employment, in 1993. This encouraged thousands of Cubans to leave for the United States in mid-1994; the large number of refugees forced the US government to re-examine its policy towards Cuban refugees, who had previously been guaranteed residence in the United States. The US government finally agreed to allow up to 20 000 Cubans into the country each year. However, the United States refused to relax its 30-year trade embargo of Cuba.

CUBA AT A GLANCE		
Official name Republic of Cuba		
Area 110 860 km² (42 793 sq miles)		
Population 10 957 090 **Per km²** 98.8 **(Per sq mile** 256)		
Capital Havana		
Government Socialist republic		
Currency 1 Cuban peso = 100 centavos		
Language Spanish		
Religion Christian (45% Roman Catholic)		
Climate Tropical, with a dry period from December to March; average temperature in Havana ranges from 22°C (71.6°F) in January to 27°C (80.6°F) in August		
Land use Cultivation 29%, meadows and pastures 23%, forest and woodlands 17%, other 31%		
Main primary products Sugar cane, rice, sweet potatoes, cassava, maize, oranges and other citrus fruits, tobacco, coffee, livestock, fish; nickel, copper, cobalt, chrome		
Major industries Agriculture, mining, food processing, tobacco products, metal refining, chemicals, cement, fertilisers, textiles, fishing		
Main exports Sugar, citrus fruits, nickel, chrome, tobacco, rum, fish		
Annual income per head (US$) 1000		
Population growth (per thous/yr) 10		
Life expectancy (yrs) Male 74 **Female** 79		

Cuba has the fourth largest nickel reserves in the world, and this is the second most valuable export. Cobalt, chromium and iron ore, of which the island has vast reserves, are also mined. The most important industry is still that of processing sugar and other foodstuffs, but factories also produce cement, farm machinery, fertilisers, paper, textiles and footwear. There is enormous potential for the development of tourism; in fact, the country's fine, sandy beaches, forested mountains and colonial architecture make Cuba a veritable tourist haven.

Just over 20 per cent of Cuba's population are mulattos – mixed-blood descendants of African slaves and Europeans. Except for a small Chinese minority, most others are descended from Spanish immigrants. Eighty per cent of the people live in towns. Although they do not share the stark poverty of most Latin American countries, Cuba's per capita gross product has dropped from about third position in Latin America (in pre-Castro days) to perhaps 23rd (in 1993).

Before 1959, foreign interests controlled the economy, owning 75 per cent of arable

Cusco *Peru* Department and city, and former Peruvian capital, in the southern Andes. The department's high pastures support sheep and alpacas; its fertile valleys are planted with maize, barley and potatoes.

The departmental capital, Cusco, stands 3500 m (11 480 ft) above sea level. Between the 11th and the early 16th century, it was the religious, political and administrative centre of the far-flung Inca empire – true to its name, 'Cusco' meaning 'navel (of the world)'. The area is rich in Inca history – at the height of its prosperity, the city and the surrounding area had a population of about 200 000 Incas.

The main square of the city was the centre for many Inca ceremonies. The ruins of Sacsahuamán and Salamancu are nearby, and to the north-west the famous 'Inca trail' leads from Ollantaytambo to Machu Picchu – 'the Lost City of the Incas'. Today, the magnificent Baroque cathedral and Church of the Compãnia, built in 1668, stand on the site of the Temple of Viracocha and an Inca palace.

The city's heritage is reflected in its festivals, which range from the Catholic Corpus Christi held on the 2nd Thursday after Whitsun, to Inti Raymi, the Inca Festival of the Sun, held at the same time.

Population (department) 1 041 800; (city) 275 000
Map Peru – Bb

Cuttack *India* City on the Mahanadi River, 435 km (270 miles) south-west of Calcutta. It is built on a spit of land between two distributaries, and is vulnerable to floods. Cuttack produces exquisite silver and gold filigree jewellery and also trades in oilseeds and rice.

Population 2 043 340
Map India – Dc

cwm See CIRQUE

Cyclades (Kikládhes) *Greece* Group of 30 islands in the centre of the Aegean Sea, lying in a circle around Delos (*kýlos* meaning 'circle'). They were vital links in the early sea-trading routes to the east, producing a race of sailors from around 2600 BC.

The islands were occupied by the Venetians in 1204, then Turks in the 16th century, before becoming part of independent Greece in 1832.

Of all the Cyclades, MÍKONOS is the most popular with tourists, NÁXOS is the largest island, and Ermoúpolis on the island of Síros is the group's capital. Soil on the Cyclades is thin, so there is little agriculture, but their coastlines and romantic settings make them attractive to tourists.

Population 95 080
Map Greece – Dc

cyclone Storm accompanied by strong winds revolving round a centre of atmospheric low pressure. There are two types. In temperate latitudes, a cyclone is known as a DEPRESSION; in the tropics, it is known as a TROPICAL CYCLONE – and is usually very violent.

cyclonic disturbance See DEPRESSION

cyclonic rain (frontal rain) Precipitation which marks the fronts and the warm sectors of a DEPRESSION in temperate latitudes, as one air mass overrides or undercuts another.

Cyprus see P 183

Cyrenaica *Libya* One of Libya's three main historical regions, covering 905 000 km² (349 340 sq miles) in the eastern half of the country and named after the ancient ruined city of CYRENE. About 23 per cent of the country's people live in the province. BENGHAZI is its chief town.

Most of Cyrenaica consists of desert, but there is fertile land along the Mediterranean coast, and the northern slopes of the coastal ridges are the only region of Libya with any forest vegetation. Also in the north lies the Jebel Akhdar, an important agricultural area.

Much of the life of the province remains nomadic or semi-nomadic, but besides herding, there is also farming and fishing. The main products are barley, wheat, grapes, olives, dates, and tuna fish. The main oil fields are in Cyrenaica.

Italy occupied Tripoli in 1912 and Italian sovereignty in Libya was recognised by Turkey. After the expulsion of the Germans and Italians in 1942 and 1943, Tripolitania and Cyrenaica were placed under British military administration. Britain recognised Mohammed Idris Al-Senussi as Amir of Cyrenaica in June 1949.

In the 19th century, the nomadic Bedouins from the region embraced the teachings of the Sanusi movement, a radical militant Muslim sect, and still remain its most numerous adherents in the region. During the Italian occupation (1912-43), they were in the forefront of the strong resistance movement.

Map Libya – Cb

Cyrene *Libya* Ancient ruined city, 224 km (139 miles) from Benghazi in the north-east of the country.

The original capital of Cyrenaica, founded in 630 BC by the Greeks, it flourished for 1000 years as the centre of a Greek colony before being devastated by an earthquake in AD 364.

▲ CUSCO The three ramparts of the Inca fort Sacsahuamán near Cusco consist of large boulders fitted into one another seamlessly. Inca masons used stones more than 5 m (160 ft) high in creating their magnificent structures.

At its height, Cyrene was a city of 100 000 inhabitants and included an important medical school. Lying a few miles inland, it was served by the port of APOLLONIA. The city's ruins include the Roman baths, the sanctuary of Apollo, the Greek theatre, the temple of Zeus, the house of the 2nd-century Jew Jason Magnus (one of the authors of the Old Testament book of Maccabees), tombs cut out of solid rock, and an early church containing fine mosaics.

Map Libya – Ca

Czechoslovakia See CZECH REPUBLIC (p 184) and SLOVAKIA

Czech Repubic see p 184.

Częstochowa (Chenstokhov; Czenstochau) *Poland* Industrial city and the country's chief pilgrimage centre. It lies in the south, 64 km (40 miles) north of the city of Katowice on the Warta River. Pilgrims flock to the richly adorned Jasna Góra ('Hill of Light') Monastery founded in 1382 which has been a symbol of Polish nationalism since the city withstood a Swedish siege in the 17th century. Its tower, measuring 105 m (344 ft) tall, dominates the city, and its treasures include the Black Madonna, a Byzantine painting of our Lady of Częstochowa, said to have miraculous powers.

The mainly modern industrial area around the monastery produces iron and steel, textiles, and paper.

Population 258 000
Map Poland – Cc

Cyprus

MEDITERRANEAN CYPRUS HAS BEEN HOST TO GODS AND CRUSADERS, SAINTS AND HEROES – AND THOUSANDS OF HOLIDAY-MAKERS

A long, long time ago, the sea foam that hisses up onto the beach of Achni on the south coast of Cyprus gave birth to golden Aphrodite, the Greek goddess of love. But mortals, too, have known the island for a very long time. Idols have been unearthed here that were carved by Stone Age sculptors 8000 years ago and more, and in the 3rd millennium BC, Cypriot mines, which are still worked, provided copper for the entire eastern Mediterranean region (the Greek name for the metal – *kypros* – and for the island are the same).

Its location on a maritime crossroads between Europe, Asia and Africa has always assured Cyprus of a variety of visitors and conquerors. Mycenaean Greeks established city-kingdoms here in about 1500 BC; thereafter, the island was in turn annexed by the Egyptians, Assyrians, Persians, Romans and Byzantines, as well as the Phoenicians. Christianity was introduced by St Paul and St Barnabas in the 1st century AD.

In 1191 Cyprus was occupied by Richard I ('the Lionheart') of England, who married Princess Berengaria of Navarre at LIMASSOL, and then handed the island over to the Knights Templar, a military and religious order founded to protect pilgrims to the Holy Land.

The French Lusignan dynasty were kings of Cyprus from the 12th century, until they were ousted by the Venetians in the 16th century. They, in turn, were expelled by the Turks who governed Cyprus until 1878, when they leased it to the British, who made the island a crown colony in 1925.

The majority of the population is of Greek descent, with a minority (one-fifth) of Turkish descent. From about 1900, there was a strong movement among Greek Cypriots for unification with the homeland of their ancestors. This escalated, in the 1950s, into the terrorist

▲ **IMPREGNABLE FORTRESS The 12th-century castle overlooking Kyrenia harbour in northern Cyprus has withstood six sieges.**

movement EOKA – *Ethniki Organosis Kypriakou Agonos*, or National Organisation of Cypriot Struggle – which carried out a campaign of assassination against both the British and the Turkish Cypriots.

It partially gained its ends in that, in 1960, the island became an independent republic within the Commonwealth, with the Greek Cypriot Archbishop Makarios III as president.

Three years later, however, civil war broke out between the Greek and Turkish Cypriots. It was quelled by a United Nations peace-keeping force, but in 1974, the situation was still sufficiently uneasy for Turkey to mount a full-scale invasion of the island under the pretext of protecting Turkish Cypriots from the Greeks. Turks occupied the northern part of the country including the towns of FAMAGUSTA and KYRENIA, and the capital, NICOSIA, which remains a divided city. The Turkish troops expelled some 200 000 Greek Cypriots from their homes to make way for settlers from Anatolia. Hundreds of Greek Cypriots were killed or injured and over 1000 are still missing.

In 1983 Turkey declared the part of Cyprus occupied by its troops to be the separate and independent Turkish Republic of Northern Cyprus. However, in terms of international law, the Turkish part of Cyprus has not been recognised by any country other than Turkey.

The island has a long, thin panhandle and is divided from west to east by two parallel ranges of mountains, which are separated by a wide central plain open to the sea at either end. In the north, the Kyrenia range reaches 1024 m (3360 ft). The south-westerly Troödos range is covered with pines, cypresses and cedars and culminates in the 1951 m (6401 ft) Mount Olympus. At that height, there is sufficient snow for skiing in January and February. The rest of the island has hot and dry summers and a warm, damp autumn and winter.

Its delightful climate contributes towards the island's astonishing variety of crops – early potatoes, vegetables, cereals, tobacco, olives, carobs, bananas, and grapes turnend into the rich, strong wines, sherries and brandies for which Cyprus has been famous since classical

times. Agriculture is, however, largely dependent on irrigation schemes.

Fishing also has some significance in the island's economy, as do copper, asbestos and chrome, though mining has deteriorated since the Turkish invasion. Additional income is provided by the British bases at DHEKELIA and AKROTIRA.

Above all, however, the island depends on its visitors – even though it is impossible to view Cyprus as a whole. Visitors who have landed in the Greek sector are not permitted to pass over into the Turkish portion, and vice versa. And though some of the most important tourist sites, such as the ruins of SALAMIS near Famagusta, are now in the Turkish zone, it is the Greek-Cypriot sector that is most popular. Indeed, tourism has led a recovery in the economy of the Greek sector since 1974, while that of the Turkish-held area has stagnated.

Cyprus has no university, but there are colleges in the Greek sector for art, forestry, engineering, nursing and business management.

CYPRUS AT A GLANCE		
Official name Republic of Cyprus		
Area (total) 9251 km² (3572 sq miles); (Greek) 5896 km² (2276 sq miles); (Turkish-held) 3355 km² (1295 sq miles)		
Population (total) 710 000; (Greek) 535 000; (Turkish-held) 175 000		
Per km² (total) 77 (**Per sq mile** 199)		
Capital Nicosia		
Government Republic		
Currency 1 Cyprus pound = 100 cents		
Languages Greek, Turkish, English		
Religions Christian (Greek Orthodox 78%), Muslim (18%), other		
Climate Mediterranean, with long, hot summers and warm, damp winters. Average temperature in Nicosia ranges from 10°C (50°F) in January to 28°C (85°F) in August		
Land use Cultivation 47%, meadows and pastures 10%, forest and woodlands 18%, other 25%		
Main primary products Copper, asbestos, pyrite; grapes, almonds, citrus fruits, potatoes		
Main exports Copper, asbestos, cement, potatoes, cigarettes, wine, citrus fruits, almonds, footwear, textiles		
Annual income per head (US$) 9820		
Population growth (Per thous/yr) 9		
Life expectancy (yrs) Male 74 **Female** 78		

Czech Republic

DIVORCED FROM ITS FORMER HALF, THE CZECH REPUBLIC IS LOOKING TO THE WEST FOR NEW PARTNERS

The Czechoslovakian Republic, comprising a Czech half and a Slovakian half, was established in 1918, after the First World War. In June 1992, it was decided to dissolve it – after more than 70 years of union. When the federation broke apart on 1 January 1993, the Czechs faced the future confidently. They were much better off than their Slovak counterparts. Small wonder – even before 1918, the Czech part of the country had more industry and infrastructure than the Slovakian section. Czechs had dominated public life, even in the Slovak region, and most political and economic decisions had been taken in Czech Prague. Despite rapid development between 1918 and 1938, and massive industrialisation after 1948, Slovakia still lagged far behind the Czech half of Czechoslovakia and never caught up with it.

GERMAN INVASION

Wedged between Germany, Austria and Poland, and protected by mountain ranges along its borders, the territory of the Czech Republic tallies with the area historically divided into BOHEMIA, MORAVIA and SILESIA. Christianity was introduced in the 9th century. PRAGUE, the capital city, was one of the leading European cities during the Middle Ages and in the 14th century was the seat of the German Holy Roman Emperor. Its countless palaces and churches are a reminder of times of wealth gone by. The university in Prague, founded in 1348, is among the oldest universities in Central Europe.

Through all their history, the Czechs cultivated strong ties – economic, cultural and political – with Western Europe, while at the same time trying to preserve their independence. For centuries, a large German population lived in the region. But after Czechoslovakia gained its independence from the former Austro-Hungarian Empire in October 1918, it strengthened its ties with France and Britain rather than Germany.

However, the young state of the Czechs and Slovaks suffered badly under Hitler's Germany. In 1938, the Munich Agreement between Germany, France, Italy and Britain allowed Hitler to march into the border regions (Sudety, or 'Sudetenland'), where the majority of the population was German-speaking – in return for a guarantee that there would be no further expansion. Six months later, German troops invaded the rest of the country and turned it into the 'Bohemian and Moravian Protectorate'. When the Second World War ended in 1945, the victorious Allies agreed that all Germans should be expelled from the country. Thus the Sudeten Germans, whose forefathers had settled in Bohemia in the 13th century and had colonised a belt along the border, were made to pay for the Nazis' war – and for their role in the violation of the 1938 Munich Agreement.

In February 1948, a coup d'état organised by the Soviets put Communists in power. From then on, the Soviet Union kept an iron grip on the country, with the Soviets exercising enormous influence in Czechoslovakia for more than 40 years.

UNDER SOVIET INFLUENCE

Two decades after the Communist seizure, new hope arose for an end to Soviet rule. Alexander Dubček, the new leader of the Czechoslovak Communist Party, introduced socialist-humanitarian reform. Dubček proposed to separate the state and the political party, and – amazingly for a Communist state – to legalise opposition parties. This encouraged Czechoslovakians to take a more active part in politics.

However, in August 1968, in answer to demands made by Dubček, and to calls for help made by staunch Stalinists inside the country, troops of the Warsaw Pact states (excepting Romania) invaded the country and despite nationwide protests, forced Dubček to resign, replacing him with pro-Soviet rulers and compelling Czechoslovakia to remain a member state of the eastern bloc military and economic alliance.

The Soviets stayed maintained control until 24 November 1989, when massive demonstrations throughout the country forced the *politburo* to hand over the reins to parliamentary democracy.

After the collapse of the Soviet Union it soon became evident that Czechoslovakia would split in two. Most Czechs resisted the demands of the Slovak nationalists – nor were all Slovaks in favour of the split – but in the end, the Slovak nationalists won. The two countries officially became independent on 1 January 1993. The currency union that they had planned to maintain was ended one month later, and the two currencies split.

Also, disagreement over property division and privatisation initially caused strained relations between the two republics.

But overall, the early signs were that the country would adjust well to the transformation from a Communist planned economy to a free market economy. Because of its versatile, well-developed industry, the Czech Republic quickly found new trading partners on the international market.

CZECH REPUBLIC AT A GLANCE	
Area 78 703 km² (30 380 sq miles)	
Population 10 389 260 **Per km²** (132) **(Per sq mile** 341.9)	
Capital Prague	
Government Republic	
Currency 1 Czech koruna = 100 haleru	
Languages Czech (official), Slovak	
Religions Christian (Roman Catholic; Protestant minority)	
Climate Continental, with hot summers and cold winters. Rain falls mainly in spring and autumn. Average temperature in Prague ranges from -3°C (26.6) in January to 18°C (64.4°F) in July	
Land use Cultivation 40%, forest 33%, grazing 11%, other 16%	
Main primary products Potatoes, sugar beet, cereals, livestock, hops; lignite, coal, uranium, iron, lead, zinc, silver, tin	
Major industries Iron and steel, transport equipment, machinery, chemicals, fertilisers	
Main exports Machinery, motor vehicles, metals, chemical products, textiles, glass, hops, beer	
Annual income per head (US$) 2370	
Population growth (per thous/yr) 0	
Life expectancy (yrs) Male 69 **Female** 76	

D layer See ATMOSPHERE

Da Nang (Tourane) *Vietnam* Chief port of central Vietnam, 600 km (about 375 miles) north-east of Ho Chi Minh City. It was an important United States military base during the Vietnam War (1965-75), and the base is now used as an international airport. Nearby is China Beach – which inspired a TV series of the same name – where American troops went for rest and recreation during the war.
Population 369 730
Map Vietnam – Bb

Da Yunhe (Grand Canal) *China* The country's longest canal, a 1200 km (745 mile) link between the Chang Jiang River and Tianjin (Tientsin). It was built mainly during the Tang (618-906 AD) and Song (960-1270) dynasties.
Map China – Hd

Dabrowa Gornicza (Dombrowa) *Poland* Southern coal and steel town, 13 km (8 miles) north-east of the city of Katowice. Its first mine was opened in 1796 to work the world's thickest known surface coal seam, 20 m (66 ft) thick. It is now preserved as a geological spectacle.
Population 136 000
Map Poland – Cc

Dacca *Bangladesh* See DHAKA

Dachau *Germany* Market town in Bavaria, about 15 km (10 miles) north and slightly west of Munich. In 1935, the Nazi rulers of Germany opened their first concentration camp here. The camp's site is preserved as a memorial to the estimated 70 000 people who died here and to the many more who were transported to Poland.
Population 35 000
Map Germany – Dd

▼ **A WHIFF OF AFRICA Dakar's Kermel market near the harbour sells mainly flowers and baskets. Like the Sandaga market with its dried fish and spices, it is heard and smelled long before it is reached.**

Daejeon *South Korea* See TAEJON

Dagestan *Russia* See MAKHACHKALA

Dahlak Islands *Eritrea* Two large and 124 smaller Red Sea coral islands off the Eritrean coast. During the Cold War, they provided an anchorage for Soviet warships in exchange for Soviet military aid to Ethiopia, which then occupied Eritrea. Dahlik Kebir, covering 900 km^2 (347 sq miles), is the largest island. Cattle raising and fishing are the main occupations of its several thousand inhabitants.
Map Ethiopia – Ba

Dakar *Senegal* The country's capital, standing on the south side of the Cape VERDE Peninsula and possessing one of the finest Atlantic harbours in the African continent.
Founded by the French in 1857 on the site of a fishing village opposite the island of Gorée, Dakar became the capital of colonial French West Africa in 1902 and of Senegal in 1958. One of West Africa's largest industrial centres, it manufactures footwear, textiles and soap. It also has a naval base and a railway yard.
Tourism is another growing industry in this town which successfully combines the French and African cultures. To emphasis this point, a Roman Catholic cathedral stands next to the old Islamic quarter in which there is a mosque with an 80 m (263 ft) high minaret.
Population 1 500 000
Map Senegal – Ab

Dal, Lake *India* Lake cradled in the folds of the Himalayas in Kashmir, about 645 km (400 miles) north and slightly west of Delhi. It became popular as a cool hill station during the British Raj, and continued to develop into a major tourist centre.
Most visitors stay on richly carved houseboats, and are attended by vendors paddling *shikara*, small vessels similar to Venetian gondolas. There are many floating vegetable beds which are constructed from reeds and cropped to take fresh food to the city. The lake shore is lined with Mogul gardens and old and new mosques.
Map India – Ba

A DAY IN THE LIFE OF A SENEGALESE BUS DRIVER

Sheikh takes his prayer mat out of the locker he has been using for 10 years, ever since he started work as a driver with Dakar's public transport company.

It is 1:20 pm, and he has just finished his seven-hour shift behind the wheel of a 75-passenger bus. He enters the depot yard, to perform his ritual ablutions, say his prayers, and have his lunch of *mafe* – meat and groundnut stew with rice.

After a short rest, he walks to Dakar's main coach station, where his afternoon job begins. Sheikh drives a taxi on a set route from the coach station to the suburb of Pikine, about 20 km (12 miles) out from the centre, and back, stopping whenever a passenger wants to get out, or whenever he is hailed by someone on the street.

Although passengers pay only a small fare, with the taxi nearly always half full to full of people, Sheikh can earn a considerable amount in just a few hours on a busy day, making the work at times highly profitable.

He has to wait only a few minutes at the coach station before his friend Abdoulaye turns up in his battered Datsun. Sheikh takes his place in the driver's seat.

At about 8 pm, after filling the tank, Sheikh returns the car and gives Abdoulaye the agreed rental for the use of the car. He keeps the rest of the money he has made and heads for home.

On his way there, he stops to give some of his earnings to his second wife, Fatou, who still lives in her parents' compound along the road from Sheikh's. She and Sheikh have a baby daughter, Awa. Fatou will continue to live with her father's family until Sheikh can afford a bigger house.

Sheikh has three children by his first wife, Aminata, who is waiting for him when he gets home. Sheikh treats both of his wives equally, although Aminata deserves more respect, socially, because of her seniority.

Sheikh's marriage to Aminata was arranged in his home town of Kaolock, and she brought him a good dowry of clothes and cattle. He met Fatou, his personal choice, in Dakar. Her dowry was straight cash.

The two-bedroomed house that Sheikh lives in is made of cement bricks, and stands in a compound of five houses, shared by five families. Altogether there are 12 adults and 15 children; the two showers and two lavatories are communal. While he sits down to his dinner of fish and rice, Aminata tells him about her day. She does some work plaiting hair and she tells him she did well at the market, where three women came to have their hair plaited. The money will go towards shoes for their eldest child, Ibrahim, aged five, who starts school next year.

As the evening draws on, friends and neighbours who share Sheikh's compound come visiting. A fire is lit in the yard, and tea utensils are brought out. Everyone chats until late into the night. The atmosphere is congenial and lighthearted.

Dalarna *Sweden* Central region covering 29 236 km² (11 288 sq miles) to the north-west of Stockholm. Largely forested, it was once an important copper-mining area and still has old established metalworking and softwood processing industries. The province's main tourist area is around Lake SILJAN.

The people of the province are noted for their traditional and colourful costumes and interesting handicrafts.

Population 286 700

Map Sweden – Bc

dale Broad, open valley. The term is especially applied to such valleys in Yorkshire, Derbyshire and the English Lake District.

Dallas *USA* Commercial and financial city in north-eastern Texas, about 110 km (68 miles) south of the Oklahoma border. It is the eastern part of the Dallas-FORT WORTH conurbation, with more than 4 million people, sharing, with Fort Worth, one of the world's busiest airports. Dallas is the second largest city in Texas and a centre of the oil, electronics and clothing industries. President John F Kennedy was assassinated here in November 1963.

Population (city) 1 006 880; (metropolitan area) 2 676 250; (conurbation) 4 037 280

Map United States – Gd

Dallols, The *Niger* Series of wide, dry valleys which were carved by rivers before the Sahara began to turn into a desert some 5000 years ago. The ancient river courses wind across the north-west from the border with Mali to the Algerian border. In places, acacias and palm trees line the former riverbanks, where water is often close to the surface.

Map Niger – Ab

Dalmatia (Dalmacija) *Croatia* Adriatic coast stretching 375 km (about 230 miles) south-east from the Kvarner Gulf to the Gulf of Kotor. The Dinaric Alps, through which there are few paths, border the region inland. Below the mountains is a narrow coastal strip on which grapes, olives, walnuts and citrus fruits are grown.

Dalmatia is a leading source of cement, marble and aluminium, as well as wine and olive oil. Its sunny beaches, coves and islands, and its charming old towns and villages – many of them once ruled by Venice – generated a booming tourist industry in the past, with RAB, TROGIR, Losinj, KRK, SPLIT and DUBROVNIK being major resorts. However, after civil war broke out in 1991, tourists stayed away.

Map Croatia – Bc-Cc

Damanhur *Egypt* Cotton processing town in the heart of the cotton-growing region of the Nile delta, 61 km (38 miles) south-east of the port of ALEXANDRIA. It stands on the Mahmudiya Canal, which links the Nile to Alexandria.

Population 216 000

Map Egypt – Bb

Damaraland *Namibia* See KUNENE

Damascus (Dimashq) *Syria* Capital of the country, situated on the fringe of the Ghouta oasis, 92 km (57 miles) east of the Mediterranean coast. It is a busy modern city, containing new high-rise hotels, office blocks, apartments, contemporary university buildings and wide, handsome boulevards. Damascus is the third most important Muslim city after Mecca and Jerusalem, and is also believed to be the oldest continuously inhabited city in the world. It was first mentioned in 2500 BC on clay tablets discovered in Mari (an ancient Mesopotamian city, now Tel-Hariri, in Syria). According to these early records, it was the town of Shem (Noah's son) and had become an important city by 1900 BC.

During its 4000-year-long history it was ruled by the Egyptians, Hebrews, Assyrians, Persians, Greeks, Romans, Arabs, Turks and finally the French. It has been the capital of independent Syria since after the end of the Second World War. Built on hills some 690 m (2264 ft) above sea level, it is now a major road and rail crossroads, linked with Amman, Baghdad and Beirut. It also has an international airport.

The city's main industries are silver and metalworking, silk weaving, glass-blowing, cement manufacture and sugar refining.

Among its antiquities are the Great Mosque with a 130 m (426 ft) long prayer hall, built in 708 AD, the tomb of the Muslim political and military leader Saladin (about 1137-93), and the ruins of a Roman temple to Jupiter. But the most important historical buildings date from the 13th century. The old city is largely ringed by ramparts built in the 13th century, and contains a number of Christian churches. In the heart of the modern city is a garden area watered by the Barada River which flows through Damascus.

The museums include the National Museum of Syria and the Museum of Epigraphy (ancient inscriptions), which are housed in a 15th-century *madrassa* (religious school).

Population 1 378 000

Map Syria – Bb

Damavand *Iran* See ELBURZ MOUNTAINS

Damietta *Egypt* See DUMYAT

Dammam *Saudi Arabia* Oil centre on The Gulf coast, situated 400 km (249 miles) east of RIYADH. Oil was found in the area in 1940; in 1953, Dammam became the capital of the easternmost province. The city centre retains a traditional Arab atmosphere with bazaars and *shisha* (pipe-smoking) houses. *Shisha* pipes, along with gold jewellery and antiques, are sold in the main bazaar, the Souk al-Hareem. Here, veiled women can be seen selling cloth – but apart from that women are seldom seen in public. Dammam is the country's second port after JEDDAH.

Population 350 000

Map Saudi Arabia – Cb

Damodar *India* River flowing 519 km (322 miles) from the north-east region of Chota Nagpur down to the plains of Bengal, where it has created its own delta west of the Ganges delta. Its valley has been made into the industrial heartland of India, and since 1948 has been developed by the Damodar Valley Authority. Large dams control floods, store water for industry and agriculture, and provide power for the states of Bihar and West Bengal.

Map India – Dc

Dampier *Australia* Iron ore port, begun in 1965, on the north-west coast of Western Australia, 1250 km (about 780 miles) north of Perth. It has a terminal for natural gas piped here from offshore deposits.

Population 1800

Map Australia – Bc

Danakil *Eritrea/Ethiopia/Djibouti* Arid lowland region, stretching about 500 km (over 310 miles) from north to south, and 150 km (nearly 100 miles) from east to west, between the Ethiopian Highlands and the Red Sea. Its only inhabitants are the Danakil, nomads tending flocks of sheep and goats.

Map Ethiopia – Ba

HARNESSING THE EARTH'S RIVERS

One of the world's earliest major dams was built across the Garawi Valley in Egypt in about 3000 BC. Used to regulate flood waters, it was some 106 m (350 ft) long and was made of packed earth with masonry walls. Egypt is also the home of one of the world's finest modern concrete dams - the Aswan High Dam, 114 m (375 ft) high and 4 km (2.5 miles) long. The world's largest concrete dam and, in fact, its largest concrete structure is the Grand Coulee on the Columbia River in Washington State, in the United States. It contains some 8 092 000 m³ (10 585 000 cu yds) of concrete.

THE WORLD'S GREATEST DAMS

Dam	Location	Statistics
Grand Coulee (largest concrete)	Washington State, USA	(length) 1272 m/4173 ft (height) 167 m/548 ft
Syncrude Tailings (biggest volume)	Alberta, Canada	540 000 000 m³/706 000 000 cu yds
Yaciretá-Apipé (longest)	Paraguay-Argentina	(length) 72 km/45 miles
Nurek (highest)	Russia	(height) 310 m/1017 ft

Danube *Europe* Second longest European river after the Volga, flowing 2860 km (1777 miles) from its source in the BLACK FOREST to its mouth on the BLACK SEA. It passes through, 10 countries – Germany, Austria, Slovakia, Hungary, Slovenia, Croatia, Serbia, Romania, Bulgaria and the Ukraine; four capitals – Vienna, capital of Austria, Budapest in Hungary, Bratislava, in Slovakia, and Belgrade, capital of Yugoslavia (Serbia and Montenegro) – stand on its banks.

The Danube, one of Europe's main water transport arteries, is navigable as far upstream as Ulm in Germany. It has some 300 tributaries and its delta covers about 2600 km² (1000 sq miles).
Map Europe – Ed-Fd

Danube Delta *Romania* See DOBRUJA

Danubian Platform *Bulgaria* Undulating region in the north, lying between the River Danube and the Balkan Mountains and stretching 450 km (280 miles) west to east from YUGOSLAVIA (Serbia and Montenegro) to the BLACK SEA. The platform's western section is crossed by rivers flowing northwards to the Danube. Situated east of VELIKO TURNOVO Province, the region is higher, drier and steppelike. Bulgaria's leading farming region, it produces wheat, maize, sugar beet, sunflowers, fruits and vines. The main towns are Vratsa, Pleven, Lovech, Razgrad, Shumen and Tol-bukhin. Less industrialised than towns in other parts of the country, they process the farm produce and service the farms. Vidin, Ruse and Silistra in the north of the platform are river ports on the Danube.
Map Bulgaria – Ab

Danzig *Poland* See GDAŃSK

Dao Phu Quoc *Vietnam* Island in the Gulf of Thailand, 40 km (25 miles) off the south-west coast. It is vital to Vietnam's claims to fishing and exploration rights in the gulf.
Map Vietnam – Ac

Daphni *Greece* Hilltop monastery dating from the 5th and the 6th century, 11 km (7 miles) west of Athens. Its 11th-century Byzantine church has some of the finest mosaics in Greece. There is a wine festival every September and October.
Map Greece – Cb

Daqing *China* Oil field of Heilongjiang Province, and the biggest field in China, which in the 1980s accounted for over a million barrels a day, almost half the country's output. Oil was discovered here in 1956.
Map China – Ib

Dar El Beida *Morocco* See CASABLANCA

Dar es Salaam *Tanzania* Chief seaport, and national capital of Tanzania – a status it shares with DODOMA. It lies 45 km (28 miles) south of the island of Zanzibar (which joined with Tanganyika to become Tanzania in 1964).

Dar es Salaam, whose name is of Arabic-Swahili origin and means 'Haven of Peace', was founded in 1862 by the Sultan of Zanzibar. From 1891 to 1916, it was the administrative centre of German East Africa. It was captured by the British in 1916, during the First World War. Today, its port is linked by rail to Zambia and handles most of Tanzania's exports. It also handles much of

land-locked Zambia's trade. The city's industries include oil refining and paint manufacture.
Population 1 657 000
Map Tanzania – Ba

Dar'a (Deraa) *Syria* Town and major road and rail junction, 96 km (60 miles) south-west of Damascus and 8 km (5 miles) north of the border with Jordan.
Population 282 000
Map Syria – Bb

Dardanelles (Canakkale Bogazi) *Turkey* Narrow strait, 72 km (45 miles) long and 1.2-8 km (0.7-5 miles) wide, linking the Aegean Sea with the Sea of Marmara. Called the Hellespont by the Ancient Greeks, it has been of great strategic and commercial importance throughout recorded history. Many battles were fought here as numerous nations struggled to gain control of the strait. The city of Troy prospered at its western entrance; the Persian king Xerxes crossed it with a bridge of boats in an unsuccessful attempt to conquer Greece in 480 BC. During the Cold War, it was an important outlet for the USSR's Black Sea fleet – and provided access to the few Soviet ports that were not frozen during winter.

The GALLIPOLI Peninsula on its northern shore was the scene of bitter fighting between Allied forces and Turks in 1915-16, during the First World War.
Map Turkey – Aa

Darfur *Sudan* Western region divided into two provinces – Northern and Southern Darfur – covering 317 809 km² (122 706 sq miles) along the Chad border. Its features include a 600-900 m (2000-3000 ft) plateau in the east, low hills in the west and north, and volcanic uplands – the Marra Mountains rising to 3024 m (9921 ft) – in the centre. Many of the inhabitants are nomadic horsemen, who raise sheep and camels.

Gum arabic, a resin used as a thickener in jellies and glues, is produced here, as are groundnuts, maize, millet, and vegetables. The regional capital is El Fasher.
Population (region) 3 094 000
Map Sudan – Ab

Darhan *Mongolia* The country's second largest city. It lies in the Haraa Gol Valley on the Trans-Mongolian Railway, 188 km (117 miles) north of the capital, Ulan Bator. Coal has been mined here since the 1960s. Cement, firebricks, textiles, and iron and steel are also produced. Darhan's coal-burning power station provides electricity for northern and central Mongolia.
Population 88 600
Map Mongolia – Db

Darién *Panama* Province occupying the eastern section of the Panama isthmus, and bordering Colombia. Its lightly populated, hot, wet forest country is the only point where there is a break in the Pan-American Highway which runs from Alaska to Chile. A road that will bridge the gap is slowly being driven through its remote jungles which are populated by a few dozen Chocó people.
Population 43 830
Map Panama – Ba

Dariya (Dir'aiyah) *Saudi Arabia* Ruined town and former capital, situated 30 km (19 miles) north-west of RIYADH. In 1818, Dariya – then the

capital of the Al Saud dynasty and refuge of the fundamentalist Islamic sect called the Wahhabis – was destroyed by Ottoman-Egyptian forces led by Ibrahim Pasha, son of Mohammed Ali, the ruler of Egypt. The Ottoman sultan had instructed Pasha to open the way to Mecca for pilgrims after the Wahhabis had closed the holy city to all other Muslims. Among the town's remains are the old fort and palace.
Map Saudi Arabia – Bb

Darjiling (Darjeeling) *India* Town and hill station in the north of the Indian state of West Bengal, close to the Nepalese border. It stands at 2134 m (7001 ft) on a mountain ridge and is linked to the plains below by a narrow-gauge railway. Darjiling became the summer capital of the government of Bengal during British rule. It is renowned for the tea estates that developed around it, which provide some of the world's finest brews. The town has unrivalled views of some of the main Himalayan peaks, including KANGCHENJUNGA.
Population 1 335 620
Map India – Db

Darling *Australia* An inland river system draining an area of southern Queensland and western New South Wales, comprising 650 000 km² (247 905 sq miles). For the major part of its 2000 km (1243 mile) course the river runs across arid plains and its flow is erratic – low in times of drought, which can last for up to three years, and flooding the surrounding countryside in times of heavy rain. It is navigable only in certain sections and at certain times. The river has several names, one of which is Barwon, which it is called before it becomes the Darling upstream from Bourke.

The Darling joins the Murray near Wentworth, and together the two rivers form the country's longest river system.
Map Australia – Ge

Darling Downs *Australia* Fertile farming district in Queensland, west of the GREAT DIVIDING RANGE to the west of Brisbane. It is drained by tributaries of the DARLING River and produces many grain crops and cattle and sheep. There are oil and natural gas deposits and the area has vast reserves of coal. The district's main towns are Warwick (population 10 390) and TOOWOOMBA.
Map Australia – Hd

Darmstadt *Germany* Former capital of the grand duchy of Hesse (the capital of Hesse was moved to Wiesbaden after the Second World War) and today the capital of South Hesse. It is 27 km (17 miles) south of FRANKFURT am Main, in the centre of a wine-producing district. Darmstadt produces machinery, chemicals and petrochemicals. It is also a cultural and scientific centre.
Population 178 900
Map Germany – Cd

Darnah *Libya* See DERNA.

Dartmoor *United Kingdom* Hilly moor in the county of Devon in south-western England. Much of it is taken in by the Dartmoor National Park, which covers 945 km² (365 sq miles). The rolling moorland and bogs are occasionally broken by rocky outcrops called TOR, the highest of which is High Willhays at 621 m (2037 ft). Sheep and wild ponies graze on the moor, but they are under threat from the increasingly heavy tourist traffic.

Dartmoor Prison, originally built in the early 19th century to house French prisoners of the Napoleonic Wars, is at Princetown, the largest town on the moor.

Map United Kingdom – De

Darwin *Australia* Capital of the Northern Territory and the main port of northern Australia. Founded in 1869, it was initially known as Palmerston. In 1911, it was renamed Darwin, after the British naturalist Charles Darwin (1809-82); the adjacent bay was then already called Port Darwin.

Darwin was the headquarters of Allied forces during the Second World War and was bombed by the Japanese in 1942. It was almost completely destroyed by a cyclone on Christmas Day 1974, but has since been rebuilt.

Population 78 140

Map Australia – Ea

Datong (Tatung) *China* Coal-mining and industrial city of Shanxi Province, about 280 km (175 miles) west of BEIJING. It is an important railway and locomotive assembly centre.

Population 1 110 000

Map China – Gc

datum level The zero from which altitudes on land and depths of the sea are measured.

Daugavpils (Dvinsk) *Latvia* Eastern Latvian town, lying 220 km (136 miles) south-east of Riga. Built on the West DVINA (or Dangava) River, it is a trading centre for grain, flax and timber. Its main industries are railway engineering, textile manufactures and food processing. Until 1893 it was the German town of Dünaburg.

Population 124 800

Map Latvia – Ba

Dauphiné *France* Historical region and former province in south-eastern France. Philip VI of France bought it in the 14th century from the count whose title was Dauphin Humbert II, on condition that the heir apparent to the French throne should adopt the title attached to the land and govern Dauphiné as a separate province.

The region's main occupations today are farming, forestry, tourism, and metal and electronic industries using hydroelectric power. GRENOBLE is Dauphiné's largest city.

Map France – Fd

Davao *Philippines* Region, comprising three provinces, and city in south-central MINDANAO. Long ruled by the sultans of Mindanao, the region became a Spanish-controlled province only in 1849.

Davao City is the Philippines' second largest urban area, and with an overall land area of 1937 km² (748 sq miles), it is one of the world's most extensive cities. It is a prosperous trade centre and port for the region, which produces much maize, Manila hemp (an industry begun by Japanese early this century), pineapples, bananas, timber and other crops, as well as fish from Davao Gulf.

At the end of the Second World War there was fierce fighting between American and Japanese forces before Davao City was liberated. About 30 km (18 miles) to the south-west is Mount Apo, a dormant volcano and the Philippines' highest mountain, at 2954 m (9690 ft).

Population (region) 1 825 000; (city) 850 000

Map Philippines – Cd

Davis Strait Broad strait, 640 km (400 miles) long and 290-640 km (180-400 miles) wide, named after the English navigator John Davis who discovered it in 1587 while searching for the northwest passage. It lies between BAFFIN ISLAND and Greenland (Kalaallit Nunaat), and is an arm of the ATLANTIC OCEAN. It is navigable in late summer and autumn, although the Labrador Current sweeps ice down its western shore.

Map Canada – Jb

▼ ROYAL FOREST Rocky outcrops dot Yellowmead Down on Dartmoor, once a 'royal forest' – a hunting preserve of Saxon kings. It was the setting for Conan Doyle's classic Sherlock Holmes thriller *The Hound of the Baskervilles.*

Davos *Switzerland* One of the largest Swiss mountain resorts, at 1560 m (5118 ft) altitude and about 115 km (72 miles) south-east of the city of Zürich. Its height and its dry, fog free climate have made it a world-famous health resort, especially for sufferers of lung diseases and allergies. Davos has also become famous as a conference and winter sports centre. It has the largest ice rink in Europe, and a pretty lake (called the Davosersee) at the end of its valley.
Population 12 680
Map Switzerland – Ba

Dawson *Canada* Settlement in the Yukon Territory at the junction of the Yukon and Klondike rivers, which was the focus of the Klondike gold rush in 1896. The population rose from a few hundred to nearly 40 000 as prospectors poured in and Dawson became a boom town. It subsequently declined, but is still a mining and tourist centre.
Population 970
Map Canada – Bb

Dayr az Zawr *Syria* Town on the EUPHRATES, 272 km (169 miles) south-east of Aleppo, and capital of the province of the same name, which covers 33 060 km² (12 762 sq miles). Dayr az Zawr is the most important town in eastern Syria, and a vital road junction between the region and southern Turkey and Iraq. Its main industries, based on the agricultural products from the province, are flour-milling, cotton textile manufacturing, and tanning. Since the completion of the Euphrates Dam in 1978, agriculture has brought new prosperity to the town and region.
Population (city) 430 000
Map Syria – Cb

Dead Sea *Israel/Jordan* Salty body of water, called Bahret Lut ('the Sea of Lot') in Arabic, lying on the Israel-Jordan border in the arid northern extension of the GREAT RIFT VALLEY. The Dead Sea covers about 1020 km² (394 sq miles), depending on the inflow, and is up to 400 m (1312 ft) deep in the north, where the River Jordan enters, but only 2 m (6.6 ft) deep in the south. There is no outlet, and water is lost only through evaporation – hence its salinity. It is said that the Dead Sea contains eight times as much salt as any of the world's oceans. It has a salt content of about 28 per cent, and is 'dead' indeed, for no fish can live in its waters. Salt, potassium, magnesium and bromine are extracted from it at a nearby Israeli chemical plant. The lake's surface is the lowest point on the earth's surface – 396 m (1299 ft) below the Mediterranean.

The Jewish sect that left the Dead Sea Scrolls lived in caves to the north-west.
Map Israel – Bb

Dean, Forest of *United Kingdom* Ancient English forest consisting mainly of oak and beech trees and covering some 110 km² (42 sq miles) between the Severn and Wye rivers south-west of Gloucester. Since ancient times, the local people, known as Foresters, have mined iron ore and coal, and grazed their livestock in the forest which is now a forest park.
Map United Kingdom – De

Death Valley *USA* Deep basin in south-eastern California and south-western Nevada. It is the lowest point in the western hemisphere, lying 86 m (282 ft) below sea level, and is a desert area of salt beds and borax formations. The valley is about 225 km (140 miles) long, and covers 8399 km² (3243 sq miles). It forms part of the largest national monument in the United States. Temperatures here have reached 57°C (134°F), the hottest recorded on the North American continent, and though rainfall is less than 51mm (2 in) annually, there is a surprising variety of fauna and flora, including over 100 bird species.
Map United States – Cc

Deauville *France* Elegant coastal resort of Lower Normandy, on the Seine estuary opposite Le Havre. It has golden beaches, a casino, a marina and a racetrack. About 15 km (9 miles) to the west is Dives, where William the Conqueror assembled his fleet for the invasion of England in 1066. In the church is a list of the noblemen who accompanied him.
Population 4380
Map France – Db

Debrecen *Hungary* Economic and cultural focus of eastern Hungary and chief town of the county of Hajdú-Bihar, situated about 190km (120 miles) east of Budapest. The largest city on the Great ALFÖLD Plain, it forms the centre of a rich crop-growing and stock-raising area, and produces foodstuffs, chemicals and various electrical goods.

Medieval Debrecen, protected by surrounding low-lying marshes and lakes, survived Turkish threats and Catholic domination by its Austrian Habsburg rulers to become Hungary's Protestant capital – the 'Calvinist Rome'. It was here that the nationalist leader Lajos Kossuth (1802-94) declared Hungary's independence from Austria in 1849.

The old city has the air of an overgrown village, with many ancient buildings, including the Old County Hall, decorated with ornaments of majolica – a type of coloured earthenware. It is also noted for its 16th-century Calvinist College, now a university, the Great Church, and the colourful Arany Bika ('Golden Bull') Hotel, built in art nouveau style.
Population 216 130
Map Hungary – Bb

Deccan *India* Name given to the Indian Peninsula, and more especially to the rocky plateau that forms most of it. Its highest point is Anai Mudi (2695 m, 8842 ft). Parts of it receive

▼ SEA OF SAND A tract of sand dunes laps the rocky walls of Death Valley – North America's hottest and driest place, where in some years there is no measurable rainfall and summer daytime temperatures often exceed 49°C (120°F)

little rain, but there are a few small dams and reservoirs which allow a limited amount of rice to be grown. The wetter districts produce grains such as millet, and cotton is also grown, especially in the south. An international research centre on semi-arid tropical crops has recently been set up in the city of HYDERABAD.
Map India – Bd

deciduous forest Forest consisting mainly of broad-leaved trees that lose their leaves at one season of the year. In a monsoon forest, as in South Asia, the leaves are shed during the hot, dry season to protect the trees against excessive loss of moisture by transpiration. In cool temperate regions, such as north-west Europe, Chile and New Zealand, leaves fall during the autumn and trees remain dormant during the winter.

189

Dedza *Malawi* Town in Central Province, 80 km (50 miles) south-east of the capital, Lilongwe, near the border with Mozambique. It is the centre of an agricultural region producing tobacco, timber, and subsistence food crops, such as maize. It stands at the foot of the beautiful, wooded Mount Dedza, which rises to 2170 m (7120 ft) and has fine views of Lake Malawi.

Population 2300
Map Malawi – Ab

deepening In meteorology, term used for decreasing atmospheric pressure at the centre of a DEPRESSION. It is the opposite of FILLING.

deforestation Removal of (virgin) forest, normally by a process of felling and clearing of wooded areas, usually to extend agriculture. It is the opposite of AFFORESTATION.

Deganya *Israel* Israel's first kibbutz, established in 1909 by Russian Jewish immigrants. It lies near the south-west shore of the Sea of Galilee (Lake Tiberias), about 25 km (15 miles) east of Nazareth. The ruined Canaanite city of Bet Yerah is nearby.

Map Israel – Ba

degradation See DENUDATION

Dehra Dun *India* Town in a valley at the foot of the Himalayas, about 236 km (147 miles) north-east of Delhi. It stands 700 m (2296 ft) above sea level and according to legend, was the home of Siva, the Hindu god of destruction and reproduction. The British established a military academy here in 1934. The Indian president's Sikh bodyguard is also based in the town.

Population 1 014 700
Map India – Cb

Delaware *USA* State on the Atlantic coast. It covers 5057 km² (1952 sq miles) on the east of the Delmarva Peninsula between Delaware and Chesapeake bays and is the country's second smallest state, after Rhode Island. Delaware was first permanently settled by Europeans in 1638, and was one of the original 13 states. Poultry, soya beans, maize and dairy produce are its agricultural mainstays, and chemicals its chief industry. The state capital is DOVER, which was part of a grant to the Quaker William Penn in 1682.

Population 622 000
Map United States – Kc

Delaware Bay *USA* Inlet, 84 km (52 miles) long and up to 56 km (35 miles) wide, between NEW JERSEY and DELAWARE on the United State's eastern seaboard, which was discovered by the English navigator Henry Hudson in 1609. It was easily approachable and thus became an important anchorage in the early days of European colonisation. The Delaware River runs into it, enabling ocean-going vessels to reach the inland ports of Wilmington and Philadelphia.

Map United States – Kc

Delft *Netherlands* City 55 km (34 miles) south-west of Amsterdam, which has been renowned for its fine glazed pottery, normally decorated in blue, since the end of the 16th century. In the Middle Ages cotton prints – indigo and white – were made here. Today, the city's manufactures include spirits, cables and penicillin.

Delft was the home town of the artist Jan Vermeer (1632-75), and much of the old city still resembles scenes in his paintings, with its canals, old churches, market place and the 1619 town hall. William the Silent, Prince of Orange (1533-84), the great-grandfather of the prince who became William III of England in 1689, lived at the Prinsenhof, which is now a museum. He was also assassinated there and is buried in the New Church at Delft.

Population 91 010
Map Netherlands – Ba

Delhi *India* Capital city of the Federal Republic of India, situated on the banks of the Yamuna River, on the borders between the northern states of Uttar Pradesh and Haryana. The city is in the strategic region linking the INDUS and GANGES valleys, bounded on the south by the Aravalli range and on the north by the Himalayas. It is just south of the town of Panipat, where many renowned battles were fought between invading powers from the North-West Frontier and the civilisations of the Ganges Plains.

The city has two distinct parts – Old Delhi and New Delhi. Old Delhi is a walled Muslim city centred on the Red Fort built by Shah Jehan between 1636 and 1658. Its streets are narrow and bustling; its beauty and serenity lie inside the courts of the main buildings.

New Delhi was proclaimed the capital of India by the British in 1911, although the newly built city was not officially inaugurated until 1931. It was designed by the British architect Sir Edwin Lutyens (1869-1944), and is graceful, tree-lined and spacious. New Delhi centres around Rajpath Avenue, which is 5 km (3 miles) long and 0.8 km (0.5 mile) wide and flanked on both sides by government buildings. Known previously as the Kingsway, it leads to the Rashtrapati Bhawan, the former Viceroy's Palace, now the residence of the president of India.

To one side is the parliament, the Lok Sabha. Just to the north is Connaught Place, an elegant ring of colonnaded shops and offices. In the southern part of New Delhi are Palam Airport, new universities and the arenas built in 1982 for the Asian Games.

Places of interest include the Red Fort, the Jama Majid Mosque, the Rashtrapati Bhawan (outside only) and Rajpath Avenue, the Qutb Minar, at 73 m (240 ft) high thought to be the highest free-standing stone tower in the world, the ruined fortified city of Tughlaqabad, and the astronomical observatory of Jai Singh, the Maharajah of Jaipur.

For all its historical buildings, Delhi has its dark side, too. In 1948, Mahatma Mohandas Gandhi, who had been trying to achieve a harmonious relationship between Muslims and Hindus, was assassinated in Delhi by a Hindu fanatic. A national shrine now marks the place of his cremation at Ram Shat. In 1984, Mrs Indira Gandhi, Prime Minister of India and the daughter of Pandit Nehru, was assassinated in Delhi by militant Sikhs.

The city's industries – many of them located in satellite townships in the outlying areas – include chemicals, textiles, rubber goods and various handicrafts.

Population 9 370 470
Map India – Bb

dell Small and secluded wooded valley.

A DAY IN THE LIFE OF AN INDIAN MONEYLENDER

Anjali stands at the doorstep of her brick house in a village near Delhi and surveys the dusty street. Breezes and passing feet keep blowing dust into her doorway. It is an ongoing battle to keep her home relatively free from dust and her servant will have to brush it out again.

Her servant is a *harijan* woman – an 'Untouchable' – one of a group of people traditionally regarded as unclean, who are the lowest members of the Hindu caste system. The use of the term 'untouchable' was declared illegal in India in 1949 and the modern constitution recognises their plight but full emancipation has not yet been achieved.

As a moneylender, Anjali's livelihood depends on the *harijans*, because they are too poor to have any possessions to pledge as security, so that banks refuse to lend them money. When the breadwinner in a local *harijan* family falls ill, the family are forced to borrow money for food from Anjali.

Often, too, they have to borrow for dowries for their daughters. And the *harijan* stonebreakers who live in Anjali's village have little chance to get out of debt because of the high prices charged by contractors to transport the stone they quarry.

Anjali feels justified in charging 25 per cent interest because of the risk in lending to people who have no security, though it means that her creditors are often in debt to her for the rest of their lives.

Three of Anjali's seven children still live at home, aged between 12 and 16. Her husband died 10 years ago, so she has to support them alone. Were it not for her moneylending, she would have to depend on her two eldest sons – one a refrigeration engineer, the other a newspaper reporter. But they, and her two eldest girls, are married, and have children of their own to look after.

Soon Anjali starts to tour the village. She checks if everyone who owes her money is working, and if anyone is ill and unable to keep up the payments (in which case she will have to raise their interest once payments are resumed).

As she does her rounds today she thinks of her latest worries. Recently, an aid project has bailed out some of her stonebreaker creditors, giving them the money, in an interest-free loan, to pay off their debts in one go.

The aid workers have also shown the stonebreakers how to form cooperatives for transporting their stone themselves. This means that they do not have to pay outside contractors at the high rates that force them to borrow. At the same time there are women from upper castes in Delhi who encourage the *harijans* to aspire beyond their traditional place as the outcasts of Indian society.

It makes Anjali nervous to think that the whole order of things might disrupt her own life. She consoles herself with the thought that change always occurs slowly in India. Her livelihood surely must be safe for some time yet.

Delos (Dhílos) *Greece* Smallest group of the Cyclades Islands and legendary birthplace of the god Apollo. Delos has the most extensively excavated archaeological site in Greece, where ancient sanctuaries, temples, colonnades and monuments in an area 1200 m (0.75 mile) long have been uncovered.

The island was a Greek political and religious centre as early as the 7th century BC. During Roman times, Delos became a centre for the slave trade; thereafter, it was almost entirely deserted. Today, this little group of islands attracts many tourists.

Map Greece – Dc

Delphi *Greece* Impressive ruins of the Temple of Apollo and seat of Apollo's oracle, built into the craggy slopes of Mount Parnassus, 166 km (102 miles) north of the capital, Athens. It was founded in the 2nd millenium BC and its oracle became the most important in classical Greece. It centred on a woman named Pythia, who prophesied to those who sought the wisdom or favour of the god. However, her communications were usually so obscured in poetry and riddle that the help of priests and other prophets were usually needed to decipher the message.

The English word 'Delphic', for an ambiguous statement, comes from the cryptic utterances of the soothsayers and priests of Delphi. One such example was 'The wooden walls will save Athens from the Persians', which is thought to have referred to the Greek fleet that defeated the invasion fleet of the Persian king Xerxes at the Battle of Salamis in 480 BC.

Later, Delphi itself sank into obscurity for many centuries and was only rediscovered in the 19th century. Since then the French have excavated and restored the site, consisting of the temple, a theatre, stadium and market place. All of these, however, are overshadowed by Mount Parnassus, which at 2457 m (8061 ft) overlooks the olive-forested ravine above the Itea Plain. A museum at the site has many findings from the ruins, including a bronze charioteer by Sotades (4th century BC).

Rugs and embroidery are made at Arákhova 9 km (5.5 miles) to the east.

Map Greece – Cb

delta An alluvial area, often triangular, at the mouth of a river. It forms where the flowing water is slowed down on contact with the sea or lake and the river's capacity to transport material is suddenly reduced. In the absence of sea – or lake – currents strong enough to carry the material away, sediments build up in the river mouth, causing repeated subdivision of the river channels as each new channel deposits its load.

Demerara *Guyana* River, 320 km (200 miles) in length, which flows into the sea at the capital, Georgetown. Farmers along its banks grow the brown cane sugar named after it.

Map Guyana – Ba

Demilitarised Zone (DMZ) *North Korea/South Korea* No-man's land boundary area between North and South Korea. The zone straddles the cease-fire line agreed to at the end of the Korean War (1950-3). It is 4 km (2.5 miles) wide and winds across the Korean Peninsula from coast to coast for 243 km (150 miles).

Map Korea – Cc

Den Bosch *Netherlands* See 'S-HERTOGENBOSCH

Den Haag *Netherlands* See HAGUE, THE

Denali National Park *USA* Area of 24 413 km² (9426 sq miles) in southern Alaska. It centres on Mount McKinley, which the local Indian Americans call Denali, and was formerly called the Mount McKinley National Park. The mountain, North America's highest peak, rises to 6194 m (20 320 ft). The park, which is open to visitors during the summer, has many lakes and glaciers, spectacular mountain scenery and tundra landscape, and wildlife including grizzly bears, mountain sheep, moose and caribou.

Map Alaska – Bb

Denbigh *United Kingdom* See CLWYD; GWYNEDD

dene Term applied to a narrow wooded valley, and also, to a sandy stretch of land (or dune) near the sea.

Denmark See p 194

Denmark Strait Sea passage, 480 km (300 miles) long and 290 km (80 miles) wide, between Greenland (Kalaallit Nunaat) and Iceland. It passes over a sea trench with a depth of 200 to 450 m (656 to 1476 ft) and contains the cold East Greenland Current.

Map Atlantic Ocean – Bc

Denpasar *Indonesia* Seaport and capital of the island of BALI.

It is a market and light industrial centre. Its museum contains many art treasures, and just outside the town is the Bali Cultural Centre with an art gallery and large, open-air theatre. Nearby are the tourist resorts of Kuta and Sanur.

Population 82 100

Map Indonesia – Ed

denudation The wearing down of the land by the combined processes of weathering, mass movement, transport of loose material, and erosion by wind, ice and water. Armed with the loose material they are transporting, they erode the surface of the ground over which they move. The process is also referred to as 'degradation'.

Denver *USA* State capital of Colorado. Because it lies at an altitude of 1610 m (5280 ft), just east of the Rocky Mountain foothills, it is called 'the Mile High City'.

Denver was founded during a gold rush in 1858; its manufactures today include aircraft, chemicals and electronics, and its new high-tech airport is expected to become the second busiest in the United States before the turn of the century. The city has the largest centre of federal government operations outside the national capital, Washington DC; they include the US mint.

Denver is also an important tourist centre, internationally famous for its ski resorts within driving distance.

Population (city) 622 980; (metropolitan area) 1 848 300

Map United States – Ec

depression (cyclonic disturbance) Near circular region of low pressure in the atmosphere in which pressure increases outwards and into which winds spiral (anti-clockwise in the north-

ern hemisphere, and clockwise in the southern hemisphere) as the depression fills. Depressions move more rapidly than ANTICYCLONES, bringing unsettled weather, stronger winds, fronts and PRECIPITATION. The winds circulate in the opposite direction to those in an anticyclone.

Dera Ismail Khan *Pakistan* Town on the west bank of the Indus River, situated at a strategic crossing point at a pontoon bridge, about 330 km (205 miles) west of the city of Lahore. It was founded in 1469 by Ismail Khan, a Baluchi ruler, but was later destroyed by floods and refounded in 1823.

The town produces fine woodwork, ivory and glassware, has a lively bazaar, and is a trading centre for grain and oilseeds.

Population 96 000

Map Pakistan – Cb

Derby *Australia* See KIMBERLEY PLATEAU

Derby *United Kingdom* City and county of north-central England. The city is 55 km (35 miles) north-east of Birmingham. It has been renowned since the 18th century for its silk and Crown Derby porcelain, and is also the headquarters of Rolls-Royce aircraft engines. It is also an important centre for the assembly of motorcars. The city's Church of All Saints was raised to cathedral status in 1927.

Derby county stretches north to the outskirts of Manchester and Sheffield, and covers 2631 km² (1016 sq miles). It includes the coal fields around Chesterfield and the hills of the southern Pennines – a major tourist area which contains the PEAK DISTRICT National Park, rising to 636 m (2086 ft) at Kinder Scout, and the spa towns of Buxton and Matlock Bath.

Population (county) 943 100; (city) 225 400

Map United Kingdom – Ed

Derna (Darnah) *Libya* Second most important port of CYRENAICA after Benghazi, situated in the north-east, about 100 km (62 miles) east of Al Bayda.

In the 15th century, Spanish Jews sought refuge here; later it became a pirates' stronghold. The town was the scene of much fighting during the Second World War when it changed hands several times.

In recent years the port has been reconstructed and modernised. Industries include cement, food processing, and Libya's first ready-to-wear clothes factory. The nearby oasis produces dates and bananas.

Population 105 000

Map Libya – Ca

Derry *United Kingdom* See LONDONDERRY

Des Moines *USA* Capital and largest city of Iowa. It stands in the middle of the state on the Des Moines River, in the heart of the feed grain and livestock belt. Des Moines is a financial and cultural centre, with a university. Its manufactures include farm machinery and tools. It also has thriving food processing, and rubber and printing industries.

Maize and dairy products are produced in the surrounding area.

Population (city) 193 200; (metropolitan area) 377 100

Map United States – Hb

desert Area where evaporation is greater than precipitation – in the form of rain or snow – with the result that only very scant vegetation is possible. See FEATURES OF THE DRY DESERTS (p 193).

Different types of deserts form in different parts of the world. There are hot and cold deserts. The SAHARA is a hot, trade wind desert – a region of high atmospheric pressure, where air which has risen and shed its moisture in the tropics descends to earth, and is warmed and dried as it is moved towards the equator by the trade winds. Rainfall here is less than 250 mm (10 in) a year.

Other hot deserts, such as the NAMIB and ATACAMA, occur on the western margins of the continents where cold currents flow. Local onshore winds are warmed on crossing the coast and retain their moisture. The GOBI, on the other hand, lies in the continental interior far from the sea.

In the hot deserts, the high temperatures cause high evaporation. Cloudless skies give rise to rapid heat loss from the earth by radiation after sunset. Temperatures may soar to over 37°C (100°F) during the day, then fall below freezing point at night. Rain occurs in isolated showers, sometimes years apart.

Cold deserts occur in polar regions such as Antarctica and Greenland (Kalaallit Nunaat) – high pressure areas of descending air and outflowing winds through the year.

desert climate (hot) A climate characterised by high temperatures, scanty, irregular rainfall and evaporation so high that the rainfall is insufficient to support vegetation typical of other regions in the same latitude.

desert pavement Desert surface of closely packed smooth pebbles, or of BEDROCK, where winds have removed fine material between them and wind-borne sand has ground and polished their tops.

desert vegetation Sparse covering of drought-resistant plants that survive in areas of very low and irregular rainfall. Drought-resistant plants are adapted in various ways to withstand prolonged dry spells. They often have small, waxy leaves and thick, fibrous bark to reduce transpiration, bulbous trunks to store water, and extensive root systems to locate GROUND WATER.

Dessau *Germany* Industrial town in Sachse-Anhalt, about 100 km (60 miles) south-west of Berlin. It stands at the confluence of the Mulde and Elbe rivers. The town was devastated by Allied bombing during the Second World War because it was a centre of the armaments industry. Machinery and beer are produced here.

Population 97 800
Map Germany – Ec

destructive plate margin Another name for a BENIOFF ZONE. See also PLATE TECTONICS.

Detmold *Germany* Town situated on the northern flanks of the TEUTOBURGER WALD, 80 km (50 miles) south-west of Hannover. Detmold's 16th-century castle was the home of the princes of Lippe, the local rulers. On a hill 6 km (4 miles) south of the town is a huge monument to Arminius, or Hermann, a local hero who annihilated three Roman legions in the district in 9 AD.

Population 69 300
Map Germany – Cc

THE ARID FACE OF THE EARTH

About one-third of the world's land surface is made up of deserts, both hot and cold, in which the land is partially or totally barren, as in the Sahara and in the polar regions. One of the world's largest and most arid hot deserts – the Rub al-Khali in Arabia – has droughts which sometimes last for years. At the other – cold – extreme are the Dry Valleys, or McMurdo Oasis, in Victoria Land, Antarctica, where it is believed that no rain has fallen for 2 million years.

THE WORLD'S LARGEST DESERTS

Desert	Location	Area (km²/sq miles)
Sahara	Africa	9 100 000/3 500 000
Great Australian	Australia	3 830 000/1 480 000
Gobi	Asia	1 295 000/500 000
Kalahari	Southern Africa	932 400/360 800
Rub al-Khali	Arabian Peninsula	650 000/251 000
Peski Karakumy (Karakum)	Turkmenistan	340 000/130 000
Taklimakan	China	327 000/126 000

Detroit *USA* Largest city in Michigan. It is the headquarters of the General Motors, Ford and Chrysler car companies and is sometimes called 'The Car Capital of the World' – or 'Motor City', or 'Motown'. It stands on the Detroit River, one of the world's busiest inland waterways, between Lake Erie and Lake St Clair, across the river from the Canadian city of Windsor.

Detroit was founded in 1701. It was captured by the British in 1760 and was ceded to the United States only at the end of 18th century. Its strategic importance led to rapid expansion soon after the Erie Canal was opened in 1825. Today, it is a busy industrial port. Much redevelopment has taken place, especially along the waterfront.

Population (city) 1 028 000; (metropolitan area) 4 665 000
Map United States – Jb

Deva *Romania* Metalworking town, regional capital and important road and rail junction on the River Mures. It lies in the west of the country, 145 km (90 miles) north of the border with Serbia and Montenegro. Deva's 13th-century fortress was held by the national hero Michael the Brave, who briefly united the provinces of Moldavia, Transylvania and Walachia before his defeat by the Turks at Deva in 600. A museum of archaeology housed in a 17th-century castle has Dacian and Roman remains. Deva manufactures building materials and processes food.

Population 82 300
Map Romania – Aa

Deventer *Netherlands* Town founded in the 8th century on the IJssel River, 87 km (54 miles) south-east of Amsterdam. It was a medieval centre of learning, and has several old churches and a library with manuscripts dating from the 10th century. Its old town is considered to be one of the oldest in the Netherlands – and one of the most beautiful. Deventer produces tapestries, carpets, bicycles and *deventerkoek* (honey gingerbread), as well as textiles and bricks.

Population 68 530
Map Netherlands – Ca

Devon *United Kingdom* County of south-west England covering 6715 km² (2593 sq miles). Its north coast, on the Bristol Channel, has steep cliffs while the south, on the English Channel, is deeply indented with many natural harbours. They include the naval base of PLYMOUTH.

Devon has a long tradition of seamanship. Sir Francis Drake (about 1540-96), who sailed round the world and later defeated the Spanish Armada, was from Devon. It also has rich farmland and is famous for its cattle and clotted cream. Tourists flock to TORBAY and to the uplands of DARTMOOR and EXMOOR, both of which are national parks. The county town is the cathedral city of EXETER.

Population 1 038 700
Map United Kingdom – De

Devonian Fourth period in the Palaeozoic era of the GEOLOGICAL TIMESCALE, named after the English county of Devon where marine rocks of the period were first studied.

dew Water droplets on the ground or on plants, which have condensed from water vapour in the air, usually during a clear night when the surfaces of the ground or plants have cooled below the dew point of the air. See CONDENSATION.

dew point The temperature at which air, on cooling, becomes saturated – the state when it is holding as much water vapour as is possible at that temperature. Any further cooling leads to CONDENSATION, or the formation of water droplets.

Dhahran *Saudi Arabia* Commercial and business centre situated 5km (3 miles) south of Dammam. Dhahran is less a town than a giant complex of oil company offices, schools, hospitals and recreational facilities. It houses the country's Petroleum and Minerals University, an eye-catching piece of contemporary architecture. It is also the site of the provincial offices of the Ministry of Petroleum and PETROMIN – the domestic oil marketing organisation – and the Oil Exhibit Centre depicting the formation, exploration and processing of oil as well as the history of Saudi Arabia's oil industry.

Population 25 000
Map Saudi Arabia – Cb

Dhaka (Dacca) *Bangladesh* The country's capital, Dhaka was the former capital of Bengal, under the Mogul emperors of India for over 100 years from 1575. Its most interesting buildings date from this period, including the Lal Bagh Fort and several mosques. Other striking buildings, including the supreme court and older parts of the university, date from 1905-11 when, under the British, the city was the capital of East

Bengal. During this period, beautifully shaded avenues were created, though some of the trees are now being felled to meet modern traffic conditions. The city grew again as capital of East Pakistan Province after India was partitioned in 1947. But its fastest growth has come since 1971, when Dhaka became the capital of Bangladesh. Its new parliament buildings and vast railway station are ultramodern in style.

The city occupies a low-lying site on the Buriganga, one of the medium-sized rivers of the Ganges delta. Much of the city and the surrounding district, which is also called Dhaka and covers 7470 km² (2884 sq miles), is prone to flooding. Even so, many flood-prone areas have been settled by squatter colonies. Industries include textiles and leather goods. Besides the University of Dhaka, the city has a University of Science and Technology .

Population (city) 5 300 000; (district) 12 989 000
Map Bangladesh – Cc

Dhanbad *India* Coal-mining city in the industrial Damodar Valley, about 290 km (180 miles) north-west of Calcutta. A college of coal mining was founded here in 1926.
Population 1 297 390
Map India – Dc

Dhaulagiri *Nepal* See ANNAPURNA

Dhekelia *Cyprus* British military base in Larnaca Bay, on the south coast, built in the 1950s. It became one of the UK Sovereign Base Areas.
Map Cyprus – Ab

Dhílos *Greece* See DELOS

Dhodhekánisos *Greece* See DODECANESE

Dhofar (Sufar) *Oman* Southern province that is the only place on the Arabian Peninsula to have reliable monsoon rains; they last from late June to September. The Dhofar Mountains, rising from a fertile plain to a height of 2000 m (6561 ft), are partly wooded and run parallel to the coast.

FEATURES OF THE DRY DESERTS

Two belts of arid desert girdle the earth on either side of the Equator. The northern belt extends through Mexico and America's south-west, across North Africa, through the Middle East and on to the Gobi Desert of central Asia. The southern belt stretches from South America's Pacific coast, across south-west Africa to central Australia. Scouring winds work on the face of the deserts, sandblasting rocky outcrops into bizarre shapes, and blowing sand dunes along in ever-changing patterns.

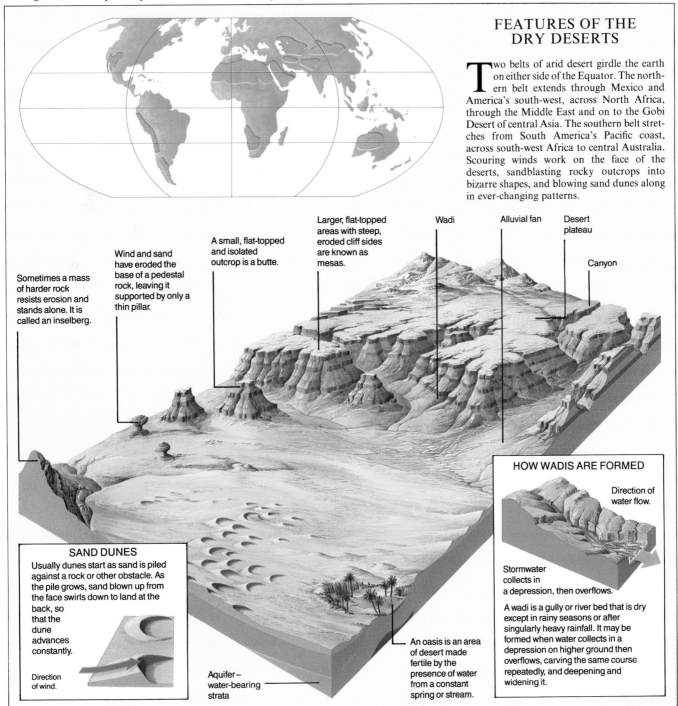

193

Denmark

A HISTORIC HOME OF THE VIKINGS, NOW FAMOUS FOR ITS DESIGNERS, ARCHITECTS, HANS CHRISTIAN ANDERSEN AND LEGO BUILDING BRICKS

Internationally, Denmark enjoys a reputation as a comfortable and contented welfare state, where people live long and happy lives. It has an egalitarian society, but retains the romantic trappings of a monarchy. It is peace-loving, yet expects all men to undertake two years' national service.

Denmark is the most dissected country in Europe and has an immensely long coast – 7300 km (about 4500 miles). No place anywhere in the country is farther than 50 km (31 miles) from the sea.

Denmark has been peopled since the end of the last Ice Age, some 10 000 years ago. Many artefacts from the Stone Age are displayed in museums, and in 1950 the body of an Iron Age man was found, perfectly preserved after 2000 years in a peat bog at Tollund in JUTLAND. The land itself is neither scenically exciting nor rich in resources. It is a homely country, with good soils lying on chalk rocks and a cool temperate climate. Rain and snow fall mostly in the south-west, where the average precipitation is about 750 mm (30 in) annually. The rest of the country has an annual rain or snowfall of 500-600 mm (19.7-23.6 in).

tales, was born on the island, in the town of Odense. Further east is the largest island, called Zealand (Sjælland), which is linked to Fünen, Jutland and Sweden by a shuttle service of train and car ferries. A bridge linking Zealand and Fünen is due for completion in 1997, and plans are underway to build another from Zealand to Sweden.

The capital, COPENHAGEN, is on the eastern shore of Zealand, as far away as is possible from the Jutland mainland, but strategically placed on the Öresund Strait between Denmark and Sweden. At the northern approach to the Öresund stands ELSINORE (Helsingør), the setting for Shakespeare's great tragedy *Hamlet, Prince of Denmark*. The most isolated Danish island, Bornholm, is several hours' ferry journey across the Baltic from the capital.

SUCCESS IN AGRICULTURE

Denmark is an agricultural country. Over 60 per cent of the land is cultivated, mostly by the rotation of grass, barley, oats, sugar beet and fodder beet. Animal husbandry has, however, been the mainstay of rural Denmark for the past century. Its best known products are bacon and butter from 4 million pigs and over 2 million cows. Farms are small, with the great majority ranging from 15 to 30 hectares (37.5-75 acres), and are mostly owned and worked by single families.

The success of Denmark's agriculture is due to cooperation, mechanisation and informed farming practices. For a century, the industry has benefited from cooperative processing and marketing of farm products, purchasing of farm supplies and financing of farm needs. Denmark made mechanisation cheaper by manufacturing its own tractors, harvesters and milking machines appropriate to smallholdings and limited purses; this blossomed into an important export industry in farm machinery.

Farmland is regarded as an important national asset. All young farmers must study to earn a 'green licence', without which they cannot own farmland. The small farmer is protected from urban investors: any property of more than 5 hectares (12 acres) must be held by someone whose principal occupation is farming. Even the development of second homes in rural areas is now strictly controlled.

INDUSTRIAL EXPORTS

Despite Denmark's rural image, only 6 per cent of the population are engaged in farming. The overwhelming majority are town dwellers engaged in manufacturing and service activities. Most factories are small, sited in the countryside, and employ fewer than 100 people. However, exports from these small factories exceed those of agriculture.

In spite of its limited range of raw materials, Denmark sells a wide range of manufactured exports, thanks largely to a strong tradition of imaginative design. This is demonstrated, for example, in furniture and kitchenware, the silverware of George Jensen, and Royal Copenhagen porcelain.

Unexpected industries have sprung up in towns on the heaths of Jutland – where cottagers once made stockings for export from the

This fortunate country, lying between the North Sea and the entrance to the Baltic Sea, is one of Europe's smallest states, being only slightly bigger than Switzerland. In medieval times (9th-12th century) it was a major European realm. So, in 990-1066, it exacted a tribute – called *danegeld* – from England in return for restraining its sea-going warriors, the Vikings. Knut the Great, the Danish king of England in 1016-35, united England and Norway (1028-35) with Denmark to form the North Sea empire of Denmark. The country embraced provinces that today form Sweden, Norway, Schleswig-Holstein and Iceland. The Faeroe Islands and the vast, ice-covered island of Greenland (Kalaallit Nunaat) are Danish, but largely self-governing.

LAND AND SEA

Denmark itself consists of a western peninsula and an eastern archipelago of 406 islands of which 90 are populated. It is a low-lying land – the highest point is Yding Skovhøs, reaching 173 m (568 ft) above sea level.

The country faces the North Sea with broad beaches which are the longest in Europe. Artificial harbours, such as ESBJERG in the south and Hanstholm in the north, had to be built because the coast offers little natural shelter for ships. Behind a coastal rampart of dunes lie flat, sandy heathlands, largely reclaimed from the sea, and peat bog, much of which has been drained. Plantations and belts of spruces and pines act as windbreaks against the strong westerly winds.

East Jutland offers a landscape of steep, low hills, with beech-clad slopes, its coast broken by broad inlets. At the heads of these inlets are old ports and market towns as well as Jutland's biggest cities, ÅRHUS and ÅLBORG.

JIGSAW OF ISLANDS

East of Jutland lies a jigsaw of islands. The second largest, FÜNEN (Fyn), is linked to Jutland by bridges. Fünen is called 'the Garden of Denmark' because of its flowery villages with their pastel-painted farmhouses. Hans Christian Andersen, famous teller of fairy

wool of their sheep, their descendants may be working in one of Europe's largest shirt factories at Herning. A carpenter's workshop created the children's constructional toy called Lego, which is sold throughout the world. An important pharmaceutical industry has developed with Denmark the world's chief exporter of insulin.

Services and ideas are also exported. Danish bridge and harbour construction, system-built factories and housing components have found wide markets, especially in the Middle East, South-east Asia, Africa and South America. Danish architects are world-famous, most notably Jørn Utzon, who designed Sydney Opera House in Australia.

All this economic success is accompanied by political stability. Coalition governments, elected by a system of proportional representation, place a high value on human rights and individual liberty. Denmark is also a strong supporter of international agencies and generous in its aid to the developing world.

MAGNIFICENT CASTLES

The happy picture is reflected in a landscape of well-manicured farmland and woodland, brick-built farmhouses with capacious barns, whitewashed churches, handsome manor houses and magnificent castles. In the towns, public gardens overflow with flowers; flags of red and white (the national colours) flutter from the main buildings, and bright shopfronts convey an air of calm prosperity. Cyclists stream through the streets; there are nearly 400 000 cycles in use, in addition to a motor-car for every two people. Danish food is another happy experience – the open sandwich (*smørrebrød*) piled high with the products of farm and fishery – extravagant pastries, akvavit, cherry brandy and lagers.

In addition to material prosperity, Denmark has a rich intellectual tradition. It is found in the influential philosophy of Søren Kierkegaard, in the atomic physics of Niels Bohr, and in the sombre music of Carl Nielsen.

For all its appeal to the outsider, however, Denmark poses its leaders with plenty of problems. Foremost is its limited range of natural resources and especially its lack of energy sources. Economic expansion since the 1950s was based largely upon cheap oil imports, and the big price increases of the 1970s have completely changed the picture. Although reserves of oil and gas have been found in the North Sea, and the first pipeline ashore went into operation in 1985, Denmark only has a meagre share in the offshore oil deposits.

Nor is agriculture able to escape from economic changes. The surpluses produced by Danish farmers have added to overproduction within the European Union (EU). Like other members, Denmark has been required to reduce its output of dairy products, and farmers' incomes have fallen. It is not surprising, therefore, that many Danes initially objected to the European Union after the Maastricht Treaty. However, after certain adaptations, and after a referendum in May 1993, the Danish government signed the treaty.

Urban and industrial expansion, and modern methods of farming, have also raised environmental problems. Denmark has a Ministry of Pollution, with water resources a matter of prime concern, and nature conservation a close second. National parks and conservation areas are preserving natural habitats, but it is too late for the storks. Only a few score of pairs now arrive on their annual migration.

Denmark was doing relatively well during the late 1980s and early 1990s compared with other countries which were hard hit by the worldwide recession. To fight the worst of its economic problems, the Danish government kept wage increases to a minimum in a bid to keep inflation down (the country had inflation of about 1% in 1993). But the Danish kept their welfare structure - one of the best in the world - and their high unemployment benefits. With the signing of the Maastricht Treaty, new investment has been encouraged, boosting growth since mid-1993. The main aim is now to adapt the country to the EU's economic criteria by 1999.

▼ GRUNDTVIG CHURCH P.V. Jensen Klint, Denmark's leading early 20th-century architect, designed this church, which stands in Copenhagen, in 1921-26. It is named after Bishop N. F. S. Grundtvig (1783-1872) who revived the Danish Lutheran Church.

DENMARK AT A GLANCE

Official name Kingdom of Denmark

Area 43 090 km² (16 633 sq miles)

Population 5 175 920 **Per km²** 120 (**Per sq mile** 311)

Capital Copenhagen

Government Constitutional monarchy

Currency 1 Danish krone = 100 øre

Language Danish

Religion Christian (91% Evangelical Lutheran)

Climate Temperate maritime; average temperature in Copenhagen ranges from 0°C (32°F) in February to 18°C (64.4°F) in July

Land use Cultivation 61%, forest and woodlands 12%, meadows and pastures 6%, other 21%

Main primary products Cattle, pigs, poultry, cereals, potatoes and other root crops, fodder

Major industries Agriculture, food processing, fishing, engineering, shipbuilding, petroleum refining, chemicals, furniture

Main exports Machinery, pharmaceutical and medical products, meat, dairy products, fish

Annual income per head (US$) 25 930

Population growth (per thous/yr) 1

Life expectancy (yrs) Male 72 **Female** 78

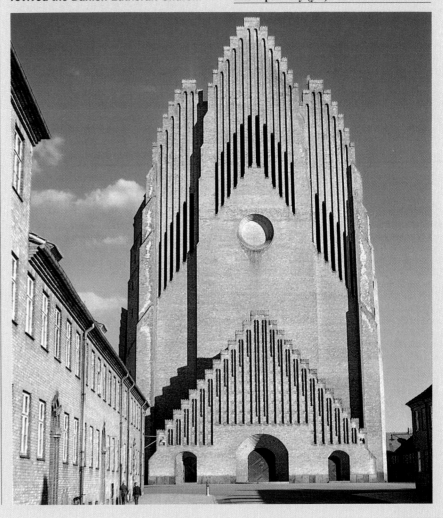

The region is a source of frankincense (an aromatic resin from trees of the genus *Boswellia*) which is used as incense. Dhofar traded in this product with the people of Yemen as far back as the 10th century BC, and today, much of the incense goes to India.

More than half of the province's inhabitants live in its capital, Salalah. From 1964 to 1975, Yemeni-backed rebels fought for Dhofar's independence from the rest of Oman.

Population 174 890
Map Oman – Ab

Diamantina *Brazil* Diamond-mining town 190 km (119 miles) north of the city of Belo Horizonte and named after the diamond industry which has flourished here since 1729. Juscelino Kubitschek (1902-76), president of Brazil in the late 1950s, and founder of the purpose-built national capital, Brasília, was born in the town.

Population 38 870
Map Brazil – Dc

diamond Crystalline form of carbon found in some igneous rocks and alluvial deposits. It is used as a gemstone, for cutting, and as an abrasive. It is the hardest known mineral and is generally colourless. However, it may be tinted yellow, orange, blue, brown or black by impurities.

diapir Body of rock formed by material piercing the earth's crust from below and forcing it upwards into a dome. The material may be hot molten rock, producing a diapir with an igneous core, or a sediment which flows under high pressure. A common type of diapir has a core of salt and is known as a 'salt dome'.

diastrophism Process of deformation of the earth's crust which produces major features such as continents, mountains and ocean basins. See PLATE TECTONICS.

diatom ooze Silica sediment consisting of the fossil remains of minute single-celled aquatic organisms known as diatoms.

diatomaceous earth Fine, powdered silicous earth, composed of the remains of single-celled organisms known as diatoms. It is used for polishing, as an absorbent in the manufacture of high explosives and as an insulating material. It is also known as 'diatomite' or 'kieselguhr'.

Dien Bien Phu *Vietnam* Small market town near the border with Laos. The defeat of French troops in a dramatic siege at the town in May 1954 effectively ended the eight-year Indo-China War and brought to an end French colonial rule in Vietnam. The battlefield is now preserved as a historical site.

Map Vietnam – Aa

Dieng *Indonesia* Volcanic plateau in central Java. It has many megalithic monuments.

Dieppe *France* Channel fishing and ferry port and seaside resort in Upper Normandy, at the mouth of the Arques River. The town was rebuilt after the Second World War, in which it was severely damaged. It now handles 500 000 ferry passengers a year.

Population 35 900
Map France – Db

Digne-les-Bains *France* Spa town 115km (72 miles) north-east of Marseilles. Fruit and lavender are grown in the countryside surrounding the town.

Population 17 500
Map France – Gd

Dijon *France* Capital of the Côte d'Or department and of the former Duchy of Burgundy, about 250 km (155 miles) south-east of Paris. It has Gothic churches and many attractive old houses, some dating from the 15th century, and is noted for its wine trade, its university and its food fair held each year in November. Celebrated local specialities include mustard, gingerbread and cassis (a blackcurrant liqueur).

Population 231 000
Map France – Fc

dike See DYKE

Dilmun *Bahrain* See QALAT AL BAHRAIN

Dimitrovgrad *Bulgaria* Industrial town on the Maritsa (Marica) River, 16 km (10 miles) north of Haskovo. Founded in 1947, the town was named after Georgi Dimitrov (1882-1949), the country's first communist leader.

It produces chemicals, especially fertilisers, manufactures cement and pottery, and processes various foods. There is some coal and lignite mining nearby.

Population 51 000
Map Bulgaria – Bb

Dinajpur *Bangladesh* Chief town of Dinajpur district near the Tista River. It stands in an isolated zone at the foot of the Himalayas. The district covers 6566 km² (2535 sq miles), much of it sandy and dry. Irrigation from undergound water supplies has helped agriculture in the district to advance considerably since independence.

Population (town) 126 000; (district) 3 200 000
Map Bangladesh – Bb

Dinant *Belgium* Tourist town about 80 km (50 miles) south of the capital, Brussels, in the gorge of the River Meuse. It stands at the foot of sheer limestone cliffs, 100 m (328 ft) high. On the very top of the cliffs stands a citadel dating from 1821, built by the Dutch on what has been for centuries a defended site. In 1914, and again in 1940, Dinant suffered great damage by the advancing German armies.

Population 12 300
Map Belgium – Ba

Dinaric Alps (Dinara Planina) *Albania/Bosnia-Herzegovina/Croatia/Slovenia/Yugoslavia (Serbia and Montenegro)* Series of fold mountain ranges covering most of the Balkan Peninsula. In the north-west they connect with the JULIAN ALPS which continue in a south-easterly direction, through ALBANIA and GREECE. Together, they form part of the geological range which traverses Europe and Asia.

Peaking at 2700 m (8858 ft), the Dinaric Alps are significantly lower than the European Alps, lying well below the climatic snowline. Only in the south do they come close to being proper high mountain ranges. The range includes the KRAS Plateau, the mountain range of Velebit and the mountains of Biokova, Lovcen and DURMITOR.

Map Bosnia-Herzegovina – Aa

Dinder National Park *Sudan* Nature reserve covering some 6475 km² (2500 sq miles) on the Ethiopian border, about 520km (320 miles) south-east of the capital, Khartoum.

Map Sudan – Bb

Dinosaur National Monument *USA* Area covering 842 km² (325 sq miles) across the northern end of the state border between Utah and Colorado.

It is a semi-arid wilderness where many fossilised dinosaur remains were found preserved in a single sandstone cliff.

Map United States – Eb

diorite Coarse-grained, intrusive igneous rock, ranging from ACIDIC ROCK to BASIC ROCK, and containing largely plagioclase feldspar and iron-magnesium silicates. See ROCK.

Diosso Gorges *Congo* Formation about 23 km (14 miles) north of POINTE-NOIRE, consisting of a series of spectacular and deeply dissected cliffs facing the Atlantic Ocean.

About 50 m (165 ft) high, the cliffs contain rocks coloured red, white and yellow by iron and other oxides. The Diosso Amphitheatre, the most impressive section, is an area which has eroded into a huge natural bowl.

Map Congo – Ab

Diourbel (Djourbel) *Senegal* Market town for a groundnut area, 140 km (87 miles) east of the national capital, Dakar, to which it was linked by rail in 1905. The town is an Islamic religious centre.

Diourbel has food processing and craft industries. There is an agricultural research station at the nearby town of Bambey.

Population 50 600
Map Senegal – Ab

dip Angle between the inclination of rock strata and the plane of the horizontal.

Dir *Pakistan* Mountainous princely state in the extreme north-west, covering 5286 km² (2041 sq miles) on the border with Afghanistan. It became part of Pakistan in 1947, but retained considerable autonomy.

In 1948, fighting broke out with the neighbouring principality of Chitral over a royal betrothal, and the Pakistani army had to intervene. The Pakistani government intervened again in 1960, on the pretext of preventing an Afghan incursion – this time deposing the country's ruler in favour of one of his sons.

The area is next to the Konar Valley in Afghanistan – scene of years of bitter resistance to the Russian invasion of 1979 – and remains semi-independent. The main settlement is the town of Dir. The state produces a variety of agricultural products.

Population 500 000
Map Pakistan – Ca

Dire Dawa *Ethiopia* Modern city about 350 km (220 miles) east of the capital, Addis Ababa, on the railway line to Djibouti. It has large textile and cement factories.

Population 85 000
Map Ethiopia – Bb

Dirschau *Poland* See TCZEW

discontinuity Boundary across which the internal character of the earth changes abruptly. The two main discontinuities are the Mohorovičić, or Moho discontinuity between the earth's crust and mantle, and the Gutenburg discontinuity between the mantle and the core. See EARTH.

Discovery Bay Bay on the south coast of Australia, 72 km (45 miles) long and 13 km (8 miles) wide, at a point where the states of SOUTH AUSTRALIA and VICTORIA meet.

It was visited in 1802 by both the French navigator Nicolas Baudin and the British explorer Matthew Flinders.
Map Australia – Gf

Disko (Qeqertarsuaq) *Greenland* Largest of Greenland's islands, covering 8580 km² (3311 sq miles) off the west coast. Its hills rise to a maximum height of 1919 m (6296 ft). The main settlement is Godhavn (Qeqertarsuaq).
Map Arctic – Cc

Disney World *USA* The largest amusement park in the world. Disney World is in central Florida, about 25 km (16 miles) south-west of Orlando.

It stands on some 113 km² (44 sq miles) of the state's lake country, and includes artificial lakes, hotels, restaurants and many different 'themelands', modelled on Disneyland on the other side of the continent in California.

One centre, Epcot (Experimental Community of Tomorrow), is a futuristic computer-controlled fantasy world, which was opened in 1982. The park also re-creates various historical landmarks from around the world.
Map United States – Je

Disneyland *USA* Popular amusement park in southern California which was opened in 1955. It is sited in Anaheim, part of the city of Los Angeles, and was designed by the film animator Walt Disney (1901-66) as a series of 'themelands', including Adventureland, Fantasyland, Main Street USA, and Frontierland. Similar parks have been opened in Florida (Disney World), France and Japan.
Map United States – Cd

divide Alternative name for a WATERSHED.

Diyarbakir *Turkey* City, and capital of the province of the same name, about 750km (465 miles) south-east of Ankara. It markets grain, tobacco, cotton, and wool and mohair produced on nearby farms, and makes textiles and leather goods. It is noted for the ancient black basalt walls, almost 5 km (3 miles) long, which surround the old town.

The largely agricultural province covers 15 355 km² (5928 miles) and produces cereals, fruits, cotton and tobacco.
Population (city) 381 140; (province) 1 095 000
Map Turkey – Cb

Djakovo *Croatia* See SLAVONIA

Djanet *Algeria* See TASSILI N'AJJER

Djelfa (El-Djelfa) *Algeria* Town in the Sahara Atlas Mountains at the end of the railway line from Algiers, on the middle route southwards to TAMANRASSET. It is an administrative centre and a market place for the nomads of the Sahara. It

Djibouti

THIS POOR DESERT COUNTRY OF NEGLIGIBLE NATURAL RESOURCES IS HEAVILY DEPENDENT UPON EARNINGS FROM ITS FREE PORT

Nomadic tribes roam the hot, inhospitable desert country of Djibouti. It is a land of 22 000 km² (8492 sq miles) with a port, also called DJIBOUTI, on the Gulf of Aden. There is a semi-arid coastal belt dotted with salt pans and surrounded by basalt plains, followed by mountain ranges rising over 1500 m (4900 ft). Djibouti became an independent republic in 1977 after 96 years as a French colony. The country still receives aid from France, and France maintains a small military force at the strategically located port of Djibouti, at the entrance to the Red Sea.

The climate is among the world's hottest, temperatures averaging 30°C (86°F) and often exceeding 40°C (104°F). Inland, rain is rare and even on the humid coast the average annual rainfall is only 130 mm (5 in). Only 10 per cent of the land can be farmed, even for seasonal grazing, and the country has great difficulty in producing food for its own modest population, let alone thousands of refugees from Ethiopia.

The native population, most of whom are Muslim, is composed of two main groups: the Afars in the north and west, and the Issas (Somalis) who live in the south and dominate public life. Both are nomadic and often cross into Ethiopia or Somalia with their sheep and goats, in search of water and grazing.

The two groups have been rivals for centuries and the Afars are demanding their own independent state, which is intended to extend beyond Djibouti's present borders into parts of Ethiopia and Eritrea.

With an unemployment rate of over 30 per cent and almost no natural resources, Djibouti is heavily dependent upon foreign aid. Crops - chiefly coffee and date palms - are grown and cattle are grazed around a few oases, but the bulk of the food for the urban population must be imported. Cattle, hides and skins are the chief exports; however, most of the country's foreign exchange is earned by the free port of Djibouti, which is also the capital and the only city. A 781 km (485 mile) railway linking the port with Addis Ababa in Ethiopia is jointly owned by Ethiopia and Djibouti, with 681 km (423 miles) of rail in Ethiopia and 100 km (62 miles) in Djibouti.

DJIBOUTI AT A GLANCE	
Map Ethiopia – Ba	
Official name Republic of Djibouti	
Area 22 000 km² (8492 sq miles)	
Population 500 000 **Per km²** 22.7 **(Per sq mile** 58.8)	
Capital Djibouti	
Government Republic	
Currency 1 Djibouti franc = 100 centimes	
Languages Arabic and French (official); Somali, Afar	
Religion Muslim (94%); Christian minority	
Climate Very hot and humid on the coast. Average temperature in Djibouti City ranges from 26°C (78.8°F) in January to 36°C (96.8°F) in July	
Land use Grazing 9%, cultivation 1%, other (desert and semi-desert) 90%	
Main primary products Sheep, goats, cattle, dates, fish; salt	
Major industries Stock rearing, processing hides and skins, fishing	
Main exports Livestock, hides and skins	
Annual income per head (US$) 971	
Population growth (per thous/yr) 31	
Life expectancy (yrs) Male 49 **Female** 49	

has dolmens (prehistoric stone burial chambers) and salt rock formations.
Population 55 000
Map Algeria – Ba

Djenne *Mali* Town 380 km (240 miles) north-east of the capital, Bamako. It is thought to have been founded by the Songhai in the 8th century and has always been an important centre of Islamic teaching. It is a market for leather goods, cloth and blankets produced by local craftsmen. Its magnificent mosque was rebuilt by the French in 1907 in the original traditional style. Djenne is the site of considerable archaeological finds been preserved in sand, including pottery, jewellery and leatherwork.
Population 7000
Map Mali – Bb

Djibouti *Djibouti* National capital and port on the Gulf of Aden. Economically, it depends largely on its role as a transit port handling trade with Ethiopia. It was declared a free port in 1981.
Population 330 000
Map Ethiopia – Ba

Djourbel *Senegal* See DIOURBEL

Dnepr *Russia* See DNIEPER

Dnepropetrovsk (Dnipropetrovs'k) *Ukraine* City and port on the DNIEPER River, called Ekaterinoslav until 1926. It is the centre of the Ukrainian wheat trade and manufactures iron, steel and machinery.
Population 1 179 000
Map Ukraine – Eb

Dnestr *Ukraine* See DNIESTER

Dnieper (Dnepr; Dniepr; Dnyapro) *Eastern Europe* Third longest European river, after the Volga and Danube, with a length of 2285 km (1420 miles). Rising in the Valdai Hills west of Moscow, it flows generally south, through Belarus and Ukraine, to a broad estuary on the Black Sea. The Dnieper is navigable in its upper course for eight months of the year and in its lower course for nine months. The water is low in minerals, but has more than 60 species of fish. It is dammed for hydroelectric power, and connected by canals with the West Dvina and Bug rivers.
Map Russia Ed; Ukraine – Ba; Belarus – Cb

Dniepr *Eastern Europe* See DNIEPER

Dniester (Dnestr) *Ukraine* River rising in the Carpathian mountain range of western Ukraine. It winds south-eastwards for 1411 km (877 miles) through the Ukraine and Moldavia, before finally entering the Black Sea about 30 km (18 miles) south-west of the city of Odesa. Between 1918 and 1940, it formed the border between Romania and the USSR.
Map Ukraine – Ab-Cb

Dnipropetrovs'k *Ukraine* See DNEPROPETROVSK

Dnyapro *Belarus* See DNIESTER

Dobreta-Turnu-Severin *Romania* Industrial city built on terraces above the Danube River, where it forms the border with north-east Serbia and Bosnia-Herzegovina. The spectacular Iron Gate – a gorge whose surging rapids feed a vast hydroelectric power station – lies 18 km (11 miles) to the north-east.
Also in the vicinity of the city are the ruins of the Roman town of Dobreta, complete with port and baths, and the remains of a huge bridge across the Danube built by the Roman emperor Trajan in the 2nd century AD. The city's main industries include food processing, shipbuilding and wood-working.
Population 83 000
Map Romania – Ab

Dobrich *Bulgaria* See TOLBUKHIM

Dobrna *Slovenia* See KARAWANKE ALPS

Dobruja (Dobrogea) *Romania* Fertile region lying between the Danube River and the Black Sea, of which CONSTANŢA is the main resort. Covering 23 258 km² (8980 sq miles), Dobruja contains huge fields of grain, sunflowers and sugar beet, as well as vineyards known for their Murfatlar grapes.
The Danube delta in the region is a major wildlife sanctuary and game reserve – a world of lakes, channels, marshes, reeds and forest. Bird-watchers can see more than 300 bird species, including herons, pelicans and waders. Fish include perch, carp and sturgeon; boar, deer and several types of wild cats are among the wildlife that can be seen. The Black Sea coast nearby has popular tourist resorts.
Map Romania – Bb

doctor Name for any of several local winds, such as the HARMATTAN of West Africa and the south easter of the Western Cape in South Africa (known locally as the 'Cape doctor'), that mitigate extremely unhealthy conditions.

Dodecanese (Dhodhekánisos) *Greece* Twelve islands in the south-east Aegean Sea, stretching from SÁMOS in the north to the largest of them, RHODES, in the south.
The islands were occupied by Italy from 1912 until the end of the Second World War. It was formally handed over to Greece in 1947. The name of the group is derived from *dodeka*, the Greek word for 'twelve'.
Population 162 440
Map Greece – Ec

Dodoma *Tanzania* Town about 400 km (250 miles) west of Dar es Salaam, with which it has shared status as national capital since 1975, largely because of its central location. A plan for a city of 400 000 people has been drawn up.
Population 203 830
Map Tanzania – Ba

Dogger Bank Large submerged sandbank in the centre of the NORTH SEA, about 180 km (110 miles) off the north-east coast of England. It is mostly 18-36 m (60-118 ft) deep, and is a famous fishing ground, especially for cod. (The word *dogger* means 'cod' in Dutch.)
A major naval battle between a British fleet under the command of Admiral Beatty and a German fleet under Admiral Hipper was fought here in 1915.
Map United Kingdom – Fc

Doha (Ad Dawhah) *Qatar* Capital, main port and commercial centre of Qatar, situated on the peninsula's east coast. A growing, modern city, it has areas named after the country's ancient families who moved in when it became the capital in 1971. Before oil was discovered, Doha was an insignificant fishing port and its inhabitants lived by fishing, harvesting pearls and piracy. The old fishing port is still used by dhows (single-masted Arab sailing vessels), while the new port handles exports and imports. An elegant corniche, or coastal road, has been built – and there is a new international airport. Doha has a desalination plant and two museums, including the ethnographic museum.
Population (Greater Doha) 350 000
Map Qatar – Ab

Doi Inthanon *Thailand* The country's highest mountain 40 km (25 miles) west of the northern city of Chieng Mai. The 2590 m (8288 ft) peak and its surrounding waterfalls have become popular with tourists since a road was built almost to the summit in the mid-1970s.
Map Thailand – Ab

doldrums A zone of sultry calms and light breezes within a few degrees of the equator, where the north-east and south-east trade winds converge. Intense heat allied with low pressure often causes strong updraughts of air, resulting in sudden squalls and heavy tropical rain. The zone was avoided by sailing ships because they could become becalmed for days.

dolerite Basic, medium-grained, intrusive igneous rock, composed mainly of FELDSPAR and PYROXENE, and sometimes olivine.

dolomite Yellowish or brownish mineral which is a double carbonate of calcium and magnesium. The name is also given to a rock, often called dolomitic limestone or magnesian limestone, which contains more than 15 per cent magnesium carbonate.

Dolomites (Alpi Dolomitiche) *Italy* Range of dolomitic limestone mountains between the upper Adige and Piave valleys of northern Italy. The limestone erodes into fantastically rugged shapes, including pinnacles, tooth-edged ridges, deep gorges and precipitous cliffs. On its southern side the highest peak, the Marmolada 3342 m (10 964 ft) which is partly glaciated, drops sheer by some 610 m (2000 ft). The range is a climber's paradise, Cortina d'Ampezzo is the main resort.
Map Italy – Ca

Dombrau *Poland* See DABROWA GORNICZA

Dominica See p 199

Dominican Republic See p 200

Domrémy-la-Pucelle *France* Village 45 km (28 miles) south-west of Nancy, where Joan of Arc was born in 1412.
Map France – Fb

Don *Russia* River rising in hills about 200 km (125 miles) south of Moscow. It flows 1870 km (1165 miles) south to the city of Voronezh, then south-east to a canal link with the Volga near Volgograd, then south-west to the Sea of Azov. It is navigable as far upstream as Voronezh, but is closed by ice for three to four months each winter. Its waters are dammed at Lake Cimbl'amskoye.
Map Russia – Fd

Donbass *Ukraine* See DONETS BASIN

Doncaster *United Kingdom* Centre of one of England's most productive coalfields, about 28 km (17 miles) north-east of Sheffield, South Yorkshire. It is also an important railway junction with workshops for building and repairing rolling stock, and since 1776 has been the venue of the annual St Leger horserace.
Population 293 300
Map United Kingdom – Ed

Dondra Head *Sri Lanka* The island's southernmost point, which is sacred to both Hindus and Buddhists. There are remains of the 7th-century Hindu shrine of Maha Vishnu Devala whose gilded roof was once a landmark for passing ships. The shrine was destroyed by the Portuguese in the 16th century.

Donegal (Dún na nGall) *Ireland* Most northerly county in Ireland, in the province of Ulster, with a wild and jagged coastline facing the North Atlantic. It covers 4830 km² (1865 sq miles), and its mountains and lakes attract tourists, especially anglers. It has good dairy farmland known as the Lagan in the east, along the border with Northern Ireland. On the poorer west coast the Gaelic language and culture are still very much entrenched. The county is renowned for its textiles, especially Donegal tweed.
The small market town of Donegal lies at the head of Donegal Bay. Lifford is the county town.
Population (county) 128 120; (town) 2190
Map Ireland – Ca

Donets Basin (Donbass) *Ukraine/Russia* Largest coal field in what used to be the Soviet Union, located in one of Ukraine's leading industrial areas.
Coal was discovered here in 1721 and mined from 1800, and became a vital fuel with the building of the railways in the 1870s. There is a vast metal industry using iron ore from the adjoining KRYVYY RIH fields. The basin's output used to be in the vicinity of 200 million tonnes of coal and over 50 million tonnes of steel a year, but production has declined since the depression in the world market price for raw materials and since the collapse of the Soviet Union. The region's chief city is DONETS'K; others include MAKIYIVKA and LUHANS'K.
Map Ukraine – Db; Russia – Ed

Dominica

FORESTED, RUGGED AND MOUNTAINOUS - THE CARIBBEAN ISLAND THAT COLUMBUS DISCOVERED ON A SUNDAY

The most northerly of the WINDWARD ISLANDS in the West Indies, Dominica is the least changed since the Caribbean was first settled by Europeans. Christopher Columbus gave the island its name because he reached it on a Sunday (in Spanish, *domingo*) in 1493. Because of its rugged terrain, colonists left it well alone at first. More than 100 years later, the British and the French began showing an interest in the island. Fierce disputes between them over the island's ownership were finally settled in 1805, when the French left the island after extracting a ransom of £12 000 (then US$53 000) from the British government.

Dominica became an independent member of the Commonwealth in 1978, with an elected president as head of state. Many of the island's place names remain French and some of the inhabitants – most of whom are descendants of African slaves – still speak a French creole.

On a reserve in the north-east of the island there are several hundred Caribs, the native Amerindians who originally inhabited the Caribbean islands and after whom the region is named. But they have lost most of their original culture and language, and live in much the same way as their black neighbours.

Dominica is the most rugged island in the eastern Caribbean. Only 3 per cent of the land is flat. The rest consists of three relatively young but inactive volcanoes, the highest of which is Morne Diablotin (1447 m, 4747 ft). There are several hot springs in the Valley of Desolation in the south. Most of the island has more than 2500 mm (98 in) of rain a year and even on the sheltered leeward coast it rains two days out of three.

The steep mountain slopes are difficult to farm and so, much of Dominica is still forested. But neither its timber nor its fishing grounds have been exploited. The island did not attract settlers as early as others in the Caribbean and still has a lower population density – as well as a poorer standard of living – than many other West Indian countries, with an average annual income per capita of US$ 2520 and an unemployment rate of 15 per cent.

BANANA EARNINGS

Agriculture employs 40 per cent of the workforce and provides almost all the exports. By far the most important export is bananas – and the island still benefits from the preferential treatment given by European countries to eastern Caribbean banana producers. The ongoing Latin American campaign to end this preferential treatment is a serious economic threat to Dominicans.

Other revenue earners are copra, citrus fruits (especially limes), cocoa, bay leaves and vanilla. Manufacturing is mostly confined to small-scale processing of agricultural products – for instance, coconuts are made into soap and cooking oil – but there is some manufacture of cement, shoes and furniture.

Dominica is not a typical tourist island, though its rugged beauty and dense forests attract adventurous travellers. The islanders are now looking to develop tourism in a bid to diversify the economy. The old fort which once defended ROSEAU has already been turned into a hotel.

Until 1982, Roseau, the small capital, was a difficult two-hour drive from the only airport; today, the new Canefield Airport is only five minutes' drive from the town. But the island still does not have an international airport, which hampers the development of tourism; ships calling at Roseau still have to anchor offshore. Portsmouth, in the north-west, has a better harbour but here, surrounding swampland has hindered growth.

DOMINICA AT A GLANCE	
Map Caribbean – Cb	
Official name Commonwealth of Dominica	
Area 750 km² (290 sq miles)	
Population 86 550 **Per km²** 115 (**Per sq mile** 298.4)	
Capital Roseau	
Goverment Parliamentary republic	
Currency 1 East Caribbean dollar = 100 cents	
Languages English (official), French creole	
Religion Christian (90%)	
Climate Tropical, with heavy rainfall. Average temperature ranges from 25°C (77°F) in January to 30°C (86°F) in June	
Land use Forest and woodlands 41%, cultivation 21%, meadows and pastures 3%, other 34%	
Main primary products Bananas, coconuts, citrus fruits, cocoa, mangoes, root crops	
Major industries Agriculture, food processing, soap, essential oils	
Main exports Bananas, copra, citrus fruits, fruit juices, cocoa, spices, bay oil, soap	
Annual income per head (US$) 2520	
Population growth (per thous/yr) 13	
Life expectancy (yrs) Male 73 **Female** 79	

Donets'k *Ukraine* City in the south east of the country not far from the border with Russia. Known as Stalino until 1961, it was founded in 1872 by John Hughes, a British industrialist, to make iron rails for the Russian railways. It is now the chief industrial centre for the DONETS BASIN, with coal, iron, steel, machinery, chemicals, cement and clothing industries.
Population 1 110 000
Map Ukraine – Db

Donga Ridge *Nigeria* See UDI PLATEAU

Dongbei (Manchuria) *China* Region in the north-east, covering 780 000 km² (310 000 sq miles). It is made up of the provinces of HEILONGJIANG, JILIN and LIAONING. From the mid-19th century, Russia showed an interest in the region; later, from 1931 to 1945, Dongbei was the Japanese puppet state of Manchukuo. In 1945, it was occupied by Soviet armies, but was only wholly returned to China in the mid-1950s. It is a land of forests and prairies with important reserves of oil, coal, iron ore and other minerals which have played a major part in the development of China's heavy industry.
Population 99 335 000
Map China – Ib

Doornik *Belgium* See TOURNAI

Dorchester *United Kingdom* County town of Dorset, about 38 km (24 miles) west of Bournemouth. It was founded by the Romans as Durnovaria in 70 AD. The massive earthworks of Maiden Castle, 3 km (2 miles) to the south-west, however, date back to the late Stone Age, around 2000 BC. This was later fortified to house a township of some 5000 people, which was overrun by the Roman 2nd Legion in 43 AD.

In 1834 six pioneer trade unionists known as the Tolpuddle Martyrs were imprisoned at Dorchester but released two years later after prolonged local protests.
Population 14 220
Map United Kingdom – De

Dordogne *France* River of western France with a length of 472 km (293 miles). It rises in the MASSIF CENTRAL, cuts through the limestone plateau of Périgord and continues west to join the Garonne north of Bordeaux, emptying into the Bay of Biscay. It crosses the department of the same name, whose capital is Perigueux and which contains the renowned LASCAUX caves.
Population (department) 378 800
Map France – Dd

Dordrecht *Netherlands* Industrial city and medieval port in the delta of the Rhine and Maas rivers, 19 km (12 miles) south-east of Rotterdam, which surpassed it in importance as a port in the 18th century. Dordrecht regained some of its value as a harbour at the end of the 19th century. Its products today include tobacco, timber, metal goods, chemicals, linen and glass.
Population (city) 112 690; (Greater Dordrecht) 212 160
Map Netherlands – Bb

Dorset *United Kingdom* County of southern England, which has been immortalised in the works of the poet and novelist Thomas Hardy (1840-1928). It covers 2654 km² (1025 sq miles) along the English Channel coast. Inland, it is a deeply rural area of chalk downs and wooded valleys, while its coast is remarkable for a variety of interesting features – stretching westwards from the resort of BOURNEMOUTH past Poole Harbour, with its narrow entrance and many islands, along the rocky and beautiful shore of the Isle of Purbeck (on which lies one of Britain's oil fields), past Lulworth Cove to the port of Weymouth and the island and naval base of Portland, and then finally along 15 km (10 miles) of Chesil Beach to Lyme Regis, with its complex of fallen cliffs. Dorset is a popular tourist area. The county town is DORCHESTER. Manufactures include microelectronics in the east.
Population 660 500
Map United Kingdom – De

Dortmund *Germany* Inland port and second largest city of the industrial Ruhr, 32 km (20 miles) east of the main Ruhr city, Essen. It is Europe's largest producer of beer and has one of its biggest steel plants. The city has been almost entirely rebuilt since the Second World War. Its Westfalenhalle is one of the largest sports halls in Europe, seating 23 000 people.
Population 599 100
Map Germany – Bc

Dortmund-Ems Canal *Germany* Canal opened in 1899 to link the expanding industries of the Ruhr with the North Sea at the port of Emden, near the Dutch border. The canal joins the River Ems at Meppen, 115 km (71 miles) north of Dortmund.
Map Germany – Bb

Douai *France* Town on the Scarpe River, 30 km (19 miles) south of Lille. The Old Testament of the Douai Bible, still used by Roman Catholics, was published here in 1609-10.
Population 41 600
Map France – Ea

Douala *Cameroon* The country's chief port and trade centre. It is a busy, densely peopled town of great contrasts. Almost anything can be bought here, from local charms and potions to Parisian *haute couture*, for there are several colourful markets and also expensive shops. The tall, modern buildings of the port contrast with a host of older, poorer buildings clustered along narrow streets. Visitors can escape the sticky heat and the flies on the slopes of Mount Cameroon or on the black volcanic sand beaches at nearby Limbe.
Population 1 211 390
Map Cameroon – Ab

Dougga (Thugga) *Tunisia* Ruined Roman city 110 km (about 70 miles) south-west of Tunis. It lies on the much older site of Thugga, founded by native Berbers and later occupied by Phoenician colonists. Built on a plateau, the town was at its height under the Romans in the 2nd century AD, although it continued to flourish well into the Arab-Islamic period. It is considered to be Tunisia's best preserved Roman site, with its theatre almost intact and its capitol one of the finest in Africa. Other monuments include a forum, temples to Jupiter, Juno, Minerva, and other Roman gods, as well as public baths, triumphal arches, private houses and streets. There is also a 2nd-century BC Carthaginian mausoleum. The surrounding plain is dotted with more Roman remains.
Map Tunisia – Aa

Douro (Duero) *Spain/Portugal* River, 895 km (555 miles) long, which rises in north-central Spain and flows westwards to cross northern Portugal and enter the Atlantic downstream from OPORTO. For part of its course it forms the border with Spain. The Portuguese section is dammed in several places to provide hydroelectric power; grapes for port wine are grown on its steep SCHIST slopes, between Pêso da Régua and the Spanish border.
Map Portugal – Cb

Dover *United Kingdom* Town in the English county of Kent, 105 km (65 miles) south-east of London. Dover has been the principal Channel port since Roman times, being the nearest to the French port of Calais, 34 km (21 miles) away. A Roman lighthouse, Saxon church and Norman castle stand on its famous white cliffs, here honeycombed with military tunnels dating back to Napoleonic times, which have inspired many, including poet Matthew Arnold and the singer Vera Lynn. Dover has a monument to Captain Matthew Webb who, in 1875, became the first man to swim the English Channel, and one to the French aviator Louis Blériot, who in 1909 became the first man to fly across it.
Population 34 300
Map United Kingdom – Fe

Dominican Republic

AFTER YEARS OF REVOLUTION AND DICTATORSHIP, DEMOCRACY HAS STILL TO BRING PROSPERITY

The state which shares the West Indian island of Hispaniola with HAITI has had its quota of poverty, revolution and dictatorships. Unlike the Negroes of French-speaking Haiti, the mixed population of the Dominican Republic share the language, religion and lifestyles of Latin America.

Sailors left behind by Christopher Columbus following the wreck of one of the ships on his first voyage in 1492 founded La Navidad, the first European settlement in the Americas. Most of the early settlers soon left to seek their fortunes in Mexico, Panama and Peru, and for three centuries the country was an unimportant Spanish colony. Within 50 years of Spanish settlement, the total Tainos Amerindian population either died of European diseases, to which they had no inherited immunity, or were slaughtered.

After a troubled and violent history, including spells of French, Haitian and again Spanish rule, the country finally gained its independence in 1865. Over the next 50 years there were 28 revolutions and 35 governments, and the US occupied the republic from 1916 to 1924. Dictators ruled the country until it became a democracy in 1966. Presidential elections in 1994 brought in President Joaquin Balaguer for his seventh consecutive term.

SANTO DOMINGO is the main port and capital. For many years it was called Ciudad Trujillo after the dictator Rafael Trujillo, who retained power, even when not holding the presidency, from 1930 until his assassination in 1961. The colonial part of the city has been rebuilt and restored. The cathedral, built between 1512 and 1540, is one of the oldest European-style buildings in the Americas.

ECONOMY BASED ON SUGAR, COFFEE AND METALS

The western part of the country is dominated by four almost parallel mountain ranges. In the Cordillera Central, Pico DUARTE (3175 m, 10 417 ft) is the highest point in the West Indies. The wide, fertile Cibao Valley lies between the two most northerly mountain ranges. The south-east, too, is occupied by fertile plains. Here, intensive farming is practised. Sugar is the most valuable crop and a mainstay of the republic's economy. It is grown mainly on plantations in the southern plains. But after the sugar price declined on the world market in the mid-1980s, production halved from 1985 to 1994. Coffee, the second export crop, is cultivated on smallholdings in the mountains; fruits and tobacco are also produced. Gold (the country has the largest operating gold mine in the Caribbean), silver, platinum, nickel and aluminium ores are mined and, in some cases, refined on the island. The government has begun to boost industrial development and encourage foreign investment. Main industries include food processing and making consumer goods for the home market. Some new industries, most of them foreign-owned, have been built in duty-free industrial zones, where they process imported, semi-finished goods for re-export.

In spite of recent industrial development, the people remain poor. In 1992 more than half the population of 7 320 000 was classified as poor, with about 25 per cent of the workforce unemployed at the beginning of 1994.

But there is hope for an upswing as the tourist industry is growing - despite the republic's proximity to troubled Haiti. The country has many fine beaches. East of La Romana, a village in 16th-century Spanish style has been built as a tourist attraction, and new resorts are now being built at La Romana, Punta Cana and Puerto Plata.

DOMINICAN REPUBLIC AT A GLANCE	
Map Caribbean – Ab	
Area 48 442 km² (18 703 sq miles)	
Population 7 320 000 **Per km²** 151 **(Per sq mile** 391)	
Capital Santo Domingo	
Government Presidential republic	
Currency 1 Dominican peso = 100 centavos	
Language Spanish	
Religion Christian (92% Roman Catholic)	
Climate Tropical; cooler in mountains. Heavy rains fall in the north and east; drier in the south. Average annual temperature in Santo Domingo ranges from 24°C (75°F) in January to 27°C (80.6°F) in August	
Land use Meadows and pastures 43%, cultivation 30%, forest 13%, other 14%	
Main primary products Sugar, coffee, cocoa, tobacco, rice, maize; bauxite, gold, silver, platinum, nickel, salt	
Major industries Agriculture, tourism, textiles, cement, food processing (sugar, rum, molasses), tobacco products, mining and metal refining, petroleum products	
Main exports Gold, silver, nickel, bauxite, sugar, coffee, tobacco, fruits	
Annual income per head (US$) 1040	
Population growth (per thous/yr) 22	
Life expectancy (yrs) Male 64 **Female** 68	

Dover *USA* State capital of Delaware, about 135 km (85 miles) east and slightly north of the national capital, Washington DC. Established in 1683 by the Quaker settler William Penn (1644-1718), who founded the state of Pennsylvania, it lies in the heart of a prosperous fruit and vegetable producing region.
Population (town) 26 630; (metropolitan) 110 990
Map United States – Kc

Dover, Strait of Busy waterway, 34 km (21 miles) wide at its narrowest point, separating England from France and connecting the ENGLISH CHANNEL to the NORTH SEA. It was called Fretum Gallicum by the Ancient Romans and is known as the 'Pas de Calais' to the French. It was first swum by Captain Matthew Webb in 1875. In 1940, it was the scene of the British evacuation from DUNKIRK. The Channel Tunnel, completed in 1994, runs beneath the Strait of Dover, connecting England to France.
Map United Kingdom – Fe

Down *United Kingdom* County covering 2448 km² (945 sq miles) in Northern Ireland's south-east corner, beside the Irish Sea. It includes the Mourne Mountains, which rise to 852 m (2795 ft). The north contains prosperous Belfast suburbs while the rest of the county is mostly good agricultural land. There is evidence of early prehistoric settlement along the coast, and monasteries at Bangor, Nendrum and Movilla date from the time of St Patrick, in the 5th century. Down Patrick (population 10110) is the reputed burial place of St Patrick.
Population 439 800
Map United Kingdom – Cc

Downs, The *United Kingdom* Two roughly parallel ranges of low chalk hills, less than 300 m (1000 ft) high, in southern England. The North Downs extend about 130 km (80 miles) in a curve from the county of Surrey across north Kent to the English Channel coast at Dover's white cliffs. The South Downs are about 110 km (70 miles) long and stretch from south-east Hampshire across southern Sussex to the cliffs of Beachy Head, east of Brighton. Between them is the rich farming country, once forested, of The Weald. The name 'downs' is also used in other parts of England for rolling, treeless grasslands.
Map United Kingdom – Ee

downthrow The vertical distance between the level of a rock on one side of a FAULT plane and that of its lower continuation on the other – the 'downthrow side'.

Drake Passage Waterway, 640 km (400 miles) wide, linking the PACIFIC and ATLANTIC oceans between CAPE HORN and the SOUTH SHETLAND ISLANDS. It was discovered by Sir Francis Drake on his voyage around the world in 1577-80.
Map Antarctica – Da

Drakensberg *South Africa/Lesotho* Mountain range stretching 1125 km (700 miles) from Eastern Transvaal, through the Free State, KwaZulu-Natal and Lesotho, to Eastern Cape Province. It is part of South Africa's GREAT ESCARPMENT, and derives its name from African legends which describe it as the home of dragons; parts of it are also known to Africans as *quathlamba* – 'Barrier of Spears'.

The highest peaks in the Drakensberg are on the border between Lesotho and KwaZulu-Natal, where Thabana Ntlenyana rises to 3482 m (11 424 ft) and Champagne Castle to 3374 m (11 069 ft). The range has several resorts and game reserves.
Map South Africa – Cc

Drava (Drave; Drau) *Austria/Slovenia/ Croatia/Hungary* Major Central European river, with a length of 718 km (447 miles). It rises in east Tyrol and flows east past Villach in Austria to enter Slovenia where it is dammed above Maribor for hydroelectric power. It crosses the northern tip of Croatia to form part of the Hungarian-Croatian border further along its course, and joins the Danube 23 km (14 miles) east of the Croatian city of OSIJEK.
Map Croatia – Ca

Drenthe *Netherlands* Farming province covering 2681 km² (1035 sq miles) in the north-east. It produces cereals and potatoes, and raises cattle. Since the 1950s, oil and natural gas have also been exploited here. ASSEN is its capital.
Population 448 300
Map Netherlands – Ca

Dresden *Germany* Historic city in the south-east, 40 km (25 miles) from the border with the Czech Republic. Once the capital of Saxony, it is now a nuclear research and medical centre and produces precision and machine tools, aircraft, instruments, electronics, office equipment, textiles and porcelain. Dresden's restored 19th-century Semper Opera House was reopened on 13 February 1985 to commemorate the 40th anniversary of the city's destruction by Allied bombs at the end of the Second World War, and to mark the completion of the painstaking reconstruction of the old Baroque city once known as 'Florence on the Elbe' and one of the most beautiful cities in Europe. Many of the city's churches, its magnificent Zwinger Palace and its Semper Gallery (containing collections of china, porcelain and paintings by old masters) have also been restored.
Population 409 600
Map Germany – Cc

drift Bank or pile of sand or snow – heaped up, for example, by currents of air or water. The term 'drift' (or sometimes, 'glacial drift') is also applied to rock debris eroded and transported by ice, then deposited either by the ice or by meltwater. In a wider sense, the term includes alluvium. In South Africa, a ford in a river is also called a 'drift', or 'drif'.

Drin *Albania/Yugoslavia (Serbia and Montenegro)* River which is 280 km (174 miles) long and empties into the Adriatic Sea from the marshy SHKODËR Lowlands in north-west Albania. Its northern mouth, the Buenë River, forms part of the border between Albania and Serbia and Montenegro. Higher up its course, it is a torrential river running through deep gorges and has many tributaries. The Drin is harnessed at several points to produce hydroelectricity.
Map Albania – Ca

Drina *Bosnia-Herzegovina/Serbia* River of 346 km (215 miles) length and running along much of the boundary between Serbia and Bosnia-Herzegovina. Formed near Sutjeska from

the Piva and Tara rivers, it flows swiftly north-east through a spectacular limestone gorge with rapids, past Foca to VISEGRAD, and on, through 'slivovica (plum-brandy) country', to the Sava. Reservoirs behind hydroelectric dams just below Visegrad and at Zvornik are used for water sports.
Map Bosnia–Herzogovina – Ca

drizzle Fine, continuous rainfall in which raindrops are less than 0.5 mm (0.02 in) in diameter, usually associated with a warm FRONT.

Drogheda (Droichead Átha) *Ireland* Port on the estuary of the River Boyne, about 40 km (25 miles) north of Dublin. It was the meeting place of Ireland's medieval parliaments. The town was sacked in 1649 after a siege by Oliver Cromwell, lord protector of England. Most of its 2000 defenders were massacred; the handful of survivors were shipped to the West Indies. Drogheda surrendered to the British again in 1690 after William of Orange defeated James II at the Battle of the Boyne. In St Peter's Church is preserved the head of St Oliver Plunkett, Archbishop of Armagh, and Primate of All Ireland from 1669. He was executed at Tyburn in London in 1681, after being implicated in a fictitious plot to murder Charles II.

Drogheda's products today include cement, chemicals, textiles, footwear, foodstuffs and beer.
Population 23 850
Map Ireland – Cb

drought A period of scanty rainfall. In the United Kingdom, the term is applied to any prolonged period of dry weather, specifically defined as 'absolute drought' – when the daily rainfall is less than 0.25 mm (0.01 in) for a period of at least 15 consecutive days, 'partial drought' – a period of at least 29 consecutive days during which the mean daily rainfall does not exceed 0.25 mm (0.01 in), and 'dry spell' – a period of at least 15 consecutive days during which the daily rainfall is less than 1.0 mm (0.04 in). However, the term takes another meaning altogether in countries with different climatic conditions.

Many countries in arid and semi-arid regions will apply the term to an absence of the rainy season, where in bad times no rainfall is measured for an entire year, and sometimes longer. Even in the USA, meteorological terms differ from English ones.

drowned valley A valley that has been filled with water from the sea or a lake, because of land subsidence or because of the rising of the sea.

drumlin Small, streamlined hill composed of glacial DRIFT, which has been shaped by ice moving over it. Drumlins are generally egg-shaped – the blunt end facing the direction from which the ice advanced, the elongated end facing away from it. Most of them are 0.4-0.8 km (0.25-0.5 miles) long and 30 m (100 ft) high.

Drvar *Bosnia-Herzegovina* Small timber processing town in western Bosnia, lying on the Una River amid forested mountains. Its full name is Titov Drvar – 'Titov' to commemorate former Yugoslavia's communist leader Marshal Tito (1892-1980), whose general headquarters during the Second World War was in the town.
Population 6420
Map Bosnia-Herzegovina – Ba

dry spell See DROUGHT

dry valley Valley which was formerly cut by running water but which now has no river running through it. Dry valleys are usually found in limestone regions; they were cut when the climate was wetter and the water table was higher, or when the subsurface was frozen and impermeable during a colder climate, the permeable limestone having since allowed the water to sink underground, leaving the valleys dry.

Dry Valleys (McMurdo Oasis) *Antarctica* Name commonly given to an ice-free area in Antarctica covering about 3000 km² (1160 sq miles). It comprises a series of extremely arid valleys near McMurdo Sound in Victoria Land, Antarctica, where no rain has fallen for at least 2 million years. The area has had no plant or animal life for about the same period of time. Several lakes in the Valleys, such as Lake Vanda, have given rise to their alternative name of McMurdo Oasis.
Map Antarctica – Jc

Duars *Bhutan* Plains region on the border with Assam, India. Much of it was annexed by the British in 1854 during a series of short wars, following which, its savannah grasslands and jungle were cleared for tea plantations. The annexed section now forms part of India. The remaining Bhutanese part is less developed, although it sustains nearly a quarter of Bhutan's population. The region takes its name from a Bhutanese word meaning 'doorway', because the torrential rivers which flow onto the plain have cut gaping ravines ('doorways') into the wall of mountains alongside.
Population 175 000
Map Bhutan – Aa

Duarte, Pico *Dominican Republic* Mountain peak, 3175 m (10 417 ft) high, in the Cordillera Central. Formerly called Monte Trujillo, it is the highest point in the West Indies.
Map Caribbean – Ab

Dubai *United Arab Emirates* Covering 3900 km² (1505 sq miles), Dubai is the second largest of the seven emirates.

Although oil has improved its wealth tremendously, Dubai has been quite prosperous since the 1830s, when the small town began turning into a flourishing trading centre. In the 1960s, the city consisted of one and two-storey houses constructed from coral limestone, some with wind towers – an architectural feature unique to this part of the world, and one of the earliest forms of air conditioning in which wind is diverted off the tower wall into the house to cool it.

There was no electricity until 1961 and no mains water until 1968. Then, during the 1970s, there was urban development including the building of many high-rise office blocks. Today, Dubai City is a busy seaport and modern capital with an international airport.

Despite its modernity, Dubai retains its links with the past. Fishing remains an important industry, and the port is still used by dhows (single-masted Arab coastal sailing vessels) and served by *abras* (water taxis). There is an early 19th-century fort and some old Arab houses.
Population (emirate) 501 000; (city) 265 700
Map United Arab Emirates – Ba

Dubbo *Australia* Country town in New South Wales, 285 km (180 miles) north-west of Sydney. It is the business centre for the surrounding sheep and cattle country and its stock markets are the largest in the state. It is the home of the Western Plains Zoo, the country's largest open-range zoo.
Population 28 060
Map Australia – He

Dublin (Baile Átha Cliath) *Ireland* Capital of the Republic of Ireland, astride the mouth of the River Liffey on the east coast. Vikings settled at the river mouth in the 9th century and remained until Ireland's warrior king, Brian Boru, drove them out after the Battle of Clontarf (1014). The Normans then made Dublin their centre for the conquest of Ireland in the 12th century. By the end of the Middle Ages, however, the area under English control had been reduced to a small area around Dublin – called the 'Pale'.

Dublin is still a largely Georgian city. The Mansion House has been the lord mayor's residence since 1715 and it was from here that Ireland's declaration of independence was made in 1919. Leinster House, where the *dáil*, or parliament, meets, was started in 1744 as a town house for the Duke of Leinster. The Custom House (1791) and Four Courts (1786) date from the same period. O'Connell Street, one of the widest and most gracious thoroughfares in Europe, runs north from the Liffey past the post office, headquarters of the insurgents during the Easter Rising of 1916. Dublin Castle, partly Norman, was the seat of British administration in Ireland for centuries. It has been restored as government offices. South of the Liffey are Christ Church and St Patrick's Cathedral.

▼ **HONOURED PAST A new mosque built in the traditional local limestone to conventional Arab design stands in a modern suburb of Dubai, the United Arab Emirate city made magnificent on oil revenues.**

The city has rich links with literature. Jonathan Swift (1667-1745), author of *Gulliver's Travels*, was born in Hoe's Court, and was dean of the cathedral for 32 years from 1713. The playwright George Bernard Shaw (1856-1950) was born in Synge Street and writer Oliver Goldsmith (about 1728-74) was a student at Trinity College. Nearby is Merrion Square, where another playwright, Oscar Wilde (1854-1900), was born. James Joyce (1882-1941), author of *Dubliners* and *Ulysses*, was born in the city, too.

In the library at Trinity College is the *Book of Kells*, an 8th-century illustrated book of the Gospels, and the harp of King Brian Boru, used as the trademark for Dublin's most famous product, Guinness stout, which is brewed near Hueston Station. Over Hueston Bridge is the road to Phoenix Park – at 7.1 km² (2.75 sq miles) the largest urban park in Europe.

Dublin's strategic position facing Britain, its harbour, airport and road network, make it the focal point of Irish trade. Electrical goods, metals, food and printed materials are produced, and it is a major financial and tourist centre.
Population (city) 478 390; (county) 1 025 300
Map Ireland – Cb

Dubrovnik (Ragusa) *Croatia* Medieval port and resort on a rocky headland in Dalmatia, about 190 km (118 miles) south-east of SPLIT. The old city, which is defended by the 16th-century Revelin Fortress to the east, is well known for its double ramparts with 20 towers and bastions, its Baroque cathedral and exquisite churches, monasteries, palaces, fountains and red or yellow-roofed houses.

Dubrovnik became fabulously wealthy from trade during the Middle Ages, and was virtually an independent city-state from 1205 to 1808 when it was conquered by Napoleon. The Congress of Vienna assigned it to Austria in 1815. Known as the 'South Slav Athens', it was renowned for its art and literature from the 15th to the 17th century. It became part of Yugoslavia in 1918 and is now part of Croatia.

The city today has several museums, a cable car up Mount Srdj behind the city, an international airport at Cilipi to the south-east, a rail link with Sarajevo and ferries to Venice, Rijeka, Split and Piraeus. However, it was badly damaged during the civil war of the early 1990s

Population	70 700
Map	Croatia – Dc

Duchcov *Czech Republic* See TEPLICE

Duero *Spain* See DOURO

Duisburg *Germany* The largest inland port in Europe, standing at the junction of the Rhine and Ruhr rivers, 18 km (11 miles) west and slightly south of Essen. Duisburg is a heavy industrial and manufacturing centre. The canals of the Ruhr Valley carry iron ore and coal to its huge steel plants, chemical works and shipbuilding yards situated alongside the port. The Flemish mapmaker Gerhard Mercator (1512-94), who invented the map projection named after him, is buried in the town.

Population	535 400
Map	Germany – Bc

Dukhan *Qatar* Oil town situated 80km (50 miles) west of the capital, Doha. It lies in the centre of the Dukhan oil field which produces about half of the country's oil output. The first well was drilled in 1940, but then the Second World War slowed production and exports did not get properly underway until 1949. The field measures about 55km (34 miles) by 9km (6 miles) and has reserves of about 2400 million barrels. The nearby Khuff natural gas field started production in 1978, and there is a liquefaction plant processing the gas for export.

Map	Qatar – Ab

Duluth *USA* Industrial port in Minnesota, at the western end of Lake Superior. It handles grain, iron ore, oil and coal, and produces steel.

Population	(city) 85 500; (metropolitan area) 239 920
Map	United States – Ha

Dumaguete *Philippines* See NEGROS

Dumbier *Slovakia* See TATRA MOUNTAINS

Dumfries and Galloway *United Kingdom* Region of south-west Scotland, covering 6397 km² (2469 sq miles) and taking in the former counties of Dumfries, Kirkcudbright and Wigtown. Its south coast, along the Solway Firth, is corrugated with bays, many of which shelter small resorts. Farther west lies the port of Stranraer on Loch Ryan, the principal crossing point to Northern Ireland. Inland, the hill country, chequered with forestry plantations, rises to 843 m (2766 ft). Dumfries (population 31 310) on the River Nith is the county town.

Population	148 400
Map	United Kingdom – Cc

Dumyat (Damietta) *Egypt* Town on the Dumyat River, the main eastern branch of the Nile delta, 13 km (8 miles) from its mouth. It makes textiles from the cotton grown in the delta region. The town has a new port complex.

Population	113 000
Map	Egypt – Bb

Dünaburg *Latvia* See DAUGAVPILS

Dunajec *Poland* River of southern Poland, 247 km (153 miles) long. It rises in the Tatra Mountains and flows north to join the Vistula River about 60 km (37 miles) north-east of the city of Cracow. The Dunajec cuts a spectacular gorge, up to 300 m (985 ft) deep, through the Pieniny Hills.

Map	Poland – Dd

Dunaújváros *Hungary* The country's first and largest post-Second World War 'new town', situated on the west bank of the River Danube, 58 km (36 miles) south of Budapest. It was formerly the small, sleepy village of Dunapentele. However, from 1950 it was transformed into the major iron, steel and cement town of Sztalinvaros, named after the Soviet leader, Joseph Stalin (1879-1953). The name was changed after 1956, when Stalin was disavowed. A museum traces the town's planning and growth.

Population	58 860
Map	Hungary – Ab

Dunbartonshire *United Kingdom* See STRATHCLYDE

Dundee *United Kingdom* Port on the east coast of Scotland, standing on the Firth of Tay, 60 km (37 miles) north of Edinburgh. The site has been occupied since before the birth of Christ, but the growth of the city came in the 19th century, with shipbuilding, engineering and jute mills. Today, the mills are largely silent, but Dundee's other industries are flourishing in the dock area and the suburbs. The port has recently come to play an important role in the North Sea oil industry.

Dundee is well known for its rich cakes, and, since the 18th century, for its orange marmalade. But the city became famous in 1879 after a rail disaster, when the rail bridge spanning the Tay River collapsed in a gale, plunging a train into the river with the loss of 75 lives. The bridge replacement was joined in 1966 by a road bridge.

Population	172 860
Map	United Kingdom – Db

dune Hill or ridge of wind-blown sand, especially one with no vegetation.

Dunedin *New Zealand* South Island port, 306 km (190 miles) south-west of the city of Christchurch. Its industries include ship repairs, beer brewing, and the manufacture of chemicals and furniture. Dunedin exports frozen meat, wool and dairy products.

The town was founded in 1848 by Scottish Presbyterians and became the country's leading settlement when gold was discovered in the area in 1861. Its wealth was used to put up a number of fine public buildings, including First Church (1868-73), the university (1878) modelled on Glasgow's in Scotland, and the elegant Gothic railway station (1902). There are also several museums depicting Maori culture and life in the days of the early European settlers.

Population	109 400
Map	New Zealand – Cf

Dunfermline *United Kingdom* See FIFE

Dungannon *United Kingdom* Market and textile town in the county of Tyrone, in Northern Ireland,

about 55 km (35 miles) west and slightly south of the capital, Belfast. It is also the administrative centre of the region. Dungannon was the seat of the O'Neills, Kings of Ulster for some 500 years until the end of the 16th century. The town's Royal School was founded by James I in 1608.

Population	(town) 8500; (district) 45 700
Map	United Kingdom – Bc

Dunkirk (Dunkerque) *France* Third largest French harbour and channel ferry port, which is 40 km (25 miles) north-east of Calais and became famous in the Second World War. In May and June 1940, some 225 000 soldiers of the retreating British Expeditionary Force and 112 000 French and Belgian soldiers escaped from its beaches in an armada of small boats in the face of the German army's advance. The town was almost completely destroyed by bombing in 1940 and has been rebuilt as one of France's most carefully planned industrial and residential towns. Major industries include steel and aluminium manufacturing, and oil refining.

Population	(city) 70 000; (metropolitan area) 196 600
Map	France – Ea

Dunkwa *Ghana* One of the largest towns of the thinly populated south-west, about 170 km (105 miles) west and slightly north of the capital, Accra. Dunkwa is the chief town of the Denkyira people and is a market for firewood, plantains, rice, cassava, vegetables and palm wine. It prospered after the Dunkwa-Awaso Railway opened in 1944 to carry bauxite from near Awaso, as well as gold from mines at Bibiani and timber.

Population	15 000
Map	Ghana – Ab

Dún Laoghaire *Ireland* Port in the south-east of the Dublin conurbation, and the car ferry terminal for the Holyhead (Wales) service. The harbour designed by John Rennie was started in 1817 as famine relief work and was completed by convicts. *Dún Laoghaire*, which is Gaelic for 'Fort of Leary', was known as Kingstown after George IV's visit in 1821. It resumed its original name with Gaelic spelling in 1920.

Population	55 540
Map	Ireland – Cb

Dún na nGall *Ireland* See DONEGAL

Durance *France* River of Provence. It flows 324 km (201 miles) through the southern part of the French Alps and is the last tributary of the Rhône, which it joins near Avignon.

Map	France – Fe

Durango *Mexico* State in northern Mexico crossed by the Sierra Madre Occidental Mountains. It has huge reserves of gold, silver, copper and iron. One of the largest iron ore deposits in the world lies just north of the state capital, also called Durango. During Spanish colonial rule, it was a wealthy religious centre.

Population	(state) 1 352 000; (city) 414 000
Map	Mexico – Bb

Durban *South Africa* Largest and busiest port on the African continent and largest city of the province of KwaZulu-Natal. Settled by white traders as Port Natal in 1823, and renamed after

Cape Governor Sir Benjamin D'Urban in 1835. Durban is a leading tourist centre on the Indian Ocean, with fine beaches and access to the resort areas of the DRAKENSBERG and the wildlife reserves to the north-east. In the harbour area are oil refineries, grain elevators and a huge sugar terminal. There are also shipyards, and factories producing processed foods, chemicals and textiles. Among the adjoining residential areas, hit by unrest in the mid-1980s and political violence in the early 1990s, are KwaMashu, Umlazi and KwaMakhutha.

Population 1 895 000

Map South Africa – Db

Durham *United Kingdom* City and county of north-east England. The city, 22 km (14 miles) south of Newcastle, is one of the most spectacular in Britain, set in a narrow loop of the River Wear, whose deep, wooded gorge separates the old and new towns. It is dominated by its massive Norman cathedral, burial place of the English historian and monk the Venerable Bede (673-735), which towers over the castle and the narrow streets of the medieval town. The castle is now part of the city's university.

The county covers an area of 2436 km² (941 sq miles). Its economy in the 18th and 19th centuries was dominated by coal, but mining has declined and the steel industry at the town of Consett has been closed down. Parts of the county were cut off in 1974 to form the new counties of CLEVELAND and TYNE AND WEAR.

Population (city) 38 110; (county) 605 800

Map United Kingdom – Ec

Durmitor *Yugoslavia (Serbia and Montenegro)* National park covering 320 km² (124 sq miles) between the torrential Piva and Tara rivers in the KARST of central Montenegro. Mount Durmitor itself, scene of the Montenegrin uprising in 1941, and considered part of the DINARIC ALPS, is a tablelike plateau rising to 2522 m (8274 ft). It is speckled with lakes, forests and in parts has snow for year-round skiing. To the east lies Zabljak, at 1450 m (4757 ft) the country's highest town, with traditional log cabins and modern hotels.

Map Serbia and Montenegro – Bc

Durrës *Albania* Port founded in the 7th century BC by the Greeks in a sheltered Adriatic bay. It was taken successively by the Illyrians (312 BC), Romans (230 BC), Ostrogoths (AD 481), Bulgars (10th century), Byzantines (12th century), Venetians (1394), and Turks (1501) who held it until 1912. It is now Albania's leading port, but retains its medieval walls and towers. Its Roman amphitheatre was built in the 2nd century BC and is the largest in Europe.

Durrës serves the country's main industrial region. Railways run to Tiranë and Elbasan, and the port is the focus for grain mills, and rubber, fertiliser, machinery and shipbuilding industries.

Population 86 900

Map Albania – Bb

Dushanbe *Tajikistan* Capital and modern industrial city, 150 km (95 miles) north of the border with Afghanistan. From 1929 to 1961, it was called Stalinabad in honour of the Russian leader Joseph Stalin (1879-1953), and grew from a village of 6000 inhabitants in the 1920s to the well laid-out city of today. It now has a university, an academy of science, and several colleges. Its

main industries are manufacturing cotton and silk textiles, engineering and food processing. Since 1992, the city has seen heavy fighting between anti-government forces and the army.

Population 604 000

Map Tajikistan – Ab

Dusky Sound *New Zealand* Inlet on the Fiordland coast of South Island, about 310 km (193 miles) west of Dunedin. It was one of the anchorages used by the English explorer Captain James Cook in 1773. Australian-based sealers arrived here in about 1792 and built the first European-style houses in the country.

Map New Zealand – Af

Düsseldorf *Germany* City on the River Rhine, about 34 km (21 miles) north and slightly west of Cologne. It is the capital of North Rhine-Westphalia and the commercial centre of the industrial Ruhr. Northern Germany's main stock exchange is situated in the city. Birthplace of the poet and essayist Heinrich Heine (1797-1856), it is also a major cultural centre, with a progressive film industry and an avant-garde art movement. Its main street, the Königsallee ('Kö'), has a lake in the middle – a leftover of the old town moat. Industries include engineering and chemicals.

Population 575 800

Map Germany – Bc

dust Small particles of matter which are fine enough to float in air. There are various sources of dust, such as volcanic activity, forest fires, and smoke from industry.

dust bowl Barren area in semi-arid regions, produced by excessive wind erosion of the soil, especially after the removal of vegetation by overgrazing or badly managed cultivation. The dust bowl that developed in the United States in the 1930s and inspired John Steinbeck's *The Grapes of Wrath* stretched from Kansas through Oklahoma, Texas, Colorado and into New Mexico.

dust devil Small, short-lived whirlwind that swirls dust, debris and sand up into the air. It varies in size and is caused by intense local heating of the earth's surface resulting in convection. Dust devils can occur in most regions, but are most common in hot deserts.

Dux *Czech Republic* See TEPLICE

Dvina *Eastern Europe* Name of two unconnected river systems in Russia, Belarus and Latvia. The North Dvina drains a large part of northern Russia. It is 1320 km (820 miles) long, and is navigable between May and November from Archangel on the White Sea to the industrial town of Kotlas, 470 km (290 miles) upstream. Once used by trappers, it is now used by forestry companies to transport timber by raft.

The West Dvina (also called 'Daugava' in Latvia) flows 1020 km (635 miles) from the Valdai Hills near Moscow to the Baltic Sea at Riga. It was part of the 'water road' used by Viking traders travelling south-east from Scandinavia to Turkey in the Middle Ages, but is now little used. Both rivers flood widely when the ice melts in April.

Map (North Dvina) Russia – Fb; (West Dvina) Russia – Dc; Latvia – Ba

Dvinsk *Latvia* See DAUGAVPILS

Dwarka *India* One of Hinduism's seven holy cities and legendary capital of the god Krishna, standing on the west coast, about 510 km (317 miles) north-west of Bombay. It has a temple of Krishna, who is said to have sought shelter there after fleeing from Mathura, and to have built the temple in one night. The temple is closed to non-Hindus.

Population 28 000

Map India – Ac

Dyfed *United Kingdom* County of south-west Wales covering 5765 km² (2226 sq miles). It was created in the 1970s from the counties of Cardigan, Carmarthen and Pembroke and named after an old Celtic kingdom.

Hill farms and open moorland of the Cambrian Mountains dominate the area, but it is the coastline for which Dyfed is famous. MILFORD HAVEN is one of Britain's finest and deepest natural harbours, and the cliffs to the west are some of the most spectacular in Britain. Aberystwyth and Tenby are the main seaside resorts.

There are two university colleges, at Lampeter and Aberystwyth, and the latter town also houses the National Library of Wales. Britain's smallest city, St David's (population 2000), with its cathedral dedicated to the patron saint of Wales, stands at Dyfed's most westerly point.

According to Welsh legend, the great Celtic wizard Merlin was born at Carmarthen, now the county town.

Population 350 000

Map United Kingdom – Cd

dyke Earth and rock embankment built to protect low-lying land from flooding. Examples include those protecting the POLDERS of the Netherlands, totalling 1300 km (over 800 miles).

The term is also applied variously to refer to a LEVEE, to a drainage ditch (or watercourse with an embankment made from the earth which was removed when digging the ditch), to an earthworks (such as Offa's Dyke on the border of Wales, originally a boundary), or, finally, to a sheet-like mass of igneous rock that has cut through the horizontal structure of the rock into which it was intruded.

Dzerzhinsk *Russia* City and centre of the Russian chemical industry, 30 km (18 miles) west of Nizhniy Novgorod. Dating from the 1940s, it was built as an overflow for Nizhniy Novgorod, and its population consists largely of workers from the factories which line the Oka River and the Moscow-Nizhniy Novgorod Railway.

Population 286 000

Map Russia – Fc

Dzhambul *Kazakstan* See ZHAMBYL

Dzierzoniow (Reichenbach) *Poland* Textile town in south-west Poland, 48 km (30 miles) south-west of the city of Wroclaw. By 1500, it was already a flourishing linen manufacturer, and with adjoining Bielawa became Lower Silesia's main textile centre after 1700.

Population 37 600

Map Poland – Bc

Dzungaria *China* See XINJIANG UYGUR AUTONOMOUS REGION

E layer See ATMOSPHERE

eagre See BORE

earth See HOW THE EARTH WAS FORMED, p 206, and JOURNEY TO THE CENTRE OF THE EARTH, p 207

earth pillar Column of soft, earthy material such as clay, capped by a hard or indurated boulder which has protected it from erosion. It can be up to 60 m (about 197 ft) tall.

earthquake Series of shock waves which travel through the earth and along its surface, caused by a sudden release of energy in the CRUST or the upper MANTLE. Some earthquakes are so slight that they can be detected only by the most sensitive instruments; others, such as the Assam earthquake of 1897, which devastated 388 000 km² (150 000 sq miles) of north-east India and was felt over an area of almost 11 million km² (4.25 million sq miles), can unleash the energy of several nuclear bombs.

Many shallow earthquakes are caused by movements of FAULTS in the earth's crust, while more severe ones occur at much greater depths related to SUBDUCTION ZONES below the earth's surface. How these deeper earthquakes are produced is not yet known. Earthquake zones or belts coincide most often with PLATE MARGINS, which suggests that the zones are closely related to PLATE TECTONICS. One of the belts, the so-called Ring of Fire, encircles the Pacific. The second belt stretches west through China along the Himalayas and across Iran and then to the north and south of the Mediterranean.

There are two ways of measuring an earthquake – by its intensity, or amount (size), and by its magnitude, or strength. Intensity is determined by assessing the degree to which shaking is felt by people in the area, the amount of damage to buildings and other artificial structures, and the extent of visible deformation of the earth. The scale most commonly used to measure intensity is the MODIFIED MERCALLI SCALE.

Earthquake magnitude is determined by the size of the seismic waves, that is, the vibrations emitted by the release of energy. The waves are measured on a RICHTER SCALE.

The centre of an earthquake is called the FOCUS, and it is from the focus that seismic waves are emitted. All foci so far located have been in the upper 720 km (440 miles) of the earth.

The earthquake's EPICENTRE is the point on the earth's surface directly above the focus. Intensity decreases outwards from the epicentre, but since buildings constructed on soft soil suffer more damage than buildings on hard rock, two separate points at the same distance from the epicentre can have a different intensity. The measurement of intensity therefore varies from place to place. Magnitude, on the other hand, is a single measurement for each earthquake.

East Anglia *United Kingdom* Area of eastern England comprising the counties of NORFOLK and SUFFOLK, and parts of CAMBRIDGESHIRE and ESSEX. Relatively flat and thinly populated, it is one of Britain's leading producers of wheat and other cereals.
Map United Kingdom – Fd

East London *South Africa* Seaport founded in 1826 at the mouth of the Buffalo River. It serves

MEASURING AN EARTHQUAKE'S MAGNITUDE

About a million earthquakes occur around the world each year, and their magnitude is measured on the RICHTER SCALE devised in 1935 by the Californian seismologist Charles Richter. The scale is logarithmic – an increase of one point on the scale means that the force of the earthquake is in actual fact 10 times greater than that of the preceding number on scale. An earthquake with a magnitude of 8, therefore, is 10 ten times more powerful than one with a magnitude of 7 and 100 times more powerful than one with a magnitude of 6.

SOME MAJOR EARTHQUAKES OF THE 20TH CENTURY

Place	Magnitude	Approximate No. of Deaths	Date
China	8.6	180 000	1920
USA (Alaska)	8.5	178	1964
Chile	8.5	4000-5000	1960
China	8.3	200 000	1927
Japan	8.3	143 000	1923
USA (San Francisco)	8.3	700	1906
China	8.2	240 000	1976
Mexico	8.1	12 200	1985
Armenia	8.0	25 000	1988
Iran	8.0	189	1977
Ecuador	7.9	600	1979
New Zealand	7.9	255	1931
Philippines	7.8	1600	1990
Japan	7.8	250	1993
Iran	7.7	10 000	1990
Algeria	7.7	3500	1980
Iran	7.7	15 000	1978
Peru	7.7	50 000-70 000	1970
Italy	7.5	83 000	1908
Pakistan	7.5	20 000-60 000	1935
Indonesia	7.5	1 300	1992
Iran	7.4	12 000	1968
Italy	7.2	3000	1980
Japan	7.2	5000	1995
USA (San Francisco)	6.9	67	1989
USA (Los Angeles)	6.6	61	1994

the Ciskei and Transkei areas of the Eastern Cape Province. Close to the city are many superb beaches, their waters warmed by the south-flowing Mozambique Current.
Population (city) 101 330; (district) 240 470
Map South Africa – Cc

East Rand *South Africa* That part of the Witwatersrand which lies east of Johannesburg and includes the towns of Benoni, Boksburg, Brakpan, Germiston and Springs, and their satellite black townships, a legacy of South Africa's apartheid era. Gold mining developed rapidly after 1923 when the mines of the central Witwatersrand became exhausted. The East Rand has become a thriving industrial area.
Map South Africa – Cb

East Siberian Sea Part of the ARCTIC OCEAN, between the coast of north-eastern SIBERIA, Wrangel Island and NOVOSIBIRSKIYE OSTROVA (New Siberian Islands). Its northern limit is the edge of the Arctic continental shelf. The East Siberian Sea is navigable only in August and September, when it is ice free.
Map Russia – Pa

Easter Island (Isla de Pascua; Rapa Nui) *Pacific Ocean* Owned by Chile, Easter Island is a tiny, remote volcanic island of 120 km² (46 sq miles) in the Pacific, 3780 km (2349 miles)

from the South American coast. 'Discovered' in 1722 by the Dutch admiral Jacob Roggeveen, who named it after the day he landed, it was annexed by Chile in 1888.

Some of the world's most fascinating and enigmatic carved monuments are to be found on the island – more than 600 strange stone faces, ranging in height from 3 m to 12 m (10 to 40 ft), which were carved by the islanders' Polynesian predecessors more than 1000 years ago. Together with other carved stonework, they are all that remain of this ancient civilisation.

The main source of income for the islanders is tourism. An airport opened in 1967 has linking flights with Tahiti and with Santiago, Chile.
Population 1800
Map Pacific Ocean – Hd

Eastern Cape *South Africa* One of nine provinces created in 1994, covering 170 616 km² (65 875 sq miles) It extends inland from the southern coast to the Free State and Lesotho and is bounded in the west by the Western Cape and in the east by KwaZulu-Natal. Home of the Xhosa people and descendants of 19th-century British and German settlers, the province is mostly agricultural, but larger centres include the industrial complex of Port Elizabeth-Uitenhage, East London and Grahamstown.
Population 6 665 400
Map South Africa – Cc

HOW THE EARTH WAS FORMED

About 4600 million years ago, a vast nebula of dust and gas surrounding the sun began to form and condense into balls of matter, which became the planets of our solar system.

There are several theories about exactly how the nebula condensed. But it is generally assumed that particles collided, or were attracted to one another by electrostatic action, to form a nucleus which became progressively more dense. As this gaseous mass grew, it began to exert a gravitational pull, which in turn attracted more material. As gravity increased, it drew heavier matter to the centre, while lighter materials collected around. The whole mass condensed into liquid form, then began to solidify.

Whether this mass was hot or cold is another subject of debate. Some scientists believe that it began cold, but heated up to a molten state as materials at the centre became more and more compressed, and radioactive decay set in.

Others believe that the primitive mass blended into a molten ball, which has been cooling ever since.

There is no doubt, however, that the earth was once extremely hot, and then cooled. As cooling occurred, some minerals started to crystallise and globules of iron and nickel sank to the centre, where a core of great density had started forming. Outside this core, the next layer of material began building up – silicates and sulphides which would later be compressed into the dense mantle of rock that now surrounds the core. Finally, the lightest materials – more silicates – rose to the surface and cooled and hardened to form the earth's crust.

By this time the lighter gases had floated off into space again and others had been absorbed, leaving almost no atmosphere. So, with a minimum of atmospheric friction to burn them up, enormous meteorites plunged unhindered through the newly forming crust to add even more materials to the mixture. But all the time, the crust was deepening and becoming more solid. Gases, including carbon monoxide and carbon dioxide, escaped from the interior furnaces to create an atmosphere of sorts. The crust in turn developed into two layers according to relative densities – a top one of lighter rocks, and a denser layer underneath.

It is thought that at this stage, a vital new ingredient was exuded into the atmosphere – water vapour. As the earth continued to cool, and the atmosphere with it, this water vapour condensed and fell as rain.

Meanwhile, surface rock was cracked and melted again and again, as the earth's seething interior erupted in volcanic spasms. But gradually, over millions of years, large, stable rafts of granitic material accumulated, floating around on the molten matter beneath them. Some collided and joined up – to become the primitive continents. Rains falling on these land masses formed lakes and streams, and flowed down onto the low-lying areas between the continents to form the first oceans.

Materials carried and dissolved into the oceans mixed in a salty soup of – among many elements – boron, bromine, chlorine, iodine and sulphur. In the seas and in the thicker soups of hot, volcanic pools, primitive living cells developed. About 2000 million years ago, plant cells capable of photosynthesis evolved. And they began adding oxygen to the atmosphere, so allowing the rapid expansion of animal life.

Dust and gas in the solar nebula condense into a molten ball. Gases are absorbed or lost into space, so there is no atmosphere.

As earth cools a crust forms, but intense volcanic activity in the super-heated regions below expels gases in large quantities through the crust, to create a primitive type of atmosphere – but one which still lacks oxygen.

Further cooling frees water vapour into the atmosphere, and finally temperatures drop sufficiently (to 100°C or 212°F) for the thick clouds to condense into rain. Great lightning storms charge the atmosphere with electricity and the rains gather new chemicals as they fall to earth. Organic compounds are formed by lightning and radiation.

The storms subside and as water evaporates organic compounds are concentrated. Self-sustaining cells evolve. Plant cells capable of photosynthesis follow, and add growing quantities of oxygen into the air. High in the atmosphere a belt of ozone forms, to shield earth from damaging solar radiation – and allow its new life forms to multiply.

Eastern Highlands *Zimbabwe* Vast granitic mountain range rising abruptly from Zimbabwe's plains, 260 km (162 miles) south-east of the capital, Harare, and close to the Mozambican border. It stretches some 300 km (186 miles) from north to south and contains mountains such as the Inyanga range and Mount Selinda, in places rising to 2500 m (8200 ft) and higher.

The Highlands, which have lakes and streams, rain forests and fascinating rock formations, offer fine views over surrounding tea and tobacco plantations. Once a refuge for mountain-dwelling peoples, the area has become a tourist resort, popular with hikers, mountaineers and trout fishers. MUTARE, one of Zimbabwe's larger and most attractive towns, lies at its centre.
Map Zimbabwe – Ca

Eastern Transvaal *South Africa* One of nine provinces created in 1994, covering 81 816 km² (31 589 sq miles). It extends east from the Witwatersrand to Swaziland and Mozambique and south to KwaZulu-Natal, and includes the former self-governing states of KwaNdebele and Lebowa. Tourist attractions in the province include the KRUGER NATIONAL PARK, which extends into Northern Transvaal, and several other game reserves, as well as the great DRAKENSBERG escarpment which separates the uplands (the HIGHVELD) from the LOWVELD. Mining, cattle ranching, forestry and fruit farming are the mainstay of its economy; Ermelo, Nelspruit, Witbank and Middelburg are its major centres.
Population 2 000 000
Map South Africa – Db

ebb tide Receding tide between high water and the succeeding low water.

Ebolowa *Cameroon* Market town 110 km (70 miles) south of the capital, Yaoundé, on the main road to Gabon. It is a collection centre for timber and cocoa for export, and its future depends largely on improving the quality of cocoa from smallholdings where trees, frequently aged and diseased, need to be replaced. The town is also a tourist centre for the surrounding forests, rich in animal and bird life, and for the spectacular mountain scenery of Ako.
Population 30 000
Map Cameroon – Bb

Ebro *Spain* Second longest river 909 km (565 miles) long, flowing in the north-east from the Cantabrian Mountains to the Mediterranean south of Tarragona. The Battle of the Ebro in 1938, in which 75 000 Republicans died, was a decisive victory for the Nationalists in the Spanish Civil War which ended a year later.
Map Spain – Eb

Ech-Chlef *Algeria* See AL ASNAM

Echmiadzin *Armenia* See EJMIADZIN

Ecrins, Parc des *France* Largest of the French national parks, covering 918 km² (354 sq miles), and situated 60 km (37 miles) south-east of Grenoble. The park contains outstanding Alpine

---

JOURNEY TO THE CENTRE OF THE EARTH

The earth's outer skin, called the CRUST, is a complex structure forming part of the LITHO-SPHERE. On the continents, the crust is 92 per cent IGNEOUS or METAMORPHIC ROCK, much of it granitic, and 8 per cent sedimentary rock. Beneath extensive flat areas of land such as the North American prairies, the crust is between 30-35 km (19-22 miles) thick. But under mountain ranges its thickness increases to as much as 80 km (50 miles). The crust under oceans, by contrast, is largely basaltic with a thin layer of sediment, and is only about 5 km (3 miles) thick. Separating the crust from the next layer, the MANTLE, is a zone known as the Mohorovicic discontinuity in which primary shock waves from EARTHQUAKES speed up dramatically. The mantle extends to a depth of 2900 km (1800 miles). The uppermost mantle is solid but between the depths of 75 and 250 km (50-155 miles), there is a semi-fluid transition zone (the ASTHENOSPHERE). The mantle below this is solid again and is called the MESOSPHERE.

Beneath the mantle lies the CORE. The outer layer – about 2100 km (1305 miles) thick – is molten and probably consists of nickel-iron. It is in constant motion, and electrical currents generated and circulating within it account for the earth's magnetic field.

The inner core of the earth, about 2740 km (1700 miles) in diameter, is thought to be a solid ball of nickel-iron squeezed to unimaginable densities by a pressure of 3800 tonnes per cm² at a temperature of around 4000°C (7200°F).

THE LAYERS OF THE EARTH

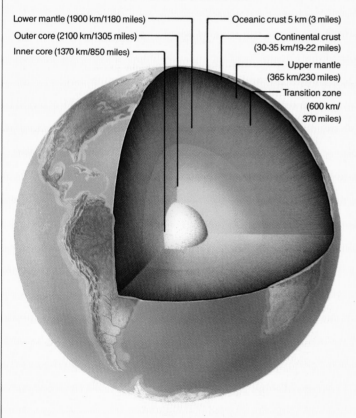

Lower mantle (1900 km/1180 miles)
Outer core (2100 km/1305 miles)
Inner core (1370 km/850 miles)
Oceanic crust 5 km (3 miles)
Continental crust (30-35 km/19-22 miles)
Upper mantle (365 km/230 miles)
Transition zone (600 km/370 miles)

SOLID AND SEMI-FLUID LAYERS

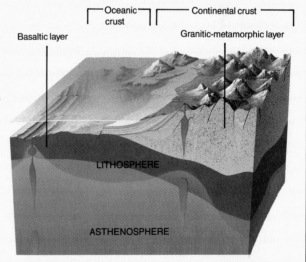

Oceanic crust
Basaltic layer
Continental crust
Granitic-metamorphic layer
LITHOSPHERE
ASTHENOSPHERE

An alternative division of earth, emphasising physical rather than chemical properties, shows the solid lithosphere (which includes the crust and part of the upper mantle) surrounding a semi-fluid asthenosphere. This in turn surrounds a solid mesosphere, which goes down to the core. The asthenosphere lies between depths of 75 km (47 miles) and about 250 km (155 miles). It is also known as the low velocity layer, because the speed at which the seismic waves of earth tremors travel – generally increasing with depth – is slowed by a few per cent in the asthenosphere. However, seismic waves speed up as they cross the Mohorovičić discontinuity, between the crust and mantle, and as they penetrate the transitional zone.

glaciers, lakes and gorges, with four peaks rising close to 4000 m (13 123 ft). They are the Barre des Ecrins, which is the highest, at 4102 m (13 458 ft), La Meije at 3983 m (13 067 ft), L'Ailefroide, reaching 3953 m (12 969 ft), and Pelvoux, at 3946 m (12 946 ft).

Map France – Gd

Ecuador See p 209

Eden, Mount *New Zealand* Highest of the volcanic cones around Auckland on North Island, rising to 196 m (643 ft) and offering panoramic views across the city.

Map New Zealand – Eb

Eder *Germany* River rising in the Rothaer hills north-west of FRANKFURT and flowing first east, then north for 175 km (110 miles) to join the Fulda River near the town of KASSEL. A dam built on the river and another on the Möhne, both providing hydroelectric power, were destroyed in 1943 by Britain's Royal Air Force in a 'Dambusters' raid, using a bouncing bomb invented by the British inventor and engineer Barnes Wallis (1887-1979).

Map Germany – Cc

Edinburgh *United Kingdom* Capital of Scotland, standing near the southern shore of the Firth of Forth. The city, the majesty of its castle, the quality of its cultural life (which led to its nickname, 'the Athens of the North') – even the bitterness of the wind when it blows off the North Sea – have inspired writers and romantic visitors alike for centuries.

Although Edinburgh has not had a resident monarch since 1603 and a Scottish parliament since 1707, it is not short of splendid buildings. Parts of its castle date back to about 1100. Holyroodhouse was the seat of the last Scottish monarchs (including Mary, Queen of Scots, who was executed in England in 1587), and is now owned by the British Crown.

The city is rich in artistic attractions, with its national gallery, Royal Scottish Academy, Royal Scottish Museum, and museum of modern art. It has three universities, three cathedrals and a number of theatres. However, it has no opera house – which means that opera lovers in the city have to go to Glasgow to enjoy this cultural activity.

Each year in August and September, the Edinburgh International Festival fills not only its theatres but every church hall, basement and, indeed, many less formal venues including pavements, with performers and audiences.

The old town's spine is the 'Royal Mile' (Lawnmarket, the High Street and Canongate), which runs down from the castle, past the 14th to 15th-century St Giles Cathedral, to Holyrood. The new town lies to the north. It dates from the 18th and early 19th century and is laid out in formal streets (notably Princes Street), squares and crescents. North again is the port of Leith. Arthur's Seat, a 251 m (823 ft) hill of volcanic rock, overlooks the city.

Population 434 520

Map United Kingdom – Dc

Edirne (Adrianople) *Turkey* Town on the Greek border, 250 km (155 miles) north-west of Istanbul. Until early this century it was called Adrianople after the Roman emperor Hadrian (AD 76-138). The Selimiye mosque with its four large minarets dominates the town. The main industries here are textiles, carpets, leather goods, soap and perfumes (rose oils).

The surrounding province of Edirne, which covers 6266 km² (2419 sq miles), produces wheat, rye, fruits and rice.

Population (city) 86 700; (province) 404 600

Map Turkey – Aa

Edmonton *Canada* Capital of ALBERTA, standing in the eastern foothills of the Rocky Mountains. Founded in 1795 as a trading post and fort of the pioneering Hudson's Bay Company, it is now the retailing outlet for a rich agricultural area. Oil entered the economy after being discovered nearby in 1947. Today, Edmonton is a major cultural, educational and financial centre.

Population (city) 616 740; (metropolitan area) 839 920

Map Canada – Dc

Edo *Japan* See TOKYO

Edward, Lake (Lake Rutanzige) *Uganda/Zaire* Lake covering about 2135 km² (824 sq miles) in the GREAT RIFT VALLEY south of the Ruwenzori Mountains, at a height of about 915 m (3000 ft). The Rwindi and Rutshuru rivers flow into it and the Semliki River drains it, flowing north into Lake ALBERT; Lake Edward is therefore one of the sources of the NILE.

The lake was explored in 1889 by the British adventurer Sir Henry Morton Stanley, who named it after Albert Edward, then Prince of Wales. Uganda's president Idi Amin renamed it Lake Idi Amin Dada, but the former name was restored when Amin was deposed in 1979.

Map Zaire – Bb

Efate, *Vanuatu* See PORT VILA

Efes *Turkey* See EPHESUS

Eger *Czech Republic* See CHEB

Eger *Hungary* Beautiful Baroque city on the southern slopes of the Bükk Mountains, 40 km (25 miles) south-west of the north-eastern city of Miskolc. It is the chief town of Heves county, and the centre of a wine-making district which produces Egri Bikavér ('Bull's Blood') wine. It is also a popular medicinal spa for people suffering from rheumatism.

The city's architecture emphasises its eventful history. Europe's northernmost surviving minaret dates from the Turkish occupation (1596-1687) of the city. Eger also has a 16th-century castle standing at the end of a street of exquisite 17th and 18th-century palaces and houses. The city's cathedral, rebuilt early this century, is one of Hungary's largest churches.

Population 63 020

Map Hungary – Bb

Egmont, Mount *New Zealand* Dormant volcano rising to 2518 m (8261 ft) on North Island, 220 km (140 miles) north and slightly west of the capital, Wellington. It is noted for its symmetry and beauty. According to Maori legend, the cone rises in splendid isolation from other volcanoes on the island because of a lovers' quarrel between Taranaki (Mount Egmont) and Mount Tongariro.

Map New Zealand – Eb

Egypt See p 210

Eifel *Germany* Volcanic highland of sparsely inhabited moors and forests between the lower Moselle River and the borders of Luxembourg and Belgium. Visitors enjoy the region's scenery and the lakes that were created when parts of the Roer River were dammed.

Map Germany – Bc

Eiger *Switzerland* Mountain, 3970 m (13 025 ft) high and 60km (35 miles) south-east of the capital, Berne. Its rugged north face – the 'Nordwand' – has been nicknamed *'Mordwand'* ('Murder Face') because of the number of climbers killed while trying to scale it – 53 between the first attempted ascent in 1935 and 1994. The first successful ascent was in 1938.

Map Switzerland – Ba

Eilat *Israel* See ELAT

Eindhoven *Netherlands* Industrial city near the Belgian border, 110km (70 miles) south and slightly east of Amsterdam. It is the headquarters of Philips, the electrical and electronics firm, founded here in 1891. The Philips Museum, the Van Abbe museum of modern art, and a technical university are located in the town.

Population (city) 195 270; (Greater Eindhoven) 390 810

Map Netherlands – Bb

Einsiedeln *Switzerland* Small town – Switzerland's chief place of pilgrimage for Catholics – situated 30 km (19 miles) south-east of Zürich. Its Black Madonna, a 15th-century wood carving, which has survived several fires and theft attempts, is kept in a Benedictine monastery church, part of an 18th-century Baroque complex. Einsiedeln is a popular centre for cross-country skiing.

Population 11 180

Map Switzerland – Ba

Eisenach *Germany* Resort town in the south-west. Eisenach lies in Thüringen State, on the edge of the Thüringer Forest. The religious reformer Martin Luther (1483-1546), leader of the Protestant Reformation, lived here during his schooldays. Composer Johann Sebastian Bach (1685-1750) was born here. Its manufactures include cars, machinery, metal, timber, textiles, electronics and chemicals.

Population 51 000

Map Germany – Dc

Eisenstadt *Austria* Town and capital of Burgenland State, which until 1921 was a part of Hungary. Situated about 40 km (25 miles) south of Vienna, Eisendstadt became the capital of Burgenland in 1925.

For centuries, the town was dominated by the wealthy Esterházy family, who owned 90 palaces across the Austro-Hungarian Empire. In 1761, the family engaged the composer Joseph Haydn (1732-1809) as their musical director, and he worked in the ornate Esterházy Castle for almost 30 years. The castle's Haydn Room is open to the public; the composer himself is buried in the nearby Bergkirche. Haydn's house is now a museum. The town is the centre of a wine and fruit region.

Population 10 500

Map Austria – Eb

Ecuador

WHERE POLITICAL FLARE-UPS ARE AS FREQUENT AS THE VIOLENT ERUPTIONS FROM THE VOLCANOES THAT MARK THE LANDSCAPE

1 CHIMBORAZO
2 BOLÍVAR
3 TUNGURAHUA

Ecuador's political scene is almost as volatile as its landscape, which includes more than 30 active volcanoes. Like eruptions in the mountains, there are regular rumblings as the military powers usurp a civilian government, hold power for a period, and are themselves replaced by another civilian government.

The main source of political strife is related to geography. In effect, Ecuador is three countries; the west, the central mountainous region of the Andes, and the east. Each has its own climate, its own terrain, and its own customs and attitudes. The west of the country, a hot coastal plain, is the seat of industry and commerce as well as an agricultural region producing the major exports of bananas, coffee, cocoa and sugar. GUAYAQUIL, Ecuador's largest city, which has been extensively modernised since oil was discovered nearby in the early 1970s, is situated in this region. Guayaquil has about 80 per cent of the country's industries.

Running down the middle of Ecuador, from north to south, are two spines of the ANDES, divided by a central plateau known as the Sierra. The plateau is broken into small, isolated river basins where most Ecuadorians live; 32 per cent of the country's workforce are still employed in agriculture, and most of the agricultural workers live here.

Life in the mountains is slower. Quito, the capital city, has the air of a Spanish colonial town, and rural Amerindian villages seem to belong to an even earlier age. The inhabitants of the Sierra are conservative by comparison to the coastal people, and the two groups consistently support different political organisations. The problem for every government is to reconcile these political differences.

The far eastern oil region is the third Ecuador. Home to just 2 per cent of the population, it is so underdeveloped that it takes hardly any part in the political dramas of the coast and the Sierra. Known as the ORIENTE, it consists of tropical savannah and rain forest around the headwaters of the River AMAZON. East of Baños lies some of the world's last really wild, unexplored territory. Since the discovery of oil here, the Peruvian-Ecuadorian conflict to define their common border has intensified. This region is valuable because, apart from oil, it offers opportunities for exploitation, both of the forest and its minerals.

DWINDLING OIL RESERVES

Ecuador's greatest problem is poverty. It has always been one of South America's poorest countries, but when oil was discovered in the 1970s salvation appeared to be beckoning. Although the oil did bring about rapid growth, especially in the cities, its benefits did not reach all of the population.

As in other parts of South America, the recession of the 1980s brought expansion to a sudden halt; in 1989, inflation in Ecuador reached a peak of 76 per cent. Then, in 1991, a cholera epidemic dealt a savage blow to the tourism industry. By the early 1990s, Ecuador was wrestling with the common Latin American diet of debt (more than US$12 000 million), financial advice and aid from the International Monetary Fund, and drastic devaluations of the currency. To compound its problems, is the prognosis that the oil reserves will be exhausted around the year 2010.

Meanwhile, those who did not profit from the oil boom while it lasted, have continued to suffer. Impoverished rural Amerindian families, who have left the countryside in the hope of making money, have built vast shanty towns on the fringes of cities such as Guayaquil.

While 1 per cent of Ecuador's landowners have large estates comprising 40 per cent of the land, two-thirds of all land holdings are smaller than 5 hectares (12 acres) – too small

▼ **WOMEN AT WORK Many women work as shepherds in the Andes of Ecuador, and children are carried on their mothers' backs. Coca leaves – a source of cocaine – are often chewed to combat weariness at high altitudes.**

for most peasant farmers to make a living. Those with plots on the steep hillsides of the Andes also have to cope with destructive soil erosion. In parts of the highlands, life is particularly hard; infant mortality among Amerindians is high (9 per cent) and life expectancy low (under 50 years).

About 30 per cent of Ecuadoreans are pure Amerindian, and although the government granted them title to 1 million hectares of land at Pastaza in May 1992, they still have little political power or wealth. Their ancestors had a great empire, but lost it to invading Incas from Peru in the 16th century. Inca rule was short-lived; the Spanish arrived in 1534 and enslaved the Amerindians and Incas for 300 years. Ecuador became part of the viceroyalty of Peru before being ceded to Greater Colombia in 1822. Independence in 1830 simply meant the transfer of power from the Spaniards to the *mestizos,* the mixed-blood descendants of the conquerors. Today, the *mestizos,* who make up 55 per cent of Ecuador's population, are still the privileged group.

ECUADOR AT A GLANCE	
Official name Republic of Ecuador	
Area 283 561 km² (109 483 sq miles)	
Population 11 280 000 **Per km²** 39.8 **(Per sq mile** 103)	
Capital Quito	
Government Presidential republic	
Currency 1 sucre = 100 centavos	
Languages Spanish (official); Amerindian languages	
Religion Christian (93% Roman Catholic)	
Climate Tropical; cooler, with less rainfall, at altitude. Average annual temperature on the coastal plains is 26°C (78.8°F); in Quito (altitude 2850 m, 9350 ft) it is 13°C (55.4°F)	
Land use Forest, woodlands 51%, meadows, pastures 17%, cultivation 9%, other 23%	
Main primary products Rice, cassava, maize, potatoes, bananas, oranges, coffee, cocoa, sugar cane, fish and shrimps; crude oil and natural gas, metal ores	
Major industries Agriculture, crude oil production and refining, cement, food processing, petrochemicals	
Main exports Crude oil, cocoa, sugar, coffee, bananas, shrimps	
Annual income per head (US$) 1070	
Population growth (per thous/yr) 28	
Life expectancy (yrs) Male 64 **Female** 68	

Egypt

AS DEPENDENT ON THE NILE AS IT WAS IN THE DAYS OF THE PHARAOHS, EGYPT IS A LAND OF ASTONISHING CONTRASTS BETWEEN RICH AND POOR, ANCIENT AND MODERN

Ancient kingdom of the Pharaohs, Egypt is the site of one of the world's oldest civilisations, with a recorded history of 5000 years. But it also has modern wonders, such as the High Dam at ASWAN on the NILE. Egypt is the doorway between Africa and Asia. With access to the MEDITERRANEAN and the RED SEA, it has always been a tempting prize for invaders. In 1956 the strategic importance of the SUEZ CANAL placed Egypt in the forefront of international affairs, when it closed the canal to shipping.

It is a land of astonishing contrasts between rich and poor, the lush Nile Valley and the surrounding desert, wretched hovels and huge monuments to the past. CAIRO is an enormous, overcrowded, sprawling city of 16 million people, one of the most cosmopolitan in the world. ALEXANDRIA, on the Mediterranean coast, has a population of nearly 4 million.

The Greek historian Herodotus, writing some 2500 years ago, called Egypt 'the gift of the Nile', for its existence depends on the waters of that great river, at 6695 km (4160 miles) the longest in the world. The rich soils deposited by flood waters along the banks of the Nile have supported large populations since history began; the Nile delta is one of the world's most fertile agricultural regions. Ninety-six per cent of the people live in the delta and the 20 km (12 mile) wide strip along the river, from Cairo to Sudan, making a population density of 1400 per km² (about 3630 per sq mile).

THE LAND AND ITS PEOPLE

Only a fraction of Egypt's 1 001 450 km² (386 900 sq miles) is settled or cultivated – the 36 000 km² (about 13 900 sq miles) along the Nile and around oases. The country consists of two deserts divided by the Nile Valley. The Western Desert – the north-eastern part of the SAHARA – is, apart from its mountainous southwest, generally low and undulating country. There are several extensive oases – BAHARIYA, Dakhla, Farafra, El Kharga and SIWA – and large, wind-scoured depressions, such as the QATTARA DEPRESSION which drops to 133 m (436 ft) below sea level.

The smaller Arabian Desert, or Eastern Desert, rises steeply from the Nile Valley to a bare, broken plateau. This slopes upwards to a high mountain range which borders the Red Sea. At the northern tip of the Red Sea, the SINAI Peninsula – a wedge of land – links Egypt and Africa with Israel and Asia. On it lies Egypt's highest peak, Gebel Katherina, 2637 m (8651 ft) high.

Egypt is a hot, dry land, with little rain except on the Mediterranean coast. The average maximum summer temperature in Cairo is 35°C (95°F); in winter it is 8°C (46.4°F). Summer temperatures in the southern and western deserts are even higher. Two winds are a curious feature of the climate. A north wind blows all year and since earliest times has enabled ships to sail up the Nile against its strong currents. But between March and June, a hot southerly wind, the khamsin, may blow, carrying sand and producing a yellow fog that may obscure the sun for days, while temperatures may rise by as much as 17°C (30°F) within two hours.

The original Egyptians were a distinctive race: a blend of African and Asian peoples of ancient lineage among whom the Arabs came as conquerors and settlers. But over the centuries, the people have become a mixture. In the desert live the Bedouin Arabs and the Berbers (a fair-skinned aboriginal people totalling between 50 000 and 80 000), and in the far south are small numbers of Nubians – a people of mixed Arabic, ancient Egyptian and Negro blood related to the Sudanese.

In recent years, Egypt has made big strides in industrial development and today, it is the second largest industrial nation on the African continent, after South Africa. But the population increases even faster than the economy – another 1.4 million people each year. If this rate is maintained, the population could well reach 70 million by the turn of the century. The country is always running to catch up with its ever-growing population – 60 per cent of food has to be imported, where not so long ago Egypt was an exporter of food. There is still a 52 per cent illiteracy rate despite great strides in education over the past 30 years.

Despite recent advances, Egypt remains one of the world's poorest countries. There are great extremes of wealth: the *fellahin*, the peasants of the rural areas, earn only one-tenth of what a craftsman earns. At the top of the scale are the businessmen, industrialists and wealthy government officials in the cities.

The *fellahin*, who live along the Nile and in the delta, make up two-thirds of the population. It is their labour across the centuries that has provided the wealth on which Egypt's achievements rest. Many have never left the villages in which they were born. Their farming methods and outlook remain much the same as centuries ago. Their hard lives, their poverty, their diseases (such as bilharzia, caused by intestinal flukes carried by water snails) have changed little, even though they may now have radio or television. They eat simply – their staple diet being rice, bread and beans.

More than 90 per cent of the people are Muslim but Egypt is not an extreme Islamic society, despite the rise of Islamic fundamentalism in the 1970s and again in the 1990s. In 1992, a planned coup by the fundamentalists failed, but by mid-1994 the situation had become serious.

Between 1992 and 1994, Islamic fundamentalists were responsible for the deaths of more than 330 people; increasingly, Coptic Christians and even tourists became targets of violence. Apart from Muslims, there are an estimated 3 to

5 million Coptic Christians who claim to be descendants of Egypt's original inhabitants before the coming of the Arabs. They are a minority, but are important politically and have been the focus of sectarian conflict.

As one might expect in the land of Cleopatra, women are more emancipated than anywhere else in the Arab world. They dress modestly but do not use the veil and can compete for a wide range of jobs – from waitresses to the highest positions in state enterprises. Polygamy, though still acceptable as part of Islam, is dying out for economic and social reasons.

LONG ROAD TO FREEDOM

Egyptians have long been used to outsiders. Foreigners come now as tourists, as business people or as technicians working for the United Nations or other international agencies, but for centuries they came as conquerors. From 525 BC, when the Persians occupied the country, Egypt was dominated by outside powers. In turn came the Greeks under Alexander the Great, the Romans, and the Arabs. In 1517, Egypt became part of the Turkish Ottoman empire. Napoleon's invasion in 1798 brought Egypt firmly to the attention of Europe.

Then, in 1854, the French engineer and diplomat Ferdinand de Lesseps was granted a concession to build the SUEZ CANAL, 160 km (99 miles) long and completed in 1869. Shortening the sea voyage from the East was not a new idea. A canal had been built by the Pharaohs around 2100 BC as far as Lake

▲ TOMBS OF KINGS The sun rises behind the pyramids at El Giza. The one built about 2570 BC for Chephren (right) was originally 143.5 m (471 ft) high and is made of limestone blocks weighing up to 15 tonnes. It stands by the great pyramid of his father, Khufu.

Timsah, and later extended by the Persian conqueror Xerxes I, but had fallen into disuse in the 8th century AD.

In 1882 Britain occupied Egypt, beginning an uneasy 75-year semi-colonial relationship; in 1914, Egypt became a British protectorate. British troops remained to defend Egypt and the canal even after Egypt officially became independent again in 1922. One of the greatest battles of the Second World War was fought in 1942 at EL ALAMEIN in the Western Desert.

In the 1950s, the nationalist cry of 'Egypt for the Egyptians' was finally translated into effective action. King Farouk was forced to abdicate in 1952. Four years later, after the British forces had left, President Gamal Abdal Nasser nationalised the Suez Canal. A triple attack by Britain, France and Israel to regain control of the vital waterway was ended by world pressure and the threat of Soviet intervention. Egypt was at last its own master.

Upon the crest of the nationalist wave, Nasser launched a huge programme of modernisation, symbolised by the Aswan High Dam. Opened in 1970, the dam created one of the largest artificial lakes in the world, Lake Nasser. By regulating the waters of the Nile,

the dam has allowed vast tracts of land to be reclaimed from the sandy desert, besides generating hydroelectric power for industry. Those people who live along the Nile no longer have to cope with annual flooding, and in some areas they can grow two crops a year by controlled irrigation.

But not all the hopes raised by the dam have been fulfilled. Large portions of reclaimed land have returned to desert through inefficient farming and lack of technical expertise – the wrong crops were grown or badly drained areas have become saline, for example.

There have been other problems – for instance, the lack of flood-borne nutrients has made it necessary to use artificial fertilisers on the land beside the river. Also, the building of the dam and the creation of Lake Nasser have necessitated moving the temples of Pharaoh Rameses II at Abu Simbel to a higher site above the flood level of the lake.

Two other major desert reclamation schemes are the New Valley project and the Tahrir (Liberation) Province project. The New Valley consists of a line of oases in the Western Desert running parallel to the Nile Valley. Supplies of subterranean water for irrigation are obtained by drilling to great depth. The Tahrir scheme is located on the western edge of the Nile delta, midway between Cairo and Alexandria, and relies on both Nile water and wells.

The land alongside the Nile produces rice, maize, barley, wheat, vegetables and beans, grown for own consumption, and sugar cane,

cotton, rice, dates, onions and vegetables, grown as cash crops. The natural vegetation, too, is abundant – there are date palms, tamarisk, mimosa, eucalyptus, cypress and jasmine. Cotton forms the main cash crop and foreign exchange earner. Some textiles are produced locally.

Cattle, buffaloes, sheep, goats and camels are the main domesticated animals. Camels are used as beasts of burden, as they are best adapted to the desert climate.

Egypt has a growing and sophisticated industrial sector producing iron and steel, aluminium, chemicals, tools, furniture, glass and textiles. Coal has been discovered on Sinai Peninsula and supplies the country's needs. Oil, too, has been discovered on Sinai Peninsula – not in large amounts, but enough to supply the country's needs and leave a surplus for export. A much bigger earner of foreign exchange is the Suez Canal which provides finance from shipping tolls. In 1990, more than 17 500 ships used the canal; the toll revenue earned in 1991 amounted to US$1770 million. At present, the canal is being enlarged to allow bigger ships to pass through.

One result of Egypt's strategic position was its ability to attract large-scale international assistance. But today, tight government control and regulation of industries (a large percentage of industrial plants are state-owned) keeps out foreign investors – already cautious because of recurring fears of political instability.

Nasser was determined that Egypt should go its own way – to be non-aligned. But the growth of the population and four wars with Israel since that country became independent in 1948 slowed economic progress. Under Nasser, the country became a one-party dictatorship but Anwar Sadat, who became president after Nasser's death in 1970, reintroduced multiparty politics in 1977. Sadat took two immensely brave steps: first, entering into negotiations with Israel and, second, signing the Camp David Agreement in 1979 under which Israeli troops withdrew from Sinai.

Others had different ideas: Sadat was assassinated by Muslim extremists and in 1981 was succeeded by the vice-president, Hosni Mubarak. The new president continued Sadat's policies of maintaining contact with Israel and seeking a peaceful solution to the Arab-Israeli problem. In May 1994, Egypt hosted the signing of the treaty negotiated between the Palestine Liberation Organization and Israel, signifying the end of Israeli occupation of the Gaza Strip. The treaty also negotiated for Palestinian rule of Jericho.

But Mubarak, too, faced enormous internal problems. Chief among these were need to attract foreign investment and to clamp down on militant Islamic fundamentalists. (During 1992-3, national income from tourism declined by as much as 20 per cent as a result of sporadic attacks by Islamic extremists,

which have scared off tourists from abroad.) His third, and possibly biggest problem, was to find ways of extending a programme of birth control. (Mubarak blamed much of the country's problems, including an unemployment rate of 20 per cent in 1992, on the rapid population growth.) The population explosion has been most evident in cities like Cairo, which has a housing shortage crisis.

The big cities have an undeniable Western facade to them. Educated Egyptians are likely to speak English or French in addition to Arabic. There is a widely varied press, often lively but periodically subject to censorship. Cairo is the second largest publishing centre in the Middle East, after TEL AVIV, and Egypt has extensive radio and television services with programmes in many languages.

There is also a substantial film industry in Egypt, though the productions are mainly unsophisticated and geared for popular consumption. The products of all these media are distributed throughout the Arab world and are very influential.

FOR THE VISITOR

Despite increasing violence in the 1990s, thousands of tourists still flocked to see Egypt's pyramids and other historical treasures. The pyramids of ancient Egypt were

royal tombs, built on the west bank of the Nile, from the 3rd dynasty of Pharaohs (about 2680 BC) to the 18th dynasty (about 1570 BC). Many of the Pharaohs had their own pyramids built, in which their mummified bodies might be preserved for eternity.

EL GIZA, or Gizeh, is the centre from which to visit the pyramids and the sphinx. The three pyramids of Gizeh are the largest and finest of their kind – the Great Pyramid of Khufu or Cheops (about 2590 BC) is one of the Seven Wonders of the Ancient World. The largest pyramid ever built, its mass of limestone blocks covers an area the size of 13 standard football pitches. It was originally 147 m (482 ft) high and 230 m (755 ft) along each side of its base. The sphinx, a mythical beast with a man's head and a lion's body, symbolised the Pharaoh Khafre, Khufu's son, as an incarnation of the sun god Ra.

From LUXOR and KARNAK, numerous temples and burial grounds can be visited, including the Valley of the Tombs of the Kings. But modern Egypt also has its charm. The sheer movement and vitality of its great cities present excitement and colour. Bargaining is the natural way to do business, whether for taxi rides or while shopping at the bazaars. Tourists who look 'fair game' may find themselves asked to pay steep prices or pestered for 'baksheesh' (alms).

Cairo is rich in mosques and museums, and Alexandria has ancient Christian catacombs (underground burial chambers). A camel ride, a Cairo shopping expedition and a trip down the Nile should be part of every visitor's itinerary, while many visit the El Alamein battlefield 110 km (70 miles) from Alexandria.

EGYPT AT A GLANCE	
Official name Arab Republic of Egypt	
Area 1 001 450 km² (386 571 sq miles)	
Population 57 396 000 **Per km²** 57	
(Per sq mile 148.4)	
Capital Cairo	
Government Republic	
Currency 1 Egyptian pound = 100 piastres	
Languages Arabic (official); Berber, Nubian, Beja	
Religions Muslim (92%, Sunni), Christian (mostly Coptic)	
Climate Hot and dry, with some winter rains along the coast. Average temperature in Cairo ranges from 13°C (55.4°F) in January to 28°C (82.4°F) in July; in Luxor, the average temperature ranges from 15°C (59°F) in January to 32°C (89.6°F) in July	
Land use Cultivation 5% (irrigated cultivation 2.6%), other (mainly desert) 95%	
Main primary products Rice, cotton, wheat, maize, sugar cane, barley, beans, lentils, millet, onions, Egyptian clover, cattle, buffaloes, sheep, goats; oil and natural gas, iron, phosphates, sea salt	
Major industries Oil refining, cement, textiles, iron and steel, fertilisers, processed foods	
Main exports Cotton, crude oil, textiles, metal ores	
Annual income per head (US$) 630	
Population growth (per thous/yr) 20	
Life expectancy (yrs) Male 62 **Female** 62	

Eisleben *Germany* Town 60 km (35 miles) west and slightly north of Leipzig. The religious reformer Martin Luther (1483-1546) was born here and eventually returned here to die; his homes are now preserved as museums.

The town's industries include textiles, furniture, cigars, clothing and copper smelting.
Population 26 000
Map Germany – Dc

Ejmiadzin (Echmiadzin) *Armenia* Ancient town and Christian religious centre, 15 km (9 miles) north of the Turkish border. Formerly called Vagarshapal, it was founded in the 6th century BC. According to local legend, the site of the town's magnificent cathedral, which was originally built in AD 301, was chosen by the Lord, who descended from heaven to strike the site with a hammer of gold. The cathedral, which was rebuilt between the 6th and the 7th century, attracts thousands of pilgrims each year.
Population 37 000
Map Armenia – Ba

Ekaterinburg *Russia* See YEKATERINBURG

El Alamein *Egypt* Desert village on the Mediterranean coast, 110 km (70 miles) south-west of the port of Alexandria. In a decisive Second World War battle in October-November 1942, British forces (under General Bernard Montgomery) beat back the German Afrika Korps (led by General Erwin Rommel) who were trying to capture CAIRO, the Middle East headquarters of the Allied forces, and to control the SUEZ CANAL. The battle was regarded as the first significant Allied victory of the war. Allied, German and Italian war cemeteries are situated near the village.
Map Egypt – Bb

El Borma *Tunisia* The country's principal oil field, situated in the far south on the Algerian border. It was discovered in 1966; by 1969, it was producing 3.5 million tonnes a year – enough to make Tunisia self-sufficient in petroleum. Crude oil from the field is piped to Sakhira, where some is exported and the rest is transported by sea to BIZERTE for refining and export.
Map Tunisia – Aa

El Faiyum (Fayum) *Egypt* Nearest oasis to the Nile Valley, situated on a plateau 100 km (60 miles) south-west of the capital, Cairo. It lies in a fertile cotton and fruit-growing stretch of the Western Desert. Nearby is Crocodilopolis, the Ancient Egyptian centre of worship of the crocodile god, Sebek. El Faiyum was a seat of the Pharaohs between 1990 and 1785 BC.
Population 244 000
Map Egypt – Bb

El Gezira (Al Jazirah) *Sudan* Cotton-growing clay plain covering about 6300 km² (2430 sq miles) between the Blue and the White Nile immediately south of Khartoum. Sudan's major irrigation scheme, the Gezira scheme, begun in 1925, is named after the region.

It consists of numerous canals supplied from the Sennar and ER ROSEIRES dams on the Blue Nile. The scheme is based on cotton crops supplemented with groundnuts, wheat and rice (alongside traditional crops of millet and beans).
Map Sudan – Bb

El Ghor Canal *Jordan* Important irrigation canal running for 98 km (61 miles) along the east bank of the River Jordan. It is fed from the YARMUK River, and has opened the Jordan Valley for extensive cultivation of fruit and vegetables.
Map Jordan – Aa

El Giza (Gizeh) *Egypt* Town on the west bank of the Nile, facing Cairo and the site of the Egyptian pyramids, built here more than 4000 years ago. It has now become a suburb of Cairo. It contains a traditional Muslim quarter alongside government offices, embassies, the University of Cairo and modern villas. Its industries include textiles, footwear and brewing; much of the country's film industry is located here.

El Giza's three main pyramids tower over the desert 15 km (9 miles) south-west of Cairo's centre. The largest is the tomb of Khufu, or Cheops, who reigned in Egypt from about 2590 to 2568 BC. It stands 137 m (450 ft) high, consists of 3 million limestone blocks, and covers an area of 5 hectares (13 acres) at its base. Its sides face directly north, south, east and west. It is the oldest of the Seven Wonders of the Ancient World and the only one still surviving.

Near it stands the sphinx, carved from a single outcrop of sandstone. It is 73 m (240 ft) long and 20 m (66 ft) high, with the body of a lion and the face of the Pharaoh Chephren (Khafre), son of Khufu, whose tomb is in another pyramid, slightly smaller than the pyramid of Cheops. The third pyramid is that of Mycernus, Chephren's son. The three huge monuments are surrounded by numerous other, smaller pyramids, temples and tombs of Egyptian nobles and court officials.
Population 2 096 000
Map Egypt – Bb

El Jadida *Morocco* Atlantic resort 100 km (60 miles) south-west of CASABLANCA. Formerly known as Mazagan, it was the site of an important Portuguese trading post in the 16th century. Imposing ramparts, which were once considered impregnable, survive from its Portuguese period. El Jadida fell to the Moroccans in 1769. Much of the old town remains, including the Portuguese underground reservoir.
Population 82 000
Map Morocco – Ba

El Jem *Tunisia* Small town halfway between Sousse and Sfax, rising abruptly from the surrounding land. Known as Thysdrus during Roman times, it is famous for its spectacular Roman amphitheatre – the third largest in the world – which could accommodate 30 000 spectators and which is better preserved than the Colosseum in Rome.
Population 12 800
Map Tunisia – Ba

El Kharga (Al-Kharijah) *Egypt* Oasis town 500 km (310 miles) south of Cairo. It lies in a green basin about 320 km (200 miles) long and 50 km (30 miles) wide on an ancient caravan route from Libya to the Nile Valley.

Around the oasis, a desert reclamation project using ground water has put a huge area under irrigation. Dates, cotton, wheat, rice, barley, bananas and grapes are produced; the town also makes cotton textiles.
Population 17 000
Map Egypt – Bc

El Mahalla El Kubra *Egypt* Egypt's fourth largest city, now merged with neighbouring Sammanud. It lies in the fertile Nile delta, 24 km (15 miles) north-east of TANTA, from where it has played a significant role in the history of Egypt. A number of temples and statues are located in and around the city, and on the nearby river banks. The reputed birthplace of Osiris, ruler of the underworld is situated to the south, at Busiris.
Population 400 000
Map Egypt – Bb

El Mansura *Egypt* City on the east bank of the Damietta River, the main eastern branch of the Nile delta, 140 km (85 miles) north of the capital, Cairo. In AD 1250, Crusaders led by Louis IX of France were defeated here by the Arab army of Turan Shah, in a battle which effectively ended the Crusades. The city takes its name from an Arabic word meaning 'triumph'. El Mansura's main industries are cotton, linen and the making of sailcloth.
Population 362 000
Map Egypt – Bb

El Misti *Peru* See AREQUIPA

El Niño Name given to the abnormal warming of the surface waters of the eastern tropical Pacific Ocean, which has had catastrophic effects on the weather, both locally and globally. Normally the highest temperatures – and consequently the lowest pressure – are over Indonesia and northern Australia, resulting in easterly trade winds piling up warm surface water in the western Pacific. When the low pressure area changes location and the trade winds weaken or reverse, warm water moves eastwards, thickening the warm surface layers off Ecuador and Peru. The phenomenon occurs every two to three years, and because it usually happens at Christmas, Peruvian fishermen call it El Niño ('the Infant' in Spanish).

The warm water displaces the cool waters of the Peruvian Current, with their wealth of microscopic plants and animals on which shoals of anchovies feed. In 1972, El Niño's warm waters were the major factor in the collapse of the Peruvian fishing industry.

Another El Niño began in early 1982 and by March 1983, the trade winds had reversed, bringing heavy rains and floods that devastated coastal Peru and Ecuador. Recent studies suggest that while flooding the west coast of South America, 1982's El Niño may also have caused major droughts in Africa. In 1992, when El Niño emerged early, there was drought in Indonesia, eastern Australia and north-east Brazil, a hurricane in Samoa, freak storms in California and months of heavy rain in Alaska and Canada.

El Obeid (Al Ubayyid) *Sudan* Capital of Northern Kordofan province, 370 km (230 miles) south-west of the national capital, Khartoum. The town is encircled by a forest reserve which gives it some protection from dust storms. It has the world's largest market for gum arabic, which is collected from trees in the region. It also trades in oilseeds, groundnuts, millet and livestock.

In 1883, an Anglo-Egyptian army was wiped out near here by followers of the Sudanese religious leader, the Mahdi.
Population 139 500
Map Sudan – Bb

El Oro *Ecuador* Province south of the Gulf of Guayaquil on the Peruvian border. Bananas, its main crop, are exported from Puerto Bolívar. The provincial capital is Machala.

Population 306 630
Map Ecuador – Ab

El Paso *USA* City in the extreme west of Texas, on the Rio Grande, and one of the main crossing points into Mexico. Its warm dry winters have made it a popular tourist destination. El Paso has large oil refineries and metal industries.

Population (city) 515 300; (metropolitan area) 591 600
Map United States – Ed

El Qahira *Egypt* See CAIRO

El Salvador See p 215

El Sumidero *Mexico* Spectacular canyon, 42 km (26 miles) long and 1800 m (6000 ft) deep, gouged out by the Río Grijalva in CHIAPAS state. In the 16th century, El Sumidero witnessed the suicide of more than 1000 Amerindians who jumped to their deaths rather than become slaves of the Spaniards. Their act shamed the conquistadores into retreating from the territory to let the indigenous people live in peace.
Map Mexico – Cc

El Tajín *Mexico* Ruined capital of the Totonac civilisation (5th to 12th centuries) in the forests of the humid coastal state of VERACRUZ. At the centre of this vast city lies the pyramid of the Nichos, with seven storeys and a niche for every day in the year. On Sundays, acrobats attached to ropes leap from a platform at the top of a 30 m (98 ft) mast beside the pyramid. They whirl to a stop, sometimes head-down, just short of the ground.
Map Mexico – Cb

El Teniente *Chile* The world's largest underground copper mine, lying 2750 m (9020 ft) up in the Andes in the Libertador region, 80 km (50 miles) south-east of Santiago. Ores have been mined here since 1912. They are refined locally and also provide molybdenum, silver and gold. The products are exported from the port of San Antonio, 130 km (80 miles) to the north-west.
Map Chile – Ac

Elat (Eilat) *Israel* Southernmost town and port, at the head of the Gulf of Aqaba, the north-eastern arm of the Red Sea. Elat is Israel's only southern outlet. It is also a seaside resort, with an airport, marina and underwater observatory. About 25 km (15 miles) to the north is the Timna Valley, which includes King Solomon's Mines, the remains of an Egyptian temple, and modern copper and manganese mines.
Population 18 800
Map Israel – Ac

Elazığ *Turkey* Market town and capital of a province of the same name, 600 km (about 370 miles) east of the capital, Ankara. It trades mostly in fruit and cotton grown in the surrounding province, which covers 9205 km² (3554 sq miles). On the Firat (EUPHRATES) River near the town is the Keban Dam, the country's largest hydroelectric scheme, opened in 1973.
Population (town) 218 120; (province) 498 230
Map Turkey – Bb

Elba *Italy* The third largest Italian island after Sicily and Sardinia, covering 223 km² (86 sq miles) and lying off the coast of Tuscany. Its capital is Portoferráio, where the French emperor Napoleon I (1769-1821) was exiled in 1814-15, after his disastrous Russian campaign. The island, now a popular tourist destination, was an ore-mining and iron-working centre for both the Ancient Greeks and the Etruscans.
Population 28 400
Map Italy – Cc

Elbe *Central Europe* River rising in the Czech Republic and flowing 1165 km (724 miles) across Germany into the North Sea just north of Hamburg. It is navigable for about 940 km (585 miles) from its mouth and is one of the most navigated rivers on earth. Large cities such as DRESDEN and MAGDEBURG stand on its banks; important rivers such as the VLTAVA, SAALE and Havel flow into it.
Map Germany – Db-Ec

Elblag (Elbing) *Poland* Baltic seaport and industrial city, 48 km (30 miles) south-east of Gdańsk. The city, which was founded in the 12th century by Teutonic Knights, was devastated during the Second World War. Only key monuments have been restored, most notably the Church of St Nicholas and its 95 m (312 ft) high tower. This and a few other remnants of the past contrast strikingly with postwar housing and shops, and factories making turbines and equipment.
Population 125 200
Map Poland – Ca

Elbrus, Mount *Russia* Highest peak in Europe, standing in the Caucasus on a northern spur from the main Caucasus range, which forms part of the boundary between Europe and Asia Minor. Mount Elbrus consists of two volcanic peaks – the western summit, which is 5642 m (18 510 ft) high, and the eastern, which is a little lower.
Map Russia – Fd

Elburz Mountains *Iran* Range of mountains running between the capital, Tehran, and the Caspian Sea. Its highest peak – and the highest point in Iran – is Damavand (5604 m, 18 386 ft), a snow-capped extinct volcano.
Map Iran – Ba

Eldoret *Kenya* One of the country's largest towns. It lies on the main road and railway to Uganda, 260 km (160 miles) north-west of Nairobi. Originally a European settler town, it declined after independence in 1963. However, the introduction of several industries in the 1970s, including textiles and clothing, gave it a boost. It has Kenya's second university, after Nairobi University.
Population 105 000
Map Kenya – Ca

Elephanta Island *India* Island in BOMBAY harbour, 10 km (6 miles) from the city centre and reached by excursion boats. Elephanta Island was named during the Portuguese rule after a rock carved into an elephant shape. The rock has since collapsed and today, the island is referred to as Gharapuri ('fortress city'). It has a complex of Hindu cave temples dating from AD 450-750, with sculptures of Hindu gods and their stories.
Map (Bombay) India – Bd

Eleuthera *Bahamas* Island in the north-central group of the Bahamas, and one of the first to be colonised by the British in the 17th century. About 160 km (100 miles) long and mostly only 3 km (2 miles) wide, Eleuthera has a narrow, fertile strip surrounded by dunes and beaches of pinkish sand.

It is one of the major resort islands of the Bahamas. Commercial farming benefits from the island's proximity to the capital, Nassau. Eleuthera was badly damaged in a hurricane in 1922, but has since been rebuilt.
Population 10 520
Map Bahamas – Ba

Elgon, Mount *Kenya/Uganda* Extinct volcano, also known as Wagagai, which rises to 4321 m (14 176 ft) on the Kenyan-Ugandan border. Its fertile lower slopes are intensively cultivated with coffee and banana plantations.
Map Uganda – Ba

Elisabethville *Zaire* See LUBUMBASHI

Ellesmere Island *Canada* Mountainous snow and ice-covered island in the Canadian arctic archipelago, separated from Greenland (Kalaallit Nunaat) by a narrow strait and covering 196 236 km² (74 591 sq miles). Only a few Inuit (formerly referred to as 'Eskimos') live here, apart from the staff of Canadian-United States research bases that are situated at Eureka and Alert.

In the north, the island peaks to heights of 3000 m (9843 ft) and is heavily glaciated. Its northern tip, Cape Columbia – the northernmost point of Canada – is just 756 km (470 miles) from the North Pole. It was from there that Robert Peary and his expedition set off to conquer the pole in 1909.
Map Canada – Ga

Ellice Islands See TUVALU

Elsinore (Helsingør) *Denmark* Town founded in the Middle Ages on the island of ZEALAND overlooking the mouth of the ÖRESUND, the main shipping channel leading to the BALTIC SEA. On the far side of the channel, which is only 4 km (2 mile) wide at this point, lies Sweden.

Elsinore is dominated by Kronborg Castle – the setting for William Shakespeare's play *Hamlet, Prince of Denmark*. The castle was built in the 16th century to guard the channel, and to extract tolls from ships heading for the Baltic.

Today, the town is Denmark's main ferry port for Sweden. It also produces machinery and textiles, and is a shipbuilding and food processing centre. Tourism is another important source of income.
Population 56 600
Map Denmark – Ca

Elvas *Portugal* Fortified hilltop town on the Spanish border, 180 km (112 miles) east of the capital, Lisbon. The Portuguese wrested it from the Moors in 1226, but in 1580, it was captured by the Spaniards.

It has a 16th-century aqueduct; its fortifications date mostly from the 17th century. The town's main manufacture today is jewellery. The area is noted for its plums.
Population 12 700
Map Portugal – Cc

El Salvador

AFTER YEARS OF CIVIL WAR, SALVADOREANS HAVE BEGUN TO REBUILD THEIR RUINED LAND AND ITS ECONOMY

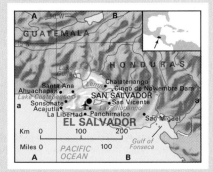

The 12-year struggle for power in the small Central American country of El Salvador resulted in the death of at least 80 000 of its people. Between 1979 and 1992, about 550 000 people were driven from their homes within the country, and more than 500 000 went into exile. Successive governments during the early 1980s were accused of abusing human rights by torture – and the disappearance of political activists and summary executions were commonplace.

Among the many victims of the violence was Roman Catholic Archbishop Oscar Romero, a leading spokesman for human rights, who was assassinated by right-wingers in March 1980 while conducting mass in the cathedral in the capital, SAN SALVADOR.

The breeding ground for the conflict was, as elsewhere in Latin America, unequal land distribution, and enormous social disparity. As recently as 1994, the country's 60 leading families owned one-third of the land – all the land which yielded coffee for export. Economically powerful, and with the military on their side, they secured a political foothold, blocking all attempts at social and economic reform. By contrast, 90 per cent of the rural population shared only one-fifth of the land. Apart from a tiny elite of white people (most of them landowners), 89 per cent of Salvadoreans are *mestizo* – of mixed Spanish and Amerindian descent; the rest are pure Amerindian.

By the late 1960s, many peasants began forming political parties. Although weak at first, their numbers grew when support came in from Christian Democrats, Socialists and Communists – especially after the victory by the socialist Sandinista movement in Nicaragua in 1979.

In the same year, Christian Democrat José Napoleon Duarte headed a provisional government; his appointment as president a year later by the military junta signalled a gradual transfer of power from the military to a civilian administration. But Duarte could not

resolve the growing social and civil conflict, and soon, the main left-wing opposition, the FMLN, called for an armed struggle to change the social inequities. The FMLN was backed by the left-wing government in Nicaragua, while the United States, fearing the spread of Communism northwards to its own borders, supported the Salvadorean military with substantial economic and military aid. Between 1982 and 1984 the ultra right-wing National Republican Alliance (ARENA) candidate, Roberto d'Aubuisson, grew increasingly powerful, and his brutality caused an outcry among Americans who objected to their government's support of him.

In 1984, the more moderate Duarte was a clear winner in El Salvador's first democratic presidential elections. However, his peace overtures were rejected by the guerrillas – and as the deadlock in the country continued, d'Aubuisson's military-dominated party, ARENA, began gaining popular support. In the 1989 elections, the new ARENA candidate, Alfredo Cristiani, was elected president.

A HISTORY OF CONFLICT

Conflict and repression were not new to Salvadoreans. After El Salvador became fully independent in 1841, internal strife and military dictatorship became a way of life for the inhabitants of the country. There were also external conflicts: for instance, in 1969, Salvadorean forces invaded neighbouring Honduras after a hotly contested World Cup soccer match between the two nations.

It was a two-leg fixture, played in the two capitals, San Salvador and Tegucigalpa. On each occasion the visiting supporters were subjected to ill-treatment and violence. This sparked the war – but the underlying cause was the illegal migration of some 300 000 Salvadoreans into Honduran territory over the previous 20 years. The Salvadorean invasion lasted for just five days, but the tensions continued for years and came to a head again in 1976, when troops clashed on the border.

In January 1992, UN-brokered negotiations brought the civil war in El Salvador to a close. The uneasy peace that followed highlighted a myriad problems, most notably economic underdevelopment and a poor transport infrastructure. Although 21 per cent of all employed Salvadoreans work in industry, most of these industries are small plants processing agricultural products.

PEOPLE, THE MAIN EXPORT

El Salvador is the most densely populated country in Central America, with 256 people per km² (662 per sq mile), and so one of its main exports is people. There is not enough work and food to support its population of 5.38 million; furthermore, civil war has damaged agriculture, industry and services. Although the United States, Germany and Japan have provided aid to help repair war damage, progress has been slow. By the early 1990s, damaged farms and industries had barely restarted production; at the same time, unemployment and inflation remained high.

Many social problems were aggravated by the civil war; a large majority of the people

today are illiterate and undernourished. In 1991, 30 per cent of the population was living in extreme poverty, another 34 per cent was classified as only slightly better off, or 'poor'.

Almost 40 per cent of the people work in agriculture, which generates some 10 per cent of the national income. Coffee accounts for about 50 per cent of the country's exports, followed by cotton (declining), sugar (declining) and cotton textiles (increasing), while the people grow rice, maize and beans to feed themselves. Some beef is also exported. El Salvador is the world's leading producer of balsam gum, which is used in patent medicines and perfumes; timber is a minor export. Soil stability and fertility have become problems since the slopes of the country's volcanic mountains have been cleared of natural forest – mostly for fuel – causing severe erosion.

Two volcanic ranges, divided by the LEMPA River valley, run from east to west across El Salvador. The river turns southwards halfway along the southern range, and then cuts through the mountains, dividing the country into western and eastern parts. Beyond the southern range, the Lempa's large, sandy delta opens into the Pacific Ocean. In the wet season, between May and October, it floods. The ranges nearest the sea form clusters of volcanic cones, several of which are active. Most of the country's towns and settlements are in the cooler, temperate highlands rather than on the hot, humid coast.

Before the arrival of the Spanish in the 16th century, the Pipils, who were related to the Aztecs, inhabited El Salvador. They called the territory Cuscatlan ('the Land of Jewels'). Monuments to their civilisation – notably the El Tazumal ruins at Chalchuapa – remain scattered all over the country. But the days when Salvadorean society was considered safely prosperous have long gone; today's population has only just begun to pick up the pieces after years of civil war.

EL SALVADOR AT A GLANCE	
Official name Republic of El Salvador	
Area 21 041 km² (8124 sq miles)	
Population 5 380 000 **Per km²** 256 **(Per sq mile** 662)	
Capital San Salvador	
Government Republic	
Currency 1 colon = 100 centavos	
Languages Spanish, Indian dialects	
Religion Christian (92% Roman Catholic)	
Climate Tropical, with the rainy season between May and October; cool in the highlands. Average temperature in San Salvador is 22-24°C (72-75°F) all year	
Land use Cultivation 35%, meadows and pastures 29%, forest and woodlands 6%, other 30%	
Main primary products Coffee, cotton, maize, rice, sorghum, sugar, beans, cattle, timber, shrimps; gold, silver, titanium	
Major industries Agriculture, textiles and clothing, food processing, chemicals, forestry, fishing, electricity	
Main exports Coffee, cotton, sugar, gold	
Annual income per head (US$) 1120	
Population growth (per thous/yr) 19	
Life expectancy (yrs) Male 63 **Female** 66	

El-Wanza *Algeria* See OUENZA

Ely *United Kingdom* English cathedral city in Cambridgeshire, 22 km (14 miles) north and slightly east of CAMBRIDGE. Ely used to stand on an island – the Isle of Ely – in the heart of the Fens, and could be reached only by boat or by three narrow causeways. After the Norman invasion of 1066, English warrior Hereward the Wake found refuge here for five years.

The island disappeared after the Fens were drained in the 17th century, leaving a flat landscape of dark and fertile soil. The cathedral, built of stone delivered by boat, is one of Britain's most beautiful. Built between the 11th and the 14th century, it has several distinct styles and architectural features, including a unique octagonal tower with angle posts carved from immense oak trees. The King's School next to the cathedral was founded in AD 970.
Population 9120
Map United Kingdom – Fd

Emden *Germany* Port in the north-west corner of the country facing Dutch territory across the estuary of the Ems River. Silting in the 16th century caused the river to change course, and Emden is now linked to the North Sea and to the Ruhr industrial region by canal. It is the site of a large Volkswagen factory.
Population 50 400
Map Germany – Bb

emerald Brilliant, transparent green form of beryl, used as a gemstone.

Emi Koussi *Chad* See TIBESTI

Emilia Romagna *Italy* Region in north-central Italy centred on its capital, BOLOGNA. Covering 22 123 km² (8542 sq miles) of the Po valley and the Apennines, it is a leading producer of apples, pears, peaches and sugar beet. Pigs and cattle are also bred here. The Italian dictator Benito Mussolini (1883-1945) was born in the village of Predappio.
Population 3 928 700
Map Italy – Cb

Emmental *Switzerland* Valley of the River Emme, especially famous for its linen cloth, its cattle and its cheese. The Emme, which is about 80 km (50 miles) long, rises east of the Swiss capital, Berne, and flows into the River Aare, near Solothurn.
Map Switzerland – Aa

Ems *Germany* River rising east of the north-western city of Münster and flowing for 370 km (230 miles) into the North Sea near the Dutch border. It is linked to the industrial Ruhr via the Dortmund-Ems Canal.
Map Germany – Bb

Enewetak *Marshall Islands* Atoll in the northern Marshalls, which in the 1940s and 1950s was used – like Bikini 320 km (200 miles) to the east – for United States nuclear weapons tests. The people were evacuated, but after a clean-up operation were allowed to return in 1980, despite fears that radioactivity remained. In 1983, they decided to make Ujelang Atoll their permanent home.
Map Pacific Ocean – Db

Engadin *Switzerland* Swiss part of the valley of the River Inn, which rises near the Italian border and flows north-east into Austria. Oberengadin, the upper valley, is dominated by lakes and the world-famous resort of St Moritz. The Swiss National Park, which covers more than 168 km² (65 sq miles), lies east of the River Inn. It consists of mainly mountainous terrain, reaching up to 3000 m (9842 ft). Unterengadin, the lower valley, is also a recreation area.
Map Switzerland – Ca

Engel's *Russia* Industrial town on the Volga River, 320 km (200 miles) north of Volgograd. It was founded in 1747 as Pokrovskaya Sloboda, and was renamed Engel's in 1932 after the German political philosopher Friedrich Engels (1820-95), who co-authored *The Communist Manifesto* with Karl Marx (1818-83).

Leather goods, chemicals, railway wagons, and machinery are manufactured here.
Population 183 000
Map Russia – Fc

England *United Kingdom* Largest constituent country of the UNITED KINGDOM, covering some 130 357 km² (50 331 sq miles). It was settled in pre-Christian times by the Celts, and later conquerors included the Romans, Danes and, finally, the Normans. Its capital is LONDON and its main industrial cities include BIRMINGHAM, LEEDS, LIVERPOOL, MANCHESTER and SHEFFIELD.
Population 48 208 100
Map United Kingdom – Ed

English Channel Waterway between England and France, which is called La Manche ('the Sleeve') by the French and is part of the ATLANTIC OCEAN. It covers 77 700 km² (30 000 sq miles) and its greatest depth is 172 m (565 ft). Its western entrance, between LAND'S END in south-western England and the Ile d'Ouessant off Brittany, is 180 km (110 miles) across.

To the north-east, the Channel narrows to 34 km (21 miles) at the Strait of DOVER, formed about 7500 years ago when ice sheets melted at the end of the Ice Age. The world's busiest sea passage (some 350 ships pass through each day), the English Channel is lined with ports and resorts on both the French and the English coast. The 50 km (31 mile) long Channel Tunnel which links England with France was opened in 1994.
Map France – Ba; United Kingdom – Ee-Fe

English Harbour *Antigua and Barbuda* Former British naval dockyard in the south of Antigua. The English admiral Horatio Nelson (1758-1805) and his fleet were based here for a time in the late 18th century; Nelson's house is now a museum. With the entire dockyard having been restored to its 18th-century appearance, English Harbour has become a tourist attraction and yachting centre.
Map Caribbean – Cb

Enna (Castrogiovanni) *Italy* City and resort known as 'the Navel of Sicily', situated 100 km (60 miles) south-east of Palermo. Enna, which stands on a plateau 1000 m (3300 ft) high, contains one of the island's most impressive medieval fortresses – the Castello di Lombardia – which rises out of sheer rock. The city trades in rock salt and sulphur.
Population 28 300
Map Italy – Ef

Ennis *Ireland* See CLARE

Enns *Austria* Small town near the confluence of the Enns and Danube rivers, 18 km (11 miles) south-east of Linz. One of the country's oldest towns, Enns received its charter in 1212. It has a Gothic parish church, a 16th-century fortress, and parts of its medieval walls still stand.
Population 10 100
Map Austria – Da

Enschede *Netherlands* Industrial town east and slightly south of Amsterdam, and only 5 km (3 miles) from the German border. It grew rapidly after 1900 – and as the centre of the country's textile industry, it became known as the 'Manchester of the Netherlands'. Since the 1960s, other industries such as electrical engineering, beer brewing and the manufacture of car tyres have replaced the textile industry.
Population (town) 147 350; (Greater Enschede) 253 560
Map Netherlands – Ca

Entebbe *Uganda* Town situated on a peninsula jutting into Lake Victoria, 33 km (20 miles) south-west of the capital, Kampala. It was Uganda's capital from 1894 until independence in 1962. It has the country's main airport where, in July 1976, Israeli commandos freed 100 Jewish hostages being held by Arab gunmen.
Population 42 000
Map Uganda – Ba

Entre Deux Mers *France* Triangular area in Aquitaine, just east of the junction of the Dordogne and Garonne rivers. It is famed for its dry white wines named after it.
Map France – Cd

Enugu *Nigeria* Capital of Enugu state, on the Udi Plateau, 440 km (275 miles) east of the Niger River. It lies on the country's only large coal field, with estimated reserves of 72 million tonnes. Most of the coal goes to the railways or power stations, but demand has fallen with increasing use of diesel locomotives and the opening of the Kainji Dam hydroelectric station.
Population 500 000
Map Nigeria – Bb

Eocene Second epoch in the Tertiary period of the Cenozoic era of the earth's timescale. See GEOLOGICAL TIMESCALE.

Eochaill *Ireland* See YOUGHAL

Eolian Islands (Lipari Islands) *Italy* Group of volcanic islands in the Tyrrhenian Sea off the north coast of Sicily, named after Aeolus, the Greek god of the four winds. The main islands are LIPARI and Salina. Since 1970, tourism has given the islands' economy a strong boost. VULCANO attracts many visitors to its volcanic mudbaths, Panarea is noted for its scenery, while STROMBOLI has an active volcano.
Population 12 500
Map Italy – Ee

Ephesus (Efes) *Turkey* Ruined city overlooking the Aegean Sea, 400 km (250 miles) south of Istanbul. The Ancient Greek city of Ephesus was one of Asia Minor's most important cities around 1000 BC. Ephesus was the site of the Temple of

Artemis (or Diana, the Greek goddess of the hunt), one of the Seven Wonders of the Ancient World. The marble temple was built in the 6th century BC by the city's immensely rich ruler, the Lydian king, Croesus. It was destroyed by invading Goths in AD 262. St Paul's Epistle to the Ephesians was directed to the citizens of Ephesus.
Map Turkey – Ab

epicentre Point on the earth's surface directly above the FOCUS, or centre, of an EARTHQUAKE.

Epidaurus (Epídhavros, Hieron Epidaurou) *Greece* Ancient site devoted to Asclepius, the Greek god of medicine. It lies 37 km (23 miles) south of Corinth. Its theatre, which dates from the 4th century BC and which is the best preserved in Greece, has a circular stage and remarkable acoustics. An international festival of drama, opera and music takes place here each summer.
Map Greece – Cc

Epídhavros *Greece* See EPIDAURUS

Epirus (Ipiros) *Greece* Region of north-west Greece covering 9203 km² (3552 sq miles), bordering Albania and the Ionian Sea, but cut off from the rest of the country by the Píndhos range of mountains. It is a hilly region with the highest rainfall in Greece. Agriculture and pasture farming thrive in the valleys. Its chief town is Yannina (Ioánnina).

In 278 BC, Epirus's king, Pyrrhus (319-272 BC), won a battle against the Romans, that was so costly in terms of troops lost, that it gave rise to the phrase 'Pyrrhic victory' – a victory that might just as well have been a defeat. Epirus was occupied by Turkey from AD 1449 until 1912.
Population 339 210
Map Greece – Bb

epoch Third category of the subdivisions of geological time – era, period, epoch, age. See GEOLOGICAL TIMESCALE.

Eptanisos *Greece* See IONIAN ISLANDS

Equateurville *Zaire* See MBANDAKA

equatorial climate Type of climate occurring in lowland areas close to the equator. It has virtually no seasonal change and brings with it 12 hours of daylight throughout the year, with monthly mean temperatures averaging 26°C (79°F). Rainfall occurs each afternoon, and more than 2000 mm (79 in) of rain fall in a year.

equatorial forest Part of the tropical rain forest found within a few degrees of the equator where there is no seasonal change. It is noted for its enormous diversity of species.

Equatorial Guinea See p 218

Er Rachidia (Ksar es Souk) *Morocco* Berber oasis town situated in the Ziz Valley on the southern flanks of the High Atlas Mountains, 240 km (150 miles) south-east of Meknès. The town lies in a huge palm grove and has a *kasbah* (old walled quarter) and fortress. It is a transport centre and market for wool, dates, olive oil and esparto grass.
Population 45 710
Map Morocco – Ba

▲ **EPHESUS Majestic ruins conjure up past images of this wealthy city in Turkey, once visited by St Paul.**

Er Roseires (Ar Rusayris) *Sudan* Blue Nile port near the Ethiopian border, handling cotton, sesame seeds and cereal exports. The Er Roseires dam to the south provides irrigation for part of the Managil extension of the Gezira scheme, and for the new Rahad scheme where cotton, millet, groundnuts and clover are produced.
Population 16 400
Map Sudan – Bb

era The major division of geological time, which is subdivided into periods. See GEOLOGICAL TIMESCALE.

Erawan *Thailand* The most productive offshore gas field in the Gulf of Thailand, 500 km (310 miles) south of Bangkok.
Map (Gulf of Thailand) Thailand – Bc

Ercolano *Italy* See HERCULANEUM

Erdenet *Mongolia* Mining town about 220 km (135 miles) north-west of the capital, Ulan Bator, between the Selenge and Orhon rivers. The mine, Erdenetiyn-Ovoo, which produces copper and molybdenum, is said to contain half the known copper deposits in Asia. Opened in 1976 with aid from the Soviet Union, it accounts for 30 per cent of Mongolia's total export earnings.
Population 58 200
Map Mongolia – Db

Erebus, Mount *Antarctica* Active volcano on Ross Island, 3743 m (12 280 ft) high, known for frequently spewing lava bombs from its 1000 m² (10 764 sq ft) lake of molten rock. In November 1979, a New Zealand airliner carrying tourists on a scenic trip crashed into its side, killing all 257 people on board.
Map Antarctica – Jc

Erfoud *Morocco* Town called 'the Gateway to the Sahara', 280 km (174 miles) south-east of Meknès. Situated at the end of the Ziz Valley, it is dominated by an old fortress. Nearby are the ruins of the 15th-century town of Sijilmassa, once the northern end of a caravan route along which gold was brought from the empire of Ghana. At Rissani, to the south, the founder of the present Moroccan royal dynasty was born in the 17th century.

The Ziz River cuts a deep, cliff-lined valley into the surrounding semi-desert, and supports palms and other brilliant green vegetation along its banks. It flows for about 280 km (174 miles) through southern Morocco.
Map Morocco – Ba

Erfurt *Germany* Tourist resort and capital of Thüringen, about halfway between Gotha and Weimar. It produces heavy machinery, typewriters, electrical goods, shoes and clothing, but it is as a historical treasure chest that the city is best known. The religious reformer Martin Luther (1483-1546) was a monk at its Augustinian monastery for five years. The city has a Scottish church built in 1200 and a 13th-century Dominican monastery, whose gardens founded the city's flower, vegetable and seed industry. The 14th-century Kramer Bridge is covered with half-timbered houses and shops.
Population 209 000
Map Germany – Dc

217

erg Term originally applied to an area of sand dunes in the Sahara, but now used with reference to any sand desert.

Erie *USA* Port on Lake Erie in Pennsylvania, 185 km (115 miles) north of Pittsburgh. It handles coal, grain, timber, oil and chemicals, and makes machinery, boilers, electric locomotives and paper products.
Population (city) 108 720; (metropolitan area) 275 570
Map United States – Jb

Erie, Lake *Canada/USA* Second smallest of the five Great Lakes, after Lake Ontario, covering about 25 667 km² (9908 sq miles). It drains north through the Niagara River (with Niagara Falls) into Lake Ontario and is linked to that lake by the Welland Ship Canal. Lake Erie is fed by the Detroit River, Lake St Clair and the St Clair River from Lake Huron. Much of its southern shore, dominated by the cities of Toledo and Cleveland, is heavily industrialised.
Map United States – Jb

Erie Canal *USA* Waterway linking the port of Buffalo on Lake Erie with the Hudson River at Albany, over a distance of about 550 km (342 miles). The original canal – some 30 km (19 miles) longer – was opened in 1825 and was a key route in the expansion westwards. It was improved in the early 20th century and became the main part of the New York State Barge Canal system, which has links with Lake Ontario.
Map United States – Kb

Eritrea see p 220

Erlangen *Germany* University town in Bavaria, 20 km (12 miles) north of NUREMBERG. Erlangen was the birthplace of George Simon Ohm (1787-1854), the physicist who gave his name to the electrical units of resistance and conductance. Appropriately, the town is noted for its electrical and electronics industries.
Population 102 400
Map Germany – Dd

erosion Wearing away of the earth's surface, principally by the slow, abrasive action of hard particulate matter carried by rivers, sea, ice and wind. Abrasive material may vary in size from boulders transported by glaciers, to dust particles blown by the wind. See THE GEOLOGICAL CYCLES: HOW ROCKS ARE FORMED AND BROKEN DOWN and THE CHANGING PATTERN OF THE COASTS, (p 219).

erratic Large piece of rock that has been transported – usually by a glacier – some distance from its source and left standing when the ice has melted. Different in composition from the rocks around it in its present location, it can be described as a geological alien.

Erzgebirge (Ore Mountains) *Czech Republic/Germany* Forested hills stretching 130 km (81 miles) along the border. Rising up to 1244 m (4081 ft) in the Czech Republic, they contain deposits of iron ore, silver, lead, copper, pitchblende and uranium. However, few mines are still worked today. Despite extensive damage caused by acid rain to the forests, the Erzgebirge with its spas and springs is a popular tourist area.
Map Germany – Ec

Equatorial Guinea

THE PEOPLE OF EQUATORIAL GUINEA ARE STILL SUFFERING THE CONSEQUENCES OF 11 YEARS OF TYRANNICAL RULE

Lying 200 km (125 miles) north of the equator on the hot, humid coast of West Africa, the Republic of Equatorial Guinea is one of the poorest countries on a poor continent. Yet before independence from Spain in 1968, this territory, which is nearly the size of Belgium, had a thriving economy based on cocoa, coffee and timber.

The economy rapidly collapsed under the first president, Francisco Macías Nguema (1968-79). Torture, forced labour, systematic murders of intellectuals, persecution of foreigners and the church, together with increasing isolationism, were features of Nguema's 11-year rule. In 1979, when he was deposed and executed after a coup, a United Nations report described the nation as 'decomposed'. After 1979, the government tried to restore the shattered economy with overseas aid from both East and West, but initially from Spain, the former colonial power. Political prisoners were released; there was even talk of restoring religious freedom.

The reality was different. President Teodoro Obiang Nguema Mbasogo continued to ban political parties until 1987 and, although a new constitution was drawn up to liberalise politics and a 'transition government' was formed in 1992, the state continued to detain and torture opposition politicians. The general elections on 21 November 1993 were boycotted by opposition parties, who described the polls as a 'farce'. Voter turnout was poor, the consensus being that nothing would change while Obiang remained in power.

After the elections, the government announced that the rebuilding of the cocoa and coffee industries to their pre-independence levels would be a priority. The fishing industry has also suffered, and the country has changed from being a fish exporter to an importer. However, forestry is gaining in importance, with the main industries being lumbering and wood products. Although the country has mineral resources, these are still largely undeveloped. Some oil and natural gas are exported.

RICH POTENTIAL

Equatorial Guinea consists of two contrasting regions. The largest, on the African mainland, is called RÍO MUNI and comprises 93 per cent of the country. It has extensive tropical forests and savannah-covered plateaus. There are few roads, and foreign visitors are almost unknown. Included in this region are three offshore islets: Corisco, Great Elobey and Small Elobey.

The second region consists of the islands of BIOKO (called Fernando Póo in Spanish times) north of the mainland off the coast of Cameroon, and Annobón, 640 km (400 miles) to the south-west. The fertile volcanic island of Bioko contains the country's capital, MALABO, which stands beside a volcanic crater flooded by the sea. The island is the centre of the country's cocoa production, and has even higher temperatures and higher rainfall than the hot, humid mainland.

The original inhabitants of Bioko, the Bubi people, speak a Bantu language, like most Equatorial Guineans. Bioko also has a few thousand Fernandinos, Creole descendants of slaves freed by Britain in the 19th century. Before 1975-6, 30 000 Nigerians lived on the island, working on the cocoa plantations. But they were evacuated following accusations of violence and maltreatment by the island's Fang rulers, the dominant Bantu-speaking group on the mainland, who, after 1968, extended their authority to Bioko.

Equatorial Guinea has considerable tourist potential, although it is difficult to reach. The natural beauty of Bioko, praised among others by the Victorian traveller Mary Kingsley, and the abundant wildlife on the mainland are potential attractions.

EQUATORIAL GUINEA AT A GLANCE	
Official name Republic of Equatorial Guinea	
Area 28 050 km² (10 828 sq miles)	
Population 466 000 **Per km²** 16.6 (**Per sq mile** 43)	
Capital Malabo	
Government Republic	
Currency 1 CFA franc = 100 centimes	
Languages Spanish (official); African languages including Fang	
Religions Christian (94% Roman Catholic), indigenous beliefs	
Climate Tropical and very humid. Average annual temperature is 25°C (77°F)	
Land use Forest and woodlands 51%, cultivation 12%, grazing 4%, other 33%	
Main primary products Sweet potatoes, cassava, bananas, coffee, cocoa, coconuts, timber; oil and natural gas	
Major industries Agriculture, forestry, fishing, food processing	
Main exports Cocoa, coffee, hardwoods and timber, oil, natural gas	
Annual income per head (US$) 330	
Population growth (per thous/yr) 22	
Life expectancy (yrs) Male 48 **Female** 48	

THE CHANGING PATTERN OF THE COASTS

The coast is where both the destructive and constructive energies of the seas come together and where the land forever dies and is perpetually born again. Some coastal landscapes feature deep indentations (rias) which are actually ancient river valleys drowned by the sea when the glaciers of the Ice Age melted; at others, where the land is slowly springing back from its Ice Age depression, the sea has retreated almost out of sight. But the swiftest and most dramatic alterations are wrought by the sea, which on one hand may carve away a cliff in a night, and on the other, lay down a new sandy beach.

Oceanic waves are the most powerful machines on earth. Driven by the wind, the wave motion that strikes western Europe, for example, begins off the tip of South America,

gathering strength on its 10 000 km (6200 mile) journey. What happens at its journey's end depends very much on the terrain. On a gently shelving coastline, the wave motion will be transformed to a current as the waves break, with the main force directed towards the headlands, while the lesser flow deposits sand and sediment, picked up from the seabed, into bays.

Rock types have a great effect upon coastal scenery. Hard rock – granite and certain limestones, for example – yields slowly, even to a storm's fury, but nevertheless, the sea undercuts it, carving caves, arches, stacks and cliffs,

which may reach up to more than 600 m (2000 ft). The effect upon softer rock is quite different. There, the sea constantly wears away rock and debris, causes landslides and hollows out bays. Cliffs are low and tumbled, and the detritus is washed along the coast to create beaches, gravel banks and pebble ridges.

Plants may take precarious root in these, but unless there is a swift build-up of sediments and protective banks, they are liable to be washed away by the next storm.

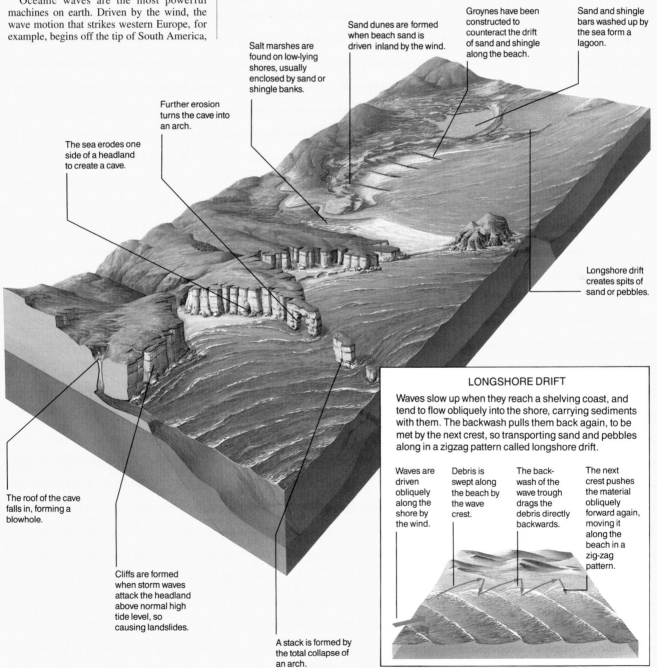

Salt marshes are found on low-lying shores, usually enclosed by sand or shingle banks.

Sand dunes are formed when beach sand is driven inland by the wind.

Groynes have been constructed to counteract the drift of sand and shingle along the beach.

Sand and shingle bars washed up by the sea form a lagoon.

Further erosion turns the cave into an arch.

The sea erodes one side of a headland to create a cave.

Longshore drift creates spits of sand or pebbles.

The roof of the cave falls in, forming a blowhole.

Cliffs are formed when storm waves attack the headland above normal high tide level, so causing landslides.

A stack is formed by the total collapse of an arch.

LONGSHORE DRIFT

Waves slow up when they reach a shelving coast, and tend to flow obliquely into the shore, carrying sediments with them. The backwash pulls them back again, to be met by the next crest, so transporting sand and pebbles along in a zigzag pattern called longshore drift.

Waves are driven obliquely along the shore by the wind.

Debris is swept along the beach by the wave crest.

The backwash of the wave trough drags the debris directly backwards.

The next crest pushes the material obliquely forward again, moving it along the beach in a zig-zag pattern.

Estonia

*THE NORTHERNMOST OF THE BALTIC
REPUBLICS THRIVES IN THE
AFTERMATH OF THE SOVIET UNION*

Independent since September 1991, the Republic of Estonia has been the most successful of all the 15 ex-Soviet republics. It has the advantage of having been under Soviet rule for a much shorter period than many of the other ex-Soviet republics; moreover, its favourable geographic position enables it to trade with Western and Nordic countries far more easily.

Of all the states that made up the former USSR, Estonia was the largest producer of consumer goods and also had the most highly developed services sector. These factors led to confident expectation that it would respond readily and efficiently to foreign tourism and capital investment.

And yet, many problems have still to be resolved: the answers to some appear to be quite straightforward, such as the need to overhaul an economic structure still partly based on trade with other members of the former Union. A more testing poser for the young republic is what to do about the large percentage of ethnic Russians that make up its population. At the end of the Second World War, only about 5 per cent of Estonia's population was Russian; in 1994, about 30 per cent of the people living in Estonia were Russian-speakers.

While the government committed itself to granting Russian inhabitants Estonian citizenship – and the rights that go with it – it set out certain conditions. By 1994 the question of citizenship rights for Russians had become more than just an internal affair, since some Russian troops continued to be stationed in Estonia. (After independence, Russian forces had remained in the country, as a type of security for the Russian residents, although most had left by the end of 1994.)

The country, situated between the Gulf of Riga and the Gulf of Finland, is a little larger than Denmark. Estonians, a Finno-Ugric people, have lived along this part of the Baltic shore for about 5000 years. Their country's long history of colonisation was set in motion at the beginning of the 13th century, when Estonians numbered no more than perhaps 100 000 people. First came Danes, in the service of a German Order of Teutonic Knights. When knightly power declined in the 16th century, Sweden took the opportunity to seize northern Estonia.

The southern section, meanwhile, became part of the Lithuanian-Polish Duchy of Courland. In 1629 all of mainland Estonia came under Swedish rule. But it has been the Russians who have been most interested in the country – and particularly its ports on the Baltic Sea, that are ice free for a large part of the year and serve as a gateway to the entire Baltic region. The Russian tsars made their first attempt at controlling the territory in the 16th century, but it was Peter the Great who acquired the Baltic states for Russia during the Nordic War (1700-21). For the next two centuries, Estonia was the northernmost of the three Russian Baltic provinces.

Then, during the confusion of the Russian revolution, the Estonians freed themselves from the grip of Russia. In February 1918, they founded the independent Republic of Estonia, which lasted until June 1940, when Stalin, during the brief period of the Soviet-Nazi Germany alliance, occupied the country and incorporated it in the Soviet Union. Thousands of inhabitants were deported, including many Balts of German origin. In 1944, Russians began immigrating to Estonia on a large scale, changing the population profile. In 1944, ethnic Estonians made up 94 per cent of the population; in 1994 this figure had dropped to just over 61 per cent.

CONTINUOUS LAND RECLAMATION

This small country with a population of just over 1.5 million has always had many different faces. Old ties still link Estonia to its neighbour Finland, and the Estonian language belongs to the Finno-Ugric group of languages. Ties also exist with neighbouring Sweden; in fact, the Estonian islands in the Baltic Sea are just a stone's throw away from that country.

Until they were expelled, Germans, too, formed part of Estonia's culture. For centuries a small upper class of German tradesmen, artisans and big landowners influenced the country's architecture, literature and religion.

Of all the Baltic republics, Estonia is the one most closely linked to and influenced by the Baltic Sea. Almost one-tenth of its total land area is made up of the islands of Saaremaa, Hiiumaa and more than 1500 smaller islands and islets. Land reclamation from the sea is a continuous process, both natural and artificial.

The Estonian mainland is rising by as much as an estimated 300 mm (11.8 in) per century. The mainland, which is bordered in the west by the Bay of Riga, in the north by the Gulf of Finland, and in the east by Lake Peipsi and Lake Pskov, seems more like a peninsula roughly split into three natural regions. Bare, smooth limestone boulders predominate in the north, moulded to their shape by Ice Age glaciers.

The central depression is made up of softer sandstone; most of this region is occupied by lakes, swamps and marshy meadows. The south is the most elevated region, culminating at Munamägi (318 m, 1043 ft) close to the Latvian border.

About one-third of the country is forested, chiefly with spruce forest; about one-fifth consists of swamps; and a further fifth is used for cultivation, principally of grains, potatoes and fodder crops.

Agriculture is well developed and though it employs only about 20 per cent of the workforce it produces enough for the country's needs and a surplus for export. The extensive forests provide close-grained timbers that, formerly widely exported to the world's construction companies, are now increasingly used in the production of paper.

Chief among the few mineral and natural resources are peat, limestone, phosphorite ore, and several million tonnes of oil shale. Estonia's oil shale deposits are among the most important worldwide. The country uses oil shale for generating electro-energy and as a raw material in the chemical industry. Even though more than half of Estonia's energy needs are produced from its oil shale, it still has to import petroleum and natural gas. Phosphorite ore is a source of phosphates used in agriculture but, unfortunately, exploitation of this resource, like oil shale, has led to unacceptably high levels of environmental pollution.

Until 1992-3, most of the country's trade – imports of fuel and other raw materials, and exports of industrial products, such as cross-country skis and electric motors – was with other ex-Soviet countries. A year later Finland had supplanted Russia as Estonia's chief trading partner, the country had a stable currency (its own), and its monthly inflation rate and unemployment figures were down to 4 and 3 per cent respectively. It had also attracted substantial foreign investment.

ESTONIA AT A GLANCE	
Official name Republic of Estonia	
Area 45 100 km² (17 409 sq miles)	
Population 1 608 470 **Per km²** 35.7 **(Per sq mile** 92.4)	
Capital Tallinn	
Government Republic	
Currency 1 Estonian kroon = 100 senti	
Languages Estonian (official), Russian	
Religions Mostly Christian (Lutheran, and Russian Orthodox)	
Climate Cool temperate climate with cold winters. Average temperature in Tallinn ranges from -6°C (21°F) in February to 16°C (60.8°F) in July	
Land use Forest and woodlands 31%, cultivation 22%, meadows and pastures 11%, other (mainly swamps and marshes) 36%	
Main primary products Oil shale, peat, phosphorites; potatoes, grains and fodder	
Major industries Hydroelectricity, electric motors, textiles, wood products, agriculture, skis	
Main exports Electric motors, textiles, cement, wood products and timber, skis and footwear, food, electricity	
Annual income per head (US$) 2750	
Population growth (per thous/yr) 5	
Life expectancy (yrs) Male 65 **Female** 75	

estuary Tidal mouth of a river where sea water mixes with fresh water. Also called a 'tidal channel', it usually supports an abundance of marine, plant and bird life.

Esztergom *Hungary* Attractive Baroque city and the country's ecclesiastical capital. It lies at the foot of the volcanic Pilis Hills, which look across the River Danube into Slovakia, 43 km (27 miles) north-west of Budapest. The Romans founded and fortified the city. In the year 1000, Hungary's first king, St Stephen, was crowned here.

Esztergom's magnificent domed cathedral, incorporating the fine 16th-century Bakócz Chapel, dates from the 19th century and is Hungary's largest church. The cardinal-primate's palace houses a superb museum of Christian art that, includes magnificent European paintings and French tapestries.

Population 29 700
Map Hungary – Ab

étang French term for a shallow pool or lake lying among sand dunes or beach gravels, applied especially to such features common along the Landes coast of France.

etesian winds Strong northerly or north-westerly summer winds blowing at intervals from the high pressure areas of the eastern Mediterranean. They strengthen in the afternoons and often die away towards evening.

Ethiopia See p 226

Etna *Italy* One of the world's most active volcanoes and the largest in Europe, rising in east Sicily to about 3323 m (10 902 ft). Its base is 200 km (125 miles) in circumference and there are almost 300 cones scattered about its slopes.

Of the 135 eruptions that have been recorded, the biggest occurred in 1329, 1669, 1910-11, 1928, 1950-1 and 1971, and the most recent in 1983 and 1992. The lower slopes on the south and east sides are densely settled, and are covered with vineyards and plantations of citrus trees and temperate fruits. The wooded upper slopes have been developed for skiing and as a summer retreat from the coastal heat.

Map Italy – Ef

Etosha Pan *Namibia* Shallow lake, about 160 km (100 miles) south of the Angolan border. Dry for most of the year, its many waterholes attract vast numbers of animals, including elephants, giraffes, kudu, leopards, lions, oryxes, springbok and zebras. The 22 270 km² (8596 sq mile) Etosha National Park, founded in 1958, includes the lake.

Map Namibia – Aa

Euboea (Évvoia; Évia) *Greece* Island in the west Aegean Sea. It is separated from the mainland by a narrow strait, which is bridged at Evripós by a 60 m (197 ft) span. The island fell into Greek hands in 1930. Its capital, Khalkís (Halkís), is an industrial centre with a yachting harbour, only 35 km (22 miles) from the mainland.

The pine-forested slopes of the inland mountain ranges are dotted with the ruins of 13th to 15th-century Venetian towers, Frankish castles from days of the Crusades and Byzantine monasteries. Charming villages such as Prokópion and Límni, and beaches at Káristos, Kími and Stíra

have become tourist attractions. Archaeologists have uncovered ancient remains, including those of a theatre and a gymnasium, at Erétria, 14 km (9 miles) south of Khalkís.

Fine marble is quarried on the island, while red wine, figs, honey and livestock are among its agricultural products; in fact, the Euboea takes its name from the ancient reputation of its cows. The word 'euboea' in Greek means 'good cattle'.

Population 209 130
Map Greece – Db

Eupen *Belgium* Manufacturing town 45 km (28 miles) east of the city of Liège. Eupen was part of Germany until 1919, when it was ceded to Belgium by the Treaty of Versailles. Many of its inhabitants are still German-speaking. Industries include brewing, food processing, tanning, wire and needle making, and the manufacture of domestic appliances.

Population 17 300
Map Belgium – Ca

Euphrates (Al Furat; Firat) *Middle East* Longest river in the MIDDLE EAST, or West Asia. It is 2736 km (1700 miles) long and has two branches, both rising in the Armenian highlands of Turkey, between the BLACK SEA and Lake VAN.

The Western Euphrates, or Kara (Karasu), has its source near Erzurum, and the Eastern Euphrates, also called the Murat (Murat Suyu), near Mount Ararat. They flow south-west and join near Keban to form the Euphrates proper, known in Turkey as the Firat. An enormous reservoir has been built at this point. The river flows

1096 km (681 miles) through Turkey before crossing the Syrian desert from north-west to south-east. Its Arabic name is Al Furat.

An ambitious water conservation scheme proposed by Turkey has caused consternation in Syria and Iraq, both of which lie downstream of the project. The construction of as many as 22 dams is visualised and one of them, the huge Ataturk Dam, has already been completed.

For much of its way through Syria, the Baghdad-Aleppo highway runs beside the river. After passing the town of Abu Kamal, near the Syrian border, the Euphrates enters Iraq, finally joining the TIGRIS River above Al Qurnah to form the SHATT AL ARAB, 64 km (40 miles) north-west of BASRA.

The river, which can be navigated only by shallow-draught vessels, is currently used for irrigation in Syria and Iraq. In addition, the extent of the effect of upstream Turkish dam-building has not yet been completely assessed or widely published.

Together with the Tigris, the Euphrates formed a fertile crescent of land that was the cradle of some of the earliest civilisations. Along its course, archaeologists have uncovered the remains of the cities of BABYLON, SUMER and UR.

Map Middle East – Ba

▼ **ANGRY MOUNTAIN** The snow-capped, fire-belching summit of Mount Etna reaches into the clouds. Despite over 130 recorded eruptions – many of them highly destructive – farmers have not budged from its fertile volcanic soils.

Europe If continents have specialities, Europe's would appear to be political and economic systems. Politics triggered the two World Wars, in 1914-18 and 1939-45. Both began in Central Europe but spread, dragging in countries as far away as the United States and Japan.

The notion of democracy was born in Europe, and so were the notions of capitalism and com-

munism. The physical line between these two 'isms', the Iron Curtain, was one of the most rigid boundaries in the world – but it came tumbling down and communications re-opened between the West and the East. In all, Europe consists of about 40 independent nations. They include such tiny, and densely populated, countries as Monaco, Andorra, Liechtenstein and San Marino.

The European continent has another claim or characteristic. Unlike the continents of America, Australia and Africa, which are each surrounded by the sea or linked to other land masses by a narrow isthmus, Europe has no natural borders. Strictly speaking, it is a subcontinent lying to the west of, and forming part of, Asia. Eurasia – that is, Europe and Asia combined – covers some

54 915 000 km² (21 197 792 sq miles) – an area that represents about one-third of the earth's land surface and one that is much larger than any other continent.

Historically, the geographic division into two separate continents is justified. The Orient, or East, is called the Asian continent, and the Occident, or West, the European continent. The two are separated by an imaginary line running, roughly, from the Urals to the Ural River, onwards across the Caspian Sea, the Caucasus, the Black Sea, Bosporus, and finally, across the Dardanelles to the Aegean Sea. The line cuts Russia and Turkey in two, leaving both countries with a smaller European and a larger Asian part.

The European continent differs vastly from the Asian continent – in fact, from all continents and islands which lie along similar lines of latitude and longitude in the northern hemisphere. Europe's average altitude is a mere 340 m (1116 ft), while Asia lies at an average 960 m (3150 ft) above sea level. Europe has a rugged coastline dotted with isles and islands – there are peninsulas jutting out to sea, and a number of large bays carved into the land.

The European average distance (340 km, 211 miles) from the sea is much smaller than that of Asia (780 km, 485 miles). This has a moderating influence on the climate. Europe, whose coasts are washed by the North Atlantic Drift, a branch of the Gulf Stream, does not experience the vast fluctuations in temperature or rainfall that are so common on the Asian continent. Exceptions are northern Sweden and Finland, whose climate is Arctic, and the North German Plain and East European countries, where it is more continental, with hot summers and cold winters.

The proximity of the sea has naturally influenced European history, and for thousands of years, the Mediterranean was a focal point of human development. Later, new impulses came from the Atlantic and North Sea coasts, especially from the Iberian Peninsula (Spain and Portugal) and the British Isles.

About 500 years ago, these were the starting points for Europe's venture into the 'New World', and for its journeys of discovery and colonisation, until the Europeans controlled almost the entire world. The 'control' they had, the battles they fought, the cruelties they perpetrated – above all, the culture and religion they introduced – and the way in which they did all these things have influenced world history for half a millennium.

Europe itself is a mosaic of landscapes and countries confined within a small space. Landscapes and different soil types follow one another in close succession, but elsewhere in the world they might be separated by many thousands of kilometres. The European continent has red soils with a Mediterranean vegetation, black soils typical of the steppe vegetation, and, nearby, polar landscapes almost devoid of plant growth. There are coniferous forests and parkland, arable farmland and endless vistas of grass.

There are great and famous rivers, too – the Elbe and the Danube, the Don, the Dnieper and the Volga – and formidable mountain ranges, from the Transylvanian Alps and the Caucasus to the Urals. Much of the continent's jagged shape, at least to the north, is due to the grinding and tearing that took place during the Ice Ages, while the other physical features were created by the upthrust of mountain chains.

Some of these are very ancient, like those formed of the Precambrian rocks in the north of Scotland that are 580 million years old. Others are more youthful, like the Alps, that are no more than 35 million years old. In between, there are plains such as those of northern France and Belgium, plateaus such as those of central Spain, and basins such as the Estonian depression and the basins in which Paris and London lie. The variety of scenery is both pleasing and extensive. The combination of natural and man-made landscapes, well-developed infrastructures and the ever-present sense of history, all combine to make Europe the prime tourist destination of the world.

Conditions are generally suitable for a broad range of farming activities. Outside Scandinavia, where little agricultural land is available, farming accounts for the use of roughly 50-75 per cent of each country's land area. Livestock, root crops, fruits, cereals, feedstuffs and industrial crops such as oilseed and rape are all-important, while to the sunny south, the vineyards of France, Italy, Spain and Portugal produce about 60 per cent of the world's wine. Olive oil and citrus fruits are also major products of the Mediterranean area.

An abundance of energy sources, raw materials, human resources and ingenuity made Europe, notably Western Europe, a great industrial region at a relatively early stage. Its Industrial Revolution, which began in Britain in the 18th century, spread outwards and, eventually, profoundly affected the lives of world's entire population.

The first, and still important, energy source was coal. It was supplemented later by hydroelectric power obtained from the region's great rivers, by imported oil and nuclear power, and latterly by North Sea and Soviet oil and natural gas. Iron ore is the most important non-fuel mineral mined, and steel production, traditionally at least, is the most important basic manufacture. Other major products include textiles, machinery, vehicles for both road and rail, aircraft, ships, foodstuffs, consumer goods and electronics.

Down the years, Europe gave much to, and took much from, the world at large. But since the Second World War, its nations have relinquished their empires, and surrendered their status as world powers to the United States.

Among themselves, they have instituted the European Economic Community (EEC) – now the European Union (EU), which is still attracting new members.

Area 10 498 000 km² (4 053 300 sq miles)

Population 693 million

Europoort *Netherlands* The western part of Rotterdam's port and industrial complex, situated at HOOK OF HOLLAND (Hoek van Holland), that city's coastal port. Europoort was created in 1958 as 'the Gateway to Europe' and has expanded enormously since the 1960s. It is accessible from the sea via a harbour entrance 850 m (2789 ft) wide and can handle the world's largest bulk carriers of oil, ore and grain. It has refineries for oil and non-ferrous metals, and shipbuilding, chemicals, engineering and food processing plants.

Map Netherlands – Bb

eustatic change Worldwide rise or fall in sea level. It is caused by the melting or growing of the polar ice caps (in which case it is called glacial-eustatic change), or by changes in the configuration of the ocean basins.

evaporite Sedimentary rock made up of minerals such as rock salt or gypsum that has been formed by recrystallisation after the evaporation of water containing dissolved salts.

evapotranspiration Combined loss of water by evaporation from the ground and by transpiration from plants.

Ethiopia

A CONSTANT VICTIM OF DROUGHT AND CIVIL WAR, ETHIOPIA IS THE OLDEST INDEPENDENT COUNTRY IN AFRICA AND ONE OF THE WORLD'S OLDEST SOCIETIES

The disastrous famine of the 1980s brought the East African state of Ethiopia to the forefront of the world stage. For although famine is not unusual in the area, this was the worst and longest of recent times – and was brought to the attention of people all over the world by television.

By mid-1985, about 300 000 people had died from starvation. Many of the dead were children, and poignant pictures of their plight appeared night after night on TV screens throughout the world. Consciences were pricked and a massive international relief operation began to gather momentum. But the relief was of short duration. At the end of 1993, Ethiopia was making renewed appeals for international food aid as a new drought began sweeping the country.

There are three main reasons for famines in Ethiopia. Firstly, while some areas are always well watered, others are regularly afflicted by severe drought, and extensive mountain terrain hinders the transport of food to stricken areas. Secondly, most Ethiopians are so poor that they cannot afford to buy food when they lose the crops they grow to feed themselves.

Thirdly, several areas been afflicted for years by civil war and guerrilla conflicts, and the government has devoted much of its energies and financial resources to these battles. By the early 1990s, however, it had started a large resettlement project aimed at depopulating the drought-stricken regions and shifting development to climatically more favoured regions in the south and east.

LIFE IN THE COUNTRY

Ethiopia is one of Africa's largest countries, with an area of 1.12 million km² (more than 470 000 sq miles) and a population of 53 million. Most of the land consists of highlands, rising to 4000 m (13 000 ft). It drops sharply towards the border with Sudan in the west and towards the DANAKIL lowlands in the north-east. The land falls more gently towards Kenya in the south and the OGADEN Desert and Somalia in the east.

The highlands are divided by the northern end of the East African GREAT RIFT VALLEY, that forms a series of lakes. The main rivers are the Abbai, which flows into Sudan and is also known as the Blue Nile, and the Awash, which drains the northern part of the Rift Valley and is used for irrigation before it disappears into the Danakil sands.

Rain falls mainly in the west, but is unreliable. Drought sometimes stops crops growing; at other times, heavy rain, often in thunderstorms, erodes soil from slopes and

reduces the area in which crops can be planted. Most Ethiopians live in the highlands in family clusters of up to 20 huts, without any amenities such as electricity and piped water, and often miles from the nearest shops or health centres. They must provide almost all their own basic needs.

Their farms generally occupy less than 1 hectare (2.5 acres), and are often fragmented. An occasional farmer may have a pair of oxen and a plough, but many – especially those living at lower altitudes – have only hoes with which to work their land. The main crops are *tef* (a local grain), barley and wheat in the

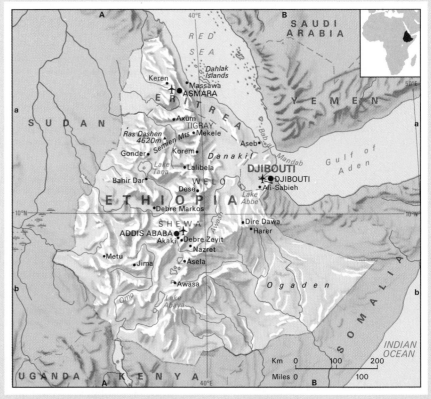

cool highlands, maize and sorghum on the lower land, and *inset* (a relative of the banana) in well-watered parts of the south. Besides producing a fruit, *insets* yield a valuable fibre, and starch is gained from their skin. But yields are usually low because the soil is poor and fertilisers are too expensive.

The lower dry areas such as the Ogaden are sparsely populated, the inhabitants being mainly nomadic tribes who move across country with their camels, cattle, sheep and goats in search of water and grazing. Some of them have permanent homes, which they occupy for only part of the year, while waiting for rain that will enable their sorghum crops to grow.

Coffee, the chief agricultural export, accounts for 60 per cent of export earnings and is grown mainly on small farms, especially in the highland regions such as Sidamo, and Kefa. A few large plantations, some financed by foreign companies, produce commercial crops such as sugar and cotton. Ethiopia has few known mineral resources, but there is a trem-

endous potential for hydroelectricity, should industrial development ever require it. At present, employment outside agriculture is confined to a small manufacturing sector covering building materials, footwear, tyres, food processing and textiles.

The towns are scattered across the country; ADDIS ABABA, with a population of some 2 million, is by far the largest. The capital city for a highly centralised national administration, it is easily Ethiopia's most important commercial and industrial centre. A number of factories are sited between the capital and Nazret, 100 km (60 miles) away, and at DIRE DAWA. Most of the

other large provincial towns have populations of 50 000 to 80 000 people and few or no industries. About 15 per cent of Ethiopians live in the towns and enjoy a higher income, more food and better housing than those who live in the rural areas.

Ethiopia is ranked among the world's 10 poorest countries; in fact, by some yardsticks it is at the bottom of the list. For instance, it is one of the world's least developed countries. It has few natural resources, but even these have not been properly developed or exploited. Thirty-eight per cent of the country's population is illiterate. There is only one doctor for every 79 000 people, which is the worst figure anywhere in the world.

The infant mortality rate is 125 per 1000 live births. Medical needs are greatly increased by famine. During the worst of the drought years, health personnel could not cope with the thousands of people needing medical care; there was an acute shortage of equipment and supplies, and clinics and health care centres were hopelessly overcrowded.

▲ FLIGHT FROM FAMINE Driven from their rural homes, thousands of starving refugees have taken shelter at the Kebre Beyah camp in the Ogaden desert region of eastern Ethiopia.

POLITICS AND RELIGION

Religion is important to most Ethiopians. Christianity was established as early as the year AD 330 around AXUM in the north, while Islam followed a few centuries later, and found a strong following in Ogaden. Today, the population is divided between the two faiths. Ethiopian Orthodox, or Coptic, Christianity was formerly the official religion.

Politically and historically, Ethiopia (formerly known as Abyssinia) is highly distinctive among African countries, for it has existed as a nation in some form for more than 2000 years and was strong enough in the late 19th century to resist Italian attempts at colonial occupation. The Italians ruled ERITREA in the north from 1890, and eventually took over Ethiopia in 1936, but they were forced out of both areas five years later by Ethiopian and Allied forces.

Emperor Haile Selassie was restored to the throne he had lost to the Italians in 1936, and this re-established a line of monarchs descended from King Solomon (972-932 BC). Britain administered Eritrea until 1952, when it was united with Ethiopia.

Haile Selassie started a programme of large-scale modernisation, but in 1974, he was overthrown by a group of army officers calling themselves the Dergue and led by Colonel Mengistu Haile Mariam. They swept away the former feudal aristocracy and adopted a Marxist-Leninist ideology, with land reform policies and a political structure based partly on peasant associations.

But by the late 1980s, Ethiopia's economic problems had increased; collective farming had many shortcomings and the military regime under Mengistu had steered off its purist socialist course. It began large-scale privatisation, but by then it was too late – Mengistu was overthrown in 1991.

At the same time, the 30 years' war against the Eritrean People's Liberation Front, which had halted internal development, ended. Ethiopian troops withdrew from Eritrea; Eritrea gained its independence in April 1993, but a year later the after effects of war were still being felt in the region. More than half its people depended on international food aid for survival.

In Ethiopia, peace remained elusive. In 1991, unrest began in pockets all over in the country, especially in the south-east and in Tigray Province. Nevertheless, in June 1994, Ethiopians went to the polls to elect a Constituent Assembly, whose members began the task of drafting a new constitution and preparing the country for multiparty elections.

ETHIOPIA AT A GLANCE	
Area 1 127 127 km² (435 083 sq miles)	
Population 60 000 000 **Per km²** 53	
(Per sq mile 138)	
Capital Addis Ababa	
Government Unitary republic, with elected transitional government	
Currency 1 birr = 100 cents	
Languages Amharic (official); English and Italian (commercial languages), and about 50 different indigenous languages	
Religions Ethiopian (or Coptic) Christian (45%), Muslim (45%), indigenous beliefs and others (10%)	
Climate Tropical on plateau, with summer rains; many regions arid and prone to drought. Average temperature in Addis Ababa ranges from 18°C (64.4°F) in March to 15°C (59°F) in August	
Land use Grazing 41%, forest and woodlands 24%, cultivation 13%, other 22%	
Main primary products Wheat, barley, maize, sorghum, millet, coffee, cotton, sugar cane, beans and peas, cattle, timber; salt, some oil	
Major industries Agriculture, food processing, textiles, cement	
Main exports Coffee, hides and skins, beans, oil products	
Annual income per head (US$) 110	
Population growth (per thous/yr) 33	
Life expectancy (yrs) Male 49 **Female** 49	

Everest, Mount (Chomolungma; Qomolangma; Sagarmatha) *Nepal/China* The world's highest mountain, rising 8846 m (29 022 ft) at the eastern end of the Great Himalayas, on Nepal's unmarked border with Chinese Tibet. Its height was first measured in 1852, and it was named after Sir George Everest (1790-1866), surveyor-general of India at the time. The Tibetans call the mountain Chomolungma – 'Goddess Mother of the World' – or, in Chinese, Qomolangma. Tibetans call it Sagarmatha.

Doubts arose later about that first measurement. A mass the size of Everest can exert its own gravitational pull, so the British survey team was not sure that their spirit levels were true – meaning that their theodolites, too, might be giving an inaccurate reading.

Because of this they took six measurements and worked out the average: exactly 29 000 ft. More recent measurements by Italian surveyors put the height at 8872 m (29 108 ft) – although for many years, the official height remained 8848 m (29 028 ft).

Measuring the mountain's exact height has proved a tricky exercise. For one thing, it's continually altered by the changing thickness of ice and snow at the summit. More recently, Everest has been measured using satellite and laser technology, and the mountain's latest confirmed height is – once again – 8846 m (29 022 ft).

Everest was first climbed in 1953 by Sir Edmund Hillary, of New Zealand, and Tenzing Norgay, a Sherpa guide. Since then there have been more than 130 successful ascents. However, more than 50 people have died attempting the climb. Today, improvements in equipment, better training and available information on the mountain have led to an increase in the number of climbers on Everest.

In 1993, 37 climbers reached the summit on a single day, and an estimated 20 000 climbers now attempt the ascent every year, despite that the fierce weather permits only three 'climbing seasons' – April-May, before the monsoons, October, after the rains, and December-January, in winter.

The increase in climbers has created a huge environmental problem, with the mountain becoming littered with garbage and human waste. Edmund Hillary himself initiated the formation of a national park in an attempt to prevent the denudation of the forests in the area.

Map Nepal – Ba

Everglades *USA* The Everglades is actually a shallow river 80 km (50 miles) wide and 0.5 m (2 ft) deep, but most people associate the name 'Everglades' with the United States' largest area of subtropical wilderness – a vast area of swamplands and forest lying in southern Florida, south of Lake Okeechobee.

Some 5668 km² (2188 sq miles) in the far south-west form a national park. The area is rich in wildlife and there are extensive areas of saw grass and mangrove.

The survival of the Everglades has in the past been threatened by water extraction for agricultural and urban use. Now, poisoning by phosphorus contained in run-off water from nearby sugar plantations is the greater threat.

It has already destroyed the indigenous vegetation in large tracts of land – and the United States Department of the Interior has taken urgent steps to save the Everglades.

Map United States – Je

Évia *Greece* See EUBOEA

Evian-les-Bains *France* Health resort with mineral springs on the south shore of Lake Geneva. It produces bottled mineral water named Evian after the resort.

Population 6200

Map France – Gc

Évora *Portugal* Capital of the district of Évora of the province of Alto Alentejo, about 110 km (70 miles) east of the national capital, Lisbon. It was captured by the Romans in 80 BC and has a Roman aqueduct and a temple of Diana. From the 14th to the 16th century, Évora was the residence of the kings of Portugal and site of a Jesuit university. Today, it is mainly an agricultural market, although there is some woollen carpet, cork and leather production. The town has a 13th-century Gothic cathedral.

Population 34 000

Map Portugal – Cc

Evpatoriya *Ukraine* See YEVPATORIYA

Evreux *France* Capital of Eure department in Upper Normandy. It stands on the Iton River, 90 km (56 miles) west of Paris. Founded in pre-Roman times, Evreux is one of the oldest towns in France. The Romans occupied it for nearly five centuries until it was seized by Clovis, king of the Franks, in the late 5th century. There are Roman ruins, an ancient Norman church and a Gothic cathedral. Since 1955, engineering and electrical firms have moved here from Paris.

Population 57 300

Map France – Db

Evros (Maritsa; Meriç) *South-East Europe* River with a total length of 514 km (319 miles). It rises in Bulgaria's Rila Mountains and flows east and then south-east past Pazardžik, Plovdiv and Svilengrad to the border with Greece and Turkey near Edirne, Turkey. It then forms the Greek-Turkish border and flows south to enter the Aegean Sea in a delta some 11 km (7 miles) wide. Its waters irrigate fields of cotton, cereals, tobacco, fruits and rice on the plain of THRACE.

Map Bulgaria – Bb, Greece – Ea

Evry *France* Capital of Essonne department and one of France's new towns, 30 km (19 miles) south-east of Paris. It was founded in 1970.

Population 29 600

Map France – Eb

Évvoia *Greece* See EUBOEA

Exeter *United Kingdom* City and port on the River Exe in south-west England, and county town of Devon, 250 km (155 miles) from London. Exeter's recorded history goes back to Roman times. It held out for two years after the Norman conquest in 1066 – only falling after an 18-day siege.

Much of the city was destroyed by German bombs during the Second World War, but a largely 13th-century cathedral, and a pre-15th century guildhall survived. The city's quay has some beautiful 19th-century buildings and a splendid maritime museum. The University of Exeter occupies a beautiful site on the north side of the city.

Population 105 400

Map United Kingdom – De

Exmoor *United Kingdom* Area of moorland pasture and heathland occupying the northern parts of the English West Country counties of Somerset and Devon. It lies near the Bristol Channel and there are steep cliffs along most of the coast.

The highest point is Dunkery Beacon, at 520 m (1705 ft). Exmoor National Park covers 685 km² (265 sq miles). The heathland is the home of red deer, ponies, sheep and grouse. Exmoor was the setting for *Lorna Doone* (1869), the best-known novel written by R.D. Blackmore (1825-1900).

Map United Kingdom – De

exosphere See ATMOSPHERE

extrusion The flow of magma, through volcanic craters and fissures in the earth's crust and out onto the earth's surface, where it forms extrusive igneous rock. Once on the surface, this cools very quickly, thus forming small crystals that give it a fine-grained or glassy texture.

extrusive rock See EXTRUSION

Exuma *Bahamas* Island group in the centre of the archipelago, consisting of Great Exuma, Little Exuma and adjacent CAYS, and covering about 335 km² (130 sq miles) in total. The islanders' main occupations are fishing, farming and tourism.

Population 3540

Map Bahamas – Bb

eye The circular area of relative calm at the centre of a CYCLONE, or HURRICANE where the atmospheric pressure may be very low. The eye, however, helps to perpetuate the raging spiral of wind and rain by drawing to itself air from surrounding higher pressures – creating winds of up to about 320 km (200 miles) per hour.

eyot Small island, especially in a river.

Eyre, Lake *Australia* Salt lake in South Australia, 700 km (430 miles) north of Adelaide. Covering about 8900 km² (3400 sq miles), it is Australia's largest lake and, at 16 m (52 ft) below sea level, the continent's lowest spot.

The lake is named after the explorer Edward Eyre, who discovered it in 1840. Most of it is barren salt flats. It has filled only four times this century, in 1950, 1974, 1984 and 1989, in each case after extremely heavy rains.

In 1964, Donald Campbell established a new world land speed record of 701.030 km/h (429.311 mph) on Lake Eyre, in his turbine-powered *Bluebird*.

Map Australia – Fd

Ezulwini Valley *Swaziland* Scenic valley between MBABANE and MANZINI. The valley, whose name means 'Place of Heaven', is Swaziland's leading tourist area, especially for visitors from neighbouring South Africa. It has a string of hotels, a casino night club and a cinema.

Outdoor attractions include the majestic Mantenga Falls on the Little Usutu River and the Mlilwane Wildlife Sanctuary, which protects animals such as white rhinoceros, giraffes, zebras, buffaloes, crocodiles, eland and other antelopes, and some 240 bird species within its 44 km² (17 sq miles). The Swazi royal residence of Lobamba is situated at Ezulwini.

Map Swaziland – Aa

F layer See ATMOSPHERE

Fada *Chad* Oasis settlement on the Ennedi Plateau, some 900 km (560 miles) north-east of the national capital, N'Djamena. It is a stronghold of the Arabic Toubou nomads. In the 1980s, Fada was drawn into Chad's civil war.
Map Chad – Bb

Faeroe Islands (Føroyar; Faeroese) *Denmark* Group of 18 islands in the North Atlantic Ocean, between the Shetland Islands and Iceland, covering 1399 km² (540 sq miles) in total. The largest island in the group is Streymoy (Strømø), on which the capital, Tórshavn (population 14 700), is located; the islands' highest point is Slaettaratindur, which rises to 882 m (2894 ft), on Eysturoy (Østerø).

The islands are largely composed of volcanic rock, which has been heavily gouged and eroded by glaciers and makes a dramatic, often forbidding, backdrop to the scattered fishing and farming settlements. Around the fringes of the islands, high cliffs are the haunt of millions of sea birds.

The principal occupations of the inhabitants are farming – mainly open-range sheep rearing – and fishing, with herring, cod and haddock as the main catch.

More than 95 per cent of exports are fish and fish products and vegetables. Imports outweigh the islands' exports by far. However, tourism now plays an increasing role as an earner of foreign currency.

The islands have had a measure of home rule since 1948. Although Danish is the official language, Faeroese, which developed out of Old Norse, is widely spoken.

The islanders also have their own flag, postage stamps and currency notes, as well as

their own by-laws; a ban on alcohol, in force since 1907, was lifted in November 1992.
Population 47 800
Map Europe – Cb

Fagatogo *American Samoa* See PAGO PAGO

Fahaheel (Al Fuhayhil) *Kuwait* Rapidly expanding port 30 km (18 miles) south of Kuwait City, of which it is now virtually a suburb. Until the 1960s, Fahaheel was no more than a Bedouin fishing village. Since then the development of a vast gas liquefaction project has brought prosperity and foreign workers to the town.
Population 50 000
Map Kuwait – Bb

Failakka Island (Faylakah) *Kuwait* Island 19 km (11 miles) from the mainland, east of Kuwait Bay. Archaeological remains show that between 3000 and 1200 BC there was a thriving community on the island, which had close trading links with the cities along the TIGRIS and EUPHRATES rivers, and with Dilmun (see QALAT AL BAHRAIN). Today, the island has become Kuwait's main leisure centre and tourist resort, with sailing clubs, museums, a hypermarket and a regular ferry service to the mainland.
Population 4900
Map Kuwait – Bb

Fair Isle *United Kingdom* Isolated island, lying between the Orkney and the Shetland islands, 37 km (23 miles) from the nearest land. Farming and fishing are the main occupations. The islanders are also noted for producing traditional Fair Isle knitwear with its intricate coloured patterns, possibly of Norse origin.
Population 75
Map United Kingdom – Gh

Fairbanks *USA* Largest inland town in Alaska, standing in the centre of the state, 190 km (120 miles) south of the Arctic Circle on the Trans-Alaska (oil) Pipeline.

Founded as a gold miners' settlement in 1902, it has become a trade, tourist and university centre. It is also a terminus and market nucleus for the Alaskan interior.
Population 30 840
Map Alaska – Cb

Faisalabad (Lyallpur) *Pakistan* City 120 km (75 miles) west of Lahore. Its prosperity dates from the opening of the Lower Chenab Canal in 1892. Although it stands in what has long been a fertile wheat- and cotton-growing area, its agricultural activities are threatened by a rising water table and increasing salinity of the soil.
Population 1 562 000
Map Pakistan – Db

Falkirk *United Kingdom* See CENTRAL REGION

Falkland Islands (Islas Malvinas) *South Atlantic* Two large and about 200 small islands, covering 12 173 km² (4700 sq miles) in all, and lying 770 km (480 miles) north-east of Cape Horn. The two main islands are East and West Falkland – bleak, rocky moorlands and bogs, swept by high winds and heavy rainfall.

▼ **DISPUTED SOUTH ATLANTIC ISLANDS** The Falkland Islands, or Islas Malvinas, are little more than rocky moorlands interspersed with bogs, and swept by the South Atlantic's icy winds. But the prospect of offshore oil deposits put them on the world map in the early 1980s.

English navigator John Davis, who landed here in 1592, is believed to have been the first European to set foot on the islands. However, the first settlers were French, who established themselves at Port Louis on East Falkland in 1764. In 1765 a British garrison was set up on West Falklands and in 1767, Spain bought the settlement at Port Louis from the French. In 1828, the newly independent Argentina established a settlement on East Falkland, which was destroyed by the United States warship *Lexington*. The US claimed that American seal hunters in the region had been attacked by settlers.

The British resumed occupation of both East and West Falkland in 1833 – and have maintained the Falklands as a British Crown colony ever since. However, Argentina continued to claim the islands, which it calls Islas Malvinas. The prospect of rich oil deposits offshore aggravated the long-standing dispute between Britain and Argentina, and in April 1982, the Argentines made a surprise attack and occupied the islands. A brief and bloody war followed, at the end of which Britain continued to control the islands.

The islanders are mostly of British stock and nearly all were born here. Their main industry is sheep farming, but the main source of revenue is from the sale of fishing licences to foreign fishing fleets operating in the Falklands Islands Conservation Zone. The sale of these licences brings in enough revenue to make the islands' government self-supporting. Stanley, on East Island, is the capital.

South Georgia, 1288 km (800 miles) to the south-east, and the South Sandwich group of small volcanic islands farther south, were Falklands Dependencies until 1985, when they became direct dependencies of the United Kingdom.

Population 2120
Map Argentina – Dd; (South Georgia and South Sandwich Islands) Antarctica – Ba

fall line The point at which a river crosses from hard rock to a softer rock is often marked by a waterfall or rapids. A line or zone linking such falls on nearly parallel rivers is known as a fall line. The rivers of the south-east United States drop from the Appalachians along such a line.

False Bay (Valsbaai) *South Africa* Inlet, 32 km (20 miles) long and 29 km (18 miles) wide, of the South ATLANTIC OCEAN, east of the Cape Peninsula. False Bay was explored in 1488 by the Portuguese navigator Bartolomeu Díaz, who was seeking a sea route to India. It was later named False Bay because homebound European navigators would mistake Cape Hangklip, at its eastern end, for the Cape of Good Hope, at its western end and, turning north too early, find themselves entrapped in the bay. Because the water in the bay is fed by warm currents of the Indian Ocean and is about 5°C (9°F) warmer than the water in Table Bay, tourists and local bathers flock to its beaches in summer. Popular bathing resorts include Muizenberg, Fish Hoek and the Strand. There are fishing harbours at Kalk Bay and Gordon's Bay, and a naval base at Simon's Town.

Map South Africa – Ac

Falster *Denmark* Island linked to the main Danish island of ZEALAND to its north by the 3.2 km (about 2 miles) long Farøbroen Bridge. The island covers 514 km² (198 sq miles) and

contains Denmark's southernmost ferry port, Gedser, which provides a ferry link with Warnemünde and Travemünde in Germany. Another bridge at Nykøbing, the main town, links the island with HOLLAND to the west.

Population 42 800
Map Denmark – Cb

Famagusta (Ammochostos; Gazi Mağusa) *Cyprus* Turkish-controlled east coast resort, and port which manufactures clothing and exports fruit and vegetables.

The 16th-century walls surrounding the old city were built by the Venetians who briefly colonised the island before being ousted by the Turks – who governed Cyprus until 1878. Famagusta's last Venetian governor, Bragadin, was murdered by the Turks after they captured the town in 1535. Bragadin's palace, as well as the 14th-century St Nicholas Church and the St Peter and St Paul's Church – both converted to mosques – can still be seen today. The palace, which now serves as a prison, is the setting of Shakespeare's play, *Othello*; its Othello Tower is a grim reminder of the tragedy that is said to have taken place here. Famagusta also has a Turkish museum, a citadel and a lively market.

Population 20 000
Map Cyprus – Ab

fan See ALLUVIAL FAN

Fanling *Hong Kong* Town in the NEW TERRITORIES, 35 km (22 miles) north of KOWLOON. Fanling is one of several new towns built to relieve population pressure in the main built-up areas. It incorporates the villages of Fanling and Sheung Shui, and has a target population of 226 000.

Population 87 900
Map Hong Kong – Ca

Fao (Al Faw) *Iraq* Port at the mouth of the SHATT AL ARAB waterway on The Gulf. It is the site of a deep-water tanker terminal, serving the country's southern oil fields. It also has a pipeline to Kirkūk in the northern fields, through which oil can flow in either direction. The pipeline was designed to provide an alternative export route via Syria for Basra, in the southern oil fields, if The Gulf were ever blocked. However, the port and its installations were damaged in 1986 during the Iran-Iraq War and again in the 1991 Gulf War.

Map Iraq – Dc

Faradofay *Madagascar* See TAOLANARO

Farafangana *Madagascar* Indian Ocean port 430km (265 miles) south of the capital, Antananarivo. It handles coffee, hides, rice, vanilla, cloves, pepper and waxes.

Population 124 000
Map Madagascar – Ab

Farakka Barrage *India* Low barrage across the River Ganges, 250 km (155 miles) north of CALCUTTA, which also carries the Assam railway line from Calcutta. The barrage diverts water into the Hooghly River, in this way keeping Calcutta open to navigation. It was opened in 1971 and became fully operational in 1975.

Map India – Dc

Faro *Portugal* Capital of Algarve, Portugal's southernmost province, 90 km (56 miles) east of

Cape St Vincent, the country's south-western tip. Faro Airport serves the many tourist resorts in the region. The town has several industries, including salt, fishing, and cork and food processing. A beach nearby is popular with holidaymakers.

Population 28 200
Map Portugal – Cd

Faroe Islands *Denmark* See FAEROE ISLANDS

Fatehpur Sikri *India* Former royal capital of the Mogul emperor Akbar, 220 km (137 miles) south of Delhi. Because of its mosques and palaces, it is a popular tourist attraction. Akbar built the town in about 1569, but it was abandoned after his death in 1605. A perfectly preserved wall 10 km (6 miles) long shelters the present-day city on three sides – the fourth side is open to a lake 32 km (20 miles) in circumference, which is usually dry but occasionally fills with water.

Population 25 000
Map India – Bb

Fátima *Portugal* Village 105 km (65 miles) north of the capital, Lisbon. A shrine in the village has been a place of pilgrimage since 1917, when the Virgin Mary is said to have appeared to three children on six occasions with messages calling for world peace. Every year around mid-October about 1 million pilgrims visit the site.

Population 6500
Map Portugal – Bc

fault Fracture of rock structures caused by stresses or strains within the earth's crust, and resulting in the displacement of the rocks on both sides of the fracture. See also THE RESTLESS EARTH.

Faya-Largeau (Faya-Abouchar) *Chad* Oasis settlement 780km (485 miles) north-east of N'Djamena. It is the capital of the Borkou-Ennedi-Tibesti (BET) prefecture, and the main centre of the northern half of the country. Faya-Largeau is the focus of several marked and unmarked desert tracks, and a trading centre where nomadic camel herders exchange dates for cereals and other goods from the south.

In the civil war that raged in Chad in the 1980s, Faya-Largeau was the stronghold of Libyan-backed anti-government forces.

Population 5500
Map Chad – Ab

Faylakah *Kuwait* See FAILAKKA ISLAND

Fayum *Egypt* See EL FAIYUM

Fdérik *Mauritania* Iron-mining town 625 km (390 miles) north-east of the capital, Nouakchott, near the border with Western Sahara. Iron ore has been mined here since 1963, and is exported via the 670 km (420 mile) railway to the port of Nouadhibou.

Map Mauritania – Ba

Fehmarn *Germany* Island in Kiel Bay, 1 km (less than 1 mile) off the coast of Holstein. Fehmarn is linked by bridge to the mainland, and visitors come over from the mainland to enjoy its fine beaches. In fact, a direct road-and-ferry route now exists between Hamburg and Copenhagen. The island covers 185 km² (71 sq miles).

Population 11 800
Map Germany – Da

THE RESTLESS EARTH

The pattern of relief, the shape and distribution of hills and valleys, mountains and lakes, is an amalgam of both the constructive and destructive forces working within the earth.

Hills, mountains and valleys are associated with FAULTS and FOLDS in the earth's crust. The folding and faulting help to construct uplands – which streams immediately begin to destroy.

Folds are waves in the structure, and faults are cracks; both result from the enormous pressures exerted by the movement of the crustal plates (see PLATE TECTONICS). The greatest pressures are exerted at plate margins where two plates meet and push the land up into mountain ranges.

Conversely, where plates separate, they pull the land apart, causing long depressions like the rift valleys of East Africa.

Simple folds usually occur in young rocks, and complex folds in older strata which have been exposed for a longer period to the earth's varied movements and may have been refolded many times.

As the distance from the plate margins increases, the folds gradually die out, as in the foothills of the Himalayas. When rocks can take no more bending, they break, so forming a fault. If this is simply stretched, a normal fault results, but if breaks are later compressed, the resulting features are known as 'reverse' or 'thrust faults'. Faults and folds frequently occur together in the same region. Such regions too are often prone to earthquakes, especially where new faulting takes place along the line of an older and buried fault.

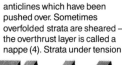

FOLDS AND FAULTS

Faulting and folding are produced by vertical and horizontal stresses in the earth's crust. Fold mountains are caused by movements in the crustal plates; examples are the Jura of France and Switzerland, the Alps and the Urals in the USSR. Fault-block mountains occur when a block between faults is uplifted and tilted, although the block may well be the remnant of an earlier fold mountain; the Sierra Nevada of California was formed in this way.

A single, simple upfold of strata is known as an anticline (1). An anticlorum (2) is a large anticline incorporating a number of smaller folds. Overfolds (3) are anticlines which have been pushed over. Sometimes overfolded strata are sheared – the overthrust layer is called a nappe (4). Strata under tension (5) may result in a normal fault. Compressed rocks produce a reverse fault (6), while horizontal movement may cause a strike-slip fault (7). A rift valley (8) occurs when land slips downwards between two parallel faults. When land is uplifted between two parallel faults (9), the formation is known as a horst or block mountain.

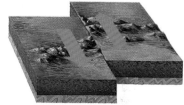

Also known, more graphically, as tear faults, strike-slip faults are dislocations of the earth's crust in which there is horizontal movement along the opposing STRIKE surfaces of the break. The faults are at least as deep as the foundations of the crust, and it is thought that some of them may form such ocean trenches as the Puerto Rico Trench in the Western Atlantic.

THE FAULT THAT THREATENS CITIES

The great majority of earthquakes that afflict the world occur round the fringes of the Pacific, where several crustal plates meet. One of these junctions, between the North American and Pacific plates, is the San Andreas Fault which cuts through the San Francisco peninsula. This makes the area prone to earthquakes, the most notorious of which was the San Francisco earthquake of 1906, which killed 700 people. Many seismologists believe that further earthquakes are imminent. In the San Andreas Fault, whose line can be clearly seen from the air, the Pacific Plate to the west moves at a rate of about 10mm a year relative to the North American Plate. In the last few million years their relative lateral movement has been several tens of kilometres, and eventually the peninsula may be entirely sheared from the continent and moved bodily north.

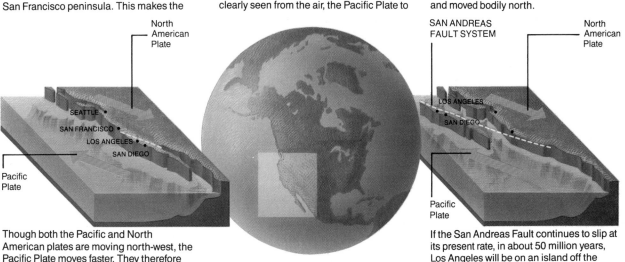

Though both the Pacific and North American plates are moving north-west, the Pacific Plate moves faster. They therefore seem to be moving in opposite directions.

If the San Andreas Fault continues to slip at its present rate, in about 50 million years, Los Angeles will be on an island off the Canadian west coast.

feldspar (felspar) A group of widely distributed rock-forming minerals consisting of silicates of aluminium combined with potassium, sodium, calcium and, rarely, barium, and probably comprising half of the earth's crust. Feldspar minerals are the main component of igneous rock. They divide into two main groups, namely potassium feldspar (silicates of aluminium combined with potassium), including orthoclase and microcline, and the plagioclase group (silicates of aluminium combined with sodium or calcium), the most common being albite and anorthite.

fell Upland stretch of open country, or moor. The term is especially applied to such landscapes in northern England.

felspar See FELDSPAR

fen Tract of low-lying swampy land in which peat accumulates beneath the surface, but in which the ground water is alkaline or neutral. The name is applied especially to the FENS region around the Wash in eastern England.

Fengfeng *China* Important coal-mining centre in Hebei Province, 250 km (155 miles) south of Beijing. It supplies high quality coking coal for the domestic steel industry as well as for export.
Map China – Hd

Fens, The *United Kingdom* Low-lying area in eastern England. It lies to the south and south-west of the Wash, mainly in Cambridgeshire, Lincolnshire and west Norfolk. It has alternated between sea and land for centuries as the sea level fluctuated. The present area, parts of which are below sea level, is the product of extensive drainage, dyking and land reclamation, begun by the Dutch engineer, Cornelius Vermuyden, in the 17th century. The land is made up of peat and silt. It is enormously fertile when drained and is devoted to arable crops.
Map United Kingdom – Fd

Fergana Valley *Uzbekistan* Irrigated agricultural region lying in the country's south-east, near the Kirghiz border. The Fergana Canal was built here in 1939 and numerous other canals have been built since then to feed the region's vast irrigation networks.

The valley is mostly a broad, fertile plain set 300-1000 m (1000-3300 ft) above sea level and surrounded by two arms of the TIEN SHAN range whose peaks reach 5500 m (18 000 ft) and higher. It produces cotton, rice, silk, and a variety of fruits – including grapes, apricots, peaches and plums. Its main cities are Kokand, the centre of the cotton industry, Andizhan, the region's rail centre, NAMANGAN, which manufactures cotton and food products, and the industrial city of Fergana (population 198 000), which lies close to a recently discovered oil field. Other products of the Fergana Valley include metals and chemicals.
Population 1 944 000
Map Uzbekistan – Ca

Fermanagh *United Kingdom* County area in the south-west of Northern Ireland. It has many forests and among its numerous lakes are Upper and Lower Lough Erne. Its main town is Enniskillen (population 11 430), 15 km (9 miles) south-west of which lies the well-known limestone cave area of Marble Arch. The area's chief industries are agriculture, forestry, and tourism – boosted especially by Fermanagh's good fishing.
Population 54 030
Map United Kingdom – Bc

Fernando Póo *Equatorial Guinea* See BIOKO

Ferrara *Italy* City situated 90 km (55 miles) south-west of Venice. Most of Ferrara still lies within its 11 km (7 mile) long medieval walls. The massive, moated Castello Estense, begun in 1385 and completed nearly 200 years later, stands in the centre. It is one of many monuments in the city, which is associated with the Dukes of Este, who ruled Ferrara from the 13th to the 16th century, making it famous for music, printing and art.

Close to the Castello Estense are two fine medieval buildings – the town hall, built in 1243, and the Romanesque-Gothic cathedral. There are also many fine Renaissance palaces, including the exquisite Palazzo Schifanoia, with 15th-century frescoes depicting the months, and the Palazzo dei Diamanti, named after its 12 600 diamond-shaped facing stones. Today, Ferrara's economy revolves around its food processing and petrochemical industries.
Population 137 340
Map Italy – Cb

Fertile Crescent *Middle East* Historical agricultural area of the Middle East. Shaped like a narrow crescent, it runs from Jordan northwards through Israel and Lebanon, passes through the west and north of Syria, then turns towards the south-east to cross Iraq, and finally continues through to The Gulf.

The Fertile Crescent receives as little as 203 to 381 mm (8-15 in) of rain a year, but since ancient times its agriculture has been greatly helped by irrigation – especially along the NILE, TIGRIS and EUPHRATES valleys.
Map Middle East – Ba

Fertöd *Hungary* Superbly restored Baroque palace built by the aristocratic Esterházy family 27 km (17 miles) south-east of the city of Sopron, near the Austrian border. Dating from the mid-18th century, the horseshoe-shaped palace with 126 rooms stands amid colourful formal gardens. The Austrian composer (Franz) Joseph Haydn who was the Esterházys' court conductor in Fertöd and EISENSTADT from 1761 to 1790, is commemorated in a museum and summer music festival. Nearby are a fine Baroque inn and Haydn's house.
Map Hungary – Ab

Fès (Fez Jedid) *Morocco* The oldest of Morocco's four imperial cities, or old royal capitals, and today a commercial city and major communications centre. It lies in a valley north of the Middle Atlas Mountains, 200 km (125 miles) south-east of Tangier. Fès remains Morocco's cultural and spiritual capital and one of the sacred cities of Islam. It contains one of the largest mosques in the world – the Qaraouyine (Qarawiyin) – which is also among the world's oldest universities, founded about AD 850.

It was chosen as the capital city in 808 by the Arab prince Moulay Idriss II. At its height in the 14th century, when Fez Jedid (as the new part of Fés was called) was built, its fame spread far beyond Africa as a place of learning, attracting students from all Muslim countries and parts of Europe. Today, it is a bustling city with many mosques, shrines, colleges, palaces, markets and gateways. Excellent brocades and silks are available here, as well as leather goods, carpets and musical instruments. The city's royal palace, Dar el Makhzen, dates from the 13th century; its medina (ancient Muslim residential quarter), with its labyrinthine streets reached through gateways in its walls, is the largest in Morocco.
Population (Greater Fès) 1 012 000
Map Morocco – Ba

Fezzan *Libya* Historical region in the south-west of the country, covering 570 000 km² (220 026 sq miles). It is very dry, consisting mostly of desert rock, gravel wastes and sand dunes. The inhabitants of this sparsely populated area live mainly in oases, where date palms grow. The chief town of the region is Sebha.
Map Libya – Bb

Fife *United Kingdom* Ancient kingdom, former county and, today, a modern administrative region in eastern Scotland. Standing on a peninsula between the firths, or estuaries, of the Forth and Tay rivers, it covers 1308 km² (505 sq miles). It contains the town of Dunfermline, capital of Scotland in the 11th century and birthplace of King Charles I (1600-49). It also contains the town of St Andrews, famous for its university (the oldest in Scotland, founded in 1411), its 12th-century cathedral ruins, and the Royal and Ancient Golf Club, one of the ruling bodies of world golf, whose Old Course was the world's first. The region has many ruined castles.

The south and west of Fife have a long history of coal mining, but only a handful of pits survive. Newer industries include marine engineering, electronics, chemicals and activities based on North Sea oil. The principal town is Kirkcaldy, where the pioneer economist Adam Smith was born in 1723. Today, a number of computer firms have made their headquarters in Kirkcaldy. The administrative centre of Fife is Glenrothes.
Population 339 280
Map United Kingdom – Db

Figuig *Morocco* Oasis on the edge of a great stretch of barren desert in the extreme east of Morocco, on the border with Algeria and linked to major desert caravan routes.
Population 14 480
Map Morocco – Ba

Fiji See p 234

Filadelfia *Paraguay* See GRAN CHACO

Filchner Ice Shelf *Antarctica* Large body of floating ice lying at the head of the Weddell Sea on the Atlantic coastline of Antarctica. It was discovered by the German explorer Wilhelm Filchner in 1912.
Map Antarctica – Cc

filling In meteorology, a term used to express increasing atmospheric pressure at the centre of a DEPRESSION. It is the opposite of DEEPENING.

finger lake Long, narrow lake occupying a U-shaped valley typically carved by a glacier, as in the LAKE DISTRICT in the United Kingdom and the Finger Lake District of New York State in the United States.

A DAY IN THE LIFE OF A MOROCCAN OASIS FARMER

With eager eyes, Mustapha scans the desert horizon as the sun rises behind him. Some time today, or perhaps tomorrow, a camel train will come into view, led all the way from Mauritania and Mali by his good friend Farouk.

It is one of the last caravans that still journey from the coastal ports to the southern Sahara; lorries now carry supplies across the desert more economically.

Mustapha's wife brings him a bowl of water and he washes his feet, hands, neck and mouth according to Muslim ritual. Prostrating himself on a small mat in the direction of Mecca, he says his prayers.

Breakfast is a substantial meal, eaten before the heat of the day. It consists of maize, dates and nuts.

Some of his children have already begun the work of breaking down the low mud walls of one of the shallow irrigation canals that run out from the lake at the heart of the oasis. Life-giving water cascades from the canal into the fields of maize chosen for today's irrigation.

At the same time, the other children are building up walls that were broken down yesterday to flood other fields.

Mustapha's wife goes out to weed the crops, and as he watches her, he notices a dark line on the horizon and can just make out the train of camels and men. He sets out to bring back his old friend and the merchants, whom he entertains in his main room, where each of them sits cross-legged upon hand-embroidered cushions.

Mustapha's wife, back from weeding, brings thick, sweet coffee in small porcelain cups with brass holders. She retreats to the kitchen when she has served it, leaving the men to talk.

The camel men would once have been Mustapha's only source of news from distant oases and from the outside world. Today he can hear all the news on his transistor radio.

Nevertheless, local gossip, and the state of the animals and distant grazing lands never reach the air waves, and talk of them occupies Mustapha and his friends for several hours.

Then Farouk, dressed in traditional loose-fitting, blue-dyed clothes, begins to put up his tent which is made of dark brown camelhair cloth.

Meanwhile, Mustapha's eldest son, Muhmoud, wearing an open-necked shirt and the latest in casual Western trousers, comes over to bid his father farewell.

Muhmoud will never be a farmer – nor indeed the leader of a camel train, and has no desire to be either. He has always wanted to be a lawyer. He is leaving for the start of another term's study in the city.

Having said his goodbyes, Muhmoud strides to his motorbike. Mustapha and Farouk watch the trail of dust he raises disappear into the distance.

Finistère *France* Windswept department in the far west. It has rugged cliffs, wide bays and many fishing harbours. BREST, on the western coast, is the largest town.
| Population 828 000 |
| Map France – Ab |

Finisterre, Cape *Spain* Rocky headland in the north-west, and the westernmost point of the Spanish mainland.
| Map Spain – Aa |

Finland See p 236

Finland, Gulf of Shallow eastern arm of the BALTIC SEA, between Finland and Estonia, which is 430 km (267 miles) long and 20-130 km (12.5-80 miles) wide. The gulf freezes over in winter; shoals and rocks also make navigation difficult. However, there are important ports along its coasts, the largest of which are HELSINKI (Finland), TALLINN (Estonia), and ST PETERSBURG and VYBORG (Russia).
| Map Finland – Cc |

Finnmark (Finmarken) *Norway* Northernmost province of Norway some 48 632 km² (18 772 sq miles). It is the largest and most sparsely populated of the Norwegian provinces. In the far north it borders on Finland and Russia. Its inhabitants include Lapps and Finns, and their main occupations are fishing, raising reindeer, farming and iron mining. Vadsø is the provincial capital.
| Population 76 000 |
| Map Norway – Ga |

fiord See FJORD

Fiordland National Park *New Zealand* Remote area along the south-west coast of South Island covering some 12 519 km² (4832 sq miles) and named after its resemblance with the fjord (fiord) landscapes of Norway. It consists of rugged mountains, precipitous valleys, roaring torrents and waterfalls, dense forests, tranquil lakes and imposing cliffs. The park is the home of the beautifully coloured takahe (*Notornis mantelli*), one of the world's rarest flightless birds, which is about the size of a small turkey.
| Map New Zealand – Af |

Firat *Middle East* See EUPHRATES

Firenze *Italy* See FLORENCE

firn (névé) Compacted snow accumulated over the years in successive layers in hollows high in the mountains, firn forms masses of granular, whitish ice, the source of glacier ice.

firth Scottish word for a long narrow arm of the sea, such as the Firth of Forth.

fissure eruption Volcanic eruption in which lava is emitted through a long crack in the earth's crust. See VOLCANO.

Fiume *Croatia* See RIJEKA

fjord (fiord) Long, narrow coastal inlet, bordered by steep hills, found especially in Norway and Alaska. Fjords are the result of a rise in the sea level, causing penetration of sea water into the U-shaped valleys eroded by former glaciers.

flagstone Hard, fine-grained sedimentary rock, usually sandstone or sandy limestone, that splits or cleaves easily along the bedding planes and is hard enough to be used for building or paving.

Flanders (Flandre; Vlaanderen) *Belgium* Coastal region in the north-west, formerly a medieval county and now two provinces covering 6115 km² (2361 sq miles). It contains the cities of BRUGES (capital of West Flanders) and GHENT (capital of East Flanders), both of which developed in the 13th and 14th centuries as centres of the European wool trade. Today, the region's prosperity is based mainly on farming the coastal POLDERS, together with the old-established textile industry and numerous modern plants manufacturing electrical goods and machinery. Flanders was the scene of some of the most vicious fighting of the First World War.
| Population 2 444 100 |
| Map Belgium – Aa |

flash flood Sudden, violent flood in a normally dry valley after heavy rain.

Flensburg *Germany* Baltic port in Schleswig-Holstein, rum-producing centre, and the most northerly city in the country, barely 2 km (1 mile) from the Danish border. In 1920, after the First World War, Flensburg voted to be part of Germany; the farming people to the north chose Denmark. In April 1945, the town was the last retreat of the Nazi government. Admiral Karl Dönitz, *Führer* for 10 days, stayed in the town's 16th-century Glucksburg Castle while negotiating the German surrender with the Allies.
| Population 87 000 |
| Map Germany – Ca |

Flinders Range *Australia* Mountains in South Australia stretching north from Spencer Gulf for more than 350 km (220 miles). Their highest point is St Mary Peak at 1168 m (3832 ft).
| Map Australia – Fe |

flint Hard form of CHERT usually found in nodules along the bedding planes of chalk. It fractures along shell-shaped planes and forms sharp edges. This, together with its hardness, made it of great importance to early humans who used flint for tools. The popular cartoon *The Flintstones* recalls its importance to Stone Age people.

Flint *United Kingdom* See CLWYD

flood plain Flat land on the borders of a river, especially its lower reaches, over which ALLUVIUM is deposited when the river floods, thus making it extremely fertile.

flood tide Rising tide between low and the succeeding high water, or the inflowing tidal stream in an inlet. It is the opposite of EBB TIDE.

Florence (Firenze) *Italy* Capital of Tuscany and one of the great artistic centres of the world. It stands on the Arno River, 230 km (145 miles) north-west of Rome. The city was founded as a military camp by the Etruscans in about 200 BC. Known as Florentia, the site was later occupied by the Romans, but it was only after the 12th century that the city grew in wealth and power. Throughout the Middle Ages, wool and silk were its staple industry.

Fiji

A VOLATILE MIX OF PEOPLES AND CULTURES GREETS VISITORS FOLLOWING IN THE FOOTSTEPS OF CAPTAIN BLIGH

The most economically advanced and racially mixed nation in its region – and one of the largest – the island state of Fiji is strategically placed at the cross-roads of the south-western Pacific. Today, Fiji allows Western navies to dock in its ports, and the most striking impression on visitors is one of broad smiles and friendliness of the locals. This is a far cry from only 200 years ago, when the Fijians had a reputation for being savage cannibals and were widely feared.

The first European visitor to the islands was the Dutch explorer Abel Tasman, in 1643; Britain's Captain James Cook called in 1774; but the first to explore the group extensively was Captain William Bligh in 1792, when he returned to the islands he had seen from an open boat after the mutiny on the *Bounty*.

Next, whaling ships and traders seeking sandalwood and *bêche-de-mer* (sea cucumbers) arrived. Guns were sold to the locals, exacerbating inter-island wars to such a degree that in 1874 Britain stepped in. Partly at the urging of European settlers, partly to ward off French and American interest in the islands, and partly motivated by its own interests, Britain accepted the offer of cession made by the chief warlord, King Cakobau, who needed money for warfare. Fiji became a colony and Britain gradually imposed order.

In 1970, the islands became an independent member of the Commonwealth, with Queen Elizabeth II as head of state; in 1987, Fiji was declared a republic, and its membership of the Commonwealth lapsed.

The native Fijians are mainly frizzy-haired Melanesians, some with Polynesian blood – but almost 50 per cent of the population are descendants of indentured sugar plantation labourers from India. Today it is this mix which sets the scene for serious racial strife. Until the late 1980s, great efforts by Britain and, later, by politicians of all nations went into avoiding this possibility.

The previous electoral system allowed no race supremacy, while giving the indigenous people a special place in their own country. Fijians own 83 per cent of the land – their name for themselves, *Taukei*, means 'the owners'. In fact, there now exists a law to protect the Fijians as the rightful landowners. The remaining land is owned by a white minority. The Indians, many of them business people, dominate commerce and the professions, as well as working on the land. The Indians also provide the political awareness that has made Fiji a regional leader.

However, in April 1987, a government dominated by Indians was voted in for the first time. But when the new government was overthrown in a military coup led by native Fijians, tension between the communities grew.

According to the new constitution of 1990, 37 of the 70 parliamentary seats are reserved for indigenous Fijians, while Indians may not occupy more than 27; five seats are reserved for other races, and a final seat for a representative from the remotest island, Rotuma.

There are some 320 islands and atolls in the group, plus another 480 islets, but only 110 are inhabited. The two main islands, VITI LEVU and VANUA LEVU, account for 87 per cent of the land area. Extinct volcanoes give them dramatic skylines, the tallest being Tomaniivi (Mount Victoria) rising to 1324 m (4344 ft) on Viti Levu, where SUVA, the capital, lies.

Coral reefs fringe most of the islands, where tropical rain forests face the south-east trade winds; grassy plains and clumps of palm-like pandanus trees occupy the leeward land. The hilly landscapes of the main islands were once heavily forested, but many trees have been felled or burnt to create grazing land, and soil erosion is now a growing problem.

Before the 1987 coup, industries revolved around tourism and the cultivation of sugar cane. Sugar and ginger are still the main export crops (25 per cent of the population depend on sugar for their livelihood), but the industries have expanded to include processing of agricultural products. Although tourism declined in the aftermath of the coup, it increased steadily again during the early 1990s.

FIJI AT A GLANCE	
Map Pacific Ocean – Dc	
Official name Republic of Fiji	
Area 18 270 km² (7052 sq miles)	
Population 780 000 **Per km²** 42.7 **(Per sq mile** 110.6)	
Capital Suva	
Government Republic	
Currency 1 Fijian dollar = 100 cents	
Languages English, Fijian, Hindustani, Urdu, Tamil, Chinese	
Religions Christian (53%), Hindu (38%), Muslim (8%)	
Climate Tropical, the main rainy season is December-March; average temperature is 25°C (77°F) throughout the year	
Land use Forest and woodlands 65%, cultivation 13%, meadows 3%, other 19%	
Main primary products Sugar, cassava, coconuts, rice, sweet potatoes, ginger, cocoa, tobacco, livestock, timber, fish; gold	
Major industries Agriculture, tourism, mining, sugar refining, food processing, clothing, forestry, fishing	
Main exports Sugar, gold, clothes, canned fish, coconut oil and copra, ginger	
Annual income per head (US$) 2010	
Population growth (per thous/yr) 19	
Life expectancy (yrs) Male 61 **Female** 64	

Florence's cultural splendour reached its peak during the 15th century after it had eclipsed rival city-states such as Pisa and Siena. After a period of turmoil, it fell under the stable rule of the wealthy Medici family. In 1865 it succeeded Turin as capital of Italy; in 1870, Rome became the capital.

Florence gave Italy its language, its art and its philosophy. The city's cultural refinement is still evident – though many artworks were destroyed, or badly damaged, when the Arno flooded in 1966. The arts developed under the patronage of the Medicis in the 15th and 16th centuries, when such artists as Giotto, Brunelleschi, Donatello, Botticelli and Michelangelo flourished here. Everywhere, works of art adorn the city's public squares, streets and buildings. The Piazza della Signoria, for example, is a veritable open-air museum of sculpture, while priceless frescoes and oil paintings are found in the city's many churches. The Uffizi and Pitti palaces have perhaps the finest collections of Renaissance paintings in the world. Michelangelo's *David* and his sculptures of slaves are in the Accademia Gallery.

Many of the city's buildings are themselves works of art, and often seem to be designed less for business or worship than for the display of art treasures and architectural skill. Among these, the Baptistry's bronze doors by Ghiberti and Pisano, was described by Michelangelo as 'the Gateway to Paradise'. Equally stunning are the *campanile* (bell tower) by Giotto and the Duomo, the domed cathedral of Santa Maria del Fiore designed by Brunelleschi.

Other historic buildings include the church of San Lorenzo, also by Brunelleschi, with its tombs of the Medicis by Michelangelo, the Palazzo Vecchio (1299-1314), which is one of the greatest Gothic palaces in Italy and the seat of the Florentine Republic until the 16th century, and the famous Ponte Vecchio (1345). The city's well-known church of Santa Croce (1294) has frescoes by Giotto as well as the tombs of Michelangelo, the political theorist Machiavelli, the scientist Galileo, and the composer Rossini.

Florence's present-day prosperity rests on its range of economic interests – banking, publishing, its university, tourism, food and drink, textiles, fashion clothing, leather goods, jewellery, furniture, metallurgy, and light engineering.

Population 408 400	
Map Italy – Cc	

Flores *Indonesia* Mountainous volcanic island of 14 250 km² (5500 sq miles) in the Lesser Sundas group, between Java and Timor. The island was once famous for its sandalwood. Today, maize, groundnuts (peanuts) and coconuts are the main crops.

Population 803 000	
Map Indonesia – Fd	

Flores Sea Indonesian sea, 240 km (150 miles) wide and 6961 m (22 838 ft) deep at its greatest known depth.

It lies between the islands of SULAWESI (Celebes), FLORES and SUMBAWA. Underneath Flores and Sumbawa – both of the Lesser Sunda group of islands – the northward-drifting Indo-Australian continental plate is sliding beneath the south-eastern part of the Eurasian plate, causing earthquakes and volcanic activity.

Map Indonesia – Ed	

Florianópolis *Brazil* Port and regional capital of Santa Catarina, 480 km (300 miles) south of São Paulo. It stands on Santa Catarina Island and is linked to the mainland by the longest bridge in Brazil, the Hercilio Luz Bridge. The natural

▲ DRAMA IN TRANQUILLITY White ibises grace a quiet pool in Florida's Everglades, but despite the apparent peacefulness, they are constantly in danger from predatory alligators and panthers, and are also threatened by water extraction for agricultural and urban use, and by pollution.

beauty of the island makes it popular with tourists. The surrounding region produces coal, textiles, wheat, apples and vegetables.

Population 256 000
Map Brazil – Dd

Florida *USA* America's 'Sunshine State', lying in the extreme south-east of the country. Because of its location, Florida is prone to hurricanes – the one which hit southern Florida in 1992 caused more than $20 billion in damages.

Florida was ceded to the United States by Spain in 1819 and became a state in 1845. It covers 139 697 km² (53 937 sq miles), and is mostly a low-lying peninsula which separates the Atlantic from the Gulf of Mexico. The state is dotted with lakes, the largest of which is Lake Okeechobee. The Florida Keys, a string of coral islands which are linked to the mainland by causeway, stretch 241 km (150 miles) just south of MIAMI BEACH, to KEY WEST.

The subtropical climate in the southern part of the state is ideal for citrus and vegetable crops, while the temperate climate in the rest of the state supports a large livestock industry. Manufacturing industries include food processing, electronics, and the making of beachwear. There is oil, and extensive deposits of phosphates which

are used in fertilisers but which are threatening the plant and animal life of the EVERGLADES. Tourism throughout the year is a boom industry. Besides the Everglades National Park, its tourist attractions include DISNEY WORLD and the John F. Kennedy Space Center at Cape CANAVERAL. The state capital is TALLAHASSEE.

Population 12 937 900
Map United States – Je

Florida, Straits of Straits, up to 145 km (90 miles) wide and 900 m (2950 ft) deep, between the Florida Keys and Cuba, and between south-eastern FLORIDA and the BAHAMAS. They contain the warm, fast-flowing Florida Current, which flows from the Gulf of MEXICO, and becomes the GULF STREAM as soon it emerges into the ATLANTIC OCEAN. Buccaneers once lurked around the western end of the straits, waiting to plunder treasure-laden galleons on their way from Mexico to Spain.

Map Bahamas – Ab

fluorite See FLUORSPAR

fluorspar (fluorite) A white or colourless mineral consisting of calcium fluoride.

Flushing (Vlissingen) *Netherlands* Port on the former island of Walcheren, 75 km (45 miles) south-west of Rotterdam, with ferry links to Sheerness in England.

It has been a Dutch naval base since the 16th century, and is also a resort and a fishing and ship-building centre.

Population 44 180
Map Netherlands – Ab

Fly *Papua New Guinea* River flowing some 1200 km (750 miles) south and east from the Victor Emanuel range, in the west-central part of the country, through swampy lowlands to the CORAL SEA at the Gulf of Papua. The mouth of the Fly River is more than 50 km (30 miles) wide and it is navigable for medium-sized craft for more than 1000 km (620 miles) of its length. Fish swarm in the muddy waters of the river, as do crocodiles, which inspire terror among the tribes living on the banks of the river.

Map Papua New Guinea – Ba

flysch Group of sediments or sedimentary rock consisting of SHALES, muds and MARLS interspersed with layers of conglomerate and coarse sandstone. Flysch is by the rapid erosion of FOLD MOUNTAINS, which occurs while the mountains are being uplifted. The sediments are deposited in adjacent troughs and are then themselves deformed by the continuing folding.

focus Centre of an EARTHQUAKE and the point from which SEISMIC WAVES are emitted.

fog Cloudlike masses of water droplets close to the ground, sometimes also containing smoke or dust particles, and often limiting visibility. Fog is formed when air is cooled below its DEW POINT. This occurs either when warm, moist air passes over a cooler sea or land surface (in which case, it is called 'advection fog'), or in calm conditions, when air cools because the land on which it rests is cooled by radiation (it is then called a 'radiation fog'), or, lastly, when warm, moist air meets cold air along a FRONT (in which case, it is called 'frontal fog').

Finland

ONCE PART OF SWEDEN, ONCE PART OF RUSSIA, THIS LAND OF LAKES AND FORESTS CLINGS TENACIOUSLY TO POLITICAL NEUTRALITY

Situated at the eastern edge of Western Europe, Finland stands where the technological 20th century meets a wilderness of forest and Arctic wasteland. Indeed, over two-thirds of Finland is covered by forest, interspersed with many thousands of lakes. The snows that fall in November do not melt until April or May. The country's austere beauty inspired the music of Jean Sibelius (1865-1957) as well as the clean lines of the architecture and furniture of Alvar Aalto (1898-1976) and Eero Saarinen (1910-61).

Most of Finland is low-lying, with large areas of swamps and peat bogs. The land rises to more than 1000 m (3300 ft) in the north in Lapland. The extensive forests, from which one-fifth of Finland's main export, paper, is produced, consist mainly of spruce where the soil is clay, and fragrant pine on sands and gravels; the birch thrives almost everywhere.

At the end of the last great Ice Age, about 10 000 years ago, when temperatures rose, the vast ice sheet retreated and left immense quantities of glacial debris on the ancient granite bedrock. In the process, thousands of lakes were formed. And as the ice melted, the sea level rose. But, relieved of the weight of ice, the land is today rising at a rate of as much as 300 mm (11.8 in) each century in HELSINKI. As a result, land is being gained from the sea at the rate of 1 km² every 10 years. Lapping the shores of this land of lakes, even the BALTIC resembles a lake and has virtually no tides.

Across the Gulf of BOTHNIA, and forming Finland's western boundary, lies Sweden; the eastern boundary of the country is shared – for 1269 km (about 790 miles) – with Russia. The Russian city of ST PETERSBURG, on the Gulf of Finland, lies only 300 km (190 miles) from the Finnish capital, Helsinki. Norway, reaching across the top of Scandinavia, forms the northern boundary of Finland. One-third of Finland lies inside the ARCTIC CIRCLE.

The Finns experience great extremes of climate between winter and summer. Summer days are long and light, but in the winter the days are short and gloomy. Inside the Arctic Circle the sun can shine at midnight for a month or two in the summer, but in winter, it does not rise at all for four to eight weeks.

Autumn is short and spring comes late; frosts in early summer are a widespread hazard. It is not surprising that Finland is sometimes referred to as 'the Land of Three Winters' – there is autumn winter, high winter, and spring winter. During the long, harsh winter everything freezes, even the Baltic Sea.

NATIONAL MINORITIES

The people of Finland probably settled in their country about 2000 years ago, coming from beyond the URAL Mountains in Russia. Their language is closely related to Estonian but bears no resemblance to any other European tongue except Hungarian.

Coastal areas were colonised in the 12th century by Swedish speakers from the west, and today, Finland is officially a bilingual country although only 6 per cent of the inhabitants speak Swedish. This figure includes all the people (about 24 000) of the province of AHVENANMAA (Åland) in the south-west archipelagoes between Sweden and Finland, who inhabit what amounts to a piece of Sweden in what is nominally Finland. Though the province voted to belong to Sweden when Finland became independent, the League of Nations insisted, for strategic reasons, that Ahvenanmaa should stay Finnish. The province was, however, granted considerable autonomy in 1920. It has its own flag and issues its own stamps; its people are exempt from Finnish military service, and Finnish speakers are not allowed to buy land in Ahvenanmaa unless they have lived there permanently for at least five years.

In 1994, the question arose as to whether Ahvenanmaans would be exempt from European Union (EU) regulations if Finland joined the European Union in 1995. The population of the archipelago were especially worried about the tax-free sales on board their ferries running the route between Finland and Sweden. Another concern was the European Union's ban on shooting wild birds during the nesting season – a tradition the population consider their right.

In the north of Finland there are some 1700 Lapps, a minority of whom are reindeer herders, their traditional livelihood. More often than not, however, they herd their reindeer on skidoos (motorised sledges) today. Most Lapps live by fishing. Lapp is taught at school and many of the Lapps still speak their native language. But the Lapps today, or Saame as they also call themselves, have become sedentary, and have replaced their tents, made of skins and hides, with houses.

Finland is not a country to which other people have emigrated to any great extent although it was for long periods ruled by others. As a result, the Finns share a physically similar appearance, and a 'family likeness' can be seen in people throughout the country.

There is not a great deal of religious diversity. The major church is Evangelical Lutheran, though there is a strong Orthodox minority

of some 53 000 owing allegiance to the patriarch in Constantinople (Istanbul).

Christianity was introduced in the 12th century, when the Swedes conquered Finland and started to convert the people. The Swedish occupation affected mainly the south-western coastal areas of the Gulf of Bothnia and the islands of Ahvenanmaa. Finland became part of Sweden in 1540. After the Russo-Swedish War of 1808-9, Finland was ceded to Russia to become a grand duchy. More than a century of Russian rule ended with the October Revolution of that country. On 6 December 1917, Finland declared its independence.

A NATION REBORN

In November 1939, the USSR invaded Finland, and although the Finns withstood the invasion, Finland was forced to give up large areas – more than one-tenth of its land in all, including most of the province of KARELIA, half of the north-eastern parishes of Salla and Kuusamo, and a corridor of land leading to the Arctic coast.

Finland tried to regain the lost territory after Hitler invaded Russia in 1941, but was forced to agree to an armistice in 1944 – an exercise which displaced some 400 000 people, besides costing huge reparation in the form of ships and machinery which Finland was obliged to pay to Russia. The pressure all this caused on Finland's economy led to a complete restructuring, from agriculture to engineering, which turned out to the country's great advantage.

After the Treaty of Paris in 1947 and until the collapse of the Soviet Union, Finland had a pact of cooperation with the USSR, which became its principal trading partner.

The country's trading links with the USSR were strengthened by its need for the USSR's energy supplies. Only one-third of Finland's energy needs can be supplied by domestic sources. The country has already developed almost all possible hydroelectric power, and readily available peat has a limited use. Finnish nuclear reactors are producing power, and the Finns today seem to have no qualms about establishing more of them. But for years, the energy gap was closed by imports of crude oil and natural gas from USSR and of coal from Poland and Germany.

The Russo-Finnish trade agreements came to a halt with the collapse of the Soviet Union. This, together with the world's economic crisis, led to a decline in the Finnish economy. Increased exports to Western markets in turn led to further estrangement with Russia. In fact, relations between Russia and Finland had already cooled over the latter's overtures to the West, even before the country agreed on the terms of entry into the EU in 1995.

Finland has always maintained strict neutrality in international affairs. It is a member of the Nordic Council of Scandinavian countries, and a member of the European Free Trade Association along with Norway, Sweden, Iceland, Switzerland and Austria.

URBAN LIFESTYLES

Since the Second World War, the scale of Finland's transformation from a largely rural to an industrial and urban country has been

▲ **DYING WAY OF LIFE A Lapp family build a tent of reindeer skins in the tundra of the far north. Few Lapps now live the 2000-year-old traditional nomadic life in which reindeer provided food, clothing, building material, herding equipment and fat for lamps. Many of those who do depend on summer tourists.**

staggering. Its economic growth rate, estimated at 3 per cent between 1983 and 1987, and at 4.9 per cent between 1987 and 1989, for a while even surpassed that of Japan. But by 1992, growth was down to -3.5 per cent and by 1993, Finland had an unemployment rate of 22 per cent.

Most people live in the south-western third of the country, where all the cities are, including Helsinki, TAMPERE and TURKU, the three biggest. One-fifth of the population live in large cities; and 80 per cent of the population in towns and settlements. But because enough people (8.5 per cent of the workforce) are still farmers, Finland continues to be largely self-sufficient in foodstuffs. Most of the rural dwellers combine agricultural work with forestry or fishing, especially in coastal Finland, although big commercial fleets now account for most of the country's catch. Some of the fishermen-farmers also breed animals such as mink for fur which is exported.

To the north, where there is so much space – only two inhabitants to the km² – much of the land is forested; it is here, too, that the reindeer are herded. Most of the crops are grown in the south-west, where the climate is mildest and where cereals, rape, sugar beet, potatoes and vegetables are main crops. Many farmers have specialised in dairy or poultry farming. Pork, too, is an important product.

Since 1900, Finland has developed a widespread network of cooperatives for marketing farm produce, for bulk buying of farm equipment, fertilisers and seed, and for organising finance and insurance. Agricultural schools and colleges have also been established, ensuring a high level of education among the rural population. Many Finns argued that membership of the EU would cause hardship among the farming community. The feeling was that local products and prices would not be able to compete with the EU's agricultural produce and price structure.

NATURAL BOUNTY

Finland's autumn and summer bring a harvest of wild berries and mushrooms, and a wealth of game birds and wildfowl; its lakes and rivers are rich in pike, perch, powan, trout and salmon, although pollution – much of it caused by wood processing, fertilisers and acid rain – has become a problem. Fishing is still a favourite sport, as is shooting and some 60 000 to 70 000 elk are shot every year, as a means of keeping down their numbers. Bears and lynxes, which were once hunted, are now protected. Wolves often make local headlines; their victims are usually the reindeer.

Finland still lives to a large extent by its forests, although the days of the seasonal employment of hordes of lumberjacks, and large-scale floating of logs down the rivers, are over. Now, skilled workers are employed in the forests all year although felling still takes place, mostly in winter; and while huge rafts of logs are sometimes floated across the lakes, most of the timber is moved by lorry.

One-third of the forests are privately owned and held in small lots – only in the north are forests state-owned. Widespread education in forestry has improved yields enormously. Continuous surveys of forest areas, both by air and on land, ensure that the timber is not cut down faster than it is replaced. Advances in forestry are matched by those in the factory.

New processes and diversification have led to new types of paper, wallboards, laminates and prefabricated timber components, as well as chemical products, all of which swell the export market.

Timber processing is not the only industry to have advanced. The complete revision of industry which followed the Second World War led to investment in metallurgy, electrical manufacturing and engineering, which now account for most of the country's industrial workers. For example, there is a new state-owned iron and steel complex at Raahe in the north. Shipbuilding has boomed, prompted in the 1950s by Finland's need for an effective fleet of icebreakers, which are are now built for other countries as well, including Russia. Shipyards in Helsinki and Turku also build ferries and cruise liners, and recently mustered the expertise of several nations to build a luxury cruise liner, the *Royal Princess*.

Entire industrial plants have been exported from Finland, including power stations, factories for processing softwoods, and oil rigs. The biggest of these undertakings has been the construction of an iron and steel complex in Russian Karelia, with a Finnish labour force who camped on site.

Skills are also exported – Finnish banking representatives work in all the world's major banking centres, and Finnish designed glass, clothing, porcelain, kitchenware, sports equipment and electronic goods have not only made for a good export trade but have influenced design elsewhere.

In the areas where seasonal unemployment is worst (the construction industries suffer especially from the long winters), new industries and public works have been set up to ease the situation. Financial support is provided on a graduated scale, so that most goes to the needy north-east and the least to the wealthier south-west.

A SPORTING LIFE

Developments are complemented by an elaborate conservation programme, with 30 national parks and more than 1000 protected areas, covering 30 000 km² (11 580 sq miles) in all, as well as controls on industrial pollution. More Finns are building summer homes, and the people have not lost their appreciation of their country's exceptional landscape.

Life for Finns follows the rhythm of the seasons. They are great sports people, with sports halls in many towns and saunas to relax in after physical exertion. Finns take pride in being fit, and some still keep the habit of combining a sauna in mid-winter with a leap into a hole in the ice. There are ski jumps on most village skylines, and schools have a skiing holiday, which coincides with the national winter sports festival in LAHTI in February.

The long summer days are to be made the most of, and the working hours are fewer then, while holidays in educational institutions run from the end of May to mid-August. Then, a great summer exodus from the towns takes place as boats and cars leave for summer cottages and family farms. Comfortable town apartments are exchanged for often primitive retreats, in response to the call of the wild.

A high point of the year is the festival of midsummer, with its bonfires. In August the Finns bid the summer farewell at crayfish parties, feasting under the last of the lingering sunshine. The Swedish-speaking people's festival of Santa Lucia in December, illuminates the dark winter with a candle-lit celebration. The long winter ends with Walpurgis Night on 30 April – a festival observed especially by students, who sing in the summer on the eve of the May Day public holiday.

For each day as well as for festivities, Finns have their own version of the Scandinavian cold table, or *smörgåsbord – voileipäpöytä* in Finnish – and their own vodka, which is sold, along with all alcohol, at shops owned by a state monopoly. Beer is popular, coming in three different strengths, and so is curdled milk, which is prepared in different regional ways by half a dozen different bacterial cultures, as is coffee.

On all festive and formal occasions, especially on Independence Day, 6 December, the blue and white Finnish flag is flown, and the music of Sibelius is not far away. A century ago the country was on the famine fringe of Europe, poverty-stricken and ruled as a dependent duchy of imperial Russia. Today, it has found a national identity, and achieved a high standard of living – despite a harsh environment with limited resources.

FINLAND AT A GLANCE		
Area 337 030 km² (130 097 sq miles)		
Population 5 057 980 **Per km²** 15 **(Per sq mile** 39)		
Capital Helsinki		
Government Republic		
Currency 1 markka, or mark = 100 penniä		
Languages Finnish, Swedish (both official), Lapp		
Religion Christian (92% Lutheran, 1.1% Orthodox)		
Climate Moderate in the south, subpolar in the north, with long, cold winters. The far north gets little rain. Average temperature in Helsinki ranges from -7°C (19.4°F) in February to 17°C (62.6°F) in July		
Land use Forest and woodlands 76%, cultivation 8%, other 16%		
Main primary products Timber, cereals, potatoes, livestock, fish; copper, zinc, iron, chromium, lead		
Major industries Forestry, timber products including wood pulp and paper, machinery, shipbuilding, clothing, chemicals, fertilisers		
Main exports Timber, timber products, paper, wood pulp, machinery, ships, metal products, chemicals, textiles and clothing		
Annual income per head (US$) 22 980		
Population growth (per thous/yr) 3		
Life expectancy (yrs) Male 71 **Female** 79		

▶ **BLUE AND GREEN JIGSAW**
Finland's lakes cover one-tenth of the country, and forests, dominated by conifers, over two-thirds. Cut logs are still sometimes floated to the timber mills and wood products – paper and paper board, timber and wood pulp – account for one-third of exports.

Foggia *Italy* City and capital of Foggia Province, 125 km (78 miles) north-east of Naples. It is set in an area rich in wheat, olives and vines; its economy depends upon food processing.
Population (city) 155 040; (province) 693 440
Map Italy – Ed

föhn (foehn) Descending warm, dry wind coming off the leeward side of a mountain range. It blows especially off the northern slopes of the Alps, where it causes rapid melting of snow in winter, with avalanches.

fold Bend in rock strata resulting from compression by forces in the earth's crust. See THE RESTLESS EARTH, p 231.

fold mountains Mountains formed by folding and uplifting of the earth's crust. They are usually associated with the BENIOFF ZONE. The Alps, Himalayas and Andes are fold mountains. See THE RESTLESS EARTH, p 231.

Fontainebleau *France* Town near the west bank of the SEINE, 55 km (35 miles) south-east of Paris. It is famous for its 16th-century château, once the home of French kings and now the official summer residence of the French presidents.
Set in formal gardens, it is visited by more than 270 000 people each year. The Forest of Fontainebleau, covering 170 km² (65 sq miles), is one of the most beautiful in France, and has inspired famous painters such as Corot and Millet.
Population 39 400
Map France – Eb

foreland Geological term applied either to a low promontory projecting into the sea, such as North Foreland in Kent, or to an ancient continental mass forming one side of a GEOSYNCLINE which mountain-building movements have pushed up into folded ranges. An area flanking a mountain range onto which glaciers moved down during the Ice Ages can also be called a 'foreland'. Examples are the Bavarian and Swiss forelands.

foreshock Minor EARTHQUAKE heralding the approach of a much larger tremor.

foreshore Area stretching from the lowest low-water line to the average high-water line.

Forli *Italy* City situated 80 km (50 miles) northeast of Florence. It boasts grandiose 1930s architecture. Its 17th-century Palazzo Paolucci was restored by Benito Mussolini (1883-1945).
Population 109 280
Map Italy – Db

Formosa See TAIWAN

Føroyar *Denmark* See FAEROE ISLANDS

Fort Dauphin *Madagascar* See TAOLANARO

Fort Hill *Malawi* See CHITIPA

Fort Knox *USA* See LOUISVILLE

Fort Lamy *Chad* See N'DJAMENA

Fort Lauderdale *USA* Atlantic resort city in Florida, 40 km (25 miles) north of Miami. Founded as a fort in 1838, it has one of the most popular beaches in the United States and one of the world's largest pleasure boat marinas. A network of more than 435 km (270 miles) of natural and artificial waterways crisscrosses the city.
Population (city) 149 380; (metropolitan area) 1 093 300
Map United States – Je

Fort McMurray *Canada* Town which stands amid vast oil-sand deposits, at the junction of the Clearwater and ATHABASCA rivers in north-east ALBERTA. It is the terminus of a water transport system through to the Arctic.
Population 31 000
Map Canada – Dc

Fort Worth *USA* City in north-east Texas. It lies in the west of the Dallas-Fort Worth conurbation, also called 'the Metroplex', with a population of more than 4 million. Fort Worth is an agricultural centre, with food processing, oil refining, aircraft and motor car manufacturing industries.
Population (city) 448 000; (metropolitan area) 1 361 030; (conurbation) 4 037 280
Map United States – Gd

Fortaleza (Fortaleja do Ceará) *Brazil* Port and capital of Ceará State, 2312 km (1435 miles) north-east of the national capital, Brasília. A population explosion between 1979-84, following a massive influx of migrants from the drought-stricken interior to its *favelas* (shanty towns), now makes it the largest city in the northern half of Brazil. It has developed a thriving tourist industry; sea food still brought in by traditional *jangada* sailing rafts provides a good local cuisine.
On either side of the city, massive sand dunes stretch along the coast. Inland, there are orchards growing cashew nuts and tropical fruits.
Population 1 825 000
Map Brazil – Eb

Fort-de-France *Martinique* Capital and port, standing on the island's west coast beside one of the best harbours in the West Indies. It was originally an administrative and military centre, but following the destruction of ST PIERRE by a volcanic eruption in 1902, it also became the chief commercial and cultural town of the island.
Population 100 000
Map Caribbean – Cc

Foshan *China* City in Guangdong Province, just south of Guangzhou. It is situated on one of the most fertile plains of southern China. It has varied light industries, including textiles. Local handicrafts can be purchased at its markets.
Population 323 000
Map China – Gf

Fouta (The Futa; Fouta Toro) *Senegal* Flood plain beside the Senegal River, and close to the Mauritanian border. It is fertilised by silt each year when the river floods. Local Toucouleur and Fulani people grow groundnuts, millet and rice on the highly prized land; however, an irrigation scheme upstream will eventually control the flooding – and alter traditional cropping patterns. Bakel and Matam are the largest towns.
Map Senegal – Ba

Fouta Djallon (Futa Jallon) *Guinea* Deeply dissected plateau in west Guinea. It rises in a steep escarpment to the west but drops more gently to the savannah plains of Upper Guinea to the east. Reaching 900 m (3000 ft) in places, it is extensively used for livestock rearing, particularly by the Fulani people, while its deep and fertile valleys support crops, including bananas and fonio (a small millet). The region contains immense deposits of bauxite, including those at Boké, Dabola, Fria, Kindia and Tougué.
Map Guinea – Ba

Fouta Toro *Senegal* See FOUTA

Foveaux Strait Strait, 24-34 km (15-21 miles) wide, which separates SOUTH ISLAND, New Zealand, from STEWART ISLAND. A ferry runs from Bluff on South Island to Halfmoon Bay on Stewart Island. Fishing and oyster cultivation are among the main activities.
Map New Zealand – Ag

Fox Glacier *New Zealand* Mountain glacier near Mount Cook in the Southern Alps of SOUTH ISLAND. Like the Franz Josef Glacier, 40 km (about 25 miles) to the north-east, it has thinned and retreated in recent years. Both glaciers are unusual in that they descend into lush forests. They move 0.5-4.5 m (1.5-15 ft) a day.
Map New Zealand – Ce

Foxe Basin Arm, 480 km (300 miles) long and 180-230 km (110-140 miles) wide, of the ATLANTIC OCEAN, between northern Canada's Melville Peninsula and BAFFIN ISLAND, named after Luke Foxe who explored it in 1631. In the south, the 320 km (200 mile) long Foxe Channel leads into HUDSON BAY.
Map Canada – Hb

France See p 242

Franceville *Gabon* See MASUKU

Franche Comté *France* Region of eastern France, comprising the agricultural plateaus of the Upper SAÔNE, the Belfort Gap, and the forests and high pastures of the Jura Mountains. It became French territory in 1674. Its main activities are dairy farming, forestry, and winter tourism. Among the region's products are cheese and watches. BESANÇON is the chief city.
Population 1 084 000
Map France – Fc

Francistown *Botswana* Centre of trade, industry and tourism in the north-east, 360 km (220 miles) from Gaborone, near the border with Zimbabwe. It was founded as the centre of the Tate Concession, a mining and farming area of more than 5200 km² (2000 sq miles) ceded to British interests in 1887 and named after the Tate River. The town itself was named after Daniel Francis, a prospector. It reached its peak as a mining town in the 1890s.
Population 65 240
Map Botswana – Cb

Frankfort *USA* State capital of Kentucky. It stands on the Kentucky River in the heart of the bluegrass country known for its thoroughbred horses. Its manufactures include textiles, shoes and whiskey.
Population 26 000
Map United States – Jc

Frankfurt (Frankfurt am Main) *Germany* City straddling the River Main, 155 km (95 miles) south-east of Cologne. Frankfurt – the name means 'Ford of the Franks' – has been a major trading city since the Romans built a bridge across the Main 2000 years ago.

In 1152, the German princes known as Electors met in Frankfurt's St Bartholomew's Cathedral and chose Friedrich Barbarossa as Holy Roman Emperor. For the next 400 years, the emperors were elected at Frankfurt – but until 1562 they were crowned at Charlemagne's old capital, Aachen.

Today, Frankfurt is a major commercial and financial centre. The country's principal stock exchange is situated here, as is the headquarters of the national bank. It was also the site from which the Rothschild family started its international banking empire.

The city also hosts many national and international fairs, including the autumn Book Fair, the largest in the world. Frankfurt operates the nation's major international airport and lies at the centre of an autobahn and railway network.

Much of the walled medieval town was destroyed by bombing in 1944, and has been replaced by tower blocks and concrete and glass skyscrapers.

Only the medieval cathedral, the 15th-century houses around cobbled Römerberg Square, and the home of the writer Johann Wolfgang Goethe (1749-1832), creator of *Faust*, have been restored. The opera house has been reconstructed as a conference centre; flower gardens now trace the path of the old ramparts.

The city is a cultural centre and has several museums and art galleries. It also boasts a renowned university.

Population	664 900
Map	Germany – Cc

Frankfurt an der Oder *Germany* Town on the Polish border, 80 km (50 miles) east of Berlin. It stands on the River Oder which forms the German-Polish border for most of its course. The town produces metal and electrical goods, and processed food.

Population	86 200
Map	Germany – Fb

Franklin *Australia* River in south-western Tasmania flowing 118 km (73 miles) before it joins the Gordon River. A combination of high rainfall, luxuriant forests and rugged scenery combine to make its course one of the most beautiful in the country.

Plans to dam the Franklin and the lower Gordon for hydroelectric power were dropped following strong worldwide protests from conservationists.

Map	(Tasmania) Australia – Hg

Frantiskovy Lázne *Czech Republic* See CHEB

Franz Josef Glacier *New Zealand* See FOX GLACIER

Franz Josef Land *Russia* See ZEMLYA FRANTSA-IOSIFA

Franzensbad *Czech Republic* See CHEB

Frascati *Italy* Resort town in the Alban Hills, some 20km (12miles) south-east of Rome.

Surrounded by vineyards, it has many beautiful 16th-century villas, and has been a retreat for the inhabitants of Rome for 2000 years.

Population	20 040
Map	Italy – Dd

Fraser *Canada* Longest river in British Columbia, which rises in the Rocky Mountains near the Yellowhead Pass and winds its way to the Strait of Georgia, south of Vancouver. It is 1370 km (850 miles) long, but only the lowest 145 km (90 miles) are navigable. The river is famous for its salmon.

Map	Canada – Cc

Fraser Island (Great Sandy Island) *Australia* Largest sand island in the world, covering 1840 km² (710 sq miles). It lies off the coast of Queensland, 190 km (120 miles) north of Brisbane. The island's sand dunes rise, in places, to 240 m (787 ft); inland they are covered by rain forest, eucalyptus trees and heath. The sands of the island also contain minerals such as titanium and zircon; plans to mine them in the mid-1920s were successfully opposed by conservationists. Much of Fraser Island is now a national park and a World Heritage Area.

Map	Australia – Id

Frauenburg *Poland* See FROMBORK

Fredensborg Palace *Denmark* See HILLERØD

Fredericton *Canada* Capital of New Brunswick. It has two universities, an art gallery, and a theatre. Chief industries are timber processing and shoe manufacturing. A Canadian forces base is situated a nearby Oromocto.

Population	46 470
Map	Canada – Id

Frederiksborg Castle *Denmark* See HILLERØD

Fredrikstad *Norway* Historic seaport and industrial town on the east shore of the Oslofjord, close to the Swedish border. Founded in 1567, it contains the remains of a fortress which was an important defence post between the 14th and the early 19th century, when Norway was united with Denmark. The old town stands on the River Glåma which drains into the Oslofjord. The port's major exports are timber, softwood products and chemicals.

Population	26 500
Map	Norway – Cd

Free State *South Africa* See ORANGE FREE STATE

Freeport City *Bahamas* Main port, industrial town and tourist centre of GRAND BAHAMA island in the north-west of the archipelago. With the adjacent Lucaya area it is the biggest tourist resort complex in the West Indies. It was founded in 1955 and grew rapidly in the 1960s with the construction of a deep-water harbour.

Population	25 000
Map	Bahamas – Ba

freestone Fine-grained, even-textured rock, usually sandstone or limestone, which can be cut in any direction without shattering or splitting.

Freetown *Sierra Leone* The country's capital, chief port and largest city. It stands on the north side of the SIERRA LEONE PENINSULA where the Sierra Leone River drains into the ATLANTIC OCEAN. Like many other West African cities, it has distinctive quarters – Kru Town, where the indigenous Kru live, Portuguese Town, quarters of the descendants of Brazilian slaves, and Maroon Town, where maroons (fugitive slaves from Jamaica) settled in the 19th century – apart from modern residential areas.

Freetown was founded by the British in 1787 as a settlement for freed slaves, which explains the city's name. In the 19th century, it was used as a base for suppression of the slave trade. The city has mills for the processing of rice and groundnuts (peanuts).

Population	690 000
Map	Sierra Leone – Ab

Freiberg *Germany* Town in the south-east, lying at the foot of the eastern Ore Mountains. It was the first of the 'free mountain cities' (*'frei'* means 'free', and *'berg'* means 'mountain') in Germany and during the Middle Ages was extremely wealthy. Freiberg has the world's oldest mining academy, founded in 1765. Lead, silver, zinc and iron pyrites were mined in the area until 1968.

Population	48 900
Map	Germany – Ec

▼ ORIENTAL EXTRAVAGANZA El Casino, built in Moorish style and now renamed Princess Casino, is a major attraction of Freeport City on Grand Bahama. Red London double-decker buses take visitors on tours of the island.

France

*THE COUNTRY OF CHAMPAGNE
AND 400 DIFFERENT CHEESES,
PUNCTUATED WITH PICTURESQUE
VILLAGES, ANCIENT TOWNS AND
BUSTLING CITIES – THIS IS THE
KALEIDOSCOPE OF FRANCE*

Picture a Parisian terrace café in the busy Champs-Elysées. Turn your head one way towards the Grand Louvre, one of the world's greatest museums of art and culture; then look in the other direction and you glimpse the Arc de Triomphe, built for the Emperor Napoleon – a symbol of centuries of French history and power. For many visitors this is France. But with all its attractions, PARIS is only the heart of a country whose language and culture flourish all over the world.

Beyond the capital, another France soon appears – vast, and largely rural. Covering nearly 551 600 km² (212 924 sq miles), including CORSICA, but not the 'overseas departments' such as Martinique, it is the third largest country in Europe after Russia and Ukraine. However, with only 57.7 million inhabitants it is one of the least densely populated, with a population density of 104.5 per km² (271 per sq mile). Wide expanses of farmland produce a rich variety of vegetables, fruit, meat, cheese and wine, all vital ingredients in the renowned French cuisine.

FROM COUNTRY TO TOWN

As recently as 1940, almost half the nation lived in the countryside and one-third of the French people worked on the farms. Things have changed rapidly since then. New jobs in factories and offices have attracted millions away from the land so that now only one-fifth of French people are country dwellers and many of these drive to work in town. One worker in three has a job in a factory and one in every two works in an office or a shop.

Only seven per cent of today's working population are farmers, and they work much more land than did the previous generation – or the one before that. Almost everywhere, farms have been enlarged, mechanised and made more productive. Many marketing co-operatives have been formed, and today there are approximately one million farms or farming cooperatives, with an average size of 28 hectares (69 acres). And the trend to enlarge is not about to stop – more than half of all farmers are 50 years or older, and many of them do not have younger family members ready to take over the farming business.

The controversial common agricultural policy introduced by the European Economic Community (EEC) – and later adopted by the European Community (EC), now the European Union (EU) – which France helped to found in 1957, aided this transformation and enriched many French farmers. But it has also contributed to 'mountains' of surplus butter and 'lakes' of milk and wine.

ONE NATION WITH MANY REGIONS

France's diverse regions are arranged for the most part like a giant amphitheatre facing west. At lower altitudes are the great basins of northern and south-western France and the peninsula of BRITTANY. At middle altitudes are the MASSIF CENTRAL and the JURA, while the country's upper tiers include the ALPS and PYRENEES – parts of the frontiers of France.

Farming is possible in all parts of France – except, of course, at high altitudes – despite variations in rainfall and temperature. The Atlantic Ocean warms the country's western shores in winter and produces moist, temperate conditions ideal for rearing livestock. Eastern France has cold winters and hot summers that are typical of Central Europe. In the south, high summer temperatures and drought make irrigation necessary for cereals and fruit. But rosemary, thyme and other aromatic herbs flourish on Provençal hillsides. In the Massif Central and the mountains the winters are much colder and agriculture more restricted than their southern location might suggest.

The Paris Basin, which contains large expanses of fertile ploughland, is crossed by major rivers, including the SEINE and MARNE. Paris grew up near their junction and has been a stimulus for agricultural improvement over the centuries. The most modern and productive farms in France are found here.

AQUITAINE, to the south-west, is the other large lowland in France, and is crossed by the broad valleys of the GARONNE and DORDOGNE. Here, farming is less modern, but the vineyards around BORDEAUX produce some of the world's finest wines.

Windswept Brittany, with its old, hard rocks, was once a backward farming area, but is now known for early vegetables (including onions and artichokes) and intensive factory farming of pigs and poultry.

▲ OPEN FOR BUSINESS Small family shops still flourish in France, and the patronne of this chemist's-cum-ironmonger's shop in Saint-Rémy, Provence, is ready to sell anything from household detergents and paint to face-creams and mineral water.

The ancient Massif Central is in the very heart of France – though it may seem alien to many urban French and has been dubbed 'the Dead Heart of France'. High plateaus, steep slopes, long dead volcanoes, deep valleys and a harsh climate combine to form a difficult environment for farming. Today, mixed husbandry here is giving way to livestock farming; uncultivated land is being planted with trees or left to revert to scrub.

Conditions are harsher still on the high ridges and in the deep valleys of the Jura, the Pyrenees and the French Alps which contain Western Europe's highest challenge to the climber, Mont BLANC, at 4807 m (15 770 ft). Here, winter sports are a main source of income, though frequently investments have to be made on very narrow profit margins in an industry that is viable for only half the year.

The valleys of the RHÔNE and SAÔNE provide a break in the ascending steps of the giant amphitheatre. For thousands of years, traders and travellers have used these natural routes which now contain motorways, railways and gas pipelines. The valley of the Rhône houses the nation's second largest city, LYONS, at the point where the river's turbulent waters meet the Saône. It flows on past terraced vineyards, powers hydroelectric works and provides irrigation before reaching its delta and the Mediterranean Sea. Here is the CAMARGUE with its rice fields and marshes, the home of flamingos and wild horses.

To the east is MARSEILLES, France's third largest city after Paris and Lyons, and the centrepiece of the Midi. This southernmost fringe of the country sweeps from the millionaires' paradise of the CÔTE D'AZUR, through rocky coves and inland valleys to PROVENCE, abundant with flowers, fruits and vegetables. It continues to the plain of LANGUEDOC, with its vines, orchards and new holiday centres.

No perfunctory visit to France will give an insight into the country's regional diversity. But regions – and their people, languages and customs – differ vastly. The diversification of the French population can in part be traced back to continuing migration from Poland, Italy and Central Europe, followed by immigrants from Spain, Portugal and the countries which France had colonised for three-quarters of a century. Today, 18 million French people – one-third of the population – have at least one foreign ancestor back to the third generation.

After decades of official disapproval, there is now a revival of regionalism. Basques,

Catalans, and Alsatians increasingly speak their own languages, while Breton and Corsican broadcasts are heard for a few hours each week. And in Corsica, a group of radical separatists occasionally draw attention to their cause through terrorist actions. Their aim is total independence from France.

Others have sought to use their cultural identity as grounds for obtaining a greater say in regional affairs. However, most groups see their dreams of cultural identity fulfilled through the broadcasts in their own language and other cultural events.

THE DELIGHTS OF FOOD

French cuisine offers a delightful way of experiencing the richness and variety of the nation's farming. Despite the appearance of

fast-food chains in large cities, the preparation of good, wholesome food is still regarded as an art. Bistros and restaurants offer excellent and often inexpensive menus which provide a good introduction to regional tastes. Many dishes are famous throughout the world, for example *bouillabaisse* (a rich fish and seafood stew from the Midi), *boeuf à la bourguignon* (tender beef, mushrooms and BURGUNDY wine), and *quiche lorraine* (a delicate blend of eggs, cream and gammon in pastry, from the north-east).

France produces a quarter of the world's wine and every restaurant has a cellar of wines, from simple *vins de table* to superior vintages. Special meals are not complete without CHAMPAGNE or COGNAC. But drink is also a national problem – nearly 38 000 French people a year die from conditions brought on by excessive

alcohol consumption: 20-40 per cent of male hospital patients, and 8-10 per cent of female patients show some alcoholic symptoms.

Shopping for the freshest ingredients, and skilfully preparing them, are essential features of French family life. One-fifth of the average household budget is spent on food. Most towns have magnificent markets crammed with colourful stalls, laden with local produce and goods brought from other regions. There are mouth-watering displays of cooked meats, pastries and locally made chocolates with temptation to buy at every turn.

The country's urban dwellers have not forgotten their rural roots – 20 per cent of French, and 25 per cent of the country's workforce, live in the region of Ile de France, which includes Paris and its environs, dotted with urban centres. Regular rush hours take place on Friday nights as Parisian families take to the motorways to enjoy *le weekend* out of town. Some have their own country cottages; and restoring an old house or barn as a second home is a serious hobby for many families.

France now has nearly 3.2 million second homes, old and new – favourite places for summer holidays (usually taken in August) as well as weekends. In addition, growing numbers of city people retire to the villages where they were born.

THE IMPRINT OF HISTORY

France's countryside and provincial towns still bear the imprint of past centuries, long before they became part of a unified nation. Provence is a good example, with its amphitheatres, triumphal arches and the best preserved aquaduct anywhere in the world, at Port du Gard – evidence of the Roman occupation. In the days of Charlemagne (c 742-814), who founded the first great French empire, feudal lords built castles and used them to defend their territory. Many of their fortifications still dot the French landscape.

The French, however, look proudly back upon a past that stretches almost as far back as the origins of the human race. In fact, the early humans known as Cro-Magnons are named after the cave of Cro-Magnon in the Dordogne. There also is evidence that hunter-gatherers walked in the territory as far back as 500 000 years ago. The legendary Gauls – considered by many to be the forebears of the French – came to France in 8 BC, bringing with them the metallurgical skills of the Hittites, which allowed them to forge tools for both agriculture and battle. The Gauls came down the Rhine and moved into the north-east of the country. Eventually they settled throughout the land, mixing with different peoples of the past who had migrated to this land of plenty.

But these ancient peoples left little behind, and France is better known for its later history. During the Renaissance of the 15th and 16th centuries, Italian craftsmen were employed to build gracious châteaux, often set in elegant gardens and parks, creating what many tourists seek when searching for the historical France of school textbooks and literature. The palaces were the homes of royalty and nobility until the French Revolution which began in 1789 and led to the execution of King Louis XVI.

The Loire Valley contains some particularly handsome chateaux such as those at Amboise, BLOIS, CHAMBORD and CHENONCEAUX.

Thousands of parish churches and scores of beautiful cathedrals were built – a sign of the wealth and power of the Roman Catholic Church and the piety of the French people. Round arches and domes are typical of the solid Romanesque churches erected during the 11th and 12th centuries; they are found across much of southern France. Tall pointed arches and large stained-glass windows are seen in the soaring 12th and 13th-century Gothic cathedrals in northern France, such as the magnificent cathedral at CHARTRES.

However, the French Revolution stripped away the Church's power. Today, 80 per cent of the people are Catholic, but only 13 per cent regularly attend mass. This is reflected in French values as well. Although the Church outlaws contraception and abortion, the state legalised the former in 1967 and the latter in 1974. One in four marriages ends in divorce. There are 22 abortions for every 100 births.

THE RISE OF INDUSTRY

Until the 20th century, many industries were rural in character and most towns were small market centres closely linked to farming. France did not experience the great wave of factory building and urban growth that transformed Victorian England. But a handful of mining districts, textile towns and iron-making centres in northern and eastern France did mushroom at that time. Their pit machinery, grim factories and workers' housing can still be seen in northern LORRAINE and the NORD-PAS-DE-CALAIS.

Paris was the exception. The city's powers were greatly enhanced when Napoleon I abolished provincial parliaments and the running of the nation's economic and cultural life was centralised in the capital. It later became the focal point of a starlike railway network which increased the city's influence. In recent years, France has developed one of the most efficient state railway systems in Europe, and the *Train à Grande Vitesse* (TGV, or high-speed train) competes with road and air travel in maintaining links between capital and provinces. High-speed trains between Paris and Lille can travel at 300 km/h (186 mph).

In 1801, Paris housed more than half a million people. During the next hundred years it attracted the poor who came in search of work, as well as scholars, artists and rich investors. New ideas were sparked off, stimulating innovations in manufacturing, fashion, culture, philosophy and science. For centuries French was the language of diplomacy and educated people throughout the world looked to Paris for inspiration. The music of Berlioz, Saint-Saëns and Bizet, the paintings of Degas, Cézanne, Monet, Renoir and Toulouse-Lautrec, the literary work of Balzac, Hugo, Zola, Flaubert and de Maupassant represented the peak of creativity in the 19th century.

By 1850, the population of Paris had trebled and two decades of frenetic expansion began. This was masterminded by Emperor Napoleon III and the city planner Baron Haussmann, who wanted the capital to be both efficient and

elegant with impressive public buildings. New markets, churches, schools and hospitals were built and parks laid out, transforming inner Paris into the beautiful city that thousands of visitors enjoy today.

Strangely, the country's population growth dipped during the 19th century. At a time when Germany's and England's population figures increased significantly, the birthrate in France decreased. This was partly due to hard living conditions among the rural French. The population consequently aged, putting a break on French industrial development – while industrial development was accelerating elsewhere.

The First World War (1914-18) cost France 1 385 000 men; during the Second World War (1939-45), a further 536 000 men, women and children were killed. Yet despite frequent changes of government, a new France grew out of the disasters of the two World Wars. The birthrate, low in the 1930s, rose after 1945 in an impressive 'baby boom' – encouraged by enormous child benefits from the state. But birthrates again tumbled from 21.1 per 1000 at the peak of the baby boom in 1948 to 13.6 in 1976, though they have risen a little since. Almost one-sixth of the population are over 65, and France is again on the way to becoming a country of the elderly. In fact, in some regions this has already happened. In the Massif Central, for instance, where young people have moved on to urban centres, some settlements appear almost deserted.

POST-WAR ECONOMIC DEVELOPMENT

National planning was introduced in 1947 to steer economic recovery and stimulate further growth. Heavy investment, including aid from the United States, enabled France to catch up on lost time and become an industrial nation almost overnight. Though it was slow with industrial development, France was at the forefront, alongside Britain and Spain, in colonising new territories – one reason why today, there are 93 million native French speakers in the world, and another 65 million people who speak French regularly as their second or third language.

Colonisation began in 1830 with the French occupation of Algeria, which was the first in a chain of colonial territories that ran along Africa's north and west coasts. France soon became a mighty power, which drew much revenue from its colonies. However, as was the case with other imperial powers, these same colonies would soon be the source of political complications, abroad and at home.

To begin with, France lost Indo-China to the Vietnamese communists in 1954; and in 1956, MOROCCO and TUNISIA were given independence. The revolt in ALGERIA in 1958 brought General Charles de Gaulle, hero of the Free French resistance to the Germans during the Second World War, to power.

De Gaulle was president for 10 years. In 1962, he gave independence to Algeria and about 1 million people of French extraction flocked to mainland France.

From 1957, free movement of workers between the EEC (now EU) countries, as well as a flowering economy at home, enabled

▲ MIRROR IMAGE The handsome Chateau de la Roche – in Poitou, in western France – is reflected in its ornamental lake. It is one of hundreds of chateaux which survived the savage pillage and destruction of the French Revolution.

France to import labour from Italy to help build new factories, homes and roads and to work in the manufacturing and service industries. Later, the French widened their policy of recruitment to attract migrants from Spain, Portugal and Algeria, and by 1975, there were estimated to be 3.5 million foreigners living and working in the country.

This increase in the working population helped the industrial economy to flourish. Old industries were brought up to date and new ones began, manufacturing motor cars, aircraft, electrical goods, electronics, and other products. However, the sudden influx of foreign workers created a housing problem which led to the rapid development of enormous residential projects, which in turn bred social problems – problems which France is still struggling to solve today.

In the early 1970s, 40 per cent of all French workers had jobs in factories. Many of the workers were housed in large blocks of flats in the towns. The first flats were cramped and poorly equipped, but conditions began to improve in the mid-1960s, when more comfortable but expensive flats were built for sale or rent to the new rising class of office workers, managers, civil servants and teachers.

Massive new office complexes, such as La Défense in western Paris and La Part-Dieu in Lyons were built. Single-family houses went up in the outer suburbs of French towns for families who wanted a house and garden of their own, rather than an apartment in a large concrete block. Today the French are a nation of townspeople and the city of Paris and its suburbs are home to more than 10 million people.

The immigration of workers from non-EEC countries was halted in the mid-1970s – some foreigners accepted 'golden handshake' payments to return home. But foreign wives and children have continued to join their menfolk in the poorer quarters of French towns. Many were illegal immigrants from North Africa. With 4.5 to 5 million foreigners living in France today, more stringent immigration laws are being introduced to stem the influx.

The government is faced with the task of creating a society which will cater for several nationalities. This is particularly difficult with unemployment rising fast. In the late 1980s and early 1990s, deep-rooted xenophobia surfaced and politicians urging the expulsion of foreigners were returned in elections. The radical conservative Jean-Marie Le Pen and his Nationalist Front gained increasing popularity until the elections of 1992 when the less conservative right won a clear majority, marginalising both the far right and the far left.

THE ADVANCE OF SOCIALISM

As early as 1968, when strikes and demonstrations disrupted everyday life throughout the country, many French people were expressing their discontent with the capitalist system and concern over social and environmental issues. Some of the fears came true after the oil crisis in the 1970s when President Giscard d'Estaing and Prime Minister Raymond Barre introduced tough measures to trim surplus manufacturing capacity, especially in the iron and steel industries. Unemployment rose and most French electors felt it was time for a change.

In May 1981, a Socialist, François Mitterrand, was elected president. He promised devolution of power, nationalisation and a generous range of social welfare schemes. In the next two years, new laws turned some of the promises into reality. Some power was handed down to elected councillors in the nation's 96 *départements* and Corsica was chosen to establish the first Regional Assembly. Nationalisation put one-fifth of the country's manufacturing capacity and most of its financial institutions under state control.

The Socialists increased family allowances, pensions and minimum wages, reduced the working week and introduced five-weeks' paid holiday and earlier retirement. Women's rights were advanced, the death penalty abolished and controversial projects for new military training grounds and extra nuclear power stations were scrapped. But money for these changes had to come from taxpayers.

Early in 1983, with the country in financial trouble, the president had to introduce an unpopular austerity programme to cut state expenditure and social benefits. Taxes and health charges were raised. Workers felt

▲ WHITE-DOMED BASILICA The beautiful late 19th-century Basilica of Sacré-Coeur is a prominent landmark on top of the hill overlooking the district of Montmartre in Paris.

betrayed and trade unions organised strikes. The president issued an ultimatum to nationalised industries to improve their efficiency. Many workers in shipbuilding and the steel, chemicals, motorcar manufacturing and mining industries faced redundancy.

This reversal of policy angered the French Communist Party, the second largest in Western Europe, and Communist members of Mitterrand's administration resigned in July 1984. Unemployment rose from 1.7 million (8 per cent) when the president came to power to 2.55 million (11.3 per cent) in early 1985. More than one million people under 25 were jobless. Young people who had never worked, and those who had interrupted periods of

employment, suffered because of the way the benefit system worked. Nearly one million people without work received 'little or nothing' – and there was talk about 'the new poor'.

Nearly all the Socialist government's policies changed in practice. Devolution ran into trouble when the Corsican experiment failed and elections for other Regional Assemblies were put off until 1986. Mitterrand's proposal to control Church schools, which educate one out of every six French children, met strong opposition and the whole idea was dropped.

In the light of these and many other problems Mitterrand had to accept that his Socialist plans could not be applied to France in the 1980s. The French people did not want wholesale changes.

In the 1986 general election the Gaullists and their right-wing allies were elected into power, and Prime Minister Jacques Chirac began undoing the socialist policies – despite the fact that Mitterrand was still president. In

May 1991, the Socialists regained the majority only to be put out of office again during the March elections of 1993, when the right-wing Gaullists took over the majority, with Mitterrand still as president.

On the whole, the French family is now healthier and better fed, housed and educated than ever before. It has more free time for holidays, a second home, hobbies, sports and other leisure activities. French men are avid supporters of sport events such as football and the annual round-France cycling race (the Tour de France). Older men spend much of their free time playing cards in cafés and *boules* (an informal version of bowls) in parks and open spaces. French women have a reputation for elegance and good dress sense.

Men and women are traditionally pictured sitting with friends under the bright umbrellas of street cafés, sipping their drinks, watching the world go by. But the picture of prosperity may be changing under the pressure of international economics. The big questions in everyone's mind focus on unemployment, racism and what will happen in national politics in the final years of the 20th century.

Another question in everybody's mind is that of the remaining French Overseas Territories. In New Caledonia, for example, the majority of locals are demanding total independence from their colonial overlords.

FRANCE AT A GLANCE
Area 551 600 km² (212 924 sq miles)
Population 57 690 000 **Per km²** 104.5 (**Per sq mile** 271)
Capital Paris
Government Republic
Currency 1 franc = 100 centimes
Languages French (official); Basque, Breton, Catalan, Corsican, Provençal, Alsatian and others (regional); Arabic and other languages among foreigners
Religion Christian (Roman Catholic 76%; Protestant), Muslim (4.5%), small Jewish community
Climate Temperate; continental in the east, with hot and dry summers on the Mediterranean coast. Average temperature in Paris ranges from 1-6°C (34-43°F) in January to 14-25°C (57-77°F) in July
Land use Cultivation 34%, meadows and pastures 23%, forest and woodlands 27%, other 16%
Main primary products Cattle, sheep, pigs, poultry, sugar beet, wheat, barley, maize, oats, potatoes, grapes, fruit, vegetables, timber, fish; iron ore, bauxite, coal, uranium, salt, oil and natural gas, potash
Major industries Iron and steel, engineering, chemicals, textiles, electrical goods, motorcars, aircraft, cement, aluminium, agriculture, perfume, forestry, fishing, food processing, oil and gas refining, tourism
Main exports Motorcars, chemicals, iron and steel, textiles and clothing, leather goods, electrical appliances, weapons, wine, cereals, processed foods, petroleum products
Annual income per head (US$) 19 480
Population growth (per thous/yr) 5
Life expectancy (yrs) Male 74 **Female** 78

Freiburg (Freiburg im Breisgau) *Germany*
Largest city of the Black Forest on the western
slopes of the Rhine Rift Valley, 110 km (68 miles)
south-west of Stuttgart. Freiburg has a beautiful
walled, old town centred on an open market. Its
13th to 16th-century cathedral is the seat of an
archbishopric; its university dates from 1457.
Population 191 000
Map Germany – Be

Fréjus *France* Small town in Provence region,
25 km (16 miles) south-west of Cannes. It was
founded in the 1st century BC by the Roman
general, and later emperor, Julius Caesar. It now
produces wine, olive oil and cork.
Population 51 500
Map France – Ge

Fremantle *Australia* See PERTH

French Guiana (Guyane) *South America* Still
living down the horrors of Devil's Island, French
Guiana is a remnant of European colonisation in
South America. Largely covered with equatorial
forest, it remains an overseas department of
France, with elected representatives in Paris and
no plans for independence.

Its economy relies on subsidies from France,
and it exports very little of its own besides
tropical hardwoods, rum, a little coffee, essences,
and shrimps – the last an industry developed
by United States companies. Most of French
Guiana's food and necessities are imported.

Ile du Diable, or Devil's Island, off the coast,
formerly a leper colony, was until 1938 a notorious penal settlement – a 'safety island' for prisoners such as Alfred Dreyfus (1859-1935) –
which inspired the novel *Papillon*. More recently,
France has established a rocket-launching station
at Kourou, expanding the mainland town from a
population of 600 to 13 500.

French Guiana has a total area of more than
91 000 km² (35 127 sq miles) and 114 700 inhabitants, of whom more than 40 000 are concentrated
in the handsome French-style capital of CAYENNE,
from which the pepper gets its name. Though few
slaves were brought to French Guiana because its
sugar industry was never greatly developed, two-
thirds of the population are descendants of
African slaves from neighbouring Guyana and
Surinam. The rest are of Amerindian, Chinese
and French descent, together with a small number
of Laotians (refugees from war-torn Indo-China
who were settled on farms in French Guiana in

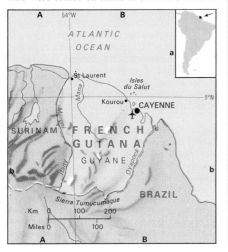

the late 1970s and have become the territory's
most successful farmers). There are also many
refugees from neighbouring Surinam.

French Guianans hope to develop tourism, as
well as kaolin and bauxite mining, and to exploit
the hardwoods of the extensive and well-watered
rain forest interior.
Population 114 700

French Polynesia *Pacific Ocean* French overseas territory consisting of some 130 islands in
the south-central Pacific, about halfway between
Australia and South America. They have a total
land area of 3941 km² (1522 sq miles) but are
spread over 4 million km² (1.5 million sq miles)
of ocean.

There are five main island groups. The
Windward Islands (including the main island of
TAHITI, which contains the capital, PAPEETE) and
Leeward Islands together make up the SOCIETY
group. These and the MARQUESAS, with steep mountains, as are most of the TUBUAIS
and the small GAMBIER ISLANDS. The TUAMOTU
group, however, are low-lying atolls.

All five groups form part of POLYNESIA. About
80 per cent of the people are Polynesians, most
of them Christians (over half of them, Protestant)
who speak Tahitian, although French is the official language. They are renowned for their beauty
and for the hip-swaying *tamure* dance. Another
10 per cent are French and some 7 per cent are
descendants of Chinese workers imported in
1865-66 by a British-owned plantation.

The first Europeans to reach the islands were
Spanish conquistadores from Peru who sighted
the Marquesas in 1595, but kept their discovery
a secret to keep out other European powers. Next,
Tahiti was reached by the English navigator
Samuel Wallis in 1767 and independently by
Louis de Bougainville, who claimed it for France,
in 1768. Each returned with stories of lush landscapes and friendly people, which created the
image of a South Sea island paradise perpetuated
by artists and writers ever since.

The tribal chiefs of Tahiti accepted a French
protectorate in 1842; the last Tahitian king,
Pomare V, ceded his kingdom entirely to France
in 1880, and by 1900 the French controlled all
the islands. In this way the islands acquired the
status of an overseas territory in 1958, but gained
increased autonomy with a Territorial Assembly,
in 1984.

There are some calls for complete independence, but the Tahitians know that their standard
of living depends on French government expenditure, especially on military facilities to support
its nuclear weapons tests at MURUROA.

Phosphate mining stopped in 1966, but the
islands export copra and coconut oil (coconut
palms grow on the lowland fringes of most
islands and on the atolls), cultured pearls, and
pearl shell, vanilla and citrus fruits. Tourism
brings in three-quarters of all non-military foreign earnings; the main tourist destinations are
Tahiti itself, nearby Moorea, and Bora-Bora on
the Leeward Islands.
Population 169 000
Map Pacific Ocean – Fc

French Southern and Antarctic Territories
Indian Ocean Territories comprising Adélie
Land in Antarctica, and several islands and island
groups in the southern INDIAN OCEAN south of latitude 37°S. The Indian Ocean islands include the

KERGUELEN group and CROZET ISLANDS, as well
as AMSTERDAM ISLAND and St Paul Island.
Administration is from Paris, and the population
consists of staff employed in scientific research
or at the meteorological stations scattered
through the territory, and at a hospital on
Amsterdam Island.
Map (Southern) Indian Ocean – Cd; Be-Ce;
(Antarctic) Antarctica – Kb-Lb

fresh breeze Force 5 on the BEAUFORT SCALE,
with a wind speed of 30-39 km/h (19-24 mph).

Fresno *USA* City in the San Joaquin Valley in
California, 265 km (165 miles) south-east of San
Francisco. It is sometimes called 'the Grape
Capital of the World' and it is certainly one of the
world's grape-growing centres, and heart of the
largest vine-growing area of the United States.
Population (city) 354 000; (metropolitan area)
755 580
Map United States – Cc

fret Mist which forms at sea and suddenly
extends a short distance inland.

Fribourg (Freiburg) *Switzerland* Capital of the
canton of the same name, 27 km (17 miles) south-
west of Berne. The town, Switzerland's centre of
Catholicism, stands dramatically in and above a
bend in the gorge of the Saane (Sarine) River. In
its medieval centre, among old churches and
spires, towers a 13th to 15th-century cathedral.

The canton of Fribourg, which covers
1670 km² (648 sq miles), straddles the country's
language frontier – about three-quarters of the
inhabitants speak French, the rest, German. The
canton is famous for its cheeses, particularly
Gruyère.
Population (town) 32 700; (canton) 218 100
Map Switzerland – Aa

Friedrichshafen *Germany* Port on the north
shore of Lake CONSTANCE. Friedrich I of Württemberg created the town, which is named after
him, in 1811 by combining the ancient town
of Buchhorn and the monastery-village of
Hofen. Friedrichshafen is the birthplace of the
Zeppelin airships, whose history is outlined in the
town's Zeppelin Museum. The main industries
are tourism and the manufacture of precision
instruments.
Population 53 700
Map Germany – Ce

Friesian Islands (Waddeneilanden; Friesians)
North Sea Fringe of islands along the North Sea
coast, which are really sandbars running parallel
to the coast and separated from it by 5-30 km
(3-20 miles) of shallows.

The islands, most of which are tourist resorts,
are divided into three groups. The West Friesian
Islands are Dutch, the largest in this group being
Texel and Terschelling. The East Friesian
Islands belong to Germany, and include Borkum
and Norderney. The North Friesian Islands
beyond the estuary of the Elbe are split up
between Germany and Denmark, and Sylt and
Heligoland (Helgoland) in this group are
German. Numerous ferries link the Friesians to
the mainland.
Population (Dutch Friesians) 22 920; (German
Friesians) 30 000; (Danish Friesians) 3750
Map Netherlands – Ba; Germany – Ca, Ab-Cb

▲ MOUNT FUJI The Japanese love affair with the country's sacred mountain goes back to earliest times. On a clear day, Mount Fuji's perfect cone can be seen from the Tokyo-Nagoya Railway; and in summer, thousands climb to the top.

Friesland *Netherlands* Northern province covering 3792 km² (1464 sq miles), including most of the West FRIESIAN (WADDEN) ISLANDS off its coast. The province owes its name to the Germanic Friesian tribe who settled here in the 1st century AD. The Friesian language is still spoken. The region has many small lakes, which make it popular with tourists. Agriculture, mainly the breeding of Friesian dairy cattle, is Friesland's principal industry. LEEUWARDEN, the provincial capital, is the only large town.

Population 604 000
Map Netherlands – Ba

Friuli-Venezia Giulia *Italy* Region covering 7846 km² (3029 sq miles) in the north-east of the country, bordering Slovenia and Austria. The region has three languages – Ladin, similar to Latin, in the north, Italian in the south, and Slovenian along the Slovenian border. Its capital is the port and shipbuilding centre of TRIESTE. The region is noted for its textiles and dairy products.

Population 1 216 400
Map Italy – Da

Frobisher Bay *Canada* Inlet of the Atlantic Ocean, which penetrates 230 km (143 miles) into the south-east coast of BAFFIN ISLAND.

Map Canada – Ib

Frombork (Frauenburg) *Poland* Town and port 64 km (40 miles) east of Gdańsk. It is dominated by a majestic 14th-century cathedral. The Polish astronomer Nicolaus Copernicus (1473-1543) was a canon at the cathedral when he put forward his belief that the earth was not the fixed centre of the universe but revolved around the sun. A tower in the cathedral courtyard now houses a museum devoted to him.

Population 2000
Map Poland – Ca

front Surface of separation between different AIR MASSES, either marking a major zone of atmospheric activity, as at the POLAR FRONT, or, on a smaller scale, occurring within a DEPRESSION as a cold, warm or occluded front.

frontal rain See CYCLONIC RAIN

frost Atmospheric condition when the temperature is at or below the freezing point of water (0°C, 32°F); also, the resultant deposit of minute ice crystals on the ground, plants and buildings, formed from frozen water vapour.

frost hollow (frost pocket) Low-lying area which may experience frost while surrounding higher areas are frost free. The frost is caused by heavy cold air sinking into the hollow and being unable to escape.

frost shattering Weathering process which causes rocks to disintegrate when water within their joints and pores freezes and expands, thus splitting them into fragments.

Frunze *Kyrgyzstan* See BISHKEK

Fuchun Jiang (Tsien Tang Kiang) *China* River of Zhejiang Province. It is the main transport waterway in the province, flowing about 400 km (250 miles). It drains into the East CHINA SEA, 50 km (30 miles) south of SHANGHAI.

Map China – He

Fuengirola *Spain* See MÁLAGA

Fujayrah *United Arab Emirates* See AL FUJAYRAH

Fuji, Mount (Fuji-yama) *Japan* Volcanic mountain 100 km (62 miles) south-west of Tokyo, and the country's best known natural feature. Fuji's majestic cone – dormant since 1707-8 – is the country's highest peak rising to 3776 m (12 389 ft) and is sacred to the Japanese. It has been celebrated for centuries in Japanese paintings and verse. The peak, which is climbed by more than 300 000 people every year, lies within the Fuji-Hakone-Izu National Park.

Map Japan – Cc

Fujian (Fukien) *China* Province in the southeast of the country, covering some 120 000 km² (46 350 sq miles). It lies beside the East CHINA SEA opposite Taiwan. It is a predominantly mountainous area, and is rich in timber. The subtropical climate allows rice, sweet potatoes, tea and citrus fruit to grow in the valleys. The provincial capital is Fuzhou.

Population 31 160 000
Map China – Hf

Fukui *Japan* Market town near the north coast of central HONSHU, 320 km (200 miles) west of Tokyo. It was badly damaged during the Second World War and again in 1948 by an earthquake. Fukui is noted for its traditional figured silk products, to which a modern rayon weaving industry has been added.
Population 253 700
Map Japan – Cc

Fukuoka *Japan* Largest city of KYUSHU Island, standing on the island's north-west coast. A commercial and industrial centre, Fukuoka is also the western terminus for the famous *shinkansen*, or 'bullet train', from Tokyo, which reaches a speed of 210 km/h (130 mph) during the 1176.5 km (735 mile) journey taking 6.5 hours.
The city was flattened in World War Two but has been rebuilt as a modern city.
The Naka River flows through its centre, separating it from the town of Hakata. During the Middle Ages, Hakata was one of Japan's major ports and a fortified town which managed to beat off an attack by Kublai Khan's Mongols in the 13th century. Today, it is famous for the Hakata dolls which it produces.
Population 1 218 300
Map Japan – Bd

Fulda *Germany* Town about 80 km (50 miles) north-east of FRANKFURT am Main. Its Baroque cathedral has the staff and relics of St Boniface, the English-born missionary whose followers founded a Benedictine monastery on the site in the 8th century.
Fulda is also the name given to the river rising 20 km (12 miles) to the south-east of the town of Fulda, and flowing 220 km (137 miles) north to join the Weser near the town of KASSEL.
Population 55 800
Map Germany – Cc

fumarole Hole or vent in the earth's surface from which issue hot gases such as steam, hydrochloric acid, sulphur dioxide and ammonium chloride. During the dry season hot springs are turned into fumaroles. Fumaroles are found in volcanoes or near volcanic areas, for instance in Alaska's Valley of Ten Thousand Smokes.

Funafuti *Tuvalu* Main atoll and capital of Tuvalu, lying 1000 km (620 miles) north of Fiji. A US military base was established here in 1943. The atoll measures 20 by 18 km (12 by 11 miles), but its actual land area is only 2.4 km² (0.9 sq miles).
Population 2600
Map Pacific Ocean – Dc

Funchal *Portugal* Capital of the Atlantic island of MADEIRA, 140 km (85 miles) off the west coast of Morocco. The town lies in a large natural amphitheatre above a bay and is surrounded by banana plantations and vineyards.
A tourist resort, especially in the winter, it has a casino and is a port of call for cruise ships. Tourist transport locally is on woven cane sledges drawn by oxen.
Population 45 600
Map Morocco – Aa

Fundy, Bay of *Canada* Bay, 160 km (100 miles) long, and 100 km (60 miles) wide at the entrance, between New Brunswick and Nova Scotia, and

Maine in the United States. It has the highest tidal range in the world, the water rising up to 17 m (56 ft) from low to high tide. Fundy National Park covers 206 km² (79.5 sq miles) on the northern coast of the bay.
Map Canada – Id

Fünen (Fyn) *Denmark* The nation's second largest island (2976 km², 1048 sq miles). It dominates the Fünen group of five large and 24 small islands which are clustered between the LITTLE BELT (Lille Bælt) and the GREAT BELT (Store Bælt) channels at the entrance to the Baltic Sea.
Eighty per cent of the land on the island is cultivated, with a large proportion under horticultural crops, such as flowers. For this reason Fünen is often called 'the Garden of Denmark' – and its hills, the highest of which reaches 131 m (430 ft), are humorously called 'the Fünen Alps'.
Almost all of its principal settlements have medieval foundations and many have well-preserved half-timbered houses.
In addition there are Stone Age burial chambers and many Viking remains, including the Ladbyskibet, the burial ship of a Viking chief (about AD 950).
On the west of the island is the medieval ferry port of Middelfart; in the east, the fortified town and ferry port of Nyborg.
To the south are pleasant resorts and towns, such as Svendborg and Fåborg. ODENSE, in the centre, is the island's main administrative, educational and industrial city.
Population 424 700
Map Denmark – Bb

funnel cloud Whirling, dark grey cloud at the centre of a tornado (or at the centre of a water-spout if the tornado touches water), sometimes

▲ MOONSCAPE IN ALASKA The Valley of Ten Thousand Smokes is covered with thick layers of pumice and ash, creating a desolate and barren scene.

extending downwards as far as the earth's (or the water's) surface.

Füred *Hungary* See BALATONFÜRED

Furka Pass *Switzerland* Mountain pass 90 km (56 miles) south of Zürich, marking the watershed between the Rhône and Rhine valleys. The pass is 2431 m (7976 ft) high.
Map Switzerland – Ba

Fushun *China* Mining city in Liaoning Province. The coal and oil shale mined here provide raw materials for the local engineering and petrochemicals industries.
Population 1 350 000
Map China – Ic

Futa, The *Senegal* See FOUTA

Futa Jallon *Guinea* See FOUTA DJALLON

Futuna *Pacific Ocean* See WALLIS AND FUTUNA

Fuzhou (Minhow) *China* Capital of Fujian Province. It is a thriving port at the mouth of the Min River on the Taiwan Strait which separates China and Taiwan. Its main industries are food processing and the manufacture of paper and chemicals.
Population 1 270 000
Map China – He

Fyn *Denmark* See FÜNEN

gabbro Coarse-grained, dark, basic intrusive rock composed largely of calcium-rich plagioclase FELDSPAR minerals and PYROXENES, sometimes also containing other minerals. Its coarse grain is due to slow cooling deep down in the earth rather than on the surface.

Gabès *Tunisia* Seaport and oasis town on the Gulf of Gabès, 320 km (200 miles) south of Tunis. It forms the entry point to the south and the desert. Apart from locally grown fruit, and hides and wool, the port exports phosphates. It is also the centre of a thriving fishing industry.

Local crafts include carpets, wicker work, wood, amber, leather and silver. The nearby Chenini oasis produces dates, oranges, peaches, apricots and pomegranates.

Population 64 500
Map Tunisia – Ba

Gabès, Gulf of Inlet, 100 km (60 miles) long and just as wide, off the MEDITERRANEAN SEA off southern Tunisia. The Ancient Romans called it Syrtis Minor to distinguish it from Syrtis Major, the larger Gulf of SIRTE to the east. Jerba Island in the Gulf of Gabès, just south of SFAX, is a popular tourist centre. GABÈS and Sfax are the chief ports on the coast.

Map Tunisia – Ba

Gabon See p 251

Gaborone *Botswana* Capital of Botswana from 1965, since when it has grown rapidly. It is now a well-planned, attractive city, containing the University of Botswana and the national museum and art gallery. It also has several hotels and a casino. Because it is only 20 km (12 miles) from the border with South Africa, it was home to a large exile community during the latter's apartheid period.

Population 138 470
Map Botswana – Cb

Gabrovo *Bulgaria* Industrial town producing textiles, footwear, furniture, timber, and leather goods. It stands on the Yantra River, 96 km (60 miles) north-east of Plovdiv. The first Bulgarian national school was opened here in 1835, but Gabrovo is better known for its international festival of comedy and satire, which is held in the House of Humour in May of every odd-numbered year. In the mountains around the town are picturesque villages such as Bozhentsi and Shipka, (with the Shipka Pass nearby), the Drjanovo Monastery and the open-air museum of Etur, depicting 19th-century life, arts and crafts.

Population 80 690
Map Bulgaria – Bb

Gabu (Nova Lamego) *Guinea-Bissau* Region and town in the wooded savannah of the northeast, about 160 km (100 miles) from Bissau. Formerly called Nova Lamego, the town was one of the final Portuguese strongholds before independence in 1974, the local Muslim Fula aristocracy having cooperated with the Portuguese in an attempt to preserve their privileges. Today, Gabu is the centre of an agricultural region which produces groundnuts, millet, guinea corn, vegetables and, in recent years, cotton and tobacco. These are transported by road to Bissau.

Population (region) 125 000
Map Guinea – Ba

A DAY IN THE LIFE OF A KIBBUTZNIK

When the alarm clock wakes Amos, its luminous hands in the dark read 4 am. Beside him, his wife Ilana stirs, turns over, and goes back to sleep. Amos dresses quickly and slips out to the cowsheds, through the kibbutz of Kfar Hanossi which has been his home for five years.

The community of 550 people lives 5 km (3 miles) from the town of Rosh Pinah near the Sea of Galilee in northern Israel, on land which has been reclaimed from a swamp. The kibbutz (Hebrew for 'gathering') was founded in 1948 and is one of over 250 such cooperative settlements in Israel – communities born of necessity in the days when teamwork was absolutely imperative to build a nation out of a wilderness.

By the time the sun rises at 6 am, Amos and his four fellow dairy workers have milked 50 cows. Israel has some of the most modern agricultural machinery – including milking machines – in the world, and the yield is high.

The dairy workers go on to clean out the byres, and then wash before going into the dairy to share out the rest of the day's chores over a cup of coffee. At 7:30 they join friends and families in the communal dining room for a buffet breakfast of eggs, cereals, salads, fruit, rolls and goat's yoghurt.

Ilana soon slips away to feed her two-year-old daughter, Rinat, in the baby house. Amos joins her there to see their child, before leaving her in the capable hands of trained childminders.

While Ilana goes to work in the kitchens, Amos takes a tractor to check the cowsheds, and then drives out to the hay fields. He and his friend Gideon load heavy bales of fodder onto the tractor trailer all morning. For both of them, it is hot work as the sun rises higher, and they are glad when it is lunchtime and the day's work is over.

After lunch, which is again taken in the communal dining room, anyone who can takes a siesta. Overhead fans whisper in the silent children's houses. At 4 pm the children wake and their parents come to collect them. Within minutes, the kibbutz is filled with noise and shouting. Amos and Ilana take Rinat home, where she can play on the lawn in front of their flat. Later, as the sun begins to sink over the wheat fields, they take her for a walk around the kibbutz. Work is progressing on new flats – the kibbutz always seems to be growing.

As he walks in the cool evening air, Amos jingles the tokens for the kibbutz shop in his pocket. He has no salary, merely an allowance, but the kibbutz provides for all his needs, so money is not a worry, nor a matter of great interest. The kibbutz belongs to everyone who lives on it.

Back at the baby house, Ilana puts Rinat to bed. Watching her, Amos knows that they made the right decision when they chose to live on a kibbutz. The children are being taken care of, there is no shortage of work or housing, and money is never a problem.

Gaeseong *North Korea* See KAESONG

Gafsa *Tunisia* Oasis town 185 km (115 miles) west of SFAX. It is a major road and rail junction on the Sfax-Tozeur Railway, and the centre of a phosphate-mining region which accounts for 85 per cent of the country's phosphate output. Founded as Capsa by the Romans, it later became a Berber stronghold and, in fact, it is one of the few places in Tunisia where traces of the Berber language and its traditions can still be found. Today its protective walls of rose-pink stone enclose a town of solid brick buildings decorated in bold geometric patterns. The local craftsmen specialise in carpets, rugs and blankets. The oasis produces dates, olives, and citrus fruit.

Population 61 000
Map Tunisia – Aa

Gaillimh *Ireland* See GALWAY

Galápagos Islands (Colón Archipiélago) *Pacific Ocean/Ecuador* Group of 15 volcanic islands on, and just south of, the equator, some 1100 km (680 miles) west of Ecuador, of which they are a province. The islands, which cover 7822 km² (3019 sq miles), were explored by the Spanish in 1535 and annexed by Ecuador in 1832. Their capital is Puerto Baquerizo on San Cristóbal Island, also called Chatham Island (population 4040). Four of the islands have permanent human populations.

Charles Darwin (1809-82), the English naturalist, visited the Galápagos Islands during his historic voyage on the *Beagle* in 1885. The animal life he found here played an important role in substantiating his theory of evolution, first expounded in *The Origin of Species* (1859). Because of their unique species of plants and animals – including the almost extinct giant tortoise – 90 per cent of the islands' land area is now a national park.

Population 10 000
Map South America – Ac

Galai (Galatz) *Romania* Major inland port on the lower Danube, 180 km (112 miles) north-east of Bucharest. As early as the Middle Ages, Galai was a booming trading centre. Today, it is an important industrial city and has steel works, shipyards and a major rail junction. Its manufactures include machinery, oil products, synthetic rubber, textiles and food. The city has a number of 17th-century churches and two fine museums.

Population 305 100
Map Romania – Ba

gale Very strong wind, varying between Force 8 and 10 on the BEAUFORT SCALE. A gale (Force 8) has a wind speed of 62-74 km/h (39-46 mph). Force 10 is a full gale, or storm, with a wind speed of 88-101 km/h (55-63 mph), and can cause considerable structural damage.

galena Soft and heavy grey mineral, essentially lead sulphide, usually found in carboniferous limestone. It is the chief source of lead.

Galicia *Spain* Region and former kingdom in the north-west, covering 29 434 km² (11 365 sq miles). It was colonised by the Goths in the 6th century and was united with CASTILE in the 11th. Bounded by the Bay of Biscay in the north and the Atlantic Ocean in the west, it has a thriving

Gabon

A POOR, FORESTED COUNTRY THAT STRUCK IT RICH WITH OIL – BUT NEVER QUITE MADE GOOD

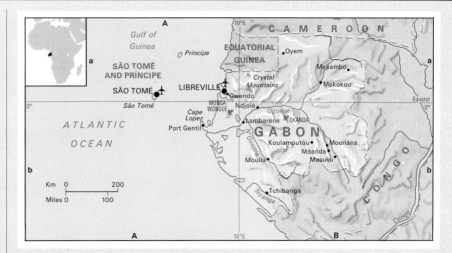

Dense tropical forest covers almost four-fifths of the hot, humid Gabonese Republic, which straddles the equator in west-central Africa. Timber was virtually its only resource until the 1960s and at independence from the French in 1960, Gabon seemed to be yet another thinly populated African country with a stagnant, near-bankrupt economy and an uncertain future. Its people suffered from acute malnutrition and endemic diseases. Then Gabon struck oil.

Immediately the country hit boom times. From 1966 to 1967, the national budget increased sixteen-fold. By the mid-1980s, Gabon was Africa's sixth largest oil producer; by 1993, it had moved up to fifth position. Other minerals are also exploited. Gabon has become the fifth most important manganese producer in the world. Uranium is mined around MOUNANA, and the manganese is mined at MOANDA, both towns on the BATÉKÉ PLATEAU in the south-east. There are huge deposits of iron ore in the north-east. With a small population and big export earnings from oil and metal ores, Gabon seemed on the brink of a miraculous transformation.

But prosperity for the nation did not bring prosperity to all its inhabitants, although there were good intentions. In 1976, an ambitious four-year plan was announced with a budget of US$32 000 million. Its aim was to bring Gabon's inadequate roads and railways up to date, and to encourage local industry. Instead, most of the money was squandered on such projects as the building of the US$120 million presidential palace and the revamping of LIBREVILLE, the national capital. Few people benefited from the country's new wealth. When the CFA franc was devalued in 1994, most Gabonese, whose earnings before had barely seen them through the month, found overnight that the value of their money had dropped by 50 per cent.

Nor is their lot likely to improve in the short term. Most of the mining industries have remained in the hands of foreign companies and too much of the capital leaves the country. Besides, there are many migrant workers in Gabon who send their earnings home. Agriculture is not well developed, only 1.7 per cent of the land area is cultivated, and there is a heavy reliance on subsistence farming. The country's tourist potential has not been developed. Wildlife flourishes in the national parks and game reserves, but elephant poaching is a continuing problem.

Because of the dense hardwood forests and the many rivers which have cut deep valleys into the interior plateaus and highlands (including the CRYSTAL MOUNTAINS and the CHAILLU MOUNTAINS in south-central Gabon), transport links are poor. However, the government has recently completed a trans-Gabonese railway project.

At first a one-party state, and now a democracy, Gabon has been comparatively stable politically since independence. However, during 1990-1 the country was thrown into turmoil by allegations of fraud in the first multiparty elections since 1964. A multiparty Government of National Unity was formed and in December 1993, President Omar Bongo, who had been in office since 1967, was elected for another term in a contested presidential election.

Gabon's history has, in many ways, been a sad one, but it has two notable monuments to altruistic and humanitarian actions. One is Albert Schweitzer's hospital at LAMBARÉNÉ, built in 1913 as part of the fight against poverty and disease, which earned him the Nobel Peace Prize in 1952. The other is the capital city, Libreville, which was founded by the French in 1849 as a home for slaves rescued from illegal slaving ships.

GABON AT A GLANCE	
Official name Gabonese Republic	
Area 267 670 km² (103 324 sq miles)	
Population 1 300 000 **Per km²** 5 (**Per sq mile** 13)	
Capital Libreville	
Government Republic	
Currency 1 CFA franc = 100 centimes	
Languages French (official); Bantu languages; Fang in the north	
Religions Christianity (Roman Catholic 72%, Protestant 8%), indigenous beliefs	
Climate Equatorial, always humid. Average temperature in Libreville is 26°C (79°F) throughout the year	
Land use Forest and woodlands 78%, grazing 18%, cultivation 2%, other 2%	
Main primary products Cocoa, coffee, palm oil, bananas, sugar cane, cassava, plantains, maize, rice, livestock, timber; petroleum, natural gas, manganese, uranium, lead, zinc, copper, niobium, titanium	
Major industries Mining, petroleum production and refining, agriculture, forestry, fishing, food processing	
Main exports Petroleum, manganese, timber	
Annual income per head (US$) 4450	
Population growth (per thous/yr) 27	
Life expectancy (yrs) **Male** 54 **Female** 54	

fishing industry. Its chief river is the Miño, which forms part of the Portuguese border; its main towns are CORUNNA, SANTIAGO DE COMPOSTELA and VIGO.

Population 2 851 000
Map Spain – Aa

Galicnik *Macedonia* See MAVROVO

Galilee *Israel* Northern region bounded by Lebanon to the north and Syria and Jordan to the east. The majority of Israel's Arab population is concentrated in this area, which is hilly in the north but has some fertile agricultural land farther south. The main towns of the region, which includes the Sea of Galilee, are TIBERIAS and NAZARETH.

Galilee is closely associated with Christ's teachings and ministry.
Map Israel – Ba

Galle *Sri Lanka* Old colonial city on a rocky headland, 105 km (65 miles) south and slightly east of COLOMBO. Formerly known as Point de Galle, it was founded as a walled seaport by the Portuguese around 1600 – although Moroccan tradesmen had already visited the site in 1344. In 1643, it was taken by the Dutch, who built a fort here which can still be seen. Galle also has many fine colonial mansions.

Locals specialise in crafts such as lace making, ebony carving and gem cutting and polishing, which are sold on the city's markets. Galle has some excellent beaches.
Population 84 000
Map Sri Lanka – Bb

gallery forest Dense tangle of trees lining the banks of a river in open savannah country. Along the smaller streams, the trees meet overhead, giving the appearance of a gallery.

Gallipoli (Gelibolu) *Turkey* Peninsula on the north shore of the DARDANELLES, 200-250 km (125-155 miles) west and slightly south of

Istanbul. It extends for about 80 km (50 miles) between the Dardanelles and the Gulf of Saros. Heavy fighting took place on its beaches in 1915 and 1916, when Allied troops – notably Australians and New Zealanders – tried unsuccessfully to take control of the strait from the Turks, who were fighting on the side of Germany. Today, the battlefields and graveyards form part of a national park.

Gallipoli is also the name of a fishing port (population 16 900) on the Dardanelles coast; it was the first place in Europe to be conquered by the Turks, in about 1356.
Map Turkey – Aa

Galloway *United Kingdom* See DUMFRIES AND GALLOWAY

Galveston *USA* Port in Texas on a barrier island just over 3 km (2 miles) offshore in the Gulf of Mexico, about 80 km (50 miles) south-east of Houston. It is linked to the mainland by a causeway. The port, which handles cotton and grain, remains a busy point of entry despite competition from the Houston Ship Canal, which was opened in 1914. Previously a pirate stronghold, the island with its 52 km (32 miles) of sandy beaches has become a popular resort.
Population (city) 62 400; (metropolitan area) 215 400
Map United States – He

Galway (Gaillimh) *Ireland* County on the north side of Galway Bay covering 5940 km² (2293 sq miles). Its county town is the city of Galway, 196 km (122 miles) west of Dublin. In the 13th century the city – then walled – became an isolated Anglo-Norman colony led by families known as 'the 14 tribes'. Despite violent opposition from the locals, they dominated the area for four centuries, and even held out for the Royalists against Oliver Cromwell until 1652.

According to local legend, an Anglo-Norman mayor named James Lynch Fitz-Stephen condemned and executed his own son for murder in 1493, and by so doing gave his name to the term 'lynch mob'. A stone near St Nicholas's Church marks the site of the gallows.

Galway today is a port and the major commercial centre for the west of Ireland; it has a modern Catholic cathedral and is the site of part of the National University of Ireland. It has a wide variety of industries and a strong tourist trade.
Population (county) 180 360; (city) 50 850
Map Ireland – Bb

Gambia *West Africa* One of Africa's major navigable rivers, with a length of some 1130 km (700 miles). It rises in the FOUTA DJALLON in Guinea, and is fundamental to the existence of the Gambia, Africa's smallest mainland country, which occupies a narrow strip of land on either side of the river and shares its name. In the dry season from December to April the river is salty from its mouth to as far as the town of Kuntaur, 200 km (125 miles) upstream, hampering irrigation projects.

The Gambia River grows from a width of about 30 m (100 ft) at Koina in the east, to an estuary 1.6 km (1 mile) wide at Banjul, the capital of the Gambia. At this point it is up to 9 m (30 ft) deep; however, a sandbar restricts access to ships of more than 25 000 tonnes.
Map The Gambia – Ab

Gambia, The See p 253

Gambier Islands *French Polynesia* Small group of islands, 1600 km (1000 miles) south-east of TAHITI. The main island is Mangaréva, a former centre of missionary activity in eastern Polynesia. It and the other small islands are enclosed by a reef forming a lagoon 26 km (16 miles) wide.
Population 600
Map Pacific Ocean – Gd

Gan *Maldives* Small coral island in the southern Maldives atoll of Addu, 644 km (400 miles) south-west of Sri Lanka. The island was an important Royal Air Force staging post during the later years of Britain's protectorate over the Maldives which ended in 1965. In the early 1990s strong efforts were being made to improve the island's tourist facilities.
Map (Maldives) Indian Ocean – Cb

Gäncä (Kirovabad) *Azerbaijan* Industrial city 290 km (180 miles) west of the Black Sea port of Baku. It was an ancient Armenian caravan town known as Gandza, which became part of Russia in 1804, when it was renamed Elizavetpol; it was called Kirovabad in 1935. Gäncä has some fine 17th-century mosques and a caravanserai – an inn with a spacious courtyard where caravans were accommodated. The mausoleum of Nizami, the 12th-century Persian poet who was born in the town, is situated nearby. Its industries include food processing, and the manufacture of textiles, agricultural machinery, cottonseed oil and soap.
Population 281 000
Map Azerbaijan – Aa

Gand *Belgium* See GHENT

Gandak *Nepal/India* River with a length of 690 km (428 miles), formed by the union of several other rivers, including the Kali Gandak and the Buri Gandak, in central Nepal. It drains the whole west-central area of that country, and then flows on into the Bihari Plains of India, occasionally causing considerable flooding, before entering the GANGES at PATNA. On its journey through Nepal, the Kali Gandak River passes between the mountains of Annapurna and Dhaulagiri, forming the world's deepest valley, which lies 5.5 km (3.5 miles) below the 8172 m (26 810 ft) high peak of Dhaulagiri.
Map Nepal – Ba

Gander *Canada* Town in eastern Newfoundland which has an airport that used to be a busy refuelling stop for transatlantic airliners. Although the airport has declined in importance because modern aircraft can now cross the Atlantic without refuelling, it is still the centre for North Atlantic air traffic control.
Population 10 340
Map Canada – Jd

Gandhara *Pakistan* Ancient kingdom and province in the area of Peshawar and Islamabad, straddling the Punjab and North-West Frontier Province. It was conquered by the Persians, before the Macedonian conqueror Alexander the Great overran it in 327-326 BC; Alexander's influence was so strong that a Graeco-Buddhist culture emerged, marked by the remains of a 2nd-century BC university at the capital, TAXILA.
Map Pakistan – Da

Gandhinagar *India* New town planned and developed since 1961, 550 km (342 miles) north of Bombay. It is the capital of Gujarat, home state of the political and religious leader Mahatma Mohandas Gandhi (1869-1948).
Population 393 470
Map India – Bc

Ganges (Ganga) *South Asia* Hinduism's holiest river, stretching for more than 2525 km (1568 miles). It rises in the Indian Himalayas near GANGOTRI Mountain (4000 m, 13 123 ft) and flows down through the Ganges plain, where it is harnessed for irrigation and power. The new low barrage at the town of Haridwar has a narrow slit in the middle, so that some of the water passes in its natural state to the sea. On its right bank it is joined by the Yamuna River; on its left bank it is joined by other great rivers flowing from the Himalayas, such as the Kosi. It then flows into Bangladesh just downstream of the FARAKKA Barrage 150 km (93 miles) north of Calcutta.

In Bangladesh the Ganges is known as the Padma until it joins the Jamuna River (Brahmaputra), after which the combined river is called the Meghna. The delta of the Ganges is the world's biggest; it is a highly active zone, with new islands emerging in the Bay of Bengal.

Along the length of the Ganges are holy shrines, most notably at VARANASI and ALLAHABAD. Hindus believe the water is so holy that it will keep clean forever. They also believe that if their cremation ashes are scattered on the river their souls will go straight to heaven. Although the river contains all kinds of pollutants – industrial and human – from the many major cities on its banks, it does have a remarkable ability to cleanse itself.
Map India – Cb-Dc

Ganges-Kobadak *Bangladesh* Irrigation scheme in western Bangladesh. Water is pumped from the Ganges to flow south through the Kushtia Canal into parts of the delta that have become silted up over the centuries, as the water flow has moved to rivers farther east. A secondary aim of the scheme is to keep salt water from intruding upriver from the Bay of Bengal during the dry season. The scheme irrigates some 90 000 hectares (220 000 acres).
Map (Ganges) Bangladesh – Ab

Gangneung *South Korea* See KANGNUNG

Gangotri *India* Mountain in the north, 40 km (25 miles) south of the Chinese border. It is 6674 m (21 896 ft) high, and its glaciers are the source of the Ganges River. Its Hindu temple, situated at an altitude of 3048 m (10 000 ft), is visited by thousands of pilgrims.
Map India – Cb

Gangtok *India* Capital of the former Himalayan kingdom of Sikkim, a state of north-east India since 1975. A road climbs 1700 m (5577 ft) to the town from the plains. And from Gangtok itself 4310 m (14 140 ft) Nathula Pass leads into Tibet. Only tracks go higher into the mountains. Visitors to Gangtok need special passes. The area is famous for its indigenous orchids; there are more than 600 species, of which 500 are on display at the Orchid Sanctuary.
Population 176 140
Map India– Db

The Gambia

A TINY, POVERTY-STRICKEN COUNTRY WITH FEW RESOURCES, DEFENDING ITSELF AGAINST ENGULFING SENEGAL

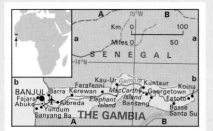

The smallest country in Africa, the Gambia pokes into the west coast of the continent like a crooked finger. The finger is 48 km (30 miles) wide at the coast, narrowing to only 24 km (15 miles) in places as it protrudes 470 km (292 miles) inland. The country is divided along its entire length by the GAMBIA River, and the division is emphasised by the lack of a single bridge over the river.

The Gambia is a little English-speaking sliver left behind by British colonists – an incongruous remnant of British rule in Africa which is a legacy of the rivalry between Britain and France for control of West African trade. The British arrived in 1664 in the wake of Portuguese traders and displaced a group of Baltic German settlers from a fort on James Island in the Gambia estuary, 30 km (19 miles) from the sea. Ivory, ebony, gold and slaves were the lure for the Europeans in those days, and these were exchanged for salt, iron, firearms and gunpowder. Fort James was the collection point for slaves shipped by the British to the Americas in the 18th century, until the slave trade was abolished in 1807.

TERRITORY UNDER SIEGE

For centuries, the Gambia was accustomed to being under siege – either by pirates or by the French (who made six attempts to take the country), or indeed by local tribal kings who cut off supplies from time to time. But Britain maintained its hold on this important trade outlet, which was finally established as a colony in 1888, while the land upriver became a protectorate in 1894. In 1889, an Anglo-French treaty defined the border between the Gambia and French-controlled Senegal.

In 1963, the Gambia was given self-government and in 1965, it became fully independent. Although the People's Progressive Party has always had an overwhelming majority in parliament, the Gambia is one of a handful of African countries that sustained a multiparty system over a long time. Its leader, Sir Dawda Jawara, was the longest serving president on the continent but was deposed in a bloodless coup in July 1994.

Several attempts have been made at a merger between the Gambia and Senegal but the Gambians continue to hang on to their separateness, because their way of life and language are so different. Indeed, one Senegalese government official remarked: 'The trouble with the Gambians is that they are so British.' Today, the Gambians are still staunchly independent. and determined to go it alone.

However, a loose confederation called Senegambia– was formed in 1981 when an attempted coup against the government led President Sir Dawda Jawara to accept the help of Senegalese troops. For years, the troops remained stationed in the Gambia as 'confederal troops' of the Senegambia Confederation; a defence alliance was established in 1982. The confederation was dissolved in 1989 and the troops moved out.

The Gambia is by far the poorer of the two countries. There are no mineral or natural resources of significance. Three-quarters of the population are employed in agriculture, which earns about 30 per cent of the gross domestic product (GDP).

In the capital, BANJUL (formerly Bathurst), which lies on a sandy island at the mouth of the Gambia River, the majority of Africans live in closely packed, corrugated iron dwellings. For most of them, the only source of water is a communal standpipe in the street. Outside the city centre, the suburbs of Cape St Mary and Fajara are a stark contrast. Here, wide, tree-lined roads where the British once lived now contain the beautifully kept offices of government departments, embassies and the houses of diplomats and foreign visitors.

RELICS OF BRITISH INFLUENCE

Evidence of British influence is apparent everywhere, from Banjul's once elegant cricket ground to the ornamental cannons that are dotted about the city centre. Even the Gambian army is still organised and drilled on British military principles. But these are only superficial relics of more prosperous days.

Today, the most luxurious existence in the Gambia is in the increasingly popular hotels on the seafront, where tourists (chiefly from Scandinavia and Britain) can step straight out of their rooms onto white, sandy beaches. It is always hot on the coast, temperatures averaging around 24°C (75°F) in January and 29°C (84°F) in June.

The local markets are geared to the tourist trade, selling carved wooden figures, ethnic masks, and brightly printed batik cloth alongside the vegetables and fish that provide the daily food of the Gambians.

THE GREAT RIVER

Boats are important in a country dominated by one of the finest waterways in Africa. There is a thriving local fishing industry using dugout canoes called *pirogues*. Most of the catch comes from the sea (about 14 000 tonnes per year), though some 3000 tonnes of freshwater fish are landed annually.

The river extends into Senegal and Guinea and is navigable over its entire distance inside the Gambia. Apart from ferries at the few main crossing points, there are, however, no passenger boat services on the river. Barges carry groundnuts and freight.

It is possible to drive the length of the Gambia on a tarmac road, skirting the edges of villages and passing alternately through cultivated land and scrub country, where cattle, sheep and goats browse. Tracks that lead off this road are hard in summer, but turn into quagmires in winter.

Most Gambians live precarious lives in the villages, tending a few animals and trying to grow enough millet and sorghum to feed themselves. If they are lucky, they vary their diet with maize, rice, vegetables and fruit. But drought and overgrazing have led to desperately low agricultural yields. Life expectancy in the villages is low – only about 45 years.

Groundnuts and groundnut products are the Gambia's main export, and account for about 85 per cent of foreign income. They are often grown by these same village farmers for cash. There is a great deal of smuggling between the Gambia and Senegal whenever there is a profit to be made; border patrols have a difficult job because the boundaries of the Gambia divide ethnic groups living on both sides of the border: the Mandingo, or Malinke – the largest group, making up 40 per cent of the population – the Fulani (or Peul), the Wolof, Dioula (or Jola), and Serer people, among others.

The Gambia's difficult times mean that large numbers of the fast-growing population are having to emigrate to find jobs. Economically, the Gambia is struggling, and depends heavily on foreign aid, mainly from Germany, China and Britain. Tourism took off after the popular television series *Roots* based on the novel by Alex Haley. Whereas in 1978, only 29 000 tourists visited the country, their number has risen to over 101 000 in recent years, and today, the country is doing much to develop tourism. But even tourism brings in less than might be hoped: most visitors come on package tours and pay for everything before they arrive.

THE GAMBIA AT A GLANCE	
Official name Republic of The Gambia	
Area 11 295 km² (4361 sq miles)	
Population 1 000 000 **Per km²** 89 **(Per sq mile** 229)	
Capital Banjul	
Government Republic	
Currency 1 dalasi = 100 bututs	
Languages English (official), African languages	
Religions Muslim (85%), Christian, indigenous beliefs	
Climate Tropical; average temperature at Banjul ranges from 24°C (75°F) in January to 29°C (84°F) in June	
Land use Cultivation 16%, forest and woodlands 20%, grazing 9%, other 55%	
Main primary products Groundnuts, cassava, millet, rice, corn, cattle, sheep, goats, fish, timber, palm kernels	
Major industries Agriculture, food processing, fishing, forestry	
Main exports Groundnuts and derived products, fish, palm kernels, hides and skins, cotton	
Annual income per head (US$) 390	
Population growth (per thous/yr) 29	
Life expectancy (yrs) Male 45 **Female** 45	

Gansu (Kansu) *China* Province in the north-west, which is shaped like a long narrow corridor and lies between Nei Mongol and the Qilian Shan mountains. It is a semi-arid region covering 450 000 km² (170 000 sq miles), taking in parts of the LOESSLANDS OF CHINA and the GOBI DESERT. It is crossed by the Silk Road, a centuries-old trade route to central Asia by which the Venetian explorer Marco Polo travelled to China in the 13th century.

Wheat, millet and *gaoliang* (a type of sorghum) are grown in the loesslands under irrigation. Although the soil is very fertile, erosion, though gradual, is a problem. The province has rich deposits of oil, iron ore, coal and copper. LANZHOU, its capital, stands on the HUENG HE (Yellow River).

Population 23 314 000
Map China – Ec

Ganvié *Benin* Picturesque lakeside village on Lake Nokoué. The houses are built on stilts, an arrangement that formerly exempted their owners from taxation. It was argued that, because they did not actually live on the ground, they could not be taxed as local residents. The village is a major tourist attraction that can be visited by dugout canoe only from the port of Cotonou.

Map (Cotonou) Benin – Bb

Gao *Mali* Town 960 km (600 miles) down the Niger River from the capital, Bamako. It was once the capital of the Songhai empire, but declined after the Moors occupied it in 1591. The town is a major market for local produce and an important river port, being the terminus of river steamers from Bamako. Its population started increasing rapidly in the early 1990s because of the vast number of nomads who have lost their herds and their traditional living in the drought that has devastated the northern parts of Mali.

Among the town's historic buildings are the Mosque of Kankan Moussa, and the tomb of the Askia dynasty who were rulers of the former Mali empire.

Population 50 000
Map Mali – Cb

Gap *France* Capital of Hautes Alpes department, 75 km (about 50 miles) south-east of Grenoble. It guards the Bayard Pass, the route taken by Napoleon Bonaparte on his return to Paris in 1815 from his exile on Elba. The town manufactures gloves and textiles.

Population 35 700
Map France – Gd

Garda, Lake *Italy* Alpine resort lake about halfway between Milan and Venice. With an area of 370 km² (143 sq miles), it is Italy's largest lake. Its beauty, its mild climate and its luxuriant vegetation and lakeside plantations of olive and citrus trees have given its shores the appearance of a miniature riviera.

Map Italy – Cb

Gargano *Italy* Mountainous promontory, 50 km (30 miles) long, extending into the Adriatic Sea 350 km (218 miles) north-east of SAN MARINO. A limestone plateau, it is not part of the Appenines. It rises to 1056 m (3465 ft) and is flanked by sandy beaches. Among its villages is the medieval pilgrimage centre of MONTE SANT'ANGELO.

Map Italy – Ed

Garmisch-Partenkirchen *Germany* Twin towns at the foot of the Bavarian Alps, 75 km (45 miles) south-west of MUNICH. Garmisch-Partenkirchen, which forms one of Germany's leading winter sports resorts, hosted the 1936 Winter Olympics. A railway climbs up to the Zugspitze which, at 2963 m (9721 ft), is Germany's highest mountain.

Population 26 700
Map Germany – De

garnet One of several widely distributed silicate minerals found in igneous and metamorphic rock, coloured red, brown, black, green, yellow or white and used as gemstones and abrasives.

Garonne *France/Spain* River, some 650 km (404 miles) long, which rises in the Spanish Pyrenees and flows north-east to Toulouse. It then turns north-west and flows on past BORDEAUX to join the DORDOGNE at the Gironde estuary, which empties into the Bay of Biscay. The middle and lower sections of the Garonne valley contain rich farmland producing wheat, maize, fruit and vegetables, as well as grapes from which the fine Bordeaux wines are made.

Map France – Cd-De

garrigue Impoverished scrub vegetation similar to maquis, found in the drier limestone areas of the Mediterranean region. It consists of stunted trees and low, drought-resistant plants, many of which are aromatic. The vegetation is sparse and interspersed with much bare earth. See MEDITERRANEAN WOODLAND.

Gascony (Gascogne) *France* Territory in the extreme south-west, whose name comes from the Vascones, a Spanish tribe who conquered it in the 6th century AD and set up the duchy of Vasconia (Gascony). It was held by the English from 1154 to 1453.

Map France – Ce

Gasherbrum *China/Pakistan* Mountain, 8068 m (26 470 ft) high, on the disputed border between Pakistani Azad Kashmir and Chinese-occupied territory in the Karakoram range.

Map Pakistan – Da

Gaspé *Canada* Peninsula projecting eastwards into the Gulf of St Lawrence in eastern Quebec. In 1534, the explorer Jacques Cartier landed here and claimed the area for France. The peninsula is mainly forested and is a summer resort. At its eastern end lies the town (and railway terminus) of Gaspé. Industries in the area include timber and tourism.

Population (town) 16 400
Map Canada – Id

Gateshead *United Kingdom* See NEWCASTLE UPON TYNE

Gatooma *Zimbabwe* See KADOMA

Gauteng (PWV) *South Africa* At 18 760 km² (7261 sq miles), the smallest of the nine provinces created in 1994 and originally called PWV (Pretoria-Witwatersrand-Vereeniging). It represents the industrial and commercial hub of South Africa with the greatest concentration of the country's population. It includes the cities of Pretoria and Johannesburg (provincial capital)

and the peripheral mining and industrial towns of Vanderbijlpark and Vereeniging, as well as those on the eastern and western extensions of the Witwatersrand, including Germiston, Benoni, Brakpan, Boksburg, Roodepoort, Randfontein and Krugersdorp.

Population 9 300 000
Map South Africa – Cb

Gävle *Sweden* Seaport and industrial city on the Baltic coast, 180 km (112 miles) north of Stockholm. It exports iron ore and softwoods, and manufactures textiles, chemicals and beer.

Population 89 200
Map Sweden – Cc

Gaza *Middle East* Mediterranean seaport 65 km (40 miles) south-west of Tel Aviv. Gaza was a Philistine city-kingdom in the 1st millennium BC; it was here that Samson pulled down the Temple of Dagon, crushing to death his Philistine captors and himself.

Population 273 000
Map Egypt – Cb

Gaza *Mozambique* See XAI-XAI

Gaza Strip *Middle East* Territory occupied by Israel since the Six Day War of 1967. It is 42 km (26 miles) long and 6-10 km (4-6 miles) wide and runs north, up along the Mediterranean coast, from the Egyptian frontier. The territory was formerly part of Palestine, and when the state of Israel was created in 1948 it was left under Arab control while its administration was handed to Egypt. However, about 45 per cent of its adult population worked in Israel. After 1948, large numbers of Palestinian refugees moved in, and it is still dominated by eight large Palestinian refugee camps. An Islamic university has been established here. In May 1994, the Gaza Strip was formally handed over to the Palestine Liberation Organisation with limited administrative powers, while negotiations with Israel continued. There is some agriculture, involving mainly citrus fruit, wheat and olives, but this is threatened by increasing shortages of water and the rising level of salinity in the water.

Population 760 000
Map Egypt – Cb; Israel – Ab

Gazankulu *South Africa* Former Bantu homeland and, from 1973, 'self-governing' state created for the Shangaan-Tsonga people of Eastern Transvaal. During its period of 'independence', it was recognised only by South Africa; after South Africa's first democratic elections in 1994, it was reincorporated into the republic. The territory, which covers roughly 7410 km² (2860 sq miles), is mainly agricultural. Occupations include cattle raising and the cultivation of cereals and sisal, much of it at subsistence level.

Map South Africa – Da

Gaziantep *Turkey* Town and provincial capital on the border with Syria, 500 km (310 miles) south-east of Ankara. Its main industries are processed agricultural produce, textiles and goatskin rugs. The neighbouring province of the same name, which covers 7642 km² (2951 sq miles), is the centre of Turkey's pistachio nut farming belt.

Population (town) 574 000; (province) 1 140 590
Map Turkey – Bb

Gbadolite *Zaire* Location of an exotic presidential village built by President Mobuto Sese Seko. It stands on the Ubangi River, in Equateur province, in the area where the president of Zaire was born. Mobutu uses it as a headquarters away from the capital, Kinshasa. Only invited guests may stay here.
Map Zaire – Ba

Gdańsk (Danzig) *Poland* The country's largest seaport and shipbuilding centre, lying 354 km (220 miles) north-west of the capital, Warsaw. It is the focus of Trojmiasto (Tri-City) – the combined urban area of Gdańsk, Sopot and Gdynia which have merged, and which stretches for 40 km (25 miles) along the Gulf of Gdańsk. Together they have a million inhabitants.

Gdańsk was already a large Polish stronghold, fishing port and crafts and amber-trading town 1000 years ago. At different times since then it has been ruled by Poles, the Teutonic Knights, and Germans. For most of the period from 1793 to 1920 it was under Prussian rule and was the capital of the province of West Prussia; from 1807 to 1815, and again from 1919 to 1939, it was the Free City of Danzig and the majority of its population was German. Hitler's claim to the city in 1939 led to the German invasion of Poland, which sparked the Second World War.

A member of the Hanseatic League made up of mostly north German towns from 1343 onwards, Gdańsk prospered from trade. Shipbuilding began only in 1850; most of its other main industries (engineering, chemicals and food processing) began after 1950.

Gothic churches, a Renaissance town hall and the 14th-century Great Mill can be seen in the oldest parts of the city. The medieval layout survives, with nine parallel streets ending at the Motlawa River. The Dominican Fair which takes place each August is a good time to buy souvenirs and antiques. Three crosses stand outside the Gdańsk Shipyard (formerly Lenin Shipyard), to commemorate workers killed in civil demonstrations in 1970, following massive increases in food prices. In 1980, the trade union Solidarity was founded here, to become the only independent Trade Union in the communist bloc. Although banned, Solidarity refused to be destroyed and, in fact, was the driving force behind the fall of communism in Poland.
Population 465 100
Map Poland – Ca

Gdańsk, Gulf of (Gulf of Danzig) Inlet of the BALTIC SEA in north-eastern Poland. It measures 64 km (40 miles) from north to south and 60 km (37 miles) across. It contains GDAŃSK and GDYNIA, two of Poland's most important ports.
Map Poland – Ca

Gdynia (Gdingen) *Poland* Baltic seaport 16 km (10 miles) north-west of Gdańsk. Founded as a fishing village in 1224, the port now handles about 18 million tonnes of cargo a year and also serves as a naval base.
Population 251 500
Map Poland – Ca

Gêba *Guinea/Senegal/Guinea-Bissau* West African river about 190 km (120 miles) long. It rises in the FOUTA DJALLON of western Guinea and flows through southern Senegal, then on, through Guinea-Bissau, to the ATLANTIC OCEAN via a wide estuary. The Gêba River is Guinea-Bissau's main transport artery. Bissau, the country's leading commercial centre and port, lies on the estuary.

Gêba is also the name of the former colonial capital of Guinea-Bissau (then Portuguese Guinea), situated near present-day Bafatá, at the river's upper limit of navigation. The town, which was abandoned as the capital because of its location in a swampy, malaria-infested area, is today virtually in ruins.
Map (river) Guinea – Aa; (Bafatá) Guinea – Ba

Geelong *Australia* Port and second largest provincial city in Victoria, 60km (38 miles) south-west of Melbourne, and once the port of entry to the Victorian gold fields. It has one of the world's largest and most modern bulk wheat terminals, and is a major wool-selling centre. It manufactures woollen textiles, motor vehicles, cement and agricultural equipment.
Population 152 800
Map Australia – Gf

Geirangerfjord *Norway* A 15 km (9 mile) long branch of Storfjord, south-east of Ålesund in west Norway. The fjord, carved by Ice Age glaciers, is claimed by many Norwegians to be the nation's most beautiful stretch of water and attracts thousands of tourists each summer. All along its length, it is edged by sheer rock walls laced with ribbons of waterfalls – among them the gauzy Bridal Veil (Brudesløret) and the multiple cascades of the Seven Sisters (Syv Søstre). The cliffs above the fjord's mirror-smooth surface are some 400 m (1310 ft) high; they extend for as much as 300 m (980 ft) below the water.
Map Norway – Bc

Gela (Terranova di Sicilia) *Italy* Town on the south coast of Sicily, 140 km (87 miles) south-east of Palermo which was originally a Greek city called Gelon. Gela has grown rapidly since a petrochemical plant was built here in the 1960s.
Population 74 800
Map Italy – Ef

▲ **LAKESIDE CITY The Swiss city of Geneva, adorned by the spectacular** *jet d'eau*, **lies at the western end of Lake Geneva. It plays host to many international organisations including the Red Cross.**

Gelderland *Netherlands* The country's largest province, covering 5129 km² (1980 sq miles). It lies in the east, along the German border, between the Maas and IJssel rivers. The region consists mainly of agricultural land, interspersed with areas of forested hills and heathland which are popular with tourists. Numerous industries have developed along the Waal and Lek branches of the Rhine, especially around Nijmegen and the provincial capital, Arnhem.
Population 1 839 900
Map Netherlands – Ba

Gelibolu *Turkey* See GALLIPOLI

Gelsenkirchen *Germany* Coal-mining town in the industrial Ruhr, 24 km (15 miles) west of Dortmund. The town produces steel, chemicals and paper.
Population 293 700
Map Germany – Bc

Genck *Belgium* See GENK

General Santos *Philippines* Port and capital of Cotabato del Sur Province, on southern MINDANAO, formerly called Dadiangas. It serves the rehousing projects of the Koronadal valley.
Population 200 000
Map Philippines – Cd

Geneva (Genève; Genf) *Switzerland* City at the western end of Lake Geneva (also known as Lac Léman), standing where the River Rhône drains the lake. Founded by the Romans, it became a powerful, independent city in the Middle Ages when it was ruled by prince bishops; its prosperity continued in the 15th and 16th centuries under its merchant rulers. During this period, it had already built up a reputation as one of Europe's most important financial centres. Surrounded on three sides by French territory, Geneva was seized by France in 1798, but joined the Swiss confederation in 1815 after the fall of Napoleon.

Geneva was one of the principal centres of the Protestant Reformation, putting the austere religious theories of John Calvin (1509-64) into political practice in the 16th century. In 1864, Henri Dunant (1828-1910) founded the International Red Cross in Geneva – and from that time onwards, the city became a centre for international organisations and multinational companies, including the United Nations and various UN agencies such as the World Health Organization (WHO) and the International Labour Organization (ILO). It has also become a centre for international negotiations, be they related to politics, trade, health, science, or peace. As a conference centre it provides five-star comfort in a scenic setting, with a 145 m (476 ft) fountain shooting up from the lake.

The canton of the same name around the city grows grapes and vegetables. It covers 282 km² (109 sq miles).
Population (city) 170 200; (canton) 383 900
Map Switzerland – Aa

Genk (Genck) *Belgium* Industrial town 30 km (19 miles) north of the city of Liège. Developed in the 1920s as a coal-mining village, Genk now also produces steel and motorcars.
Population 62 000
Map Belgium – Ba

Genoa (Genova) *Italy* Seaport in the north-west, 120 km (75 miles) south-east of TURIN. During the Middle Ages it was a wealthy and extremely powerful city-state. It has retained its position as an important trading centre and naval base, and it is also Italy's leading port. The explorer Christopher Columbus (1451-1506) was born here. In the 19th century, Giuseppe Mazzini (1805-72) and Giuseppe Garibaldi (1807-1882), leaders of the Risorgimento (or 'Resurgence') movement, which paved the way for Italian unification, made the city their headquarters.

Today, Genoa's economy revolves around heavy industries, principally steel and shipbuilding, although these are declining. A scarcity of flat land around the city prevents the expansion of new industries.
Population 760 300
Map Italy – Bb

Genoa, Gulf of Part of the LIGURIAN SEA, washing the shores of north-western Italy. The port of GENOA, once a great maritime power, occupies a central position on the coast of the gulf. The Riviera di Levante, east of Genoa, and the Riviera di Ponente, extending to the French border to the west, are dotted with resorts.
Map Italy – Bb

Genova *Italy* See GENOA

Gent *Belgium* See GHENT

geo (gio) Long, narrow, steep-sided inlet running inland from the sea along the edge of a cliff. It is probably caused by the collapse of the roof of a cave worn by the waves along a line of weakness such as a major JOINT or minor FAULT.

GEOLOGICAL TIMESCALE

Contrary to the postulations of the 17th-century Archbishop James Ussher, the earth was not created in 4004 BC – but a great deal earlier. Most geologists and physicists, with the aid of radioactive dating, are agreed that our planet evolved about 4600 million years ago. As to how it evolved, or assumed its present form with crust, mantle and core, there is considerably less unanimity.

That some rocks are older than others was discovered in the late 18th century when a few people of inquiring mind began to study rock strata and the fossils embedded in them. These gave some idea of the relative ages of rocks, and names were assigned to distinguish them; but their ages in years was incalculable. With the discovery in 1896 of radioactivity, however, and the subsequent discovery that radioactivity is present in rocks, a 'time ruler' was established. Each radioactive element in a rock decays and transforms itself into a new element at a constant, measurable rate. By measuring the amount of original element left in rock, and the amount of new element, it is possible to calculate the date at which the rock was formed. Thus it has been discovered that the oldest rocks, part of the original crust, are about 3900 million years old.

THE TIMESCALE AND EARTH'S DEVELOPMENT

Era	Period – and epoch		Beginning (in millions of years ago)	Activity
Precambrian	Archaeozoic		4600	Origin of the earth. Formation of crust, continents and oceans.
			3900	Oldest known rocks.
	Proterozoic		3300	Origin of life. Formation of present-day atmosphere.
Palaeozoic	Cambrian		570	
	Ordovician		500	
	Silurian		430	Life comes ashore from the seas.
	Devonian		395	Caledonian mountains raised.
	Carboniferous		345	
	Permian		280	Appalachians and Central European mountains raised.
Mesozoic	Triassic		225	Urals raised. Pangaea starts to break up (200-180 million years) and North Atlantic starts to open.
	Jurassic		190	South Atlantic starts to open (140-130 million years).
	Cretaceous		135	India splits from Antarctica (105-100 million years).
Cenozoic	Tertiary	– Palaeocene	65	Formation of the Rocky Mountains.
		– Eocene	53	Australia splits away from Antarctica (45 million years).
		– Oligocene	37	India collides with Asia (30 million years). Formation of the Himalayas and the Alps.
		– Miocene	26	
		– Pliocene	5	Earliest forms of hominid appear.
	Quaternary	– Pleistocene	2	
		– Holocene		Started 15 000 years ago.

geode A small, usually spheroidal hollow in a rock, which is lined with crystals. It is common in sedimentary rock, especially limestone.

geographic poles The north and south poles, or the points at which the axis of rotation (the imaginary line about which the earth rotates) reaches the earth's surface. The geographic poles do not coincide with the GEOMAGNETIC POLES.

geological cycle See ROCK CYCLE

geological timescale See panel above

geomagnetic poles The points at which the axis of the earth's magnetic field reaches the earth's surface. The axis lies at an angle of 11 degrees to the axis of rotation (see GEOGRAPHIC POLES), which is why the geomagnetic poles do not coincide with the geographic poles. At present, the north geomagnetic pole lies at 79°N, 70°W, and the south geomagnetic pole at 79°S, 110°E, but both are moving very slowly. For example, although the latitudes of the poles have remained fairly constant over the past 150 years, the poles have moved westward by about 6 degrees of longitude. As well as varying in location, the axis of the earth's magnetic field has also undergone reverses of polarity. The effect of these reversals on the alignment of magnetic particles in rocks is important evidence supporting the theory of PLATE TECTONICS.

geomagnetism The magnetism of the earth.

George *South Africa* Cathedral town and commercial centre at the foot of the Outeniqua Mountains, 440 km (275 miles) east of CAPE TOWN and 8 km (5 miles) from the Indian Ocean. It was founded in 1811 and named after Britain's King George III. The town, noted for its setting of mountains and forests, is on the Garden Route, a scenic highway linking Cape Town with PORT ELIZABETH and DURBAN.
Population 62 300
Map South Africa – Bc

George Town (Pinang) *Malaysia* Port and capital of Pinang Island, on a headland 300 km (186 miles) north-west of the capital, Kuala Lumpur. The city grew beside Fort Cornwallis, founded by Captain Francis Light of the British East India Company in 1786 to guard the island's harbour. Today, most of the harbour trade has moved to the nearby mainland port of Butterworth.
George Town, with its noisy mixture of East and West, Malay and Chinese, is now primarily a tourist centre. It has Buddhist temples, streets of Chinese shops, and easy access to the island's hills and beaches.
Population 250 600
Map Malaysia – Ba

Georgetown *Guyana* National capital and the country's largest city. It lies on the Atlantic coast and the mouth of the Demerara River. Most of it lies below sea level and is protected by dykes. The city was founded by the French in 1781, then developed by the Dutch, who drained its marshy site with a canal system, and seized by the British in 1812. The remaining dykes and timber houses on stilts are a reminder of Dutch colonial days. Today, the city is an administrative and commercial centre, with small-scale industry producing beer, soft drinks and timber products. Trade winds bring relief from the tropical climate.
Population 200 000
Map Guyana – Ba

Georgetown *USA* Residential section of Washington, DC. It lies on the north bank of the Potomac River, and was already a thriving town when the national capital was created in 1790-1800. Georgetown still retains many of its fine terraced houses and cobbled streets and much of its old, small-town elegance. Georgetown University was founded here in 1789.
Map United States – Kc

Georgia See p 258

Georgia *USA* South-eastern state, just north of FLORIDA, covering 152 490 km² (58 876 sq miles). It was founded as an English colony, named after King George II, in 1733, and became one of the original 13 states after the United States declared independence from England. The largest state east of the Mississippi River, Georgia's terrain rises from the Atlantic Ocean in the south-east to the Blue Ridge in the Appalachian Mountains in the north.
With mild winters and warm (but hurricane-prone) summers, the state formerly produced tobacco and cotton on large plantations. Today, its products also include poultry, peaches, pecans, peanuts, soya beans, food products, and wood-products from the extensive forests, as well as

textiles, and transport equipment. Tourism is expanding. ATLANTA is the state capital, and SAVANNAH the chief port.
Population 6 478 200
Map United States – Jd

Georgia, Strait of Channel about 240 km (150 miles) long and 30-64 km (19-40 miles) wide between south-western BRITISH COLUMBIA and VANCOUVER ISLAND, in Canada. It forms part of the scenic inland water route called 'the Inside Passage', which extends from SEATTLE, in the American state of Washington, to ALASKA.
Map Canada – Cd

geosyncline Major elongated depression in the earth's crust, sometimes several hundreds of kilometres long, which sinks as it is filled with sediment worn away from the landmasses on either side of it. Geosynclines are often uplifted to produce FOLD MOUNTAINS.

geotherm A surface which links all the points in the earth's interior, or in part of the earth, at which the temperature is the same.

Gerlach Peak (Gerlachovsky Stít) *Slovakia* See TATRA MOUNTAINS

German Bight Part of the NORTH SEA, hugging the north-western coast of Germany. The Germans call it the 'Deutsche Bucht', and it is also known as the Heligoland Bight, after the island of HELIGOLAND (Helgoland).
Map Germany – Ca

Germany See p 259

Germiston *South Africa* Industrial and gold-mining city on the EAST RAND, 13 km (8 miles) south-east of Johannesburg. The Rand Refinery, established here in 1921, is the world's largest and handles more than 70 per cent of the western world's gold output.
Population 166 400
Map South Africa – Cb

▲ YELLOWSTONE NATIONAL PARK Castle Geyser, so called because it is slowly raising a 'castle' of deposits around its rim as it shoots up to 30 m (100 ft) into the air, is one of about 200 geysers in the United States' oldest national park.

Gerona *Spain* Cathedral town 88 km (55 miles) north-east of BARCELONA. It manufactures chemicals, textiles, electronic equipment and soap. The province produces cereals, fruit and wine.
Population (town) 68 660; (province) 509 630
Map Spain – Gb

Gettysburg National Military Park *USA* Site of the most decisive battle in the American Civil War. It lies in southern Pennsylvania, 105 km (65 miles) north and slightly west of Washington, DC. Here, on 1-3 July 1863, Union forces led by General George Meade defeated the Confederates under General Robert Lee and stopped Lee's advance into the North. Nearly 50 000 men were killed or badly wounded in the clash. It was in the cemetery four months later that President Abraham Lincoln delivered his famous Gettysburg Address.
Map United States – Kc

Geyre *Turkey* See APHRODISIAS

geyser Natural hot spring that intermittently ejects a column of water and steam into the air. Geysers occur in active or recently active volcanic areas, and a well-known example is GEYSIR in Iceland.

Geysir *Iceland* One of the largest geysers in the world, which has been the object of scientific investigation for more than two centuries. It lies at the centre of Iceland's major group of hot springs, 85 km (53 miles) north-east of Reykjavík. Every 80-90 minutes, it spouts steam and water into the air, up to a height of 100 m (330 ft). The English word 'geyser' comes from its name.
Map Iceland – Bb

Georgia

THE TRANSCAUCASIAN REPUBLIC THAT THREATENED TO SPLIT APART JUST A FEW YEARS AFTER INDEPENDENCE

Georgia or 'Grusinia', as the Russians call the country on the far side of the Great Caucasus, is the westernmost of the three Transcaucasian republics between the Black Sea and the Caspian Sea – Armenia and Azerbaijan being the other two. Unlike most of the ex-Soviet republics, the Republic of Georgia did not become a member of the Commonwealth of Independent States (CIS) immediately after proclaiming its independence in November 1990.

After 1990, Georgia strove resolutely to go its own way. It started developing a private sector and began looking for trading partners in the West. The future of the republic, which is more than twice as large as Belgium and has a population a little over half the size of Belgium's, looked very favourable – Georgia has a diverse economy with much potential for development.

However, like its neighbours to the south, Armenia and Azerbaijan, large regions in this country became involved in an armed struggle for independence. Abkhazia in the north-west along the Black Sea was one region; the autonomous region of the South Ossets bordering Russia in the north was another. It wanted to reunite with the North Ossets in the North Ossetian Autonomous Republic in Russia; finally, the Adzhar Autonomous Republic with its capital city of BAT'UMI also began claiming independence.

On independence, Zviad Gamsakhurdia was elected president of Georgia in May 1991, but it was not long before he was forced to flee the capital and a military council took power. Former USSR Minister of Foreign Affairs, Eduard Shevardnadze, became the country's next head of state in October 1992. Until Gamzakhurdia's death late in 1993, however, there was a faction fighting in support of reinstating the deposed president, which added to the country's fragmentation.

The worst fighting took place in Abkhazia, which declared its independence from Georgia in June 1992. Russia, which continued to keep troops in the region, began negotiating for a cease-fire between the government and Abkhazian troops. But relations between Georgia and Russia became increasingly strained after the Georgian government blamed Russia for the region's instability, and for the rebels' capture of Abkhazia's capital, SOKHUMI, in September 1993. In 1994, a treaty was signed between Georgia and Russia, permitting the latter to establish troops on Georgia's frontier with Turkey. Giving in to economic pressure, Georgia became a member of the CIS.

The country is a patchwork of autonomous regions – and it consists of vast contrasts of landscapes. The Great Caucasus in the north forms an almost impassable wall of mountains, culminating in ELBRUS, an extinct volcano which boasts a height of 5642 m (18 510 ft). The Lesser Caucasus in the south is much lower, at 3322 m (10 899 ft). Deep, narrow valleys incise it. Between the two ranges lies a vast region of depressions and basins crossed by the rivers Rioni and Kura. The Rioni Valley in the west opens up like a funnel towards the Black Sea and Europe. This region has been populated for at least 2500 years and was the centre of historical Georgia, known as Kolkhida in Greek and Roman times. Georgia was the land of the Golden Fleece in Greek mythology, and was at times under Roman rule.

It has a warm, subtropical climate with a high annual rainfall of 1000-2500 mm (39-98.4 in). Citrus fruits and other subtropical trees and plants grow here – 97 per cent of the oranges that were harvested in the former Soviet Union came from this region. The figure for tea was almost as high, the Georgian tea crop accounting for 93 per cent of the USSR's total harvest.

One-quarter of the country's population lives off the land, growing sugar beet, vegetables, grains, potatoes, and rearing pigs and poultry. In the east, the climate becomes increasingly arid and continental; and on the far side of the watershed between the Rioni and Kura rivers, in the mountainous regions of the upper Kura, winters are cold with little snowfall and summers are hot. Here, grapes, cereals and vegetables are harvested.

Industries, such as food processing, heavy industries, and mechanical and auto engineering, are concentrated in the capital, T'BILISI, and in other cities such as K'UT'AISI and Bat'umi. About 30 per cent of the country's workforce is employed in industrial and construction industries.

Georgia is an important mining country and an exporter of ferrous and non-ferrous metals. Its manganese deposits in the Chiatura region are among the largest in the world and have in the past supplied over half of world demand. There are some deposits of oil, natural gas, and coal but they are small and Georgia has to import most of its fuels from neighbouring countries. Most of the oil comes from Azerbaijan via a pipeline and is refined locally.

Dependence on oil and gas is the country's most vulnerable point; since independence, conflicts in both Abkhazia and South Ossetia have disrupted fuel supplies. Azerbaijan has used its vital oil supplies to pressure Georgia to assist it in its war against Armenia; and the Georgian government has had to tread carefully to keep on good terms with the other ex-Soviet republics that supply it with its fuel.

RULED BY FOREIGNERS

The Georgians have shared the fate of most Caucasian peoples: much of their past was spent under foreign rule, as the country passed successively through the hands of the Greeks, Romans, Turks, Persians, Arabs and Russians.

Georgia gained independence in the 12th century, but was conquered by Monguls in 1236 and later divided into several principalities which preserved a nominal independence under foreign rulers. Occupation by the Persians and later the Turks, introduced Islam in some parts of the country. The Christian faith had been introduced to the region in the 4th and 5th centuries – from this time, too, stems the Georgian literature.

When the Russian tsars claimed territory on the far side of the Caucasus, they were welcomed as protectors against the Persians and the Turks. But all too soon they showed their true colours – they were yet another imperial power which ruthlessly exploited the mineral resources in its occupied territories. They introduced Russian as the official language and suppressed the Georgian Church in favour of the Russian Orthodox Church, quelling any resistance by force.

Later, Joseph Stalin, himself a Georgian, brutally suppressed any rebellion of the Georgian peoples. His was a policy of arbitrary delimitation, which was to lead to so much ethnic conflict in the region.

GEORGIA AT A GLANCE		
Official name Republic of Georgia		
Area 69 700 km² (26 905 sq miles)		
Population 5 634 296 **Per km²** 81		
(Per sq mile 209)		
Capital T'bilisi		
Government Republic		
Currency 1 rouble = 100 copeks		
Languages Georgian (official); Russian, Armenian		
Religions Christian (65% Georgian Orthodox, 10% Russian Orthodox, Armenian Orthodox 8%); Muslim (11%)		
Climate Moderately warm and humid in the west; little rainfall in the east. Average temperature in T'bilisi ranges from 1°C (33.8°F) in January to 24°C (75.2°F) in July		
Land use Forest and woodlands 30%, pastures and meadows 30%, cultivation 11%, other 29%		
Main primary products Manganese, copper, tungsten, arsenic, molybdenum, some oil, natural gas, and coal; grapes, citrus fruits, tea		
Major industries Mining, machinery, vehicles, textiles, food processing		
Main exports Machinery, vehicles, textiles, ferrous and non-ferrous metals; citrus fruits, wine, tea		
Annual income per head (US$) 4410		
Population growth (per thous/yr) 9		
Life expectancy (yrs) Male 69 **Female** 76		

Germany

*EAST AND WEST GERMANY HAVE
BROKEN DOWN THE BERLIN WALL AND
CREATED A NEW UNITED EUROPEAN
POWER WHICH IS NOT WITHOUT ITS
STRESSES AND STRAINS*

What the Germans refer to as their 'economic miracle' (*Wirtschafts-wunder*) was followed, 40 years later, by a second miracle – the reunification of their severed land, in November 1989. The process began when East Germany opened its borders to West Germany and to West Berlin. The Iron Curtain which had separated the two Germanys for more than 40 years was lifted; the Berlin Wall, built in 1961 to divide the city of Berlin in two, was broken down – a year later, on 3 October 1990, Germans were officially celebrating their reunification.

The events leading to the fall of the German Democratic Republic (GDR) are complicated, but during the months preceding it they were accelerated. When Hungary opened its borders with Austria in the summer of 1989, hundreds of East Germans left their country via this gateway to the West. The exodus was a direct challenge to the head of state, Erich Honecker. Inside the country, opposition parties formed and began demonstrating for democratic reform. Though somewhat alarmed, the government responded with well-tried East German tactics. Demonstrations were brutally suppressed – but nothing would now stem the political tide.

Honecker eventually stepped down; the government was radically reformed, almost overnight. Free elections and an amnesty were announced, and the order to shoot on sight, previously issued to all East German border guards, was repealed.

Lifting its terror rules did not save 'the first socialist state on German soil' from being doomed. In LEIPZIG, DRESDEN, POTSDAM and other centres, people were demanding democratic power, and, increasingly, for the two Germanys to be reunited into one united 'fatherland'.

At this juncture, it was only the victorious Allies of the Second World War (the United States, Britain, France and the Soviet Union) who could still stop the free elections in East Germany which were scheduled for 18 March 1990. However, they agreed to the reunification of the two countries and on 1 July 1990, the two severed nations formed a currency, economic and social union; the political union followed shortly after, on 3 October, and on 2 December 1990 the united Germans went to the election polls to elect a new united German parliament.

Longed-for though it had been, some say that the reunification came too fast – not least for the European neighbours who now have a German state of 81 million people in their midst. These nations feared Germany not so much as a political power but as an influential economic giant.

However, their fears have proven quite unnecessary. The two Germanys, with their vastly different economic and social systems, were united without transition, and Germany is now facing innumerable problems brought about by this dramatic political step. The most challenging problem was to integrate the two economies; besides, the two new partners are split over questions of environmental protection, law, transport and telecommunication. The most optimistic prognosis for Germany's future is that existing differences will be smoothed out by the mid-1990s – but at tremendous expense.

A MULTITUDE OF STATES

Germany today is made up of 16 states – eleven in the West and five in former East Germany. The country occupies 356 910 km^2 (137 771 sq miles) in Central Europe. For the most part, it lacks natural borders – possibly one of the reasons why Germany has for centuries been the European battlefield. In its 1000-year-long history, Germany was almost always split politically – even during periods of so-called political 'union'.

German history began around 911-19, when the East Frank kingdom was for the first time referred to as the '*Reich* of the Germans'. Soon after the German tribes were united under Henry, the Saxon king, and the first German emperor was crowned, inner strife erupted. While the German emperors – coveting the title of Holy Roman Emperor – came into conflict with the popes in Italy, the actual power in the north was held by the knighthood. After the 16th century, individual towns, which had become rich through trade, also laid claim to power, and finally established themselves, after much fighting, as 'free cities'.

The year 1517 – when Martin Luther published his 95 theses against the malpractices of the Roman Catholic Church – marks the beginning of the second epoch in German history. The reformation and the counter-reformation caused a rift in the church which had until then united the German states. Political conflicts were from then on aggravated by denominational difference.

The Thirty Years' War (1618-48) followed, which began as a religious war, but soon grew into a power struggle between the major European powers. In the end, the German emperor (*Kaiser*) lost large chunks of his *Reich*, as well as political power to the German dukes. Consequently, during the 17th and 18th centuries the country consisted of no fewer than 1700 small duchies, earldoms, free cities and principalities. Only Austria and PRUSSIA remained outside this mosaic of statelets and developed into great European powers.

In fact, the competition between these two continued to shape German politics well into the mid-19th century. In 1866, Prussia gained a decisive victory over the Austro-Hungarian empire. The Prussian-dominated Second German Reich was founded in 1871.

The 120-odd years since then have put the German nation through many diverse state systems. Their *Kaiser* first ruled over a confederation of monarchies and republics. After the First World War, there was the Weimar Republic, a parliamentary democracy which was supplanted in 1933 by the Third Reich and the Nazis. In 1949, the country was split in two – the socialist German Democratic Republic (GDR) in the east, and the parliamentary democracy of the Federal German Republic in the west – to be reunified in 1990.

Since Germany has always been divided into so many little states, a federal republic today remains the most workable form of government. Indeed, all through its history, the country has been something of a federation. However, unlike Switzerland where each canton has its own legal system, the German government does not grant its federated states a lot of freedom, and federal law overrides provincial law.

FRIESIANS, SAXONS, HESSIANS – AND JAPANESE

The Second and the Third Reich ended in humiliating defeats which cost the German nation large tracts of land. Today, the reunited Federal Republic of Germany covers an area which is only about two-thirds of its prewar size. Germany is slightly smaller than Japan, and its population is about two-thirds that of Japan. Both Japan and Germany are highly industrialised, and are competing for markets the world over. In fact, similarities between the two are such that the Japanese have been called the 'Asian Prussians', with reference to the perception of their being regimented, organised, martial and bureaucratic, as well as to their blind obedience and incorruptible sense of duty.

Prussia was once a melting pot of many different peoples, with their own dialects, mentality and traditions. They were the Selesians, Friesians, Saxons, Hessians, Westphalians and many more. Even today, the German nation is a mix of peoples. indeed, the German dialects differ so vastly that people from different *Bundesländer* (states) sometimes have difficulty communicating.

The cuisine and customs of the various states also show marked differences. While COLOGNE and MAINZ celebrate the carnival, North Friesians welcome spring with a festival of bonfires. Celebrations are more serene in the Saxon Ore mountains on *Hutzen* evenings, or during the pre-Christmas period in Vogtland nearby.

Many festivals are rooted in religious traditions and still reflect the split in the Church. Most of the north is Protestant; the south is Catholic; an influx of foreigners into the country during recent years has, however, introduced foreign religions.

Germany with its 81 million people has an average population density of 226 people per km^2(586 per sq mile). Some regions, however, are fairly sparsely populated, such as the LÜNEBURGER HEIDE or the EIFEL, which average 40 to 80 people per km^2 (100 to 200 per sq mile). At the opposite end of the scale, there are two regions with up to 1000 people per km^2 (2500 per sq mile). One of these stretches from the Dutch border, along the Rhine, and south

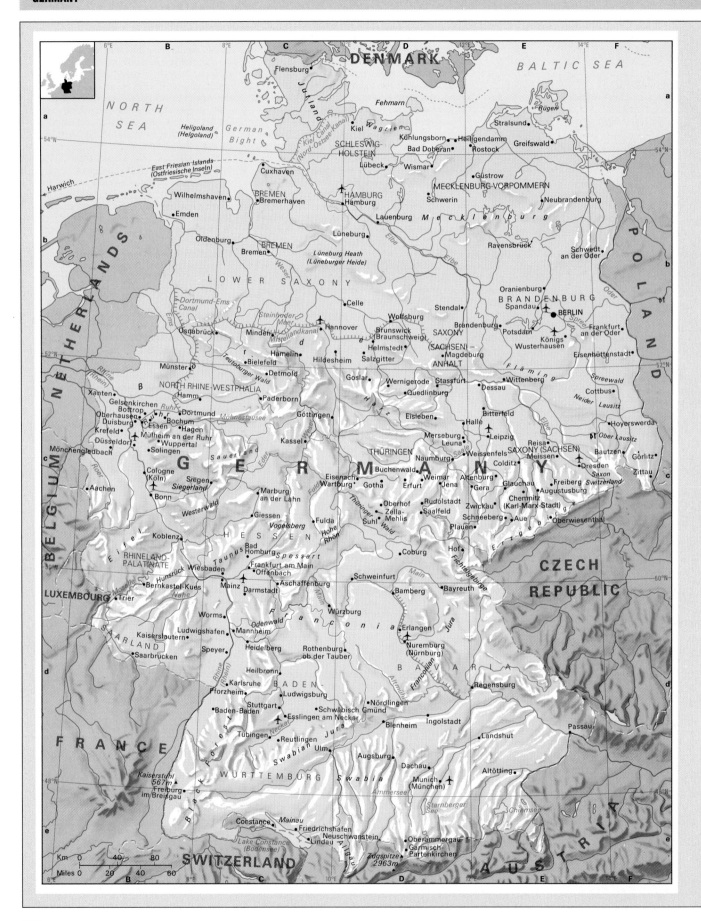

to FREIBURG; the other runs along the northern edge of the Mittelgebirge and then across to Thüringen and Saxony. Both cross in the RUHR, a conurbation of many cities crowded together in a small space.

The three largest cities of Germany, Berlin, HAMBURG and MUNICH, are not part of these densely populated regions, however. They are insular pockets of conurbation surrounded by sparsely populated agricultural land. About five out of six Germans live in urban centres.

Most German towns have a long tradition, some with histories going back as far as Roman times or even to Celtic times. The oldest German cities are Cologne, Trier, Augsburg and Kempten. Bonn was 1000 years old when the capital, Berlin, was founded.

CULTURAL CENTRES

The 'nation of poets and thinkers', as the Germans like to think of themselves, has a multifaceted culture which has absorbed aspects of virtually every other European culture. Independence in cultural and educational matters, however, is unique to Germany, where each *Bundesland* decides on all matters pertaining to culture and education without any interference from the federal government.

The cultural independence of states, though enriching, also has its disadvantages. Families who move from one state to another, for instance, face many problems because the educational systems differ significantly, and not all previous school courses are credited. Professional people, too, may face education and training-related discrimination when looking for work in another state.

Germany has never been dominated by one cultural metropolis, unlike France, whose cultural focus is clearly Paris, and England, which has been dominated by London. A few German cities have therefore grown up to be cultural centres, and these are scattered all over the country. Most of them were founded in the Middle Ages, when German princes moved from one residence to another, taking with them their entourage, including musicians and artists. Among their princely centres are cities such as FRANKFURT am Main, AACHEN, Magdeburg, Mainz, Cologne and Nuremberg.

A second generation of German cultural centres developed in the 16th to 19th centuries, when the *Reich* consisted of many small countries, and the ruler of each was ambitious to build his own residence complete with theatres, libraries and museums. Residential towns dating from this period are Weimar and Würzburg. Today, most large German cities have several universities, research institutes, and colleges.

AN AGEING POPULATION

Every year, about 2 to 3 million Germans move from one *Bundesland* to another; before reunification, there were an additional 6 million immigrants from East Germany and other Eastern European countries. Bavaria, Schleswig-Holstein and Baden-Württenberg have experienced a significant increase in population over the past few years; other *Bundesländer* have shown a decrease.

During the 1980s, the population figures in both East and West Germany seemed to be declining and there was a low birthrate.

However, German-speaking immigrants from Eastern Europe have raised the population figures. In the long term, the demographers expect the country's population to decrease to 50 million people by the year 2030. Overall, the German population is ageing – a problem which the government has to address urgently.

Lack of labour power led the Germans to import migrant workers during the 'economic miracle' of the 1950s. The percentage of foreigners was about 1.5 per cent in East Germany at the time of reunification – but almost 8 per cent of the West German population were foreigners, and West Berlin had a record of 12 per cent non-German native speakers. By far the largest group of foreigners were Turks, with 1.5 million of them living in Germany. Other groups are Italians and emigrants from former Yugoslavia.

FROM RÜGEN TO THE ALPS

Germany is also becoming ever more popular as a holiday destination, and rows upon rows of cars with foreign registration plates are using the autobahns. Each year, tens of thousands of visitors come from the United States, Japan and elsewhere to see the cultural and historical sights such as the dome in Cologne, the house where Johann Goethe was born in Frankfurt, the castle in Heidelberg and

▼ **MAD KING'S DREAM The castle of Neuschwanstein perches on a lofty crag above the Lech Valley in the foothills of the Bavarian Alps. It was built for King Ludwig II of Bavaria.**

▲ **A COUNTRY REUNITED** Berlin's famous Brandenburg Gate, long the symbol of a divided city, was the scene of exuberant celebration in October 1990 as the Berlin wall tumbled.

also the Baltic hinterland, with countless lakes and attractive cities such as Lübeck, Schwerin, Wismar, Stralsund and Greifswald. Weimar, Dresden and Potsdam of former East Germany have now been added to the list. The autoroutes through Germany usually bypass the large conurbations and cross, instead, the country-side strewn with picturesque villages.

Despite unfavourable, cool weather conditions, the coastal regions, from East Friesia to Heligoland (Helgoland), and the marshy landscapes around Dithmarschen and Eider-stedt right up to the North Friesian Islands, receive their fair share of tourists each year.

Another popular attraction is the wine-growing region in the south-west, where romantic villages, a glass of the good wine, and sunny days encourage a stay-over. The central mountain ranges still have landscapes that come closest to what the whole of Germany must once have been – though there is some deforestation here and the acid rains have left their mark. Many holidaymakers come here to hike in the mountains or to ski down the mountain slopes.

However, the southernmost regions – the Chiemsee, Starnberger See and Lake Constance – attract the greatest number of tourists. Nowhere else in Germany do anglers, yachtspeople, wind surfers, mountaineers and ski enthusiasts meet on so small a land area.

AGRICULTURE, MINING AND INDUSTRIES

More than three-quarters of the land area is cultivated or forested. The remainder is mostly taken up by urban centres and by the many transport routes that crisscross the country. Nature reserves and wasteland account for the remaining two per cent.

Agriculture contributes only 2 per cent to the national budget. Many farming concerns (mostly family businesses) are dependent on state subsidies – one reason being that they have to compete against the highly mechanised 'agricultural factories'.

Germany meets most (about 75 per cent) of its food needs, and besides, produces more sugar and butter than it can consume. Surpluses are exported, and Germany is the fifth largest exporter of agricultural products in the world. However, some of the surplus sugar and butter piles up, costing the government money instead of earning revenue. Germany is also the second largest importer of food, after the United States, in the world. Its imports include exotic fruit and other delicacies from all over the world.

The country also imports minerals. Though much of its industry, especially in the Ruhr, was based on coal mining during the 19th century, Germany today has few mineral deposits besides coal and lignite, rock salt and potash, and some natural gas and oil. Most of its coal close to the surface has been mined, and shaft mines make the extracted coal ever more expensive, the deeper the shafts. In Westphalia, where coal is still mined, the mining is done at a world record depth of 1417 m (4649 ft) – in many other places in the world, the mineral is extracted in open-cast mines, and so it is cheaper for the Germans to import the raw material.

Germany is still the world's largest miner of lignite, together with the United States. Indeed, the country takes first place in consumption of lignite. Its four large deposits, at Cologne-Aachen, Helmstedt, in the Halle-Leipzig region and at Niederlausitz, are extensive enough to yield lignite for another few centuries, at the present rate of extraction. However, lignite has a high sulphur content, and was one of the main contributors to pollution in former East Germany, since it was the East Germans' chief source of fuel.

The central mountain ranges, such as the HARZ and the Ore mountains, are rich in minerals and have been deeply carved and excavated by human hands. Today, most of the deposits are exhausted and ore mining in this region has come to a gradual halt. What has taken its place, instead, is mining of gravel and building sand for road construction and the building industry.

▼ **HISTORIC PORT Beside St Pauli Quay, a naval sail training vessel is a reminder that Hamburg's greatness as a port goes back to the 12th century. The stone Pegelturm once recorded the tidal levels of the Elbe river.**

Other major industries include the motor vehicle industry, machinery, electronics, chemicals, optics, and processed foods. Germany is well known for its industrial products, and they are exported worldwide.

INTERNATIONAL POLITICS

In the post-Second World War years, West Germany experienced an economic upswing which was likened to a miracle. It is doubtful, though, whether this kind of economic boom can be repeated for East Germany in the years after reunification. The world is a different place. Simply in terms of manufacturing competitors, the rising Asian powers are a major new factor. Indications are that for a transition period lasting several years, Germany will continue to be split economically. Although the western part of the country will continue to have many companies that do extremely well on the global market, the enterprises of former East Germany have recently lost their guaranteed markets in Eastern Europe and will have to find markets on the world market, when many branches of industry are unable to compete internationally.

But the west, too, will not be able to maintain its phenomenal growth. Asian and South American countries – as well as the newly free Eastern European countries, such as Poland – with cheaper labour have begun competing with the Germans internationally and have made heavy inroads on the German share of the global market.

At home, unemployment is rife (at the end of 1993, there were more than one and a half million unemployed) and the once well-known steel and shipbuilding industries cannot survive without state subsidies.

Other areas that have been hit by the economic slump are the automobile industry, mechanical engineering, as well as chemical

and electrotechnical industries; in the past, they were the mainstay of the German economic miracle.

On the whole, however, Germany is still earning far more in exports than it is spending in imports. A lot of money leaves the country with migrant workers sending their savings home and with Germans themselves travelling abroad, spending billions of Deutschmark outside their country.

With the emergence of the European Union (EU), expectations that Germany would play a key role are now being realised. Only a few years ago the prospect of a German dominating Europe would have sent shock waves around the globe. It is significant that with the re-election in 1994 of German Chancellor Helmut Kohl, Europeans actually started looking towards Germany to take the lead in negotiating and defining the final EU.

Kohl succeeded in reuniting Germany through vigorous negotiation with the victorious Second World War Allies – which, with the exception of the United States, all had reservations about the creation of the new federal republic. Kohl's often expounded principle 'What's good for Europe is good for Germany, and what's good for Germany is good for Europe', appears to have allayed European fears. It seems that the Chancellor's political ambition to go down in history as the man who united Europe is within reach, and that he has ensured for Germany an ever-larger role on the world political stage.

At the insistence of other UN countries, the German Constitutional Court in May 1994 ruled that German troops may participate in foreign peace-keeping operations. This was a major departure from policies designed to restrict German militarisation that had been followed since the Second World War .

GERMANY AT A GLANCE	
Official name Federal Republic of Germany	
Area 356 910 km² (137 771 sq miles)	
Population 80 767 590 **Per km²** 226	
(Per sq mile 586)	
Capital Berlin	
Government Federal republic	
Currency 1 Deutschmark = 100 pfennig	
Languages German (official); some Danish in the north of Schleswig-Holstein; Sorbisch and other European languages	
Religions Christian (45% Protestant, 37% Roman Catholic), Muslim 2%, some Greek Orthodox, Jewish and Buddhist minorities, and approximately 15% unaffiliated	
Climate Cool temperate, moist climate, maritime in the west and continental further east. Average temperature in Berlin is –1°C (30°F) in January, and 19°C (66°F) in July	
Land use Cultivation 35%, forest and woodlands 30%, meadows and pastures 16%, other 19%	
Main exports Motor vehicles, machinery, machine tools, electronics, chemical products, precision mechanics, optics, textiles, wines, processed food; coal, lignite, rock salt, potash, iron ore, copper, nickel, oil, natural gas	
Annual income per head (US$) 20 500	
Population growth (per thous/yr) 4	
Life expectancy (yrs) Male 73 **Female** 79	

Gezhouba *China* Site of the country's largest dam, on the Chang Jiang River. It took 18 years to complete, from 1970 to 1988. Today, the dam supplies 14 000 million kilowatt-hours of electricity a year.
Map (Chang Jiang) China – Ed-Ee

Ghadames *Libya* Oasis town in the west of the country, 760 km (472 miles) south-west of Tripoli. It has been the crossroads of ancient caravan routes across the desert since pre-Roman times. Also known as 'the pearl of the desert', Ghadames is one of the oldest Berber settlements in the Sahara. It was captured by the Romans in 20 BC, when it became known as *Cydamis*. Later, it became a citadel of Islam, and was closed to outsiders until the present century.

The oasis, which is famed for its hot springs, produces apricots, dates and figs. Ghadames has an airfield.
Population 52 200
Map Libya – Aa

Ghana See p 266

Gharb *Morocco* See RHARB

Ghardaia *Algeria* The capital and most impressive of the five towns in M'ZAB. It was settled by the Mozabites, an austere Muslim sect of Berber origin who built this 11th-century town on a hill to make it look like a pyramid; they also dug the wells and planted date palms. It contains a remarkable mosque with an imposing 22 m (72 ft) high minaret; most of the houses and mosques in the town are painted white.

Situated 500 km (310 miles) south of Algiers, Ghardaia has become an important road junction on the Route du HOGGAR across the Sahara. It has an airport, several hotels, and an active, mixed market dealing in produce brought by caravan from all over the Sahara and the deep south, making it an attractive stopping place for tourists. Nearby, at the holy town of Beni Isguen, the gates are closed to strangers while prayers are said from midday to 3:30 pm. (Strangers are also forbidden to stay overnight.)
Population 65 000
Map Algeria – Ba

ghat In India, a mountain pass or mountain chain; in the West Indies, 'ghat' is a term for a ravine.

Ghats *India* Two mountain ranges – the Eastern Ghats and the Western Ghats – in the south of the country. Between the two ranges lies the DECCAN Peninsula.

The Eastern Ghats have an average height of 457-609 m (1500-2000 ft), and run for about 800 km (495 miles) along the south-east and east coast from just south of Cuttack in the north-east to Mount Anai Mudi in the south. Set among them is the Holy Shrine of TIRUPATI.

The Western Ghats, with an average height of 914-1524 (3000-5000 ft), extend for 1287 km (800 miles) along the south-west and west coast from the River Tapti to Mount Kanyakumari.
Map India (Eastern Ghats) – Cd; (Western Ghats) – Bd-Be

Ghawar *Saudi Arabia* The world's largest oil field, lying inland along The GULF, to the west of Qatar. It stretches for 241 km (150 miles) from north to south, and is up to 35 km (22 miles) wide. Ghawar can produce up to several million barrels of oil per day.
Map Saudi Arabia – Bb

Ghazni *Afghanistan* Town and trading centre for wool and fruit, 120 km (74 miles) south-west of Kabul. It became Muslim in the 9th century AD; in 977, it became the capital of the Ghaznavid dynasty. It is renowned for its bazaar and the handicrafts sold there – and also for the delicious Alubekhara plums which grow in the area.
Population 32 000
Map Afghanistan – Bb

Ghent (Gent; Gand) *Belgium* Medieval city 60 km (35 miles) north-west of Brussels. It stands on a number of islands connected by bridges over the Lys and Schelde rivers. The capital of East Flanders Province, Ghent flourished on the wool trade during the 13th and 14th centuries, and still has a remarkable collection of beautiful old buildings, including the moated castle of 's Gravensteen – once the seat of the counts of Flanders, dating from the 9th century and rebuilt in its present form in 1180.

Ghent is the centre of the flower seed and bulb market, and the country's second largest port, linked by the 29 km (18 miles) long Ghent-Terneuzen canal, across Dutch territory to the North Sea. John of Gaunt (Ghent), son of Edward III of England, was born here in 1340.
Population (city) 230 400; (metropolitan area) 1 490 000
Map Belgium – Ba

Giant's Causeway *United Kingdom* Northern Ireland's most famous tourist attraction, situated on the north coast, 80 km (50 miles) north-west of the capital, Belfast. It is a spectacular complex of hexagonal basalt columns, stretching 275 m (900 ft) along the coast and 150 m (500 ft) into the sea. The tallest is the Giant's Organ, 12 m (39 ft) high. The most valuable treasure recovered from the Spanish Armada of 1588 was found here in 1967 on the wreck of the galleon *Girona*. The gold and silver hoard, including coins and jewels, is kept in the Ulster Museum in Belfast.
Map United Kingdom – Bc

gibber An Australian term for a stone or rock, especially one polished by the wind.

gibber plain Type of gravel plain in arid parts of Australia.

gibli (ghibli) Local name for the hot southerly sirocco blowing across Libya and Tunisia.

Gibraltar See p 265

Gibraltar, Strait of Strategically important strait linking the MEDITERRANEAN SEA with the ATLANTIC OCEAN. At its narrowest point it is a mere 14 km (9 miles) wide, at its widest, 43 km (27 miles). It is 58 km (36 miles) long. A strong surface current, about 80 m (260 ft) deep, flows through it from the Atlantic into the Mediterranean, replacing water lost by evaporation. At greater depths a counter-current of dense, salty water flows outwards from the Mediterranean. During the Second World War, German U-boats used this undercurrent to carry them noiselessly past the Allied blockade into the Atlantic.
Map Spain – Ce

Gibson Desert *Australia* Mainly arid area in central Western Australia, extending about 400 km (250 miles) from north to south and 800 km (500 miles) from east to west and covering 320 000 km² (123 523 sq miles). It includes the Gibson Desert Nature Reserve, which covers 186 000 km² (71 798 sq miles). The terrain consists largely of parallel gravelly ridges up to 15 m (59 ft) high. Annual rainfall is approximately 150 mm (5.9 in) and there is sparse grass and scrubby vegetation.

The desert, which is virtually unpopulated, was crossed by the explorer Ernest Giles in 1874, and named by him after a member of the expedition who died on the crossing.
Map Australia – Cc

Giessen *Germany* University town on the Lahn River, 50 km (30 miles) north of FRANKFURT am Main. It manufactures precision instruments.
Population 73 600
Map Germany – Cc

Gifu *Japan* Castle town and regional capital on central HONSHU Island, 270 km (170 miles) west of Tokyo. It is a well-known manufacturer of paper parasols and lanterns. It is also popular for its night-time spectacle of fishing with cormorants on the Nagara River.
Population 410 000
Map Japan – Cc

Gijón *Spain* Seaport and industrial town on the Bay of Biscay, producing machinery and hardware, chemicals, cement, glass, and pottery. It also exports coal. Much of the town was destroyed during the Spanish Civil War (1936-9) and has subsequently been rebuilt. Few of its original buildings remain.
Population 259 070
Map Spain – Ca

Gilbert Islands See KIRIBATI

Gilgit *Pakistan* District, river and town in the far north of the country, at the centre of some of the highest and most beautiful mountain scenery on earth. The Silk Road, a trade route used by the Venetian adventurer Marco Polo on his way to China in the 13th century, ran through the area. The Karakoram Highway into western China passes here, winding among peaks of more than 8000 m (26 250 ft).
Map Pakistan – Da

gill (ghyll) A swift-flowing mountain stream; also, a ravine.

Gillingham *United Kingdom* See MEDWAY TOWNS

Gimcheag *North Korea* See KIMCH'AEK

Gippsland *Australia* Region of south-eastern VICTORIA, stretching from Melbourne to the border with New South Wales. Brown coal is mined by the open-cast method in its Latrobe Valley and used for power generation in the region. There are major oil fields in Bass Strait, to the south. Cattle are raised and crops – mainly corn, oats and sugar beet – are grown. Along the south-east coast is the Ninety Mile Beach with, behind it, an extensive lake system.
Map Australia – Hf

Gibraltar

THE STRATEGIC ROCK BASTION IN THE STRAIT BETWEEN EUROPE AND AFRICA IS ONE OF THE REMNANTS OF BRITISH COLONIAL POWER

On the south coast of Spain a sheer limestone rock rises to 425 m (1394 ft) from a promontory overlooking the entrance to the Mediterranean Sea. The promontory is the British Crown Colony of Gibraltar, a strategic naval and air base and communications centre for monitoring naval traffic .

Since the ancient Greeks settled here, Gibraltar has seen many invasions. In 711, the Moors who arrived from North Africa gave the rock the name Jebel Tariq, ('the Rock of Tariq') from which the name Gibraltar is derived. The peninsula remained under the Moors almost continuously until 1462, when the Spanish drove them out. It was captured by the British and Dutch in 1704, during the War of the Spanish Succession, and was recognised as a British colony in 1713 by the Treaty of Utrecht. Over the years it has been besieged by the Spanish and the French. Today, Spain still lays claim to it.

In a United Nations referendum in 1967 all but 44 of more than 12 000 voters stated that they would rather be ruled by Britain than by Spain. Britain's respect for the wishes of the Gibraltarians about their nationality is written into the country's constitution.

In 1969 General Francisco Franco, the Spanish head of state, closed the border between Spain and Gibraltar, so that visitors from one country to the other had to travel via Tangier in Morocco, and planes to Gibraltar had to make a tricky landing to avoid Spanish airspace. In 1982, Spanish and Gibraltarians pedestrians were allowed to walk through the border. Then, in 1985, as Spain prepared to join the European Economic Community (later the EC, and then the EU), the road was opened completely. Many tourists now pass through both ways, although sometimes subjected to deliberate delays, but especially into Gibraltar where duty-free goods are available.

But some Gibraltarians are worried about the effects of the new agreement between Britain and Spain which will allow Spanish workers, previously excluded from Gibraltar, to find jobs here. Unemployment in Gibraltar is about 5 per cent; in Andalucia, across the border, it is 30 per cent.

Gibraltarians are a blend of Moorish, British, Maltese, Asian, Genoese and Spanish descent, and they speak both English and Spanish. The majority are Roman Catholic. Since Gibraltar has no agriculture or mineral resources, most of the inhabitants depend for their living on the port, dockyards and NATO bases. Although the British naval presence in Gibraltar is much reduced from its peak before the Second World War, the Strait of Gibraltar is one of the world's busiest waterways, with a ship passing every six minutes.

The Rock of Tariq is well known for its colony of wild barbary apes – which has grown from only five in 1923 to 53 in 1985. Some see the opening of the border as the first stage in a gradual handover of Gibraltar to the Spanish – but folklore relates that the British will stay in Gibraltar as long as the apes do.

GIBRALTAR AT A GLANCE	
Map Spain – Cd	
Area 6.5 km² (2.5 sq miles)	
Population 31 508 **per km²** 4847 **(Per sq mile** 12 603)	
Capital Gibraltar Town	
Government Self-governing British crown colony	
Currency 1Gibraltar pound = 100 new pence	
Languages English, Spanish	
Religion Christian (74% Roman Catholic, Protestant 11%), Muslim 8%, Jewish 2%	
Climate Warm temperate; westerly winter winds bring rain. Average temperature ranges from 12°C (54°F) in February to 24°C (75°F) in August	
Land use Cultivation 0%, pastures 0%, forests 0%, other 100%	
Main primary products Negligible	
Major industries Ship repairs, ship bunkering, beverages and food processing, tourism, and service industries	
Main exports Re-exports of petroleum products, tobacco and other manufactures, wines and spirits	
Annual income per head (US$) 9500	
Population growth (per thous/yr) 5	
Life expectancy (yrs) Male 76 **Female** 79	

Gir Forest *India* Game reserve 40 km (25 miles) north of the western seaport of Diu on the Gulf of Khambhat. It is the main home of the Indian lion, the only Asian lion species.
Map India – Ac

Girna *India* Sacred mountain about 450 km (280 miles) north-west of Bombay. It is 1118 m (3668 ft) high and contains several temples, including the Amba Mata temple where newly wed couples pray for their marriage to be blessed. Rock-carved decrees issued during the rule of Asoka (from 272 BC), a Buddhist emperor of India, can also be seen here.
Map India – Ac

Gironde *France* Estuary, 10 km (6 miles) wide near its mouth and 75 km (47 miles) long, into which the DORDOGNE and GARONNE rivers empty. It extends from the junction of the two rivers, 20 km (12 miles) north of BORDEAUX, to the sea at the Bay of Biscay.

Covering an area of 10 000 km² (3860 sq miles), the largest department in France, in the Aquitaine region, is also called Gironde. It contains many of France's best known winelands, such as Bordeaux. Two-thirds of the department's population live in the Bordeaux metropolitan area. The main industries in the region are electricity, chemical works and textiles.
Population 1 213 000
Map France – Cd

Gisborne *New Zealand* Seaport and resort on Poverty Bay, about 320 km (200 miles) south-east of Auckland. It is a centre for tourists wishing to explore the surrounding area, of forested mountains, which rise to 1754 m (5754 ft), at Mount Hikurangi. The region, also called Gisborne, exports meat, dairy produce, and wool produced locally.
Population (town) 31 480; (region) 44 390
Map New Zealand – Gc

Giuba *Ethiopia/Somalia* See JUBBA

Giulie, Alpi *Slovenia* See JULIAN ALPS

Gizycko *Poland* See MASURIA

Gjirokastër (Argyrokastron) *Albania* Picturesque town hugging the slopes of a mountain on which towers a fortress built by the Ottoman Turks under Ali Pasha in 1811. It lies at 350 m (1148 ft) in Albania's southern highlands and is linked by road with Yannina, in Greece. The town has a mixed population of Albanians and ethnic Greeks.

Gjirokastër was the scene of a Greek victory over the Italians in 1940-1, during the Second World War.
Population 23 800
Map Albania – Cb

Glace Bay *Canada* Coal-mining town, on the Atlantic coast of Nova Scotia, where the Italian inventor Marconi set up his first transatlantic wireless service on 17 October 1907.
Population 19 500
Map Canada – Jd

glacial lake A body of water contained by the walls of a valley and the margin of a glacier. Also, a lake formed in a depression cut by ice or held up by deposits of MORAINE left by ice. See HOW GLACIERS CHANGED THE LANDSCAPE, p 269

glaciation Term referring to the formation, movement and retreat of glaciers and ice sheets, as well as to the overall effects on a landscape – both erosion and deposition of debris – brought about by glacial action. See HOW GLACIERS CHANGE THE LANDSCAPE, p 269

glacier A huge mass of ice, originating from compacted snow, forming FIRN, moving slowly away from the zone of accumulation towards an area of lower ground. A glacier can also be an ice sheet which has spread out from a central mass and covers a large part of an entire continent – as, for example, in Antarctica. See HOW GLACIERS CHANGE THE LANDSCAPE, p.269

Glacier *Canada* National park with magnificent glacial and alpine scenery, covering 1350 km² (521 sq miles) in the Selkirk Mountains of southern British Columbia.
Map Canada – Dc

Glacier National Park *USA* Area of jagged peaks, ice fields, glaciers and waterfalls in the Rocky Mountains in north-western Montana, on the Canadian border. The park covers 4101 km² (1583 sq miles). Its U-shaped valleys and most of the lakes were formed during the last Ice Age.
Map United States – Da

265

Ghana

COCOA IS STILL THE BIGGEST SOURCE OF INCOME FOR THE FORMER GOLD COAST

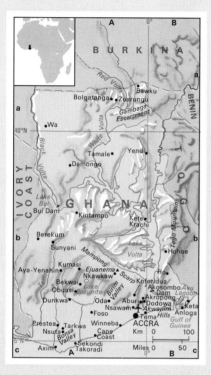

I n 1957, when Ghana became the first black African state to achieve independence from British colonial rule, waves of excitement rippled through the entire continent. The atmosphere in the newly independent state was electric. Colourful celebrations in the towns and in the countryside, singing, drumming and dancing expressed the ecstasy of a people who had lived under foreign colonial rule for decades.

As an additional boost to morale, the country's name was changed. The European appellation 'Gold Coast' had always recalled a history painful to the indigenous people. Trade in gold with the Portuguese in the 15th century had developed by the 17th century into the slave trade, involving – besides the Portuguese – the Dutch, the British, the Danish, the French and the Spanish.

The new name, Ghana, was borrowed from the great African empire which flourished in Mali and Senegal between 700 and 1100, and was intended to inspire a comparable greatness in the new state on the southern coast of West Africa.

But the euphoria and the expectation that Ghana would become a prosperous industrialised nation soon turned to disappointment. The new government's control was undermined by opposition from traditional chiefs and big farmers, while repressive laws passed

to contain the opposition tarnished the idealistic image of the ruling Convention People's Party. Its leader, Kwame Nkrumah, kept out of the fray, and continued to inspire other nationalist movements abroad with his denunciations of imperialism. But money was spent on prestige projects which ignored the country's real needs, debts to the West went unpaid, and Nkrumah was deposed by a military coup in 1966. He was overthrown partly because of his dictatorial and self-indulgent style of government – in 1966, for instance, he expanded his personal guard into a regiment. Nkrumah went into exile in Guinea and died in 1972.

A series of coups followed, which swung Ghana from one political extreme to another, and from long periods of military rule interrupted by short stretches of civilian rule – and, through them all, corruption and mismanagement damaged the economy. The production of cocoa, the country's main export product, declined and the unfavourable economic climate abroad made things worse. From 1983, a successful economic recovery programme began to turn the downslide around. Inflation came down, real income per capita rose, budgets began to show surpluses, and the country began meeting its foreign debt obligations. However, Ghana remains heavily reliant on foreign aid and government measures to stimulate the production of cocoa, and other crops have not yet produced the desired results. In contrast, the mining sector has increased investment and output.

A new democratic constitution was approved by referendum in April 1992; the ban on political parties was lifted in May that year. The first multiparty elections since 1979 were held in late 1992 and saw the former military ruler, President Jerry Rawlings, who stood as a candidate, continuing in office. The Fourth Republic of Ghana was inaugurated on ·7 January 1993.

WHITE BEACHES AND BLUE LAGOONS

Visitors to the country will find many attractions, including palm-fringed beaches of white sand along the Gulf of GUINEA in the south. There are peaceful blue lagoons where the great VOLTA River and the smaller rivers to the west of it meet the sea.

Much of the coastal region is farmed. Around Keta Lagoon, in the east, shallots are intensively cultivated. On the flat coast farther west, near Elmina and around Cape Coast and Apam, the locals build shallow pools with low mud walls, which trap the sea water at high tide. When the water evaporates, it leaves a crust of salt, for which there is a ready market.

The coast is humid, its stickiness relieved slightly by the dry and dusty harmattan wind blowing from the north in December and January. The beaches reach northwards to steaming tropical evergreen forests of hardwoods and silk-cotton trees, so called because their fruit, the kapok, yields a silky fibre used for stuffing cushions and upholstered furniture.

The steaming forests with their brilliant birds and flowers and winding creepers give way to deciduous woodland farther north, where the climate is hotter and drier. In this

region, parts of the original forest were chopped down by itinerant farmers who cultivated a piece of land until its fertility was exhausted, then moved on to clear another piece of forest. The areas thus cleared are now being turned into garden areas where cocoa and palm oil are produced. In the west, original rain forest on the Ashanti Plateau has been cleared to grow cocoa; in the east, more areas of the old forests have been destroyed by fires in periods of drought.

Still farther north, woodland gives way to the tropical savannah of the upper Volta basin. This is the home of antelopes and buffaloes, elephants, lions and wild hogs. At the northern end of Lake VOLTA, the landscape becomes harsh and barren; near the northern frontier with Burkina (once Upper Volta), it becomes an arid plain relieved by shea trees, whose nuts yield an edible yellow fat, and acacias and baobabs with their swollen water-storing trunks. The nuts, gums and resins of these trees are collected and sold by the local people. But in some parts of this country almost all trees have been chopped for fuel. Billowing clouds of dust are stirred up by herds of cattle, sheep and goats on the move in search of pasture. The land is overgrazed and soil erosion is a growing problem. In the north-east, years of drought have added to agricultural problems. The crops that people live on, as well as those they might sell – including millet, sorghum and groundnuts – have given poor harvests.

LIFE IN THE VILLAGE

About two-thirds of Ghanaians live in villages. In the north, their homes tend to be round huts with flat or conical roofs made of mud, sticks or any other available local materials. The huts of an extended family are grouped around a yard and fenced or walled. The yard is used for keeping poultry, storing grain and tethering animals overnight. Ancestor worship is still practised in these villages, involving ceremonies for the departed spirits of the dead.

In the south are the tribal lands of the Akan and the Fante who once traded with the Europeans in slaves captured from other tribes. Here, too, are the Ashanti, whose kingdom dominated the area in the 17th century. The dwellings in this region are more sophisticated than those in the north, often rectangular with gable roofs of corrugated iron or, near the coast, of split bamboo and palm leaves. As in the north, family homes are grouped together around a courtyard.

The ethnic groups of the north, which include the Gonja, Wala, Dagomba and Mamprussi, adhere most closely to their traditional way of life. They were the least exposed to European influence, as the colonists virtually ignored the area. As a result, the level of economic development in southern Ghana is far higher than in the north, where poverty is rife. In early 1994, the north was the scene of clashes between the Dagombas, Nanumbas and Gonjas on the one side and the Konkombas on the other. The Kokombas have been spreading west and south from neighbouring Togo into Ghana and have laid claims to land traditionally owned by the local tribes.

The British promoted cocoa, rubber, palm oil and coffee production in the southern forests. These crops are still grown, side by side with food for the local people, so almost anywhere in the forest zone, women (rarely men) can still be seen planting, hoeing or weeding plots of yams, cassava, maize and vegetables under cocoa, rubber or palm trees.

Twenty years ago, Ghana's cocoa crop was the biggest in the world. But drought and bush fires hit production and at the same time prices in the world market fell. Other factors today discourage cocoa growers. The roads badly need repair, and the harvested crop sometimes does not reach a sales depot for many months – and even when it does, payment is extremely slow. For all these reasons, producers have been reluctant to replant cocoa, preferring to grow food crops such as maize and cassava which can be sold locally. Ghana now produces about one-tenth of the world's cocoa,

and the government is trying to redevelop the country's ailing industry to boost its export earnings. Cocoa today accounts for 45 per cent of all Ghanaian exports.

Ghana has important sources of wealth in manganese and bauxite, as well as gold and diamonds, which were mined even before the colonial era. Gold accounts for 17 per cent of export earnings. The Ashanti Goldfields are among the largest and most productive in the world. In the BONSA VALLEY the local people, pan for diamonds in the gravels deposited by the river. There is bauxite (aluminium ore) mining at Kibi; the metal supplies the aluminium smelter at TEMA.

Ghana's most ambitious project since independence was Lake Volta, made by damming the Volta River. It is one of the largest artificial lakes in the world, covering 8480 km² (3251 sq miles), or 3.5 per cent of Ghanaian land. The project involved rehousing 76 000

gather noisily to board the ferries. And in the crush on board, as the boat sails past the canoes of silent fishermen, the passengers converse noisily – invariably about politics.

The rate of urbanisation is alarming. Most of Ghana's towns lie in the south, the majority of them on the coast. They combine wealth and poverty. Relics of the colonial era remain – two-storey buildings with broad verandahs at the front and back, supported by tall wooden or stone pillars. But in cities such as ACCRA, dramatically rapid growth has turned the outskirts into an unplanned sprawl.

All the Ghanaian towns and cities have market days – cheerful and noisy occasions when business meets with pleasure. In the larger towns the markets sell anything from vegetables to love potions and elaborate medicines based on the folklore of primitive indigenous beliefs. Dominating the market are the formidable 'market mamas' – women whose business acumen is rarely equalled, and whose political clout at local, regional and, in some cases, national level is considerable.

None of the varied governments of Ghana's chequered political career since independence have managed to come to grips with Ghana's social and economic problems. In the early 1990s, under the regime of President JJ (formerly 'Flight Lieutenant Jerry') Rawlings, concerted efforts were made to restore the responsibility for local development to the local people. In 1983, Ghana started implementing an economic rebuilding programme, which included some privatisation and less stringent government control. Some progress was made, but with too much reliance still on its raw materials, and with the cocoa production in trouble, the country's economy needed to make much greater headway to justify the optimism with which Ghana celebrated independence.

▲ ON WITH THE NEW Awnings and brightly coloured umbrellas have replaced the traditional thatched roofs of the stalls in Kumasi market. Most men now wear 'Western' dress, and many women have discarded the 'kente', the wax-painted cloth worn over the shoulder.

people. Now rice is grown successfully around the lake on irrigated land.

Roads and railways are very poor in Ghana, which hinders meaningful economic development. But ferries run to and fro across Lake Volta. Crowds of people, their luggage tied in brightly coloured bundles, their goats, chickens and in some cases their furniture, too,

GHANA AT A GLANCE		
Official name Republic of Ghana		
Area 238 540 km² (920 79 sq miles)		
Population 17 351 000 **Per km²** 73		
(Per sq mile188)		
Capital Accra		
Government Republic		
Currency 1 cedi = 100 pesewas		
Languages English (official), Akan, Ga, Ewe, and other West African languages		
Religion Indigenous beliefs (38%), Christian (42%), Muslim (20%)		
Climate Tropical, with the main rainy season in December/January and August; less rain falls in the north. Average temperature in Accra ranges from 25°C (77°F) to 28°C (82.4°F)		
Land use Forest and woodlands 37%, grazing 15%, cultivation 12%, other 36%		
Main primary products Cassava, taro, maize, yams, bananas, cocoa, sorghum, timber; gold, diamonds, manganese, bauxite		
Major industries Agriculture, bauxite refining, steel, oil refining, food processing, cement, vehicle assembly, forestry, mining		
Main exports Cocoa, palm oil, diamonds, gold, hardwoods, manganese ore, bauxite		
Annual income per head (US$) 450		
Population (per thous/yr) 31		
Life expectancy (yrs) Male 55 **Female** 55		

Gladstone *Australia* Port in Queensland, 140 km (275 miles) north-west of Brisbane. The port is a base for tourists visiting the Great Barrier Reef resort of Heron Island. To the south the coast around Bustard Bay is relatively unspoilt.

Coal mined in the hinterland is the main export, and bauxite (aluminium ore) is brought in to be processed into alumina at the world's largest aluminium plant.

Population 23 460
Map Australia – Ic

Glåma (Glomma) *Norway* The country's longest river, draining an area of 42 000 km² (16 213 sq miles). It rises near the border with Sweden in the province of Sør-Trøndelag. From here, it flows towards the south for 598 km (372 miles), through Østerdalen in the south-east, and then into Oslofjord at Fredrikstad. In its lower reaches, below Lake Øyeren, it is known as the Glomma. Timber is floated down the river and its waters are harnessed for hydroelectricity at various locations.

Map Norway – Cc

Glamorgan *United Kingdom* Former county of south Wales and the most densely settled and industrialised part of Wales. In the 1970s, it was split into three administrative counties. Mid Glamorgan is the biggest, covering 1019 km² (393 sq miles). It takes in what are known as the Valleys, which extend, like the fingers of a hand, north from the coast into the formerly coal-mining flanks of the hills. Names such as the Rhondda and Merthyr Tydfil are synonymous with Welsh mining the world over.

South Glamorgan, covering 416 km² (161 sq miles), centres on the city of CARDIFF and the port of Barry. West Glamorgan, with an area of 815 km² (315 sq miles), includes the industrial complex of SWANSEA, Port Talbot and Neath, and the beautiful Gower peninsula with its beaches and holiday homes.

Population (Mid Glamorgan) 541 800; (South Glamorgan) 408 600; (West Glamorgan) 371 000
Map United Kingdom – De

Glarus *Switzerland* Capital of the canton of the same name on the Linth River, 55 km (34 miles) south-east of ZÜRICH. The town was largely destroyed in 1861 by a fire fanned by a violent FÖHN wind. The canton is an alpine region, rising to 3614 m (11 857 ft). Most of it is forests and 'alps' – high pastures where cattle and sheep are reared. However, in terms of its total population, Glarus is the most heavily industrialised canton in Switzerland. Industries include machinery, electrical works and textiles.

Population (town) 5 700; (canton) 38 500
Map Switzerland – Ba

Glasgow *United Kingdom* Scotland's largest city, lying on the River Clyde, and the centre of the administrative region of Strathclyde. Glasgow started growing in the 12th century shortly after the cathedral was built, in what is now the east end of the city, on the remains of a 6th-century church. From there, the city grew steadily westwards, following the Clyde toward the sea. Eventually, Glasgow grew downstream past Govan and Clydebank with their shipyards, overran the separate burgh of Paisley with its abbey, cathedral and textile mill, and created a built-up area for 25 km (16 miles) along the river.

Glasgow's great prosperity began with 18th-century overseas trade, and shipbuilding soon followed. 'Clyde-built' became a mark of quality for steamships the world over from about 1820; so, too, in the field of rail transport, did the name of Springburn. The stately ocean liners known as the Cunard 'Queens' were built at Clydebank. Since the Second World War these major industries have fallen into a steep decline.

Central Glasgow today is mainly 19th century in appearance, with George Square and the City Chambers at its heart, and its streets laid out in a formal grid pattern. West of the centre are the higher-priced residential areas of the past century and the cultural institutions and facilities that went with them – the University of Glasgow (founded in 1451 but on its present site since 1870), the botanical gardens, the Kelvin Hall exhibition centre, and the art gallery with a fine collection of British and European paintings. A second university, the University of Strathclyde, was founded in 1964.

Population 654 540
Map United Kingdom – Cc

glass Hard, shiny volcanic rock which has cooled too quickly to have a crystalline structure.

Glastonbury *United Kingdom* Town in the county of Somerset in western England, 185 km (115 miles) from London. It stands at the foot of a steep, isolated hill, and is renowned for its Glastonbury Abbey. A small chapel built on the site of the abbey in early times was replaced by a monastic church which burned down in 1184. This was replaced in turn by a huge monastery, which was largely destroyed in 1539. Its chief remains are a fine Lady Chapel and a superb late 14th-century kitchen.

After the late 12th century, two legends grew up around Glastonbury. According to the first, St Joseph of Arimathea, who buried Christ, built a wattle church here and housed in it the Holy Grail – the chalice used by Christ at the Last Supper. The second legend told that King Arthur of the Round Table was buried in the monastery in the 6th century.

Population 6810
Map United Kingdom – De

Glatz *Poland* See KLODZKO

Gleiwitz *Poland* See GLIWICE

Glendalough (Gleann dá Loch) *Ireland* Valley in the Wicklow Mountains, 39 km (24 miles) south of DUBLIN, where the hermit St Kevin established a monastery and seat of learning in the 6th century. Remains include a round tower, a 10th-century cathedral, and St Kevin's Kitchen, an early Irish oratory, now a museum.

Map Ireland – Cb

Glittertind *Norway* The nation's second highest mountain, at 2465 m (8087 ft). It is in the Jotunheimen region in south Norway, 112 km (70 miles) north-west of Lillehammer.

Map Norway – Cc

Gliwice (Gleiwitz) *Poland* Southern manufacturing city 23 km (14 miles) west of Katowice. Founded by German settlers as a trading centre, it soon grew into a wealthy beer-brewing city. Gliwice produces beer, textiles, iron, steel and chemicals. Despite its industrialisation, the old town retains much of its beauty, with medieval walls and many buildings, such as the town hall ('Rathaus') and the Allerheiligen Church.

Population 214 200
Map Poland – Cc

Glomma *Norway* See GLÅMA

Gloucester *United Kingdom* City and county of western England. The city was the Roman town of Glevum. It stands on the River Severn, 150 km (93 miles) west and slightly north of London. Its cathedral, said to be one of the most beautiful buildings in Britain, was founded in the 11th century and completed in the 16th; its cloisters are particularly impressive. Gloucester is one of the venues, with Hereford and Worcester, of the annual Three Choirs Festival.

The county covers 2638 km² (1019 sq miles). It includes the Forest of DEAN on the Welsh border, the beautiful 18th-century spa town of Cheltenham – birthplace of the composer Gustav Holst in 1874 – the COTSWOLD HILLS, and Cirencester, which was the second largest town in Roman-occupied Britain.

Population (county) 539 400; (city) 104 700
Map United Kingdom – De

Gnesen *Poland* See GNIEZNO

Gniew *Poland* Medieval northern town, 62 km (39 miles) south of the port of Gdańsk. Its 13th-century red-brick castle, which still towers over the town, was one of the strongholds of the Order of Teutonic Knights in 1282. Attractive arcaded houses surround the market square.

Map Poland – Cb

Gniezno (Gnesen) *Poland* Ancient town in central Poland, 45 km (28 miles) north-east of the city of POZNAŃ. It was the country's first capital, in the 11th century, but its roots go back further. Remains of an 8th-century fort have been found on the town's Gora Lecha ('Hill of Lech'). According to legend, the fort was founded by Lech, the leader of the Polan tribe which founded the Polish state and gave the country its name. Gniezno gets its name from the word *gniazdo*, meaning 'the nest' – a reference to the white eagle, symbol both of the Polan tribe and of the country Poland. The massive cathedral, standing where a church has stood since 973, contains the remains of St Adalbert (Wojciech), Poland's first patron saint.

Population 70 000
Map Poland – Bb

Goa *India* State on the west coast, covering 3702 km² (1429 sq miles) and lying about 550 km (342 miles) south of Bombay. Together with Daman (on the Gujarat coast, 112 km, or 70 miles, north of Bombay) and the small island of Diu (off Kathiawar peninsula to the north-west), it used to form the Union Territory of India. Goa, Daman and Diu were formerly Portuguese possessions; Goa was seized by the naval officer Alfonso de Albuquerque (later to become the Portuguese East India viceroy) in 1510, Diu in 1534 and Daman in 1559. All three were annexed by India in 1961. Goa became a state in 1987, and the remaining Union Territories of Daman and Diu are administered by the governor of Goa. Today's capital is the port of PANAJI (Panjim)

HOW GLACIERS CHANGE THE LANDSCAPE

A glacier is like a river in slow motion – it flows downhill, it erodes the ground it passes over and carries the eroded material with it, then finally deposits the debris when it melts and its force is spent.

A glacier is a major tool in sculpting the surface of the land. At its head, as it moves downhill from a mountain top, it scoops out a steep-sided amphitheatre-like hollow, or CIRQUE. On some mountains, several glaciers may have carved a group of cirques around the summit, separated from one another by sharp ridges called ARÊTES.

As the glacier moves, it picks up rock fragments and carries them along with it. The fragments act like grit in glasspaper and erode the underlying rocks, gouging them with parallel scratches, or striations. These marks show the course of a glacier long after it has vanished.

The scouring action of a glacier transforms V-shaped river valleys into U-shaped ones by carving away the valley sides and levelling the floors. The ends of spurs between tributary valleys are removed, leaving HANGING VALLEYS with their outlets high above the main valley floor. Hillocks and outcrops of rock over which a glacier has passed are often left with gently sloping sides facing the glacier's advance, and rugged slopes on the lee side where the glacier has plucked away material. They are called roches moutonnées – or, in English, 'sheeplike rocks' – because of their resemblance to recumbent sheep.

Unlike a river, a glacier does not deposit the debris it carries until it melts or recedes. The eroded material, the TILL, ranges from boulders to fine dust. It may be transported many hundreds of kilometres by the glacier. If a glacier has melted gradually and steadily, the till will be spread over a wide area and is called ground MORAINE. But if it has remained stationary for a long time before retreating, the till builds up along its leading edge. When the ice melts, the till is left as a ridge called a terminal moraine.

Glaciers also produce lateral moraines. These are composed of materials that fall or are scraped from the valley sides and build up along the outer edges of a glacier. When two glaciers meet, their two inside lateral moraines merge to form a medial moraine down the centre of the combined glacier.

KAMES – long, low hills of fine sediment – are left where streams issued from the edge of an ice sheet. KETTLES are ice-filled depressions covered by till that remain after an ice sheet has retreated. The ice eventually melts and leaves a bowl-shaped hollow lined with debris. DRUMLINS are small hillocks of fine till, which may rise to 100 m (330 ft). They are shaped like half an egg, with the thick, steep end facing the flow of ice. Long, winding ridges of sand and gravel, known as ESKERS, are deposited by streams flowing beneath an ice sheet.

A valley glacier is not a smooth sheet of ice. Its surface is covered by debris that falls from the sides of the valley down which it moves. Cracks, or crevasses, appear in the surface when the glacier is 'stretched' as it moves over a hummock in the ground or turns a sharp corner. A *bergschrund* is a crevasse formed around the upper rim of the ice within a cirque – the huge weight of the glacier tears the ice away from the steep rock face. The scouring action of the glacier as it moves down the valley leaves many depressions in the ground. After the glacier has melted, some of these may fill with water and form cirque lakes, or tarns, in the mountains, and ribbon lakes where the valley is dammed by terminal moraines.

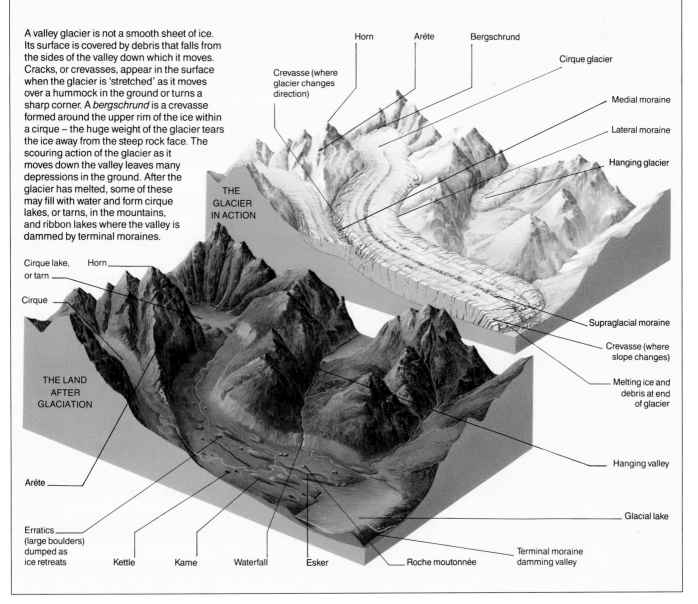

THE GLACIER IN ACTION

Horn — Aréte — Bergschrund — Cirque glacier — Medial moraine — Lateral moraine — Hanging glacier — Crevasse (where glacier changes direction) — Supraglacial moraine — Crevasse (where slope changes) — Melting ice and debris at end of glacier — Hanging valley — Glacial lake — Terminal moraine damming valley — Roche moutonnée

THE LAND AFTER GLACIATION

Cirque lake, or tarn — Horn — Cirque — Aréte — Erratics (large boulders) dumped as ice retreats — Kettle — Kame — Waterfall — Esker

Old Goa, the abandoned former capital, is noted for its Baroque cathedrals. St Francis Xavier, the Apostle of the Indies, who died in 1552, was buried here. But despite four centuries of Portuguese rule, Goa's population is predominantly Hindu.

Goa has now developed into a major tourist destination – partly because of its superb beaches which occur along the entire coastline. Coconut palms line the beaches and coconuts are one of Goa's major export products, as are fruit, spices, manganese and iron ores, bauxite, fish and salt. It manufactures fertilisers, sugar and textiles.

Population 1 168 600
Map India – Bd

Gobi Desert *Mongolia* Wasteland covering 1 295 000 km² (500 000 sq miles) – over half the country's area. It extends south into the Inner Mongolian Autonomous Region of China.

Surrounded by mountains, the desert is between 1000 and 2000 m (3300-6600 ft) above sea level, and ranges from sand in the west to stone in the east. Its climate is one of extraordinary extremes – temperatures can rise to 45°C (113°F) in summer and fall to -40°C (-40°F) in winter. Rain falls in summer but is sparse and never brings more than 150 mm (6.2 in) in one year. Yet there is vegetation which is much like that of a desert steppe. The recent discovery of coal at Tavan-Tolgoyt, in the middle of the Gobi Desert, has encouraged hopes that other minerals may be present.

The main route across the desert, followed by a road and by the Trans-Mongolian Railway, runs south-east from the capital, Ulan Bator. It takes about two days to cross the desert by train.

Map Mongolia – Db

Godavari *India* River, 1465 km (910 miles) long, rising in the Western GHATS near Bombay, forming a constantly growing delta in the Bay of Bengal. The delta is laced with irrigation canals, and the region is becoming prosperous through improved rice farming, although the coastal areas are at risk from cyclones and tidal waves (tsunamis). Lake Kolleru, on the delta between the Godavari and Krishna rivers, is a bird sanctuary. The Godavari is one of the seven sacred rivers of the Hindus.

Map India – Cd

Godthåb (Nuuk) *Greenland* Largest town and capital of Greenland, on the south-west coast. Founded by the Danes in 1721, it is the oldest Danish settlement in the country. Gothåb is the seat of the national council and the supreme court. It has fish canning and freezing plants.

Population 12 700
Map Arctic – Cc

Godwin Austen *China/Pakistan* See K2

Golan Heights *Syria/Israel* High, bare hills rising to 2225 m (7300 ft) in south-west Syria and overlooking the Sea of Galilee and Israel. Because of their location, they are of great military and strategic importance. The western slopes and highest crest were captured by the Israelis during the 1967 Arab-Israeli Six Day War, and were annexed by Israel in 1981.

Although Israel acknowledged Syrian sovereignty of the Golan Heights in 1994 as part of the negotiating process to resolve the Middle-East conflict, the area continued to be the subject of dispute and international negotiation.

Map Syria – Ab

Golconda *India* Ruined town and fortress 8 km (5 miles) north-west of the city of Hyderabad. It stands on a rocky mountain 120 m (393 ft) above the plain and was the capital of the kingdom of Qutb Shahi (1518-1687). Its fortified remains are well preserved and visitors can also see the tombs of the Qutb Shahi kings which lie just outside the fort. The tombs, some of which have their own attached mosques, and surrounding gardens gradually fell into disrepair but were restored in the late 19th century.

Golconda once protected the nearby diamond mines, from which, some claim, came the magnificent Koh-i-noor diamond, though its history is controversial before 1849 when it was acquired by the British and placed with Queen Victoria's crown jewels. In 1937 it was incorporated into the queen's crown for the Coronation of Queen Elizabeth, consort of George VI.

Map India – Cd

Gold Coast *Australia* A series of beach resorts stretching 32 km (20 miles) along the coast of southern Queensland from the New South Wales border to within 60 km (38 miles) of Brisbane. Its main centres include Coolangatta and Surfers Paradise. Developed from a series of small settlements since the late 1940s, it is now Australia's seventh biggest urban region.

Population 274 200
Map Australia – Id

▼ **SILK ROAD Nomads make their way along this ancient trade route below the Altun Shan at the edge of the Taklimakan Desert – an extension of the Gobi in Xinjiang, China.**

Golden Triangle *South-East Asia* Name given to the largely mountainous border area where Thailand, Laos and Myanmar (Burma) meet. It is inhabited mainly by hill people whose chief occupation is growing the opium poppy. The Golden Triangle is the source of about 70 per cent of all opium produced worldwide. Refined morphine, which is also manufactured here, is controlled by Chinese and Shan warlords who run private armies to protect their interests.
Map Myanmar – Cb

Gomel' *Belarus* See HOMYEL'

Gomera *Spain* See CANARY ISLANDS

Gonder *Ethiopia* City located about 400 km (250 miles) north and slightly to the west of the capital, Addis Ababa. It was the royal town of the Emperor Fasiladas (1632-67) and was the capital of Ethiopia until the end of the 19th century. Several rocky outcrops within the city are topped by the castles of successive emperors. It is the centre of a predominantly Christian community and has seven impressive churches.
Population 80 000
Map Ethiopia – Aa

Gondwana Southernmost of the two ancient continents into which Pangaea, an earlier supercontinent, divided about 200 million years ago. The northern continent was Laurasia. See CONTINENTAL DRIFT.

Gongju *South Korea* See KONGJU

Good Hope, Cape of *South Africa* Stormblown southernmost tip of the Cape Peninsula, first recorded in 1488 by the Portuguese navigator Bartolomeu Díaz, who named it Cape of Storms. Later, King John II of Portugal renamed it Cape of Good Hope for 'the hope it gave of finding [a sea route] to India'. The area at the peninsula's tip is now the Cape Point Nature Reserve.
Map South Africa – Ac

Goodwin Sands *English Channel* Stretch of shoals and sandbanks, 16 km (10 miles) long, off the coast of KENT in south-eastern England. Although they are marked by lightships and buoys, numerous ships have been wrecked on the sands, which are partly exposed at low water.
Map United Kingdom – Fe

Góra Kalwaria *Poland* Pilgrimage town in eastern Poland. Set amid orchards 34 km (21 miles) south-east of the capital, Warsaw, it was founded in 1672 by a bishop of Poznań in western Poland, and laid out in the shape of a cross. Its name translates as 'Mount Calvary'.
Population 10 600
Map Poland – Dc

Gorakhpur *India* City in the north, 80 km (49 miles) south of the Nepalese border. Traditionally, it is the place of Buddha's death and cremation in about 483 BC. However, Buddha died in Kushinagar, 55 km (34 miles) away. Gorakhpur has numerous old Buddhist temples and ruins.
The city manufactures textiles and dyes and has railway workshops.
Population 3 067 280
Map India – Cb

Gordon *Australia* River rising in the highlands of south-west Tasmania and flowing 200 km (124 miles) north-west through ancient forests of Huon pine – now protected – to Macquarie Harbour. It has been dammed to form Lake Gordon and Lake Pedder, and generates hydroelectric power. Plans for further dams downstream were abandoned in 1983 after worldwide protests by conservationists.
Map Australia – Hg

Gorée *Senegal* Volcanic islet and town off the coast beside Dakar, the capital. It was captured in turn by the Portuguese (15th century), the Dutch (16th century), and the French (in 1677). The town was a flourishing base for the slave trade and a provisioning port on the route to India. Today, it is a quiet fishing port, tourist resort and UN-protected World Heritage site. Visitors can still see the dungeons in which slaves were kept before they were shipped across the Atlantic. Three museums in the town trace the island's life and history.
Map Senegal – Ab

Gorgan (Gurgan) *Iran* Town in the former province of Gorgan, 50 km (30 miles) east of the south-eastern shore of the Caspian Sea. Cotton, wheat and salt are traded here.
A further 30 km (19 miles) to the north of the town lie the remains of Alexander's Wall, a line of earthworks named after Alexander the Great by historians who believed that they had been constructed by the King of Macedonia. Once 70 km (43 miles) long, Alexander's Wall was, in fact, built during the Sassanid dynasty (3rd to 7th centuries AD). Later it was largely disassembled for its stones.
The Gorgan River, 240 km (150 miles) long, which drains into the Caspian Sea just north of the town of Bandar-e Torkeman, is one of the few Iranian rivers that never dries up.
Population 162 000
Map Iran – Ba

gorge Deep, steep-sided rocky river valley, formed where river erosion cuts down more rapidly than WEATHERING and MASS MOVEMENT can wear back the sides.

Gori *Georgia* See T'BILISI

Gor'kiy (Gorky) *Russia* See NIZHNIY NOVGOROD

Görlitz *Germany* Industrial city straddling the Neisse River which forms the Polish border, 80 km (50 miles) east of Dresden. Its products include railway vehicles, machinery, precision instruments, textiles and furniture.
That part of the city which lies east of the Neisse – it has belonged to Poland since 1945 and is called Zgorzelec – has a ruined 12th-century castle and many 16th-century houses. On the German side, Görlitz has an old town with some beautiful Renaissance houses.
Population 72 700
Map Germany – Fc

Goroka *Papua New Guinea* Capital of Eastern Highlands Province, 400 km (250 miles) north-west of Port Moresby. The fertile valleys of the region, discovered by outsiders only in the 1930s, have extensive coffee plantations. At the biennial Goroka Show, villagers from the surrounding area celebrate at a 'sing-sing', wearing traditional dress and make-up.
Population 17 770
Map Papua New Guinea – Ba

Gorongosa National Park *Mozambique* See SOFALA

Góry Swietokryskie (Holy Cross Mountains) *Poland* Range of hills in the south, about halfway between Warsaw and Cracow. Thick larch and fir forest clothe its ridges. Many summits are topped by churches or monasteries, including the Benedictine Abbey of the Holy Cross on Lysa Gora, at 612 m (2008 ft) the highest hill. The main town is KIELCE.
Map Poland – Dc

Goslar *Germany* Former imperial capital at the foot of the Harz Mountains, 70 km (43 miles) south-east of Hannover. Lured by the wealth of the lead and silver mines in the Harz Mountains, the Holy Roman Emperor Heinrich II built a palace here early in the 11th century. All mining operations in the area stopped in 1988.
Today, Goslar has become an administrative, tourist and industrial centre. Its cathedral, rebuilt in 1820, still contains Heinrich's bronze throne.
Population 46 300
Map Germany – Dc

Göta Kanal *Sweden* Waterway, using rivers, lakes and a canal, which links the city of Gothenburg on Sweden's south-west coast with the Baltic. The waterway follows the Göta River to Lake VÄNERN, then runs on, via canals and stretches of river, to Lake VÄTTERN. It emerges on the east coast near Söderköping. It is about 560 km (350 miles) long. Opened in 1832, it was initially used to transport industrial goods, but today it is used almost exclusively by tourists. The Göta River section, 90 km (56 miles) long, is the only part navigable by ocean-going ships.
Map Sweden – Bd

Gothenburg (Göteborg) *Sweden* The country's second largest city after Stockholm and its main seaport. It lies on the Kattegat strait at the mouth of the Göta River. It was founded in 1624 near the site of a medieval settlement. The city plan – with its defensive canal system – was based on the Dutch model. Many Dutch and, later, British immigrants settled here, and up to today Gothenburg has maintained close trade links with Britain. Its harbour, ice free throughout the year, is a base for ferries serving Denmark, and across the North Sea, Britain. The main industries are motor car and ball-bearing production, foodstuffs, textiles and clothing.
Gothenburg is an important seat of learning. Gothenburg University was founded in 1891, and the country's oldest technical university – Chalmers' University of Technology – was founded in 1829 thanks to a donation from an English engineer named William Chalmers. The city also has an oceanographic institute, a medical research centre, a botanical garden, museums and many theatres.
Population 433 800
Map Sweden – Ad

Gotland *Sweden* The largest island in the Baltic, covering 3001 km² (1158 sq miles), 150 km (93 miles) south of Stockholm. It has been part

271

of Sweden since 1645 (except for a few short periods of foreign rule). During Viking times, it was an important trade centre, reaching the peak of its prosperity in the Middle Ages. It was once a haunt of pirates.

Its ports include the walled city of VISBY, Slite and a number of fishing harbours. The island has 91 medieval churches and has a rich selection of archaeological remains, including some 200 runic stones – stones inscribed with ancient Scandinavian symbols. The island's limestone, the basis of a modern cement industry, has been exported throughout the Baltic. Fishing, sheep, and sugar beet also contribute to its economy. However, thanks to its mild, sunny climate, Gotland's main industry is tourism.

Each summer the island hosts the Stånga Games. Among the events are an early form of tennis, a throwing contest in which the competitors hurl a heavy stone disc, and a contest resembling the traditional Scottish Highland event of tossing the caber.

Population 57 600
Map Sweden – Cd

Göttingen *Germany* University town 90 km (55 miles) south of Hannover. Twenty-nine Nobel prize winners have studied, lectured or researched at its university, which specialises in the sciences and medicine and was founded in 1737 by England's George II, as Elector of Hannover. The physicist Max Planck (1858-1947), who devised the quantum theory, one of the foundations of modern atomic physics, was a student at Göttingen and is buried here.

Population 121 800
Map Germany – Cc

Gouda *Netherlands* Cheese-making town 42 km (26 miles) south and slightly west of Amsterdam. Much of the medieval town centre with its encircling canals (originally defensive moats) and market place survives. The town's famous cheese market is held from mid-June to mid-August each year. It still retains its traditional character and attracts huge crowds.

Population 67 410
Map Netherlands – Ba

Governador Valadares *Brazil* Mining town 324 km (201 miles) north-east of the city of Belo Horizonte. Its mines produce gemstones including emeralds, amethyst, topaz, agate, aquamarine and tourmaline.

Population 236 000
Map Brazil – Dc

Gozo (Gozzo) *Malta* Second largest island of the Maltese archipelago, separated from the main island of Malta by a channel 5 km (3 miles) wide which contains the 3 km² (1.1 sq mile) islet of Comino. Gozo is still substantially free of tourist trappings and presents the restful, unpretentious nature of Mediterranean island life at its best. The people are modest, friendly and dignified. Most of the male population are involved in farming and fishing, while many women are employed in the lace-making industry. The island, which covers 67 km² (26 sq miles), is noted for its oranges, potatoes, onions, carrots and broad beans. Its rough wines have a legendary potency, as many visitors have found to their cost. Ferries to Malta run from the main port, Mgarr, 8 km (5 miles) by road from the capital, Victoria.

Most of Gozo's dozen or so villages are strung out on rocky ridges overlooking broad, carefully cultivated valleys. Huge Baroque churches dominate the landscape. The one at Xewkija is large enough to hold 12 000 people – even though the entire village population is less than a quarter of that number. Ramla Bay has a large, sandy beach on its north coast; there are smaller beaches at Marsalforn, Xlendi and Dwerja, which is the scene of the phenomenon of Qawra, the 'Inland Sea', where water rushes through a natural archway into a bay encircled by cliffs. Other places of interest include the impressive prehistoric Ggantija temples and the pilgrimage church of Ta Pina, built between 1920 and 1936.

Population 25 200
Map Malta – Ca

Graaff Reinet *South Africa* Beautifully preserved old town lying within a loop of the Sundays River in the Eastern Cape, 190 km (120 miles) north-west of PORT ELIZABETH. In 1795, local burghers proclaimed themselves independent of the Cape Town-based rule of the Dutch East India Company and set up a short-lived republic. The town is noted for its many beautifully preserved 19th-century buildings and its proximity to a strange, eroded landscape of basalt towers and cliffs known as the Valley of Desolation. The town and the surrounding district produce wool, angora hair and furniture.

Population 34 440
Map South Africa – Bc

Gracanica *Slovenia* See PRISTINA

Gradec (Slovenj Gradec) *Slovenia* See KARAWANKE ALPS

Grahamstown *South Africa* Cathedral and university town in the Eastern Cape, 130 km (80 miles) north-east of PORT ELIZABETH. Its many churches have earned it the nickname of 'the City of Saints' and it is regarded as the spiritual home of English-speaking South Africans, especially the descendants of the 4000 British settlers who arrived here 1820. Grahamstown was founded as a fortified frontier post in 1812 and became the principal town of the district of Albany where British settlers were allocated farms. Today, the town is noted for an annual festival of the arts.

Population 19 780
Map South Africa – Cc

Grampian *United Kingdom* Administrative region in east-central Scotland covering 8698 km² (3357 sq miles) and taking in the former counties of Aberdeen, Kincardine and Banff, as well as part of Moray. It is named after the Grampian Mountains, which extend into neighbouring regions and rise to 1344 m (4409 ft) at Ben Nevis, Britain's highest peak, in the Highland region; they also include the ski resorts of the Cairngorm massif. In the east they descend gradually to a coastal fringe of fertile land, the home of the Aberdeen Angus breed of cattle. The region's best known product is whisky and its best known inhabitants are the royal family, whose castle at Balmoral is 70 km (43 miles) west of the North Sea oil port of ABERDEEN, the main administrative centre. The region also has industries associated with offshore oil technology.

Population 493 160
Map United Kingdom – Db

Gran Canaria *Spain* See CANARY ISLANDS

Gran Chaco *Paraguay / Argentina / Bolivia* Wilderness covering some 240 000 km² (92 660 sq miles) – an area about the size of Britain – on the borders of Argentina and Bolivia. Territorial disputes over the area with Bolivia led to the Chaco War of 1932-5, and though Paraguay was victorious, victory came at the cost of 40 000 lives. The north and west of Chaco is largely arid thorn scrub. Towards the east and south there are cattle pasture and forests famous for the quebracho, or 'axe-breaker tree', a source of tannin. The Chaco's population is sparse, with some Amerindians and scattered and isolated settlements of Mennonites, a Christian fundamentalist sect originally from Germany. The area's main town is Filadelfia.

Population 100 000
Map Paraguay – Ab

Gran Pajatén *Peru* Magnificent ruined city of a pre-Inca civilisation, discovered by the American explorer Gene Savoy in 1964 in the forest of the Urubamba River valley. It has many circular stone buildings, temples and blocks of carved stone depicting human figures and condors.

Map Peru – Ba

Gran Paradiso National Park *Italy* Oldest of the country's national parks, founded in 1922, 50 km (30 miles) north-west of Turin. It covers 620 km² (240 sq miles) of the Graian Alps, rising to the Gran Paradiso peak at 4061 m (13 323 ft). The park protects rare animals and birds such as the ibex, chamois and royal eagle.

Map Italy – Ab

Gran Sasso *Italy* Part of the APENNINE Mountains 120 km (75 miles) north-east of Rome. The highest point in the range is the Monte Corno, reaching 2912 m (9554 ft).

Map Italy – Dc

Granada *Nicaragua* The country's oldest Spanish settlement, founded in 1524 at the northern end of Lake Nicaragua, 50 km (30 miles) south-east of the capital, Managua. During the 19th century, Granada formed the nucleus of the country's independence movement. Many of its colonial buildings survive, including the church of La Merced ('Church of Mercy'), completed in 1781. The town's products include furniture, clothing, cotton, rum, and soap.

Population 88 600
Map Nicaragua – Aa

Granada *Spain* Tourist resort and city 195 km (120 miles) north-east of Gibraltar. It is famous for its Moorish architecture, most notably the Alhambra, a magnificent palace built between 1248 and 1354, and the adjacent Generalife – the summer palace of the Moors – with its exquisite gardens. There is also a 16th-century cathedral which contains a royal chapel with the tombs of the Catholic monarchs Ferdinand and Isabella.

Granada produces textiles, paper and liqueurs; it also caters for tourists, one of its attractions being the flamenco dancing of the gypsy community living in caves in the nearby mountain of Sacromonte. Granada Province produces cereals, sugar cane and tobacco.

Population (city) 255 210; (province) 790 520
Map Spain – Dd

Grand Bahama *Bahamas* One of the largest islands in the Bahamas, covering 1370 km² (530 sq miles). It lies in the north-west of the group, only 100 km (62 miles) off the coast of Florida. The interior is heavily forested, and the island was largely undeveloped, with a population of about 8000, until the early 1960s. It hit a boom period after a group of American, Bahamian and European business people began to develop it as a tax-exempt tourist, industrial and commercial area, mainly in and around the town of FREEPORT CITY and its suburb, Lucaya. Today, Grand Bahama has the second largest population in the Bahamas.

Population 41 040
Map Bahamas – Ba

Grand Banks *North Atlantic* Part of the continental shelf off south-eastern NEWFOUNDLAND where the warm GULF STREAM and the cold Labrador Current meet. The mixing of the warm and cold waters creates one of the world's leading fishing grounds, although overfishing has led the Canadian government to introduce restrictions in recent years. Cod fishing in particular has been badly affected. Fogs, particularly in the north-east, and Arctic icebergs are shipping hazards, especially in spring. The banks measure 680km (422 miles) from east to west and 560 km (348 miles) from north to south, and vary in average depth from 90 m to 180 m (295-590 ft). Oil was discovered here in the late 1970s.

Map Canada – Jd

Grand Canal *China* See DA YUNHE

Grand Canyon *USA* Spectacular gorge, the world's largest, cut into the earth's crust by the Colorado River. It lies in north-western Arizona, is 350 km (220 miles) long and up to 1870 m (6135 ft) deep and is cut in multicoloured rock strata laid down over more than 1000 million years. The most beautiful stretch of the canyon walls lies within a national park covering 4931 km² (1904 sq miles). More than 3 million people visit the canyon each year.

Map United States – Dc

Grand Popo *Benin* Port, market and tourist resort 22 km (14 miles) from the border with Togo. Founded in 1727 by refugees fleeing the slave traders in Ouidah, it is one of two places along the coast where a channel links the coastal lagoons and the sea.

Map Benin – Ab

Grand Rapids *USA* City in western Michigan, 40 km (25 miles) east of Lake Michigan. It has been named after the rapids on the Grand River. The city was founded as a fur trading post in 1826 but soon developed as a centre of furniture manufacture, and held its own furniture fair annually from 1887 until the early 1960s. Its main museum contains streets recreating the period 1870-1900. Today, Grand Rapids has substantial iron and steel industries.

Population (city) 183 000; (metropolitan area) 626 500
Map United States – Ib

Grand St Bernard Pass *Switzerland/Italy* See GROSSER ST BERNHARD and ST BERNARD PASSES

Grand Terre *Guadeloupe* See GUADELOUPE

Grand Teton National Park *USA* Mountainous area covering 1241 km² (479 sq miles) in western Wyoming, immediately south of Yellowstone National Park. It is the most spectacular part of the Teton Range, rising to 4196 m (13 766 ft) at the Grand Teton, and is dotted with lakes, streams, extensive forests and areas of sage brush. Declared a national park in 1929, it has a great variety of game and attracts millions of visitors every year.

Map United States – Db

Grande Comore *Comoros* See NJAZÍDJA

granite Coarse-grained igneous rock consisting largely of quartz, feldspar minerals and micas.

granitic layer Upper, less dense of the two layers into which the continental portions of the earth's CRUST are divided.

Grasse *France* Manufacturing town and resort on the Côte d'Azure, 12 km (7.5 miles) north of Cannes. It is set among orange groves and fields of flowers from which essences are extracted to make perfumes. The French artist Jean Fragonard (1732-1806) was born here. A museum in the town commemorates his life and work.

Population 42 100
Map France – Ge

grassland Region where the vegetation consists mainly of grass, the rainfall being too light for forest growth. The main types of grassland are tropical grassland, or savannah, mid-latitude (temperate) grassland which includes the STEPPES, PRAIRIES and PAMPAS, and mountain grasslands, a zone above the treeline on mountains. Grasslands are mostly natural, though usually modified and extended by humans over thousands of years, particularly by the use of fire.

Graubünden (Grisons) *Switzerland* Largest of the 26 cantons and home to most of Switzerland's

▲ **GRAND CANYON The Colorado River carved its vast, 6-29 km (4-18 mile) wide canyon as the land slowly rose during earth movements, and still erodes huge amounts of debris daily.**

Romansch speakers, covering a mountainous area of 7106 km² (2744 sq miles) in the south-east. The main occupations in the canton are tourism, construction, cattle breeding and dairy farming.

Population 179 300
Map Switzerland – Ba

gravel Mass of rock fragments ranging in diameter from 2 to 4 mm (0.08 to 0.16 in).

Graz *Austria* The country's second largest city, and capital of Styria State, 140 km (87 miles) south-west of VIENNA. Graz started as a settlement on the trade route from Italy to Hungary, and became a fortress town in the 12th century guarding the northern route through the Alpine foothills. Later Graz played an important part in Habsburg resistance against the Turks, who were pushing westward. The city has an 11th-century castle, a 15th-century Gothic cathedral, a 17th-century arsenal, and a university dating from 1596. Its manufactures include paper, shoes, clothing and chemicals. It also has steel foundries, ironworks and railway workshops.

The German mathematician and astronomer Johannes Kepler, who confirmed that the planets revolve around the sun, lived in Graz from 1594 to 1600.

Population 232 200
Map Austria – Db

Great Artesian Basin *Australia* The world's largest underground natural reservoir, lying beneath an area of about 1.7 million km² (656 000 sq miles). It stretches from the Gulf of Carpentaria into South Australia and New South Wales, and gathers its water from the porous rocks west of the GREAT DIVIDING RANGE. About

4700 flowing boreholes have been constructed to tap the basin's water for irrigation, livestock and domestic use. Only about half of these are still flowing, with a total discharge rate of 1500 million litres (330 million gallons) a day. Without them, however, it would be impossible to farm in Australia's dry interior.
Map Australia – Gd

Great Australian Bight Arm of the INDIAN OCEAN lying off southern Australia, and extending 1160 km (720 miles) from east to west and 350 km (218 miles) from north to south. It is partly bordered by the inhospitable NULLARBOR PLAIN. The bight was first explored by a Dutch navigator, Captain Thyssem, in 1627. However, he turned back before reaching the more fertile part of SOUTH AUSTRALIA. The coast was charted by the English sailor and explorer Captain Matthew Flinders in 1802.
Map Australia – De

Great Australian Desert *Australia* Vast arid heart of the island continent, covering some 3 830 000 km² (1 480 000 sq miles) and comprising such distinct areas, as the GIBSON DESERT, GREAT SANDY DESERT, GREAT VICTORIA DESERT, NULLARBOR PLAIN and SIMPSON DESERT.
Map Australia – Cd-Dd, Fc

Great Barrier Island *New Zealand* Main island in the Hauraki Gulf, 100 km (60 miles) north-east of Auckland on North Island. It covers an area of 285 km² (110 sq miles) and is a mixture of pasture, scrub and forest. Tryphena is the island's main town and centre of activity.
Population 1140
Map New Zealand – Eb

Great Barrier Reef *Australia* Largest coral reef system in the world, stretching about 2000 km (1250 miles) along the coast of Queensland in the Coral Sea. The entire reef actually consists of more than 3400 individual reefs. It is up to 500 m (1650 ft) thick, extending up to 300 km (190 miles) from land in the south, but much less in the north. Its waters are dotted with more than 600 islands, some of which have been developed as tourist resorts. The reef, which has about 350 species of corals and is rich in colourful fish, is a marine park covering 348 700 km² (134 602 sq miles) and a World Heritage Area.
In February 1991, it was designated the world's first 'particularly sensitive area' by the UN International Maritime Association. This means that the Australian government has the power to insist that ships passing through the reef carry a pilot.
Map Australia – Ga

Great Bear Lake *Canada* Fourth largest lake in North America, covering 31 328 km² (12 093 sq miles). It lies in the Mackenzie district of NORTHWEST TERRITORIES and drains into the Mackenzie River. The lake, which is rich in fish, has a maximum depth of 413 m (1356 ft) and is frozen for more than eight months of the year.
Map Canada – Db

Great Belt (Store Bælt) *Denmark* Central channel through the Danish archipelago, with the islands of FÜNEN and LANGELAND to the west and ZEALAND and LOLLAND to the east. It is 115 km (71 miles) long, 11-35 km (7-22 miles) wide and

20-58 m (66-190 ft) deep. Contracts for one of the world's longest suspension bridges, with a suspended section of 1624 m (5328 ft), were signed in 1991 to link its western and eastern shores. The bridge is due to be in use by 1997.
Map Denmark – Bb

Great Bitter Lake *Egypt* See BITTER LAKES

Great Britain *United Kingdom* The island comprising ENGLAND, SCOTLAND and WALES.

Great Dividing Range *Australia* Mountain chain running almost the full length of eastern Australia from Queensland to Victoria. The range includes the SNOWY MOUNTAINS, BLUE MOUNTAINS and the ATHERTON TABLELAND. The highest peak on both the range and the continent, at 2228 m (7309 ft), is Mount Kosciusko .
Map Australia – Gb-Id

Great Dyke *Zimbabwe* A Precambrian intrusion of igneous rock forming a ridge stretching more than 500 km (310 miles) across the high plains of central and southern Zimbabwe and containing many important minerals including chrome, asbestos and gold.
Map Zimbabwe – Ba

Great Escarpment *South Africa* The uptilted rim of the broad plateau forming most of southern Africa. It rises steeply, reaching up to 1500 m (4920 ft) in the east. Much of the escarpment is formed by the DRAKENSBERG range.
Map South Africa – Cc

Great Glen *United Kingdom* See NESS, LOCH

Great Lakes *North America* Series of five lakes centrally located in the north, drained by the St Lawrence River to the Atlantic Ocean. Four of them – HURON, SUPERIOR, ERIE and ONTARIO – straddle the Canadian-United States border. LAKE MICHIGAN, the fifth, lies within the United States.
Map United States – Ia; Canada – Gd-Hd

Great Plains *North America* Vast area of undulating grassland stretching south-east from the Mackenzie Mountains in north-west Canada through the west-central United States to west Texas; it includes the PRAIRIES. Bordered by the foothills of the Rocky Mountains, it stretches eastwards almost to the Great Slave Lake and Lake Winnipegosis in Canada, and to the Missouri River in the United States. The Great Plains are up to 1000 km (625 miles) wide in the United States. They w ere the domain of buffalo and native American peoples until the middle of the 19th century, at which point wagon routes, followed by railways, led to the colonisation of the area by Europeans. Devoted to farms and ranches, the area is now North America's granary, with wheat as its main product. Other important products are livestock, oil, gas and coal.
Map United States – Ea-Fc

Great Rift Valley *Middle East/East Africa* Great depression running about 6400 km (4000 miles) from the Jordan Valley through the Red Sea to Mozambique in south-east Africa (where it is sometimes known as the East African Rift Valley). It consists of a series of geological faults caused by huge movements of the earth. For much of its length, its traces have been lost by

erosion. In some parts, however, such as in southern Kenya, cliffs rise thousands of metres. The Great Rift Valley floor reaches its highest levels in parts of central Kenya, where it rises to more than 1830 m (6000 ft); its lowest point is the DEAD SEA, whose surface lies 396 m (1299 ft) below sea level – the lowest point on the earth's surface.
Map Middle East – Bb; Africa – Ff-Gd

Great Salt Lake *USA* A shallow lake in central Utah, west of the Wasatch Range and just northwest of Salt Lake City, capital of the Mormon state. Its area fluctuates according to the flow of the rivers emptying into it, reaching a maximum of 6476 km² (2500 sq miles). Its average depth ranges from 4 m to 7.3 m (13-24 ft).
Map United States – Db

Great Sandy Desert *Australia* Arid region in the north of Western Australia, stretching inland from the coast towards the state boundary with the Northern Territory. This is the hottest part of Australia.
Map Australia – Cc

Great Sandy Island *Australia* See FRASER ISLAND

Great Slave Lake *Canada* The deepest lake in North America, with a depth of 614 m (2015 ft), covering 28 570 km² (11 030 sq miles) in the southern Northwest Territories. It is fed by the Slave River and drained by the MACKENZIE River. Gold was discovered on its northern shore in the 1930s, when the mining town of YELLOWKNIFE was established.
Map Canada – Db

Great Smoky Mountains *USA* Range in the Appalachian Mountains, straddling the border of North Carolina and Tennessee. Named after the blue haze that always envelops them, the mountains rise to 2025 m (6643 ft) at Clingmans Dome. The range is enclosed in a 2092 km² (807 sq mile) national park. With the area receiving plentiful rainfall, few places in the United States have such varied plantlife.
Map United States – Jc

Great Victoria Desert *Australia* Huge area of parallel sand dunes straddling the state borders of Western Australia and South Australia.
Map Australia – Dd

Great Wall *China* The longest fortification in the world, winding a total of 3460 km (2150 miles) from Shanhaiguan, close to the Gulf of Bo Hai, to Yumen in the Gobi Desert, with branches adding a further 2865 km (1780 miles). It was originally built of earth and stone by an army of conscript labour, many of whom died constructing it. Most of the wall was completed during the Qin dynasty (221-206 BC), when it measured some 10 000 km (6215 miles); however, it was strengthened by later emperors. It was intended to protect China against nomads from the north, though it did not entirely prevent later invasions.
The present wall was largely constructed during the Ming dynasty (1368-1644). The best preserved section is at Nankou Pass, 80 km (50 miles) north of Beijing. In parts the Great Wall is 9 m (30 ft) high and wide enough for a column of soldiers to march on. United States astronaut Alan Bean exploded the myth that it is

visible from the moon, during a United States lunar mission in 1969.
Map China – Ec-Hc

Great Yarmouth *United Kingdom* Port and seaside resort at the mouth of the River Yare, 175 km (110 miles) north-east of London in the English county of Norfolk. It thrived as a port for 100 years before receiving a new boost as a centre for developing the southern North Sea natural gas fields in the 1960s. It is now a major base for some 400 companies involved in gas production and exploration.
Population 62 430
Map United Kingdom – Fd

Great Zimbabwe *Zimbabwe* Historic site 300 km (185 miles) south of the capital, Harare, and 28 km (17 miles) south-east of the town of Masvingo, the main town of the province of the same name. Great Zimbabwe is the largest of about 150 dry-stone village sites in the country, which were occupied by the Shona-Karanga civilisation that flourished between about AD 1200 and 1450. Its buildings include a hilltop fort and a circular temple with a conical tower 11 m (36 ft) high. The country takes its name from the site.
Map Zimbabwe – Cb

Greece See p 276

greenhouse effect Term used to describe the heating of the earth's surface by the retention of infrared radiation. When the ground gives off heat as infrared radiation, some of it is absorbed by water vapour and carbon dioxide in the air and is radiated back to the ground, warming it further. Some scientists think that the greenhouse effect could eventually lead to the melting of the polar ice caps and the inundation of the low-lying – and most heavily populated – parts of the world.

Greenland See p 280

Greenland Sea Southern arm of the ARCTIC OCEAN lying north of Iceland and north-east of Greenland. Pack ice covers large areas of it and the East Greenland Current carries ice southwards from the sea.
Map Arctic Ocean – Ab

Greenwich *United Kingdom* See LONDON

Greenwich meridian Line of LONGITUDE through Greenwich, England, taken as the prime or zero meridian – 0° east and west – for worldwide navigation and timekeeping.

gregale Strong blustery north-east wind, sometimes accompanied by showers, which blows in the central Mediterranean during the winter.

▲ **GREAT WALL OF CHINA** Under China's first emperor, the Great Wall kept out barbarians and was also a communications link; messages were passed by smoke signals in the day and by fire at night.

Greifswald *Germany* Baltic port and university town north of Berlin, founded by Dutch merchants in 1240. To the east is Peenemünde, the site where V-rockets were developed during the Second World War. Boat trips to the island of Rügen leave from Greifswald harbour.
Population 66 700
Map Germany – Ea

Grenada See p 281

Grenadines *West Indies* Group of about 600 small islands – the largest being CARRIAOU – in the WINDWARD group of islands between ST VINCENT and GRENADA. The group was 'discovered' by Christopher Columbus in 1498. Its inhabitants are largely descendants of African slaves. The northern part of the group belongs to St Vincent, the southern part to Grenada. The Grenadines are a paradise for yachting enthusiasts, and tourism has brought some prosperity.
Population 14 500
Map Caribbean – Cc

Greece

LAND OF LEGENDARY BEAUTY AND OLD GLORIES, STRUGGLING TO MAKE ITS WAY IN THE MODERN WORLD

A 17th-century travel writer said of Greece: 'The mind is fascinated, the body faints, the eyes become moist and are delighted.' It is a legendary country which attracts lyrical prose. It has great natural beauty, a rich architectural heritage in monuments which are famous throughout the world, and priceless art treasures.

Some 2500 years ago, Greece was the centre of the Western world, where European civilisation and the concept of democratic government were born. Modern Greece has suffered bloody civil war, dramatic political upheavals and economic hardship. Between these extremes, Greece has had a tumultuous history. Progressive and successful as the ancient Greek cities may have been, they were too competitive amongst themselves to unite into a single state. ATHENS and SPARTA, especially kept on striking at each other's throat in a struggle for power. This played into the hands of their smaller rivals. Nonetheless, Greece flourished even under the Roman Empire, which ruled Greece from 146 BC, and even more in the Byzantine period from AD 330. But by 1460, seven years after the fall of Constantinople to the Turks, most of Greece had become a province of the Ottoman Empire and remained under Turkish rule for nearly 400 years – a period that has coloured Greek attitudes and foreign affairs to this day. After a nine-year war (1821-30), Greece became an independent kingdom, but its present frontiers were not stabilised until after the Second World War.

AFTERMATH OF WAR

Greece sided with the Allies in the First World War, and won further territory from the Turks, who had sided with Germany. Many Turkish residents in Macedonia and THRACE consequently left the country and moved further east. Italian and German troops occupied Greece during the Second World War, but were harassed by Greek resistance fighters and withdrew finally in November 1944 after British troops landed. For years after the war, communist guerrillas fought in the mountains against nationalist troops in a bloody civil war that cost 120 000 lives. American and British aid helped to ensure defeat of the guerrillas, as did President Tito's quarrel with Moscow, which led to the withdrawal of Yugoslav support for the guerrillas.

The civil war ended in 1949, but political instability returned some years later, and in 1967 Colonel George Papadopoulos led a successful military coup. The young king, Constantine II, was driven into exile in Rome, and in 1973 Papadopoulos declared a republic, with himself as president. Within months, however, he was deposed in another coup and a new military junta took power.

Meanwhile, relations with Turkey worsened over the problem of CYPRUS. Turkey invaded northern Cyprus in 1974 and the regime in Greece fell in a bloodless countercoup. Democracy had returned to the country that invented the word.

Greece rejoined NATO (after a six-year gap) in 1980, and became a member of the European Economic Community in 1981, which has given it the opportunity to market many of its agricultural products, in addition to a significant budgetary subsidy.

Mountain ranges, resembling giant ribs, and deep incisions by the sea, creating peninsulas and islands are the country's dominant physical features. The PÍNDHOS mountains divide Greece from the Albanian border in the north to the Gulf of CORINTH in the south, peaking at Smolikas (2637 m, 8651 ft). Another rib runs southwards and then east, soaring to Mount Olympus (2917 m, 9570 ft) and extending to the Aegean Sea.

About 70 per cent of the land is hilly, with harsh mountain climates and poor soils, but agriculture is the chief activity. Until 1950 most of the people were involved in subsistence farming, producing enough food for themselves but with no surplus for sale. There has since been a steady migration to the towns, but 25 per cent of the working population still live in the countryside.

Forests cover large areas in the highlands of Macedonia, Thrace and Píndhos. Woodland in lower areas has been reduced by forest fires during long, dry summers and by felling to provide fuel and clear land for pasture.

Dry Mediterranean vegetation covers southern and central Greece, with bushes and open woods. The island of CRETE has an astonishing variety of wild plants and flowers; there are more than 100 species with medicinal properties. Back on the mainland, Attiki is renowned for herbs and wild flowers which provide the nectar for excellent honey. Firs are common in the mountains; oaks, beeches and chestnuts lower down.

In coastal waters, sea sponges are an important crop in the DODECANESE islands, and sea anemones and urchins are a tourist attraction among the submerged rocks. Dolphins sometimes follow ships along the coasts.

FARMERS AND SAILORS

Large-scale agriculture is concentrated on the fertile plains on the eastern coasts. Crops include tomatoes, cotton and cereals in THESSALY, fruit trees in the river valleys of AXIOS and ALIÁKMON in Macedonia, tobacco in the Strymon valley in Thrace, and vineyards, orange and olive groves in the PELOPONNESE and Crete.

Communications between most regions are poor because of distance and mountains. The sea unifies the country, and the country has almost 14 000 km (8700 miles) of coastline. More than 89 per cent of imports and exports are carried by water. On land, the coastal roads are the main means of communication between the principal towns and the countryside. Greece was, in 1869, one of the last European countries to start building railways.

The IONIAN SEA in the west and the Aegean in the east contain 2000 Greek islands. Fewer than 200 are inhabited, but they contain 11.3 per cent of the population. The main tourist islands, such as CORFU, Crete, MÍKONOS and RHODES, have international airports, while many others handle domestic flights only. The rest of the islands use PIRAEUS, the port of Athens, as their link with the outside world.

As the southernmost part of the BALKANS, the Greek peninsula is the most south-easterly extension of Europe – cut off from West European countries by 1300 km (800 miles) of road through former communist republics. Again, sea and air provide important links; Italy is less than 200 km (125 miles) away by sea.

The Greek economy had slumped alarmingly by the end of the civil war in 1949. Industry consisted of small processing plants for tobacco, leather, cotton and foodstuffs, and there was little money to develop natural resources. Over the next 25 years an ambitious development programme was pushed through, financed largely from Western Europe and the United States. Large irrigation schemes helped agriculture, and new power stations laid the foundations for industries such as shipbuilding, aluminium, steel, plastics, chemicals and electrical engineering.

But oil price rises and the world recession of the mid-1970s slowed development. Much manufacturing work is still carried out manually, and technology lags behind that of many other countries. So Greece remains relatively poor, with an income per head about half that of most of its partners in the European Union (EU). As before, the country must continue to import many industrial products.

While tourism brings a steady income, the economy needs foreign exchange from other sources. In the past communist countries were some of the biggest customers for Greece's agricultural products and the ship repair yards in SIROS, in exchange for electricity, telecommunications equipment and oil. After the collapse of the Soviet Union, Greece lost many of its former East-bloc trading partners, and had to start paying hard cash for oil on the world market.

Merchant shipping has enjoyed mixed fortunes in the past 40 years, but with a seafaring tradition stretching back to the legendary Odysseus, Greece today has the third largest merchant fleet after Liberia and Panama, with 6 per cent of the world's tonnage. There are some very rich Greek shipowners, but many of them register their ships in foreign countries. If these ships flying 'flags of convenience', such as those of Liberia, Panama and Cyprus, are included, Greek-owned tonnage is believed to be the world's largest.

PEOPLE AND POLITICS

Greek political views are intense, and electoral campaigns divide the nation. Patronage is rife, and victorious politicians are often expected to give their supporters rewards such as appointments in the civil service.

Domestic politics are concerned mainly with urban problems. Industrial development since 1950 has caused a rapid move to the towns and an explosion of housebuilding. The capital, Athens, has increased its population

▲ CROWING GLORY The Parthenon
stands on the Acropolis or 'high city'
in Athens. It was built as a temple to
the goddess Athena in 438 BC, and
was destroyed when Venetians
besieged the city in 1687.

and education are not always adequate; there
are not enough hospitals and schools.

Many young people have emigrated, mostly
to the United States, Canada, Australia, Africa
and as migrant workers to European countries
(especially Germany). Some return to Greece,
having saved enough to buy a house, or a car,
or perhaps to start a small business.

Greece attracts more than 5 million tourists
a year, and three out of every four come from
Western Europe. Most visitors go to Athens or
the major islands. In some historic towns,
handsome mansions and houses have been
replaced by featureless tourist apartments, but
most coasts have not been disfigured by high-
rise flats. In recent years formerly unspoilt
islands in the Ionian Sea, such as LEVKÁS,
Paxoí and ZÁKINTHOS, and others in the
Aegean (Amorgos, Kea, PÁTMOS, Skopelos)
have also been invaded by tourists during the
summer months.

ISLANDS FOR TOURISTS

The Greek islands have a variety of rocks
which give them distinctive characters: limestone in the Ionian islands, marble in Skiros,
PÁROS, ÍOS, Anafi and TÍNOS, granite in
Mikonos, pumice and lava in SANTORINI. Some
islands are waterless and bare like Pátmos,
others have picturesque wooded hills like
AEGINA or sheltered and watered valleys like
CEPHALONIA.

The Ionian islands of Corfu and Levkás are
known for their green landscape, Zákinthos for
its fragrant gardens and ITHAKA for its miniature harbours. Most of the CYCLADES are
barren, with white houses and windmills.
CHIOS, LESBOS, SÁMOS and Ikaria are fertile
with abundant springs and thick woods.
Amorgos and Santorini have dramatic and
colourful cliffs. Set in transparent seas,
Rhodes and Crete are sprinkled with beautiful
historic towns surrounded by orchards, vineyards, fields of wheat and tree-clad mountains.

Rural settlements are dotted over much of
the mainland and the islands. Most houses are
built of stone and timber, with tiled roofs,
wooden pillars and balconies hanging over
narrow, sloping streets. Inland villages develop around the plateia, the lively central square;
the centre of most coastal towns is the harbour.

The mountain ranges of central and southern Greece form a mass of jagged, rocky
prongs, and descend through craggy ridges to
sea-girt peninsulas and promontories, as in the
southern Peloponnese and CHALCIDICE. Deep
gulfs and inlets penetrate the land, so that no
place in Greece is more than 70 km (45 miles)
from the sea. The narrow coastal plain meets
the sea in a fretwork of bays, some of which
are inaccessible from land. Elsewhere, gulfs
are bounded by mountains cradling tiny
harbours and fishing villages, or bordered by
pine-clad slopes.

Each season lends its own charm to the
landscape. Spring comes early in April and
myriads of Mediterranean plants burst into a
blaze of colour. The hot season starts in May
and builds to an average summer temperature
of 27°C (81°F), with heatwaves reaching 40°C
(104°F). Cool breezes may temper the heat on
the coasts and the islands, but the clear air and

sixfold since 1945, and now has more than
30 per cent of the total population, 60 per cent
of manufacturing capacity, and practically all
government and business administration.
More than half of the 10 million population
lives in the six largest cities: Athens, SALON-
ICA, PATRAS, HERAKLION, VÓLOS and Larisa.

People have moved from the rural areas and
from the islands, except those where tourism
has become a source of income, as in Míkonos,
Kos, Rhodes, Corfu and Crete. Many of the
city immigrants are youngsters escaping from
a life of hard labour on poor farms. There is
not enough work for them all in the towns, but
the hardships of unemployment are cushioned
because Greeks live in extended families of
two or three generations, all helping one
another (in 1994 the official unemployment
rate stood at just over 10 per cent). Towns are
becoming overcrowded, with many suburbs
lacking sewerage and piped water. Health care

bright sunshine bring out all the beauty of the landscape. Winters are warm and wet in the lowlands, but there is snow in the mountains.

While the Greek summer is a great tourist attraction, the country's rich history brings visitors all year. The grandeur of ancient civilisations is recalled by spectacular ruins. Most are in the central and southern parts of Greece and the islands. The Bronze Age (2000-1100 BC) produced the splendid Minoan palaces of KNOSSÓS, Faistos and Mallia in Crete, the citadels of MYCENAE and Tirins, and the city of ancient Thíra on Santorini. Athens contains the most important group of architectural landmarks of the Classical period (500-300 BC) at the Acropolis.

The ancient sites of DELPHI and OLYMPIA and the ruins of the sacred city of DELOS are great tourist attractions. The vast ancient theatre of EPIDAURUS, which overlooks the Argolic plain,

and the Roman Odeon of Herodes Attikos at the foot of the Acropolis are used for international festivals of the arts.

Byzantine monuments include elegant churches and monasteries decorated with mosaics. Salonica, which was the second city of Byzantium, has a rich heritage of these stylish churches, and the monasteries of DAPHNI and Osios Loukas are famous for both their architectural beauty and their art treasures.

Venetian and Frankish occupation early in the 13th century enriched Greece's capes and islands with picturesque castles and walled cities such as MONEMVASIA, NÁXOS, Koroni and, most beautiful of all, Rhodes.

COUNTRY AND TOWN LIFE

Greeks are traditionally associated with folk dancing and singing, often in picturesque national dress. Their biggest folk festivals are

linked with the Greek Orthodox religion, which is followed by 98 per cent of the people. Easter Sunday, especially, is a day of rejoicing when families gather in town and village squares to sing and dance, and roast lambs on spits. (It is often a different day from that celebrated by the Western churches.) On saints' days, fairs are held outside monasteries and churches. There are two national holidays in honour of the Virgin Mary – on 25 March (coinciding with Independence Day) and 15 August, when there is a national pilgrimage to the monastery on the island of Tinos.

The people of Greece are united by religion and by the modern Greek language known as Demotic (or dhemotike), derived from the ancient language and spoken by almost everyone. A more formal, literary version of Greek, called katharevousa, customarily used in government documents and by officials, is dying

▲ VIOLENT BIRTH A chapel over-looks the tranquil harbour of the island of Santorini (Thira) in the Aegean Sea. The bay formed when the volcanic island blew apart about 1400 BC. The resulting shock wave, sea wave and hail of ash devastated much of Crete, 120 km (75 miles) to the south.

out. Dialects survive in many rural areas, but Ancient Greek is incomprehensible to the layman, and is taught only in high schools and universities.

Rural life is simple, and recreation is mostly confined to Sundays, feast days and weddings. Older people pass the time sitting and talking in the village square. Traditional costume is still worn by older people on festive occasions. There are great regional variations in embroidery, colours and patterns; old hand-woven costumes, coloured with plant dyes, are valued by collectors. Handwoven carpets and blankets are still made in Thrace, Crete and central Greece, assisted by state subsidy. Villagers are generally of the older generation, and many villages are ageing rapidly.

In the cities and towns, social life for older people is centred on the cafés and eating out at tavernas is still a widespread custom for those who can afford it. In the few traditional tavernas that survive, one can still see men

dancing the zeimbekiko by themselves. Poorer city families eat at home, and the meal is an important family gathering.

When people from rural areas move to the towns, their values and family behaviour change little. A village girl may use her parents' savings to buy a flat in Athens as a step towards marriage. But she will probably marry a country boy who is using his family's resources to study in Athens and seek a job in the civil service. Many families deprive themselves to send their children to live in Athens. In return, children tend to work hard at school, and are disciplined and obedient at home. Young people usually need their parents' permission to go out in the evenings, take a job or get married. The father is the head of the family and makes decisions for all its members. This is a rural lifestyle, but it is only slowly dying out in urban environments.

Women suffer discrimination in many ways. A bride is expected to provide a dowry, perhaps a flat or the equivalent cash. After marriage, a wife must obey her in-laws as well as her husband. Husbands rarely help with domestic chores, even if the wife is working. In many rural areas, single girls are not expected to appear in public unless accompanied by a male relative. Married women are not supposed to go to work, have guests at home, or meet friends outside, without their

husband's permission. Many small businesses are family-run with female family members – unpaid – helping in numerous jobs.

It is said to be of the Greek character to 'watch the world go by'. The country has faced a huge task in trying to create a modern industrial economy in the face of slowly dying traditional attitudes.

GREECE AT A GLANCE	
Official name Hellenic Republic	
Area 131 940 km² (50 930 sq miles)	
Population 10 470 460 **Per km²** 79 (**Per sq mile** 206 sq mile)	
Capital Athens	
Government Parliamentary republic	
Currency 1 drachma = 100 lepta (not in use)	
Language Greek	
Religions Christian (98% Greek Orthodox), Muslim (1%)	
Climate Mediterranean; average temperature in Athens is 9°C (48°F) in January and 28°C (82°F) in July	
Land use Cultivation 31%, grazing 40%, forest and woodland 20%, other 9%	
Main exports Clothing and textiles, cotton, citrus fruits, sultanas, vegetables, olive oil, petroleum products, metal ores and metals, tobacco	
Annual income per head (US$) 6000	
Population growth (per thous/yr) 8	
Life expectancy (yrs) Male 62 **Female** 71	

Greenland

THE WORLD'S CHILLIEST NATION, KNOWN TO ITS INHABITANTS AS KALAALLIT NUNAAT, HAS GONE OUT INTO THE POLITICAL COLD TO CARVE ITSELF A NEW POLAR ROLE

In February 1985, the European Common Market lost half its territory when Greenland became the first nation to leave the EEC since the community's foundation in 1958. The gesture, by the people of the world's largest island, summed up the determination of Greenlanders to go it alone economically as well as politically.

The concertina-playing Lutheran priest who steered the island out of the EEC – Prime Minister Jonathan Motzfeldt – also wanted Greenland to cut its remaining political ties with Denmark, ending an association that dates back to the 14th century, when Norway and its possessions, including Greenland, came under Danish rule, and Danes founded the first trading posts.

The island, which its inhabitants call Kalaallit Nunaat (Land of the People), has had home rule since 1979 but is still nominally part of Denmark. The people are a mixture of Inuit ('Eskimo') and Danish extraction. 'We just don't belong to Europe,' said Motzfeldt. 'We are a polar people. Our ties are with the Inuit of Canada, Alaska and Siberia. We want to make our own future.'

The task is not easy. The rocky and mountainous island is nearly 50 times the size of Denmark – with only one-hundredth of its population. All but one-tenth of its vast area is permanently covered with an ice cap whose average depth is more than 1500 m (nearly 5000 ft), leaving a 20 km (12 mile) wide coastal belt. Crop growing on any scale is impossible, and the only livestock farming of any commercial consequence is sheep farming. All but the southern end is inside the Arctic Circle. And even near the southern tip, at the capital GODTHÅB, (or Nuuk, as the Greenlanders prefer to call it), the sun rises in winter at 10 am and sets at 2 pm.

Gothåb, with a population of 12 200 – about the size of a small British town – is home to nearly one Greenlander in five. Like the other main towns on the island, it is on the west coast, warmed by southerly currents from the Atlantic. By contrast, a northerly current from the Arctic Ocean chills the east coast so that pack ice makes it virtually totally inaccessible for most of the year.

PROBLEMS OF MODERN SOCIETY

The capital's inhabitants, like other urban Greenlanders, live uneasily between the modern and the primitive. Secure jobs in fish factories – Greenland's only major manufacturing industry – have drawn many Inuits to the towns. Families that would once have wandered across the ice as nomadic hunters

▲ **SILENT BEAUTY Huge, desolate icefields sweep down to seas around Greenland. Sparkling icebergs like this were once the 'snout', or seaward end of a glacier, which has broken off in a process called 'calving'.**

with harpoons, kayaks, huskies and rifles now get their food from supermarkets kept stocked by freezer ships from Denmark. But these well-meant developments have carried a price. Greenland's suicide rate is the highest in the world, about one per thousand, and alcoholism is frighteningly common. Moreover, a high degree of sexual freedom – partly a legacy of Inuit tradition – has given the capital a high rate of venereal disease.

Although there is heavy dependence on Danish aid, the island's leaders are confident that Greenland will be able to survive on its own. It has fishing grounds producing abundant catches of cod and shrimp. In 1990, fish and fish products made up 83 per cent of its exports – and because fishing is so important, Greenlanders view environmental issues very seriously. It is believed that there may be oil reserves beneath the eastern coast. In addition, Greenland benefits economically from the American base at THULE, part of the United States' nuclear early warning system. (Another former US base at SØNDRE STRØM-FJORD was transferred into the hands of Greenland's government in September 1992.) There is also a growing tourist trade. Holidaymakers are flown by helicopter to villages along the west coast (there are few roads in Greenland outside the towns) to sample the sea fishing, or to explore on foot and sledge the fringes of the island's empty interior.

There is some precedent for the islanders' confidence. The first European to visit the island was a Norseman named Eirikr Raudi (Eric the Red), who sailed from Iceland in AD 982. He was sufficiently confident about its future to make his home here; his farmstead,

at Brattahlid, is now a tourist attraction. Eirikr also named the island Greenland, as a means, he admitted later, of making it sound more attractive to potential settlers – although in fact it was much greener then and it even had some forest, the climate being much better than today. His optimistic strategy worked. In 986, 35 small shiploads of pioneers followed him to Greenland, founding a Norse colony which lasted for 500 years.

Since 1985 Greenland has had its own flag; it is white and red with the figure of a rising sun.

GREENLAND AT A GLANCE	
Map Arctic – Cb	
Official name Kalaallit Nunaat	
Area 2 175 600 km² (840 000 sq miles), of which about 341 700 km² (131 930 sq miles) are ice free	
Population 58 000 **Per km²** 37.5 **(Per sq mile)** 14.5	
Capital Godthåb (Nuuk)	
Government Self-governing part of Denmark	
Currency 1 Danish krone = 100 øre	
Languages Greenlandic (Inuit, or 'Eskimo') and Danish (official)	
Religion Christian (95% Evangelical Lutheran)	
Climate Polar; along the south and west coast, tundra climates. Average temperature in Nuuk ranges from 7°C (46°F) in July to -7°C (20°F) in January. The interior is much colder; the average annual temperature is 0.7°C (0.5°F).	
Land use Grazing 1%, no forest, no cultivation; 85% covered with ice	
Main primary products Fish, sheep; zinc, lead	
Major industries Fishing, mining, fish and food processing, hides and skins	
Main exports Fish (cod, shrimps) and fish products, cryolite	
Annual income per head (US$) 8780	
Population growth (per thous/yr) 11	
Life expectancy (yrs) Male 63 **Female** 69	

Grenoble *France* Commercial and manufacturing city and tourist centre, and birthplace of Stendhal (1783-1842), author of the novel *Le Rouge et le Noir (Scarlet and Black),* 100 km (60 miles) south-east of Lyons.

It manufactures textiles, sweets, chemicals and electrical goods, and has a large university. In the 19th century, industrialists from Grenoble pioneered the generation of hydroelectricity. The city has various old churches, including a cathedral built in the 12th to the 13th centuries.
Population (city) 150 800; (metropolitan area) 400 000
Map France – Fd

greywacke Dark, very hard sandstone consisting of angular rock particles, up to 2 mm (0.08 in) in diameter, cemented together.

Grijalva *Mexico* Major river, some 480 km (300 miles) long, flowing from the highlands of CHIAPAS State northwards across the humid lowlands of TABASCO State into the Gulf of Campeche. It is also known as the Tabasco. Several hydroelectric power projects have been built along its course – including the Nezahualcóyotl Dam in Chiapas, the largest in Mexico. The vast river basin is a rich agricultural area.
Map Mexico – Cc

grikes (grykes) Deep fissures or cracks caused by the widening of joints by solution. They develop on horizontal limestone surfaces between CLINTS and are commonly found in limestone areas such as the Pennines in northern England, and the eastern Alps.

Grimsby *United Kingdom* Fishing port on the south shore of the Humber estuary, 24 km (15 miles) south-east of Kingston upon Hull in the English county of Humberside. Flemish fishermen began using the port in the 15th century, but new harbour works and railway links to the rest of the country greatly boosted it in the 19th century. It became one of the world's great fishing ports, with a fleet of about 250 vessels landing up to 200 000 tonnes of fish each year. However, the industry has recently been badly hit by diminishing stocks and fishing restrictions.

Today, Grimsby is a major city for food production, and has acquired the title of 'Europe's Food Town'. The National Fishing Heritage Centre, opened in 1991, is the central tourist attraction.
Population 91 900
Map United Kingdom – Ed

Grindelwald *Switzerland* Resort town lying at 1393 m (4570 ft) altitude and 55 km (35 miles) south-east of the capital, Berne. Its winter sports and its closeness to the Bernese Oberland – the EIGER, Wetterhorn and Schreckhorn tower over it – have made it popular with tourists for more than a century.
Population 3800
Map Switzerland – Ba

Grisons *Switzerland* See GRAUBÜNDEN

grit (gritstone) Coarse sedimentary rock, usually sandstone, composed of angular grains with a diameter of 0.06 mm-2 mm (0.0025 in-0.08 in), and occurring widely – in the Pennines in northern England, for instance.

Grenada

A BEAUTIFUL ISLAND OF SPICES WHERE MARXISTS FELL OUT BEFORE AMERICAN-LED TROOPS WERE SENT IN

The spice island of Grenada in the Caribbean made world headlines in 1983, when United States marines and troops from other Caribbean countries invaded the country to topple a Marxist government. Grenada had moved to the left after a bloodless coup in 1979. The new prime minister, Maurice Bishop, had immediately established links with the Soviet Union, Cuba, Libya and Algeria. The invasion came after he and three Cabinet ministers were murdered by rival Marxists who then seized power. Since then, much foreign aid has been directed to the island in an effort to ensure both a stable elected government and a sound economy. The first elections for eight years were held in 1984 and thereafter again in 1990 – in accordance with the 1974 constitution.

The most southerly of the Windward Island chain, Grenada is one of the most beautiful in the West Indies, with an attractive wooded landscape. The island consists of the remains of extinct volcanoes, the youngest and highest being Mount St Catherine (840 m, 2756 ft). Two of the old volcanic craters have become the Grand Etang and Antoine lakes. The only flat land is along the lower courses of the rivers and along part of the coast.

Grenada was sighted by Christopher Columbus in 1498. The first serious attempt at European settlement in 1609 was prevented by the Carib Indians who lived here. The French established a settlement in 1650, wiping out the Caribs, but Grenada was captured and ceded to Britain in 1783. Together with the small neighbouring islands of Carriacou and Petit Martinique, it became an independent member of the Commonwealth, with the Queen as head of state, in 1974.

Agriculture is the island's main industry, with cocoa, nutmegs bananas and mace the chief exports – the island supplies nearly one third of the world's nutmeg. Most manufacturing is concerned with the processing of crops, but there are factories which brew beer and make clothes and furniture. About 25 per cent of the workforce have no full-time job.

Tourism, which came almost to a standstill during Bishop's rule and the subsequent coup, is today an important source of foreign income. The wooded hillsides, the long coral beach of Grand Anse and the beautiful setting of the capital, ST GEORGE'S, are the main attractions. Much of the town was rebuilt after a hurricane destroyed it in 1955. Its harbour has berths for ocean-going ships.

Cruise ships call at St George's but the number of visitors used to be restricted by the lack of an international airport. The Cubans built a long runway for the new Point Salines airport, causing fears that it would be used by Communist military aircraft. This was one reason for the invasion in 1983. The airport was opened in 1984 and can handle long-haul flights, increasing tourist potential.

Of the other islands belonging to the GRENADINE group north of Grenada itself, Carriacou, most of which belongs to ST VINCENT, is the largest. Tourism and boatbuilding are the main activities. Petit Martinique is uninhabited.

GRENADA AT A GLANCE	
Map Caribbean Cc	
Area 344 km² (133 sq miles)	
Population 93 343 **Per km²** 271	
(Per sq mile 701)	
Capital St George's	
Government Parliamentary democracy	
Currency 1 East Caribbean dollar = 100 cents	
Languages English, Creole	
Religion Christian (64% Roman Catholic)	
Climate Subtropical, with a dry season from January to May. Average temperature in St George's hovers around 27°C (79°F) all year	
Land use Cultivated 41%, pastures 3%, woodland 9%, other 47%	
Natural resources timber	
Main primary products Cocoa, nutmegs, mace, bananas	
Major industries Agriculture, food processing, tourism	
Main exports Cocoa, bananas, nutmegs, mace, textiles	
Annual income per head US$ 2310	
Population growth (per thous/yr) 2.4	
Life expectancy (yrs) Male 68 Female 73	

Grmeč Planina *Bosnia-Herzegovina* Wild, forested limestone massif in western Bosnia between the Sana and Una rivers, rising to 1604 m (5262 ft) at Crni Vrh. A winding road links the ancient fortified towns of Bihac at the north-west end and Kljuc to the south-east, running through an area rich in monuments commemorating the partisans who fought here during the Second World War.
Map Bosnia-Herzegovina – Ba

Grodno *Belarus* See HRODNA

Groningen *Netherlands* Regional centre for the north-east, 145 km (90 miles) north-east of Amsterdam. Among its attractions are medieval moats and fortifications, which have been converted into canals and parks, museums and a university dating from 1614. The city's manufactures include book printing and sugar.

Groningen is the capital and only town of any size of the farming province of Groningen, which has a land area of 2611 km² (1007 sq miles). Chief farming occupations are the rearing of horses and the production of sugar beet, cereals, potatoes, and dairy goods.
Population (city) 169 390; (Greater Groningen) 208 470; (province) 553 900
Map Netherlands – Ca

Grootfontein *Namibia* Cattle-rearing district covering 26 520 km² (10 239 sq miles) around a town of the same name. After independence, it was absorbed into the larger Omaheke region. The town was founded in the 1890s around a German garrison. It lies 450 km (280 miles)

281

north-east of the capital, Windhoek. The world's second largest known meteorite, a mass of iron and nickel 2.75 m (9 ft) by 2.43 m (8 ft), and weighing an estimated 59 tonnes, was found in 1920 on Hoba West farm, 20 km (12 miles) west of the town.

Population (district) 34 900; (town) 18 000
Map Namibia – Aa

Grosser St Bernhard (Grand St Bernard Pass; Col du Grand St-Bernard) *Switzerland/Italy* Alpine pass between Switzerland and Italy 25 km (15 miles) east of Mont Blanc. The area is famous for its St Bernard dogs, which were used to rescue travellers from snowdrifts. The successors of the monks who bred the dogs run a hospice, founded in the 11th century on the 2469 m (8100 ft) high summit. Today's travellers cross the mountain through a 5.8 km (3.6 mile) tunnel.
Map Switzerland – Ab

Grossglockner *Austria* See HOHE TAUERN

grotto A natural or artificial cave.

ground frost Term used to indicate a ground level temperature of 0°C (32°F) or below, while the air temperature remains above freezing point.

ground water Term used, generally, for any water beneath the earth's surface, and more specifically, for a region of subsurface water beneath the WATER TABLE, including underground streams, where it forms the saturation zone in which all spaces and fissures are filled with water.

groyne Narrow jetty or wall built out into the sea from the shoreline to arrest the movement of waterborne sand or gravel along the shore and thereby to protect beaches from erosion.

Groznyy *Russia* Oil city and capital of Chechenya (formerly the Chechen-Ingush Republic). It lies in the north Caucasus foothills, 145 km (90 miles) west of the Caspian Sea. Groznyy was founded in 1818 as a fort and developed after the discovery of oil in 1893. It has oil pipelines to the Caspian port of Makhachkala and to Tuapse on the Black Sea. Its other industries include petrochemicals, textiles and processed foods. Groznyy was devastated in 1994-5 when Russian troops invaded the country to end Chechen's proclaimed independence.
Population 401 000
Map Russia – Fd

Grudziadz *Poland* Industrial town 100 km (60 miles) south of the port of Gdańsk. It was developed as a fortress by the Teutonic Knights in 1291. Today its factories make metals, beer, footwear, rubber and farm machinery, and centuries-old riverside granaries are still in use.
Population 100 900
Map Poland – Cb

Grünberg *Poland* See ZIELONA GÓRA

Grünwald (Tannenberg) *Poland* Battlefield near the village of Stębark in northern Poland, 140 km (85 miles) south-east of the port of Gdańsk. On 15 July 1410, a Polish-Lithuanian Tatar force led by King Wladyslaw Jagiello defeated the Teutonic Knights here. And in August 1914, during the First World War,

German troops defeated Russian forces at the same spot. A granite obelisk and a museum mark the site of the battle.
Map Poland – Db

Gruziya See GEORGIA

Gstaad *Switzerland* Fashionable skiing resort 1050 m (3444 ft) up in a valley, 55 km (35 miles) south and slightly west of the capital, Berne. It is also a popular summer resort for tourists visiting the Bernese Oberland. The resort hosts an annual international tennis tournament in July and a music festival from end-July to mid-September every year.
Population 2500
Map Switzerland – Aa

Guadalajara *Mexico* Capital of the state of Jalisco. It lies on a plateau 1580 m (5190 ft) above sea level, and is the main industrial, commercial and agricultural centre of the western highlands. In colonial times it was reputed to be 'more Spanish than Spain itself' and much of its colonial heritage survives, including parks, fountains, squares, churches, a majestic 16th-century cathedral and the very ornate government palace, as well as an 18th-century university. The city still has a gracious, provincial atmosphere, despite its skyscrapers and its factories which manufacture goods such as shoes and textiles. It is also noted for its many crafts such as leatherwork, embroidery, glass, weaving and ceramics.
Population 2 300 000
Map Mexico – Bb

Guadalajara *Spain* Market town about 50 km (30 miles) north-east of Madrid, which trades in cereals, wine and olives produced on farms in the surrounding province of the same name.
Population (town) 63 650; (province) 145 600
Map Spain – Db

Guadalcanal *Solomon Islands* Island in the south of the group on which stands the Solomon Islands' capital, HONIARA. During the Second World War, it was the scene of bitter fighting between Japanese and United States forces. The island has extensive rice paddies and plantations of oil palms.
Population 50 400
Map Pacific Ocean – Cc

Guadalquivir *Spain* River, 657 km (408 miles) long, flowing south-west into the Gulf of Cádiz on the southern Atlantic coast. Its water is tapped for irrigation and there are several dams along its course.
Map Spain – Cd

Guadalupe Mountains *USA* Range in west Texas on the southern border of New Mexico with Texas, reaching 2667 m (8751 ft) at Guadalupe Peak, the highest point in Texas. Guadalupe Mountains National Park contains significant Permian limestone fossil beds.

Guadeloupe *West Indies* Group of islands in the Leeward chain in the eastern Caribbean. Christopher Columbus reached Guadeloupe in 1493, but the indigenous Carib people (who called the main island Karukera) resisted Spanish would-be colonists. Eventually, in 1635, French settlers occupied the islands and apart from short

periods of British occupation (and one of Swedish control), Guadeloupe has remained under French rule. In 1946 it became an overseas department of France, with full representation in the National Assembly.

The main island of Guadeloupe is made up of two sections separated by a narrow strait, Rivière Salée ('Salt River'), 6 km (3.5 miles) long. The western section, Basse Terre, is one of a chain of volcanic islands in the eastern Caribbean. The eastern section, Grande Terre, is part of an arc of limestone islands. Basse Terre contains Soufrière, an active volcano 1467 m (4813 ft) high that is the highest point in the eastern Caribbean. The other islands in the group are nearby MARIE GALANTE, La Désirade and the tiny Iles des Saintes – as well as ST BARTHÉLÉMY and ST MARTIN, more than 200 km (125 miles) further north-west. In all, the islands total 1702 km² (657 sq miles).

Agriculture is important, with the main crops being bananas, sugar cane and aubergines. The chief industries are sugar refining and rum distilling. Year-round sunshine, good sea bathing and beautiful mountain scenery attract tourists – especially from France and the United States.

Most of the islands' inhabitants are descended from African slaves. However, the populations of St Barthélemy and Terre de Haut in the Iles des Saintes are descended from 17th-century settlers from northern France. French is the official language throughout the islands, but many people speak a Creole dialect.

The seaport of BASSE TERRE is the capital of Guadeloupe, but the largest town is the commercial centre and port of POINTE-À-PITRE on Grande Terre.
Population 422 000
Map Caribbean – Cb

Guadiana *Spain/Portugal* River, 778 km (483 miles) long, which rises south-east of the Spanish capital, Madrid. It forms parts of the southern Spanish-Portuguese border and flows into the Atlantic at the Gulf of Cádiz.
Map Spain – Cc

Guaíra Falls *Paraguay* See ITAIPÚ

Guajira *Colombia* Department with a thinly populated, largely desert peninsula jutting into the Caribbean Sea in the extreme north of the country. The capital is Ríohacha (population 126 000), a treeless city and port 220 km (140 miles) north-east of BARRANQUILLA. Marijuana is widely, though illegally, grown in the area and is then smuggled out for sale in North America. Deposits of coal and natural gas are being developed.
Population 347 540
Map Colombia – Ba

Guam *Mariana Islands* Largest and southernmost of the Mariana Islands, covering an area of 549 km² (212 sq miles) and lying about 2250 km (1400 miles) south-east of Okinawa, Japan. When it was 'discovered' by Fernaoã Magellan in 1521, it had already been inhabited by Micronesian people for at least 3000 years. The Spanish claimed it in 1565. Half of its present population is Chamorro – people of mixed, mainly Filipino, Indonesian and Spanish descent; 15 per cent are immigrants from the United States. Guam became an American territory after the Spanish-

American War in 1898 and, apart from Japanese occupation in 1941-4, has remained so. It has internal self-government but a 1982 referendum favoured Commonwealth status like that of the neighbouring NORTHERN MARIANAS. Its present status remains as unincorporated US territory, which means the populace are United States citizens but do not take part in the United States' elections.

More than one-third of the land is a US military base, and this dominates the local economy, ensuring a far higher standard of living than elsewhere in Micronesia. Nevertheless, in 1979 voters overwhelmingly demanded a return of much of their land to civilian use. The second major source of income is tourism, with almost a million people visiting the island every year – 80 per cent of them from Japan. Guam is a transport hub of the western Pacific and is second only to Hawaii as a Pacific tourist destination;

Although parts of the island are highly developed, 18 per cent of the land area is dense rain forest (in part of which a Second World War Japanese soldier was found still hiding in 1972). It also has beautiful beaches (some with crashing surf) and sleepy South Sea island villages – as well as architectural relics of Spanish rule. Some copra is produced, but most food is imported.

Population 146 000
Map Pacific Ocean – Cb

Guanajuato *Mexico* Capital of Guanajuato State, and one of the most striking colonial towns of central Mexico. Its picturesque houses and winding, cobbled streets nestle at the bottom of a narrow gorge in the mountains of the Sierra Madre. Founded in 1554, Guanajuato soon became a boom town after silver was discovered in the surrounding area. By the end of the 17th century it had become the richest city in Mexico. The legacy of its prestigious past now includes the gleaming white university, the ornate Juarez Theatre, and the Alhondiga de Granaditas – an old grain warehouse converted into a museum dedicated to the heroes of the independence movement. Guanajuato provides a charming backdrop to the annual Cervantes Festival – a three-week international event of the performing arts held twice a year, in April/May and October.

Population (town) 45 000; (state) 3 980 000
Map Mexico – Bb

Guangdong *China* Mountainous province in the south covering 210 000 km² (81 000 sq miles). Part of it lies within the tropics, and the growing season in the valleys of the Xi, Bei and Dong rivers and in the delta of the Zhu Jiang lasts all year. Farmers can grow three crops of rice annually, as well as sugar cane, sweet potatoes and tropical fruits. The population is concentrated in the cities, of which the capital, GUANGZHOU, is the largest.

Guangdong has become an increasingly prosperous province. Its economy has benefited from recent changes in Chinese agricultural policy allowing farmers to sell their surplus at a profit instead of growing it only for the state, and from increased trade with Hong Kong and Macau.

Population 65 250 000
Map China – Gf

Guangxi-Zhuang *China* Autonomous region on the north of the Gulf of Tongking, bordering Vietnam. The interior is hilly and agriculture is confined to narrow river valleys and the coastal belt. The region covers 230 000 km² (88 800 sq miles) and straddles the Tropic of Cancer. Two crops of rice are grown each year, as well as sugar cane, tobacco and tropical fruits. Several ethnic minorities closely linked to the peoples of South-East Asia live here, including 12 million Zhuang, who are related to the Thais and who were the original inhabitants of the area. The regional capital is NANNING.

Population 43 800 000
Map China – Ff

Guangzhou (Canton) *China* Capital of Guangdong Province and the country's sixth largest city. It lies in the delta of the Zhu Jiang River and rose to prominence as a maritime trade centre during the Han dynasty (206 BC-AD 220). Portuguese traders arrived in 1514 and were followed by British merchants in the 17th century. Later, the city became a centre for the opium trade, encouraged by Britain but outlawed by the Chinese government, whose attempt to suppress the traffic in 1839 led to the so-called Opium War with Britain. After three years the Chinese were defeated and opium traders flourished. Today, the city is one of China's most cosmopolitan cities.

Population 5 669 000
Map China – Gf

guano The dried droppings of seabirds or bats, which have accumulated along dry coastal areas or in caves. Guano is a rich source of phosphate of lime used in fertiliser.

Guantánamo *Cuba* City in the extreme southeast, 65 km (40 miles) east of Santiago de Cuba. It is the commercial centre and capital of an agricultural province of the same name, with sugar milling, coffee-roasting, tanning and confectionery industries. Guantánamo Bay, on which the city stands, is the site of a United States naval base. An area of 112 km² (43.3 sq miles) was leased to the United States in 1903. In recent years a fence and two mine fields – one Cuban and one American – separated it from Fidel Castro's communist domain. In 1994 its barren, rock-strewn hills and beaches, contrasting sharply with the lush semi-tropical island across the mine fields, became the site of a tent town housing up to 65 000 Cuban refugees from the communist regime.

Population (city) 200 000; (province) 500 000
Map Cuba – Ba

Guatavita la Nueva *Colombia* Modern town 75 km (45 miles) north-east of Bogotá. It was built in colonial style to replace an older town submerged by a hydroelectric scheme in 1967. Near the town, in the crater of an old volcano, is Lake Guatavita, source of the legend of El Dorado – 'the Gilded One'. According to legend, the leader of the Chibcha Indians would coat himself in resin and gold dust once a year and leap into the lake from a raft as an offering to tribal gods. He would emerge cleansed of the gold.

Gold objects were also thrown into the lake as offerings, and several unsuccessful attempts have been made to drain it to find the treasure. It was this story that lured the conquistadores into the heart of South America. A gold model of a raft, found in the lake, is in the museum at Bogotá.

Population 15 000
Map Colombia – Bb

Guatemala See p 284

Guatemala City *Guatemala* National capital and the largest city between Mexico and Colombia. It sprawls across a plateau 1500 m (4920 ft) high, 75 km (47 miles) inland from the Pacific Ocean. It is 10 times larger than any other town in Guatemala, and is responsible for half of the country's industrial output.

The city was founded as Guatemala's third capital in 1776 after earthquakes had destroyed the earlier capitals of Antigua and Ciudad Vieja. It was itself virtually destroyed in 1917-18, by earthquakes which lasted on and off for six weeks, and was again seriously damaged by an earthquake in 1976.

Today Guatemala City is a modern, commercial city with good shopping and a colourful market. It houses the national archaeological museum which has an impressive collection of finds from Mayan sites. The city's University of San Carlos is one of the oldest in the Americas, founded in 1676. The city produces textiles, foodstuffs and silverware.

Population 2 000 000
Map Guatemala – Ab

Guayaquil *Ecuador* Capital of Guayas Province, standing on the Guayas River near its mouth on the Gulf of Guayaquil. It is the country's largest city and chief port, whose exports include bananas, cocoa, sugar, coffee and cattle; among its manufactures are textiles, leather goods, cement and iron products.

The original European settlement was founded just east of the present site by the Spanish conquistador Sebastián de Belalcázar in 1535. In 1537, the Spanish explorer Francisco de Orellana established the present town but it was destroyed by earthquake in 1942. Much of the town has been rebuilt.

In 1822, Guayaquil was the scene of the fateful meeting between Simón Bolívar and José de San Martín, the principal leaders of the struggle for South American independence from Spain. The two leaders disagreed on policy and San Martín retired from active service in the wars of liberation, leaving Bolívar to complete the struggle alone.

Population 1 508 400
Map Ecuador – Bb

Guayaquil, Gulf of *Ecuador* Large bay, 180 km (112 miles) wide at the mouth, on the Pacific coast. It was visited in 1526 by the Spanish conquistador Francisco Pizarro (1478-1541) who found a temple covered with gold in the Amerindian town of TUMBES on the south shore. This confirmed his belief that South America was rich with treasure, and led to the Spanish conquest of the region. The two main ports on the bay are Guayaquil and Puerto Bolívar.

Map Ecuador – Ab

Guayas *Ecuador* Coastal province which produces more than half of the country's bananas as well as rice, sugar cane, cotton and, farther inland, coffee and cocoa. Most of the produce is taken to the provincial capital, GUAYAQUIL, by road or rail, but some is carried by small boats on the Guayas River. There are substantial oil and natural gas deposits in the river basin.

Population 2 515 100
Map Ecuador – Ab

Guatemala

A COUNTRY OF BEAUTIFUL MOUNTAINS AND LAKES, WHOSE TRANQUILLITY MIGHT BE REGAINED THROUGH DEMOCRACY

The mountainous land of Guatemala is regularly shaken by earthquakes and by political unrest. A ridge of volcanoes parallel to its Pacific coast marks a line of physical instability which has devastated successive earlier capitals – first ancient Ciudad Vieja in 1542, then colonial Antigua in 1773, and the present capital, GUATEMALA CITY, in 1917, 1918 and 1976.

▲ **LIFE AMONG THE RUINS A market is held in the remains of a Jesuit monastery in Antigua, the old capital which was wrecked by an earthquake in 1773. Vegetables, bags and brightly coloured women's tunics are on sale.**

Most of the remainder are *mestizos* (of mixed Amerindian and Spanish blood). The population is less varied than elsewhere in Central America because few Spaniards settled in the country and no slaves were brought in.

The indigenous population is divided culturally into *ladinos*, who are Westernised, and *indígenas*, who have kept their old customs and languages. The culture of the *indígenas* is threatened by the civil conflict. Villages have been destroyed and villagers displaced or killed by government and rebel forces.

The Amerindians are extremely poor. Roman Catholic missionaries who tried to improve the lot of the peasants were regarded as subversives by the government and had to flee the country. In their place came North American evangelists who supported the government. Thirty per cent of the people are now born-again Christians.

Although there is widespread poverty, Guatemala is one of the richest countries in Central America. But only a tiny minority – the powerful land barons, the coffee-trading merchants and US companies – have a share in the country's riches, so that 3.7 per cent of the agricultural companies own more than two-thirds of the country's arable land.

Coffee made the country prosperous and still dominates the economy. It is grown on the lower slopes of the volcanic highlands; the heights, though cool and inhabited by 65 per cent of the people, are almost uncultivated. Here, people plant small patches of maize, beans and potatoes for their own use. A 50 km (30 mile) strip of the hot Pacific coast produces sugar, cotton, cattle and bananas.

Civil conflict has almost destroyed tourism, and has dragged the economy down. The land could generate more wealth; nickel mining, once important, has been suspended but there

is oil, zinc, lead and copper, as well as some antimony, tungsten, sulphur and marble, and hydroelectric power potential. The volcanic subsoil is fertile, and although one-third of the country is uninhabited and very little cultivated, crops could be grown almost everywhere in the country.

Guatemala has the largest and one of the most developed manufacturing sectors in Central America. This took a dip in the 1980s but showed signs of recovery by the early 1990s. Guatemala City is growing, with a drift of people from the countryside over the past 35 years swelling the population to just over 2 million – nine times the size of any other town in the country. It is the seat of industry – textiles, paper and pharmaceuticals – and the core of economic life. Although education is free, 55 per cent of adults are illiterate.

Politics are also unstable: governments have mostly been established by armed coups, and their regimes have been unjust and cruel. The military handed power over only in 1985; and in 1986 an elected civilian government made a start in dismantling the apparatus of repression, such as the secret police. But political killings continued (between 1983 and 1985 there had been up to 100 political assassinations and 40 abductions a month), price rises led to labour unrest, and the army remained the power behind the scenes. The war between government troops and the leftist guerrillas also continued. However, in January 1994 a civilian government called a national referendum to pave the way at last to peaceful resolution. By introducing amendments to the constitution, President Ramiro de Léon Carpio hoped to bring about democracy and an improved economy, as well as fight human rights violations and corruption. Talks with the guerrillas at the beginning of 1994 also promised a peaceable solution to the country's long-standing internal conflict.

Fifty per cent of Guatemalans are pure-blood Amerindians, descendants of the Mayans whose ruined cities, built 2000 years ago, still stand in remote lowlands in the north.

GUATEMALA AT A GLANCE		
Official name Repulic of Guatemala		
Area 108 889 km² (42 042 sq miles)		
Population 9 746 000 **Per km²** 89.5		
(Per sq mile 231)		
Capital Guatemala City		
Government Republic		
Currency 1 quetzal = 100 centavos		
Languages Spanish, local native American languages		
Religions Christian (75% Roman Catholic), tribal religions		
Climate Tropical, temperate in highlands. Average annual temperature in the mountainous regions is 20°C (68°F); in the lowlands it is more than 25°C (77°F)		
Land use Forest 34%, cultivation 17%, grazing 13%, other 36%		
Main primary products Coffee, bananas, cotton, sugar, maize, cattle, tobacco, timber; crude oil, nickel, lead, zinc		
Major industries Crude oil refining, pharmaceuticals, food processing		
Main exports Coffee, sugar, cotton, bananas, cardamom		
Annual income per head (US$) 980		
Population growth (per thous/yr) 29		
Life expectancy (yrs) Male 55 **Female** 60		

Gubbio *Italy* Well-preserved medieval town 160 km (100 miles) north of Rome. It has a church and cathedral, both of the 13th century, and an elegant 18th-century ducal palace. Its chief product is ceramics.
Population 30 540
Map Italy – Dc

Guernica *Spain* Town 19 km (12 miles) east of the northern port of Bilbao and known as the 'holy city' of the Basque people. Their parliament met here, under an oak tree, from the Middle Ages until the 19th century. The town was destroyed in 1937 by German bombers supporting the Nationalists in the Spanish Civil War, and became a symbol of the war's horrors when it inspired Pablo Picasso's painting, *Guernica,* now at the Reina Sofia Museum in Madrid. Today the town makes metal goods and furniture.
Population 18 130
Map Spain – Da

Guernsey *English Channel* See CHANNEL ISLANDS

Guerrero *Mexico* Mountainous state with a long Pacific coastline in south-west Mexico. It contains the famous beach resorts of ACAPULCO and ZIHUATANEJO, and the beautiful 18th-century silver-mining town of TAXCO near its northern border. Guerrero's capital, Chilpancingo, is principally an agricultural centre, although it has a prestigious university and some industry.
Guerrero is also the name of a town on the Conchos River in Chihuahua State. Horses are raised here and cereals and fruit are grown in the surrounding districts.
Population (state) 2 622 000; (town) 35 630
Map Mexico – Bc

Guerrero Negro *Mexico* Marshes on the BAJA CALIFORNIA peninsula which are the largest salt flats in the world. Oil and natural gas have been discovered here.
Map Mexico – Ab

Gui Jiang *China* River of the Guangxi-Zhuang Autonomous Region, flowing 350 km (220 miles) to join the Xi Jiang at Wuzhou.
Map China – Gf

Guiana Highlands *South America* Vast, thinly populated area south of the Orinoco River, extending eastwards from eastern Venezuela across northern Brazil, Guyana, Surinam and French Guiana. Its forested hills contain many minerals, including bauxite, iron ore and manganese. Their highest point is Roraima, 2810 m (8565 ft), a flat-topped mountain which lies at the intersection of the boundaries of Brazil, Venezuela and Guyana. The highlands are the source of many rivers including those of the Amazon system to the south and of the Orinoco system to the north.
Map Venezuela – Bb

Guildford *United Kingdom* See SURREY

Guilin *China* City in the Guangxi-Zhuang Autonomous Region, 400 km (250 miles) north-west of Guangzhou and close to some of the most spectacular mountain scenery in China. To the south of the city, an extraordinary landscape of natural domes and towers, some rising almost vertically 100 m (330 ft) and more from shimmering paddy fields, stretch on either side of the Gui Jiang River. The hills are riddled with caves and garlanded with trees, orchids and vines. Their haunting beauty has frequently been the inspiration for Chinese landscape paintings and attracts many tourists.
Population 686 000
Map China – Gf

Guimarães *Portugal* Town 40 km (25 miles) north of the city of Oporto. It was the birthplace of Portugal's first king and is therefore often referred to as 'the cradle of the nation'. It was also the ancient capital of Portucale, the heartland from which Portugal was reconquered from the Moors between the 11th and the 13th centuries. The castle, with a 10th-century tower, and the 15th-century palace of the dukes of Braganza survive. The city has textile and craft industries, notably several scissors and knife manufacturers.
Population 22 000
Map Portugal – Bb

Guinea See p 286

Guinea *West Africa* Name derived from an ancient African kingdom, and at one time applied to the whole coastal region of West Africa from present-day Senegal to Angola. Much of this 6250 km (3900 mile) coast is hot, humid rain forest. The region today includes the states of EQUATORIAL GUINEA, GUINEA and GUINEA-BISSAU.

Guinea, Equatorial See EQUATORIAL GUINEA

Guinea, Gulf of Arm of the ATLANTIC OCEAN off the south coast of West Africa extending about 1900 km (1180 miles) from Cape Palmas, in Liberia, to Cape Lopez, in Gabon. It includes the bights of BENIN and BIAFRA.
Map Africa – Cd

Guinea Highlands *West Africa* Mountainous region in Guinea extending into Sierra Leone, Liberia and Ivory Coast. The mountains are generally rounded; thanks partly to their inaccessibility, they are heavily forested and contain much wildlife.
The peaks include some of West Africa's highest points, including Guinea's Mount Nuon, 1825 m (5987 ft), and Mount Nimba, 1752 m (5748 ft). There are coffee, tea and kola plantations in some areas. Rice is the staple crop of the local Kissi people. The Nimba Mountains have iron ore deposits.
Map Guinea – Cb

Guinea-Bissau See p 287

Guipúzcoa *Spain* One of the three BASQUE REGION provinces, on the Bay of Biscay. At 1997 km² (771 sq miles), it is Spain's smallest province, though densely populated.
Population 676 490
Map Spain – Da

Guiyang *China* Capital of Guizhou Province, about 400 km (250 miles) north of the border with Vietnam. It is an important railway junction, with factories manufacturing rolling stock, and an iron and steel works.
Population 1 530 000
Map China – Fe

Guizhou *China* Mountainous and largely forested province in the south, covering 170 000 km² (65 600 sq miles). It is part of a great highland plateau stretching into neighbouring Yunnan Province. Its climate is mild and wet. Rice and wheat are grown in the fertile, narrow valleys and livestock is reared in the mountains. Many of Guizhou's inhabitants are not ethnic Chinese: the Buyi and Zhuang, for example, who live in the mountains, are related to the Thais of South-East Asia. Guizhou is one of the most remote, backward and sparsely inhabited provinces of China. Its capital is GUIYANG.
Population 33 610 000
Map China – Fe

Gujarat *India* Western state covering 196 024 km² (75 665 sq miles), which includes the Kathiawar Peninsula and the adjoining fertile Indus flood plain in the north and stretches to the Gulf of Khambhat in the south-east. It was formed in 1960 when Bombay state was divided between the Gujarati-speaking north and the Marathi-speaking south.
It is overwhelmingly Hindu, although a significant number of Muslims live here too, especially in the city of AHMADABAD. The new city of Gandhinagar is the state capital. Named after Mahatma Mohandas Gandhi, construction of Gandhinager began in 1965.
Population 41 174 000
Map India – Ac

Gujranwala *Pakistan* Town in the Punjab, 65 km (40 miles) north of LAHORE. The greatest of the Sikh kings, Ranjit Singh, was born here in 1780. A memorial to him commemorates his cremation in Lahore in 1839, at which his concubines threw themselves on to his funeral pyre.
Population 1 076 000
Map Pakistan – Da

Gulag Archipelago *Russia* See NORIL'SK

gulch Deep rocky ravine or valley, especially one cut by a torrent, in the western United States.

Gulf For physical features where a name begins with 'Gulf', see main part of the name.

Gulf, The Variously called the Arab, Arabian or Persian Gulf, it is linked to the ARABIAN SEA by the Strait of HORMUZ and the Gulf of OMAN. The Gulf, which covers 233 100 km² (90 000 sq miles) and reaches a depth of 91 m (300 ft), has become an important shipping lane for tankers and other cargo vessels because of the tremendous wealth of the oil-producing nations that surround it. In many places the water is so shallow that ships of more than 5000 tonnes cannot get within 8 km (5 miles) of the shore, and the shoals and reefs on the Arabian coast, where famous pearl oyster beds lie, are hazards to shipping. Several of the Gulf states have constructed artificial 'islands' offshore that are linked to the mainland by submarine pipelines. This enables tankers up to 500 000 tonnes deadweight to load crude oil from the inland oil fields.
Oil pollution as a result of Iraqi attacks on oil installations in Kuwait during the Gulf War (1991-92) caused considerable damage to the environment in and around The Gulf, particularly in the northern areas.
Map Saudi Arabia – Cb

Guinea

LONG ISOLATED THROUGH FRENCH RETRIBUTION AND ITS LEADER'S PARANOIA, THE FIRST INDEPENDENT STATE OF FRENCH-SPEAKING AFRICA STRUGGLES FOR SELF-SUFFICIENCY

▲ **POVERTY WITHOUT LIBERTY A poster of Guinea's first president, Sékou Touré, overlooks a street in Conakry. At independence in 1958 he claimed: 'We prefer poverty in liberty to wealth in slavery.'**

A green, lush and beautiful country with perhaps the greatest agricultural potential of all the former French territories in West Africa, Guinea was for 20 years after independence hidden from the outside world by its own isolationist policies – a reaction to the abrupt withdrawal of French aid. At the same time, its people suffered extreme poverty and political persecution meted out by its socialist leader, Ahmed Sékou Touré.

European involvement was restricted to coastal trading and slave raiding. Trading posts were set up, and in the 19th century France gradually assumed control.

In a referendum in 1958, Guineans chose total independence against membership of a French-African community. France's reaction was to break off diplomatic relations, stop all aid and assistance, withdraw investment and close its market to Guinean produce. Instantly crippled, Guinea took aid from wherever it could – mainly the former USSR, which offered it way below world market prices for its most important export, bauxite (aluminium ore). More recently loans have been obtained from the World Bank and the International Monetary Fund (IMF), and have gone into developing the infrastructure and boosting the depressed economy.

Guinea's main source of income – and its main hope for the future, apart from revitalising agriculture – is still its mining industry. The country has some of the world's biggest reserves of high-grade bauxite, and in 1991 ranked as the world's second largest bauxite producer. It is also rich in iron ore, mines uranium and gold (since 1988), and its diamonds are its second most important export after bauxite. But development has been hampered by a poor transport system.

From 1978 the dire need for aid forced the government to adopt a more 'open door' policy, and after Touré died in 1984 the new military administration began reforms that could at last speed development, and set up a Transitional Committee for National Recovery, consisting of both military and civilian members. In 1992 a multiparty electoral system was introduced.

Situated on the bulge of West Africa, Guinea is about the same size as the United Kingdom. Dense mangroves edge the deeply indented 320 km (199 mile) Atlantic coastline, but parts of the swamps and the forested coastal plains have been cleared to grow crops such as rice, cassava, yams, maize and vegetables. Sheer sandstone escarpments rising from the plains mark the edge of the FOUTA DJALLON, where level plateaus – often rising to more than 900 m (2950 ft) – are cut by fertile valleys. Despite the often bare red dusty surface, the plateaus provide grazing for the unique dwarf N'Dama cattle, while crops such as *fonio* (a small millet), cassava, vegetables and fruit (especially pineapples and bananas) are cultivated in the valleys.

East of the Fouta are the savannah plains of Upper Guinea, where livestock and cereals are the main products. To the south-east rise the rounded, forested hills of the GUINEA HIGHLANDS, reaching peaks of more than 1800 m (5900 ft); here, coffee and kola nuts are important cash crops and rice is also grown.

Guinea has a number of ethnic groups. The Susu predominate near the coast, the Kissi in the Guinea Highlands and the Tenda in the east. The Fulani are pastoralists of the Fouta Djallon and Upper Guinea, while the Malinke (Mandinka) are traders who live throughout the country. As in much of West Africa, early

GUINEA AT A GLANCE	
Official name Republic of Guinea	
Area 245 857 km² (94 926 sq miles)	
Population 6 400 000 **Per km²** 26 (**Per sq mile** 67)	
Capital Conakry	
Government Republic	
Currency 1 Guinean franc = 100 centimes	
Languages French (official), various local languages	
Religions Predominantly Muslim; indigenous beliefs, Christian minority	
Climate Tropical; wet season from April to November. Average annual temperature in Conakry is 27°C (80.6°F)	
Land use Forest 59%, grazing 25%, cultivation 3%, Other 13%	
Major industries Agriculture, mining, bauxite refining, fishing	
Main exports Bauxite, diamonds, gold, coffee	
Annual income per head (US$) 510	
Population growth (per thous/yr) 28	
Life expectancy (yrs) Male 44 **Female** 44	

Guinea-Bissau

EARNING LESS IN A YEAR THAN MOST WESTERNERS DO IN A WEEK, THE PEOPLE OF THIS SMALL COUNTRY STRUGGLE TO FEED THEMSELVES

Guinea-Bissau (formerly Portuguese Guinea) is one of the smallest and poorest nations in West Africa, and one of the poorest in the world. Yet it is a peaceful country with stunning scenery. It rises gently from the deeply indented and island-fringed coastline (with a dense cover of forest, interspersed with marshes and deltas) to wooded savannah on the low inland plateau. Three main rivers form natural highways into the interior. The 15 ethnic groups include the Balanta (the biggest group at 30 per cent), Fulani, Manjaka, Mandinga and Pepel.

In 1963, the Partido Africano da Independencia da Guiné e Cabo Verde (PAIGC) began a war of independence for Guinea-Bissau and the nearby Portuguese colony of Cape Verde Islands (aiming to combine them as a single independent state).

Guinea-Bissau proclaimed its independence in 1973; Portugal finally recognised its independence in 1974 and that of the Cape Verde Islands a year later.

Eleven years of war after a century of Portuguese rule left the country impoverished. The new PAIGC government concentrated on agricultural development, aiming for self-sufficiency in food, but progress was set back by a drought in 1977, another in 1979-80 and in 1983. Later, it introduced a four-year plan

(1988-91) to further develop the agricultural sector, which employs 90 per cent of the workforce and accounts for almost 100 per cent of all exports. The country has some mineral resources but lack of funds has hampered their development.

In 1991, the PAIGC approved the introduction of a multi-party democracy. Multi-party national elections scheduled for November and December 1992 were postponed until March 1993, and again, under much criticism by opposition parties, until late in 1994.

GUINEA-BISSAU AT A GLANCE

Map Guinea – Aa-Ba	
Official name Republic of Guinea-Bissau	
Area 36 120 km² (13 943 sq miles)	
Population 1 078 000 **Per km²** 29.8 **(Per sq mile** 77.3)	
Capital Bissau	
Government Republic	
Currency 1 Guinean peso = 100 centavos	
Languages Portuguese, Criolo (Portuguese Creole), African languages	
Religions Tribal (65%), Muslim (30%), Christian (5%)	
Climate Tropical, heavy rains May to October; average temperature is 27°C (81°F) throughout the year	
Land use Grazing 43%, forest 38%, cultivation 12%, other 7%	
Main primary products Cereals, coconuts, groundnuts, cashew nuts, rice, palm kernels, timber, fish; bauxite, phosphates	
Major industries Agriculture, fishing, forestry, beverages	
Main exports Oil seeds, processed fish, groundnuts, palm oil, timber, cashew nuts	
Annual income per head (US$) 210	
Population growth (per thous/yr) 21	
Life expectancy (yrs) Male 45 **Female** 48	

Gunnbjørn Fjeld *Greenland* Highest mountain in Greenland, at 3702 m (12 146 ft). It rises on the east side of the island in King Christian IX's Land.
Map Arctic – Bc

Gunsan *South Korea* See KUNSAN

Gunung Agung *Indonesia* Sacred Hindu mountain of Bali, rising to 3142 m (10 308 ft). A volcanic eruption in 1963 killed more than 1500 people during an important religious festival.
Map Indonesia – Ed

Gunung Kerinci (Kerintji) *Indonesia* Active volcano and the highest mountain on Sumatra. It rises to 3805 m (12 484 ft) in the centre of the island.
Map Indonesia – Bc

Gunung Merapi *Indonesia* Volcanic peak of central Java, one of the most active on the island. It rises to 2911 m (9550 ft) some 30 km (18 miles) north of Yogyakarta.
Map Indonesia – Cd

Gurgan *Iran* See GORGAN

Guri Dam *Venezuela* Hydroelectric scheme on the Caroní River south of Ciudad Guayana. It is Venezuela's largest producer of hydroelectric power, supplying 70 per cent of its needs.
Map Venezuela – Bb

Guru Sikhar *India* Outlying mountain of the Mount Abu massif in the Aravalli range, on the Rajasthan border with Gujarat, rising to 1722 m (5648 ft). It is noted for its 11th to 13th-century Dilwara Jain temples, which contain some of the finest marble stone carvings in India. The mountain is one of the haunts of the sacred monkeys called Hanuman langurs. Revered by Hindus,

▼ HOLY MOUNTAIN The cone of Bali's Gunung Agung, locally called 'the Navel of the World', towers over the rice fields. Besakih, Bali's main Hindu temple, stands halfway up.

Gulf Stream *Atlantic Ocean* Warm ocean current of the North Atlantic, flowing from the Gulf of Mexico north-eastwards up the east coast of the United States. After passing Newfoundland it turns more to the east and is properly known as the NORTH ATLANTIC DRIFT, which tempers the climate of north-western Europe and helps to ensure copious rainfall. The Gulf Stream moves at an unusually high speed of 9 km/hr (6 miles/hr) and reaches depths of up to 1000 m (3280 ft). It carries 55 million m³ (1943 million cu ft) of warm water, with a high salt content, every second. Coasts even within the Arctic Circle benefit by being ice free most of, if not all, the year round.

Gullfoss *Iceland* Waterfall, 50 m (164 ft) high, on the Hvitá River 100 km (60 miles) east of REYKJAVÍK, the capital. The word 'foss' is Icelandic for 'waterfall'.
Map Iceland – Bb

gully Small water-worn channel on a hillside, which has been cut by a rapidly flowing stream. Gullies also form on poorly farmed agricultural land during so-called 'gully erosion'.

Gulu *Uganda* Town 270 km (168 miles) north of the capital, Kampala, and the regional centre for the Acholi people. It stands in a mainly subsistence farming and cattle-herding district, but

some cotton is produced commercially. There is a large military base nearby.
Population 42 000
Map Uganda – Ba

they are found in forests on the mountain and allowed to roam through the temples and outlying villages, stealing food from general stores and raiding crops at will. They live in groups of 20 to 30, and are noted for their slender bodies.
Map India – Bc

Gusau *Nigeria* Northern industrial town and market 730 km (450 miles) north-east of the capital, Lagos. It deals in hides and skins, cotton, groundnuts, tobacco and kola nuts (which are chewed as a mild stimulant), and makes textiles.
Population 111 400
Map Nigeria – Ba

Güstrow *Germany* Farming town 160 km (100 miles) north-west of Berlin. It has a fine range of 16th-century buildings and is a tourist centre for both the Baltic coast and the lakes of MECKLENBURG.
Population 39 100
Map Germany – Eb

gut Narrow channel opening into the sea or into a large estuary. The term is mostly used to refer to such phenomena in the eastern United States.

Gutland *Luxembourg* The southern, and more fertile, two-thirds of the grand duchy. The northern third – Ösling – is part of the Ardennes plateau.
Map Belgium – Cb

Guwahati *India* North-eastern town in Assam, on the Brahmaputra River, 70 km (43 miles) south of the border with Bhutan. It is the centre of a flourishing tea industry established in 1838. It also trades in cotton, jute and rice. It has flour mills and oil refineries.
Population 1 987 660
Map India – Eb

Guyana See p 289

Guyane *South America* See FRENCH GUIANA

Guyenne *France* South-western former province covering some 41 000 km² (15 800 sq miles) between the Gulf of Gascony and the south-western part of the Massif Central. BORDEAUX is the chief city.
Population 2 660 000
Map France – Cd

guyot Flat-topped submarine mountain, possibly a volcano truncated by waves, rising from the ocean floor. Guyots are found mainly in the Pacific Ocean.

Gwalior *India* City 319 km (198 miles) south-east of Delhi. On a sandstone plateau overlooking Gwalior is a medieval Hindu fort, with palaces and temples, some of which are more than 1000 years old. Nearby are some impressive rock sculptures of Jain saints. The statues, which were carved around AD 1450, are some 18 m (59 ft) high.

The city's main industries are textiles, footwear, leather goods and pottery. There are also flour and oilseed mills.
Population 1 414 950
Map India – Cb

Gwangju *South Korea* See KWANGJU

A DAY IN THE LIFE OF A GUYANAN CATTLE RANCHER

From the large window of the ranch manager's office, which forms part of his home, Duane Mendes surveys the 40 500 hectares (100 000 acres) of gently undulating savannah grassland where his cattle roam. Far into the distance he can just make out the dark line of the edge of the jungle inhabited by Amerindians, many of whom have worked from time to time for Duane's team of several dozen employees, making excellent *vaqueros* or cowhands.

But Duane has business to attend to. Beside his office stands a new four-wheel-drive Jeep. But as this is the wet season he is well aware that it could get bogged down in the mud in minutes. So Duane walks to the corral to get his horse.

As he mounts, he is joined by *vaqueros* from the southern part of the ranch, who have come to report that a few of the cattle are suffering from worms. Duane issues instructions for the medicines they are to be given, and then gallops down to the airstrip to meet the plane from the capital, Georgetown, which will carry away a cargo of Duane's beef.

The ranch lies in an area of Guyana where there are no suitable roads for transporting the beef to Georgetown's market. Fortunately, Duane has managed to persuade the owners of the company he works for that the most efficient way to transport beef is by air. The beef cattle are killed in the ranch's abattoir, and this way the meat can still be fresh when it gets to market.

Duane's grandparents were born in Portugal, but he feels fully Guyanan himself; he speaks English as fluently as do most of the population of this former British colony. However, he was educated outside the country: he went to college in Brazil and in Canada, where he learned the modern methods of ranch management. Now he has sent his son, Andrew, aged 20, to college, to ensure that he has the same advantages.

Late in the afternoon Duane returns to his office, having seen off the last planeload of beef. He is eager to check if a letter has arrived from his son. There is a glow of red light in his office, cast by the fire burning over part of the ranch's grassland – one of Duane's modern methods designed to encourage the regrowth of the grass. It seems to symbolise the new hope that has characterised Guyana since the country achieved its independence from Britain in 1966 – the hope that has spurred ranchers like Duane to make the most of their land.

But there is no letter from Andrew. Duane hopes it is because his son is working so hard; he knows himself how much there is to learn in this adventurous but demanding career. Thinking of his boy, he prepares to sit down to dinner with his wife and three teenage daughters. Perhaps the girls will marry young men who can help on the ranch too. The future he contemplates, as he beams at his family, is a rosy one.

Gwent *United Kingdom* County in south-east Wales, formerly known as Monmouthshire, which was renamed in the 1970s after an ancient Welsh principality. It covers 1376 km² (531 sq miles) and centres on the industrial city of Newport at the mouth of the River Usk. Gwent's spectacular Wye Valley forms much of the county's border with England.
Population 448 500
Map United Kingdom – De

Gweru *Zimbabwe* Formerly known as Gwelo, Gweru, 275 km (171 miles) from Harare, is the country's third largest city and the industrial centre of an area rich in gold, chrome, iron, asbestos, limestone and coal. It takes its name from the Shona word for 'dry' which refers to the river which flows through the city intermittently.
Population 85 000
Map Zimbabwe Ba

Gwynedd *United Kingdom* County of north-west Wales, created out of the former county of Caernarfon, plus parts of Denbigh and Merioneth, and named after the ancient Celtic kingdom that dominated the country after the Romans left in the 5th century. It covers 3868 km² (1493 sq miles) and includes the island of ANGLESEY, Lleyn peninsula and SNOWDONIA National Park. There are superb medieval castles at Conwy, Caernarfon and Harlech, and many seaside resorts, of which the largest is Llandudno. Tourists flock to the narrow-gauge mountain railway at Ffestiniog and the rack railway to the summit of Snowdon, at 1085 m (3560 ft) Wales's highest mountain. Holyhead on Holy Island is the principal port for crossing to the Irish Republic. Gwynedd has a university at Bangor and a nuclear power station at Wylfa.
Population 239 300
Map United Kingdom – Cd

Gyeongju *South Korea* See KYONGJU

Györ *Hungary* Regional capital of the Little ALFÖLD and administrative centre of Györ-Moson-Sopron county, 108 km (67 miles) west of Budapest. The city is a busy river port at the confluence of the Rába and Danube rivers. It is a major industrial centre, whose many products include textiles, spirits, foodstuffs, and Ikarus buses made in the former Rába railway works.

The old town at the heart of the city clusters around Kaptalan Hill, site of the Roman fortress of Arrabona, which guarded a route along the Danube Valley. Its elegant Baroque buildings include a cathedral which was built in the 12th century and contains the elaborate golden reliquary of St Ladislas, the 13th-century bishop's palace, an 18th-century Carmelite church, and the house where the emperor Napoleon lived in 1809, now turned into a museum.

Pannonhalma Abbey, a large Benedictine foundation 21 km (13 miles) south-east of Györ, which is famous for its library and grammar school, is also largely Baroque in style, but retains its 10th-century church.
Population 130 290
Map Hungary – Ab

gypsum A white EVAPORITE mineral, calcium sulphate, found in clays and limestones, and used in the manufacture of cements and plasters, especially plaster of Paris.

Guyana

A PIECE OF THE BRITISH WEST INDIES ON THE CONTINENT OF SOUTH AMERICA REMAINS BEDEVILLED BY RACIAL PROBLEMS

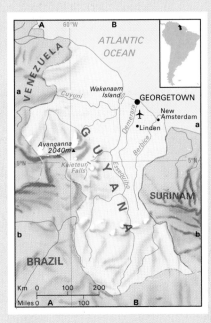

The Spanish adventurers turned up their noses at Guyana as a land with muddy harbours and no gold. The Dutch, French and English were all interested in the area, which continued to change hands until 1816 when they split the area up among themselves. The western part, which became Guyana, went to the British (then known as British Guyana), while the central region was controlled by the Dutch and the east by the French. Independent since 1966, Guyana is physically part of South America, but is culturally linked with the West Indies. It is the only English-speaking country on the South America continent.

Gold was found in small quantities, but the real wealth of Guyana is its bauxite (aluminium ore), on which the economy depends, and the sugar and rice it grows. The land could be much more productive, but large areas of the country are unexploited.

The south-west is high savannah, from which the jungle drops down towards the coast. The land is intersected by the numerous rivers which gave Guyana its name – it means 'land of many waters'. Four major rivers cross the coastal plain – the Demerara, Berbice, Essequibo and Courantyne rivers. Except for the Demerara River, all rivers can only be crossed by ferry.

The jungle and rivers could produce minerals, hardwood and hydroelectric power, but 90 per cent of the people live and work on the narrow coastal plain. Here, often on land reclaimed from tidal marshes and mangrove swamps, rice is grown and vast plantations produce sugar, as they have done since the Dutch brought slaves here from Africa in the 17th century.

The capital city of GEORGETOWN is the best-preserved town of wooden architecture in the Caribbean. Its civic buildings are all made of white-painted wood, and its cathedral, rising to 39.6 m (132 ft), is one of the tallest wooden buildings in the world. The city is known for its beautiful flowering trees.

LAND OF SIX PEOPLES

The descendants of the African slaves are one of the six racial groups that make up Guyana's diverse population. When slavery was abolished in 1833, the British colonists were deprived of their labour force. Foreign labourers were offered five-year contracts to work on the plantations, with the opportunity to stay in Guyana afterwards. This brought an influx of Portuguese and Chinese workers, who still form small communities in Guyana.

A steady flow of poor Asian Indians arrived during the second half of the 19th century, and they are now the largest racial group in the country, forming just over 50 per cent of the population. They still provide the labour for the sugar farms and rice paddies. People of mixed race make up a fifth group.

A sixth group are the original inhabitants, now swamped by the tide of immigrants; only 30 000 Carib Indians are left, some living as nomads or hunter gatherers in the rain forests.

The country is deeply divided, racially and politically. The two main opposing parties are the black-dominated People's National Congress (nominally socialist), which held power from 1964 to 1992, and the (Asian) Indian-dominated People's Progressive Party (which propounds a Marxist doctrine and which came to power in 1992). Their rivalry has caused economic and social problems. Antagonism between the Asians, who control a large proportion of the nation's economy, and blacks led to clashes and bloodshed in the 1960s.

Bauxite companies were nationalised in 1971, and the state took over the sugar industry in 1976. But political competition diminished the benefits this might have brought. The People's National Congress used these industries to increase its political support by employing many more people but without improving productivity. Very little was done to promote economic growth, and the economy plummeted into crisis.

Unemployment increased to 40 per cent, and the three main exports of sugar, bauxite and rice were produced at well below their full potential. Foreign investment was drastically curtailed, and a desperate shortage of foreign exchange arose.

The government also placed severe restrictions on imports so that spare parts for machinery had to be smuggled in from Surinam and Brazil. Some imports such as wheat flour were classified as luxuries and banned, and although there is no shortage of food, there is a limited variety. Morale in the country sank to an all time low. The Asians complained of discrimination and the state grew more repressive. Health care and education, which were once good by South American standards, went into decline. Many people were emigrating, mostly to Britain and the United States, and new uncertainties followed the death in 1985 of President Forbes Burnham, who led the country for 21 years.

With the first free elections in 28 years held in 1992 and the new government, the economy has improved significantly. GDP growth in 1992 was reported to be 7.5 per cent and expected to increase during 1993. During the same period, rice production increased by 23 per cent, bauxite and mining went up by 50 per cent and gold production increased by 300 per cent. New loans were negotiated with the IMF and (notably) India, marking an important change in the prospects for this divided land.

GUYANA AT A GLANCE	
Area 215 000 km² (83 000 sq miles)	
Population 798 000 **Per km²** 3.7 (**Per sq mile** 9.6)	
Capital Georgetown	
Government Presidential republic	
Currency 1 Guyana dollar = 100 cents	
Languages English (official), Creole, and various ethnic languages	
Religions Christian (52%), Hindu (33%), Muslim (10%)	
Climate Tropical. Average temperature is 27°C (81°F)	
Land use Forest 79%, grazing 6%, cultivation 2.5%, other 12.5%	
Major industries Agriculture, mining, forestry, sugar refining, timber milling	
Main exports Sugar, rice, molasses, rum; shrimp; bauxite, diamonds, gold	
Annual income per head (US$) 370	
Population growth (per thous/yr) -6.8	
Life expectancy (yrs) Male 65 **Female** 68	

gyre Circular flow of water in each of the great ocean basins of the world which is produced by the combined effects of prevailing winds and the rotation of the earth. In the northern hemisphere, each gyre flows in a clockwise direction, while in the southern hemisphere the flow is anticlockwise. The Gulf Stream is part of the North Atlantic gyre. See HOW OCEAN CURRENTS ARE CAUSED.

Gyula *Hungary* Town on the Great ALFÖLD near the Romanian border, 90 km (55 miles) northeast of the city of Szeged. It has a fine medieval castle. Since 1960, its spa has become popular.
Population 36 150
Map Hungary – Bb

Gyumri (Leninakan) *Armenia* City, formerly called Aleksandropol or Alexandropol, and under Soviet rule known as Leninakan. It lies on the flanks of Mount Aragats, which reaches 4090 m (13 418 ft), 15 km (9 miles) from the Turkish border. Gyumri was founded in 1837 by Armenian refugees fleeing Turkish persecution. Its manufactures include textiles, carpets, copper products and machinery.
Population 123 000
Map Armenia – Aa

Ha Khotso *Lesotho* Village 40 km (25 miles) east of the capital, Maseru. Nearby is a rock shelter called Ha Baraona, translating as 'home of the little Bushmen', that contains a superb collection of rock paintings.
Map South Africa – Cb

Ha'apai *Tonga* Large group of low-lying coral islands 160 km (100 miles) north of NUKU'ALOFA. The main town is Pangai and it lies on Lifuka Island. The notorious mutiny on HMS *Bounty* took place on 28 April 1789 at Tofua in the west of the group.
Population 8550
Map Pacific Ocean – Ec

haar Local name for a cold mist or fog off the east coast of England or Scotland, usually occurring in spring or early summer.

Haarlem *Netherlands* Provincial capital, 18 km (11 miles) west of Amsterdam, and the centre of the Netherlands' tulip-growing region. Haarlem's 15th-century Church of St Bavo is one of the largest in the country; its museums include one devoted to the painter Frans Hals (about 1580-1666) who lived and died in the city.
Population (city) 149 320; (Greater Haarlem) 213 690
Map Netherlands – Ba

Haast Pass *New Zealand* Mountain pass, 560 m (1840 ft) high, through the Southern Alps of South Island. A road was built through the pass in 1965. It provides a low-level link between Queenstown in the central Otago region and the west coast. As both areas are of very little economic significance, the route is primarily an all-weather tourist one.
Map New Zealand – Bf

Habbaniyah *Iraq* Lake just south of the Euphrates River, 100 km (60 miles) west of the capital Baghdad. It is 20 km (12 miles) long and between 5 and 10 km (between 3 and 6 miles) wide. A British airfield, which was built on its north shore in the Second World War, has been converted into a tourist village.
Map Iraq – Bb

Hachinohe *Japan* Fishing port in the north of HONSHU Island. It is also an industrial centre specialising in fish products, steel, chemicals, paper and pulp. Local craftsmen carve *yawata-uma* – traditional, brightly painted wooden horses.
Population 241 400
Map Japan – Db

hade Angle of inclination from the vertical of a VEIN, FAULT or LODE.

Hadhramaut *Yemen* Long, wide valley running parallel with the coast of the Gulf of ADEN and between 160 km and 320 km (100 and 200 miles) inland. It is 560 km (348 miles) long and is surrounded by limestone rockface up to 300 m (984 ft) high. Grain, dates and sesame are grown in the upper and middle parts of the valley, which have torrential, seasonal rainfall.

The barren eastern section is largely uninhabited. Many of the valley's inhabitants have emigrated to form communities in other Middle East countries, in East Africa and in Indonesia. The capital, Shibam, which has a population of about 7000, consists of about 500 mud-brick tower houses – all 5 to 7 storeys tall and all crammed into less than 1 km² (0.4 sq miles).
Map Yemen – Aa

Hadrian's Wall *United Kingdom* Roman fortification built on the orders of Emperor Hadrian, who ruled from AD 117 to 138 and visited Britain in 122. It stretches 117 km (73 miles) from the Solway Firth near Carlisle in the west to Wallsend near the east coast, and for centuries formed the north-west frontier of the Roman empire and the border between England and Scotland.

Originally, it was about 4.5 m (15 ft) high and had a milecastle housing troops every Roman mile (1480 m, 4860 ft), with watchtowers in between. There were also 17 forts along the wall. Though it fell into disrepair after the Romans left Britain in the 5th century, many stretches still remain, mainly in the county of Northumberland.
Map United Kingdom – Dc

Hagen *Germany* Ruhr steel town, 35 km (22 miles) south-east of Essen. It manufactures machinery, has food processing plants and is the seat of Germany's correspondence university.
Population 214 400
Map Germany – Bc

Hague, The ('s-Gravenhage; Den Haag) *Netherlands* The Dutch seat of government since 1580, and also capital of Zuid-Holland Province. It lies 55 km (35 miles) south-west of Amsterdam and includes the seaside resort of Scheveningen.

The city's name 's-Gravenhage ('the Counts' Private Domain') derives from the former hunting lodge of the Counts of Holland in a wooded area known as the Haghe ('hedge'). In 1248 Count William II built a castle here, and around it grew a palace, known as the Binnenhof. An artificial lake, the Hofvijer, was constructed beside the palace in 1350. Surrounded by government buildings and embassies, these now lie at the heart of The Hague. Beyond is a spacious city, with canals, parks and woods.

Prominent buildings include the Palace of Peace (which houses the International Courts of Justice, established here in 1922), the Netherlands Congress Centre (which has helped to earn the city a reputation as a conference centre), the Grote Kerk (Great Church) dating from 1399, the town hall (1565), the Mauritshuis Royal Art Gallery, the royal residence called 'The House in the Wood' (1645) and its pleasure gardens, and the Madurodam model village. Many multinational business firms, including Royal Dutch Shell, have their headquarters here.
Population (city) 444 660; (Greater Hague) 694 320
Map Netherlands – Ba

Hahotoé *Togo* Phosphate-mining region in the south, just north of Lake Togo. In the early 1990s it had estimated reserves of 100 million tonnes.
Map Togo – Bb

Haifa *Israel* Capital of Haifa district, and the country's chief port, on the Mediterranean coast 85 km (55 miles) north of Tel Aviv. It lies at the foot of Mount Carmel, the summit of which can be reached by a cable car. Among the many places of interest are the university, technical college, zoological gardens, maritime and art museums, a Bahai temple and a shrine where the founder of the Bahai religion is buried. Haifa's industries include textiles, chemicals, glass, soap, oil products and electrical equipment.
Population 251 000
Map Israel – Aa

hail (hailstones) Precipitation in the form of pellets of ice. Droplets of water that condense in the lower part of a thundercloud may be carried by rising air currents to a level where they freeze into ice pellets. The strong convection currents prevent the hailstones from falling until they have grown much larger. In a turbulent cloud current, small hailstones can be tossed up and down between the layers of moisture and ice many times, growing larger all the time. Hailstones more than 700 grams (1.5 lb) in weight have been recorded.

Hail *Saudi Arabia* Town and oasis situated 250 km (155 miles) north-west of Buraydah. It lies on the old caravan and pilgrim route from Iraq to MEDINA and MECCA. Once the capital of the Ibn Rashids, who were the main opponents of the Saud family, and the Wahhabis, the fundamentalist Muslim sect that now predominates in Saudi Arabia, it was besieged and taken by King Abdul-Aziz in 1921 and incorporated into the new kingdom of Saudi Arabia.

Because of its long history, Hail has many antiquities, including a number of forts, castles and palaces that date back to the Ottoman era. The city is growing rapidly and many new industries have been established here. Fruits are grown in the oasis and most of the country's wheat crop comes from this area.
Population 92 000
Map Saudi Arabia – Bb

Hainan *China* Tropical island in the South China Sea off the coast of Guangdong Province. Hainan covers 33 670 km² (13 000 sq miles) with more than 1000 km (620 miles) of largely unspoilt coastline. Coffee and rubber plantations cover parts of the interior. Oil exploration is being carried out offshore and oil shale is mined. The island's chief city is Haikou.
Population 6 860 000
Map China – Gg

Hainaut (Henegouwen) *Belgium* Industrial province covering 3787 km² (1462 sq miles) along the south-west border with France. From 1850 to 1950, the coal field that crosses the area from east to west fuelled metalworking, textile and chemical industries around the main towns of Charleroi, Mons and Tournai, but the economy has become depressed since the worked-out coal mines closed in the 1970s. Today, other industries, such as the petrochemical and electrical industry and the manufacture of machinery, are gaining in importance. Cereals and root crops are grown in the countryside.
Population 1 278 000
Map Belgium – Aa

Haiphong *Vietnam* Port and third largest city of the country. It lies on the Red River delta, 90 km (55 miles) east of the capital, Hanoi. Founded in 1874 by the French, it was heavily damaged by bombing during the Vietnam War (1965-75).
Population (city) 456 000; (metropolitan area) 1 400 000
Map Vietnam – Ba

Haiti

VOODOO AND BOGEY MEN SUPPORTED THE DICTATORS OF THE WORLD'S FIRST BLACK REPUBLIC

Dr François 'Papa Doc' Duvalier ruled the West Indian republic of Haiti from 1957 to 1971, using voodoo (a religious cult from Africa based on sorcery and fetishism) and his special police, the *Tontons Macoutes* (bogey men), to subjugate the population. When Papa Doc died, he was succeeded as president-for-life by his 19-year-old son, Jean-Claude ('Baby Doc').

His regime was little less oppressive than his father's; it collapsed in January 1986, and he and his family fled the country, to live in a French chateau. He was succeeded by army leader General Henri Namphy, but unrest and instability continued. He in turn was deposed two years later, as were a long line of military governments, the one succeeding the other for periods of about two years.

Throughout these transitions the streets of PORT-AU-PRINCE and other towns and villages would frequently be stained with the blood of civilians (or police) hacked to pieces, shot or burned, or all three. Finally the opposition called for internationally monitored elections, and a former priest, Jean-Bertrand Aristide, was elected president in 1991. Another coup was staged eight months later and Aristide became the president-in-exile (living in Washington DC from where he sputtered vituperative condemnations at his enemies).

The military junta's rule continued a brutal and savage tradition, changing Haiti from the poorest nation in the western hemisphere to the most callously ruled, poorest nation – with the highest incidence of AIDS. Most of the population do not have ready access to clean water, nor can they count on adequate medical services. Nearly all Haitians are subsistence farmers and rarely have sufficient food. Because of these difficulties people left their country in droves – often on boats that could hardly manage the short distance to the shores of Florida in the United States where it was hoped a better life awaited them.

The small nation occupies the western part of the island of HISPANIOLA (the eastern part is taken up by the Dominican Republic). The island was sighted by Christopher Columbus in 1492 and settled by the Spanish. But in 1697 French pirates in the western part of the island forced Spain to cede this area to France. The French sugar planters imported thousands of slaves from Africa, and made the island a prosperous colony.

In 1791 the 500 000 slaves revolted – inspired by the French Revolution of 1789 – and in 1804 they declared their independence. They formed the world's first black republic. Haiti also ruled Santo Domingo, now the Dominican Republic, from 1822 to 1844. Haiti had 22 dictators between 1843 and 1915, when the United States moved in to bring about stability. The US occupation ended in 1934.

More than 90 per cent of Haitians are of African descent and the rest are mostly of mixed race. French is the official language, but 90 per cent of the people speak Creole, a French patois. Catholicism is the official religion, but voodoo is widely practised. About 80 per cent of the people live in rural areas and, although education is free, fewer than half the children go to school. Yet the country has a strong artistic and literary tradition.

Hispaniola is the second largest West Indian island after Cuba, and the most mountainous. The mountain ranges are separated by deep valleys and plains; the highest point in Haiti is La Selle (2680 m, 8793 ft).

Agriculture is the chief occupation but there is much rural unemployment. The average holding is only 1.5 hectares (3.7 acres) and many farmers just barely grow enough to feed their own families. Large estates grow coffee, sugar and sisal for export (that were boycotted under UN sanctions until September 1994 in efforts to force the military junta from power and into exile).

Most of the forests have been cleared, with severe soil erosion and declining crop yields. Malnutrition is widespread and the country holds the record for infant mortality (10 out of every 100 die before their first birthday).

The supply of cheap labour has attracted US investment in assembly plants producing goods for export, including textiles, sports and electrical goods, but foreign aid has remained vital to the country's economy.

Before the total collapse of government, tourism was the largest source of foreign income after farm exports. American visitors were attracted by the scenery, the culture and almost complete absence of street crime. Most stayed in the capital, Port-au-Prince, and its hillside suburb of Pietonville. The Sans Souci palace and the spectacular Citadelle near CAP HAITIEN in the north were the chief attractions. Both were built by a former waiter who became King Henri Cristophe of northern Haiti (1806-20).

In September 1994, United States-led forces (20 000 US troops plus other police and soldiers from Caribbean countries and various UN member states) invaded Haiti to restore the government of Aristide, who returned as president a few weeks later.

HAITI AT A GLANCE	
Map Caribbean – Ab	
Area 27 750 km² (10 714 sq miles)	
Population 6 384 880 **Per km²** 230	
(Per sq mile 595)	
Capital Port-au-Prince	
Government Republic	
Currency 1 gourde = 100 centimes	
Languages French, Creole	
Religions Christian (80% Roman Catholic, 16% Protestant), most also follow voodoo	
Climate Tropical. Average temperature in Port-au-Prince ranges from 25°C (77°F) in January to 29°C (84°F) in July	
Land use Cultivation 33%, pastures 18%, forest 1.3%, other 47.7%	
Main primary products Bananas, cassava, maize, rice, sorghum, coffee, sugar cane, sisal, timber, bauxite, copper (unexploited)	
Major industries Agriculture, mining, textiles, food processing, cement, assembly of imported parts, forestry, tourism	
Main exports Textiles, toys, electrical parts, coffee, bauxite, sugar	
Annual income per head (US$) 370	
Population growth (per thous/yr) 16	
Life expectancy (yrs) Male 44 **Female** 47	

Hajar Mountains *Oman/United Arab Emirates* Range running parallel with the Gulf of Oman and forming the spine of the MUSANDAM Peninsula. In its mid-point, the mountains rise to Jebel Akhdar ('the Green Mountain'), crowned by the 3107 m (10 194 ft) high peak Jebel Shams.

Below Jebel Akhdar is the long oasis of Rustaq, containing numerous villages and five watercourses. The Sumail Gap separates the Western Hajar from the Eastern Hajar mountains.
Map Oman – Aa

Hakodate *Japan* Port and gateway to HOKKAIDO on that island's southernmost coast. Open to foreign trade since the 19th century, it has the country's only western-style castle (in fact, its name translates as 'box castle').

Hakodate is linked to the island of HONSHU by the Seikan rail tunnel, which was completed in 1988, and at 54 km (33.5 miles) the world's longest underwater tunnel. In parts, the tunnel runs 100 m (330 ft) below the seabed.
Population 307 000
Map Japan – Db

Hakone *Japan* Mountainous region of central HONSHU Island lying south-west of the capital, Tokyo. Forming part of the Fuji-Hakone-Izu National Park, scattered with volcanoes and hot springs, it is a popular summer holiday destination. The main town is Odawara in Sagami Bay.
Map Japan – Cc

Hakusan *Japan* Mountainous national park on HONSHU Island, covering 474 km² (83 sq miles) and facing the Sea of Japan, 235 km (145 miles) west of the capital, Tokyo. One of the park's most popular attractions is its *gassho-zukuri* – houses topped by steep, thickly thatched roofs.
Map Japan – Cc

Halas *Hungary* See KISKÚNSÁG

Halicarnassus *Turkey* See BODRUM

Halifax *Canada* Capital and largest city of NOVA SCOTIA, and Canada's main ice-free Atlantic port. Its strategic position made it an important wartime naval base in 1759, when James Wolfe, the British general, sailed from here to capture Quebec; in both World Wars, convoys set out from here.

The first trans-Atlantic steamship service between Canada and Great Britain started at Halifax in 1840. Today, the city is important for bulk cargoes and container traffic, and is an educational, cultural and industrial centre. Industries include iron and oil foundries, and factories making soap and shoes.
Population (city) 295 920; (metropolitan area) 330 380
Map Canada – Id

Halifax *United Kingdom* Town in West Yorkshire, that has been an important English wool trade centre since the 15th century. It lies between Bradford and Huddersfield. Piece Hall, rebuilt in 1770 and now converted for craft shops and galleries, was a manufacturers' trade hall where numerous weavers sold their 'pieces' (lengths) of cloth.
Population 77 350
Map United Kingdom – Ed

halite (rock salt) A white EVAPORITE mineral, sodium chloride (common salt), found in sedimentary rocks, such as sandstone and shale. It is a source of salt and chlorine.

Halland *Sweden* South-western farming province covering 5448 km² (2103 sq miles). It was ceded by Denmark to Sweden in the 17th century. Its chief city is HALMSTAD.
Population 266 000
Map Sweden – Bd

Halle *Germany* Town and inland port on the Saale River, 30 km (20 miles) north-west of LEIPZIG. It was founded as a fortress in AD 806 and later became a salt-mining centre. Modern industries include printing, machinery, rubber, cement, paper and petroleum products. The composer George Frederick Handel (Georg Friedrich Händel, 1685-1759) was born here.
Population 310 200
Map Germany – Dc

Halmstad *Sweden* Seaport and industrial city on the Kattegat coast, 150 km (93 miles) south-east of Gothenburg. It has a ruined medieval castle. Its industries include shipbuilding, steel, paper, textiles and brewing.
Population 81 100
Map Sweden – Bd

halo Circular band of light, sometimes coloured, seen around the sun or moon, caused by the refraction of light by ice particles in CIRROSTRATUS clouds. Halos herald wet weather.

Hälsingborg *Sweden* See HELSINGBORG

hamada Rock desert that has been swept clear of sand and dust by winds. The term is usually used to refer to such areas in the Sahara.

Hamadan (Hamedan) *Iran* City 300 km (185 miles) south-west of the capital, Tehran. It produces carpets, rugs and leather goods, as well as copperware and raisins. It also trades in wool, hides and skins. Hamadan was the capital of the Medes in the 6th century BC and of the Achaemenid Persian empire (559-330 BC). It was sacked by the Mongols in 1220.

Today it is a provincial capital and seat of a university. The city has a large, colourful bazaar and numerous monuments, including the tomb of the Persian physician and philosopher Avicenna (AD 980-1037), and the mausoleum of the Biblical characters Esther and her cousin and guardian, Mordecai.
Population 350 000
Map Iran – Aa

Hamah *Syria* Commerical and industrial city on the Orontes River, 125 km (78 miles) south of Aleppo. The centre of Syria's steel industry,

Hamah also has tanning and weaving industries and trades in cereals and fruit. It lies 308 m (1010 ft) above sea level and is surrounded by orchards and market gardens.

Hamah has been continuously inhabited since the late Stone Age and was known in ancient times as Epiphaneia. In AD 638 it was captured by the Arabs. The city is noted for the huge wooden water-powered wheels, first used in ancient times, that lift water from the Orontes for irrigation. The largest is 22 m (72 ft) in diameter. Most of them have now been replaced by electric pumps. Among the town's places of interest are the Great Mosque with its 14th-century water wheel.
Population 237 000
Map Syria – Bb

Hamamatsu *Japan* Industrial city near the southern coast of HONSHU Island, 215 km (135 miles) south-west of the national capital, Tokyo. Its main products are pianos and motorcycles. In May each year, the city holds a 'battle of the giant kites' festival, in which rival teams try to sever the strings of one another's kites.
Population 535 000
Map Japan – Cd

Hamar *Norway* Market town on the west shore of Lake Mjøsa in south-central Norway. The town was founded in 1152 by the only Englishman to ever become a pope, Adrian IV (1154-9), while he was a cardinal bishop in Scandinavia. The town's industries include processing of wood and farm produce.
Population 26 000
Map Norway – Cc

Hamburg *Germany* Second largest city in the country, after Berlin, and its biggest port, straddling the Elbe River 110 km (68 miles) from its mouth. Founded by the Holy Roman Emperor Charlemagne in 810, it was a leading member of the Hanseatic League, the medieval trading association of north European towns. Its modern docks cover an area of more than 100 km² (39 sq miles) on both sides of the Elbe, and it has petrochemical, electronics and food processing industries.

For centuries, Hamburg was an independent city and today it keeps some autonomy as a *Bundesland* (state) of the Federal Republic. It has a long musical tradition. The composers Johannes Brahms (1833-97) and Felix Mendelssohn (1809-47) were born here and George Frederick Handel (Georg Friedrich Händel, 1685-1759) was a violinist in the opera orchestra before he moved to England. The Second World War song 'Lili Marlene' was written by a Hamburg man, Hans Leip.

In St Pauli district, west of the city centre, is the Reeperbahn, ('Rope Walk'), an infamous street of nightclubs and brothels. Hamburg has two universities and several colleges.
Population 1 652 400
Map Germany – Db

Hamedan *Iran* See HAMADAN

Hämeenlinna (Tavastehus) *Finland* Capital of the province of Hame, 98 km (60 miles) north-west of the capital, Helsinki. It was the birthplace of Jean Sibelius (1865-1957), composer of *Finlandia*, the symphonic tone poem

he wrote in 1900 which became Finland's unofficial national anthem. The town's 13th-century castle was built by Swedes when Finland was part of Sweden. The town now manufactures textiles and plywood. Aulanko, a large pleasure park, is situated near the city.
Population 44 000
Map Finland – Bc

Hamelin (Hameln) *Germany* Carpet-making town on the Weser River, 49 km (30 miles) south-west of Hannover, and site of the legend of the Pied Piper. In AD 1284, a rat catcher is said to have lured with his pipes a plague of rats from the town into the Weser to drown. But when the piper was not paid, he piped away the town's 128 children, leaving one crippled boy, and they were never seen again. The basis of the legend is long forgotten; the so-called 'Rat Catcher's House' in the town dates only from 1600.
Population 58 300
Map Germany – Cb

Hamersley Range *Australia* Mountains in the PILBARA region of Western Australia, rising to 1235 m (4052 ft) at Mount Bruce. The range contains rich deposits of iron ore. Two of the peaks – Mount Whaleback and Mount Tom Price – are composed almost entirely of high-grade ore and are being systematically demolished. Most of the central section of the range constitutes the Hamersley Range National Park, which was established in 1969.
Map Australia – Bc

Hamhung (Hamheung) *North Korea* Industrial port and provincial capital on the east coast, 180 km (112 miles) north-east of the national capital, Pyongyang. Heavily bombed during the Korean War (1950-53) and since rebuilt, it now manufactures machinery, chemicals and fertiliser.
Population 670 000
Map Korea – Cc

Hamilton *Bermuda* Capital and main port of Bermuda. Founded in 1790, it was made the capital in 1815 and became a free port in 1956. Today, large cruise ships dock alongside The Front, the main shopping and commercial street.
Population 3000
Map (Bermuda) Atlantic Ocean – Bc

Hamilton *Canada* Large industrial city on the shore of Lake Ontario. The city extends from the lake shore onto the Niagara Escarpment 100 m (328 ft) high. It is Canada's steel centre. Other main products are motor vehicles and chemicals.

Hamilton was first settled in 1778 and became a city in 1846. It is the home of the Royal Botanical Gardens, historic Dundurn Castle and McMaster University.
Population (city) 318 500; (metropolitan area) 599 760
Map Canada – Hd

Hamilton *New Zealand* Market and industrial town on North Island, 120 km (75 miles) south of Auckland. Its industries include food processing, the manufacture of agricultural equipment, meat freezing and electronics. The University of Waikato was founded here in 1964; the town has several research institutions.
Population 148 630
Map New Zealand – Eb

Hamilton *United Kingdom* See STRATHCLYDE

Hammamet *Tunisia* Resort town and port on the southern coast of the peninsula ending in Cape Bon. It has long been renowned as a beauty spot patronised by the rich and famous, and has splendid beaches and many hotels that cater for European visitors.

It is also the venue for an annual international cultural festival held in July. There is a small medina (the old Muslim quarter), a souk (market) for brassware and carpets, a 15th-century fort, the Great Mosque, and a museum containing a collection of traditional Tunisian costumes.

Population 30 500
Map Tunisia – Ba

Hamma Salahine *Algeria* See BISKRA

Hammerfest *Norway* Fishing port and one of the most northerly towns in the world, on Kvaløy Island, 95 km (59 miles) south-west of North Cape. It lies 460 km (285 miles) north of the Arctic Circle and has midnight sun from mid-May to the end of July. The town was badly damaged by fire in 1890. It was rebuilt then, and again rebuilt after the Second World War, after the Germans destroyed most of the province of FINNMARK. Hammerfest manufactures cod-liver oil and exports fish and furs.

Population 9500
Map Norway – Fa

Hampshire *United Kingdom* County covering 3772 km² (1456 sq miles) of southern England on the English Channel coast. It used to include the Isle of Wight, now a separate county. Its main centres are the major ports of SOUTHAMPTON and PORTSMOUTH. Aldershot, in the north, is the home base of the British army. The middle of the county is a rural haven of chalk downland and quiet trout streams, focused on the ancient city of WINCHESTER. The old royal hunting grounds of the NEW FOREST, now a popular holiday region, are in the south-west.

Population 1 581 900
Map United Kingdom – Ee

Han Shui *China* River of central China, flowing about 1130 km (700 miles) south through Hubei Province to join the Chang Jiang River at the city of Wuhan.

Map China – Gd

Hangchow *China* See HANGZHOU

hanging valley Tributary valley whose floor is much higher than that of the main valley, so that its river enters the main valley by a waterfall or rapids. It is usually caused by deeper glacial erosion of the main valley.

Hangzhou (Hangchow) *China* Capital of Zhejiang Province, 160 km (100 miles) south-west of Shanghai. It was the capital of southern China from 1127 until the Mongol conquest in 1276. It is now an administrative and industrial centre. The Grand Canal runs from Hangzhou northwards, linking all the great cities of the Chinese lowland.

Population 1 340 000
Map China – Ie

Hankow *China* See WUHAN

Hannover (Hanover) *Germany* Capital of Lower Saxony, 100 km (62 miles) south-east of Bremen. It was the city of the Electors of Hannover – one of whom, George I, a grandson of James I, became England's king in 1714. The British astronomer Sir William Herschel (1738-1822), who discovered the planet Uranus, was born in Hannover. Today the city is a cultural and educational centre. It manufactures chemicals, textiles, rubber and steel, and holds several fairs each year, among them a trade fair in spring.

Population 513 000
Map Germany – Cb

Hanoi *Vietnam* Capital city in the north of the country on an arm of the Red River, 130 km (80 miles) from the Gulf of Tongking. It was chosen as the capital in 1010, when it was called Thang Long ('City of the Soaring Dragon'). Its present name was first chronicled in 1831 and means 'City inside the River'.

The heart of the modern city was laid out by the French during the colonial period when it was the region's administrative centre from 1887 to 1946. Hanoi was heavily bombed during the Vietnam War (1965-75). Its main manufactures are cars, machine tools, cement, textiles and chemicals. Hanoi is expanding its tourist facilities rapidly to cope with an influx of tourists.

Population (city) 1 089 000; (metropolitan area) 3 056 150
Map Vietnam – Ba

Hanover *Germany* See HANNOVER

Hanshin *Japan* Industrial zone on the north-east shore of the Inland Sea, around the cities of OSAKA and KOBE. It accounts for about one-sixth of the country's manufactures. Its main products include iron and steel, chemicals and machinery.

Map Japan – Cd

Haora (Howrah) *India* Large industrial city on the west bank of the River Hooghly, opposite Calcutta. It has railway workshops, jute and cotton mills, paper factories and iron and steel rolling mills.

Population 3 718 910
Map India – Dc

Har Horin *Mongolia* See KARAKORUM

Har Karmel *Israel* See CARMEL, MOUNT

Harappa *Pakistan* Site of an ancient city, 80 km (50 miles) south-west of LAHORE, that flourished from 2500 to 1500 BC. It covered an area 5 km (3 miles) in diameter and had a heavily fortified castle. The city was built of brick, but much of it had been ransacked for railway ballast and local buildings before proper archaeological excavations began in the 1920s. Harappa was the second largest city of the Indus Valley Harappan civilisation; the largest was MOHENJO DARO

Map Pakistan – Db

Harare *Zimbabwe* National capital, formerly called Salisbury after the 19th-century British prime minister, Lord Salisbury. Founded in 1890 on a site 1470 m (4822 ft) above sea level, it is now a modern city, many of whose streets are lined with jacaranda trees that flower in spring, summer and early autumn. Harare has one of the world's largest tobacco markets and the city

produces a wide range of products, including beer, building materials, furniture and sugar. It also processes other foods.

It is the home of the national parliament and the University of Zimbabwe. The satellite city of Chitungwiza, with a population of at least 300 000, is part of the Greater Harare urban area.

Population 1 458 000
Map Zimbabwe – Ca

Harbel *Liberia* Headquarters and industrial centre of the American Firestone Plantation Company, 45 km (28 miles) east of the capital, Monrovia. It lies in one of the world's largest and most modern rubber plantations, which covers 290 km² (110 sq miles).

Harbel has one of the world's biggest rubber-processing factories, and most of the services of a city – including a hydroelectric power station, a radio station, schools, churches, hospitals, a brickworks and a soft-drink bottling plant.

Map Liberia – Aa

Harbin (Charbin) *China* Capital of Heilongjiang Province on the south bank of the Songhai Jiang River. It is a major railway junction and a centre for food processing and heavy engineering industries.

Population 2 800 000
Map China – Jb

Hardangerfjord *Norway* One of Norway's longest fjords, measuring 145 km (90 miles) and running east and south-east of BERGEN. It is the site of several hydroelectricity plants and of some of Norway's best-known tourist resorts, as well as one of the fruit-growing areas of Norway, since it has a mild climate inland. To the east is Hardangervidda, an extensive plateau.

Map Norway – Bc

hardness Relative resistance of a mineral to scratching, as measured by the MOHS SCALE OF HARDNESS.

Hardwar *India* See HARIDWAR

hardwood Timber from broad-leaved trees, as opposed to conifers. The temperate varieties include oak, beech, cherry, maple and walnut, while from tropical forests come ebony, ironwood, mahogany, rosewood and teak. Most temperate hardwood trees are deciduous, shedding all their leaves at one time. In equatorial forests, there is no seasonal leaf fall – leaves persist for a period and are shed irregularly.

Hargeysa *Somalia* Northern market town, 130 km (80 miles) south-west of the port of BERBERA. Nomadic herders, who form the majority of the population, trade livestock here. It was the capital of British Somaliland from 1941 to 1960 and is today the 'capital' of the self-styled independent state of Somaliland.

Population 400 000
Map Somalia – Ab

Harghita Mountains *Romania* Ridge of volcanic mountains in eastern TRANSYLVANIA, 168 km (104 miles) south of the border with Russia. Stretching for 50 km (30 miles) from north to south, the Harghita range rises to 1800 m (5905 ft).

Map Romania – Ba

Hari Rud *Afghanistan/Turkmenistan* River, 1125 km (about 700 miles) long, that rises in the Kuh-e-Baba mountains, 150 km (93 miles) west of Kabul, from where it flows west through the fertile Herat Valley. It then curves north to form part of the border with Iran, and ends at the remote oasis of Tedzhen in Turkmenistan.

Map Afghanistan – Ba; Turkmenistan – Bb

Haridwar (Hardwar; Gangadwara) *India* Holy town on the Ganges River, 263 km (163 miles) north-east of Delhi. Each year more than 2 million Hindu pilgrims come here to bathe, and to spread the ashes of cremated relatives. The bathing ghat (area) contains a footprint said to be that of the Hindu god Vishnu.

Population 189 000
Map India – Cb

Harkány *Hungary* See MOHÁCS

harmattan Dry, dusty north-east wind blowing from the high atmospheric pressure areas of the Sahara towards humid low pressure areas over the West African coast.

Harper (Maryland) *Liberia* Port in the southeast of the country, 25 km (16 miles) from the Ivory Coast border. Founded in 1834 as an independent colony, it became part of Liberia in 1857 and was named after Robert C. Harper, a member of the American Colonisation Society that founded Liberia in the early 19th century as a home for freed slaves from the United States.

The town's modern development began after the establishment of a United States-owned rubber plantation near Gedetarbo on the Cavally River, 25 km (16 miles) to the north. The port's exports are rubber, cocoa and timber.

Population 14 000
Map Liberia – Bb

Harrisburg *USA* State capital of Pennsylvania, 150 km (93 miles) north-west of the city of Philadelphia. Standing on the Susquehanna River, it was founded in the early 18th century as a trading post. Nearby coal mines were the basis of its industries then; today, Harrisburg produces machinery, aircraft parts and timber products. In 1979, it was the scene of a near-disaster when a fault in the nearby Three Mile Island nuclear power plant caused a partial meltdown of a reactor's core.

Population (city) 52 380; (metropolitan area) 587 990
Map United States – Kb

Harrogate *United Kingdom* Town in North Yorkshire, 30 km (18 miles) west of York, that became one of England's most renowned spas in the 19th century. As demand for taking 'the cure' – drinking of and bathing in its sulphur, iron and magnesia spring waters – declined, Harrogate developed conference and exhibition facilities.

Population 64 920
Map United Kingdom – Ed

Hartford *USA* State capital of Connecticut, on the Connecticut River, 150 km (95 miles) north-east of New York City. Hartford was one of the earliest and strongest of the colonial cities and was a military supply depot during the American War of Independence.

Today, it is an industrial, commercial and insurance centre. Manufactures include firearms, aircraft engines, office equipment and domestic appliances.

Population (city) 139 740; (metropolitan area) 1 157 590
Map United States – Lb

Haryana *India* Agricultural state to the north-west of the capital, Delhi, covering 44 212 km² (17 066 sq miles). It was established in 1966. In 1986, its capital, CHANDIGARH, was to have been transferred to the adjacent state of Punjab, but the plans were delayed after protests – although Haryana was promised a new capital.

The state's wheat farming depends upon new 'green revolution' technology – the use of high-yielding seeds, fertilisers and new cultivating techniques – and irrigation from a canal system built since 1947.

Population 16 317 700
Map India – Bb

Harz *Germany* Range of forested mountains between the German states of Lower Saxony, Thüringen and Saxony-Anhalt. The mountains, whose highest point is the Brocken at 1142 m (3747 ft), have mines and many summer tourist resorts. The region was immortalised in the poetic drama, *Faust*, by the German writer Johann Wolfgang von Goethe (1749-1832).

Map Germany – Dc

Hasselt *Belgium* Industrial town, 66 km (41 miles) east of Brussels, and capital of the province of Limburg. The medieval heart of the town is surrounded by modern suburbs and an industrial belt along the Albert Canal, which runs past the town on the northern side. Main industries are brewing, distilling and the manufacture of agricultural machinery. The estate of Bokrijk, with a large open-air museum of Flemish rural life, lies 6 km (4 miles) east of Hasselt.

Population 63 800
Map Belgium – Ba

Hassi Messaoud *Algeria* Centre of the Algerian oil industry, situated in the middle of the Sahara, some 320 km (about 200 miles) south of Biskra. Pipelines begin at Haoud El Hamra and take the oil to the northern ports of ARZEW, BEJAIA and SKIKDA. The field is also linked by pipeline to Edjele and Tiguentourine, and to the HASSI R'MEL gas field. Since oil was discovered in the 1950s, water has also been found and the area now has swimming pools and tree plantations.

Map Algeria – Ba

Hassi R'Mel *Algeria* One of the world's largest natural gas fields with reserves estimated in the early 1990s at about 1.2 million m³ (42.38 million cu ft). The field lies in the Sahara, 100 km (62 miles) north of GHARDAIA, and pipelines take the gas to the ports of ARZEW, 416 km (258 miles) to the north-west and SKIKDA, 590 km (366 miles) to the north-east. There is a pipeline link to Hassi r'mel and another takes the gas to Europe via Sicily.

Map Algeria – Ba

Hastings *New Zealand* Town on North Island, 250 km (155 miles) north-east of Wellington. It stands inland from Hawke Bay and is the centre of a rich fruit-growing and pastoral region. Settled in 1864 on land leased from local Maoris, it was named after Warren Hastings (1732-1818), Britain's first governor-general of India. Like NAPIER 20 km (12 miles) to the north-east, it was rebuilt after an earthquake in 1931.

Population 57 750
Map New Zealand – Fc

Hastings *United Kingdom* Town on the Sussex coast and one of the CINQUE PORTS. It lies 85 km (53 miles) south-east of London and is best known for the battle in 1066 when the invading Normans under William the Conqueror defeated the English under King Harold. The battle actually took place near the town of Battle, 10 km (6 miles) inland.

The old part of Hastings, with narrow winding streets, half-timbered houses and, on the beach, tall wooden 16th-century huts for drying nets, preserves the atmosphere of a medieval fishing village. Hastings is overlooked by the remains of the castle built by William I in 1067. Industrial estates at Churchfields and Castleham are an important source of employment.

Population 76 680
Map United Kingdom – Fe

Hat Yai *Thailand* Town in the far south, 40 km (25 miles) from the Malaysian border. It is a rubber processing town that is also popular with Malaysian tourists.

Population 261 400
Map Thailand – Bd

Hatteras, Cape *USA* Southern tip of Hatteras Island, one of a 160 km (100 mile) string of long, narrow islands known as the Outer Banks along the stormy Atlantic coast of North Carolina. Because of its violent storms and shallow waters, it has become notorious for shipwrecks along its shores. Much of Hatteras Island is preserved as a National Seashore.

Map United States – Kc

Hattusas *Turkey* See BOĞAZKALE

Hauraki Gulf Bay on the north-eastern coast of NORTH ISLAND, New Zealand, stretching 55 km (34 miles) from east to west, and 42 km (26 miles) from north to south. The port of AUCKLAND stands on an isthmus between the MANUKAU and Waitemata harbours, which are two inlets of the Hauraki Gulf.

Map New Zealand – Eb

Hausaland *Nigeria* See NORTHERN NIGERIA

Hautes Fagnes *Belgium* See ARDENNES

Haut-Zaire *Zaire* The country's largest administrative region, covering 503 239 km² (194 300 sq miles) in the north-east. It lies mostly in the humid Zaire River basin, with highlands overlooking the Great Rift Valley system in the east. Forests blanket the centre and south, with savannah in the north and mountain vegetation in the east. The economy depends mainly on plantation crops, including cotton. With the exception of a few scattered groups of Pygmies, most of the inhabitants of the region are members of various Bantu or Sudanese farming groups. The capital is Kisangani.

Population 5 119 750
Map Zaire – Ba

Havana (La Habana) *Cuba* Seaport and capital of Cuba, and the largest city in the West Indies, situated on the island state's north-west coast. The whole metropolitan area forms the province of Cuidad de la Habana. The harbour, one of the best in the western hemisphere, was fortified by the Spanish in the 16th century. An entrance canal, guarded by the 16th-century Castillo del Morro, leads into the broad interior haven. Opposite lies the Castillo de la Fuerga Real, guarding the entrance to the old city. To the west of the harbour, the Río Almendares gently runs through the city and down to the sea.

The old city, with its 16th and 17th-century palaces, churches and monasteries, which reflect its Spanish past, is the commercial centre of the country. The new city is mainly residential, and reflects American rather than Spanish influence. In 1898, the harbour was the scene of the sinking of the United States battleship *Maine*, that precipitated the Spanish-American War, and resulted in US military occupation of Cuba until 1902, when it became independent.

Until Cuba's communist revolution, Havana was considered the economic, cultural and entertainment capital of the Caribbean. Havana today exports sugar, tobacco products, coffee and tropical fruit, and handles most of the country's imports. Industries include oil refining, cigar making (of the world-famous Havanas), textile mills, clothing and packaging plants.

Havana is also the name of a Cuban province lying near the west end of the island. Most of its economy is based on agriculture – mainly tobacco, sugar cane, citrus fruits, cattle, rice and coffee – while copper and iron are mined in the highlands. The capital is Güira de Melena (population 32 650).

Population (metropolitan area) 2 096 000; (Havana Province) 633 400
Map Cuba – Aa

Hawaii *USA* State of the United States in the North Pacific Ocean. Its 122 islands were settled by Polynesian people (originally from Southeast Asia) between AD 500 and 900.

The English navigator Captain James Cook (1728-79) reached the islands in 1778 on his second Pacific voyage and named them the Sandwich Islands. However, they remained an independent kingdom throughout the 19th century. At the request of the Hawaiian people, they were annexed to the United States in 1898. They became the 50th US state in 1959. The population today is made up of many nationalities, mainly from Japan and Europe. Polynesian-descended locals account for just under 20 per cent of the inhabitants.

The coral and volcanic islands, of which 14 are inhabited, lie more than 3380 km (2100 miles) from the city of San Francisco on the American mainland. There are many volcanoes, including MAUNA KEA, reaching 4206 m (13 796 ft) and MAUNA LOA, a little lower, at 4169 m (13 667 ft), both on the largest island, also called Hawaii. Agriculture is subtropical, with sugar cane, pineapples and coffee being the chief products.

Hawaii has served as a United States naval and military base – PEARL HARBOR is now a national historic landmark. The state, which has a total land area of 16 705 km² (6450 sq miles), is popular with tourists. HONOLULU on the island of OAHU is the state capital.

Population 1 159 610

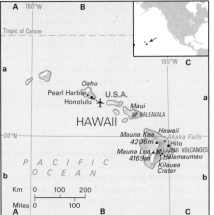

Hawaii Volcanoes National Park *USA* Park covering 990 km² (382 sq miles) on the island of Hawaii. It includes the volcano MAUNA LOA, which towers to 4169 m (13 677 ft), with its active craters of Kilauea and Mokuaweoweo. The park has much tropical vegetation and birdlife, and fantastic lava formations.
Map Hawaii – Bb

Hawke Bay *New Zealand* Bay, 80 km (50 miles) long and 56 km (35 miles) wide, on the east coast of NORTH ISLAND, named in 1769 by Captain James Cook after the British first lord of the admiralty, Sir Edward Hawke (1705-81). Its southern tip, Cape Kidnappers, was so named because here, Maoris seized a boy from Cook's ship. (The boy escaped.)
Map New Zealand – Fc

Hawke's Bay *New Zealand* District of North Island centred on Hawke Bay, 320 km (200 miles) south-east of Auckland. It is a farming area covering 11 289 km² (4358 sq miles) and produces sheep, fruit, vegetables and wine.
Population 139 480
Map New Zealand – Fc

Hazarajat *Afghanistan* Bleak and forbidding mountain region in the centre of the country, to the west of GHAZNI. Its fertile valleys are inhabited by the Hazaras, a Mongol people believed to have settled there after the invasion of Genghis Khan in the early 13th century.
Map Afghanistan – Bb

▲ SLEEPING GIANT Heleakala, a dormant volcanic crater of Kolekole, rises to 3055 m (10 023 ft) in the national park on Maui Island in Hawaii. The crater covers 49 km² (19 sq miles).

haze Mass of atmospheric water droplets mixed with dust, smoke and salt, that restricts visibility.

Heard Island *South Indian Ocean* Uninhabited island about 4000 km (2484 miles) south-west of the Western Australian capital of Perth. It is 43 km (27 miles) long and 20 km (12 miles) wide, and has an active volcano, the 2745 m (9006 ft) high Mawson Peak, named after the Australian scientist and Antarctic explorer Douglas Mawson (1882-1958). Heard Island is one of the last Antarctic habitats left that has not been subjected to introduced organisms. Together with the McDonald Islands, about 40 km (25 miles) to the west, it is administered as an Australian external territory and is inhabited intermittently by scientific research parties.
Map Indian Ocean – Ce

heat equator Zone of maximum heating on the surface of the earth that migrates seasonally, following the position of the overhead sun. It moves farther north and south over areas of land than it does over the oceans, since the land heats up quicker than water.

heath (heathland) Extensive tract of open, uncultivated land, often on sandy soil and covered with shrubby plants of the heather family.

heave Horizontal displacement of rock strata by an inclined FAULT.

Heaviside-Kennelly layer (E Layer) See ATMOSPHERE

Hebei (Hopeh) *China* Province in the north of China, covering 195 000 km² (75 272 sq miles) in the north of the North China Plain. It surrounds Beijing and includes part of the mountainous region between China and the Mongolian plateau. In the west, it borders on the Great Wall which forms the boundary between the autonomous region of Nei Mongol and China. Before artificial dykes were built, the region was prone to extensive flooding; however, since their introduction, much of the soil has become salinated. The climate, too, is one of extremes. The winters are bitterly cold; the summers are uncomfortably hot and sometimes bring disastrous droughts. Despite its harsh environment, Hebei is densely populated, particularly in the lowland areas, where wheat and cotton are grown. Abundant coal reserves are used to fuel iron and steel works and other heavy industries.

Population 62 750 000
Map China – Hc

Hebrides *United Kingdom* See WESTERN ISLES

Hebron *West Bank* Town which was originally part of Jordan but was occupied by the Israelis during the Six Day War in 1967. However, in terms of a 1994 agreement, Israel agreed to relinquish control of the WEST BANK, including Hebron, to the Palestinian Liberation Organisation (PLO). Situated 35 km (21 miles) south-west of Jerusalem, the town lies in the centre of a rich agricultural area. Its main industry is glass-making. Hebron is believed to be one of the world's oldest cities; the nearby Cave of Machpelah is the traditional burial place of Abraham, the father of the Hebrew people. Other Biblical figures said to be buried here are Isaac, Jacob, Leah, Rebecca and Sarah. Later, the Holy Mosque of Harem el-Khalil was built on the site. Hebron hit the world headlines in February 1994 when an Israeli settler shot and killed 29 Palestinians.

Population 100 000
Map Jordan – Ab

Heerlen *Netherlands* Former coal and steel-producing town in Limburg, in the extreme south-east just 5 km (3 miles) from the German border. Since 1975, when the Dutch closed all their coal mines and switched to mainly locally produced natural gas, Heerlen has been hard pressed to create replacement jobs for its redundant miners.

Population 95 350; (Greater Heerlen) 269 420
Map Netherlands – Bb

Hefei *China* Capital of Anhui Province, 450 km (280 miles) west of SHANGHAI. The city's industries include steel and aluminium production, and engineering.

Population 1 541 000
Map China – He

Heidelberg *Germany* Town on the Neckar River, 18 km (11 miles) south-east of Mannheim. Its main tourist attraction is a 13th-century castle whose ruins overlook the town.

From the 13th to the 18th centuries, Heidelberg was the residence of the Electors of the Palatinate. Heidelberg has the country's oldest university, founded in 1386. The chemist Robert Bunsen (1811-99), who invented the Bunsen burner, was a professor here.

Population 136 800
Map Germany – Cd

Heilbronn *Germany* Wine-making town in the valley of the Neckar River, 40 km (25 miles) north of Stuttgart. It has leather, paper, chemical, tool-making and ironware industries.

Population 115 800
Map Germany – Cd

Heilong Jiang *Russia/China* See AMUR

Heilongjiang *China* The country's most northerly province, covering nearly 460 000 km² (178 000 sq miles) and bordering on Russia. Its climate and soil are similar to those of Alberta, Canada, and, like the Canadian prairie province, the Chinese province is a major wheat-growing region. Because of this, farms are larger and more mechanised than in the rest of the country.

Other crops include soya beans and sugar beet, and sunflowers grown for oilseed. The province has much timber and some minerals (notably iron ore), but its most important natural resource is oil. Its capital is HARBIN.

Population 36 080 000
Map China – Ib

Hejaz (Hijaz) *Saudi Arabia* Red Sea coastal region and former province (now renamed Western Province), stretching from the Gulf of Aqaba in the north to Asir in the south. Once an independent kingdom, it was conquered by Ibn Saud and became part of the dual kingdom of Nejd and Hejaz, formed in 1926, that became Saudi Arabia in 1932.

The Hejaz Mountains run parallel to the coast, rising to more than 2000 m (6562 ft). The area is mainly arid, but in oases on the coastal belt wheat, millet and dates are grown. JEDDAH on the Red Sea and MECCA, focus of Muslim pilgrimage, are the two most important towns.

Map Saudi Arabia – Ab

Hekla *Iceland* Active volcano in the south of the country, 112 km (75 miles) east of REYK-JAVÍK. Its crater is 5 km (3 miles) across and the mountain is nearly 1491 m (4892 ft) high. There have been 26 recorded eruptions, the last in 1991, that have left huge lava fields covering 420 km² (162 sq miles).

Map Iceland – Cb

Helder, Den *Netherlands* Port and the country's main naval base. It lies on the north-western tip of the mainland, and its ferries provide a link to the Friesian island of Texel.

Population 61 150
Map Netherlands – Ba

Helena *USA* State capital of Montana. It lies in the foothills of the Rocky Mountains, on the eastern slope of the continental divide. Gold was discovered in 1864 in what was dubbed 'the Last-Chance gulch' – now the city centre. Today, Helena is a commercial and trading centre of a ranching and mining area.

Population 24 570
Map United States – Da

Heligoland (Helgoland) *Germany* Island 2.1 km² (1 sq mile) in the North Sea, some 70 km (44 miles) from the mouth of the River Elbe. It was held by the Danes and then by the British from 1814 until 1890, when it was handed over to Germany in exchange for the island of Zanzibar off the East African coast. It was used as a naval base during the two World Wars. Heligoland is now a duty-free zone.

Population 1800
Map Germany – Ba

Heliopolis *Egypt* North-eastern suburb of Cairo, on the site of an ancient holy city (referred to in the Bible as 'Ob'). It was dedicated to the sun god Ra, and its temple later became an important library. Two of its stone obelisks, which are called 'Cleopatra's Needles', now stand in London and in New York City. Cairo's international airport is at Heliopolis, which is 15 km (9 miles) from the centre of Cairo.

Map Egypt – Bb

Helmand *Afghanistan* The country's longest river, 1100 km (683 miles) long, rising in the HINDU KUSH west of Kabul. It flows south-west across the country, through the Hazarajat region and on to the Iranian border, where it turns north and drains into the swamps of Lake Harmum.

Map Afghanistan – Bb

Helsingborg *Sweden* Main ferry port to Denmark, at the narrowest part of ÖRESUND, 4.5 km (3 miles) wide, opposite the Danish town of ELSINORE. Its manufactures include rubber, cotton, sugar, pottery and beer.

Population 110 600
Map Sweden – Bd

Helsingfors *Finland* See HELSINKI

Helsingør *Denmark* See ELSINORE

Helsinki (Helsingfors) *Finland* Capital of Finland, founded in 1550 on a peninsula in the Gulf of Finland. It was made capital in place of Turku after 1812, when Russia annexed Finland from Sweden, of which Finland had been a part since the 12th century. It is the most northerly of continental European capitals.

Helsinki, planned upon a grid-iron pattern of streets, rapidly outstripped other Finnish towns in size to become the largest port and manufacturing centre. The city's industries include ship-building (one of the city's major shipyard docks is completely enclosed to protect workers from the bitter winter weather), printing, sugar refining and the manufacture of textiles and porcelain. The Arabia porcelain factory is the largest ceramic plant in Europe.

In addition to the neo-classical buildings of the city centre, there are a number of other attractive public buildings, among them the neo-classical cathedral, the railway station (1916), the grey granite parliament building, the Olympic stadium (built for the 1952 Olympic Games) and the marble-faced Finlandia concert and congress hall. The city also has an international airport and an expanding network of underground trains.

Many of the city's inhabitants live in a semi-circle of suburban developments set in the forested hinterland and along the coves and headlands of the waterfront. Helsinki has Finland's oldest and biggest university and many museums and galleries.

Population 501 500
Map Finland – Cc

hematite Red, grey or black mineral, ferric oxide, found mostly in igneous rock and veins. It is an important source of iron.

Henan (Honan) *China* Populous province covering 167 000 km² (64 464 sq miles) in central China and forming the heart of China, both geographically and historically. Eastern Henan is low-lying and is drained by tributaries of the Huang He and Huai He rivers. The western half is hilly and includes part of the LOESSLANDS OF CHINA as well as the foothills of the Daba Shan. Henan is a major wheat-growing area and also produces tobacco, sesame, rapeseed and peanuts. Flooding of the land, once a major problem, has been stopped by river control schemes. ZHENGZHOU is the capital.

Population 88 610 000

Map China – Gd

Henegouwen *Belgium* See HAINAUT

Henzada *Myanmar* Town on the Ayeyarwady (formerly Irrawaddy) delta in the south of the country, 125 km (78 miles) north-west of Yangon. An important rice trading centre, it is linked by ferry to the country's rail network.

Population 283 600

Map Myanmar – Bc

Heraclea Lyncestis *Macedonia* See BITOLA

Heraklion (Candia; Iráklion) *Greece* Main port and capital of the island of CRETE. In ancient times, it was the port of the Minoan kingdom. Minoan vases, statues and frescoes, many from the excavated palace of KNOSSÓS, 5 km (3 miles) to the south, are in the city museum. In 900 AD Heraklion was used as a base by Arab pirates and slave traders. The Venetians took over in 1204. Heraklion fell to the Turks in 1669 after a siege of 22 years. It became part of Greece in 1913. Its tourist attractions include Venetian fortifications dating from the 14th to the 17th century, Venetian churches and a Venetian castle commanding the city's twin harbours. Heraklion was the birthplace of the painter Domenikos Theotokopoulos (El Greco) (1541-1614), and of the novelist Nikos Kazantzakis (1885-1957), author of *Zorba the Greek*.

Population 117 000

Map Greece – Dd

Herat *Afghanistan* The country's third largest city, standing on the Hari Rud River, 220 km (137 miles) north of Farah. Known as 'the Gateway to Afghanistan', Herat lies 914 m (2998 ft) above sea level, and is enclosed by 15th-century walls. The city is dominated by its spectacular citadel, and noted for its mosques, palaces and ancient tombs. Its main products are textiles, silk and carpets.

Population 177 000

Map Afghanistan – Ba

Hercegovina See BOSNIA-HERZEGOVINA

Herculaneum (Ercolano) *Italy* Graeco-Roman archaeological site on the Bay of Naples at the foot of Mount VESUVIUS. It was smothered by a volcanic eruption in AD 79 which buried it in a sea of mud and ash up to 25 m (82 ft) thick. Another eruption again covered it in mud and ashes in 1631. The excavations, entered from the modern town of Ercolano, lie well below ground level. The site contains several fine black and white mosaic floors and many well-preserved public buildings, shops, houses and statues.

Map Italy – Ed

Hercynian Belonging to a phase of mountain building in the late Palaeozoic era, between 200 and 250 million years ago. It is also called Variscan and ARMORICAN.

Hereford *United Kingdom* City (and former county) of western England, 75 km (47 miles) south-west of BIRMINGHAM. Its beautiful sandstone cathedral shares the Three Choirs Festival with its neighbours, Worcester and Gloucester.

The city stands on the River Wye among hills and fertile farmland where Hereford cattle originated and where the region's cider is made. In 1994, a local government commission recommended that the county of Hereford be re-established.

Population 51 000

Map United Kingdom – Dd

Hereford and Worcester *United Kingdom* County of the west of England on the Welsh border, combining the two former counties of its name – although it has been recommended that Hereford be re-established as a county in its own right.

Hereford and Worcester covers 3927 km² (1516 sq miles), stretching north to the edge of the Black Country near the city of Birmingham and south-east into the Vale of Evesham, one of Britain's leading fruit-growing areas. Tourism is centred on the valley of the River Severn and on the Malvern Hills, which rise to 425 m (1394 ft) and were the home of the composer Edward Elgar (1857-1934).

Population 685 400

Map United Kingdom – Dd

Hermon, Mount *Lebanon/Syria* Peak 44 km (27 miles) south-east of the port of Sidon, standing 2814 m (9232 ft) above sea level on the borders of Lebanon and Syria. Known to the Arabs as Jebel esh Sheikh, it is also called 'the Old Man Mountain' and 'Snow Mountain'.

Its summit is snow-covered for most of the year, and its three peaks tower over the ruins of the Biblical town of Dan. Mount Hermon's lower slopes are well wooded, and it is surrounded by the ruins of ancient temples. The mountain is the source of the River Jordan.

Map Lebanon – Ab

Hertford *United Kingdom* Town and county of southern England. The town lies 32 km (20 miles) north of London and has the ruins of a castle founded in the 10th century by the son of the Saxon King Alfred the Great. Queen Elizabeth I (1533-1603) spent her childhood at Hatfield House. Nearby are the garden cities' of Letchworth and Welwyn, which were built in the 1920s as an experiment in planned living. In the centre of the county, which covers 1634 km² (631 sq miles), lies the Roman city of Verulamium, today the cathedral city of ST ALBANS. Aircraft, equipment for space exploration, computers, electronics and pharmaceuticals are manufactured in the county. It also has a large printing industry.

Population (county) 988 700; (town) 21 600

Map United Kingdom – Ee

Herzegovina See BOSNIA-HERZEGOVINA

Hessen *Germany* Hilly state covering an area of 21 114 km² (8150 sq miles) of forested land between the Neckar River in the south and the Weser and Diemel in the north, and between the Rhine Valley in the west and the former East German border to the east. Its capital is the spa town of WIESBADEN.

Population 5 791 300

Map Germany – Cc

Hex River Valley *South Africa* A relatively narrow valley of the Western Cape, north-east of Cape Town, where table grapes are intensively cultivated. The surrounding mountains are snow-clad in winter, providing both sport and water for summer irrigation.

The largest settlement is the town of De Doorns, 210 km (130 miles) from Cape Town.

Map South Africa – Ac

Hidalgo *Mexico* Small mining state, mining mainly silver, gold and mercury, north of Mexico City. It has impressive Toltec remains, dating from about the 12th century, at Tula.

Population 1 881 000

Map Mexico – Cb

Hieron Epidaurou *Greece* See EPIDAURUS

high An area of high atmospheric pressure. See ANTICYCLONE.

high water Highest point reached by the sea in any one tidal cycle.

Highland *United Kingdom* Administrative region covering 25 398 km² (9804 sq miles) of northern Scotland – the largest in the country. It includes the former counties of Inverness, Nairn, Ross and Cromarty, Sutherland and Caithness, as well as parts of Argyll and Moray.

Its sparsely populated highlands have suffered through a turbulent history of depopulation and clearance, but although they have been regenerated – by forestry, salmon farming, tourism, and industries based on the North Sea oil fields – both they and the islands lying off the coast have been designated one of the European Union (EU)'s poorest regions. Industries based on North Sea oil have also suffered a decline.

Ben Nevis, the highest mountain in Britain at 1344 m (4409 ft), which is part of the Grampian Mountains, falls within the Highland region. The region's main town is INVERNESS.

Population 209 420

Map United Kingdom – Cb

Highlands, The *United Kingdom* The part of Scotland north and west of an approximate line joining the Firth of Clyde and the town of Stonehaven on the east coast at the foot of the Grampian Mountains. It includes the most rugged of Scotland's mountains but also some low-lying country, especially near the coast.

Its symbols are heather and tartan, kilt and clan. The area to the south of the line – known as the Lowlands, although it contains some substantial hills – has long been culturally distinct.

Map United Kingdom – Cb

highveld Term used in southern Africa for open grassland found at altitudes of about 1500 to 1850 m (4900-6100 ft) above sea level. Trees are few because of the altitude, the large range of temperatures each day (especially in winter, when frosts occur), and the high frequency of droughts. Corresponding to the pampas, prairies and steppes in other continents, the highveld is used to rear cattle and for arable farming.

Hildesheim *Germany* Cathedral town 30 km (19 miles) south-east of Hannover. The cathedral was started in the 9th century but was mostly rebuilt after being damaged during the Second

297

World War. Its bronze doors and a chandelier remain from the 11th century.

In the cloisters is a rose, which, according to folklore, has been growing there since 815, when a chapel was first built on the site.

Population 105 300
Map Germany – Cb

hill fog Low cloud enveloping high ground.

Hillerød *Denmark* Small tourist town that has grown up around the magnificent Frederiksborg Castle, on a lakeside 36 km (22 miles) north-west of Copenhagen. The Renaissance Castle, where the kings of Denmark used to be crowned, was built by Christian IV in the 17th century. It was gutted by fire in 1859, but was restored by private subscription in 1860-75 and is now Denmark's national history museum. Fredensborg Palace, the spring and autumn home of the Danish royal family, lies 9 km (6 miles) north-east of Hillerød. The palace is open to the public in July; the park is open all year round.

Population 25 200
Map Denmark – Cb

Hilversum *Netherlands* Town 25 km (15 miles) south-east of the capital, Amsterdam. It has the appearance of a garden city, and has grown in the 20th century as a commuter town for the capital and as the headquarters of the Dutch broadcasting system.

Population (town) 84 560; (Greater Hilversum) 102 430
Map Netherlands – Ba

Himachal Pradesh *India* Mountainous state bordering on Tibet and covering 55 673 km² (21 490 sq miles). It was created as a territory from 30 former hill states in 1948. In 1966, parts of Punjab were added to it, and it became a state in 1971. Tea, rice, maize and wheat are grown. Its capital is the hill town and resort of SHIMLA.

Population 5 111 100
Map India – Ba

Himalayas *Southern Asia* The world's mightiest mountain range, curving across southern Asia for some 2500 km (1550 miles). It stretches between the Indian states of Jammu and Kashmir in the west and Assam in the east, and along the way sweeps right across northern Nepal.

The range is divided into four sections – the Trans-Himalayas in Tibet, and the Great Himalayas in the north, that include Mount EVEREST (8846 m, 29 022 ft), the Lesser Himalayas in the centre, and the Outer Himalayas in the south, that include the Siwalik range. The Himalayas have an average height of 6100 m (20 000 ft), and of the 109 peaks in the world rising to more than 7315 m (24 000 ft) above sea level, 96 are in this range.

The name 'Himalayas' comes from the combined Sanskrit words of *him*, meaning 'snows', and *alya*, meaning 'home of'. Some of the Western World's favourite flowering shrubs were introduced from the Himalayas, including many fine species of rhododendrons in the 18th and 19th centuries.

The Himalayas are an important climatic divide between the tropical southern areas with monsoon climates and the dry mountain deserts and steppes of Central Asia.

Map Asia – Fe

Himeji *Japan* Steel-making and oil-refining city and port on the INLAND SEA, 75 km (47 miles) west and slightly north of the city of OSAKA. It is overlooked by the beautiful White Heron Castle dating from the 14th century, with its five-storey main tower, 38 separate buildings and 21 gates.

Population 454 000
Map Japan – Bd

Hims *Syria* See HOMS

Hindu Kush *Central Asia* The western extension of the Himalayas stretching 600 km (373 miles) from the PAMIRS and Karakoram range in the north-east of the region to the KUH-E BABA Mountains in the west. Many of its passes are more than 2500 m (8200 ft) above sea level and it has more than 20 peaks that reach heights of 7000 m (22 966 ft) and more. The highest peak is Tirich Mir (7690 m, 25 229 ft), lying in the east in Pakistan. To the west, the mountains form a great almond-shaped range. There, the average height is 3200 m (10 500 ft). Since ancient times the lower passes have been used by armies heading south to invade India. The name Hindu Kush means 'Indian Destroyer'.

Map Afghanistan – Ba

Hirmand *Afghanistan/Iran* Huge arid basin, 240 km (150 miles) long, to the east of the Dasht e Lut Desert, on the border between Afghanistan and Iran. The basin contains Lake Hamun, which also straddles the border. Several rivers flow into it, but none flow out.

Map Iran – Ca

Hirosaki *Japan* University town of northern HONSHU Island, 35 km (20 miles) south-west of the port of AOMORI. It is dominated by the 17th-century Hirosaki Castle, which is famous for the cherry blossoms in its gardens – among the most beautiful in Japan.

Population 175 000
Map Japan – Db

Hiroshima *Japan* Regional capital in western HONSHU Island straddling the Ota River delta at the western end of the Inland Sea. It is now Japan's leading car manufacturing centre. Other industries include shipbuilding, petrochemical works and fish canning.

The Peace Memorial Park and the ruined dome of the Municipal Industrial Promotions Hall are reminders of the world's first atomic bomb that was dropped on the city on 6 August 1945, killing 200 000 people and destroying 80 per cent of the city's buildings. August 6 is commemorated each year with religious ceremonies and rallies.

Population 1 086 000
Map Japan – Bd

Hispaniola *West Indies* Second largest island in the Caribbean after Cuba, covering about 76 200 km² (29 400 sq miles). It lies between Cuba and Puerto Rico, and comprises HAITI in the west and the DOMINICAN REPUBLIC in the east. On it stands Pico Duarte, formerly Monte Trujillo, at 3175 m (10 417 ft) the highest point in the West Indies. The island was noted by Christopher Columbus on his first voyage to the Americas in 1492 and settled by the Spanish a year later.

Population 12 486 000
Map Caribbean – Ab

Hispur Glacier *Pakistan* Longest glacier in the Himalayas and, at 50 km (30 miles), one of the longest in the world outside the polar regions. It lies in the Karakoram Range in Azad Kashmir.

Map Pakistan – Da

Histria *Romania* Extensive Greek, Roman and Byzantine archaeological site overlooking the Black Sea, 50 km (30 miles) north of Constanța. Histria, which dates from the 7th century BC, contains ruins of a temple to Aphrodite, the Greek goddess of love, as well as Greek baths and Roman and Byzantine homes and shops.

Map Romania – Bb

Hitachi *Japan* City on the east coast of HONSHU Island, 125 km (78 miles) north-east of the capital, Tokyo. Originally a copper-mining area, the city now produces heavy electrical machinery.

Population 202 000
Map Japan – Dc

Hluhluwe Game Reserve *South Africa* Reserve in northern KwaZulu-Natal, 280 km (175 miles) north-east of Durban. It is noted for its big game, including both black and white rhinoceroses.

Other species in this 230 km² (89 sq miles) reserve include lions, elephants, buffalos, leopards, hippos, several species of antelope, baboons, crocodiles and many bird species.

There are facilities for day visitors as well as for those wishing to stay longer. Hluhluwe (pronounced 'shloo-shloo-wee' with the accent on the second syllable) was founded in 1897.

Map South Africa – Db

Ho Chi Minh City (Than Pho Ho Chi Minh; Saigon) *Vietnam* The country's largest city, lying in a rich plain on the River Saigon, 50 km (30 miles) from the southern coast. The city was once the French colonial capital and the shopping and trading centre for the people of wealthy rubber and rice plantations of the plains and the Mekong River delta. Later, it was the capital of South Vietnam. Its layout reflects its colonial past – it has broad, tree-lined boulevards and many pavement cafés.

It kept the name Saigon until the collapse of the United States-backed South Vietnam regime in 1975, when the two Vietnams were unified under the Hanoi government. Then it was renamed Ho Chi Minh City after the communist leader Ho Chi Minh (1890-1969) who was president of North Vietnam from 1954 until his death.

In the late 1960s, a huge influx of people from the war-torn countryside streamed into the city, and many of its former open spaces have now been redeveloped with commercial and residential development.

Industry still prospers in the Chinese town of Cholon beside the city. The deep-water port (still called Saigon) on Khanh Ho Island in the delta handles almost all the trade of the southern part of the country.

Population 3 169 140
Map Vietnam – Bc

Ho Chi Minh Trail *South-East Asia* Complex of roads, trails and paths cutting through the hills of the Truong Son and leading from the northern part of Vietnam through Laos to southern Vietnam.

Some of the routeways date from French colonial times, but the trail gained importance

during the Vietnam War as the supply route for the insurgent movement in the south. Heavily bombed by the United States airforce from bases in Thailand, it was nevertheless maintained and today has been upgraded into a drivable road along large sections of its length.
Map (Truong Son) Vietnam – Bb-Bc

hoar frost (white frost) Frozen dew or water vapour that forms a white surface on the ground and on objects near the ground. It forms instead of dew when the dew point is below freezing point (0°C, 32°F).

Hobart *Australia* Capital of Tasmania, and one of the country's oldest cities, lying at the mouth of the Derwent River. It was founded in 1804 as a port, penal colony and whaling centre, and was made the island's capital in 1812. Many early 19th-century buildings remain. The Tasmanian museum and art gallery in the city depicts the culture of the Tasmanian Aboriginal people, who were nearly driven to extinction by white people and their diseases. Today the port handles fruit, grain, wool minerals and timber. Industries include flour mills, sawmills and iron foundries.
Population 181 840
Map Australia – Hg

Hodeida (Al Hudaydah) *Yemen* The country's main industrial and commercial port situated 144 km (87 miles) south-west of SAN'A. It is the largest export harbour after ADEN. Its principal exports are coffee, dates, senna, myrrh and sesame seeds. It has an oil refinery, an oil-blending plant, an oxygen plant, a textile mill and two cotton-cleaning plants; there are also dairy and fruit juice factories. Other products include cigarettes, soft drinks, and aluminium products. Hodeida's harbour was completed with Soviet aid in 1962, and has been expanded since then. The city also has an international airport.
Population 155 110
Map Yemen – Ab

Hofuf *Saudi Arabia* Capital of Al Hasa, the world's largest oasis, situated 300 km (186 miles) east of the national capital of RIYADH. It is an

▲ **PEDAL POWER** Two wheels are overwhelmingly the most popular form of transport in crowded Ho Chi Minh City, formerly Saigon, now recovering from the wounds of the Vietnam War.

important trading centre and deals in locally grown cereals, fruits and dates. The city is a charming patchwork of alleyways, mud-plastered buildings, and Arabian forts and walls. Bedouin women sell newly spun camel and sheep wool at a weekly camel market.
Population 102 000
Map Saudi Arabia – Bb

hogback (hog's back) Sharp ridge with steeply sloping sides, produced by the eroded broken edges of highly tilted strata on one side and the dip of the rocks on the other. It is a steep form of a CUESTA.

Hoggar (Ahaggar) *Algeria* Mountain massif of great beauty. Lying in the far south-east of the country, it is known for its moonlike landscapes. Its highest point is Mount Tahat, reaching 2918 m (9573 ft) near Tamanrasset; its valleys are now under irrigation. The region is the home of the nomadic Tuareg people, noted for their excellent basketwork.
Map Algeria – Bb

Hohe Tauern *Austria* Range of mountains in the eastern Alps, that runs through the states of Salzburg, Carinthia and Tyrol. The range's highest point is the Grossglockner, at 3797 m (12 460 ft) Austria's highest mountain. Farther east is the Niedere Tauern range, which reaches 2863 m (9393 ft).

The Hohe Tauern National Park, opened in 1971 and the country's largest, covers 2000 km² (770 sq miles). It contains the Pasterze Glacier, 10 km (6 miles) long and covering 24 km² (9 sq miles). The Grossglockner Hochalpenstrasse toll road, which begins at Zell-am-See, runs through some of the finest scenery in Austria. It reaches a height of 2505 m (8218 ft) at the Edelweiss Spitze before descending to Lienz. It is usually blocked by heavy falls of snow from November to May. The 8.5 km

(5.3 mile) long Tauern railway tunnel, opened in 1909, was for a long time the only major route through the eastern Alps, but today the Felbertauern Tunnel also carries motorway traffic. There is a major hydroelectric power station in the Kaprun valley, south of Zell-am-See.
Map Austria – Cb

Hohes Venn *Belgium* See Mardennes

Hohhot (Kueisui) *China* Capital of the Nei Mongol Autonomous Region (Inner Mongolia). It lies 400 km (250 miles) west of Beijing. Its industries include food processing and the manufacture of chemical fertilisers. Called 'the Blue City', it was founded in the 16th century as a settlement for nomads of the steppes.
Population 1 206 000
Map China – Gc

Hokitika *New Zealand* Port on the west coast of South Island, 160 km (100 miles) north-west of CHRISTCHURCH. It developed during the gold rush of the 1860s. The town has a nephrite factory that produces trinkets for tourists. Kaniere, 5 km (3 miles) to the east, was the site of the country's last surviving gold dredge.
Population 3340
Map New Zealand – Ce

Hokkaido *Japan* The nation's second largest island, after Honshu. It is rugged and largely forested, and covers 78 509 km² (30 312 sq miles) in the far north, where winters are bitterly cold. Hokkaido is sparsely populated, and conditions for agriculture are less favourable than on Japan's other islands. Its main products are paper, pulp, butter, cheese, sugar and fish. Hokkaido is home to the distinctive Ainu people – but there are only about 17 000 of them left. SAPPORO is the chief town, and the highest peak is Mount Asahi, at 2290 m (7513 ft).
Population 5 644 000
Map Japan – Db

Holguín *Cuba* Capital of Holguín Province. It lies in a fertile plateau region, 105 km (65 miles) north-west of SANTIAGO DE CUBA. It manufactures furniture and tiles, and exports tobacco, coffee, sugar, cereals, beef and hides through the port of Gibara, 30 km (19 miles) to the north.
Population (city) 228 000; (province) 927 700
Map Cuba – Ba

Holland *Netherlands* Name commonly given to the Kingdom of the Netherlands, but more accurately applied to two north-western provinces – Noord-Holland, including the Friesian island of Texel (2948 km², 1138 sq miles), and Zuid-Holland (3363 km², 1298 sq miles).

Most of the land lies below sea level and is protected by dykes and coastal dunes. Holland is the heart of the country, housing 38 per cent of its people, and containing Randstad – a ring of the largest Dutch cities including Amsterdam, Rotterdam, and the Hague, as well as Haarlem and Utrecht. Industries in the ever-growing cities compete for space with market gardeners and bulb growers.
Population (Noord-Holland) 2 440 200; (Zuid-Holland) 3 295 500
Map Netherlands – Ba

Hollókö *Hungary* See MÁTRA

Hollywood *USA* Northern suburb of Los Angeles, California. For a long time it was the centre of the country's film industry and still remains the symbolic centre for the motion picture and television industry even though many of the studios and executive offices have moved to outlying areas. Many film and television stars still live in the suburb of Beverly Hills nearby.
Map United States – Cd

holm Term referring to a small island in a river, or to low-lying land near a stream.

Holocene Second epoch of the Quaternary period of the Cenozoic era of the earth's timescale. See GEOLOGICAL TIMESCALE.

holt Small wood, or wooded hill.

Holy Cross Mountains *Poland* See GÓRY SWIETOKRZYSKIE

Holy Island *United Kingdom* See GWYNEDD; NORTHUMBERLAND

Homburg *Germany* See BAD HOMBURG

Homonhon *Philippines* See VISAYAN ISLANDS

Homs (Al Khums) *Libya* Seaport and market town 104 km (65 miles) east of TRIPOLI, at the foot of the Jebal Nafusah escarpment. Founded by the Turks, it became important after 1870 for its exports of esparto grass, which is used for rope, shoes and paper. Today, the town is a tourist base for the ruins of LEPTIS MAGNA, a Roman walled city lying 3 km (2 miles) to the east. Its principal activities are olive oil refining, cement, tuna fish processing and esparto processing.
Population 180 000
Map Libya – Ba

Homs (Hims) *Syria* Industrial city and important road and rail centre standing on the Orontes River, 140 km (87 miles) north of DAMASCUS. The city's industries include oil refining, fertilisers, sugar beet processing, textiles, jewellery and metallurgical works.
 In ancient times, Homs (then called Emesa) was noted for the worship of the sun god Baal. It was the birthplace in AD 204 of the Roman emperor Heliogabalus, and the Church of St Elian commemorates a Roman governor of Emesa who became a Christian martyr. Today there are a number of churches for the Christian community, and in 1957 several catacombs were discovered. Another interesting monument is the Tomb of Khaled Ibn Al-Walid, the warrior who brought the Islamic religion to Syria in AD 636.
 Near Homs is the even more ancient site of Kadesh, where the Hittites and the Egyptians fought a battle in 1288 BC.
Population 481 000
Map Syria – Bb

Homyel' (Gomel') *Belarus* Industrial city 300 km (186 miles) south-east of MINSK. It was first a Lithuanian and then a Polish possession until 1772, when Russia took control. After being devastated during the Second World War, the city was rebuilt to become a centre for timber products, machinery, textiles and footwear.
Population 500 000
Map Belarus – Bb

Honduras See p 301

Honduras, Gulf of (Bay of Honduras) Inlet of the CARIBBEAN SEA, that washes the shores of Honduras, Guatemala and Belize and was explored by the Spanish in the 1520s.
Map Honduras – Aa

Hong Kong See p 302

Hong Kong Island *Hong Kong* One of the three territories that make up the Crown Colony of Hong Kong, the other two being the peninsula of KOWLOON and the NEW TERRITORIES. It covers 80 km² (31 sq miles). Britain acquired the island under the Treaty of Nanking in 1842 as a base from which to provide protection for British traders operating in Chinese waters.
 Separated from the mainland by Victoria Harbour, Hong Kong Island is very hilly, its highest point being Victoria Peak, at 551 m (1808 ft). On the northern shore stands the city of VICTORIA, capital of Hong Kong. The southern part of the island with its deep bays and inlets contains scattered towns such as ABERDEEN and STANLEY.
Population 1 184 000
Map Hong Kong – Cb

Honiara *Solomon Islands* National capital on the north coast of Guadalcanal Island. It was developed after the Second World War near the site of the wartime air base of Henderson Field, now the country's main airport.
Population 33 750
Map (Guadalcanal Island) Pacific Ocean – Cc

Honolulu *USA* State capital and chief port of Hawaii as well as main airport and seaport which the Pacific Ocean area. It stands on the south coast of the island of OAHU, just east of the US naval base of Pearl Harbor – the bombing of which by the Japanese brought the United States into the Second World War in December 1941. The renowned Waikiki Beach with its tourist hotels is in the eastern part of the city, which is backed by mountains of the Koolau Range.
Population (city) 365 300; (metropolitan area) 836 200
Map Hawaii – Ba

Honshu *Japan* Largest of the Japanese islands, covering 230 988 km² (89 185 sq miles) in the centre of the country. It contains about two-thirds of Japan's land area and three-quarters of its population. The country's political, economic and cultural life revolves around its great cities of TOKYO, the national capital, KYOTO, YOKOHAMA, NAGOYA and OSAKA – all on or near the south coast. A mountainous backbone divides the island; the Pacific Ocean side enjoys hot summers, while the west on the Sea of Japan gets heavy winter snows.
Population 99 254 000
Map Japan – Cc

Hooghly (Hugli) *India* One of the arms of the GANGES River in the Ganges delta. It splits off as the Bhagirathi from the Ganges just above the Bangladesh border and then flows for 193 km (120 miles) into the Bay of Bengal. The banks north of Calcutta – known as Hooghlyside – are built up with jute and other industrial factories. A barrage at FARAKKA aids the river's flow.
Map India – Dc

Hook of Holland (Hoek van Holland) *Netherlands* Rotterdam's coastal port, at the mouth of the Nieuwe Waterweg ('New Waterway'), 25 km (15 miles) west of the city centre. It is the terminus for North Sea ferries.
Map Netherlands – Ba

Hoorn *Netherlands* Former seaport, now a town on the west side of IJSSELMEER, 30 km (20 miles) north-east of Amsterdam. It is popular with tourists, for much of the prosperous port of the 17th century, when the Netherlands built a vast empire in the East Indies, survives intact.
 Hoorn was the birthplace of the explorers Willem Schouten (1580-1625), who named Cape HORN after the town in 1616, and Abel Tasman (1603-1659), who reached Tasmania (named after him) in 1642.
Population 59 960
Map Netherlands – Ba

Hopetown *South Africa* Town near the south bank of the Orange River in the Northern Cape, 120 km (75 miles) south of KIMBERLEY.
 The first South African diamond was found near Hopetown in 1867, but today the surrounding area is better known for its production of livestock and wool.
Population 5060
Map South Africa – Bb

Hormuz, Strait of (Strait of Ormuz) Strait that is 60-100 km (37-62 miles) wide and that separates Oman and the United Arab Emirates from Iran. It was the notorious haunt of pirates from the 7th century BC until the 19th century AD.
 Today, its significance lies in its strategic position at the southern outlet to the the GULF which is rich in oil. About 17 per cent of the world's oil passes through the Strait of Hormuz from the Gulf to the Arabian Sea.
Map United Arab Emirates – Ca

horn Pyramidal peak formed when several CIRQUES develop back to back, leaving a central mass with prominent faces and ridges. The most famous horn is the Matterhorn in Switzerland.

Horn, Cape (Cabo de Hornos) *Chile* Most southerly point of South America. It is a rocky headland 424 m (1391 ft) high on Horn Island in the Magallanes region of Tierra del Fuego. It was sighted by the English adventurer Sir Francis Drake in 1578 and was named by the Dutch explorers Jakob le Maire and Willem Schouten in 1616, after the latter's home town of Hoorn. As an alternative route to the Strait of Magellan to the north of Tierra del Fuego, it played a major role in the world's seafaring history until the opening of the Panama Canal in 1914 – despite its heavy seas and gale-force winds.
Map Chile – Be

horse latitudes Zones of high air pressure around latitudes 30 degrees north and 30 degrees south, that move north and south with the overhead sun, with trade winds and westerlies blowing outwards from them.
 On land, these zones are associated with deserts, while at sea they are associated with areas of calms. In the days of sail, horses (if part of the cargo) were often thrown overboard from becalmed ships in order to save drinking water – and this practice may be the origin of the name.

horst Block of the earth's crust upthrust between two parallel FAULTS. It is the opposite of a graben (see RIFT VALLEY).

Hortobágy *Hungary* Europe's only surviving area of steppe or prairie outside the former USSR countries. It covers some 10 000 km² (about 3900 sq miles) west of the city of Debrecen in eastern Hungary. The drainage of its salt marshes and shallow lakes has provided farmland for crops, including rice and cotton. Even so, the Hortobágy remains a traditional *puszta* (Hungarian for 'grassland') of the Great ALFÖLD. It has the flavour of the American West and of central Asia with its seemingly endless spaces roamed by shepherds and their flocks, splendid horses and haughty *csikós* (horsemen). The Nagycsárda (Great Inn), dating from 1699, stands at the centre of the lonely region by the nine-arched bridge built in the 1830s over the Hortobágy canal, and is noted for its cuisine.
Map Hungary – Bb

hot spring (thermal spring) Natural spring continuously discharging water that is hotter than 37°C (99°F). Hot springs are usually associated with volcanic action.

Houston *USA* Largest city in Texas. It is linked to the Gulf of Mexico by the 92 km (57.3 mile) long Houston Ship Canal, and is the second busiest port in the United States and a major oil refining and petrochemical centre.
Recently it has become a centre for finance and has leading medical research facilities. Houston also has the Lyndon B. Johnson Space Center, which was established in 1961, a World Trade Centre and the Astrodome that was the prototype for domed stadiums.
Population (city) 1 630 600; (metropolitan area) 3 322 030; (conurbation) 3 731 130
Map United States – Ge

Hovenweep National Monument *USA* Area of 318 hectares (785 acres) in the extreme south-east of Utah and south-west of Colorado. It contains the remains of six groups of cliff dwellings, towers and pueblos (communal villages of some native Americans), some dating back to prehistoric times.
Map United States – Ec

Howrah *India* See HAORA

Hradec Králové (Königgrätz) *Czech Republic* Regional capital of Vychodocesky (eastern Bohemia), 105 km (65 miles) east of Prague. The city, which stands on the River Elbe, has a fine castle and a 14th-century Gothic cathedral of red brick. It manufactures musical instruments, machinery, chemicals and timber products.
Population 163 600
Map Czech Republic – Ba

Hrodna *Belarus* Industrial city, formerly called Grodno, in the lowlands of Belarus beside the Polish border. It dates from the 12th century and was the capital of Lithuania in the 13th century. Its main products are electrical equipment, textiles, leather, tobacco and food.
Population 239 000
Map Belarus – Ab

Hsin-kao Shan *Taiwan* See YÚ SHAN

Honduras

THIS POOR 'BANANA REPUBLIC' FACES A BRIGHTER FUTURE EXPORTING SHRIMPS AND LOBSTERS

From the 1920s until the late 1940s Honduras was the world's leading banana exporter – the original banana republic. Foreign-owned companies – particularly the United Fruit Company of the United States – ran the industry and the government, and had the power to determine national policy. Dependent on a single crop of limited value whose profits went abroad, Honduras did not prosper. Produce is now more diverse, though bananas are still the second most important export after coffee. There might be a brighter economic future for Honduras, with the export, mainly to the USA, of shrimps and lobsters.

Honduras, which became a fully independent republic in 1838, is sparsely populated. Four-fifths of the country is mountainous, indented with river valleys, marshes and lagoons running down to the short Pacific coast in the south and the longer banana-growing Caribbean coast in the north.

Mountainous terrain and tropical forest make agricultural development difficult; 29 per cent of the land is forested, but the valuable hardwoods, including mahogany, guayacán, rosewood and walnut, have not been effectively exploited and deforestation is becoming a problem. Cattle raising was developed in the 1970s, peaked in the 1980s, but lost importance by the early 1990s.

The capital, TEGUCIGALPA, is surrounded by mountains and many of them rise to more than 2000 m (6560 ft). Its name means 'silver hills' in a language of the Indians (Mayas) who lived there before the Spanish arrived in the 16th century.

Over 90 per cent of the 5.4 million Hondurans are *mestizos* – of mixed Spanish and Indian blood. Seven per cent of the population is of Indian descent; the rest are *mulattos*, whites and blacks. Most are peasants producing maize, rice and beans for their own use; only a few find jobs harvesting bananas on mechanised plantations. Forty-four per cent of the people live in towns. Education is free, but 27 per cent of the people cannot read.

By the early 1990s, Honduras had continued to escape the serious political unrest experienced by Nicaragua, Guatemala and El Salvador (although a World Cup football match sparked a brief war with El Salvador in 1969). It was a two-leg fixture, played in the two capitals. On each occasion the visiting supporters complained of ill-treatment and violence, including rape. The allegations led to military action by both countries, who were already at daggers drawn over the migration of Salvadoreans into rural areas of Honduras.

However, in September 1992 El Salvador and Honduras signed a historic border agreement. Earlier that year the two countries and Guatemala agreed to create a free trade zone.

This century's major domestic disturbance was a strike in 1954 of 40 000 workers of the United Fruit Company, that secured improved conditions and reduced the company's political power.

HONDURAS AT A GLANCE		
Area 112 088 km² (43 267 sq miles)		
Population 5 418 000		
Capital Tegucigalpa		
Government Presidential republic		
Currency 1 lempira (peso) = 100 centavos		
Language Spanish, Amerindian dialects		
Religion Christian (86% Roman Catholic)		
Climate Tropical on coast, with high rainfalls; temperate in the south, with a dry season from November to April. Average temperature in Tegucigalpa ranges from 19°C (66°F) in January to 24°C (75°F) in May		
Land use Forest 29%, grazing 23%, cultivation 16%, other 32%		
Mineral resources Zinc, gold, silver, lead		
Main primary products Bananas, coffee, maize, beans, rice, sugar cane, tobacco, timber, shellfish, cattle, fruits; lead, zinc, silver, gold, tin		
Major industries Agriculture, mining, forestry, cement, fishing		
Main exports Coffee, bananas, sugar, hardwoods, fish and shellfish		
Annual income per head (US$) 580		
Population growth (per thous/yr) 31		
Life expectancy (yrs) Male 62 **Female** 66		

Huai He (Hwai-Ho) *China* River flowing for about 1100 km (690 miles) over a vast alluvial plain south across Jiangsu Province, and joining the Chiang Jiang east of ZHENJIANG.
Map China – Hd

Hua-lien *Taiwan* One of the few cities on Taiwan's east coast, 120 km (75 miles) south of T'ai-pei. It is the largest port on the Pacific coast.

Its industries include the production of cement and fertilisers, sugar refining and pineapple canning; it is famous for its marble and jade products. T'ai-lu-ko Gorge, one of the most scenic spots in Taiwan, lies 20 km (12 miles) to the north. A section of the East-West Highway runs through the gorge.
Population 3 575 000
Map Taiwan – Bc

301

Hong Kong

THE INHABITANTS OF THIS TINY BRITISH COLONY ARE HAVING MISGIVINGS ABOUT THE FUTURE – IN 1997 THEY WILL BE HANDED OVER TO THEIR GIANT NEIGHBOUR, THE PEOPLE'S REPUBLIC OF CHINA

The time: a summer midnight. The place: a street corner on Nathan Road in downtown KOWLOON, heart of Hong Kong. The air is sticky and warm on the skin. Lights blaze from every window. Huge neon advertising signs compete with dazzling shopfronts against a background of high-rise blocks hung with washing. Construction workers swarm across floodlit towers of scaffolding made of bamboo canes as thick as a man's leg. Buses bulging with passengers snarl past, bumper to bumper with hooting trucks, delivery vans and limousines. Motorcycles, bicycles and handcarts – rarely a rickshaw – clatter beside and between them.

In a hole-in-the-wall café, tucked beside a six-storey emporium, mah-jong players slap down their ivory tiles, the staccato clacks punctuating the syncopated rhythm of flip-flop sandals on the pavements outside. On a side street, an elderly Chinese man in vest and shorts lies asleep beneath an awning, oblivious to the pedestrians eddying around his wooden camp bed. Overhead, a few stars twinkle palely through the damp, exhaust-laden air – vanishing from time to time as another jet thunders over the rooftops on its climb out of the international airport.

WORKING ROUND THE CLOCK

This bewildering, electrifying bustle goes on round the clock in Hong Kong. For the city never sleeps. Thousands of flats have small workshops in the back, so that moonlighting workers can go on making money long into the night. Order a made-to-measure shirt or suit in Hong Kong one afternoon, and it is ready for a final fitting the next morning. Much of the city's clothing, however, is made specially for export to Europe, North America and Australia; for home consumption many cheap but serviceable garments are imported from China. Food, water and electricity, too, have to be imported from China. At the other end of the scale, another popular line in imports is expensive French wines and perfumes for sale to tourists – who flood in at the rate of more than 7 million a year.

Many international business empires have luxury offices in this flourishing business centre. A glass-fronted skyscraper completed in 1985 for the Hong Kong and Shanghai Banking Corporation is reported to have cost US$615 million, and others have since put up similar buildings.

The colony's remarkable history began in 1840 when the British occupied a largely barren rock at the mouth of the ZHU JIANG (Pearl River) south of GUANGZHOU (Canton) because of its protected natural harbour. Before the British occupation, Hong Kong was inhabited only by a small fishing population and was the haunt of pirates. Later evidence has shown that human habitation of the land dates from Neolithic times, with strong influence from the northern Chinese Stone Age cultures, notably the Lungshan.

Britain's illegal opium trade in the 19th century led to its acquisition of Hong Kong. In 1839 an anti-opium campaign was launched in Canton by the Chinese government, resulting in a siege of the British factory there and confiscation of a huge number of chests containing vasts amounts of opium.

The British retreated to Macau and demanded either a commercial treaty or a small island where operations could be conducted safely. Fighting in the Hong Kong waters expanded into the first Opium War. After defeat in the Wars, China ceded this rocky island to Britain. In 1860, after more fighting, a mainland peninsula opposite the rock was ceded to Britain as well. In 1898, a much larger area on the mainland and numerous outlying islands were leased to Britain for 99 years.

Thereafter, Hong Kong's history has been vibrant to say the least. Quickly developing into a flourishing trading centre, it attracted large numbers of Chinese immigrants, especially after the 1911 Revolution, then during the Japanese occupation of China and most recently after the Chinese Civil War (1945-9). After 1945 it was realised that economic recovery was tied to increased industrialisation, and much energy has been expended to this end.

The 'rock' is now HONG KONG ISLAND, a hill fringed by golden beaches and studded with luxurious blocks of flats. Twenty-seven per cent of the population lives on Hong Kong Island. The peninsula is now Kowloon, where the population density reaches up to 250 000 per km² (96 500 per sq mile) in some places. These are now the most densely populated places on earth.

The third part, known as the NEW TERRITORIES, forms more than 90 per cent of the colony's land. It includes the island of LANTAU, and more than 200 smaller outlying islands, as well as the mainland part – where the steep-sided valleys are dotted with factories and new towns interrupted with much smaller fields used for cultivation.

Beyond the New Territories is the vastness of Communist China. Between the two is a river not much wider than a country lane, and a high barbed-wire fence erected by the British to keep refugees from communism out – an ironic reversal of the European Iron Curtain built by communist governments to keep would-be refugees in. Many such refugees swarmed to the area especially in the time of Mao's Cultural Revolution.

WINNING COMBINATION

China is vital to Hong Kong's prosperity. All but 2 per cent of its people are Chinese, coming from the neighbouring Chinese province of Kwantung, and lesser numbers from Fukien and other provinces. The most numerous of the regional groups are the Cantonese who inhabit both rural and urban areas. Much of the raw material that Hong Kong transforms into a fountain of products for the world comes from China. Hong Kong itself has no natural resources – no minerals and no fuels. Even its water comes largely from reservoirs across the Chinese border.

Hong Kong has only three assets: its position, close to the main trading routes of the Pacific; its magnificent natural harbour; and its hard-working people. This combination has proved to be a winner. Hong Kong exports more radios than any other country in the world – including even the United States and Japan. It handles more international trade than far larger countries, outranking Brazil, India, South Africa and China. And its economy has been growing, after allowing for inflation, at around 10 per cent a year – mainly financed by reinvestment of profits. This rate of growth exceeds that of the world's industrial giants. As a result of this success, Hong Kong is almost as important to China's prosperity as China is to Hong Kong's.

Hong Kong's economic achievements may be the best guarantee of its political future. In 1997, Britain's 99-year lease on the New Territories runs out – and with it Britain's sovereignty over the entire colony. On 1 July 1997, under an agreement signed in Beijing in 1984, China takes over. But it will not be a simple handover. Under the agreement, China has guaranteed to preserve Hong

Kong's freewheeling capitalist economy for 50 years: a political recognition of Hong Kong's economic strength. But China's own need for cash for development may well be enough to restrain communist politicians until well into the 21st century from disturbing a goose that lays so many golden eggs.

Hong Kong's people, however, view post-1997 Hong Kong with some trepidation. Almost 20 per cent Chinese and non-Chinese residents (which include those from the Commonwealth countries, the US, Portugal and Japan) many of them highly qualified, plan to emigrate before the handover.

Along with them, much capital will leave, and investors have been pulling out funds for some years. The British and Chinese have also been at loggerheads over the financial arrangements for the new US$21 billion airport at Chek Lap Kok to replace the over-burdened Kai Tak airport.

Many fear that China will not honour its international commitment to let Hong Kong continue along its present path – despite the Sino-British declaration that the rights and freedoms that Hong Kong people enjoy today will be protected in the future. The future of the 2 per cent of non-Chinese people living in Hong Kong also remains uncertain. In 1997, they will lose their status of British Dependent Territory citizens but will not be granted Chinese citizenship.

HONG KONG AT A GLANCE

Area 1078 km² (416 sq miles)

Population 6 019 900 **Per km²** 5583 (**Per sq mile** 436)

Capital Victoria

Government British Crown Colony

Currency 1 Hong Kong dollar = 100 cents

Languages Chinese (Cantonese), English

Religions Buddhist, Christian, Taoist, Confucian, Muslim, Hindu

Climate Tropical; monsoon from May to September. Average temperature ranges from 16°C (61°F) in February to 29°C (84°F) in July

Land use Settlement 78%, forest 11%, cultivation 7%, grazing 1%, other 3%

Main primary products Pigs, poultry, fruit, vegetables, fish

Major industries Textiles, clothing, toys, electronic goods, clocks and watches, plastic products, ship repair, aircraft engineering, iron and steel rolling, fishing

Main exports Textiles, clothing, electronic goods, computers, clocks and watches, toys, plastic products, jewellery

Annual income per head (US$) 11 490

Population growth (per thous/yr) 11

Life expectancy (yrs) Male 75 **Female** 81

▶ HUB OF TRADE Junks and sampans add to the bustle of Hong Kong harbour, one of the best and busiest in the world. Some are home to families of Hong Kong Chinese; many act as lighters to load and unload ships at anchor. The largest junks – many from China – are trading vessels in their own right.

Huambo *Angola* Railway town and provincial capital founded in 1912, 520 km (320 miles) south-east of the national capital, Luanda. It was formerly known as Nova Lisboa.

The savannah-covered province, also called Huambo, has Angola's highest peak, Mount Moco, which rises to 2620 m (8595 ft). The main products are coffee, grain and cattle.

Both the town and the province became a focus of fighting during the Angolan civil war, with the population of the town increasing dramatically as refugees swarmed in from the devastated countryside. During 1993-5, the entire population was living on food relief.

Population (town) 400 000; (province) 1 500 000
Map Angola – Ab

Huancavélica *Peru* South-central department in the Andes. Mercury was discovered here in the 16th century and was used in the production of silver amalgam at Potosí.

The department's capital, also called Huancavélica, 240 km (150 miles) south-east of LIMA, was founded in 1570 by Francisco de Toledo, Spanish viceroy of Peru, under the name Villa Rica de Oropezza. It retains much of its Spanish colonial heritage in its cathedral and its 16th and 17th-century churches.

Population (department) 375 700; (town) 27 400
Map Peru – Bb

Huang He (Hwang Ho; Yellow River) *China* The country's second longest river, after the CHANG JIANG. It rises in the mountains of the west and flows 4667 km (2900 miles) into the Gulf of Bo Hai (the Yellow Sea), 320 km (200 miles) from Beijing. Its name, which means 'Yellow River', comes from the huge amount of silt it carries – so much, in fact, that the river has raised the channel it flows in well above the surface of the surrounding North China Plain.

Much of the silt ends up in the Gulf of Bo Hai, but the delta is not particularly fertile owing to the accumulation of salt in the soil. The river was formerly nicknamed 'China's Sorrow' because of its tendency to flood; however, the construction of artificial dykes since 1949 has greatly reduced this danger.

Map China – Fc, Hd

Huang Shan (Yellow Mountains) *China* Beautiful range in Anhui Province 30 km (19 miles) south of Jiuhua Shan. The mountains form giant towers, with gnarled pine trees growing out of rock crevices, and are frequently swathed in mists and cloud. Huang Shan's peaks, up to 1860 m (6102 ft) high, are a favourite subject of Chinese landscape painters.

Map China – He

Huánuco *Peru* Central department producing sugar cane and tea. The departmental capital, called San León de Huánuco, is an attractive Spanish colonial town standing 1812 m (5945 ft) above sea level. The archaeological sites of the Inca city of Huánuco Viejo and the Temples of Kotosh built during the almost 3000-year-old Chavín culture are nearby.

Population (department) 609 200; (city) 86 300
Map Peru – Ba

Huaráz *Peru* Capital of Ancash department on the Río Santa, lying at an altitude of 3060 m (10 040 ft). The city was severely damaged by an earthquake in 1970 that left 20 000 people dead. It has now been completely rebuilt. The archaeological museum has a fine collection of antiquities from the Chavín culture which flourished in the area between 900 and 200 BC.

Population 65 000
Map Peru – Ba

Huascarán (Nevado Huascarán) *Peru* The country's highest mountain, reaching 6768 m (22 205 ft) and overlooking the Río Santa Valley. It is one of 20 peaks above 6000 m (19 685 ft) in the White Cordillera range. The Huascarán National Park, established in 1975, covers the whole of the White Cordillera Mountains above 3900 m (12 795 ft). In 1962 and 1970, avalanches in the Huascarán tore down into the cities of Yungay and HUARÁZ, killing about 23 000 people.
Map Peru – Ba

Hubei (Hupeh) *China* Province of central China straddling the CHANG JIANG River and covering 187 500 km² (72 377 sq miles). The Chang Jiang has cut deep gorges through the western mountains, part of a series of spectacular canyons beginning upstream in Sichuan Province.

The climate is moist and warm and the soil fertile, making Hubei an important producer of rice, soya beans, wheat, maize, tea and cotton. It has large reserves of iron ore, coal, magnesium, manganese and copper and heavy industries centred in the provincial capital, Wuhan.

Population 85 800 000
Map China – Ge

Huddersfield *United Kingdom* West Yorkshire town, 24 km (15 miles) south-west of LEEDS, that prospered on wool in the 19th century. Hand-woven cloth had already been produced in cottages and farms for centuries, but mechanisation brought the huge steam-powered mills.

Many of these survive, but some have been converted to other uses and their chimneys smoke no longer. The market hall, town hall and grand Victorian railway station reflect the former prosperity of the textile industry.

Population 148 540
Map United Kingdom – Ed

Hudson *USA* River rising in the Adirondack Mountains of New York State and flowing 492 km (306 miles) south to the Atlantic at New York City. It is named after the English explorer Henry Hudson (about 1550-1611) who was the first European to travel up the river in 1609. It is navigable by ocean-going vessels as far as Albany and connects with the Great Lakes, Lake Champlain and the St Lawrence River through divisions of the New York State Barge Canal.

Map United States – Lb

Hudson Bay A vast bay of some 1 233 000 km² (475 952 sq miles) in east-central Canada that has been part of Canada since 1869. The bay, and the 724 km (450 mile) long Hudson Strait, which links it with the ATLANTIC OCEAN, were discovered by the English navigator Henry Hudson (about 1550-1611) in 1610. CHURCHILL, in Manitoba, is the bay's main port. Here, ships take on cargoes of livestock products and cereals brought by rail from central Canada. The bay is frozen over from November to July each year. Its greatest depth is 259 m (850 ft).

Map Canada – Gb

Huê *Vietnam* Ancient imperial capital, 80 km (50 miles) north-west of the port of Da Nang on the Song Huong River. It was the seat of conquering Chinese warlords in 200 BC, and 400 years later became the capital city of invading Cham tribes from the region of present-day Cambodia. The heart of the present city, built in the early 19th century, is a moated enclave beside the River of Perfumes. It covers an area of 7 km² (3 sq miles) and includes palaces, tombs, temples, and the Forbidden City, once the seat of Vietnamese emperors.

Huê declined after the French colonists moved their capital to Saigon, now Ho Chi Minh City, in 1883. The city was bombed during the Vietnam War (1965-75) and its grandeur was allowed to deteriorate even further after the communist takeover in 1975.

Population 260 490
Map Vietnam – Bb

Huelva *Spain* Port and provincial capital 95 km (60 miles) north-west of Cádiz on the south Atlantic coast. It exports copper, iron and manganese mined in its adjacent province of the same name. It also has a tuna and sardine fishing fleet. Christopher Columbus planned his 1492 voyage of discovery to the West while staying at the nearby La Rabida Monastery. Much of the town was destroyed by an earthquake in 1755.

Population (town) 142 550; (province) 443 480
Map Spain – Bd

Huesca *Spain* Mountain town and capital of Huesca Province, 225 km (140 miles) north-west of the Mediterranean port of Barcelona, marketing fruit, wine and cereals and making agricultural machinery, cement, pottery, cloth and leather. The province rises from the Ebro Valley to the peaks of the Pyrenees and contains Pico de Aneto, on the French border, the highest point in the Pyrenees at 3404 m (11 168 ft).

Population (town) 44 170; (province) 207 810
Map Spain – Ea

Hugli *India* See HOOGHLY

Huíla *Angola* Hilly province whose capital, Lubango, is 680 km (420 miles) south of the national capital, Luanda. There are large iron deposits at Cassinga, 280 km (175 miles) east and slightly south of Lubango.

Population 869 000
Map Angola – Ab

Huila *Colombia* Rice-growing and cattle-rearing department in the south-west. Its capital is Neiva, 240 km (150 miles) south-west of Bogotá. Huila is named after the volcanic peak of Nevado de Huila, which reaches 5750 m (18 865 ft).

Population 776 880
Map Colombia – Bb

Hull *Canada* City across the Ottawa River from Ottawa, to which it is connected by five bridges. It manufactures pulp, paper and newsprint, and many of its inhabitants are employed in federal government offices. More than 88 per cent of the population are French Canadians.

Population (city) 60 710; (metropolitan area with Ottawa) 201 540
Map Canada – Hd

Hull *United Kingdom* See KINGSTON UPON HULL

Humber *United Kingdom* Estuary of the Ouse and Trent rivers on the east coast of England, 225 km (140 miles) north of London. It is 60 km (35 miles) long and is crossed by a suspension bridge 1410 m (4626 ft) long, opened in 1981, to the west of KINGSTON UPON HULL.
Map United Kingdom – Ed

Humberside *United Kingdom* County of 3512 km² (1356 sq miles) in eastern England, taking in parts of Yorkshire and Lincolnshire around the Humber estuary. It centres on the port of KINGSTON UPON HULL and includes the steel town of Scunthorpe and the ports of GRIMSBY and Immingham. Important industries are oil refining, chemicals, processed foods and steel.
Population 877 300
Map United Kingdom – Ed

Humboldt Glacier *Greenland* Largest glacier in Greenland (Kalaallit Nunaat). It is 114 km (71 miles) long and 95 km (59 miles) wide, in the north-west of the country. The glacier is named after the German explorer and naturalist Alexander von Humboldt (1769-1859), a well-known writer on America, sometimes called the founder of modern physical geography.
Map Arctic – Db

humidity Water vapour content of a mass of air, given either as absolute humidity, expressed in grams of water per cubic metre of air, or as relative humidity, in which case the water vapour content is expressed as a percentage of the total amount of water vapour required to saturate the air at a stated temperature.

Hunan *China* Province covering 210 500 km² (81 255 sq miles) in central China. It lies south of the great Lake Dongting on the rich alluvial plains of the Xiang Jiang River with hills in the south-west. It was settled by the Chinese in the early 15th century, and because of its warm, humid climate and fertile soils it developed rapidly into a major rice-growing area.
Today, it supplies one-sixth of the country's rice, mainly from the densely settled lowlands around Lake Dongting and the Xiang floodplain. The province also produces tungsten and other metal ores. The capital is CHANGSHA.
Population 62 670 000
Map China – Ge

Hunedoara *Romania* Iron and steel town in the west of the country. It has a massive medieval fortress built largely by the Romanian nobleman and ruler of Transylvania, Iancu Hunedoara, who in 1456 defeated the Turks at Belgrade. The castle was badly damaged by fire in the early 1850s but has since been restored.
Population 85 800
Map Romania – Aa

Hungary See p 306

Hungnam *North Korea* East coast fishing port 200 km (125 miles) north-east of the capital, Pyongyang. Since the 1920s it has become an industrial centre making machinery, chemicals and has been supplied with coal from local deposits and hydroelectric power from the nearby Bujon River.
Population 143 600
Map Korea – Cc

Hunter Valley *Australia* Wine, sheep and coal-producing area in New South Wales. Lying 100 km (60 miles) north of Sydney, the valley covers about 30 000 km² (11 600 sq miles).
The Hunter River, which flows through the valley, is 462 km (287 miles) long, rises in the GREAT DIVIDING RANGE and flows east to the Pacific at Newcastle.
Map Australia – Ie

Huntingdon *United Kingdom* See CAMBRIDGE

Huntly *New Zealand* Mining town on North Island, about 80 km (50 miles) south of Auckland. It is the centre of the country's most important coal field.
Population 6500
Map New Zealand – Eb

Huntsville *USA* City in northern Alabama, about 30 km (20 miles) from the Tennessee border. It is an old agricultural town with textile industries, but its main employer today is the Redstone Arsenal and its associated rocket and space equipment plants.
Population (city) 159 790; (metropolitan area) 293 050
Map United States – Id

Huron, Lake *Canada/USA* Second largest of the Great Lakes, lying between Ontario, Canada, and the US state of Michigan. It is supplied by both Lake Superior and Lake Michigan, and drains via the St Clair River, Lake St Clair and the Detroit River into Lake Erie. About 60 per cent of its approximately 59 610 km² (23 010 sq mile) area is in Canada.
Map Canada – Gd; United States – Jb

hurricane Name used for a CYCLONE in the North Atlantic (especially in the Gulf of Mexico and West Indies). 'Hurricane' is also the term used for the highest wind speed on the BEAUFORT SCALE, in excess of 119 km/h (73 mph).

Hutt Valley *New Zealand* Industrial centre and commuter area for the capital, Wellington. Settlers first settled here in 1839, but later moved to Wellington because of regular flooding by the Hutt River. Hutt Valley was an agricultural area until residential overspill from the capital made it effectively a part of the city. Its industries include cars, heavy engineering and textiles.
Population 131 630
Map New Zealand – Ed

Hvar (Lesina) *Croatia* Rocky, forested Dalmatian island lying off the tourist resorts of the Makarska riviera. Covering 289 km² (112 sq miles), it is Croatia's Madeira, producing wine, honey, grapes, olives, figs and dates, as well as marble. Hvar town has a Baroque cathedral, several fine 16th-century fortresses and a Franciscan monastery. Starigrad is the island's main ferry port.
Population 11 400
Map Croatia – Cc

Hwang Ho *China* See HUANG HE

Hwange *Zimbabwe* Coal-mining town, formerly called Wankie, near the country's western tip, more than 500 km (310 miles) west and slightly south of the capital, Harare. Hwange National

Park, formerly known as Wankie National Park, covers 14 651 km² (5657 sq miles). It lies south-east of the Victoria Falls and has 107 recorded animal species, including buffalos, eland, elephants, giraffes, kudus, wildebeest and zebras. It also has some 400 species of birds.
Population 39 000
Map Zimbabwe – Ba

Hyderabad *India* Walled city founded in 1589, and capital of the state of Andhra Pradesh, about 500 km (310 miles) north-west of Madras. The former capital of the Nizam (ruler) of Hyderabad's princely state, it was forcibly taken over by India in 1948.
The city has an 11 km (7 mile) long city wall with 13 gates. There are many fine Muslim buildings including the Charminar Tower, a 16th-century civic monument standing 46 m (150 ft) high, and mosques, tombs and museums. Hyderabad is a thriving centre of crafts, including metal inlay work, carpets and silks. It has many major colleges and research institutions.
Population 3 005 000
Map India – Cd

Hyderabad *Pakistan* City 160 km (100 miles) north-east of Karachi, standing at the head of the Indus delta. In its modern form the city dates from 1782, when it was planned by the Afghan prince Sarfaraz Khan. Its fort is now a ruin, but there are several surviving royal tombs.
The city, which is renowned for its crafts – mainly embroidery and hand-woven silk, is home to the University of Sind. A few kilometres to the north-west is the site of a battle fought in 1843 in which the British under Sir Charles Napier defeated Mir Sher Mahomed Khan and annexed Sind Province. The site, on the outskirts of Miani village, is marked by a monument.
Population 928 000
Map Pakistan – Cc

Hydra (Ídhra) *Greece* Island resort in the Saronic gulf, 71 km (41 miles) south-west of PIRAEUS. Hydra is barren but unspoilt, and has no motor traffic.
Its harbour town has 18th-century mansions, small hotels and a marina. The mansions and white houses climb high above the harbour to create a classic Greek island scene. Today, Hydra has become a popular tourist resort.
Population 2700
Map Greece – Cc

hydrological cycle (water cycle) Endless series of changes that water undergoes between the sea, air and land. The water in the sea evaporates and becomes water vapour, which condenses into clouds and falls back to earth in the form of rain, running off into rivers and so back into the sea.

hydrosphere All the waters of the earth, including oceans, rivers, lakes, streams, snow, ice, water vapour and underground water.

Hyères, Iles d' *France* A group of three small islands off the coast of Provence. Ile de Porquerolles has some tourist facilities, Ile de Port Cros is a 16 km² (6 sq mile) national park which is both marine and terrestrial, and Ile du Levant is a naval base and a holiday resort for naturists.
Map France – Ge

Hungary

THE LAND OF THE MAGYARS WAS THE FIRST OF THE FORMER EASTERN BLOC STATES TO CHOOSE DEMOCRACY

Hungary declared its independence in October 1989 and had free elections in the spring of 1990 bringing a conservative coalition, led by the Democratic Forum, into power. However, after decades under a command economy, the country faced high unemployment (by 1993 close to 13 per cent), inflation (22 per cent per annum), a mounting budget deficit (around US$3.5 billion), and social

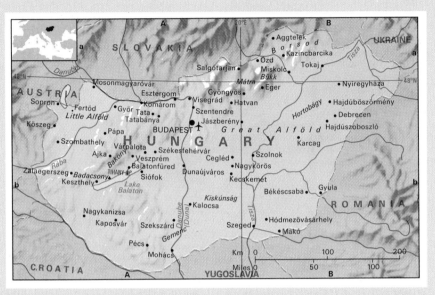

In 1956, Russian troop reinforcements were sent into Hungary to quell a national uprising that threatened to take the country out of the Eastern Bloc. The streets of the capital, BUDAPEST, echoed to the crackle of gunfire as Hungarian freedom fighters (later joined by some Hungarian troops) faced Soviet armoured divisions. The rebels were demanding the withdrawal of all Soviet troops from the country, the withdrawal of Hungary from the Warsaw Pact, and free elections.

The tragic events of 1956 cost much bloodshed – about 4000 Hungarians died and a further 200 000 fled to the West. The new leader who was installed with Soviet support, János Kádár, gradually led Hungary to a degree of liberalisation linked with economic progress. In 1968 he began a series of reforms which aimed to decentralise planning, encourage private enterprise and run an economy based on world market trends.

The result was a measure of prosperity and freedom unknown elsewhere in the Eastern Bloc, and repeatedly denounced by the left as 'goulash communism'. The country attracted foreign investment long before other Eastern Bloc countries dreamed of free market economies. In 1991, it received 60 per cent of all foreign investment flowing into Eastern Europe; a year later, it still received the largest share.

After Kádár stepped down in 1988, Hungary began taking giant strides in the direction of becoming a parliamentary democracy. It was the first country to lift the Iron Curtain and Hungarians were allowed to leave their country as migrant workers from 1983 – thus many East Germans used Hungary as their escape route to the West in 1989.

tension between supporters of the new government and of the former communists. The result of dissatisfaction with the economic performance was that elections in May 1994 put the Socialist Party, led by Gyula Horn, who had been foreign minister and part of the reform group in the final period of communist rule, back in control.

But the new government asserted its allegiance to the ideas of Western Europe socialists rather than old-style communism. Earlier that year the country became an associate member of the European Union (EU).

Landlocked in the heart of Europe, Hungary is dominated by the Great ALFÖLD (Nagyalföld, or Great Plain) to the east of the DANUBE, that runs north to south across the country. Most of the plain is fertile agricultural land. Two more lowlands lie west of the Danube. The Little ALFÖLD (Kisalföld, or Little Plain) in the north-west is separated from the Transdanubian plains by the Central Highlands. In the south of the country, the Mecsek Mountains rise to 681 m (2234 ft). In the shadow of the BAKONY Mountains, which dominate the Central Highlands, lies Lake BALATON – the largest lake in Central Europe, covering 595 km^2 (230 sq miles). The highlands continue in the north of Hungary, sweeping round to the east beyond the 300 m (980 ft) deep Danube gorge. Here they form an extension of the CARPATHIAN Mountains, and include the MATRA range, with Hungary's highest point, at 1015 m (3330 ft).

Hungarian winters are severe caused by cold air flowing down from the surrounding mountains. However, in summer the mountains shield the plains and temperatures average

22°C (72°F) in July, though May and June are often cold and wet. Western Hungary is wetter than the east, where summer droughts occur.

TROUBLED PAST

The country's history goes back more than a thousand years to the foundation of the state. In 896, the Magyars (pronounced 'modjars'), nomadic horsemen from the Russian steppes, swept into the Danube Valley and, under their legendary leader Árpád, occupied territory that is now south-western Hungary. The first Hungarian kingdom was established under the Magyar King (later Saint) Stephen (1000-1038). The kingdom expanded but there were frequent wars with the Turks, who finally conquered the central plain and ruled for 150 years during the 16th and 17th centuries.

The Turks were challenged by the Austrian Habsburgs, who took over the country in 1699. A war of independence against Austria was crushed in 1849 with Russian help, but it was in the interests of the Magyars to make common cause with the Austrians, and in 1867, a dual monarchy was established and Franz Joseph I was made King of Hungary. The country became an autonomous partner in the Austro-Hungarian empire, which also included Czechs, Slovaks and Serbs.

The empire finally collapsed after the First World War, and Hungary was left with one-third of its former territory – the rest going to Czechoslovakia, Romania and Yugoslavia. In the 1930s, Hungary came under German influence, and in the Second World War the Hungarians fought with Germany against the Russians. After the war, the country became a Soviet satellite. It became a people's republic in 1949 and was a founder member of the Warsaw Pact in 1955.

ECONOMIC BOOM

Hungary enjoyed more of an economic boom and less foreign indebtedness in the 1970s and 1980s than many of its then-communist neighbours (especially Poland, Romania and Yugoslavia). The secret of success lay in the Hungarian government's own blend of policies since the mid-1960s. It combined progressive incentive schemes for management and workers with the kind of improvements made by the Bulgarians in agriculture and by the East Germans in industry, and allowed elements of private production – it is estimated that in 1992-3, the private sector produced one-third of national output.

When Hungary became communist in 1948, land ownership was more feudal than elsewhere in Eastern Europe. One per cent of those who gained a living from farming owned 57 per cent of the land. The largest landowners were the Roman Catholic Church and the leading Magyar noble family – the Esterházys.

The estates became state farms or were divided among 640 000 landless labourers and smallholders. In 1950 a start was made in collectivising all peasant land but this was reversed after the change of regime and a great part of the land was re-privatised. Today, 17 per cent of cultivated land is farmed by peasants who produce about one-third of the total potato, vegetable and wine production.

After initial neglect, money has been poured into agriculture since the mid-1960s. Enormous strides were made in mechanising farm work. The use of fertilisers soared from 13 kg per hectare (about 12 lb per acre) in 1938 to 300 kg per hectare (about 270 lb per acre) in the 1980s, supported by large-scale expansion of the chemical industry, but in 1990 it dropped again. Irrigation, extended from 14 000 hectares (about 35 000 acres) in 1945 to 450 000 hectares (more than 1.1 million acres), brought new land into cultivation along the Danube, TISZA and Körös rivers and the most arid areas of the Great Plain, the HORTO-BÁGY and the Hajdúság. Here, irrigation even made rice cultivation possible.

As a result, yields of cereals used for making bread increased in many areas. The policy has been to expand output and improve the quality of specialised produce in which Hungary is competitive in both Western and Eastern European markets. The cultivation of grain for animal fodder has often displaced bread cereals to ensure big increases in the output of ham, pork, beef, salami, milk, skins and wool. Hungary today also exports processed foods; it is the largest exporter of goose livers (mainly to France) worldwide, and Hungarian salami is well known.

Large areas, especially in the KISKÚNSÁG between the Danube and Tisza rivers, are now devoted to a variety of crops whose Hungarian harvest time falls between those of Mediterranean and northern Europe, or for which Hungary has environmental advantages. They include such vegetables as cabbages, lettuces, onions and red and green peppers; grapes for wine; and fruits such as melons, apricots, peaches, plums, and strawberries.

In recent years, Hungary's industrial path has been radically altered. Under the first five-year plan (1950-54) Rakosi tried to turn Hungary into what he termed 'a country of iron and steel'. There was, however, not enough iron ore and coal to support a large-scale steel industry, and these had to be imported from the Ukraine. The plan was an expensive failure. Later five-year plans were more realistic; industries have been developed where adequate natural resources exist – for example, food production from central, eastern and southern areas, and aluminium production in the Bakony Mountains. Greater use has been made of traditional skills in textiles, leather and ceramics. Newer industries have been promoted, such as the manufacture of electrical and electronic equipment; other products include machinery and motor cars.

SOCIAL UPHEAVAL

Yet Hungary is not without its problems, most of which are historically based. In the past, orthodox communists in some other Eastern European countries opposed its liberal management, banking and foreign trade policies. With poor domestic energy resources, Hungary depended on these neighbours to supply oil, gas, coal and electricity in exchange for food and manufactured goods.

Another problem was the huge cost of providing essential services in the Great Plain. Here, scattered towns – often no more than

large villages such as Hódmezövásárhely – and remote farmsteads require modern amenities, cultural facilities and social services.

Lack of essential services caused many to move to the cities, leaving farming to the old. To stem the tide and retain a productive workforce on the land, Hungary in the 1960s provided pensions and other welfare rights for farm workers in cooperatives equal to those enjoyed by state employees – the first East European country to do so.

Most rural migrants flocked to Budapest, which houses more than 2 million people (one-fifth of Hungary's population) and provides nearly 30 per cent of the country's jobs; it has a serious housing shortage. Most people there live in high-density areas of flats ranging from 19th-century inner city art nouveau to 1960s suburban socialist utilitarian.

Today, car ownership has soared, giving Budapest problems of parking, congestion, noise and pollution. Many managerial, administrative, professional, scientific and industrial workers own second homes. This has been made possible because of rising real incomes, cheap housing and public utilities and free medical care and education.

Hungarians are hearty eaters and many drink heavily. Paprika is present in most national dishes. Tokaj and 'Bull's Blood' are among the country's excellent wines. Cafés and restaurants everywhere are lively.

The Hungarians have managed to retain their culture through a turbulent history, including subservience to and revolt against Turks and Austrians. Their Magyar language – a unique tongue related only to Finnish – exotic food, ancient folk music and unusual rhythms immortalised in the works of Liszt, Bartók

and Kodály sharply distinguish them from their German, Slav and Romanian neighbours. These were shaped 15 centuries ago when the Magyars still roamed the Ural-Altai steppes of southern Russia.

Picturesque rolling country in Transdanubia in the west is dotted with old towns, spas, orchards and vineyards, and merges eastwards into the Bakony Mountains. These thrust their spurs and remarkable shapes, including extinct volcanoes, towards Lake Balaton. An international tourist playground in Central Europe, the lake has splendid beaches, modern holiday resorts, old towns with cool wine cellars, and the beautiful countryside of the Tihany Peninsula. Northern Hungary is different again – a chain of thickly forested hills and mountains that cradle spas, attractive villages and the vineyards of EGER and TOKAJ.

HUNGARY AT A GLANCE		
Area 93 036 km² (35 913 sq miles)		
Population 10 375 000 **Per km²** 111.5		
(Per sq mile 289)		
Capital Budapest		
Government Republic		
Currency 1 forint = 100 filler		
Language Hungarian		
Religions Christian (65% Roman Catholic, 22% Protestant), Jewish (1%), Atheist (13%)		
Climate Continental; average temperature in Budapest ranges from -1.1°C (34°F) in January to 22.2°C (72°F) in July		
Land use Cultivation 57%, forest 13%, grazing 13%, other 17%		
Main primary products Cereals, potatoes, sugar beet, fruit and vegetables, livestock, timber; bauxite, coal, lignite, oil, natural gas		
Major industries Agriculture; iron, steel, textiles, chemicals, machinery, transport equipment; foresty, timber products, mining		
Main exports Motor vehicles, machinery, iron and steel, bauxite, chemicals; meat, wheat, wine, textiles and clothing		
Annual income per head (US$) 3400		
Population growth (per thous/yr) Declining (–0.2)		
Life expectancy (yrs) Male 65 **Female** 74		

▼ **THE BLUE-GREY DANUBE The vast 19th-century neo-Gothic Parliament Building stands beside the river in Budapest. In 1944, the Nazis blew up the nearby Margaret Bridge at the height of a rush hour, with much loss of life. It has been rebuilt.**

Iaşi *Romania* City, university town and capital of Moldavia until 1862. It lies 11 km (7 miles) west of the Moldavian border. Although it was ravaged by Tatars (1513), Turks (1538) and Russians (1686), Iaşi retains many ancient buildings, including the magnificent Trei Ierarhi Church ('Church of the Three Hierarchs'), built in 1639, the 17th-century Golia Monastery, and the neo-Gothic Palace of Culture, built at the turn of the century.

In front of it is the statue of Stephen the Great (1457-1504), the Moldavian prince most successful in resisting the Turkish advance. These buildings contrast markedly with the modern high-rise districts created with the growth of Iaşi's electrical, chemical, textile, tobacco and engineering industries.

Population 334 400
Map Romania – Ba

Ibadan *Nigeria* Capital of Oyo State and the country's second city after Lagos, some 120 km (75 miles) to the south-west. Ibadan, with its world-renowned university, is Nigeria's intellectual heart. It is also a market for cocoa, palm oil and kola nuts.

Population 1 600 000
Map Nigeria – Ab

Ibagué (San Bonifacio de Ibagué) *Colombia* Market town and capital of Tolima department, 140 km (85 miles) west of Bogotá, to which it is linked by road and rail. It is the centre of a cattle-rearing and coffee-growing district; its industries include flour milling, brewing, and the making of leather goods. Renowned as the musical capital of Colombia, it has a conservatory of music. It holds a folk festival each June.

Population 334 080
Map Colombia – Bb

Ibar *Serbia and Montenegro* Serbian river, 276 km (172 miles) long. It rises north of the market town of Peć and flows east, then north through a scenic gorge to the Western Morava River (locally called 'Zapadna') at the city of Kraljevo. It is rich in carp and ideal for canoeing.

Map Serbia and Montenegro – Cc

Ibarra *Ecuador* Capital of Imbabura Province on the Pan-American Highway, 90 km (55 miles) north-east of QUITO. It lies in a deep fertile basin in the Andes and is the market centre for the surrounding agricultural region, trading in coffee, sugar cane and cotton. It manufactures textiles and furniture. It is also a tourist centre, noted for its wood carvings.

Population 80 990
Map Ecuador – Ba

Ibb *Yemen* Small town 60 km (37 miles) north of TAIZ. Ibb is the capital of the Green Valley, an area noted for the superb scenery of contrasting mountains and valleys.

Population 19 700
Map Yemen – Ab

Ibiza (Iviza) *Spain* One of the BALEARIC ISLANDS, with a port of the same name that exports figs, raisins, pine timber and salt. Tourism is an important source of revenue; the town and island are fashionable tourist resorts.

Population (island) 80 500; (town) 25 300
Map Spain – Fc

Iboland *Nigeria* See SOUTH-EAST REGION

Içá *Brazil* See PUTMAYO

Ica *Peru* South-western department that produces cotton and grapes for wine and brandy making. The area is rich in pre-Columbian history, particularly at Paracas, Nazca and Tambo Colorado. The departmental capital, also Ica, is located on the banks of the Ica River, 273 km (170 miles) south-east of LIMA. It has a university and a fine archaeological museum.

Population (department) 542 900; (city) 152 300
Map Peru – Bb

ice ages See THE GRIP OF THE ICE AGES, p 309

ice fall Broken, tumbled mass of ice where a glacier steepens and crumbles.

ice floe Level expanse of floating ice. It is also called a 'floe'.

ice sheet (ice cap) Vast, continuous expanse of land ice, such as that covering the Antarctic.

ice shelf Thick, floating ICE SHEET attached to a coastline.

iceberg Massive floating body of ice that has broken away from a glacier or ice sheet. Icebergs move under the influence of currents and winds, and can be hazardous to shipping. The largest ever seen, 335 km (208 miles) long and 97 km (60 miles) wide, was sighted in the south Pacific Ocean in 1956. The tallest, 167 m (550 ft) high, was recorded off western Greenland in 1959.

Iceland See p 310

icing Formation of ice on exposed objects. The term is especially applied to ice formed on an aircraft or ship from moisture in the atmosphere.

Idaho *USA* Rocky Mountain state covering 213 497 km² (82 412 sq miles). It is mountainous, with the rugged Bitterroot Range of the Rockies in the east and the Snake River crossing the plateau in the south. The area was not permanently settled until the 1860s, when gold was discovered. Today, silver, lead, zinc and phosphates are mined in what has appropriately been named 'the Gem State' because of the more than 70 kinds of precious and semi-precious stones found here. Agriculture is mostly limited to the cultivation of potatoes and wheat.

Its unspoilt beauty and the growth of the winter sports industry have helped to turn Idaho into a leading tourist state. Its tourist attractions include the Yellowstone National Park and the national monument Craters of the Moon. BOISE is the state capital.

Population 1 006 700
Map United States – Ca

Idfu *Egypt* Town on the Nile River, 123 km (76 miles) north of ASWAN. It is the site of a temple dedicated to the falcon-headed god Horus. Built in about 250 BC, the temple is one of the best preserved of ancient Egypt's monuments. The people of Idfu trade in cotton, cereals and dates, and make pottery.

Population 27 000
Map Egypt – Cc

Ídhra *Greece* See HYDRA

Idi Amin Dada, Lake *Uganda/Zaire* See EDWARD, LAKE

Idlib *Syria* Commercial town 60 km (37 miles) south-west of ALEPPO. It is situated in a fertile agricultural region in which fruit, olives, cereals and tobacco are produced.

Population 428 000
Map Syria – Bb

Ieper *Belgium* See YPRES

Ife *Nigeria* Town 170 km (105 miles) north-east of Lagos in Osun State. It is probably the oldest town of the Yoruba people, who can trace their ancestry back to Oduduwa, a tribal chief who is said to have founded the town between the 7th and the 10th century.

By the 13th century, Ife was producing exceptionally fine bronzes and terracotta heads. Today, it is a cultural and religious centre with a university, museum, and, it is said, more than 400 gods. It also markets cocoa and timber.

Population 215 100
Map Nigeria – Ab

Ifni (Santa Cruz de Mar Paqueña) *Morocco* Fishing is the main industry of this arid former Spanish province on Morocco's Atlantic coast, occupying 1502 km² (580 sq miles) just opposite the Canary Islands. Spain established a fortified trading, slaving and fishing settlement here in 1476, abandoned it half a century later, then reclaimed it from Morocco in 1860. Ifni was finally returned to Morocco in 1969. Besides fishing, the mainly Berber population of about 46 000 raise camels, sheep and goats. About 16 000 Berbers live in Sidi Ifni, a small port engaged in fishing and trading with the Canary Islands. The town is a hand crafts centre for the region.

Population 46 000
Map Morocco – Bb

Igaraçu *Brazil* Town 30 km (18 miles) west of the city of RECIFE, near the country's north-eastern tip. Much of the town is protected as a national monument. Its church of São Cosme y São Damião, dating from 1535, is the oldest in Brazil. The town's main products are sugar, coconuts, yams, cassava and paper.

Population 74 000
Map Brazil – Eb

igneous rock Crystalline rock that has been formed by the solidification of magma (molten matter formed within the earth's crust and upper mantle). Common examples of igneous rock include BASALT, DIORITE and SERPENTINE .

Iguaçu Falls (Iguazú Falls) *Brazil/Argentina* Waterfall on the border of Brazil and north-eastern Argentina, and 19 km (12 miles) from the Paraguayan border. The Iguaçu River is fed by 30 tributaries on its way from its source near the Brazilian city of Curitiba, and is 4 km (2.5 miles) wide just before it plunges 82 m (269 ft) down a crescent-shaped cliff.

At times of peak flow, enough water tumbles over the falls to fill more than six Olympic-sized swimming pools every second. But in some years, rainfall is so slight that the river can dry up to form a mere trickle, as it did for a month

THE GRIP OF THE ICE AGES

During the past 2 million years, earth has been deep-frozen nine times, and through the study of glacial scarring on ancient rocks, it is apparent that similar Ice Ages have occurred at intervals over something like 930 million years. Tentative theories have been advanced to explain these phenomena – periodic wanings of the sun's heat, volcanic dust blocking the sun's rays, a decrease in the atmosphere's carbon-dioxide content accelerating the heat loss to outer space, and wobbles of the earth as it orbits the sun. Earth's most recent icy blanket covered Antarctica, parts of New Zealand, Patagonia and the southern Andes, the Caucasus and Himalayas, and large parts of northern Eurasia and North America. The ice sheet covering the North Sea and Britain down to the line of London's northern suburbs retreated some 10 000 years ago.

Today the considerable remains of this last Ice Age are the ice caps of Antarctica and Greenland. The results of this glaciation may be seen in ice-carved valleys, and in uplifted coastlines as the land slowly rose after being relieved of the weight of the ice.

The vast ice sheets locked up enormous quantities of water, so much that sea levels stood at least 135 m (442 ft) below present tidal marks. What are now continental shelves were then dry land, and many offshore islands, including Britain, were part of adjacent land-masses. Russia and North America were joined at what is now the Bering Strait. At these times, lower sea levels allowed the movement of people and animals between the land-masses.

Whether earth has seen its last Ice Age is unknown; most likely we live in an inter-glacial period, one of many in earth's history, and in ten, or twenty, or a hundred thousand years, the ice will return. But a warmer planet would be equally disastrous. The polar ice caps would melt, raising the sea level by some 60 m (200 ft) and submerging many major cities, including New York, Rio de Janeiro and London.

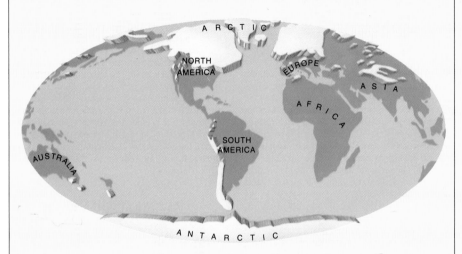

During the most recent Ice Age, which ended some 10 000 years ago, much of the Northern Hemisphere, and part of the Southern, was covered by glaciers and ice sheets. Their courses can be traced from the debris and scarrings they left on the land.

---

in 1978. The largest of the 275 cascades in the falls, the Devil's Throat, is on the border line in mid-river. It can be reached by boat from the Brazilian side. The Iguaçu River flows into the Paraná River at the Paraguayan border.
Map Brazil – Cd

Ijebu Ode *Nigeria* Market town in Ogun State, 75 km (47 miles) north-east of LAGOS. It handles cocoa, palm oil, rubber, kola nuts (chewed as a mild stimulant) and timber, and makes wood products and bicycles.

In addition, various goods are manufactured in its approximately 1000 small craft workshops.
Population 113 000
Map Nigeria – Ab

IJmuiden *Netherlands* See NOORDZEEKANAAL

IJssel *Netherlands* River, 115 km (70 miles) long, that acts as a mouth of the Rhine. It flows out of the Lower Rhine near the city of Arnhem and runs north into IJSSELMEER.
Map Netherlands – Ca

IJsselmeer *Netherlands* Former arm of the North Sea, 95 km (60 miles) long and up to 60 km (37 miles) wide. Known as the Zuiderzee, it was formed by floods in the 13th century. Land reclamation began in 1920, and the arm was finally sealed off by the Afsluitdijk (dam) in 1932 to form IJsselmeer, a freshwater lake fed by the IJSSEL River.

Four POLDERS covering about 1650 km² (637 sq miles) are now farmland. Work on a fifth polder, the Markerwaard (or 'West Polder')

was planned, but because of ecological objections, these plans were postponed.
Map Netherlands – Ba

Ikopa *Madagascar* River rising south of the capital, ANTANANARIVO, and flowing 400 km (250 miles) north-west into the Betsiboka River. The river's source is the Mantasoa reservoir, completed in 1939, that supplies water to the capital and the surrounding rice fields.
Map Madagascar – Aa

I-lan *Taiwan* City in north-eastern Taiwan, 40 km (25 miles) south-east of T'ai-pei. The coastal plain on which it lies is one of the main rice-growing areas of eastern Taiwan, and I-lan acts as its commercial and distribution centre.
Population 462 000
Map Taiwan – Bb

Ile (Iles) For physical features whose names begin with 'Ile' or 'Iles', see main part of the name.

Ilebo *Zaire* Town, formerly called Port Francqui, 580 km (360 miles) east of Kinshasa, the capital. It is located on the Kasai River just above its confluence with the Sankuru.

The terminus of a railway from the mineral-rich Shaba region, it transports copper and other products to Kinshasa by the Ilebo-Kinshasa river route, Zaire's business waterway. It is also an important commercial, industrial and agricultural centre.
Population 40 000
Map Zaire – Bb

Ile-de-France *France* Historically at the centre of the Paris basin from which the king ruled the country, today it is a region of 12 100 km² (4671 sq miles) around Paris, and contains seven departments, 20 per cent of the nation's population and 25 per cent of its economic potential. It has a rich variety of farmland and forests, and many châteaux are dotted within its confines, including those at FONTAINEBLEAU and VERSAILLES.
Population 10 661 000
Map France – Eb

Ilesha *Nigeria* Cocoa-marketing town in Osun State, 200 km (125 miles) north-east of LAGOS. Missionary influence here was strong and, unlike other towns in the state, Christians outnumber Muslims by about six to one.
Population 266 700
Map Nigeria – Ab

Iligan *Philippines* Industrial centre on the north coast of MINDANAO, 50 km (31 miles) south-west of CAGAYAN DE ORO. The capital of Lanao del Norte Province, it has chemical, steel and other industrial plants fed with hydroelectric power from the 97 m (318 ft) high Maria Cristina Falls on the Agus River.
Population 227 000
Map Philippines – Bd

Illampu *Bolivia* Spectacular mountain with a permanent snowcap, rising to 6550 m (21 490 ft) above sea level over the town and valley of Sorata, 80 km (50 miles) north-west of the Bolivian capital, LA PAZ.
Map Bolivia – Bb

Iceland

THE LAND IS STILL YOUNG AND THE MOUNTAINS ARE STILL MOVING, BUT THE GOVERNMENT IS A MODEL OF STABILITY

When American astronauts wanted to simulate walking on the moon, they went to the lunar-like landscapes of Iceland. It is mostly a desolate country of black rock, lava fields and glaciers, described as 'a desert of the ocean' by early missionaries who visited the country. The tourist industry highlights the drama of volcanoes, geysers and pools of boiling mud, and calls it 'The Land of Ice and Fire'.

In human terms this strange island just south of the Arctic Circle has a long history, but it is geologically young. If the aeons of time in which the earth was formed were compressed into one day, Iceland would be born at five minutes to midnight.

The country forms the peak of the mid-Atlantic ridge. It is an unstable part of the earth's crust – the two plates are still drifting apart and the activity which formed the country continues. The ocean deep still produces a constant flow of lava. Iceland has more than 200 volcanoes (best known are HEKLA and Askja), about 30 of which are active, and at least one of which can be expected to erupt every five years; new lava flows cover the old lava, and change the face of the island. In 1963 a new volcano, SURTSEY, emerged from the sea off the south coast; in 1973 lava from Helgafell on the island of Heimaey, in the WESTMAN ISLANDS group, virtually destroyed one of Iceland's largest towns.

DESOLATE LANDSCAPE

During each of the successive ice ages, Iceland lay completely hidden under ice, and one-tenth of the country is still covered in ice and snowfields. The glaciers include VATNA-JÖKULL (9000 km², 3474 sq miles), the largest in the northern hemisphere outside Greenland. The ice cover waxes and wanes and its retreat over the past century has exposed the relics of farms that were covered in the Middle Ages.

There are more than 700 hot springs, which provide three-quarters of Iceland's population

– and the island's hothouses – with central heating. At Haukadalur the springs can be heard grumbling underground.

The base-rock of Iceland is basalt, which has weathered to a bluish-black colour and produced a central plateau with forbidding cliffs. Although it does not have high mountain peaks, Iceland is a high country – more than one-third lies above 600 m (1970 ft) – with limited coastal lowlands. Lava flows have been eroded into fantastic shapes by water and weather. The country is treeless except for some birch scrub and a few coniferous plantations.

The sea makes Iceland habitable. The chief lowland, the most densely populated area, lies in the south-west corner where the North Atlantic Drift warms the shores and keeps them free from ice. The capital, REYKJAVÍK, lies here and its harbour is ice-free. In winter, average temperatures in this area hover around freezing point; in July and August they rise to about 11°C (52°F). The summer days have short nights and the winter days are short and gloomy.

The economy of Iceland is heavily dependent on its sea fishing industry, which accounts for 75 per cent of exports and employs 12 per cent of its workers. Over the past 50 years fishing has grown from a small-scale family activity to a large-scale commercial enterprise with trawlers and processing plants owned by big companies or cooperatives. The fishing industry in 1994 still accounted for half of Iceland's gross domestic product (GDP) and 70-80 per cent of its export earnings. However, falling world prices for fish and government cuts in the fishing quotas in 1992 to maintain stocks, contributed to a deepening recession, bringing unemployment by the beginning of 1993 to 5 per cent, the highest since the Second World War. The worrying indications of overfishing have been the main reason why many Icelanders mistrust the increasing degree of economic interaction with other countries in Europe – which gives foreigners the right to fish legally off Iceland's coasts.

Only 21 per cent of the country is suitable for farming and only 1 per cent for cultivation,

mainly of fodder and root crops, the remaining 20 per cent being used for sheep and cattle.

A quarter of the labour force works in energy production and manufacturing. The people are voracious readers and their needs are met by a lively graphics and printing industry.

Until a few years ago, Iceland was thought to have first been settled by Vikings from Norway in 874. Recent archaeological discoveries suggest, however, that there were earlier Scottish and Irish settlements. In 930 the settlers created the world's first parliament (the *Althing*). Iceland united with Norway in 1262 and in 1380 both countries came under the Danish crown. During the Second World War, Iceland was occupied by Allied troops. In 1944 it became independent.

Icelanders enjoy remarkably high standards of living, education, social security and health care. Life expectancy is high. The people are proud of their egalitarian society and their cultural heritage. They have ambivalent views about the American base on their island at KEFLAVÍK, the more so because although Iceland is a member of the NATO alliance, it has no military forces of its own.

ICELAND AT A GLANCE	
Official name Republic of Iceland	
Area 103 000 km² (39 768 sq miles)	
Population 261 270 **Per km²** 2.5 **(Per sq mile** 6.6)	
Capital Reykjavík	
Government Parliamentary republic	
Currency 1 Icelandic krona = 100 aurar	
Language Icelandic	
Religion Christian (93% Evangelical Lutheran)	
Climate Cold maritime; warmed by the North Atlantic Drift. Average temperature in Reykjavík ranges from -0.3°C (32.5°F) in January to 11.4°C (52.5°F) in July	
Land use Grazing 20%, cultivation 1%, forest 1%, other 78% (mainly icebound)	
Main primary products Fish, potatoes, sheep, dairy cattle, poultry	
Major industries Fishing, fish and food processing, agriculture, cement, aluminium	
Main exports Fish, fish products, aluminium, wool and sheepskin products	
Annual income per head (US$) 23 670	
Population growth (per thous/yr) 10	
Life expectancy (yrs) Male 76 **Female** 81	

▼ **HOT AND COLD Ice-cold glacial meltwaters snake among hot springs at Landmannalaugar in southern Iceland.**

Illimani (Nevado de Illimani) *Bolivia* Triple-peaked mountain rising to 6882 m (22 579 ft) and towering over the capital, La Paz. It has the largest glacier in tropical Latin America.
Map Bolivia – Bb

Illinois *USA* Midwestern state bordered by the Mississippi River in the west and Lake Michigan in the far north-east. It covers 144 286 km² (55 696 sq miles) and maize, wheat, soya beans, pigs and cattle are produced on its gently undulating lowlands. It is one of the largest industrial centres in the nation. Coal is mined and some oil produced. CHICAGO, the chief city, is a leading rail centre, has one of the world's busiest airports, and is famous for its huge grain mills and elevators. It is also a leader in the printing and publishing industry. SPRINGFIELD is the state capital.
Population 11 430 600
Map United States – Hb

ilmenite Lustrous brownish-black mineral, titanium iron oxide, found in basic IGNEOUS ROCK, in VEINS and in SEDIMENTS derived from rocks bearing this ore.

Ilo *Peru* Major fishing port at the mouth of the Moquegua River in the south-west, with a fish-meal processing industry. It exports avocados, wine, and copper mined at Toquepala and Cuajone in the river valley. A smelter 8 km (5 miles) north of Ilo produces copper ingots.
Population 31 700
Map Peru – Bb

Ilocos *Philippines* Densely populated narrow coastal region of north-western LUZON, between the Cordillera Central and the South China Sea. It produces mainly tobacco and rice.
The coast is rugged and scenic, rising to 3000 m (9843 ft) almost immediately; it also has fine beaches. Bauang, south of the port of San Fernando, is the Philippines' most important beach resort area. Ilocos has many fine old stone churches, especially in Vigan. Laoag, in the north, is a lively commercial centre.
Population 3 551 000
Map Philippines – Bb

Iloilo *Philippines* Capital of Iloilo Province on the south-west coast of PANAY. A regional centre and major port of the western VISAYAN ISLANDS, it has many fine Spanish colonial churches and other colonial buildings. Rice and sugar are the main agricultural products.
Population (city) 310 000; (province) 1 300 000
Map Philippines – Bc

Ilorin *Nigeria* Capital of Kwara State, 260 km (160 miles) north-east of LAGOS. It is a fascinating mixture of Hausa-Fulani, Yoruba and other African cultures. Traditional industries include pottery, weaving, dyeing, shoe making and carpentry. Modern factories refine sugar and make cigarettes and matches.
Population 335 400
Map Nigeria – Ab

Imatong Mountains *Sudan/Uganda* Range, 150 km (93 miles) long, which straddles Sudan's southern border with Uganda. The mountains' highest point, Kinyeti, is Sudan's highest mountain at 3187 m (10 456 ft).
Map Sudan – Bc

Imatra *Finland* Industrial town in south-east Finland, 35 km (20 miles) north-east of the town of Lappeenranta and near the border with Russia. It is located close to some of the country's largest sawmills, and manufactures cellulose, as well as chemicals, copper and iron.
The tumbling waters of the nearby Imatra cataract on the Vuoksi River used to be a tourist attraction, but since the construction of a hydro-electric dam on the river, the rapids have almost dried up. On Sundays, when the sluice-gates are opened to the old river bed, the rapids can be seen as before.
Population 32 900
Map Finland – Dc

Imbabura *Ecuador* North-central province lying mainly in the Andes but with a narrow westward extension into the coastal plain. It is crossed by the Pan-American Highway and dotted with volcanic crater lakes. Most of the population, comprising mainly Amerindians and descendants of African slaves, live in the Ibarra basin, where there is some light industry. The province produces maize, wheat, barley, potatoes, sugar cane and cotton. Dairy cattle, sheep and llamas are pastured. The provincial capital is IBARRA.
Population 265 500
Map Ecuador – Ba

Imjin *North Korea/South Korea* River, 160 km (100 miles) long, rising in North Korea's southern mountains. It crosses the DEMILITARISED ZONE, and flows into the Yellow Sea west of the South Korean town of Munsan via an estuary shared with the Han. It was the scene of a heroic stand by British troops – the First Gloucesters – in 1951 during the Korean War.
Map Korea – Cc

Imperia *Italy* Seaport and resort on the Gulf of Genoa, 40 km (25 miles) east of the French border. It is noted for its olive oil and flowers.
Population 40 170
Map Italy – Bc

impermeable rock The opposite of PERMEABLE ROCK

impervious rock The opposite of PERVIOUS ROCK.

Impfondo *Congo* Capital of Likouala region in the far north of the country. Formerly Desbordesville, it is a landing point for steamboat passengers on the UBANGI River, and serves as a trading and transport centre for the forested region surrounding it.
Population 30 000
Map Congo – Ba

Imphal *India* City and capital of the north-east state of Manipur, 60 km (37 miles) west of the border with Myanmar. It is situated in a beautiful valley tucked in a fold of mountains. The Japanese attacked and laid siege to it in 1944, during the Second World War. The lifting of the siege after three months forced the Japanese to drop plans to invade India. Today, foreigners need special permission to visit Imphal.
Population 707 180
Map India – Ec

In Aménas (In Amnas) *Algeria* Important staging post and road junction, near the Libyan border, on the route southwards to the TASSILI N'AJJER mountain region. It has an airport and is the centre of an oil and gas region, with an oil pipeline running to Sakhira in Tunisia.
Population 7500
Map Algeria – Bb

Inagua *Bahamas* Two islands, Great Inagua and Little Inagua, in the extreme south of the archipelago. On Great Inagua, the third largest island of the Bahamas, salt is produced by evaporating sea water. The chief town is Matthew Town in the south-west. Little Inagua is uninhabited.
Population 990
Map Bahamas – Cb

Inari (Enare) *Finland* The largest of the lakes in Finnish Lappland, with an area of 1400 km² (540 sq miles), and dotted with hundreds of islands. It is fed by the Ivalo River from the south, and drains by way of the Paats River to the Barents Sea. It is a popular tourist attraction.
Map Finland – Ca

Inch'on (Incheon) *South Korea* Second largest port in the country, after Pusan. It lies beside the Yellow Sea, 39 km (24 miles) west of the capital, Seoul. Inch'on (formerly Chemulpo) was opened to international commerce in 1876. It produces iron and steel, glass, plywood, cars and chemicals.
Population 1 818 000
Map Korea – Cd

incised meander River MEANDER which has been cut deeply into its original valley floor. The river's power to erode is increased, and it maintains the same pattern of the meander at progressively lower levels. Examples include the Wye near Chepstow and the Dee near Llangollen in Wales, and the Moselle near Trier in Germany.

India See p 312

Indian Desert *India/Pakistan* See THAR

Indian Ocean Smaller than the PACIFIC and ATLANTIC oceans, the Indian Ocean covers 74 120 000 km² (28 611 133 sq miles) and lies between Asia, Australia, Antarctica and Africa. It includes the RED SEA, The GULF, the ARABIAN SEA, the Bay of BENGAL and the ANDAMAN SEA, but only in two places – at the Red Sea and The Gulf – does it deeply incise the land mass of the continents.
The Indian Ocean was formed in the last 170 million years as the ancient continent of GONDWANA broke up and the Indian subcontinent, which was formerly joined to Africa, Australia and Antarctica, drifted north and collided with the Eurasian continental plate. Apart from Madagascar, Sri Lanka and the Seychelles, none of the islands in the Indian Ocean is older than 150 million years – in fact, most of them are no more than a few million years old.
CONTINENTAL DRIFT is still changing the shape of the ocean. OCEAN FLOOR SPREADING is occurring along the central ocean ridges on the sea floor, which run across the length of the ocean floor, from north-west to south-east, and south-west to north-east, crossing just east of the island of Mauritius. Ocean floor spreading occurs most noticeably along the ocean's north-eastern boundary, on and around the Sunda Islands, where the Indo-Australian *(continued on p 317)*

(continued on p 317)

311

India

IN A VAST LAND WITH A 4000-YEAR HISTORY OF CIVILISATION, POPULATION PRESSURES, POVERTY AND CONFLICTS OF RELIGION AND CASTE POSE A MOUNTING CHALLENGE TO THE WORLD'S LARGEST DEMOCRACY

More than 900 million people live in India, and there will be 1000 million by the end of the century if present trends continue. It is a country with huge problems, centred mainly on poverty and

▲ CROWNING GLORY The Maharajah's Palace, rebuilt in 1897, stands on Chamundi Hill in the centre of Mysore City. It contains a gold and ivory throne – a gift of the Mogul Emperor Aurangzeb (1658-1707).

population growth. Before the problem of population is remedied, no other problem can be effectively tackled. There is bitter poverty, as well as underdevelopment in agriculture and rural infrastructure, high unemployment, and hopeless overcrowding in the cities.

The country in itself is a vast contradiction: although it is one of the poorest countries in the world, it is among the largest industrial nations, manufacturing airplanes, communications satellites and nuclear power stations. For years a stable democracy and a model for other developing nations, it is today beset by internal unrest and religious rivalries as its millions

seek political power and its government has turned its back on controlled-market socialism in favour of a free-market system.

In its landscape, too, it is a country of contrasts. The HIMALAYAS lie in the north, forming a barrier between India and Tibet. The Himalayas are the world's youngest and highest mountains. They run into two other mountain chains to the west; the KARAKORAM, which lie in KASHMIR, and the HINDU KUSH, in Afghanistan.

The lower slopes of the mountains in West Bengal are covered in lush vegetation, and the area around DARJILING is one of the world's best tea-growing areas. Here, rainfalls of up to 10 000 mm (390 in) a year have been recorded. The cool north-eastern monsoon, which blows from January to March, brings rain only to the far south-east.

The climate in the rest of the country is almost tropical, with plenty of sunshine and an abundance of water from the monsoons and from numerous rivers flowing from the snows of the Himalayas. The monsoon rains last from June to September, bringing high humidity and slightly cooler days.

At the foot of the Himalayas the land changes abruptly into a huge plain, drained by three great rivers – the INDUS, BRAHMAPUTRA and the GANGES – and their tributaries. This is the most densely populated part of India. It fills one-third of the country and extends 3500 km (2200 miles) from the PUNJAB in the west to BENGAL and Assam in the east.

Although it receives enormous amounts of water from the monsoon-fed rivers and the occasional flooding caused by melting Himalayan snows, irrigation is still needed. However, it is one of the most fertile areas in the world, producing rice and sugar cane. The plains run south to the Vindhya Mountains, which border the DECCAN plateau – the oldest

part of India with rocks more than 600 million years old. The Deccan extends to the southern tip of the country.

CASTES AND UNTOUCHABLES

The people of India have varied origins. Before 2000 BC they were mostly dark-skinned tribes, including the Dravidians, most of whom live in the south today. The course of history was changed when fair-haired, blue-eyed Aryan tribes moved down from the north-west. The beliefs of these invaders – recorded in religious texts called the Vedas – formed the basis of the great Hindu religion. Four thousand years later in modern India the Vedic hymns are still recited by Hindu priests.

The Aryan peoples had a three-tier social system of priests, warriors and common people. Later they became four tiers, – the priests (*brahmins*), warriors (*kshatriyas*), creators of wealth (*vaishyas*), and the labourers and peasants (*shudras*). These were the origins of the caste system in Hinduism. Caste dictated what work you did, whom you married and even with whom you could eat, and the same restrictions applied to your descendants. Outside the main castes were the people whose occupations made them the 'Untouchables'. They were responsible, for instance, for the removal of dead bodies, or of human excrement. It was believed that these *pariah*, as they were called, were being punished for ill deeds performed in a previous life. They have been renamed *harijan* ('children of God'), or *dalits* ('the oppressed').

In the modern Republic of India, caste has little legal significance – Untouchability is outlawed by the constitution – but it is still a paramount influence on society. A majority still marry within their own castes and caste plays an important role at election times. Even today, India's bureaucracy is dominated by the upper castes. The *dalits* have to fight for political representation; in Uttar Pradesh where *dalits* and Muslims formed a coalition and secured power in the provincial elections in 1994, caste clashes have soared.

MUSLIM WEALTH AND ARTISTRY

The rival Muslim religion was established in India in 1192 by a Muslim prince from Afghanistan, Muhammad Ghuri. The prince conquered the territory when he defeated the Hindu ruler, Prithviraja, in a historic battle near DELHI. Ghuri founded a sultanate which he ruled from Delhi and expanded. Though most Indians remained Hindu, the ruling class were Muslim until the British took control.

Under the Mogul emperors (1526-1761), Muslim art flourished and gave the world priceless paintings and buildings, including arguably the most beautiful building on earth, the TAJ MAHAL (today threatened by pollution). Europeans were fascinated by the wealth and products of this empire, which attracted traders from Britain, the Netherlands, Portugal and France. The British East India Company was established in 1600 and during the next 100 years held considerable power over the country.

By supporting rival Indian principalities, the British soon controlled large parts of India. When the Mogul empire collapsed from within, riven by wars of succession, the British took sides in the conflict and became a governing power after Robert Clive defeated the Nawab of Bengal at the Battle of PLASSEY in 1757. The British presence was challenged by the Indian Mutiny in 1857-58, which was savagely repressed. After that, the British took complete control and remained in power for another 90 years, developing railways and irrigation canals, expanding the port cities of CALCUTTA, MADRAS and BOMBAY, and exploiting the country's raw materials. This period brought little change for the rural population. In states where Britain ruled indirectly through the Indian princes, it was as if time stood still.

BIRTH PANGS OF FREEDOM

Indian nationalists, led by Jawaharlal Nehru (1889-1964) and Mohandas Gandhi (1869-1948), began demanding freedom from the British in the early 1920s. Gandhi had risen to power and earned the title of Mahatma ('Great Soul') through his 50-year struggle to help the poor of India, especially the Untouchables.

A famous advocate of passive resistance, he insisted that the independence campaign be non-violent and he was largely successful in this aim until the last terrible birth pangs of India's freedom. Independence struck the fetters from two Indian nations: the Hindus, who

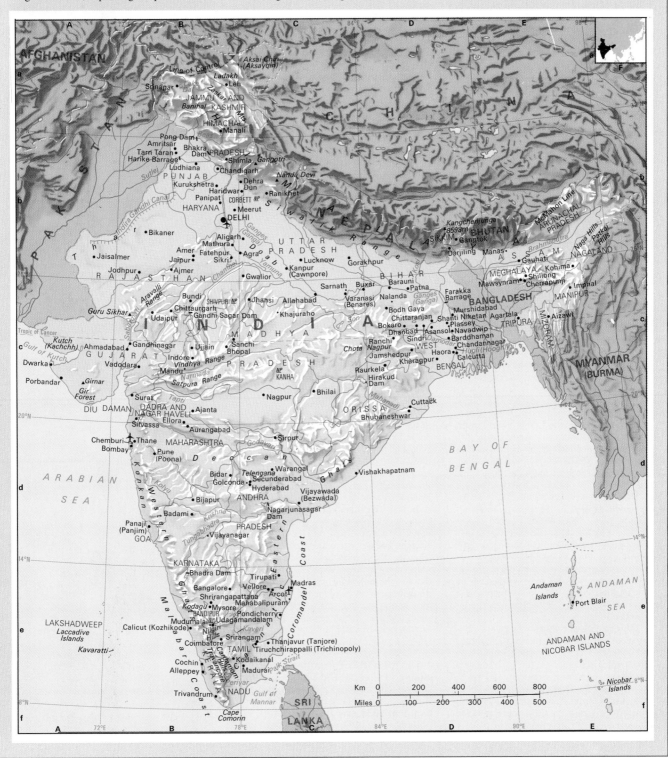

had clamoured for self-rule for 750 years, and the Muslims. To meet these different ambitions, the subcontinent was partitioned.

The Muslims, led by Muhammad Jinnah, were given a new state called Pakistan. It was a country with two 'wings' – one in the Indus Valley, the other in Bengal – separated by more than 1600 km (1000 miles) of Indian territory.

As these two parts were torn from India in August 1947, tensions between Hindus and Muslims erupted into large-scale massacres. Millions of people fled their homes – the Muslims found refuge in PAKISTAN, the Hindus fled to India. One of the princely states, Kashmir, wavered between joining India or Pakistan. The two countries immediately went to war over the issue, which is still unresolved. Kashmir today is claimed by both India and Pakistan. Their conflict has widened and in recent years India and Pakistan have been involved in an arms race compared by many to that of the Cold War.

Mahatma Gandhi was assassinated by a Hindu fanatic in 1948. The first Prime Minister of India after independence was Nehru, who was committed to modernising his country, developing agriculture to provide food for all, and implementing social justice. He held power from 1947 to 1964. He was succeeded by Lal Bahadur Shastri, whose period as Prime Minister brought war with Pakistan in 1965.

Nehru's daughter, Indira Gandhi (no relation to the Mahatma), became Prime Minister in 1966 and maintained a stable government for 18 years. In 1971 her government fought another war with Pakistan that led to the formation of BANGLADESH from the eastern wing of Pakistan. In 1975 Gandhi declared a state of emergency – said by critics to have been characterised by brutal suppression of opposition. Her government, known for its controversial measures, lost the 1977 elections but returned to power in 1980.

The next violent crisis involved the Sikhs, a religious minority whose beliefs blend those of Hinduism and Islam. Living mainly in the wealthy Punjab State, they wanted their own separate nation. Sikh terrorists used Sikh temples as their strongholds, and on 5 June 1984, the Indian army stormed the Sikhs' most sacred place, the Golden Temple at AMRITSAR.

The government said that 500 terrorists and civilians died in the Golden Temple – including the Sikh leader, Jarnail Singh Bhindranwale. Ninety-two soldiers were also killed. Unofficial sources said that about 1200 people died. Four months later Mrs Gandhi was assassinated, and two Sikh officers in her personal security guard were accused of her murder. Since June 1984 bloody clashes have occurred repeatedly between radical Hindus and other religious groups.

Mrs Gandhi's son Rajiv was appointed as her successor and his position was confirmed by a sweeping victory at a general election soon afterwards. He encouraged the use of modern technology, especially computers. However, his period of office was marked by such political issues as terrorism in Kashmir, Tamil separatism in Sri Lanka and allegations that he and his close associates accepted bribes for defence purchases. In the 1989 elections he and his

Congress Party were defeated. Two years later he was assassinated in the midst of elections that restored his party to power. P V Narasimha Rao took over. Savage communal conflict between Muslim and Hindu extremists and violent secessionist movements in Kashmir and the Punjab intensified during his term of office. On the economic front, however, the government's structural reform programme continued, with bolder liberalisation measures such as the lifting of price controls on some raw materials, the relaxation of currency restrictions, the opening of capital markets and the privatisation of oil and gas fields.

POOR PEASANT FARMERS

The drive for modernisation begun by Nehru and continued in six successive five-year plans, has made India a major industrial power in the world. With its abundant coal reserves the country at first generated electricity from coal. After the discovery of offshore oil in the 1970s the emphasis shifted and in recent years about 33 per cent of India's power supply has come from oil and from nuclear and hydroelectric plants, with coal still providing the rest. The government has progressively relaxed its control of industry, and most consumer goods are produced by private enterprise.

But despite the great industrial revolution, the India that Mahatma Gandhi loved, based on village life, is still dominant. About 75 per cent of India's people live in the villages, and 70 per cent of employment is still in agriculture. More than one-third of the farmers have smallholdings of less than 0.2 hectare (0.49 acre) – the average Indian smallholding is 2.63 hectares (6.48 acres). Others own no land at all. They till the soil, but rent the land, paying the landlord either in cash or with part of their harvest. This sharecropping system is blamed for the low levels of investment and productivity in much of Indian agriculture and is one reason for the widespread rural poverty.

Official sources say that 40 per cent of villagers – more than 200 million people – live below the poverty line. Nevertheless, there has been a 'Green Revolution' in Indian agriculture in recent years. This has given the government a reserve stock of grain, held against the inevitable bad harvest – but it is the big landowners who have benefited most from it.

Besides the often exploited sharecropper, there are landless peasants, employed day by day. They are extremely poor. Some will work for food and a place to sleep; others for cash when the farmers need help, as at transplanting time. Many of them migrate to cities where they join the millions of homeless unemployed. Those who stay in the country often do not even have access to village land on which to put a small mud hut. Recent estimates put the total backlog of housing at 31 million units, of which

▶ SILVER STREET Merchants sell cloth and silver in Old Delhi's Chandni Chowk (Silver Street). It leads to the Red Fort, the palace of Shah Jehan begun in 1638, and its balconied houses were built when the street was widened to take his great processions.

10.4 million were needed in urban areas. Requirements for the last decade of this century were put at 79 million dwelling units to meet the backlog and the growth in population.

Most families cook on dried dung or wood fires, and there is no longer enough of either fuel. Women and children forage for hours for a few sticks or in the hopes of finding an unclaimed cowpat. The alternative is to own an animal, but it is only the farmer with some hectares of land who has enough rice-straw and other fodder to keep a cow. Fodder is so scarce that stray cows are shooed out of the countryside and find their way into the towns, where they are allowed to wander virtually at will and graze on rubbish tips.

WHY THE COW IS SACRED

The cow was proclaimed a sacred animal of Hinduism by India's holy men centuries ago to save it from slaughter. Cows and bullocks are an essential part of the rural economy. They pull the ploughs, help to puddle rice fields and pull carts to market. In some places they even grind corn as they did in Biblical times. Cows also provide a little milk, but their yield is small by Western standards.

India has more cattle (198 million) and buffaloes (77 million) than any other country in the world. But there is barely enough food for them and their quality is correspondingly low.

Tractors could take over much of the cows' work but are economical only on large farms and are still quite rare in India. They are concentrated in prosperous areas that have good water supplies and lower population densities – such as the Punjab and Haryana.

To improve agricultural productivity the government has developed irrigation, but little more than half of the irrigation potential has been used so far; the aim is to develop the rest by the early 21st century. Some of the irrigation schemes are massive: the Indira Gandhi Canal project, for instance, includes a canal hundreds of kilometres long, and thousands of kilometres of distribution channels that take water to the THAR Desert.

Rice is the best known of India's food grains, but in the north and west, wheat is the staple, the basis for chapatti and other unleavened bread that is an important food. In much of the Deccan plateau, coarse grains such as sorghum or millet are grown.

MARKET DAY IN THE VILLAGE

Nearly all of India's rural people go regularly to weekly markets, even though they may have little surplus to sell, and will probably have to carry their wares on their heads for many miles. On market day a township of stalls will sprout around a small village or town. It is a day for gossip, perhaps for sounding out the marriage market for one's children. It is a time to seek medicines, many of them of a traditional herbal kind.

Some women will go with their husbands to a jeweller and spend thousands of rupees on gold and silver bangles. Even families who are nearly destitute will buy jewellery – as an investment to be worn until the terrible day when it has to be sold to finance a marriage. Towns as well as villages face great pressure on

resources. There is not enough public transport and rickshaws are too expensive for the poor so everything has to be within walking distance. Each quarter of the town will have its own bazaar, where vegetables can be bought daily. Many small traders and artisans live above their shops; in one part of town the metal workers, in another the grain traders. Buildings are low, the streets narrow, yet the density of human settlement is higher than in Western skyscraper cities.

The houses of the well-to-do often turn their backs on the outside world: all the rooms open on to an internal courtyard. More often than not, they are the home of an extended family, including grandparents and uncles, brothers, sisters and cousins.

Poverty is worst in the countryside but much more visible in towns, where poor people are concentrated. Towns become more and more crowded as people stream in from the country, seeking any opportunity to make money. Overcrowding is increased by the number of animals that join the crush in the streets – monkeys, goats, chickens and the sacred cows.

The urban poor build shacks out of mud, tin and scraps of wood, usually on land which they occupy illegally. Some will pay rent to racketeers just to sleep on the pavement. Sometimes a group of people organise a swoop on a building site, then defy eviction. Such squatters are adept at stealing power from overhead cables. In one year, more than 30 per cent of Calcutta's electricity was reputed to have been stolen.

But squatters cannot steal sewers that should have been laid before they raised their makeshift shacks, so sanitation is the worst problem. Sewage runs down the streets in open canals. The shanty people cook their evening meals on fires fuelled by coal dust, dung, oily rags and rubbish, and the smog that results is the worst in the world. Besides the air pollution, the water is heavily polluted. Still, practising Hindus perform cleansing rites all along the rivers, next to drinking cattle and women washing clothing. India regularly has cholera epidemics and leprosy is common.

THE ROAR OF THE TRAFFIC

Indian cities are cacophonous and overcrowded. Many lack proper planning. The roads are congested with every type of vehicle, mainly cycles and rickshaws. Every truck and car repeatedly blasts its horn. On festive occasions, street corner loudspeakers blare music at maximum volume. Hundreds of shouting sellers of lottery tickets add to the din.

The buses, leaning at crazy angles, with people squeezed inside like sardines or hanging outside from the window bars, add their contribution to noise and smog, the engines belching black diesel clouds in the faces of rickshaw pullers and motor scooter riders. The educated classes seek salaried jobs with big companies or with government. The poor who live in shanty towns may work in factories, but often they have their own workshops, turning waste materials into useful items.

Long-distance travel between cities is still mostly by train, although state-owned coaches connect small villages to major district towns. Hundreds of different languages are spoken in

India. English is used in the legislative and judiciary but Hindi is the official language.

Primary education is free and theoretically compulsory. Many children do not attend school, however, particularly in rural areas where they are needed on the farms. Equipment standards are very low and the government does not provide books. About 40 per cent of the population is illiterate.

GOVERNMENT

Despite poverty and low levels of literacy and education, and despite political upheavals, India has never suffered an army coup, nor a transfer of power between parties except by general election.

There are two major tiers of government, at federal and state level. Some states are bigger than any nation in Europe – Uttar Pradesh, the most populous, has nearly 140 million people.

Life is easier for the central government if its own party holds power at state level. The centre can also hold power through the administration: the senior civil servants of each state belong to the nationwide Indian Administrative Service, and they are always chosen from a different state from that in which they serve. In the army, the local commanders will also be out-of-state people. This makes it easier to exercise impartial authority.

Despite all its problems, modern India must be acclaimed as a major triumph – a synthesis of Eastern spiritualism and tradition, and Western political liberalism.

INDIA AT A GLANCE	
Official name Republic of India	
Area 3 287 590 km² (1 269 046 sq miles) excluding parts of Jammu and Kashmir occupied by Pakistan and China	
Population 897 443 000 **Per km²** 273 (**Per sq mile** 707)	
Capital New Delhi	
Government Federal republic	
Currency 1 Indian rupee = 100 paisa	
Languages Hindi (official) and English; 17 other languages used, including Urdu	
Religion Hindu (80%), Muslim (11%), Christian (2.4%), Sikh (2%), Buddhist and Jain	
Climate Tropical; monsoon from June to September. Subtropical in the north, and desert climate in the north-east. Average temperature in New Delhi ranges from 13.9°C (57°F) in January to 40°C (104°F) in May	
Land use Cultivated 51%, forest 20%, grazing 4%, other 25%	
Main primary products Rice, wheat, sugar cane, barley, sorghum, millet, potatoes, tea, groundnuts, cotton, jute, pulses, vegetables, fruit; tobacco, rubber; coal, iron ore, oil and gas, bauxite, copper, manganese, gemstones	
Major industries Textiles, iron and steel, transport equipment, chemicals, fertilisers, machinery, oil refining, agriculture, cement, coke, food processing, beverages	
Main exports Textiles, machinery, tea, coffee, iron and other metal ores, gemstones, leather goods, cotton, spices	
Annual income per head (US$) 310	
Population growth (per thous/yr) 20	
Life expectancy (yrs) Male 57 **Female** 58	

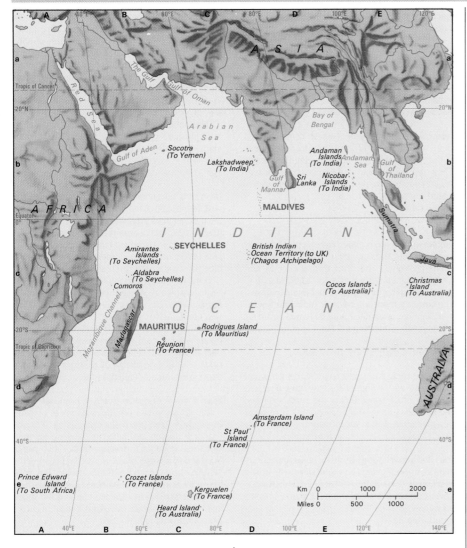

continental plate is sliding under the European plate, causing quakes and volcanic eruptions.

The Indian Ocean also contains the deep JAVA TRENCH, where the Indian Ocean reaches its greatest known depth at 7455 m (24 459 ft).

South of the equator, the surface water circulates in an anticlockwise direction, through the West Australian, South Equatorial and Mozambique currents. North of the Equator, the monsoon causes seasonal changes in the directions of surface currents. In summer, the south-west monsoons blow, often reaching gale force, while in winter, they reverse, to blow in a north-easterly direction.

With the exception of that part of the ocean that lies south of the Tropic of Capricorn, the Indian Ocean is a tropical ocean with temperatures of more than 25°C (77°F) and consequently has a high evaporation rate. The fact that few rivers empty into it means that the ocean has a high salt content in comparison with the Atlantic and Pacific oceans.

Fishing is less developed in the Indian Ocean (again in comparison with the Atlantic and Pacific oceans). The main fishing grounds are in the Arabian Sea and the Bay of Bengal. Oil is extracted in The Gulf and manganese nodules are scattered over large areas of the ocean floor, especially in the Red Sea.

Arab, Indian and Chinese trading ships have sailed across the Indian Ocean since early times. European trade increased after the Portuguese explorer Vasco da Gama discovered the sea route around Africa to India in 1498. The development of oil resources in The Gulf region has greatly increased the ocean's strategic importance.

Map See above

Indian summer Spell of unusually fine, sunny and dry weather, especially in late autumn.

Indiana *USA* Midwestern state with an area of 92 925 km² (35 870 sq miles) east of ILLINOIS, between Lake Michigan and the Ohio River. Wheat, barley, maize, soya beans, pigs and cattle are produced. Coal is mined in the south-west and some limestone is quarried. Electrical and transport equipment and steel are the main manufactures. INDIANAPOLIS is the state capital.

Population 5 544 200
Map United States – Ic

Indianapolis *USA* State capital of INDIANA. The city is an agricultural and industrial centre, and venue of the annual Indianapolis 500 car race.

Population (city) 731 330; (metropolitan area) 1 380 500
Map United States – Ic

Indonesia See p 319

Indore *India* Cotton textile city lying 806 km (501 miles) south-west of Delhi. The former capital of the princely State of Indore, it is now one of the major cities of the State of MADHYA PRADESH. It has many buildings from the colonial period of the British Raj, when it was the focus of British political activity in central India. Daly College was established in the 19th century to educate the sons of Indian princes. Today, Indore has a university as well as several scientific institutions.

Population 1 830 870
Map India – Bc

Indus (Sindh) *South Asia* One of the great rivers of Asia, rising north of the Himalayas at 5200 m (17 000 ft) on Mount Kailas in Tibet (where it is called the Shiquan He) and flowing 2897 km (1800 miles) between mountains, through gorges up to 3000 m (10 000 ft) deep and across blistering deserts to the Arabian Sea south of KARACHI. The river and its tributaries, such as the Jhelum, Chenab and Sutlej, are vital sources of irrigation water for both Pakistan and India, who signed an agreement in 1960 on how to share their flow.

A great civilisation, centred on MOHENJO and dating back 4000 years, once flourished along the river's banks. Today, the Indus Valley is an important transport route, especially between Sind and the Punjab. Notorious in the past for its flooding, the Indus is dammed at TARBELA for hydroelectric power, and numerous irrigation barrages have been constucted where it crosses the Punjab and Sind. The river's flow is diminished once it reaches the plains because of evaporation and irrigation, and because of the sponge-like effect of its alluvial deposits, which are up to 1.6 km (1 mile) deep. On irrigated lands, increasing salinity of the soil is becoming a problem.

Map Asia – Ef; Pakistan – Cb-Da

Infanta *Philippines* See QUEZON CITY

Inga Dam *Zaire* One of the world's largest hydroelectric projects, on the INGA RAPIDS in the Zaire River 230 km (145 miles) south-west of KINSHASA. Begun in 1968, two of the project's several phases have been completed. Inga Dam has a potential of 30 million kilowatts, and has already made possible an expansion of industry in the Bas-Zaire region.

Map Zaire – Ab

Inga Rapids (Livingstone Falls) *Zaire* A series of 32 cataracts, previously called the Livingstone Falls, between KINSHASA, the capital, and MATADI on the lower Zaire River. The rapids fall 260 m (853 ft) over 354 km (220 miles) as the river rushes through the Crystal Mountains.

Map Zaire – Ab

Ingolstadt *Germany* Industrial town on the Danube River, 72 km (45 miles) north of MUNICH. It is the pipeline terminus for oil supplies from the Mediterranean Sea, and has oil refineries, a car industry (it is the seat of the Audi works), and electronics and textile factories.

Population 105 500
Map Germany – Dd

Inhambane *Mozambique* Seaport on the site of an ancient Arab settlement at the tip of a peninsula, some 400 km (249 miles) north-east of

MAPUTO. The port is the capital of Inhambane Province, which covers 68 615 km² (26 485 sq miles). Most people in the province belong to the Tsonga and the Chopi ethnic groups. The province's chief products are cashew nuts, cotton, rice, sugar and, in the drier interior, cattle.
Population (town) 27 000; (province) 1 167 000
Map Mozambique – Ac

inland sea Extensive isolated expanse of water having no link with the open sea, and often associated with distinctive species of animals that have bred there in isolation. Lake Baikal in Russia is an example.

Inland Sea (Seto Naikai) *Japan* Almost completely enclosed area of sea between the islands of HONSHU, SHIKOKU and KYUSHU. It stretches nearly 500 km (310 miles) between the Honshu cities of Osaka in the east and Shimonoseki in the west, and forms the Seto Naikai National Park. The sea, though in places heavily polluted by coastal industries, has many areas of outstanding beauty, including thousands of pine-clad islands with small fishing villages under steep cliffs.
Map Japan – Bd

Inle *Myanmar (Burma)* Lake in Shan State, 200 km(125 miles) south of Mandalay. A major tourist attraction, the lake lies in a valley close to the state capital, Taunggyi. Buddhist temples line its banks; and 'leg-rowers' – boatmen who row standing up, with a paddle strapped to one leg – fish the waters.
Map Myanmar – Cb

Inn *Switzerland/Austria/Germany* River rising near Sils in south-east Switzerland. It flows for 510 km (317 miles) into south-west Austria, passing through Innsbruck, and on through the Bavarian Alps into Germany where it joins the River Danube at the city of PASSAU. More than 20 hydroelectric plants have been constructed along its course.
Map Austria – Bb

Inner Mongolia *China* See NEI MONGOL AUTONOMOUS REGION

Innsbruck *Austria* Tourist resort, manufacturing city and the capital of Tyrol, on the Inn River, 140 km (87 miles) south-west of SALZBURG. In 1809, it was the headquarters of Andreas Höfer, the Austrian who led an unsuccessful revolt of Tyrolese peasants against Napoleon's Confederation of the Rhine, which Austria had joined after capitulating to the French. A statue of Höfer in the city marks the event.
Badly damaged during the Second World War, Innsbruck has been extensively rebuilt. Its major buildings include the 18th-century imperial palace (the 'Hofburg'), and a 16th-century Franciscan church (the 'Hofkirche'), in which stands the magnificent tomb of the Holy Roman Emperor Maximilian I (1459-1519), though the emperor is actually buried at Wiener Neustadt.
The city's main industries include metal products, textiles, bookbinding and bell casting. Innsbruck is a transport hub, a cultural centre, with a university founded in 1669, and a winter sports resort (it hosted the 1964 and 1976 Winter Olympics).
Population 115 000
Map Austria – Bb

inselberg Isolated domed hill left by erosion in a desert or semi-arid region.

insolation weathering The WEATHERING process that occurs when rocks or rock minerals expand and contract as a result of heating by the sun during the day and cooling at night. Different minerals expand and contract at different rates, causing accelerated weathering.

Interlaken *Switzerland* Tourist centre in the Bernese Oberland, 42 km (26 miles) south-east of the capital, Berne. It lies between the lakes of Thun and Brienz, and is famous for its view of the JUNGFRAU. Textiles, clocks and watches are manufactured in the town.
Population 5100
Map Switzerland – Aa

intertropical convergence zone Zone of ascending, moist air and low pressure where the trade winds converge. This zone migrates north and south of the equator with the HEAT EQUATOR. It is characterised by thick clouds and thunderstorms. The summer monsoon rains of India and Southeast Asia are associated with it. It was formerly known as the 'intertropical front'.

intrusion A body of INTRUSIVE ROCK.

intrusive rock An IGNEOUS ROCK formed when magma flows or forces its way into a crack, cavity or other space below the earth's surface and solidifies here. The body of rock formed in this manner is called an intrusion; BATHOLITHS, DYKES and SILLS are common examples.

Inuvik *Canada* Administrative, communications and fur-trading centre on the eastern side of the Mackenzie delta in the NORTHWEST TERRITORIES. It was laid out as a model Inuit ('Eskimo') centre between 1954 and 1962.
Population 3270
Map Canada – Bb

Invercargill *New Zealand* Town on an inlet of the Foveaux Strait in the extreme south of South Island. It is the centre of a sheep and dairy-farming area, and the bulk of its trade goes through the port of BLUFF, 25 km (15 miles) to the south. Its industry is based on the manufacture of fertiliser and meat-freezing works.
Population 51 980
Map New Zealand – Bg

Inverness *United Kingdom* Northernmost major town of Scotland, on the River Ness, 130 km (80 miles) north-west of ABERDEEN. It is the administrative, transport and supply centre for the whole of the Highland region; it is also a major tourist centre, lying to the north-east of Loch NESS.
At Inverness, Duncan, the Scottish king, was murdered in 1039 by Macbeth – a deed immortalised in Shakespeare's play *Macbeth*. Nothing remains of the old castle of Inverness where the deed took place. About 8 km (5 miles) east of the town is CULLODEN, where an English army defeated Bonnie Prince Charlie's forces in 1746.
Population 38 200
Map United Kingdom – Cb

inversion layer A layer of the atmosphere in which temperature increases with height, contrary to the usual pattern. As a rule, temperature decreases significantly with height since the atmosphere is not normally warmed directly by the sun, but indirectly by the radiation of solar energy from the earth's surface. The exception is the stratosphere where ozone filters out ultra-violet radiation, holding the heat, and producing an inversion layer with a temperature of 80°C (176°F). The air below this layer has a temperature of -5°C (23°F).
The condition is often associated with major high pressure systems. The inversion layer prevents dust escaping from the lower layers of air, so causing haze. See THE DIFFERENT LAYERS OF THE ATMOSPHERE, p 53

inversion temperature Increase of temperature with height above the earth's surface, contrary to the normal conditions of temperature falling with height. The increase may occur just above the earth's surface (in which case, it is called a 'surface inversion') or at altitudes far above the surface (this is called a 'high altitude inversion').
A surface inversion may be the result of rapid heat loss from the ground by irradiation at night, especially when the air is calm and the sky clear, or by warm air being cooled over a cold surface. A high altitude inversion may occur when a cool air mass undercuts a warm one at a cold front, or a warm air mass overrides a cold one ahead of a warm front in a depression; this is called an OCCLUSION.

Iona *United Kingdom* Scottish island covering 8 km² (3 sq miles) off the west coast of Mull Island. In AD 563, the Irish evangelist St Columba founded a monastery here, from where his missionaries spread Christianity throughout pagan Scotland. An abbey was built on the island in the Middle Ages and 60 Scottish, Norse and Irish kings are said to be buried here. A religious brotherhood founded in 1938 has restored the abbey and seeks to re-create the island's spirit of piety.
Map United Kingdom – Bb

Ionian islands (Eptanisos) *Greece* The seven main islands of the Ionian Sea – CORFU, CEPHALONIA, LEVKÁS, ITHACA, ZÁKINTHOS, KÍTHIRA and PAXOÍ. They have a mild climate, lush vegetation, sandy beaches, fishing ports and timeless villages.
The islands have been settled since the Homeric age (11th to 12th century BC). They were ruled in turn by the Byzantines, Romans, Venetians (1203-1797), a Russo-Turkish force (1798-9), the French (1807-15) and the British (1815-64), after which they were returned to Greece. Tourism, olive and fig production, and fishing are the main activities. The Greek name for the islands, Eptanisos, comes from Greek words meaning 'seven islands'.
Population 191 000
Map Greece – Ab

Ionian Sea Lying between southern Italy and Greece, the sea includes some of the deepest parts of the MEDITERRANEAN, with a greatest known depth of 5121 m (16 800 ft). The sea's greatest width is 676 km (420 miles). Named after Io, mistress of the god Zeus, it is rich in classical and mythological associations. Zeus changed his mistress into a white cow to conceal her from Hera, his jealous wife. However, Hera sent a gadfly to torment Io, and Io wandered around the seashores until Zeus finally returned her to human form.
Map Europe – Ee

Indonesia

A SPRAWLING MASS OF VOLCANIC ISLANDS THAT IS OCCUPIED BY ONE OF THE MOST DIVERSE AND EXOTIC NATIONS ON EARTH

Scattered across the Indian and Pacific oceans in a vast emerald crescent, Indonesia is made up of 13 677 islands, forming by far the world's biggest island chain. Other statistics are nearly as impressive. With more than 197 million people, it is the world's fourth largest country. Nearly 90 per cent of its population follow the Islamic faith, making it also the world's largest Muslim country.

From the tip of SUMATRA in the north-west to IRIAN JAYA in the south-east, the islands of Indonesia curve through more than 5000 km (about 3200 miles) of ocean – farther than the distance across the continental United States. It has almost as many tribal and ethnic groups as there are days in the year (360 in all), and between them they speak more than 250 languages and dialects.

Indonesia has more than 100 active volcanoes, and the turbulence of nature has been echoed by the turbulence of the country's political life. Hundreds of thousands of Indonesians were massacred after an attempted communist coup in the mid-1960s that led to President Raden Suharto seizing power; and more than 200 000 people have been killed in East TIMOR since 1975 when the Indonesian army occupied that small island country after the Portuguese left.

Maintaining the unity of this sprawling, many-peopled nation of dense jungle, smouldering volcanoes and stunning scenery is one of the most vital tasks facing Indonesia's leaders. Small wonder that the nation's official motto is *Bhin-neka tunggal ika* ('Unity in diversity'). Since independence was won from the Dutch in 1949, two men, President Achmed Sukarno and President Suharto, have in turn striven to unite this vast republic. At times it meant restricting the press, but with the new world order emerging in the 1990s, and with rising prosperity, reins on the local press have, on occasion, been loosened.

WHERE THE LAND TREMBLES

Indonesia's land area is nearly 2 million km² (about 750 000 sq miles), but its islands are scattered over 8 million km² (more than 3 million sq miles). The biggest landmass is KALIMANTAN, Indonesia's share of the island of Borneo, followed by Sumatra, Irian Jaya (formerly West New Guinea) and the irregularly shaped SULAWESI (Celebes). JAVA, though fifth in size, is the dominant island and by far the most heavily populated (with 813 people per km², 2106 per sq mile). It contains more than 91 million people, and in some regions population reaches 2000 per km² (5180 per sq mile). This emerald chain of mountainous islands is a series of high points on undersea extensions of mainland Asia and Australia. More than 10 000 years ago the water levels rose, and most of the land was submerged. The high points, which are the islands today, had been formed at about the same time as the Himalayas. Kalimantan, Sumatra, Java, LOMBOK and BALI are all peaks on the SUNDA SHELF, a largely submerged extension of Asia. Irian Jaya and its attendant islands are on the SAHUL SHELF, which links New Guinea and Australia. Sulawesi and the MOLUCCAS are on a third extension – from Japan and the Philippines.

All the larger islands have a volcanic mountainous area flanked by coastal plains. Steep-sided mountain ranges face the deep seas, and sloping lowlands fall away gently to the shallow seas covering the continental shelves. The highest mountains are PUNCAK JAYA (5030 m, 16 498 ft) and Gebel Daam (4922 m, 16 148 ft) on Irian Jaya. There are peaks above 3000 m (9 840 ft) on Sumatra, Java, Sulawesi, Lombok and Bali.

The Indonesian archipelago is among the most restless zones in the earth's crust. This is where the Eurasian and the Indo-Australian plates meet, where ridges and trenches open up, accompanied by volcanic eruptions and earthquakes. About one third of Indonesia's volcanoes are active, and some have been exceedingly destructive. In 1883 KRAKATAU, an island volcano in the SUNDA STRAIT between Java and Sumatra, exploded after 200 years of inactivity killing 36 000 people and forming new islands with its lava and scattering debris as far as Madagascar. The sound of the explosion, the most violent ever recorded, was heard nearly 5000 km (about 3000 miles) away.

In the 10 years to 1983, there were 21 eruptions. Galunggung in west Java, for instance, blew a dense cloud of ash into the upper atmosphere and nearly brought down a passing airliner. Earthquakes are frequent in the southern islands. In 1976 and 1977 more than 6300 people were reported killed or missing and 4160 injured by earthquake motion.

These massive volcanic eruptions have produced a jagged and fragmented landscape – a feature that is accentuated by the geological action known as fracturing in some mountain areas. Here, huge movements within the earth's crust have created plateau-like uplands and rift valleys – such as the one that slices through the northern Barisan Mountains on Sumatra. Because the volcanic soil is easily eroded, the heavy rains wash masses of material off the highland areas, resulting in a rapidly changing landscape in terms of geological time. For example, in southern Kalimantan, northern Sumatra and southern Irian Jaya, coastlines swollen by debris washed from the hills push steadily seawards, creating ever-increasing swamplands.

Rainfall, often falling as torrential downpours, reaches more than 3000 mm (118 in) a year, while temperatures average 27°C (81°F). Because it is so hot and humid throughout its huge area, Indonesia contains more equatorial rain forest than any other country. However, in many areas, sections of the dense forest have been extensively cut for valuable hardwoods, leaving exposed fragile soils that are easily washed away. Only on Java and the LESSER SUNDA ISLANDS – which lie in the path of dry air moving north from Australia, creating a relatively dry season from April to October – is there any variation in the dominant rain forest. The lower humidity allows mixed deciduous forest to grow – with evergreen teak a major component. But even here, more and more timber has been cleared to provide land to resettle and feed the population.

The islands have been inhabited since long before historic times. In 1891, near Trinil village in east Java, Dr Eugene Dubois found fossils of Java Man (*Homo erectus*) dating back 500 000 years. Stone Age people spread from mainland Asia in several waves between 2500 and 1000 BC, and, in the 1st century AD, Indian traders and Hindu and Buddhist priests from the Indian subcontinent arrived and introduced their culture. The remains of their many temples can be seen today.

Muslim traders, possibly from the state of Gujarat in India, introduced Islam to the country in the 14th century; the new religion spread rapidly but Bali remained an outpost of Hinduism. Eventually, in 1509, the first Europeans – the Portuguese – established a foothold in Indonesia and its spice trade. Spain, Holland and England also competed for the lucrative spice business but the Dutch East India Company triumphed over its rivals and gained control of JAKARTA, which it called Batavia, in 1619. Dutch government rule followed, apart from a brief period when Britain's Sir Thomas Stamford Raffles took over the islands during the Napoleonic Wars.

FORGING NATIONAL UNITY

Indonesia was occupied by the Japanese in the Second World War, but the Dutch returned in 1945 to reclaim the colony. Open fighting broke out with the independence movement in 1947 and, after a two-year war, the Dutch conceded and left the islands. Ahmed Sukarno became president.

Sukarno gave Indonesia a national identity but his ambitious plans brought economic failure. He also sent paratroopers into Dutch New Guinea, which was eventually to become Irian Jaya; and he openly waged war with Malaysia, the formation of which in 1963 Sukarno strongly opposed.

Under Sukarno's leadership, the country was initially tolerant of communism, but in 1965 an attempted communist coup, which involved Indonesian communists and in which six generals of the Indonesian army were assassinated, was quelled by the army. The people rose up and in three months massacred an estimated 500 000 communists and their supporters; whole villages were wiped out. A new strong man emerged in the turmoil, Lieutenant-General Suharto, who replaced Sukarno as president in 1967.

Suharto ended the confrontation with Malaysia, restored links with the West and dealt with inflation of 600 per cent – bringing it down eventually to around 5 per cent. He launched a series of five-year plans to increase food production and tackle unemployment. He also extended Indonesia's territory in 1975 by

invading and annexing East (Portuguese) Timor where civil war was raging. However, the East Timor independence movement, Fretilin, has continued to fight a guerrilla war against Indonesian troops into the 1990s. The suppression of the independence movement has been brutal. Many of the thousands killed have been civilians, including six foreign journalists, and relations with the United States have been strained by concern over Indonesia's abuse of human rights.

President Suharto has faced political unrest at home, too, with opposition from labour activists and fundamentalist Muslims, leading to

violent clashes and riots. One of the government's greatest problems is overpopulation.

This is being dealt with by migration schemes and family planning campaigns. Already 3 million people have been moved to less densely populated areas, particularly Sumatra and Irian Jaya – a futile programme, since each year hundreds of thousands of people migrate in the opposite direction, away from the underdeveloped island to the main island of Java, aggravating conditions there.

Only 6000 of the 13 677 islands are inhabited. Overcrowded Java, with only 7 per cent of the land, contains 60 per cent of the population.

▲ BENEATH THE VOLCANO Smoke funnels from Merapi – one of 35 active volcanoes on the island of Java – as workers toil in the outlying rice fields. The mountain rises to some 2911 m (9550 ft) and its last major eruption was in 1867.

Thanks mainly to the Dutch, who kept down local tribal warfare and improved productivity, the population has been growing for almost two centuries. Improvements in health care have accelerated this growth, and put increasing pressure on the land available for cultivation.

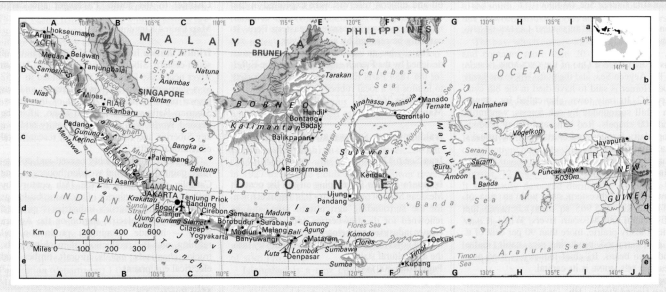

More than one-third of rural families are landless and the average holding is only half a hectare (just over an acre).

Despite irrigation schemes and the development of high-yielding varieties of rice, Java's fertile rust-coloured soil, enriched by volcanic residues, is in danger of becoming exhausted. In 1980, 47 per cent of rural Javanese lived below the poverty line, and many had left for the cities. The capital, Jakarta, is the largest city in South-east Asia, with 7.8 million people, but it only has housing, jobs and proper sewerage and other services for fewer than 1.5 million people. The rest are crammed into ramshackle shanty towns, especially along the railway tracks. Regional centres such as SURABAYA and BAN-DUNG have also grown rapidly and have populations of more than 2 million and 1.5 million respectively.

In contrast, parts of Kalimantan and Irian Jaya remain wildernesses with fewer than 10 people per km² (26 per sq mile). Here the forests are difficult to clear and are unhealthy for settlement, with malaria a problem. Apart from scattered forest clearance for logging and shifting cultivation, settlement of the outer islands has been traditionally restricted to the coasts. The people make a living from fishing and trading in local forest produce.

The main problem for Indonesia's leaders since independence has been to draw together the different lands and people which the country inherited from the Dutch. The population falls into two main groups, Malayan and Papuan, with a transitional area in central Indonesia. The main ethnic groups include the Javanese – who make up 60 per cent of the population – and Sundanese in Java; the Acehneses, Bataks and Minangkabaus in Sumatra; the Madurese in Madura; the Sasaks in Lombok; the Ambonese in the Moluccas; the Timorese on Timor; the Menadonese and Buginese in Sulawesi; the Dyaks in Kalimantan; and the Irianese in Irian Jaya.

There are also about 3 million Chinese, as well as minority groups of Indians and Dutch. The Chinese have managed to secure for themselves an influential position in the islands' commercial life. Attempts to bring unity to these many different groups have included the adoption of one language – Bahasa Indonesia, similar to Malay – as the country's official tongue. Agreement has been reached with Malaysia on a common Roman script for both countries. Modern technology also has helped to unify the nation. Indonesia was the fourth country in the world to operate a domestic communications satellite system, which can carry TV and radio to all Indonesia's 27 provinces.

Oil was discovered in Indonesia in 1885 but only in the 1970s and 80s did the country become a major producer. Large deposits of natural gas have also been found in Aceh and eastern Kalimantan. Indonesia is now the world's biggest exporter of liquefied natural gas.

Tin has long been mined in the RIAU archipelago, and Indonesia's reserves remain the largest in the world. There is nickel in Sulawesi and the Moluccas, copper in Sumatra and Irian Jaya, bauxite in the Riau islands and coal in Kalimantan.

Indonesia is one of the world's leading rubber producers and timber is also an important export. There are plantations of sugar cane, coffee, tea, tobacco, palms and spices. Fields of rice provide the main food crop, and maize, cassava, sweet potatoes and soya beans are also grown for the home market. There is an abundance of tropical fruits, including bananas, pineapples, papayas and mangoes. Fishing has great potential from the well-stocked seas around the islands, but the methods used are often primitive and not geared to a modern export market.

These resources, as yet not fully developed, make Indonesia a storehouse of energy and valuable raw materials, and offer the country a solid base for economic development. There is still a sharp contrast between the abject poverty of the vast majority of the population and the vast riches that slumber under the Indonesian soil. So far, only a small part of the country's mineral resources has been tapped. Earnings from oil and gas exports have already been invested in irrigation in Java and in expanding the plantations of rubber, tea and oil palms on the outer islands. The country has begun to build up its heavy industry, using its huge mineral resources and powered by its natural gas and the hydroelectric power generated in Sumatra. Technical knowledge provided by the other members of the Association of South-east Asian Nations (Malaysia, the Philippines, Singapore and Thailand) should help Indonesia to develop more advanced industries such as electronics and machine-tool manufacture in the future.

INDONESIA AT A GLANCE

Official name Republic of Indonesia

Area 1 919 440 km² (741 095 sq miles)

Population 197 232 430 **Per km²** 103 (**Per sq mile** 266)

Capital Jakarta

Government Presidential Republic

Currency 1 Rupiah = 100 sen

Languages Bahasa Indonesia (official), also many other languages and dialects

Religions Muslim (86%), Christian (10%), Hindu (2%), Buddhist and Confucian

Climate Tropical/monsoonal. Only the Lesser Sunda Islands have a dry season. Average year-round temperature in Jakarta is 27°C (81°F)

Land use Cultivation 11%, grazing 7%, forest and woodland 67%, other 15%

Main primary products Rice, maize, cassava, sweet potatoes, soya beans, sugar, bananas, palm oil, copra, rubber, coffee, tobacco, tea, groundnuts, fish, timber; oil and natural gas, tin, nickel, copper, bauxite, coal

Major industries Agriculture, oil and gas production and refining, mining, forestry, fishing, textiles, transport equipment assembly, food processing, paper, cement, matches, tyres, glass, chemicals, fertilisers

Main exports Crude oil, natural gas, refined petroleum products, metal ores, tin, hardwoods, coffee, tea, spice, tobacco

Annual income per head (US$) 560

Population growth (per thous/yr) 16

Life expectancy (yrs) Male 59 **Female** 62

Ireland

FIFTEEN CENTURIES OF FAITH AND POETRY OVERLAID BY EIGHT CENTURIES OF STRIFE HAVE LEFT THEIR MARK ON THE LAND AND ON ITS PEOPLE

Not for nothing is Eire – to use Ireland's ancient, Celtic name – called the Emerald Isle. Its stunning, shamrock green is the product of the mild, humid and ever-changing climate created by Atlantic weather systems washing over one of Europe's most westerly islands.

Warm fronts sweep in from the ocean at any time of the year, drenching the countryside beneath sheets of low cloud. At such times a well-soaked stranger seeking solace in a snug bar may be surprised to be told that 'it's a fine, soft day, praise be to God'. Then a cold front will arrive, bringing bright skies and drifting showers that endow the landscape with a sparkling brilliance of colour.

The Republic of Ireland occupies fourfifths of an island that bears roughly the shape of a gigantic saucer, with a broken chain of mountains forming an irregular rim around the coast. The central plain is largely limestone, covered mainly with boulder clay, providing good farmland and pasture.

In the mountainous west, full face to the Atlantic, the rainfall varies from 1016 mm (40 in) on the coast to twice that or more in the mountains. There, in the far south-west of KERRY, are Ireland's highest peaks. CARRAUNTOOHIL, tallest of all, rears to 1041 m (3414 ft) and Brandon Mountain near Dingle falls sharply to coastal cliffs from its 953 m (3127 ft) summit.

Winter temperatures average between 4°C and 7°C (39°F and 45°F) with the highest figures in the west, which has little frost. Summer temperatures range from 14°C (57°F) in the north to 16°C (61°F) in the south.

The country's history is a good deal less temperate. The stormy story of its centurieslong fight for independence ended only in 1922 and the Republic ranks among Europe's youngest free nations. Yet its contribution to Western culture is immeasurable. From Swift and Sheridan to Yeats and Shaw, from Oscar Wilde and James Joyce to Samuel Beckett, Ireland's literary giants have enthralled readers and theatregoers in many countries.

Eight bitter and often bloody centuries have left the island divided into two parts – the Republic, which is predominantly Roman Catholic, and Northern Ireland, which is twothirds Protestant and is part of the United Kingdom. Although DUBLIN politicians have advocated a united Ireland, apart from a small minority neither they nor their supporters have condoned terrorism – and the Irish Republican Army (IRA) has been an illegal organisation in both the Republic and Northern Ireland.

Ireland's tragedy has been that it has had more than its share of history. The land became a centre of Christian faith and learning, a realm of kings and poets, after it was converted by St Patrick in AD 432-465. An extensive monastic life developed that shaped the Irish church. Many Irish missionaries were sent out over all of Europe.

But three centuries later its prosperity lured Vikings, who came to raid and stayed to settle. One of Ireland's first national heroes, Brian Boru, defeated the Vikings at CLONTARF in 1014, but this taste of freedom did not last long. In 1171 the Norman conquest began for Ireland when Henry II of England sent his knights across the Irish Sea to take possession.

THE TROUBLED LAND

Henry proclaimed himself overlord of the entire island and secured recognition of his actions from church leaders. The independence of Gaelic Ireland had come to an end. This was the beginning of a turbulent relationship with Ireland's powerful neighbour, marked by misunderstanding, religious intolerance, rebellion, repression, evictions, atrocities, retaliations – and rare outbreaks of goodwill.

The list of Irish grievances and misfortunes is long. In 1541 the island was firmly tied to England. Henry VIII made himself King of Ireland. His attempts to reform the Irish church failed but in the 16th and 17th centuries, Irish lands were confiscated, and 'plantations' of Protestants from England and Scotland were settled in ULSTER, in the north-east – a policy that was to produce a permanent Protestant majority in Northern Ireland.

Oliver Cromwell left a name that is still detested for the ruthlessness with which he crushed a Catholic and Royalist rebellion in 1649. Catholics were forbidden to exercise their religion after their defeat by William III's forces at the Battle of the BOYNE in 1690.

An abortive uprising by Wolfe Tone and his United Irishmen in 1798 was the curtain-raiser to another troubled century. In 1801 an Act of Union came into force under which Ireland's Parliament was abolished and Irish MPs were elected instead to the British House of Commons at Westminster.

The temper of the times in England was anti-Catholic, and Catholics were not allowed to hold public office. But in 1828 the Catholic Daniel O'Connell was elected Member of Parliament (MP) for County CLARE, an event which forced the British Government to accept Catholic emancipation the following year.

The harsh facts of economics took over from politics in the middle of the century when Ireland's potato crop fell victim to blight – a fungal disease that rotted the tubers in the ground or in store.

Before the Great Famine, which began in 1845, the land supported a population of more than 8 million people, half of them living on home-grown potatoes. By 1848 Ireland had

lost more than a million people to disease, starvation and emigration. In the single year 1847, 250 000 died and 3 million were being supported by public funds. Many landlords sacrificed fortunes trying to save their tenants, but others simply evicted them. The famine years launched a decline in population, at first rapid then more gradual from the 1890s. But more than 2 million have emigrated since 1900, many to North America.

Today there are over 10 million people of Irish descent living in the United States alone – almost three times the number living in Ireland itself. The departures have etched clear marks on the Republic's countryside. Throughout the province of CONNAUGHT and the counties of KERRY and DONEGAL, ruined and deserted hamlets, abandoned fields and forgotten peat diggings tell the tale.

After the famine, a revolutionary movement known as Fenianism sprang up. Its efforts resulted in some Parliamentary reforms. On the eve of the First World War, the British Government seemed at last ready to grant Home Rule – despite the many Protestants in Ulster who were prepared to take up arms rather than submit to Catholic rule.

'Ulster will fight, and Ulster will be right' was the slogan of their leader, the Protestant southern Irishman Sir Edward Carson. And fight they did – but not against other Irishmen. Many of the southern Irish laid aside their grievances on the outbreak of the World War

and 250 000 of them volunteered to fight alongside Ulstermen in the British army. Even so, there was enough discontent for Germany to think it worthwhile supporting an insurrection against British rule – the Easter Rising of April 1916. It was crushed, and 90 rebels were condemned to death, of whom 15 were executed. Eamon De Valera (1882-1975), who was born American, was one of those reprieved.

In 1918 the political movement called *Sinn Féin* ('Ourselves Alone') proclaimed an Irish Parliament, the *Dáil Eireann*, with De Valera as president. The IRA (the *Sinn Féin's* military wing) declared war on the British, and Britain responded by drafting ex-servicemen into the Royal Irish Constabulary. These auxiliaries were the notorious Black and Tans (named after the colour of their uniforms) who fought terror with terror.

In 1921 the IRA leader Michael Collins (1890-1922) signed a treaty with Britain which gave Ireland independence – with the exception of six of the nine counties of Ulster. 'Early this morning I signed my death warrant,' he wrote to a friend. And he was right, for before a year had passed, he had been assassinated in a bitter civil war between those *Sinn Féiners* who accepted the treaty and those who did not.

The Irish Free State came into formal existence in 1922, at first as a dominion within the British Commonwealth. De Valera, who became president in 1932, set about demolishing the country's last links with Britain. Finally, in

▲ **CITY OF MEMORIES The sun sets on Dublin and the shimmering River Liffey. On the right stands the dome of the Four Courts, headquarters of the judiciary. Time's imprint is everywhere in the city, whose mixture of elegance and shabbiness casts a potent spell, as many writers have testified.**

1949, the Free State was proclaimed a republic, no longer within the Commonwealth. (De Valera retired as prime minister in 1959 and was elected to the presidency, serving until 1973.) In 1955 came acceptance into the United Nations. There were setbacks in attempting to join the European Community, but in 1973, Ireland joined the organisation.

The demand for unification of all Ireland is written into the country's constitution, and inexorably, the Republic has been drawn in to share the agony of Northern Ireland. In 1985 a pact between the British and Irish governments gave the Republic a consultative role in running Northern Ireland. In December 1993 the two governments issued the Downing Street Declaration setting out general principles for holding peace talks on Northern Ireland.

Political parties, church representatives and private individuals and groups had also tried to find solutions. But the terrorist campaign of the IRA (and other nationalist groups) and Protestant Northern Irish para-militaries (who favoured Northern Ireland remaining in the UK)

had continued, with atrocities in the Republic, mainland UK and Europe, as well as in Northern Ireland – where more than 3000 were killed and more than 36 000 wounded in the 15 years following 1969.

In August 1994, after behind-the-scenes negotiations begun by the British government, the *Sinn Féin,* which had previously rejected the Downing Street Declaration, announced 'a complete cessation of military operations' by the IRA. This was followed two months later by a similar declaration by the Protestant equivalent, the Combined Loyalist Military Command.

HARNESSING ITS WATER POWER

The country's limited energy resources and remoteness from major population centres and markets in the developing Europe of the 19th century slowed industrial progress. Coal – the essential ingredient of the times – had to be imported and most of it came from Britain. An electrification programme began in the 1920s with a hydroelectric scheme on the River SHANNON. Later schemes harnessed the power of many other rivers so that today at least 75 per cent of the country's water-power potential is used. Peat, created by a moist climate acting on ill-drained tracts of glacial clays, covers one-seventh of the Republic's 70 283 km² (27 130 sq miles), and is

the country's chief indigenous fuel source. Five million tonnes a year are cut by a government-sponsored corporation; 15 major peat works and briquette factories process it and seven peat-fired power stations generate about 14 per cent of the country's electricity.

However, present reserves may last only 20 years. Two gas fields off the south coast began operation in 1989 and have reserves for about 10 years. A gas pipeline from Scotland will boost supplies. But most electricity comes from imported coal and oil.

Government policies have been directed towards industry since 1932, partly to reduce emigration. This was running at 40 000 people a year in the 1950s but then declined. Hardly any left in the early 1980s, when unemployment soared in Britain, while many European and Japanese companies, attracted by Ireland's cheap labour force, opened Irish branches, so stimulating domestic employment. More recently, renewed economic problems once again brought an increase in emigration.

Dublin and its suburbs attracted many of the foreign firms and the capital, the country's leading commercial, industrial and administrative centre, remains a magnet for migration within Ireland. The region's population of about one million is roughly one-third of the national total. Its industries include brewing, food production, light engineering, clothing and glass making.

The capital is also the main financial, publishing and printing centre and the focus of radio, television and other communications industries. Despite efforts to distribute industry more widely, Dublin still has half the country's manufacturing output. The Republic ranks among the least industrialised countries in Europe. It is also one of the poorest, with a per capita income about half that of France. Its unemployment rate, that remained stubbornly at 15 per cent or more in the early 1990s, has caused much concern. In October 1993 the government unveiled a six-year US$28 billion plan to provide 200 000 more jobs.

Some of Ireland's most ambitious enterprises remain state operated; the Aer Lingus airline, the Peat Board and the CIE (*Coras Iompair Eireann*) road-rail system are major companies. An Industrial Development Board was set up in 1950, using tax concessions to attract foreign firms. In 1992 a new agency combining industrial development, training and science technology was announced.

Perhaps the most striking venture was the creation of a whole new industrial town beside Shannon International Airport on the west coast – a customs and tax-free zone, which encouraged industries producing a wide range of goods, from radios to machine tools and textiles. Unfortunately the development of long-range jets meant that transatlantic flights no longer needed to refuel at Shannon, which is now more of a flight-control centre.

However, the country profits handsomely from a thriving import trade in tourists. An estimated 17 million people of Irish descent live abroad, and many dream of returning to or visiting the land of their forebears. Thousands do so every year. They have included two American presidents – John F Kennedy and Ronald Reagan – in recent times. Others go just for the wonderful, unspoilt beauty of the land and the lakes, the friendly, hospitable people, the relaxed atmosphere and change of pace. In the early 1990s the Republic welcomed about 1.75 million visitors a year from Britain alone, by far the largest proportion (47 per cent) of the total from all countries. Another 700 000 (19 per cent) were from Northern Ireland and 9 per cent came from the USA in a total of nearly 3.7 million.

Visitors find a land where the Catholic faith remains a conservative influence. Its citizens voted overwhelmingly in 1986 against the acceptance of divorce and in 1992 against the legalisation of abortion – but the referendum was in favour of allowing Irish women to travel outside the country for such operations.

Homosexual acts between consulting adults were legalised in June 1993. However, there have been recent advances in the status of women in Irish society and the first woman head of state, President Mary Robinson, elected in 1990, welcomed another woman in 1993 as head of one of the political parties. In 1990 the (Anglican) Church of Ireland voted to ordain women.

The land is divided into provinces that were once its ancient kingdoms. To the north and west is wild, beautiful Connaught, including the counties of SLIGO, MAYO and GALWAY and the blue lakes and matching blue mountains of CONNEMARA; south, in MUNSTER, the long Atlantic-probing peninsulas of CORK and Kerry are backed by towering peaks of the Caha and Slieve Mish mountains, and MACGILLICUDDY'S REEKS.

At the centre LEINSTER, MEATH, LAOIS (or Leix, both pronounced 'Leesh'), CARLOW and WICKLOW have rolling hills and prosperous farmlands, full of small towns and hamlets. On the mild east coast, between Dublin and WEXFORD, small harbours are interspersed with sandy beaches. At an anglers' paradise called Pontoon, between loughs Conn and Cullin in Mayo, a licence is required to fish for salmon – but only after you have caught one. And that, like a fine, soft day, is very Irish.

◀ **LAKELAND MAGIC Lush green countryside and high mountains encircle Killarney's Upper Lake. Tourists who throng the little town of Killarney can travel to the lake in pony-drawn jaunting cars.**

IRELAND AT A GLANCE	
Area 70 283 km² (27 136 sq miles)	
Population 3 525 700 **Per km²** 50	
(Per sq mile) 130	
Capital Dublin	
Government Parliamentary republic	
Currency 1 Punt (Irish pound) = 100 pighne	
Languages English, Irish	
Religion Christian (93% Roman Catholic, 3% Church of Ireland, 1% Presbyterian)	
Climate Temperate maritime, warmed by the North Atlantic Drift. Average temperature ranges from 4-7°C (39-45°F) in January to 14-16°C (57-61°F) in July	
Land use Meadow and pasture 66%, cultivation 13%, forest 5%	
Main primary products Cereals, potatoes, sugar beet, vegetables, livestock, fish; peat, natural gas, lead, zinc, barytes, gypsum	
Major industries Agriculture, machinery, food processing, chemicals and fertilisers, textiles, clothing, tourism	
Main exports Livestock, meat, textiles, dairy products, chemicals, machinery, beverages	
Annual income per head (US$) 8500	
Population growth (per thous/yr) -2.5	
Life expectancy (yrs) Male 72 **Female** 78	

Issyk-Kul' (Rybach'ye) *Kyrgyzstan* Mountain lake in the central Asian republic of Kyrgyzstan, 105 km (65 miles) from the Chinese border. Its deepest point is 702 m (2303 ft) and it covers an area of 6280 km² (2424 sq miles). Its surface is 1609 m (5279 ft) above sea level. The hot springs make it a fishing paradise, since it is rich in perch, pike and carp. The lake also has medicinal springs used to treat rheumatism.
Map Kyrgyzstan – Ba

Istanbul (Constantinople) *Turkey* The country's largest city, straddling the Bosporus at the entrance to the Black Sea. It is one of the most exciting and lively cities in the world, with an exotic nightlife and excellent tourist hotels. Its skyline is studded with the domes and minarets of the mosques that crown Istanbul's seven hills. The port handles 60 per cent of Turkey's trade, and the railway terminus links European and Turkish networks.

Founded by the Greeks in about 660 BC, it is the only city in the world to be built on two continents – Europe and Asia. Formerly called Byzantium and then Constantinople, it was the capital of the Byzantine empire for 1000 years, and of the Ottoman empire for 500 years. In 1923 Kemal Ataturk, modern Turkey's founder, moved the capital 350 km (220 miles) east to ANKARA.

The city, which remains a commercial centre, consists of three historic parts – Stambul on the peninsula between the Marmara Sea and the arm of the Bosporus called the Golden Horn, Beyoglu, which includes the former suburbs of Galata and Pera, and Üsküdar, which was formerly called Scutari and stands on the Asian side of the Bosporus. The hospital where the British nurse Florence Nightingale tended the wounded during the Crimean War (1853-6) is at Üsküdar.

Stambul is the original centre, and most of the city's tourist sights are found here, including the Hagia Sofia (in turn a church, a mosque, and now a museum), the Blue Mosque, the Mosque of Suleiman the Magnificent, the Topkapi Palace, and the Great Bazaar. The city's main industries are shipbuilding, textiles, pottery, tobacco, cement, glass and leather goods.
Population 6 620 240
Map Turkey – Aa

Istanköy *Greece* See KOS

Istra (Istria) *Italy/Slovenia/Croatia* Peninsula on the north Adriatic coast, stretching 97 km (60 miles) south from Trieste, Italy, on its western neck, and 80 km (50 miles) from Rijeka, Croatia, on its eastern side. PULA is the chief town. Most of the peninsula was ceded by Italy to Yugoslavia in 1947. The north-west section was divided between Italy and Yugoslavia in 1954. About 50 per cent of the peninsula is cultivated, and wheat, maize, grapes and olives grown.
Map Croatia – Ab

Itaipú *Brazil/Paraguay* Dam on the Paraná River, straddling the Brazil-Paraguay border, 720 km (450 miles) west of the city of São Paulo. Completed in 1982, it has 18 giant turbines, the last of which was installed in 1991, and can generate 12 600 MW of electricity, making it the world's largest power station.

The lake behind the dam has, however, submerged the homes of 200 000 people and drowned the Guairá Falls 190 km (118 miles) upstream. The falls, now marked only by an underwater cliff, were once the world's largest body of falling water. The flow over them was more than double that over the Niagara Falls.

The dam was built as a joint venture between Brazil and Paraguay, the agreement having been sealed by the Itaipú Treaty of 1973.
Map Brazil – Cd

Italy See p 335

Ithaka (Itháki) *Greece* One of the IONIAN islands – the kingdom of Homer's legendary hero, Odysseus, the main harbour town of Vathí lies on a sheltered inlet.
Population 3650
Map Greece – Bb

Itsukushima *Japan* See MIYAJIMA

Ituri Forest *Zaire* Named after its chief river, Ituri Forest covers between 30 000 and 54 000 km² (11 580-20 850 sq miles) in north-east Zaire. It is the largest primeval forest outside the Amazon basin and its dense canopy of leafy foliage blocks out sunlight from the dank forest floor.

In 1887, the British adventurer Sir Henry Morton Stanley became the first European to cross this twilight world from east to west. The journey took five months, and the expedition, which suffered 180 deaths in its final stages, was frequently attacked by Pygmies using poisoned arrows. Pygmies (Mbuti) are still the region's main inhabitants. They have preserved their traditional hunting and gathering lifestyle.

Much of the Ituri Forest remains unexplored, but tourism has begun in the region. Visitors who approach the forest from Kisangani or Goma can sometimes catch a glimpse of such rare animals as okapis or mountain gorillas.
Map Zaire – Ba

Ivanovo *Russia* Administrative centre of the Ivanovo region, and a major textile and clothing city in the centre of a rich flax-producing area, 240 km (150 miles) north-east of Moscow. It was formed as Ivanovo-Voznesensk in 1871 by joining two villages. By 1905, it had more than 20 textile mills. That year, 80 000 mill workers, striking for more money and better conditions, formed Russia's first workers' councils, or 'soviets'. The city also produce machinery and chemicals.
Population 482 000
Map Russia – Fc

Iviza *Spain* See IBIZA

Ivory Coast See p 340

Ivrea *Italy* City at the foot of the Alps, 50 km (30 miles) north of TURIN. It is the headquarters of the Olivetti office machinery empire. The higher, old town has a 10th-century Romanesque cathedral and a 14th-century castle with a cylindrical tower.
Population 24 550
Map Italy – Ab

Iwaki *Japan* City on the east coast of HONSHU Island, 175 km (110 miles) north-east of the capital, Tokyo. Created in 1966 out of the former mining towns of the Joban coalfield, it is now a centre of the chemicals industry.
Population 356 000
Map Japan – Dc

Iwo Jima *Japan* Largest of the Volcano islands, covering 21 km² (8 sq miles) in the western Pacific, about 1250 km (776 miles) south-east of Tokyo. In 1945 it was the scene of bloody fighting between United States and Japanese forces. The United States forces occupied Iwo Jima and, after the war, America administered the Volcano islands. They were returned to Japan in 1968.
Map Pacific Ocean – Ca

Iztaccíhuatl (Ixtaccíhuatl) *Mexico* Spectacular snow-capped volcano, 5286 m (17 342 ft) high, that stands beside its twin, Popocatépetl, in the Anahuac Valley south of MEXICO CITY.

Izhevsk (Ustinov) *Russia* Industrial city and capital of the Udmurt Republic, it is situated near the western edge of the Ural Mountains, 225 km (140 miles) south-west of PERM'. It was founded in 1760 and has been making steel products since then. Its other main products are machine tools, motor vehicles, pianos and furniture.
Population 635 000
Map Russia – Gc

Izmir (Smyrna) *Turkey* Port and provincial capital on the Aegean coast, 335 km (210 miles) south-west of Istanbul. Founded by the Greeks more than 5000 years ago, it was captured by the Ottoman Turks in 1424. Invading Greeks occupied the city in 1919.

It was retaken by the Turks in September 1922, and a few days later was almost totally destroyed by fire. Izmir produces copper, dyes, cotton, wool, carpets, textiles, silk, leather goods, tobacco, figs, raisins, olive oil, pharmaceuticals, soap and sponges. The neighbouring province of the same name, that covers 12 825 km² (4952 sq miles), is rich in silver, mercury and iron.
Population (city) 2 319 190; (province) 2 694 770
Map Turkey – Ab

Izmit *Turkey* Naval base on the Sea of Marmara, 90 km (55 miles) south-east of Istanbul. It is the capital of Kocaeli Province, that covers 3986 km² (1539 sq miles) and where tobacco and olives are grown. Its main industries are oil refining, shipbuilding, and the production of chemicals, cement and paper.
Population 257 000
Map Turkey – Aa

Izu *Japan* Peninsula on the south coast of HONSHU Island, between the bays of Sagami and Suruga, in the south of the Fuji-Hakone-Izu National Park. Its rugged coastline and the hot springs at the town of Ito attract day-trippers from the capital, Tokyo, 100 km (60 miles) to the north-east. Ito, on the west coast of the peninsula, was the home of Will Adams, an English seaman who was shipwrecked there in the early 17th century. He was the inspiration for the hero in James Clavell's novel, *Shogun*.
Map Japan – Cd

Izumo *Japan* Town near the north coast of western HONSHU Island, 115 km (70 miles) north-east of the city of HIROSHIMA. It is the site of the country's oldest Shinto shrine, the Izumo Taisha, dating from the 1st century AD. In legend the Shinto gods gather here each October to discuss marriage arrangements for single people.
Population 80 700
Map Japan – Bc

Israel

THE PROMISED LAND OF THE JEWS AND BIRTHPLACE OF CHRISTIANITY HAS BEEN THE FOCUS OF CONFLICT IN THE MIDDLE EAST FOR MORE THAN FOUR DECADES

Born in strife, the modern state of Israel has been almost constantly at war with its powerful Arab neighbours, many times its size, ever since its establishment in 1948. The triumphant return of Jews to their historic homeland ended 18 centuries of exile, but the land they returned to had long been the home of the Palestinian Arabs, who made determined attempts to strangle the tiny state at birth. Five wars and countless acts of terrorism followed as Arab countries and Palestinians rose against Israel.

The country occupies a long narrow stretch of land at the south-eastern corner of the Mediterranean. This hilly country has as its natural eastern boundary the GREAT RIFT VALLEY, the great fault that split the earth's surface from Syria to Africa millions of years ago. The valley contains the River JORDAN, which flows first through subtropical vegetation, then arid wasteland to the DEAD SEA, 396 m (1299 ft) below sea level. The triangular wedge of the NEGEV Desert ends at the Gulf of AQABA, at the head of which is Israel's southern outlet, the small port of ELAT.

Israel's northern half is temperate and fertile, while the south is arid and barren. Most of the country's 4.9 million citizens live on the coastal plain bordering the Mediterranean, where TEL AVIV is the leading commercial city. The greenest part is GALILEE, though here, as in all Israel, no rain falls in summer. The Sea of Galilee, lying among the northern hills, is Israel's main reservoir of fresh water. This is piped to the Negev – the desert and semi-desert home of wandering Bedouin Arabs – and it is the Israelis' proud boast that they have 'made the desert bloom' with sunflowers, sugar beet and other crops.

The Jews are a clever, restless, vibrant people. In the years since modern Israel was founded, large numbers of them have left Europe, North America, the Middle East and Africa to make new lives in their historic homeland. About 825 000 Arabs lived here before the Jewish state was founded. The Jewish population was 650 000 in 1948; today it is more than 4 million, forming one-fifth of the world's Jews. Non-Jews – Arabs, Druse and others – number 900 000.

It is a young nation. About one-quarter of the population is under 14. The people are highly educated. Many of those who flocked to Israel brought with them a wealth of skills and it is not uncommon for, say, a truck driver to have a degree. It is also one of the most polyglot societies on earth. There are, for instance, Iraqi, Ethiopian, Moroccan and Russian Jews, and many others. Besides Hebrew or Arabic, most of the people speak English, the commercial and official language.

FIGHT FOR ECONOMIC SURVIVAL

Israel's economic and military survival has been largely made possible by the massive help it has received from the United States, both at government level and from the powerful American Jewish community. Despite the fact that much of it lies in desert or semi-desert regions, Israel is virtually self-supporting in food. The most modern farming methods are used and almost one-tenth of the country's exports are agricultural produce.

Much of its agriculture is based on methods employed at the *kibbutz* (collective settlement) or *moshav* (cooperative village). The *kibbutzim,* with their organised communal life, are designed as agricultural collectives but are also geared, especially in border areas, for rapid defence in times of war. There are 270 *kibbutzim,* with a total population of 129 300, scattered around the country. They produce 40 per cent of the country's agricultural output and there is hardly one without a factory as well – making electronic or irrigation equipment, processed foods, farm machinery or furniture.

A 1951 law gave Israeli women equal status with men, in theory at least. Men and women serve side by side in the Israeli defence forces; the men do three years' national service, single women two. Married women are exempt.

There are 179 000 Israelis in the armed forces and nearly 18 per cent of the national budget is spent on defence, which puts a heavy strain on the economy. There are constant, huge balance of payments deficits, partly made good by aid from the United States. Nevertheless, the inflation rate reached a staggering 1000 per cent a year in 1984. By 1992, however, it had been brought down to 10 per cent.

THE ARAB-ISRAELI CONFLICT

The reason for Arab hostility is rooted in history. For more than 2600 years there was no independent Jewish state. From the conquest of ancient Israel by the Assyrians in 722 BC until the emergence of modern Israel in 1948, the land was part of a succession of empires. Jewish revolts against the Romans in 66 BC, AD 70-73 and 132-135 led to the devastation of the land and the dispersal (Diaspora) of the Jews, leaving them without a homeland for 1800 years. The result was that, though never losing their identity – in spite of frequent persecution – they took on the flavour of the countries of their adoption.

Israel forms part of the region which the Romans named PALESTINE and which had diverse rulers down the centuries. It eventually became part of the Turkish Ottoman empire, until the British occupied it during the First World War. In 1917 the British foreign secretary, Arthur Balfour, promised British support for a Jewish national home in Palestine. But Britain had also agreed to back the creation of independent Arab states, which the Arabs claimed included Palestine. The League of Nations authorised a British mandate in 1922.

During the 1920s and 1930s Jews from many parts of the world moved to Palestine, much to the displeasure of the Arabs living there, who were rapidly becoming a minority. During and after the Second World War, Jewish survivors of Nazi persecution in Europe flocked to Palestine, and Zionists (Jewish nationalists) adopted terrorist tactics against the British, who were trying to restrict immigration. Following a United Nations vote to establish separate Arab and Jewish states in Palestine, the British left in 1948.

Egypt, Jordan, Lebanon, Syria and Iraq attacked at once; the Israelis lost the old part of Jerusalem but gained territory elsewhere. In 1956, provoked by invasion threats and a blockade of its RED SEA port of Elat, Israel struck at Egypt and occupied the Gaza Strip and the Sinai Peninsula, but eventually withdrew. In the Six Day War of 1967, Israel was attacked by Egypt, Syria and Jordan.

Again, the Israelis took Sinai, as well as the GOLAN HEIGHTS, from Syria, and the WEST BANK area and old Jerusalem from Jordan. In 1973 Egypt and Syria hit back on the Day of Atonement, Yom Kippur. Egyptian and Israeli forces both crossed the Suez Canal, making gains; the Syrians were driven back. The UN negotiated a ceasefire and some withdrawals followed, but Israel continued to hold Sinai until peace moves by President Anwar

Sadat of Egypt resulted in the Camp David Agreement with Israel in 1979 under which Sinai was handed back to Egypt. However, the problem of Arab Palestinian refugees in the areas surrounding Israel was unresolved.

The Palestine Liberation Organisation (PLO), recognised by Arab countries, made many terrorist attacks on Israel. Israel retaliated by bombing Palestinian refugee camps with fighter planes or helicopter gunships. Finally, PLO forces, led by Yasser Arafat, were driven out of Lebanon in 1982 by an Israeli invasion aided by right-wing Lebanese Christian militia.

The PLO remained active elsewhere though. The operation, involving a lengthy stay in Lebanon and the massacre by right-wing Lebanese of more than 700 Palestinian refugees in Israeli-occupied BEIRUT, brought worldwide criticism and political controversy in Israel; so did the settlement of Jews on the West Bank. By 1985 the Israelis had withdrawn from Lebanon. In September 1993 it became apparent that Israel was willing to negotiate about its occupation of West Jordan and the Gaza Strip.

As late as the beginning of 1990s, Jewish settlements were still being built there on a large scale. Even Israel's allies condemned this move, and unrest among the Palestinians living in the West Bank became frequent because they feared losing their land to the Israelis permanently. The situation was improved after Israel signed an historic peace accord with the PLO in May 1994 agreeing to grant Palestinians self rule in the Gaza Strip and in Jericho on the West Bank as the first stage of a phased return of the rest of the towns and villages in the occupied territories. Israel also moved towards peace settlements with Jordan and Syria over the West Bank and Golan Heights. But unrest continued. To add to the uncertainty, the PLO was accused by the more radical Palestinian activists of selling out to the Israelis.

LIFE IN ISRAELI CITIES

About 90 per cent of Israelis live in urban areas, such as Jerusalem, TEL AVIV-JAFFA and HAIFA. Until about 1875, Jerusalem – special to people of differing beliefs – consisted only of a walled city with four distinctive quarters: Jewish, Christian, Muslim and Armenian. Then the growing Jewish population began to establish new neighbourhoods outside the walls.

After Israel's war of independence in 1948, Jerusalem was divided for nearly 20 years. The Old City was in the Jordanian sector and Israelis were barred from it. After the 1967 Yom Kippur War, the city was reunited under Israeli rule and the walled sector extensively renovated. Here is the seat of government, with the Knesset (parliament) of 120 members elected by proportional representation.

Although it has no significant mineral resources, Israel has grown into a dynamic industrial nation over the past decade. Machinery, weapons, electronics and other technical products make up the exports. Diamond polishing and finishing is a valuable industry and there is growing emphasis on research and development in the high technology fields. The country is geared to tourism – in part a deliberate propaganda exercise to show Israel

to best advantage despite its hostile environment. In 1992, Israel was host to 1.6 million tourists. Today the Israelis, backed by thousands of years of faith, represent one of the oldest cultures in the world.

Yet there is also a younger, almost brash nationalism, a pride in the achievements of this small state surviving against great odds. Prices are high but there are few small countries that have so much to offer the visitor.

ISRAEL AT A GLANCE
Official name State of Israel
Area 20 700 km² (7992 sq miles) as defined in the 1949 armistice agreements
Population 4 918 950 **Per km²** 238 (**Per sq mile** 615)

▼ HOLY CITY Behind the onion domes of St Mary Magdalene, a Russian church in Jerusalem, rears the Dome of the Rock, built on what Muslims revere as the site from which Muhammad rose to heaven. In a city of holy places, the main focus of Jewish devotion is the Wailing Wall.

Capital Jerusalem	
Government Parliamentary republic	
Currency 1 new Israel shekel = 100 agorot	
Languages Hebrew, Arabic, English	
Religions Jewish (82%), Muslim (14%), Christian and others (4%)	
Climate Mediterranean and desert climates. Average temperature in Jerusalem 9°C (48°F) in January and 24°C (75°F) in August	
Land use Cultivation 22%, grazing 40%, forest and woodlands 6%, other 32%	
Main primary products Livestock, citrus fruits, grapes, wheat, sugar beet, olives, figs, vegetables, potash, crude oil, phosphates, bromine, natural gas	
Major industries Agriculture, mining, food processing, textiles, clothing, leather goods, transport equipment, aircraft, chemicals, metal products, machinery, diamond cutting	
Main exports Finished diamonds, textiles, fruit, vegetables, chemicals, machinery, fertilisers, electronics, military hardware, metals	
Annual income per head (US$) 13 230	
Population growth (per thous/yr) 30	
Life expectancy (yrs) Male 76 **Female** 80	

Italy

FROM SOUTH TO NORTH, OLIVES TO OLIVETTIS, THE LANDSCAPES AND LIFESTYLES PRESENT A HARMONY OF CONTRASTS – ALL UNMISTAKABLY ITALIAN

In Italy, history is all around you. Its cities are so venerable that they seem to have existed since time began. In fact its network of towns and roads was formed mostly in the times of Imperial Rome. The glories of Rome, its styles and ideas, were revived in the early years of the Italian Renaissance, an unprecedented flowering of the arts and sciences that began in the 14th century and continued for 250 years. Between them, Rome and the Renaissance made an incalculable contribution to Western civilisation, a contribution that continues to be provided by today's Italy, especially in the areas of style and design.

Yet, despite its antiquity, Italy as a political unit is little more than 125 years old. After the fall of the Roman Empire in AD 476, the peninsula fragmented politically. The Lombards overran the north, established their capital at PAVIA and gained footholds in the south at SPOLETO and BENEVENTO. Much of the rest of Italy stayed under the Eastern Roman Empire, which was ruled from Byzantium. Around the 8th century new powers emerged: the Papal States along the ROME-RAVENNA axis; the trading city of VENICE, as well as the republic of Genoa, both of which became leading powers in the trade with the Orient; the Frankish empire, which absorbed much of the north; and the Arabs in the south.

From the 12th century, 200 or more city-states flourished between Rome and the Alps. This was the great communal age which led into the Renaissance. Meanwhile the south, after a brief spell of splendour under the Normans, sank into colonial obscurity under Spanish rule. The 16th and 17th centuries were periods of economic decline, the sea route opening new trade shifting to the Atlantic, where other powers (England, Spain, Portugal and the Netherlands) reigned supreme, and with Austria emerging as the key imperial power in the north. Thus, virtually all parts of Italy changed hands several times between the fall of the Roman empire and unification.

After 15 centuries of political chaos, along came Napoleon, whose brief period of supremacy struck a spark of nationalism which gave rise to the *Risorgimento*, the unification movement. Unification took 50 years. It was spearheaded by three of Italy's most famous figures, Giuseppe Mazzini the theorist, Giuseppe Garibaldi the buccaneer, and Camillo Cavour the politician.

The first all-Italian parliament met in TURIN in 1861. In 1865 the capital moved to FLORENCE and in 1870 to Rome. Except for the Vatican City, that last vestige of the once powerful Church State, Rome was the last major piece of Italian territory absorbed into the new kingdom.

A century or more of unity has seen Italy make steady, and sometimes spectacular, economic progress. The more shameful aspects of Italy's recent history were its attempts at creating an empire in Africa, and the inter-war Fascist period, when Benito Mussolini ruled for more than 20 years. He was responsible for Italy's involvement – on the German side – in the Second World War.

In 1946, Italy became a republic and the 1948 elections ushered in the Christian Democrats who have dominated Italian politics mainly ever since. However, governments have not lasted long. Italy has had more changes of government than any other European country. And to some foreigners it may seem that Italians are always on strike.

A GLITTERING MOSAIC

The cities of the north, such as Turin and MILAN, have an industrial and commercial mentality that rivals any in Europe. Living standards are high; the rates of car and second home ownership, for example, are higher than in Britain. Excellent railways – with notably cheap fares – and first-rate motorways provide swift, convenient transport links throughout the north. The south – the region called the

▲ **HEART OF A CITY** A pavement artist draws religious pictures – including 'The Heart of Mary' – outside Milan's vast cathedral, the Duomo. Made of pink marble, it can hold 20 000 people.

Mezzo-giorno, 'land of the midday sun' – is a complete contrast. Here medieval concepts of honour and behaviour dictate a fierce loyalty to family and a lingering suspicion of outsiders.

In SICILY, NAPLES and CALABRIA secret societies such as the Mafia and Camorra are social plagues. Peasant farmers wrest a living from eroded hillsides. Here there is resignation and a degree of hopelessness. Waiting for someone else to do something is an all too prevalent attitude in this sunbleached land. Most of the land belongs to a few wealthy landowners who invest their southern profits into the north and are little interested in the economic well-being of southern Italy. Young Italians, bitter about the hard lot of the older generation, refuse to look almost certain unemployment in the eye. Many people have left the land since the 19th century and moved to northern urban centres.

Governments have long ignored the *Mezzogiorno* problem, and so they have gained few followers and little influence in the south – an important reason why secret organisations such as the Mafia (in Sicily) and the Camorra (in Naples) have been allowed to become so powerful. Italians have a long tradition of turning to the protection of the powerful and influential members of society. In the past these were rich, often ruthless landowners; in recent times the *mafiosi* have taken over the same patriarchal role.

The character of many southern cities has also been changed as a result of post-1960 policies of industrial development, as in Naples (Alfa Romeo cars), Taranto (steel) and Syracuse (petrochemicals). In further efforts to raise the south's affluence, governments have built large industrial plants in coastal towns – steel works, chemical plants and oil refineries.

However, small industries are still lacking and financial aid channelled to institutions or construction projects has regularly vanished into invisible pockets. During the early 1990s a spate of arrests and suicides hit the Italian upper class, contributing to the downfall of the leading parties and ushering in the right wing (which was soon also inculpated). Investigating magistrates arrested politicians and leading industrialists, some of whom preferred death to the humiliation of public exposure.

But despite differences between north and south, it is the legacies of history, the culture, the works of civilised and sophisticated people, that are the abiding glories. In the south, at PAESTUM, Selinunte and AGRIGENTO, are Greek temples as fine as those of Greece, as well as irrigation systems introduced by Arab colonists a thousand years ago. In central Italy stands Florence, cradle of the Renaissance, and Rome itself. Around them stretch landscapes redolent of Leonardo da Vinci, Dante and St Francis, where hilltop towns and hillside terraces bear the fruits of centuries of labour.

With the exception of GENOA, most of the major northern towns are on the margins of the great plain drained by Italy's largest river, the PO. One line of towns is strung like beads along the Via Emilia, the arrow-straight Roman road that stretches from Milan to the ADRIATIC coast: PIACENZA, PARMA, REGGIO NELL'EMILIA, MODENA, BOLOGNA, Imola, Faenza and FORLI. The road meets the coast at the popular resort of RIMINI. Bologna, chief city of this group, disputes with Capua in Campania the distinction of being the oldest town in Italy: people have lived here continuously for nearly 3000 years.

On the northern side of the plain is another row of towns at the mouths of Alpine valleys. They, too, are ancient in foundation, but many have developed important industries over the past century. Turin (Fiat cars), IVREA (Olivetti typewriters and electronics), Biella (wool), VARESE (engineering), COMO (silk), and the engineering towns of Lecco, BERGAMO and BRESCIA are examples. A third row, between the 'high' and 'low' plain, includes Vercelli, Novara, Milan, PADUA and TREVISO.

Then there are the cities on or near the Po River itself: PAVIA, CREMONA, MANTUA and FERRARA. Finally, sited on the fringes of the delta, VENICE and RAVENNA are unique: Venice, the Queen of the Seas, because of its imperial history, artistic heritage and island site, Ravenna because of its role as a Byzantine capital.

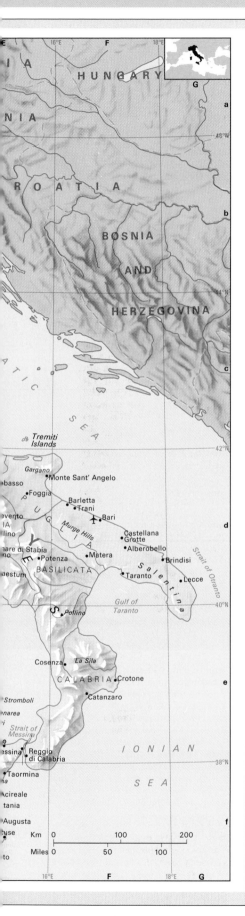

Apart from a group of famous valley towns in northern TUSCANY (Florence, PISTOIA, PISA, LUCCA), hill settlements are characteristic of central Italy. Most of them are in southern Tuscany, UMBRIA and MARCHE. Star-shaped URBINO, spreadeagled along its ridges, or triangular SIENA are distinctive forms. Others, such as PERUGIA and ASSISI, cling picturesquely to their verdant slopes.

The south, too, has its hill towns, the most striking being ENNA and Erice in Sicily. But most of its major cities are coastal and have Graeco-Roman origins. Among them are Naples, BRINDISI, TARANTO, REGGIO DI CALABRIA, MESSINA, CATANIA, SYRACUSE and Agrigento. Byzantine rule continued the classical traditions of many southern cities but, later, feudalism caused widespread stagnation of urban and rural life.

VILLAGE LIFE, ITALIAN-STYLE

Italy's urban, and urbane, way of life – at its best in cities such as Bologna and Florence that have not become too industrialised – extends also into rural areas where about 30 per cent of the population still live and about 10 per cent of the labour force works on the land. Most farmers do not live in isolated farmsteads or hamlets, but cluster in bustling 'town-villages' that may, especially in the south, contain as many as 60 000 people.

Life in these villages tends to follow a pattern, as do the villages themselves. At the village centre is a large, open, paved space, the *piazza*. Facing on to this at one end will be the façade of a church; near by will be the *municipio*, the town hall. Other public buildings include the *carabinieri* (police) station. The *carabinieri* are the traditional police force and are sometimes even present in small villages. The more recently instituted *polizia* are found mainly in provincial capitals and larger towns. The bank and a few bars and shops are also likely to be set around the *piazza*, along with the large houses of the local landowning and professional families.

The changing scene in the *piazza* has a precise daily rhythm which presents the observer with a pageant of everyday life Italian style. At dawn the farmers appear, snatching a cup of coffee on the way to their fields; some still ride donkeys or horses, especially in the south. Next, the children pass through. School in Italy starts between 8 am and 8.30 am. The younger children are on their way to the local elementary school; the older ones may take a bus to a secondary school or college in a neighbouring town. Shops, too, are open early – before 8 am – by which time the noise of craftsmen and mechanics is added to the cacophony. Once a week, a market fills the square. Later on the professional workers – doctors, lawyers and so on – take the stage, along with office workers.

Conservative attitudes continue to prevail – especially in the south, where, for example, there is still prejudice against women being out too much on their own. For teenagers and young unmarried adults, contact between the sexes still depends, to a certain extent, on the rituals of the village: carnival, churchgoing and the *passeggiata* (evening stroll). The women's domain lies in the streets leading off the *piazza*, where the food shops are found. Around 1 pm the *piazza* fills up again with children returning from school, office workers going home and shopkeepers leaving their counters. But they do not linger long in the heat of the sun; besides, lunch awaits. During the next few hours the *piazza* is deserted, save perhaps for a few children. For after lunch comes the siesta.

Some offices and most shops open again from late afternoon to mid-evening. Then around 7 to 8 pm (earlier in winter) the *piazza* and the main streets come alive again. This is the time for the *passeggiata*, the traditional strolling up and down, arm-in-arm with family and friends; a time for relaxed chatting and greeting, for showing off one's best clothes, and for teenagers and young people to exchange playful banter. Only wives and mothers miss the *passeggiata* – they are at home preparing the evening meal.

Later, after the meal, the *piazza* may return to life – especially on summer evenings, as men gravitate towards the bars and young people and engaged couples stroll about.

A SCENIC JOURNEY

A journey through Italy is a scenic idyll. The northern borderlands embrace spectacular Alpine scenery culminating in the Mont BLANC massif on the French frontier and the MATTERHORN on the Swiss. Small glaciers survive in places, and splendid ski slopes have popularised winter resorts such as COURMAYEUR, in the AOSTA valley, and CORTINA D'AMPEZZO in the DOLOMITES.

Flat-bottomed valleys seam the Alpine massif. Glacial action has deepened and then dammed many of these valleys to produce a series of lakes – MAGGIORE, COMO, GARDA.

Some widely differing Italians live there too – 100 000 French-speakers in the VALLE D'AOSTA, 300 000 German-speaking people in BOLZANO Province and a few thousand Slovenes and Croatians around Trieste and Gorizia near the Croatian border.

The Alps merge abruptly into the great northern plain, a 45 000 km² (17 000 sq mile) triangle of lowland which gradually sinks and widens towards the Adriatic Sea. This is the nation's economic heartland, containing most of Italy's larger cities, industries and productive farmland.

Somewhere behind Savona, a port on the LIGURIAN SEA, the Alps give way to the APENNINES, which then wind southwards down the entire length of the country.

And as you go south, so the climate seems to grow perceptibly warmer, the sky bluer and the people more outgoing and gregarious. The trees change from the pines of the Alps and poplars of the plain to more Mediterranean species – evergreen oaks and cypresses – then, continuing southwards, olives. Farther south still, almonds, oranges and lemons reflect the influence of the southern summer drought and mild winter.

The Apennines are mainly sandstone and limestone – young mountains where the earth's crust is still active, as the frequent tremors and earth movements testify. Italy's most recent major earthquake, in November

1980, struck in the heart of the southern Apennines east of Naples.

The western flank of the Apennines is also volcanic, from the extinct volcanoes and volcanic plateaus of southern Tuscany and LAZIO to the cone of VESUVIUS (at present quiescent but terrifyingly active in the past as in AD 79, when it destroyed POMPEII) and then, via the volcanic LIPARI ISLANDS, to Sicily's ETNA, bonfire of Europe.

On their eastern, Adriatic flank the Apennines are less dramatic, with smooth, rounded outlines. Landslides and soil erosion tear great gashes in the clay soils and the rural economy in regions such as MOLISE and BASILICATA is among the poorest in the country.

Finally there are the two island regions: Sicily, an exaggerated microcosm of everything Italian; Sardinia, a law unto itself –

isolated and boulder-strewn, where shepherds carry on their fiercely traditional way of life.

DEEP SOCIAL CHASMS

Italy is a rather rigidly stratified society, with serious conflict between classes and little opportunity for climbing the social ladder. Centuries of foreign rule, the hereditary power of Italy's aristocratic families and chronic unemployment have been significant factors in perpetuating the deep chasms separating social groups. There is a long history of class conflict, from the brigandage of the past to the modern urban terrorism of the Red Brigades, who in 1978 kidnapped and murdered the former prime minister, Aldo Moro.

In the past, Italy's educational system did much to preserve social inequalities, with its emphasis on classical languages and history,

and with learning by heart taking precedence over imaginative project work. Teachers tended to come from middle-class backgrounds, and to impart a culture that was alien to the working classes, from the south or from areas of relatively closed and traditional culture.

But since 1968 there have been extensive changes – best demonstrated by the fact that by 1985 half the students at university were from working-class or peasant backgrounds. Italy also has one of the highest proportions of university students among EU countries.

BOOM TIME IN THE CITIES

Many social problems resulted from large-scale destruction during, and the great expansion of cities after, the Second World War. Rapid urban growth has taken place in most parts of the country, but has been more evident

◀A GATHERING OF GONDOLAS Their prows curved like the necks of proud black swans, some of Venice's 300 or so gondolas line up in the San Marco Canal – with the Salute domes in the background. They have been painted black since 1562, when a city law was passed prohibiting too much pomp.

Florence or the late 19th-century textile and engineering industries of LOMBARDY, had made little difference on a national scale.

In the immediate postwar period, the economic prospects again looked dreary. Armies had tramped over the countryside and bombardment had destroyed towns and industries; 2 million unemployed – among them many former soldiers and landless peasants – threatened revolution. Living standards were abysmally low compared with those of northern European countries. Yet, astonishingly, Italy underwent a modern renaissance, and rose phoenix-like from the ashes of war.

There were two main developments. First came the industrial boom of the 1950s and 1960s, with a rate of expansion second only to that of Japan. Italy's gross national product (GNP) doubled from 1950 to 1962 and seemed well on the way to repeating the performance in subsequent years. Then recession struck in the early 1970s. World oil prices quadrupled, administering a huge shock to an economic system founded largely on imported fuels.

Italy produces only small quantities of coal and oil, although it generates considerable hydroelectric power in the Alps and extracts natural gas in the lower Po Valley. The second development was Italy's entry into the European Common Market (of which it was a founder-member). This enlarged the markets open to its expanding industries – especially for cars, electrical goods, clothing and shoes. Also, large numbers of Italian workers could cross frontiers freely in search of jobs.

Two great population movements began: some workers emigrated to other Common Market countries – especially to France and West Germany – during the 1950s and 1960s, and at the same time more than 2 million people moved from rural areas – especially the south – to Rome and the 'Industrial Triangle' of Milan, Turin and Genoa.

Not until the 1980s did these migrations come to an end. Emigrants began returning from abroad to their home villages. Cities such as Milan, Turin, Genoa, Venice and TRIESTE stopped growing as people chose to remain in the countryside or settle in smaller towns.

UNSTABLE GOVERNMENTS

It is difficult to reconcile the dynamism of Italy's economy for most of the postwar period with the stagnation of its political system.

There are eight main national parties: Communist, Socialist, Social Democrat, Republican, Christian Democrat, Liberal, Monarchist and Neo-Fascist. Over a period of more than 40 years no party, except the Christian Democrats in 1948, won a clear majority of votes in any election and the governments have been characterised by unstable alliances between the parties of the centre and the left.

It was a classic case of multiparty political stalemate. This eventually led to the rise of new political entities, such as the Federalist *Lega Nord* (Northern League) and *Forza Italia*, a movement created by Silvio Berlusconi, owner of the most important private television stations. In the elections of March 1994 – thanks to new electoral rules and an electorate wanting radical change – the right-wing Freedom Alliance, led by Berlusconi, won an outright victory. His triumph was shortlived however, as he was forced to resign later that year.

An important postwar political trend has been the rise of regional governments. This has taken place in two phases. First, four regions – Sicily, SARDINIA, VALLE D'AOSTA and TRENTINO ALTO ADIGE – received special status between 1946 and 1949. They were joined by FRIULI-VENEZIA GIULIA in 1963.

This act was a recognition of their cultural distinctiveness as islands or frontier regions. In each, excepting Sicily, a second official language is spoken beside Italian – Ladin and German in Trentino, Friulian in Friuli-Venezia Guilia, and French in the Aosta Valley. The second phase of decentralisation came in the early 1970s when individual governments were created in the other 15 regions.

The hope was that regional devolution would lessen the inertia inherent in a centralised political system, and allow more local participation in dealing with issues closer to the hearts of the people.

With one or two exceptions, this hope was not initially realised. Regional administrations struggled unsuccessfully with the central government to define their respective roles and to gain true independence of action.

ITALY AT A GLANCE	
Official name Italian Republic	
Area 301 230 km² (116 304 sq miles)	
Population 58 018 540 **Per km²** 187 (**Per sq mile** 498)	
Capital Rome	
Government Parliamentary republic	
Currency Lira	
Languages Italian, some French, German and Slovene	
Religion Christian (100% Roman Catholic)	
Climate Mediterranean; average temperature in Rome is 7°C (44°F) in January and 25°C (77°F) in July	
Land use Cultivation 40%, forest 22%, meadows and pastures 16%	
Main primary products Cereals, potatoes, sugar, vegetables, soft fruits, citrus fruits, grapes, olives, livestock, fish; oil and natural gas, asbestos, potash, iron ore, zinc, mercury,	
Major industries Iron and steel, motor vehicles, chemicals, oil and gas refining, machinery, textiles, clothing, agriculture, food processing, wine, fishing, tourism	
Main exports Machinery, chemicals, petroleum products, motor vehicles, clothing, food, textile yarns and fabrics, iron and steel, footwear	
Annual income per head (US$) 16 850	
Population growth (per thous/yr) 2	
Life expectancy (yrs) Male 74 **Female** 81	

in the north because of the migration of job-seekers from the south. The proportion of Italians living in towns of at least 100 000 people increased from 35 per cent in 1951 to 50 per cent by 1981.

The outer suburbs of large cities seem particularly soulless, consisting of large, tightly packed blocks of flats with little public open space and few communal services. Municipal housing schemes are few, and poorer people live in dilapidated tenements, seen at their worst in central Naples. Often, the most basic services are not provided and many of the new settlements lack adequate sanitation.

This city growth has resulted from the remarkable industrialisation of Italy since 1950. Before then Italy had been mainly a rural economy. Earlier industrial phases, such as the medieval textile industries of Venice and

Ivory Coast

A LONG PERIOD OF PROSPERITY AND STABILITY FOR THIS AFRICAN COUNTRY IS FOLLOWED BY A SERIES OF ECONOMIC CRISES

From the 15th to the 19th centuries those few Europeans who landed on the Ivory Coast's tropical shore came only to trade in elephants' tusks – which gave the land its name – and slaves. Meanwhile, in the rain forest-clad and savannah interior, those people who escaped the slavers continued to farm or wage war on one another, far removed from Western influences.

But the French arrived in 1830 and established a colony in 1893. After the Second World War, the liberation movement gained momentum, leading to independence in 1960. However, the French influence remained strong. About 50 000 French people still live here and French is the official language.

Ivory Coast's 60 ethnic groups have no common language apart from French, and their loyalties to their groups are still stronger than their loyalty to the country as a whole. They have a very rich and varied culture: the Baoul in the centre are skilled woodcarvers and goldsmiths, living in homes with intricately carved doors. Even more elaborate doors guard the homes of the Senouto.

The first European visitors were discouraged by the rocky cliffs of the south-west coast, along the Gulf of GUINEA; but farther east are coastal plains that today are the country's most prosperous region. The former capital city of ABIDJAN lies here.

A tourist centre has grown up by the lagoons where vast sandy beaches are lined with palm trees – an area now known as the African Riviera. The interior of the Ivory Coast is less affluent, but has its oases of prosperity, such as BOUAKÉ, on the central plateau of the country, and Ferkessedougou in its more sparsely forested north. In 1984 the town of YAMOUSSOUKRO became the new seat of government. It boomed into a prestigious capital boasting Africa's largest cathedral and many palaces.

For years the République de la Côte d'Ivoire, as the country has officially called itself since January 1986, was one of the richest places in West Africa. It still is one of the most Westernised. The general prosperity and particularly the boom of Yamoussoukro were the result of the policies of the government in power for the 13 years after independence.

It was dominated by one man, President Félix Houphouët-Boigny, known for his pro-Western outlook. Until his death in late 1993, he pursued a policy of liberal capitalism, encouraging Western investors to put their money into the Ivory Coast, by placing few restrictions on the import or export of capital.

The devaluation of the CFA franc in early 1994 more or less coincided with the president's death and the country was plunged into an unknown future. Houphouët-Boigny was succeeded by Henri Konan Bédié. Economically, there were hard times ahead. But even during the times when the French supported the country economically, the wealth from foreign investment benefited the few more than the many. With Abidjan being one of the most expensive cities in the world, life was even harder for those who profited least from the country's 'economic miracle'.

Some improvements, though, have been for the good of all. Roads are well kept and the Ivory Coast has the greatest length of metalled road in West Africa. There is an efficient rail service, too, from Abidjan up to the northern border with BURKINA.

Riding the railway northwards, the visitor passes through the humid tropical rain forest of the south, where teak and mahogany grow, along with iroko trees, whose wood was once used as a substitute for teak. Felling is now prohibited. Oil palms and raffia palms provide useful products. The Ivory Coast is an agricultural land that has nudged back the natural forest to grow cocoa, coffee, cotton, bananas and pineapples. The country is among the world's largest producers of cocoa, coffee, and palm-kernel oil. Together, these crops bring in more than 35 per cent of the country's export revenue.

TRADITIONAL EXISTENCE

But beyond the plantations nestle villages that have seen little of that revenue, where the people still lead a largely self-sufficient, traditional existence. They grow their own food, usually cassava – the sturdy root from which tapioca is made – vegetables and rice. Most people at some time in their lives contract at least one of many debilitating diseases such as malaria, yellow fever and bilharzia.

North of the humid lowland plain, the land rises gradually to a savannah plateau and becomes increasingly dry. The far north is grassy and treeless. Here the rainy season is shorter than in the south, so there is only one harvest a year. Towards the northern border with Mali, around KORHOGO, the dry months bring intense heat. In the west lies quite a different land – the forested MAN Mountains, the eastern flank of a range centred in neighbouring Guinea.

Despite its roads, dams and hydroelectric power stations, its oil refinery and deepwater ports at Abidjan and San Pedro, the country is still underdeveloped. Drought in the 1970s and early 1980s reduced harvests and depressed the economy.

Although the rains returned in 1983 and brought an upturn in the economy, many people face economic hardship. The government has encouraged farmers to diversify their crops and has embarked on several economic reform programmes.

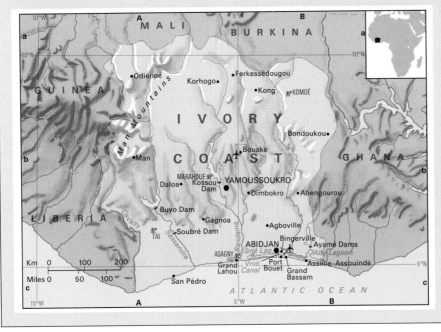

IVORY COAST AT A GLANCE		
Official name République de la Côte d'Ivoire		
Area 322 463 km² (124 503 sq miles)		
Population 14 200 000 **Per km²** 44		
(Per sq mile 114)		
Capital Yamoussoukro		
Government Republic		
Currency 1 CFA franc = 100 centimes		
Languages French; African languages		
Religions Indigenous beliefs (60%), Christian (19%, mostly Roman Catholic), Muslim (20%)		
Climate Tropical; average temperature in Abidjan is 27°C (80.6° F) through the year		
Land use Grazing 9%, forest 26%, cultivation 13%, other 52%		
Main primary products Cocoa, coffee, cotton, bananas, pineapples, rubber, sugar, rice, cassava, yams, timber; crude oil, diamonds, cobalt, uranium, iron ore, copper, manganese		
Major industries Agriculture, food processing, textiles, leather goods, forestry, mining, petroleum refining		
Main exports Cocoa, coffee, timber, pineapples, bananas, cotton, oil		
Annual income per head (US$) 670		
Population growth (per thous/yr) 35		
Life expectancy (yrs) Male 53 **Female** 53		

Jabal ad Duruz (Jebel al-Arab) *Syria* Mountainous district situated in the extreme south of the country near the border with Jordan. Its highest point is the extinct volcano of Jabal ad Duruz, 1735 m (5691 ft) high, lava from which has formed a series of sheets, cones and caves. Cereals are grown in the relatively fertile west side of the district; the east side is a barren wilderness. Excursions are made to the ruins of Kanatha from Es Suweidiya, which is the district's main settlement.
Map Syria – Bb

Jabal al Nusayriyah *Syria* Mountain range running parallel to the coast for 113 km (70 miles). The mountains, which rise from about 1200 m (3937 ft) to about 1500 m (4921 ft), peak at 1562 m (5125 ft). A number of ruined Crusader forts are dotted along the range.
Map Syria – Bb

Jablonec *Czech Republic* See LIBEREC

Jachymov (Sankt Joachimsthal) *Czech Republic* Former mining town on the German border north-west of Prague. Silver mining began here in 1516. From 1519, the town – then called Sankt Joachimsthal – minted silver *Joachimsthalers,* or 'thalers' (from which the word 'dollar' originated).

The pioneering physicists Marie and Pierre Curie used pitchblende (uranium ore) from Jachymov for their investigations into radioactivity. Marie (1867-1934) went on to isolate the elements radium and polonium from it after Pierre's death in 1906. Production of the ore for radium began in 1908, and after the Second

World War much of the East Bloc's uranium came from here. Jachymov is also a spa with radioactive water.
Map Czech Republic – Aa

Jackson *USA* State capital and largest city of MISSISSIPPI, located in the centre of the state. Its origins go back to the late 18th century, when the French established a trading post here known as Le Fleur's Bluff. Later, it was named after Andrew Jackson, hero of the war of 1812 (against Britain) and seventh president of the United States. For a long time the town was regarded as the crossroads of the South. Today it is an industrial, educational and distribution centre. Nearby is an 89 hectare (220 acre) model of the Mississippi River basin used to study flood control.
Population (city) 196 600; (metropolitan area) 395 400
Map United States – Hd

Jacksonville *USA* Port and major transport centre in north-east Florida, 40 km (25 miles) south of the Georgia border. It is the largest city in Florida and straddles the St Johns River, 45 km (28 miles) from the Atlantic Ocean. In 1901, a devastating fire destroyed most of Jacksonville. In addition to being an exporter of forest and agricultural products, the city has emerged as an important manufacturer of paper, chemicals and cigars.
Population (city) 635 230; (metropolitan area) 906 700
Map United States – Jd

jade Either of two distinct minerals, JADEITE and NEPHRITE, used as gems or in carved ornaments. Of the two, jadeite is the more highly prized.

jadeite Green or white mineral, sodium aluminium silicate, found in IGNEOUS and METAMORPHIC rocks. It is a form of JADE.

Jadotville *Zaire* See LIKASI

▼ **ONE OF THE MOST BEAUTIFUL INDIAN CITIES** The palace within the fort at Amber, the old Rajasthani capital near Jaipur, acquires a rosy magic at sunset each day.

Jaén *Spain* Town of Roman origins, 65 km (40 miles) north of GRANADA. It is famous for its olive groves, producing olive oil as well as brandy, chemicals and leather goods. Its province of the same name has several lead mines.
Population (town) 103 260; (province) 637 630
Map Spain – Dd

Jæren *Norway* Peninsula covering 1700 km² (656 sq miles) in the extreme south-west of Norway, and often referred to as 'a little bit of Denmark attached to Norway' because of its absence of hills. It has been farmed intensively since prehistoric times and now specialises in dairy products and the cultivation of vegetables.
Map Norway – Bd

Jaffna *Sri Lanka* Port in the north-west, exporting fruits, cotton, tobacco and timber, situated 300 km (185 miles) north of COLOMBO. It is located on a peninsula less than 70 km (45 miles) from the Indian coast. Tamils from India established a flourishing kingdom here in the 13th century; it has been the headquarters of the island's Hindi Tamil population ever since. The Portuguese annexed the port in 1617, but were ousted about 40 years later by the Dutch, who built several forts here (one of which has been preserved as the city's police station).

Although Jaffna has fine sandy beaches, its tourism industry has been non-existent since the 1980s, when the peninsula on which it is located became a base for Tamil guerrillas seeking an independent Tamil state. Bitter fighting has taken place in the vicinity.
Population 129 000
Map Sri Lanka – Ba

Jaipur *India* City and capital of the north-west state of Rajasthan, about 260 km (162 miles) south-west of Delhi. One of the most beautiful cities in India, it was founded in 1727 by Maharajah Jai Singh II, who ordered it to be painted pink, the colour that his astrologers declared auspicious for him.

One frontage of the royal palace – 'the Palace of the Winds' – at nearby Amber is renowned for its many stone filigree windows. There are numerous other magnificent buildings inside the 6 m (20 ft) high city walls, including the finest of Maharajah Jai Singh's open-air observatories. Here, the heavens were charted on stone gnomons designed by Jai Singh himself – some of which are up to 30 m (98 ft) tall.

The spacious government headquarters as well as the city's modern suburbs are located outside the city walls. Jaipur is noted for its carpets, jewellery and paintings, including some exquisite miniature work.
Population 1 455 000
Map India – Bb

Jajce *Bosnia-Herzegovina* Former Bosnian capital situated amid forested hills, 53 km (33 miles) south of the city of BANJA LUKA. Its chalets, Turkish-style houses and mosques, and *campanile* (bell tower) cluster around a hilltop fort, near where the Pliva River tumbles into the Vrbas via a spectacular waterfall. Yugoslavia's former communist leader Marshal Tito (1892-1980) proclaimed Yugoslavia a federal socialist republic at Jajce on 29 November 1943.
Population 9100
Map Bosnia-Herzegovina – Ba

Ják *Hungary* See SZOMBATHELY

Jakarta (Batavia) *Indonesia* National capital, in western Java, and the most heavily populated city in South-east Asia. The city was the centre of the Dutch East India Company's trading empire from 1619, when it was known as Batavia. Its old city centre is marked by typical Dutch architecture, with numerous bridges, pleasant squares and elegant public buildings. However, Jakarta has grown beyond this central area into a sprawling conurbation surrounded by shanty towns. Today, it is one of Indonesia's key ports, exporting tea and rubber. Its industries revolve around its textile mills, iron foundries and saw-mills.
Population 7 880 000
Map Indonesia – Cd

Jalalabad (Djelalabad) *Afghanistan* Town on the Kabul River, 120 km (74 miles) east of Kabul, commanding the head of the KHYBER PASS. It is the capital of the country's most densely populated province, Nangarhar. Founded in 1560, the town has become an important trading centre for fruit, sugar and timber. Bitter fighting took place here during the 19th-century Afghan Wars.
Population 57 800
Map Afghanistan – Ba

Jalapa (Jalapa Enríquez; Xalapa) *Mexico* Capital of VERACRUZ State, in the Sierra Madre Oriental mountains, 130 km (80 miles) from the Gulf of Mexico. It is a typical colonial town, criss-crossed with many cobbled streets, and dotted with squares and gardens. It has become well known for its interest in the performing arts. It has a university (opened in 1944) and an anthropological museum with precolonial artefacts.

Sugar refining, flour milling and coffee roasting are important industries. Jalapa gave its name to a purgative produced here from a local plant.
Population 288 300
Map Mexico – Cc

Jalisco *Mexico* Principal state in western Mexico, whose capital is GUADALAJARA. It is the heartland of Spanish culture in Mexico and is renowned for its conservatism and traditionalism.

Silver and copper are mined in the mountains inland, while a variety of industries (textiles, leatherware, chemicals, building materials) and agricultural products such as cereals, coffee, tobacco and cotton (grown in the valleys and on the coast) make up the bulk of the economy. Some cattle are reared in the area.

Jalisco is the land of the blue maguey cactus – the source of tequila and mescal, Mexico's best-known alcoholic drinks. Puerto Vallarta is the most popular of a number of beaches.
Population 5 279 000
Map Mexico – Bb

Jamaica See p 343

James Bay *Canada* Shallow southern arm, 440 km (273 miles) long and 160 km (100 miles) wide, of HUDSON BAY, which was discovered by Henry Hudson in 1610. It was here in 1611 that Hudson's crew mutinied and cast him adrift with his son John and seven men. They were never heard of again. The bay was named after Thomas James, who explored the area in 1631 and 1633, while searching for the NORTHWEST PASSAGE.
Map Canada – Gc

Jamestown *USA* First permanent English settlement in the United States, founded beside the James River in VIRGINIA in 1607. The village is about 185 km (115 miles) south of Washington, DC, in the Colonial National Historical Park.
Map United States – Kc

Jammu and Kashmir *India* Former princely state in the north-west, bordering on Chinese Tibet. The majority of the population are Muslim, but in 1947 the ruling Hindu maharajah refused to cede the state to either India or Pakistan. War between the two countries resulted, followed in 1949 by a UN-negotiated ceasefire agreement dividing the area between Pakistani AZAD KASHMIR and Indian Kashmir.

India claims all 222 236 km² (85 783 sq miles) of the state, although it occupies only about half of it. There was further fighting in 1965 and 1971; by the early 1990s the dispute had still not yet been resolved. From the mid-1950s, China occupied a portion of the disputed territory.

SRINAGAR, Jammu and Kashmir's summer capital, is popular with tourists. Jammu (population 943 400) is the winter capital. The main products produced here are wool, fruit, maize, wheat and rice. However, this remains one of the most backward regions on the subcontinent.
Population (Indian Kashmir) 7 718 700
Map India – Ba

Jamshedpur *India* Steel centre in the north-east, 308 km (191 miles) west of Calcutta. Its development coincided with the founding of the first steel plant here in 1907-11 by Jamshetji Tata, the first private businessman to build such a plant in India. At that time the site, rich in iron ore, was just a remote valley of rain forests.

Today Jamshedpur is a hive of blast furnaces, coke ovens, machine factories, metalworks, and steel and iron concerns.
Population 834 000
Map India – Dc

Jamuna *Bangladesh* River forming the main stream of the BRAHMAPUTRA from the Tista to the Ganges. The Ganges and Jamuna together form the Padma, which flows into the Meghna River.
Map Bangladesh – Bb

Jan Mayen *Norway* Island in the Arctic Ocean, 540 km (335 miles) north-east of Iceland. Discovered in 1607 by the British navigator Henry Hudson (1550-1611), it was annexed by Norway in 1929. It covers 380 km² (146 sq miles) and is dominated by the Beerenberg, an extinct volcano 2277 m (7470 ft) high and partly glaciated. The island is the site of a weather station, whose operators are the only inhabitants.
Map Arctic – Ab

Janakpur *Nepal* Lowland town and religious centre in the south-east of the country, 10 km (6 miles) north of the Indian border. Janakpur is centred on the Hindu Janaki temple (after which it is named) and contains many tanks used for ritual bathing. The town is visited by Hindu pilgrims from all over India as well as Nepal.
Population 143 000
Map Nepal – Ba

Japan See p 344

Japan Alps *Japan* See CHUBU SANGAKU

Japan, Sea of Arm of the PACIFIC OCEAN, between Japan, Korea and Russia, covering 1 008 000 km² (389 200 sq miles), and which is up to 4225 m (13 862 ft) deep. The Tsushima branch of the warm Japanese Current flows northward through the sea, while a cold counter-current brings ice to the mainland shore. The sea contains important commercial fishing grounds for anchovy, herring and mackerel.
Map Japan – Bb

Jarash (Jerash) *Jordan* Town containing well-preserved remains of a provincial Roman settlement, situated 38 km (24 miles) north of AMMAN. Jarash's paved Roman streets, in which the ruts made by chariots can still be seen, are lined with hundreds of carved columns. Under the town, the 2000-year-old drainage system still works.

There are also theatres, temples, plazas, baths and public buildings and a triumphal arch commemorating a visit by Emperor Hadrian in AD 129. The HIPPODROME has stone polo posts erected after the Persian invasion of 614.

Founded in 322 BC by soldiers of Alexander the Great, the city was abandoned in AD 747 after a series of earthquakes. The modern town was founded in the late 19th century.
Population 29 000
Map Jordan – Aa

Jaroslaw (Jaroslau; Yaroslav) *Poland* Agricultural town in the south-east, standing on the San River, about 30 km (18 miles) from the Ukrainian border. During the Middle Ages, it was a flourishing market place at the crossroads of the Rhineland-to-Russia and Baltic-Black Sea trade routes. The town still has Renaissance and 18th-century Baroque buildings.
Population 41 400
Map Poland – Ec

Jarrow *United Kingdom* See NEWCASTLE UPON TYNE

Jars, Plain of *Laos* Central part of the TRAN NINH plateau, characterised by narrow river valleys and limestone and sandstone hills rising to about 2000 m (6560 ft) above sea level. The name of the region is derived from the prehistoric stone funerary jars found here in the 19th century by the French. In the early 1960s fierce fighting occurred here between government and communist Pathet Lao troops during the Laotian civil war.
Map (Tran Ninh Plateau) Laos – Aa-Ab

Java (Jawa; Djawa) *Indonesia* Island covering 132 187 km² (51 026 sq miles) and home to about 60 per cent of the republic's people, although it covers only 7 per cent of the country's land area. Besides agriculture there are textile mills, car factories and smaller manufacturing industries. The national capital, JAKARTA, is located on the island. Java has 112 volcanoes, many of which are active.
Population 107 514 000
Map Indonesia – Cd

Java Sea Area covering 310 800 km² (120 000 sq miles) of the PACIFIC OCEAN, lying between Java and Borneo, with Sumatra in the west and the FLORES SEA in the east. In the Battle of the Java Sea on 27 February 1942, the Japanese navy defeated an Allied fleet to open the way to invade Java.
Map Indonesia – Cc

Jamaica

VIOLENCE AND POVERTY LIE BEHIND THIS CARIBBEAN ISLAND'S IMAGE OF A LUSH AND PEACEFUL PLAYGROUND FOR TOURISTS

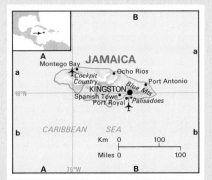

More than a million tourists a year visit Jamaica. They sip rum punches in the blazing Caribbean sun, ride bamboo rafts down lazy rivers, and dance through balmy nights to the throbbing beat of reggae music. Along the way, they spend some US$870 million a year.

These prosperous visitors stay in luxurious hotel complexes near palm-fringed beaches and find it easy to imagine that Jamaica is an island of peace and lotus-eating plenty. But they see only one face of this turbulent island.

Behind the facade of the beaches is Jamaica's less happy face: one lined by hard work, and lean from the struggle to survive. Among the hills that rise in the centre of the island from the mostly narrow coastal flatlands, farmers' wives rise before dawn to walk – sometimes for hours – to reach the nearest road where they can get a bus to market. Maids in the plush hotels of MONTEGO BAY, OCHO RIOS, PORTANTONIO and Negril earn in a week about the same as the price of a glass of whisky from the hotel bar. Just over one in six of the rapidly expanding workforce may be without a job.

Although tourism has been booming since the 1980s and has replaced sugar as the second largest earner of foreign exchange, the growth has not been enough to offset the decline in the principal export: bauxite and alumina, the ore and mineral from which aluminium is made. Jamaica is the world's third largest producer of bauxite after Australia and Guinea. If the ore were processed into aluminium, it would command higher prices; but the operation would be too expensive, because Jamaica has no cheap supplies of energy to power the smelters. Production of bauxite and alumina dropped by nearly 40 per cent in the early 1980s, but has been gradually recovering.

The government has encouraged foreign investment and agricultural diversification. The aim is to raise the Jamaicans' standard of living and diminish the country's massive foreign debts – which, by 1992, had reached US$3700 million. At the same time, the country has become one of the largest recipients of aid from the United States.

New crops, such as winter vegetables, flowers, fruit and honey, are being developed for export alongside the more traditional products: sugar and bananas, molasses and rum, peppers and ginger, cocoa and coffee. The bulk of the coffee harvest is Blue Mountain coffee, rated by some gourmets as the world's best.

Agriculture suffered a huge setback as a result of Hurricane Gilbert in 1988. Particularly badly affected were the 150 000 subsistence farmers. But the island's agricultural sector has now fully recovered.

There are less orthodox crops, too. Cannabis – known locally as *ganja* – is widely, though illegally, grown in the inland hills and hawked in bars and on beaches. In addition, nearly 1900 tonnes of it is smuggled into the USA annually, bringing Jamaican growers and dealers US$175 million and making the drug one of Jamaica's most lucrative products.

Despite economic measures taken by the government, life for most islanders is an unhappy contrast with the beauty of their land. In the ghettos of West KINGSTON, crowded with people from rural areas looking for work, tiny houses of wood and corrugated iron sag beside litter-strewn streets – and gang leaders rule their run-down territories with guns. Frustrations sometimes flare into political violence. In 1976, rioting became so grave that the government declared a year-long state of emergency. During a general election campaign in 1980, nearly 700 people were killed; and tensions rose again before the 1993 elections.

Jamaica gets its name from its original inhabitants, the Arawak Indians, who called it Xaymaca, meaning 'land of wood and water' Christopher Columbus, who in 1494 claimed the island for Spain (it remained Spanish until the British captured it in 1655, and was British until independence in 1962), described it simply as 'the fairest isle that eyes have beheld'.

JAMAICA AT A GLANCE	
Area 10 990 km² (4242 sq miles)	
Population 2 529 980 **Per km²** 230 **(Per sq mile** 596)	
Capital Kingston	
Government Parliamentary monarchy	
Currency 1 Jamaican dollar = 100 cents	
Language English, Creole	
Religions Christian (Church of God 18%, Baptist 10%, Anglican 7%, Seventh-Day Adventist 7%, Roman Catholic 5%, and others), Rastafarian	
Climate Tropical at sea level; in the south, dry season from November to April. Average temperature in Kingston ranges from 25°C (77°F) in January to 29°C (84°F) in August	
Land use Cultivation 25%, meadows and pastures 18%, forest and woodlands 28%, other 29%	
Main primary products Sugar, bananas, coffee; bauxite	
Main industries Mining, agriculture, tourism, sugar refining, molasses, rum, tobacco products, cement, textiles, oil refining	
Main exports Bauxite, alumina, sugar, bananas, other fruit and vegetables, rum	
Annual income per head (US$) 1340	
Population growth (per thous/yr) 10	
Life expectancy (yrs) Male 72 **Female** 76	

Java Trench Deep trench, 7455 m (24 459 ft) at its deepest point, in the INDIAN OCEAN off southern Indonesia. Here the Indo-Australian continental plate is sliding under the Eurasian plate (see PLATE TECTONICS). As it descends, earthquakes occur.
Map Indonesia – Bd

Jaworzno *Poland* Industrial town some 25 km (15 miles) east of the city of KATOWICE. One of the country's largest coal-mining and electricity-generating centres, it also produces lead and zinc.
Population 93 900
Map Poland – Cc

Jayapura *Indonesia* Town in Irian Jaya Province, situated on the north coast of New Guinea. It is the administrative centre of the province and was formerly known variously as Sukarnapura, Kota Baru and Hollandia.
Population 45 800
Map Indonesia – Jc

Jbail *Lebanon* Historic seaport 29 km (18 miles) north of the capital, BEIRUT. Site of the Biblical Gebal, it claims to be the oldest continuously inhabited town in the world. The ancient Greeks called it Byblos, and initially also called papyrus by the name 'byblos' because it was exported from Egypt to the Aegean through the port of Jbail. Trade in papyrus brought wealth to the town, which went through its first boom in the 3rd millennium BC.

The massive walls of the ancient city date from 2800 BC and nearby traces of civilisation – including late Stone Age and Bronze Age remains – go back 9000 years.

A necropolis dating from about 3500 BC contains some 1450 burial jars, and excavations have uncovered the remains of a Greek temple. The Romans added a theatre, and the Crusaders a citadel, to the town's architecture.
Population 1200
Map Lebanon – Aa

Jeddah (Jidda) *Saudi Arabia* Major port on the RED SEA, situated 70 km (44 miles) west of MECCA. The main point of arrival for pilgrims to Mecca and MEDINA, it has an ultra-modern airport that can handle thousands of pilgrims a day.

Jeddah is also the commercial and diplomatic capital of Saudi Arabia, and its shore is lined with high-rise office blocks. Among the most notable constructions is the spectacular Corniche, which is lined with sculptures by a number of internationally acclaimed artists.

A walled city some 300 years old, Jeddah has a well-preserved old town containing coral-fronted houses with wooden balconies, merchants' houses with elaborate carvings, and the House of the Tree, dating from the 1850s and built next to a huge tree, a local landmark.

Until 1918, the city was ruled by Turkey; the Saudis took control in 1925.
Population 1 200 000
Map Saudi Arabia – Ab

Japan

AS THE PACE OF THE ECONOMIC MIRACLE SHOWS SIGNS OF SLACKENING, JAPAN ADAPTS TO A CHANGING POLITICAL LANDSCAPE

Strung across the eastern margin of Asia, the island nation of Japan has transformed itself in little more than a century from an isolated feudal backwater with no natural resources into the second biggest economy in the world after the United States. After four decades of economic success, and annual growth rates of up to 10 per cent, Japan is the dominant economy of East Asia, and will remain so for years to come.

A plain recital of the statistics indicates the strength of the foundations that were laid during the postwar expansion. With less than 3 per cent of the world's population, Japan enjoys some 10 per cent of the world's income as measured by gross domestic product. It vies with South Korea to be the world's biggest shipbuilder and with the United States to be the world's biggest car manufacturer.

sophisticated chips, a field led by the USA. In the late 1980s, the Americans had been shocked by claims that Japan's prowess in chip manufacture gave it a stranglehold over weapons technology. By the mid-1990s the idea had come to seem like a nationalist fantasy.

Other worrying trends are emerging. Big Japanese manufacturers – household names all over the world – have begun to make their sophisticated products outside Japan, in countries such as the US and the UK and in low-wage economies such as China and Malaysia. As jobs move abroad, Japanese industry is being 'hollowed out'. Companies are taking on fewer high-school and university graduates and there is concern that a shrinking workforce will have to support more and more unemployed.

By Western standards, joblessness is low – less than 3 per cent compared with twice that in the US and three or four times that in Europe. But the Japanese figure is artificially reduced by the number of people who are in jobs with little or no work to do. The ranks of the under-employed – people who, in Japanese parlance, are 'sitting by the window' – strains Japan's jobs-for-life employment tradition.

Big car makers give workers unpaid holidays and shut down production to save costs. Privately, some also suggest that Japan cannot for ever avoid mass lay-offs, which have been virtually unknown even in economic down-

sumption rather than on income for a major part of its revenue. The controversy over how, and to what extent, this can be done is a central preoccupation in the final decade of the century and has contributed to the disintegration of the once-stable political system.

Nevertheless, the achievements are still remarkable. In 1945, Japan became the first nation to experience a nuclear holocaust when atomic bombs were dropped on HIROSHIMA and NAGASAKI. Emperor Hirohito told his people that the war had developed 'not necessarily to our advantage'. Japan's economy lay in ruins, her people malnourished, demoralised and suffering for the first time in their history the humiliation of occupation by a foreign power.

Today, by contrast, Japan is, in theory, a modern liberal democracy, its people are among the world's healthiest – industrial pollution has been drastically reduced since the 1970s – and it has the lowest infant mortality rates of anywhere. A fiercely competitive education system has produced a 99 per cent literacy rate and 90 per cent of people consider themselves middle class. In the process of modernisation, Japan has sacrificed little of its cultural identity, which it cherishes above all things.

The price of individual failure, however, can be heavy – suicide is not infrequent, especially among the young. And the price of international success has been a mounting distrust of Japan among foreign governments. Relationships with the US, in particular, are soured by perennial disputes over trade. America's deficit with Japan hit a record $60 billion in 1993 and the Americans argue that this is largely due to unfair protectionism.

LIFE ON THE PLATE EDGE

The Japanese archipelago (consisting of about 3900 islands) sits across the so-called Pacific 'ring of fire'. It owes its existence to the cataclysmic forces that are unleashed wherever the giant plates making up the earth's crust move against one another. Six chains of steep, serrated mountains form the broken skeleton of the country. Studded with volcanoes (more than 60 of them active), the mountains endow Japan with a rare scenic beauty, but little else.

There are few workable deposits of minerals or fuels, and the silts washed down by tumbling mountain streams have rarely created more than a narrow and fragmented coastal plain. Less than a quarter of the land is flat enough for agriculture or human settlement, though in the south spectacular flights of mountain terraces are cultivated.

Much of the land upon which the modern cities and factories stand has been painstakingly won from the sea, at the risk of damage from storm or *tsunami* – giant waves up to 30 m (100 ft) high caused by earthquakes. Frequent and widespread quakes, on average 5000, are registered each year and, though most of them are only slight, the threat of catastrophe is ever present. In 1923, a huge quake killed 150 000 people in and around Tokyo and destroyed 60 per cent of the city and 80 per cent of Yokohama. In 1995, an earthquake destroyed much of the port city of Kobe and killed more than 5000 people.

Nevertheless, economists believe that the deep recession of the early 1990s may prove a watershed for the Japanese economy. In the last decade of the century the first hints were emerging that Japan might eventually be overshadowed by China. And there have been symbolic reversals. It was once possible to boast that one in every three television sets in the world was made in Japan, but in 1993 for the first time Japan imported more television sets than it exported.

Similarly, its dominance in computer chip and microprocessor manufacture has been shaken. Under central government guidance, Japanese manufacturers bet heavily in the 1980s on mass-market chips – only to find in the mid-1990s that the most valuable market was in

▲ **MOUNTAIN-GIRT SEA The fine gateway to a Shinto shrine stands on the shore of Miyajima (Itsukushima), an island in the Inland Sea near the city of Hiroshima. The island is popular with tourists and has many ancient temples.**

turns. The situation is exacerbated by the fact that Japan's birth rate has fallen to 1.5 children per family. On current trends, 27 per cent of the population will be aged 65 or over by the year 2025. This, like the hollowing out of industry, implies fewer wage-earners supporting even more people out of work.

An important consequence of this is that the government will have to look to taxes on con-

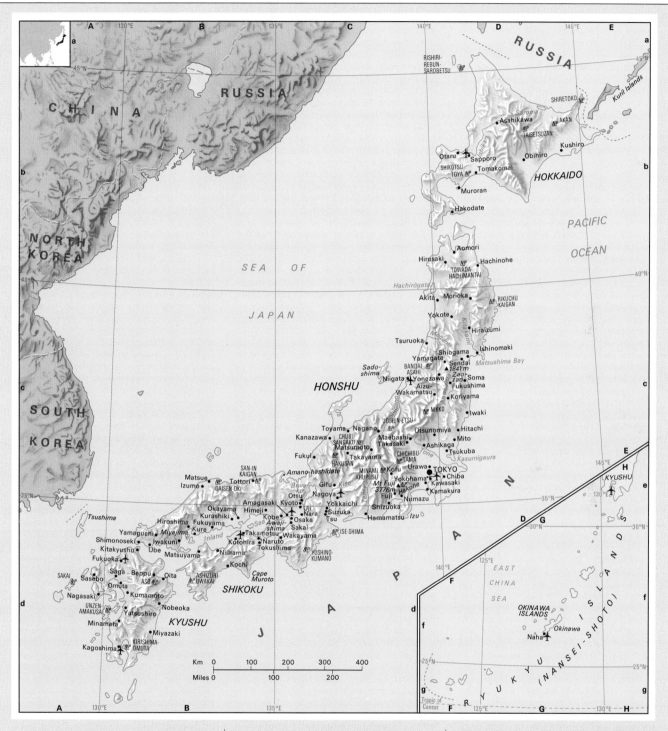

Heavy rains and hot summers allow rice to grow on most lowlands, although chill breezes can damage crops in the north. The north-eastern 'snow country' is heavily blanketed in winter, and hurricanes in summer and early autumn can devastate town and countryside anywhere.

Within this naturally hazardous environment live almost 125 million people; and although overall population density is not quite the highest in the world, the ratio of people to habitable land certainly is. Before

the modernisation of industry, most of them worked on the land and as recently as the mid-1950s many still lived in rural areas. However, the rapid growth of industry and commerce after the Second World War generated an enormous drift away from the country and 76 per cent of the people now live in cities.

Some 40 million people, nearly a third of the population, are settled in greater Tokyo, within 50 km (30 miles) of the centre of the capital. Modern department stores, hotels, shops and offices have transformed city centres all over the

country, but the suburbs beyond can be chaotic. Few streets bear names and the custom of numbering houses according to when they were built rather than their location adds to confusion, not just for visitors but for local people as well.

Japan's modern history can be said to have started in the late 16th century, when a long civil war exposed the people to foreign influences. The victorious Tokugawa shoguns (military dictators) set out to check the spread of alien ideas and of Christianity in particular. Almost any form of contact with the outside world was forbidden. Christianity was suppressed and the country remained in virtual seclusion for more than two centuries.

Thus isolated, the Japanese had time to perfect many elements of their unique culture. By the middle of the 19th century, however, when the arrival of an American fleet under Commodore Matthew Perry finally awoke Japan to the renewed threat from outside, it was technologically backward and quite unable to defend itself. Only a concerted effort to emulate the foreigners offered any hope of escape from domination.

At first, Japan was compelled to sign treaties on poor terms – now termed the 'unequal treaties' – but this bought time in which to modernise the country. In 1868 a coup overthrew the last shogun, an imperial government was established and a crash programme of modernisation began.

Foreign experts advised on all aspects of government, economy and defence, and bright young men were sent overseas to learn all they could from the West (as they still do). Educational reforms led to a rapid spread of literacy and fostered a deepened sense of nationalism based upon reverence for the emperor and respect for ancient Confucian ideals. Despite having few natural resources, Japan was soon powerful enough to defeat both China (1894-5) and Russia (1904-5) and began to build an overseas empire.

By the 1920s, Japan had gained considerable international prestige and influence but the great depression of the 1930s ruined the economy. Trade slumped as overseas markets collapsed and the government drifted into the embrace of military extremists.

A Japanese invasion started a war with China in 1937 – imperial soldiers massacred at least 150 000 people in Nanking – and in December 1941, attacks upon Pearl Harbor, in Hawaii, and on British, French and Dutch colonial outposts throughout South-east Asia signalled Japan's entry into the Second World War in alliance with Nazi Germany.

Brilliant initial successes stoked dreams of a 'Greater East Asia Co-prosperity Sphere' but these were shattered when Japan lost control of the sea lanes vital for the import of raw materials. Despite fierce (and often suicidal) resistance, defeat was inevitable. The war ended with the atomic bombs. By August 1945 the goal of 'Rich country, strong army' had brought mass destruction and ruin.

Japan remained occupied until 1952 and General Douglas MacArthur ensured that the new constitution banned military expeditions overseas. Nevertheless, Japan now has the sixth largest military force in the non-communist world and is also supported by substantial American military bases.

The so-called 'peace constitution' became controversial in the 1990s as Japan sought to acquire a more important diplomatic role on the world stage in line with its economic power. The aim was to secure a permanent seat on the United Nations Security Council but other permanent members, notably Britain and France, argued that Japan could not expect to join the elite group with veto powers of peace-keeping and other UN operations unless it was

prepared to put its own troops into war zones. This led to fierce debate between those who wanted to amend the constitution and unshackle the armed forces, and those who demanded that Japan should remain pacifist and accept its lesser world role.

The postwar constitution adopted ideas from both the American congressional system and Britain's parliamentary system. But for nearly four decades, from 1955 onwards, it produced not a Western two-party system but one-party rule by the conservative Liberal Democrats. Their governments and the electoral system became increasingly unrepresentative and finally came undone in July 1993 when voters, tired of years of corruption scandals, voted a centrist coalition to power.

The old, moribund political system began to disintegrate, with new parties and splinter groups repeatedly forming, splitting apart and forming in new configurations. In early 1994, Prime Minister Morihiro Hosokawa forced electoral reforms through a reluctant parliament, introducing a measure of proportional representation.

The move was intended to bring two benefits: first, votes would be redistributed to reflect the population's postwar drift to the cities; and second, it would create a system in which parliamentary candidates would be less likely to buy votes.

By shattering the assumptions on which so many politicians had built their careers, the reforms forced MPs to re-think strategy and alliances. Japan appeared set for political turmoil for several years into the future.

COMPANY LOYALTY

Japan's remarkable post-war economic performance can be attributed to many factors, first among which is the quality of its workforce – more reliable, conscientious and better educated than those of most competitors. Although education is compulsory only from six to 15 years of age, more than 90 per cent of children finish high school to sixth-form standard.

A third of them go to university, where many obtain degrees in subjects immediately relevant to industry's needs. Training continues throughout a worker's career and university graduates destined to manage in industry must first perform even the most basic jobs.

Workers are given responsibility for the quality of their work and take pride in what they do. Large firms, which have traditionally offered lifetime employment, try hard, even during deep recession, not to lay off workers. However, as the economic slump of the early 1990s continued, many employees even of top companies were forced into early retirement.

There remains far greater loyalty between employers and employees in Japan than in most other countries and there is often little friction when companies move staff from one job to another. But gradually, rising disenchantment with their work-dominated lives has prompted more 'salarymen' to refuse to live away from their families.

Each factory or office is built up from small 'sections', which act both as production units and as a focus of social activity – a vital func-

tion in a culture where 'belonging' is important. Loyalty with sections extends outwards to the company as a whole and finds expression in collective holidays and excursions – even in singing the company anthem.

Investment in new industrial technology has been stimulated by high levels of personal savings, low company dividends and a willingness by all to consider long-term gains at the expense of short-term performance. The government, through its powerful Ministry of International Trade and Industry, guides industry, with mixed results, into business sectors which it thinks important.

Although the Japanese have recently tried to improve their ability to develop technology from fundamental research, they remain reliant on foreign, chiefly American, inventiveness in several areas, particularly communications. Some 80 per cent of the computer software used inside Japan is pirated, much of it from packages written in the US.

Japan is bereft of most industrial raw materials – a little copper, iron ore, lead, zinc, natural gas and sulphur are extracted. Being the second largest oil consumer in the world, the country has to import oil, as well as coking coal and metallic ores worth several billion US dollars a year, and so benefited greatly from low commodity prices during the 1960s. Economic growth was halted briefly in 1974 after the rise in world oil prices, but Japan adapted well to an era of expensive energy.

Energy efficiency was improved and the sources of supply were diversified to include nuclear power. Emphasis shifted from heavy industries such as steel and shipbuilding to new high-technology sectors in which the qualities of its workforce gave Japan an advantage over competitors. The switch also brought improved processes to older industries, leading to a cut in environmental pollution and a productivity boost that has helped to protect industries from Third World rivals.

However, its economic success and its burgeoning current account surpluses have pushed its currency, the yen, so high that many of its products can no longer compete in world markets. The high yen did much to perpetuate the 1990s recession. It also contributed to what is almost certainly a permanent change in the character of the economy. It pushed Japan finally into the elite club of the world's fully developed economies.

This means that it will probably never again be able to stimulate growth by investing yet more deeply in capital equipment, boosting production and finding new markets abroad. In future, Japanese economic cycles will depend on domestic consumer demand. This is inextricably linked to demands for more representative government. In the mid-1990s, economics made the consumer, the general public, a much more powerful force; it is to this that politicians are scrambling to adjust.

COUNTRYSIDE AND TOWN

Fewer than 5 million households remain on the land and two-thirds of these earn more from part-time work in factories and offices than from the fields. The average farm of 1.4 hectares (3.5 acres) may be uneconomic, but tax concessions and high rice prices (about five times the world market price) ensure a degree of rural prosperity. The Japanese are gradually eating less rice, as incomes have risen and preferences have shifted to alternative foods. Meat, like bread, is a relatively new commodity in Japan – the first time a Tokyo restaurant ever put a meat dish on its menu was in 1869. Today meat is a major import.

Few Japanese farmers have switched from rice growing to animal rearing or vegetable growing, despite government urging. On high ground, rice gives way to fruit and vegetables such as apples, oranges, cabbages, potatoes, radishes, and mulberries for silk production.

Fishing communities have always given Japan most of its protein foods. Large ports send ships thousands of kilometres across open sea to fish off the coastal waters of such countries as Canada and Peru. There are also fish farms producing oysters and edible seaweeds as well as fish. Japan has traditionally been a major whaling nation. It accepted the international whaling ban imposed in 1986 but has fought to have it lifted for minke whales, which it claims are not in danger of extinction.

In the city, breakfast is as likely to be bacon and eggs as it is to be a bowl of rice and fish. White-collar workers wear suits and cram into commuter trains from the suburbs. Offices are usually open-plan. Electronic word processors have eased the problems of handling Japanese writing, with its thousands of characters. Like other developed countries, Japan has a rapidly expanding service sector.

For the factory hand, conditions relate closely to the size of the company. Workers in large, automated assembly plants often spend the day tending robots that do all the hard labour. But those working for small subcontractors turn out components in tiny, cramped workshops, earning less and enjoying less job security. A substantial number of these are Japan's so-called burakumin (village people) who are descended from the outcastes who did the dirty work of tanning, undertaking and sweeping in previous centuries.

There are few career opportunities for women. Tradition holds that they should retire and look after the home as soon as they marry. Many ambitious women tell of bosses putting them under pressure if they do not conform to the national stereotype. Others tell of virtual ostracism by neighbours if they do not stay at home doing their 'wifely duty'.

In these circumstances many women pour all their energies into their children's education. These 'education mamas' can be merciless in their efforts to ensure that vital examinations are passed, for doing well is still virtually the only passport to a successful future. Many children go to at least one cramming school in addition to daytime school. Failure to pass an entrance examination may bring disgrace upon the child's family.

A typical Japanese home is small by European or American standards but they are getting bigger as incomes rise. Building regulations prevent many tall apartment blocks being built in cities and this contributes to the crush on commuter trains. Even in modern homes, at least one room is likely to be in

traditional style, with reed *tatami* mats on the floor and a small alcove where a favourite painting or calligraphy scroll can be hung. Here, the ancient tradition of tea ceremonies continues. It is this room that keeps the household in contact with its cultural roots.

The typical Japanese is both a Shintoist and a Buddhist. Christians account for 1.4 per cent of the population. Few Japanese are ardently religious – codes of behaviour derive from social mores rather than doctrine or scripture. Nevertheless, religious festivals are popular. Shrines and temples are kept in perfect order. Austere Zen gardens are a particular attraction.

AFTER A DAY'S WORK

Despite the scarcity of free time, the Japanese have a wide variety of leisure pursuits. An office worker returning home might, for example, pop into the public bath house for a scalding dip, or go to a Pachinko parlour to immerse himself in pinball, a national obsession. Soccer, baseball and sumo wrestling are all wildly popular spectator sports. Golf courses are regarded as close to heaven and, though they are prohibitively expensive for all but the very rich, people can be seen practising their swings on the pavement, and even on rooftops.

Instead of returning home, a 'salaryman' may spend an entire evening in town with colleagues, exploring the 'pleasure quarter', a noisy and garish maze of narrow streets lined with bars, sex shows, burlesques and tiny restaurants. Here it is possible to let off steam, secure in the knowledge that whatever happens all will be forgiven in the morning.

In the main shopping arcades, amid fashion boutiques and expensive department stores, coffee bars with exotic Western names offer relaxation for tired office workers, or a rendezvous for young lovers – important now that fewer marriages are arranged. A few people drop out of this conformist society. Juvenile

▲ SHORT BREAK City-dwellers relax beside a bathing pool in Tokyo. The average Japanese worker takes only 101 days off a year, including weekends, compared with a European's 140.

crime, though slight by Western standards is rising. Other people – up to 3 million burakumin and 700 000 ethnic Koreans – suffer adverse discrimination.

JAPAN AT A GLANCE
Area 372 313 km² (143 750 sq miles)
Population 124 330 000 **Per km²** 334 **(Per sq mile)** 865)
Capital Tokyo
Government Constitutional monarchy
Currency Yen
Language Japanese
Religions Shintoism and Buddhism (94%), Christianity (1.4%)
Climate Temperate; in the north, winters bring cold north-western winds, subtropical in south. Average temperature in Tokyo ranges from 3.7°C (38.7°F) in January to 26.4°C (79.5°F) in August
Land use Forest 66%, cultivation 12%, meadows and pastures 11%, other 11%
Minerals Coal, iron ore, copper, zinc, lead, sulphur, natural gas
Primary products Rice, potatoes, sweet potatoes, tea, fruit, vegetables, edible seaweeds, timber, poultry, fish
Industries Iron and steel, nonferrous metals, ships, cars, motorcycles, consumer electronics, textiles, petrochemicals, fertilisers, cement, optical equipment, watches, food processing
Exports Cars. consumer electronics, industrial and construction machinery, chemicals, textiles
Annual income per head US$ 27 826
Population growth (per thous/yr) 2.4
Life expectancy (yrs) Male 76 **Female** 82

Jefferson City *USA* State capital of Missouri, on the banks of the Missouri River, about 170 km (105 miles) west of St Louis. The tourist drawcard, the Lake of the Ozarks, lies to the south-west. Jefferson City is the seat of Lincoln University.

Population	35 480
Map	United States – Hc

Jejudo *South Korea* See CHEJU DO

Jena *Germany* Town 80 km (50 miles) south-west of LEIPZIG and site of a battle in which the French under Emperor Napoleon I decisively defeated Prussia in 1806. The philosopher Georg Hegel (1770-1831) and the poet and dramatist Friederich von Schiller (1744-1805) were on the staff of the town's university (founded in 1557); political theorist Karl Marx was a student here. In 1846, the optician Carl Zeiss opened a workshop here, starting what became Europe's biggest optical instruments industry. Precision instruments, optical products, cameras, chemicals and pharmaceuticals continue that tradition today.

Population	105 500
Map	Germany – Dc

Jenolan Caves *Australia* Extensive limestone cave system in the Great Dividing Range, 115 km (72 miles) west of SYDNEY. Open to visitors, the caves have been known to Europeans since the 1840s when a settler found them while hunting a bushranger who used them as a hide-out.

Map	Australia – Ie

Jeonju *South Korea* See CH'ONGJU

Jerba *Tunisia* Island in the Gulf of Gabès, 67 km (42 miles) south of SFAX and connected to the mainland by a causeway. It is said to be the mythical land of the Lotus Eaters visited by Ulysses. The island covers 514 km² (198 sq miles). Its resort area, which is served by an airport, has fine beaches, good hotels, colourful souks and some Roman remains, including two forts and a triumphal arch. Dates, olives and oranges thrive in the fertile soil; wine is also produced here. The growth of tourism has encouraged a revival of traditional crafts, such as weaving and pottery.

Population	92 300
Map	Tunisia – Ba

Jerez de la Frontera *Spain* Town situated just north of Cádiz, on the south-west Atlantic coast, from whose name 'Jerez' the word 'sherry' derives. It produces and stores the wine, liqueurs and brandies in numerous *bodegas* (cellars), and makes bottles and casks.

Population	183 310
Map	Spain – Bd

Jericho (Yeriho; Ariha) *West Bank* Oasis town situated 22 km (14 miles) north-east of Jerusalem. At 251 m (825 ft) below sea level, it is the world's lowest city. It is also the site of one of the world's oldest cities, dating back to about 7800 BC. Jericho came under the control of the Palestinian Liberation Organisation in 1993 as the first step in the agreement that Israel would hand back the entire occupied West Bank, which it had occupied since 1967, to the PLO.

Population	15 000
Map	Jordan – Ab

Jersey *English Channel* See CHANNEL ISLANDS

Jerusalem (Yerushalayim; El Quds esh Sherif) *Israel/Jordan* Capital of Jerusalem district, and capital of Israel since 1950 (though not accepted in international law), lying 30 km (20 miles) west of the northern end of the DEAD SEA. A holy city for Jews, Christians and Muslims, it was divided between Israel and Jordan after 1948. Israel took control after the Six Day War (1967). Prior to this, Jordan held the eastern part ('the Old City'). Jerusalem's Arabic name is El Quds esh Sherif.

In 1980 the Israeli government declared the city, including the eastern section, the 'eternal capital of Israel'; however, this decision was partially reversed in 1994 when Israel recognised Jordanian sovereignty over East Jerusalem. The controversy surrounding the future of East Jerusalem is not over however, as the PLO has also laid claim to this part of the city.

Conquered from the Philistines by David in 1000 BC, Jerusalem is the site of Solomon's First Temple, destroyed by the Babylonian king Nebuchadnezzar in about 586 BC, and of the Second Temple, destroyed by the Roman emperor Titus in AD 70. The city's Arab captors built the Dome of the Rock (Mosque of Omar) on the site of the First and Second temples in AD 632. Jerusalem was taken by the Crusaders in 1099, then recaptured by the Muslim leader Saladin in 1187. Ottoman Sultan Suleiman built the existing wall of the Old City in the 16th century.

Inside the Old City is the Western (Wailing) Wall. Here are also the El-Aqsa Mosque, the Via Dolorosa (Sorrowful Way, or Way of the Cross), and the Franciscan Church of the Holy Sepulchre. The Old City includes Jewish, Christian, Muslim and Armenian quarters, and a colourful market.

Just outside the Old City wall, to the south-west, is Mount Zion, with the traditional tomb of David, and the Room of the Last Supper. To the north-west is Mea Shearim (the ultra-orthodox Jewish quarter), and the tombs of the kings of Judah, where King David is now believed to be buried. The Jewish quarter has been almost entirely rebuilt since 1967. To the east of the Old City is the Mount of Olives, with its ancient Jewish cemetery, and the site of the Garden of Gethsemane, where Judas betrayed Jesus.

At En Karem, on the outskirts of Jerusalem, are John the Baptist's birthplace and the new Hadassah Medical Centre, remarkable for the stained-glass windows, designed by the Russian-born Jewish painter Marc Chagall (1887-1985), of its synagogue.

Among the city's many museums are the Israel Museum, the national museum with the Shrine of the Book, which houses the Dead Sea scrolls, and Yad Vashem, the memorial to Jewish victims of the Nazi Holocaust. Public buildings in the modern part of the city include the Knesset (parliament), the president's residence, the Hebrew University, and the Jewish Agency (World Zionist Organisation) complex.

Population	542 500
Map	Israel – Bb

Jervis Bay Territory *Australia* An area of 70 km² (27 sq mile) at Jervis Bay on the New South Wales coast. It lies 195 km (121 miles) south of Sydney and was given to the Commonwealth government in 1915 as a separate area of the Australian Capital Territory, to allow access to the sea. Four-fifths of Jervis Bay Territory is occupied by a national park.

Population	800

Jessore *Bangladesh* Chief town of Jessore district, which covers 6573 km² (2538 sq miles) in the west. Jute, rice, tamarind, tobacco and sugar cane are grown. The indigo dye industry was based in the town before being moved to India at the time of Pakistan's partition in 1947. As one of the principal garrison bases of the Pakistani army and situated only 20 km (12 miles) from the Indian border, it was one of the first targets of the Indian army during the war of 1970-1.

Population	(town) 160 000; (district) 4 020 000
Map	Bangladesh – Bc

jet Dense black lignite, capable of taking a brilliant polish, used in jewellery.

jet stream Fast-moving air current at high altitude. It is usually several thousand kilometres long, several hundred kilometres wide, and only a few deep. Jet streams occur at the boundary level (tropopause) between the troposphere and the stratosphere (see THE DIFFERENT LAYERS OF THE ATMOSPHERE p53). With winds of more than 160 km/h (100 mph), they are sometimes used by easterly flying aircraft to conserve fuel.

Jezzine *Lebanon* Once a popular resort, this small stone-built town perches almost 1000 m (3280 ft) up in the southern Lebanon Mountains, about 16 km (10 miles) inland from the port of SIDON. Magnificent walks amid splendid mountain scenery once drew tourists, and the town also had a reputation for making fine silver plate cutlery. But now, with Jezzine at the centre of a Christian enclave in the continuing Lebanese civil war, and with its modest population swollen by thousands of refugees and militiamen, the tourists stay away.

Map	Lebanon – Ab

Jhansi *India* City with large railway works, 420 km (261 miles) south-east of Delhi. Its early 17th-century Mogul fort, still standing though modernised in the 19th century, was the scene of a massacre in 1857, when the Sepoys (Indian soldiers under British command) rebelled. The British regained the fort after fierce fighting and exchanged it for Gwalior in 1966.

Population	284 100
Map	India – Cc

Jhelum *India/Pakistan* One of the five main Punjabi rivers, flowing some 720 km (450 miles) through the Vale of Kashmir and the northern Punjab to join the Chenab River about 200 km (125 miles) west of LAHORE. It is dammed for irrigation at Mangla.

Map	Pakistan – Da

Jiangsu (Kiangsu) *China* One of the country's richest provinces, covering 107 300 km² (41 419 sq miles) of low-lying land on the northern side of the fertile Chang Jiang (Yangtze) delta. It is studded with lakes and crisscrossed by canals and waterways. Jiangsu is only about half the size of Britain, yet its population by far exceeds that of the UK. With a population density of 564 people per km² (21 per sq mile), it is one of the most densely populated regions of China.

It is a major rice and cotton-growing area, but the most striking feature of its recent economic development has been the growth of its small rural industries, producing mainly textiles and agricultural machinery and processing foods.

These and established urban industries, have made it China's second largest industrial hub after SHANGHAI. Jiangsu's capital is NANJING.

Population 69 110 000
Map China – Hd

Jiangxi *China* Province in the south, covering 164 000 km² (63 306 sq miles) and centred on the basin of the Gan Jiang River. Its capital is NAN-CHANG. To the north, Jiangxi opens onto the valley of the Chang Jiang (Yangtze) River and Poyang Hu, China's largest lake.

Jiangxi's climate is warm and moist and its fertile soils are used to grow rice, tea and cotton. Coal is mined at Pingxiang and clay at Gaoling mountain near JINGDEZHEN.

Population 39 130 000
Map China – He

Jičin *Czech Republic* Bohemian market town, founded in 1302, 76 km (47 miles) north-east of PRAGUE. Jičin and Turnov, to the north-west, are the main tourist centres for an area called 'the Bohemian Paradise' (Český Ráj), with pictur-esque wooded hills and outcrops of basalt and sandstone eroded into fantastic shapes.

Population 17 000
Map Czech Republic – Ba

Jih-yüeh T'an *Taiwan* See SUN MOON LAKE

Jilin (Kirin) *China* Province covering some 187 000 km² (72 184 sq miles) of central Dong-bei (Manchuria). Its climate is extreme – winter temperatures fall far below freezing point while summers are warm. In contrast with the neigh-bouring provinces of Heilongjiang and Liaoning, mineral resources are scant in Jilin. Agriculture is the main occupation – the province is China's leading producer of soya beans. Its capital is CHANGCHUN.

Population 25 320 000
Map China – Ib

Jima *Ethopia* Market town about 230 km (145 miles) south-west of the capital, Addis Ababa. It is the centre of a coffee-growing area.

Population 70 000
Map Ethiopia – Ab

Jinan (Tsinan; Litsheng) *China* Capital of SHANDONG province, about 360 km (225 miles) south of Beijing. Its famous silk industry is sup-plemented today by food processing, petrochem-ical and heavy industries.

Population 3 376 000
Map China – Hd

Jingdezhen *China* City famous for ceramics. Situated in Jiangxi province, it is the country's leading centre for the manufacture of porcelain. China clay comes from the Gaoling Mountain, from which the English word 'kaolin' is derived.

Population 581 000
Map China – He

Jinhae *South Korea* See CHINHAE

Jinja *Uganda* Town on the Nile, beside the north-ern shore of Lake VICTORIA, 80 km (50 miles) east of the capital, Kampala. It was a cotton centre in the early 20th century, and became the country's second largest town, after Kampala, in the 1950s with the building of the Owen Falls Dam just to the north-west. The dam was East Africa's first major source of hydroelectricity. Today, textiles and light engineering products are manufactured by Jinja's factories.

Population 100 000
Map Uganda – Ba

Jiujiang *China* Ancient city founded by the Han dynasty (206 BC-AD 220) on the south bank of the Chang Jiang River. It is the site of the beauti-ful ancient Yanshui Pavilion, built on an island in the middle of Lake Gantang.

Population 390 000
Map China – He

Joal-Fadiout *Senegal* Tourist resort formed from two settlements 115 km (70 miles) south-east of DAKAR. The ancient port of Joal, first visited by Europeans in the 15th century, was the birthplace of Léopold Sédar Senghor, the country's first president (from 1960-80). Fadiout village stands on a large bank of shells built up by the sea.

Map Senegal – Ab

João Belo *Mozambique* See XAI-XAI

Joáo Pessoa *Brazil* Capital of Paraíba State, university city and seaside resort about 100 km (62 miles) north of the city of RECIFE. Named after a governor of the state who was assassinated in 1930, the city lies close to the eastern end of the Trans-Amazonian Highway. Its main industries revolve around textiles, plastics and paper. The city has several churches dating from the 16th cen-tury. Nearby are a number of beautiful beaches lined with coconut palms. The beach of Tambaú sweeps round to the cliffs of Cabo Branco – the easternmost point of South America.

Population 460 000
Map Brazil – Eb

Jodhpur *India* Walled city on the edge of the THAR DESERT, 604 km (375 miles) south-west of Delhi. Formerly the capital of the princely state of Jodhpur, it was founded by Rao Jodha, its first maharajah, in 1459.

The fort, built on rocky cliffs 120 m (394 ft) above the city, is reached by a zigzag ascent through numerous gateways. At the last gate are the handprints of 15 Hindu maharanis who per-formed suttee – burning themselves to death on the funeral pyres of their dead husbands.

The city is a centre for equestrian sports, and the close-fitting riding breeches called jodhpurs take their name from here.

Population 648 600
Map India – Bb

Jogjakarta *Indonesia* See YOGYAKARTA

Johannesburg *South Africa* Largest city in the country, commercial centre, and capital of GAUTENG Province. It is known to most Africans as 'eGoli' – 'Place of Gold'. It encloses the spot where gold of the Main Reef was found in 1886 and contains many of the early deep-level mines.

Huge dumps of spoils alongside the city centre have been landscaped and planted with vegeta-tion. Johannesburg is also a centre for the manu-facture of chemicals, foodstuffs, pharmaceuticals, textiles, metal goods and machinery. The city has many museums and open spaces. It is home to the University of the Witwatersrand and the Rand Afrikaans University. In 1994 Johannesburg and its vast peripheral towns, including Soweto, Alexandra and Lenasia, were earmarked for incorporation into a single metropolis. Their establishment as separate 'townships' was the result of laws that required Africans and Asians to live in segregated areas.

Population (metropolitan) 5 336 000
Map South Africa – Cb

John o'Groats *United Kingdom* Small Scottish settlement in the north-east, traditionally re-garded as the northernmost point of the British mainland, although it is about 4 km (2.5 miles) farther south than Dunnet Head, 20 km (12 miles) to the west.

Map United Kingdom – Da

Johnston Atoll *Pacific Ocean* Atoll 1430 km (888 miles) south-west of Honolulu, HAWAII. It became a US naval base in 1934. It was used as a base for nuclear weapons tests until 1962 and is now used for the disposal of chemical weapons waste. It now falls under the US Defense Nuclear Agency and the Department of the Interior.

Population (mainly government) 1200
Map Pacific Ocean – Eb

Johor (Johore) *Malaysia* Most southerly state of peninsular Malaysia, extending over 18 986 km² (7330 sq miles). It was an independent sultanate until European colonisation in the early 19th century. Palm oil, rubber and pineapples are the region's chief products.

Population 2 106 500
Map Malaysia – Bb

Johor Baharu *Malaysia* Capital of Johor State situated on JOHOR STRAIT. It is rapidly developing as a port, exporting palm oil from new plantations to the north. The older part of the town has a palace of the sultans of Johor. The ruins of the ancient capital of the sultanate lie 30 km (19 miles) north at Johor Lama.

Johor Baharu has benefited from the close proximity of Singapore just across the strait, whose trade and industries have spilled over, turning it into a thriving industrial centre and boosting its exports.

Population 249 900
Map Malaysia – Bb

Johor Strait (Tebrau Strait) Narrow arm of the sea between the southern tip of the Malay Peninsula and Singapore Island. It is 48 km (30 miles) long and 1-5 km (0.6-3 miles) wide, and is crossed by a road and rail causeway.

Map Singapore – Aa

joint Fracture or crack in a rock mass along which no displacement has occurred. Joints are caused by rocks contracting on cooling or drying out, or by folding.

Jokkmokk *Sweden* Small settlement on the Arctic Circle, 825 km (515 miles) north of STOCKHOLM. It is the centre of an area of 18 143 km² (7003 sq miles) and serves a large community. Every Feb-ruary, Lapps (or Saame, as they sometimes refer to themselves) come here to trade at its open market, which attracts thousands of tourists. A centre of Lapp culture, Jokkmokk contains the Lapp People's College and the Lappland Museum.

Population 6800
Map Sweden – Cb

Jolo *Philippines* Town and administrative centre for the SULU ARCHIPELAGO, on the north coast of the agriculturally rich Jolo Island. A walled Islamic stronghold, the town is a port of entry and a busy fishing and pearling centre, although much of this activity has been disrupted in recent years by the activities of Muslim nationalists.

Jolo Island, which covers 894 km² (345 sq miles), produces manioc and a wide range of fruits. It contains many extinct volcanoes, the highest of which is Mount Tumatangas, at 812 m (2664 ft).
Population 52 500
Map Philippines – Bd

Jonglei Canal *Sudan* Canal, about 360 km (223 miles) long, that cuts across a loop of the White Nile about 800 km (500 miles) south of the capital, Khartoum. It was being built to reduce evaporation and to conserve the waters of the river from loss in the SUDD swamps. Work started in 1978, but has been disrupted by the war in southern Sudan. The project has been criticised by environmentalists who are concerned about its effects on the ecology of the swamps.
Map Sudan – Bc

Jönköping *Sweden* Industrial city, dating from the 13th century, at the south end of Lake VÄTTERN, about 130 km (80 miles) east of GOTHENBURG. The mechanised production of safety matches started here in 1845; a match museum traces this aspect of the city's history. Jönköping also produces sewing machines, bicycles, chemicals, and footwear.
Population 112 800
Map Sweden – Bd

Jordan See p 351

Jordan *Middle East* The longest and largest river of both Israel and Jordan. It rises on the slopes of Mount HERMON and flows southwards through the Sea of Galilee (Lake Tiberias) into the Dead Sea. It is 330 km (205 miles) long and is used for hydroelectric power and for irrigation. The section between the Sea of Galilee and the Dead Sea forms the Jordan-Israeli border.
Map Israel – Ba

Jordan Valley *Jordan/Israel* Fertile valley in which wheat and barley have been grown since 8000 BC. Watered by the River JORDAN and the EL GHOR CANAL, it is 100 km (62 miles) long and contains more than 30 farming villages on the Jordanian side of the river.

Here, agricultural development has been greatly aided by the building of the King Talal Dam on the ZARQA River. Grains, grapes and olives, as well as tomatoes and bananas are grown in the valley. Some aqua-farming is also carried out in its dams and reservoirs.
Map Israel – Ba; Jordan – Aa

Jos *Nigeria* Capital of Plateau state, on the JOS PLATEAU, about 710 km (440 miles) north-east of LAGOS. It was founded by British colonists at the beginning of the 20th century, and developed as the centre of the country's tin-mining industry. It has also become a popular hill resort. Fine sculptured heads and terracotta figures from the civilisation that flourished at nearby Nok from 500 BC to AD 200 are exhibited in its museum.
Population 145 400
Map Nigeria – Bb

Jos Plateau *Nigeria* Upland covering some 7780 km² (3003 sq miles) in the centre of the country, with an average height of about 1260 m (4135 ft). It is littered with huge granite outcrops, extinct volcanoes and waterfalls such as the Assob and Kuru falls on the steep plateau edge, about 50 km (30 miles) south of the town of JOS. The striking scenery makes the plateau a leading resort area. It is also a source of tin and columbite.
Map Nigeria – Ba-Bb

Joshin-etsu Kogen *Japan* National park, covering 1889 km² (729 sq miles), just north-west of the town of NAGANO in central HONSHU. Its alpine landscape rises to 2446 m (8025 ft) at Mount Myoko. The park is a popular summer resort for the inhabitants of Tokyo, about 155 km (95 miles) to the south-west.
Map Japan – Cc

Joshua Tree National Monument *USA* Area of 2266 km² (875 sq miles) in California, about 160 km (100 miles) east of Los Angeles. It contains Joshua trees (tree-like species of yucca), cacti and other desert plants, and strikingly eroded granite outcrops.
Map United States – Cd

Jostedalsbreen *Norway* Glacier on mainland Europe, covering 486 km² (188 sq miles). It lies just north of Sognefjord, about 150 km (93 miles) north-east of BERGEN.
Map Norway – Bc

Jotunheimen *Norway* Mountain region with an impressive cluster of peaks in south-central Norway. Its name means 'the home of the giants'. Glittertind, at 2470 m (8104 ft), and Galdhøpigga, at 2469 m (8100 ft), are the two highest peaks in the region.
Map Norway – Cc

Juan de Fuca Strait *North America* United States-Canadian sea passage, between VANCOUVER ISLAND and the WASHINGTON and BRITISH COLUMBIA mainland, giving SEATTLE, VICTORIA and VANCOUVER access to the PACIFIC OCEAN. It was reputedly named after a Greek navigator who used the name Juan de Fuca while serving in the Spanish navy in 1592. The waterway is 160 km (100 miles) long and 18-27 km (11-17 miles) wide.
Map Canada – Cd

Juan Fernández Islands *Pacific Ocean* Pacific archipelago named after the Spaniard who landed here in 1574. The islands, whose main industry is lobster fishing, belong to Chile. The main islands – Más Afuera (or Alejandro Selkirk), Más á Tierra (or Robinson Crusoe), and Santa Clara (or Goat Island) – lie 650 km (400 miles) west of the Chilean port of VALPARAÍSO. The English writer Daniel Defoe (1660-1731) based his novel *Robinson Crusoe* on the real-life adventures of a Scottish seaman, Alexander Selkirk, who lived on Más á Tierra in 1704-9.
Population 1000
Map Pacific Ocean – Id

Juba *Sudan* Market town on the Upper Nile, about 1200 km (720 miles) south of the capital, KHARTOUM. It is the capital of Sudan's autonomous southern region, which stretches along the national borders of Kenya, Uganda and Zaire. The townspeople are mainly African and non-

Muslim, unlike the Arab people of the north. Most hold to indigenous beliefs although there are some Christians.
Population 83 800
Map Sudan – Bc

Jubail *Saudi Arabia* Port and growing industrial complex on The Gulf, situated 85 km (53 miles) north of DAMMAM. A former fishing village, Jubail began expanding in the late 1970s, when the government started work on the complex. Today it has oil refineries, petrochemical industries, a steel mill, fertiliser plant, an aluminium smelter and other factories.
Population 50 000
Map Saudi Arabia – Bb

Jubba (Giuba) *Ethiopia/Somalia* River rising in the Ethiopian highlands and flowing southwards for 1200 km (750 miles). It reaches the Indian Ocean near the southern Somali port of KISMAAYO. Its year-round water is used to irrigate farms along its valley.
Map Somalia – Ab

Júcar *Spain* River, 505 km (314 miles) long, entering the Mediterranean Sea at Cullera just south of the port of VALENCIA.
Map Spain – Ec

Judaea *Middle East* The southern part of ancient Palestine, that was ruled at various times by Persians, Greeks and Romans. Stretching from the Mediterranean to the River Jordan and the Dead Sea, it was bordered in ancient times by Sinai to the south and Samaria in the north – an area divided today between Israel and the WEST BANK. The highest point is Mount HEBRON, at 1013 m (3323 ft), about 30 km (19 miles) south-west of Jerusalem.
Map Israel – Ab

Jujuy *Argentina* See SAN SALVADOR DE JUJUY

Julian Alps (Alpi Giulie; Julijske Alpe) *Slovenia* Mountain range in the north-west, forming the Italian-Slovenian border. West of it lies the Tagliamento Valley; to the north and east, the Sara forms a natural boundary. The range has 250 peaks exceeding 2000 m (about 6560 ft), including Triglav, Slovenia's highest mountain.

It is crossed by the forested valleys of two main rivers – the upper Soca, and the Trenta, which cradles Lake BLED and the wilder Lake Bohinj to the west. A major skiing area, its resorts include Bled and Bohinjska Bistrica to the west, and Mount Vogel above Lake Bohinj. Triple-peaked Triglav (meaning 'the three-headed') rises to 2864 m (9396 ft) above the Triglav National Park. Established in 1924, the park protects 780 km² (301 sq miles) of forests and pastures noted for wild flowers, and seven blue lakes.
Map Slovenia – Aa-Ba

Jumla *Nepal* Town in the west of the country, about 350 km (215 miles) north-west of KATHMANDU. Until the 18th century, Jumla was the capital of Nepal. The town leads to some of the best mountain forest trekking areas in Nepal and to some of the finest Himalaya scenery.
Population 10 000
Map Nepal – Aa

Jumna *India* See YAMUNA

Jordan

AN OASIS OF STABILITY WITHIN THE TROUBLED MIDDLE EAST, JORDAN HAS BECOME THE MIDDLE-MAN IN THE CONFLICTS OF ITS NEIGHBOURS

The development of modern Jordan has been dominated by one man for more than 40 years – its ruler King Hussein. His courage has seen the country through constant political crises. Jordan remains a frontline state in relation to its neighbour Israel and among the Arab countries no other has shown a greater interest in negotiating a lasting settlement to the Arab-Israeli conflict. In 1994 it seemed that these efforts would finally bear fruit.

The area was long ruled by the Ottoman Turks. The emirate of Transjordan, created in 1923, remained under British protection until 1946 when it achieved full independence as the Kingdom of Jordan. Almost immediately, the new state became embroiled in the 1948-9 Arab war against Israel. In the armistice that followed, the West Bank (of the River Jordan), which had been occupied by Palestinian Arabs for centuries, was annexed by Jordan. At the same time, Jordan received Palestinian refugees fleeing from Israel. Later, in the Six Day War of 1967, Jordan lost the West Bank to Israel and gained even more refugees.

King Hussein allowed refugee militants from the Palestine Liberation Organisation (PLO) to raid Israel from bases in Jordan. However, the guerrillas defied the government's authority, and Hussein's army acted against them in 1971. The bitter conflict that ensued ended with the expulsion or imprisonment of the guerrillas until Hussein declared a general amnesty for all political prisoners and exiles in 1973. The following year, Jordan acknowledged the PLO as the sole legitimate representative of the Palestinians. The kingdom recognised the PLO's right to the Israeli-occupied West Bank and in 1988 King Hussein formally ceded the territory to the PLO.

Many Palestinians continued to live in Jordan's 10 UN refugee camps. A March 1993 estimate put the number at 1 057 300 compared with only 476 300 living in the West Bank. All have been granted Jordanian citizenship and take a full part in the political and economic life of the kingdom. Recently, nationalist sentiment surfaced urging that Palestinians be encouraged to 'go home', in sharp contrast to the king's policy of national reconciliation.

The rich Arab states, such as Kuwait and Saudi Arabia, gave substantial economic aid to strengthen Jordan as an Arab country bordering Israel. Unfortunately King Hussein, and the Palestinians, chose to back Saddam Hussein of Iraq during the Gulf War, thus earning the ire of the Gulf states who were directly threatened. On the other hand, the continuing civil war in Lebanon allowed Jordan's capital AMMAN to replace the Lebanese capital, BEIRUT, as the financial centre of the region. Jordanians have a reasonably high standard of living and 80 per cent of adults are literate. Although they have no oil, they refine it and have an efficient system of agriculture.

Only one-fifth of Jordan is fertile enough to be farmed; the rest is desert. The fertile areas include the disputed West Bank and the valley of the Jordan. To the east lie the hills and mountains of Biblical history which are also fertile. All these areas have cool winters with moderate rainfall.

▲ ROCK CITY The city of Petra, carved in the red sandstone walls of a desert gorge, prospered on a caravan route for more than 1000 years from 300 BC.

Though Jordan may continue to suffer as an unfortunate middle-man in Israeli-Arab conflicts, it is the one state that has tried hardest in its efforts of genuine rapprochment with Israel and the West. In 1994, Israel acknowledged Jordan's sovereignty over the West Bank and its role in East Jerusalem, to the alarm of the PLO. It may be hoped that while the Gulf states may have reduced their aid to this small kingdom, its goodwill will be rewarded by Israel and the West. As one of the most advanced Middle Eastern states, it stands to benefit most from any assistance, technical or financial.

JORDAN AT A GLANCE	
Official name Hashemite Kingdom of Jordan	
Area 97 740 km² (37 730 sq miles), including about 5880 km² (2270 sq miles) on West Bank	
Population (East Bank) 3 823 636; (West Bank) 1 404 114 **Per km²** 53 **(Per sq mile** 138)	
Capital Amman	
Government Constitutional monarchy	
Currency 1 Jordanian dinar = 1000 fils	
Languages Arabic; English widely understood among upper and middle classes	
Religions (East Bank) Sunni Muslim (92%), Christian (8%)	
Climate Hot and dry; cool in winter; only the west receives winter rain. Average temperatures in Amman: 8°C (46°F) in January to 25°C (77°F) in August	
Land use Cultivation 4.5%, grazing 1%, forest and woodland 0.5%, other 94%	
Mineral resources Phosphates, potash, shale oil	
Main primary products Vegetables, fruits, olives, wheat; phosphates, potash	
Major industries Agriculture, mining, fertilisers, chemicals, cement, natural gas	
Main exports Phosphates, chemicals; fruit and vegetables, tobacco; potash, cement	
Annual income per head (US$) 1120	
Population growth per (thous/yr) 35	
Life expectancy (yrs) Male 70 **Female** 74	

Juneau *USA* Capital of Alaska, situated on the Gastineau Peninsula in the south-eastern panhandle of the state, roughly 400 km (250 miles) north-west of the Alaska's southern boundary. It was a gold rush town in the 1880s and continued yielding gold until the 1940s.

Today it is a port and fishing centre linked to the rest of the state only by sea and air. Because of its inaccessibility it was decided to move the capital, and a site for a new one at Willow, north of Anchorage, was chosen in 1974. However, for now, Juneau will remain Alaska's capital.
Population 26 800
Map Alaska – Dc

Jungfrau *Switzerland* Magnificent mountain in the Bernese Oberland, 60 km (37 miles) south-east of the capital, Berne. It is 4158 m (13 642 ft) high. It was first climbed in 1811 by two Swiss brothers, Rudolf and Hieronymus Meyer.

It was crossed in 1912 by a railway, which took 16 years to build. Passing mostly through tunnels, the 9 km (5.5 mile) long railway starts in Kleine Scheidegg, from where it climbs to the Jungfraujoch, a mountain saddle with a station at 3450 m (11 320 ft) – the highest in Europe. From here, a cable car rises a further 112 m (367 ft) towards the sphinx-like summit below the Jungfrau.
Map Switzerland – Aa

Junggar Pendi (Dzungarian Basin) *China* Vast area of grassland and desert covering about 880 000 km² (340 000 sq miles) in the Xinjiang Uygur Autonomous Region. It is remote and inhospitable but has rich oil deposits, which are being exploited at the city of Karamay.
Map China – Cb

jungle Term loosely applied to mean dense tropical forest, or rain forest, but never used scientifically. It is derived from a Sanskrit and Hindi word meaning 'uncultivated land'.

Junín *Peru* Central department, which produces half of the country's wheat. Its capital, Huancayo (population 164 950), lies on the banks of the Mantaro River, 195 km (120 miles) east of Lima, to which it is linked by the highest railway in the world. The railway reaches 4817 m (15 806 ft) above sea level at La Cima, on the Morococha line, and 4781 m (15 688 ft) in the Galera tunnel, on the main line.

The town of Jauja, 40 km (25 miles) to the north of Huancayo, was the capital of the Spanish conquistador Francisco Pizarro in 1534 before he founded Lima the following year.
Population 1 113 600
Map Peru – Bb

Jura *France/Switzerland/Germany* Crescent-shaped mass of limestone plateaus, valleys and mountain ridges, extending 240 km (150 miles) on either side of the Franco-Swiss border. The Jura's southern end is east of the city of LYONS. Its highest peak is the Crêt de la Neige (1723 m, 5653 ft), just north-west of GENEVA. Dairy farming, especially for cheese, forestry, winter sports, and watch and clock making are the region's main industries. The French department of Jura occupies the western slopes.

The Jura continue north-eastwards into Germany on the other side of the BLACK FOREST. The Swabian Jura and its continuation, the Franconian Jura, reaching a height of about 900 m (2950 ft), run on for more than 320 km (200 miles), ending east of Nuremberg.
Population (department) 249 000
Map France – Gc

Jura *Switzerland* Watch-making canton which was created as recently as 1979. It lies at the north-western end of the country and covers 840 km² (325 sq miles) around the cantonal capital at Delémont (population 11 700), some 30 km (19 miles) south-west of BASLE.
Population 68 300
Map Switzerland – Aa

Jurassic Second period of the Mesozoic era in the earth's timescale. See GEOLOGICAL TIMESCALE.

Jurong *Singapore* District in the south-west of Singapore Island containing the country's main industrial area. There are more than 2270 factories in just 20 km² (8 sq miles), employing 135 280 people – approximately 40 per cent of the country's industrial labour force. The district's products include ships, steel rods and pipes, chemicals, pharmaceuticals, plastics, cement, bricks, cables, textiles and tyres. The area is a free trade zone and has a major port.

A massive town development that surrounds the industrial area includes parks, gardens and a bird sanctuary. The Botanic Gardens are now used for recreation, but in the early 20th century the centre supplied millions of rubber plant seeds to farmers establishing rubber plantations.
Map Singapore – Ab

Jutland (Jylland) *North-west Europe* Peninsula covering 29 620 km² (11 434 sq miles) between the North Sea and the Baltic Sea. It is more than 400 km (250 miles) long and most of it comprises the mainland of Denmark, where it is called Jylland; the southern section forms part of the German state of SCHLESWIG-HOLSTEIN (Danish before the Danish-Prussian War of 1864). The border, which crosses the peninsula between Flensburg and the North Sea coast, was settled after a referendum in 1920.

The peninsula tapers northwards to the northernmost tip, called the Skaw (or Grenen). On the western (North Sea) side, the peninsula lacks natural harbours and contrasts with the deeply indented and sheltered Kattegat coast in the east. In the south-west, the remains of a much older coastline, breached largely by storms in medieval times, are marked by the North FRIESIAN ISLANDS. Large areas of salt marsh between the islands and the modern coastline have been reclaimed.

When Denmark ceded the twin duchies of Schleswig and Holstein to Prussia following the war of 1864, it lost one-third of its territory. By way of compensation, the Danes began a vigorous campaign to reclaim barren heathlands in central Jutland, and have turned some 10 000 km² (3860 sq miles) into productive farmland.

Jutland gave its name to the only major naval battle between the British and German fleets in the First World War. It was fought off the peninsula's west coast in 1916 and resulted in Britain retaining control of the North Sea.
Map Denmark – Ba

Jwaneng *Botswana* Southern town 110 km (68 miles) west of Gaborone. Its diamond mine is owned jointly by Botswana and De Beers, the South African mining company, and has been said to be probably the most important kimberlite pipe (diamond deposit) discovered anywhere in the world since the original discoveries in KIMBERLEY, South Africa.

The mine, which went into full production in 1982, is a source of high-quality gemstones accounting for almost half of Botswana's exports; its expected life has been put at about 40-45 years.
Population 11 190
Map Botswana – Bb

A DAY IN THE LIFE OF A SINGAPOREAN FINANCIER

Robin Lee will make a fortune before he is 50, some 14 years from now. Like many of his fellow 2.5 million Singaporeans, he tells himself this daily as he drives to meetings, or takes a train to Malaysia, or jets to Bangkok and Hong Kong.

Singapore's energetic and ambitious people, three-quarters of whom are Chinese like Robin, have transformed a marsh and jungle island no bigger than a big Western city into one of the world's most dynamic states. Its people are young – almost half were not born when it became independent in 1965. The growing finance firm where Robin is a rising star funds the construction of oil rigs and factories – and Robin handles a large number of clients.

Like 77 per cent of Singaporeans, he lives in a government-built flat, with his wife Rose and two children, James and Susan, aged six and seven. He bought his modern three-bedroomed flat in Jurong six years ago from Singapore's public housing authority, the Housing and Development Board. He paid for it with part of his savings in the Central Provident Fund, the government's compulsory savings scheme. A quarter of a Singaporean's income goes into the fund but he gets it back with interest on retirement.

Robin studied finance at the National University and at work speaks English, the language of administration. He also speaks three of Singapore's six Chinese dialects plus Malay, the national language, and has some experience of engineering, thanks to his two years' military service. He looks forward to a fat year-end bonus on top of his salary and to the day when he will start his own consultancy. Rose works as a well-paid secretary while the children are at school. The couple earn enough to go gambling occasionally. Robin likes to put money on the horses at the Singapore Turf Club. He also gambles a little with friends at mahjong, a game with four players that resembles cards, but with the motifs on small, hard tiles.

At mid-morning, Robin drives back to Jurong estate to see a factory site. After his tour he takes his client out to a Chinese lunch. He spends the afternoon on figures and paperwork and on the telephone, forgetting the time until seven, when Rose calls and reminds him they – and the children – have a dinner appointment at eight with friends in a Western restaurant. There is plenty of choice – cosmopolitan Singapore is the culinary capital of South-east Asia.

K2 (Godwin Austen; Qogir Feng) *Pakistan/China* Second highest mountain in the world, after Mount EVEREST, rising to 8611 m (28 250 ft) in the KARAKORAM, on a part of the disputed border between Azad Kashmir and China. It was first climbed by an Italian team in 1954.
Map Pakistan – Da

Kabah *Mexico* Ruined city of the Mayan civilisation which flourished between the 7th and 10th centuries. It lies in southern YUCATÁN, 19 km (12 miles) from Uxmal. Its main attraction is the Palace of the Masks, which is faced with hundreds of stone reliefs of the rain god Chac.
Map Mexico – Db

Kabalega Falls *Uganda* Waterfalls on the River Nile, north-east of Lake ALBERT. Formerly known as the Murchison Falls, they are situated in a national park which is rich in wildlife.
Map Uganda – Ba

Kabul *Afghanistan* The country's capital, largest city and main cultural and commercial centre. It is also the capital of Kabul Province. It stands on the Kabul River, 190 km (118 miles) west of the KHYBER PASS.

Situated almost 1830 m (6000 ft) above sea level, it commands the mountain passes to the north and west – and has been on the route of several major invasions, including those of Alexander the Great in the 4th century BC and the Indian emperor Babur, who made Kabul the capital of the Mogul Empire in 1504. Babur's tomb lies in the western outskirts of the city. In 1842, during the first Afghan War, Kabul was occupied and partially destroyed by British troops. The city was rebuilt in the late 19th century by Emir Abder-Rahman Khan.

Kabul's most colourful street is the Jodi Maiwand, site of the bustling main bazaar. Its main manufactures include textiles, footwear, vehicle components and timber products.

The city has also borne the brunt of the Afghan civil war; at the beginning of 1994, the city came under siege for months, during which time it was relentlessly pounded by rockets which reduced much of it to rubble. Nearly 400 000 people were forced to flee their homes; many others had to rely on food relief to survive.
Population 1 500 000
Map Afghanistan – Ba

Kabul *Afghanistan/Pakistan* River rising in the HINDU KUSH west of the Afghan capital, KABUL, and flowing generally eastwards for about 480 km (300 miles) to join the INDUS between Peshawar and Islamabad, the Pakistani capital. It is dammed at Warsak, north-east of the Khyber Pass, to provide hydroelectric power and some water for irrigation. Although it is not navigable, its lower reaches are used for logging.
Map Afghanistan – Ba; Pakistan – Ca

Kabwe (Broken Hill) *Zambia* The country's oldest mining town, 100 km (62 miles) north of the capital, LUSAKA. Its mine, which was opened in 1902, produces high-quality lead vanadium (a metallic element used in rust-resistant high-precision tools) and zinc.
Population 210 000
Map Zambia – Bb

Kachchh *India* See KUTCH

Kachin *Myanmar (Burma)* State that covers 89 042 km² (34 379 sq miles) in the far north of Myanmar, between the Indian and Chinese borders. Myanmar's highest peak, Hkakabo Razi, which rises to 5881 m (19 294 ft) is situated here.

The thickly forested hills are inhabited largely by the Kachin people, who have been fighting a war of independence for many years with the central government. The Shans and Bamars of the lowlands have developed a jade-mining industry near the town of Mogaung, and sugar cane growing around the state capital, MYITKYINA. Opium poppies are also grown in Kachin.
Population 904 000
Map Myanmar – Ca

Kadhimain (Al Kazimiyah) *Iraq* Holy city of the Shiite Muslims, which has been assimilated into Baghdad as one of its northern suburbs. An elaborate mosque, the third most important in Iraq, has became a shrine to former Shiite imams. It was built in 1515 and has domes and minarets coated with gold.
Map Iraq – Cb

Kadoma *Zimbabwe* Formerly a gold-mining town called Gatooma, Kadoma has become the centre of the country's cotton textile industry. Situated 140 km (87 miles) south-west of the capital, Harare, it produces goods from locally grown and imported cotton.
Population 67 000
Map Zimbabwe – Ba

Kaduna *Nigeria* Capital of the state of the same name, on the Kaduna River, some 625 km (390 miles) north of LAGOS. Founded by the British in 1917, it became the capital of the former Northern Region. Situated in the heart of Nigeria's main cotton-growing area, it produces cotton textiles, beer, soft drinks, bicycles, ammunition, and shoes.
Population (town) 202 000; (state) 3 969 250
Map Nigeria – Ba

Kaesong (Gaeseong; Kaegyong) *North Korea* City 150 km (93 miles) south-east of the capital, PYONGYANG. It was the capital of Korea during the Koryo dynasty (AD 918-1392), but its prosperity waned after Seoul, now in South Korea, was made the capital in 1394. Textiles and porcelain are its chief industries.
Population 310 000
Map Korea – Cd

Kafr Kanna *Israel* See CANA

Kafue *Zambia* River, 970 km (600 miles) long, which rises close to the border with Zaire and flows into the Zambezi south-east of the capital, LUSAKA. Lake Kafue, just south of Lusaka, was created when the river's gorge was dammed at the confluence with the Zambezi; a hydroelectric station was opened here in the 1970s. The river supports some fishing, but navigation is restricted by rapids.

A national park named after the river and covering 22 400 km² (8649 sq miles) was opened here in 1951. Its wildlife includes elephants, lions, rhinoceroses, crocodiles and the fish eagle, Zambia's national emblem.
Map Zambia – Bb

Kagan Valley *Pakistan* Beautiful mountain area, north of ISLAMABAD, which is renowned for its lakes and fishing. It is linked to the Indus Valley by the 4200 m (13 780 ft) high Babusar Pass.
Map Pakistan – Da

Kagera *Rwanda/Tanzania/Uganda* River which rises in south-east Rwanda and which forms the border with Tanzania for more than 150 km (93 miles). It then turns eastwards to form part of the Ugandan-Tanzanian border before flowing through Tanzania to Lake Victoria at the border with Uganda. With a length of 530 km (330 miles), it is the longest headwater of the Nile. The Kagera National Park on the river's west bank, which covers an area of 2500 km² (965 sq miles) in north-east Rwanda, is well stocked with big game, crocodiles and water birds.
Map (river) Tanzania – Ba; (national park) Rwanda – Ba

▼ **RIVER OF LIFE AND DEATH The Kabul River valley is a precious green oasis in a dusty, dry land, but it has seen heavy fighting during the Afghan Wars and the civil war in Afghanistan.**

Kagoshima *Japan* City and port on Kagoshima Bay in southern KYUSHU Island. It lies in the shadow of an active volcano, Sakurajima, which reaches 1118 m (3668 ft). The city's inhabitants carry umbrellas even in fine weather to protect themselves from showers of fine ash thrown out by the volcano. The mountain was a small island in Kagoshima Bay until 1914, when a tremendous eruption hurled out enough lava and ash to link it with the mainland.
Population 537 000
Map Japan – Bd

Kahuzi-Biega National Park *Zaire* See BUKAVU

Kaieteur Falls *Guyana* Spectacular waterfalls on the Potaro River, 230 km (145 miles) south-west of the capital, Georgetown. More than 90 m (295 ft) wide at the brink of the falls, the river plunges down a cliff as a solid wall of water, with an uninterrupted drop of 226 m (741 ft) – more than four times the height of North America's Niagara Falls. The waterfalls, whose existence became known to the outside world after a British geologist, C. Barrington Brown, visited them in 1870, have become the centrepiece of Guyana's only national park.
Map Guyana – Bb

Kaifeng *China* Ancient city in Henan Province, about 650 km (400 miles) south of Beijing. Xiangguosi monastery, a Buddhist sanctuary founded in the 6th century AD, is situated in the city. Kaifeng is noted for its handicrafts.
Population 690 000
Map China – Hd

Kaikoura Ranges *New Zealand* Two mountain chains (Kaikoura Range and Seaward Kaikoura Range) on South Island, 175 km (108 miles) north of Christchurch. Their highest point is Mount Tapuaenuku at 2885 m (9465 ft). Cattle and sheep farming are the main occupations here.
Map New Zealand – De

Kaingaroa Forest *New Zealand* One of the largest planted forests in the world (some 200 000 hectares, 495 000 acres), lying about 200 km (120 miles) south-east of Auckland on North Island's VOLCANIC PLATEAU. The area was planted – mostly with radiata pines – during the Depression years of the 1930s as part of the government's relief work scheme.
Map New Zealand – Fc

Kainji Reservoir *Nigeria* One of Africa's largest artificial lakes. It is on the Niger River in western Nigeria and covers 1295 km² (500 sq miles), stretching 137 km (85 miles) downstream from the town of Yelwa to the Kainji hydroelectric dam. After the dam was completed in 1968, 150 villages were flooded and 50 000 people had to be resettled. The project supplies irrigation water and aids flood control. The reservoir is well stocked with fish.
Map Nigeria – Aa

Kairouan (Qairawan) *Tunisia* The holiest Muslim city in Africa and the fourth holiest Muslim city in the world, after MECCA, MEDINA and JERUSALEM, situated 57 km (35 miles) west of SOUSSE. Seven trips to Kairouan for the faithful are equivalent to one visit to Mecca. Kariouan was founded in AD 671 by Okba Ibn Nafaa, an Arab warrior who had known Muhammad. Built originally as a strategic military outpost it blossomed into the greatest Islamic cultural centre in the MAGHREB, producing Tunisia's finest romantic poets and experts in religious law.

Kairouan's buildings reflect the clean, unfussy classic lines of the early Islamic architectural style. Its imposing city walls and ramparts date from 1052 and the Medina – the old Muslim quarter – has many Islamic shrines, mosques and minarets, and is a busy place of cafés and rug-weaving shops. The city is famous, too, for its fascinating souks – markets selling rugs, carpets and leatherware.

Among other places of interest are the beautiful 9th-century Great Mosque (the oldest in Africa), the 9th-century Thleta Bibane Mosque, the splendid 13th-century *zawia* (courtyard) of the house of Abid El Ghariani and the Sidi Sahbi Mausoleum (according to legend, Sidi Sahbi was a holy man who preserved three hairs from the beard of the prophet Muhammad, and so was called 'the Barber').
Population 72 300
Map Tunisia – Ba

Kaiserslautern *Germany* Industrial and university city 100 km (62 miles) south-west of FRANKFURT am Main. It produces machinery and sewing machines, cars, textiles, tobacco, and beer.
Population 98 000
Map Germany – Bd

Kaiserstuhl *Germany* Group of volcanic hills in the Rhine rift valley near the Black Forest town of FREIBURG. The name of this prominent, abrupt feature, which rises to 557 m (1827 ft), translates as 'Emperor's Chair'. Its slopes, few of which are still forested, are terraced for vine growing.
Map Germany – Bd

Kajaani *Finland* City founded in 1651 beside rapids south-east of Lake Oulu in the centre of the country. A trade and transport centre, its industries process and saw timber, pulp wood, and produce paper and cellulose. The remains of a late medieval castle are situated near the city.
Population 36 700
Map Finland – Cb

Kakadu National Park *Australia* Australia's largest national park and World Heritage area, covering nearly 20 000 km² (7720 sq miles). It lies in the NORTHERN TERRITORY, 200 km (124 miles) east of Darwin. Aboriginal rock art spanning more than 20 000 years can be seen painted and carved on hundreds of cave walls and rock faces. The park also has a wide variety of animals and plants, many of which have not yet been scientifically described. It has a 600 km (370 mile) long escarpment and vast flood plains.
Map Australia – Ea

Kalaallit Nunaat See GREENLAND

Kalahari *Southern Africa* Arid, semi-desert area of sand and scrubby vegetation covering 932 400 km² (360 800 sq miles) in South Africa, Botswana, Zimbabwe and Namibia. Its remoter reaches are inhabited mostly by San (Bushmen), but human population throughout is sparse.

Nights are cold and days searingly hot, with the annual rainfall varying from 175 mm (7 in) in the south to 575 mm (23 in) on the Kavango River. Almost none of the run-off reaches the sea, and the abundant wildlife is dependent on the many pans in which water collects. The principal rivers are the Kavango and the Kunene, both of which rise in northern Angola. Wildlife reserves include the Kalahari Gemsbok National Park, Etosha National Park, Moremi Wildlife Reserve and others within the area of the vast inland delta of the Kavango River, the Okavango Swamps.
Map South Africa – Ba

Kalávrita *Greece* Rail terminal and modern city on the northern Peloponnese, 75 km (47 miles) from CORINTH. The clock of the Metropolitan Church stands permanently at 2:34, the time on 13 December 1943 when the building was set on fire by German soldiers who then massacred the town's 1436 males, as retaliation for anti-Nazi guerrilla activities in the region.

The monastery of Ayía Lávra, founded in AD 961, lies 6 km (4 miles) to the south. In 1821, the revolution that eventually led to Greece's independence from Turkish occupation began here. Kalávrita is the terminus of a late 19th-century narrow-gauge railway line, which runs from Dhiakoptón, on the Gulf of Corinth, through gorges, mountain passes, bridges and tunnels.
Map Greece – Cb

Kalawewa Tank *Sri Lanka* Reservoir in central Sri Lanka, 140 km (85 miles) north-east of COLOMBO. Formed in AD 460 when the Kala River was dammed, it originally covered 18 km² (7 sq miles). The tank is now used as part of an irrigation scheme designed to encourage settlers to move into the region.
Map Sri Lanka – Ba

Kalémié *Zaire* Town founded in 1891 as Albertville. It is a transport centre on the west shore of Lake TANGANYIKA, and linked by rail to Kabalo on the Lualaba River. Lake steamers operate between Kalémié in Zaire and Kigoma in Tanzania, which in turn is connected by rail to Dar es Salaam. Fishing is important while the chief manufacturing industries are textiles and cement.
Population 61 000
Map Zaire – Bb

Kalgoorlie *Australia* Gold and nickel-mining town in Western Australia, 540 km (340 miles) east of PERTH. Kalgoorlie and its twin town, Boulder, have produced more than 1000 tonnes of gold since the discovery of the metal here in 1893. Mining is still carried out; the historic Hainault mine is preserved as a tourist attraction.

At Coolgardie, another 19th-century gold-mining town situated 40 km (25 miles) to the south-west of Kalgoorlie, the population has dropped from 15 000 to a little over 1000. Kambalda (population 4300), to the south, is enjoying a new lease of life as a nickel-mining centre.
Population 25 000
Map Australia – Ce

Kalimantan *Indonesia* The Indonesian part of BORNEO, covering 538 718 km² (208 000 sq miles) of the mainly mountainous island which is notorious for its huge forest fires. Forestry, oil and natural gas are its economic mainstays. With a population density of just 12 people per km² (41 per sq mile), Kalimantan has become the focal point of a government resettlement scheme.
Population 8 677 000
Map Indonesia – Dc

Kálimnos (Calino) *Greece* Island in the DODE-CANESE group in the east Aegean Sea. The island is known for its natural sponges, which are gathered off the North African coast at the beginning of summer and transported for processing to Pothiá, the chief town. Kálimnos lies just south of the island of Léros, where the Roman commander Julius Caesar (100-44 BC) was captured by pirates in 75 BC. Caesar raised his ransom, raised a naval force, captured his captors and crucified them.
Population 26 900
Map Greece – Ec

Kalinin *Russia* See TVER'

Kaliningrad (Königsberg) *Russia* Commercial port and naval base near the Baltic Sea, 30 km (18 miles) north of the Polish border and isolated from the rest of Russia by the Baltic States of Lithuania, Latvia and Estonia. Founded as Königsberg in 1255 by Teutonic Knights, it became the capital of the German state of East Prussia in the Middle Ages.

As the cultural, commercial and political centre of East Prussia, it played an important role in the trade between Germany and the Baltic states. During the Second World War it fell to the Russian army after a long, devastating siege. It was ceded to Russia at the 1945 Potsdam Conference involving Britain, the United States and the USSR. The following year it was renamed Kaliningrad after the Soviet politician Mikhail Kalinin (1875-1946).

Built on the Pregel (Pregolya) River, it is linked to the Baltic port of Baltiysk by a canal 42 km (26 miles) long. Most of the postwar city is situated north of the original town (of which little remains). Kaliningrad, the centre of a district of the same name, is an important fishing port; it also makes ships, paper and machinery. The German philosopher Immanuel Kant (1724-1804) was born in the town and lectured here for a while. He was buried in the 13th-century cathedral, which was destroyed during the Russian siege.
Population (city) 408 000; (district) 887 000
Map Russia – Dc

Kalisz (Kalisch) *Poland* Industrial city 105 km (65 miles) south-east of the city of POZNAŃ. It has textile, metal, engineering and piano-making industries.
Population 106 000
Map Poland – Cc

Kalmar *Sweden* Seaport on the south-east coast, 390 km (242 miles) south of STOCKHOLM. It lies opposite the isle of Öland, to which it is linked by a 6 km (4 mile) long bridge. The Union of Kalmar, the agreement which united Sweden, Denmark and Norway under the Danish royal house, was signed in Kalmar's medieval castle (rebuilt in the 16th and 17th centuries). Besides shipyards and railway workshops, Kalmar has match factories and paper mills. It also manufactures motorcars and processes food.
Population 56 900
Map Sweden – Cd

Kalmyk Republic *Russia* Autonomous republic of the Russian Federation, situated north-west of the Caspian Sea and covering 76 100 km²

▲ SHIPS OF THE DESERT Camels cross the shifting sands of the Kalahari in Botswana. A party led by the missionary Dr David Livingstone abandoned a crossing in 1849 because the wagons sank into the sand.

(29 375 sq miles). Consisting mainly of arid lowland and desert, it is the home of the nomadic Kalmyk people, who drive their flocks of merino sheep, pigs, cattle and camels between winter and summer pastures.

The republic, which is partly irrigated by the Volga River, yields fodder and melons. The capital, Elista (population 120 000), produces leather and woollen textiles and processes food from the region's livestock. The already reduced amount of fish caught in the Caspian Sea is decreasing even further due to pollution.
Population 322 000
Map Russia – Fd

Kalocsa *Hungary* City in the centre of the country, 110 km (70 miles) south of BUDAPEST. It lies 6 km (4 miles) east of the River Danube in a particularly fertile part of the Great ALFÖLD, which specialises in growing the red seasoning spice, paprika. Kalocsa has a wealth of fine Baroque buildings, including an 18th-century cathedral and archbishop's palace, with a library containing many rare manuscripts. This city of brightly painted houses is also noted for its vividly coloured embroideries.
Population 18 190
Map Hungary – Ab

Kaluga *Russia* Industrial city and railway junction on the Oka River, 160 km (100 miles) south-west of Moscow. Iron and steel products, railway equipment, matches, glass, perfume, footwear, clothing and food are manufactured here.
Population 314 000
Map Russia – Ec

Kamakura *Japan* Town on Sagami Bay, 48 km (30 miles) south-west of the capital, Tokyo. The seat of Japan's shogun chiefs in the 12th and 13th centuries, it is now a centre of Zen Buddhism. Since 1900 it has become a prosperous commuter town for Tokyo.
Population 174 000
Map Japan – Cc

Kambui Hills *Sierra Leone* Low plateaus and hills rising to 320 m (1050 ft) in the extreme south-east astride the Moa River. The hills are rich in minerals and precious stones, especially diamonds, which are found in the river gravels.
Map Sierra Leone – Ab

Kamchatka *Russia* Peninsula in the far northeast of Russia, separating the Sea of Okhotsk from the Bering Sea and the Pacific Ocean. It is about 1200 km (750 miles) long, up to 560 km (350 miles) wide and covers some 350 000 km² (135 000 sq miles) – about the size of Japan.

It contains 22 active volcanoes, of which the highest, Klyuchevskaya Sopka, rises to 4750 m (15 580 ft). The main occupation is fishing – for crabs, salmon, herring and cod. The largest port on the peninsula is PETROPAVLOVSK-KAMCHATSKIY on the south-east coast.
Population 422 000
Map Russia – Qc

kame Sand and gravel deposited as a cone where a stream derives from an ice sheet, and left either as a hummock or a steep-sided ridge as the ice retreats.

Kamina *Zaire* Transport centre in Shaba region, 420 km (260 miles) north-west of the city of Lubumbashi, and linked by railway to ILEBO. Kamina is the centre of a farming region producing cotton, tobacco and vegetables, some of which are processed locally.
Population 57 000
Map Zaire – Bb

Kamloops *Canada* City, founded in 1812, situated at the junction of the North and South Thompson rivers at the heart of a gold and copper-mining area and the centre of British Columbia's cattle industry.
Population 67 660
Map Canada – Cc

Kamnik *Slovenia* See KARAWANKE ALPS

Kampala *Uganda* National capital since independence in 1962. It developed next to the site known as Mengo, where the *kabaka* (king) of Buganda had his palace. The British built a fort here in 1890, and the town that grew around it became the country's commercial centre because of its position near Lake VICTORIA, in the country's most prosperous region.

The city is built on a series of hills, two of which are occupied by the cathedral, hospital and schools of the Protestant and Roman Catholic churches. Another is the site of the main mosque, and a fourth has the university.

Most of the city's factories, which produce goods for local consumption, are built on drained swampland. Many of Kampala's poorer inhabitants live in shanty towns around and between the main buildings.
Population 773 000
Map Uganda – Ba

Kamphaeng Phet *Thailand* Market town 320 km (200 miles) north-west of the capital, BANGKOK. Many parts of the original town, which was founded in the 13th century, can still be seen. Farmers in the surrounding province of the same name grow mainly rice, sugar cane and mung beans (used in Chinese cooking as bean sprouts).
Population (town) 19 600; (province) 634 900
Map Thailand – Ab

Kampong Ayer *Brunei* See BANDAR SERI BEGAWAN

Kampuchea See CAMBODIA

Kananga *Zaire* City founded as Luluabourg in 1894, situated 750 km (465 miles) west of Lake Tanganyika. It is a centre of the Luba people and capital of Kasai Occidental region. Its location on the Lulua River and on the Lubumbashi-Ilebo railway, has seen it grow into an important commercial and transport centre. The city serves the diamond-producing region around Tshikapa to the south-west. Coffee and cotton are also processed here.
Population 938 000
Map Zaire – Bb

Kanazawa *Japan* Port near the north-west coast of HONSHU Island, 295 km (185 miles) north-west of the capital, Tokyo. A beautiful castle town largely untouched by modernisation, its chief attractions are the traditional Japanese gardens of lakes, waterfalls and azalea groves in the castle grounds. Mount Hakusan (2702 m, 8865 ft), one of Japan's three holy mountains, towers to the south.

Kanazawa has a university and is an important cultural centre. The nearby Myoryuji Temple, built in 1643, was once the house of the Ninja, a secret sect widely employed, because of their skills in concealment, as spies and assassins.
Population 443 000
Map Japan – Cc

Kanchanaburi *Thailand* Town about 120 km (75 miles) west of the capital, Bangkok. The surrounding province, which lies along the border with Myanmar (Burma), is an important area for growing sugar cane. Its main tourist attractions are the war cemeteries and the railway bridge over the River KWAI, built by Allied prisoners of war during the Second World War for the occupying Japanese; the bridge was rebuilt after being destroyed in air raids.
Population (town)139 100; (province) 612 750
Map Thailand – Ac

Kanchenjunga *India/Nepal* See KANGCHENJUNGA

Kandahar *Afghanistan* Capital of Kandahar Province and the country's second city, 470 km (292 miles) south-west of KABUL. It has long been an important crossroads town, and is the only stopover on the road between Kabul and Herat, 440 km (273 miles) to the north-west. Kandahar is thought to have been founded by Alexander the Great in the 4th century BC when it was known as Alexandrai Arachosiorum. It changed hands many times and was destroyed more than once.

In the 16th century, it was conquered by Babur (about 1482-1530), founder of the Mogul empire of India. A reminder of those days is the throne of the emperor, hewn out of a rock, at the top of a flight of 42 steps. In 1748, Kandahar became the first capital of Afghanistan. The old city is dominated by the mosque and tomb of Ahmed Shah, who created the first Afghan federation in 1747. Kandahar's main industries today are textiles and fruit canning.
Population (city) 226 000
Map Afghanistan – Bb

Kandy *Sri Lanka* Former royal capital of the Sinhalese kings. It stands beside a lake constructed 465 m (1525 ft) above sea level in the central mountains, 95 km (60 miles) north-east of COLOMBO. The Sinhalese kings were established in Kandy by the 14th century, and from the 16th century until 1815, when the last king ceded the city to Britain, they had held off Portuguese, Dutch and British invaders.

Today Kandy is one of the world's most sacred Buddhist sites and a symbol of Sri Lankan nationalism. A tooth believed to be from Siddartha Gautama, who founded Buddhism in the 6th century BC, is kept in the 16th-century Dalada Maligawa ('Temple of the Tooth') and is carried in procession through the city at an annual festival in July or August. The tooth is kept concealed in seven jewel-encrusted caskets, each nested like a Russian doll inside a larger casket. The country's national museum is situated near the temple in what were once the quarters of the royal concubines. The remains of the Sinhalese royal palace contain an archaeological museum. Sri Lanka's largest university and the royal botanical gardens are at Peradeniya, 6 km (4 miles) west of Kandy.
Population 104 000
Map Sri Lanka – Bb

Kanem *Chad* Region, and present-day prefecture, in western Chad, north-east of Lake CHAD. A powerful kingdom flourished here between the 10th and 15th centuries; the people of Kanem were converted to Islam in the 11th century. The kingdom later joined with the Nigerian kingdom of Bornu to form Kanem-Bornu, an empire which lasted until it was conquered by Europeans in the 19th century. Ruled as a separate French protectorate, Kanem became part of Chad only in 1958. Mao is the capital of the prefecture.
Population 268 200
Map Chad – Ab

Kanem-Bornu *Nigeria/Chad* See KANEM; MAIDUGURI

Kangaroo Island *Australia* Island, 145 km (90 miles) long and 50 km (30 miles) wide, off the South Australian coast south-west of Adelaide. Flinders Chase at its western end is a national park. There is a number of other smaller reserves on the island containing wildlife which has become quite tame. Barley and sheep are the island's main agricultural products.
Population 4100
Map Australia – Ff

Kangchenjunga *India/Nepal* Mountain on the border between Nepal and Sikkim, which at 8598 m (28 208 ft) is one of the highest peaks in the world. The head waters of the River TISTA and one of the major tributaries of the KOSI River flow from its glaciers. The name 'Kangchenjunga' is Tibetan and translates literally as 'the Fire Treasures of the Great Snow'. The mountain was first climbed in May 1955 by a British team led by Charles Evans.
Map India – Db

Kangerlussuaq *Greenland* See SØNDRE STRØM-FJORD

Kangnung (Gangneung) *South Korea* Rail terminus on the north-east coast, almost due east of the capital, SEOUL. Just north of the city lies Lake Kyongpo, a bathing resort in summer and used by skaters in winter.
Population 153 000
Map Korea – Dd

KaNgwane *South Africa* Former self-governing state (1984-1994), created for the Swazi people of the Eastern Transvaal and recognised only by South Africa. In the early 1980s, an offer by the South African government to cede the area to neighbouring Swaziland was dropped after opposition from KaNgwane and KwaZulu leaders.
Map South Africa – Db

Kankan *Guinea* Main town of Upper Guinea. It stands on the banks of the Milo River, a tributary of the Niger, 485 km (300 miles) north-east of the capital, CONAKRY. Kankan is the eastern terminus of the railway from Conakry and the focus of the main roads linking Conakry with Mali, Liberia and Ivory Coast. The town itself resembles an enormous village, with huts and fields of crops situated next to the roadside. Kankan is a trading centre for the region's produce, especially rice, and has a lively walled market.
Population 265 000
Map Guinea – Ca

Kano *Nigeria* Capital of the state of the same name, on the border with Niger. One of Nigeria's largest cities, Kano is a Muslim religious and educational centre, as well as a tourist centre with a lively modern nightlife.
Kano is one of the largest of the medieval Hausa cities. By the 12th century it was a thriving trading centre at the end of a Saharan caravan route. The old section of the town, with its bustling 15th-century market selling traditional leather goods, baskets, cloth and jewellery, survives. A rail link, opened in 1912, connects Kano to Lagos on the Atlantic. Modern factories produce peanut oil, corned beef, textiles, shoes and soft drinks.
Population (city) 580 000; (state) 5 632 000
Map Nigeria – Ba

Kanpur (Cawnpore) *India* Industrial city on the River Ganges, in north-central India, 79 km (49 miles) south-west of LUCKNOW. Wool, cotton, leather and sugar are produced here. In July 1857, 1000 British soldiers and their families were murdered here during the Indian Mutiny after being promised safe conduct on surrender. A memorial garden commemorates the event.
Population 2 485 490
Map India – Cb

Kansas *USA* One of the Great Plains states, covering 211 864 km² (81 782 sq miles) of undulating prairie. Before the coming of the railways, the Santa Fe Trail crossed the state, bringing traders and settlers to the area. Later, cattle drives from Texas ended at railway depots in Abilene and Dodge City. In the 1930s, a 'dust bowl' spread over the state after widespread drought had turned the topsoil into a fine powder. Many settlers were forced to leave their homes.
Kansas has hot, dry summers and cold winters, and is the nation's chief wheat-producing state. Other farm products include sorghum, maize and beef and dairy cattle. The state has gas fields and some oil; its industries include insurance, the assembly of light aircraft and motor vehicles, and food processing. TOPEKA is the state capital.
Population 2 477 600
Map United States – Gc

Kansas City *USA* Industrial city in the north-east corner of Kansas straddling the Kansas River on the border with Missouri. Called Kansas City, Kansas, to distinguish it from nearby Kansas City, Missouri, it is situated on a river crossing which made it a gateway to the West. Its location helped to make it the principal market for wheat and livestock in the southern Great Plains. Grain milling, meat packing, agricultural equipment, vehicle assembly, chemicals and engineering are its main industries. The University of Kansas Medical Center is its single largest employer.
Kansas City, Missouri, is a commercial and industrial city in the north-west of Missouri at the junction of the Kansas and Missouri rivers, just east of Kansas City, Kansas. Originally a fur trading post, it now has the second largest population in Missouri, after St Louis, and is one of the busiest rail centres in the United States. It is the nation's largest winter wheat market with grain elevators second in size only to those at Minneapolis and St Paul in Minnesota.
Limestone caves below the metropolitan area are used as warehouses, especially for frozen foods. Wholesale retail trade makes up a large part of the city's economy, along with printing and publishing, and food processing.
Population (Kansas City, Kansas) 684 900; (Kansas City, Missouri) 435 150
Map United States – Hc

Kansk *Russia* Industrial city in central Siberia, 800 km (500 miles) north-east of NOVOSIBIRSK. Founded as a fort in 1628, it became a station on the Trans-Siberian Railway in the early 1890s. The present city grew from the development of the nearby coal mining industry after 1950. Other industries revolve around machinery, chemicals, food processing and textiles.
Population 110 000
Map Russia – Kc

Kao-hsiung (Takao) *Taiwan* Major industrial city and leading seaport, on the south-western coast. Taiwan's second largest city, it has a busy international airport, a large iron and steel works, an oil refinery, an aluminium processing plant, one of Asia's biggest shipyards and one of the largest shipbreaking yards in the world. Kao-hsiung imports crude oil, iron ore and timber, and exports manufactured goods and agricultural products. The city's manufactures range from electronics and precision instruments to plastic goods and toys. Despite its industrial character, the city (called 'Takao' until 1920) has numerous temples, shrines and pagodas.
Population 1 404 000
Map Taiwan – Bc

Kaolack (Kaolak) *Senegal* The country's third largest city after Dakar, which lies some 150 km (95 miles) to the north-west. Kaolack is the commercial and cultural centre of Senegal's main groundnut-producing region.
Population 181 000
Map Senegal – Ab

kaolin See CHINA CLAY

kaolinite White or grey mineral, hydrated aluminium silicate, formed when granite decomposes. It is the principal constituent of kaolin or CHINA CLAY.

Kaposvár *Hungary* City and administrative centre of the south-western county of Somogy, 50 km (30 miles) south of Lake BALATON. It is a market centre for the picturesque fruit-growing Transdanubian hills to the east and for the Plains to the west where livestock is farmed. The development of postwar light industries set around textiles, shoes, foodstuffs and electrical goods has led to the considerable expansion of the town.
Population 71 210
Map Hungary – Ab

Kaptai Dam *Bangladesh* See KARNAFULI

Kapuni *New Zealand* Gas field 200 km (120 miles) north-west of the capital, WELLINGTON, and linked to Auckland and Wellington by pipeline. Since natural gas was discovered here in 1959, many New Zealand cars have been converted to run on compressed natural gas.
Map New Zealand – Ec

kar See CIRQUE

Kara Kum *Turkmenistan* See PESKI KARAKUMY

Kara Sea Shallow sea with a depth of mostly less than 200 m (656 ft), forming part of the ARCTIC OCEAN between the NOVAYA ZEMLYA and SEVERNAYA ZEMLYA islands, and covered by ice all year except for a few weeks in late summer. The OB', YENISEY and other Siberian rivers flow into it. In 1597, the Dutch explorer Willem Barents died here after leading an expedition into the area.
Map Russia – Ha

karaburan Strong, hot, dust-laden north-east wind blowing in the Tarim Basin of central Asia during spring and summer.

Karachi *Pakistan* Port and industrial city on the Arabian Sea, capital of Sind Province, and Pakistan's largest city. In 1800, Karachi was a mere fishing hamlet on a desert shore. It expanded in the 19th century after the British built a railway along the Indus River valley.
Because it is situated far enough from the Indus delta to avoid silting up, it became Pakistan's leading port. Besides exporting cotton and wheat, the city has metal and engineering industries. Karachi became the national capital when Pakistan was created in 1947. In 1967, the new city of Islamabad became the capital. Although it has often been described in the past

as a 'city of superlatives', Karachi's most recent claim to fame is not one of which it can be proud: it is now described as one of the most dangerous cities in Asia.

Spiralling unemployment (30 per cent in 1994), inadequate power, water, health care and educational systems, as well as the plentiful supply of arms, a legacy from the days when the United States used Karachi as a transshipment point to supply anti-Soviet guerrillas in Afghanistan, have led to this state of affairs.

Population 8 445 000
Map Pakistan – Cc

Karaganda *Kazakhstan* Industrial and mining city in the transition zone between the steppe and Kazakhstan's desert, 750 km (466 miles) north of ALMA-ATA.

Founded in 1857 as a mining village with 150 inhabitants, the city's growth coincided with the development of the local coal industry. Since the Second World War, Karaganda has become Kazakhstan's second largest city. Its other main industries are iron and steel.

Population 614 000
Map Kazakhstan – Db

Karak *Jordan* Busy commercial centre, market town and road junction situated 22 km (13 miles) from the south-east edge of the Dead Sea, and 90 km (60 miles) south of the capital, Amman.

Originally an Italian Crusader town, Karak has a castle overlooking the Dead Sea valley, high town walls and winding, medieval streets. The castle has been partially restored. An Italian hospital in the town is run by nuns.

Population 169 000
Map Jordan – Ab

Karakoram *China/India/Pakistan* Mountain range in northern Kashmir, stretching some 400 km (250 miles) across disputed Indian and Pakistani territory along the border – also disputed in parts – with Chinese Tibet. It is linked with the Himalayas to the south and south-east.

Together with the Pamirs to the north-west, the Hindu Kush to the west and the Kunlun Shan range to the east, it forms a mountain knot known as 'the Roof of the World'.

It has 33 peaks higher than 7300 m (23 950 ft), including the world's second highest peak, K2, and several passes above 5500 m (18 050 ft). The Karakoram also has one of the longest glaciers – the HISPUR – outside the polar regions.

Map China – Ad; Pakistan – Da

Karakorum (Har Horin) *Mongolia* Historic site 322 km (200 miles) west of the capital, ULAN BATOR, in the valley of the Orhon River. In 1235, Ogodai Khan, the son of the conqueror Genghis Khan, built his headquarters here.

After becoming the capital of the Great Khans, the settlement was abandoned in the 16th century as Mongol power collapsed. Today, only a few scattered ruins remain.

Map Mongolia – Db

Karamoja *Uganda* Arid, sparsely inhabited region in the north-east, on the Kenyan border. The Karamojong and Jie peoples who live here are primarily cattle herders, although they do grow some sorghum.

Population 300 000
Map Uganda – Ba

Karawanke Alps (Karawanken; Caravanche) *Austria/Slovenia* East-west mountain range rising to 2238 m (7343 ft) at Grintavec in Slovenia. The range extends about 370 km (230 miles) eastwards through Slovenia as the Pohorje Spur in the north and the Savinjske Spur in the south. The fertile Savinja basin between the spurs centres on the prosperous town of Celje. A spa since Roman times, it makes beer, textiles, zinc and machinery. The quickest route across the Karawanke Alps is via the Loibl Pass, at 1368 m (4488 ft).

Map Slovenia – Aa

Karbala (Kerbela) *Iraq* Regional capital and Shiite place of worship 90 km (55 miles) south-west of the national capital, Baghdad. The tombs of Hussein bin Ali and his brother Abbas – grandsons of Muhammad, the founder of Islam and who were killed here in the Battle of Tuff in AD 680 – are in the town. Karbala trades in dates, hides and wool, and provides facilities for pilgrims visiting the tombs.

Population 185 000
Map Iraq – Cb

Karelia *Finland/Russia* Home of a Finnish people who were first recorded in the 9th century and, historically, a buffer province between Finland and the former USSR, Karelia has at different times been part of both countries. When the frontier was last adjusted, in 1944, Finland had to cede most of the province and was left only with what is now the province of North Karelia (Pohjois-Karjala) and its capital Joensuu. The rest forms the KARELIAN AUTONOMOUS REPUBLIC of the Russian Federation.

Population (North Karelia) 770 000
Map Finland – Dc

Karelian Autonomous Republic (Karjala; Karelia) *Russia* Autonomous republic of the Russian Federation, lying just south of the Arctic Circle between Finland, the White Sea (an inlet of the Barents Sea) and Lake LADOGA. It covers 172 400 km² (66 500 sq miles) and consists largely of dense forest, marshes, rivers and some 50 000 lakes. The eastern part of the region has belonged to Russia since the 12th century. The western part, which belonged to Finland until

▲ **HEIGHT OF DESOLATION The bleak Skigar Valley lies beyond the pass of Strangdokmola in the Karakoram in northern Kashmir. Perpetually snow-capped, its Koser Gunge rises to 6400 m (21 000 ft).**

1940, was occupied by the Russians after the Russo-Finnish War of 1939-40, and was ceded to them in 1944. Today, Finnish Karelians are a minority, making up 10 per cent of the population.

Timber is the republic's main industry, followed by fishing, hunting, and some agriculture. Copper and iron are mined. Transport is aided by a canal from the White Sea to Lake ONEGA, which links with the Volga-Baltic Waterway and the Murmansk-St Petersburg railway line. The republic's main towns are its capital and ship-repairing centre of PETROZAVODSK, and the ports of Kem' on the White Sea and Sortalava on Lake Ladoga.

Population 799 000
Map Russia – Eb

Karen *Myanmar (Burma)* Hilly state formerly called Kawthulei and covering 30 383 km² (11 731 sq miles). It lies in the south-east of the country, bordering Thailand. The Karen people (Kayens) of this isolated region have been fighting for separation from Myanmar since the country gained independence in 1948. Pa-an, the state capital, is linked to the outside world by the Thanlwin (Salween) River. Most Karens are subsistence farmers, growing mainly rice.

Population 1 057 500
Map Myanmar – Cc

Kariba *Zimbabwe/Zambia* Hydroelectric dam on the ZAMBEZI River, 150 km (93 miles) south of the Zambian capital, Lusaka. At 5230 km² (2019 sq miles), the lake created by the dam is the largest artificial expanse of water in Africa, after Lake Volta in Ghana. Lake Kariba, whose southern shoreline is largely in Zimbabwe, reached its present size in 1961 after Operation Noah rescued thousands of wild animals trapped by the rising waters. The area surrounding the lake still has plenty of wildlife, and parts of it have been declared a nature reserve.

Map Zambia – Bb; Zimbabwe – Ba

Karjala *Russia* See KARELIA

Karl-Marx-Stadt *Germany* See CHEMNITZ

Karlovy Vary (Carlsbad; Karlsbad) *Czech Republic* Health spa 115 km (70 miles) west of the capital, Prague. It is named after its founder, King Karl (Charles) I of Bohemia (1347-78), who was crowned Holy Roman Emperor Charles IV in 1355. The waters of the spa are said to cure digestive disorders.

Karlovy Vary's present luxurious hotels, sanatoria, baths and colonnades date mostly from its heyday in the 19th century when it was patronised by royalty, aristocrats and famous people from all over Europe, including the German composer Johannes Brahms (1833-97), the Polish composer Frédéric Chopin (1810-49) and the Russian poet Alexandr Pushkin (1799-1837). Karlovy Vary hosts an international film festival each year. Its manufactures include fine porcelain and crystal objects.
Population 59 200
Map Czech Republic – Aa

Karlskrona *Sweden* The country's main naval base, and a tourist centre on the Baltic coast, 220 km (136 miles) east of HELSINGBORG. The layout of its streets and dockyard – built on the mainland and five neighbouring islands – dates from the late 17th century; its dry docks are hewn out of solid granite rock.

In the 18th century, Karlskrona was Sweden's second largest city. Today, many Swedish cities have outgrown it. The city's main manufactures include telephones, light bulbs, porcelain goods and canned fish.
Population 59 400
Map Sweden – Bd

Karlsruhe *Germany* University and industrial city in the Rhine Valley, 65 km (40 miles) northwest of STUTTGART. It was laid out in the shape of a wheel in 1715 as a new capital city for Karl Wilhelm, ruler of Baden. His palace, which is now a museum, stood at the hub, and the avenues and streets radiated outward like spokes.

Today, Karlsruhe has become Germany's second largest inland port and a major industrial centre, with oil refineries, nuclear research laboratories, and cosmetics, electronics, tyre, bicycle and motorcycle industries.
Population 275 100
Map Germany – Cd

Karlstad *Sweden* Provincial capital of Varmland, situated on the north shore of Lake VÄNERN, 305 km (190 miles) west of Stockholm. The treaty ending the union of Sweden and Norway was signed here in 1905. Softwood products, chemicals, textiles and clothing are manufactured in the city.
Population 77 300
Map Sweden – Bd

Karnafuli *Bangladesh* River draining the Chittagong Hill Tracts and reaching the sea at CHITTAGONG. Its estuary provides Chittagong's principal harbour area. The river has been dammed upstream at Kaptai to provide power and increase the water depth available at Chittagong in the dry season. Boats travelling up and down the river have to be lifted over the dam by a crane.
Map Bangladesh – Cc

Karnak *Egypt* Temple site on the east bank of the Nile, 500 km (310 miles) south of CAIRO. The ruins at Karnak, the most extensive group of surviving ancient buildings in the world, represent the creative peak of the brilliant New Kingdom, whose pharaohs ruled Egypt from Thebes for more than 500 years. Most of the buildings date from about 1560 to 1090 BC, and are temples to Amon, the supreme god of the 18th dynasty.

The colonnaded hall at Karnak was built mainly for the warrior King Rameses II who reigned for 66 years in the 13th century BC; it is probably the largest single room in a religious building and could hold Paris's Notre Dame Cathedral with space to spare. It has 134 pillars, many bearing inscriptions praising Rameses. Beyond, in an avenue of pillars stretching more than 400 m (1300 ft), stands an inscribed obelisk, shaped from a single block of granite quarried at Aswan. It was erected by Queen Hatshepsut, the dynasty's only queen, who ruled around 1500-1480 BC.
Map Egypt – Cc

Karnataka *India* State in the south-west, covering 191 791 km² (74 031 sq miles). It was established in 1956-60 from parts of the older states of Mysore, Hyderabad and Bombay. It includes the distinctive coastal Konkan area and parts of the Western GHATS and the DECCAN plateau. Its capital is the city of BANGALORE.
Population 44 977 200
Map India – Bd

Kärnten *Austria* See CARINTHIA

Karoo (Karroo) *South Africa* Arid and semi-arid region of the Northern, Western and Eastern Cape provinces, extending into the Free State. It is divided into the Great Karoo and the Little Karoo. The arid Great Karoo extends north from the Swartberg Mountains into central South Africa, while the semi-arid Little Karoo lies between the Swartberg range in the north and the mountains that mark the inland limit of the coastal plateau to the south.

The typical landscape consists of flat plains with low-growing scrub, dotted with characteristic hills, or *koppies*. Fertile when irrigated, the Karoo is used mainly for grazing sheep and goats.
Map South Africa – Bc

Karroo *South Africa* See KAROO

Kars *Turkey* Fortified town near the northeastern border with Armenia, about 1000 km (620 miles) east of ANKARA. It was occupied by Russia three times and was ceded to the Russians with the surrounding province of the same name in 1878, but returned to Turkey in 1921. The town produces dairy goods, textiles and carpets, and the mountainous province, which covers 17 379 km² (6710 sq miles), has large salt deposits.
Population (town) 70 400; (province) 662 120
Map Turkey – Ca

karst Barren limestone or dolomite region in which erosion has produced rock PAVEMENTS, GRIKES, SINKHOLES, underground streams, and caverns. See HOW CAVES ARE FORMED, p 141

Karst *Slovenia* See KRAS

Karun *Iran* River rising in the Zagros Mountains south-west of the capital, Tehran, and flowing

848 km (527 miles) to the Shatt al Arab waterway near the tip of The Gulf. It is navigable for 150 km (95 miles), from the Shatt al Arab waterway to the inland port of AHVAZ.
Map Iran – Aa

Kasai (Cassai) *Angola/Zaire* River, 2150 km (1350 miles) long, which rises in the eastern plateaus of Angola, flowing first east and then north to form part of the border between Angola and Zaire. It continues north and west to its confluence with the Zaire River. The upper course contains a number of waterfalls and rapids, but a navigable waterway runs for 772 km (479 miles) below the river port of ILEBO, providing an outlet for the mineral wealth of the Shaba region.
Map Zaire – Ab

Kasai Occidental (Western Kasai) *Zaire* Administrative region covering 156 967 km² (60 589 sq miles) in south-central Zaire, and producing gem diamonds and gold. Besides mining, the main activity is farming, and the region produces coffee, cotton, palm products and livestock. Most of Kasia Occidental's population belong to the Lulua group. The capital is Kananga; other towns include Tshikapa and Ilebo.
Population 3 465 760
Map Zaire – Bb

Kasai Oriental (Eastern Kasai) *Zaire* Administrative region covering 168 216 km² (64 931 sq miles) in south-central Zaire, with its capital at Mbuji-Mayi. Inhabited mainly by members of the Luba group, the region produces most of the world's industrial diamonds. Coffee, cotton, palm products, rubber and livestock are the main agricultural products.
Population 2 859 220
Map Zaire – Bb

Kasama *Zambia* Market town situated 700 km (435 miles) north-east of the capital, Lusaka. It is the regional capital of Northern Province, and has a coffee processing plant.
Population 9000
Map Zambia – Cb

Kashan (Keshan) *Iran* City at the edge of the Dasht-e-Kavir Desert, 175 km (110 miles) south of the capital, Tehran. Its chief products are carpets, cotton, silk and copperware. The city's skyline is dominated by a 13th-century minaret 45 m (148 ft) high.

Just to the west of Kashan, Shah Abbas I (ruled 1587-1629), who is buried in the city, laid out his magnificent *Bagh-i-Shah* ('Garden of the King'), where cypress trees more than 300 years old, fountains and tiled canals remain.
Population 155 000
Map Iran – Ba

Kashi (Kashgar) *China* City in the Xinjiang Uygur Autonomous Region, 200 km (125 miles) from the border with Kyrgyzstan. It was formerly an important oasis on the Silk Road trade route to Central Asia; now it is the economic centre of southern Xinjiang.

With most of its inhabitants being members of the Muslim faith, the city occasionally experiences tension between the Uygur majority and the Han rulers; in 1990, riots led to the city being closed to foreigners for a time.
Map China – Ac

Kashmir *India/Pakistan* See AZAD KASHMIR; JAMMU AND KASHMIR

Kassala *Sudan* Market town 400 km (250 miles) east of the capital, Khartoum (to which it is linked by road, rail and air). It is the capital of a cotton-growing province of the same name, which became home to thousands of refugees who fled the Ethiopian civil war in the 1980s.
Population (town) 143 000
Map Sudan – Bb

Kassel *Germany* Commercial and industrial city on the Fulda River, 155 km (95 miles) north-east of FRANKFURT am Main. It has machinery, precision instruments, electronics and textile industries. The brothers Jakob and Wilhelm Grimm, who worked in the city library, lived in Kassel from 1798 to 1828, during which time they published their fairy tales. The 'Documenta', an international arts festival and modern art exhibition takes place here.
Population 194 300
Map Germany – Cc

Kasserine (Al-Qasmayn) *Tunisia* District capital and major road and rail junction, 216 km (134 miles) south-west of Tunis. It is situated at the foot of Jebel Chambi and close to the El Douleb oil field. The rolling land to the south and north of Kasserine and the narrow valleys hedged in by steep, rocky cliffs, which form the Kasserine Pass, was the site of a furious series of attacks and counter-attacks in 1943 between Rommel's Afrika Korps and the Allied Forces.
The German defeat at Kasserine made it one of the turning points in the Allied push for victory in North Africa. Today, the town is the centre of an irrigated agricultural area in which wood pulp and cellulose are manufactured.
Population 47 700
Map Tunisia – Aa

Kastoría *Greece* Town on a peninsula in Lake Kastoría, Macedonia, 221 km (137 miles) west of SALONICA. Since 200 BC, Kastoría has been occupied successively by Romans, Bulgars, Serbs and Turks (1385-1912), who made it a centre for the Middle Eastern trade in bear, fox and beaver furs. Still a fur trading centre today, it holds an annual fur fair in March. It is noted for its beautiful 18th-century houses and 72 Byzantine churches with medieval frescoes, of which the oldest – Anargyroi and Varlaam – date from the 10th century. The nearby village of Árgos Orestikón is famed for its woollen carpets.
Population 20 700
Map Greece – Ba

Kastrup *Denmark* See AMAGER

Kasvin *Iran* See QAZVIN

katabatic wind Valley wind blowing downhill during the night, caused by the flow downward of dense air chilled by radiation on the upper slopes. See also ANABATIC WIND.

Katanga *Zaire* See SHABA

Katherine Gorge *Australia* Magnificent gorge of the Katherine River, situated in the Northern Territory, 275 km (170 miles) south-east of Darwin. It is about 12 km (7.5 miles) long and up to 100 m (330 ft) deep, but its water level may rise 18 m (60 ft) during the wet season. An oasis in an arid region, the Katherine Gorge teems with amphibians, reptiles (including snakes and crocodiles), birds, kangaroos and wallabies.
Map Australia – Ea

Kathmandu (Katmandu) *Nepal* The country's capital and also the name of the fertile valley around the city. Kathmandu lies in central Nepal, some 90 km (55 miles) from the Indian frontier. Founded in AD 723, it became the national capital in the late 18th century. In 1934 it was badly damaged by an earthquake.
In the 1960s, it became a favourite destination for hippies, who travelled here from developed countries to sample the cheap and readily available drugs. However, the implementation of tough new drug laws in the 1980s and 1990s put the trade into decline.
The city centre – around Durbar Square – is a maze of narrow streets, houses with wooden balconies, markets and Hindu and Buddhist temples. Further out, there are modern European-style buildings, including the royal palace with some 1700 rooms, built around 1900. (The royal palace lies in the old part of the city.) The Buddhist temple and monastery of Swayambhu, whose terraces are the haunt of monkeys, are on a hill overlooking the city. The temple is topped by a four-sided *stupa*, or shrine, of which each face is painted with an eye to symbolise the all-seeing wisdom of the Buddha.
Population (city) 235 160; (district) 400 000
Map Nepal – Ba

Katowice (Kattowitz) *Poland* Southern industrial city in Upper Silesia, 72 km (45 miles) north-west of the city of CRACOW. Katowice, which has developed into a cultural centre, has four technical colleges, a mineralogical research institute, several theatres and museums as well as the region's library. Its main industries pivot around coal, zinc, steel and engineering. In the city centre, a striking three-winged monument commemorates the Silesian Uprisings of 1919-21, when the city's Polish majority rebelled successfully against German rule.
The Bledow Desert (also called Pustynia Bledowska), the only true sand desert in Europe, lies 40 km (25 miles) to the east. During the Second World War, the German General Erwin Rommel (1891-1944), who became known as the 'Desert Fox', trained his Afrika Korps on the 32 km² (12 sq miles) of shifting dunes.
Population 366 800
Map Poland – Cc

Katsina *Nigeria* Northernmost town in Nigeria and capital of the state of the same name, situated only 20 km (12 miles) from the border with Niger. Founded in the 11th century at the end of a Saharan caravan route, it flourished from 1591 onwards after the fall of the Songhai empire to the west which resulted in an influx of Songhai scholars and craftsmen, especially leather workers. At the same time, caravans brought salt, slaves, leather, kola nuts and European goods into the town. However, Katsina was sacked during a religious war in 1807. Today it markets camel hides, leather goods, cotton, and groundnuts. It is a Muslim seat of learning.
Population (town) 145 500; (state) 3 878 340
Map Nigeria – Ba

Kattegat (Cattegat) *Denmark/Sweden* Strait covering 25 000 km² (9650 sq miles) between south-western Sweden and the east coast of Denmark's JUTLAND Peninsula, and forming part of the sea passage between the NORTH and BALTIC seas. Its name, Swedish and Danish for 'cat's throat', might be a reference to its shape, since it is 220 km (nearly 140 miles) long, and only 34-160 km (21-100 miles) wide; it is 124 m (407 ft) deep at its deepest point. Ferries run from Frederikshavn and Grenå in Denmark to Varberg and Gothenburg in Sweden. The area is a popular sailing centre in summer.
Map Denmark – Ba; Sweden – Ad

Kattowitz *Poland* See KATOWICE

Katyn *Russia* Forest and village some 20 km (12 miles) west of SMOLENSK. In 1940, during the Second World War, some 4500 Polish army officers were massacred nearby shortly after they and about 11 000 others (whose whereabouts are unknown) had been taken prisoner by the Russians.
Although their graves were discovered in 1941, the discovery did not become generally known until two years later. The Russians finally admitted responsibility in 1990 for the massacre carried out by the Soviet secret service (NKVD), and claimed that thousands of Soviet citizens had also been disposed of in a similar way at Katyn.
Map Russia – Ec

Kaunas (Kovno) *Lithuania* Industrial city about 100 km (62 miles) south-east of KLAIPEDA. From 1918 to 1939 it was the capital of independent Lithuania before Lithuania was annexed by the USSR. Under Soviet rule, Kaunas (then renamed Kovno) flourished as a university town, river port, railway junction and industrial centre.
Today it is a cultural centre with several technical colleges, museums and theatres. Its main products are metal goods, electrical motors, textiles, processed foods and paper.
Population 433 000
Map Lithuania – Ba

Kaválla *Greece* City and port of MACEDONIA, 125 km (78 miles) east of SALONICA. A Roman fleet led by Brutus and Cassius landed here after the assassination of Julius Caesar and before the Battle of Philippi, which took place another 15 km (9 miles) to the north-west in 42 BC.
Under Byzantine rule it was called 'Christopolis'. Its present name derives from the horses (*cavalli*) left behind by Crusaders who boarded ships to Asia Minor in the port. Kaválla has a well-preserved Byzantine citadel with an aqueduct, a mosque and old Turkish houses, baths and shops. It exports tobacco, clothing, wood, furniture and plastics; offshore oil wells operate from Prinos. St Paul preached for the first time in Europe at Philippi and wrote an epistle to the Philippians.
Population 56 400
Map Greece – Da

Kavango *Namibia* Region, and formerly a homeland called Okavango, set up in 1970 for the Kavango people on the Angolan border, and bordered in the north by the KAVANGO River. A subsistence farming area, covering 43 417 km² (16 759 sq miles), it is centred on the regional capital of Rundu (population 19 000).
Population 136 590
Map Namibia – Aa

Kavango (Cubango) *Southern Africa* River, formerly known as the Okavango, which rises in central Angola, where it is called the Cubango. It flows for about 1600 km (1000 miles) in a generally south-easterly direction, along part of the border of Angola and Namibia, and on into the OKAVANGO DELTA in Botswana.
Map Angola – Ab; Namibia – Aa

Kaveri (Caveri; Cauvery) *India* River in the south, rising in the DECCAN Plateau and flowing first south-east, then east, for 764 km (475 miles) to the Bay of Bengal. The Mettur Dam, completed in 1934, is 1615 m (5300 ft) long and 54 m (176 ft) high. It was the largest in the world when built and was the first combined power and irrigation dam in India.
Map India – Ce

Kavkaz *Asia/Europe* See CAUCASUS

Kawasaki *Japan* Industrial city and port in the Tokyo-Yokohama conurbation. It has oil refineries, chemical plants, shipbuilding plants and steel works built on land reclaimed from Tokyo Bay.
Population 1 174 000
Map Japan – Cc

Kawthulei *Myanmar (Burma)* See KAREN

Kayah *Myanmar (Burma)* Tiny eastern border state of 11 733 km² (4530 sq miles) which straddles the Thanlwin (Salween) River. A hydroelectric power station at Lawpita Falls on the Baluchaung River supplies electricity for most of southern Myanmar. The inhabitants, mostly Karens (Kayens), are largely subsistence farmers. Loikaw is the state capital.
Population 168 400
Map Myanmar – Cc

Kayes *Mali* Town 100 km (60 miles) from the border with Senegal. It was the terminus of the rail link between the Niger and Senegal rivers until the railway was extended to the Senegalese capital, Dakar in 1923.
Population 45 000
Map Mali – Ab

Kayseri *Turkey* Textile town and carpet-trading centre 230 km (140 miles) south-east of ANKARA. It is the capital of an agricultural province of the same name. The town lies at the foot of an extinct volcano, Erciyas Daği, that rises to 3916 m (12 848 ft). Its main industry is rug and carpet weaving.
The province, which covers 16 917 km² (6532 sq miles), produces fruit, raisins, millet and rye.
Population (town) 461 415; (province) 943 480
Map Turkey – Bb

Kazakhstan See p 362

Kazan' *Russia* Capital of the Tatar Republic, port and industrial city, 700 km (434 miles) east of MOSCOW. Founded as Bolgar in the 13th century by the Golden Horde, the Mongol army that swept over Eastern Europe, it was captured and destroyed by Tsar Ivan the Terrible (1530-84) in 1552. During Ivan's reign, the city's kremlin (fortress), which is still standing, was built.
Its university was founded in 1804. Leo Tolstoy (1828-1910), the author of *War and Peace*, and the revolutionary leader Vladimir Lenin (1870-1924) are recorded as being among the students at the university.
Set on a dam and tributary 5 km (3 miles) from the Volga, Kazan' is the industrial centre of the booming Volga-Urals region. Its traditional industries producing leather, silk and fur goods have been joined by chemicals, electrical equipment, machinery, tools, and paper as well as oil refining and shipbuilding plants.
Population 1 107 000
Map Russia – Fc

Kazanluk (Kazanlâk) *Bulgaria* Town, 47 km (29 miles) south of Gabrovo on the Sofia-Burgas and Ruse-Kurdzhali rail and road routes. It is the capital of the Valley of Roses (between the Balkan Mountains and Sredna Gora Mountains) and the centre of the attar of roses industry. Its rose petals are harvested in May and June, each day's crop being gathered before the sun robs the petals of their oil (an important base ingredient for many perfumes).
The Georgi Dimitrov hydroelectic power station, beneath which lies the ancient Thracian capital of Sevthopolis, is situated 16 km (10 miles) to the south-west on the Tundzha River.
Population 61 000
Map Bulgaria – Bb

Kebnekaise *Sweden* Peak in the Kjölen Mountains in north-west LAPPLAND, and, at 2111 m (6945 ft), the highest point in Sweden.
Map Sweden – Cb

Kecskemét *Hungary* City in the centre of the country on the Great ALFÖLD, 84 km (52 miles) south-east of Budapest. It is the market centre of a rich agricultural area producing fruit, a potent apricot brandy known as *barackpálinka*, livestock and leather goods. The city is also the regional capital of KISKÚNSÁG – the area between the Danube and Tisza rivers.
Kecskemét was the birthplace of the composer Zoltán Kodály (1882-1967), and there is a music institute named after him. The city centre is popular with artists, and contains buildings in the distinctive Hungarian art nouveau style developed in the 1890s by the architect Ödön Lechner. The city also has two fine Baroque churches and a restored synagogue.
Population 104 560
Map Hungary – Ab

Kedah *Malaysia* State on peninsular Malaysia's north-west coast, covering 9425 km² (3639 sq miles). It has the country's largest area of rice lowland, in the irrigated plain of the Muda River near the state capital, Alor Setar. Rubber is grown on hilly slopes north of Kedah Peak.
Population 1 412 000
Map Malaysia – Ba

Kediri *Indonesia* Town in East Java and the centre of a rice-growing region. It is famous for its handicrafts, such as basketwork, silverwork, rattan and batik.
Population 200 000
Map (Java) Indonesia – Cd-Dd

Keeling Islands *Indian Ocean* See COCOS ISLANDS

Keelung (Chi-lung) *Taiwan* Major seaport and industrial centre on the north coast, 30 km (20 miles) north-east of T'ai-pei. One of the oldest settlements in Taiwan, it was occupied by Spanish and Dutch colonists in the 17th century and became a flourishing port during the period of Japanese colonial rule over Taiwan (1895-1945).
Parallel with Taiwan's national economy, it has grown rapidly since the mid-1960s and now handles more than 85 million tonnes of cargo a year. Exports include rice, mushrooms, canned agricultural products and, increasingly, industrial products. The coastline to the north-west is famed for its sculptured rock formations.
Population 362 000
Map Taiwan – Bb

Kefallinia *Greece* See CEPHALONIA

Kefar Nahum *Israel* See CAPERNAUM

Keflavík *Iceland* Town 35 km (22 miles) south-west of REYKJAVÍK, and site of the country's main civil airport. The airport, part of which is a United States military base, was built in 1941 as a wartime staging post for the Allies. There is a fishing harbour nearby.
Population 7300
Map Iceland – Bb

Kei (Great Kei) *South Africa* River in the Eastern Cape, flowing 640 km (397 miles) from the southern Drakensberg range to the Indian Ocean. Historically a frontier between European and Xhosa territories, it later formed the western border of the so-called 'independent' Republic of Transkei (1976-1994).
Map South Africa – Cc

Keihin *Japan* The country's prime industrial zone. It lies on south-east HONSHU Island and includes the cities of Tokyo and Yokohama. It grew rapidly in the 1930s as the country expanded its armaments industry ahead of the Second World War. Since 1945, it has switched to car manufacturing, printing and publishing, shipbuilding, oil refining and steelmaking.
Map Japan – Cc

Kékes *Hungary* See MÁTRA

Kelang *Malaysia* River rising in central peninsular Malaysia and flowing through the national capital, Kuala Lumpur, into the Strait of Malacca at Port Kelang (formerly Port Swettenham). This modern container port has been expanded to reduce Malaysia's dependence on Singapore. The town of Kelang, 19 km (12 miles) upstream from the port, is a commuter town for the capital.
Population (town) 192 100
Map Malaysia – Bb

Kelantan *Malaysia* Densely populated rice-growing state in north-east Malaysia on the Thai border, covering 14 943 km² (5769 sq miles). Its main town is Kota Baharu. Development of the region – one of Malaysia's poorest – has been slow, in part because of its isolation, and in part because Kelantan, heartland of Islamic fundamentalism in Malaysia, did not vote for the ruling government coalition in the 1960s.
Recently, however, the government has implemented several agricultural reform programmes and developed communications to improve the lot of the inhabitants.
Population 1 220 000
Map Malaysia – Bb

Kazakhstan

A MINERAL-RICH LAND BETWEEN THE URAL AND THE ALTAI MOUNTAINS, AND THE ARAL AND THE CASPIAN SEAS

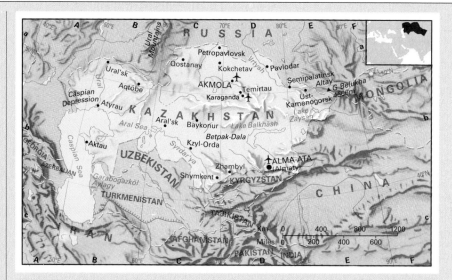

A Central Asian republic, Kazakhstan is more than 10 times the size of the United Kingdom and ranks among the 10 largest countries of the world. It is the second largest of all the former Soviet republics, stretching 3000 km (1860 miles) across from the Caspian Sea in the west to the Chinese region of Xinjiang Uygur in the east.

In the south-east, its boundaries with Kyrgyzstan mostly follow the course of the mountain ranges of the TIEN SHAN, reaching heights of more than 4000 m (12 120 ft). The borders with Uzbekistan and Turkmenistan are almost straight, running through the middle of Kizil Kum Desert, the ARAL Sea and the hilly plains on the shores of the Caspian Sea. These artificial boundaries were created in the 1920s by the Soviet government. About 1700 km (1060 miles) to the north of the Kizil Kum, the Kazakh steppes merge into the Siberian plains.

The land of the Kazakhs consists for the most part of lowland, which includes the Caspian plains and the Aral lowland; only about one-fifth lies at an altitude of more than 500 m (1640 ft). The vastness of the land is enhanced by endless steppe and desert landscape. High rainfalls in the extreme north encourage scattered tree growth but rain decreases further south and so the wooded tracts of wetland steppe merge into desert.

Kazakhstan has a continental climate, with vast fluctuations in temperature. In the capital, ALMA-ATA, temperatures of more than 40°C (104°F) have been recorded in the summer; winters bring temperatures of –50°C (–90°F).

The steppe originally provided grazing for the herds of the nomadic Kazakh and other Central Asian peoples. But now the economy has changed radically, and so has the lifestyle of the population. Farming land was gained, after 1954, when vast areas of the steppe were ploughed and fertilised. Kazakhstan soon became one of the largest wheat-growing regions of the Soviet Union, accounting for about one-fifth of the USSR's total arable land and about one-seventh of its wheat production. Agriculture today accounts for almost 40 per cent of the country's output. Besides wheat, wool and meat are major agricultural exports.

Kazakhstan's agricultural reforms in the 1950s were successful, but unpredictable rainfalls still cause crop failures once every 10 years. There has been wind erosion making agricultural yields set to dwindle in the future.

One of the ecologically most endangered areas is the Semipalatinsk region where radioactive contamination due to atomic testing has occurred. Another environmental problem area is the Aral Sea, which is drying up, causing high concentrations of chemical pesticides and natural salts.

Industrial pollution, too, is becoming all too evident. The country began its rapid industrialisation in the 1940s. During the Second World War the Soviet government dismantled important factories in the west and reassembled them further to the east, away from the war front. Kazakhstan became an industrial nation almost overnight. Rich deposits of coal, petroleum, natural gas, chrome, molybdenum, nickel, cobalt, lead, zinc, gold, silver and bauxite turned the Karaganda region and the Altai and Ural foothills into giant industrial centres.

Today, industry and construction employs more than 30 per cent of the country's workforce. Industrial development attracted many foreigners. Beside the Muslim Kazakhs, who account for about 42 per cent of the population, there are Russians (making up 37 per cent of the total population), Ukrainians (5.2 per cent), German speakers (4.7 per cent), Uzbeks (2 per cent) and other groups.

RUSSIAN MINORITY

Some ethnic tensions have broken out in Kazakhstan since independence. The society is Muslim-dominated, but there are large Christian (mainly Russian Orthodox) groups. In December 1992, relations with the country's Russian minority became strained, when Russian demonstrators in Ust-Kamenogorsk demanded a revision of the citizenship and language laws. When the new constitution was adopted in January 1993, it made Kazakh the state language. Although it gave special status to Russian, it specified that the country's president had to be a speaker of Kazakh.

The Republic of Kazakhstan proclaimed its independence in December 1991 – it had formed part of the region administered by the Soviet Union's General Government of the Steppes since 1731. Since independence, it has kept close ties with its former colonial master. Despite its vast oil, coal and agricultural resources it is still dependent on trade with Russia. Although most of its exports go to the former Soviet states, with which it formed the Commonwealth of Independent States (CIS), it has begun to steer its own course towards political and economic independence.

One of the obstacles to real independence is the fact that Kazakhstan is landlocked. Initially, oil trade with the West was hindered by the fact that all the pipelines run across former Soviet republics. But in 1993 a 40-year joint United States-Kazakh oil venture was signed to develop and exploit the Kazakh Tengiz and Korolev oil fields. Oil deals with France and other Western countries followed.

In 1992 the Kazakh government freed many prices and embarked upon economic reforms. However, in that year national output dropped by 15 per cent. A national privatisation programme for 1993-5 was launched at the beginning of 1993.

With promises of foreign aid, its vast mineral resources, and its privatisation scheme, Kazakhstan has been able to make the transition from a planned to a market economy.

KAZAKHSTAN AT A GLANCE	
Official name Republic of Kazakhstan	
Area 2 717 300 km² (1 048 907 sq miles)	
Population 16 156 370 **Per km²** 6 **(Per sq mile)** 15	
Capital Alma-Ata (Almaty)	
Government Republic	
Currency 1 Russian rouble = 100 copeks (transitional)	
Languages Kazakh (official), Russian	
Religions Muslim (47%), Russian Orthodox (15%), other (38%)	
Climate Continental climate, with hot summers and cold winters. Arid in the west, with more rainfalls in the east. Average temperature in Alma-Ata range from –8.8°C (–16°F) in January to 22.2°C (72°F) in July	
Land use Grazing 59%, cultivation 15%, forest and woodlands 4%, other 22%	
Main primary products Wheat; ferrous and non-ferrous metals, coal, oil	
Major industries Mining industries, iron and steel production, agricultural machinery, chemicals, construction materials	
Main exports Grains, wool, meat, cotton; oil, ferrous and non-ferrous metals, coal, chemicals	
Annual income per head (US$) 1680	
Population growth (per thous/yr) 7	
Life expectancy (yrs) Male 63 **Female** 73	

Kelibia *Tunisia* Charming fishing port and tourist centre on the south-eastern shore of the Maouin Peninsula, which ends at Cape Bon. It is dominated by a great Roman castle. Tomatoes, oranges and grapes are grown in the surrounding country and muscadel wine is made here. The Cape Bon lighthouse stands along the coast.

Population 24 600
Map Tunisia – Ba

Kells (Ceannanus Mór) *Ireland* Ancient town, situated 58 km (36 miles) north-west of DUBLIN. St Columba founded a 6th-century monastery here which became a refuge for Iona monks. It was dissolved in 1551, but a round tower, high crosses and other buildings survive. The 8th-century *Book of Kells*, an illuminated edition of the Gospels made by the monks at Kells, and one of the most precious books in the world, is now kept in the library at Trinity College in Dublin.

Population 2180
Map Ireland – Cb

Kelsty *Poland* See KIELCE

Kemerovo *Russia* Industrial city in southern Siberia, 195 km (120 miles) north-east of NOVO-SIBIRSK. It developed mainly after 1928 with the opening of the nearby Novokuznetsk coal field. Its industries include coal, coke chemicals, fertilisers, pharmaceuticals, and agricultural machinery.

Population 521 000
Map Russia – Jc

Kemi *Finland* Town and seaport founded in 1867 on the north coast of the Gulf of Bothnia, 25 km (15 miles) from the Swedish border. It is situated at the mouth of the Kemi River, Finland's richest source of hydroelectric power. The town is one of the main centres of Finland's timber industry.

Population 25 100
Map Finland – Cb

▼ **FANTASTIC FLORA** Giant plants such as 9 m (30 ft) tall groundsels grow on the upper slopes of Mount Kenya and other high peaks in equatorial Africa.

Kempenland (Campine) *Belgium* Heath and forest-covered plateau region, 45 km (28 miles) north of the city of Liège. It extends just across the Dutch border, covers about 6000 km² (2300 sq miles) and contains Belgium's only working coalfield. Its largest town is GENK.

Population 200 000
Map Belgium – Ba

Kempton Park *South Africa* Residential and industrial Witwatersrand town 24 km (15 miles) east of JOHANNESBURG and 40 km (25 miles) south of PRETORIA. It is the site of South Africa's first aircraft factory; adjoining it is Jan Smuts international airport. Negotiations preceding the country's first democratic election in 1994 were held in its World Trade Centre.

Population (town) 106 610; (district) 354 790
Map South Africa – Cb

Kenana *Sudan* Vast, state-run sugar plantation, covering 336 km² (130 sq miles). It lies in southern El Gezira, just east of the White Nile and 290 km (180 miles) south of the capital, KHARTOUM. Officially opened in 1951, it is one of the world's largest sugar complexes with finance from the Sudanese, Kuwaiti and Saudi Arabian governments, and international groups. Kenana draws its irrigation water from the White Nile.

Map (Khartoum) Sudan – Bb

Kenema *Sierra Leone* Eastern regional capital in the Kambui Hills, 225 km (140 miles) south-east of the national capital, Freetown. Kenema is a market for locally produced cocoa, palm oil, coffee, ginger and timber, and for diamonds which are sifted from local river gravels.

The town has a saw-mill and a furniture factory. It has government and missionary secondary schools, a Catholic teacher training college for women, a technical institute, as well as a government library, and a hospital.

Population 14 000
Map Sierra Leone – Ab

Kengtung *Myanmar (Burma)* Town in Shan State, one of the areas forming the opium poppy-growing GOLDEN TRIANGLE. The town is reputed

to be the centre of the illegal opium trade carried out in the surrounding hills east of the Thanlwin (Salween) River.

Population 174 000
Map Myanmar – Cb

Kénitra (El Qenitra; Mina Hassan Tani) *Morocco* Port, formerly known as Port Lyautey, on the Sebou River, 40 km (25 miles) north of RABAT. Founded by the French in 1913, it is the main outlet for the RHARB Plain and the agricultural areas of FÈS and MEKNÈS.

Its industries include textiles, processed fish, fertilisers and tobacco. Nearby is Mehdiya, a former Carthaginian trading post and site of the 1st-century Roman port of Thamasada.

Population 189 000
Map Morocco – Ba

Kennedy, Cape *USA* See CANAVERAL, CAPE

Kent *United Kingdom* County in the south-east corner of England, covering 3732 km² (1441 sq miles) and crossed by the chalk hills of the North DOWNS. The centre of the county, called the Weald, was once covered in dense wood. Kent is known as 'the Garden of England' because of its orchards and spring flowers. The pastures of Romney Marsh are noted sheep grazing areas.

Paper making and cement manufacturing industries line the Thames and Medway rivers. Margate and Ramsgate are popular tourist resorts, while the ports of DOVER and Folkestone handle most of Britain's cross-Channel traffic. Kent has become known as 'the Gateway to Europe' – not only because of its ports, but as a result of the major developments associated with the Channel Tunnel. Inland, tourists flock to the cathedral city of CANTERBURY. The county town is Maidstone.

Population 1 536 100
Map United Kingdom – Fe

Kentucky *USA* South-central state covering 102 930 km² (39 732 sq miles) of mostly rolling Blue Grass hills linking the Appalachian Mountains in the east to the Mississippi River in the west. The Ohio River borders the state to the north. Thoroughbred horses have long been symbolic of Kentucky, and the famous Kentucky Derby is run annually at Churchill Downs.

The state is also the world's largest producer of bourbon and the country's largest producer of bituminous coal. Other main products include tobacco, whiskey, machinery, foodstuffs and cigarettes. FRANKFORT is the state capital.

Population 3 685 300
Map United States – Ic

Kenya See p 366

Kenya, Mount (Kirinyaga) *Kenya* Second highest African mountain after Kilimanjaro, this long-extinct volcano rising to 5199 m (17 057 ft) is high enough to be glaciated between its jagged peaks even though it lies almost on the equator. The slopes between 4600 and 3200 m (15 100 and 10 500 ft) consist of moorland dotted with giant heather, groundsel and lobelia.

Below them there is forest, including a zone of bamboo – and lower than about 1700 m (5600 ft), the southern and eastern slopes are farmed by Kikuyu, Embu and Meru people. The main crops are maize and beans, with coffee and tea as cash crops. The drier western slopes yield grains and provide grazing. The first recorded ascent of Mount Kenya, or Kirinyaga (Mountain of Whiteness), as it is often called today, was by the British geographer Sir Halford Mackinder in 1899.

Map Kenya – Cb

Kerala *India* South-western state bordering on the Arabian Sea, and covering 38 863 km² (15 005 sq miles). It has a high rainfall, in many places more than 2997 mm (118 in) annually, and much of it is covered by tropical rain forest. The coastal area is dotted with lagoons and crisscrossed with channels; not surprisingly, most of the travelling here is done by small boats.

Fishing and farming are the main occupations. Agricultural products include spices (mainly pepper), rice, timber, rubber and tea. There are some textile and chemical industries and aluminium mines. Kerala's capital is the port of TRIVANDRUM.

Population 29 011 200

Map India – Be

Kerch *Ukraine* Crimean port between the Sea of AZOV and the BLACK SEA. One of the region's oldest cities, it was known as Pantikapaion when it was founded by the Greeks in the 6th century BC. It was later ruled by Rome, Genoa, Turkey and, from 1771 to 1991, Russia.

Many important relics have been discovered in the area, including the remains of a Greek acropolis and Roman burial mounds and catacombs; articles from these are kept in the city's history museum. Kerch is one of the Crimea's chief fishing bases, and exports iron ore. It has iron works and chemicals and shipbuilding industries.

Population 176 000

Map Ukraine – Db

Kerguelen *Indian Ocean* Largest of a group of about 300 islands, islets and rocks, Kerguelen covers 3414 km² (1318 sq miles) and forms part of the FRENCH SOUTHERN AND ANTARCTIC TERRITORIES. It lies 5310 km (3300 miles) south-east of Cape Town, in the southern Indian Ocean, and was aptly named 'Desolation Island' by the French navigator Kerguélen-Trémarec, who discovered it in 1772.

Much of its western half is a huge glacier; snowfields carpet the centre, in places giving way to peat marshes and glacial lakes. The island is mountainous, rising to 1850 m (6070 ft) at Mount Ross in the south. Kerguelen has belonged to France since 1893, but was not occupied effectively until 1949. A permanent base and scientific centre, Port-aux-Français, which was built in 1950, now has a staff of about 75.

Map Indian Ocean – Ce

Kericho *Kenya* Highland town 195 km (120 miles) north-west of NAIROBI. It is the centre of Kenya's leading tea-growing district.

Population 40 000

Map Kenya – Cb

Kérkira *Greece* See CORFU

Kerkouane *Tunisia* Impressive ruins of a Carthaginian town which fell to the Romans in the Punic Wars (264-146 BC), situated at the eastern end of Cape BON.

Founded in about the 6th century BC, the town was a retreat for Carthaginian aristocrats, and appears to have been hastily abandoned before the Roman takeover. The old site remains intact,

A DAY IN THE LIFE OF A KENYAN FARMER'S WIFE

Under the pure blue Kenyan skies, another long day is beginning for Wamboi Muthoka, a 25-year-old farmer's wife from Kenya's largest tribe, the Kikuyu. She and her husband, Kitau, 27, own a coffee farm on the slopes to the south of Mount Kenya. They have six children, all aged under eight. Wamboi's daily work is not easy. The young woman labours from dawn to dusk on the farm. As well as planting, weeding or harvesting, she has to prepare food for her family and take care of them.

About half of the 1.6 hectare (4 acre) farm is given over to coffee and the rest is planted with maize, bananas, beans and other vegetables, which make up the family's staple diet. The coffee crop brings in hard cash, equivalent to around US$100 a year. Coffee and tea are Kenya's leading exports, and fluctuations in world prices have a profound effect on the economy and ultimately on the wealth of Wamboi's family.

The first task of Wamboi's day is to feed and milk the family's two cows. Mostly they are fed on maize stalks, supplemented by any available vegetation. Then Wamboi packs off her two eldest children, aged six and seven, to the nearby primary school, a timber hut that all the local people helped build a few years ago.

Although primary school education is free in Kenya, the cost of uniforms and textbooks is a constant strain on family finances. (Fortunately, shoes are not obligatory.) But despite the cost and the fact that the national educational system is noncompulsory, few people question the need for an education – which can earn their children a prized government job in Nairobi – even if they have had no formal education themselves. Wamboi had none, and Kitau only a few years of primary schooling.

Wamboi's four other children are not yet old enough for primary school. The latest addition to the family, a six-month-old girl, goes everywhere strapped tightly to her mother's back, while the others spend the day doing odd jobs or running around the farm. Wamboi's mother and her husband's parents also live on the farm. The whole family is housed in two grass-roofed huts made of wood and mud. Kitau has just one wife – some of the older Kikuyu have several. When her day's work in the fields is done, Wamboi goes down to the stream, 20 minutes away, to bring back water in a huge jar which she balances on her head.

She also has to go out in search of firewood – an increasingly difficult task, as nearby trees are felled for fuel, and collecting wood these days can take as much as an hour and a half. Wamboi carries the firewood in a large bundle on her back supported by a leather strap around her forehead.

She and her family will then spend the evening either chatting with their neighbours or listening to the government-owned 'Voice of Kenya', broadcast in Swahili or English, on their large old radio.

with excavations revealing luxurious houses and shops as well as workshops for the production of the blueish dye known as Tyrian purple.

Map (Cape Bon) Tunisia – Ba

Kermadec-Tonga Trench *Pacific Ocean* Ocean trench in the south-western PACIFIC, extending from just north of New Zealand to south of Western Samoa. It is flanked in the west by the volcanic Kermadec Islands and, farther north, by the Tonga Islands.

Along the trench, the westward-drifting Pacific plate is sliding under the Indo-Australian plate (see PLATE TECTONICS), causing earthquakes and the volcanoes which make this area part of the Pacific 'Ring of Fire'. The trench was discovered in 1895. The greatest known depth of the Kermadec Trench is 10 047 m (32 963 ft); the depth of the Tonga Trench is 10 822 m (35 499 ft).

Map Pacific Ocean – Ed

Kerman (Kirman) *Iran* Capital of Kerman Province, 820 km (510 miles) south-east of the capital, TEHRAN. Founded in the 3rd century AD, the town has numerous ancient buildings, including the Friday Mosque (built in 1348) and the Pamenar Mosque.

Its products include beautiful shawls, carpets and brassware. The surrounding province covers 220 150 km² (85 000 sq miles).

Population (town) 312 000; (province) 1 623 000

Map Iran – Ba

Kermanshah *Iran* See BAKHTARAN

Kerry (Ciarraí) *Ireland* County covering about 4700 km² (1815 sq miles) on Ireland's jagged south-west coast with offshore islands and mountainous peninsulas jutting out into the Atlantic Ocean. The county town, Tralee (population 17 230) stands at the neck of Dingle Peninsula, which rises to 953 m (3127 ft) at Brandon Mountain. Sandy beaches, cliffs, lakes and lush, unspoilt countryside have helped to make Kerry popular with tourists.

Population 121 890

Map Ireland – Bb

Keszthely *Hungary* Medieval town, cultural centre, and modern resort with fine sandy beaches on the west shore of Lake BALATON in western Hungary. Besides the great 18th-century Baroque Festetics Palace, now a venue for musical performances, its attractions include collections of folk art in the Balaton Museum and numerous facilities for water sports.

The town's prestigious Agricultural University, originally founded as the Georgikon College in 1797, is the oldest of its type in Europe.

Population 22 180

Map Hungary – Ab

Ketrzyn (Rastenburg) *Poland* Market town in the north-east. Ketrzyn lies in the former German province of East Prussia, 30 km (20 miles) south of the Russian border. Near it is Poland's largest state industrial farming complex. The remains of the 'Wolf's Lair' – the massive concrete bunker headquarters of Adolf Hitler's general staff, where senior officers made an abortive attempt to assassinate the Nazi leader on 20 July 1944 – are situated 8 km (5 miles) from the town.

Population 27 200

Map Poland – Da

kettle (kettle hole) Depression left in a mass of glacial DRIFT, formed by the melting of an isolated block of ice.

Key West *USA* Southernmost city in the continental United States. It lies in the state of Florida, at the end of the chain of coral islands known as the Florida Keys, 210 km (130 miles) south-west of MIAMI. It is a fishing port and tourist centre connected to the mainland through the Keys. The home of writer Ernest Hemingway (1899-1961), who lived here in the 1930s, is a museum.
| Population 24 300 |
| Map United States – Jf |

Khabarovsk *Russia* Second largest city in Russia's Far East, after Vladivostok. Situated 35 km (22 miles) from the north-eastern tip of China on the Amur River, it has an embankment walk with views, across a treeless plain, into China. Khabarovsk, which was founded as a trading station for trappers and hunters, has become the main centre for the local fur trade. Its industries include oil refining, shipbuilding and the production of machinery, furniture and timber.
| Population 613 000 |
| Map Russia – Od |

Khabur *Turkey/Syria* River formed from tributaries in south-east Turkey, flowing south for 320 km (200 miles) through Syria. It irrigates the desert area in the Syrian north-east and ends in the Euphrates just below Dayr az Zawr. The most important town on the river is Al Hasakah. The Abbasid Bridge, built across the Khabur in the 8th century, is considered a perfect example of early Arab architecture.
| Map Syria – Cb |

Khajuraho *India* An important temple site 150 km (93 miles) south of the city of KANPUR in north-central India. Eighty-five Hindu and Jain temples dating from the 10th and 11th centuries, known, to outsiders, for their highly erotic sculptures were scattered over 20 km² (8 sq miles); only about 20 of them have survived.
| Map India – Cc |

Khalkidhikí *Greece* See CHALCIDICE

Khamis Mushait *Saudi Arabia* Rapidly expanding military and airforce town in the Asir Mountains, situated 30 km (19 miles) east of ABHA. It is the home of the King Faisal Military City (a major army base and training centre) and the headquarters of the Saudi Arabian airforce. Many expatriates work here and there is a strong European feel to the place.
| Population 54 000 |
| Map Saudi Arabia – Bc |

Khammouan *Laos* Province of central Laos, stretching from the Mekong River to the Vietnamese border. It is situated on the main route to Vietnam. Two passes through the Truong Son chain of mountains join the two countries.
| Population 259 000 |
| Map Laos – Bb |

Khaniá (Chania) *Greece* Port on the north-west coast of CRETE. Known in ancient times as Cydonia, the town was occupied from 1252 to 1645 by the Venetians, who called it Canea; the Turks held the town from 1645 to 1898. Khaniá's

old fortified quarter with a 14th-century rampart can still be seen, as well as a number of Venetian churches and a Turkish lighthouse. Kastéllí, a 16th-century Venetian town with a fortress, stands on a beach 42 km (26 miles) to the west.
| Population 62 000 |
| Map Greece – Dd |

Kharg Island (Khark) *Iran* Small island, 3 km (2 miles) long, in The Gulf. It lies off the Shatt al Arab waterway, about 30 km (20 miles) from the coast. Iran's largest oil terminal on the island was repeatedly bombed during the Iran-Iraq War (1980-8).
| Map Iran – Bb |

Kharkiv (Khar'kov) *Ukraine* The second largest city in the Ukraine, after KIYEV, 640 km (400 miles) south of Moscow. After 1880, it expanded rapidly as an industrial centre and market city serving Ukraine's fertile farming region and the Donets Basin coalfield. Today, its factories produce tractors, combine harvesters, mining machinery, tools, railway wagons, electrical equipment and foodstuffs.
| Population 1 618 000 |
| Map Ukraine – Da |

Khar'kov *Ukraine* See KHARKIV

Khartoum *Sudan* National capital, situated at the confluence of the Blue and White Nile. It is one of the world's hottest capital cities, with average temperatures reaching 41°C (106°F) in June and 32°C (90°F) in January.

The city was founded by the Egyptian ruler Muhammad Ali in 1823, destroyed by the religious leader, the Mahdi, in 1885 after a siege in which the former governor of Sudan, General Charles Gordon, was killed – after which the Mahdi declared OMDURMAN, across the river, to be the new capital.

Khartoum was, however, rebuilt by the British after Lord Kitchener defeated the Mahdi's followers in 1898. Kitchener laid out the new city centre in the form of a Union Jack. In the present city, many fine examples of colonial architecture stand side by side with mosques and bazaars. For tourists, the city has several hotels, and a national museum. Khartoum today consists of three towns – the modern city centre, Khartoum North and

▲ **LONELY HEADLAND** Constant waves of Atlantic rollers break below lofty Slea Head at the tip of the Dingle Peninsula in the Irish county of Kerry.

Omdurman. The city centre is situated between the two rivers on a curving strip of land shaped like an elephant's trunk; indeed, its name, from the Arabic *Ras-al-hartum*, translates as 'end of the elephant's trunk'.

Khartoum North, across the Blue Nile, is a modern city of factories and offices and the country's industrial capital producing processed foods, pharmaceuticals, leather goods and textiles. Omdurman across the White Nile retains the characteristics of the old Arab town with many tiny shops and a colourful souk.
| Population (conurbation) 2 477 000 |
| Map Sudan – Bb |

Khaskovo *Bulgaria* Industrial town 56 km (35 miles) south of Stara Zagora in fertile THRACE. Many refugees from Macedonia and Grecian Thrace settled here after the two World Wars. The town's industries include silk, textiles, tobacco products, processed foods, wine and metal goods.
| Population 88 700 |
| Map Bulgaria – Bc |

Khawr Fakkan *United Arab Emirates* Small port set against a backdrop of mountains on the east coast of the Gulf of Oman, 25 km (15 miles) north of AL FUJAYRAH. It is part of the enclave belonging to the Emirate of SHARJAH on The Gulf. Its natural harbour is used by dhows, and has facilities for the handling of bulk and ferry ships.
| Population 2000 |
| Map United Arab Emirates – Ca |

Kherson (Cherson) *Ukraine* Port and shipbuilding city on the Dnipro (Dnieper) River, 25 km (16 miles) from the Black Sea. It was founded in 1778 as a fort and naval base by the Russian soldier and statesman Prince Gregory Potemkin (1739-91) for Empress Catherine the Great. Much of the original fort – including its walls, gates and arsenal – survives. Besides shipbuilding, Kherson produces textiles and machinery, refines oil and processes foods.
| Population 361 000 |
| Map Ukraine – Cb |

Kenya

THE CRADLE OF HUMANITY MILLIONS OF YEARS AGO NOW HAS TO DEAL WITH ETHNIC CONFLICT AND A POPULATION EXPLOSION

For a country whose history as a single entity extends back less than 100 years and whose life as a nation is just over a quarter century old, it may seem paradoxical that Kenya is often described as the 'cradle of humanity'. It was here – on the remote shores of Lake TURKANA (formerly Lake Rudolf) in the country's north-west – that the skull and leg-bones of '1470 Man' (the number is a museum catalogue reference), who probably lived some 2 million years ago, were found by Dr Louis Leakey. Anthropologists believe that today's human beings descended from such a creature.

Before the 19th century, the only part of Kenya known to the outside world was its coast – more than 500 km (310 miles) of white beaches, fringed by palm trees, fanned by the trade winds and protected from the Indian Ocean by a continuous flank of coral reef. From medieval times onwards, traders from Europe, Asia and Arabia came here in search of slaves and ivory. Some of them settled and today's coastal population, the Swahili, are a racial blend of Arabs and black Africans.

It was not until the late 19th century that British explorers ventured into the interior and even then their interest was not so much in Kenya itself as in charting a route from the coast to the British protectorate in Uganda. To underline this point, the railway that was subsequently built between MOMBASA on the coast and KISUMU on Lake VICTORIA was called the Uganda Railway; NAIROBI, now Kenya's capital and one of the largest cities in East Africa, was the site chosen for the railway workshops.

The British were, however, quick to realise that some of the land they had come upon, almost by accident, was in fact extremely fertile. By 1914, several thousand British farmers had settled in the south-western region. They appropriated large tracts of well-watered and volcanic soil from the Kikuyu and Masai peoples and, in what became known as the 'White Highlands', employed cheap African labour to produce crops such as coffee. The resentment caused by this colonisation erupted in 1952 when the Mau Mau, a guerrilla organisation, began its campaign of violence.

To this day, the south-west is Kenya's most densely inhabited region, accounting for the

▶ GLORIOUS DAWN The flamingos of Lake Nakuru awaken at first light. The lake – protected as a national park – lies in the Great Rift Valley, a vast geological fault which slashes 6400 km (4000 miles) across Africa, cutting through Kenya's highlands.

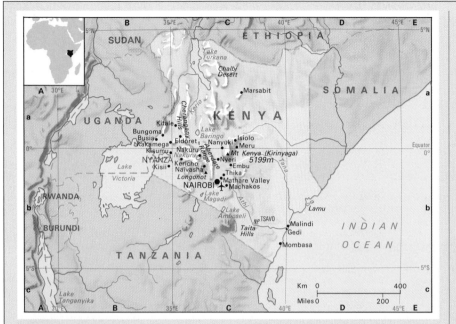

bulk of the country's population and almost all its economic production. In fact, it comprises one of Africa's most successful agricultural production regions. Maize is the main food crop. In contrast, the rest of Kenya is very arid. In the east is a wilderness of dry scrubland and desert, dissected by the country's longest river, the TANA, and inhabited by a small nomadic population. The north-west is similarly dry, although it does include Lake Turkana.

As a single entity Kenya is entirely a colonial creation, first under the control of the British East Africa Company, then, from 1895, as a British protectorate. It was proclaimed a colony by the British in 1920 and granted its independence on 12 December 1963. Its first president, Jomo Kenyatta, who had himself been released from prison only two years before, succeeded in keeping the tribal tensions well beneath the surface. He founded the Kenya African Union (KANU) which eventually became the sole legal party until 1991 when opposition parties were legalised.

Despite an attempted coup by junior air force officers in 1982, Kenyatta's successor, President Daniel arap Moi, from 1978, continued to steer the country into the 1990s. Many KANU members of parliament left that party in early 1992 and joined the Forum for the Restoration of Democracy (FORD), the main opposition party. The government blamed FORD for many of the conflicts that broke out during 1992. However, the last KANU victory in elections, in 1992, was overwhelming.

KANU has been successful in keeping relative peace for all the Kenyans, but the country is a collection of different peoples closely identified with specific territories, rather than an integrated nation. About 1000 years ago, Bantu-speaking migrants entered Kenya from the south up the coastal belt. Then, 500 years later, other peoples, including the nomadic Masai, entered Kenya from the north. Later, the Luo entered from the west. Finally, there are still some 78 000 Asians in the towns.

The long dominance of the Kikuyu in Kenyan national affairs reflects not only their numbers – although only 21 per cent of the population, they are by a good margin the biggest ethnic group – but also their relatively high levels of education and income and their occupation of the national heartland which includes the capital.

The second largest group are the Luhya (14 per cent) followed by the Luo (13 per cent). President Moi is a member of the Kalenjin group which have an 11 per cent share of the population, and the Kikuyu no longer control the government. This and other factors have led to power tensions which broke out in 1992 and again briefly in 1993, affecting the Kikuyu and the Kalenjin people and leading to the death of about 365 people, while more than 10 000 were made homeless.

Although it directly affects a very small proportion of the population, tourism is Kenya's third most important source of foreign exchange after coffee and tea. But the tourists who visit Kenya are exposed to only selected aspects of the country.

POPULATION GROWTH

One of Kenya's most pressing problems is its population growth, which has been increasing at well over 3 per cent a year, one of the highest rates in the world. If this trend continues, the population of more than 28 million will double every 20 years. Despite a more active family planning programme than in most parts of Africa, there has been no drop in the birthrate. Half of Kenya's population is under 15 years old and, unless enough jobs can be created, the people of Kenya will soon outstrip the country's economic capacity. Already only 14.8 per cent of the population is employed.

Not only is the population growing too fast, it is also unevenly distributed. A bare 4 per cent of the land is arable. Some 75 per cent of the population is concentrated on about 20 per cent of the land area in the south-west. The remaining 25 per cent, mainly pastoralists,

inhabit the semi-arid north and north-east covering 80 per cent of the country. Small wonder then that the south-west is experiencing a severe shortage of land and, as people attempt to cultivate more intensively, there are bound to be environmental problems.

In the countryside there are wide disparities of income. During the colonial period cultivated land consisted of large European-run farms, some producing coffee and tea for export, and small plots where African farmers scratched enough to feed their families. In the 1950s, there developed smaller, African-run coffee and tea farms. Then, after 1963, most of the settler farms changed hands: some were bought outright by the new African elite; others were used in resettlement schemes for the landless; and many began to be farmed by consortia and cooperatives.

It is for this reason that many Kenyans migrate to the towns which are growing rapidly although they still account for only 20 per cent of the population. The cities are still centres of administration and commerce but industry is growing. Though agriculture still employs 19 per cent of the working population and accounts for 65 per cent of exports, the industrial sector already employs more than 26 per cent. Kenya is poorly endowed with minerals and the country has only very limited hydroelectric potential; for most people the only available fuel is wood.

The country's greatest tourist attraction and a major natural resource is its wildlife. It was therefore a far-sighted move when, in 1977, the government banned hunting. Today, more than 250 000 tourists a year visit AMBOSELI, TSAVO and other national parks and game reserves.

KENYA AT A GLANCE	
Official name Republic of Kenya	
Area 582 650 km² (224 909 sq miles)	
Population 28 700 000 **Per km²** 49	
(Per sq mile 127.6)	
Capital Nairobi	
Government Republic	
Currency 1 Kenya shilling = 100 cents	
Languages Swahili (official), English (official), indigenous languages	
Religions Christian (54%), indigenous beliefs (18%), Muslim (6%)	
Climate Tropical; hot and humid on the coast, temperate inland, dry to the north. Average temperature in the highlands is 15-17°C (59-63°F) through the year; temperatures in Mombasa range from 24.4°C (76°F) in July to 28°C (82°F) in February	
Land use Cultivation 4%, grazing 7%, forest and woodland 4%, other 85%	
Main primary products Maize, millet, beans, cassava, potatoes, sweet potatoes, sugar cane, coffee, tea, cotton, pineapples, sisal, cattle, pyrethrum (flower used in insecticides)	
Major industries Agriculture, oil refining, cement, food processing, tourism	
Main exports Coffee, tea, petroleum products, fruit and vegetables, sugar, chemicals, cement, hides and skins	
Annual income per head (US$) 330	
Population growth (per thous/yr) 35	
Life expectancy (yrs) Male 52 **Female** 56	

Khíos *Greece* See CHIOS

Khiva *Uzbekistan* Town on the banks of the Amudar'ya River, 250 km (160 miles) south of the ARAL Sea. Once a major centre of the slave trade, it remains a largely medieval city, with its original mosques, minarets, bazaars, gates and shaded courtyards still intact.

Many of its wooden houses are exquisitely carved; other buildings are painted or faced with ceramic tiles in vivid floral or geometric designs. Khiva produces silk, cotton and carpets.
Population 24 200
Map Uzbekistan – Ba

Khodzhent *Tajikistan* See KHUDZHAND

Khojak Pass *Pakistan* Pass 2200 m (7220 ft) in the mountains of BALUCHISTAN Province, on the road between the provincial capital QUETTA and the Afghan city of Kandahar. There is a railway tunnel underneath the pass.
Map Pakistan – Cb

Kholm *Poland* See CHELM

Khon Kaen *Thailand* Regional capital and university city 400 km (250 miles) north-east of BANGKOK in one of the country's poorest regions. Most farmers in the surrounding province eke out a living by growing rice, cassava and sugar cane.
Population (city) 215 800; (province) 1 666 670
Map Thailand – Bb

Khone Falls *Laos* Complex group of waterfalls on the Mekong River in the extreme south of the country, which forms a formidable barrier to navigation on the river. Before the Laotian civil wars of the 1960s and 1970s, the falls were bypassed by a railway portage system, but this has been abandoned, leaving Laos with no outlet to the sea along the river.
Map Laos – Bc

Khor (Al Khawr) *Qatar* The state's second city, situated 40 km (25 miles) north of the capital, Doha, on the shore of a narrow inlet. Khor used to be a major fishing harbour, and dhows still land their catches here. Today, the seashore is being transformed into a leisure area. Khor is also a major archaeological site; discoveries that have been made here include burial chambers and bones dating back to 7000 BC. The Museum of Archaeology contains information on the sites.
Map Qatar – Ab

Khorasan *Iran* Mostly arid north-eastern province with an area of 313 337 km² (120 980 sq miles). Cereals, cotton, tobacco, sugar beet, fruit and nuts are grown around oases such as Birjand. It also manufactures rugs and processes foods. Its capital is the holy city of MASHHAD.
Population 5 280 600
Map Iran – Ba

Khorramabad (Khurramabad) *Iran* Market town for wool and fruit, that is about 400 km (250 miles) south-west of the capital, TEHRAN. Its massive citadel was the stronghold of the Lurs, a fiercely independent people who raided the surrounding area for several centuries until Shah Abbas I captured the citadel in the 17th century.
Population 249 000
Map Iran – Aa

Khorramshahr *Iran* Inland port on the Shatt al Arab waterway, near the border with Iraq. The terminus of the trans-Iranian railway, it normally handles almost half of Iran's international trade, although this activity was severely disrupted by the Iran-Iraq War (1980-8). Exports include cotton, dates, hides and skins.
Population 151 000
Map Iran – Aa

Khouribga *Morocco* Centre of phosphate mining and associated industries, 105 km (65 miles) south-east of CASABLANCA.
Population 128 000
Map Morocco – Ba

Khudzhand (Leninabad) *Tajikistan* One of Central Asia's oldest towns, 145 km (90 miles) almost due south of TASHKENT, in Uzbekistan. The town grew around a fortress founded by the Macedonian conqueror Alexander the Great (356-323 BC). Originally known as Khodzhent, it was renamed Leninabad after the communist leader Vladimir Lenin (1870-1924); following the collapse of the Soviet Union, it was again renamed. It produces traditional ceramics, wood carvings, embroidery and silks.
Population 165 000
Map Tajikistan – Ba

Khulna *Bangladesh* Coastal district in the south-west bordering India. Although most of its people were Hindu, Khulna was awarded to predominantly Muslim Pakistan after the partition of East and West Pakistan in 1947. Many of the Hindus fled to India during the riots that followed. In 1970, the district was struck by one of the worst cyclones this century, killing at least 250 000 people. The port of Khulna is the chief town.
Population (district) 4 329 000; (town) 877 000
Map Bangladesh – Bc

Khumbu *Nepal* Himalayan Mountain region which includes Mount EVEREST. Khumbu is also the name of a glacier which flows from a shoulder of Mount Everest over the dangerous Khumbu icefall.
Map Nepal – Ba

Khunjerab Pass *China/Pakistan* See MINTAKA PASS

Khurramabad *Iran* See KHORRAMABAD

Khuzestan (Khuzistan) *Iran* South-western province, covering 64 651 km² (24 962 sq miles) between the Zagros Mountains, the Iraqi border and The Gulf. It is the main source of Iran's oil; dates, melons, vegetables, rice and cotton are also grown here. Most of Iran's Arab minority, who are Sunni Muslims, live in the province. AHVAZ is the capital. SUSA, capital of the Elamite kingdom some 4000 years ago, also lies between the Zagros Mountains and The Gulf.
Population 2 681 980
Map Iran – Aa

Khyber Pass *Afghanistan/Pakistan* Most important pass between the two countries, carrying a modern road and railway line and an ancient caravan route across the Safed Koh Mountains between Peshawar in Pakistan and the Afghan capital, Kabul. It is 50 km (30 miles) long and reaches a height of about 1070 m (3520 ft). The Khyber Pass has been a major trade and invasion route

▼ **NORTH-WEST FRONTIER A motor road snakes up the Khyber Pass in Pakistan, which has been called 'the Gateway to India'. Its great strategic importance has led to many foreign forces trying to control it.**

since ancient times. It may have been used by the Macedonian conqueror Alexander the Great in 327 BC and the Mongol Tamerlane in 1398 AD.

It was certainly of great importance to the British during the Afghan Wars in the 19th century. Since the late 1970s, the pass has been used by millions of Afghani refugees fleeing into Pakistan. The pass was briefly closed after the Russian invasion of Afghanistan in 1979.

Map Afghanistan – Bb; Pakistan – Ca

Kicking Horse Pass *Canada* Route through the Rocky Mountains between Alberta and British Columbia. The pass reaches a height of 1627 m (5338 ft), and carries the highest stretch of the Canadian Pacific Railway.

Map Canada – Dc

Kiel *Germany* Port on an estuary opening into Kiel Bay in the extreme north and capital of SCHLESWIG-HOLSTEIN, Germany's northernmost state. It has the country's largest shipyards and is an international yachting centre. Its university conducts most of Germany's marine research.

The Kiel Canal, opened by Kaiser Wilhelm II in 1895, is 45 m (148 ft) wide and 14 m (46 ft) deep, and runs for 96 km (60 miles) from Kiel to the mouth of the Elbe River, north of Hamburg. It provides a passage for ocean-going ships between the Baltic and the North Sea.

Kiel was the German Reich's leading naval base and has remained the country's naval port, though much of the city was destroyed towards the end of the Second World War.

Population 245 600
Map Germany – Da

Kielce (Kelsty) *Poland* Industrial city 145 km (90 miles) south of the capital, WARSAW. Kielce has copper mines and marble quarries, and its factories produce electrical goods, machinery and processed foods.

Population 213 000
Map Poland – Dc

Kiev *Ukraine* See KIYEV

Kigali *Rwanda* National capital situated in the centre of the country. Before independence in 1962, Belgium administered Rwanda from Bujumbura, the present capital of Burundi. After becoming Rwanda's capital, Kigali's population soared from some 5000 during the colonial period to about 300 000 at the end of 1993.

However, the city was badly damaged during the civil war which raged in 1994. Many of Kigali's Tutsi inhabitants were murdered by Hutus, who subsequently deserted the city when the predominantly Tutsi Rwandan Patriotic Front took control.

Population (pre-1994) 300 000
Map Rwanda – Ba

Kigoma *Tanzania* Capital of Kigoma region, on the east shore of Lake Tanganyika, 50 km (30 miles) south of the border with Burundi. It is the terminus of the railway from Dar es Salaam and the main transit port for Burundi. It ships timber, cotton and tobacco. The old fishing settlement of Ujiji, where Henry Morton Stanley found the missing Scottish missionary and explorer Dr David Livingstone on 28 October 1871, is situated 8 km (5 miles) to the south.

Population (town) 50 000
Map Tanzania – Aa

Kikládhes *Greece* see CYCLADES

Kildare (Cill Dara) *Ireland* County covering 1694 km² (654 sq miles) in the south-east, in the flattest part of Ireland. Glacial deposits cover much of the surface of the county, and soils are varied. The Grand Canal, built in the 18th and early 19th centuries to carry trade between DUBLIN and the River Shannon, passes through the county, and now carries holiday boats. Naas (population 11 140) is the county town.

The market town of Kildare, 48 km (30 miles) south-west of Dublin, is the heart of Ireland's horse-racing industry, and home of the National Stud and Horse Museum. The Irish Derby is held each June at the Curragh just to the east.

Population (county) 122 650; (town) 4200
Map Ireland – Cb

Kilimane *Mozambique* See QUELIMANE

Kilimanjaro *Tanzania* Africa's highest mountain. It lies in the north-east of Tanzania near the Kenyan border. Kilimanjaro has two peaks – Kibo, which rises to 5895 m (19 340 ft), and Mawenzi, rising to 5353 m (17 564 ft). Both were once active volcanoes.

Although it is a mere 3 degrees south of the equator, the summit of Kibo is permanently covered in snow. Lower down, on the fertile slopes of the mountain below 2000 m (6560 ft), bananas, coffee and maize are grown.

Kilimanjaro was first climbed in 1889 by two Germans, Ludwig Purtscheller and Hans Meyer, who scaled Kibo. Now, many thousands of climbers reach the summit each year.

Map Tanzania – Ba

Kilkenny (Cill Chainnigh) *Ireland* City 102 km (63 miles) south-west of DUBLIN. It lies in the middle of a hilly county of the same name which covers 2062 km² (796 sq miles). The city was an Anglo-Norman stronghold; later, parliaments met here, passing enactments such as the Statute of Kilkenny (1366), which made it illegal for the English to marry local inhabitants. Kilkenny's medieval castle, round tower and two cathedrals still dominate the closely packed streets.

Black marble was once quarried in the surrounding hills. Today, the city processes food and makes textiles and shoes.

Population (city) 8520; (county) 73 640
Map Ireland – Cb

Killarney (Cill Airne) *Ireland* Tourist centre and market town in the county of Kerry, 77 km (47 miles) north-west of CORK. It is surrounded by mountains and lakes, and jaunting cars – open two-wheeled carts, with back-to-back seats, drawn by ponies – carry visitors on trips through the fine valleys to the south of the town.

Ireland's highest range, Macgillicuddy's Reeks, lies to the south-west, topped by the country's loftiest peak, Carrauntoohil (1041 m, 3414 ft) whose slopes in spring are covered with thousands of flowering rhododendrons.

Population 7280
Map Ireland – Bb

Kilmain *Mozambique* See QUELIMANE

Kilmarnock *United Kingdom* See STRATHCLYDE

Kilwa Kisiwani *Tanzania* See KILWA KIVINJE

Kilwa Kivinje *Tanzania* Small market town on the coast, 210 km (130 miles) south of DAR ES SALAAM. It was established in the 1830s by residents of Kilwa Kisiwani, who had abandoned their island settlement 30 km (20 miles) to the south for a mainland site. After Kilwa Kisiwani itself was founded in the 11th century by people from Shiraz in Persia, it became the centre of a sea trading area stretching south into Mozambique.

In 1505, it was largely destroyed by the Portuguese. It then became an Arab slave-trading centre until it was abandoned by its inhabitants. Its impressive remains include the Great Mosque, a palace and a Portuguese fort, dating from the 11th to the 14th century. Just north of Kilwa Kisiwani is Kilwa Masoko, a new administrative centre with a small airstrip.

Population 10 000
Map Tanzania – Ba

Kimberley *South Africa* Capital city of the Northern Cape Province. Also referred to as 'the Diamond Capital of the World', Kimberley has the world's largest excavation, known as the 'Big Hole', where open-cast diamond mining was carried out between 1871 and 1915. The hole is 1.6 km (1 mile) in circumference and about 1200 m (4000 ft) deep; it produced some 3 tonnes of diamonds for the removal – mostly by pick and shovel – of 20 million tonnes of earth. It is now the centre of the Kimberley Mine Museum.

Financier and imperialist Cecil John Rhodes endured a four-month siege in the town during the Anglo-Boer War of 1899-1902. Kimberley has both Anglican and Catholic cathedrals, art galleries, museums and memorials. It still has active diamond mines.

Population 167 060
Map South Africa – Bb

Kimberley Plateau *Australia* Region – also known simply as the Kimberley – of low, rugged hill blocks cut by gorges, covering about 420 000 km² (162 000 sq miles) in the far north of Western Australia. Monsoon rains fall from November to April, turning the Fitzroy and ORD rivers into raging torrents. The rains are followed by dry, hot winter months. Kimberley is the traditional home of several thousand Aborigines.

The region's highest peak is Mount Ord, reaching 936 m (3071 ft) in the King Leopold Range. Derby (population 3000) is the main town. Beef cattle are raised in the Kimberley; diamonds are mined at Argyle. Cultured-pearl farming is carried out at Cygnet and Kuri bays.

Population 36 390
Map Australia – Db

kimberlite The greenish-blue intrusive igneous rock in which diamonds are found, as at KIMBERLEY in South Africa.

Kimch'aek (Gimchaeg) *North Korea* East coast fishing port, 350 km (218 miles) from the capital, PYONGYANG. It has fish oil and fish processing factories, and iron foundries.

Population 281 000
Map Korea – Db

Kincardineshire *United Kingdom* See GRAMPIAN

Kindia *Guinea* Hill town 120 km (75 miles) north-east of the capital, CONAKRY, at the edge of the Fouta Djallon, about 400 m (1300 ft) above

sea level. It was developed as a health resort for Europeans after the opening of the railway to Conakry in 1904, and still has many colonial-style buildings, including offices, prisons and schools. Nearby are numerous streams and waterfalls. Kindia sells locally grown fruits, including bananas; a fruit research centre and a branch of the Pasteur Institute are situated in the town.
Population 79 900
Map Guinea – Ba

Kindu *Zaire* Port on the Lualaba River, 370 km (230 miles) west of the border with Burundi. It is the terminus of a railway which extends southwards to the mineral-rich Shaba region and has a branch line running eastwards to Kalémié on Lake Tanganyika.

Below Kindu, the river is navigable as far as Ubundu (Ponthierville), which is linked by rail to the river port of KISANGANI.
Population 50 000
Map Zaire – Bb

Kings Canyon National Park *USA* Californian national park covering 1863 km² (719 sq miles) in the SIERRA NEVADA, 300 km (190 miles) north of Los Angeles. It has two canyons – each between 762 and 1525 m (2500 and 5000 ft) deep – on the Kings River, and giant sequoia trees.
Map United States – Cc

Kingston *Canada* City on Lake Ontario at the mouth of the Cataraqui River. Founded in 1783 by loyalist refugees from the American Revolution, Kingston was the capital of Canada from 1841 to 1844. Today, it is an important grain processing and shipping port, and has industries making locomotives, ships, diesel engines and aluminium products. Queen's University and the Royal Military College of Canada are based here.
Population 56 600
Map Canada – Hd

Kingston *Jamaica* Capital and chief port, in the island's south. It has a deep, landlocked harbour, which is enclosed by the Palisadoes, a 13 km (8 mile) long sandspit extending from the coast east of the city. The island's main international airport lies in the middle of the spit, and PORT ROYAL is situated at its western end.

Founded by the British after Port Royal was destroyed in an earthquake in 1692, Kingston became the commercial capital in 1703 and the island's political capital in 1872. An earthquake and fire in 1907 destroyed many buildings in the town. However, a few architectural relics remain among the modern buildings, most notably Headquarters House (formerly the seat of government), Devon House, a restored mansion, and Gordon House, which houses Jamaica's legislature.

Rockfort, a 17th-century moated fortress stands on the city's eastern outskirts; on its northern outskirts lies New Kingston, the new commercial and residential suburb and the centre of the Jamaican coffee trade.

Today, the city is a busy commercial centre whose industries include oil refining, food processing, as well as the manufacture of tobacco, furniture, clothing and shoes. A government-owned railway connects the city with most of the island's 14 parishes.
Population 641 500
Map Jamaica – Ba

Kingston upon Hull (Hull) *United Kingdom* One of the country's leading seaports situated on the north shore of the Humber estuary in the county of Humberside. The town has been associated with ships ever since its docks were founded by Edward I in 1293. Whaling and coastal fishing have come and gone, and most of today's maritime business is connected with traffic across the North Sea.

Devastated by German bombs during the Second World War, the city has been largely rebuilt; many new industries, mainly in the fields of engineering and chemicals, have been established. One of the buildings to survive the German bombing was the home of William Wilberforce, champion of the campaign to abolish slavery, who was born in Hull in 1759. There is also a university. The Humber suspension bridge links the city with the river's south bank. Opened in 1981, it has a span of 1410 m (4626 ft).
Population 245 000
Map United Kingdom – Ed

Kingstown *Ireland* See DUN LAOGHAIRE

Kingstown *St Vincent* Seaport and capital, which lies on the south-west coast at the head of Kingstown Bay. Its botanic gardens, established in 1765, contain a breadfruit tree descended from one brought here from Tahiti by Captain William Bligh in 1792.

Bligh's first mission to bring breadfruit to the island ended in the mutiny on board the *Bounty* in 1787. The objective of planting breadfruit was to provide food for African slaves working on West Indian sugar plantations.

This venture failed, however, because the slaves preferred other foods, which grew here in abundance. Today Kingstown exports arrowroot, bananas and coconuts. Its chief industries are the processing of arrowroot, copra, fruit and sugar.
Population 26 540
Map Caribbean – Cc

Kinross *United Kingdom* See TAYSIDE

Kinshasa *Zaire* Capital of Zaire and the largest city in Central Africa. It is situated on the south bank of Malebo (formerly Stanley) Pool, a broad stretch of the Zaire River, 400 km (250 miles) from the ATLANTIC OCEAN.

Before the arrival of the British adventurer Sir Henry Morton Stanley in 1877, the site of the city contained two riverside villages: Kintambo and Kinshasa. In 1881, Stanley acquired the rights to Kintambo and named it Léopoldville after his patron Léopold II of Belgium. In 1898, a railway was completed from Léopoldville to the seaport of Matadi, by-passing the Inga Rapids (Livingstone Falls) and enabling Léopoldville to develop quickly as a commercial centre. It became the national capital in the 1920s.

After Zaire became independent in 1960, the city grew rapidly from a population of 400 000 to more than 3 million in 1990. Renamed Kinshasa in 1966, it has become the focus of a network of communications that extends far inland along the rivers and railways. Kinshasha also has an international airport.

The city is a major industrial centre and its food processing, wood working, textile, chemical and other industries have grown rapidly since the opening of the Inga Dam hydroelectric project on the nearby INGA RAPIDS. The appearance

of the city centre, once an area of attractive modern buildings, spacious parks and boulevards, has deteriorated and many shops, hotels and restaurants have been closed. However, the colourful central market and a national museum are well worth a visit. The Beaux Arts Academy and the Université National du Zaire, one of the country's three universities, are in Kinshasa.
Population 3 562 000
Map Zaire – Ab

Kirghizia See KYRGYZSTAN

Kiribati See p 372

Kirin *China* See JILIN

Kirinyaga *Kenya* See KENYA, MOUNT

Kiritimati *Kiribati* See CHRISTMAS ISLAND

Kirkcaldy *United Kingdom* See FIFE

Kirkcudbright *United Kingdom* See DUMFRIES AND GALLOWAY

Kirkenes *Norway* See VARANGERFJORD

Kirkūk *Iraq* Regional capital and oil town, some 240 km (150 miles) north of the capital, BAGHDAD. It has the largest oil field in Iraq and several pipelines run from here to the Mediterranean Sea and The Gulf.

However, international sanctions imposed on Iraq after the 1991 Gulf War have significantly reduced oil production. Kirkūk is also an agricultural and market centre, and trades in salt, sheep, grain and fruit.
Population 650 000
Map Iraq – Cb

Kirov (Vyatka) *Russia* Industrial city on the Vyatka River, 790 km (490 miles) north-east of MOSCOW. It was founded in 1174 by traders. In 1934, the city – formerly Khlynov, then Vyatka – was renamed in memory of Sergei Kirov (1886-1934), a locally born communist leader who was assassinated in that year. Kirov's industries include agricultural equipment, chemicals, metal working and heavy machinery.
Population 491 000
Map Russia – Fc

Kirovabad *Azerbaijan* See GÄNCÄ

Kirovograd *Ukraine* See KIROVOHRAD

Kirovohrad (Kirovograd) *Ukraine* City in the south, 250 km (155 miles) north-east of Odesa; formerly called Elizavetgrad. The centre of a rich farming – particularly hop-growing – region, it produces agricultural equipment and clothing.
Population 296 000
Map Ukraine – Cb

Kirşehir *Turkey* Carpet and rug-making town, which is particularly well known for its prayer rugs, 135 km (85 miles) south-east of ANKARA. It is the capital of a heavily forested province of the same name. The province, with an area of 6570 km² (2537 sq miles), produces grain, linseed oil, cereals, fruits and mohair.
Population (town) 65 900; (province) 256 860
Map Turkey – Bb

Kisalföld *Hungary* See ALFÖLD, LITTLE

Kisangani *Zaire* Capital of Haut-Zaire region. It straddles the Zaire River just below the BOYOMA (formerly Stanley) FALLS, 530 km (330 miles) west of the Ruwenzori Mountains on the border with Uganda.

Formerly called Stanleyville, it was established in 1898 on the site of a trading post that had been set up by the British adventurer Sir Henry Morton Stanley in 1883.

It is now a busy commercial, fishing, manufacturing and tourist centre, and a transshipment point for goods arriving by rail from the south or by river boat from KINSHASA.

Population 557 000
Map Zaire – Ba

Kishinev *Moldova* See Chişinău

Kiskörös *Hungary* See KISKÚNSÁG

Kiskunhalas *Hungary* See KISKÚNSÁG

Kiskúnság *Hungary* The section of the Great ALFÖLD (Plain) which lies between the Danube and Tisza rivers in central Hungary. A flat, sandy region, it produces table wines, apricots, peaches and other tree fruits, tobacco, and poultry and livestock. KECSKEMÉT is the regional capital.

The town of Kiskunhalas, or Halas (population 31 260), is situated 55 km (34 miles) south of Kecskemét on the plain amid the Bugac grasslands and is noted for its traditional lace making.

The small town of Kiskörös (population 14 790) is situated 48 km (30 miles) south-west of the regional capital. The cottage birthplace of the Hungarian poet and nationalist Sándor Petöfi, one of the leaders of the 1848 Hungarian rebellion against Austrian rule, is preserved here as a monument.

Map Hungary – Ab

Kismaayo (Kismayu) *Somalia* Southern port between the capital, MOGADISHU, and the Kenyan border. The town exports bananas and has meat-canning factories.

Population 200 000
Map Somalia – Ac

Kistna *India* See KRISHNA

Kisumu *Kenya* One of the nation's largest towns, on the north-east shore of Winam (formerly Kavirondo) Gulf off the east side of Lake Victoria, which lies 260 km (160 miles) north-west of NAIROBI. Kisumu was the original terminus of the Uganda Railway, and the busiest port on the lake until the break-up of the East African Community in the 1970s reduced lake traffic with Tanzania and Uganda. Today, the town is a minor manufacturing centre.

Population 200 000
Map Kenya – Bb

Kitakyushu *Japan* Industrial city at the northern tip of KYUSHU island. Established in 1963 by the merger of five towns, it has become Japan's leading iron and steel producer; it also manufactures chemicals and machinery.

Kokura-jo Castle is now a folk museum. The original structure dates from 1602.

Population 1 026 000
Map Japan – Bd

Kiribati

THESE FAR-FLUNG PACIFIC ISLANDS STRADDLE BOTH THE INTERNATIONAL DATE LINE AND THE EQUATOR

The flag of Kiribati shows a frigate bird flying over a sun rising from a blue sea. It is a fitting symbol for these sunny coral islands in the middle of the Pacific Ocean. They were known as the Gilbert Islands until their independence in 1979. Their new name is pronounced 'Kirri-bass' – the closest the Gilbertese language can get to 'Gilberts'.

They consist of 33 islands and atolls, including BANABA (formerly Ocean Island), the Phoenix Islands and some of the Line Islands. They were named the Gilbert Islands in the 1820s. The waters around the islands were a favourite sperm-whaling ground for American and British vessels from the early 1800s until the 1870s. The first permanent European settler arrived in 1837. Together with the Ellice Islands (now Tuvalu), most of the islands were a British protectorate from 1892 and a colony from 1916. The Ellice Islands broke away in 1975. Altogether, the republic of Kiribati spreads over some 5 million km² (2 million sq miles) of the Pacific. The Japanese occupied the islands during the Second World War, but they were recaptured after the battle of TARAWA in November 1943.

The indigenous people are Micronesian (unlike those of Tuvalu, who are Polynesian) and Christian following the influence of European missionaries. Fish is the staple food. The soil is poor and the most common vegetable is the coarse *babai* (taro), which is grown in deep pits. Coconut palms grow prolifically on most of the islands. From 1900 to 1980, Banaba yielded vast amounts of phosphates for export, but this resource has now been exhausted. Kiribati has negotiated fees for the fleets of other countries, including Russia, to fish in its waters. It is heavily dependent on overseas aid.

KIRIBATI AT A GLANCE	
Map Pacific Ocean – Db	
Area 717 km² (277 sq miles)	
Population 63 000	
Capital Tarawa	
Government Parliamentary republic	
Currency 1 Australian dollar = 100 cents	
Languages Gilbertese, English	
Religion Christian (50% Roman Catholic, 45% Protestant)	
Climate Tropical; average temperature in Tarawa is 29°C (84°F) all year	
Main primary products Breadfruit, pandanus, coconuts, vegetables, fish	
Major industries Copra processing, handicrafts, fishing, fish farming	
Main exports Copra, handicrafts, fish, postage stamps	
Annual income per head (US$) 480	
Population growth (per thous/yr) 16	
Life expectancy (yr) Male 50 **Female** 54	

Kitchener-Waterloo *Canada* Twin cities in southern Ontario, 100 km (62 miles) west of TORONTO. Their industries include meat packaging, distilling, and the manufacture of furniture, cars, clothing and various wood products. Kitchener was settled in 1805 by Mennonites from Pennsylvania, and later by Germans.

Until 1916, it was known as Berlin and it still celebrates an Oktoberfest each year. Waterloo has two universities.

Population (metropolitan area) 356 420
Map Canada – Gd

Kíthira (Cerigo) *Greece* Remotest of the IONIAN ISLANDS, lying 14 km (9 miles) south-east of the Peloponnese. According to classical mythology, Kíthira is the birthplace of the goddess of love, Aphrodite. In 1537, its entire population was sold as slaves by the Turkish pirate Barbarossa.

The main town has a Venetian castle and there are large caverns nearby. The chief sources of revenue today are tourism and agriculture; honey is an important export. However, with increasing numbers of islanders leaving for the mainland, many of Kíthira's villages lie deserted.

Population 3350
Map Greece – Cc

Kitwe-Nkana *Zambia* Copper-mining city which was founded in the 1930s, 290 km (180 miles) north of the capital, LUSAKA. The town receives its power supply from the KARIBA DAM.

Population 495 000
Map Zambia – Bb

Kitzbühel *Austria* Ski resort in the Kitzbühel Alps, 80 km (50 miles) north-east of INNSBRUCK. In the Middle Ages, it was a centre for iron, lead, copper and silver mining. However, the last mine closed in 1770. Skiing began in the area in the latter part of the 19th century.

Population 8200
Map Austria – Cb

Kivu *Zaire* Administrative region covering 256 662 km² (99 072 sq miles) in eastern Zaire. Its western boundaries lie within the Zaire River rain forest zone, but to the east, mountains border the GREAT RIFT VALLEY, which encloses lakes Edward, Kivu and Tanganyika.

Eastern Kivu, tourist area, contains the Kahuzi-Biega, Virunga and part of the Maiko national parks. The chief towns are the capital, Bukavu, Goma (the site of huge Rwandan refugee camps in 1994), and Kindu.

Population 4 714 000
Map Zaire – Bb

Kivu, Lake *Rwanda/Zaire* Lake covering an area of 2850 km² (1100 sq miles) in the GREAT RIFT VALLEY. With a surface 1460 m (4790 ft) above sea level, it is Africa's highest lake. The only outlet is the Ruzizi River, which drains into Lake Tanganyika. Unlike other East African lakes, Lake Kivu has no crocodiles or hippopotamuses, and contains few fish.

Map Zaire – Bb

Kiyev (Kiev, Kyiv) *Ukraine* The capital and the country's largest city, standing on the banks of the Dnipro (Dnieper) River, 400 km (249 miles) north of ODESA. The Russians call it 'the Mother of Cities' because it is one of the oldest cities in the region, dating from the 6th century. Kiyev became the capital of the powerful Kiev-Rus principality (the infant Russian state) in 882, and its grand prince, having been converted, made Christianity the state religion in 988. The Mongols sacked the city in 1240, and it was part of Lithuania and then Poland before it was restored to Russia in 1654. It became Ukraine's national capital after that country's independence from the Soviet Union in 1991.

The city grew rapidly once the rail links with Moscow (1863) and Odessa, now Odesa (1870), were forged. It was the scene of bitter fighting in the aftermath of the Russian Revolution of 1917, and was severely damaged during the Second World War, when 200 000 people died and the city centre was razed. It was rebuilt with wide, tree-lined boulevards and parks which enhance relics such as the 11th-century St Sophia's Cathedral, now a museum. Kiyev has engineering, electrical, food-processing and precision tools industries. Its university, founded in 1834, is now an academy of science.

In April 1986, the world's worst recorded civil nuclear accident occurred at a power station at Chernobyl', 90 km (55 miles) north of the city. The roof of one of the plant's four reactors blew off in a huge explosion, releasing radioactive material into the atmosphere. The fallout contaminated much of the Ukraine, parts of Poland and Scandinavia and affected countries as far away as Britain. The immediate danger was not contained until the following month, by which time thousands of people had been moved from the area.

A number died of radiation exposure – notably firemen who fought the blaze that followed the explosion – and there has been a high incidence of cancer and a high rate of birth deformities among the population. More than 120 000 people were evacuated, all villages within a 30 km (18.5 mile) radius were razed to the ground and a large area of farmland made unusable. In the early 1990s, old people were allowed to return to their farms, but the crops which they grew were found to be heavily contaminated.

Population 2 616 000
Map Ukraine – Ca

Kizil Irmak *Turkey* The country's longest river. It flows for 1130 km (700 miles) from its source near Sivas, some 340 km (210 miles) east of ANKARA, to the Black Sea, which it enters at Cape Bafra, 350 km (215 miles) north-east of Ankara.
Map Turkey – Ba

Kladno *Czech Republic* Bohemian industrial town 25 km (15 miles) north-west of PRAGUE. Coal has been mined here since 1720. The town also produces coke and steel.

Lidice lies 8 km (5 miles) to the west. Roses cover the remains of the old village, destroyed in 1942, in reprisal for the assassination in Prague of Reinhard Heydrich, the German SS deputy protector of Bohemia and Moravia. The Nazis shot the men of Lidice, sent the women and children to camps, and razed the village to the ground.
Population 72 700
Map Czech Republic – Ba

Klagenfurt *Austria* Capital of CARINTHIA State, lying 100 km (62 miles) south-west of GRAZ. Much of the city was destroyed by fire in 1514, and its few remaining buildings of note date from the mid- to late-16th century.
Population 89 500
Map Austria – Db

Klaipeda (Memel) *Lithuania* The country's chief port and naval base on the Baltic coast, 100 km (62 miles) north-west of KANNAS. Ice-breakers are used between December and March to keep the port open. A canal links it to the mouth of the Neman (Memel) River nearby.

The city, founded as a fort in 1252, later became a member of the Hanseatic League – a trading federation of mostly north German and other Baltic Sea towns. It was German territory (as Memel) for much of its history, but was independent from 1924 to 1939; in 1945, it was occupied by Russian troops. Today, it is an industrial city producing plywood, cellulose, paper, textiles, foodstuffs and canned fish. Oil carried by pipeline from the Volga-Ural oil field is one of its important exports. Crafts people in the town carve jewellery from amber, a fossil resin found along the Baltic coast.
Population 204 000
Map Lithuania – Ba

Klis *Croatia* See SPLIT

Kljuc *Bosnia-Herzegovina* See GRMEC PLANINA

Klodzko (Glatz) *Poland* Manufacturing city and tourist centre lying 84 km (52 miles) south-west of WROCLAW. It was established in 981 AD to defend the Cracow-Prague route over the Sudety Mountains, and lies in 'the Valley of Health' – a fertile basin of rolling farmlands and forests. The valley has many spas – known in Polish as *zdroj*. From the 12th century, the city attracted many German immigrants, but after the Second World War it was returned to Poland.

Today Klodzko produces metals, paper, textiles, timber and foodstuffs. The Skull Chapel (1776-8) at Czermna, some 2 km (about 1 mile) to the north, is lined with the skulls and bones of 3000 people, most of them victims of the Thirty Years' War (1618-48).
Population 30 300
Map Poland – Bc

Klondike *Canada* River, 160 km (100 miles) long, in Yukon Territory, flowing west to join the Yukon River at DAWSON. It gave its name to the gold-producing region on Bonanza Creek, a tributary, where gold was discovered in 1896.

Over the next two years, nearly 30 000 prospectors joined the gold rush and, within the first seven years, Can$100 million worth of gold was found. However, by 1910, when the gold became more difficult to extract, only 1000 prospectors remained. Some mining continued along the river until 1966.
Map Canada – Bb

kloof In South Africa, the name given to a gorge, ravine or pass.

Knock (Cnoc) *Ireland* Village 60 km (37 miles) north-east of the city of GALWAY, where a vision of the Virgin Mary was seen on a church wall in 1879. A huge modern basilica has been built near the site, which Pope John Paul II visited in 1979. The village is close to an international airport that was built in the 1980s.
Population 440
Map Ireland – Bb

Knokke *Belgium* Coastal resort 18 km (11 miles) north of the city of BRUGES. One of a string of former fishing villages, it attracts visitors for its magnificent scenery, proximity to Bruges, and its horse-riding facilities, golf courses and casino.
Population 28 000
Map Belgium – Aa

knoll Small, rounded hill or mound.

Knossós *Greece* The largest excavated palace of the Minoan civilisation, 5 km (3 miles) south-east of the present-day capital of Crete, HERAKLION (Iráklion). It was first built in 1950 BC, rebuilt in 1700 BC and destroyed by fire between 1400-1380 BC. Knossós was the cultural and economic centre of one of the first great Mediterranean civilisations.

The ruins comprise a stone-built rectangular complex, laid out on four levels around a square covering 1325 m^2 (1584 sq yds). The British archaeologist Sir Arthur Evans, who began excavations here in 1899, spent large sums of his own money on the project. An area of 20 000 m^2 (24 000 sq yds) has been uncovered, revealing many chambers, workshops, and a throne room with the alabaster throne of King Minos – the master, according to Greek mythology, of the bull-headed Minotaur, which wandered the underground labyrinth below the palace, feeding on human flesh. Many of the artefacts found at Knossós are on display at Heraklion.
Map Greece – Dd

Knoxville *USA* Commercial and industrial city in eastern Tennessee, 280 km (174 miles) east of NASHVILLE. The centre of the tobacco-growing eastern Tennessee River valley, it also produces textiles, clothing, furniture and large amounts of marble. It is the headquarters of the Tennessee Valley Authority.
Population (city) 125 100; (metropolitan area) 604 800
Map United States – Jc

Koba National Park *Senegal* See NIOKOLO-KOBA NATIONAL PARK

Kobe *Japan* Port on HONSHU Island. Kobe lies on Osaka Bay, at the eastern end of the Inland Sea. The site of one of the world's largest container terminals, the city also produces ships, steel, and some of the country's best *sake*, or rice wine.

On 17 January 1995, Kobe suffered a devastating earthquake measuring 7.2 on the Richter Scale. Over 50 000 buildings were damaged or destroyed, both in the quake and the subsequent fires, 25 000 people were injured and more than 5000 died. The road system and port facilties suffered severe damage.
Population 1 477 000
Map Japan – Cd

København *Denmark* See COPENHAGEN

Koblenz (Coblenz) *Germany* Wine-making city at the junction of the Rhine and Moselle rivers, 53 km (33 miles) south-east of Bonn. In Roman times and all through the Middle Ages, it was a fortress town. Later, the German publisher Karl Baedeker (1801-59), now famous for his series of guidebooks, had his offices in Koblenz. The city produces aluminium, paper and parts for the car industry.
Population 108 000
Map Germany – Bc

Kodagu (Coorg) *India* Mountainous region on the south-west coast, 1292 km (802 miles) south of BOMBAY. It contains the Nagarhole Wildlife Sanctuary, known for its wild elephants, panthers and deer. Its capital is Madikeri.
Map India – Be

Koforidua *Ghana* Capital of the Eastern Region, and market town at the foot of the Kwahu uplands, 55 km (35 miles) north of the capital, ACCRA. It markets agricultural produce of the region, including kola nuts, which are chewed as a mild stimulant.
Population 60 300
Map Ghana – Ab

Kofu *Japan* City, provincial capital and university town in the mountains of Jamanashi Province on HONSHU Island. It lies west of the capital, Tokyo, between Mount Fuji and the Japan Alps. Kofu produces silk and wines.
Population 201 000
Map Japan – Cc

Kohima *India* Hill town and capital of the state of Nagaland, 110 km (68 miles) west of the border with Myanmar. It is situated 1500 m (4920 ft) above sea level and was the farthest that the Japanese drove into India during the Second World War. They captured the town after a fierce battle in June 1944, but it was recaptured by Anglo-Indian forces later that month.
Population 394 180
Map India – Ec

Kokkola (Gamlakarleby) *Finland* Seaport established in 1620 on the Gulf of Bothnia, 112 km (70 miles) north-east of the port of VAASA. It exports timber and manufactures chemicals, metal goods, clothing and sports goods.
Population 34 900
Map Finland – Bc

Kokoda Trail *Papua New Guinea* Jungle track across the Owen Stanley Range north-east of PORT MORESBY. It was the site of bitter fighting during the Second World War, when Japanese troops, having followed the trail from their landing bases on the north coast of New Guinea, threatened to capture the capital Port Moresby. Several thousand Japanese, Australian and American troops died in the subsequent battle for control of the area.
A war memorial stands at the village of Kokoda at the northern end of the trail.
Map (Port Moresby) Papua New Guinea – Ba

Kola (Kol'skiy) Peninsula *Russia* Arctic peninsula and an important strategic area with major military bases near the Norwegian border in the far north-west of Russia, between the Barents Sea and the White Sea. Lapps herd reindeer in the tundra of the north or work in the great coniferous forests of the south. Apatite (calcium phophate) is mined at APATITY and Kirovsk.
The ice-free port of MURMANSK is the main town and capital of the region incorporating the peninsula, which covers some 129 500 km² (50 000 sq miles).
Map Russia – Eb

Kolarovgrad *Bulgaria* See SHUMEN

Kolberg *Poland* See KOLOBRZEG

Kolding *Denmark* Town in east JUTLAND, lying at the head of a 10 km (6 miles) long inlet of Little Belt (Lille Bælt). A market town since the Middle Ages, it exports fish, grain and cattle. The ruin of a large castle, Koldinghus, which was built in 1208 and destroyed by fire in 1808, is a well-known landmark.
The town's Geographical Garden contains about 2000 plants from all over the world, among which are more than 200 varieties of roses.
Population 57 300
Map Denmark – Bb

Köln *Germany* See COLOGNE

Kolobrzeg (Kolberg) *Poland* Resort on the Baltic coast, 100 km (60 miles) from the German border. The old town, a Hanseatic port, grew wealthy from trading salt and has a fine Gothic cathedral and a fortress dating from the 10th century.
Population 40 200
Map Poland – Aa

Kolonia *Federated States of Micronesia* See POHNPEI

Kol'skiy Peninsula *Russia* See KOLA PENINSULA

Kolwezi *Zaire* Leading mining and processing centre for the Shaba region, Zaire's chief mining area. It is located 240 km (150 miles) north-west of the city of LUBUMBASHI.
Population 200 000
Map Zaire – Bc

Kolyma *Russia* River of north-east Siberia, 2513 km (1562 miles) long. Rising in the gold-bearing Kolyma Mountains, it flows north-east into the East Siberian Sea. From October to June, Arctic ice holds back its water, causing widespread flooding in summer and vast ice jams in winter. In summer and early autumn it is navigable for 2000 km (1243 miles) of its course and is an important transport route through Siberia (which has few roads).
Map Russia – Qb

Komárno *Slovakia* The country's southernmost town, lying at the junction of the Danube and Váh rivers. River boats and small sea-going vessels are built at this busy river port. It is linked by a road bridge to the Hungarian town of KOMÁROM.
Population 35 400
Map Slovakia – Bc

Komárom *Hungary* Port on the River Danube in north-west Hungary opposite the Slovakian town of KOMÁRNO. The road bridge linking the two is the only bridge across the Danube which connects Hungary and Slovakia. Komárom has aluminium and chemical plants.
Population 19 620
Map Hungary – Ab

Komi Republic *Russia* Autonomous republic of the Russian Federation covering 415 900 km² (160 600 sq miles) on the north-western slopes of the Urals. Its southern two-thirds is covered with coniferous forests, and its northern third by Arctic tundra. The area once supported little more than fur-trappers and hunters. However, since it became an autonomous republic in 1936, Komi has become a supplier of raw materials to the factories of Leningrad (now St Petersburg), the Volga valley and the Ural Mountains. Timber rafted down the Sysola and Vychegda rivers (both tributaries of the North Dvina) is processed at Syktyvkar, the republic's capital, and the towns of Ukhta and Sosnogorsk are expanding producers of oil and natural gas which are piped to Western Europe.
However, the old pipelines are prone to leakages – and a major leak in 1994 in the region of Kolva and the Kolva River caused consternation among environmentalists. Coal is mined at Vorkuta. The Komi people, a nation of East Finns, make up almost one-quarter of the population; the other inhabitants are a mixture of people from Russia and the other ex-Soviet states who immigrated to the Komi Republic to exploit its mineral wealth.
Population 1 263 000
Map Russia – Gb

Kommunizma, Pik *Tajikistan* See COMMUNISM PEAK

Komodo *Indonesia* Island of some 520 km² (200 sq miles) in the Lesser Sundas group, west of FLORES. It is the home of a giant monitor, the 3 m (10 ft) long Komodo dragon, which is the world's largest lizard.
Map Indonesia – Ed

Komoé *Ivory Coast* Most easterly of the country's three major rivers – the other two are the BANDAMA and the SASSANDRA. It rises near the border with Burkina and flows some 600 km (373 miles) to the sea at Grand Bassam. It is not navigable because the water level is frequently too low. However, its waters are used for irrigation and a hydroelectric dam was built on it in 1979, at Taabo. Rice is grown on its lower reaches, which also form the western boundary of the area occupied by the Agni peoples.
Map Ivory Coast – Bb

Komoé National Park *Ivory Coast* The country's largest park, which covers 1 150 000 km² (444 000 sq miles), in the north-east, around the headwaters of the KOMOÉ River. Elephants, buffalo, deer and wild boar roam the wooded savannah. The park, which was established in 1926, has hills reaching 640 m (2100 ft).
Map Ivory Coast – Bb

Kompong Cham *Cambodia* Central province astride the Mekong River. Rubber-growing plantations in the north were badly hit by the war in South-east Asia (1965-75), but rice yields in the southern lowlands have been consistently high.
Population 820 000
Map Cambodia – Bb

Kompong Chhnang *Cambodia* Central province between the Tonle Sap River and the capital, PHNOM PENH. It is a mainly rice-growing area, but there is some fishing along the river.
Population 273 000
Map Cambodia – Ab

Kompong Som *Cambodia* The country's main seaport, situated on the Gulf of Thailand. It has grown rapidly since road and rail links over the Elephant Mountains were established with the country's heartland in the 1960s. Cement works and an oil refinery operate nearby.
Population 53 000
Map Cambodia – Ab

Komsomol'sk-na-Amure *Russia* One of the largest cities of Russia's Far East. A port on the Amur River, some 880 km (550 miles) north-east of VLADIVOSTOK, Komsomol'sk-na-Amure lies on a branch of the Trans-Siberian Railway from Khabarovsk (the Baikal-Amur railway line).

The city was founded in 1932 by Komsomol (Young Communist League) pioneers in an area of remote forests and swamps. It grew rapidly, with factories making heavy machinery and equipment, foodstuffs, clothing, footwear and paper. Ship-building, oil refining (oil is pumped via pipeline from the oil fields in Sakhalin) and fishing are also important industries.

Population 318 000
Map Russia – Oc

Kongju (Gongju) *South Korea* Ancient town, and capital of the Paekche kingdom from AD 475 to 538. It lies 120 km (75 miles) south of the capital, SEOUL. Much of the city is being restored, and today it is virtually an open-air museum.

Population 17 000
Map South Korea – Cd

Kongsberg *Norway* Town in south Norway, 70 km (44 miles) south-west of OSLO. It was founded in 1624 as a silver-mining town, but the silver had been completely mined by 1957; the mines are now a museum and tourist attraction. The town has a magnificent 18th-century church. Kongsberg manufactures small arms, electronics and – as the site of Norway's mint – money.

Population 21 300
Map Norway – Cd

Kongsvinger *Norway* Town in south-east Norway, lying in the valley of the Glåma River, 75 km (47 miles) north-east of OSLO. Its ruined fortress dates from 1683. A special commercial relationship exists between Kongsvinger and the Swedish town of Arvika. The Norwegian and Swedish governments have promoted the twinning of the towns, with financial assistance for trade and special arrangements for the exchange of goods across the border.

Population 17 400
Map Norway – Dc

Königsberg *Russia* See KALININGRAD

Königsgrätz *Czech Republic* See HRADEC KRÁLOVÉ

Konin *Poland* Coal-mining and aluminium smelting city, 89 km (55 miles) south-east of the city of Poznań. During the Second World War, the Nazis first used mobile gas-chamber lorries to murder Jews at the town of Chelmno-nad-Nerem, 25 km (16 miles) east of Konin; 360 000 Jews died here between 1941 and 1944.

Population 73 000
Map Poland – Cb

Konkouré *Guinea* The main west-flowing river of Guinea. It rises in the Fouta Djallon and flows through a narrow, deep valley for about 250 km (160 miles) to enter the Atlantic Ocean just north of the capital, CONAKRY. There are power stations at Souapiti and Amaria on the lower reaches of the river.

Map Guinea –Ba

Konstanz *Germany* See CONSTANCE

Konya *Turkey* Carpet-making town lying 235 km (145 miles) south of the capital, ANKARA. It is the capital of a province of the same name. The town was the cultural centre of the Anatolia region during the 13th century when the sect of the Whirling Dervishes was founded here by the Sufi Muslim mystic Jalal ad-Din Rumi. Although the sect is officially banned, it still holds its main festival in the town each December, when devotees literally whirl themselves into a mystic trance.

The province covers 47 420 km² (18 309 sq miles) and its farmers raise sheep, goats, horses and camels, and grow grain, tobacco, cotton, poppies, fruit and vegetables on irrigated lands.

Population (town) 513 350; (province) 1 750 300
Map Turkey – Bb

Kootenay *Canada* National park covering 1378 km² (522 sq miles) on the western slopes of the Rocky Mountains in south-east BRITISH COLUMBIA. It contains snow-capped peaks, tree-covered valleys, hot springs and geysers.

Map Canada – Dc

Kopar *Slovenia* See KOPER

Kópavogur *Iceland* Town lying just south of the capital, REYKJAVÍK. Since the Second World War, it has developed into the country's second largest town, after the capital, mainly because it is an outlier of the capital in a favoured part of the country.

Population 15 900
Map Iceland – Bb

Koper (Kopar; Capo d'Istria) *Slovenia* Adriatic seaport and resort on the Istra Peninsula, 15 km (9 miles) south-west of Trieste, in Italy. It was conquered by Greeks, Romans and Byzantines before the Venetians held it for more than 500 years. It was ruled by Austria from 1815 to 1910, when the Italians occupied it. Made part of the Free Territory of Trieste in 1945, it was ceded to Yugoslavia in 1954, giving Slovenia – independent since 1991 – a port. Since the 1950s, a new harbour as well as car, lorry and motorcycle plants have been built here.

The Venetian old town, once an island, has massive fortifications, a Gothic-Renaissance cathedral and palace, and several *campanili* (bell towers). The little salt-producing port of Piran, 9 km (6 miles) to the west, with a web of arched alleyways and piazzas, contains a copy of St Mark's *campanile* in Venice.

Population 45 000
Map Slovenia – Ab

koppie (kopje) In southern Africa, the term applied to a small, isolated rocky hill.

Korcë *Albania* The country's largest mountain town, situated 855 m (2805 ft) above sea level. It is the centre of an extensive wheat and sugar beet-growing basin in the south-east highlands and Macedonian lakes region. Its population is mixed, and includes Romanian-speaking Vlachs. Albania's only pre-war hydroelectric power station is situated here. Since 1950, Korcë has grown into a modern town, with its main industries revolving around sugar refining, brewing, and the making of leather goods and glass. Copper and lignite are mined nearby.

Population 67 100
Map Albania – Cb

Korčula (Curzola) *Croatia* Lush Dalmatian island, covering 273 km² (105 sq miles), which commands the Dubrovnik-Split sea route. Known in ancient times as Corcyra Nigra, Korčula was held by Venice until 1815. It was at the 13th-century sea battle of Curzola between Venice and Genoa that the Venetian explorer Marco Polo, who was born on the island, was captured by the Genoese. Polo wrote his account of his travels to the East while he was being held in a Genoese jail.

Long a community of sailors, shipbuilders and fishermen, Korčula also produces grapes, olives, fruit and fine white marble from the ancient quarries at Vrnik. Blato, the island's largest settlement,

▼ **MUSLIM RELIC IN ALBANIA** The former mosque built by a local warrior – a Muslim convert who fought at Constantinople in 1453 – still manages to dominate old Korcë.

is famous for its *Kumpanjinja*, a traditional battle dance performed each April. A similar ceremony is staged each week in summer at the old walled port of Korčula. Prior to the civil war, which broke out in the region in 1991, tourism was an important source of revenue.

Population 19 600
Map Croatia – Cc

Korea, North See p 378

Korea, South See p 379

Korea Strait Sea passage between the south coast of South Korea and Japan, linking the CHINA SEA to the Sea of Japan. The island of TSUSHIMA divides the strait into two parts: the western section which is 97 km (60 miles) wide and the eastern section, with a width of 64 km (40 miles), and which is sometimes called the Tsushima Strait.

In 1905, during the Russo-Japanese War, the Japanese nearly annihilated the Russian Baltic Fleet in the Tsushima Strait, the Russians having sailed halfway around the world to confront the Japanese navy.

Map Korea – De

Korem *Ethiopia* Mountain town some 400 km (250 miles) north of the capital, ADDIS ABABA. Its population grew rapidly after 1982 as refugees from the drought-hit countryside arrived in their thousands. It was television reports from Korem which prompted the Irish singer Bob Geldof to establish the pop music appeals, Band Aid and Live Aid, in 1984-5.

Population 30 000
Map Ethiopia – Aa

Korhogo *Ivory Coast* Capital of the northern savannah region, and main town of the Muslim Senoufo people, lying 190 km (118 miles) north-west of BOUAKÉ. Millet, sorghum, rice and yams are the staple crops, and cotton the cash crop.

Population 65 000
Map Ivory Coast – Ab

Kórinthos *Greece* See CORINTH

Koror *Palau* Capital and main port, scenically situated on a lagoon facing the eroded formations of coral limestone called the Floating Garden Islands. Its main industries are processing copra and fish. From 1914 to 1945, Koror was the capital of Japanese-administered Micronesia; in 1935, it had a Japanese population of more than 25 000.

Population 8000
Map (Palau) Pacific Ocean – Bb

Kortrijk *Belgium* See COURTRAI

Korup National Park, *Cameroon* Nature reserve covering some 126 000 hectares (311 220 acres) of rain forest in the north-west, bordering Nigeria. Declared a national park in 1986, it has the largest number of plant species found in any African forest – 1200 species, including 400 different kinds of trees. Extensive research is being carried out to tap the forest's vast resources for pharmaceutical purposes.

Map Cameroon – Ab

Kos (Cos; Istanköy) *Greece* One of the larger DODECANESE islands, lying in the eastern Aegean Sea, close to the the Turkish coast (it is called Istanköy by the Turks). Kos was the home of Hippocrates (about 460-380 BC), the 'Father of Medicine', who founded a medical school on the island. Its harbour, Mandraki, is guarded by a 14th-century Frankish castle. Mandraki also has a mosque dating from 1786 and, in the same square, a gigantic plane tree with a trunk 13.5 m (44 ft) in girth and some of its branches held up by scaffolding. There is a fountain playing beneath it. A Doric temple to Asclepius, the god of medicine, lies 6 km (4 miles) south-west of the town.

The volcanic island of Nísiros, with natural sulphur springs, lies south of Kos. According to mythology, the island is said to have been a piece of Kos which the sea god Poseidon flung at Polybotes, the Titan.

Population 20 300
Map Greece – Ec

Kosciusko, Mount *Australia* The country's highest mountain. It rises to 2228 m (7310 ft) in the Snowy Mountains near the state border between New South Wales and Victoria. The Kosciusko National Park, of which the mountain forms a part, covers 6469 km² (2497 sq miles).

Map Australia – Hf

Kosi *Nepal/India* River flowing for 480 km (300 miles) through eastern Nepal and northern India. It is formed by the confluence of two minor rivers – the Sun Kosi and the Arun. The Sun Kosi, which drains much of eastern Nepal, plunges through some rugged gorges before emerging as the Kosi on the plains of northern India, where it soon joins the Ganges River. Although it is dammed for hydroelectric power and flood control in Nepal, it still causes massive flooding in India during heavy monsoons.

Map India – Db; Nepal – Ba

Košice *Slovakia* Regional capital of Vychodoslovensky (eastern Slovakia), lying in a wine-producing area in the south-east, 25 km (15 miles) from the Hungarian border. Since 1950, Košice's population has quadrupled; coinciding with the expansion of metal, engineering, chemical, and food industries.

The city also has a university and other cultural facilities, as well as a 14th-century Gothic cathedral and the macabre Miklus's medieval prison, complete with instruments of torture.

Population 238 300
Map Slovakia – Cb

Kosovo (Kosmet) *Yugoslavia* (*Serbia and Montenegro*) Autonomous province in south-west Serbia covering 10 887 km² (4202 sq miles). It is situated on the Albanian border, and three in four of its people are ethnic Albanians.

The area has long been an economic backwater despite its well-watered, fertile lowlands, and the rich deposits of silver, chrome, lead and zinc in its mountains. However, since the 1950s, determined attempts have been made to improve the province's agriculture and industry in an effort to address the problem of unemployment.

Tourists are now being encouraged to visit Kosovo for its skiing and its beautiful mountain scenery. A 14th-century sultan's tomb at Kosovo Polje, and two medieval mosques at PRIŠTINA are reminders that Kosovo was held by the Turks from 1389 to 1912.

However, the civil war that broke out in the 1990s in parts of former Yugoslavia has not helped to encourage tourism. Priština is the capital of the province.

Population 1 584 000
Map Yugoslavia – Cc

Kosovo Polje *Yugoslavia* See PRIŠTINA

Kosrae *Federated States of Micronesia* Most easterly state of the federation, consisting of the single atoll of Kosrae which lies some 500 km (310 miles) south-east of POHNPEI. It was a whaling port in the last century, but most of its working population are now subsistence farmers or fishermen.

Population 5600
Map Pacific Ocean – Db

Kossou *Ivory Coast* Dam 1.6 km (1 mile) long and 60 m (196 ft) high on the BANDAMA River near the town of YAMOUSSOUKRO. The government had to rehouse 75 000 people in order to build the dam, which was designed to produce enough hydroelectric power to supply all of the country's needs. However, production has been hit by droughts which have reduced water levels.

Map Ivory Coast – Ab

Kostroma *Russia* One of the oldest cities in Russia, founded in AD 1152, 315 km (195 miles) north-east of the capital, MOSCOW. It began as a fortress built to guard trade routes along the Kostroma and Volga rivers; its kremlin (fortress) and cathedral still survive.

Kostroma has been a centre of the Russian linen industry since the 16th century, and most of the city's workers are employed in its linen textile mills, which use locally grown flax. Other industries include engineering, shipbuilding, and tobacco and wood processing.

Population 280 000
Map Russia – Fc

Kota Baharu *Malaysia* Port, and capital of Kenlantan State, on the north-east coast of peninsular Malaysia, just south of the Thai border.

Population 170 600
Map Malaysia – Ba

Kota Kinabalu *Malaysia* Capital of Sabah State on the north-west coast of Borneo. The town, formerly called Jessleton, was badly damaged during the Second World War, but has been rebuilt. Mount Kinabalu, 55 km (34 miles) to the east, is 4101 m (13 455 ft) high.

Population 108 800
Map Malaysia – Ea

Kotka *Finland* Seaport established in 1878 on the Gulf of Finland, on the south-east coast 50 km (30 miles) west of the Russian border. It is Finland's busiest port after the capital, Helsinki, and exports timber and timber products while importing fuel and machines. Its products include machinery, paper, pulp, sugar and processed foods.

In 1789, during a naval battle between Russia and Sweden near here, more than 100 ships foundered, yielding a rich harvest for underwater archaeologists. In 1855, the port was set on fire by the British navy and the only building to remain was an 18th-century Orthodox church. Nearby is Langinkoski, once a summer residence of the Russian royal family.

Population 56 500
Map Finland – Cc

Kotor, Gulf of (Boka Kotorska) *Yugoslavia (Serbia and Montenegro)* Bay on the coast of Montenegro. Enclosed by high mountains, it is one of the world's finest natural harbours. The small medieval seaport of Kotor (Cattaro), on the south side of the gulf, was ruled by Venice from 1420 to 1797. It has a fine 12th-century cathedral, 14th-century mint and a seamen's guild founded in the early 9th century. It was badly damaged by earthquakes in 1979 and 1981. The road inland from the port to the former Montenegrin capital of CETINJE twists through 32 hairpin bends up the side of the Lovcén Mountain range, offering breathtaking views over the gulf.
Map Yugoslavia – Bc

Kotte *Sri Lanka* Town being developed as a new seat of government, 11 km (7 miles) east of central COLOMBO, the present capital. When completed, it will be renamed Sri Jayawardhanapura after the 15th-century Sinhalese name of the old capital at Kotte. Colombo will remain the country's commercial capital.
Population 109 000
Map Sri Lanka – Ab

Koudougou *Burkina* Market town for export crops on the edge of the Mossi Highlands, 70 km (45 miles) west of the capital, OUAGADOUGOU. It is situated on the country's only railway and has some industry, including a shea butter mill, which extracts edible fat from the fruit of the shea tree, a cotton ginnery, and a textile complex.
Population 51 670
Map Burkina – Aa

Kouilou (Kwilu) *Congo* River, some 320 km (200 miles) long, rising on the Batéké Plateau in the south-west, near the border with Gabon. Its upper section, called the Niari, flows south, then west, then north-west, before turning south-west at Makabana and flowing, as the Kouilou, to the ATLANTIC Ocean. Gorges along its path provide potential sites for hydroelectric power stations. Large plantations in the Niari Valley, Congo's most fertile region, produce export crops including cocoa, coffee, palm products, sugar and tobacco, which are shipped to Pointe-Noire via the Congo-Océan railway.
Map Congo – Ab

koum (kum) In Turkestan, the name given to a sandy desert with dunes and stretches of sand, much like the ERG of the Sahara.

Koumba *Guinea* See CORUBAL

Kourou *French Guiana* Coastal town 56 km (35 miles) west of the capital, CAYENNE. Until 1968, when it became the European Space Agency's rocket-launching site, it was a fishing village. In the 19th century, it was colonised by 15 000 French settlers, under the illusion that it was the legendary El Dorado; all died.

Today, the space centre covers some 120 km² (46 sq miles) beside the coast. The site was chosen because it is only 5 degrees north of the equator, where the earth is moving fastest as it spins. The extra speed acts like a slingshot, enabling rockets to get into space with less fuel.

It is used mainly to launch weather and telecommunications satellites.The European space shuttle *Hermes* was due to be launched here in 1995.
Population 13 960
Map French Guiana – Bb

Kovno *Lithuania* See KAUNAS

Kowloon *Hong Kong* City and one of the three constituent territories of the British Crown Colony of Hong Kong, the others being HONG KONG ISLAND and the NEW TERRITORIES. The acquisition in 1860 of the small peninsula gave the British control over Victoria Harbour, which separates Kowloon from Hong Kong Island.

Kowloon developed in the 19th century as a harbour settlement. Today it is an industrial region and an important commercial centre. It is also one of the most crowded areas on earth – the average population density being 26 180 people per km² (67 824 per sq mile) – although many of its inhabitants have resettled in the New Territories over the last decade.

Toys, cotton goods, footwear and rope are manufactured here. Kowloon has an ocean terminal with the world's third largest container port, the colony's Kai Tak international airport, and a railway link to Guangzhou in China.
Population 2 000 000
Map Hong Kong – Cb

Kpalimé (Palimé) *Togo* Market town in the Togo Highlands, 105 km (65 miles) north-west of the capital, LOMÉ. It is the centre of Togo's main cocoa-producing area, and also handles coffee and palm oil. The town is being developed as a tourist resort.
Population 31 000
Map Togo – Bb

▼ **CHINA'S BACKDOOR** The skyscrapers of Hong Kong's Kowloon City stand on a tip of the Chinese mainland. Victoria Harbour separates it from Hong Kong Island; Victoria, here in the foreground, is the colony's business capital.

North Korea

AN ISOLATED STATE, WITH THE FIRST COMMUNIST DYNASTY IN HISTORY, NORTH KOREA LAGS ECONOMICALLY FAR BEHIND ITS VITAL NEIGHBOUR TO THE SOUTH

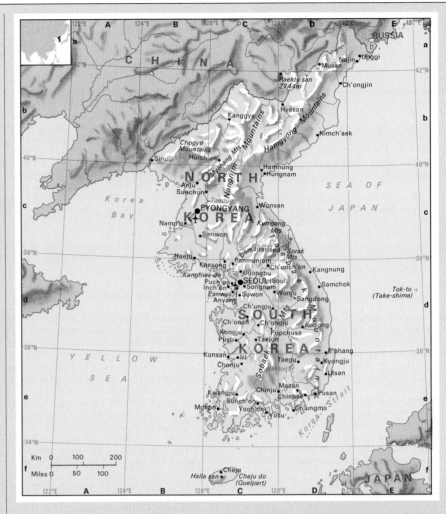

Remote, closed and hostile to outside contact, North Korea has continued as a virtual dictatorship under its new leader Kim Jong Il, son of the late 'Great Leader' Kim Il Sung. It is a communist state of the Stalinist school, seemingly dedicated to two main aims: reunification (under communist control) with its neighbour SOUTH KOREA, and the glorification and perpetuation of the Kim dynasty.

Since the armistice that ended the Korean War on 27 July 1953, little has been achieved towards the first aim; abusive propaganda and the occasional incident in the DEMILITARISED ZONE along the ceasefire line being punctuated by more conciliatory spells. With regard to the second, it seems that North Koreans – who mourned the death of their 'Great Leader', Kim Il Sung in 1994, like the passing of a benevolent king – have happily accepted the leadership of the son, Kim Jong Il.

With North Korea, however, it is difficult to be sure of anything. Few Western visitors are allowed entry, and even contacts with China, its last remaining friend, are kept to a minimum.

Whatever its shortcomings, North Korea undoubtedly has provided its population with adequate housing, basic education and health care, but the economy has been stagnant. In most respects, life in this rigid, authoritarian and disciplined society is drab and unexciting.

North Korea has some 55 per cent of the Korean peninsula's land area, stretching from the YALU and Tumen rivers, which form its border with CHINA and, for an 18 km (11 mile) stretch in the east, with Russia, south to the Demilitarised Zone, which runs mostly just north of the 38th Parallel – forming its border with South Korea. It is a mountainous country and is especially rugged in the far north, where one of its peaks exceeds 2500 m (8200 ft). Three-quarters of the land is forested highlands or scrubland and only 19 per cent is cultivated.

The climate is also similar to that of South Korea, but slightly cooler and less humid in summer and even colder in winter. There is generally less rainfall than in the south.

In contrast to its southern neighbour, however, North Korea is quite well endowed with fuel and mineral resources. Although it has no oil, coal mines can supply 90 per cent of its energy needs. Other mineral resources include iron, copper, lead, zinc, tungsten, graphite and 50 per cent of the world's magnesite.

Much North Korean industry was developed by the Japanese, who occupied the peninsula from 1910 to 1945. (For earlier history, see SOUTH KOREA.) With the defeat of Japan in

1945, Russian occupying troops opened the way for a communist state to be formed in 1948 under Kim Il Sung, a former partisan. The Korean War began with an attack by North Korea on South Korea in June 1950. South Korea was defended by United Nations troops (mainly American) and North Korea's cities were devastated. One million North Koreans were killed or fled to the South.

The war ended in 1953, after which reconstruction centred on heavy industry and collective farms. The country's prosperity peaked in the mid-1960s, but then declined to well below that of the South. All North-South Korean contact was cut after the war. Meanwhile, Kim Il Sung built up a personality cult that made those of Stalin seem modest by comparison.

Throughout 1993 and into 1994 there were successive confrontations between North Korea and the international community over its refusal to allow inspection by the International Atomic Energy Agency following US suspicions that it was planning to develop nuclear bombs. UN sanctions against North Korea seemed imminent.

After the death of Kim Il Sung, Kim Jong Il appeared to take a more conciliatory stance. An accord signed with the United States three months later opened the way to commercial and diplomatic relations between the two countries in exchange for the dismantling of North Korea's

nuclear capability. This raised prospects of an end to the cold war between north and south.

NORTH KOREA AT A GLANCE

Official name Democratic People's Republic of Korea

Area 120 540 km² (46 530 sq miles)

Population 22 645 800 **Per km²** 188 **(Per sq mile** 487)

Capital Pyongyang

Government People's republic

Currency 1 North Korean won = 100 chon

Language Korean

Religions Mostly atheist, some Shamanist, Chundo Kyo, Buddhist, Christian

Climate Moderate to cool, with cold winters and warm, humid summers. Average temperature in Pyongyang ranges from –8°C (18°F) in January to 24°C (75°F) in August

Land use Cultivation 19%, forest and woodlands 74%, other 7%

Main primary products Rice, coal, iron ore

Major industries Agriculture, iron and steel, engineering, cement, chemicals

Main exports Metals and metal ores (mainly iron), chemicals

Annual income per head (US$) 1685 (estimated)

Population growth (per thous/yr) 19

Life expectancy (yrs) Male 66 **Female** 73

South Korea

INSPIRED BY ONE NEIGHBOUR, THREATENED BY ANOTHER, SOUTH KOREA HAS A VIGOROUS, BOOMING ECONOMY OVERSHADOWED BY FALTERING PROGRESS IN POLITICAL LIBERALISATION

Korea used to be called 'Land of the Morning Calm', but there is precious little tranquillity today in this dynamic industrial country of East Asia. From the capital SEOUL, in the north-west, to the main port of PUSAN in the south-east, factories belch smoke into the sky, while bulldozers and excavators grind across the landscape, clearing the way for further industrial development and preparing sites for motorways and houses.

The beauties of rural Korea can still be found – especially in the mountainous east, where vast stretches of forest are home to a diverse animal life – and there are historical sites, notably at KYONGJU, dating back for more than 1000 years. But signs of dramatic change are apparent everywhere, for South Korea's economy has been expanding at a rate that makes even Japan's growth seem sluggish by comparison.

Like that of Japan, South Korea's economic miracle has been achieved on the basis of very slender natural resources. A country slightly larger than Portugal, it occupies the southern half of the Korean Peninsula, stretching some 400 km (250 miles) from the KOREA STRAIT to the DEMILITARISED ZONE bordering NORTH KOREA. It is predominantly mountainous. The highest mountain ranges run north and south along the east coast. Only 22 per cent of its surface area is cultivated, the lowlands being mainly in the south and west. Here also are the main areas of settlement, where population densities are extremely high.

Domestic coal and hydroelectric power help to meet the burgeoning demand for energy, but South Korea has no oil and relies on imported fuels for three-quarters of its energy needs. Farmers have to cope with a monsoonal climate that veers between the extremes of icy winters and hot, humid summers. Nevertheless, the growing season is long and wet enough for good yields of rice, the staple crop. Fishing has also emerged as an important foreign exchange earner.

BETWEEN TWO POWERFUL NEIGHBOURS

As a political entity, South Korea dates from the end of the Second World War. From 1910 to 1945 it was simply the southern – primarily agricultural – part of a Japanese colony comprising the whole Korean peninsula. For most of its history, China has been the dominant neighbour but the ancient Korean culture itself has proved remarkably resilient through the centuries.

Korea was first united under the Silla emperors, based at Kyongju, in the 7th century AD when the kingdoms of Kogyuro (in the north) and Paekche (in the south) were conquered with Chinese help. Culture, art and the Buddhist religion flourished under the Sillas.

In 1392, the Yi (or Choson) dynasty, allied to the Ming rulers of China, re-established self-rule; their capital was Seoul. Again there was a cultural blossoming, including the introduction of Confucianism and the invention of a phonetic alphabet. But in the 17th and 18th centuries, internal Chinese strife and Japanese expansionism brought foreign domination.

After the Sino-Japanese war of 1894-5 and Russo-Japanese war of 1904-5, Korea became a Japanese protectorate in 1905 and was formally annexed in 1910. The Japanese ruled the peninsula until American troops landed in the south in August 1945. Russia occupied the north and Korea was divided by the 38°N line of latitude – the 38th Parallel – into two zones of occupation. This split was meant to be temporary but the deepening of the cold war led to two rival states – one communist, the other pro-Western – being established, fixing the division. In June 1950, the armies of North Korea launched a massive surprise attack on the South, to begin the Korean War (1950-3).

The North Koreans were repelled by United Nations (mainly American) forces but were reinforced by Chinese troops. The eventual outcome was an armistice and a ceasefire line near the starting point.

BUILDING AN INDUSTRIAL NATION

Its economy being largely agricultural and its cities and communications devastated by the war, South Korea in 1953 was exhausted and impoverished. American aid enabled the country to survive but it was only in the 1960s that its fortunes began to improve. An export promotion policy directed energies towards developing light industries such as textiles and clothing, and cheap electrical and electronic products. From the early 1970s, heavy industry grew rapidly. Japanese investment and know-how aided industrial development and in many ways the South Korean industrial miracle reflected and was inspired by postwar Japanese expansion.

As in Japan, the population is well educated and hardworking. In the recent years, wages have increased tremendously and South Korea is one of the most expensive producers, after Japan, in Asia. As in 'the Land of the Rising Sun', the people of South Korea still expect to be utterly disciplined and to fit into the niche which society has defined for them, at least when it comes to work ethics. Politics is another matter. An ever-increasing part of the population expressed dissent with the government's authoritarian rule. The South Korean government has not been noted for its respect for human rights, a fact which has drawn international censure.

THE TWO KOREAS

Government planning has helped to steer the economy towards success, but perhaps the most important factor of all has been the people's attitude towards their North Korean rivals. Not only were South Koreans determined to outperform the North on all fronts, but the ever-present threat of invasion instilled an unusually strong sense of discipline and seriousness of purpose.

To counter North Korean ambitions, South Korea maintains a well-equipped army of 633 000, backed by 35 000 American troops. But all has not always been well on the home front. Although theoretically a democracy, South Korea was ruled by a succession of powerful and authoritarian military leaders from 1961 to 1993. Political opponents were harassed and persecuted. Violent student demonstrations led in July 1987 to the resignation of the president, the regime's acceptance of direct elections to choose his successor and to an amnesty for opposition leaders. So it came about that demands for a more liberal form of democracy bore fruit in 1993 when the first civilian president was appointed after 32 years of military rule.

Large overseas debts and the threat of protectionist measures against South Korean goods in foreign markets could undermine its apparently robust economy. For the present, however, the country offers a spectacular example of rapid transformation from rags to riches. Its people enjoy a level of prosperity undreamt of a generation ago.

Relations between North and South Korea reached fever pitch once again in 1994 with the refusal of North Korea to allow inspectors of the International Atomic Energy Agency to visit its nuclear power plants where it was suspected the North Koreans were preparing the manufacture of their first atomic bomb.

The US Pentagon began to make plans to win the next Korean War, while the south stood by nervously, its troops in position, watching the clock tick. Much of the tension was relieved, however, after the death of North Korea's 'Great Leader', Kim Il Sung, and the more amenable attitude adopted by his son and heir.

SOUTH KOREA AT A GLANCE	
Official name Republic of Korea	
Area 98 480 km² (38 014 sq miles)	
Population 44 614 000 **Per km²** 453 **(Per sq mile** 1173)	
Capital Seoul	
Government Republic	
Currency 1 South Korean won = 100 chon	
Language Korean	
Religions Christian (48%), Buddhism (47%), Confucian, Chondogyo	
Climate Far south warm, temperate; rest continental. Average temperature in Seoul ranges from –5°C (23°F) in January to 25°C (77°F) in August	
Land use Cultivation 22%, pastures 1%, forest and woodlands 67%, other 10%	
Main primary products Rice, fish, tungsten	
Major industries Agriculture, textiles and clothing, iron and steel, shipbuilding, chemicals, electrical and electronic goods	
Main exports Clothing and textiles, electronics equipment, iron and steel, ships, food products	
Annual income per head (US$) 6790	
Population growth (per thous/yr) 10	
Life expectancy (yrs) Male 67 **Female** 74	

Kra *Myanmar (Burma) /Thailand* Isthmus linking peninsular Malaysia with mainland Southeast Asia. Some 200 km (125 miles) long, it is only about 50 km (30 miles) wide on average. Kra is also the name of a river, 60 km (37 miles) long, flowing into the Andaman Sea.
Map Thailand – Ad

Kragujevac *Yugoslavia* (*Serbia and Montenegro*) Commercial town lying 97 km (60 miles) south-east of BELGRADE. It was the capital of Serbia from 1818 to 1939, and later became the site of Serbia's first schools, theatre, newspaper, and printing press. Today, the town produces Zastava cars.
   A monument and park commemorate 7000 men and boys aged 14 to 70 who were shot by the Nazis on 21 October 1941 as a reprisal for partisan resistance.
Population 146 610
Map Yugoslavia – Cb

Krak des Chevaliers *Syria* See CRAC DES CHEVALIERS

▼ KRUGER NATIONAL PARK A herd of elephants comes down to a river for a drink. Elephants do particularly well in the park and have to be culled regularly to protect the balance of the environment.

Krakatau (Krakatoa, Rakata) *Indonesia* Volcanic island in the SUNDA STRAIT between Java and Sumatra. It is the remains of a much larger volcano which collapsed on 27 August 1883 in what is regarded as the world's greatest natural explosion. The shock wave of the explosion was felt as far away as California, and giant waves created by the blast killed about 36 000 people.
Map Indonesia – Cd

Krakow *Poland* See CRACOW

Kraljevo *Yugoslavia* (*Serbia and Montenegro*) Large industrial and market town in Serbia, situated at the junction of the Ibar and Western Morava (or Zapadna) rivers, 124 km (77 miles) south of BELGRADE. Its factories produce machinery, railway vehicles and metal goods.
   Serbia's medieval kings were crowned at Kraljevo (in fact, the name 'Kraljevo' means 'the Kings' Place'). In 1945 the town was renamed Rankovicevo after Aleksandar Rankovic, a leading Serbian communist. However, Rankovic fell into disgrace in 1962 and Kraljevo regained its old name. Nearby are the stately ruins of Maglic, a 15th-century archbishop's castle on the hilltop overlooking the Ibar, and Kalenic monastery, famous for its fine 15th-century frescoes.
Population 124 600
Map Yugoslavia – Cc

Kranjska Gora *Slovenia* See KARAWANKE ALPS

krans In South Africa, an overhanging, sheer wall of rock or precipice.

Kras (Karst; Carso) *Slovenia* Bare, dry limestone plateau in the DINARIC ALPS near the Italian border, north of Trieste and east of Gorizia on the Soca (Isonzo) River. Rising to 1495 m (4908 ft), it is noted for its caves, such as those around POSTOJNA, its waterfalls and underground rivers.
   Karst, the German name for the plateau, is also applied to the much larger limestone area stretching the length of western Slovenia and Croatia. The word 'KARST' is used by geographers to describe any deeply eroded limestone area with perhaps a very thin layer of topsoil.
Map Slovenia – Ab

Krasnodar *Russia* City in the Caucasus, 80 km (50 miles) north-east of the Black Sea. Founded in 1793 as the Cossack fort of Ekaterinodar, it is now the hub of the rich Kuban' River plains, where melons, tomatoes, tobacco, sunflowers, sugar beet and wheat are grown, and of the west Caucasian wine-producing and cattle-rearing area. Krasnodar also produces machinery, textiles and chemicals.
Population 631 000
Map Russia – Ed

Krasnoyarsk *Russia* One of the largest cities in Siberia and capital of the administrative territory of the same name. Cossacks founded it in 1628 as a fort beside the crossing of the Yenisey River, about 620 km (385 miles) east of NOVOSIBIRSK.

The city lies on the Trans-Siberian Railway, and since the 1930s has become the focus of a huge territory which is rich in gold, timber, copper, iron ore and uranium. To the south is the Krasnoyarsk Dam (capacity 6000 MW), one of the world's largest hydroelectric plants.

Population (city) 924 000
Map Russia – Kc

Kratie *Cambodia* Northern rubber-growing province. The provincial capital, which is also called Kratie, lies on the Mekong River, at its highest navigable point.

Population (province) 136 000; (town) 12 000
Map Cambodia – Bb

Krefeld *Germany* Industrial town and port on the River Rhine, 20 km (12 miles) north-west of DÜSSELDORF, and the centre of the country's silk and artificial fibre industry. It produces hosiery and textiles and has a textile museum.

Population 244 000
Map Germany – Bc

Krishna (Kistna) *India* River rising 160 km (100 miles) south-east of BOMBAY. It flows generally south-east, for 1300 km (800 miles), into the Bay of Bengal, and is a major source of power. It has a large and very fertile delta and, although not navigable, it provides water for irrigation.

Map India – Bd

Kristiania *Norway* See OSLO

Kristiansand (Cristiansand) *Norway* Main port in south Norway, on the SKAGERRAK at the mouth of the Otra River. It has a spacious, ice-free harbour, protected by offshore islands.

It was founded in 1641 by King Christian of Denmark and later fortified. In 1892, it was destroyed by fire. The town's industries include shipbuilding, and the production of metal goods, timber, wood pulp and chemicals; ferry services run from here to Denmark and England.

Setesdal, one of Norway's most picturesque valleys, can be found nearby.

Population 67 000
Map Norway – Bd

Kríti *Greece* See CRETE

Krivoy Rog *Ukraine* See KRYVYY RIH

Krk (Veglia) *Croatia* Fertile island covering 408 km² (158 sq miles), lying off Rijeka in the northern Adriatic. It is the largest of the Croatian islands. Vines, olives and figs grow on its sheltered western side and it is renowned for its red wines. Its main town, also called Krk, is a medieval, walled seaport with a Romanesque cathedral, and a magnificent bishop's palace. It lies on a hill above the sheltered Bay of Krk. Until civil war broke out, the tourist industry here was developing rapidly.

The island was settled by Croats in 1059. The Bascanska Ploca, a tablet of stone dating from the 11th century, which was found in the town of Baska, is inscribed with the oldest surviving example of the Croatian language. The original stone is now in Zagreb, but there is a copy of the important relic in the town church.

Population 15 800
Map Croatia – Bb

Kronshtadt *Russia* Naval base on Kotlin Island in the Gulf of Finland, 25 km (16 miles) west of ST PETERSBURG. The Russians under Tsar Peter the Great (1682-1725) captured the island from the Swedes (who called it Kronslott) in 1703 and began the construction of its fortress seven years later. Its defences played a decisive part in the October Revolution in 1917, and a key role in the Siege of Leningrad during the Second World War, when the city held out against German forces from September 1941 to January 1944.

Map Russia – Db

Krosno *Poland* Town lying 150 km (95 miles) south-east of the city of CRACOW. It has fine 16th to 18th-century pastel-painted houses set around a central square. Krosno lies at the centre of Poland's Carpathian oil field. The town has a museum of kerosene lamps, which were invented in Poland in 1853.

Population 42 900
Map Poland – Dd

Kruger National Park *South Africa* State-run wildlife reserve in the Eastern Transvaal and Northern Transvaal lowveld, dating from 1898 and covering 19 485 km² (7523 sq miles). Probably southern Africa's single most important tourist attraction, it is home to more than 130 species of mammals, 114 of reptiles, 450 of birds and 48 species of fish. It also covers five main botanical divisions.

The park has more than 2000 km (1243 miles) of roads; other facilities include rest camps, caravan sites, shops and restaurants.

Map South Africa – Da

Krugersdorp *South Africa* Mining and industrial centre, and principal town of the West Rand, 32 km (20 miles) north-west of Johannesburg. Its mines produce gold, manganese, uranium, lime, iron, asbestos and dolomite.

The town was established in 1887 and is named after Paul Kruger (1825-1904), Boer leader and president of the Transvaal Republic. The Kromdraai Palaeontological Reserve is nearby and there is a 150 km² (58 sq miles) game reserve on the town's outskirts.

Population 196 210
Map South Africa – Cb

Krung Thep *Thailand* See BANGKOK

Kruševac *Yugoslavia (Serbia and Montenegro)* Industrial and market town on the Western Morava River (also called the Zapadna) in Serbia, 153 km (95 miles) south-east of BELGRADE. Its restored castle dates from the 14th century.

Population 164 800
Map Yugoslavia – Cc

Kruszwica *Poland* See KUJAWY

Krym *Ukraine* See CRIMEA

Kryvyy Rih (Krivoy Rog) *Ukraine* City in the Donets Basin of southern Ukraine, 250 km (155 miles) north-east of ODESA. Founded by Cossacks in the 17th century, it remained a small village until high grade iron ore was discovered nearby in the 1880s. Since then it has grown into one of the world's biggest iron- and uranium-mining centres. Its other products include steel, coke, mining machinery, diamond drills, cement and processed foods.

Population 713 000
Map Ukraine – Cb

Ksar es Souk *Morocco* See ER RACHIDIA

Kuala Lumpur *Malaysia* National capital which stands at the confluence of the Kelang and Gombak rivers, 30 km (19 miles) upstream from the Strait of Malacca. It was founded in 1857 as a tin-mining camp, and its name comes from Malay words translating as 'Muddy River Mouth' – a reference to the river bottom from which tin ore was scooped by giant dredges. The mining camp grew rapidly to become the economic centre of Selangor State, Malaysia's most developed region. In 1895 it was chosen as the capital of the Federated Malay States.

Since independence in 1963, the city's population has more than doubled. Its modern government complex, surrounded by lake gardens, mixes uneasily with old temples and mosques, Oriental shopping streets, a magnificent, white-domed Victorian railway station, new housing blocks and the Putra World Trade Center, one of the largest in South-east Asia. However, Kuala Lumpur, or KL, as it is called, also has its share of squatter colonies.

Population 937 900
Map Malaysia – Bb

Kuala Terengganu *Malaysia* Capital of Terengganu State on Malaysia's east coast, 140 km (85 miles) south of the Thai border. It has no rail links, but the east coast highway has been improved, opening up communications. Hydrocarbons have been discovered offshore.

Population 186 700
Map Malaysia – Ba

Kuantan *Malaysia* Capital of Pahang State on the east coast, 200 km (125 miles) east of KUALA LUMPUR, the national capital. A new trans-Peninsular road now links Kuantan with the industrialised west. A port has been built to export palm oil, produced on plantations inland.

Population 136 700
Map Malaysia – Bb

Kuanza *Angola* See CUANZA

Kuching *Malaysia* Town and riverside port in the extreme south-west of the State of Sarawak. Kuching was the headquarters of Sarawak's remarkable ruling family of 'white rajahs'.

The dynasty was founded by James Brooke, an Englishman formerly employed by the East India Company. Brooke helped the Sultan of Brunei to suppress a revolt and as a reward, the Sultan made him Rajah of Sarawak in 1841. Brooke used Kuching as his base for a campaign against local piracy in the 1840s and his family made their home here. After three generations, the third and last of the 'white rajahs', Sir Charles Vyner Brooke, relinquished all his rights, in favour of Britain, in 1946. In 1963, Sarawak became part of newly independent Malaysia.

Population 120 000
Map Malaysia – Db

Kufra (Kufrah) *Libya* Group of five oases, 48 km (30 miles) long and 20 km (12 miles) wide, in the Cyrenaican Desert, 920 km (572 miles) south-east of BENGHAZI. They have always been an important caravan centre on the trans-Saharan route from Benghazi to Chad. In 1895, they became the headquarters of the Sanusi, a Muslim religious fraternity. They strongly resisted the Italian colonial invasion in 1912, and it was not until 1931 – and only after prolonged bombing – that the Italians were able to occupy the oases.

Kufra produces barley and the country's best dates, as well as traditional hand-woven textiles. The area has been selected for major agricultural improvements using pumped water (there are large underground water reservoirs).

Some 10 000 hectares (about 25 000 acres) of irrigated land produce fodder for sheep rearing, and another 15 000 hectares (37 000 acres) have been put under irrigation. Iron ore and low-grade manganese deposits have been found nearby.

Population 25 000
Map Libya – Cb

Kuh-e Baba *Afghanistan* Massive mountain range separated from the Hindu Kush by the Bamian Valley. Its spectacular passes include the Unai, reaching 3100 m (10 170 ft), and the Hajigak, at 3250 m (10 660 ft).

The highest peak is Shah Fuladi, which reaches 5600 m (18 370 ft). The range is the source of the Hari Rud and Kabul rivers.

Map Afghanistan – Ba

Kuito (Bié Silva Porto) *Angola* Market town 550 km (342 miles) south-east of the capital, LUANDA. It is the capital of Bié Province. Lying at the heart of one of the country's most fertile regions, the town normally handles coffee, meat, rice, beeswax and cereals.

However, it suffered greatly when placed under siege by UNITA forces in February 1993 – 60 000 people, many of whom had fled to Kuito from the countryside, were trapped here, and more than 25 000 people died in the fighting. By 1994 the town had been reduced to a ruin, with its inhabitants surviving on food relief.

Population (Bié Province) 1 125 000
Map Angola – Ab

Kujawy *Poland* Ancient province straddling the Noteć River in central Poland. Kruszwica, the regional capital, 105 km (65 miles) north-east of the city of Poznań, was the cradle of the Polish state. The city's Romanesque church – built between 1120 and 1140 – stands on the original site of a 10th-century wooden cathedral built by King Mieszko I, Poland's first ruler, who embraced Christianity.

Map Poland – Cb

Kukës *Albania* Main town of north-east Albania, situated at the confluence of the Black and White Drin rivers. It lies in a basin beneath the high, dazzling white gypsum and marble Korab Mountains. Surrounding the town are several chrome and copper mines.

Kukës has ore processing plants built by the Italians before the Second World War and expanded with Czech and Soviet aid after 1948, in the wake of the communist takeover of Albania in 1946.

Population 12 500
Map Albania – Ca

Kulm *Poland* See CHELMNO

Kumamoto *Japan* City and administrative centre near the west coast of KYUSHU Island, 75 km (47 miles) east of the port of NAGASAKI. A former producer of swords, Kumamoto manufactures small electrical components for other factories on Japan's 'Silicon Island'.

Its spectacular 17th-century castle has been rebuilt and is now a museum. Kumamoto's 300-year-old park is a prime example of Japanese landscape gardening.

Population 575 300
Map Japan – Bd

Kumasi *Ghana* The country's second largest city, after ACCRA. It was founded as the capital of the Ashanti people in the 17th century and was one of the first towns to develop in Ghana's southern forest, with the Ashanti taking control of the surrounding area so that they could channel as much trade as possible through their capital. Even today, Kumasi is known as 'the City of the Golden Stool' – the golden stool being a symbol of the Ashanti royal line – and the Ashanti king still has his court at Manhiya Palace.

The town, which is the centre of the cocoa-growing belt, has Ghana's largest market. Kumasi has been modernised with paved streets, parks and gardens and is dominated by the Kumasi Central Hospital.

It also has a university, a museum, several theatres, including an open-air theatre, a crafts workshop, model farm and zoo. It is linked by road and rail to the capital, Accra, which lies 200 km (125 miles) to the south-east.

Population 1 000 000
Map Ghana – Ab

Kumana *Sri Lanka* See RUHUNA

Kundelungu National Park *Zaire* Founded in 1970 in Shaba region, south-west of Lake Mweru on the Zambian border, the park includes Lofoi Falls, some forest country, and savannah vegetation on the Kundelungu plateau. Wildlife includes antelopes, cheetahs, elephants, and zebras.

Map Zaire – Bc

Kunene *Angola* See CUNENE

Kunene *Namibia* Region in the north-west of the country, covering 36 804 km² (14 207 sq miles). It stretches from the Kunene (Cunene) River in the north, along the Angolan border to Namibia's Atlantic coast in the west, and includes what was known before independence, under apartheid, as 'Damaraland' – a region to which the Damara people were restricted.

Today, the unique Skeleton Coast Park, a large nature reserve (16 900 km², 6524 sq miles) also falls within the borders of Kunene. Numbers of visitors to the park are strictly controlled to protect its fragile and unusual desert ecology. Khorixas is the regional capital.

Population 56 500
Map Namibia – Ab

Kunlun Shan *China* Mountain range with peaks of more than 7000 m (23 000 ft). It stretches for almost 4000 km (2485 miles) eastwards from the Karakoram Range, forming the northern edge of the Xizang Gaoyuan (Tibetan Plateau).

Map China – Bd

Kunming *China* Capital of Yunnan Province. It is an important railway junction and trading centre. It has iron and steel plants, engineering works and truck assembly plants.

Population 1 976 000
Map China – Ff

Kunsan (Gunsan) *South Korea* Port on the west coast, standing on the south bank of the Kuan estuary. Paper, plywood, rubber, chemicals and textiles are made here.

Population 218 000
Map Korea – Ce

Kuopio *Finland* City and tourist resort founded in 1782 on the west coast of Lake Kallavesi, 340 km (210 miles) north-east of the capital, HELSINKI. It is the seat of the Orthodox archbishopric and has a museum devoted to the Orthodox church in Finland.

Population 82 300
Map Finland – Cc

Kupang *Indonesia* Capital and main port of East Nusa Tenggara Province at the western tip of the Island of TIMOR.

Population 123 400
Map Indonesia – Fe

Kurashiki *Japan* City on HONSHU Island, 160 km (100 miles) west of OSAKA. Old Kurashiki, an important port during feudal times, is carefully preserved. Its quiet canal, spanned by 'willow-pattern' bridges, is shaded by weeping willows and flanked by solid white warehouses with heavy black-tiled roofs, many of which have been converted into museums.

New Kurashiki, which sprawls across land reclaimed from the Inland Sea, is an important industrial centre specialising in oil refining, and the production of petrochemicals, iron and rayon. Kurashiki also houses the astronomical observatory from which Honda's Comet was first sighted in 1949.

Population 415 000
Map Japan – Bd

Kurdistan *Middle East* Area of Central Asia inhabited by the highly independent Kurds – a nomadic, stock-raising Muslim people. Kurdistan comprises south-east Turkey, north-east Iraq and north-west Iran and covers an area of some 200 000 km² (77 202 sq miles).

There are immense variations in the estimates of the numbers of Kurds, ranging from about 14 million to around 24 million – 8 to 12 million in south-east Turkey, 3.7 to 7.3 million in north-eastern Iran, and 2.5 to 4.4 million in northern Iraq. An independent Kurdistan was proposed in the 1920 Treaty of Sèvres, between the First World War Allies and Turkey, but the idea has come to nothing.

Since then, Turkey, Iraq and Iran have repressed the Kurds, giving rise to periodic disturbances and rebellions. In the early 1970s there was a major uprising of Kurds in Iraq. This led to the creation of the Kurdish Autonomous Region, recognising the special status of the Kurds. However, fighting continued sporadically over the following years. During the 1991 Gulf War, the Kurds in Iraq once again rose up and rebelled against Saddam Hussein. Their rebellion was brutally suppressed, and many fled to Iran and Turkey.

Map Middle East – Ba

Kurdzhali (Kârdžali) *Bulgaria* Main regional town of the eastern Rhodope Mountains, 55 km (34 miles) north of the Greek border in a broad valley on the Arda River. Since 1950, when it was a quiet town of 9000 inhabitants, most of whom were involved in tobacco farming and processing, it has grown into a major lead and zinc-mining and processing centre. Power for the new industries is supplied by two hydroelectric dams close to the town.
Population 59 000
Map Bulgaria – Bc

Kure *Japan* Town on the INLAND SEA, 20 km (12 miles) south-east of HIROSHIMA. A former naval base with an excellent natural harbour, it now has modern steel mills and builds some of the world's largest ships. During the Second World War, its shipyards produced the *Yamato*, one of the largest battleships ever built.
Population 217 500
Map Japan – Bd

Kuril Islands (Chishima; Kurile Islands) *Japan/Russia* Chain of 56 volcanic islands, many uninhabited, stretching from the Japanese island of HOKKAIDO north-east to Russia's KAMCHATKA Peninsula. They have at least 100 volcanoes of which 35 are still active.They have a total land area of 15 600 km² (6020 sq miles). The southern islands (Kunashir, Etorofu, Shikotan and Habomai) were seized from Japan by the USSR in 1945.
Map Japan – Eb

Kuril Trench *Japan/Russia* An extension of the Japan Trench in the Pacific Ocean floor adjacent to the KURIL ISLANDS and Russia's KAMCHATKA Peninsula, which form part of the area known as the Pacific 'Ring of Fire'. Volcanoes in the area erupt and earthquakes are caused by the Pacific plate meeting and descending beneath the Eurasian plate (see PLATE TECTONICS). The Kuril Trench reaches a depth of 10 498 m (34 442 ft).
Map Russia – Pd

Kursk *Russia* Industrial city 450 km (280 miles) south of MOSCOW. It was established on the Seym River in the 9th century, but was destroyed by Tatars in 1240 and not rebuilt until 1586. The biggest tank battle of the Second World War, involving 5700 tanks, took place around Kursk in 1943. It ended in a Russian victory, but most of the city, except for the centre, was destroyed.
The sections that survived include the 18th-century St Serge Cathedral, which has a finely carved 17 m (46 ft) high screen. Kursk has one of the world's richest iron ore deposits, containing 160 000 million tonnes.
Population 433 000
Map Russia – Ec

Kurukshetra *India* Holy Hindu town in Karnal district, Haryana, 120 km (74 miles) north of Delhi. Here, Lord Krishna is said to have revealed himself to Prince Arjuna before a mighty battle (about 1000 BC) in which many died. The battle, and the moral discourse in duty and obligation, and right and wrong that resulted from it, are related in the *Bhagavad-Gita* ('the Song of the Lord'), Hinduism's most sacred and philosophical book.
Population 635 000
Map India – Bb

Kushiro *Japan* Fishing port on the south-east coast of HOKKAIDO Island. A pulp and paper industry has developed here.
Population 206 000
Map Japan – Db

Kuta Beach *Indonesia* See BALI

K'ut'aisi *Georgia* The country's second largest city, after T'BILISI. It stands on the Rioni River in the south-west, 85 km (53 miles) east of the Black Sea and 70 km (40 miles) from the Turkish border. It produces motor vehicles, mining machinery, chemicals, silk, clothing, footwear, furniture, and canned foods. K'ut'aisi contains the remains of a fortress built on the site in the 1st century AD and of an 11th-century cathedral.
Population 238 000
Map Georgia – Bb

Kutch (Kachchh) *India/Pakistan* Coastal district some 550 km (340 miles) north-west of BOMBAY, covering 45 612 km² (17 611 sq miles). It borders the Gulf of Kutch and consists largely of a vast, swampy depression, the Rann, which floods from both the sea and the rivers during the monsoon season. At other times it is dry, salty and inhospitable.
Part of it is claimed by Pakistan, which waged a tank campaign here against India in 1965. Though the campaign was unsuccessful, Pakistan was granted about 10 per cent of the region in 1968.
Population 1 245 970
Map India – Ac

Kutna Hora *Czech Republic* Former silver-mining town in Bohemia, situated 65 km (40 miles) south-east of PRAGUE. Mining began here in the 13th century and the royal mint was established in the 14th-century Vlassky Dvur ('Italian Court'), where craftsmen brought in from Florence, designed and struck silver coins. The process is now recreated in a coin museum.
The sumptuous Gothic cathedral dedicated to St Barbara – the patron saint of gunners and miners – has a chapel with 15th-century wall paintings of the craftsmen plying their trade. The mines were gradually exhausted and the mint closed in the 18th century. Today, the town has engineering, metal, textile, and food processing industries.
Population 21 400
Map Czech Republic – Bb

Kuwait See p 385

Kuwait City *Kuwait* Situated at the southern entrance to Kuwait Bay, the country's capital stands on the site of a 16th-century Portuguese fort. In the early 1960s, it consisted of one main street and a few thousand people living on the edge of the desert. During the next 20 years it was transformed into a wealthy, expanding city of high-rise buildings, large office blocks, boulevards, parks and gardens. The modern city has been laid out with roads radiating from its centre.
Kuwait's principal buildings include the Seif Palace (the Emir's headquarters), built in 1896, and the Kuwait Towers, built in the 1970s, which contain restaurants, an indoor garden, and an observatory from which there are panoramic views of the city. The city's museums include the Kuwait Natural History Museum and the Kuwait Museum (which was looted and set ablaze by Iraqis during the 1991 Gulf War).

A DAY IN THE LIFE OF AN ASIAN EXPATRIATE IN KUWAIT

Economic hardship in Maria Rodriguez's native Philippines brought her to oil-rich Kuwait in 1991 to work like thousands of other Asian women in The Gulf as maids for wealthy Arabs. The Gulf War reduced the emirate's intake of foreign labour by 30 per cent but experienced domestic workers like Maria who are well educated and speak English fluently are still in demand.

She comes from a large Catholic family in her home province near Manila where her three young children are being looked after by relatives. Her husband works as a driver in Saudi Arabia and they see each other every two years when on home leave. The money they earn ensures an education for their offspring and helps their extended family.

In Kuwait City, Maria is effectively on duty 24 hours a day but sleeps between midnight and sunrise and for a few hours after lunch when the Arab world traditionally enjoys a long break as temperatures outside get as high as 45°C (113°F). She looks after the entire running of the house except gardening and shopping.

Maria's day begins at around 5.30 am as the nearby mosque calls male Muslims for the first of five daily prayer sessions. She must have breakfast prepared – usually bread, cheese, olives, jam, milk, yoghurt, tea and coffee – by 6.30 am, although some family members do not eat until around 10 am.

By late morning, lunch, an extravagant spread of salads, fish, pitta bread and fresh fruits, must be prepared and Maria stands by in her black uniform until the family of five is satisfied and has retired to their rooms for a couple of hours' sleep. The remainder of the day goes quickly as she washes clothes, compiles daily shopping lists for the chauffeur and cleans and dusts the spacious seven-room double-storey white-washed villa in a quiet suburb on the outskirts of Kuwait City.

Sometimes, extra duties such as rearranging furniture or entertaining the children are expected of her. She usually rests between 3 pm and 6 pm. By sunset the family has often gone out shopping, visiting or working at their nearby store. Her cousin in a neighbourhood household often phones her at this time to chat and gossip.

She occasionally baby-sits on Friday – the Muslim holy day – when the couple go out together. Her day off arrives about every two weeks but not always on the same day. She does very little on this day except think of her family, pray, write letters and sleep as she is always very tired.

As women – especially non-Muslims – have very few human rights in Kuwait, she stays in to avoid being harassed on the streets by local men. She gets paid in US dollars tax-free and, because her two-year contract includes accommodation and food, she sends her entire salary home where she is saving for a small plot of land on which to build a home for her retirement when she and her husband can be together.

Damage to the entire city during the war was considerable. However, a massive reconstruction programme has now been completed. Building materials and chemicals are manufactured, and there is a fishing industry.
Population (Kuwait and suburbs) 360 000
Map Kuwait – Bb

Kuybyshev *Russia* See SAMARA

Kuznetsk *Russia* See NOVOKUZNETSK

Kwai (Khwae) *Thailand* Two tributaries of the Mae Khlong River, the Kwai Yai ('Big Kwai'), also called the Mae Khlong, and the Kwai Noi ('Little Kwai') both about 150 km (93 miles) long. The bridge over the River Kwai, famous from the film of the same name, was built between 1942 and 1943, during the Second World War, over the Kwai Yai near KANCHANABURI, 120 km (75 miles) west of Bangkok.

The railway it carried, which continued into Burma (now Myanmar), was completed in October 1943, but only part of it is still in use. The first bridge no longer exists, but the cemeteries filled with those who died here remain to attract visitors. An estimated 110 000 Allied prisoners of war and slave labourers died during the construction of the bridge and work on the 415 km (257 mile) stretch of railway line.
Map Thailand – Ac

Kwajalein *Marshall Islands* The world's largest atoll. It lies in the Ralik chain about 450 km (280 miles) north-west of MAJURO. Its 283 km (176 mile) long reef encloses a lagoon covering 2850 km² (1100 sq miles). The atoll is leased by the United States under a 30-year agreement from 1982 for use as a test target for missiles fired some 8000 km (5000 miles) away in California.
Population 6600
Map Pacific Ocean – Db

KwaNdebele *South Africa* Former self-governing state created in 1981 for the Ndebele people and covering 3410 km² (1317 sq miles) north-east of PRETORIA. With meagre natural resources in the territory, most inhabitants were obliged to commute to areas in South Africa in which they could sell their labour. KwaNdebele was only recognised by South Africa, and after that country's first democratic elections in 1994, it was re-integrated into South African territory.
Map South Africa – Cb

Kwangju (Gwangju; Chonnam) *South Korea* Provincial capital and university town 275 km (170 miles) south of the national capital, SEOUL. The city, which is more than 1000 years old, is the centre of South Korea's textile industry. It lies on fertile lowlands – producing mainly cotton, rice and silk – in the once remote province of Cholla. Modern communications have ended Kwangju's isolation, but its tradition of independence manifested itself in violent student riots against the Seoul government in 1980. It has many historical remains, and there are old temples and tombs in the surrounding hills.
Population 1 145 000
Map Korea – Ce

Kwango (Cuango) *Angola/Zaire* River, 1100 km (680 miles) long, which rises in the central plateaus of Angola, and flows northwards to form part of the Angolan-Zairian border before reaching its confluence with the Kasai River near BANDUNDU. Most of its course in Zaire is navigable.
Map Zaire – Ab

Kwanza *Angola* See CUANZA

KwaZulu *South Africa* From 1977 to 1994, self-governing state which was based on the historic kingdom of Zululand (founded in about 1785) but recognised only by South Africa. Combined in 1994 with the former province of Natal to form KWAZULU-NATAL, the region produces livestock, cereals and sugar cane. Its scenic landscape of hills and mountains and rivers attracts many tourists. Its capital, ULUNDI, was the site of the last battle of the Anglo-Zulu War of 1878-9.
Map South Africa – Db

KwaZulu-Natal *South Africa* Province of the Republic of South Africa covering 91 481 km² (35 320 sq miles) and created in 1994 by joining the province of NATAL with the self-governing territory of KWAZULU. Thousands of people in the province have died through political violence in the last few years. The region was originally named Terra Natalis ('Land of the Nativity') by the Portuguese navigator Vasco da Gama who reached its shores on Christmas Day, 1497.

Once inhabited mainly by Zulus, Natal was settled by voortrekkers (emigrant Afrikaners) in the 1830s but was annexed to the (British) Cape Colony in 1843. Indentured Indian workers were introduced from 1860 to work on sugar estates. Mohandas 'Mahatma' Gandhi (1869-1948), who led India's struggle for independence, lived in Natal and practised law here between 1893 and 1914, during which time he fought for the civil rights of Indians. Today, Indians make up almost one-third of the province's population.

Sugar cane is the province's chief crop; its natural resources include coal and timber. Beaches and mountain scenery make KwaZulu-Natal a major holiday destination. Its principal towns include PIETERMARITZBURG, DURBAN and ULUNDI, while Richard's Bay has developed as a major export harbour since 1967, handling ores and coal.
Population 8 549 000
Map South Africa – Db

Kwilu *Congo* See KOUILOU

Kwinana *Australia* Port and industrial town in Western Australia, situated on the natural harbour of Cockburn Sound, about 20 km (12 miles) south-west of PERTH. An important industrial centre, Kwinana handles bauxite and nickel.
Population 13 520
Map Australia – Be

Kyaukse *Myanmar* (*Burma*) Town in central Myanmar, 50 km (30 miles) south of MANDALAY. It is the centre of an intensively cultivated area producing rice, cotton, chillies and groundnuts. Kyaukse has well-developed irrigation systems which are 900 years old and still in use.
Population 115 700
Map Myanmar – Cb

Kyiv *Ukraine* See KIYEV

Kyoga, Lake *Uganda* Area of shallow water and swamps fed by the River Nile, 110 km (69 miles) north of Lake Victoria. It stretches more than 200 km (125 miles) from east to west, forming a barrier to communications.
Map Uganda – Ba

Kyongju (Gyeongju) *South Korea* Southeastern city situated 80 km (50 miles) north of the port of PUSAN. It was the capital of the Silla kingdom from 57 BC to AD 935 and still contains many buildings from that period. South-east of the city lies Pulguk-sa Temple, built in 751.
Population 142 000
Map Korea – De

Kyoto *Japan* City lying 45 km (28 miles) north-east of OSAKA near the southern end of Lake Biwa-ko in central HONSHU. The cultural treasure house of Japan, it was the imperial capital for more than 1000 years until 1868. The city's three palaces are a reminder of past days of wealth and glory. By the 9th century BC, Kyoto already had more than 500 000 inhabitants.

The city was the only major Japanese city to escape damage in the Second World War. It still retains the atmosphere of traditional feudal Japan and abounds with temples, shrines and museums. The Zen Buddhist temple of Ryoanji, on the city's outskirts, contains one of the world's oldest stone gardens. Laid out in 1499, the garden consists of 15 large stones set in a sea of grey-white gravel. The garden exists as an aid to meditation.
Population 1 461 000
Map Japan – Cc

Kyrenia *Cyprus* North coast port and market town for the neighbouring farming district. It has been under Turkish occupation since 1974. Founded by the Ancient Greeks, the town has a 12th-century castle and Byzantine chapel, and the surrounding countryside is dotted with the remains of medieval abbeys and fortresses.

The Kyrenia Mountains stretch for 160 km (100 miles) along the coast just inland from the town. The mountains' highest point is Kyparissovouno, at 1024 m (3360 ft).
Population 23 000
Map Cyprus – Ab

Kyrgyzstan See p 386

Kyushu *Japan* The country's southernmost island and its third largest, after HONSHU and HOKKAIDO. Kyushu covers 43 065 km² (16 627 sq miles) and is dotted with many active volcanoes. It enjoys a mild climate, and rice paddies yield two harvests per year. But agriculture is declining as more of the islanders move to the major cities, the largest of which are FUKUOKA, KITAKYUSHU and KUMAMOTO. Kyushu has now become a centre of Japan's electronics industry, earning itself the nickname of 'Silicon Island'.
Population 13 296 000
Map Japan – Bd

Kyustendil (Kjustendil) *Bulgaria* Spa town situated about 65 km (40 miles) south-west of the capital, SOFIA. It trades in wine, fruit and tobacco, and is the market and processing centre for Bulgaria's major fruit-growing area. The spa's hot sulphur springs were used by the Roman Emperor Trajan in the 1st century AD, and have drawn visitors ever since. To the north of the town lies Zemen Monastery.
Population 55 500
Map Bulgaria – Ab

Kuwait

THIS SMALL EMIRATE RAPIDLY RECOVERED FROM THE GULF WAR AS RESTORED OIL PRODUCTION LEVELS PAID FOR RECONSTRUCTION

When on 2 August 1990, the tiny, oil-rich state of Kuwait on The GULF was invaded by the Iraqi army, it was taken completely by surprise. Although Iraq's President Saddam Hussein had long threatened the Kuwaitis with forcible seizure of a shared oil field that he claimed was rightfully Iraq's, most Middle-East observers had thought the territorial dispute could be settled peaceably.

Hussein's ruthless act of aggression caused shock waves around the world and started the Gulf War – at the cost of scores of thousands of Arab lives and billions of dollars' worth of military aid provided by the Western powers who sought to free Kuwait and contain Hussein.

Within days, Hussein imposed a provisional government in Kuwait and claimed the invasion had been requested by Kuwaitis opposed to the ruling house of Al Sabah. Any resistance was brutally crushed; billions of dollars worth of gold and cash were stolen from banks and Islamic art treasures looted from museums. Hussein declared Kuwait the 19th province of Iraq, and the southern Iraqi province of Basra was enlarged to annex Kuwait's part of the Rumalya oil field and also give Iraq its long desired access to the Persian Gulf.

The Gulf War would involve many countries, including the United States, Britain and France. What must have been inconceivable to most Arabs only days earlier had also come true: hundreds of thousands of Western troops were on sacred Saudi soil, and Arab nations, from Riyadh to Cairo (with an approving nod from Damascus), were united in a coalition with the West to fight an Arab brother. The war lasted six months; in that time much of Baghdad's infrastructure was destroyed by cruise missiles, while American B-52s carpet-bombed tank positions in the desert. Iraq had to cower back into its lair, humiliated; Hussein had gambled and lost when he had taken on the military might that the United States, Britain and France could throw at it, plus the best that the rich Gulf states could afford to buy.

Now recovering, Kuwait maintains its tradition of being one of the most comprehensive welfare systems in the world, with the Al Sabah government setting out to distribute its oil wealth to its people. The average income is equivalent to US$11 100 a year (1992) and one of the world's highest. Education and health care are free to all nationals; food, water, electricity and transport are subsidised.

But a middle-class suburban house costs the equivalent of more than US$1 million, and flats are expensive to rent. However, the government has provided homes for young couples at nominal rents. Meanwhile, revenue from oil exports has decreased drastically with the fall in prices and the government has been running at a deficit.

The country, of only 17 820 km² (6879 sq miles), consists of the city of KUWAIT at the southern entrance of Kuwait Bay and a small wedge of undulating desert between Iraq and Saudi Arabia. It was founded in the 18th century by Anaizah tribesmen, driven from the Arabian desert by drought. The ruling house of Al Sabah, which came to power in 1756, still provides the country's emir, who is advised by a partly elected national assembly.

At the end of the 19th century, Kuwait came under British protection, which lasted until 1961. Both neighbouring Saudi Arabia and Iraq have claimed the territory and even invaded it – the latter in 1973 and 1990. Although Iraq, after its recent defeat, accepted the validity of Kuwait's existing borders, its leadership continued to speak of their legitimate claims to Kuwait – and its rich oil wells.

Kuwait is a commercial crossroads, alive with development. Pearls and camel hides were once the main exports. However, since oil was found in 1938, Kuwait's reserves have been estimated at 94 000 million barrels, representing 10 per cent of the world's total.

Its growing range of industries, mostly oil-related, are highly dependent on foreign technicians and imported skills. Of the almost 1.7 million people, nearly 60 per cent are immigrants, who provide 86 per cent of the work force but do not share the state's oil wealth.

It is an arid hot land in summer; winters are cool and sometimes there is frost. Little rain falls and Kuwait has the world's largest salt-water distillation plants, producing as much as 977 million litres (215 million gallons) a day. Surprisingly, there is a variety of animals and flowers, and a total of 373 desert flower species has been listed.

Only 8 per cent of the land can be farmed. Cultivated land is found around oases but there is some pasture for sheep and goats. Although most food is imported, the government has financed wheat growing and other schemes. Hunting has reduced the desert wildlife but there are gazelles, the ratel (a badger-like animal), the fennec fox and some wolves, jackals, hares and sand rabbits.

Kuwaiti men traditionally meet in the evening in *diwaniyas* (large tents) to sit and talk and drink bitter coffee flavoured with cardamom seeds. Alcohol is banned, even for embassies. Soccer is popular and the national team have been Gulf champions on a number of times. Women wear Western dress, take jobs and drive cars but are not allowed to vote or hold parliamentary posts.

Parliament has 75 members, 50 of whom are elected; there are free elections but no political parties. Only 10 per cent of citizens are able to vote, and only half of these do so. Kuwait also has the most independent Press in the Arab world. However, during the Gulf War greater distrust grew between Kuwaitis and their government. Now, more Kuwaitis demand democratisation and a larger share in political life.

Not much of the old Kuwait is left, although the city has its ancient gates, museums and the royal palace. The shopper is more likely to find the latest Western goods in the souk, or bazaar, than anything traditional. FAILAKKA ISLAND, in Kuwait Bay, has magnificent beaches; and excavations here have revealed 3000-year-old Bronze Age dwellings.

These same beaches were heavily mined and littered with military obstructions by the Iraqis during the war. As the invaders left, they destroyed oil-well heads and set the black jets ablaze. The sky turned black for months – almost as black as the sea where tons of oil had been spilled. In the desert, large patches of the fragile ecology were smothered in oil lakes.

Repairs have been completed. Yet the future of this emirate still looks uncertain. Kuwaiti power structures are complex, the country's economy is too focused on oil and its lack of democracy has been exposed. It is possible that Kuwait's turbulent history has not ended and there may be further difficulties to come.

KUWAIT AT A GLANCE	
Official name State of Kuwait	
Area 17 820km² (6879 sq miles)	
Population 1 698 080 **Per km²** 95 **(Per sq mile)** 247)	
Capital Kuwait (Kuwait City)	
Government Emirate	
Currency 1 Kuwaiti dinar = 1000 fils	
Languages Arabic (official), English (commercial language)	
Religion Muslim (mostly Sunni)	
Climate Desert climate with humidity. Average temperature in Kuwait City ranges from 13°C (55°F) in January to 35°C (95°F) in July	
Land use Pastures 8%, other 92%	
Main primary products Sheep, goats, dates and other fruit, vegetables, fish, shrimp; oil and natural gas	
Major industries Oil and gas production and refining, boat building, fishing, processed foods, chemicals, ammonia fertilisers, cement, seawater desalination	
Main exports Crude oil and oil products, natural gas	
Annual income per head (US$) 11 100	
Population growth (per thous/yr) 86	
Life expectancy (yrs) Male 72 **Female** 77	

Kyrgyzstan

POVERTY HAS MARRED THE FIRST YEARS OF INDEPENDENCE, BUT THIS SMALL STATE IS MOVING RAPIDLY TOWARDS ECONOMIC REFORM

Kyrgyzstan is one of the smallest of the former Soviet republics, covering 198 500 km² (76 600 sq miles) on the border with north-western China. Its land area is slightly smaller than the state of South Dakota in the United States, and has a population slightly greater than Detroit's. In the west are the republics of Kazakhstan, Tajikistan and Uzbekistan.

The name Kyrgyzstan means 'land of the nomads'. A member of the Commonwealth of Independent States (CIS), it has called itself by that name since independence in December 1990. To have called it 'land of mountains' might have been more fitting, since three-quarters of the region consists of high mountain ranges. Traditionally, much of it was pasture land for sheep, goats and cattle, but the Kirghiz peoples who used to roam about in search of pasture for their livestock have long since settled down.

In places, the land rises to spectacular peaks such as Pik Pobedy (Tomur Feng), at 7439 m (24 406 feet), in the TIEN SHAN Mountains on the Chinese border. The range is the source of several large streams and also the point from which several mighty, glaciated mountain ranges, such as the Alexander, Talas Alatan and Altai, extend westwards. In the east, they straddle the Issyk-Kul', a large mountain lake of 6280 km² (2424 sq miles); further west, two vast valleys, the Naryn and FERGANA, lie embedded in the heart of the mountains.

The Fergana Valley cuts deep into the surrounding ranges, but Kyrgyzstan has no other low-lying regions, besides a few narrow foothills. Agricultural activities are therefore limited by altitude. The country's dry, continental climate is a further impediment to large-scale agriculture. Whereas the western escarpment gets an average of between 400 and 800 mm (between 16 and 32 in) of rain a year, the valleys and basins receive only between 100 and 150 mm (between 4 and 6 in) and consist of semi-arid grass and scrub steppe, blooming into life here and there around an oasis.

Sparse walnut groves can be found along protected slopes. At higher altitudes, the steppe merges into alpine meadows, which have been used for summer pasture since time immemorial. At 3400 to 4000 m (11 150 to 13 100 ft) begins the world of glaciers, whose melt water in high summer is tapped by farmers to irrigate crops.

Since 1926, cultivation on irrigated land has produced fruit, cereals, sugar and cotton; a total of 10 320 km² (3984 sq miles) is now under irrigation.

ON THE ROAD TO ECONOMIC INDEPENDENCE

The republic of the Kirghiz people is a poor country. Under Soviet rule, its gross national product per capita lay far below the average for the Soviet Union. Today, the country's economy is still largely oriented towards agriculture. Apart from livestock rearing, other main products are tobacco, silk, coal, agricultural machinery and textiles. Industry, concentrated around the capital of Bishkek, produces electric motors, washing machines, livestock feeding equipment, cement and furniture. The country is also an exporter of electricity. There is some mining, mainly of coal, rare earth metals and gold. Until independence, petroleum and natural gas were subsidised by the Soviet government.

Independent Kyrgyzstan has been in the process of large-scale privatisation. In 1992, many prices were freed but a lack of spare parts and fuel shortages brought about a 20 per cent decline in that year. In 1993, the country embarked upon agricultural and land reforms and introduced its own currency in the place of the Russian rouble.

But its agricultural products have not fetched high prices on the world market and it has to import many of its foods, as well as fuel and parts and, as it is landlocked, all of its trade has to be transported through other former Soviet states. Subsidies from Russia were cut during the first year of independence.

Nevertheless, Kyrgyzstan's economy has suffered less than was to be expected and prospects brightened during 1994 when president Askar Akayev initiated a policy of strict adherence to market liberalisation, tightened money supply and reduced state spending, as prescribed by the International Monetary Fund. Inflation was brought down from a peak of 40 per cent to low single digits and he sought Western investment to help exploit the country's untapped mineral wealth – which includes large gold deposits. Its most important export now – hydroelectric energy – is a valuable commodity. Another potential source of income is tourism, which could in future take its place beside agriculture and mining.

There has been some political fragmentation based on ethnic or religious groups. The largest of these, accounting for about 52 per cent of the population, are the Kirghiz Muslims who are closely related to the Kazakhs; their language belongs to the Siberian Turkic group. Their ancestors were driven out of Siberia by Mongolians in the 10th century and they found a new home in the Tien Shan. Around the mid-19th century, they were conquered by the Russians; and after the October Revolution their country became an autonomous region of the Soviet Union.

During the 1920s they developed considerably in cultural, educational and social life. Literacy was greatly improved and a standard literary language was introduced. Economic and social development was also notable. Land reforms were carried out which resulted in the settlement of many of the nomadic Kirghiz. From 1936, it was known as the Kirghiz Republic of the south-central USSR.

Some leading members of the Kirghiz Communist Party (KCP) attempted to increase the role of ethnic Kirghiz in the government of the Republic, but these so-called 'national Communists' were expelled from the KCP and often exiled or imprisoned, particularly during the later 1930s. Despite the suppression of nationalism during the Stalin era, many aspects of Kirghiz national culture were retained.

Besides the Kirghiz, there is a large number of Russian residents – the second largest population group, making up almost 22 per cent. Until recently, Russians filled many important public positions (Kirghiz did not replace Russian as the official language until 1989) whereas the Central Asian groups – the Kirghiz, Uzbek (accounting for 12.9 per cent of the population), Ukrainians (almost 3 per cent), German, Kazakhs and Tajik – have had to fight for political power. For instance, there was controversy, in early 1993, over opening a Kirghiz embassy in Israel. Many Muslims in the country feared that this would mar foreign diplomatic relations with the Arab and the Muslim world.

KYRGYZSTAN AT A GLANCE	
Official name Kyrgyz Republic	
Area 198 500 km² (76 623 sq miles)	
Population 4 625 950 **Per km²** 23 (Per sq mile) 60)	
Capital Bishkek	
Government Republic	
Currency 1 som = 100 tyin	
Languages Kirghiz (official), Russian	
Religions Muslim (70%), Russian Orthodox	
Climate Continental climate, with cold winters and hot summers and an average low rainfall. Average temperature in Bishkek ranges from -5°C (23°F) in January to 25°C (77°F) in July	
Land use Grazing 42%, cultivation 7%, forest 5%, other 46%	
Mineral resources Oil, natural gas, coal, mercury, antimony, lead, zinc	
Main primary products Ferrous and non-ferrous metals, cotton, livestock, tobacco	
Major industries Mining, machinery, chemicals, food processing	
Main exports Ferrous and non-ferrous metals; wool, cotton, tobacco, machinery, chemicals, electricity	
Annual income per head (US$) 810	
Population growth (per thous/yr) 16	
Life expectancy (yrs) Male 63 **Female** 72	

La Ceiba *Honduras* North-western port which exports bananas and pineapples grown inland.
Population 68 900
Map Honduras – Aa

La Condamine *Monaco* The town centre of the principality – including its main business area, and residential area where many Monégasques live – on the north-western side of the port.
Population 12 500
Map (Monaco) France – Ge

La Coruña *Spain* See CORUNNA

La Goulette (Halq al-Wadi) *Tunisia* Seaport town and beach resort, 11 km (7 miles) north-east of the capital, TUNIS, to which it is connected by a deep-water channel. It was the original harbour for the capital before the channel was dug in 1893, which enabled Tunis to take ocean-going vessels. A new port to serve the capital was built at La Goulette in 1968. However, many port services have remained at the old port of Tunis, close to the heart of the city. La Goulette has a massive *kasbah* (Muslim residential quarter), and its main street contains many inexpensive seafood restaurants.
Population 61 700
Map Tunisia – Ba

La Guaira *Venezuela* Main port, some 13 km (8 miles) to the north of the capital, CARACAS, which can be reached by train along a 37 km (23 miles) winding mountain railway line. A road (completed in the 1950s) follows a more direct route. Founded in 1567, La Guaira's old town remains largely intact. Ships are built in the port, which also has dry dock facilities and a good harbour.
Population 30 000
Map Venezuela – Ba

La Libertad *Peru* North-western department, producing most of the country's sugar cane. The crop is exported mainly through the port of Salaverry, 500 km (310 miles) north of Lima. Salaverry also handles copper from mines 120 km (75 miles) inland. The ruins of the pre-Inca Chimú city of CHAN CHAN lie 5 km (3 miles) north of the department's capital, TRUJILLO.
Population 1 243 500
Map Peru – Ba

La Mancha *Spain* Arid treeless plateau, 680 m (2230 ft) high and covering some 4800 km² (18 530 sq miles) about 160 km (100 miles) south of Madrid. It provides the setting for the fictional exploits of the knight Don Quixote in the 17th-century novel by Miguel de Cervantes. Grapes and cereals are the principal crops. La Mancha has a number of large villages with populations of more than 10 000.
Map Spain – Dc

La Oroya *Peru* Major lead, tin and silver-mining and smelting centre, lying in a deep narrow valley at the fork of the Mantaro and Yauli rivers, 187 km (116 miles) east of Lima.
Population 36 000
Map Peru – Bb

La Pampa *Argentina* See PAMPAS

La Paz *Bolivia* National capital – and also the highest capital city in the world, lying at 3627 m (11 900 ft) above sea level. La Paz is also the name of the surrounding department. The city lies in a valley at the foot of multiple-peaked ILLIMANI, sheltered from the winds of the surrounding ALTIPLANO plateau.

Founded in 1548 by the Spanish adventurer Alonso de Mendoza, it flourished as a town on the route from the silver mines in the south-east to the Peruvian capital, Lima. During the War of Independence (1809-25), it was one of the main centres of resistance. After victory, it was named La Paz de Ayacucho ('the Peak of Ayacucho') to commemorate the decisive battle at Ayacucho in Peru. It has been the seat of government since 1898, although the constitution nominated Sucre as the national capital.

Steep streets climb from the business and wealthier residential areas at the bottom of the canyon to the terraced heights where most of the Amerindian population live. Woollen garments made from alpaca wool, together with embroidered shawls and leather goods are sold in the city's markets, which are crowded with bowler-hatted Amerindian women. The Plaza Murillo forms the centre of the old city, where the church and monastery of San Francisco (1743-84) are among the few good remaining examples of colonial architecture in Bolivia.

Textiles, electrical appliances, chemicals, furniture and glass are manufactured in the city. La Paz also handles lead and silver for export.
Population (city) 1 126 000; (department) 1 913 200
Map Bolivia – Bb

La Plata *Argentina* Port situated on the Río de la Plata (also called the River Plate, or the 'Silver Stream'), 56 km (35 miles) south-east of BUENOS AIRES. Its main cargoes are oil and frozen meat. The huge estuary, covering 35 000 km² (13 510 sq miles), where the rivers Plate, Paraná and Uruguay enter the Atlantic Ocean, was the scene of the Second World War Battle of the River Plate.
Population 644 200
Map Argentina – Db

La Rochelle *France* Seaport, tourist centre and capital of Charente-Maritime department, 120 km (75 miles) south of NANTES. A leading port from the 14th to the 17th century, it was a stronghold of the Protestant Huguenots during the 17th-century Wars of Religion, until it was captured by the Catholics in 1628 after a 14-month siege.

It now manufactures fertilisers, plastics and cement; other industries include shipbuilding, fish canning and saw-milling. The port of La Pallice is 5 km (3 miles) to the west.
Population 104 700
Map France – Cc

La Serena *Chile* Capital of Coquimbo region, some 400 km (250 miles) north of Santiago. It was founded by the Spaniards in 1543 and named after the Spanish birthplace of the conquistador Pedro de Valdivia. Throughout the 16th and 17th centuries it was a frequent target of Amerindian and pirate attacks.

La Serena stands at the mouth of the Elqui River, in an area renowned for the quality of its peaches and oranges.
Population 118 000
Map Chile – Ab

La Skhirra *Tunisia* See SAKHIRA

La Spezia *Italy* Major naval base and port, some 75 km (47 miles) south-east of GENOA. It has the country's most important naval arsenal and, for tourists, a naval museum.
Population 115 200
Map Italy – Bb

La Unión *El Salvador* Department on the south-east coast, growing cotton and sugar cane. Its capital, also called La Unión (population 36 930), is

▼ MARKET DAY Garments as colourful as the balloons enliven the scene in La Paz, Bolivia's capital city, where half the population are Amerindian.

a deep-water harbour on the Gulf of Fonseca near the border with Honduras and Nicaragua.
Population 251 140
Map (Gulf of Fonseca) El Salvador – Ba

laagte In South Africa, the term used for broad hollows, or dips, which are less well defined than valleys, between wide rises in a comparatively flat landscape.

Laatokka *Russia* See LADOGA, LAKE

Laâyoune (El Aaiún) *Western Sahara* Capital town of the territory, situated in the north-west, some 20 km (12 miles) from the Atlantic coast. Its name means 'the Springs' in Arabic. It is served by a small port, and an airport.

During the war with Polisario guerrillas fighting for independence, the Moroccans established a 34 km (21 mile) system of perimeter defences around the town, as well as building new homes in the town to induce nomadic tribesmen to settle.
Population 38 000
Map Western Sahara – Aa

Labasa *Fiji* See VANUA LEVU

Labé *Guinea* Town in the Fouta Djallon, lying 260 km (160 miles) north-east of the capital, CONAKRY, on the road running north from Mamou towards Senegal. Its cool climate and mountain scenery made it a popular resort among French colonists. Bitter oranges, jasmine and coffee are grown nearby; livestock and other agricultural produce are marketed in the town.
Population 79 700
Map Guinea – Ba

Labrador *Canada* The mainland section of the province of Newfoundland, covering an area of 295 800 km² (112 826 sq miles) and separated from Newfoundland's island section by the Strait of Belle Isle. It is a plateau area of the CANADIAN SHIELD and varies in height from 200 m to more than 1000 m (656 ft to 3280 ft). Its north is tundra and dense coniferous forests cover the south. Its several mineral deposits include iron ore. Native Canadians and Inuit (formerly called 'Eskimos') engaged in fishing and trapping live in small communities along the coast. Goose Bay Airport and the mining towns of Labrador City and Wabush are further inland.
Map Canada – Ic

Labuan *Malaysia* Island of 75 km² (35 sq miles) lying 55 km (35 miles) off the coast of BRUNEI. The excellent Victoria Harbour, once a key port in trade with China, is now a free port.
Population 26 400
Map Malaysia – Ea

Laccadive Islands *India* See LAKSHADWEEP

laccolith (laccolite) Mushroom-shaped body of IGNEOUS ROCK intruded between the layers of sedimentary rock which it has pushed up.

Lacq *France* Village in the extreme south-west where oil and natural gas were discovered in 1951. Sulphur, extracted from the gas, is also produced here. The nearby town of Mourenx (population 7500) was built in 1957-61 to house the oil field's workers.
Map France – Ce

Ladakh *India* District of JAMMU AND KASHMIR in the western Himalayas. The area known as AKSAI CHIN has been occupied by the Chinese since the 1950s.
Map India – Ba

Ladoga, Lake (Ladozhskoye Ozero; Laatokka) *Russia* The largest lake in Europe, covering 18 390 km² (7100 sq miles), including its approximatly 500 islands, and lying just northeast of ST PETERSBURG. It is fed by the Volchov, Svir and Vuoksi rivers and drained by the Neva. Its north and west shores belonged to Finland from 1919 to 1940. Lake Ladoga was a major trading route even in Viking times. In the 18th century, a canal system was started along its southern shore to protect ships from the frequent storms that lashed the lake. Today the canal is part of the Volga-Baltic waterway, which links it to the Baltic-White Sea Canal. From November to April each year the lake freezes over.
Map Russia – Eb

Ladysmith *South Africa* Town in KwaZulu-Natal, lying 190 km (120 miles) north-west of DURBAN. Founded in 1850, it is named after the wife of Sir Harry Smith, the then governor of the Cape Colony. Ladysmith was besieged for 118 days during the Anglo-Boer War (1899-1902). Many relics of this war as well as other conflicts can be seen in both the town and the countryside. In fact, Ladysmith lies on the 'Battlefield Route' that includes sites of the two Anglo-Boer wars (1880-1, 1899-1902) and the Anglo-Zulu War (1878-9). Cattle, horses, cereals and various dairy products are produced in the district.
Population 7530
Map South Africa – Cb

Lae *Papua New Guinea* The country's second largest town, standing on Huon Gulf, 320 km (200 miles) north of the capital, PORT MORESBY. Its port handles gold, timber, plywood, cattle, cocoa and coffee. Devastated by bombing in the Second World War, Lae had to be completely rebuilt. Today, it is the country's major industrial centre and the site of Papua New Guinea's University of Technology. Its tropical heat and heavy rainfall – 4570 mm (180 in) a year – enable flowers to grow everywhere.
Population 88 170
Map Papua New Guinea – Ba

Laghouat *Algeria* Administrative centre, and staging post on the Route du Hoggar, 400 km (about 250 miles) south of ALGIERS. It is the first extant oasis in the desert after the Saharan Atlas Mountains. Founded in the 11th century by the Hilians, a Berber group which had fled from the north, the town has become the main market and commercial centre for the Ouled Nail Mountains Confederation of Peoples.
Population 45 000
Map Algeria – Ba

lagoon Body of salt water separated from the sea by sand or shingle BARS or CORAL REEFS.

Lagos *Nigeria* Former national capital and the capital of Lagos State, situated on an island in the south-west. Lagos is the largest West African city and port. It was founded by the Yoruba people in the 13th century as a refuge from their enemies. Until it was occupied by the British in the mid-

19th century, it was a centre of the West African slave trade; after the abolition of the slave trade, the British converted it into the country's trade, administrative and industrial centre. Today it is the most congested city in West Africa. Overcrowded slums with open sewers and squalid winding streets contrast with modern houses and new main roads. Many parts of the city have become notorious for muggings and theft.

The administrative and business heart of the city is on Lagos Island where there are large shops (although not always well stocked) and high-rise office blocks. Two road bridges link the island to the mainland and to the expensive and elegant residential areas of Ikoyi Island and Victoria Road. Because most people live on the mainland, there is commonly appalling traffic congestion on the bridges. All efforts to ease the congestion have failed, despite the fact that the administrative capital has been moved to ABUJA.

Most industry is located on modern estates, and products include foodstuffs, metal goods, chemicals, textiles, furniture and glass. Tourist attractions include the 18th-century Oba's (king's) Palace on Lagos Island, good beaches, polo, golf, fishing and sailing. Local craftsmen produce dyed cotton, handwoven cloth, leather goods and jewellery.
Population (metropolitan area) 5 700 000
Map Nigeria – Ab

Lahej *Yemen* Town, and district of the same name, 10 km (6 miles) north-west of ADEN. The district is one of the country's richest agricultural and cotton-growing areas.
Population (town) 35 000
Map Yemen – Ab

Lahore *Pakistan* City once known as 'the Queen of British India', when it was capital of the imperial Punjab. Since Pakistan achieved independence in 1947, the city has continued to grow although it has lost much of its surrounding land to India in border changes. The Mogul emperors built many fine palaces, gardens and mosques here in the 16th to the 18th century, and it prospered both under the Sikh kings of the Punjab and after the British takeover in the 1840s.

Lahore retains many of the tree-lined malls of fine public buildings from imperial days. The old city has a 9 m (30 ft) high city wall and 13 gates, a 16th-century fortress and a 17th-century mosque. Outside the museum stands the great cannon beside which Rudyard Kipling's fictional hero Kim played. In 1940, Muhammad Ali Jinnah, leader of the Muslim League, proclaimed the foundation of Pakistan as the League's goal in a speech here.

Lahore, which is about 270 km (170 miles) south-east of the national capital, Islamabad, remains the capital of the Pakistani province of Punjab. It is the country's second biggest city after KARACHI, and its largest industrial centre. Numbered among its many industries are metalworking, engineering, and the production of chemicals, textiles and leatherware.
Population 2 952 700
Map Pakistan – Db

Lahti *Finland* Modern industrial town, situated 100 km (62 miles) north-east of HELSINKI. Most of its factories are involved in the processing of timber and the manufacture of wood products, including furniture. The town is also the venue for Finland's annual winter sports competitions. Its ski jumps, located on the great ridge called SALPAUSSELKÄ, command magnificent views of the country's central lake district.

Population 93 800

Map Finland – Cc

Laibach *Slovenia* See LJUBLJANA

lake Inland body of fresh or salt water, varying in extent from a few tens of square metres to thousands of square kilometres, and in depth from a few metres to more than 1000 m (about 3300 ft). Lakes occupy impermeable basins formed by earth movements, erosion, deposition or volcanic activity. The lake with the largest area in the world is the CASPIAN SEA, covering an area of 371 000 km² (143 210 sq miles). The deepest lake, with a depth of more than 1620 m (5315 ft), is Lake BAIKAL in Siberia.

Lake District *United Kingdom* Scenic area of lakes and mountains in the north-west English county of Cumbria; it stretches about 40 km (25 miles) in each direction. The Cumbrian Mountains, which form its core, have many peaks, including England's highest, Scafell Pike, at 978 m (3208 ft). The major lakes of the district – WINDERMERE and Coniston to the south, ULLSWATER to the north-east, and Derwent Water and Bassenthwaite Lake to the north – radiate out from this core.

The Lake District has rich literary associations, notably with the 19th-century Lakeland Poets, of whom William Wordsworth (1770-1850) is the best known. For over a century the Lake District has been one of Britain's major tourist areas, even though it lies in one of the country's wettest areas. It has been declared a national park, but remains threatened by the growing number of visitors each year and by the over-exploitation of its water resources.

Map United Kingdom – Dc

Lake of the Woods *Canada/USA* Extremely irregular lake dotted with some 17 000 islands and covering about 4390 km² (1695 sq miles). It lies on the United States-Canadian border, mostly in the Canadian province of ONTARIO, with a small part in MANITOBA and another part in the US state of MINNESOTA. It drains to the north-west into Lake Winnipeg. It is a popular tourist area.

Map Canada – Fd

Lakshadweep *India* Territory in the south Arabian Sea consisting of three groups of islands, the Laccadive, Minicoy and Amindivi islands. They have a land area of 32 km² (12 sq miles) and consist of low coral atolls enclosing lagoons. Lakshadweep's inhabitants are mainly fishermen and coconut farmers. The majority of its population are Muslim, but in the Minicoy Islands people speak Mahal, which is written in Divehi script.

Population 51 700

Map India – Ae

Lalibela *Ethiopia* Mountain town situated 320 km (200 miles) to the north of the capital, ADDIS ABABA. It is renowned for its 11 Christian churches carved out of solid rock during the 12th and 13th centuries.

Map Ethiopia – Aa

Lalita Patan *Nepal* See PATAN

LAKES AND INLAND SEAS

There are two main types of natural lake: salt and freshwater. Inland seas – such as the Caspian – have no outlet and water is lost only by evaporation so that salinity gradually builds up. Freshwater lakes have outflowing streams and the amount of water entering them usually exceeds that lost by evaporation, so they do not become salty.

LARGEST INLAND WATERS OF THE WORLD

Lake	Location	Area (km²/sq miles)
Caspian Sea (salt)	Azerbaijan/Russia/Kazakhstan/ Turkmenistan/Iran	371 000/143 210
Superior (freshwater)	USA/Canada	82 400/31 800
Victoria (freshwater)	Kenya/Uganda/Tanzania	69 500/26 800
Huron (freshwater)	USA/Canada	59 610/23 010
Michigan (freshwater)	USA	57 454/22 178
Aral Sea (salt)	Kazakhstan/Uzbekistan	37 000/14 280
Tanganyika (freshwater)	Burundi/Zaire/Tanzania/Zambia	32 900/12 700

Lalitpur *Nepal* See PATAN

Lambaréné *Gabon* Capital of Moyen-Ogooué Province, situated on a large island in the wide Ogooué River, 160 km (100 miles) east of PORT GENTIL en route from Libreville to Brazzaville. Lambaréné became famous as the site of a hospital established in 1913 by the German-born philosopher, musician, missionary doctor and 1952 Nobel Peace Prize-winner Albert Schweitzer (1875-1965). The hospital still stands, although Lambaréné now has another, modern hospital. The town is a trading centre; its saw-mills process timber for shipping downriver by barge.

Population 26 260

Map Gabon – Bb

Lambayeque *Peru* Department in the north-west. It is the country's largest rice producer and a major producer of sugar cane, particularly in the irrigated sandy lowlands near Chiclayo, the department's capital. Irrigation is turning the Sechura Desert on the western coastal plain into cultivable land. The Chancay and Leche valleys in the department have several pre-Inca sites; the Bruning Archaeological Museum in Lambayeque town contains historical ceramics and artefacts.

Population (department) 935 300; (town) 27 000

Map Peru – Aa

Lambert Glacier *Antarctica* The world's longest known glacier, flowing over the Prince Charles Mountains to the Amery Ice Shelf in east Antarctica. It is at least 400 km (250 miles) long and is fed by several smaller glaciers. Where it reaches the Amery Ice Shelf it is 200 km (125 miles) wide and flows at about 750 m (2460 ft) per year. It was discovered by an Australian aircraft crew in 1956.

Map Antarctica – Oc

Lampang *Thailand* Market and timber town, and former royal residence, lying some 535 km (332 miles) north of the capital, BANGKOK. The town's timber yards store teak cut from nearby forestland before it is transported south by truck. Elephants used in the logging camps are trained at a school near the town. Rice is the main crop in the surrounding province of the same name.

Population (town) 215 800; (province) 734 100

Map Thailand – Ab

Lamphun *Thailand* Northern province and town situated about 585 km (365 miles) north-west of BANGKOK. The town was the capital of a kingdom ruled by the Mon people who lived in the area more than 1000 years ago. One of their town moats can still be seen.

Population (province) 395 200; (town) 154 300

Map Thailand – Ab

Lampung *Indonesia* Province of southern Sumatra, covering 35 367 km² (13 655 sq miles) and producing, amongst others, coffee, pepper, rubber, tea and peanuts.

Population 4 624 800

Map Indonesia – Bc

Lanao *Philippines* Volcanic plateau comprising the provinces of Lanao del Norte and Lanao del Sur, on north-central MINDANAO. It has several volcanic peaks, one of which, Mount Ragang, is 2815 m (9236 ft) high; the second largest lake of the Philippines, Lake Lanao, is also situated on the plateau. Possibly formed by the collapse of another volcano, the lake is some 340 km² (131 sq miles) in area. Much of the area is now farmed for maize and other crops.

Population (Lanao del Norte) 461 000; (Lanao del Sur) 405 000

Map Philippines – Bd

Lanark *United Kingdom* See STRATHCLYDE

Lancashire *United Kingdom* County of north-west England, formerly including Furness (now part of CUMBRIA) and the areas around LIVERPOOL and MANCHESTER, but today covering an area of 3043 km² (1175 sq miles) to the north of these cities. Lancashire was a focus of development during the Industrial Revolution in the 18th and 19th centuries, and became the world's major cotton centre before the First World War in mill towns such as PRESTON (now the county town), ROCHDALE (now part of Greater Manchester), BLACKBURN and Burnley.

Today, most of the great mills are closed or have been converted to other businesses. The county's seaside resorts, such as BLACKPOOL and the PENNINE Hills, are tourist attractions.

Population	1 409 900
Map	United Kingdom – Dd

Lańcut *Poland* Small south-eastern town, 160 km (100 miles) east of the city of CRACOW. It contains a magnificent Baroque palace, built between 1629 and 1641 for the powerful and aristocratic Lubomirski family. The ivy-clad building, which is topped by onion domes, is now a museum, with fine collections of paintings, sculpture, clocks, furniture, tapestries, carriages and coaches.

Population	15 600
Map	Poland – Ec

land breeze A wind blowing from the land towards the sea or a lake at night or in the very early part of the day, following cooling of the land by radiation during the night. It is the opposite of a SEA BREEZE.

Landes *France* Coastal department of some 9237 km² (3566 sq miles) in the south-west. The name also applies to the wider area of sandy coastal plain. It was thought to be suitable only for rearing sheep until 19th-century landowners planted pine trees and drained coastal marshes. In the 20th century, some pines were cleared for farming, and the main activities are now forestry, farming and tourism; the fine, sandy beaches at Arcachon are particularly popular.

Population	(department) 311 500
Map	France – Cd

Land's End *United Kingdom* Most westerly point of mainland England, in the county of Cornwall. Its treacherous granite cliffs are visited by thousands of tourists each year.

The Wolf Rock lighthouse, about 13 km (8 miles) offshore, and Longships lighthouse, some 2.4 km (1.5 miles) away, were built in the 19th century to warn ships away from the dangerous waters around the headland. Over the centuries, these treacherous waters have claimed scores of ships and hundreds of lives.

Map	United Kingdom – Ce

landslide (landslip) Sudden fall of masses of rock and soil down a cliff or hillside due to heavy rain acting as a lubricant or to the tremors of an earthquake.

Langa *South Africa* African residential area some 10 km (6 miles) from the centre of Cape Town, established in the 1920s as one of the first model 'townships' for migrant workers and designed for a population of 5000.

By 1967 additions included barracks for 20 000 single men out of a total population of 32 000. Langa today is a sprawling township.

Population	75 700
Map	(Cape Town) South Africa – Ac

Langeland *Denmark* Island in the FÜNEN group, covering 283 km² (109 sq miles) and linked by bridge to Fünen. Langeland's beaches are popular holiday spots; its chief town is Rudkøbing.

Population	17 700
Map	Denmark – Bb

Langkawi *Malaysia* Island in the Straits of Malacca beside the Thai border. It is a tourist resort and has a number of fishing villages.

Population	28 350
Map	Malaysia – Aa

Languedoc-Roussillon *France* Wine-producing region noted particularly for its red table wines. It stretches behind the Mediterranean coast west of the Rhône River to the Pyrenees. However, the historical province of Languedoc occupied a much larger area than the present region. Besides wine, the region's industries include farming and tourism. MONTPELLIER is the main city.

Population	2 115 000
Map	France – Ee

L'Anse-aux Meadows *Canada* Historic site near St Anthony in northern Newfoundland. Vikings from Greenland, travelling in *knorrs* (heavy long-ships), landed here in the late 10th century to establish possibly the first European colony in North America. Since the ruins of their settlement were discovered in the 1960s, eight turf houses – open to view – have been excavated and restored.

Map	Canada – Jc

Lansing *USA* State capital of Michigan. It lies 130 km (80 miles) north-west of DETROIT and is, like Detroit, a motorcar-manufacturing centre where, in 1897, one of the first petrol-powered motorcars was built. Its other products include electrical motors, tools and chemicals; the city is also the main trade and processing centre of the surrounding agricultural area. Michigan State University is at East Lansing.

Population	(city) 127 320; (metropolitan area) 432 700
Map	United States – Jb

Lantau *Hong Kong* Largest of the many small islands of the NEW TERRITORIES which belong to the British Crown Colony of Hong Kong. It covers 150 km² (58 sq miles) and lies to the west of Victoria Harbour. A mountainous but tranquil island, fringed by beautiful beaches, it contains the Po Lin Buddhist monastery and the world's largest outdoor statue of the Buddha – 32 m (112 ft) high and weighing 528 tonnes.

Population	(including associated islands) 17 100
Map	Hong Kong – Bb

Lanzarote *Spain* See CANARY ISLANDS

Lanzhou *China* Capital of Gansu Province, on the Huang He River, 1200 km (750 miles) south-west of Beijing. It is an important communications centre for western China and contains oil refining, chemicals and woollen textiles plants.

Population	2 260 000
Map	China – Fd

Laoag *Philippines* See ILOCOS

Laois (Leix; Laoighis) *Ireland* Farming county covering 1719 km² (664 sq miles), about 65 km (40 miles) south-west of DUBLIN. It was formerly known as Queen's County, because during the reign of Mary I, there was an unsuccessful attempt to start an English community here. The county town is Port Laoise (population 3620), once known as Maryborough.

Population	52 310
Map	Ireland – Cb

Laon *France* Capital of the Aisne department, 130 km (81 miles) north-east of Paris, where it overlooks the plain of Champagne. Old 17th-century houses and narrow streets cluster around the town's 12th-century cathedral.

Population	28 700
Map	France – Eb

Laos See p 391

Lappeenranta (Villmanstrand) *Finland* Lakeside port, founded in 1649 on the south shore of SAIMAA Lake in south-east Finland, some 25 km (15 miles) from the Russian border. It has one of Finland's largest concentrations of softwood processing plants. Chemicals, cellulose and machinery are manufactured here.

Population	55 600
Map	Finland – Dc

Lappland *Scandinavia* Region lying largely inside the Arctic Circle, but having no formal boundaries. Traditionally, it is inhabited by the Lapp people, but today Lapps make up only about 10 per cent of its population. It stretches across the north of Norway, Sweden and Finland, and covers some 320 000 to 350 000 km² (123 500 to 135 100 sq miles). Mining, (mainly of iron ore), forestry, fishing and reindeer herding are the main activities. Lappland has several national parks.

Map	Finland – Cb

Laptev Sea *Russia* Shallow part, with a depth of no more than 100 m (328 ft), of the ARCTIC OCEAN, formerly called the Nordenskjold Sea. It is named after the 18th-century Russian navigators and explorers Dmitri and Chariton Laptev. It lies between the SEVERNAYA ZEMLYA and NOVOSI-BIRSKIYE OSTROVA islands, with eastern Siberia and the Lena River delta to the south and the edge of the continental shelf to the north. The Laptev, or Dmitri Laptev, Strait connects it to the EAST SIBERIAN SEA. The Laptev Sea is icebound and not navigable for 10 months every year.

Map	Russia – Ma-Oa

Larache (El Araïche) *Morocco* Spanish-style fishing port on the Atlantic coast, some 70 km (45 miles) south-west of TANGIER. It exports wool, hides, cork, wax, fruit and vegetables. The 17th-century Stork Castle, so called because of the birds that nest on its ramparts, lies in the town. Lixus, an ancient settlement containing various megalithic, Phoenician and Roman remains, is situated on a hillside to the north.

Population	64 000
Map	Morocco – Ba

Laramie *USA* Market town in a timber, mining and stock-raising area of south-east Wyoming, 35 km (22 miles) north of the Colorado border.

Laos

A LANDLOCKED AND UNDERDEVELOPED COUNTRY THAT HAS BEEN FORGOTTEN BY THE WORLD BECAUSE OF ITS OWN ISOLATIONISM

This small country has long suffered the twin fates of isolation from economic development and interference by outside forces. Progress was also hindered by centuries of internal political division. The French colonial period brought peace, but little development came with it.

Laos, abandoned by the French in 1954, faced crippling problems and is one of the least developed countries in the world today. Ruggedly mountainous, the country possesses a rudimentary communications system – besides having no industry and old-fashioned agricultural techniques. To enable significant development it needed peace, national unity and foreign aid. Today Laos still suffers from the after effects of civil war which broke out in 1960 and which was escalated by foreign aid to rival factions.

The communist Pathet-Lao regime, with Vietnamese support, won political control of the country in 1975. It ousted the king and proclaimed Laos a people's republic. But the ravages of war, the withdrawal of American economic aid, the new regime's doctrinaire socialist policies and a crippling drought hindered development. More than 300 000

Laotians fled the country. After 1980 more liberal government policies coupled with huge Soviet and Vietnamese aid produced a change in fortune. But Laos remains dependent on the weather. Good monsoon seasons create self-sufficiency in rice, the staple diet. Drought or floods can mean disaster – and the diversion of development funds to pay for rice imports.

The country's mountainous terrain broadens in the north to a series of wide plateaus, some 2000 m (6560 ft) above sea level, which contains the Plain of JARS. In the south, the ANNAM Mountains reach 2500 m (8202 ft) on the Vietnam border. On the western side are the country's only lowlands, the alluvial plains of the MEKONG. This is Laos's principal river and its main communications artery, since there are no railways and few roads. It is also the country's main rice-growing region. On its north bank, on the border with Thailand, lies VIENTIANE, the capital and largest city. It is the country's main trade outlet, via Thailand.

Laos exports timber, coffee and tin, as well as electricity. The black market also exports opium and gold. All manufactured goods must be imported and Laos has a constant foreign trade deficit. The country's principal crop is rice, mostly still grown on small peasant plots. Little goes to market, since many farms are too small to produce a surplus, though many farmers have joined cooperatives. Some cassava, maize and potatoes are grown for domestic consumption and a small crop of coffee and tea is grown for export. However, the country still cannot grow enough food for its needs and has to rely on imports. Subsistence farming accounts for 60 per cent of the GDP and more than 85 per cent of total employment.

The main ethnic group are the Lao, who make up about 50 per cent of the population. They are Buddhists, live along the Mekong Valley, and their language is the country's only official language. The rest of the population is made up of hill-dwelling groups, such as the Tai (about 20 per cent of the population), Phoutheung (15 per cent), Meo, Hmong, Yao and other peoples.

Tai tribes from the north settled Laos in the 10th and 11th centuries. By the 14th century a powerful kingdom, Lan Xang, had developed, centred on today's LUANG PRABANG, its capital. But feudal rivalries and intervention by Burma (Myanmar), Siam (Thailand) and Vietnam weakened it. By the 18th century it had split into two rival states, both of which were dominated by Siam in the 19th century.

After French colonial expansion in Vietnam and Cambodia, Laos was established as a French protectorate in 1893, and remained under French rule, with the French rejecting its

demands for independence and establishing the Kingdom of Laos under French control.

Meanwhile the exiled Souphanouvong formed an alliance with Vietnam's Ho Chi Minh. The forces of the former (the Pathet Lao) and of Ho (the Viet Minh) united in Laos to set up a Pathet Lao government, controlling much of the east by 1954, when the French left.

After 1955 the royal government received massive American support. In 1975 the Pathet Lao forced the abdication of the king and Souphanouvong became president.

The new government has made some progress recently, having acknowledged that their attempts to impose a socialist system on the country's agriculture and trade have been disastrous. Since 1990 private trade has come back. And although Laos is still one of the world's poorest nations, foreign investment has reportedly been flowing in at a pace almost alarming to the Laotians (who do not look forward to the noise and cultural pollution which comes with development). French villas have been renovated for private use or as businesses and traffic has increased to such an extent that for the first time in recent memory traffic police officers are needed on the streets. New telephone cables have been laid and the Laotians have already chosen which type of tourists they should let into their country – opting for the more controlled, expensive package tours for the select few.

LAOS AT A GLANCE	
Official name Lao People's Democratic Republic	
Area 236 800 km² (91 400 sq miles)	
Population 4 569 330 **Per km²** 19 (**Per sq mile** 50)	
Capital Vientiane	
Government Communist republic	
Currency 1 kip = 100 at(t)	
Languages Lao (official), French, also indigenous languages	
Religions Buddhist (85%), indigenous religions, Christian	
Climate Tropical monsoon, with most rains in summer. Average temperature in Vientiane ranges from 21°C (70°F) in January to 28°C (82°F) in May	
Land use Cultivation 4%, pastures 3%, forest and woodlands 58%, other 35%	
Main primary products Rice, maize, vegetables, tobacco, coffee, cotton, tin	
Major industries Agriculture, mining and ore processing, forestry and timber milling	
Main exports Timber, coffee, electricity	
Annual income per head (US$) 250	
Population growth (yr thous/yr) 28	
Life expectancy (yrs) Male 50 **Female** 53	

Settled in 1868 as a depot for the Union Pacific Railroad, it is now the seat of the state's university and a tourist centre offering good skiing, hiking and fishing in the surrounding mountain ranges.
Population 26 690
Map United States – Eb

Larap *Philippines* See BICOL

Laredo *USA* Border town in south Texas, on the Rio Grande 475 km (295 miles) south-west of

HOUSTON. It was settled by Spaniards in 1755 and occupied by the Texas Rangers in 1846; the Mexican War (1846-8) established it as part of Texas, but it still retains a Mexican flavour. It has economic ties with its sister city Nuevo Laredo in Mexico across the border, to which it is linked by three bridges over the Rio Grande. Laredo thrives on export-import trade and tourism; the surrounding area produces cattle, oil and gas.
Population 122 900
Map USA – Ge

Larnaca *Cyprus* South-east coastal resort town and port, trading in local farm produce such as cereals, potatoes and livestock, making brandy and soap, and processing tobacco. The town contains the Hala Sultan Mosque and a shrine to Umm Haran, a female relative of the Prophet Muhammad. According to Christian legend, Lazarus, who was raised from the dead by Jesus, lived in Larnaca.
Population 53 500
Map Cyprus – Ab

Larne *United Kingdom* Manufacturing town and port in Northern Ireland, 34 km (21 miles) north-east of BELFAST. Car and passenger ferries link the town to Cairnryan and Stranraer in Scotland, 60 km (37 miles) away. Larne manufactures generators and communications equipment, as well as machinery and paper products.

| Population 17 580 |
| Map United Kingdom – Cc |

Las Palmas de Gran Canaria *Spain* Largest city and chief port of the CANARY ISLANDS and capital of the autonomous region of the Canary Islands. It exports sugar, tomatoes, almonds and bananas. It is a popular year-round resort.

| Population (Gran Canaria Island) 767 970 |
| Map Morocco – Ab |

Las Vegas *USA* Gamblers' paradise in the desert in the south-east corner of Nevada. It began as a Mormon settlement in 1855-7 and at the same time was a trading post along the Salt Lake City-Los Angeles railway line. The city grew rapidly when the state legalised gambling in 1931, the same year that construction began on the Hoover Dam, a hydroelectric dam, 40 km (25 miles) south-east along the Colorado River.

Las Vegas has become world-famous for its glamorous 'Strip' of 24-hour casinos, nightclubs and entertainment. The Strip contains nine of the world's 10 biggest and gaudiest hotels. Visitors swell the coffers of the city, which is the 12th richest in the United States, by US$4.4 billion a year.

| Population (city) 258 300; |
| (metropolitan area) 852 740 |
| Map United States – Cc |

Lascaux, Grottes de *France* Caves in the Vézère Valley near the Dordogne River in the south-west. Discovered by four boys out walking with their dog in September 1940, the caves contain what is now regarded as the world's finest collection of prehistoric art. The paintings on the walls of the caves depict bulls, bison, deer, horses and hunters. Their Stone Age creators lived between 20 000 and 15 000 years ago. The caves have, however, been closed to the public since 1963 to preserve them from damage. A facsimile, Lascaux II, was opened nearby in 1983; it was created over a period of 10 years by artists using similar techniques and materials to those employed by the Stone Age painters.

| Map France – Dd |

Lassen Volcanic National Park *USA* Area of 433 km² (167 sq miles) at the southern end of the Cascade Mountains of California, 300 km (185 miles) north-east of SAN FRANCISCO. It contains Lassen Peak, at 3187 m (10 457 ft) the largest plug dome volcano in the world, and fine lava flows, hot springs, steam vents and lakes.

| Map United States – Bb |

Latakia (Al Ladhiqiyah) *Syria* Principal port and capital of the district of the same name, 100 km (62 miles) north-west of HAMAH. Most of the country's exports and imports pass through Latakia, whose main industry is processing tobacco. The bustling modern city, now the centre of a rich agricultural area, dates from Roman times, when it was known as 'Laodikeia'. Among its Roman remains is a triumphal arch.

| Population (city) 267 000 |
| Map Syria – Ab |

laterite Hard, reddish-brown earth formed by the LEACHING of weathered rock material high in iron content. It is found in tropical areas with marked wet and dry seasons.

latitude Angular distance of a point north or south of the equator. It is the angle, at the centre of the earth, between a line to this point and a line to a point on the equator due north or south of it. See also PARALLEL.

Latium *Italy* See LAZIO

Latvia See p 393

Lauenburg *Poland* See LEBORK

Lauis *Switzerland* See LUGANO

Launceston *Australia* North Tasmanian port at the head of the Tamar River estuary, lying some 160 km (100 miles) north of HOBART. It is Tasmania's second largest city.

It was founded in 1805 by Colonel William Paterson and was originally named Patersonia, but was later renamed Launceston after Governor King's birthplace in Cornwall, England.

Today, it is often called 'the Garden City' because of its beautiful parks and gardens. Launceston exports agricultural produce and timber. Its industries include hydroelectricity, textiles, and flour milling.

| Population 66 750 |
| Map Australia – Hg |

Laurasia The northern continent of the two ancient continents into which Pangaea, an earlier supercontinent, divided about 200 million years ago. Its southern counterpart was GONDWANA. See CONTINENTAL DRIFT, p 171

Laurentian Mountains *Canada* Range situated in southern Quebec Province, which rises to 1166 m (3825 ft) at Mount Raoul-Blanchard. In addition to being an important area for winter sports, it also has a number of major resorts.

| Map Canada – Hd |

▲ **SUNSEEKERS' CITY Tourists and small boats crowd the beach at Las Palmas. Founded in 1478, it developed as a modern port from 1883 and as a luxury resort after 1950.**

Lausanne *Switzerland* French-speaking city on the northern shore of Lake Geneva, and seat of the International Olympic Committee and of the Federal Court of Justice. A centre for international fairs and conferences, it also has a university, libraries and numerous professional schools, theatres and museums. The cathedral, consecrated in 1275 (construction started in 1173 on the site of an older church) is one of the few notable medieval buildings preserved in Lausanne; the 17th-century town hall is one of the most remarkable in Switzerland. Ouchy, once a hamlet, is now a suburb and Lake Geneva's busiest quay. The city is the capital of the French-speaking Protestant canton of Vaud (population 593 000).

| Population 117 600 |
| Map Switzerland – Aa |

Lautoka *Fiji* The country's second largest city and port, lying on the north-west coast of Viti Levu Island. Its exports – mainly sugar – exceed those of the capital, SUVA. It is the centre of an extensive sugar cane-growing region, and has one of the largest sugar mills in the southern hemisphere. A number of tourist resort coral islands are situated offshore from Lautoka. Nadi (pronounced 'Nandi'), Fiji's main international airport and one of the busiest in the South Pacific, lies 20 km (12 miles) to the south.

| Population 28 730 |
| Map (Viti Levu) Pacific Ocean – Dc |

lava Molten rock that has issued from a volcano or fissure in the earth's surface.

lava tube Tube formed by the cooling and solidification of the outer surface of a lava stream while the molten lava inside has flowed away.

Laval *Canada* City in Quebec Province which is separated from neighbouring MONTREAL by the

Latvia

*LATVIA STILL FACES THE CHALLENGE OF
BREAKING WITH ITS SOVIET PAST*

During most of its 1000-year history, Latvia has remained fragmented and controlled by foreign rulers. The Latvians have inhabited their green country on the Baltic since the 9th century, but have only twice had their own independent state: from 1918 to 1940 and since the collapse of the Soviet Union.

Latvia was first mentioned in historical records in 1185 when an Augustinian monk founded the first Christian church close to today's capital, RIGA. From the 13th to the 17th century, most of the country was under the rule of the order of the Augustinians. Then it was conquered in turn by the Poles and Swedes. From the 18th century until the fall of tsardom it was ruled by the Russians.

The country's two provinces, Livonia and Courland, were united after the October Revolution and proclaimed the Republic of Latvia in November 1918. Their independence was short-lived: in 1940, when Hitler and Stalin divided the Baltic countries up between them, Latvia went to the Soviets. It became the Latvian Socialist Soviet Republic in 1940 after occupation by the Red Army.

The Republic of Latvia declared full independence from the Soviet Union in May 1990. But returning to its pre-Soviet lifestyle was not all that easy. Economically the country was still dependent on the former Soviet republics, especially for its supply of raw materials and energy. There has been some privatisation but unlike Estonia, its northern neighbour, Latvia has been struggling hard to compete with the free markets of the West. A high unemployment rate, rocketing inflation

and frequent bottlenecks in the supply of goods, for instance foodstuffs, have stirred up dissatisfaction among the people.

It is among the Russian population group especially that the situation has become volatile. Russians make up about 34 per cent of the population but are now, in terms of new immigration laws, 'second-class citizens' – so much so that they refer to themselves as victims of an unofficial apartheid.

Latvians, on the other hand, view the stringent immigration laws as an effective means of preventing any further 'Russianisation' of their country. The Latvian percentage of the total population has dropped to just over 50 per cent, while the Russian portion has more than tripled over the past 50 years, mainly owing to the immigration of Russian industrial workers to the bigger urban centres, such as Riga and Daugavpils. Here, Latvians have become a minority. Most attempts to mix the different ethnic groups have been unsuccessful as each group lives on in isolation.

BRIDGING EAST AND WEST

The country which lies between Estonia in the north, Lithuania in the south, and Russia and Belarus in the east, could ideally serve as a bridge between Eastern and Western Europe. For centuries the bulk of Russia's trade with the West was handled through Latvia's usually ice-free port of Liepaya. Important trade routes started here, then followed the course of the Daugava (Dvina) River and continued eastwards along the Dnieper and Volga rivers. The country is hoping to resume its traditional role in trade; however, today it is a transshipment centre for illicit drugs from Central and South-west Asia to Western Europe.

Besides its favourable position with regard to trade and traffic, the country has several other valuable commodities. Among these are amber, found along the Baltic coastline, and peat, which is used as fuel in thermal power stations. The country at present produces only 10 per cent of its energy needs.

Almost two-fifths of the land is covered by deciduous and coniferous forests, but timber is a resource that has barely been tapped. Cattle raising still takes an unchallenged first place in agriculture; other products include grains, sugar beet and potatoes. There is some fishing and fish packing and canning.

The industrial sector employs about 33 per cent of Latvia's workforce and manufactures a variety of products, such as motor vehicles and agricultural machinery, synthetic fibres, pharmaceuticals, electronics and processed foods. However, it is dependent on importing all raw materials and most of its energy.

In their landscapes the two historical provinces of Livonia (south of the Daugava River) and Courland (east of the Bay of Riga and extending into Estonia) are very similar to the German morainal country along the Baltic Sea. Fertile and loamy soils alternate in quick succession with leached, sandy soil, bogs and marshes, interspersed with high morainic hills. The region has been noted for its natural beauty and is often compared to Switzerland, with its abundance of rivers and lakes, of which there are more than 3000.

However, air and water pollution are a growing problem. Already the Gulf of Riga and the Daugava River are heavily polluted.

During the last Ice Age, two huge glaciers covered the land. One left a depression in the east, which is dotted today with lakes and swamps; the other advanced from the north-west, grinding away the Bay of Riga and the adjacent depression through which flows the Daugava River. This depression, fringed by moraine hills, forms the nucleus of Latvia. In it lie the biggest Latvian cities – Riga, which houses over one-third of the population and is the main industrial area, and further inland, close to the Lithuanian and Belarus borders, Daugavpils, an important agricultural trading centre and industrial town producing textiles and railway equipment.

LATVIA AT A GLANCE	
Official name Republic of Latvia (Latvijas)	
Area 64 100 km² (24 743 sq miles)	
Population 2 735 570 **Per km²** 43	
(Per sq mile 110)	
Capital Riga	
Government Republic	
Currency 1 lats = 100 satims	
Languages Latvian (official); Lithuanian, Russian	
Religions Christian (Lutheran, Russian Orthodox, Catholic)	
Climate Maritime, with wet, moderate winters; further east, increasingly continental. Average temperature in Riga ranges from 5°C (23°F) in January to 17°C (63°F) in July	
Land use Forest 39%, cultivation 27%, meadows and pastures 13%, other 21%	
Main primary products Peat, amber, timber, fish, beef and dairy products, grains, sugar beet, potatoes, vegetables	
Major industries Processed foods, vehicles, machinery, fertilisers, pharmaceuticals, electronics, synthetic fibres	
Main exports Electrical equipment and appliances, vehicles, chemicals, textiles, food	
Annual income per head (US$) 6740	
Population growth (per thous/yr) 7	
Life expectancy (yrs) Male 64 **Female** 75	

Rivière des Prairies. The city was created in 1965 from 14 communities on Ile Jésus.

Population 314 400

Map Canada – Hd

Lazio (Latium) *Italy* Region covering an area of 17 203 km² (6642 sq miles) in central Italy. Lazio's richest agriculture area is the reclaimed coastal lowland 50 km (30 miles) south of ROME.

Population 3 076 000

Map Italy – Dc

Le Havre *France* The nation's second largest port, after Marseilles, and its most important transatlantic port, situated on the north bank of the Seine estuary on the coast of the English Channel. Founded in 1517 by Francis I, it rose to prominence in the 19th century, when it handled transatlantic and tropical cargoes. It was largely destroyed in 1944, during the Second World War. In the reconstruction which began in 1946, the city acquired wide boulevards, modern buildings and a greatly enlarged port. Shipping, banking,

oil refining, chemicals, engineering and the assembly of cars are its main industries.

Population 254 000

Map France – Db

Le Mans *France* City about 185 km (115 miles) south-west of PARIS. A 24-hour race for sports cars is held here each year. The city contains a fine cathedral and an 11th to 12th-century church.

Population 145 400

Map France – Db

A DAY IN THE LIFE OF A LEBANESE SCHOOL TEACHER

Sanaa Falloush returned to her Lebanese village in 1993 after an 18-year absence brought on by the country's civil war. She spent her time in exile teaching Arabic in Paris, but after all those years she felt she wanted to return home to help rebuild her country. She could only vaguely remember the country of her childhood days.

Now she is a teacher of French – the language of Lebanon's former colonial masters – in the same southern village that she and her family abandoned in 1975 when, along with 500 000 others, she fled the region at the outbreak of war. Qana is set in a land of wadis, caves and valleys, with Israel some 30 km (18.6 miles) to the south and the Mediterranean Sea closer still. Sanaa lives in a five-roomed house with her family.

After preparing a breakfast of pitta bread, cheese, jam and thick strong coffee for her retired dentist father and two elder brothers who run a small import-export business, she catches a communal taxi to school. Her first lesson begins at 7:30 am; her students are girls aged between 11 and 18 who have lived in Lebanon all their lives and are eager to hear of Sanaa's exciting life in exile. Their own short lives have been made up only of bombs and bloodshed.

A mid-morning break gives Sanaa the chance to relax and chat to other teachers at the 200-pupil school. As elsewhere in the teaching profession, the teachers complain about their students' laziness and lack of concentration; but the delinquency and disrespect which Sanaa witnessed in Paris are fortunately missing here. Education is a prized commodity in Lebanon and most of the children seem to really appreciate the opportunity to learn.

School finishes at 1 pm. Sports follow in the afternoon, but most children and staff go home instead and help their families with household chores or working in the family business. A teacher's monthly salary of about £L50,000 – not quite $US30 – is hardly enough to live on, and many are forced to supplement their income by waitressing, assisting in a shop or working on an agricultural plot.

Sanaa normally gets home by 2 pm after stopping at the local market to buy fresh vegetables and fruit. She prepares a salad lunch for her family and then spends her afternoon shopping or doing the books for her brothers' business.

A late dinner of mutton or fish and vegetables gives the family time to sit down quietly together to discuss the day's events. Television supplies them with foreign films and documentaries to keep in touch with the outside world.

Sanaa misses France and her French friends. But she feels she belongs in Lebanon, although as a Maronite Christian she realizes that she is in the minority in a largely Muslim country.

Le Puy (Puy-en-Velay) *France* Manufacturing city situated among eroded volcanic hills, 110 km (70 miles) south-west of LYONS. Its 12th-century Notre Dame Cathedral contains a statue of the Black Madonna, the object of an annual pilgrimage on the Feast of the Assumption (15 August). Le Puy's church of St Michel-d'Aiguilhe stands on an 80 m (260 ft) rock pinnacle.
Population 22 000
Map France – Ed

Le Touquet-Paris-Plage *France* Fashionable seaside resort on the English Channel coast, 50 km (30 miles) south of CALAIS. It has sandy beaches, casinos, a racecourse and a harbour.
Population 5600
Map France – Da

leaching Removal of soluble mineral salts from soil or rock by percolating water. Soils become progressively less fertile with leaching.

Lebanon See p 396

Lebanon Mountains (Jebel Lubnan) *Lebanon* Range of mountains running for about 160 km (100 miles) parallel to the Mediterranean coast from TRIPOLI in the north to TYRE in the south. The highest point is Qornet es Saouda, reaching 3087 m (10 125 ft) at the northern end of the range. The range receives a high rainfall of more than 1000 mm (39 in) a year and there are large springs some 1200 to 1500 m (3930 to 4921 ft) up, which are the source of a number of small rivers. Small communities of Druses, Maronites and other religious minorities have settled near the springs, where they have farms.
The western slopes facing the Mediterranean are particularly well watered, and farming is possible as high as 1525 m (5000 ft).
Map Lebanon – Aa-Ab

Lebombo (Lobombo; Lubombo) Mountains *Southern Africa* Low range, about 660 m (2000 ft) high, running north-south along the boundary of South Africa and Mozambique. It continues southwards through eastern Swaziland, where its flat-topped, grassy surface is used mainly for cattle rearing and farming. The climate is cool and pleasant. In Swaziland, three rivers – the Usutu, Mbuluzi and Nkalashane – have cut scenic gorges through the mountains. One of the gorges, on the Mbuluzi River, is the route used by the Swaziland-Maputo railway which links the country's interior with the Mozambican coast.
Map Swaziland – Ba

Lebork (Lauenburg) *Poland* Market town in Pomerania, known for its 15th-century castle. It lies about 65 km (40 miles) north-west of the Baltic port of Gdańsk.
Population 31 000
Map Poland – Ba

Lebowa *South Africa* Former self-governing state, covering 22 476 km² (8678 sq miles), created in 1972 for the North Sotho people in the Northern Transvaal and recognised only by South Africa. In 1994, it was fully reintegrated into South African territory. Lebowakgomo was its capital. The main activities consist of subsistence farming and the raising of livestock.
Population 2 740 590
Map South Africa – Ca

Lecce *Italy* Beautiful Baroque city in the heel of Italy, 12 km (8 miles) from the Adriatic Sea. Lecce was important in Roman and Norman times, but reached its zenith under Spanish influence in the 16th to the 18th century. During this period its best buildings – from the cathedral and church of Santa Croce to palaces and ordinary homes – were built from honey-coloured limestone. The town also has a Roman amphitheatre, a tall Roman column brought from the Appian Way and a 16th-century castle. It trades in olive oil, wine and tobacco.
Population 102 300
Map Italy – Gd

Ledenika Cave *Bulgaria* See VRATSA

Leeds *United Kingdom* City of West Yorkshire, England, about 60 km (37 miles) north-east of Manchester. It rose to prominence in the 19th century as one of the country's leading producers of ready-made clothing, textile machinery and railway equipment. It also became known for badly overcrowded slums which have, however, been improved or cleared since the Second World War. The city centre is now a modern business area mostly reserved for pedestrians. The huge town hall, built in 1858, reflects Leeds' 19th-century wealth; similarly, its imposing civic centre, television studios and university mirror its 20th-century growth.
Population 445 240
Map United Kingdom – Ed

Leeuwarden *Netherlands* Provincial capital of FRIESLAND, 110 km (70 miles) north-east of Amsterdam. Its canal-ringed centre has several old churches; among them is an interesting landmark consisting of a leaning tower of a church left standing when the rest of the church building was demolished in 1595. Leeuwarden is home to the Friesian Museum. The town manufactures dairy products, textiles and paper.
Population 85 100
Map Netherlands – Ba

Leeward and Windward Islands *Caribbean* The Leeward Islands and the Windward Islands together make up the Lesser Antilles. The Leeward Islands – so called because, as they are the northern group of the two, they are more sheltered from the prevailing north-easterly winds – include island states such as ANTIGUA AND BARBUDA and ST KITT AND NEVIS.
Among the southern group, the less sheltered Windward Islands, are ST VINCENT AND THE GRENADINES, DOMINICA and ST LUCIA. The Leewards run south-east from the Virgin Islands to Guadeloupe; the Windwards in a crescent from Dominica to Grenada, near the north-eastern coast of Venezuela.
Map Caribbean – Cb

Leeward Islands *French Polynesia* See SOCIETY ISLANDS

Legaspi *Philippines* See BICOL

Leghorn (Livorno) *Italy* Port and industrial city 150 km (93 miles) south-east of GENOA. Its industries include shipbuilding, engineering, electrical equipment, chemicals, marble, copper and glass.
Population 175 300
Map Italy – Cc

Legnica (Liegnitz) *Poland* Industrial city 63 km (39 miles) west of the city of WROCLAW. In 1241, at Legnickie Pole (Field of Legnica), 4 km (2.5 miles) to the east, advancing Tatar hordes fought the Poles, killing the Polish leader Prince Henryk the Pious. Despite being defeated, the Polish army managed to halt the westward advance of the Tatars. A Gothic chapel is now the battlefield museum. Legnica is a copper smelting and sulphuric acid-producing centre.
Population 104 200
Map Poland – Bc

Leicester *United Kingdom* City some 140 km (87 miles) north-east of London. The county of the same name covers 2553 km² (986 sq miles) of the eastern Midlands of England. It stretches from the hills of Charnwood in the west, with their ancient forests, to the uplands of Rutland in the east, the site of Britain's largest constructed lowland lake, Rutland Water, with a shoreline of 38 km (24 miles).

Formerly the Roman town of Ratae, the city still contains the remains of Roman baths. In 1264 Leicester made another historical mark when the Earl of Leicester, Simon de Montfort, forced his brother-in-law, Henry III, to hold the nation's first parliament.

Today, Leicester's main industries consist of hosiery, knitwear, footwear and engineering. The city's Church of St Martin was elevated to cathedral status in 1926, and a local college to university status in 1957. Loughborough, about 16 km (10 miles) north of Leicester, boasts a leading technological university.
Population (city) 284 700; (county) 894 000
Map United Kingdom – Ed

Leiden *Netherlands* University city and cultural centre about 35 km (20 miles) to the south-west of AMSTERDAM. The city was the birthplace of the painter Rembrandt (1606-69); its university dates from 1575. Pilgrims, seeking to escape religious persecution in England, spent 11 years in Leiden before sailing to America in 1620.
Population (city) 113 840;
(Greater Leiden) 192 230
Map Netherlands – Ba

Leinster (Cuige Laighean) *Ireland* The most fertile of Ireland's four historic provinces (the others are Connaught, Munster and Ulster), which occupy the south-east of the country.
Population 1 860 950
Map Ireland – Cb

Leipzig *Germany* Southern city which the 18th-century German writer Johann Wolfgang von Goethe (1749-1832) called 'Paris in miniature'. It is situated some 180 km (112 miles) south-west of BERLIN and is a major cultural centre. The religious reformer Martin Luther (1483-1546) was active here; Goethe and his fellow poet Johann Schiller (1759-1805) were both students at Leipzig University, which was founded in 1409 (making it the second oldest in Germany, after Heidelberg University).

Music thrived here through the genius of Johann Sebastian Bach (1685-1750), who lived and worked in the city for the last 27 years of his life. The opera composer Richard Wagner was born here in 1813. Today, the Leipzig Gewandhaus Orchestra is one of the finest in the world and the Thomanerchor, renowned as a choir of world standing, is based here. Leipzig is also a publishing, printing and industrial centre, producing agricultural machinery, chemicals and paper. Its first international trade fair was held more than 700 years ago.
Population 511 000
Map Germany – Ec

Leitrim (Liatroim) *Ireland* Thinly populated county in the north-west, with a short Atlantic coastline. An area of bogs and mountains, it covers 1525 km² (589 sq miles) and is split by Lough Allen. The county town is Carrick on Shannon.
Population 25 300
Map Ireland – Ba

Leix *Ireland* See LAOIS

Lek *Netherlands* See RHINE

Lelystad *Netherlands* New town some 40 km (25 miles) north-east of AMSTERDAM, on the East Flevoland POLDER, an area of low-lying land reclaimed from the IJSSELMEER in 1957. The town, named after Cornelius Lely, the statesman-engineer who conceived the reclamation project begun in 1920, is the administrative centre of the IJsselmeer polders.
Population 60 170
Map Netherlands – Ba

Léman, Lac *Switzerland* See GENEVA

Lemberg *Ukraine* See L'VIV

Lempa *El Salvador* The country's longest river, at 260 km (160 miles).
Map El Salvador – Ba

Lena *Russia* East Siberian river, some 4300 km (2672 miles) long. It rises at 1810 m (5940 ft) in the mountains along the west shore of Lake BAIKAL, and tumbles northwards over rapids to the small town of Kachug, where it becomes navigable. It then flows north-east and north past the city of Yakutsk to enter the Laptev Sea via a large delta. It is navigable for 3500 km (2175 miles) of its course but is frozen from October until May.
Map Russia – Nb-La

Lenin Peak (Pik Lenina) *Kyrgyzstan/Tajikistan* The former Soviet Union's third highest peak after Communism Peak and Pik Pobedy. Named after the Russian revolutionary leader Vladimir Lenin (1870-1924), it soars to 7134 m (23 405 ft) in the Pamir Mountains.
Map Tajikistan – Bb

Leninabad *Tajikistan* See KHUDZHAND

Leninakan *Armenia* See GYUMRI

Leningrad *Russia* See ST PETERSBURG

Lens *France* Industrial city in northern France, surrounded by pit villages, slag heaps and coal mines, which have been closed since the 1970s. The city produces chemicals, car parts, electrical and metal goods, and clothing.
Population (town) 35 000; (conurbation) 323 000
Map France – Ea

lenticular cloud Lens-shaped cloud, often associated with squally winds over hills and mountains.

León *Mexico* Commercial city in GUANAJUATO province on the fertile plain of the Gòmez River in the central highlands of Mexico. The river floods regularly and León suffered a terrible flood in 1888 when it was almost totally destroyed. The city now thrives as Mexico's shoe-making capital.
Population 872 500
Map (Guanajuato Province) Mexico – Bb

León *Nicaragua* The country's second largest city, after the capital, Managua. Lying near the north Pacific coast, León has Nicaragua's oldest university, founded in 1804, and Central America's largest cathedral, built between 1746 and 1846. The city was the national capital for 300 years until 1858, when the seat of government was moved to Managua. Cotton, timber, processed foods, and fertilisers and insecticides are its main industries.
Population (city) 158 600; (department) 248 700
Map Nicaragua – Aa

León *Spain* Ancient city 290 km (180 miles) north-west of MADRID, founded by the Romans and capital of the Spanish kingdom of ASTURIAS – later named León – which flourished between the 10th and 12th centuries. Today it is the centre of an agricultural province of the same name; it manufactures textiles, pottery and leather goods.
Population (town) 144 020; (province) 525 900
Map Spain – Ca

Léopoldville *Zaire* See KIUSHASA

Lepanto *Greece* See NÁVPAKTOS

Leptis Magna *Libya* Ruined ancient port near HOMS, 123 km (76 miles) east of Tripoli. Leptis Magna, which was founded by Phoenicians in about 600 BC, was a leading Carthaginian city by the 2nd century BC. Unlike CARTHAGE, it was not destroyed by the Romans during their conquest of the Carthaginians in the Third Punic War (149-146 BC). Annexed by the Romans in 46 BC, it became important as a trade centre and port for goods coming across the Sahara. The city reached the pinnacle of its influence in the 2nd century AD. Its well-preserved remains include forums, baths, temples, walls, arches and an impressive theatre.
Map Libya – Ba

Lérida *Spain* Town 160 km (100 miles) west of the Mediterranean port of BARCELONA. Former residence of the King of Aragón, it has a number of old buildings, including a huge 12th-century fort and cathedral. Its industries include tanning and the manufacture of textiles and paper. The surrounding province of the same name produces wine and olive oil. It was the scene of a nine-month battle in 1937-8, during the Spanish Civil War.
Population (town) 112 090; (province) 353 460
Map Spain – Fb

Lerma (Río Grande de Santiago) *Mexico* Mexico's longest river, flowing west from the Sierra Madre range to Lake Chapala, where it becomes the Río Grande de Santiago, and onwards to enter the Pacific in northern JALISCO State. The Lerma irrigates the central highlands for agriculture, and since 1951 its headwaters have supplied Mexico City's huge population with water.
Map Mexico – Bb

Lebanon

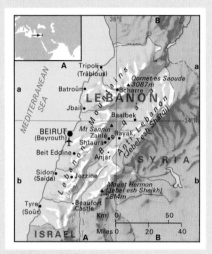

Once the jewel of the eastern Mediterranean, the playground of the rich with a distinctly French flavour, this sad country was under violent assault for about 16 years. The Palestine Liberation Organisation (PLO), a Syrian army, an Israeli occupation force, Western and United Nations peace-keeping forces have all been involved. Lebanese fought Lebanese, and strangers entering a town had, for their own safety, to find out which local militia was in control. A tentative peace agreement was finally signed in October 1990 and these resilient people prepared to use their astute business skills to finance the re-building of their once spectacularly lovely country.

Lebanon's cosmopolitan people were until recently among the most prosperous in the Arab world. The Lebanese are shrewd business people and so adaptable that for many years the country was a kind of middleman in a sea of Middle Eastern trouble. Lebanon's main ports, the capital BEIRUT and TRIPOLI, acted as transit centres for the countries of the interior – Syria, Jordan and Iraq. Beirut was the banking centre of the Middle East and the home of entrepreneurs who could arrange every kind of business deal.

All this has changed. The country is now heavily in debt, inflation has risen steeply and the people's main aim for almost two decades has been simply to survive – in a land where sniping, bombing, kidnapping and assassination were the daily norm. They carried on as traders and business people in circumstances in which most others would have given up in despair.

Lebanon covers an area of 10 400 km² (4015 sq miles), with a 225 km (140 mile) coastline on the Mediterranean. Inland from a narrow plain, spectacular mountains, snow-covered in winter, rise to a height of 3086 m (10 125 ft). In the east of the country these shelter the fertile BEQA'A. Still farther east, the land rises to the Anti-Lebanon, a range that reaches up to 2629 m (8625 ft).

The coast has a warm Mediterranean climate with dry summers. In the mountains the snow lies on the ground from December to May so that in winter it is possible to swim in a warm sea and then go no more than a few kilometres inland to ski. Evergreen forests grow in the mountains and the oak is a familiar tree in much of the country. But only a small number of Lebanon's famous giant cedars remain – carefully protected in groves more than 900 m (2950 ft) above sea level.

Lebanon, with about 3.5 million people, is truly a country of minorities. Apart from the Arabs, who make up 95 per cent of the population, there are also Armenians, Kurds, Greeks, Turks and Jews. More significant however, are the religious classifications, since these govern political affiliations. Christians, once just more than half the population, now form 30 per cent and comprise four main sects. Most powerful are the Maronites, who are in union with the Roman Catholic Church but have their own Eastern rites and Patriarch; there are also Catholic, Orthodox and Armenian Christians.

There are five Muslim sects: Shiites (the majority), Sunnis (who formerly outnumbered the Shiites), Druse, Alawite, Isma'ilite and Shi'a. The Druse believe in reincarnation, do not fast during the Muslim holy month of Ramadan nor go on pilgrimage to MECCA, and show equal hostility to both Christians and other Muslims.

This mixture of races and beliefs led inexorably to civil strife. The country's position adjoining Syria and Israel, the two most bitter opponents in the Middle East, meant also that Lebanon was bound to become involved, however much it might try to avoid trouble.

About one-fifth of the employed people work in agriculture, near the coast and in the Beqa'a Valley. The country produces a variety of fruits and vegetables – olives, citrus, grapes, figs, apricots, apples, bananas, onions and potatoes – as well as cotton, tobacco and sugar beet. The coast is fertile and the steep valleys are suitable for growing olives and vines. Industry is on a small scale. Manufactures include cement, fertilisers, jewellery, sugar and tobacco. There are oil refineries at Tripoli and SIDON.

Thirty-six per cent of the population are under 15 years old. The Lebanese are better educated than most Arabs: free primary education was introduced in 1960 and literacy is about 80 per cent. In addition to Arabic many people speak French or, increasingly, English.

A HISTORY OF STRIFE

In past centuries most major armies in the area passed through Lebanon, while those who fled before them often sought refuge in the mountains. The ancient Phoenicians reached their peak of power in the 12th to 9th centuries BC, leaving the towns of TYRE and Sidon as their monuments. Assyrians, Babylonians, Egyptians, Persians, Greeks and Romans fought in turn for the rich land. Later came the Arabs, who introduced Islam in the 7th century AD, followed by the Crusaders and the Turks.

In the 1860s the Maronite Christians appealed for aid after a series of massacres by the Druse. Britain and France prevailed on the Turks to set up a Christian government. The French established autonomy for Beirut and the region surrounding it, which was then ruled by a Christian governor. In 1918 British troops invaded the country but left when the French created the state with its present borders. When the Ottoman empire broke up after the First World War Lebanon came under the control of the French, who left their influence in language, food and customs. During the Second World War, in 1941, the Free French proclaimed Lebanon an independent state. In 1943, however, a new parliament came into conflict with the French authorities and it was only in 1945 that France transferred the country to Lebanese control and agreed in 1946 to withdraw its troops.

Under an unwritten agreement of 1943, government posts were shared out among the various religious groups. Thus the president was always a Maronite, the premier a Sunni and the speaker of the Chamber of Deputies a Shiite; other government posts and seats in the Chamber reflected the sectarian make-up of Lebanese society. However, these proportions did not remain static and by the mid-1960s the now predominant Muslims – and particularly the Shiites – were demanding more power, as befitted their numbers; the Maronites resisted change. Added to this was the arrival of Palestinian refugees and, more seriously, the PLO after it was expelled from Jordan in 1970.

PRIVATE ARMIES

Lebanon took little part in the Arab-Israeli wars of 1948 and 1956 and none in the wars of 1967 and 1973. But clashes between private armies of Muslims, Druse, Christians and militant Palestinians led to civil war in 1958 and again in 1975. Attacks by the PLO across the Israeli border in 1969 and 1973 brought reprisals from the Israelis and from Lebanese forces led by Christians of the right-wing Falangist Party.

From 1975 the fragile political concord in the country began to collapse in the face of Christian-Muslim rivalries and mistrust. Lebanon began to slip into a more or less permanent state of civil war. In 1977 40 000 Syrian troops entered the country to support the Muslims. The next year Israel invaded southern Lebanon to establish a buffer zone between itself and PLO groups who frequently raided over the Israeli border. The United Nations sent a temporary peace-keeping force and the Israelis withdrew.

By 1980 there were 33 private armies or militias in the country. In 1982 the Israelis, supported by Falangists in the south, mounted a full-scale invasion of southern Lebanon to eject the PLO forces under Yasser Arafat, first from Beirut to Tripoli and then from the country altogether. The same year the Falangist president, Bachir Gemayal, was murdered after three weeks in office and succeeded by his brother, Amin – who was elected by an overwhelming majority.

▲ **FERTILE LAND A white cap of snow crowns the mighty range of the Lebanon Mountains in the north of the country. Fruits and vegetables are grown on terraced slopes rising behind Bcharre; some of the fields are irrigated. Little remains of the forests that once occupied the mountain slopes.**

A Western peace-keeping force from the United States, Britain, France and Italy was withdrawn after suicide attacks by Muslim extremists on French and American troops. Until 1984 the Israelis and Syrians had each controlled 40 per cent of the country with 20 per cent in the hands of various factions and militias, leaving the legal Lebanese government virtually powerless. Beirut was a divided city: the Christians held the east and the Muslims the west of the capital. Sunni Muslims were strong in Tripoli, Shiites in the Beqa'a and the south, and Druse in the Chouf Mountains. The government was an uneasy coalition, excluding Walid Jumblatt, the Druse leader, but including Nabih Berri, the Shiite militia chief, who was then the Minister of Justice.

The civil wars were fought with savagery. Falangists massacred Palestinians in refugee camps; Shiites murdered Christians in their village strongholds. The country became an astonishing patchwork of no-go areas. Towns were divided into sections where particular Christian or Muslim or Druse militants held sway; once prosperous places became ghost towns. People became used to a life of sudden violence followed by business as usual. Visitors from abroad usually made sure that trusted contacts made arrangements for them.

In 1985 the Israelis withdrew their forces, except from a 'security zone' in the south. The Syrians remained, still holding on to their 40 per cent share.

Then, after the Ta'if Accord in 1990, the Lebanese established a more equitable political system which recognised a greater role for Muslims in the political process. In December of the same year legislative elections were held for the first time in 20 years. Most militias were disbanded or significantly weakened when the government extended its rule over 50 per cent of the country and seized large quantities of weapons from the various contestants.

FOR THE VISITOR

Lebanon's climate makes it a delightful place for a holiday and it has much for tourists to see and enjoy. In more peaceful times one-fifth of the country's income came from tourism. Visitors were offered a mixture of Arab and French cuisines, beach resorts, old Crusader castles at Sidon and Byblos (now JBAIL) and ancient Phoenician (Byblos) and Roman sites (the chariot circus at Tyre), superb scenery (especially the view across the Beqa'a Valley from the Dah el Beidar Pass in the Lebanon range) and winter sports in the mountains.

Among the places of historic interest are the two Phoenician cities of Tyre and Sidon, BAAL-BEK with its Roman temple to Jupiter, and the cedars at BCHARRE and QORNET ES SAOUDA.

Beirut also had a reputation for arranging most kinds of pleasure – at a price.

Will the Lebanese succeed in regaining their earlier prestige, or has their history been irrevocably changed?

LEBANON AT A GLANCE	
Official name Republic of Lebanon	
Area 10 400 km² (4015 sq miles)	
Population 3 552 370 **Per km²** 342	
(Per sq mile 885)	
Capital Beirut	
Government Republic	
Currency 1 Lebanese pound = 100 piastres	
Languages Arabic (official), French, English	
Religions Muslim (70%), Christian (30%), Judaism	
Climate Mediterranean, cool in highlands. Average temperature in Beirut ranges from 14°C (57°F) in January to 28°C (82°F) in August	
Land use Cultivation 30%, grazing 1%, forest and woodland 8%, other 61%	
Main primary products Citrus fruits, apples, grapes, potatoes, sugar beet, wheat, olives, cotton goats, sheep, some iron ore	
Major industries Trade and banking; cotton yarn, textiles, agriculture; tobacco, cement, food processing, manufacturing, fertilisers	
Main exports Fruit and vegetables, chemicals, machinery, metals, textiles and clothing, cement, tobacco, precious and semi-precious stones and jewellery	
Annual income per head (US$) 1400	
Population growth (per thous/yr) 18	
Life expectancy (yrs) Male 67 **Female** 72	

Lesotho

WITH FEW NATURAL RESOURCES, LESOTHO IS DEPENDENT ON ITS GIANT NEIGHBOUR, SOUTH AFRICA

Snow-capped mountains and treeless uplands cut by spectacular gorges and isolated valleys cover two-thirds of the small kingdom of Lesotho (formerly Basutoland), perched fortress-like on southern Africa's HIGHVELD. Landlocked Lesotho – about the size of Belgium – is entirely surrounded by the Republic of South Africa, on which it is economically dependent.

Characteristic of the people of Lesotho, the Basotho (or Basuto), are the coloured blankets they wear – acting as a shield against both the cold winters and the frequent summer hail and thunderstorms.

The Basotho were forged as a nation in the early 19th century. Groups of smaller tribes fleeing from the relentlessly expanding Zulu nation gathered under the outstanding leadership of King Moshoeshoe I in the sheltering mountains, where they made the flat-topped, impregnable THABA BOSIU (Mountain of Night) their citadel. However, when they clashed with the Boers, who settled on much of their land, Moshoeshoe was forced to turn to Britain and ask for help.

The country became a British protectorate in 1868, but remained under its indigenous chiefs. Full independence finally came in 1966 with King Moshoeshoe II as head of state, although two months after independence he was forced to accept a purely ceremonial role. The power lay essentially with a military council which, early in 1986, deposed Chief Leabua Jonathan, prime minister since independence, who had maintained his position by suspending the constitution during 1970-3. In 1990 Moshoeshoe was also deposed and went into exile, to be replaced by King Letsie III. Elections in March 1993 – the first since 1970 – brought a landslide victory to the Basotho Congress Party (BCP) which won all 65 parliamentary seats. Although the election results were disputed by the Basotho National Party (BNP), the BCP leader, Ntsu Mokhele, was sworn in as prime minister in April 1993. In the following year government was disrupted by a drawn-out struggle between the king, supported by the military, and the new government. In August 1994 the king

suspended parts of the constitution, dissolved parliament and dismissed the government. Disturbances which followed brought the diplomatic intercession of the South African, Botswana and Zimbabwe governments, who threatened to impose sanctions against Lesotho. The outcome was an agreement signed in September 1994 reinstating Mokhele and parliamentary rule and providing for Letsie's abdication in favour of his deposed father. Moshoeshoe II was reinstated in 1995.

As a result of the influence of 19th-century French Protestant missionaries, Lesotho has one of the highest literacy rates in Africa – 68 per cent for women and 44 per cent for men (more girls attend school because boys often herd livestock). Yet it is a poor country. There are few industries and the mountainous terrain allows only about one-tenth of the land to be cultivated, the chief crop being maize. Yields are low, partly because of soil erosion caused by steep slopes and overgrazing by large herds of sheep, goats and cattle. Counter-measures now been taken, including reducing the number of animals and promoting mixed farming. Wool, mohair and manufactured products are exported. A major blow to the economy came when the largest of Lesotho's diamond mines closed in 1982. Although some diamonds are still mined, most of the country's foreign exchange now comes – and has done for the past two decades – from remittances sent home by Lesotho workers in South Africa.

More than 100 000 of the population of 1.9 million migrate to South Africa to earn a living, chiefly in the gold and coal mines of the WITWATERSRAND; in 1991, a total of US$439 million was earned from remittances by labourers employed in South Africa – more than 70 per cent of the country's domestic output that year. But ever since South Africa's foreign labour cutbacks in the late 1980s, providing alternative employment for its growing labour force and alternative sources of foreign exchange have become the government's most pressing tasks. Industrialisation has increased: in 1982 industry earned 6 per cent of the gross domestic product (GDP); by 1991, it earned 13 per cent. However, most of the population (60 to 70 per cent) continues to live off subsistence farming and migrant labourers' earnings.

As a result of its economic dependence on South Africa, Lesotho was until recently under enormous political pressure from its large neighbour, particularly after the Lesotho government began to oppose apartheid. While the African National Congress (ANC) was banned in South Africa, Lesotho found itself regularly pressured over its support for the ANC. In 1986 South Africa instituted an

economic blockade, thereby forcing Lesotho to expel the ANC. South Africa also helped in the overthrow of the Jonathan government and maintained cordial relations with Lesotho's military rulers from 1986 to 1992. While Jonathan was still in power, South Africa had supported the military wing of the opposition Basotho Congress Party. When, in 1993, the BCP became the government, Lesotho's friendly relations with South Africa continued.

The area in which Lesotho's fortunes have been improving is tourism. Many of the tourists are South Africans attracted by the casinos in the capital, MASERU. Skiing at Oxbow in the mountains is also popular. Another potential source of income is the ORANGE River, which rises on the Lesotho side of the DRAKENSBERG range.

The first phase of the joint Lesotho-South African Lesotho Highlands Development Project, damming the river's headwaters to provide power needed by South Africa's industrial Witwatersrand, will be completed in 1995-96. It will supply 70 m³ (2474 cubic ft) of water per second to South Africa and provide Lesotho with its own electricity. This is one of the largest engineering projects in the world, requiring the construction of six major dams, four pumping stations and 225 km (140 miles) of tunnels over a period of 25 years. The Lesotho government expects that the project will also attract tourism to the area.

LESOTHO AT A GLANCE	
Map South Africa – Cb	
Official name Kingdom of Lesotho	
Area 30 350 km² (11 715 sq miles)	
Population 1 900 000 **Per km²** 630	
(Per sq mile 162)	
Capital Maseru	
Government Constitutional monarchy	
Currency 1 loti = 100 lisente	
Languages Sotho, English (both official)	
Religions Christian (92.8%), indigenous beliefs (6.2%)	
Climate Continental; temperature in Maseru ranges from 9°C (48°F) in July to 21°C (70°F) in January; temperatures are much lower in the mountains	
Land use Grazing 66%, cultivation 10%	
Mineral resources Diamonds	
Main primary products Wheat, maize, sorghum, pulses, livestock; diamonds	
Major industries Agriculture, textiles, tourism	
Main exports Wool, mohair, textiles, leather goods, diamonds	
Annual income per head (US$) 590	
Population growth (per thous/yr) 27	
Life expectancy (yrs) Male 60 **Female** 64	

Lesbos *Greece* Third largest island in the Aegean Sea, 10 km (6 miles) from the coast of Turkey. It is the birthplace of Sappho (about 612-580 BC), classical Greece's greatest female poet. Her passionate poetry – which seems to have been directed at her female admirers – has given us the word 'lesbian', after the name of the island.

Lesbos is a fertile land of hot springs, olive groves and sandy beaches. The chief town, Mitilíni, and its harbour, on the eastern side of the island, are overlooked by a huge castle dating

from 1373. The castle was built by a Genoese family who ruled the island from AD 1355 until the Turks occupied it in 1462. South of Sigri, on the west coast of Lesbos, is a petrified forest formed when conifers and sequoias were buried in volcanic ash. A rare species of salamander and blind mice live here.

The island has several Byzantine churches, monasteries and an archaeological site at Eressós.
Population 103 700
Map Greece – Eb

Lesina *Croatia* See HVAR

Leskovac *Yugoslavia* See MORAVA

Lesotho See above

Lesse *Belgium* River, 80 km (50 miles) long, which flows underground for part of its course. At the village of Han about 60 km (35 miles) southwest of LIÈGE, it has created spectacular limestone caves – the Grottes de Han. Visitors

enter the caves through one of the river's older and now dry channels, and leave, farther along its course, by boat. The largest of the caverns is the Salle du Dôme, a vaulted chamber 150 m (490 ft) across and 130 m (425 ft) high. A local museum contains relics from the caves.
Map Belgium – Ba

Lesser Sunda Islands *Indonesia* See NUSA TENGGARA

leste Hot, dry southerly to easterly wind blowing from the Sahara and experienced in Madeira. It is often dust-laden and precedes a DEPRESSION.

Lésvos *Greece* See LESBOS

Leszno (Lissa) *Poland* Beer-brewing and china-making town 66 km (41 miles) south of the city of Poznań. It was settled by Protestants from the Czech province of Moravia in the 16th century, and became the centre of Polish Calvinism. The Czech theologian and educational reformer Comenius (Jan Komensky, 1592-1671) lived in the town. Cereals and cattle are marketed here.
Population 52 900
Map Poland – Bc

Lethbridge *Canada* City in southern Alberta on the Oldman River, 177 km (110 miles) south-east of CALGARY. Established in 1870 as a coal-mining settlement, it is now the centre of a large ranching and irrigated agricultural area, whose products it processes. The city's university opened in 1967.
Population 69 420
Map Canada – Dd

Lettland See LATVIA

Leuven *Belgium* See LOUVAIN

levanter (levante) Strong easterly or east-north-easterly wind blowing across south-east Spain, the Balearic Islands, and the Strait of Gibraltar.

leveche Hot, dry southerly wind, similar to the SIROCCO, blowing from Morocco across to the coast of southern Spain. It precedes an advancing DEPRESSION and often carries a great deal of dust. Under certain conditions the dust is capable of reaching northern Europe and has been known to cause pink 'blood' rain.

levee Bank of a river built up above the flood plain by the deposition of sediment close to the channel during flooding. Levees are formed naturally, but may be strengthened artificially as protection against flooding. Wholly artificial levees are also built for the same purpose.

Levkás *Greece* Island in the Ionian Sea, which is separated from the mainland only by a canal first cut in ancient times and reopened in 1905. In the south of the island, at Cape Doukáton, is the Leucadian Rock from which the poet Sappho is said to have leapt to her death in about 580 BC in despair at being rejected in love.
Population 20 000
Map Greece – Bb

Lexington *USA* City in north-central Kentucky. It lies in the heart of the bluegrass country, 120 km (75 miles) south of Cincinnati, Ohio. It

is an important tobacco market and thoroughbred horse-breeding centre.
Population (city) 210 200; (metropolitan area) 405 940
Map United States – Jc

Leyte *Philippines* Large, irregular-shaped island covering 7213 km² (2785 sq miles) in the eastern VISAYAN ISLANDS, just south-west of Samar, to which it is linked by a bridge. Rugged and undeveloped, it is often lashed by heavy rains and hit by typhoons from the Pacific. Rice, maize, sugar and other crops are grown on its coastal lowlands. Leyte was the landing place, just south of the provincial capital, Tacloban, of United States invasion forces on 20 October 1944; over the next week, the Americans won a decisive air-sea battle against the Japanese fleet in Leyte Gulf, to the east.
Population 1 480 000
Map Philippines – Bc

Lhasa (Lasa) *China* Capital of the Xizang Autonomous Region (formerly Tibet) and one of the remotest cities in the world. It lies in a broad valley on a tributary of the Yarlung Zangbo (Brahmaputra) River, at an altitude of 3800 m (12 450 ft) between the Himalayas and the Nyainqên-tanglha Range, and is separated from eastern China by the rugged wilderness of the Xizang Gaoyuan (the Tibetan Plateau). Lhasa, which translates as 'the Seat of the Gods', contains many Buddhist temples and the Bodala (Potala) Palace, the former residence of the Dalai Lama, who fled to India in 1959 after a Tibetan rebellion against Chinese rule had been crushed. Since Tibet was ceded to China the pilgrims have stayed away.
Population 343 200
Map China – De

Lhotse *Nepal/China* Twin-peaked mountain in the Himalayas on the Nepalese frontier with Tibet, just south-west of Mount Everest. Lhotse I rises to 8501 m (27 890 ft) and is some 118 m (387 ft) higher than Lhotse II.
Map Nepal – Ba

Liaoning *China* Province in the north-east covering about 140 000 km² (540 000 sq miles). Bordering on North Korea, it is the southernmost of the three provinces of Dongbei (Manchuria). The fertile Liao River basin forms the centre of the province with low hills on either side. Tobacco, apples and pears, soya beans, sugar beet and cotton are grown in the province, which is also unusually well endowed with minerals. The great coal fields of Fushun, Fuxin and Benxi are among the largest and richest in China. Iron ore is mined at Anshan, Benxi and Liaoyang; the province also has valuable reserves of manganese and molybdenum, as well as bauxite, oil shale, lead, zinc and copper. Liaoning is China's leading producer of iron and steel, aluminium and heavy machinery. The capital is SHENYANG.
Population 40 160 000
Map China – Ic

Liatroim *Ireland* See LEITRIM

Libau *Latvia* See LIEPAJA

Liberec (Reichenberg) *Czech Republic* Industrial city lying 20 km (12 miles) from both the German and Polish borders. It produces Liaz

lorries and diesel engines, textiles, processed foods, clothing, textile machinery, and furniture. To the south and west lies Bohemia's main glass-making area, which exports its renowned glass and glass jewellery around the world. The area's main towns are Jablonec, Novy Bor and Železny Brod.
Population 99 600
Map Czech Republic – Ba

Liberia See p 402

Libertador (O'Higgins) *Chile* Region just south of Santiago taking in Cachapoal and Colchagua provinces. The economy is based on copper mining at EL TENIENTE and pastoral and stock farming around RANCAGUA, the regional capital, and San Fernando. Rodeo championships are held in the towns. There is a hydroelectric plant on the Rapel River.
   The region's full name, El Libertador General Bernardo O'Higgins, commemorates Chile's first president and liberator from Spanish rule.
Population 688 390
Map Chile – Ac

Libreville *Gabon* National capital and capital of Estuaire Province. Founded in 1849 as a settlement for freed slaves, it stands on the north shore of the Gabon River estuary, on a vast indented bay. Libreville has expanded rapidly since the exploitation of Gabon's mineral resources in the 1960s.
   It is now a major port (see OWENDO), exporting hardwoods, rubber and palm products, and an industrial centre. It is also the educational centre of the country, and seat of the Université Omar Bongo, founded in 1970 and named after Gabon's second president. The city retains a French character, though many modern structures now tower above the old French colonial buildings.
Population 308 000
Map Gabon – Aa

Libya See p 404

Libyan Desert *Libya* See SAHARA

Lidice *Czech Republic* See KLADNO

Lidköping *Sweden* Small manufacturing town on the south-east shore of Lake VÄNERN, 130 km (80 miles) north-east of Gothenburg. It is situated in fertile farmland amid prehistoric remains, medieval churches and old villages. Rörstrand porcelain is made in the town.
Population 36 300
Map Sweden – Bd

Liechtenstein See p 406

Liège (Luik; Lüttich) *Belgium* City situated on the Meuse River, some 95 km (60 miles) east of Brussels. It gives its name to a Belgian province 3862 km² (1491 sq miles) in area, and is also claimed by French-speaking Belgians as the capital of French-language culture in Belgium, since Brussels, the national capital, is a bilingual city.
   Owing to frequent attacks on Liège through its history, few of its medieval buildings survive. The great cathedral of St Lambert was destroyed in the spreading fires of the French Revolution in 1792, but the Palace of the Bishops (built in the 1530s, re-faced in the 1730s) survived and today serves as the law courts.

Industrial zones containing much of Belgium's iron and steel industry are situated both upstream and downstream from the city centre. The downstream zone also comprises the port of Liège, the entrance to the ALBERT CANAL and the town of Herstal. Coal mines and slag heaps were formerly features of the landscape around the city, but mining has now ceased along the Meuse. The space formerly taken by the coalfield has been occupied by new industrial estates and the motorway junctions that make Liège a focus of trans-European routes.

Population (city) 212 000; (metropolitan area) 589 670; (province) 1 000 000
Map Belgium – Ba

Liegnitz *Poland* See LEGNICA

Lienz *Austria* Main town of the east Tyrol at the junction of the Isel and Drava valleys, 80 km (50 miles) west of Villach. The area was cut off from the rest of Austria after the First World War, when south Tyrol was ceded to Italy. However, in 1967, the Felbertauern Tunnel connected the east Tyrol with the rest of the Austrian Tyrol. Despite improved transport and communications, and despite the growth of winter sports, its remoteness has resulted in people leaving the region.

Population 11 700
Map Austria – Cb

Liepaja (Lepaya; Libau) *Latvia* Baltic port, naval base and industrial city, situated 195 km (120 miles) west of RIGA. Liepaja's harbour is ice free all year, and during winter it handles Riga's traffic, for the Latvian capital is on the icebound Gulf of Riga. Liepaja has shipbuilding industries; its products include iron and steel, agricultural machinery, leather and spirits.

Population 114 900
Map Latvia – Aa

Lietuva See LITHUANIA

Liffey (An Life) *Ireland* River which rises in the Wicklow Mountains and runs 80 km (49 miles) in a loop northwards to Dublin, passing under the city's 10 bridges and into the Irish Sea. It feeds the Lacken reservoir and hydroelectric scheme in the mountains.

Map Ireland – Cb

lightning Large-scale natural electrical discharge in the atmosphere in the form of a visible flash of light. The discharge, which results from a build-up of opposing electrical charges in two regions, may be between two clouds, between cloud and air, within a cloud, or from cloud to ground. The commonest forms of lightning are fork lightning, a zigzag type of lightning, often with several branches, and streak lightning, with fewer branches. Sheet lightning is a discharge which gives a cloud a white sheet appearance. Ball lightning is a rare cloud-to-ground discharge in which a luminous ball appears near the point of impact on the ground, moves about for a few seconds and then may explode or simply disappear.

Lightning Ridge *Australia* Opal field in New South Wales about 570 km (360 miles) north-west of Sydney. It produces some of the world's finest stones, including the rare black opal.

Population 1500
Map Australia – Hd

lignite (brown coal) Brown carbonaceous rock, intermediate between peat and the more familiar black coal. It is formed in the process by which vegetable matter is converted into coal.

Liguria *Italy* Region in the north-west covering 5415 km² (2091 sq miles) around the Gulf of Genoa. The coast is sheltered by mountains, and because of the mild climate and beautiful scenery many fashionable resorts have opened up since the 19th century – among them are Ventimiglia, BORDIGHERA, SAN REMO, ALASSIO, Albenga, RAPALLO and Santa Margherita. Industry, centred in between these on the ports of GENOA, Savona and LA SPEZIA, includes steel, shipbuilding and oil refining and creates a serious pollution problem.

Population 1 719 200
Map Italy – Bb

Ligurian Sea *Italy* Branch of the MEDITER-RANEAN named after the Italian province of LIGURIA, which borders it in the north. TUSCANY is to the east and CORSICA to the south. The Ligurian Sea, which reaches a depth of more than 2830 m (9285 ft), includes the Gulf of GENOA. Parts of the sea have been badly polluted.

Map Italy – Ac

Likasi *Zaire* Mining, industrial and commercial city in Shaba region in the south-east, 115 km (70 miles) north-west of the city of Lubumbashi. Formerly called Jadotville, it is situated on the Lubumbashi-Ilebo Railway. Its industries include copper and cobalt refining and metalwork.

Population 194 000
Map Zaire – Bc

Likoma *Malawi* Island of 18 km² (7 sq miles) in Lake MALAWI, just 10 km (6 miles) from the Mozambican shore. It was the site of a Church of England mission founded in 1885, and the island contains an impressive cathedral, built in 1911. The village of Likoma (population 7900) is a fishing port. Its twin isle, Chisumulu, is 10 km (6 miles) to the west.

Population 9160
Map Malawi – Ab

Li[lle Bælt *Denmark* See LITTLE BELT

Lille-Roubaix-Tourcoing *France* Conurbation near the Belgian border, 210 km (130 miles) north of PARIS. Lille produces textiles, chemicals, metal goods and food products. Roubaix and Tourcoing, north-west of Lille, produce woollen yarn, carpets and textiles. Many of the workforce commute daily across the international frontier.

Population (conurbation) 950 000
Map France – Ea

Lilongwe *Malawi* National capital since 1975, chosen to replace ZOMBA as capital because of its position in the centre of Malawi. It is now the country's second largest city after BLANTYRE-LIMBE, and is an attractive, well-planned city which is growing quickly.

Lilongwe is a trading centre, serving a fertile area which produces maize, peanuts and tobacco. It also serves as the centre of a 445 000 hectare (1 100 000 acre) rural development project launched in 1977 to improve crop production and provide social services.

Population 350 000
Map Malawi – Ab

Lima *Peru* National capital and capital of Lima department, standing on the Rímac River, 13 km (8 miles) east of the Pacific port of Callao. It was founded in 1535 by the Spanish conquistador Francisco Pizarro, and is now the economic and cultural centre of Peru. The University of San Marcos, founded here in 1551, is the continent's oldest university. Although Lima has been hit by numerous earthquakes, the old city still has several Spanish colonial buildings, most of which lie close to the Plaza de Armas, now dominated by modern skyscrapers. Historic buildings include the cathedral, the Torre Tagle Palace and

▼ **DUAL PURPOSE** An enterprising citizen of Lima combines a shoe-shine booth with a news-stand so that his customers can polish up their knowledge of news and sports.

the House of the Inquisition. Pizarro's home, too, was built here, on a site now occupied by the Palacio de Gobierno ('Government Palace').

Lima has spread to meet its port, Callao, and together they form a vast conurbation housing about 70 per cent of the country's manufacturing industry, including textiles, clothing, pharmaceuticals, and food processing.

Population (conurbation, with Callao) 6 404 000
Map Peru – Bb

Limassol *Cyprus* The island's main port and its second largest town after the capital, Nicosia. It is a major tourist centre and produces wine (said to be the island's best), brandy, cigarettes and perfume. Wine, fruit, vegetables, asbestos, copper and iron ore exports are handled at Limassol port. Since the Turkish occupation of northern Cyprus begaan in 1974, the town's economy has expanded considerably.

Richard the Lionheart, King of England, from 1189 to 1199, used Limassol as his base for the Third Crusade in 1191. In 1815 an earthquake killed all but 150 of the town's inhabitants; Limassol did not fully recover until the 1870s.

Population 132 100
Map Cyprus – Ab

Limbang *Borneo* River, some 150 km (95 miles) long, in northern Borneo. Its lower course and estuary, which is extremely swampy, form the enclave of Malaysian territory that divides the two wings of Brunei.

Map Brunei – Ab

Limburg (Limbourg) *Belgium* Flemish-speaking province lying between the city of Liège and the Dutch border, with an area of 2422 km² (935 sq miles). The capital is Hasselt. Its main industries – chemicals, non-ferrous metals, car assembly, electrical goods – are based on the province's Kempenland coalfield, but there is some forestry among the heathlands.

Limburger cheese was originally made in the town of Limbourg, which now lies in the province of Liège about 110 km (70 miles) east and slightly south of Brussels. The town was the seat of the former Duchy of Limbourg; the area was split up in 1839 to form the Belgian and Dutch provinces of the same name.

Population (province) 727 000
Map Belgium – Ba

Limburg *Netherlands* Hilly province which covers 2209 km² (853 sq miles) in the extreme south-east. It contained the country's only coalfield, until this was closed in 1975. The province produces ceramics, cement and chemicals; tourism is another important source of income. The capital is MAASTRICHT.

Population 1 119 900
Map Netherlands – Bb

Limerick (Luimneach) *Ireland* City and port on the south bank of the River Shannon, 87 km (54 miles) north of the southern city of Cork. Viking invaders founded it in 922, but it was later sacked by the medieval Irish king Brian Boru. Still later, the Anglo-Normans occupied the city and King John (1199-1216) built the castle here which still carries his name. Modern Limerick is mostly Georgian, and built on a grid system. It has a wide variety of agricultural industries, and is noted for its ham and bacon. Limerick is also

a commercial centre and university town. It is the main town of the surrounding county of the same name covering some 2686 km² (1037 sq miles), and includes the west of the Golden Vale, which produces livestock and dairy products.

Population (city) 52 020; (county) 162 960
Map Ireland – Bb

limestone SEDIMENTARY ROCK, consisting mainly of calcium carbonate, and found in the form of CALCITE, although DOLOMITE may also be present. It is formed from the skeletons of small marine organisms, or by deposition of calcium carbonate dissolved in water. It is used as building stone, for making lime and cement, and in smelting iron.

Limfjorden *Denmark* A broad channel across north JUTLAND, extending from the Kattegat in the east nearly all the way to the North Sea in the west. The 1 km (0.6 mile) long Thyborøn Canal links its western end with the North Sea. Denmark's largest cement plants are located on the channel's shores, and its waters yield a rich harvest of oysters.

Map Denmark – Ba

Limoges *France* Town on the Vienne River, 180 km (110 miles) north-east of BORDEAUX. It is renowned for both its enamel work and its fine porcelain. Limoges enamel was well established in the 12th and 13th centuries, and reached its peak in the 16th century, but it never properly recovered after being devastated in the late 16th-century Wars of Religion. The porcelain industry which began in the 18th century, continues to be an important employer. Other industries include tanning, shoe making, vehicle manufacture and aeronautics; uranium is mined nearby.

The town has a 13th to 16th-century cathedral, a museum for porcelain art works, and an art gallery containing works by Auguste Renoir (1841-1919), who was born in Limoges.

Population 133 500
Map France – Dd

limon Fine-grained deposit in France and Belgium from which brown, loamy soils have developed. It is of wind-blown origin, deposited under humid conditions, and possibly redeposited by running water. See also LOESS.

Limón *Costa Rica* Agricultural province which extends along the country's entire Caribbean coast. It grows bananas, cacao (used to make cocoa), abaca (a Manila hemp) and sugar cane. Its capital is the town of Limón, the country's main Caribbean port, built on the site of the Amerindian village of Cariari which the explorer Christopher Columbus visited in 1502.

Population (province) 219 490; (town) 67 780
Map Costa Rica – Ba

Limousin *France* Region of rocky plateaus in the north-west of the Massif Central. It rises to a height of 978 m (3208 ft) at the Millevaches ('the Thousand Cows') plateau. Limousin bulls – used by stock-breeders all over the world – are reared in the region. LIMOGES is the main city.

Population 723 000
Map France – Dd

limpo Type of SAVANNAH in Brazil, consisting mainly of tall, coarse grasses with relatively few bushes or trees.

Limpopo *Southern Africa* River, some 1610 km (1000 miles) long, also known as the Crocodile River in its upper reaches. The 'great grey-green, greasy Limpopo', as the writer Rudyard Kipling (1865-1936) described it, rises near JOHANNESBURG in South Africa and flows first north-west, then north-east to form the country's borders with Botswana and Zimbabwe. In its lower course, it flows sluggishly through Mozambique to the Indian Ocean, 85 km (53 miles) to the north-east of Maputo, Mozambique's capital. The river's length is navigable for 100 km (60 miles) into Mozambique up to a dam near the town of Guija.

Map Mozambique – Ac; South Africa Ca-Da

Lincoln *United Kingdom* City and county of eastern England. The city has a fine 11th-century castle and stands on a ridge above the Witham River, 190 km (120 miles) north of LONDON. Its cathedral, one of the largest in Britain, has been rebuilt several times since the 11th century because of fire and earthquake damage; a storm in the 16th century cost it its main spire. The modern city is an industrial centre specialising in farm machinery and agricultural supplies. The first military tank was built in Lincoln during the First World War.

The county – 5885 km² (2272 sq miles) in area and largely flat apart from The WOLDS – has rich farmland and a host of seaside resorts, including Skegness. Part of north Lincolnshire was incorporated in the new county of HUMBERSIDE in 1974.

Population (city) 84 800; (county) 591 000
Map United Kingdom – Ed

Lincoln *USA* Capital of Nebraska, lying in the south-east of the state, 75 km (47 miles) south-west of Omaha. It is an agricultural, marketing and servicing city, and a financial centre, as well as an educational centre, with two universities. Its most famous landmark is the Capitol, with a tower 120 m (390 ft) high.

Population (city) 192 000; (metropolitan area) 213 600
Map United States – Gb

Lindau *Germany* Bavarian tourist town centred on a little island on the north shore of Lake Constance near the Austrian border. It has a casino, boating facilities and opulent architecture. It also has textile, machinery, electronics and food processing industries. Lindau is the meeting place each year of the Nobel Prize laureates.

Population 24 100
Map Germany – Ce

Lindisfarne *United Kingdom* See NORTHUMBERLAND

line squall Band of extremely stormy weather bringing with it sudden strong, gusty winds along a line sometimes as much as 500 km (about 310 miles) in length. Line squalls are associated with COLD FRONTS and are characterised by dark rolls of cloud and downpours of rain and hail.

Line-of-Rail *Zambia* Densely populated strip of land which runs for 700 km (435 miles) along the railway linking the capital, Lusaka, with Maramba in the south and the northern copper-mining towns. Built at the beginning of the 20th century, the railway has drawn people to it leaving the remainder of the country sparsely populated.

Map Zambia – Bb

Liberia

IN A LAND OF FREED SLAVES RULED BY AN ELITIST SOCIETY, A MILITARY COUP SET OFF CHAOTIC FIGHTING AMONG RIVAL FACTIONS

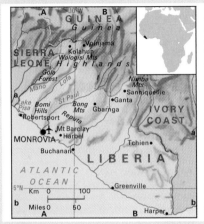

Africa's oldest independent republic and the only African country never ruled by a foreign power, Liberia was created by freed slaves returning from America from 1822 onwards. It was recognised as an independent state in 1847. With the help of philanthropic societies, among them the American Colonisation Society, more than 11 000 Africans were repatriated to Liberia by 1860. Blacks from America have continued to migrate intermittently, but the Americo-Liberians have remained a minority of the country's population.

From the start, life was difficult for them. The beautiful but treacherous coast, with rocky cliffs and lagoons enclosed by sand bars, offered few landing places. Even when ships found one, the new settlers had much to endure. Torrential rain, flies and stinging insects, and mysterious and deadly diseases took their toll. The bright red soils were not as fertile as they looked. Those who stuck it out formed the core of an elite who for long dominated the 30 or so other ethnic groups – made up mostly of the Kpelle, Bassa and Gio, and the Kru, Mano, Gola and Kissi peoples.

It was that elite, the descendants of the Americo-Liberians, that for many years formed strong governments which ensured political stability within a one-party framework. However, the Afro-Americans' treatment of the indigenous people left much to be desired. Their forced labour policies were strongly criticised in the West. So too was their extensive flag-of-convenience merchant navy which exercised inadequate control over its 1370 or so foreign-owned ships.

In 1980 a bloody military coup led by Master Sergeant Samuel Doe brought an end to the era of minority rule. Doe became president. Then, in late 1989, a revolt marked the beginnings of a protracted civil war between army factions and other rebel groups. The opposition was spearheaded by the National Patriotic Front of Liberia (NPFL) under Charles Taylor and the rival Independent National Patriotic Front led by Prince Johnson. In the ensuing chaos many thousands of Liberians fled to neighbouring countries.

In August 1990 a military peace-keeping force assembled by the Economic Community of West African States (ECOWAS) moved in. Soon afterwards Doe was murdered by Prince Johnson's supporters. The ECOWAS group sponsored a ceasefire agreement between the government forces and the two rebel factions and installed a provisional government headed by Dr Amos Sawyer. However, Charles Tayler and the former vice-president, Harry Moniba, simultaneously declared themselves president, and the civil war continued with new rebel groups emerging to fight the others.

Eventually, in July 1993, a peace agreement was signed and an ECOWAS Monitoring Group was deployed throughout the country. A transitional government formed in early 1994 prepared the way for elections in September 1994, but these were postponed after renewed fighting. A fresh peace agreement signed by the leaders of the three principal rival factions that month was greeted within days by outbreaks of fighting in various parts of the country, during which Taylor was ousted as leader of the NPFL and tens of thousands of refugees fled to Ivory Coast and Guinea.

Altogether the civil war was estimated in 1994 to have cost Liberia at least 150 000 lives, displaced more than a third of the population and ruined the economy.

To visitors in more peaceful times, Liberia's roots, once nurtured in America, were clearly evident, with buildings in early American styles, churches of many denominations and mostly American cars. In towns American food was popular, and shops, boutiques and supermarkets were well stocked.

The climate is warm throughout the year and has marked wet and dry seasons. Rainfall is high on the coast, decreasing substantially in areas farthest inland. Roughly 40 per cent of the country consists of tropical rain forest. The narrow coastal strip is a constructed mosaic of tropical rain forest and savannah. The Kru and Vai peoples who live here cleared areas of woodland to grow crops. Rainfall from April to October ensures that it is green all year. Inland, the land rises to a densely forested plateau, dissected by deep, narrow valleys. The beaches are beautiful and swimming is excellent in the lagoons, such as at Lake PISA. But in the inland rivers and lakes there is a risk of bilharzia – a disease carried by a tropical, parasitic worm.

Farther inland the scenery becomes more spectacular. There are magnificent waterfalls, such as St John's; the NIMBA MOUNTAINS, rising to 1752 m (5748 ft) which are rich in iron ore; and the 1380 m (4525 ft) Wologisi Range. Towards the border with Guinea, where the temperature range is greater and the humidity lower than on the coast, the forest is broken by large areas of savannah. Only about 4 per cent of Liberia is considered arable.

In the forest and animal reserves there is a rich variety of wildlife including the rare dwarf hippopotamus, several species of antelopes and anteaters, elephants, crocodiles, monkeys and chimpanzees.

From 1926, a short time after the Firestone Rubber Company of America was given a 99-year concession to plant rubber in return for revenue, Liberia's governments pursued an 'open door' policy, inviting foreign investment. Rubber plantations developed rapidly in the forest zone and as Firestone's interests grew, other multinational companies followed them in. Eventually rubber was replaced as the country's main export by iron ore.

However, the image of erstwhile prosperity was deceptive. Most Liberians remained poor, the government was in debt and the country's exports, such as rubber, iron ore, timber and diamonds, were sold mainly in their raw, unprocessed state, which limited their value. Apart from that, most industries were in the hands of foreigners. A thriving timber industry – there were 25 000 km² (9650 sq miles) of forest, the source of some valuable tropical hardwoods – was entirely owned by overseas logging companies.

Around 70 per cent of the 2.8 million population eke out a living from agriculture, mostly on poor land which is left over from foreign concessions and large state farms. Rice is grown on the flood plains of the many rivers. Cassava, yams, maize, rice and vegetables are cultivated under small stands of rubber, cocoa and banana trees. This diet is supplemented with fish, meat and fruit.

Money is earned by women selling surplus fruit and vegetables in markets and along the roadside. Liberia could, indeed, be self-sufficient in basic crops; but vast amounts of scarce foreign exchange has been spent on imports of food and consumer goods to satisfy the tastes of townspeople, many of whom were foreign workers.

LIBERIA AT A GLANCE	
Official name Republic of Liberia	
Area 111 370 km² (43 000 sq miles)	
Population 3 000 000 **Per km²** 27	
(Per sq mile 70)	
Capital Monrovia	
Government Republic	
Currency 1 Liberian dollar = 100 cents	
Languages English (official), various African languages	
Religions Indigenous (70%), Christian, Muslim	
Climate Tropical, with high summer rainfalls; average temperature in Monrovia ranges from 24°C (75°F) to 27°C (81° F)	
Land use Forest and woodlands 22%, cultivation 4%, grazing and other 74%	
Main primary products Rice, cassava, bananas, rubber, palm kernels, maize, coffee, cocoa, timber; iron ore, gold, diamonds	
Major industries Mining, forestry, agriculture	
Main exports Iron ore, rubber, hardwoods, coffee, cocoa, diamonds	
Annual income per head (US$) 450	
Population growth (per thous/yr) 30	
Life expectancy (yrs) Male 55 **Female** 60	

Lingayen *Philippines* Trading port and capital of Pangasinan Province, standing on Lingayen Gulf at the mouth of the Agno River in LUZON. The centre of a rich farming area, it was first visited by Chinese, Japanese and other barter traders during the 14th century. Spanish and, some time later, American colonists developed the town. Lingayen was badly damaged during the Second World War when it was the first landing point on Luzon by Japanese forces in 1941, and by American forces in 1945.

Population 56 300

Map Philippines – Bb

Linköping *Sweden* Industrial city some 198 km (123 miles) south-west of STOCKHOLM. It was an important religious centre in the Middle Ages and has a magnificent 12th to 15th-century cathedral. Today, it manufactures cars, aircraft and rolling railway stock.

Population 126 400

Map Sweden – Bd

Linz *Austria* The country's third largest city and capital of Upper Austria, 160 km (100 miles) west of VIENNA. Straddling the Danube, Linz has been an important river port and trading centre since the Middle Ages. In 1939 Hermann Göring (Goering), the head of Nazi Germany's *Luftwaffe* (airforce), established a major steel works in the city; as a result it was heavily bombed during the Second World War. Besides iron and steel, Linz manufactures chemicals, textiles, tobacco and fertilisers. The city has one of the oldest Austrian churches, the 8th-century Martinskirche, a castle, and numerous other historical buildings.

Population 202 860

Map Austria – Da

Lion, Golfe du (Gulf of Lions) *France* Section of the MEDITERRANEAN Sea washing a large bay that stretches from the Franco-Spanish border to the French port of TOULON. The MISTRAL, a cold, northerly wind which is funnelled through the Rhône Valley and sweeps across southern France to the Mediterranean, reaches a great speed over the gulf, sometimes making boating hazardous. The French city of MARSEILLES is the chief port on the gulf.

Map France – Ee

Lipari *Italy* Principal island of the EOLIAN group, covering 38 km² (15 sq miles) off the north-east coast of Sicily. About half its inhabitants live in Lipari Town, and the rest in four other villages and on many scattered farms. The islanders' main occupations are fishing, agriculture (chiefly viniculture), and tourism. Lipari Island has 12 extinct volcanoes, which were the source of obsidian (black volcanic glass), mined here in the Stone Age and traded throughout the Mediterranean region.

Population 10 700

Map Italy – Ee

Lisbon (Lisboa) *Portugal* Capital, port and the country's largest city standing on the north bank of the Tagus River, about 15 km (9 miles) from the sea. It was from Lisbon that the Portuguese navigator Vasco da Gama sailed for India in 1497.

According to legend, the city was founded by the Homeric hero Ulysses (also known as Odysseus). Most modern authorities, however, attribute its foundation to the Phoenicians around

1200 BC. It was conquered in turn by a variety of peoples, including the Greeks, Carthaginians, Romans and Visigoths, before the Moors overran the region in AD 714.

In 1147, Alfonso I, the first King of Portugal, captured the city for the Portuguese, and in 1255 Alfonso III made it his capital in place of the city of Coimbra. In the 14th and 15th centuries, Lisbon was one of the most magnificent cities in Europe – with much of the wealth of the New World passing through it. However, on All Saints' Day in 1755 large parts of the city were destroyed by an earthquake and associated tidal wave (tsunami), followed by fire. More than 30 000 people died in the disaster. The city was rebuilt by the Marquis de Pombal, and the 18th-century town plan is still evident today. A fire again destroyed large parts of Lisbon in 1988.

One of the best views of Lisbon is from the renamed Ponte 25 de Abril, the city's first bridge across the Tagus; some 2278 m (7474 ft) long, the bridge was completed in 1966. The modern centre of the city is Praça Dom Pedro IV, known as 'The Black Horse Square' because of the equestrian statue standing in it.

To the east lies the Alfama, or medieval town, with its 12th-century cathedral and narrow streets. Above these is the Castelo de São Jorge, fortified by the Visigoths in the 5th century and modified first by the Moors, then by Alfonso I. Among other places of interest are the Jeronimos Monastery, the Tower of Belem (with carved balconies and domed turrets), the Museum of Ancient Art, and the Monument to the Discoveries which commemorates the great Portuguese explorers.

The city's main industries produce textiles, soap, flour, and steel, refine sugar and oil, and build ships.

Population (city) 807 900;
(metropolitan area) 2 500 000

Map Portugal – Bc

Lisburn *United Kingdom* Market town in Northern Ireland, 10 km (6 miles) south-west of the capital, BELFAST. French Huguenots settled here and organised the fledgling linen industry into what is now a centre of the country's linen manufacture. Christ Church Cathedral, in the market square, dates from the 17th century.

Population 40 000

Map United Kingdom – Bc

Lissa *Poland* See LESZNO

Litani *Lebanon* River in southern Lebanon which rises near BAALBEK and flows south between the Lebanon Mountains and the Anti-Lebanon Mountains. Near the town of Marjayoun, it turns sharply west, plunging through a deep gorge in the Lebanon Mountains, and enters the Mediterranean Sea about 10 km (6 miles) north of the port of Tyre.

Map Lebanon – Ab

lithification Process by which loose sediment is consolidated into hard rock.

lithosphere Solid outer layer of the earth lying above the semi-fluid asthenosphere. It includes the earth's crust and upper mantle. See JOURNEY TO THE CENTRE OF THE EARTH, p 207

Lithuania See p 407

Little Belt (Lille Bælt) *Denmark* The channel between the Danish peninsula of JUTLAND and the FÜNEN island group. It is about 130 km (80 miles) long, and between 30 km (19 miles) and 600 m (nearly 2000 ft) wide. The channel was bridged for the first time in 1935.

Map Denmark – Bb

Little Rock *USA* Capital and largest city of Arkansas, on the Arkansas River in the centre of the state. It was founded in 1820 and is now a financial and industrial centre producing textiles, electronics and processed foods. The largest bauxite reserves of the United States lie nearby. The city achieved worldwide notoriety in 1957 when President Dwight David Eisenhower sent in federal troops to enforce a supreme court ruling outlawing racial segregation in schools.

Population (city) 175 800;
(metropolitan area) 513 100

Map United States – Hd

littoral current See LONGSHORE CURRENT

Litva See LITHUANIA

Liverpool *United Kingdom* Port city of north-west England situated on the north side of the Mersey estuary, formerly in Lancashire but now the main centre of the county of Merseyside. It flourished in the late 17th and 18th centuries on trade with the New World and Africa, first in slaves and sugar, then in cotton.

In the 19th century, it became the main port for emigrants to the United States, not only from Britain but also from Scandinavia and Germany. Its docks and quays stretched along the Mersey for 12 km (7 miles) while, on the south shore, the town of Birkenhead grew to provide further docks and, after 1824, shipbuilding yards. Liverpool and Birkenhead were linked by a 4.8 km (3 mile) long road tunnel in 1934; a second one was opened in 1971.

However, the city has declined seriously since the Second World War. Although it possesses two large cathedrals, many fine civic buildings and a university, the transatlantic passenger traffic now goes by air, cotton is no longer king, unemployment is high, and the docks are all but empty and have been partly reclaimed for a park. The Albert Dock area has been transformed and is now the site of a maritime museum and Tate Gallery. For many, the city's greatest claim to fame is that the Beatles pop group – Paul McCartney, John Lennon, George Harrison and Ringo Starr – were all born here. But Liverpool is also the home of two major orchestras and is regarded as 'the football Mecca' of Britain.

Population 480 700

Map United Kingdom – Dd

Livingstone *Zambia* See MARAMBA

Livingstone Falls *Zaire* See INGA RAPIDS

Livingstonia *Malawi* Town 400 km (250 miles) north of the capital, LILONGWE, and named after the explorer David Livingstone. It has splendid views of Lake Malawi, which Livingstone reached in 1859. The town was initially founded as a mission in 1875 and stands about 1370 m (4500 ft) above sea level. It has a museum and craft shops. Coal has been found nearby.

Map Malawi – Ab

Libya

AN ANCIENT DESERT COUNTRY WHICH WAS CATAPULTED INTO THE 20TH CENTURY BY OIL MONEY AND A CONTROVERSIAL COLONEL

The Socialist People's Libyan Arab Jamahiriyah and its leader, Colonel Moammar Gaddafi, have often played a provocative role on the world stage since 1969 – and have usually been cast as villains by the nations of the West.

Libya was one of the militant members of the Organisation of Petroleum Exporting Countries (OPEC) which pushed through big price rises that shocked Western economies in the 1970s. Libyan money has been used to support guerrilla movements around the world. The country has been criticised as a training ground for guerrilla warfare and terrorism and blamed for a number of outrages. It has in the past sent arms and money to help Muslims and leftist militias in Lebanon, to Eritreans fighting the Ethiopian government, to anti-American forces in Nicaragua and to the Irish Republican Army fighting the British in Northern Ireland.

American anger against Gaddafi and his support of guerrilla groups reached a climax when United States bombers attacked 'terrorist-related' targets in Libya on 15 April 1986. Many countries expressed strong disapproval of the attack. The Lockerbie air disaster of 1988, which Libyan terrorists were suspected to have caused, was another outrage, which led to United Nations sanctions of Libya in the early 1990s.

Gaddafi's power at home and abroad rests on oil. Libya was one of the world's poorest countries until oil was discovered in 1958. Today an annual 70 million tonnes of oil is produced, most of it being exported to the European Union (EU) countries.

Crude and refined oil accounts for 99 per cent of exports, giving Libyans the highest income per head on the African continent. At the present rate of drilling, oil reserves will last until about 2045 but already the government has begun preparing for post-oil years, for instance by investing worldwide.

Gaddafi, who led an army coup to end the monarchy in 1969, has been seen by many of his people as a folk hero and by others as a dangerous tyrant. Because of assassination attempts, he has made his home in a fortified barracks outside TRIPOLI, the capital. While he has used oil revenues to improve life for Libyans in his Socialist Cultural Revolution, he has also financed political, military and guerrilla forays worldwide.

NATION OF YOUNG PEOPLE

Inside Libya, the revolution has produced a decentralised form of democracy with many powers devolved to people's congresses in each community. This gives representation to everyone in a diverse society of townspeople, farmers and nomads – 'Jamahiriya' means 'the state of the masses'. But despite a system of nationwide representation, Libya remains in effect a dictatorship. Nevertheless, there is increasing political awareness, especially among young people, who predominate in numbers. More than 65 per cent of the population is under 24. Thanks to a high birth rate and improved medical care, the population has been increasing at a rate of 3.7 per cent a year.

This young land of young people has a very ancient history. There were ancient Egyptian and Greek settlements on the north-east coast; the Phoenicians and then the Carthaginians settled in the north-west. Then came the Romans, who adapted some of the Greek buildings. The finest Roman ruins in Libya, and perhaps in Africa – walls, baths, arches, temples and forums – are on the coast at LEPTIS MAGNA. The Romans were followed by the Vandals and the Byzantines, before the Arab invasions and migrations began in the 7th century AD. The Arabs brought Islam and converted the native Berbers. In the mid-16th century the Turks brought the area under Ottoman rule. The final colonisation, by the Italians, began in 1911, but it took 30 years to suppress fierce opposition by the desert tribes. By 1939 Libya was ruled by Italy as a province.

During the Second World War the British Eighth Army fought the Italians and Germans over the coastal region. TOBRUK, in the north-east, was a major battlefield. After the War Britain and France occupied the country until the United Kingdom of Libya was established under King Idris in 1951.

The Sahara covers much of Libya. Apart from desert oases, the only green areas are the Mediterranean scrublands of the north-west and the forested hills of the north-east. Between these two, where 90 per cent of the people live, the desert extends as far as the shore of the Mediterranean. It extends southwards for 800 km (500 miles).

The interior of Libya has had some of the highest recorded temperatures of anywhere in the world, ranging up to 50°C (122°F) in the shade, although the desert nights can often be bitterly cold. The north-east and north-west enjoy a Mediterranean climate of hot, dry summers and cool, moist winters. The highlands in these areas receive much of the rain and provide the country's most fertile soils.

About 20 per cent of the people work on the land, although agriculture contributes no more than 5 per cent to the gross domestic product (GDP). The main agricultural region lies in the triangle formed by the towns of Tripoli, HOMS and Gharyan, in the north-west. The main crops are barley, dates, fruit, olives, groundnuts and wheat. There are about 5.5 million sheep in Libya, as well as goats and cattle. Leather, skins and hides are exported.

Most of the Libyan people are Arabs, although there still remain distinct North African native Berber groups and Berber-speaking villages. Among the minority groups are the Tuaregs, a nomadic Berber people who herd their animals in the desert.

GADDAFI'S DREAM

Gaddafi has pursued a dream of ending poverty in Libya. Huge irrigation projects are creating fertile land from the desert, with wells tapping underground reservoirs. A total of more than 20 000 hectares (49 420 acres) are under irrigation. There is also a huge scheme to build 'the great man-made river', a system of pipelines to transport water from these reservoirs to the towns and farms of the coast. Phase 1 of the project was completed in 1991. The country has rapidly been industrialised, with new factories producing petrochemicals, steel and cement.

Inevitably the rapid changes have brought problems. The growth of the oil and other industries prompted a drift to the towns by people seeking work, which caused over-

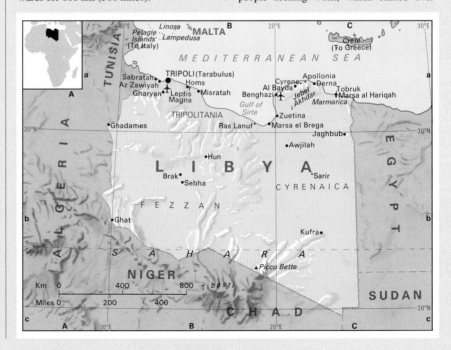

▲ **TWENTY-FIVE YEARS IN POWER**
Colonel Moammar Gaddafi leads a
military parade through the streets of
Tripoli in September 1994.

crowding and a shortage of housing. But social service programmes, including free education and free medical services, were launched.

The housing programme has been very important to Gaddafi. Born of a nomadic Arab family, he grew up in tents in the desert. When he proclaimed himself the Leader of the Revolution, he vowed that he would not move his mother out of her tent and into a house until every Libyan had a home. No one quarrels with housing for the poor, but parts of Libya's revolution have been more controversial.

FINDING AN IDENTITY

Libyans suffered centuries of foreign rule, which has prevented the nation from finding its own identity. Upon independence, Gaddafi abolished savings accounts and most of the legal system. He nationalised urban property, foreign trade and the retail trade. He also nationalised all foreign oil companies.

He chose the Muslim religion as the national religion, although he is perhaps not a strict fundamentalist. At first, his policies were contrary to Muslim traditions. For instance, there were indications that he believed in the liberation of women. But from the late 1980s there was a

rise of Islamic fundamentalism in Libya. Hundreds of people were held in detention and gross violations of human rights were reported. In 1993 Gaddafi announced the introduction of Islamic laws (*sharia)* which relate to forms of punishment, for instance for thieves (who get their hands severed), adulterers (who are given 100 strokes of the whip) and adulteresses (who are stoned to death). In 1994 he announced that the application of *sharia* would be extended.

Arab nationalism, anti-imperialism and anti-Israeli feeling have been the dominant themes in Gaddafi's foreign policy. Over the years he has made repeated bids to unite his state with other Arab countries, but so far his attempts to found a Federation of Arabic Republics (1971), to unite with Egypt (1973) and Tunisia (1974), or to lay territorial claims to Chad (1980) have been unsuccessful.

In 1992 the United Nations imposed sanctions on Libya after the government refused to extradite two Libyans believed to have been involved in the Lockerbie air bombing of 1988. Thus the country became increasingly isolated and its oil revenues declined. Gaddafi, too, seemed less in control than formerly. In October 1993 there was an attempted coup, linked to the Lockerbie affair. So Gaddafi's days of being the national hero seem to have passed. Despite privatisation and economic reform, economic growth has slowed and the country, which has to import 75 per cent of its

food needs, has learnt the feel of international isolation, with shortages affecting many areas of the economy.

LIBYA AT A GLANCE	
Official name Socialist People's Libyan Arab Jamahiriya	
Area 1 759 540 km² (679 358 sq miles)	
Population 4 872 600 **Per km²** 2.8 **(Per sq mile** 7)	
Capital Tripoli	
Government Islamic socialist republic	
Currency 1 dinar = 1000 dirhams	
Language Arabic	
Religions Muslim (99%), Christian (1%)	
Climate Hot and dry in the south; Mediterranean on the coast. Average temperatures in Tripoli: 12° C (53° F) in January; 26° C (79° F) in August	
Land use Grazing 8%, cultivation 2%, other (desert) 90%	
Main primary products Wheat, barley, olives, dates, citrus fruits, groundnuts, tobacco; oil and natural gas	
Major industries Oil and gas production and refining, agriculture, food processing, textiles, handicrafts, cement, fishing	
Main exports Oil, natural gas and petroleum products; olive oil, leather, skins and hides	
Annual income per head (US$) 8300	
Population growth (per thous/year) 37	
Life expectancy (yrs) Male 61 **Female** 65	

Liechtenstein

THE PRINCIPALITY THAT COUNTED ITS TROOPS HOME FROM THE WAR AND FOUND IT HAD GAINED AN EXTRA MAN

Among the microstates of Europe, Liechtenstein is still most people's favourite. Andorra is too remote, and Monaco too expensive. Liechtenstein is just right – a Swiss motorway runs within a stone's throw of the border, while the main railway from Zurich to Austria crosses it.

The Principality of Liechtenstein, whose history dates back to 1342, lies on the east bank of the Upper RHINE River. It uses Swiss currency and belongs to the Swiss customs union but is otherwise independent. The population speaks German as well as local dialects. There is a one-chamber parliament in VADUZ, the capital.

Many foreign companies have their head-quarters in Liechtenstein so that they can profit by the country's lenient tax laws. This is good for the state's revenue, as is the trade in its decorative postage stamps. Nearly 80 000 visitors a year also mean big business. Beautiful scenery and fine skiing are tourist attractions.

Its territory is roughly 25 km (16 miles) from north to south, with an average width of just over 6 km (3.7 miles). Agriculture continues, but the farming population has been reduced from 40 per cent in 1930 to 1.7 per cent in 1991 – a period during which Liechtenstein moved into industry, producing quality metal goods, machinery, processed food, pharmaceuticals and ceramics; 45 per cent of its exports go to countries of the European Union. Unemployment is low (about 1.2 per cent in 1994), a factor contributing to the large proportion of foreigners (38 per cent of the population) living and working here.

The number of immigrants began building up only in the 1920s. But one of the earliest recorded population gains in recent times was made in 1866, when Liechtenstein sent men to fight for Austria against Italy in the Tyrol – the contingent is said to have returned with one man more than when it set out.

LIECHTENSTEIN AT A GLANCE

Map Switzerland – Ba	
Official name Principality of Liechtenstein	
Area 160 km² (61 sq miles)	
Population 29 790 **Per km²** 186 **(Per sq mile** 488)	
Capital Vaduz	
Government Parliamentary monarchy	
Currency 1 Swiss franc = 100 centimes	
Languages German, Alemannic dialects	
Religion Christian (81% Roman Catholic)	
Climate Temperate. Average temperature in Vaduz ranges from 0.6°C (33°F) in January to 18.7°C (66°F) in July	
Land use Pastures and meadows 16%, cultivation 24%, forest 35%	
Main primary products Dairy cattle	
Major industries Ceramics, metalworking, electrical instruments, clothing, processed foods, pharmaceuticals, machinery, business services, tourism	
Main exports Machinery, metal and other industrial manufactures, chemicals, processed foods, postage stamps	
Annual income per head (US$) 30 270	
Population growth (per thous/yr) 12	
Life expectancy (yrs) Male 74 **Female** 81	

Livorno *Italy* See LEGHORN

Liw *Poland* One of eastern Poland's oldest settlements, 83 km (52 miles) east of the capital, WARSAW. Rising from poppy fields, its massive, 14th-century brick fortifications, now a museum, are the remnants of the third stronghold to be built on the site.
Map Poland – Eb

Lizard Point *United Kingdom* Most southerly point on the British mainland, jutting into the English Channel from the county of Cornwall. The cliffs soar to 55 m (180 ft) on its rugged coast, which is corrugated with coves and harbours. A lighthouse overlooking the treacherous waters around the headland was built here in 1619. Goonhilly Downs, located inland, is a space satellite communications station.
Map United Kingdom – Cf

Ljubljana (Lyublyana; Laibach) *Slovenia* The capital of Slovenia, lying on the Ljubljanica, a tributary of the Sava River. It is also a road and rail junction, a tourist centre and an industrial city producing machinery, textiles, porcelain, paper, furniture, shoes, chemicals and leather goods. Ljubljana grew around the 12th-century hilltop Grad fortress, which commands fine views of the surrounding basin and the Karawanke and Julian Alps to the north and west respectively. The town has many beautiful Baroque buildings, such as the 18th-century Ursuline Church and the 17th-century St Francis Church.
Population 323 300
Map Slovenia – Ba

Llangollen *United Kingdom* Small town on the Dee River in the Welsh county of Clwyd, 50 km (30 miles) south of Liverpool. It has become renowned since 1947 for its annual summer eisteddfod: an international festival of folk music, dancing and poetry that attracts thousands of performers from many lands.
Population 2630
Map United Kingdom – Dd

llanos Term originally meaning 'plains', now applied chiefly to the tropical grassland or savannah of the Orinoco Basin and the Guiana Highlands in South America. In particular, it is applied to the Llanos of central Venezuela, lying between the Cordillera de Mérida range and the Orinoco River. This is a rich grazing area, watered by the Orinoco and its tributaries.
Map (Llamos) Venezuela – Ab-Bb

Llanquihue *Chile* Largest lake in the country, covering 878 km² (339 sq miles). It is set in the southern region of Los Lagos in a spectacular setting of snow-capped mountains, including the 2660 m (8727 ft) high Osorno volcano. It is now a popular fishing and water sports resort. On its western shore, near Puerto Varas, is one of South America's oldest archaeological sites.
Map Chile – Ad

loam Soil containing sand, silt and clay in roughly equal proportions.

Loanda *Angola* See LUANDA

Lobamba *Swaziland* Town lying some 16 km (10 miles) south-east of MBABANE. It is the site of the Swazi parliament and the seat of the queen mother, called the *Ndlovukazi* (Swazi for 'the Great She-Elephant'). Lobamba is famed for two ceremonies, both of which take place over several days. The reed dance, which is held in August-September, honours the queen mother, while the *incwala* kingship ceremony held in December-January honours the king. These ceremonies include spectacular dancing in traditional costumes, singing and feasting. Visitors are welcome for most of the time, but they must leave during certain parts of the ceremony. On the sixth day of the kingship ceremony, warriors build a fire with wood collected from the hills. Dancing and singing around the fire, they call on the spirits of their ancestors to send rain to douse the fire as a sign of favour for the coming year.
Population 5800
Map Swaziland – Aa

Lobito *Angola* Port 390 km (240 miles) south of the capital, LUANDA. Founded in 1834, it grew rapidly after the completion of the Benguela Railway in the 1920s. In normal times it handles much of the country's agricultural exports, and minerals from Zaire and Zambia.
Population 59 000
Map Angola – Ab

Lobombo Mountains *Southern Africa* See LEBOMBO MOUNTAINS

Locarno *Switzerland* Popular resort in the Italian-speaking part of Switzerland, lying at the northern end of Lake Maggiore. The town's cultural highlight is the annual international film festival on the Piazza Grande in August. The pilgrimage church of the Madonna del Sasso, founded in 1480 and enlarged in 1616, stands above the town.
Population 14 100
Map Switzerland – Ba

loch In Scotland, a lake or a long narrow arm of the sea.

Lod (Lydda) *Israel* Town with an international airport, about 15 km (10 miles) south-east of TEL AVIV. Its factories produce aircraft parts as well as electronic equipment. Lod was the site of a biblical miracle; here, St Peter told Aeneas to get up and walk as Jesus had cured him of paralysis. In the 12th century the city was destroyed by the Muslim leader Saladin, but it was rebuilt soon afterwards under King Richard I (Richard the Lionheart) during the Third Crusade.
Population 41 400
Map Israel – Ab

lode Mineral deposit that is contained in hard rock, usually occurring in a VEIN. In East Anglia, an area of England, 'lode' is the term given to an artificial watercourse.

Lithuania

*THE SOUTHERNMOST OF THE THREE
BALTIC REPUBLICS IS ONLY A TINY
SECTION OF A LITHUANIAN STATE
WHICH ONCE EXTENDED AS FAR
AS THE BLACK SEA*

U nlike its Baltic neighbours, Latvia and Estonia, which have almost exclusively been ruled by foreigners, the land along the Neman (Memel) River was once a great European power. The Lithuanians who lived along the banks of the middle and upper Neman were able to withstand the attacks of the knights of the German Order in the Middle Ages. In the mid-13th century, their various tribes united under King Mindaugas and went on to conquer large parts of Eastern Europe. By the 14th century the Grand Duchy of Lithuania extended all the way through Russia and the Ukraine to the Black Sea.

A new epoch in Lithuanian history began in 1386 with the marriage of Jagiello, the Lithuanian Grand Duke, to the successor to the Polish throne, Jadviga: with the two countries united, they counted among the largest European powers.

At the end of the 18th century, however, the country was divided between Russia, Prussia and Austria. The Russian tsars did everything in their power to merge the Baltic country in their realm. They suppressed all Lithuanian culture, persecuted any visible followers of the Lithuanian nationalist movement and drove thousands of Lithuanians from their country in the 19th century. As a result there are about twice as many Lithuanian descendents living in the United States today as there are living in Lithuania.

The October Revolution of 1917 gave the Lithuanian people the opportunity to break free from their Russian rulers. They founded their own independent state in February 1918. But soon they found themselves facing one crisis after another. Polish troops occupied the area around the present capital of VILNIUS; in response, Lithuanians annexed the Memel region. (The border squabbles with Poland were ended by treaty as late as 1992.) Lands were expropriated and the democratic government made way for a dictatorship. In August 1940 the Red Army invaded the country. Lithuania eventually gained independence from the Soviet Union in September 1991.

Lithuania's economy is based mainly on agriculture, which employs about 20 per cent of its workforce, producing meat and dairy products, grain, vegetables, potatoes and sugar beet. Its coast is sandy and prone to silting up. The only large port is KLAIPEDA (Memel), which is ice-free throughout the year. Vilnius is a railway and road intersection, providing land communication with Eastern Europe.

NUCLEAR POWER CONCERN

The country has few mineral resources, besides some deposits of amber, which have been largely depleted, and mineral water. There are large quantities of peat – a cheap but highly polluting fuel suitable for thermal power stations. Electricity is exported from a large nuclear power station (Ignalina) in the country's east – the same type of power station as at Chernobyl' in Ukraine which was responsible for a nuclear accident in 1986. The Lithuanians are concerned about the environmental implications if another Chernobyl'-type accident should occur, but cannot afford to shut down the power station. The country also has the only oil refinery of the Balkan states (at Mažeikiai), which supplies oil to the other former Soviet republics.

Industrialisation came late, although some factories went up during Soviet rule in Lithuania's four major cities – Vilnius, KAUNAS, Klaipeda and Šiauliai. Today about 25 per cent of the country's workforce is employed in industry, processing food, manufacturing textiles, and producing metal-cutting machine tools, refrigerators, electric motors, and television sets. Other industries include shipbuilding and furniture making.

▼ABANDONED BARRICADES
Children play on the barricades which were erected by thousands of Lithuanians to defend the Parliament buildings in Vilnius in January 1991.

Although Lithuania still lags behind Latvia and Estonia in industrial development – and economically – one of the advantages of late development has been that the country has not inherited large numbers of industrial workers from Russia. Unlike Latvia and Estonia, where local people feel they have become minorities in their own country, Lithuanians make up almost 80 per cent of the population, forming a strong majority. As a result, Lithuania has more lenient immigration laws and there is almost total absence of ethnic tension.

The country's economy is dependent on its retaining good relations with Russia and the other Commonwealth of Independent States (CIS) members. It imports most of its raw materials from the former Soviet republics. It also needs the markets in the East, although it is looking to the West for new markets and investment. Despite its dependence on its former partners, the country ultimately aims to become an independent privatised economy.

LITHUANIA AT A GLANCE	
Official name Republic of Lithuania	
Area 65 200 km² (25 168 sq miles)	
Population 3 819 640 **Per km²** 58.6	
(Per sq mile 151.8).	
Capital Vilnius	
Government Republic	
Currency 1 litas = 100 centas	
Language Lithuanian (official); Polish, Russian	
Religion Christian (mainly Roman Catholic; some Lutheran and Russian Orthodox)	
Climate In the west, moderate maritime; towards the east, increasingly continental. Average temperature in Vilnius ranges from –5°C (23°F) in January to 18°C (64°F) in July	
Land use Cultivation 37%, forest 25%, meadows and pastures 17%, other 21%	
Main primary products Peat, amber, cattle, dairy products, vegetables, sugar beet	
Major industries Electronics, machinery, oil refining, shipbuilding, textiles, food processing	
Main exports Electric motors, machinery, textiles (especially linen), foods	
Annual income per head (US$) 5880	
Population growth (per thous/yr) 8	
Life expectancy (yrs) Male 66 **Female** 76	

lodestone (loadstone) Strongly magnetised, pure form of MAGNETITE.

Lodz (Lódź) *Poland* The country's second largest city after the capital, WARSAW. Founded in 1423, it is situated in central Poland, 121 km (75 miles) south-west of Warsaw.

In the 19th century occupying Russians turned the town into a polluted agglomeration of spinning, weaving, dyeing and clothing industries, with unsanitary housing, especially in the Jewish ghetto, which was razed by the Nazis in 1944.

Today Lodz's modern factories manufacture textiles, clothing, chemicals, electrical, engineering and photographic goods. The town has a university as well as several technical colleges.

Population 848 300
Map Poland – Cc

Loei *Thailand* Province and town near the Laotian border, 450 km (280 miles) north of the capital, BANGKOK. There are substantial lignite and iron ore reserves. Cotton is the province's main crop.

Population (province) 507 700; (town) 102 100
Map Thailand – Bb

loess (löss) Deposit of fine-grained, friable, porous siltlike dust, yellowish to grey in colour, generally deposited by the wind after being winnowed from deserts, or left by the retreat of the last ice sheets. LIMON in France and Belgium and brick-earth in Britain are substantially the same material, possibly redeposited to some extent by water. Loess covers about 10 per cent of the Earth's land surface.

Loesslands of China Region of northern China, forming a great plateau stretching through the provinces of SHANXI, SHAANXI, and eastern GANSU and on, through the NINGXIA-HUI AUTONOMOUS REGION. Its yellowish LOESS soil, derived from the windblown sands of the Gobi Desert, is fertile but erodes easily. Most of the eroded soil is carried by rain into the HUANG HE, giving the river its name, which means 'yellow river'. Drought is a perennial threat to crop production throughout the region. In parts of Shanxi and Shaanxi, the local farmers used to live in caves dug into the soft loess earth beneath their fields.
Map China – Fd

Lofoi Falls *Zaire* Waterfalls on the Lofoi River, about 90 km (55 miles) north of the city of LUBUMBASHI. It has a sheer drop of 340 m (1113 ft). The Lofoi Falls are now protected in the Kundelungu National Park.
Map Zaire – Bc

Lofoten *Norway* Scenic Arctic island group off the north-west coast of Norway. The group consists of four main islands (Austvågøya, Vestvågøya, Moskenesøya and Ilakstadøya) and several smaller islands and rocky islets. Covering 1227 km² (474 sq miles), the Lofoten stretch 190 km (118 miles) along the north side of the Vestfjorden Channel. Their cod fishing grounds are among the richest in the world. Every year hundreds of fishing boats converge on the islands at the height of the cod season from January to April, so much so that catches have declined alarmingly due to overfishing. The main ports for fish processing are Svolvær and Henningsvær.

The Lofoten Islands are notorious for the powerful tidal current which sweeps through the strait between the southern tip of Moskenes – the most southerly island in the group – and a nearby islet. Imaginative descriptions of the current's effects by writers such as Edgar Allan Poe (1809-49) and Jules Verne (1828-1905) have made its name – 'Moskenesstraumen' (Moskenes Current) or, more familiarly, the 'Maelstrom' – almost synonymous with a malevolent whirlpool capable of swallowing entire ships. Usually, however, even small boats can cross it safely, although some combinations of wind and tide can make the Maelstrom a terrifying hazard to navigate.
Population 26 200
Map Norway – Da

Logan, Mount *Canada* The country's highest peak – and the second highest, after Mount McKinley, in all North America – at 5959 m (19 550 ft). It constitutes part of the Saint Elias Mountains in south-west Yukon.
Map Canada – Ab

Logroño *Spain* Walled town 250 km (149 miles) north of Madrid, on the Ebro River. It was taken from the Moors by Alfonso VI in 1095. During the Middle Ages it was an important station on the pilgrims' route to Santiago. Today, it is the

▼ **FAIRYTALE CASTLE** The medieval Château de la Roche towers over the Loire River at Roanne. Most later chateaux of the region were designed more for beauty than for defence.

capital of La RIOJA Province. The town's main products are textiles, fruit preserves and the equipment for winemaking.

Population 122 250

Map Spain – Da

Loire *France* Longest river in France, covering 1020 km (635 miles). It rises in the southern MASSIF CENTRAL and flows north through gorges to Orléans, before sweeping westward through a wide valley to the Bay of Biscay west of Nantes. The middle valley – known as 'the garden of France' – is lined by many elegant châteaux, including those located at Amboise, Blois and Chambord. Several nuclear power stations are sited along the middle stretch of the river. Grapes, other fruit and vegetables are grown here, and mostly white wines such as Muscadet, Saumur, Pouilly and Vouvray are produced.

Loire is also the name of a department in the Rhône-Alpes region drained by the Loire River, used principally for agriculture and livestock. Industries in Saint Etienne-Roanne are currently being diversified.

Population (department) 746 000

Map France – Cc, Fd

Loire, Pays de la *France* Region of western France, covering 32 083 km² (12 387 sq miles). It is centred on the city of NANTES. The main activities are agriculture and, in the cities, food processing, electronics, and transport.

Population 3 060 000

Map France – Cc

Loja *Ecuador* Capital of the most southerly province of the same name. It stands on the Pan-American Highway and the Zamora River, 200 km (125 miles) south-east of GUAYAQUIL. It is the commercial centre for the surrounding agricultural and dairy farming region. Loja is noted for its university, which was founded as a law school in 1897. It was for a time, during Spanish colonial rule, a world centre for cinchona (a source of quinine) production.

Population (city) 94 310; (province) 384 700

Map Ecuador – Bb

Lolland (Laaland) *Denmark* Third largest Danish island which covers an area of 1234 km² (476 sq miles). Its broken, indented coastline consists of forested land situated in the north and east, while the marshy south is protected from flooding by dykes and large sand dunes.

Its main crops are sugar beet, oilseeds and tobacco. Places of interest include the 12th-century royal palace, Alholm Castle, in Nysted, which has the largest collection of veteran cars in Europe. The island boasts various beautiful medieval buildings including manor houses and churches. Its principal towns are Nakskov, Maribo and Saskøbing. Lolland forms an important part in the system of bridges linking Central Europe with Scandinavia.

Population 73 600

Map Denmark – Bb

Lomami *Zaire* River of central Zaire, 1450 km (900 miles) long. It rises in the south near KAMINA, some 450 km (280 miles) north-west of Lubumbashi, and flows north to the Zaire River which is near Isangi. Its lower 400 km (250 miles) are navigable.

Map Zaire – Bb

Lombardy (Lombardia) *Italy* The country's wealthiest and most densely populated region covering some 23 854 km² (9210 sq miles) of the Italian Alps and the North Italian Plain. Lombardy is Italy's industrial powerhouse and is prominent in all main sectors of manufacturing. Agriculture and dairy farming are also highly productive, due to careful control of water in the low plains of the great Po River.

The region's cities include the regional capital MILAN, BERGAMO, BRESCIA, COMO, CREMONA, MANTUA, MONZA, PAVIA, and VARESE. At Magenta, 25 km (16 miles) west of Milan, French and Italian armies defeated Austrian troops in 1859, during the campaign to unify Italy. About 9000 men died in the battle, which was so bloody that Magenta gave its name to a shade of red.

Population 8 939 400

Map Italy – Bb

Lombok *Indonesia* Belonging to the Lesser Sunda group of islands, Lombok is an island of 5435 km² (2098 sq miles) lying across a 35 km (22 mile) wide strait from BALI. Through the strait runs the imaginary Wallace's Line, which was suggested by the British biologist Alfred Russel Wallace (1823-1913) as dividing the Asian biogeographical realm from the Australasian Realm. Whereas Lombok has marsupial creatures related to kangaroos, Bali to the west does not.

Lombok has a mountainous landscape, with the highest point the peak of the Gunung Rinjahi volcano, which reaches 3726 m (12 224 ft). Rice and coffee are the island's main crops.

Population 1 300 200

Map Indonesia – Ed

Lomé *Togo* The country's capital and by far its largest town. Lomé, which is situated on the coast beside the border with Ghana, was founded as the colonial capital of German Togoland in 1897. It now has an international airport, and a modern deep-water port which exports phosphates, cocoa, coffee, cotton and palm oil.

It also handles much of the transit trade from the landlocked states of Mali, Burkina and Niger. Some colonial buildings in the old town, such as the governor's palace on the Grand Marché and a cathedral (1910), are a reminder of days under German rule. The city has an attractive palm-lined seafront and hotels offering entertainment and sports facilities. Lomé's university was founded in 1965. The local double storey market stocks cloth from all over West Africa, gold and silver jewellery, and local craftwork.

Population 580 000

Map Togo – Bb

Lomond, Loch *United Kingdom* Lake in Scotland, about 28 km (17 miles) north-west of GLASGOW. It covers 70 km² (27 sq miles) and stretches 35 km (22 miles) from Ardlui in the north to Balloch. The loch lies in the shadow of 973 m (3192 ft) high Ben Lomond and is dotted with islands. It is commemorated in the folk song 'The Bonnie, Bonnie Banks o' Loch Lomond', and is a popular tourist attraction.

Map United Kingdom – Cb

London *Canada* Industrial and financial centre of south-western Ontario. London was planned as a regional capital in 1793, but was not settled until 1826. A university was established here in 1878. Processed foods, paper products,

locomotives, textiles and refrigerators are among the goods manufactured in London.

Population 381 520

Map Canada – Gd

London *United Kingdom* National capital and major metropolitan area, with 32 boroughs plus the City of London itself spreading over 1580 km² (610 sq miles) on both banks of the River Thames. The Romans founded Londinium in AD 43, in what is now the 'square mile' of 'the City', the business heart of the capital. They also built the first London Bridge. In the 11th century a royal palace and then a minster were built some 3 km (2 miles) to the west at what became known as Westminster. In this way twin cities were created which did not really merge until the 17th century; the City itself which became the commercial capital and Westminster, the centre of royal and, later, political power. The movement westwards was accelerated by the Great Fire of 1666 which destroyed most of the City.

Today the western part of central London contains the Houses of Parliament, government departments, Buckingham Palace and other royal homes, the main national museums, concert halls and art galleries, and, in the 'West End', the major theatres, restaurants and shops. In the west, too, are the principal parks, of which Hyde Park with Kensington Gardens is the largest.

The City, developed over the centuries as an international centre of banking and commerce, had to its east the roots of its wealth – the docks from which ships traded all over the world. The active docks have now moved downriver, to be replaced by modern industrial, leisure and tourist developments, but the City retains its role, if not its appearance. Redevelopment after the bombing in the Second World War has all but submerged its greatest glory – Sir Christopher Wren's St Paul's Cathedral, built after the Great Fire between 1675 and 1710 – in a sea of office blocks.

The Thames snakes its way through London in graceful, sweeping curves and is spanned by six road bridges between the Tower of London – built next to the City by William the Conqueror after 1066 – and the Houses of Parliament alone. It was the capital's main transport artery, for local as well as international trade, before the railways arrived. It has long posed a threat of surge-tide flooding. This risk was eliminated only in 1984 with the completion of a movable barrier downstream at Woolwich.

Today, travellers in London have the choice of moving on often congested streets, many of them rather narrow compared with those of other major cities, or of using one of the world's most comprehensive networks of surface and underground railways; the river is seldom used for commuter transport. The new 16 km (10 mile) extension of London Underground's Jubilee line will provide a fast link from the Docklands into the west of London and boost the further development of new residential and industrial areas, such as at Canary Wharf.

The rail network was built to serve an ever-growing capital from the mid-19th century, and in turn encouraged urban growth as workers moved out to the suburbs. As it grew, London swallowed communities that were once outlying country towns and villages. Many, such as Greenwich (with its old Royal Observatory buildings, its Royal Naval College, National Maritime Museum and the clipper ship called

Cutty Sark), are full of historic interest. Others have merely become commuter suburbs. But some former villages with literary or artistic associations, such as Chelsea, Hampstead and Highgate, have retained their local identity.

Few major capitals so dominate the cultural and commercial life of their nation as does London, and this helps to explain the vast influx of visitors – tourists and business people alike – who arrive each year. With more than 10 million overseas tourists and more than 14 million Britons visiting London each year it may claim to top the international tourist league.

Visitors discover a population that is cosmopolitan and lively to a degree – who are pleased to help strangers but not afraid to tell the world their opinions from atop a soapbox at Speakers' Corner in Hyde Park. They find traditions that have changed little over the centuries – from the pageantry of the changing of the guard at Buckingham Palace to the humble street markets – alongside a vitality that keeps London a trend-setter in fashion and the arts. They observe a vivid cultural life that has few equals, including one of the world's great opera houses at Covent Garden, more than 40 major theatres and many smaller ones, and concerts by five full-time symphony orchestras and countless other groups which perform music from madrigals to pop.

Above all, however, they come to see London's physical heritage – the great buildings, art collections and museums. These include great religious buildings – St Paul's Cathedral, Westminster Abbey and the Roman Catholic Westminster Cathedral. They also include Westminster Hall and the Houses of Parliament,

and those associated with the royal family, such as the Tower of London (where, among many other things, the crown jewels can be seen), St James's Palace (1532), Kensington Palace (1605) and the 19th-century Buckingham Palace (opened to the public for the first time in 1993).

Major museums include the British Museum (with its superb collection of antiquities and associated British Library), Victoria and Albert Museum (with its exhibits of fine and applied arts of all kinds), the Bethnal Green Museum of Childhood, and specialised museums such as the museums of science, natural history and geology. There are numerous art galleries, of which the most famous are the National Gallery, the Tate Gallery, the National Portrait Gallery, the Royal Academy, the Wallace Collection and the Courtauld Institute. And there are countless lesser buildings of note, statues and public gardens.

Population (city) 4 100 000; (metropolitan area) 6 378 00

Map United Kingdom – Ee

Londonderry *United Kingdom* Northern Ireland's second largest city after the capital, BELFAST. The city is situated on the Foyle River, 100 km (60 miles) north-west of Belfast. The city's original name, Derry – from the Irish *Doire*, meaning 'a place of oaks' – is derived from the wood where a monastery was founded by St Columba in AD 546. The building is now a ruin.

Derry was granted its first charter in 1604 and in 1613 was given to the city of London as a plantation – an area to be colonised. Funds for the plantation were raised by the London livery

▲ BUSTLING HEART Amid the busy traffic of London's Piccadilly Circus, people stand in the sunshine at the base of the century-old figure of Eros. Like the statue, London's red buses are recognised around the world as a symbol of Britain's capital city.

companies, and as a result, the town's name was changed to Londonderry. The massive walls, which still form a complete circuit of the old town and date from the period of the plantation, were also funded by the London companies.

Londonderry was besieged in 1641 and 1649 by Royalist forces during the English Civil War, and again in 1688-9, when it held out for 105 days against an army supporting the deposed Catholic king, James II. At one stage, its Governor Lundy was prepared to let the king's troops into the city, but the gates were closed by 13 apprentices and Londonderry stood firm. Nationalists usually call the city Derry, leaving off the British prefix. In recent years, this group has controlled the council whose office name is now Derry City Council. The city has a Church of Ireland cathedral, which was completed in 1633, and a Roman Catholic cathedral, which stands on a hilly site opposite it.

The county of Londonderry covers 2076 km² (801 sq miles) in north-west Northern Ireland. Much of it is given over to farming, especially dairying and cereal crops. Traces of what may be the oldest dwellings in the British Isles have been found in the lower valley of the Bann River in the east of the county.

Population (city) 72 340; (county) 202 030

Map United Kingdom – Bc

Londrina *Brazil* Modern, prosperous city about 400 km (250 miles) west of SÃO PAULO. It manufactures soft drinks, coffee and vegetable oils, among other things. A British company was responsible for much of its early development, which explains the city's name.
Population 540 980
Map Brazil – Cd

Long Island *USA* Island forming part of New York State, covering 4463 km² (1723 sq miles) and extending 190 km (118 miles) east from New York City. The city boroughs of Queens and Brooklyn lie at its western end. The rest is given over largely to commuter communities, farms and summer homes, although industry has grown rapidly since the Second World War. The southern coast has long sandy beaches on the Atlantic Ocean, and there is good sport and commercial fishing.
Map United States – Lb

Long Island Sound Busy inlet of the ATLANTIC Ocean between CONNECTICUT and LONG ISLAND, New York State. Measuring 145 km (90 miles) in length and 5 to 32 km (3 to 20 miles) across, it is linked to New York Bay by the East River.

The Dutch navigator Adriaen Block explored the inlet and surrounding valley in 1614. He sailed up the Connecticut River which flows into the sound, passing an island he named Block Island, and was impressed by the area's beauty and potential. His enthusiastic report and hand-drawn maps encouraged the establishment of Dutch settlements in the region. Today the sound is lined with residential areas, resorts and leisure boating facilities.
Map (Long Island) United States – Lb

Longford (Longfort) *Ireland* Market town 109 km (68 miles) north-west of DUBLIN. It is the county town of a cattle-raising county of the same name covering 1044 km² (403 sq miles). The writer Oliver Goldsmith (about 1728-74) was born in the county, at Pallas, about 20 km (12 miles) south of Longford. His poem 'The Deserted Village' is set in the area.
Population (town) 6390; (county) 30 300
Map Ireland – Cb

longitude Angular distance of a point east or west of the GREENWICH MERIDIAN. It is the angle at the centre of the earth, as measured at the equator, between the Greenwich meridian and the meridian through the point.

longshore current (littoral current) Current that flows close to and parallel with the shore, produced by the waves moving into the coast at an angle.

longshore drift Movement of material along a beach by the action of waves breaking at an angle to the shoreline. The rush of water up the beach (the swash) pushes sand and gravel obliquely up the beach, then the backwash pulls it straight down, producing a sideways drift of the beach deposits. See the CHANGING PATTERN OF THE COASTS, p 219

Lop Buri *Thailand* Town some 130 km (80 miles) north of the capital, BANGKOK, whose history dates back to the 8th century AD when it was ruled by the Khmer people of Cambodia. Today the town has a large Thai military base.

The chief crops of the surrounding province, also called Lop Buri, are maize and cotton.
Population (town) 36 600; (province) 680 000
Map Thailand – Bc

Lop Nur (Lob Nor) *China* Remote salt lake covering roughly 2000 km² (770 sq miles) of the TARIM PENDI. It lies in the empty desert country of central Xinjiang Uygur Autonomous Region and has been used by China as a testing ground for nuclear weapons. Depending on the amount of water that feeds into it and the changing course of the rivers that flow into it, Lop Nur constantly changes shape, and its direction – so much so that it has been called 'the Wandering Lake'.
Map China – Dc

lopolith Saucer-shaped body of intrusive igneous rock lying between rock strata.

Lord Howe Island *Australia* Forested island, covering 16 km² (6 sq miles) in the Pacific Ocean, about 800 km (500 miles) north-east of SYDNEY. Administered as part of New South Wales, it is classified as a World Heritage area, and is a very popular tourist destination.
Population 300
Map Pacific Ocean – Cd

Loreto *Italy* The country's most celebrated pilgrimage town, 20 km (12 miles) south-east of ANCONA. Each year thousands of people visit a house here in which Jesus is said to have grown up. It was supposedly flown in from Nazareth in Israel by angels in 1294. The house lies inside the Santuario della Santa Casa – the 'Sanctuary of the Holy House'.
Population 10 600
Map Italy – Dc

Loreto *Peru* Northernmost department bordered by Ecuador, Colombia and Brazil. It is a rain forest region, crossed by many tributaries of the Amazon and separated from coastal departments by the Andes. It produces rubber, timber and palm oil. It also has deposits of gold, iron and oil. Ocean-going vessels travel up the Amazon from the Atlantic 3680 km (2285 miles) away to Loreto's commercial centre and capital city, IQUITOS.
Population 654 100
Map Peru – Ba

Lorraine *France* Region in the north-east of the country, alongside the German border. With ALSACE, most of Lorraine was annexed by Germany during the Franco-Prussian War (1870-1) but restored to France in 1919, after the end of the First World War. The region, which covers an area of 23 547 km² (9089 sq miles), includes several First World War battlefields, including VERDUN. The double-barred Cross of Lorraine was an important symbol of the Free French forces during the Second World War. NANCY and METZ are Lorraine's main cities.
Population 2 305 000
Map France – Fb

Los Angeles *USA* City in southern California, about 200 km (125 miles) from the Mexican border. It has one of the largest urban areas in the world, sprawling over much of southern California and connected by an extensive network of freeway systems. It is home to more

than 14 million people, giving the city and metropolitan area population figures second in the United States only to those of New York City. However, it is almost entirely a city of suburbs, with no real centre.

The original settlement, founded in 1781, was near the present downtown business district beside the Los Angeles River, 24 km (15 miles) from the Pacific Ocean. It expanded when the Spanish Franciscans built a mission here in 1822. As the city prospered, it grew beyond the surrounding hills, although it is still contained by the San Gabriel, San Bernardino and Santa Ana mountains, which help to trap the bane of the city – smog.

Rich oil fields, film-making surrounding the suburb of Hollywood, and tourism are its main livelihoods. It has several universities, an aircraft industry, as well as electrical and chemical manufacturing. Los Angeles is one of the country's busiest ports. Like San Francisco,

Los Angeles lies on the San Andreas Fault and has been prone to earthquakes. The North Ridge earthquake early in 1994 sent highrises, bridges and expressways toppling. Large complexes were evacuated and Los Angelenos began to re-examine safe structural design.
Population (city) 3 485 400; (metropolitan area) 8 863 160; (conurbation) 14 531 530
Map United States – Cd

Los Baños *Philippines* See BAY, LAGUNA DE

Los Lagos *Chile* Region constituting the far southern limits of the unbroken settled area of central Chile, which stretches from Santiago. Some of the most beautiful of the country's lakes lie inland in OSORNO Province, and there are ski resorts on the slopes of the Andes. The economy is based on tourism. The regional capital, PUERTO MONTT, is a fishing port.
Population 953 330
Map Chile – Ad

Lot *France* River, 481 km (299 miles) long, which rises in the southern MASSIF CENTRAL and flows westward to join the Garonne 30 km (19 miles) west of Agen.

Lot is also the name of a department of the arid Midi-Pyrénées region in the south-east of the country. Tobacco and vegetables are its main crops. Tourism is an important source of revenue.
Map France – Dd

Lothian *United Kingdom* Scottish region based on the country's capital, EDINBURGH, which covers 1756 km² (678 sq miles). The name is taken from an ancient kingdom that was also known as Lyonesse in the Middle Ages. West Lothian has a number of small industrial towns struggling to develop new industries following the decline of coal mining. Barley, grown mainly for whisky distilling and beer brewing, is the chief cash crop.
Population 723 680
Map United Kingdom – Dc

Lötschberg *Switzerland* Mountain pass 60 km (37 miles) south-east of the capital, BERNE. Although it has been used since the Bronze Age, it was difficult to cross until 1913 when a railway between Kandersteg and Brig was built through the Lötschberg Tunnel, 14.6 km (9 miles) long.
Map Switzerland – Aa

411

lough In Ireland, the name given to a lake or a long narrow arm of the sea; equivalent to the Scottish LOCH.

Louisbourg *Canada* Small port on the east coast of CAPE BRETON ISLAND. The remains of an imposing fortress built by the French in 1720-45 and destroyed by the British in 1758, is the focal point of a national historic park.
Population 1260
Map Canada – Jd

Louisiana *USA* Fertile southern state covering 115 334 km² (44 520 sq miles) mainly between the Mississippi River, the Gulf of Mexico and Texas. Most of the state consists of lowland, and the Mississippi, which is flanked by LEVEES to reduce flood risks, is above the level of the land on either side. The climate is subtropical in the south and temperate in the north. Sweet potatoes, rice, sugar cane, cotton and soya beans are grown. Louisiana heads the United States in the production of rock salt and sulphur and ranks second in the production of crude oil. Other products include natural gas, chemicals and foods.

The name Louisiana – after the French King Louis XIV – originally applied to the whole Mississippi River basin, which was settled by French colonists in 1699. After first losing the territory to England and Spain, then regaining the Spanish (western) part in 1800, the French sold the remainder to the United States in 1803 for US$15 million in a transaction known as 'the Louisiana Purchase'. In 1812 the southern part of this territory (called 'the Territory of Orleans') became the state of Louisiana. To this day, many French and Spanish influences can be seen in the state. Louisiana's main city is NEW ORLEANS. BATON ROUGE is the state capital.
Population 4 220 000
Map United States – Hd

Louisville *USA* City in north-west Kentucky, on the Ohio River where it forms the northern border of the state. It is an agricultural marketing centre; it manufactures domestic appliances and has some of the nation's largest whiskey distilleries and cigarette factories. It is also the home of Churchill Downs, site of the Kentucky Derby horse race. Fort Knox, the country's gold bullion vault, is situated about 40 km (25 miles) south-west of the city.
Population (city) 289 800;
(metropolitan area) 962 600
Map United States – Ic

Lourdes *France* Centre of pilgrimage on the fringes of the Pyrenees, 130 km (80 miles) south-west of TOULOUSE. The shrine was founded after a peasant girl named Bernadette Soubirous said she had seen the Virgin Mary there 18 times between February and July 1858. Today, more than 4 million pilgrims, some of them seriously ill or handicapped, visit the shrine each year, and since 1858, more than 300 have been cured, apparently by miracle. Lourdes has a vast underground basilica and several hospitals.
Population 13 600
Map France – Ce

Lourenço Marques *Mozambique* See MAPUTO

Louth (Lú) *Ireland* Smallest county in Ireland, covering an area of 823 km² (318 sq miles),

40 km (25 miles) north of DUBLIN. Carlingford Lough lies to the north beyond the Cooley Mountains, which extend to a height of 590 m (1963 ft). Today cereals, textiles and footwear are produced here. The county town is DUNDALK.
Population 90 720
Map Ireland – Cb

Louvain (Leuven; Löwen) *Belgium* City on the Dijle River, 24 km (15 miles) east of BRUSSELS, and the seat of the country's best known university, founded in 1426. Despite serious damage to the city during the First and Second World Wars, Louvain retains its medieval, perfectly circular form. However, because it lies north of the language line (see BELGIUM), Louvain has become Leuven and the university wholly Flemish. A new French university, Louvain-la-Neuve, has been set up 25 km (16 miles) to the south, near Wavre. Louvain's main industries are brewing, flour milling and light engineering.
Population 84 900
Map Belgium – Ba

low See DEPRESSION

Löwen *Belgium* See LOUVAIN

Lower Austria (Niederösterreich) *Austria* The country's largest state, known as 'the cradle of the nation', and covering an area of 19 172 km² (7402 sq miles) in the hilly north-east. It produces timber, livestock, grain and wine, as well as oil and natural gas, and industrial products such as cars, rolling stock, aircraft and textiles. It contains VIENNA, which, though administered as a separate state, is both the national capital of Austria and capital of Lower Austria State.
Population 1 480 930
Map Austria – Da

Lower Egypt *Egypt* Northern Egypt including Cairo, the Nile delta and the Mediterranean coast. Southern, or Upper, Egypt by contrast covers the whole region south of the Nile delta up to the Sudanese border. After the two parts were united in about 3100 BC, 30 dynasties of pharaohs ruled Egypt until Alexander the Great's conquest of the country in 332 BC.
Map Egypt – Bb

Lower Saxony (Niedersachsen) *Germany* State covering an area of 47 348 km² (18 276 miles) which stretches from the North Sea coast southwards to the borders of Hessen and North Rhine-Westphalia. Except for the south-east, it is part of the North European Plain, a mosaic of fertile farmland, sandy heaths, marshy fenland and peat bog. Some of Germany's best agricultural land – producing wheat, rye, potatoes and fodder crops – lies between the cities of BRUNSWICK and MINDEN. The city of BREMEN, a leading Hanseatic town during the Middle Ages, is another centre.

Lower Saxony produces lignite, peat, oil and natural gas. Industries revolve mainly around car assembly, engineering, shipbuilding, and the production of electronics and chemicals. HANNOVER is the state capital.
Population 7 423 700
Map Germany – Cb

Lowicz *Poland* Market town 76 km (47 miles) south-west of the capital, WARSAW. Nieborow, a magnificent Baroque palace (1695-7), stands in

fine formal gardens 10 km (6 miles) to the south-east. Owned by the Radziwill family from 1774 to 1945, it is now part of the national museum.
Population 26 900
Map Poland – Cb

Lowlands, The *United Kingdom* See HIGHLANDS, THE

lowveld Also known as bushveld, lowveld is found in parts of subtropical southern Africa at an altitude of between 300 and 900 m (990 and 2950 ft). It includes regions of dry wooded savannah and scrub, as found in South Africa's KRUGER NATIONAL PARK.

Loyalty Islands *New Caledonia* Major outlying islands of the territory, consisting of three main islands, themselves 100 km (62 miles) east of New Caledonia. Most of the population are Melanesian, but there are some Polynesians living on Ouvéa. Their principal crop is coconuts.
Population 15 500
Map Pacific Ocean – Dd

Lo-yang (Luoyang) *China* Ancient city in Henan Province. Previously known as Luoyang, the city dates from 1200 BC and with Anyang and Zhengzhou was one of the main centres of the Shang civilisation, the earliest in China. From 770 to 221 BC it was the capital of the Eastern Zhou kingdom – one of the successor states to the Shang. Many important archaeological sites have been opened up in the area. Now known as Lo-Yang, it is a centre for the manufacture of bearings, tractors and mining machinery.
Population 386 000
Map China – Gd

Lú *Ireland* See LOUTH

Lu Shan *China* Mountain range in Jiangxi Province, south of the Chang Jiang River; its highest peak reaches 1474 m (4836 ft). The Lu Shan botanical garden, on the peak of the Nine Strange Things, is one of the finest in China. Lu Shan is a popular summer holiday destination.
Map China – He

Lualaba *Zaire* River rising near the Zambian border, 120 km (75 miles) west of the LUBUMBASHI, and flowing 1800 km (1120 miles) northwards. It is navigable between Bukama and Kongolo, Kasongo and Kibombo, and KINDU and Ubundu. Below Ubundu are the Boyoma Falls, where the Lualaba becomes the Zaire River.

The Scottish missionary and explorer David Livingstone (1813-73) thought that the Lualaba was a source of the Nile, but in 1876-7 the British-American adventurer, Sir Henry Morton Stanley, (1841-1904) followed its course and proved that it was a headstream of the Zaire River.
Map Zaire – Bb

Luanda (Loanda) *Angola* National capital and port founded by Portuguese settlers in 1576 and, before the civil war that began in 1975, one of the most beautiful cities on the African coast. One of its landmarks is the castle of St Michael, built in 1638. Before the war, Luanda was a busy port and manufacturing centre; with food processing, paper, woodworking and textile plants, and an oil refinery. Pipelines link it to the oil fields of the Cuanga River delta on the Atlantic coast. Its population expanded rapidly during the civil war as refugees fled to the city from the various areas of conflict. The neighbouring province, also called Luanda, produces coffee, cotton, palm products and sisal.

Population (city) 1 700 000; (province) 2 000 000
Map Angola – Aa

Luang Prabang *Laos* Province and town on the Mekong River in northern Laos, 220 km (140 miles) north of the capital, VIENTIANE. Once the seat of the Principality of Luang Prabang, the town was the nation's royal capital during the French colonial period and until 1975; it was also the heart of Laotian Buddhism and contains numerous fine temples, palaces and pagodas. However, since the communist takeover in 1975, the monarchy has been abolished and Buddhism discouraged.

Population (province) 450 000; (town) 45 000
Map Laos – Ab

Luapula *Zambia* North-eastern province which borders Zaire, covering an area of 50 567 km² (19 524 sq miles) and named after the Luapula River, which marks the Zaire border. Its capital is the town of Mansa, about 450 km (280 miles) north-east of the national capital, Lusaka. The main occupations of the thinly populated area are arable and livestock farming and fishing.

Population 526 000
Map Zambia – Bb

Lubango *Angola* Market town some 680 km (420 miles) south of the national capital, LUANDA. The surrounding province of Huíla produces coffee and cattle, but farming was disrupted in the 1980s by raiding parties of rebel Unita guerrillas and by South African troops pursuing Swapo guerrillas across the Namibian border.

Population 31 700
Map Angola – Ab

Lübeck *Germany* Port on the Trave River, lying 15 km (9 miles) from its mouth on the Baltic Sea, and 60 km (38 miles) north-east of HAMBURG. Lübeck was founded in 1157 by Henry the Lion, Duke of Saxony, and rapidly became a thriving Baltic port, trading in timber. In the 13th century, it was one of the leaders of the Hanseatic League.

Today its shipping trade is mostly to and from Scandinavia; it handles exports of paper, timber, cellulose, iron ore, steel, and motor vehicles. Its industries include shipbuilding, engineering, the production of textiles, chemicals, and medical instruments as well as the processing of food, in particular marzipan. Tourism is another major source of income.

Willy Brandt (1913-92), the West German Chancellor between 1969 and 1974, was born in Lübeck; so was the writer and Nobel Prize winner Thomas Mann (1875-1955), author of *Death in Venice* (1913) and *Dr Faustus* (1947).

Population 214 800
Map Germany – Db

Lublin (Lyublin) *Poland* Eastern Poland's largest city and cultural centre, situated on the Bistritz River, 153 km (95 miles) south-east of WARSAW. It is the market place for a fertile farming region which produces meat, wheat, flour and sugar; besides beer, tobacco products and vegetable oils, the city also produces agricultural equipment and other machinery. It has a total of

five academic institutions, including the Catholic University – the only one of its kind in Poland – and the Marie Curie-Sklodowska University, named after the Polish-born scientist and Nobel laureate (1867-1934) who discovered radium.

Lublin was the seat of two independent Polish governments. The first was set up on 7 November 1918 as a temporary socialist government; the second was established on 22 July 1944 as the German army retreated.

The Museum of Martyrology and a monument in the south-eastern suburb of Majdanek are grim reminders of Nazi terror. They stand on the site of an extermination camp where 370 000 Poles, Russians, Jews and people of 17 other nationalities were murdered between 1941 and 1944.

Population 351 400
Map Poland – Ec

Lubombo Mountains *Southern Africa* See LEBOMBO MOUNTAINS

Lubumbashi *Zaire* The country's second largest urban area, after KINSHASA. Founded in 1910 as Elisabethville, it lies only 30 km (20 miles) from the Zambian border. It is the capital of Shaba Region in the south-east, Zaire's chief mining region, and has copper and cobalt refining, as well as processing industries. Four railway lines connect the city of Lubumbashi to the Atlantic and Indian oceans.

Population 700 000
Map Zaire – Bc

Lucca *Italy* Beautiful walled Tuscan city and market centre 60 km (38 miles) west of Florence. Its early wealth derived from silk, wool and banking, and is expressed in the city's many fine 12th to 15th-century Romanesque churches. The narrow streets are still rich in medieval palaces, towers and attractive tenements. Surrounding the old town are 16th-century fortifications, with beautiful avenues of trees facing a grassy moat. The operatic composer Giacomo Puccini (1858-1924) is Lucca's most renowned native son. The surrounding region is noted for its olive oil.

Population 86 700
Map Italy – Cc

Lucerne, Lake (Vierwaldstätter See; Lac des Quatre-Cantons) *Switzerland* Alpine lake covering 114 km² (44 sq miles) in central Switzerland. Surrounded by mountains and fringed with woodland, it is one of the country's most popular and beautiful tourist spots. It is often known simply as Lake Lucerne, after the city of LUCERNE on its north-west tip.

It was on the shores of the lake, in 1291, that representatives of the three original Swiss cantons – Uri, Schwyz and Unterwalden – signed a pact of unity against the country's Austrian rulers. The pact set up 'the Everlasting League', which was the embryo of the modern state of Switzerland.
Map Switzerland – Ba

Lucerne (Luzern) *Switzerland* Lakeside city 65 km (40 miles) north-east of the capital, BERNE. Large sections of the town's medieval walls and watch towers are still standing. Unfortunately, its famous 14th-century covered Kapellbrücke, the oldest footbridge in Europe, burnt down in 1993; it was rebuilt in 1994.

Other tourist attractions are Lucerne's Renaissance town hall, its 17th-century cathedral

and its giant Lion of Lucerne (1821). The dying lion, 9 m (30 ft) long and 20 m (65 ft) high, was carved on a wall of rock as a monument to the Swiss Guards of the French king Louis XVI, more than 700 of whom were killed defending the Tuileries Palace in Paris in 1792 during the French Revolution.

Today, the agriculturally important Canton of Lucerne stretches across 1492 km² (576 sq miles) to the north, south and west of the city.

Population (city) 59 600; (canton) 331 800
Map Switzerland – Ba

Lucknow *India* Capital of the northern State of Uttar Pradesh. It lies in the Ganges River valley, 569 km (354 miles) south-east of DELHI. Its old buildings include the Pearl Palace of the rulers of the former Mogul Province of Oudh, and the Residency, where in 1857 during the Indian Mutiny a British force withstood a siege from July until mid-November. The city is a centre for Indian handicrafts, including *bidri* – the ancient art of silver inlay on gunmetal – which is now being revived.

Population 2 744 580
Map India – Cb

Lüda (Dalian) *China* City and port in Liaoning Province. It stands on the Liaodong Peninsula and is a centre for heavy engineering.

Population 4 619 000
Map China – Ic

Lüderitz *Namibia* Lobster-fishing port on the arid Namibian coast, 230 km (145 miles) from the South African border. It was named after Franz Lüderitz, a German merchant who bought the site and persuaded Germany to take it under imperial protection in 1884. This was Germany's first step in colonising German South West Africa. The port also serves the nearby diamond-mining zone, to which access is restricted.

Population 17 000
Map Namibia – Ab

Ludhiana *India* Agricultural trading town in the north-western State of Punjab, situated 310 km (193 miles) north-west of DELHI. Its agricultural university is recognised throughout the world for its research on wheat cultivation.

Population 1 012 000
Map India – Bb

Ludlow *United Kingdom* English town situated in the Midland county of Shropshire, about 3 km (2 miles) from the Welsh border. It is rich in Tudor and Georgian buildings and has a fine 11th to 14th-century castle. Many of its houses were built between the 16th and the 18th century by wealthy landowners and merchants.

Population 7500
Map United Kingdom – Dd

Ludwigshafen *Germany* Town and port on the Rhine River opposite Mannheim, some 80 km (50 miles) south of FRANKFURT am Main. It grew around the BASF chemical works, and today also produces glass and steel.

Population 162 200
Map Germany – Cd

Lugano *Switzerland* Lakeside tourist resort near the Italian border in the extreme south of the country. The economic and cultural centre of the

Canton of Ticino, it is overlooked by the hills of Monte San Salvatore and Monte Brè, both more than 900 m (nearly 3000 ft) high.

Population 25 170
Map Switzerland – Ba

Lugansk *Ukraine* See LUHANS'K

Lugo *Spain* Town noted for its Roman walls and towers. It lies in the north-west, 80 km (50 miles) south-east of CORUNNA. Lugo trades in cattle and has tanning, flour-milling and textile industries. The surrounding province of Lugo produces fish and timber products.

Population (town) 83 000; (province) 384 370
Map Spain – Ba

Lugou Bridge (Marco Polo Bridge) *China* Bridge, 15 km (9 miles) south-west of Beijing, which was the site of the outbreak of war between China and Japan in 1938. Also known as the Marco Polo Bridge, it is said to be the modern version of a bridge crossed by Marco Polo during his travels in China in the 13th century.

Map China – Hc

Luhans'k (Lugansk; Voroshilovgrad) *Ukraine* City 275 km (170 miles) south-east of KHARKIV, at the heart of the Donets coal-mining basin. Luhans'k was known as Lugansk before 1935 and again from 1958 to 1970, and thereafter as Voroshilovgrad until 1991.

The city was founded in 1795, when an iron foundry making arms for the Russian Black Sea fleet was established here. It later specialised in railway equipment and by 1941 produced half of the USSR's locomotives. It also became an engineering and research centre, and today produces textiles, processed foods, and chemicals.

Population 501 000
Map Ukraine – Db

Luik *Belgium* See LIÈGE

Luimneach *Ireland* See LIMERICK

Luleå *Sweden* Seaport at the head of the Gulf of Bothnia, about 946 km (588 miles) north-east of STOCKHOLM. Its harbour is icebound for most of the winter, but when it is open – on average, 150 days of the year – it exports about one-third of Lappland's output of iron ore, as well as wood pulp and timber.

It has a large iron and steel mill; its Norrbotten Museum has a notable collection of Lapp handicrafts and costumes.

Population 66 600
Map Sweden – Db

Luluabourg *Zaire* See KANANGA

Lumbini *Nepal* Grove in the modern village of Rummindei, formerly Kapilavastu, which was the birthplace of Prince Siddhartha Gautama (about 563-488 BC), the founder of Buddhism, who came to be known as the Buddha – a Sanskrit title meaning 'the Enlightened One'. His birthplace – near the Indian border 200 km (125 miles) south-west of KATHMANDU – is marked by a stone pillar erected by the Buddhist Indian emperor Ashoka in about 250 BC and is now a shrine. The site has a museum and a library.

Map Nepal – Aa

Lund *Sweden* City in the extreme south of Sweden, 16 km (10 miles) north-east of Malmö. It was founded in the early 11th century by the Danish king, Canute, who also ruled England between 1016 and 1035. It has a fine Romanesque cathedral, and its university, established in 1668, is the oldest in the country, after Uppsala.

Population 92 000
Map Sweden – Be

Lüneburg *Germany* Canal port and industrial centre 40 km (25 miles) south-east of HAMBURG. In the Middle Ages it became rich by trading in salt. On Lüneburg Heath, a sandy, partly forested area south of the town, the British field marshal Bernard Montgomery received the surrender of German troops in May 1945, ending the Second World War in Europe. Most of what remains of the original heath now forms part of a nature reserve, which attracts thousands of tourists each year. Lüneburg has mineral springs and a health spa.

Population 61 500
Map Germany – Db

Luoxiao Shan *China* Range of mountains in Jianxi Province. It was used as a refuge by communist rebels under the late Mao Zedong from 1927 until they broke out of a siege by General Chiang Kai-Shek's Nationalist forces in 1934. This was the first step of the Long March, when the rebels fought their way 9500 km (6000 miles) to safety in Yan'an.

Map China – Ge

Luoyang *China* See LO-YANG

Lurgan *United Kingdom* Market town in Northern Ireland, 30 km (20 miles) south-west of the capital, BELFAST. It has been a centre of the linen industry since the weaving of damask (a reversible material with a pattern made by different weaves) began in 1691. The small lanes leading from the main street contained weaving shops in the 18th century. Lurgan merged with Portadown in 1973 to form the city of Craigavon.

The small community of Waringstown, 5 km (3 miles) south-east of Lurgan, is named after Samuel Waring who, by using methods similar to those he had seen in Holland and Belgium, introduced damask weaving to the district.

Population 21 900
Map United Kingdom – Bc

Lusaka *Zambia* The country's capital and centre of the north-south railway and east-west road network. Founded in 1905 to serve the lead mine at Kabwe, it became the capital in 1935. The city's main buildings include the National Assembly, the president's State House and an Anglican cathedral. Lusaka is set in the centre of a fertile agricultural region rich in cotton and tobacco.

Population 1 000 000
Map Zambia – Bb

Lüshun (Port Arthur) *China* Naval base at the tip of the Liaodong Peninsula. It was Chinese territory until 1898, when it was taken over by Russia on a 25-year lease which was cut short by Japan's victory in the Russo-Japanese War of 1904-5. The lease was transferred to the Japanese authorities who held on to the base even after the lease expired. It was restored to China in 1955 after Soviet Union occupation from 1945.

Map China – Lc

Luton *United Kingdom* Industrial town in the English county of Bedfordshire, 50 km (30 miles) north of London, which produces mainly cars and trucks. Luton has one of London's airports.

Population 174 600
Map United Kingdom – Ee

Luxembourg See p 415

Luxembourg *Belgium* South-eastern province adjoining the independent grand duchy of the same name. It is mostly covered by the forested upland of the ARDENNES.

Population 232 700
Map Belgium – Cb

Luxembourg *Luxembourg* Capital of the grand duchy, consisting of the old city, standing on a cliff above the Alzette River, and the newer city to the east and south. The city's two halves are linked by viaducts. The old city, which is centred on the remains of the old fortress, whose foundations are Roman and Frankish, dates from the 10th century and contains the ducal palace (1572), the town hall where the parliament sits, and the gothic cathedral of Notre Dame (1613-23).

The new city, founded in the 19th century, and its north-eastern suburb of Kirchberg, house several European Union (EU) institutions such as the Court of Justice, the Investment Bank and the Coal and Steel Community. Numerous conference and exhibition centres as well as Europe's oldest commercial radio station, Radio Luxembourg, which broadcasts to Britain, France, Germany and the Netherlands are situated here.

Population 78 000
Map Belgium – Cb

Luxor *Egypt* Site of temple monuments and cenotaphs to the pharaohs of the New and Middle Kingdom, on the south side of their ancient capital, Thebes. The site is on the east bank of the River Nile, 500 km (310 miles) to the south of the present capital, CAIRO. Luxor temple is 260 m (853 ft) in length, making it one of the longest temples of ancient times. Dating mostly from about 1400 BC, it is linked to the temples at Karnak, 3 km (2 miles) to the north, by an avenue of sphinxes, each with a ram's head. The name Luxor is derived from the Arabic word for 'palaces'. There is a museum at the site, and hotels for tourists.

Population 65 000
Map Egypt – Cc

Luzern *Switzerland* See LUCERNE

Luzon *Philippines* The largest island of the Philippines, some 800 km (500 miles) long and covering some 104 688 km² (40 420 sq miles). It accounts for about one-third of the country's land area and more than half of its people. It is the Philippines' most northerly major island, and has an irregular coastline with several natural harbours. In the south-east it extends into a narrow peninsula, Bicol.

Luzon has a monsoon climate and receives copious rainfall, as well as being exposed to typhoons between August and October. It is geographically diverse, containing high mountains (notably Mount Pulag, at 2928 m, 9606 ft), mountain ranges (notably the Cordillera Central, the Sierra Madre, and the Zambales) and volcanic peaks, as well as rich rice-growing lowlands and

Luxembourg

*EUROPE'S LAST INDEPENDENT DUCHY
IS A SMALL STATE THAT PLAYS A BIG
ROLE IN INTERNATIONAL AFFAIRS*

This state is small, yet it has been so active in West European affairs that it is always represented on committees – even though it seldom rates more than a single delegate.

Luxembourg became a Grand Duchy in 1354 and is the only one of the hundreds of independent duchies that once made up west-central Europe to have survived to become a founder member of today's United Nations (1945). It was a founder member of the North Atlantic Treaty Organisation (1949) and of the European Economic Community (1957). It is the seat of the European Court of Justice, the Secretariat of the European Parliament. It is a major international banking centre, with about 190 different banks. In 1958 it joined with Belgium and the Netherlands to form the Benelux economic union – a totally free international goods and labour market.

The northern part of Luxembourg is hilly, picturesque country. It includes the southern edge of the ARDENNES and EIFFEL uplands and is drained by the Sûre (Sauer) River. The southern part, which is physically part of French LORRAINE, is lower and more fertile. Here beds of iron ore form the basis of a substantial steel industry. On the east, Luxembourg is bordered by the MOSELLE River and the lower Sûre in whose valleys wines are produced.

Luxembourgers speak both French and German but, for preference, they use their own Germanic dialect they call *Letzeburghish*. Their heavy industry, together with their international institutions, have attracted many thousands of foreign workers, who make up about a third of the population.

In May 1940, during the Second World War, Nazi Germany overran Luxembourg and made it part of the Third Reich. About 10 000 of its inhabitants were forced to fight for Hitler, while others joined the Resistance against the occupying Germans. In 1986 West

Germany finally agreed to pay Luxembourg a sum of Dm12 million (about US$5.2 million) in reparation.

Europe's first commercial radio station, Radio Luxembourg, went on the air in 1934 and is still broadcasting today. Its studios stand in scenic gardens near the old town in the capital, Luxembourg.

▲ FORTRESS CITY The fortifications of the old city of Luxembourg still stand above the steep walls of the Alzette gorge, and are now topped by a spectacular corniche road.

LUXEMBOURG AT A GLANCE	
Map Belgium – Bb	
Official name Grand Duchy of Luxembourg	
Area 2586 km² (998 sq miles)	
Population 390 000 **Per km²** 151 **(Per sq mile** 391)	
Capital Luxembourg	
Government Constitutional monarchy with a unicameral parliament	
Currency 1 Luxembourg or Belgian franc = 100 centimes	
Languages Letzeburghish; French and German widely spoken; some English	
Religion Christian (95% Roman Catholic)	
Climate Temperate maritime; average temperature in the capital ranges from 0.3°C (32.5°F) in January to 17.4°C (63°F) in July	
Land use Forest 32%, meadows and pastures 27%, cultivation 22%	
Main primary products Cereals, potatoes, grapes, dairy cattle; iron ore	
Major industries Iron and steel, chemicals, banking, food processing, metal products and engineering, agriculture, wine making	
Main exports Machinery, metal goods, steel and steel products, plastics and chemical products, textiles	
Annual income per head (US$) 28 770	
Population growth (per thous/yr) 4	
Life expectancy (yrs) Male 71 **Female** 78	

barren uplands where people can barely scrape a living. Luzon also contains the bulk of the country's industry, centred on the vast metropolitan area of the capital, MANILA, and has important mineral and timber resources.

The official language of the Philippines, Pilipino, is based on Tagalog, spoken by people of the Manila region, but several other languages are spoken in more isolated areas on the island.

Population 34 232 000
Map Philippines – Bb

L'viv (Lwow; Lwiw; Lemberg; L'vov) *Ukraine* Industrial city in the west, about 470 km (290 miles) west of KIYEV. Founded in 1256, it was conquered by the Poles in 1340. The city was part of Austria from 1772 until 1918, when it was again returned to Poland until 1939. Its mixture

of Ukrainian, Polish and Austrian building styles, side by side with modern architecture, reflects its colourful past. Among its oldest buildings are the remains of a 13th-century fort, a Roman Catholic cathedral which dates from the 14th to the 17th century, an Armenian cathedral built in 1370-1493, and some old Russian Orthodox churches.

Among L'viv's principal activities today are food processing, beer brewing, manufacturing paints and agricultural machinery.

Population 790 000
Map Ukraine – Ab

L'vov *Ukraine* See L'VIV

Lyallpur *Pakistan* See FAISALABAD

Lydda *Israel* See LOD

Lyons (Lyon) *France* The country's second largest city after Paris, at the junction of the Rhône and Saône rivers. It is a major trade centre with the oldest bourse (stock exchange) in France, dating from 1595; it is also regarded as the gastronomic centre of France. Its main industries are metallurgy, motor vehicle manufacture, chemicals and textiles. Among the city's historic buildings are the cathedral, built between the 12th and the 15th century and the Basilica of Notre Dame de Fourvière, opened in 1894.

Population (town) 415 000; (metropolitan area) 1 262 000
Map France – Fd

Lyublin *Poland* See LUBLIN

Lyublyana *Slovenia* See LJUBLJANA

Ma'an *Jordan* Market town and road and rail junction, situated about 115 km (71 miles) south of KARAK. It is the administrative centre of southern Jordan; it is also a commercial centre trading mainly in grain. Its main industry is tobacco processing.

Formerly a staging post for pilgrims to Mecca, Ma'an's main tourist attractions are the remains of an 18th-century Turkish fort and a rest-house. Most of its old town is built of clay; an irrigation system allows gardens to be cultivated. It also has an airport.

Population (town) 12 000; (region) 148 000
Map Jordan – Ab

Maas *Western Europe* See MEUSE

Maastricht *Netherlands* Provincial capital of LIMBURG, on the Maas (Meuse) River 170 km (105 miles) south-east of Amsterdam. Lying in a narrow strip of the Netherlands between Belgium and Germany, it is a multilingual city with a well-respected university. The Netherlands' oldest church, St Servatius, founded in the 6th century, is situated on its main square.

Maastricht manufactures ceramics, cement, paper and glassware, as well as clothing and chemicals. Its greatest claim to fame, however, is the Maastricht Treaty – a key document in the formation of the European Union – which was signed here by European Community heads of state in December 1991.

Population 118 290
Map Netherlands – Bb

Macau See p 417

MacDonnell Ranges *Australia* Central Australia's major mountain system. The ranges form parallel ridges in the Northern Territory, running east and west from ALICE SPRINGS. Their highest point is Mount Zeil at 1510 m (4954 ft). Gold was once mined at Arltunga to the east.

Map Australia – Ec

Macedonia See p 418

Macedonia *Greece* Largest region of Greece, bounded in the north by Albania, the former Yugoslavian republic of Macedonia, and Bulgaria, and covering 34 177 km² (13 192 sq miles). It is a geographically varied region of forested mountains (including PÍNDHOS, OLYMPUS and Vérmion), lakes, broad fertile plains, and three long rivers – the Aliákmon, Axios (Vardar), and Strimón. Wildlife includes brown bears, boars, jackals, wildcats, deer, wild goats, wolves, vultures and hawks. Agriculture in the region is the most varied and productive in Greece. Maize, beans, vegetables, rice, fruit, tobacco and timber are grown and livestock raised.

In addition to being Greece's major power-producing and oil-refining region, Macedonia also has an abundance of minerals, including copper, gold, iron, lead, magnesite, silver and chrome. The main cities are SALONICA, KAVÁLLA, Sérrai and Dráma.

Until the 4th century BC, Macedonia was on the fringe of Hellenic culture, but under Philip II it began to play a dominant and unifying role in Greek society. Philip's son, Alexander the Great (356-323 BC), used Macedonia as a springboard from which to build his world empire. Later, the region was ruled by Romans, Goths, Huns, Slavs

and Crusaders; from 1371 until the Balkan Wars (1912-13) it was held by the Turks. It was then split between Greece and Serbia, with a small area going to Bulgaria. The Greek region is now a southern bastion of the North Atlantic Treaty Organisation (NATO).

Population 2 122 000
Map Greece – Ba

Maceió *Brazil* Port and capital of Alagoas state, 160 km (100 miles) south of the coastal city of RECIFE. Its lagoons and sandy white beaches fringed by coconut palms have seen it develop into a popular tourist area. The chief industries in the area revolve around sugar, but production of salt and the manufacture of machinery also contribute to the economic well being of the city. At 27 731 km² (10 704 sq miles), Maceió is the smallest state in Brazil.

Population 628 210
Map Brazil – Eb

Macerata *Italy* City in the Marche region, 175 km (almost 110 miles) north-east of ROME. Founded in the 11th century, its old town is enclosed within walls built in the 14th century; in addition to several late medieval and Baroque buildings, it has a university dating from 1290.

Population 43 600
Map Italy – Dc

▼ CITY OF 3000 STEPS The ruins of Machu Picchu – the Inca city which the plundering Spanish conquistadores failed to find – cascade down 13 km² (5 sq miles) of terraces built high up in the Peruvian Andes.

Macgillicuddy's Reeks (Na Cruacha Dubha) *Ireland* Highest mountain range in Ireland, lying in the south-west, 105 km (65 miles) west of the town of CORK. It rises to 1040 m (3414 ft) at Carrauntoohil.

Map Ireland – Bb

Machu Picchu *Peru* Ruined Inca city in Cusco department. It is located 2280 m (7480 ft) above the Urubamba River on the saddle of a mountain known as the Machu Picchu ('old peak'). This complex of palaces, houses, plazas and terraces, linked by stairways, was one of the last strongholds of the Inca people before they were destroyed by the Spaniards in the 16th century. And yet the Spanish never found this settlement. Machu Picchu lay abandoned and unknown until 1911, when an American archaeologist, Hiram Bingham, found it.

Map Peru – Bb

Mackay *Australia* Port on the east coast of Queensland, 800 km (almost 500 miles) north-west of BRISBANE. The centre of a sugar cane-growing area, it processes and exports one-third of the country's total crop. The port also ships some of the coal mined 200 km (124 miles) inland to Japan, and is a base for exploring the Great Barrier Reef.

The Whitsunday Islands, of which many are tourist resorts, are 100 km (62 miles) further north, off Proserpine (population 3030).

Population 40 300
Map Australia – Hc

Mackenzie *Canada* River which rises in the Great Slave Lake in the Northwest Territories and flows 4251 km (2640 miles) in a north-

Macau

ONCE A SEEDY OUTPOST OF EMPIRE, PORTUGAL'S FORMER CASINO COLONY IS BOOMING

The tiny Portuguese enclave of Macau on the China coast offers some bizarre contradictions. Although Chinese make up 95 per cent of its population, Portuguese cultural influence shows everywhere. Sometimes Chinese and Portuguese traditions blend, with curious results.

Macau consists of a peninsula bordering the Chinese mainland province of GUANGDONG, and two small islands, Taipa and Coloane, which are joined by a causeway and bridge. There is no airport, and most visitors arrive by ferry or jetfoil from HONG KONG, which lies 64 km (40 miles) to the east across the ZHU JIANG (Pearl River) estuary.

The Portuguese first used Macau as a trading post in 1521. They occupied it in 1557 and used it as a base for trading and missionary activities in China. Portuguese sovereignty over the territory, formally recognised by China in 1885, lasted until 1974. Macau was then redefined as a Chinese territory under Portuguese administration. In 1987, Portugal agreed to return the territory to China on 20 December 1999. Provisions were made to ensure the autonomy of Macau, including its right to elect local leaders and the right of its residents to travel freely. The Chinese also agreed to respect Macau's culture and socio-economic systems for 50 years after the start of Chinese rule.

Macau was never able to compete commercially with the dynamism of Hong Kong, and for many years it remained a seedy and neglected backwater, its economy dominated by gambling. However, since the mid-1970s Macau has emerged as an industrial and commercial centre. It manufactures clothes, toys, artificial flowers and fireworks – all produced cheaply because wages are poor for thousands of refugees from mainland China.

MACAU AT A GLANCE	
Official name Macau	
Map Hong Kong – Ab	
Area 16 km² (6 sq miles)	
Population 477 850 **Per km²** 29 865 **(Per sq mile** 79 641)	
Capital Macau City	
Government Chinese territory administered by Portugal, with internal autonomy	
Currency 1 Pataca = 100 avos	
Languages Cantonese, Portuguese (official)	
Religions Buddhist and Taoist (45%), none (45%), Christian (8%), Confucian (2%)	
Climate Subtropical monsoon. Average temperature ranges from 16°C (60°F) in January to 28°C (82°F) in July	
Land use Urban area	
Main primary products Fruit, vegetables, livestock, fish	
Major industries Textiles, light manufacturing, tourism, gambling	
Main exports Clothing, textile yarns and fabrics, artificial flowers, toys	
Annual income per head (US$) 6700	
Population growth (per thous/yr) 14	
Life expectancy (yrs) Male 77 **Female** 82	

westerly direction to enter the Arctic Ocean in a 16-pronged delta. It was traced to the Arctic by Sir Alexander Mackenzie, the Canadian explorer. The river is navigable to the Great Slave Lake from June to October. The economy of the area is dominated by its large oil and gas fields.
Map Canada – Cb

Mackenzie Basin *New Zealand* Sheep-farming region in the centre of South Island, about 200 km (124 miles) south-west of the city of CHRISTCHURCH. Overlooked by Mount Cook, it covers 2000 km² (772 sq miles).
Map New Zealand – Be

mackerel sky Sky covered with cirrocumulus or altocumulus clouds resembling the markings on a mackerel. It is usually seen in summer during unsettled weather.

Mâcon *France* Market centre of the Mâconnais wine area 70 km (44 miles) north of LYONS. It has medieval wine cellars, which are open to visitors.
Population 37 200
Map France – Fc

Mactan *Philippines* See CEBU

Madagascar See p 419

Madain Saleh *Saudi Arabia* Important archaeological site situated 800 km (497 miles) north of JEDDAH. It consists of about 100 tombs, many with ornate façades, hewn out of the face of sandstone cliffs. They date from about 100 BC to AD 100, when the town of Al-Hejr stood on the site. It was the main stopping-place on the caravan route between the now ruined city of PETRA, in south-west Jordan, and the Arabian Peninsula.
Map Saudi Arabia – Ab

Madang *Papua New Guinea* Port on the north coast of New Guinea, 500 km (310 miles) north-west of PORT MORESBY. Its main exports are copra, coconuts, coffee and cocoa. Industries include

timber milling, tobacco processing and light engineering. A lighthouse on the coast is a memorial to the 'Coastwatchers' – volunteers who reported Japanese troop and ship movements to the Allies from behind enemy lines during the Second World War.
Population 27 180
Map Papua New Guinea – Ba

Madara *Bulgaria* See SHUMEN

Madeira *Brazil* One of the Amazon River's principal tributaries. The Madeira River, which is fed by tributaries from the Andes, flows 2013 km (1251 miles) to join the Amazon east of the city of MANAUS. Near the Bolivian border, prospectors still pan gold from the river.
Map Brazil – Bb

Madeira *Atlantic Ocean* Group of volcanic islands controlled by Portugal as an autonomous region. They lie 900 km (almost 560 miles) south-west of the Portuguese capital, Lisbon, and 100 km (62 miles) off the Moroccan coast. Probably known to the Genoese from the early 14th century, the islands consist of the main island of Madeira, covering 728 km² (281 sq miles), Porto Santo, covering 42 km² (16 sq miles), and the small uninhabited and barren island groups of the Desertas (Dezerte) and the Selvagens. FUNCHAL on Madeira Island is the capital.

A popular winter resort, the island of Madeira has a mild climate, subtropical vegetation and picturesque scenery. The island rises to 1816 m (5958 ft) at Pico Ruivo. Grapes (for the world-famous Madeira wine), bananas and sugar cane are the main crops.
Population (Madeira) 271 400; (Porto Santo) 4400
Map Morocco – Aa

Madhya Pradesh *India* The largest Indian state, covering 443 446 km² (171 170 sq miles) in the centre of the country. It includes the hills south of the Ganges plains and much of the Narmada

River basin. About one-third of the area is still forested. Farming, the mainstay of the economy, revolves around the cultivation of wheat, rice, maize, sugar cane, oilseeds and cotton. Mineral resources include diamonds, coal, iron and manganese ores. The state has a diverse, though mainly Hindi-speaking population, including Bhil people who continue to cling to a tribal way of life. BHOPAL is the capital.
Population 66 135 900
Map India – Cc

Madina do Boé *Guinea-Bissau* Small town 150 km (93 miles) east of Bissau. In 1980, it was designated to become the country's administrative capital, but by 1994 this had not happened yet.
Map Guinea – Ba

Madison *USA* State capital of Wisconsin, standing on an isthmus between Lake Mendota and Lake Monona, 190 km (almost 120 miles) north-west of CHICAGO. Founded in 1836 and named after the United States' fourth president, James Madison, the city lies in the middle of a rich dairy-farming area. It is the seat of the University of Wisconsin, which was also established in 1836.
Population (city) 191 280; (metropolitan area) 367 080
Map United States – Ib

Madras *India* Capital of Tamil Nadu State and the country's main east coast port. Because it lies on the sandy Coromandel Coast, its artificial harbour is protected by a breakwater started in the late 19th century. Founded as Fort St George by the British East India Company in 1639, the city prospered on cotton and textiles. Later, however, its status as the most important town of British India, came under threat from the rapidly growing Calcutta. Indian independence in 1947 gave it a new boost, and its industries today include large engineering works and car plants; it is also the centre of the Tamil film industry.

Madras is India's fourth largest city, after Calcutta, Bombay and Delhi. It has the country's

Macedonia

*A COUNTRY WHOSE NEIGHBOUR
CONTESTS ITS RIGHT TO USE A NAME
RESONANT OF AN ANCIENT EMPIRE*

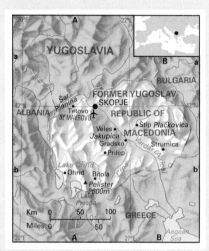

This mountainous country in the heart of the Balkan Peninsula has laid claim to a name that is rich in historical associations, for it was from his base in the adjoining Greek region of Macedonia that Alexander the Great built a vast empire in the 4th century BC. Inevitably, the choice of name has led to disagreement with Greece. It also caused some initial difficulties in obtaining international recognition.

Consisting of the southernmost portion of the former republic of Yugoslavia, Macedonia held a referendum in September 1991 and declared its independence from Yugoslavia on 20 November 1991. Other European Union countries recognised the new state under the name of the Former Yugoslav Republic of Macedonia, but Greece refused to do so and imposed a blockade when the new government rejected its conditions. The Greeks agreed to recognise the new country only if it dropped the name of Macedonia, changed its flag –

which included a sun symbol recalling the dynasty of Alexander the Great – and its constitution – which stated that the 'Republic cares for the rights and status of those persons belonging to the Macedonian people in neighbouring countries' – and forfeited all claims to former territories.

The Greeks were convinced that the Former Yugoslav Republic of Macedonia had designs on their northern province, also called Macedonia. A large tract of land, always the focal point of squabbles, Macedonia was divided up only in 1913, after the Balkan wars and after having been under Turkish rule for more than five centuries. The northern part went to Serbia, the south to Greece. From 1919, until the declaration of the independent state of Macedonia in 1992, the north formed part of Yugoslavia.

Today's inhabitants of Macedonia are only distant relations of the ancient Macedonians of the time of Alexander the Great – if indeed there is any relation at all. As far back as the 6th and 7th centuries AD, Slav peoples came to settle in the mountainous land on either bank of the Vardar. During the Middle Ages, they were ruled alternately by Serbs and Bulgarians. Both in terms of language and in their Orthodox faith, the present-day Macedonians are related to the Serbs and Bulgarians.

There have been tensions over demands for political and educational autonomy by the large Albanian minority. The Albanians are significant because they live in three countries, Albania, Serbia and Macedonia, and have the sympathy of the most important regional power, Turkey. Several hundred American troops were sent to Macedonia as part of a United Nations peacekeeping force, although America had refused to commit ground forces to other parts of the former Yugoslavia,

By the nature of its position, Macedonia is the link between Greece and the Slav countries of south-eastern Europe. The most important international traffic arteries of the Balkan Peninsula lead through the valley of the Vardar. The river, which flows into the Aegean Sea, has carved a deep course through a mountainous region continually shaken by earthquakes. North of Skopje, Macedonia's capital – which was hit by a severe earthquake in the summer of 1963 – it has eroded an easy course all the way into Serbia. There are several basins left in the mountainous terrain, which

in the vicinity of Lake Ohrid and Lake Prespa are often flooded.

All of these are surrounded by high mountain ranges reaching more than 2000 m (6562 ft). In the rain shadow of the mountain chains there is a continental climate with hot summers and very cold winters. Through the Vardar Valley, though, the more temperate Mediterranean climate advances a little inland.

The Macedonian economy is principally based on cereal crops. Where the climate is less harsh and there is irrigation, cotton, rice, citrus fruit and tobacco do well. The republic mines zinc, chromium, iron ore, lignite and asbestos. After the end of the Second World War, industries were developed in Skopje and in a few other centres.

But despite industrialisation and despite its natural resources, Macedonia is still one of the poorest Balkan states, with an unemployment rate of about 30 per cent and little technological development.

With sanctions imposed against Serbia throughout the early 1990s and the Greeks blockading its southern border, Macedonia's future seemed bleak indeed, especially for a newly formed republic which desperately sought the change to a market economy after decades of floundering under a centrally controlled system.

MACEDONIA AT A GLANCE		
Official name Former Yugoslav Republic of Macedonia		
Area 25 856km² (9981 sq miles)		
Population 2 193 950 **Per km²** 85		
(Per sq mile) 220)		
Capital Skopje		
Government Republic		
Currency 1 Denar = 100 Deni		
Languages Macedonian 70%, Albanian 21%, Turkish 3%, Serbo-Croat 3%, other 3%		
Religions Eastern Orthodox 59%, Muslim 26%, Catholic 4%, Protestant 1%, other 3%		
Climate Hot and dry summers, with cold winters and heavy snowfalls		
Land use forests 30%, pastures 20%, cultivation 10%, other 40%		
Traditional exports Machinery and transport equipment, raw materials, food, live animals, tobacco and chemicals		
Annual income per head (US$) 1400		
Population growth (per thous/yr) 9		
Life expectancy (yrs) Male 71 **Female** 75		

oldest town charter (granted in 1688). It also has India's oldest English church, St Mary's (founded in 1678), in which the British general and statesman Robert Clive (known as 'Clive of India', 1725-74) was married in 1753.

Population 3 795 030

Map India – Ce

Madre de Dios *Peru* South-eastern department which is crossed by several rivers. Its capital, Puerto Maldonado (population 12 700), stands at the confluence of the Tambopata and Madre de Dios rivers. Its economy is fairly diverse: gold is mined in the region – and although the forest here continues to yield timber and brazil nuts, it is cattle ranching that is making the most progress. The Madre de Dios River, which is 960 km

(600 miles) long, rises in the south-east and flows first north, then east into Bolivia to join the Beni.

Population 49 000

Map Peru – Bb

Madrid *Spain* Capital and largest city, in the centre of the country. Its name may be derived from the Arabic name *medshrid*, which comes from *materia*, meaning 'timber' – a reference to the forests that grew here when the Moors built a fortress on the site in the 9th century.

Today Madrid is Spain's second largest industrial centre, after Barcelona. Chief among its manufactures are aircraft, electrical equipment, agricultural machinery, and leather goods. The *Palacio Real*, home of the Spanish royal family, stands west of the *Plaza Mayor* with its many

old buildings and its maze of streets and alleys. Madrid has many museums, including the Prado Museum which has an outstanding art gallery containing over 5000 painting.

Madrid was captured from the Moors by Castilians in 1083 and made the capital of Spain in place of Valladolid by Philip II in 1561. Because of its remoteness, however, it became Spain's largest city only after the coming of the railways in the 19th century. The city came under heavy fire during the Spanish Civil War (1936-9). Wheat, vines and olives are grown in the surrounding province (also known as Madrid).

Population (city) 3 010 490; (province) 4 947 550

Map Spain – Db

Madagascar

AN AFRICAN STATE, THE FOURTH LARGEST ISLAND IN THE WORLD, IS ENDOWED WITH ABUNDANT NATURAL RESOURCES – BUT IS EXTREMELY POOR

Although it lies only about 400 km (250 miles) off the East African coast, the island of Madagascar is more closely related in its people, culture and language to the islands of Indonesia some 5000 km (about 3000 miles) eastwards across the Indian Ocean than to Africa. The islanders probably owe the Asian element in their ancestry to migrants who began settling here about 2000 years ago. Madagascar is unusual in that all 18 different ethnic groups speak the same language – Malagasy. There are local dialects, but they are mutually intelligible.

Madagascar is the world's fourth largest island (after Greenland, New Guinea and Borneo, but excluding Australia), stretching about 1600 km (1000 miles) from north-east to south-west and up to 580 km (360 miles) across. Portuguese explorers landed here in 1500 but for three centuries the islanders resisted all attempts at colonisation. Madagascar finally came under French rule from 1896 until 1960.

The island is distinctive not only in its people but also in its wildlife. Its animals (and plants) evolved in long isolation after the island broke away from the African continent some 150 million years ago. The lemurs are probably the best known example of the unique Madagascan animal species. To protect them and other rare plants and animals more national parks are now being developed. The Madagascan government has a policy of combining environmental issues with development projects for the community.

Madagascar is one of the world's poorest and least developed countries. Most of its more than 13 million people work on the land and over 80 per cent grow only enough to feed themselves. The staple food is rice. A decline in the rice crop in recent years, caused partly by persistent bad weather and soil erosion, and exacerbated by an expanding population, led to rice rationing in 1980 and the economy's collapse. In 1986, the government introduced a five-year programme to boost rice production and become self-sufficient in overall food production. But political upheavals in 1990-1 disrupted the programme. Also, in 1990-2 the failure of rains virtually destroyed the rice crop. In February 1994, the island was hit by a cyclone – the worst since 1927 – which destroyed the homes of 150 000 people, as well as the island's main port and oil refinery, and left large parts of agricultural land devastated.

A series of high savannah-covered plateaus occupy the centre of the island, reaching up to 2886 m (9468 ft). To the east, the forested mountains fall steeply to the hot, humid coast. To the west and south-west, the land falls more gradually through drier grassland and scrub.

Cassava, a starchy root, is almost as widely grown as rice, but only about 5 per cent of the total land area is cultivated. Some 58 per cent is pasture, but the huge number of cattle has led to over-grazing. The cattle, numbering over 10 million, are of poor quality. Coffee is the main export earner, with cloves, vanilla – Madagascar is the world's largest exporter of vanilla – and sugar. Industries, apart from mining, include sugar refining.

Most of the people live on the central plateau and east coast. The largest ethnic group, the light-skinned Merina (or Hova) of Indonesian origin, form nearly a quarter of the population. They inhabit the central plateau. The Betsimisaraka are *côtiers*, or coast dwellers. They have controlled the government since 1960, and there is antagonism between them and the plateau people. The coastal groups are of mixed Indonesian, African and Arab ancestry.

Upon independence in 1960, Madagascar became known as the Malagasy Republic. However, the country's constitution was changed by referendum in 1975, shortly after the socialist military government of Lieutenant Commander Didier Ratsiraka took power, and it became the Democratic Republic of Madagascar. President Ratsiraka was re-elected for a second seven-year term in 1982.

In 1989, he was re-elected for yet another term and after pressure for reform, he issued a decree in 1990 announcing a form of multi-party government. In August 1991, thousands of Madagascans took to the streets of Antananarivo to demand true democracy, and by October that year a transitional government had been formed. In the November 1992 and February 1993 elections, President Ratsiraka's 17-year rule came to an end and Dr Albert Zafy, leader of the *Comité de Forces Vives* party, took over as president.

The Madagascans now face a better future. Deposits of bauxite, oil, chromite and iron ore have been discovered and prospects for tourism are bright – after declining in 1991, a record number of tourists visited in 1993.

▲ **HIGH CAPITAL Antananarivo lies on a ridge 1249 m (4094 ft) above sea level. Much of the city dates from the French occupation of the island from 1896 to 1960.**

MADAGASCAR AT A GLANCE	
Official name Republic of Madagascar	
Area 587 040 km² (226 603 sq miles)	
Population 13 500 000 **Per km²** 23 **(Per sq mile** 60)	
Capital Antananarivo	
Government Republic	
Currency 1 Madagascar franc = 100 centimes	
Languages Malagasy and French (official), local languages	
Religions Indigenous religions (52%), Christian (41%), Muslim (7%)	
Climate Tropical; cooler at altitude. Average temperatures in Antananarivo range from 13.3°C (56°F) in July to 20°C (68°F) in February	
Land use Grazing and other 69%, forest 26%, cultivation 5%.	
Main primary products Rice, cassava, mangoes, vanilla and other spices, bananas, potatoes, sugar cane, maize, coffee, pepper, cattle, timber; graphite, chromite, mica, semi-precious stones	
Major industries Agriculture, oil refining, forestry, food processing, mining, cement, paper, textiles, glassware	
Main exports Coffee, vanilla, cloves, sugar, graphite	
Annual income per head (US$) 230	
Population growth (per thous/yr) 32	
Life expectancy (yrs) Male 51 **Female** 55	

Madura *Indonesia* Island covering about 5290 km² (2042 sq miles) off the north-east coast of Java. Cattle, teak, copra and coconut oil are Madura's main products; revenue from tourism is also increasing.

Population	1 860 000
Map	Indonesia – Dd

Madurai *India* Manufacturing city lying 215 km (134 miles) north-east of the country's southern tip and an ancient Tamil city and seat of learning. For more than 14 centuries, from the 5th century BC, it was the capital of the Pandya dynasty, which had conquered northern Ceylon (Sri Lanka). The city has several palaces; its Great Temple, part of a complex covering 6 hectares (15 acres), has the Hall of 1000 Pillars. Madurai's economy is centred on the production of cotton and textiles.

Population	952 000
Map	India – Ce

Mae Mo *Thailand* Town situated 550 km (about 340 miles) north of the capital, BANGKOK. It lies in the centre of a large deposit of lignite (brown coal), the mining of which began in the mid-1980s to reduce Thailand's dependence on imported oil. The Mae Mo River, which runs through the town, is 60 km (37 miles) long, and drains into the Wang River south of Lampang.

Population	28 700
Map	Thailand – Ab

Mae Nam Khong *Thailand* See THANLWIN

Mae Suai *Thailand* Remote northern town lying about 700 km (435 miles) north of the capital, BANGKOK. Its economy is largely based on black market trade with nearby Myanmar. In this way, consumer goods from Thailand pass into Myanmar in return for cattle, jade and precious stones. Rice is the chief crop of the mainly agricultural district of the same name.

Population (town)	65 500
Map	Thailand – Ab

Maebashi *Japan* City in central HONSHU, lying 100 km (62 miles) north-west of the capital, Tokyo. Once an important silk town, it now depends largely on local commerce.

Population	286 000
Map	Japan – Cc

Maelstrom *Norway* See LOFOTEN

maestrale Cold northerly to north-westerly wind blowing in northern Italy, corresponding to the MISTRAL of France.

Mafikeng (Mafeking) *South Africa* Town in North-West Province, 260 km (162 miles) west of Pretoria. Formerly known as Mafeking, the town survived a seven-month siege during the Anglo-Boer War (1899-1902). Until 1965 it was the seat of the British administration of Bechuanaland, now Botswana. In this way it was unique, in that it was situated outside the country being administered. Mafikeng adjoins the urban area of Montsiwa and the new town of Mmabatho; in 1994, the combined conurbation, under the name of Mafikeng, was proclaimed the capital of the newly founded North-West Province.

Population	29 400
Map	South Africa – Cb

Mafraq *Jordan* Town and road junction of routes from Iraq and The Gulf, situated 50 km (31 miles) north-east of AMMAN. It also lies on the rail route from Amman to Syria and it has a small airport. A Turkish castle dating from the 16th century is a tourist attraction.

Population	26 000
Map	Jordan – Ba

Magadan *Russia* Port and industrial city on the north shore of the Sea of Okhotsk in Russia's Far East. It was founded in 1933 on a fine harbour ringed by saw-toothed mountains. Besides being a busy naval base, it is also the administrative centre of a huge region covering 1 200 000 km² (463 000 sq miles) – more than five times the size of Britain. It is situated in the middle of a rugged landscape, noted for its bitter winters. The majority of Chukchi, Evenky and Yakut peoples, who inhabit the region, are reindeer herders. Magadan has large fish-canning and engineering plants; tin and gold are mined in the vicinity.

Population (city)	143 000; (region) 534 000
Map	Russia – Qc

Magadi, Lake *Kenya* Lake in the Great Rift Valley, 85 km (53 miles) south-west of NAIROBI. It is covered by a white layer of trona, a sodium-based mineral used in the manufacture of glass, which is pumped ashore and exported through Mombasa as soda ash. The small town of Magadi is a railhead on the east side of the lake.

Map	Kenya – Cb

Magallanes *Chile* Southernmost region of the country, stretching across the Strait of MAGELLAN to Cape HORN, and including the TIERRA DEL FUEGO in the south. Its major industries are sheep farming and oil. All Chile's oil fields are in either Tierra del Fuego or in an area just north of the Strait of Magellan. They supply the country with 25 per cent of its fuel needs. The discovery of coal at Pecket and the development of tourism around the capital, PUNTA ARENAS, have added to the region's economic importance.

Population	143 060
Map	Chile – Ae

Magat *Philippines* A tributary of the Cagayan River, in northern LUZON, which has been developed for hydroelectric power.

Map	Philippines – Bb

Magdalena *Colombia* The country's longest river. Rising in the Andes south of Popayán, it flows north for 1550 km (963 miles) between the eastern and central ranges of the Andes, before draining into the Caribbean Sea near Barranquilla. Before a network of roads was built during the 20th century, the Magdalena was the main artery of communication between the coast and the interior. Goods were carried to and from Barranquilla by boat.

Map	Colombia – Bb

Magdeburg *Germany* Industrial city, first chronicled in the 9th century, 120 km (75 miles) south-west of Berlin. It stands on the Elbe River and has become a busy river port. Although the city was badly damaged during the Second World War, its cathedral, dating from the 13th century, has been carefully restored. The city's factories produce machinery, chemicals and precision instruments, among other goods. The city has

produced many famous people, including the German physicist Otto von Guericke (1602-86) after whom a technical college has been named, and the composer Georg Philipp Telemann (1681-1767); a Telemann festival in is held in Magdeburg every three years.

Population	278 800
Map	Germany – Db

Magellan, Strait of *South America* Strait, 560 km (nearly 350 miles) long and 3-32 km (2-20 miles) wide, separating mainland South America from TIERRA DEL FUEGO, and linking the ATLANTIC and PACIFIC oceans. Overlooked by snow-capped peaks and strewn with islands, the strait has great scenic beauty. It was discovered in 1520 by the Portuguese navigator Ferdinand Magellan (Fernão Magalhães, 1480-1521), who wept with joy when his three ships emerged safely into the Pacific (after five weeks sailing the strait).

Navigation difficulties faced by sailing ships were compounded by the high tidal range (resulting in fast currents) and frequent fogs. However, the importance of the strait as a shipping route increased when steamships came into use – until the opening of the PANAMA CANAL in 1914 provided a short cut between the Atlantic and Pacific and greatly reduced the importance of the strait. It is now an international waterway, even though it lies within Chile's territorial waters.

Map	Chile – Be

Maggiore, Lake *Italy/Switzerland* Alpine lake covering 212 km² (82 sq miles) in northern Italy and southern Switzerland. Shaped by an Ice Age glacier, it has a basin 372 m (1220 ft) deep. Many lakeside villas and hotels are dotted along its shores; resorts such as Verbania, Stresa and Pallanza stand on the western shore. LOCARNO is the main town on the Swiss part of the lake. Although the surrounding area is mountainous, it is extensively used for farming.

Map	Italy – Ba

Maghreb (Maghrib) *North Africa* Region of north-west Africa comprising the states of ALGERIA, TUNISIA and MOROCCO. The name is derived from the Arabic *Djezira el Maghreb*, which translates as 'Island of the West'. This island of land between the Sahara and the Mediterranean is the western extremity of the Arab world. It has a cultural as well as geographical unity, since all three countries were once ruled by France.

Map (Algeria, Tunisia and Morocco)	Africa – Ca-Cb, Da-Db, Ea

magma Molten matter formed within the earth's crust or upper mantle, which may rise towards the earth's surface and consolidate on cooling, to form IGNEOUS ROCK.

magma chamber Reservoir in the earth's crust where magma that has risen from greater depths accumulates. It is believed that volcanoes are connected to magma chambers, and that these are the source of the material that the volcanoes ultimately eject onto the surface.

magnesian limestone See DOLOMITE

magnetic poles See GEOMAGNETIC POLES

magnetite Black mineral, iron oxide, often occurring with titanium or magnesium, and an important ore of iron. A magnetically polarised piece of the mineral is called a LODESTONE.

magnetosphere Outermost part of the earth's atmosphere, which blends into interplanetary space, and in which charged particles are trapped and dominated by the earth's magnetic field. See THE DIFFERENT LAYERS OF THE ATMOSPHERE, p 53

Magnitogorsk *Russia* Industrial city on the south-eastern slopes of the Ural Mountains, 385 km (240 miles) south of Yekaterinburg. It was established in 1929 as a new town for workers brought in to the local iron mines and steel works. Apart from iron and steel, the city manufactures military vehicles and weapons, mining and transport equipment, and glass.

Population 443 000
Map Russia – Gc

Magwe *Myanmar (Burma)* Market town on the Ayeyarwady (Irrawaddy) River, which lies 230 km (143 miles) south-west of Mandalay, and capital of the division of the same name covering 44 820 km² (17 305 sq miles). The town is situated in one of the country's driest areas, near Myanmar's most important gas and oil fields, at Man and Yenangyaung.

Population (division) 3 241 000
Map Myanmar – Bb

Mahabalipuram (Mamallapuram; 'Seven Pagodas') *India* Ancient port and modern tourist centre on the east coast, 60 km (37 miles) south of MADRAS. Its Hindu religious complex dating from the 7th to the 8th century includes caves, temples, a rock relief depicting 'the Penance of Arjuna' (Arjuna was a legendary prince who fought at KURUKSHETRA), and the remains of seven pagoda-like temples made of single, huge blocks of stone. Mahabalipuram is renowned for its Shore Temple washed by the ocean's tides.

Map India – Ce

Mahajanga *Madagascar* Port situated 400 km (nearly 250 miles) north-west of the capital, ANTANANARIVO. Founded in about 1700 by Arab traders, it was formerly called Majunga. It manufactures processed foods, textiles and tobacco.

Population 121 970
Map Madagascar – Aa

Maharashtra *India* West coast, mainly Marathi-speaking state, covering some 307 690 km² (118 768 sq miles) on the dry, western Deccan Plateau. Conditions in the area are generally poor for farming, although grains such as millet are grown in some areas, and cotton and sugar cane are cultivated in parts that are under irrigation. Rice is grown on the wetter coastal strip below the Western Ghats.

Manufacturing industries – mainly textiles and machinery, and food processing – are concentrated in the cities of Thane, Pune and Bombay, the capital. The area is susceptible to earthquakes; one that occurred in 1993 killed more than 12 000 people.

Population 78 706 700
Map India – Bd

Mahaweli Ganga *Sri Lanka* The island's longest river. It rises in the wet central mountains and flows for 330 km (205 miles) generally northwards across a dry plain to Koddiyar Bay, near

Trincomalee. Because the Mahaweli Ganga has more water than can be used by the farmers along its lower valley, an irrigation scheme transfers water to neighbouring rivers, such as the Kala Oya and Maduru Oya, and to ancient reservoirs, known as tanks. The Victoria Dam is the largest dam of the scheme, which also generates electricity.

Map Sri Lanka – Bb

Mahé *Seychelles* Chief island of the group, situated in the western Indian Ocean. Covering 153 km² (59 sq miles), it makes up half the land area of the Seychelles. Its main town and the island state's capital is the seaport of VICTORIA. Mahé is mountainous, its highest point reaching some 905 m (2969 ft).

Population 59 000
Map (Seychelles) Indian Ocean – Bc

Mahilyow (Mogilev) *Belarus* Industrial city situated on the banks of the Dnyapro (Dnieper) River, 180 km (112 miles) east of MINSK. The city was badly damaged during the Second World War, when nine in 10 of its population were killed. It has since developed into a key industrial centre making metal goods, chemicals, synthetic fibres, leather goods, beer and wine.

Population 363 000
Map Belarus – Cb

Mahón *Spain* Seaport and capital of Minorca in the BALEARIC ISLAND, and, during the 18th century, a British naval base. It exports wine, cheese, brandy and agricultural produce, and manufactures shoes and jewellery.

Population 21 540
Map Spain – Hc

Mahore *Indian Ocean* See MAYOTTE

Mähren *Czech Republic* See MORAVIA

Maiduguri *Nigeria* Capital of Borno State, 100 km (62 miles) south-west of Lake Chad. Since the arrival of the railway in 1964, it has developed into an attractive town, with wide tree-lined avenues and a bustling daily market.

A busy industrial centre, its factories produce shoes, cotton products and palm oil. It is linked

▲ TEMPLE ON THE SHORE The many intricately carved Hindu shrines at Mahabalipuram in India were built for the powerful Pallava princes. They include the wave-washed Shore Temple.

by road to Cameroon and Chad. Although it is the country's largest state, at 116 400 km² (44 931 sq miles), Borno is sparsely populated. Most of its inhabitants are farmers – and the chief crops grown here are millet, sorghum, groundnuts and vegetables; cattle are also reared. The Bornu plain, beside Lake Chad, was the centre of the Bornu Empire, which flourished in the 10th and 11th centuries.

Population 225 100
Map Nigeria – Ca

Main *Germany* River rising near BAYREUTH in north-eastern Bavaria, and flowing for 524 km (325 miles) through Würzburg and Frankfurt to join the Rhine at MAINZ. It is navigable along most of its course. Since 1992, the Main-Danube Canal has provided an important link between the Rhine and Danube rivers.

Map Germany – Cd

Maine *USA* Most north-easterly of the country's states, covering 86 176 km² (33 265 sq miles). In addition to its mountainous landscape – it rises to 1605 m (5268 ft) at Mount Katahdin – it has more than 2500 lakes. Offshore of Maine's jagged coastline on the Atlantic Ocean are more than 2000 islands.

Forestry is the main industry, but farming (blueberries and potatoes), fishing (especially lobster), and tourism are other important sources of income. AUGUSTA is the state capital and Portland is its largest city.

Population 1 227 900
Map United States – Ma

Mainz *Germany* Capital of Rhineland-Palatinate State and port at the meeting point of the RHINE and the MAIN rivers, 30 km (19 miles) south-west of FRANKFURT am Main. A market town since Roman times, Mainz hosts Germany's biggest wine market in August-September each year. Machinery, glassware, sparkling wine, toothpaste and cement are manufactured at its factories. In 1452, Johann Gutenberg (c1398-1468), a resident of Mainz, revolutionised printing when he became the first person to use movable type to print a book (the Bible). Mainz's Museum of Printing has a Gutenberg Bible of 1452-5 and a recreation of Gutenberg's printing works. Mainz has a university that was founded in 1477. The city is also known for its annual carnival.

Population 179 500
Map Germany – Cd

Majorca (Mallorca) *Spain* Largest of the BALEARIC ISLANDS, covering some 3639 km² (1405 sq miles). It is one of the most popular tourist destinations in the Mediterranean.

Population 296 750
Map Spain – Gc

Majunga *Madagascar* See MAHAJANGA

Majuro *Marshall Islands* Atoll in the Ratak Chain in the east of the territory. Three linked islands at its eastern end – Dalap, Uliga and Darrit – form the Marshall Islands capital, also called Majuro; the islands were badly damaged by a huge wave in 1979.

Population 8700
Map Pacific Ocean – Db

Makarska *Croatia* Ancient Dalmatian port and, before the civil war that shattered the region during the 1990s, a popular tourist resort, 56 km (35 miles) south-east of SPLIT. Its palm-lined streets and seafront lie beneath the magnificent crags of Biokova Mountain which rises to 1962 m (6435 ft). The town has retained its old-world charm by keeping modern hotel developments at a distance – at Tucepi, Podgora, Zivogosce, Baska Voda, and at Brela, a former fishing hamlet just to the north.

Population 7000
Map Croatia – Cc

Makassar *Indonesia* See UJUNG PANDANG

Makassar Strait (Macassur; Makasar Strait; Ujung Pandang Strait) *South-east Asia* Broad strait, 960 km (595 miles) long and 130-370 km (81-230 miles) wide, running between Borneo and SULAWESI (Celebes) in Indonesia. It links the CELEBES SEA to the JAVA and FLORES seas. In January 1942, a Japanese convoy fought American and Dutch air and naval forces here. Although the Allies put up stiff resistance for five days, they could not stop the Japanese invasion of Borneo.

Map Indonesia – Ec

Makedonia *Greece* See MACEDONIA

Makiyivka (Makeyevka) *Ukraine* Industrial city in the south, situated near the city of Donets'k in the Donets coal-mining basin. It was founded as a steel town in 1899; besides steel, it produces machinery and chemicals.

Population 427 000
Map Ukraine – Db

Makhachkala *Russia* Port and industrial city on the west coast of the Caspian Sea, north of the Caucasus range. It was rebuilt following a devastating earthquake in 1970.

Makhachkala is also the capital of Dagestan, an autonomous republic of the Russian Federation

covering 50 300 km² (19 400 sq miles). This mountainous region, which was annexed from Persia in 1723, is home to some 30 nationalities. It was the setting of the novel *Hadji Murad* by Leo Tolstoy (1828-1910). Its factories produce textiles, footwear and cement.

Population (republic) 709 000; (city) 269 000
Map Russia – Fd

Ma-kung *Taiwan* See PESCADORES

Makurdi *Nigeria* Capital of Benue State, 580 km (360 miles) north-east of Lagos. It is situated on the rail route that links Kaduna to Port Harcourt, and on the road between Jos and Enugu. The Benue River steamers use the town as their headquarters during the flood season. The town is a market centre; its manufactures include cigarettes and soft drinks.

Population 86 800
Map Nigeria – Bb

Malabar Coast *India* The coastal strip of the south-western State of Kerala. It was known to Greek, Roman and earlier Middle Eastern voyagers, who sailed here and back on the seasonally changing winds. With more than 3000 mm (120 in) of rain a year, its natural vegetation is tropical rain forest.

Map India – Be

Malabo *Equatorial Guinea* National capital and chief port of Bioko Island. It stands on the north of the island, on the rim of a volcanic crater which has been breached by the sea, in this way creating a natural harbour. Cocoa and hardwood are its main exports. Malabo has a cathedral that was built in 1916.

Population 37 200
Map Equatorial Guinea – Aa

Malacca *Malaysia* See MELAKA

Malacca, Strait of Sea passage, 800 km (nearly 500 miles) long and 50-180 km (31-112 miles) wide, between the west coast of peninsular Malaysia and Sumatra, linking the ANDAMAN SEA and the South CHINA SEA. With Singapore at its southern end, it is one of the world's busiest shipping lanes, where liners and cargo ships sail alongside traditionally rigged Arab dhows, Chinese junks and Malaysian praus.

Map Malaysia – Ab

malachite Green mineral, copper carbonate used as a source of copper and for ornamental stoneware.

Málaga *Spain* Port and somewhat polluted tourist resort on the south Mediterranean coast. The adjacent province, also called Málaga, includes the resorts of Fuengirola, Torremolinos and Marbella. Building materials, metal goods, foodstuffs, fertilisers, textiles, beer and wine are produced here. The painter and sculptor, Pablo Picasso (1881-1973) was born in Málaga.

Population (town) 522 110; (province) 1 160 840
Map Spain – Cd

Malanje (Malange) *Angola* Market town and railhead linked to the national capital, Luanda, 350 km (217 miles) to the west. Malanje is the capital of a coffee- and cotton-producing province of the same name. However, because of the

civil war in Angola, little farming took place here during the early 1990s – and the town's population had to rely on relief food.

Population (town) 31 600; (province) 892 000
Map Angola – Aa

Mälaren *Sweden* Lake stretching some 110 km (about 70 miles) westwards from Stockholm, and covering an area of 1140 km² (440 sq miles). Many historic ruins, castles, palaces and residential villas dot its shores and the more than 1000 islands that are scattered across its waters.

Many of the buildings date back to the 16th and 18th centuries, including Drothingholm Summer Palace, the residence of the Swedish royal family. The lake is a popular weekend destination for Stockholm families.

Map Sweden – Cd

Malaspina Glacier *USA* One of the world's largest ice sheets. It lies in the St Elias Mountains in St Elias National Park, just west of Yakutat Bay on the south coast of Alaska. The glacier is more than 300 m (1000 ft) thick and covers some 3900 km² (1500 sq miles).

Map Alaska – Cc

Malawi See p 423

Malawi, Lake (Lake Nyasa; Niassa) *East Africa* Third largest of East Africa's lakes, after Lake VICTORIA and Lake TANGANYIKA, covering 23 300 km² (9000 sq miles). It is about 80 km (50 miles) across at its widest point; its greatest known depth is 678 m (2224 ft). Most of the lake lies within Malawi, but part of it is in Mozambique and part borders Tanzania.

Today, fishing vessels and tourist boats crisscross the lake. When the Scottish missionary and explorer David Livingstone first visited it in 1859, he named it 'Lake Nyasa' after being told the word 'nyasa' when he asked the locals what the lake was called. However, 'nyasa' was simply the local word for 'lake'.

The level of the lake varies. In 1915 it dropped to a record low of 469 m (1540 ft) above sea level; in 1980 it reached its highest level – 475 m (1560 ft) above sea level.

Map Malawi – Ab

Malaysia See p 424

Malbork (Marienburg) *Poland* Industrial town and trading centre 58 km (36 miles) south-east of the Baltic port of Gdańsk. Its restored castle, originally built by the Teutonic Knights between 1274 and 1400, and consisting of three turreted strongholds, is one of Europe's largest medieval fortresses. Malbork's Amber Museum, in one of the turreted strongholds, has a display of jewellery and other items made in the town from fossil resin found on the nearby Baltic coast.

Population 38 900
Map Poland – Ca

Maldives See p 426

Malé *Maldives* Principal atoll covering 2.6 km² (1 sq mile), lying in the centre of the Maldives. It is the country's capital, administrative centre and the only urban area on the islands. It trades in fish (bonito and tuna), copra, and other coconut products. But it is tourism that is becoming increasingly important to the Maldives' economy, and the

Malawi

WHERE LAKES, RIVERS, WATERFALLS, WOODS AND MOUNTAINS PROVIDE A PARADISE FOR THE ADVENTUROUS TOURIST

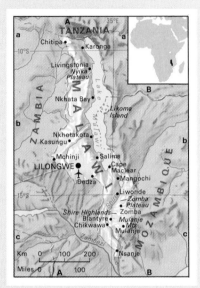

The name Malawi means 'the lake where sun-haze is reflected in the water like fire', and it is the vast and splendid Lake MALAWI – the third largest lake in Africa – from which the country takes its name. Nyasa, the name originally given to the lake by the explorer David Livingstone after he discovered it in 1859, is actually a local word meaning 'lake', or 'broad water'. As Nyasaland, the country was under British rule from 1891 to 1964.

Malawi's great natural beauty lies not only in its shimmering lake but in its park-like grasslands and woodland scenery. In the south, the SHIRE River with its spectacular waterfalls flows through a valley overlooked by wooded, towering mountains.

For all its scenic diversity, Malawi lies in a narrow strip never more than 160 km (100 miles) wide, and the country's share of Lake Malawi covers one-fifth of the lake's area. The lake, renowned for its golden beaches and teeming fish and bird life, lies in a deep valley. Crocodiles have been removed from resort areas such as Salima, making them safe for water sports and swimming. There are several national parks, including Nyika, Kasungu and Lengwe.

Malawi is a tourist's paradise, but until recently, tourists and the local people had to follow strict dress rules: women were not allowed to wear trousers or skirts above the knee, and men were forbidden to wear their hair longer than collar length. Like the country's rigid censorship, the Dress Act of 1973, repealed in 1993, reflected the attitudes of Malawi's first president, Dr Hastings Kamuzu Banda.

Banda became president when the country obtained its independence in 1964 and had himself declared president for life by his parliament in 1971. This was not difficult since he was also head of the Malawi Congress Party, the only party represented in the national assembly. Besides deciding what the Malawians should wear and be permitted to read, Banda was not known for being a champion of human rights. During his 30 years as president, many thousands of political prisoners were detained without trial, some tortured and many harassed.

By 1992, criticism of his regime led to much foreign aid being frozen and growing pressure finally forced Banda to legalise opposition parties. Malawi had its first multiparty elections in 1994, which brought victory for the United Democratic Front. Banda, who was in his nineties and in failing health, was replaced by Bakali Muluzi as the new president. Early in 1995, Banda and his close associate, the former Minister of State, John Tembo, were charged with the murder in 1983 of four politicians believed to have opposed Tembo's appointment as acting president in Banda's absence.

Malawi depends on foreign aid for more than 80 per cent of its development budget. It is one of the 10 poorest countries in the world; more than 70 per cent of its population cannot read or write and there is an infant mortality rate of more than 14 per cent. The government's priority has been to tackle poverty and improve health and education services. Agriculture is still the predominant occupation, employing 90 per cent of the population and accounting for 40 per cent of gross domestic product (GDP).

Many Malawians live off their own crops. The government has encouraged wealthy Malawians to invest in plantation farming to increase agricultural exports. This has resulted in more and more land being held by an elite group and has caused an acute land shortage for small-scale peasant farmers.

Apart from maize, crops also include groundnuts, coffee, beans, cassava, millet, and rice on the lake shores, as well as tobacco, tea and sugar cane – the main exports – grown on the large, privately owned estates. Most tea estates are on lush, terraced hillsides in the south. Tobacco is grown mainly on the fertile central plateau.

Exporters rely heavily on the two railway lines across Mozambique to the coast, but for years these were disrupted by the civil war in Mozambique which finally ended in 1993. Another prominent trading partner is South Africa; some Malawians are migrant workers in the South African mines.

Malawi has deposits of bauxite and coal but little is mined because these lie in inaccessible areas. Manufacturing, using hydroelectricity, has been developed rapidly in recent years, but imports of manufactured goods are still large.

Most people live in rural areas. Nearly all (98 per cent) belong to one of nine groups – the Chewa, Nyanja, Lomwe, Yao, Tumbuko, Sena, Tonga, Ngoni and Ngonde. More than two-thirds of Malawi's inhabitants live in the central and southern regions – and, in fact, the 1994 elections were won on a regional-ethnic basis, with most voters from the south voting for Muluzi; his two opponents, Chakufwa Chihana and Hastings Banda, won the northern and central regions respectively.

MALAWI AT A GLANCE	
Area 118 480 km² (45 735 sq miles)	
Population 10 055 000 **Per km²** 85 **(Per sq mile)** 220)	
Capital Lilongwe	
Government Republic	
Currency 1 kwacha = 100 tambala	
Languages Chichewa, English	
Religions Christian (75%), Muslim (20%), indigenous beliefs	
Climate Tropical; cooler in the highlands. The rainy season is from November to March. Average temperature in Lilongwe ranges from 15.5°C (60°F) in July to 23°C (73°F) in November	
Land use Forest 50%, cultivation 25%, grazing and other 25%	
Main primary products Maize, rice, cassava, sorghum, millet, beans, groundnuts, sugar, tea, tobacco, cotton	
Major industries Agriculture, food processing, brewing, cement, tourism	
Main exports Tobacco, sugar, tea, groundnuts, cotton, textiles, cereals	
Annual income per head (US$) 210	
Population growth (per thous/yr) 33	
Life expectancy (yrs) 44	

authorities have extended Malé's airport, Hulule, to take jumbo-sized jets. Although there are some hotels and guesthouses in the capital, most visitors tend to travel on to one of the many resort islands.
Population 55 000
Map Indian Ocean – Cb

Mali See p 427

Malines (Mechlin; Mechelen) *Belgium* Cathedral city and Belgium's religious capital since the 15th century, lying 32 km (20 miles) south of the city of ANTWERP. The cathedral of St Rombout, which dates from the 13th century, has a tower nearly 100 m (330 ft) high, and a painting of the Crucifixion by the Flemish artist Anthony van Dyck (1599-1641). Once famous for its lace, the city now makes tapestries, furniture, and beer.
Population 77 000
Map Belgium – Ba

mallee scrub Dense, scrubby thicket of low-growing eucalyptus bushes, found in parts of the semi-arid, subtropical regions of south-east and south-west Australia where the rainfall is only 200-350 mm (8-14 in) a year.

Malmédy *Belgium* Tourist town in the Ardennes hills, about 40 km (25 miles) south-east of LIÈGE. Formerly a part of Germany, it was ceded to Belgium in 1919. It achieved notoriety in 1944, during the Second World War, when about 100 United States prisoners of war were shot there by German soldiers.
Population 10 000
Map Belgium – Ca

Malaysia

DIVIDED BY THE SOUTH CHINA SEA, AN UNLIKELY UNION OF LANDS MORE THAN 700 KM (435 MILES) APART

Jungle-clad mountains, dense and humid forests, vast plantations of rubber trees and oil palms and miles of golden beaches make up Malaysia. In a combination of heat and high humidity, a thousand different kinds of orchid bloom. Lush and tropical, Malaysia is one of the most successful of the developing countries: the world's largest producer of rubber, palm oil, pepper and tin. It is a major source of hardwoods and has recently begun to export oil and gas, as well as machinery and motor vehicles.

All this has resulted from the unlikely union of West (now Peninsular) Malaysia and East Malaysia (SABAH and SARAWAK on the island of Borneo). They are separated by 700 km (435 miles) of the South China Sea; they had no ties with each other, apart from the accident of British colonial rule; and the west was far more developed than the east.

Malaysia's success has been achieved by nearly 19 million people who are a potentially explosive cocktail of races: Malays (50 per cent), Chinese (32 per cent), Indians (9 per cent) and other indigenous peoples. The Malays are politically dominant while the Chinese are economically powerful and thus tensions between them have cast an uneasy shadow over the country's success story.

Much of Malaysia is mountainous and forested: some mountains rise to 2189 m

(7182 ft) at Gunung Tahan on the Malay peninsula and to 4101 m (13 454 ft) at Gunung Kinabalu in Sabah. Rainfall is high everywhere, averaging 2540 mm (100 in) a year.

The ancestors of the Malays came from Yunnan in southern China, and moved into the peninsula between about 2000 and 1000 BC. Arab traders from The Gulf converted the Malays to Islam in the 15th century. MELAKA (Malacca) was made the sultan's seat of government and, because of its strategic position en route to India and China, became a flourishing trading centre.

Soon after, the Europeans arrived. The Portuguese (1509) and the Dutch (1641) in turn occupied Melaka on the south-west coast, before the British took over in 1824. Earlier, the British had established trading settlements on PINANG Island in 1786 and at SINGAPORE in 1819. In 1826 the three were combined as the Straits Settlements and in 1867 they were proclaimed a Crown Colony. In 1874, Britain established control over PERAK and SELANGOR and by 1930 had absorbed the whole Malay peninsula.

The two territories of East Malaysia, however, did not come under formal British rule until after the Second World War. The Briton, James Brooke, put down a revolt against the Sultan of Brunei in 1841; he was rewarded by being made the Rajah of Sarawak. His successors ruled there until the Japanese invasion in the Second World War, at the end of which the rajah gave the country to the British Crown. In the case of Sabah, it was a British company – the British North Borneo (Chartered) Company – that administered the country from 1882 to 1942.

Anti-colonial movements had grown in the Malay Peninsula during the 1920s and 1930s, influenced by the pressures for independence in India. The Japanese occupation of Malaya

(1942-5) taught the Chinese Communist guerrillas that Westerners were not invincible, so giving a new impetus to the independence movement. After the war, groups of guerrillas took to the jungle again, waging a terrorist campaign against the British. The Emergency, as it was called, lasted from 1948 until 1960 when the rebels were finally suppressed.

Britain realised, however, that the situation could not be resolved by military means alone. In 1957, the Malay Peninsula became the self-governing Malayan Federation. This was expanded in 1963 to include Singapore, Sarawak and Sabah in the Federation of Malaysia. But tension soon led to Singapore's departure and the loss of the federation's principal port and industrial centre.

Nine of Malaysia's 13 states have a sultan or hereditary head of state, the other four have governors appointed by the king. Every five years the nine hereditary rulers elect a king: in 1994 they elected King Tunku Jaafar. There is a House of Representatives with 180 members and a Senate of 69 members, and each state has its own constitution and state government.

Malaysia's economic success since independence – an industrial growth rate of 8 to 9 per cent a year between 1987 and 1992, and one of the world's strongest currencies – has been achieved by developing its wealth of natural resources.

Apart from coastal oil and small timber handling ports, the Borneo territories were undeveloped when independence was first attained. Tropical rain forest was everywhere, including large tracts of swamp forest along the coast of Sarawak; much of the interior was inaccessible and populated sparsely by farmers who dwelt communally in longhouses. Along the coast lay the scattered villages of the Sea Dayaks, or Iban, who lived off the forest and by fishing, but whose ancestors

were pirates. Today, Sabah and Sarawak have grown rich on an expanded oil industry, and on logs and sawn hardwoods, sold to Japan and the American west coast. Felling and the rapid expansion of the road network have cut great swathes through the forest.

Peninsular Malaysia, on the other hand, at the time of independence already had a thriving commercial economy based on rubber growing and tin dredging. Rubber was introduced into the Malay Peninsula in 1877 after Sir Henry Wickham, a British explorer had smuggled seeds from Brazil (which up to then had monopolised the rubber trade) to England. About 20 seedlings were sent to Malaya and vast plantations developed from them.

The British, taking advantage of good communications along the railway lines already used for tin exports, rapidly expanded rubber cultivation to cover the western foothills of the peninsula. Here much of the tropical rain forest was largely replaced by a plantation of rubber trees and, later, of oil palms. Today, two-thirds of rubber plantations are in the hands of the Malaysians themselves.

But even during the colonial era, the local population had followed the lead of the British planters and investors. The typical Malay *kampongs* (villages), especially in the centre of the peninsula, line the roadway between a narrow belt of riceland in the valley and rubber smallholdings on the hillsides. The eastern side of the peninsula remained largely untouched. But since independence, the government has opened up vast tracts, mainly for oil-palm growing. Blocks of smallholdings have been organised on estate lines by the Federal Land Development Authority, which provides service vehicles and processing plants.

Tranquillity remains in the small *kampongs* along the river valleys, where rice is grown and cattle are reared, and in the fishing villages of the TERENGGANU coast, where *perahu* – outrigger canoes with triangular sails – line the shore. But irrigation projects have begun to modernise rice growing, and local fishermen face competition from larger fishing boats. Tourists in their chalets, too, have begun to disturb the giant turtles known for their nocturnal egg laying – one of the area's attractions.

RACIAL HOTCHPOTCH

In the eastern villages the Muslim Malays dominate: the largest building is invariably the mosque. But in the older estate areas of the west coast, rubber tapping is carried out mainly by Indians, descendants of workers originally brought to the country by the British. In the ports are Hindu temples and a variety of Tamil shops.

In the states of Pinang, Perak, Selangor and Melaka, the Chinese make up the majority. The earliest Chinese came to the peninsula to trade in the 17th century, settling particularly in Melaka, then controlled by the Dutch. But most arrived two centuries later when Chinese workers were brought in to dredge for tin. The industry also attracted business interests from China and Europe.

Although some of the larger enterprises around the capital KUALA LUMPUR and in the Kinta Valley depend on foreign capital, in general the Chinese control the tin industry. This control has increased since the richest deposits, which made Malaysia undisputed leader of the world tin industry, have begun to give out.

With their tin profits, the Chinese moved on to control other parts of the Malaysian economy. In South-east Asia, most 'Chinatowns' are enclaves within the indigenous cities, but in Malaysia many towns are Chinese. The people live in shophouses – three or four-storey buildings in terraces which are used as home, factory and shop by the hard-working Chinese residents. In their midst are Buddhist and Taoist temples.

Only in recent years has the economic power of the local Chinese been challenged by the Malays. In 1969, racial tensions led to riots, and the Malay parties' vote dropped at elections. The then prime minister, Tunku Abdul Rahman (a Malay), concluded that the economic gap between the *bumiputras* (indigenous Malays) and the richer communities had to be reduced.

In 1970, the Malays owned 2 per cent of the shares in limited companies. The government's aim was to increase the Malays' share to 30 per cent by 1990. The intention was that the Chinese would own and manage 40 per cent, and others (including foreigners) the remaining 30 per cent. By 1985, the Malays' share had grown to 21 per cent. However, later policies had to deal more subtly with issues relating to community interests in order to alleviate racial tensions. In addition, the government decided more urgently to modernise an export-oriented industry and encourage foreign investment.

The 'new economic policy' brought many younger Malays into the towns to seek jobs in Malaysia's growing industries, especially on the island of Pinang, as well as in the KELANG Valley leading from Kuala Lumpur to the country's main port, Port Kelang. Kuala Lumpur is now a modern maze of skyscrapers and highways, industrial estates and suburbs. Towards Kelang, new towns house workers in electronics and other industries. But the

▲ VANISHING TRANQUILLITY
Motorised as well as traditional boats ply from this village in Terengganu in eastern Peninsular Malaysia, and increasing numbers of tourists are visiting the area's fine beaches.

migration to the towns has been frowned on by Muslim traditionalists, anxious for the younger generation to maintain strict Muslim values.

MALAYSIA AT A GLANCE
Official name Malaysia
Area 329 790 km² (127 331 sq miles)
Population 18 845 340 **Per km²** 57 **(Per sq mile** 148)
Capital Kuala Lumpur
Government Parliamentary monarchy
Currency 1 Ringgit (Malaysian dollar) = 100 sen
Languages Malay (official), Chinese, English, Tamil and several indigenous languages
Religions Muslim (54%), Buddhist (18%), Hindu (7%), Christian (7%), Confucian and others
Climate Tropical. Monsoon from October to February in the east and from May to September in the west with high rainfalls throughout the year. Average temperature in Kuala Lumpur is 27°C (81°F) all year.
Land use Cultivation 13%, forest and woodland 63%, other 24%
Main primary products Rice, palm oil, rubber, timber, cocoa, pepper, pineapples, fish, crude oil, natural gas, tin, bauxite, iron ore, copper
Major industries Agriculture, crude oil production, mining, food processing, forestry, fishing, rubber, tyre manufacture, cement, electronics, chemicals, textiles
Main exports Crude oil, petroleum, liquefied natural gas, timber, electronic components, palm oil, rubber, tin, pepper, cocoa, pineapples, bauxite, iron ore, copper
Annual income per head (US$) 2790
Population growth (per thous/yr) 23
Life expectancy (yrs) Male 66 **Female** 72

Maldives

A SCATTERING OF JEWEL-LIKE ISLANDS ACROSS THE INDIAN OCEAN

Travellers who visit the Maldives find them an unspoilt utopia of waving palms, glittering white coral sand and translucent blue-green seas. The republic has no direct taxation, no political parties, and almost no crime.

All these attractions may, paradoxically, prove a mixed blessing, for, in the age of modern communications, travellers are soon followed by too many tourists. But isolation – the 1200 or so atolls that comprise the republic lie some 640 km (400 miles) south-west of Sri Lanka – may postpone the day when the Maldives become just another group of holiday islands.

Only about 210 of the atolls, which are all fairly low-lying, are inhabited. The people, mostly of far-off Indian or Sri Lankan descent, were ruled by sultans of the Didi family from about 1100 until 1968, when a republic was proclaimed. Once Buddhists, the islanders were converted to Islam by Moroccans and Persians in the 12th century. From 1887 to 1965, the Maldives were a British protectorate, but thereafter gained full political independence.

The main occupations are fishing, which produces 60 per cent of all exports, farming and tourism. The Maldivians were always great seafarers, and used to carry cowrie and tortoise shells, dried fish and ambergris to India. Now, the chief export is canned or frozen tuna, most of which goes to the United States, United Kingdom and Sri Lanka. Rice, the islanders' staple food, has to be imported. The capital, MALÉ, was founded by Portuguese traders in the 16th century.

MALDIVES AT A GLANCE

Official name	Republic of Maldives
Map	Indian Ocean – Cb
Area	300 km² (116 sq miles)
Population 243 100	**Per km²** 810 **(Per sq mile** 2096)
Capital	Malé
Government	Republic
Currency	1 Rufiyaa = 100 laari
Languages	Divehi (a form of Sinhalese), English
Religion	Muslim (Sunni)
Climate	Tropical. Average temperature is 27°C (80°F) throughout the year
Land use	Cultivation 10%, meadows and pastures 3%, forest and woodland 3%, other 84%
Main primary products	Fish, fruit, vegetables, coconuts
Major industries	Fishing, agriculture, tourism
Main exports	Canned, processed fish, clothing
Annual income per head (US$)	500
Population growth (per thous/yr)	36
Life expectancy (yrs)	Male 63 Female 65

Malmö *Sweden* Port on the coast of the Öresund channel which separates Sweden from Denmark. The country's third largest city after STOCKHOLM and GOTHENBURG, Malmö is situated opposite the Danish capital, Copenhagen, with which it is linked by ferry. A trading post since the 12th century, the city was ruled by the Danes until 1658, when Charles Gustav X seized it for Sweden. The Earl of Bothwell – the third husband of Mary, Queen of Scots – was held in protective custody in its 16th-century castle from 1567 to 1573. He had fled to Malmö in 1567 after being implicated in the assassination of Mary's second husband, Lord Darnley.

Historic buildings include a 14th-century church, which is a fine example of Baltic Gothic architecture, and a fortress that is now a museum. Its main industries are food processing, shipbuilding and the production of cement, chemicals and textiles.

Population	236 700
Map	Sweden – Be

Malolos *Philippines* See BULACAN

Malopolska *Poland* See POLAND, LITTLE

Malta See p 428

Maluku (Moluccas) *Indonesia* Archipelago, covering 74 504 km² (28 766 sq miles), south of the Philippines. Between the 16th and the 19th century, the islands were known to Europeans as the Spice Islands; they include Halmahera, Ambon, Seram, Buru, and the Aru group. Their main exports are timber, spices (including cloves, mace and nutmeg), fish and copra.

Population	1 814 000
Map	Indonesia – Gc

Mammoth Cave National Park *USA* Park covering 205 km² (79 sq miles) on the Green River in western Kentucky, 125 km (78 miles) south of LOUISVILLE. Its main attraction is a spectacular system of huge underground caverns and passages, with a total known length of 560 km (348 miles) – the most extensive known cave system in the world. See also BENEATH THE SURFACE OF THE EARTH, p 140

Map	United States – Ic

Mamry, Lake *Poland* See MASURIA

Man *Ivory Coast* Administrative and commercial capital of the country's western region, 460 km (286 miles) north-west of Abidjan. The town is located in a lush valley of the Man Mountains near the borders of Guinea and Liberia. It is best reached by air, although road links are rapidly improving.

A jagged rock called the Dent de Man ('Tooth of Man'), a bridge made out of liana vines at Danané, and a luxury hotel designed to resemble an African village at Gouessesso, some 50 km (31 miles) to the north, are its main attractions.

Population	80 000
Map	Ivory Coast – Ab

Man, Isle of *Irish Sea* Island, midway between England and Ireland, which is a dependency of the British Crown but is not part of the United Kingdom. It covers 585 km² (226 sq miles) and rises to 621 m (2036 ft) at Snaefell. Douglas is the principal town and ferry port. The island has

its own parliament, the Court of Tynwald, which dates from the 9th century and is one of the oldest in the world; it includes the House of Keys, a Representative Assembly. Its own Manx language, a Celtic tongue, is hardly ever spoken these days.

The island has low taxes, and banking and finance contribute one-third of its income. Its main industries are agriculture and tourism. Tourists come especially during the TT (Tourist Trophy) motorcycle races, held each summer on a 60 km (37 mile) circuit. Historic structures include Neolithic monuments from 2000 BC, and old Norse memorial stones. Tailless Manx cats, which originated on the island, are prized by cat fanciers all over the world.

Population	69 790
Map	United Kingdom – Cc

Manabí *Ecuador* Coastal province to the west of COTOPAXI. The region, which is subject to periodic droughts, produces the toquilla palm from which leaves panama hats are made. Its capital is PORTOVIEJO, just west of which lies the seaport of Manta, a processing centre for vegetable oils and cotton. Manta is also an important tuna, and game-fishing port and an export harbour handling coffee and bananas.

Population	1 031 930
Map	Ecuador – Ab

Manado *Indonesia* Provincial capital of North Sulawesi, standing at the north-east tip of the island. It is a fishing port and a processing centre for coconuts, cloves and nutmeg.

Population	129 910
Map	Indonesia – Fb

Managua *Nicaragua* National capital city, standing on the southern shore of Lake Managua, only 55 m (175 ft) above sea level. Despite its humid climate and its record of earthquake disasters – 8000 people died in one in December 1972 – Managua contains one-fifth of the country's population.

Established as capital in 1858 as a compromise choice between conservative Granada, to the south, and liberal León, to the north, it is now a commercial and industrial city producing some 60 per cent of the nation's goods. These include cotton products, soft drinks, processed foods and building materials. Maize, beans, rice, sugar, bananas and sesame seeds are grown in the surrounding volcanic lowlands.

On the shoreline of Lake Managua (1049 km², 405 sq miles) are 1000-year-old footprints preserved in lava and probably made by Amerindians fleeing an eruption of the nearby Masaya volcano.

Population	820 000
Map	Nicaragua – Aa

Manapouri *New Zealand* Lake covering some 142 km² (55 sq miles) in the Fiordland National Park in the south-west of South Island. At the western end of the lake, a tunnel almost 10 km (6 miles) long draws water from the lake to generate hydroelectricity.

According to Maori tribal stories, the hills around Manapouri lake contain the grave of a legendary princess who died trying to rescue her sister. The lake's Maori name means 'Lake of the Sorrowing Heart'.

Map	New Zealand – Af

Mali

THE SAHARA IS ADVANCING IN THIS DROUGHT-STRICKEN COUNTRY, WHICH ONCE BOASTED SOME OF THE RICHEST CIVILISATIONS IN AFRICA

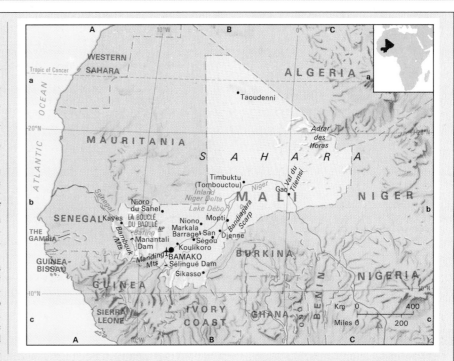

Great empires once flourished on the plains of Mali, where now there are scattered villages in a parched, dusty wilderness. The mighty kingdom of Ghana ruled these lands as long ago as the 4th century AD and the Mali empire which succeeded it reached its peak of wealth and culture in the 13th and 14th centuries.

Mali was renowned for its gold, and was a great trading nation. The name of the legendary Tuareg city of TIMBUKTU (Tombouctou), founded in the 11th century, retains its mystique to this day. But now there is neither gold nor glory. The SAHARA is encroaching and Mali is one of the most unfortunate countries in the world suffering famine, disease and abject poverty.

▲ POT MARKET Pottery traders' canoes – hollowed out tree trunks known as pirogues – line the bank of the Niger River at the town of Mopti.

After Moroccan occupation in the 16th century, the French conquered the country in 1893. They called it Soudan; others knew it as French Sudan; the name Mali came back with independence. In April 1959, Mali and neighbouring Senegal formed the Mali Federation which became independent on 20 June 1960, but exactly two months later the federation was dissolved and the two countries continued their own separate ways.

By 1991, having spent three decades under single-party and military rule, the Malians demanded democratic reform. The army joined the popular revolt and overthrew the dictatorial government of President Moussa Traore. After the first multiparty elections in the following year, Alpha Oumar Konaré became president.

The land is mostly vast and monotonous plains and plateaus. The few high spots in this unvaried landscape are far apart. In the south some rain falls and the plains are covered by grassy savannah with scattered trees. But this peters away to the north where the barren and virtually rainless Sahara covers half the country.

Mali's people include many ethnic groups, most of whom are farmers or animal herders. Life revolves around the vast inland delta of the NIGER River. Almost every family keeps cattle, goats or sheep. Camels and oxen are used for transport. Tensions erupted with the nomadic Tuareg in 1990 but since 1993, they have been granted limited autonomy.

Much worse are the implications of drought and desertification. A prolonged drought that first swept across West Africa in the 1970s and 1980s has brought grief to Mali's nomads and farmers alike. It has extended the desert southwards into southern Mali, turning land that used to provide grazing into useless dust.

Today, about 70 per cent of the country consists of desert and semi-desert. Agriculture is confined to the area on both sides of the river – only 50 km² (19 sq miles) is under irrigation. Industries are small and concentrate almost entirely on food processing. Gold is mined only in small quantities. Iron ore, bauxite, uranium, tin and copper have been discovered.

MALI AT A GLANCE

Official name Republic of Mali

Area 1 240 000 km² (478 654 sq miles)

Population 9 766 000 **Per km²** 8 (**Per sq mile** 20)

Capital Bamako

Government Republic

Currency 1 CFA franc = 100 centimes

Languages French (official), Bambara, Fulani, and other indigenous languages

Religions Muslim (90%), indigenous beliefs (9%), Christian (1%)

Climate Hot and dry; some summer rain in the south. Average temperature at Bamako ranges from 25°C (77°F) to 30°C (86°F)

Land use Grazing 25%, cultivation 2%, forest and woodlands 7%, other 66%

Main primary products Millet, maize, vegetables, rice, groundnuts, cotton, sorghum, sugar cane, livestock, fish

Major industries Agriculture, processed foods, tanning, fishing

Main exports Cotton, livestock (cattle), skins and hides, groundnuts, rice, dried fish

Annual income per head (US$) 300

Population growth (per thous/yr) 29

Life expectancy (yrs) Male 44 **Female** 47

427

Malta

AN ISLAND FORTRESS FACED A NEW CHALLENGE WHEN THE BRITISH SAILED AWAY – NOW THE BRITONS ARE FLYING BACK, SEEKING SUN AND TAX RELIEF

In the time of Classical Greece, about 2500 years ago, Malta was known as 'the Navel of the Inland Sea'. Its position, standing sentinel across the narrows between Sicily and Tunisia which divide the Mediterranean, has always governed its destiny.

British naval and strategic interests largely moulded the island's character, developing the Grand Harbour area beside VALLETTA, the capital, far beyond the capacity of the island's meagre resources. The Valletta conurbation, crowded on to finger-like peninsulas and spreading inland, has more than 200 000 people – two-thirds of the total population. There are 1342 people per km² (3475 per sq mile) living on the main island of Malta, one of the world's highest population densities. The second largest island, GOZO, has only 355 people per km² (920 per sq mile).

Overcrowding has not obliterated the fascinating vestiges of Malta's far-distant past. The oldest are the cave dwellings at Ghar Dalam in the south which are 5800 years old.

Inevitably, as a small, strategically placed island, Malta has a history of foreign domination, successively under the Phoenicians, Carthaginians, Romans, Arabs, Normans and Spaniards. Then followed a period of more formal colonial history, first under the Knights of St John (1530-1798), then fleetingly under the French (1798-1800), and finally under the British (1800-1964).

Throughout the past four centuries, Malta has been used as an island fortress, first as the southern bastion of Christian Europe, later as a guardian over the short sea route to India after the SUEZ CANAL was opened in 1869. During the Second World War it was an Allied base controlling the central Mediterranean and threatening enemy supply routes to North Africa. The island's two great events in history were first in 1565, when it resisted the Turks

▲ VENETIAN AIR Boats resembling Italian gondolas line a quiet quayside of Grand Harbour. The 16th-century houses show both European and Middle Eastern influences.

during the Great Siege, and then in the Second World War when it suffered many months of fierce German and Italian bombardment. The islanders' bravery in withstanding this latter onslaught earned the whole island the George Cross, the UK's civilian award for valour.

Until the late 1950s, the British military base was the mainstay of the Maltese economy, accounting for more than half of Malta's foreign exchange earnings and a quarter of its civil employment. Then Britain began running down the base and by 1979 had withdrawn altogether. Those 20 years saw great changes in Malta's economy. Light industries were encouraged to develop and tourism boomed – from 20 000 visitors in 1960 to 730 000 in 1980; the island also became a haven for retired Britons seeking sunshine and low taxes. Nevertheless, nearly half of the working population is still engaged in agriculture.

At the same time there were political changes. After Britain rebuffed a suggestion that Malta should be made an integral part of the United Kingdom, it became an independent member of the Commonwealth in 1964 and a parliamentary republic in 1974. Under the mercurial Dom Mintoff, prime minister from 1971 until his retirement in 1984, Malta adopted a

fiercely independent stance, which included economic and trade agreements with China and Libya. Mintoff's Labour Party and the politically powerful Catholic Church (aligned with the Nationalist Party) have been the two main players in the island's postwar political scene. Elections in 1987 returned a Nationalist government which sought to restore links with the West. Under Edward Fenech Adami's leadership, Nationalists worked towards re-instating free trade and joining the European Union (EU).

Many Maltese are passionate Catholics and see their religion as a fundamental link with European culture. But their way of life is highly individualistic, linked to the special character that comes from being islanders. They have their own language, Maltese, universally spoken but not written until the present century.

The landscape is flattish, rocky and bare, save for a sprinkling of evergreen carob and olive trees, although rather untidy. The shortage of water and porous limestone bedrock account for the sparseness of the vegetation and pose problems for farming, industry and the hotel trade.

MALTA AT A GLANCE	
Official name Republic of Malta	
Area 320 km² (124 sq miles)	
Population 361 800 **Per km²** 1131 **(Per sq mile)** 2918	
Capital Valletta	
Government Parliamentary republic	
Currency 1 Maltese lira = 100 cents = 1000 mils	
Languages Maltese, English	
Religion Christian (97% Roman Catholic)	
Climate Mediterranean, with dry summers and mild winters; average temperature in Valletta is 12°C (54°F) in January and 26°C (79°F) in August	
Land use Cultivation 41%, other 59%	
Main primary products Vegetables, fruit, pigs, poultry, wheat	
Major industries Agriculture, shipbuilding and repairs, textiles, clothing, electronic equipment, foodstuffs, tourism	
Main exports Textiles, clothing, footwear, ships, flowers, synthetic materials	
Annual income per head (US$) 6630	
Population growth (thous/yr) 1	
Life expectancy (yrs) Male 74 **Female** 79	

▲ BOOMING CITY IN BRAZIL'S RAIN FOREST Traditional houses on stilts line a quiet creek in Manaus. The city is enjoying a 20th-century boom as the key centre and port in the opening up of the vast Amazon basin.

Manaus *Brazil* Capital of Amazonas State, on the Rio Negro, just upstream from where it flows into the Amazon River, 1440 km (900 miles) north-west of the national capital, Brasília. It is now a free port, and offers duty-free shopping. Although much of the city is modern, there are traditional houses on stilts along the waterside. A 19th-century floating dock is a reminder that Manaus has always been a major port, even though it is 1600 km (1000 miles) from the sea.

The city is home of the Teatro Amazonas opera house, which numbers among past performers the French actress Sarah Bernhardt and the Italian tenor Enrico Caruso. Its magnificent building was financed by rubber merchants who had made fortunes after 1839, when the invention of vulcanisation made rubber indispensable to wheeled transport all over the world. The opera house was completed in 1896, shortly before the Amazon rubber market collapsed in the wake of the development of Malaysia's rubber plantations which were cultivated from seeds smuggled out of Brazil by an English botanist, Sir Henry Wickham. The recent discovery of oil nearby has launched Manaus into a second boom era.

The surrounding Amazonas State (population 2 088 680) is Brazil's largest state, which at 1 564 445 km² (603 875 sq miles), is about three times the size of France. Consisting mainly of equatorial rain forest, its main products are timber, rubber, jute, nuts and vegetable oils.
Population 635 000
Map Brazil – Bb

Manawatu *New Zealand* River on North Island, rising 130 km (80 miles) north-east of the capital, WELLINGTON. It flows 182 km (113 miles) through the Manawatu Gorge to the Tasman Sea.
Map New Zealand – Ed

Manche, La See ENGLISH CHANNEL

Manchester *United Kingdom* City of north-west England, formerly in Lancashire but now headquarters of the county of Greater Manchester, which covers more than 1286 km² (497 sq miles). The Romans were first to make their mark here, building a fort which they named Mancunium in AD 79; in the 14th century, Flemish weavers settled here. The city's rise as one of the major cities of Great Britain coincides with the growth of the mechanised cotton industry in the 18th century.

This transformed a rural area into one dotted with mill towns, such as BOLTON, ROCHDALE, Bury, OLDHAM and BLACKBURN. Manchester dominated them all, overrunning Salford, on the opposite bank of the Irwell River, and merging with STOCKPORT, 10 km (6 miles) to the southeast. In 1894, the Manchester Ship Canal linked the city with the sea, 55 km (35 miles) away.

With the notable exception of its 18th-century cathedral, most of Manchester's public buildings date from the 19th century; these include the magnificent town hall and the Free Trade Hall. The Royal Exchange (the city's stock exchange) was opened in 1921 – just in time, ironically, to preside over the decline of the cotton industry that had sparked the city's growth. Today, Manchester is a banking centre, second in England only to London. It has two universities (three, if another at Salford is included) and two of the country's finest libraries, the Rylands Library and the Central Reference Library. It is also home to two of Britain's most famous soccer teams – Manchester United and Manchester City – and the Hallé orchestra.
Population (district) 440 000; (Greater Manchester) 2 570 500
Map United Kingdom – Dd

Manchuria *China* See DONGBEI

Mandalay *Myanmar (Burma)* Major city and traditionally the cultural heart of the country, Mandalay is situated on the Ayeyarwady River, 560 km (348 miles) north of YANGON . It is the capital of the division of the same name covering 37 024 km² (14 295 sq miles) in the centre of Myanmar. Mandalay's royal palace, built between 1853 and 1878, was destroyed during

the Second World War. In 1982, much of the old city around it was burnt down. But the commercial hub around Zegyo market is still a centre for the gem trade – particularly rubies.

The most important of Mandalay's many temples are the Arakan pagoda – named after the principal Buddha image which was brought from Arakan State in western Myanmar – and Kuthodaw, with its mass of small pagodas carved with the texts of Buddhist scriptures. Mandalay Hill, the site of numerous Buddhist temples, can be climbed via a shrine-lined stone staircase. The views of the city from the top of the hill are described as magnificent.
Population (city) 417 300; (division) 4 581 000
Map Myanmar – Cb

Mandara Mountains *Cameroon/Nigeria* Ranges of rounded granite hills and extinct volcanic peaks rising to 1500 m (about 4920 ft) on the northern part of the border between Cameroon and Nigeria. Erosion on the Kapsiki Plateau in Cameroon has left a stunning landscape of rock pinnacles, where Kirdi people, known for their unique traditional dancing, live in hanging cliffside villages. These and the area's forests, streams, waterfalls and wildlife are attracting increasing numbers of tourists.
Map Cameroon – Ba

Mandu *India* Ruined fortified city in central India, standing on a rocky promontory of the Vindhya Range, 705 km (438 miles) north-east of BOMBAY. From the early 15th century it was the capital of the Muslim state of Malwa. Its main mosque is one of the finest examples of Afghan architecture in India.
Map India – Bc

manganese nodule Irregularly shaped, crumbly fragment of rock, containing on average 20 per cent manganese, 6 per cent iron and 1 per cent nickel and copper. They are found on the ocean floors and are likely to become an important commercial resource in the future.

mangrove swamp Tidal mud flat covered chiefly by mangroves – tropical evergreen trees and shrubs with stiltlike roots – occurring mainly near the Amazon and Niger deltas and along the coasts of Sumatra and Borneo.

Manica *Mozambique* See CHIMOIO

Manila *Philippines* National capital, the Philippines' chief commercial, financial and industrial centre and one of the major cities of Asia, lying on the eastern shore of Manila Bay in westcentral LUZON, at the mouth of the Pasig River. The bay (area 2000 km², 770 sq miles) is a superb natural harbour – a major reason why the Spanish colonists chose it as the site for their capital of the islands in 1574. The old walled district of Intramuros beside the Pasig dates back to this era, but most of the old Spanish buildings (and much of the rest of Manila) were destroyed either by earthquakes or by bombs dropped on it during the Second World War.

Over the past few decades, Manila has expanded widely north and south of the river. Today, the urban area of so-called Metro Manila covers 635 km² (245 sq miles) and incorporates several cities on its fringes, such as Pasay, Makati and Mandaluyong in the south, QUEZON CITY (the

429

capital from 1950 to 1976) and San Juan in the east, and Caloocan in the north, as well as the 14 municipal districts of Manila proper.

Each district has its own character, reflecting the Philippines' diversity and extremes of wealth and poverty. In the inner northern district of Tondo, World Bank finance was used to try to alleviate conditions in a squalid, densely populated slum; nearby Binondo is Manila's Chinatown, while the Ermita and Malate areas, behind the sweep of Roxas Boulevard along the bay front south of Intramuros, form the main tourist zone of hotels, restaurants and nightclubs.

Sampaloc is primarily a centre of higher education, with several private universities, while Malcañang, the presidential palace, originally a Spanish aristocrat's mansion is in nearby San Miguel. Makati, is the city's financial district.

Nearby, but outside Metro Manila itself, are the Second World War battlefields of BATAAN and CORREGIDOR, the CAVITE naval base, the large lake of Laguna de BAY, and the Taal volcano in BATANGAS province.

Population (city) 1 599 000; (Metro Manila) 7 832 000
Map Philippines – Bc

Manipur *India* State covering 22 327 km² (8618 sq miles) of hill country on the border with Myanmar (Burma). Most of its inhabitants still lead a tribal way of life. Imphal is its capital and boasts a university and numerous colleges.
Population 1 837 149
Map India – Ec

Manisa *Turkey* Market town 300 km (186 miles) south-west of the capital ISTANBUL. Farmers in the

▼ **TAXI! American army surplus jeeps – gaudily painted and bedecked with signs, stickers and accessories – ply as taxis (or jeepneys) in Manila's streets.**

surrounding province of the same name produce mainly grain, olives, raisins and tobacco. The province covers 13 810 km² (5332 sq miles).
Population (town) 158 430; (province) 1 154 420
Map Turkey – Ab

Manitoba *Canada* Most easterly of Canada's three prairie provinces, covering 650 087 km² (250 998 sq miles). Two-thirds of Manitoba lies in the CANADIAN SHIELD area, and one-sixth of its area is covered by lakes. Manitoba is a notable producer of grain crops, and beef and dairy cattle. Minerals include oil, nickel, copper and zinc; industries centre on processed food, agricultural equipment, clothing, and furniture. Its capital is WINNIPEG.

The first agricultural colony was established in the province by Scottish settlers in the RED RIVER Valley in 1812. Today, almost 20 per cent of the population are of East European extraction, of which the majority have a Ukrainian background. A Ukrainian festival is held every year at Dauphin.
Population 1 091 940
Map Canada – Fc

Manizales *Colombia* City 175 km (110 miles) north-west of Bogotá. It is perched on both sides of an Andean ridge at a height of 2126 m (6975 ft) above sea level. Manizales is the business heart of Caldas, Colombia's biggest coffee-growing region; it hosts a coffee festival each year in January.
Population 327 160
Map Colombia – Bb

Mannar, Gulf of (Gulf of Manaar; Manar) *Sri Lanka/India* Inlet, 160 km (100 miles) long and 130-270 km (80-168 miles) wide, of the INDIAN OCEAN between Sri Lanka and the Indian state of TAMIL NADU. Its northern boundary is Adam's Bridge which, according to legend, was built by

the mythological Hindu hero Rama. It is, in fact, a chain of sandbanks between Pamban Island in India and Mannar Island in Sri Lanka.
Map Sri Lanka – Aa; India – Ce

Mannheim *Germany* Port and industrial city at the junction of the Rhine and Neckar rivers, 70 km (43 miles) south of FRANKFURT am Main and across the river from LUDWIGSHAFEN. Like Ludwigshafen, it is a major commercial centre.

The city centre, built by the electors of the Palatinate Province in the 18th century, is laid out in a fan-shaped pattern radiating from the Baroque palace, now a university. At Mannheim in 1885, the engineer Karl Benz (1844-1929) built a petrol-driven, three-wheeled carriage – the first true motorcar. Today, Mannheim produces machinery, motor vehicles and electronics.
Population 310 400
Map Germany – Cd

mantle Middle layer of the earth, between the CRUST and the CORE. See JOURNEY TO THE CENTRE OF THE EARTH, p 207

Mantua (Mantova) *Italy* City built on the marshy land in the central North Italian Plain, 120 km (75 miles) south-east of MILAN. During the rule of the Gonzaga dynasty in the 15th and 16th centuries, it became a renowned centre of art and learning. It has many fine buildings, including the vast Palazzo Ducale and, opposite the palace, the cathedral, replanned by the architect Giulio Romano (1492-1546) after a fire had damaged the original in 1545.

The Roman poet Virgil (70-19 BC) was born nearby and lived in the city for most of his life. Mantua's industries include tourism, printing, tanning, and processing farm products.
Population 54 800
Map Italy – Cb

Manukau Harbour *New Zealand* Largest of the two harbours on which the port of AUCKLAND stands. Its name is Maori for 'the Place of Wading Birds' and indeed, it is largely surrounded by mud flats and sandbanks which attract waders. A sandbar at its entrance restricts shipping. A narrow isthmus separates Manukau Harbour, which covers 394 km² (152 sq miles), from the economically more important Waitemata Harbour, which covers 200 km² (77 sq miles).
Map New Zealand – Eb

Manzini *Swaziland* Town situated some 41 km (about 25 miles) south-east of MBABANE on the Mzimneni River – Manzini means 'at the water'. Founded as a trading post in 1885, it was called Bremersdorp until 1960. Together with the nearby industrial township of Matsapa, it has grown into a leading agricultural, commercial, industrial and trading centre. Its industries include brewing, cotton ginning, and the processing of meat and dairy products.
Population 32 700
Map Swaziland – Aa

Maoming *China* City of Guangdong Province, lying 320 km (nearly 200 miles) south-west of GUANGZHOU. Its proximity to oil-shale deposits and other minerals has seen it emerge as an important mining centre.
Population 450 000
Map China – Gf

Maputo *Mozambique* Capital of Mozambique. A major seaport formerly known as Lourenço Marques, it became the national capital in 1907, even though it lies in the extreme south of the country, only 100 km (62 miles) from the South African border. Maputo has a large natural harbour, and its modern development based on transit trade began in 1895 when it was linked by rail with Johannesburg in South Africa.

The city is also Mozambique's leading manufacturing centre, with food processing, textile and iron industries. However, its tourist industry collapsed in the wake of a guerrilla war waged by right-wing Renamo rebels in the early 1980s; to compound matters, most of Mozambique's 250 000 Europeans left the country. Soon afterwards, a peace agreement was signed between the government and the rebels, and attempts were made to clean up the city and reopen hotels.

Maputo is the capital of the country's smallest and southernmost province, also called Maputo, which covers 25 756 km² (9941 sq miles), and produces cotton, maize, sugar and rice.
Population (city) 1 070 000; (province) 1 138 700
Map Mozambique – Ac

maquis (macchia) Drought-resistant scrub of small myrtle and wild olive trees and aromatic shrubs, usually found on acid soils in parts of the Mediterranean region. On limestone soil, it is replaced by GARRIGUE. Maquis corresponds to the CHAPARRAL of the south-western United States and Mexico. See MEDITERRANEAN WOODLAND.

Mar del Plata *Argentina* Beach resort 400 km (almost 250 miles) south-east of Buenos Aires. Famous in South America for its nightlife and for being the home of one of the world's largest casinos, the city attracts more than 3 million visitors during summer (December to March).
Population 520 000
Map Argentina – Db

Maracaibo *Venezuela* Oil city, port and provincial capital of Zulig, situated on the west bank of the strait linking Lake Maracaibo with the Caribbean, 500 km (310 miles) west of the capital, CARACAS. Maracaibo has become the country's second largest city after the capital since one of the richest oil fields in the world, the Bolívar Coastal Field, was discovered nearby in 1917.
Population 1 400 640
Map Venezuela – Aa

Maracaibo, Lake *Venezuela* Shallow inland sea opening into the Caribbean, on the extreme north-west tip of Venezuela. It covers 13 280 km² (5127 sq miles), but is nowhere more than 35 m (115 ft) deep. The lake contains the country's largest oil fields, which provide 70 per cent of Venezuela's oil.
Map Venezuela – Ab

Maracay *Venezuela* Capital of Aragua State, 50 km (31 miles) west of the country's capital, CARACAS. It was the favourite city of General Juan Vicente Gómez, the dictator who ruled Venezuela from 1908-35. Several of his grandiose architectural follies still stand, including an opera house, a bull ring modelled on the one in Seville, Spain, and the huge triumphal arch of the Gómez mausoleum.

Gómez also encouraged investment in the city and supported local industry so that today, the city is one of the country's most important industrial centres. Its industries include cigarette production, meat packaging, textiles, paper and chemicals.
Population 956 660
Map Venezuela – Ba

Marajó *Brazil* Island situated at the mouth of the Amazon covering 47 964 km² (18 514 sq miles) – big enough for Switzerland to fit into, with room to spare. It has a rich wildlife, including birds and crocodiles, but is prone to flooding during the rainy season from December to June. Much of its flat plains are used for cattle and buffalo ranching.
Population 50 000
Map Brazil – Db

Maralinga *Australia* Area in the Great Victoria Desert of South Australia, which was used to test British atomic bombs in the 1950s.
Map Australia – Ee

Maramba (Livingstone) *Zambia* Tourist town on the border with Zimbabwe, some 380 km (236 miles) south-west of the capital, Lusaka. The capital of Northern Rhodesia from 1911 to 1935, it has a museum devoted to the life of the Scottish medical missionary David Livingstone (1813-73), after whom the town was named.

Today, it is the capital of Zambia's Southern Province, producing cotton, maize, tobacco and sugar cane. The VICTORIA FALLS are situated 11 km (7 miles) to the south-west of the town.
Population 72 000
Map Zambia – Bb

Maramures *Romania* North-west province on the borders of Russia and Hungary. Noted for its rich and varied peasant culture, it consists mainly of well-wooded mountains, fertile plains and unspoilt villages. Maramures represents the best of 'old' Romania. Richly decorated peasant costumes are a daily sight here, and village houses have ornately carved wooden porches and gateways.
Population 517 000
Map Romania – Aa

Maranhão *Brazil* North-eastern state covering 328 660 km² (126 860 sq miles). Despite a booming aluminium industry, the state remains one of the poorest in the country. Babaçu palms (nuts, oil) are grown along the coastal belt; the interior is savannah where livestock are reared. The state capital is the port of São Luís, founded by the French in the 17th century as St Louis; many of its tile-faced colonial buildings survive.
Population 5 182 000
Map Brazil – Db

Marañón *Peru* Headstream of the Amazon, some 1650 km (1025 miles) long. It rises in the Andes at Lake Huaiuash and flows first north and then north-east across Loreto Department's lowlands to join the Ucayali to become the Amazon.
Map Peru – Ba

Marathón *Greece* Small town, 5 km (3 miles) north of the battlefield where the Athenians routed the Persians in 490 BC. After the battle, a courier named Pheidippides is said to have run the 42 km (26 miles) to Athens without stopping. On arrival, he gasped out his message, 'Rejoice, we conquer!' – and dropped dead, possibly from heat stroke. His feat gave the name 'marathon' to the long-distance race which is a major feature of the modern Olympic Games.
Map Greece – Cb

Marbella *Spain* See MÁLAGA

marble METAMORPHIC ROCK formed by recrystallisation of limestone under great heat and pressure. Marble is quarried for construction, and is also used in sculpture and for decorative purposes.

Marble Bar *Australia* See PILBARA, THE

Marburg *Slovenia* See MARIBOR

Marburg (Marburg an der Lahn) *Germany* University town on the Lahn River, some 80 km (50 miles) north of FRANKFURT am Main. Timbered houses climb the side of a hill beneath its medieval castle. In 1527, the residing court of Philip the Generous founded the first German Protestant university here – and the university in its turn was responsible for founding the pharmaceutical company, Behring, based at Marburg.

Each hour during daytime, the figure of a herald on the town-hall clock blows a 'Trumpet of Justice'. Field-Marshal Paul von Hindenburg (1847-1934), President of the Weimar Republic from 1925 to 1933, is buried in Marburg's 13th-century cathedral of St Elizabeth.
Population 73 200
Map Germany – Cc

Marche (Marches) *Italy* Rural region covering 9694 km² (3743 sq miles) from the eastern flank of the Apennines to the Adriatic coast, and extending about 160 km (100 miles) south-east from SAN MARINO. It is covered with well-tended farms, a patchwork of vines, olives, cereals and livestock. Many of its inhabitants are employed in the crafts industry, making shoes, paper, and musical instruments. The capital is ANCONA.
Population 1 435 600
Map Italy – Dc

Marco Polo Bridge *China* See LUGOU BRIDGE

Mardan *Pakistan* Agricultural town containing one of Asia's largest sugar factories, 120 km (75 miles) north-west of ISLAMABAD. Nearby are Buddhist monuments, and one of several carved stone pillars erected by the Buddhist Indian emperor Ashoka in about 250 BC.
Population 193 000
Map Pakistan – Da

Mareb *Yemen* See MARIB

mare's tail Drawn-out, wispy cirrus cloud, a sign of strong winds in the upper atmosphere, and often a sign of approaching bad weather.

Margarita (Isla de Margarita) *Venezuela* Largest island in the Nueva Esparta group which occupies 920 km² (355 sq miles) in the Caribbean, about 290 km (180 miles) east of the capital, CARACAS. The island used to be famous for its pearl fishing. Today it is a duty-free zone with many beaches. Its capital is La Asunción (population 14 100), and its largest town is Porlamar (population 22 500).
Map Venezuela – Ba

Maria Cristina Falls *Philippines* See ILIGAN

Mariana Islands *Pacific Ocean* Chain of mountainous islands 2400 km (nearly 1500 miles) east of the northern Philippines. The islands consist of the United States territory of GUAM along with the Commonwealth of the NORTHERN MARIANA ISLANDS.
Map Pacific Ocean – Cb

Marianas Trench Curving trench on the seabed near Guam in the north-western PACIFIC OCEAN. It is a PLATE MARGIN and contains Challenger Deep which, at 11 033 m (36 198 ft), is the deepest known spot in the oceans. It was discovered by the British survey ship *Challenger* in 1951.
In January 1960, the United States Navy bathyscaphe *Trieste*, crewed by Dr Jacques Piccard of Switzerland, and Lieutenant Donald Walsh of the United States, reached a depth of 10 915 m (35 810 ft). Echo soundings by a Russian survey ship during the same year confirmed the current record depth.
Map Pacific Ocean – Cb

Marianske Lazne (Marienbad) *Czech Republic* Bohemian spa town 12 km (7 miles) from the German border. Developed in 1812, it was patronised by, among others, England's Edward VII (reigned 1901-10), the German composer Richard Wagner (1813-83), and the German writer Johann Wolfgang von Goethe (1749-1832).
Today, visitors come to the spa in search of relief from respiratory, nervous and skin disorders, or to play golf, fish, walk in the surrounding hills, attend conferences or see the annual international film festival.
Population 18 400
Map Czech Republic – Ab

Maria-Teresiopel *Yugoslavia* See SUBOTICA

Mariazell *Austria* Winter and summer resort in the state of Styria and the country's main pilgrimage centre, 95 km (59 miles) south-west of VIENNA. Founded by Benedictine monks in 1157, it has become famous for its torchlight processions. A 12th-century wooden statue of the Virgin Mary ('Our Lady of Mariazell') is kept in the Gnadenkirche (pilgrimage church). The statue has been the object of veneration probably since the 16th century, when a Hungarian victory over the invading Turks was attributed to the intervention of the Virgin Mary.
Population 1900
Map Austria – Db

Marib (Mareb) *Yemen* Ruined city situated 140 km (87 miles) east of SAN'A. Marib was the capital of the kingdom of Sheba in about 1000 BC. Among its remains are the Temple of Bilquis (the Muslim name for the Queen of Sheba) and parts of its 5th-century BC city walls.
A temple near the walls has been converted into a mosque. To the south of the town are the ruins of a 550 m (1800 ft) stone and masonry dam, also dating from the 5th century BC. With irrigation reviving agriculture, and Yemen's first oil well coming on stream in 1986, a new town has been built at Marib.
Map Yemen – Aa

Maribor (Marburg) *Slovenia* Major industrial city on the Drava River, close to the Austrian border. Founded in the 10th century around a castle and a citadel, it prospered on Austrian trade with Venice. It manufactures cars, railway equipment, steel and non-ferrous metal products, leather goods, tobacco, textiles, and foodstuffs. An attractive town at the foot of a spur of the Karawanke Alps, Maribor is a tourist centre for Alpine ski resorts and nearby spas. Ptuj, 24 km (15 miles) to the south-east, has a fine medieval castle. There are large wine cellars at the nearby towns of Ljutomer and Ormoz.
Population 185 700
Map Slovenia – Bc

Marie Galante *Guadeloupe* Island situated 25 km (16 miles) south-west of Grande Terre Island. It is the largest of Guadeloupe's dependent islands, covering 157 km² (61 sq miles). Its capital is Grand Bourg on the south-west coast. Sugar cane is the chief crop, although production has declined since the 19th century.
Population 13 460
Map Caribbean – Cb

Marienbad *Czech Republic* See MARIANSKE LAZNE

Marienburg *Poland* See MALBORK

Marigot *Guadeloupe* See ST MARTIN

marin Warm, moist south-easterly wind that blows across the coast of southern France, especially in spring and autumn, as a result of a depression in the Golfe du Lion (Gulf of Lions).

Marinduque *Philippines* Island covering an area of 958 km² (370 sq miles) in the Sibuyan Sea between MINDORO and southern LUZON, and containing large iron ore mines. A re-enactment of the crucifixion of Jesus is staged each year on Good Friday in Boac, the capital of Marinduque, and in the towns of Gasan and Mogpog.
Population 163 000
Map Philippines – Bc

Maritsa *South-east Europe* See EVROS

Mariupol' (Mariupol; Zhdanov) *Ukraine* Port on the north shore of the Sea of Azov. It was established in 1779 on the site of an ancient Greek colony. Originally called Mariupol, it became Zhdanov during the Soviet era (1948-91) but reverted to Mariupol' when Ukraine became independent. Mariupol' is the country's second most important Black Sea port, after Odesa. Its main industries are iron and steel, engineering and chemicals, and it exports coal, grain and salt.
Population 520 000
Map Ukraine – Db

marl A clay or mudstone with a lime content. It is placed on sandy soils to improve texture.

Marlborough Sounds *New Zealand* Rugged coastline at the north-eastern tip of South Island. The English explorer Captain James Cook anchored here on his visits to New Zealand in the 1770s. Its beaches, woods and pastures make it a popular tourist attraction. Its longest inlet, Pelorus Sound, stretches on for more than 55 km (34 miles).
Map New Zealand – Dd

Marmara, Sea of (Marmara Denizi) *Turkey* Known to the Turks as Marmara Denizi, the sea is part of the strategically important waterway that links the BLACK SEA (via the Bosporus) with the MEDITERRANEAN (via the Dardanelles), with Istanbul commanding the entrance to the Bosporus. The sea covers about 11 140 km² (4300 sq miles); its greatest depth is some 1370 m (4495 ft). A number of fishing villages, romantic ruined castles and gleaming Turkish tombs line the coasts.
Map Turkey – Aa

Marmarica *Libya* North-east coastal district of CYRENAICA. Most of it is a desert plateau of ridges no more than 200 m (656 ft) high, which extends into Egypt parallel to the Mediterranean coast. Through the centuries, many armies have fought here – Romans, Egyptians, Libyans, Arabs and modern Europeans. During the Second World War, much of the fighting of 1942-3 took place in Marmarica. TOBRUK is the chief town.
Map Libya – Ca

Marne *France* River, 525 km (326 miles) long, of north-central France. It rises in the Plateau de Langres 70 km (43 miles) north of DIJON and flows north, and then west, to join the Seine in eastern Paris.
The department of the same name, covering 8163 km² (3151 sq miles), lies in the region of Champagne-Ardennes, to the north-east of Paris. Although this is one of the most densely populated regions of France, 10 per cent of its workforce are engaged in agriculture, producing cereals on large and highly mechanised farms, and, of course, champagne. Most people in the region, however, are employed in service industries, working in the region's major towns. CHÂLONS-SUR-MARNE is the chief town.
Population (department) 559 000
Map (river) France – Eb; (Châlons-sur-Marne) France – Fb

Maroua *Cameroon* Chief town (and main Islamic centre) of the far north. Cattle, cotton, groundnuts, millet and sorghum are marketed from here; it also has a cattle-breeding research station. Maroua's buildings, constructed mostly of mud, are in the traditional style. The peaceful, relaxed atmosphere of the town is in marked contrast with that of the modern cities farther south.
Population 143 000
Map Cameroon – Ba

Marquesas Islands *French Polynesia* Group of about 10 rugged islands with deep, fertile valleys lying some 1400 km (875 miles) north-east of TAHITI. Tai-o-haé on Nuku Hiva Island is the capital. The islands were 'discovered' by the Spanish in 1595. They had a population of 50 000 in 1800 but this number had dropped to only 2000 by the 1920s as a result of 'blackbirding' (kidnapping by Europeans for forced labour) and disease. The French painter Paul Gauguin was buried on Hiva Oa Island in 1903.
Population 7540
Map Pacific Ocean – Gc

Marrakech *Morocco* One of Morocco's four royal cities, and a former capital. Marrakech, from which Morocco takes its name, was founded in the 11th century by the Berber Almoravid dynasty. Surrounded by a vast oasis of palm groves, the town lies on a major crossroads at the foot of the High ATLAS MOUNTAINS 220 km

(140 miles) south of CASABLANCA. Berber, Arab and African cultures meet at this point in a dazzle of colour, sound and activity.

The Djemaa el Fna Square in particular is a hive of noise and movement – and among the attractions here are snake charmers, acrobats and storytellers. In the souks (bazaars), craftsmen produce leatherwork, silver, brass, pottery and jewellery. Weavers abound and the city is renowned for its carpets and rugs.

Other places of interest include the 12th-century city walls, complete with gateways, the 70 m (230 ft) high minaret of the 12th-century Koutoubia mosque, the Saadian (Sovereign's) Mausoleum, and the Ben Youssef madrassa (religious school).

Population	440 000
Map	Morocco – Ba

Marsa el Brega (Mersa el Brega) *Libya* Oil port on the Gulf of Sirte. The first Libyan oil terminal to be built, it was opened in 1961, and is the country's premier petrochemical centre. It has a refinery, gas liquefaction plant, and ammonia and urea plants. It is also the site of Libya's first technical university.

Population	7100
Map	Libya – Ba

Marsa Matruh *Egypt* See MATRUH

Marsala *Italy* Wine-making and fishing town at the western tip of Sicily. Marsala wine was first produced here in the 18th century. The Italian patriot Giuseppe Garibaldi (1807-82), together with 1000 red-shirted volunteers, landed at Marsala in 1860 at the beginning of his campaign to unite Italy.

Population	79 900
Map	Italy – Df

Marseilles (Marseille) *France* The country's premier port and third largest city, after Paris and Lyons, on the Mediterranean coast east of the Rhône River delta. Commerce has flourished here since Greek traders arrived in the 7th century BC. Today, the city with its cosmopolitan mix of Mediterranean people including many Arabs, sprawls up the slopes above its small natural harbour, to the south of the modern port.

The city centre is a jumble of narrow streets and alleyways. Farther out are broad streets and

▲ **PARIS IN THE CARIBBEAN Balata Church, a replica of Sacré Coeur in Paris, overlooks Martinique's capital, Fort-de-France, and Les Trois Ilets, where Napoleon's Joséphine was born.**

modern buildings; among its better known buildings are the Unité d'Habitation, an apartment block designed by the Swiss-born architect Le Corbusier (1887-1965) and completed in 1952, which is regarded as a landmark of modern architecture. The Château d'If – a castle built in the early 16th century on an island in the harbour – was the real prison in which the fictional Count of Monte Cristo was held in the novel by Alexandre Dumas (1802-70).

The French national anthem *La Marseillaise* was named after a troop of volunteers from the city who marched to Paris in 1792 to join the revolution against France's old aristocratic regime. The Marseilles men sang the song as they entered the capital, and the stirring words and music caught on.

Population	(city) 801 000; (conurbation) 1 087 000
Map	France – Fe

marsh Area of temporarily flooded land beside a river or a lake, characterised by water-loving vegetation such as rushes and reeds, and often silty. The term 'marsh' is also used loosely to describe any area of low-lying, wet land such as a fen, swamp or bog.

Martaban *Myanmar (Burma)* Town on the Gulf of Martaban, an inlet of the ANDAMAN SEA to which Martaban Town has given its name. It lies 160 km (100 miles) east of YANGON (Rangoon) and has prospered from vast deposits of natural gas which have been discovered in the gulf.

Martaban is the terminus of the railway from Yangon. A ferry plies across the estuary of the Thanlwin River transporting passengers to the city of Mawlamyine (Moulmein) and the railway system southwards.

Population	15 000
Map	Myanmar – Cc

Martinique *West Indies* One of the Windward group of islands in the eastern Caribbean, made up of three groups of volcanoes and covering 1100 km² (425 sq miles). The highest peak, Mont PELÉE, reaches 1397 m (4583 ft). This tropical,

Marshall Islands

ONCE THE SITE OF NUCLEAR TESTS, THE ISLANDS NOW DEPEND ECONOMICALLY ON AN AMERICAN MILITARY BASE

There are some 1250 islands, atolls and reefs in the Marshalls, part of MICRONESIA in the western Pacific Ocean, yet they total just 181 km² (70 sq miles) and have a population of not quite 52 000. The islands lie 3200 km (2000 miles) south-west of Hawaii, and form two chains: Ralik and Ratak. The largest atoll is KWAJALEIN, 120 km (75 miles) long, but the most famous are BIKINI and ENEWETAK where the United States performed 64 nuclear tests from 1946 to 1958.

An internally self-governing republic since 1979, the island's economy is dependent on aid from the US, totalling about US$40 million a year, in exchange for the 30-year use of a military base on Kwajalein. The base is used to recover missiles launched over the Pacific in test flights from California, 6700 km (4200 miles) away. Most (92 per cent) of the Marshallese, a Micronesian people, live in villages – farming and fishing for their food, as they have always done, and growing coconuts for export. Rent is received from the USA for its missile base on Kwajalein atoll.

The islands were named after John Marshall, a British sea captain who visited them in 1788, although they had been sighted by a Spanish expedition in the 1520s. They became a German protectorate in 1884, but were occupied by the Japanese during the First World War. The Marshalls were Japanese until US forces captured them during the Second World War. From 1947 until 1990 they were part of the US Trust Territory of the Pacific. The islands became a full member of the United Nations in September 1991.

MARSHALL ISLANDS AT A GLANCE	
Map Pacific Ocean – Db	
Official name Republic of the Marshall Islands	
Area 181 km² (70 sq miles)	
Population 51 990 **Per km²** 287 (**Per sq mile**) 742)	
Capital Majuro	
Government Parliamentary republic	
Currency 1 US dollar = 100 cents	
Languages Marshallese, English	
Religion Christian	
Climate Tropical, with the rainy season from May to November; average temperature is 27°C (81°F)	
Main primary products Coconuts, cassava, sweet potatoes, fish	
Major industries Agriculture, fishing, coconut processing	
Main exports Copra, coconut oil	
Annual income per head (US$) 1500	
Population growth (per thous/yr) 39	
Life expectancy (yrs) Male 61 **Female** 64	

humid island was once the home of the Carib people, who called it Madinina. It was reached by Christopher Columbus in 1493 but he did not land until his fourth voyage in 1502. Determined resistance by the Caribs discouraged colonisation until the French settled there in 1635. Apart from short periods of British occupation, Martinique has remained under French rule. In 1946, it became an overseas department of France with local self-government and full representation in the National Assembly in Paris.

Bananas, sugar cane – originally produced with slave labour – and pineapples are the main export crops. Sugar refining, the distilling of rum, cement manufacture, construction and petroleum products are the main industries. Although tourists, mainly from France and the United States, bring in additional revenue (in fact, more than the revenue earned from agricultural exports), the island continues to depend heavily on financial support from the French government. FORT-DE-FRANCE is the group's capital and main port.

Population 360 000
Map Caribbean – Cc

Maryland *Liberia* See HARPER

Maryland *USA* East coast state covering some 25 484 km² (9837 sq miles). It straddles Chesapeake Bay and continues in a north-westerly direction to the Appalachian Mountains; its southern boundary follows the Potomac River. Maryland was first settled by Europeans in 1634 and was one of the original 13 states. With its old, picturesque towns, and with some of the suburbs of the national capital, WASHINGTON DC, lying in the state, it attracts many tourists. ANNAPOLIS is its capital and BALTIMORE its largest city.

Chickens, dairy goods, soya beans, maize, coal, metals, electronics, processed foods and steel are its chief products. Fishing and crabbing are important economic activities, even though their importance has declined in recent times.

Population 4 392 000
Map United States – Kc

Masada (Mezada) *Israel* Hilltop fortress 55 km (35 miles) east of BEERSHEBA, overlooking the Dead Sea. In AD 73 it was the scene of the last stand of the Zealots, an extreme Jewish sect, in the Jewish revolt against the Romans. After being put under siege for two years, the 960 remaining defenders committed suicide rather than surrender to Rome.

Many of Masada's buildings have been restored, including the palace of Herod the Great (who ruled from 37 to 4 BC), the synagogue and the storehouses. A cable car carries visitors to the 49 m (160 ft) summit.

Map Israel – Bb

Masan *South Korea* City and port 60 km (35 miles) west of the southern port of PUSAN. Developed during Japanese colonial rule (1905-45), it is now the centre of a free trade zone set up by the government to promote exports. Its industries include flour milling, and the manufacture of textiles and electrical machinery.

Population 497 000
Map Korea – De

Masbate *Philippines* Island of the VISAYAN group, covering some 3269 km² (1262 sq miles) between PANAY and the southern tip of LUZON.

Consisting mainly of rolling hills, it is best known as a cattle raising area, although maize is also grown.

Population 585 000
Map Philippines – Bc

Maseru *Lesotho* National capital, situated about 1500 m (4921 ft) above sea level. It has light industries, manufacturing, among other things, candles, carpets, sheepskin goods and pottery. The royal palace, a Catholic and an Anglican cathedral and the country's national museum are located in the city. Maseru is linked to South Africa by rail. Maize, sorghum, wheat, beans and peas are the main crops of the district of Maseru, which covers 4279 km² (1652 sq miles).

Population (city) 120 000; (district) 292 000
Map South Africa – Cb

Masherbrum *Pakistan* Mountain rising to 7821 m (25 660 ft) in the Karakoram Range in Azad Kashmir, just south of K2.

Map Pakistan – Da

Mashhad (Meshed) *Iran* Holy city and capital of Khorasan Province, situated about 750 km (470 miles) east of the national capital, TEHRAN. Situated on the ancient trade and military route between India and Persia, it trades in carpets and cotton goods, and produces leather, tinned fruit, sugar, and rugs. The city is a place of pilgrimage for Shiite Muslims, because the Shiite religious leader Imam Ali Reza was buried here in 817.

Population 1 759 000
Map Iran – Ba

Mashonaland (Shonaland) *Zimbabwe* Northern region consisting of three provinces (Mashonaland West, Mashonaland Central, and Mashonaland East) covering 112 089 km² (43 278 sq miles) around the national capital, HARARE. It is the home of the Shona people, the country's largest language group.

Population 2 837 400
Map Zimbabwe – Ca

Masjed Soleyman (Masjid Suleiman) *Iran* Town 250 km (155 miles) south-west of the capital, TEHRAN. It lies at the centre of Iran's oldest oil field, which began production in 1908. The surrounding area has numerous archaeological remains, including a huge stone terrace, which was built by the Parthians (200-100 BC) and is approached by a flight of steps 25 m (82 ft) wide.

Population 108 000
Map Iran – Aa

Mason-Dixon Line *USA* The southern boundary of PENNSYLVANIA. It was surveyed in the 1760s by two Englishmen, Charles Mason and Jeremiah Dixon, in order to settle a dispute between the Pennsylvania and Maryland colonies. Before and during the American Civil War (1861-65) it represented the boundary between so-called slave states and free states, and subsequently, in a general way, the boundary between the North and South of the United States.

Map (Pennsylvania) United States – Kb; (Maryland) United States – Kc

mass movement (mass wasting) The downward movement of weathered material on a slope under the influence of gravity, usually lubricated by rain or melting snow.

Massa *Italy* Town 40 km (25 miles) north-west of PISA. It produces chemicals and marble is also quarried here.

Population 66 600
Map Italy – Cb

Massachusetts *USA* New England state covering 20 269 km² (7824 sq miles) in the north-east, on the Atlantic coast. In 1620, the Pilgrims landed at nearby Cape Cod Bay, and established the colony of PLYMOUTH. In 1775, the American War of Independence started in the Massachusetts town of Lexington. The state has 106 colleges and universities, including the renowned Harvard and Yale universities.

Most of Massachusetts is situated in lowland, which rises to 1064 m (3491 ft) in the Berkshire Hills to the west. Like most New England states, it has few mineral riches. The once lucrative textile, clothing and footwear industries have all declined and have been overtaken by those making electrical equipment, electronic instruments and machinery. Other sources of income are agriculture, involving mainly dairy produce, market gardening and cranberries, and tourism, especially along the coast and in the Berkshires. BOSTON is the state capital and largest city.

Population 6 016 400
Map United States – Lb

Massachusetts Bay *USA* Inlet of the ATLANTIC OCEAN measuring 105 km (65 miles) from northwest to south-east. It lies between Cape Ann and CAPE COD, and has two arms – Boston Bay, and Cape Cod Bay, where the Pilgrims landed in 1620 and founded PLYMOUTH. The Puritans who followed the first group founded SALEM and BOSTON on Massachusetts Bay.

Map United States – Lb

Massawa *Eritrea* Red Sea port situated on an island linked by a causeway to the mainland. Its prosperity was profoundly affected during the 1970s and 1980s by famine and civil war.

Population 35 000
Map Ethiopia – Aa

massif Large plateau-like mass of uplands, usually with clearly defined margins, which are often formed by FAULTS. The term is also used to describe a compact group of connected mountains which form a distinct portion of a mountain range.

Massif Central *France* Largely pastoral upland area in central France, covering 85 000 km² (32 800 sq miles), or about one-sixth of the country. Its highest peak is the Puy de Sancy, at 1885 m (6184 ft).

Map France – Ed

Masuku (Franceville) *Gabon* Capital of the south-eastern Province of Haut-Ogooué, situated alongside a tributary of the Ogooué River. Founded in 1880 by the Italian-born explorer Pierre Savorgnan de Brazza, and initially called Franceville, it is now the main town and trading centre in a region rich in manganese and uranium.

It is also the terminus on the Trans-Gabon Railway, which runs from Libreville along the coast. The line was officially opened in 1989. The Supernational Centre for Medical Research of Franceville is located in the town.

Population 75 000
Map Gabon – Bb

Masuria (Mazuria; Masuren; Mazury; Pojezierze Mazurskie) *Poland* Northern lakeland, situated east of the Vistula River. Since the Second World War, it has been divided between Poland, Russia (or until 1991, the USSR), and Lithuania (although by far the largest part is in Poland). Its ridges, hills of gravel and long, winding lakes were left by retreating glaciers at the end of the last Ice Age some 10 000 years ago. The lakes include Lake Sniardwy (Spirding), Poland's largest, which covers some 106 km² (41 sq miles) and is a nesting place for grey herons.

If the Suwalki lake district to the north-east is included, the region has more than 1000 lakes, a labyrinth of connecting rivers and canals and about 90 nature reserves. This makes it a superb holiday area. Wild European bison are protected in the Borecka Forest on the Russian border, and Europe's largest herd of elk can be seen in the Red Swamp (Czerwone Bagno) reserve near AUGUSTÓW. The resort of Giżycko (Lötzen; population 26 300) lies on an isthmus near Lake Mamry, Poland's second largest lake, which covers 105 km² (about 40 sq miles). During the day, the lake is alive with cormorants; at night, tourists hunt for crayfish by torchlight.
Map Poland – Db

Matabeleland (Ndebeleland) *Zimbabwe* Southwestern region consisting of two provinces (Matabeleland North, and Matabeleland South) covering a total of 132 087 km² (51 000 sq miles) around the city of BULAWAYO. During the 1970s, it was a centre of guerrilla activity against the government of Ian Smith and his Rhodesian Front Party. The region is the home of the Matabele (or Ndebele) people.
Population 1 379 400
Map Zimbabwe – Ba

Matadi *Zaire* Capital of Bas-Zaire region, and Zaire's principle seaport. It is located on the Zaire River estuary at the upper limit of navigation for ocean-going ships. Built on a hillside at the base of the Crystal Mountains, it was established by the English adventurer Sir Henry Morton Stanley as a trading station in 1879. It is now a busy commercial city with a variety of industries, including pharmaceuticals and printing. The Matadi-Kinshasa railway skirts the Inga Rapids and Inga Dam, which are located just above Matadi.
Population 162 000
Map Zaire – Ab

Matale *Sri Lanka* Town situated in the central mountains, 20 km (12 miles) north of the royal Sinhalese city of KANDY. Aluvihara, an ancient Buddhist rock temple, lies 3 km (2 miles) to the north. Monks are still copying the story of the Buddha's many incarnations on to palm leaves – a project that was started almost 150 years ago. The project is designed to replace a set of leaves written in 43 BC, but destroyed in 1848 during a skirmish between the Sinhalese and British troops.
Population 29 700
Map Sri Lanka – Bb

Matanzas *Cuba* Seaport, resort and capital of Matanzas Province, situated in a fertile sugar cane region 104 km (64 miles) east of HAVANA. It exports sugar and the coarse, red fibre of the henequen plant, which is used in making rope and twine; factories here also make rayon, shoes and fertilisers. The beach resort of VARADERO lies 35 km (22 miles) farther east.
Population (city) 113 700; (province) 599 200
Map Cuba – Aa

Matera *Italy* Town lying some 200 km (about 125 miles) east of NAPLES. Its most outstanding feature is its numerous cave dwellings cut into the side of a limestone gorge and known as *Sassi*.
Population 52 200
Map Italy – Fd

Mathura (Mattra; Muttra) *India* Northern city on the Yamuna River, 130 km (80 miles) south-east of DELHI. It is revered by Hindus as the birthplace of Krishna, the legendary Hindu deity, and has relics from local Hindu, Buddhist and Muslim kingdoms. Its Buddhist artists produced some of India's finest sculptures from the 2nd century BC to the 6th century AD.

Today, cotton and fertilisers are the main products made in the city.
Population 1 923 920
Map India – Bb

Mato Grosso *Brazil* Remote plateau region bordering Bolivia and Paraguay, covering some 1 232 000 km² (475 700 sq miles) – an area larger than France and Spain combined – and containing the states of Mato Grosso and Mato Grosso do Sul. The region's main towns are Cuiabá and Campo Grande. In addition to gold (which sparked a gold rush to the area in the 17th century), the region contains deposits of iron and manganese ore, as well as diamonds.

However, agriculture – mainly rice, soya beans, wheat and livestock – is still the chief source of income. The name of the region is derived from its characteristic landscape: *mato grosso* translates as 'thick scrub'.
Population 3 524 000
Map (region) Brazil – Cc; (Mato Grosso State) Brazil – Cc; (Mato Grosso do Sul) Brazil – Cd

Matopo Hills *Zimbabwe* Granite range south of the city of BULAWAYO. Rising to 1554 m (5098 ft), the hills have been settled for about 40 000 years. The tomb of Cecil Rhodes (1853-1902), the British-born South African statesman and founder of Rhodesia (now Zimbabwe), is at World's View in Matopos National Park.
Map Zimbabwe – Bb

Mátra *Hungary* Group of volcanic mountains near the Slovakian border north-east of BUDAPEST. They reach 1015 m (3330 ft) at Kékes, Hungary's highest peak, and drop steeply to the Great ALFÖLD to the south. Vineyards clothe their lower slopes; extensive beech and oak forests grow just above the vineyards.

A popular holiday area for winter skiing and summer hikes, the mountains have several fine resorts, including Mátrafüred, situated 400 m (about 1300 ft) above sea level to the north of Gyöngyos, the regional centre.

The Börzsöny Hills, which rise to 936 m (3071 ft), and which run westwards to the River Danube, are an extension of the Mátra. The region has numerous ruined castles and popular summer resorts above the great Danube bend; among a number of scenic villages in the Nógrád Valley, is Hollókö, home to the Palócs – one of Hungary's most distinctive ethnic groups.
Map Hungary – Aa

Matruh (Marsa Matruh; Mersa Matruh) *Egypt* Harbour town on the Mediterranean coast 300 km (186 miles) west of the port of ALEXANDRIA. In a cave near the town, General Erwin Rommel, is said to have had his headquarters while vainly trying to stem the Allied advance after the battle of El Alamein in October-November 1942. The cave is now a war museum.
Population 27 900
Map Egypt – Ab

Ma-tsu Tao (Matsu Islands) *Taiwan* Group of islands lying 19 km (12 miles) off the coast of the Chinese Province of Fujian, near FUZHOU. Like Chin-men Tao the islands are heavily fortified by the Chinese Nationalist government.
Map Taiwan – Ba

Matsue *Japan* City near the north-west coast of HONSHU Island, 130 km (80 miles) north of HIROSHIMA. Known as 'the City of Water', it stands on the Ohashi River which links Lake Shinji to the Naka lagoon. In July each year, thousands of candle-lit paper lanterns, commemorating the souls of the dead, are set adrift on the waters of Shinji during the Toro Nagashi festival.
Population 143 000
Map Japan – Bc

Matsushima *Japan* Bay on Honshu's east coast just north-east of the city of SENDAI. Its name translates as 'Islands of Pines' – a reference to more than 260 forested rocky islets that are a feature of the area. Seaweed and oysters are farmed in the waters of the bay.
Map Japan – Dc

Matsuyama *Japan* Port and city on the north-west coast of SHIKOKU island overlooking the Inland Sea. It has oil and petrochemical plants. Dogo Spa is a hot spring resort with a popular public bathhouse.
Population 443 000
Map Japan – Bd

Matterhorn (Mont Cervino; Mar Cervin) *Italy/Switzerland* One of the world's most spectacular and distinctive mountain peaks, situated on the frontier, 5 km (3 miles) south of the Swiss town of ZERMATT. It was eroded by ice to the shape of a steep, bent pyramid, and has a particularly sheer east face.

The mountain, 4477 m (14 688 ft) high, was first climbed in 1865 by a party of seven led by the English mountaineer Edward Whymper (1840-1911). On the way down, four of the party fell to their deaths. Whymper and two of the guides were saved when the rope broke.
Map Switzerland – Ab

Maui *USA* Second largest island in the United States Pacific Ocean state of Hawaii, after Hawaii Island. It rises to 3055 m (10 023 ft) at Kolekole Peak. Maui has a flourishing tourist industry.
Map Hawaii – Ba

Maule *Chile* Noted wine-, wheat- and livestock-producing region in the basin of the Maule River. It embraces some of the country's lake district and most spectacular mountain scenery. The summer resort of Constitución lies some 90 km (55 miles) west of the regional capital, TALCA, where Chilean independence was declared in 1818.
Population 834 050
Map Chile – Ac

Mauna Kea *USA* See HAWAII

Mauna Loa *USA* See HAWAII and HAWAII VOLCANOES NATIONAL PARK

Mauritania See p 437

Mauritius See p 438

Mavrovo *Macedonia* National park covering some 686 km² (265 sq miles) in the north-west, on the Albanian border. It includes the craggy SAR PLANINA, and Korab Mountain, rising to 2764 m (9068 ft), and contains beech and fir forests harbouring bears, lynxes, foxes and deer.

The spectacular Radika River gorge lies in the south of the park. Among several villages that are rich in folklore is Galicnik, noted for its traditional costumes and a multiple wedding ceremony held each year in July.
Map Macedonia – Ab

Mawlamyine (Moulmein) *Myanmar (Burma)* City, port and capital of Mon state, 120 km (75 miles) south-east of the national capital, Yangon. It is a market place for sugar cane and rubber grown along the coastal strip at the mouth of the Thanlwin River. Timber, particularly teak, is floated down the Thanlwin River from logging sites in the rain forest and processed here.
Population 203 000
Map Myanmar – Cc

Mawsynram *India* The wettest place on earth, Mawsynram lies high up in the hilly north-eastern state of MEGHALAYA. Lying on the south side of the hills, it gets an average of 11 873 mm (467 in) of rain per year. Farming and the processing of forest products are the main occupations.
Map India – Ec

Mayagüez *Puerto Rico* Seaport, industrial city and cultural centre on the west coast, 115 km (72 miles) from SAN JUAN. Sugar, coffee, pineapples and citrus fruit are processed and exported from here. Mayagüez is the centre of the island's knitwear industry, which has found a lucrative market in the United States.

The city has a university college of agriculture and mechanics, founded in 1763.
Population 100 370
Map Caribbean – Bb

Maymyo *Myanmar (Burma)* See PYIN OO LWIN

Mayo (Meigh Eo) *Ireland* Rugged county covering 4831 km² (1865 sq miles) on the Atlantic coast. Its west coast is wild and broken with many inlets and fringed by islands; inland, it is dotted with lakes. Although the land is generally poor, the county's main activity is agriculture. Some textiles are produced. The county town is Castlebar (population 6070).
Population 110 710
Map Ireland – Bb

Mayon *Philippines* Active volcano 2462 m (8077 ft) high on the Bicol Peninsula of southern LUZON, just north of Legaspi. Symmetrically shaped, its name is derived from a local dialect word for 'beautiful'. Mayon has erupted nearly 40 times since 1800. After torrential rains, mud slides are an additional hazard.
Map Philippines – Bc

Mayotte (Mahore) *Indian Ocean* French dependency in the COMOROS island group covering 375 km² (144 sq miles) and valued by France as a naval base. A French possession since 1843 and one of the group's four main islands, Mayotte voted to remain French when the other islands in the Comoros group voted for independence in 1974. Its main town is Dzaoudzi. Mayotte's main crops are ylang-ylang (flowers for perfume oil), vanilla, coffee and rice; livestock is also raised.
Population 89 983
Map Madagascar – Aa

Mazar-e Sharif *Afghanistan* Manufacturing town 304 km (nearly 190 miles) north-west of KABUL. Its industries include flour milling, brick making and textile manufacturing, and it trades in carpets and skins. It contains the tomb of Ali, the son-in-law of Muhammad, and is a place of pilgrimage for Shiite Muslims. The town is known for the game of bazkashi, which is played on horseback – with the riders trying to gain possession of the body of a dead goat or lamb.
Population 113 000
Map Afghanistan – Ba

Mazatlán *Mexico* Fast growing beach resort and port at the mouth of the Gulf of California, in the State of Sinaloa. It was once an important port for trade with the Philippines; now it is a resort for tourists, drawn here by excellent swimming, surfing beaches and deep-sea fishing contests.
Population 314 200
Map Mexico – Bb

Mazovia (Mazowsze) *Poland* Ancient, semi-independent principality in eastern Poland, which joined the Polish kingdom in 1529. It is a flat lowland, with the capital, WARSAW, at its centre. Vast pine forests punctuated by sand dunes and peat bogs rich in bilberries and cranberries are features of the area. Traditional Mazovian thatched cottages, often crowned by storks' nests, with gardens full of sunflowers and hollyhocks, can still be found in certain areas. A lively dance that developed locally gets its name from the Polish word *mazurka*, meaning a Mazovian woman.
Map Poland – Db

Mazuria *Poland* See MASURIA

Mbabane *Swaziland* Capital of the country, situated in west-central Swaziland. It was founded as a trading station in the late 19th century, and became the capital of the British Protectorate of Swaziland in 1902. Although parliament meets in nearby Lobamba, Mbabane has remained the national capital and the country's chief commercial centre. Occupying an attractive position overlooking the Ezulwini Valley about 1150 m (3770 ft) above sea level, it lies on Swaziland's main east-west road.
Population 52 000
Map Swaziland – Aa

Mbala *Zambia* North-eastern market town near the Tanzanian border. Formerly known as Abercorn, it serves a region producing coffee, fruit, maize and tobacco. The 221 m (725 ft) high Kalambo Falls plummet over a cliff at the southern end of Lake TANGANYIKA, 30 km (19 miles) west of the town.
Population 5200
Map Zambia – Ca

Mbale *Uganda* One of Uganda's largest towns, 190 km (nearly 120 miles) north-east of the capital, KAMPALA. Established in 1902, it is the regional centre for the east of the country, and handles coffee grown on the slopes of Mount Elgon to the east.
Population 54 000
Map Uganda – Ba

Mbandaka *Zaire* Capital of Equateur region, 580 km (360 miles) north-east of KINSHASA. Formerly called Equateurville and later Coquilhatville, it is a commercial and industrial city, producing coffee, rice, rubber and timber. It has food processing and printing industries.
Population 294 000
Map Zaire – Aa

M'Banza-Congo *Angola* Historic town 290 km (180 miles) north-east of the capital, LUANDA. Formerly the capital of the powerful Kongo kingdom, it was a flourishing place when Portuguese settlers arrived in the 15th century and named it São Salvador.
Population 12 700
Map Angola – Aa

Mbanza-Ngungu *Zaire* Commercial and industrial town, formerly called Thysville, lying some 120 km (75 miles) south-west of the capital, KINSHASA. It has railway workshops.
Population 84 000
Map Zaire – Ab

Mbeya *Tanzania* Regional capital in the south-west, 90 km (55 miles) north-west of Lake MALAWI. It markets tobacco, tea, coffee, cattle and goats.
Population 100 000
Map Tanzania – Ba

Mbomou *Equatorial Africa* See BOMU

Mbuji-Mayi *Zaire* Capital of Kasai Oriental region, 615 km (382 miles) west of Lake TANGANYIKA. Formerly called Bakwanga, it is the centre of a major diamond-mining region, which produces, by weight, about 75 per cent of the world's industrial diamonds.
Population 625 000
Map Zaire – Bb

McKinley, Mount *USA* See DENALI NATIONAL PARK

McMahon Line *China/India* Part of the boundary between Tibet in China and what is now the Indian territory of Arunachal Pradesh, running in a north-easterly direction from the Bhutanese border to the Brahmaputra River. The line was negotiated by Tibet (then an independent country) and Great Britain at Shimla in 1914; it was named after Sir Henry McMahon, the chief British negotiator. China disputed the agreement then, and again after the communist takeover in 1949. In fact, it was one of the causes of the war between China and India in 1962.
Map India – Eb

McMurdo Oasis *Antarctica* See DRY VALLEYS

McMurdo Sound *Antarctica* A branch, 148 km (92 miles) long and 37-74 km (23-45 miles) wide, of the ROSS SEA, forming a channel between ROSS ISLAND and Victoria Land in Antarctica. It was

Mauritania

*A MUSLIM LAND WHERE TENSIONS
HAVE ERUPTED BETWEEN MOORS
AND BLACK AFRICANS*

Prolonged droughts have devastated Mauritania's traditional economy. It can produce little of the food it needs, and relies heavily on foreign, particularly Arab, aid. In 1991, when Mauritania supported Iraq during The Gulf War, its two main donors, Kuwait and Saudi Arabia, withdrew their aid.

Mauritania is nearly twice the size of France, yet it has only 2.25 million people. It links the Arab MAGHREB with black West Africa. The SAHARA covers the north, and here the only settlements are those around oases such as Atar, where millet, dates, melons and vegetables are grown. Temperatures rise well over 30°C (86°F) during the day, but can fall to freezing point at night. In the south is the main agricultural region, the fertile valley of the SENEGAL River; between the two lie the drought-stricken SAHEL grasslands.

About two-thirds of the people are of Moorish (Arab/Berber) descent or of mixed Moorish and black descent. The *bidan* (white Moors) have pale skins, and legend has it that the traditional pale blue robe is worn to enhance their colouring. The *harattin* (black Moors) have the same features, but much darker skins, and were slaves of the *bidan* in times past. The remaining Mauritanians are black Africans, who live in the south and include the Peul and Wolof peoples.

Although Islam is a unifying force, for the faith is almost universal and Islamic law is applied, tensions between the Moors and the blacks broke out in 1989 when members of the minority black population were persecuted – many of them arrested and tortured to death – by members of the armed forces; thousands fled across the border into neighbouring Senegal. The presence of the refugees in Senegal has led to deterioration of diplomatic relations, already strained, between the two countries. And inside the country, the tension between the two population groups continues.

The majority of the people – the Moors and Peul – are traditionally nomadic or partly nomadic herders of goats, sheep, cattle or camels. However, severe drought in the 1970s and the 1980s killed 50 to 70 per cent of the country's animals.

In 1963, 83 per cent of Mauritanians were to some extent nomadic, but by 1980 the figure had fallen to 25 per cent. The rest had settled along the Senegal River, near remaining oases, or in vast shanty towns in urban areas – particularly NOUAKCHOTT, the capital – where unemployment has reached 20 per cent. The World Bank has now given aid to help rebuild the livestock herds so that the nomadic population can sustain themselves once more.

Iron ore makes up nearly half of the country's exports, but the world price for iron ore has declined in recent years. It is mined near Zouérate; copper is mined near Akjoujt and has provided the other main export; despite the fact that two-thirds of the country is virtually desert, Mauritania is one of the Sahel's richer countries. Development of its mineral resources and fishing industry is the best hope for its future – however, its rich fishing grounds have been depleted by foreign fleets. The country now has accumulated enormous foreign debt – exacerbated by the recent droughts, by rising energy prices, and falling prices for its primary resources.

Originally a Berber kingdom, Mauritania became a French protectorate early in the 20th century and a colony, forming part of French West Africa, in 1920. It emerged as an independent republic on 28 November 1960, but a military coup occurred in 1979. Links with France have diminished as the country strives for recognition as a true Arab nation.

In 1979, Mauritania withdrew its claim to Western Sahara. A referendum held in 1991 endorsed a new constitution, which paved the way for multiparty civilian rule. Elections were held a year later, and President Moaouia Ould Sidi Mohammed Taya was returned to office, although the election results were disputed by the opposition.

Relations have meanwhile improved with Saudi Arabia – which has made a renewed offer of assistance – and Senegal. But the government has continued to face accusations of human rights violations perpetrated against its black minority.

MAURITANIA AT A GLANCE	
Official name Islamic Republic of Mauritania	
Area 1 030 700 km² (397 953 sq miles)	
Population 2 252 000 **Per km²** 2.2 **(Per sq mile** 5.6)	
Capital Nouakchott	
Government Republic	
Currency 1 ouguiya = 5 khoums	
Languages Arabic (official), African languages, Wolof (official)	
Religion Muslim (100%)	
Climate Hot and dry, summer rains near coast. Average temperature in Nouakchott ranges from 21°C (70°C) in January to 30°C (86°F) in September	
Land use Grazing 38%, cultivation 1%, forest and woodlands 5%, other 56%	
Main primary products Livestock, millet, sorghum, dates, fish; iron ore, copper, phosphates, salt	
Major industries Agriculture, mining, fishing	
Main exports Iron ore, copper, processed fish	
Annual income per head (US$) 530	
Population growth (per thous/yr) 27	
Life expectancy (yrs) Male 45 **Female** 50	

discovered in 1841 by the British explorer Sir James Clark Ross, who named it after one of his officers. The British explorers Ernest Shackleton (1874-1922) and Robert Falcon Scott (1868-1912) and the Commonwealth trans-Antarctic Expedition of 1957-8 established their bases alongside the sound.

The American McMurdo Base, on Ross Island, is the largest Antarctic base; it is staffed by about 750 people during summer.

Map Antarctica – Jc

Mdina (Notabile) *Malta* Walled town in the west of the island, which as Citta Vecchia was the capital until it was superseded by Valletta in 1570. It is less than 500 m (550 yards) square, but has many fine palaces and religious buildings which date mainly from the 18th century.

Population 930

Map Malta – Db

Mead, Lake *USA* Constructed lake covering some 600 km² (230 sq miles) on the Arizona-Nevada border. It was built in 1930-6, at the same time as the Hoover Dam on the Colorado River.

Map United States – Dc

mean sea level Average level of the sea, calculated from a long series of continuous records of tidal oscillations, and used as a basis from which to measure the height of land masses.

meander One of a series of sweeping curves in the course of a river across its FLOOD PLAIN.

Mauritius

NEW INDUSTRIES ARE GROWING ON A TROPICAL ISLAND WHERE SUGAR IS NO LONGER THE BIGGEST EXPORT

An overpowering odour of molasses surrounds the island of Mauritius. It used to be the sweet smell of success when the cane sugar industry and its by-products of molasses and rum kept the island in relative prosperity. But changing world markets have affected the economy.

Sugar is still a major export – the second biggest after textiles. The cane plantations cover 90 per cent of the cultivated land and use one-fifth of the labour force. But new industries have been developed and sugar accounts for only 40 per cent of the island's exports compared with 98 per cent in former days.

The government's strategy has been to encourage industrialisation, and Mauritius now has a manufacturing sector which employs 100 000 people, mainly making textile goods and fertilisers. From 14 per cent in the mid-1980s, the unemployment rate was down to 2.4 per cent by the mid-1990s. In 1992, when the world economy was still at a low mark, Mauritius had an impressive growth rate of more than 6 per cent – all the more impressive because, during the 1980s, the development costs of the industrial sector pushed up foreign debt.

Besides sugar and manufacturing, the tourist industry which was developed in the 1980s, has made a significant contribution – it is now the third largest source of foreign exchange, after sugar and clothing exports. In 1991, almost 300 000 tourists visited the island which is almost completely surrounded by coral reefs, making scuba diving a main attraction. There is a choice of superb tropical beaches along 177 km (110 miles) of coast.

This volcanic island in the Indian Ocean has many craters with spilt lava flows. The undulating northern plain rises gradually to the central plateau at one point 823 m (2700 ft) high, before dropping sharply to the south and west coasts. The climate is varied, but mainly hot and humid, and the prevailing south-east winds bring an annual rainfall of 5000 mm (195 in) to the windward slopes. Fast-flowing rivers have been harnessed to produce hydro-electric power.

The well-watered, fertile soil was once covered with luxuriant natural vegetation but much of this has given way to the sugar plantations. Yet every island paradise has a negative side: between November and April, Mauritius is battered by raging cyclones, which can cause extensive damage to crops and human settlements.

SUCCESSION OF SETTLERS

Mauritius was known to 10th-century Arabs, but the first recorded settlement was by the Dutch in 1598, who named the island after their ruler, Prince Maurice of Nassau. The Dutch abandoned the settlement in 1710, the French arrived in 1715, and the British seized the island in 1810. Mauritius, with the small island of RODRIGUES, became an independent Commonwealth state in 1968 and a republic in 1992. The island state has a multiparty system and is ruled by a coalition of parties, headed by Sir Aneerood Jugnauth, who became prime minister in 1982.

Despite its history of European colonialism, two-thirds of the population are of Indian extraction – descendants of contract workers who were brought in to work the plantations after the abolition of slavery in 1835. The rest are mostly Creole descendants of African slaves and French settlers.

The population has doubled since the 1960s, this is mainly because the eradication of malaria has reduced the death rate. Population density exceeds 600 per km² (1560 per sq mile) and the capital city of PORT LOUIS is growing to form a conurbation with the towns of Beau Bassin and CUREPIPE.

MAURITIUS AT A GLANCE	
Map Indian Ocean – Bd	
Official name Republic of Mauritius	
Area 1860 km² (718 sq miles) excluding Rodrigues and dependencies	
Population 1 121 000 **Per km²** 603 **(Per sq mile** 1561)	
Capital Port Louis	
Government Republic	
Currency 1 Mauritius rupee = 100 cents	
Languages English (official), French, Creole, Hindi, others	
Religions Hindu (52%), Christian (28%), Muslim (17%), other 3%	
Climate Tropical; average temperature is 23°C (73°F) throughout the year	
Land use Cultivation 58%, forest 31%, grazing and other 11%	
Main primary products Sugar cane, tea, tobacco, potatoes, fish	
Major industries Agriculture, sugar processing, molasses, rum distilling, textiles, clothing, fertilisers, electronics, tourism	
Main exports Sugar, molasses, rum, textile yarns, clothing, and fabrics, tea, tobacco	
Annual income per head (US$) 2700	
Population growth (per thous/yr) 10	
Life expectancy (yrs) Male 66 **Female** 74	

Meath (An Mhí) *Ireland* Fertile grassland county covering 2336 km² (902 sq miles) north-west and north of DUBLIN. It is known as Royal Meath because it includes Tara, ancient seat of the Irish kings. Navan, the county town (population 3420), has textile and engineering industries and Ireland's largest metal-mining complex.
Population 105 370
Map Ireland – Cb

Meaux *France* Chief town of the Brie cheese-making district on the MARNE River, some 40 km (25 miles) east of Paris. Meaux is also an important producer of mustard.
Population 48 300
Map France – Eb

Mecca (Makkah, Al Makkah) *Saudi Arabia* Capital of the Western Province (formerly HEJAZ), situated 64 km (40 miles) east of JEDDAH. It lies in a narrow valley overlooked by hills crowned with forts. Mecca was the birthplace of Muhammad, the founder of Islam, and is the religion's most holy city. All Muslims try to make the *haj* (pilgrimage) to Mecca at least once in their lives, and 2 million pilgrims visit it each year.

Muhammad was born here in about the year 570 and it was his home until 622, when he fled to Medina – 350 km (217 miles) to the south – to avoid religious persecution. Mecca contains the Kaaba, a shrine which houses the sacred Black Stone, towards which Muslims face, wherever they are in the world, when they pray. The Black Stone is said to have been given to the prophet Abraham by the archangel Gabriel. The city is also the site of the University of Ummul Qura. Like Medina, Mecca is protected on behalf of the Muslim world by the Saudi royal family.

Non-Muslims are forbidden to enter Mecca; any caught within its walls face heavy penalties and deportation. The city's commercial life revolves almost entirely around the pilgrims who find accommodation in enormous tent cities on Mecca's eastern outskirts.
Population 618 000
Map Saudi Arabia – Ab

Mechelen *Belgium* See MALINES

Mechlin *Belgium* See MALINES

Mecklenburg *Germany* Lakeland region covering 26 694 km² (10 307 sq miles) between the capital, Berlin, and the Baltic Sea. It is a popular area for tourists.
Population 1 106 900
Map Germany – Eb

Medan *Indonesia* Provincial capital of North Sumatra. Local produce includes coffee, tea, palm oil, latex, cinnamon and tobacco. Orang-utans, members of one of the world's most endangered animal species, live in the mountains to the west of Medan.
Population 2 378 000
Map Indonesia – Ab

Médéa (Lemdiyya) *Algeria* Administrative centre in the middle of a vine- and fruit-growing district in the Tell Atlas Mountains, some 24 km (15 miles) south-west of BLIDA. It lies 914 m (3000 ft) above sea level at the foot of the Djebel Nador. Plastic goods are manufactured here.
Population 90 000
Map Algeria – Ba

Medellín *Colombia* Second largest city of the republic, situated 240 km (150 miles) north-west of the capital, Bogotá. The capital of Antioquia department, it is known as the city of eternal spring, because the combination of its altitude (1486 m, 4875 ft) and its tropical location give it a warm climate throughout the year.

Founded in 1616 in the Aburrá Valley, Medellín was the country's first city to develop a modern industrial base. It began producing textiles in the late 19th century, and still produces 66 per cent of all Colombian textiles.

It also produces plastics, and employs one in four of the country's manufacturing workforce.

Medellín is a city of flowers and is the self-proclaimed orchid capital of the world. However, it is also the chief centre of the international trade in illicit cocaine. Its drug barons are among the richest people in the world, with assets valued in billions of United States dollars.

Population	1 639 000
Map	Colombia – Bb

Medias *Romania* Industrial town in central TRANSYLVANIA, 190 km (118 miles) south of the Russian border. Medias, which grew from an 11th-century craftsmen's village, retains many of its old streets, a fortified church built in the 14th and 15th centuries, and the remains of other medieval buildings. The town manufactures, among other things, machinery, shoes, glass, china, porcelain and wine.

Population	68 400
Map	Romania – Aa

Medicine Hat *Canada* Main industrial and commercial city of south-eastern Alberta, situated on the South Saskatchewan River, some 265 km (165 miles) south-east of CALGARY. Its industries revolve around rich gas reserves, flour and timber mills, grain elevators and iron foundries.

According to legend, the town stands on the spot where a Cree medicine man lost his hat in the river while fleeing from Blackfoot braves who had slaughtered his tribe.

Population	43 630
Map	Canada – Dc

Medina (Al Madinah) *Saudi Arabia* The country's second holiest city, situated some 350 km (217 miles) north of the holiest one – MECCA. In AD 622, Muhammad, the founder of Islam, fled here from his birthplace Mecca, where he was being persecuted for his religious beliefs.

His flight (known as the Hegira), which took place on 22 September, was later marked as the beginning of the Islamic calendar. Muhammad's tomb is in the city's main mosque. The Islamic University was established in Medina in 1962. Thousands of pilgrims visit the city each year.

Population	500 000
Map	Saudi Arabia – Ab

Mediterranean climate Warm temperate climate on the western margins of continents, and around the Mediterranean Sea, characterised by hot, dry, sunny summers and moist, mild winters.

Mediterranean Sea *Africa/Europe* Covering 2 505 000 km² (967 230 sq miles) between the continents of Europe and Africa, the Mediterranean Sea is strictly an arm of the ATLANTIC OCEAN to which it is linked by the Strait of GIBRALTAR. In the north-east it is connected to the BLACK SEA. A submerged ridge between Sicily and Tunisia divides the Mediterranean into two parts. In the west are the TYRRHENIAN and LIGURIAN seas and the BALEARIC ISLANDS Basin. In the east are the ADRIATIC, IONIAN and AEGEAN seas. Its average depth is 1370 m (4495 ft) and its greatest depth of 4846 m (15 900 ft) is in the Hellenic Trough, off south-western Greece.

Many of the Mediterranean's sea-floor features are the result of compression caused by the northward drift of the African plate which has pushed against the Eurasian plate (SEE PLATE TECTONICS). These plate movements have also thrown up the Alps, the Apennines and other fold

mountains, and still cause earthquakes and volcanic activity in the region.

Evaporation and an undercurrent of dense, salty water to the Atlantic removes water from the Mediterranean three times as fast as it is replaced by rain or by rivers. This makes the Mediterranean saltier than the Atlantic and also, to maintain the level, results in a constant inflow of surface water from the Atlantic.

Because the Mediterranean is almost tideless, toxic materials tend to accumulate in the coastal waters where they have been discharged. The severe pollution this has caused in places has affected both tourism and the fishing (which is already suffering as a result of overfishing).

The Mediterranean has been a busy shipping lane for at least 5000 years, and the opening of the SUEZ CANAL in 1869 placed it on the main sea route from northern Europe to Asia.

Map	Africa – Da

Mediterranean woodland Wooded areas of drought-resistant species of evergreen trees, such as the wild olive, cork oak and conifers, together with flowering, often aromatic, shrubs. This type of vegetation is found around the Mediterranean and on the western margins of continents in subtropical latitudes where similar climates prevail. Grazing and deforestation by fire has caused degeneration into areas of scrub such as MAQUIS and GARRIGUE.

Médoc *France* One of the world's greatest wine-producing areas, on the west bank of the Gironde River, some 30 km (20 miles) north-west of BORDEAUX. It is particularly famous for its red wines, named after the region. Among the area's renowned vineyards are those of the Châteaux Lafite, Latour, Margaux and Mouton-Rothschild.

Map	France – Cd

Medway Towns *United Kingdom* The riverside towns of Rochester, Chatham and Gillingham in the south-east English county of Kent, 50 km (30 miles) from LONDON. They are situated on the River Medway, which is 112 km (70 miles) long and flows first north and then north-east into the Thames estuary.

Rochester was a Roman walled city at the Medway River's lowest bridge point; some parts of its city walls remain, as well as a splendid 12th-century Norman castle and a cathedral founded in 604 and rebuilt between the 12th and the 15th century.

Rochester is Britain's second oldest bishopric, after Canterbury, 40 km (25 miles) to the east. Until the 1980s, Chatham and Gillingham shared a naval dockyard that had been opened during Elizabeth I's reign in the 16th century; ironically, it was closed in the 1980s, during the reign of Elizabeth II. Both towns depended heavily on the dockyard for employment and income, and its closure created an urgent need for new jobs.

Population	210 000
Map	United Kingdom – Fe

Meerut *India* Northern industrial city, situated 66 km (41 miles) north-east of DELHI. In 1806, the British established a military base in Meerut, and it was here that the Indian Mutiny broke out in 1857. The city's factories produce textiles, sugar, chemicals, and leather goods.

Population	536 600
Map	India – Bb

Meghalaya *India* North-eastern state covering 22 445 km (8664 sq miles) and comprising the Garo, Khasi and Jaintia Hills, which are inhabited mainly by Garo, Khasi and Jaintia tribespeople. The south side of the hills is one of the world's wettest places, where the city of Cherrapunji has more than 10 300 mm (406 in) of rain a year and MAWSINRAM has an annual 11 873 mm (467 in).

Although most of Meghalaya's inhabitants are farmers, coal and corundum (a hard metal used in abrasives) are mined and forest products are processed. Its capital is SHILLONG.

Population	1 760 600
Map	India – Ec

Megiddo *Israel* Archaeological site in the Jezreel Valley, 15 km (10 miles) south-west of NAZARETH. Remains have been found here of a fortress built by King Solomon in the 10th century BC, and of stables built by King Ahab in the 9th century BC. According to Christian tradition, it is at Megiddo that the Armageddon will be fought on the Day of Judgment.

Map	Israel – Ba

Meigh Eo *Ireland* See MAYO

Meiktila *Myanmar (Burma)* Textile town in central Myanmar, 130 km (80 miles) south of MANDALAY. A textile mill built in 1958 and a series of irrigation reservoirs excavated in the 1960s have boosted the development of the town as the centre of a cotton-growing region.

Population	229 600
Map	Myanmar – Bb

Meissen *Germany* Town on the River Elbe, 20 km (12 miles) north-west of DRESDEN. It is world-renowned for its porcelain, commonly known as Dresden china. The town is dominated by a fine Gothic cathedral and by Albrechtsburg Castle (1471-85) where the delicate tableware was made between 1710 and 1863. It is now a museum devoted to the display of the product that made it famous.

Population	38 200
Map	Germany – Ec

Mekambo *Gabon* Small town in the north-east, 177 km (129 miles) north-east of Makokou. It is a minor trading centre for rubber gathered in the forests which surround the town. The region around Mekambo contains some of the world's richest iron ore deposits, which were discovered in 1971, but which have hardly been exploited due to lack of funds.

Population	2000
Map	Gabon – Ba

Mekele *Ethiopia* Chief town of Tigray Province, about 500 km (310 miles) north of the capital, ADDIS ABABA. It is the site of the castle of Yohannes IV, Emperor of Ethiopia in the 1870s and the 1880s.

Population	50 000
Map	Ethiopia – Aa

Meknès *Morocco* Historic city and one of the country's four royal cities (the other three being FÈS, MARRAKECH and RABAT). It lies some 58 km (36 miles) south-west of Fès and was founded by the Berbers in the 11th century. Many of the surviving buildings date from the 13th and 14th centuries. In the 17th century, Sultan Moulay Ismail made Meknès his capital and built his magnificent Dar El Kebira Palace (which was damaged by an earthquake in 1755).

Meknès lies within a huge triple wall comprising 40 km (25 miles) of battlements and towers and containing the medina (old Muslim residential quarter) and palace. In the palace complex are the ruins of stables that once lodged 12 000 horses. Other places of interest include the Bab Mansour gateway, the Kouba El Khiyatine (reception place for foreign ambassadors), the present palace, called the Dar El Makhzen, and the mausoleum of Sultan Moulay Ismail. Today, the town is a rich market centre trading in fruits and vegetables, wine and olives, as well as ceramics, leatherwork and carpets produced locally.

Population 750 000

Map Morocco – Ba

Mekong (Lancang Jiang; Mae Nam Khong) *South-East Asia* South-east Asia's largest river, with a total length of 4185 km (2600 miles) and with a drainage basin covering 310 000 km² (nearly 12 000 sq miles). For almost half its length, from its origin in Chinese Tibet as far as the mountains of northern Laos, the Mekong flows through a series of deep gorges.

Although the valley opens up south of Luang Prabang, it is still interrupted by rocky structures and major waterfalls. It is only below Kratie in Cambodia that the Mekong forms an extensive alluvial plain. Following its confluence with the Tonle Sap River near PHNOM PENH, the Mekong enters its extensive delta which forms the whole of the extreme south of Vietnam, flowing out to sea in two main branches.

Fed by melting snows from the Tibetan mountains, the Mekong is a river with wide fluctuations in flow and its floods have been destructive throughout history, even making colonisation of the swampy delta difficult. Since the mid-1950s there have been plans to control the river and put its resources to work for the benefit of industries and agriculture, but the Mekong is an international river, forming the border of Thailand and Laos for much of its middle course, before flowing through Cambodia and Vietnam. Political divisions have prevented any development of the main stream, although the United Nations Mekong Committee, set up to coordinate its development, has done useful work on tributaries, especially in Thailand and in Laos.

Map Asia – If; Laos – Ab; Cambodia – Bb

Mekong Delta (Mien Nam) *Vietnam* The southern plain of Vietnam from the south coast to the foothills of the Truong Son (Annamite) mountain chain. It was colonised by the French (who called it 'Cochin China') in 1862; large, prosperous rice and rubber plantations were set up. Its buoyant economy – sharply different from the small-scale farming of the north – was one of the factors that led to Vietnam's partition in the mid-1950s and to the subsequent Vietnam War, in 1965-75.

Map Vietnam – Bc

Melaka (Malacca) *Malaysia* Tourist resort on the Strait of MALACCA, 120 km (75 miles) south-east of KUALA LUMPUR. Founded in about 1400, Melaka became one of the Far East's chief trading centres. It was colonised in turn by the Portuguese, Dutch and, in 1824, the British; the

buildings erected by each of the colonial powers still survive. The town lost its main source of income when trade shifted to Singapore farther south, and when silting blocked up the entrance to the harbour.

The state of Melaka, whose capital is Melaka Town, covers 1657 km² (640 sq miles). Its main crops are rice and rubber.

Population (town) 87 500; (state) 583 500

Map Malaysia – Bb

Melanesia *Pacific Ocean* One of the three main regions of the Pacific islands (the other being Micronesia and Polynesia) which are divided in three, along ethnic lines. Melanesians are mostly dark-skinned, with frizzy hair. They arrived in the Pacific from South-east Asia in waves of migration beginning some 10 000 years ago, and spread from New Guinea (now IRIAN JAYA and PAPUA NEW GUINEA, where even earlier people, the Papuans, were already settled) as far east as FIJI and south to NEW CALEDONIA. Melanesia (which means 'the black islands') also includes the SOLOMON ISLANDS and VANUATU. The region covers a land area of 1000 km² (3860 sq miles).

Traditional Melanesian society is based on village units, each often with its own language (on the Solomon Islands, for instance, more than 87 different languages are spoken), and generally without hereditary kings or chiefs. Elaborate indigenous beliefs involve initiation ceremonies, sacred places and objects, and ancestor worship.

Population 7 145 780

Map Pacific Ocean – Cc

Melbourne *Australia* Australia's second biggest city and state capital of VICTORIA, standing astride the Yarra River at the head of Port Phillip Bay. It was founded in 1835 by settlers from Tasmania and grew as a result of the gold rush during the 1850s, when Victoria became a separate colony. Many buildings, including the State Parliament House, the Treasury (1853) and Melbourne University (1855) date from this period. Melbourne was the country's seat of parliament until 1927 when parliament was moved to Canberra.

The city centre is laid out on a grid pattern, with spacious avenues. The botanical gardens, founded in 1845, cover 43 hectares (106 acres) and contain more than 12 000 plant species. Melbourne is one of Australia's main financial, commercial and industrial centres. It exports

▲ RICE BOWL A vast patchwork of glittering green rice paddy-fields covers the valley of the Mekong and Tonle Sap rivers in Cambodia.

wheat, wool, meat, fruit and dairy products. Its industries are based on electronics, engineering, car assembly plants, metallurgy, chemicals, textiles, and food processing. Monash (1958) and La Trobe (1964) are two modern universities. The city hosted the 1956 Olympic Games.

Population 3 153 500

Map Australia – Gf

Meleda *Croatia* See MLJET

Melilla (Mliya; Tamlit) *Spanish enclave in Morocco* Seaport 220 km (137 miles) east of CEUTA on the Mediterranean coast, and in Spanish hands since 1497. It is administered as part of the Spanish province of Málaga. The enclave, which covers only 12.5 km² (about 4.5 sq miles), exports iron ore from the nearby Beni bu Ifrur mines. The main industry is fishing. Melilla's old walled town stands on a huge rock rising from the sea.

Population 63 670

Map Morocco – Ba

Melnik *Bulgaria* See BLAGOEVGRAD

Memel *Lithuania* See KLAIPEDA

Memphis *USA* City on the Mississippi River in the extreme south-west of Tennessee; in fact, its metropolitan area extends across the river into Arkansas and south into the state of Mississippi. Memphis rose to prominence as a centre for marketing cotton during the steamboat era. Although it contains many reminders of the pre-American Civil War plantation era, it is very much a city of the 1990s and has become a pace-setter in industry, commerce, agriculture and medical research.

The Nobel Peace Prize-winner and civil rights leader Dr Martin Luther King Jr was assassinated in Memphis in 1968. Rock-and-roll and ballad singer Elvis Presley (1935-77), born nearby in Mississippi, spent most of his life in the city. Graceland, his Memphis mansion, attracts many thousands of fans and tourists each year.

Population (city) 610 300; (metropolitan area) 1 007 300

Map United States – Hc

Menai Strait *United Kingdom* Strait, 22.5 km (14 miles) long and 180-1100 m (590-3600 ft) wide, separating Anglesey from mainland Wales. It is crossed by a graceful road-carrying suspension bridge built by the Scottish engineer Thomas Telford in 1826. The present rail bridge was built on the surviving stone and brickwork of Robert Stephenson's Britannia Tubular Bridge (1850), which burnt down in 1970.
Map United Kingdom – Cd

Mendip Hills *United Kingdom* See SOMERSET

Mendoza *Argentina* City located some 1000 km (620 miles) west of BUENOS AIRES, lying at the foot of the Andes Mountains. Founded in 1561, it was destroyed by an earthquake in 1861, and has been largely rebuilt since. A huge oil field nearby has turned the city into a regional centre. It is also the centre of Argentina's principal wine-producing region and its red wines are well known internationally. Nevertheless, Mendoza faces stiff competition from Chile, whose wine exports outstrip those of Argentina's.
Population 775 000
Map Argentina – Cb

Mergui (Myeik) *Myamar (Burma)* Port in the Tenasserim (Tanintharyi) Division, in the extreme south of the country. The townspeople earn much of their income from the pearling grounds among the Mergui archipelago offshore, while cargo boats use Mergui's docks to collect rubber and coconuts from nearby plantations.
Population 199 800
Map Myanmar – Cd

Meriç *South-east Europe* See EVROS

Mérida *Mexico* Gracious colonial capital of the State of YUCATÁN, built by the Spanish conquerors in 1542 on the site of a Mayan city, after a violent encounter between the Spanish and the Mayas. Its colonial buildings, many of which were built of stone taken from torn-down Mayan temples, include a beautiful cathedral, the richly ornamented Casa de Montejo, and the chapel of Cristo de las Ampollas. Mérida is the gateway to the magnificent Mayan ruins of Yucatán.
Population 557 300
Map Mexico – Db

Mérida *Spain* Market town 265 km (165 miles) south-west of MADRID. Two irrigation dams built by the Romans in the 2nd century AD are still in use. The only major maintenance work the dams have needed in 1800 years has been the renewal of their stone facings, carried out in the 1930s.
Population 49 280
Map Spain – Bc

Mérida *Venezuela* Capital of Mérida State high in the Andes, about 650 km (403 miles) south-west of CARACAS. Founded in 1558, its buildings, a mixture of colonial and modern architecture, are arranged around 21 parks.
These include the Parque de las Cinco Repúblicas, which contains soils from the five countries liberated by the Venezuelan nationalist Simon Bolívar (1783-1830).
Population (city) 275 360; (state) 570 220
Map Venezuela – Ab

meridian A line of LONGITUDE, an imaginary north-south line drawn between the poles and crossing the equator at right angles.

Merioneth *United Kingdom* See CLWYD; GWYNEDD

Meroe *Sudan* Ruined city of ancient Nubia. It lies some 280 km (nearly 175 miles) north of the capital, KHARTOUM, on the east bank of the River Nile. Meroe was the capital of an Ethiopian kingdom between about 700 and 300 BC, and of the kingdom of Cush between 300 BC and AD 350.
Excavations have revealed streets and buildings that indicate a heavily populated and well-developed city. Its remains include a riverside quay, palaces, a temple, and a copy of a Roman public bath.
Map Sudan – Bb

Mers El Kébir (El-Marsa El-Kebir) *Algeria* Former fortified seaport and French naval base to the north-west of ORAN, and overlooked by the 16th-century Santa Cruz fortress perched on the hills inland. It was under Spanish control until it was taken by the French in 1830. In 1940, after the fall of France to Germany during the Second World War, the harbour was the scene of Operation Catapult, in which the British destroyed most of the French warships to prevent them from falling into enemy hands. Today, Mers El Kébir is an Algerian naval base.
Population 7800
Map Algeria – Aa

Mersa el Brega *Libya* See MARSA EL BREGA

Mersa Matruh *Egypt* See MATRUH

Mersey *United Kingdom* River, about 110 km (70 miles) long, of north-west England. It is formed by the confluence of the Etherow and

▼ CITY AT SUNSET Skyscrapers in Melbourne are reflected in a lake in the city's public gardens. During the gold rush of the 1850s, Melbourne was the fastest growing city in the world – by 1873 it already had 207 000 people.

Goyt rivers at Stockport in Greater Manchester, and flows into the Irish Sea downstream of Liverpool. Its estuary is of great commercial importance, allowing ocean-going ships to reach the ports of Liverpool, Birkenhead and, via the Manchester Ship Canal, Manchester.
Map United Kingdom – Dd

Merseyside *United Kingdom* County of north-west England, formed in 1974 out of parts of Lancashire and Cheshire, and covering 652 km² (252 sq miles). It is centred on the ports of LIVER-POOL and Birkenhead on either side of the River MERSEY. It also includes the town of St Helens, with its glass industry dating from the 18th century.
Population 1 449 700
Map United Kingdom – Dd

Mersin (Içel) *Turkey* The country's largest port on the Mediterranean Sea, 400 km (250 miles) south-east of the capital, ANKARA. It handles cotton, wool, chrome, fruit, cereals and timber, and refines oil from nearby Batman.
Population 422 000
Map Turkey – Bb

mesa Flat-topped elevation with one or more clifflike sides, common in the south-western United States.

Mesa Central *Mexico* Vast mountain plateau in the centre of Mexico, fringed by the Sierra Madre Oriental range in the east, and the Sierra Madre Occidental in the west. It is about 1130 km (690 miles) long, and has an average height of between 1220 and 2440 m (4000-8000 ft). Its highest point is the Citlaltépetl volcano, Mexico's tallest peak at 5747 m (18 855 ft). Mining was once the mainstay of the region's economy, but irrigated agriculture in the fertile basins has now superseded it.
Map (Sierra Madre Occidental) Mexico – Bb; (Sierra Madre Oriental) Mexico – Bb

Meseta *Spain* Central plateau covering about 350 000 km² (135 000 sq miles) – roughly three-quarters of the country.
Map Spain – Ca

Meshed *Iran* See MASHHAD

Mesolóngion *Greece* Town standing on a large lagoon west of Stereá Ellás on the north side of the Gulf of Patras. Between 1822 and 1826, the town withstood three sieges by the Turks during the Greek struggle for independence.

Many of its defenders broke out in 1826; those who remained lit the arsenal and blew themselves up rather than surrender. The English poet Lord Byron arrived at the town in January 1824 to help the Greeks in their fight for independence, but died of fever three months later.
Population 10 200
Map Greece – Bb

Mesopotamia *Middle East* Oil-rich region between the Euphrates and Tigris rivers, stretching from the Armenian Mountains in the north to The Gulf. Most of it lies in modern Iraq. The region's ancient civilisations – such as those of UR, BABYLON and NINEVEH – are credited with the development of writing, mathematics and astronomy, and with developing the first wheel.
Map Iraq – Bb

mesosphere A layer, 30 km (20 miles) deep, in the atmosphere, between the ionosphere and the stratosphere. See THE DIFFERENT LAYERS OF THE ATMOSPHERE, p 54

The solid part of the earth's mantle, between the 'soft' middle mantle – the asthenosphere – and the core, is also called the mesosphere. See JOURNEY TO THE CENTRE OF THE EARTH, p 207

Mesozoic era Third great era in the evolution of life, when the dinosaurs and other reptiles were dominant. See GEOLOGICAL TIMESCALE, p 256

Messina *Italy* Seaport near the north-east tip of Sicily, facing the Italian mainland, 6 km (4 miles) away across the Strait of Messina. It handles most of Sicily's exports of citrus fruits, and most of the traffic crossing the strait.

The city was founded in the 8th century BC, but its period of greatest prosperity as a trading centre began with the Crusades in the 12th century and lasted until the 14th. Cholera, earthquakes and war have killed many thousands of its citizens over the centuries; the 12th-century church of Annunziata dei Catalani is one of only a handful of ancient buildings to have survived the earthquakes. The city's cathedral is a 20th-century reconstruction of the original Norman building. Messina's biggest earthquake, in 1908, killed 83 000 people.
Population 274 800
Map Italy – Ee

Messina, Strait of (Gulf of Messina) *Italy* Lying between the Italian mainland and Sicily, the strait is 56 km (36 miles) long and 6-48 km (4-30 miles) wide, linking the IONIAN and TYRRHENIAN seas. Its strong currents and numerous whirlpools gave rise to legends about monsters, including the Greek legend about Charybdis, who was turned into a whirlpool by Zeus, and Scylla, who had six heads, each with three rows of teeth, which were ideal for plucking rowers from passing ships.
Map Italy – Ee

Mestre *Italy* See VENICE

Meta *Colombia/Venezuela* River, some 1000 km (620 miles) long, rising in the eastern Andes, near BOGOTÁ. It flows north-east, then east, forming part of the Colombia-Venezuela border, to Puerto Carreno, where it joins the Orinoco.
Map Colombia – Bb

metamorphic rock Igneous or sedimentary rock which, in the chemical composition or texture, or both, have been altered by heat, pressure or the chemical action of migrating fluids – or by a combination of any of these. Examples include granulite (which resembles granite but is a metamorphosed form of sandstone), MARBLE (metamorphosed limestone), and SLATE (which was once shale).

metamorphic zone Zone or region where META-MORPHISM has taken place.

metamorphism The process by which rocks in the earth's crust are altered, by heat, pressure or chemical action, or a combination of these.

Metéora *Greece* Eroded valley in THESSALY, on the eastern side of the Píndhos Mountains, which is noted for its monasteries. From the valley floor,

huge iron-grey towers of rock rise almost perpendicularly to 300 m (950 ft). In the 14th century, Byzantine monks built 24 monasteries on these precarious but impregnable towers. Today, five remain, the largest being Great Metéoron.

The monasteries were once accessible only by windlasses, ropes and cut steps, but now narrow paths wind up the towers past caves where hermits once lived. The monasteries contain Byzantine artefacts, including rare books; some are open to the public.
Map Greece – Bb

meteorite Piece of solid material from space which passes through the atmosphere without being completely burnt up and reaches the surface of the earth.

Metropolitana (Santiago) *Chile* Central region taking in the capital, SANTIAGO, and the provinces of Melipilla and Maipo. At Maipú, just south-west of Santiago, monuments commemorate a victory over the Spaniards in 1818 by Chilean nationalists led by José de San Martin. Wheat, fruit and livestock are the main products of the countryside, supporting the industries of Santiago and its port, Valparaíso. The ski resort of Farellones in the Andes is popular with tourists.
Population 5 170 290
Map Chile – Ac

Metz *France* City straddling the Moselle River near the German border in LORRAINE, 329 km (204 miles) east of PARIS. A military strongpoint since Roman times, it was fortified in the 17th century and again when it was under German rule from 1871 to 1918; parts of the walls and fortifications are still standing. The city lies in a fertile area famous for its Moselle wines. It is also the main centre of Lorraine's iron and steel region.
Population (town) 120 000; (conurbation) 193 000
Map France – Gb

Meuse (Maas) *France/Belgium/Netherlands* River, 950 km (590 miles) long, which rises in Lorraine, eastern France. From here, it flows north through the edge of the Ardennes upland, into Belgium and on to the Netherlands, where it is known as the Maas. It shares its delta with the RHINE, with which it is linked by canals.

It is navigable from the sea up into France, where it is linked by canals eastwards with the Rhine and westwards with the rivers of the Paris basin. It carries heavy barge traffic, especially through Liège, where it meets the Albert Canal from Antwerp. Its valley is lined with towns such as Verdun, Sedan, Namur, Liège, and Maastricht, whose names recur in Europe's wars.
Map Belgium – Ba

Mexico See p 444

Mexico (Estada de México) *Mexico* State covering 21 414 km² (8266 sq miles), west of MEXICO CITY's Federal District, in the mountainous centre of the Republic of Mexico. A quarter of Mexico's urban inhabitants live here and many of Mexico City's industries have spilled over into Mexico State. The state capital, TOLUCA, 56 km (35 miles) south-west of Mexico City, is a busy commercial centre in its own right and is popular with tourists.
Population 9 816 000
Map Mexico – Cc

Mexico, Gulf of The world's largest gulf, linked to the ATLANTIC OCEAN by the Straits of FLORIDA and to the CARIBBEAN SEA by the YUCATÁN CHANNEL. Its warm waters were teeming with pirates and adventurers in the late 18th and early 19th centuries. Its broad continental shelves contain important oil deposits. Warm waters from the Gulf of Mexico contribute to the GULF STREAM via the Florida Current. The gulf covers about 1 544 000 km² (596 140 sq miles); its greatest depth is 4377 m (14 360 ft).
Map South America – Ab; Mexico – Cb; United States – He-Ie

Mexico City (Ciudad de México) *Mexico* Capital of the Republic of Mexico and of Mexico City Federal District. It lies about 2200 m (7350 ft) above sea level on the southern end of Mexico's high central plateau, where the air is exceptionally thin, and is surrounded by mountains and volcanoes rising to some 5000 m (16 404 ft) and higher.

The homes of Mexico City, many of them single-storey houses or shanty dwellings, sprawl over a vast area across the plateau, beyond the Federal District and into the state of Mexico. With more than 20 million inhabitants, Mexico City is one of the world's largest urban areas – and it is constantly growing. More than half of the Mexicans employed in industry work here, in metallurgical, chemical, food-processing, electronics, tyre-making and textile factories, in cement plants and paper mills. But in spite of this activity, large numbers of the population live in abject poverty. Many try to make a living as street vendors, rubbish pickers, scavengers, or prostitutes. There are countless beggars in the city, and numerous people live in squalid housing that they built themselves.

The excessive concentration of industry has added to the air pollution and the city is often veiled in a haze of pink smog. The city has also been vulnerable to natural disasters. In September 1985, an earthquake measuring 8.1 on the RICHTER SCALE killed more than 7000 people and destroyed hundreds of buildings in the city.

In spite of this depressing picture, Mexico City has many pockets of great beauty and historical interest. The city was built on the ancient Aztec site of Tenochtitlan, the ruins of which are being excavated near the central plaza.

The ancient culture centred on the Zocalo, a 240 m² (2583 sq ft) place of Aztec worship. The plaza itself is surrounded by magnificent colonial buildings. Another fine square, the Plaza de Las Tres Culturas (Square of the Three Cultures), contains examples of Aztec, Baroque and modern architecture – Mexico's three main cultural influences. The imposing modern buildings of the university campus are covered with colourful murals depicting Mexico's struggle for independence and the subsequent revolution.

Evidence of the nation's rich early history is preserved in the city's famous Museum of Anthropology, which has more than 60 000 exhibits of ancient treasures. Mexico City's university, which dates from 1551, is the oldest university on the American continent.

Some of the city's residential areas are notable, too. For instance, the houses around Chapultepec Park, once the summer residence of the Aztec rulers, are among the most luxurious in the country. They are situated close to the equally imposing shopping area of the *Zona Rosa* ('Pink Zone'). The neighbourhood of San Angel in the south of the city has a different kind of charm: it has a village atmosphere, especially on Saturdays when its craft bazaar is held.
Population (city) 8 257 000; (metropolitan area) 20 500 000
Map Mexico – Cc

Mezada *Israel* See MASADA

Miami *USA* City in south-east Florida, separated from the Atlantic Ocean by Biscayne Bay and a fringe of barrier islands, including MIAMI BEACH. At the beginning of the 20th century, it had fewer than 500 inhabitants; however, thanks to the skills of property developer and promoter Henry Flagler, the population began increasing rapidly, jumping from 30 000 to 110 000. Today, Miami has become a major tourist, industrial, financial and business centre.

Miami is the southernmost major city of continental United States, and its proximity to the Caribbean has resulted in it being a haven for Cuban refugees. A large community of Cuban exiles from the Castro regime have settled here, many of whom live or shop in the Little Havana area of the city.
Population (city) 358 500; (metropolitan area) 1 937 100
Map United States – Je

Miami Beach *USA* Booming year-round resort in southern Florida, on an island 5 km (3 miles) off the east coast. It lies between the Atlantic Ocean and Biscayne Bay, is connected by four causeways to the mainland city of Miami, and is a concentration of hotels, apartment blocks and holiday homes, which stretches on for miles.
Population 92 640
Map United States – Je

mica Any of a group of complex silicate minerals occurring as thin flaky sheets in igneous and metamorphic rock. The two main types are muscovite, which contains aluminium and potassium, and biotite, containing aluminium, potassium, magnesium and iron.

Michigan *USA* State made up of two peninsulas in the heart of the GREAT LAKES area, separated by the MacKinac Straits. Lower Peninsula is flanked by lakes Huron, St Clair, Erie and Michigan, while Upper Peninsula lies between Lake Superior, Lake Michigan and Lake Huron. The latter is more hilly, forested and less populated, and was the setting for Henry Wadsworth Longfellow's poem *The Song of Hiawatha*. Iron and copper ores are mined here.

Dairy produce, wheat, maize, barley, beef, fruit and vegetables are the main agricultural products of Lower Peninsula. Iron ore, oil and gas are also produced. Manufacturing industries revolve around machinery, cars and accessories – especially in and around DETROIT, the largest city. Michigan covers 147 545 km² (56 954 sq miles); the state capital is LANSING.
Population 9 295 300
Map United States – Ia-Ib, Jb

Michigan, Lake *USA* The only one of the five Great Lakes wholly in the United States. It covers 57 454 km² (22 178 sq miles). Gary, CHICAGO and MILWAUKEE are its chief ports.
Map United States – Ib

Michoacán *Mexico* State in central Mexico, on the Pacific coast, consisting of lush woodland, volcanoes, hot springs, lakes and rivers. It is the home of the Tarascans who still practise traditional customs and produce traditional crafts including leather, wood carvings, and pottery.

Lakeside towns such as Pátzcuaro, the old Amerindian capital of Tzintzuntzan, and the picturesque colonial capital of Morelia are popular with tourists. Farming and iron mining are important, and Michoacán's pottery and lacquerware are famous.
Population 3 534 000
Map Mexico – Bc

Micronesia *Pacific Ocean* Geographically one of the three main regions of the Pacific islands, the other two being Melanesia and Polynesia, each having its own ethnic group. The name 'Micronesia' means 'small islands', and indeed, the region consists of thousands of tiny islands and atolls, most of them situated in the North Pacific.

It stretches from PALAU in the Caroline Islands to the MARSHALL ISLANDS and KIRIBATI (which straddles the equator); also included are GUAM, the Federated States of MICRONESIA, NAURU and the NORTHERN MARIANA ISLANDS. Micronesians, who were great seafarers, first arrived here from Asia about 3000 years ago.
Population 453 530
Map Pacific Ocean – Cb

microseism Faint, recurrent tremor of the earth's crust, due to natural causes such as wind and waves, or traffic and machinery.

Middelburg *Netherlands* Provincial capital of Zeeland and resort on the former island of Walcheren, 130 km (80 miles) south-west of AMSTERDAM. Most of the oldest part of the city had to be rebuilt after being destroyed during the Second World War.
Population 40 110
Map Netherlands – Ab

Middle East As a geographical definition, the term 'Middle East' has undergone a change of meaning. Before the First World War, Western Europe distinguished the Near East from the Middle East. The Near East meant the old Ottoman empire – Turkey and the countries bordering the eastern Mediterranean; while the Middle East was understood to be Afghanistan and the lands around The Gulf.

When the Ottoman empire ceased to exist in the 1920s, the Near East stopped being a definite unit, and the countries that once comprised it are now generally included in the Middle East. Today this collective label covers Turkey, Syria, Jordan, Israel, Lebanon, Afghanistan, Iran, Iraq, Saudi Arabia, Yemen, Oman, the United Arab Emirates, Qatar, Bahrain and Kuwait. As a political concept, the term also includes Egypt and Cyprus.

The Middle East covers some 7 916 058 km (3 056 405 sq miles). The northern and southern extremities of the Middle East are two great peninsulas: the Turkish peninsula between the Black Sea and the Mediterranean, and the Arabian peninsula bordered by the Red Sea, the Arabian Sea and The Gulf.

However, the term 'Middle East has been used loosely to include other (continued on p 448)

Mexico

*VIOLENCE BUILT MEXICO, WHERE EVEN
THE GROUND TREMBLES. ITS PAST
WAS TRAGIC – AND ITS HOPES OF
PROSPERITY AND PEACE
REMAIN IN THE BALANCE*

More than 20 million people live in MEXICO CITY, capital of Mexico. It is the fastest growing city on earth. So rapidly is Mexico's current population expanding that its people are flooding into the United States – either legally or illegally. More than 10 million Mexicans are estimated to be legally resident in the United States. No one knows how many more have crossed the border unofficially.

Mexico can be a wonderful place to visit – as booming tourist figures prove. It can also be a wonderful place to live. Sadly for many, it is also a country to get out of. For despite social and political reforms won by centuries of blood and tears, the blood and tears are by no means over, and the reforms have still not benefited all Mexicans while the population growth is outstripping the country's resources. Every year more than 2.5 million newborn Mexicans must be provided for.

Mexicans joke that in crossing the US border they are just winning back California – which used to be part of their country, along with Texas, Arizona, New Mexico and other US states. America has responded with strong border patrols – and counter-invasions by millions of *gringo* tourists bound for ACAPULCO and other superb resorts on the Pacific coast. Tourist attractions include more than 10 000 awesome monuments to ancient civilisations and the Latin attitude that makes for maximum enjoyment of the many feast days and fiestas throughout the calendar.

Nowhere are these festivals celebrated with more enthusiasm than in Mexico City itself. Appalling slums it may have; air made barely breathable by millions of smoking exhausts; whole areas destroyed in the 1985 earthquake or subsiding gently into the spongy ground. But it does have style, exuberance, treasures of art and architecture, and a talent for living.

The city's 1500 km² (580 sq miles) embraces the last great temple of Montezuma's Aztec capital, Tenochtitlán, the palace of his Spanish conqueror Cortés, Olympic sports stadiums and the world's biggest bull ring, seating 60 000. In the south, there is the modern university building. Other modern buildings decorated by Diego Rivera and fellow Mexican masters of the monumental mural live with Spanish-colonial churches and the quiet, elegant plazas of secluded neighbourhoods that could be in another time, another place.

Mexico was literally born of violence, for it lies on a particularly sensitive part of the earth's crust and remains prone to earthquakes. It is a land of volcanic mountain ranges and high plateaus, with lowlands only along the Pacific coastline, the Gulf of Mexico and the YUCATÁN Peninsula. The spectacular 5452 m (17 887 ft) volcano POPOCATÉPETL, dormant since 1928 looms south of the capital; beyond it rises 5699 m (18 697 ft) CITLALTÉPETL (Pico de Orizaba), the highest peak in Mexico and perpetually snow-capped. About 70 per cent of the country is more than 500 m (1640 ft) above sea level.

In the north-west the SIERRA MADRE OCCIDENTAL Mountains lie between the coast and the great central plateau, an area of high plains and great deserts, merging into broad valleys, lakes and dormant volcanoes towards the south. The SIERRA MADRE ORIENTAL range rises between the eastern edge of the plateau and the coast. South of the capital, the SIERRA MADRE DEL SUR extends to the Guatemalan border, broken by the narrow isthmus of Tehuantepec. Here the Atlantic and the Pacific oceans are separated by a mere 216 km (134 miles) of land – and here the North American continent ends.

ANCIENT LEGACIES

The first recorded civilisation in the territory was that of the Olmecs, who lived near the eastern coast from about 1100 BC and left many carvings and sculptures. Next came those great empire builders, the Mayas of Yucatán, who flourished from about AD 300 and spread through the central uplands and south into Guatemala, Honduras and El

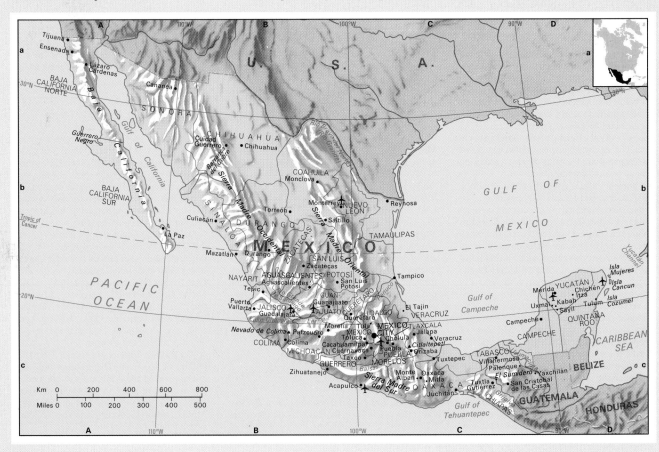

Salvador. Famous Mayan sites are at Chichén Itzá and Palenque. In the early 13th century, the Aztecs swept down from the north to dominate much of the area. Their astonishing cities, fabulous wealth and horrific religious ceremonies amazed and appalled even the invading Spaniards of the 16th century.

Blood ran down the steep steps of the Aztecs' pyramid temples as grotesquely masked priests wielding obsidian knives slashed open the chests of those to be offered up to the gods and tore out their still-beating hearts. Countless thousands died to propitiate hundreds of gods and goddesses. In 1487, when the Aztecs dedicated their Great Temple of Tenochtitlán, they are said to have sacrificed people four at a time from dawn to sunset for four days.

But one of their most potent gods, the plumed serpent Quetzalcóatl, helped in their downfall. The god embodied the spirit of a once all too human ruler, a fair-bearded dedicated drunk who fled his throne in disgrace but was expected back some day, from the east. When a bearded white-skinned man in a shiny metal suit turned up in 1519, the Aztec Emperor Montezuma II thought Quetzalcóatl had come home. The blond stranger turned out to be the Spaniard Hernán Cortés, who with his soldiers and priests would kill, torture, enslave and 'convert' the Aztecs and other peoples of Mexico with a zeal ferocious enough to match their own.

It took Mexico three centuries to cast off the iron grip of Spain and almost another one to free the people of other assorted adventurers and oppressors. Even now, foreign interests loom large in the economy.

Little remains of Montezuma's magnificent island capital on Lake Texcoco – or even of the lake itself – which has been submerged, built over and lost in the growing sprawl of today's capital. The site of the Great Temple is being excavated, but the most important building in the vicinity today is a Spanish-built cathedral, which was begun in 1573.

Not only the works of the Aztecs survive, so too does their language Náhuatl – as well as the monuments and languages of the Maya, Zapotec, Otomí and Mixtec peoples of pre-Columbian Mexico. From these tongues are derived the 59 different dialects spoken among more than 5 million of the country's 26 million or so Amerindians, who form about 29 per cent of the population; another 55 per cent are *mestizo* – of mixed European and Amerindian descent. The rest (16 per cent) are *criollos* – of European stock.

OLD COUNTRY OF YOUNG PEOPLE

Mexico has been called an old country of young people: 37 per cent are under 14 years old and the population has an average age of 20 years. Throughout the present century the birthrate has been high; the population is growing at 2 per cent a year, compared with 0.9 per cent in the United States and 0.3 per cent in the United Kingdom. Rapid economic growth in a half century of relative calm is a major cause of this growth.

Extensive oil discoveries, increasing industrialisation and penury among the landless peasants have caused a huge shift of people from the countryside to towns and cities. About 73 per cent now live in urban areas, one-fifth of them in or around the capital. The population of GUADALAJARA City is nearing 3 million; MONTERREY and PUEBLA have long since passed the million mark. Many millions have simply exchanged rural poverty for urban deprivation, but despite low incomes and poor housing they are in many ways better off: most houses have electricity, most children go to school, and social services and health care are more efficient.

The great oil fields are on the east coast around the city of TAMPICO and in the state of TABASCO and offshore in the Mexican Gulf itself. Mexico became one of the world's major oil producers, with reserves second only to those of Saudi Arabia. However, great industrial and public service projects financed by investments and foreign loans geared to the oil boom turned sour in the late 1970s and early 1980s as oil prices began falling and interest rates soared. In 1982, inflation zoomed from a previously 'normal' 10-20 per cent to 70 per cent a year, and foreign debts equalled a staggering 82 per cent of the gross national product. In a bid to stem the tide of cash pouring out of the country, the government nationalised Mexico's 54 private banks. Further economic measures, based on International Monetary Fund (IMF) proposals and geared to

▼ **NATURE'S HARSHNESS Barren granite mountains stretch inland from the Pacific coast on the Baja (Lower) California peninsula. Sparse sage scrub is the only natural vegetation that can survive on the poor desert soils, which in some years receive no rain at all.**

repay national debt, reduced inflation to 56 per cent by 1984, and 12 per cent by 1992. Tightened controls on spending and some international financial diplomacy in 1990 won a breathing space for interest repayments, but controls were further tightened in 1992, although banks were re-privatised. Another serious financial crisis hit the country late in 1994, with effects spilling over into other Latin American countries. The peso fell by 40 per cent against the United States dollar in 10 days, jeopardising loan repayments and causing a 50 per cent drop in the Mexican stock exchange and sharp falls in the Argentinian and Brazilian markets.

Mexico seems better equipped than most developing countries to win through economic turbulence. Industrial output in 1989 included 5.8 million tonnes of steel, about 24 million tonnes of cement and more than 700 000 motor vehicles. The country still turns out most of its own consumer durables. Oil production in 1990 was 155 million tonnes – more than twice the production of OPEC countries like Libya or Algeria – and Mexico remains, as it has been for centuries, the world's leading producer of silver, with more than 2000 tonnes a year from 70 mines.

Copper and sulphur are also mined in large quantities. Significant mining of lead, copper, zinc and other metals is being done. Despite industrialisation, agriculture continues to be a major factor in the economy, although now employing only about one-third of the workforce (whereas industry employs about a quarter). By January 1994, Mexico was ready to implement the North American Free Trade Agreement (NAFTA), and so combine the capital, technology and expertise of Canada and the United States with its own natural resources and labour, thus further boosting its economy.

LANDLESS FARMERS

Mexico still grows most of its own food and is supplying the United States on an increasing scale. Maize, sorghum, wheat and beans are the staples, sugar cane is another major crop, and there is large-scale cattle ranching; vegetables, fruit, cotton and coffee are grown in the north-west and are an important export.

That the land employs such a large part of the workforce but contributes such a small part of the national income (in 1990, it employed one-third of the working population but contributed only 8.9 per cent of the gross domestic product) reflects both the natural difficulties that beset farmers on much of the cultivable land and the negative outcome of numerous attempts to redistribute the land since the end of Spanish rule. Apart from the western highlands and the jungles of the southeast, the country gets little rainfall – and almost none at all across the whole of the great central plateau. Most of the cultivable land needs irrigation and vast government projects have almost increased irrigated areas to 6 million hectares (8.5 million acres) since the 1950s.

However, the government clearly did not do enough for the peasant population – as the Chiapas uprising in January 1994 demonstrated. The Zapatista National Liberation Army, named after Emiliano Zapata, a popular hero of Mexico's 1910-17 revolution who fought to recover communal land seized by rich land owners, demanded better health services and education, an end to discrimination and oppression, and a share in the nation's improving economy. Above all, they wanted land reforms and redistribution. Despite independence, revolution and the stirring exhortations and promises of a succession of charismatic leaders, most of the Mexican people never did win their promised lands.

Agricultural laws today limit the size of privately owned land, but many large ranches – called *hacienda* – were merely 'divided up' on paper, among the owners' family and friends. Right up to the revolution of 1910, just 1 per cent of the population owned 97 per cent of the land, 96 per cent of the people owned 1 per cent of the land, and 88 per cent of the peasant farm workers had no land. Today, 37 families – most of them 'barons' *(caciques)* with strong links to the ruling party – hold about a quarter of the nation's wealth.

COSTLY WAR OF INDEPENDENCE

The Chiapas' uprising had nation-wide repercussions, with other poverty-stricken states joining in the protest; and though the conflict forced the PRI, the Institutional Revolutionary Party, to concede to most of the Zapatistas' demands, it was not before bombings took place in Mexico City, a leading Mexican financier was kidnapped and the presidential candidate, Luis Donaldo Colosio, was assassinated.

And so, history repeats itself: in 1928 a president elect was killed, thus ending an epoch of violence and bringing in the present government, the PRI, which has controlled Mexican politics since 1929. But the Mexican nation has a history of violence and tremors – Napoleon's invasion of Spain in 1808 started unrest within the colony of Mexico, which scented independence and won it in 1821 after a war that cost 600 000 lives.

In 1823, the country was proclaimed a republic. There followed a period of constant changes of government, insurrection and civil war. In 1836, the cattle and cotton barons of Texas, angered by the abolition of slavery, rebelled and declared their own independence. During the Mexican War against the United States (1846-8), Mexico was heavily defeated and had to cede Utah, Arizona, California and New Mexico – almost half of its land area – as well as parts of Colorado and Wyoming to the USA for little more than US$28 million. It had already lost Florida in 1819; Texas became part of the United States in 1845.

Conflict continued as Benito Juárez introduced reforms during two spells as president. His presidency was interrupted by a regime headed by Archduke Maximilian of Austria, who was installed in 1864 as emperor by a French invasion force supported by conservative Mexicans. This lasted until 1867.

A paternalistic but autocratic dictator, Porfirio Díaz, ruled from 1876 to 1910 and saw huge foreign investments pour in, 24 000 km (nearly 15 000 miles) of railway track laid, industries and mineral resources developed and a civil service created. Unfortunately, the resultant prosperity did not extend to the peasant masses; their land was systematically stolen and many of them were reduced to semi-slavery in mines and on plantations. A law passed by Díaz in 1883 resulted in 55 000 km² (214 285 sq miles) of 'national' land being sold off at bargain prices to rich *criollos*, foreigners and his personal friends. One wealthy family ended up with 50 ranches covering 30 000 km² (11 583 sq miles).

There was increasing unrest and finally revolution. When it came in 1910 it was confused, horrific and long drawn-out, as peasants, smallholders, bandits, Amerindians and even middle-class ranchers rose in fury. Díaz fled to exile in Paris and a succession of revolutionary leaders came to the fore, among them Zapata and the teenage bandit Pancho Villa.

At appalling cost, the revolution triggered real social upheavals. A constitution of 1917 heralded major land reforms, advanced social welfare schemes, compulsory free education, minimum wages and the right to strike. Many of these aims would not be achieved until President Lázaro Cárdenas held office, between 1934 and 1940. Only then were the great estates parcelled out into *ejidos*, or common lands, and basic education made more widely available.

Oil fields and railways were nationalised; power stations and the major airlines followed. Land reform remains a continuing process and agricultural development in some areas – such as the north-west – has been spectacularly successful. However, this has benefited mainly those companies or middle-class families which have sufficient capital to invest in irrigation schemes, fertilisers and mechanisation. Outside the growth areas, rural life continues to be a desperate struggle.

The country's economy is still too greatly dependent on the export of raw materials. More than 75 per cent of Mexico's foreign revenue comes from metals and oil exports – in fact, oil fetches 30 per cent of all foreign income earned.

AMERINDIAN CULTURE

Perhaps because of its exceptionally violent history, Mexico has not attracted the mass immigration from Europe that has done so much to shape the rest of the continent. And given the large native population present when Cortés landed – estimated at 7 or 8 million – the country has retained a distinctively Amerindian identity. Indeed, ever since the Zapotec-born President Juárez's time, there has been an increased national bias in favour of the Amerindian culture and race. How deeply this feeling really permeates is hard to assess, since Mexico retains a highly stratified, class-based society in which pale faces seem more often than not to occupy the positions of real power. It may also be more than a coincidence that the majority of people in the poorest regions of the south are Amerindian. (The Amerindians of Chiapas certainly felt they were being discriminated against when they rose up in protest on 1 January 1994.)

Indeed, the socialist ideology proclaimed by successive revolutionary and elected governments is embarrassingly at odds with the glaring contrasts in income and wealth that is the reality of life in Mexico. The reason that relative political stability was maintained for so long was due largely to the skill of the majority party in co-opting opposition groups while retaining the support of its traditional allies. The PRI's power base was a triple alliance with labour, the peasantry and state employees, and for years it was so closely intertwined with the government apparatus that it was difficult to untangle them: thus the party supported the state, and vice versa.

Since opposition parties were banned until 1983, policies until recently were determined by the results of power struggles within the PRI. The most vital of these conflicts was about choosing a party candidate for the presidency. In the past, the PRI candidate always won, and the president had almost absolute power, so the choice was critical both to the country and the party, for the successful candidate would hold office for six years. However, he was then barred from ever seeking re-election – one of the few rules in Mexican politics.

In the August 1994 elections the PRI was again returned to power with Ernesto Zedillo Ponce de León as president. But soon after, the party's Secretary-General, José Francisco Ruiz Massieu, was assassinated. The killing was initially attributed to revenge for a crackdown on drug-traffickers. But two months later José's brother, Deputy Attorney General Mario Ruiz Massieu, resigned, alleging that the killing of his brother was instigated by a hard-line PRI group that was opposed to José's political reform programme.

Culturally, the country is very rich. It has produced not only Diego Rivera but two other great masters of the mural – José Clemente Orozco and David Alfaro Siqueiros; internationally famous writers such as Octavio Paz and Carlos Fuentes; modern architects such as Luis Barragán and Pedro Ramírez Vázquez. It possesses a wealth of Amerindian cultures which contribute to contemporary song and dance, as performed by Mexico's renowned folklore ballet company. And, as in the beginning, it has its numberless treasures of other, unknown masters – the sculptors, artists and architects of the early civilisations.

MYSTERY OF THE MONUMENTS

To many visitors the great ruined cities, spectacular pyramids and enormous, mysterious monuments of ancient peoples are the abiding glories of Mexico. They survive in great numbers, but perhaps the most striking are the huge pyramids at TEOTIHUACÁN, 48 km (30 miles) north-east of Mexico City. This was the cultural centre of a trading empire that flourished between 300 BC and AD 750.

It is believed that as many as 200 000 people lived in its grid network of streets covering 21 km² (8 sq miles) – making the city larger than Imperial Rome. The Pyramid of the Sun once soared in four massive terraces to a height of 65 m (216 ft); its neighbouring Pyramid of the Moon, though smaller, is still 42 m (138 ft) high. The city was destroyed by fire and had been deserted for a thousand years when the Spaniards first saw it.

▲ NATURE'S FURY Rescue workers and survivors gather beside a ruined hotel in Mexico City after the major earthquake of 1985. Over 7000 lives were lost, but more than a week later a number of victims, including several newborn babies, were still being rescued from the debris.

Maya ruins include the magnificent city-complexes of CHICHÉN ITZÁ and Uxmal, in the jungle of the Yucatán Peninsula; PALENQUE, in CHIAPAS, is another. Maya beginnings go back to 1000 BC, but their golden age was AD 250-900. They were sophisticated mathematicians and astronomers, with an accurate calendar calculated back more than 3000 years, which took leap years into account. An observatory survives in Chichén Itzá; in a carved crypt at Palenque was found the skeleton of a 7th-century ruler covered with jade jewellery.

The Mayas' hieroglyphic writing is still, unfortunately, only partly understood. MONTE ALBÁN, in OAXACA, was the capital of the Zapotec culture. Its huge main square was formed by levelling an entire hilltop. Pyramids, terraces, sculptures, tombs and staircases remain to show how the place looked at the height of Zapotec power, in the 8th century AD.

Mexico's ancient cities once blazed with colour – vast murals, vividly decorated statues, the feathered capes and plumed headdresses of rulers, priests and warriors. Some of that colour survives in peeling frescoes, on pottery and on the fierce images of the gods they worshipped – faded now but bright enough to show why the modern Mexican still loves a bold stroke of the paint brush.

MEXICO AT A GLANCE	
Official name United Mexican States	
Area 1 972 550 km² (761 426 sq miles)	
Population 90 500 000 **Per km²** 46 **(Per sq mile** 118)	
Capital Mexico City	
Government Federal republic	
Currency 1 Mexican peso = 100 centavos	
Languages Spanish, Náhuatl, other Amerindian languages	
Religion Christian (93% Roman Catholic)	
Climate Tropical and temperate according to altitude. Average temperature in Mexico City ranges from 12.2°C (54°F) in January to 18.9°C (66°F) in May; in the coastal regions the range is from 17°C (63°F) to 24°C (75°F); in the arid regions from 5°C (41°F) to 40°C (104°F).	
Land use Grazing 38%, forest 22%, cultivation 13%, desert 21%, other 6%.	
Main primary products Maize, sorghum, wheat, barley, rice, cotton, sugar, coffee, beans, fruits, cattle; uranium, copper, manganese, iron, coal, lead, zinc, tin, silver, gold, oil and natural gas, aluminium, phosphates, sulphur, mercury, fluoride	
Major industries Oil and natural gas production and refining, agriculture, mining, iron and steel, aluminium refining, vehicles, cement, machinery, textiles, pottery	
Main exports Oil and natural gas, oil products, metals, sulphur, textiles, cotton, sugar, chemical products, machinery, consumer electronics	
Annual income per head (US$) 3470	
Population growth (per thous/yr) 20	
Life expectancy (yrs) Male 62 **Female** 66	

Federated States of Micronesia

A NATION OF HUNDREDS OF ISLANDS STRETCHING OVER THOUSANDS OF KILOMETRES

Spread over a vast area of the western Pacific to the north of Papua New Guinea are the 600 or so islands and atolls of the Federated States of Micronesia, or FSM. Their name, Micronesia, means the 'World of Little Islands'. After a succession of colonists – from Spain (1529), Germany (1899) and Japan (1914) – followed, in 1947, by United States rule as part of the Trust Territory of the Pacific Islands and years of negotiations, the US Congress in 1986 approved a 15-year Compact of Free Association for the FSM (as it did for the neighbouring Marshall Islands).

The agreement gives self-government to the islands while safeguarding American defence interests. In December 1990, the United Nations recognised the termination of the trusteeship and on 17 September 1991, Micronesia became a full UN member.

The island nation stretches some 3200 km (2000 miles) from end to end, yet the total land area – 702 km² (271 sq miles) – is only a quarter of Rhode Island, America's smallest state. One island – POHNPEI, on which the capital, Kolonia, stands – accounts for almost half the land. The country consists of four states – from east to west, KOSRAE, Pohnpei, Chuuk (formerly known as TRUK) and YAP.

The Federated States of Micronesia comprise all Caroline Islands with the exception of the Palau group which is farther west. It decided in 1978 to become independent.

The name Caroline Islands was given to the FSM and Palau by the Spanish who landed here in the 16th century and gained sovereignty over the islands in 1885-86 but sold them to Germany in 1899. The forebears of the present native Micronesian population arrived here from Asia possibly as early as 3000 BC. And some 600 years ago, an unknown civilisation on Pohnpei built a community – Nan Madol – of stone buildings on artificial islands separated by a system of canals. Today, the normal home for many islanders is a wooden hut.

The FSM has received (and will continue to receive) substantial American economic aid. In the bigger towns, many people follow an Americanised way of life, but elsewhere the lifestyle has been touched remarkably little by centuries of colonial rule. Here, the people are subsistence farmers and fishermen. Copra is the main cash crop and there are rich tuna fishing grounds offshore, exploited mainly by American, Japanese and Korean fleets. However, the FSM has established a 200 nautical mile (370 km) economic zone, bringing in revenue from fishing. The nation's main economic prospects lie in the growing development of fishing and tourism.

FEDERATED STATES OF MICRONESIA AT A GLANCE	
Map Pacific Ocean – Cb	
Area 702 km² (271 sq miles)	
Population 117 590 **Per km²** 168 (**Per sq mile** 434)	
Capital Kolonia	
Government Federal republic	
Currency 1 US dollar = 100 cents	
Languages English, Micronesian dialects	
Religion Christian (mainly Roman Catholic)	
Climate Tropical; average temperature at Kolonia is 27°C (81°F) all year	
Land use Cultivation 33%, forest 22%, grazing 13%, other 32%	
Mineral resources Phosphates	
Main primary products Coconuts, yams, cassava, tropical fruits, fish	
Major industries Copra, agriculture, fishing, food processing	
Main exports Copra, coconuts, fish, handicrafts	
Annual income per head (US$) 1500	
Population growth (per thous/yr) 30	
Life expectancy (yrs) Male 65 **Female** 69	

Muslim and Arab countries such as Algeria, Djibouti, Egypt, Libya, Mauritania, Morocco, Somalia, Sudan and Tunisia in North Africa, the former Soviet republics of Kazakhstan, Kyrgyzstan, Tajikistan and Turkmenistan, as well as the island of Cyprus, and Pakistan.

The region includes some of the world's highest and most spectacular mountains. In Turkey, two great ranges run from west to east – the Pontus Mountains in the north and the Taurus Mountains in the south. They meet in the mountainous tangle of the Armenian Knot, whose greatest peak is Ararat, at 5165 m (16 945 ft). From the Knot, the Elburz Mountains stretch across northern Iran to merge with the western and lower extremity of the high ranges of Afghanistan.

The most spectacular of these, the Hindu Kush, rises to over 7500 m (24 600 ft) and forms the western extremity of the Himalayan and Pamir system of central Asia. Some of the passes through these Afghan ranges soar to 5300 m (17 000 ft). Finally, the Zagros Range, rising to 4548 m (14 921 ft) at Zardeh Kuh, runs south-east from the Armenian Knot through Iran to form a barrier that faces Iraq and The Gulf.

The names of the region's two great rivers, the Tigris and the Euphrates, ring through human history, for it was in the land between them, about 10 000 to 11 000 years ago, that agriculture, the foundation of modern civilisation, was developed. The Euphrates, 2736 km (1700 miles) long, rises in Turkey, then flows through Syria and Iraq, irrigating the farmlands of the three countries. The Tigris, 1840 km (1143 miles) long, also rises in Turkey, but for much of its course it flows through Iraq. A number of rivers in Afghanistan and Iran, the Helmand for example, have no outlet to the sea; a number of others seasonally dry out. The vast Arabian Peninsula, however, possesses virtually no rivers.

The feature that gave rise to the early Middle Eastern civilisations is the FERTILE CRESCENT. Extensive irrigation schemes based on the Tigris and the Euphrates, and reinforced by a moderate rainfall, have made this one of the most productive regions in the world. Long ago, it provided perfect conditions for the agricultural communities that grew into the city-kingdoms of Sumer, Babylon and Assyria.

But the climate of much of the Middle East is harsh. The region has some of the hottest places on earth. In the Rub al-Khali (Empty Quarter) of southern Arabia, for example, or the Dasht-e-Kavir, the central desert region of Iran, temperatures as high as 55°C (130°F) in the shade have

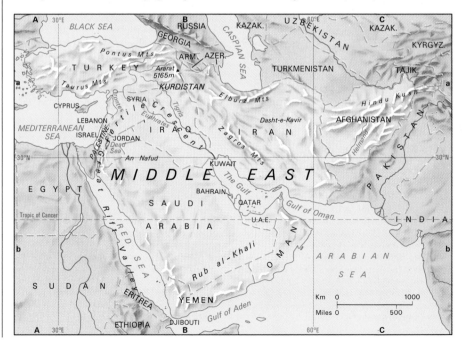

been recorded. The climate of central Turkey and much of Iran and Afghanistan is continental, with searing heat in summer and temperatures as low as -37°C (-35°F) in winter. Vegetation conforms to the climate – Turkey, northern Iran and Syria have conifer and pine forests, the Mediterranean coastal countries are typified by scrubland. In The Gulf and Arabia, most of the land is desert.

Until very recently the economy of these arid regions was not unlike that described in the Old Testament – a world of peasant farmers and nomadic herdsmen. But the discovery of oil and, later, of other minerals has completely transformed the Middle East. The Gulf contains the greatest oil reserves in the world, which have brought vast wealth and modern development to the region.

Population 200 million
Map See p 448

Middle West *USA* See MIDWEST

Middleback Ranges *Australia* Hills rich with iron ore, lying 260 km (about 160 miles) north-west of ADELAIDE in South Australia. Ore from the mines such as Iron Knob and Iron Monarch is sent some 80 km (50 mile˜) by rail to the port of Whyalla on Spencer Gulf for smelting.
Map Australia – Fe

Middlesbrough *United Kingdom* Iron and steel-making town and port on the River Tees, 70 km (45 miles) north of YORK; formerly in North Yorkshire, it is now in the county of Cleveland.

When the Stockton and Darlington railway reached it in 1830, it was a tiny village. Thanks to the shipping of coal from the south Durham coalfields, the working of iron ore in the Cleveland Hills to the south of the town, and the establishment of an iron and steel industry (now severely depressed), its population had grown to 40 000 by 1870. The 19th-century town is linked to the older market town of STOCKTON-ON-TEES over the Tees via the Transporter Bridge, unique in Britain, with its travelling suspended roadway.
Population 146 400
Map United Kingdom – Ec

middleveld In southern Africa, a region of undulating grassland and scrub-covered RANDS and KOPPIES which occur between about 900 and 1500 m (2950 and 4900 ft) above sea level.

Midi-Pyrénées *France* The nation's largest region, covering 45 348 km² (17 647 sq miles) and centred on the southern city of TOULOUSE. It stretches from the Massif Central to the Pyrénées. About 25 per cent of French cattle are raised in this region.
Population 2 431 000
Map France – Dd

Midlands, The *United Kingdom* Central region of England which is generally considered to include the counties of DERBY, LEICESTER, NORTHAMPTON, NOTTINGHAM, STAFFORD, WARWICK and WEST MIDLANDS. It is Britain's industrial heartland, and includes manufacturing centres such as BIRMINGHAM, COVENTRY, Derby, Leicester, Nottingham, and STOKE-ON-TRENT.
Map United Kingdom – Ed

Midlothian *United Kingdom* See BORDERS; LOTHIAN

▲ **ARTISTS' RETREAT** A gleaming white windmill stands over the resort of Míkonos in Greece's Cyclades group of islands. Like the houses, it contrasts with the blue sea, providing stunning scenes for painters to capture.

mid-ocean ridge Line of submarine mountains rising from the ocean floor.

Midway Islands *Pacific Ocean* Atoll containing two small islands, about 1850 km (1150 miles) north-west of Honolulu, in Hawaii. An American possession since 1867, it is a United States naval air base. The Battle of Midway in June 1942, when a Japanese invasion fleet was stopped, was the turning point of the Second World War in the Pacific.
Population (mainly military) 450
Map Pacific Ocean – Ea

Midwest (Middle West) *USA* The country's fertile heartland around the upper Mississippi, Missouri and Ohio rivers. The area includes the states bordering the Great Lakes (Ohio, Indiana, Michigan, Illinois, Wisconsin and Minnesota), and Iowa and Missouri to the west, with Kentucky, Kansas, Nebraska and North and South Dakota sometimes included as well. About 20 per cent of the United States' land area lies in the Midwest.

In the east, the growing of maize ('corn'), on which cattle and pigs are fattened, is the main farming activity (the area is sometimes called the 'Corn' or 'Feed Belt'). Wheat, soya beans and tomatoes are also grown here. Chicago is the main city.
Map (North Dakota) United States – Fa; (Missouri) United States – Hc; (Ohio) United States – Jb

Mien Bac (Bac Bo) *Vietnam* Northern region of Vietnam, including Hanoi and the whole of the Red River valley plain, stretching from the northern slopes of the Truong Son chain to the Chinese border. In French colonial times it was called Tonkin, from Dong Kinh (meaning 'Eastern Capital'). It was the heart of communist-led North Vietnam during the Vietnam War (1965-75).
Map Vietnam – Aa-Ba

Mien Trung *Vietnam* Hilly central region, called Annam in French colonial days, situated between the plain of the Mekong River delta and the valley of the Red River, south of the capital, HANOI. It consists almost entirely of the Truong Son (Annamite) chain, about 1100 km (700 miles) of rain-forest covered mountains stretching along the Laotian border. The chain's highest point is Ngoc Linh, reaching 2598 m (8524 ft) west of the coastal town of Quang Ngai.

The hills are inhabited by people sometimes called *montagnards*, of mixed Malayo-Polynesian and Mon-Khmer origins, who grow rice and tea on burnt-over forest land. Their houses are usually raised above the ground on pilings.
Map Vietnam – Bb

migmatite Composite rock body formed when an existing rock, usually a METAMORPHIC ROCK, is invaded by another rock material, such as MAGMA. The reaction between them changes the structure and texture of the existing rock.

Mihintale *Sri Lanka* Sacred hill about 310 m (1020 ft) high, situated 16 km (10 miles) east of the ancient Sinhalese royal city of ANURADHAPURA. In 247 BC, Mahinda, son of India's Buddhist Emperor Ashoka, converted the Sinhalese King Devanamplyatissa to Buddhism on the hill. Since then, Mihintale has been a home for monks and has attracted pilgrims to its many rock shrines, which are linked by a stone staircase of 1840 steps. Since the 10th century, Sinhalese laws have preserved the hill's forest and wildlife, making it probably the world's longest surviving wildlife park.
Map Sri Lanka –Ba

Mikinai *Greece* See MYCENAE

Míkonos *Greece* Island in the CYCLADES group, lying north of Náxos. The island resort of the same name, which can be reached by boat from Athens, is a town of winding lanes, white-washed houses and windmills. It has a double-storey, domed church. Apart from tourism, its economy revolves around wines, figs and manganese.
Population 5500
Map Greece – Dc

Milan (Milano) *Italy* The country's leading industrial and commercial city, standing on the north-western part of the North Italian Plain. It was capital of the Western Roman Empire from the 4th to the 5th century, then flourished as a city-state from the 11th to 13th century. During the Renaissance period (14th to 16th century) its reputation as a centre for art, culture and politics spread throughout Europe. Finally, it developed through the industrial revolution in the 19th century to become Italy's most prosperous city.

With three universities and a polytechnic, several newspapers and publishing houses, a host of theatres, museums and art galleries, offices, a stock exchange, a trade fair and many industries including cars, machinery, chemicals, textiles, clothing and printing, Milan is the only Italian city to be fully integrated into the network of European cultural, industrial and financial capitals. Rome is Italy's administrative capital, but Milan creates its wealth.

Milan has a medieval centre surrounded by successively more recent suburbs. The Piazza Duomo marks the middle of the city where the cathedral explodes skywards in a Gothic riot of pinnacles and spires. It is Italy's largest Gothic building and the third largest cathedral in the world; it was started under Gian Galeazzo Visconti in 1386 and not completed until the 19th century.

The Galleria, a huge iron and glass arcade built in 1865, leads to the renowned La Scala Opera House; the Brera, the city's main art gallery, is a little to the north. The Sforza Castle, second of Milan's great public monuments and the major legacy of the Renaissance period, fronts a large park to the west. The convent of the Renaissance church of Santa Maria delle Grazie, built between 1466 and 1492, contains Leonardo da Vinci's fresco, *The Last Supper*.

Population 1 432 200
Map Italy – Bb

Milford Haven *United Kingdom* Town and fine natural harbour at the south-west corner of Wales, in the county of Dyfed. Since 1960 it has been developed as a major oil port, with several refineries and petrochemical factories.

Population 13 930
Map United Kingdom – Ce

Milford Sound *New Zealand* Glacier-sculpted inlet of the Tasman Sea on South Island, 260 km (160 miles) north-west of DUNEDIN. The spectacular cone of Mitre Peak, rising 1692 m (5551 ft) from the sea, dominates the entrance. High rainfall – around 2650 mm (246 in) a year – in the area feeds hundreds of tumbling streams and waterfalls, among them the 158 m (518 ft) high Bowen Falls. The Sound is a major tourist attraction, as is the 55 km (34 mile) long Milford Track, described as the most beautiful walk in the world.

Map New Zealand – Af

Mílos *Greece* One of the CYCLADES Islands. Mílos is an arid, volcanic island with hot springs. It was renowned in ancient times for its obsidian, which was exported as far afield as Egypt. It was in the sea near the ruined 700 BC city of Melos that the ancient Greek sculpture known as the Venus de Milo was found in 1820. The statue is now in the Louvre museum in Paris. The island's main port is Adhámas, 4 km (2 miles) from the biggest town, Pláka, which has a 13th-century church and a Venetian castle. There are cata-

combs at nearby Trypití; a small, hillside Roman theatre can be seen near old Mílos.

Population 4550
Map Greece – Dc

Milwaukee *USA* Chief port and industrial centre of Wisconsin. It stands on the west shore of Lake Michigan. Many German immigrants settled here after 1848 and made the city a beer-brewing and meat-packaging centre. Other industries include production of steel and electrical equipment, printing, engineering, and the manufacture of transport equipment.

Population (city) 628 100; (metropolitan area) 1 432 150
Map United States – Ib

Mina al-Ahmadi *Kuwait* The country's main oil terminal, situated 40 km (25 miles) south of KUWAIT City. It handles some of the world's largest tankers and can load 2 million barrels of oil a day. The Kuwait Oil Company's refinery here is the largest in the country, and can process up to 300 000 barrels a day.

Map Kuwait – Bb

Mina Hassan Tani *Morocco* See KENITRA

Minamata *Japan* Fishing port and industrial town in Yatsushiro Bay on the west coast of KYUSHU Island. The town is notorious for Minamata disease which has claimed about 200 lives since the early 1950s. The disease was caused by methyl mercury wastes discharged into the bay. The mercury poisoned the fish, and in turn, many of the people who ate the fish.

Population 36 500
Map Japan – Bd

Minami Arupusu *Japan* National park also known as the Southern Alps National Park, on central HONSHU Island, 130 km (80 miles) west of the capital, Tokyo. It contains Mount Shirane, at 3192 m (10 472 ft) the second highest peak in Japan, after Mount Fuji.

Map Japan – Cc

Minas Gerais *Brazil* See BELO HORIZONTE

Minch, The *United Kingdom* Channel, 56 km (35 miles) long, separating mainland Scotland from the Outer Hebrides, also called the North Minch. This is to distinguish it from the Little Minch, 24-40 km (15-25 miles) long, between North Uist and Skye further south. Both channels have many associations with the Jacobite Rebellion of 1745 and the subsequent wanderings of Bonnie Prince Charlie.

Map United Kingdom –Bb

Mindanao *Philippines* Second largest island covering 94 631 km² (36 537 sq miles) – about one-third of the Philippines' total land area. Lying to the south of the VISAYAN ISLANDS, it is the most southerly major island of the archipelago. The island's longest river, flowing about 320 km (200 miles) through COTABATO region, is also called Mindanao.

The island has an indented coastline, extending into the peninsula of Zamboanga in the west. Largely mountainous, it has the Philippines' tallest peak, Mount Apo near DAVAO; it also has swampy lowlands in the Agusan Valley and the Cotabato region. It gets more rain than most other parts of

the Philippines, making year-round cultivation possible. Most of the Philippines' pineapples, maize and coconuts are grown here. Coffee, cacao and Manila hemp are also produced, and the Philippines' most important mineral deposits, including iron, nickel, and copper, are on the island.

Mindanao's rich resources have attracted many migrants from the other islands since the 1930s. This has led to tension, and even occasional violence, between the newcomers (mainly Christians) and the predominantly Muslim local people. Development continues, however, and as a result Davao is now the Philippines' third largest city, with a population of 850 000.

Population 14 100 000
Map Philippines – Cd

Minden *Germany* Port, cathedral city and industrial town situated on the Weser River, 60 km (37 miles) south-west of HANNOVER. Minden's cathedral dates from the 13th century; its manufactures include machinery, textiles, glass, ceramics, and furniture. At Minden in August 1759, British and Hanoverian armies defeated the French in the Seven Years' War (1756-63).

Population 75 900
Map Germany – Cb

Mindoro *Philippines* Island covering 9736 km² (3759 sq miles) south-west of LUZON. It comprises two provinces, Mindoro Occidental and Mindoro Oriental. One of the least developed of the Philippines' islands until recent years, it remains heavily forested and infested with malaria. Because of its mountainous central area, most of the inhabitants live on its eastern and western coastal plains, where rice is the main crop. Sugar cane is grown near San Jose in the south. Mindoro's forests yield valuable timber, but parts have been protected as nature reserves.

Population (Mindoro Occidental) 145 700; (Mindoro Oriental) 329 000
Map Philippines – Bc

Mindoro, Sea of See SULU SEA

mineral Inorganic chemical element or compound that occurs naturally in the earth. Minerals are distinguished from one another by their chemical formula, crystalline structure, colour, lustre, hardness, and degree of transparency. Rocks are aggregates consisting of a mixture of minerals.

mineral spring Spring of water that contains an appreciable amount of mineral salts, such as iron compounds, magnesium chloride and sodium chloride. Spring water is often used for medicinal purposes.

Minho *Portugal* Coastal province covering 4928 km² (1902 sq miles), to the north of the city of OPORTO. Its capital is the town of Braga. Minho's main crops are maize, beans and grapes; vinho verde wines are produced here.

Population 965 700
Map Portugal – Bb

Minho (Miño) *Spain/Portugal* River which rises in north-western Spain and flows in a south-westerly direction to the Atlantic. Some 275 km (170 miles) long, it forms part of the northern border between Portugal and Spain.

Map Portugal – Ba; Spain – Ba

Minicoy Islands *India* See LAKSHADWEEP

Minneapolis *USA* City on the west bank of the Mississippi River, and the business, cultural, financial and industrial centre of Minnesota. The nearby Falls of St Anthony – a series of rapids with a total fall of 25 m (82 ft) – determined the site of the city because they provided power for the original timber and flour mills. The city is still the market centre for a large cattle and grain area and has a large meat-packaging industry. Minneapolis lies across the river from its twin city of ST PAUL.
Population (city) 368 500; (metropolitan area) 2 538 900
Map United States – Hb

Minneriya Giritale *Sri Lanka* See POLON-NARUWA

Minnesota *USA* North-central state situated on the Canadian border west of Lake Superior. Although it was explored in the 17th century, it did not become a state until 1858. It covers some 206 077 km² (79 548 sq miles) and its northern part is wooded and dotted with lakes that have earned it the nickname, 'state of 10 000 Lakes'.
Besides iron ore, which is mined in the north, Minnesota now produces dairy goods, maize, sugar beet, wheat and cattle. Its industries include machinery and food processing. The state capital is ST PAUL.
Population 4 375 000
Map United States – Ga

Miño *Spain/Portugal* See MINHO

Minorca (Menorca) *Spain* Second largest of the BALEARIC ISLANDS, after Majorca, covering 702 km² (271 sq miles). Mahon is its capital and main port.
Population 20 870
Map Spain – Gb-Hc

Minquiers, Les *Channel Islands* Group of un-inhabited islets and reefs in the Gulf of St Malo.
Map Channel Islands – Bc

Minsk *Belarus* The country's capital and a leading industrial centre, 260 km (160 miles) from the Polish border. It was founded in 1067 as Menesk, became part of Lithuania in 1326, was ceded to Poland in the 16th century, then to Russia in 1793. By 1939, 40 per cent of its people were Jewish, but virtually all of them were killed after invading Germans transferred them to concentration camps during the Second World War.
Since 1945, new printing and timber plants have been set up, along with factories making lorries, tractors, machinery, radios and television sets, clothing, and textiles.
Population 1 612 000
Map Belarus – Bb

Mintaka Pass *China/Pakistan* Pass 4709 m (15 500 ft) up in the mountains on the border of Azad Kashmir and western China. It lies on a spur of the Karakoram Highway. The principal Karakoram Highway runs over the 4602 m (15 100 ft) high Khunjerab Pass to the east.
Map Pakistan – Da

Miocene Fourth epoch of the Tertiary period in the Cenozoic era of the earth's timescale. See GEOLOGICAL TIMESCALE, p 256

mirage Optical phenomenon, particularly in hot deserts, in which an image, often inverted, is produced as a result of refraction of light by layers of air with differing densities. The most common form involves an image of the sky, producing the illusion of water.

Miranda *Venezuela* Northern state which includes part of the metropolitan area of the national capital, CARACAS. It covers 7950 km² (3069 sq miles); its capital is Los Teques.
Population 1 871 100
Map Venezuela – Ba

Misiones *Argentina* North-eastern province tucked away between Paraguay and Brazil. The provincial capital, Posadas (population 220 000), on the Paraguayan border, is 840 km (520 miles) north of BUENOS AIRES; the port of Posadas stands on the Paraná River.
The province's name is derived from the Jesuit missions set up in the region in the 17th and 18th centuries to convert the local Guaraní people to Christianity and teach them handicrafts and agricultural techniques. The missions also protected the indigenous people from the slave trade until 1767 when the Jesuits were driven out of the area. The province, which was settled by German and Italian immigrants after the First World War, is subtropical and heavily forested. It produces yerba maté (a type of tea), tobacco, citrus fruits, timber and fish.
Population 723 000
Map Argentina – Da

Miskolc *Hungary* The country's second largest city, after the capital, Budapest, situated 137 km (85 miles) north-east of the capital. Miskolc is the administrative capital of Borsod-Abaúj-Zemplén County, and is a major industrial city with iron and steel, engineering, food processing and textile plants. It spreads across the Sajó River valley and up into the Borsod upland to the north. The southern suburb of Tapolca is a popular spa.
Population 192 360
Map Hungary – Ba

Misratah *Libya* Oasis town and seaport 140 km (87 miles) east of TRIPOLI, and the capital of the Misratah Division in the north-west. It produces dates and cereals; other products are iron and steel, carpets and traditional handwoven textiles.
Population 103 400
Map Libya – Ba

Mississippi *USA* Southern, largely forested state covering 122 354 km² (47 230 sq miles) between the Mississippi River, which forms most of its western boundary, and the state of Alabama to the east. Everywhere the land is below 246 m (806 ft).
Agriculture, which once centred mainly on cotton, has now diversified to include rice, soya beans and broiler chickens. Gas and oil are produced and there are refining and engineering industries. JACKSON is the state's capital and largest city.
Mississippi is also the name of the river rising in the northern state of Minnesota and flowing 3779 km (2348 miles) south to the Gulf of Mexico. It is the second largest river in the United States, after the MISSOURI, which is its main tributary. Together, the Mississippi-Missouri form the world's third longest river system, flowing a total length of 6019 km (3740 miles) and draining some 3.25 million km² (1.25 million sq miles). Other major tributaries of the Mississippi include the ILLINOIS, ARKANSAS and OHIO rivers.
The river has long been a major transport artery, carrying about 60 per cent of the country's waterway freight, and has been navigated by steamboats since about 1820.
It is, however, prone to serious flooding. In 1993, for example, 32 383 km² (12 500 sq miles) were badly flooded and another 48 575 km² (18 751 sq miles) were too wet to produce crops. Nevertheless, massive control schemes have now stabilised the river significantly.
Population 2 5753 200
Map (state) United States – Hd; (river) United States – Hb, Hd

▼ DOWN THE MISSISSIPPI A century ago, cargo was transported on stern-wheeler steamboats. The *Creole Queen*, a modern reproduction of a steamboat, carries tourists exploring nearby New Orleans and the United States' second largest river.

Missolonghi *Greece* See MESOLÓNGION

Missouri *USA* Midwest state bounded by the MISSISSIPPI River in the east and crossed by the Missouri River, which joins the Mississippi near ST LOUIS. The state covers some 180 557 km² (69 697 sq miles) and stretches south into the Ozark Plateau, where a series of lakes, springs and caves have turned it into a popular tourist area.

Livestock, dairy produce and soya beans are the chief sources of farm income. Lead is the chief mineral, providing most of the country's supply. Manufactures now include transport and aerospace equipment, and chemicals. The state capital is JEFFERSON CITY, but the largest cities are KANSAS CITY and St Louis. Some 150 km (95 miles) north-west of St Louis, on the Mississippi, is the city of Hannibal, the boyhood home of the writer Samuel Clemens (1835-1910), better known as Mark Twain.

The Missouri River, which has given the state its name, is the longest river in the country, with a length of 4088 km (2540 miles). It rises in Montana and flows first east then south-east into the Mississippi just above St Louis. It has been dammed in places for flood and navigation control. Despite these measures, the worst flooding on record occurred in 1993, leaving some 70 000 people homeless and causing damage estimated at US$12 billion.
Population 5 117 100
Map (state) United States – Hc; (river) United States – Ea, Hc

mist Mass of water droplets in the lower layers of the atmosphere caused by the condensation of water vapour in the air, reducing visibility. At lower visibility and greater density the mist becomes a FOG.

mistral Strong, cold, dry northerly to north-westerly wind blowing from the central plateau of France towards a low pressure system over the Mediterranean Sea, experienced especially in the Rhône Valley and the Golfe du Lion (Gulf of Lions). It corresponds to the MAESTRALE of Italy.

Mistras *Greece* Ruined Byzantine town on a hilltop beside Mount Taíyetos, 7 km (4 miles) west of Sparta. It was ruled by Byzantine princes from the 13th to the 15th century and then became an intellectual centre and meeting place of the Western and Byzantine cultures. A palace, churches, monasteries, houses and Turkish baths of the period survive, with fine stone and brickwork, frescoes, paintings and icons.

The Byzantine philosopher Gemistos Plethon (about 1355-1452), who advocated a return to the Ancient Greek religion and influenced many writers in Renaissance Europe, lived here.
Map Greece – Cc

Mitchell, Mount *USA* Highest peak in the country east of the Mississippi River. It rises to 2037 m (6684 ft) in the Black Mountains range of the Appalachians in western North Carolina.
Map United States – Jc

Mitilíni *Greece* See LESBOS

Mitla *Mexico* Ancient ruined city of the Zapotecs, in the south-eastern state of Oaxaca, 400 km (248 miles) south-east of MEXICO CITY. The former residence of the Zapotec and Mixtec rulers, Mitla was built more than 2000 years ago over catacombs consecrated by the Zapotecs. Stone fretwork and geometrical mosaics adorn its long, low palaces, and its Hall of Columns has six enormous pillars, each one a single stone.
Map Mexico – Cc

Mito *Japan* City near the east coast of HONSHU Island, lying 100 km (62 miles) north-east of the capital, Tokyo. Its fine landscape garden, the Kairakuen, is noted for its plum blossoms and a pavilion where poetry readings are held regularly.
Population 235 000
Map Japan – Dc

Miyajima (Itsukushima) *Japan* Small island in the Inland Sea, 19 km (12 miles) south-west of the city of HIROSHIMA, one of the most beautiful parts of Japan's coastline. It is dotted with shrines – the main shrine of Itsukushima has given the island its other name. Miyajima is the venue of many colourful festivals, including Tamatori Ennen in August, in which young men fight through water to retrieve a sacred ball.
Map Japan – Bd

Miyazaki *Japan* Market town on the east coast of KYUSHU Island. The town is known as 'the Hawaii of Japan' for its sunny climate. A Shinto shrine at Miyazaki commemorates Jimmu Tenno, the legendary first emperor of Japan, who reigned in the 7th century BC.
Population 287 000
Map Japan – Bd

Mizoram *India* Territory covering 21 081 km² (8137 sq miles) of wet, rain forest-clad hills on the border with Myanmar. It is inhabited by the Mizo and Lushai people, whose main activity is rice farming.
Population 689 760
Map India – Ec

Mjøsa *Norway* The country's largest natural lake, covering 368 km² (142 sq miles), situated 55 km (34 miles) north of OSLO. It is surrounded by an extensive cultivated area, in which the main crops are grain and potatoes, and small manufacturing towns.
Map Norway – Cc

Mljet (Meleda) *Croatia* Southernmost Dalmatian island, lying north-west of Dubrovnik. Mljet may have been the Biblical site of Melita where St Paul was shipwrecked. Covering some 98 km² (38 sq miles), it is now a pine-forested national park. Mongooses have been introduced on the island to control the large number of snakes. Mljet has three lagoons, on one of which, the 46 m (151 ft) deep Veliko Jezero ('Great Lagoon'), is an islet with a 12th-century monastery, now a hotel.
Map Croatia – Cc

Mo i Rana *Norway* Industrial town, some 30 km (20 miles) south of the Arctic Circle, where a major state-owned iron and steel works was built in the 1950s to combat local unemployment. The nearby Dunderland Valley has been mined intermittently for a century, but today most of the ore for the works is brought from the Sydvaranger mine in Finnmark.

Energy is provided by the nearby Røssaga hydroelectric power station and from imported coal. By-products of the works include zinc, copper and sulphur. The town also has port facilities. Being closest to the Svartisen Glacier, Mo i Rana is used by many tourists as a stopover.
Population 20 700
Map Norway – Db

Moanda *Gabon* Modern town with an airport, situated in the south-east, 50 km (31 miles) north-west of MASUKU. It lies in an area containing some of the world's largest manganese deposits, which were first mined in 1951. The manganese is refined at Moanda and then taken to Brazzaville in neighbouring Congo by cable railway. From there it is carried by rail to the port of Pointe-Noire.
Population 22 910
Map Gabon – Bb

Mobile *USA* Port on the Gulf of Mexico in Alabama. The city, which was founded in the early 17th century by the French and was once the capital of Louisiana, stands on the Mobile River. Shipbuilding, production of chemicals and bauxite processing are the basis of its industries.
Population (city) 198 300; (metropolitan area) 476 900
Map United States – Id

mobile belt Any long, narrow region of the earth's crust in which structural or mountain-building activities are currently taking place; for example, the zone along which two lithospheric plates are colliding, or a GEOSYNCLINE which is still subsiding and receiving more sediment.

Mobuto Sese Seko, Lake *Uganda/Zaire* See ALBERT, LAKE

Moçambique *Mozambique* Ancient Arab port some 580 km (360 miles) south of the Tanzanian border. It lies on a small coral island and was visited by the Portuguese explorer Vasco da Gama in 1498. The Portuguese established a stronghold here in 1507 and Moçambique soon became a provisioning port for Portuguese fleets. It served as the national capital until 1907.
Population 12 200
Map Mozambique – Bb

Moçamedes *Angola* See NAMIBE

Moche Pyramids *Peru* Largest pre-Columbian structures in South America. Made of adobe bricks, they stand in a valley 5 km (3 miles) south of Trujillo. They are relics of the Mochica culture, which flourished between the 3rd and 8th century AD. The biggest, Huaca del Sol (temple of the sun), is 48 m (158 ft) high, 228 m (748 ft) long and 135 m (443 ft) wide. The smaller pyramid is called Huaca de la Luna (temple of the moon).
Map Peru – Ba

Modena *Italy* Industrial city 38 km (24 miles) north-west of BOLOGNA. It has a handsome Romanesque cathedral, built between 1099 and 1148, and a colossal 17th-century palace, now a military academy.

The town is a noted manufacturer of high-performance sports cars, agricultural machinery, machine tools, other metal goods, and leather.
Population 177 500
Map Italy – Cb

Modified Mercalli Scale A 12-point scale designed in 1931 to measure the intensity of an earthquake, replacing the earlier 10-point Rossi-Forel Scale. It measures the amount of damage at a particular point, and decreases with distance from the earthquake's focus, unlike the absolute RICHTER SCALE of magnitude.

Mogadishu (Mogadiscio; Muqdisho) *Somalia* National capital and chief port, on the south-east coast. It exports livestock and sugar and was an important trade centre in antiquity, with Indian, Arabian and Chinese merchants using its port facilities. In the 10th century, Persian and Arabian merchants founded a settlement here, which was conquered by the Portuguese in the 16th century. The old quarter, known as Hammar Wein, is a remnant of the Arab settlement.

The town was seized by the Sultan of Zanzibar in 1871, sold to the Italians in 1905, and taken by the British in 1941 during the Second World War. It was a ready-made capital when, in 1960, Somalia became independent.

Population 750 000
Map Somalia – Ab

Mogami *Japan* River 230 km (143 miles) long in the north-west of HONSHU Island. It flows north-west to enter the Sea of Japan at the town of Sakata.
Map Japan – Dc

Mogilev *Belarus* See MAHILYOW

Mogok *Myanmar (Burma)* Mining town on the edge of the eastern hills, some 125 km (75 miles) north-east of Mandalay.

The town is the centre of Myanmar's ruby-mining industry, which – despite the primitive mining methods currently used – makes an important contribution to the country's exports.
Population 79 500
Map Myanmar – Cb

Mohács *Hungary* City and port on the west bank of the River Danube in the south of the country, some 12 km (7 miles) from the Croatian and the Serbian border. Here on 29-30 August 1526, King Louis of Hungary and his 20 000-strong army were heavily defeated by forces of the Turkish sultan, Suleiman I (the Magnificent, 1520-66). The Ottoman victory resulted in the Turkish domination of Hungary for more than 150 years, and paved the way for the Turkish assaults on Vienna in Austria (1529 and 1683).

The event is commemorated by an annual carnival in the city. However, the tables were turned 162 years later, in 1687, when the retreating Turks were defeated at the Second Battle of Mohács fought at Harkány to the south-west.
Population 20 310
Map Hungary – Ab

Mohammedia *Morocco* Port and seaside resort 25 km (16 miles) north of CASABLANCA. It was formerly settled by the Portuguese, and some of their buildings can still be seen in the fedala kasbah ('old town'). In 1942, Mohammedia was one of the landing places for United States troops in the North African campaign. Today, the port is famed for its fish and fine fish restaurants. It is also the site of a major oil refinery.
Population 105 100
Map Morocco – Ba

Mohave Desert *USA* See MOJAVE DESERT

Mohéli *Comoros* See MWALI

Mohenjo Daro *Pakistan* One of the world's greatest archaeological sites, 200 km (125 miles) north of HYDERABAD. Excavations of this 'City of the Dead' began in 1922. A city of the Indus civilisation, it lasted 1000 years from 2500 BC. In fact, some of the remains are 4000 years old.

It had wide streets, good sewers and large houses, as well as a huge public bath, 12 m by 7 m (39 ft by 23 ft), which is faced with gypsum mortar and backed with bitumen to make it watertight. Most buildings were built with fired bricks of uniform size and quality. A public granary was built with ventilation passages below the floors to prevent the grain from becoming mouldy.

It seems that the city traded with China and the Middle East. But it was abandoned, apparently suddenly, in about 1500 BC for reasons that are still uncertain. Military conquest by invading Aryans, economic decline and an earthquake have all been suggested. The exodus may have been accelerated by a river flood. Today, the site faces the risk of a second destruction because of the rise in the water table and soil moisture brought about by irrigation canals, and the United Nations agency UNESCO has launched a rescue plan to counteract this.
Map Pakistan – Cb

Möhnestausee *Germany* Lake created by damming the Möhne River in the hills 30 km (18 miles) east of the Ruhr industrial complex. The dams on the Möhne, and on the Eder River to the south-east, were destroyed in a British 'Dambusters' raid by the RAF in 1943, but had been repaired by the end of the war.
Map Germany – Cc

Moho Abbreviation for the Mohorovičić discontinuity, the boundary which separates the earth's crust from the mantle. See JOURNEY TO THE CENTRE OF THE EARTH, p 207

Mohokare *Lesotho/South Africa* See CALEDON

Mohole Hole which scientists planned to drill through the earth's crust and into the mantle to obtain samples of rocks. The project, which was conceived in the 1950s, was not completed, and the idea was abandoned in 1966 because of rising costs and disagreements among the scientists involved. It did help, however, to stimulate more limited projects to drill into the basaltic crust of the ocean floors, such as the Deep Sea Drilling Project.

Mohs Scale of Hardness A 10-grade scale used to measure and specify the hardness of a mineral, devised in the 19th century by the German mineralogist Friedrich Mohs. Each mineral in the scale can be scratched by minerals having higher numbers, and can in turn scratch those that are lower down on the scale. The hardest mineral, with a 10 rating, is the diamond; the softest, with a 1 rating, is talc.

Móin Aluine *Ireland* See BOG OF ALLEN

Mojave Desert (Mohave Desert) *USA* Wilderness covering about 38 850 km² (15 000 sq miles) in south-eastern California, between DEATH VALLEY and the city of LOS ANGELES. It is one of the hottest regions in the world and receives

an annual average rainfall of less than 100 mm (3.94 in). It is the site of Edwards Airforce Base, main landing site for space shuttle missions.
Map United States – Cc

Mokanji Hills *Sierra Leone* Upland region west of the Jong River, some 176 km (110 miles) south-east of the capital, FREETOWN. Most of the country's bauxite, an important export, is mined in the hills.
Map Sierra Leone – Ab

molasse Soft, predominantly sandstone sediment deposited in the Alpine foothills by streams bringing eroded material down from the rising Alps. The term is also used for similarly formed deposits of much greater age – such as the Newark Sandstone of the eastern United States, derived from the Appalachians, and the Old Red Sandstone of Britain – derived from the Caledonian Mountains.

Moldau See MOLDOVA

Moldau *Czech Republic* See VLTAVA

Moldavia See MOLDOVA

Moldavia *Romania* Fertile north-west region containing the rugged eastern CARPATHIANS and noted for its beautifully decorated monasteries and vividly painted houses. Established as a principality in the 14th century, largely under Hungarian control, it was taken over by the Turks in the 16th century and, with WALACHIA, became the core of Romania in 1862. The region consists mainly of rich farmland.
Map Romania – Ba

Moldova See p 454

Molise *Italy* Rural region covering 4438 km² (1714 sq miles) immediately south of ABRUZZI, between the Adriatic Sea and the Apennines. Its mainly peasant inhabitants make a living farming grapes, wheat, vegetables, sheep and goats.
Population 332 900
Map Italy – Ed

Molotov *Russia* See PERM'

Molucca Sea *Pacific Ocean* Indonesian sea, an arm of the PACIFIC OCEAN between SULAWESI (Celebes) and the MALUKU Islands. It is linked in the south to the BANDA SEA and in the south-east, to the SERAM SEA. The first expedition to circumnavigate the globe sailed into the Molucca Sea in 1521, soon after the death of its commander, Ferdinand Magellan.
Map Indonesia – Fc

Moluccas *Indonesia* See AMBON; MALUKU

Mombasa *Kenya* Second largest city of Kenya, after the capital, NAIROBI, and the country's main port, as well as the main export-import harbour for neighbouring Uganda and Rwanda. Mombasa stands on an island with a natural harbour in two sheltered creeks. Kilindini Harbour, built after the Second World War, has deep-water berths for ocean-going vessels.

The town was occupied by medieval Arab traders, and later on by the Portuguese, before becoming part of a British protectorate in 1895.

453

Moldova

ONE OF THE POOREST FORMER SOVIET
STATES, LANDLOCKED MOLDOVA IS
TORN BY NATIONAL MINORITIES
SEEKING INDEPENDENCE

Moldova is the second smallest of the former Soviet countries after Armenia, but it is the most densely populated. In its past it was repeatedly sliced up by neighbouring states and then pieced together, like a jigsaw – and this history of fragmentation has shaped its politics since it became the Republic of Moldova.

The region was settled by the Romans in the 1st century AD and by Slavic peoples in the 9th and 10th centuries. In the 14th century it was founded as a principality and parts of it have since been ruled by Romania and Turkey. It was part of Romania when the Soviet Union annexed it in 1944. The Moldovan language is very similar to Romanian and there was a strong move towards reunification with Romania after Moldova gained its independence from the Soviet Union in August 1991. In a plebiscite in March 1994, however, more than 90 per cent of the population voted against reunification.

A FARMING COUNTRY

Bessarabia, that area of Moldova which lies between the Prut River (which forms part of the Romanian border) and the DNIESTER (Dnestr) River, rises to 429 m (1407 ft) and is cut by steep-sided ravines. But the Dniester plains of Moldova have a warm climate and a fertile soil. They produce grapes, tree fruits, strawberries, walnuts and honey. The farm crops include wheat, maize, sunflowers, sugar beet and tobacco. There is also illicit production of opium and cannabis, which is exported to the other former Soviet states.

Agriculture is important, employing more than 33 per cent of the workforce. But there is progressive deforestation and extensive soil erosion caused by poor farming methods, as well as badly contaminated soils and groundwater. Agricultural methods put a heavy emphasis on using chemicals, including DDT – which has been banned elsewhere.

Food processing, distilling, wine making and engineering industries, as well as footwear and textile industries, are concentrated in the towns of Belítsy, Tiraspol and Bendery, and in Chişinău (Kishinev), the capital, which was largely destroyed in 1944, but has been rebuilt.

Moldova joined the Commonwealth of Independent States (CIS) in December 1991. During Soviet times its living standard had ranked poorly among the lowest of the Soviet republics, its output accounting for just over 1 per cent of the Soviet Union's total output. Today, with its dependence on raw materials from Russia and other CIS countries, there is little chance that it will rise to become a major producer. In 1991, and again in 1992, its output fell by more than 20 per cent, due to a drop in energy supplies from Russia.

Russo-Moldovan relations became strained during the first year of independence. Russia still kept 7000 troops of its notorious 14th Army in the Dniester region of Moldova, most of which lies east of the Dniester River. The region, which has a largely Russo-Ukrainian population, declared its sovereignty in September 1990, but was never recognised. After independence, the Russians living here again began to clamour for their independence – or else for reunification with their 'Motherland'. Moldova spent most of 1992 fighting over the Dniester region; at the end of 1993 it was finally granted a high degree of autonomy – but no federal status.

Relations between Russia and Moldova were further strained in 1992 when Moldova refused to sign the draft charter of the CIS, which was designed to provide Russia with a new central authority over all the CIS states. By refusing to sign, Moldova did not become a full member. In retaliation, the Russian government cut oil exports to Moldova by 10-20 per cent at the end of 1992, crippling Moldova's industries.The country now imports most of its fuels from Azerbaijan.

Politics in Moldova are in most ways related to the fact that the country consists of three main regions. Firstly, there is the west, which is home to the Moldovians, who make up just under two-thirds of the total population. Secondly, there is the 'Dniester republic'; and thirdly, there are separatist movements in the 'Gagauz republic' in the south, situated around the city of Comrat, which has a large Turkish population.

Another concern of the government has been to remove all Russian influence. After independence, the Communist Party was banned, although it was again unbanned in 1993. Legislative elections in February 1994 brought an overwhelming victory to the ruling Agrarian Democratic Party, which under President Mircea Snegur freed prices on most goods and moved towards privatisation.

In August, a new constitution came into force establishing the country as a presidential parliamentary republic and allowing special autonomy for the Dniester and Gagauz regions; talks were also held with Russia on the phased withdrawal of its troops from the Dniester region.

MOLDOVA AT A GLANCE	
Official name Republic of Moldova	
Area 33 700 km² (13 009 sq miles)	
Population 4 455 650 **Per km²** 132 **(Per sq mile** 343)	
Capital Chişinău (Kishinev)	
Currency Lei	
Languages Moldovan (official), Romanian, Russian	
Religions Christian (mainly Eastern Orthodox), some Jewish	
Climate Temperate continental, with long, hot summers and fairly cold winters, and little rainfall. Average temperature in Chişinău ranges from -4°C (25°F) in January to 21°C (70°F) in July	
Land use Cultivation 50%, meadows and pastures 9%, forest 8%, other 33%	
Main primary products Lignite; maize, vegetables, fruit, tobacco	
Major industries Agricultural machinery, tools, electronics, shoes, textiles, food processing	
Main exports Maize, vegetables, fruit, wine, tobacco, meat	
Annual income per head (US$) 1260	
Population growth (per thous/yr) 4	
Life expectancy (yrs) Male 64 **Female** 72	

In the 1890s, Mombasa was linked by rail with inland Kenya and Uganda. Today there is an oil refinery on the mainland beside the harbour, and tourist hotels line the coast north and south of the city. Tourists enjoy its attractive beaches.

Population 500 000
Map Kenya – Cb

Mon *Myanmar (Burma)* State in the south of the country, covering 12 297 km² (4748 sq miles). Its capital is MAWLAMYINE (Moulmein).

Population 1 682 000
Map Myanmar – Cc

Møn *Denmark* Island covering some 218 km² (84 sq miles) in south-eastern Denmark. Most of it is low-lying but at its eastern end the land rises above the Baltic to form a line of chalk cliffs, a famous landscape some 7 km (4 miles) long and in places more than 120 m (390 ft) high. One of these yellow and white cliffs is known as 'the Weeper' because water runs down it from a spring. Although Møn attracts visitors, tourism takes second place to agriculture. Root crops and wheat are grown here, and cattle are reared.

Population 10 200
Map Denmark – Cb

Monaco See p 455

Monaco-Ville *Monaco* Ancient capital of the small principality, occupying a rocky headland projecting into the Mediterranean Sea south of Monaco's harbour. It contains the 16th-century royal palace of the Grimaldi family, the 19th-century cathedral and Monaco's noted oceanographic museum. The commune itself has narrow streets lined with attractive old houses.

Population 12 500
Map France – Ge

▲ CITADEL BY THE SEA Phoenician traders colonised Monaco-Ville's rocky base. Genoese settlers fortified it in 1215, and the ruling house of Grimaldi gained control in 1297.

Monaghan (Muineachán) *Ireland* Rural county covering 1291 km² (498 sq miles) on the border with Northern Ireland. It is part of the Province of Ulster and, with nearly 200 lakes, is a popular area for coarse fishing. The county town of the same name lies 6 km (4 miles) from the border.
Population (county) 51 290; (town) 5750
Map Ireland – Ca

Monastir *Tunisia* Small coastal town and tourist centre south of SOUSSE. It was the birthplace in 1903 of President Habib Bourguiba, president of Tunisia from 1956 to 1987. Among the places of interest are an 8th-century Arabic fortress, the oldest in Africa, in which *son et lumière* performances are held in summer, an Islamic museum, the Great Mosque dating from the 9th to the 11th century, the Habib Bourguiba Mosque (an excellent example of modern Islamic architecture), the presidential palace and the old, walled town. Since independence, the town has become a showpiece for urban development.
Population 39 000
Map Tunisia –Ba

Monastir *Macedonia* See BITOLA

Mönchengladbach *Germany* Industrial city some 25 km (16 miles) west of DÜSSELDORF, to the south-west of the Ruhr industrial region. The town grew as a textile centre around a 10th-century Benedictine abbey and expanded rapidly in the 19th century with the growth of Germany's chemical and engineering industries.
Population 259 400
Map Germany – Bc

Monclova *Mexico* Important industrial town in the northern state of Coahuila. Together with MONTERREY, in neighbouring Nuevo Léon, and SALTILLO it forms a triangle that constitutes one of the most productive mining and industrial regions in Mexico. The area yields zinc, copper, silver and lead, and Monclova has a huge steel mill, iron foundries, railway workshops and a range of manufacturing industries.
Population 178 020
Map Mexico – Bb

Moncton *Canada* City on the Petitcodiac River in south-east NEW BRUNSWICK. It is the transport hub of the maritime provinces (New Brunswick, Nova Scotia and Prince Edward Island). The town has meat-packaging plants and manufactures steel, chemicals and woollen goods. It has Canada's only French university outside QUEBEC.
Population 65 280
Map Canada – Id

Mondego *Portugal* Longest river flowing entirely within Portugal. It rises in the Serra da Estrêla, in the Beira Alta Province, and after 220 km (137 miles) reaches the Atlantic at the town of Figueira da Foz, about 150 km (95 miles) north of the capital, LISBON.
Map Portugal – Bb

Monemvasía (Napoli di Malvasia) *Greece* Venetian walled town built on a high crag rising from the sea, with a single entrance on a causeway linking it with the mainland. It lies on the south-east coast of the Peloponnese.

Founded in the 8th century by the Byzantines, it flourished as a wine-making centre, exporting Malmsey wines (named after the town's medieval name, Malmesia) all over Europe.
Map Greece – Cc

Mongolia See p 456

Monaco

THE WHEEL OF FORTUNE KEEPS ON TURNING FOR A FAIRYTALE PRINCIPALITY WHICH THRIVES ON FUN AND GAMES

Sunlight sparkling on the blue Mediterranean, luxurious yachts bobbing in the marinas, and bronzed people basking on the beaches – these are the main ingredients of the heady cocktail that is Monaco. To which must be added the risks and romance of the casino, the thrills and spills of the Monte Carlo Rally and Monaco Grand Prix, the fairytale marriage, in 1956, of Prince Rainier and Grace Kelly, the beautiful American film star, and various alliances and marriages (copiously recorded) of the next generation of Grimaldis.

There are only 4960 Monégasque citizens but another 26 040 people, mainly French and Italian, live in this tiny country in the south-east corner of France, with which it has a customs union. Members of the Grimaldi dynasty – originally a Genoese family – have ruled Monaco since the 13th century. But their country attracted little interest until the second half of the 19th century, when the casino and first hotels were built.

There are four distinct districts, all of them almost entirely urban. The old town of MONACO-VILLE, set on a rocky promontory, houses the royal palace and the cathedral. MONTE CARLO retains its world-famous casino, but has changed drastically as stately houses have given way to high-rise apartments and hotels. Between the two is LA CONDAMINE, with thriving businesses, shops, banks and attractive residential areas. The fourth district is Fontvieille. The Monégasques pay no taxes in this rich little country, but neither do they share all the fun – they are not allowed to gamble in the casino.

MONACO AT A GLANCE	
Map France – Ge	
Official name Principality of Monaco	
Area 1.9 km² (0.73 sq miles)	
Population 31 000 **Per km²** 16 316 **(Per sq mile** 42 466)	
Capital Monaco-Ville	
Government Constitutional principality	
Currency 1 French franc = 100 centimes	
Languages French (official), the Monégasque dialect, Italian and English	
Religion Christian (Roman Catholic 95%)	
Climate Mediterranean, with mild winters. Average temperature ranges from 10°C (50°F) in January to 23°C (73°F) in July	
Land use Urban	
Major industries Tourism, gambling, banking, perfumery, plastics	
Main exports Cosmetics, electronics, plastics, tinned food	
Annual income per head (US$) 16 000	
Population growth (per thous/yr) 9	
Life expectancy (yrs) Male 74 **Female** 81	

Mongolia

A CENTURIES-OLD NOMADIC LIFESTYLE IS VANISHING AS CITY LIFE CALLS THE HERDERS IN FROM THE HILLS

The traditional circular domed tents *(yurts)* of Mongolia's nomadic herders can still be seen, but they are likely to be equipped with television sets and transistor radios. There are fewer *yurts* because a growing number of Mongolians live in cities: the capital, ULAN BATOR, alone, holds more than 25 per cent of this country's sparse population.

Most of Mongolia is more than 1500 m (4900 ft) above sea level. The north-western area consists of remote mountain ranges, including the Hangayn Mountains and the ALTAI, which includes the country's highest peak, Taban Bogdo Ula (4356 m, 14 290 ft).

To the south-east, mountains give way to grass-covered steppes and the desert wastes of the GOBI. Most of Mongolia's territory is pasture – cultivated land accounts for 1 per cent of the surface. The environment is harsh

and demanding – some land is permanently frozen and some is sandy desert – and the climate is extreme. Long and cold winters are followed by short, hot summers. Rainfall is low and unreliable.

The country was once part of China and was divided into Inner Mongolia and Outer Mongolia. Inner Mongolia remains Chinese; Outer Mongolia's emergence as an independent country dates from the time of the Russian-backed revolution in 1921 which established the Communist Mongolian People's Republic in 1924 – only the second country in the world to become a People's Republic. Relations with big brother Russia in the north are still close after the disintegration of the Soviet Union. Mongolia needs to market its products to the other CIS members.

Until 1990, all power had been vested firmly in the hands of the Mongolian People's Revolutionary Party. After severe cutbacks in aid from the Soviet Union, and local demonstrations, the entire Politburo resigned and political opposition was legalised. A new constitution introduced democracy in 1992 and all Russian troops were withdrawn by September of the same year.

For many centuries, the herding of horses, cattle, sheep, yaks, and camels has been

the mainstay of the economy. But since the Second World War there have been significant changes.

The mining industry – which was developed with lavish Russian aid at centres such as ERDENET and DARHAN – rivals animal husbandry as a major source of national income. Today, however, the nomadic life of the herders has given way to raising livestock on cooperative ranches. But with the reduction in Russian aid, the country has been slowly and surely returning to privatisation and a capitalist economy.

MONGOLIA AT A GLANCE	
Official name Mongolia	
Area 1 566 500 km² (604 825 sq miles)	
Population 2 367 100 **Per km²** 1.5 **(Per sq mile 4)**	
Capital Ulan Bator	
Government Republic	
Currency 1 Tugrik = 100 mongo	
Languages Mongolian (official), Kazakh	
Religions Lamaist Buddhist (predominant), Muslim (4%)	
Climate Dry and cold. Average temperature in Ulan Bator ranges from -26°C (-15°F) in January to 17°C (63°F) in July	
Land use Grazing 79%, forest 9.7%, cultivation 1%, other 10.3%	
Main primary products Livestock, cereals, coal, lignite, petroleum, copper, molybdenum, gold, fluorspar	
Major industries Agriculture, food processing, textiles, mining, processing hides and skins	
Main exports Minerals, livestock, wool, hides, meat	
Annual income per head (US$) 660	
Population growth (per thous/yr) 26	
Life expectancy (yrs) Male 64 **Female** 68	

Mongu *Zambia* Market town on the edge of the ZAMBEZI River floodplain, 550 km (340 miles) west of the capital, Lusaka. Local farmers earn a living from maize, millet, goat and sheep farming. Lealui, a village 18 km (11 miles) west of Mongu, is the home of the chief of the Silozi people. To escape river floods during the January-April rainy season, the chief moves his family by royal barge to higher ground at Limungula in a ceremony called Ku-omboka.
Population 24 900
Map Zambia – Bb

Monmouth *United Kingdom* See GWENT

Monrovia *Liberia* The country's capital, largest city and major seaport. It lies in the west, 75 km (45 miles) from the border with Sierra Leone, at the mouth of the Mesurado River, which almost encircles the built-up area. The city was founded by the American Colonisation Society in 1822 as a home for freed slaves from the United States. It gets its name from James Monroe, United States president from 1817 to 1825.

The American influence in Monrovia could be felt for years and was evident in the city's broad avenues lined by modern buildings, as well as stone houses of the earliest settlers, which simulate the architectural styles of the American South. But there are also shanty towns, mud

huts and slums and much of the city has been destroyed since civil war broke out in 1990. Monrovia has a high unemployment rate which has soared since 1990.

A museum of memorabilia and African crafts, and an amphitheatre where African music and dancing are staged, are situated on Providence Island, the settlers' first landing place.
Population 668 000
Map Liberia – Aa

Mons (Bergen) *Belgium* Medieval town 55 km (34 miles) south-west of the capital, BRUSSELS. It has a Gothic cathedral, and six important museum collections. It is located on a coal field, and mining attracted other industries; in the 19th century, it was an important textile centre. Today, however, the mines are closed and industries are of minor significance.

On 23 August 1914, Mons was the scene of the first encounter in the First World War between the advancing Germans and the British army. The battle gave rise to the legend that heavenly forces – 'the Angels of Mons' – had been seen coming to the aid of the British, standing between them and the Germans. However, the British were forced to retreat from Mons almost all the way to Paris before they could counter-attack
Population 95 000
Map Belgium – Aa

monsoon Regular seasonal reversal of the wind system bringing summer rainfall to India, other parts of Asia, northern Australia and Africa. The reversal occurs because of atmospheric pressure differences over large land masses and over the oceans. In winter the land masses are dominated by anticyclonic systems and winds blow from the continent, but by early summer the heat from the sun warms the land, thins the air above and produces low pressure over the land masses. Winds blowing from the sea into the low pressure area bring moisture with them that falls as the monsoon rains, often as torrential downpours.

monsoon forest Form of tropical forest found in areas that have a marked dry season, such as the monsoon regions of Myanmar, Indonesia, Thailand, India and northern Australia. The rainfall is sufficient during the wet season to support the growth of trees, but they shed their leaves during the dry season. New foliage appears when the next monsoon rains begin.

montana Forested slopes of the tropical eastern Andes, particularly in Peru.

Montana *USA* Sparsely populated north-western state along the Canadian border. Almost half its 381 087 km² (147 138 sq miles) lie within the Rocky Mountains and the rest is in the Great

Plains to the east. Wheat, cattle, sheep, barley and sugar beet are the main crops. Minerals found here include coal, oil, natural gas, and some copper and non-ferrous metals. HELENA is the state capital.

Population 826 000

Map United States – Da

Monte Albán *Mexico* Site of an ancient Zapotec and Mixtec religious centre, known as 'the White Acropolis', on a steep hill overlooking Oaxaca City. Occupied originally from 800 BC to AD 1200, it was the ruling city of south-central Mexico until the 10th century. Later, in the last few centuries before Spanish rule, the Zapotecs and Mixtecs only buried their dead here.

Palaces, pyramids, observatories, sanctuaries, dwellings and terraces cover 39 km² (15 sq miles). In the centre, lies a great plaza 300 m (1000 ft) long and 200 m (660 ft) wide. Tombs here have yielded priceless treasures of funeral urns, jewels and goblets.

Map Mexico – Cc

Monte Bello Islands *Australia* Remote and uninhabited archipelago in the Indian Ocean off the coast of Western Australia. The first British atomic bomb was tested here in 1952.

Map Australia – Bc

Monte Carlo *Monaco* Home of the famous casino and main tourist area of the tiny principality, containing numerous luxury hotels. The casino, facing the Mediterranean on one side and lovely gardens on the other, was opened in 1856; it has gaming rooms and an ornate theatre. Nearby are conference rooms and exhibition centres.

Population 13 500

Map France – Ge

Monte Sant'Angelo *Italy* Town and pilgrimage centre standing at nearly 800 m (2025 ft) on the Gargano Peninsula, about 170 km (105 miles) north-east of NAPLES. It has grown up around the Sanctuary of the Archangel Michael, which was founded in the 5th century.

Population 16 500

Map Italy – Ed

Montecatini Terme *Italy* Health resort 40 km (25 miles) north-west of FLORENCE. It is noted for its baths and thermal springs.

Population 21 300

Map Italy – Cc

Montego Bay *Jamaica* Second largest town and seaport, after KINGSTON, and the island's best known tourist centre. It stands on the north-west coast, 130 km (80 miles) from Kingston.

Originally the site of a large Arawak village, it was visited by the famous explorer Christopher Columbus in 1494. It has exceptionally fine bathing beaches, and the town's prosperity is largely derived from the luxury tourist trade. The port's chief export is bananas.

Population 100 000

Map Jamaica – Ba

Montélimar *France* Town lying about 130 km (80 miles) south of LYONS. It is best known for the manufacture and sale of nougat made from locally grown almonds.

Population 30 000

Map France – Fd

▲ IN THE RAINY SEASON Workers near Bombay take a break from transplanting rice during the annual torrential monsoon rains. It is back-breaking work, and their straw protectors shield them when bending to the task.

Montenegro See YUGOSLAVIA

Monterey *USA* Pacific coastal resort situated in California, 135 km (85 miles) to the south of SAN FRANCISCO. Popular with writers and artists, it hosts a jazz festival each year.

Population 32 000

Map United States – Bc

Monterrey *Mexico* The country's third largest city, after MEXICO CITY and GUADALAJARA. It is the capital of Nuevo León State and lies 690 km (430 miles) north-west of the national capital. It was captured by American forces during the Mexican War (1846-8); today it is a major industrial centre, producing 75 per cent of Mexico's iron and steel, as well as lead, chemicals and a wide range of other manufactured goods.

Population 2 800 000

Map Mexico – Bb

Montevideo *Uruguay* Capital city and port on the northern bank of the River Plate, 100 km (62 miles) across the estuary from Argentina. The city was founded in 1726 by the Spanish, and expanded with the country's growing meat industry. Later, it grew as a financial centre. Successive waves of Spanish and Italian immigrants have made Montevideo's population – which today represents about half of the national total – almost exclusively European.

The city has a Mediterranean climate and its shady squares, parks and gardens (notably the El Prado with 850 varieties of roses), its stately 19th-century buildings, its tea rooms and well-kept old cars give it an atmosphere of pre-Second World War Europe. About 90 per cent of Uruguay's trade passes through the port, where the ship's bell of HMS *Ajax* commemorates the Battle of the River Plate, in which the German pocket battleship *Graf Spee* was sunk offshore in 1939.

A late 18th-century fort, which now contains a military museum on a hill situated west of central Montevideo, has commanding views over the city and river.

Population 1 500 000

Map Uruguay – Ab

Montgomery *United Kingdom* See POWYS

Montgomery *USA* Capital of ALABAMA, situated on the Alabama River in the centre of the state, 220 km (135 miles) from the Gulf of Mexico. The city is a business and market centre which once handled large amounts of cotton. New industries such as the manufacture of fertilisers have been attracted here because of reliable sources of hydroelectric power.

Population (city) 187 100; (metropolitan area) 292 500

Map United States – Id

Montpelier *USA* Capital of the north-eastern state of VERMONT. It lies 80 km (50 miles) from the Canadian border, and is a centre of granite quarrying, skiing, and fishing.

Population 8200

Map United States – Lb

Montpellier *France* Busy commercial centre 125 km (78 miles) north-west of MARSEILLES. It trades in wine, and manufactures electronic equipment and computers. Its botanical garden, dating from 1593, is the oldest in France.

Population 208 000

Map France – Ee

Montreal (Montréal) *Canada* Canada's second largest city, after TORONTO, and one of the world's

largest French-speaking cities (although English is also spoken). Settled by the French in 1642, two-thirds of its population are Roman Catholics of French origin today.

Montreal gets its name from Mount Royal, 231 m (758 ft) high, which dominates the city. The city stands at the junction of the St Lawrence and Ottawa rivers and, though it is 1600 km (1000 miles) from the ocean, is a major port.

The heart of the city is an island, but development has spread northwards to LAVAL and the North Shore, and south and eastwards across the St Lawrence. Montreal has four universities, a major financial centre, and industries producing electrical goods, aircraft, railway equipment, clothing, and chemicals.

Vieux Montreal, the old part of the city near the harbour, has been declared a historic area. Other attractions for the visitor include large, temperature-controlled underground shopping areas, Mary Queen of the World Cathedral, Notre Dame church, the 1976 Olympic stadium and a variety of museums and parks.

Population (city) 2 862 290; (metropolitan area) 3 127 240
Map Canada – Hd

Mont-Saint-Michel *France* Vast granite rock, 78 m (256 ft) high and 900 m (2953 ft) around its base, set amid sandbanks just off the Normandy coast 35 km (22 miles) east of Saint-Malo. It is topped by a 10th to 13th-century Benedictine abbey on a site where St Michael the Archangel is said to have appeared in AD 708.

Around the base are medieval walls and towers below a village of close-packed houses. There is a panoramic view of the bay from the walls. A causeway leads to the rock through partially reclaimed marshland. The abbey, whose spire rises 150 m (492 ft) above sea level, is visited by more than 700 000 tourists each year.

Population 72
Map France – Cb

Montserrat *Spain* Mountain 1236 m (4055 ft) high and about 50 km (30 miles) north-west of the Mediterranean port of BARCELONA. On its foothill is a 10th-century Benedictine monastery, where pilgrims come to worship at a black wooden statue, known as Santa Imagen, or more familiarly as the 'Black Madonna', the patron saint of Catalonia.

Map Spain – Fb

▲ **MONTREAL SKYLINE The city of Montreal rises across the St Lawrence River. The river port is actually 1600 km (1000 miles) from the open sea and is only able to operate all the year round because icebreakers keep the waterway free of solid ice.**

Montserrat *West Indies* British Crown Colony in the Leeward Islands, south-west of ANTIGUA. It was reached in 1493 by Christopher Columbus, who named it after the Santa Maria de Montserrat monastery in Spain. British settlers under Sir Thomas Warner colonised the island in 1632; 20 years later, a large contingent of Irish settlers arrived from neighbouring St Kitts. Partly because of the Irish connection and partly because of the vivid colour of its tropical landscape, it is sometimes called the Emerald Isle. It became internally self-governing in 1960. The small town of PLYMOUTH is the capital.

Montserrat is a small rugged island covering 102 km² (39 sq miles) and made up of the remains of three groups of volcanoes, the highest of which, Chance Peak, rises to 914 m (2999 ft). About a quarter of the land is cultivated, and vegetables and fruit are sold to neighbouring islands, particularly Antigua. Nine in 10 of the islanders are descended from African slaves brought to the island to work on the sugar plantations that were abandoned in the 19th century.

Some Sea Island cotton is grown and woven locally into a high quality cloth which is sold to tourists or exported. Cotton was once the island's principal industry. More recently, some light industry has moved to the island, attracted by tax concessions, and tourists now come in increasing numbers, drawn by the island's lush scenery and relaxed atmosphere.

Population 12 600
Map Caribbean – Cb

Monza *Italy* Cathedral town 12 km (8 miles) north-east of MILAN. Today its manufactures include textiles and carpets; the Italian Formula One Grand Prix motor race is held here each year.
Population 122 500
Map Italy – Bb

moor (moorland) Broad tract of open land, often elevated but poorly drained, that has patches of heather, coarse grass, bracken or similar vegetation, or acid peat bogs.

Moose Jaw *Canada* Town situated in south Saskatchewan, 70 km (43 miles) west of REGINA, which serves a large farming area. It has meat-packaging and dairy plants, flour and timber mills, grain storage and stockyard facilities.
Population 33 590
Map Canada – Ec

Mopti *Mali* Town 460 km (285 miles) north-east of the capital, BAMAKO, on the Bani River near its meeting place with the Niger. It is built on three islands joined by dykes and is sometimes known as 'the Venice of Mali'. Its market deals in cattle and fish, and is a centre for the river trade with the Ivory Coast.
Population 54 000
Map Mali –Bb

Moquegua *Peru* Southern department stretching inland from the Pacific Ocean. Copper is mined at Cuajone, and avocados, olives and wine are exported. The department capital, also called Moquegua, stands on the Río Moquegua, 75 km (47 miles) north-east of its port, Ilo. It was founded in 1541 and has many splendid Spanish colonial houses along its winding, cobblestoned streets, which have survived several earthquakes.
Population (department) 134 100; (city) 31 500
Map Peru – Bb

moraine Accumulation of rocks, stones or other debris, carried and deposited by a glacier or ice sheet. See GLACIATION.

Morang *Nepal* See BIRATNAGAR

Morat *Switzerland* See MURTEN

Morava *Czech Republic* See MORAVIA

Morava (March) *Yugoslavia* (*Serbia and Montenegro*) Main Serbian river system with three major arms. The Western (Zapadna) Morava, 218 km (135 miles) long, rises in the west Serbian mountains. It flows eastwards to join the 318 km (198 mile) long Southern (Juzna)

Morava at Stalac, forming the Great (Velika) Morava. The Great Morava flows some 221 km (137 miles) north to join the Danube near the city of Smederevo.

With the Axios (Vardar) River, the Morava valleys have always provided easy access between the Aegean coast and the Danube plains, and major road and rail routes follow their paths. Orchards and tobacco fields surround Vranje in the upper Southern Morava Valley. Lower down the river's course, below the Grdelica Gorge, lie the industrial towns of Leskovac and Nis surrounded by fertile farmlands growing wheat and maize. The Western Morava has carved the spectacular Ovcarsko-Kablarka Gorge.
Map Yugoslavia – Cb

Moravia (Morava; Mähren) *Czech Republic* Region in the south-east, between Bohemia in the west, Sudety in the north and the Carpathians in the east, and crossed by the Morava (March) River. Its main products include cereals, sugar beet, fruit, vegetables, wine, and *slivovitz* (plum brandy) – the Czech national drink. The region consists of South and North Moravia and covers 26 095 km² (10 073 sq miles). It has deposits of coal and small reserves of oil and gas. Its industrial towns include BRNO and OSTRAVA.
Map Czech Republic – Cb

Moray *United Kingdom* See GRAMPIAN; HIGHLAND

Moray Firth *United Kingdom* Scottish inlet of the NORTH SEA with INVERNESS at its head. With Inverness being the centre of the Highlands region, and with LOCH NESS and CALLODEN nearby, the area attracts many visitors.
Map United Kingdom – Db

Mordovia *Russia* See SARANSK

Morelia *Mexico* Market town, and capital of the south-western state of MICHOACAN, set in scenic mountains. It was founded in 1541 by Mexico's first viceroy, Antonio de Mendoza, who named it Valladolid after his home town in Spain. In 1828 the town was renamed after the leader of Mexican independence, José María Morelos. Morelia is noted for its 18th-century aqueduct, Baroque cathedral, courtyards and rose-tinted, carved façades.
Population 489 760
Map Mexico – Bc

Morelos *Mexico* Small state lying directly south of the Federal District around MEXICO CITY. It was originally the territory of the Tlahuicas. At Oaxtepec resort near the state capital of CUERNAVACA, the Aztec emperor Montezuma I (1466-1520) had his botanical gardens, which have now been converted into a public recreation centre.

Many wealthy Mexicans have weekend homes in Morelos. The main crops grown here are sugar cane, cereals, rice, coffee, fruit and vegetables for the markets of Mexico City.
Population 1 195 000
Map Mexico – Cc

Morena, Sierra *Spain* Mountain range in the south-west, dividing the Meseta Central from the Guadalquivir Valley. Its highest peak is Cerro Angulosa, at 1301 m (4267 ft).
Map Spain – Cc

Morioka *Japan* City in north-east HONSHU Island, producing traditional cast-iron tea kettles and dyed cotton fabrics. It is the northern terminus for the *shinkansen*, the 210 km/h (130 mph) 'bullet train' from Tokyo, which lies 460 km (285 miles) to the south.
Population 235 000
Map Japan – Dc

Morocco See p 462

Morogoro *Tanzania* Regional capital 175 km (110 miles) west of DAR ES SALAAM. It is the centre of Tanzania's sisal industry. Its other products include tobacco and sugar.
Population 90 000
Map Tanzania – Ba

Morondava *Madagascar* Port on the Mozambique Channel, some 375 km (235 miles) south-west of the capital, ANTANANARIVO. Morondava exports farm produce, including beans, maize, rice and raffia. It has food processing and woodworking industries.
Map Madagascar – Ab

Moroni *Comoros* Capital of the island group, set on the west coast of NJAZIDJA. It is a fascinating mixture of modern government buildings and old Arab houses and mosques, of broad squares and winding alleys.
Population 24 000
Map Madagascar – Aa

Morphou *Cyprus* North-western town near the bay of the same name. It is at the centre of an area growing citrus fruits, strawberries and tulips. The Church of St Mamas is dedicated to the Cypriot saint who lived in Byzantine times. Mamas is revered here for having refused to pay taxes, and for further showing his contempt for authority by taming a lion and riding on it into the presence of the governor, who was so impressed that he exempted Mamas from taxation for life.
Population 10 180
Map Cyprus – Ab

Morris Jesup, Cape *Greenland* Most northerly point of the mainland of Greenland. Just north of the cape lies the world's northernmost land, a 30 m (100 ft) wide speck called the Islet of Oodaaq, off Peary Land. It is 706 km (439 miles) from the north pole and the sea around it is frozen for most of the year.
Map Arctic – a

mortlake An OXBOW lake.

Moscow (Moskva) *Russia* Capital of Russia and of the Russian Federation – and until the collapse of the Soviet Union, the city from which all power flowed in one of the world's most centralised nations.

It is the home of the glittering and elegant Bolshoi Opera and Ballet Theatre, of the special department stores which in the past were open only to foreigners and high Communist Party officials, of thinly stocked state shops where housewives queue, of forest suburbs dotted with luxury weekend cottages, or *dachas*, and of bleak workers' suburbs lined with high-rise blocks of flats.

The city straddles the Moskva River, after which it is named. Since the time of Ivan the

▲ DOMED CATHEDRAL The Cathedral
of St Basil the Blessed, in Red Square,
Moscow, is characterised by its onion-
shaped domes. In October 1990 the first
service was held here since 1924.

Great (1462-1505), it has been the cultural and
educational capital of the Russian people. Even
after 1712, when Peter the Great (1672-1725)
moved his court and the country's government
to his new capital of St Petersburg (called
Leningrad in Soviet days) on the Baltic, the tsars
were still crowned in Moscow's Kremlin. In

1918, the leaders of the Russian Revolution
restored Moscow as the capital.

The city is first mentioned in Russian chronicles in 1147. In 1156, the Russian prince Yuri
Dolgoruky built a wooden *kremlin* here – the
word means 'town fortress', and the city that
grew around it prospered at the hub of the river
routes across Russia. By 1400, it was already
the capital of a thriving principality, often known
as Muscovy, founded in 1263 by Alexander
Nevsky (1220-63). Having survived Tatar raids,
it gradually absorbed neighbouring principalities, such as Kiyev and Novgorod. Grand Duke

Ivan III (Ivan the Great) consolidated the state in
1480, finally defeated the Tatars and made himself tsar of all the Russias. However, the city of
Moscow was captured by Tatars in 1571; in 1610
it was partly destroyed by Poles. In the 19th and
20th centuries, it survived its two greatest sieges
damaged but unconquered: the siege by the
French general Napoleon in 1812 and by Adolf
Hitler's German army in 1941.

Moscow grew rapidly as the country's centralised administrative centre after 1918; large-
scale industrial growth took place in the 1930s,
with steel, machinery and vehicle plants being
built. Postwar rebuilding since the 1940s, new
suburbs of workers' apartments, and industries
producing chemicals, paper, textiles, processed
foods, furniture and electronics pushed the city's
limits out to a ring motorway which is 110 km
(68 miles) long. In 1969, Moscow had 6.6 million inhabitants; 20 years later its population
numbered almost 9 million. The city's limits
have now outgrown the ring motorway – today
there are more than 13 million people living in
Greater Moscow.

The heart of the city is still the Kremlin – the
symbol of Russian power and authority, with
former royal palaces and the glittering onion
domes of its cathedrals rising above red-brick
walls beside Red Square.

The Kremlin's walls, which overlook the
Moskva River, are 2.3 km (nearly 1.5 miles)
long, with 19 towers. They enclose Kremlin
Square, a magnificent assemblage of 15th- to
18th-century Russian architecture, whose four
cathedrals and church are reminders that
Moscow has been the headquarters of the
Russian Orthodox Church since 1326.

The 19th-century Grand Palace, for years the
headquarters of the Supreme Soviet of the USSR,
and the 18th-century Council of Ministers building – where the revolutionary leader Vladimir
Lenin (1870-1924) lived and worked from 1918
to 1924 – also stand on the square.

Away from the Kremlin, avenues radiate from
the centre and are crossed by ring roads. Moscow
has more than 120 museums and galleries,
35 theatres and concert halls, including the
Moscow Arts Theatre, and 75 cinemas. Farther
out, in the nearby Lenin Hills, is the Olympic
village built for the 1980 Olympic Games. Still
a centre of world politics, and Europe's largest
city, Moscow has many technical colleges and
research institutes, as well as the well-known
Lomonossow University.

Population (city) 8 802 000; (metropolitan area)
13 150 000
Map Russia – Ec

Moselle (Mosel) *Western Europe* River, some
550 km (340 miles) long, that rises in the Vosges
Mountains of eastern France. It flows north
through Lorraine, passing Nancy and Metz, to the
German border. From here it forms part of the
frontier between Luxembourg and Germany, and
then descends through a narrow, winding valley
to join the Rhine at Koblenz.

Locks between Nancy and Koblenz allow 1500-
tonne barges to use the river. The sides of its valley,
particularly in Germany, are planted with vines
that produce the fruity Moselle white wine. Its
chief tributaries are the Meurthe and Saar.
Map France – Gb

Moskva *Russia* See MOSCOW

moss Name given to an acid bog, especially in northern England and Scotland, such as Chat Moss in Lancashire.

Mossamedes *Angola* See NAMIBE

Mossel Bay (Mosselbaai) *South Africa* Fishing port and resort on the south coast of the Western Cape, first recorded by the Portuguese navigator Bartolomeu Díaz in 1488. The town is the land base for the exploitation of an offshore gas field.

Population 59 170
Map South Africa – Bc

Most *Czech Republic* Coal-mining town in north-west Bohemia, 15 km (10 miles) from the German border. In 1975, its late Gothic cathedral was moved, complete, to a nearby site to make way for new mining operations.

Population 62 100
Map Czech Republic – Aa

Mosta *Malta* Small yet bustling town in the middle of the island. It remains dominated by the huge dome of its 19th-century neoclassical church, built largely with donations from local emigrants. On display near the altar is the now defused bomb which crashed through the roof during a crowded service in 1942 and miraculously failed to explode.

Population 9030
Map Malta – Db

Mostaganem (Mestghanem) *Algeria* Trading and fishing port, and administrative capital of the Mostaganem Department – a rich agricultural area also designated for industrial development. The town lies some 80 km (50 miles) north-east of ORAN, to which it is linked by rail. Founded in the 11th century, it reached the height of its prosperity under the Turks in the 16th century. Today, it exports citrus fruits, wine, wool, grain and vegetables. An 11th-century citadel is the feature of the older part of town.

Population 101 600
Map Algeria – Ba

Mostar *Bosnia-Herzegovina* Former capital city of Herzegovina, 80 km (50 miles) south-west of Sarajevo. Surrounded by barren, pitted mountains, the city, whose name translates as 'old bridge', is spread around a fine 16th-century bridge, which spanned the blue Neretva River by a single arch, 29 m (95 ft) wide and rising 20 m (66 ft) above the river.

In 1993, the bridge was destroyed during fighting between Muslims and Croats. Much of the city was also badly damaged, including its Muslim architecture and many of its factories. Mostar – Herzegovina's largest city – with an ancient bazaar and handicrafts quarter, is the centre of a small but rich basin producing high quality tobaccos, fine *blatina* and *zilavka* wines, fruit, walnuts, and vegetables.

Before the civil war, the city's aluminium plant, which used local bauxite, and its cotton mills were powered by the nearby Jablanica hydroelectric dam, opened in 1957.

Population 126 100
Map Bosnia-Herzegovina – Bb

Mosul (Al Mawsil) *Iraq* City on the Tigris River, situated some 350 km (220 miles) north-west of the capital, BAGHDAD. It is the second largest Iraqi city, after Baghdad, and in the Middle Ages was the centre of Indo-European trade. Today, Mosul has an oil refinery and is a market place for the surrounding oil-producing and farming region, trading in grain, fruit, livestock, and wool. Its other industries include flour milling and tanning hides. A mosque in the city is dedicated to the Biblical prophet, Jonah, who is said to be buried here. The remains of the ancient Assyrian capital, NINEVEH, lie on the opposite bank of the Tigris.

Population 1 571 000
Map Iraq – Ba

Motherwell *United Kingdom* See STRATHCLYDE

Moulay Idriss *Morocco* Town, 50 km (30 miles) west of FÈS, lying on the edge of rocky hills among olive groves and cacti. It contains the tomb of Moulay Idriss I, the founder of the Islamic kingdom of Morocco, who died in AD 791. A *moussem* (religious festival) takes place each September in his memory. The 1800-year-old remains of VOLUBILIS are nearby.

Map Morocco – Ba

moulin Circular sinkhole in the surface of a glacier, or in the bedrock beneath, worn by swirling meltwater falling down a crevasse.

Moulins *France* Town and former capital of the Duchy of Bourbon, about 145 km (90 miles) north-west of LYONS. Moulins produces leather, hosiery and furniture, and is an important road and railway junction.

Population 25 550
Map France – Ec

Moulmein *Myanmar (Burma)* See MAWLAMYINE

Mount Gambier *Australia* City in the extreme south-east of South Australia, 380 km (240 miles) south-east of ADELAIDE. It is situated at the foot of an extinct volcano which has three crater lakes. The largest, Blue Lake, is 500 m (1640 ft) across and 80 m (260 ft) deep; almost circular in shape, the lake is a brilliant blue in summer but fades to a dull grey in winter.

Population 21 150
Map Australia – Gf

Mount Hagen *Papua New Guinea* Capital of Western Highlands province, standing at more than 1700 m (5600 ft) above sea-level and situated 1000 km (620 miles) north-west of PORT MORESBY. It takes its name from a mountain peak, an extinct volcano which rises to some 3777 m (12 392 ft), 24 km (15 miles) to the north-west.

Mount Hagen was first visited by Europeans in 1933, but had long been a trading centre for locals. The Europeans established it as a small patrol post around an airstrip; it now has factories for processing locally grown coffee, tea and pyrethrum (used as an insecticide). A show and 'sing-sing' are held here every second year; every other year they are held at GOROKA in the Eastern Highlands.

Population 17 860
Map Papua New Guinea – Ba

Mount Isa *Australia* Mining town in harsh, remote country in Queensland, about 1500 km (930 miles) north-west of BRISBANE. It is a major mining centre, producing some 80 per cent of Australia's copper in the 1990s, as well as large quantities of lead, zinc and silver.

Population 23 670
Map Australia – Fc

Mount Rainier National Park *USA* Area of 976 km² (377 sq miles) in the Cascade Range in western Washington State, 110 km (70 miles) north of the Oregon border. The park centres on Mount Rainier, a 4392 m (continued on p 464)

▼ PAPUAN FESTIVAL Highlanders in Papua New Guinea don their finery and gather every second year at Mount Hagen for a 'sing-sing' – two days of parades, gift exchanges, ritual pig killings, feasting and dancing.

Morocco

THIS GATEWAY TO NORTH AFRICA STANDS HEIR TO THE MAGNIFICENT MOORISH KINGDOM

Strategically placed at the western entrance to the Mediterranean Sea, Morocco is but a step away from Europe and yet is unmistakably Middle Eastern in flavour. It has long attracted Europeans, from traders to tourists. But the traffic has been far from one-way.

Across the Strait of GIBRALTAR – a mere 13 km (8 miles) wide at one point – the Moors of North Africa poured into Spain in the 8th century, bringing their religion, their architecture and their culture – proud monuments that still stand today. From 1085, the rulers of the newly established Berber Kingdom of Morocco played an important part in the Muslim domination of Spain.

It was a domination that did not last, of course. In time Spain, and particularly France, came to rule Morocco, although never so completely as some other North African countries.

Today Morocco, one of only three kingdoms remaining in Africa, is a conservative state that tries to maintain close links with other Arab countries yet is firmly pro-Western. However, its claim to the sparsely populated neighbouring territory of WESTERN SAHARA has embroiled it in years of costly warfare.

With a land area of some 446 550 km² (17 000 sq miles), Morocco has a 1400 km (875 mile) coastline on the Atlantic but only 400 km (250 miles) on the Mediterranean. It is a land of vivid contrasts: high rugged mountains which are snow-capped in winter, the fierce arid SAHARA, and the green and cultivated Atlantic and Mediterranean coasts. The country is split from south-west to north-east by the ATLAS MOUNTAINS, which continue across Algeria and into Tunisia. To the north of the mountains are fertile coastal plains along the Atlantic; to the south, the desert with scattered oases.

The RIF MOUNTAINS in the north rise to 2456 m (8058 ft). They are separated from the Middle Atlas by the TAZA Gap. Farther south is the High Atlas range. In the south-west, the Anti Atlas Mountains form an elevated rim at the edge of the Sahara plateau.

The north has a pleasant Mediterranean climate of hot, dry summers with mainly clear blue skies and mild, moist winters. Average temperatures at TANGIER range from 11°C (52°F) in winter to 29°C (84°F) in summer, with 810 mm (32 in) of rain annually; in MARRAKECH the winters are warmer and the summers even hotter. In winter, snow often covers the High Atlas Mountains.

The natural vegetation varies from Mediterranean scrub in the north to mountain forests in the Atlas ranges; about 12 per cent of the country is wooded. In spring the mountain slopes become carpeted with beautiful alpine wild flowers. On the drier southern flanks of the Atlas, sparse vegetation merges into desert. Among the wild animals are macaque monkeys – the so-called Barbary apes.

Morocco contains more than 28 million people. Most are something of a mixture. Arabs, Berbers and, later, Negroes, French and Spanish have intermingled. The Berbers are descendants of the original inhabitants who were in this area thousands of years ago and at one time controlled all the land between Egypt and Morocco.

Arabic is the official language, spoken by most people except the Berbers, whose ancient tongue still predominates in the central mountains. French is the business language of the cities, but Spanish is also spoken. The foreign population of some 60 000 includes small French, Spanish and Italian communities. Nearly all Moroccans are Muslims, but they are more tolerant than in many Arab states.

Morocco is still mainly a farming country – 50 per cent of people work on the land and agricultural products account for 30 per cent of export earnings. Wheat, barley and maize are the main food crops and Morocco is one of the world's chief exporters of citrus fruits. Olives, grapes and many varieties of vegetables are grown. However, the standard of living varies greatly and there is much poverty, with one out of three people living below the poverty line.

Most of the colonial settlers who left were replaced by Moroccan landowners, but the big estates were not broken up – they became, instead, huge agri-business operations. Most farmers have only small plots from which they try to feed themselves and their families. As production has not kept pace with the rising population, much food has to be imported.

In the past, Morocco's main wealth was derived from phosphates, used for fertiliser manufacture. The country is the third-largest producer in the world after the United States and Russia and its reserves are the world's largest – about half of the world's known total. But with environmentalists' pressures to use organic fertilisers instead of phosphates and fluctuating prices on the international market, Morocco has embarked on a campaign to lessen its reliance on this product.

Morocco is Africa's largest producer of silver. It has other minerals as well – iron, zinc, manganese, lead, cobalt and the only sizeable coal stocks in North Africa – but these are still comparatively undeveloped.

The country's mixed economy is one of the most complex and advanced in Africa. Morocco is almost self-sufficient in textiles and it cans most of its fruit. It has car assembly plants and soap and cement factories, while its sea fishing industry is the largest in Africa. The country is now looking across the Mediterranean to the European Union as a future trade partner; but so far talks have led only to an extension of bilateral trade.

Tourism, a major source of revenue, is growing in importance. In 1989, the country attracted 3.5 million visitors from Europe and other Arab countries. It is in a unique position to offer unspoilt beaches close to Europe – and never far from cities with Middle Eastern charm.

ANCIENT KINGDOM

Like other countries of North Africa, Morocco came in turn under Carthage, Rome and then the Vandals. Byzantium ruled it in the 6th century. Around AD 685, Arabs from the east swept across the country, bringing the Muslim religion. From 711, many Moroccans joined the Muslim, or Moorish, invasion of Spain, led by Tariq ibn Zaid, which stamped the Arab influence on Andalucía and even briefly reached south-western France. The name Moor itself is derived from Mauri, a North African Berber people. The Almoravid dynasty, which ruled from Marrakech in the 11th century, established a Berber kingdom

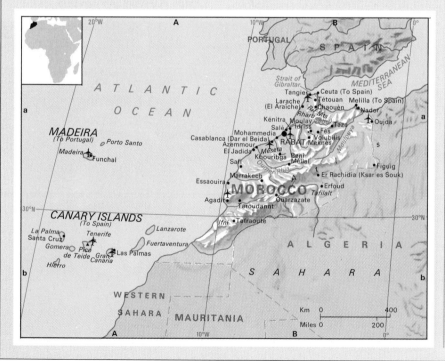

stretching from central Spain to Senegal. But conflict arose between the Arabs and Berbers. The Christian kingdoms of northern Spain and Portugal took advantage of the divisions in Moorish power and began to obtain footholds in northern Morocco. In 1492, the Christian armies drove the Moors from their last Spanish stronghold, GRANADA, and proceeded to conquer several cities on the Moroccan coast. The Moors had regained most of them by 1578, but CEUTA and MELILLA remain Spanish enclaves to this day.

For more than 300 years, Morocco remained independent. Then, in the late 19th century, the country became the centre of European imperialist rivalries. France and Spain wanted to divide it, and Germany also tried to intervene. These rivalries were resolved eventually and most of the country became a French protectorate in 1912. Spain was left in possession of a small protectorate in the north; Tangier became an international zone.

By and large, the following 45 years were successful and prosperous ones for Morocco. The French administration in particular, led until 1925 by General Louis Lyautey, did much to modernise the country. The sultan remained, but with little power, and the only serious opposition came from Berber groups in the hills. Moroccan troops fought alongside French in the First World War and with General Francisco Franco in the Spanish Civil War (which began with a revolt at Spanish army bases in Morocco).

In the Second World War, Allied forces landed in Morocco and Algeria in 1942 to drive the Germans out of North Africa and Allied leaders – Winston Churchill, Franklin Roosevelt and Charles de Gaulle – met at CASABLANCA in January 1943. After the war, Sultan Sidi Mohammed ben Youssef became the focus of a popular and unified independence movement. He was ordered to leave the country but Moroccans brought him back and he was able to negotiate Morocco's independence in 1956 without the bitter guerrilla war that convulsed Algeria.

Tangier was returned to Morocco but Spain retained the two small enclaves at Ceuta and Melilla as well as the Mediterranean islands of Al Hoceimas, Chafarinas and Albóran; Spanish IFNI was ceded to Morocco in 1969. The sultan became king in 1957 and was succeeded in 1961 by his son, Hassan II, an astute though conservative political leader. Violent unrest had to be quashed in 1965 and Hassan declared a state of emergency which lasted until 1971. In 1972, the restrictions on political parties were lifted. A 1992 referendum resulted in a new constitution, with the number of parliamentary seats increased from 306 to 333.

In contrast to most other constitutional monarchies, the Moroccan king wields important executive powers. King Hassan has at times run the country without parliament. In 1975 he organised the peaceful 'green march' in which 350 000 unarmed Moroccans entered the mineral-rich Spanish Sahara (now Western Sahara). Spain, the former colonial power, agreed that the country should be divided between Morocco and Mauritania. But the indigenous Sahrawis, with Algerian support,

▲ DYER'S MARKET Skeins of wool hang out to dry after dyeing in a market in Marrakech. Morocco's *souks* (markets or bazaars) enchant nearly 2 million tourists who visit the country each year.

formed a movement for self-government, the Polisario Front, and in 1976 they proclaimed a new independent Saharan Arab Democratic Republic (SADR). By 1982, the SADR was recognised by more than 70 countries worldwide – but Morocco still remained in the region.

In 1976, Polisario began a guerrilla war against the occupation by Morocco and Mauritania. Mauritania gave up its claim to the region in 1979. Morocco then annexed the whole of Western Sahara. The war, though nationally popular, led to bad relations with Algeria – until 1992, when Algeria agreed to stop supporting Polisario. Although Moroccan troops remain stationed in the Western Sahara, Polisario and Morocco reached a UN-negotiated ceasefire in 1991 and the two parties met for talks for the first time in mid-1993.

It is fairly well off in comparison with other African states, but still faces enormous social problems. Not least of these is its rapid population growth – between 1970 and 1990 the population grew by 10 million. Education is compulsory between the ages of 7 and 13 but half the adult population is illiterate. However, newspaper and radio campaigns have started

to improve the literacy rate. Unemployment is high, causing Moroccans to flood to the cities and enter Europe illegally to find work.

MOROCCO AT A GLANCE	
Official name Kingdom of Morocco	
Area 446 550 km² (172 373 sq miles)	
Population 28 500 000 **Per km²** 64 **(Per sq mile** 165)	
Capital Rabat	
Government Constitutional monarchy	
Currency 1 dirham = 100 centimes	
Languages Arabic (official), Berber dialects, French, Spanish	
Religion Muslim (almost 99%)	
Climate Warm on coast, hot inland. Average temperature in Rabat ranges from 12.5°C (55°F) in January to 23°C (73°F) in August	
Land use Grazing 47%, cultivation 21%, forest 12%, other 20%	
Main primary products Cereals, pulses, citrus fruits, vegetables, olives, grapes, sheep, goats, poultry, almonds, dates, timber, fish; phosphates, iron, lead, manganese	
Major industries Agriculture, food processing, textiles, leather goods, cement, wine, fertilisers, forestry, fishing, mining	
Main exports Phosphates, metal ores, citrus fruit, vegetables, wine, tinned food, carpets, textiles, cotton	
Annual income per head (US$) 1040	
Population growth (per thous/yr) 26	
Life expectancy (yrs) Male 65 **Female** 69	

(14 410 ft) volcanic peak from which a spectacular series of glaciers radiate. The area has dense forests, wide expanses of alpine meadow and hosts of wild flowers.
Map United States – Ba

mountain Mass of land with steep slopes projecting well above its immediate surroundings. A mountain differs from a PLATEAU in that the area of its summit is much less than that of its base, and it generally has steep sides with a large proportion of bare rock. There is no precise definition of the height which distinguishes a hill from a mountain. (For the world's highest mountains, see below.)

mountain grasslands Area lying above the treeline in most mountainous regions of the world where a short growing season and moisture from melting snow or summer rain support the growth of grasses. See PAMIR and ALP.

Mourne Mountains *United Kingdom* Range in County Down, Northern Ireland, rising to 852 m (2795 ft) at Slieve Donard. The mountains have several reservoirs and are a popular tourist attraction. As described in an old song, the Mourne Mountains 'sweep down to the sea' close to the seaside resort of Newcastle.
Map United Kingdom – Bc

Moxico *Angola* Farming province whose capital, LUENA, is 810 km (503 miles) south-east of the national capital, LUANDA. Copper is mined in the province, but the economy was disrupted by civil war from 1975 into the 1990s. Subsistence crops include maize, millet, beans, and groundnuts.
Population 213 000
Map Angola – Bb

Mozambique See p 468

Mozambique Channel *Mozambique/Madagascar* Strait between Madagascar and the African mainland, which is more than 1600 km (1000 miles) long and between 400 and 970 km (250 and 600 miles) wide. An important route for East African shipping, the warm Mozambique Current flows through it.
Map Mozambique – Ab

Mtskheta *Georgia* See T'BILISI

Mtwara *Tanzania* Regional capital and port lying 40 km (25 miles) north of the Mozambican border. It handles local exports such as cashew nuts and other crops, and imports such as petroleum products for its hinterland.
Population 60 000
Map Tanzania – Cb

Mu *Myanmar (Burma)* Central river draining the area between the Ayeyarwady (Irrawaddy) and Chindwinn valleys.
Map Myanmar – Bb

Muara *Brunei* Deep-water port on the promontory north of the Limbang estuary, some 16 km (10 miles) north-east of BANDAR SERI BEGAWAN. It handles most of Brunei's imports, including foodstuffs, consumer goods and machinery.
Map Brunei – Ba

mud Apart from describing wet, sticky, soft earth, geologists use the term to define a fine-grained sediment of clay and silt, which usually contains a high percentage of water, found, for instance, where lakes and rivers have been drained.

mud flat Land covered at high tide and exposed at low tide.

mud pot Pool of boiling mud, usually sulphurous and often brightly coloured, found in areas of volcanic activity, such as in the Yellowstone National Park in Wyoming in the United States.

Muda *Malaysia* River rising in the hills of north-west Malaysia on the border with Thailand. It flows about 320 km (200 miles) to the south-west, entering the Andaman Sea near Sungei Petani. The river has been dammed near the town of Alor Setar to irrigate the Kedah plain, enabling farmers to grow two rice crops a year.
Map Malaysia – Ba

mudstone Fine-grained, compact rock similar to SHALE, but without its capacity to cleave.

Mudumalai *India* Game reserve in the Nilgiri Hills of the south, 560 km (348 miles) south-west of the east coast city of MADRAS.
Map India – Be

Mufulira *Zambia* Copper-mining town 300 km (185 miles) north of the capital, LUSAKA. In 1970 a flood at the mine killed 89 men.
Population 206 000
Map Zambia – Bb

ON THE ROOF OF THE WORLD

The Himalayas, 2400 km (1500 miles) long and varying between 200 and 400 km (125 and 250 miles) wide, contain many of the world's highest peaks – 96 of them are more than 7315 m (24 000 ft) high. The massive range started to form 38 million years ago when the Indian Plate of the earth's crust collided with and moved under the Eurasian Plate and forced the land upwards. This movement continues still.

THE WORLD'S HIGHEST MOUNTAINS

Mountain	Location	Height (m/ft)
Everest	Nepal/China	8846/29 022
K2 (Godwin Austen)	Pakistan/China	8611/28 250
Kangchenjunga	Nepal/India	8598/28 298
Lhotse I	Nepal/China	8501/27 890
Makalu	Nepal/China	8481/27 824
Dhaulagiri	Nepal	8172/26 810

HIGHEST POINTS OF THE CONTINENTS

Continent	Mountain	Height (m/ft)
Africa	Kilimanjaro (Tanzania)	5895/19 340
Antarctica	Vinson Massif	4897/16 066
Asia	Everest (Nepal/China)	8846/29 022
Australia	Kosciusko (New South Wales)	2228/7310
Europe (excluding Russia)	Mont Blanc (France)	4807/15 770
Europe (including Russia west of the Urals)	Elbrus (Russia)	5642/18 510
North America	McKinley (Alaska)	6194/20 320
South America	Aconcagua (Argentina)	6959/22 831

Muharraq Island *Bahrain* Second largest of the 33 islands which make up Bahrain State, lying 3 km (2 miles) north-east of Bahrain Island. It is connected to Al Manamah on the main island by a causeway. In the old quarter of the main town, Al Muharraq, many of the houses are built around enclosed courtyards.

They date from a pearl-fishing boom in the 18th and 19th centuries. Al Muharraq is now a centre for building dhows, single-masted coastal sailing vessels. Its main buildings include the Bahrain Museum and a fine wooden mosque.

Population 74 250
Map Bahrain – Ba

Muineachàn *Ireland* See MONAGHAN

Mukallah (Al Mukallah) *Yemen* Port lying 480 km (298 miles) east of ADEN. It is a beautiful town with white buildings reflecting an unusual architectural mix of Yemeni, Arabic and Indian influences. The centre of Yemen's fishing industry, it has a fish-canning factory and a fish-freezing plant.

Population 100 000
Map Yemen – Ab

mulga scrub Type of scrub consisting mainly of a species of acacia known as mulga, which covers parts of central and western Australia.

Mülheim an der Ruhr *Germany* Port on the Ruhr River, about halfway between DUISBURG and ESSEN in the Ruhr industrial area. About 100 years ago, Mülheim was the second largest mining and steel town in the Ruhr, but its last coal mine closed in 1966. Today it has steel and engineering plants.

Population 177 700
Map Germany – Bc

Mulhouse *France* Industrial city near the border with Germany and Switzerland. It was a cotton town in the 19th century. Today its industries include engineering, car assembly, and chemicals and textile industries.

Population 224 000
Map France – Gc

Mull *United Kingdom* Mountainous island off the west coast of Scotland, covering 925 km² (357 sq miles) and separated from the mainland by the Sound of Mull. Its highest point is Ben More, at 966 m (3169 ft). Its economy centres on sheep, fish and tourism. The chief settlement is Tobermory.

The islands of Iona and Staffa lie off the western shore. Staffa's series of extraordinary basalt columns and caverns, including Fingal's Cave, inspired the Mendelssohn overture *The Hebrides*.

Population 2700
Map United Kingdom – Cb

▲ **MOUNTAIN PANORAMA The view from 7900 m (about 25 900 ft) on the south-west face of Mount Everest shows the West Ridge of the mountain (right foreground) and Pumon lying below (centre).**

Multan *Pakistan* Industrial town in the Indus Valley, 300 km (185 miles) south-west of LAHORE and described by some of its inhabitants as 'the Manchester of Pakistan' because of its cotton industry. It was captured by the 4th-century BC Macedonian conqueror Alexander the Great, and was a major Hindu shrine before Muslims destroyed its temples in AD 950; they were restored in 1138.

Population 950 000
Map Pakistan – Cb

Munich (München) *Germany* Capital of Bavaria, situated 60 km (37 miles) from the Austrian border. Munich, site of the 1972 Olympic Games, has expanded in the past century to become the country's third largest city.

It is also one of Germany's most industrialised cities, making cars, machinery, chemicals, pharmaceuticals, optometrics, cigarettes, and beer. At its heart it keeps some of its character as the city of the Wittelsbachs, the princely family that ruled Bavaria as dukes and, later, kings from 1180 to 1918. Their palace, built in Renaissance style

465

during the 17th century and rebuilt after it was heavily damaged during the Second World War, has its own theatre, where the 18th-century Austrian composer Wolfgang Amadeus Mozart first produced his opera *Idomeneo* which he wrote in Munich. The Baroque Nymphenburg Palace, summer home of the Wittelsbach family, lies north-east of the city. It was built in the 17th century and has a French-style park.

Munich was the birthplace of Adolf Hitler's Nazi movement in the 1920s. About 120 km (75 miles) to the south-east is the Alpine resort town of Berchtesgaden; Hitler had a mountain retreat above the town, and it was here, in September 1938, that he met the British Prime Minister Neville Chamberlain to discuss peace in Europe before signing the Munich agreement.

Munich, which calls itself 'the Beer Capital of the World', holds an international beer festival in October ('the *Oktoberfest'*) each year, and its *Fasching* (the German shrovetide carnival) is held each year in January-February. Its oldest and largest beer cellar is the Hofbräuhaus, founded by Duke Wilhelm V in 1589, and which is now the state brewery. The city is also noted for its 15th-century cathedral, university, art galleries – particularly the Pinakothek, with paintings by Dürer, Rembrandt, El Greco and Rubens – and theatres, including the Bavarian State Opera.
Population 1 229 000
Map Germany – Dd

Münster *Germany* Port on the Dortmund-Ems Canal, 70 km (45 miles) north-east of ESSEN. Built on the bank of the River Aa, it is the cultural and commercial centre of Westphalia. Its university, in a Baroque palace, is one of Germany's largest; its 13th-century cathedral is the largest in Westphalia.
Population 259 400
Map Germany – Bc

Munster (Cuige Mumhan) *Ireland* One of Ireland's four historic provinces – the other three being Connaught, Leinster and Ulster – in the south-west of the country, with Cork as its chief town.
Population 1 009 530
Map Ireland – Bb

Muqdisho *Somalia* See MOGADISHU

Murchison Cataracts *Malawi* See SHIRE

Murchison Falls *Uganda* See KABALEGA FALLS

Murcia *Spain* University city 180 km (110 miles) south of Valencia. It trades in cereals, almonds and citrus fruits, and manufactures silk, textiles, flour, aluminium and leather goods. A former Moorish capital, it is now a mining centre for iron and zinc. The surrounding province of the same name has one of Spain's largest areas under irrigation and produces fruit.
Population (city) 328 100; (province) 1 045 600
Map Spain – Ed

Murmansk *Russia* The world's largest city north of the Arctic Circle. It was founded in 1916 along the Murman-St Petersburg railway line on an estuary of the Kola River, 50 km (30 miles) from the Barents Sea in the extreme north-west of Russia. Thanks to the warm ocean current, known as the North Atlantic Drift, it is ice-free all year

round and is a major fishing and cargo port, as well as an industrial centre manufacturing ships, beer, canned fish, and timber. Nearby Polyarnyy is an important naval base.
Population 472 000
Map Russia – Eb

Murray *Australia* One of the country's longest rivers, with a length of 2570 km (1600 miles). With its extensive system of tributaries, including the Lachlan and Murrumbidgee, the Murray and its main tributary, the DARLING (which in turn has many tributaries), drain an area of some 910 000 km² (351 000 sq miles) – one-eighth of the entire country, and an area nearly twice the size of France.

The Murray rises in the Snowy Mountains in the GREAT DIVIDING RANGE and forms the state boundary between New South Wales and Victoria before entering South Australia and flowing into the sea south of Adelaide. In some places, restored paddle steamers re-create for tourists the Murray's transport heyday. The waters of both the Murray and the Murrumbidgee are widely used for irrigation.
Map Australia – Ge

Murree Hills *Pakistan* Beautiful forested foothills of the Himalayas situated north-east of ISLAMABAD. They rise to 2150 m (7050 ft), and attract many tourists. The small town of Murree was originally a small British hill station.
Map Pakistan – Da

Murshidabad *India* Eastern town some 180 km (110 miles) north of CALCUTTA. It was founded in 1704 as the capital of the Muslim Kingdom of Murshidabad (Bengal). In 1757, when the British took over Bengal soon after the Battle of PLASSEY nearby, the town became the administrative centre of British India and the seat of the governor-

▲ MUNICH'S TRIUMPH Leopoldstrasse runs in front of the Triumphal Arch at the heart of Munich. Beyond it is Ludwigstrasse, and on the left is the church spire of Ludwigskirche. To the right are the two tall cupolas of Frauenkirche.

general from 1773 to 1834. The magnificent palace of the nawabs (Muslim rulers) survives.
Population 356 000
Map India – Dc

Murten (Morat) *Switzerland* Fortified town on Lake Murten, about 25 km (15 miles) west of the capital, BERNE. It still has its 12th to 15th-century town walls and its 13th-century castle.
Population 4790
Map Switzerland – Aa

Mururoa *French Polynesia* Atoll situated in the TUAMOTU Archipelago, 1500 km (932 miles) south-east of Tahiti. It has been used by the French as a nuclear weapons testing ground since the 1960s. An underground test in 1985 sparked worldwide controversy when French secret service agents sank the Greenpeace conservation organisation's ship *Rainbow Warrior* in New Zealand (killing one of its crew members) as it prepared to sail to the atoll to lead a protest.
Map Pacific Ocean – Gd

Musandam *Oman* Spectacular mountainous peninsula jutting into the Strait of HORMUZ and separating The Gulf from the Gulf of Oman and the Arabian Sea.

Though part of Oman, it is divided from the rest of the sultanate by a 70 km (43 mile) wide stretch of the United Arab Emirates (UAE). The isolated settlements on the peninsula are mainly fishing villages. Musandam's main town is

Khasab, which has a shallow harbour and a 17th-century Portuguese fort. Until recently the peninsula had few proper roads, and most local travelling was done by sea. However, in 1982 a road was built, linking Khasab to Musandam's two other towns, Bakha and Bayah, and on into the UAE.
Map Oman – Aa

Musay'id *Qatar* See UMM SAID

Muscat *Oman* Capital, main port and seat of government of the sultanate in the east of the Arabian Peninsula. The neighbouring town of MUTTRAH is the trading and residential area while Ruwi, a few kilometres inland, is the commercial centre of the area.

Muscat, with its fine natural harbour ensconced between sheer face rock, grew in influence and power during the Middle Ages, when it gained importance on the trade route between Arabia, India and Africa. It was a thriving city when the Portuguese captured it in 1507, remaining until the middle of the 17th century.

The city walls and gates built during the Portuguese occupation, as well as the many houses in the old town, with beautifully carved doors, are still mostly intact. The city was held by the Persians from 1650 to 1741, after which it became the capital of an independent sultanate. In 1798, the sultan signed a treaty establishing economic and political links with Britain, which lasted until the late 1960s.

Muscat's places of interest include several fine old Arab *bait*s (houses) such as Bait Nadir (now a museum), Bait Graiza (converted to state apartments), Bait Fransa (occupied by the French embassy), Bait Zawawi (the United States embassy), as well as three 16th-century forts and the royal marble palace, built on the seafront in the 1970s.
Population (including Muttrah) 380 000
Map Oman – Aa

muscovite Most common form of MICA, which is found in acid IGNEOUS and METAMORPHIC ROCKS. It ranges in colour from colourless or pale brown to green, has a vitreous lustre and is used as an insulator. It is also called 'isinglass'.

Muscovy *Russia* See MOSCOW

muskeg Swamp largely filled with sphagnum moss in the subarctic zone and coniferous forest of northern Canada.

Mustique *St Vincent* Privately owned island in the Grenadines forming part of the state of St Vincent. It has been developed as an exclusive tourist resort and provides goods and services for yachting people in the area.
Population 200
Map Caribbean – Cc

Mutare *Zimbabwe* Town, formerly called Umtali, 220 km (137 miles) south-east of the capital, HARARE, near the Mozambican border. It is the main town of the Manicaland province and the gateway to the tourist resorts of the eastern highlands. The town lies in the foothills of the Vuma and Nyanga mountains, which rise in places to 2500 m (8200 ft) and more.
Population 69 600
Map Zimbabwe – Ca

Muttra *India* See MATHURA

Muttrah (Matrah) *Oman* New port and residential area of greater MUSCAT and the sultanate's trading centre, situated immediately to the north of the capital. For centuries, Muttrah remained a separate commercial town, and until 1929 it was connected to Muscat by only donkey tracks. The terminus of the caravan routes from the interior, it traded in pearls, dates and fruit. Today the town's numerous souks (bazaars) deal in silver, gold, cloth and grains. The seafront is dominated by a 16th-century fort and there are some fine mosques and old Arab merchant houses.

Just to the north-west, in Muttrah Bay, is Mina Qaboos, the sultanate's main modern port. Completed in 1974, it has nine deep-water berths for handling large container ships. Nearby, the old fishing harbour and fish market still thrive.
Population (including Muscat) 380 000
Map Oman – Aa

Mwali (Mohéli) *Comoros* Smallest island of the Comoros group, covering 290 km² (112 sq miles). Before independence, Mwali was known as Mohéli. Its main town is Foboni.
Population 25 000
Map Madagascar – Aa

Mwanza *Tanzania* Regional capital and the country's second largest town, after DAR ES SALAAM. Mwanza stands on the south shore of Lake Victoria; it makes and exports textiles.
Population 170 000
Map Tanzania – Ba

Mweru, Lake *Zaire/Zambia* Lake lying 930 m (3051 ft) above sea level, on the border between south-east Zaire and Zambia. It covers about 4920 km² (1900 sq miles). The lake is fed by the Luapula River from the south and drained by the Luvua River, a tributary of the LUALABA. Swamps border its south and east shores.
Map Zaire – Bb

My Lai (Song My) *Vietnam* Village in former South Vietnam, which was the scene of a massacre in March 1968, during the Vietnam War (1965-75). American troops killed more than 100 villagers, including women and children, during a 'search and destroy' operation against Viet Cong guerrillas. An American officer was later found guilty of murder, and the case helped to turn public opinion against the war. A memorial has been erected on the site.
Map Vietnam – Bb

Myanaung *Myanmar (Burma)* Small town on the Ayeyarwady (Irrawaddy) River in southern Myanmar, about 180 km (110 miles) north-west of Yangon (Rangoon). Natural gas was found locally in the late 1960s, and the town grew rapidly during the 1980s as a producer of chemicals synthesised from the raw material, which is also used to fuel power stations.
Population 183 200
Map Myanmar – Bc

Myanmar (Burma) See p 470

Mycenae (Mikínes) *Greece* Excavated, fortified city of Homer's king Agamemnon on a hill some 30 km (18 miles) south-west of CORINTH on the Peloponnese. Mycenae was occupied from 1580 to 1100 BC, when it was destroyed by the invading Dorians, and was the heart of a rich trading kingdom now referred to as the Mycenaean civilisation, named after the site.

It was excavated from 1874 by the German archaeologist Heinrich Schliemann, who uncovered a palace, temples, a citadel and the remarkable royal chamber tombs – including the largest, the Tomb of Agamemnon – containing gold death masks. Most of the finds date from 1600-1100 BC, and many are now in the Athens museum.
Map Greece – Cc

Myitkyina *Myanmar (Burma)* Town and capital of the state of Kachin in Myanmar's far north – and the end of the railway line which runs up the centre of the country from Yangon (Rangoon). It lies in a largely agricultural area of rice and sugar cane fields, and has a sugar refinery.
Population 103 300
Map Myanmar – Cb

Mykolayiv (Nikolayev) *Ukraine* Black Sea port 110 km (70 miles) north-east of Odesa. Founded as a Russian naval base at the mouth of the Southern Bug River in 1788, it soon became a major grain-exporting harbour. Today it has grown into a shipbuilding centre as well.
Population 503 000
Map Ukraine – Cb

Mymensingh *Bangladesh* Chief town of Mymensingh District on the old course of the Brahmaputra River. The district covers 9668 km² (3733 sq miles) and produces jute.
Population (town) 186 000; (district) 6 568 000;
Map Bangladesh – Cb

Mysore *India* Southern industrial city 470 km (290 miles) south-west of MADRAS. It was a capital of the wealthy, progressive Hindu state of Mysore from the late 16th century. The palace of its ruling maharajas, a spectacular pavilion of granite, incorporates cast-iron pillars specially made for it in Glasgow (Scotland). A summer palace, now a hotel, stands on a cliff overlooking the city. Mysore is an educational and commercial centre, and produces textiles, fine silk saris, leather goods, chemicals and cigarettes.
Population 479 000
Map India – Be

Mytilene *Greece* See LESBOS

Mývatn *Iceland* Lake covering some 37 km² (14 sq miles) in the north-east of the island. Sometimes called 'Mosquito Lake', it is popular with scientists and tourists because of the hot springs and volcanic craters around its shores, and its bird life. Europe's principal deposits of the mineral diatomite are found in the area.
Map Iceland – Ca

M'Zab *Algeria* Stony, barren region on a plateau to the north-east of the Great Western Erg in the northern Sahara. The area was settled at the beginning of the 11th century by the Mozabites, an austere Muslim sect of Berber origin, who dug wells, planted date palms and built five towns united in a confederation. These towns, of which the best known is GHARDAIA, are built on small hills and look like coloured pyramids. They have markets and are centres of traditional arts crafts.
Map Algeria – Ba

Mozambique

A COUNTRY STRUGGLES TOWARDS RECOVERY FROM A DEVASTATING CIVIL WAR, DROUGHT AND FLOODS THAT HAVE BROUGHT IT TO THE BRINK OF STARVATION

When independence came to Mozambique on 25 June 1975, its Portuguese former masters walked out, leaving a black African nation with hardly anyone trained to run the farms and factories.

The Africans had received virtually no education – 99 per cent of them were illiterate. Those who could read and write were mainly youngsters, and in desperate moves to avert economic chaos young people were appointed to key positions. While a 19-year-old boy became head of a cotton research centre, an 18-year-old was helping to run one of the country's largest secondary schools.

Mozambique was the last European colony in south-east Africa. It is largely a tropical country with mainly savannah vegetation ranging from open grassland to thickly wooded areas. A coastal plain covers most of the southern and central territory, giving way to the western highlands and to a plateau in the north which includes the Nyasa Highlands.

The Portuguese explorer Vasco da Gama visited the territory in 1498 and the first Portuguese settlers arrived early in the 16th century, conducting a lucrative trade from their coastal towns with countries in India and East Asia.

Little was done to improve the lot of the Africans – or, indeed, to develop the country. As late as 1970, when most colonial powers had freed their foreign territories, the Portuguese still dominated Mozambique. About 250 000 Europeans held virtually every professional, managerial and technical post in a country with 8 million Africans.

African discontent had already flared into fierce guerrilla warfare waged from 1964 by the Mozambique Liberation Front (Frelimo). This brought some moves towards economic development, but it was the revolution of 1974 in Portugal that gave the final push to independence and the creation of a one-party Marxist state under President Samora Machel (1933-86). A great exodus of Portuguese, who feared that the Africans would exact revenge for past oppression, followed; by 1977 all non-Mozambicans had left.

The new government was expecting communist countries to help Mozambique, but in fact the former Soviet Union and Bulgaria showed more interest in the possibility of exploiting the country's natural resources than in aiding development. Mozambique has a variety of mineral resources. As well as large reserves of coal, tantalite and titanium there are offshore gas, bauxite, gold, iron ore, tin, graphite and fluorite.

STARTING FROM SCRATCH

The new Mozambique had to start from scratch, training managers and technical staff while somehow keeping agriculture and industry going. The people, having struggled hard for independence, were enthusiastic and determined. Mozambicans with any ability were soon helping to run the new society but the lack of expertise caused many problems. Some of the worst were in agriculture, which produces mainly cotton, groundnuts and cashew nuts – of which the country is the world's largest exporter – and grain subsistence crops.

In the chaos following independence, the government decided that huge state farms were the answer to a desperate shortage of food. These farms failed to produce efficiently, partly because they were short of tractors, and fertilisers and pesticides, and partly because of mismanagement. But the new farm managers eventually began to learn the necessary skills and the small peasant farms began to make a greater contribution in a country where more than 85 per cent of the people are farmers.

But as agriculture began to emerge from crisis, nature took a hand, and the country was struck by the drought that affected much of Africa in the early 1980s. Drought was interrupted by the other extreme of weather in 1984 when Maputo Province in the south suffered its worst floods in 30 years. More than 350 000 farmers lost their crops. Added to the years of drought, this disaster brought famine, which led to more than 100 000 deaths. In 1992, the region was stricken with yet another drought.

Generally, conditions are fairly good for agriculture. Rivers water the plains where most of the cultivation and population are concentrated and rain falls – normerly as thunder showers – from November to April. It is heaviest in the highlands, where altitude gives cooler temperatures; the north-west highlands receive more than 1500 mm (59 in) a year. Temperatures on the central coast average 30°C (86°F) all the year.

Industry as well as trade brought ideological problems for the Marxist government, which was forced to intervene in hundreds of manufacturing firms abandoned by the Portuguese in efforts to keep them producing. But although it was opposed to capitalism, it was not equipped to take over large sectors of the economy. Over the years of crisis, it learnt to take a lenient view of private capital in some areas.

Industries include food processing, beverages, and tobacco, as well as chemicals, petroleum products, textiles, and building materials – although the industrial sector has now been operating at only 20-40 per cent of its 1981 level.

THE ENEMY WITHIN

Since 1983, the Frelimo government has implemented some economic reform. Peasants are allowed to produce food to meet demand and though agricultural output is still only at 75 per cent of what it was in the early 1980s, the food situation has improved somewhat. But the country is still heavily dependent upon foreign assistance – and its record of civil strife has put off most foreign investors.

While coping with natural and human-made disasters, the Mozambican government also had to fight right-wing guerrillas, known as the Mozambique National Resistance (MNR) – also called Renamo. Support for the MNR came from neighbouring South Africa, which was angered by Frelimo's backing for the African National Congress (ANC) in South Africa, and from Rhodesia (now Zimbabwe) during the period of illegally declared Rhodesian independence from Britain.

The MNR tried to sabotage the new state. They destroyed thousands of shops, blew up railways and power lines running from the hydroelectric CAHORA BASSA DAM on the ZAMBEZI River. Large areas were unsafe because of their activity – it was even dangerous to travel between major cities – and in 1992, when most of the country was dependent on food aid, they hampered its distribution.

A new constitution was approved in 1990, paving the way for multiparty democracy and liberalisation of the economy. In 1992, Renamo and Frelimo fighters signed a UN-negotiated peace accord, most troops were disarmed and the two former enemies were unified into one army. Altogether, it is estimated that 600 000 people died in the civil war, and 4.5 million were made homeless, either fleeing to other parts of the country, or to neighbouring states.

After repeated delays in the UN-supervised peace process, free elections were eventually held in October 1994 when Machel's successor, President Joaquim Alberto Chissano, was returned to power.

THE STRUGGLE CONTINUES

While embroiled in civil war and battling with a wide range of problems, Mozambique has made great progress in some areas. Huge vaccination programmes have led the way to improving health standards. An extensive education programme is replacing the skills taken out of the country by the retreating Portuguese. Today, 33 per cent of the population can read and write – compared with 1 per cent at the time of independence.

The new society has also greatly improved the standing of its women, who traditionally held low status despite their crucial roles as agricultural workers, water carriers, mothers and cooks. New laws were introduced to stop these injustices.

Women's participation in political and non-agricultural economic roles has also increased, although strange double standards still operate. For instance, some women have been expelled from cities because they were single and it was assumed that they must therefore be prostitutes. An important political official lost her job on the grounds of adultery, although she had already separated from her husband; much worse behaviour from male officials rarely causes any public action.

For women, Mozambique's slogan of 'A luta continua' (the struggle continues) is doubly significant. The struggle has continued: now

▼ SCHOOLING IN THE SUN Outdoor 'classrooms' are not uncommon in Mozambique, and these young students squat informally in the shade of a tree to learn their numerals – written in Arabic and Roman.

lost skills must be replaced and the economy has to get up and get going.

The most urgent priorities have been to rehabilitate returning refugees and boost agricultural production. Highly dependent on international food aid, the country was referred to, by the World Bank, as the world's poorest country in 1992. Tete Province in the north-west, for instance, was almost entirely deserted during the war – although it had once been the granary of the country. The refugees have returned to a pillaged countryside, where most of their villages are burnt and the land strewn with landmines – an estimated 1 million landmines are still in the ground. Some refugees have been equipped with farming implements, but it will take some years before production will reach anywhere near pre-war level.

However, significant progress has been made to boost the country's economy. There has been an influx of foreign expertise to help industry and education.

Also, tourism is again picking up, although slowly. The government has encouraged investment in this sector. Hotels have been rebuilt and other tourist facilities have been restored. An increasing number of foreign visitors are coming to the national parks – which are rich in zebra, buffalo, rhinoceros, elephant, giraffe, and lion – and to the beaches. The country has 2 470 km (1535 miles) of beautiful, island-speckled coastline.

MOZAMBIQUE AT A GLANCE	
Official name Republic of Mozambique	
Area 801 590 km² (309 423 sq miles)	
Population 16 341 800 **Per km²** 20 **(Per sq mile** 53)	
Capital Maputo	
Government Republic	
Currency 1 metical = 100 centavos	
Languages Portuguese (official), indigenous languages	
Religions Indigenous beliefs (60%), Christian (30%), Muslim (10%)	
Climate Humid, tropical; rainy season from November to April. Average temperature in Maputo ranges from 18°C (64°F) in July to 26°C (79°F) in February	
Land use Cultivation 4%, grazing 56%, forest and woodlands 20%, other 20%	
Main primary products Cotton, cashew nuts, tea, sugar, cassava, cereals, bananas, sisal, groundnuts, coconuts, rice. sunflower seeds; coal	
Major industries Agriculture, textiles, chemicals (fertiliser, soap and paints), food processing, petroleum products, cement, mining, steel, engineering, asbestos, glass docks and railways	
Main exports Cashew nuts, shrimps, cotton, sugar, copra, citrus fruits	
Annual income per head (US$) 60	
Population growth (per thous/yr) 28	
Life expectancy (yrs) **Male** 46 **Female** 50	

Myanmar

IN FORMER BURMA, DETENTION IS THE PENALTY A NOBEL PEACE PRIZE WINNER PAYS FOR HER EFFORTS TO ESTABLISH DEMOCRACY

On 18 June 1989, the country's name was changed from Burma to the Union of Myanmar. At the same time the English spellings of many places were changed. The capital, Rangoon, for example, became YANGON, and the Burmans became Bamars. These changes reflected the new government's isolationist sentiments – and hinted at many other changes going on behind the scenes.

Not much remains of the Burma under British colonial rule when the author Rudyard Kipling wrote a ballad about the road to MANDALAY. That was a road that ran through the mainstream of Burmese village life. It was also to reach that road that the British 14th Army, nicknamed 'the Forgotten Army', fought the Japanese in the Second World War.

From as early as 1612, the British made progressive inroads into Burma and, after three border wars, annexed it in 1886. It was made a province of India in 1923. In 1937, it was separated from India to become a Crown Colony and was given an elected assembly. During the Second World War, Burmese forces under the nationalist leader General Aung San, fought first for the Japanese, who had promised them independence, then for the British.

Independence came in 1948, but Aung San was assassinated in 1947 before he could take office as Burma's first prime minister. Burma left the Commonwealth; a military dictatorship followed in 1962 and a one-party state in 1974.

The one party was the Burma Socialist Programme Party, headed until 1981 by Ne Win. The party's blend of socialism, Buddhism and isolationism made Burma one of the ten poorest countries in the world. This was despite great natural resources in teak forests, coal, oil, gas, iron, tin, tungsten, lead, zinc, silver, copper, nickel, rubies, sapphires and emeralds. Rigid economic controls caused great shortages of all goods and there was a thriving black market.

When Ne Win stepped down, to be replaced by San Yu, more liberal policies were implemented and foreign aid and investment were encouraged. As a result, the economy began to grow much more rapidly, with gross domestic product (GDP) expanding at 6 to 7 per cent a year. But the shortages and black market remained, and when promises of an improved economy were only partly fulfilled, the armed forces under the leadership of General Saw Maung assumed power.

After mass pro-democracy protests in mid-1988, multiparty elections were eventually held in 1990, but the military refused to accept the results, which gave victory to the National League for Democracy. A period of ruthless suppression of opposition groups ensued. The military have also had a terrifying record of human rights violations – ethnic minorities, for instance, were dealt with mercilessly – in response to which, the United States imposed trade sanctions in August 1991.

Opposition leader Aung San Suu Kyi, who was placed under house arrest in 1989, was still in detention in late 1994 despite repeated world-wide appeals for her release. In 1991 she was awarded the Nobel Peace Prize for her efforts to establish democracy in her country.

HORSESHOE OF MOUNTAINS

Myanmar, with an area of 678 500 km² (261 968 sq miles), is the second largest country in South-east Asia after Indonesia. The heartland of the country is the Irriwaddy Valley (now called Ayeyarwady), home of the Bamars who comprise 70 per cent of the country's 43.5 million population. Tribal people live in the foothills of a giant horseshoe of mountains which forms a barrier between Myanmar and its neighbours.

The Ayeyarwady, 2000 km (1243 miles) long, flows south through gorges strewn with rapids to MANDALAY and Yangon. Its delta, extending into the ANDAMAN SEA, provides ideal swampy land for rice cultivation. Until 1964, Burma was the world's largest rice exporter, and rice still accounts for nearly half of the country's export earnings. Rice and salted fish are the staple foods.

Farther upstream, the valley is sheltered from the monsoon by the surrounding hills, and in this 'dry zone' cotton, sesame and beans are interspersed with irrigated ricelands, particularly around Mandalay. At KYAUKSE an irrigation system ordered by King Anawratha, who established Burmese supremacy over the delta 900 years ago, is still in use.

King Anawratha built his ceremonial capital, Pagan (now renamed Bagan), in the 'dry

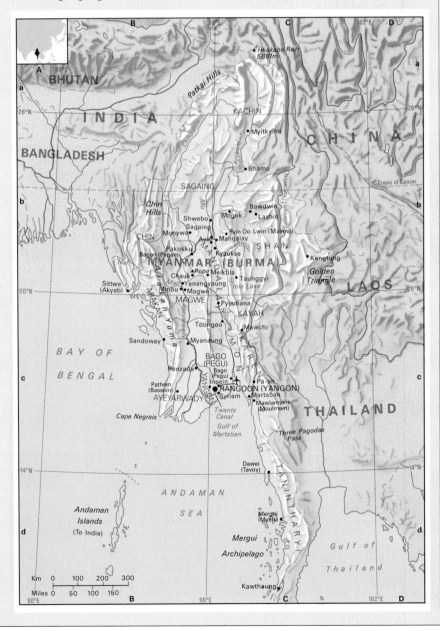

1993 and the DAB announced it was willing to enter peace talks, but various private armies still exist. In the north and east, feudal warlords battle with the authorities to protect their profitable and deadly trade in opium, grown in the infamous GOLDEN TRIANGLE, which supplies 70 per cent of the world's production. In north-western Arakan, thousands of Muslims were driven into neighbouring Bangladesh early in 1992; but many were repatriated in 1993.

The Burmese Way to Socialism was launched as a do-it-yourself economic system which rejected foreign aid despite the ravages of the Second World War. Agriculture, vital to Burma's prosperity, was government-planned and the price of rice was controlled. There was a foreign policy of neutrality; after the coup in 1962, widespread nationalisation followed, in mining, forestry, manufacturing and retailing.

The strategy was unsuccessful. Farmers were reluctant to invest in crops for inadequate returns and held back their produce from the government. Industries were unable to get enough raw materials from agriculture. The rebel movements denied the government the timber and minerals of the uplands areas.

Rice exports fell from 1.7 million tonnes a year to 170 000 tonnes, leaving the country with insufficient capital to restore much of its war damages. In 1975, when food shortages brought riots, Burma was forced to seek aid. Not only has foreign aid brought improvements, under more liberal economic policies there has been some growth. Prices for crops have been increased and high-yielding varieties of rice have raised production by 60 per cent in eight years. New oil and gas finds may succeed in making the country self-sufficient in these commodities.

The government has been putting more money into the mining and timber industries for development. However, with the amount of felling done in bumper years such as 1990, and no reforestation, all teak will be exhausted by the year 2005. Education has been reorganised and adult literacy is put at 81 per cent. But the people of Myanmar still have little freedom.

▲ 'CITY OF A THOUSAND TEMPLES' Buddhist temples surround the old Burmese capital of Pagan (now renamed Bagan), founded in AD 849. The city, once a leading centre of Buddhist learning, never recovered from sacking by Mongols in 1287.

zone' in the 11th century. It was the Pagan dynasty which introduced Buddhism to the country. By 1550 the Burmese controlled the present territory and by 1800 had extended their influence into Assam and Thailand. The last Burmese king, Thibaw, was displaced by the British in 1885.

The 'dry zone' was also the home of Burma Oil, containing the small oil fields which once made Burma the second largest producer in the British Empire. Today, it still holds enough oil and natural gas for most of the country's needs. Near Mandalay, the CHINDWINN River joins the Ayeyarwady; to the north are found most of Myanmar's precious stones.

About half of Myanmar is covered by trees. As well as teak forests, there are rubber plantations, some oil palms, and other woods. In the north, mountains rise to 5881 m (19 294 ft) at Hkakabo Razi, and great rivers such as the Chindwinn and Ayeyarwady rise to form torrents, flowing through deep gorges. In the west, the Chin and NAGA hills (reaching 3826 m, 12 511 ft) extend towards the isolated Arakan coast, bordering the Bay of Bengal.

In the east, the hills between Myanmar and Thailand rise to more than 3000 m (10 000 ft),

broadening northwards into the SHAN plateau. There is still no way of crossing these hills other than by foot, mule, horse or elephant; the infamous 'Death Railway', built across the THREE PAGODAS pass by prisoners of war of the Japanese, has been abandoned.

Yangon has 2.5 million people; the division of Yangon has more than 4 million. Arriving by air, the traveller's first sight of the capital may be the sun glinting on the golden pinnacle of the Shwe Dagon Pagoda, 100 m (330 ft) high. There are countless monasteries and shrines: Yangon claims to have the world's largest reclining Buddha, 78 m (256 ft) long and 17 m (56 ft) high.

Away from Yangon, the road to Mandalay remains in part a narrow, rutted track. Mandalay contains the Arakan Pagoda and the ruins of King Thibaw's finely carved and gilded palace, destroyed by shellfire in 1945.

There are about 10 main ethnic groups in the country. The largest, after the Bamars (68 per cent of the population) are Shans (7 per cent), followed by the hill-dwelling Karens (now renamed Kayins, 6 per cent), and the Kachins (2.3 per cent). After independence in 1948, the hill tribes objected to domination by the Burmans and some tribesmen took to arms.

The Karen National Union and its Liberation Army, which has waged a guerrilla war since 1965, is the dominant member of the Democratic Alliance of Burma (DAB), and comprises about 20 ethnic and pro-democracy factions. Kachin guerrillas signed a peace agreement with the military government in late

MYANMAR AT A GLANCE

Official name Union of Myanmar (Burma)
Area 678 500 km² (261 969 sq miles)
Population 43 456 000 **Per km²** 64 **(Per sq mile)** 165)
Capital Yangon (Rangoon)
Government Republic (ruled by military council)
Currency 1 kyat = 100 pyas
Languages Burmese, indigenous
Religion Buddhist (89%), Christian (4%), Muslim (4%), other (3%)
Climate Tropical; monsoon from May to September. Average temperature in Yangon ranges from 25°C (77°F) in January to 31°C (88°F) in April
Land use Cultivation 16%, meadows and pastures 1%, forest and woodlands 49%, other 34%
Main exports Rice, teak, rubber, crude oil, metal ores, opium (illegal)
Annual income per head (US$) 250
Population growth (per thous/yr) 18
Life expectancy (yrs) Male 57 **Female** 62

Na Cruacha Dubha *Ireland* See MACGILLI-CUDDY'S REEKS

Naarden *Netherlands* Former seaport on the south side of the IJSSELMEER, 19 km (12 miles) south-east of Amsterdam. The town lies within fortifications built in a perfect star shape by the French military engineer Sébastien de Vauban in the late 17th century. Modern developments, including chemical plants, are outside the moat in neighbouring Bussum.

Population 16 520
Map Netherlands – Ba

Nabadwip *India* See NAVADWIP

Naberezhnyye Chelny (Brezhnev) *Russia* Industrial town on the Kama River, 900 km (nearly 560 miles) east of Moscow. From 1984 until the collapse of the Soviet Union it was called Brezhnev after President Leonid Brezhnev (1906-82) to commemorate its rapid growth during his leadership into one of the world's largest truck-manufacturing centres. In 1991 it reverted to its original name.

Population 510 000
Map Russia – Gc

Nabeul *Tunisia* Administrative capital of Cape BON, situated 61 km (38 miles) south-east of TUNIS on the Gulf of Hammamet. It is a tourist centre with hotels, restaurants, night clubs, a theatre and good beaches. The manufacture of ceramics has been the main industry since Roman times. Hundreds of potteries still use traditional designs. Other local crafts include embroidery, needlework and lace making. Perfume distilleries working from ancient formulae use, among other natural ingredients, orange-tree blossom and the flowers of jasmine.

Population 50 000
Map Tunisia – Ba

Nablus (Nabulus) *West Bank* Largest WEST BANK town and home of the university of Al-Abjah, situated 48 km (30 miles) north of JERUSALEM. The market and centre of an important agricultural region that has fallen into decline under Israeli occupation, it manufactures oil and soap.

Population 44 000
Map Jordan – Aa

Nacala *Mozambique* Natural harbour considered to be the finest on Africa's east coast, about 100 km (62 miles) north of Mozambique Island. The port was constructed as an outlet for landlocked Malawi in the 1970s and is connected to that country by rail via Nampula; however, war has retarded development.

Map Mozambique – Bb

Nadi *Fiji* See LAUTOKA

Nadia *India* See NAVADWIP

Nador *Morocco* Port on the Mediterranean coast, 13 km (8 miles) south of MELILLA, handling iron ore from the Beni bu Ifrur mines. The town trades in livestock (especially sheep), cereals and fruits, and is the site of an integrated iron and steel complex.

Population 63 000
Map Morocco – Ba

Næstved *Denmark* Town in the south of ZEALAND, linked to the sea by a canal. It originated around a Benedictine monastery founded in 1135, near the site of which now stands Herlufsholm, a private residential school. The town developed as a market centre for southern Zealand. Its industries now include paper, glass, pottery, textiles and fishing. It has the remains of a medieval hospital (the Helligåndshuset, now a museum), as well as a Gothic church (St Morten's), and St Peder's (St Peter's) Church, which is the only surviving monastic building.

Population 45 300
Map Denmark – Bb

Nag Hammadi *Egypt* Industrial town on the Nile, 480 km (nearly 300 miles) south of the capital, CAIRO. Situated on Egypt's main north-south railway line, Nag Hammadi handles produce from the oases to the west, and has textile, sugar and aluminium industries. It is best known for the discovery in the nearby Jabal al-Tarif Mountain in 1945 (two years before the finding of the Dead Sea Scrolls at Qumram) of 13 papyrus volumes of gnostic scriptures dating from AD 130.

Population 19 800
Map Egypt – Cc

Naga *Philippines* See BICOL

Naga Hills *Myanmar (Burma)/India* Ranges rising to 3826 m (12 553 ft) on the border between India and Myanmar. Much of their tropical forest cover has been replaced by dense stands of bamboo as a result of centuries of shifting cultivation by the hill peoples.

Map India – Eb

Nagaland *India* Eastern state with a mainly Naga population and English as well as Naga as its official languages. It covers 16 579 km² (6399 sq miles) in the southern NAGA HILLS on the border with Myanmar. Missionary activity has been intense in the area and as a result, about 45 per cent of the inhabitants are Christians. The people have little affinity with predominantly Hindu India, and some have in recent years waged a separatist war in densely wooded hills.

Population 1 209 550
Map India – Eb

Nagano *Japan* City in central HONSHU Island, 175 km (110 miles) north-west of Tokyo, and deep in the Japan Alps. Its industries include electrical equipment, foodstuffs and printing. A revered Buddha statue in the city's temple of Zenkoji is put on public display once every seven years; 1994 was a display year.

Population 347 000
Map Japan – Cc

Nagasaki *Japan* Port and industrial city in the west of KYUSHU Island, specialising in shipbuilding and heavy engineering and dominated by the Mitsubishi company. It was the target of the second atomic bomb, which was dropped on 9 August 1945 – three days after the HIROSHIMA bomb. The blast killed more than 50 000 people and destroyed 40 per cent of the city, though many old buildings survived.

Nagasaki has been a centre of Japanese Christianity since the 16th century, and the Oura Cathedral, built by a French missionary in 1865, is the oldest Gothic church building in Japan; the nearby Site of the Martyrdom of the 26 Saints is a mural dedicated to the Christians who were crucified here after their religion had been banned in the 17th century. For centuries the town was virtually the only point of contact between Japan and the outside world. However, from 1637 all foreigners (with the exception of Dutch traders) were banned from entering Japan. Chinese merchants were allowed to trade through the small island of Dejima in Nagasaki Bay. The port was not reopened until 1859.

Population 445 900
Map Japan – Ad

Nagoya *Japan* Manufacturing and commercial city and port on HONSHU Island, 270 km (nearly 170 miles) south-west of TOKYO. It was extensively rebuilt after severe wartime bombing and its products today range from cars to china. The Atsuta Shrine in the city houses an ancient sword which is part of the sacred imperial regalia.

Population 2 155 000
Map Japan –Cc

Nagpur *India* Textile-producing, administrative and commercial city on the northern Deccan plateau of central India, 860 km (534 miles) north-east of the west coast city of BOMBAY. The Gonds, an ethnic group who are largely outside the mainstream of Indian (Hindu) life, live in the district, which is noted for its sweet oranges.

Population 3 274 970
Map India – Cc

Nagyalföld *Hungary* See ALFÖLD, GREAT

Nagybecskerek *Yugoslavia* See ZRENJANIN

Nagycenk *Hungary* See SOPRON

Nagyvázsony *Hungary* See BALATONFÜRED

Naha *Japan* Capital and main port of OKINAWA Island. It was once the seat of the Ryukyu dynasty who ruled the Ryukyu Islands from 1429 to 1879. In May and June 1945, it was the scene of a fierce battle between United States and Japanese forces.

Population 305 000
Map Japan – Gf

Nahuel Huapí *Argentina* See BARILOCHE

Nairn *United Kingdom* See HIGHLAND

Nairobi *Kenya* The national capital, once a leisurely colonial town, now a bustling metropolis whose services have been strained by a rapidly growing population. It lies some 440 km (nearly 275 miles) north-west of MOMBASA, on the edge of the highlands at a height of 1700 m (about 5580 ft).

Originally a construction camp on the Uganda Railway in the 1890s, in an unoccupied strip between Kikuyu tribal country and grazing land used by the nomadic Masai people, Nairobi grew quickly, until it came to replace Mombasa as the colonial capital in 1907.

Tourism has grown in recent years to become one of the city's – and Kenya's – chief foreign currency earners. The regional offices of a host of United Nations agencies are located in Nairobi. The City Square, which forms the centre

A DAY IN THE LIFE OF A NAIROBI WATCHMAN

In a long, shed-like room on the outskirts of Nairobi 18-year-old Timbau, who rents the room with four other young Masai, warms a paste of maize meal and chopped cabbage over a wood fire on the earth floor. It is late afternoon, and Timbau works at night in the city as a guard and watchman (or askari) at the suburban home of a middle-class family.

He always eats before he goes to work. But the pattern of the day and even the food he eats still seem strange and new to him. When he lived with the Masai, who herd cattle over the grassy plains of southern Kenya, his diet was mostly milk and a little cow's blood.

His home then was a hut in one of the settlements along the route the Masai travel every year in search of water and grazing. The huts, or *bomas*, are made of mud plastered over a frame of poles, and a group of *bomas* are fenced around with acacia thorn branches, to make a *manyatta*.

Timbau abhors the thought of changes in the way of life of his people, who are under pressure to become settled cultivators. But grazing land is becoming more scarce; some is being fenced off for ranching or farming, and traditional grazing grounds have been made into a national park. Droughts that have devastated so many parts of Africa have killed cattle, so food is short and the Masai have to supplement their diet with maize meal.

If they do not grow crops they have to buy maize in the markets. Besides, their lives as warrior-nomads have been affected by Western culture – and there is now pressure on them to have an education and own Western goods. All these things have to be paid for and it was the need to make money that sent Timbau away from his people and into the city. Nairobi attracts thousands of other migrants from rural Kenya for the same reason. It is the principal industrial centre of Kenya and one of the largest cities in tropical Africa and this is where it is hoped jobs will be found and money earned.

When Timbau arrived in Nairobi he could speak only Masai, but within a week he had learned enough Swahili to buy things in the local market. Still he feels he is an outsider. Other Africans treat him as a primitive because he wears his red cloak, or *shuka* – and especially because he has gaping holes in his ear lobes, made by the traditional decorative plates used by his people. He has now laid aside the jewellery and beads worn by the Masai, but still resists Western clothes. His employers seem content to have a red-robed guard armed with a spear and club in the sentry box at the gate of their home.

As he walks the several kilometres from his shanty-district home to work, Timbau thinks that what he most needs to buy out of his first wage packet is a charcoal stove for his room. After that he can send any spare money home. In a few years, perhaps, he will have earned enough to return to what remains of his old life.

of the city, is lined with government and administrative buildings. Kenyatta Avenue is a busy main road with rows of office blocks and hotels.

Many of the adult Nairobians today are migrants – rural people seeking work in the city's light industries and commerce in the south of Nairobi. Most of these people keep close ties with their rural homes and eventually return to them.

Population 1 500 000
Map Kenya – Cb

Najran *Saudi Arabia* Oasis town situated 220 km (nearly 140 miles) south-east of ABHA. An unusual feature of the town is a number of five- to eight-storey mud houses, some of which have stained-glass windows. The palace of the former emirs in the Najran Valley, is also built of mud. The remains of Ukhdud, a town which flourished in the early centuries of Christianity, lie a few kilometres south of Najran. Parts of its 4 m (13 ft) stone walls are still standing; among its carvings are two intertwined snakes.

Population 47 500
Map Saudi Arabia – Bc

Nakhodka *Russia* Port in the far east of Russia, situated on the Sea of Japan. Built in the 1950s to relieve overcrowded facilities at VLADIVOSTOK, 90 km (55 miles) to the west, it is now the Trans-Siberian Railway terminal for passengers and freight containers going to and from Japan or Hong Kong. Fishing and shipbuilding are its main industries.

Population 163 000
Map Russia – Od

Nakhon Pathom *Thailand* Province and town 60 km (37 miles) west of the capital, BANGKOK. Farmers here grow rice, sugar cane, fruit and vegetables for the capital. Thailand's tallest Buddhist pagoda (127 m, 417 ft) is located in the town (which is regarded as the oldest in the country).

Population (province) 603 700; (town) 211 000
Map Thailand – Bc

Nakhon Phanom *Thailand* Agricultural town on the Laotian border, 630 km (390 miles) northeast of BANGKOK. It lies in a rice-growing province of the same name. During the Vietnam War (1965-75), the United States airforce maintained an airbase here.

Population (town) 124 600; (province) 582 700
Map Thailand –Bb

Nakhon Ratchasima *Thailand* Engineering and agricultural market city 270 km (168 miles) northeast of BANGKOK. It is the gateway to the Khorat plateau of the north-east. The surrounding province of the same name grows maize, cotton, fruits and rice. Some of the most important Khmer monuments in Thailand, notably the 9th-century temple city of Phimai, can be seen in the city.

Population (city) 366 200; (province) 2 360 800
Map Thailand – Bc

Nakhon Sawan *Thailand* Province, and riverside city some 240 km (150 miles) north-west of BANGKOK. Bung Boraphet, the country's largest lake, covering 120 km² (46 sq miles), is situated near the city; it contains many crocodiles. The province's farmers grow mainly maize, cotton, soya beans and sorghum.

Population (province) 1 081 500; (city) 223 300
Map Thailand – Bb

Nakhon Si Thammarat *Thailand* Province and city on the coast of the Gulf of Thailand, 600 km (almost 375 miles) south of BANGKOK. Rubber, tin, tungsten, rice and fruits are the province's chief products. The Wat Phra Mahathat is the most notable of several ancient Buddhist temples that can be seen in the city. Local crafters make traditional Thai puppets out of leather and patterned silver nielloware (silver inlaid with a black alloy).

Population (province) 1 411 970; (city) 102 120
Map Thailand – Ad

Nakuru *Kenya* One of the country's largest towns, 140 km (87 miles) north-west of NAIROBI. It lies on the highest part of the Great Rift Valley floor, just north of Lake Nakuru, a soda lake occupied by vast flocks of flamingos. Nakuru was the business centre of the Kenya Highlands when that region was being developed as farmland by European settlers. Today it is a regional market town with food processing industries.

Population 163 000
Map Kenya – Cb

Nalanda *India* Remains of a massive Buddhist monastic centre or 'university' on the Ganges River plain near the town of Bihar Sharif, 470 km (about 290 miles) north-west of CALCUTTA. It flourished from the 8th to the 12th century, when it had 10 monasteries, three libraries and numerous *stupas* (shrines).

It was already thriving in the 7th century AD when a Chinese pilgrim named Hsuan-tsang visited the centre: it had 10 000 student monks from throughout Asia as well as 1500 teachers, he reported. Nalanda was sacked by Muslims in about 1200. Relics gathered from the ruins – including stone and bronze sculptures – are kept at the site museum.

Population (district) 2 003 310
Map India – Dc

Nal'chik *Russia* City on the northern flank of the Caucasus Mountains. A tourist resort, spa and university town, it is the capital of the Kabardin-Balkar Republic, an autonomous region of the Russian Federal Republic covering 12 500 km² (4800 sq miles). Nearby, the Baksan River valley leads to the twin summits of volcanic Mount ELBRUS which, at 5642 m (18 510 ft), is Europe's highest peak.

Population 237 000
Map Russia – Fd

Namangan *Uzbekistan* Town some 200 km (125 miles) east of TASHKENT. One of the main towns in the fertile, irrigated and densely populated Fergana Valley, it produces cotton textiles, silk, processed foods, wine, and beer.

Population 312 000
Map Uzbekistan – Ca

Namaqualand *Namibia/South Africa* Arid region rich in copper and diamonds, situated along the Atlantic coast, astride the Northern Cape/Namibian border. It is traditionally the home of the Nama people. Today, the term Namaqualand refers to the South African region only. Following the brief rainy season just before spring (September and October), the deserts of Namaqualand are transformed into a blazingly colourful carpet of flowers.

Map South Africa – Ab

Namen *Belgium* See NAMUR

Namib Desert *Namibia* One of the world's bleakest and most ancient deserts, with sand dunes soaring to 275 m (900 ft). It stretches the length of the Namibian coast – about 1900 km (1180 miles) – and on, into southern Angola, and is some 50 to 140 km (30 to 87 miles) wide. It relies almost solely on sea mist for moisture, and many living things inhabiting its hostile environment have developed remarkable survival methods. One species of beetle, for example, stands on its head in the mists so that condensation will drip down its shell to its mouth.

Plants of the Namib include the fantastically shaped *Welwitschia mirabilis*, which uses its frayed leaves, shaped like the tentacles of some strange animal, to collect moisture from the air. The plant, whose two leaves are contantly growing, can live for more than 1000 years.

Map Namibia – Ab

Namibe (Moçamedes; Mossamedes) *Angola* Port lying 700 km (435 miles) south of the national capital, LUANDA. It is the capital of a province of the same name. Founded in the 1840s by the Portuguese, Namibe developed as a fishing and cargo port after the construction early in the 20th century of a railway to the town of Menongue, about 620 km (385 miles) to the east.

Namibe Province has an arid coastal strip below the Serra da Chela Mountains which rise to over 2100 m (6900 ft). Its main crops are sisal, maize, millet and beans.

Population (town) 115 000; (province) 233 150
Map Angola – Ab

Namibia See p 476

Namp'o (Chinnampo) *North Korea* The country's largest port, situated on the Taedong River, 37 km (23 miles) from its mouth on the west coast. The harbour, which was opened to international trade in 1897, now serves the port of the capital, PYONGYANG.
Population 130 000
Map Korea – Bc

Nampula *Mozambique* Trading centre 150 km (93 miles) inland on the railway linking the seaports of Moçambique and Nacala to the interior. It is the capital of the northern agricultural province of the same name, which covers 81 606 km² (31 500 sq miles).
Population (city) 126 100; (province) 2 498 400
Map Mozambique – Ab

Namur (Namen) *Belgium* City situated 60 km (37 miles) south-east of the national capital, BRUSSELS, and itself the capital of Namur Province. Strategically positioned at the confluence of the Meuse (Maas) and Sambre rivers, it was the scene of battles and sieges from the 17th century right up to 1914. Its main industries are food processing and machinery.

The province, which covers some 3665 km² (1413 sq miles), consists mainly of the hills and forests of the Ardennes. Livestock farming and tourism are the chief sources of income.
Population (city) 100 000; (province) 421 200
Map Belgium – Ba

Nan *Thailand* Mountainous northern province on the Laotian border, named after the Nan River which flows through it. A town of the same name in the province lies 585 km (363 miles) north of BANGKOK. Both the town and province have a large population of so-called hill people, such as the Yao, Hmung, Khamu, Htin and Phitong Luang, who cultivate rice and maize.
Population (province) 414 000; (town) 104 650
Map Thailand – Bb

Nan Ling *China* Mountain ranges between the Chang Jiang River basin and the tropical lowlands of the far south. They stretch for almost 1000 km (almost 625 miles) from the eastern flanks of the Yungui Plateau to northern Zhejiang Province, reaching 2100 m (6900 ft).
Map China – Gf

Nanchang *China* Capital of Jiangxi Province. It is a centre for the processing of agricultural products and for the manufacture of aircraft, chemicals and diesel trucks.
Population 1 330 000
Map China – He

Nancy *France* Capital of Meurthe-et-Moselle Department, and once the capital of the Duchy of Lorraine. An elegant city, it was enlarged and beautified in the 18th century by Stanislas Leszczynski, the deposed King of Poland and father-in-law of the French King Louis XV. Leszczynski made Nancy his capital when Louis 'gave him' part of Lorraine. Place Stanislas, the city's magnificent central square, is named after him.
Population 99 400
Map France – Gb

Nanga Parbat (Diamir) *Pakistan* One of the world's highest peaks, rising 8126 m (26 660 ft) and marking the western end of the Great Himalaya Mountain range. It was first climbed by a German-Austrian expedition in 1953.
Map Pakistan –Da

Nanjing (Nanking) *China* Capital of Jiangsu Province in east-central China. The city, whose history can be traced back to the 8th century BC, stands on the banks of the Chang Jiang River, 250 km (155 miles) from SHANGHAI. Nanjing was the capital of Chiang Kai-Shek's Nationalist government between 1928 and 1937. Its capture by the Japanese in December 1937 was accompanied by savage killing and brutality – many thousands of people were killed. Today, the city is an important industrial and trading centre.
Population 2 470 000
Map China – Hd

Nanning *China* Capital of GUANGXI-ZHUANG Autonomous Region, some 160 km (100 miles) from the Vietnamese border. It is a centre for sugar refining.
Population 1 070 000
Map China – Gf

Nansei-shoto *Japan* See RYUKYU ISLANDS

Nantes *France* Port and capital of Loire-Atlantique department, 50 km (31 miles) from the mouth of the Loire River. In the 18th century, Nantes was France's most important port for trade with the Caribbean and Africa. It was at Nantes that Henry IV of France signed the Edict of Nantes in 1598, guaranteeing a large measure of religious and civil liberty to the Protestant Huguenots. The edict was revoked in 1685.
Population 492 000
Map France – Cc

Nan-t'ou *Taiwan* City in the western foothills of the island's central mountains, in a region producing rice, sugar, bananas and pineapples. It is a good base for exploring the mountains to the east, as well as SUN MOON LAKE.
Population 543 000
Map Taiwan – Bc

Napier *New Zealand* Port on North Island, 270 km (nearly 170 miles) north-east of the capital, WELLINGTON. It stands on Hawke Bay and

▼ **RICHES FROM DESOLATION The barren, shifting sands of the southern Namib Desert, which in some years has no rainfall, yield gem-quality diamonds.**

exports fruit, wool, frozen meats and dairy and timber products. Like HASTINGS, 20 km (12 miles) to the south-west, most of it was destroyed by an earthquake in 1931, when more than 250 people were killed. The town was rebuilt in Art Deco style.

Population 52 470
Map New Zealand – Fc

Naples (Napoli) *Italy* The country's third largest city, after ROME and MILAN, capital of the south as well as of the Campania Region, and a seaport and tourist centre 185 km (115 miles) south-east of Rome. The glittering Bay of Naples and the smouldering volcano VESUVIUS provide a dramatic backdrop to the city, which has some of the best museums in the world, as well as some of the worst slums in Europe, and an appalling crime rate.

Naples was founded as a Greek colony in the 7th to the 6th century BC and flourished during the Roman Empire and subsequently under Byzantine, Norman, Swabian and Spanish rule. The street plan of the old city reflects the Roman, medieval and Spanish periods. The Roman part, between the church of Santa Maria of Constantinople and the Capuano Castle, contains the cathedral. The medieval parts lead down to the docks and the Castel Nuovo. A rabbit warren of densely populated narrow alleys and small squares, this section of the city was badly damaged by an earthquake in 1980. The Spanish segment is located to the west of the Via Toledo.

The historic buildings of Naples include Palazzo Reale, the imposing Castel Nuovo (built in 1279-82 and remodelled in the 15th century), the Floridiana (a splendid early 19th-century villa), the San Carlo Theatre (one of the world's largest opera houses, built in 1737), the vast Albergo dei Poveri (a former Bourbon poorhouse) and the Museo Nazionale with its Graeco-Roman remains, including many statues, mosaics and murals from POMPEII.

Industries spread through the urban sprawl around Naples centre on steel, oil refining, chemicals, clothing, leather and food processing. Apart from Naples itself, many tourists visit nearby centres such as CAPRI, ISCHIA and SORRENTO.

Population 1 206 000
Map Italy – Ed

Napoli di Malvasia *Greece* See MONEMVASIA

nappe Almost horizontal overfold, normally of mountainous proportions, that has been forced some distance from its site of origin by earth movements. See THE RESTLESS EARTH, p 231

Naqsh-i-Rustam *Iran* See PERSEPOLIS

Nara *Japan* City on south HONSHU Island, 34 km (21 miles) south of KYOTO. Nara was the national capital from 710 to 784, and is second only to Kyoto as a storehouse of national treasures. One of its many temples, Todai-ji, founded in AD 752, contains the world's largest bronze Buddha. The statue is 16.2 m (53 ft) high and weighs more than 450 tonnes. Most of the city's temples are in Nara Park; others – such as the 8th-century Toshodaji Temple and the well-known Horyuji Temple – are on the outskirts of the city. Nara is a well-known manufacturer of calligraphy brushes and ink.

Population 349 000
Map Japan – Cd

Narbonne *France* Market city near the Mediterranean coast about 80 km (50 miles) from the Spanish border. Founded in 118 BC, it is said to have been the first Roman colony beyond the Alps. A favourite tourist haunt is its fortified cathedral, which was started in the 13th century, but which remains unfinished. Narbonne also contains many elegant public buildings and squares.

Population 46 000
Map France – Ee

Narenta *Bosnia-Herzegovina/Croatia* See NERETVA

Narew (Narev) *Poland/Belarus* River, 484 km (301 miles) long. It rises just inside Belarus, 85 km (55 miles) north-east of the border city of Brest, and joins the Bug River north of WARSAW.

Map Poland – Db

Naruto *Japan* Town and port on north-east SHIKOKU Island. It lies at one end of the Onawato Bridge, which was completed in 1986 and which links Shikoku, via the island of Awaji, to HONSHU Island across the Naruto Strait. Huge whirlpools, up to 25 m (82 ft) across, form in the strait with each turn of the tide as water passes between the Pacific and the Inland Sea.

Population 64 300
Map Japan – Bd

Narvik *Norway* Port in northern Norway on Ofotfjord, east of the Lofoten and Vesterålen islands. Iron ore from the Swedish mines at Kirunavaara and Luossavara is exported from Narvik's ice-free harbour. A rail link from the mines was built before the First World War; the port is also linked by road to Sweden. In 1940, while the port was being used by the Germans as a naval refuelling base and as a supplier of raw materials, a British fleet headed by the battleship *HMS Warspite* sailed into the fjord and sank nine German destroyers.

Population 18 900
Map Norway – Ea

Nashville *USA* State capital of Tennessee, standing on the Cumberland River, 240 km (150 miles) east of the Mississippi River. Founded in 1779, it is now called 'Music City USA' and is also known as the 'capital' of country and western music. Apart from recording studios its industries include glass, tyres and motor cars.

Population (city) 510 800; (metropolitan area) 985 000
Map United States – Ic

Nassau *Bahamas* Seaport, largest town and capital of the Bahamas, situated in the north of NEW PROVIDENCE Island. Founded by the British in the 17th century, it has many old colonial-style buildings. Its trade increased during the American Civil War (1861-5) when it was a base for ships defying the North's blockade to supply the South. From 1920-33 it served as a base for bootleggers during the Prohibition period. Nassau's main industries today are tourism and banking. Moreover, it is now a tax haven for many international companies which have registered offices here. It is also a tourist haven, with over one million tourists visiting each year.

Population 191 500
Map Bahamas – Ba

Nasser, Lake *Egypt* See ASWAN HIGH DAM

Natal *Brazil* City, port and state capital of Rio Grande do Norte on the country's north-east tip. Its name means 'birth', and it was founded on Christmas Day (the birth of Christ), 1597. Its 16th-century fort is still in excellent condition. Brazil's experimental rocket base is nearby at Barreira do Inferno. Rio Grande do Norte covers 53 015 km² (20 469 sq miles), and its people are called *Potiguares* ('shrimp eaters'), derived from the Amerindian name for the powerful Tupis, who occupied the area before the arrival of European settlers. The state's main products are salt, which is harvested from salination pans at the seaside, cement, furniture, cotton and sugar cane. All these products are exported through the port at Natal.

Population (city) 606 300; (state) 2 319 000
Map Brazil – Eb

Natal *South Africa* Former British colony (1843-1910) and then province (1910-1994) of South Africa. See KWAZULU-NATAL.

Natitingou *Benin* Chief town of the ATAKORA HIGHLANDS, 450 km (280 miles) north of the Atlantic coast and only 27 km (17 miles) from the border with Togo. It is the centre of a subsistence farming region producing guinea corn, sorghum, yams, and vegetables and maize.

Population 50 800
Map Benin – Aa

natural vegetation Wild plant communities that have survived in, or have colonised, an area and have not been planted by people. However, few places have remained unaffected by human activities and the term can rarely be applied in the strictest sense, except in the remotest regions. See also CLIMAX VEGETATION.

Nauplia (Návplion) *Greece* Port on the Gulf of Argos, 45 km (28 miles) from CORINTH in the north-eastern Peloponnese. According to legend, it was founded by Nauplios, son of the sea god Poseidon; in fact, it was most likely originally colonised by sea traders.

Known at one time as Napoli di Levanti, it was a major stronghold of the Venetians during their period of domination of the Mediterranean (AD 1387-1715), and a vital centre of the Greek revolt against Turkish rule from 1821. It was the capital of Greece from 1828 until 1834, when the newly elected King Otto moved to liberated ATHENS. In 1831, Count Capodistrias, the first president of modern Greece, was shot dead in the town by rival politicians. Nauplia has a museum, the twin Venetian rock fortresses of Palamídi and Íts Kalé, as well as Byzantine fortifications and two mosques.

Population 10 600
Map Greece – Cc

Nauru See p 478

Nausori *Fiji* See SUVA

Navadwip (Nabadwip) *India* Eastern town on the Bhagirathi River, 95 km (about 60 miles) north of CALCUTTA, and formerly called Nadia. It is the 'Benares of Bengal', a Hindu place of pilgrimage, and a seat of Sanskrit learning.

Population 129 500
Map India – Dc

Namibia

ADVANCED TECHNOLOGY IS USED TO EXPLOIT THE WORLD'S OLDEST DESERT IN ONE OF ITS NEWEST REPUBLICS

For millennia the Namib Desert along the Atlantic coast protected the peoples of Namibia from invaders, explorers and adventurers. Then in the 18th century the first foreigners penetrated the land from the Cape Colony, now in South Africa. They encountered harsh, lunar landscapes beneath which were concealed rich reserves of diverse mineral resources.

Along the forbidding southern coast diamonds have been exploited since 1908. By the beginning of the First World War, 5 million carats had been lifted from the desert floor and present-day estimates are that between 1.5 and 3 billion carats may still be encrusted in the pre-Cambrian bedrock, much of it in submarine layers off ancient beaches that were shaped with the succeeding ice ages over tens of millions of years. Sophisticated ships are required to vacuum the sea-bed to bring the booty of valuable gems to world markets – but in the same country, 1000 km (620 miles) inland to the north, the Ovahimba, and to the east the Khoisan, still live much the way their ancestors did 10 000 to 30 000 years ago.

The arid plains and jagged mountain ranges of Namibia also have some of the world's largest uranium deposits and vast supplies of copper, tin, lead, zinc and vanadium. Diamond and uranium sales account for about 60 per cent of the country's foreign exchange earnings.

Nearly 1.6 million people live in this newly born state. They include about 80 000 white people and many African tribes, such as the Khoisan (known also as Bushmen or simply San), who are believed to have created the magnificent prehistoric rock paintings found recently in the Namib Desert and other parts of the country.

Few people live in Namibia, but many have coveted it. As European powers carved up Africa in the late 19th century, Germany declared the territory a protectorate. The discovery of diamonds in 1908 caused the European population to increase to 15 000 by 1913. South African troops ousted Germany in 1915 during the First World War and after the war the League of Nations gave South Africa a mandate to govern the country. The mandate stressed the paramountcy of human dignity and development and the League retained primary responsibility for the country.

When the League was replaced by the United Nations after the Second World War, South Africa refused to recognise the UN trusteeship that succeeded the mandate and proceeded to apply its policy of apartheid – separate development of races – in Namibia. Black nationalists, led by the South-West African People's Organisation (SWAPO), demanded independence. Following South Africa's continued refusal, they began a guerrilla war and brought as much international pressure to bear against South Africa as they could muster over a period of 25 years.

In 1971 the International Court of Justice ruled South Africa's occupation illegal – but to no avail. Although the UN declared the elections held under South African supervision in 1978 null and void, elected multiracial governments were installed from 1979 onwards.

Ten years later, after intensive international negotiations had taken place during 1988, South Africa and the UN embarked on a joint programme to prepare Namibia for free, multiparty elections and independence. SWAPO, having boycotted all previous elections, now agreed to participate and thousands of exiles, including the SWAPO leaders, returned home in 1989.

On 21 March 1990 Namibia finally gained independence after SWAPO had won the UN-supervised elections held in November 1989. At midnight the South African flag was lowered and a new republic was born. The first president of Namibia, Sam Nujoma, was sworn in by Javier Péres de Cuellar, then Secretary-General of the United Nations. By March 1994 Namibia had reintegrated Walvis Bay (an enclave disputed with South Africa) and introduced its own currency. In December that year President Nujoma and his government were re-elected

DESERTS AND MOUNTAINS

Discrepancies between the rich (which now comprise many whites and a new black élite) and the poor (including some whites) are vast. About 95 per cent of Namibians live in abject Third World conditions, while the other 5 per cent live in First World style (there are about 15 000 private swimming pools in the capital, WINDHOEK, a city of 165 000 inhabitants).

Statistically, the country has a living standard significantly higher than most African nations, while poverty pervades the rural areas where communal subsistence farmers eke out a meagre existence in a harsh terrain. Mineral industries account for approximately 25 per cent of the gross national product (GNP) and are based mainly on diamonds, uranium and copper.

Apart from about 30 000 Khoisan, the black population of Namibia includes about 50 000 Caprivians, 100 000 Damara, 100 000 Herero, 124 000 Kavango, 665 000 Ovambo, 64 000 Nama and 7000 Tswana. Today the Khoisan people live predominately in or near the KALAHARI DESERT.

One of the driest countries on earth, Namibia has three main regions running from north to south across the country. The Namib Desert is a 50-140 km (30-85 mile) wide strip running along the entire Atlantic coastline. The Central Plateau, east of the Namib, rises to 1690 m (5550 ft) above sea level, with rugged outcrops, sandy valleys and plains of poor scrub and grasslands. To the east is the Kalahari, with sand, patchy scrub, coarse grass and dried-up salt flats.

The Namib has only 50 mm (2 in) of precipitation a year, most of which comes in the form of fog rolling in from the sea at night and condensing to form dew as the desert sands cool. Rainfall in Windhoek in the highlands is higher but still meagre at 200-250 mm (8-10 in) a year.

Traditional farmers raise cattle and grow sorghum and maize in the northern highlands; sheep are farmed in the southern highlands, where 'Persian lamb' skins from the karakul breed are produced for export. On the central plateau are the unique Rehoboth Basters, of mixed Dutch, Nama and German descent, who speak Afrikaans. They number more than 30 000 among about 90 000 people of mixed descent in Namibia, many of the others being fishermen in Walvis Bay and LÜDERITZ, or artisans in the towns.

The white population is mostly of Afrikaner stock, about a quarter of them being descendants of the German settlers who arrived in 1892. The whites live mainly in the towns, although some remain on the land, often responsible for running large farms. Much of the small and large local industry is dominated by German Namibians.

Whites still enjoy a relatively advantaged way of life, in contrast to the blacks, many of whose lives were shattered by the guerrilla war, causing them to flee south to overcrowded towns. This process intensified after independence with the pressures to succeed and the high expectations that inevitably followed. Many blacks continue to exchange the traditional life in rural areas for unemployment in the capital.

Offshore, the cold Benguela Current, which flows north from Antarctica, feeds one of Africa's richest fishing grounds. Mackerel, tuna, pilchards and snoek are among the catches processed at Walvis Bay and Lüderitz or frozen on the many foreign trawlers and factory ships which come from as far afield as Spain, Japan and Russia.

If the fish resource is properly conserved and regulated it could provide a substantial source of revenue for this small nation. It is estimated that the fishing industry alone could lead the way to prosperity. But economic miracles do not happen overnight and the new government – though clearly market oriented

– must still prove its ability to manage both democracy and the potential wealth that remains unexploited.

Tourism has also increased significantly with independence. Game viewing is most popular but many tourists also come for the unique landscapes, and to experience the desolation of a world where nature is still supreme. Other popular attractions include the extraordinary Fish River Canyon, 900 m (3000 ft) deep and 60 km (37 miles) long, with hot springs and a spa called Ais-Ais (meaning 'very hot') in the most rugged section.

In the north is the ETOSHA PAN ('the Great White Place'), a huge 4000 km² (1737 sq mile) shallow depression that floods briefly after rains, attracting vast herds of game – elephants, giraffes, wildebeest, zebras and even some rhinoceroses. Lions, leopards and cheetahs arrive to prey on them and the many species of buck which also congregate around the waterholes. Great flocks of flamingos gather in the muddy shallows. The dry season is the best time to visit because then animals converge on the few water holes. The Etosha National Park, which encompasses the Etosha Pan, is one of the largest conservation areas in the world, and covers some 22 270 km² (8598 sq miles).

A variety of rare plants grows in the desert, including the welwitschia, which has two leaves, each up to 3 m (10 ft) long. The oldest known living example is estimated to be 1500 years old. And strange as it may seem, this desert country boasts a prolific bird life, with more than 600 species.

Although many Namibians are weary of their country being seen as the world's largest game park, the government is under pressure to develop a healthy balance between ecotourism and the necessary industry to urge a poor African nation towards prosperity and the 21st century.

NAMIBIA AT A GLANCE
Official name Republic of Namibia
Area 824 290 km² (318 258 sq miles)
Population 1 541 300 **Per km²** 2
(Per sq mile 5**)**
Capital Windhoek
Government Republic
Currency 1 Namibia dollar = 100 cents
Languages English (official), Afrikaans, Oshivambo, Nama/Damara and various other Bantu languages, German
Religions Christian (90%) remainder indigenous
Climate Temperate and subtropical, very dry. Temperatures in Windhoek range from 6-20°C (43-68°F) in July to 17-29°C (63-84°F) in January
Land use Cultivation 1%, grazing 64%, forest and woodland 22%, other 13%
Main primary products Cattle, sheep, fish; maize, millet, sorghum; diamonds, copper, lead, zinc, tin, uranium, vanadium
Major industries Mining, stock rearing, food processing, textiles, ore smelting, fishing
Main exports Diamonds, uranium, copper, lead, zinc; fish, cattle, karakul fur pelts
Annual income per head (US$) 1610
Population growth (per thous/yr) 30
Life expectancy (yrs) Male 59 **Female** 64

Navajo National Monument *USA* Three of the largest and most elaborate American Indian cliff dwellings found in the country. Situated in northern Arizona, 30 km (20 miles) from the Utah state border, they date back more than seven centuries. The monument covers 146 hectares (360 acres). Although these cliff dwellings of *Anasazi* (the Navajo word for 'the ancient ones') are in the federal Navajo reservation, which covers 64 800 km² (25 000 sq miles) across Arizona, New Mexico and Utah, these ancient people were not ancestors of the Navajo, but of the Pueblo groups.
Map United States – Dc

Navan *Ireland* See MEATH

Navarino *Greece* See PÍLOS

Navarra (Navarre) *Spain* Mountainous northeastern region on the French border, once known as the Kingdom of Navarra, now an autonomous region covering 10 421 km² (4023 sq miles). Sugar beet, cereals and grapes are grown here; its capital is PAMPLONA.
Population 519 280
Map Spain – Da-Ea

Návpaktos (Lepanto) *Greece* Harbour town 170 km (106 miles) west of ATHENS on the north side of the Gulf of Corinth. The town overlooks the straits where in 1571 a Christian fleet mobilised by Spain, Venice, Genoa, Sicily, Naples and the Papacy routed the Ottoman navy in the Battle of Lepanto, the last great sea battle fought by oared galleys. Miguel Cervantes, the Spanish author of *Don Quixote*, was a participant in the fight, in which 200 of the 230 Turkish naval ships present were sunk or captured. Návpaktos is now a tourist centre, and holds a sports and arts festival each July.
Map Greece – Bb

Návplion *Greece* See NAUPLIA

Náxos *Greece* Largest of the CYCLADES islands, lying in the south Aegean Sea. According to Greek mythology Theseus, slayer of the Minotaur, abandoned the Cretan Princess Ariadne at Náxos. The legend suggests that the island may have been an outpost of the Minoan Empire of Crete. Náxos has been settled since at least 2000 BC, but its sought-after marble was being used for sculptures and receptacles as far back as 3000 BC. From AD 1207 to 1344, Náxos was the seat of a Venetian duchy. It has a 13th-century cathedral, a French convent and a museum. The island has a fertile soil, a hilly interior and sandy beaches which are popular with tourists. Red and white wines, honey, olive oil, fruits, figs and pomegranates are its main products.
Population 14 000
Map Greece – Dc

Nayarit *Mexico* One of Mexico's smallest states, on the Pacific coast north of JALISCO. Its mineral-rich mountains are inhabited by the Huichol and Coras people. Tepic, the capital, lies in a piece of rain forest in Nayarit's humid south-west corner.
Population 816 110
Map Mexico – Bb

Nazareth (Nazarat) *West Bank* Capital of the Northern District, 30 km (19 miles) south-east of HAIFA. Jesus spent his childhood here with his parents Mary and Joseph. The Basilica of the Annunciation, built in 1965 on the site where archaeologists believe the Archangel Gabriel appeared to the Virgin Mary, is one of the town's many shrines and churches.

Other notable buildings include the Church of St Joseph, on the traditional site of Joseph's carpentry shop, and the Greek Orthodox Church of the Assumption, built over the well where Mary is said to have drawn water. Having changed hands many times during the Crusades, Nazareth was captured by the Turks in 1517 and held until the British took it in September 1918 during the First World War. As part of the West Bank it will fall under PLO administration.
Population 44 800
Map Israel – Ba

Nazca *Peru* Small colonial town in Ica Province, on the Pan-American Highway. It is a centre for those wishing to visit the Nazca Lines, a series of huge geometric and animal patterns, some more than 2 km (1.25 miles) long, etched into the surrounding stony desert. Scientists believe that the patterns may have been an astronomical calendar created by the people of the Paracas, Nazca and Ayacucho cultures between 900 BC and AD 800.
Map Peru –Bb

naze See NESS

Nazwa *Oman* See NISWA

Ndebeleland *Zimbabwe* See MATABELELAND

N'Djamena (Fort Lamy) *Chad* National capital and river port, lying on the Chari River near its junction with the Logone, 80 km (50 miles) south of Lake Chad. Founded as Fort Lamy by the French in 1900, it is a mixture of elegant colonial architecture, mud-brick houses and modern buildings. The museum contains relics of ancient civilisations of the region. Much of the city was badly damaged in the civil war that ravaged the country in the 1980s.

N'Djamena is Chad's communications centre, with overland routes to outlying parts of the country as well as to Nigeria and Cameroon. During the 1979-83 drought it was the centre for the distribution of relief aid. When rainfall is good, groundnuts are grown in the surrounding area and sent to N'Djamena for processing. Hides and skins are also processed and cattle on the hoof are exported to Nigeria.The city also has a meat-chilling plant. N'Djamena has a university.
Population 728 000
Map Chad – Ab

Ndola *Zambia* Copper-mining town 270 km (nearly 170 miles) north of the national capital, LUSAKA. Ndola is the regional capital of the Copperbelt Province. Founded in 1904, it now has copper and cobalt refineries and steel works.
Population 467 000
Map Zambia – Bb

Neagh, Lough *United Kingdom* Largest freshwater lake in the British Isles, covering 381 km² (147 sq miles) in Northern Ireland, some 20 km (12 miles) west of the capital, BELFAST. Eight rivers flow into the lake but only one, the Bann, leaves it. The lake is Europe's largest source of freshwater eels.
Map United Kingdom – Bc

Nauru

THE PROSPERITY OF THIS TINY ISLAND REPUBLIC COMES FROM ITS HUGE PHOSPHATE DEPOSITS, WHICH WILL SOON BE EXHAUSTED

The smallest republic in the world is the 21 km² (8 sq mile) Pacific island of Nauru, halfway between Australia and Hawaii, just 40 km (25 miles) south of the Equator. Its inhabitants, totalling about 10 000, pay no taxes or import duties, no school fees or health charges and yet the tiny nation – and particularly the families that own the land – is rich.

This prosperity comes from a treasure chest in the shape of a plateau that rises about 60 m (200 ft) above sea level and is bounded by coral cliffs. The plateau contains rich deposits of high-quality phosphate rock which is sold for fertiliser to Australia, Japan and South Korea. But the deposits are expected to run out by the year 2000, and although the government is now investing overseas many islanders may have to emigrate.

The island has known foreign occupation several times since it was discovered by the British in 1798. The Germans annexed it in 1888, the Australians took it in 1914, and the Japanese occupied it in 1942. The island was retaken by the Australians in 1945 and in 1947 became a United Nations trust territory. It became an independent republic in 1968, retaining a special relationship with the Commonwealth. Two years later the government took control of the phosphate industry from the British Phosphate Commissioners, a joint Australian-British-New Zealand body.

A narrow band of fertile land between the coast and the uninhabited plateau is cultivated. The climate is hot and wet, but the rains have been known to fail. The government subsidises imports, so food is cheap, although almost all the county's requirements have to be brought in – even water.

The 5000 or so native Nauruans are of mixed Micronesian and Polynesian origin. There are also several thousand migrant phosphate workers from other Pacific islands, as well as a small Chinese community.

NAURU AT A GLANCE	
Map Pacific Ocean – Dc	
Official name Republic of Nauru	
Area 21 km² (8 sq miles)	
Population 10 000 **Per km²** 476 **(Per sq mile** 1250)	
Government centre Yaren	
Government Republic	
Currency 1 Australian dollar = 100 cents	
Languages Nauruan, English	
Religion Christian (Protestant 50%; Roman Catholic 30%)	
Climate Tropical, with varying rainfall; average temperature range is 23-32°C (73-90°F) all year	
Land use Coconut palms 30%, wasteland and phosphate mining 60%, other 10%	
Main primary product Phospate	
Major industry Phosphate mining	
Main export Phosphates	
Annual income per head (US$) 10 000	
Population growth (per thous/yr) 14	
Life expectancy (yrs) Male 64 **Female** 69	

neap tide Tide of lowest range, occurring twice a month, when the sun's gravitational pull is at right angles to that of the moon. See TIDES: HOW TIDES ARE PRODUCED.

Neblina, Pico da *Brazil* The country's highest mountain, rising to 3014 m (9888 ft) near the Venezuelan border. Perpetually veiled by mist, the mountain was only discovered in 1962, and until 1965 it was not clear whether its summit was in Brazil or Venezuela.
Map Brazil – Ba

Nebraska *USA* Midwest state covering some 200 018 km² (77 227 sq miles) between the fertile lowlands of the Missouri River and the high grasslands of Colorado and Wyoming. One of the four richest agricultural states, it produces cattle, pigs, maize, wheat, sorghum and soya beans. LINCOLN is its capital and OMAHA its largest city.
Population 1 578 400
Map United States – Fb

neck The vent of an extinct VOLCANO, which is filled with solidified lava.

Neckar *Germany* River rising in the BLACK FOREST in the south-west and flowing north for some 370 km (230 miles) to join the Rhine at MANNHEIM. Its valley produces wine.
Map Germany – Cd

Needles, The *United Kingdom* Line of chalk stacks in the English Channel off the western tip of the Isle of WIGHT.
Map United Kingdom – Ee

Nefud (Nafud) *Saudi Arabia* Desert of red sand in the north of the country. It covers about 56 000 km² (21 600 sq miles) and extends into southern Jordan. Isolated water wells and slight rainfall enable nomadic herders to survive.
Map Saudi Arabia – Bb

Negeri Sembilan *Malaysia* State extending to some 6643 km² (2565 sq miles) and lying in Peninsular Malaysia, south-east of the national capital, KUALA LUMPUR. Rice and rubber have replaced the declining tin-mining industry as the state's main source of revenue.
Population 723 800
Map Malaysia – Bb

Negev *Israel* Southern part of Israel, covering 14 107 km² (5446 sq miles) and bordered by Jordan and Egypt. Assigned to Israel at the partitioning of Palestine in 1948, the region's main towns are Beersheba in the north and the Red Sea port of ELAT, at Israel's southernmost tip. Although much of the Negev is desert, irrigation has made agriculture possible.
Population (Southern district) 478 800
Map Israel – Ab

Negombo *Sri Lanka* Fishing port and holiday resort some 30 km (19 miles) north of COLOMBO. Situated between the sea and a lagoon, it has a fine sandy beach, old colonial buildings and modern hotels. It is noted for its seafood from Negombo Lagoon. Colombo's international airport is at nearby Katunayaka.
Population 61 400
Map Sri Lanka – Ab

Negro *Brazil/Uruguay* River, some 800 km (500 miles) long, rising in Brazil and flowing across Uruguay from the north-east to the town of Soriano in the south-west, where it joins the Uruguay River. The Negro is dammed near the town of Paso de los Toros to form one of South America's largest constructed lakes, Embalse del Río Negro, covering 10 360 km² (4000 sq miles).
Map Uruguay – Ab

Negro, Rio (Río Guanía) *Colombia/Brazil* Largest tributary, 2253 km (1400 miles) long and draining 670 000 km² (258 600 sq miles), on the north bank of the Amazon River. It rises in the Río Guanía Department of Colombia and for part of its length is known as Río Guanía. It forms the Colombian-Venezuelan border for several hundred kilometres, before veering off into Brazil where its name changes to Rio Negro (Black River). From the Brazilian border it flows on for another 1370 km (850 miles), 1000 km (620 miles) of which are navigable, before it joins the Amazon at MANAUS.
Map Colombia – Cc

Negros *Philippines* Fourth largest island of the archipelago, lying in the VISAYAN ISLANDS between Panay and Cebu. It covers 12 704 km² (4905 sq miles), and has a volcanic mountain core that separates the two sides of the island, which are administered as separate provinces: Negros Oriental and Negros Occidental. The eastern part of the island is mainly mountainous and little development has taken place.

This part has strong cultural links with Cebu. To the north and west, plains of rich volcanic soil are covered with sugar plantations. These brought the area – and its capital Bacolod, 'the sugar capital of the Philippines' – considerable prosperity until world sugar prices slumped in the 1970s and 1980s. By the 1990s, Negros Occidental had become one of the Philippines' most depressed areas.
Population (Negros Oriental) 819 400; (Negros Occidental) 1 930 300
Map Philippines – Bd

Nei Mongol Autonomous Region (Inner Mongolia; Nei Monggol Zizhiqu) *China* Region stretching along the borders of China and the Mongolian People's Republic. It covers an area of 1 177 500 km² (454 528 sq miles) including part of the Gobi Desert. The Great Chinese Wall forms part of its border. It is a thinly populated land of vast arid wastes and empty grasslands.

Most of the population are ethnic Chinese. Mongolians, many of whom are herders of oxen, horse and sheep, make up about one-tenth of the population; some small minorities, including Manchus, make up the rest. The region is rich in coal, iron ore, asbestos, talc, mica, and rare earth oxides. Mining is being developed at centres such as Bayan Obo. HOHHOT is the region's capital.
Population 21 457 000
Map China – Fc

A DAY IN THE LIFE OF
A NEBRASKAN FARMER

On an arrow-straight road that stretches over rolling plains from one horizon to the other, Stephen Oman's station wagon heads from his isolated Nebraskan farmstead to the small town of North Platte, an hour away, for the Sunday morning Baptist Church service.

His wife Sue sits next to him stiffly and respectable in her Sunday best, and in the back of the car the children, Glen (12) and Peggy (10), are scrubbed and smart. Even the towering aluminium grain silos they pass beside the road seem to be polished for the occasion.

The silos are monuments to the success of Stephen and his fellow farmers, on land that was once known as the Great American Desert. Stephen's grandfather, an immigrant from Germany, had to gamble his whole future on a good season's rain here. Now, new strains of wheat and new farming techniques have made harvests, and farmers' incomes, more secure.

The landscape the Omans cross is chequered black and gold, with alternate strips of waving wheat and bare earth, each strip planted only once in two years in order to conserve moisture in the soil. And as a safeguard against soil erosion, the surface of the earth is disturbed as little as possible. It is not ploughed, but only harrowed and then rolled before planting.

North Platte's New Hope Baptist Church, a neatly kept bungalow, is surrounded by the sedans, station wagons and trucks of many of Stephen's neighbours. The congregation is large and enthusiastic. Many of the congregation members, including Stephen, give 10 per cent of their income towards the running of the church.

After the hymns have been sung and the sermon given, a baptismal service is conducted. The minister immerses a new believer into a sunken pool in the centre of the church. Stephen and Sue both underwent the same ceremony at around the age of 16. Only then can a Baptist be taken fully into the church; children younger than that age are thought to lack the necessary understanding of the ceremony.

Stephen regards himself as a good Christian. He has strong moral values is patriotic, supports the Republican Party, and is a great traditionalist. He values discipline.

For the rest of the day, after church, he takes the children to a picnic lunch on a bluff near Laramie, overlooking the North Platte River. They say grace before they tuck in to Sue's delicious homemade pies, and sit with their legs dangling over a deep trench in the limestone rock. Stephen explains that it was carved by the iron rims of countless wagon wheels, as the pioneers ground their way westward along the Oregon Trail.

Stephen still farms this land in the spirit of a pioneer, steadily conquering it with new agricultural techniques.

Neisse (Neiße; Lausitzer Neisse; Nysa Luzycka) *Eastern Europe* River rising above the Czech town of LIBEREC in northern Bohemia and flowing generally north for some 255 km (159 miles) to join the Oder River 155 km (96 miles) south of the Polish port of Szczecin. From the Czech border the Neisse forms part of the 'Oder-Neisse Line', which is the frontier between Germany and Poland.
Map Poland – Ac

Nejd *Saudi Arabia* Region and former province (now renamed Central Province), covering central and eastern Saudi Arabia. Before the First World War it was nominally under Turkish rule, and from 1905 it was ruled by Ibn Saud. In 1926 it was united with HEJAZ as a dual kingdom; in 1932 it became the single kingdom of Saudi Arabia. The Nejd plateau consists of ridges and shallow vales, many of them covered with sand desert. Its chief cities are RIYADH and BURAYDAH, and its main settlements are found around oases.
Map Saudi Arabia – Bb

Nelson *New Zealand* Port on South Island, at the head of Tasman Bay, about 120 km (75 miles) west of the capital, WELLINGTON. First settled in 1842, it was named after the British naval hero Lord Horatio Nelson (1758-1805). Today its chief industries revolve around timber, food processing and tourism.

The Nelson Lakes National Park, a tourist attraction situated 80 km (50 miles) south-west of the town, contains hundreds of red deer and chamois goats, which were introduced from Europe by settlers.
Population 47 390
Map New Zealand – Dd

Nepal See p 480

nephrite Mottled, often pitted variety of jade with an oily lustre.

Neretva (Narenta) *Bosnia-Herzegovin/Croatia* Major river, 218 km (135 miles) long, rising some 60 km (37 miles) south of SARAJEVO. From here the river flows north-west, then through the

145 km² (56 sq mile) lake behind the Jablanica hydroelectric dam, before sweeping south, past the medieval Turkish towns of Pocitelj and Mostar, to the Adriatic. It is well stocked with trout in its upper reaches, and with eels and mullet farther down. It has cut some spectacular gorges through the deeply eroded limestone mountains of the KRAS region.
Map Bosnia-Herzegovina – Bb-Cb

Nesebâr *Bulgaria* Black Sea resort situated 30 km (19 miles) north-east of Burgas, which was developed more than 2500 years ago as the Greek trade port of Mesembria. Sitting on small rocky peninsula linked by a causeway to the mainland, it has become an artists' retreat highlighted by attractive old houses and workshops overhanging cobbled streets. Its churches, built in a variety of Orthodox styles, date from the 5th to the 17th century.

Other features of the town are its extensive salt pans, medicinal mud lake and, on the northern side of the town, the popular tourist spot of Sunny Beach (Slunchev Bryag). Pomorie (population 20 000), lying 15 km (9 miles) to the south, is the centre of the local wine-growing area.
Population 12 000
Map Bulgaria – Cb

Ness, Loch *United Kingdom* Lake in the Great Glen, the valley that bisects northern Scotland between INVERNESS and Fort William. It is 37 km (23 miles) long and forms part of the Caledonian Canal, built between 1803 and 1847.

The lake is famous because of repeated claims that a monster – a dinosaur-like creature – lives in its depths. Many of those who believe the claims suggest that 'Nessie' is a survivor from prehistoric times. However, all efforts to track down the monster, using the most modern equipment available, have failed.
Map United Kingdom – Cb

▼ NESSIE'S LAIR The placid waters of Scotland's Loch Ness became in 1933 the site of the legend of the monster 'Nessie' believed to lie in its depths.

Nepal

CLOSED TO THE OUTSIDE WORLD UNTIL THE EARLY 1950s, THIS MOUNTAIN KINGDOM IS TODAY A POPULAR TOURIST DESTINATION

Nepal is a land tilted on its side. It is a long, narrow rectangle on the flanks of the eastern HIMALAYAS, with its northern border running along the mountain tops. The border area is known as 'the Roof of the World' and includes the world's highest mountain, EVEREST (8846 m, 29 022 ft).

The rectangle is only 160 km (100 miles) wide; yet in that short distance it drops from the dizzy heights of its northern frontier with Chinese Tibet to an altitude of only 100 m (330 ft) on its southern border with India.

In the TERAI, the flatlands of the south, tigers, elephants and the rare Indian rhinoceroses prowl the thick jungle, and rice, sugar cane, jute and oranges grow. Farther north in the spring the foothills blaze with azaleas and rhododendrons, both of which are native to Nepal. North again, yaks browse the high pastures below the eternal snows of the great peaks of Everest, KANCHENJUNGA, ANNAPURNA and Dhaulagiri.

Nepal is one of the world's poorest and least developed nations. Nine out of every 10 members of the population eke out a living from the land as peasant farmers and labourers. Only about one person in five can read or write, and the average income per head is about US$170 a year.

This poverty owes as much to the geography of the country as to past politics. Landlocked and largely cut off by the mountains from the trade routes of the Indian plains, Nepal has had little opportunity to develop a manufacturing economy. In addition, it has no significant mineral production with which to pay for imports from either of its two giant neighbours – India and China.

Politically, the country developed as a cluster of feudal valley kingdoms, separated from each other by the hills. It still consists of a number of ethnic groups and continues to be divided between Hindu (90%) and Buddhist (5%) faiths. Although the Gurkhas conquered the kingdoms and established modern Nepal in 1769, the leaders of each subject kingdom showed more loyalty to their own clans than to the nation.

As a result, the political history of Nepal is a web of plots and counter-plots, assassinations and palace intrigues, with noble families conspiring against one another to gain influence at the expense of the central throne.

From 1846 these families, chiefly the Ranas, one of whom always held the office of prime minister, reduced the kings to mere figureheads – until another palace revolution in 1951 restored the power of the monarchy and began the process of bringing Nepal into the 20th century.

However, successive kings became more autocratic and political parties were banned. In the spring of 1990, mass pro-democracy demonstrations led to the introduction of a multiparty system and amnesty for all political prisoners. Under a new constitution the king, Birenda, relinquished his absolute

▼ TRADE AND RELIGION Bicycle rickshaws bring buyers to a colourful vegetable market in Kathmandu. The pagoda-studded city is the home of Kumari, the girl whom Nepalis worship as a living goddess.

powers and elections were held in May 1991.

With Chinese and Indian aid, roads have been built from the northern and southern borders to KATHMANDU and along the central valley to link the major towns.

Until 1951, the country was completely closed to foreigners but it is now known as a tourist paradise, with over 200 000 visitors a year. It is also famous as the home of the Gurkha warrior. About 80 000 of these fearless fighters serve in the British army, having fought alongside the British since the early 19th century, when Nepal was one of Britain's protectorates. Another 70 000 serve in the present-day Indian army.

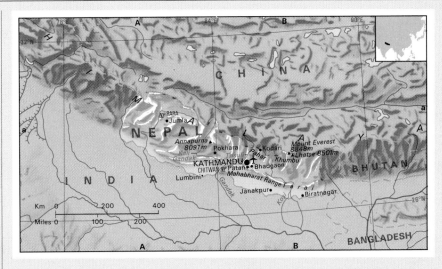

NEPAL AT A GLANCE

Official name Kingdom of Nepal

Area 140 800 km² (54 362 sq miles)

Population 20 535 466 **Per km²** 146 **(Per sq mile** 377)

Capital Kathmandu

Government Constitutional monarchy

Currency 1 rupee = 100 paisa

Languages Nepali (official) and other Tibetan languages divided into various dialects

Religions Hindu (90%), Buddhist, Muslim

Climate Tropical monsoon with abundant summer rains; drier in north. The average temperature in Kathmandu is 10°C (50°F) in January and 24°C (75°F) in July.

Land use Cultivation 17%, grazing 13%, forest and woodland 33%, other 37%

Main primary products Rice, wheat, barley, sugar cane, cattle, medicinal herbs, jute, pepper, tobacco, fruit, timber

Major industries Agriculture, jute spinning, sugar milling, textiles, forestry, tourism

Main exports Cattle, rice, jute, goat skins, textiles, carpets, hardwoods, spices

Annual income per head (US$) 170

Population growth (per thous/yr) 24

Life expectancy (yrs) Male 52 **Female** 52

ness (naze) Headland jutting into the sea.

Netanya *Israel* Mediterranean resort, some 30 km (20 miles) north of TEL AVIV. Netanya is the centre of the country's diamond cutting and polishing industry.
Population 139 700
Map Israel – Aa

Netherlands, The See p 484

Netherlands Antilles and Aruba *West Indies* Dutch territory in the Caribbean consisting of two groups of islands more than 800 km (500 miles) apart. ARUBA, and CURAÇAO and BONAIRE lie off the north coast of Venezuela, while ST MARTIN (only partly Dutch), ST EUSTATIUS and Saba are in the Leeward Islands. They cover about 800 km² (309 sq miles). The islands were settled by Dutch colonists in the 1630s; in 1954 they became internally self-governing and constitutionally equal with the Netherlands and Surinam.

Queen Beatrix of the Netherlands is the head of state and is represented by a governor. (Aruba adopted separate status within the Kingdom in 1986.) The capital is WILLEMSTAD on Curaçao. Most inhabitants are of African and Dutch descent. Although Dutch is the official language, many people speak Papiamento, a mixture of Dutch, Spanish, Portuguese and other languages. The economy revolves around tourism and the refining of oil imported to Curaçao from Venezuela.
Population 191 300
Map (Aruba, Curaçao and Bonaire) Caribbean – Ac-Bc; (St Martin, St Eustatius and Sabra) Caribbean – Cb

Neuchâtel (Neuenburg) *Switzerland* City on the north-western shore of Lake Neuchâtel, 40 km (25 miles) west of the capital, BERNE. It is the capital of the canton of the same name, which covers 803 km² (310 sq miles). The city manufactures electrical appliances, watches and chocolates.

Its church, which dates from the 12th and 13th centuries, has Gothic and Romanesque elements. It also has a castle built in the 12th to the 15th century, a university and other educational institutions. Lake Neuchâtel, the largest lake entirely within Switzerland, covers 218 km² (84 sq miles).
Population (city) 31 700; (canton) 162 600
Map Switzerland – Aa

Neusander *Poland* See NOWY SACZ

Neusatz *Yugoslavia* See NOVI SAD

Neuschwanstein *Germany* See MUNICH

Neva *Russia* River, 74 km (46 miles) long, which drains Lake LADOGA in north-west Russia, flowing through the city of St Petersburg to the Gulf of Finland. The Neva is frozen from about December to April, and after the thaw, spring floods can raise its height by up to 5 m (16 ft). It is navigable by ocean-going ships, and is a key section in waterways linking the Baltic Sea to the White Sea and the Volga River.
Map Russia – Ec

Nevada *USA* Seventh largest state, covering 286 298 km² (110 540 sq miles) to the east of California, between the SIERRA NEVADA and the Rocky Mountains. Most of the state consists of desert and mountains, which explains its low population density of about 31 per km² (12 per sq mile). Cattle and sheep ranching are the main types of farming. The United States' richest known deposit of silver is situated in Nevada; other minerals mined here include gold, mercury, manganese and diatomite.

However, Nevada's principal industry is neither farming nor mining, but gambling. More than 35 million hopefuls a year visit the gaming tables, particularly in LAS VEGAS and RENO. The state capital is CARSON CITY. The United States' nuclear weapons test site is situated some 100 km (60 miles) north-west of Las Vegas.
Population 1 201 800
Map United States – Cc

Nevado de Colima *Mexico* Smoking, active volcano, 4328 m (14 200 ft) high, in Jalisco State. It erupted in 1941, causing many deaths.
Map Mexico – Bc

Nevado de Huila *Colombia* See HUILA

Nevado de Toluca *Mexico* Snow-capped extinct volcano rising to a height of 4565 m (14 977 ft) above Toluca, capital of MEXICO State. Two of its craters have filled up with meltwater, forming the Lake of the Sun and the Lake of the Moon.
Map (Toluca) Mexico – Cc

Nevado del Ruiz *Colombia* See ARMERO

névé See FIRN

Nevers *France* Capital city of Nièvre Department and of the former Duchy of Nivernais. It stands on the Loire River, 140 km (90 miles) south-east of Orléans. It has a huge 13th-century cathedral and a 15th-century ducal palace. Today it is a noted manufacturer of china, earthenware and aircraft parts.
Population 42 000
Map France – Ec

Nevis See ST KITTS AND NEVIS

New Amsterdam *USA* See NEW YORK (city)

New Britain *Papua New Guinea* Largest island of the Bismarck archipelago, covering some 36 500 km² (11 125 sq miles), and lying off the north-east coast of New Guinea. It is mountainous, with active volcanoes, hot springs, and peaks reaching 2438 m (8000 ft).

Annual rainfall is between 4000 mm (157 in) and 6000 mm (236 in). The soil, particularly in the north and north-east, is extremely fertile, and produces coconuts, cocoa, copra, rubber, timber and palm oil. RABAUL is the main town.
Population 315 650
Map Papua New Guinea – Ba

New Brunswick *Canada* Small maritime province covering 73 436 km² (28 354 sq miles) in eastern Canada, between the state of Maine in the United States, Quebec to the west and north-west, and Nova Scotia to the south-east.

With the French being the first to settle here in 1604, the territory was in turn part of the French province of Acadia and part of the British province of Nova Scotia before being made a colony in 1784. About 35 per cent of its population, which is largely rurally based, are French-speaking. New Brunswick has a long coastline and consists mainly of rolling, forested hills. Its economy is based chiefly on forest products and the processing of foods; coal, oil products and tourism are also important sources of revenue. FREDERICTON is its capital and SAINT JOHN its largest urban centre.
Population 723 900
Map Canada – Id

New Caledonia *Pacific Ocean* Group of islands about 1400 km (875 miles) to the north-east of BRISBANE, Australia, which in the 1980s and early 1990s was in the throes of a sometimes bloody journey towards independence. The assassination of pro-independence leader Pierre Declercq in 1981 sparked widespread riots; these were repeated in 1985, when police shot another liberation leader.

Leading the opposition to independence for the islands (French-owned since 1853) were a powerful European minority, who make up more than 33 per cent of the population. Ranged against them were indigenous Melanesian people or 'Kanaks', who make up a little less than half the total. In July 1989, France discontinued direct rule (although the islands remain a French Overseas Territory) under an agreement which provides for a territorial referendum on full independence, to be held in 1998. Political and interracial tensions continued, however, and in 1992, the commercial centre of the capital, NOUMÉA, was extensively damaged during rioting.

New Caledonia covers an area of 19 103 km² (7376 sq miles) and is made up of one large hilly, forested island called New Caledonia (like the group), or La Grande Terre, and covering some 16 750 km² (6466 sq miles) and several smaller ones. The latter include the Isle of PINES and the LOYALTY ISLANDS. Almost 50 per cent of the population live in Nouméa. The main island has some of the world's largest known nickel deposits; most of the ore is processed before being exported. Large deposits of iron ore, chrome and other minerals have also been found on the island. Less than 10 per cent of the land can be cultivated, but substantial quantities of maize and coffee are grown; large areas are grazed by cattle, and forestry is important.
Population 164 170
Map Pacific Ocean – Dd

New England *Australia* Plateau region in northeastern New South Wales, stretching south for 300 km (186 miles) from the Queensland border. Forming part of the GREAT DIVIDING RANGE, the plateau is mostly 1000-1500 m (3300-5000 ft) high, rising to some 1608 m (5275 ft) at Round Mountain. Much of it is rich sheep and cattle-grazing country, but some areas are covered by thick forests of eucalyptus trees.

The region has several spectacular waterfalls, including Wollomombi Falls, the highest in Australia at 445 m (1460 ft). The main towns are Tamworth (population 31 720) and ARMIDALE.
Map Australia – Ie

New England *USA* Region comprising the north-eastern states of Maine, New Hampshire, Vermont, Connecticut, Rhode Island and Massachusetts. The latter three are part of the heavily populated industrial belt stretching along the Atlantic coast from the national capital, Washington DC, through New York City to BOSTON. The other three are more rural and less industrialised. New England was settled mainly

by English people in the 17th century. Their Puritan Protestant system of democratic local government and the attention given to education have remained strong traditions in the area.
Map United States – Lb

New Forest *United Kingdom* Ancient English royal hunting forest covering some 364 km² (141 sq miles) in south-west Hampshire, between Southampton and Bournemouth. It was placed under forest laws more than 1000 years ago, but many of its great oak trees were felled from the 16th to the 18th century to build ships. As a result, large areas of the forest are now open heath or grassland. In 1992, the New Forest was given national park status to protect it from urban spread.
Map United Kingdom – Ee

New Guinea See IRIAN JAYA; PAPUA NEW GUINEA

New Hampshire *USA* One of the original 13 states and first settled in 1623. It covers some 24 038 km² (9279 sq miles) in north-eastern NEW ENGLAND, between Massachusetts and Maine; its western boundary with Vermont is the Connecticut River. It consists mainly of woodland. The White Mountains rise to 1917 m (6288 ft) at Mount Washington in the north, and it has several lakes, of which the largest is Winnipesaukee. Agriculture centres on dairy farming, fruits and vegetables; the main manufactures are electronics and computers, paper and pulp. Tourism makes an important contribution to its economy. As well as lakes, there are pleasant beaches, facilities for winter sports and opportunities for hiking and fishing. CONCORD is its capital, Manchester its biggest city. Portsmouth, its only port, is also a naval shipyard.
Population 1 109 300
Map United States – Lb

New Haven *USA* Port city in Connecticut. Situated on an inlet of Long Island Sound, it was laid out in 1638 by Puritan settlers around nine squares. One of them, 'The Green', still remains an open public space. The city is the seat of Yale University, one of the oldest and most prestigious universities in the United States, dating from 1701. Some of Yale's libraries and collections – notably the Peabody Museum of Natural History – are open to visitors.
Population (city) 130 500; (metropolitan area) 530 200
Map United States – Lb

New Hebrides See VANUATU

New Ireland *Papua New Guinea* Mountainous island, 320 km (200 miles) long and up to 60 km (37 miles) wide, in the BISMARCK ARCHIPELAGO. Copra and cocoa are its principal crops, and its tuna-fishing industry is an important source of revenue. Its chief town is the port of Kavieng. The Melanesians who live here are famous for their wooden carvings and masks which they wear during *malanggan* (death) celebrations.
Population 87 000
Map Papua New Guinea – Ca

New Jersey *USA* State covering 20 295 km² (7836 sq miles) in the north-east, on the mid-Atlantic coast between New York State and Pennsylvania. It is part of the commuter belt for New York City and Philadelphia, and has the highest population density of all the United States, with more than 95 per cent of its people living in towns or cities. Nevertheless, the state's farms produce considerable quantities of fruit, vegetables, poultry, eggs and dairy products. Industry is diverse, with shipyards, refineries and petrochemical plants along the coast, and New Jersey's electronics, pharmaceutical companies, food processing plants and print works inland. Tourism is the second largest single industry, and casinos at Atlantic City and the Meadowlands entertainment complex are among the attractions. TRENTON is the state capital and NEWARK and Jersey City are its main industrial areas.

New Jersey's first European settlers were Dutch, but the area was ceded to England in 1664. It was one of the original 13 states and the site of nearly 100 battles of the American War of Independence. Princeton, which is the fourth oldest university in the United States (1746), is situated in the centre of the state.
Population 7 730 200
Map United States – Lc

New Mexico *USA* South-western state situated on the Mexican border. It was settled by the Spanish in the 16th century and was part of Mexico until it was ceded to the United States in 1848. It became a state in 1912. New Mexico covers some 314 895 km² (121 553 sq miles) and rises to more than 4012 m (13 161 ft) at Wheeler Peak in the Sangre de Cristo range of the Rocky Mountains. The state has the country's largest number of American Indian inhabitants (almost 10 per cent of the total number). There are 24 federal reserves in New Mexico and it is a cultural, research and tourist centre for native American lore and artefacts. Almost 40 per cent of the population are of Hispanic origin, a third of them being descendants of the original Spanish settlers.

Farming in this largely arid and semi-arid state centres on cattle and sheep ranching and on crops such as chilli peppers, pinto beans and pecans for the local native American-Hispanic cuisine. Gas, oil, uranium and coal are the principal minerals and there are large copper reserves.

The state capital is SANTA FE; the largest city is ALBUQUERQUE. The first atom bomb test was carried out near Alamogordo in today's White Sands Missile Range.
Population 1 515 000
Map United States – Ec

New Orleans *USA* Major historic port city on the Gulf of Mexico situated at the mouth of the MISSISSIPPI River in the southern state of Louisiana. Founded in 1718, it was under French or Spanish control until 1803. Its chief exports are cotton, rice, oil and petrochemicals, and its manufactures include aerospace equipment, chemicals, and processed foods.

New Orleans is also a busy tourist centre, best known as the birthplace of jazz and for the Mardi Gras – two weeks of revelries and parades which culminate on Shrove Tuesday with a major parade through the Vieux Carré, the French Quarter. This area is distinctive for its St Louis Cathedral, its narrow streets and distinctive houses with iron-trellis balconies. Riverboat gambling, which was legalised in 1992, has become another tourist attraction.
Population (city) 496 900; (metropolitan area) 1 285 300
Map United States – Hd

New Plymouth *New Zealand* Market town on North Island, 260 km (160 miles) north-west of the capital, WELLINGTON. Founded in 1841, it was one of the earliest European settlements in the country. It stands in a rich dairy-farming area and is the supply base for the offshore Maui gas fields.
Population 48 420
Map New Zealand – Ec

New Providence *Bahamas* Island in the central group of the Bahamas archipelago. Its chief town is NASSAU, also the capital of the Bahamas. The island has some farming, fishing and light industry, but the main occupations are the service industries based in Nassau. The hub of the nation, New Providence has an international airport. Some 60 per cent of Bahamians live on the island.
Population 171 540
Map Bahamas – Ba

New Siberian Islands *Russia* See NOVOSI-BIRSKIYE OSTROVA

New South Wales *Australia* The country's most populous state, covering some 801 600 km² (309 426 sq miles) and bordered by the Pacific coast in the east, VICTORIA to the south, and QUEENSLAND to the north. It was originally the name of the British colony which covered the entire eastern half of mainland Australia when it was founded in 1788. Various sections were used to form other colonies and states during the 1800s.

The state capital, SYDNEY, stands in the fertile coastal plain which lies to the east of the Great Dividing Range. The plains of the far west are semi-arid grazing land, but there is richer wheat and intensive grazing country in the central region and along the coast. Dairy products, wheat and other grains, fruits, wine, sheep, and cattle are the main agricultural products; the state produces more than a third of the country's wool. Large coal deposits have been discovered in the Hunter Valley near the port of Newcastle and around Wollongong, and at Broken Hill in the far west there are rich reserves of silver, lead and zinc. About one-third of Australians live in New South Wales, of whom just under two-thirds live in and around Sydney.
Population 5 901 400
Map Australia – Ge

New Territories *Hong Kong* The most extensive of the three parts of the Crown Colony of Hong Kong, the other two being HONG KONG ISLAND and KOWLOON. Acquired by Britain from China in 1898 on a 99-year lease, the New Territories extend from the peninsula of Kowloon northwards to the Shum Chun River – which marks the border between Hong Kong and China – and includes LANTAU and more than 200 other outlying islands. Originally inhabited by a fishing and farming population, the once rural landscape of the mainland section has rapidly given way to massive urban and industrial growth.

Many of Hong Kong's largest factories are now in the New Territories, whose southern fringe is often referred to as New Kowloon, in recognition of the steady expansion into the New Territories of the built-up area centred on Kowloon Peninsula. Elsewhere in the New Territories are eight new towns – each of them an existing or planned city of considerable size.
Population 1 812 000
Map Hong Kong – Ca

The Netherlands

A WEALTHY, DENSELY POPULATED NATION, WHICH HAS THE WORLD'S LARGEST PORT, PROVIDES A GATEWAY TO MODERN EUROPE

The Kingdom of the Netherlands exists by the whim of the sea and the will of the Dutch who live there. More than one-fifth of the country is below sea level, including all the main centres of population.

Over the centuries the sea has sometimes taken from the land and sometimes given back. In 1421 the sea claimed a huge area of land at the mouth of the RHINE River known ever since as the Hollands Diep. In the 17th century, drainage and dyking added some 1120 km² (432 sq miles) to the land area. In the 18th century another 500 km² (193 sq miles) of new land and in the 19th century another 1170 km² (452 sq miles) were gained.

But the sea, too, has had its gains since that first disaster in 1421 – notably in the flooding of 1953. Only in the past 50 years has the bal-ance between gain and loss tipped decisively towards the inhabitants. With modern machinery, Dutch reclamation works have so far this century added 2500 km² (965 sq miles) to the country's land.

This is what comes from establishing the core area of a nation in the delta of three great rivers – the Rhine (with two main branches, the Lek and Waal), the Maas (MEUSE) and the SCHELDE. As in all deltas at all times, whatever land there is exists through the uneasy equilibrium of three forces – the work of the river, the work of the sea, and the intervention of people in restraining the other two. In 1995 severe flooding of the Rhine, Waal and Maas rivers caused 250 000 people in the low-lying provinces of Gelderland and Limburg to be evacuated from their homes as floodwaters weakened the dykes.

The Dutch have had plenty of practice at turning back the sea – enough to tackle huge reclamation schemes in recent years. One is the IJSSELMEER (Zuiderzee) where, after the building of an enclosing dam in 1932, they have drained four large areas to give themselves an extra 1650 km² (637 sq miles) of former seabed to cultivate, not to mention an overspill town for AMSTERDAM called Almere.

Another major project was prompted by the disastrous storm floods of 1953, when 1835 people drowned and 200 000 head of cattle were lost. In the Rhine-Maas-Schelde delta there were a number of islands with coasts open to attack by the sea. The Delta Plan has sealed these islands from the sea by building barrages across the delta channels, leaving open to wind and tide only the shipping lanes up to ROTTERDAM in the north and the Belgian port of ANTWERP in the south. This has not only made the islands secure but has also linked them to the mainland and attracted tourists to the area's constructed lakes.

Another project, also intended to attract tourists, will – paradoxically – obliterate centuries of dyke building, dams and drainage. The province of Groningen's 'Blue City' plan, proposed early in 1995, aims to reduce the amount of farmland in the Oldambt region in favour of a lake for recreation.

In contrast to the low-lying and reclaimed areas there is another Netherlands – the 'high' (relatively) country containing the province of Gelderland, Overijssel, Noord-Brabant and Limburg. It lies east and south of the delta and much of it is low hills of sands and gravels. In the east, these materials were brought in by ice from Scandinavia in the Ice Ages; in the south they are river terraces of the Rhine and Maas. Further south, in LIMBURG, an area of chalk rises to 321 m (1053 ft) and is underlain by coal.

That landscape depends very much on the material beneath it. The sands and gravels are infertile. They are largely covered by heath and forest, much of it now protected in nature reserves. The soils of the river valleys are cold and heavy clays, which lie under water meadows or orchards. The peat of the delta lands is almost all under grass, with a high water table to prevent shrinkage; in fact, 57 per cent of Dutch farmland is under grass. The sea clays of the coast are areas of tillage; they are lighter than the river clays, and are generally cropped.

THE FAMOUS DUTCH BULB FIELDS

The most famous Dutch landscapes of all are the bulb fields and the areas of horticulture on the line of old sand dunes which lie just behind the present coastal dune barrier from the HOOK OF HOLLAND north to ALKMAAR. Today, the most notable feature of this landscape is the expanse of glass under which salad vegetables, fruit and flowers are raised.

The name Netherlands (or Low Countries) originally applied to the area now occupied by the modern state as well as Belgium and Luxembourg. In the 19th century, Belgium and then Luxembourg gained independence, leaving a Netherlands Kingdom with a Protestant majority and a Protestant monarchy, but a strongly Catholic south.

The Netherlands today are wealthy, even by European standards. But in their first century of independence from Spain, between 1600 and 1700, they led the Continent. The young republic attracted Protestant refugee artists and craftsmen from all parts of Europe. It had military and, especially, naval power, and it went in search of the world's wealth, to found a great empire overseas. While the Dutch East India Company made the money, the great cartographers – Mercator, Ortelius, Blaeu – provided the maps, and Rembrandt, Hals and Vermeer painted the daily scenes.

Things were never quite so good again. The 18th century brought French armies. The loss

▲ CITY OF CANALS One of Amsterdam's finest attractions is its network of canals, with their trim bridges and banks of tree-lined streets. Some of the anchored barges serve as floating homes – and visitors can travel on special canal-boats by day and night.

of Belgium in 1830 was a blow. Four wars fought against France and three against Britain heralded the end of the colonial era.

Life became desperate in 1940, when Germany invaded the Low Countries and wiped out the centre of Rotterdam in the first few hours of its attack. Four years of German occupation were ended for the southern Netherlands in autumn 1944, but the Rhine and Maas, which had once protected the Dutch republic, now protected the Germans: In the north, German occupation continued through a further terrible 'hunger winter' and by 1945 few parts of Europe were more denuded of every kind of removable resource than the Dutch heartland. It was from this nadir that the country's postwar recovery had to proceed.

The Netherlands, with 449 inhabitants to every square kilometre (1163 per sq mile), is the most densely populated country in Europe, but this is only an average figure. The east and north of the country are quite sparsely populated, while the density in South Holland province is more than 1148 per km² (2973 per sq mile). The Dutch population is overwhelmingly concentrated in the cities of the so-called Randstad Holland, a horseshoe of urban areas which starts with DORDRECHT in the south and includes Rotterdam, DELFT, The Hague, LEIDEN, HAARLEM, Amsterdam and HILVERSUM, before it reaches the other tip of the horseshoe at UTRECHT. The diameter of the

horseshoe is about 50 km (31 miles). Here, a congested road system, lack of housing and serious pollution are inducing an increasing number of city dwellers to move out to the surrounding countryside, which has some of the world's most productive farmland and market gardens. The decentralisation phenomenon is gradually chewing away at the valuable farmlands and it is a continuing struggle to preserve agricultural land and recreational areas within and around the horseshoe.

Almost the whole of the Randstad is below sea level, so the drainage must be planned and controlled before anything can be built – let alone a building like the 112 m (367 ft) Dom Tower at Utrecht or the underground railway at Rotterdam. Other parts of the country, too, are not empty of opportunity. Boom cities exist in the south, especially EINDHOVEN with its great electrical industry. There is Limburg, with its industries formerly based on local coal, and now on chemical products and tourism. There are the former textile towns of OVERIJSSEL Province and there is the natural gas field of the far north-east to attract development. But the Randstad dominates the Dutch economy.

Its harbours, especially Rotterdam, the largest in the world, handle more than one-quarter of the European Union's total sea freight. Its coastal industries provide refined oil, steel (5-6 million tonnes per year at IJmuiden), ships and processed tropical produce. It also has highly developed chemical and electrotechnical industries. It is, finally, the administrative and commercial heartland.

Dense settlement demands a dense network of transport links. The Dutch have created the most concentrated motorway network in the world. Their electric railways operate frequent, regular services, although they carry little freight. The canals and, above all, the

Rhine are the great arteries of freight traffic – traffic that converges, over oceans, by river and canal, on Rotterdam. With the mouth of the Rhine under their control, the Dutch are the gatekeepers of Western Europe.

THE NETHERLANDS AT A GLANCE
Official name Kingdom of the Netherlands
Area 41 526 km² (16 029 sq miles); (land area 33 940 km², 13 101 sq miles)
Population 15 239 180 **Per km²** (of land area) 449 (**Per sq mile** 1163)
Capital Amsterdam (seat of government The Hague)
Government Parliamentary monarchy
Currency 1 guilder = 100 cents
Languages Dutch, Frisian (minority); English and German widely spoken
Religion Christian
Climate Temperate, maritime, with cool summers and mild winters. Average temperature in Amsterdam is 1.7°C (35°F) in January and 17°C (63°F) in July
Land use Pasture 26%, cultivation 22%, forest 8%, inland water 19%, other 25%
Main primary products Cereals, potatoes, sugar beet, fruit, vegetables, livestock, fish; oil and gas, salt
Major industries Agriculture, iron and steel, food processing, chemicals, textiles, clothing, printing, shipbuilding, tobacco processing, diamond cutting, fertilisers, oil and gas production and refining, fishing
Main exports Petroleum products, chemical products, electrical appliances, metalware, machinery, motor vehicles, sea vessels and aircraft, dairy products, meat, vegetables, flowers, chocolate, tobacco, flower bulbs
Annual income per head (US$) 21 800
Population growth (per thous/yr) 6
Life expectancy (yrs) Male 74 **Female** 80

New Valley Scheme *Egypt* Development scheme for the revival of a discontinuous line of oases – EL KHARGA, Dakhla, Farafra, BAHARIYA and SIWA – in Egypt's Western Desert. The scheme involves tapping deep artesian water in the Nubian sandstone lying some 900 m (3000 ft) below. Apart from increasing food production, the main objective of the scheme (in operation since the 1960s and expected to continue into the next century) is to ease population pressure on the Nile Valley and delta regions.

Sealed roads have been constructed between newly sited villages, and health and social services have been provided for the villagers. The aim of the scheme is to recover up to 20 per cent of approximately 30 000 km² (11 580 sq miles) covered by the oasis depressions.
Map (El Kharga, Dakhla and Farafra) Egypt – Bc; (Bahariya) Egypt – Bb; (Siwa) Egypt – Ab

New York *USA* Second most populous state, after California, lying on the mid-Atlantic coast and covering 127 219 km² (49 108 sq miles). It includes LONG ISLAND and the HUDSON River Valley, stretching from New York City harbour north-west to the Appalachian highlands, lakes Erie and Ontario, and to the Canadian border. The Adirondack Mountains rise to 1629 m (5344 ft) in the north-east; in the south lie the Catskill Mountains. The state has more than 2000 lakes.

New York was the Dutch colony of New Netherlands until 1664, when it was annexed by the British and renamed after the Duke of York, brother of King Charles II. It was one of the 13 original states of the Union. The state has more than 800 museums and 326 institutions of higher learning, including the military academy of West Point.

The state is the second largest manufacturing state in the United States, its main industries being printing, the manufacture of electronic and electrical equipment, machinery and scientific instruments, and the production of chemicals, clothing and processed foods. The chief agricultural products of the state are cattle, dairy goods, fruits, vegetables, and wine. NIAGARA FALLS and New York City are the main tourist attractions. ALBANY is the state capital.

New York is also the name of the country's most populous city, situated at the southern tip of New York State. Known affectionately as 'the Big Apple', it is the centre of a conurbation spreading into neighbouring parts of Connecticut and New Jersey that contains almost 20 million people – one of the biggest, if not the biggest, in the world.

New York City lies on the east coast, and started out in 1612 as a Dutch trading post on Manhattan Island at the mouth of the Hudson River. Later, with its fine natural harbour attracting a flourishing trade, it developed into the settlement of New Amsterdam. Even before Britain captured and renamed it in 1664, it drew many immigrants from England, Scotland, Ireland, Germany and Scandinavia. However, its greatest period of growth was between 1880 and 1920, when the Statue of Liberty in the harbour marked the gateway to the New World for millions of European immigrants. Indeed, one of its boroughs, Brooklyn, which had 4000 inhabitants in 1810, had a population of 1 600 000 a hundred years later. Today the city, still a magnet for people from almost every part of the world, has become home to more than 150 ethnic groups.

New York City consists of five boroughs – Manhattan, the Bronx, Brooklyn, Queens and Staten Island. The metropolitan area spreads north into Westchester County (of New York State) and Connecticut, east to Long Island (a continuation of Queens), and west and south into New Jersey.

The city is the country's – and in many aspects the world's – leading financial, trade, manufacturing, communications, fashions, entertainment and tourist centre. And Manhattan is still its heart. The financial sector is based in and around Wall Street in Lower Manhattan. The name Wall Street – equated worldwide with the New York Stock Exchange – comes from a wooden pallisade, or wall, built by Dutch settlers in 1653 to protect themselves from raiding American Indians. Many other businesses – notably publishing and printing, television and radio (with Rockefeller Centre as their focal point), and advertising and public relations (synonymous with Madison Avenue) – have their headquarters in midtown Manhattan, extending from the southern end of Central Park. Clothing manufacture and fashion houses are based mainly south of midtown, in an area known as the 'Garment Centre'.

The city is rich in centres of entertainment, the arts, education and culture, with a wide selection of theatres, concert halls, universities and museums. Broadway is the focal point of theatre (although few theatres are actually on that street, most being in side streets near Times Square). Many plays are also presented in 'off Broadway' and 'off-off-Broadway' theatres in Manhattan and Brooklyn.

Music has as its focal point the Lincoln Center for the Performing Arts on Upper West Side. Many concerts are also presented at Carnegie Hall near Central Park. The New York Philharmonic Orchestra and the New York City Ballet are other world-famous attractions.

Pre-eminent among the many museums are the Metropolitan Museum of Art on 5th Street,

▲ **NEW YORK, N.Y.** Two police officers patrol 52nd Street in the heart of midtown Manhattan, near the main offices of several radio and television networks, including CBS and NBC.

with almost 3 million works of art representing nearly every culture of the past 5000 years, and the Museum of Modern Art on 53rd Street.

Dominant features of the city's famous skyline are the 381 m (1250 ft) high Empire State Building and the twin towers, 414 m (1360 ft) high, of the World Trade Center. The United Nations headquarters stand beside the East River.

The city's landmarks and cultural centres attract millions of visitors each year, but many tourists also come to savour the life of the city's vast mosaic of nationalities and ethnic groups – and to sample their cuisines.
Population (state) 17 990 500; (city) 7 322 600; (metropolitan area) 8 546 800; (conurbation) 19 549 600
Map (state) United States Kb; (city) United States – Lb

New York State Barge Canal *USA* See ERIE CANAL

New Zealand See p 488

Newark *USA* Industrial port in New Jersey. It lies 16 km (10 miles) west of NEW YORK City and forms part of the New York conurbation. Industries include pharmaceuticals, chemicals and jewellery. Newark's international airport serves the whole New York metropolitan area.
Population (city) 295 200; (metropolitan area) 1 915 900
Map United States – Lb

Newcastle *Australia* Industrial city and port in New South Wales, standing at the mouth of the Hunter River, some 120 km (75 miles) north of

A DAY IN THE LIFE OF A NEW YORK COP

Roll call at the Todt Hill police station, Service Area No 1A, Staten Island takes place at 7.45 am: Officer Brian Patrick Byrnes, 24, steps out of his car outside the station at 7.40. Inside, a sergeant issues instructions to a dozen officers for their eight-hour shifts. Byrnes is allotted a routine patrol of a multistorey tenement in Stapleton, a mainly black area that has seen better days.

Byrnes rides a three-wheel motor scooter to Stapleton. There, he checks into the room in the run-down apartments which is assigned to the police. In a slum block like this, a careless cop can soon be a dead cop. Byrnes, every nerve alert, takes the lift to the top floor, and, hand on revolver, climbs the stairs to the roof. Nothing suspicious up there. On the way down, by the stairway, he looks and listens at each floor for signs of trouble.

Back on the street, he starts his foot patrol, keeping an eye on upper windows for bottles or other missiles that might be thrown at him. But Byrnes has a quiet day. Local shopkeepers wave hello or share some news or gossip. Passers-by ask him for directions. He takes the number of a speeding car, and radios ahead to colleagues in a squad car – he cannot give chase on his 50 km/h (30 mph) scooter.

Even his quiet days are tense. He knows that when there is trouble it happens suddenly. He has to think fast, keeping in mind the strict rule book that dictates his response to every emergency. Happily, he has never yet had to fire the gun he always carries on duty.

In the three years since he began his job, Byrnes' salary has risen about 75 per cent to a level comparable to what many junior executives earn. He gets 29 days' vacation a year, unlimited sick leave, free medical and dental care. And in 20 years, at the age of 44, he can retire on a pension of half his finishing salary and, if he chooses, take another job.

Byrnes graduated in sociology from a Florida college and went on to the New York College of Criminal Justice and the New York Police Academy. Now he lives in a Staten Island bungalow with his father, a retired fireman, and his mother, a retired schoolteacher.

He is the youngest of three children, but the only one still at home. His sister, a nurse, is married; his brother, a chemical engineer. Byrnes thinks he will probably stay at home until he marries.

His girlfriend, Patty, is from New Jersey. He likes to spend his days off with her, and perhaps drive his sports car to a mountain lake just over two hours away in Pennsylvania, where he keeps a 5 m (18 ft) motorboat. They can spend whole afternoons adrift on the lake, sitting in the inflated tubes of old car tyres, sipping beer, listening to the radio, and being spun round dizzily in the wake of other boats.

SYDNEY, and founded as a penal settlement in 1804. It exports coal, wool, dairy produce and wheat; its industries include iron and steel, chemicals, shipbuilding, metalworking and textiles.

Population 432 600
Map Australia – Ie

Newcastle *South Africa* Coal-mining and steel-manufacturing town at the foot of the Drakensberg in northern KwaZulu-Natal. Iron ore and steel are railed for export from Richards Bay.

Population (town) 38 770; (district) 53 590
Map South Africa – Cb

Newcastle upon Tyne *United Kingdom* City of north-east England in the county of Tyne and Wear, 400 km (about 250 miles) from LONDON. There was a bridge over the River Tyne in the 2nd century AD when the Romans built Hadrian's Wall, whose eastern end was at Wallsend, now a Newcastle suburb. A castle from which the city got its name was built in Norman times.

Although the port flourished in the 16th century, the real boom for the city came with the increased mining and export of coal in the 17th century and with the rise of engineering industries, especially shipbuilding, in the 19th century. At the height of the area's prosperity, the Tyne was lined with yards from Newcastle, past Jarrow – abbey home of the Venerable Bede (AD 673-735), biblical commentator and father of English historical writing – to Tynemouth with its 11th-century priory. When the Great Depression of the 1930s hit Britain, no area of the country suffered more than the Tyneside ship towns, especially Jarrow and Gateshead.

Since then, life has been a struggle for the area, but few cities have a better record of urban renewal. Newcastle has a spacious 19th-century city centre with grand civic buildings and a medieval church turned cathedral. It has a university, industrial estates on both banks of the river, a light railway metro system, and some shipbuilding. Perhaps the most impressive of the city's buildings is the railway station (1850-65), a reminder that the area produced engineer George Stephenson (1781-1848) who built the first steam locomotive for a public railway.

Population 280 000
Map United Kingdom – Ec

Newcastle-under-Lyme *United Kingdom* See POTTERIES, THE

Newfoundland *Canada* The country's most easterly province, covering some 372 000 km² (143 634 sq miles) and comprising a triangular island at the mouth of the Gulf of St Lawrence and Labrador on the mainland. Archaeological finds indicate that Vikings established a colony here in about AD 1000. Newfoundland was rediscovered in 1497 by an English expedition led by the Italian-born explorer John Cabot; the first settlers arrived from south-west England in the early 17th century. Much of Newfoundland Island is a plateau, rising to 842 m (2762 ft) in the Long Range Mountains. The coastline is heavily indented. Offshore, the Newfoundland Banks provide rich catches of cod for the many Newfoundlanders living in small fishing villages.

Besides fishing, the province's economy depends on forestry and mineral resources, including copper and gypsum. The development of mining and pulp and paper production has

created large centres at Corner Brook and Grand Falls. The province's capital and main city is ST JOHN'S on the east side of the island.

Population 568 420
Map Canada – Ic

Newgrange *Ireland* See BRUGH NA BOINNE

Newport *USA* Port and naval base on Rhode Island, at the mouth of Narragansett Bay, in the state of Rhode Island. Settled by Europeans in 1639, it quickly became a prosperous trade centre. Although it was partly destroyed during the American War of Independence (1776-9) it has more than 300 colonial buildings, including the oldest Jewish synagogue (1763) in the United States. It was a fashionable resort in the 19th century, and many millionaires' mansions are now tourist attractions.

A major yachting centre, it has hosted numerous America's Cup races. Jazz, folk and classical music festivals are held in Newport every year.

Population 28 200
Map United States – Lb

Newport News *USA* Major Atlantic port and part of a United States naval base at the mouth of the James River in Virginia. It handles coal, oil, tobacco, grain and ores and has one of the world's largest privately owned shipyards.

Population (city) 154 600; (metropolitan area) 1 443 200
Map United States – Kc

Newry *United Kingdom* Port and market town in Northern Ireland, 55 km (35 miles) south-west of the capital, BELFAST.

It grew up around a Cistercian abbey founded in the mid-12th century. The Newry Canal, built between 1730 and 1741 to link the Irish Sea and Lough Neagh via the Bann River, was one of the earliest ship canals in the British Isles. Its main industries are food processing and the manufacture of linen and clothing.

Population 21 660
Map United Kingdom –Bc

Neyshabur *Iran* See NISHAPUR

Ngaliema, Mount *Uganda/Zaire* See RUWENZORI

Ngami Depression *Botswana* Marsh rich in bird life, which was once a huge lake that was described by explorer David Livingstone as an inland sea. Fed by the OKAVANGO DELTA, it probably covered 52 000 km² (20 070 sq miles) when Livingstone saw it. Since the 1890s, papyrus reeds have blocked the main inflow and the marsh has gradually shrunk. A drought in 1965-6 dried it up completely for a time.

The Ngamiland district, which encompasses both the Ngami Depression and the Okavango Delta, covers 109 300 km² (42 200 sq miles). Its chief town is Maun.

Map Botswana – Bb

Ngaoundéré *Cameroon* Town situated on the ADOUMAOUA MASSIF, 450 km (280 miles) north-east of the capital, Yaoundé, to which it is linked by rail. Ngaoundéré is an Islamic centre dominated by the pastoral Peul people.

Population 47 510
Map Cameroon – Bb

New Zealand

AN URBAN PEOPLE IN A BEAUTIFUL WILD LAND, OUTNUMBERED 15 TO 1 BY THEIR LIVESTOCK, ARE FINDING NEW WAYS TO TRADE AND A UNIQUE NATIONAL IDENTITY

Bonds of trade and kinship tied New Zealand so firmly to Britain that it was long considered as the Britain of the South Pacific. Although it ceased to be a colony in 1907, another 40 years passed before it set up its own foreign ministry and developed an independent foreign policy. Moreover, Britain continued to provide a guaranteed market for New Zealand's primary produce.

Not until Britain joined the European Economic Community in 1973 did New Zealand's economic dependence on Britain really begin to decline and the country make a serious effort to diversify its trade. Success in that effort has made New Zealand an affluent modern state with prospering trade, with not only Europe but also Asia, the Americas and the Middle East. Its main trading partners today are, in order of predominance, Australia, Japan, the United States, the United Kingdom, Germany, Korea, Taiwan and China. Trade is now focusing increasingly on the markets of the East.

At the same time, New Zealand has become a leader in the South Pacific region. In 1971 it was host to the first meeting of the South Pacific Forum, a political grouping of 13 independent countries in the region. The country has been a vocal opponent of French nuclear testing on Mururoa Atoll in the Pacific, and in 1985 its government barred nuclear-armed or nuclear-powered ships from its ports – a policy that strained relations with the United States and jeopardised ANZUS, its defence treaty with the US and Australia. The same year, French agents sank the anti-nuclear protest ship *Rainbow Warrior* at AUCKLAND.

TWO MAJOR ISLANDS

New Zealand consists of two major islands (NORTH ISLAND and SOUTH ISLAND) and several smaller ones, including STEWART ISLAND and the CHATHAM ISLANDS; the Antipode Islands lie at a point in the world diametrically opposite Britain. Four out of five New Zealanders are of European origin (mainly British) and English is the most widely spoken language. New Zealanders' customs and social structure remain essentially British.

The country lies about the same distance as Italy from the Equator but the climate is more temperate, with regular rainfall all year and no extremes of heat or cold. Its terrain also resembles Italy in the nature and diversity of its features. The North Island does not have only fertile grazing land but also active volcanoes, such as the conical Mount EGMONT on the west coast, the majestic snow-capped Mount RUAPEHU (2797 m, 9177 ft) – a major skiing

area – and Mount NGAURUHOE, which forms an impressive backdrop to Lake TAUPO, a mecca for trout fishermen. Geysers, hot springs and boiling mud attract tourists to ROTORUA.

The mountainous South Island is fringed in the south-east and on the central east coast by extensive plains where cereals are grown and huge flocks of sheep graze. Fruit trees grow along the sloping foothills. Running the whole length of the island and rising from the plains with astonishing suddenness are the snow-capped peaks of the Southern Alps. Evergreen beech forests clothe the slopes and water abounds. Clean, clear lakes such as WAKATIPU and MANAPOURI, lying in ice-scooped hollows, and rushing mountain torrents create a breathtaking landscape.

Eighteen of the peaks here exceed 3000 m (10 000 ft) – the highest is Mount COOK at 3754 m (12 316 ft). At the higher levels, where the snow is permanent, there are ice fields and glaciers. Some are very large – the Franz Josef and FOX glaciers, dropping to the nearby west coast, and the Tasman, heading east from the main divide, are the most accessible. The area was heavily glaciated 2 million years ago and

the south-western coast is cut by fjords. Among the most striking is MILFORD SOUND, where sheer cliffs rise for hundreds of metres.

New Zealand is slightly larger in area than the United Kingdom but has a population of only 3.4 million, 74 per cent of whom live on the North Island. Eighty per cent of all the people are town dwellers – nearly half of them live in the area around Auckland, the boom city of North Island. Much of the South Island, which has greater areas of mountains, forests and farmland than the North Island, is very sparsely populated.

Farming and farm-related industries are still the basis of New Zealand's wealth but they now employ only about 10 per cent of the population. Most of the highly educated and urbanised New Zealanders work in service industries, such as administration, teaching and banking. The small fraction that work in agriculture tend some 55.2 million sheep and nearly 8.1 million cattle (beef and dairy).

THE MAORI PEOPLE

An important feature of New Zealand's new society is the recent increase in the number of

Maoris – the Polynesians who inhabited the islands for about 700 years before the arrival of the first Europeans. Dutch explorer Abel Tasman and his crew, saw them in 1642. Legend has it that the first Polynesian immigrants to New Zealand came from across the seas in seven canoes. Although there was a substantial Maori population when the Europeans arrived, two centuries later it seemed that the Maoris might die out. Their lands had been taken from them and the European way of life had weakened their culture. Tribal warfare with European-introduced firearms, and European diseases to which the Maoris had no resistance had reduced their numbers. However, improved health care and a cultural resurgence have changed that.

Now almost 15 per cent of New Zealand's population are at least half Maori and many others have some Maori ancestry. There is much intermarriage between Polynesian and European (or *pakeha*) New Zealanders. Calls have been made for more widespread teaching of the Maori language in schools and today there is a Maori pre-school programme (*kohanga reo*) teaching the language, traditions and values. Maori immersion (*kura kaupapa*) and bilingual schools are rapidly gaining popularity. The authorities have also recognised Maori land claims. In October 1992 the Maoris were given a 35 per cent share in the country's fishing industry. This resolved all outstanding fishing claims.

Early Maori artefacts – tools, weapons and ornaments – are mainly seen in museums now, but their population group's traditions of fine carving and weaving continue. The Maoris, like other New Zealanders, take great interest in European art forms; the Maori soprano Dame Kiri Te Kanawa is a world-renowned opera singer. The Maoris, converted to Christianity nearly 150 years ago, have developed their own forms of the religion, incorporating their own traditions. They are also active in the conventional churches and a Maori was Anglican primate of New Zealand until 1985, when he was appointed governor-general – the representative of the British sovereign, who is also sovereign of New Zealand.

The political and cultural resurgence of the Maori coincided with a wave of immigrants from the Pacific islands, notably Western Samoa, Fiji and the Cook Islands; most of them settled in Auckland. Relations between the races, while not as harmonious as previously, are still very good by international standards. The urbanisation of the Maori and Pacific islander populations since the Second World War has caused some social problems; nevertheless, New Zealand remains at heart a caring society and extreme views are exceptional.

INNOVATION AND ADAPTATION

Immigration from Europe, which swelled the population during the 1960s and 1970s, slowed in the 1980s but picked up in the 1990s; but the birthrate has fallen, bringing a decline in population growth. However, the population is still young (though getting older) – the median age is 31.6 years – and well educated (20 per cent of school leavers go to university), making for a high-calibre workforce. New Zealanders have always been innovators and are quick to adopt new technologies. Their sheep and cattle rearing benefited from the early use of refrigerated transport, crop spraying and mechanised shearing and milking.

Young and new attitudes have also changed national politics over the past decade. The conservative National Party had been in government for all but two three-year terms in the previous 35 years when the Labour party swept them aside in the 1984 election.

The October 1990 election was again won by the National Party led by Jim Bolger.

During the 1980s, the government faced economic difficulties created partly by the limitations on exports to Britain since 1973 and partly by the increased costs of imported oil. These factors, combined with the baby boom of the 1960s (which put 30 000 new workers on

▼ **ALPINE CREST Fed by heavy snowfalls brought by westerly winds, Fox Glacier stands at the crest of the Southern Alps on New Zealand's South Island. It falls steeply by about 200 m per km (1000 ft per mile).**

the labour market each year in a country whose labour force numbered less than 1.5 million), resulted in worsening unemployment in the early 1980s. Unemployment in the mid-1990s was still high, at about 9 per cent. In addition, surplus production of meat and dairy produce in Europe and the United States over several years led to oversupply and price cutting. Despite selling everything it produces, New Zealand has seen its agricultural profits shrink.

To help compensate for these problems, the country has developed several small-scale, but efficient secondary industries, for instance aluminium and methanol production and the prefabrication of houses for export. It has also diversified its dairy products to include new types of cheese, skimmed milk, casein and baby milk powders; and by aggressive selling it has won new markets in the Middle East, Eastern Europe, Japan and other Eastern countries, and Australia in particular.

An agreement on closer economic relations with Australia has in effect enlarged New Zealand's domestic market from 3 million to 21 million. Australia is the most important market for the new manufactured products.

AGRICULTURE AND INDUSTRY

Despite innovations, agriculture still holds the key to the country's future prosperity. New Zealand has some of the finest pastures in the world, thanks to its moist, warm climate. Well-managed farms with sheep, beef and dairy cattle produce extremely high yields of wool, meat and milk. The country's tiny population means that demand by the home market is small and an exceptionally large proportion of the produce is exported. In fact, New Zealand is the world's largest exporter of lamb and is second only to the European Union (EU) in the export of dairy products and second to Australia in wool exports.

Moves have been made to widen the basis for the country's wealth. For example, the timber industry has been developed to produce not only sawn timber (radiata pine, Douglas fir and Corsican pine), but also such products as veneer, chipboard, fibreboard, pulp and paper. Additional land has been planted with vines to support a growing wine industry in the HAWKE'S BAY, MARLBOROUGH SOUNDS and Auckland areas.

Exports have also been given a fillip with increased production of oranges, lemons, grapefruit, kiwi fruit (also known as Chinese gooseberries) and vegetables from the Bay of PLENTY and NORTHLAND, as well as apples from the orchards of NELSON and Hawke's Bay.

Moeover, the seas are yielding increased riches. New Zealand has declared an exclusive economic zone extending for some 370 km (200 nautical miles) beyond its shores, covering much of an extensive submerged continental landmass. This is a rich source of seafood, and its seafood exports have consequently increased in value.

Although it has little oil, the country is well endowed with other sources of energy. There are abundant reserves of coal; volcanic areas on North Island produce geothermal energy from the hot springs, and the mountain streams have been harnessed to produce 75 per cent of the country's eletrical needs. Offshore natural gas reserves have helped cut down oil imports, but are expected to last only until 2005.

Major industrial developments in recent years have included the building of an iron and steel works using iron sands from the west coast of the North Island, and an aluminium smelter, worked by hydroelectric power, that uses imported bauxite.

Manufactured goods now account for 29 per cent of the country's exports, compared with only 4 per cent in the mid-1960s.

CITIES AND EVERGREEN FORESTS

Much of the manufacturing is done in Auckland, which sprawls along and beyond an isthmus separating the Waitemata and Manukau harbours. The land is dominated by dozens of low volcanic cones, the most striking being Mount EDEN, near the city centre, and Rangitoto, an island in the Hauraki Gulf. Aucklanders, like other New Zealanders, live mainly in detached houses, mostly American or Scandinavian in style. Four out of five New Zealand dwellings are separate houses occupied by one household. Half of these have timber walls and nearly two-thirds are covered with galvanised steel roofs.

Timber houses are the rule in windy and earthquake-prone WELLINGTON, which sits astride a fault in the earth's crust (see PLATE TECTONICS, p 542) at the southern end of the North Island. The houses perch on steep hillsides overlooking the harbour and the high-rise offices that form the nation's financial and administrative centre. Although less than half the size of Auckland, Wellington is the capital and seat of New Zealand's parliament.

The principal town of the South Island is CHRISTCHURCH, lying on the CANTERBURY PLAINS below the lava flows of a huge extinct volcano; the city's port is in the flooded crater. Christchurch, one of the world's first garden cities, with weeping willows that trail their leaves in rivers meandering through a grid of streets, reminds British visitors of such English towns as Cambridge or Stratford and Americans of their own older East Coast towns. Its air of spacious, languid good living suggests an idealised image of the past.

DUNEDIN, also in the South Island, is still regarded as New Zealand's fourth main centre, although its population (about 110 000) has been surpassed by the new North Island city of HAMILTON. Dunedin was founded by Scots, but 19th-century gold rushes brought the city a flood of settlers who soon outnumbered the Scots. Gold wealth gave the city a fine heritage of Victorian buildings.

The isolation of New Zealand for some 70 million years has preserved creatures that exist nowhere else, especially flightless birds such as the long-beaked kiwi, which is the national symbol. The inhabitants of the country have taken its name and refer colloquially to themselves as 'Kiwis'.

In many parts, the most characteristic trees today have been brought into the country – the radiata or Monterey pine and the Monterey cypress, both from California. Vast forests of introduced conifers on the volcanic plateau of the North Island include one of the world's largest plantations. But in the great national parks the primeval evergreen forests of native conifers and broad-leaved trees are still dense and dark. In places they reach to broad, empty beaches lapped by the sparkling blue of the Pacific Ocean or Tasman Sea.

The splendid scenery of glaciers, mountains, forests and lakes, the empty beaches, clear air and lush pastures continue to justify the title of 'God's own country', bestowed by an early prime minister. The attractions are many and visitors come in increasing numbers; tourism has become a major earner of foreign exchange.

The local people, also, relish the space and beauty. Swimming, surfing and sailing are notable pastimes in a sport-mad country, where no place is more than 110 km (70 miles) from the sea.

New Zealanders remain especially proud of their national rugby team, the All Blacks, but consider their most famous achiever to be Sir Edmund Hillary who, with Sherpa Tensing Norgay, climbed Mount Everest in 1953.

In all its interests, the country knows no class or colour barriers. Ever since the days of the first settlers, when New Zealanders found that social divisions did not help to farm the land, society has been fair and perhaps uniquely classless.

This was the first country to give women the vote – in 1893 – and the Treaty of WAITANGI in 1840 gave Maori and Europeans equal rights as citizens. This exceptional equity may be the foundation of the new national identity which New Zealanders are building.

NEW ZEALAND AT A GLANCE	
Area (including all islands) 270 986 km² (104 604 sq miles)	
Population 3 449 700 **Per km²** 13 **(Per sq mile** 33)	
Capital Wellington	
Government Parliamentary monarchy	
Currency 1 New Zealand dollar = 100 cents	
Languages English, Maori	
Religion Christian (more than 50% Protestant)	
Climate Temperate maritime, with abundant rainfall. Rain falls throughout the year on South Island and during the winter on North Island. Average temperature in Wellington ranges from 8.1°C (47°F) in July to 16.2°C (61°F) in January	
Land use Grazing 50%, forest 27%, cultivation 1.4%, other 21.6%	
Main primary products Sheep, cattle, fruits, vegetables, timber, cereals, fish; natural gas, coal, iron, sand	
Major industries Natural gas processing, paper, iron and steel, aluminium smelting, fertilisers, cement, glass, transport equipment, machinery, agriculture, wool and textiles, meat and dairy processing, wine making, timber milling and wood processing, tanning, fishing, tourism	
Main exports Meat, dairy products, wool, timber, wood pulp and paper, timber products, fish, fruit and vegetables, hides and skins, aluminium	
Annual income per head (US$) 12 060	
Population growth (per thous/yr) 5	
Life expectancy (yrs) Male 72 **Female** 78	

A DAY IN THE LIFE OF A NEW ZEALAND SHEEP FARMER

Geoff Williams wakes late after the excitement of the previous day. It is Sunday and he did not get home until well after midnight after going to the races yesterday at the nearby small country town of Paeroa near Thames on North Island. His neighbour, Brian Macinroe, had a horse running which won its race and the celebrations went on through the afternoon and evening.

Geoff looks over at his sleeping wife, Jeannette, as he rises to make a cup of tea. The morning peace and quiet makes him aware of his worries over the farm and his family. Jeannette is pregnant. The thought of another child to bring up and pay for is a troublesome one. The price of wool and meat from his 4000 sheep has been gradually declining and the future looks uncertain.

He already has three children. His two sons, Dick and Jack, are at boarding school not far away in Auckland, and his daughter, Susan, is studying on a home science course at Otago University over in Dunedin on South Island.

Some of Geoff's friends have turned their sheep farms over to cultivation. But his land is not good enough. He has seriously thought of selling his 400 hectare (988 acre) hill property and trying his hand at something else, such as running a motel, perhaps at a beach resort somewhere in New Zealand, or even across the sea in Queensland, Australia.

Geoff's thoughts are broken by the loud yapping of his sheepdogs outside. After striding out into the open air, he lets his dogs loose, kick-starts his Honda motorcycle, and roars off into the hills behind his house to move a small herd of beef cattle on to fresh grazing. Geoff has bought the cattle to fatten and then resell.

When Geoff gets back, Jeannette is cooking a big breakfast of bacon, eggs and bubble-and-squeak.

Geoff spends a relatively easy day setting up electric fences in preparation for moving more stock later. He tours his boundary fences, checking where he will have to make repairs, searches out his shotgun for the duck-hunting season, and digs thoroughly over the vegetable patch. It is autumn and he has some precious time to catch up on farm maintenance before his busiest times of lambing in the spring and shearing in the summer.

Jeannette delivers a large box of jams and preserves to the local church bazaar and in the afternoon heads for Miranda Beach on Hauraki Gulf, about half an hour's drive away, where friends keep a boat for water-skiing. This time, Jeannette does not take her turn; she has had to give up the sport until after the baby is born. Geoff joins the group in his car before sunset for a quick swim and a few beers. He and Jeannete stay for a barbecue on the beach.

Ngauruhoe, Mount *New Zealand* Active volcano rising to 2291 m (7516 ft) in the Tongariro National Park on North Island, some 250 km (155 miles) north-east of the capital, WELLINGTON. For most of the time it emits only smoke, but occasional violent outbursts send molten lava flowing down the mountainside. The last big eruption was in 1978.
Map New Zealand – Ec

Ngorongoro Crater *Tanzania* Volcanic crater in the north of the country, about 230 km (145 miles) south-east of Lake VICTORIA. The crater, which is 22 km (14 miles) across, is one of the largest in the world. Nomadic Masai herders graze cattle on its floor. The crater floor is also home to more than 30 000 wild animals, including wildebeest, lions, leopards, elephants, and rhinoceroses. Extinct for several million years, the crater is now part of the Ngorongoro Conservation Area, which includes OLDUVAI GORGE.
Map Tanzania – Ba

N'Gounié *Gabon* River, 440 km (273 miles) long, rising in the Chaillu Mountains and flowing past Mouila, Fougamou and Sindara to its confluence with the OGOOUE River, 19 km (12 miles) above LAMBARÉNÉ. Its valley is Gabon's leading food-producing region. The river is navigable between Sindara and Lambaréné and between Mouila and Fougamou for much of the year.
Map Gabon – Bb

Nha Trang *Vietnam* Port and beach resort, some 300 km (nearly 190 miles) north-east of HO CHI MINH CITY. It is also a market town for farmers growing rice, sugar cane, rubber and coffee in the hills inland.
Population 263 090
Map Vietnam – Bc

Niagara Falls *Canada/USA* Spectacular falls providing a great tourist attraction on the Niagara River on the Canada-United States border. The river, which flows from Lake Erie to Lake Ontario, divides into two falls at Goat Island. The Horseshoe (Canadian) Falls are 54 m (176 ft) high and the crest of the falls is 675 m (2215 ft) wide.

The American Falls are 56 m (184 ft) high and 325 m (1066 ft) wide. The river drops 98 m (320 ft) in 48 km (30 miles). The gorge is spanned by bridges and the waters have been harnessed for hydroelectricity. Since the last Ice Age, erosion has caused the falls to move upstream at a rate of 1 m (39 in) a year; structures have now been put up to reduce the rate to 3 cm (1.18 in) a year.
Map Canada – Hd

Niagara-on-the-Lake *Canada* Popular resort at the northern end of the Niagara River, where it enters Lake Ontario near the Canada-United States border. It was the first capital of Upper Canada (1792-6), the present-day province of Ontario. The annual Shaw Festival of plays is held here in the summer.
Population 12 950
Map Canada – Hd

Niamey *Niger* National capital, about 100 km (62 miles) east of the border with Burkina (formerly Upper Volta). Until 1926, when it replaced ZINDER as the capital, Niamey was no more than a cluster of villages. Today, its tree-lined boulevards contain elegant French colonial buildings and modern office blocks. Two markets – the Small Market, which specialises in food, and the Great Market, which sells colourful cloth as well as leather, iron and copper ware – are features of the city. The trans-Sahara road and the road running west-east across the Sahel region meet here.

The magnificent Great Mosque – a large, white, modern building in traditional style – overlooks the Niger River, which flows past the city and which, like the country, is said to get its name from the Tuareg word *n'eghirren*, which translates as 'flowing water'. Nearby is the national museum, which is set in a 24-hectare (60-acre) park complete with zoo and botanical gardens.
Population 583 000
Map Niger – Ab

Niari *Congo* See KOUILOU

Niassa, Lake *East Africa* See MALAWI, LAKE

Nicaragua See p 492

Nicaragua, Lake *Nicaragua* Largest freshwater lake in Central America, covering 8264 km² (3191 sq miles), and the only freshwater lake containing marine life such as sharks, swordfish and sawfish (which reach it from the Caribbean Sea via the San Juan River). Many rare birds are found in the surrounding vegetation.
Map Nicaragua – Aa-Ba

Nice *France* The capital of Alpes-Maritimes department, situated on the Côte d'Azur, 25 km (16 miles) from the Italian border. Although it was annexed by France from the Kingdom of Sardinia in 1860, it has retained an Italian flavour. The Italian soldier and nationalist Giuseppe Garibaldi (1807-82), whose redshirt supporters helped to unify Italy, was born in the city.

The old town around the harbour is a network of narrow streets and squares with open-air markets. The new town is spread along the beach of the Baie des Anges, and the famous Promenade des Anglais with its grand hotels extends for more than 6 km (4 miles) to the airport. In addition to being a popular resort, Nice has important industries, including electronics and food processing.
Population 476 000
Map France – Ge

Nicosia (Lencosia; Lefkoşa) *Cyprus* Capital and largest city. It lies near the centre of the island, and has been divided since the island was partitioned into Greek and Turkish zones in 1974-5, after the Turkish invasion of northern Cyprus. Local industries include footwear, textiles, leather goods, cigarettes and tourism. The city is named after Nike, the Greek goddess of victory.

Its old town centre – founded before the 7th century BC and known then as Ledra – is encircled by 8 km (3 miles) of 16th-century walls and includes, in the northern (Turkish) sector, St Sophia Cathedral, now a mosque, and a 14th-century Armenian church. The Cyprus Museum, which has a collection of archaeological treasures found on the island and dating back to the Bronze Age, is in the southern (Greek) sector; also in this sector are the St John's Cathedral and the palace of the Greek Orthodox Archbishop of Cyprus.
Population 168 800
Map Cyprus – Ab

Nieborow *Poland* See LOWICZ

Nicaragua

ALMOST DESTROYED BY CIVIL WAR, NICARAGUA IS MAKING ITS WAY SLOWLY TOWARDS RECONCILIATION AND RECONSTRUCTION

There have been wars of one sort or another in Nicaragua, which became an independent republic in 1838, for more than a century. Indeed, the country has been troubled with conflicts and foreign interference ever since it was conquered by the Spanish in the early 16th century. Until 1990, little had changed, and its left-wing Sandinista government was fighting a guerrilla army which survived on aid and armaments sent by the United States.

Nicaragua lies between the Pacific and the Caribbean on the isthmus of Central America, bordered to the north by Honduras and to the south by Costa Rica. The Caribbean coast on the east is forested lowland and the country's wettest part. Here, malaria is rife – it is called 'Mosquito Coast'. Behind this lies a triangle of volcanic mountains, rising in the north to 2438 m (7998 ft). West of this, a belt of savannah lowland runs parallel to the Pacific coast, south-east from the Gulf of Fonseca. This region, containing lakes Managua and Nicaragua, is where nine in 10 Nicaraguans live.

Between the lowlands and the Pacific coast is another range of volcanic mountains, and the whole area is subject to devastating earthquakes. In both 1931 and 1972, MANAGUA, the capital city, was almost totally destroyed by two disastrous earthquakes. But this is the most fertile region in the country; volcanic ash from seven active volcanoes enriches the soil, and cotton, coffee, sugar cane, bananas, rice and tobacco are grown.

The east coast was colonised by the British in the 18th century and Jamaican slaves were brought in to work on plantations. The British also signed peace treaties with the local people, the Miskitos. Today, the east coast population is largely descended from the black slaves and from the original Amerindian inhabitants – the Niguiranos, Chorotegas, Chontales and

Caribs. The rest of the population (nearly 70 per cent) are *mestizo* – of mixed white and Amerindian descent. The north-east is the home of the Miskitos, who are descended from a mixture of black slaves and Caribs. These were the people most threatened by the conflict that racked Nicaragua until recently – it was in the north-eastern hills that government troops and guerrillas were fighting.

THE SANDINISTAS AND THE 1980s

The Sandinista government came to power after a revolution in 1979, which overthrew a repressive military regime that had been dominated by one family – the Somozas – since the 1930s. The Sandinista National Liberation Front took its name from General Augusto César Sandino, a patriot who was assassinated by the Somozas in 1934.

Before the revolution, the United States backed the Somoza military regime against the left-wing guerrillas. It trained the National Guard – the force of police and soldiers by which the Somozas kept their hold on power. Later, the United States, fearing the spread of communism through its 'backyard', supported the counter-revolutionaries, or Contras, a number of whom were former members of the National Guard. The Contras were trained by US advisers in Honduras and it was with US help that Nicaragua's ports were mined. The United States, which was formerly Nicaragua's main trading partner, also introduced a trade embargo in 1985, refusing to accept imports of Nicaraguan sugar.

This was not the first time in Nicaragua's history that the US had intervened. American adventurer, William Walker, invaded it in 1855, declared himself president and stayed in power until the US Navy ousted him in 1857. In 1912, US marines entered a country torn by conflict between the two major cities of Granada and Léon and stayed in occupation until 1925, and again from 1927 until 1933.

WAR AND PEACE

The Nicaraguans greeted the fall of the Somozas in 1979 with enthusiasm, but the war that had brought it about cost them dearly. Extensive damage had already been caused to the economy under the General Anastasio Somoza. Much of the country's agriculture had been destroyed, and continued to be damaged by subsequent fighting. Under the leadership of President Daniel Ortega, the Sandinistas reformed the country along socialist principles. They nationalised industries, commerce and banking.

The Somozas had also possessed large tracts of Nicaraguan land, which the Sandinistas redistributed more evenly among the people – in small landholdings worked and owned by themselves, or in collective farms. Health and education reforms were impressive under the Sandinistas. The Sandinistas tried to silence all resistance. Although the constitution guaranteed pluralism and basic democratic rights, newspapers were shut down, members of opposition parties were arrested and the Catholic church was suppressed.

A change came with the 1990 elections, which showed a clear victory for the oppos-

ition. During the weeks after the elections, the political parties, until then at war with each other, negotiated a ceasefire and agreed an amnesty for all political prisoners. The long civil war was officially ended by June 1990. But that was not the end of all the problems.

Unrest continued. When President Violeta Barrios de Chamorro came to power and a new constitution was approved in 1990 it was generally believed that the conflict was finally over. The country was at last at rest and an attempt at national reconciliation could be made. Money formerly wasted on a civil war could be better spent. Foreign investors were again beginning to show an interest in Nicaragua – the fact that the country has a wealth in mineral deposits helped attract foreign capital. Nicaragua also has sufficient resources to produce enough food, not only to feed its own population but also to become a major exporter of agricultural products.

But in 1993, left-wingers (now renamed *recompas*) and rightists (*recontras*) clashed again and Chamorro was unable to negotiate. A rift in her government and more political upheavals once again scared off foreign investors, with United States loans frozen and other investors ready to follow suit if the situation did not improve dramatically. Early in 1994, after a realignment of political blocs and some constitutional reforms, right-wing guerrillas in the north announced a ceasefire.

During most of that year, however, the peace process remained fragile. The struggle for power between the right and left of the coalition government continued, as did a dispute with the US over the property of US citizens confiscated during Sandinista rule.

Former Sandinistas still dominated banking and held important government positions and the combined unemployment and underemployment rate was almost 60 per cent; 70 per cent of people were living in abject poverty.

NICARAGUA AT A GLANCE	
Official name Republic of Nicaragua	
Area 129 494 km² (49 986 sq miles)	
Population 3 987 000 **Per km²** 31 **(Per sq mile** 80)	
Capital Managua	
Government Presidential republic	
Currency 1 córdoba = 100 centavos	
Language Spanish (official), some English, and Amerindian dialects	
Religion Christian (88% Roman Catholic)	
Climate Tropical – only the Pacific coast has dry winters; temperature throughout the year in Managua is 27°C (81°F); the mountains are much cooler	
Land use Grazing 42%, forest 26%, cultivation 10%, other 22%	
Main primary products Coffee, cotton, rice, maize, sugar, bananas, cattle, timber; gold, silver, copper	
Major industries Agriculture, chemicals, textiles, cement, mining, food processing, forestry	
Main exports Cotton, coffee, meat, bananas, sugar, metals, timber	
Annual income per head (US$) 830	
Population growth (per thous/yr) 34	
Life expectancy (yrs) Male 60 **Female** 66	

Niedersachsen *Germany* See LOWER SAXONY

Nieuw Amsterdam *Surinam* See PARAMARIBO

Nieuwe Waterweg (New Waterway) *Netherlands* Deep-water channel, 28 km (17 miles) long, cut between 1866 and 1872 from Rotterdam to the North Sea. It replaced the shallow channels of the Rhine delta, thus enabling ocean-going vessels to reach Rotterdam. The HOOK OF HOLLAND and EUROPOORT are at its seaward end.
Map Netherlands – Bb

Niger *West Africa* River, 4170 km (2590 miles) long, whose basin is the largest in West Africa, and the third largest in Africa, after the Nile and the Zaire (Congo) Rivers. The Niger rises on the Sierra Leone-Guinea border and flows north-east across Guinea into Mali. Here, just below the city of TIMBUKTU (Tombouctou), it makes a great curve south-eastwards, and continues across south-west Niger. It then forms part of the Niger-Benin border and continues into Nigeria, where it sweeps south to the Bight of Benin to empty into the Gulf of Guinea.

The vast Niger delta, an area of mud flats and mangrove swamps covering some 20 000 km² (7720 sq miles), stretches for 300 km (185 miles) north-west from Port Harcourt. Several rivers, besides the Niger, cross the delta, including the Benin, Escravos and Forcados, and they are increasingly blocked by sandbars created by longshore currents. However, new prosperity has come to the delta with the exploitation of its vast oil deposits, which were first discovered in 1956, at the town of Oloibiri.

The Niger is an ancient highway of trade. The traffic in palm kernels (which yield palm oil) from the delta was so great that in the 18th century the British called the delta rivers 'the oil rivers'. The Niger is navigable from the sea to the Nigerian town of Jebba, and for about 1610 km (1000 miles) above the Nigerian town of Yelwa. Several hydroelectric and irrigation schemes have been set up along its length, including the lake at KAINJI covering 1295 km² (500 sq miles).
Map Africa – Dc

Niger See p 494

Nigeria See p 496

Nijmegen *Netherlands* City situated on the Waal River (Lower Rhine) south-east of AMSTERDAM and only 7 km (4 miles) from the German border. In September 1944, Allied and German forces clashed here for the possession of the city's bridge over the river. Although Nijmegen suffered extensive damage, its 16th-century town hall and 15th-century Church of St Stephen still stand in the market place. The preserved foundations of a palace of the Holy Roman Emperor, Charlemagne (742-814), can also be seen. Today, its products include machinery and paper.
Population (city) 146 990; (Greater Nijmegen) 246 880
Map Netherlands – Bb

Nikko *Japan* Tourist town set on the edge of a beautiful mountainous national park, 1407 km² (543 sq miles) in area, in central HONSHU Island, 120 km (75 miles) north of TOKYO. The town contains a large complex of temples and shrines, including the Toshogu, the magnificent maus-

oleum of the Shogun (warlord) Ieyasu (1544-1616). Ieyasu, who ruled Japan from 1603 to 1605, was the founder of the Tokugawa dynasty of shoguns, which kept Japan in feudal seclusion for more than 250 years until 1853.

The mausoleum's main building, covered in vermilion lacquer, is said to decorated with more than 2 million sheets of gold leaf. It is the setting for an annual May pageant in which men parade in Japan's traditional costume of samurai warriors.
Population 23 900
Map Japan – Cc

Nikolayev *Ukraine* See MYKOLAYIV

Nilaveli *Sri Lanka* See TRINCOMALEE

Nile (Al-Bahr; Bahr el Nil) *Africa* The world's longest river, flowing 6695 km (4158 miles) northwards from its farthest headstream, the Luvironza in Burundi, to the Mediterranean via its delta in north-eastern Egypt.

One of its major sources is Lake VICTORIA, from where it drops, as the Victoria Nile, through lakes Kyoga and Albert, and then flows on, as the Albert Nile, to the Sudanese border at Nimbule. Winding through the swamps of the SUDD it is called the Bahr el Jabel, it then continues to Khartoum as the Bahr el Abiad, but both sections are known in English as the White Nile. At Khartoum it joins the Blue Nile (Abay Wenz; Bahr el Azraq), which rises at Lake TANA in the Ethiopian highlands and is about 1370 km (850 miles) long. From Khartoum the river is known simply as the Nile.

The Nile system drains some 2 870 000 km² (1 108 000 sq miles) – about one-tenth of Africa. Summer rains and melting snows in Ethiopia cause the annual flooding of the river – now controlled by the ASWAN HIGH DAM – which enabled the flourishing civilisation of ancient Egypt to develop in its valley. The dam has now disturbed the natural rhythm of the flooding of the Nile valley and has made it necessary to use artificial fertilisers on the lands in the valley. The Nile is navigable for most of the way through Sudan and Egypt, except in the low-water season between Khartoum and Lake Nasser (formed by the Aswan High Dam), where there is a series of rapids, the Cataracts.
Map Africa – Gc-Ga

Nilgiri Hills (Nilgiris) *India* Compact group of plateaus rising to 2636 m (8647 ft) at Doda Betta, a precipitous peak about 400 km (248 miles) north-west of the mainland's southern tip. Much is open, rolling country, and before independence in 1947, the British maintained an English-style hunt at UDAGAMANDALAM. About half the area is forested, and plantations in the area produce coffee, tea, rubber and medicinal plants.
Map India – Be

Nimba Mountains *West Africa* Mountain range in south-east Guinea and north-east Liberia, along the border with Ivory Coast. It forms part of the Guinea Highlands and rises to 1752 m (5748 ft) at Mount Nimba. The mountains contain high grade iron ore, and diamonds are also exploited. Its wildlife includes hippopotamuses, crocodiles, boars, jackals, various members of the cat family, monkeys and baboons.
Map Guinea – Cb; Liberia – Ba

nimbostratus Low thick cloud formation of a uniform dark mass from which rain or snow falls.

nimbus General name for a thick cloud from which rain falls, often used in combination with other cloud names, as in cumulonimbus.

Nîmes *France* City overlooking the Rhône River, 100 km (62 miles) north-west of Marseilles. It was an important Roman settlement from 120 BC until the 5th century AD, and many Roman remains survive. These include the Pont du Gard, a three-storey aqueduct across the Gardon Valley, and an amphitheatre that can hold 21 000 people.

Today, the manufacture of textiles and clothing is one of the city's chief industries, with denim getting its name from a locally manufactured cloth known as *serge de Nîmes*.
Population 128 500
Map France – Fe

Nin *Croatia* See ZADAR

Nineveh *Iraq* Remains of an ancient Assyrian capital standing on the Tigris River near the city of Mosul, 350 km (217 miles) north-west of the capital, BAGHDAD. Mentioned often in the Bible, the city was the hub of an empire in the 8th and 7th centuries BC under the Assyrian kings Sennacherib and Assurbanipal. Excavations here have revealed palaces and a library containing Sumerian and Assyrian dictionaries, giving scholars an invaluable key to these languages.
Map Iraq – Ba

Ningbo *China* City of Zhejiang Province, 60 km (37 miles) east of HANGZHOU. It is a major seaport and an important centre for engineering
Population 1 090 000
Map China – Ie

Ningxia-Hui Autonomous Region (Ninghsia-Hui) *China* Region of northern China between Nei Mongol and the province of Gansu, stretching south from the Huang He River. The region is named after the Muslim Hui people who constitute about 30 per cent of the population. The southern part of the region lies in the LOESSLANDS OF CHINA. Ningxia-Hui covers some 60 000 km² (23 160 sq miles). Its extensive pastureland is the home of the argal sheep, known for their high-quality wool. Irrigation has opened up some of the semi-arid land for the cultivation of rice, wheat and cotton, particularly along the Huang He. Ningxia-Hui also has deposits of coal.
Population 4 870 000
Map China – Fd

Niokolo-Koba National Park *Senegal* Wildlife reserve in the south-east, covering about 9000 km² (3475 sq miles). The reserve, which surrounds the village of Niokolo, some 55 km (35 miles) from the Guinea border, consists mainly of savannah (grassland of tropical and sub-tropical Africa) with forest along the Gambia River and its tributaries. It is home to 70 species of mammals, including elephants, hippopotamuses and antelopes, 300 species of birds and 60 species of fish and reptiles.
Map Senegal – Bb

Nios *Greece* See fos

Nirinyaga *Kenya* See KENYA, MOUNT

Niger

*THE SAHARA IS ADVANCING IN A
COUNTRY WHERE WATER FROM
THE NIGER RIVER HAS BEEN THE
SOLE LIFEBLOOD THROUGH
YEARS OF DROUGHT*

In the 15th century, the landlocked African state of Niger was the site of two rich kingdoms: the Songhai Empire in the south-west and the Hasusa States in the south-east. These two corners of the country are still the most densely populated.

In the extreme south-west, the NIGER River and its tributaries bring life to the landscape and the south-east has Lake Chad, but the rest of the country is terribly short of water. In a country where 90 per cent of the population are engaged in agriculture and stock raising, more than half of the land in the north is covered by the encroaching SAHARA – with the wasteland of Ténéré probably the most hostile – while to the south lies the drought-stricken SAHEL. Only 3 per cent of the country is arable, and even this small portion is threatened by increasing soil erosion and overgrazing.

The capital, NIAMEY, stands on the Niger in the south-west. It is an elegant town with wide, tree-lined roads and a modern centre. Busy markets add African colour to the remnants of French colonial rule, pavement cafés and French restaurants. Colonial rule began in the 1890s and ended with independence in 1960.

A military regime came to power in 1974 but in 1989 the country's ruling body, headed by Ali Saibou, began paving the way to democracy and multiparty politics. The constitution, rewritten in 1989, was amended in 1991 by a national conference and again rewritten in 1992. Transitional government began in 1991. In 1993 the first multiparty elections since independence brought victory to a nine-party coalition government and President Mahamane Ousmane. In the next elections, in Janaury 1995, opposition parties narrowly defeated the government.

TRADITIONAL GRAZING GROUNDS

Near the south-western border, the Niger River passes through the 'W' NATIONAL PARK – so called because the river runs in a 'W' shape. The elephants, buffaloes, antelopes and lions which roam the park's wooded savannah are protected. Before reaching the park, the river runs past villages of square, mud houses with flat roofs and tiny windows. Every settlement has its own mosque, with a tall minaret, elegant in spite of it being made of mud.

The people are mostly Muslim. Mainly members of the Hausa, Djerma and Songhai tribes, they fish and farm their own food, growing rice and vegetables on land flooded by the river, and scattering millet and sorghum seed farther afield. In addition, cotton, cowpeas for fodder, and groundnuts are grown for sale. Harvests away from the river depend on the erratic rains between June and October, and since there has been intermittent drought from 1968, and a devastating one in 1983-4, the millet and sorghum crops have failed again and again. Food supplies have been decreasing for years. Hunger has driven people to the towns, especially Niamey, in search of work, money and food, and many have travelled on into neighbouring Nigeria, or even to the Ivory Coast. The same pattern of drought and migration has affected all the southern region of Niger, eastwards to Lake Chad.

Along the southern border with Nigeria, in places like ZINDER, the people are Hausa, while the Kanouri or Beriberi (of mixed Hausa and Kanouri stock) live in the Lake Chad area. All of them, except for splinter groups who have retained their traditional religious beliefs, are Muslims. The movement of population threatens to destroy this culture, too, and towns such as Zinder and Maradi are almost as over-crowded as the country's capital.

The population in the north is concentrated around the AÏR Mountains, whose jagged peaks rise to 1944 m (6378 ft) from the desert flatness. Valleys are traditional grazing grounds for the camels, cattle, horses and goats of nomadic Tuareg people, of Berber origin. A few caravans are still taken into the desert by the Tuareg people. One of their main trails runs from AGADEZ to the oasis of Fachi, where they collect dates, then to Bilma, where they load their camels with salt. However, since 1973, whole clans have been wiped out by drought. In the mountains there are hot springs and caves with magnificent rock drawings dating back 5000 years.

Uranium has been mined in the Aïr Mountains since the 1970s and is now Niger's main export, superseding groundnuts. The mines have taken metalled roads into the desert where once there were only trails. Uranium mining created economic growth at first, but since the 1980s, when the world price dropped, the country has again been immersed in severe poverty.

A challenge for the government has been to appease the Tuareg, who have demanded their share of uranium earnings and organised an armed resistance movement in support of their claim to autonomous regions. (Some progress was made in peace talks in mid-1994.)

Another challenge has been to deal with the effects of the CFA franc's devaluation in January 1994, which led later that year to riots and a three-day workers' strike in response to the fall in the franc's purchasing power. But the overriding challenge has been to feed all of Niger's people after years of drought – relieved only by catastrophic floods in the east and centre of the country in August 1994, when 20 000 people were left homeless and the country appealed for international aid.

NIGER AT A GLANCE	
Official name Republic of Niger	
Area 1 267 000 km² (489 191 sq miles)	
Population 8 962 000 **Per km²** 7	
(Per sq mile 18)	
Capital Niamey	
Government Republic	
Currency 1 CFA franc = 100 centimes	
Languages French (official), Hausa, Djerma, and other indigenous languages	
Religions Muslim (83%); animist, Christian	
Climate Hot and dry in the north; the extreme south has a tropical climate with summer rainfalls. Average temperature in Niamey ranges from 23.8°C (75°F) in January to 34°C (93°F) in May	
Land use Grazing 7%, cultivation 3%, forest 2%, desert 88%	
Main primary products Millet, sorghum, groundnuts, cotton, cowpeas, rice, livestock; uranium, tin, phosphates	
Major industries Agriculture, textiles, food processing, mining	
Main exports Uranium, groundnuts, livestock, hides and skins, meat, cotton	
Annual income per head (US$) 300	
Population growth (per thous/yr) 31	
Life expectancy (yrs) Male 42 **Female** 45	

Nis (Nish) *Yugoslavia (Serbia and Montenegro)*
Ancient east Serbian city 201 km (125 miles) south-east of BELGRADE, situated on the Nišava River near its confluence with the Southern Morava. A strategic Roman town commanding the major trade routes between the central Danubian plains and the Aegean and Bosporus, it was the birthplace of Constantine the Great (about AD 274-337), Rome's first Christian emperor. Attacked in turn by Goths, Huns, Bulgars, Hungarians and Byzantines, it never-theless prospered under the Turks from 1456 to 1877 as a key caravanserai between Constantinople and Hungary. Nis is still a hub of central European road and rail routes, and an important manufacturing centre producing machinery, electrical appliances, ceramics and processed foods.
Population 247 900
Map Yugoslavia – Cc

Nishapur (Neyshabur) *Iran* Town 700 km (435 miles) east of the capital, TEHRAN. Omar Khayyam, the Persian poet, mathematician and astronomer (about 1050-1123), was born here. His tomb, set in a garden, is 3 km (2 miles) to the south-east.

Craftsmen in the town make pottery and in the district around it, fruits, cereals and cotton are grown. Turquoise, which is mined in the mountains to the north, is cut and polished in workshops in the town.
Population 59 200
Map Iran – Ba

Niswa (Nazwa) *Oman* Provincial capital of the interior, Niswa is situated 140 km (87 miles) south-west of the capital, MUSCAT. Lying behind the Hajar Mountains, it is linked to the coast by an asphalt road that was built through the Sumail Gap in 1976. The old part of Niswa dates from the 6th and 7th centuries. The modern town that has sprung up in recent years contains offices, a police academy, a hospital and an agricultural college. Alfalfa, leather, silverware, spices and animals are traded and goats are auctioned at the city's weekly market.

Population 56 230

Map Oman – Aa

Niterói Bridge *Brazil* Bridge linking the city of Niterói with RIO DE JANEIRO, and also, more formally, known as the Ponte Costa é Silva. At 14 km (8.5 miles), it is one of the world's longest bridges and was completed in 1974. At its highest point, it soars 72 m (236 ft) above the waters of Guanabara Bay.

Map Brazil – Dd

Nitra *Slovakia* Market city 75 km (47 miles) north-east of BRATISLAVA. It is situated among rich farmlands, orchards and vineyards, and has food-processing, brewing, winemaking, engineering and electrical industries. Nitra's cathedral incorporates Slovakia's oldest church, consecrated about 11 centuries ago.

Population 82 000

Map Slovakia – Bb

Niue *Pacific Ocean* The world's largest raised coral island, covering 259 km² (100 sq miles) in the South Pacific, 2100 km (1304 miles) northeast of New Zealand. Its population is Polynesian; however, population figures are dropping as people leave to settle in New Zealand.

British explorer Captain James Cook landed here in 1774; it became a British territory in 1900. It was then transferred to New Zealand, but was granted self-governing status in 1974. As with the Cook Islands, New Zealand handles its foreign affairs and defence, and contributes aid. Niue's main exports are passion fruit, copra, honey and limes. Its capital is Alofi.

Population 2500

Map Pacific Ocean – Ec

Nizhniy Novgorod (Gor'kiy; Gorky) *Russia* Major industrial city, regional administrative centre and port on the Volga River, 380 km (235 miles) east of MOSCOW. It was founded as a fort in 1221 and grew into a riverside market centre. In the 19th century it was the site of Russia's largest trade fair. It was renamed in 1932 in honour of the Russian author Maxim Gorky (1868-1936), who was born and bred in the city's slums and described it in works such as his autobiographical *Childhood*, published in 1913-14.

In 1991, it reverted to its original name. For years during Soviet rule it was a place of internal political exile and closed to foreigners. The Soviet scientist and dissident Andrei Sakharov (1921-89) was banished here in January 1980. Volga motor vehicles, as well as riverboats and hydrofoils, are made in the city. Other industries include oil refining and the manufacture of aircraft, diesel motors, machine tools, paper and agricultural equipment.

Population 1 445 000

Map Russia – Fc

Nizhniy Tagil *Russia* Industrial city in the Ural Mountains, about 125 km (78 miles) north of YEKATERINBURG. Founded in 1725 and close to gold, copper and iron ore mines, its main products are chemicals, machinery, iron and other metals.

Population 439 000

Map Russia – Gc

Nizina *Bulgaria* See THRACE

Njazídja *Comoros* Largest island within the Comoros group, covering an area of 1147 km² (443 sq miles), and formerly known as Grande Comore. The capital of the island, MORONI, is situated on the west coast.

Population 255 800

Map Madagascar – Aa

Nkongsamba *Cameroon* Picturesque town in a narrow valley 220 km (137 miles) north-west of the capital, YAOUNDÉ. It is a busy coffee market, and the home of some of the country's richest people. Fog, which is caused by nearby active volcanoes, is common.

The area is popular with tourists, who visit Mount Manengouba and the waterfalls at Ekom.

Population 86 870

Map Cameroon – Bb

Nong Khai *Thailand* Mekong River town beside the Laotian border, 575 km (357 miles) north-east of BANGKOK. The town, which is only 25 km (15 miles) downstream from the Laotian capital, VIENTIANE, on the opposite bank of the river, is a railhead for goods being transported to and from Laos. Tobacco and rice are the main crops in the surrounding province of Nong Khai.

Population (town) 118 100; (province) 749 800

Map Thailand – Bb

Nonthaburi *Thailand* Province and town just north of BANGKOK, which are rapidly being swallowed by the expanding capital. The province was once renowned for its durians – a fruit said to have the most exquisite taste, but which even its devotees agree has the foulest smell. Today, only a few durian orchards have survived the metropolitan sprawl.

Population (province) 373 000; (town) 31 000

Map Thailand – Bc

Noord-Brabant *Netherlands* Southern province covering 5106 km² (1971 sq miles) on the Belgian border. Much of the south is infertile heath and forest, suitable only for rearing sheep and cattle. However, recent industrial development, especially around EINDHOVEN, has brought an increase in population and prosperity.

The province was part of the old Duchy of Brabant, which was divided into three parts under French rule. The southern part now forms Belgium's northern provinces, including Belgian Brabant. The capital is 's-Hertogenbosch.

Population 2 243 500

Map Netherlands – Bb

Noordzeekanaal *Netherlands* Canal opened in 1876 between AMSTERDAM and the NORTH SEA at IJmuiden. It is 27 km (17 miles) long, and has shortened the sailing route to Amsterdam, formerly reached via the Zuiderzee. Because the land which the canal crosses is below sea level, locks lower rather than raise vessels at the entrance.

Map Netherlands – Ba

Nordkapp *Norway* See NORTH CAPE

Nördlingen *Germany* Medieval town in Bavaria, 80 km (50 miles) south-west of NUREMBERG. Its ancient buildings are encircled by fortified walls and gates. Each summer, the town holds a fancy-dress horserace, Germany's oldest festival.

Population 18 300

Map Germany – Dd

Nord-Pas-de-Calais *France* Densely populated industrial and farmland region in the far north of the country. Its growth and urbanisation were linked to coal mining, steel making, engineering and textiles, but these industries have now declined. New industries include the manufacture of car parts, plastics, chemicals and electronics. The region centres on the conurbation of LILLE-ROUBAIX-TOURCOING.

Population 3 970 000

Map France – Da

Norfolk *United Kingdom* Eastern English county of 5355 km² (2068 sq miles) occupying the northern bulge of East Anglia. Between the coast and the county town, NORWICH, lie the Norfolk Broads, a series of river-linked lakes, partly natural but largely created by medieval peat diggers. The lakes are a popular venue for water sports. Norfolk's Breckland area has some 1030 km² (400 sq miles) of heathland, pine and conifer plantations. The coast is lined with holiday resorts, the largest being GREAT YARMOUTH, a fishing port and supply base for North Sea gas rigs.

Population 714 000

Map United Kingdom – Fd

Norfolk *USA* Atlantic port in the southern state of Virginia at the mouth of Chesapeake Bay. It has a deep natural harbour and is the home of the world's largest naval base, the command centre of the United States' Atlantic Fleet.

Population 279 700

Map United States – Kc

Norfolk Island *Pacific Ocean* Island, covering 35 km² (13 sq miles) and lying about 1400 km (869 miles) east of BRISBANE, Australia. Reached by the British explorer Captain James Cook in 1774, it served as a penal colony between 1788 and 1855 and was settled by families of mutineers from HMS *Bounty* (who moved here from PITCAIRN) in 1856. Islanders celebrate a public holiday, Bounty Day, on 8 June. The population is divided into 'mainlanders' (later immigrants) and 'islanders', descendants of Bounty mutineers, who have preserved their identity. The island is administered as an Australian external territory with a considerable degree of self-government.

Population 2000

Map Pacific Ocean – Dd

Noril'sk *Russia* Mining town in the Siberian Arctic, 80 km (50 miles) east of the YENISEY River. It was founded in 1935 as part of what became known as the Gulag Archipelago – the chain of prison camps described in the novel *The Gulag Archipelago* by Russian author Alexander Solzhenitsyn (1918-). Now populated by workers earning above average wages, rather than by exiles and convicts, it lies 380 km (236 miles) north of the Arctic Circle.

Population 173 000

Map Russia – Jb

Nigeria

A COUNTRY WITH A MIXTURE OF DIFFERENT PEOPLES AND MANY ECONOMIC PROBLEMS PAINFULLY ENCOUNTERS THE REALITIES OF MODERN NATIONHOOD

Known as 'the giant of West Africa', Nigeria is the region's largest country and Africa's most populous nation. Changes have been forced at breakneck pace, partly due to political upheaval and partly to the discovery of oil and natural gas which have come to dominate the nation's economy. The former capital LAGOS, a teeming city of skyscrapers, flyovers, slums and almost 6 million people, is a testament to Nigeria's riches, and the city has tended to colour the country's image in the eyes of foreigners. But despite the roar of cars and taxis and the blare of hooters, Nigeria is a nation where the brash, bustling present coexists with a very persistent past.

Nigeria is a collection of different peoples and separate cultures thrown together by the British, without consent, and welded into a nation. Although many Nigerians are not keen to admit it, ethnic tensions lie beneath the surface and have in the past erupted in terrible conflicts, such as the Biafran War of 1967-70. Religious differences, expecially between the Christians and the Muslims, cause tensions, too.

The population, believed to be over 100 million, consists of some 250 ethnic groups. Of these, four – the Hausa and the Fulani in the north, the Yoruba in the south-west, and the Ibo in the south-east – account for around 65 per cent of the population.

VARIED LANDSCAPES

The landscapes that these peoples inhabit are equally varied: from the coastal lands of the south through the tropical forest belts of the interior to the mountains and savannahs of the north. The southern coast is fringed by miles of idyllic sandy beaches, broken only by mangroves at those points where watercourses trickle into the sea. Back from the coast, a maze of creeks and lagoons fans out on either side of the NIGER delta. It was these inland waterways which provided a means of transport into the interior before roads were built.

Today, several channels are silted up; but dugout canoes and small boats still carry people and goods between villages and towns such as Badagri and Calabar. Farther north, the forest belt which produces Nigeria's tree crops (palm oil, rubber and hardwoods) as well as bananas and cocoa, gradually gives way to grassy parkland. The land rises, past the junction of the two great rivers, the Niger and BENUE, towards the mountains and holiday resorts of the central JOS PLATEAU and, in the east, towards the steep valleys and wooded slopes of the Cameroon highlands. North of the Jos Plateau, the savannahs become increasingly dry and

grassy; and the presence of the skeletons of animals are a constant reminder that the drought of the SAHEL can reach this far south.

Inevitably the contrast in landscape between northern and southern Nigeria is reflected in the livelihoods of the country's people. The heavy rains that fall in the south encourage the growth of yams, cassava, maize and a wide range of vegetables. Rice is grown on tiny patches of land beside rivers and streams which flood in the wet seasons, and on large stretches of irrigated land.

The north, however, is not so self-sufficient. It depends on one rather unreliable rainy season, whereas the south has two. Cereals, such as millet, sorghum and maize, are the staples of the northern economy, with groundnuts and cotton also important. Drought in the early 1990s severely affected agricultural activities in the north and there is progressive desertification in the region. However, unlike the south, the north is not plagued by tsetse fly because it does not have the swamps and vegetation that are the fly's natural habitat. Livestock can therefore play a vital part in the economy.

Before the arrival of the European powers in the 15th century, economic activity in West Africa – including what is now Nigeria – focused on the Sahel. City-states, such as SOKOTO, KATSINA and KANO, grew up along the fringes of the SAHARA and thrived on trading northwards across the desert. The Europeans shifted this focus to the coast; and there it has remained ever since. At first they traded in gold and ivory, but it was not long before they switched to slaves.

When the British outlawed the slave trade in 1807, European merchants were forced to turn to other commodities, particularly palm oil. This was used in soap, a product suddenly much in demand by workers in the grimy

factories of Britain's Industrial Revolution. Cocoa, rubber, timber, groundnuts and cotton soon followed as leading export products, transforming the economy of Nigeria from one in which land was farmed for subsistence into one in which land was farmed for profit. The British flag followed trade, and in 1914 – purely for administrative convenience – the northern and southern parts of the country were merged into what is now Nigeria.

The British, who from the start intended to return Nigeria to the Nigerians, adopted a system of indirect rule for its colony. This involved the local chiefs in administration, in gathering raw materials for export, and in tax collection – an arrangement which disrupted, but did not totally destroy, the traditional system. Roads and railways were constructed to transport produce to the coast, and today Nigeria has a denser transport network than any other country in West Africa.

INDEPENDENCE ACHIEVED

After years of negotiation, Nigeria was granted independence in 1960 as a federal state. The nation consisted of three distinct (and hostile) regions: north, south-west, and south-east. National politics degenerated into a vicious struggle for power between the Hausa-Fulani, the Yoruba and the Ibo. In June 1967, the eastern region, in which the oil fields were situated, declared its secession from Nigeria as the independent Ibo state of Biafra.

A violent civil war broke out and, in one of the most tragic events in the history of modern Africa, more than one million people died as the Ibos battled against the federal forces – before succumbing to a famine which laid waste their land. In the following years many more regions were created, as autonomous component states of the federation, in an

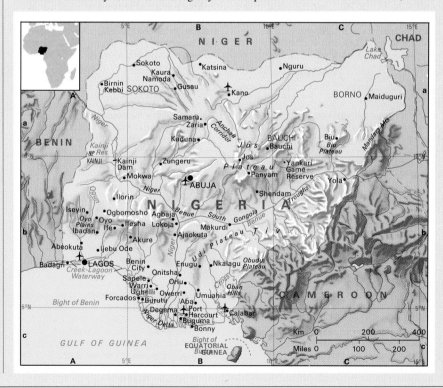

attempt to break the cohesion and power of the three principal ethnic groups while, at the same time, devolving more autonomy to some of the other ethnic groups. Since 1991 the Nigerian Federation has been composed of 30 states and a federal capital territory.

Although Nigeria was bequeathed a Westminster model of government by the British, military juntas ruled for 24 out of the first 34 years after independence. There was general rejoicing, therefore, when in October 1979 the country returned to civilian rule under President Shehu Shagari. But the jubilation was short-lived. On 31 December 1983, another military takeover put an end to four years of corruption, economic mismanagement and the undermining of the electoral process through violence and fraud.

Two years later this was followed by yet another military coup, headed by General Ibrahim Babangida, who stayed in power until 1993. His so-called 'attempts at returning to democracy', with a new constitution drafted in 1989, in reality meant that the two political parties which were allowed were both government-run. Elections in 1993 were annulled after the presumed victory by pro-democracy presidential candidate Moshood Abiola, a Yoruba multi-millionaire. After countrywide and international protests, Babangida stepped down and appointed a civilian interim president, but remained in control.

After less than three months – a period of civil and industrial turbulence in which pro-democracy movements struggled vainly to end military dominance – the interim government was ousted in a bloodless coup by General Sani Abacha. Abacha dissolved parliament and banned all political activity while insisting that he would convene a constitutional conference and restore civilian rule.

Political turmoil continued throughout 1994, during which Abacha assumed an increasingly repressive stance. Abiola was arrested and charged with treason, the oil industry was brought to a standstill by a two-month pro-democracy strike and Abacha decreed 'absolute power' for his military government. (It was also, incidentally, a year during which Nigerian swindlers became increasingly notorious for their successful multi-million US dollar international frauds.)

Today, Nigeria, potentially a rich country, has enormous national debt, 60 per cent inflation, and a high level of corruption that pervades the system and has effectively ruined the economy. Under the British, its economy depended on the profitable export of cash crops. As a result, until 1973 Nigeria was, after Ghana, the world's second largest producer of cocoa. Although it is the country's only agricultural export of any significance, cocoa production has been declining steadily since 1965.

Production of groundnuts, cotton and palm oil dropped sharply in the 1970s and 1980s and although it has now begun to pick up again, production today is much lower than in the early 1970s. Several factors have conspired to produce this decline. Drought and low prices to the producer fixed by the Nigerian Federal Marketing Board in the early 1970s were an important cause of low output. Initially, how-

ever, the drop in revenue from crops was masked by the impact of crude oil. Between 1973 and 1981, Nigeria was transformed from a rural country with an agricultural economy to one in which oil accounted for 95 per cent of its revenues.

Very little of Nigeria's new wealth has been channelled back into the country's ailing rural economy. Oil money has until recently been paid straight into central government coffers – with the oil-producing states receiving only 3 per cent of earnings. Large Western-style development projects, such as the construction of the new capital at Abuja, have sapped revenue, and progress in economic reform has been painfully slow, despite these recent development projects. Market reforms begun in 1986 were abandoned by 1994.

Nigeria's dependence on oil inevitably links the country's fortunes with the fluctuations of the world oil market. In 1981, for example, when the oil price slumped, the country was precipitated into a recession from which it has still not fully recovered. In the early 1990s, the government tried to reduce the level of dependence on oil exports and shift the focus to non-inflationary growth. However, corruption and the lack of finance prevented any significant development. Living standards in 1994 were still far below the level of 1982.

Many Nigerians cannot accept the fact that the government has spent billions on the construction of Abuja – on its presidential palace, hotels, conference centres, its gold-domed mosque – when their villages do not have running water or electricity, roads, schools or hospitals.

▲ CLOSE-KNIT COMMUNITY Many Nigerians still follow a tribal way of life in mud-hut villages. They live in large, extended family groups, many of which produce only enough food for their own needs.

NIGERIA AT A GLANCE	
Official name Federal Republic of Nigeria	
Area 923 770 km² (356 585 sq miles)	
Population 100 000 000 **Per km²** 108 **(Per sq mile** 280)	
Capital Abuja	
Government Federal republic under military government	
Currency 1 naira = 100 kobo	
Languages English (official), Hausa, Ibo, Yoruba and other African languages	
Religions Muslim (50%), Christian (40%), indigenous beliefs	
Climate Tropical; the wettest month is June. Average temperature in Abuja is 27°C (81°F) throughout the year	
Land use Cultivation 34%, grazing and other 51%, forest and woodlands 15%	
Main primary products Groundnuts, cotton, cocoa, rubber, maize, millet, sorghum, cassava, rice, yams, palm kernels, livestock, timber; crude oil and natural gas, coal, iron ore	
Major industries Crude oil, natural gas production, refining; agriculture, mining, petrochemicals, fertilisers, vehicles, cement	
Main exports Crude oil and refined products, natural gas, tin, timber and hides, rubber	
Annual income per head (US$) 320	
Population growth (per thous/yr) 28	
Life expectancy (yrs) Male 54 **Female** 56	

Normandy *France* Region in the north-west of the country, between Brittany and Picardy, which borders the English Channel. It was the site in June 1944 of the D-Day invasion of France, which began the liberation of German-occupied western Europe. The province is noted for its cream, cheeses, cider and calvados (an apple brandy). Its major cities are CAEN, LE HAVRE and ROUEN.
Population 3 006 000
Map France – Cb

Norrköping *Sweden* Industrial seaport on a Baltic inlet 157 km (97 miles) south-west of STOCKHOLM. Founded in the 14th century, it was burnt by the Russians in 1710 during the Great Northern War of 1700-21. Today it is Sweden's seventh largest city. Its manufactures include paper, woollen textiles, chemicals, rubber, and radio and television sets.
Population 120 800
Map Sweden – Cd

North Cape (Nordkapp) *Norway* One of Europe's most northerly points, jutting out to sea 500 km (310 miles) north of the Arctic Circle. It is a promontory rising 300 m (1000 ft) on Magerøya (Mager Island), 95 km (59 miles) north-east of HAMMERFEST. Contrary to popular belief, it is neither Europe's northernmost cape nor its most northerly point; the northernmost cape is the nearby, but less accessible, Knivskjellodden, while Europe's most northerly point lies 70 km (44 miles) farther east. North Cape is a popular tourist destination for visitors wanting to see the midnight sun in summer.
Map Norway – Ga

North Carolina *USA* State covering some 136 444 km² (52 669 sq miles) on the Atlantic Ocean south of Virginia. The land rises to the Appalachian Mountains, which reach 2037 m (6684 ft) at MOUNT MITCHELL in the far west. The state is the country's leading producer of tobacco and cigarettes. Its farming products include livestock, maize and peanuts. North Carolina was one of the original 13 states. RALEIGH is its capital and Charlotte its biggest city.
Population 6 628 600
Map United States – Kc

North Dakota *USA* North-central plains state on the Canadian border. Covering 183 160 km² (70 702 sq miles), it stretches westwards from the Red River to the high plains and BADLANDS on the border of Montana. Agriculture is the main occupation here, and among the crops grown are wheat, animal feed, and sunflower and flax seeds; livestock is also kept.
Oil and lignite are the state's chief minerals and food processing and farm machinery its main industries. It was originally the home of various American Indian peoples, including the Dakota. BISMARCK is its capital and Fargo its largest city.
Population 638 800
Map United States – Fa-Ga

North Field *Qatar* Natural gas field in The Gulf, one of the world's largest, covering 2590 km² (1000 sq miles), and lying some 70 km (43 miles) north-east of the northern tip of the Qatar Peninsula. Discovered in 1972, it has estimated reserves of 4.2 trillion m³ (148 trillion cu ft) of gas, and may be in production for the next 100 years.
Map Qatar – Aa

North Island (Te Ika-a-Maui) *New Zealand* Smaller of the country's two principal islands, covering 114 685 km² (44 281 sq miles). Its population, however, is almost three times that of the larger South Island. Although the centre of North Island is mountainous, it also has rich pastureland which feeds some 37.5 million sheep – nearly half the country's sheep population. The main cities of North Island are Auckland, Hamilton and the national capital, WELLINGTON.
Population 2 553 410
Map New Zealand – Eb

North Land *Russia* See SEVERNAYA ZEMLYA

North Pole Point at the northern extremity of the earth's axis of rotation, located in the shifting ice packs of the ARCTIC. Most authorities agree that the first person to reach the Pole was a United States naval officer, Robert Peary, on 6 April 1909, but some dispute his claim. In 1958, a United States nuclear submarine surfaced through the pack ice here. Thousands of passengers fly over or near the Pole on intercontinental flights every day.
Map Arctic – a

North Rhine-Westphalia *Germany* Westernmost and most populous state of the republic, established in 1946 by the union of two ancient provinces on the borders of Belgium and Holland. The union placed the huge industrial conurbation of the Ruhr under a single administration. North Rhine-Westphalia's capital is DÜSSELDORF, although COLOGNE, ESSEN and DORTMUND have larger populations. The state, which covers 34 066 km² (13 150 sq miles) and is crossed by the Rhine River, produces coal and steel, machinery, tools, chemicals, textiles and processed foods.
Population 17 414 900
Map Germany – Bc

North Sea *United Kingdom/Europe* Shallow sea, also once called 'the German Ocean', lying between Britain and mainland Europe. It covers 575 000 km² (222 000 sq miles) and reaches a depth of 661 m (2170 ft). It is a major shipping lane, a rich fishing ground and a major source of oil and natural gas, the main beneficiaries of these fuels being Britain and Norway. By 1994, some £39 million worth of oil and gas from Britain's sector was gushing every day.
The North Sea's highest tidal range is 4-6 m (13-20 ft) near The WASH. High tides combined with storm winds occasionally raise water levels to dangerous heights, causing flooding of British and Dutch coastal areas.
Map United Kingdom – Dc; Germany – Ba

North Slope *USA* Region between the Brooks Range and the Arctic Ocean in northern Alaska. It has enormous reserves of oil and natural gas, and its crude oil is carried south by the trans-Alaska pipeline to the port of Valdez on Prince William Sound off the Pacific coast.
Map Alaska – Bb

North West Cape *Australia* Promontory on the coast of Western Australia, 1125 km (about 700 miles) north of PERTH. It is the site of a United States naval communications base. The greater part of the cape, about 506 km² (195 sq miles), is occupied by the Cape Range National Park.
Map Australia – Ac

Northampton *United Kingdom* Town and county of central England. The county, covering 2367 km² (914 sq miles), straddles the south Midlands and is rich in ironstone, which was used to feed the blast furnaces at the steel town of Corby until they were closed. It now produces shoes. The town, which lies 105 km (65 miles) north of London, has become a residential area for the capital. It has a splendid 18th-century market square, one of the biggest in England, and a round church founded in the 12th century.
Northampton was the scene of a battle during the Wars of the Roses in 1460, which led to the capture of Henry VI and the Yorkists' proclamation of Edward IV as king.
Population (town) 164 000; (county) 547 000
Map United Kingdom – Ed

norther Sudden cold, dry north wind, bringing low temperatures to Texas, the Gulf of Mexico coast and other parts of southern North America, usually following a DEPRESSION. It often blows with great ferocity, reaching up to 100 km/h (more than 60 mph), may be accompanied by severe thunderstorms and hail, and temperatures may fall by as much as 20°C (36°F) in 24 hours.

Northern Cape *South Africa* Province created in 1994, extending eastwards from the Atlantic coast to the borders of Free State, Eastern Cape and North-West provinces. It covers 363 389 km² (140 272 sq miles) and consists of semi-arid and sparsely settled KAROO-type countryside. KIMBERLEY, the provincial capital, is the biggest city.
Northern Cape has vast reserves of iron ore at Sishen while the recovery of diamonds and semi-precious stones and stock rearing are the mainstays of the economy. Several irrigation schemes have been set up, especially along the Vaal and Orange rivers.
Population 763 900
Map South Africa – Bb

Northern Ireland *United Kingdom* Constituent province of the UK, covering some 14 121 km² (5452 sq miles). It was established in 1921 when IRELAND divided, and consists of six counties – ANTRIM, ARMAGH, DOWN, FERMANAGH, LONDONDERRY and TYRONE – of the ancient province of ULSTER. Since 1972, it has been governed directly from London following increased political violence by the Irish Republican Army (IRA) seeking the formation of a united Ireland independent of Britain and paramilitary loyalists fighting for the retention of Home rule. In 1994, both the IRA and their paramilitary opponents announced a cessation of hostilities. The capital is BELFAST.
Population 1 610 000
Map United Kingdom – Bc

Northern Mariana Islands See p 500

Northern Nigeria *Nigeria* Geographical region inhabited by the Hausa peoples, and centred on the Sokoto River basin in the north-west and the great plateau to the east. It is mostly dry and flat, and rises from 600 m (1970 ft) in the west to 1050 m (3445 ft) near the adjoining Jos Plateau to the south-east. One of Nigeria's main cattle-producing areas, its crops include millet, sorghum, groundnuts, cotton, vegetables and, in the river valleys, rice. The main towns are Kano, Sokoto, Kaduna and Zaria.
Map Nigeria – Ba

Northern Province *Zambia* Province bordering Malawi and Tanzania. The provincial capital, KASAMA, is 660 km (410 miles) north-east of the national capital, Lusaka. Home of the Bemba-speaking people, it covers about 147 820 km² (57 076 sq miles) of largely infertile soil. The most productive farming area is around MBALA, where coffee, fruit and tobacco are grown.

Population 867 000

Map Zambia – Cb

Northern Territory *Australia* Largely self-governing territory covering some 1 346 200 km² (519 770 sq miles) in the central north of the country. It was originally part of New South Wales and then South Australia before becoming a Commonwealth (federal) government territory in 1911. Just under one-quarter of its inhabitants are Aboriginal and almost two-fifths of the territory is now Aborigine land.

The chief town is DARWIN, where nearly half of the population live. The semi-arid scrubland that covers most of the territory is used mainly for cattle ranching and mining. Groote Eylandt, an island in the Gulf of Carpentaria, contains Australia's largest manganese deposit. Bauxite is found along the gulf shore, and there are huge uranium reserves in the Alligator River. Uranium, aluminium, manganese, copper and gold have been found at Tenant Creek.

Population 167 800

Map Australia – Eb

Northern Transvaal *South Africa* Province created in 1994, extending southwards from the international boundaries of Botswana and Zimbabwe to the borders of Eastern Transvaal, Gauteng and the North-West Province. Much of the province, that extends over 119 606 km² (46 169 sq miles), consists of grassy highveld divided by the GREAT ESCARPMENT from the lowveld, which extends to the boundary of Mozambique. Forestry, mining, cattle rearing, tourism and the commercial raising of crops as diverse as tea, citrus fruit and tobacco form the basis of the economy. About half of the KRUGER NATIONAL PARK lies within the province. Northern Transvaal's capital is PIETERSBURG.

Population 5 100 000

Map South Africa – Ca

Northland *New Zealand* Peninsula on North Island, stretching from just north of AUCKLAND to Cape Reinga at the island's northern tip and covering some 12 649 km² (4883 sq miles). Tourism, agriculture and oil refining (there is a refinery at Marsden Point), are the main industries. The largest city is WHANGAREI.

The Waipou State Forest, which lies on the western side of the peninsula, is one of the few remaining kauri pine forests in the country. It contains the largest tree in New Zealand, 52 m (170 ft) tall and 14 m (46 ft) in girth. Known to the Maoris as *Tane Mahuta* ('God of the Forest'), the tree is estimated to be older than 1200 years.

Population 131 620

Map New Zealand – Da

Northumberland *United Kingdom* County of north-eastern England, on the border with Scotland, extending over 5033 km² (1943 sq miles). Centuries of conflict between the two countries have given the county a frontier spirit and a wealth of medieval castles, such as those at Alnwick and BERWICK-UPON-TWEED. It also has some of the emptiest but most beautiful landscapes in England – particularly the Cheviot Hills, which rise to 815 m (2674 ft), and the Kielder Forest. A favourite spot with visitors is Lindisfarne, or Holy Island, with its ruined 11th-century Benedictine priory; the area teems with wildfowl and wading birds. The county town is Morpeth.

Population 306 700

Map United Kingdom – Ec

North-West Frontier Province *Pakistan* Rugged, mountainous province covering 74 521 km² (28 773 sq miles) along the northern part of the Afghan border. It was used by the British as a protective buffer zone for the Indus Valley and was ruled indirectly through local tribal chiefs. Most of the inhabitants are Pathans, renowned for their skills as guerrilla fighters, whose homeland is split by the international border. The most famous of its many passes is the KHYBER Pass.

Population 11 061 000

Map Pakistan – Ca

Northwest Passage *Canada* Northern sea passage joining the Atlantic and Pacific oceans and running through Canada's Arctic archipelago in the waters north of Alaska, north of BAFFIN Island and south of Victoria and Banks islands. First sought by 16th-century English explorers, including Sir Martin Frobisher (1576-8), looking for a short sea route to the Orient, it was not discovered until the 19th century.

It was first crossed by sledge, from west to east, by Sir Robert McClure in 1850-3 and first navigated, from east to west, by Norwegian Roald Amundsen in 1903-6. The first commercial ship to navigate the passage was the United States oil tanker *Manhattan* in 1969.

Map (Baffin Island) Canada – Ha-Hb; (Victoria Island) Canada – Da-Ea

North-West Province *South Africa* Province created in 1994, its boundaries being Botswana and the provinces of Northern Transvaal, Northern Cape, Free State and Gauteng. It covers 118 710 km² (45 823 sq miles), with its arid north-western parts lying close to the KALAHARI Desert. The capital is Mmabatho. Commercial farming revolves around maize, sheep and cattle. Manganese, nickel and diamonds are mined.

Population 3 500 000

Map South Africa – Cb

Northwest Territories *Canada* Northern region of Canada, including all the land north of 60°N, except for YUKON and the northern tip of Labrador and Quebec. The region covers 3 426 320 km² (1 322 597 sq miles) – nearly one-third of Canada – but over 133 000 km² (51 340 sq miles) of this vast region is covered by lakes and rivers. Mountain ridges rise above 2700 m (9000 ft) in the west and more than 2600 m (8000 ft) on the eastern islands of BAFFIN and ELLESMERE.

Tundra extends over much of the north and east, and agriculture is limited. About 60 per cent of the population are Inuits (formerly referred to as 'Eskimos'), native Canadians and Métis (mixed French and native Canadian) and, as in the past, most are engaged in fur trapping and fishing. In June 1993, the Canadian government signed an agreement granting self-government to Inuits in their own territory, to be created from part of the Northwest Territories. Minerals, including lead, gold, zinc, cadmium and silver, are the main source of income. YELLOWKNIFE is the region's capital.

Population 57 650

Map Canada – Db

Norway See p 502

Norwegian Sea *Norway* Fishing ground for cod and herring in the North ATLANTIC OCEAN, lying off Norway. Despite its northerly latitude, it is usually ice free because the relatively warm Norwegian Current (the most northerly extension of the GULF STREAM) flows through it. Its greatest depth is 3667 m (12 030 ft).

Map Atlantic Ocean – Ea

Norwich *United Kingdom* River port in eastern England, about 160 km (100 miles) north-east of LONDON. It is the county town of Norfolk; during the medieval heyday of the East Anglian wool trade, it was the country's second city, after London. It stands on the Wensum River, 30 km (20 miles) from the North Sea. Its riches and reputation attracted many skilled immigrants from Europe between the 14th and 16th centuries.

A large 12th-century castle – now a museum – overlooks the city, and some sections of a 14th-century city wall remain. It has many beautiful streets, such as cobbled Elm Hill, and churches. The Anglican cathedral with its spectacular nave roof dates mainly from the 15th century. Its spire, rising to 96 m (315 ft), is the country's second tallest, after Salisbury. The city is the home of the modern University of East Anglia.

Population 127 200

Map United Kingdom – Fd

Notabile *Malta* See MDINA

Nottingham *United Kingdom* City and county of the northern Midlands of England. The city, 70 km (45 miles) north-east of BIRMINGHAM, grew around an 11th-century castle on a crag above the River Trent. Charles I raised his standard here in 1642 at the start of the English Civil War. The city's real growth came in the 19th century, and Nottingham now produces lace, hosiery, bicycles and tobacco. It was one of the first towns in Britain to ban private motor vehicles from its centre, diverting them to a ring road.

The historic Goose Fair, which dates back to the Middle Ages, is held in Nottingham every October. The city and the remains of Sherwood Forest near it are inextricably linked with the legendary 12th-century outlaw Robin Hood.

The county covers 2164 km² (836 sq miles). Before the collapse of the once dominant coal industry, Nottingham's coal field was one of the most productive in Britain. Between 1961 and 1992, 26 of the county's 39 pits were shut down; another nine were closed in the following years. Clothing and textiles have become the main hope for the revival of the local economy.

Population (city) 280 900; (county) 1 020 200

Map United Kingdom – Ed

Nouadhibou *Mauritania* Port, formerly called Port Etienne, on Cape Blanc Peninsula, on the border with Western Sahara. It exports iron ore from the mines at Fdérik, 670 km (420 miles) away by rail, and is a fishing port.

Population 60 000

Map Mauritania – Ba

Northern Mariana Islands

IN A REMOTE CHAIN OF THE PACIFIC THE PEOPLE ARE FULL CITIZENS OF THE UNITED STATES OF AMERICA

When the Trust Territory of the Pacific Islands – those far-flung specks in the western Pacific entrusted to American care by the United Nations in 1947 – was negotiating its future in the 1970s, one group of islands opted for close American ties rather than full independence. Thus, in 1978, the Northern Mariana Islands became an internally self-governing state with its foreign affairs and defence controlled by the United States.

A closer association was forged in 1986 when the islands became a Commonwealth in political union with the US. This gave the Mariana people full US citizenship and opened the way to full statehood in the future, perhaps in conjunction with GUAM, the southernmost of the Mariana Islands, which is a separately administered US territory. The United Nations terminated the islands' trust status in 1990.

There are 16 mountainous islands and islets in the group, which stretch in a curving chain about 725 km (450 miles) long, and 2400 km (1500 miles) east of the northern Philippines. They are volcanic, Agrihan (the highest) reaching to 959 m (3146 ft), and Pagan having hot springs and an active volcano that erupted violently in 1981.

The most important islands are SAIPAN – which contains the capital, Susupe, and 86 per cent of the population – TINIAN and Rota. The first two are well known for the fierce Second World War battles, when United States forces took them from the Japanese in June and July 1944, and for their subsequent role as military bases. After the war, Saipan became the administrative centre for the entire trust territory, and the Mariana people benefited economically as a result.

Most of the people are Chamorros of mixed ancestry, including Spanish and Filipino. The islands, ruled by Spain from 1668 to 1889, are named after a regent of Spain, Mariana of Austria. Before the Second World War, Saipan and Tinian grew much sugar cane but today there are only a few full-time farmers and fishermen. Japanese and Korean boats fish the Mariana waters.

About 50 per cent of the workforce is employed in tourism which provides most of the islands' income. The tourists, who number around 450 000 a year, are mainly Japanese, many of whom come to see wartime relics. It was from Tinian that US B-29 bombers flew to drop the atomic bombs on HIROSHIMA and NAGASAKI in August 1945.

NORTHERN MARIANAS AT A GLANCE

Map Pacific Ocean – Cb
Official name Commonwealth of the Northern Mariana Islands
Area 477 km² (184 sq miles)
Population 43 300 **Per km²** 90 **(Per sq mile** 235)
Capital Susupe
Government Self-governing commonwealth in political union with the USA
Currency 1 US dollar = 100 cents
Languages English, Chamorro, Carolinian
Religion Christian (mainly Roman Catholic)
Climate Tropical; average temperature at Susupe is 26°C (79°F) through the year
Land use Cultivation 5% (on Saipan island), grazing 19%, other 76%
Main primary products Coconuts, vegetables, fruit, fish, cattle
Major industries Tourism, construction, light industry
Main exports Clothing, manufactured goods, processed food, concrete blocks
Annual income per head (US$) 11 500
Population growth (per thous/yr) 30
Life expectancy (yrs) Male 66 **Female** 69

Nouakchott *Mauritania* The country's capital, standing on the Atlantic coast. Built in the 1950s, it is an attractive city of flat-roofed houses, wide sandy streets and beautiful mosques. It was enlarged during the droughts of the 1970s and 1980s by a massive 'tent city' housing about 300 000 refugees from rural areas.

A small industrial area near the harbour contains a power station and Africa's first water desalination plant. The port, 260 km (160 miles) to the north-east, was enlarged in the 1970s, and a deep-water harbour was opened in 1986.
Population 850 000
Map Mauritania – Bb

Nouméa *New Caledonia* Capital and main port of the territory, standing near the southern end of New Caledonia Island. It is the largest French-speaking city in the Pacific. There are tourist hotels on a peninsula to the south of the city, while on the northern outskirts is the smelter that processes nickel ore.
Population 65 110
Map (New Caledonia) Pacific Ocean – Dd

Nova Lamego *Guinea-Bissau* See GABU

Nova Scotia *Canada* One of the Maritime Provinces in eastern Canada, comprising a mainland peninsula and CAPE BRETON ISLAND. It gets its name of 'New Scotland' from the charter given to the Scottish poet and politician William Alexander by England's James I (James VI of Scotland) in 1621 for the establishment of a colony in Canada.

Nova Scotia has wooded hills and valleys, containing numerous rivers and lakes, and a very picturesque coastline. Its economic wealth is provided by agriculture, forestry, the mining of coal and gypsum, and industries such as wood processing, chemicals, steel and engineering. Nova Scotia had the first Canadian newspaper (*The Halifax Gazette*, 1752) and Canada's first university (King's College, 1788). HALIFAX is the capital.
Population 899 940
Map Canada – Id

Novaya Zemlya *Russia* An almost uninhabited pair of islands separated by a narrow channel and lying in the Arctic Ocean between the Barents and Kara seas. They extend from north-east to south for more than 1000 km (625 miles), and cover 83 400 km² (32 193 sq miles). Fishermen and hunters visit the tundra islands in the summer.

Most of the islands' permanent inhabitants work at weather, radar or geological research stations. Samoyeds, a Mongolian people, herd reindeer and collect the down of eiders (large Arctic Sea ducks). The Samoyed dog breed was originally developed here by them.
Population 400
Map Russia – Ga

Novgorod *Russia* City 155 km (95 miles) south-east of ST PETERSBURG. Founded in AD 859 on the banks of the Volkhov River, Novgorod is now generally regarded as the cradle of Russian civilisation. It became an important trading post on the Viking route from Scandinavia to Constantinople (present-day Istanbul), and in the Middle Ages it rivalled Moscow as the country's most influential and prosperous city trading in the Baltic region.

Among its historic buildings are the cathedral of St Sophia, built in the mid-11th century, and its kremlin (citadel), built in 1044 and enlarged in 1116. It has several light industries, including furniture, fish canning, clothing and chinaware.
Population 232 000
Map Russia – Ec

Novi Pazar (Novibazar) *Yugoslavia (Serbia and Montenegro)* West Serbian town on the Raška River, 185 km (115 miles) south of BELGRADE. It was a cradle of Serbian culture in the Middle Ages, the period in which the 9th-century Church of St Peter, the brilliant 12th-century frescoes in the ruined Djurdevi Stubovi (St George's Pillars) monastery, and the superb monastery of Sopocani, built in 1265, originated. The Turks captured Novi Pazar (the name means 'new bazaar') in 1456 and ruled here until 1913.
Population 85 600
Map Yugoslavia – Cc

Novi Sad (Ujvidék; Uvidék; Neusatz) *Yugoslavia (Serbia and Montenegro)* Capital of the Vojvodina autonomous province of Serbia, standing on the banks of the Danube, some 75 km (47 miles) north-west of BELGRADE. The pre-1914 town centre is dominated by the Gothic cathedral's multicoloured roof and spire, while modern apartment blocks, and factories manufacturing farm machinery, textiles, electrical goods and chemicals stand in village-like suburbs. A university was founded in the city in 1960.
Population 178 900
Map Yugoslavia – Bb

Novokuznetsk *Russia* Industrial city in south-central Siberia, 300 km (185 miles) south-east of NOVOSIBIRSK. Novokuznetsk (formerly called Kuznetsk, then Stalinsk) was founded in 1617, and had only about 4000 inhabitants until the late

1920s, when it became the site of the Soviet Union's largest iron and steel works. Since then, more industries have been established, among them plants producing chemicals, aluminium, alloys and heavy machinery. Today, Novokuznetsk is the largest city of the Kuznetsk basin industrial and coal-mining region.

Population 602 000
Map Russia – Jc

Novosibirsk *Russia* One of the largest industrial centres of the former Soviet Union and of Russia today, straddling the Ob' River in Siberia, more than 2800 km (1738 miles) east of MOSCOW. It was founded in 1893, when the Trans-Siberian Railway needed a base there. It grew rapidly in size and importance – especially during the Second World War, when engineering and electrical factories were moved here after the Germans occupied the Ukraine and Belarus.

The city's products now include machinery, trucks, aircraft, ships, tools, electrical goods, plastics, pharmaceuticals and other chemicals, textiles, knitwear, shoes, timber products and furniture. It is Siberia's largest scientific research centre; just to the south is the 'science town' of Akademgorodok.

Population 1 476 000
Map Russia – Jc

Novosibirskiye Ostrova (New Siberian Islands) *Russia* Group of islands in the Arctic Ocean between the East Siberian and Laptev seas. They are covered with snow and ice for nine months of the year. Also named Anjou Islands after polar explorer Admiral Anjou, the islands became famous after fossilised mammoths and other Ice Age animals were found here.

In 1925, the Soviet Academy of Sciences started running research stations here.

Map Russia – Pa

Nowa Huta *Poland* Europe's largest new town outside the former Soviet Union, now a suburb of the city of CRACOW, separated from it by a green belt. Nowa Huta, or 'New Foundry', was begun in 1949 to serve the Lenin Steelworks, built at the same time. Besides iron and steel, Nowa Huta produces chemicals.

Population 224 100
Map Poland – Dc

Nowy Sącz (Neusander) *Poland* Medieval city 74 km (46 miles) south-east of the city of CRACOW. It was founded in the 13th century as a fort on the Dunajec River, then extensively used as a trade route.

The old quarter, with its medieval, 17th- and 18th-century buildings with gleaming metal roofs, now attracts artists and tourists. The city is the centre of an apple-growing district.

Population 68 300
Map Poland – Dd

Nu Jiang *China* See SALWEEN

nuée ardente Rapidly moving, turbulent, incandescent cloud of gas, ash and rock fragments that flows close to the ground after violent ejection from a volcano, destroying all life in its path – as Mont PELÉE on Martinique island did in 1902.

Nuevo Léon *Mexico* Important northern industrial state containing iron, steel, lead and manu-

facturing industries concentrated in its capital, MONTERREY. It is also a citrus growing region.
Population 3 086 470
Map Mexico – Bb

Nuku'alofa *Tonga* Chief port and capital, lying on the north coast of Tongatapu Island, between the lagoon and the ocean. Ships tie up at wharves projecting beyond the coral reef. The main industry is copra processing. Nuku'alofa's royal palace, a handsome timber building, was constructed in 1867; nearby are the tombs of former Tongan kings.
Population 27 740
Map (Tonga) Pacific Ocean – Ed

Nullarbor Plain *Australia* Arid, flat and largely uninhabited limestone region along the southern coast of Western and South Australia, which is, as its name ('no trees') suggests, treeless.

For much of its length of 1200 km (745 miles), sheer cliffs drop from the plain into the Great Australian Bight. The plain is crossed by a highway and the transcontinental railway. One stretch of the line runs straight for 478 km (297 miles), the longest straight track in the world.
Map Australia – De

nunatak Isolated mountain peak or hill that projects through the surface of surrounding glacial ice.

Nuremberg (Nürnberg) *Germany* Medieval city in Bavaria, some 150 km (93 miles) north of MUNICH. It was the home of Tannhäuser, the legendary lyrical poet of the 13th century, and of several famous artists, such as Albrecht Dürer (1471-1528). It was also the setting for the opera *Die Meistersinger von Nürnberg* by the 19th-century composer Richard Wagner. Several kilometres of the medieval walls, with rampart walks, still stand, and the city's 11th-century castle and its 14th-century churches and houses, many destroyed or badly damaged during the Second World War, have been restored.

The city was used for mass rallies by Adolf Hitler's Nazi party during the 1920s and 1930s, and for the trials of Nazi leaders and war criminals in 1945.

Today it is an important industrial centre making electrical equipment, machinery and motor vehicles. It also has thriving printing and publishing industries. The well-known Nuremberg sausage and *Lebkuchen* (spicy cake) are produced here. Also known for its toys, the city has a toy museum and an annual toy fair.
Population 493 700
Map Germany – Dd

Nuristan *Afghanistan* Well-forested and mountainous region bordering on Pakistan. It produces 80 per cent of Afghanistan's timber, and is noted for its intricate wood carvings.
Map Afghanistan – Ba

Nürnberg *Germany* See NUREMBERG

Nusa Tenggara (Lesser Sunda Islands) *Indonesia* Chain of islands extending east from JAVA. They include Bali, Lombok, Sumbawa, Sumba, Flores and Timor. The islands are mainly mountainous and volcanic.
Population 6 688 000
Map Indonesia –Ed

Nuuk *Greenland* See GODTHÅB

Nuwara Eliya *Sri Lanka* Town standing at 1990 m (6530 ft) altitude in the central mountains, 100 km (62 miles) east of COLOMBO, in a region that produces Sri Lanka's best tea. The British established it as a hill station – a refuge from the summer heat of the lowlands – in 1825, but its present buildings are mostly in the later Edwardian English style.

The beautifully landscaped town has a lake, park, club, race track, golf course and well-stocked trout streams, and is a good centre for hill walking. Pidurutalagala, Sri Lanka's highest peak at 2524 m (8280 ft), overlooks the town from the north.
Population 21 300
Map Sri Lanka – Bb

Nyasa, Lake *East Africa* See MALAWI, LAKE

Nyeri *Kenya* Prosperous market town and district capital between the Aberdare Mountains and Mount Kenya, some 95 km (60 miles) north of NAIROBI. Coffee and tea are grown in the surrounding Nyeri district. Most of its inhabitants are Kikuyu people.
Population 89 000
Map Kenya – Cb

Nyíuregyháza *Hungary* Regional capital of the Nyírség, the country's most important apple-growing region, which lies in the north-east on the Slovakian, Ukrainian and Romanian borders. The city has grown rapidly since 1965 with the establishment of light industries and colleges, and as a tourist centre for nearby Sóstó, a spa with hot springs.
Population 114 960
Map Hungary – Bb

Nyköping *Sweden* Seaport and industrial city, situated some 98 km (61 miles) south-west of STOCKHOLM. Founded in the 13th century on the site of an old trading town, it was destroyed by a fire in 1665 and rebuilt on a gridiron plan.

It manufactures margarine and tobacco, and also has engineering works and electrical and chemical factories.
Population 48 100
Map Sweden – Cd

Nysa *Poland* River, 193 km (120 miles) long, which rises on the Czech border and flows north-east to join the Oder River, 60 km (35 miles) south-east of the city of WROCLAW.

Nysa is also the name of a city which straddles the river, 75 km (45 miles) south-east of Wroclaw. Founded as a trading town in the 13th century, it has since become the home of many Jesuit and other Catholic monasteries. It produces bricks, timber, foodstuffs and metal goods.
Population 43 500
Map Poland – Bc

Nysa Luzycka *Eastern Europe* See NEISSE

Nzwani *Comoros* Second largest of the Comoros Island group, covering 425 km² (164 sq miles), and formerly known as Anjouan. Its main export is ylang-ylang oil, which is used to make perfume. The chief town is Mutsamudu.
Population 197 900
Map Madagascar – Aa

Norway

THE NEW HARVEST FROM THE SEA IS OIL AND GAS – AND IT PAYS MUCH BETTER THAN FISHING

The Viking warriors of the land that is now Norway were once a much-feared people. Between the 9th and the 11th centuries the British Isles suffered their invasions, while places as far afield as North America experienced their impact. Today, although Norwegians are still strong and resilient in the face of a harsh environment, they have a far gentler image abroad.

Norway is a prosperous and picturesque country, welcoming 4 million or more tourists every year to its attractive villages and spectacular scenery of fjords, cliffs, rugged uplands and forested dales.

It is a narrow, very long country, stretching 1752 km (1100 miles) from Lindesnes, in the south, to Nordkin, near the NORTH CAPE. Norwegians like to mention the fact that the northern frontier of Italy is as close to Oslo as the North Cape – and less expensive to reach.

In the south, Norway is about 400 km (250 miles) across, but near NARVIK in the north the land narrows to only 6.5 km (4 miles). It has a Siamese twin – Sweden, to which it is joined back-to-back along a 1619 km (1017 mile) border. In the northern province of FINNMARK, the frontier is shared with Finland and Russia. Even farther north, the islands of SPITSBERGEN in the SVALBARD archipelago, lie within 1100 km (700 miles) of the North Pole. The coastline is dotted with islands – about 150 000 of them.

The land is rugged and mountainous. Two-thirds of it is more than 300 m (1000 ft) above sea level, and a quarter of the country lies at 1000 m (3280 ft) or higher. Massive erosion during the last great Ice Age scoured the fjords of the west coast which are among the deepest in the world. For instance, the longest fjord, Sognefjord, some 204 km (127 miles), reaches a depth of 1303 m (4275 ft). Inland, the ice scooped out deep dales which are now ribboned with lakes, one of which, Hornindalsvatn, is the deepest in Europe at 514 m (1686 ft).

The ice also sculpted a giant dragon's tail of northern islands, the LOFOTENS, extending 190 km (118 miles) in the Norwegian Sea. Between two of the Lofoten Islands flows the strong and treacherous tidal current known as the Moskenstraumen. It is known in literature and legend as the Maelstrom and is the subject of exaggerated stories of a destructive whirlpool that sucks ships down.

Such dangerous natural phenomena, together with avalanches, landslides, floods and storms at sea, partly explain Norway's rich folklore of the supernatural – with trolls and giants inhabiting its lakes, caves and fjords, and stirring up catastrophes. The Norse mythology from before the 10th century also had its destructive gods, such as Thor, wielding thun-

derbolts, and Woden, snatching warriors from battle to an afterlife of feasting in Valhalla – the Hall of the Slain. The Norwegian imagination was perhaps all the more stimulated by the long dark winters, which bred storytellers.

Although one-third of the country is north of the Arctic Circle, the climate is modified by the North Atlantic Drift (known also as the Gulf Stream) off the west coast. This warm current ensures that almost all the coastal waters are ice-free in winter and helps to support a rich variety of fish. Wet west winds bring heavy rain and snow to most parts of the country, particularly the west coast; BERGEN in the south-west, for example, receives nearly 2000 mm (79 in) a year.

As a compensation for dark winters, northern Norway is one of the best places from which to see the AURORA Borealis (Northern Lights). In summer, Arctic Norway enjoys the beauty of the midnight sun – at TROMSO, for as long as two months.

Much of Norway is heavily wooded, but as the climate grows colder farther north, the

forests give way to rocky wastes where only mosses and lichens grow, overlooked by snow-clad peaks. Norway has mainland Europe's largest glacier, JOSTEDALSBREEN, with an area of 486 km² (188 sq miles).

About 75 per cent of the more than 4 million people live in towns, a quarter of them in or around Oslo, TRONDHEIM (the old capital), BERGEN and, since the early 1970s, in the oil boom town of STAVANGER. The large-scale migration of the last century was due to the harsh conditions of the countryside.

Just under 3 per cent of the land is cultivated and agriculture, as well as fishing, is heavily state-subsidised. The average holding is fairly small, although many farmers also own forest land, which provides additional income. About three-quarters of all farms are dairy farms and cattle are reared in all parts of the country, even the far north. Sheep farming is widespread – there are at least 2 million sheep in Norway. Pig and poultry rearing are also important. Some goats are kept in hilly areas and reindeer are herded in the north above the timber line.

The most productive agricultural areas are around Oslofjord and Lake MJØSA in the south-east, where the main crops are hay, grain, and vegetables. In the south and along the west coast, apples, cherries and soft fruits are grown.

Up to 150 years ago, Norway was a poverty-stricken country, and at the beginning of this century it was still one of the poorest countries in Europe. Today, Norway has one of the highest per capita incomes in the world. Much of this new wealth comes from the oil and gas discovered in the North Sea in 1971. Oil and gas products account for more than half of Norway's exports; the industry is largely gov-ernment-controlled through state enterprises.

Norway is the highest per capita producer of hydroelectric power, tapped from the rivers flowing down the high plateaus – and only about half of its rivers have been harnessed. There are coal mines in Arctic Spitsbergen and a major iron and steel plant at MO I RANA.

The growth of Norway's industry started long before the recent energy boom, however. Aluminium smelting based on hydroelectric power and imported bauxite was first devel-oped in the 1920s.

SEAGOING TRADITION

Traditionally, Norwegians have looked to the sea for their livelihood. Indeed, 85 per cent of them live within sight of a coastline which, even disregarding Norway's thousands of islands, measures 21 900 km (13 608 miles). The fishing industry has always been impor-tant. For many centuries, fish was the country's chief export. Today, Norway is Europe's lead-ing producer of fish. The sea-going tradition is also maintained by Norway's modern shipping companies. With a total of 2000 ships, the country can boast the fourth largest merchant fleet in the world.

The earliest known people of Norway were hunters and gatherers who left rock carvings 5000 years ago. However, the state of Norway is young. During the 11th century, Denmark occupied Norwegian territory. From the 14th century, Norway formed part of the Danish-Norwegian union, which later included neighbouring Sweden.

From 1814 until 1905, Norway was united with Sweden. In 1905, Norway became inde-pendent, when the Danish Prince Charles was instated as King Håkon VII. German forces occupied Norway from April 1940 to the end of the Second World War.

The young nation has a strong sense of national identity, not affected by the fact that it has two dominant languages. In the mid-19th century, a new literary language, Nynorsk (or Landsmål) was fostered to challenge the offic-ial language of Bokmål (or Riksmål), which had lingered from the days of Danish rule. More than 80 per cent of the people speak Bokmål and it is also used by the press and TV stations. Nynorsk is used mainly in literature. The two variants burden the schoolroom, com-plicate signposts and confound mapmakers; but they do not differ much and Norwegians understand both. A third language, spoken in northern Norway, is that of the 25 000 Lapps.

Norway has had some internationally famous heroes such as the Arctic explorers

Fridtjof Nansen (1861-1930) and Roald Amundsen (1872-1928), who beat a British party led by Captain Robert Scott in the race to the South Pole in 1911. More recently, the explorer and anthropologist Thor Heyerdahl sailed his raft *Kon-Tiki* from Peru across the Pacific Ocean to the TUAMOTU islands in 1947.

The reputation of Norwegian literature was enhanced by the playwright Henrik Ibsen

tiations have kept on deadlocking over the Norwegian fishing industry and the country's territorial waters.

However, the country is a member of the European Free Trade Association and the Nordic Council (with Denmark, Sweden, Finland and Iceland). Norway was a founder-member of NATO, but remains opposed to the stationing of foreign troops on its soil.

▲ LAND OF THE VIKINGS
Aurlandsfjord branches south from Sognefjord towards the permanent snows of Storeskavlen (1729 m, 5673 ft). The name Viking derives from the word *vik,* an Old Norse term for a creek or inlet.

(1828-1906), whose plays portrayed Norwe-gian life. The composer Edvard Grieg (1843-1907), whose music is part of the international concert repertoire, was inspired by the folk tunes of his native Norway.

Today, Norwegians have a high standard of living. Despite a recent unemployment rate of about 6 per cent, there are no very poor people, thanks to the extensive social security system. The welfare state provides services even to the most isolated of the scattered communities. There is space for everyone and most people own their own houses. Most also have leisure homes on the shores of lakes, on the seaside or in the mountains. There is leisure fishing for everyone – the country's salmon fishing is renowned. Norway rivals Switzerland for skiing and mountaineering – indeed, Norwegians invented skiing. Woodlands offer elk hunting.

A referendum in Norway voted against membership of the European Economic Com-munity in 1972 and again, in late-1994, against membership of the European Union. Between 1993-4, 67 per cent of Norway's exports went to EU countries, but EU-Norwegian nego-

NORWAY AT A GLANCE

Official name Kingdom of Norway
Area 324 220 km² (125 153 sq miles)
Population 4 299 200 **Per km²** 13
(Per sq mile 34)
Capital Oslo
Government Parliamentary monarchy
Currency 1 Norwegian kroner = 100 øre
Languages Norwegian (Bokmål and Nynorsk), some Lapp and Finnish
Religion Christian (88% Evangelical Lutheran)
Climate Temperate maritime in coastal regions; cold in the north and east. Average temperature in Oslo ranges from -4.7°C (23.5°F) in January to 17.3°C (63°F) in July
Land use Forest 27%, cultivation 3%, grazing, other, 70%
Main primary products Barley, oats, potatoes, livestock, apples, timber, fish; crude oil and natural gas, coal, iron, lead, zinc, copper, nickel, titanium, quartz (silicon)
Major industries Mining, crude oil and natural gas refining, mineral refining, chemicals, shipbuilding, food processing, fish-ing, forestry, timber products
Main exports Crude oil and natural gas, aluminium and other non-ferrous metals, chemicals, ships, machinery, fish, petroleum products, paper, timber products
Annual income per head (US$) 23 120
Population growth (per thous/yr) 4
Life expectancy (yrs) Male 73 **Female** 80

Oahu *USA* Island covering about 1549 km² (598 sq miles) in the central Pacific US state of Hawaii. The most highly developed of the Hawaiian islands, it contains the capital, HONOLULU, and the naval base of PEARL HARBOR.
Map Hawaii – Ba

Oakland *USA* Port on San Francisco Bay in California, and part of the San Francisco-Oakland-San Jose conurbation. It has prospered as the western end of the transcontinental railway and is an industrial and business centre, noted as a producer of computers, electrical equipment, cars, ships, food, glass and chemicals.
Population (city) 372 200; (conurbation) 6 253 300
Map United States – Bc

oasis Area in a desert where sufficient water is available to support plant life. It varies in size from a few palm trees around a spring to an area of several hundred square kilometres with a large agricultural population.

Oaxaca *Mexico* Mountainous rural state in southern Mexico. Its highland villages are the home of descendants of the ancient Zapotec and Mixtec peoples, of whom about one-fifth speak only their native language. Although coffee is grown and the state is rich in minerals, Oaxaca has a poorly developed agricultural and mining sector. In fact, it is one of Mexico's poorest states.

Its capital, also called Oaxaca, is located on a plateau 1640 m (5000 ft) above sea level, surrounded by the peaks of the southern Sierra Madre. Amerindians flock into this colonial city on Saturdays to sell their tooled leather, embroidered blouses and woollen ponchos at the market. The ruins of Monte Albán are nearby.
Population (state) 3 022 000; (city) 212 910
Map Mexico – Cc

▼ FIREWALKERS Fijians on the island of Mbengga in Oceania heat stones for the firewalking ceremony. The walkers parade around a 4 m (13 ft) circular pit of white-hot stone, amazingly suffering no pain or injury.

Ob' *Russia* Siberian river which rises in the Altai Mountains near the Mongolian border, and flows generally north-west and north, before emptying into the Kara Sea (a part of the Arctic Ocean). Together with the Irtysh River, which it joins in western Siberia, it runs for 5570 km (3460 miles), which makes it the fifth longest in the world after the Nile, Amazon, Chang Jiang (Yangtze) and Mississippi-Missouri. Below the city of Tomsk, the Ob' is the main means of transportation in western Siberia – although roads and airstrips are now being built to serve the booming Siberian oil industry. It is also a major source of hydroelectric power. The river drains an area of 2975 km² (1148 sq miles); when in flood it measures up to 40 km (25 miles) across near its mouth.
Map Russia – Hb-Jc

Oban *United Kingdom* Fishing and ferry port of western Scotland, standing on the Firth of Lorn in Strathclyde region, 100 km (62 miles) north-west of Glasgow. Boats connect it with many of the Western Isles.
Population 8000
Map United Kingdom – Cb

Oberammergau *Germany* Alpine town 70 km (43 miles) south-west of Munich, where, every 10 years since 1634, the inhabitants have performed a Passion play in May to celebrate their deliverance from the plague. The performance has become a major international event, attended by thousands of visitors.

The play is not Oberammergau's only attraction: tourists come throughout the year to see its colourful houses, wood carvings and 18th-century Rococo church. They are also attracted by the Alpine scenery and winter sports.
Population 5200
Map Germany – De

Oberhausen *Germany* Industrial town of the Ruhr between DUISBURG and ESSEN. The region's first iron ore was mined and smelted here in 1758; coal mining followed 100 years later. Today Oberhausen produces coal and manufactures steel and machinery.
Population 223 800
Map Germany – Bc

obsidian Acid-resistant, lustrous, black volcanic glass. It was used in early times for weapons and tools, and is now used principally for ornaments and jewellery.

Obuasi *Ghana* Gold-mining town 170 km (105 miles) north-west of ACCRA. Some of the gold for which the Ashanti kingdom was renowned was mined in the vicinity. The railway from Sekondi reached Obuasi in 1902 amid speculation that the region would turn out to be as rich as the gold-bearing reefs of the Witwatersrand in South Africa. This, however, proved to be unfounded, although Ghana is the third biggest producer of gold in Africa after South Africa and Zimbabwe. Today, Ghana's gold exports – 75 per cent of it originating around Obuasi – make up 17 per cent of export earnings.
Population 47 400
Map Ghana – Ab

occlusion (occluded front) In an atmospheric DEPRESSION, the overtaking of a warm front by a cold front, which ultimately lifts the warm sector completely off the surface of the earth and forms an occluded front. If the overtaking cold air is colder than the air mass in front, it is called a cold occlusion; if, however, the air is warmer it is a warm occlusion. If there is no marked difference in temperature, it is a neutral occlusion.

ocean Any of the large bodies of salt water that cover approximately two-thirds of the earth's surface. Most geographers count four: the ARCTIC, ATLANTIC, INDIAN and PACIFIC; some consider the ANTARCTIC OCEAN to be a fifth. See also HOW OCEAN CURRENTS ARE CAUSED, p 505

ocean basin Low-lying part of the earth's crust that is filled by the salt water of the oceans. It does not include the CONTINENTAL SHELF.

ocean floor spreading Part of the geological process in which ocean floors are created and destroyed. Sections of the oceanic crust (called plates) move apart and basaltic magma, or molten lava, erupts from the mantle and solidifies, adding new crust to the diverging plates. The zones of separation are known as constructive plate margins and are marked by OCEAN RIDGES. See PLATE TECTONICS, p 542

Ocean Island *Kiribati* See BANABA

ocean ridge Continuous system of mainly submarine mountains, extending into all the major oceans and running approximately parallel to the ocean margins. The mountains rise to thousands of metres above the ocean floor, some forming islands such as Jan Mayen Island in the Arctic Ocean and Bouvet Island in the South Atlantic. A feature of the mountain ranges is a central rift valley – a zone of earthquake and volcanic activity. Outpourings of basaltic lava create the ridges.

ocean trench Long, narrow and steep-sided depression in the ocean crust, which may be up to 2 km (about 1 mile) deeper than the surrounding ocean floor. Trenches are usually situated parallel to the adjacent coastlines and lie above SUBDUCTION ZONES where oceanic plates are descending into the earth's mantle. They are associated with subduction zones and island arcs. See PLUMBING THE DEPTHS, p 507

Oceania *Pacific Ocean* About 50000 years ago, people from Asia began the colonisation of the Pacific islands of Oceania. The last great migrations in this area were made by voyagers dubbed by some historians 'the Vikings of the Sun' (believed to be Polynesians), who went as far as EASTER ISLAND, which is 2500 km (1553 miles) from the nearest neighbouring island group.

Oceania encompasses tens of thousands of islands scattered over a vast area of the Pacific Ocean, from Palau in the west 12500 km (7762 miles) to Easter Island in the east, and from Midway in the north 7500 km (4657 miles) to New Zealand in the south. Only about 3000 of the islands are large enough to have names. The term may or may not include Australia, New Zealand and New Guinea which may be separated and called AUSTRALASIA.

Unquestionably included are three main island groups: MELANESIA ('black islands', from the inhabitants' skin colour), an arc of islands curving south-east from New Guinea, in which lie the Solomon Islands, Vanuatu and Fiji; MICRONESIA ('small islands'), another arc of small coral islands including the Mariana, Marshall and Caroline island groups, and the tiny countries of Kiribati and Nauru; and, in mid-ocean, the great triangle of POLYNESIA ('many islands'), with New Zealand, Hawaii and Easter Island at its corners, and encompassing Samoa, Tonga and Tahiti.

Many of the islands are volcanoes which have risen from the sea-bed of the Pacific close to where two plates of the earth's crust have collided (see PLATE TECTONICS, p 542). This phenomenon is still occurring; in Tonga two volcanoes periodically rise from the sea, only to be eroded by the waves again. Many islands are subject to eruptions and earthquakes. Most are surrounded by coral reefs which are made up of accumulations of the skeletons of tiny creatures.

Explorers came from Europe in the 16th and 17th centuries. They included the Portuguese, such as Ferdinand Magellan (1480-1521), who was killed by Philippine islanders; the Dutch, such as Abel Tasman (1603-59), who was the first European to sail to New Zealand, New Guinea and Tasmania; and the Spanish. There was no established link with Europe until the 18th century, after the voyages of English explorers such as Samuel Wallis (1728-95), who reached Tahiti in 1767, and James Cook, who was killed by Hawaiians in 1779. By 1850, many of the major island groups had become colonial outposts of France, Britain, the United States and the Netherlands, and missionaries who had sewn the seeds of a deep-rooted Christianity which is a striking aspect of island life today.

Physical separation meant that island groups developed distinct local cultures, which are now often overlaid with the (continued on p. 507)

HOW OCEAN CURRENTS ARE CAUSED

There are two main generators of ocean currents. The first is wind. Surface winds whip huge masses of water along with them in drift currents. The second is density difference. Temperature and salinity changes alter the density of the water, making it rise to the surface or sink to the depths, creating vertical circulatory currents. The earth's spin and the shape and position of landmasses affect the direction currents take.

Deflection of moving bodies due to the earth's rotation is known as the Coriolis Effect.

Currents in the Northern Hemisphere veer right, and those in the southern veer left. The effect is more marked as distance from the Equator increases, and results in ocean currents circulating clockwise north of the Equator, and anti-clockwise south of it.

The density of seawater increases as it becomes colder or saltier. The hot sun warms equatorial waters, making them less dense, but at the same time evaporation increases their salinity. Ice and cold air chill polar seas. Rain, melting ice and rivers dilute the salinity.

An example of how a landmass can affect a current is the Gulf Stream. Warm Atlantic and Caribbean currents merge off the coast of Florida, are turned north-east by the US coastline and are then launched across the ocean as the North Atlantic Drift, arriving still warm on the western coastlines of Europe.

Where winds are offshore, waters are driven away from the coast. Deep water rises to the surface, bringing with it many nutrients from the bottom. This explains the rich fisheries, for example, of the Peruvian and Namibian coasts.

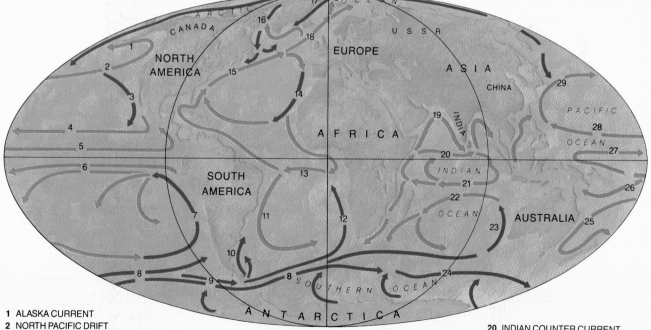

1 ALASKA CURRENT		
2 NORTH PACIFIC DRIFT		
3 CALIFORNIA CURRENT	**12** BENGUELA CURRENT	**20** INDIAN COUNTER CURRENT
4 NORTH EQUATORIAL CURRENT	**13** GUINEA CURRENT	**21** NORTH EQUATORIAL CURRENT
5 EQUATORIAL COUNTER CURRENT	**14** CANARIES CURRENT	**22** SOUTH EQUATORIAL CURRENT
6 SOUTH EQUATORIAL CURRENT	**15** GULF STREAM	**23** WEST AUSTRALIAN CURRENT
7 PERU (HUMBOLDT) CURRENT	**16** LABRADOR CURRENT	**24** WEST WIND DRIFT
8 WEST WIND DRIFT	**17** EAST GREENLAND CURRENT	**25** EAST AUSTRALIAN CURRENT
9 CAPE HORN CURRENT	**18** NORTH ATLANTIC DRIFT	**26** SOUTH EQUATORIAL CURRENT
10 FALKLAND CURRENT	**19** SOUTH-WEST AND NORTH-EAST	**27** EQUATORIAL COUNTER CURRENT
11 BRAZIL CURRENT	MONSOON DRIFT	**28** NORTH EQUATORIAL CURRENT
		29 KURO SHIO

THE FLOORS BENEATH THE OCEANS

The hidden scenery that lies beneath the seas has intrigued humanity since classical times. The insatiably curious Alexander the Great had himself lowered to the sea-bed in a barrel – though to no great depth – while the legend of the submerged continent of Atlantis has continued to fascinate even to the present time.

Deep-diving chambers and instruments and techniques developed over the decades have given us a reasonable idea of the structure of the deep ocean floors. It is a younger scene than that of the continents, since much of it is being perpetually renewed (see PLATE TECTONICS, p542), and less varied. Nevertheless, it does possess numerous features, such as mountain ranges, chasms, volcanoes and the equivalent of vast, sandy deserts.

The submarine world begins in the shallows at low water mark on the beach and plunges down to such depths as the Pacific's Mariana Trench (11033 m, 36197 ft). Continental shelves, part of the continental landmasses, slope gently down from the shore, split here and there by ravines, down which sediments, washed to the coast by rivers, slide into the depths. At a depth of about 200 m (656 ft), the edge of the continental shelf is reached, and the sea-bed plunges abruptly downwards at an angle of 3 to 6 degrees to the horizontal on the continental slope. The less steep slope of the continental rise – at a gradient of between 1 in 100 to 1 in 700 – follows, reaching an average depth of some 5 km (3 miles) below the surface. There begin the abyssal plains, vast desert-like stretches, inhabited by a few fish, worms and molluscs adapted to these lightless depths and whose ultimate food source must come from above. In places, nodules of manganese lie scattered over the ocean bed.

From the plains rise seamounts – volcanoes, some of which break the surface as islands. Hawaii Island is the exposed top of a mountain more than 9150 m (30000 ft) high – over 200 m (656 ft) higher than Mount Everest. In addition, there are eroded, extinct, flat-topped volcanoes called guyots, together with the ridges that are found in all oceans. These are rugged underwater mountain chains, up to 1000 km (620 miles) across, whose individual peaks sometimes rise as high as several kilometres above the ocean floor. In a few places, such as Iceland and Tristan da Cunha, the ridges rise above sea level, and appear as islands. They are frequently divided and traversed by transform faults. Immensely long and deep oceanic trenches, often flanked by volcanic islands, border some oceanic regions and cut deep down into the oceanic crust. Through them, the rocks that compose the oceanic plates slide deeper to blend there with sub-crustal magma along a DESTRUCTIVE PLATE MARGIN, the whole region being a Benioff Zone. A blend of magma and metamorphosed rock re-emerges at the crests of the oceanic ridges and spreads over the ocean floor to maintain the cycle of destruction and renewal.

LANDMARKS OF THE OCEAN FLOOR

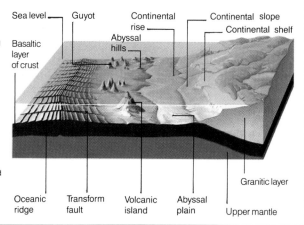

About two-thirds of the earth's crust is seabed. All the great oceans show a similar topography, in which continental shelves, slopes and rises drop down from coastal waters to the abyssal plains, from which rise submarine hills, volcanoes and mountain ranges. Until quite recently, less was known about this hidden scenery than the geography of distant planets. The topography of the great depths has been revealed by echo soundings and seismic probes, while some idea of its appearance and composition has been obtained from dredged rock samples, deep-sea drilling and comparison with ancient oceanic rocks now upthrust into the land.

Sea level — Guyot — Continental rise — Continental slope — Continental shelf — Abyssal hills — Basaltic layer of crust — Oceanic ridge — Transform fault — Volcanic island — Abyssal plain — Granitic layer — Upper mantle

1	East Pacific Ridge	9	Mid-Indian Ridge
2	Mid-Atlantic Ridge	10	Indian-Antarctic Ridge
3	Southwest Indian Ridge	11	Kermadec-Tonga Trenches
4	Walvis Ridge	12	Philippines Trench
5	Middle American Trench	13	Marianas Trench
6	Puerto Rican Trench	14	Aleutian Trench
7	Peru-Chile Trench	15	Kuril-Japan Trenches
8	South Sandwich Trench	16	Sunda-Java Trenches

cultures of European colonists. Tahiti and New Caledonia have a French flavour; Hawaii is American. Japan has become an increasingly prominent participant in trade in Oceania, but New Zealand also plays an important economic role. Australia has wielded the greatest economic influence. Although many Australians once worked on the islands for trading companies or for Australian banking and insurance companies, local employment policies of island governments have greatly reduced their numbers.

Copra (dried coconut flesh) and tourism form the basis of the economies of several small island states. The larger islands have other resources: for instance, Fiji sends timber to Australia and New Zealand and sugar to Europe; New Caledonia has nickel and Papua New Guinea has enormous resources of gold, copper, oil and timber.

The small scale of operations and the distance from major markets are problems for the tiny emerging nations of the area. Tourism provides a partial, but two-edged solution: the price of financial benefits has been some destruction of culture and environment. A South Pacific Forum promotes economic cooperation between the separate islands. Recently, considerable economic change has come to several island states with exploitation of their fishing resources: Fiji has its own fishing industry; the Solomon Islands are in partnership with the Japanese; Tuvalu licenses fishing fleets from Taiwan and South Korea giving them access to its waters. Kiribati, with a small fishing industry of its own in cooperation with the Solomon Islands, licenses United States, South Korean and Japanese boats to operate within one of the world's largest fishing zones, some 3 million km² (1.2 million sq miles).

Although most of the indigenous people are still subsistence farmers, there has been a drift towards towns in recent years. Erosion of traditional village life, with its extended families, has resulted in the breakdown of old systems of support for the ill and aged; few island states have the infrastructure or resources to meet this new need. European and Asian immigrants tend to live in the towns and are often a significant section of the population. In Fiji nearly half the people are of Indian origin. French Polynesia has some 20 000 residents from mainland France out of a population of 200 000, while in New Caledonia more than half are non-Melanesians.

During the Second World War, Oceania was a battleground for the Americans and Japanese. During the Cold War, the Super Powers vied for influence in the region. At the end of the Cold War, economic relationships began replacing security and strategic alliances. This is not surprising: in the early 1990s, the islands' economic bargaining power was almost non-existent, with their gross domestic product having grown by merely 0.1 per cent since the 1980s.

By the mid-1990s, the 14 independent or self-governing states faced an uncertain future. Ideologies may take second place to personalities in these small new nations, but internal politics can be as strident as in other parts of the world, particularly in Papua New Guinea, where the formation of secessionist movements has resulted in serious bloodshed since the Second World War. One of the major colonial forces in the area, France, while reiterating its determination to maintain its presence in the area, entered an accord that provided for a planned referendum in 1998 on independence for New Caledonia.

Over all these developments hang population pressures and environmental concerns such as damage from mining in Nauru, Kiribati, Fiji and Papua New Guinea, and the possible rise in the sea level as a result of global warming. This may now threaten the existence of many thousands of low islands and islets: all but one of Kiribati's 33 islands are low-lying atolls. On the positive side, however, France declared a moratorium on nuclear testing in its territory after years of protest from neighbouring states, particularly New Zealand.

Map Pacific Ocean

Ocho Rios *Jamaica* Seaport and resort town on the north coast, 90 km (56 miles) east of MONTEGO BAY. Many cruise ships visit the port and it serves as a tourist base for visits to the many waterfalls and cascades, as well as the limestone scenery. One of the most notable features is the Dunn's River Falls, where a clear mountain stream cascades 180 m (about 600 ft) down a wooded gorge to the sea. Prospect Plantation offers a view of an agricultural lifestyle long past.

Population 10 500
Map Jamaica – Ba

Odawara *Japan* See HAKONE

Odense *Denmark* Third largest city of Denmark and the administrative centre for the island of FÜNEN. An 8 km (5 mile) long ship canal, built in 1797-1804, links the city to Odense Fjord, the site of a major shipyard at Lindø. The city has been a bishopric since 1020, and its 11th-century Cathedral of St Knud (rebuilt in the 13th century after being destroyed by a fire) is one of the finest Gothic buildings in Denmark.

Most visitors, however, come to the city because it is the birthplace of Hans Christian Andersen (1805-75), the author of many of the world's best-loved fairytales. Several museums display mementos of his life. Among the places of particular interest are Hans Christian Andersen's childhood home and the house in which he later lived, a railway museum and Fünen Village – which is an open-air museum of traditional, local country life.

Odense, Denmark's second largest industrial centre, specialises in heavy industries and electrotechnical manufacturing. It also has sugar refining and food processing plants.

Population 177 600
Map Denmark – Bb

Odenwald *Germany* Range of hills east of Mannheim, situated between the Neckar and Main rivers. Because its western slopes, along the Bergstrasse ('mountain road'), are sheltered from the east winds, spring comes exceptionally early. Tobacco, grapes, peaches, apricots, cherries, almonds and walnuts are among the produce grown; the show of early blossoms is a great tourist attraction.

Map German – Cd

Oder (Odra) *Central Europe* River, 912 km (567 miles) long. It rises in the Oder Mountains in the Czech Republic 60 km (35 miles) south-west of the city of OSTRAVA. From Raciborz, just inside Poland, it is navigable all the way to the Baltic, 710 km (440 miles) downstream. The Oder transports coal, building materials, timber and chemicals. With its tributary, the Neisse, it forms the 'Oder-Neisse Line', which is the border between Poland and Germany.

Map Poland – Bc

Odesa (Odessa) *Ukraine* Black Sea port and tourist resort in the Ukraine, 400 km (249 miles) south of KIYEV. It has five harbours and has been a naval base since 1794. During the Soviet era, Russia's Antarctic whaling fleet had its base here. The city, which stands on a group of hills overlooking the Black Sea, was built around a 14th-century Tatar fort. Much of its rich cultural life revolves around its handsome 19th-century opera house. Its historic buildings and seaside location attract thousands of tourists. It is also an industrial centre, producing steel, ships, machinery, chemicals, petroleum products, and foodstuffs. It handles much of Ukraine's imports and exports.

In 1905 a mutiny aboard the battleship *Potemkin* ended in the killing of thousands of inhabitants of the city. The incident was sparked off when the crew of the ship rebelled after being served maggot-infested meat. A civil uprising followed in the city, which was then bombarded by government forces; 6000 people died.

Population 1 115 000
Map Ukraine – Cb

Odra *Eastern Europe* See ODER

PLUMBING THE DEPTHS

The lowest recorded point in the world – the Challenger Deep in the western Pacific – has not been visited by man. Part of the Mariana Trench, it was discovered in 1951 by the British survey ship Challenger – and in 1960 a manned American bathyscaphe descended to 10 915 m (35 810 ft). But the record depth of 11 033 m (36 197 ft) was established later that year by a Russian survey ship using echo soundings.

THE DEEPEST POINTS OF THE OCEANS

Ocean	Location	Depth (m/ft)
Pacific	Mariana Trench (Challenger Deep)	11 033/36 197
Atlantic	Puerto Rico Trench	9200/30 200
Indian	Java (Sunda) Trench	7455/26 400
Arctic	Angara Basin	5440/17 850
Antarctic	(Southern) South Sandwich Trench (Meteor Deep)	8428/27 651

Offaly (Uíbh Fhailí) *Ireland* County covering 1998 km² (771 sq miles) in central Ireland, 50 km (about 30 miles) west of DUBLIN. It consists mainly of grassy plain, with the Slieve Bloom Mountains in the south. Offaly was formerly known as King's County after Philip of Spain, husband of Mary I, Queen of England (1553-8). The county town is Tullamore (population 8620).
Population 58 490
Map Ireland – Cb

Offenbach *Germany* Town on the southern outskirts of FRANKFURT am Main with large tanning and engineering industries. It has a leather museum and runs an annual leather fair.
Population 115 000
Map Germany – Cc

Ogaden *Ethiopia* Desert region, some 600 km (373 miles) east to west and 400 km (248 miles) north to south, which is claimed by neighbouring Somalia because it is inhabited mainly by Somali nomadic herders. The two countries fought a full-scale war over it in 1977; the Western Somali Liberation Front is still active here. The Ethiopian government is unwilling to give up the territory, particularly because it may be rich in oil.
Map Ethiopia – Bb

Ogbomosho *Nigeria* One of Nigeria's largest cities in Oyo State, 200 km (124 miles) north-east of Lagos. Founded in the 17th century as a traditional Yoruba tribal town, it has become a busy industrial centre producing foodstuffs. It also has a tobacco market.
Population 514 400
Map Nigeria – Ab

Ogooué (Ogowe) *Congo/Gabon* River, 1200 km (745 miles) long, rising in Congo south of the town of Zanaga, and flowing through the rain forests of Gabon to the Atlantic. It drains some 222 700 km² (85 984 sq miles), and its upper course is marked by rapids and waterfalls, including the Poubara Falls. But it is navigable for part of the year below Ndjolé and throughout the year below LAMBARÉNÉ. The latter section is Gabon's busiest waterway especially for timber exports. Near the Atlantic the river divides into several channels, and has built up a large delta south of Port Gentil.

The Ogooué was explored separately by the French Paul du Chaillu, the Italian-born Pierre Savorgnan de Brazza (who also found its source), and by the Victorian traveller Mary Kingsley, who braved its rapids in a canoe. She described the Ogooué as 'the greatest river between the Niger and the Congo'.
Map Gabon – Bb

O'Higgins *Chile* See LIBERTADOR

Ohio *USA* Midwestern industrial state covering 107 070 km² (41 330 sq miles) of gently undulating country southwards from Lake Erie to the Ohio River. It was first settled in the 1780s as part of the Northwest Territories, and joined the Union in 1803. Ohio's farmers produce livestock (cattle, pigs and poultry), soya beans and maize. Coal, oil and gas are the main mining products while manufactures include machinery, transport equipment, rubber goods, metals and food products. The largest cities are CLEVELAND, CINCINNATI and the state capital COLUMBUS.

The Ohio River was used as a key route by settlers travelling to the Midwest. It is formed from the Allegheny and Monongahela rivers, which join at the city of PITTSBURGH in Pennsylvania, and flows 1575 km (980 miles) south-west to the Mississippi River at Cairo in southern Illinois. The Ohio forms the boundary between the states of Ohio, Indiana and Illinois to the north and West Virginia and Kentucky to the south. Today, the river is regulated by a series of 46 locks and dams, and drains one of the most populous and industrialised regions of the United States.
Population (state) 10 847 100
Map (state) United States – Jb; (river) United States – Ic-Jc

Ohrid, Lake (Ohridsko Jezero) *Albania/Macedonia* The Balkan's second largest lake (after Lake Shkodër) covering 348 km² (134 sq miles) astride the Albanian border in south-west Macedonia (approximately one-third lies in Albania). It is up to 286 m (938 ft) deep, and its abundant trout and eels, fine beaches, mild winters, warm summers, clear water and superb scenery make it Macedonia's leading tourist centre. Fed by underground springs, the lake empties via the Drin into the Aegean Sea.

The ancient town of Ohrid on the east shore has picturesque white-painted houses with upper floors jutting out over quaint paved alleys. Several fine Byzantine churches can also be seen here. The town hosts a Folk Song and Dance Festival each summer.
Map Macedonia – Ed

oil and natural gas See THE STORY OF OIL AND NATURAL GAS, p 509

oil shale Fine-grained black or dark grey rock rich in kerogen (fossilised organic remains) which, when heated, produces hydrocarbons that can be distilled to produce oil.

Oileáin Arann *Ireland* See ARAN ISLANDS

Ojos del Salado *Argentina/Chile* After ACONCAGUA, this is the highest mountain in South America rising to a height of 6880 m

▲ RIVER HARVEST Young Botswanans fish with baskets in the Kavango River, which rises in Angola and flows into the Okavango Delta in Botswana.

(22 572 ft). It stands on the northern border between Argentina and Chile.
Map America, South – Bc

Ok Tedi *Papua New Guinea* Copper and gold mine in the Star Mountains of central New Guinea, near the border with Irian Jaya. Named after a local river, its reserves of ore have been estimated at 350 million tonnes.
Map Papua New Guinea – Ba

Oka *Russia* Tributary of the Volga River which rises in the uplands south of Moscow and flows generally north and east before joining the Volga at NIZHNIY NOVGOROD. The Oka, which is 1480 km (920 miles) long, is used largely for transporting grain and timber.
Map Russia – Fc

Okanagan Valley *Canada* Fruit-farming area and popular holiday resort in southern BRITISH COLUMBIA. Tourist areas include Lake Okanagan and the towns of Penticton and Summerland.
Map Canada – Cd

Okavango *Southern Africa* See KAVANGO

Okavango Delta *Botswana* Swampy delta covering 10 360 km² (4000 sq miles) in the north-west of the country. It is supplied with water by the KAVANGO RIVER, which rises in Angola. Vast numbers of animals and birds, including hippopotamuses, buffaloes, storks and pelicans, live within its braided channels. About one-fifth of the delta is protected in the unfenced Moremi Game Reserve (1800 km², 69 sq miles) where elephants are a common sight. Plans for commercial use of the water for irrigation and for livestock ranches have been restricted by the area's remoteness and the high evaporation rate of the delta. During floods, water drains east through the Botletle River into the Makgadikgadi Pans.
Map Botswana – Ba

THE STORY OF OIL AND NATURAL GAS

Of all the ages of humankind, it seems probable that the Oil Age will be the shortest lived. A few limited uses apart, neither petroleum nor natural gas played much of a part in human affairs until almost within living memory. They now provide more than 70 per cent of world energy, but it is variously prophesied that all reserves will be exhausted by the turn of the century, or at the most, in 40 to 50 years' time. The first well specifically drilled for petroleum was sunk in 1859 by Colonel E L Drake in Pennsylvania to exploit the process invented by the Scots chemist James 'Paraffin' Young, who had found a means of distilling paraffin, or kerosene, from crude oil. For both Drake and Young the object of the exercise was the recovery of the by-product paraffin wax, a cheaper material for candle making than beeswax or tallow.

People have been aware of petroleum for a very long time. More than 5000 years ago, the Assyrians, Babylonians and Sumerians used bitumen – a semisolid form of petroleum oil

occurring in surface seepages – as mortar, in road making, and as waterproofing in irrigation channels. It was also used to create mosaics, to caulk ships, and in liquefied form as a liniment and a laxative. The ancient Chinese, drilling down to 140 m (450 ft) in search of brine, lit associated natural gas deposits to evaporate the water content from the salt. Oil and gas seepages, probably originally ignited by lightning, at BAKU in what is now Azerbaijan, fuelled the 'eternal fires' of the fire-worshipping early Persians.

Like coal, oil and gas are for all practical purposes non-renewable resources and, like coal too, they are the remains of ancient life forms. But while coal evolved out of the trees and plants of long-ago tropical forests, petroleum had its beginnings in the seas. It is derived from the remains of simple planktonic life forms, minute plants and animals, although the exact means by which plankton has been converted to petroleum is not completely clear. Plankton were first abundant in

the warm seas some 570 million years ago and have continued to be abundant ever since. Over a long period, dead plankton have rained down to form a layer on the sea-bed which, if near coasts, has become covered by sediments. Over aeons, the sediments and the organic matter have become pressed ever more deeply into the earth's crust by new strata forming above. Millions of years of pressure, combined perhaps with heat from the planet's interior, have forced structural changes upon the organic layer, converting it to liquid and gas.

The sediments above have since evolved into sedimentary rocks, such as sandstones and limestones, and because of their low density, the liquid oil and gas have moved upwards and sideways to penetrate porous sedimentary rocks. If the deposits have reached the surface in this manner, they were, and remain, lost as seepages. But if the upward movement of the oil and gas has been halted by a layer of impervious rock, then subterranean pockets and reservoirs are created, which can be tapped.

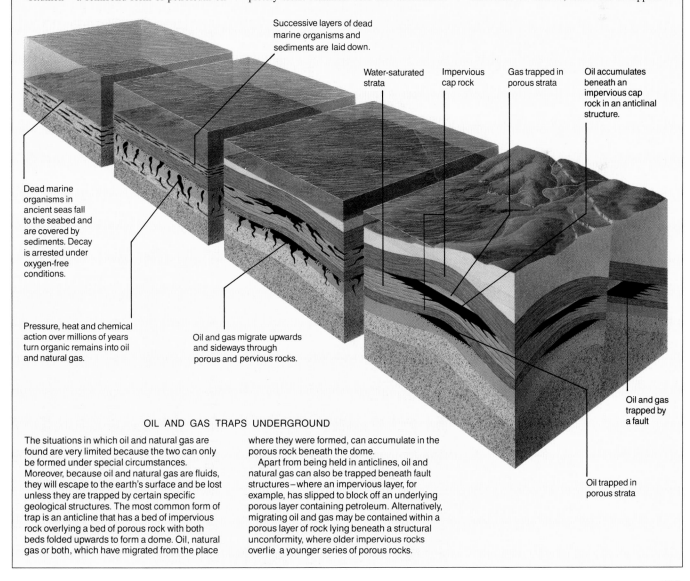

OIL AND GAS TRAPS UNDERGROUND

The situations in which oil and natural gas are found are very limited because the two can only be formed under special circumstances. Moreover, because oil and natural gas are fluids, they will escape to the earth's surface and be lost unless they are trapped by certain specific geological structures. The most common form of trap is an anticline that has a bed of impervious rock overlying a bed of porous rock with both beds folded upwards to form a dome. Oil, natural gas or both, which have migrated from the place

where they were formed, can accumulate in the porous rock beneath the dome.

Apart from being held in anticlines, oil and natural gas can also be trapped beneath fault structures – where an impervious layer, for example, has slipped to block off an underlying porous layer containing petroleum. Alternatively, migrating oil and gas may be contained within a porous layer of rock lying beneath a structural unconformity, where older impervious rocks overlie a younger series of porous rocks.

Okayama *Japan* Commercial centre in south-west HONSHU Island, 140 km (87 miles) west of the city of Osaka. Its 'Black Crow Castle' was built in black in 1573 to contrast with the snow-white castle at nearby Himeji. The Korakuen landscaped gardens, with a tiny rice field, tea plantation, ancient trees, lawns, ponds and water-falls are rated among Japan's finest. Industries vary from chemicals and machinery to the pro-duction of traditional *tatami* mats of woven straw.

Population 591 300
Map Japan – Bd

Okhotsk, Sea of *Russia* Arm of the north-western PACIFIC OCEAN, covering 1 528 000 km² (589 961 sq miles), and bounded by the KAMCHATKA Peninsula and the KURIL ISLANDS. Its 100 km to 200 km (62 mile to 125 mile) wide continental shelves are major fishing grounds. It is also a source of oil for Russia. While it has an average depth of 971 m (3186 ft), its greatest recorded depth is 5210 m (17 093 ft). In summer it provides an outlet for eastern Siberia, but fogs and ice are a winter hazard. MAGADAN is the chief port.

Map Russia – Pc

Okinawa *Japan* Main island of the RYUKYU ISLANDS, covering 1176 km² (454 sq miles) in the East China Sea, 500 km (310 miles) south-west of Japan. During the Second World War, fierce fighting took place here between the United States and Japan. Bombed from March until June 1945, it was invaded and finally overrun by the United States who used it to establish air bases close to the Japanese mainland. After the war, the group was put under the control of the United States. The northern section was returned to Japan in 1954, and the entire group of islands (covering 2388 km², 922 sq miles) was restored to Japanese rule in 1972.

Population 1 222 000
Map Japan – Gf

Oklahoma *USA* South-western Great Plains state covering 173 320 km² (66 919 sq miles) of gently undulating country between the Red River and Texas to the south and Kansas to the north. From 1828 to 1845 thousands of native Ameri-cans who were driven from their homes in the south-eastern United States migrated here along 'Trail of Tears'. It remained a reservation until the United States government opened it for set-tlement in 1889. It became a state in 1907; but it still has the largest native American population (250 000) of all the US states.

During the 1930s farming was badly hit when a long drought, combined with unscientific farm-ing methods, created a dustbowl in the state. Much of the soil was blown away by winds. Many Oklahomans moved to California, where they were often exploited by ruthless farmers. Their plight was highlighted in the moving novel, *The Grapes of Wrath*, by the 1962 Nobel Prize winner for literature, John Steinbeck (1902-68).

Today, Oklahoma's agriculture centres mainly on cattle, pigs, poultry, wheat and cotton. It has ample reserves of coal, and also rates as one of the country's largest producers of natural gas. Its industries include the production of petroleum products, together with transport and mining equipment. OKLAHOMA CITY, the state capital, and TULSA are the biggest cities.

Population 3 145 600
Map United States – Gc

Oklahoma City *USA* State capital of OKLAHOMA, situated in the centre of the state on the North Canadian River, directly over an oil field. It was founded on 22 April 1889 when surrounding ter-ritory (previously reserved for native Americans) was opened to white settlement. Claims were staked in true cowboy fashion: at the shot of a gun, 10 000 people raced from the site to stake out their land. Another 40 000 people were similarly involved elsewhere in the territory. A cowboy hall of fame is situated in the city.

In addition to being a cattle market and agri-cultural service centre, Oklahoma City also has a number of important industries, including petro-chemicals, electronics and food processing.

Population (city) 444 700; (metropolitan area) 958 800
Map United States – Gc

Öland *Sweden* Long and low-lying island covering 1345 km² (520 sq miles), off the south-east Baltic coast. It is linked by a 6 km (4 mile) long bridge to the mainland port of Kalmar. Its fishing fleet provides an important source of revenue; farming centres on the rearing of sheep and the cultivation of wheat and root crops. Limestone is also quarried.

Chief among Öland's numerous archaeo-logical features, is a reconstructed Iron Age settlement. The island, a popular destination for tourists, is well known for its many windmills.

Population 25 400
Map Sweden – Cd

Olbia *Italy* Port on the north-east coast of Sardinia, 190 km (118 miles) north of Cagliari. It is the point of entry for most ferry passengers from mainland Italy.

Population 32 600
Map Italy – Bd

Oldenburg *Germany* Northern river port and market town on the Hunte River, some 45 km

▼ HEADS IN THE DESERT The Aboriginal name for the Olgas is Katatjuta – 'many heads' – because of their uncanny resemblance to a group of skulls conferring in the desert.

(28 miles) north-west of Bremen. It has a castle, part of which has been converted into a museum of art and cultural history. The town is also known for the mud-brick tower (known as Lappen) of the Holy Ghost Hospital (1468), one of its few remaining medieval buildings.

Population 143 100
Map Germany – Cb

Oldham *United Kingdom* Mill town on the north-eastern outskirts of Greater Manchester. In the 17th and 18th centuries a thriving woollen cloth industry was established here. However, the real boom period for the town came in the 19th century, with the development of the cotton industry. Today, the manufacture of cotton goods has been replaced by a wide range of other activi-ties, including vehicle assembly, construction of boats, textile machinery and electronics.

Population 107 830
Map United Kingdom – Dd

Olduvai Gorge *Tanzania* Prehistoric site in the north of the country 30 km (19 miles) north-west of the NGORONGORO CRATER. Between 1959 and 1961 the British palaeontologist Louis Leakey (1903-72) and his family discovered fossil skulls here that were thought to be between 1.5 and 2 million years old, and to represent a stage of human evolution prior to *Homo sapiens*. The gorge is up to 150 m (490 ft) deep, and stone tools and the bones of many extinct animals have also been found here.

Map Tanzania – Ba

Oléron, Ile d' *France* Island 30 km (19 miles) long in the Bay of Biscay to the south of LA ROCHELLE. At 175 km² (68 sq miles), it is France's second largest island after Corsica. It has a mild climate, good fishing, fine beaches and breeding grounds for mussels and oysters.

Population 18 400
Map France – Cd

Olga, Mount *Australia* One of a cluster of 36 dome-shaped peaks in central Australia, 30 km (19 miles) west of AYERS ROCK in the Northern Territory. At 1066 m (3497 ft), Mount Olga is the tallest of the group, informally called the Olgas, or Katatjuta ('many heads') by their aboriginal

custodians. The first European to sight the peaks was the explorer Ernest Giles in October 1872. He is said to have named them after Queen Olga of Württemberg – at the request of his patron.
Map Australia – Ed

Oligocene Third epoch of the Tertiary period of the Cenozoic era of the earth's timescale. See GEOLOGICAL TIMESCALE, p 256

Ólimbos *Greece* See OLYMPUS

Olinda *Brazil* Historic city 7 km (5 miles) north of the city of RECÍFE, and today part of greater Recífe. Founded in 1535 by the Portuguese, it is protected by the United Nations Educational, Scientific and Cultural Organisation (UNESCO) as part of the world's cultural heritage. It is so named because early settlers are said to have exclaimed '*O linda!*' ('How beautiful') on their arrival at the site.

One of the first major urban settlements in Brazil, it has beautiful colonial churches – many with distinctive blue and white tiles. The most notable among these are da Sé (1537), da Misericórdio (1540) and São Francisco (1577).

Olinda hosts one of the most exciting carnivals in Brazil, just before Lent. Unlike the celebrated carnival in Rio de Janeiro, anyone can join in, between the set-piece groups – and each year about 250 000 people do. They dance the frenetic *frevo* (beside which the samba seems sedate) through the narrow hilly streets of the old town for at least four days and nights.
Population 389 200
Map Brazil – Eb

olivene Group of silicate minerals, consisting of compounds of iron silicate and magnesium silicate, occurring in basic and ultrabasic igneous and some metamorphic rocks. One dark green variety, also known as peridot, is valued as a gem.

Ollantaytambo *Peru* Site of an Inca tambo (posting station) and fortress, on the Urubamba River, 70 km (43 miles) north of CUSCO. The ruins include flights of terraces, the temple and palace, the Baío de la Ñusta (Bath of the Princess) and a solar observatory. The last Inca leader, Manco Capac, and his soldiers used the fortress in 1537 when they retreated from the Spanish army after the siege of Cusco.
Map Peru – Bd

Oloibiri *Nigeria* See NIGER

Olomouc *Czech Republic* Market town and capital of Moravia in the 11th and 12th centuries, 64 km (40 miles) north-east of BRNO. It is situated in the heart of a fertile grain-growing plain; its main industries are food processing and engineering. The old town has a Gothic cathedral and town hall, and a university founded in 1566.
Population 104 300
Map Czech Republic – Cb

Olympia *Greece* Best known sanctuary of the god Zeus, at the foot of Mount Kronos near the Ionian coast of the Peloponnese, and the original site of the Olympic Games. The sanctuary was a tree-lined, sacred grove in which a flame, relit each spring in a stone enclosure, burned in honour of Zeus. All-male foot races, which later evolved into the Olympic Games, were held

in a stadium in the grove from 776 BC. They lasted for more than 1000 years until the Roman emperor Theodosius banned them. The temples were destroyed in the 5th and 6th century AD. Excavations by German archaeologists from 1875 revealed the Temple of Zeus, the Altis, or sacred olive grove, the stadium, the hippodrome, where chariot races were held, a gymnasium and baths. Excavation of the almost complete village led directly to the revival of the Olympic Games by a Frenchman, Baron Pierre de Coubertin, at ATHENS in 1896.
Map Greece – Bc

Olympia *USA* Capital of the Pacific state of Washington and port at the head of Puget Sound, on the Olympic Peninsula, more than 250 km (155 miles) from the Pacific Ocean. It is a gateway to the OLYMPIC NATIONAL PARK's coastal and mountain wilderness and forests. Local products include Olympia oysters and beer.
Population (city) 33 800; (metropolitan area) 161 200
Map United States – Ba

Olympic National Park *USA* Mountain and coastal wilderness covering an area of 3735 km² (1442 sq miles) in Washington State. It includes Mount Olympus (2428 m, 7965 ft), glaciers and 81 km (50 miles) of undisturbed Pacific coastline. The park, which has the finest remains of the Pacific north-western rain forest, is rich in wildlife.
Map United States – Ba

Olympus (Olympos, Ólimbos) *Greece* Highest mountain massif in Greece on the borders of Thessaly and Macedonia, and legendary home of the gods. The massif stretches from the Thessaly coast for 40 km (25 miles) into Macedonia, peaking at Mítikas (2917 m, 9570 ft), Greece's highest mountain. Its precipitous, crystalline rocks are broken by thickly wooded ravines. The massif is snow-capped for most of the year, and is now the home of a number of ski resorts for holiday-makers as well as a military ski school.
Map Greece – Ca

Omaha *USA* Market city on the MISSOURI River on the eastern border of Nebraska. In 1846-7 Mormons on their way to Salt Lake City used it as their winter quarters. Two years later it became a busy trading post during the California gold rush. Its connection with cattle ranching goes back a long way – to the late 1800s of the big cattle drives from Texas. Omaha carries on that tradition today: it has one of the largest livestock markets and meat processors in the world.

Omaha's industries include the manufacture of farming equipment, the production of fertiliser and food processing.
Population (city) 335 800; (metropolitan area) 639 800
Map United States – Gb

Oman, Gulf of Arm of the ARABIAN SEA, leading to the Strait of HORMUZ and the oil-rich GULF. It is 560 km (350 miles) long and 320 km (200 miles) wide, and its greatest known depth is 4153 m (13 625 ft). Despite the monsoon winds which sweep through the gulf for three months of the year, it was an important shipping lane in ancient times, when cargoes of ivory, jewels, silks, spices and teak from China and South-east Asia were transferred between

ships at Omani ports and then taken on to Babylon. Today huge oil tankers are the most common ships in The Gulf.
Map Oman – Aa

Omdurman *Sudan* City of flat-roofed, baked clay houses and narrow streets situated across the Nile from the capital, Khartoum, and the seat of the Sudanese parliament. It specialises in the production of jewellery, carvings and the delicate working of gold and silver. It also has markets which are best known for their trade in camels, skins and souvenir ostrich eggs.

The city was founded on the site of an insignificant riverside village by the Sudanese religious leader, the *Mahdi*, in 1884. His tomb is in the city. In 1898, after the Battle of Omdurman, in which the British defeated the *Mahdi*'s followers, Khartoum again became the capital.
Population 526 300
Map Sudan – Bb

Omo *Ethiopia* River rising in the central highlands and flowing southwards for about 800 km (500 miles) into Lake TURKANA (Lake Rudolf) on the Kenyan border. Archaeological finds in the lower Omo Valley since 1967 point to the presence of an early prehistoric human settlement here. One of the oldest discoveries was the jawbone of a hominid, estimated to be 2.5 million years old. Older remains, dating back about 4.4 million years, have since been found in the Awash Valley.
Map Ethiopia – Ab

Omsk *Russia* Industrial city in western Siberia about 2200 km (nearly 1370 miles) east of Moscow where the Trans-Siberian Railway crosses the IRTYSH RIVER. Founded in 1716 as a strategic military fort at the confluence of the Irtysh and Om' rivers, it is now an important transshipment centre and river port.

During the Second World War, Omsk began to manufacture machinery and fertilisers for farms on the plains around it. Its industries now include oil refining and petrochemicals, and the production of cars, flour, footwear and textiles.
Population 1 167 000
Map Russia – Ic

Onega, Lake (Onezhskoye Ozero) *Russia* Second largest lake in Europe after Lake Ladoga, covering 9720 km² (3750 sq miles) in KARELIA, in north-western Russia. It is linked by canal to the Baltic Sea, White Sea and Volga River. Its main settlement is the industrial city of Petrozavodsk on the west shore. From here, boats ply to Kizhi Island, which has a magnificent open-air museum of wooden churches, including the 22-domed Church of the Transfiguration. The lake, which in places is 100 m (328 ft) deep, is frozen from December to May.
Map Russi – Eb

Onitsha *Nigeria* Trading town on the east bank of the Niger River, 360 km (224 miles) east of Lagos, in Anambra State. Situated where the rain forests of the south begin to give way to savannah, it has marketed river-borne goods from both areas for centuries. Today, Onitsha handles palm oil, dried fish, vegetables, rice, cassava, and manufactured goods.
Population 262 100
Map Nigeria – Bb

Oman

*IN LESS THAN 20 YEARS AN
ENLIGHTENED RULER BROUGHT
THIS ONCE BACKWARD STATE
INTO THE 20TH CENTURY*

Ruled by a sultan with absolute power, this country – third largest on the Arabian Peninsula, after Saudi Arabia and Yemen – has a population which is 88 per cent Arab. Minority groups are the Baluchis, from Iran and Pakistan, and Africans descended from former slaves. There are also 250 000 immigrant workers from India, Pakistan, Iran, South Korea and Europe.

Oman is divided in two by the United Arab Emirates. A small mountainous area in the north at the tip of the MUSANDAM Peninsula overlooks the strategic Strait of HORMUZ, which controls the entrance to The GULF. The main part of the country consists of a fertile coastal plain (BATINAH) in the far north, the limestone HAJAR MOUNTAINS rising to more than 3000 m (9840 ft), and a largely barren plateau with small fishing harbours to the south. The DHOFAR coast in the far south is fertile and has reliable monsoon rain. There are good pastures in Dhofar's mountains. Inland the plateau merges into the largely unexplored and arid Empty Quarter in Saudi Arabia.

The old Sultanate of Muscat and Oman was one of the most backward Arab states when oil production began in 1967. But in 1970 the present ruler, Sultan Qaboos Bin Said, deposed his father, and since then he has directed oil revenues into making massive strides in social development. Today, in the renamed Oman, education is free (but not compulsory) and there are 741 schools – compared with three in the former sultan's time.

Oil, found mainly in the north, provides 95 per cent of exports; there are also copper and natural gas. In recent years, looking ahead to the depletion of its oil reserves, Oman has embarked on an ambitious programme to diversify its economy, encourage foreign investment and enter into joint ventures overseas. However, 70 per cent of the people continue to live by farming fruit and vegetables or fishing. With high temperatures and low, generally unreliable rainfall, less than 2 per cent of the country's land is cultivated. This area includes the Batinah, a potentially rich garden area, land around mountain villages which have ancient irrigation systems, and sheltered mountain valleys.

The port of MUSCAT was a flourishing trade centre by the 5th century BC. It was a Portuguese base from 1508 to 1648, and has been the seat of the present dynasty since 1741.

In the late 18th century Oman established a close treaty connection with Britain. It became fully independent in 1971 but has military links with Britain and the United States. Its relatively uneventful political scene was marked in mid-1994 by the arrest of about 200 soldiers, university lecturers and students who were alleged to be members of a secret network of Islamic militants.

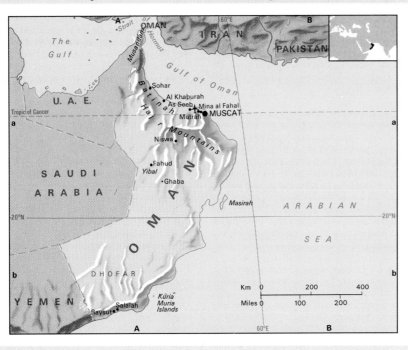

OMAN AT A GLANCE	
Official name Sultanate of Oman	
Area 314 000 km² (121 207 sq miles)	
Population 2 225 000 **Per km²** 7 (**Per sq mile** 18)	
Government Monarchy (Sultanate)	
Currency 1 Omani rial = 1000 baiza	
Languages Arabic (official), Persian, English, Baluchi, Urdu and other Indian dialects	
Religions Muslim (Ibadhi 75%), Hindu (small minority)	
Climate Desert climate, with hot summers and mild winters. Average temperature in Muscat ranges from 22°C (72°F) in January to 34°C (93°F) in June	
Land use Cultivation 2%, pastures 5%, other 93%	
Main primary products Dates, fruits, vegetables, fish, crude oil and natural gas	
Major industries Agriculture, mining, oil refining, copper smelting, fishing, cement	
Main exports Crude oil, processed copper, fish, textiles	
Annual income per head (US$) 6490	
Population growth (per thous/yr) 35	
Life expectancy (yrs) Male 65 **Female** 69	

Ontario *Canada* The country's second largest province, covering an area of some 1 068 582 km² (412 580 sq miles) south of Hudson Bay. Northern Ontario is a rocky CANADIAN SHIELD area dotted with lakes, and rich in abundant forests and minerals (such as nickel, copper, gold, silver and platinum) at the mining centre Sandbury. The southern part of the province has gently undulating areas of clays and sands. The southern shoreline stretches 3700 km (2290 miles) along four of the Great Lakes. Agriculture is important in the mild south-west, where crops include wheat, barley, sugar beet, grapes and tobacco. About half of Canada's manufactures come from Ontario, mainly from a southern crescent between Hamilton and Oshawa. Products include steel, cars, electrical goods and chemicals. Ontario is Canada's most populous province. It contains the national capital OTTAWA, but the largest city is the provincial capital, TORONTO. Tourist attractions include the NIAGARA FALLS and 250 000 lakes.
Population 10 084 890
Map Canada – Fc

Ontario, Lake *Canada/USA* Smallest and most easterly of the GREAT LAKES, covering about 18 960 km² (7321 sq miles). It is supplied with water from the higher-lying Lake Erie by the Niagara River, via the Niagara Falls, and drains to the Atlantic by the St Lawrence River.
Map United States – Kb; Canada – Hd

Onyx River *Antarctica* Longest of the very few rivers in Antarctica. It flows in summer, carrying meltwater from the Lower Wright Glacier about 50 km (30 miles) to Lake Vanda.

oolite Sedimentary rock, usually a limestone, composed of tiny rounded grains embedded in a fine matrix.

Oostende *Belgium* See OSTEND

ooze Layer of mudlike sediment covering the floors of oceans and lakes, and composed principally of the remains of microscopic animals. Oozes are named after the organism whose remains are dominant in them, such as diatom ooze and radiolarian ooze.

opal Variety of silica used as a gemstone. Vivid colours are produced by impurities or by light catching on minute cracks in the stone. Opals range from pearly white, through yellows and reds, to the most valuable black or blue gems.

Oporto (Porto) *Portugal* The country's second largest city after Lisbon, lying some 270 km (168 miles) north of the capital on the steep banks of the lower DOURO River. Today a vibrant commercial and industrial centre, Oporto started out as a Roman settlement known as Portus Cale, from which the medieval principality of Portucalia – and the country today – derived their names. Among several interesting bridges over the Douro, are the metal Maria Pia railway bridge designed by the French engineer, Gustave Eiffel, the designer of the Eiffel Tower in Paris, and completed in 1877; the Dom Luis I bridge built in 1886, with two layers of roads; and the Arrábida road bridge, completed in 1963. Two further modern bridges span the river. Before the mouth of the river silted up, the city was a busy port. Nevertheless, it still exports wine, including port wine, which takes its name from the city.

Oporto is at the heart of a busy industrial region producing tyres, chemicals, metals, electrical equipment, car parts, textiles, leather goods, silk, shoes and soap.

Among the city's many places of interest are the 12th-century cathedral; the Clérigos Tower, a Baroque church with a granite tower 75 m (246 ft) high; the Gothic and Baroque Church of St Francis; the Soares dos Reis Museum containing pottery, paintings and sculptures; the Ethnographic Museum; and the Palácio da Bolsa, now the stock exchange.
Population (metropolitan area) 1 500 000
Map Portugal – Bb

Oran (Wahran) *Algeria* Port and second city of the country, situated on the north-west coast. The city, overlooked by the 16th-century fort of Santa Cruz on a nearby hill and surrounded by hills planted with orange, lemon and olive groves, was founded in 903 by Arabs from southern Spain. Later it was expanded by the Spaniards who ruled the region from 1509 until it fell to the Turks in 1708. In 1732 it reverted to Spain, and Spanish influence can be seen in a number of buildings and monuments. However, most of the town was devastated by an earthquake in 1790. In 1831, Oran was occupied by the French, who developed it as a modern port. Industries include cement, chemicals, fertilisers, agricultural machinery, iron smelting, fruit and fish canning, as well as the manufacture of textiles, glass, footwear and cigarettes. The port exports wheat, vegetables, wine, wool and esparto grass.

Places of interest include the Kasbah (the old Spanish and Arab city); the 18th-century mosque of the Pasha Sidi El Houari; Fort Lamorrue, built by the Spaniards in 1742; the Demaeght Museum with prehistoric exhibits and paintings; and many Moorish and Spanish gates.
Population 700 000
Map Algeria – Aa

Orange *France* Historic town in Vaucluse Department, 105 km (65 miles) north-west of MARSEILLES. Its majestic Roman theatre, seating 10 000, is the best preserved in Europe. The principality of Orange was annexed by France in the 17th century. It was previously the seat of the Nassau dynasty, members of which were princes of the Netherlands and one of whom, William of Orange, also became King William III of England in 1689.
Population 27 500
Map France – Fd

Orange *Southern Africa* Longest river of Southern Africa (2090 km, 1299 miles) rising in the Drakensberg range in Lesotho and forming part of the border between South Africa and Namibia before entering the Atlantic Ocean. The river has been dammed in several places as part of a huge water-conservation and hydroelectric power project largely financed by South Africa.
Map South Africa – Ab-Cc

Orange Free State *South Africa* Landlocked province of 129 437 km² (49 964 sq miles), with borders based on those of the Orange Free State Republic of 1854-1900, the Orange River Colony (1900-1910) and those of the post-1910 province of the Union (later Republic) of South Africa. With its sister Boer state – the South African Republic (later known as Transvaal) – the Orange Free State went to war with Great Britain in the Anglo-Boer War of 1899-1902, both states being annexed to the Crown in 1900. In 1994 the name of the province was changed to Free State.

The province lies on the high plateau between the Orange and Vaal rivers and has its capital at BLOEMFONTEIN. The economy is based increasingly on mining – including gold and coal – and, traditionally, on livestock and maize farming.
Population 2 800 000
Map South Africa – Cb

Oranjestad *Aruba and Netherlands Antilles* Seaport and capital of the island of ARUBA. The town, which takes its name from the Dutch Royal House of Orange, grew around a fort built by the Dutch to protect the island from Spanish and British invasion. The chief industry is oil refining.

Oranjestad is also both the port and capital of ST EUSTATIUS in the Leeward Islands.
Population (Aruba) 21 300: (St Eustatius) 1500
Map Caribbean – Ac; Caribbean – Cb

Orapa *Botswana* Site of the world's second largest diamond mine (after Mwadui in Tanzania), discovered in 1967 about 260 km (162 miles) west of FRANCISTOWN. When mining began in 1971, it was expected to continue for nearly 30 years. Most of the diamonds are now used for industrial purposes.
Population 8830
Map Botswana - Cb

▲ PORT WINE AND ROSES Prince Henry the Navigator (1394-1460), national patron of Portugal's early voyages of exploration, was born in Oporto – now renowned for port and fine gardens of roses and camelias.

Orcadas Base *Antarctica* The oldest permanently inhabited scientific base in Antarctica, on Laurie Island in the South Orkney group. It was established by a Scottish expedition in 1903 and is now run by Argentina.
Map Antarctica – Cb

Ord *Australia* River rising in the Kimberley Plateau of Western Australia and flowing 480 km (300 miles) east then north into Cambridge Gulf, 80 km (50 miles) from the Northern Territory border. It has the greatest flow of any Australian river, but this is almost entirely in the summer 'wet'. It is dammed for irrigation to form Lake Argyle which lies south-west of Darwin.
Map Australia – Db

Ordesa *Spain* Game reserve and national park in the Pyrenees, on the French border about 180 km (110 miles) south-east of the port of SAN SEBASTIÁN. The park is in a glacier-sculpted valley discovered by a Frenchman – Ramond de Carbonniers – in 1802. This, one of Spain's most beautiful spots, is dominated by Monte Perdido (Lost Mountain), 3355 m (11 004 ft) high.
Map Spain - Ea

ordnance datum The MEAN SEA LEVEL, calculated from tidal observations made at Newlyn, Cornwall between 1915 and 1921, from which all heights on official British maps are derived. The term is abbreviated to OD.

Ordovician Second oldest period in the Palaeozoic era of the earth's geological timescale, characterised by primitive types of fish. See GEOLOGICAL TIMESCALE, p 256

Ordzhonikidze *Russia* See VLADIKAVKAZ

ore Rock or naturally occurring MINERAL or minerals from which metals or other constituents can be economically extracted.

Örebro *Sweden* Market town situated on Lake Hjälmaren, some 196 km (122 miles) west of STOCKHOLM. Founded in the 11th century, Örebro has a 13th-century church and a well-preserved 16th-century castle. It is the centre of Sweden's shoe industry, and it also manufactures paper.
Population 123 200
Map Sweden – Bd

Oregon *USA* Mountainous state covering 251 477 km² (97 073 sq miles) on the Pacific coast north of California. It rises to 3424 m (11 235 ft) at Mount Hood in the Cascade Range in the north-west. A trading post for furs was set up in 1811, but large-scale immigration by settlers began in 1842 via the Oregon Trail. This wagon route, some 3200 km (2000 miles) long, began at Independence in western Missouri and followed the Platte and North Platte rivers westwards, crossed the Rocky Mountains at South Pass, and then ran between the Snake and Columbia rivers, over the Blue Mountains and then either to Walla Walla or to the Columbia River and Fort Vancouver (now the city of Vancouver) in Washington State, across the Columbia from PORTLAND.

Agriculture in Oregon is concentrated in the Willamette and Columbia river valleys, and cattle, wheat, dairy produce, fruit and vegetables are the chief products. Oregon also has one of the world's largest salmon fisheries. Its single most important industry is timber and wood processing, which accounts for 40 per cent of all industries. Oregon's timber reserve, the largest in the United States, covers nearly half the state. Food processing (including winemaking), instrument making and tourism are other important industries.

In addition to the famous CRATER LAKE NATIONAL PARK, it has 13 national forests and more than 200 state parks. As a result, the state is considered to be a recreational paradise. SALEM is the state capital and Portland the major city.
Population 2 842 300
Map United States – Bb

Orenburg *Russia* River port and industrial city (formerly Chkalov) on the Ural River 360 km (224 miles) south-east of SAMARA. It was founded as a fort in 1743, and expanded when rail links were established here in the early 1870s. It lies in a fertile agricultural district which is also rich in copper, iron ore, nickel, coal and salt; its vast oil and gas resources have been piped to Eastern and Western Europe. Its industries include metal refining, chemicals and food processing.
Population 557 000
Map Russia – Gc

Öresund (The Sound) Deepest and most easterly channel connecting the NORTH SEA and the KATTEGAT to the BALTIC SEA. It is about 140 km (90 miles) long. It runs between the Danish island of Zealand and Sweden. At its narrowest point it is only 4 km (2.5 miles) wide.
Map Denmark – Cb

Oriente *Ecuador* Region to the east of the Andes, incorporating the provinces of Morona Santiago, Napo, Pastaza and Zamora Chinchipe. Its northern and eastern borders (with Colombia and Peru respectively) were the subject of a long dispute, which was only settled at a conference in 1942. Border skirmishes continue, the latest in 1995.
Map Ecuador – Bb

Orinoco *Venezuela* The country's principal river, about 2200 km (1370 miles) long. It rises in the Sierra Parima in the south, flows north along the Colombian border, then eastwards right across the country into the Atlantic through a wide delta of innumerable banks and channels covering some 22 500 km² (8700 sq miles).
Map Venezuela – Bb

Orissa *India* Eastern state covering 155 707 km² (60 103 sq miles) on the Bay of Bengal. Besides Oriya, which is the official language, numerous other languages are spoken, especially in the Eastern Ghats. Rainfall tends to be unreliable, and periodic famines once swept Orissa. It remains underdeveloped, 80 per cent of its people are dependent on rice growing, and 40 per cent of its area is still covered with forests. However, the government has set up more than 50 large industrial projects since 1978. These include an iron and steel plant at Raurkela. Coal and manganese are also mined in the area. CUTTACK is the largest city and BHUBANESHWAR the capital.
Population 31 512 100
Map India – Cc

Orizaba *Mexico* Town in Veracruz State, 240 km (149 miles) east of MEXICO CITY. It grew around natural springs between the high plateaus of the Sierra Madre Oriental and the humid lowlands of the Gulf Coast. Although severely damaged by an earthquake in 1973, it continues to produce beer, rum, textiles, sugar, coffee and tobacco. Orizaba is also the name of the highest peak in Mexico, better known as CITLALTÉPETL.
Population 610 000
Map Mexico – Cc

Orkney *United Kingdom* Group of about 90 islands and islets, totalling 976 km² (377 sq miles), off the north-east tip of mainland Scotland. The remains of a Stone Age settlement dating from 2000 BC has been found at Skara Brae on Mainland Island. Settled by Vikings in the 9th century, the islands are still rich in Scandinavian remains. During the two World Wars the islands were used as an anchorage by the Royal Navy. At the end of the First World War the trapped German fleet scuttled itself in Scapa Flow – between the islands of Mainland, Hoy and South Ronaldsay – rather than fall into British hands. The capital of the islands is Kirkwall on Mainland, which has a 12th-century cathedral. The group is linked by air and sea transport (a two-hour journey by ship) to the Scottish mainland. Fishing, cattle rearing and oil industries are the chief sources of revenue.
Population 19 450
Map United Kingdom – Gi

Orlando *USA* City in central Florida, 330 km (205 miles) north-west of Miami. It is in a citrus fruit-growing area and is a tourist base for DISNEY WORLD and Cape CANAVERAL.
Population (city) 164 700; (metropolitan area) 1 224 900
Map United States – Je

Orléans *France* Capital city of the Loiret department, situated on the River Loire 110 km (65 miles) south-west of PARIS. In 1429 Joan of Arc, then just 17, forced the English to abandon their siege of the city – a feat which earned her the name of Maid of Orléans. Many old buildings were destroyed during the Second World War but

the city, now rebuilt, retains much of its medieval character. Among its modern industries are the manufacture of car parts and chemicals.
Population (city) 105 100; (metropolitan area) 243 000
Map France – Dc

orogeny, orogenesis Process of mountain building, especially by folding and faulting of the earth's crust.

orographic precipitation Precipitation that is caused by cooling of moisture-laden air as it rises over a mountain range.

Orontes *Middle East* Unnavigable river, 384 km (238 miles) long, rising in the BEQA'A Valley in Lebanon and flowing north-east into Syria, where it remains for most of its course. It is used for irrigation near the cities of Homs and Hamah. The river then crosses the Turkish frontier, turns west and enters the Mediterranean north of the Syrian port of Al Ladhiqiyah.
Map Middle East – Ba

orthoclase Whitish, red or green potassium aluminium silicate (a feldspar) which is found in acid IGNEOUS and METAMORPHIC rocks, and used in the manufacture of glass and ceramics.

Orumiyeh (Urmia) *Iran* City near the Turkish border, 125 km (78 miles) south-west of the city of TABRIZ. It is said to be the birthplace of Zoroaster (about 628-551 BC), the founder and spiritual leader of the Zoroastrian religion. The population of the surrounding area is made up of Armenians, Turks and Kurds. Lake Urmia, which at 4700 km² (1815 sq miles), is Iran's largest lake, lies about 10 km (6 miles) east of the city. The water, too salty for fish, is considered to have curative properties.
Population 300 750
Map Iran – Aa

Oruro *Bolivia* Departmental capital and city 200 km (125 miles) south-east of the capital, LA PAZ. The city, founded in 1606, is an important mining and rail centre, and the seat of a technical college. Oruro plays host to the country's most famous carnival, La Diablada, which begins on the Saturday before Ash Wednesday each year with a procession of dance groups led by ornately costumed and masked figures representing Satan, his wife and St Michael the Archangel. The festival lasts for eight days.

About half of Bolivia's tin is mined around Oruro. Other minerals include silver, tungsten and antimony from the western hills and slopes of the Cordillera Real range.
Population (city) 208 300; (department) 437 300
Map Bolivia – Bb

Orvieto *Italy* Well-preserved medieval city set on a cliff 95 km (nearly 60 miles) north of Rome. Orvieto, which attracts many tourists each year, is known principally as a producer of quality white wines and for its Romanesque-Gothic cathedral, which is built in curiously striped stone layers of dark green and white.
Population 22 800
Map Italy – Dc

Osaka *Japan* The country's third largest city after Tokyo and Yokohama. It straddles the delta

▲ **EXACT COPY** It took more than 600 000 men to build Osaka's original castle, which was destroyed by fire on three separate occasions in 1615, 1665 and 1867. The present building is a reproduction completed in 1931.

of the Yodo River at the eastern end of the INLAND SEA, 400 km (248 miles) south-west of Tokyo. More than 1000 bridges span the many arms of the Yodo River. The city features the world's first offshore airport, built on an artificial island. Osaka was severely affected by the earthquake that devastated nearby Kobe in January 1995, although the damage was relatively slight. The city also suffers from severe urban congestion, air pollution and land subsidence caused by the extraction of groundwater for its wide variety of manufacturing industries. It does, however, have Japan's largest and most sumptuously decorated castle, originally built in the 16th century by the warlord Toyotomi Hideyoshi. Its arrival signalled the beginning of the transformation of the city into feudal Japan's commercial and financial capital. Osaka's merchants dominated trade and became paymasters to Japan's Samurai class

Although still a major trade centre, the city also has a reputation for hedonism, reflected in its fine food, active nightlife, superb department stores of the Shinsaibashi shopping area and theatres devoted to Japan's traditional but popular *kabuki* dramas and *banraku* puppet plays.

The city's two main religious sites are the ancient Shinto shrines of Sumiyoshi dating from 202 AD, which are surrounded by hundreds of carved stone lanterns, and the 10th-century Temmangu shrine, famous for the decorated boats which are paraded at festival times.

Population 2 624 000
Map Japan – Cd

Oshawa *Canada* Industrial city and port on Lake Ontario, 53 km (33 miles) east of Toronto. It is an important car manufacturing centre and also produces glass, textiles, drugs and furniture.

Population (city) 186 450;
(metropolitan area) 240 100
Map Canada – Hd

Osijek (Esseg) *Croatia* City overlooking the Drava River about 30 km (20 miles) from the Hungarian border. It developed around a medieval fort, and has a Gothic cathedral and both Catholic and Orthodox churches. As Esseg, the city was part of Hungary from 1699 to 1918, and was the birthplace of Bishop Strossmayer (1815-1905), champion of Croatian independence from the Austro-Hungarian Empire.

A major road and rail junction, Osijek is the largest market town of Slavonia, processing cereals, fruits, livestock and timber (which is rafted down the Drava). It is also the nation's chief producer of matches. Until the outbreak of civil war in the early 1990s, the city was also Slavonia's educational and cultural centre.

Population 164 600
Map Croatia – Db

Ösling *Luxembourg* See GUTLAND

Oslo *Norway* Capital, largest city, main port and chief industrial centre in the country, lying at the head of Oslofjord on Norway's south-east coast.

The city was founded in 1048 by King Harald Hardråde III of Norway, who ruled from 1046 to 1066, and prospered as a trading centre and port, becoming the Norwegian capital in 1299. Norway's principal fortress was built here at Akershus in the Middle Ages. After a disastrous fire in 1624, the city was rebuilt and renamed Christiania (later Kristiania) after Christian IV, then king of both Norway and Denmark, but it reverted to its original name in 1925.

The early town's industries developed along the banks of the Aker River, whose successive waterfalls could be used to drive water wheels.

Around them, the houses of workers, merchants and government officials spread outwards through the lakes, forests and marshes behind the harbour. As a result of the sprawl, Oslo now covers more than 450 km² (173 sq miles), making it one of the world's largest cities in area. However, it retains a rural atmosphere outside the much smaller city centre because of large 'green belt' areas used today by Oslo residents for hiking, skiing and outdoor recreation.

The city's spine is the avenue known as Karl Johansgate. It is named after King Karl XIV Johan, formerly Jean (Johan) Baptiste Jules Bernadotte, one of Napoleon's marshalls who was elected Crown Prince of Sweden in 1810 and became King of Norway under the dual monarchy established at the end of the Napoleonic Wars. His palace in central Oslo is still the official residence of Norwegian monarchs. The avenue named after him links the palace to the city's other main buildings: the cathedral (built in 1697), in front of which is a statue of King Christian IV; the central market; the original university buildings, founded in 1811; and the Stortinget, the nation's parliament.

Around this core are theatres, museums and galleries, including a gallery devoted to the work of the Norwegian expressionist painter Edvard Munch (1863-1944). Nearby Frogner Park is filled with the works of the Norwegian sculptor Gustav Vigeland (1869-1943). Farther inland are the soaring ramps of the 19th-century ski jumps at Holmenkollen, where an annual jumping championship has been held since 1892.

The island of Bygdøy in the harbour houses a magnificent maritime exhibition which includes restored Viking ships from the 9th century; the *Fram*, the vessel used by Norwegian explorer Fridtjof Nansen for his voyage through the Arctic ice pack in 1893-6; the *Gjøa*, the ship used by Norway's Roald Amundsen when he became, in 1903, the first man to sail through the North-west Passage; and the *Kon Tiki*, the balsawood raft on which the Norwegian anthropologist Thor Heyerdahl crossed the Pacific from South America to Polynesia in 1947.

Oslo's manufactures include electrical equipment, metal goods, timber, dairy products, machine tools, chemicals, textiles and ships.

Population (city) 473 300;
(metropolitan area) 748 000
Map Norway – Cd

Osnabrück *Germany* Industrial town situated 125 km (78 miles) west of Hanover. It has been a bishopric since the 8th century and has a 13th-century cathedral. It was a member of the cities confederation known as the Hanseatic League, and its greatest historical moment dates to 1648 when the Thirty Years' War came to an end here. Its many factories produce steel, cars and car parts, cables, textiles, paper, and food products.

Population 163 200
Map German – Cb

Osorno *Chile* Province in the south-central region of Los Lagos. It has some of the country's most beautiful lakes, including Puyehue and Rupanco, both of which have thermal springs, and Todos Los Santos (All Saints), also known as Esmeralda because of its emerald-green water. The snow-covered slopes of the volcanic Mount Osorno, which rises to 2660 m (8727 ft), form part of a thriving ski centre. The provincial capital, also called Osorno, was founded in the 1550s but was destroyed by Amerindians half a century later. It was re-established in 1796, and later settled mainly by German immigrants. It is the site of a 16th-century Spanish fortress.

Population (province) 186 000; (town) 136 000
Map Chile – Ad

Ostend (Ostende; Oostende) *Belgium* Flemish-speaking port and ferry terminal for the Dover-Ostend ferry, 20 km (12 miles) west of BRUGES. It is also a naval base, fishing port and resort. On the seafront, spa baths stand alongside a casino and villa used by the Belgian royal family.

Population 72 000
Map Belgium – Aa

Ostia Antica *Italy* Finest surviving Roman city in the country after Herculaneum and Pompeii, remarkable for its public buildings and beautiful black and white mosaic floors. Standing near the mouth of the Tiber River, it served as a port for Rome, which lies 20 km (12 miles) to the northeast. But after being regarded as Rome's most important suburb and commercial harbour, it was destroyed by invading Barbarians and gradually disappeared under mud and wind-blown sand.

Excavation of the city started in the early part of the 20th century. The Via delle Tombe leads to the main street of the city, the Decumanus Maximus. Public buildings include the Theatre, the House of Diana (the Roman goddess of hunting) and, facing the Forum (market place), the Temple of Vulcan (the god of fire) – most of them built in the 2nd to the 4th century AD. The area is also the site of a 15th-century castle.

Map Italy – Dd

Ostrava *Czech Republic* Coal-mining and manufacturing city in Czech Silesia, the country's main industrial area. The city, which is regional capital of Severomoravský (northern Moravia), lies 250 km (155 miles) east of PRAGUE, and 15 km (10 miles) south of the Polish border. It is situated near the Czech Republic's largest coal deposits. Steel, chemicals, metal and heavy engineering goods, as well as building materials and foods, are produced in the city.

Several prominent people were born in the Carpathian foothills, south-east of the city: the Austrian-Jewish psychiatrist Sigmund Freud (1856-1939) was born in Pribor; the Czech composer Leos Janácèk (1854-1928) in Hukvaldy; and Emil Zátopek (born 1922), the Czech long-distance runner who won three gold medals at the 1952 Helsinki Olympic Games, in Koprivnice.

Population 331 500
Map Czech Republic – Db

Oswiecim *Poland* See AUSCHWITZ

Otago *New Zealand* District of South Island north-west of the port of DUNEDIN. Its features are flat-topped mountains and glacier-dug lakes. Gold was discovered here during the 1860s, but today sheep and fruit are the economic mainstays. The main town is Alexandra.

Population 186 067
Map New Zealand – Bf

Otjozondjupa *Namibia* Area located in central and eastern Namibia covering some 105 327 km² (40 657 sq miles). It consists mostly of shrub savanna and is partly wooded, especially in the

A DAY IN THE LIFE OF A CANADIAN BUSINESS COUPLE

By 6 am, career wife Moniquc Webster is wakening twins Daniel and Diane, and her husband, Andrew, 40, is frying bacon and eggs. The Ottawa family is up an hour earlier than usual this frosty October morning; Andrew has to be in Calgary, Alberta – over 2900 km (nearly 1800 miles) away – within six hours. '*Dépêchez-vous*' ('Hurry up'), Monique urges her dawdling three-year-olds. But 'eat slowly', Andrew admonishes later over breakfast. The couple are among some 3 million bilingual Canadians who are fluent in English and French.

Andrew's four-hour Air Canada flight from the local airport will cross two time zones before landing at 10 am, Calgary-time. A middle-level civil servant with Tourism Canada, he is due to meet officials of the Calgary Stampede, the world-famous rodeo which is held each July.

Breakfast over, Andrew picks up his overnight bag, kisses Monique and the twins goodbye and drives off in his small car which he will leave at the airport. Once he is on his way, 38-year-old Monique dresses the twins, stacks plates and coffee mugs in the dishwasher, vacuums the carpeted, two-bedroom, 10th-floor apartment, and goes down to do the family's laundry in the coin-operated machines in the basement.

At 12.30 pm she buckles the youngsters in hcr ageing, Japanese-made stationwagon for a 30 minute drive to the *Garderic Paradis des Petits,* a day-care centre she owns with Marie-Claire Archambault, a friend from her university days when they studied history together. The centre yields a modest income, yet allows Monique to be with her children and gives her time to do household tasks before the weekend.

Monique's earnings (the equivalent of US$7000 a year after taxes) are all saved. Soon she will have the $10 000 down payment for a three-bedroom $70 000 house – modest by Ottawa standards – and a further $10 000 for furniture and appliances. The Websters' total income, which is well above the Canadian average of $30 000 for a family of four, is commonplace in Canada's capital. Taxes take a quarter of Andrew's $40 000 salary, leaving $2500 in monthly take-home pay. After fixed expenses, $500 for rent and $800 for payments on their two cars, there is $1200 left for food (which claims about $600 monthly) and for clothes and outings.

On her way home, Monique buys a frozen *tourtière* at the supermarket. She heats the spicy pork pie while the twins watch their favourite television programme, *Polkadot Door.*

Once the children are asleep, Monique relaxes. She is watching the Johnny Carson Show on her bedroom TV set when Andrew telephones. He will be home tomorrow afternoon. Monique tries to share his excitement about the one million spectators and 600 cowboys expected at the Calgary Stampede. But she is weary and soon falls asleep.

east. Most of the inhabitants are involved in subsistence farming, while a small percentage live and work in small market towns.

The area, formerly known as Bushmanland, is a popular tourist destination, its main attractions being game parks, for example, the recently proclaimed Khaudom Game Reserve where 64 different mammal species, particularly buck, can be seen. The area also has many private game farms.

Population 85 000
Map Namibia – Aa

Otranto, Strait of Sea passage, 69 km (43 miles) long, connecting the ADRIATIC and IONIAN seas, and separating the heel of Italy from Albania.

Map Albania – Bb

Ottawa *Canada* Capital of Canada, at the junction of the Ottawa and Rideau rivers in south-east Ontario. Established in 1809, it was initially called Bytown after Lieutenant-Colonel John By, the engineer who built the Rideau Canal. It was renamed in 1854 when it was incorporated as a city; in 1857 it was chosen by Queen Victoria as Canada's future capital. The first session of its parliament was held 10 years later.

Ottawa contains many government offices and the skyline is dominated by its parliament buildings, rebuilt after a fire in 1916. The capital contains the national museum, national arts centre (with a theatre, concert hall and the national opera), national library and public archives. It has many parks and walkways maintained by the National Capital Commission.

The Ottawa River, which flows through Quebec and Ontario provinces for 1271 km (790 miles), was once an important route for explorers, missionaries and fur traders.

Population (city) 678 150; (metropolitan area) 920 860
Map Canada – Hd

Ouagadougou *Burkina* The country's capital since 1954 when it was linked by rail with ABIDJAN in Ivory Coast. It has been the capital of the Mossi people since the 15th century, and its buildings contain a mixture of native African and colonial French influences.

Buildings worth seeing include the palace of the Moro Naba, the Mossi emperor; a neo-Romanesque cathedral; and the former French governor's residence. More recently, a huge sports stadium was built. Tourist attractions are limited, but Ouagadougou boasts one of the country's few wooded areas, called the Bois de Boulogne after the Parisian park of the same name.

Population 442 220
Map Burkina – Aa

Ouargla (Wargla) *Algeria* Large oasis town and administrative centre, 560 km (nearly 350 miles) south-east of Algiers, on the cross route that links the central and eastern trans-Saharan routes to the south. It is also a pipeline junction in the middle of the HASSI MESSAOUD oil region, and is being developed as an industrial area. Water from deep artesian wells has given the town a new lease of life over the past 30 years.

Population 49 000
Map Algeria – Ba

Ouarzazate *Morocco* Oasis and former colonial French fortress town 130 km (80 miles) southeast of MARRAKECH on the southern slopes of the

High Atlas Mountains. The town is noted for its Ouazguita carpets, which are woven by women in traditional geometric designs of orange-red on a black background. To the east of the town stands Taourirt, one of the most impressive *kasbahs* (old walled quarters) in Morocco.

Population 29 000
Map Morocco – Ba

Oudenaarde (Audenarde) *Belgium* Cotton textile town 30 km (19 miles) south of GHENT. In 1708 British troops under the command of the Duke of Marlborough defeated a French force here during the War of the Spanish Succession.

Population 25 000
Map Belgium – Aa

Oudtshoorn *South Africa* Town about 64 km (40 miles) north of MOSSEL BAY in southern Western Cape Province, noted for the large-scale commercial breeding of ostriches. Although the ostrich feather industry collapsed in 1914, it has shown signs of revival in recent years, partly as a result of the tourist trade. Marketing of ostrich meat, claimed to be cholesterol-free, appears set to create a new boom. The famous CANGO CAVES are 26 km (16 miles) from Oudtshoorn.

Population (town and district) 68 000
Map South Africa – Bc

Ouenza (El-Wanza) *Algeria* Centre of iron ore production, on the railway 120 km (75 miles) south of Annaba near the Tunisian border. Together with neighbouring Bou Khadra, it has the oldest iron mines in Algeria; between them they produce 75 per cent of the country's ore, which is exported through Annaba.

Population 35 000
Map Algeria – Ba

Ouidah (Whydah; Ajuda) *Benin* Administrative centre and port 50 km (30 miles) west of the port of COTONOU. It has always been a stronghold of the voodoo cult and became a principal centre of the slave trade after the Portuguese constructed a port here in 1580.

In the 17th century the slave trade in these parts was mainly conducted by the French. In the 19th century, liberated Portuguese-speaking slaves, also known as Creoles, returned to Ouidah from Brazil and founded a distinct community. British and Danish forts nearby are a grim reminder of the hundreds of thousands of slaves shipped to the New World from Ouidah.

Population 35 000
Map Benin – Bb

Oujda *Morocco* The eastern gateway to Morocco, and an important road and rail junction near the Algerian border, 300 km (186 miles) east of FÈS. The town, founded in AD 944, was fought over for centuries by Arabs, Berbers and Turks, earning it the name Medinet el Haira (City of Fear). Little is left of the old town today apart from traces of its ancient walls and a fine gateway, the Bab Sidi Abdelwahad.

Oujda is a thriving commercial centre, trading in sheep, wool, cereals, fruit and wine. Nearby is the Sidi Yahya oasis, the legendary burial place of John the Baptist and the site of the Battle of Isly where the French defeated the Moors in 1844.

Population 260 000
Map Morocco – Ba

Oulu (Uleåborg) *Finland* Seaport located on the Gulf of Bothnia, 100 km (62 miles) from the Swedish border, and capital of the province of the same name covering 61 580 km² (23 776 sq miles). Founded in 1605, it is now the largest city in northern Finland, has a reputable university and is the country's leading timber-processing centre. It is also established as the major gateway for Finnish exports. The city is located at the mouth of the 108 km (67 mile) long Oulu River, which drains the lake of the same name south-east of the city.

Population 103 500
Map Finland – Cb

Ouro Prêto *Brazil* Historic town some 70 km (43 miles) south-east of the city of Belo Horizonte. Founded in 1711, and declared part of the world's cultural heritage by the United Nations agency UNESCO. A national monument since 1933, it is an 18th century time capsule, with cobbled streets and 13 Baroque churches, fountains, mansions and gardens.

Population 67 550
Map Brazil – Dd

overburden Layer of rock that covers a more useful material, such as a coal seam, and which has to be removed in open-cast mining. The term is also used to describe any loose material overlying solid rock.

overcast Term used to describe the sky when more than 75 per cent of it is cloud-covered.

Overijssel *Netherlands* Eastern province covering an area of 3925 km² (1575 sq miles) 'over' or east of the IJssel River. It is principally dairy farmland, but Enschede, Almelo, Hengelo and Oldenzaal, in the east, used also to be the country's largest textile manufacturing towns. The provincial capital is Zwolle.

Population 1 039 100
Map Netherlands – Ca

Oviedo *Spain* Northern steel-making city, some 210 km (130 miles) east of CORUNNA. The city's 14th-century cathedral contains a famous cross that was carried into the Battle of Covadonga in 718 by Don Pelayo, first king of Asturias, at which the Christians began the reconquest of Spain from the Moors

Oviedo is the capital of the present-day province of Asturias. Its industrial products include glass, chemicals and pottery.

Population (city) 196 050; (province) 1 134 600
Map Spain – Ca

Owen Falls *Uganda* See VICTORIA, LAKE

Owen Stanley Range *Papua New Guinea* Mountains stretching 950 km (590 miles) in the extreme south-east of New Guinea. The highest point is Mount Victoria (4073 m, 13 363 ft).

Map Papua New Guinea – Ba

Owendo *Gabon* Libreville's deep-water port which lies just south of the city and the country's main shipping point for general cargo.

Map Gabon – Aa

Owerri *Nigeria* Town to the north and east of the Niger River delta, 410 km (255 miles) east of Lagos. It has grown rapidly since it was made the

capital of Imo State in 1976. A twin town has been established across the Nwaori River.

Population 150 000
Map Nigeria – Bb

oxbow Horseshoe-shaped lake formed from a U-bend, or meander, in a river, when the river cuts through the neck of the meander and shortens its course. See HOW WATER SHAPES THE LAND, p 571

Oxford *United Kingdom* City and county lying in the upper basin of the River Thames. The county stretches as far downstream as the outskirts of Reading and covers an area of 2611 km² (1008 sq miles). It is mainly agricultural with industries concentrated round the city and the towns of Banbury, Didcot and Abingdon.

The city of Oxford is 80 km (50 miles) northwest of LONDON. Its university is the oldest in England, dating back to at least the 12th century, and has one of the greatest libraries in the world, the Bodleian. Its greatest area of learning has long been the humanities, although it excels also in the sciences. Its beautiful colleges, mostly of honey-coloured stone, are scattered throughout the inner city, and every narrow street reveals some new architectural treasure.

The industrial suburb of Cowley manufactures cars, and is still the city's largest employer. Printing works and publishers – notably the Oxford University Press – are also based here.

Population (county) 580 900; (city) 130 300
Map United Kingdom – Ee

Oxus *Central Asia* See AMUDAR'YA

Oymyakon *Russia* See VERKHOYANSK

Oyo *Nigeria* Yoruba town, situated in the state of the same name, 170 km (105 miles) north-east of LAGOS. The old town stood 130 km (80 miles) to the north, but was destroyed by Fulani people in a war in the 1830s.

Nigeria's first road, built in 1905, linked Oyo with the city of Ibadan, some 50 km (30 miles) to the south. Today Oyo markets tobacco, cotton and indigo dye, together with its traditional craft products which include beads, embroideries, leather goods and cloth. It lies on the Oyo Plains, which cover most of south-west Nigeria, and rise from 120 m (395 ft) in the south (the main cocoa-producing area) to 400 m (1310 ft) in the north.

Population 250 000
Map Nigeria – Ab

ozone Faintly blue gas with a pungent odour, occurring naturally in the layer of the ATMOSPHERE known as the stratosphere, or ozone layer. It is a form of oxygen in which each molecule contains three atoms of oxygen, unlike the two-atom oxygen in the air we breathe.

Ozone is highly poisonous, but plays a vital part in maintaining conditions suitable for life on earth, since it shields out harmful ultraviolet radiation from the sun, and produces a warm covering, or inversion layer, which limits convectional movements of air to lower atmospheric layers. Recently, widening holes in this protective layer have been detected over the Arctic and Antarctic, apparently due to increasing chemical pollution from earth. The main culprits are believed to be chlorofluorocarbons (CFCs), which are used in aerosols and refrigeration units, but which are now gradually being replaced.

Paarl *South Africa* Town at the centre of a major wine-making region 56 km (35 miles) north-east of CAPE TOWN, in the Western Cape Province. French Huguenots introduced the vines to the area soon after they arrived here in 1688. A tall monument on one of three granite domes outside the town commemorates the development of the Afrikaans language (a variant of Dutch). The *Genootskap van Regte Afrikaners* ('Fellowship of True Afrikaners') was founded in the town in 1875 to promote Afrikaans as a formal language, through the development of its literature.

Population 136 120

Map South Africa – Ac

Pacific Ocean Covering about 16 000 000 km² (6 176 175 sq miles), a greater area than all the continents put together, the Pacific is the world's largest ocean. Together with all its seas, straits and adjacent channels, it covers 181 000 000 km² (69 687 980 sq miles). It was named by the Portuguese navigator Ferdinand Magellan, following a particularly calm voyage from the tip of South America to the Philippines in 1520-1. However, the Pacific is by no means always peaceful. In fact, the highest wave in an open sea, measuring 34 m (112 ft) was recorded there during a hurricane in 1933.

This great ocean includes the BERING SEA and the Sea of OKHOTSK in the north, and the Sea of

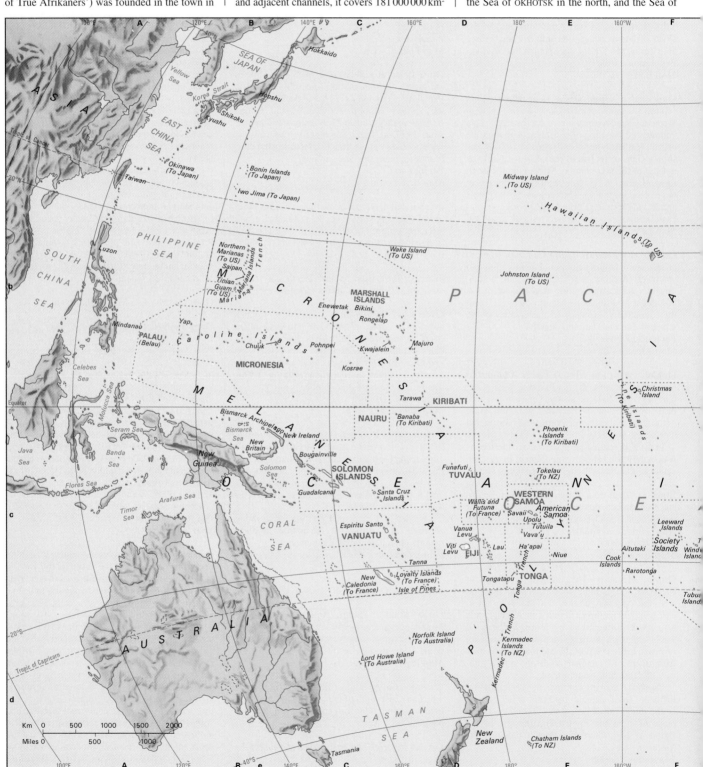

JAPAN and the CHINA SEA in the west. It washes the shores of the north and south American continents, Antarctica, Australia and Asia. The Philippine and Indonesian island groups enclose the BANDA, CELEBES, SERAM, FLORES, JAVA and MOLUCCA seas. The ARAFURA, CORAL and TASMAN seas are in the south-west. The islands in the western Pacific are part of Asia. However, the many volcanic and coral islands of Micronesia, Melanesia and Polynesia are included in OCEANIA.

With an average depth of 4200 m (13 780 ft), the Pacific is the world's deepest ocean. Its greatest known depth is 10 924 m (35 840 ft). Because of its depth, scientists believe, a smaller proportion of the water evaporates, and therefore the ocean remains less salty than the shallower Atlantic.

The main currents in the North Pacific – the North Equatorial Current, the Japan or Kuroshio Current and the California Current – rotate in a clockwise direction, north up the coast of Asia and south down the coast of North America. The main currents in the South Pacific – the South Equatorial Current, the East Australia Current, the Antarctic Circumpolar Current, and the Peru (Humboldt) Current – move in an anti-clockwise direction, north up the coast of South America and south past Australia.

The Pacific provides half the world's catch of fish and shellfish. The leading fishing grounds cover the continental shelves in the north-west. The next most abundant waters are off the coast of South America. Oil is extracted near the Asian and North American coasts, while manganese nodules (as yet unexploited) are scattered over large areas of the deep ocean floor – the ABYSS.

Map See map alongside

Paczkow *Poland* Town situated 75 km (47 miles) south of the city of WROCLAW. It is surrounded by medieval walls, which stand complete with towers and bastions.

Population 10 000
Map Poland – Bc

Padang *Indonesia* Provincial capital of West SUMATRA. It produces cement and rubber.
Population 480 900
Map Indonesia – Bc

Paderborn *Germany* Town situated 100 km (62 miles) south-west of HANNOVER. It is named after the Pader River which rises here, bubbling out of more than 100 underground springs. The town became a bishopric in 806. Its university was established in 1614. Paderborn's church tower is about 100 m (328 ft) high.
Population 120 700
Map Germany – Cc

Padma *Bangladesh* See JAMUNA

Padua (Padova) *Italy* City lying some 35 km (22 miles) west of VENICE. One of Italy's most important artistic centres, it preserves among its cultural treasures many medieval palaces, gold-domed churches and priceless works of art.

Known as Patavium in Roman times, it grew into one of the most prosperous and influential towns in the Empire. The classical historian Livy (59 BC-17 AD) was born here.

Galileo Galilei (1564-1642), the astronomer and mathematical physicist, lectured at Padua's medieval university (founded in 1222 by Emperor Frederick II) for 18 years.

The city centre is built around a number of fine piazzas surrounded by medieval palaces. The Palazzo della Ragione, which was built in the 13th and 14th centuries, overlooks the Piazza della Frutta and the Piazza della Erbe, the ancient fruit and vegetable markets which are still flourishing. Padua's imposing cathedral dates from the 9th century, but most of it was rebuilt in 1552.

The city's economy centres on food and drink, agricultural machinery, bicycles and motorcycles, electrical goods and textiles.
Population 218 200
Map Italy – Cb

Paektu san (Baegdu Son) *China/North Korea* Extinct volcano and, at 2744 m (9003 ft), North Korea's highest peak. It stands astride the Chinese border and its crater contains a deep lake.
Map Korea – Db

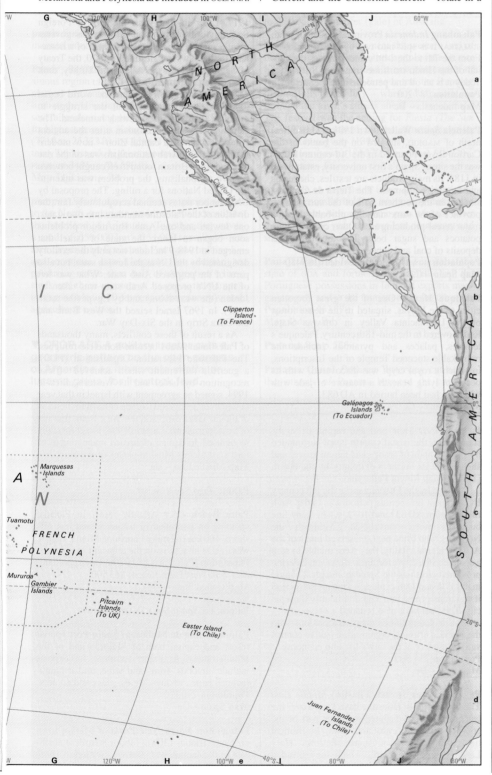

▲ **CAPTURED IN OILS An artist on the Left Bank of the Seine in Paris paints the Cathedral of Notre Dame, with its graceful 14th-century spire, just across the Pont de l'Archevêché.**

hoods remain, and Parisians often say that their city is really a collection of 100 villages. Its eastern districts still have small workshops, making furniture and clothing. The Latin Quarter, on the left (or south) bank of the Seine, is dominated by the Sorbonne (the University of Paris) and by bookshops and publishers. The right bank has the *haute couture* fashion houses.

Elsewhere in the city, there are distinctive Arab, African, Indo-Chinese and Jewish quarters. The presence of immigrant groups adds further diversity to the city, which houses some 1 250 000 foreigners, one-third of all those residing in France. Portuguese, Algerians, Moroccans and Spaniards are currently the most numerous among the city's immigrants.

Population (city) 2 152 000; (Greater Paris) 9 060 000

Map France – Eb

Paris Basin *France* Vast saucer-shaped depression in northern France, containing the capital. It covers 120 000 km² (46 000 sq miles) – almost a quarter of the area of country, but a much larger proportion of its population. The basin, much of which is richly fertile farmland, is drained by the Seine, Somme, Marne and Loire rivers.

Map France – Db

Parma *Italy* City bisected by the Parma River, 120 km (75 miles) south-east of MILAN. Founded by the Romans in 183 BC, Parma flourished as a trade centre until it was destroyed by the Ostrogoth King Theodoric (AD 455-526). It was rebuilt in the Middle Ages.

The old working-class districts and the partly ruined Palazzo Ducale of the former Dukes of Parma lie to the west, while to the east, the Romanesque cathedral has a beautiful fresco, *The Assumption of the Virgin*, by Correggio (1494-1534). The home of the famous Parmesan cheese and Parma hams, the city's economy is based mainly on the processing of foods, as well as textiles and shoe manufacturing.

Population 174 300

Map Italy – Cb

Páros *Greece* Island of 194 km² (75 sq miles) lying to the west of Náxos in the CYCLADES, in the south Aegean Sea.

Inhabited since about 3000 BC, the island was renowned in ancient times for its snow-white marble, quarried near its present-day capital, Paríkia. Gentle slopes drop down to quiet beaches and the harbour town of Páros, which has a ruined castle dating from AD 1266, and a 10th-century Byzantine cathedral.

Population (island) 7900

Map Greece – Dc

Parramatta *Australia* Originally a farming town in New South Wales on the Parramatta River, now a western suburban satellite of SYDNEY. Founded in 1788 as the country's second settlement, it contains some of Australia's oldest buildings, such as Elizabeth Farm House (1793) and Old Government House (1799). It has become a major business centre manufacturing textiles, machinery, cars and flour.

Population 133 500

Map Australia – Ie

Pasadena *USA* City in south-west California. Lying at the foot of the Sierra Madre Mountains, it forms a north-eastern suburb of Los Angeles. Pasadena has a spectacular Tournament of Roses parade each New Year's Day when it hosts the famous college football game at the Rose Bowl stadium. The Mount Wilson Observatory stands above the city.

Population 131 600

Map United States – Cd

Pasargadae *Iran* Ruined city situated 100 km (62 miles) north-east of the southern city of SHIRAZ. The remains include the palace and tomb of Cyrus the Great (ruled 559-530 BC), who made it his capital. Cyrus's tomb – a mammoth sarcophagus – rests on an imposing pyramid of stepped stone blocks.

Map Iran – Ba

Pasay *Philippines* See MANILA

Pasco *Peru* Central department in the high Andes and one of the country's main mining regions. Iron ore and manganese are mined in the valley of the Rió Perené. Copper, zinc, lead bismuth and silver come from the department capital, Cerro de Pasco.

Population 282 900

Map Peru – Bb

Pascua, Isla de *Pacific Ocean* See EASTER ISLAND

Pas-de-Calais *France* Northern department, covering parts of Artois and Picardy. Its chalk cliffs are part of the same geological formation as the white cliffs of Dover, across the Channel in England. The principal towns are ARRAS, BOULOGNE-SUR-MER and CALAIS.

Population 1 412 000

Map France – Da

Passau *Germany* University town and bishopric on the German-Austrian border. It is situated on the confluence of three rivers, including the Danube, and is known as the *Dreiflüssestadt* (Three-River-City). Its 17th-century Baroque cathedral has the world's largest church organ, with more than 17 000 pipes and five keyboards.

Population 49 900

Map Germany – Ed

Passchendaele (Passendale) *Belgium* Village 60 km (37 miles) west of the city of GHENT. Monuments have been erected to more than half a million Allied and German soldiers who died here in heavy fighting in October and November 1917, during the First World War.

Population 4000

Map Belgium – Aa

Pasto *Colombia* City on a high, dry plateau in the south-western mountains, 90 km (56 miles) from the border with Ecuador. The surrounding countryside is inhabited mainly by Amerindian farmers. The city is flanked by the still active volcano of Galeras (4276 m, 14 028 ft). In August 1991 the area had to be rapidly evacuated when the volcano began shooting out ash.

Population 303 400

Map Colombia – Bb

Patagonia *Argentina/Chile* Cold, high desert covering the four southernmost provinces of Argentina and part of MAGALLANES region in Chile. A vast plateau of some 770 000 km² (300 000 sq miles) south of the Río Negro and the Limay River, it is virtually barren, except around the river valleys, where grapes and other fruits are grown. Welsh-speaking farmers, whose descendants emigrated to Argentina in the mid-19th century and who established settlements such as Trelew and Esquel, also raise sheep in the region. Patagonia's two principal towns are the Argentinian port of Comodoro Rivadavia (population 124 000) and PUNTA ARENAS in Chile. The region's natural resources include oil, iron, copper and uranium.

Map Argentina – Cc

Pataliputra *India* See PATNA

Patan (Lalita Patan; Lalitpur) *Nepal* Oldest town of the Kathmandu Valley, which lies just

5 km (3 miles) south of the capital, into which it has gradually been absorbed. Although chiefly a Buddhist temple town, it also has Hindu and Muslim buildings. There are 55 major temples and 136 monastries in the area.

Population 79 000
Map Nepal – Ba

Pathein (Bassein) *Myanmar (Burma)* City in the Ayeyarwady (Irrawaddy) delta, some 140 km (87 miles) west of the capital YANGON (Rangoon). It is the centre of the country's rice trade and has rice mills and storage silos.

Population 355 600
Map Myanmar – Bc

Pátmos *Greece* Volcanic island lying among the northernmost of the DODECANESE Islands, in the eastern Aegean Sea. St John the Evangelist was exiled here in AD 81-96. The battlement-walled monastery of St John, standing on a crag overlooking the town, was founded in AD 1088, and has some richly carved icons and fine medieval frescoes. In the Convent of the Apocalypse is the cave where St John is reputed to have written the Book of Revelation.

Population 2250
Map Greece – Ec

Patna *India* Capital of Bihar state, on the south bank of the River Ganges 460 km (285 miles) north-west of CALCUTTA. The modern city is just downstream from Pataliputra, the capital of Hindu and Buddhist empires in the 5th to 3rd centuries BC. The city's gigantic beehive-shaped granary (25 m, 82 ft) tall, was constructed by an Englishman in 1786 to prevent famine, but has hardly ever been used. Patna has many academic institutions, including a university and library containing rare Persian and Arabic manuscripts; it also has Hindu and Mogul paintings.

Population 3 623 230
Map India – Dc

Patos, Lagoa dos *Brazil* Lagoon in the southern coastal state of Rio Grande do Sul, with PORTO ALEGRE lying at its northern tip. At 10 144 km² (3915 sq miles) it is the country's largest salt-water lagoon. Two nearby lagoons to the south, Mirim and Mangueira, extend to the Uruguay border.

Map Brazil – Ce

Patras *Greece* Largest city on the PELOPONNESE, situated on the north-west coast facing the Ionian Sea. Founded in 1100 BC, Patras is the main gateway to Greece from the west. Under Roman, and later, Venetian, domination it grew into a busy Mediterranean port. Today it is an industrial centre, processing textiles, leather and olive oil, and exporting wine and currants. A reliquary at the church of Ayios Andreas is said to contain the skull of St Andrew, patron saint of both Greece and Scotland.

Population 155 000
Map Greece – Bb

Pattani *Thailand* Town on the Gulf of Thailand, 110 km (70 miles) from the country's southern tip. Like the surrounding province of the same name, the town is now dominated by Malay-speaking Muslims. The province's main crops are rice, cashew nuts and fruits.

Population (town) 81 000; (province) 478 700
Map Thailand – Bd

patterned ground Clusters of stones that are arranged in stripes or polygons in the surface layer. The larger rock fragments are probably arranged into these patterns following intensive freezing and thawing of the ground.

Pátzcuaro *Mexico* Picturesque town on the south shore of Lake Pátzcuaro, deep in the Michoacán Mountains of central Mexico. Its native name means 'Place of Delights' – possibly because the area has more than 250 species of humming birds. The lake, which measures 50 km (31 miles) in circumference, is dotted with seven islands, one of which, Janitzio, has a popular all-night vigil for All Souls' Day (1-2 November), when Amerindians gather to illuminate the cemetery with candles and adorn the graves with flowers.

Population 30 000
Map Mexico – Bc

Pau *France* Pyrenean resort, commercial and industrial town and capital of the Pyrénées-Atlantiques Department, 90 km (56 miles) from the Atlantic coast. It was the birthplace of Jean Baptiste Jules Bernadotte (1763-1844), a lawyer's son who became a Revolutionary marshal under Napoleon and was adopted in 1810 by the ageing and childless King Charles (Karl) XIII of Sweden. In 1818 he ascended the throne as Charles (Karl) XIV, founding the present Swedish ruling house.

Population 135 700
Map France – Ce

pavement Smooth ground surface of rock formed by wind erosion, weathering or glacial scouring. See DESERT PAVEMENT and CLINT and HOW CAVES ARE FORMED, p 141

Pavia *Italy* University city, 32 km (20 miles) south of MILAN. The ancient city centre preserves the gridiron street plan of the original Roman fort called Ticinum. Pavia's cathedral was started in 1488 and completed in 1936. The university was founded in 1361.

Pavia has food industries and manufactures agricultural machinery, textiles, electrical goods and sewing machines.

Population 80 700
Map Italy – Bb

Pavlodar *Kazakhstan* Town on the Irtysh River, 385 km (239 miles) south-east of OMSK. It lies in fertile farmland and its main industry is food processing – the steady growth of which has led to the trebling of the population since 1960.

Population 337 000
Map Kazakhstan – Da

Pays Basque *France* See BASQUE REGION

Paz de Rio *Colombia* See BOYACÁ

Pazardzhik (Pazardźik) *Bulgaria* Town on the Maritsa River, 100 km (62 miles) south-east of SOFIA. It serves a fertile agricultural and vine-growing area. The town's chief industries are textiles, and rubber and leather goods. Its underground Church of the Holy Virgin has a magnificent carved icon screen.

Population 79 500
Map Bulgaria – Bb

Peace River *Canada* A major tributary, 1923 km (1195 miles) long, of the MACKENZIE River. It rises

in the Rockies in western Canada, and flows north-east to enter into the GREAT SLAVE LAKE, which is drained by the Mackenzie. Its valley, in which cereals are grown, is the most northerly arable farming area in Canada.

Peace River is also the name of a town on the Peace River, in an oil, gas and farming area in north-west ALBERTA. It is the centre of an Anglican bishopric extending to the Arctic Circle.

Population (town) 8 200
Map Canada – Dc

Peak District *United Kingdom* Upland area of moorland, crags and dales at the southern end of the Pennines in Derbyshire. It is popular with hikers and potholers, who explore its numerous underground caves. Most visitors come from the nearby industrial areas of LANCASHIRE, SOUTH YORKSHIRE and The POTTERIES.

The area's highest point, Kinder Scout ('The Peak'), is 636 m (2088 ft) high. Despite most of the Peak District being a national park, fluorite – including the prized coloured form, Blue John – is mined in some places.

Map United Kingdom – Ed

Pearl Harbor *USA* Deep-water harbour and naval base on OAHU Island in the US Pacific Ocean state of Hawaii. It was bombed without warning by the Japanese on 7 December 1941 – an event which prompted the United States to enter the Second World War.

Map Hawaii – Ba

Pearl Islands *Panama* See CONTADORA, ISLA DE

Pearl River *China* See ZHU JIANG

peat Partially decomposed, compacted vegetable matter found in bogs and fens, and used as a soil conditioner and fuel.

pebble Rock fragment 4-64 mm (0.16-2.52 in) in diameter. It is larger than GRAVEL but smaller than a COBBLE, and usually rounded.

Peć (Pech; Petch; Ipek) *Yugoslavia (Serbia and Montenegro)* Ancient market town situated some 20 km (12 miles) from the north-east tip of Albania. It lies astride the torrential Bistrica River on the western edge of a fertile mountain basin. Its old Turkish quarters, with narrow streets, bazaars and mosques, contrast sharply with the apartment blocks and factories that have been built in the post-1960 period.

Industries revolve around food and tobacco processing, leather goods and textiles. Peć was the headquarters of the Serbian Orthodox Church from the 14th to the 18th century; the imposing Patriarchate, with three churches, lies at the mouth of the spectacular Rugovo Gorge, just west of the town.

Decani, medieval Serbia's largest monastery, is situated 16 km (10 miles) south of Peć. It was built in the 14th century and has fine frescoes.

Population 111 000
Map Yugoslavia – Cc

Pech *Yugoslavia* See PEĆ

Pechenga (Petsamo) *Russia* Ice-free port situated in the north close to the Finnish border on the Barents Sea. It has a large fishing fleet and refines copper and nickel from local mines.

The area belonged to Finland from 1920 to 1944, when it was annexed by the USSR.
Population 51 000
Map Russia – Eb

Pechora *Russia* River which rises in the Ural Mountains and flows 1809 km (1124 miles) north-west to the Barents Sea. Its basin contains coal, oil and gas.
Map Russia – Gb

Pécs *Hungary* Chief city of south-west Hungary, and administrative centre of Baranya County, 171 km (106 miles) south-west of BUDAPEST. It lies near the old town of Villány on the vine-clad southern flanks of the densely forested Mecsek Mountains, which reach 681 m (2234 ft). A major manufacturing centre whose products include leather goods, Zsolnay porcelain and Villányi red and white wines, Pécs has two 'new town' satellites: Uránváros, where uranium is mined, and Komló, a producer of lignite.

Established by the Celts and later developed by the Romans, the city prospered as a trading centre in the Middle Ages; Pécs' early Christian catacombs, its fine cathedral with four graceful spires, and its university, founded in 1367, date from this period. The Gazi Kassim Pasha and Jakovali Hassan mosques date from the Turkish occupation (1543-1686), while the Baroque bishop's palace survives from the 18th century.
Population 170 540
Map Hungary – Ab

pedalfer Soil, rich in iron and aluminium, found in wet, and often warm regions.

pediment Gently sloping, eroded surface, which may be partially covered with ALLUVIUM, found at the base of mountains in dry areas.

Peebles *United Kingdom* See BORDERS

Pegasus Bay *New Zealand* Bay on the east coast of New Zealand's South Island, north of Banks Peninsula, which was named after the British botanist Sir Joseph Banks who accompanied Captain James Cook on his circumnavigation of New Zealand in 1769-70. The bay is 66 km (41 miles) long, and measures 24 km (15 miles) from east to west. The port of Lyttelton on the bay's south side serves the city of CHRISTCHURCH.
Map New Zealand – De

pegmatite Coarse-grained IGNEOUS ROCK, with crystals at least 100 mm (0.5 in) in diameter.

Pegu *Myanmar (Burma)* See BAGO

Pekanbaru *Indonesia* Provincial capital of RIAU in Sumatra. It is a prosperous oil port and a processing centre for rubber.
Population 186 300
Map Indonesia – Bb

Peking *China* See BEIJING

pelagic sediment Ocean-floor sediment formed from dead organisms that inhabit the open ocean, rather than the shoreline areas. These organisms extract minerals such as lime or silica from the ocean water.

Pelagonia *Macedonia* See BITOLA

Pelée, Mont *Martinique* Active volcano, 1397 m (4583 ft) high, in the northern part of the island. Its eruption on 8 May 1902 destroyed the town of St Pierre on the west coast, 20 km (12 miles) north-west of FORT-DE-FRANCE. More than 30 000 people were killed in just three minutes by a cloud of ash and hot gas which escaped from the side of the volcano. The eruption created a column of lava 360 m (1180 ft) high which collapsed in 1903.

Only one person survived the 1902 eruption – a stevedore who had been jailed for brawling. He was protected by the thick stone walls of the jail in which he had been locked up. Today St Pierre is a small village (population 6000), and the ruins of the old town are a tourist attraction.
Map Caribbean – Cc

Pelion *Greece* See PÍLION

Pella *Greece* Site of the birthplace of Alexander the Great (356-323 BC), who made it his capital. It lies 40 km (25 miles) west of SALONICA, and was rediscovered only in 1957. A number of streets, some with mosaics, have since been uncovered.
Map Greece – Ca

Peloponnese (Pelopónnisos) *Greece* The southernmost part of mainland Greece connected to northern Greece by the isthmus of CORINTH. It gets its name from Pelops, the mythical king who was chopped up by his father and fed to the gods.

The region consists of hills and rugged coastline, bounded by the Aegean Sea to the east and the Ionian Sea to the west. Barren mountains, among them Taíyetos (2404 m, 7885 ft) and Zereia (2377 m, 7798 ft), fall to narrow coastal plains. The Peloponnese was the cradle of the Mycenaean civilisation, the forerunner of classical Greece. From AD 267 until the 11th century, it was ravaged by Goths, Huns, Slavs and pirates; later, it was ruled by Franks (after 1210) and Turks (after 1453). It was united with Greece in 1828. The Peloponnese produces red and white wine, olives and fruit. It makes textiles and paper, and processes food. The main city is PATRAS.
Population 1 012 500
Map Greece – Bc

Pelotas *Brazil* Agricultural city at the southern end of the lagoon of Lagoa dos Patos, 210 km (130 miles) south of PORTO ALEGRE. It processes locally grown rice, fruit and soya beans.
Population 294 000
Map Brazil – Ce

Pemba *Tanzania* Island in the Indian Ocean, about 60 km (35 miles) north-east of the island of ZANZIBAR, with which it was a British protectorate from 1890 to 1963. The islands constituted the independent country of Zanzibar for four months in 1963-4 before joining mainland Tanganyika to form the Republic of Tanzania. Pemba, whose chief export is cloves, measures 985 km² (380 sq miles). Its main town is Chake Chake on the west coast.
Population 270 000
Map Tanzania – Ba

Pembroke *United Kingdom* See DYFED

Penang *Malaysia* See PINANG

P'eng-hu Lieh-tao (Penghu Islands) *Taiwan* See PESCADORES

A DAY IN THE LIFE OF A FRENCH AGRICULTURAL SALESMAN

Pierre Dufour gets up at 7.30 in the morning. He has a busy day ahead and only has time to glance at the newspaper as he sips a strong, black coffee. Pierre, 38, is an agricultural salesman in the south of France. He sells fertilisers, pesticides and machinery to farmers in the Languedoc-Roussillon region who grow wine grapes, fruit trees, lettuces and tomatoes.

Pierre likes his job and is good at it. Once a week he calls his manager in Bordeaux to place his orders. He hopes one day for promotion – to extend his region over a larger agricultural area.

This morning, Pierre has two clients to see, one in the town of Carcassonne and one in the village of Elne, close to home. The Dufours live in a small but comfortable house in the village of Toreilles, only a few miles from Perpignan. Pierre had the house built a few years ago. The mortgage takes about a quarter of his salary.

Pierre spends a lot of time on the road, driving to see clients. Then he often thinks about his wife Françoise and how they met. He had just finished his compulsory year in the army and was working in his father's small grocery shop, in a poor district of Perpignan. Françoise lived with her mother and would often come into the shop. The couple fell in love and they married in St Jean's Cathedral in Perpignan.

Their children, Alain (14) and Anne-Marie (12), go to a private Catholic school. Pierre has high hopes that Alain will go to university, and perhaps become a doctor or even a lawyer.

After seeing his clients, Pierre returns home at around noon for a tasty lunch of fried aubergines and fillet of sole. The children eat at school. This gives Pierre some free time. He can watch the 1 pm news on television and also have a nap in peace before setting off again at around 2 pm.

He generally does not set out for home until 7 or 8 in the evening, but before he gets in he often likes to go to the nearby café to drink a couple of glasses of Pernod and have a chat. Conversations vary, but most frequently they are about politics, especially if there is an election coming up in the near future. Pierre, who is a moderate centre-rightist, does not profess to strong political beliefs, but he does like exchanging ideas with his friends.

When he reaches home, he changes into casual clothes before sitting down to dinner with the family. This is time for all of them; the children talk about their day at school and Pierre can catch up with family news.

After the meal the whole family watch television, usually a film, but sometimes variety shows, political debates or quizzes. At about 11pm Pierre goes to bed and reads a few pages of a thriller before falling asleep. His life may not be as adventurous as the hero's, but he enjoys it.

peninsula Long projection of land into a sea or a lake.

Pennines *United Kingdom* Range of hills forming the backbone of England. They stretch 225 km (140 miles) north from the River Trent in the Midlands to the Cheviot Hills on the Scottish border, rising to 636 m (2087 ft) at Kinder Scout in the south and 893 m (2930 ft) at Cross Fell in the north. Rough grass, peat bog and forestry plantations cover their slopes, while there is good farmland in the dales in the eastern flanks. The range includes the Northumberland, Yorkshire Dales and PEAK DISTRICT national parks.
Map United Kingdom – Dc

Pennsylvania *USA* State granted as a colony in 1681 to the English Quaker William Penn (1644-1718), and later one of the original 13 states. It lies in the north-east, and covers 117 412 km² (45 322 sq miles) between Lake Erie and the Delaware River, a large part of it being in the (Allegheny) Appalachian plateau. Pennsylvania is a coal, oil and iron mining centre; its industries include the production of steel, machinery, electrical equipment and chemicals, and food processing. Farming is based on dairying, livestock, vegetables and fruit.
PHILADELPHIA is the principal city and port, PITTSBURGH the main centre of the coal and steel industry and HARRISBURG the state capital.
Population 11 881 600
Map United States – Kb

Pentland *Firth* Channel some 23 km (14 miles) long and 10-13 km (6.2-8 miles) wide, situated between the northern tip of the Scottish mainland and the ORKNEY Islands. It can be difficult to navigate and many people have drowned in its seas.
Map United Kingdom – Gj

Penzance *United Kingdom* Tourist resort and most westerly town in England, standing on the south coast of the county of Cornwall, 12 km (8 miles) from LAND'S END. It is a rail terminus and seaport for services to the Isles of Scilly, and has an airport. St Michael's Mount, a small island just offshore, has an imposing ruined 11th-century monastery and a 15th-century fort.
Population 19 210
Map United Kingdom – Ce

Peradeniya *Sri Lanka* See KANDY

Perak *Malaysia* West-coast state which covers an area of more than 21 005 km² (8110 sq miles) and embraces the basin of the Perak River. It is the country's major tin-mining area. There are rubber estates on the Main Range foothills inland.
Population 2 222 000
Map Malaysia – Bb

perched block A boulder, or an ERRATIC which has been carried by a glacier or ice sheet and left standing in a precarious, exposed position.

Pereira *Colombia* Market town and capital of Risaralda Department, 190 km (120 miles) west of BOGOTÁ. It trades in coffee and cattle, and also manufactures textiles, clothing and paper.
Population 335 960
Map Colombia – Bb

Peremyshl *Poland* See PRZEMYSL

Pergamum *Turkey* See BERGAMA

peridotite Any of a group of coarse-grained ULTRABASIC ROCKS composed mainly of OLIVINE, and various PYROXENES and amphiboles, but with less than 10 per cent of FELDSPAR.

Périgord *France* Low limestone plateau in south-western France, 100 km (62 miles) to the east of BORDEAUX. Caves in the region were inhabited in prehistoric times – as is shown by the LASCAUX cave paintings. The plateau is also noted for its truffles – edible fungi that grow underground and are traditionally collected by using pigs to sniff them out. The region has a mild climate which makes it a popular weekend retreat from the cities.
Map France – Dd

period Unit of geological time, longer than an epoch and shorter than an era. See GEOLOGICAL TIMESCALE, p 256

Perister, Mount *Macedonia* See BITOLA

Perlis *Malaysia* The country's smallest state, covering 795 km² (307 sq miles) on the north-west coast bordering Thailand. Rice and some sugar cane are grown here.
Population 187 500
Map Malaysia – Ba

Perm' (Molotov) *Russia* Major industrial city in the western Ural Mountains, about 280 km (175 miles) north-west of YEKATERINBURG. Perm' began its industrial life as a copper-smelting centre in 1723. It grew rapidly after the First World War, when chemical, engineering, motor vehicle and aircraft industries began to be set up there. It later began to produce steel, paper, timber, petroleum products and textiles. The city lies on three main transport routes: the navigable Kama River, the Great Siberian Highway and the Trans-Siberian Railway across the Urals.
From 1940 to 1957 the city was called Molotov in honour of Vyacheslav Molotov (1890-1986), the Soviet foreign minister under Joseph Stalin. But with Stalin's death in 1953, Molotov's influence gradually declined and in 1957 the city reverted to its earlier name of Perm'.
Population 1 091 000
Map Russia – Gc

▲ ROYAL CITY Sunlight throws into relief the carved gift-bearers lining the King's Gateway near the towering Apadana, or great audience hall, at ancient Persepolis in Iran.

permafrost Permanently frozen ground, found only in cold regions at high latitudes, generally within 2000-3000 km (about 1200-1900 miles) of the north pole. The surface may thaw for a short time in summer.

permeable rock Rock that allows water or other liquids to pass freely through it. It may be either a PERVIOUS ROCK or a POROUS ROCK.

Permian Sixth period in the Palaeozoic era of earth's history. See GEOLOGICAL TIMESCALE, p 256

Pernambuco *Brazil* See RECIFE

Perpignan *France* Market town lying near the Mediterranean coast, 25 km (15 miles) from the Spanish border. It has a cathedral, a château and the palace of the Kings of Majorca, all dating from the 14th century.
Population 106 000
Map France – Ee

Persepolis (Takht-e-Jamshid) *Iran* One of the world's greatest archaeological sites, 670 km (416 miles) south-east of TEHRAN. The Achaemenid Emperor, Darius I (ruled 522-486 BC), intended Persepolis to be the ceremonial capital of Persia, but before the planned city could be completed it was sacked and burnt down by Alexander the Great in 330 BC. Nevertheless, the pillars and other parts of many of the buildings are still standing.
Among the principal remains are the Great Terrace built by Darius, the Gate of Xerxes, the palace of Darius and Xerxes, Darius's treasury, the Apadana (great audience hall), the Hall of One Hundred Columns, the tomb of Artaxerxes II and III and that intended for Darius III, who died in the war against Alexander.
The tombs of Darius I and II, Xerxes I and Artaxerxes I, hollowed out of a cliff high above the ground, are situated at Naqsh-i-Rustam, 5 km (3 miles) north of the city.
Map Iran – Bb

Persian Gulf *Saudi Arabia/Iran* See GULF, THE

Perth *Australia* State capital of Western Australia. Founded in 1829 on the Swan River, it grew slowly until it became a convict settlement in 1850. Perth and its port, Fremantle, developed rapidly after gold was discovered 547 km (340 miles) to the east at KALGOORLIE in 1893. The port handles refined oil, wheat and wool. Rottnest Island, 20 km (12 miles) off the coast, is a wildlife sanctuary where the quokka, a dwarf ratlike wallaby, lives.
Population 1 143 270
Map Australia – Be

Perth *United Kingdom* Scottish city 55 km (34 miles) north of EDINBURGH, standing on the salmon-rich River Tay, and former county now divided between CENTRAL REGION and TAYSIDE. Perth was a strategic fort for the English kings of the Middle Ages in their repeated attempts to conquer Scotland; it was seized by the Scottish leader Robert Bruce in 1311 in his battle against them. The city was the Scottish capital until the murder of James I of Scotland in 1437.

The Scottish Reformation began in Perth in 1559 when the Calvinist John Knox preached his sermon against 'Catholic idolatry' in the Kirk of St John, which still stands in the city centre.

Perth has a small port for coastal shipping, but much of the traffic to the oil fields of the North Sea passes through Perth by road to ports farther north. The main industries of the area include agriculture, tourism, glass manufacturing, and whisky distilling and bottling.
Population (Perth and Kinross) 125 770
Map United Kingdom – Db

Peru See p 534

Perugia *Italy* Beautifully preserved medieval city set in the Apennines 135 km (84 miles) north of Rome. Perugia was a centre of Renaissance painting, and was the home of Pietro Perugino (1445-1523), teacher of Raphael (Raffaello Santi, 1483-1520) and Bernardino Pinturicchio (1454-1513). Works by these painters are scattered through the city's churches and public buildings. The Palazzo dei Priori (1243-1443), a beautifully preserved medieval palace, is the home of the Umbrian National Gallery. The richness of Perugia's cultural life continues today, with two universities – one for foreigners – and many other educational and artistic institutions.

The economy of the city, which is the capital of Umbria, is centred on tourism and chocolate making, with a few textile and furniture factories.
Population 150 600
Map Italy – Dc

pervious rock A PERMEABLE ROCK that has no pores. It allows liquids pass through it via cracks and fissures, joints and bedding planes. Limestone and chalk are common examples of pervious rocks. See also HOW CAVES ARE FORMED, p 141

Pesaro *Italy* Adriatic port and holiday resort 130 km (80 miles) east of FLORENCE. The operatic composer Gioacchino Rossini (1792-1868) was born here. Tourism, agricultural machinery and pottery are the town's main industries.
Population 88 500
Map Italy – Dc

Pescadores (P'eng-hu Lieh-tao) *Taiwan* Group of about 64 islands in Taiwan Strait, 45 km (28 miles) off the west coast of Taiwan. Low-lying and mainly formed from coral, the islands have a total land area of 127 km² (49 sq miles). Only 24 of the islands are inhabited, and the main occupations here are fishing and farming. The main town, Ma-kung (population 60 000), was a Japanese naval base during the Second World War; it lies on P'eng-hu, the largest island.
Population 94 000
Map Taiwan – Ac

Pescara *Italy* Tourist resort and seaport on the Adriatic coast, 155 km (96 miles) north-east of ROME. Its industries are fishing, shipbuilding, and manufacture of machinery, textiles and furniture.
Population 121 370
Map Italy – Ec

Peshawar *Pakistan* Historic town that guards the foot of the KHYBER PASS in North-West Frontier Province. Standing on the lower Kabul River, the city has a university and agricultural research institute. It produces machinery, textiles and canned fruit. One of its Buddhist shrines contained relics of the Buddha himself until 1909, when they were sent to Burma (now Myanmar). The city also has a number of Mogul, Sikh and British buildings. Its bazaars have traded for thousands of years, dealing in various goods brought down the pass from central Asia by camel caravans. In the early 1990s, the majority of its inhabitants were refugees from Afghanistan.
Population 1 114 000
Map Pakistan – Ca

Peski Karakumy (Kara Kum, Karakumy) *Turkmenistan* The country's largest desert, covering some 340 000 km² (130 000 sq miles) along the Iranian and Afghan borders east of the Caspian Sea. Consisting mostly of sand, it includes much of Turkmenistan. Some areas receive as much as 100-150 mm (4-6 in) in rainfall per year – sufficient to provide grazing for herds of karakul sheep.

The Karakumskiy Canal runs for 1400 km (870 miles) from the Amudar'ya River, east of the desert, to Krasnovodsk on the Caspian Sea.
Map Turkmenistan – Ab

Petch *Yugoslavia* See PEĆ

Petén *Guatemala* Most northerly department of the country, lying between Mexico and Belize, south of Mexico's Yucatán Peninsula. It covers about 36 000 km² (13 900 sq miles) and consists mainly of rain forest and grasslands dotted with lakes. The region's Amerindian farmers practise slash-and-burn agriculture, moving from one patch of rain forest to another every few years as their crops exhaust the thin soil.

Flores, the main town of the region, is situated on an island in the Petén Itzá Lake, 260 km (160 miles) north of the capital, GUATEMALA CITY. The nearby forest contains extensive Mayan ruins dating from before the 10th century AD.
Population 253 000
Map Guatemala – Aa

Peterborough *United Kingdom* Cathedral city in Cambridgeshire, situated 120 km (75 miles) north of LONDON. The cathedral, which was built largely in the 12th and 13th centuries, is one of the finest surviving Norman buildings in Britain.

The modern city of Peterborough has agricultural industries and is a residential overspill for London. Much of the capital was built from bricks produced near Peterborough.
Population 154 000
Map United Kingdom – Ed

Petra *Jordan* Ruined rock city and the country's main tourist attraction, situated 98 km (61 miles) south-west of KARAK. It has been called 'the rose-red city' after a line in the poem *Petra* by the Victorian writer, the Reverend John Burgon (1845-88), because of the startling effect of sunrise and sunset on its pinkish-red sandstone. The original name, Sela, and the present name both mean 'rock'.

Petra was founded in about 1000 BC by the Edomites, said to be the descendants of Esau, the son of Isaac and Rebecca. In about 300 BC the Edomites were driven out by the Naboteans, who made Petra their capital. All that remains of the Nabotean capital are the rockfaces with numerous dwellings cut into them. The Street of Façades – great crags with buildings cut into them – contains the striking treasury of Khazneh. There are many other cave dwellings and a Roman theatre.

The city was once on the crossroads of caravan routes between Arabia, Egypt and Assyria, but by the 5th century they had fallen into disuse and the town became deserted soon after.

It is ringed by high mountains and is approached through the Siq, a steep gorge more than 3 km (2 miles) long, whose sheer sides rise to 100 m (about 330 ft). The tomb of Aaron, the elder brother of Moses and high priest of the Hebrews, is on a hill overlooking the city.
Map Jordan – Ab

Petrikau *Poland* See PIOTRKÓW TRYBUNALSKI

Petrograd *Russia* See ST PETERSBURG

Petrokov *Poland* See PIOTRKÓW TRYBUNALSKI

petroleum (crude oil) A natural, thick, yellow to dark-coloured liquid hydrocarbon mixture which is found beneath the earth's surface. The word 'petroleum' is derived from the Greek, and means 'rock oil'. See THE STORY OF OIL AND NATURAL GAS, p 509

Petropavlovsk *Kazakhstan* Timber town on the Trans-Siberian Railway in northern Kazakhstan, 265 km (165 miles) west of OMSK. Its industries now produce processed food, clothing and timber and wood pulp.
Population 245 000
Map Kazakhstan – Ca

Petropavlovsk-Kamchatskiy *Russia* Pacific port and naval base used by Russian whalers and fishing boats situated on the south-east coast of the Kamchatka Peninsula. Its population has trebled since the 1960s with the expansion of 'factory-ship' canning, and the servicing and repair of the Russian fleet.
Population 271 000
Map Russia – Qc

Petrovgrad *Yugoslavia* See ZRENJANIN

Petrozavodsk *Russia* Capital of the Karelian Republic, on the western shore of Lake Onega. It

was originally founded in 1703 as an ironworking centre, with metalworking, ship-repairing and timber-based industries. It has shut down its ironworking factories, but remains an industrial centre.
Population 274 000
Map Russia – Eb

Pforzheim *Germany* City lying halfway between KARLSRUHE and STUTTGART. It is the centre of Germany's optical industry and is also a leading watchmaking and jewellery trade centre.
Population 112 900
Map Germany – Cd

phacolith (phacolite) Body of intrusive IGNEOUS ROCK shaped like an upturned saucer and lying in the upfold of beds or layers of rock. It is formed by the injection of hot molten rock either during the folding or after it.

Phaistos *Greece* Large excavated Minoan palace on a hill, 60 km (37 miles) south of HERAKLION (Iráklion) in Crete. It was built in about 1700 BC on the site of an earlier palace. Parts of both palaces have been uncovered. The Phaistos Disc, a pottery disc stamped on both sides with hieroglyphics (as yet undeciphered), was found in the new palace and is in a museum in Heraklion.
Map Greece – Dd

Phayao *Thailand* Rice-growing province and town in the north, about 630 km (390 miles) north of BANGKOK. The originally 13th-century, largely wooden, town lies beside Lake Khwan Phayao.
Population (province) 477 900; (town) 141 700
Map Thailand – Bb

phenocrysts Large crystals set among a mass of smaller crystals in an IGNEOUS ROCK. An example is PORPHYRY. The two sizes result from the rock solidifying in two stages. Slow cooling produces the larger crystals, then fast cooling produces the ground mass of smaller crystals.

Philadelphia *USA* City and port situated in Pennsylvania, some 160 km (100 miles) up the Delaware River from the Atlantic Ocean. It was settled in 1681 by colonists sent by the English Quaker William Penn. The American Declaration of Independence was signed here in 1776. Today it has engineering and shipbuilding industries and is an educational and arts centre, with three universities. The city's centre retains a flavour of old Philadelphia and its buildings are protected.
Population (city) 1 646 700; (metropolitan area) 4 768 400
Map United States – Kc

Philae *Egypt* Island in the Nile River just below the ASWAN HIGH DAM, 700 km (435 miles) south of CAIRO. The island was dedicated to the fertility goddess Isis from about 600 BC until the time of Christianity. Its ancient temples were submerged when the old Aswan Dam was built in 1907. They were revealed again in 1970 when the new Aswan High Dam held back the river farther upstream. The Temple of Isis has since been reconstructed on the nearby island of Agilkia.
Map Egypt – Cc

Philippines See p 536

Philipsburg *Netherlands Antilles and Aruba* See ST MARTIN

Phitsanulok *Thailand* Agricultural province and prosperous market town 355 km (220 miles) north of the capital, BANGKOK. The principal crops grown here are rice, maize and beans.
Population (town) 215 900; (province) 729 350
Map Thailand – Bb

Phlegrean Fields (Campi Flegrei) *Italy* Volcanic area to the west of NAPLES formed by collapsed cones and craters and which remains partially active. It contains the crater lake of Averno, the legendary entrance to the underworld, and Mount Nuovo, an ash cone which is 140 m (460 ft) high created by an eruption in 1538.
Map Italy – Ed

Phnom Malai *Cambodia* Hilly region north of the CARDAMOM MOUNTAINS in western Cambodia. It became the main refuge for the Khmer Rouge guerrillas after they had been removed from government by the Vietnamese army in 1978.
Map Cambodia – Ab

Phnom Penh *Cambodia* National capital on the Mekong River just north of its confluence with the Tonle Sap. It has remained the capital almost continuously since 1434, but recent wars have scarred the city badly. It has only partly recovered from forced depopulation imposed by the communist Pol Pot regime in 1975. In 1974, for instance, the city had a population of more than 2 million; five years later, only 23 000 were left.

In 1979 Phnom Penh was occupied by the Vietnamese who had defeated the Pol Pot regime in December 1978; the last of their troops left the Cambodian capital in 1989. The city's industries revolve around textiles and food processing.
Population 800 000
Map Cambodia – Ab

▼ **TRADITIONAL STYLE** Houses on stilts with boats alongside line the shore at a quiet fishing village on the island of Phuket in Thailand.

Phoenicia *Middle East* See LEBANON; SYRIA

Phoenix *USA* Capital of Arizona and one of the fastest growing metropolitan areas in the United States. It lies in the centre of the state, 200 km (about 125 miles) north of the Mexican border, and is a retirement and tourist centre. It has high-technology research and production industries.
Population (city) 938 000; (metropolitan area) 2 238 500
Map United States – Dd

Phoenix Islands *Kiribati* Group of eight small scattered islands with no permanent population, south-east of the Gilbert Islands. Some of them, including Kanton and Enderbury, were claimed by the United States until that country signed a treaty of friendship with Kiribati in 1979, abandoning its claim on condition that they were not used by another country for military purposes.
Map Pacific Ocean – Ec

Phou Bia *Laos* Mountain which rises to 2820 m (9252 ft) immediately south of the Plain of Jars. It is the country's highest peak.
Map Laos – Ab

Phuket *Thailand* The country's largest island covering 801 km² (309 sq miles). It lies in the ANDAMAN SEA, 675 km (about 420 miles) south-west of the capital, Bangkok, and is joined to the mainland by a causeway. The centre of a group of 37 beautiful islands, Phuket has mushroomed as a holiday destination since the Bay of Phangnga nearby was chosen as a setting for the 1974 James Bond film *The Man With The Golden Gun*. It also has tin and rubber industries. The island's main town is also called Phuket.
Population (island) 146 400; (town) 86 800
Map Thailand – Ad

Piacenza *Italy* City on the Po River, 60 km (35 miles) south-east of MILAN. The centre preserves its Roman chequerboard plan, and there are about 5 km (3 miles) of *(continued on p 539)*

Peru

ONCE THE CENTRE OF THE RICH AND MIGHTY INCA EMPIRE, PERU IS NOW STRUGGLING AGAINST POVERTY

Just over 500 years ago the greatest Amerindian civilisation to exist in South America – that of the sun-worshipping Incas – had its heartland in Peru. The well-organised Inca Empire, rich in gold and silver, stretched to the north into Ecuador and south into present-day Chile. The spectacular remains of the ancient Inca cities at CUSCO and MACHU PICCHU are among the continent's most amazing monuments.

But Peru today is scarred by poverty and violence. Most of its people are the victims of an unequal social structure and the struggle to survive in shanty towns around the cities or win a meagre living from the soil in a countryside that until 1993 was racked by civil war.

THREE LANDS

Although the climate and geography of the country have helped to preserve Peru's ancient treasures, they have also been obstacles to prosperity. There are three distinct regions west to east: the coast, the high sierra of the Andes and the tropical rain forest.

The narrow coastal belt west of the Andes is mainly desert, created by the influence of the cold Peruvian Current (once called the Humboldt Current) flowing north from Antarctica along the Pacific coast. The current cools the air, preventing the formation of rain.

The dry coastal air has helped to preserve ruins and has saved many of the relics of Amerindian civilisations thousands of years older than the Inca civilisation.

The heights of the Andes, where the Incas made their last stand against the Spaniards, are wet, mist-soaked solitudes with peaks of more than 6000 m (19 600 ft). Here, the vegetation varies with altitude, from dense forest to mountain scrub. In the south, the two main Andean ranges separate and diverge, to embrace between them a massive plateau, the ALTIPLANO. Lake TITICACA lies here, the highest navigable sea of water, which has been inhabited for centuries by Amerindians living on reed islands. The climate on its shores is mild – and like the Incas long ago, today's population cultivates the ground and grazes sheep and llamas close to the lake.

To the east of the Andes is the Montaña region, the Amazon lowland covered in *selvas* (tropical forests). The region covers some 60 per cent of the country, yet has only 3 per cent of its people – some parts can still be reached only by river. The boom in wild rubber in the early 20th century gave rise to a few towns such as IQUITOS, which boasts a town hall designed by the same Gustave Eiffel who built the Eiffel Tower; the materials for the building were imported from France to this improbable location by a rubber baron. Archaeologists are still finding ancient settlements in the jungles of Montaña, where they lie buried in the fast growing foliage.

The combination of desert, mountain and rain forest limits the areas that can be lived in or cultivated. The Incas grew crops on hillside terraces and irrigated areas. Today subsistence farming still continues in the mountains, but most large-scale agriculture is in the oases and fertile, irrigated river valleys that cut across the coastal desert. Forty per cent of all agricultural products come from this region, including sugar and cotton (which are exported), and rice and bananas. In the mountains large areas are unfarmable and some farmers who grow food on the slopes are faced with the problem of soil erosion. Furthermore, the zone is subject to earthquakes, which have repeatedly brought destruction. In 1970 an earthquake dislodged a glacier which buried the town of Yungay and 18 000 of its people.

The Peruvian Current has in the past brought great wealth from the sea in the form of vast shoals of anchovies. Coastal settlements developed around industries based on fish canning and the production of fishmeal. When it appeared that the shoals had become depleted the government imposed a fishing ban in 1983 to allow stocks to recover. By 1990 Peru was again the fourth biggest fish producer in the world, with 6.8 tonnes harvested.

Fish today is Peru's second largest revenue earner after copper. In the past the Peruvian economy was based mainly on the huge, thick layers of sea-bird droppings (*guano*) in the desert which were collected and exported to be sold as valuable fertiliser, but *guano* is no longer an important income earner.

UNYIELDING LAND

Nature is not always bountiful to the Peruvians. Occasionally, there is a warm marine current, known as EL NIÑO, which reverses the cooling effect of the Peruvian Current. Then the anchovies disappear and devastating storms and floods on land destroy crops, bringing hunger.

The rain forests in the east are still impenetrable – the few existing tracks and roads are impassable in the rainy season. The rain forest metropolis of Iquitos, which grew up thanks to the rubber boom, has long ago sunk into oblivion since the demand for rubber subsided. The red soils of the region yield little else. But recently the discovery of oil has once more meant boom time for the *selvas*. Today the source of wealth is oil – and some exploitation is beginning to intrude into an area previously the preserve of missionaries working with Amerindian tribes. Oil contributes 8 per cent of Peru's exports, but new oil discoveries are now needed; present reserves are likely to be exhausted by the end of the century.

Peru's economy relies heavily on its mineral wealth – 50 per cent of exports come from copper, zinc, silver, gold, lead and iron ore mined in the Andes. Then, of course, there is the illegal export of coca leaves – Peru is the world's largest coca leaf producer.

TWO NATIONS

More than 25 per cent of people in Peru speak only the Amerindian languages of Quechua or Aimara, and not the official Spanish. There are some 50 Amerindian tribes in the country. These people have shared very little in what limited benefits economic growth has brought Peru and live in conditions of great hardship.

In the countryside many peasants scratch a living from tiny patches of land. And in the towns they sell food, vegetables and clothing

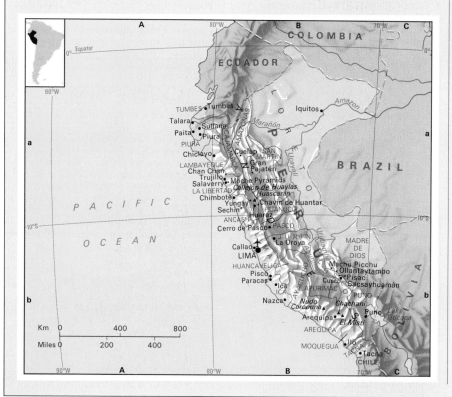

▲ MARKET DAY The open-air Amerindian market in the hill town of Chincheros, in south-central Peru, attracts both natives and tourists. The wares on offer range from wine jugs to colourful Amerindian shawls.

in markets which are often picturesque but yield a poor return. The diet and general health of these people is equally poor; 7.6 per cent of children die before they are one year old, 40 per cent of children suffer from malnutrition. Peru is one of the countries with the lowest average daily per capita intake of food. Ninety-nine per cent of the rural population and 60 per cent of town dwellers have no running water or drainage. Most of the people of the capital city of LIMA live in shanty-town settlements called *barriadas* which they build illegally on government land, and which are only gradually being supplied with water and sanitation.

The widespread poverty results from the combined effects of climate, terrain and natural disasters. The unequal distribution of wealth – 1 per cent of the population earns about 30 per cent of the national income – has also resulted in social discontent. This reached new heights in the early 1980s, especially where the rural poor had the opportunity of an education. A movement, based on Chinese Maoist doctrine and called the Shining Path (*Sendero Luminoso*), began in the University of AYACUCHO, in the middle of an impoverished mountain region. Throughout the 1980s the *Senderistas* guerrillas bitterly clashed with government forces – and the inhabitants of the mountain villages were caught in between.

Although largely quashed, this rebellion was the greatest threat to the government, which is headed by a democratically elected president. Fernando Belaúnde Terry was ousted by a military coup in 1968, but re-elected in 1980. Alan García Pérez succeeded him in July 1985 and introduced some ineffective reforms. Five years later, with terrorism rampant and inflation up to 8000 per cent a year, Alberto Fujimori was elected president. In April 1992 Fujimori suspended the constitution, dissolved the parliament and sacked half of the judiciary in a presidential coup which was much criticised both internally and abroad.

However, his government managed to bring inflation down to below 20 per cent by 1994 and lift the country's growth rate to around 7 per cent. Fujimori also put up a determined fight against the *Senderistas* and captured most of their political leadership.

Still, Peruvian business people claimed that Fujimori's reforms were not sufficient and the rural population of Amerindians, whose vote brought in the president in the first place, had not seen any improvement in their lot.

CONQUERED BY 200 MEN

Like its recent politics, Peru's history has witnessed a series of upheavals and its share of violence. Before the Inca Empire was established in the 15th century, several great Amerindian civilisations flourished. Over a period of 1400 years these rose in turn until finally the highland Inca tribe absorbed the remains of the others into its culture.

The Inca civilisation originated in the Cuzco region; Cusco itself became the empire's capital and the centre ('navel') of the nation's religion, which was based on sun worship. The empire, for all its splendour, only survived for 300 years before it fell to the Spanish conquerors in 1532. Parts of the Inca city walls of Cusco survive today, but much of the Inca capital was lost beneath the Baroque buildings of the Spanish colonisers. The empire was rich in gold and silver, but treated these metals as ornamental, for the Incas used no money.

When the Spanish invaded under Francisco Pizarro, the Inca nation was split apart by an internal crisis. The last great emperor, Huayna Cápac, had divided his territory between two sons, who established rival capitals at Cusco and QUITO, and went to war. Both sides hoped for help from the Spanish in their own fight but were betrayed by the conquerors. The empire fell to a force of only 200 men.

But one isolated Inca citadel remained undiscovered: Machu Picchu, straddling a high mountain in the Andes above the Urubamba River, about 50 km (30 miles) north of Cusco. Here, the Inca 'Virgins of the Sun', the temple priestesses, took refuge from the violence of the Spaniards. The city remained undisturbed for four centuries.

Pizarro needed a base from which to exploit Peru's treasures, so Lima, the capital, was founded. To handle exports to Spain, the port of Callao was developed on the coast, 13 km (8 miles) to the north-west.

Lima is said to be a different world from the rest of Peru, but it might be truer to say that every part of Peru is different from the rest. On the Altiplano near the Bolivian border, for instance, the bowler-hatted Amerindians still herd llamas and alpacas, make clothes from the wool of these creatures, eat their meat and keep warm by using their dried dung for fuel.

PERU AT A GLANCE	
Official name Republic of Peru	
Area 1 285 220 km² (496 109 sq miles)	
Population 23 210 400 **Per km²** 18 **(Per sq mile** 47)	
Capital Lima	
Government Presidential republic	
Currency 1 new sol = 100 céntimos	
Languages Spanish (official); Quechua, Aimará	
Religion Christian (more than 90%)	
Climate Temperate on coast, with a desert climate; cooler in Andes, tropical in *selvas*, with summer rains. Average temperature in Lima ranges from 15°C (59°F) in August to 22°C (72°F) in February	
Land use Forest 53%, grazing 21%, cultivation 3%, other 23%	
Main primary products Fish, cotton, sugar cane, coffee, rice, potatoes, fruit, vegetables, livestock; crude oil, copper, silver, zinc, lead, iron ore, tungsten	
Major industries Fishing and fish processing, agriculture, mining, oil production and refining, textiles, cement, leather goods, plastics, chemicals, metal refining	
Main exports Copper, fish and fish products, oil, silver, iron ore, zinc, cotton	
Annual income per head (US$) 1160	
Population growth (per thous/yr) 29	
Life expectancy (yrs) Male 63 **Female** 67	

Philippines

MORE THAN SEVEN THOUSAND ISLANDS WHERE TYPHOONS BLOW AND VOLCANOES ERUPT, THE PHILIPPINES IS A LAND OF PASSION AND BEAUTY – A PIECE OF LATIN AMERICA IN THE PACIFIC OCEAN

The fury of the typhoons that hit this beautiful island chain in the western Pacific and the destructive power of its active volcanoes are matched by the passions of the people in religion and politics.

The Philippines is the only predominantly Christian country in Asia. At Easter, Roman Catholic penitents simulate the crucifixion; yet when Pope Paul visited the capital, MANILA, in 1970, an attempt was made to assassinate him. And so too in politics: Benigno Aquino, the chief rival of former President Ferdinand Marcos, was shot as he stepped off his plane at Manila airport on his return from exile in 1983.

Named after Philip II of Spain (1527-98), the 7107 islands of the Philippine archipelago stretch for 1850 km (1150 miles) between Taiwan and Borneo. Together they have a coastline longer than that of the United States. Spain colonised the Philippines in 1565 and ruled here until the Americans took over in 1898. When the Americans left in 1946, a cynic said that the Filipinos had spent '400 years in a convent followed by 50 years in Hollywood'. Filipinos today are the least oriental of the orientals; the republic has been called 'a piece of Latin America in the Pacific'.

In 1941 the Philippines was invaded by the Japanese, then liberated by the Americans in 1945. The greatest sea battle in history was fought off the Island of Leyte in 1944, when American naval forces destroyed the might of the Japanese fleet.

Healing the scars of war and re-establishing economic and political life were complicated by a rebellion in central LUZON by the communist-led Hukbalahap guerrillas (Huks). The Filipino name means 'People's Anti-Japanese Army' – an accurate description of its activities during the occupation. After the war, the Huks openly declared their communist orientation and launched an armed revolt against the government, seeking sweeping rural land reforms, which attracted many peasants to their ranks, and political power. Military pressure, some progress on a fairer distribution of agricultural land and economic reforms led to the collapse of the revolt in 1954, when its leader Luis Taruc surrendered. The need for further land reforms continued and the Huks became active again in the 1960s. They surrendered in 1972, but a breakaway Maoist group – the New People's Army – maintained its terrorism into the early 1990s.

The total land area of the Philippines is 300 000 km² (115 830 sq miles) – about the size of Italy – but the islands are scattered over an expanse three times as great. They form four main groups. To the north are Luzon, on which stands the large, sprawling city of Manila, and MINDORO. The VISAYAN ISLANDS are situated in the centre, and include SAMAR, Leyte, CEBU, NEGROS and PANAY. In the south are MINDANAO and the SULU ARCHIPELAGO, while isolated in the south-west is PALAWAN.

NATURE'S VIOLENCE

Most of the seas around the Philippines are shallow and easily navigable. The archipelago is dotted with small ports and anchorages. These are usually on the western coasts and interior seaboards, where they are sheltered from the prevailing east winds which blow from the Pacific. From June to December there is a risk of typhoons – in some years there are 200 or more. In 1984 two typhoons with winds of up to 298 km/h (185 mph) sank 11 ships and caused more than 1600 deaths.

The central and western parts of the islands are partly sheltered from the worst of the Pacific storms by mountains on the eastern coasts. On Luzon, the SIERRA MADRE mountains form the eastern barrier, while Mindanao is protected by the Diuata range. On both islands there are some more mountains farther west; Mindanao also has the Philippines' highest peak, Mount Apo (2954 m, 9690 ft).

Through several of the smaller islands, such as Cebu and Palawan, run hilly backbones flanked by narrow coastal plains, with little land for settlement. The main lowland areas are the central plain and the CAGAYAN Valley on Luzon, and the valleys of the Agusan and Mindanao rivers on Mindanao.

The Philippines were created by a disturbance of the earth's crust between 50 and 60 million years ago. There are more than 30 active volcanoes and some have erupted violently in recent times. In 1984 the MAYON volcano on Luzon hurled boulders as big as cars down its slopes as terrified villagers fled their land. The eruption sent a cloud of ash 10 km (6 miles) high over southern Luzon. In June 1991 Mount Pinatubo erupted several times, covering an extensive area with millions of tons of ash, sand and rock, and claiming 555 lives. In February 1993, Mayon erupted again, spewing forth deadly ashes and causing devastation in the surrounding villages. Earthquakes and seaquakes are not uncommon either. One in east Mindanao in 1976 sent a huge wave (tsunami) inland and 8000 people were reported dead or missing.

Off the east coast of the Philippines, the movement of the earth's crust has created one of the deepest ocean trenches in the world, the Mindanao Trench, which plunges to 10 670 m (35 007 ft) near Mindanao.

SEASONS AND PEOPLE

The islands have three seasons: rainy from June to October, cool from November to February and hot from March to May. In the lowlands the climate is hot and humid – the average daily temperature is 28°C (82°F) – but in the highlands temperatures can drop to 16°C (61°F). Manila has 2080 mm (82 in) of rain a year, western Luzon 1000 mm (39 in) and north-east Mindanao 3800 mm (150 in). More than 35 per cent of the Philippines is covered by dense tropical rain forest which contains many hardwoods, including teak. The rain forest of Mindanao is one of the most productive in the world, and timber is an important foreign exchange earner. But too much cutting and too little replanting has taken its toll. Several of the central Visayan Islands are suffering from erosion, and much of the previously forested island of MASBATE is now rolling tropical grassland of little value. The country's minerals include chrome, copper, gold, iron ore, silver, nickel and zinc.

Of the 7107 islands only some 400 are inhabited. The first known settlers, 25 000 years ago, were the *negritos*, or pygmies, who are believed to have arrived by a land bridge from mainland Asia that existed at the time. This primitive group still hunt with spears in the upland jungles of Negros, Samar and Leyte, remote from modern life. Between 2000 and 3000 years ago, the Ifugaos arrived, probably from Taiwan, and began building rice terraces which survive today on the island of Luzon. The terraces, chiselled out of mountainsides and then filled with soil and gravel, depend on a complex system of sluices and canals to water the rice.

The next wave of immigrants were Malays from Indonesia in the 3rd century BC. They were the ancestors of the main present-day population, and the local administrative unit is still called the *barangay* after the small sailing vessels that carried the Malays to the islands. In this mountainous archipelago, life was concentrated in the coastal lowlands. Each new wave of settlers fished the rivers and grew rice along the banks, pushing earlier inhabitants back into the hills; this is one reason why the rice terraces became so extensive.

The famous Portuguese explorer Ferdinand Magellan, while on a Spanish round-the-world expedition via South America, brought the first Europeans to the islands in 1521. Although Magellan was killed in a clash between rival indigenous groups, in 1565 Miguel Lopez de Legaspi formally claimed the Philippines for Spain.

SPANISH AND AMERICAN HERITAGE

In the northern two-thirds of the Philippines, the Roman Catholic religion which the Spanish invaders brought is today practised with fervour. Catholics make up 83 per cent of the population, and Protestants 9 per cent. Many Protestants are members of the Nationalistic Philippine Independent Church. The extreme south was never subdued by the Spanish, and almost 5 per cent of the total population – mainly in western Mindanao and the Sulus – are Muslims. Among the primitive farmers and fishermen, animism is still practised.

Under Spanish occupation, the coconut palms which are a feature of almost every village were cultivated commercially for their copra; it can be seen drying on the roadside throughout the archipelago. In areas such as the ILOCOS region of Luzon, tobacco cultivation was introduced in the 18th century and many smallholders grow little else. Galleons plied between Manila and South America,

Chinese junks sailed in with silk, jade and spices, and a prosperous Chinese merchant community grew up.

In some parts of the new Spanish colony, much of the old village organisation was retained. But in the more central areas, such as the fertile central plain of Luzon and the island of Negros, sweeping changes took place. Large estates were given up by the Spanish administration to Spanish settlers and religious orders. The Filipinos had to be content

with small tenanted holdings; the size of these is a major social problem in rural areas today and has brought about demands for land reform. Sugar cane and pineapples are extensively grown, but the slump in the world sugar market has caused much hardship on Negros.

Most of the small towns in the Philippines resemble their Spanish equivalents, with a central square which has the local church on one side and secular buildings on the other. In education, the Dominican and Franciscan friars

▲ NATURE TAMED Rice terraces begun more than 2000 years ago climb the steep hillsides of northern Luzon. It is estimated that if all the terraces were laid end to end they would reach halfway round the world.

have left many institutions of learning: one Filipino in four goes on to higher education, and the adult literacy rate is 90 per cent.

In 1898 the four-month Spanish-American

War began after the Spanish sank the USS *Main* in the harbour of Havana, Cuba. An American squadron destroyed the Spanish fleet in Manila Bay, and Manila was occupied with the help of Filipinos, who were already in revolt against Spanish rule. However, Filipino hopes of immediate independence were dashed: the United States 'bought' the islands from Spain for US$20 million.

Nationalist resistance continued into the early part of the 1900s, while the Americans gradually handed over more and more administrative power to Filipinos. In 1934 President Roosevelt signed the Tydings-McDuffie Act, which made the Philippines a Commonwealth, with a guaranteed independence in 1946. During the Second World War Filipinos fought alongside American soldiers against the Japanese invaders and in 1946 full independence came, as promised.

Under American influence most of the urban Filipinos developed an assertiveness and informality unusual in South-east Asia. Close ties with the United States – including military bases – remain, and English is widely used. There are two official languages: Filipino, which is based on Tagalog (the main language of central Luzon), and English. Spanish is still commonly spoken, together with more than 80 regional languages and dialects.

Filipino men wear sports shirts and slacks by day, and a *barong tagalog* (a loose, almost transparent shirt worn outside the trousers) for formal wear. Filipinas wear light summer frocks, and *ternos* (formal dresses with butterfly sleeves) on special occasions.

PROBLEMS OF INDEPENDENCE

Since independence the Philippines has faced many problems reminiscent of Latin America: inflation, government debt, high unemployment, social inequalities, rising crime and corruption. Social unrest and austerity increased as the country's industrial expansion in textiles, chemicals and food processing slowed. Ambitious projects such as finding alternative sources of energy to imported oil have caused increasing debt. The hill peoples are opposed to long established homelands being flooded for hydroelectricity. The Muslims resent control from the Christian north, especially to the migration of land-hungry northerners to parts of Mindanao.

For 12 years Muslims of the Moro National Liberation Front (MNLF) waged guerrilla warfare in the south, demanding autonomy. Some 60 000 died in the mid-1970s, but the rebellion faltered through internal dissension and lack of outside support.

Another long-standing threat to security came from the Maoist New People's Army (NPA). The NPA's 12 000 troops attacked army camps, convoys and towns, and in 1984 they killed nearly 3000 people, among whom were 800 soldiers and 70 police. By 1994, however, good progress was made in peace talks between the government and both the MNLF and NPA.

Many observers felt that insurgency in the Philippines had been fuelled mainly by the repressiveness of government forces under Ferdinand Marcos. Ferdinand Marcos, aided by his wife Imelda, was the dominant political

figure in the Philippines for two decades after his election in 1965. He imposed martial law from 1972 to 1981, when he was re-elected. His most prominent opponent, Benigno Aquino, was killed in 1983.

However, in the elections of 1986, Corazon (Cory) Aquino – widow of the assassinated Benigno – made sweeping gains. Both Marcos and Aquino claimed victory, but after allegations of widespread election fraud, pressure from overseas, and the defection of most of the armed forces to the Aquino cause, the ailing Marcos and his family fled to exile (and reportedly a vast fortune) in the United States, where he eventually died in Hawaii.

Cory Aquino became president amid a wave of popular rejoicing. The government released many political prisoners and set about the tasks of reconciliation, land and political reforms, and the rebuilding of the country's shaky economy. No one believed that it would be an easy job, especially when the NPA ignored appeals to surrender, and when, in 1986-7, elements of the armed forces mounted coup attempts.

When Aquino's term of office came to an end in 1992, she was succeeded by Fidel Ramos, an army general who had remained loyal to her government through the various failed coups. He gained international and national recognition as a stronger ruler.

The economy picked up, with exports increasing, and the perennial problem of electricity shortages was resolved. At the beginning of 1995 the Philippines looked forward to a brighter future as it anticipated

joining the growing prosperity of the other ASEAN (Association of South-east Asian Nations) countries.

PHILIPPINES AT A GLANCE	
Official name Republic of the Philippines	
Area 300 000 km² (115 830 sq miles)	
Population 68 464 400 **Per km²** 228 **(Per sq mile)** 591	
Capital Manila	
Government Republic	
Currency 1 Philippine peso = 100 centavos	
Languages Filipino (official), English, Spanish and local dialects	
Religions Christian (90%), Muslim (5%)	
Climate Tropical; winters are drier in the west, but always humid in the north-east. Average temperature in Manila ranges from 25°C (77°F) to 28°C (82°F) through the year	
Land use Cultivation 37%, meadows and pastures 4%, forest and woodland 40%, other 19%	
Main primary products Rice, maize, coconuts, sugar cane, abaca (Manila hemp), rubber, tobacco, pineapples, bananas, coffee, timber, fish; copper, chrome, gold, iron, nickel	
Major industries Agriculture, food processing, textiles, chemicals, forestry, fishing, mining	
Main exports Electrical goods, clothing, metal ores, coconut oil, sugar, fruit and vegetables, timber, Manila hemp	
Annual income per head (US$) 770	
Population growth (per thous/yr) 19	
Life expectancy (yrs) Male 63 **Female** 68	

16th-century walled fortifications still standing. The city's economy centres on the agricultural products of its fertile hinterland.

Population 103 500
Map Italy – Bb

Piatra Neamţ *Romania* Tourist centre on the Bistrita River in the north-east of the country, 120 km (75 miles) west of the Moldovan border. The city contains a late 15th-century church and belfry, and the Carpathian natural history museum. Nearby are the 17th-century Agapia Monastery, the 16th-century Bistrita Monastery, the spectacular Ceahlau Mountains, Lake Izvoru Muntelui, Lacu Rosu, and the Bicaz gorge. Piatra Neamt contains a papermill, an oil refinery, and chemicals, food processing and textile plants.

Population 92 300
Map Romania – Ba

Piauí *Brazil* See TERESINA

Picardy (Picardie) *France* Northern region of 19 399 km² (7490 sq miles) between PARIS and FLANDERS. It is composed of sweeping chalk plateaus drained by the Oise and Somme rivers. Parts were badly damaged by shelling during the First World War, and some villages and towns have since been completely rebuilt. Large mechanised farms now work the region's loamy soils to grow wheat and sugar beet; AMIENS is Picardy's largest town. The region has some light industries.

Population 1 810 000
Map France – Eb

Pichincha *Ecuador* North-central province that stretches across the large fertile Quito basin and into the rain forest of the north-west lowlands around Santo Domingo. Mount Pichincha, at 4794 m (15 728 ft) overlooks the provincial and national capital of QUITO; it was the scene in May 1822 of a decisive battle in which the Spaniards were defeated by forces led by the South American liberator Antonio José de Sucre.

Population 1 756 200
Map Ecuador – Bb

Pico Spanish and Portuguese term for 'mountain' or 'peak'. For the names of geographical features beginning with Pico, refer to the main part of the name.

Pidurutalagala *Sri Lanka* See NUWARA ELIYA

piedmont Formed or lying at the foot of a mountain or mountain range. For example, a piedmont glacier lies at the foot of a mountain range. The term 'piedmont' is derived from the Piedmont Region at the foot of the Alps in northern Italy.

Piedmont (Piemonte) *Italy* Region covering 25 399 km² (9807 sq miles) in the north-west. Its capital is TURIN; other major towns include ASTI, Alessandria, Cuneo, Novara and Vercelli. Piedmont is divided into three by the great sweep of the Alps which all but encircle the region: the flat plain of the upper Po Valley where most of the towns are situated, and by the Monferrato and Langhe hills.

Piedmont, through its statesman Camillo Cavour (1810-61), was the driving force behind the unification of Italy, which was largely completed in 1860. Turin was the nation's first capital, losing out later to Florence and then to Rome. The Piedmont region is rich in farmland, iron mines and sawmills.

Population 4 356 200
Map Italy – Ab

Pieniny *Slovakia/Poland* See BESKIDY

Pierre *USA* Capital of South Dakota, situated on the Missouri River in the centre of the state. It stands across the river from the original fur trading post of Fort Pierre, founded in 1817. Today it is a distributor and processor of livestock for the region.

Population 12 900
Map United States – Fb

Pietermaritzburg *South Africa* Former capital of Natal Colony (1843-1910), and of Natal Province (1910-1994). Founded in 1839 and named after the Voortrekker leaders Pieter Retief and Gerrit Maritz, Pietermaritzburg lies 70 km (45 miles) north-west of DURBAN.

Much attractive Victorian architecture has survived and, with well-planted open spaces and parks, creates a period atmosphere of elegance. However, by the mid-1990s, surrounding black townships continued to be the scene of ongoing violence between supporters of the African National Congress (ANC) and the largely KWAZULU-based Inkatha Freedom Party.

Population (town) 156 470; (district) 228 550
Map South Africa – Db

Pietersburg *South Africa* Capital, administrative centre and principal town of Northern Transvaal Province, situated 300 km (186 miles) north-east of Johannesburg on the main highway to Zimbabwe. The town was established in 1886 south of the Soutpansberg Mountains and, during the Anglo-Boer War (1899-1902), served briefly as the capital of the South Africa Republic (later, Transvaal).

About 90 mines in the district produce precious and semi-precious stones, as well as silica, corundum and tin. Farming includes stock rearing, while tobacco, maize, sorghum and sunflower seeds are among the crops.

Population 64 210
Map South Africa – Ca

Pigg's Peak *Swaziland* Village, situated 60 km (37 miles) north of MBABANE, named after a prospector, William Pigg, who found gold nearby in 1884. Today the mine is defunct, but the settlement is the centre of a timber-producing and popular tourist area; several magnificent waterfalls, including the Malolotsha Falls on the Malolotsha River, are situated in the area.

Population 3200
Map Swaziland – Aa

Pigs, Bay of (Bahia de Cochinos) *Cuba* Bay on the south coast, about 175 km (110 miles) south-east of HAVANA. On 17 April 1961, it was the scene of an attempted invasion by exiled anti-Castro Cubans, trained and equipped by the American CIA. The 1600 invaders were quickly overcome and many were killed or captured.

Map Cuba – Aa

Pikes Peak *USA* Snow-capped mountain rising to 4301 m (14 110 ft) in the Rocky Mountains in Colorado, near COLORADO SPRINGS. It is one of the state's principal skiing and resort areas. Gold used to be mined nearby.

Map United States – Ec

Pilbara, The *Australia* Iron-rich area in Western Australia, centred on the HAMERSLEY RANGE 1200 km (nearly 750 miles) north of Perth. It was inhabited only by Aborigines until the first iron ore mines were opened in the 1960s. These mines now produce 97 per cent of Australia's iron ore. Marble Bar, 160 km (100 miles) south-east of the main outlet, Port Hedland, is one of the hottest places in the country. Temperatures here often soar to 49°C (120°F).

Map Australia – Bc

Pílion (Pelion) *Greece* Heavily forested mountain range rising abruptly on the east coast of THESSALY, and stretching in a peninsula into the Aegean Sea north of Euboea. Mount Pílion itself – the main peak in the range – rises to 1548 m (5078 ft). In legend, it was known as the land of Centaurs and its foothills, washed by abundant streams and blessed with mild winters, was a wild garden of herbs.

The district, ideal for camping and walking, is dotted with villages of timber-framed houses, such as Portariá, Makrinítsa, Zagorá, Miléai, Khania and Tsangarádha. There are ski slopes and, on the Aegean coastline, several holiday resorts, including Makryrakhi and Áyios Yiannis.

Map Greece – Cb

pillow lava Form of lava that has solidified underwater, consisting of a series of closely fitting pillow-shaped, or globular, masses, usually about 500-1000 mm (20-40 in) across.

Pílos (Navarino) *Greece* Harbour town on Navarino Bay, on the south-west coast of the PELOPONNESE. In AD 1827 at the naval Battle of Navarino, a combined French, English and Russian fleet defeated the Turks during the War of Greek Independence.

The remains of 58 vessels sunk during the battle can be seen under the water in calm weather. Nearby, parts of the palace of the Homeric King Nestor have been excavated.

Map Greece – Bc

Pilsen *Czech Republic* See PLZEŇ

Pinang (Penang) *Malaysia* State on the west coast, 300 km (185 miles) north-west of the capital, KUALA LUMPUR. It includes Pinang Island and the mainland port of Butterworth, covering a total of 1033 km² (399 sq miles). Pinang Island is a popular holiday destination, where tourists can choose between the tranquillity of Malay fishing villages and the bustle of Chinese-dominated GEORGE TOWN.

Population 1 142 200
Map Malaysia – Ba

Pinar del Río *Cuba* Capital of the Pinar del Río province, some 160 km (100 miles) south-west of HAVANA. It manufactures tobacco products, pharmaceuticals and furniture.

Population (city) 122 000; (province) 681 000
Map Cuba – Aa

Pinatubo, Mount *Philippines* Volcano, 1780 m (5840 ft) high, which erupted in June 1991, killing more than 500 people and damaging the nearby United States Clark Air Base, which was subsequently closed.

Map Philippines – Bc

Píndhos *Greece* Mountainous backbone of Greece, a massive range extending from Albania in the north and falling away into the Gulf of Corinth in the south. Its peaks include Smólikas (2637 m, 8651 ft), Grámmos (2503 m, 8210 ft), Gióna (2510 m, 8235 ft) and the sacred Parnassus (2457 m, 8061 ft) with its twin summits, one dedicated to Apollo, the god of light, the other to Bacchus, the god of wine.

The range, which is mainly metamorphic and volcanic with a gently rolling plateau, has a pass at the village of Metsovon near Katara. Deep ravines are clothed in forests of beech, oak, chestnut and fir. Five major rivers of Greece rise in the range: the AKHELÖÖS, the Arakhthos and the Aóös, which drain into the Ionian Sea; and the Aliákmon and Piniós, which lead to the Aegean Sea.

Map Greece – Bb

Pines, Isle of *New Caledonia* Beautiful tourist resort island, covering 152 km² (59 sq miles), situated 45 km (28 miles) off the south-eastern tip of New Caledonia. The 30-45 m (100-150 ft) tall 'pines', are actually related to the monkey-puzzle tree and are unique to the area.

Population 1470

Map Pacific Ocean – Dd

pingo Dome-like mound of earth found in Arctic regions, enclosing a core of ice, where ground water is trapped by PERMAFROST. The water freezes and expands, forcing up the mounds like earth blisters. Pingos are up to 90 m (295 ft) high and 800 m (2625 ft) across. Some have craters with circular ponds, where the ice has melted and caused the earth to collapse.

P'ing-tung *Taiwan* Industrial and commercial city in the south of the island, situated 22 km (14 miles) east of Kao-Hsiung. It lies in a region that produces rice, sugar, bananas and sisal, and has food processing industries.

Population 905 000

Map Taiwan – Bc

Pinsk Marshes *Belarus/Ukraine* See PRIPYAT' MARSHES

Piotrków Trybunalski (Petrikau, Petrokov) *Poland* Industrial city 45 km (nearly 30 miles) south of the city of LODŹ. Because of its central position in the country, it was the seat of the Crown Tribunal, Poland's highest court (from which it gets its name), between 1578 and 1792. Today textiles, glass, timber and machinery are manufactured here.

Population 77 200

Map Poland – Cc

Pipestone National Monument *USA* Prairie area covering 114 hectares (282 acres) in southwest Minnesota, 10 km (6 miles) from the South Dakota border. It has red pipestone quarries where native Americans dug the material for their ceremonial peace pipes.

Map United States – Gb

Piraeus (Pireás) *Greece* The country's largest port, standing on a promontory beside ATHENS. Its several industries include engineering, cement, chemicals, textiles, oil refining and shipbuilding.

Piraeus has been the port of Athens since 482 BC, the main harbour being a natural basin surrounded by land on three sides. There are two smaller harbours: Pasalimani, known in ancient times as Zea, where ancient boat sheds can be seen under water, and Microlimano. A museum on Pasalimani Quay traces the history of the port.

Population 170 400

Map Greece – Cc

Piran (Pirano) *Slovenia* Fortified Venetian-built port situated 32 km (20 miles) south-west of the Italian border. A beautiful town of steep, narrow streets and red-roofed houses set around a 14th-century cathedral and bell tower, it has a piazza named after the violinist and composer Giuseppe Tartini, who was born here in 1692. One of the fine palaces located on the square, built by a rich merchant from Venice for his mistress, carries the inscription *'Lassa pur dir'* ('Let them talk'). The merchant is said to have ordered the inscription as a defiant riposte to townsfolk who disapproved of his amorous visits from across the Adriatic.

Population 5500

Map Slovenia – Ab

Pisa *Italy* City on the Arno River, 70 km (44 miles) west of FLORENCE. It was once a Roman port but is now 10 km (6 miles) from the sea because of silting in the estuary.

Its renowned bell tower, 55 m (180 ft) high and leaning at an angle of 5°30′, can be reached by climbing 294 steps. It was started in 1173 and completed more than 200 years later; in recent years, work has been carried out to prevent any further settling of its foundations. The summit, used by the scientist Galileo Galilei (1564-1642) to conduct gravity experiments, has fine views over the city. Pisa's civic museum boasts a fascinating display of early Pisan sculptures and paintings. There are many Romanesque and Gothic churches and richly ornamented 16th-century palaces such as the Palazzo dei Cavalieri.

The city's industries are based on glass, engineering, clothing and pharmaceuticals.

Population 171 300

Map Italy – Cc

Pisa Lake (Lake Piso) *Liberia* Shallow saltwater lake, covering nearly 600 km² (230 miles) about 70 km (some 45 miles) to the north-west of MONROVIA, the capital. Hills covered with thick tropical forest descend to the wide sandy beach along the lake's shore, and extend along the two peninsulas that form the lake's narrow outlet to the sea. Fishing villages line the shores, but prior to the civil war that raged in the country in the early 1990s, Lake Pisa had become popular with tourists, who were attracted by its calm waters and the wildlife on the island of Massating – which includes monkeys, boars, porcupines, jackals and a wide variety of birds.

Map Liberia – Aa

Pisac *Peru* Magnificent ruins of an Inca fortress in the Urubamba Valley, 32 km (20 miles) north of CUSCO. They consist of extensive terraces and towers. The temples of the Sun and Moon face a large burial ground.

Map Peru – Bb

Pisco *Peru* Coastal town, founded in 1640, which stands in Ica Department south of LIMA and is made up of two districts, Pisco Puebla and Pisco Puerto. Pisco Puerto has Spanish colonial houses and a statue of the country's liberator General José San Martín, who landed with his troops on 7 September 1820 at nearby Paracas Bay. Pisco Puerto is the commercial port and the site of a distillery producing Pisco brandy; other industries revolve around fishmeal, cotton-seed oil and cotton textiles.

Population 82 300

Map Peru – Bb

Piso, Lake *Liberia* See PISA, LAKE

Pistoia *Italy* City situated 32 km (20 miles) northwest of FLORENCE. It has a fine range of art treasures, including works by the early sculptors of Pisa and Como and by outstanding Renaissance artists such as Andrea del Verrocchio (1435-88). The city's Piazza del Duomo is one of Italy's finest medieval squares. Pistoia's magnificent cathedral, midway in style between Pisan and Florentine Romanesque, is dominated by a large 13th-century bell tower.

The economy of the city is based on vegetables and fruit, railway industries, textiles, shoes, beds and furniture.

Population 89 900

Map Italy – Cc

Pitcairn Islands *Pacific* Ocean Group of four small islands in the South Pacific about halfway between New Zealand and Panama, comprising Pitcairn itself – covering 4.5 km² (1.75 sq miles) – and the uninhabited islands of Henderson, Ducie and Oeno which together cover an area of 38 km² (15 sq miles). It is administered by the British High Commission in New Zealand and Adamstown is its settlement.

Pitcairn is renowned for its association with Fletcher Christian and eight other mutineers from HMS *Bounty*, who settled here with a group of Tahitian women in 1790, and whose descendants still live here. Nothing was known of their existence until 1808. By 1856 there were too many people (194) on the island, so some moved to NORFOLK ISLAND, about 5900 km (3664 miles) to the west; 43 subsequently returned.

Population 60

Map Pacific Ocean – Gd

pitch (plunge) Angle made by the axis of a fold in geological strata with the horizontal.

Pitch Lake *Trinidad and Tobago* Natural deposit of BITUMEN (pitch), extending over more than 46 hectares (114 acres) in the south-west of Trinidad, near SAN FERNANDO. The lake is hard around the edges and viscous towards the centre, and is thought to be formed by the seepage of crude oil, which on evaporation has left a residue of bitumen. The supply is replenished from underground, and has yielded millions of tonnes since its discovery in the 16th century; it is the world's principal source of natural pitch. Fossils of prehistoric animals have been found fully preserved in the pitch.

Map Trinidad and Tobago – Aa

Piteşti *Romania* Industrial city on the Argeş River, some 105 km (65 miles) north-west of BUCHAREST. Oil was discovered here in the early 1960s, and since then the former market town has expanded rapidly. It is also the country's

▲ INHOSPITABLE PACIFIC COAST
A long boat approaches the precipitous, volcanic island of Pitcairn, near Adamstown, the island's sole village.

main car manufacturing centre. Other products include electric motors, chemicals and shoes.
Population 162 800
Map Romania – Ab

Pitons *St Lucia* Twin conical mountains, Gros Piton (798 m, 2619 ft) and Petit Piton (750 m, 2460 ft), rising sheer from the sea near the village of Soufrière on the south-west coast. They are the laval plugs, now exposed by erosion, of extinct volcanoes which erupted about 15 000 years ago; and they are one of the most dramatic sights in the Caribbean. Soufrière is named after another volcano situated nearby, which is surrounded by numerous hot sulphur springs.
Map Caribbean – Cc

Pittsburgh *USA* Major industrial and financial centre in western Pennsylvania, some 175 km (110 miles) from Lake ERIE. It was founded by the French in 1754, but they were soon ousted by the British. It lies in a coal-mining area.
By the 19th century it had developed into a flourishing iron-making town, and it became known as Steel City under the influence of the industrialist and philanthropist Andrew Carnegie (1835-1919). It is still a steel centre, but since the 1960s it has switched increasingly to the finance and service sectors.
Population (city) 369 900; (metropolitan area) 2 394 800
Map United States – Kb

Piura *Peru* City and department on the border with Ecuador. Rice and high grade cotton are grown under irrigation on the department's broad coastal plain. Founded in 1532 by the Spanish conquistadores, the city is the birthplace of Miguel Grau, hero of the War of the Pacific with

Chile (1879-84). To the north lie the oil fields of Lobitos and Negritos.
Population (city) 324 500; (department) 1 494 300
Map Peru – Aa

placer deposit Surface material deposited by a river, ice, the wind or waves, and containing heavy, resistant minerals that have accumulated by weathering. The minerals may include gold, platinum, cassiterite (tin ore) and diamonds.

plain An extensive low-lying area of land of gentle variations in height.

Planina *Slovenia* See POSTOJNA

Plassey *India* Historic village in West Bengal, 150 km (93 miles) north of CALCUTTA. In June 1757, 3000 British troops commanded by Robert Clive (later Baron Clive of Plassey) fought an epic battle here against 50 000 Indians. Clive's victory transformed the British from merely a trading company into the ruling force in India.
Map India – Dc

Plate, River *South America* See LA PLATA

plate margin The edge of one of the rigid, adjoining plates that make up the earth's LITHOSPHERE. See PLATE TECTONICS, p 542

plate tectonics The theory that the earth's crust is composed of a series of rigid plates and that movement of these plates in relation to one another is responsible for the major structural features of the earth's surface.

plateau Elevated, and comparatively level, expanse of land.

Platte *USA* River in Nebraska, formed by the confluence of the North and South Platte rivers which rise in Wyoming and Colorado. The Platte

flows some 500 km (310 miles) eastwards into the Missouri. The OREGON Trail, used in the 19th century by settlers on their way westwards, followed the Platte and North Platte valleys.
Map United States – Gb

Plenty, Bay of *New Zealand* Wide inlet, some 260 km (160 miles) long, on the north-east coast of North Island, 200 km (125 miles) south-east of AUCKLAND. The region around the bay produces citrus fruits, dairy products and sheep. Timber is its main industry and is exported from the port of Mount Maunganui near TAURANGA. There is a large wood processing factory on the shores of the bay at Whakatane.
The bay was named by the explorer Captain James Cook in 1769. It was here that friendly Maoris supplied him with food and water.
Map New Zealand – Fb

Pleven *Bulgaria* Town situated some 137 km (85 miles) north-east of SOFIA. A major market, and industrial and service centre for the fertile Danubian Platform, it currently produces textiles, machinery, cement and ceramics. In 1877, Pleven fell to Russian troops after a 143-day siege during a Russo-Romanian invasion which eventually forced the Turks to recognise Bulgarian independence. The town has a mausoleum dedicated to those who died in the battle, and a museum depicting the siege.
Population 138 000
Map Bulgaria – Bb

Pliocene Fifth, and final, geological epoch of the Tertiary period in the Cenozoic era of the earth's timescale. See GEOLOGICAL TIMESCALE, p 256

Pliska *Bulgaria* See VARNA

Plitvice *Croatia* Thickly forested national park in Croatia, 20 km (12 miles) north-west of Bihać. It has a string of 16 lakes that are linked by cascades and waterfalls, one of them 76 m (249 ft) high. The area is noted for its bears and wolves.
Map Croatia – Bb

Plock (Plozk) *Poland* City situated about 95 km (60 miles) north-west of the capital, WARSAW. It was founded in the 10th century and has a 12th-century cathedral. Today the city's postwar factories process food and manufacture farm implements; the city also has an oil refining and petrochemicals complex. Each June, Plock hosts the National Festival of Folklore and Folk Art, and a fair where traditional wooden carvings, ceramics and embroideries are sold.
Population 122 000
Map Poland – Cb

Ploieşti *Romania* Major oil city in the Carpathian foothills, 60 km (35 miles) north of BUCHAREST. Oil production and refining began here in 1857.
The industry expanded rapidly this century and during the Second World War was so crucial for Germany's war machine that the Allies subjected the city to heavy air attacks. Intensive exploitation which continued after the war severely depleted the oil fields. But the installations were rebuilt and, today, refineries and petrochemical works light up the sky at night, although the fields yield only one-third of locally produced oil.
Population 248 700
Map Romania – Bb

PLATE TECTONICS

Most geophysicists now agree that the earth's outer crust, the lithosphere, is made up of at least 15 separate but adjacent plates which constantly move in relation to one another. Where plate margins adjoin, massive forces are unleashed which manifest themselves in three different ways, the products of which are ocean ridges, ocean trenches and transform faults. The study of this movement is known as plate tectonics.

Ocean ridges occur at the constructive plate margins where molten rock wells up from the earth's interior, as the plates move apart, and cools to make new crust. Oceanic trenches mark destructive plate margins or subduction zones, where the edges of some plates slide down into the earth's hot interior as two plates collide. Along transform faults, plates slide past each other and crust is neither created nor destroyed, as along the San Andreas Fault in California. However, most transform faults occur at the ocean ridges, cutting across the ridges at right angles to the plate margins.

All types of plate margin are characterised by active volcanoes, earthquakes or a combination of both. As a result, all the plates are continuously changing their size and shape, growing along constructive plate margins and being consumed along destructive margins.

WHEN CONTINENTS COLLIDE

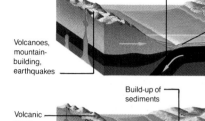

1
When continental blocks move together, ocean plate between is destroyed as it slides beneath the continental plate to form a trench. Earthquakes, volcanoes and mountain-building occur on the land.

2
Sediments are scraped off the ocean floor by the continental plate and build up thickly. Sediments on the continental margin are also compressed and raised. Oceanic plate continues to be consumed.

3
When the ocean floor has been completely consumed, the landmasses collide. Accumulated sediments are compressed, folded and uplifted to form mountain ranges. Older volcanic ranges are also compressed and raised.

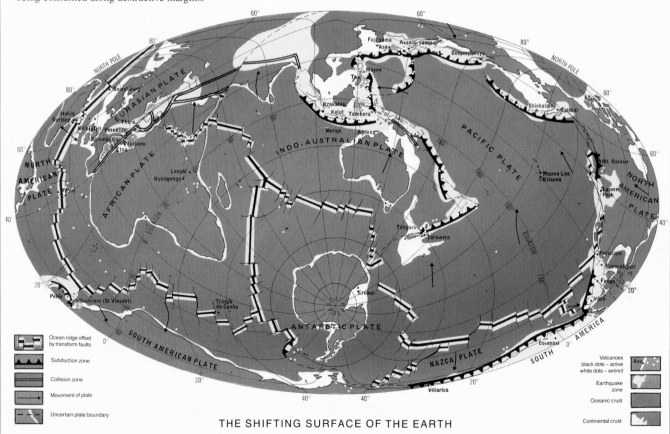

THE SHIFTING SURFACE OF THE EARTH

Ocean ridge offset by transform faults
Subduction zone
Collision zone
Movement of plate
Uncertain plate boundary

Volcanoes
black dots – active
white dots – extinct
Earthquake zone
Oceanic crust
Continental crust

The mosaic of lithospheric plates that makes up the earth's crust floats on semi-molten rock – the asthenosphere. Each plate moves at about 12.5 mm (0.5 in) each year, but what causes the plates to move is not yet known. Some scientists believe that convection currents in the asthenosphere may be responsible. It is visualised that pancake-like convection cells rise beneath the ocean ridges and spread outwards with a return current coming back only a few hundred kilometres below the upper current. Another theory suggests that gravity is the driving force, pulling the denser lithospheric plate into the less dense asthenosphere at the subduction zones. Alternatively, new material injected at the ocean ridges forces the plates apart.

Plovdiv *Bulgaria* Town standing on the Maritza River, about 140 km (87 miles) south-east of the capital, SOFIA. It is a market town in a region which produces rice, fruits, wines and tobacco. Industries include machinery, electrical goods, textiles, footwear and food processing. Buildings of note include an agricultural college, the ruins of a 13th-century fortress, a number of Roman remains, including a theatre, and many charming 19th-century houses.

Plovdiv was founded in the 1st century BC as Trimontium and became the capital of Thrace in 46 BC. It was frequently sacked during the Middle Ages, fell to Turkey in the 14th century, and became Bulgarian in 1885. It is now the country's second largest city after Sofia.

Population 379 000
Map Bulgaria – Bb

Plozk *Poland* See PLOCK

plug See NECK

pluton (plutonic rock) Large mass of IGNEOUS ROCK formed beneath the surface of the earth by the consolidation of magma.

Plymouth *Montserrat* Chief port and capital, which stands on the island's south-west coast. Georgian houses on the main street were built of stone brought from England as ship's ballast.

Population 3500
Map Caribbean – Cb

Plymouth *United Kingdom* Port and naval base some 300 km (nearly 190 miles) south-west of London and situated in the English county of Devon. It was one of the bases from which the English fleet set out to defeat the Spanish Armada in 1588, and was the final departure point for the voyage of the pilgrims to North America in 1620. The naval dockyard at neighbouring Devonport on the estuary of the River Tamar was built 1691.

The city of Plymouth was badly damaged by Second World War bombing, but parts of the old port survived, including the statue of the naval hero Sir Francis Drake (1540-96), whose home was at nearby Buckland.

Population 255 000
Map United Kingdom – Ce

Plymouth *USA* Historic seaport situated in Massachusetts about 55 km (35 miles) south-east of BOSTON. It was the site of the first permanent European settlement in New England – the Plymouth Colony – set up by the Pilgrims in 1620. They had sailed from Plymouth, England to escape religious persecution.

The town is now a major tourist attraction, and among its draw cards are Plymouth Rock (the supposed site of debarkation from the pilgrim's vessel, the *Mayflower*), and nearby 'Plymouth Plantation', which is a reconstruction of the Pilgrim village.

Population 45 600
Map United States – Lb

Plzen (Pilsen) *Czech Republic* Regional capital of western Bohemia and industrial city some 84 km (52 miles) south-west of PRAGUE. The old town has a 14th-century cathedral and 16th-century town hall. It expanded after 1850 with heavy industries producing Skoda armaments, locomotives, turbines and vehicles. Plzeň's older products include pilsner lager beers, which originated there in 1842.

Population 175 000
Map Czech Republic – Ab

Po *Italy* Longest river in Italy, flowing 652 km (405 miles) from the Cottian Alps to the Adriatic, where its growing delta, formed by the river as it deposits layer upon layer of sediment, has shifted 70-80 km (43-50 miles) out into the sea.

The Po drains 75 000 km² (28 950 sq miles) of terrain and is prone to flooding. The risk of flooding over the years has meant that few towns are located on its banks (notable exceptions are TURIN and Piacenza). However, its fertile plains are densely settled. The river is navigable for much of its length and has been an important trade route since Roman times.

Map Italy – Bb-Cb

Pobedy, Pik (Tomur Feng) *Kyrgyzstan/China* Mountain situated in the Tien Shan range on the Kyrgyzstan-Chinese border. It reaches a height of 7439 m (24 406 ft).

Map Kyrgyzstan – Ca

Podgorica (Titograd) *Yugoslavia (Serbia and Montenegro)* Capital of Montenegro situated some 20 km (12 miles) north of Lake Shkodër. Named Titograd in 1946 after the former Yugoslav leader Marshal Tito (1892-1980), it was renamed Podgorica after the collapse of the Yugoslav federation. From 1474 to 1879 it was a Turkish stronghold. Most of old Podgorica was destroyed during the Second World War and by an earthquake in 1979.

The modern city of Podgorica has a busy airport, a university, and factories manufacturing tobacco products and aluminium.

Population 152 000
Map Yugoslavia – Bc

Podgoritsa *Yugoslavia* See PODGORICA

Podhale *Poland* Southern region around the upper valley of the DUNAJEC River. It is now a popular holiday area. ZAKOPANE is its main resort.

Map Poland – Cd

P'ohang *South Korea* Industrial city and port on the east coast, 120 km (75 miles) north-east of PUSAN. South Korea's largest iron and steel works was built here on land reclaimed from the sea. A large fishing fleet is based here. The city is a noted producer of tinned fish. Grapes are grown in the vicinity.

Population 309 000
Map Korea – Dd

Pohjois-Karjala *Finland* See KARELIA

Pohnpei (Ponape) *Federated States of Micronesia* State consisting of one main island, Pohnpei itself, and eight outlying atolls. The national capital, Kolonia (population 5500), is situated on the north coast of Pohnpei Island, which covers about 300 km² (116 sq miles). A new federal capital is being built at Palikir. The ruins of Nan Madol, a group of stone buildings erected on constructed islands more than 600 years ago, are in the south-west.

Population 30 000
Map Pacific Ocean – Cb

point Promontory or cape projecting into the sea.

point bar Low, curving sandy or gravelly spit that forms on the inside of a river MEANDER when the mainstream channel shifts towards the outer bank of the bend.

Point Lisas *Trinidad and Tobago* Industrial complex on the west coast of Trinidad, 10 km (6 miles) north of SAN FERNANDO. Lying on former mangrove and sugar cane land, the complex makes use of the abundant supply of oil and natural gas for the production of petrochemicals, fertilisers and steel. It has a deep-water harbour, a plant for producing liquid oxygen and nitrogen, and the eastern Caribbean's largest power station.

Map Trinidad – Aa

Pointe-à-Pitre *Guadeloupe* Main commercial centre and port of Guadeloupe, standing beside a large and protected bay in the south-west of Grande Terre Island. Together with its suburb of Abymes, it forms the largest town in the French overseas department of Guadeloupe. Today its industries include sugar refineries, distilleries and tobacco processing plants.

Population (Pointe-à-Pitre) 26 100; (Abymes) 56 240
Map Caribbean – Cb

Pointe-Noire *Congo* The country's main seaport, situated on the Atlantic coast at the end of the Congo-Océan Railway from BRAZZAVILLE, 370 km (230 miles) away. Opened in 1939, the city grew rapidly and was established as the capital of the Middle Congo region of French Equatorial Africa from 1950 to 1958.

Its exports include cotton, cocoa, hardwoods, various minerals, palm products and rubber. Manufacturing is increasing and the city has a refinery to process oil from coastal oil fields. It also has an international airport. Besides its own trade, the port handles the foreign trade of Gabon, Chad and the Central African Republic. Exports of oil from offshore oil fields pass through the port of Rivière Ronge close by. The former port of Loango, where Congo's first Roman Catholic mission was founded in 1833, is situated 20 km (12 miles) north-west of Pointe-Noire.

Population 576 000
Map Congo – Ab

Poitiers *France* Capital of Vienne department, situated 95 km (nearly 60 miles) south of TOURS. In 1356, during the Hundred Years' War, Edward the Black Prince and his English longbowmen defeated the French here and captured the French king, John II. Poitiers has several medieval churches, the remains of Roman amphitheatres and a university founded in the 15th century. It lies in a wine, wheat and livestock producing area.

Population (town) 79 000; (metropolitan area) 105 000
Map France – Dc

Poitou-Charentes *France* Largely rural region, which mainly lies to the south and west of the city of POITIERS. The first French dairy-farming cooperatives were established here in the 1880s. Butter, cheese, cereals and brandy from the Cognac area are the main products of the region. Poitiers and LA ROCHELLE are the chief cities.

Population 1 595 000
Map France – Cc

543

Pojezierze Mazurskie *Poland* See MASURIA

Pokhara (Pokhra) *Nepal* Town of wood and plastered houses in the centre of the country, 140 km (87 miles) north-west of KATHMANDU. It is set in a valley bottom at the relatively low altitude, for Nepal, of about 1000 m (nearly 3300 ft). Because of its much lower altitude, the area's climate is warm enough to grow rice and citrus fruits. The ANNAPURNA massif, to the north, towers over the town. Tourists wishing to trek into the Himalayas use Pokhara as a base.

Population 50 000
Map Nepal – Aa

Pokrovskaya Sloboda *Russia* See ENGEL'S

Pola *Croatia* See PULA

Poland See p 545

Poland, Greater (Wielkopolska) *Poland* Historical heartland of the Polish state, where in the 10th century Prince Mieszko I began uniting the tribes which later called themselves Poles. The region lies west of the city of LODZ, south of the

Vistula and Noteć rivers, and north and east of the Oder River. Cereals, sugar beet and timber are produced in its flat-bottomed, fertile valleys. The Wielkopolski National Park south-west of Poznań, the regional capital, preserves 100 km² (39 sq miles) of low, forested hills and lakes, and is popular with tourists.

Map Poland – Bb

Poland, Little (Malopolska) *Poland* The region of southern and south-eastern Poland, called 'Little' to distinguish it from Greater Poland. It includes the city of CRACOW and the Carpathian Mountains, as well as the Cracow Jura Upland, the GÓRY SWIETOKRZYSKIE MOUNTAINS (Holy Cross Mountains) and the Lublin Plateau.

Map Poland – Dc

polar air mass Cold air which has originated in middle latitudes, either over an ocean, in which case it is called a 'polar maritime', or over a continental interior, in which case it is called a 'polar continental'. See AIR MASS, p 53

polar front A frontal zone in the North Pacific and North Atlantic oceans, along which polar maritime and tropical maritime air meet, in which DEPRESSIONS are formed. See AIR MASS, p 53

polar wandering Movement of the so-called GEOMAGNETIC POLES away from, towards or around the GEOGRAPHIC POLES.

Polatsk (Polotsk) *Belarus* Market and industrial town 170 km (106 miles) north-east of MINSK. Polatsk is one of the country's oldest towns; in fact, it was first mentioned in national chronicles in AD 862. Its industries include oil refining, timber processing and the production of textiles, flax, leather goods and beer.

Population 78 000
Map Belarus – Ba

polder An area of low-lying reclaimed land, especially in the Netherlands, which is protected by a DYKE.

poles See GEOGRAPHIC POLES; GEOMAGNETIC POLES

Poles'ye *Belarus/Ukraine* See PRIPYAT' MARSHES

Polonnaruwa *Sri Lanka* Ruined city in the east of the country, 170 km (105 miles) north-east of COLOMBO. It was the Sinhalese capital in the 11th and 12th centuries, when the city and the surrounding plain were irrigated by a dammed lake and vast waterway system. The lake, Minneriya Giritale, was originally built in the 3rd century; it was restored in 1903 and is still in use today. The city itself, which was once protected by three concentric walls, had many beautifully carved and frescoed palaces, temples and shrines. The walls of one temple still rise 22 m (72 ft) around a statue of a 14 m (46 ft) high Buddha.

Map Sri Lanka – Bb

▼ **RURAL NEPAL** A farmer's wife in a mountain village near Pokhara turns precious millet to dry it after the harvest in late November.

Poland

THROUGH CENTURIES OF OPPRESSION, THE POLISH SPIRIT FLOWERED UNDAUNTED AND LED THE WAY TO THE COLLAPSE OF THE EASTERN BLOC AND THE SOVIET UNION

Like a granite reef in a stormy ocean for most of its thousand-year history, Poland has been battered, overwhelmed and not infrequently submerged. Yet always it appears again in the sunlight, sturdy, unbroken and unchanged. Its beginning was in the 10th century, when the Slav peoples who lived between the Vistula and Oder rivers united under a single Christian ruler, Mieszko I.

Thereafter, Poland's borders contracted or expanded according to the whims of powerful neighbours and the efficiency of successive monarchs. Its great age began in the 14th century, in the reign of Casimir the Great, who made Poland into a major European power, and continued until the 1680s, when King John Sobieski drove the Turks from the gates of Vienna and saved Christian Europe from the Muslim onslaught.

But Poland's weakness has always been the vulnerability of its borders. Situated on the wide North European Plain, its frontiers are open to invaders on almost every side. The Mongols came from the east, and the Teutonic Knights from the west, followed by Russians and Germans down the centuries to our own time. The gentle, sandy dunes of the Baltic coast, too, were difficult to defend against Swedish invaders in the 17th century; and only in the south, where the CARPATHIAN Mountains climb up to almost 2500 m (8200 ft), is there any degree of natural protection.

In the second half of the 18th century, Poland became progressively weaker and was partitioned between Prussia, Russia and Austria. The last Polish king, Stanislaw II, abdicated in 1795. From then until 1918 Poland virtually disappeared from the map.

But throughout the years of occupation, the fires of Polish freedom still burned; flaring up occasionally into open rebellion. The most desperate of these were the risings of 1830 and 1863 against the Russians. The reprisals that followed led to the Great Emigration, which began in 1831. Leading national figures such as the composer Frédéric Chopin, the poet Adam Mickiewicz and Marie Curie – discoverer of radium – gravitated to Paris, while poorer folk left in droves to work in the coal mines of the Ruhr and northern France. Later, in ever-increasing numbers, they departed for the United State ; today, some 10 million people of Polish descent live abroad.

In 1918, at the end of the First World War, Poland once again achieved statehood, but its independence was short-lived. Two decades later, the armies of the Nazi dictator Adolf Hitler (1889-1945) crossed the frontier – an act that precipitated the Second World War.

In the years that followed, Poland suffered more perhaps than any other nation. Large areas of the country were devastated and some 6 million Polish citizens, including 3 million Jews, were exterminated. Most of the Jews died in notorious Nazi death camps such as AUSCHWITZ (Oswiecim), near Cracow. The experience of the Poles in the east was also bitter. The Soviets annexed large parts of eastern Poland in 1939, in a single stroke turning 1 million people into refugees or deporting them to Central Asia and Siberia. About 14 000 Polish officers, taken prisoner by the Russians, were executed, along with many others, in the KATYN Forest on the orders of the Russian secret police, the NKVD.

Once again, the Poles were caught between two fires, and once again they fought back with incredible bravery. The Jewish rising of the Warsaw Ghetto and the 1942 fight of the Polish Home Army against impossible odds stand among the great legends of the war. In other parts of the world, soldiers, sailors and airmen of the Free Polish Forces fought gallantly to bring about an Allied victory – in the Battle of Britain, at Tobruk, El Alamein and Cassino, and in dozens of other campaigns.

It availed them little. In 1944, while the Home Army fought to the death through the ruins and sewers of Warsaw, the Russians paused on the far bank of the Vistula. While German forces crushed the resistance, the Russians were laying the foundations of Communist Poland. The Soviet-supported Polish United Workers' Party, known simply as 'The Party', ruled Poland until 1989, when

▼ **CULTURE CAKE Warsaw's Palace of Culture and Science, erected in the 1950s, towers above snow-covered streets and parkland. The vast 'wedding cake' skyscraper contains theatres, museums and sporting facilities – an official symbol of the new Poland.**

the Polish initiative began the reaction against communism that led eventually to the collapse of the Eastern Bloc and the Soviet Union. They overthrew The Party, drafted a new constitution and declared the country a republic.

CITIES REBORN

Whatever their political persuasion, the new rulers, after the Second World War, had some fearful problems to overcome – to restore a land laid waste, and to feed and house a population that was virtually destitute. The United Nations Relief and Rehabilitation Association (UNRRA) and war reparation funds levied from the Germans provided some relief. But under Soviet pressure, and in common with all other Eastern Bloc countries, Poland refused Marshall Aid from the United States and looked to its own resources, with some help from the Soviet Union, for recovery.

Ancient cities, such as the capital Warsaw and Gdańsk (formerly Danzig), rose again, lovingly remodelled, so far as was possible, upon their former images. Some things, of course, could never be replaced; where the Warsaw Ghetto once stood, for example, there is now only a dark monument of bronze and marble. Fortunately, the Germans were unable to destroy Cracow before they were driven out by the Russians, and the city's medieval, Renaissance and Baroque buildings are unscathed.

Taken all round, it might have been thought that Poland was at last emerging into the sunlight; but once again, the storm clouds were gathering. The new rulers were intent on remodelling Poland according to the Soviet pattern. The nationalisation of agriculture and industry was coupled with grandiose investment projects and wasteful management of the economy. The cost of living soared, the political opinions of the population were ignored and the secret police became all-powerful.

Between 1939 and 1946, Poland's population shrank from 36 million to less than 24 million. Many of those lost were war casualties, but part of the decline was due to one of those

boundary changes all too familiar to the Poles. This one, agreed upon by the major Allies, was drastic, in that the country's area was reduced from 388 600 km² (150 038 sq miles) to a mere 312 683 km² (120 727 sq miles). The land was lost to the Soviet Union, although at the same time Poland's western border was pushed over to the Oder-Neisse Line in what had been called East Prussia.

About 4 million German settlers fled these territories, but several million former Polish citizens who now found themselves in the Soviet Union were not so fortunate; unless they could prove Polish birth they had to stay – or were deported, a course of action which cost another 900 000 lives. These changes, coupled with war deaths, changed Poland's multinational population structure. The land that had once held Poles, Jews, Germans, Lithuanians, Belarussians and Ukrainians was now almost exclusively Polish.

Native Poles emigrating from the territories lost to the Soviet Union moved to the cities or to the lands of the west and north regained from Germany. This, and a general drift from the countryside to the towns, brought about a further demographic shift in the Polish kaleidoscope. Although the rural population remained steady at 15 million, the population as a whole increased from 24 million in 1946 to 38.5 million in 1992.

Poland has made good use of its living space, both old and new. From 1919 to 1939 the only coast it possessed was the 140 km (86 miles) strip of the 'Polish Corridor', containing GDYNIA, the country's only major port. The 560 km (350 miles) that it gained in 1945 included not only the major ports of Gdańsk and SZCZECIN (Stettin) but numerous smaller harbours as well. Upon these outlets, Poland based shipbuilding industry, a thriving merchant marine, a deep sea fishing fleet, ferry links with Sweden and East Germany (and later Germany) and a liner service to Montreal.

North-eastern Poland has poor soils and a cool climate. Agricultural production is limited, even though nationalisation of former German estates provided ready-made large farms suitable for mechanisation and scientific management. By contrast, west-central and south-west Poland have fertile soils and a better climate; these are the nation's 'breadbasket' and 'sugar bowl', producing wheat and sugar beet. These regions also have a dense conglomeration of towns linked by railways, and considerable natural resources such as coal, Europe's largest copper deposits, salt and water power.

Central and eastern Poland, the 'old' 1918-39 areas remaining within the present boundaries, are more rural, with landscapes rather like those of the eastern shires of England. Urban development is concentrated mainly in Warsaw and LODZ.

During the 1960s and 1970s the government made particular efforts to build up industry in the predominantly agricultural regions of central and eastern Poland. These are the chief producers of the national 'staples' – rye,

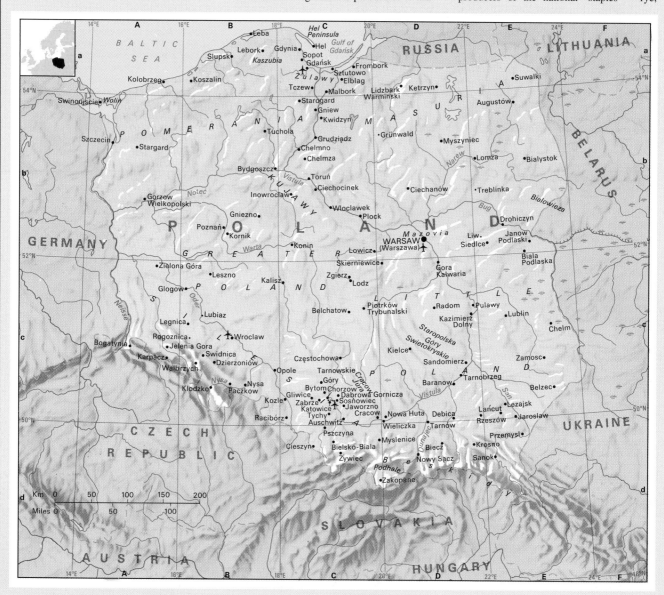

potatoes, cabbage – and of specialities such as tobacco, fruits and flowers. There are now food, textile and shoe factories, and plants to process local resources such as sulphur, gas and coal, and convert imported oil into petrochemicals. Skill-intensive industries such as pharmaceuticals, electronics and engineering were established around Warsaw and Lodz.

Poland's most valuable acquisition in 1945 was the Silesian coalfield, united within Polish borders at last, after having been variously shared over two centuries or so with Austria and Prussia. Silesia contains the richest coal reserves in Europe and the Poles made good use of them, as well as of the other mineral resources in the area. From 1945 to 1979 coal output rose from 47 million to 240 million tonnes a year and steel from 1 million to 20 million tonnes. Coal, steel, lead and zinc were exported to Hungary, Romania, East Germany, Yugoslavia and other Eastern Bloc countries, and besides, fuelled the growth of Polish shipbuilding, heavy engineering and lively automobile and chemical industries.

Altogether, Poland's industrial progress has been dazzling, although bought at a price – some of the worst environmental despoliation and air pollution in Europe, with cities such as Katowice and Gliwice recording unusually high incidences of respiratory disorders.

The country's path since 1945 has been rough and uphill. The cities, with their ballooning populations, were hardly able to cope, even after reconstruction. Improvements were slow in coming and consumer goods even slower. As the queues outside almost every shop grew longer with the years, and expenditure on the military and other government departments increased, it occurred to many of the workers that they had been striving for Soviet, not Polish, aspirations.

THE PEOPLE REBEL

These were the resentments that led to the founding of the Solidarity trade union in the Gdańsk shipyards, and to the union's subsequent and rapid spread throughout the country. It began in the late 1970s with falling standards of living. Escalating government debts to Western banks led to the steady disappearance from the shops of consumer goods – household appliances, cars, cigarettes, canned foods, and imported drinks – although many were obtainable at government-run stores for US dollars or other Western currencies.

Poland's black market in foreign currency – the most blatant in Eastern Europe – exploded and the value of the zloty dropped from 150 to the US dollar in the 1970s to 800 in the early 1980s. In 1981, food vouchers were introduced for even the most basic foods. Workers and farmers watched the purchasing power of their labour dwindle away and resentfully contrasted their lot with that of the Party elite, state officials and media people. But the issue was not simply that of wages and shortages; the workers wanted to gain some influence over their daily lives and the government of their country. Coupled with this was the government's anticlerical attitude, deeply offensive to a devout Roman Catholic

nation. Peasants, students and workers rioted in protest against communist and Soviet control in 1956, 1968, again in 1970 when the police fired upon the shipyard workers of Gdańsk, and yet again only six years later. Concessions won on these occasions usually gave way to further repressions.

In 1980 there were widespread strikes, which gave birth to Solidarity, the first free – that is, not state-controlled – trade union in postwar Eastern Europe. But the period of freedom generated by Solidarity lasted only 16 months. The new prime minister and Party boss, General Wojciech Jaruzelski, reacted strongly. He declared martial law, ordered mass arrests and banned the union. Martial law was repealed in 1983 and most of the prisoners released, but relations between workers and government remained uneasy. In 1983 Lech Walesa, leader of Solidarity, was awarded the Nobel Peace Prize as a salute to the union.

In 1989, Walesa and Solidarity won the first free parliamentary elections. A year later he was elected president. The transition from a planned economy to a market economy has been, however, a painful process.

At first Poland banked heavily on American and German support and managed to negotiate a deal with its Western creditors to have half of its US\$ 48 billion debt written off by 1994. Although Poland could not bring itself to comply with its IMF programme and stalled discussions with its creditors in the early 1990s, it proved to be the most rapidly changing Eastern Bloc country, rising most easily from the ashes of the former Soviet empire.

The Catholic Church remains an important unifying force among the Polish people. Even during the communist days, when no church was given official recognition, the Poles insisted on getting married before the altar, as well as perpetuating other Christian traditions. In 1993 the Poles passed new legislation making abortion far more difficult to obtain than in other communist (or formerly communist) countries. Whatever its political affiliations, this is one of the few countries in Europe where church services on Sundays are filled to over-flowing. Although its roots are deep in the villages, the Church's influence on towns and cities is just as great. Each year on 4 December, the miners of Silesia celebrate *Barburka*, the day of St Barbara, patron saint of miners. On special feast days, as many as half a million people converge on CZĘSTO-CHOWA, a city over-looked by the Jasna Gora (Hill of Light) monastery, a place of international pilgrimage that contains the famous Byzantine painting of the Black Madonna.

Over the years the communist authorities had tried, unsuccessfully, to limit the influence of the Church. In the 1950s, the then primate was imprisoned, as were many of his priests. Until elections in 1989, the building of churches was restricted and Mass could not be broadcast. Many Poles considered Roman Catholicism to be their own faith, while communism was seen as an alien creed. Every Pole rejoiced with fierce pride when Cardinal Karol Wojtyla of Cracow was elected as Pope in 1978, for the act seemed to confirm the

nation's oneness with the Church. It was an ancient emotion.

Down the centuries, through partition and foreign oppression, it was the Church that kept the fires of Polish nationalism aglow. Its priests went to concentration camps and death rather than retract their message of spiritual and national faith.

The last martyr in such a cause was Father Jerzy Popieluszko, a supporter of Solidarity who, in 1984, was kidnapped, beaten, strangled and thrown into a reservoir near Warsaw. Some 250 000 mourners appeared at his funeral and riots in many parts of the country forced the authorities to bring four members of the security police to trial for his murder. The lengthy sentences awarded quietened the riots, but the trial did little to convince the population that high-ranking state and party officials were not involved.

The Poles said of themselves that their communism was like the radish, red on the surface, but white within – not a bad summing-up of so proud and tenacious a nation. Perhaps this is what made it possible for the Democratic Left Alliance (a coalition between two 'post-communist' parties that were successors to the ruling parties of the communist period) to form a government after elections in September 1993. However, it was Lech Walesa who remained president, while the new left-wing government affirmed that it was committed to the development of a market-based economy and dispelled suggestions that it would revive pre-1989 policies.

During 1994, tension between the government's two component parties emerged over the rate of privatisation. There was also conflict between the president and the government over several issues – including Walesa's veto of legislation passed by parliament to liberalise the strict abortion law of 1993.

POLAND AT A GLANCE	
Official name Republic of Poland	
Area 312 683 km² (120 727 sq miles)	
Population 38 519 500 **Per km²** 123	
(Per sq mile 319)	
Capital Warsaw	
Government Republic	
Currency 1 zloty = 100 groszy	
Language Polish	
Religions Christian (95% Roman Catholic, 1% Polish Orthodox, some Lutheran)	
Climate Continental. Average temperature in Warsaw ranges from -4°C (25°F) in January to 19°C (66°F) in July	
Land use Cultivated 47%, forests 28%, pastures 13%, other 12%	
Main primary products Cereals, sugar beet, oilseed, potatoes, livestock, fish, timber; coal, sulphur, copper, zinc, lead, iron ore	
Major industries Machinery, iron and steel, mining, chemicals, shipbuilding, food processing, agriculture, petroleum refining, fishing, forestry	
Main exports Machinery, coal, foodstuffs, chemicals, non-ferrous metals, ships, vehicles, clothing, iron and steel	
Annual income per head (US\$) 1960	
Population growth (per thous/yr) 3.5	
Life expectancy (yrs) Male 68 Female 77	

Polotsk *Belarus* See POLATSK

Polyarnyy *Russia* See MURMANSK

Polynesia *Pacific Ocean* Largest of the three main regions of the Pacific Islands (the other two being Melanesia and Micronesia) which are delimited along ethnic lines. Polynesians are brown-skinned people distinguished by Asian features, and generally with wavy black hair. They include the Maoris of New Zealand.

Their ancestors probably came from parts of India and Indonesia more than 3500 years ago; they were fine sailors, and made long voyages in huge canoes guided by winds, wave patterns and stars. They settled in TONGA in about 1140 BC and SAMOA by 1000-500 BC. They later spread further to the SOCIETY ISLANDS (AD 200), the Marquesas and other parts of present-day French Polynesia, and by AD 500 had reached as far east as Easter Island and north to Hawaii. Within the next 500 years they had reached the Cook Islands and New Zealand. Other Polynesian islands include Tuvalu, the Tokelaus and Niue. Despite this huge area and long periods of separation, all Polynesian languages are closely related.

The ancient Polynesians developed a strict caste system with hereditary chiefs. They worshipped many gods and built large temples on platforms called *marae*, on which they made offerings and sometimes human sacrifices. Many sacred objects and places were *tabu*; our word 'taboo', meaning forbidden, is derived from this word.

Population (Polynesians only) 780 000
Map Pacific Ocean – Dd-Fb

Pomerania (Pomorze; Pommern) *Germany/ Poland* Former Prussian province stretching all along the Baltic Sea from STRALSUND to the western parts of Gdańsk (Danzig) Bay and covering some 38 401 km² (14 823 sq miles). The Oder River formed the border, in Prussian days, between a smaller region known as Vorpommern in the west, and a large region called Hinterpommern, in the east. Today the same boundary separates the German region (once Vorpommern) of Pommern from Poland's Pomorze (the former Hinterpommern). Pomorze lies in the north-west of Poland, on the Baltic coast, between the Oder and Vistula rivers. It is a low lakeland region popular with holidaymakers.
Map Germany – Eb; Poland – Ab

Pomorie *Bulgaria* See NESEBUR

Pomorze *Poland* See POMERANIA

Pompeii *Italy* Ancient ruined city near the foot of the volcanic Mount VESUVIUS, about 24 km (15 miles) south-east of NAPLES. Pompeii was struck by a severe earthquake in AD 63; in AD 79 it was destroyed by a cloud of choking ash and hot lava falling from the sky when Vesuvius erupted. At the time of the disaster there were about 20 000 people living here. The ash has grotesquely preserved the postures and even the facial expressions of many fleeing or sheltering citizens at the moment of death.

By the early 1990s, more than 60 per cent of this huge Roman city has been excavated, including several kilometres of sunken streets fronted by blank walls of houses. Although many of the more important individual exhibits have been removed to museums, the archaeological site is

an incredibly well-preserved record of urban life in ancient times. The forum is justly famous for its beautiful setting against the backdrop of Vesuvius. Other highlights are the amphitheatre, the large and small theatres, the Basilica, the House of Loreius Tiburtinus – a reconstruction of a Pompeian courtyard garden – the House of the Vettii with its painted walls and, outside the main city, the Villa of Mysteries, which contains the finest of all ancient paintings, including scenes of religious celebrations.

There are lesser relics, too, that bring Pompeii's tragedy down to a human scale, such as political slogans scratched on the walls and notices for public entertainment.
Map Italy – Ed

Ponape *Federated States of Micronesia* See POHNPEI

Ponce *Puerto Rico* Seaport and major city of the south coast, 75 km (47 miles) south-west of San Juan, to which it is linked by the Las Américas Expressway, opened in 1975. Its port exports sugar; the city's industries include the production of cement, textiles, footwear and paper products, as well as rum distilling and oil refining. A university was established here in 1948.
Population 187 750
Map Caribbean – Bb

Pondicherry (Puducheri) *India* Port and territory on the Bay of Bengal 169 km (105 miles) south of MADRAS. It was founded by the French East India Company in 1673, and later became the capital of French India. From 1816 to 1954, the French assumed 'permanent' control. Today it exports groundnuts and manufactures cotton textiles and toys.
Population 789 400
Map India – Ce

Ponta Delgada *Portugal* Capital and main port of the Azores, a group of islands lying 1290 km

▲ HAUNTING IMAGES On Easter Island in Polynesia, some of the fallen 'living faces', 3-12 m (10-40 ft) tall, have been re-erected. They were carved to give the islanders the protection of the ancestors they represent.

(about 800 miles) off the west coast of Portugal. Ponta Delgada is on the island of Sõ Miguel.
Population 21 200
Map (Azores) Atlantic Ocean – Dc

Pontevedra *Spain* Port near the border with northern Portugal, trading in grain and fruits, and with shipbuilding and fishing industries. A Roman bridge with 12 arches spans the Lerez River on which the town stands. The province beside the town is also called Pontevedra.
Population (town) 71 490; (province) 896 850
Map Spain – Aa

Pontine Marshes *Italy* See LAZIO

Poole *United Kingdom* See BOURNEMOUTH

Poona *India* See PUNE

Popayán *Colombia* Colonial town and capital of Cauca Department, situated 1536 m (5039 ft) above sea level, about 375 km (230 miles) south of CALI. It was founded in 1537 by Sebastián Belalcázar, a Spanish conquistador seeking Eldorado. Colonists set up sugar plantations in the damp valley below and made their homes in the cool, tree-lined tranquillity of the town.

It became a religious town and centre of learning noted for its splendid, Spanish-style churches and monasteries. Pilgrims come from all over Latin America to participate in the annual Good Friday processions. In 1983, an earthquake shattered the town, leaving only one church intact.
Population (town) 203 770; (department) 933 640
Map Colombia – Bb

Popocatépetl *Mexico* Volcano, 65 km (40 miles) south-east of Mexico City, joined by a ridge to its twin, Ixtaccíhuatl, which reaches 5286 m, (17 343 ft). The higher of the two peaks, 5452 m (17 887 ft), Popocatépetl is easier to climb, and has three marked trails. It contains a crater 1 km (0.5 mile) in circumference and 75 m (250 ft) deep. Its name is the Amerindian word for 'smoking mountain' and, although dormant since 1928, it occasionally emits vast clouds of smoke

According to legend, Popocatépetl was an Aztec warrior who was falsely reported to have been killed in battle. His lover, Ixtaccíhuatl, died of a broken heart after she heard the news. When Popocatépetl returned to find her dead, he watched over her until he too died.

Map Mexico – Cc

Porbandar *India* Town on the Arabian Sea 960 km (596 miles) north-west of BOMBAY. It was the birthplace of the nationalist leader Mahatma Mohandas Gandhi (1869-1948). Textiles and cement are manufactured in the town.

Population 160 000
Map India – Ac

Pordenone *Italy* Town situated 65 km (40 miles) north-east of VENICE. The older part of the town contains a wealth of Gothic, Renaissance and Baroque buildings including the Palazzo Comunale (1291-1365) and the late Gothic cathedral. Pordenone is Italy's leading producer of washing machines and refrigerators.

Population 49 750
Map Italy – Db

pores See POROUS ROCK

Pori (Björneborg) *Finland* Port in the south-west. It is situated on the Gulf of Bothnia 250 km (145 miles) north-west of the capital, HELSINKI. It has metallurgical, softwood and food-processing industries, with an outport at Mäntyluoto.

Population 76 300
Map Finland – Bc

porous rock Rock with pores – cavities between its mineral grains – which will soak up water or other liquids like a sponge. Some porous rocks, such as sandstone, allow the water to pass through them; they are called 'PERMEABLE ROCKS'. Others, such as clay, are porous, but not permeable.

porphyry IGNEOUS ROCK in which large crystals, known as phenocrysts, are embedded in smaller crystals of a different composition.

Port Antonio *Jamaica* Seaport and one of the earliest tourist resorts in Jamaica, lying on the north-east coast, 40 km (25 miles) north-east of KINGSTON. Its first hotel was built by the United Fruit Company at the turn of the century, and the port is still a major shipping port for bananas.

Population 12 500
Map Jamaica – Ba

Port Arthur *Australia* Former penal settlement in Tasmania, 50 km (about 30 miles) south-east of HOBART. It was built in 1830 and abandoned in 1877, after the shipping of convicts from Britain ceased in 1853. But some buildings remain, including a church which was designed and built by the convicts. The settlement receives many thousands of visitors each year; but there is no permanent population except for custodial staff.

Map Australia – Hg

Port Arthur *China* See LÜSHUN

Port Blair *India* Seaport and the capital of the ANDAMAN AND NICOBAR ISLANDS. It lies on the south-east coast of South Andaman Island. Founded in 1789, Port Blair became a penal colony in 1858. During the Second World War it was occupied by the Japanese, who used it as a naval base. Today it exports copra, timber, rubber and coconuts. It is popular with tourists for its beaches, fishing and for reef diving.

Population 74 960
Map India – Ee

Port Bouet *Ivory Coast* Deep-water port beside ABIDJAN, standing on the Ebrie Lagoon. It is open to the sea via the VRIDI CANAL. The first wharf was built at Port Bouet after a pilot canal was cut between 1905 and 1907, but the port was left abandoned when the canal silted up. Rebuilt in 1932, the wharf expanded rapidly after 1950, when the deep-water canal was opened.

Map Ivory Coast – Bb

▼ **SMOKING MOUNTAIN The smouldering fury of snow-capped Popocatépetl volcano can be seen from the national capital, Mexico City.**

Port Cros, Ile de *France* See HYÈRES, ILES D'

Port Elizabeth *South Africa* Main seaport city of the Eastern Cape, with industries that include motor vehicle assembly, tyre manufacturing and food processing. Named in 1820 in memory of the wife of a Cape Governor, Sir Rufane Donkin, Port Elizabeth still has many interesting old buildings, museums and memorials – including a large horse memorial which is probably unique. It is close to popular beaches and coastal resorts.

Population (city) 303 350; (district) 670 650
Map South Africa – Cc

Port Etienne *Mauritania* See NOUADHIBOU

Port Gentil *Gabon* Capital of Ogooué-Maritime Province, and Gabon's chief port situated on an island between two mouths of the OGOOUÉ River. It was founded to handle the export of slaves, but in the late 19th century it boomed as an export harbour for okoumé, a valuable hardwood shipped downriver from Ndjolé and LAMBARÉNÉ. Today Port Gentil exports refined petroleum products from an installation built in 1957.

Population 164 000
Map Gabon – Ab

Port Harcourt *Nigeria* Capital of Rivers State and the country's second port, after LAGOS. It stands on firm land beside the Bonny River on the east side of the Niger River delta. Work on the deep-water port was started in 1912; in 1916 it was connected by rail to the city of Enugu.

Its exports include coal from Enugu, palm oil and groundnuts. Port Harcourt is also the centre for the oil fields of the eastern delta, as well as a thriving industrial centre producing refined mineral oil, aluminium sheets, bottles, tyres and paints.

Population 336 000
Map Nigeria – Bc

Port Hedland *Australia* Port found in Western Australia, about 1300 km (nearly 810 miles) north of PERTH. It handles iron ore from mines in the Pilbara region, which is exported mainly to Japan. Salt is another major export.

Population 11 340
Map Australia – Bc

Port Jackson *Australia* Inlet on the east coast of Australia, also called Sydney Harbour. One of the world's finest anchorages, it is spanned by the elegant Sydney Harbour Bridge. The inlet is a drowned river valley with many coves and bays and is navigable by the largest vessels, although at weekends it is alive with smaller leisure sailing craft.

The harbour contains Sydney Cove – now the centre of SYDNEY – where Captain Arthur Phillip set up Australia's first convict settlement on 26 January 1788. Today Port Jackson covers 52 km² (22 sq miles) and is 2.4 km (1.5 miles) wide at the mouth.

Map Sydney, Australia – Ie

Port Kembla *Australia* Port in New South Wales, 88 km (55 miles) south of SYDNEY. Part of the Wollongong industrial area, it boasts the country's biggest steel works, as well as factories producing tin plate, copper and chemicals.

Population 5060
Map Australia – Ie

Port Láirge *Ireland* See WATERFORD

Port Louis *Mauritius* Capital, largest town and chief port of the island. It stands on the north-west coast, sheltered from the prevailing trade winds, at the head of a deep inlet which provides an excellent harbour. Founded in 1736 by the French and named after Louis XV, the town is dominated by an early 19th-century fortress, the Citadel. The island's main commercial centre, Port Louis, exports the national sugar crop and has factories processing sugar and producing a variety of goods, including clothing.

Population 158 000
Map (Mauritius) Indian Ocean – Bd

Port Moresby *Papua New Guinea* Port, university town and capital of Papua New Guinea, standing at the south-east end of the Gulf of Papua. Although it has a relatively dry climate, it is isolated from much of the country, relying mainly on sea and air transport. An Allied base during the Second World War, it was severely damaged by Japanese bombing. It exports rubber, coconut products, coffee and timber, and has food processing and light manufacturing industries.

Population (metropolitan area) 170 000
Map Papua New Guinea – Ba

Port of Spain *Trinidad and Tobago* Main seaport and national capital, situated on the north-west coast of Trinidad. After KINGSTON, Jamaica it is the largest city in the English-speaking West Indies, with a continuous line of suburbs which stretch eastwards as far as Arima, some 25 km (16 miles) to the east. It produces rum, beer, plastics, timber and textiles. Among its tourist attractions are a national museum, an art gallery, and a 27 hectare (65 acre) botanic garden dating from 1820. On the western edge of the city, on a hill 335 m (1100 ft) high, stands the restored Fort George, from which there are splendid views of the city. A new cruise ship terminal was completed at Port of Spain in 1989.

Population (city) 46 010; (metropolitan area) 200 000
Map Trinidad and Tobago – Aa

Port Phillip Bay *Australia* Large inlet of BASS STRAIT on the coast of southern VICTORIA, Australia, with MELBOURNE at its head. It was surveyed in February 1802 by a young explorer, Lieutenant John Murray. He named it Port King after the governor of New South Wales who declined the honour and named it after Captain Arthur Phillip, founder of the first Australian convict settlement at Sydney Cove. The bay is 48 km (30 miles) long and 40 km (25 miles) wide.

Map Australia – Gf

Port Pirie *Australia* Port situated in South Australia, standing on Spencer Gulf, some 200 km (124 miles) north of ADELAIDE. It is a lead-smelting centre using ore from BROKEN HILL, and exports lead and wheat.

Population 14 110
Map Australia – Fe

Port Royal *Jamaica* Small fishing village at the western end of the Palisadoes (the sand spit enclosing KINGSTON harbour). Its fortified harbour was once a haven for pirates; later, the British navy chose it as their Caribbean headquarters. In 1692, much of the town fell into the sea during an earthquake and subsequent tidal wave (tsunami). Parts of it were rebuilt. St Peter's Church, dating from 1725, stands on the site of the first cathedral in the New World, built by the Spanish in 1523. The old Naval Hospital (1819) is now an archaeological museum. There is a maritime museum at Fort Charles, which was built in 1655 and expanded to become the best protected fort in the Caribbean.

Map Jamaica – Bb

Port Said *Egypt* City and port situated at the Mediterranean entrance to the Suez Canal, 170 km (105 miles) north-east of CAIRO. It was opened as a port in 1868, the year before the Suez Canal was opened, and named after Sa'id, the viceroy of the time.

It quickly became the world's largest coaling station, and the headquarters of the canal operations and maintenance. In 1956, when Egypt chose to nationalise the canal, the city was badly damaged during unsuccessful attempts by Anglo-French forces to win back control of the canal. Port Said experienced hard times in 1967-75, while the Suez Canal was closed.

Since 1975, however, it has flourished and today it has chemical, tobacco and cotton industries, and is a free trade zone.

Population 461 000
Map Egypt – Cb

Port Stanley *South Atlantic* See STANLEY

Port Sudan *Sudan* The country's largest port, built early in the 20th century on the Red Sea coast some 640 km (398 miles) north-east of the capital, KHARTOUM and situated opposite the Saudi Arabian port of Jeddah. It replaced the ancient Arab port of Suakin, which had been abandoned because it was choked with coral. Port Sudan has modern docks and exports cotton, sesame seeds, gum arabic, hides and skins.

Population 205 000
Map Sudan – Bb

Portalegre *Portugal* Weavers' town 160 km (100 miles) north-east of the capital, Lisbon, at the foot of the Serra de Sào Mamede. The town, noted in the 16th century for its tapestries, later became important for its silk industry, which was established in the 17th century.

Population 14 800
Map Portugal Cc

Port-au-Prince *Haiti* Principal port and capital city, standing on the south-east shore of the Gulf of Gonâve. The city, founded by the French in 1749, has been destroyed by fire many times, as well as by an earthquake in 1751 and 1770.

French colonial buildings of the 19th century dominate the town's architecture, although the cathedral dates from the 18th century. Among the city's modern buildings are the university (1944) and the technical institute (1962). Haitian history is depicted in the National Pantheon Museum, and the art and culture of the nation in the Museum of Haitian Art. The city refines sugar, flour and cottonseed oil and manufactures textiles.

Port-au-Prince attracted the world's attention in 1994 when United States troops landed here in order to reinstate Haiti's president, Jean-Bertrand Aristide, who had been toppled in a coup in 1991.

Population 1 000 000
Map Caribbean – Ab

Portimão *Portugal* Fishing port and boatbuilding and fish-canning town in the Algarve, about 25 km (16 miles) east of Cabo de SÃO VICENTE, at the country's south-western tip. The neighbouring resort of Praia da Rocha, which is rich in unusual rock formations carved by the sea from the limestone cliffs, lies at the mouth of the Arade River, guarded by two medieval forts.
Population 19 600
Map Portugal – Bd

Portland *USA* Pacific port and industrial and trade centre in Oregon on the Willamette River near its junction with the Columbia, 100 km (62 miles) inland from the ocean. Portland is a timber, metal and food processing centre and a hub of foreign trade with Asia. It holds an annual Rose Festival (June) which is extremely popular, and every autumn hosts the International Livestock Exposition.
Population (city) 437 300; (metropolitan area) 1 515 500
Map United States – Ba

Porto *Portugal* See OPORTO

Porto, Golfe de *France* See CORSICA

Porto Alegre *Brazil* Port and capital of the southernmost state, Rio Grande do Sul which covers 282 184 km² (108 920 sq miles) and is bordered in the south by Uruguay and in the west by Argentina. The city lies at the confluence of five rivers which flow into the Patos Lagoon. Its main products today are food, textiles, chemicals and leather. Exports include rice, wheat, meat, hides, wool and wine.
The city has many business and financial institutions and is also an educational centre, being the seat of the Catholic University of Rio Grande do Sul. Points of interest include the Governor's Palace and the Nossa Senhora da Dores church.
Population (city) 1 254 600
Map Brazil – Ce

Porto-Novo *Benin* Port and administrative capital of Benin. It stands on Ouémé Lagoon, which is open to the sea at LAGOS in Nigeria. Porto-Novo also has an outlet to the sea via Lake Nokoué and the port of COTONOU nearby. However, this route is subject to silting. The town was made the national capital in 1894, but its influence has declined recently, mainly because its road and rail links with the interior are inferior to those of Cotonou. Because Cotonou has superceded Porto-Novo as the biggest city and business centre, many government departments and embassies have moved there.
Porto-Novo was the seat of King Joffa of the Goun people who accepted French protection in 1882. An attractive town, its narrow streets are lined with the red-earth walls of merchants' homes built in the traditional style of the Yoruba people. Many of its larger buildings date from the early 18th century, when the French made it a centre of their trade in slaves. The town's ethnic museum has various displays devoted to the history of its African kings.
Population 213 000
Map Benin – Bb

Porto Velho *Brazil* Market town and capital of the western Amazonian state of Rondônia, situated some 750 km (465 miles) south-west of the city

A DAY IN THE LIFE OF A BRAZILIAN COLONIST

This is not the first morning that Floriano has looked at his small field in despair. He stands in front of his small house, built of trees he cleared from the land, and can hardly summon the energy to start weeding.

He can remember when he was full of hope, listening to the speech of the governor of the newly created state of Rondônia, as he offered farmers free land in the western rain forest. At the time, Floriano lived in the state of Matto Grosso, where he was a share-cropper, a tenant-farmer who handed over a large part of his crops as rent to his landlord. Deciding to chance his luck, he – like tens of thousands of other peasants and their families – followed the new road into the forest.

He had been given his land grant of 50 hectares (124 acres), and had begun to clear the trees enthusiastically. But since then he faced disappointment. Whatever crops he tried to grow failed. And all around him the forest seemed to shout its success. 'Why,' he asked himself, 'if giant trees can grow here in such profusion, can't I get tiny maize plants to survive?'

Wearily, Floriano begins his task of weeding. He clambers over the scattered charred remains of tree trunks he felled and burnt as they lie crazily across his field like huge used matchsticks.

Floriano searches out each maize seedling from among the fallen timber. He was not able to plant in proper rows and there is no easy water supply for the seedlings. Floriano reflects bitterly on the need for water in the midst of a rain forest, with the nearest stream more than a kilometre away.

Today is an important day in Floriano's struggle to beat the forest. He spent most of his savings in the nearby town on some coffee plants in one last bid for success. He will plant them today and if this crop fails, he and his family will have to abandon this patch of ground and return to work on someone else's land once again.

Floriano's wife and two sons help him to make holes in the earth for the coffee plants, working vigorously with a hoe and their bare hands. They untie the sacking from the roots of the plants, and spread the roots gently across the base of each hole. The soil is replaced, leaving a little hollow to gather water. Their future rests on the success of their planting methods.

At the end of the day Floriano and his family survey the result of their labour – hardly able to make out the plants among the forest debris. One cannot just depend on nature. Perhaps a prayer will ensure that the work will not have been in vain.

But not far away another family is packing their meagre belongings and leaving their land, defeated. No one told them, just as no one told Floriano, that rain-forest soils, without the life-giving dead remains of the forest, are infertile. The coffee plants have little hope of surviving.

of MANAUS. The state covers an area of about 243 044 km² (93 814 sq miles), and its principal products today are timber, rubber, livestock, cocoa, coffee, gold and tin.
Population 286 000.
Map Brazil – Bb

Portoviejo *Ecuador* Capital of Manabí Province on the east bank of the Portoviejo River, 135 km (85 miles) north-west of GUAYAQUIL. It is an important commercial centre with vegetable oil and cotton processing plants. Irrigation schemes have been established in the river valley after a series of severe droughts.
Population 132 940
Map Ecuador – Ab

Portsmouth *United Kingdom* Port situated in the southern English county of Hampshire, situated 100 km (62 miles) from LONDON, on the Spithead Channel which separates the Isle of Wight from the mainland.
It was granted a charter in 1194 by Richard I (the Lion Heart), and established as a naval dockyard in 1540 by Henry VIII – a role it has retained ever since. It has both an Anglican and a Roman Catholic cathedrals, an old town of quaint, narrow streets and a resort area, Southsea, with 3 km (about 2 miles) of beaches.
Portsmouth's principal tourist attractions, however, are two ships: HMS *Victory*, on which Horatio, Lord Nelson died in 1805 at the Battle of Trafalgar, and Henry VIII's great ship the *Mary Rose,* which sank off Portsmouth in 1545. It was raised in 1982, to be restored and put on display. The novelist Charles Dickens (1812-70) was born in the city.
Population 184 100
Map United Kingdom – Ee

Port Vila *Vanuatu* Capital and main port of Vanuatu, on the island of Efate in the centre of the group and also known as Vila.
It has a good picturesque natural harbour which is provided with a natural breakwater, Ifira Island, lying just outside the harbour entrance. Port Vila used to export manganese ore – mined at Forari in the east of Efate Island – to Japan; but mining stopped in 1978 because of a fall in manganese world prices.
Population 19 400
Map (Vanuatu) Pacific Ocean – Dc

Portugal See p 552

Porvoo (Borgå) *Finland* Picturesque cathedral town founded in 1346 on the south coast, 45 km (28 miles) east of HELSINKI. The Russian tsar, Alexander I, formally became Grand Duke of Finland here in 1809 when Sweden ceded Finland to Russia.
Population 20 600
Map Finland – Cc

Posen *Poland* See POZNAŃ

Positano *Italy* Picture-postcard resort situated 30 km (19 miles) south-east of NAPLES on the Gulf of Salerno. It was once an ancient fishing village, and its spectacular position among mountains rising to 1500 m (5000 ft) has proved an irresistible tourist lure.
Population 3600
Map (Naples) Italy – Ed

Portugal

A LAND FAVOURED BY TOURISTS HAS KNOWN GLORIOUS DAYS OF EXPLORATION AND EMPIRE, AND ANCIENT LIFESTYLES STILL PERSIST

The word Portugal usually brings to mind sunny, sandy beaches and luxury hotels, or the famous port wine, which has been produced for 400 years in the DOURO Valley. In many ways the country, its people and its landscapes remain largely unknown.

Today, Portugal is one of the poorest countries in Western Europe, but it had the first great European overseas empire. Prince Henry the Navigator (1394-1460) inspired the Portuguese discoveries in West Africa and beyond in the 15th century. Vasco da Gama sailed round Africa to reach India in 1498 and the Portuguese were the first Europeans to reach China and Japan by sea. By 1550 they had claimed Brazil.

Today, about 200 million people around the world speak Portuguese – a figure that includes seven Portuguese-language countries and the many emigrants living abroad.

Portugal has modern cities and ancient lifestyles. The hustle and bustle of the capital, LISBON, is a world removed from the oxcarts creaking their way up hillside tracks in the MINHO region. In the north-east are the gorse-clad, steep-sided valleys and mountains of Trás-os-Montes, cold and damp in winter, and still the haunt of wolves and wild boars.

The coastal plain of the north-west, with its woodlands of pine and eucalyptus, has a patch-work landscape of small fields of maize and beans, surrounded by arbours of vines. The grapes are used to make the sparkling white *vinho verde* wines.

The Douro River is the artery of port wine – in fact, the main artery of the north, the route down which countless barrels have been shipped. It runs from the steep slopes of the upper valley between the Spanish border and Peso da Régua, to the lodges at Vila Nova de Gaia, opposite OPORTO.

Moving south, across the MONDEGO River, are the high granite peaks and glacial valleys of the SERRA DA ESTRÊLA, Portugal's winter sports area, which peaks at 1991 m (6532 ft). Farther south are lands increasingly hotter and drier in summer: the vast expanses of the ALENTEJO, with its wheat fields and cork and olive plantations. High-walled towns such as EVORA, ESTREMOZ, ELVAS and Marvão stand on the skyline like guardians watching over the toil of generations of peasant farmers.

In the west, at the mouth of the TAGUS River, lies Lisbon, built on seven hills, surrounded by wealthy suburbs and the tourist haunts of the Serra da Arrabida and the Serra de Sintra.

The Alentejo continues to the hinterland of the ALGARVE, with its beautiful groves of almond, fig and olive trees. Here, the climate is warm and equable even in winter. Along the southern coast of the Algarve are the old fishing harbours of PORTIMÃO, FARO and Tavira – centres of an extensive tourist region along the sandy beaches washed by the Atlantic Ocean. The fishing fleets still operate, landing catches of sardines and tuna.

POLITICAL INSTABILITY

A republic since 1910, the country was ruled from 1932 to 1968 by a single prime minister, António de Oliveira Salazar. A right-wing dictator, Salazar, though at first sympathetic to Germany, kept Portugal stable and out of the Second World War. During the war, spies from both sides haunted Lisbon.

From the mid-1950s the country became more and more divided politically: the north remained strongly religious and conservative, but in the south, a large landless peasantry began to express its discontent.

Salazar was incapacitated by a stroke and succeeded by Marcello Caetano in 1968, a time when the costs of colonial wars against nationalist movements in Angola, Portuguese Guinea (now Guinea-Bissau) and Mozambique had begun to place a heavy burden on the country, both in manpower and finance. In 1974 the Armed Forces Movement staged a bloodless coup, and a national hero, the monocled General Antonio de Spinola, briefly became president. The African colonies were freed but internal political confusion followed.

Many short-lived governments (mainly left-wing and socialist alliances) followed in quick succession, but they all foundered due to economic problems. After Salazar, the 1974 coup continued to hinder foreign investment. Subsequently Portugal's low labour costs attracted some investment, but this fell away as real wages dropped by 27 per cent between 1983-5, giving rise to strikes and economic turmoil. From 1980 some privatisation was allowed, which improved the situation somewhat, and the trading benefits from membership of the European Economic Community (Portugal joined the EEC in 1986 and the European Union in 1988) began to attract a growing number of foreign investors. This trend was further encouraged by the victory of Cavaco Silva's social-democrats in the 1987 elections and again in 1991.

Portugal's industry is focused mainly in the districts of Lisbon and SETUBAL situated in the centre-west, and Oporto, AVEIRO and Braga in the north. Around Lisbon, metallurgy and engineering are the dominant industries, whereas in the north, textiles take first place. Food processing is also of some economic importance, and no one can visit coastal towns such as Matosinhos or Peniche without

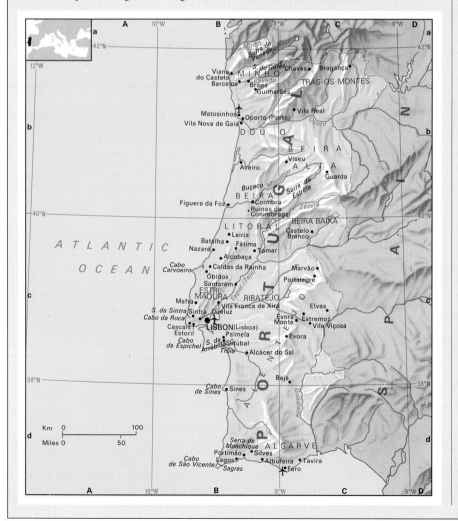

noticing the all-pervading smell of fish canning.

Most companies employ no more than 20 workers, and industry is mainly labour-intensive, but increasing investment is leading to the modernisation of some sectors, such as petrochemicals and shipbuilding.

Although 18 per cent of Portugal's work force is employed in agriculture, only few agricultural products are produced for export, such as cork, vegetables, wine and olive oil. What is more, the country has to import more than half of its food. There is a continuing, large balance of payments deficit.

Low yields are due to droughts and often to inferior soils, but also to bad structuring. Farms are small in the north – often less than 2 hectares (5 acres). Here oxen are used instead of tractors to plough the land; sowing and picking are done by hand. Many of the farmers are subsistence farmers, growing maize, beans, potatoes and vegetables for their own use.

Farms are larger in the south. After 1974 most of the large southern estates were turned into farm cooperatives, but subsequent falls in grain production led to the government leasing land back to individual farmers, and most of the 600 cooperatives were broken up.

In the north, farming continues to be an old-fashioned business; most of the cattle are stall-fed, and a common sight in the evenings is to see children taking their family's two or three cows to the collective milking parlour in the village. The women, often dressed in black, may spend the morning washing clothes at the communal tubs by the stream.

After lunch, the men meet in the village café to drink coffee and watch the latest soap opera on television. An older son may be working abroad, probably in France or Germany, spending his holidays back in his home village, perhaps with his foreign car, and using

much of his income to build a new house. Foreign currency brought back by migrant workers contributes greatly to a healthier balance of payments.

The whole rhythm of town life is different. Cars blare their horns in traffic jams, people in brightly coloured clothes view the latest fashions in the shops; men in dark suits go briskly about their work. Yet in all but the largest towns the cafés and bars continue to be very much the preserve of men. Traditions are changing, but the visitor will be surprised at the continuance of old customs and ways of life.

Since 1960 tourism has played an important role in Portugal's economy. Almost 10 million tourists visit the country each year, bringing in millions in foreign exchange. A large percentage of the tourism is based on the coastal resorts, but Portugal also has striking inland landscapes and a rich cultural heritage.

Much of the pleasure in visiting Portugal lies in exploring underdeveloped parts of the country and staying in state hotels called *pousadas*. These are often buildings of great historical interest, such as the castles at Obidos and Palmela. There are many old castles, such as those at Elvas, Evoramonte and Monsaraz. Beautiful countryside can be found in the national park of Peneda-Gerês and the Serra da Estrêla National Park.

Splendid churches and cathedrals are a common feature of the towns and cities. The most eye-catching architectural style is the early 16th-century Manueline, in which dramatic use is made of curved and twisted columns, ribs and corbels. One of the best examples is the famous window at the Convent of Christ at Tomar. Other spectacular religious architecture can be found in the fine Gothic cathedral at Évora and the 15th-century monastery at Batalha.

▲ VALLEY OF WINE Port wine grapes grow on terraces above the Douro Valley in north-east Portugal. The wine is shipped down the Douro to lodges where it matures for up to 20 years.

Portugal's 860 km (530 miles) of coastline and its rivers and streams offer some excellent fishing and there are many fine golf courses, for example, at ESTORIL and Penina.

PORTUGAL AT A GLANCE	
Official Name Portuguese Republic	
Area 92 082 km² (35 553 sq miles), including the Azores and Madeira	
Population 9 858 000 **Per km²** 107 (**Per sq mile** 277)	
Capital Lisbon	
Government Parliamentary republic	
Currency escudo	
Language Portuguese	
Religion Christian (94% Roman Catholic)	
Climate Hot dry summers, mild moist winters; rainfalls decrease from north to south. Average temperature in Lisbon ranges from 10.8°C (51°F) in January to 22.5°C (72.5°F) in July	
Land use Cultivation 34%, forest 32%, grazing 9%, other 25%	
Main primary products Cereals, olives, timber, rice, grapes, citrus fruits and vegetables, cork, fish; copper, iron, tungsten	
Major industries Agriculture, textiles, machinery, food processing, wood products, chemicals, wine, fishing, tourism	
Main exports Textiles, footwear, machinery, chemicals, cork, wine, canned sardines, wolfram	
Annual income per head (US$) 8550	
Population growth (per thous/yr) 8	
Life expectancy (yrs) Male 71 **Female** 78	

Postojna (Postumia; Adelsberg) *Slovenia* Attractive Slovenian town in the Julian Alps, 40 km (25 miles) south-west of LJUBLJANA. Nearby, the KRAS limestone is pitted by 18 km (11 miles) of spectacular, stalactite-hung caves, halls and tunnels carved by the Pivka River. Graffiti on the walls show that the caves were known as early as 1213.

Today, an electric railway, 2 km (about 1 mile) long, takes visitors to the start of a conducted underground tour through grottoes known as 'the Ballroom' and 'Paradise'. The highlight of the caverns is an enormous vaulted chamber called 'the Concert Hall', in which the Italian conductor Arturo Toscanini (1867-1957) once held a full orchestral concert. The chamber, which covers about 2800 m² (3350 sq yds), can hold 1000 visitors at a time. Planina is another caving centre nearby.

Population 5000
Map Slovenia – Bb

Potchefstroom *South Africa* University town situated in North-West Province, about 105 km (65 miles) to the south-west of JOHANNESBURG. Founded by Voortrekkers in 1838, it was the first capital of the South African Republic (Transvaal) and remained such until 1854.

Today Potchefstroom is the commercial centre of the fertile Mooi River valley which produces maize, vegetables and fruit as well as livestock.

Population (town) 46 530; (district) 185 550
Map South Africa – Cb

Potenza *Italy* Market town 130 km (80 miles) east of NAPLES, some 820 m (2690 ft) up in the Lucanian Apennines. It is the capital of the impoverished Basilicata Region, but which today has several new state-aided industries in food, clothing and construction.

Population 65 600
Map Italy – Ed

pothole Hole in the rocky bed of a stream, formed by the grinding effect of pebbles as they are whirled around by the eddies. The term is also a popular name for a SINKHOLE.

Potosí *Bolivia* City and department bordering Chile and Argentina. It is best known for the Cerro Ricode Potosí, the mountain reaching 4829 m (15 843 ft), in which vast deposits of silver were discovered by the Spanish.

The city, founded in 1546 by the Spaniards Juan de Villarroel and Diego Centeno, lies south-west of Sucre, at the foot of the hill at an altitude of 3978 m (13 050 ft). By 1650 Potosí had become the largest city in the Americas, with 160 000 inhabitants. It still retains much of its colonial heritage, with some splendid 17th and 18th-century Baroque churches and public buildings such as the mint (the Casa Real de Moneda), rebuilt in 1759.

Although the silver deposits have long been exhausted, the mines still produce tin, zinc, copper and lead.

Population (city) 120 000; (department) 841 200
Map (city) Bolivia – Bb; (department) Bolivia – Bc

Potsdam *Germany* City on the Havel River, 25 km (16 miles) south-west of BERLIN. It has many fine palaces – notably the Cecilienhof Schloss and Sanssouci – and was the summer retreat of the Prussian king Frederick the Great

A DAY IN THE LIFE OF A BOLIVIAN PEASANT

More than 60 per cent of Bolivia's 7.5 million people are Amerindians. Among these, the largest single group is the Quechua who live high up in the mountainous Altiplano area of south-eastern Bolivia. These were the original inhabitants of Bolivia who were forced up to the plateau by the arrival of the Spaniards.

Mariano Guallpa, his wife Augustina and their five children are Quechuas. They live together in a one-room house made of sun-baked earth near the town of Potosí, almost 4000 m (more than 13 000 ft) above sea level. Although Potosi is an active tin-mining centre, this has little impact on the daily life of Mariano, who ekes out a living as a subsistence farmer on the poor, thin soil where vegetation is sparse and there is little water for irrigation.

Mariano owns a few sheep and cattle, but he still follows the Quechua tradition of communal land ownership and farms the surrounding area with the other people of his village. Potatoes, maize and quinou, a kind of millet, are the main crops, and form the staple diet.

Mariano's 13-year-old son, Tito, usually helps out in the fields, where the work is hard because of the lack of oxygen at such an altitude and because of the absence of any but the simplest agricultural tools.

Augustina often lends a hand, although much of her time is spent at home looking after the younger children and cooking or weaving. She weaves sheep, llama and alpaca wool, on an old loom handed down from her mother, into brilliant striped cloth. Her dyes, by-products of Bolivia's mining industry, are bought from merchants or at the market in Potosi. They can use some of Augustina's cloth to trade with neighbouring tribes, or with fellow merchants. Some is made up into the cloaks or ponchos that Mariano wears over a thick woollen shirt and trousers – together with a domed black hat, embroidered in bright colours, which comes down over his ears like the helmets once worn by the Spanish conquistadores.

Like others of his tribe, Mariano speaks Spanish, the official language of Bolivia, as well as the native Quechua language, and he has carefully educated his children to be similarly bilingual. However, there are no formal schools for the Guallpa children to attend, and like Mariano, they remain unable to read or write.

Although he is nominally a Roman Catholic, in practice Mariano's faith owes much to ancient myths and to him the mountains, lakes and caves are the homes of spirits which can harm or help him in his everyday life.

In the evenings he turns to his most prized possession – a small transistor radio. This is his only contact with the outside world and it brings the news of the distant capital, La Paz, cracklingly to life in the remote windy uplands.

(1712-86). Allied leaders came together here in July and August 1945 to discuss the administration of a defeated Germany in the aftermath of the Second World War.

Today it is an important centre of education with academies of political science, law, finance, film, agriculture, and medicine. Its factories produce locomotives and textiles and process food.

Population 139 800
Map Germany – Eb

Potteries, The *United Kingdom* Industrial area in the English Midland county of Staffordshire. It is centred on the towns of STOKE-ON-TRENT, Tunstall, Burslem, Hanley and Longton, and was the setting for the 'Five Towns' portrayed in a series of novels by Arnold Bennett (1867-1931).

The area has produced pottery since Roman times. The modern industry, based on local coal (used to fire the kilns) and canal transport, dates from the mid-18th century; its porcelain manufacturers include the famous family names of Wedgwood, Minton, Royal Worcester, Spode and Doulton. Newcastle-under-Lyme, to the west, produces bricks and tiles.

Map United Kingdom – Dd

Poubara Falls *Gabon* Falls on the upper Ogooué River, harnessed since 1975 to produce the electricity for the fast-developing mining and industrial region near MASUKU.

Map (Masuku) Gabon – Bb

Poverty Bay *New Zealand* Bay on the east coast of NORTH ISLAND, 320 km (nearly 200 miles) south-east of Auckland, where the British explorer Captain James Cook made his first landing in New Zealand in October 1769. After a skirmish with Maoris, Cook named it Poverty Bay 'because it afforded no one thing that we wanted'. The seaside resort of GISBORNE lies at the head of the bay, which is 10 km (6.2 miles) long and 6 km (3.7 miles) wide

Map New Zealand – Fc

Powys *United Kingdom* Sparsely populated county of central Wales, which was created from the old counties of Brecon (Brecknock), Radnor and Montgomery, and named after an ancient Celtic principality.

It covers some 5077 km² (1960 sq miles) and stretches from the English border almost to Cardigan Bay. It is mostly hill country, with the Brecon Beacons and Black Mountains rising to 886 m (2907 ft) and 811 m (2660 ft) respectively in the south, and the Plynlimon ridge – source of the Severn and Wye rivers – to 752 m (2467 ft) in the west.

Today the Brecon Beacons are a major tourist attraction and much of the area is used for sheep farming. The county town is Llandrindod Wells.

Population 118 600
Map United Kingdom – Dd

Poyang Hu *China* One of the country's largest freshwater lakes, joined in the north to the Chang Jiang (Yangtze) River. Poyang Hu once covered more than 5000 km² (1900 sq miles) but is now only some 3600 km² (1400 sq miles) because of the accumulation of silt and land reclamation. It is impossible to measure its area precisely as the difference between flood- and low-water levels can be as much as 7.5 m (25 ft).

Map China – He

Poza Rica *Mexico* Oil-producing boom town in VERACRUZ State. Oil was first found here in 1930. Today most of the oil produced is taken elsewhere for refining as Poza Rica has only one small refinery.
Population 151 000
Map (Veracruz State) Mexico – Cb

Poznań (Posen) *Poland* Historic city 269 km (167 miles) west of the capital, WARSAW. It was founded in the 10th century on an island in the Warta River, and became Poland's first bishopric and the focus of Greater Poland, the heartland of the Polish state. The city, halfway between Gdańsk and Prague, and Berlin and Warsaw, thrived as a market and crafts centre.

After 1922, the medieval St John's Fair – held annually on 24 June – became the International Trade Fair for crafts and is now for industrial products. The city's products include textiles, foodstuffs, agricultural equipment, marine and railway engines, machine tools, electrical goods, rubber and paper.

One of Europe's best preserved prehistoric swamp towns, Biskupin, was discovered in 1933, 84 km (52 miles) north-east of Poznań; today it is a museum. The settlement, situated on a peninsula in Lake Biskupin, covers 40 hectares (100 acres) and dates from about 550 BC. It was built by the Lusatians, the forerunners of Slav civilisation. An earth-and-timber rampart, incorporating more than 35 000 stakes, encloses 12 timber-paved streets lined with 105 wooden huts and workshops.
Population 590 100
Map Poland – Bb

Pozzuoli *Italy* Coastal town situated 12 km (8 miles) west of NAPLES. It has extensive Roman remains, including the amphitheatre of Flavius, one of the largest Roman theatres in Italy. Pozzuoli's economy is based upon fishing and iron and steel production.
Population 75 710
Map Italy – Ed

Prague (Praha) *Czech Republic* The country's capital and largest city, which is also capital of Středočeský region in central Bohemia, and covers an area of some 11 000 km² (4247 miles). Prague straddles the Vltava River in the heart of Bohemia and, like Rome, covers seven hills. It was founded in the 9th century, when the Premysl family of princes established themselves at Vysehrad, a high rock beside the river in the south of the modern city.

Wenceslas, or Vaclav (about 907-29), an early Christian member of the royal family, who was murdered by his heathen brother, later became the patron saint of the country. The 19th-century English carol *Good King Wenceslas* is loosely based on legends about the saint. The massive Palace of Culture (built in 1981) stands near Vysehrad. Across the river is Bertranka, the house (now a museum) where the Austrian composer Wolfgang Amadeus Mozart (1756-91) stayed during his many visits to Prague.

The old town, on the east bank of the river, preserves the Gothic Tyn Church, the 14th-century town hall (with its fine astronomical clock), Charles University (founded in 1348), and the Jewish ghetto (dating from the 10th century). The Charles Bridge (1357), embellished with 18th and 19th-century statues, links the old town to the 'Lesser Town'. This maze of streets, with ornate craftsmen's signs, old taverns, and beautiful palaces, including the Wallenstein Palace, was founded in 1257.

Hradčany Hill, above the city, is crowned by both the royal castle and St Vitus Cathedral (1344-1929), the present home of the Bohemian crown jewels. In 1618, during the Thirty Years' War, an event known as 'the Defenestration of Prague' occurred in Hradcany: two deputies of the (Catholic) Holy Roman Emperor were thrown out of a window in the royal castle by angry Bohemian Protestants. Since 1918 the castle has been the official home of the country's president.

The old town and the Lesser Town were merged into one city in 1784. The new town to the south-east of the old dates mostly from the 19th century, although its ground plan was already laid out by Emperor Charles IV in 1348. It includes the elegant boulevard known as Wenceslas Square.

Chemicals, engineering, electrical, food, clothing, film and pharmaceutical industries have been established in Prague. The city's post-1948 suburbs, with massive apartment blocks, are now linked to the centre by an underground railway system, which was expanded in the 1980s.
Population 1 215 100
Map Czech Republic – Ba

Praia *Cape Verde* Port and capital of CAPE VERDE. It stands on the main island of São Tiago in the Atlantic Ocean, about 600 km (370 miles) west of the coast of Senegal, in Africa. Praia exports sugar cane, castor beans, coffee and bananas.
Population 64 000
Map Africa – Ac

prairie Flat, temperate grassland in the North American interior east of the Rocky Mountains, similar to STEPPE. Most of the prairies are now ploughed for cereal and fodder crops. The term is used for similar areas elsewhere.

Praslin *Seychelles* Second largest island of the Seychelles group, situated in the western Indian Ocean, some 43 km (27 miles) to the north-east of MAHÉ, and covering 18 km² (7 sq miles). A tourist spot today, it is renowned for its attractive sandy beaches and for the coco-de-mer palms of the Valée de Mai.
Population 5000
Map (Seychelles) Indian Ocean – Bc

Prato *Italy* Industrial town 16 km (10 miles) north-west of FLORENCE. Wool textiles are its main industry.
Population 165 360
Map Italy – Cc

Precambrian period Oldest and largest division of geological time, covering the period from the formation of the earth up to about 570 million years ago. See GEOLOGICAL TIMESCALE, p 256.

precipitation Moisture that is deposited from the atmosphere in the form of dew, frost, rain, sleet, hail or snow.

Prešov *Slovakia* Industrial town situated some 36 km (22 miles) north of KOŠICE. Its various factories produce machinery and process food and textiles. The Gothic old town has an 18th-century Greek Catholic cathedral. Today Prešov is the centre of Slovakia's Ukrainian community, which is some 45 000 strong.
Population 78 200
Map Slovakia – Cb

Prespa, Lake (Prispansko Jezero) *Albania/Greece/Macedonia* Lake covering an area of 274 km² (106 sq miles) in Macedonia, now divided between the three countries. Up to 54 m (177 ft) deep, it lies cradled in mountains at an altitude of 853 m (2799 ft). The lake drains by underground streams to Lake OHRID to the north-west. It is noted for its carp, the fine sandy beaches along its eastern shore and the pelican and cormorant colonies on the lake island of Golem Grad.
Map Macedonia – Ab

Pressburg *Slovakia* See BRATISLAVA

pressure See ATMOSPHERIC PRESSURE

pressure gradient (barometric gradient) Rate at which the atmospheric pressure changes on the earth's surface, as indicated by the spacing of the isobars on a weather chart. If the isobars are close together, there is a considerable difference of pressure between two points on the earth's surface and a steep gradient, resulting in strong winds. If the isobars are far apart, the gradient and the winds are slight.

pressure ridge Ridge of floating ice formed as two ice floes push against each other. A similar pressure ridge may also form on the surface of a glacier where the flow is impeded by the shape of the valley.

Preston *United Kingdom* County town of Lancashire, situated at the head of the Ribble estuary 45 km (28 miles) to the north-west of MANCHESTER. It received a charter in 1179, making it one of the oldest boroughs in England, and has had a member of parliament since the 13th century. In 1648, in a decisive battle of the English Civil War, in Preston, the Roundhead army of Oliver Cromwell routed 20 000 Scottish supporters of Charles I.

Today Preston is an industrial centre producing aircraft, trucks and buses, textiles, chemicals and machinery. It is the centre of a new town planned to house 285 000 people eventually. The Preston Dock, which closed in 1981, is being redeveloped as a 182 hectare (450 acre) estate for industrial, recreational and residential use.
Population 168 400
Map United Kingdom – Dd

Pretoria *South Africa* Administrative capital of South Africa since 1910. Between 1854 and 1900, it was the capital of the South African Republic and, from 1900 to 1910, of Transvaal Colony. The city lies in GAUTENG Province, its centre 70 km (44 miles) from that of Johannesburg. Although listed as the national capital, Pretoria was one of three capitals nominated after formation of the Union of South Africa (1910-1961). The others are Cape Town (now seat of the legislature) and Bloemfontein (the judicial capital and seat of the Appeal Court). A stately city, Pretoria has many public parks and streets lined with jacaranda trees. The city's industries include motor vehicles, steel and cement.
Population 667 700
Map South Africa – Cb

prevailing wind Wind that blows most frequently from a specific direction in a certain place. For example, the south-westerly winds are the prevailing winds over much of western Britain. The prevailing wind is not necessarily the dominant wind.

Prince Albert *Canada* Town on the North Saskatchewan River, 145 km (90 miles) northeast of SASKATOON. It was founded in 1866 as a Presbyterian mission to the Cree people. The town serves the surrounding grain and cattle area, and manufactures timber products.

The Prince Albert National Park, covering 3875 km² (1496 sq miles), lies 55 km (34 miles) north of the town. It contains coniferous forests (black and white spruce), parklands and prairies, grasslands and hundreds of lakes.

Population (town) 37 520
Map Canada – Ec

Prince Edward Island *Canada* Island in the Gulf of St Lawrence, and Canada's smallest province, covering 5660 km² (2185 sq miles). Separated from the mainland by the Northumberland Strait, the crescent-shaped island consists of red sandstone lowlands with a maximum height of 142 m (466 ft). Two-thirds of the area is cultivated, with the main crops being vegetables (especially potatoes) and tobacco.

Manufacturing centres mainly on food processing. Tourism is an important source of revenue; chief among the island's attractions is the Prince Edward Island National Park, 18 km² (7 sq miles) of dunes, salt marshes and fine bathing beaches. Deep-sea fishing is a popular pastime. CHARLOTTETOWN is the island's capital and its main town.

Population 129 770
Map Canada – Id

Prince George *Canada* Town in central BRITISH COLUMBIA situated at the junction of the Fraser and Nechako rivers. It is the commercial and industrial centre for a timber, mining and agricultural area. The town is an important road and rail junction.

Population 69 650
Map Canada Cc

Prince Rupert *Canada* Important port on the northern coast of BRITISH COLUMBIA. It is a tourist centre and a fishing port, and exports grain, ores and timber.

Population 16 620
Map Canada – Bc

Príncipe *São Tomé and Príncipe* Second largest island of the republic, covering some 142 km² (55 sq miles). It is situated in the Gulf of Guinea, 200 km (nearly 125 miles) west of Equatorial Guinea. Volcanic, tropical and largely forested, Príncipe rises to some 948 m (3110 ft) at Pico de Príncipe. The principal town is Santo António. Cocoa, coconuts and coffee, which are grown on plantations, are the main exports.

Population 6700
Map Gabon – Aa

Pripyat' Marshes (Palyessye; Pinsk Marshes; Poles'ye) *Belarus/Ukraine* Europe's biggest marsh and peat bog. It lies in Belarus and Ukraine along the Pripyat' River, a tributary of the Dnyapro (Dnieper). Almost uninhabited, densely forested, and frozen in winter, it was divided between the USSR and Poland from 1920 to 1939; in 1945 the Russians assumed sole control over it, but in 1991 it was returned to Belarus and Ukraine upon their independence.

Map Ukraine – Bb; Belarus – Bb

Pristina *Yugoslavia (Serbia and Montenegro)* Capital of the autonomous province of Kosovo in Serbia, 80 km (50 miles) north of SKOPJE. It was a capital of the medieval Serbian Empire and lies in the plain known as Kosovo Polje, or 'the Field of Blackbirds', where the Turks inflicted a crushing defeat on the Serbs in 1389. The town stagnated under the Turks and became a backward region. However, since 1950 it has become part of a modern administrative, educational, communications and industrial centre, with cotton spinning, metal and electrical industries.

Most of the townsfolk are ethnic Albanians and, in the mid-1990s, tension between the Muslim Albanians and the Serb residents ran high. The beautiful 14th-century monastery of Gracanica, with cupolas, a Byzantine dome and fine frescoes dating from the 14th to the 16th century, lies some 10 km (6 miles) to the south-east.

Population 210 000
Map Yugoslavia – Cc

Prizren *Yugoslavia (Serbia and Montenegro)* Market town at the mouth of a valley on the northern flank of the Sar Mountains, about 300 km (186 miles) south of BELGRADE. Many of its inhabitants are Albanian Muslims who were encouraged to settle here by the Turks.

The old Turkish town in the valley has some fine painted houses, mosques, bazaars and 16th-century Turkish baths. The town's Orthodox churches and monasteries stand on the forested slopes above; while the post-Second World War town, with food and tobacco processing, cotton-spinning and textile factories, lies on the plain. The ski resort of Brezovica is above the town.

Population 134 500
Map Yugoslavia – Cc

Prokop'yevsk *Russia* Coal-mining city in southern Siberia, about 270 km (170 miles) south-east of NOVOSIBIRSK. Despite modern machinery and the fact that Prokop'yevsk's mines have coal seams up to 40 m (130 ft) thick, some pits have been closed by the government. As a result, the population in the city has been declining.

Population 274 000
Map Russia – Jc

promontory High ridge of land or rock jutting out into a sea or other expanse of water, or into low-lying land.

Proterozoic era The more recent Precambrian era, when various primitive plant and animal life forms evolved from single-celled organisms.

Provence *France* Historical Mediterranean region covering 26 135 km² (10 091 sq miles) between the Rhône River and the Italian border. Settled by the Romans, it became a kingdom in AD 855, and part of France in 1481. The main cities today are MARSEILLES, NICE and TOULON. The modern metropolitan region of Provence-Côte d'Azur covers 31 400 km² (12 125 sq miles) and is administered from Marseilles. The historic capital was AIX-EN-PROVENCE. The Provençal language, which had a thriving literature in the Middle Ages, was revived in the 19th century by a group of writers including the poet Frédéric Mistral (1830-1914). Provence's scenery and brilliant light inspired many painters of the Impressionist school.

Population (Provence-Côte d'Azur) 4 258 000
Map France – Fe

Providence *USA* Port and state capital of Rhode Island, at the head of Narragansett Bay, 65 km (40 miles) south-west of BOSTON. Founded in 1636 as a refuge from religious persecution, it was the state's first European settlement. Brown University (founded 1764) moved to its present site in the city in 1770. Providence's traditional textile industry has been outpaced by jewellery making and silverware. Other industries include metalworking and engineering.

Population (city) 160 700; (metropolitan area) 1 134 400
Map United States – Lb

Providenciales *Turks and Caicos Islands* Most developed tourist island in the group. An international airport was opened on Providenciales in 1984 to serve a newly built holiday resort village.

Population 4960
Map Caribbean – Aa

Provincetown *USA* Tourist resort at the northern tip of Cape Cod, Massachusetts. It was the first stopping place of the Pilgrims in 1620, one month before they founded the PLYMOUTH Colony on the other side of Cape Cod Bay. It was at Provincetown that they drew up the Mayflower Compact, the first 'constitution' in the New World.

Population 3600
Map (Cape Cod) United States – Mb

Prussia *Northern Europe* Former state of north Germany whose capital was BERLIN. Its heartland was Brandenburg, but it eventually covered 295 000 km² (114 000 sq miles) from the Rhine far along the Baltic coast into present-day Poland and Lithuania, and included the industrial cities of the RUHR.

Prussia was formed as an independent duchy in the 17th century. It became the dominant power of northern Germany and led to the formation of the *Zollverein* (customs union of German states) in the 19th century. The centre of German militarism, it played a leading role in the unification of Germany at the end of the Franco-Prussian War (1870-1). Its king, Wilhelm I, became the first *Kaiser* (emperor) of Germany in 1871 and its leading statesman, Prince Otto von Bismarck, the German chancellor. Prussia lost its independence under the Nazis and was dissolved by the 1945 POTSDAM Conference.

Map (Berlin) Germany – Eb

Prut *Romania/Ukraine/Moldova* River forming Romania's eastern frontier with Moldova. It rises in south-west Ukraine and flows some 900 km (560 miles) south to join the Danube 200 km (124 miles) north-east of BUCHAREST.

Map Romania – Ba; Moldova – Aa

Przemysl (Peremyshl) *Poland* City in the southeast 10 km (6 miles) from the Ukraine border. It was founded as a fortress beside the San River, and flourished from 981 AD as a market on the trade route from Central Europe to the Black Sea.

The city is now part of a huge 'dry port' marshalling complex, where freight is switched between the Polish standard-gauge and the broad-gauge railways of the former Soviet Union.
Population 64 100
Map Poland – Ed

Pskov (Pleskau) *Russia* City 250 km (155 miles) south of ST PETERSBURG. It is one of Russia's oldest cities, being mentioned in AD 903. In 1240 it became a significant trade power after being occupied by Teutonic Knights. After being annexed by Moscow in 1510, it withstood 26 sieges from Poles, Germans and Swedes. In 1917 it was the scene of the abdication of the last tsar, Nicholas II (1868-1918).

During the Second World War, Pskov was occupied by German troops (1941-4) and was badly damaged. Its most notable buildings and monuments were later painstakingly restored. They include the kremlin (citadel), the remains of the old town and the fine, silver-domed 12th-century Trinity Cathedral.
Population 206 000
Map Russia – Dc

Puch'on *South Korea* City in the industrial region stretching between SEOUL, the capital, and the port of Inch'on.
Population 668 000
Map Korea – Cd

puddingstone See CONGLOMERATE

Puebla (Puebla de Zaragoza) *Mexico* Capital of Puebla state, 120 km (75 miles) south-east of MEXICO CITY. Surrounded by a ring of volcanoes, it was founded in 1531 by Spaniards from the Talavera region of Spain, who brought their local tile-making skills to adorn Puebla's colonial red-brick houses.

The city has 360 churches and the largest church in Mexico, Santa Domingo, built in the 16th and 17th centuries and ornamented with gold leaf. The city is also famous for its wicker-work and pottery, and for objects made of local onyx. Industries have overflowed from the national capital into Puebla, including a major car factory. The ancient Toltec site of CHOLULA is 8 km (5 miles) to the west.

Puebla State, which is mountainous and forested, is important for mining and produces textiles, sugar cane, cotton, cereals, tobacco, rice, fruits, vegetables and chillies.
Population (city) 1 055 000; (state) 4 118 000
Map Mexico – Cc

Puerto Cortés *Honduras* The country's chief port, situated in the north-west corner close to the Guatemalan border. It exports mostly bananas, and has a large oil refinery.
Population 43 300
Map Honduras – Aa

Puerto La Cruz *Venezuela* Oil port about 280 km (175 miles) east of CARACAS. Formerly a fishing hamlet, it now has two oil refineries. It is also being developed as a tourist resort.
Population 220 000
Map Venezuela – Ba

Puerto Montt *Chile* Seaport and capital of Los Lagos region, on the northern edge of the Gulf of Ancud, some 900 km (nearly 560 miles) south

Puerto Rico

THESE CARIBBEAN ISLANDS COULD CHOOSE TO BECOME THE 51st STATE OF THE UNITED STATES

The people of Puerto Rico are United States citizens who speak Spanish and live 1600 km (1000 miles) from the US mainland. Spain lost this group of Caribbean islands after the Spanish American War in 1898. Puerto Ricans were given US citizenship in 1917, but they cannot vote in American elections unless they live on the mainland. Although there is some pressure for independence, the people voted in 1952 to become a self-governing commonwealth in association with the United States.

The question that continues to dominate the political scene is whether Puerto Rico should keep its present political status or not. The alternatives would be to accept full independence, or else become the 51st state of the United States. In a non-binding referendum in 1993 a narrow majority voted for maintaining commonwealth status rather than applying for inclusion into the US.

The option of independence gained little support. However, legislation passed in April 1991 made Spanish the sole official language – as opposed to both English and Spanish as previously.

This mainly Spanish-speaking, Roman Catholic, West Indian country, with its people mostly of Spanish or African descent, has more in common culturally with the neighbouring Dominican Republic, Cuba and mainland Central America, than with the nation to which it belongs. Puerto Rico (Spanish for 'Rich Port') is the main island of the territory, which also includes the two small islands of Vieques and CULEBRA, and a fringe of uninhabited smaller islands and islets. Puerto Rico Island has a backbone of mountains, the Cordillera Central, which reach to a height of 1338 m (4390 ft) at the peak of Cerro de Punta. Agriculture has cleared most of the island's natural forest, but trees in the Sierra de Luquillo in the northeast are protected in a national park. Cooling winds temper the tropical heat.

AGRICULTURE AND INDUSTRY

Dairy farming is the most important agricultural activity. Sugar is the main crop, followed by tobacco, coffee, pineapples, bananas and coconuts; but the agricultural sector has been overtaken by industry in recent years. The change was forced by a rapid growth in

population, despite a drop in the birthrate from 1.7 per cent in the mid-1980s to 0.3 per cent a year in the mid-1990s. The need to support more people put pressure on farmland, and slopes were cleared that should have been left forested. This caused serious soil erosion which consequently reduced agricultural output. Eventually farming could not provide enough food and export income for the population, many of whom emigrated to the United States.

The US government encouraged the growth of industry in order to give Puerto Rico another source of income. Tax reliefs and cheap labour encouraged a variety of American businesses to invest there, and today manufacturing has taken over from agriculture as the main export earner. Products include textiles, clothing, electrical and electronic Sgoods, plastics and chemicals, made mainly by subsidiaries of big US companies. Despite considerable growth, unemployment has risen, peaking at 18 per cent in 1992.

Puerto Rico attracts many tourists to its mountain scenery, excellent beaches, and game fishing. There are many relics of Spanish colonial days, including 16th and 17th-century houses in the old town of SAN JUAN, the capital and chief port, and two great 16th-century clifftop forts which dominate the harbour of San Juan Bay.

PUERTO RICO AT A GLANCE	
Map Caribbean – Bb	
Official name Commonwealth of Puerto Rico	
Area 8897 km² (3435 sq miles)	
Population 3 600 000 **Per km²** 405 **(Per sq mile** 1048)	
Capital San Juan	
Government Self-governing commonwealth in association with the United States	
Currency 1 US dollar = 100 cents	
Languages Spanish (official), English	
Religion Christian (80% Roman Catholic)	
Climate Tropical maritime; with most rain falling in summer. Average annual temperature is 25.5°C (78°F)	
Land use Grazing 38%, forest 20%, cultivation 14%, other 28%	
Main primary products Sugar cane, livestock (dairy cattle, pigs, goats, poultry), tobacco, pineapples, bananas, coconuts, fish	
Major industries Oil refining and petrochemicals, clothing, agriculture, sugar refining, electrical equipment, machinery, fishing, food processing	
Main exports Chemicals, petroleum products, clothing and textiles, pharmaceutical products, metal goods, sugar, alcohol, bananas	
Annual income per head (US$) 6610	
Population growth (per thous/yr) 3	
Life expectancy (yrs) Male 72 **Female** 77	

of the national capital, SANTIAGO. It is a centre for agriculture, fishing and tourism, and is also the terminus of the southern railway.

The city was founded by German immigrants in 1853, and was devastated by an earthquake after little more than a hundred years in 1960.
Population 125 000
Map Chile Ad

Puerto Presidente Stroessner *Paraguay* See CIUDAD DEL ESTE

Puerto Vallarta *Mexico* Fastest growing beach resort in Central America, situated on the Bay of Flags (Bahía de Banderas), in Jalisco State. It exports bananas, hides, coconut oil and wood. Despite its luxury hotels, it conveys the charm of

a fishing village, with red-tiled, white-walled houses, scattered along the Cuale River.

Population 70 000

Map Mexico – Bb

Puglia (Apulia) *Italy* Region covering an area of 19 250 km² (7500 sq miles), in south-eastern Italy. It is known as 'the Kingdom of Drought and Stone' because of its hot, dry climate.

In the past, malaria was rife in the area, which was also economically backward. Moreover, the limestone which forms the backbone of the Murge Hills, seamed with gorges and caves. Today, with an improved water supply, Puglia has become a flourishing agricultural region. Grapes are its main crop, but olives, almonds, wheat, tobacco and vegetables are also grown.

There are engineering and food processing plants at the regional capital, BARI, petrochemicals at BRINDISI and steel making at TARANTO.

Population 4 081 500

Map Italy – Ed

Pukaskwa *Canada* National park protecting the wilderness of western Ontario and its wildlife. Established in 1971, it covers an area of 1885 km² (728 sq miles), south of Marathon on the north shore of Lake Superior and is accessible only by boat. Wildlife includes beaver, muskrat, timber wolf and black bear.

Map Canada – Gd

Pula (Pulj; Pola) *Croatia* Modern resort, naval base, major shipbuilding centre and chief town of Istra, on the Adriatic peninsula just south of the Italian border. The Romans knew the town in the 2nd century BC as Piestas Julia.

Parts of its monumental amphitheatre, built and enlarged between 30 BC and AD 72 to hold more than 20 000 spectators, can still be seen; its walls rise 32 m (105 ft) above 72 arches. Just north-west of Pula lie the Brijuni (Brioni), a group of small islands where Yugoslavia's former leader Marshal Tito (1892-1980) had a villa.

Population 47 200

Map Croatia – Ab

pumice (pumice stone) Lightweight, acid volcanic rock full of vesicles (cavities that once held hot gases). It is often used as an abrasive.

puna High, dry plateau situated at 3000-5000 m (9800-16 400 ft) between the West and East Cordilleras of the Andes in Peru and Bolivia, with a sparse covering of coarse grasses and drought-resistant shrubs.

Punakha *Bhutan* A former capital, standing beside the Sankosh River at 1200 m (3950 ft), some 24 km (15 miles) north-east of the present capital, THIMPHU. Its large monastery, which was constructed in 1637 in the temperate valley, is one of the best surviving examples of Bhutanese architecture. The coronation of former Bhutanese kings took place here.

Map Bhutan – Aa

▲ **MAJESTIC RETREAT** Monks from Bhutan's capital, Thimphu, spend the winter at the dzong, or fortified Buddhist monastery, at Punakha.

Puncak Jaya *Indonesia* The nation's highest mountain, rising to 5030 m (16 500 ft). It lies in IRIAN JAYA on the island of New Guinea.

Map Indonesia – Ic

Pune (Poona) *India* City situated some 164 km (102 miles) south-east of BOMBAY. It stands on top of an escarpment of the Western GHATS. Pune was the summer capital of Bombay State during the British Raj; before that time it was the Marathi capital. It has many beautiful palaces and Hindu temples, and some magnificent public gardens and government buildings. It came under British rule in 1818, and since India's independence in 1947 has developed as an industrial centre and a seat of learning, with many research institutions. Its principal products include pharmaceuticals, machinery, paper, soap and textiles.

Population 5 494 340

Map India – Ed

Punjab *India* North-western agricultural state covering 50 362 km² (19 440 sq miles). It was part of the pre-independence Indian province of Punjab, which in 1947 was divided between Pakistan and India. A new capital for the Indian state of Punjab was built at CHANDIGARH. In 1966 the state was split in three, into a Sikh-dominated

Punjab in the west, and what are now the Hindu-dominated states of HARYANA and HIMACHAL PRADESH. Punjab, with its lush, rolling expanse of well-irrigated wheatlands, is known as 'the Granary of India'.

The name comes from Persian words – 'punj' meaning 'five' and 'aab' meaning 'water' – referring to the five rivers, the Jhelum, Chenab, Ravi, Beas and Sutlej, which cross the historic Punjab region; all but the Beas cross the border into the Pakistani Punjab and all enter the Indus River.

Population 20 281 000
Map India – Bb

Punjab *Pakistan* Province covering 205 344 km² (79 283 sq miles), comprising the larger western part of the British Indian province of Punjab which was divided up between India and Pakistan in 1947.

Economically and politically it is the most important region in the country. It relies on irrigated agriculture using water from the great rivers that cross it – the Indus, Jhelum, Chenab, Ravi and Sutlej. The region produces wheat, cotton and rice. The provincial capital, LAHORE, was once dominated by Sikhs, many of whom still want to unite with the Indian Punjab to form an independent Sikh state.

Population 62 199 000
Map Pakistan – Cb

Puno *Peru* City and department on the border with Bolivia. The highlands are home mainly to Amerindians who depend on the pastures to support their sheep, llamas and alpacas. The city, on the north-west shore of Lake TITICACA, South America's largest lake, is widely renowned for its fiestas, its music and its carnival.

A 17th-century cathedral stands in the centre of the city. The nearby town of Pucará is famous for its pre-Inca remains – the pre-Columbian tall burial towers called *chullpus* at Sillustani, and the floating islands of the Urus people situated on Lake Titicaca.

Population (city) 99 600; (department) 1 023 500
Map Peru – Bb

Punta Arenas *Chile* Capital of MAGALLANES Region and a major port, standing on the north shore of the Strait of Magellan. Today it is the centre of the sheep-farming industry and exports both meat and wool.

Population 125 000
Map Chile – Ae

Punta del Este *Uruguay* Popular resort on a peninsula, 140 km (86 miles) east of the capital, MONTEVIDEO, at the mouth of the River Plate. Beaches on either side offer visitors the breakers of the Atlantic or the shelter of the estuary.

Inland are pine, mimosa and eucalyptus woods near the colonial town of Maldonado, and offshore is Isla de Lobos, a protected island reserve inhabited by some half a million seals.

Population 10 000
Map Uruguay – Bb

Puntarenas *Costa Rica* Fishing port and tourist resort situated on the Pacific coast some 80 km (50 miles) west of the capital, SAN JOSÉ. The coastal province of the same name produces bananas, cocoa and palm oil.

Population (port) 92 360; (province) 338 380
Map Costa Rica – Ba

Pusan *South Korea* The country's second largest city after the capital, SEOUL. Lying on the southeast tip of South Korea, it is also the country's leading port. Pusan was developed during Japanese colonial rule (1905-45). It became overpopulated with repatriates from overseas when Korea gained independence in 1945 and again with refugees during the Korean War (1950-53) when it was the temporary capital of the Republic of Korea. Its products include machinery, chemicals and pharmaceuticals, textiles and ships. In addition, tourists enjoy the nearby beach resort of Haeundae and the hot springs of Tongnae. The city's United Nations cemetery contains the graves of soldiers who fell in the Korean War.

Population 3 798 000
Map Korea – De

Pustynia Bledowska *Poland* See BLEDOW DESERT

puszta Temperate, almost treeless grassland in the plains of Hungary.

Putumayo *Colombia/Ecuador/Peru/Brazil* River rising in the Andes of south-west Colombia near the Ecuadorean border and running south-east for about 1900 km (1180 miles). It forms part of the Colombian border with Ecuador and almost all of that with Peru before crossing into Brazil – where it is known as the Içá. Here it joins the Amazon some 200 km (124 miles) from the Colombian-Brazilian border.

Map Colombia – Bc

Puys, Chaine des *France* Mountains of volcanic origin in the AUVERGNE. About 70 summits rise above the main plateau of the Massif Central which has an average height of 800-1000 m (2625-3280 ft), the highest being the Puy de Sancy, at 1885 m (6184 ft).

Map France – Ed

PWV *South Africa* See GAUTENG

Pyatigorsk *Russia* Town and health resort in the northern Caucasus Mountains, 140 km (85 miles) south-east of STAVROPOL'. Mineral waters at the town's numerous sanatoriums are used to treat ailments such as rheumatism and arthritis.

The famous Russian poet and writer Mikhail Lermontov (1814-41) was killed in a duel in a glade just outside Pyatigorsk; the town has a museum devoted to him.

Population 130 000
Map Russia – Fd

Pyeongyang *North Korea* See PYONGYANG

Pyinmana *Myanmar (Burma)* Town in central Myanmar, lying halfway between MANDALAY and YANGON (Rangoon). Surrounded by a major sugar cane-growing area, it has an agricultural university which opened in 1975.

Population 172 700
Map Myanmar – Cc

Pyin Oo Lwin, *Myanmar (Burma)* Hill town, formerly called Maymyo, 50 km (30 miles) east of, and 800 m (265 ft) above, Mandalay. Its cool climate attracted the British, who developed it after 1886 as Burma's summer capital.

Population 100 800
Map Myanmar – Cb

Pyongyang (Pyeongyang) *North Korea* National capital, reputed to be the oldest city in Korea, beside the Taedong River, 87 km (54 miles) from its estuary on Korea Bay. Chinese trading colonists founded the city, then called Lolang, in 108 BC. Renamed Pyongyang, it became capital and cultural centre of the Korean peninsula for centuries. After the Japanese occupation of Korea (1905-45), the city became the capital of communist North Korea in 1948. It was heavily bombed during the Korean War (1950-3).

The new city built since then has broad, tree-lined avenues with many parks and open spaces. There is little motor traffic. Tourists are shown the Museum of the Korean Revolution, the Tombs of the Revolutionary Heroes, and the 60 m (200 ft) high Arch of Triumph, built in marble and granite. Dominating the skyline is the uncompleted Willow Hotel – with 105 storeys, it was supposed to become the tallest building in Asia, but it now stands abandoned and unfinished.

Pyongyang is the seat of a medical school as well as a communist university for training party leaders. Education is encouraged through classes taught at factories.

The arts are strongly supported with many institutions dedicated to opera, theatre and dance. Pyongyang produces more than a quarter of the country's industrial output – mainly iron and steel, machinery, railway locomotives, rolling stock and textiles.

Population 2 000 000
Map Korea – Bc

Pyrenees (Pirineos; Pyrénées) *Western Europe* Mountain chain which extends for about 500 km (311 miles) from the Bay of Biscay as far as the Mediterranean, and forms a natural boundary between France and Spain. Within the eastern part of the mountains lies the tiny independent state of ANDORRA, whose official language is Catalan (which is also spoken in north-east Spain and the adjoining area of France).

The shape of the mountain was largely carved by Ice Age glaciers, but only 34 km² (13 sq miles) of permanent snowfields remain – all at heights above 3000 m (9840 ft). The highest point is Pico d'Aneto, reaching up to 3404 m (11 170 ft) in the central Pyrenees on the Spanish side of the border.

Maize, cereals, and fruits are grown in the western valleys, and olives and vineyards are commonplace in the eastern valleys of the range.

Map France – Ce

pyrite (iron pyrites) Brassy yellow mineral sulphide of iron. It is an important source of sulphur and is also used to produce sulphur dioxide for the manufacture of sulphuric acid. Its metallic yellow lustre has led to it being mistaken for gold, which explains its other name – 'fool's gold'. The name comes from the Greek word *pyr* which means 'fire'.

pyroclastic rock Rock formed from fragments of volcanic material that are blown into the air by a volcanic explosion. On landing, the multitudinous fragments become consolidated over a wide area around the VOLCANO

pyroxene Any of a group of complex silicate minerals found in basic IGNEOUS ROCK and some METAMORPHIC ROCK. Pyroxenes include augite (calcium, magnesium, iron and aluminium silicate) and JADEITE.

Qaanaaq *Greenland* See THULE

Qalat al Bahrain *Bahrain* Historic site, 5 km (3 miles) west of AL MANAMAH. The remains consist of a series of cities, the earliest being 5000 years old, which were part of the Bronze Age state of Dilmun, as Bahrain was then known. Excavations have revealed the remains of palaces, broad streets and a harbour. The remains of three temples, the oldest dating from 2500 BC, are situated 6 km (4 miles) west along the coast of Barbar. A third site of the same period lies 2 km (just over 1 mile) west of Barbar at Ad Diraz, where traces of village buildings have been discovered. The Dilmun people were buried in single-chamber grave mounds, 100 000 of which lie in north-west Bahrain Island.
Map Bahrain – Ba

Qam'do *China* Geographical region of about 270 000 km² (104 250 sq miles), which covers eastern XIZANG (Tibet), and north-west YUNNAN and west SICHUAN provinces. It is a remote, sparsely populated region of parallel mountain ranges, separated from one another by the deep gorges of the upper Nu Jiang (Salween), Lancang Jiang (Mekong) and Chang Jiang rivers. Maize, rice and winter wheat are cultivated in the region.
Map China – Ee

Qatif *Saudi Arabia* Coastal oasis on The Gulf, situated 15 km (9 miles) north of DAMMAM. Numerous underground springs irrigate the surrounding land; in fact, the oasis has been a centre of trade, fishing and agriculture for centuries. Although Qatif is rapidly being modernised, it still is a maze of narrow lanes and old houses linked by overhead bridges.
Population 95 000
Map Saudi Arabia – Bb

Qattara Depression *Egypt* Huge, sunken area of the Western Desert, midway between Cairo and the Libyan border. It is 320 km (nearly 200 miles) long and 120 km (75 miles) wide, and in parts it is up to 134 m (440 ft) below sea level. It consists almost entirely of soft sand, brackish ponds and salt marshes. During the Second World War, it was a vital, impassable flank to the British defences at the Battle of EL ALAMEIN.
Map Egypt – Ab

Qazvin (Kasvin) *Iran* Market town founded in the 4th century, about 150 km (95 miles) north-west of the capital, TEHRAN. It manufactures carpets, textiles and soap, and also trades in pistachios, raisins and other agricultural products. During the 16th century it was made the country's capital for a brief period.

In the Elburz Mountains above Qazvin is the ruined citadel of Alamut, one of the strongholds of the Assassins, a secret Muslim order founded in about 1090, who set out to kill their religious enemies – notably Crusaders. The citadel was destroyed in 1256, and the Assassins were exterminated in 1264, by the Mongol leader Hulagu, grandson of Genghis Khan.
Population 279 000
Map Iran – Aa

Qatar

THIS TINY ARAB KINGDOM HAS COME FROM POVERTY TO WEALTH IN A FEW DECADES – AND IMMIGRANTS OUTNUMBER ITS LOCAL INHABITANTS

The little emirate of Qatar, halfway along the western coast of The GULF, presents a classic rags-to-riches story. Before 1949, when the first oil was exported, it was a tribal society which existed on pearl diving, fishing and camel breeding. Within 45 years it had a per capita income of US$16 240, one of the highest in the world.

The Arabs of this stony desert land have a proverb: 'Hitch your camel first and then depend on God.' They have made it one of the most progressive states in the Arab world, although traditional Muslim values have been respected. The oil money has been used to build up social services, with free education and health care, and develop heavy industry so that prosperity will not vanish overnight when the oil runs out. In 1947 there was only one hospital and one resident doctor in Qatar. Today there is a wide variety of social services.

Qatar consists of a peninsula and a few small islands, which cover a total area of only 11 000 km² (4246 sq miles). The peninsula thrusts out into The Gulf from the Saudi Arabian coast like a fat thumb. The summers are long, hot and uncomfortably humid. A north-west wind, the *shamal*, helps to cool the land during the winter months. It rains in the north in the winter; otherwise Qatar is an arid country, and most fresh water comes either from natural springs and wells or, increasingly today, from desalination plants producing fresh water from sea water.

Arab writers first mentioned Qatar in the 10th century. The people today are descended from migratory Bedouin tribes who arrived from central Arabia in the 1730s. They founded DOHA – for centuries a little village, now the capital. Qatar was invaded by the Persians in 1783 and later became a dependency of its neighbour BAHRAIN. In 1868 it came under the influence of the British, who installed the present ruling family. The Turks took nominal control at the end of the 19th century. In 1916 Qatar became a British protectorate. Independence was achieved in 1971.

OIL AND GAS RESERVES

The discovery and exploitation of oil and natural gas have transformed modern Qatar. The oil field at DUKHAN, at current levels, is expected to continue production till about 2020, and the big offshore North Field has reserves of natural gas estimated to be the world's largest. It began production in 1991 and is believed to have reserves of 4.2 trillion (million million) cubic metres (148.4 trillion cubic ft) of gas.

The government, in its efforts to diversify the economy, offers attractive incentives such as tax breaks. A new range of heavy industry includes iron and steel, cement, fertiliser and petrochemical plants. The fishing industry has been modernised and frozen shrimps are exported. Dates are also exported.

These modern developments have attracted large numbers of immigrants to work in Qatar, so that they now outnumber the native population. Only about 120 000 of the country's people are Qataris. The rest come from other Arab countries and from India and Pakistan; the white-collar workers include Egyptians, Lebanese and Syrians. It is now a highly urbanised society; 350 000 people live in Doha, the capital.

The emir, Sheik Khalifa bin Hamad Al-Thani, is head of state and prime minister, ruling with the help of an appointed Consultative Council of 30 members. There are no elections. Despite all the economic changes, Qatar remains fundamentally an Islamic society (most Qataris are Sunni Muslims), in which traditional Arab dress is worn, women are enjoined to modesty, and floggings are still carried out for some crimes – although not, according to a government spokesman, with great severity.

QATAR AT A GLANCE		
Official name State of Qatar		
Area 11 000 km² (4246 sq miles)		
Population 499 115 **Per km²** 45 **(Per sq mile** 118)		
Capital Doha		
Government Absolute monarchy		
Currency 1 Qatar riyal = 100 dirhams		
Languages Arabic (official), English		
Religion Muslim (95%)		
Climate Desert climate with high humidity and mild winters. Average temperatures range from 19°C (66°F) in January to 35°C (95°F) in August		
Land use Pastures 5%, desert 95%		
Main primary products Livestock, fodder, vegetables, fruits, fish, oil and natural gas		
Major industries Oil and natural gas production and refining, cement, iron and steel, chemicals, agriculture, fishing		
Main exports Oil and natural gas, frozen shrimps, dates, steel, chemicals		
Annual income per head (US$) 16 240		
Population growth (per thous/yr) 28		
Life expectancy (yrs) Male 70 **Female** 75		

Qena *Egypt* Market town on the east bank of the Nile, about 450 km (280 miles) south of the capital, CAIRO. It lies on an ancient route from northern Egypt to the sacred Muslim city of MECCA, in Saudi Arabia. Pottery is made in Qena.
Population 119 790
Map Egypt – Cc

Qesaria *Israel* See CAESAREA

Qilian Shan *China* Mountain range rising from the extreme north-east of the XIZANG GAOYUAN (Tibetan Plateau). Its towering snow-capped peaks exceed 4000 m (13 000 ft), the highest reaching 6346 m (20 819 ft).
Map China – Ec

Qin Ling *China* Mountain range stretching 1500 km (900 miles) across central China and rising to 4205 m (13 474 ft). It forms a watershed between the Huang He and Chang Jiang river basins. North of the range, rainfall is light and winter temperatures fall below freezing. To the south, rainfall is abundant and, because of the sheltering effect of the mountains, the winters are relatively mild.
Map China – Fd

Qingdao (Tsingtao) *China* City of Shandong (Shantung) Province on the Yellow Sea (Huang Hai). It was a German treaty port (a city where a Western power – in this case, Germany – enjoyed privileges obtained through treaties with China, such as immunity for its nationals from trial under Chinese law) from 1898 to 1914 and is now Shandong's biggest city. Locomotives and railway rolling stock are manufactured in the town.
Population 2 000 000
Map China – Id

Qinghai *China* Province in the west covering 720 000 km² (280 000 sq miles). It takes its name from a salt lake and includes the Qilian Shan, the Qaidam Pendi (Tsaidam basin) and part of the plateau of Tibet. The climate is one of extremes: winter temperatures fall far below freezing and rainfall is sparse; summers are hot. Cultivated land is confined largely to the valley of the Huang He River. Elsewhere, the breeding of yaks, sheep and horses is the mainstay of the economy. The province is rich in coal, oil, potash and bromine.
Population 3 720 000
Map China – Dd

Qinghai Hu *China* Salt lake of QINGHAI Province. It covers 5957 km² (2300 sq miles) and lies 3000 m (10 000 ft) above sea level.
Map China – Ed

Qiqihar (Tsitsikar; Lungkiang) *China* City of HEILONGJIANG Province, 1000 km (620 miles) north-east of BEIJING. It manufactures locomotives, railway rolling stock and heavy machinery.
Population 1 209 200
Map China – Ib

Qom (Qum) *Iran* Holy city, 145 km (90 miles) south-west of the capital, TEHRAN. Fatima, the sister of a Shiite religious leader, Imam Reza, died here in 816, and the 17th-century sanctuary built over her tomb is now a place of pilgrimage for Shiite Muslims. Qom was also the home of Ayatollah Ruholla Khomeini (1900-89), the Shiite Muslim leader whose followers ousted the last shah and established an Islamic republic in Iran in 1979.
Population 681 000
Map Iran – Ba

Qornet es Saouda (Qurnet es-Sauda) *Lebanon* The country's highest mountain at 3087 m (10 128 ft). It lies in the north of the Lebanon Mountains. For much of the year, the peak of the mountain is snow-covered.
Map Lebanon – Ba

Quang Yen *Vietnam* Coal town lying 100 km (62 miles) east of the capital, HANOI. The surrounding coal field was developed by the French in the late 19th and early 20th centuries; by the mid-1970s it was producing 6 million tonnes of anthracite a year. However, expansion faltered as a result of the emigration of skilled Chinese miners following the Chinese invasion of Vietnam in 1979.
Map Vietnam – Ba

quartz Hard, crystalline form of silicon dioxide, commonly occurring as colourless and transparent crystals. Quartz is the world's most common mineral, frequently making up the majority of sand grains in sand and SANDSTONE. It occurs in the form of hexagonal columns surmounted by pyramids. There are coloured varieties, such as rose, milky and smoky quartz. Amethyst is a purple quartz, and citrine is yellow. These and other coloured quartzes are prized as semi-precious stones. The crystals are widely used in electronics.

quartzite Metamorphic rock, consisting almost entirely of silica, formed by the recrystallisation of sandstone.

Quaternary Second of the two periods in the Cenozoic era of the earth's timescale. The term is sometimes applied to the fourth era in the geological timescale, beginning at about the time of the onset of the last Ice Age. See GEOLOGICAL TIMESCALE, p 256

Quatre-Cantons, Lac des *Switzerland* See LUCERNE, LAKE

Quebec *Canada* Largest province, covering 1 358 000 km² (524 300 sq miles) in the east between Hudson Bay, Labrador, and the Gulf of St Lawrence. The capital is QUEBEC CITY and the largest city is MONTREAL. The province includes three geographic regions: the CANADIAN SHIELD in the north, the ST LAWRENCE Lowland Valley, where most of the people live, and the Appalachian region, composed of undulating plateaus which rise to more than 1000 m (3280 ft) near the United States border. The Shield region is densely forested and has mines producing copper, iron, zinc, silver and gold. The Lowland farms yield sugar beet, tobacco, market garden and dairy products. Abundant hydroelectricity has led to the development of metallurgical industries. Manufactured goods include transport equipment, machinery, textiles and clothing. The people of Quebec are mostly French-speaking and increasing numbers are in favour of Quebec Province seceding from the (predominantly English) Canadian union.
Population 6 895 960
Map Canada – Hc-Ic

Quebec City *Canada* Capital of Quebec Province, university town and Canada's oldest city, standing on the St Lawrence River at its junction with the St Charles River. Quebec City was founded in 1608 by the French explorer Samuel de Champlain. In 1759, it was taken from the French by British troops under General James Wolfe. It is the only walled city in Canada. The upper town, built on cliffs, has narrow streets and old houses, and contains the main public buildings. The lower town has an excellent harbour and contains the business and industrial areas. The city retains much of its old French atmosphere; 95 per cent of its inhabitants are native French-speakers and 99 per cent are Roman Catholics.
Population 583 920
Map Canada – Hd

Queen Charlotte Islands *Canada* Group of islands 160 km (100 miles) off the west coast of BRITISH COLUMBIA. The main islands are Graham, covering 6452 km² (2491 sq miles) and Moresby, with an area of 2567 km² (991 sq miles). Most of the islands are covered by forest. Coal, copper and gold have been mined, but timber and fishing are more important. The majority of the inhabitants are Haidas, a native American people who survive only on the Queen Charlotte Islands and on Prince of Wales Island off Alaska.
Population 5620
Map Canada – Bc

Queen Charlotte Sound *New Zealand* Inlet of the COOK STRAIT, 40 km (25 miles) long, in the north-eastern corner of New Zealand's SOUTH ISLAND. The British explorer Captain James Cook wrote that it 'is a collection of some of the finest harbours in the world'.
Map New Zealand – Ed

Queen Charlotte Strait *Canada* A scenic, sheltered waterway running between the north-eastern coast of VANCOUVER ISLAND and the Canadian mainland. It is 97 km (60 miles) long and 26 km (16 miles) wide. It is part of the inland water route (the Inside Passage) from SEATTLE in WASHINGTON State, to ALASKA.
Map Canada – Cc

Queensland *Australia* Most north-easterly state of Australia, covering 1 727 200 km² (666 718 sq miles). The mountains of the GREAT DIVIDING RANGE split the state in two, with the well-soaked, very hot and monsoonal coastal plain in the north, and arid lands covering much of the interior. The Great Barrier Reef lies off the east coast. There are huge deposits of coal, bauxite, copper, lead, silver and zinc, while the state's primary agricultural products are cattle, sugar cane, pineapples, bananas and cereals. Queensland has an overall population density of 1.7 people per km² (4.5 per sq mile). More than 45 per cent of the population live in the state capital, BRISBANE.
Population 2 999 900
Map Australia – Gc

Queenstown *New Zealand* Tourist resort on Lake Wakatipu, 160 km (100 miles) north-west of DUNEDIN on South Island. It looks across to the 2286 m (7500 ft) high mountains of the Remarkables Range. One of the country's finest ski resorts – Coronet Peak – lies to the north.
Population 2930
Map New Zealand – Bf

Quelimane (Quilimane; Kilimane; Kilmain) *Mozambique* Seaport on the mouth of the Quelimane River, about 300 km (185 miles) north-east of the port of BEIRA. Its coconut plantation, one of the world's largest, covers 202 km² (78 sq miles). During the 18th and 19th centuries, Quelimane was a notorious slave port; today, it exports copra, cotton, maize, sisal, sugar, tea and tobacco. It is the capital and only major port of Zambézia Province, which covers 105 008 km² (40 533 sq miles).

Population 71 800
Map Mozambique – Ab

Quelpart *South Korea* See CHEJU DO

Quemoy *Taiwan* See CHIN-MEN TAO

Quercy *France* Former province, covering 5215 km² (2013 sq miles), on the south-west of the Massif Central. Its limestone plateaus are cut by steep-sided valleys. Land in the large valleys is utilised, vines are grown on the sunny slopes and the plateaus are used for grazing and forestry. Cahors is the main town.

Map France – Dd

Querétaro *Mexico* State on Mexico's central plateau, covering 11 480 km² (4431 sq miles), between the Sierra Madre Occidental and the Sierra Madre Oriental. In the north, the mountain slopes are used for cattle grazing, and cotton, maize, coffee and beans are grown in the fertile valleys. Many industries have been encouraged in the lower-lying south of the state by a government plan to establish industry outside the Federal District around Mexico City. Mercury and gemstones such as agates, opals, amethysts and topaz are mined.

The capital of Querétaro State is also Querétaro and is located 175 km (110 miles)

▼ IDYLLIC LAKELAND New Zealand's Queenstown nestles besides Lake Wakatipu at the foot of Queenstown Hill – a superb viewpoint. Sheep graze on the green hills dotted with lakes and farmsteads.

north-west of MEXICO CITY. It was founded by the Otomís and Chichimecas in 1445 and was taken by the Spanish in 1532. Emperor Maximilian was tried and shot here in 1867. The city produces cotton textiles and pottery, and processes crops from the surrounding area. Its cotton factories are among the most important in Mexico, and new industries encouraged by the government have increased the population fivefold in the last 25 years. The city has an 18th-century aqueduct, a 16th-century cathedral, and 55 churches.

Population (state) 1 044 000; (city) 454 810
Map Mexico – Bb

Quetta *Pakistan* Capital and largest city of Baluchistan Province, 80 km (50 miles) from the Afghan border. The name Quetta means 'fort' and the city, tucked in among the arid hills between the Bolan and Khojak passes, was founded to control the route between the Indus Valley and the Afghan city of KANDAHAR. The British built a military college in Quetta which is still used by the Pakistani army.

The town was destroyed by an earthquake in 1935, but has since been rebuilt.

Population 589 000
Map Pakistan – Cb

Quetzaltenango *Guatemala* Second largest city in the country, after Guatemala City, situated 110 km (70 miles) west of the capital, in the heart of a rich agricultural area. Founded in 1524, it lies 2335 m (7660 ft) above sea level on the slopes of Santa María Volcano (3768 m, 12 362 ft), which destroyed it in 1902.

The city's name means 'Place of the Quetzal', after the rare bird of the rain forests which was worshipped by the Mayas. Quetzal is also the name of the national currency. The city has two universities. It also has textile and brewing industries.

Population 252 910
Map Guatemala – Ab

Quezon *Philippines* Province occupying the narrow isthmus that connects the Bicol Peninsula with the rest of LUZON; it includes the isolated Bondoc Peninsula. Most of Quezon remains remarkably undeveloped for an area so close to

the capital, MANILA. This is largely because much of it is mountainous and forested; part of it is a national park. The main product is coconuts.

Population 1 116 000
Map (Bicol and Bondoc peninsulas) Philippines – Bc; (Luzon) Philippines – Bb

Quezon City *Philippines* Administrative capital of the country from 1948 to 1976, but now merely a suburb of MANILA. It contains embassies and the huge University of the Philippines.

Population 1 165 000
Map Philippines – Bc

quicksand Bed of loose, round-grained sand saturated with water. The water acts as a lubricant and the round grains reduce friction, causing heavy objects to sink easily.

Quilimane *Mozambique* See QUELIMANE

Quimper *France* Capital of Finistère Department at the end of the Brittany Peninsula. It manufactures pottery, known as Quimper or Brittany ware. The other main industries are sardine fishing and tourism. In July a festival of Breton folklore, with bagpipes and dancing takes place here.

Population 59 500
Map France – Ab

Quintana Roo *Mexico* Mexico's newest state, created in 1974 on the north-east coast of the Yucatán Peninsula, an area of swampy tropical lowland and white beaches. The capital, Chetumal, a free port, lies in the extreme south near the Belize border. But the real centre of Quintana Roo is the purpose-built Caribbean resort of CANCUN in the north.

Population 256 000
Map Mexico – Dc

Quito *Ecuador* National capital and capital of Pichincha Province, lying almost on the equator 2820 m (9250 ft) up in a narrow valley amid the Andes. It was founded in 1534 by the Spanish conquistador Sebastián de Belalcázar on the site of a ruined Amerindian town of the same name. In 1822 the South American liberator Antonio José de Sucre (1795-1830) defeated the Spanish on Pichincha volcano just to the north-west, thus gaining Ecuador's independence. The cathedral in the central square contains Sucre's tomb.

One of the country's two major industrial centres (the other being GUAYAQUIL), Quito manufactures textiles, clothing, footwear and pharmaceuticals. It also produces handcrafts from leather, wood, gold and silver. Its churches and old mansions contain some of the finest art treasures in South America. Among them are the church and monastery of San Francisco and the Jesuit church of La Compañía. There are two universities and an international airport.

Population 1 100 800
Map Ecuador – Bb

Qum *Iran* See QOM

Qwaqwa *South Africa* Former 'homeland' known as Witsieshoek, and after 1974, a self-governing state in terms of the South African apartheid policies, situated in the south-eastern Free State. Its capital was Phuthaditjhaba.

Map South Africa – Cb

Rab (Arbe) *Croatia* Island covering 93 km² (36 sq miles) in the Adriatic Sea off the north Croatian coast. It is noted for its marble quarries and silk industry. The town of the same name on the island's south coast is a former Venetian fortified settlement dating from the 15th century; it has a fine cathedral, several attractive palaces built by wealthy Venetians, and four Romanesque bell towers.

Population 3500
Map Croatia – Bb

Rabat *Malta* Town, 10 km (6 miles) west of the capital, VALLETTA. In Roman times, Rabat and the adjoining smaller Mdina formed the capital city, then called Melita, where St Paul landed after his shipwreck (Acts 28:1). He is supposed to have lived in a grotto adjoining the church named after him. This was the first Maltese church to have been built on the grand scale following the prosperity brought by the Knights of St John, who made the island their headquarters in 1529. Many other churches in Rabat are built over or within early Christian catacombs.

Population 12 000
Map Malta – Db

Rabat *Morocco* National capital and royal residence, Rabat stands on the site of a 10th-century stronghold, on the Atlantic at the mouth of the Bou Regreg River. Founded in the 12th century, it remained small and unimportant until 1913 when the French made it the capital in place of FÈS.

The old city is surrounded by 12th-century battlements, and has many souks (bazaars) offering a range of handicrafts. It also has a 12th-century mosque, and the Oulayas Kasbah, a citadel which looks out over the river to the ancient corsair town of SALÉ. The new city is spacious with many good hotels and restaurants; there are fine beaches close by. When the king is in residence at the royal palace there is a regal procession as he goes to the mosque to pray.

Outside the walls is Chella, the site of a former Roman town, which the Arabs turned into a cemetery in the 14th century. An ancient ruined mosque stands over the tombs, which are beautifully carved in Arabic script and surrounded by gardens.

Population 1 472 000
Map Morocco – Ba

Rabaul *Papua New Guinea* Port and capital of Bismarck Archipelago, situated on the northern tip of NEW BRITAIN ISLAND. It was founded in 1910 as a German colonial headquarters, and was the administrative centre of the Australian mandate territory after the First World War. Allied bombers destroyed Rabual after the Japanese had occupied it during the Second World War. It is now Papua New Guinea's biggest town off the New Guinea mainland, and among its industries are the manufacture of furniture and building materials. Its main exports are coconut products, timber and cocoa.

Rabaul's magnificent harbour is a sea-filled crater. Two active volcanoes overlook the town; whose inhabitants had to be evacuated in 1937 after a violent eruption. In 1984, residents again fled after experts had warned that another big eruption was imminent. Nothing happened, however, and the relieved islanders gradually began to return to their homes.

Population 17 050
Map Papua New Guinea – Ca

race Term applied to a strong, swift flow of sea water through a restricted channel, produced by marked tidal differences at either end of the channel. The channel created by such a current may be called a race. The term 'race' is also used to refer to the canal or channel used to tap a stream and lead water to drive the wheel of a water mill.

Racibórz (Ratibor) *Poland* Industrial city near the Czech border, 65 km (40 miles) south-west of the city of KATOWICE. It was the capital of an independent Silesian Duchy from 1172 to 1336, and is now a busy rail junction. Its factories produce metals, electrical goods and processed food.

Population 62 800
Map Poland – Cc

radiation fog FOG formed inland by radiation cooling of the earth's surface, which in turn cools the air near the ground to below dew point. Favourable conditions for the formation of radiation fog are a long night with clear skies, moist air and little wind.

radiogenic heat Heat produced in the earth's CRUST and upper MANTLE by radioactive decay of minerals.

radiolarian ooze Sediment of red clay and organic remains, primarily skeletons of small organisms of the order *radiolaria*, found on the ocean bed in deep tropical seas, particularly the mid-Pacific and Indian oceans. See OOZE.

Radnor *United Kingdom* See POWYS

Radom *Poland* Industrial city, 98 km (61 miles) south of the capital, WARSAW. It flourished in the 14th and 15th centuries as a cloth-making centre at the crossroads of two major trade routes – from Greater Poland to Silesia, and from Lithuania to the Ukraine. From 1795 to 1809 it was ruled by the Austrians and from 1815 to 1918 by Russia.

Today, Radom is still an important manufacturing centre, lying on a railway junction, with engineering and machine-building industries. It also processes fruit and vegetables as well as producing glass, textiles, chemicals, electrical and leather goods.

Population 226 300
Map Poland – Dc

rag (ragstone) Any of various hard, coarse, sedimentary stones which break into thick slabs, such as the Kentish Rag. Ragstones are widely used as stones for building.

Ragang *Philippines* See LANAO

Ragusa *Croatia* See DUBROVNIK

Ragusa *Italy* Town in the south-east corner of Sicily. Its economy is traditionally based on agricultural, but there are now oil, asphalt and chemical plants.

Population 67 300
Map Italy – Ef

rain forest See TEMPERATE RAIN FOREST; TROPICAL RAIN FOREST

rain shadow Lee side of a mountain or hill range where precipitation is low, such as east of the Rockies in North America.

rain spell In Britain the term is generally applied to a raining period of 15 consecutive days, each with a rainfall of at least 0.25 mm (0.01 in). Throughout the world, the term varies depending on the aridity of a region.

rainbow An arc of colours – red, orange, yellow, green, blue, indigo, violet – appearing in the sky opposite the sun. It is seen when sunlight falls on rain and is caused by the reflection and refraction of the light in the water droplets which act as prisms. Occasionally some of the sunlight falling on the water droplets is reflected twice and a second rainbow is formed, less distinct than the first and with the colours in the reverse order. In the primary rainbow, red is on the outer side of the arc and violet on the inside.

rainfall Total quantity of water deposited on a given area in a given time, as measured by a rain gauge. It may be condensed or precipitated as rain, SNOW, HAIL, DEW, HOAR FROST, RIME or SLEET. (See also BONE DRY and SOAKING WET under WEATHER.)

raised beach Wave-cut platform covered with ancient beach deposits, sometimes backed by old cliffs, that has been raised above the shoreline by earth movements, or left behind by a fall in the sea level or the waterline of a lake. The term is not applied if the beach deposits are left behind when the lake dries up or is drained.

Rajang *Malaysia* The country's longest river. It flows 564 km (350 miles) from the hills of central Borneo westwards through Sarawak's swampy forest land into the South China Sea west of the town of Sibu.

Map Malaysia – Db

Rajasthan *India* North-western state covering some 342 267 km² (132 149 sq miles) from the Aravalli Range to the Pakistani border in the Thar Desert. Its name is derived from the Rajput tribes, who settled here in the 9th century and founded various noble lineages. During the British Raj, the area contained many princely states. Rajasthan is noted for its arts and crafts, particularly its paintings, many of which are miniatures in Mogul style; others are painted on silk.

One of the world's longest irrigation canals, the Indira Gandhi Canal flows for some 170 km (106 miles) across the Punjab without distributing water, then has a working length of 479 km (298 miles) and a total length of minor distributaries of 4510 km (2802 miles). Although the land in the state is mainly arid, irrigation has now made crop farming possible alongside the traditional rearing of sheep, goats and camels. The state capital is the city of JAIPUR.

Population 43 880 700
Map India – Bb

Rajshahi *Bangladesh* Chief town of the Rajshahi District, which covers some 9456 km² (3651 sq miles) in western Bangladesh. It stands beside the Ganges River and is the home of a university and centre of the silk industry.

Population (town) 517 000; (district) 5 270 000
Map Bangladesh – Bb

Raleigh *USA* State capital of North Carolina, lying in the centre of the state, about 185 km (115 miles) from the Atlantic Ocean. Laid out in

1792, it was named in honour of the English adventurer Sir Walter Raleigh, who had helped open up the New World two centuries earlier.

Raleigh is located in the heart of tobacco-growing country, although nearby Durham is referred to as 'tobacco city'. Raleigh has developed into an important research and educational centre and is one of the United States' fastest growing cities.

Population (city) 208 000; (metropolitan area) 855 600

Map United States – Kc

Ramat Gan *Israel* Industrial town lying to the east of Tel Aviv, of which it is a suburb. Founded in 1921, it has a diamond exchange and produces textiles, processed foods and furniture.

Population 122 700

Map Israel – Aa

Ramillies *Belgium* Village situated some 50 km (31 miles) south-east of the capital, BRUSSELS, where in 1706 the British under the Duke of Marlborough defeated the French, led by François de Neufville, Duc de Villeroi (1644-1730), during the War of the Spanish Succession.

Population 4200

Map Belgium – Ba

Rancagua *Chile* Capital of Libertador region, lying in a fertile agricultural area in the Central Valley, 80 km (50 miles) south of SANTIAGO. In 1814, troops led by the country's liberator, Bernardo O'Higgins, fought a pitched battle here against the Spaniards.

Population 200 000

Map Chile – Ac

Rance *France* River, 100 km (62 miles) long, in northern Brittany. It drains into the Gulf of Saint-Malo via a 19 km (12 mile) long estuary. The world's first tidal power station was completed on the estuary in 1968.

A dam 750 m (2460 ft) long was built across the river to take advantage of the unusually large rise and fall in the estuary's tides. Special turbines in the dam can tap the power of both the incoming and the out-going water.

Map France – Bb

Ranchi *India* Industrial town, administrative centre, and former summer capital of Bihar State. It is located on the Chota Nagpur plateau, some 450 km (280 miles) north-west of CALCUTTA. Among its heavy engineering products are machine tools. It has a university.

Population 2 205 030

Map India – Dc

rand In South Africa, a term applied to a ridge of hills or an extended area of high ground, often covered with scrub.

Rand, The *South Africa* See WITWATERSRAND

Randstad *Netherlands* See HOLLAND

range A term applied to an extended group of mountains or hills. In the United States the term is also used to refer to an extensive area of land which is covered by shrubby vegetation and where animals, usually cattle, roam in search of food.

Rangoon *Myanmar (Burma)* See YANGON

Ranong *Thailand* Province and town on the Andaman Sea. Both lie beside the southern tip of Myanmar (Burma), about 475 km (295 miles) south-west of the Thai capital, BANGKOK. The main industry is tin mining, and many mine workers are immigrants from Myanmar (Burma).

Population (town) 50 100; (province) 96 600

Map Thailand – Ad

Rapa Nui *Pacific Ocean* See EASTER ISLAND

Rapallo *Italy* Seaport and tourist resort 30 km (20 miles) south-east of GENOA. It developed from a fishing village to a resort in the late 19th century. Today it produces olive oil and wine.

Population 29 300

Map Italy – Bb

rapids Fast flowing rocky section of a river, usually caused by a steepening of its bed.

Rarotonga *Cook Islands* Largest and most important island of the group, covering 67 km² (26 sq miles) and lying about 2750 km (nearly 1710 miles) north-east of New Zealand. It is mountainous, rising to several sharp peaks; Te Manga, at 652 m (2139 ft) is the highest. Fruit and vegetables are grown on a narrow coastal plain; the main industry is a fruit and fruit juice canning factory. Avarua, the chief town and capital of the Cook Islands, lies on the north coast; it has much of the atmosphere of a 19th-century South Seas trading post. By law, no building on Rarotonga may be higher than a coconut palm.

Population 10 910

Map Pacific Ocean – Fd

Ras al Khaymah *United Arab Emirates* The most north-easterly and most fertile of the small emirates, covering 1700 km² (656 sq miles) on the west coast of the MUSANDAM Peninsula, next to the Strait of Hormuz. It is regarded as one of the most beautiful spots in the United Arab Emirates with carefully irrigated, abundant greenery.

Its main industries are the production of building materials, and fishing and agriculture. About half of the United Arab Emirates' supply of fresh food comes from here. The emirates' growing town of Ras al Khaymah is divided in two by a lagoon. It is a favourite weekend getaway for people from Dubai and includes water sports and camel racing among its recreational activities.

Population (emirate) 130 000; (town) 5300

Map United Arab Emirates – Ba

Ras Dashen *Ethiopia* Highest peak in the country, rising to 4620 m (15 160 ft) in the Semien Mountains of the north.

Map Ethiopia – Aa

Ras Lanuf *Libya* Oil terminal built on the Mediterranean coast (on the Gulf of Sirte) in the 1960s to handle oil from fields to the south and east which are connected to it by pipelines.

Map Libya – Ba

Ras Tanura *Saudi Arabia* Port and major oil terminal and gas processing plant on The Gulf, situated on a peninsula, 40 km (25 miles) north of Dammam. Ras (Cape) Tanura was built shortly before the Second World War and is the site of the country's first oil refinery, which came on stream in 1946.

Map Saudi Arabia – Cb

Rashid (Rosetta) *Egypt* Town and port on the west bank of the Rashid (Rosetta), the main western river of the Nile delta, 55 km (34 miles) north-east of the port of ALEXANDRIA. In 1799, one of Napoleon Bonaparte's soldiers, digging a trench near the town, found a black basalt slab measuring about 1 m² (1 sq yd).

The slab, now known as the Rosetta Stone, was taken to Britain in 1801, and is now in the British Museum in London. It is inscribed with a decree issued by Ptolemy V in 196 BC, in Greek, Egyptian hieroglyphics and demotic characters. In 1822, the brilliant French Egyptologist Jean François Champollion (1790-1832) used it as the key to decipher the hieroglyphics.

Population 43 000

Map Egypt – Bb

Rasht (Resht) *Iran* Industrial city near the Caspian Sea, 250 km (155 miles) north-west of the capital, TEHRAN. Its chief manufactures are textiles, hosiery and carpets. Rasht is the capital of the province of Gilan. The inhabitants of the area, many of whom are Armenian, produce silk, and process rice and tea.

Population 341 000

Map Iran – Aa

Rastenburg *Poland* See KETRZYN

Ratibor *Poland* See RACIBORZ

Ratnapura *Sri Lanka* Gem-mining town on the south-west flank of the central mountains, 65 km (40 miles) south-east of COLOMBO. The area around the town is dotted with deep pits from which soil is excavated, washed and sieved for sapphires, rubies, amethysts, aquamarines, tourmalines and cat's-eyes. A gem museum in the town is overlooked by a large statue of the Buddha.

Population 37 400

Map Sri Lanka – Bb

Rauma (Raumo) *Finland* Town founded in 1442 on the Gulf of Bothnia, 215 km (nearly 135 miles) north-west of the capital, HELSINKI. Its industrial plants include timber processing factories and shipyards. Fine old frame and timber dwellings line the streets in the older part of the town.

Population 38 200

Map Finland – Bc

Ravenna *Italy* Historic city, some 115 km (about 70 miles) south of VENICE. It was the base of the Roman Adriatic fleet and served as capital of the Western Roman Empire from AD 402, after the fall of Rome. However, due to silting up of the Po River delta, the coast shifted 9 km (5.5 miles) forward into the Adriatic Sea and Ravenna is today connected to the Adriatic Sea by a canal 10 km (6 miles) long.

Ravenna was the Ostragoth capital from 493 until 540, when it became the Byzantine capital of Italy. It is noted for its rich array of Roman and Byzantine remains, including several well-preserved churches founded by the Romans in the 5th century, many containing fine mosaics. The tomb of the Ostragoth King Theodoric (455-526) stands in a 10-sided stone building with a 250-tonne block of solid stone serving as its roof. At about the start of the 6th century the banker Julianus Argentarius built the baptistry and church of San Vitale which, with its superb mosaics, is Ravenna's most widely renowned monument. The remains of the poet Dante Alighieri, who died in Ravenna in 1321 at the age of 56, lie in a small 18th-century temple; nearby is the Dante museum.

Industrial development took off in the 1950s, after natural gas was discovered in the area. A new canal port was built and huge refining and petrochemicals complexes were set up; today rubber, cement and fertilisers are also produced.

Population 136 700
Map Italy – Db

Ravensbrück *Germany* Site of a Nazi concentration camp for women, about 80 km (50 miles) north of the capital, BERLIN.
Map Germany – Cb

ravine Narrow, steep-sided depression, which is smaller than a VALLEY or CANYON but larger than a cleft or GULLY.

Rawalpindi *Pakistan* Ancient town in the northern Punjab on a trans-Indian highway known in the days of the British Raj as 'the Grand Trunk Road'. The national capital from 1958 to 1960, Rawalpindi was the military headquarters for British control of the North-West Frontier Province; it is now the headquarters of the Pakistani army. Its industries centre on engineering, oil refining and the manufacture of chemicals.
Population 1 076 000
Map Pakistan – Da

Rayong *Thailand* Province and fishing port on the Gulf of Thailand, 145 km (90 miles) southeast of the capital, BANGKOK. Since 1980, the town has developed rapidly as a gas processing centre for offshore gas fields. The main crop grown in the province is cassava.
Population (town) 129 100; (province) 410 600
Map Thailand – Bc

Ré, Ile de *France* Flat island, 28 km by 5 km (17 miles by 3 miles), just west of LA ROCHELLE on the Bay of Biscay. It is noted for its beaches, oyster beds, vineyards and market gardens.
Population 14 000
Map France – Cc

reach Unbroken stretch of water, between two bends or locks, especially on a river or canal.

Reading *United Kingdom* County town of Berkshire, England, standing on the River Thames, 60 km (37 miles) west of LONDON. It has a university and a wide range of industries, including electronics and engineering. The town's prison was made famous by the Irish writer Oscar Wilde, who composed *The Ballad of Reading Gaol* in 1898, after he had served two years' hard labour here for homosexual offences.
Population (city) 136 200 ; (metropolitan area) 198 341
Map United Kingdom – Ee

Recent An alternative name for HOLOCENE period. See GEOLOGICAL TIMESCALE, p 256

Rechna Doab *Pakistan* Region between the Ravi and Chenab rivers in the Punjab. It grew in importance after the building of irrigation canals in the 19th century, and it remains an important area for

agriculture – especially the cultivation of wheat, cotton and rice.
Map Pakistan – Db

Recife (Pernambuco) *Brazil* State capital of Pernambuco near the country's north-eastern tip. Of all South America's major cities and ports, Recife lies nearest to Europe and Africa. Because the city is crisscrossed by canals and the channels of the Capiberibe River, it is frequently referred to as 'Brazil's Venice'. From 1629 to 1654, Recife was the centre of Dutch Brazilia; the Dutch architectural influence can still be seen in the villas and houses in the city.

The surrounding state, which covers some 98 281 km² (37 936 sq miles), was one of the earliest areas settled by the Portuguese. The brazilwood tree, after which the country is named, grew in the original Atlantic rain forest of the coast around Recife. (The trees were felled for their highly prized red dye, brazilin.)

The land is ideal for the cultivation of sugar cane; in the 19th century, when it was the centre of the country's sugar-cane region, many African slaves were brought here to work on the cane plantations. Today cotton, tobacco and fruits are grown in addition to the sugar cane.
Population (city) 1 335 700; (state) 7 225 000
Map Brazil – Eb

red earth Non-scientific term applied to tropical soil produced by intensive chemical weathering under conditions of high temperatures and humidity, and markedly seasonal rainfall. The result is a loamy mixture of clay and quartz coloured by iron compounds. It can be up to 15 m thick. Red earth should not be confused with LATERITE.

red mud Mud derived from erosion of the land surface, and found on the CONTINENTAL SLOPE near the mouth of certain rivers. The colour is due to iron oxides.

Red River *Canada/USA* River, some 877 km (545 miles) long, which rises in the United States in Lake Traverse on the South Dakota-Minnesota border and flows north into Lake Winnipeg in south Manitoba. Its valley, first settled in 1812, is one of Canada's best farming areas.
Map Canada – Fd

Red River (Song Hong; Yuan Jiang) *China/Vietnam* River that rises in the Chinese province of YUNNAN and flows for about 1200 km (nearly 750 miles), past the Vietnamese capital, Hanoi, and on into the Gulf of TONGKING through a delta up to 80 km (50 miles) wide.

The river carries into the delta thousands of tonnes of the red soil that gives it its name, flooding surrounding flatlands with a rich silt. In parts, it runs between banked-up levees with the water level as much as 12 m (40 ft) above the plain. This makes irrigation simple. However, flooding, when it happens, is severe.
Map China – Ef-Ff; Vietnam – Aa

Red Sea *Middle East* Branch of the INDIAN OCEAN, covering some 438 000 km² (169 073 sq miles) between north-east Africa and the Arabian Peninsula, occupying part of the GREAT RIFT VALLEY system. The Strait of Bab al Mandeb and the Gulf of Aden link the Red Sea to the Indian Ocean. In the south there are numerous volcanic islands – which signify tectonic activity, respon-

sible for the formation of the Great Rift Valley and for OCEAN FLOOR SPREADING which is occurring along a valley in the centre of the sea and pushing the Arabian plate north-east. The Red Sea owes its name to a species of algae which imparts a red bloom to the water. Its greatest known depth is 2246 m (7370 ft).

The Red Sea lies in a hot and humid region, in which water temperatures in summer reach 29°C (84°F). In 1963, rich deposits of zinc, copper and other minerals were discovered on the seabed in the north and central parts.

The sea was a major shipping lane in antiquity, but its importance declined after the discovery of the sea route to India around southern Africa. However, after the opening of the SUEZ CANAL in 1869, it once again became a major transit route.
Map Saudi Arabia – Ab

Redwood National Park *USA* Forest and shore area covering 424 km² (164 sq miles) along the Pacific coast in the extreme north of California between Eureka and Crescent City. The park was established in 1968 to preserve one of the few remaining areas of California redwood trees. The tallest known tree, at 112 m (368 ft), is in the park.
Map United States – Bb

reef Strip or ridge of rocks, sand or soil rising from the seabed and extending to just below the surface of the water, and sometimes exposed at low tide.

re-entrant Marked indentation into an upland area or into a steep slope.

reg Flat, stony desert plain, especially in the Sahara, where tightly packed sheets of wind-scoured gravel and pebbles cover the surface.

Regensburg *Germany* River port and tourist centre on the confluence of the Regen and the Danube, 110 km (68 miles) north-east of Munich. The town was founded by the Celts; in AD 150, the Romans built a military camp called Castro Regina beside the Celtic settlement. From the 6th century to the 13th century, when Bavarian princes resided here, Regensburg became an important trade centre.

It stands as a living museum of German architecture from the Middle Ages to the 20th century, with more than 20 Gothic or Romanesque churches and a 13th-century cathedral. The first German parliament, or Diet, met in the town hall in 1663, and Diets were held here until 1806, when Napoleon, the French emperor, abolished them. The city has a university and several other educational institutions, including an academy of music. Its industries include textiles, metal goods, chemicals and electronics.
Population 121 700
Map Germany – Ed

Reggio di Calabria *Italy* Seaport and tourist resort at the end of the toe of Italy, founded in the 8th century BC. It was rebuilt after being destroyed by an earthquake in 1908. Its industries include olive oil, pasta and fruit canning, and its port handles agriculture products such as citrus fruits, grapes and fruits grown in the province of Calabria. There is a ferry link, via the Strait of Messina, with Sicily.
Population 178 500
Map Italy – Ee

Reggio nell'Emilia *Italy* Town situated 60 km (37 miles) north-west of BOLOGNA. It was a Roman settlement and the Roman Via Emilia (built in 187 BC) runs through the heart of the city. Its industries include electrical equipment, cement and processed food.

Population	130 800
Map	Italy – Cb

Regina *Canada* Capital of SASKATCHEWAN on the Trans-Canada Highway, and centre of one of the world's largest wheat-growing plains. The university city has been extensively landscaped and the provincial government buildings are set in 920 hectares (2270 acres) of lakes (notably Wascana Lake) and woods. Regina was founded in 1882 on the Canadian Pacific Railway as this progressed westwards.

Regina was the headquarters of the North-west Mounted Police from 1892 to 1920 when it became the western headquarters of the renamed Royal Canadian Mounted Police. Today its industries revolve around oil, gas, steel, chemicals, electrical equipment and dairy products.

Population	191 180
Map	Canada – Ec

regolith Term that is applied to the layer of loose material (including soil, sand, rock fragments, volcanic ash and glacial DRIFT) which covers the solid bedrock, and which is subject to weathering.

Rehovot (Rehovoth) *Israel* Town on the Judaean plain, 21 km (13 miles) south-east of TEL AVIV. It is the centre of an area that grows citrus fruits and produces fruit juice and dairy products, and metal goods, glass and pharmaceuticals. The tomb of Chaim Weizmann, Israel's first president (1949-52), and the Weizmann Institute of Science are at Rehovot.

Population	70 000
Map	Israel – Ab

Reichenbach *Poland* See DZIERZONIOW

Reichenberg *Czech Republic* See LIBEREC

Reims (Rheims) *France* Historic city on the Marne River, and centre of the Champagne wine industry. It is famed for its massive 13th to 14th-century cathedral, which is longer and higher than the Cathedral of Notre Dame in Paris. Until 1830, the kings of France were crowned here. In 1429, Joan of Arc, then 17 years old, led an army of 12 000 to Rheims to have the Dauphin crowned as Charles VII.

Most of the town was destroyed in the First World War, and the cathedral was not completely restored until 1938. The city's technical college contains a famous room which General Dwight Eisenhower (1890-1969), the Supreme Allied Commander, used during the Second World War, as his headquarters. The formal German surrender was signed in the room on 7 May, 1945.

Population	(city) 181 000; (metropolitan area) 206 000
Map	France – Fb

Reinga, Cape *New Zealand* Headland at the northern tip of North Island. According to Maori folklore, it is the place from where the souls of the dead depart on their journey back to their legendary Pacific homeland, Hawaiki.

Map	New Zealand – Da

rejuvenation The term literally means to become young again, and is applied to the development of younger surface forms, such as the renewed down-cutting of a river caused by a lowering of sea level or an uplift of the land through which the river flows.

relative humidity See HUMIDITY

relief The shape of the earth's surface; the variations in its height and steepness of slopes.

Renfrew *United Kingdom* See STRATHCLYDE

Rennes *France* Industrial city and the capital of Ille-et-Vilaine Department, 100 km (62 miles) north of NANTES. It was severely damaged by fire in 1720 and rebuilt with spacious squares and elegant houses. The 17th-century parliament house was restored. Rennes' industries, several of which have moved to the city from Paris, include vehicle assembly, electrical goods, clothing and chemicals.

Population	(city) 197 600; (metropolitan area) 245 800
Map	France – Cb

Reno *USA* Gambling resort in western NEVADA, 16 km (10 miles) from the California state border. It used to be a place for quick and easy marriages and divorces, but since divorce laws have been liberalised in most states it has lost this appeal. Today, Reno is a base for winter and water sports in the surrounding mountains and lakes.

Population	(city) 133 900; (metropolitan area) 254 700
Map	United States – Cc

Resht *Iran* See RASHT

Réthimnon *Greece* Town on the north coast of Crete, 60 km (37 miles) from HERAKLION. Many buildings, including the loggia of the old governor's palace and the church of San Francesco, remain from its days as a Venetian provincial capital. The monastery of Arcadi lies 23 km (14 miles) to the south. In 1866, several hundred Cretans who had fled here to escape invading Turks, blew themselves up with explosives rather than be taken prisoner.

Population	17 700
Map	Greece – Dd

Réunion *Indian Ocean* A small piece of France lying east of Madagascar, in the south-west Indian Ocean. The island of Réunion, formerly Bourbon, has been a French overseas department (administered on the same basis as departments in mainland France) since 1946, although it was first colonised in 1638 and belonged to the British from 1810-15.

Control of Réunion is expensive, and some argue that it brings France little reward; the inhabitants, too, are divided about the potential advantages of political independence. The present arrangement at least gives the young people access to jobs in France; unemployment is a growing problem in Réunion.

The island covers 2512 km² (970 sq miles). It is volcanic like its nearest neighbour, MAURITIUS, and has one active volcano and nine dormant cones. Its two mountain masses, with peaks reaching 3069 m (10 069 ft), provide spectacular scenery. They are divided by a depression and

skirted by coastal plains. As in Mauritius, the economy is based on sugar, which is grown on nearly 70 per cent of the cultivated land. Sugar and its by-products, molasses and rum, are the main exports through the chief port and capital city of SAINT-DENIS. Vanilla, tea, maize, potatoes and plants grown for perfume oil are also cultivated. Livestock is raised on the high pastures.

Tourists are attracted by Réunion's tropical climate and superb beaches with their offshore coral reefs, but the remoteness of the island limits the tourism potential. About a quarter of the population is of French origin. Others are a mixture of Indian descendants of the workers once brought to farm the plantations (who retain their Hindu culture), Malays, Indo-Chinese and the Creole descendants of French settlers and African slaves.

Population	598 000
Map	Indian Ocean – Bd

Reutlingen *Germany* Textile-making city amid the vineyards of the Neckar valley, some 30 km (19 miles) south of STUTTGART. Some parts of its medieval walls and gates, as well as a 13th-century church, remain.

Population	96 300
Map	Germany – Cd

Reval (Revel) *Estonia* See TALLINN

Reykjanes *Iceland* Peninsula in the south-west of the island. The town of KEFLAVÍK is on the peninsula's northern side. Hot springs on the peninsula are used to provide domestic heating.

Map	Iceland – Bb

Reykjavík *Iceland* Capital and principal port in the south-west of the island on Faxaflói Bay. According to tradition, the city was founded by Ingolfr Arnarsson, a Viking settler, in AD 874. It became a trading post in the 18th century, when a cathedral was built. Reykjavík became the seat of government in 1845; a university was founded in 1911.

Industries include fishing, fish processing, metalworking, shipbuilding, clothing manufacture and printing. Nearby are cement and fertiliser plants. Reykjavík handles about 70 per cent of Iceland's trade; its airport is at KEFLAVÍK.

Many of the city's buildings are heated by water piped from nearby volcanic hot springs. An open-air swimming pool in the city is heated so effectively by this method that it remains in use all year round. The steam heat has also given the city its name: Reykjavík means 'smoking bay'.

Population	97 600
Map	Iceland – Bb

Reynosa *Mexico* Dusty town on the Rio Bravo (Río Grande) in the eastern state of Tamaulipas on the border with Texas. Founded by the Spanish in 1749, it burgeoned in the 20th century following the discovery of oil and natural gas. Now the Gulf Plain around Reynosa produces much of Mexico's gas. Reynosa has become a major centre for commerce, textiles, agriculture (mainly cotton) and stock breeding.

Population	281 800
Map	Mexico – Cb

Rharb (Gharb) *Morocco* Fertile plain north of Rabat, extending for 80 km (50 miles) along the Atlantic coast and 110 km (70 miles) inland. Crossed by the Oued Sebou River, it is an

important agricultural region, producing wheat and citrus fruits. During French colonial times the land was greatly improved by irrigating dry areas, draining swampland and introducing modern agricultural methods. Rharb is also a source of phosphates, rock salt and petroleum.

Map Morocco – Ba

Rheims *France* See REIMS

Rhine (Rhein; Rhin; Rijn) *Western Europe* Major European river, some 1320 km (nearly 825 miles) long. It rises in the Swiss Alps, and proceeds to follow a mountain course as far as Lake CONSTANCE. Its waters traverse the lake and plunge over the Rhine Falls at SCHAFFHAUSEN westwards to BASLE. Here the river abruptly turns to the north, and enters the Rhine rift valley. The rift, let down between parallel faults, has a broad, flat floor, with the VOSGES Mountains of France rising on the west side and the BLACK FOREST massif on the east.

The rift continues for 300 km (nearly 190 miles); at its end, in Hessen, the Rhine turns west, then north-west, to cut through the Middle German upland in its famous gorge. It emerges at BONN into the North German Plain, flows down past COLOGNE and the RUHR and, just after reaching the Dutch frontier, divides into three, forming the Waal, the most southerly and the main channel, which reaches the sea via Nijmegen, the Lek, which runs farther north, past Arnhem, and the IJSSEL, which flows into IJsselmeer.

The Rhine today is fully navigable up to Basle. Large-scale work was done between 1817 and 1876 to shorten its winding course, remove obstacles and rapids, reduce flooding and reclaim farmland. Since the Second World War, the greatest changes have been made by the French, between Strasbourg and Basle, with the building of the Grand Canal d'Alsace. In 1994 and again in 1995, the Rhine flooded causing severe damage to towns, cities and farmland – the 1995 floods were reported to be the worst this century.

Freight transport on the Rhine dwarfs that carried on all other West European waterways. The heaviest freight movements occur between the Ruhr industrial region and the Dutch port of Rotterdam. Iron ore, coal and oil products, and grains are principal cargoes, but so are cars and building stone. Inland from Rotterdam, the largest ports on the river are DUISBURG and LUDWIGSHAFEN-MANNHEIM.

To the tourist, the Rhine evokes vineyards and legends. The vineyards are in the rift valley and on the steep sides of the gorge. The legends are everywhere. The Lorelei rock is named after the Siren whose singing lured sailors to their doom. The Rhine is badly polluted. One incident in 1986 resulted in 30 tonnes of poisonous waste going into the river from a Swiss chemical works. Efforts are under way to cut down the pollution levels.

Map Germany – Cd

Rhineland-Palatinate (Rheinland-Pfalz) *Germany* State stretching along the west bank of the RHINE, south of Bonn. It borders on Luxembourg, Belgium and France. Rhineland-Palatinate was ruled from the 10th to the 19th century by the Counts Palatinate of the Holy Roman Empire. Its wooded hills include the EIFEL and the wine-growing region of the lower Moselle.

Population 3 785 100

Map Germany – Bc

▲ OLD TOWN, YOUNG WINES The lovely town of Bacharach, founded here before AD 900, stands in the Rhine gorge below Stahleck Castle. The surrounding vineyards are noted for their pungent, full-bodied wines, best drunk while they are young.

rhinn In Scotland, a rugged ridge.

Rhode Island *USA* The country's smallest state, covering some 3140 km² (1212 sq miles) on the Atlantic coast between Connecticut and Massachusetts. It was established in 1636 by Roger Williams, a Baptist dissenter from the religious rule of the Puritans in Massachusetts Bay Colony. Rhode Island was one of the original 13 states and became a haven of religious freedom. The oldest synagogue in the United States is also on Rhode Island.

Agriculture centres chiefly on greenhouse and nursery products, potatoes, chickens, eggs and dairy produce. The principal manufactures are machinery, textiles and metals. Fishing and tourism thrive along the coast. PROVIDENCE is the state capital and biggest port, with a yachting marina in NEWPORT.

Population 1 003 500

Map United States – Lb

Rhodes *Greece* Largest of the DODECANESE Islands, covering 1399 km² (540 sq miles) in the south-eastern Aegean Sea, and the most developed of the Greek island tourist centres. Rhodes was settled in about 1400 BC and later became a successful maritime trading nation, forming the first rules of the sea, the Rhodian Laws, in about

900 BC. The town of Ródhos, the island's capital, was founded in 408 BC. In 305 BC, after beating off an attack by the Macedonians, the islanders erected one of the Seven Wonders of the World, a 32 m (105 ft) high statue of the sun god Helios, at Rhódos harbour. The Colossus of Rhodes, as it was called, crumbled in an earthquake in 224 BC, as did the town.

The Knights of St John, a multinational European military society who had their headquarters at Rhodes from the 13th to the 16th century, made the island a place of pilgrimage before moving to Malta. After the Knights had departed, the island came under Turkish rule. In 1912, however, it again changed hands, this time coming under Italian rule. The Italianate Grand Master's Palace and the Gothic Governor's Palace still stand near the Street of the Knights. Rhodes was ceded to Greece in 1947. Today it has many modern hotels; and produces olives, citrus fruits and wines.

Population 88 500
Map Greece – Ec

Rhodope Mountains (Rodopi Planina) *Bulgaria/Greece* Forested mountain system, extending 290 km (180 miles) south-east from the Maritsa River into north-eastern Greece. During the Roman Empire, it marked the boundary between Thrace and Macedonia. Today its many rivers are harnessed for hydroelectric power. The region has many spas.

Map Bulgaria – Ac

Rhódos *Greece* See RHODES

Rhône *Switzerland/France* River, some 812 km (505 miles) long, 522 km (324 miles) of which are located in France. It rises in the Swiss Gotthard massif, flows through Lake Geneva and enters France through the Jura Mountains.

The Rhône then flows south and west to Lyons where it is joined by the Saône, and on south, to the Mediterranean. South of Lyons, 12 major projects built since the 1930s tap the river for hydroelectric power and irrigation, and tame it for navigation.

Map France – Fd

Rhône-Alpes *France* Large region of eastern France, covering 43 698 km² (16 872 sq miles). It consists of the eastern fringe of the Massif Central, the valleys of the Rhône and Saône rivers, and the northern Alps. Economically it is the most developed region in the country, after the area around Paris. The main city is LYONS.

Population 5 351 000
Map France – Fd

rhyolite Fine-grained, acidic IGNEOUS ROCK consisting largely of quartz, feldspar and mica. It is the extrusive, or volcanic, equivalent of granite.

ria Long, narrow branching inlet caused by drowning of a narrow, steep-sided river valley through a rise in sea level. The River Fal in Cornwall is typical. Examples are also found in south-west Ireland and north-west Spain. Unlike a FJORD, a ria deepens towards the sea.

Riau *Indonesia* Province covering the east of Sumatra opposite Singapore. It is the centre of the nation's oil industry and is also a tin-mining area.

Population 2 168 500
Map Indonesia – Bb

Ribe *Denmark* One of Denmark's most picturesque market towns, situated in south-western JUTLAND, where Denmark's first church was built in 854. The church is gone, but the Romanesque cathedral, dating from about 1130, is one of the earliest brick-built structures in the country.

The town also has Denmark's oldest provincial museum. Beside the small medieval port, now silted up, are grassy mounds that mark the site of the Riberhus, a medieval royal castle. The Treaty of Ribe, in terms of which Schleswig and Holstein were united, was signed here in 1460.

Population 17 900
Map Denmark – Bb

Riccione *Italy* See RIMINI

Richard Toll *Senegal* Market town located in the Senegal River delta in the north-west. The town is named after Monsieur Richard, a French horticulturalist who founded an experimental garden (*toll* means 'garden') in 1830. Outside the town lie the remains of a mansion built by Baron Roger, Governor from 1822 to 1827. Lying to the east of Richard Toll is an integrated sugar complex which has done much to revive the local economy. The Delta Irrigation Scheme was opened at the head of the delta in 1947, in an attempt to increase the area's rice production.

Map Senegal – Aa

Richard's Bay *South Africa* Developing industrial town and harbour, primarily for the export of mineral ore. It is situated on the KwaZulu-Natal coast, about 160 km (100 miles) north-east of DURBAN. Although the tonnage of cargo landed is relatively small, Richard's Bay handles by far the greatest tonnage of cargo exported from any African port.

Population 23 000
Map South Africa – Db

Richmond *Australia* Town in Tasmania, 18 km (11 miles) north-east of HOBART. It was founded in 1824 and has many historic buildings, including St John's Church (1836-7), Australia's oldest Catholic church. On the mainland, 50 km (about 30 miles) north-west of SYDNEY, lies another town called Richmond, adjacent to WINDSOR.

Population (Richmond, Tasmania) 750;
(Richmond, New South Wales) 9 310
Map (Richmond, Tasmania) Australia – Hg;
(Sydney) Australia – Ie

Richmond *USA* State capital of VIRGINIA. It stands on the James River in central Virginia, 160 km (about 100 miles) south-west of Washington DC. The city is a commercial, educational and historical centre. It was founded in 1637 – one of the United States' earliest English settlements – and became the capital of Virginia in 1780. It was the capital of the Confederacy of the South during the Civil War (1861-5). Richmond's industries today include tobacco, paper making, publishing and food processing.

Population (city) 203 100; (metropolitan area) 865 600
Map United States – Kc

Richter scale Logarithmic scale ranging from 1 to 10, which was devised in 1935 by the American seismologist Charles Richter, and is still used today for measuring the magnitude of EARTHQUAKES.

Rideau Canal *Canada* Canal, some 198 km (123 miles) long, which links the capital, OTTAWA, with KINGSTON, Ontario. It was built between 1826 and 1832 as a British military route secure from American attack. The canal is now used by pleasure craft.

Map Canada – Hd

ridge Long, narrow, steep-sided upland.

ridge of high pressure Elongated extension of high ATMOSPHERIC PRESSURE between two areas of low pressure, responsible for short periods of fine weather during a rainy period.

Riding Mountain *Canada* National park covering 2976 km² (1149 sq miles) in west Manitoba. It is an area of forested highland with several lakes and a wildlife sanctuary for bison, elks, moose, wolves, black bears and deer.

Map Canada – Ec

Riesengebirge *Czech Republic/Poland* See SUDETY

Rif Mountains *Morocco* Coastal mountain range in the north-east of the country, extending 400 km (248 miles) from GIBRALTAR and TANGIER over CEUTA to MELILLA, and separated from the Middle ATLAS MOUNTAINS by the TAZA Gap. The area, which was originally forested but is now heavily eroded with deep valleys and rugged peaks, is difficult to penetrate and has been the retreat of pastoral Berber people threatened by invaders from the 7th century to modern times. In the east, the great bulge of the mountains reaches a height of 1800 m (5900 ft); in the centre the highest point is Tidirhine, at 2456 m (8058 ft).

Map Morocco – Ba

rift valley Long, narrow depression in the earth's surface, formed when the land sinks between two fairly parallel faults, a structural formation known as a 'graben'. One of the most spectacular is the GREAT RIFT VALLEY, extending from the valley of the River Jordan south through East Africa.

Riga *Latvia* Capital and port on the Baltic Sea, 290 km (180 miles) north of the border with Poland. Riga was founded in 1190 and became a major trading centre in the 13th century. It was the capital of independent Latvia from 1918 until 1940. During the First and Second World Wars it was occupied by German troops.

The medieval old town on the Dvina River is a charming jumble of streets, parks, river banks and a canal; its most imposing buildings are the early 13th-century Doma Cathedral, and Pils Castle, built between 1328 and 1515.

Riga's main industries include marine engineering, shipbuilding, chemicals, glass, cement, textiles, electrical machinery and electronics.

Population 910 000
Map Latvia – Aa

Rijeka (Fiume) *Croatia* Seaport at the mouth of the Rijeka River, 129 km (80 miles) south-west of ZAGREB. It was a focus of international dispute during the First World War because of its mixed Italian and Croat population, and was the Free City of Fiume from 1920 to 1939. The city reverted to Croatia after the war and grew rapidly as the country's leading port and shipbuilding centre. Its position on Bakar Bay, its good road and rail communications, and plentiful timber and hydroelectric power from the Gorski Kotar Mountains to the east, have fostered engineering, paper and textile manufacturing, food processing and oil-refining industries.

Population 206 000
Map Croatia – Bb

Rijn *Western Europe* See RHINE

Rikuchu Kaigan *Japan* National park, covering 124 km² (48 sq miles), on the wild Pacific coast of northern HONSHU Island, about 460 km (285 miles) north-east of the capital, Tokyo. The region, which has spectacular cliffs and white beaches, has been repeatedly devastated by tsunamis – giant waves generated by underwater earthquakes. One such wave, the Great Sanriku Tsunami of 1896, killed 23 000 people.
Map Japan – Dc

rime Accumulation of ice crystals on the windward side of exposed objects, such as trees or telegraph poles. It is formed when water droplets in a fog or mist are driven by a slight wind and freeze on contact with the exposed objects.

Rimini *Italy* Seaside resort on the Adriatic Riviera, some 155 km (96 miles) south of VENICE. Medieval Rimini is preserved in the old town centre. Standing here is the Malatesta Temple, a huge Renaissance mausoleum; the town also has the oldest Roman triumphal arch still standing in Italy (it was built in 27 BC).

Rimini has a stretched-out sandy beach and its many hotels, villas and recreational facilities attest to the fact that it is a popular tourist destination. The resort of Riccione (population 31 300), which lies 10 km (6 miles) south-east of Rimini, has annual film, drama and art festivals.
Population 129 900
Map Italy – Db

Rîmnicu Vîlcea *Romania* City on the River Olt, some 150 km (93 miles) north-west of BUCHAREST. Rîmnicu Vîlcea stands at the neck of a wooded valley and produces timber, leather and fur goods.
Population 78 000
Map Romania – Aa

Río Bravo *USA/Mexico* See RIO GRANDE

Rio de Janeiro *Brazil* Metropolis, port and the national capital from 1763 to 1960, when the purpose-built city of BRASÍLIA took over. Rio de Janeiro – 'Rio' for short – is one of the most dramatically sited cities in the world. The Portuguese, who discovered the spot on New Year's Day 1502, named it 'January River' because they believed, wrongly, that the entrance to the huge Guanabara Bay – on which Rio now stands – was the mouth of a great river.

The bay is dotted with rocky, palm-covered islands and around it are steeply domed mountains, some, like the celebrated Sugar Loaf, reaching to the sea. Another domed peak – Corcovado (translating as 'Hunchback'), which rises to 704 m (2310 ft) – is topped by a statue of Christ, 35 m (115 ft) tall, with arms outstretched over the city.

Rio de Janeiro itself is crowded onto a narrow shelf of land around the bay, and has also grown into the mountain slopes and the hills. Nowhere are the sharp contrasts of Brazilian wealth and

A DAY IN THE LIFE OF A MEXICAN GARAGE OWNER

When a Chevrolet with US licence plates pulls up to the Frontera Garage in Nuevo Laredo at 11 am, garage-owner Luis Alvarez says a silent prayer of thanks to the picture of the Virgin Mary above the doorway.

Business has been bad the past week and he has sold only a couple of tyres and fixed two engines – half a normal week's work. And an American customer means dollars, not inflated pesos. This customer – like most who come to this border-town mechanic's shop – is a Chicano, or Mexican-American, crossing the Rio Grande from Texas for a few hours to buy cheaper clothing, food, maybe even to visit the bullring, and, in this case, to buy a new fan belt.

Luis, like many of Nuevo Laredo's 200 000 residents, stakes his livelihood on the thousands of tourists, day-shoppers and migrant workers who cross the border. He and his cousin Pepe bought the property five years ago. With its cattle and natural gas production, and greatly improved irrigation schemes, the city had blossomed since the 1950s and the tourists poured in. But since the economic crisis caused by the tailing off of the oil boom, along with high interest on Mexico's foreign debt, Luis's income has been eroded by inflation and the slide in the peso's exchange rate against the US dollar.

Fortunately, his wife Lupe earns money working part-time in a supermarket. And Nuevo Laredo lies in the 26 km (16 mile) wide duty-free zone, which means Luis can import US car parts without paying import taxes.

Luis hopes his 14-year-old son Carlos will attend engineering school, and that one of his three young daughters will marry a man who could take over the running of the garage. Then he could retire at 60.

Luis's musings are interrupted by a honking car. A fair-haired American couple in a Toyota have their radiator replaced and ask for directions to Mexico City. Luis practises the English he learned while working for four years in California as a mechanic and night security guard.

By the time the car leaves, the temperature has climbed to 32°C (90°F) and Luis realises it is 2 pm: time for lunch.

The three-hour lunch break is the only period that Luis devotes entirely to his family, which includes Lupe's mother. Today, the women serve consommé, rice, refried kidney beans (*refritos*) and white flour tortillas, or flat pancake-like bread.

Luis reopens the garage at 5 pm and works on repairing the axle of his best friend Paco's Buick. When Paco collects the car three hours later, he reminds Luis that several friends are waiting at the El Presidente *cantina* (bar) down the road. There the men end the day with jokes, gossip and argument as they drink tequila with lime and salt before finally making their way home.

poverty more apparent. The sleek beachfront suburbs of Copacabana and Ipanema – whose names have become familiar the world over through popular songs – have as their backdrop, wherever the slopes are too precipitous for normal housing, the huge *favelas*, or shanty towns, of the poor. Rio has magnificent museums, beautiful parks and, in the Maracāna Stadium, the world's largest soccer arena, with space for 205 000 spectators.

The city is a commercial, financial and industrial centre and a major tourist attraction. Today its main industries include clothing, furniture, chemicals, shipbuilding, glass, tobacco products and processed food. Its main tourist attractions are the Copacabana and the carnival.
Population 5 487 400
Map Brazil – Dd

Rio Grande *Guinea/Guinea-Bissau* See CORUBAL

Rio Grande (Río Bravo) *USA/Mexico* River, 3078 km (1885 miles) long, rising in the San Juan Mountains in south-west Colorado, United States. It flows south through central New Mexico, then meanders south-east from El Paso in Texas, and Juarez in Mexico, forming the border between the two counties. It finally empties into the Gulf of Mexico between Brownsville (its chief port) in the USA and Matamoros in Mexico.

The river is dammed in many places for irrigation and flood control. It is unnavigable, except near its mouth. In Texas, it flows north-east through the canyons of the Big Bend National Park. In Mexico, from which illegal immigrants have entered the United States by swimming the river, it is known as the Río Bravo.
Map Mexico – Bb

Río Grande de Santiago *Mexico* See LERMA

Rio Grande do Norte *Brazil* See NATAL

Rio Grande do Sul *Brazil* See PORTO ALEGRE

Río Muni *Equatorial Guinea* Mainland part of the country, covering 26 017 km² (10 043 sq miles) and at one time called Mbini. Timber, coffee and cocoa are its chief products. The main population group are the Bantu-speaking Fang.
Population 241 000
Map Equatorial Guinea – Bb

Riobamba *Ecuador* Capital of Chimborazo Province, lying at a height of 2700 m (8860 ft) on a flat plain between the Chimborazo and Sangay volcanoes. Its attractive colonial buildings include the Convento de la Concepción, which houses the Museum of Sacred Art. The original town, founded in 1534 on the site of Cajabamba 25 km (15 miles) to the south-east, was destroyed in 1797 by a landslide which killed more than half of its 9000 inhabitants. Riobamba manufactures textiles, carpets and footwear.
Population 94 500
Map Ecuador – Bb

Rioja, La *Spain* Autonomous region south of the Basque country. It produces Spain's best table wines as well as cereals, asparagus, peas and apricots. Textiles are produced in the region, and lead, copper and coal are extracted from its mines. La Rioja's capital is LOGRONO.
Population 263 430
Map Spain – Da

rip (rip current; rip tide) Turbulence and agitation in the sea (or sometimes a river), caused either by the meeting of tidal streams or a tidal stream suddenly entering shallow water, or by the return of water piled up on the shore by strong waves.

Ripon *United Kingdom* Cathedral city in the English county of North Yorkshire, some 35 km (22 miles) north-west of YORK. The cathedral was built between the 13th and the 16th century, but has an earlier Saxon crypt. In the market place, a civic official known as the Wakeman blows his horn at 9 o'clock each evening, a custom going back more than 1000 years, to announce the setting of the watch for the night.

The spectacular ruins of Fountains Abbey, the greatest Cistercian remains in Britain, lie to the south-west of the city. It was founded in 1132 and grew rich and splendid on the proceeds of the Cistercians' skill as sheep farmers. It was dissolved by Henry VIII in 1539 and gradually fell into decay.

Population 13 230
Map United Kingdom – Ec

river A natural stream of water flowing, regularly or intermittently, in a definite channel towards a sea, lake, marsh, another river or a depression in a desert basin. See NATURE'S WATERY HIGHWAYS; HOW WATER SHAPES THE LAND, p 571

river terrace Stretch of level land covered with ALLUVIUM and lying along a river valley above the level of the river's present FLOOD PLAIN. A river terrace develops when the river cuts through the old flood plain into a deeper valley. Each pair of terraces in a paired series marks the level of a former flood plain.

Riyadh *Saudi Arabia* Capital and commercial centre, Riyadh is situated almost in the centre of Saudi Arabia in an oasis some 680 km (about 420 miles) west of MEDINA. Its old walls were pulled down in the 1950s to make way for new buildings, and it is now a city of sharp contrasts. The poor live in central Riyadh and the better-off in the modern suburbs. In the centre there are traditional souks (bazaars) dealing in gold, silver and ornamental coffee pots. Looming over the city, the tall, modern water tower has become a landmark, dwarfing the still surviving clay forts.

Until recently, Riyadh was one of the most isolated and least known cities in the world. Extremely puritanical, Riyadh is the centre of Wahhabism, the Muslim sect founded in the 18th century by Muhammad Abdul Wahhab and noted for its strict observance of the original words of the Koran.

Due to its position high in the central NEJD, Riyadh has a pleasant and healthy climate. There is an international airport and also railway links with DAMMAN on The Gulf. Until 25 years ago, the city was officially the capital though it was not in practice. The last foreign embassies left Jeddah only in the 1970s. Since then, Riyadh, which is both modern and clean, has expanded rapidly, from covering 85 km^2 (33 sq miles) in 1950 to 1600 km^2 (618 sq miles) today.

Riaydh's places of interest include the king's palace, numerous mosques and Saudi Arabia's first museum – the Museum of Archaeology and Ethnology.

Population 1 300 000
Map Saudi Arabia – Bb

Rizal *Philippines* Province of central LUZON bordering on Laguna de BAY. Those parts that had become heavily built up and industrialised (including Caloocan, Makati and Quezon City) were made part of Metro MANILA in 1976, leaving the rest as a mainly farming and fishing area.

Population 3 754 000
Map (Laguna de Bay) Philippines – Bc

Rjukan *Norway* Town in south Norway, 120 km (75 miles) west of OSLO, beside the 104 m (341 ft) high waterfalls known as Rjukanfoss. It is the site of the original chemical plant of Norsk Hydro, which began making nitrate fertilisers from 1906. One chemical plant – used during the wartime German occupation to make 'heavy water' (a vital component of atomic bombs) – was destroyed by a team of Norwegian commandos in 1943.

Since the war, the area's chemical industry has largely shifted to the towns of Notodden, 60 km (40 miles) to the south-east, and Herøya

Population 7000
Map Norway – Cd

Road Town *British Virgin Islands* Capital and main port of the Caribbean colony, standing on the island of TORTOLA. It is mainly a port of call for cruise ships.

Population 6330
Map Caribbean – Cb

Roanoke *USA* City in the Roanoke valley of southern VIRGINIA, about 80 km (50 miles) north of the border with North Carolina. It is a railway and manufacturing centre which produces electronics, furniture and plastics. Roanoke stands near the Blue Ridge National Parkway, a 750 km (466 mile) scenic road through the Appalachian Mountains.

Population (city) 96 400; (metropolitan area) 224 500
Map United States – Kc

Roaring Forties Sailors' name for the region between latitudes about 40° and 50° south, where the prevailing westerly winds blow unimpeded and with considerable force across the oceans. Depressions bring rainy, but generally mild weather to this zone of squally, stormy seas – notoriously rough off Cape HORN.

Roca, Cabo da *Portugal* Cape lying 35 km (22 miles) west of the capital, LISBON. It ends in a cliff 150 m (490 ft) high. The cape is the westernmost point of continental Europe.

Map Portugal – Bc

Rochdale *United Kingdom* English town 16 km (10 miles) north of MANCHESTER, which was formerly a major centre of the Lancashire cotton industry. Today, the importance of cotton has dwindled. The town, which is now part of Greater Manchester, was the birthplace of the Cooperative Movement in 1884 and of the popular music hall singer and recording artiste Gracie Fields (1898-1979).

Population 97 940
Map United Kingdom – Dd

Rochester *United Kingdom* See MEDWAY TOWNS

Rochester *USA* Industrial city about 10 km (6 miles) south of Lake ONTARIO in western New York State. It is a centre for producing precision instruments and electronics, and is the home of Xerox Corporation as well as the world's largest photographic manufacturer, Eastman Kodak. Rochester has a university; and the renowned Eastman School of Music is also located here.

Population (city) 231 600; (metropolitan area) 1 062 500
Map United States – Kb

rock crystal Transparent colourless QUARTZ.

rock cycle See THE GEOLOGICAL CYCLE: HOW ROCKS ARE FORMED AND BROKEN DOWN, p 573

NATURE'S WATERY HIGHWAYS

Although the world's longest river, the Nile, has an official length of 6695 km (4158 miles), it is actually somewhat shorter than that. This is because it lost part of its watercourse with the creation of Lake Nasser behind the Aswan High Dam in Egypt in 1971. However, geographers do not take this into account when they measure the river's length.

THE WORLD'S LONGEST RIVERS

River	Location	Length (km/miles)
Nile	Africa	6695/4158
Amazon	South America	6440/4000
Mississippi-Missouri	North America	6019/3740
Chang Jiang (Yangtze)	Asia	5980/3716
Ob'-Irtysh	Asia	5570/3460
Zaire (Congo)	Africa	4670/2902
Huang He (Hwang Ho)	Asia	4667/2900
Amur (Heilong Jiang)	Asia	4510/2800
Lena	Asia	4300/2672
Parana	South America	4264/2560
Mackenzie-Peace	North America	4251/2640
Mekong	Asia	4185/2600
Niger	Africa	4170/2590
Yenisey	Asia	4092/2543
Murray-Darling	Australia	3750/2330
Volga	Europe	3688/2292

HOW WATER SHAPES THE LAND

Water is the true genius of landscape, that paints and moulds the basic rough rock of earth's crust. It carves valleys, spreads plains, brings seed, nurtures growing things; after aeons of erosion, it reduces the loftiest peaks down to gentle hills. Even then, its work is not finished, for the destruction of the mountains changes climate and vegetation, and the hills may transform into rich, arable plains. The transition is slow, but eternal, as may be seen by tracing the path of any of the world's great rivers. A typical example is born high on a mountain range, of melted ice and snow. Trickles and streams infiltrate and tear the surrounding rock, then unite into a torrent that grinds pebbles and fragments into sediments which are carried into the quieter reaches of the river below. There, some of the eroded material is deposited as gravel bars and as embankments at river bends, while lighter sediments may be spread by floods to create wide flats of rich, alluvial soil. Throughout its course, the river picks up and lays down material, until it reaches its mouth where the bulk of the sediments are dumped, sometimes creating a delta. The remaining debris is carried out to sea, where it eventually evolves into sedimentary rock.

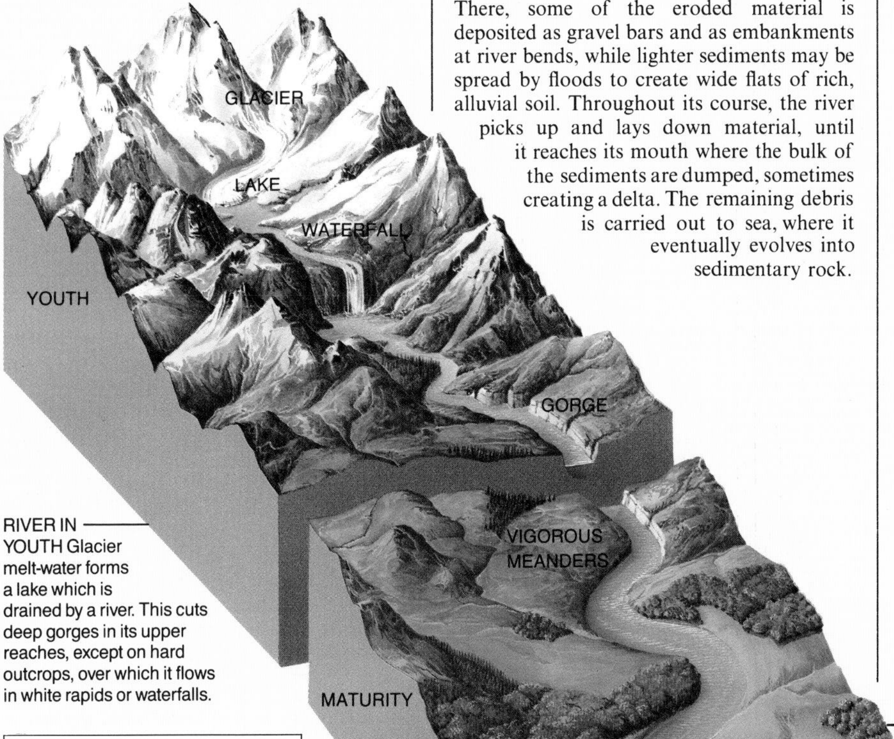

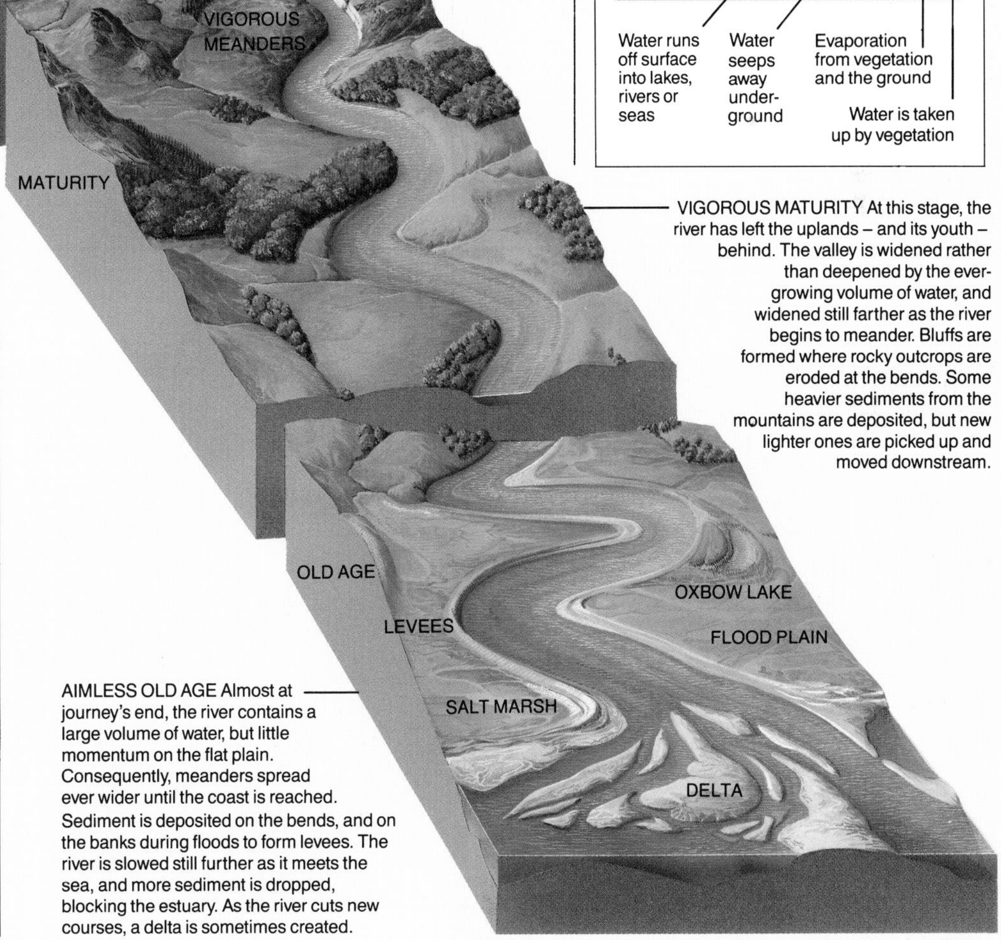

THE HYDROLOGICAL CYCLE

Virtually all the water in the world and in its atmosphere has existed for at least 3000 million years; almost no 'new' water at all has been created in that time. About 97 per cent of it is in the oceans, and a further 2 per cent is locked into the polar ice caps. The remaining 1 per cent is the mainstay of all life on earth, and is driven by the sun in an eternal cycle of evaporation from the sea, precipitation, run-off from the land, and return to the sea via the rivers. Precipitated water is replaced by evaporation every 12 days.

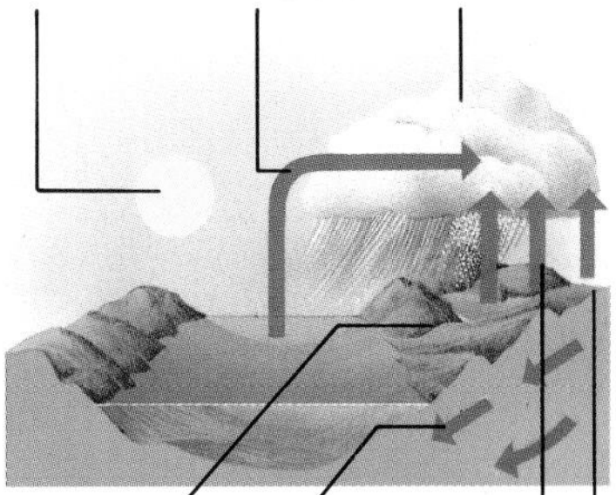

Sun evaporates water from the oceans

Wind moves clouds towards the land

Water precipitated as rain or snow

Water runs off surface into lakes, rivers or seas

Water seeps away underground

Evaporation from vegetation and the ground

Water is taken up by vegetation

GLACIER

LAKE

WATERFALL

YOUTH

GORGE

VIGOROUS MEANDERS

MATURITY

RIVER IN YOUTH Glacier melt-water forms a lake which is drained by a river. This cuts deep gorges in its upper reaches, except on hard outcrops, over which it flows in white rapids or waterfalls.

OXBOW ABANDONED

Meanders or bends in a river's lower course occasionally become so pronounced that the river cuts through the land at the neck of the bend. The stream follows the new course, and the meander becomes a stagnant 'oxbow' lake.

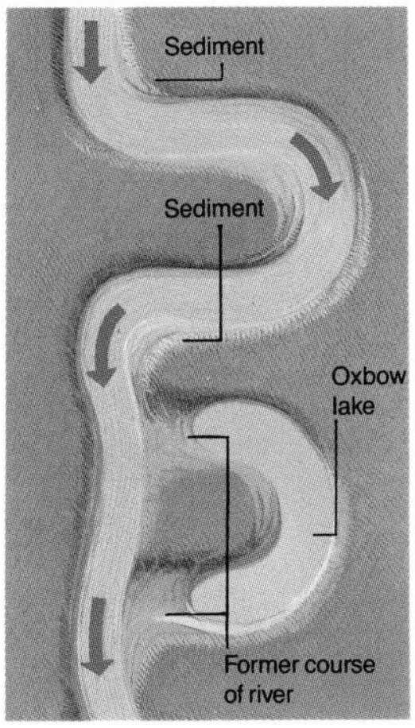

Sediment

Sediment

Oxbow lake

Former course of river

VIGOROUS MATURITY At this stage, the river has left the uplands – and its youth – behind. The valley is widened rather than deepened by the ever-growing volume of water, and widened still farther as the river begins to meander. Bluffs are formed where rocky outcrops are eroded at the bends. Some heavier sediments from the mountains are deposited, but new lighter ones are picked up and moved downstream.

OLD AGE

LEVEES

SALT MARSH

OXBOW LAKE

FLOOD PLAIN

DELTA

AIMLESS OLD AGE Almost at journey's end, the river contains a large volume of water, but little momentum on the flat plain. Consequently, meanders spread ever wider until the coast is reached. Sediment is deposited on the bends, and on the banks during floods to form levees. The river is slowed still further as it meets the sea, and more sediment is dropped, blocking the estuary. As the river cuts new courses, a delta is sometimes created.

rock flour Powdered rock produced when rocks are ground together – for example, along the faces of a moving fault, or during movement of a glacier when pieces of rock carried by the ice abrade each other or the underlying rock.

rock salt See HALITE

Rockall *North Atlantic* Small and uninhabited island of rock about 400 km (nearly 250 miles) to the north-west of Ireland. It is claimed by the United Kingdom.
Map Europe – Bc

Rockhampton *Australia* City situated on the Fitzroy River in Queensland, lying on the Tropic of Capricorn 520 km (nearly 325 miles) north-west of BRISBANE. It is the centre of a coal mining and agricultural region. Rockhampton is known as the nation's beef capital, and has two of the country's largest meat processing plants.
Population 55 800
Map Australia – Ic

Rocky Mountain National Park *USA* Area of 1061 km² (410 sq miles) in the Rocky Mountains to the north-west of DENVER in central Colorado. It contains Ice-Age features, among which are glacial terrains and 155 lakes, as well as alpine valleys and much tundra vegetation. There are more than 100 peaks above 3148 m (10 000 ft) in the park.
Map United States – Eb

▼ REALM OF MISTS Several frost-sculpted peaks rise to more than 3050 m (10 000 ft) among the rivers of ice and lakes of Glacier National Park in the Rocky Mountains of Montana.

Rocky Mountains (Rockies) *North America* Largest mountain system of the continent, extending some 4800 km (3000 miles) from central New Mexico, north-westwards, through the United States and Canada, into Alaska. It approaches the Pacific coast in the north, but for much of its extent it is more than 1000 km (620 miles) inland. The Southern Rockies rise to more than 4000 m (13 132 ft) in the states of Wyoming, Colorado and New Mexico. Mount Elbert in Colorado is the system's highest peak at 4399 m (14 431 ft).

The Rockies, renowned for their spectacular scenery, contain several national parks. They have rich natural resources, including vast forests and minerals such as uranium, gold, silver, lead, zinc, copper, coal, oil and natural gas.
Map Canada – Cc; United States – Da

Rodnei Mountains *Romania* Rugged mountain area in the north of the country, rising to 2305 m (7564 ft) above dense forests. It is noted for its many unspoilt villages rich in folklore. Tourist attractions include the Izvorul Tausoarelor Cave.
Map Romania – Aa

Rodopi Planina *Bulgaria* See RHODOPE MOUNTAINS

Rodrigues Island *Mauritius* Volcanic island, surrounded by fringing coral reefs, lying in the Indian Ocean, 600 km (about 370 miles) east of MAURITIUS. Covering 104 km² (40 sq miles), it has a predominantly Creole farming population. Its main town is Port Mathurin. Rodrigues is in the cyclone-hurricane zone of the Indian Ocean, and is prone to being struck by cyclones in the summer and autumn months.
Population 34 200
Map Indian Ocean – Cc

Romania See p 575

Rome (Roma) *Italy* National capital and the country's largest city, standing on the River Tiber, 25 km (16 miles) from the west coast. Rome, the so-called 'Eternal City', has grown twelvefold since 1870 when it was made the capital – an explosive rate of growth which largely accounts for two of Rome's greatest problems: lack of planning and chaotic traffic.

According to legend, the city was founded by Romulus, the mythical son of Mars, the Roman god of war; according to historians, the city dates from 753 BC. It was built on seven hills and surrounded by the Aurelian Wall. Rome is now more than 10 km (6 miles) wide, not counting shanty towns that have sprung up around it. Its character and architectural splendour derive from two widely separated periods: the first three centuries of the Roman Empire (AD 27-300), and the three centuries of the Renaissance and the Baroque (about 1450-1750).

Although many of the remains have been plundered, the grandeur of the imperial city was such that probably never in history have so many magnificent public buildings been assembled in one place. After the fall of the Roman Empire, there followed a thousand years of neglect and devastation, tempered only by the building of some early Christian churches, and the vast Roman catacombs – an underground system of passages and tombs later used for refuge by persecuted Christians.

In the 15th century, the popes brought the full vigour of the high Renaissance into the city. The notable features of this period are the golden-brown façades of the plasterwork of many buildings, and the beautifully sculpted monuments of Santa Maria del Popolo. The influence of Michelangelo (1475-1564), *(continued on p574)*

THE GEOLOGICAL CYCLE: HOW ROCKS ARE FORMED AND BROKEN DOWN

The surface of the earth is forever changing, subtly but inexorably, in a cycle that takes tens of millions of years to complete before beginning all over again. All of the planet's most powerful forces are concerned in the cycle which begins, or at least recommences, with magma, material that lies beneath the surface crust in a near-molten state and which becomes molten when pressure is released.

Then the magma is expelled through fissures and volcanoes to solidify as IGNEOUS ROCKS. These, hard as they are, when exposed to weathering by wind, rain and frost over many years, are broken down into tiny fragments that are washed down by streams and rivers as sediments to the sea.

In the past 150 million years, for example, the Mississippi has deposited a 12 km (7 mile) thick layer of sediments on to the floor of the

Gulf of Mexico. With the pressure of overlying sediments, deeper layers harden into SEDIMENTARY ROCKS. Some are so deep that the pressure from above and thermal energy from below change them into METAMORPHIC ROCKS.

Where the crustal plates of the earth (see PLATE TECTONICS, p 542) collide, rocks may be pushed up into fold mountains, and these are subjected to weathering. Deep within the mountain core, pressure and heat may turn the rocks to magma, which intrudes into the rocks above and may reach the surface. At ocean trenches, igneous rocks produced at oceanic ridges are returned to the earth's interior. These also replenish the magma supply – and the cycle can begin again.

Weathering can occur in three ways – chemically, mechanically and biologically. In chemical weathering, the oxygen and carbon

dioxide contained in rain and the humic acid in the soil attack rocks, especially in warm climates, dissolving their components.

In mechanical weathering, the rocks become split by frost, or by the varying expansion and contraction of their components during temperature changes. In biological weathering – sometimes regarded as a type of mechanical weathering – rocks are broken down by the roots of plants, or by the action of animals.

Weathered fragments, transported by ice, wind and water, are ground smaller still against other stones. Ice, wind and water, armed with rock fragments, wear away more of the earth's land surface producing more debris. All the debris is carried down to the lowlands and seas. The combined processes of weathering and transport of its products are known as erosion.

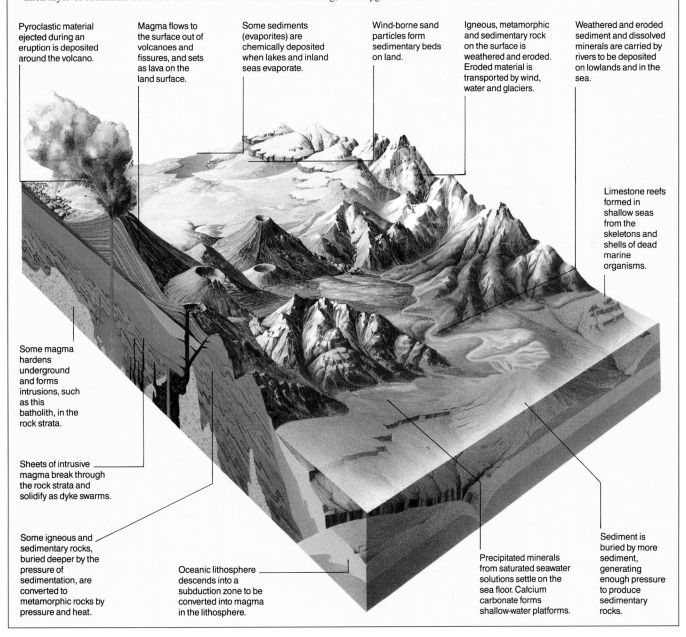

Pyroclastic material ejected during an eruption is deposited around the volcano.

Magma flows to the surface out of volcanoes and fissures, and sets as lava on the land surface.

Some sediments (evaporites) are chemically deposited when lakes and inland seas evaporate.

Wind-borne sand particles form sedimentary beds on land.

Igneous, metamorphic and sedimentary rock on the surface is weathered and eroded. Eroded material is transported by wind, water and glaciers.

Weathered and eroded sediment and dissolved minerals are carried by rivers to be deposited on lowlands and in the sea.

Limestone reefs formed in shallow seas from the skeletons and shells of dead marine organisms.

Some magma hardens underground and forms intrusions, such as this batholith, in the rock strata.

Sheets of intrusive magma break through the rock strata and solidify as dyke swarms.

Some igneous and sedimentary rocks, buried deeper by the pressure of sedimentation, are converted to metamorphic rocks by pressure and heat.

Oceanic lithosphere descends into a subduction zone to be converted into magma in the lithosphere.

Precipitated minerals from saturated seawater solutions settle on the sea floor. Calcium carbonate forms shallow-water platforms.

Sediment is buried by more sediment, generating enough pressure to produce sedimentary rocks.

573

painter, sculptor and architect, was crucial. His architectural genius is clearly illustrated in the Farnese Palace. The city has more than 500 churches – and most of them were built during the Baroque period.

The 19th century saw rapid expansion of the city and the break up of many of the large estates, although the Villa Borghese and many of the major archaeological features such as the Forum and the Caracalla Baths survived. The dictator Benito Mussolini (1883-1945) tried to re-create many of the glorious views of Ancient Rome by clearing away old quarters such as the Borgo, and laying out new triumphal avenues such as the Via della Conciliazione, which leads to St Peter's Square, and the 1932 Via dei Fori Imperiali, which runs from Piazza Venezia to the Colosseum.

Rome is not an industrial city. Apart from the construction industry and those industries automatically part of any large city – such as food, clothing and furniture – there is no manufacturing tradition. Rome is, perhaps above all, a city of bureaucrats working for the national, regional, provincial and communal governments, and for international organisations based here. The city is a major centre of learning – Rome University has 100 000 students – and of publishing, libraries and museums. Cinecittà, the centre of Italy's cinema industry, helps to provide the fashionable Via Veneto with a flow of celebrities.

Places of interest include the VATICAN, the Piazza Campidoglio (designed by Michelangelo in 1536 and containing a 2nd-century AD bronze statue of Emperor Marcus Aurelius), the Spanish Steps (1725) and the Trinitá dei Monti Church (1495-1585), the 17th-century Villa Borghese, enlarged in the 18th and 19th centuries, the Via Appia Antica, the main Roman road to southern Italy opened in the 3rd century BC, the Colosseum (AD 72-80), the Foro Romano (the Roman Forum or square) and the Palatine Hill, the Pantheon (the only building of the Ancient World to have survived virtually complete), and the Arches of Titus (AD 81) and Constantine (AD 315). There are numerous other churches and monuments, including San Giovanni in Laterano, the so-called Cathedral of Rome, rebuilt in 1649, and the Trevi Fountain (1732-62), into which people throw a coin to ensure their return one day to the city.

Population 2 791 400
Map Italy – Dd

▲ ETERNAL CITY The majestic, Baroque colonnade designed by Gianlorenzo Bernini (1598-1680) all but encloses St Peter's Piazza in Rome. It is a fusion of Christian inspiration with classical themes in the city's art.

Roncesvalles *Spain* West Pyrenean village 35 km (21 miles) north-east of PAMPLONA. It is celebrated in literature as the scene, in AD 778, of an Arab victory over Charlemagne, king of the Franks, and the death of his commander, Roland, which inspired the epic poem *Chanson de Roland*.
Population 50
Map Spain – Ea

Ronda *Spain* Historic town, situated 65 km (40 miles) west of the southern Mediterranean beach resort of MÁLAGA. Set on a steep-sided plateau, the town is one of Spain's prime tourist attractions. The old town is Moorish, the 'new' town dates from the 15th century. Ronda's bull-ring is claimed to be the country's oldest.
Population 33 900
Map Spain – Cd

Rondane *Norway* Mountainous area covering 575 km² (222 sq miles) in south-eastern Norway. It was made Norway's first national park in 1962. Rondane is the setting for the legends of Peer Gynt, which were adapted by the Norwegian dramatist Henrik Ibsen and used for his play *Peer Gynt*, written in 1867, which in turn inspired the incidental music by Norwegian composer Edvard Grieg (1843-1907).
The area contains remains of early hunting pits and Iron Age sites for smelting bog ore – an iron ore found in marshy areas.
Map Norway – Cc

Rondônia *Brazil* See PORTO VELHO

Rongelap *Marshall Islands* Atoll in the northern Marshalls, some 160 km (100 miles) east of BIKINI. It was badly contaminated by nuclear fallout after the United States tested a hydrogen bomb at Bikini in 1954. The inhabitants, who were only evacuated two days later, showed many symptoms of radiation sickness. Despite contamination remaining, they were allowed to return in 1957.
Population 235
Map Pacific Ocean – Db

Rønne *Denmark* Medieval seaport and main town of BORNHOLM Island. It has ferries to Copenhagen and the Swedish port of Ystad.
Population 15 200
Map Denmark – Cb

Roraima *South America* See GUIANA HIGHLANDS

Rorke's Drift *South Africa* Mission village and arts centre on the Buffalo River in northern KwaZulu-Natal, 170 km (105 miles) north of Durban. On 22-23 January 1879, about 130 men of the 24th Regiment of Foot (2nd Warwickshire, later South Wales Borderers) held an improvised fort against 4000 Zulu warriors, many of them exhausted after their victory at ISANDHLWANA.
Eleven Victoria Crosses were awarded, the greatest number of these medals ever won in a single action.
Map South Africa – Db

Røros *Norway* Town in south Norway, 110 km (68 miles) south-east of TRONDHEIM. It was the country's chief source of copper until the 1950s, when the ore ran out. Industrial archaeologists have converted the town into a virtual museum piece. Røros now has a thriving tourist industry, and has become a major winter ski resort.
Population 6000
Map Norway – Cc

Ros Comain *Ireland* See ROSCOMMON

Ros Láir *Ireland* See ROSSLARE

Rosa, Monte *Italy/Switzerland* Mountain rising to 4634 m (15 203 ft) on the Italian-Swiss border, east of the MATTERHORN. The Swiss tourist resort of ZERMATT lies to the north-west.
Map Italy – Ab

Rosario *Argentina* River port on the Paraná, 280 km (nearly 175 miles) north-west of BUENOS AIRES. Although the settlement only named Rosario in 1725, it was actually established in 1689 when the first house was built.
The river allows ocean-going ships to reach the city, and Rosario is now the main outlet for grain, meat and timber from central and northern Argentina. Its industrial products include steel and agricultural machinery.
Population 1 096 300
Map Argentina – Cb

Roscommon (Ros Comain) *Ireland* Market town about 145 km (90 miles) north-west of DUBLIN. It is the county town of the surrounding county of the same name which covers some 2460 km² (950 sq miles).
Population (town) 1310; (county) 51 900
Map Ireland – Bb

Roseau *Dominica* Capital, situated on the southwest coast near the mouth of the Roseau (or Queen's) River.
Formerly Charlotte Town, it was burned by the French in 1805, the year that marked the end of the dispute between Britain and France over ownership of DOMINICA. The French left, after extracting a handsome reward from the British, and Roseau was subsequently rebuilt. Today, its port, an open roadstead, exports tropical produce.
Population 21 000
Map Caribbean – Cb

Romania

A NATION OF LATINS IN A SEA OF SLAVS STRUGGLES TO REVIVE ITS ECONOMY AFTER DECADES OF AUTOCRATIC RULE WITHIN THE SOVIET BLOC

The Romanians are a Latin people who have tenaciously retained their language and customs despite being surrounded by Hungarians and Slav-speaking Bulgars, Serbs and Ukrainians. In turn they fought against Hungarian, Turkish and Habsburg occupation. Heroic leaders resisted Turkish onslaughts for five centuries; one of them, Vlad Tepes (1456-75), known as the Impaler, was immortalised later as Dracula ('The Order of the Dragon'), the vampire count of the famous horror story.

Romania became an independent state in 1878, fought against Germany, Austria and Hungary during the First World War and with Germany in the Second. Although it joined the Allies in 1944, its postwar fate was decided by the Soviet advance into central Europe. In 1947, with the abdication and exile of King Michael I, it became a one-party communist state and a member of the Warsaw Pact. After the accession to power in 1965 of Nicolae Ceausescu as head of state, it was virtually a dictatorship until his overthrow in 1989.

On the map, Romania resembles a Catherine wheel. In the south-east, the 'fuse' comprises the reed jungle, river channels and wildlife paradise of the DANUBE delta, and the dry rolling farmlands of DOBRUJA separating the river from the BLACK SEA coast. Inland is a ring of rich agricultural plains, flat in the south (WALACHIA) and west (Banat), and hilly in the east (MOLDAVIA).

These plains enclose the sweeping arc of the forested CARPATHIAN MOUNTAINS that rise to 2663 m (8737 ft). On the south-western frontier with Yugoslavia, the Danube breaches the mountains in the magnificent 160 km (100 mile) IRON GATES gorge. The circular 'core' of Romania is TRANSYLVANIA, within the Carpathian arc.

Today, traditional Romanian culture can be seen most strongly in MARAMURES, Moldavia and Walachia. In addition to Romanians, the country has sizable minorities of Hungarians (8.9 per cent), Germans (0.4 per cent), as well as Ukrainians, Serbs, Croats, Russians, Turks, Jews and Gypsies. With Ceausescu's rise to power, the national minorities lost their special rights and ever-increasing numbers of Hungarians and Germans emigrated during his years of dictatorship.

In December 1989, Romania's people rose up in arms. The military sided with the people and conflict with the pro-government secret police troops (the Securitat) ensued. The fighting cost many lives before Ceausescu was taken prisoner and executed. The new government under the leadership of President Ion Iliescu announced free elections and embarked on harsh economic reforms.

Ceausescu's communist economic policies during the 1970s focused entirely on industrialisation and for more than two decades Romania pushed through five-year development plans concentrating investment in heavy industry, at the expense of consumer goods. From 1960 to 1970 it had the world's third fastest growing industrial production; and from 1970 to 1976, the third fastest growth in agricultural output. The state owned virtually everything, apart from peasant farmland and housing. These reforms, however, eventually led the country into nearly devastating economic problems. Rising international debt led to import restrictions. A major setback was the drying up of the PLOIESTI oil fields.

Today, about 20 per cent of the workforce is employed in agriculture. While the large-scale industrialisation turned Bucharest, the capital, and Ploiesti into manufacturing centres, most of the country is still farmland. Horse-drawn carts lit by flickering oil lamps used to jingle through the dusk at harvest time, carrying farm workers home by the dozen. Today, there are processions of huge, single-driver combine harvesters, with headlamps ablaze.

Industry depends on electricity generated by coal-fired power stations and the rivers flowing from the Carpathians. The hydroelectric dams at Bicaz and CURTEA DE ARGEŞ are impressive, yet they are dwarfed by the huge Iron Gates station jointly developed by Romania and Yugoslavia. But over-ambitious planning has plunged the country into a serious energy crisis. Coal and iron ore support steel factories at Resita and HUNEDOARA, the rich mineral deposits in the APUSENI and Maramures mountains supply copper, lead, zinc, gold, silver and bauxite refining industries.

With the fall of the communist government, new economic policies were introduced. The first free elections in 53 years were held in May 1990, and brought an overwhelming presidential victory for Ion Iliescu, who was narrowly re-elected in 1992. The years 1993 and 1994 were marked by repeated labour unrest and demonstrations by the opposition as the minority left-leaning coalition government appeared uncommitted to reform and became increasingly unpopular.

However, early in 1994 an austerity programme agreed to with the International Monetary Fund was accepted. By mid-1994, at least basic foodstuffs were again available and primary health care – which was severely set back in the 1980s – had improved, largely assisted by international aid agencies.

ROMANIA AT A GLANCE		
Area 237 500 km² (91 700 sq miles)		
Population 23 172 400 **Per km²** 98 **(Per sq mile)** 253)		
Capital Bucharest		
Government Republic		
Currency 1 Leu = 100 bani		
Language Romanian (official), Hungarian, German		
Religions Christian (70% Romanian Orthodox, 6% Roman Catholic, 6% Protestant), other 18%		
Climate Continental, with little rainfall in the interior. Average temperature in Bucharest ranges from -3°C (27°F) in January to 23°C (73°F) in July.		
Land use Cultivated 46%, meadows and pastures 19%, forests 28%, other 7%		
Main primary products Cereals, potatoes, sugar, fruit and vegetables, grapes, livestock, timber, fish; oil and natural gas, coal, iron ore, salt		
Major industries Agriculture, iron and steel, mining, forestry, chemicals, shipbuilding, machinery, motor vehicles. textiles, fertilisers, coke, food processing		
Main exports Machinery, transport equipment, metal goods, steel, petroleum products, foodstuffs, chemicals, clothing		
Annual income per head (US$) 1640		
Population growth (per thous/yr) 2		
Life expectancy (yrs) Male 64 **Female** 74		

Rosetta *Egypt* See RASHID

Roskilde *Denmark* Town on the Danish island of ZEALAND, some 30 km (19 miles) west of Copenhagen. It was one of Denmark's major towns from the 10th century into the Middle Ages, and was the scene of the signing of the Peace of Roskilde in 1658 between Denmark and Sweden. Its Gothic cathedral, dating from 1170, has been the burial place of Danish kings for some 600 years. There is a university (1972), and a museum with the remains of six Viking ships from a nearby fjord.
Population 49 600
Map Denmark – Cb

Ross and Cromarty *United Kingdom* See HIGH-LAND; WESTERN ISLES

Ross Ice Shelf *Antarctica* The world's largest body of floating ice, lying at the head of the Ross Sea on the Pacific coast of Antarctica. It is estimated to be as large as France and it covers Roosevelt Island entirely. The ice increases in thickness from about 400 m (1300 ft) in the north to more than 750 m (2460 ft) in the south. It moves seawards at a rate of about 1000 m (3280 ft) a year. Huge icebergs break off, leaving ice cliffs up to 70 m (230 ft) high in the Ross Barrier.
Map Antarctica – Id-Jd

Ross Island *Antarctica* Island, 75 km (47 miles) long, situated in the Ross Sea beside MCMURDO SOUND. It is dominated by several volcanic peaks, including the 3794 m (12 447 ft) high Mount EREBUS, which is active, and Mount Terror; both were named after ships commanded by Sir James Clark Ross, their discoverer. The American McMurdo and the New Zealand Scott research bases are on the island.
Map Antarctica – Jc

Ross Sea *Antarctica* Deep inlet in the coast of Antarctica, discovered in 1841 by the British explorer Sir James Clark Ross. The inlet was the departure point for the shortest route across Antarctica from the sea to the south pole.
Map Antarctica – Hc

Rössing *Namibia* Uranium mine founded in 1971. It lies 250 km (155 miles) west of WIND-HOEK. Until the decline of the uranium market it was one of the largest open-pit mines in the world.
Map Namibia – Ab

Rosslare (Ros Láir) *Ireland* Resort on the east coast, 121 km (75 miles) south of DUBLIN. The harbour, 8 km (5 miles) south-east of the town, is a ferry terminal, and is the major arrival port for visitors from Fishguard, Wales, and Le Havre, France.
Population 850
Map Ireland – Cb

Rostock *Germany* Chief port between LÜBECK and SZCZECIN on the Baltic Sea, about 200 km (120 miles) north-west of Berlin. It is also a shipbuilding and marine engineering centre and has a marine engineering faculty at its university. Other products include electrical equipment, machinery and textiles. Although Rostock was badly damaged during the Second World War, many beautiful medieval buildings remain.
Population 248 100
Map Germany – Ea

Rostov (Rostov-Yaroslavskiy) *Russia* One of the oldest towns in Russia, first mentioned in AD 862 and formerly called Rostov-Velikiy; it is 195 km (about 120 miles) north-east of MOSCOW. In the Middle Ages, it grew into a flourishing centre for handmade enamelware and linen goods (in which it still specialises). Rostov has a magnificent lakeside kremlin (citadel), a 13th-century cathedral and some fine 17th-century churches.
Population 31 000
Map Russia – Ec

Rostov-na-Donu (Rostov-on-Don) *Russia* Industrial city and port on the River Don near the north-east tip of the Sea of AZOV. Founded as a customs post in 1750 when Turkey still ruled the coast of the Sea of Azov 50 km (about 30 miles) downstream, Rostov-na-Donu became Russia's second largest grain export harbour during the 19th century. Today, it produces ball bearings, road-building and agricultural machinery, electrical equipment, chemicals, barges, glass, wine and cigarettes. It is also a cultural centre, with a university, museum, theatres and art galleries.
Population 1 020 000
Map Russia – Ed

Rothenburg ob der Tauber *Germany* Walled medieval town on the River Tauber, 65 km (40 miles) west of NUREMBERG. Guard towers and gates survive from the Middle Ages, and inside are galleried and timbered houses, Gothic and Renaissance churches and cobblestoned streets. Tourists crowd it in summer.
Population 11 400
Map Germany – Dd

Rotherham *United Kingdom* Industrial town on the River Don, 10 km (6 miles) north-east of SHEFFIELD in the English county of South Yorkshire. It has been a market centre since Saxon times, but grew with the 19th-century development of coal mining and the spread of steel making from Sheffield.
Rotherham is also the chief destination for tourists visiting the Peak District, the Dukeries and the moors of West and North Yorkshire.
Population 123 310
Map United Kingdom – Ed

Rotorua *New Zealand* Tourist city and health resort about 160 km (100 miles) south-east of AUCKLAND. It lies beside Lake Rotorua on North Island, in an area steaming with hot springs, mud pools and geysers. Some of the volcanically warmed springs are used to heat nearby buildings and homes in winter, and to power refrigeration units in summer. Pohutu, the most powerful geyser, sometimes shoots water 30 m (100 ft) into the sky. Whakarewarewa, on the edge of the city, is a Maori village whose inhabitants use the hot springs for swimming, washing and cooking.
Population 53 700
Map New Zealand – Fc

Rotterdam *Netherlands* The country's largest city and the world's busiest port. Rotterdam handles some 32 000 ships per year, and a greater tonnage of cargo than any other port. It was founded in the 13th century as a fishing village where the Rotte River flowed into the Rhine-Maas delta.
The Rhine is the 'main street' of Europe, and after the NIEUWE WATERWEG gave the city deep-water access to the North Sea in 1872, ocean cargo vessels were quick to take advantage of Rotterdam, where river and ocean traffic could meet. The port has expanded all the way to the sea; docks and oil refineries line the waterway, and EUROPOORT stands at its seaward end. However, the building problems were immense in an area where most land is below sea level and made of delta mud.

Rotterdam's city centre, wiped out by German bombs in May 1940, has been rebuilt, with new residential and shopping areas, conference and concert facilities, and several museums, including one devoted to maritime history, and the Boymans art museum, with a splendid art collection. The docks, warehouses, shipyards and plants processing imported raw materials, such as chocolate, coffee and tobacco, provide most of the jobs in the city.
Population (city) 596 023; (Greater Rotterdam) 1 069 365
Map Netherlands – Bb

Roubaix *France* See LILLE-ROUBAIX-TOUCOING

Rouen *France* Port and capital of Seine-Maritime Department, standing on the Seine, 123 km (76 miles) north-west of PARIS. The French national heroine Joan of Arc, captured by the Burgundians and sold to the English, was tried as a heretic and sorceress and burned at the stake here in 1431.
Rouen was the medieval capital of Normandy, and has many medieval buildings, including a beautiful 13th-century cathedral (restored after damage in the Second World War) and the Abbey of St Ouen, where Joan of Arc was imprisoned. It is now one of the country's largest ports, exporting manufactured goods, and importing coal and oil.
Population (town) 103 000; (metropolitian area) 380 000
Map France – Db

Roussillon *France* Former department covering some 2000 km² (about 770 sq miles) beside the Mediterranean and the Spanish border, now part of LANGUEDOC-ROUSSILLON. Irrigated areas on the plain produce wheat, fruit, wines and vegetables. PERPIGNAN is the main town.
Map France – Ee

Rouyn-Noranda *Canada* Centre of a gold and copper-mining area in western Quebec, 386 km (240 miles) north-east of OTTAWA. Other industries include dairy products and timber.
Population 26 450
Map Canada – Hd

Rovaniemi *Finland* City often called the capital of Finnish Lappland. It lies just south of the Arctic Circle, 160 km (100 miles) north of the port of OULU on the Gulf of Bothnia, at the broad confluence of the Kemijoki and Ounasjoki rivers.
It was virtually destroyed during the Second World War and reconstructed on the basis of a plan by the Finnish architect, Alvar Aalto. It is now a commercial and tourist centre.
Population 34 300
Map Finland – Cb

Rovinj (Rovigno) *Croatia* Small port on the Istran coast, 55 km (35 miles) from the Italian border. The old town, a pastel-painted former Venetian settlement of narrow, winding alleys and colourful piazzas, lies on a promontory surrounded by pine-clad islands. It has a renowned institute of marine biology and an aquarium.

Population 20 400
Map Croatia – Ab

Roxburgh *United Kingdom* See BORDERS

Rožňava *Slovakia* Medieval gold-mining town in eastern Slovakia, 13 km (8 miles) from the Hungarian border. Its main square stands above the former mines. The town is a base for tourists visiting the Slovak Ore Mountains to the north, and the caves and gorges of the Slovak Karst (Slovensky Kras), which extends into Hungary. Iron ore is mined nearby.

Population 19 500
Map Slovakia – Cb

Ruapehu, Mount *New Zealand* Active volcano 230 km (142 miles) north-east of the capital, WELLINGTON, in Tongariro National Park. At 2797 m (9176 ft), it is the highest point on North Island; it is popular with hikers in summer and skiers in winter. Near the summit is Crater Lake, its cloudy green waters strongly acidic and warmed by volcanic steam.

On Christmas Eve 1953, an eruption sent a torrent of mud, water and rocks from the lake smashing down the volcano's flanks and thundering through a river gorge. The torrent swept away the railway bridge at Tangiwai ('Weeping Waters') just minutes before the Wellington-Auckland express was due to pass on it. In the darkness the speeding train plunged into the gorge, killing 151 people.

Map New Zealand – Ec

Rub al-Khali *Saudi Arabia/Yemen/Oman* Desert known as 'the Empty Quarter', covering some 650 000 km² (251 000 sq miles) on the Arabian Peninsula. It extends from the NEJD into Yemen, and from Yemen to Oman.

One of the world's most arid and hostile deserts, and in fact the world's largest continuous stretch of sand, with dunes up to 300 m (984 ft) high, Rub al-Khali has yet to be fully explored. In 1930, the British traveller and writer, Bertram Thomas, followed the one well-watered route from the west coast north-east to the Qatar Peninsula. The Shaybah oil field lies in the north-east of the Rub al-Khali.

Map Saudi Arabia – Bc

Rudolf, Lake *Kenya* See TURKANA, LAKE

Rugby *United Kingdom* Town in the English Midlands county of Warwickshire, some 45 km (28 miles) east of BIRMINGHAM. It has a number of engineering and car assembly plants and is the home of the public school where the game of rugby football originated in 1823, when a soccer player picked up the ball and ran with it.

Population 59 720
Map United Kingdom – Ed

Rügen *Germany* The country's largest island, covering 926 km² (357 sq miles) in the Baltic Sea. It is a popular summer resort, fringed by chalk cliffs and dotted with beechwoods, sandy coves, fishing villages and prehistoric remains. The islanders earn a living from agriculture, fishing, extraction of chalk, and tourism.

Population 85 000
Map Germany – Ea

Ruhr *Germany* River in the north-west flowing 235 km (146 miles) west to the Rhine at DUISBURG. Centred on its lower valley is Western Europe's largest coal field and industrial region, with mines and steel and engineering plants stretching from DORTMUND in the east to Duisburg.

The Ruhr region, one of the biggest industrial areas on earth, is known as 'the Forge of Germany'. The mining of top-grade coking coal began here in the 1850s. But since the 1970s, recession and the closure of a number of the mines have forced the eastern Ruhr to diversify into lighter and computer-based industries. Petrochemical industries flourish in the west, based on the Rhine ports and pipelines, and soft coal is produced west of the Rhine.

Population 6 000 000
Map Germany – Bc

Ruhuna (Yala) *Sri Lanka* The country's second largest national park, after Wilpattu. It is 980 km² (378 sq miles) in extent and lies near the island's south-east coast, about 180 km (110 miles) to the south-east of COLOMBO. Many species of animals, including elephants, leopards, bears, buffaloes, deer and wild boars roam freely over its thorny scrub and open woodlands. The nearby Kumana bird sanctuary is a refuge for ibis, pelicans, parrots, peacocks and storks.

Map (Colombo) Sri Lanka – Ab

Rumaila *Iraq* One of the largest oil fields in the world, about 50 km (30 miles) south-west of The Gulf port of BASRA. It started production in 1972. The oil field has led to the creation of the nearby industrial towns of Az Zubayr, Shu'aiba and Umm Qasr.

Map Iraq – Cc

run In America, a fast flowing stream or brook, such as Bull Run, Virginia.

runnel (swale) Long, narrow hollow on a beach between two ridges running parallel to the coastline which are exposed at low tide. The term is also applied to a small stream, or the channel in which such a stream flows.

Runnymede *United Kingdom* Meadow on the south bank of the River Thames, 8 km (5 miles) south-east of WINDSOR. King John sealed the Magna Carta – the Great Charter setting out the rights of citizens to justice and security – at Runnymede on 15 June 1215.

There are three memorials on the site: the first to the Magna Carta; the second to the 20 000 Commonwealth airmen who died in the Second World War but who have no known grave; and the third to the American President John F Kennedy, who was assassinated in Dallas, Texas, in 1963.

Map United Kingdom– Ee

run off Amount of water, originating as precipitation, that reaches streams and rivers, and flows away to the sea. It consists of the water which flows off the surface, as well as some of the water that sinks into the ground and later reaches the streams.

Ruse (Russe) *Bulgaria* River port on the Danube, 244 km (152 miles) north-east of SOFIA, standing near the only bridge which links Bulgaria and Romania across the Danube. Founded as a Roman

fortress in the 2nd century AD, it expanded after 1878 and is now a major industrial centre with shipbuilding and tanning industries. Among its manufactures are rubber goods, agricultural machinery, electrical equipment and plastics.

Population 192 400
Map Bulgaria – Bb

Rushmore, Mount *USA* Granite mountain rising to 1707 m (5600 ft) in the Black Hills of South Dakota, and site of the Mount Rushmore National Memorial, covering 5.2 km² (2 sq miles). The memorial features giant carvings of the heads of four presidents of the United States, up to 18 m (60 ft) tall, cut out of the face of the mountain. The heads are of Washington, Jefferson, Lincoln and Theodore Roosevelt and were carved between 1927 and 1941, in six years of work.

Map United States – Fb

Russia See p 578

Rustenburg *South Africa* Mining town at the foot of the Magaliesberg range in North-west Province, 100 km (62 miles) north-west of PRETORIA. It is the site of one of the world's largest platinum mines; chrome and nickel are also mined. The district produces tobacco and maize.

Population 125 310
Map South Africa – Cb

Rutanzige, Lake *Uganda/Zaire* See EDWARD, LAKE

Rutland *United Kingdom* See LEICESTER

Ruwenzori *Uganda/Zaire* Mountain range, just north of the equator, between lakes Edward and Albert, thought to be the legendary Mountains of the Moon, where the Ancient Greeks believed the Nile River rose. Its highest point is Margherita, at 5109 m (16 761 ft), one of the peaks of Mount Stanley (Mount Ngaliema). Dense forests skirt the mountains, but above 1650 m (5410 ft) the forests give way to successive bands of mountain grassland, bamboo thicket, tree heathers, giant senecios, and finally tundra, with snow fields and glaciers at the top.

Map Uganda – Aa; Zaire – Ba

Rwanda See p 583

Ryazan' *Russia* Industrial city on the Oka River, some 175 km (110 miles) south-east of MOSCOW. Ryazan' was founded in 1095 but destroyed by the Tatars in 1237. It was rebuilt, and in the late 18th century became a grain-milling centre.

The city began refining oil in the early 1960s; its other chief products now include electrical appliances, machine tools, chemicals, leather goods and farm equipment.

Population 527 000
Map Russia – Ec

Rybinsk (Andropov) *Russia* City on the Volga River at the south end of the Rybinsk reservoir. Formerly called Shcherbakov, then Rybinsk, it was named Andropov in 1984 after President Yuri Andropov (1914-84), and renamed Rybinsk after the collapse of the Soviet Union. It has engineering, boat-building, timber-machinery, food processing and leather industries.

Population 252 000
Map Russia – Ec

Russia

AFTER THE COLLAPSE OF THE SOVIET UNION, THE WORLD'S LARGEST COUNTRY IS SHAKEN BY POLITICAL CONFLICT AND ECONOMIC CRISES

With an area of over 17 million km² (6.56 million sq miles), the Russian Federation, as the former Russian Federal Soviet Socialist Republic is now called, has the dimensions of a continent. Russia is the largest continuous stretch of land in the world. About one-ninth of all land globally belongs to this giant nation. A country the size of the United Kingdom could fit into it almost 70 times – and even an island continent the size of Australia would fit twice into Russia's vast territory and still not fill it.

More than 10 000 km (6200 miles), as the crow flies, lie between the Baltic Sea in the west and the Bering Strait in the east.

The land stretches from MURMANSK near the Finnish border to VLADIVOSTOK on the Sea of Japan, and from the BLACK SEA to well north of the ARCTIC CIRCLE. Its climate ranges from vast frozen wastes to subtropical deserts. Russia has Europe's highest mountain, ELBRUS (5642 m, 18 510 ft), and its longest river, the VOLGA (3688 km, 2292 miles). The country also has the world's deepest lake, BAIKAL, which plunges to 1620 m (5315 ft).

Despite its gigantic dimensions, most of the country's landscapes are quite monotonous, for it consists largely of seemingly endless steppe and gently undulating, open countryside. Lowlands form the western part of Russia, dotted in the north with lakes. An enormous area – altogether 1 360 000 km² (525 000 sq miles) – of the lowlands is drained by the Volga. South of the 55th latitude are Russia's fertile black soils, on which most of the country's grains are grown.

The extreme south is sandwiched between the Black Sea and the Caspian Sea and the glaciated ranges of the Caucasus; the Black Sea provides access, via the Bosporus and the Dardanelles, to the Mediterranean Sea. In the north, on the Baltic and the Arctic Ocean coast, harbours are blocked by ice for months. Here, the exception is the port of Murmansk on the Barents Sea, which is kept ice-free all year round thanks to the warm ocean current known as the North Atlantic Drift.

To the east, the forested Ural Mountains stretch for more than 2000 km (1240 miles), dividing the whole county into west and east. Beyond the Urals, the endless Siberian plains

▶**HISTORIC CAPITAL At the heart of Moscow, citizens stroll across Red Square, in Soviet times the scene of displays of military might. Above them tower the domes of Ivan the Terrible's Cathedral of St Basil the Blessed, a unique combination of nine chapels.**

begin. The west Siberian lowlands – mainly tundra, bogs and coniferous forests – stretch on to the Yenisey River. Farther east, the central Siberian Mountains rise above the plains and continue eastward to the River Lena, beyond which the east Siberian mountain ranges tower to heights of more than 3000 m (9800 ft).

The Sea of Okhotsk, the Bering Sea and the Sea of Japan wash around the country's east coast. In winter, when the sun goes down over the maritime region of this coast, it is just rising above the horizon in the capital, Moscow.

Russia's climate is continental, with short, hot summers, and long, bitterly cold winters. Only near the Baltic and Barents seas, and on the Russian Black Sea coast, can the moderating influence of water on the climate be felt. The average temperature during the coldest month is -10.3°C (13.5°F) in Moscow, but in Yakutsk, it is -43.2°C (-45.8°F).

Arctic air lowers winter temperatures in the region's 'pole of cold', at OYMYAKON, to -50.1°C (-58°F). This is the coldest region on earth inhabited by human beings. In 1938, the temperature plummeted to an all-time low of -77.8°C (-108°F); although this was an exception, temperatures of around -70°C (-94°F) are not unusual. Despite this cold, amazingly the area is vegetated.

Summers can be warm, with temperatures rising to 30°C (86°F). These vast fluctuations in temperature are the largest recorded anywhere in the world.

BRATSK in the southern taiga has only 120 frost-free days a year. Modern flats, shops, offices and factories in towns such as Yakutsk are supported on concrete stilts to allow freezing air to circulate underneath and so preserve the permafrost – otherwise the ground would melt and buildings would sink into the mud. At temperatures of -45°C (-49°F) or lower, truck engines and diesel motors must be left running day and night in winter. Yet summer brings heat, enabling people in Yakutsk to raise cattle.

RISE OF A NATION

Seas that freeze for up to eight months of the year in the north and east and the rugged mountains in the south provide natural defences for many of the frontiers. Russia's northern and eastern borders are the seas; in the south-east the country is cut off from its neighbours by high mountain ranges; in the south-west the Black Sea provides a natural boundary.

But there is a critical 2000 km (1250 mile) stretch of lowland on the border with China, in the south-east. And the vast plains lying west of the URAL Mountains which divide Europe from Asia are easily crossed by land or by the river systems of the DNIEPER and the Volga. The Baltic Sea provides an additional gateway. In the times of Peter the Great it was called Russia's window to Europe and the West.

Between the 9th and 16th centuries there were invasions by Vikings from the north, Bulgarians and Turks from the south-west, Mongols and Tatars from the east, and Poles and Lithuanians from the west. Napoleon reached MOSCOW in 1812 but was forced into a humiliating retreat by the bitter Russian winter. In turn, the Russians took land from the invaders. The Tsars Ivan III (1462-1505) and

Ivan IV, 'The Terrible' (1533-84), pushed out the borders as they ruled from their kremlin fortress in Moscow.

From 1703, Peter the Great (1682-1725), Catherine the Great (1762-96) and Alexander I (1801-25) directed the country's expansion from the Imperial capital of St Petersburg. The largest territory was won in the 19th century when vast parts of Central Asia were acquired and Siberia was incorporated in the tsardom. Russia gained access to the Black Sea, annexing large areas of Poland and Belorussia, the Baltic, Bessarabia, Moldavia and Romania. In 1809, the tsardom annexed all of Finland. Some of these areas were later relinquished by the

Russian government, while others, such as the northern half of East Prussia and a few islands in the Sea of Japan, were surrendered after the Second World War. But Russia remains in dispute with Japan over the return of the southern Kurils, a group of islands it occupied at the end of the Second World War, as well as with China over sections of their common boundaries. Almost half of Asia stayed in the hands of the Soviets until the collapse of their Union.

FORMS OF GOVERNMENT

Russia's historic openness to attack is the root of its apparently paranoid concern for strong military defence and centralised gov-

ernment. In the time of the tsars, controls were already deeply rooted in the Russian Empire. The tsars' rule was authoritarian, mostly reactionary and intolerant of opposition. Serfdom was not abolished until 1861. Peter the Great, who brought much West European civilisation to Russia, sent the first political prisoners to Siberia in 1710.

After an uprising in 1825 the numbers exiled to Siberia or to south-eastern deserts swelled. They included Moscow University professors, writers such as Fyodor Dostoyevsky and, later, revolutionaries such as Lenin. Paul I (1796-1801) and Nicholas I (1825-55) built up a secret police force and

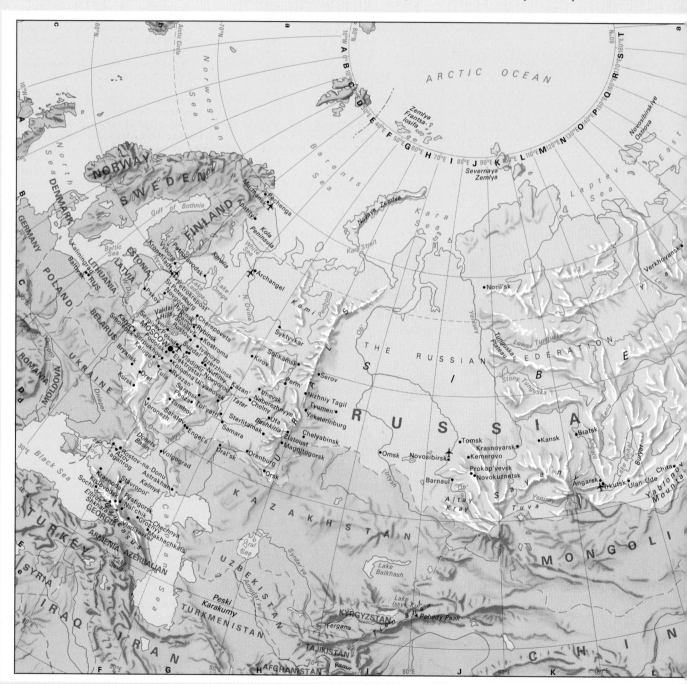

introduced censorship, with bans on foreign books dealing with reform or the plight of oppressed European peasantry. Restrictions on movement prevented intellectuals like the poet Alexandr Pushkin (1799-1837) from travelling abroad. People were already in the process of being isolated from the outside world.

Many people made the revolution possible, but two names stand out: Vladimir Ilich Ulyanov, or Lenin (1870-1924), and Lev D Bronstein, who was better known as Trotsky (1879-1940). Both were Marxists, intellectuals, visionaries, orators and organisers.

The socialist doctrine born in the minds of 19th-century philosophers such as Karl Marx

became political reality in 1917 when Lenin and Trotsky took over a war-weary nation and bent it to their will. The tsar and the royal family were executed, the well-to-do were hounded, imprisoned or killed, and the state became all powerful. Lenin died in 1924. Trotsky, having lost a power struggle, was finally murdered in Mexico in 1940.

After Lenin's death, Joseph Dzhugashvili, later known as Stalin, used ruthless purges, executions and labour camps to maintain a position of absolute, unquestioned power as Secretary-General of the Communist Party. Terror – fear of the knock on the door by secret agents in the night – achieved social obedience. It enabled Stalin to push through huge plans for industrialisation, collective farms, canal and railway building as part of successive five-year plans from 1928 onwards.

Backed by a reign of terror in which many millions died, Stalin reigned supreme for 30 years. During this period, the Soviet Union annexed eastern Poland and the three Baltic states which had gained independence from Russia in 1918. In 1941, Hitler's invading armies were driven back across the border after years of bloody war – an operation which cost about 20 million Russian lives and left large parts of the country devastated. Allies in war, the victorious Russians and Americans soon fell out and so began the Cold War.

The once backward Soviet Union became a mighty military power. The Soviets were always on the spot, ready to deploy their troops and material, wherever in the world a nation rose up against capitalism, or where, stirred up by Western allies, anti-communist feelings rose. At home, industry and agriculture were developed on a large scale. Depite this, the promised material gains of communism seemed to fade farther and farther from realisation. The economy could meet only a small part of the population's needs.

REFORMS OF GORBACHEV

Stalin's successors obstinately stuck to the communist line, in spite of the impending political and economic bankruptcy. But in 1985, the rising star of communism was Mikhail Gorbachev, elected Secretary-General of the Communist Party at 54. With the Soviet defence budget reaching US$18 000 million a year, talks with the United States on reducing the number of missiles and avoiding a possible space war slowly got under way.

In a new policy of 'openness', Gorbachev proposed electoral reforms and relaxed censorship on films, books and pop music. Some well-known dissidents were released and dancers, musicians and writers who had defected to the West were invited back to the USSR. Not only did Gorbachev begin the restructuring of the Russian economy and society – he also released the Eastern European countries from their communist fetters, allowing them to take their fate into their own hands.

Gorbachev tried to prevent the disintegration of the Soviet Union, but in the end had to accept its dissolution and the interim formation of 15 sovereign states, including Russia, in the place of the former Union. The republics which became independent were the Baltic

republics (ESTONIA, LATVIA and LITHUANIA), the three Slav republics (Belorussia, now BELARUS, Moldavia, now MOLDOVA, and the UKRAINE), the trans-Caucasian republics (GEORGIA, ARMENIA and AZERBAIJAN), KAZAKHSTAN, and four Central Asian republics (TURKMENISTAN, UZBEKISTAN, TAJIKISTAN and Kirghizia, today called KYRGYZSTAN).

Most have retained some of the former ties with Russia by becoming members of an eastern trade community, the Commonwealth of Independent States (CIS), which is mainly kept alive by the former Soviet republics' dependence on Russia's oil at a price slightly lower than the world market price.

AGRICULTURE AND INDUSTRY

The harsh climatic and soil conditions found in most of Russia are not conducive to much crop growing. Climatic extremes increase and conditions for crop and cattle farming become increasingly unfavourable from west to east. Whereas in the west, the agricultural belt is about 1500 km (930 miles) wide, it narrows in the east until finally, in Siberia, it peters away between the taiga and the desert. Cold temperatures in the north and east put an end to all agricultural activity; the isolated warmer pockets have limited cultivation of potatoes, flax and rye.

Land to the east and west of the lower Volga was gained for cultivation in agricultural reforms during the 1950s and 1960s. Although farm output quadrupled from 1922 to 1985, food shortages were still the order of the day. The Virgin Lands scheme, launched in 1953, opened up the Altay territory in southern Siberia. Large areas were put under irrigation. New grains which were better adapted to the short growing season were cultivated, specialised farming techniques were introduced, and agriculture was mechanised.

But the agricultural reforms were only partly successful and although mechanisation and fertilisers have substantially raised yields in the west, drought will always be a factor, resulting in periodic crop failures. Also, Russia's consumption has risen because of population growth – between 1922 and 1985, the population doubled – and better diets.

After the collapse of the Soviet Union in 1991, giant strides were made towards privatising agriculture. Countless farming operations were started, especially in the south and north Caucasian steppe. Where the soils are less productive, however, farmers hesitated to take the leap into private ownership. If these trends continue, it might still be years before Russia will be self-sufficient in cereals and fodder. A large percentage of the nation's food needs are imported from the United States, Canada, Argentina and the European Union

Industrially, however, Russia has made enormous progress since the 1920s. The country has had to accommodate the biggest rural-to-urban migration in the developed world this century. In 1926, only 18 per cent of the population lived in towns – today, some 75 per cent live in urban centres.

After 1922 there was commitment to technical modernisation, military supremacy and self-reliance. The Soviet Union became the

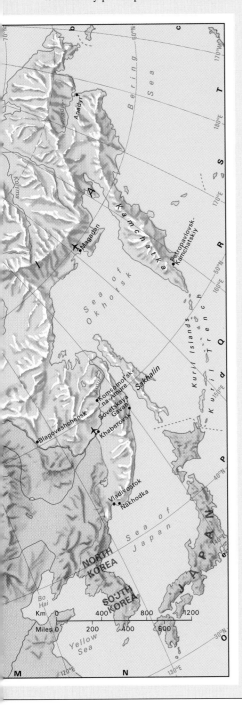

world's leading producer of petroleum, coal, hydroelectric power, iron ore, steel, some non-ferrous metals and timber. Railways were improved, and enormous wilderness frontiers tamed. New towns were built.

A great part of the vast urbanisation has been due to Russia's mineral wealth – enormous resources in iron ore, copper, zinc, gold, diamonds, uranium, and also coal, oil and natural gas have been discovered. The vast, thinly populated and wilderness regions of Siberia and the Russian far east are a rich storehouse of fuel, energy, minerals and timber. They are difficult and costly to develop and to live in because of the weather. Wells drilled between TYUMEN' and Surgut since 1965 now yield most of Russia's oil, and in the eastern Siberian plateaus, forests cover huge mineral deposits, including massive coal reserves. Western Siberia has vast natural gas reserves; a gas pipeline runs from Siberia into Western Europe.

The country may be rich in mineral resources, plenteous wood and rivers that can be harnessed for hydroelectric energy, but harvesting its wealth creates enormous problems. Despite large-scale urbanisation, Russia's industries are concentrated in a few isolated districts. Vast distances lie between the resources – most of them in Siberia – and the industrial regions in the more populated European Russian west. Transport therefore plays a vital, but costly, role.

The main rivers flow from north to south, or south to north, and have limited use for transporting minerals across the width of the country. The Trans-Siberian Railway which links Moscow with the port of Vladivostok on the Sea of Japan 9302 km (5780 miles) away was built in 1904 and has since then been the main means of cross-country transportation. It has also opened up Siberia for development.

In the north, the Baikal-Amur-Magistral (BAM) railway, extending some 3200 km (2000 miles), from Taishet on the Trans-Siberian Railway to the Pacific port of Sovetskaya Gavan', was completed in 1985. The most important post-Second World War transport project, it was intended to make south-east Siberia, with its timber, coal, copper, iron ore, asbestos and other minerals, one of the world's major development areas.

But it has not had the desired results. Fuel depots along the line are regularly plundered; a decade after its completion some of its rolling stock depots were in disrepair; large sections of track were in poor condition and not being repaired because of a lack of rails and sleepers. However, it did bring about a measure of urbanisation. For decades, settlement had been restricted to land alongside the Trans-Siberian Railway. Since 1975, 70 boom towns have emerged on the BAM route beyond Bratsk.

The most important industrial centres today are still St Petersburg, Moscow, the Volga-Kama region, the Southern Urals, the DONETS BASIN (most of which lies in the Ukraine) and Kuzbas in west Siberia. Large industrial complexes produce non-ferrous metals and mine iron ores in the Urals; timber working and paper making are scattered along the north, food processing in the south. Today, the country is highly industrialised but the transition

from a planned economy to a free market system has been difficult. Russia's industry has inherited big concerns from Soviet times – which are not well able to adapt to the new market situation. In 1990, the entire Soviet Union had a total of 46 700 industrial concerns; by comparison, the small German state of Baden-Württemberg has today about 18 800 concerns, each employing 10 or more workers.

Under President Boris Yeltsin, who came to power in 1991, giant strides were made in 1992 towards a market economy. A major privatisation programme was begun. Most prices were freed, and the defence budget slashed. Foreign trade, however, soon declined sharply – within a year, exports were down to 25 per cent, imports to 21 per cent. Trade with other former Soviet countries also declined due to the unstable rouble and escalating prices in the region.

By the end of 1992, wages had started to lag behind inflation, and Russians soon found themselves unable to buy the kinds of foods that they were used to, and they were eating increasing amounts of bread and potatoes instead of proteins. Industrial production slumped and was still decreasing in 1993. Many companies found that they could not compete with imported goods – even vodka, traditionally Russian, was being imported at a lower price than that of the local product. Unemployment soared. Many Russians began asking themselves whether they had not been better off under communism.

THE PRESENT AND THE FUTURE

Although Russia's internal restructuring was begun with much enthusiasm, it soon slowed. The government under Yeltsin had a tough battle to fight at home, especially against a parliament opposed to his style of leadership. In October 1993, after Yeltsin had ordered the suspension of parliament, an armed revolt of parliamentarians led by Vice-President Alexander Rutskoi was quelled by forces loyal to the president after shelling of the parliament building, which the rebels had occupied.

The subsequent elections in December under a new constitution, the first free parliamentary elections since 1917, brought an unexpectedly large share of the votes to the candidates of the extreme left and right. Prominent among them was the flamboyant leader of the Liberal Democratic Party, Vladimir Zhirinovsky, who won the largest proportion of votes after campaigning for an expansionist militaristic policy, playing on Russian nationalistic sentiments.

Zhirinovsky's popularity underlined the struggle for power between Yeltsin's supporters and those on both the left and right, which throughout 1994 continued to give concern to the international community – and particularly the country's foreign aid donors and trading partners. That concern was accentuated by a slump in the rouble in October and by events in Chechnya, the Caucasian republic which had declared its independence from Russia in November 1991 after the Chechen nationalist Dzakhar Dudayev took power in a military coup and installed himself as president.

In late 1994, fighting between Dudayev's supporters and pro-Russian forces escalated,

leading to intervention from Moscow. In December, a Russian force of some 4000 troops advanced on Chechnya's capital, Grozny. The Russians encountered fierce opposition and casualties mounted rapidly in the fighting that followed.

The bungled military intervention in Chechnya was a severe blow to Yeltsin's position and added to fears that others of the 21 components of the Russian Federation might seek to break away. Once the core of a mighty empire, Russia has in effect been experiencing the end of its colonial era – much later than the other former colonial powers, such as Great Britain and France. Their empires collapsed and their former constituents experienced political, economic and social turmoil decades earlier.

After granting independence to the former Soviet republics, Russia now comprises 76 per cent of the area it had during Soviet times. Thanks to the colonisation policies of the tsars, it is still a country with about 100 nationalities whose members continue to speak their own languages. There are 21 autonomous republics, regions and districts – from the Karelian Republic in the far north-west to the Jewish autonomous region in the Russian Far East.

Dissatisfied with Russian dominance, the nationalistic feelings of the ethnic minorities have gained in strength. By the mid-1990s the Caucasus was rumbling with unrest. As the Siberian people began demanding their rightful share of the natural resources in their region Russia's future seemed anything but rosy.

RUSSIA AT A GLANCE	
Official name Russian Federation	
Area 17 075 200 km² (6 591 214 sq miles)	
Population 149 300 400 **Per km²** 8.7	
(Per sq mile 22.6)	
Capital Moscow	
Government Federation	
Currency 1 rouble = 100 kopeks	
Languages Russian (official), and more than 100 others	
Religions Christian (Russian Orthodox); some Muslim and Jewish	
Climate Continental, with little rainfall and enormous fluctuations in temperature. Extremely cold winters, arctic in the north. Summers warm, and hot in the south. Average temperature in Moscow ranges from -10.3°C (13.5°F) in January to 17.8°C (64°F) in July	
Land use Forest 41%, cultivation 8%, grazing 5%, other 46%	
Main primary products Cereals, livestock, fruit and vegetables, timber, fish; oil and natural gas, coal, lignite, iron ore, diamonds, bauxite, copper, lead, zinc, salt, uranium	
Major industries Iron and steel, cement, transport equipment, engineering, armaments, electronic equipment, chemicals, fertilisers, oil and gas processing, fishing, shipbuilding, mining and mineral refining, agriculture, forestry, and timber processing	
Main exports Oil and refined products, natural gas, wood and wood products, metals, coal, chemicals, machinery, transport equipment, timber, paper, textiles	
Annual income per head (US$) 3240	
Population growth (per thous/yr) 2	
Life expectancy (yrs) Male 64 **Female** 74	

Rwanda

*ANCIENT RIVALRIES HAVE ERUPTED IN
A GENOCIDAL BATTLE FOR POWER
ACROSS ONE OF THE MOST
DENSELY POPULATED RURAL, AND
LANDLOCKED, NATIONS ON EARTH*

In the heart of Africa, the Republic of Rwanda is set in one of the world's most densely populated rural regions which for four centuries has been riven by hatred between the Tutsi and Hutu peoples.

The country is almost wholly composed of steep hills, whose slopes are intensively cultivated. Rwanda and the neighbouring country of BURUNDI are almost identical twins. Until independence they were linked as Ruanda-Urundi and were administered by Belgium.

In the past, the richer Tutsi in the east, making up a small minority of the population, were a ruling aristocracy. However, by independence in 1962, the Hutu, vastly outnumbering the Tutsi, had gained political control.

In the years that followed, ancient hatreds exploded into wholesale slaughter. Hutu extremists were accused of embarking on the systematic genocide of their Tutsi rivals. In 1973, a military coup led to a Hutu government which banned party politics and created the sole ruling party, the National Revolutionary Movement for Development (MRND). Periodic outbreaks of violence with mass killings continued. In October 1990, the Rwandan Patriotic Front (FPR), a Tutsi-led movement opposed to the military government, invaded from Uganda.

After repeated international efforts to make peace, a new constitution was approved and a multiparty system introduced in 1991. United Nations-monitored steps towards transitional government began in 1992 which were to take the country to multiparty elections scheduled for 1995, but delays and divisions within parties slowed down the process and a new eruption of violence intervened.

A peace accord to end the civil war was signed in August 1993. However, in April 1994, the Hutu presidents of Rwanda and Burundi were killed when their aircraft was shot down. The response by Hutu militia, who blamed the Tutsi rebels, plunged the country into an orgy of slaughter. Between April and July 1994, tens of thousands were massacred and hundreds of thousands from both groups fled into neighbouring Uganda and Tanzania.

However, the FPR was rapidly gaining control over large parts of the country, and on July 18, after securing the capital, KIGALE, claimed victory and announced the formation of a coalition government headed by a Hutu president and prime minister. By then, according to relief agencies, up to 5 million people had been displaced within the country and in refugee camps in neighbouring territories.

Estimates in the mid-1980s put the Tutsis at about 15 per cent of the population against 85 per cent for the Hutu (there were also a small number of pygmy Twa people). A decade later it was estimated that the Tutsis formed only 9 per cent. Between 1956 and 1994, the conflicts in Rwanda and Burundi were believed to have cost at least 350 000 lives. In 1995, reprisals by the Tutsi-dominated army against the majority Hutus recommenced the bloodbath.

MOUNTAINOUS COUNTRY

Rwanda has a central spine of highlands from which streams flow west to the Zaire River, and east to the Nile. In the north are active volcanoes rising to about 4500 m (14 760 ft). The land falls gently eastwards towards Tanzania, and more sharply westwards to Lake KIVU, which forms most of the boundary with Zaire. Fertile soils are limited, and rainfall is less than might be expected in such mountainous country – ranging from 750 to 1750 mm (30-70 in) a year – but the rain is more reliable than in much of Africa and serious droughts are rare. Yet deforestation, overgrazing, and soil erosion are increasing.

An estimated 8 million population gives a density of 326 people per km² (846 per sq mile) – which is very high for a country where 95 per cent of the people live in rural areas. At independence in 1962 there were hardly any towns; the country had been governed by the Belgians from BUJUMBURA in Urundi. The settlement of Kigali, selected as the Rwandan capital, had a mere 5000 inhabitants. A rapid influx increased Kigali's population to about 300 000, but there were only three other towns, and they each had fewer than 35 000 inhabitants. Most Rwandans live not even in villages, but in homes dispersed throughout the countryside, generally family units of three or four huts built of mud or *adobe* (mud bricks).

Rwanda is one of the world's least developed countries. Agriculture – which employs 90 per cent of the workforce – is mainly for subsistence, but even the people's self-sufficiency is declining. Most families farm tiny plots, producing just enough bananas, cassava roots, sorghum, sweet potatoes and pulses to feed themselves. A few grow coffee for sale and this accounts for 85 per cent of Rwanda's total exports, while some rice is grown commercially on valley floors. Tea, which is state-subsidised, and chrysanthemums are two other cash crops. Many families have a few goats and chickens, and there are some cattle, but these are kept mainly by the richer Tutsi in the east.

Rwanda's economic development has been hindered by isolation from world trade routes. The main access to the sea is by road through Uganda and Kenya, and transport costs are extremely high. More important, the economy remains dependent on coffee exports and foreign aid. But between 1986 and 1994, the international price for coffee was low; and the civil war caused the economy to go into rapid decline.

RWANDA AT A GLANCE
Official name Republic of Rwanda
Area 26 340 km² (10 167 sq miles)
Population 8 600 000 **Per km²** 326 (**Per sq mile** 846)
Capital Kigali
Government Republic
Currency 1 Rwanda franc = 100 centimes
Languages Kinyarwanda (official), French (official), Kiswahili
Religions Christian (74%), indigenous beliefs (25%), Muslim (1%)
Climate Tropical modified by altitude; with two main rainy seasons. Average temperature in Kigali is 19°C (66°F) through the year
Land use Cultivation 40%, grazing 18%, forest and woodlands 10%, other 21%
Main primary products Sweet potatoes, cassava, potatoes, beans and peas, sorghum, bananas, groundnuts, coffee, tea, pyrethrum, cattle, sheep, goats, pigs, timber; tin, tungsten
Major industries Agriculture, mining, forestry, processing hides and skins, textiles, food processing, furniture, shoes, plastic goods
Main exports Coffee, tea, metal ores, hides and skins, pyrethrum
Annual income per head (US$) 250
Population growth (per thous/yr) 29
Life expectancy (yrs) Male 40 **Female** 42

Rybinsk Dam *Russia* Large hydroelectric dam on the Volga River, situated 270 km (170 miles) north of MOSCOW. Completed in 1941, it flooded the nearby lowland to create a vast reservoir covering some 4550 km² (1756 sq miles).

The inhabitants of some hundreds of drowned villages were rehoused in the neighbouring city of RYBINSK.
Map Russia – Ec

Rysy, Mount *Poland* See TATRA MOUNTAINS

Ryukyu Islands (Nansei-shoto) *Japan* Chain stretching some 1200 km (750 miles) south-west from KYUSHU Island to Taiwan. It includes the OKINAWA group of islands. The islands – in particular the Okinawa – were the scene of heavy fighting during the United States invasion of Japan in 1945. The Okinawa group were only returned to Japan in 1971, but the United States retains military bases on some islands.
Population 1 222 600
Map Japan – Gf

Rzeszów *Poland* Modern industrial city, 153 km (95 miles) east of the city of CRACOW. The medieval town at its heart was frequently looted, by Tatars, Swedes and Austrians. However, a fine 17th-century Cistercian church survives. Since 1945, new factories have been set up here, making railway equipment, machinery, chemicals and clothing, and the city's population has more than quadrupled.
Population 151 000
Map Poland – Dc

Sa'da (Sa'dah) *Yemen* Town and road junction 190 km (118 miles) north-west of the capital, SAN'A. It was first settled in AD 901 on the frankincense trade route and has a mosque built in the 12th century. A road built with Chinese aid and opened in 1977 connects Sa'da with San'a and a 79 km (49 mile) long road links it with Zahran in Saudi Arabia.
Population 4500
Map Yemen – Aa

Saale *Germany* River rising north-east of BAYREUTH and flowing some 426 km (265 miles) through the towns of Jena and Halle into the Elbe River north of Leipzig. Its upper course is dammed for reservoirs and hydroelectric power.
Map Germany – Dc

Saarbrücken (Sarrebruck) *Germany* Iron ore-mining and industrial city and capital of Saarland. It stands on the Saar River near the French border, and both French and German are spoken here. Steel was the main industry, but the world steel recession dating from the 1970s has seriously affected production. Saarbrücken has a university and many colleges and academies, as well as theatres and museums.
Population 191 700
Map Germany – Bd

Saarland *Germany* Small state covering some 2570 km² (992 sq miles) on the French border opposite Lorraine. Since the 17th century the region has been fought over by France and Germany, but in 1935 the inhabitants voted to become part of Germany – and the integration was completed in 1957. Saarland is a hilly, forested region of coal fields, heavy industry and vineyards. SAARBRÜCKEN is the capital.
Population 1 074 600
Map Germany – Bd

Šabac *Yugoslavia (Serbia and Montenegro)* Major north Serbian market town and river port on the Sava River, 65 km (40 miles) west of Belgrade. It clusters around the ruins of a 15th-century fortress built by Sultan Mehmet II as the base for the Turkish conquest of Croatia.
Population 119 700
Map Yugoslavia – Bb

Sabah *Malaysia* Most easterly state of Malaysia, occupying about 73 700 km² (28 450 sq miles) on the island of BORNEO. The state was known as North Borneo until it joined Malaysia in 1963. Rapid growth of its hardwood industry since then has made it the richest state in the country. There is concern, though, that logging is upsetting the ecology of the hilly interior. In parts, the cleared rain forest has been replaced by commercial tree crops such as rubber and oil palm. Other parts are being protected; Kinabalu National Park, for instance, has been set up on the slopes of Mount Kinabalu – at 4101 m (13 455 ft) Malaysia's highest peak.
Sabah's capital is KOTA KINABALU. The island is populated along the coast by Malays and Chinese; Dayaks live in the mountainous areas.
Population 1 034 000
Map Malaysia – Ea

Sabrahta (Sabrata) *Libya* Ancient city with foundations dating back to the 5th century BC. It is 67 km (42 miles) west of TRIPOLI, and once stood at the end of the caravan route bringing gold, slaves, ivory and ostrich feathers from the south. As a Roman city it was of importance, exporting olive oil. A Graeco-Roman theatre facing the sea has been restored and seats 5000 people. Other remains include baths, temples, a basilica, forum, shops and houses with mosaics.
Population 14 500
Map Libya – Ba

Sacramento *USA* State capital of California, 137 km (85 miles) north-east of San Francisco, and market centre for agricultural produce of the Sacramento River valley. It grew out of an 1839 trading post, near Sutter's Mill, where gold was discovered in 1848. Fort Sutter, founded in 1841, today houses a museum of the gold-rush days. A canal provides a link with the Pacific Ocean, making Sacramento a port. Processed foods, aerospace equipment, weapons and defence systems are manufactured here.
The Sacramento is also California's longest river, rising near Mount Shasta in the north of the state and flowing some 615 km (382 miles) southwards through the Sacramento Valley (the northern section of the Central Valley) into Suisun Bay, an extension of San Francisco Bay. Many towns along the river were established during the California gold rush of 1848-9.
Population (city) 369 400; (metropolitan area) 1 340 000
Map (city) United States – Bc; (river) United States – Bb

Sacsayhuamán *Peru* Inca fortress above CUSCO whose name means 'Coloured Falcon'. There are three massive zigzag walls containing 21 bastions, made of enormous stones weighing up to 300 tonnes and up to 5 m (16 ft) long. The walls flank a parade ground and an Inca throne carved out of solid rock. The fortress was razed to the ground by the Spaniards after they had defeated the Inca Manco Capac in the 16th century. They used the stones for their own buildings in Cusco.
Map Peru – Bb

saddle Saddle-shaped depression in the crest of a ridge.

Safad (Zefat) *Israel* Hill town 12 km (7 miles) north-west of the Sea of Galilee. It was a stronghold of resistance against the Romans in the 2nd century, and the remains of a Crusader castle built in the 12th century still survive. In the 16th century it became a centre of Jewish mysticism. Today, as well as being a summer resort, Safad is a base for skiing on Mount Hermon, and there is an artists' colony in the Arab quarter.
Population 17 200
Map Israel – Ba

Safajah *Egypt* See BUR SAFAGA

Saffaniyah *Saudi Arabia* The world's largest offshore oil field, situated in The Gulf 230 km (143 miles) north-west of DAMMAM. It is twice the size of the Forties field in the North Sea and can produce up to 1 million barrels of oil a day.
Map Saudi Arabia – Bb

Safi *Morocco* Atlantic port 220 km (140 miles) south-west of CASABLANCA that handles phosphates mined at Youssoufia, 100 km (62 miles) to the east. It is also the world's biggest exporter of sardines, caught off the Moroccan coast and processed in Safi. Although it is a growing industrial town, with a chemical plant that processes the phosphate rock, it retains its medina (old Muslim residential quarter), an old Portuguese fortress, a palace, and potters' workshops.
Population (Greater Safi) 198 000
Map Morocco – Ba

Safid Rud *Iran* Name given to the lower course of the Qezel Owzan River, whose source is near Sanundaj. The whole course runs for 800 km (500 miles) and flows into the Caspian Sea. Its delta, to the east of the city of RASHT, is a rice-growing area.
Map Iran – Aa

Sagaing *Myanmar (Burma)* Town and capital of Sagaing Division, which stretches far to the north, covering 94 626 km² (36 535 sq miles) between the Ayeyarwady (Irrawaddy) and Chindwinn rivers. It produces wheat, cotton and beans. Standing beside the Ayeyarwaddy River opposite the city of MANDALAY, the town is over-looked by pagoda-lined hills.
Population (town) 190 000; (division) 3 856 000
Map Myanmar – Bb

sagebrush Scrub vegetation in the semi-desert regions of the south-western United States and on the Mexican plateaus.

Sagres *Portugal* Small port on the windswept Cabo de São Vicente Peninsula on the country's south-western tip. It was at Sagres in about 1416 that Henry the Navigator, the Portuguese prince and renowned patron of explorers, founded a school of navigation, where Vasco da Gama's voyages of discovery to Africa and India were planned.
Map Portugal – Bd

Ságvár *Hungary* See SIÓFOK

Sahara *Africa* The world's largest desert, occupying nearly one-third of the African continent. Its size is estimated at 9 065 000 km² (3 500 000 sq miles), stretching from the Atlantic Ocean to the Red Sea; it is bordered by the Mediterranean and the Atlas Mountains in the north and by the SAHEL belt to the south.
Most of the Sahara is a low plateau, averaging about 460 m (1500 ft) above sea level; its lowest point is the Qattara Depression in Egypt which falls to 133 m (436 ft) below sea level. Its highest point is 3415 m (11 204 ft) at TIBESTI in Chad. Only about one-ninth of the desert is sandy – notably the Libyan Desert, Egypt's Western Desert and, in Algeria, the Great Eastern Erg and Great Western Erg.
Map Africa – Cb

Saharan Atlas *Algeria* See ATLAS MOUNTAINS

Sahel *Africa* Semi-arid region stretching from Senegal to Sudan and forming a belt dividing the Sahara Desert from Africa's tropical forests. The terrain is mostly grass, thorn and scrub SAVANNAH, and there is some poorly cultivated cropland watered by the Niger and Senegal rivers. Cattle, goats and camels are reared. Drought in this region caused widespread famine in the 1970s, in the mid-1980s and again in the early 1990s.
Map Africa – Cc

Sahul Shelf *Timor Sea* The undersea extension of the continental platform off north-west Australia. The shelf is at relatively shallow depth and is a source of petroleum. Tax revenue from production is shared with Indonesia.
Map Australia – Ca

Saida *Algeria* Administrative centre in the Atlas Mountains, 200 km (124 miles) south of MOSTAGANEM on the main western rail and road route across the mountains to the SAHARA.

It stands in the middle of an important agricultural region producing mainly wheat, where the country's 'agricultural revolution' – a plan to make Algeria self-sufficient in food – was launched in the 1970s.
Population 60 000
Map Algeria – Ba

Saigon *Vietnam* See HO CHI MINH CITY

Saimaa *Finland* Finland's biggest lake, covering 4380 km² (1690 sq miles) in the south-east of the country. Its rugged shoreline extends for a total of about 15 000 km (9320 sq miles).

Saimaa is also a canal completed in 1856 between the lake and Vyborg Bay at the head of the Gulf of Finland. It is 45.5 km (28 miles) long, including the passage through several lakes along its route. The southern part belonged to the USSR after the Second World War, but in 1968 the two countries signed a treaty giving Finland control of the canal for 50 years.
Map Finland – Dc

St Albans *United Kingdom* English cathedral city in Hertfordshire, 32 km (20 miles) north of London. It was founded nearly 2000 years ago as Verulamium and was one of the largest towns of Roman Britain. Remains of the Roman baths, walls, theatre and a street are preserved near the city centre. The cathedral, 168 m (550 ft) long, dates from the 11th century, with 13th and 14th-century additions. It stands on a hill where

Alban, a Roman soldier, was executed in AD 209 for harbouring a Christian priest; he became Britain's first saint and gave his name to the Saxon town that replaced Verulamium.
Population 77 190
Map United Kingdom – Ee

St Barthélémy *Guadeloupe* Small island, commonly called St Barts, situated 210 km (130 miles) north-west of the main islands of GUADELOUPE, of which it is a dependency. The inhabitants are descended from Norman immigrants, and speak a 17th-century Normandy dialect. The economy depends upon fishing and tourism, and the harbour of Gustavia, the capital, is a favourite anchorage for yachtsmen.
Population 5040
Map Caribbean – Cb

St Bernard Passes *France/Italy/Switzerland* Two Alpine passes used since Roman times. The Grand St Bernard, 25 km (16 miles) east of Mont Blanc, is 2469 m (8100 ft) high and joins the Valle d'Aosta in Italy with Martigny and the Rhône Valley in Switzerland. The modern road dates from 1905 but is bypassed by a tunnel which was opened in 1964.

The Little St Bernard, 2199 m (7215 ft) high, is the pass through which, some scholars believe, the Carthaginian general Hannibal and his elephants invaded Italy in AD 218. It is about 18 km (11 miles) south of Mont Blanc, joining the Valle d'Aosta with the Isère Valley and CHAMBÉRY in France. The present road was built in 1871.
Map Italy – Ab

Saint Catharines *Canada* Industrial centre in a fruit-growing region at the outlet of the WELLAND CANAL in southern Ontario. Industries include shipbuilding, food processing, and the manufacture of automobile parts and electrical equipment.
Population 129 300
Map Canada – Hd

St Croix *US Virgin Islands* Largest and most populous island of the group, covering 215 km² (83 sq miles), containing the territory's only extensive area of lowland. The island is the site of the huge Amerada Hess oil refinery and has factories which make rum, and a variety of goods, especially watches, from imported materials. The

two small towns of Christiansted and Fredriksted contain attractive old Danish-style buildings.
Population 50 100
Map Caribbean – Cb

Saint-Denis *France* Industrial suburb of northern Paris. It manufactures chemicals, plastics, diesel engines, leather goods and fireworks. The abbey of Saint Denis, founded in 626, was the home of Peter Abelard, the 12th-century French philosopher whose tragic love affair with his pupil, Héloïse, has become part of romantic legend. Abelard became a monk and Héloïse a nun. They are both now buried in a single sepulchre in the Père Lachaise cemetery in Paris. Most of the kings of France are buried in the abbey.
Population 91 300
Map France – Eb

Saint-Denis *Réunion* Capital of the island, lying on the north coast and backed by volcanic peaks in the interior. It has attractive 18th-century French and Creole buildings in the centre, but modern development is spreading west to the island's main shipping port, Le Port, and east to the airport.
Population 121 670
Map (Reunion) Indian Ocean – Bd

Saint-Emilion *France* Fortified town lying near Libourne in the lower DORDOGNE Valley. It is surrounded by vineyards that produce the red wine named after the town.
Population 2800
Map France – Cd

Saint-Etienne *France* Industrial city situated about 50 km (30 miles) south-west of LYONS. With coal mines, metalworking and textile factories, it was the cradle of the Industrial Revolution in France in the early 19th century. However, coal mining decreased during the 1970s and ceased in 1981. Its dominant industries now include precision engineering, clothing and plastics.
Population 313 000
Map France – Fd

St Eustatius *Netherlands Antilles and Aruba* Small island, also known as Statia, in the Caribbean Leeward chain, consisting of two volcanoes joined by an area of lowland. There is a little farming and pumice mining and some tourism, but the island is largely undeveloped.

St Eustatius was settled by the French in 1625, and became Dutch in 1632. For much of the 18th century it was a rich duty-free trading centre. Indeed, it was the first foreign government to recognise the fledgling United States and was an important supply source for the revolutionaries.

Incensed by this, Britain invaded and captured the island in 1781, sacking warehouses and homes. The invaders continued to fly the Dutch flag and captured several American and other merchant ships by this ruse. The capital ORANJESTAD stands 90 m (295 ft) above the sea on cliffs on the southwest coast. Also known as Upper Town, it contains the remains of a 17th-century Dutch fort.
Population 1870
Map Caribbean – Cb

St Gall (Sankt Gallen; St Gallen) *Switzerland* Town situated in the economic heart of eastern Switzerland, 65 km (40 miles) east of Zürich. It

▼ SAHARAN SEA Sand dunes reach out towards the Atlas Mountains, part of the Djezira el Maghreb – 'the Western Isle'– between the desert and the Mediterranean Sea.

585

is named after an Irish monk, Gallus, who settled here in 612 and has an 8th-century Benedictine monastery (now a UNESCO world heritage site) that made the town an important cultural centre in the 9th and 10th centuries.

Textiles were the main industry of the canton of the same name until the middle of the 20th century and its embroidery ('St Galler Stickerei') is still popular.

Population (town) 72 000; (canton) 432 800
Map Switzerland – Ba

St George's *Grenada* Port and capital city of the island, lying on the south-west coast. It is built around a horseshoe-shaped bay, the brightly painted houses rising up the wooded hillsides.

The city was founded in 1650 by the French and became the seat of the British governor after Grenada changed hands in 1783. Point Salines Airport is 6 km (3.5 miles) south of the town and there is a modern deep-sea port which handles the island's exports of cocoa, bananas, nutmeg, fish, rum and sugar.

Population 32 000
Map Caribbean – Cc

St Gotthard (St Gothard) *Switzerland* Pass through the central Alps, 55 km (35 miles) south-east of Lucerne, near the Italian border. A 15 km (9 mile) tunnel, built in 1872-80, carries a railway line under the pass. In 1980 a 16 km (10 mile) road tunnel replaced the old winding road over the top, which is usually open from June to mid-October.

Map Switzerland – Ba

St Helena *South Atlantic* Mountainous island, 1930 km (1200 miles) west of Africa, where Britain held the exiled Napoleon Bonaparte, following his final defeat at the Battle of Waterloo, for six years until he died in 1821. The island covers only 122 km² (47 sq miles). About 1500 people live in the capital and port, Jamestown. The climate is mild. St Helena is a British colony, with ASCENSION ISLAND and TRISTAN DA CUNHA as dependencies.

Population 5650
Map Atlantic Ocean – Ee

St Helens, Mount *USA* Volcano, 2549 m (8364 ft) high, in the Cascade Range in western Washington State. It erupted in 1980, destroying surrounding forests, killing more than 60 people and losing some 400 m (1312 ft) in height. The mountain lies within the 583 km² (225 sq mile) Mount St Helens National Volcanic Monument established in 1982.

Map United States – Ba

Saint John *Canada* Seaport city on the Bay of Fundy at the SAINT JOHN River mouth in southern New Brunswick. The French had a trading post here in 1630 but the first major settlement was made in 1783 by a group of New Englanders. In 1785 it received Canada's first city charter.

Saint John has an excellent ice-free harbour and large dry docks; both trans-Canadian railway lines have their termini in the city. Industries include shipbuilding, paper making and oil refining. Tourist attractions include the 18th-century courthouse and the Reversing Falls rapids.

Population (city) 74 970; (metropolitan area) 124 980
Map Canada – Id

Saint John *USA/Canada* River, some 673 km (418 miles) long, which rises in Maine, USA, and forms part of the US-Canadian border before flowing across New Brunswick into the Bay of FUNDY at the port of SAINT JOHN.

The river is navigable for small vessels as far as FREDERICTON, New Brunswick. At Grand Falls, on the US border, the river drops 23 m (75 ft) in a great cataract; and at Reversing Falls rapids, near the mouth of the river, the flow is reversed by high tides surging up the Bay of Fundy.

Map Canada – Id

St John *US Virgin Islands* Smallest of the territory's three main islands, covering 52 km² (20 sq miles). Expensive but uncrowded tourist resort. About two-thirds of the island is national parkland, including offshore waters and islets.

Population 2400
Map Caribbean – Cb

St John's *Antigua and Barbuda* Capital and chief port of Antigua, which forms part of the Lesser Antilles. The town is laid out in a formal grid plan and is a mixture of attractive 18th and 19th-century buildings, which survived an earthquake in 1843; several new developments have been started to attract visitors.

Population 36 000
Map Caribbean – Cb

St John's *Canada* Capital of NEWFOUNDLAND and the easternmost city of Canada. It is a major fishing port and one of the oldest settlements in North America, having been colonised by the British under Sir Humphrey Gilbert in 1583. The city extends across a hillside overlooking the harbour.

Twice during the 19th century fire destroyed much of the old town and only a few of the old wooden buildings remain. Alcock and Brown made the first non-stop transatlantic air flight from here in 1919, and Marconi received the first transatlantic wireless message at the Cabot Tower on Signal Hill in 1901. Industries include fish processing, shipbuilding, textiles and paper.

Population (city) 95 770; (metropolitan area) 171 860
Map Canada – Jd

St Kitts and Nevis See p 587

St Lawrence *Canada/USA* River, some 3058 km (1900 miles) long, that flows out of Lake Ontario through Quebec and Ontario into the Gulf of St Lawrence. It is the main trade artery of Canada and a major source of hydroelectricity. It forms part of the boundary between Canada and the USA, and is open to US shipping. Shoals, rapids and islands (the Thousand Islands) slow navigation but the scenery is a magnet for tourists.

Map Canada – Id

St Lawrence, Gulf of *Canada* Major waterway and fishing ground in eastern Canada, and the outlet to the ATLANTIC OCEAN for ships using the ST LAWRENCE SEAWAY. The first European to sight the gulf was the French explorer, Jacques Cartier, in 1535. Fogs and drifting ice are shipping hazards from mid-December to mid-April. It covers 259 000 km² (100 000 sq miles).

Map Canada – Id

St Lawrence Seaway and Great Lakes Waterway *Canada/USA* A system of canals, dams and locks linking the Great Lakes through the St Lawrence River with the Atlantic Ocean. The system, some 3800 km (2350 miles) long, was opened in 1959.

It permits ocean-going vessels up to 221 m (725 ft) long, 23 m (75 ft) wide and with up to 7.9 m (26 ft) draught to sail from Montreal, Quebec, to the west end of Lake SUPERIOR. The system provides the water that drives hydro-electric plants in Ontario and New York State.

Map Canada – Hd

St Louis *Senegal* Seaport and regional capital of the north, founded by the French in 1659 on an island in the mouth of the Senegal River. It was the first French settlement in West Africa.

In the 17th and 18th centuries, the French, Dutch, Portuguese and British grew rich on the trade in slaves and gum arabic, but St Louis was superseded by the deep-water port at Dakar as the mouth of the river silted up. It was the capital of French West Africa from 1895 to 1904, however, and remained Senegal's capital until 1958.

Population 180 000
Map Senegal – Aa

St Louis *USA* Major transport and industrial centre in eastern MISSOURI, on the Mississippi River, 16 km (10 miles) south of its confluence with the Missouri River.

In the 19th century St Louis was a key trading post and departure point for settlers heading west, and it remains a busy commercial city. Industries include transport equipment, electronics, beer, processed food and chemicals. Its suburbs extend across the river into Illinois.

Population (city) 396 700; (metropolitan area) 2 492 500
Map United States – Hc

St Lucia See p 588

St Lucia *South Africa* Conservation area of lake, lagoon and wetlands on the coast of northern KwaZulu-Natal, about 260 km (160 miles) to the north-east of Durban.

Map South Africa – Db

St Maarten *Netherlands Antilles and Aruba* See ST MARTIN

Saint-Malo *France* Fishing and ferry port in northern Brittany, 70 km (44 miles) north-west of RENNES. It was founded in the 6th century as a monastic settlement, which was later fortified. Today it is a popular tourist resort with sandy beaches nearby. Only the medieval ramparts and some granite houses survived destruction in 1944 during the Second World War, but the town has been rebuilt in the old style.

Population 48 100
Map France – Cb

St Martin *Guadeloupe/Netherlands Antilles and Aruba* Island in the Leeward group, just south of ANGUILLA. The northern section is administered by France as part of GUADELOUPE, which lies 240 km (150 miles) to the south-west.

The southern section, known as St Maarten, is part of the Netherlands Antilles, whose capital is on CURAÇAO, 900 km (560 miles) to the south-west. There are no customs or other barriers between the two sections; the French capital of Marigot and the Dutch Philipsburg – both with

St Kitts and Nevis

ISLANDS FORMED BY EXTINCT VOLCANOES REMAIN THE 'MOTHER COLONY OF THE WEST INDIES'

The islands of St Kitts (also called St Christopher), covering some 168 km² (65 sq miles), and Nevis, covering 93 km (36 sq miles) lie in the LEEWARD ISLAND group in the eastern Caribbean. They were sighted by Christopher Columbus in 1493 but there were no settlers until the British came to St Kitts in 1623 and to Nevis in 1628. St Kitts became a base for the British colonisation of neighbouring islands and it is proud of its title 'Mother Colony of the West Indies'.

For a time St Kitts was settled by the French and British simultaneously, and ownership was initially disputed, but in 1783 it became solely British. Administered for many years with ANGUILLA, St Kitts and Nevis became an independent Commonwealth country in 1983 with the British sovereign as head of state. The inhabitants of both islands are descended from African slaves and are English speaking.

St Kitts consists of three young but extinct volcanoes which are linked by a narrow sandy isthmus to much lower volcanic remains in the south. Mount Liamuiga (formerly Mount Misery) at 1156 m (3792 ft) is the island's highest point. Around most of the island the gentle slopes which rise from the coast are covered with fertile soil on which sugar cane is grown; sugar is the chief export. Market gardening and livestock rearing are being expanded on the steeper lands above the cane fields. Agriculture accounts for 7 per cent of the country's gross domestic product (GDP).

There are some light industries which use imported materials to make electronic equipment, footwear and batik for export. Some new factories manufacture clothing and television sets for export. Older industries continue, notably sugar processing, brewing, distilling and bottling. Another major source of income is, and has been for years, remittances from overseas workers. However, in the 1990s the government began encouraging development of industries and infrastructure.

The white sands and relatively dry climate of southern St Kitts led to a major tourist development at Frigate Bay, and the government is keen to build up this money-spinner. Many of the older buildings of historical interest in BASSETERRE, the capital and deepwater port, have been carefully restored.

The partly constructed fort of Brimstone Hill, built on a clifftop in the 17th and 18th centuries and the scene of many past battles between the British and French, has been restored and provided with a museum.

Nevis lies 3 km (2 miles) south-east of St Kitts, with which it has daily links by boat and light aircraft. The island is an extinct volcano which rises to 985 m (3232 ft) at Nevis Peak. The cloud-covered peak, which reminded Columbus of snow-capped mountains in Spain, gave the island its name – the Spanish word for snow is *nieves*.

Farming has declined on Nevis, but the government has encouraged more food production. Small factories process locally grown copra and sea-island cotton seed into oil. There are small boat-building and fishing industries, but tourism is still the main source of income, thanks to the fine beaches and clear waters. Visitors stay in cottages in the grounds of old plantation houses. CHARLESTOWN is the island capital.

ST KITTS AND NEVIS AT A GLANCE

Map Caribbean – Cb	
Official name Federation of St Kitts and Nevis	
Area 261 km² (101 sq miles)	
Population 40 620 **Per km²** 156 **(Per sq mile)** 402)	
Capital Basseterre	
Government Parliamentary monarchy	
Currency 1 East Caribbean dollar = 100 cents	
Language English (official)	
Religion Christian (mostly Anglican)	
Climate Tropical and humid. Average temperature is 26°C (79°F) all year	
Land use Cultivation 39%, meadows and pastures 3%, forest and woodlands 17%, other 41%	
Main primary products Sugar, cotton, coconuts, vegetables, livestock	
Major industries Agriculture, textiles, sugar processing, electronic equipment, tourism	
Main exports Sugar, clothing, electronic equipment, footwear	
Annual income per head (US$) 3990	
Population growth (per thous/yr) 6	
Life expectancy (yrs) Male 63 **Female** 69	

good harbours – are only 10 km (6 miles) apart. At one time, the production of salt through evaporating sea water was the chief source of income, but today tourism is the major industry.

Population 28 520
Map Caribbean – Cb

St Moritz *Switzerland* Fashionable resort and spa in the south-east, 15 km (9 miles) north of the border with Italy. It has thermal baths – one of which has been used for 3000 years – and is noted for its winter sports and summer mountaineering.

St Moritz hosted the Winter Olympics in 1928 and in 1948. It is the site of the Cresta Run for toboggans. In 1992 Christian Bertschinger of Switzerland set a record for the course, which is 1212 m (3977 ft) long, with a drop of 157 m (515 ft). His time was 50.41 seconds.

Population 5070
Map Switzerland – Ba

Saint-Nazaire *France* Shipbuilding town at the mouth of the Loire River, developed in the 19th century as a deep-water port for the city of NANTES, 50 km (30 miles) to the east. During the Second World War it was an important German submarine base for U-boats attacking Allied merchant shipping convoys on their way across the Atlantic, and the British bombed the base several times. Today, there is a viewing terrace over the remains of the submarine pens.

Population 132 000
Map France – Bc

St Paul *USA* Twin city of MINNEAPOLIS and the state capital of Minnesota. It was founded in 1840 at the navigable limit of the Mississippi River. Most of the city is laid out on three terraces – the lowest for the railroad and factories, the middle for businesses and the top for residences. The period since the 1960s has seen major urban renewals. It is a business, government and tourist centre and has a popular annual winter carnival.

Population (city) 272 200; (metropolitan area) 2 538 800
Map United States – Hb

St Petersburg *Russia* Former Russian capital (from 1712 to 1918) and the country's second largest city after Moscow. It was known as St Petersburg until 1914, then as Petrograd until 1924, when it was renamed Leningrad in honour of Vladimir Lenin (1870-1924), one of the leaders of the Russian Revolution. With the collapse of the Soviet Union the city regained its old name of St Petersburg.

Tsar Peter the Great (1682-1725) founded the city as a gateway to the west at the head of the Gulf of Finland on about 100 marshy islands in the delta of the Neva River. The marshes are extremely treacherous and St Petersburg was built at great cost in human lives.

Today the flood-prone city, sometimes called the 'Venice of the North', is linked by nearly 700 bridges. The tsar's architects, engineers and artisans – mostly from Western Europe – gave central St Petersburg an air of grandeur, with well-planned squares, parks and wide avenues between neo-classical Baroque buildings, often pastel-painted, with gilt spires and domes.

St Petersburg is considered to be one of Europe's most beautiful cities. The Peter and Paul Fortress, the old heart of the city beside the Neva, encloses a cathedral and was once used as a political prison. The author Maxim Gorky (1868-1936) was imprisoned here in 1905.

The old city also includes the sumptuous Winter Palace of the tsars – now the Hermitage Museum, with a superb collection of paintings – and the Kirov Theatre, home of the Kirov Ballet. The main shopping street, Nevsky Prospekt, is named after a national hero, Alexander Nevsky (1220-63), who defeated invading Swedes on the Neva in 1240.

The city was the scene of the Decembrist uprising against the tsar in 1825, and the Russian Revolution of 1917 also began here. In the Second World War, German forces besieged it for almost 900 days, from September 1941 to January 1944, when more than 1 million of its inhabitants died – 640 000 of them from starvation. More than 500 000 were buried in the city, in the world's largest cemetery.

St Petersburg is, with Moscow, Russia's most important cultural centre and is a popular tourist destination. It is one of the country's leading manufacturing cities and its main export port. Its major industries include engineering, shipbuilding and food processing, and the production of chemicals, paper, textiles and clothing.

Population 6 705 000
Map Russia – Dc

St Lucia

*CARIBBEAN ISLAND THAT IS A
REGULAR PORT OF CALL FOR OIL
TANKERS AND CRUISE LINERS
BRINGING VALUABLE REVENUE*

One of the WINDWARD ISLANDS in the eastern Caribbean, St Lucia was settled briefly by the British in 1605. Over the next 200 years control passed some 10 to 14 times between them and the French before the island was finally ceded to the British in 1814. English is the island's official language, but many of the people, mostly descendants of African slaves, speak a French patois. St Lucia became a fully independent member of the Commonwealth, with the British sovereign as head of state, in 1979.

It is an island of extinct volcanoes. Mount Gimie is the highest peak, at 950 m (3117 ft). To the south there lies the horseshoe-shaped Qualibou which contains 18 lava domes and seven craters; in the west are the impressive peaks of the PITONS – Gros Piton (798 m, 2619 ft) and Petit Piton, which rise directly from the sea to more than 750 m (2461 ft).

Bananas are the main crop and chief export, but production is often affected by hurricanes, drought, disease and, more recently, falling prices which have meant hardship for banana farmers. In October 1993 emotions ran high when a three-day strike demanding higher pay for workers in the industry ended in the death of two banana farmers; several others were injured, and the government was forced to dismiss the parastatal banana association on the grounds of corruption. Coconuts are the second most important crop.

The free-port and trade zone at Vieux Fort in the south of the island has attracted industries producing textiles and electrical components for export.

Grande Cul de Sac Bay, to the south of the capital, CASTRIES, is one of the deepest and most modern tanker ports in the Americas for oil transshipment. The beautiful Castries harbour is a regular calling point for cruise liners – and tourism is increasing, with new facilities, such as those at Hewanorra Airport, being developed with an eye to the future.

ST LUCIA AT A GLANCE

Map Caribbean – Cc	
Area 620 km² (239 sq miles)	
Population 144 340 **Per km²** 233 **(Per sq mile** 604)	
Capital Castries	
Government Parliamentary monarchy	
Currency 1 East Caribbean dollar = 100 cents	
Languages English, French patois	
Religion Christian (Roman Catholic 90%, Protestant 10%)	
Climate Tropical. Average temperature is 27°C (81°F) throughout the year	
Land use Cultivation 28%, meadows and pastures 5%, forest and woodlands 13%, other 54%	
Main primary products Bananas, cocoa, coconuts, vegetables, fruits, spices	
Major industries Food processing, rum distilling, clothing, electrical components, fishing, tourism	
Main exports Bananas, cardboard boxes, clothing, cocoa, coconut products, fruits, vegetables	
Annual income per head (US$) 2900	
Population growth (per thous/yr) 5	
Life expectancy (yrs) Male 67 **Female** 72	

St Pierre *Martinique* See PELÉE, MONT

St Pölten *Austria* Capital, since 1986, of Lower Austria (Niederösterreich) and industrial town 56 km (35 miles) west of Vienna. It was known in Roman times as Aelium Cetium and during the Middle Ages was renowned as a centre for the leather and textile industries. In the 1850s it became one of the country's earliest industrial towns. Its manufactures include turbines, paper, machinery, furniture and rolling stock.
Population 49 800
Map Austria – Da

Saint-Raphael *France* See SAINT-TROPEZ

St Thomas *US Virgin Islands* Main tourist island of the territory, covering some 83 km² (32 sq miles). Its fine harbour, beside which stands the capital and duty-free port, Charlotte Amalie, was long the haunt of pirates, but now attracts cruise ships and luxury yachts.
Population 44 000
Map Caribbean – Bb, Cb

Saint-Tropez *France* Fashionable tourist resort on the Mediterranean, 100 km (62 miles) east of Marseilles, with excellent beaches and yacht marinas. It has been popular with artists since the 1890s, mainly because of its fine, clear light. The resort of Saint-Raphael is just to the north.
Population 5800
Map France – Ge

St Vincent and the Grenadines See p 589

Saint Vincent, Cape *Portugal* See SÃO VICENTE, CABO DE

St Vincent, Gulf *Australia* Deep inlet of the Indian Ocean south of Australia, about 145 km (90 miles) long and some 72 km (45 miles) wide. ADELAIDE stands on its eastern shore and Kangaroo Island at its mouth. The south-western outlet, Investigator Strait, was named after the *Investigator*, the ship used by the British navigator Matthew Flinders, who discovered the gulf in 1802 during his circumnavigation of Australia.
Map Australia – Fe

Saintes *France* Industrial town on the Charente River, about 60 km (37 miles) south-east of LA ROCHELLE. It produces iron goods, pottery and brandy. Known as Mediolanium in ancient times, it contains the remains of a Roman amphitheatre. There is also a fine 15th to 16th-century church.
Population 25 900
Map France – Cd

Saintes, Iles des *Guadeloupe* Group of eight tiny volcanic islands situated 10 km (6 miles) south of BASSETERRE. The only inhabited islands are Terre de Haut and Terre de Bas. Terre de Haut's fishermen are descended from 17th-century settlers from northern France, while the fishermen of Terre de Bas are mostly black.
Population 2900
Map Caribbean – Cb

Saintonge *France* Historic province on the Bay of Biscay, north of the GIRONDE estuary. It was a fief of AQUITAINE and, as such, part of England from 1154 to 1375. The chief town is SAINTES.
Map France – Cd

Saipan *Northern Mariana Islands* Largest and most populous island of the territory, covering some 122 km² (47 sq miles) and lying 200 km (125 miles) north-east of GUAM. It has high cliffs except on the western side, where the capital, Susupe (population 8000), is situated. The capture of Saipan by US forces in June 1944 was a step towards the recapture of the Philippines and the defeat of Japan in the Second World War.
Population 38 800
Map Pacific Ocean – Cb

Sakai *Japan* Port and industrial city on south HONSHU Island, just south of Osaka. It was a leading city from the 14th to the beginning of the 17th century – independent, with its own militia. The city's economy collapsed when Japan outlawed most links with the outside world in the 17th century. Today, however, Sakai is a leading commercial and oil and steel centre.
Population 808 000
Map Japan – Cd

Sakhalin *Russia* Island in the Russian Far East, separated from the Japanese island of HOKKAIDO to the south by the 40 km (25 mile) wide La Pérouse Strait. Long, narrow and mountainous, with thick forests in the north, Sakhalin covers about 76 400 km² (29 500 sq miles). The island was originally settled by Russian fishermen in 1853 and later became a convict colony. The southern part was taken over by Japan in 1905. However, Russia took it back in 1945 after the defeat of Japan in the Second World War.

The inhabitants' main occupations are fishing, canning, coal mining, oil drilling, paper making and hunting. The main town and capital is Yuzhno-Sakhalinsk (formerly Toyohara).
Population 660 000
Map Russia – Pd

Sakhira (La Skhirra) *Tunisia* Port on the Gulf of Gabès, 100 km (about 62 miles) south-west of SFAX. It is the oil terminal for the pipeline from the Algerian Edjele oil field and from Tunisia's El Borma field.
Map Tunisia – Ba

Sal (Ilha do Sal) *Cape Verde* Desert island in the CAPE VERDE archipelago and the site of the international airport. The only town is ESPARGOS. The island's name derives from the salt that was once exploited here.
Population 6000
Map (Cape Verde) Africa – Ac

St Vincent and the Grenadines

ISLANDS THAT HARBOUR EXCLUSIVE YACHTING RESORTS – AND THE HOME OF A FAMOUS BREADFRUIT TREE

Situated in the beautiful botanic gardens of KINGSTOWN, capital of St Vincent in the WINDWARD ISLANDS, grows a breadfruit tree descended from one brought from Tahiti in 1792 by Captain William Bligh. It was Bligh's second attempt to introduce it; the first ended in 1789 with the famous mutiny aboard HMS *Bounty*.

The aim was to provide a source of food for the African slaves working on the West Indian sugar plantations. This plan failed – the slaves preferred other foods such as bananas and plantains – but St Vincent is still sometimes called the 'Tahiti of the Caribbean', because like Tahiti, the island has productive soils and luxuriant vegetation.

Christopher Columbus may or may not have spotted the island in 1498 – there is some controversy – but he was certainly not the first person to see it. The Ciboney people are said to have lived here from about 4300 BC, followed by the Arawaks and, in the 16th century, the Caribs. Only in the 17th century did Europeans – French and English – really attempt to settle. The Caribs resisted them, but after the last Carib War (1795-7), most of them

were deported to the Bay Islands, off Honduras. Meanwhile, control switched many times between Britain and France, the island being finally ceded to Britain in 1783.

With the Grenadine Islands of MUSTIQUE (a much-favoured yachting resort), Bequia, Canouan, Mayreau and Union to the south, St Vincent became an independent country and member of the Commonwealth, with the British sovereign as head of state, in 1979. Two-thirds of the population are descended from African slaves and most of the rest are of mixed race. Some descendants of the Carib live on the slopes of Mount SOUFRIÈRE in the north of St Vincent. English is the main language.

A chain of volcanoes makes up the backbone of St Vincent. Soufrière, 1234 m (4048 ft) high, is one of the two active volcanoes in the eastern Caribbean (the other is Mont PELÉE on Martinique). Its last violent eruption was in 1979. In 1812 it killed 56 people and in 1902 more than 1000 died from breathing the hot dust, from burns or as victims of falling stones.

St Vincent is one of the poorest islands in the eastern Caribbean and unemployment is high (35-40 per cent in 1992). Much of the island is forested. Farming is the main occupation, employing 60 per cent of the workforce, and bananas are the main export, although banana earnings declined in the early 1990s due to low world prices and unreliable weather. The island is the world's leading producer of arrowroot starch (used to make medicines, fine flour and paper for computers). Some copra, coconut oil and spices are exported, but sugar is now grown only for local use. The fishing industry is being developed.

There is very little manufacture; the few factories process agricultural products and make cement, furniture and tennis racquets. A flour mill serves all the Windward Islands. The

government has launched a huge programme to develop tourism to provide employment – a large new tourist complex on Union Island in the southern Grenadines has been built, and the government has been improving roads and airports for tourism. Cruise ships call at the deep-water harbour of Kingstown; about 160 000 tourists visit the islands each year.

The seas around the tiny Grenadines and the many bays are some of the best boating waters in the world. Luxury tourism and yachting have made the inhabitants of Bequia and the exclusive island of Mustique more prosperous than those on the main island of St Vincent.

ST VINCENT AT A GLANCE	
Map Caribbean – Cc	
Area 349 km² (135 sq miles)	
Population 114 560 **Per km²** 328 **(Per sq mile** 849)	
Capital Kingstown	
Government Parliamentary monarchy	
Currency 1 East Caribbean dollar = 100 cents	
Languages English, French patois	
Religion Christian (Anglican, Methodist, Roman Catholic)	
Climate Tropical; dry season January to May. Average temperature is 27°C (81°F) throughout the year	
Land use Cultivation 50%, forest and woodlands 41%, meadows and pastures 6%, other 3%	
Main primary products Bananas, arrowroot, eddoes (taro), coconuts, spices	
Major industries Agriculture, flour milling, food processing, clothing, cement manufacture, forestry	
Main exports Bananas, arrowroot starch, taro, copra, tennis racquets	
Annual income per head (US$) 1990	
Population growth (thous/yr) 8	
Life expectancy (yrs) Male 70 Female 73	

Salalah *Oman* Port and capital of DHOFAR Province, situated about 100 km (62 miles) east of the border with Yemen. It has a temperate climate because it is positioned to catch the Indian summer monsoon. Salalah is a modern city set among coconut groves and banana plantations. Parts of the old city remain, including the former residence of the sultan, some traditional Arab houses and an old mosque.

In addition to its own harbour, Salalah is served by the new port of Raysut, 20 km (12 miles) to the east. It has an airport that was built in 1977.

Population 15 000
Map Oman – Ab

Salamanca *Spain* University town about 170 km (105 miles) north-west of the capital, Madrid. The whole town has been declared a national monument because of its magnificent 17th and 18th-century buildings. The town's two universities date from the 13th century. Salamanca is the commercial centre for the surrounding agricultural province of the same name.

Population (town) 162 890; (province) 357 800
Map Spain – Cb

Salamis *Cyprus* Site of a ruined Greek city, founded in 1180 BC on the east coast. It was the scene of a naval victory for the Greeks in 306 BC

when Demetrius I defeated Ptolemy I of Egypt. It has the remains of an amphitheatre, a temple of Zeus and a forum (main square).

Map Cyprus – Ab

Salamís (Salamina) *Greece* Island in the Saronic Gulf, 13 km (8 miles) west of PIRAEUS. In 480 BC the Athenians destroyed a Persian fleet in the narrow strait between the island and the coast of Attica. Salamís is now a major Greek naval base.

Population 28 600
Map Greece – Cc

Salang Tunnel *Afghanistan* Tunnel through mountains 100 km (62 miles) north of Kabul. Opened in 1964, it cuts the distance from Kabul to the northern plains by 200 km (125 miles). Lying 3360 m (11 030 ft) above sea level, the tunnel is 3400 m (11 155 ft) long and 8 m (26 ft) wide. It is closed at night during winter as the temperature can drop to -20°C (-4°F).

Map Afghanistan – Ba

Salaverry *Peru* Modern port 8 km (5 miles) south of TRUJILLO. It exports copper from Quiruvilca, 120 km (75 miles) inland, and sugar from the valleys and plains of La Libertad.

Population 5250
Map Peru – Ba

Saldanha *South Africa* Town and fishing harbour on Saldanha Bay in the Western Cape, 110 km (70 miles) north-west of Cape Town. It is the site of a naval and military academy. The harbour was extended in the 1970s to export iron ore mined at Sishen. South of the town is the West Coast National Park, which includes the shallow Langebaan lagoon, measuring 16 by 4.5 km (10 by 2.5 miles), a paradise for seabirds and a graveyard of countless fossilised oysters.

Population 33 970 (including the nearby town of Vredenburg)
Map South Africa – Ac

Salé *Morocco* Atlantic seaport and suburb of RABAT, to which it is linked by bridge and ferries across the Bou Regreg River. Founded on the site of an ancient Roman port, it was a major commercial centre from the Middle Ages. It has a 12th-century mosque, a Koranic college dating from 1341, the monumental Bab Mrisa gateway and many examples of 13th and 14th-century Moroccan architecture. Its markets specialise in basketwork, jewellery and embroidery, and there are streets for cabinet makers, stone carvers, blacksmiths and brass workers. In the 17th century it was the home of the Barbary pirates.

Population 450 000
Map Morocco – Ba

589

Salem *USA* Two cities in the United States. The first is a historic city in Massachusetts, on the Atlantic 20 km (12 miles) north-east of Boston. It is one of the oldest settlements in the country, founded in 1626. The city became notorious after a series of witchcraft trials in 1692, which ended in the execution of 20 people, mostly women. It later became a prosperous port, and today it is also an industrial centre and tourist attraction. Salem still has several 17th-century buildings, including the House of the Seven Gables, immortalised in the novel of the same name by Nathaniel Hawthorne (1804-64).

Salem is also the capital of Oregon, on the Willamette River about 70 km (45 miles) south-west of PORTLAND. It is the centre of a rich farming and livestock region, and has numerous food and wood processing plants and a papermill. It was founded in 1840 by a Methodist missionary, and has a university founded two years later.
Population (Massachusetts) 38 600; (Oregon) (city) 107 800
Map (Massachusetts) United States – Lb; (Oregon) United States – Bb

Salentina *Italy* Area forming the heel of Italy. Grapes and tobacco are the main crops. The towns, especially LECCE, are noted for their ornate Baroque architecture.
Map Italy – Fd

Salerno *Italy* Seaport about 45 km (28 miles) south-east of Naples. It was renowned in the Middle Ages as a centre of medical science and has a fine 11th-century cathedral. Its industries include cement, machinery and textiles.
Population 155 900
Map Italy – Ed

Salford *United Kingdom* See MANCHESTER

salina In the south-western United States, a SALT PAN or a marsh, spring, pond or lake containing salt water.

Salinas Valley *USA* Fertile lettuce, artichoke and fruit-producing region between two parallel ridges of the Coast Ranges in California. The Salinas River runs 240 km (150 miles) north-west into Monterey Bay south of San Francisco. The writer John Steinbeck was born and is buried in the city of Salinas; his novel *East of Eden* is set in the valley.
Map United States – Bc

salinity Measure of the total quantity of salts dissolved in water, especially sea water, expressed in parts per thousand by mass. Sea water's average is 35 parts per thousand.

Salisbury *United Kingdom* Cathedral city in Wiltshire, some 32 km (20 miles) north-west of Southampton.

It is best known for its beautiful cathedral, begun in 1220, whose slender spire – which at 123 m (404 ft) is the tallest spire in Britain – was completed in 1334. It replaced an earlier cathedral lying 3 km (2 miles) to the north, on a hilltop site now known as Old Sarum, which is the Norman name of the town.
Population 37 830
Map United Kingdom – Ee

Salisbury *Zimbabwe* See HARARE

Salonica (Thessaloníki) *Greece* The country's second largest city after Athens, and Macedonia's chief port, lying on the Thermaic Gulf. It is a university town and a major industrial centre with oil refineries, engineering and textile plants.

The city was founded in 315 BC. Little remains of the original buildings, but some Roman ruins – the arch and palace of the 4th-century Roman Emperor Galerius – and Byzantine ramparts are still visible. An archaeological museum containing some of the treasures of ancient Macedonia, including most of the finds from the tomb of King Philip (383-336 BC), is at Vergina.

In 1430 the town was seized by the Turks and ceded to Greece only in 1912. The city stages an annual international trade fair in September.
Population 706 200
Map Greece – Ca

Salpausselkä *Finland* A terminal moraine – a line of hills formed from debris left by retreating ice sheets during the Ice Age. The hills extend across southern Finland from the province of KARELIA in the south-east to the Hanko Peninsula in the south-west of the country.
Map Finland – Cc

Salt *Jordan* Small town and administrative centre set in fertile hill country some 29 km (18 miles) north-west of Amman. Salt has been inhabited since 3000 BC. It has the remains of some Roman tombs and the mosaic of an early Byzantine church.
Population 35 000
Map Jordan – Aa

salt dome (salt plug) DIAPIR, or rock dome with a core of salt. The rock surrounding a salt dome is heavily folded and faulted and may act as a reservoir for oil and gas. The salt in a dome can be up to 1000 m (3300 ft) deep.

salt flat Wide, level stretch of country that has very salty soil or a surface covering of salt, representing the bed of a dried-up salt lake.

Salt Lake City *USA* State capital of Utah and the headquarters of the Mormon Church (the Church of Jesus Christ of Latter-Day Saints). The site, 22 km (14 miles) east of the GREAT SALT LAKE, was chosen in 1847 by Brigham Young (1801-77), who had led the Mormons more than 1800 km (1100 miles) from Illinois and Missouri, through the Rockies to their new home.

It was laid out around the Mormon Temple and Tabernacle with wide, tree-lined avenues and numerous parks. It is now a manufacturing, cultural, financial and tourist centre.
Population (city) 159 900; (metropolitan area) 1 072 200
Map United States – Db

salt marsh Marsh alongside a low-lying shore that is frequently flooded by the sea.

salt pan Basin containing an evaporating salt lake, bordered by a deposit of salt.

salt plug See SALT DOME

Salta *Argentina* City about 1280 km (795 miles) north-west of Buenos Aires. Founded in 1582, it has suffered a number of earthquakes although some of its colonial buildings – among the best

in the country – have survived. It stands at an altitude of 1190 m (3900 ft) just east of the Andes in a region that is rich in Inca and other pre-Columbian archaeological sites. Salta's main industries are tourism, mining of iron, zinc, lead, tin and silver, and cattle raising.
Population 370 000
Map Argentina – Ca

Saltillo *Mexico* Capital of the northern state of Coahuila, and a thriving manufacturing and commercial town making machinery, silverware, pottery and textiles. It is known for its *sarapes* – brightly coloured woollen over-garments. From 1824 to 1836, Saltillo was the capital of Texas State. Just outside Saltillo is the Angostura battlefield, scene in 1847 of one of the bloodiest battles of the Mexican-American War.
Population 440 800
Map Mexico – Bb

salting Slightly higher area of a SALT MARSH where there is grass and little bare mud.

Salto *Uruguay* Second largest city in the country after the capital, Montevideo. It is on the Argentinian border about 430 km (270 miles) north-west of Montevideo. Salto is a river port, in a citrus fruit-growing area and makes wine, processed meat and boats.
Population 80 000
Map Uruguay – Ab

Salut, Iles du *French Guiana* Atlantic islands with a grim past. At one time the whole of French Guiana was a penal colony. The most infamous of the prisons were these islands – they include Devil's Island, which became particularly notorious. Captain Alfred Dreyfus (1859-1935) was sent here in 1894 after being court-martialled and wrongly sentenced as a German spy. The 20th-century writer and former inmate Henri Charrière wrote about his term here in the book *Papillon*, which was also made into a film.

The rocky islands lie 16 km (10 miles) offshore. Palm-covered Devil's Island measures only about 1200 by 400 m (3940 by 1300 ft). Some of the prison buildings still stand; the former wardens' mess hall is now a hotel.
Map French Guiana – Ba

Salvador *Brazil* Port and capital of Bahia State, about 1200 km (745 miles) north-east of Rio de Janeiro. Traditionally, it was the first European landfall in Brazil – by the Portuguese admiral Pedro Alvares Cabral in 1500 – and the country's first capital. Much of the Alta or Upper City – on a bluff overlooking the bay – is a national monument; it contains more than a hundred 17th and 18th-century churches and civic buildings. Today Salvador has a university and several technical institutions.

Bahia – once the name of the city and still the name of the state – means 'bay' and refers to the city's huge natural harbour. Shipbuilding is an important industry. The port's main exports are cigars and other tobacco goods, petrochemicals, coffee, hides and vegetable oils. The main product of the state, which covers 561 026 km^2 (216 556 sq miles), is cocoa. But Salvador is also the centre of a vast sugar cane, tobacco and banana-growing area.
Population (city) 2 075 400; (state) 11 738 000
Map Brazil – Ec

Salween (Thanlwin; Nu Jiang) *China/Myanmar (Burma)/Thailand* River, 2900 km (1800 miles) long, rising in eastern Tibet in China. It flows east in a deep valley, then south into eastern Myanmar (where, since Burma's change of name it is known as the Thanlwin) across the hilly Shan and Kayah states. It continues southwards, forming the Myanmar-Thai border for about 125 km (80 miles), and then crosses lowlands to the Gulf of Martaban at the city of Mawlawyne in Myanmar. Rapids, shoals and other dangers confine navigation to the section below Kayah State.
Map Myanmar – Cb; China – Ee

Salzburg *Austria* Celebrated city of music and mountainous central state renowned for its Alpine spas and resorts. At one time an independent state ruled by prince-archbishops, Salzburg became part of Austria in 1816, following the Napoleonic Wars.

Its capital is Salzburg City, on the Salzach River, 255 km (160 miles) west of Vienna. It was the birthplace in 1756 of the composer Wolfgang Amadeus Mozart and a festival is held in his honour each August. Mozart was born in the old town, where his house is now a major tourist attraction.

The city, which lies between two high and craggy hills, was built by successive bishops in the 16th and 17th centuries. Its narrow streets include the Getreidegasse, lined with six-storey buildings, each bearing a wrought-iron shop sign depicting its wares. The city was badly bombed during the Second World War; almost half of it had to be rebuilt afterwards. Among its older buildings are the 17th-century cathedral and an 11th-century castle.
Population (state) 483 800; (city) 143 970
Map Austria – Cb

▼ HIGH FORTRESS Hohensalzburg towers over Salzburg's Old City. The 11th-century castle on the tree-lined Mönchsberg was a home of the prince-bishops who ruled the city for more than 1000 years from AD 798.

Salzkammergut *Austria* Mountain and lake region in the eastern Alps, crossed by the Traun River. Once important for salt production, it is now a tourist region of spas, with BAD ISCHL as its chief centre. There are more than 40 lakes, of which the largest is the Attersee. The highest peak is Dachstein at 2995 m (9826 ft).
Map Austria – Cb

Samanala *Sri Lanka* See ADAM'S PEAK

Samar *Philippines* Third-largest island of the Philippines, covering 13 080 km² (5050 sq miles). Most easterly of the VISAYAN ISLANDS, it is exposed to Pacific typhoons and receives up to 3500 mm (136.5 in) of rain a year. The island consists mainly of a rugged dissected plateau, with dense forests and few roads, except along the west coast. It has few mineral resources other than copper mined at Bagacay, and the main occupation is subsistence rice farming.

The main towns include Calbayog, which has an airport, and Catbalogan, a seaport, both on the west coast. Tacloban on LEYTE (which is linked to Samar by a bridge) serves as its main trading port.
Population 1 100 000
Map Philippines – Cc

Samara (Kuybyshev) *Russia* Industrial city and Volga River port 860 km (535 miles) south-east of Moscow. It was the wartime capital of the Soviet Union between October 1941 and August 1943, during the German invasion of Russia. Founded as Samara in 1586, it was renamed Kuybyshev in 1935 after the Revolutionary leader Valerian Kuybyshev (1888-1933) but reverted to Samara after the break-up of the Soviet Union. It produces precision machinery, transport equipment, petrochemicals, timber, processed food and consumer goods. A dam on the Volga River near the city forms the largest reservoir in Russia, at 6450 km² (2490 sq miles).
Population 1 251 000
Map Russia – Gc

Samaria *Israel/Jordan* A region of ancient Palestine extending from the Mediterranean to the Jordan River. It was made a division of the province of Judaea by the Roman Emperor Augustus in AD 6. The region receives high rainfalls, compared with the rest of Palestine and produces olives, vines, fruits, grains and mutton.

Its main city, founded in 887 BC, was also called Samaria, and was the holy city of the Samaritans. Its ruins lie in the West Bank, the part of Jordan which was occupied by Israel in 1967.
Map Jordan – Aa

Samarkand *Uzbekistan* Ancient city 250 km (155 miles) north of the Afghan border. In 329 BC, when the Macedonian conqueror Alexander the Great captured it, it was known as Maracanda – and it was then already an established city. In the 7th century AD it became a stopping point on the Silk Road from China to Europe, and it was visited by the Venetian explorer Marco Polo in the 13th century. The city developed into a flourishing centre in the 14th and 15th centuries, and again when it became a Russian city in 1868.

Modern blocks of flats now surround the mosaic-filled Muslim buildings of the city centre. The city's factories produce cotton and silk textiles, carpets, leather, shoes, tea, tobacco, canned fruit and wine.
Population 370 000
Map Uzbekistan – Bb

Samarra *Iraq* Holy city on the Tigris River, 100 km (62 miles) north-west of the capital, Baghdad. Built in about AD 838, it became a capital city under the Abbasid caliphs who ruled the region from 838 to 883. Its 17th-century, copper-domed mosque is revered by Shiite Muslims and its ruined Great Mosque, built in about AD 852, was once the largest in the world. Nearby, a dam built on the Tigris River regulates the river's flooding and helps irrigate the fields.
Population 62 000
Map Iraq – Bb

Samoa, American *Pacific Ocean* Group of five islands and two atolls covering some 197 km² (76 sq miles) to the east of WESTERN SAMOA and midway between Sydney, Australia, and Hawaii. They have been American territory since 1900, and there is little pressure for reunion with independent Western Samoa whose standard of living is much lower.

American Samoa has an elected legislature and elected executive governor. It receives substantial grants from the United States government, which also buys most of its exports, mainly canned tuna fish. Taro, breadfruit, yams, coconuts and bananas are grown for local consumption. Most of the islands (including TUTUILA, the main one) are hilly and densely forested.
Population 46 800
Map Pacific Ocean – Ec

Samoa, Western See WESTERN SAMOA

Sámos *Greece* Wooded island in the eastern Aegean Sea, only 2 km (1.2 miles) from the coast of Turkey. It was the birthplace of the mathematician and philosopher Pythagoras (about 582-507 BC), who gave his name to the famous theorem which is taught in schools all over the world. The island is known for its Sámos wines, which are a popular export product.
Population 31 600
Map Greece – Ec

Samothrace (Samothráki) Greece One of the northern Aegean islands, 40 km (25 miles) from the coast of Thrace. In Homeric legend the sea god Poseidon watched the battle for Troy from the island's Mount Fengari (1600 m, 5250 ft). A marble sculpture of the winged goddess of victory was found during excavations of the Sanctuary of Great Gods, near Kamariótissa.
Population 2900
Map Greece – Da

Samsun Turkey Black Sea port about 420 km (260 miles) north-east of Ankara, and capital of the province of the same name. In May 1919 Kemal Ataturk (1881-1938), founder of modern Turkey, launched a revolution in the area to rid the country of foreign control.

The port's main cargo today is tobacco grown in the fertile and heavily populated province, which covers 9579 km^2 (3698 sq miles).
Population (town) 303 980; (province) 1 158 400
Map Turkey – Ba

Samut Prakan Thailand Industrial province and town, both of which have become a southern suburb of Bangkok. The main industries are steel, glass, electrical goods, and pharmaceuticals.
Population (province) 646 800; (town) 247 900
Map Thailand – Bc

San Agustín Colombia Town, some 300 km (190 miles) south-west of Bogotá. It lies beside the San Agustín National Park, which contains remarkable stone carvings of a pre-Columbian farming society. Hundreds of rough-hewn figures of men, animals and birds line the forest paths and small streams of the Valley of Statues. The earliest of these date from about AD 600.

Archaeologists have discovered little about the people other than that they lived in bamboo huts and grew crops. Many of the stone animals guard elaborate tombs. Pottery, gems and necklaces found in some of the tombs can be seen at the museum in the town.
Population 16 850
Map Colombia – Bb

San Andreas Fault North America See
CALIFORNIA; FAULT; SAN FRANCISCO

San Antonio Chile Seaside resort and port 113 km (70 miles) west of Santiago. It handles the output from EL TENIENTE, one of the world's largest underground copper mines.
Population 60 800
Map Chile – Ac

San Antonio USA Major commercial, financial and industrial city in southern Texas, midway between Houston and the Mexican border. It has several military bases.

San Antonio is one of the country's most historic cities. It was the capital of culture and finance in the era before oil supplanted cattle as Texas's source of wealth. Four 18th-century Spanish missions in the city have been preserved. Another mission, the Alamo, was made into a fort in 1793, and 187 Texans – including the folk hero Davy Crockett – held out there against the Mexican army from 23 February to 6 March 1836; none of the Texans survived.
Population (city) 935 900; (metropolitan area) 1 324 700
Map United States – Ge

San Bernardino USA Market city and county in California, 90 km (55 miles) east of Los Angeles, forming part of the Riverside-San Bernardino metropolitan area and also part of the Los Angeles-Riverside-Orange County conurbation. It was founded by the Spanish in 1816 and sold in 1851 to the Mormons who settled the area between the San Bernardino and San Gabriel mountains. It is a fast-growing residential centre and the hub of a citrus fruit and dairy-producing area. All or part of seven military bases are in San Bernardino county – the largest county in the US.
Population (city) 164 200; (metropolitan area) 1 810 900
Map United States – Cd

San Cristóbal Venezuela Capital of Táchira State, high in the Andes, and about 650 km (400 miles) south-west of the capital, Caracas. It was founded by the Spanish in 1561 and retains much of its old colonial appearance.
Population 364 730
Map Venezuela – Ab

San Diego USA Southernmost port in California and industrial, commercial and cultural centre on the Mexican border. It was founded as a Spanish mission in 1769 and today is one of the fastest growing cities in the United States. Its metropolitan area embraces 15 other cities and it receives 29 million visitors a year.

It is a major naval base, administrative centre of the Pacific Fleet, and a commercial tuna fishing port. Its industry specialises in aerospace and defence equipment, electronics and shipbuilding, and it is a centre of scientific and medical research. The largest zoo in the US is located here.
Population (city) 1 110 500; (metropolitan area) 2 498 000
Map United States – Cd

San Fernando Trinidad and Tobago West coast port and Trinidad's second largest city, after Port of Spain, which lies 40 km (25 miles) to the north. An important market and administrative centre, it serves the southern sugar-growing districts.
Population 28 580
Map Trinidad and Tobago – Aa

San Francisco USA City in California on a hilly neck of land between San Francisco Bay and the Pacific Ocean, and the centre of one of the USA's major conurbations, including OAKLAND and SAN JOSE. It grew around a Spanish mission and fort founded in 1776 and enjoyed a boom during the gold rush of 1849. Much of the city was destroyed by earthquake and fire in 1906 – it lies on the San Andreas Fault – and has been rebuilt and extended on to land reclaimed from the bay. The last major earthquake was in 1989, killing more than 60 people and causing about US$5.6 to 7 billion damage to property.

San Francisco is one of the country's main ports and is a major financial, educational and cultural centre. It remains a Mecca for about 2.5 million tourists each year, who come to ride its old-fashioned cable-car trams that climb the steeply sloping streets, as well as to sample the fish at Fisherman's Wharf and the cuisines of its large and varied Asian population from China, Japan, the Philippines, Korea and Vietnam.
Population (city) 724 000; (metropolitan area) 1 603 700; (conurbation) 6 253 300
Map United States – Bc

A DAY IN THE LIFE OF A CALIFORNIAN COMPUTER TECHNOLOGIST

Roger Gerski sinks back into his armchair, nightcap in hand, as his wife Anne waves goodbye to their dinner guests. It has been a long day.

At 6.30 am he was up and jogging round the block at his favourite time of day when the air is refreshingly crisp, and the sun just filters through the leaves of the garden palms before the scorching heat of the day.

He then returned for a swim and a shower followed by a breakfast of grapefruit, muesli, fresh orange juice and decaffeinated coffee enjoyed by the pool.

As he left for the office, where he works as a software designer for a computer company, the children were only just getting up for school and he barely had a chance to greet them. He had wanted to talk to them – especially to his son Peter, for whom he has high hopes; already he is a computer whizz at school and there is plenty of scope for a career in his father's footsteps.

At 7.30 Roger backed his car out of the garage. His office is on El Camino Road, on the northern outskirts of San Jose, to the south of San Francisco. He slipped through the quiet streets of suburban Palo Alto and together with thousands of other vehicles filtered on to the busy freeway.

It was a tricky morning. The company was launching a new computer software product that Roger had designed and he had to attend a marketing meeting. Marketing did not really fall into his area of expertise, but as a company director he had to keep a close watch on the 'machine' – the network of sales and marketing staff who sold and generated publicity for the company's product – and take responsibility if anything went wrong.

The most successful entrepreneurs in new high-tech industries such as his need a mixture of skills, some of which, when he is perfectly honest with himself, he admits he does not have. Behind those high, tinted glass walls in the bright modern air-conditioned office world, life is harshly competitive and very often precarious.

He rushed his lunch, which was nothing unusual, snatching a hamburger at the McDonald's drive-through. Roger's afternoon was spent on paperwork. But he managed to find a spare 30 minutes which he devoted to new thoughts on his latest software project. He regarded this as the only real work of the day.

He got home just in time to welcome his dinner guests – an old friend and his boss from a rival transnational software company, with their wives. They had vaguely talked of a new job for Roger – more or less the same work but with a higher salary than the generous one he earned already.

Roger wonders for a minute, before he heads for bed, if the extra money would be worth it. He suspects the job might only be a faster way to get an ulcer, or worse.

San Francisco Bay *USA* Largest bay on the coast of California. It is about 95 km (60 miles) long and up to 20 km (12 miles) wide. The bay is fed by the Sacramento and San Joaquin rivers, and opens to the Pacific Ocean through the Golden Gate, a strait spanned by the Golden Gate Bridge – one of the longest suspension bridges in the world, spanning 1280 m (4200 ft) and completed in 1937. The nearby Oakland Bay Bridge, linking San Francisco with Oakland and Berkeley, has a total length of 13 km (8 miles).
Map United States – Bc

San Gimignano *Italy* Beautifully preserved medieval town about 40 km (25 miles) south-west of Florence. Set amongst the vineyards and gentle hills of southern Tuscany, its skyline is dominated by the towers of palaces and churches.
Population 7400
Map Italy – Cc

San Joaquin *USA* River much used for irrigation in California. It rises in the Sierra Nevada Mountains, some 300 km (185 miles) east of San Francisco and flows about 560 km (350 miles) north-west through the southern Central Valley, a rich fruit, vegetable and wine-producing region, to the Sacramento River near San Francisco Bay.
Map United States – Bc

San José *Costa Rica* National capital founded by Spanish settlers from the nearby city of CARTAGO in 1736. It replaced Cartago as the capital in 1823 because it has a pleasant climate all year round thanks to its altitude of 1170 m (3840 ft).

San José has an attractive mix of Spanish and modern buildings, with wide avenues and a number of parks. The city's main industries are clothing manufacture, food processing, textiles

▼ SILICON VALLEY The electronics centre of San Jose in California stretches into the distance like a board of giant computer circuitry.

and pharmaceuticals. San José Province produces much of Costa Rica's sugar cane and coffee on its rich, deep volcanic soil.
Population (city) 296 300; (province) 1 105 800
Map Costa Rica – Bb

San Jose *USA* Industrial and market city in California 20 km (12 miles) south-east of San Francisco. Renowned for its wine grapes and other fruits, it is the main town of 'Silicon Valley', one of the biggest centres of the computer industry in the US. The city's population has quadrupled since 1960 and it is now bigger than San Francisco.
Population (city) 782 200; (metropolitan area) 1 497 600
Map United States – Bc

San Juan *Argentina* Wine-making city about 1000 km (620 miles) north-west of the national capital, Buenos Aires. It is the capital of the province of the same name. Founded in 1562, the city was flattened by an earthquake in 1944 but has since been rebuilt. The province extends into the Andes to the Chilean border.
Population (city) 353 000; (province) 528 840
Map Argentina – Cb

San Juan *Puerto Rico* Seaport and capital city on the north coast. The oldest part of the city, founded by the Spaniards in 1508, is built on a small island connected to the mainland by a causeway and bridges. Buildings such as the mighty Castillo de San Felipe del Morro attest to the fact that San Juan was once under constant threat from pirates and rival colonial powers.

The School of Tropical Medicine, part of the University of Puerto Rico, is here; the rest of the university is at Rio Piedras, 11 km (7 miles) away. San Juan exports sugar, tobacco, pineapples, bananas, oranges and cacao, manufactures clothing, cigars and cigarettes, and refines sugar. The chief industry is tourism.
Population 437 750
Map Caribbean – Bb

San Luis Potosí *Mexico* Industrial and agricultural state in north-central Mexico, which is rich in gold and silver. Both the state and the capital, also called San Luis Potosí, were inhabited by the semi-nomadic Chichimeca peoples before they were displaced by Spanish speculators seeking gold and silver.

The city, an elegant colonial mining town of pink sandstone buildings, was formally founded in 1592 by the Spanish and named after the Franciscan mission of San Luis, established here in 1590, and the Bolivian silver town of Potosí (Potosí means 'The Place of Great Wealth'). Lead, silver, zinc and iron are still mined, and flour, beer and leather goods manufactured.
Population (state) 2 200 000; (city) 525 820
Map Mexico – Bb

San Marino See p 594

San Marino *San Marino* The towers and battlements of this tiny, picturesque capital of Europe's smallest republic are perched more than 610 m (2000 ft) high on the craggy western slope of Monte Titano, some 100 km (62 miles) east of Florence. Triple walls guard it, only one road enters it, and the narrow winding streets are mostly closed to vehicles.

The city is on the site of a hermitage built by a stonemason named Marino in about AD 300: his bones now rest in the 16th-century Basilica di San Marino. The Rocca Fortress, with three linked towers on the three peaks of Monte Titano, offers magnificent views – including one across the plain below to Rimini, 23 km (14 miles) to the north-east, on the Adriatic coast.

Several art museums, six churches, a Gothic government house and other fine public buildings also draw tourists to the prosperous city.
Population 4500
Map Italy – Dc

San Miguel *El Salvador* Manufacturing town 136 km (84 miles) east of the capital, San Salvador, producing textiles, pharmaceuticals and plastics. It was founded in 1530 and has an 18th-century cathedral.

Coffee, sisal and cotton are grown in the surrounding department of the same name; there are gold and silver mines. San Miguel volcano, overlooking the town, is 2130 m (6990 ft) high. It last erupted in 1976.
Population (town) 182 820; (department) 380 440
Map El Salvador – Ba

San Miguel de Allende (Allende) *Mexico* Charming colonial town of narrow cobbled streets, and houses with hand-carved wooden doors, in GUANAJUATO State, central Mexico. Once important for silver mining, the town is now a meeting place for artists, actors, writers and intellectuals.
Population 50 000

San Miguel de Tucumán (Tucumán) *Argentina* Capital of Tucumán Province, about 1090 km (680 miles) north-west of the national capital, Buenos Aires. It was founded by the Spanish in 1565, and was the site of Argentina's declaration of independence in July 1816. It is the centre of Argentina's sugar-cane industry.
Population (city) 622 000; (province) 1 134 310
Map Argentina – Ca

San Pedro Sula *Honduras* North-western town near the Guatemalan border. It is the country's second largest city and one of Central America's fastest growing towns. Good, paved roads link the city to the capital, TEGUCIGALPA, and to Caribbean ports, and there are direct flights to Tegucigalpa, Miami and New Orleans. It is the country's industrial centre, with many new industries, such as small steel rolling mills, textiles and clothing, plastic ware, food processing, zinc roofing, cement and furniture.
Population 461 000
Map Honduras – Aa

San Remo *Italy* Seaport and resort 115 km (70 miles) south-west of Genoa and 20 km (12 miles) from the French border. It abounds in sedate villas and subtropical plants. The medieval town was developed as a resort by aristocratic British tourists in the 19th century.
It is the region's principal market for flowers, which are grown on protected terraces on either side of the town. In February the annual Italian pop music festival is held here.
Population 62 100
Map Italy – Ac

San Salvador *Bahamas* Small island in the east central group of the archipelago, of historical interest as the first land on which Columbus set foot in the New World on 12 October 1492. The day is now a national holiday. There are three monuments marking the site because the exact landing spot is disputed.
Population 540
Map Bahamas – Cb

San Salvador *El Salvador* National capital and largest city, producing about one third of the country's industrial output. Its products include textiles, tobacco, processed foods, plastics, pharmaceuticals and dairy produce, though the basis of its wealth is coffee, grown on the volcanic soils of the nearby mountains.
The city was founded by the Spaniard Diego de Alvarado in 1525, but has been rebuilt 12 times as a result of earthquakes. The modern city is laid out in the form of a cross, with low buildings designed to withstand new tremors.
The 1960 m (6430 ft) high San Salvador volcano overlooks the city. The summit crater is 15 km (9 miles) across and in places 1000 m (3280 ft) deep. Its wooded inner slopes sweep down to a small, black cone – the remnant of a major eruption in 1917. A smaller eruption in 1986 cost the lives of about 1500 people.
Population (city) 422 570
Map El Salvador – Ba

San Salvador de Jujuy *Argentina* Ancient city and capital of Jujuy Province, 1255 km (780 miles) north-west of BUENOS AIRES. Founded in 1565, the city lies at an altitude of 1260 m (4130 ft), and is surrounded by the windswept *altiplano*, or 'high plateau'.
Before the Spanish conquest, the whole area was part of the Inca Empire. There are reserves of iron, zinc, lead, tin and silver which are being exploited, but agriculture is limited by the altitude. Nearby is the Amerindian town of Humahuaca, standing at 2940 m (9650 ft), and the Valle Grande National Park.
Population (city) 183 000; (province) 502 700
Map Argentina – Ca

San Marino

EUROPE'S SMALLEST REPUBLIC HAS BEEN INDEPENDENT FOR CENTURIES WHILE BEING SURROUNDED BY ITALY

The smallest European independent republic lies in the eastern foothills of the Apennines. It has wooded mountains, pasture land, citadels – notably La Guaita in the heart of the state – and medieval villages, clustered around the peaks of Monte Titano, which rises to some 743 m (2438 ft).

Monte Titano was settled in about AD 300 by Marino, the stonemason saint after whom the republic is named. The country became an independent commune in the 13th century and has remained so – its first constitution was written in 1263. It is governed by the Great and General Council consisting of 60 elected members; coalitions have ruled since 1957, when 12 years of communist control ended. Most citizens between 16 and 55 can be called on by the local militia in an emergency, but the police are hired from Italy. San Marino entered a customs union with Italy in 1862. The republic draws some 3.5 million tourists a year – almost 150 to each inhabitant. One attraction is the capital, the tiny fortified city of SAN MARINO.

Much of the republic's revenue comes from the sale of its own stamps, coins, postcards, souvenirs and duty-free liquor. A less important contribution to the economy is made by light manufacturing industries, and a long tradition of pottery making and silk weaving. Most Sanmarinese, however, still work on the land and in forestry. Wine, milk, cheese and wool are the main products.

Despite the inevitable influence of Italy, San Marino still values its separate administration and identity.

SAN MARINO AT A GLANCE	
Map Italy – Dc	
Area 61 km² (24 sq miles)	
Population 23 860 **Per km²** 391 **(Per sq mile)** 994)	
Capital San Marino	
Government Parliamentary republic	
Currency Italian lira	
Language Italian	
Religion Christian (95% Roman Catholic)	
Climate Mediterranean. Average temperature ranges from 4°C (39°F) in January to 24°C (75°F) in July.	
Land use Arable 17%; other 83%	
Main primary products Wheat, olives, vines, dairy cattle; building stone	
Major industries Agriculture, tourism, food processing, textiles, quarrying, ceramics, forestry	
Main exports Postage stamps, wine, wool, skins, hides, stone, ceramics	
Annual income per head (US$) 17 000	
Population growth (per thous/yr) 10	
Life expectancy (yrs) Male 77 **Female** 85	

San Sebastián (Donastia; Izurum) *Spain* Seaport and industrial city on the Bay of BISCAY near the French border, and capital of the province of Guipúzcoa. Its moderate climate and sandy beaches make it one of Spain's most sought-after tourist resorts. Its industries include fishing, chemicals, cement and metal goods.
Population 171 440
Map Spain – Ea

San'a *Yemen* Capital and leading city situated in the middle of the main plateau, 2286 m (7500 ft) above sea level. Surrounded by high, thick stone walls containing eight gates, San'a is said by the Arabs to be the world's oldest city and the birthplace of the Arab nation. Its founder is claimed to be Shem, the eldest son of Noah, and it was first called Medinet Sam, or 'City of Shem'. The name San'a comes from the period of Abyssinian rule in the 6th century AD and means 'fortified'.

San'a has long been a centre of handicraft industries, such as weaving and jewellery making; and in recent years larger industries have been set up. These include a textile plant, an iron foundry, plants to make building fixtures and fittings, tools and water pumps, as well as water and electricity projects.

Places of interest include the Medina (the old city and Arab quarter with the most important of the many mosques); the dairy market or *suq*; the citadel (Al Qasr); the ruined fort of al-Birash; and the Mutawakil, which formerly contained the palace of the imams and is now the Dar as Sa'd ('House of Good Luck') Museum.

San'a is connected by good roads to HODEIDA on the coast; SA'DA in the north; and TAIZ in the south. There is a university (founded in 1970).
Population 500 000
Map Yemen – Aa

Sanchi *India* Village in the north, 810 km (503 miles) south of Delhi. It is the site of India's biggest Buddhist temple complex, which lay hidden under the jungle until uncovered in the early 19th century. Sanchi prospered under the great Indian emperor Asoka (273-232 BC), when the Great Stupa – a stone cairn more than 16 m (52 ft) high – was built to house holy relics, and it thrived for 1000 years. Like SARNATH, it was abandoned, but the monks' living quarters and many *stupas* survive.
Map India – Bc

sand Loose, rounded particles of disintegrated rock, especially quartz. Fine sand ranges from 0.02 to 0.2 mm (0.0008 to 0.008 in) in diameter. Coarse sand is between 0.2 and 2 mm (0.008 and 0.08 in) in diameter. Sand is finer than gravel and coarser than silt. If the particles are angular, it is called grit.

Sandakan *Malaysia* Port in east Sabah on the north-east coast of the island of Borneo. It handles most of the exports of the state's profitable hardwood industry.
Population 113 500
Map Malaysia – Ea

sandbank Submerged ridge of sand in a sea or river, formed by currents and often exposed at low tide.

sandstone Variously coloured SEDIMENTARY ROCK consisting of compressed or cemented particles of sand. Among the most important sandstone deposits in Europe are the Old Red Sandstone, laid down in the DEVONIAN period, and the New Red Sandstone of the PERMIAN and TRIASSIC periods.

Sandwich Islands *Pacific Ocean* See HAWAII

Sandwip *Bangladesh* An exposed island off the south-east coast in the delta of the Ganges, north-west of CHITTAGONG. Many thousands of people died when huge cyclones hit the island in 1970 and 1985.
Map Bangladesh – Cc

Sankt Gallen *Switzerland* See ST GALL

Santa Ana *El Salvador* The country's second largest city after the capital, San Salvador. It has textile and food-processing industries, and coffee is grown in the surrounding department of the same name. The city retains some colonial architecture, including a neo-Gothic cathedral.
Santa Ana volcano, the country's highest mountain at 2365 m (7760 ft), broods near the city above Lake Coatepeque, which is fed by hot springs warmed by the volcano.
Population (city) 202 340; (department) 451 620
Map El Salvador – Ba

Santa Anna Hot, dry, FÖHN-like wind blowing from the north and north-east, descending from the Sierra Nevada mountain range across the deserts of southern California.

Santa Barbara *USA* Attractive resort and residential city on the Californian coast 129 km (80 miles) north-west of Los Angeles. It grew up around a beautiful Spanish mission established in 1786. Electronics and aerospace equipment are its main industries, and there are oil fields and offshore oil wells. However, tourism is the city's main income earner.
Population (city) 85 600; (metropolitan area) 369 600
Map United States – Cd

Santa Catarina *Brazil* See FLORIANOPOLIS

Santa Clara *Cuba* Capital of Villa Clara Province, on the plateau 300 km (186 miles) south-east of Havana. Founded by the Spanish in 1689, it has become an important crossroads and commercial centre in an agricultural region producing sugar, tobacco and coffee. Industries include tobacco manufacture and leather processing. Since the 1959 revolution, its university, founded in 1949, has become a major educational centre for the island.
Population (city) 194 000 (province) 788 800
Map Cuba – Ba

Santa Cruz (Santa Cruz de la Sierra) *Bolivia* City and the largest and potentially the richest department in the country, covering the central and southern areas of the Oriente eastern lowlands. Santa Cruz is Bolivia's major centre of new colonisation from the Andean region.
A wide variety of crops are grown in the central area, especially sugar cane, rice, coffee and cotton. Oil fields have been developed at Caranda and Colpa, and natural gas at Río Grande. Large iron ore and manganese deposits have been discovered in the south-east. The city, founded in

1561 by the Spaniard Ñuflo de Chávez, originally on a site 322 km (200 miles) east of the present location, is the fastest growing in the country.
Population (city) 669 000; (department) 1 314 600
Map Bolivia – Bb

Santa Cruz de la Palma *Spain* Chief town of La Palma Island in the CANARY ISLANDS.
Population 16 780
Map Morocco – Ab

Santa Cruz de Tenerife *Spain* Seaport on Tenerife Island in the CANARY ISLANDS and also the name of one of the island's provinces. The town exports bananas, tomatoes and potatoes. The province covers 3442 km² (1329 sq miles) and consists of Tenerife Island and the smaller Canary Islands of La Gomera, La Palma and Hierro.
Population (town) 200 170; (province) 725 820
Map Morocco – Ab

Santa Fe *USA* Capital of New Mexico, and tourist resort at the foot of the Sangre de Cristo Mountains in the north of the state, about 145 km (90 miles) south of the Colorado border. The area was occupied by the ancestors of the Pueblo people some 2000 years ago and settled by Spanish colonists in 1609-10.
The city became a key trade centre on the route south and west from Missouri before the railways were constructed. About half of its population is Spanish speaking. The nuclear weapons research centre at Los Alamos, where the first atomic bomb was made, is 56 km (35 miles) to the north-west.
Population (city) 55 900; (metropolitan area) 117 000
Map United States – Ec

Santa Isabel *Equatorial Guinea* See MALABO

Santa Marta *Colombia* Port resort and capital of Magdalena Department, on the Caribbean coast 96 km (60 miles) east of BARRANQUILLA. Tourists are attracted to its long, sandy beaches and warm climate, and to the Tairona Nature Park, 35 km (22 miles) to the east, an unspoilt stretch of wooded coastline.
Santa Marta's deep-water dock exports bananas, coal from reserves discovered in the nearby Guajira Peninsula and oil from fields near the Venezuelan border. Colombia's liberator, the Venezuelan-born Simón Bolívar, died penniless near Santa Marta in 1830, aged 47.
Population 350 000
Map Colombia – Ba

Santander *Colombia* North-east Andean department about 310 km (190 miles) north of Bogotá. The capital is BUCARAMANGA. Most of Columbia's oil reserves lie to the west of the department. The country's largest refinery is at Barrancabermeja beside the Magdalena River.
Population 1 642 580
Map Colombia – Bb

Santander *Spain* Seaport, industrial city and resort on the Bay of Biscay and capital of the province of Cantabria. It exports minerals, wine and wheat, and its industries include shipbuilding, oil refining, tanning, chemicals, cables and machinery. At Altamira, 27 km (17 miles) to the south-west, are caves containing Stone Age

paintings between 13 000 and 20 000 years old. Entrance is now restricted.
Population 180 300
Map Spain – Da

Santiago (Santiago de Chile) *Chile* Capital and main industrial centre of the country, in the Central Valley about 90 km (55 miles) from the coast. It was founded by the Spanish conquistador Pedro de Valdivia in 1541 at Santa Lucía Hill, now a city park. The city was destroyed several times in turn by Amerindians, earthquakes and floods. Its main street is the Avenida O'Higgins (the Almeda), which is 100 m (330 ft) wide and more than 3 km (2 miles) long, and separates the new city from the old. The old city contains statues of Chile's liberator, Bernardo O'Higgins, and José de San Martín and a monument commemorating the Battle of Concepcíon in 1879.
Near the main square, the Plaza de Armas, are the cathedral built in 1558 and a museum in what was formerly the Casa Colorado, built in 1769 for the Spanish governor. A cable car runs 228 m (750 ft) to the top of the highest vantage point in the city, San Cristóbal Hill, which is crowned by an enormous statue of the Virgin Mary. Industries in the new part of the city include food processing, textiles, pharmaceuticals, chemicals, clothing and leather goods.
Population 5 343 000
Map Chile – Ac

Santiago de Compostela *Spain* Pilgrimage and university city in the north-west of the country, 56 km (35 miles) south of CORUNNA. Its cathedral, built between the 11th and 13th centuries, contains the remains of the Apostle St James the Great, discovered nearby in AD 813. The town is named after the saint, who is known in Spanish as *Sant' (Santo) Yago*. In the Middle Ages Santiago was the most popular place of pilgrimage in Christendom, after Jerusalem.
Population 87 800
Map Spain – Aa

Santiago de Cuba *Cuba* Seaport and capital of Santiago de Cuba Province, on the south coast 145 km (90 miles) from the western tip of the island. Founded in 1514, it was the island's capital until 1589, and is now the second largest city after Havana. It was the scene, in July 1953, of Fidel Castro's first revolutionary strike against the Batista government – a raid on an army barracks – which resulted in his capture. He left the island under an amnesty in 1955 and returned a year later to launch a new uprising. The city's industries refine oil, produce rum and textiles and process sugar. Its port ships iron, manganese, nickel and copper ore, sugar, rum and tobacco.
Population (city) 405 000; (province) 974 100
Map Cuba – Ba

Santiniketan *India* See SHANTI NIKETAN

Santo *Vanuatu* See ESPIRITU SANTO

Santo Domingo (Ciudad Trujillo) *Dominican Republic* Main port and capital city, lying on the south coast, at the mouth of the Río Ozama. It was founded in 1496 by Bartholomew, the brother of Christopher Columbus, and is the oldest continuous European settlement in the Americas. The cultural and commercial centre of the country, it has two universities and an early

16th-century cathedral containing the tomb in which Christopher Columbus was buried before his remains were taken to Seville Cathedral in Spain. Formerly Ciudad Trujillo, it was renamed after President Rafael Trujillo was assassinated in 1961. Sugar is exported, and tourism is a principal industry.

Population 1 600 000
Map Caribbean – Bb

Santorini (Thíra) *Greece* Island of the CYCLADES group, north of Crete. It is thought by some archaeologists to be the site of the lost and legendary land of Atlantis. A volcanic eruption in about 1500 BC sank a crater in one side of the island, and today the remains of the crater's ring wall form an arc of almost vertical cliffs which rise some 300 m (980 ft) from the sea.

Excavations since 1967 at Akrotíri on the island have revealed evidence of a flourishing civilisation before the eruption. The main contemporary town, Thíra, stands spectacularly on the volcanic clifftop.

Population 7100
Map Greece – Dc

Santos *Brazil* The country's largest port, 60 km (38 miles) south-east of the city of São Paulo. It is 5 km (3 miles) from the open sea and approached through the winding Santos Channel. More than 40 per cent of Brazil's imports, and half of its exports, pass through Santos, which started as a coffee-exporting port but today handles mainly industrial products. It also has oil refineries and chemical plants.

Population 546 600
Map Brazil – Dd

São Francisco *Brazil* River rising in the hills west of the city of BELO HORIZONTE, and flowing some 2900 km (1800 miles) into the Atlantic, about halfway between the cities of Salvador and Recife. Hydroelectric dams along its length provide electricity for the whole region. The dams also provide irrigation waters, allowing the cultivation of crops such as grapes in an area of perpetual sunshine but little rain.

Map Brazil – Dc

São Miguel *Portugal* Largest island (770 km², 298 sq miles) in the AZORES group, in the Atlantic, about 1290 km (800 miles) off the west coast of Portugal. It has been the scene of numerous volcanic eruptions since the 15th century, and the landscape is pitted with volcanic cones and lake-filled craters. The soil is extremely fertile and maize, pineapples, vines, oranges, figs and tea are grown. The capital is Ponta Delgada.

Population 131 900
Map Atlantic Ocean – Dc

São Paulo *Brazil* Industrial city and capital of São Paulo State, 360 km (225 miles) south-west of Rio de Janeiro and almost exactly on the Tropic of Capricorn. It vies with Mexico City as one of the fastest growing and largest conurbations in the world.

The state, covering an area of 247 900 km² (95 690 sq miles), is about the size of Britain. It was transformed by the coffee boom of the late 19th century which brought in a huge number of immigrants: one million people from Italy, 500 000 from Portugal and Spain, 250 000 from Japan and many from Germany. By 1936 the total had reached 2 850 000. This magnetic attraction continues today, with considerable internal migration, especially from the poor north-east of Brazil.

Apart from its industrial pre-eminence – almost 60 per cent of the country's industry is concentrated in the state – it is Brazil's leading producer of sugar and livestock.

Population (city) 9 700 100; (metropolitan area) 18 000 000; (state) 33 070 000
Map Brazil – Dd

São Tomé and Príncipe See p 597

São Tomé *São Tomé and Príncipe* National capital and main port, on an inlet on the north-eastern coast of São Tomé Island. It was founded in the late 15th century by the Portuguese and still retains many old colonial buildings. Its main exports are cocoa and coffee.

Population 51 000
Map Gabon – Aa

São Vicente *Cape Verde* The island with the second largest population in the CAPE VERDE archipelago. The second largest town and cultural centre of the islands, Mindelo, is located here.

Population 42 000
Map (Cape Verde) Atlantic Ocean – Dd

São Vicente, Cabo de (Cape St Vincent) *Portugal* Cape at Portugal's south-western tip, ending in cliffs 75 m (245 ft) high. In 1797 an English fleet under Sir John Jervis (later Lord St Vincent) defeated a Spanish fleet off the cape. A powerful lighthouse is situated here.

Map Portugal – Bd

Saône *France* River, about 480 km (300 miles) long, in east-central France. It rises in the Vosges Mountains and flows south past the vineyards of Burgundy to merge with the Rhône at Lyons.

Map France – Fc

Sapporo *Japan* Capital of HOKKAIDO, in the south-west of the island. The city was founded in 1871 as a springboard for the colonisation of the island. Now a thoroughly modern city with a gridiron street plan and a university, it hosted the 1972 Winter Olympics. Its nearby ski and hot-spring resorts make it a winter sports paradise.

Population 1 672 000
Map Japan – Db

Saqqara *Egypt* Village alongside an ancient necropolis and site of the 'Step Pyramid' 23 km (14 miles) south-west of Cairo. In February 1986 the tomb of Maya, treasurer to the boy king Tutankhamun, who died in about 1352 BC, was found here. The discovery was made by two archaeologists, Dr Geoffrey Martin of University College, London, and Dr Jacobus Van Dijk of the Leiden Museum in the Netherlands, who had spent 10 years excavating the tomb of Maya.

Map Egypt – Bb

Sar Planina *Yugoslavia (Serbia and Montenegro)* Mountain range separating the Serbian autonomous province of Kosovo from Macedonia. It rises to 2702 m (8865 ft) at Titov Vrh. A region of dense forests, lakes and pastures, it is also a fine skiing area, served by the cities of Tetovo and Prizren.

Map Yugoslavia – Cc-Cd; Macedonia – Aa-Ab

Saragossa *Spain* See ZARAGOZA

Sarajevo *Bosnia-Herzegovina* National capital, lying on the Miljacka River, 150 km (93 miles) from the Adriatic coast. Much of the city was destroyed in the civil war that racked Bosnia-Herzegovina in the 1990s between Muslims and Bosnian Serbs, who besieged the city for 22 months from 1992 to 1994.

Before the siege, the city had some 80 mosques, and the old Turkish quarter below the city's citadel contained some of the finest Ottoman architecture in the region.

The city's history is marred by many catastrophes, including fire, floods and the plague. Footprints on the pavement near Princip's Bridge (Principov Most) mark the spot where a 19-year-old Bosnian nationalist, Gavrilo Princip, assassinated Archduke Franz Ferdinand, heir to the throne of the Austro-Hungarian Empire, on 28 June 1914.

Bosnia was then under Austrian rule, and the assassination unleashed the First World War.

Population 448 500
Map Bosnia-Herzegovina – Cb

Saransk *Russia* Industrial town 240 km (150 miles) south of NIZHNI NOVGOROD (Gor'kiy). Saransk is capital of Mordovia, an autonomous republic of the Russian Federation, covering 26 200 km² (10 100 sq miles).

Population (city) 312 000; (republic) 964 000
Map Russia – Fc

Saratov *Russia* Industrial city and river port in the south-west, sprawling for about 50 km (30 miles) along the west bank of the Volga River, 320 km (200 miles) north of VOLGOGRAD. Saratov was founded on a neighbouring site in 1590 and it grew in the 1870s after the building of a railway from Moscow, 715 km (445 miles) to the north-west.

For three centuries it held an important position in the trade between Russia and Central Asia. Set amid rich farmland, Saratov's main products are tractors, farm machinery, fertilisers, processed food and furniture.

Population 911 000
Map Russia – Fc

Sarawak *Malaysia* The country's largest state, covering 124 449 km² (48 039 sq miles) and lying south-west of SABAH on the island of Borneo. From 1841 until the Japanese invasion in the Second World War, it was ruled by the 'white rajahs' – descendants of Sir James Brooke, to whom the territory was ceded by the Sultan of Brunei – and was a British protectorate from 1888. It became part of Malaysia in 1963.

Sarawak's forbidding highland interior, bordering Indonesia, and its swampy plain have limited human settlements to the river deltas and valleys. The Iban people – cannibals in former times, now mostly slash-and-burn farmers who live in communal long-houses – and the Sea Dayak people, coastal fishermen related to the Iban, make up most of the population. The population in the centres, such as Kuching, the capital, is largely Chinese.

Offshore oil discoveries since the 1960s and a spreading road network have challenged the traditional way of life.

Population 1 669 000
Map Malaysia – Db

São Tomé and Príncipe

*COCOA REMAINS A VITAL FACTOR
IN THE ECONOMY OF THIS GROUP OF
SMALL ISLANDS LYING OFF THE
COAST OF WEST AFRICA*

Straddling the equator, the volcanic islands of São Tomé and Príncipe are Africa's second smallest independent nation after the Seychelles. It was the Portuguese who discovered the islands in 1471 and they ruled them from 1740 until independence in 1975.

A colonial plantation economy – the main crop is cocoa – still dominates life today. However, production has dropped as a result of drought and mismanagement. In the first 12 years after independence, cocoa production fell by 50 per cent. Most food is imported, and the country relies on foreign aid, notably from France.

The country is mainly composed of the two islands in its title. São Tomé covers some 845 km² (326 sq miles). Its highest point is a peak of 2024 m (6640 ft); the many lower summits are all extinct volcanic cones. Sparkling streams rush down steep hillsides and surge through rain forests on their way to the hot, humid coastlands. Príncipe is another craggy island, reaching 948 m (3110 ft). In addition, there are a number of islets, including Tinhosa Pequena and Tinhosa Grande.

The native islanders, known as *filhos da terra* (Portuguese for 'sons of the land'), are mostly descendants of imported slaves, mulattos, contract labourers and Portuguese settlers. About 46 per cent of the workforce is employed in small manufacturing industries, commerce and services, and 54 per cent on the land. A high proportion of the plantation workers are itinerant and live on the plantations, now state-owned, although in the mid-1990s the government began reversing the socialist trend which it followed after independence.

In the long term, the government plans to develop tourism and become less dependent on the only major export, cocoa.

SÃO TOMÉ AT A GLANCE

Map Gabon – Aa
Official name Democratic Republic of São Tomé and Príncipe
Area 1001 km² (370 sq miles)
Population 133 225 **Per km²** 133 **(Per sq mile** 360)
Capital São Tomé
Government Republic
Currency 1 dobra = 100 centimes
Languages Portuguese (official), Creole, and local African languages
Religions Christian (more than 90%, mainly Roman Catholic), indigenous beliefs
Climate Tropical and very humid; average temperature in São Tomé is 26°C (79°F) throughout the year
Land use Cultivation 21%, grazing 4%, forest and woodlands 75%
Main primary products Cocoa, coconuts, palm kernels, bananas, coffee, timber
Major industries Agriculture, food processing, timber products
Main exports Cocoa, copra, coffee, palm kernels and nuts
Annual income per head (US$) 370
Population growth (per thous/yr) 26
Life expectancy (yrs) Male 63 **Female** 61

Sardinia (Sardegna) *Italy* Second largest island in the Mediterranean after Sicily, covering 24 089 km² (9300 sq miles), separated from neighbouring CORSICA by the Strait of Bonifacio. It is sparsely populated because of its rugged landscape and isolation. There are few towns apart from CAGLIARI, the capital, and the provincial capitals of Nuoro, Oristano and SASSARI. Much of it is covered by granite mountains, culminating in the Gennargentu massif which rises to 1834 m (6016 ft). The main lowland, the Campidano, runs between Cagliari and Oristano.

Sardinia's economy is mainly pastoral, based on sheep and goats. Cork oaks grow in the northwest and wheat, vines and vegetables are grown in the east. Petrochemical industries have been developed at Porto Torres and Cagliari since the Second World War. Coal, lead, zinc and iron mining have a long tradition in the south-west, but this activity has been reduced because of dwindling reserves. Tourism flourishes at ALGHERO and along the COSTA SMERALDA. Sardinians have a tradition of hospitality, but this is also the land of vendetta (blood-feud).

During the Bronze Age (4000-2000 BC) the shepherds built megalithic fortress-villages of squat, cone-shaped towers. The remains of more than 7000 of these *nuraghi* still dot the landscape.
Population 1 664 400
Map Italy – Bd

Sargasso Sea Region of comparatively calm water in the north ATLANTIC Ocean between the North Equatorial Current, the Gulf Stream and the Canaries Current. Covering an area of about 5 200 000 km² (2 007 000 sq miles), it contains many patches of floating sargassum weed, from which it gets its name, and is the breeding place of European and North American eels.

However, because there is little movement of water, the Sargasso Sea is deficient in nutrients and plankton, so it contains comparatively few large marine animals. Sailors once believed that the Sargasso was a graveyard for wrecked ships lured there by gods or devils. Many thought that ships trapped in the weed would forever sail in circles. In fact there is not enough weed to trap the smallest ship.
Map Atlantic Ocean – Bc

Sarh *Chad* Trading and industrial town on the River Chari about 500 km (310 miles) south-east of N'DJAMENA, near the border with the Central African Republic (CAR). It lies on a busy route used for exporting cattle to Bangui in the CAR. The Sarh area is one of the richest and most densely populated parts of Chad, where good-quality cotton is produced. Sarh is also one of the country's main industrial centres, with cotton textile mills, an abattoir and light industries.
Population 65 000
Map Chad – Ac

Sark *English Channel* See CHANNEL ISLANDS

Sarnath *India* Site just north of the city of VARANASI where the founder of Buddhism – Siddhartha Gautama (about 563-483 BC) – preached his first sermon. A monastery was founded here, and its ruins survive with those of Buddhist shrines, including one built by the Emperor Asoka (273-232 BC). The site was deserted after the decline of Buddhism in India from the 6th century AD, partly because of the revival of popular Hinduism and later because Muslim invaders desecrated it. A modern temple contains Japanese Buddhist paintings.
Map India – Cc

Sarnia *Canada* Lake port on the Ontario shore of the St Clair River at the southern end of Lake Huron. It is an oil refining and petrochemicals centre and is connected to Port Huron, across the US border, by tunnel and bridge.
Population 74 870
Map Canada – Gd

Sárospatak *Hungary* See BORSOD

sarsen Large sandstone boulder, dating from the EOCENE or OLIGOCENE epochs, found mainly on the chalk lands of southern England. Sarsens were used in the construction of the outer circle of STONEHENGE.

Saskatchewan *Canada* Central prairie province covering 651 900 km² (251 640 sq miles), and the country's biggest wheat producer. The northern third forms part of the CANADIAN SHIELD; the rest is plains sloping from 1000 m (3280 ft) above sea level in the west, down to 500 m (1640 ft) in the east. It is a rural province with a relatively small population and many farms of 850 hectares (2100 acres) or more. The territory was settled by Assiniboine, Algonquin and Athabasca people. Europeans settled here in the 20th century, after the railways came. Mineral resources being exploited include coal, oil, gas, zinc and copper. The provincial capital is REGINA.

The Saskatchewan River, which is 1939 km (1205 miles) long, flows from the Rocky Mountains into Lake Winnipeg. Its upper part is divided into the North and South Saskatchewan rivers, which join near Prince Albert.
Population (province) 988 930
Map Canada – Ec

Saskatoon *Canada* City situated on the South Saskatchewan River, 240 km (150 miles) north-west of REGINA. It is a major service centre for the surrounding wheat-growing area, and has industries which mill flour and make dairy products. It also has art galleries, museums and a university. In 1882 a temperance colony settled on the east bank of the river; the following year the railway put its station on the west bank. The two communities were amalgamated in 1906.
Population (city) 186 060; (metropolitan area) 210 020
Map Canada – Ec

Sasolburg *South Africa* Industrial town built on a vast coal field in the Free State, about 80 km (50 miles) south of Johannesburg. It is the site of South Africa's first oil-from-coal plant, set up in 1954. The name comes from SASOL, the state-subsidised South African Coal, Oil and Gas Corporation. Apart from oil and petrol, other chemicals are produced as by-products.

Population (town) 33 310; (district) 89 080
Map South Africa – Cb

Sassandra *Ivory Coast* River rising in the Guinea Highlands, and flowing 563 km (350 miles) to the Atlantic Ocean at the town of Sassandra, 65 km (40 miles) along the coast from the port of San Pedro. A hydroelectric dam at Buyo was completed in 1980.
Map Ivory Coast – Ab

Sassari *Italy* Second town of Sardinia after the capital, CAGLIARI, and about 170 km (105 miles) to the north-west of it. Its university was founded by the Jesuits in the 16th century. The town produces wine and a wide range of foods, including pasta, olive oil and cheese.
Population 119 700
Map Italy – Bd

Satpura Range *India* East-west range of hills rising to 1325 m (4347 ft) in central India. It lies between the trough-like valleys of the westward-flowing Tapti and Narmada rivers and, with the Vindhya Range to the north, marks the northern end of the Deccan plateau.
Map India – Bc

Sattahip *Thailand* Port on the east coast of the Gulf of Thailand. It has been developed since 1980 as a deep-water alternative harbour to Bangkok, 130 km (80 miles) to the north-west.
Population 83 600
Map Thailand – Bc

Satu Mare *Romania* Manufacturing town set in rich farmland in the north-west corner of the country, 10 km (6 miles) east of the border with Hungary. Satu Mare was founded at least 750 years ago beside the Somes River by boatmen taking salt to central Romania. Its products include mining and transport equipment.
Population 113 600
Map Romania – Aa

saturation State reached when air is holding as much water vapour as is possible at a given temperature.

saturation zone See WATER TABLE

Sau *Slovenia / Croatia / Bosnia - Herzegovinia/ Yugoslavia* See SAVA

Saudi Arabia See p 600

Sault Sainte Marie *Canada* Port on the north bank of St Mary's River between Lake Superior and Lake Huron in Ontario. It is a centre for tourism. Manufacturing includes chemicals, lumber, pulp, paper and steel. Opposite, on the south bank of the river, is Sault Sainte Marie, Michigan. Five ship canals, four in the USA and one in Canada, connect the lakes.
Population 81 480
Map Canada – Gd

Sava (Sau; Szava) *Slovenia/Croatia/Bosnia-Herzegovina/Yugoslavia (Serbia and Montenegro)* River, 945 km (587 miles) long, part of which forms the border between Croatia and Bosnia-Herzegovina. It has two sources in Slovenia – in the Karawanke Alps above Planica (near the Austrian border) and in the Julian Alps – and flows south-east to meander sluggishly across the Croatian-Slavonian plains and into Serbia to join the Danube at Belgrade.
Map Slovenia – Ba-Bb

Savai'i *Western Samoa* Largest island of Western Samoa, lying to the west of Upolu and covering 1820 km^2 (702 sq miles). The volcanic peak of Silisili is 1858 m (6096 ft) high and much of the north-east of the island is covered by black lava flows. After eruptions in the early 20th century, many people migrated to Upolu. Savai'i has been inhabited by Polynesians for 3000 years and has several ancient temples.
Population 44 930
Map Pacific Ocean – Ec

savannah Open grassland with tall grasses and scattered trees and bushes, covering large areas of Africa, South America and northern Australia. See CAMPOS.

Savannah *USA* Major port and historic city, the oldest in GEORGIA, near the mouth of the Savannah River, which forms the Georgia-South Carolina border. Founded in 1733, it is laid out in a series of squares leading to Forsyth Park. It played an important part in the American War of Independence and the Civil War. The city's industries include oil refining, paper products, cotton and wood processing and marine supplies. It is popular with tourists.
Population (city) 137 600; (metropolitan area) 258 100
Map United States – Jd

Savannakhet *Laos* Town and rice-growing province in southern Laos. The town is on the Thai border about 210 km (130 miles) north-west of the town of PAKSE.
Population (town) 51 000; (province) 543 600
Map Laos – Ab

Savoie *France* Former independent duchy covering some 10 416 km^2 (4022 sq miles) in the northern Alps. It became French territory in 1860, and now forms two departments – Savoie and Haute-Savoie. Hydroelectric power and winter tourism are important to its economy. CHAMBERY is the main town.
Population 310 000
Map France – Gd

Savonlinna *Finland* Town, founded in the 17th century, on an island in SAIMAA Lake in south-east Finland. It is dominated by the ruined medieval castle of Olavinlinna, one of the largest in northern Europe. It is a summer resort and the setting of an annual opera festival in July.
Population 28 700
Map Finland – Dc

Sayaboury *Laos* Province and town in north-western Laos between the MEKONG River and the Thai border. Once part of Thailand, the province was annexed by Laos' French colonial rulers in 1909. A hilly and underpopulated area with only narrow lowlands along the Mekong, it has been the source of border disputes between Thailand and Laos.
Population (province) 223 600; (town) 14 000
Map Laos – Ab

Sbeitla *Tunisia* Small town on an important road junction on the eastern edge of the Tell Atlas, close to the El Douleb oil field and some 30 km

▼ BLACK AND WHITE The lovely Romanesque Church of Santa Trinita de Saccargia stands on the outskirts of Sassari, a city of dazzling white limestone surrounded by olive groves.

(19 miles) north-east of KASSERINE. It was the site of the 2nd-century Roman town of Sufetula, whose impressive remains include a triumphal arch, a forum, baths and three finely preserved temples built in red sandstone. They are believed to have been dedicated to the Roman god Jupiter and the goddesses Juno and Minerva.
Population 12 100
Map Tunisia – Aa

Scafell Pike *United Kingdom* See LAKE DISTRICT

scalded flat In Australia, a plain whose soil is of little use because of a high salt content.

Scapa Flow *United Kingdom* Natural anchorage, extending over 130 km² (50 sq miles), and enclosed by the Scottish ORKNEY ISLANDS. It was a base for British fleets in both World Wars. In 1919 the defeated German navy scuttled 71 of its ships in Scapa Flow, where they had been interned. In 1939 a German submarine slipped past the British defences and sank the British battleship *Royal Oak*. Churchill Barrier, a series of concrete causeways built between the islands, blocks the eastern approaches.
Map United Kingdom – Gj

scar Steep, bare rock face or crag. The name is chiefly used in northern England to indicate a limestone cliff.

Scarborough *Trinidad and Tobago* Main port and capital of Tobago, on the south-east coast. Remains of a French fort stand on a 130 m (430 ft) hill above the town.
Population 3000
Map Trinidad and Tobago – Aa

Scarborough *United Kingdom* English seaside resort and port in North Yorkshire, 58 km (36 miles) north-east of YORK. It developed as a spa in the mid-17th century. It has a ruined Roman signal station and a Norman castle, and is a popular angling and conference centre.
Population 38 050
Map United Kingdom – Ec

scarp See ESCARPMENT

scarth Bare rock face, especially in the English LAKE DISTRICT.

Scebeli *Ethiopia/Somalia* See SHABEELLE

Schaffhausen *Switzerland* Town about 40 km (25 miles) north of Zürich and capital of the canton of the same name. The main tourist attraction is the nearby Falls of the Rhine, the largest Central European waterfalls. Tourists also come for the international Bach festival, which is held every three years, and to see the painted 16th to 17th-century houses that bear witness to the town's prosperous past. Main industries in the canton are metal and electrical goods, machinery and watch making.
Population (town) 33 900; (canton) 73 000
Map Switzerland – Ba

Schelde (Scheldt; Escaut) *France/Belgium/Netherlands* River which rises in north-east France and flows for 435 km (270 miles) through the Belgian cities of Tournai, Ghent and Antwerp. It used to empty into the North Sea

▲ ATLANTIC ISLES Temperatures rarely fall below 7°C (45°F) on the Isles of Scilly – England's westernmost point, warmed by the North Atlantic Drift.

through two estuaries, both in the Netherlands, but dyking has cut off the Oosterschelde (East Schelde) estuary, leaving the Westerschelde (West Schelde) as the only outlet.
Map Belgium – Aa

schist Any of a variety of medium to coarse-grained metamorphic rocks composed of parallel, often wavy and flaky, bands. The rock breaks easily along the bands, and is usually named after its dominant mineral, for example, mica schist.

Schlesien *Poland* See SILESIA

Schleswig-Holstein *Germany* Northernmost state of the country occupying the southern part of the Danish peninsula between the North Sea and the Baltic Sea. Two duchies – Schleswig in the north and Holstein to the south – were united in 1773 to form a province, which became part of Prussia in 1864. It is agricultural country with dairy farming, sugar beet, wheat and vegetables. Its only industries lie around the ports of KIEL, the capital, the old Hanseatic town of LÜBECK, and FLENSBURG, all three on the Baltic coast.
Population 2 634 100
Map Germany – Da

Schönbrunn *Austria* See VIENNA

Schottegat *Netherlands Antilles* See CURAÇAO

Schwarzwald *Germany* See BLACK FOREST

Scilly, Isles of *United Kingdom* Group of islands some 40 km (25 miles) south-west of LAND'S END in England. The largest island is St Mary's, followed by St Martin's and Tresco. St Mary's is served by ship and helicopter from PENZANCE in Cornwall. The mild climate is ideal for spring

flowers and early vegetables, which are exported to the mainland, and it also attracts many tourists. Palms and other exotic plants flourish in Tresco's Abbey Gardens.
Population 2000
Map United Kingdom – Bf

scoria (cinders; slag) Rough, angular fragments of rapidly cooled lava containing cavities caused by bursts of escaping gas. Scoria is darker and more cindery than pumice.

Scotland *United Kingdom* Northern constituent country of the UK, covering some 78 762 km² (30 410 sq miles). The kingdom of Scotland – peopled by native Picts, immigrant Scots from Ireland and Scandinavians – was formed in the 9th century AD. King James VI of Scotland became James I of England in 1603, and the kingdoms joined in 1707. Scotland's capital is EDINBURGH and its chief industrial cities are GLASGOW, ABERDEEN and DUNDEE.
Population 4 957 290
Map United Kingdom – Cb

Scott Base *Antarctica* Headquarters of the New Zealand Antarctic research programme, situated on ROSS ISLAND on the shores of McMurdo Sound. It was founded in 1957.
Map Antarctica – Jc

scree (talus) Loose, usually angular fragments of rock debris which have fallen and collected at the foot of a weathered steep slope or cliff.

scrub Vegetation consisting of stunted trees and low-growing shrubs with occasional taller trees. It is found where the soil is poor and rainfall scanty.

scud Ragged mass of very low cloud driven by a strong wind, below the main rain clouds.

Scutari *Albania* See SHKODËR

Scutari *Turkey* See ÜSKÜDAR

Saudi Arabia

A STRICTLY MUSLIM OIL-RICH STATE AND FOCAL POINT OF THE ISLAMIC FAITH, TO WHICH MILLIONS OF PILGRIMS FLOCK EVERY YEAR

The largest Arab country (in terms of area, though not population) and the birthplace of the Prophet Muhammad and of the Islamic faith, Saudi Arabia contains more oil than any other nation on earth. In the early 1990s, its reserves were estimated at more than 252 billion barrels. It is also the largest exporter of oil. With the quadrupling of oil prices after the 1973 Arab-Israeli War, enormous wealth poured into the desert kingdom – more than US$100 000 million a year at its peak. A nation of desert nomads became one of the richest in the world.

Using its money from oil, Saudi Arabia launched a huge development programme. Suddenly towns were bustling with foreigners: contractors, immigrant workers and business representatives from all parts of the world. Towns and ports were transformed into major commercial centres with modern office blocks, factories and housing estates. RIYADH, the capital, JEDDAH, the main port on the RED SEA, and DAMMAM, the oil centre on The GULF, were among the major centres of change. Yet despite radical changes Saudi Arabia remains a strict Muslim society, with laws based on the Koran.

THE LAND AND ITS PEOPLE

Saudi Arabia, with an area estimated at 2 250 000 million km² (868 525 sq miles), covers most of the great Arabian Peninsula. It is thinly populated, with an estimated 17.6 million people, and it contains some of the hottest, bleakest desert in the world: temperatures often rise to 45°C (113°F). In the north, the NEFUD Desert stretches into Jordan and Syria. In the west, the narrow, humid coastal plain along the Red Sea is backed by steep mountains, reaching 2600 m (8530 ft) in the HEJAZ to the north and 3133 m (10 278 ft) at Jebel Abha in the ASIR highlands.

Beyond the mountains, the land slopes gradually down to the Ad Dahna Plain, about 240 km (150 miles) wide, bordering The Gulf. Fierce winds and sandstorms sweep the deserts and rocky regions in the centre of the country. In the south-east is the forbidding desert of the RUB AL-KHALI (Empty Quarter). The largest continuous sand desert on earth, it is waterless, uninhabited and almost featureless. Large areas of the interior are still unexplored and the southern boundaries with Kuwait, Yemen and Oman remain vague.

The people of the Peninsula are the descendants of the original Arabs who swept out of Arabia across the Middle East and North Africa in two centuries of conquest after Muhammad proclaimed the new faith of Islam in the 7th century. On the Red Sea coast, the descendants of African slaves have inter-

mingled with the Arabs, although as a rule their population remained in total isolation – until the oil boom attracted a flood of immigrants to the country.

Today, because of rapid industrialisation, there are about 4.6 million foreign workers in addition to 13 million Saudis. As in Kuwait, Palestinians were evicted during The Gulf War of 1990-1 (for siding with the Iraqis). But the Jordanians, Sudanese, Indians, Syrians and Egyptians still remain. At the managerial and technical levels Americans, Britons and other Europeans do tours of duty.

Despite the overall low population density of the country, there are centres where urbanisation and crowded conditions are part of everyday life. Altogether 77 per cent of the population live in cities – one quarter alone in Riyadh, the capital; the arid regions are, of course, virtually uninhabited.

As in the past, many inhabitants still lead a nomadic life, following their sheep, goats and camels in search of grazing. But today's Bedouins may look for fresh pasture in a fast car, drive their herds to new areas in trucks, or bring up water by road tanker.

The Ottoman Turks held sway over the area from the 16th to the early 20th century. During the First World War they were ousted by the Arabs in a daring guerrilla campaign organised by a British colonel, T. E. Lawrence (Lawrence of Arabia). But the creator of modern Saudi Arabia was Ibn Saud, who in tribal warfare between 1902 and 1924 united the four main regions of NEJD, AL-HASA, the Hejaz and Asir. In 1932, as King Abdul Aziz Ibn Saud, he proclaimed the new kingdom of Saudi Arabia.

The king died in 1953. The present ruler, King Fahd Ibn Abdul Aziz, is the fourth of his sons to govern the country. Saudi Arabia is an absolute monarchy; the king is also prime minister and owns all the undistributed land in the country. There are no elections, no political parties are allowed and there is no constitution. The country is ruled according to Shari'a law.

There are about 5000 princes in the royal family and many important government offices are in their hands. Enlightened help for the people has gone hand in hand with royal extravagance: possibly 10 per cent of the oil wealth has gone to the royal family.

However, in 1993 King Fahd kept his undertaking to create a *majlis-al-shura* (a non-elected consultative assembly), with 60 members. It is intended that eventually half of the members will be elected through a system of provincial assemblies. With no official opposition, the only publicised dissent has come from militant Muslim radicals and from human rights activists; in September 1994 the authorities said 110 people were in detention, but unofficial reports claimed that more than 1000 dissidents were being held.

DEVELOPMENTS BASED ON OIL

Oil was first discovered in the 1930s, but it made little difference to the country until well after the Second World War; pilgrims visiting the holy cities of MECCA and MEDINA remained the most important source of income until then. However, from 1973 Saudi Arabia led OPEC (the Organisation of Petroleum Exporting Countries) in drastically cutting oil production and pushing up its price – a step which had far-reaching effects on the Western, and world, economies. It also nationalised the oil reserves owned by foreign oil companies.

Since 1980, in contrast, the country has had a moderating influence on world markets. It boosted production to stabilise price increases when the Iran-Iraq War (1980-8) again disrupted supplies, and it led in cutting output when overproduction and a fall in demand in the West pushed oil prices down again.

However, by 1985 the Saudis had tired of seeing their income dwindle while other OPEC nations exceeded agreed quotas. They stepped up output, with the result that oil prices tumbled to below US$10 a barrel in 1986. In 1987 Saudi Arabia led a new move by OPEC to cut output and raise the price to US$18 a barrel. They were, however, unable to stop the decline during the 1990s which led to some of the lowest oil prices ever (in real terms). Added to this was the cost of The Gulf War, reported to have cost Saudi Arabia US$64 000 million (including US$16 800 million in cash to the United States, which provided more than 400 000 troops).

In 1994 the Cabinet promised some drastic spending cuts in an effort to balance the budget and announced a restructuring of debt to United States arms companies. Plans for

▲ ISLAM'S HOLIEST PLACE Pilgrims throng around the *Kaaba,* in the centre of Mecca. The building contains the sacred Black Stone, said to have fallen from Paradise with Adam.

the partial privatisation of state concerns, after government contractors had reported delays in payments of up to 15 months, were also announced.

The oil fields that are the source of nearly all the country's wealth lie in the eastern region, near and under The Gulf. GHAWAR, the largest oil field in the world, is 241 km (150 miles) long and 35 km (22 miles) wide and can produce 5 million barrels a day. There are 14 other oil fields.

Development plans using the oil riches began with housing, schools and water supplies. At one stage three schools a day were being opened. Education is free but illiteracy remains high – about 60 per cent. Free land and interest-free loans are provided for housing. Between 1980 and 1995, some 2000 factories were built and much government spending has gone into agriculture. Although

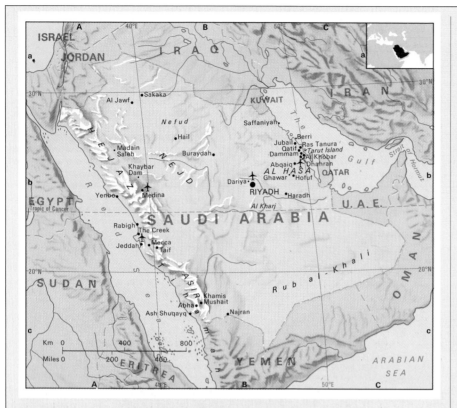

drink water, coffee or tea with a tribe or kiss their tent to secure hospitality and protection for three-and-a-third days without question.

Although a desert country, Saudi Arabia is not without vegetation. Wild flowers appear in the Asir Mountains in the spring and TAIF is famous for its roses and pomegranates. Date palms grow in oases and the higher regions support acacias, junipers and tamarisk trees.

Camels and asses, herded by their nomadic owners, are a familiar sight. The wildlife includes leopards, wolves, foxes, hyenas, the rare oryx (a large antelope) and three types of gazelle; baboons, sand cats, ratels (honey badgers), hyraxes (the 'coney' referred to in the Bible), and a variety of snakes. Migrant birds from Europe rest along the Red Sea coast, and falcons are highly prized and trained for hunting. The rich underwater life of the Red Sea has been made famous by the French film maker Jacques Cousteau.

Jeddah, Saudi Arabia's commercial and diplomatic capital on the Red Sea, is also the entrance for the pilgrims (more than 1 million from abroad) who come to visit Mecca and Medina every year.

The holy city of Mecca is 70 km (44 miles) from Jeddah in a narrow valley overlooked by hills and castles. It contains the Great Mosque, the *Haram*, which encloses a small, sacred building, the *Kaaba* (House of God), where pilgrims kiss the venerated Black Stone, a meteorite that fell from the sky in the forgotten past. Medina, 180 km (112 miles) from the Red Sea, is surrounded by double walls with nine gates. Its mosque contains the tombs of Muhammad, his daughter Fatima and the Caliph Omar. The city also has an Islamic University. Muslims try to visit Mecca and Medina at least once in their lifetime. Non-Muslims are banned. In certain months up to 2 500 000 pilgrims were officially estimated to have arrived in the country for the annual *haj*.

only 1 per cent of the country's total area is used for cultivation, lavish subsidies raised grain output from 130 000 tonnes in 1980 to 3.9 million tonnes in 1992. Saudi Arabia is now virtually self-sufficient in grain. Large artificially watered market gardens sheltered beneath huge plastic sheets have been built. At one time there was even a proposal to tow Arctic icebergs into The Gulf to provide fresh water, but this idea was abandoned.

GUARDIANS OF ISLAM

The Saudis see themselves as the guardians of Islam. They have one of the strictest Muslim societies in the world. In AD 622 the Prophet Muhammad, driven out of his birthplace, Mecca, began his teaching in the city of Medina. Within a hundred years the Arabs had spread Islam as far as Spain in the west and India in the east. Then, in 750, the centre of Islamic power moved to BAGHDAD and Arabia lapsed into obscurity. But in the 18th century a new prophet arose – Muhammad Ibn Abdul Wahhab. Preaching an austere Islamic doctrine, his followers became members of the Wahhabi sect. This is the basis of religion and law in Saudi Arabia today.

The possession of alcohol, drugs and pornographic material (which may include fashion magazines) is strictly prohibited, even for foreigners. Detection of alcohol on a person's breath may result in up to two months' imprisonment and a lashing. Drug offences carry up to 15 years' imprisonment and even someone with prescribed drugs may be detained while the drug is analysed. Traffickers are executed.

Theft is punished by imprisonment, and repeated theft by amputation of the right hand. Murder, rape, adultery and renouncing the

Islamic faith carry the death penalty. This is carried out in public after Friday prayers – by stoning for adultery and beheading in most other cases. Up to 60 people may be kept in a large cell awaiting trial, and 200 after conviction; they pay for food and sleep on the floor.

Women still go veiled in public, may not drive cars and can work only in all-female establishments. Girls receive primary education, but in higher education their numbers drop – although 73 per cent of men are literate, only 48 per cent of women can read and write. At universities women can watch lectures by men only on closed-circuit television. But there has inevitably been questioning of the restrictions, especially by women who have travelled to the United States and Europe.

FOR THE VISITOR

Saudi cuisine has widened as a result of immigration from other Arab countries, but meat and rice still form the staple dish. When entertaining friends, Saudis usually prepare dishes containing mutton, chicken, goat or camel; the meat may be stuffed with rice, nuts and herbs and will be placed on a large bed of rice. The guests eat from the communal dish with the fingers of the right hand. Soups and salads may be flavoured with chives, coriander or dill.

A variety of fish is available in coastal regions; *shami* or *samul* bread is eaten with every meal. Dates are a traditional sweet, but there is also *muhallabiyah*, a milk pudding, and the sticky *kneifah* or *bakhlava*, a kind of pastry filled with honey and nuts. Ground coffee spiced with cardamom seeds completes the meal. Bedouin hospitality is legendary. Among the nomads a stranger had only to

SAUDI ARABIA AT A GLANCE	
Official name Kingdom of Saudi Arabia	
Area 2 250 000 km² (868 525 sq miles)	
Population 17 615 400 **Per km²** 8	
(Per sq mile 20)	
Capital Riyadh	
Government Absolute monarchy	
Currency 1 Saudi rial = 20 quirsh = 100 hallalas	
Language Arabic	
Religion Muslim (85% Sunni, 15% Shiite)	
Climate Hot and dry, mild in winter. Only the highlands in the south-west receive summer rains. Temperature in Riyadh ranges from 14°C (57°F) in January to 37°C (99°F) in July	
Land use Cultivation 1%, pastures 39%, forest and woodland 1%, other 59%	
Main primary products Cattle, goats, sheep, poultry, alfalfa, dates, grapes, water melons, wheat, sorghum; crude oil and natural gas	
Major industries Crude oil and natural gas production and refining, cement, chemicals, fertilisers, steel	
Main exports Crude oil and refined products, natural gas	
Annual income per head (US$) 7940	
Population growth (per thous/yr) 33	
Life expectancy (yrs) Male 67 **Female** 69	

sea A part or branch of an ocean, especially if partly surrounded by land, or a large inland stretch of salt water. For geographical features whose name begins with 'Sea of', see under main part of name.

sea breeze Breeze blowing during the afternoon and evening from the sea towards an area of low atmospheric pressure over the land which has heated up more quickly than the sea, particularly in tropical regions. It is the opposite of a LAND BREEZE.

sea level See MEAN SEA LEVEL

seamount Isolated volcanic peak rising from the ocean floor.

season One of the periods into which the climatic year is divided, according to variations in temperature, duration of daylight and rainfall. In latitudes outside the tropics a seasonal rhythm of temperature is imposed by the large differences in the angle of elevation of the sun that occur as the earth follows its orbital path around the sun. In the summer the angles are at their highest and the days longest, which combine to give the highest temperatures of the year. The converse applies in winter. The transitional seasons are spring and autumn. In tropical areas, which are always hot, the rainfall regime is the factor that distinguishes the seasons from one another (see MONSOON).

Seattle *USA* Port in Washington State which handles much trade with the Far East and Alaska. It stands on Puget Sound, about 160 km (100 miles) south of the Canadian border, and is the main financial, market, industrial and cultural centre for the Pacific North-West region. The city is the home of Boeing Aircraft and also has shipbuilding, electronics, food and wood processing, fishing and tourist industries.
Population (city) 516 300; (metropolitan area) 2 033 200
Map United States – Ba

Sebenico *Croatia* See ŠIBENIK

Sechin *Peru* One of the most important pre-Columbian ruins, standing on the coast 365 km (nearly 225 miles) north of the capital, Lima. There are many adobe and stone buildings decorated with carvings of warriors.
Map Peru – Ba

secondary depression Small concentrated area of low atmospheric pressure on the margins of a main DEPRESSION.

Secunda *South Africa* Town in Eastern Transvaal, about 65 km (40 miles) east of Johannesburg, where the second and third oil-from-coal plants were built by the government in the 1970s and 1980s. The first such plant was in SASOLBURG.
Population 31 350
Map South Africa – Cb

Secunderabad (Sikandarabad) *India* Twin city just north of HYDERABAD in southern India. Its cantonment (permanent military base) was one of Britain's largest in India. Here in 1898, while in the Indian Medical Service, the British scientist Sir Ronald Ross (1857-1932) discovered that the

malaria parasite is carried by mosquitoes. He won a Nobel Prize for his work in this field.
Population 136 000
Map India – Cd

Sédhiou *Senegal* Market town on the Casamance River, 45 km (28 miles) from the Guinea-Bissau border. It lies at the heart of a farming area producing rice and bananas and, to a lesser extent, groundnuts, maize, vegetables and tropical fruits.
Population 150 000
Map Senegal – Ab

sediment Material comprising weathered particles of rocks, or particles of chemical or organic origin, deposited by wind, water or glacial ice. Sediment is the raw material of SEDIMENTARY ROCK.

sedimentary rock Type of rock formed by the consolidation of particles weathered from older rock, or from particles of chemical or organic origin. Rocks composed of hardened sediments or particles abrased from mountains and hills by wind and weather are called clastic or detrital rocks; examples are sandstone, shale and conglomerates. Organic sedimentary rocks, such as limestone, are made up of the shells and other remains of innumerable long-dead sea creatures; while chemical sedimentary rocks, like gypsum and rock salt, had their beginnings in naturally occurring chemical solutions.

Sedlez *Poland* See SIEDLCE

Sedom *Israel* Town, 6 km (4 miles) from the south end of the DEAD SEA. Established in 1934 by the Palestine Potash Company of Kaliya, it is now the headquarters of chemical and fertiliser industries using salt and minerals extracted from the Dead Sea. The biblical town of Sodom (from which Sedom takes its name), which was destroyed with Gomorrah, is believed to lie buried beneath the waters of the Dead Sea.
Map Israel – Bb

Ségou *Mali* Town on the Niger River, some 250 km (155 miles) downstream from the capital, BAMAKO. It was the capital of the Bambara civilisation from 1660 to 1861. Today it is a stopping place on the route to the towns of Mopti and Gao. The headquarters of the Office du Niger are here and some of its attractive office buildings are built in traditional style in spacious grounds. The organisation, which is concerned with improving agriculture in the Niger Delta, is one of the town's main employers.
Population 99 000
Map Mali – Bb

Segovia *Spain* Ancient fortified town 68 km (42 miles) north-west of Madrid, whose Roman aqueduct still supplies water. Its industries include flour milling, tanning and pottery, while the surrounding province of the same name grows cereals and raises sheep.
Population (town) 54 380; (province) 147 190
Map Spain – Cb

seiche An oscillation of the water in lakes, inland seas, bays and channels, resembling a tide, caused by SEISMIC WAVES or atmospheric disturbances such as pressure changes and wind. The rise and fall may be a few centimetres or several metres.

seif dune Ridge of sand, often many kilometres long, crossing a desert parallel to the direction of the prevailing wind. The steep-sided dune may form when a line of BARCHANS joins up. Its height and width are increased by crosswinds.

Seine *France* River of northern France 775 km (482 miles) long, which rises on the Plateau de Langres, near DIJON, and flows north-west through Paris. The stretch between Paris and its mouth, at Le Havre, is navigable.
Map France – Db

seismic energy Energy that is released by an EARTHQUAKE.

seismic focus Centre of an earthquake, below the earth's surface, from which the SEISMIC WAVES are emitted.

seismic waves The vibrations emitted by an EARTHQUAKE or man-made explosion.

seismograph Instrument for detecting and measuring SEISMIC WAVES.

seistan Strong, northerly wind sometimes exceeding 110 km/h (nearly 70 mph), blowing in the Sistan (Seistan) region of eastern Iran during the four months of summer. It is sometimes known as 'the wind of 120 days'.

Sekondi *Ghana* Coastal town some 170 km (105 miles) west of the capital, Accra. It was the country's first railway terminus, serving a hinterland rich in gold. It also handled most of Ghana's import and export trade until 1923, when it was superseded by adjoining Takoradi. The two were united in 1963; Sekondi remains a residential and trade centre. Sekondi-Takoradi is the capital of Ghana's Western Region.
Population (Sekondi-Takoradi) 254 500
Map Ghana – Ac

Selangor *Malaysia* The country's most developed state, on the west coast of Peninsular Malaysia surrounding the capital, Kuala Lumpur. Selangor, which covers 7962 km² (3974 sq miles), produces much of Malaysia's rubber and 40 per cent of the federation's industrial output. Its capital is Shah Alam.
Population 1 978 000
Map Malaysia – Bb

Selenge *Mongolia/Russia* River rising in north-west Mongolia and flowing 2120 km (1317 miles) north-east into Lake BAIKAL in Russia. Most of Mongolia's arable land is concentrated in the lower Selenge Valley.
Map Mongolia – Cb

Selkirk *United Kingdom* See BORDERS

Selous Game Reserve *Tanzania* Africa's largest wildlife reserve, covering 54 600 km² (21 076 sq miles) and the deep, narrow Stiegler's Gorge. The reserve lies beside the Rufiji River about 200 km (125 miles) south-west of Dar es Salaam.
Map Tanzania – Ba

selva Commonly, the equatorial forest of the Amazon basin in South America. The term is also applied to similar forests in other regions.

Senegal

IN A COUNTRY WITH A BUSTLING MODERN CAPITAL ON THE COAST, EIGHT OUT OF TEN PEOPLE STILL LIVE IN FARMING VILLAGES

As France's first and most favoured colony in West Africa, Senegal is still very French in character, although it has been independent since 20 June 1960. It was the only colony where French citizenship was granted to Africans and where conscious efforts were made to educate Africans to become 'black Frenchmen'. A

multiparty system has been in place since 1977 but politics are dominated by President Abdou Diouf's Socialist Party, which has consistently controlled the bulk of national assembly seats.

The largest ethnic group in Senegal is the Wolof people, accounting for about 45 per cent of the population. Traditionally farmers on the savannah, this group now includes wealthy, educated townspeople who control Senegalese economic and political life. The economic and political nerve centre of Senegal is the capital, DAKAR, on the volcanic CAPE VERDE Peninsula which juts into the Atlantic and is the most westerly point in Africa.

More than 90 per cent of Senegalese are Muslims – indeed, the name of the country comes from the Zenaga Berbers of Mauritania, who invaded the region in the 11th century, and converted the inhabitants to Islam. The strictest rules of the faith are not always observed: south of the Sahara, for instance, women do not customarily wear the veil.

Most of Senegal consists of low plains, covered with sand or dry savannah of the SAHARA and SAHEL. The lowest, coolest and wettest part is the south-west, which is also the most densely populated.

The rural areas are inhabited and farmed by 80 per cent of the Senegalese. In the villages, the huts of extended families (parents, children, and other close relatives living together) are enclosed in fenced compounds. Work on the land for the whole family begins in June or July when the rains bring relief from the oppressive heat, and almost overnight the flat, brown countryside turns green. The men go to the fields early and return late; the women help there, as well as doing domestic chores. Tribal customs determine whether women are allowed to work in the millet fields, the main source of food for the family. However, the women work with the crops for sale.

The main cash crop is groundnuts; these are farmed on collective farms and production is largely state-controlled. Senegal is one of the world's leading producers of groundnut oil. The women also have their own plots of vegetables, and if they are near rivers which can be used to flood the fields, they grow rice too. The vegetables and rice supplement the people's meagre diet. Surpluses are sold to bring in additional income. Cattle, goats and sheep are kept by most families, and they are the symbols of a farmer's wealth and prestige.

Phosphate, mined near Thies, accounts for 9 per cent of Senegal's exports. World demand for phosphates has been declining, however, and the country has had to produce other goods for export. Major exports today are manufactures (30 per cent of exports), fish products (23 per cent), groundnuts (12 per cent), and petroleum (16 per cent). A film industry based in the capital is thriving and tourism is a growing industry.

Hard times have come to Senegal with the drought that has afflicted the Sahel across the whole of West Africa. In normal years there is just enough food to go around – but now grain stores are low and food does not always last the year; many go hungry for two or three months before the annual October harvest. More irrigation is needed to make the food supply secure against the erratic rains, and agricultural techniques in the dry areas need to be improved. A bad drought occurred during the 1970s and again in 1983-4, causing widespread famine.

The country depends on food imports and international aid – today Senegal is close to bankruptcy, with a huge trade deficit. In 1993 the government introduced an emergency plan to cut national spending, but the situation was further exacerbated by the devaluation of the CFA franc in 1994.

Politically, Senegal has been relatively stable in recent years, with President Diouf having been re-elected for a third term in 1993, though with a reduced majority. However, there have been violent clashes between government troops and the Movement of Democratic Forces of Casamance (MDFC), which demands independence for the southern province. Other disturbances occurred early in 1994 when a radical Islamic youth movement, the *Moustarchidine Oua Moustarchidine*, was banned after it called for the overthrow of President Diouf in violent demonstrations in Dakar in which six people were killed.

SENEGAL AT A GLANCE	
Official name Republic of Senegal	
Area 196 190 km² (75 750 sq miles)	
Population 8 463 300 **Per km²** 43 **(Per sq mile** 112)	
Capital Dakar	
Government Republic	
Currency 1 CFA franc = 100 centimes	
Languages French (official), African languages, mainly Wolof	
Religions Muslim (94%), Christian (5%), indigenous beliefs (1%),	
Climate Tropical; average temperature in Dakar ranges from 21°C (70°F) in January to 28°C (82°F) in October	
Land use Cultivation 12%, grazing 33%, forest and woodlands 54%, other 1%	
Main primary products Rice, millet, sorghum, maize, groundnuts, fish, timber; phosphates	
Major industries Agriculture, petroleum refining, phosphate fertiliser manufacture, textiles, beverages, fishing, forestry, cement and processed food	
Main exports Manufactures, including building materials and textiles; fish products, groundnuts, petroleum and petroleum products, phosphates	
Annual income per head (US$) 780	
Population growth (per thous/yr) 30	
Life expectancy (yrs) Male 55 **Female** 57	

Semarang *Indonesia* One of the oldest cities in Indonesia, it is now a textile and cigarette-producing city and port on the north coast of Java, about 400 km (250 miles) east of Jakarta. It has a large Chinese community.
Population 503 200
Map Indonesia – Dd

Semendria *Yugoslavia* See SMEDEREVO

semi-desert Semi-arid region between true desert and grassland, characterised by scrubby, thorny plants, coarse grass and bare patches of sand or stony waste.

Semipalatinsk *Kazakhstan* Eastern industrial city and port on the Irtysh River, 900 km (562 miles) north of ALMA-ATA. It was founded in 1718. It has one of Central Asia's largest meat processing plants. Atomic testing near here has caused widespread radioactive contamination.
Population 339 000
Map Kazakhstan – Ea

Semmering Pass *Austria* Pass carrying road and rail routes south-west from Vienna to the Mur and Mürz valleys and the town of Bruck-an-der-Mur, 135 km (84 miles) south-west of the capital. The village of Semmering, a popular winter sports resort, lies near the 985 m (3232 ft) high summit of the pass.
Map Austria – Db

Sendai *Japan* City near the east coast of HONSHU Island, about 300 km (185 miles) north-east of Tokyo. It is the home of many central government offices and a dozen universities and colleges. Industrially it specialises in oil refining, timber goods, electrical goods and foods. The old town was severely damaged during the Second World War but has since been rebuilt.
Population 918 000
Map Japan – Dc

Senegal *Guinea / Mali / Mauritania / Senegal* Important West African river, which is about 1790 km (1110 miles) long. It rises as the Bafing River in the Fouta Djallon uplands of central Guinea, and flows north into Mali, where it is joined by the Bakoy River to form the Senegal proper. The enlarged river marks the border between Senegal and Mauritania, and reaches the Atlantic at St Louis, Senegal.

The Senegal, which flows all year, is used for local transport, despite shifting sandbanks and an erratic flow – forming a shallow stream in the dry season (October-May) and, usually, flooding in the wet season (May-October). However, river control schemes in Senegal and Mali even out the flow and reduce the dependence of farmers on the region's erratic rains.

Map Senegal – Ba

Sensuntepeque *El Salvador* See CABAÑAS

Seoul (Soul) *South Korea* Capital city and teeming metropolis on the Han River, 60 km (37 miles) upstream from the YELLOW SEA.

The city became the capital of unified Korea under the Yi dynasty in AD 1393. Royal palaces were built in its northern part below Mount Pukak, and remnants of defensive walls survive. Until the end of the Second World War it was called Hansong. In 1948, after Japanese colonial rule (1905-45), it became the capital of the Republic of Korea. During the Korean War (1950-3) the city was occupied by invading communist armies from the north, and severely bombed. Its population shrank from more than 1 million to 50 000.

Reconstruction began in 1953 and since the 1960s the city's expansion has been dynamic, fuelled by wholesale migration from the countryside. Today the city accounts for almost half of South Korea's industrial output. High-rise blocks tower over the business centre; in the suburbs houses are crammed together; and traffic congestion blocks the river bridges daily.

Although the city is ultra-modern, it still has royal palaces, temples, pagodas and stone gateways. A few art treasures are held in the National Museum, and the Toksugung Palace offers a glimpse of Yi architecture. The city was host to the 1988 Olympic Games; the Olympic city is in the south-east, on the left bank of the river.

Population 10 915 000

Map Korea – Cd

Sepik *Papua New Guinea* River rising in the Victor Emanuel Range and flowing first north and then east for 1200 km (750 miles) into the Bismarck Sea. It is navigable for some 475 km (300 miles) from its mouth. People of the region are renowned for their wood carving.

Map Papua New Guinea – Ba

Sept Iles (Seven Isles) *Canada* Port city on the St Lawrence River in Quebec, founded as a trading post in 1651. It boomed in the 1950s when iron ore was mined and processed in the area.

Population 29 300

Map Canada – Ic

Sequoia National Park *USA* Area covering 1629 km² (630 sq miles) in the Sierra Nevada Mountains of California, 260 km (160 miles) north of Los Angeles. It has groves of giant sequoia trees, the biggest of which – 'General Sherman' – is 11 m (37 ft) in diameter at its base, 84 m (275 ft) tall, weighs about 1400 tonnes and is more than 3000 years old.

The park also contains Mount Whitney (4418 m, 14 495 ft), the highest peak in the country outside Alaska. The national park was founded in 1890.

Map United States – Cc

sérac Ice pinnacle between intersecting crevasses in an ice fall on a glacier.

Seraing *Belgium* Industrial town situated 5 km (3 miles) south-west of the city of Liège. It is the centre of the country's steel and heavy engineering industry, and headquarters of the steel-making and shipbuilding firm of Cockerill, whose English founder, John Cockerill, built a foundry and machine factory at Seraing in 1817.

Population 68 000

Map Belgium – Ba

Seram (Ceram) *Indonesia* Mountainous island of 17 148 km² (6619 sq miles) in the Maluku (Molucca) group between Sulawesi and New Guinea. Sago palms are grown on the island, and timber is exported.

Map Indonesia – Gc

Seram Sea (Ceram Sea) *Pacific Ocean* An arm of the PACIFIC OCEAN in Eastern Indonesia which was formerly called Pitt Passage. It lies between Halmahera and western New Guinea to the north and Seram Island to the south.

Map Indonesia – Gc

Serbia (Srbija) See YUGOSLAVIA

Seremban *Malaysia* Capital of the state of Negeri Sembilan, 65 km (40 miles) south-east of the national capital, Kuala Lumpur. It is the centre of the shrinking tin-mining industry of the state.

Population 136 300

Map Malaysia – Bb

Serengeti *Tanzania* Savannah plain in the northeast of the country, stretching from Lake Eyasi northwards to the border with Kenya. It contains many buffaloes, elephants, giraffes, lions, leopards and thousands of gazelles, wildebeest and zebras. The Serengeti National Park – one of Africa's most spectacular game reserves – occupies 14 500 km² (5600 sq miles) of the plain.

Map Tanzania – Ba

Sergiyev Posad (Zagosrk) *Russia* Town, known as Zagosrk until the collapse of the Soviet Union, and place of pilgrimage 60 km (37 miles) north-east of MOSCOW. It has a fortified 14th-century monastery, Trinity-St-Sergius, and a seminary founded in 1742. The monastery complex contains the 15th-century Trinity Cathedral and the 16-century Cathedral of the Assumption.

The modern town has engineering and chemical industries and produces electrical goods.

Population 115 000

Map Russia – Ec

Seria *Brunei* Major oil-producing town 70 km (44 miles) south-west of BANDAR SERI BEGAWAN. Oil was first discovered here in 1929 and, although the field (which is half onshore and half offshore) was almost exhausted at one time, new reserves have been discovered.

The town is the administrative and service centre for the Sultanate's oil industry.

Population 23 420

Map Brunei – Ab

series Subdivision of a system that represents the rocks formed during an epoch. For example, Carboniferous Limestone is a series within the Carboniferous system, and Chalk is a series within the Cretaceous system.

▼ ELEPHANT COUNTRY The permanent snows of Oldeani (3188 m, 10 459 ft) rise above the torrid grasslands of Tanzania's Serengeti Plain.

Seringapatnam *India* See SHRIRANGAPATTANA

serir Stony desert of the central and eastern Sahara within Egypt and Libya which is covered with sheets of angular gravel. It is similar to the REG of the western Sahara.

Serowe *Botswana* Chief village of the Bamangwato people, 140 km (90 miles) south-west of FRANCISTOWN. It was founded in 1902 in a fertile area around a tree-covered hill, which became the burial ground for the Bamangwato royal family.
Population 30 260
Map Botswana – Cb

serpentine Dark green or brownish rock, often variegated or mottled like the skin of a snake, which is composed mainly of minerals of the serpentine group (hydrated magnesium silicates). The rock is a source of magnesium and used as a decorative stone in building. Asbestos is a fibrous mineral of the group.

Serra da Estrêla *Portugal* Range of mountains about 120 km (75 miles) south-east of the city of OPORTO. It is the highest range in mainland Portugal. Estrêla, the main summit, is 1991 m (6532 ft) high, with a skiing centre nearby at Penhas da Saúde.
Map Portugal – Cb

Serra do Mar *Brazil* Mountain range running along the south-east coast between the cities of Rio de Janeiro and Porto Alegre. Historically a considerable obstacle to the settlement of the interior, the range reaches its highest point at Serra dos Orgãos, a 2200 m (7215 ft) peak overlooking Rio de Janeiro.
Map Brazil – Dd

Sétif *Algeria* Market town 112 km (70 miles) west of CONSTANTINE on the road and rail route to Algiers. Lying 1067 m (3500 ft) above sea level in the Kabylie Mountains, it is the administrative centre for the district and serves as an important agricultural region dealing mainly in grain and cattle. Sétif was founded by the Romans in the 1st century AD, and remains from that period can still be seen. The town's chief industry is flour milling.
Population 195 000
Map Algeria – Ba

Seto Naikai *Japan* See INLAND SEA

Setúbal *Portugal* City and port about 30 km (20 miles) south-east of the capital, Lisbon. Its industries include cement, chemicals, shipbuilding and food processing. The area around Setúbal produces salt, fish, oranges and wine.
Population 78 000
Map Portugal – Bc

Sevastopol' *Ukraine* Black Sea port on the south-west coast of the Crimea. Founded in 1784 close to the antique settlement of CHERSONESOS, it became the main base of the Russian Black Sea fleet in the early 19th century and has large dockyards and arsenals. It is also a spa and has food processing, engineering and clothing factories.

Sevastopol' has twice been besieged: in 1855 during the Crimean War between Russia and the forces of England, France and Turkey (when it stood firm); and in 1941-2 during the Second World War, when the Germans took possession of it. The Russians recaptured the city in 1944.

There are some 400 monuments in the city commemorating its turbulent past. The coastal village of Balaklava – the scene of the disastrous charge of the Light Brigade in 1854 during the Crimean War, when 270 out of a force of 673 British cavalrymen rode to their deaths – is 15 km (9 miles) to the south-east.
Population 361 000
Map Ukraine – Cc

Seven Pagodas *India* See MAHABALIPURAM

Severn *United Kingdom* The country's longest river, rising on the slopes of Plynlimon in central Wales and flowing some 350 km (220 miles) into England and out into its estuary below the city of GLOUCESTER. The funnel shape of the estuary constricts the incoming tide and produces a BORE (high wave). The Severn is rich in salmon.
Map United Kingdom – Dd

Severnaya Zemlya (North Land) *Russia* Group of ice-covered islands in the Arctic Ocean off Cape Chelyuskin, the most northerly point of Asia. The islands, which cover some 37 000 km^2 (14 300 sq miles), contain weather, radar and research stations.
Map Russia – Ka

Seville (Sevilla) *Spain* Port and industrial city near the country's southern tip. It is also capital of the province of Seville and the administrative centre of the autonomous region of Andalusia. Some of Europe's highest temperatures – up to 48°C (118°F) – have been recorded here. It was the birthplace of the Spanish artists Murillo (1617-82) and Velazquez (1599-1660). It once had the monopoly of Spain's fabulously wealthy New World trade, but now exports fruit (particularly oranges) and wine from its fertile province.

The Old City contains the world's largest Gothic building – a cathedral begun in 1402; other buildings include the Alcázar fortress and palace started by the Moors on the bank of the Guadalquivir, and the fascinating old Jewish quarter of Santa Cruz where Murillo lived. No other Spanish city knows how to celebrate traditional festivals with the panache displayed by Seville. Its best known are the *Eria de Abril*, continuing over six days and six nights, and the *Semana Santa* (Holy Week) during which long processions wind through the streets.
Population (city) 683 030; (province) 1 619 700
Map Spain – Bd

Sèvres *France* Suburb of south-western Paris near VERSAILLES, and the site of the renowned state-owned porcelain factory transferred here by Louis XV in 1756. Previously the factory had been in the Parisian suburb of Vincennes.
Population 22 000
Map France – Eb

Sewa *Sierra Leone* Central river flowing some 320 km (200 miles) to the Atlantic. Diamonds are found in its valley gravels.
Map Sierra Leone – Ab

Seychelles See p 607

Sfax *Tunisia* The country's second city and a major port and industrial centre, situated on the northern shore of the Gulf of Gabès. It grew from two ancient Phoenician settlements and became an early trading centre. However, Sfax was of little importance before the 12th century when the Sicilians took it, after which it began to prosper as an olive-exporting centre.

There is a huge old medina (Muslim residential area) and numerous souks (markets) selling metalwork, woodwork and embroidery. The Grand Mosque, with its finely decorated, three-tiered minaret, dates from AD 849. The city's industries include chemicals, leather, food processing, olive oil, soap and fishing, especially for sponges and octopuses. Together with Gabès it is the main port for the export of phosphates.
Population 278 000
Map Tunisia – Ba

▼ FESTIVAL OF REPENTANCE Hooded, bare-foot penitents follow carvings of the Stations of the Cross through the streets of Seville during Holy Week.

Seychelles

'PARADISE ISLANDS' SOCIETY REMAINS DIVIDED BETWEEN THOSE WHO OWN THE LAND AND THOSE WHO WORK IT

The Garden of Eden lay in the Seychelles, according to some islanders. Indeed, it is in many ways an island paradise, with a superb climate and idyllic palm-fringed beaches. Its isolation – 1200 km (750 miles) from the east coast of Africa – has given rise to unique species of plants and animals, including rare birds and the endangered hawksbill and green turtles, which in the past were slaughtered for their tortoise shell.

The two main islands, MAHÉ and PRASLIN, are partly granite and reach heights of 905 m (2969 ft). Granite is also the bedrock of 30 more of the 115 or so islands; the rest are coral, low-lying and largely uninhabited.

About 90 per cent of the people live on Mahé, site of the capital, VICTORIA. The population is descended from French settlers, who came in 1770, and African slaves from MAURITIUS. The islanders' diet is mainly cassava, sweet potatoes, yams, imported rice and fish; coconut is the cash crop. The fishing industry is expanding, but tourism accounts for more than 70 per cent of foreign exchange earnings. The tourist industry boomed when Mahé got its international airport in 1971 and a number of international hotels were built. But at the same time the government has tried to move away from the country's dependence on tourism by developing agriculture, fishing and some industries.

The islands became a republic in 1976 after 162 years of British rule. The first president,

James Mancham, was replaced by France-Albert René after a 1977 coup, and from 1979 to 1991, the Seychelles People's Progressive Front (SPPF) was the sole legal party. A law was passed in 1991 legalising other political parties – and now, ex-president Mancham leads the opposition in parliament.

The 1993 elections brought victory for René and the SPPF. René's social welfare policies and economic reforms seem to be popular with most of the electorate. However, society is still divided between the landowners (mostly white) and the landless (usually of mixed race), employed in agriculture, fishing and tourism.

SEYCHELLES AT A GLANCE

Map Indian Ocean – Bc	
Official name Republic of Seychelles	
Area 455 km² (176 sq miles), including Aldabra	
Population 71 500 **Per km²** 157 **(Per sq mile** 406).	
Capital Victoria (on Mahé)	
Government Republic	
Currency 1 Seychelles rupee = 100 cents	
Languages English, French, Creole (all official)	
Religion Christian (90% Roman Catholic, 8% Anglican)	
Climate Tropical and very humid. Average temperature is 26°C (79°F) throughout the year	
Land use Cultivation (mostly permanent crops) 22%, forest and woodlands 18%, other 60%	
Main primary products Fish, vegetables, fruit, coconuts	
Major industries Tourism, fishing, brewing, cigarette manufacture	
Main exports Fish, copra, cinnamon bark, vanilla	
Annual income per head (US$) 5480	
Population growth (per thous/yr) 8	
Life expectancy (yrs) Male 66 **Female** 73	

's-Gravenhage *Netherlands* See HAGUE, THE

Sha Tin *Hong Kong* City in the NEW TERRITORIES on the south-eastern side of a sea inlet known as Tolo Harbour. One of eight new towns built to relieve congestion in Hong Kong's built-up areas. Sha Tin consists largely of soaring apartment blocks built on land reclaimed from the sea.

The city also contains the campus of the Chinese University of Hong Kong, and is well equipped with recreational facilities, including the racecourse of the Royal Hong Kong Jockey Club and the Jubilee Sports Centre.
Population 300 000
Map Hong Kong – Ca

Shaanxi (Shensi) *China* Province in the north covering 190 000 km² (73 000 sq miles) of the central LOESSLANDS OF CHINA. The great Huang He River forms the eastern boundary of Shaanxi, and the southern boundary follows the crests of the Qin Ling. Shaanxi contains the valley of the Wei River, a tributary of the Huang He. This was the heartland of Chinese civilisation. The dusty loess plateau is prone to drought but produces wheat and millet. Farmers used to live in caves dug in the loess, with chimney holes rising into

the fields above. The area is rich in coal reserves, much of them undeveloped, and has valuable deposits of molybdenum and mercury. The capital is XI'AN.
Population 34 050 000
Map China – Fd

Shaba *Zaire* Administrative region, formerly called Katanga, covering some 496 965 km² (191 828 sq miles) in the south-east. It lies on a plateau between 900 and 1900 m (2953 and 5900 ft) high. The region has great mineral wealth, including copper, cobalt, manganese, platinum, silver, uranium and zinc. It also includes two national parks: Kundelungu and Upemba. The plateau, which is mostly wooded savannah and mountainous in the east, is drained by headstreams of the Zaire River. Cattle are reared in the south and west, and crops grown on plantations include citrus fruits. The capital of the region is Lubumbashi.

When Zaire (then the Congo-Kinshasa) became independent in 1960, Katanga seceded and remained separate until 1962, when its rebellion was quelled with the aid of United Nations troops. In late 1993 the region made an unsuccessful attempt to secede again. The

mining industry, which was controlled by European companies before independence, is now substantially state-owned.
Population 4 452 620
Map Zaire - Bb

Shabeelle (Shebelle; Scebeli) *Ethiopia/Somalia* River, more than 1700 km (1055 miles) long, which rises in central Ethiopia and flows roughly southwards through the Igaden desert region. It comes within 30 km (18 miles) of the Indian Ocean near the capital, MOGADISHU, then turns to join the Jubba River 350 km (220 miles) farther south-west. The Shabeelle and the Jubba are the only two rivers in Somalia that never dry up, and the water of both is used for irrigation.
Map Somalia – Ab

shake-hole See SINKHOLE

shale Fine-grained sedimentary rock composed mainly of clay minerals, with well-defined narrow strata which readily split apart.

Shan *Myanmar (Burma)* State covering some 155 801 km² (60 155 sq miles) in the eastern part of the country bordering China, Laos and Thailand. It is named after the Shan people, the largest ethnic minority in Myanmar, who are closely related to the Thais and who in the mid-1990s were engaged in a bitter fight for independence. The state capital is TAUNGGYI.

Much of Shan consists of hilly, forested plateaus. Rice, potatoes and onions are the main food crops of the Shans, but hill dwellers in the east of the state, including the Wa and Palaung peoples (formerly headhunters) grow opium in regions largely beyond the control of the government. The crop is sold to opium warlords operating in the area known as the GOLDEN TRIANGLE.
Population 3 719 000
Map Myanmar – Cb

Shandong (Shantung) *China* Province in the north, covering 153 300 km² (59 174 sq miles) on the Yellow Sea coast. It takes in much of the North China Plain and is one of the country's most populous provinces, with about 550 people per km² (1375 per sq mile). The Huang He River reaches the sea in the north of the province.

Shandong's climate is more humid and temperate than that of the northern interior of China, but farming is often hampered by the build-up of salt in the soil. The coastal belt and lower-lying areas with fertile soil yield two, and sometimes three, harvests a year.

Shandong has long been famous for its fine quality silk cloth – known in English as 'shantung'. The province also contains the Shengli oil field. The Tai Shan, one of China's holiest mountains, 1524 m (5000 ft) high, is one of the province's major attractions. The capital is JINAN.
Population 84 393 000
Map China – Hd

Shanghai *China* China's most populous city and the site of its greatest concentration of manufacturing industries. It is also the nation's chief port. It stands in a 6185 km² (2400 sq mile) municipality on the densely settled and fertile delta of the Chang Jiang River. Shanghai grew rapidly after the arrival of the Western imperial powers in China in the 19th century. Foreign merchants and diplomats enjoyed special privileges and also rights of residence in the city which became notorious for the stark contrast it presented between European prosperity and Chinese poverty. By the 1930s, Shanghai had developed into one of the great seaports of Asia, with opulent banks and office buildings flanking the harbour on the Huangpu River. Today, overseas trade is returning rapidly. During the first half of the 1990s, foreign companies had already invested many times more money than they had been allowed to invest during the 1980s.

The city's industries, which include shipbuilding, iron and steel, chemicals and textiles, are increasingly being moved to a rapidly developing suburbia, while the city becomes a centre of regional headquarters for international companies.
Population (city) 7 800 000; (metropolitan area) 13 450 000
Map China – Ie

Shannon (An Tsionainn) *Ireland* Longest and largest river in the British Isles, with a length of 386 km (240 miles) including its 90 km (56 mile) estuary. The Shannon rises in north-west County Cavan near the Northern Ireland border, and runs south and west through Lough Ree and Lough Derg. It enters its estuary at LIMERICK and empties into the Atlantic. The Shannon is a salmon-fishing river, and at Ardnacrusha in County Clare it powers a hydroelectric scheme which was built in the 1920s.
Map Ireland – Bb

Shanti Niketan (Santiniketan) *India* Rural retreat and university in West Bengal, 130 km (80 miles) north-west of CALCUTTA. It was founded at the turn of the century by Sir Rabindranath Tagore (1861-1941), India's best known poet and leading educator, and a Nobel Prize winner, to help regenerate India's villages and to widen international understanding with the aid of a multicultural university.
Map India – Dc

Shantou *China* City and port on the coast of GUANGDONG Province. It is one of several areas designated as 'special economic zones' where overseas investment is encouraged.
Population 860 000
Map China – Hf

Shanxi (Shansi) *China* Province in the north of the country, reaching into the LOESSLANDS OF CHINA. Although it lies in a fertile area covering about 150 000 km² (58 000 sq miles), its soil erodes easily and needs careful management. Moreover, rainfall is light and variable, and Shanxi is vulnerable to drought. Its chief crops are maize, wheat and cotton. The province is also China's leading coal producer. Though coal has been mined here for centuries, enormous reserves remain to be exploited. The capital is TAIYUAN.
Population 29 790 000
Map China – Gd

Shaoshan *China* Village in Hunan Province, 104 km (65 miles) south-west of CHANGSHA. It is the birthplace of Mao Zedong (1893-1976), leader of communist China from 1949 to 1976. Mao's family house and the nearby museum attract thousands of Chinese visitors every year.
Map China – Ge

Shari *Equatorial Africa* See CHARI

Sharjah *United Arab Emirates* Third largest of the seven emirates, covering 2600 km² (1003 sq miles). Its main town, also called Sharjah, lies on The Gulf, 13 km (8 miles) north-east of DUBAI. The emirate has a small oil field, which lies in an area which Iran also claims. Revenue is shared between the two countries.

However, Sharjah concentrates mainly on commerce, encouraging new companies to establish themselves in its territory. Manufactures include paint, plastic pipes and cement. There is a container service at Port Khaled. The town has impressive new buildings, including a post office and the Sharjah New Souk (bazaar), as well as new port facilities. An old, covered souk, and a few traditional Arab houses can also be seen.
Population (emirate) 314 000; (town) 126 000
Map United Arab Emirates – Ba

Shark Bay *Australia* Inlet, 240 km (150 miles) long and 100 km (62 miles) wide, in Western Australia, 670 km (416 miles) north of PERTH. Dirk Hartog Island, at the mouth of the bay, was visited in 1616 by the Dutch navigator Dirk Hartog, who left behind an engraved pewter plaque to mark his visit. The plaque was rediscovered in 1697 and is now in Amsterdam.
Map Australia – Ad

Shatt al Arab *Iran/Iraq* Waterway formed by the union of the Euphrates and Tigris rivers at the Iraqi city of Al Qurnah, 400 km (nearly 250 miles) south-east of BAGHDAD. From Al Qurnah, it flows south-east for some 170 km (105 miles) to The Gulf. The lower course forms the disputed Iran-Iraq border. Along the waterway are the largest continuous palm plantations on earth.
Map Iran – Aa; Iraq – Cc-Dc

Shayib el Banat *Egypt* Highest mountain, at 2187 m (7175 ft), in the Eastern Highland which runs the length of Egypt's Red Sea coast and stretches from the Gulf of Suez to the Sudanese border.
Map Egypt – Cc

shear The movement of one body of rock relative to another along a fault or thrust plane, parallel to their plane of contact. See also THE RESTLESS EARTH, p 231.

Sheffield *United Kingdom* City in the English county of South Yorkshire, 55 km (35 miles) east of MANCHESTER. Its name has for centuries been synonymous with cutlery, steel and Sheffield plate, a fusion of silver and copper. Its industries are fuelled by the coalfields surrounding the city. Sheffield's guild of craftsmen, known as the Cutlers' Company, was founded in the 17th century under James I. The city has a cathedral dating from the 15th century and a university founded in 1905.

Today the city of Sheffield is a tourist and conference centre, with extensive leisure and sporting facilities.
Population 521 000
Map United Kingdom – Ed

Shenandoah *USA* River in northern Virginia, which flows some 90 km (55 miles) north-east to the Potomac at Harpers Ferry in West Virginia. Its valley, between the Allegheny and Blue Ridge mountains, includes the Shenandoah National

Park, and was the scene of several American Civil War engagements. In 1859, the anti-slavery rebel John Brown, with 21 men, captured the Federal arsenal at Harpers Ferry. He was hanged for treason.
Map United States – Kc

Shenyang *China* Capital of LIAONING Province. It is the largest city in the north-east and stands at the centre of the railway network of DONGBEI (Manchuria). It has enormous coal and iron ore reserves and produces machine tools, metals, aircraft and textiles.
Population 4 500 000
Map China – Ic

Shenzhen *China* City in GUANGDONG Province just across the border from Hong Kong. It is the largest of the country's 'special economic zones' set up to attract investment from overseas.
Population 335 000
Map China – Hf

's-Hertogenbosch (Den Bosch) *Netherlands* Capital of the province of Noord-Brabant, situated some 80 km (50 miles) to the south-east of AMSTERDAM. Its name (meaning 'duke's wood') probably derives from Henry I, Duke of Brabant, who in the 12th century had a hunting lodge nearby. The town's cathedral of St John, built from the 13th to the 16th century, is the country's largest Roman Catholic church.
Population (city) 94 340; (greater city) 196 200
Map Netherlands – Bb

Shetland Islands *United Kingdom* Group of more than 100 islands about 160 km (100 miles) north-east of the Scottish mainland. They were settled by Norse invaders in the 9th century and became Scottish only in 1469. The festival of Up-Helly-Aa each January, when a longship is ceremonially burnt, recalls the Viking connection. Before the discovery of North Sea oil, fishing and sheep-rearing were the main occupations of the islanders; the oil industry is now one of the major employers, notably at the Sullom Voe terminal.

The industry has, however, also caused some pollution to the region. In January 1993 an oil tanker, the *Braer*, ran aground on the southern tip of the Shetland Islands, damaging the environment and the local salmon industry.

The capital of the Shetlands, Lerwick, is a ferry port linking the islands to Aberdeen, the Orkneys and Norway.
Population 28 000
Map United Kingdom – Gg

Shewa *Ethiopia* Region covering 85 500 km² (33 003 sq miles) around the capital, ADDIS ABABA. It has formed the political and economic heartland of the country for more than 100 years. Its main products include barley, maize, sorghum, skins and hides.

There is no large-scale industry, though there are diverse small industries in Addis Ababa and other main towns.
Population 7 000 000
Map Ethiopia – Ab

shield See CRATON

Shijiazhuang *China* Capital of HEBEI Province, 250 km (155 miles) south of China's capital,

Beijing. It is a centre for the manufacture of woollen textiles.
Population 1 320 000
Map China – Hc

Shikoku *Japan* Smallest of the country's four main islands, the others being HONSHU, HOKKAIDO and KYUSHU. Shikoku covers some 18 787 km² (7252 sq miles) and has a rugged, mountainous core flanked by small alluvial plains. There is a slow, traditional flavour to life on the island, despite bridge links to the rest of Japan.

The north coast is dry and warm, while the south is exposed to wet summer winds which allow two crops of rice to be grown each year. Tomatoes, cucumbers and melons are the main winter crops. The island is prone to heavy typhoons which damage autumn and winter crops.
Population 4 195 000
Map Japan - Bd

Shillong *India* Capital of MEGHALAYA State, 1251 km (777 miles) north-east of Calcutta. It lies nearly 1500 m (4930 ft) up in the Khasi Hills, and with its golf course and racetrack, was a popular hill resort for the British colonial families of Bengal.
Population 174 700
Map India – Ec

Shimizu *Japan* See SHIZUOKA

Shimla *India* Hill resort and capital of HIMACHAL PRADESH State, about 280 km (174 miles) north of DELHI. It stands on wooded Himalayan ridges some 2000-2430 m (6600-8000 ft) above sea level, where temperatures rarely exceed 25°C (80°F). The British began the resort in about 1820 and made it the summer capital of British India. The town has a bizarre profusion of Victorian styles of architecture.
Population 614 890
Map India – Bb

Shimonoseki *Japan* Deep-sea fishing port and city on the south-west tip of HONSHU Island. An April festival here commemorates the sea battle of Dannoura in 1185 in which the fleet of the Taira clan was destroyed and the seven-year-old Emperor Antoku drowned.
Population 263 000
Map Japan – Bd

Shipka (Sipka) Pass *Bulgaria* Mountain pass, 1326 m (4350 ft) high in the Balkan Mountains between the towns of GABROVO in the north and KAZANLUK in the south. It was the scene of major battles in 1877 during the Russo-Turkish war for Bulgarian independence.
Map Bulgaria – Bb

Shiraz *Iran* Provincial capital of Fars, about 700 km (435 miles) south-east of TEHRAN. Shiraz, which may have been founded before the 6th century BC, was by the 12th century AD the literary capital of Persia. It was the birthplace of the poets Saadi (1184-1292) and Hafiz (1300-88), both of whom were also buried there. From 1750 to 1786 it was the capital of Persia.

Celebrated for many centuries for its wine, silver filigree and silk rugs, Shiraz today is a regional metropolis with an airport and a university. Its products include textiles, carpets, sugar, fertilisers and cement, and it is the commercial

centre of a farming region producing cereals, sugar beet and grapes. Shiraz's industries were targeted by the Iraqi airforce during the Iran-Iraq War (1980-88).
Population 965 000
Map Iran – Bb

Shire *Malawi/Mozambique* River flowing 400 km (nearly 250 miles) south from Lake Malawi to join up with the ZAMBEZI River in Mozambique. The upper course, through the shallow Lake Malombe, has a gentle gradient. By contrast, the middle course below Matope drops 383 m (1256 ft) in 80 km (50 miles). Here the river descends through scenic gorges in a series of cataracts, called the Murchison Cataracts by the missionary and explorer David Livingstone after a president of the Royal Geographical Society.

There are hydroelectric power stations at the Nkula Falls and, 8 km (5 miles) down-stream, at the Tedzana Falls. The cataracts end north of Chikwawa and the river then flows across a broad swampy plain before entering the Zambezi.
Map Malawi – Ac

Shire Highlands *Malawi* Fertile plateau in Southern Province, rising to 1700 m (5600 ft). It stretches about 80 km (50 miles) from south of BLANTYRE to ZOMBA, and was the chief area of European settlement in the colonial period. Its main products are tea and tobacco.
Map Malawi – Bc

Shizuoka *Japan* City on central HONSHU Island, 145 km (90 miles) south-west of Tokyo. It has a university, pharmaceutical college and several renowned schools. Shizuoka is also the cultural and administrative centre of Suruga Bay. With the nearby city of Shimizu on the bay, it forms an extensive industrial zone processing tea and producing canned fruit, petrochemicals and ships.
Population 472 000
Map Japan – Cc

Shkodër (Scutari) *Albania* Quiet, largely undeveloped riverside town in a beautiful setting near the lily-covered Lake SHKODËR and backed by an arc of the ALBANIAN ALPS. It is overlooked by the ruins of a 14th-century Venetian citadel – a legacy of the days when the port was regularly visited by traders from Italy. Now its river is silted up and shallow and there is no railway to replace it. The inadequate transport facilities and the town's closeness to Yugoslavia (Serbia and Montenegro) have deterred greater development, other than a few light industries, since Albania's closed frontier policy has deprived Shkodër of its natural hinterland across the border.
Population 83 700
Map Albania – Ba

Shkodër, Lake (Lake Scutari; Lake Skadar; Skadarsko Jezero; Liqen i Shkodrës) *Albania/Yugoslavia (Serbia and Montenegro)* The largest Balkan lake, covering 391 km² (143 sq miles) astride the Albanian-Montenegrin border. Bare limestone mountains descend abruptly into the lake, where summer mirages are common. Water chestnuts, a local delicacy, and fish, including carp, are plentiful. Virpazar in Yugoslavia and Shkodër in Albania are the main towns beside the lake. Outside Crnojevica (known as 'the Venice of Montenegro') near the

lake's northern shore are the remains of Obod monastery, where the first books in the Cyrillic alphabet were printed in 1493.
Map Yugoslavia – Dc

shoal Ridge of sand, mud or pebbles just below the surface of the sea or a river, creating a hazard to navigation.

Shonaland *Zimbabwe* See MASHONALAND

shore Term used to refer to the area between the lowest low tide mark and the highest point reached by storm waves. Loosely applied, it refers to land immediately next to a sea or any large body of water.

shott (chott) Shallow temporary salt lake in a hot desert, or the hollow in which such a lake lies. The term is used especially in North Africa in the areas bordering the Atlas Mountains.

shoulder Gentle slope above a steep-sided valley in a mountainous area.

Shreveport *USA* Industrial port at the navigable limit of the Red River in north-west LOUISIANA. It is an oil, gas, timber and cotton processing and shipping centre. Manufactures include metal, glass products and telephone equipment.
Population (city) 198 500; (metropolitan area) 376 300
Map United States – Hd

Shrewsbury *United Kingdom* County town of Shropshire, England. It stands on a wide loop of the River Severn 65 km (40 miles) north-west of BIRMINGHAM, and was an English stronghold through centuries of conflict with the Welsh.

It has many fine half-timbered buildings, one of the handful of round churches in Britain, and a public school whose pupils included the naturalist Charles Darwin (1809-82), originator of the theory of evolution by natural selection.
Population 59 170
Map United Kingdom – Dd

Shrirangapattana (Seringapatnam) *India* Fortress town in the south, standing on the Kaveri River, 485 km (301 miles) south-west of the east coast city of MADRAS. It was the capital of Tipu ('the Tiger'), Sultan of Mysore.

A successful siege by the British resulted in Tipu's death in 1799, and the establishment of British power in southern India. The city's breached walls survive.
Map India – Be

Shropshire *United Kingdom* County of west-central England, bordering Wales. It lies just west of the industrial West Midlands and stretches from the fringes of the industrial north to the placid countryside around the historic town of Ludlow in the south – a total area of 3490 km² (1347 sq miles). There are many hills, such as the Wrekin, which at 407 m (1335 ft) is topped by a 2000-year-old fort and from which, it is said, 17 counties can be seen on a clear day. The county's fertile Severn Valley provides rich agricultural land.

Shropshire was a cradle of Britain's Industrial Revolution. The world's first iron bridge was built at Iron-Bridge in 1778, from girders produced in a foundry at Coalbrookdale nearby. The new town of Telford has attracted some of Britain's most promising high-technology industries. The county town is SHREWSBURY.

Population 411 000

Map United Kingdom – Dd

Shumen (Šumen; Kolarovgrad) *Bulgaria* City and capital of Shumen Province, 80 km (50 miles) west of VARNA. It trades in grain and wine, and manufactures farm machinery, beer, furniture and enamelware.

A strategically important stronghold, Shumen surrendered to the Russians in June 1878 during the Russo-Turkish war for Bulgarian independence. About 10 km (6 miles) to the east is the ruined ancient city of Madara whose remains date from the 5th century. Carved in the nearby cliffs is the figure known as the Horseman of Madara, which dates from the 8th century.

Population 110 750

Map Bulgaria – Cb

Shush *Iran* See SUSA

Shuwaikh *Kuwait* The country's largest port, lying in the western suburbs of KUWAIT City. About 40 per cent of the goods which pass through it are re-exports – the result of the country's expanding manufacturing sector. There is a special industrial zone, which offers land at a nominal price to attract factory development. Power and water plants have already been established.

Shuwaikh is one of the campuses of the University of Kuwait, which has over 12 000 students. It takes some 60 per cent of its students from around The Gulf and other Arab states.

Population 9100

Map Kuwait – Ab

Shwebo *Myanmar (Burma)* Town situated 75 km (45 miles) north-west of MANDALAY. Shwebo was the national capital during the reign of the warrior king, Alaungpaya (1752-63), who is buried here. Today the town is a market for cotton and beans grown in irrigated fields along the Mu River.

Population 167 000

Map Myanmar – Bb

sial General name given to the rocks of the earth's upper continental CRUST, which are rich in silica and alumina. The word is formed from the first two letters of these two minerals.

Sialkot *Pakistan* Garrison town in the PUNJAB, 100 km (62 miles) north of LAHORE. It was the scene of a bloody insurrection during the 1857 Indian Mutiny against British rule. The town now manufactures sports goods, particularly hockey sticks and squash and tennis racquets.

Population 487 000

Map Pakistan – Da

Siam See THAILAND

Šibenik (Sebenico) *Croatia* Seaport and fishing and naval base at the mouth of the Krk River in Dalmatia, 50 km (about 30 miles) north-west of SPLIT. The heavily fortified old town, containing one of Dalmatia's finest cathedrals (built 1431-1536) and local coves and beaches, is popular with tourists.

Divers can explore the underwater world of corals and sponges around the nearby islands of Zlarin and Krapanj.

Population 29 600

Map Croatia – Bc

Siberia (Sibir) *Russia* Vast North Asian region lying between the Ural Mountains and the Pacific, covering 13 800 000 km² (5 300 000 sq miles), mostly in the Russian Federation. It is rich in timber, metals and wildlife, and the steppes of western Siberia are fertile agricultural land if irrigated. But the extremely harsh climate and the inhospitable land in the north and east – where tundra, swamps and swampy forests are frozen, snow-covered or flooded for much of the year – have greatly restricted settlement. Even so, the Siberian region now has a population of more than 30 million.

Originally peopled largely by Tatars, most of Siberia was conquered by Russian Cossacks between 1582 and 1633. It was colonised by settlers in search of furs, timber, gold and silver; later, political prisoners and criminals were exiled here to work in the camps and mines. During the 19th century about 1 million exiles, many of them from the Baltic region, were sent to Siberia, and many more followed after the Russian Revolution – especially in the 1930s and 1940s, during the iron rule of Joseph Stalin (1879-1953).

The TRANS-SIBERIAN RAILWAY, built between 1891 and 1905, linked Siberia with Russia proper – and opened up the region. From the mid-1920s, development spread eastwards, with the growth of large industrial complexes using local resources such as coal, zinc, lead, copper and iron ore, and hydroelectric power. This led to the growth of some of the region's chief cities, including IRKUTSK, KRASNOYARSK, NOVOSIBIRSK and OMSK. Since 1945, discoveries of huge gas and oil reserves have boosted the region's industrial development.

Population 32 100 000

Map Russia – Hb

Sibiu *Romania* Major industrial city and tourist centre in southern TRANSYLVANIA, 210 km (130 miles) north-west of BUCHAREST. Originally a Roman colony, Sibiu was settled by Germans in the 12th century. Some ethnic Germans still live here and much of the city's architecture is Germanic. Its 14th-century walls and fortifications are still standing, and they surround and are surrounded by narrow streets, small squares, and red-roofed houses with dormer windows.

The old town radiates from Republic Square, near which is the 18th-century Brukenthal Museum, which houses a fine collection of Romanian paintings – as well as canvases by the Flemish artists Peter Rubens (1577-1640) and Anthony van Dyck (1599-1641). Sibiu's industrial products include footwear, textiles, building materials, machinery and chemicals.

Population 161 000

Map Romania – Aa

Sichuan (Szechwan) *China* The nation's most populous province, lying in the west of the country. Sichuan covers about 570 000 km² (220 000 sq miles) between XIZANG AUTONOMOUS REGION (Tibet), and the provinces of SHAANXI and Sansu. Most of it belongs to the Red Basin, a vast region surrounded almost entirely by high mountains. West of the basin is the Wolong nature

reserve which is the last refuge of the giant panda. There are many spectacular gorges in the east, created by the Chang Jiang River as it cuts down through the mountains.

The climate in Sichuan is humid and temperate and its soils are unusually fertile. Rice, wheat, millet, sugar cane and citrus fruits are grown here. The region is noted for its spicy cuisine.

Population 109 980 000

Map China – Ee

Sicily (Sicilia) *Italy* The Mediterranean's largest and most populous island, covering 25 708 km² (9926 sq miles).

Its capital and largest city is PALERMO; other major towns and cities include the provincial capitals of AGRIGENTO, CATANIA, ENNA, MESSINA, RAGUSA, SYRACUSE and TRAPANI. A mountain chain crosses the north between Palermo and Messina. Mount ETNA, standing at 3323 m (10 902 ft) and Europe's largest active volcano, towers in the east.

The central and southern parts of the island are a jumble of open hill country, with widespread sulphur outcrops. Agriculture in this inland area is poor and backward, but in the south-east, along the north coast and around the lower slopes of Etna, where Sicily's lemons grow, it is rich and flourishing. Industry is limited to recently established oil refining, petrochemical and vehicle assembly plants. Tourism focuses on the major archaeological sites and on resorts such as TAORMINA and Cefalù.

The island was named after the Ancient Siculi tribe and was colonised in the 7th and 8th centuries BC by the Greeks, who left the remains of several splendid city-states, notably at Agrigento, GELA and Syracuse. The Romans conquered Sicily, exploiting it for its wine, olive oil, wheat and sulphur. Later the island became the centre of an Islamic empire in the central Mediterranean, with Palermo the great Arab capital. Arab and Byzantine influences are found in the characteristic Norman architecture of the 11th and 12th centuries, best seen in the splendid cathedrals of Palermo, Cefalù and Monreale. Catalan Gothic influences were introduced by Spain in the 15th century when the island was an impoverished viceregal colony.

Unification with Italy in 1860 brought few benefits to the island; indeed, the bitter joke about the Italian boot kicking Sicily dates from this time. Sicilian society is highly stratified, and the criminal Mafia organisation, which the Italian authorities are trying to eradicate, has spread from Sicily to many other countries, particularly the United States, to which many Sicilians have emigrated during this century.

Population 5 196 800

Map Italy – Df-Ef

Sidi Barrani *Egypt* Coastal village 400 km (nearly 250 miles) west of ALEXANDRIA. During the Second World War, it was captured by the Italians in September 1940, the British in December 1940 and the Germans in 1941; it was finally recaptured by the British during Erwin Rommel's retreat in November 1942.

Map Egypt – Ab

Sidi Bel Abbés *Algeria* Walled town situated 80 km (50 miles) south of ORAN. It grew up around a French camp built in 1843 and was the

A DAY IN THE LIFE OF A SICILIAN DOCTOR

Maria Ferro, 30, graduate in medicine from the University of Catania, is, unusually for a Sicilian, blonde and blue-eyed – which means that she is a true descendant of the Normans who came to the island around AD 1000.

Jobs for doctors are scarce in Italy. There are simply too many of them – more than twice the number of doctors in Britain or the United States, proportional to population, and incomes are relatively low. Maria does not have a full-time job. But every day she drives her Lancia to the hospital in Catania to practise on a voluntary basis.

Sometimes she earns a little as a stand-in for other doctors. She earns a fee for every private patient she sees and is also paid for the days she works in a health centre when the regular doctor is away. But money is not a major worry for her; she lives with her parents, who own a shoe shop and are comfortable enough financially to support their only child.

Maria has a knack for gaining people's confidence and for providing psychological and emotional comfort for those of her patients – and there are many in Sicily – who go to the doctor as much for emotional help as for the diagnosis and cure of their physical illnesses.

With her male patients, there is sometimes a certain shyness to be overcome, as it is still unusual in this society for Sicilian women to have a profession. Maria's fiancé, Giovanni d'Antona, whom she met at the hospital, has a salaried position as an assistant doctor. They will be married in a few months' time, but it will be a small, quiet Catholic ceremony attended only by members of the family.

Marriage will not mean the end of work for Maria. She wants at least two children, but intends to have a hospital career as well. Like Giovanni, she would like to reach at least the rank of primario, the head of a specialised department. But the prospects for both of them are poor.

If Maria has children, she is entitled to six months' paid maternity leave. If she has to work nights, Giovanni will mind the children. Her parents will always be available to look after them during the day.

Maria snatches a quick lunch at the hospital cafeteria; her main meal of the day will be in the evening, at home with her parents. Her mother mostly cooks lots of pasta, in many different ways, or *schiaccata* which is Maria's favourite. This is a kind of Sicilian pizza with lots of onion and cheese sprinkled on top. Sometimes her mother cooks rice and meat dishes.

Tomorrow Maria is going out to eat with Giovanni and will dress up. Her wardrobe consists of stylish casual clothes for work, and very elegant dresses for evening outings. Afterwards, they may go for an evening stroll, or *passegiata*, along the seafront.

headquarters of the French Foreign Legion until Algerian independence in 1962.

The town is the administrative centre of the district as well as an important road and rail crossroads. Situated in a major agricultural region, it trades in cereals, wine, olives, livestock, tobacco and esparto grass. Local manufactures include flour, cement and furniture.

Population 158 000
Map Algeria – Aa

Sidon *Lebanon* Port and third city of Lebanon after the capital, Beirut, and Tripoli. Situated 40 km (25 miles) south-west of Beirut, it is the Mediterranean terminus of the pipeline from the Saudi Arabian oil fields and has an oil refinery. The city grows and exports citrus fruits – and its residential areas have spread beyond the citrus orchards and banana groves which formerly marked its boundaries. In recent years the city's business and social life has been disrupted by civil disturbances and war.

The site of Sidon – on a promontory facing an island – was inhabited as early as 4000 BC, and possibly earlier; later, it became one of the main cities of the Phoenicians and was noted for the manufacture of glass and purple dyes. From the 16th to the 19th century, it flourished under the Ottoman Turks as the chief port for Damascus. Decline set in with the expulsion of French traders by the Ottoman Turks in 1791 and an earthquake in 1837.

Sidon's places of interest include the 13th-century Crusader castle on the promontory, and the harbour with Roman remains; the Great Mosque, whose walls date from the 13th century; the Castle of St Louis built by Crusaders on the acropolis; and the citadel, to the south of which lies Murex Hill. This is formed by a mound of crushed shells of the snail *Murex trunculus* which provided the purple dye known as Tyrian purple, made and traded by the Phoenicians and renowned throughout the ancient world.

Population 24 700
Map Lebanon – Ab

Sidra, Gulf of *Libya* See SIRTE

Siegen *Germany* Iron and steel town, and former iron-mining town, on the River Sieg, 75 km (45 miles) south-east of COLOGNE. The artist Peter Paul Rubens (1577-1640) was born in the town while his parents were in exile from religious persecution of Protestants in Antwerp. The town has a university and two castles.

Population 109 200
Map Germany – Cc

Siegerland *Germany* Hilly and largely forested holiday region east of BONN, where the River Sieg rises. It was once a major source of the country's iron, when charcoal from the nearby forests was used for smelting. After the discovery of coking coal in the Ruhr in the 1850s, its industries declined, but steel is still made at SIEGEN.

Map Germany – Bc

Siem Reap *Cambodia* Western province north of Tonle Sap Lake. It is the site of the ANGKOR ruins of the Khmer civilisation. The town of the same name, in the province, has an airport, which was built in the 1960s to encourage tourism.

Population (province) 313 000; (town) 20 000
Map Cambodia – Ab

Siena *Italy* Superb medieval Tuscan town built on red earth hills, 50 km (30 miles) south of FLORENCE. It gives its name to burnt sienna – a warm reddish-brown pigment made from the earth and used in painting.

Siena was an independent city-state from the 12th to the 16th century and evolved its own brilliant culture which made it one of Europe's major capitals of Gothic art. The 13th and 14th centuries also saw the development of a distinctively Siennese school of sculptors, of whom the greatest was Jacopo della Quercia (1374-1483).

The city's university was founded in 1247, and its medieval heritage is evident in, among other things, its popular festivals, such as the *Palio delle Contrade*, a thrilling city-centre horserace held twice each summer between teams in costume representing the city's medieval quarters.

Siena's shell-shaped Piazza del Campo is one of the finest medieval squares in Europe. It is completely surrounded by ancient houses and fronted by the Palazzo Pubblico, with its graceful tower, the 102 m (335 ft) Torre del Mangia. The palace is now Siena's town hall and civic museum, and has a fine collection of local art. The city's vast Gothic cathedral dates from the 13th century.

Population (city) 59 970; (province) 249 940
Map Italy – Cc

sierra Term applied to a rugged range of mountains having a serrated or irregular profile, particularly in Spain, Spanish America and the United States. The term is often used in English simply to mean 'mountain' or 'mountains'.

Sierra de Guadarrama *Spain* Forested mountain range, 180 km (110 miles) in length, north-west of MADRID. The mountains' highest point is Pico Peñalara, at 2430 m (7970 ft).

Map Spain – Cb

Sierra Leone See p 612

Sierra Leone Peninsula (Freetown Peninsula) *Sierra Leone* Forested and steeply-sided peninsula that is some 40 km (25 miles) long and 16-20 km (10-12 miles) wide.

It rises to 888 m (2913 ft) at Picket Hill. The national capital, FREETOWN, lies on the peninsula's north side.

Map Sierra Leone – Ab

Sierra Madre *Philippines* Mountain range extending more than 500 km (310 miles) from the northern tip of LUZON Island down its eastern flank to Quezon Province. It reaches 1850 m (6068 ft) at its highest point.

Descending steeply to the Pacific Ocean, the range bears the brunt of the powerful trade winds and destructive typhoons that regularly strike the island, and is among the wettest places in the Philippines.

Map Philippines – Bb

Sierra Madre del Sur *Mexico* Mountain range stretching southwards from the Río Balsas along the Pacific coasts of GUERRERO and OAXACA states to the isthmus of Tehuantepec. It reaches a height of 3850 m (12 630 ft) in central Oaxaca.

Map Mexico – Cc

Sierra Madre Occidental *Mexico* Rugged mountain range stretching for some 1600 km

Sierra Leone

A PLACE CREATED AS A HOME FOR RESCUED SLAVES IN THE 19th CENTURY, SIERRA LEONE IS NOW A LAND OF PEOPLE SHACKLED BY POVERTY

The creation of Sierra Leone's capital, FREETOWN, in 1787 as a refuge for liberated slaves from the United States marked the beginning of the end of the Atlantic slave trade – a major milestone in

Africa's history. From then through the 19th century the British navy intercepted slave ships in the Atlantic, released their West African prisoners and brought them to settle here.

The liberated slaves acquired the cultural ideals of the British and became one of the most highly educated and influential elites in West Africa. But conflict arose with the native tribes of Sierra Leone, predominantly the Mende in the south and the Temne in the north who together make up 65 per cent of the population. They resented the superior attitude of this elite. The power of the Creoles (the descendants of the liberated slaves originating from many African countries) diminished after independence in 1961, in favour of political parties dominated by one or other of the two main indigenous groups.

From 1978 to 1992 Sierra Leone had a one-party government under the Temne-backed All People's Congress. A new, multiparty constitution was approved in a referendum in 1991 and opposition parties were allowed. Before an election could be held, the country was disrupted by its involvement in the conflict in neighbouring Liberia. The army, unable to halt the expanding guerrilla warfare in the south-east of the country bordering on Liberia,

became dissatisfied with the government's handling of the war and took over the government in April 1992. A 27-year-old army captain, Valentine Strasser, became president. A year after assuming power, he indicated that he was considering a gradual return to democracy and promised elections in 1996.

The name Sierra Leone ('Lion Mountain') was conferred on the country in about 1462 by Portuguese sailors – either because of the lion shape of one of the forested mountains of the peninsula, where Freetown stands, or, some say, because they mistook the thunder which accompanies the frequent torrential rains of this coast for the roaring of lions in the mountains. British slave traders followed the Portuguese and there was some British settlement in the 17th and 18th centuries before the arrival of the former slaves. The Freetown area became a British Crown Colony in 1808 and in 1895 a British protectorate was declared over the interior.

The rains, which fall between May and October, swell the dense mangrove swamp forests and flood the grasslands along the coast. The coastal belt is swampland where rice, the main food of Sierra Leoneans, is grown, as well as yams, cassava and millet. Freetown Bay – one of Africa's finest natural harbours – is lined with beautiful, palm-fringed beaches which attract tourists. But life for Sierra Leoneans is not easy. In the capital, and in the smaller inland towns such as KENEMA, unemployment is extremely high.

Industries such as timber production are few and underdeveloped. Many people live under the poverty line, yet Sierra Leoneans have still to pay for education as well as for medical treatment. In towns such as Bo the hospitals are merely huts where patients supply their own blankets and food, buy their own medicines, and even pay for kerosene for the hospital generators. About 80 per cent of adults are illiterate – the country has the second highest illiteracy rate in the world – and disease is rife, especially tropical diseases that are left untreated.

Around the towns of Bo, Kenema and Moyamba, the coastal swamps give way to tropical forest. Farmers make a living from small plantations of coffee, cocoa and oil palm. Under the trees, or on nearby plots, yams, cocoyams, cassava, vegetables and ginger are grown; along with rice, these comprise a monotonous diet for the local people. If life for the urban Sierra Leoneans is not easy, in the villages it is even more precarious. Almost everything depends on the harvest. There is no running water, no electricity, no sanitation, no emergency health care and little motor transport.

Beyond the wide, forested coastal plains the land rises to a mountainous plateau in the east, which is an extension of the GUINEA HIGHLANDS. Here, much of the forest has been haphazardly cleared for agriculture, although the government has tried to stem this by creating large reserves. On the drier plateaus and mountains of the far north, the forest gives way to open savannah, where millet and sorghum are grown for food and groundnuts for cash.

Although most of the people make a living from farming, the country has a wealth of minerals, such as diamond and iron ore deposits in the Sula Mountains, as well as bauxite (aluminium ore), rutile (titanium ore) and some gold. Diamonds are panned from the rivers Moa and Sewa, and mined at Koidu. But mismanagement and poverty have meant that money to develop mining is lacking. Indeed, the output of the mines is declining – the iron ore mines at Sula recently closed down.

Added to its problems have been falling prices for export crops, large trade deficits, growing dependence on aid, and the crippling cost of the guerrilla war in the south-east of the country – which by late-1994 had claimed more than 20 000 lives and displaced about 500 000 people who found refuge in neighbouring Guinea. However, observers saw some hope in the stable currency, falling inflation and the renewal in 1994 of International Monetary Fund lines of credit, which had been suspended since 1988.

SIERRA LEONE AT A GLANCE	
Official name Republic of Sierra Leone	
Area 71 740 km² (27 691 sq miles)	
Population 4 700 000 **Per km²** 66 **(Per sq mile** 169)	
Capital Freetown	
Government Military government	
Currency 1 leone = 100 cents	
Languages English (official), Krio, Mende, Temne, and other African languages	
Religions Indigenous beliefs (50%), Muslim (40%), Christian (10%)	
Climate Tropical, with summer rains. Average temperature in Freetown is 27°C (81°F) throughout the year	
Land use Grazing and other 44%, forest and woodlands 29%, cultivation 27%	
Main primary products Rice, cassava, groundnuts, coffee, cocoa, timber, palm nuts, fish; diamonds, gold, bauxite, rutile	
Major industries Agriculture, mining, fishing, forestry	
Main exports Diamonds, bauxite, rutile, coffee, cocoa, palm kernels	
Annual income per head (US$) 170	
Population growth (per thous/yr) 25	
Life expectancy (yrs) Male 43 **Female** 49	

(1000 miles) along western Mexico, parallel with the Pacific coast.

It has several peaks which rise above 3000 m (10 000 ft), and canyons up to 2000 m (6560 ft) deep, and is the source of several rivers, including the LERMA and Yaqui.
Map Mexico – Bb

Sierra Madre Oriental *Mexico* Mountain range in eastern Mexico, about 1130 km (700 miles)

long, running roughly parallel with the Gulf of Mexico. The highest point is the CITLALTÉPETL volcano, which at 5700 m (18 700 ft), is also the highest point in Mexico. The headwaters of the Papaloapán and Pánuco rivers rise in the eastern Sierras.
Map Mexico – Bb-Cb

Sierra Maestra *Cuba* Mountain range in the extreme south of the island, rising to 2005 m

(6578 ft) at Pico Turquino. It is from here that Fidel Castro, who had been exiled in 1955, began the communist uprising in 1956.
Map Cuba – Bb

Sierra Nevada *Spain* Andalucian mountain range, some 95 km (60 miles) long just south of GRANADA.

It is named after the snow which covers its peaks from autumn to late spring. Mulhacén

tops the range and is mainland Spain's highest peak at 3478 m (11 411 ft). (The highest peak in the entire country is Pico de TEIDE in the Canary Islands.)
Map Spain – Dd

Sierra Nevada *USA* Mountain range that runs through eastern CALIFORNIA and forming most of the western border with the state of Nevada. It rises to 4418 m (14 495 ft) at Mount WHITNEY and is the largest mountain range west of the Rockies.
    The Sierra Nevada Mountains are almost always covered with snow and are a popular all-year-round vacation region.
Map United States – Bc

Sierra Nevada de Santa Marta *Colombia* The highest range of mountains in the country, rearing out of the Caribbean coastal plain north of the main Andes range. It reaches 5800 m (19 029 ft) in the twin peaks of Cristóbal Colón (Christopher Colombus) and Simón Bolívar.
Map Colombia – Ba

Sighetu Marmației *Romania* Border town in the north-west of the country, overlooking the River Tisza and Russia on the river's far bank. The area to the south of the town is a centre of traditional peasant culture, with unspoilt villages, beautiful wooden churches, people in national costumes, and wood carving and embroidery workshops.
Population 42 000
Map Romania – Aa

Sighișoara *Romania* Town in the heart of the country, some 230 km (145 miles) north-west of BUCHAREST. It was the birthplace, in about 1430, of the bloodthirsty Prince Vlad Tepes, upon whom the fictional Count Dracula is thought to have been based.
    Founded by German settlers in the late 13th century, Sighișoara was built around a hilltop castle. The citadel is the best preserved of its kind in Transylvania. There are several medieval towers along the city wall, the most notable of which is the 14th-century clock tower. A 13th to 14th-century church dominates the old town.
Population 33 100
Map Romania – Aa

Sigiriya *Sri Lanka* Spectacular palace fortress on a 180 m (600 ft) high granite block in central Sri Lanka, situated 150 km (93 miles) north-east of COLOMBO.
    It was built during the reign of King Kasyapa (477-495), and its 2 hectare (5 acre) summit is reached by a staircase that passes between the paws and through the throat of a colossal carved lion. Fifth-century frescoes depicting 21 voluptuous half-naked women survive in a grotto in the side of the rock.
Map Sri Lanka – Bb

Sikandarabad *India* See SECUNDERABAD

Sikasso *Mali* Town 280 km (174 miles) south-east of the capital, BAMAKO, near the borders with Ivory Coast and Burkina.
    Sikasso stands in the part of the country that has escaped the ravages of drought and is an important centre for cotton. It is also the clearing house for trade – both legal and illegal

– with the neighbouring countries of Burkina and Ivory Coast.
Population 60 000
Map Mali – Bb

Sikkim *India* State covering nearly 7100 km² (2740 sq miles) astride one of the main Tibetan-Indian trade routes, and bounded by China, Nepal and Bhutan. The valleys of the Tista River – a tributary of the Brahmaputra – and its tributaries which originate in the Himalayas, occupy most of the state.
    The land varies in altitude from 244 m (800 ft), at its lowest point, to some 8585 m (28 165 ft), and in landscape from tropical forest in the south to glaciers in the north. The people of Sikkim call it *Denjong* (translating as 'the Land of Rice'). Most of the inhabitants are emigrants from Nepal.
    In 1890 the British made the Kingdom of Sikkim a protectorate, which passed to India at independence in 1947. The monarchy was abolished and Sikkim became India's 22nd state in 1975. GANGTOK is the capital.
Population 403 610
Map India – Db

Siklós *Hungary* See PÉCS

Silesia (Slezsko; Schlesien; Slask) *Poland/ Czech Republic/Germany* Region of some 40 000 km² (15 440 sq miles) in east-central Europe which now lies mostly in Poland, and includes the upper valley of the Oder River.
    During the Middle Ages the region consisted of a number of Polish principalities, which fell into Habsburg hands after the 14th century.

▼ SERRATED MOUNTAIN PEAKS Mount Whitney in the Sierra Nevada of California, United States, provides one of the longest routes in the world for rock-climbers.

Later, the Treaty of Breslau divided Silesia into a Prussian and an Austrian part. After 1945, the Prussian province of Silesia (Schlesien) was ceded to Poland.
    Coal-rich Upper Silesia contains Poland's industrial core – a congested 19th-century agglomeration of 12 cities grouped around KATOWICE, with a total of 2.2 million inhabitants. The area supplies electricity, steel, lead, zinc, heavy metal goods and chemicals.
    Lower Silesia – north-west of Upper Silesia – is a warmer, fertile and partly forested area which produces wheat, sugar beet and oilseeds. Copper is mined and refined around the cities of LEGNICA and Glogow.
Map Poland – Ac

silica Crystalline compound of silicon and oxygen, which are the two most abundant elements in the earth's CRUST and MANTLE.
    Silica takes many forms, occurring as minerals such as quartz, agate and flint, as a common constituent of sedimentary rocks such as sandstone, or in combination with other compounds to produce the rock-forming minerals called SILICATES.

silicates Most abundant group of minerals found in rocks. Silicates are composed of SILICA in complex chemical combinations with oxides of aluminium, potassium, calcium, sodium, magnesium and iron. Examples are amphiboles, feldspar minerals and pyroxenes.

Silicon Valley *USA* See SAN JOSE

Silistra *Bulgaria* River port on the Danube, 112 km (70 miles) north-east of RUSE.
    Silistra port handles grain, and the town manufactures bricks, furniture and cotton goods. Lake Sreburna Nature Reserve is about 20 km (12 miles) to the east of the town.
Population 58 300
Map Bulgaria – Ca

Siljan *Sweden* Lake covering an area of 355 km² (137 sq miles) in southern-central Sweden, about 250 km (155 miles) north-west of STOCKHOLM. Its shores are lined with tourist resorts, including Leksand, Mora and Rättvik. Every year some 10 000 contestants take part in the Wara long-distance cross-country ski tournament held to the north of the lake during winter.
Map Sweden – Bc

Silk Road *China* Ancient route leading from north China through central Asia to Bukhara, Samarkand and ultimately, Europe. The road, on which Chinese silk was transported west as early as Roman times, leaves the city of YUMEN (meaning 'the Jade Gate') in Gansu and skirts the GOBI and TAKLIMAKAN deserts.

It took caravans six to eight years to complete the journey of over 20 000 km (12 400 miles). The Silk Road flourished from the 2nd century BC to about AD 200. After silk worms were introduced to Europe in the 6th century, trade via the Silk Road decreased significantly. The 13th-century Venetian explorer Marco Polo journeyed to China on this well-used trail. Modern roads, partly metalled, also follow the line of the old Silk Road.
Map (Yumen) China – Ec; (Gobi Desert) Mongolia – Db; (Taklimakan Desert) China – Bc

Silkeborg *Denmark* Health resort in the hill and lake district of east-central JUTLAND. In the town museum is the mummified head of a 2200-year-old man, known as Tollund man, which was found in 1950 in a Jutland peat bog.
Population 34 500
Map Denmark – Ba

sill Sheet-like body of intrusive IGNEOUS rock formed by the injection of hot molten rock between the beds or layers of existing rock. Sills can vary in thickness from a few centimetres to hundreds of metres. One of the best known of these is the Great Whin Sill that runs across northern England and southern Scotland.

silt Deposit of fine particles, between 0.002 and 0.02 mm (0.00008 and 0.0007 in) in diameter, laid down by rivers and streams. Silt is coarser than CLAY but finer than SAND. The term also applies to particles of this size in soils, whether waterborne or windborne.

Silurian Third of the six defined periods in the Palaeozoic era of the earth's timescale. See GEOLOGICAL TIMESCALE, p 256

sima General name given to the rocks of the earth's oceanic CRUST, which are rich in silica and magnesia. The word is formed from the first two letters of these two minerals.

Simferopol' *Ukraine* City and capital of the Autonomous Republic of Crimea, set in a fruit-growing area, some 55 km (35 miles) north-east of SEVASTOPOL'.

It was founded in 1784 on the site of a Tatar town that was destroyed in battles between the Russians and the Crimean Tatars. Nearby are the sites of the ancient settlements of Ak-Mechet (which were once the home of Tatars) and Neapolis (occupied in the 3rd century BC by the Scythian people). Today the city's industries produce machinery, tools, textiles, perfume, cigarettes and processed food. Many tourists travel to the Crimean beaches via Simferopol'.
Population 349 000
Map Ukraine – Cb

Simon's Town *South Africa* Picturesque harbour town and principal base of the navy, situated on the False Bay shore of the Cape Peninsula. It lies 36 km (22 miles) south of CAPE TOWN.

Simon's Town was developed by the British Royal Navy after 1806. The town became known as 'the Gibraltar of the South' because of its strategic position. There remain many interesting old buildings, among which are churches and museums for tourists to visit, and there are beaches and angling spots nearby.
Population 6300
Map South Africa – Ac

simoom (simoon) A swirling, intensely hot, suffocating, sand-laden wind, which is the result of intense local heating and convection, blowing across the northern Sahara desert in summer. At times there may be so much sand in the air that visibility is reduced to zero.

Simplon (Simplone) Pass *Switzerland* Alpine pass, 2005 m (6578 ft) high, on the route between Brig, in Switzerland, and Domodossola, Italy. The road, built in 1801-5 by Napoleon, has been less used since the 19.8 km (12.3 mile) long Simplon Tunnel was opened in 1906. The pass, however, remains a spectacular and popular route with its hairpin curves winding into upper Italy.
Map Switzerland Ba

Simpson Desert *Australia* Uninhabited arid region covering an area of about 130 000 km² (50 000 sq miles) in the NORTHERN TERRITORY, SOUTH AUSTRALIA and QUEENSLAND, and known as Arunta Desert before 1939. The explorer Charles Sturt entered the region in 1845, but it was not until 1939 that the desert was crossed by Cecil Thomas Madigan, who explored its endless sea of dunes. Madigan called the region 'the Dead Heart of Australia'; the desert is named after the Australian geographer A. A. Simpson.
Map Australia – Fc

Sinai *Egypt* Peninsula, consisting mostly of desert and mountains and stretching from the Mediterranean Sea between the Suez Canal and the Israeli border to the Red Sea. At its southern end, sharp peaks culminate in Gebel Katherina, at 2637 m (8651 ft) Egypt's highest mountain.

The Sinai is the land bridge between Africa and Asia, and has been used by traders and armies since the time of the pharaohs. Since 1948, when the state of Israel was established, it has been a focus of conflict between Arab nations and Israel. It was occupied by Israel after the 1967 Six Day War, but returned to Egypt in 1982 under a treaty signed by the two countries in 1979. Oil, manganese and iron have been found in the barren El Tih Plateau of the south.
Map Egypt – Cb

Sinaloa *Mexico* Large coastal state in the north-west, crossed in the south by the Tropic of Cancer. Once the territory of the nomadic Chibcha people of Colombia, it is now a region of silver mining and settled farming – of livestock, cereals, sugar cane, rice, tomatoes, cotton and groundnuts. Shrimp fishing and tourism, centred around the resort of MAZATLÁN, are also economically important.
Population 2 210 770
Map Mexico – Bb

Sind *Pakistan* Province situated on the lower reaches of the Indus, covering some 140 914 km² (54 407 sq miles). It is mostly a scorching desert plain, relying on regular irrigation for crops (mainly cotton and rice). Water supplies to the area depend on how much is taken upstream in the Punjab, and how much is stored in dams such as Mangla and Tarbela on the edge of the Himalayas. The increasing salinity of the soil is also a problem.

The province was annexed from Afghanistan by the British in the 19th century. The main cities are KARACHI, the provincial capital and Pakistan's largest city, and HYDERABAD.
Population 28 013 000
Map Pakistan – Cb

Singapore See p 615

Singapore *Singapore* Island located off the southern tip of the Malay Peninsula which forms the main part of the Republic of Singapore. It covers 570 km² (220 sq miles) and is separated from the mainland by the Johor Strait which is less than 1 km (0.6 mile) wide at this point, and is spanned by a causeway.

The city of Singapore lies on the island's southern shore. It is the country's capital, as well as a major international port on the Strait of Singapore, on the south side of the main island. It is home to more than 90 per cent of the people in the country.

Orchard Road lies at the heart of the city's commercial and tourism area, with many international hotels and tax-free shopping complexes.

Kalang is one of the oldest parts of the port area, dating from the 19th century, where the Kalang River has since been widened into a tidal basin, now a marina. The western suburb of Queenstown and People's Park in the centre of the old city are two of the first housing estates built to rehouse people from the overcrowded Chinese city centre.

Ang Mo Kio and JURONG are more recent new towns. Jurong was founded as part of a state-supported industrialisation programme.
Population (city) 2 826 400
Map Singapore – Ab

Singapore, Strait of *Singapore* Important shipping lane, 105 km (65 miles) long and 16 km (10 miles) wide, running between the south coast of Singapore island and the Indonesian Riau Islands. It links the Strait of MALACCA to the South CHINA SEA.
Map Singapore – Ab

Singora *Thailand* See SONGKHLA

Sinj *Croatia* See SPLIT

sinkhole (sink) Natural depression in a limestone KARST land surface which communicates with a subterranean passage into which surface water disappears (a 'swallow-hole') or used to disappear. Sinkholes are formed by solution below the soil, or the collapse of a cavern roof (a 'shake-hole'). They are popularly known as 'potholes'. See HOW CAVES ARE FORMED, p 141

Singapore

TINY SINGAPORE'S WELL-ORGANISED STATE HAS CREATED PROSPERITY AND HARMONY – BUT RULES AND REGULATIONS ABOUND

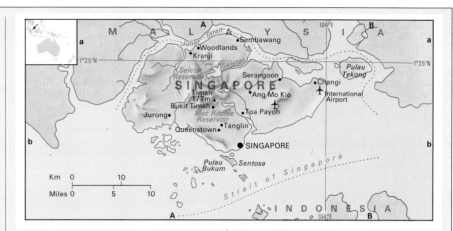

Densely populated Singapore – there are almost 4472 people living in every square kilometre (11 582 per sq mile) – is one of the world's smallest yet most successful countries. It comprises 60 islands at the foot of Malaysia just north of the Equator, but most of its 2.8 million population live on the main island of Singapore.

Once a British naval base, Singapore is now a sleek, efficient international business centre at the crossroads of Asia, and one of the world's busiest ports. On any day there are some 600 ships in port. Its airport too is one of the world's busiest, and also home to an expanding airline which has become world famous. The people of Singapore are among the richest in South-east Asia – income per head is US$15 750. Unemployment, poor housing and illiteracy have been almost wiped out.

The architect of Singapore's 'miracle' was the austere Lee Kuan Yew (premier from 1959 to 1990), whose People's Action Party holds a majority of the seats in parliament. The price is a strictly regulated society; the prize is the cleanest, most efficient, most corruption-free state in South-east Asia.

The government sees itself as the guardian of the people's interests in the mould of the Chinese imperial past; rules and regulations abound. Jay-walking, litter offences and smoking in public places are punished with high fines equivalent to about US$225. Islanders are admonished not to chew gum and not to waste water when cleaning their teeth. But a 1987 Population Policy encourages early marriage, three-child families and the immigration of 'talented' Asians to increase the population to 4 million by the 21st century.

There is a heavy tax imposed on cars to reduce traffic jams and a new electronic system will restrict movement of vehicles in various parts of the city state. In addition, cars carrying fewer than four people are forbidden to enter the centre of Singapore City during peak traffic hours without a special pass.

To enforce these restrictions, the government employs a large secret police and the press is heavily censored, as are films and books. The circulation of foreign publications which are critical of Singapore can be legally restricted. A security act also allows the government to detain 'subversives' without trial.

LEASED FROM THE SULTAN

Singapore (the name means 'Lion City' in Sanskrit) was part of a Sumatran trading empire in the 13th century. It was destroyed by the Javanese in the 14th century and was being reclaimed by the rain forest when Sir Stamford Raffles, of the British East India Company, arrived in 1819. Raffles proceeded to lease it from the Sultan of Johore and seven years later it became a tax-free part of the new colony of the Straits Settlements.

Singapore has few resources of its own – it grows very little of its own food and even has to get water from Malaysia, to which it is linked by a causeway. Its success is due to its position and its industrious, highly disciplined people. It dominates the narrow waterways which lead through South-east Asia to link the Indian and Pacific oceans, and has a natural deep-water anchorage, Keppel Harbour. The British wiped out piracy in the South-east Asian waters and set the stage for economic development. Within its first four years Singapore's population grew from 150 to about 10 000 – largely through Chinese (and some) Malaysian immigration. By 1881 the settlement had a population of 180 000 and in 1867 became a British Crown Colony. When the Suez Canal was opened in 1869, European-East Asian trade expanded; with the start of development in Malaya at the beginning of the 20th century, Singapore became one of the world's leading ports for tin and rubber exports.

As Singapore grew, increasing numbers of Chinese arrived from the south China coast to work in the port and form the link between European enterprises and local Eastern communities. Today 77 per cent of the population is Chinese. About 15 per cent of the people are Malays and 6 per cent Indians or Pakistanis. There are four languages – Malay, Chinese (mainly Mandarin), Tamil and English.

From 1924 to 1938 the British built their biggest Far East naval base at Sembawang, and Singapore became an important marine servicing centre. The base was vaunted as an impregnable fortress. However, its guns were pointing the wrong way. The island was fortified against attack from the sea; in the Second World War the Japanese attacked through Malaya and the British garrison were forced to surrender. After the war Singapore once again came under British rule, until it became self-governing in 1959. It joined briefly in the Federation of Malaysia in 1963 but became an independent republic in 1965.

In the past 30 years Singapore has become a busy metropolis of soaring glass and concrete offices, attracting banks, finance houses and industrial companies. Orchard Road is a vibrant commercial centre of hotels and duty-free shops which attract about 2 million tourists annually. There are good telecommunications and airline services, and the airport at CHANGI is the best in South-east Asia.

New industrial and housing developments are springing up all over the islands, and land is being reclaimed from the sea to add to the area available for development. Most people work in industry on the outer edges of the main island, producing a vast range of goods, including computer equipment, rubber and rubber products and a variety of telecommunications equipment for export to the United States and Asian countries.

With its vast reserves, Singapore decided to go 'regional' in the mid-90s, with large-scale development projects earmarked for China and India. Despite a slump during the world-wide recession, Singapore, with its increasing living standards, improved technology, per capita output and labour discipline, was well on its way to becoming a developed country.

SINGAPORE AT A GLANCE	
Official name Republic of Singapore	
Area 632 km² (244 sq miles)	
Population 2 826 400 **Per km²** 4472	
(Per sq mile 11 582 sq mile)	
Capital Singapore	
Government Parliamentary republic	
Currency 1 Singapore dollar = 100 cents	
Languages Malay, Chinese, Tamil, English	
Religions Buddhist, Muslim, Hindu, Christian, Taoist, Confucianist	
Climate Equatorial; rain falls throughout the year, the main rainy season being from October to January. Average daily temperature is 27°C (81°F) throughout the year	
Land use Cultivation 11%, forest and woodland 5%, other 84%	
Main primary products Rubber, coconuts, fruit, vegetables, livestock, fish	
Major industries Oil refining, oil-drilling equipment, chemicals, textiles, processed foods, printing, shipbuilding and repair, electronic equipment, international trade	
Main exports Petroleum products, machinery, rubber, electrical and electronic equipment, chemicals, vegetable oils	
Annual income per head (US$) 15 750	
Population growth (per thous/yr) 12	
Life expectancy (yrs) Male 73 **Female** 79	

Sinop *Turkey* Black Sea port about 300 km (186 miles) north-east of the capital, ANKARA. Founded in the 8th century BC by colonists from the Greek Ionian city of Miletus, it was the birthplace of the Greek philosopher Diogenes (about 400-325 BC), who taught the virtues of the simple life. Today the port handles mostly tobacco, fruit and timber.

Population 25 000
Map Turkey – Ba

sinter Crust of porous SILICA, deposited around a hot spring or geyser when its water is over-saturated with silica. Sinters occur around many geysers in Iceland and in the Yellowstone Park in the United States.

Sintra *Portugal* Town situated 25 km (15 miles) west of the capital, LISBON, in the Serra da Sintra. It was the favourite summer resort of Portuguese royalty and has two palaces and a ruined Moorish castle. It was celebrated by the English poet Lord Byron (1788-1824) in his poem *Childe Harold*.

Population 20 200
Map Portugal – Bc

Siófok *Hungary* The largest holiday resort on the south shore of Lake BALATON, 100 km (62 miles) south-west of BUDAPEST. It is the lake's most modern tourist centre, with a complex of hotels, sanatoria, holiday homes and camp sites behind a fine sandy beach. Sagvar, 9 km (6 miles) to the south-east, has the remains of the Roman fort of Tricciani, which dates from the 3rd century AD. Several resorts lie to the south-west along the lake, including Zamardi, a picturesque village with traditional cottages and old wine cellars, Balatonendred, noted for lace making, Szántód-Rév, at the narrowest part of the lake and with a ferry link to Tihany, and Fonyod, a fashionable older resort amid picturesque hills.

Population 22 590
Map Hungary – Ab

Sion (Sitten) *Switzerland* Town about 80 km (50 miles) south of BERNE, and the political and religious centre of the canton of Valais. The city is dominated by two hills with medieval strong-holds, built to guard trade routes along the Rhône Valley. The town hall, with an astronomical clock, was built in the 17th century and the cathedral dates from the Middle Ages. Largely destroyed by a fire in 1788, Sion is today the capital of the mainly French-speaking, wine-producing canton.

Population (town) 25 300; (canton) 262 400
Map Switzerland – Aa

Sioux City *USA* Farming and livestock centre on the MISSOURI River in western Iowa, near the South Dakota and Nebraska borders. It has stock-yards, abattoirs and meat-packaging facilities and processes food, animal feed, and fertiliser.

Population (city) 80 500; (metropolitan area) 115 000
Map United States – Gb

Sioux Falls *USA* Market city for livestock and grains, and the meat-packaging centre in south-east South Dakota. It lies near the Minnesota and Iowa borders.

Population (city) 100 800; (metropolitan area) 139 200
Map United States – Gb

Sipka Pass *Bulgaria* See SHIPKA PASS

Siracusa *Italy* See SYRACUSE

sirocco (scirocco) Hot, south or south-easterly wind of North Africa, southern Italy, Sicily and the other Mediterranean islands, originating in the Sahara as a dry, dusty wind, but becoming moist and oppressive as it passes over the Mediterranean. Its dust gives rise to 'blood rains' which occasionally reach northern Europe.

Síros *Greece* Island in the centre of the CYCLADES group in the western Aegean Sea. It is a tourist centre, though not as well developed as other islands, also producing textiles and leather goods.

Population 19 700
Map Greece – Dc

Sirte (Sidra), Gulf of *Libya* Arm of the MEDITERRANEAN SEA, 440 km (273 miles) wide, between the towns of MISRATAH and BENGHAZI. It is bordered mainly by arid desert and oil terminals connected by pipelines to inland oil fields.

Map Libya – Ba

Sitra *Bahrain* Small island just off the north-east coast of Bahrain Island with a causeway linking the two. Sitra has an oil refinery, Bahrain's main power station, and a desalination plant which turns 90 million litres (20 million gallons) of sea-water into freshwater every day.

Population 36 760
Map Bahrain – Ba

Sitten *Switzerland* See SION

Sittwe (Akyab) *Myanmar (Burma)* Port situated in Arakan State. It grew in importance during the British colonial period between 1824 and 1948, and is now a major rice market. It is also the state capital.

Population 143 200
Map Burma – Bb

Sivas *Turkey* Industrial and commercial town 440 km (nearly 275 miles) east of the capital, ANKARA. A former Roman city, it stood around the point where caravan routes from Persia and Baghdad joined on their way to Europe.

It is now a market centre making cement, metal goods, carpets and textiles, as well as dealing in agricultural produce.

In the 14th century, when the town was sacked by the Tatar warlord Tamerlane, 4000 Armenian soldiers were buried alive, Christians were strangled or drowned and children were trampled to death by his Mongol cavalry.

Population 219 900
Map Turkey – Bb

Siwa *Egypt* Oasis town in northern Egypt, situated between the Libyan border and the Qattara Depression, about 470 km (293 miles) to the south-west of ALEXANDRIA.

Siwa lies on an ancient North African caravan route; the uninhabited houses of the old abandoned caravan village stand on a craggy outcrop near modern Siwa. Ancient Siwa was the site of the oracle of the god Amun (later renamed Zeus Amon by the Greeks), and is said to have been visited by the legendary Greek hero, Heracles, and by Alexander the Great.

Plans were announced in the mid-1990's to restore the Oracle Temple, in disrepair for a long time. Dates, olives and grapes are grown in the vicinity of the town.

Population 3600
Map Egypt – Ab

Sjælland *Denmark* See ZEALAND

Skadar, Lake (Skadarsko Jezero) *Albania/Yugoslavia* See SHKODËR, LAKE

Skagen *Denmark* Northernmost town situated in JUTLAND and one of the country's two chief fishing ports, the other being ESBJERG. The town was first recorded in 1299. To its north lies the Skaw, a dangerous sandy spit at the point where the waters of the Skagerrak and the Kattegat meet at the tip of Denmark. Skagen attracted a colony of artists and authors at the end of the 19th century; today it is a popular tourist resort.

Population 11 500
Map Denmark – Ba

Skagerrak *Norway/Denmark* Channel between Norway and Denmark, 240 km (150 miles) long and 130-145 km (80-90 miles) wide.

It links the NORTH SEA to the KATTEGAT, the ÖRESUND and the BALTIC SEA. Navigable throughout the year, the water is deepest, reaching about 809 m (2654 ft), near the Norwegian coast but quite shallow near JUTLAND, Denmark, where the shore is lined with sandbanks. Frequent ferry services operate to connect a number of ports between the two countries.

Map Sweden – Ad

Skåne *Sweden* Province in the south-west of the country, covering 10 914 km² (4213 sq miles). Its chalk rocks form the basis of a thriving cement industry, and its fertile plains are rich in grain and root crops, particularly sugar beet. It contains the ports of MALMÖ, HELSINGBORG and TRELLEBORG, and the university of LUND.

Population 1 084 800
Map Sweden – Be

Skeleton Coast *Namibia* Stretch of arid, often fog-bound coast and shifting sands in the north-west. It is littered with old and new shipwrecks and gets its name from the human and whale bones that are periodically uncovered by the unpredictable, shifting sands together with the wrecks. The 16 000 km² (6178 sq miles) Skeleton Coast National Park protects this eerie region.

Map Namibia – Aa

Skellefteå *Sweden* Port on the Gulf of BOTHNIA, 812 km (505 miles) north-east of Stockholm. It lies at the mouth of the Skellefte River, along which logs are transported from inland forests. Besides timber, Skellefteå exports a variety of mineral products such as copper from the nearby Boliden mining area.

Population 75 500
Map Sweden – Db

skerry Low, rocky island, chiefly in Scandinavia and northern Scotland.

Skíathos *Greece* Most westerly of the Sporades islands in the Aegean Sea, lying only 5 km (3 miles) from mainland Greece. It is a pine-forested resort with some fine beaches, including Koukounariés, west of the harbour. The home of the Greek novelist Alexandros Papadiamandis (1851-1911) is near the port.

Population 4200
Map Greece – Cb

Skien *Norway* Central town in an industrial area 100 km (62 miles) south-west of OSLO, and the capital of Telemark Province. The conurbation includes the smaller towns of Porsgrunn and Brevik, and has a combined population of some 100 000. Skien, which has had an iron industry since the 16th century, became the first town in Norway to install electric lighting in 1885.

Population 48 500
Map Norway – Cd

Skikda (Philippeville) *Algeria* Seaport on the Mediterranean coast, 80 km (50 miles) west of ANNABA. Once an ancient Phoenician trading centre and then a Roman settlement, it is now chiefly important as an oil and gas terminal at the end of pipelines from the Sahara. It also manufactures petrochemicals and plastics, and has a large natural gas liquefaction plant. Other exports include agricultural products, wood, esparto grass, iron ore and marble. The modern town was founded by the French in 1838, and is today the administrative capital of the district.

Population 146 000
Map Algeria – Ba

Skopje (Skoplje, Üssküb) *Macedonia* Capital of Macedonia, standing on the Vardar (AXIOS) River. It grew as a communications hub on the Belgrade-Aegean and Dubrovnik-Constantinople routes, and is an industrial centre producing textiles, tobacco, steel, ceramics and metalwares.

In the summer of 1963 much of Skopje was destroyed by an earthquake, which left seven in ten of its people homeless. International aid helped to rebuild the city, and its modern buildings are now designed to withstand fairly strong earthquakes. Much of the Turkish old town, parts dating from 1392, survived the 1963 earthquake or was rebuilt. It contains some fine mosques, including that of Mustapha Pasha, bazaars, *hans* (caravanserais), streets of craftsmen's shops, and the small 17th-century Orthodox church of the Holy Saviour (Sveti Spas).

Population 563 300
Map Macedonia – Aa

sky cover (cloud cover) Amount of sky obscured by cloud, measured in Britain on a scale of 0 (cloudless) to 8 (entirely covered). Internationally it is measured in tenths of the sky covered.

Skye *United Kingdom* Largest island of the Inner Hebrides off the west coast of Scotland. It covers 1417 km² (547 sq miles) and measures about 80 km (50 miles) from north to south, yet its many inlets mean that no point is more than 10 km (6 miles) from the sea. It is covered by rock, moorland and bog, and is used mainly for sheep farming. The Cuillin Hills, which rise to a height of 993 m (3258 ft), are generally considered to provide the finest climbing in the British Isles. Today, the only major town is the fishing port of Portree which is situated on the east coast. In 1994, construction began on a bridge linking the island to the mainland.

After his defeat by the English at the Battle of Culloden in 1746, Charles Stuart, 'Bonnie Prince Charlie', crossed to Skye disguised as a

---

BUILDINGS THAT SCRAPE THE SKY

The world's first metal-frame skyscraper, the ten-storey Home Insurance Building (52 m, 171 ft) was erected in Chicago in 1884-5. It was then the tallest office block in the world – a title now held by the Sears Tower, also in Chicago. What was the world's tallest structure – the Warszawa Radio Mast (646 m, 2120 ft) near Plock in Poland – was completed in 1974 but collapsed in 1991.

SOME OF THE WORLD'S TALLEST STRUCTURES

Name	Location	Height (m/ft)
CN Tower	Toronto, Canada	553/1814
Petronas Towers (completion 1996)	Kuala Lumpur, Malaysia	450/1476
Sears Tower (without spires)	Chicago, USA	442/1454
World Trade Center	New York, USA	412/1350
Empire State Building (without spire)	New York, USA	381/1250
John Hancock Centre (without spire)	Chicago, USA	343/1127
Eiffel Tower	Paris, France	320/1050
Central Plaza (without spire)	Hong Kong	309/1015

---

woman, an episode that is commemorated by the *Skye Boat Song*.

Population 8000
Map United Kingdom – Bb

skyscrapers See BUILDINGS THAT SCRAPE THE SKY (above).

slash A term applied in the south and south-east United States to wet or swampy ground. More generally it may be applied to the debris of felled trees, or the area in a wood or forest strewn with such debris.

Slask *Poland* See SILESIA

slate Fine-grained METAMORPHIC ROCK formed by the effects of heat and pressure on shale. Slate splits easily into thin, smooth-surfaced layers, long used for roofing.

Slavkov (Austerlitz) *Czech Republic* Small town in southern Moravia, situated some 19 km (12 miles) east of BRNO. It is the site of the Battle of Austerlitz, at which on 2 December 1805, Napoleon and his French army of 70 000 men routed an allied army of 86 000 Russians and Austrians under the Russian general Kutuzov. The allies lost a total of 18 500 men, Napoleon 900. A museum and chapel mark the battlefield.

Population 6320
Map Czech Republic – Cb

Slavonski Brod *Croatia* Major road and rail junction as well as a port on the Sava River, some 170 km (105 miles) south-east of ZAGREB. It has large railway engineering works.

Population 106 400
Map Croatia – Db

sleet Rain which freezes, or snow or hail that melts, as it falls. In the United States, the term is also applied to a thin sheet of ice which forms when sleet or rain freezes on cold surfaces.

Slezsko *Poland* See SILESIA

slickenside Polished and grooved rock surface caused by one rock mass sliding past another under pressure, as along a fault.

Sliema *Malta* The main resort and biggest town on the island, lying just across Marsamxett Harbour from the capital, VALLETTA. It is totally different in character from the 16th-century Baroque harmony of the capital. Hotels, modern buildings and a vigorous nightlife are the main attractions. It stands on the sea and has a yacht marina but no beach.

Population 20 200
Map Malta – Eb

Sligo (Sligeach) *Ireland* Market town located 217 km (135 miles) north-west of DUBLIN. The poet William Butler Yeats (1865-1939), some of whose work describes the surrounding countryside, spent much of his childhood at his grandfather's home in Sligo. He is buried at Drumcliff, 6 km (4 miles) to the north. Just south-east of the town lies Lough Gill, containing Yeats's 'Lake Isle of Inisfree'. Sligo's port now caters mainly for tourists. The town is the county town of Sligo county which covers 1796 km² (693 sq miles).

Population (town) 17 300; (county) 54 760;
Map Ireland – Ba

Sliven *Bulgaria* Town on the main road and rail line, situated 95 km (60 miles) west of BURGAS. Its chief industries are silk, woollen textiles and woodwork. Bulgaria's first factory – a textile mill – was set up here in 1834.

Population 112 200
Map Bulgaria – Cb

Slovakia See p 618

Slovenia See p 619

Slunchev Bryag *Bulgaria* See NESEBUR

Slupsk (Stolp) *Poland* Industrial city 100 km (62 miles) west of the Baltic port of GDAŃSK. Once a fortified market town and a member of the Hanseatic League of mostly north German towns, it now processes food and produces timber and metal goods.

Population 90 600
Map Poland – Ba

Småland *Sweden* Historic province in the south of the country surrounding Lake VÄTTERN, south-

Slovakia

AFTER CENTURIES OF SUBORDINATION TO THEIR NEIGHBOURS, THE SLOVAKS HAVE REALISED THEIR LONG-HELD DREAM OF AN INDEPENDENT STATE

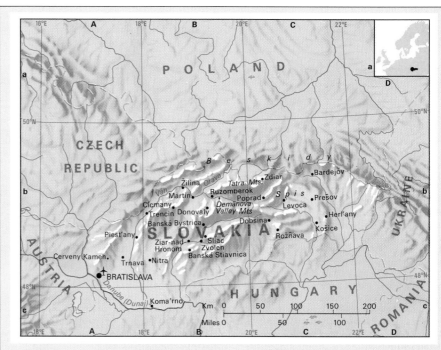

Slovakia has been an independent state since 1 January 1993, when the complete dissolution of the federal state of Czechoslovakia took full effect. The countries of the Czechs and Slovaks had been united from 1918 to 1939, and again from 1945 to 1992 – but it was Slovakia which demanded independence and democracy.

At first, the Slovaks' demands for independence met with strong resistance from the Czechs. But the Slovaks insisted that it was time to dissolve the union. They were unhappy that most important government positions were filled by Czechs, who made up only 1 per cent of Slovakia's population but dominated the region's public life. Furthermore, the economy of the eastern – Slovakian – part of Czechoslovakia was lagging far behind the western region. But the main reason was simply that after centuries of foreign rule, they wanted self-determination and independence.

A CHEQUERED PAST

The Slovaks, a West Slav people, have been living in the mountainous region between the Danube and the West Carpathians since the 7th century, and perhaps even before that. Natural boundaries kept them isolated – and over the ensuing centuries they developed and preserved their own culture, free from outside influence. Then, towards the end of the 9th century, the Hungarians expanded deep into Slovak territory, and between 1038 and 1918 Slovakia was part of the Hungarian Empire.

Poverty forced many Slovaks to emigrate to America in the 19th and early 20th century. It was among a group of exiled Czechs and Slovaks that the idea of a united country took shape, and in 1918 representatives of these two nationalities signed the Treaty of Pittsburgh, which unified the countries. Both regions were, however, to retain a certain amount of autonomy. The reality proved to be somewhat different – and so, through the years, dissatisfaction among the Slovaks grew.

In 1939 the Czech government in Prague, weakened by Germany's claims to Czech territory, could not hold together the union. Bohemia and Moravia, in the west, was made a 'protectorate' of the German Reich. A German-backed Slovak government proclaimed Slovakia independent and its troops fought on the German side in the Second World War. At the end of the war in 1945, after liberation by the Soviet Army, the two countries were again united. A communist government came to power and the Slovaks' nationalistic aspirations grew; this resulted in their decision in 1992, after the collapse of communism, to form a separate state.

Although most of the country is mountainous and little cultivated, Slovakia has some fertile lowland along its borders with Hungary. Here, in the far south-west, lies BRATISLAVA, the capital, on the River Danube. The Danube flows for some 170 km (106 miles) along the country's southern border. Its fertile valley is traversed by the VÁH, Nitra and Hron rivers. The lowlands produce grains, potatoes, sugar beet, hops, fruit and poultry; pigs and cattle are raised for meat and dairy products. Slovakia is largely self-sufficient in food production.

The northern and central parts of Slovakia are covered by mountains, which include the FATRAS, the TATRAS, High and Low Tatras and the Ore Mountains. The Tatras Range rises high above the tree line, reaching over 2600 m (8530 ft). While the foothills are densely populated, the outlying mountainous regions have been preserved as national parks, where bears and lynx still roam. The parks attract large numbers of tourists and provide good skiing facilities in winter. Timber from the mountains feeds a number of paper and furniture-making plants. But industry in the narrow valleys, together with the resultant acid rain, has begun to leave scars on the forests.

The area once had enormous deposits of gold, silver, copper and iron ore, which have been mined since the Middle Ages. At the start of the 19th century, the Ore Mountains were considered to be the second most important European iron ore deposit. But today resources are almost exhausted. KOGICE, which is the second largest city of Slovakia, processes iron ore imported from the Ukraine. Mining of other ores has also slowed down. Some brown coal is still mined and exported..

There was a negative growth rate after independence, due partly to efforts to decentralise and split from the Czech Republic. Although at disbandment the two countries intended to remain within a currency union, this ceased in 1993, at a time when Slovakia experienced an unemployment rate of nearly 14 per cent and an inflation rate between 25 and 30 per cent. During 1994 a new coalition government was formed, led by the pro-Western politician Jozef Moravcik, which reduced state spending, cut the budget deficit, brought inflation under control, halted the rise in unemployment and reversed the decline in production.

The political situation remained uncertain. In an October general election in 1994 the nationalistic, anti-reform party, led by the previous prime minister, Vladimir Meciar, again received popular majority support. Months of manoeuvring to form a new coalition produced inconclusive results.

SLOVAKIA AT A GLANCE

Official name	Slovak Republic
Area	49 035 km² (18 928 sq miles)
Population	5 375 600 **Per km²** 110 (**Per sq mile** 284)
Capital	Bratislava
Government	Republic
Currency	1 Slovak krone = 100 haleru
Languages	Slovak (official), Hungarian, Czech
Religion	Christian (mostly Roman Catholic)
Climate	Temperate; most rain falls during summer. Average temperature in Bratislava ranges from 1.6°C (35°F) in January to 20.1°C (68°F) in July
Land use	Forest 41%, cultivation 31%, grazing 17%, other 11%
Natural resources	brown coal and lignite; small amounts of iron ore, copper and manganese ore; salt; gas
Major industries	brown coal mining, chemicals, metal-working, consumer appliances, plastics, armaments
Main exports	Machinery, chemicals, textiles, minerals, metals, agricultural products, timber
Annual income per head (US$)	2900
Population growth (per thous/yr)	5
Life expectancy (yrs)	**Male** 68 **Female** 77

Slovenia

THE FIRST BREAKAWAY YUGOSLAV STATE IS ALSO THE WEALTHIEST

A little republic situated on the south-eastern boundary of the Alps, Slovenia was the first of the Yugoslavian republics to declare its independence – on 25 June 1991 – when it broke away from the multinational Balkan state.

Although the Yugoslav Federal Army was sent in, fighting was brief. A ceasefire was concluded in July and by January 1992 the country was officially recognised by the

international community as a sovereign state.

Among the six former Yugoslav republics, the land of the Slovenes was always seen as an outsider, not merely due to its geographical situation, but because of its history and population. Today it is the most prosperous of these countries, with a per capita income nearly twice that of its former confederates, and not far below that of its neighbours to the west, Austria and Italy. This fact probably facilitated the split from the socialist federal republic, in contrast with neighbouring Croatia and Bosnia-Herzegovina, where civil war broke out soon after the declaration of independence.

Slovenia is the second smallest of the former Balkan republics. It is just larger than New Jersey, USA, and lies in the far north-west of the Balkan Peninsula, bordering Austria, with which it was united for centuries. More than 90 per cent of Slovenia's population is Slovene. Croats and Serbs are small minorities, each forming only 2 to 3 per cent of the

overall population. Slovenia's long road to independence was therefore made easier by its ethnic uniformity. In the 6th and 7th centuries AD the Slovenes and other Slav peoples came from the north-east to the river valleys and mountain ranges between the Adriatic Sea and the Danube and settled here. Slovenia was one of the founder members of Yugoslavia in 1918. However, in 1920 the Slovene people had to cede large parts of their country to Italy. And in 1941 Slovenia was divided up among Germany and Italy. It was only after the Second World War that it was united with the People's Republic of Yugoslavia.

For more than 1000 years, from the reign of Charlemagne until the fall of the Austro-Hungarian Empire, Slovenia was a part of the German Empire and later of the Danube monarchy. Resistance against political suppression and cultural domination by the Austrians and Hungarians began stirring towards the end of the 18th century. However, it would take another century for the national rebirth to take place. As they had been under the Austro-Hungarians, the Slovenes were dominated by outsiders in the 'Kingdom of Serbs, Croats and Slovenes' – later to be Yugoslavia. This time it was Serb domination they came to resent.

In a referendum in December 1990 the Slovenes voted overwhelmingly in favour of independence. They demanded that Slovenia leave the Yugoslavian People's Republic. Not only were they dissatisfied with Belgrade's communist government, but they also fervently wanted national independence and democracy. Slovenia's new constitution is modelled on Western European democracies.

Geographically, Slovenia belongs to the Alpine countries more than it does to the south-east European states. The north of the country consists of the steep, rugged limestone mountain ranges of the southern Alps. The sheer rock faces of the Karawanke Alps rise from the left bank of the SAVA River; on its right bank are the mountain ridges of the JULIAN ALPS, including Triglav (2864 m, 9396 ft), the highest peak. From here, they slowly veer to the south-east and continue on as karstic mountain ranges of the DINARIC ALPS. The abundant rainfall that occurs in spring and autumn over the Dinaric Alps, drains off into the fissured limestone. Over millennia, huge caves have been washed out underground, and limestone stalactites formed. Some spectacular caves, such as those at POSTOJNA, are among the most popular tourist attractions in Slovenia. The mountainous, largely forested regions, called Gorenjsko in Slovenian, are sparsely populated. Most of the country's 2

million people live in the river valleys, where all the larger cities are situated. The capital city of LJUBLJANA, and the second largest city of MARIBOR are together home to more than one-quarter of the population.

The Dolenjsko, the Slovenian Alpine foothills, constitute the economic hub of the republic. The river valleys, with their lower-lying pastures, produce mainly cereals and root crops. The strong Alpine streams are harnessed for hydroelectric power. More important, though, the country's industrial centres are situated at the intersection of the busy traffic routes from the Adriatic coast to Hungary and from central Europe to south-eastern Europe. Timber is one of the major products, since almost half of the country is forested. But there are other natural resources, including coal and – of major importance – the mercury deposit at Idria, one of the largest in Europe. This diversity – boosted by tourism – makes Slovenia the wealthiest region of former Yugoslavia.

However, the loss of markets in the neighbouring republics has added to its economic problems. At first there was a drop in industrial output due to Yugoslavia's civil war and the economic sanctions imposed against it. By 1994, however, privatisation was progressing and inflation had begun to fall, although Western companies were still reluctant to invest in a country with so small a market.

SLOVENIA AT A GLANCE	
Official name Republic of Slovenia	
Area 20 296 km² (7834 sq miles)	
Population 1 967 660 **Per km²** 97 **(Per sq mile** 251)	
Capital Ljubljana	
Government Republic	
Currency 1 tolar = 100 stottin	
Languages Slovenian, Serbo-Croatian	
Religions Roman Catholic (96%), Muslim (1%)	
Climate Mediterranean on the coast, and continental, with mild to hot summers inland	
Land use Forests 45%, pastures 20%, cultivated 12%, other 23%	
Natural resources lignite, coal, lead, zinc, mercury, uranium, silver	
Major industries ferrous metallurgy, aluminium reduction, lead and zinc smelting, electronics, trucks, wood products, textiles	
Main exports Machinery and transport equipment, manufactures, chemicals, food, raw materials, tobacco	
Annual income per head (US$) 6330	
Population growth (per thous/yr) 2	
Life expectancy (yrs) Male 70 **Female** 78	

west of Stockholm, and covering an area of 29 322 km² (10 163 sq miles). In the 19th century it was one of the country's poorest farming areas, and many of its inhabitants emigrated to North America. Subsequently it became renowned for its innovative industrialists and thriving small manufacturing towns.
Population 698 800
Map Sweden – Bd

Smederevo (Semendria) *Yugoslavia (Serbia and Montenegro)* Serbian city on a bluff over-

looking the Danube, 40 km (25 miles) south-east of BELGRADE. Smederevo was built around the walls of a moated fortress, with 20 towers and an inner citadel and palace – the remains of which still survive. The city was founded in 1429-30 and was Serbia's capital until 1459, when the fortress fell to the Turks, who held the city until 1809. Smederevo is the centre of a vine-growing area producing fine red and rosé wines, and is also a steel and heavy engineering centre.
Population 107 400
Map Yugoslavia – Cb

smog Thick, yellow, polluted fog over a built-up area where smoke particles promote condensation and sulphur dioxide contributes to the acridity. The term 'smog' is coined from the words 'smoke' and 'fog'. Since pollution control laws were introduced in many industrialised countries, there has been a considerable decrease in this atmospheric killer, which has been responsible for many deaths from bronchitis and pneumonia.

Smolensk *Russia* Industrial city on the Dnieper River, some 370 km (230 miles) south-west of

Soufrière *St Vincent* Active volcano, 1234 m (4048 ft) high, in the north of the island. During a violent eruption in 1979, the crater lake disappeared and a lava dome rose in its place. Some 150 000 tonnes of fine ash from the eruption fell on BARBADOS, 160 km (100 miles) away, and ash destroyed one-third of St Vincent's banana crop.
Map Caribbean – Cc

Soul *South Korea* See SEOUL

sound Narrow passage of water, generally wider than a STRAIT, connecting two larger bodies of water. The term is also applied to a channel between the mainland and an island and to a coastal inlet.

Soúnion, Cape *Greece* Cape situated on the south-eastern corner of Attica, 60 km (37 miles) south-east of ATHENS. A white Doric temple to the sea god Poseidon dating from 444 BC stands on its precipitous tip, which offers panoramic views over the Aegean Sea.
Map Greece – Dc

Sour *Lebanon* See TYRE

Sourou *West Africa* See BLACK VOLTA

Sousse *Tunisia* Seaport and one of Tunisia's three major tourist areas, built around an old walled city on the Gulf of Hammamet on the east coast. Sousse lies on the site of ancient Hadrumetum, a key Carthaginian city which Hannibal, the Carthaginian general, used as a base in his campaign against Rome during the Second Punic War (218-201 BC). It became an important Arab city between the 9th and 11th centuries and was taken by the Sicilians in the 12th century. It was recaptured by Arabs in the 13th century.

The old walled town, built on the side of a hill facing the sea, contains an ancient medina (Muslim residential area) and kasbah (citadel). Among the historic buildings are the 9th-century Ribat fortress, with a high tower and seven bastions. The local museum has a fine collection of Roman, Byzantine and Christian mosaics. Just outside the town are a series of 2nd to 4th-century Christian catacombs containing 10 000 graves. Apart from tourism, the town's industries are textiles, engineering and car assembly.
Population 100 000
Map Tunisia – Ba

South Africa See p 624

South Australia *Australia* State covering some 984 000 km² (379 834 sq miles) in the central south. Founded in 1836, it was the fourth British colony established in Australia.

Only the south-east has significant rainfall; inland and westwards, farmland gives way to scrub and, finally, desert which covers two-thirds of the state. More than 70 per cent of the population live in the state capital, ADELAIDE. The state has rich deposits of iron ore, uranium, silver, lead and gold. Wheat, livestock and, along the Murray River, fruit are the agricultural mainstays. Grapes are produced for wine in the BAROSSA VALLEY.
Population 1 456 700
Map Australia – Ed

South Carolina *USA* South-eastern state covering 80 601 km² (31 113 sq miles) between the Blue Ridge Mountains, the Atlantic Ocean, and the Savannah River, which forms its south-western border with the state of GEORGIA. The state was first permanently settled in 1670 and was one of the original 13 states. It was officially separated from North Carolina in 1729. It played a key role in the American Civil War, since it was the first state to break away from the Federal Republic of the United States after Abraham Lincoln's election to the presidency in December 1860. As the Confederate State of South Carolina, it is credited with starting the war by attacking Fort Sumter, in Charleston harbour.

Agriculture in the state concentrates on soya beans, maize, peaches, tobacco, cattle (both dairy and beef), poultry and cotton. The principal manufactures include textiles, clothing, paper, various wood products, chemicals and machinery. Tourism is the second largest money earner, after textiles. COLUMBIA is the state capital.
Population 3 486 900
Map United States – Jd

South China Sea See CHINA SEA

South Dakota *USA* North-central state covering some 199 776 km² (77 116 sq miles) of the Great Plains. It is a sparsely populated area, with only 3.5 people per km² (9 per sq mile). It is bisected by the Missouri River. The Black Hills rise to 2207 m (7242 ft) in the south-west and gold is mined here. South Dakota is an agricultural region, and livestock (including cattle, sheep and pigs), wheat, maize and dairy farming are the mainstays. Manufacturing is mainly concerned with food processing and production of industrial machinery, as well as medical instruments. PIERRE is the state's capital, while SIOUX FALLS and Rapid City are its biggest cities.
Population 696 000
Map United States – Fb

South-east Region (Iboland) *Nigeria* One of the most densely populated regions of Nigeria. Palm oil is its main product, but cocoa and rubber, as well as maize, cassava, rice and vegetables for local consumption, are also grown. Enugu is the main town, in the heart of the Ibo cultural region.

In 1967 the Ibo people seceded from Nigeria as the independent republic of Biafra, which included the entire south-east region. In the civil war which followed, lasting until 1970, famine killed more than a million people.
Population 10 000 000
Map Nigeria – Bb

South Georgia *South Atlantic* Mountainous, barren island 1300 km (more than 800 miles) east of the FALKLAND ISLANDS, of which it is a dependency. It is 160 km (100 miles) long and up to 30 km (19 miles) wide, and has an area of 3750 km² (1450 sq miles). It rises to a peak at Mount Paget, 2934 m (9626 ft) high.

The island was claimed for the British by James Cook in 1775. The British explorer Sir Ernest Shackleton died on South Georgia in 1922 and is buried here. Formerly a base for whaling and sealing, it now has a British Antarctic research base.
Map Antarctica – Ba

South Island (Te Wahi Ponamu) *New Zealand* Larger of the country's two principal islands, covering some 151 484 km² (58 488 sq miles). Although the Southern Alps Range extends for

Solomon Islands

THE JAPANESE, WHO INVADED THE ISLANDS IN THE SECOND WORLD WAR, ARE NOW COMMERCIAL PARTNERS

The Solomon Islands, discovered by the Melanesians in 1000 BC or earlier, were named after the Biblical king by Alvaro de Mendana, a Spanish conquistador who sighted them on a voyage from Peru in 1568. But he found no gold, let alone the fabled King Solomon's mines.

The nation – an independent member of the Commonwealth since 1978, having been a British protectorate from the 1890s – consists of six large islands and a myriad of smaller ones in the Pacific, 1600 km (1000 miles) north-east of Australia's Queensland coast.

The climate is hot and wet and prone to typhoons. A typhoon in 1986 killed more than 100 people and made 90 000 homeless. The land is mountainous and forested. The trees cover the scars of the battles in 1942-3 between the Japanese and Americans, especially on the main island of GUADALCANAL, where nearly 24 000 Japanese died in seven months of fighting. Today, the Japanese are partners in developing the Solomons' fishing industry, and the islands' tropical vegetation has been partly cleared for oil and coconut palm plantations and cattle raising.

The country also has resources of lead, zinc, gold and nickel. However, for most Solomon Islanders, who are Melanesians, the way of life has not changed for centuries. They grow fruit and vegetables and catch fish.

SOLOMON ISLANDS AT A GLANCE	
Map Pacific Ocean – Dc	
Area 28 450 km² (10 982 sq miles)	
Population 372 750 **Per km²** 13 **(Per sq mile** 34)	
Capital Honiara	
Government Parliamentary monarchy	
Currency 1 Solomon Is dollar = 100 cents	
Languages English (official), Pidgin, 87 tribal languages	
Religions Christian (34% Anglican, 17% South Sea Evangelical, 19% Roman Catholic), indigenous beliefs	
Climate Equatorial; temperature in Honiara averages 27°C (81°F) all year	
Land use arable land 13%, permanent crops 1%, meadows and pastures 1%, forest and woodland 85%	
Main primary products Sweet potatoes, rice, taro, yams, coconuts, livestock, palm kernels, timber, fish, gold,	
Major industries Agriculture, fishing, forestry	
Main exports Fish, timber, copra, palm oil, cocoa	
Annual income per head (US$) 710	
Population growth (per thous/yr) 34	
Life expectancy (yrs) Male 67 **Female** 72	

Somalia

THIS DROUGHT-STRICKEN COUNTRY WITH A NOMADIC POPULATION IS TORN APART BY WARRING CLANS

The rich cultural heritage of Somalia – strategically situated on the Horn of Africa – contrasts with a poverty that is extreme even by African standards. The country is one of the poorest in the world, and one of the least developed. Life expectancy is a mere 33 years and the infant mortality rate is 162 per 1000 live births. Most Somalis are nomads with few possessions other than their sheep, goats and camels. When drought kills their animals they quickly face starvation, as happened in the mid-1970s and again in the early 1980s and 1990s. Years of clan wars, which have created a state of anarchy in large parts of the country, have made things worse and more than half the population has been unable to survive without substantial food aid.

The country is arid; most of it consists of

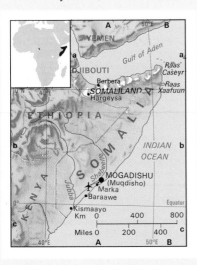

low plateaus with scrub vegetation. There are mountains near the northern coast and two rivers, the JUBBA and the SHABEELLE, which are used to irrigate crops. The settled population is concentrated in the mountains and river valleys and in a few coastal towns. In the past, the few industries that there were concentrated

on food processing – but it is probable that almost all industries have been destroyed or have shut down as a result of the civil war.

About 65 per cent of exports consist of live animals, meat, hides and skins, sold mainly to Arab countries. A few large-scale banana plantations by the rivers provide the second export. Priority in irrigated areas – a total of some 1 600 km² (1000 sq miles) – is now being given to basic food crops, such as sorghum, corn and sugar, to reduce dependence on aid and imported food. However, soil erosion, overgrazing and periodic drought remain serious obstacles.

Somalia is rare among African states in that it had a homogeneous population, speaking the same language, long before it became an independent republic on 1 July 1960. Its disparate clans began forming an Islamic group centuries ago, when it developed trading links with the Arab world and far across the Indian Ocean. In colonial days Britain occupied the north (1884) and Italy the south (1905). These two territories were joined on independence.

The years since then have been turbulent and politically chaotic. Apart from the country's own internal rivalries, there was a bitter conflict with Ethiopia, which resisted a Somali invasion of the OGADEN region where the population is mainly Somali. Conflict and drought have led to a huge inflow of refugees from the Ogaden into Somalia, as well as an exodus of Somali refugees into neighbouring Kenya.

In 1969 Mohammed Siad Barre seized power in a military coup, making the Somali Revolutionary Socialist Party the country's sole legal political party. As self-proclaimed president he began years of tyrannical rule. In the late 1980s an armed rebellion, led by the United Somali Congress (USC), the other major group among about 14 factions, led to civil war. Barre was ousted in January 1991; USC leader Ali Mahdi Mohammed was made interim president and elections were promised.

But fighting broke out among the various factions, and there was large-scale looting of international aid supplies – and even the kidnapping and murder of relief workers.

In May 1991 the Somali National Movement proclaimed independence for Northern Somalia, once British Somaliland, but their secession under the name of the Somaliland Republic was not recognised by the south. The tragedy of mass starvation, the continuing civil war and looting by the warring militia led to an international UN peacekeeping force, consisting largely of United States troops, being deployed in the region in

December 1992. Its aim was to bring an end to civil war, disarm the warring factions, and enable the food aid to reach about 4.5 million famine victims. The UN intervention – during which billions of dollars were spent and the casualties among the peacekeepers included 36 US soldiers and Marines, 24 Pakistani soldiers, nine Indian soldiers and three Indian doctors – failed to end the fighting.

Repeated UN-brokered peace pacts between the leaders of the two most prominent of the warring factions, one under Ali Mahdi Mohammad, the other under General Muhammad Farrah Aideed, were signed – and broken. In March 1994 these two leaders once again signed a peace agreement at a conference in Nairobi and by the end of that month all but a few of the international troop contingents had been withdrawn. The agreement had little immediate effect. It was followed by repeated postponements of a scheduled national reconciliation conference, while fighting in and around the capital, MOGADISHU, resumed.

However, a regional peace conference in the south in June led to a ceasefire between the two largest warring factions there. In March 1995 the last of the UN peacekeepers left the country, while both Aideed and Mohammad were reported to be hosting conferences of their allies, each hoping to form an interim government that could claim power.

SOMALIA AT A GLANCE	
Official name Somali Democratic Republic	
Area 637 660 km² (246 144 sq miles)	
Population 9 000 000 **Per km²** 14	
(Per sq mile 37)	
Capital Mogadishu	
Government Republic	
Currency 1 Somali shilling = 100 centesemi	
Languages Somali (official), Arabic, English and Italian (both commercial)	
Religion Muslim (99%)	
Climate Hot and dry, with a very short rainy season in the south. Average temperature in Mogadishu is 27°C (81°F) through the year	
Land use Grazing 68%, forest and woodlands 14%, cultivation 2%, desert 16%	
Main primary products Sheep, goats, camels, bananas, sugar cane, cotton, maize, millet	
Major industries Stock-rearing, oil refining, sugar refining, flour milling	
Main exports Meat, hides and skins, livestock, bananas, cotton	
Annual income per head (US$) 120	
Population growth (per thous/yr) 30	
Life expectancy (yrs) Male 33 **Female** 33	

most of its length, it also has pastureland for some 32.5 million sheep, almost half of the country's total. The Canterbury Plains near the east coast are the chief wheat-growing area. The principal cities are CHRISTCHURCH, DUNEDIN, INVERCARGILL and TIMARU.

Population 881 357

Map New Zealand – Cd

South Magnetic Pole *Antarctica* The point on the earth's surface at the southern end of the axis of the earth's magnetic field. It is the point to which the southern end of a magnetic compass

points. Its location changes by about 10-15 km (6-9 miles) a year; in 1995, its estimated position was 65.1°S, 139.5°E – about 145 km (90 miles) off the coast of Adélie Land.

Map Antarctica – Lb

South Orkney Islands *South Atlantic* Barren and uninhabited, this largely glaciated group of islands lies in the South ATLANTIC Ocean, nearly 1360 km (850 miles) north-east of the Antarctic Peninsula. The islands are part of the British Antarctic Territory in an area also claimed by Argentina, and were once used by whalers and

sealers. The United Kingdom has a scientific base here (Signy on Coronation Island). Argentina also has a base – Orcadas on Laurie Island.

Map Antarctica – Cb

South Polar Plateau *Antarctica* Desolate, ice-covered core of Antarctica. Although the ice here can be 4800 m (15 700 ft) thick, the average thickness is some 2100 m (6900 ft).

Map Antarctica – d

South Pole *Antarctica* Southern end of the earth's axis, lying about *(continued on p 628)*

South Africa

AFTER MORE THAN 40 YEARS OF APARTHEID, 40 MILLION BLACK PEOPLE AND 5 MILLION WHITES MEET FOR THE FIRST TIME AS EQUAL CITIZENS

South Africa, a large, rich and beautiful country, is emerging at last from the legacy of long years of internal dissent and international condemnation.

It was the insistence of the South African government on maintaining white supremacy through *apartheid* – the separate and unequal development of racial groups – that alienated and isolated the country. As colonialism died in Africa, so South Africa and its former satellite, Namibia, stood out as the only countries on the entire continent still ruled by whites.

Strategically, the country straddles the sea route from the Middle East to Europe and the United States around the Cape of GOOD HOPE, taken by oil tankers too large for the SUEZ CANAL. To add to its importance for the Western world's oil supply and for world peace, South Africa has immense mineral wealth, from gold (it is the world's biggest producer), diamonds, uranium and silver, to coal and natural gas, asbestos, chrome, copper, iron, magnesium, nickel and platinum.

All these treasures, however, proved to be no protection against world recession, the increasingly militant mobilisation of disadvantaged masses at home and international sanctions that included cultural exchanges, trade and finance, weapons and transport and – perhaps most painful to many white South Africans – a strict boycott of participation in international sporting events. All these factors contributed to a new pragmatic approach by the government which led to the surprisingly sudden evolution of a new South Africa in the early 1990s.

THE SHAPE OF THE COUNTRY

The country occupies a huge oval, saucer-shaped plateau surrounded by a belt of land, 55 to 240 km (35 to 150 miles) wide, which drops in steps to the sea. The plateau averages 1200 m (3950 ft) above sea level, and the rim of the saucer, the GREAT ESCARPMENT, rises in the east to 3482 m (11 425 ft), just over the Lesotho border in the DRAKENSBERG.

In 1994 South Africa's traditional four provinces – Cape of Good Hope and Natal, (historically former British colonies), and Transvaal and Orange Free State (historically former Boer republics) – were further divided into nine provinces, each with its own local government and independent administrative system: WESTERN CAPE, EASTERN CAPE, NORTHERN CAPE, KWAZULU-NATAL, FREE STATE, EASTERN TRANSVAAL, NORTH-WEST, GAUTENG and NORTHERN TRANSVAAL.

Landlocked within South African territory is the kingdom of LESOTHO (formerly the British High Commission Territory of Basuto-land). Another landlocked country is the kingdom of SWAZILAND, also a former British High Commission Territory, which shares borders with South Africa and MOZAMBIQUE.

Most of South Africa's northern borders are formed by rivers. From east to west, these are the 'great, grey-green, greasy Limpopo' that Rudyard Kipling wrote about, the Marico, Molopo, Nossob and ORANGE. The Orange is the country's longest river, rising in the highlands of Lesotho and running west for some 2090 km (1299 miles) to reach the Atlantic. Together with its tributaries it drains most of the plateau. Several other rivers follow relatively short courses from the escarpment to the sea, but the country as a whole is short of water and suffers from periodic droughts.

In the southern and south-western parts of the Western Cape, rain falls mainly in the winter months, from May to October, but most of the rest of the country receives summer rainfall, from November to April. In some areas, especially along the eastern and southern coastal plains, annual rainfall may be as high as 1500 mm (60 in). North of the Cape's folded mountains and west of the Drakensberg, the country lies in a rain shadow that receives less than 500 mm (20 in) annually, and rainfall decreases steadily towards the NAMIB and KALAHARI deserts in the west and north.

Temperatures are fairly uniform over large sections of the country, with a mean annual temperature of about 17°C (63°F). The warm Mozambique current, flowing south along the eastern seaboard, keeps the coastal temperatures several degrees higher than those on the west coast, which is cooled by the BENGUELA CURRENT flowing from the Antarctic.

VARIED NATURAL REGIONS

Despite the overall pattern, there are great regional differences in climate. In winter, while much of the Western Cape experiences wet weather with temperatures of about 12°C (54°F), the centre of the country has dry conditions and temperatures may fall to -12°C (10°F). In summer on the central plateau or HIGHVELD, huge formations of thunderclouds bring brief torrential storms.

Vegetation, too, is varied. Along the Western Cape coast and extending towards KwaZulu-Natal, a low-growing, small-leaved scrub known as fynbos includes an extraordinary variety of species, particularly of ericas (heaths) and proteas (evergreen shrubs with large flower heads). These plants are astonishingly resistant to drought – and even to periodic fires. Pockets of temperate evergreen forests are found along this coast.

North of the Cape's folded mountains with their tortuously flexed and folded strata is the semi-arid KAROO region. These dusty flatlands, relieved by characteristic, bare hillocks or KOPPIES, are sparsely dotted with grasses and low-growing scrub that support goats and large flocks of sheep. Small villages – the typical South African *dorp* – supply the wants of farmers whose homesteads, dams and probing wind pumps dot the countryside. By day the Karoo bakes under relentless heat, but winter brings a cuttingly cold wind and freezing nights.

The temperate grasslands of the Free State have mostly given way to vast fields of maize and sunflowers, extending to the thorn-bush and baobab country of the far Northern Transvaal. Much of the low-lying eastern area of the Northern and Eastern Transvaal was once covered with subtropical thorn forests, but these were cut down to provide timber for building, and for shoring the tunnels of gold and coal mines. The former forests have been replaced with extensive pine and eucalyptus plantations, while remnants of indigenous evergreen forests survive in some mountainous areas. From the craggy Drakensberg to the eastern coast, the hilly landscape is covered with low-growing, evergreen subtropical bush.

THE PEOPLE

The recorded history of settlement in South Africa dates from the late 15th century and the sporadic landfalls of Portuguese explorers seeking a sea route to India. African settlement in the subcontinent, however, was more than 1000 years older. When the first Europeans arrived, the pastoralist Khoikhoi and hunter-gatherer San (Bushmen), speaking the Khoisan group of languages, inhabited mainly the south. Bantu speakers, whose origins are thought to have been between the Niger and Congo deltas and who shared a common heritage with many groups speaking some 300 related languages, were distributed over a wide area of southern and central Africa. They were established in the Transvaal and Natal by AD 300 and in the Transkei area by AD 700. By the 15th century two main groups had emerged – the Nguni and the Sotho – and two smaller groups – the Venda and Tsonga. Among the Nguni the most active sub-groups politically today are the Xhosa and Zulu.

South Africa's former racial policy, which was universally known as *apartheid* ('apartness'), separated people not only into blacks and whites, but also into coloured people (about 3 million) and Asians (about 1 million). Coloured people have diverse origins that date from the establishment of the Dutch East India Company outpost at what is now CAPE TOWN, in 1652. They were created by marriage and informal liaisons between European settlers on the one hand and, on the other, indigenous Khoikhoi and slaves imported principally as farm labour from Africa, the Indian subcontinent and the East Indies. They are spread throughout the country, but by far the greatest concentration is in the Western Cape.

The Asian population is descended mainly from Indians brought as indentured labourers to the sugar-cane fields of Natal for several decades from the mid-19th century. Most of South Africa's Asians are to be found in KwaZulu-Natal and GAUTENG provinces.

South Africa's white population is a mixture of the descendants of the original Dutch and French (Huguenot) settlers who later called themselves Afrikaners or Boers (a Dutch and South African Afrikaans word for farmers), English and Germans who arrived later, and other more recent European influxes. Afrikaans-speaking groups comprise about 60 per cent of the white population, the rest being predominantly English-speaking. Divisions within the white group have not been solely

along the lines of language or religion, however, but based on broad political affiliations, often inherited from earlier generations. As a result of extensive intermarriage, many grow up speaking both languages.

Ethnic divisions are clearer among black South Africans, almost 20 per cent of whom are Zulu. Preservation of specific black identities through cultural and political affiliations has been a frequent source of friction leading to violent and bloody clashes.

Animosity between the Afrikaners and the English – more recently in the political field – had its origins in the British occupation of the Cape, initially from 1795 to 1803 and permanently from 1806. In the 1700s Dutch-descended colonists had expanded eastwards from Cape Town, most of them living the life of the *trekboer* or itinerant grazier rather than that of settled farmers.

THE GREAT TREK NORTH

As Afrikaner Boers moved away to the east they clashed with the Xhosas, who were moving southwards. A series of wars set off by land hunger and cattle theft began in 1779 and continued – later waged also by the British – for the next 100 years

British rule was not to the Boers' liking. According to some Boers, inadequate compensation after their slaves were freed was a final provocation, sparking a mass exodus to the north in search of a new Promised Land.

This became known as the Great Trek, which in the years 1835-8 took 10 000 Boers and their ox wagons across vast tracts of land. Their most fiercely fought engagements with black Africans were those against the Zulus, who were then occupied with their own era of imperial expansion from a power base located in present-day northern KwaZulu-Natal.

A famous battle was fought on the banks of the Ncome River in KwaZulu-Natal in 1838, where 12 000 Zulus attacked 470 Boers in a well-sited laager. The Boers prayed for victory and promised that their deliverance would be remembered forever as a day of solemn thanksgiving. Clubs and spears could not stand up to massed firepower, even from the unreliable muzzle-loaders of the time. The battle ended with some 3 000 Zulus killed and three Boers slightly wounded. The Ncome River ran red, and is known still as BLOOD RIVER. The day – 16 December – became the Day of the Vow and was kept as such annually until 1994. Still a public holiday, it is known now as the Day of Reconciliation.

In 1840 the victorious Boers proclaimed their Republic of Natalia in KwaZulu-Natal. But this was annexed by Britain two years later. The Boers trekked again, to found the landlocked republics of Orange Free State and *Zuid-Afrikaansche Republiek* (Transvaal). These, too, were annexed as the fortunes of South Africa started to rise with the discovery of diamonds and gold.

▲ FAIREST CAPE Table Mountain forms an unforgettable backdrop to Cape Town – as majestic a sight as when Sir Francis Drake described the Cape Peninsula as 'the fairest cape we saw in the whole circumference of the earth' in the 16th century.

Independence was restored to the Transvaal after the Anglo-Transvaal (first Anglo-Boer) War when a badly led British force was defeated at Majuba Hill in 1881.

Only five years later, the world's largest gold strike was made on the Witwatersrand. This was to lead to the South African (second Anglo-Boer) War which began in 1899 with a string of humiliating defeats for Britain. But imperial numbers and unlimited resources began to tell and the Boers' capitals – first BLOEMFONTEIN, then PRETORIA – were quickly occupied. Several thousand diehards – the '*bittereinders*' – served the Boers' cause in a guerrilla war that dragged on for two years before peace was concluded in May 1902. By then the term 'concentration camp' had found a firm place in the language. More than 40 000 people, black and white and mostly women, children and old people, died in epidemics in these camps which the British set up to receive civilian refugees from their scorched earth policy that devastated the land.

Under the terms of the relatively generous peace settlement, the defeated Boer republics became British colonies that in 1910 were merged with Natal and the Cape of Good Hope to form the Union of South Africa, a self-governing dominion of the British Empire. With the exception of limited voting rights in the Cape, blacks were totally excluded from participation in government, as they had been in the individual states before union.

REPRESSIVE LEGISLATION

Many of the Boers' wartime leaders – 'the generals' – came to the fore in politics. Most of them, in fact, were admired by the British, and included Louis Botha, first prime minister of the Union of South Africa, and Jan Smuts, one of the principal contributors to the new country's constitution and later to the formation of the League of Nations after the First World War. Smuts, famed also for creating the concept of holism, was prime minister from 1919 to 1924 and again from 1939.

In 1912 the South African Native National Congress was formed. (This was later to become the African National Congress [ANC], victor at the polls in 1994.) Its stated objectives included African unity, social and economic advancement and the extension of political rights. But almost immediately came the Natives Land Act of 1913, which firmly established in law the principle of segregated areas and granted blacks – who then made up 67 per cent of the population – the use of just 8.5 per cent of the country's rural land.

In 1948 Smuts, who led a moderate party composed of both Afrikaans and English-speaking members, was swept from power by the National Party, formed almost exclusively by Afrikaners. Sternly authoritarian and conservative, the new government retained power for 46 years. Most of the basic *apartheid* laws were introduced by the first National Party government under Dr Daniel Francois Malan, a former Dutch Reformed clergyman. But the chief architect of apartheid was regarded as Hendrik Frensch Verwoerd, Malan's Minister of Native Affairs.

Verwoerd became Prime Minister in 1958 and dominated the political scene until his death at the hand of an assassin in 1966. During that time Afrikaner republicanism enjoyed a resurgence when a whites-only referendum produced a narrow swing in favour of republican status. Since its domestic policies were unacceptable to what was now the British Commonwealth, the new Republic of South Africa celebrated its own creation in 1961 by withdrawing from that body.

Despite an initial and impressive economic boom, South Africa faced increasing international opposition to *apartheid*. Among the cornerstones of this policy were the Mixed Marriages and Immorality acts which banned marriages and sexual relations between whites and 'non-whites'. The Group Areas Act ensured that white, black, Asian and coloured people all lived separately, usually with whites occupying the suburbs conveniently close to city centres and blacks and coloured people forcibly removed to outlying 'townships' or 'locations'. As part of the system of 'influx control', blacks residing in urban areas were required to carry a passbook at all times, giving

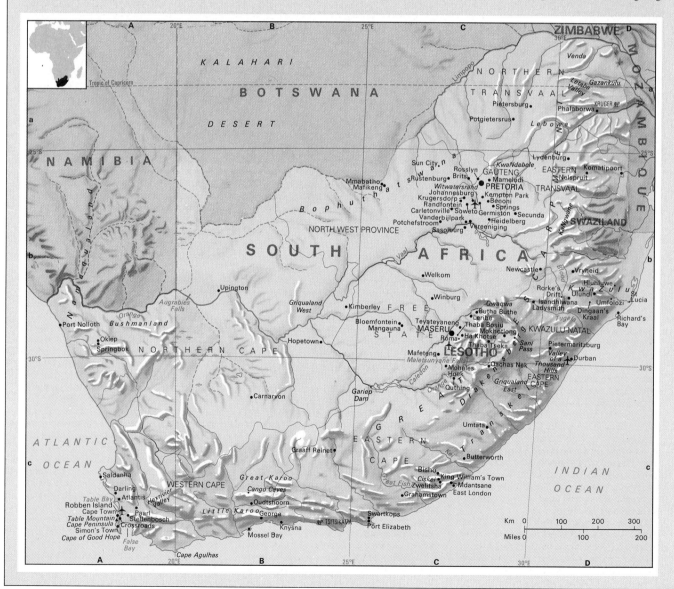

details of the legal basis on which they were entitled to be there. Job reservation was also reinforced. The law did not recognise black trade unions, nor were their members allowed to take part in structured negotiations, nor to go on strike. The leaders of extra-parliamentary opposition were hounded and imprisoned. Their organisations were banned.

The proponents of apartheid aimed to create a white-dominated South Africa in which almost no blacks would be able to claim citizenship. This was to be achieved by a system of ethnically based reserves or 'tribal homelands', derisively referred to as 'bantustans'. When black people were removed from cities or other areas designated 'white', they were resettled in the appropriate homeland of which, technically, they were citizens. In due course, homelands might apply to the South African government to become self-governing and, eventually, to become 'independent'. By the time the system collapsed in 1994, there were no fewer than 10 such states, of which four – BOPHUTHATSWANA, TRANSKEI, CISKEI and VENDA – were 'independent', although only their leaders and South Africa recognised that independence. In total, these internal states occupied about 14 per cent of the country.

Homelands were seen as labour pools and special government concessions were granted to 'border industries' – those concerns that sited their factories either in the homelands or close to their borders. Blacks had to fulfil complex requirements to qualify to live in cities outside the homelands, and most existed as contract workers whose rights of residence depended on the possession of a valid contract of employment, usually renewable annually.

ALL-WHITE REFERENDUM

In the hope of showing movement towards a more democratic system of government, a referendum was held – only among the white electorate – in 1983. It resulted in the approval of a new, tricameral system of parliament, in which there were separate chambers for whites, coloureds and Asians, but with power remaining firmly with the whites and, in particular, with the state president who, for the first time, possessed full executive powers. Blacks had no meaningful franchise in this new South African constitution. The entire system was based on the racial division of interests into 'own affairs' and 'general affairs'. Blacks were dissatisfied at having received no real advantage. Ultra-conservative Afrikaners considered that whites had given away too much, and split from the National Party to form their own Conservative Party.

The constitution failed, as many had predicted it would. Worse, it failed during a time of alarming economic decline brought on, in large measure, by international trade sanctions and the flight of foreign capital.

Active, long-term military involvement in the destabilisation of neighbouring countries such as Angola and Mozambique came full circle as the intensity of the wars declined and South Africa was flooded with cheap, illegal and readily available automatic weapons – in particular the AK47 assault rifle. Murders and other violent crimes increased alarmingly, the

great majority of them being perpetrated by blacks on blacks.

A change in the state presidency occurred in 1989 when Pieter Willem Botha was replaced by Frederik Willem de Klerk, who began negotiations between the government and the (ANC), many of whose leaders – some of them imprisoned for more than 25 years – were released from prison.

On 2 February 1990, de Klerk announced in parliament a series of fundamental moves that changed forever the pattern of South African politics. Banned organisations such as the ANC, the Pan-Africanist Congress and the South African Communist Party were declared legal and the imminent release of Nelson Mandela, the world's most famous prisoner after 27 years of incarceration on ROBBEN ISLAND, was announced.

During long periods of deliberation to draw up an interim constitution and to provide for the election of a government of national unity by all adult South Africans, bloodshed increased. Much of it occurred in townships and rural areas as a result of the intense rivalry between the ANC and the Zulu-based Inkatha Freedom Party, although there were also attacks on isolated farmers and congregations in churches. Widespread strikes – part of a campaign described by the ANC and its ally, the South African Communist Party, as 'rolling mass action' – hit the economy hard.

Reincorporation of the homelands into South Africa and the reaction of right-wing activists also produced sporadic outbreaks of violence, but fewer than had been feared.

Meanwhile, representatives from all interested parties proceeded with negotiations towards an equitable, if temporary, solution – the creation of a new, interim constitution and the installation of a government of national unity to administer it. Election day itself, in April 1994, passed surprisingly peacefully. The results, though, were no surprise.

The ANC won close to 63 per cent of the votes and, thereby, 252 seats in a 400-seat National Assembly – falling short of the two-thirds majority that would have enabled it to rewrite the constitution unilaterally. The National Party, monolithic governing party for four decades and more, took 82 seats, the Inkatha Freedom Party 43 seats and the mainly Afrikaner Freedom Front just nine seats. It was a new face for South Africa, one that many people had not expected to see in their lifetime

White rule formally ended on 10 May 1994 with the swearing-in of Nelson Rolihlahla Mandela as the new president. South Africa was swiftly welcomed back into the world community and to membership of the many international organisations from which it had been excluded – among them the Non-Aligned Movement, the Organisation of African Unity and, after an absence of more than 30 years, the British Commonwealth.

But the road to personal prosperity, especially for the majority of newly enfranchised blacks, is steep and uncertain. The gaps between prosperity and poverty are to be seen everywhere, especially in the contrasts between the established urban areas and the

ragged squatter camps that have grown – and are still growing – in widening rings around them. But the country, if well managed, can provide abundance for all its people.

CITY OF GOLD

South Africa's huge mineral wealth overshadows all its other natural resources. Mining accounts for more than 10 per cent of the total wealth (gross domestic product) produced by the country every year.

The first major discovery was diamonds, in Griqualand West (now part of the Northern Cape), in 1866. More were found along the VAAL RIVER and then buried in the blue clay, or kimberlite, that took its name from the diamond-rush city of KIMBERLEY, where the Big Hole mine is the largest hole ever dug by hand. The hole covers an area of 15.5 hectares (38 acres) and is 365 m (1200 ft) deep.

In 1886 the world's richest gold-bearing ore was discovered in an outcrop of the main gold reef on a Transvaal ridge known as the Witwatersrand. This brought about the birth of JOHANNESBURG, South Africa's largest city and known to most Africans as *eGoli* – place of gold. A few years after this vital gold-strike, a start was made with the large-scale mining of coal, of which there are vast fields in the Eastern Transvaal and KwaZulu-Natal as well as in the Free State. Coal has now become an important export; it fuels thermal power-stations and is the source of oil extracted by advanced techniques devised in South Africa.

South Africa is the most economically developed country in Africa, with its economy underpinned by mining, agriculture and, increasingly, by manufacture.

Policies instituted by the interim government have included its Reconstruction and Development Programme, chiefly concerned with raising the living standards of the impoverished majority. The redefining of priorities and redirection of finance have resulted in the end of compulsory military service. Research in scientific fields such as medicine has been increasingly seen as secondary to the provision of basic care and amenities.

Agriculture, although it provides less than 10 per cent of the gross domestic product, still produces – in years of adequate rainfall – enough food for the country's needs and for export. Cape wines and apples and Outspan oranges are among familiar items in the shops of many countries. Without irrigation, much of the country is suited only to livestock farming. The largest national irrigation scheme is known as the Orange River Project and includes impressive storage dams, transfer tunnels and major works in neighbouring, landlocked Lesotho, which is the source of the Orange River.

Commercial activity today is assisted by extensive road and railway systems and internal air-routes. There are international airports at Johannesburg, DURBAN and Cape Town.

CITY AND COUNTRYSIDE

The events – both tragic and hopeful – that have absorbed observers the world over, have taken place against a background of great natural beauty. South Africa has some of the world's most varied and spectacular scenery, from idyllic, sun-splashed beaches to beautiful plains and rugged, soaring mountains.

The most distinctive architecture is that of the Western Cape Province, where there evolved the domestic style known as Cape Dutch, and where whitewashed, symmetrical facades have curling gables and wide, wood-framed sash windows, roofs of mellow thatch and verandahs, or stoeps, of slate or tile.

Cape Town, the legislative capital of the country and seat of parliament, lies in a natural amphitheatre at the foot of TABLE MOUNTAIN, which is often wreathed in a summer cloud known as the 'tablecloth'. This harbour city extends along the CAPE PENINSULA, its attractive suburbs set between sea and mountain.

The administrative capital of Pretoria is dominated by massive civic architecture, including the Union Buildings (1910) on a ridge overlooking the city. More solemn is the granite Voortrekker Monument which commemorates the Dutch-descended pioneers who journeyed by wagon into the subcontinent. A light touch is given to the city by its jacaranda trees which colour the streets purple with fallen blossoms in October and November.

To the south is Johannesburg, still retaining something of its brashness as a gold-strike town. An earlier version has been recreated as Gold Reef City where tourists can absorb a bygone atmosphere free from the risks of pioneering in Africa. Today Johannesburg is the commercial and industrial heartbeat of South Africa, a magnet alike for entrepreneurs and for the unemployed. Its topographical twin city is SOWETO, from where hundreds of thousands of blacks commute daily to their places of work in formerly 'white' Johannesburg.

KwaZulu-Natal's beaches, mountains and game reserves are the holiday destination of those who live inland, especially in the Transvaal provinces and in the Free State. DURBAN has Africa's largest harbour and is also its holiday city supreme, while a colonial aura clings more closely to Pietermaritzburg, with its atmospheric shaded streets and dignified architecture overlooking parks and cricket grounds.

African communities provide picturesque attractions for tourists. In the Northern and Eastern Transvaal, villages of the Ndebele people dot the landscape, the walls of their homes painted with striking geometric designs. Throughout the South African countryside, the traditionally shaped huts – the cone-on-cylinder and the beehive – are still to be seen, albeit in diminishing numbers.

Game conservation is an important concern, game being a major tourist attraction. The largest reserve is the KRUGER NATIONAL PARK, some 20 700 sq km (8000 sq miles) along the border of Mozambique, renowned for its abundant big game and other species.

Other reserves include Umfolozi, where the white rhino, almost extinct elsewhere in Africa, has been preserved from oblivion and can now be shipped to suitable destinations around the world. There is wildlife outside the reserves, too. In many rural areas the grunts of leopards may still be heard at night, and a wary eye must be kept open for crocodiles. The climate encourages an outdoor life, both for daily recreation and annual holidays.

South Africa, in its transition from apartheid to democracy, has overcome problems formerly thought insuperable. It is clear that still more problems lie ahead, although both the country's government and the mass of its people, in a spirit of reconciliation and hope for the future, show a determination to complete the process of reform.

SOUTH AFRICA AT A GLANCE	
Official name Republic of South Africa	
Area 1 220 088 km² (470 967 sq miles)	
Population 45 million **Per km²** 37	
(Per sq mile 95.5)	
Capitals Cape Town (legislative), Pretoria (administrative)	
Government Parliamentary republic	
Currency 1 rand = 100 cents	
Languages English, Afrikaans, Zulu, Xhosa, Sotho and six other indigenous languages (all official)	
Religions Christian, Hindu, Muslim, Jewish	
Climate Mostly semi-arid, Mediterranean in south-west, sub-tropical on north-eastern coast. Average temperatures range from 8°C (46°F) in July to 23°C (73°F) in January	
Land use Grazing 65%, cultivation 10%, forest and woodland 3%, other 22%	
Main primary products Cereals, sugar, grapes, livestock, potatoes, citrus fruits, apples, pineapples, tobacco, cotton, timber, fish; gold, coal, iron ore, diamonds, copper, manganese, limestone, chrome, silver, nickel, phosphates, asbestos, tin, zinc, vermiculite	
Major industries Gold, coal, diamond and other mining, iron and steel, food processing, motor vehicles, mineral refining, machinery, chemicals, petroleum refining, agriculture and viticulture, tobacco products, clothing, paper, textiles, forestry, fishing	
Main exports Gold, coal, metal ores and products, diamonds, foods, machinery and transport equipment	
Annual income per head (US$) 2670	
Population growth (per thous/year) 26	
Life expectancy (Yrs) Male 62 **Female** 67	

480 km (300 miles) south of the Ross Ice Shelf. The south pole lies 2992 m (9816 ft) above sea level, where ice is over 2500 m (8200 ft) thick.

It was first reached by the Norwegian explorer Roald Amundsen on 14 December, 1911, then by the British explorer Robert Scott 34 days later. Scott perished on the journey home. It is the site of the United States AMUNDSEN SCOTT STATION.
Map Antarctica – d

South Sandwich Islands *South Atlantic* Uninhabited group of 16 barren, volcanic islands in the South Atlantic Ocean, situated north of the WEDDELL SEA. They constitute a dependency of the FALKLAND ISLANDS, 1100 km (685 miles) to the north, and cover 340 km² (130 sq miles). The South Sandwich Islands are administered by Great Britain and claimed by Argentina.
Map Antarctica – Ba

South Shetland Islands *Antarctica* Barren, uninhabited, volcanic islands. They cover some 4662 km² (1800 sq miles), and lie north of the Antarctic Peninsula, between the Atlantic and Pacific oceans, providing the geological link between the Antarctic Peninsula and South America. Formerly a sealing and whaling centre, the islands are now a site for scientific surveys. They are part of the British Antarctic Territory – an area claimed also by Chile and Argentina.
Map Antarctica – Db

Southampton *United Kingdom* Port city at the head of Southampton Water on the south coast of England, some 110 km (68 miles) south-west of LONDON. It is the largest city in the county of Hampshire, and has been a port since Roman times. Some parts of its medieval walls remain.

The rise of the modern city dates from the 19th-century development – by a railway company – of the docks that made it, in the era of the great transatlantic liners, the principal British port for passengers to North America and South Africa. As liners declined in the 1960s, oil refineries and container traffic took over, and parts of the docklands have been transformed into opulent housing and business complexes at Ocean Village and Town Quay. Developments include a variety of tourist attractions focusing on maritime history.
Population 207 300
Map United Kingdom – Ee

Southend-on-Sea *United Kingdom* See ESSEX

southerly buster Strong, dry, cold wind bringing low temperatures to New South Wales, Australia, when a mass of polar air pulled northwards behind a DEPRESSION bursts suddenly into warmer areas. It corresponds to the PAMPERO of Argentina and Uruguay.

Southern Ocean See ANTARCTIC OCEAN

Southland *New Zealand* Southernmost, mainly sheep-farming, region of South Island, covering 29 624 km² (11 437 sq miles). The town of BLUFF has the country's only aluminium smelter. INVERCARGILL is the main town.
Population 103 440
Map New Zealand – Af

Southport *United Kingdom* Seaside town in the English county of Merseyside, 28 km (17 miles) north of LIVERPOOL. Founded in the late 18th century as an elegant residential and holiday resort, it still maintains this exclusive reputation in contrast to nearby BLACKPOOL.
Population 90 960
Map United Kingdom – Dd

South-west Nigeria (Yorubaland) *Nigeria* Traditional homeland of the Yoruba people, south-west of the Niger River. It is fertile and densely populated and produces cocoa, yams, tapioca and kola nuts (chewed as a mild stimulant). The area includes the ancient towns of IFE and OYO and the urban agglomerations of LAGOS, IBADAN, OGBOMOSHO, ILORIN and Oshogbo.
Map Nigeria – Ab

Sovetskaya Gavan' *Russia* Port and naval base on a fine natural harbour on the Tatar Strait, opposite SAKHALIN in the far east of the country. Completion in 1984 of the Baikal-Amur Railway line, linked to the Trans-Siberian Railway, greatly boosted its role in the shipping of freight between Japan and Europe.

Population 80 000
Map Russia – Pd

Soweto *South Africa* Acronym for Southwestern Townships, a contiguous group of townships inhabited by black people, situated about 20 km (12 miles) outside Johannesburg, in GAUTENG province. These townships were set up as dormitory townships to house black migrant workers. The inhabitants of Soweto were in the forefront of the anti-*apartheid* campaigns from the 1970s to the 1990s. Although *apartheid* was officially abolished after the election of 1994, Soweto seems certain to keep its predominantly black profile for a considerable time.

Population 2 000 000
Map South Africa – Cb

Spa *Belgium* Resort 32 km (20 miles) southeast of the city of LIÈGE. It gave its name to all resorts which, like the town, have mineral or medicinal springs.

Population 9600
Map Belgium – Ba

Spain See p 630

Spalato *Croatia* See SPLIT

Spanish Town *Jamaica* Town in the south-east, 20 km (12 miles) west of KINGSTON. Founded by Diego Columbus (the son of the famous explorer Christopher Columbus) in about 1520-6, it was the national capital until superseded by Kingston in 1872. It contains several historic buildings,

including St Catherine's Cathedral (1714), the Court House, Eagle House (dating from the 17th century), and the ruined Old King's House (1762) which was the official residence of Jamaica's governors until 1870. The town now processes sugar, coffee, cocoa and fruit.

Population 92 380
Map Jamaica – Bb

Sparta (Spárti) *Greece* Town in the south-east Peloponnese at the foot of Taíyetos Mountain. Sparta rose to power in the 7th century BC as a ruthless and mighty militaristic city-state which soon expanded its territory. Though it did exceptionally well on the battlefield, Sparta lagged far behind other Hellenic cities in its culture and intellectual development. It fought beside Athens against the Persians at the Battles of Thermopylae and Salamís in 480 BC, then against Athens in the Peloponnesian War (431-404 BC). Much of Sparta was destroyed and its population perished in 464 BC in a catastrophic earthquake. The city later fell under Roman rule and was sacked by Visigoths in AD 395. The modern town was founded in 1834 near the ancient ruins. Olives and oranges are its main products.

Population 14 400
Map Greece – Cc

Speightstown *Barbados* Small fishing port on the north-west coast. After BRIDGETOWN, it is the island's most important town. In the 18th century, it was a major sugar port. Some of its old colonial buildings are still standing. Close to the town is the new Heywoods Beach tourist development, a luxury complex built to provide jobs in northern Barbados and to relieve the tourist pressure on the south. The beautiful beach here is about 1.6 km (1 mile) long.

Population 3000
Map Barbados – Ba

Spencer Gulf *Australia* Large inlet in the Indian Ocean, between Eyre and Yorke peninsulas, in southern Australia. It is 320 km (200 miles) long

and 130 km (80 miles) wide. It was explored in 1802 by the British navigator Matthew Flinders, during his circumnavigation of Australia. He named it after the 2nd Earl Spencer (1758-1834), the First Lord of the Admiralty who had approved Flinders' voyage.

Map Australia – Fe

Speyer *Germany* City in the Rhine Valley, 20 km (12 miles) south-west of HEIDELBERG. It has one of the largest Romanesque cathedrals in Europe. The building was started in the 11th century; eight Holy Roman Emperors were buried in it. In 1529 a protest in the town by princes and local officials against the Roman Catholic Church's treatment of the reforming priest Martin Luther (1483-1546) gave the reform movement the name Protestantism. Today the city produces electronics, aircraft and building materials.

Population 46 000
Map Germany- Cd

Spice Islands *Indonesia* See MALUKU

spit Narrow ridge of sand or shingle built up by longshore drift and stretching from the shore out to sea or across a bay. If it extends right across the mouth of a bay it is called a BAR. See THE CHANGING PATTERN OF THE COASTS, p 219

Spithead *United Kingdom* Deep channel off the entrance to PORTSMOUTH Harbour, between the Isle of WIGHT and the English mainland. It is a busy shipping lane that copes with naval traffic as well as supertankers, bulk carriers and ferries.

Map United Kingdom – Ee

Spitsbergen *Arctic Ocean* Island group in the SVALBARD archipelago in the Arctic Ocean, some 580 km (360 miles) north of Norway and about 1000 km (621 miles) from the North Pole. Spitsbergen is also the name of the main island in the group, formerly called West Spitsbergen.

The archipelago covers an area of 62 521 km² (24 134 sq miles), of which almost 80 per cent is covered by glaciers. The highest peak on the group is Newton's Peak at 1717 m (5633 ft). The islands were discovered in 1596. Spitsbergen Island, covering 39 000 km² (15 060 sq miles), has extensive national parks and nature reserves, as well as both Norwegian and Russian coal-mining camps. These are a legacy of the mining rights given to all 40 signatories of the 1920 convention that recognised Norway's sovereignty over the islands. The chief Norwegian settlement is at Longyearbyen, named after an American, J. M. Longyear, who in 1905 was the first person to mine coal there. Besides its coal miners and a few people employed in research, the islands have no permanent inhabitants.

Population (Norwegian) 1000; (Russian) 2500
Map Arctic – Rb

spitskop (spitzkop) In South Africa, a hill with a sharply pointed top, caused by the presence of hard rock, such as dolerite, which has resisted erosion.

Split (Spljet; Spalato) *Croatia* Largest city in Dalmatia, 50 km (about 30 miles) south-east of ŠIBENIK. It is a major Adriatic port, fishing and naval base, cultural centre, and an industrial area producing ships, cement, plastics, timber, wine and processed food. Until (continued on p 633)

▼ APPALLING REFUGE Men of the British polar expedition, led by Sir Ernest Shackleton, spent 105 days on Elephant Island in the South Shetlands in 1916 after the *Endurance* sank.

Spain

EUROPE'S FAVOURITE TOURIST DESTINATION, SPAIN STILL RETAINS THE MYSTERY AND ROMANCE OF ITS ONCE GLORIOUS HISTORY

Each year more than 10 per cent of all international travellers – more than 50 million – go to Spain in search of the sun. The figure may or may not include the movements of the tens of thousands of older people living in Spain who have discovered that pensions go farther here than in their native lands. Housing is cheaper, almost everything costs less than anywhere else in Europe and, provided a suitable settlement is chosen, there is hardly any need to learn Spanish at all.

However, relatively few visitors, residents or tourists, venture away from the golf courses, tower blocks and hastily built urbanisations on the Mediterranean coast to discover one of the most fascinating countries in Europe – a country of such diversity that it would take several lifetimes to know it properly.

One reason is its isolation. It is sealed off from the main body of the European continent by the PYRENEES, which rise to some 3404 m (11 170 ft) at their highest point, Pico d'Aneto. Spain itself is a mountainous country, much of which is a great plateau slashed by valleys and gorges. The western Pyrenees are part of the BASQUE REGION, which in Spain extends westwards along the north coast through NAVARRA. Here the mountains of CANTABRIA begin, and stretch along the Bay of Biscay to GALACIA, the corner of Spain which lies on the Atlantic coast north of Portugal. ANDALUCIA in southern Spain also has mountains, the SIERRA NEVADA range which includes Spain's highest mountain, at 3478 m (11 411 ft), and a short Atlantic coast, between Gibraltar and Portugal.

But Spain's longest shoreline is the one that borders the Mediterranean, from Gibraltar to the frontiers of France. This is the coast whose hot dry summers and mild winters lure visitors. Inland the story is rather different – baking heat in summer and savage cold in winter – although it is here, away from the resorts, that the true Spain is revealed.

The greatest days of Christian Spain followed almost immediately upon the expulsion in 1492 of the Moors of North Africa from GRANADA, their last European stronghold, and the unification of the country under King Ferdinand of ARAGON and Queen Isabella of CASTILE. It was these far-sighted monarchs who dispatched Columbus on his journeys to the New World, then recouped their investments a million times over. Beginning with the conquest and sacking of Mexico by Hernán Cortés in 1521 and of Peru by Francisco Pizarro in 1533, riches poured into Spain.

In the 17th and 18th centuries the power of Spain declined. It lost Gibraltar to the British during the War of the Spanish Succession (1701-13), and when the country was later occupied by Napoleon's troops from 1808 to 1814, most of its colonies took advantage of the situation to gain their independence.

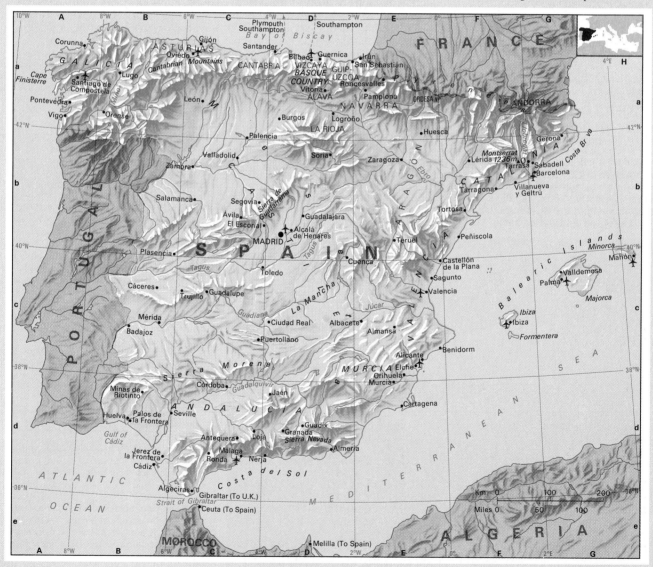

Spain's history in the first half of the 20th century is not a happy one. It was backward industrially, poverty was rife and after the First World War – in which Spain was neutral – anarchists, trade unionists and Basque nationalists united in violent clashes against the government. With the backing of the king, Alfonso XIII, a military dictatorship under General Primo de Rivera was established in 1923. But the required economic reforms were beyond him, and he resigned in 1930; elections were held in 1931, a republic was declared and the king driven into exile. In 1936 pressure from the Popular Front – consisting of left-wing republicans, socialists, communists and anarchists – caused the government to totter. Meanwhile, the right-wing General Francisco Franco took advantage of the situation to make his own bid for power, and landed an army of crack Moroccan troops at CADIZ.

So began the Spanish Civil War. The opponents were the Loyalists – a mingling of the old Republic and the Popular Front – and the Nationalists, a blending of the regular army, conservatives and much of the Catholic Church. The Loyalists were supported by international brigades of left-wing idealists from many parts of the world. The Nationalists' allies were regular army and air force units, thinly disguised and well equipped by Hitler and Mussolini. Victory went to the Nationalists' big battalions. In 1939 Franco became Chief of State, and virtually a dictator.

FASCISTS AND THE GUARDIA CIVIL

Franco's regime was conservative in the extreme, and repressive. Although he was sympathetic to the Axis cause, Franco kept Spain out of the Second World War. His main supporters were the industrialists and landowners, the Church and the army, making for a moralistic and traditional society.

It was a long time before Spain began to recover from the ravages of the Civil War. Already backward agriculturally and industrially, there was much ground to be made up if the country were to regain its place among the European nations. In the 1950s, the tourist trade burgeoned. At the same time the National Institute for Industry was set up. Much foreign capital, attracted by cheap labour, began to flow in and new industries developed. Spain's largest car manufacturer, Seat, was established in 1950 with Italian backing. In 1951 the Ensidesa steelworks was set up with British collaboration. In the 1960s industry continued to grow with help from the United States – in exchange for naval and air bases.

LIBERALISATION AND GROWTH

The rapid growth of tourism opened up the country to outside influences and changed social attitudes and tastes, particularly among the young; but it was not until Franco's death, in 1975, that true liberalisation could begin.

In 1947, Franco had promised to restore the monarchy, and, in accordance with his wishes, Juan Carlos I, grandson of Alfonso XIII, succeeded him and took the oath as head of state and king of Spain. No sooner was he in power than the young king began to institute reforms. Political prisoners were freed, opposition parties were permitted (as was divorce), censorship was relaxed, and in June 1977 the first general election in 40 years was held. In 1981 the king's firmness defeated a coup attempt by right-wing army officers – and the democratic credentials of Juan Carlos were confirmed. A constitution was approved and national elections were held in October 1982.

Since Franco, urban Spain at least has become more cosmopolitan – and less distinctively Spanish – with crowds of commuters, fast-food chains and fashionably dressed young people. Signs of a post-Franco freedom are everywhere: political graffiti, satirical journalism and topless bikinis on the beaches. But beneath this modern veneer the old Spain lives on. The flamenco guitar holds its own against its electric counterpart, there are more *tapas* bars selling the traditional snacks of anchovies and spiced sausage than hamburger counters, and a top *torero* (bullfighter) is as big a hero as any footballer.

Most of the central part of Spain consists of a great plateau, the MESETA (tableland). It rises to about 600 m (1970 ft) and is partly covered with a low scrub of thyme, rosemary and lavender – the very scent of Spain.

The capital, MADRID, lies in the centre of the plateau, at the heart of Castile. A diagonal line across the plateau from the south-west to north-east divides Old Castile (Castilla y Léon), to the north-west, from New Castile (Castilla La Mancha) in the south. Two ranges of mountains follow the diagonal: the Sierra de Gredos, rising to 2592 m (8504 ft), and the slightly lower Sierra de Guadarrama. Lying north of Madrid, the mountains are loved by *madrileños* and tourists alike for their clear air.

The Meseta is largely arid; its principal agricultural activities are sheep rearing on the mountain slopes, and the growing of wheat and saffron which stretch in vast fields across the wide landscape. The area was more productive in the days of the Moors who irrigated it; later, during the Middle Ages, it was impoverished by sheep grazing. Today, large wheat and sunflower fields have supplanted the sheep, although the yields are meagre. In the extreme south of the Meseta, vines are grown in the valley of the GUADALQUIVIR, which cuts Castile off from southern Spain, dividing the plateau from the mountains of the Sierra Nevada. A similar river valley borders Castile to the north.

Geographically and historically, Castile is at the heart of Spain. Its dialect, Castilian Spanish, is the official language of the nation. As the country's industrial and commercial centre, focused on Madrid, it is in some ways one of the most modern parts of Spain. All Castile proclaims a fervent Catholicism in a splendid array of churches and monasteries. There, too, are fairy-tale castles, especially in the province of SEGOVIA.

The Meseta is mostly hot and dry. Contrary to the words of the song, the rain falls mainly on BILBAO in the north. The wet north-west, though cool, is nevertheless warmed by the North Atlantic Drift in the Bay of Biscay. This part of Spain could be – and constantly expresses a wish to be – another country. Along the steep north coast of Navarra, into the Pyrenees, and over into the south-west corner of France, is the land of the Basques – who defended themselves against the Moors in the 8th and 9th centuries when most of Spain capitulated. They call themselves Euskaldunak. They have lived here, possibly, since the Stone Age and speak a language, Euskera, which may be the oldest in the world.

The Basques' chief characteristic is a yearning for separate nationhood, especially in Spain, where there are more than 2 million of them. Another quarter of a million live in the Basque departments of France, but there the Euskera-speaking proportion is much smaller. While the French Basques are mostly farmers and fishermen, the Spanish Basques include a vociferous, city-dwelling intelligentsia, whose 'capital' is Bilbao. Among these people a movement has grown up which seeks to create an independent Basque state.

In the north-east, CATALONIA, too, has its separatist movements. The region also bridges the Pyrenees, between France and Spain, from French Perpignan to the Ebro delta, and has its own language, Catalan, which resembles Provençal, the old tongue of southern France.

Catalonia and the BALEARIC ISLANDS in the Mediterranean – MAJORCA, MINORCA, IBIZA and the many smaller islands – share a common history and tongue. Catalonian seafarers have been famous since the Middle Ages, and their capital, BARCELONA, is still a great seaport. Textile manufacture, a traditional industry, also thrives, as do new branches of multinational industries – while vineyards, olive groves and orchards provide Catalonia's main agricultural wealth.

Spain's principal tourist area is the COSTA BRAVA, running south from the French-Spanish border to just north-east of Barcelona. Much of the coast is built up. But it is still striking with its red-brown headlands, cliffs and pine forests that reach down to sandy beaches. South of Barcelona is the Costa Dorada where the old Roman city of TARRAGONA remains unspoilt. Farther south still, the pine forests give way to orange and almond groves, and Catalonia to VALENCIA. The port of Valencia, Spain's third largest city, after Madrid and Barcelona, is surrounded by steelworks and cement factories.

In the southern part of Valencia begins the Costa Blanca – the White Coast, a name derived from its clear white light. Here is the specially created resort town of BENIDORM. Here, too, the warm, sunny south begins.

The south coast provides most foreigners' images of Spain, although in fact it is the area in which the Moorish heritage is strongest of all. This is Andalucia, the land of flamenco dancers, of gypsies and toreadors, a land of romance and blood.

In SEVILLE, where magnificence is commonplace, the Moorish tower (La Giralda) is still outstanding, even though its neighbour is the largest and most opulent cathedral in Spain, lavishly ornamented with Baroque flamboyance, while within, paintings, precious metals and jewels glow about the tomb of Columbus.

The conjunction of Moorish and Baroque is not unusual in Andalucia: CÓRDOBA has a splendid, simple mosque, so beautiful that even the victorious Christians forbore to destroy it. Instead, on the orders of Charles V,

▲ HEART OF THE NATION The magnificent 12th-century castle at Alarcón stands on the arid Meseta plateau of Castile, 165 km (100 miles) south-east of the capital, Madrid. The castle is now an exclusive hotel.

a beautiful Baroque cathedral was built inside.

The famous staccato flamenco is the dance of the Andalucian gypsies. Many foreigners take the women's costume for the dance – tight, flounced skirts, high combs and lace mantillas – to be the traditional dress of all Spain. Gypsies still live in Andalucia; in the Moorish quarter of Granada, for instance, where some of them live picturesquely, but quite comfortably, in caves bristling with TV aerials. The old quarters of Andalucian towns are perfection; blinding-white with lime wash, the air jasmine-scented and the courtyards hung with vines and bird cages, brightly splashed with ceramics and geraniums, and cooled by fountains. Seated by outdoor tables, the Andalucians take their evening sherry, whose English name is derived from the wine-producing town of JEREZ DE LA FRONTERA, and watch the world go by.

No wonder Andalucia appeals to tourists, even if for the most part they seldom leave the gravelly beaches of the sunny coast – the Costa del Sol – and its miles-long backdrop of hotels, villas, golf courses and high-rise apartments that run from MÁLAGA all the way to the fashionable resorts of Marbella and San Pedro de Alcántara. Here, the descendants of the Moors have begun a modern invasion, as is apparent from a splendid new mosque and the palaces of Arab royalty.

Away from the beaches, however, in the inland areas, Andalucia is poverty-stricken. There is little work outside agriculture, and the villages are growing old. The young generation has left in search of work.

Galicia is the north-west corner of Spain. Like the provinces which lie to the west of Castile, near the Portuguese border, it is one of the least visited areas of the country. Until the discovery of America, Galicia was known as Finisterre – the end of the earth. It is this very isolation that has preserved the Spanishness of the north-west. This region successfully held out against the Moors for longer than the rest of the country, and the Moors never conquered the mountain fastnesses of ASTURIAS, on the north coast.

The countryside of Galicia and Asturias is lush and green, rather like the landscape of southern Ireland, and is still ideal for growing cereals. The west coast of Galicia, on the Atlantic, is deeply indented with *rias*, submerged river valleys. The region's riches are principally agricultural, with a valuable addition of timber from its pine and eucalyptus forests. Galicia is also known as a holy place. Throughout the Middle Ages, millions of pilgrims from all parts of Europe journeyed to the town of SANTIAGO DE COMPOSTELA and its cathedral, in which are kept the relics of St James the Apostle.

An appreciation of Spain's past is vital to an understanding of the country. Yet it is clear that a new Spain is emerging, and one that is very much part of the modern age. Spain today is part of Europe in many more ways than simply imitating its social habits. It joined the European Community (EC) in January 1986, which spurred economic growth, and the European Union two years later.

SPAIN AT A GLANCE	
Official name Kingdom of Spain	
Area 504 750 km² (194 839 sq miles)	
Population 39 804 000 **Per km²** 79 **(Per sq mile** 204)	
Capital Madrid	
Government Parliamentary monarchy	
Currency 1 peseta = 100 centimos	
Languages Castilian Spanish (official), Catalan, Galician, Basque	
Religion Christian (99% Roman Catholic)	
Climate Mediterranean in the southern and eastern coastlands, temperate elsewhere. Average temperature in Madrid ranges from 5°C (41°F) in January to 24°C (75°F) in July	
Land use Cultivation 41%, meadows and pastures 21%, forest 31%, other 7%	
Main primary products Cereals, fruit, vegetables, sugar cane, grapes, tobacco, cotton, almonds, olives, timber, cork, fish; coal, lignite, iron, lead, copper, tin, zinc, mercury	
Major industries Tourism, agriculture, food processing, wine making, yarns and textiles, iron and steel, mining, petroleum refining, chemicals, engineering, transport equipment, forestry and timber products, fishing, cement manufacture, metal processing	
Main exports Machinery, motor vehicles, iron and steel, chemicals, petroleum products, metal ores and small metal manufactures, sulphur, copper, zinc, fruit and vegetables, wine, olive oil, shoes, textiles	
Annual income per head (US$) 14 020	
Population growth (per thous/yr) 2	
Life expectancy (yrs) Male 74 **Female** 81	

the civil war which broke out in the early 1990s, the city was an important tourist resort serving Dalmatia. In November 1991, it came under Serbian fire but little damage was done.

The old city is an architectural treasure set on a promontory below a ridge which rises to some 1330 m (4363 ft). Split grew up around the retirement home of the Roman emperor, Diocletian, who was born nearby in AD 245. He retired in 305 to the magnificent palace he had built, and was buried here in 313. Afterwards the fortified palace – a 185 by 220 m (about 600 by 720 ft) rectangular complex of luxury apartments, public halls, temples, garrison quarters and catacombs – became a factory producing Roman uniforms, and the nucleus of a city. Early in the 7th century, after the nearby Roman town of Salona (Solin) was sacked by the nomadic Avars from Central Asia, refugees turned the palace into a fortified city with a warren of narrow streets; Diocletian's mausoleum became the city's cathedral. The city prospered under the Byzantines (812-1089) and Venetians (1420-1797), and spread beyond the walls, gaining a graceful *campanile* (bell tower) and many fine buildings. Split has several ancient churches, including Sv Nikola and Gospa od Zvonika, fine museums and a gallery devoted to the works of the Yugoslav sculptor Ivan Mestrovic (1883-1962), which is housed in the sculptor's own villa; a number of Mestrovic's works adorn the city.

Klis, inland from Salona, has the remains of a 5th-century hilltop fortress. The ancient textile town of Sinj, 30 km (19 miles) north-east of Split, is famous for the Sinjska Alka – a medieval tournament held annually on 15 August to commemorate a battle in 1715, when the townsmen, though heavily outnumbered, defeated the Turks.

Population 207 000
Map Croatia – Cc

Spokane *USA* City in eastern WASHINGTON State, 25 km (16 miles) from the Idaho border. It is a transport centre and a regional market for timber, agriculture and mining. Spokane produces food, wood and metal goods.

Population (city) 177 200; (metropolitan area) 361 400
Map United States – Ca

Spoleto *Italy* Town about 90 km (55 miles) north of ROME. It is noted for its Roman theatre, its Arch of Drusus, its vast aqueduct built by the Lombard dukes in the 10th century, and a Gothic cathedral consecrated by Pope Innocent III in 1198. Its industries include leather and textiles. An international music and drama festival is held here each summer.

Population 38 000
Map Italy – Dc

Spratly Islands *South China Sea* Scattered group of reefs midway between Vietnam and north Borneo. Japan seized them for a submarine base during the Second World War, but they are now claimed by a number of countries, including China, Malaysia, the Philippines and Vietnam.

Map Vietnam – Cd

spring Natural fountain or flow of water from the ground. A spring is formed where the water table intersects the surface of the ground. Its position is related to the structure and type of the local rocks, the water table's level and the shape of the

land. For example, a spring is formed when water sinks through a porous rock layer until it meets the saturated rock above an impervious layer. Water within the saturated rock will flow laterally to the spring. Spring water may be hot or cold, hard or soft, depending on its origin underground. See GEYSER; HOT SPRING; MINERAL SPRING.

spring tide Tide of highest range which occurs twice a month about the time of the new moon and the full moon when the sun, moon and earth are approximately aligned. See HOW TIDES ARE PRODUCED, p 670

Springfield *USA* The name of two cities in the United States. The first, the state capital of Illinois, lies 280 km (nearly 175 miles) southwest of CHICAGO in a rich agricultural and coal-mining area. President Abraham Lincoln (1809-65) lived in Springfield for 24 years and was buried here.

The second Springfield, in Massachusetts, lies on the Connecticut River, 145 km (90 miles) west of BOSTON. Founded in 1636, it was the site of the first national arsenal from 1794 to 1968. Manufactures include electronic equipment, clothing, machinery and chemicals. Basketball was invented here in 1891, at Springfield College, the site of the Basketball Hall of Fame.

Population Springfield, Illinois (city) 105 200; (metropolitan area) 189 600;
Springfield, Massachusetts (city) 156 900; (metropolitan area) 587 900
Map (Springfield, Illinois) United States – Ic; (Springfield, Massachusetts) United States – Lb

Springs *South Africa* Mining town on the East Rand, founded in 1885 to supply coal to JOHANNESBURG, 40 km (25 miles) to the west. Gold was discovered here in 1908, uranium followed later.

Population 78 700
Map South Africa – Cb

spruit In South Africa, a term applied to a small river or rivulet.

spur A prominent ridge projecting from a mountainside, mountain range or valley side.

▲ SQUATTERS' PALACE Diocletian's vast palace overlooks the harbour at Split. In the 7th century, the disused building was taken over by refugees fleeing Barbarian invaders.

squall Sudden, violent, short-lived burst of wind, often accompanied by a shower. See LINE SQUALL.

Srbija See YUGOSLAVIA (SERBIA AND MONTENEGRO)

Sri Jayawardhanapura *Sri Lanka* See KOTTE

Sri Lanka See p 634

Sri Padastanaya *Sri Lanka* See ADAM'S PEAK

Srinagar *India* North-western city, in the Vale of Kashmir, some 890 km (552 miles) north-west of Delhi. It is cradled by the Himalayas at the end of Lake DAL which has floating gardens and traditional houseboats.

Srinagar is the capital of JAMMU AND KASHMIR state, and is a busy tourist centre, with palaces, mosques, museums and a university. Among the vast variety of craft goods available in its bazaar are carpets of silk and wool, furs, silver thread-work, jewellery and semi-precious stones, and carved wooden furniture.

Population 708 330
Map India – Ba

stack Pillar of rock rising from the sea, which has been isolated from the land through erosion caused by wave action. See THE CHANGING PATTERN OF THE COASTS, p 219

Stafford *United Kingdom* Borough and county in the Midlands of England, which takes in the POTTERIES and the moorlands that fringe the Peak District National Park. The county, which covers 2716 km² (1049 sq miles), lies north and north-west of the industrial West Midlands and is centred on the valley of the River Trent and the town of Lichfield with its 13th-century cathedral. Burton upon Trent, in the east of the county, has a major brewing industry. (*continued on p 636*)

Sri Lanka

A BEAUTIFUL ISLAND, RENOWNED FOR ITS TEA, ITS SCENERY AND ITS BUDDHIST MONUMENTS, HAS BEEN TORN BY YEARS OF CIVIL CONFLICT

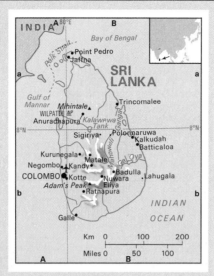

Shaped like a teardrop that has fallen from the cheek of India, the lush tropical island of Sri Lanka has been marred by persistent civil conflict that might well make India – which did briefly become involved in the conflict – weep. Over the centuries many people have traded with Sri Lanka (the name means 'Resplendent Land'), colonised it or settled here. The Romans called it Taprobane, the Arab merchants Serendip, the Portuguese Ceilas and the British Ceylon. It was from Serendip that the word 'serendipity' – the making of happy discoveries by accident – was coined by the English writer Horace Walpole in 1754.

Members of two of the oldest races on the island – the Tamils and the Sinhalese – have not, however, found each other's presence here serendipitous.

The first inhabitants of Sri Lanka, the Veddas, lived here from about 3000 BC. They were conquered around 600 BC by the Sinhalese, who came by sea from north India. The Sinhalese were probably Aryans, since their language is of Aryan origin. They settled mostly in the dry areas of the east and centre of the island, establishing great civilisations which relied on elaborate irrigation systems, with dams and water tanks. Their capital city was ANURADHAPURA, and their religion was, and for most Sinhalese still is, Buddhism.

FOOTPRINT ON A MOUNTAIN

Buddhists say that Buddha left his footprint at the top of a mountain in the south central highlands known as ADAM'S PEAK (2243 m, 7359 ft). Hindus claim this 'footprint', 1.6 m (5.25 ft) long, was made by the god Shiva, but Muslims attribute it to Adam. The spot is a point of pilgrimage for all these religions.

All over the island today there are Buddhist temples and statues, attended by saffron-robed monks, who also conduct wedding ceremonies or lead funeral processions, banging on their drums of mourning.

The Veddas dwindled in number, and only a few small groups survive in the remote interior of Sri Lanka, but another people were to become the lasting enemy of the Sinhalese. Between the 11th and 12th centuries, Hindu Tamils invaded from south India, a mere 29 km (18 miles) away across the Palk Strait. They drove the Buddhist Sinhalese southwards, forcing them to establish a new kingdom in the mountainous interior, around the city of KANDY. The city still contains temples dating from the 11th century when it was the Buddhist capital. The Tamils are still mostly Hindu, although some today are Christian.

Spice-seeking Arab merchants and Portuguese and Dutch colonisers were the next people to arrive in Sri Lanka. Then, in 1796 the British took the island and ruled it until its independence in 1948. The Kandy area, however, held out as an independent Sinhalese kingdom until 1815. The British developed coffee plantations in the mountains, but the crops were devastated by disease in the 1870s and were replaced by Ceylon tea. Rubber and coconut plantations were established on the coast – and in the 19th century more Indian Tamils were brought in to work on the British plantations.

This influx of people changed the distribution of the population – the wet southwest became heavily settled, and the city of COLOMBO became the new capital. The dry east, however, lost all importance, although rice is still cultivated here.

The descendants of the Tamil immigrants now form about 18 per cent of the population, living mainly in the north around JAFFNA, and down the east coast. In the central mountains Tamils still pluck the tender top leaves from the tea bushes, and live in harmony with their Sinhalese neighbours, but in the north trouble between Tamils and Sinhalese led to martial law and a shoot-on-sight curfew which quickly emptied the streets of Jaffna at dusk.

The problems that caused the conflict have surfaced since independence from Britain in 1948. Under British rule, the Tamils, although a minority, held the majority of responsible white-collar jobs. Ceylon was restyled the Republic of Sri Lanka in 1972, when Buddhism was made the state religion. Since then the Sinhalese majority has reclaimed the top jobs. Both communities are well-educated – Sri Lanka has the highest literacy rate (88 per cent) in Asia, after the twin leaders Japan and the Philippines – and the Sinhalese felt that they had simply redressed a numerical imbalance. The Tamils argued that they were being discriminated against politically and in higher education and employment.

As unemployment among educated classes rose in Sri Lanka, what began as a non-violent campaign against discrimination developed into a violent guerrilla fight for an independent homeland for Tamils, to be called Eelam, in the northern and eastern provinces. The government regarded the separatist movement as a communist threat, and the mainly Sinhalese army took on the Tamil guerrillas – sometimes acting violently against ordinary Tamil citizens in retaliation for terrorist attacks on soldiers. Street fighting in Colombo in 1983 left 400 dead – most of them Tamils. Tamil guerillas conquered Jaffna in 1987 and began attacking Sinhalese villages in the east. The government retaliated by bombing Tamil strongholds in Jaffna. India, feeling responsible for the Tamils, sent peace-keeping troops in July 1987 and forced the government to stop the bombings. A treaty was signed with India guaranteeing autonomous status to the Tamils, and introducing Tamil as a national language. Conflict, however, persisted and government forces began to suffer heavy losses against the guerrillas, known as the Liberation Tigers of Tamil Eelam, in late 1993. Earlier that year the president of Sri Lanka, Ranasinghe Premadasa (elected in 1988), had been assassinated. Some progress was made in peace talks during 1994 but these were interrupted by the killing in a suicide bomb attack of a presidential candidate, Gamini Dissanayake, and about 50 other people at a political rally preceding a general election in November. Early in 1995, however, hopes rose again for a resolution of the 11-year conflict. Winner in the elections, Chandrika Kumaratunga, who became both president and prime minister, announced the signing of a truce with the Tigers to be monitored by international observers.

AMBITIOUS SCHEME

At independence in 1948, the population had one of the highest living standards in Asia: income per head was twice as high as that in India. However, while agriculture, forestry and fishing dominate exports, high levels of unemployment persist and the country still has to import food. An ambitious scheme to dam the Ganga River at MAHAWELI provides both irrigation and hydroelectric power. The irrigated region will eventually provide a rice surplus for export. Small farmers already grow rice on the tropical lowlands which make up 80 per cent of the island, but mainly to feed their families. Additionally, power from the dam will provide for the large energy needs of planned heavy industry.

Since 1948 successive governments have continued to provide free education for all, which has given the country its excellent literacy record. They have also given food subsidies to poorer people, and health care is good. These factors, however, have led to a large rise in population (running until recently at a rate of 2 per cent a year – now, with family planning campaigns, down to 1.1 per cent). The population growth, combined with the heavy fall in export prices of coconut fibre (coir), rubber and tea has resulted in a standard of living that is little above India's. The government is trying to encourage development and growth, by improving domestic conditions and encouraging foreign investments.

The export trade of the plantations and gem mines is also the mainstay of the capital and

▲ **A KING'S BEAST** A giant lion, hewn from granite, fronts the magnificent palace fortress of Sigiriya, in Sri Lanka's Central Province. In the 5th century it was the citadel of King Kasyapa.

port of Colombo, a city of 615 000 people. The population of Sri Lanka is young (50 per cent under 25), and Colombo swarms with poor urchins who hawk plastic combs, polish shoes, and work as houseboys and market bearers.

Attempts are being made to increase revenue from tourism. It is a beautiful island and is rich in wildlife. Animals include elephants, leopards, bears and wild boars. But the years of conflict have discouraged many tourists.

SRI LANKA AT A GLANCE

Official name Democratic Socialist Republic of Sri Lanka

Area 65 610 km² (25 332 Sq miles)

Population 17 838 190 **Per kmm²** 271 **(Per sq mile** 704)

Capital Colombo

Government Parliamentary republic

Currency 1 Sri Lankan rupee = 100 cents

Languages Sinhala (official and national), Tamil (national), English

Religions Buddhist (69%), Hindu (15%), Muslim (8%), Christian (8%)

Climate Tropical monsoon, with afternoon breezes on the coast and cooler air in the higher areas. Average temperature in Colombo is 27°C (81°F) throughout the year

Land use Cultivation 33%, meadows and pastures 7%, forest and woodland 37%, other 23%

Main primary products Rice, coconuts, tea, rubber, cassava, fruit, spices, timber, fish, gemstones and semiprecious stones (sapphire, ruby, beryl, topaz, spinel, garnet, moon stone), iron ore, graphite

Mineral resources Limestone, graphite, mineral sands, gems, phosphates, clay

Major industries Agriculture, textiles, mining, forestry, fishing, oil refining

Main exports Tea, rubber, coconut products, textiles, gemstones, petroleum products

Annual income per head (US$) 540

Population growth (per thous/yr) 11

Life expectancy (yrs) Male 69 **Female** 74

The borough of Stafford itself, about 32 km (20 miles) north-west of BIRMINGHAM, dates back to Anglo-Saxon times and was the birthplace in 1593 of the angler and author Izaak Walton. Today it is an industrial town specialising in electrical engineering. South-east of the town is the 120 km² (46 sq mile) heath known as Cannock Chase; in medieval times, this was a royal hunting forest.

Population (Borough) 120 000; (county) 1 049 800
Map United Kingdom – Dd

stalactite Icicle-shaped deposit of calcite (calcium carbonate) hanging from the roof of a cave. See HOW CAVES ARE FORMED, p 141

stalagmite Icicle-shaped mass, similar to a stalactite, but growing upwards from the floor of a cave. See HOW CAVES ARE FORMED, p 141

Stalin Peak *Tajikistan* See COMMUNISM PEAK

Stalinabad *Tajikistan* See DUSHANBE

Stalingrad *Russia* See VOLGOGRAD

Stalino *Ukraine* See DONETS'K

Stalinsk *Russia* See NOVOKUZNETSK

Stanley *Hong Kong* Fishing village situated on a peninsula on the southern coast of HONG KONG ISLAND. Although there are many shops selling goods to tourists, Stanley remains less affected by modernisation than most other settlements on Hong Kong Island, and is a convenient base from which to explore the coves and beaches along the south-eastern coast of the island.

Population 15 000
Map Hong Kong – Cb

Stanley *South Atlantic* See FALKLAND ISLANDS

Stanley, Mount *Uganda/Zaire* See RUWENZORI

Stanley Falls *Zaire* See BOYOMA FALLS

Stanley Pool *Zaire/Congo* See ZAIRE

Stanleyville *Zaire* See KISANGANI

Stara Planina *Bulgaria* See BALKAN MOUNTAINS

Stara Zagora *Bulgaria* City on the south slopes of the Balkan Mountains, some 80 km (50 miles) north-east of PLOVDIV. It is an agricultural trading centre dealing in wheat, wine and attar of roses. Its chief industries are brewing, distilling, tanning, textiles, fertilisers and flour milling.

In 1877 it was the scene of a Turkish victory over the Russians, during the Russo-Romanian invasion which resulted in the formation of the principality of Bulgaria (1878).

Population 164 550
Map Bulgaria – Bb

Stargard (Stargard in Pommern) *Poland* Medieval fortress town in Pomerania, 32 km (20 miles) south-east of the port of Szczecin. Much of its 13th-century walls and a Renaissance town hall survive. The town produces metals, processed food, soaps and detergents.

Population 63 300
Map Poland – Ab

Starnberger See *Germany* Bavarian lake 25 km (16 miles) south-west of MUNICH. It covers some 57 km² (22 sq miles). The tourist town of Starnberg lies at its northern end. A cross on its eastern shore marks the spot where in 1886 the Bavarian king, Ludwig II, confined in a château on the lakeshore because of insanity, drowned his doctor and himself while out for a walk.

Map Germany – De

Staropolska *Poland* The country's oldest heavy industrial area, 130 km (80 miles) south of the capital, WARSAW. The Romans began smelting ores here; the first charcoal-iron furnace began production in 1598. Foundries and works which produce Star trucks continue the metal-working tradition at Ostrowiec, the area's main town.

Map Poland Dc

Statia *Netherlands Antilles and Aruba* See ST EUSTATIUS

Stava *Italy* See TRENTINO ALTO ADIGE

Stavanger *Norway* City in the extreme southwest of the country, on Stavanger Fjord. It is one of Norway's largest ports, now focusing on the construction of drilling rigs and tankers for liquefied petroleum gas. Close to Norway's North Sea oil and natural gas fields, it is the headquarters of the Norwegian Oil Board, and of many supply and charter firms. Founded in 1125, when building started on its cathedral, Stavanger also has a theatre and several museums.

Population 101 500
Map Norway – Bd

Stavropol' *Russia* The name of two Russian towns. One, in the northern foothills of the Caucasus Mountains, 235 km (145 miles) east of KRASNODAR, was founded in 1777 as a fort, and became a trading centre on the main Russia-Georgia road. Its products today include canned meat, fruit juice, wine and leather goods. The other, a town on the Volga River, was drowned when the Volga was dammed. A new town called TOL'YATTI was built on the shoreline.

Population 324 000
Map (Stavropol') Russia – Fd; (Tol'yatti) Russia – Fc

Stein am Rhein *Switzerland* One of the country's best preserved medieval towns, with its castle, churches, market place and town gates all intact. It also has some fine, old timbered houses with elaborately painted façades. It stands beside the western tip of Lake CONSTANCE, near the German border.

Population 2900
Map Switzerland – Ba

Stellenbosch *South Africa* The second oldest European settlement, after Cape Town, which lies 30 km (20 miles) to the south-west. Founded in 1679 and named after Governor Simon van der Stel, Stellenbosch shares the status of cultural heartland of the Afrikaner people with PAARL. Its university has strong Afrikaner affiliations, and six out of eight of South Africa's prime ministers between 1910 and 1983 were former students. Stellenbosch lies at the centre of a rich fruit- and wine-farming district.

Many of the district's wine estates – a number boasting classic Cape Dutch dwellings – are open

to visitors. Some attractive architecture has survived in the oak-shaded older parts of the town.

Population 73 840
Map South Africa – Ac

Stelvio National Park *Italy* Alpine park covering 1316 km² (508 sq miles) in a remote area near the Swiss frontier, 150 km (95 miles) north-east of MILAN. It has a number of peaks over 3500 m (11 500 ft) high, including Ortles (3905 m, 12 812 ft), and also takes in Italy's biggest glacier, the dei Forni, which is 5 km (3 miles) long.

Stelvio, founded in 1935, is the home of deer, chamois, lynx and bears, and of rare sub-alpine forest, tundra moss, lichen and wild grasslands.

Map Italy – Ca

Stelvio Pass *Italy* Alpine pass between Italy and Switzerland, 150 km (95 miles) north-east of MILAN. It was developed in 1825 to provide the Austrian rulers with an easier route to Milan. It rises to 2758 m (9048 ft) and leads through the spectacular scenery of the STELVIO NATIONAL PARK.

Map Italy – Ca

Stendal *Germany* Former weaving town and rail junction, 110 km (68 miles) west of BERLIN. It is rich in Gothic churches. Nearby, on the banks of the River Elbe, is a nuclear power station.

Population 49 400
Map Germany – Db

steppe Mid-latitude grassland similar to PRAIRIE, consisting of level plains which extend across Eurasia from the UKRAINE to MANCHURIA.

Stereá Ellás *Greece* Mountainous central region south of Thessaly and Epirus, bounded by the Aegean Sea in the east and the Ionian Sea in the west. It includes the capital, ATHENS, and the classical centres of DELPHI, Orehomenos and THEBES. Covering 24 500 km² (9572 sq miles), it contains some of Europe's largest reserves of bauxite. Its main industries are agriculture, fishing, tourism, engineering and the manufacture of chemicals, textiles and aluminium.

Population 4 127 200
Map Greece – Cb

Stettin *Poland* See SZCZECIN

Stewart Island *New Zealand* Largest of the country's offshore islands, covering 1746 km² (674 sq miles), off the southern tip of South Island. It is a haven for some of New Zealand's rarest birds, such as the kakapo, a type of owl-like, nocturnal parrot which lives on the ground.

Population 680
Map New Zealand – Ag

Steyr *Austria* Industrial city on the confluence of the Steyr and Enns rivers, 28 km (17 miles) south of LINZ. The old city dates from the 13th century and has many fine houses, particularly in the main square. Its manufactures include lorries, tractors, mopeds, iron goods and sporting guns. It is an iron and steel industrial centre. Its goods are shipped via the Enns River.

Population 39 540
Map Austria – Da

Stirling *United Kingdom* City of central Scotland, some 35 km (22 miles) north-east of

GLASGOW. The rocky outcrop above the city, which guards the valley of the River Forth, has probably been fortified since Roman times. The castle dates back to the 11th century and is a smaller version of Edinburgh's. To the south of the city, a monument commemorates the Battle of Bannockburn in 1314, when the Scots under Robert Bruce defeated the English and won independence from Edward II. To the north of Stirling is its university, founded in 1967. The former county of Stirling was divided in the 1970s between CENTRAL REGION and STRATHCLYDE.

Population 39 000
Map United Kingdom – Db

stock See BATHOLITHS

Stockholm *Sweden* The Swedish capital, set on some 20 islands and peninsulas between Mälaren Lake and the Baltic Sea. Stockholm grew up in the mid-13th century around a stockaded island fortress. Its name derives from the Swedish words *stock,* which translates as 'pole' and *holm* which translates as 'island'. The three main islands are those of Stadsholmen, Helgeands-holmen and Riddarholmen.

Today the historical heart of the city, now linked to nearby islands, is known as 'the Town between the Bridges'. It contains the houses of parliament, the royal palace and Riddarholm Church, the burial place of Sweden's kings. Many of the medieval streets are now pedestrian precincts, crammed with antique shops; the 17th-century burghers' houses have been turned into offices and private homes.

To the east of the old town is a group of islands whose buildings face waterfronts where motorboats, yachts and schooners are moored. On Djurgården Island the capital's maritime past lives on in a museum housing the *Wasa,* a 60 m (200 ft) long warship that sank on her maiden voyage in Stockholm harbour in August 1628 and was raised to the surface again in 1961.

Over to the north, much of the 19th-century shopping and residential district has been replaced by the high-rise Högtorget quarter. The city's garden suburbs and newer districts are linked by underground railway to central Stockholm. The open-air museum – a living reminder of Sweden's rural past – was a pioneer museum of its type. In total contrast is the royal summer palace of Drottningholm.

The city is Sweden's largest manufacturing centre, producing machinery, chemicals, textiles, rubber and beer – and it also has flourishing electrical, engineering and printing industries.

There is a frequent ferry service to Finland, and the main international airport is at Arlanda, 40 km (25 miles) north of the city. Each December the Nobel Prizes – founded by Alfred Nobel (1833-96), the Swedish chemist and pacifist who invented dynamite and gelignite – are awarded in Stockholm, except for the Peace Prize which is awarded in Oslo, Norway.

Population (city) 684 580; (metropolitan area) 1 600 000
Map Sweden – Cd

Stockport *United Kingdom* Town on the southeastern outskirts of MANCHESTER. Originally a market town, it developed as a cotton-spinning centre in the 19th century. Today its principal industries are in engineering. A railway viaduct more than 30 m (100 ft) high, with 22 arches, spans the valley of the River Mersey, which runs through the town.

Population (town) 136 790; (district) 288 300
Map United Kingdom – Dd

Stockton *USA* Industrial city and inland seaport in California, standing on the San Joaquin River about 100 km (62 miles) east of SAN FRANCISCO. It grew during the gold rush of 1849, and is the distributor and processor of agricultural produce and wines of the Central and San Joaquin valleys.

Population (city) 211 000; (metropolitan area) 480 600
Map United States – Bc

Stockton-on-Tees *United Kingdom* Town on the north bank of the River Tees in the English county of Cleveland, 50 km (30 miles) south of NEWCASTLE UPON TYNE. Originally a quiet market town – with the widest main street in Britain – it developed as an industrial centre after the opening in 1825 of the world's first public railway, the Stockton & Darlington. This brought coal from mines inland to wharves on the Tees. Today there are enormous chemical works in the northern suburb of Billingham.

Population (town) 87 220; (district) 172 540
Map United Kingdom – Ec

Stoke-on-Trent *United Kingdom* Principal city of The Potteries in the central English county of Staffordshire. It lies 65 km (40 miles) north of BIRMINGHAM and is the home of some of the world's largest china and porcelain firms, among which are Wedgwood and Spode.

Population (district) 253 100
Map United Kingdom – Dd

Stolp *Poland* See SLUPSK

Stonehenge *United Kingdom* Prehistoric stone monument near Amesbury in the English county of Wiltshire, 120 km (75 miles) south-west of LONDON. It was built between about 2750 and 1300 BC, possibly as a religious or ceremonial centre, and the remains are made up of some 150 boulders – the tallest 6.7 m (22 ft) high and weighing about 45 tonnes – in two main circles. There are traces of other earth and stone circles.

Although Stonehenge was reputedly used by Celtic Druids as a temple for sun worship, the Druids did not arrive in Britain until the 3rd century BC – at least 1000 years after Stonehenge was completed. Some stones were transported over 320 km (nearly 200 miles) from Wales.

Map United Kingdom – Ee

Store Bælt *Denmark* See GREAT BELT

storm Violent atmospheric disturbance with strong winds accompanied by rain, snow or other precipitation, and often by thunder and lightning. The term is also applied to a strong wind that is rated force 10 and force 11 (violent storm) on the BEAUFORT SCALE, with speeds between 88-101 km/h (55-63 mph) and 102-119 km/h (64-73 mph) respectively.

storm beach Accumulation of coarse material such as shingle and cobbles formed above the foreshore during a period of powerful storm waves at the highest tides.

storm surge Rapid rise of the sea level above predicted tidal levels when water is piled up against a coast by powerful onshore winds.

strait (straits) Narrow passage of water, narrower than a sound, joining two larger bodies of water. For geographical features whose name begins with 'Strait', see the main part of the name.

Stralsund *Germany* Port on the Baltic Sea opposite the island of RÜGEN. It has many medieval buildings and is a growing fishing, shipbuilding, manufacturing and tourist town.

Population 73 100
Map Germany – Ea

▼ MYSTERIOUS STONES A storm sweeps over Stonehenge, a lonely sentinel on Salisbury Plain. To preserve the monument, access is restricted.

Strasbourg *France* Industrial city and port near the place where the Ill River flows into the Rhine on the German border, almost due east of Paris. It is the capital of ALSACE, and has many canals.

The city's Gothic cathedral – dating from the 11th to the 15th century – is built of red sandstone from the VOSGES; many of the city's 17th-century houses have timbered exteriors. The Council of Europe and the European Parliament of the Common Market hold their meetings in the city.

Population 378 500
Map France – Gb

strata See STRATUM

Stratford *Canada* Town 160 km (100 miles) west of TORONTO and the home of Canada's summer Stratford Shakespearean Festival, started in 1953. It is also a railway engineering centre.

Population 27 670
Map Canada – Gd

Stratford-upon-Avon *United Kingdom* English town in the county of Warwickshire, 40 km (25 miles) south-east of BIRMINGHAM. It was the home town of the playwright and poet William Shakespeare (1564-1616). His birthplace, a beautiful half-timbered house, the school he probably attended and, at nearby Shottery, the family home of his wife, Anne Hathaway, have been well preserved. The Memorial Theatre overlooking the River Avon is now the Royal Shakespeare Theatre. Adjoining it is the Swan Theatre. The town has given its name to many other towns throughout the English-speaking world, for example, STRATFORD in Ontario, Canada.

Population 21 730
Map United Kingdom – Ed

strath In Scotland, the name given to a steep-sided, flat-floored valley that is wider than a glen.

Strathclyde *United Kingdom* Administrative region in western Scotland, named after the valley of the River Clyde, which is centred on GLASGOW. It covers 13 503 km² (5212 sq miles) and takes in the counties of Ayrshire, Bute, Dunbartonshire, Lanarkshire and Renfrewshire, and parts of Stirlingshire and Argyllshire, so that it extends from within 50 km (30 miles) of the English border to the southern Highlands and islands. Most of the north is mountain and moorland. The south was heavily industrialised, but most industries have closed down. Hamilton and Coatbridge developed as iron towns in the 19th century, providing materials and machines that, until the 1960s, supplied the world with ships and railway locomotives. The last surviving steel plant in Scotland, Ravenscraig, closed in 1992. However, the computer industry is rapidly expanding, as are industries in electronics, information technology and financial services.

Population 2 218 230
Map United Kingdom – Cc

stratocumulus Low cloud formation of large, soft grey rolls frequently covering the whole sky, usually associated with dry, dull weather.

stratopause See THE DIFFERENT LAYERS OF THE ATMOSPHERE, p 53

stratosphere See THE DIFFERENT LAYERS OF THE ATMOSPHERE, p 53

stratum Distinct layer of sedimentary rock. One stratum is usually called a bed. The plural, strata, applies to several beds.

stratus Low cloud formation of white, foggy sheets, often bringing drizzle.

strike Direction of a horizontal line in the bedding planes of inclined strata. It lies at right angles to the direction of DIP.

Stromboli *Italy* Active volcano island in the EOLIAN ISLANDS to the north of Sicily. Its summit towers about 926 m (3038 ft) above sea level. The volcano's nightly fireworks are spectacular, though rarely dangerous; however, the volcano erupted strongly in 1966.

Population 300
Map Italy – Ee

Stuttgart *Germany* State capital of Baden-Württemberg, situated 150 km (93 miles) south of FRANKFURT am Main. In the 10th century, a German duke set up a *stutgarten*, or stud farm, in this natural basin at the north of the Swabian Jura. The city takes its name from this stud and keeps a horse in its coat of arms today. It also lives up to its name as a *garten* (the German for 'garden') – more than half of the land in the city centre is filled with parks and woods.

Stuttgart's port lies on the Neckar River, north-east of the city, and industrial plants line the river's banks. Among them is the Daimler-Benz factory, the world's oldest car plant, founded in 1890, which has a museum and the original workshop of Gottlieb Daimler (1834-1900), as well as the Porsche car plant. Other products include electronic goods (the Bosch company operates from here), machinery, precision instruments and textiles. The city is also an important educational, cultural and publishing centre with more than 100 publishers based here.

Population 580 000
Map Germany – Cd

Styria (Steiermark) *Austria* Forested, mountainous *Bundesland* (state) in the south-east, that covers an area of 16 387 km² (6326 sq miles), crossed by the rivers Mur, Mürz and Enns. It is noted for its mining and metal industries and is rich in iron ore. Another natural resource is its timber, and there are paper, cellulose and timber industries. The state is also well known for its agricultural products. The capital is GRAZ.

Population 1 184 600
Map Austria – Db

subduction zone Long, narrow zone along which an oceanic plate is moving down into the earth's MANTLE. It is also known as a 'Benioff zone'. See PLATE TECTONICS, p 542

Subotica (Subotitsa; Szabadka; Maria-Teresiopel) *Yugoslavia (Serbia and Montenegro)* Second largest city in the Serbian autonomous province of VOJVODINA after Novi Sad. It is a centre for the fertile Bačka Plains which stretch south to the Danube; its factories produce heavy machinery, farm equipment, processed food, furniture and chemicals. Subotica lies 160 km (100 miles) north-west of BELGRADE, 30 km (19 miles) from the Hungarian border. Its inhabitants are largely ethnic Hungarians and there are many ornate buildings in the

Hungarian style. The resort of Palić on the shore of Lake Palić, surrounded by acacias, vineyards and orchards, is 8 km (5 miles) to the east.

Population 150 300
Map Yugoslavia – Ba

subsidence Sinking of part of the earth's surface relative to its surroundings due either to local factors, such as mining activities, or to major processes affecting the earth's crust, such as the formation of a RIFT VALLEY.

subtropical climate Warm climate in low, temperate latitudes, with a distinct seasonal rhythm. There are two types: dry subtropical, known as Mediterranean climate, which has long, dry summers and is found on the west side of continents; and wet subtropical, with higher rainfall and wet summers, found on the east side of continents.

Sucre *Bolivia* City and the seat of the judiciary, a university and an archbishopric. Sucre is the country's constitutional capital. It was founded in 1538 as Charcas by the Spaniard Pedro de Anzures and renamed in 1826 after Antonio José de Sucre, the Venezuelan who freed parts of Peru and Bolivia from Spanish rule and was president from 1826 to 1828. It has some fine colonial buildings, including the legislative palace where Bolivia's independence was declared in August 1825. Local laws decree that all buildings be painted in the original colonial white.

Population 95 690
Map Bolivia – Bb

Sudan See p 640

Sudbury *Canada* Mining town some 370 km (230 miles) north-west of TORONTO. It is the centre of one of the world's richest nickel-mining areas and also has copper and platinum mines.

Population 150 000
Map Canada – Gd

sudd Floating mass of compact vegetation that is found on the White NILE, where it often obstructs navigation. It consists chiefly of plants from swamps lying alongside the river. The mass may sometimes form a barrier over 30 km (19 miles) long and up to 5 m (16 ft) thick.

Sudd *Sudan* Swampland in the south, covering some 129 500 km² (50 000 sq miles) – about the size of England. Evaporation from the swamp leads to major water loss from the White Nile.

Map Sudan – Ac

Sudetenland *Central Europe* See SUDETY

Sudety (Sudetes; Sudeten; Sudetenland) *Czech Republic/Poland* Series of mountain ranges between the Czech Republic and Poland. The Riesengebirge (Karkonosze; Krkonoše; Giant Mountains) are the highest. They reach 1602 m (5257 ft) at Mount Sniezka (Sněžka) on the border, and are a well known skiing area.

The Czech Sudety was part of Austria-Hungary until 1919 and had a significant proportion of German-speaking inhabitants. Called 'Sudetenland', it was handed over to Nazi Germany by the Munich agreement of 1938. However, together with similar areas of north-west and south-west Czechoslovakia, also called

Sudetenland, it was returned to Czechoslovakia in 1945, and its German-speaking people were later expelled.
Map Czech Republic – Ba-Ca

Suez (El Suweis) *Egypt* Town at the southern end of the SUEZ CANAL, for which it is a key control point. Suez has a fuelling station for ships; it has two oil refineries and manufactures fertilisers. To the south are long stretches of sandy beaches.
Population 195 000
Map Egypt – Cb

Suez, Gulf of *Egypt* Horn-like north-western arm of the RED SEA, which is 290 km (180 miles) long and 32 km (20 miles) wide. It is an important shipping lane because of the SUEZ CANAL which links it to the MEDITERRANEAN SEA. It also has offshore oil wells. This branch of the GREAT RIFT VALLEY system is bordered by Egypt's Eastern Desert in the west and the SINAI in the east.
Map Egypt – Cb

Suez Canal *Egypt* Ship canal built by the French engineer Ferdinand de Lesseps to link the Mediterranean with the Red Sea. The accepted length is 173 km (107 miles). It runs from Port Said on the Mediterranean coast through Lake Timsah and the Bitter Lakes to the port of Suez.
    The canal was opened in November 1869, shortening the journey to India by more than 11 000 km (7000 miles) by cutting out the trip around the Cape of Good Hope. It was run by a French company, which Britain took over in 1875.
    Britain remained a major shareholder until 1956 when the then Egyptian president, Gamal Abdel Nasser (1918-70), nationalised the canal, sparking the Suez conflict with Britain, France and Israel, during which the Egyptians blocked it with scuttled ships. Since then it has been entirely controlled by Egypt. The canal was blocked again in 1967 during the Six-Day War between Arab states and Israel, and reopened in June 1975. Today, Israeli shipping is allowed through. More than 17 000 vessels a year pass through the canal.
Map Egypt – Cb

Suffolk *United Kingdom* County of eastern England covering 3800 km² (1467 sq miles) and forming part of EAST ANGLIA. The south-east has the estuaries of the Stour, Orwell and Deben, the county town of IPSWICH and the ports of Harwich and Felixstowe. In the west are sandy heathlands known as the Breckland, north of BURY ST EDMUNDS, and the grasslands of the horse-racing town of Newmarket. In between is a gentle agricultural land immortalised by the paintings of the Suffolk artists John Constable (1776-1837) and Thomas Gainsborough (1727-88).
Population 653 800
Map United Kingdom – Fd

Sukhothai *Thailand* Agricultural province and market town about 385 km (240 miles) north of the capital, BANGKOK. The old town was the country's first capital, in the 13th and 14th centuries, and has been restored. The crops of the province include rice, mung beans, soya beans and cotton.
Population (province) 562 800; (town) 11 500
Map Thailand – Ab

sukhovey Hot dry wind, blowing mainly from the east during summer in Ukraine, Moldova, south-western Russia and Kazakhstan.

Sukhumi *Georgia* See SOKHUMI

Sulawesi (Celebes) *Indonesia* Curiously shaped volcanic island of 179 370 km² (69 255 sq miles) between BORNEO and New Guinea. It has six active volcanoes and is largely mountainous, but has some good farmland, producing rice, sugar cane, coffee, cassava beans and maize.
Population 12 508 000
Map Indonesia – Ec

Sulu Archipelago *Philippines* Group of some 400 named and many more unnamed islands covering 982 km² (379 sq miles) between MINDANAO and BORNEO, and forming the southern boundary of the SULU SEA. It consists of two main island chains – the northern Pangutaran group, mainly coral reefs, and the southerly main group of larger, mostly volcanic islands, which include BASILAN and JOLO. It is a distinctive region whose inhabitants – the Moros – make a living by fishing and raising crops such as rice.
    The islands became part of the Philippines in 1940. Long the home of pirates, they were ruled by the sultans of Jolo until the late 19th century, when they came under Spanish and later American control. The population is staunchly Muslim.
Population 360 600
Map Philippines – Bd

Sulu Sea (Sea of Mindoro) *South-east Asia* Lying between the south-western Philippines and BORNEO and covering an area of 260 000 km² (100 386 sq miles), the sea was once dominated by Moro (Philippine Muslim) pirates. Moro boats with their vividly coloured sails are still much in evidence, but fishing has now replaced piracy.
Map Philippines – Ad

Sumatra (Sumatera) *Indonesia* Second largest of the Greater SUNDA ISLANDS, covering an area of 473 607 km² (182 860 sq miles) off the west coast of Peninsular Malaysia. The centre of Indonesia's oil industry, it also produces rubber and timber, as well as coffee, coconuts, rice and spices.
Population 36 882 000
Map Indonesia – Ab-Bc

▲ **INTERNATIONAL ARTERY** Oil tankers make their way through the Suez Canal. The waterway can now take all but the very largest supertankers, and it is still being enlarged.

Sumba *Indonesia* Island covering 11 153 km² (4306 sq miles) in the Lesser SUNDA ISLANDS, south of FLORES. It produces maize, tobacco, rice, coconuts and fruit, but receives less rainfall than the other Indonesian islands. Once forested, Sumba is now deforested and overgrazed.
Population 251 000
Map Indonesia – Ed

Sumbawa *Indonesia* Island covering 15 448 km² (5965 sq miles) in the Lesser SUNDA ISLANDS, west of FLORES. Its highest volcano, Tambora, standing at 2821 m (9258 ft), is the remnant of a much larger volcano which devastated the island in 1815, killing 50 000 people and causing many thousands to starve; 35 000 people emigrated. Tropical products are grown, including rice, sweet potatoes and soya beans.
Population 195 600
Map Indonesia – Ed

Sumer *Middle East* Ancient kingdom in the region of Mesopotamia, founded about 5000 BC. The Sumerians are credited with inventing the earliest form of writing in about 3500 BC. It consisted of cuneiform (wedge-shaped) symbols and pictures inscribed on damp clay tablets which were then dried. Some 50 000 tablets, depicting the Sumerian way of life, have been discovered at Nippur, the greatest city of Sumer, 100 km (62 miles) south of BABYLON. Sumer was conquered in about 1950 BC by the Babylonians; today, the area forms part of Iraq.
Map (Babylon) Iraq – Cb

Sun City *South Africa* Modern holiday complex with a staff complement of over 4500, created in 1979 in the former 'independent' Republic of BOPHUTHATSWANA, some 150 km (93 miles) north-west of Johannesburg. A vast new section, known as 'the Lost City' (continued on p 642)

Sudan

AFRICA'S LARGEST COUNTRY, THOUGH RAVAGED BY DROUGHT AND WAR, IS A LAND OF GREAT POTENTIAL

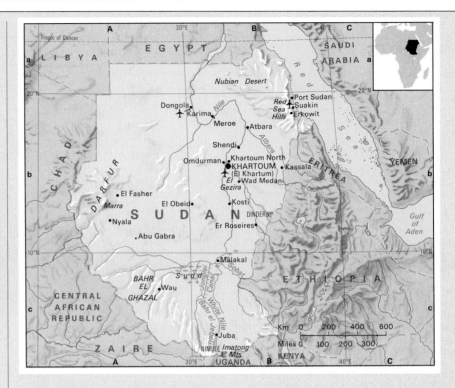

The Republic of Sudan is the largest country in Africa, and also has some of its biggest problems – drought, famine, economic crisis, political instability and civil war. The famine has grown steadily worse as annual rainfall has decreased since the mid-1970s. The crisis-ridden economy has for years suffered from a lack of investment and an inability to earn foreign currency. In 1992 inflation stood at 150 per cent; moreover, the economy was stagnant.

The fragile political scene, initially held together by the dictatorial President Jaafar Nimeiri – there were 12 attempted coups from the time he seized power in 1969 – faced new uncertainties when an army coup ousted him in April 1985. From 1986 several coalition civilian governments held power until 1989, when a military coup was led by Lieutenant-General Omar Hassan Ahmad al-Bashir.

A civil war between the north and south, caused by differences in race and religion, has been on and off the boil since 1955. Sudan is a melting pot of races. The people of the north are Arab and Muslim; their affinities tend to be with the Middle East. The people of the south are black Africans – some Christian, although most have indigenous beliefs – speaking their own languages. Southern peoples include the Dinka, Nuer (some of the tallest people on earth, with many men reaching more than 2 m [6 ft 7in]), Shilluk, Bari and Azande. Each has its own customs and beliefs, as well as music, dances and handicrafts.

Arabic is the main language of Sudan and English is often understood, but altogether 115 languages are spoken. About a quarter of the people live in towns; however, Sudan still has a long way to go in its development. For instance, about one child in every ten dies before the age of one.

Sudan occupies much of the Upper NILE basin, from the foothills of the East African plateau to the SAHARA. The river flows some 3475 km (2160 miles) from the Uganda border to Wadi Halfa on the border with Egypt. The White Nile drains the Ugandan highlands with their year-round rain; the Blue Nile drains the mountains of Ethiopia where rain comes only in summer, causing the great annual flood so important to Egypt. Sometimes flooding can have catastrophic consequences – as in 1988, when large parts of KHARTOUM, where the rivers join, were under water and more than 1 million people made homeless.

In the north the Nile winds through the Libyan and Nubian deserts, extensions of the Sahara, its palm-fringed valley a narrow strip of habitable land. Towards the RED SEA in the east is broken hill country where irregular rain gives scant pasture for sheep and camels of the Beja nomads. Across central Sudan stretches the savannah which produces most of the country's food; 250-750 mm (10-30 in) of rain normally falls in summer, while east of the White Nile there is large-scale irrigation. The DARFUR Highlands in the far west rise to about 3071 m (10 073 ft) at Jebel Gimbala.

In southern Sudan the flat, well-watered clay plains are flooded in the summer wet season, and cash crops can be grown. Areas of permanent swamp, papyrus beds and floating vegetation block navigable channels in the SUDD region. During the dry season, grass fires sweep the area, so few trees grow.

Sudan is a country of tropical heat. Summer temperatures range from 27°C (81°F) in the north to 46°C (115°F) in the south, winter temperatures from 16°C (61°F) to 27°C (81°F).

The economy is largely dependent on agriculture: farming – mainly subsistence farming – employs 80 per cent of the workforce. Many northern people are nomads whose way of life has changed little in generations.

Although only 15 per cent of the arable land is cultivated, the country is a leading African producer of some agricultural products, such as cotton, groundnuts and sesame seed. But a large percentage of agricultural export earnings are spent on importing grains.

Commercial farming has expanded with several large-scale irrigation projects. Cotton is produced under the long established EL GEZIRA cooperative scheme, and cotton and cotton products normally account for about half of exports. The similar KENANA scheme aims to make Sudan an exporter of sugar. In the south, the JONGLEI CANAL, 360 km (224 miles) long, will drain much of the vast Sudd swamp.

The country has some 2 million hectares (4.9 million acres) of agricultural land under irrigation, most of this lying between the courses of the White and Blue Nile in east-central Sudan. Although these areas account for less than half of Sudan's potential for irrigation, they represent 60 per cent of Africa's total irrigated land.

But a constant problem has been to find foreign exchange to pay for the farm machinery, fertilisers and oil that these huge developments need. An ambitious scheme in the 1970s had envisaged using Arab oil money to develop Sudan as a 'breadbasket' for the Middle East, but the petrodollars never materialised.

Sudan is the world's greatest source of gum arabic, used in medicines and inks. Oil has been discovered, and development of a large oil field near Bentiu in the Sudd would provide all the country's energy needs. But its exploitation has been delayed by guerrilla warfare.

A small manufacturing sector concentrates on food processing, building materials, cigarettes, petroleum products, sugar and textiles. Many skilled Sudanese work in The Gulf states. The money they send home helps the economy, but their absence is felt in a country which urgently needs a skilled workforce.

In ancient times, two civilisations flourished in Sudan: Nubia and Kush. Three Sudanese kingdoms developed in the 4th century; in the 6th century, they were converted to Christianity, and resisted the spread of Islam from Egypt until the 14th century. Then, in 1820 Egypt occupied much of the north and developed Sudan's trade in ivory and slaves, as well as the country's cotton industry.

In 1881, the *Mahdi* ('divine guide') Muhammad Ahmed led a revolution to reform Islam and drive out foreigners from Sudan. General Charles Gordon, who had been the British Governor-General of Sudan from 1877-9, a time when Britain was the colonial power in Egypt, was sent back with orders to withdraw the Egyptian army. But he was killed when Khartoum fell in January 1885 after a ten-month siege – two days before a relief force arrived. In 1898 the army of the *Khalifa*, the *Mahdi's* successor, was defeated at OMDURMAN by an Anglo-Egyptian army under General Kitchener. The two countries then established a condominium over Sudan, but in practice the country was run by the British.

Civil war broke out in 1955 as independence approached. The black African south tried to break away from the Arab north. Independence on 1 January 1956 led to further political instability. A multiparty democracy was followed by a military coup in 1958 and by a return to civilian rule in 1964. In 1969 Colonel Nimeiri led a second coup, and three years later ended the civil war by giving regional autonomy to the three southern provinces. The war had lasted 17 years and cost nearly 500 000 lives.

Nimeiri established a leftist government and nationalised industries and banks. In 1973 he made Sudan a one-party state. In 1977, however, he expelled his Soviet military advisers and began to turn to the West for aid. By the mid-1980s, Sudan was receiving more than US$700 million of foreign aid each year – more than any other African country except Egypt.

In 1983 Nimeiri tried to consolidate his support among the fundamentalist Muslims by introducing *Sharia* law, with its harsh punishments of whippings, maimings and hangings. Alcohol was banned, thieves maimed, and couples unable to prove their marriage were lashed.

This sparked rebellion by non-Muslims in the south, and civil war broke out afresh. The Sudanese People's Liberation Army sank a Nile steamer, killing hundreds and halting river traffic. They attacked foreign installations and mined roads and railways. By 1986 they were controlling much of the south.

Sudan has in fact waged war against itself since independence – a war that has claimed about 1 million lives and displaced more than 3 million people, mainly from the south. Some of them fled to the north, others into neighbouring countries – Uganda, Zaire, Kenya, Ethiopia and the Central African Republic.

By the 1990s the Sudanese government began to clamp down heavily on the country's southern population. Not only did it send armed forces into the region and bomb its villages and towns, but it also introduced discriminatory and often inhumane laws, such as crucifixion, and group punishment for an individual's crimes. The West, outraged at human rights violations committed by the Sudanese government against its southern population, and enraged at the government's continued support of terrorism worldwide, cut all aid, and imposed sanctions. Sudan found itself in international isolation. War has drained its coffers, as have the years of drought. Added to this, drastic cuts in foreign aid have been made – from US$1 billion in 1989 to less than $50 million in 1993 – and the country faces bankruptcy.

Successive years of drought have hit the Sudan, a country which is prone to dust storms and desertification. The north-west and northeast have been continuously affected, and large tracts of land have been left parched and spoiled. Rivers run dry; no green foliage grows; the carcasses of animals lie in the sun.

About 300 000 Sudanese died of starvation in a long drought in 1988, when horrifying scenes of famine brought international aid organisations to the rescue – but they were hampered by poor administration and serious transport difficulties.

In early 1994, drought occurred once again in the war-torn south, exhausted after years of civil war. The southern rebel forces were split along religious and tribal lines, and had spent their entire resources. Simultaneously, the government launched a carefully planned offensive against the south, and the UN made an urgent appeal for humanitarian aid for people displaced due to the fighting.

Despite the onslaught on the south, and the tightening of laws, the country was putting on a semblance of political liberalisation. The first democratic elections for 18 years, held in April 1986, brought a coalition of northerners led by Sadiq al Mahdi to power. However, the democratic government was shortlived. After the military coup in 1989, all political parties were banned, and the country was ruled by a 15-member Revolutionary Command Council (RCC). In January 1992 a transitional parliament was formed as a step towards democracy. In October 1993 the ruling council was dissolved and a civilian president appointed to lead a transitional government that was to usher in elections within two years. During 1994 peace talks were held with the southern rebels after a redivision of the country into 26

▼ **WEEKLY CELEBRATION** The Dinka gather near the southern capital of Juba for a tribal dance every Friday. They and their herds migrate from villages in the savannah to riverside pastures in the dry seasons.

states and the appointment of a Christian southerner as a vice-president. Later that year the IMF, which had earlier decided to withdraw Sudan's membership because of non-cooperation on a debt repayment schedule, agreed to allow the country more time to reform its economy. However, tensions in Sudan's relations with Egypt over the disputed area border region of Halaib grew, with Egypt accusing Sudan also of fomenting Islamic fundamentalist terrorism.

SUDAN AT A GLANCE	
Official name Republic of the Sudan	
Area 2 505 810 km² (967 270 sq miles)	
Population 29 000 000 **Per km²** 11 **(Per sq mile** 30)	
Capital Khartoum	
Government Republic	
Currency 1 Sudanese pound = 100 piastres	
Languages Arabic (official), English, African languages in the south	
Religions Muslim (70%), indigenous beliefs (25%), Christian (5%)	
Climate Arid in the north; tropical in the south, with summer rains. Average temperature in Khartoum ranges from 22.8°C (73°F) in January to 33.9°C (93°F) in June	
Land use Grazing 24%, cultivation 5%, forest and woodlands 20%, desert 51%	
Main primary products Cotton, dates, groundnuts, sesame seed, gum arabic, sugar, sorghum, wheat, beans, livestock; crude oil	
Major industries Agriculture, oil production and refining, food processing, textiles, cement	
Main exports Cotton, sesame seed, groundnuts, cereals, gum arabic, livestock, petroleum products	
Annual income per head (US$) 440	
Population growth (per thous/yr) 28	
Life expectancy (yrs) Male 53 **Female** 55	

(1992), has an architectural theme suggestive of a rather fanciful ancient civilisation. Major attractions are the gambling casinos – illegal before the 1994 democratic elections in South Africa – and entertainment centres. Next to Sun City is the 50 000 hectare (23 550 acre) Pilanesberg National Park, which has elephants, lions, hippos, rhinos and many other species of game.

Map South Africa – Cb

Sun Moon Lake (Jih-yüeh T'an) *Taiwan* Largest lake in Taiwan, covering an area of 8 km² (3 sq miles) in the central mountain ranges, some 45 km (28 miles) south-east of T'AI-CHUNG. A temple on its shore houses remains of Hsuan Chuang, a Tang dynasty monk who was largely responsible for introducing Buddhism to China. The lake, which is a major tourist attraction, today supplies hydroelectric power from a plant built by the Japanese in the 1930s.

Map Taiwan – Bc

Sunch'on (Suncheon) *South Korea* Railway junction and market town, lying in a vegetable-producing area 145 km (90 miles) south-west of the port of PUSAN. It is the tourist gateway to the beaches and islands of the south-western coast.

Population 167 000

Map Korea – Ce

Sunda Islands *Indonesia* Collective name for islands of the Malay Archipelago. The Greater Sunda Islands consist of JAVA, SUMATRA, BORNEO, SULAWESI and adjacent islands. They cover some 1 333 677 km² (514 953 sq miles). The Lesser Sunda Islands (NUSA TENGGARA) comprise the chain of islands east of Java, from Bali to Timor. Their total area is 76 249 km² (29 441 sq miles). The Sunda Islands are grouped along the oceanic Java Trench, where the Indo-Australian continental plate is moving in below the European plate (see PLATE TECTONICS, p 542), resulting in frequent earthquakes and volcanic eruptions.

Population (Greater Sunda Islands) 170 000 000; (Lesser Sunda Islands) 6 688 000

Map Indonesia – Cc-Ed; (Timor) Indonesia - Fd-Gd

Sunda Shelf *South China Sea* Submarine extension of the South-east Asian mainland which underlies much of the South China Sea. It is one of the widest areas of continental shelf in the world, and its relatively shallow waters make it an attractive place to search for oil.

Map Asia – Ih

Sunda (Soenda) Strait *South-east Asia* Strait between JAVA and SUMATRA, Indonesia. The Sunda Strait is 26-110 km (16-68 miles) wide and 27-183 m (90-600 ft) deep. It contains many small volcanic islands, including KRAKATAU which erupted in 1883 with such force that its explosions were heard 4700 km (2930 miles) away. Explosions of another kind occurred in 1942, when the United States fleet suffered heavy losses here during the Battle of the JAVA SEA.

Map Indonesia – Bd

Sundarbans *Bangladesh* Dense mangrove swamp and rain forest of the Ganges delta. The area teems with wildlife, including snakes, deer and crocodiles, but it is plagued by leeches and mosquitoes. There is some development of a timber industry, but much of the rain forest is now a wildlife sanctuary and one of the last refuges of the Bengal tiger.

Map Bangladesh – Bd

Sunderland *United Kingdom* English industrial city at the mouth of the River Wear, 16 km (10 miles) south-east of NEWCASTLE UPON TYNE in the county of Tyne and Wear. It thrived in the 19th century on coal mining and shipbuilding and was once the greatest shipbuilding centre in the world. However, coal mining has declined and shipbuilding has ceased. Today light industries have grown up, with many foreign companies investing here; the port has been developed to handle containers. The city's St Peter's Church dates from AD 674, but was partly rebuilt in the 1870s.

Population (district) 296 400

Map United Kingdom – Ec

Sundsvall *Sweden* Timber port on the Gulf of BOTHNIA, 406 km (about 250 miles) north of Stockholm. It manufactures and exports wood pulp and paper.

Population 94 300

Map Sweden – Cc

Sungei Kolok *Malaysia/Thailand* River rising close to the Thai border in the northern Malaysian state of KELANTAN. For much of its length it forms the border between Malaysia and Thailand. It turns north-west into Thailand before joining the South China Sea just north of the border.

Map Malaysia – Ba

Sunset Crater National Monument *USA* Volcanic cinder cone in northern ARIZONA, some 180 km (110 miles) south of the Utah border, formed by an eruption in about 1065. The colours of the 300 m (1000 ft) cone are reminiscent of sunset, ranging from yellow to orange to red.

Map United States – Dc

Sunyani *Ghana* Largest town and capital of Brong-Ahafo region, 320 km (200 miles) north-west of the capital, ACCRA. It is also a busy market for the cocoa and kola nuts which are intensively grown in the surrounding area.

Population 25 000

Map Ghana – Ab

Superior, Lake *Canada/USA* Highest of the Great Lakes and the world's largest sheet of fresh water, covering 82 400 km² (31 800 sq miles). The Soo (or Sault Sainte Marie) Canals enable ships to bypass the 6 m (20 ft) difference in water level between Lake Superior and Lake HURON.

Map United States – Ia

Suphan Buri *Thailand* Market town set in a prosperous agricultural province of the same name, 80 km (50 miles) north-west of the capital, BANGKOK. The province's farmers grow rice and sugar cane.

Population (town) 134 500; (province) 761 500

Map Thailand – Bc

Surabaya *Indonesia* The nation's second largest city after the capital, JAKARTA, and the provincial capital of East Java. It is an industrial port with shipyards, an oil refinery, textile mills and a fishing fleet, and exports sugar, rubber, tobacco, copra, coffee and other agricultural products.

Population 2 470 000

Map Indonesia – Dd

Surat *India* West-coast port at the mouth of the Tapti River, 308 km (190 miles) north of BOMBAY. It was India's chief port under early Mogul emperors (1573-1650), and once was the main western port of British India, surpassing Bombay for a time in the value of goods handled. No longer a great port, it still produces traditional gold and silver threadwork, and handwoven cottons.

Population 1 496 900

Map India – Bc

Surat Thani (Ban Don) *Thailand* Coastal province and town 560 km (350 miles) south of the capital, BANGKOK, which is developing as an agricultural area. The nearby island of Ko Samui is a growing tourist resort. The province's major crop is the oil palm from which palm oil is made.

Population (province) 669 600; (town) 35 200

Map Thailand – Ad

Surin *Thailand* Province and town situated some 335 km (208 miles) north-east of the capital, BANGKOK. The province, which lies along the Cambodian border, is the site each November of a spectacular round-up of elephants. At the round-up, the elephants, which have been trained to transport logs in the teak forests, are paraded, raced and pitted against 70 to 100 humans at a time in tugs-of-war. The elephants usually win. Many of the province's farmers are Khmers from across the border. Rice is their main crop.

Population (province) 1 272 560; (town) 33 000

Map Thailand – Bc

Surinam See p 643

Surrey *United Kingdom* English county of 1655 km² (639 sq miles) on the southern border of LONDON, now partly used for residential over-spill from the capital. However, conservation laws have protected most of the North DOWNS, the chalk hills that cross the county. The county town, Guildford, 45 km (28 miles) south-west of London, has a modern cathedral and is the seat of the University of Surrey.

Population 1 033 600

Map United Kingdom – Ee

Surtsey *Iceland* Volcanic island which rose from the sea off the south coast of Iceland, 30 km (19 miles) south-west of the WESTMAN ISLANDS, in November 1963. Within a few months the island, named after Surtr, the god of fire in Norse mythology, was 171 m (561 ft) high and covered about 2.5 km² (1 sq mile). Within months after the eruptions grumbled to a halt, plants – whose seeds were carried in by wind and birds – were colonising the new land. A second wave of eruptions occurred in 1966-7, pushing up another two islets close by, but they have submerged again.

Map Iceland – Bb

Susa (Shush) *Iran* Archaeological site on the Susiana Plain, 100 km (62 miles) from the Iraqi border. Clay tablets and pottery dating from before 3500 BC have been found here. For centuries the capital of the Elamite Kingdom, it was destroyed in the 7th century BC by the Assyrian King Assurbanipal. Cyrus the Great (ruled 559-530 BC) rebuilt it as his first capital before he conquered all of Persia. It was destroyed by the Macedonian conqueror Alexander the Great (356-323 BC) and again in the 4th century AD. It was eventually abandoned towards the end

Surinam

A HOTCHPOTCH OF CULTURES IS IN THE PROCESS OF BUILDING A NATION

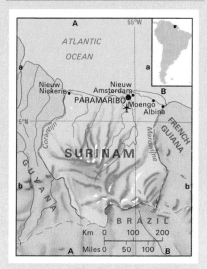

In 1667 the Dutch gave a North American settlement to the British in exchange for a territory on the northern coast of South America. The settlement, then called New Amsterdam, became New York. The territory became Dutch Guiana and is today the independent state of Surinam. The Dutch had occupied it in the 16th century, but European

wars led to treaties which alternately took away and returned the territory to them. The country became an official Dutch colony in 1814. Independence was declared in 1975. The Dutch brought slaves from Africa in the 17th century and, after the abolition of slavery in 1863, indentured labourers from India, Indonesia and China to work on the sugar plantations. Other marks left by the Dutch are the enormous dykes which they built.

The population is now a mixture of Creole descendants of African slaves, who make up most of the 200 000 people in the capital city of PARAMARIBO; Indians, Chinese and Indonesians, who mostly farm rice and sugar on the coastal plains; Europeans (most of whom live near the coast); and native Amerindians, small numbers of whom live mainly in the dense rain forest and savannahs of the interior which rise towards the Guiana Highlands in the south.

The Amerindian tribes have been nudged deeper into the rain forest by settlements of 'Bush Negroes' who are the descendants of escaped slaves. Nevertheless, all the races live in harmony. On the eve of independence, a large proportion of the population – many of them well educated – emigrated, mainly to the Netherlands, so that they would not lose their Dutch citizenship. The exodus has continued, for, with high unemployment and soaring inflation, the future of the country is uncertain.

The economy is dependent on bauxite and the production of aluminium, which comprises 70 per cent of exports. Surinam has under-exploited resources of timber – planned large-scale timber extraction will, however, pose an environmental problem. Great potential exists

for agriculture and tourism, too. The economy, however, desperately needs an influx of money.

A military junta seized power in 1980, and again from 1990 to 1992, an intervening period of democratic government being more ostensible than real. Continuing political instability, guerrilla activity and nationwide strikes have discouraged foreign investment.

SURINAM AT A GLANCE	
Official name Republic of Surinam	
Area 163 265 km² (63 037 sq miles)	
Population 416 320 **Per km²** 2.5 (**Per sq mile** 6.5)	
Capital Paramaribo	
Government Republic	
Currency 1 Surinam guilder = 100 cents	
Languages Dutch (official), English, Sranang Tongo (Surinamese, or Taki-Taki), Hindi, Javanese, Chinese	
Religions Christian (Protestant 25%, Roman Catholic 23%), Hindu (27%), Muslim (20%)	
Climate Tropical and very humid. Average temperature in Paramaribo is 27°C (81°F) throughout the year	
Land use Forest and woodlands 97%, other 3%	
Main primary products Rice, sugar, timber, bananas, citrus fruits, coconuts, coffee, cocoa, shrimp and fish; bauxite	
Major industries Agriculture, bauxite mining, alumina refining, aluminium smelting, forestry, lumbering, food processing	
Main exports Bauxite, alumina and aluminium, rice, bananas, timber, shrimp and fish	
Annual income per head (US$) 3700	
Population growth (per thous/yr) 15	
Life expectancy (yrs) Male 67 **Female** 72	

of the 14th century, after the Mongol invasion and devastation by Tamerlane. Four *tells*, or mounds, have been excavated since the mid-19th century, uncovering citadels, palaces and other buildings.
Map Iran – Aa

Susquehanna *USA* River rising in NEW YORK state and flowing 715 km (444 miles) southwards across Pennsylvania and into Chesapeake Bay, 50 km (30 miles) north-east of BALTIMORE.
Map United States – Kb

Sussex *United Kingdom* Twin counties – East and West Sussex – of southern England on the English Channel. Their combined area is about 3811 km² (1471 sq miles). The hills of the South DOWNS were once covered by forests, providing timber for shipbuilding and charcoal for a primitive iron industry. Today the downs are mostly open pastureland. The coast is lined with resorts, of which the largest is BRIGHTON. Inland, the chief towns are the county town of West Sussex, CHICHESTER, Horsham and Crawley, which has grown with the development of nearby Gatwick airport. Lewes, with its Norman castle, is the county town of East Sussex, and HASTINGS is situated near the site of the battle in 1066 that established Norman rule in England. Herstmonceux Castle became the home of the Royal Greenwich Observatory in the 1950s, but there are plans to move the observatory to Cambridge.
Population 1 427 900
Map United Kingdom – Ee

Susupe *Northern Marianas* See SAIPAN

Sutherland *United Kingdom* See HIGHLAND

Sutjeska *Bosnia-Herzegovina* National park west of the upper Drina River in Bosnia-Herzegovina. It includes the lovely mountain of Maglić, which reaches 2387 m (7831 ft) near the Montenegrin border, as well as rare virgin forests around Perucica, and the Sutjeska Gorge west of Maglić. A striking white marble memorial complex at Tjentiste commemorates the month-long Battle of Sutjeska in 1943 when 19 700 partisans were surrounded by 127 000 Nazi troops and 7356 of them were killed. The partisan commander, Marshal Tito (1892-1980), Yugoslavia's subsequent leader, was wounded in the battle.
Map Bosnia-Herzegovia – Cb

Sutlej (Satluj) *China/India/Pakistan* River which rises in the Himalayas of south-west Tibet, China, and flows some 1370 km (850 miles) west into India, then south-west through the Punjab to join the Chenab River in Pakistan. It is part of the Indus River system, and is dammed at BHAKRA. Its headwaters were allocated to India following a treaty between India and Pakistan in 1960.
Map India – Bb; Pakistan – Db

Suva *Fiji* Capital and major port of Fiji, standing on the south-east coast of the island of VITI LEVU. Almost 20 per cent of Fiji's population live in and around Suva. Cargo and cruise ships use its fine

natural harbour. The city is the home of the University of the South Pacific, founded in 1968 and financed by 11 Pacific states. Copra and food processing (coconuts and soya beans), and tourism are the main industries. The botanical garden has a number of rare tropical plants.
Population (city) 74 000; (metropolitan area) 133 000
Map (Fiji) Pacific Ocean – Dc

Suwon (Suweon) *South Korea* City, founded in the 18th century, situated 27 km (17 miles) south of SEOUL. It produces textiles and chemicals.
Population 645 000
Map Korea – Cd

Suzdal' *Russia* Small city and tourist centre 190 km (118 miles) north-east of MOSCOW. It dates from the 11th century, and has an imposing 12th to 13th-century kremlin (citadel), several fortified monasteries, a 13th-century cathedral, several churches and a market square.
Population 10 000
Map Russia – Fc

Suzhou (Soochow) *China* City of Jiangsu province south of the Chang Jiang delta, 80 km (50 miles) west of SHANGHAI. Dating from the 6th century AD, it is an attractive city crisscrossed by canals, and is sometimes called 'the Venice of China'. It is famous for fine silk embroidery.
Population 900 000
Map China – Ie

Svalbard *Norway* Arctic Ocean archipelago whose name means 'the cold coast'. It covers 62 700 km² (24 203 sq miles) due north of mainland Norway. Discovered by Vikings in 1194, and then rediscovered by the Dutch navigator William Barents in 1596, Svalbard became a centre for international whaling during the 17th century; by the mid-17th century, some species were near extinction. Later, Russian and Norwegian hunters visited the islands, decimating the wild animal population. Norwegian sovereignty was recognised internationally in 1920 and became effective in 1925. The islands are now demilitarised, but signatories of the 1920 convention have the right to mining concessions – which is why there is a Russian coal-mining operation on the main island, SPITSBERGEN.

Population 3500
Map Arctic – Rb

Sverdlovsk *Russia* See YEKATERINBURG

Swabia (Schwaben) *Germany* Former province of south-west Germany, south of STUTTGART. Mountainous and forested, it is now part of Baden-Württemberg and south-western Bavaria.
Map Germany – Cd

Swakopmund *Namibia* Resort at the mouth of the Swakop River, 360 km (224 miles) west of Windhoek. Once Namibia's chief port, it was superseded by Walvis Bay 32 km (20 miles) to the south. Many workers at the nearby Rössing Uranium Mine live here.
Population 20 000
Map Namibia – Ab

swale See RUNNEL

swallow-hole See SINKHOLE

swamp A permanently waterlogged lowland region and its vegetation – often reeds. Swamps form when a lake basin fills and the surface is so flat that the run-off of rainwater is very slow. Some permanently waterlogged areas have become home to certain highly adapted trees – for example, swamp cypresses in Florida's EVERGLADES, and mangroves on tropical coasts.

Swansea (Abertawe) *United Kingdom* Port and the second city of Wales after Cardiff, which lies 55 km (35 miles) to the east.

It stands at the mouth of the River Tawe in the Welsh county of West Glamorgan. Swansea rose to prominence in the 19th century with the export of coal and the growth of industries such as tin plating in the Tawe Valley. In 1992 a barrage was completed across the river which enabled an industrial and leisure area to be developed 3.2 km (2 miles) upstream. The splendid scenery and beaches of the Gower Peninsula have proved popular with tourists and lie just to the southwest. The poet Dylan Thomas (1914-53) was born in and grew up in Swansea.
Population 175 170
Map United Kingdom – De

swash Another term for a BAR of sand which is washed by the waves. The term is also applied to a channel of water that cuts through or runs around a sandbank. Commonly, the word 'swash' is used to describe the rush of seawater up a beach following the breaking of a wave.

Swat *Pakistan* Former tribal state ruled by the Wali of Swat. It covers 7511 km² (2900 sq miles) in the mountains north-east of PESHAWAR. The region was self-governing until 1960.
Population 900 000
Map Pakistan – Da

Swaziland See p 646

Sweden See p 647

swell Regular undulating movement of waves out in the open sea, with no breaking and with a considerable distance between successive crests.

Switzerland See p 650

Sydney *Australia* Largest, brashest and oldest city in the country, and the state capital of NEW SOUTH WALES. It is centred on the magnificent harbour, which was named Port Jackson (but never explored) by the English navigator James Cook in 1770, and now spreads north, west and south over 1735 km² (670 sq miles).

The city was settled in 1788 by 1487 people – 759 of them convicts – who had sailed in a fleet of 11 ships from Britain. Its historical buildings include St James's Church and Hyde Park Barracks, both designed by the convict architect Francis Greenway in the early 19th century. But its most spectacular features are more modern – the Centrepoint Tower, complete with revolving restaurant; the shell-roofed Opera House, opened in 1973; and the Sydney Harbour Bridge, opened in 1932. The 49 m (161 ft) wide bridge, which carries two railway tracks, eight lanes of roadway, a cycle path and footway across a span of 503 m (1650 ft), is the world's widest longspan bridge. It links the city centre, to the south of the harbour, with the rapidly expanding second commercial centre of North Sydney, formerly a residential suburb. A 2 km (1.2 mile) long tunnel under the harbour was opened in 1992.

The city draws thousands of tourists who flock to its famous beaches, such as BONDI. Sydney is also Australia's busiest port, has four universities and, since the Second World War, has developed an easy-going cosmopolitan flavour, with large numbers of immigrants from all over the world.

▲ IMPERIAL RIVER The Southern Grand Canal, 400 km (250 miles) long, glides through Suzhou. It joins the Da Yunhe ('Grand Canal') with the city of Hangzhou, and was built by the Zhou emperors in the 4th century BC.

Its industries include shipbuilding, timber processing, chemical works and car assembly.
Population 3 698 500
Map Australia – Ie

Sydney *Canada* Port on Cape Breton Island, Nova Scotia. A coal, iron and steel-producing centre, its industries include mining equipment, marine and offshore engineering and electronics. The port is designated a free trade zone.
Population 33 320
Map Canada – Id

Syedlets *Poland* See SIEDLCE

Sylhet *Bangladesh* Town and district in the east near the Indian border south of SHILLONG. The district, covering 12 718 km² (4910 sq miles), is the home of Bangladesh's tea industry.
Population (town) 166 800; (district) 5 656 000
Map Bangladesh – Cb

Sylt *Germany* See FRIESIAN ISLANDS

syncline A downfold in the BEDROCK, in which rocks incline together from opposite sides.

Syracuse (Siracusa) *Italy* Seaport on the east coast of Sicily, 50 km (30 miles) from its southeastern tip. It was founded by Corinthian Greeks in 734 BC and grew rapidly to become the most notable city of the Hellenic Age outside Greece itself. At its peak it probably had more inhabitants than any other city in the world. Greek Syracuse covered not only the island of Ortygia – its centre today – but also large areas of the Sicilian mainland. Archimedes, the mathematician, born here in 287 BC, was killed by the Romans when they seized the city in 212 BC.

Ortygia is the site of an extraordinary cathedral, moulded around a Greek temple and with an imposing Baroque façade. It has the Castello

A DAY IN THE LIFE OF A SYDNEY BUS DRIVER

In the 1950s Jim Arnold was among the thousands of young Britons who seized the chance of a new life 'down under' in Australia. Jim had taken some savings and invested what was then a week's wages in the price of his assisted sea passage, and was full of high hopes.

The reality of Australia was, at first, disappointing. There was stiff competition for the best jobs and Jim tried a series of trades, including working on a building site, driving a long-distance truck and being a cook on a cattle station.

Then he met Anne Carter at a barbecue in Sydney. They were engaged within a few months and he settled down in his present job driving a Sydney bus.

His wife's great-great-grandfather was a convict transported from Britain to Australia in the 1840s, a heritage of which she is quite proud. He was convicted for stealing a handkerchief.

The Arnolds live in a three-bedroomed bungalow in the eastern Sydney suburb of Bondi. Many of their neighbours are also immigrants. Jim became a naturalised Australian citizen more than 20 years ago. Now his parents are dead, he has no desire to retire to Britain. He doesn't even want to visit Britain again.

The climate and relaxed lifestyle have won Jim over completely to his adopted country. With his salary and overtime pay, and Anne's earnings from working part-time in a shop, they can live comfortably with very little financial stress.

Their son Greg studied engineering, and is now an engineer on oil-exploration projects in Queensland.

Jim's bus route begins at Watson's Bay, an affluent suburb with a splendid view up Sydney Harbour to the skyscrapers of the city centre. The 11 km (7 mile) route ends there at Circular Quay, a terminal for buses, trains and ferries. This spot is alongside the dock where Jim himself first landed in Australia.

Jim would like to move out to more spacious surroundings – but not if it means going too far from the beach. Unless he's on a very early shift, Jim takes a swim before breakfast at the world-famous Bondi beach. And, like many of his outdoor-loving companions, he intends to continue this habit well into his old age.

Now approaching retirement, he hopes to buy a house near one of the superb beaches that stretch for 25 km (16 miles) north of the harbour to the wealthy suburb of Palm Beach, and settle into a leisurely life of fishing, swimming and basking in the sun.

On Saturday evenings Jim and Anne often go to the South Sydney Leagues club, a huge social centre, for a meal, to watch the cabaret and put a few dollars in the 'pokies' – the poker machines or one-armed bandits. Jim, like most Australians, likes to gamble.

'We'd move to Palm Beach tomorrow if we won the lottery,' he says.

Maniace (which is dated about 1239), the Palazzo Bellomo (13th and 15th centuries), the Arethusa Fountain and the national museum, which contains some of the finest Greek exhibits outside Athens. The greatest wonders of ancient Syracuse on the Sicilian mainland include the Roman amphitheatre, the Greek theatre, the *latomie* or ancient quarries, now planted with gardens, and the catacombs of San Giovanni. The Castello Euralio, constructed by the tyrant Dionysius II between 402 and 397 BC for defence, lies 9 km (5.5 miles) to the north.

Apart from the nearby petrochemical plants, tourism is the city's main industry today, and the port exports local wine, olive oil and citrus fruits.

Population 124 600
Map Italy – Ef

Syracuse *USA* Industrial city in central New York State, 230 km (145 miles) east of Lake ERIE, to which it is linked by canal.

There are salt springs nearby and the city was once a leading salt producer. Its manufactures today include electronics, electrical equipment and pharmaceuticals. The city is the home of Syracuse University.

Population (city) 163 900; (metropolitan area) 742 200
Map United States – Kb

Syrdar'ya *Kazakhstan* Central Asian river which is 2860 km (1780 miles) long.

Known at its source, in the Tien Shan Mountains near the Chinese border, as the Naryn, it flows west between the cities of TASHKENT and SAMARKAND, then continues north-west to the Aral Sea. Along the way it is dammed in order to provide electricity, and also tapped for irrigation purposes. The river was known to the ancient Greeks as the Iaxartes.
Map Kazakhstan – Cb

Syria See p 652

Syriam *Myanmar (Burma)* Oil town on the southern bank of the Yangon (Rangoon) River, opposite the capital, Yangon (Rangoon).

Formerly a Portuguese trading post, it is now the site of the country's main oil refinery.
Population 102 000
Map Myanmar – Cc

system Succession of rocks or rock STRATA formed during a geological period such as the Cambrian or the Triassic.

Szabadka *Yugoslavia* See SUBOTICA

Szántód-Rév *Hungary* See SIÓFOK

Szava *Slovenia* See SAVA

Szczecin (Stettin) *Poland* Port in the north-west, standing on the ODER River 55 km (34 miles) from the open sea. It rivals Gdańsk as Poland's leading Baltic port and shipbuilding centre. Other industries include engineering and chemicals. Between 1720 and 1945 it was the main port for the German city of Berlin. About two-thirds of the city was destroyed in the Second World War and it has been extensively rebuilt. A few old buildings remain as a reminder of the city's prosperous Hanseatic days, and of when it served as residence for the Pomeranian counts. The 15th-century church of St Peter and St Paul and part of the Renaissance castle are among the surviving buildings.
Population 414 300
Map Poland – Ab

Szechwan *China* See SICHUAN

Szeged *Hungary* River port on the Tisza, some 10 km (6 miles) from the Yugoslav border, and the economic and cultural focus of the southern Great ALFÖLD. The old town, which was extensively damaged by river floods in 1879, was rebuilt with concentric boulevards linked by avenues radiating from the inner city. The heart of the city is dominated by a neo-Romanesque

▼ EVENING SKYLINE Skyscrapers, Sydney Harbour Bridge and the Opera House are symbols of Sydney.

cathedral (1912-29), one of the country's largest churches. An open-air festival of the performing arts has been held in the main cathedral square virtually every July-August since 1913. Szeged has two universities, several scientific institutes and a theological college.

Long famous for its fiery paprika spice, fine restaurants and food processing factories, the city now also has chemical plants, which use local and Romanian oil and natural gas, and engineering and electrical works. Ujszeged, a resort across the Tisza, is noted for its water sports.

Population	177 700
Map	Hungary – Bb

Székesfehérvár *Hungary* Principal city of the central county of Fejér, 65 km (40 miles) south-west of BUDAPEST on the motorway to Lake Balaton. The kings of Hungary were crowned in the city from 1000 to 1527, and several were buried here. Székesfehérvár is a beautiful city with a splendid Baroque cathedral (1758-68), churches, palaces and houses set among parks and gardens. Long a market and commercial centre for a wine and fruit-producing district in the Bakony foothills, it has expanded rapidly since 1960 with the building of factories making aluminium, Ikarus buses, and electrical and telecommunications equipment. Lake Velence, 8 km (5 miles) to the east, has summer resorts.

Population	109 310
Map	Hungary – Ab

Szekszárd *Hungary* Chief town of the central county of Tolna, 137 km (85 miles) south of BUDAPEST on the road to PÉCS. Over the ruins of an 11th-century castle stands the old county hall, an impressive palace built in the early 19th century. It lies in a region producing Szekszárdi red wine, and close to the Gemenc wildlife reserve.

Population	37 140
Map	Hungary – Ab

Szolnok *Hungary* City on the Tisza River at the heart of the Great ALFÖLD, 89 km (58 miles) south-east of BUDAPEST. It has a fine Baroque church, but is best known for its water sports and river regattas. The city is the administrative centre of Jász-Nagykun-Szolnok County and a busy communications and commercial centre at the heart of a prosperous farming area. It has food processing and fertiliser factories.

Population	79 460
Map	Hungary – Bb

Szombathely *Hungary* City in the west of the country, 18 km (11 miles) east of the Austrian border. As Savaria, it was a capital of the Roman province of Pannonia, which covered the central Danubian Plains. The city has two large squares, with superb Baroque houses, a cathedral and a bishop's palace. It is the chief city of Vas County, a prosperous wine-producing area. New industries include chemical plants using local oil and natural gas as raw materials. The grounds of the ruined 2nd-century Roman temple of Isis to the south are used for the Savaria Festival of music each summer. Ják, 11 km (7 miles) south of Szombathely, has a fine 13th-century triple-aisled Romanesque basilica, restored in 1896.

Population	85 740
Map	Hungary – Ab

Sztálinváros *Hungary* See DUNAÚJVÁROS

Swaziland

ALMOST SURROUNDED BY SOUTH AFRICA, SWAZILAND HAS BECOME INCREASINGLY DEPENDENT ON ITS GIANT NEIGHBOUR

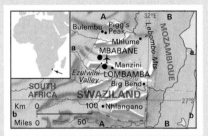

One of the three monarchies remaining in Africa, Swaziland is faced by an unusual mix of political and economic problems, as the country tries to balance the old and the new. Swaziland gained independence from Britain in 1968 and went back to its pre-colonial state of absolute monarchy in 1973. Although political parties keep offices in the country, elections are held on a non-party basis.

Surrounded on three sides by South Africa and to the north-east by Mozambique, this landlocked hilly enclave is well watered, has a pleasant climate, and contains splendid scenery. The land drops in three great steps from west to east. The first, the highveld (more than 1000 m, about 3280 ft), consists of rugged mountains clothed in forestry plantations and cut by streams and waterfalls. The middle-veld (400-850 m, 1310-2790 ft), planted with orange groves and pineapple fields, falls away to the undulating, park-like eastern lowveld (150-300 m, 490-980 ft), where sugar cane plantations flourish on irrigated lands. The extreme eastern boundary is formed by the Lebombo Mountains (825 m, 2700 ft), which are indented by gorges and divide Swaziland from Mozambique and the Indian Ocean.

The death in 1982 of King Sobhuza II, who had reigned for 61 years, brought into the open the struggle between traditional politics and modern developments. For a time, the nominal head of the country was *Ndlovukazi*, Queen Regent Dzeliwe, one of the late king's elder wives. However, she was simply a figurehead – behind her lay the *liqoqo*, the supreme tribal council of senior princes and chiefs, who in 1983 deposed her and the prime minister Mabandla. Queen Ntombi, one of Sobhuza's younger wives whose 15-year-old son, Makhosetive, had meanwhile been chosen by the royal counsellors as the future king, then became *Ndlovukazi*. In April 1986, after reaching the age of 18, Makhosetive, then a pupil at a British public school, succeeded to the throne as Mswati III. Dissension within the *liqoqo* led to its dissolution and the appointment of a new prime minister after the king's inauguration.

Just over half the land is held in trust for the 906 940-strong nation by the king. It supports peasant farmers who produce chiefly maize and cattle to feed themselves and grow a few crops to sell for cash, such as cotton and tobacco. Much of their land is overgrazed and eroded. Drought in 1992 destroyed 80 per cent of the harvest, which meant extreme hardship and starvation for 42 per cent of the population – there was no international food aid to provide for the Swazi nation.

The rest of the land is worked by commercial companies, who grow all the main export crops – sugar (the most valuable), timber, and citrus fruits – and mine the country's mineral resources, chiefly asbestos from Havelock mine at Bulembu in the north-west, and industrial diamonds. Iron ore, once a major export, is now exhausted. Asbestos exports, too, have decreased since health concerns cut the world demand. Some gold, quarry stone and kaolin are mined. The only coal mine closed down in July 1992, but another is being developed.

Manufactures include fertilisers, textiles, leather and tableware. Tourism also thrives, most tourists being from South Africa.

Swaziland is overshadowed by the Republic of South Africa, with which it is in a customs union. It receives 75 per cent of its imports from South Africa and sends 50 per cent of its exports to its giant neighbour. Indeed, Swaziland is so closely linked to the South African economy that when the latter became depressed, so did its own economy.

It was this close relationship with its large neighbour that put the Swazi government in a difficult position during South Africa's *apartheid* period. Wholly dependent upon that country, Swaziland had to tread carefully between keeping relations with South Africa amiable and yet not offending the Organisation of African Unity (OAU).

SWAZILAND AT A GLANCE		
Official name Kingdom of Swaziland		
Area 17 400 km² (6705 sq miles)		
Population 906 940	**Per km**	52
(Per sq mile 135)		
Capital Mbabane		
Government Constitutional monarchy		
Currency 1 lilangeni = 100 cents		
Languages Siswati (official), English (official)		
Religions Christian (60%, mostly Protestant), indigenous beliefs (40%)		
Climate Subtropical moderated by altitude; average temperature in Mbabane ranges from 12°C (54°F) in June to 20°C (68°F) in January		
Land use Grazing and other 84%, cultivation 10%, forest and woodlands 6%		
Main primary products Maize, cotton, rice, sugar, citrus fruits, tobacco, cattle, timber; asbestos, coal, diamonds		
Major industries Agriculture, forestry, mining, timber processing, chemicals, tourism		
Main exports Sugar, citrus fruits, soft drink concentrate, timber and wood pulp, meat, asbestos, diamonds		
Annual income per head (US$) 1080		
Population growth (per thous/yr) 34		
Life expectancy (yrs) Male 52 **Female** 60		

Sweden

SOPHISTICATED TECHNOLOGY AND EFFICIENT INDUSTRY EXIST SIDE BY SIDE WITH THE WILD BEAUTY OF GREAT FORESTS AND COUNTLESS LAKES IN THIS ENVIABLE COUNTRY

Now far into its second century without war, with a high living standard, one of the most extensive social welfare programmes on earth, and endless stretches of untouched nature, Sweden is in many respects the envy of Europe.

But earthly paradises do not come cheaply; the Swedes carry the world's heaviest tax burden (although their earnings are high) and the cost of their welfare services is proportionally one of the world's highest. And since the recession of the late 1980s and early 1990s, even Sweden can no longer offer the type of social security its citizens have been used to. In 1992 the new centre-right government introduced large-scale spending cuts, tax increases and reduced social security provisions. Although the previous Social Democratic government was returned to power in 1994, its 1995 budget also provided for stringent curbs on the high rate of public spending, which was almost 70 per cent of gross national product (GDP).

After voting for the Social Democrats, Swedes also went along with that party's pro-European stance. In a referendum held in November 1994, they agreed to membership of the European Union. Earlier that year, a project to construct a 17 km (10 mile) road and rail link across the ØRESUND to Denmark by bridge and tunnel, due to be completed in the year 2000, also received a formal go-ahead.

COLOUR IN THE TUNDRA

Sweden is a large, handsome, northerly country whose southernmost tip is on roughly the same latitude as the border between England and Scotland. Its northernmost tip, almost 1600 km (1000 miles) away, lies well within the Arctic Circle. Much of the south is flat, much of the north mountainous, and the coastline is a bewildering scatter of 20 000 and more islands and islets. But most of Sweden is an undulating plain, whose chief elements are coniferous forest and dark lakes scoured out of granite during the ice ages. There are, however, some startling variations, such as beech and oak forests surrounded by a patchwork of fields and meadows in the south, while the northern latitudes are treeless. Moving north, the coniferous forests gradually give way to tundra-like landscape which in summer at least is a colourful wilderness of bog, lichens and mosses illuminated by cotton grass, wild flowers and banks of cloudberries, bilberries and wild strawberries.

On the whole, the climate is extreme. Summers can bring heatwaves, but they are brief. Winters in the north are seven to eight months long and in the south, three; to all areas they bring deep snow and ice. In the Arctic Circle the sun does not go down for weeks in June and July, but winters are bitter and dark.

About half the country is covered by forest; spruce and pine, with an admixture of birch – the national tree – being the usual pattern. In the past, Swedish woodlands were endangered by overcutting for charcoal, tar, timber and fuel. But today they are among the best maintained in the world, with more trees being planted annually than are cut – a policy that is enhanced by draining and planting some of the extensive marshlands. Huge areas of wilderness remain, however, where flocks of migrant birds move in and out with the swing of the seasons, and bears, elks, roe deer and reindeer roam. In addition, an estimated 270 000 domesticated reindeer migrate between the fells in summer and the forests in winter. With such large numbers, there is grave danger of overgrazing and serious damage to the forests.

In Sweden land and water are everywhere intermingled. Off the coasts there are archipelagoes, great and small – the most extensive and varied being in the Stockholm area. The bald, storm-lashed archipelago of the west coast presents a striking contrast, while the shoreline of the Gulf of Bothnia, still rising from the compression of the last Ice Age, produces new islands, skerries and shoals every two or three generations. There are islands too in the lakes of the interior, some of which – including VÄNERN and VÄTTERN – are among Europe's largest. The rivers that drain them have wild falls, torrents and rapids. The most spectacular rivers are found in Norrland, the northern two-thirds of the country. They issue from high-level lakes along the Norwegian border and race down parallel valleys to the Gulf of Bothnia, plunging over majestic waterfalls, most of which have been harnessed to produce electricity. The fish in the lakes, rivers and surrounding seas include such delicacies as the North Sea and Baltic herring, salmon, lobsters, crabs and river trout.

FARMS AND MINES

Just over 8.5 million people live in this large country, but whereas the southern third has always been the most populous, vast areas remain thinly occupied. Thus the Stockholm

▼ **THE BIG COUNTRY The sun glints on a tumbling river in Jamtland, in central Sweden. This is a crisp, clear-eyed 'away from it all' region of lakes, rivers, mountains, forests – and very few people.**

region has a population density of about 252 inhabitants per km² (4032 per sq mile), while there are areas in the north with fewer than five people to a square kilometre.

Agriculture was the most important source of income around the turn of the century, but today it employs no more than 3 per cent of the workforce. Most Swedish farmers are owner-occupiers, with the family providing the labour force; produce is sold through co-operatives. Dairy farming in northern Sweden and cattle farming in the south predominate, even though all cattle have to be housed during the winter. About 6.8 per cent of Sweden is cultivated, with the emphasis on fodder crops,

grain and sugar beet. Southern farms are larger than those in the north, but northern forests, intensively worked for timber products, are far greater in extent than the woodlands of the south. The number of farms is about half of what it was in 1955, but production has risen considerably. Today Sweden is self-sufficient in food production.

Timber is a major export and forestry is now a separate industry, although traditionally it was a winter occupation for farmers. So too was mining, particularly around Bergslagen in southern Sweden. There, iron ore was extracted, smelted with charcoal and forged with the aid of water power. The introduction

of new processes in the mid-19th century rendered many small foundries obsolete and iron and steel production were concentrated instead upon a few towns in the north-east. These new processes later protected Sweden from the worldwide glut which affected many other steel producers. Sweden is now one of the world's leading producers of iron ore, most extracted from the LAPPLAND MINES of Küranavaara, Luossavaara and Svappavaara. Lapland ores are processed at LULEA in the north, while southern Sweden's principal steel plant is at Sandviken. Mines around Boliden on the Bothnian coast produce zinc, silver, gold and copper, although Europe's longest worked copper mine at Falun expired, like the small foundries, in the 19th century.

The 19th-century industrial boom that transformed the steel industry also dragged wood processing from tiny plants powered by upriver waterfalls to large factories sited on navigable waterways. These large-scale enterprises now produce a wide range of papers, wallboards, laminates, prefabricated units, chemicals and, of course, matches. The story of the match industry is told in the museum at JÖNKÖPING.

Sweden has very few deposits of coal or oil, so its industrial development has depended very largely on the development of hydro-electric techniques. The demand for energy has increased phenomenally – a demand not least due to the entirely electrified state railway system. The trouble is that most of the chief sources of hydroelectric power are in the far north of the country, while the greatest demand is in the south. Faced with such a problem, Swedish engineers led the world in developing methods of transmitting electricity. Later, the installation of nuclear energy plants and imports of cheap oil provided the basis for a final phase in energy expansion; then public opinion swung against nuclear reactors at just about the same time as the mid-1970s oil crisis and the reactors will continue working only until 2010. Sweden has therefore had to find alternative sources of energy – its oil comes mainly from the North Sea and coal from Germany and Poland.

INDUSTRIAL EXPERTISE

Sweden came rather late to the industrial field. But having adopted British and German methods of steel production in the late 19th century, together with mechanical and chemical processes for paper and pulp production, native Swedish inventiveness began to assert itself, although often in matters surprisingly warlike for such a peace-loving nation. Alfred Nobel invented dynamite and established an industrial empire before making amends with his prizes, while Bofors, a munitions factory in central Sweden, was known the world over for the light automatic cannon it produced. More praiseworthy, perhaps, was Sweden's impact upon the design of furniture and household utensils – clean, unfussy lines in glassware, porcelain, textiles and kitchenware.

Heavy industry also benefited from the well-coordinated energy programme, with electrical and engineering firms and their products acquiring world stature. Major shipbuilding yards grew up at the great ports of

GOTHENBURG, MALMÖ and Uddevalla (although the latter two are no longer in production). The Husqvarna company pioneered sewing machines and motorcycles in the early 20th century, while the leap to international fame in car design was achieved in not much more than a generation. The success of the industry is apparent in the immense assembly plants of Volvo, Saab and Scania. The Swedish aircraft industry (Saab), too, has gained a name for itself, especially for its single-seater fighters. As in most countries, specialised industries have become associated with particular places – ESKILSTUNA is best known for cutlery; Orrefors for glass; VÄSTERÅS for electrical components; and BORÅS for textiles. The nation's craftsmen have built a reputation for skilled workmanship, although this might seem to be somewhat endangered by the new Swedish expertise in robotics.

Since the Swedish home market is limited, large-scale industrial developments have been possible only by looking to the outside world. There, many major Swedish concerns have established subsidiary plants, some of which employ larger labour forces than the parent company at home. The country also sells its skills abroad – banking and insurance, statistical advisory services and engineering. Swedish engineers are very active in the Third World, building dams, hydroelectric power plants, high voltage transmission systems, timber processing plants, and silos for grain storage, as well as airfields, hospitals and public buildings. Not infrequently, revenue derived from such projects exceeds that generated by Swedish companies at home.

Sweden's late arrival on the industrial scene had one distinct advantage: its towns were able to avoid the horrors of the slums and satanic mills that beset the first generation of manufacturing towns in other parts of Europe. Many Swedish towns are of ancient origin, and usually have at least one or two historic buildings, but most down the years have been devastated by fire, the plague of all timber-built settlements. When they came to be reconstructed, the Swedish love of order manifested itself in gridiron street plans. The width of streets, the size of residential lots and of public spaces were, and are, strictly regulated. In the larger cities, apartment blocks predominate and there are some tower blocks; save in the city suburbs, small terraced or detached, privately owned houses with gardens are less common. A considerable number of towns have only a few thousand inhabitants and centre on a single enterprise – a mine, a small processing plant, a specialised factory.

In the north, where rural life is harder, the drift to the towns continues, especially from northern towns to those of the south. Unemployment is consistently higher in the north. Heavy investment in communications, subsidies for industry, the establishment of growth centres and salary adjustments ease, but do not cure, the situation.

To some extent, the problem is one of a small population in a very large country. On the other hand, it is the abundance of space that enables Swedes to enjoy their particular lifestyle, and to create, for example, such great

national parks as the 1970 km² (760 sq mile) Sareks in Lappland, the largest such park in Europe. Just about every family has a car; many have a motor or sailing boat as well, and there is easy access to the countryside. Many families have a second home, whether a large villa or a prefabricated unit put together by a do-it-yourself buff. If winterised – that is, with water, fuel and sewage pipes buried beneath the reach of frost – these leisure homes can also be used for autumn hunting and winter skiing. Like the Norwegians, the Swedes are enthusiastic cross-country skiers. The highlight in winter sports is the annual Wasa cross-country skiing event, with a 85 km (53 mile) course which always attracts enthusiasts.

A SEASONAL LIFESTYLE

Life is geared very much to the seasons. Swedes make the most of the short summers, with school holidays running from early June until late August; during this period, business activities start and finish earlier in the day. The National Day, June 6, is a major summer event; but midsummer eve, when every community erects a tall cross decked with fresh greenery, flowers and ribbons, and throws an all-night party, is summer's climax. At this time, especially, Swedish people like to fly their blue and yellow flag. As the nights continue to lengthen, there are August crayfish parties to herald summer's close.

Gloom begins to descend in November and deepens until the mid-winter solstice, when the south of Sweden has about six hours of daylight, and the north not much more than three. In Arctic Sweden, street lighting is never extinguished during the winter months and snow ploughs, snow blowers, defrosting machines and icebreakers are in regular daily use for four months.

Each year a brave challenge is thrown into the face of darkness – the mid-winter festivals of light. On the fourth Sunday before Christmas, Advent Sunday, candelabra burn in every window and Christmas trees are already aglow in town squares. On December 13, St Lucia's Day, girls put on crowns of lighted candles, representing St Lucia, the patron saint of light, and boys wear hats decorated with stars. At this time the festive drink is *glögg*, mulled wine which is aflame with brandy. Gingerbreads are baked and tables are decorated with the red-capped trolls and straw goat of Yuletide. The festive fare is the Christmas cold table – the most extravagant of the varied national *smörgåsbord*.

Temperatures drop to their lowest at about the end of February, although March can be radiant with sunshine. Schools have skiing holidays, while skiing competitions, local and national, are held as the days lengthen. Meanwhile, the students begin to look forward to May Day Eve – Walpurgis Night – when they put on their white caps and everybody celebrates the arrival of spring.

Although Sweden is a byword for neutrality, the country has had its share of military adventure – the Protestant king, Gustavus Adolphus, was the greatest commander of the Thirty Years' War (1618-48). But the country has not fought a war since 1814.

On the whole, Sweden is about as near to a model state as the 20th century is likely to see. It is technically advanced, efficiently organised, affluent, egalitarian, politically neutral. In practice, it is not without its problems and paradoxes. Some Swedes voice objections to the standardisation, state interference in private life, and high level of taxation that are necessary adjuncts to the welfare state. They feel that social featherbedding has led to over-government, and are uneasily aware of a blandness and uniformity in the world about them. Others fear that standards cannot be maintained and that the generous donation of 0.7 per cent of GNP that Sweden contributes to international aid may have to be cut back.

There are those too who are disturbed about possible racial tensions. A post-Second World War influx of guest workers and refugees from Finland, the Baltic states, Italy, Greece and Yugoslavia has meant that one in ten of Swedish citizens has been either born abroad, or born of immigrant parents.

A large majority of citizens clearly supports Sweden's neutral stance, although the country remains an armed neutrality with a large defence budget, military conscription and a highly remunerative arms industry. Perhaps it is only in religious matters that the Swedes have ceased to conform. No more than a small minority are still active churchgoers, although most citizens happily pay the tithes asked by the state church as an aid towards maintaining its splendid, historic buildings. Nevertheless, the stern Lutheran ethic is firmly entrenched in the Swedish soul. From it stem the national virtues of discipline and hard work, and the high value placed upon achievement.

SWEDEN AT A GLANCE	
Official name Kingdom of Sweden	
Area 449 964 km² (173 665 sq miles)	
Population 8 730 300 **Per km²** 19 **(Per sq mile)** 50)	
Capital Stockholm	
Government Parliamentary monarchy	
Currency 1 Swedish krona = 100 öre	
Languages Swedish; small Lapp and Finnish-speaking minorities	
Religion Christian (94% Evangelical Lutheran)	
Climate In the south and central parts, temperate maritime, with short, hot summers and long, cold winters; the north has subpolar conditions. Average temperature in Stockholm ranges from –3.8°C (25°F) in February to 17.8°C (64°F) in July	
Land use Forest 62%, cultivation 6%, grazing 1%, other 31 %	
Main primary products Dairy products, cereals, potatoes, sugar beet, rapeseed, timber; iron ore, uranium, copper, lead, zinc, gold	
Major industries Engineering and electrical goods, motor vehicles, mining, timber including wood pulp and paper, furniture	
Main exports Engineering and electrical goods, motor vehicles, timber, timber products, paper and wood pulp, chemicals, iron and steel	
Annual income per head (US$) 23 680	
Population growth (per thous/yr) 5	
Life expectancy (yrs) Male 73 **Female** 81	

Switzerland

INDEPENDENCE AND NEUTRALITY, THE CORNERSTONES OF THE NATION'S POLICY, HAVE BROUGHT PEACE AND PROSPERITY TO THE SWISS

To the Swiss, it seems that everybody else wants to be Swiss too. Not, perhaps, permanently, but long enough to take advantage of the benefits provided by the country. The Italians want to work there. The French and many others want to keep their money there. The Americans want to set up international institutes there. And the British want their daughters to learn French there.

Meanwhile the Swiss work steadily at creating for themselves the world's highest standard of living. National annual income per person is US$36 320, and the gross national product (GNP) has been increasing by an average of around 1 per cent annually. Although primarily an industrial economy, with machinery, chemicals, watches, instruments and textiles among the main exports, Switzerland has huge earnings from international finance and tourism. It has become a leading centre for international banking and insurance, with some 530 banks and financial institutions. Apart from the country's economic stability, the great attraction for foreign customers has been Swiss banking secrecy – although that has been diminished in some respects recently.

Being Swiss is a less definite state of affairs than being, for instance, French or Italian. For Switzerland is a confederation of 23 cantons, or sovereign states (three of which are divided into 'half cantons'), and people are citizens first and foremost of their canton. There is even a word for this cantonal loyalty – *Kantönligeist*. It is the canton which issues a residence permit or runs a university. 'Swiss' refers to a ski team, to a railway system; above all (and

this is probably the greatest unifying force in the nation) to an army – indeed, one of the world's most remarkable armies, in which every Swiss male serves actively or on the reserve until he is 42 (52 for high-ranking officers). In order to be prepared, every Swiss soldier keeps all his equipment, arms and ammunition at home.

Nor do the Swiss intend to be unprepared for nuclear war. Underground nuclear shelters will be provided for the entire population by the year 2010.

There is no Swiss language by which to identify Swiss people. Three major languages are in use: the most widespread is German (64 per cent of the population), followed by French in the west (19 per cent) and Italian in the south (8 per cent). There is even a fourth official language, called Romansch and spoken by 1 per cent, mainly in the canton of Graubünden. Most Swiss people will respond politely if addressed in any of the three main languages, but they will talk to one another in a quite different language – Swiss German.

Such are the attractions of living in Switzerland that more than 18 per cent of the population are foreigners. A further 155 000 workers cross the frontier each day, mainly from France, to work in Switzerland. In addition, another 130 000 seasonal workers are admitted for a few months at a time, mainly to work in hotels, restaurants and cafés. To bring in his family, however, a worker must have at least an annual work permit and must have lived in Switzerland at least 12 months. After a stay of ten years, foreigners can obtain a residence permit and, after long probation, can eventually become Swiss. In 1993 the total number of foreigners working in Switzerland was 950 000 out of a population of 6.9 million. However, the Swiss are becoming increasingly reluctant to let in foreigners. As matters stand, with a population density of 167 people per km² (433 per sq mile), and being mountainous at that, Switzerland is fairly densely populated – and although for many years unemployment was around only 1 percent, it exceeded 5 per cent for the first time in 1993.

Switzerland began in 1291 as a defensive alliance between three cantons – Unterwalden, Uri and Schwyz (from which the country takes its name). The other 20 joined the league over succeeding centuries, to form what became in 1848 a federal republic. GENEVA, VALAIS and NEUCHÂTEL joined only in 1815; Jura, a new canton, was established in 1979.

The system of government is federal: the capital is BERNE (Bern), and it houses the national council, the council of states and a seven-man federal council, one member of which acts as president of the Swiss Confederation (the nation's official name) each year. But the cantons cling to their individual rights, and only such central functions as the control of the army, the railways or the postal service are left to the federal government.

INDEPENDENCE AND NEUTRALITY

The nation was born out of resistance to Austria and its Habsburg rulers. It began as a rather loose alliance, destined to defend the independence of its members, but later it engaged in an aggressive expansionism, which ended with defeat at Marignano (1515). Under the influence of the French Revolution, the old regime broke down and was replaced by the short-lived centralised Helvetic Republic. Napoleon pacified Switzerland in 1803 by giving it a constitution which reconciled both the new centralistic and older federalistic interests. During the Second World War Switzerland was surrounded on all sides by countries at war. But it survived unscathed the boast by Germany's Adolf Hitler that after his armies had conquered Europe he would take Switzerland with the Berlin fire brigade.

Independence and neutrality – guaranteed in perpetuity by the international treaty in 1815 – are the essentials of Swiss policy. To defend that neutrality the Swiss have a formidable army, although Swiss neutrality has been far too useful to other nations for any country, however aggressive, to find it worth violating. For example, Swiss or Swiss-based agencies have in the past acted as go-betweens in the exchange of prisoners-of-war and release of hostages, and, in times of war, to protect and help military and civilian victims.

Probably no other nation has made quite such a business out of neutrality as the Swiss. Ever since 1864, when the International Red Cross was initiated here, Switzerland – in particular, Geneva – has become the natural place for international organisations to set up their headquarters. One of these, the ineffective League of Nations (1919), provided Geneva with a splendid white elephant of a building which, after the Second World War, was filled with branches and offices of the League's successor organisations – many of them associated with the United Nations.

Insistence on neutrality was exemplified in 1994 when a federal referendum rejected a proposal which would allow Swiss participation in United Nations peace-keeping troops.

To many people, Switzerland is the country of the ALPS, and although not all of it is mountainous, 60 per cent of the country is Alpine. Northern Switzerland, like neighbouring regions of eastern France and south-west

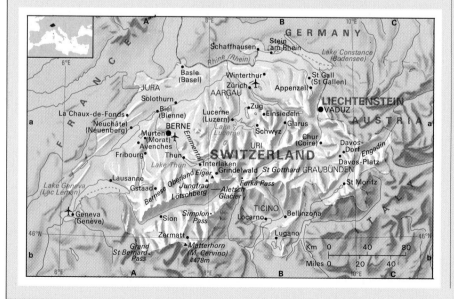

Germany, is mainly a land of hills and woods, but also of cities and industries. BASLE is world famous for pharmaceuticals and ZURICH and its suburbs for electrical engineering and machinery. It is non-Alpine Switzerland that produces the cheese, the chocolate, the clocks and the watches for which the country is renowned.

The Alps occupy the southern half of the country. They form two main east-west chains, divided by the straight line of the upper valleys of the RHÔNE and RHINE.

The northern chain of the Alps is wholly in Switzerland and contains such spectacular peaks as the EIGER (3970 m, 13 025 ft) and JUNGFRAU (4158 m, 13 642 ft). The southern chain contains the highest summits, but these Switzerland shares with its neighbours. The highest, most imposing peak of Mont BLANC (4807 m, 15 771 ft) lies just inside France, and the MATTERHORN (4478 m, 14 690 ft) lies on the Italian border.

In Alpine Switzerland, snowfields feed numerous glaciers, of which the largest is the Grosse (Great) ALETSCH (23.6 km, 15 miles long), and winter sports traffic transforms small localities like ZERMATT into large ones during the season.

▼ HAPPY VALLEY Prosperity beams like the sun on Lauterbrunnen, cosily enfolded in its valley in the Bernese Oberland. The Breithorn (3782 m, 12 408 ft), in the background, is only 11 km (about 7 miles) from the mighty Jungfrau, Mönch and Eiger.

The Alps also influence Swiss life through their impact on climate. They divide the Mediterranean world from the central European world. Going south, to emerge from the Gotthard or Simplon rail tunnels is to enter a new environment. Italian Switzerland is a land of lakes and subtropical vegetation, distinctive in architecture as well as speech. Going north there is less sunshine, and also an uncomfortable hot, dry south wind – the Föhn – which blows periodically and causes headaches.

There are many regional variations in climate. But generally the mountain air is clear and clean, a factor that has resulted in Switzerland attracting invalids from many parts of the world to its hospitals and clinics.

North of the Alps lies the Swiss Plateau, a strip of relatively low land, some 10-70 km (6-44 miles) broad, with its many lakes, its agriculture (here are the homes of Gruyère and Emmenthal cheese) and the bulk of the Swiss population. The plateau extends from Lake Geneva to Lake Constance. The region lying between Basle, Zurich and WINTERTHUR is heavily industrialised.

The third and smallest section of Switzerland is formed by the mountains of the JURA, which it shares with France. This region lies north-west of Lakes Geneva and Neuchâtel and comprises a number of steep, forested ridges running from south-west to north-east up to Schaffhausen. The Jura peaks are much lower than those of the Alps. In the heart of the Jura lies LA CHAUX-DE-FONDS, centre of Swiss watchmaking.

Helping to look after the needs of tourists, who arrive by the million every year, are more than 190 000 workers employed in the hotel and catering trades.

SWITZERLAND AT A GLANCE	
Official name Swiss Confederation	
Area 41 284 km² (15 936 sq miles)	
Population 6 908 000 **Per km²** 167	
(**Per sq mile** 433)	
Capital Berne	
Government Federal republic	
Currency 1 Swiss franc, or franken = 100 centimes or rappen	
Languages German, French, Italian, Romansch	
Religion Christian (46% Roman Catholic, 40% Protestant)	
Climate Temperate, mild in the valleys to the south; average temperature in Zurich ranges from –1.1°C (30°F) in January to 16.8°C (62°F) in July	
Land use Pastures and meadows 15%, forest 30%, cultivation 10%, other 45 %	
Main primary products Dairy, cattle, corn, fruits, grapes, vegetables, potatoes	
Major industries Tourism, machinery, chemicals, pharmaceuticals, banking, clock and watchmaking, instruments, textiles, forestry, metal goods, graphics industry, food	
Main exports Machinery, electrical appliances, chemical and pharmaceutical products, instruments, clocks and watches, metals, textiles, timber, cheese, chocolate	
Annual income per head (US$) 36 320	
Population growth (per thous/yr) 7	
Life expectancy (yrs) Male 74 **Female** 81	

Syria

FOR THOUSANDS OF YEARS THE CROSSROADS OF TRADE, VARIOUS CULTURES AND POWER SPHERES HAVE SHAPED THE HISTORY OF THIS EASTERN MEDITERRANEAN COUNTRY

During the 1980s Syria was armed with Russian missiles and revolutionary fervour, and it still holds the key to any lasting settlement between Israel and her Arab neighbours. Although Egypt and Jordan have made peace with Israel and the Palestinians have been in the process of doing so, Syria has shown great reluctance to follow suit. Its troops, who lost the GOLAN HEIGHTS in the Six Day War with Israel in 1967, have for years occupied parts of northern Lebanon. Here, Syria has manipulated the rival political and religious factions to weaken its adversaries.

Syria, always on the crossroads between the three continents since the days of the ancient world, has for thousands of years been the intermediary between the cultures of North Africa, Europe and Asia. All too often foreign powers tried to lay their hands on this strategically located land – and often it was the scene of bloody battles. Not always innocent, and often willing to maintain its autonomy at any cost, Syria was from 1979 into the early 1990s embargoed by the United States as a state sponsoring international terrorism. Indeed, Syria was one of the most nationalist and radical of Arab states.

People have lived in the capital, DAMASCUS, in unbroken succession since 2500 BC. It is the oldest continuously inhabited city in the world. From the city's coast the Phoenicians (9th-4th centuries BC) set out to trade in their ships, built from the cedars of Lebanon. St Paul was converted to Christianity on his way to Damascus and his refuge, Hanania's House, still stands off the Via Recta.

Agriculture still flourishes in the Syrian part of the rich FERTILE CRESCENT formed by the ORONTES and EUPHRATES rivers, but now there is oil (although not a great amount), which has become the nation's leading export.

THE LAND AND ITS PEOPLE

Much of Syria is mountainous. Behind the narrow, fertile coastal plain, the mountain range of JABAL AL NUSAYRIYAH, scored by deep valleys, drops eastwards to the GREAT RIFT VALLEY which leads to the Red Sea and into Africa. Farther south, the ANTI-LEBANON range rises to Mount HERMON (2814 m, 9232 ft). Most of the eastern part of the country is desert or semi-desert, a stony and inhospitable land. Although the mountain ranges make up only a part of the country, it is here that most of the population is concentrated. The arid east of the country is largely steppe and desert – only the Euphrates and Kabur valleys allow for irrigation farming and are densely populated.

The coast has warm dry summers and cool wet winters; inland, the climate becomes hot and there is little rain. In complete contrast, snow falls on the mountains in winter. Vegetation varies from Mediterranean scrub – tamarisks, camel thorns and acacias – to coniferous forests in the highlands; much of the country behind the coastal range is steppeland that turns green after the sparse rains encourage grasses to grow. Wildlife includes gazelles, wolves and wild cats; eagles, buzzards, kites and falcons are found in the mountains.

Ninety per cent of Syria's people are Arab, but the country presents a fascinating mixture of diverse communities, languages, religions and customs. Some of the minorities can trace their histories back long before the coming of the Arabs in the 7th century.

The Kurds of the north live along the Turkish border and make up 6 per cent of the population, with their own language, culture and national dress. There are Armenians in ALEPPO, Circassians (mountain people like the Kurds) in the south-west, Assyrians from Iraq, Turkmans from central Asia and the nomadic Bedouins of the desert. Many Arab Palestinians live in camps around Damascus, and there is a small Jewish community in the capital.

Ninety per cent of all Syrians are Muslims. The orthodox Sunni Muslims outnumber the Shiites (the sect that holds sway in Iran) by about six to one; but it is the small Alawite sect, an offshoot of Shiism, that controls the government. Although President Hafez al-Assad has been Iran's ally, he has not tolerated Islamic militancy: alcohol is available and Western dress is common among men and women in the cities.

In the south, the secretive Druse sect practises a religion containing Muslim, Christian and gnostic elements. There are about a dozen Christian communities, the strongest being the Greek Orthodox.

One-third of Syria's workforce is involved in agriculture; 31 per cent of the land today is cultivated and another 46 per cent provides some form of pasture, although this becomes increasingly poor on the edges of the desert.

Agriculture produces 27 per cent of Syria's income. Barley, wheat, tobacco, fruit, grapes and vegetables are grown on the Mediterranean coast – while in the irrigated valleys cotton is the main crop; sheep, goats and cattle are raised; mules and camels are beasts of burden. The Euphrates Dam at Al-Thaura, built with the help of Russian technology, feeds a vast reservoir, Lake ASSAD, which has enabled large-scale irrigation and supplies 50 per cent of the country's electricity needs.

MINERAL WEALTH

Syria's oil reserves are small by Middle East standards – only about 1700 million barrels. Production is small, and although it is by no means enough to finance large-scale industrial expansion, it is enough to make the country self-sufficient and provide half of the nation's export earnings.

Most of the oil comes from fields around Karachuk in the north-west, while the Al Jozirah region in the north-east yields natural gas – in the early 1990s, reserves were nearly 157 000 million m³ (5 548 000 cu ft). Other minerals found include limestone, phosphates (around Al Shargiya and at Khneifis), salt, manganese, iron, gypsum, marble and asphalt.

In the past 30 years Syria has diversified and greatly expanded industries such as textiles (using home-grown cotton), leather, chemicals and cement with the help of its oil revenues. Most of the industrial growth has taken place in the major towns, such as Damascus, HOMS, HAMAH and Aleppo. These have drawn people from all parts of the country. Indeed, half the population is urban-based and there has been a rapid growth of slums and shanty towns, especially in Damascus and Aleppo.

Nearly half the population is under 15, so there is great pressure on schools; Syria's population increase (3.7 per cent a year) is one of the world's highest. Education is free and about one-third of university students are women. The adult literacy rate is 64 per cent.

Health facilities have improved in recent years: there is now one doctor to every 2300 people, a figure comparable with several other Middle East states but lower than that in the developed Western countries. However, even the remotest arid regions are served by mobile clinics. The average life expectancy is 66 years, but 4.3 per cent of children still die in their first year.

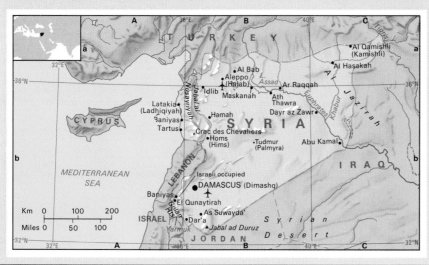

Syria has been a trading crossroads since ancient times. The Phoenicians were sea-faring traders and overland trade routes extended to Arabia, India and China. Even today oil pipelines across Syria follow ancient trading routes. Syrian traders are found throughout the Middle East and in parts of Africa, where the word 'Syrian' is a synonym for 'merchant'.

After the Phoenicians, Syria came in turn under the Greek, Roman and Byzantine empires. Then, three years after the death in AD 632 of the founder of the Muslim faith, the Prophet Muhammad, Arab armies took Damascus. It became the most important political and military base in an Islamic empire ruled by descendants of Umayya, the Prophet's great-great uncle. From 661 until 750, some 14 successive caliphs of the Umayyad dynasty held sway in Damascus.

BATTLEGROUND OF THE CRUSADES

In 1095 Christians in Europe launched the first of their Crusades. Crusader knights, sworn to recover the Christian holy places from the Islamic infidels, landed in Asia and in 1099 took Jerusalem. For two centuries, wars and battles were sparked off intermittently, during which period the Crusaders built a defensive chain of superb castles along the coast, many of which are still standing. The most important of these is the CRAC DES CHEVALIERS, built on a mountain west of Homs.

The Syrians fought back. In 1114 Damascus beat off an attack and Sultan Nur ad-Din (1146-74) captured many of the Christians' northern strongholds. In 1169 he took Cairo. His son Salah ad-Din (the famous Saladin) recaptured Jerusalem in 1187. Five years later peace came and Saladin returned to Damascus in triumph. But he had taken no share of the spoils and died in poverty; his friends had to borrow money to pay for his funeral.

For two centuries Syria was united with Egypt under the rule of the Turkish Mamelukes, until it was absorbed into the Turkish Ottoman Empire in the 16th century. At that time, Syria included the territory that is now Lebanon. When Turkey sided with Germany in the First World War, the Allies invaded Syria, and after the war the Levant (Syria and Lebanon) became a French mandate. In 1926 Lebanon became a separate state, taking with it the ports of BEIRUT and TRIPOLI. Twenty years later Syria became independent.

LEADING THE ARAB REVIVAL

Since then, Syria has played a leading role in the Arab revival that has marked recent Middle East politics. In 1958 it joined Egypt to form the United Arab Republic, but broke away after a coup in 1961. Other coups then followed until Hafez al-Assad came to power in 1970 – and set about consolidating his Ba'ath Party as the prominent polical force.

Under President Assad, Syria achieved reasonable stability despite war with Israel, a bitter feud with Iraq, hostility towards Jordan (for that country's more pragmatic policies towards Israel) and intervention in Lebanon.

Syria has always been hostile to Israel. The Six Day War was followed by another war in 1973. The government sent 20 000 troops into

Lebanon as a 'peace-keeping force' in 1976; they remained in a commanding position there for nearly two decades. Syria's pro-Soviet stance also strained relations with the United States, especially after attacks on the US embassy and American marines in Lebanon were attributed to Syrian extremists.

In recent years, however, with the dismantling of the Soviet Union, tensions between Syria and Western countries appear to have eased. By late 1994 President Assad had shown a notably more conciliatory attitude towards Israel after the progress in Israel's peace-making with the Palestinians and Jordan, and after a meeting between Assad and Israeli Prime Minister Itzhak Rabin.

FOR THE VISITOR

Syria has fostered tourism and encouraged more hotels to be built, together with other facilities, in a bid to make visitors welcome. In a good year, 1.2 million people, mainly from the Middle East, may come in order to enjoy the climate, the scenery, the historic sites and the modern places of interest.

Damascus is mentioned in early Egyptian records, and legend says it was the town of Shem, named after a son of Noah. The walled city lies in an oasis where the Barada River descends like a torrent from the Anti-Lebanon Mountains. Today, with a population of 1.3 million, it is a blend of the old and the new, with the Great Mosque, the Citadel and Saladin's tomb to see. The site of the Umayyad Mosque has been a place of worship for at least 3000 years, serving Aramaeans, Romans (who built a temple to Jupiter on the site), Byzantines and Arabs in turn. The present building was erected as a Byzantine church for the Romans under Emperor Theodosius I in AD 379, but was rebuilt as a mosque in 705 after Arabs had taken Damascus.

Aleppo, with a population of 1.3 million, was a key town on trade routes for thousands of years. It has a citadel, nearly 200 minarets, a 13th-century royal palace and Syria's biggest souk (market). Tourist villages are being

▲ **ONCE PROUD CITY The Grand Arch stands amid the ruins of Palmyra. The city grew in an oasis on the edge of the Syrian desert, and prospered from trade under the Romans. However, the conquerors destroyed Palmyra in AD 273 after its queen, Zenobia, defied them.**

developed on the Mediterranean at Ras Al-Basit, near the port of LATAKIA. There are mountain resorts at Zebadani and Bloudan, near Damascus.

SYRIA AT A GLANCE	
Official name Syrian Arab Republic	
Area 185 180 km² (71 498 sq miles)	
Population 14 338 530 **Per km²** 77	
(Per sq mile 199)	
Capital Damascus	
Government Republic	
Currency 1 Syrian pound = 100 piasters	
Languages Arabic (official), Kurdish, Armenian, Aramaic, Circassian, French (widely understood)	
Religions Muslim (90%) Christian (8%), Druse (2%), Jewish (tiny communities in Damascus)	
Climate Mediterranean on the coast; hot and dry inland, cold winters in highlands. Average temperature in Damascus ranges from 7°C (45°F) in January to 28°C (82°F) in August	
Land use Cultivation 31%, pastures 46%, forest and woodlands 3%, other 20%	
Main primary products Sheep, goats, cattle, cotton, fruit, potatoes, sugar, wheat, barley, vegetables, crude oil, natural gas, crude asphalt, phosphates, salt, manganese	
Major industries Agriculture, oil and gas production and refining, textiles (wool and cotton), cement, flour, soap, leather goods, glass, metal goods	
Main exports Oil products, natural gas, phosphates, cotton, fruit, vegetables, wool, wheat	
Annual income per head (US$) 2300	
Population growth (per thous/yr) 38	
Life expectancy (yrs) Male 65 **Female** 67	

Taal *Philippines* See BATANGAS

Taban Bogdo Ula *Mongolia* See ALTAI

Tabasco *Mexico* Swampy and forested state covering 25 267 km² (9756 sq miles) in the southeast, on the Gulf of Campeche. Rains are exceptionally heavy, the annual rainfall being up to 5000 mm (197 inches), and vast areas are often flooded by swollen rivers. Once only a source of timber, hardwood, dyes, coffee, sugar and bananas, the state was opened up for oil exploration in the late 1960s. Now Tabasco produces a large percentage of Mexico's oil, much of which is refined in Villahermosa, the state capital. When the rain forest was cleared for the oil industry, huge carved stone heads were uncovered. They were made by the Olmecs who settled in Tabasco in about 1200 BC.

Population 1 494 000
Map Mexico – Cc

Tabgha *Israel* Traditional site of Christ's miracle of the Feeding of the 5000, 10 km (6 miles) north of TIBERIAS, on the west shore of the Sea of Galilee. The 6th to 7th-century Church of the Multiplication has a Byzantine mosaic depicting a basket containing loaves and two fishes.

Map Israel – Ba

Table Bay *South Africa* Inlet in south-western Africa, which is 5 km (3 miles) long and 10 km (6.2 miles) wide. Overlooked by TABLE MOUNTAIN, it was 'discovered' around the end of the 15th century by Portuguese navigators seeking a sea route to India. In 1652 the Dutch founded a settlement on the bay, which eventually became known as CAPE TOWN, today called 'the mother city' of South Africa.

Map South Africa – Ac

Table Mountain *South Africa* Flat-topped sandstone mountain reaching 1086 m (3563 ft) and overlooking CAPE TOWN. It is visible to ships some 150 km (almost 95 miles) away, and is often seen swathed by its 'tablecloth' – a layer of cloud that rolls over the top when a south-east wind blows. A cableway built in 1929 runs to the top.

Map South Africa – Ac

Tábor *Czech Republic* Town in Bohemia 76 km (47 miles) south of PRAGUE. It was founded in 1420 by followers of John Huss (Jan Hus, about 1369-1415), the Czech religious reformer who lived for several years at nearby Kozí Hrádek Castle. The town was the Hussite capital during the wars in which Huss's followers sought to conquer Bohemia.

Population 33 800
Map Czech Republic – Bb

Tabor, Mount (Har Tavor) *Israel* Mountain, 10 km (6 miles) east of NAZARETH, overlooking the Jezreel Valley. It is the traditional site of the miracle of the Transfiguration, when Christ's face 'shone like the sun and his garments became white as the light' (Matthew 17:2). A Franciscan church and hospice and a Greek Orthodox church are situated at the summit, 588 m (1929 ft) high. There are also remains of Crusader fortifications.

Map Israel – Ba

Tabora *Tanzania* Regional capital situated some 340 km (about 210 miles) east of Lake TAN-GANYIKA. Founded by the Arabs in 1820, it trades in millet, groundnuts and cotton.

Population 214 000
Map Tanzania – Ba

Tabriz *Iran* Provincial capital of Eastern Azerbaijan, 550 km (342 miles) north-west of the national capital, TEHRAN. Set among high mountains, and at the foot of the Kuh-i-Saland volcano, Tabriz possibly dates back to Assyrian times; it is known to have existed in the days of the Macedonian conqueror Alexander the Great (356-323 BC). Earthquakes have destroyed it several times.

Today it is a commercial and industrial city, manufacturing textiles, carpets, leather and soap. Farms in the surrounding province produce almonds and dried fruit. Tabriz is the cultural and commercial centre of north-western Iran.

Population 971 500
Map Iran – Aa

Ta-chia *Taiwan* River of west-central Taiwan. It rises in the northern part of the central mountains and flows 140 km (nearly 90 miles) westwards to the Taiwan Strait. The fast flowing waters of its upper reaches have been harnessed for hydro-electric power.

Map Taiwan – Bb

Tacloban *Philippines* See LEYTE

Tacna *Peru* City and department on the border with Chile. Its arid lands are now being irrigated to produce crops of grapes and olives. The city lies in the shadow of the Campo de la Alianza, the site of a decisive battle between Peru and Chile in 1880 during the War of the Pacific (1879-84), and was under Chilean rule until 1929, when its inhabitants voted for it to be returned to Peru. The fountain and cathedral situated on the Plaza de Armas in Tacna were designed by the French engineer Gustave Eiffel (1832-1923), who built the Eiffel Tower in Paris. There is a railway museum and local museum exhibiting pottery.

Population (department) 209 800; (city) 150 200
Map Peru – Bb

Tacoma *USA* Port on Puget Sound, about 40 km (25 miles) south of Seattle, in Washington State. It has timber and mineral processing industries. Tacoma is the gateway to the MOUNT RAINIER NATIONAL PARK.

Population (city) 159 400; (metropolitan area) 515 800
Map United States – Ba

Taegu *South Korea* The country's third largest city after Seoul and the port of Pusan. Taegu lies 95 km (60 miles) north-west of Pusan. It is the country's chief textile centre, making silk, cotton, woollen and synthetic fabrics. It also has the country's highest temperatures and the surrounding farmland yields two crops of rice per year, as well as apples, tobacco, mulberries and cotton.

In 1950, during the Korean War, the advance of the North Korean armies was checked just north of Taegu.

Population 2 286 000
Map Korea – De

Taejon (Daejeon) *South Korea* Modern city 150 km (nearly 95 miles) south of the capital, SEOUL. Destroyed during the Korean War (1950-3), it has been rebuilt as a manufacturing centre for textiles, ceramics and furniture.

Population 1 062 000
Map Korea – Cd

tafelberg See TAFELKOP

tafelkop In South Africa, the term for a mountain with a flat top. Large tafelkops are known as tafelbergs.

Tafilalt (Tafilet) *Morocco* Largest Saharan oasis in Morocco, comprising 1375 km² (530 sq miles) of fortified villages and groves of date palms stretching for 50 km (30 miles) along the Ziz Valley, with ERFOUD as the main settlement. In AD 757 the Berbers established an independent kingdom here, with its capital at Sijilmassa.

Sijilmassa was destroyed by Arab invaders in 1363, refounded in the 17th century by Emir Moulay Ismail (from whom the present Moroccan king descends), and devastated by nomadic tribesmen in 1818. Tafilalt is noted for its dates, which it exports worldwide.

Population 70 000
Map Morocco – Ba

Taganrog *Russia* Seaport and industrial town on the Sea of AZOV, east of the Crimea. It ships grain and has a steel industry, shipyards and commercial fisheries. Founded as a fortress and naval base in 1698 by Tsar Peter the Great (1672-1725), it was occupied by the Turks in 1730-69. The Russian playwright Anton Chekhov (1860-1904) was born here; his birthplace is now a museum.

Population 293 000
Map Russia – Ed

Tagaytay City *Philippines* See CAVITE

Tagus (Tajo; Tejo) *Spain/Portugal* Longest river on the Iberian Peninsula, rising in eastern Spain and flowing some 1010 km (628 miles) southwestwards through the Portuguese capital, LISBON, to the Atlantic. The Tagus drains an area of 80 000 km² (30 881 sq miles).

Map Portugal – Bc

Tahiti *French Polynesia* Largest island of the territory, covering 1042 km² (402 sq miles), and main island of the SOCIETY ISLANDS. It is mountainous, reaching 2237 m (7339 ft) at Orohena, and covered in dense vegetation and tropical flowers. Sometimes termed 'the original South Seas paradise island', Tahiti is fringed by coral reefs. Its beauty has been immortalised by the painter Paul Gauguin, who lived here in 1895-1901, and such writers as Pierre Loti and Herman Melville. It is a popular tourist destination, and it produces coconuts, vanilla and tropical fruits. Its capital is PAPEETE. The smaller but equally beautiful island of Moorea lies 16 km (10 miles) to the west.

Population 96 000
Map Pacific Ocean – Fc

Tahoe, Lake *USA* Lake in the Sierra Nevada, 1898 m (6228 ft) high, covering some 500 km² (193 sq miles). It lies astride the Nevada-California state border, 250 km (155 miles) northeast of SAN FRANCISCO. It is a popular tourist area, with gambling on the Nevada side, notably at nearby RENO.

Map United States – Bc

Tai Po *Hong Kong* City in the NEW TERRITORIES, stretching westward from a sea inlet known as Tolo Harbour. Originally a small market centre for the surrounding region, it has been developed, much of it on reclaimed land, since 1973 as one of the new towns aimed at relieving population pressure in the built-up areas of the colony.
Population 116 000
Map Hong Kong – Ca

Tai Shan *China* Mountain in Shandong Province, 85 km (52 miles) south of JINAN City. It rises on the Shandong Peninsula and is the highest peak in the peninsula, at about 1545 m (5100 ft). The Tai Shan is the most sacred of the five sacred mountains of the Taoist religion.
Map China – Hd

T'ai-chung (Taizhong) *Taiwan* City and port, 125 km (nearly 80 miles) south-west of T'AI-PEI. Developed as a planned settlement during the Japanese colonial period (1895-1945), it has emerged as an expanding industrial city since the mid-1960s. Construction of its port, 20 km (12 miles) west of the main built-up area, began in 1971; although incomplete, it handles a high volume of cargo. T'ai-chung has an export processing zone where foreign companies manufacture products ranging from electronics and optical goods to plastics.
Population 814 000
Map Taiwan – Bb

Taif *Saudi Arabia* Mountain resort situated 60 km (37 miles) south-east of MECCA. Some fine old Arab houses remain near the souk (market), which sells traditional Arab clothes made locally. Its streets are lined with pepper trees. Set about 1700 m (5577 ft) above sea level, the roads leading up to Taif provide spectacular views of brown, wind-sculpted rocks overlooking fertile green valleys, in which figs, peaches, apricots, pomegranates, almonds and vegetables are grown. Taif itself is noted for its roses.
Population 205 000
Map Saudi Arabia – Bb

taiga Coniferous forest region of northern Europe bordered on the north by the TUNDRA and on the south by deciduous forest or STEPPES. The term is also used for similar forest in North America.

T'ai-lu-ko gorge *Taiwan* See HUA-LIEN

T'ai-nan *Taiwan* City of south-western Taiwan, about 50 km (30 miles) north of KAO-HSIUNG. The 'City of a Hundred Temples' (in fact there seem to be more like 200 temples) is Taiwan's oldest city and dates from the late 16th century. The site was occupied briefly by the Dutch in the early 17th century, and some of their fortifications survive, such as Fort Zeelandia. The city was the capital of Taiwan until superseded by T'ai-pei in 1885. Although now industrial, T'ai-nan retains many buildings of historical interest. The Confucian Temple, over 300 years old, is one of the best examples of traditional Chinese architecture on the island, as is the Buddhist Kai-Yuan Temple.
Population 699 000
Map Taiwan – Bc

T'ai-pei *Taiwan* Capital and largest city of Taiwan, lying near the northern end of the island on the east bank of the Tan-shui River at its confluence with the Keelung. The capital since 1885, it grew rapidly during the Japanese colonial administration of Taiwan (1895-1945), but even more dramatically after the Second World War. Its population expanded from 335 000 to more than 2.6 million in just over 40 years.

A busy, bustling city of broad avenues, T'ai-pei contains government offices and parliament buildings as well as the headquarters of many of Taiwan's industrial and commercial companies. Industries include engineering, manufacture of textiles, and food processing and printing.

Little of old T'ai-pei remains, but the nightly market, with open-air foodstalls, survives. The most imposing buildings are recent structures, many uneasily combining the traditional Chinese and 20th-century Western architectural styles. More historic structures include the late 19th-century city wall and gates, and the ornate and colourful Lungshan Temple, the most famous of Buddhist temples in T'ai-pei, some 250 years old, but largely rebuilt after the Second World War. The National Palace Museum, 8 km (5 miles) from the centre, in the Wai-shuang-hsi Hills, houses a spectacular collection of Chinese art treasures – some 250 000 items, many removed from mainland China on the eve of the communist revolution in 1949. They include carved jade, enamel and lacquerware, porcelain, bronze items, paintings, tapestries and books. T'ai-pei has several parks, a zoo and a botanical garden containing some 700 species of trees, shrubs and palms.
Population (city) 2 656 000; (metropolitan area) 4 200 000
Map Taiwan – Bb

Taiwan See p 656

Taiwan Strait *Taiwan/China* Body of water 160 km (100 miles) wide, and 70 m (230 ft) deep, separating Taiwan from the Chinese mainland. Formerly known as the Formosa Strait, the Taiwan Strait is subject to typhoons which sweep in from the central PACIFIC OCEAN.
Map Taiwan – Ab

Taiyuan *China* Capital of Shanxi Province, 400 km (nearly 250 miles) south-west of BEIJING. It is a centre for the manufacture of iron and steel, heavy machinery and chemicals.
Population 1 960 000
Map China – Gd

Taiz *Yemen* The country's third largest city situated some 80 km (55 miles) north-east of Mocha. A hill resort at 1300 to 1400 m (4265 to 4593 ft) altitude, Taiz offers panoramic views over the surrounding land and is a popular local holiday centre mainly because of its pleasant climate. It lies at the foot of the Saber Mountains, which are more than 3000 m (9843 ft) high. It has government offices, several mosques, of which the most important are the Ashrafiyah Mosque and the Mudhafar Mosque, and an international airport. The city is the centre of a coffee-growing area, and its industries include cotton weaving, tanning, jewellery making and plastics. New water and electricity projects have been set up.
Population 178 040
Map Yemen – Ab

Taj Mahal *India* See AGRA

Tajikistan See p 657

Takamatsu *Japan* City on north-east SHIKOKU Island, which was badly damaged by Second World War bombing. Today it is a ferry port and industrial centre. Ritsurin Park has a graceful bridge, tranquil ponds and a forest backdrop.
Population 330 400
Map Japan – Bd

Takasaki *Japan* City on central HONSHU Island, 100 km (62 miles) north-west of the national capital, Tokyo. It is a gateway to the Japan Alps and is noted for its *daruma*, red roly-poly dolls.
Population 236 000
Map Japan – Cc

Taklimakan *China* The country's largest desert, covering over 327 000 km² (126 000 sq miles) in the Tarim Pendi basin of western China. In the language of the Uygurs, the main ethnic group in and around the desert, Taklimakan means 'no escape after entry' – an apt description of the desert's immense, trackless wastes. Some 85 per cent of the desert consists of drifting sand dunes, some as high as 300 m (1000 ft) – which explains why the old caravan routes, such as the Silk Road, as well as today's roads, skirt it.
Map China – Bc

Takoradi *Ghana* Town just west of SEKONDI, lying 170 km (105 miles) west of the capital, Accra. It grew only after the country's first deep-water harbour was built here in 1923, cutting the cost of exports. The port was improved in 1953 and is now fully mechanised, handling about 2 million tonnes of cargo a year, mainly timber, cocoa and mineral exports. Takoradi has a thriving industrial area producing cocoa butter and powder, plywood and veneer, cigarettes, hardware and cement. It was linked with Sekondi as a city in 1963, and Sekondi-Takoradi is now the capital of Ghana's Western Region.

The nearby village of Nkroful was the birthplace of the independent country's first president, Dr Kwame Nkrumah (1909-72), who held power from 1957 until 1966. He is also buried here.
Population (Sekondi-Takoradi) 254 500
Map Ghana – Ac

talc Fine-grained, white, grey, brown or pale green mineral, magnesium silicate, having a soft, soapy texture, and rating 1 on the MOHS SCALE of hardness – the lowest value. It is used in talcum powder and as an electrical insulator.

Talca *Chile* Province and city in the Maule region of the Central Valley. It is a major area for growing wheat and wine and for stock-farming. The city, which is also capital of the province, lies 240 km (nearly 150 miles) south of SANTIAGO. Founded in 1692, it was destroyed by earthquakes twice. Its industries include flour and paper mills.
Population (province) 262 200; (city) 138 000
Map Chile – Ac

Tallahassee *USA* Capital of FLORIDA, situated in the north-west of the state, between the Gulf of Mexico and the border with Georgia. Tallahassee was the territorial capital in 1823 when southern Florida was still undeveloped and unexplored. Computers, food, paper and timber products are its economic mainstays today.
Population (city) 124 800; (metropolitan area) 233 600
Map United States – Jd

Taiwan

MOST COUNTRIES NO LONGER RECOGNISE THE ISLAND STATE OF TAIWAN – AND YET THEY CONTINUE TO NEED ITS TRADE

Taiwan is a political anomaly – a major trading country whose very existence few governments recognise officially. Virtually none today supports its claim to be the sole legitimate Republic of China. The United Nations and other international organisations recognise only the Communist People's Republic, ruled from BEIJING since 1949. Yet Taiwan – or Formosa, as the island

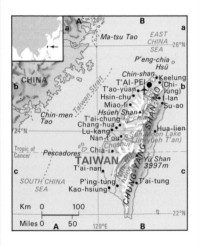

was once known – cannot be ignored. In the past 30 years, it has become one of the world's most dynamic newly industrialised nations.

Sixteenth-century Portuguese navigators called Taiwan *Ilha Formosa* – the beautiful island – and despite modern cities, factories and motorways, Taiwan still contains some spectacular landscapes.

Lying about 160 km (100 miles) off the south-east coast of China, the island is predominantly mountainous, with many majestic peaks; the tallest is YÜ SHAN, at almost 4000 m (13 123 ft). Cultivation is largely restricted to the western and northern coastal plains, which also contain most of Taiwan's population and industry. Apart from the main island, which is about 375 km (235 miles) long and 145 km (90 miles) wide, the country includes the PESCADORES Islands and CHIN-MEN TAO and MA-TSU TAO island groups.

Taiwan lies astride the Tropic of Cancer, giving it a warm and humid climate for most of the year. Winters are mild, with average lowland temperatures rarely falling below 10°C (50°F); summers can be oppressive, July temperatures in T'AI-PEI, the capital, often exceeding 30°C (86°F). But a wide range of crops – tea, sugar cane, bananas, pineapples and especially rice – thrive in the high temperatures, abundant rainfall and fertile soil, particularly along the coast. Intensive farming on the coastal belt has enabled the Taiwanese to become self-sufficient in food production and even to produce a little extra for export.

On the other hand, fuel and mineral resources remain meagre. Domestic coal and hydroelectric power are not sufficient to meet important industrial needs, so oil and other raw materials must be imported. Forty per cent of Taiwan's electricity currently comes from nuclear power.

Japan annexed the island from the Chinese Empire in 1895, after the Sino-Japanese War. After the Second World War, Taiwan reverted to China, then controlled by the right-wing Nationalist Party. Routed by Mao Zedong's Communist forces, some 2 million Nationalists, under their leader Chiang Kai-shek, fled from the mainland to Taiwan in 1949 and set up a government in T'ai-pei, calling their state the 'Republic of China'.

With American help, the government began to develop the island's economy from its agricultural base, beginning with a successful land reform programme and industrial expansion. The first aim was to replace imports, but from the mid-1960s the emphasis switched to promoting exports. Taiwan's export-processing zones are among the most successful of their kind in the world, accommodating domestic and overseas companies. Exports, including

electronic goods and computerware, as well as clothing and footwear, today help to drive the Taiwanese economy forward, contributing towards a spectacular rise in living standards. Cheap labour has also ensured that prices stay competitive on the international market.

Overshadowing the future of this prosperous and politically stable island has been the question of its relationship with China. Two major blows were its expulsion from the United Nations (in favour of mainland China) in 1971, and the switch of American diplomatic recognition from T'ai-pei to Beijing in 1979. Relations between mainland China and Taiwan improved during the early 1990s and by late 1994 talks on reunification were being held between the two countries. Nevertheless, many Taiwanese began to fear a deterioration in the political situation once Hong Kong reverts to China in 1997, and applications for emigration rose sharply.

TAIWAN AT A GLANCE	
Official name Republic of China	
Area 35 981 km² (13 892 sq miles)	
Population 21 100 000 **Per km²** 586 (**Per sq mile** 1519)	
Capital T'ai-pei	
Government Republic	
Currency 1 new Taiwan dollar = 100 cents	
Language Chinese (Mandarin – official), Hokkien, Hakka	
Religions Confucian and Taoist (48%), Buddhist (24–30%), Christian	
Climate Tropical to subtropical monsoon climate; in T'ai-pei, temperatures range from 14.8°C (58.6°F) in February to 28.2°C (83°F) in July	
Land use Forest 60%, cultivation 25%, other 15%	
Main primary products Rice, sugar cane, sweet potatoes, bananas, pineapples, citrus fruits, pigs, poultry, fish, timber; coal, sulphur	
Major industries Electronic and electrical goods, textiles, chemicals, fertilisers, paper, cement, glass, plastics, iron and steel, mining, agriculture, forestry, fishing	
Main exports Electrical and electronics equipment, machinery, clothing, metals and metal goods, textiles, food, toys, plastics	
Annual income per head (US$) 4989	
Population growth (per thous/yr) 11	
Life expectancy (yrs) Male 71 **Female** 76	

Tallinn (Reval; Revel) *Estonia* Baltic Sea port, industrial centre and national capital, lying on the Gulf of Finland opposite Helsinki. It began as a fortified trading post in the 10th century and was captured by the Danes in 1219. It was subsequently occupied by Germans, Swedes and, from 1710 to 1918, Russians. During the Second World War it was damaged by Russian bombing, and has since been substantially rebuilt. Tallinn's industries include shipbuilding, machinery, electrical equipment, textiles, paper and chemicals.
Population 484 400
Map Estonia – Ab

talus See SCREE

Tamale *Ghana* Largest town in the Volta River basin, 440 km (nearly 275 miles) north of the

capital, ACCRA. It is the capital of Ghana's Northern Region, and has its own airport. It is a market for rice, groundnuts and cotton.
Population 219 200
Map Ghana – Ab

Tamanrasset (Fort-Laperrine) *Algeria* Market town and administrative centre, 1400 m (4600 ft) above sea level, located on the edge of the Hoggar Mountain region in the deep south. It is the key starting place for desert travel to AGADEZ in Niger and GAO in Mali. It is also the capital and meeting place of the Tuaregs, desert nomads who graze their livestock in the Hoggar Mountains.
Population 30 000
Map Algeria – Bb

Tamatave *Madagascar* See TOAMASINA

Tambov *Russia* Town on the plain of the Oka and Don rivers, 420 km (260 miles) south-east of MOSCOW. Its industries include engineering, chemicals and food processing. Tambov was founded as a fort in 1636 to defend Moscow against the Tatars.
Population 307 000
Map Russia – Fc

Tamil Nadu *India* Rural state in the southeast, formed from the Tamil-speaking southern 130 058 km² (50 202 sq miles) of the former state of Madras. The main commercial crop of the coastal plain, where myriad 'tanks' (small artificial lakes) provide water for irrigation, is rice. MADRAS is the state capital.
Population 55 858 950
Map India – Be

Tajikistan

THIS IMPOVERISHED ASIAN COUNTRY HAS BEEN MADE POORER STILL BY A POST-INDEPENDENCE CIVIL WAR

Of all the former Soviet states, land-locked Tajikistan, independent since September 1991, has always had the lowest standard of living – and today it faces the bleakest future. About 87 per cent of its population live below the poverty line as defined by the United Nations.

The country is dominated by the snow-capped Pamir Mountains. Only 7 per cent of its 143 100 km² (55 250 sq miles) land area is below 1000 m (3280 ft) and more than half is over 3000 m (9850 ft). The country is crossed by several major rivers. The climate is sharply continental and the land is somewhat arid. The topographic and climatic range has given it an extremely varied plant life, with more than 5000 different kinds of flowers.

The lowland area is now irrigated, mainly in the FERGANA and AMUDAR'YA valleys, and cotton and fruit crops are harvested. Today, 38 per cent of the workforce in employed in agriculture, and besides fruit and cotton, which are exported, the country's agricultural products include grains, mulberry trees (for silkworms), wheat and vegetables; pigs, goats and cattle are raised. There is some illicit production of cannabis and opium, which are exported to other member countries of the Commonwealth of Independent States.

Hydroelectricity from tributaries of the Amurdar'ya powers cotton and silk mills, food factories, mines and smelting works. Industries are concentrated in the cities – mainly in the capital, DUSHANBE, and in KHUDZHAND and Ura-Tyube – and include food processing, as well as chemicals and fertilisers, cement, electricals, and aluminium. The republic is rich in minerals such as coal (the major mineral fuel), zinc, lead, molybdenum, uranium, radium, arsenic, bismuth, oil and natural gas, but as yet they have not been developed. Neither is the transport system well structured or developed.

All of Tajikistan's trade is with former Soviet states, and in the first quarter of 1993, further trade and military treaties were entered into with Russia. A trade and aid agreement was also signed with the country's giant neighbour to the east, China. However, short-term prospects for an upswing remain poor for an economy ravaged by civil war.

ETHNIC MIX

Like most Central Asian countries, Tajikistan is named after its largest population group, whose members speak a language that is closely related to Persian. Many Tajiks also live in neighbouring Afghanistan, China, Kyrgyzstan and Uzbekistan. But unlike the Kyrgyz, Uzbeks or Kazakhs, the Tajiks are not related to other Turkic peoples of Central Asia. Tajikistan was invaded by the Persians in the 6th to 7th centuries BC, and later by the Greeks under Alexander the Great, the Arabs, Tatars and Mongols. From the 15th to the mid-18th century the country was ruled by the Uzbek khanate Bukhara. It was then taken over by Russia in the late 19th century. Its people are predominantly Muslim.

The country is divided along ethnic, religious and regional lines. The main ethnic source of conflict has been between the Tajiks, of Persian origin, and the Uzbeks, of Turkic origin. Although the Tajiks clearly form a majority, there is a total of 12 minority groups. Politically the conflict has been mainly between former communists and an opposition formed of Muslims and democrats.

Like many of the neighbouring republics, Tajikistan has a high rate of population increase; half the population is under 20 years of age and children under nine years old constitute almost a third. Women form over two fifths of the non-farm work force. Most of the people live in rural villages called *qishlaqs*.

In the Pamirs in the east lies the autonomous region of Gorno-Badakhshan, which constitutes 45 per cent of the country's land area but only about 3 per cent of its population. The so-called 'autonomy' of the Badakh is limited; they do not have a self-elected government.

THOUSANDS DISPLACED

While it was a Soviet republic, no political parties with religious objectives were permitted in Tajikistan; however, a few weeks after independence the first Islamic party was founded. There followed a scurry nationwide to form new political parties and to obtain political representation.

The Communist Party had remained in power after independence, with 58 per cent of the national vote, but in early 1992 the newly elected president was ousted. This triggered civil war between the communists based in Dushanbe and Islamic militants in the central and eastern regions – a war which displaced many thousands of people and disrupted food imports, causing terrible shortages in a number of regions.

By the end of 1992 about 30 per cent of the country's industries had shut down; agricultural output, too, had declined, with the cotton harvested being only half of that produced during the previous year. By early 1993 the

inflation rate had shot up to more than 30 per cent a month. For instance, in February 1993 the price of bread was increased overnight by 150 per cent.

Since independence, two presidents have been forced to resign; there have been several coup attempts, and a minister has been assassinated. New national elections which were scheduled for December 1992 had to be cancelled due to the war. Fearing the spread of violence across the country's borders, peace-keeping forces from Uzbekistan, Russia, Kazakhstan and Kyrgyzstan were deployed in the country. By early 1993 a state of emergency had been declared, which brought a relative calm.

The areas most affected by the strife have been Garm and Gorno-Badakhshan, where the ousted government has taken refuge. By the end of 1992 more than 200 000 people were reported homeless inside the country; many more had fled to Afghanistan. From there, the anti-government Rastokezh People's Front operated across the border.

Another focal point of the conflict was the capital, Dushanbe, where several politicians were taken hostage and the presidential palace was captured by anti-government forces. After the recapture of the capital by the army, reports of serious human rights violations were made.

Late in 1994 peace talks held in the Iranian capital, Tehran, led to a ceasefire agreement between the government and the Muslim democratic opposition; however, there was a renewed outbreak of fighting with Tajik rebels in Gorno-Badakhshan. In November 1994 presidential elections were held, when the acting head of state, Imamoli Rakhmmanov, won 58 per cent of the vote. In a simultaneous referendum more than 90 per cent of the electorate supported a new constitution

TAJIKISTAN AT A GLANCE	
Official name Republic of Tajikistan	
Area 143 100 km (55 250 sq miles)	
Population 5 836 200 **Per km²** 41	
(Per sq mile) 106)	
Capital Dushanbe	
Government Republic	
Currency 1 Russian rouble = 100 kopeks	
Languages Tajik (official), Russian	
Religions Sunni Muslim (80%), Shi'a Muslim (5%)	
Climate Continental, with cold winters and hot, dry summers; warm and mild at lower altitude. Average temperature in Dushanbe ranges from 1.4°C (34.5°F) in January to 28.2°C (83°F) in July	
Land use Grazing 23%, cultivation 6%, forest and woodlands 4%, other 67%	
Mineral resources Oil, natural gas, lignite, lead, uranium, zinc, gold, tungsten, molybdenum, antimony, mercury	
Main primary products Cotton, fruit; ferrous and non-ferrous metals	
Major industries Mining, textiles, food processing	
Main exports Ferrous and non-ferrous metals, textiles; fruit, vegetable oil, cotton	
Annual income per head (US$) 480	
Population growth (per thous/yr) 27	
Life expectancy (yrs) Male 66 **Female** 71	

Tammerfors *Finland* See TAMPERE

Tampa *USA* Port on the west coast of FLORIDA, situated at the head of Tampa Bay, on the Gulf of Mexico. It attracts many Cuban immigrants. A noted producer of cigars, Tampa also processes and distributes local vegetables and seafood (especially shrimp).

Population (city) 280 000; (metropolitan area) 2 068 000

Map United States – Je

Tampere (Tammerfors) *Finland* Third largest city in Finland after the capital, HELSINKI, and ESPOO. Tampere was founded in 1779 in the south-west of the country, 165 km (nearly 105 miles) north-west of Helsinki. It lies beside a waterfall between lakes Näsijärvi and Pyhäjärvi, which provided energy for a cotton-mill established by a Scot, James Finlayson, in the 1820s. The city grew up around the mill to become Finland's leading textile centre. Its industries now also include locomotive and rolling stock production, engineering, timber processing, plastics and footwear manufacture and printing. The city also has an open-air theatre with a revolving auditorium and several museums and art galleries.

Population 174 900

Map Finland – Bc

Tampico *Mexico* Oil-refining city and port in the state of Tamaulipas, near the mouth of the River Pánuco on the Gulf of Mexico. Summers here are hot and humid, and torrential rains in late September often cause floods. Despite its trail of refineries and storage tanks along the river, Tampico is a popular resort.

Population 271 600

Map Mexico – Cb

Tana, Lake *Ethiopia* The country's largest lake covering 2849 km² (1100 sq miles) in the highlands, about 350 km (217 miles) north-west of ADDIS ABABA, in the Amhara region. It is the source of the Abay, or Blue Nile. More than 60 rivers flow into the lake, which is nowhere more than 14 m (46 ft) deep.

Map Ethiopia – Aa

Tana *Kenya* The country's longest river, flowing 800 km (nearly 500 miles) from the Aberdare Mountains south-eastwards into the Indian Ocean, 90 km (55 miles) north-east of Malindi. The upper section is used to generate electricity via a series of dams built in the 1960s and 1970s. In the coastal region, irrigation supports cultivation of bananas, cassava and rice.

Map Kenya – Cb

Tananarive *Madagascar* See ANTANANARIVO

Tanga *Tanzania* Regional capital and port on the Indian Ocean, about 50 km (30 miles) south of the Kenyan border. It exports sisal, coffee and copra, and has a fertiliser industry.

Population 200 000

Map Tanzania – Ba

Tanganyika, Lake *Burundi / Tanzania / Zaire / Zambia* Africa's second largest lake, after Lake Victoria, covering 32 900 km² (12 700 sq miles) in the western arm of the East African section of the GREAT RIFT VALLEY. With depths up to 1480 m (4755 ft), it is the world's second deepest fresh-water lake, after Lake BAIKAL in Russia. The first Europeans to reach the lake were the British explorers Richard Burton and John Hanning Speke in 1858. Contrary to earlier belief, Lake Tanganyika is not a source of the Nile. In fact, the lake has only one outlet, the Lukuga, which is a tributary of the Zaire River. There are two main feeders, the Ruzizi and Malagarasi rivers. Lake Tanganyika is navigable and has some fine beaches, fishing and wildlife. Its chief ports are Bujumbura (Burundi), Kalémié (Zaire) and Kigoma (Tanzania).

Map Zaire – Bb; Tanzania – Ba

Tangier (Tanger) *Morocco* Seaport at the south-western end of the Strait of GIBRALTAR; on the part of Africa closest to Europe, it has been the gateway between the two continents for centuries.

Greeks, Phoenicians and Romans all built cities on the site. The Roman city, Tingis, fell successively to the Vandals (in the 5th century), Byzantines (6th century) and Arabs (8th century). In 711, it became the springboard for the Arab invasion of Spain. Later occupants of Tangier included the Portuguese (1471-1580 and 1656-62), Spanish (1580-1656), and English (1662-84), before the city again became part of Arab Morocco. It had its heyday from 1912 to 1956 when it had international status and was a booming trade centre. Today it is a free port and a busy trade and industrial city. The city has a distinctly international flavour with numerous banks, a famous casino and exciting nightlife.

The centre of Tangier is the busy Grand Socco market, alongside which is the old city dominated by its *kasbah* (citadel) containing the Dar al Makhzen (the Sultan's Palace), the apartments of which have been transformed into a museum of Moroccan arts, displayed region by region. The Museum of Moroccan Antiquities lies in the Dar Shorfa Palace. Also in the old city is the 17th-century Great Mosque built by Moulay Ismail (1672-1727), who pacified Morocco's warring tribes and drove out European interlopers.

To the east and west of Tangier lie Cape Malabata and Cape Spartel, both with lighthouses. Between them are some 22 km (14 miles) of fine beaches and rocky bays with views across the straits to Spain.

Population 554 000

Map Morocco – Ba

Tangshan *China* Coal-mining city of Hebei Province, situated some 150 km (95 miles) east of BEIJING. It was devastated in 1976 by an earthquake measuring 7.8 on the RICHTER SCALE; more than 500 000 people were killed. Tangshan's mines have now been repaired and the city, with more than 200 new factories, has been rebuilt.

Population 1 500 000

Map China – Hc

Tanintharyi (Tenasserim) *Myanmar (Burma)* Hilly division, covering an area of 43 344 km² (16 735 sq miles) in the south of the country. It stretches into the narrow isthmus which forms the northern part of the Malay Peninsula, and is swept by the south-west monsoon, which brings an annual 4000 mm (158 in) of rainfall between May and October. Tanintharyi's scattered population works largely on rubber and fruit plantations. Tavoy (population 102 000) is the capital.

Population 918 000

Map Myanmar – Cd

Tanjore *India* See THANJAVUR

Tanna *Vanuatu* Island in the south of the group whose active volcano, Yasur, is a popular tourist attraction; its crater continuously issues smoke, steam and lava fragments. Tanna is a stronghold of the 'cargo cult' (see MELANESIA).

Population 18 530

Map Pacific Ocean – Dc

Tannenberg *Poland* See GRUNWALD

Tanta *Egypt* Town in the centre of the Nile delta, 83 km (51 miles) north of CAIRO and midway between the Dumyat (Damietta) and Rashid (Rosetta), the river's two main branches.

Population 372 000

Map Egypt – Bb

Tanzania See p 659

Taolanaro *Madagascar* Indian Ocean port in the extreme south-east, formerly called Fort Dauphin and later, Faradofay. It exports beans, cattle, timber and rice. The first French settlement on the site was founded in 1643. Its governor was Etienne de Flacourt, whose *History of the Great Island of Madagascar* (1658) remains a major source of information about the island.

Population 13 000

Map Madagascar – Ab

Taormina *Italy* Well-preserved medieval town and resort on Sicily's east coast, 60 km (37 miles) from its north-east tip. The old town's position, 250 m (820 ft) high, has protected it from tourist development, and hotel building is restricted on the beaches below. But even in Roman times this was a popular stopover for holidaymakers, as it affords views of the active volcano, Mount ETNA. There are many strikingly attractive buildings from the later Middle Ages. The view of Mount Etna from the Graeco-Roman theatre is the most breathtaking in Sicily.

Population 10 300

Map Italy – Ef

T'ao-yüan *Taiwan* City 20 km (12 miles) west of T'AI-PEI, on Taiwan's northern plain. It lies at the centre of a prosperous agricultural region. Nearby is Taiwan's principal airport, Chiang Kai-shek International Airport, opened in 1979.

Population 1 446 000

Map Taiwan – Bb

Tapiola *Finland* See ESPOO

Tapti (Tapi) *India* River rising on the northern Deccan in central India, flowing some 702 km (436 miles) west to the Gulf of Khambhat at the city of SURAT. Its flow varies with the seasons.

Map India – Bc

tar pit Region where natural BITUMEN has accumulated and is exposed at the surface, such as at Rancho La Brea in California, United States.

tar sand (oil sand) Sandstone in which crude oil has been trapped and the lighter components have evaporated, leaving behind a residue of asphalt or bitumen that fills the pores or holes in the rock. An example is the tar sand of the ATHABASCA Valley in northern Canada. See also THE STORY OF OIL AND NATURAL GAS, p 509

Tanzania

ECONOMIC REFORM SEEKS TO EASE DECADES OF POVERTY, BUT THE COUNTRY'S GREATEST WEALTH IS IN SCENERY AND WILDLIFE

With some 800 km (500 miles) of untouched palm-fringed coast bordering the warm seas of the Indian Ocean, Tanzania abounds in scenic superlatives. Mount KILIMANJARO is Africa's highest mountain and Lake TANGANYIKA the continent's deepest and longest freshwater lake. Lake VICTORIA, shared with Kenya and Uganda, is the world's third largest lake and the volcano NGORONGORO contains its second largest crater. One of the longest rivers in the world – the NILE – rises in Lake Victoria

In the 1950s the country – then known as Tanganyika – was very much the Cinderella of British colonial territories, especially by comparison with its East African neighbours, Kenya and Uganda. But by the 1970s, Tanzania had become one of the best known of post-colonial African countries, largely because of the personality of its president, Julius Nyerere, and his innovative approach to social and political problems. His policies were hailed as a shining example to other poor countries, but have since been widely criticised as failures.

The truth lies somewhere between. Incomes have hardly risen in real terms, and the state of the national economy is critical. On the other hand, social welfare has improved greatly and far more has been achieved in nation-building than in most parts of Africa.

Tanzania is entirely a colonial creation. Occupied by the Germans in the 1880s and colonised as German East Africa, the area became British-administered Tanganyika after the First World War. Independence was obtained in 1961, but Tanzania was not formed until 1964, when the small state of ZANZIBAR joined the vastly larger mainland territory.

Mainland Tanzania consists mostly of plateaus, broken by mountainous areas and the East African section of the GREAT RIFT VALLEY. The south-eastern plateau, which is covered by dry grasslands, rises behind the narrow, reef-fringed coastal plain, and is bordered on the west by the southern highlands. Between them and the Kilimanjaro volcanic region in the north-east is the Masai Steppe, a plateau that is covered by grass and thorn bush. Mount Kilimanjaro itself rises to 5895 m (19 340 ft). Although only three degrees south of the equator, its peak is snow-capped all year.

Between the eastern and western arms of the Rift Valley is the vast interior plateau, about 1200 m (4000 ft) above sea level. It is dry, infertile and covered by poor grassland. In the north it descends to the more fertile Lake Victoria basin. Tanzania's western border follows the western arm of the Rift Valley.

Height makes the land much cooler than it would otherwise be, so close to the equator. The mountains are temperate and the central areas warm and dry. Only the coast is very hot, with temperatures averaging 25-35°C (77-95°F) through the year. It is dry almost everywhere, but none of the country is desert. Most areas have just enough rainfall for agriculture. In contrast, the islands of Zanzibar and PEMBA are hot and humid.

The country's 28.5 million people are widely scattered, with the main population concentrations occurring along the borders and on the narrow coastal plain. The heartland is relatively empty. Of the 120 different tribes – most of which are of Bantu origin and Swahili-speaking – the largest is the Sukuma. On the Masai Steppe, some nomadic Masai people still herd their humpbacked cattle.

Over the past 1000 years, Arabs from The Gulf countries have settled and intermarried with the local people. There are immigrants from the Indian subcontinent also, who have settled here during the past 300 years.

Despite the harsh environment, agricultural products account for 85 per cent of exports and 90 per cent of Tanzanians make a living from the land. Productivity is low, however, and there is no surplus from the crops they grow – principally maize – after they have fed themselves. Only a few farmers earn a regular income from cash crops such as cotton and coffee. Food for the towns has to be imported.

The islands of Zanzibar and Pemba, with higher and more reliable rainfall, are more successful agriculturally than the mainland. The western half of Zanzibar in particular is very fertile. Copra and coconut fibre are their leading exports, and the two islands are among the world's leading suppliers of cloves.

With the exception of one diamond mine, a small gold mine and some coal and kaolin mining, Tanzania's mineral resources have in the past been considered either too low grade or too isolated to be worth working. Manufacturing is limited. However, in 1992 both the manufacturing and the mining (especially gold mining) sectors began picking up.

FAILURE OF 'FAMILYHOOD'

Tanzania began independence as one of the poorest countries in Africa, and so it remains today. In 1967 there were hopes of changing the position. President Nyerere's objectives included raising incomes through increased agricultural output and industrial growth, as well as improved welfare and education for everyone, and a fair distribution of wealth.

To achieve these objectives, two main courses of action were taken. One was the nationalisation of a large part of the country's economy. The other, more radical, move was the rearrangement of Tanzania's communities. Based on the principle of *ujamaa*, a Swahili word meaning 'familyhood', 13 million people were moved from rural settlements and concentrated in 7700 new villages. The new communities were to manage themselves on democratic lines, and were to be supplied with everything necessary for communal agriculture. At first people joined the new villages willingly, but later coercion was used.

In the event, resources and the resettled farmers both fell far short of the government's hopes. Some people preferred to cultivate their own small plots, rather than take part in larger communal schemes. Corruption plagued the system, and funds were often 'misdirected'. Gradually people drifted back to the lands they had left. Overall, the upheaval of resettlement led to a decline in agricultural productivity from which the country has not recovered. There are still food shortages.

Tanzania in the 1970s received more aid per head of population than any other African country. In early 1994, additional food aid was provided by the United Nations and the Southern African Development Community (SADC) countries to avert nationwide famine due to drought.

Nyerere retired in 1985, after 24 years as president. His successor, President Ali Hassan Mwinyi, introduced an economic reform

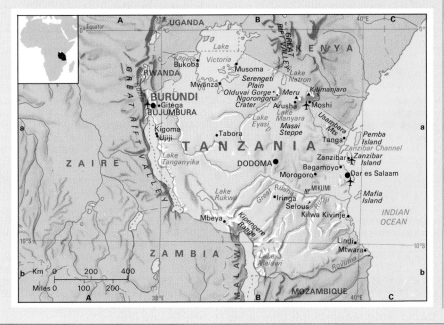

programme in 1986, strongly encouraging the principles of private initiative and a free market economy. Tanzania remained a single-party state until July 1992 when restrictions on opposition parties were lifted. Multi-party elections were due late in 1995.

TOURIST WONDERLAND

Tanzania has all the necessary ingredients for successful tourism. One of the country's greatest attractions is its abundant wildlife, including elephants, rhinoceroses, lions, leopards, buffaloes, hippopotamuses, zebras, giraffes, cheetahs, elands, impalas, wildebeest, warthogs and monkeys. There are 17 national parks and game reserves, in total occupying some 100 000 km² (nearly 40 000 sq miles), more than one-tenth of the country's total area. Among the most celebrated of these are the SERENGETI Plain in the north and the SELOUS GAME RESERVE in the south, covering more than 50 000 km² (19 000 sq miles).

Tanzania's wonders also include archaeological sites such as OLDUVAI GORGE on the western edge of the Serengeti Plain, where the remains of early man – some 2 million years old – were discovered in 1964 by the renowned British archaeologist and anthropologist Louis Leakey. Further evidence of early man includes the 30 000-year-old Stone Age rock paintings of wild animals and hunters around Kondoa, east of the Masai Steppe.

Remains of mosques and palaces dating from as early as the 12th century can be found along the coast, especially on Zanzibar and Pemba, also known as the Spice Islands.

▲ SNOWS OF KILIMANJARO Though only 3 degees south of the Equator, a thin layer of snow permanently outlines the summit of Tanzania's Mount Kilimanjaro, once an active volcano and at 5895m (19 340ft) Africa's highest mountain.

TANZANIA AT A GLANCE	
Official name United Republic of Tanzania	
Area 945 050 km² (364 886 sq miles)	
Population 28 500 000 **Per km²** 30	
(Per sq mile 78)	

Capital Dar es Salaam, due to be replaced by Dodoma in the late 1990s, by which time all government offices will have moved there.
Government Republic
Currency 1 Tanzanian shilling = 100 cents
Languages Swahili (official), English, African languages, Arabic
Religions Christian (40%), Muslim (33%), indigenous beliefs, Hinduism

Climate Hot and humid coast; dry inland with temperatures governed by altitude. Average temperature in Dar es Salaam ranges from 24°C (75°F) in July to 28°C (82°F) in January
Land use Forest and woodland 47%, grazing and other 47%, cultivation 6%
Main primary products Maize, cotton, coffee, sisal, cloves, coconuts, tobacco, cassava, beans, tea; diamonds, gold, nickel, oil, phosphates
Major industries Agriculture, food processing, textiles, cement, oil refining
Main exports Cotton, coffee, tobacco, sisal, cloves, coconut products, tea, diamonds
Annual income per head (US$) 110
Population growth (per thous/yr) 30
Life expectancy (yrs) Male 51 **Female** 55

Tara *Yugoslavia (Serbia and Montenegro)* Montenegrin river, which unites with the Piva to form the DRINA. The Tara rises near the Albanian border, and flows north-west, cutting a spectacular gorge below Mojkovac. East of the town of Zabljac it is spanned by a 365 m (1197 ft) bridge, 150 m (490 ft) above the churning river.
Map Yugoslavia – Bc

Tara Planina *Bosnia-Herzegovina* See VIŠEGRAD

Tarai (Terai) *Nepal* The southern lowlands of Nepal, containing one-third of the country's population – over 5 million people. The area at the foot of the Himalayas is humid and thick with tropical forest and elephant grass. Growing up to 5 m (16 ft) high, the grass is almost impenetrable. Much of it has been cleared for the cultivation of rice, jute and tobacco. Nepal's main industrial centre is at BIRATNAGAR, in the south-eastern Tarai near the border with India.
Map Nepal – Ba

Taranaki *New Zealand* District covering 9720 km² (3753 sq miles) on North Island, some 160 km (100 miles) north-west of WELLINGTON. It has some rich dairy pastures, around the volcanic cone of Mount Egmont (called Taranaki by the Maoris), and dairy processing factories

which produce butter and cheese. The inland hill country is mostly given over to sheep. It is also an oil and gas-producing centre.
Population 107 220
Map New Zealand – Dc

Taranto *Italy* Seaport and naval base on the 'instep' of Italy, situated 250 km (155 miles) south-east of NAPLES. Founded by the Spartans in 706 BC, there are few ancient remains. The city's medieval core is on a rectangular island connected to the mainland by two bridges which separate the inner and outer harbours. This tightly packed district is the site of a castle built by the Spaniard Ferdinand of Aragon in 1480, a cathedral, built in the 10th and 11th centuries and adorned with a 1713 Baroque façade, and the 14th-century Gothic church of San Domenico. Modern Taranto is a sprawling city, reaching to the mainland where many new industries, including one of Europe's largest steel works, have been established.
Population 244 600
Map Italy – Fd

Taranto, Gulf of *Italy* Arm of the IONIAN SEA some 140 km (90 miles) long and just as wide, occupying the 'instep' of Italy. The main port, TARANTO, sheltered much of Italy's fleet during

the Second World War, though British carrier-borne aircraft inflicted much damage on 11-12 November 1940. Today, Taranto is a NATO base. The gulf yields shellfish.
Map Italy – Fd

Tarawa *Kiribati* Most important atoll of the group, containing nearly 40 per cent of Kiribati's population as well as its administrative centres. It lies towards the north-western end of the Gilbert Islands chain. It was the site of a fierce Second World War battle when, in November 1943, the first major American amphibious landing took the Japanese positions. In four days of fighting, only 17 Japanese troops survived out of 4700; there were 3300 American casualties.
Population 28 790
Map Pacific Ocean – Db

Tarbela Dam *Pakistan* Largest dam in the country, built in the 1970s on the Indus River 50 km (30 miles) north-west of Islamabad to irrigate and power large areas of the Indus Valley. It is 148 m (485 ft) high, 2750 m (9022 ft) long, and contains a huge quantity, 121 000 000 m³ (158 000 000 cu yds), of rock and earth. Its projected life is less than 80 years; after that, it will be irretrievably choked with silt.
Map Pakistan – Da

Tarbes *France* Capital of the Hautes-Pyrénées Department, situated 124 km (77 miles) south-west of TOULOUSE. Developed by the Romans, Tarbes was occupied by the Saracens for a time in the 8th century. It still retains echoes of the Arab period in the form of stud farms, which have won an international reputation since the early 19th century for the horses they have bred from English and Arabian stock.
Population (town) 48 000; (metropolitan area) 75 000
Map France – De

Tarija *Bolivia* Department and city in the south, on the borders of Argentina and Paraguay. The department possesses flourishing vineyards and orchards and is rich in oil, particularly around Sanandita and Bermejo near the Argentine border. The city, founded in 1574 by the Spaniard Luis de Fuentes, is noted for its colourful *niño* (child) processions held in September and for a flower festival which is held each October during the southern hemisphere spring.
Population (department) 293 220; (City) 68 430
Map Bolivia – Bc

Tarim Pendi *China* Basin covering 530 000 km² (205 000 sq miles) in the west, bounded by the massive mountains of the Altun Shan, Karakoram, and Tien Shan, and named after the River Tarim. The Taklimakan Desert lies at the heart of the basin. Wheat, maize, some rice, millet, cotton and fruit are grown in oases on its rim. There are oil reserves, but they have not yet been exploited.
Map China – Bc

Tarn *France* River in the south-west, 375 km (233 miles) long, which rises on Mont Lozère, about 70 km (45 miles) north-west of Nîmes, and flows generally west to join the Garonne River near Montauban. In its upper reaches, the Tarn has cut spectacularly beautiful canyons, known as Les Gorges du Tarn. The Tarn Department is in the Midi-Pyrénées. It is a land of pastures and forests; but traditional industries are in decline, resulting in mass immigration to the cities.
Population 339 345
Map France – De

Tarnopol *Ukraine* See TERNOPIL

Tarnów *Poland* Industrial city 72 km (45 miles) east of the city of CRACOW. Its products include nitrates, textiles, timber and machinery.
Population 120 000
Map Poland – Dc

Tarnowskie Góry (Tarnowitz) *Poland* The hub of Upper Silesia's industry, 26 km (16 miles) north-west of the city of KATOWICE. Its mines were already producing lead and zinc by the 13th century. The first smelter was established 300 years later, and Poland's oldest mining school opened here in 1803. A square of buildings dating from the 16th to the 18th century lies at the heart of the old town. Today, the city makes boilers, mining machinery and safety equipment, as well as clothing, chemicals and meat products.
Population 65 300
Map Poland – Cc

Taroudannt *Morocco* Ancient town with 17th-century battlements, lying 80 km (50 miles) east of AGADIR in the fertile Sous Valley, with the High and Anti ATLAS MOUNTAINS rising north and south. Founded in the 11th century, the town rose to prominence in the late 16th century as the capital of the country under the Saadid dynasty. Its past glories have faded away and now Taroudannt is no more than a sleepy, sun-baked market town.
Population 17 100
Map Morocco – Ba

Tarragona *Spain* Ancient city situated on the Mediterranean, about 95 km (60 miles) south-west of the port of BARCELONA, and province. The city was the centre of Roman activities on the Iberian Peninsula between the 3rd century BC and the 5th century AD and it has many Roman and medieval remains. Pontius Pilate was once its governor. It has a 12th to 13th-century cathedral surrounded by a city wall partly built in the 6th century BC. It exports wine and manufactures chemical products and electrical equipment.
The surrounding province of Tarragona produces wine and fruit, and has deposits of lead, copper, silver and marble.
Population (town) 112 700; (province) 542 000
Map Spain – Fb

Tarsus *Turkey* Agricultural market town 380 km (235 miles) south-east of the capital, ANKARA. It is the birthplace of St Paul.
Population 168 650
Map Turkey – Bb

Tartu *Estonia* University town, formerly called Yur'ev, about 275 km (170 miles) south-west of ST PETERSBURG. The university was founded by the Swedish King Gustav II Adolf in 1632, when Estonia was ruled by Sweden and Tartu was called Dorpat. Today, the town manufactures machinery and footwear as well as processing food and timber products.
Population 115 300
Map Estonia – Bb

Tartus *Syria* Coastal fishing port and agricultural centre, situated some 64 km (40 miles) south of LATAKIA. Crops grown include cereals, fruit, olives and cotton. In the Middle Ages Tartus – originally a Phoenician settlement – was one of the arrival points for pilgrims making their way to the Holy Land and it soon became a major supply port for the Crusaders. Tartus fell to the Muslims in 1291. Its medieval Cathedral of Notre Dame de Tortose is now a museum.
The port is the terminal for the pipeline from Karachuk, the country's first oil field. There has been much development since the discovery of oil; the city also trades in agricultural products.
Population 475 000
Map Syria – Ab

Tashkent *Uzbekistan* National capital, lying 725 km (450 miles) south-east of the Aral Sea. It is also the centre of an irrigated area producing cotton, rice, fruit and tobacco. Tashkent was the fourth largest city of the USSR; today it is a major industrial centre producing machinery, cotton, silk textiles, electrical equipment, chemicals, furniture and processed food.
In 1966 it was hit by an earthquake in which 300 000 people were made homeless; it has since been rebuilt with a modern centre.
Population 2 094 000
Map Uzbekistan – Ba

Tasman Bay *New Zealand* Inlet on the north coast of SOUTH ISLAND, in New Zealand. It covers 72 km (45 miles) from east to west and 48 km (30 miles) from north to south. In 1770, Captain James Cook recorded evidence of widespread Maori cannibalism in the area. After landing, he saw a Maori with the bone of a human forearm picked clean of flesh. His journal recorded, 'They gave us to understand that but a few days before they had taken, killed, and eaten a boat's crew of their enemies or strangers.' The town of NELSON lies on the bay's south-eastern shore.
Map New Zealand – Dd

Tasman Sea Arm of the South PACIFIC OCEAN, in the ROARING FORTIES between south-eastern Australia, Tasmania and New Zealand. It stretches for 1600 km (1000 miles) between Australia and New Zealand, covering an area of 2 300 000 km² (888 030 sq miles); its greatest depth is more than 5994 m (19 665 ft). Discovered in 1642 by the Dutch navigator Abel Tasman, it was named after him in 1890. The Tasman Sea is often stormy.
Map Pacific Ocean – Cd

Tasmania *Australia* Island state off the south-east coast, covering an area of some 67 800 km²

▼ WILDERNESS COUNTRY Mount Ruby at 771 m (2529 ft) towers over Port Davey and the forested ranges of Tasmania's Southwest National Park.

(26 172 sq miles). The capital is HOBART. The first European to discover Tasmania was the Dutch navigator Abel Tasman in 1642. It was originally called Van Diemen's Land after the governor-general of the Dutch East Indies, on whose order Tasman sailed to Australia. The British took over the island in 1803 and used it for penal settlements. It became a separate colony in 1825 and was renamed Tasmania in 1855.

Tasmania lies in the moist, temperate zone, and produces apples, berries, pears, potatoes, other vegetables, sheep and cattle. Forestry and paper making are major industries, as is the mining of tungsten, tin, copper, silver, lead and zinc. The heavy rainfall and mountainous terrain enables hydroelectric plants to supply 90 per cent of Tasmania's electricity. The island's stretches of beach and national park and its colonial buildings make it a popular tourist destination.

Population 460 500
Map Australia – Hg

Tassili N'Ajjer *Algeria* Sandstone plateau, situated 2000 km (about 1240 miles) south-east of ALGIERS, resembling a moon landscape of great beauty with gigantic canyons, deep gorges, petrified forests and sandstone features fashioned by erosion. The highest point of the Tassili N'Ajjer is 2254 m (7400 ft) above sea level. The region is one of the richest in prehistoric art in the world. There are thousands of impressive rock paintings, dating back to around 6000 BC, which provide a record of the life of the area when it was a grassland region supporting animals such as elephants and giraffes. At the foot of the Tassili N'Ajjer in the east is the oasis town of Djanet. This is the main centre for the plateau.

Map Algeria – Bb

Tatar Republic *Russia* Autonomous republic of the Russian Federation around the Volga River centred on the city of KAZAN', its capital. It covers 68 000 km² (26 250 sq miles) and is populated mainly by Muslim Tatars – descendants of the Mongol horde and the peoples they conquered when they galloped across Russia in the 13th century and held the land in thrall for 250 years. The inhabitants' main occupations are agriculture and dairy farming. There are oil and gas fields – accounting for one-third of the Commonwealth of Independent States' (CIS) production.

The region's industries include engineering, tanning and timber as well as the production of chemicals. The Tatar Republic made a unilateral declaration of independence in 1992.

Population 3 679 000
Map Russia – Gc

Tatra Mountains (Tatry) *Slovakia/Poland* Three ranges of the western Carpathians. Along the border lie the Western Tatras (Zapadné Tatry; Tatry Zachodnie) and, to the east, the High Tatras (Vysoké Tatry; Tatra Wysoke).

The Western Tatras reach 2250 m (7382 ft) at Mount Bystrá in Slovakia. The High Tatras rise to about 2655 m (8711 ft) at Gerlach Peak (Gerlachovský Štít), which is the highest point in Slovakia and in the Carpathians, and to 2499 m (8199 ft) at Rysy on the border, one of whose peaks is Poland's highest mountain. Tourist resorts include Zakopane and Podhale in Poland, and Strbské Pleso and Tatranská Polianka in Slovakia. The Low Tatras (Nízké Tatry) to the south, within Slovakia, rise to some 2043 m

(6703 ft) at Dumbier peak and include the lovely Demanova Valley.

Map Slovakia – Bb-Cb

Tatung *China* See DATONG

Tauern Mountains *Austria* See HOHE TAUERN

Taunggyi *Myanmar (Burma)* Tourist resort and capital of Shan state. It lies near Inle Lake some 175 km (110 miles) south-west of MANDALAY. It is a centre of the Shan handicraft industry.

Population 149 000
Map Myanmar – Cb

Taunus *Germany* Range of mainly forested hills just north-west of FRANKFURT am Main. The hills, the highest of which rise to 880 m (2887 ft), are mostly flat-topped, with steep-sided valleys.

Map Germany – Cc

Taupo, Lake *New Zealand* The largest lake in the country. It covers 606 km² (234 sq miles) on North Island, midway between Auckland and the capital, Wellington. It is rich in trout.

Today the lakeside town of Taupo (population 18 370) is used by tourists as a base for exploring the volcanic plateau in summer and for trout fishing throughout the year.

Map New Zealand – Ec

Tauranga *New Zealand* Market town on North Island, situated 160 km (100 miles) south-east of AUCKLAND on the Bay of Plenty. It is the centre of a prosperous fruit-growing and farming area. The nearby port is Mount Maunganui.

Population 70 800
Map New Zealand – Fb

Taurus Mountains *Turkey* Mountain range which runs east to west, separating the Anatolian Plateau from the Mediterranean Sea. The western end rising from the Gulf of Antalya has peaks of over 3000 m (9840 ft), the highest of which is Mount Erciyas Daği, at 3916 m (12 848 ft). In the eastern Taurus, peaks rise to more than 4000 m (13 123 ft), and the highest is Cilo Dagi, at about 4168 m (13 675 ft). The Cilician Gates, a pass in the eastern Taurus, has been used for centuries by armies and traders on the highway from Europe to Syria and Persia.

Map Turkey – Bb

Tavastehus *Finland* See HÄMEENLINNA

Taveuni *Fiji* See VANUA LEVU

Taxco *Mexico* Delightful hillside town situated in GUERRERO State, where the twin-spired Santa Prisca Church towers over red-tiled roofs and small plazas with fountains and flowers. Silver mines, some worked as early as 1521, made Taxco prosperous, and today the town's narrow, winding streets are crowded with shops selling jewellery and silverware. Taxco has been declared a national monument to restrict any new building which might spoil its colonial character.

Population 60 000
Map Mexico – Cc

Taxila *Pakistan* Town with the excavated remains of three ancient cities covering 25 km² (10 sq miles), some 36 km (22 miles) north-west of RAWALPINDI. The first city, Bhir Mound, dates

from about the 7th century BC and was the capital of the kingdom of Gandhara, which became a province of the Persian Empire of Darius I. It was then overrun by Alexander I in 326 BC and the Indo-Greek site – the second city – flourished as Sirkap under a succession of rulers, the last of which were the Parthians. The city subsequently fell to the Kushans in the 1st century AD, who founded the third city, Sirsukh, on the site. It became a famous seat of learning throughout the Buddhist world. The domestic buildings were multi-storeyed and there were several Buddhist monuments. Eventually, in the 5th century, the city was sacked by the Huns. By the 7th century, it was ruined. A museum in Taxila displays some of the artefacts found at the site.

Population 48 000
Map Pakistan – Da

Tayside *United Kingdom* Administrative region of east-central Scotland, which takes its name from the River Tay. It covers some 7942 km² (3066 sq miles) and includes the former counties of Angus, Kinross and most of Perthshire. The population is concentrated in the cities of DUNDEE and PERTH. The south has Scotland's finest agricultural land, producing soft fruits, potatoes and livestock. The north rises to the Highlands and draws tourists to its mountains and ski slopes.

Population 385 270
Map United Kingdom – Db

Taza *Morocco* Eighth-century Arab town on an ancient site, about 88 km (55 miles) east of FÈS on the Taza Gap, a pass that separates the RIF MOUNTAINS from the Middle ATLAS MOUNTAINS. In the 12th century its strategic position, 550 m (1800 ft) above sea level, led the Almohad conquerors to construct its fortifications. Today it manufactures footwear and building materials, and is famous for its fine carpets. The town has an interesting madrassa (religious college) and a mosque with a beautifully crafted Hispano-Moorish bronze lantern.

Population 56 000
Map Morocco – Ba

T'bilisi *Georgia* National capital and industrial centre which is situated in the CAUCASUS Mountains near the Turkish border; formerly known as Tiflis. Its manufactures include machine tools, electrical equipment, locomotives, chemicals and petroleum products.

Founded in AD 455 or 458, when the capital of the Georgian kingdom was transferred from Mtskheta, T'bilisi was for centuries an important centre on trade routes between Europe and Asia. Captured in turn by Persians, Byzantines, Arabs, Mongols and Turks, the city came under Russian rule in 1801.

T'bilisi is one of the oldest remaining cities in the world. Many of the older parts are built on the slopes of Mount Mtatsminda. They include medieval buildings and courtyards, a 6th-century cathedral (rebuilt in the 16th century), a basilica dating from the 6th and 7th centuries, and a 13th-century castle.

Gori, 65 km (40 miles) to the north-west of the city, was the birthplace of Joseph Stalin (1879-1953), the Soviet leader from 1924 to 1953. Mtskheta, the Georgian capital from the 2nd century to the 5th century, lies 20 km (12 miles) north of T'bilisi. The tiny town has several old churches, including an 11th-century cathedral

which incorporates a 6th-century church and contains the tombs of Georgian kings.
Population 1 264 000
Map Georgia – Cb

Tczew (Dirschau) *Poland* Vistula River port 32 km (20 miles) south-east of the Baltic port of Gdańsk. Its 890 m (2920 ft) long bridge, which was built across the river in 1857, was at the time Europe's longest iron bridge.
Population 49 900
Map Poland – Ca

Tébessa (Theveste) *Algeria* Road and rail junction situated near the Tunisian border, and the administrative capital of the district, which is an important phosphate-mining region. Tébessa manufactures some of the best carpets in Algeria. Founded by the Romans in AD 71, the town has some of the finest Roman remains in Africa, including a basilica, an arch and an enormous circus. Then called Theveste, it was one of the first towns in Africa to adopt Christianity.

Tébessa was destroyed by the Vandals in the 5th century AD, but restored by the Byzantines in the 6th century, and the modern city lies within Byzantine walls. It was strategically important to the Allies in February 1943, when it was captured from the Germans and successfully defended against one of the last counter-thrusts in the North African campaign made by the German general, Erwin Rommel (1891-1944).
Population 68 000
Map Algeria – Ba

tectonics See PLATE TECTONICS, p 542

Tegucigalpa *Honduras* National capital since 1880, when the president of the time moved from Comayagua because the citizens there resented his marriage to an Amerindian.

Founded as a gold-mining camp in the mid-16th century, the city still has some fine Spanish colonial architecture, including the Church of St Francis, completed in 1592 and now rebuilt. Today, Tegucigalpa produces textiles, timber, chemicals and processed food, which are mostly for domestic consumption.
Population 678 700
Map Honduras – Ab

Tehran (Teheran) *Iran* National capital 100 km (62 miles) south of the Caspian Sea, at the foot of the ELBURZ MOUNTAINS, which tower 5600 m (18 373 ft) east of the city. Agha Muhammad Shah, founder of the Qajar dynasty (1794-1925), moved his capital from Shiraz to Tehran in 1794, and during the 19th century the city was enlarged and developed by his descendants. By the mid-1800s, it already had a population of more than 100 000. However, although the southern part of the city retains its 18th-century character, the centre of Tehran was rebuilt by Reza Shah (ruler from 1926 to 1941), who changed the face of the capital by demolishing the old fortifications and adopting a geometrical layout with broad avenues, open parks and modern buildings.

In 1943, Tehran was the scene of the first wartime meeting (the Tehran Conference) between the Allied leaders Winston Churchill, Franklin Roosevelt and Joseph Stalin. Today, the city is one of the largest cities in the Middle East. Its industries include chemicals, textiles, car assembly, tanning and glass making.

Places of interest include the Pahlavi Palace, the Majlis (parliament building) and the 19th-century King's Mosque, as well as the archaeological museum (which includes items from the ruined city of Persepolis), the huge bazaar and the Sepahsalar Mosque, with its eight minarets.
Population 6 476 000
Map Iran – Ba

Tehuacán *Mexico* Spa town in the south-eastern state of Puebla, with a mild, pleasant climate, and mineral springs where visitors bathe. Its mineral water is bottled and sold throughout Mexico.
Population 112 000
Map (Puebla State) Mexico – Cc

Teide, Pico de *Spain* Volcanic mountain, some 3718 m (12 198 ft) high, on TENERIFE in the Canary Islands. It is the highest point on Spanish territory and in the Atlantic.
Map Morocco – Ab

Tejo *Portugal* See TAGUS

Tel Aviv-Jaffa (Tel Aviv-Yafo) *Israel* Capital of Tel Aviv district, situated on the Mediterranean coast, about 50 km (30 miles) west of JERUSALEM. Founded in 1909, Tel Aviv was the country's capital until 1950 and is still the site of foreign embassies, since the majority of foreign countries do not recognise Jerusalem as the Jewish state's permanent capital.

Tel Aviv-Jaffa is the nation's second largest city, after Jerusalem, and the main industrial centre. Its products include chemicals, textiles, metal goods and processed foods.

▼ **SHORT-LIVED CELEBRATION This grandiose edifice at the heart of Tehran was opened as the Shahyad Monument by Iran's last shah in 1971 to mark 2500 years' continuity of successive Persian and Iranian royal states.**

In 1950 the neighbouring area of Jaffa (Yafo) – the Biblical port of Joppa (Jappho) – became part of Tel Aviv. The old city has been rebuilt and now has a colourful artists' colony, a marina and cafés, shops and restaurants.

Among the city's museums are the Museum of the Diaspora, telling the story of the Jewish people in exile from AD 70 up to the founding of the state of Israel in 1948, the municipal museum and the Helena Rubinstein Pavilion (both art museums), and the Ha'aretz museum complex, which contains exhibitions of science and technology, glass, folklore and archaeology.

The city's airport is situated at Lod (Lydda), 15 km (10 miles) to the south-east.
Population 353 200
Map Israel – Aa

Telemark *Norway* Mountain and lake province covering an area of 15 315 km² (5912 sq miles) in the south, west of OSLO. Almost one-third of the region stands at 1100-1200 m (3610-3937 ft) in altitude; its highest point is Gausta, at 1883 m (6178 ft). The region is dotted with lakes and contributes one-fifth of Norway's hydroelectric energy. The 'telemark', a type of turn made in cross-country skiing, was perfected in the area. Skien is the capital.
Map Norway – Bd

Tell Atlas *Algeria* See ATLAS MOUNTAINS

Tema *Ghana* The country's chief port since 1961, some 30 km (19 miles) east of the capital, ACCRA. It has a well-planned modern town with attractive residential areas and excellent hotels – including one on the GREENWICH MERIDIAN. There is an industrial area near the railway, with an aluminium smelter, an aluminium utensil works, an oil refinery and textile mills. The port specialises in handling cocoa. The AKOSOMBO Dam across the Volta provides electricity for the factories.
Population 54 000
Map Ghana – Ab

temperate climate An ambiguous term which is sometimes used to define a climate without extremes. Correctly, however, the term denotes climates of temperate latitudes (mid-latitudes), while 'equable' defines climates with a lack of extremes. Many temperate latitude climates are far from equable. See COOL TEMPERATE CLIMATE

temperate rain forest Dense forest of tall trees – often up to 30 m (100 ft) tall, but lower and less dense than in TROPICAL RAIN FOREST – found in areas with an equable climate and heavy rainfall, such as north-western North America, southern Chile, the west coast of New Zealand, south-east Australia and southern Japan. The trees often have leathery leaves and form a dense canopy; ferns and other plants needing little light grow on the forest floor.

temperature See WEATHER, p 730 and FROM BOIL-ING HOT TO FREEZING COLD, p 731

Temuco *Chile* Capital of the Araucanía region, 610 km (380 miles) south of SANTIAGO. It was founded in 1881 and is the centre of a rich agricultural area producing cereals, fruit and timber. A peace treaty with the warring Araucanían Mapuche people was signed in 1881 under a tree on the nearby peak of Cerro Nielol.

| Population 250 000 |
| Map Chile – Ac |

Tenasserim *Myanmar (Burma)* See TANIN-THARYI

Tenerife *Spain* Largest of the CANARY ISLANDS, at 2057 km² (794 sq miles).

| Population 663 000 |
| Map Morocco – Ab |

Tennant Creek *Australia* Gold, silver and copper mining centre in the Northern Territory, 450 km (280 miles) north of ALICE SPRINGS.

| Population 3480 |
| Map Australia – Eb |

Tennessee *USA* South-eastern state – and river – covering 109 178 km² (42 144 sq miles) between the Mississippi River and the Appalachian Mountains. Livestock, animal feed, tobacco, cotton and soya beans are its main farm products. Coal and zinc are its chief minerals, and chemicals, transport equipment, machinery, textiles and foodstuffs the principal manufactures. NASHVILLE is the state capital and MEMPHIS the largest city.

The Tennessee River rises near Knoxville, Tennessee, and flows some 1050 km (652 miles) south-west, through Tennessee and Alabama, then north through Tennessee and north-west through Kentucky, into the Ohio River, which is a tributary of the Mississippi.

In 1933 the federal government set up the Tennessee Valley Authority (TVA) to provide flood control and navigation improvement on the Tennessee and lower Mississippi rivers. It also aimed to improve conditions in a depressed area by encouraging better agricultural, mining and forestry practices and reducing soil erosion, and by providing hydroelectric power for domestic and industrial use (although only 16 per cent of electricity needs are met today by hydro-electricity, the rest being provided by coal-powered stations). The TVA covers 106 214 km² (41 000 sq miles) in seven states; its headquarters

are at Knoxville. In 1985, the Tennessee-Tombigbee Waterway, which begins near Memphis, connected the Tennessee River to the Mississippi and its outlet to the Gulf of MEXICO.

| Population (state) 4 877 000 |
| Map (state) United States – Ic-Jc; (river) United States – Id |

Teotihuacán *Mexico* Ruined capital, situated 55 km (34 miles) north-east of Mexico City, of the Teotihuacán culture, which extended its influence over central Mexico between the 2nd and the 6th century AD. It was named Teotihuacán ('the City of the Gods') only after the original inhabitants had vanished. The site comprises 21 km² (8.1 sq miles) of palaces, temples and courtyards and is dominated by the 65 m (213 ft) high Pyramid of the Sun and the 42 m (138 ft) high Pyramid of the Moon. The pyramids are connected by the broad Street of the Dead. The site's most famous building, the Temple of Quetzalcóatl, is adorned with elaborate carvings.

| Map (Mexico City) Mexico – Cc |

Teplice *Czech Republic* Bohemia's oldest spa, which grew around a monastery built next to a spring in 1158. It lies north-west of PRAGUE, 11 km (7 miles) from the German border. The small town of Duchcov (Dux) to the south-west has made fine porcelain for more than 300 years, including the type known as Royal Dux. Dux Castle was the home of the Italian adventurer and rake Giovanni Giacomo Casanova (1725-98) in his retirement. He died there at the age of 73.

| Population 53 500 |
| Map Czech Republic – Aa |

Tequendama Falls *Colombia* See BOGOTÁ

terai Area of marshy rain forest along the foothills of the Himalayas, found chiefly in Uttar Pradesh State in India and in Nepal. See TARAI.

Terengganu (Trengganu) *Malaysia* East coast state peopled predominantly by Malays. It covers 12 955 km² (5002 sq miles) and is largely hilly rain forest with a densely populated coastal region. Most of the people are rice farmers, but offshore oil is bringing new prosperity and new roads have boosted coastal tourism. On the beaches north of the town of Kuantan, leather-back turtles, some more than 2 m (6.5 ft) long and weighing almost 1 tonne, waddle ashore at night from May to September to lay eggs in the sand.

| Population 752 000 |
| Map Malaysia – Bb |

Teresina *Brazil* Market town and regional capital 800 km (nearly 500 miles) west of the country's north-eastern tip. It is said to be Brazil's hottest city, after Manaus – with an average temperature of 30°C (86°F). The surrounding state of Piauí, covering 250 934 km² (96 886 sq miles), is the poorest in the country. Its main products are vegetable oils, tropical fruits and livestock.

| Population (city) 598 450; (state) 2 581 050 |
| Map Brazil – Db |

Terni *Italy* City 75 km (47 miles) north of ROME. It is one of central Italy's most important industrial centres, specialising in steel, engineering, chemicals and hydroelectricity.

| Population 111 400 |
| Map Italy – Dc |

Ternopil (Ternopol') *Ukraine* West Ukrainian industrial town, formerly known as Tarnopol. It lies 120 km (75 miles) south-east of L'VIV. Its products include concrete, leather goods, footwear and canned foods.

| Population 212 000 |
| Map Ukraine – Bb |

terrace See RIVER TERRACE

Terschelling *Netherlands* See FRIESIAN ISLANDS

Tertiary First of the two periods in the Cenozoic era. See GEOLOGICAL TIMESCALE, p 256

Teschen *Poland* See CIESZYN

Tesin *Poland* See CIESZYN

Tessin *Switzerland* See TICINO

Tete *Mozambique* Capital of Tete Province, which covers 100 724 km² (38 879 sq miles). It stands on the ZAMBEZI River and is the trade centre for west-central Mozambique, exporting cattle, cotton, hides and skins. Tete has the potential to become a major industrial centre based on power from the CAHORA BASSA DAM, coal from Moatize and other minerals found locally.

| Population (town) 39 000; (province) 864 200 |
| Map Mozambique – Ab |

Tethys Sea which was situated between the ancient continents of Laurasia and Gondwana. See CONTINENTAL DRIFT EXPLAINS EARTH'S ANCIENT MYSTERIES, p 171

Tétouan *Morocco* The capital of Tétouan Province, situated some 64 km (40 miles) south-east of TANGIER and 10 km (6 miles) from the Mediterranean. It was founded in the 14th century and was the capital of former Spanish Morocco. There is a fine medina (old Muslim residential quarter) and a museum with antiquities from the ancient site of Lixus (see LARACHE).

| Population 856 000 |
| Map Morocco – Ba |

Tetovo *Macedonia* City, 40 km (25 miles) west of SKOPJE. An old Turkish settlement, it has colourfully painted houses and mosques, and textile and carpet industries. It is a centre for the ski resorts of the Sar Planina Mountains.

| Population 162 400 |
| Map Macedonia – Ab |

Teutoburger Wald (Teutoburg Forest) *Germany* Chalky ridge of wooded hills in the north-west. It is an extension of the highlands south of the North German Plain, and lies in an arc from OSNABRÜCK to PADERBORN, where it rises to some 468 m (1535 ft). In AD 9, Arminius, a local tribal chief, ambushed the main Roman expeditionary force in the forest and virtually annihilated three legions. The disgraced Roman general, Varus, committed suicide, and the Romans never again made any great effort to conquer the tribal territories to the east of the Rhine.

| Map Germany – Cb |

Tevere *Italy* See TIBER

Texas *USA* Second largest state, after Alaska, covering some 691 190 km² (266 807 sq miles) between the Rio Grande (which forms the Mexican border), the Red River and the Gulf of MEXICO. It was first settled in 1659 by Europeans, and was ruled by Spain and later Mexico until it won independence as the Republic of Texas in 1836. It became a US state in 1845, but its boundary with Mexico was only agreed upon after the Mexican-American War (1846-8). Today, more than 25 per cent of Texans are of Mexican origin.

The climate ranges from subtropical in the south-east to continental in the interior, and in places irrigation is necessary. Beef and dairy cattle, and wheat are the main farm products and Texas is the nation's leading cotton-grower. It is also the main producer of oil, natural gas and petroleum by-products. Principal manufactures include chemicals, processed food, electronics, computers, aerospace equipment and machinery. The state capital is AUSTIN and the largest cities are HOUSTON, DALLAS and SAN ANTONIO.

Population 16 986 500
Map United States – Fd

Texel *Netherlands* See FRIESIAN ISLANDS

Thaba Bosiu *Lesotho* Steep-sided natural fortress occupied by the Basotho king, Moshoeshoe I, in 1824 as a refuge for his people during the conflict between African, Boer and British interests. It stands 19 km (12 miles) east of the capital, MASERU, and contains the grave of Moshoeshoe I, who died in 1870.

Map South Africa – Cb

Thai Binh *Vietnam* Province with an area of 1531 km² (591 sq miles), 80 km (50 miles) east of the capital, HANOI. It is one of the most densely populated rural areas of South-east Asia, with 1064 people per km² (2756 per sq mile). Rice is cropped twice a year from its fertile plain.

Population 1 632 000
Map Vietnam – Ba

Thailand See p 666

Thailand, Gulf of Arm, 720 km (450 miles) long and 480-560 km (300-350 miles) wide, of the South CHINA SEA. It lies between the Malay Peninsula, Thailand, Cambodia and the south-western tip of Vietnam. The Gulf of Thailand's shallow coastal waters are rich fishing grounds; the west and north coasts, in Malaysia and Thailand, are lined with beautiful resorts. Some offshore oil is produced.

Map Thailand – Bc

Thai Nguyen *Vietnam* Steel-producing town about 50 km (30 miles) north of the capital, HANOI. Its integrated steel plant, built during French rule and fuelled by the nearby Quang Yen coalfield, has been modernised with Russian and Chinese aid.

Population 171 810
Map Vietnam – Ba

Thames *United Kingdom* English river, 335 km (210 miles) long, on which LONDON stands. It rises in the Cotswold Hills in Gloucestershire, and flows generally eastwards past the city of OXFORD, through the Chiltern Hills, and on through London to the North Sea. The river has been an important trade and transport route since prehistoric times, with many large houses and estates on its banks, including seven past or present royal residences – at Windsor, Hampton Court, Richmond, Kew, Westminster, the Tower of London and Greenwich.

Ocean-going ships used to sail up to the Pool of London in the heart of the city, or to docks a little downstream, but today most ships berth at modern cargo ports, such as Tilbury, on the river's lower reaches. The Thames used to be liable to flood at exceptionally high water, but an anti-flood barrier – completed at Woolwich in 1984 – now protects the capital. Above London, boating has become a popular tourist attraction, and many riverside towns, notably Henley, hold regattas in summer.

Map United Kingdom – Ee

Thames, Firth of *New Zealand* Inlet, 35 km (20 miles) long and 16 km (10 miles) wide, on the NORTH ISLAND of New Zealand, forming the south-eastern arm of the HAURAKI GULF. Captain James Cook explored it in 1769 and named the river that flows into it the Thames. The river is now known by its Maori name, Waihou, but the name Thames is still used for the Firth and for the small industrial town on its shores (population 6366), situated 65 km (40 miles) south-east of AUCKLAND, whose industries today include iron and steel foundries, a car assembly plant and a factory producing prefabricated houses.

Map New Zealand – Eb

Thane (Thana) *India* Town at the head of the Bombay estuary, and now a north-eastern suburb of BOMBAY City. It has grown rapidly since 1960, and is today one of the leading western Indian industrial centres. Its products include chemicals and cars. Thane also refines oil from the offshore 'Bombay High' field, which supplies almost one-quarter of India's oil.

Population 797 000
Map India – Bd

Thanjavur (Tanjore) *India* Silk-producing city in the Kaveri River delta in the south-east, some 334 km (208 miles) south-west of the city of MADRAS. It was the capital of several empires, including that of the Cholas in the 10th century. Among the city's places of interest are its Great Fort and palace, and its 11th-century temple of the Hindu god Siva, which has adjoining libraries of Sanskrit and Tamil manuscripts.

Population 184 000
Map India – Ce

Thar (Indian Desert) *India/Pakistan* Area of sand desert and rocky hills covering about 200 000 km² (77 200 sq miles) astride the India-Pakistani border. It receives up to 510 mm (20 in) of rain a year, but this is erratic, and temperatures reach 50°C (122°F) in May and June. In places the underground water is salty and useless for irrigation, although scant crops of wheat are grown in some hollows. However, the massive Indira Gandhi Canal project will bring new agriculture to the area. The desert scrub and grassy dunes support herds of cattle and goats, and camels provide transport. Several cities, including JODHPUR and Bikaner, lie in the Thar.

Map India – Ab

thaw Period of relatively warm weather when snow and ice melt. Thaw occurs each spring in high latitudes, freeing icebound seas, lakes and rivers.

A DAY IN THE LIFE OF A THAI RICE FARMER

Byapha looks out of the doorway of his thatched bamboo hut and surveys Nong Tong, 20 huts in a clearing in the forest in Thailand's Nan Province, the homes of his friends and his clansmen who have chosen him as their headman. He must speak for their interests to representatives of a government that has until recently paid little heed to the wishes of the hill peoples of northern Thailand.

But now the government has started to take notice, as Byapha and others like him have learned the national language and have ventured among the urban dwellers, who no longer regard them as mere curiosities, the 'embroidery people' in their distinctive costumes. The pink and blue embroidered waistcoat and green pantaloons that Byapha wears were woven on a foot loom and hand-sewn by his wife before their marriage.

Byapha has been willing to adapt to the expectations of outsiders, but he still has a duty to preserve the traditions of his ancestors whose spirits are present around him and whose ashes, brought from the village his clan recently left, lie in the village's new spirit house. His own father's ashes are kept on a shelf in his hut.

It is more than loyalty to tradition that leads Byapha to defend the way of life of his clansmen. He knows that it is the one best adapted to survival. Byapha and his fellow farmers clear land from the forest in order to grow the rice on which they live. For two or three years the land is fertile enough to produce a good harvest. Then the villagers must move on; it takes a dozen years of lying fallow for the worked land to regain its former fertility.

But if the government forbids his people to clear more land elsewhere, they are obliged to farm their existing slopes until the soil's goodness is used up and the yields are low. Then there is not enough food for everyone. It is this that in the past has caused many of his people to devote land to white opium poppies, which survive on quite infertile soil for 10 years or more. They also provided a crop which fetched consistently high prices both in Thailand and, of course, abroad. A new government clampdown on the growing of opium has, however, made it even harder for the people to buy the rice they are short of – not to mention luxuries of Western culture such as radios, plates and plastic washing bowls and tinned meat.

It is not only the granting of new forest land that Byapha must argue for when he visits the government offices in town. He must resist the influx of impoverished lowland Thais into these hills. There are many reasons why the influx should be stemmed, but chief amongst them is the fact that the methods of farming of the lowland Thais speeds up the erosion of the soil which is already a problem.

He will return in the evening and report to his fellow villagers – and to his dutifully waiting wife.

Thailand

SUBTLE DIPLOMACY EXERCISED BY ITS KINGS SAVED THE FORMER SIAM FROM EUROPEAN COLONISATION – NOW WESTERN TOURISTS FLOCK TO SEE THIS ORIENTAL GEM

Like a scintillating jewel that flashes a myriad colours, Thailand sparkles with a special magic unlike any other country of the Orient. Perhaps it is because this country, alone in South-east Asia, has never been a European colony and had its traditions diluted by Western influences; even the word Thai means 'free'. Perhaps it is because Thailand's culture – the gilded temples and devotion to the Buddhist religion, the houses on stilts, the sinuous classical dances, the adulation of the royal family – goes hand in hand with a welcome to the West.

But there is another, less attractive face of Thailand. The country's capital, BANGKOK, has a reputation as the world's sex capital, where the 'bar girls' of Patpong Road – often little more than children, many of them sold into prostitution by their families – entertain tourists in seedy establishments once frequented by American servicemen on 'rest and recuperation' from Vietnam.

Some other parents take the barely preferable alternative of sending children to work in 'sweatshop' factories; while the whole family may inhabit a tiny dwelling perched above a filthy canal – regarded as 'picturesque' by tourists, yet in reality a squalid slum. Again largely for economic reasons, for thousands of farmers in the golden triangle – the isolated area spanning the borders of Thailand, Myanmar (Burma) and Laos – the main crop was until recently the opium poppy. But largely due to government efforts, Thailand is no longer a major producer of heroin or cannabis, although it remains an important traffic route from other countries to the outside world.

Yet Thailand has much genuine beauty. It is a tropical country of mountains and rain forests and emerald green plains, *klongs* (canals) and rice fields, huge statues of the Buddha and miles of unspoilt beaches. Bordered by Myanmar, Laos, Cambodia and Malaysia, the country is about the size of France. Within it live 58.7 million people, the main group – the Thais proper – being descended from tribes that came south from China about 1000 years ago. There are minorities of Chinese, Malays and Indians, while in the hills live the Meo, Yao and Karen tribes.

Most Thais are Buddhists, but practise a Buddhism of their own which borrows from Hinduism and other religions. (For example, the walls of a Buddhist temple may be decorated with scenes from the great Hindu epic the Ramayana.) There are 27 000 *wats* (temples), many festivals and a quarter of a million monks, for almost all young Thai men spend time as monks as well as in national service.

THE HEART OF THE COUNTRY

Bangkok, which the Thais call Krung Thep, the 'city of angels', is a sprawling, metropolis of 5.8 million people, notorious for its hours-long traffic jams, aggravated by the lack of a mass transport system. However, some relief of its congestion is promised by an elevated highway to the airport and plans for the building of a skytrain system.

Its modern shopping plazas are loud with Western pop music, yet Bangkok retains a flavour of traditional Thailand. In the 300 temples the visitor can sense the inner tranquillity induced by the Buddhist religion.

The Grand Palace, built in 1782 by Rama I, symbolises the reverence which the Thai people feel for King Bhumibol (Rama IX) and Queen Sirikit. The *klongs* which traverse the city, and the River CHAO PHRAYA on which it stands, gave Bangkok its unofficial title, the 'Venice of the East'. Many of the canals have been filled in to make roads, but on those that remain, stilted houses, some draped with fishing nets, are visited daily by market vendors in boats piled high with fruit and vegetables.

The heartland of the country is the alluvial central plain, north of Bangkok, formed by the River Chao Phraya. Here, there are three seasons – cool, hot and monsoon – all of them sweltering. A 'cool' maximum temperature between November and February is 26°C (79°F); a 'hot' maximum between February and May is 36°C (97°F). In the monsoon, the weather is hot, very humid and wet – more than 1500 mm (59 in) of rain may fall between May and October and the average temperature is 32°C (90°F). The central plain contains a vast expanse of paddy fields, for Thailand is the world's fifth largest producer of rice, and since 1981 the leading exporter. The Chao Phraya is crisscrossed by transport and irrigation canals which serve the paddy farmers who live on their banks.

In this rice bowl, about 70 km (43 miles) from Bangkok, lies AYUTTHAYA, the old capital of Siam. Once the centre of a large empire, it was sacked in 1767 by the Burmese, who left its proud buildings in ruins. Today it is a quiet provincial town, but the remains of its former grandeur can still be seen.

In the far north of the country, thickly forested hills form a series of north-south ridges, separated by wide flat valleys and rising to 2590 m (8497 ft) at DOI INTHANON. Here the climate is much cooler than in the

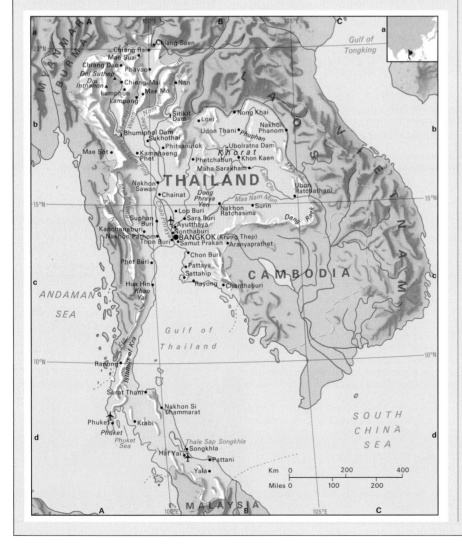

south. Teak is hauled to the water by elephants and floated to the timber mills down-river. Rice is grown in the valleys, as are such temperate crops as garlic and onions.

At festival times, many people from the surrounding hills flock into the regional capital of CHIENG MAI, Thailand's second largest city. But for most of the year the hill tribes remain in their small villages on the forested ridges, up to 1000 m (about 3300 ft) above sea level. The tribes can be distinguished from one another by their style of costume, mainly in black and white or red and blue. They practise shifting agriculture, clearing forest and brush and abandoning the land after two or three years.

The north-east, known as the Isan region, is a low undulating sandstone plateau, cut off from the rest of the country by steep slopes. In the dry season from November to April, a red dust is thrown up from the roads. It pervades the atmosphere, settling on trees, on houses, on everything everywhere. The people wear checked cloths, akin to the cowboy bandana, around their heads and mouths. This is the mark of the Isan, or Thai-Lao.

If the north-east is dry Thailand, the narrow southern peninsula is wet Thailand. Here, in the upland massifs and coastal plains that lead to the Malaysian border, the landscape is more luxuriant. This is rubber country: Natural rain forest blends with the dark green of the plantation of rubber trees, grown mainly by individual farmers.

Much of the landscape resembles Malaysia, and this is where Thailand's Malay-speaking minority live. Rice is grown in small plots lined with rubber trees. A wide variety of fruit – rambutans (red and oval, with soft spines), oranges, grapefruit and limes – is also grown. Muslim mosques are to be seen, and there are fewer houses on stilts than elsewhere in Thailand. The Thai speech becomes more clipped and singsong, the sarong replaces the shorter *phakama* of the northern areas, and shadow plays are an important cultural feature.

THAILAND TODAY

Thailand did not resist Japanese occupation in the Second World War, and under pressure even declared war on the US. America refused to accept the declaration, saying it could not regard Thailand as an enemy. Since the war Thailand has adopted a pro-Western stance and has achieved stability despite several coups and long spells of military rule; the monarchy has been a stabilising force, respected by both military and civilian governments, although it lost absolute power in 1932. The military last ruled from February 1991 to June 1992. Thailand aided the US during the Vietnam War, providing bases and airfields, and has received large grants from the US.

The aftermath of the war still lingers. Thailand took in some 80 000 refugees from Vietnam and more than 100 000 Cambodians are housed in camps along the border. Fighting between the Vietnamese in Cambodia and the Khmer Rouge guerrillas frequently spilled over into Thailand.

Since 1960 Thailand's economic growth has been rapid. Traditionally the economy was dominated by four major commodities: rice,

rubber, tin from the south and teak from the northern hills. In the 1960s highways were built across the country, opening up vast tracts of land. Farmers in the uplands began to grow maize, cassava, jute and sugar cane. Irrigation schemes and storage dams extended the cultivation season, high-yielding rice increased crops and tractors replaced the water buffalo.

In 30 years the country's forest area shrank from 60 per cent of the land to 20 per cent, leading to soil erosion in some areas, and a ban on all commercial logging in Thailand's rain forest. There are other problems: the Gulf of Thailand is being overfished, tin resources in the ANDAMAN sea are being drained by hundreds of suction boats, and Bangkok itself has a growing flood problem. The city is sinking a little each year because its heavily used underground water supplies are drying up.

Today, more country people are seeking their fortunes in the towns, only to find that opportunities in industry are limited. However, as Hong Kong and Taiwanese industry produces more sophisticated products, there is scope for Thai firms – whose labour is cheaper – to export a growing proportion of simple manufactured goods, led by textiles and electronics. Natural gas discoveries in the Gulf of Thailand and a small onshore oil field are spawning a small petrochemical industry.

A potential boost to trade and foreign investment has been provided recently by the government's new financial liberalisation programme, which included the relaxation of

foreign exchange controls and allowed a wider scope to foreign banks to open branches in the country. Also promising greater development has been the opening of the first bridge across the MEKONG River between Thailand and Laos, a key link in a planned road network which will extend from Beijing, capital of China, all the way to Singapore.

For many young people, even the educated, the only opportunities lie in the service sector – above all, in the tourist trade. More than 2 million people visit Thailand each year, some for its more dubious pleasures, but most for the exotic attraction of its sights.

THAILAND AT A GLANCE	
Official name Kingdom of Thailand	
Area 514 000 km² (198 410 sq miles)	
Population 58 722 500 **Per km²** 114	
(Per sq mile 295)	
Capital Bangkok	
Government Constitutional monarchy	
Currency 1 baht = 100 satang	
Language Thai	
Religions Buddhist (95%), Muslim (4%), Christian, Hindu	
Climate Tropical, monsoon, rains from May to October. Average temperature in Bangkok ranges from 26°C (79°F) in December to 30°C (86°F) in April	
Land use Cultivation 26%, pastures 4%, forest and woodland 28%, other 42%	
Main primary products Rice, rubber, maize, sugar cane, cassava, pineapples, bananas, timber, fish, tin, tungsten, iron, manganese	
Major industries Agriculture, light manufacturing, processed foods, cement, paper, textiles and clothing, mining, forestry	
Main exports Rice, tapioca, rubber, sugar, tin, tungsten, textiles, maize, manufactures	
Annual income per head (US$) 1840	
Population growth (per thous/yr) 36	
Life expectancy (yrs) Male 54 **Female** 59	

▼ GLITTERING DEVOTION A host of colourful Buddhas and shrines fills the compound of the Grand Palace in the old walled city in Bangkok. The palace was home to the court when the country was called Siam, and was forbidden to more lowly mortals.

Thebes *Egypt* Ruins of the ancient city of the New Kingdom pharaohs, and capital of Egypt for almost 1000 years from about 1600 BC. It lies beside the Nile, about 500 km (310 miles) south-east of CAIRO.

Ancient Thebes was a walled city described by Homer, the Greek poet, as 'hundred-gated Thebes'. Huge temples to the gods and temple tombs for the pharaohs were built at the city's twin religious sites, LUXOR and KARNAK, which are linked by an avenue of sphinxes. To stop looting, which had emptied the pyramid resting places of earlier pharaohs, the Thebans buried their royal dead in great chambers carved out of the sandstone cliffs in the Valley of Kings across the river. The largest of these, the tomb of the teenage king Tutankhamun, who died in about 1352 BC, was discovered in 1922 by the British archaeologist Howard Carter. It was crammed with an astonishing collection of gold furniture, masks and jewelled ornaments, most of which is now in the Cairo Museum.

Thebes declined after it was sacked by the Mesopotamian ruler Assurbanipal and his Assyrian hordes in the 7th century BC. Today it is one of Egypt's main tourist attractions.
Map Egypt – Cc

Thebes (Thívai; Thíva) *Greece* Modern city situated 50 km (31 miles) north-west of ATHENS on the site of ancient Thebes, the city of seven gates and the legendary capital of Oedipus, the mythical king who according to legend unwittingly killed his father and married his mother.
Population 18 700
Map Greece – Cb

thermal Local, rising CONVECTION CURRENT of warm air. It occurs because some land surfaces, such as built-up areas and bare sand, are heated by the sun much more rapidly than others, such as grassland and forest. Thermals assist the ascent of birds and gliders.

thermal equator See HEAT EQUATOR

thermal spring See HOT SPRING

thermocline Temperature gradient in an ocean or a lake, in which there is a marked decrease of temperature with depth. It occurs because the surface water is warmed by the sun and becomes less dense, and therefore remains on the surface. The water below is colder and more dense, and so does not rise.

In oceans, as only a thin layer of the surface is heated, and there is no large-scale convection mixing, the surface water cools off quickly in winter. Seasonal changes in the relative temperature of the ocean surface and land cause changes in wind patterns, such as the monsoons.

thermosphere See ATMOSPHERE

Thessalía *Greece* See THESSALY

Thessaloníki *Greece* See SALONICA

Thessaly (Thessalía) *Greece* Central rural region of Greece covering some 12 939 km² (5382 sq miles) and bounded by Macedonia in the north, the Aegean Sea in the east and STEREÁ ELLÁS in the south. It contains the largest plain of Greece, drained by the Piníos River, and is the granary of Greece. The region, almost always under foreign rule, was also known in ancient times for breeding horses. Its capital is Lárisa.
Population 695 700
Map Greece – Bb

Thetford Mines *Canada* Centre of an asbestos-mining area in southern QUEBEC. The town is the largest producer of asbestos in the Western world. There is skiing nearby.
Population 20 360
Map Canada – Hd

Thiès *Senegal* The country's second largest city, about 80 km (50 miles) east of DAKAR, at the railway crossing to northern Senegal (St Louis) and neighbouring Mali to the east. Tapestries produced at a local factory are renowned and sold abroad.
Population 320 000
Map Senegal – Ab

Thimphu (Thimbu) *Bhutan* Capital of the Himalayan kingdom since 1960 (construction was begun in 1955) and home of its ruler, the 'Dragon King'. The town stands in a broad valley in western Bhutan at an altitude of some 2370 m (7775 ft), and is linked to Assam in India by one of the country's few roads. The *dzong* (monastery-fortress) of Tashi Chho dominates the town; it was built in 1641 but largely reconstructed (without using nails) in the 1960s. It houses government offices and the king's throne room; his palace is on the opposite side of Thimphu River. About 5 km (3 miles) from the town, at the entrance to the Thimphu Valley, is Bhutan's oldest *dzong*, Simtokha (1627).
Population 15 000
Map Bhutan – Aa

Thionville *France* Former iron-mining town on the Moselle River near the southern tip of Luxembourg. It was a favourite residence of Charlemagne in the late 8th century. It grew rapidly in the early 20th century with the Lorraine steel industry, but was badly hit in the 1980s by cutbacks in steel.
Population (town) 39 700; (metropolitan area) 132 000
Map France – Eb

Thíra *Greece* See SANTORINI

Thíva *Greece* See THEBES

Thívai *Greece* See THEBES

Thon Buri *Thailand* Twin city of BANGKOK, on the west bank of the Chao Phraya River. It was, for five years, the national capital after the Burmese sacked the royal capital of AYUTTHAYA in 1767. Traditionally, many of its inhabitants – a large part of whom are Chinese – live on boats moored in the river, and many still live a canal-side existence. The city's most famous building is Wat Arun, or 'the Temple of the Dawn', which was built between 1824 and 1851.
Population 919 000
Map Thailand – Bc

Thorn *Poland* See TORUŃ

thorn forest Dense thickets of thorny scrub growing where the rainfall is too scanty or unreliable for the growth of normal forests. See CAATINGA

Thousand Islands *Canada/USA* Group of about 1500 islands and islets extending some 128 km (80 miles) along the St Lawrence River from Gananoque to Brockville, Ontario. Those on the west side, which include the largest, Wolfe, covering 127 km² (49 sq miles), Howe and Grenadier, are mostly Canadian. Those on the east, including Grindstone and Wells, are in the US. Some islands are large enough to have villages.

The St Lawrence Islands National Park, covering 405 hectares (1000 acres), includes 18 of the Canadian islands, 80 islets and part of the Ontario coastline. The five-span Thousand Islands International Bridge, opened in 1938, crosses the river between Coelins Landing in New York State and Ivy Lea, in Ontario.
Map Canada – Hd

Thrace (Thráki; Thraki; Trakija; Trakiyska) *South-east Europe* Historical region spanning the borders of modern Bulgaria, Greece and Turkey. It stretched at one time from Macedonia to the Black Sea and Sea of Marmora, and from the Aegean to the Danube. Inhabited by various tribes, it became a Roman province in AD 46 and, from 1453, part of the Turkish Ottoman Empire. The present frontiers date from the 1920s.

The Bulgarian part of Thrace (known as Trakiyska or Nizina) north of the Rhodope Mountains – centred on PLOVDIV – became known as Eastern Rumelia and was annexed to Bulgaria in 1885. Turkish Thrace corresponds to European Turkey, stretching from the Evros (Meriç) to the Bosporus and Dardanelles straits, an area of 23 973 km² (9256 sq miles). Apart from ISTANBUL, the largest town is EDIRNE.

The modern Greek region of Thrace (locally called Thráki; and sometimes called 'Western Thrace' to distinguish it from the Turkish Eastern Thrace), covers 8578 km² (3312 sq miles). The largest towns are Komotiní and Xanthi; there are excavated ruins at Abdera, birthplace of the philosopher Democritus (about 460-370 BC).

In ancient times, Thrace was renowned for its gold and silver mines; today it is mainly agricultural, producing tobacco, wheat and fruits.
Population (Turkish) 4 325 300; (Greek) 345 200
Map Bulgaria – Bc; Greece – Da; Turkey – Aa

Thráki *Greece* See THRACE

Three Pagodas Pass *Myanmar (Burma)/Thailand* Mountain pass linking the south of Myanmar's Karen State with Thailand – though it is not used, except illegally, for cross-border transit. It carries the road from the port of MOULMEIN south-east to BANGKOK, the Thai capital. The pass was the route for the infamous Second World War 'death railway' constructed by Allied prisoners of war for the Japanese. The bulk of the line on the Thai side of the pass has now been abandoned, and the line on the other side of the border, though planned, was never built.
Map Myanmar – Cc

Three Rivers *Canada* See TROIS RIVIÈRES

throw (throw of fault) Term used to refer to the vertical distance between a rock bed on one side of a FAULT and its displaced continuation on the other side. The distance may vary from a few millimetres to hundreds of metres. See THE RESTLESS EARTH, p 231

thrust Force of compression in the earth's crust, resulting in a low-angle reverse FAULT. The term is also used for such a fault. See THE RESTLESS EARTH, p 231

Thule *Greenland* Northernmost settlement in the country, lying on the north-west coast, 1530 km (950 miles) from the North Pole. It has an airport, the Dundas weather station and a large American military base. It was Kund Rasmussen's (1879-1933) point of departure and base during several of his Arctic exhibitions. During the Second World War, all the inhabitants were moved to New Thule (Qaanaaq) at Inglefield Bay.
Population 800
Map Arctic – Db

Thunder Bay *Canada* Port at the western end of Lake SUPERIOR, created in 1970 by the merger of Fort William and Port Arthur. It is a centre for hunting, fishing and winter and water sports.
Population (city) 113 950; (metropolitan area) 124 430
Map Canada – Gd

thunderhead Term, used principally in North America, for the swollen upper part of a CUMULONIMBUS CLOUD with shining white edges. Thunderheads usually herald storms.

thunderstorm Storm in which intense heating induces air to rise rapidly and form vast CUMULONIMBUS CLOUDS, with heavy rain, lightning, thunder, and sometimes hail.

Thüringer Wald *Germany* Mountains in the south-west, stretching 60 km (37 miles) south-east from the Werra River and EISENACH. They rise to 984 m (3228 ft) on the Great Beerberg, and are popular with skiers.
Map Germany – Dc

Thysville *Zaire* See MBANZA-NGUNGU

Tiahuanaco *Bolivia* Ruined city near the southern end of Lake TITICACA, thought to date from about AD 500 to 1000. These pre-Inca ruins, which are spread over a large area, are the remnants of one of the oldest civilisations in South America. The major surviving structures on the site are the Gate of the Sun, the Acapana Pyramid, which is 15 m (49 ft) high, and the Temple of Kalasasaya, which is covered with hieroglyphics that are still to be deciphered. The main statues are housed in the national museum, the Museo Arqueológico Tiwanacu, in La Paz.
Map Bolivia – Bb

Tianjin (Tientsin) *China* Third largest city in China after Shanghai and Beijing. Founded as a garrison town where the Da Yunhe (Grand Canal) joins the Hai He River, it grew in the 19th century as a port for the capital, 100 km (62 miles) away. From 1860 to 1949 foreigners enjoyed rights of residence in the city and many of its buildings are in the Western style.
Tianjin is an important centre on the communications network of northern China and has a wide range of industries, including textiles, steel and chemicals. Oil from the Ziyang basin is refined here. Most overseas trade is now handled by nearby XINGANG.
Population 5 700 000
Map China – Hc

Tiber (Tevere) *Italy* Third longest river in the country after the Po and Adige. The Tiber rises near the 1407 m (4616 ft) high Mount Fumaiolo to the east of FLORENCE and flows 405 km (252 miles) south through Umbria to ROME and its mouth at Ostia. The towns of Todi, Perugia and Città di Castello overlook the Tiber Valley.
Map Italy – Dc

Tiberias (Tevarya) *Israel* City, market town and health resort, with hot springs, about 50 km (30 miles) east of HAIFA, and some 207 m (680 ft) below sea level.
It stands on the west shore of the Sea of Galilee (now also known as Lake Tiberias), where Jesus walked on the waters. The city was founded by Herod Antipas in the 1st century AD, and named after the Roman emperor, Tiberius. In the 11th to the 16th centuries, it was captured in turn by the Crusaders, the Arabs and the Turks. The first Jewish settlers from Poland arrived here in the mid-1700s. Tiberias has many ancient synagogues and tombs of Jewish scholars, and is a centre for pilgrims to the Galilee area. It is among the four holiest Jewish sites, the other three being Jerusalem, Hebron and Safad.
To the south, at Hammat, there is a 1st-century synagogue with a fine mosaic floor depicting the signs of the Zodiac. At the nearby spa of Hammat Gader, Roman antiquities can be seen and there are hot mineral springs which feed a natural pool.
Population 29 500
Map Israel – Ba

Tibesti *Chad* Chain of extinct volcanic mountains in the Sahara of north-western Chad, in the prefecture of BORKOU-ENNEDI-TIBESTI near the Libyan border. The range has dramatic cliffs, curiously shaped rocks and deep ravines; its highest peak is Emi Koussi, at 3415 m (11 204 ft).
Fossil finds in the area include the skull of a protohuman primate called *Chadanthropus*. Ancient rock drawings of wildlife suggest that the climate used to be much wetter than it is now. Still, the area receives more rainfall than the surrounding lowland, and after the rains have fallen, sheep and goats are grazed here.
Map Chad – Aa

Tibet *China* See XIZANG AUTONOMOUS REGION

Tibetan Plateau *China* See XIZANG GAOYUAN

Ticino (Tessin) *Switzerland* Only canton in Switzerland where Italian is the main language. It lies on the southern slopes of the Alps near the Italian border.
Ticino covers 2812 km² (1086 sq miles) and includes parts of Lake Lugano and the northern end of Lake MAGGIORE. It has a warm climate and because of the considerable change in altitude throughout the canton, a varied and rich flora. Ticino's capital is Bellinzona with a population of 17 100. The Ticino River rises in the St Gotthard massif and enters the Po River in Italy.
Population 294 100
Map Switzerland – Ba

tidal current Movement of tidal water into and out of a bay, estuary, harbour or other body of water at the flood and ebb tide.

tidal flat Area of sand or mud which is uncovered at low tide.

tidal wave Informal name for a TSUNAMI

tide Daily rise and fall of the sea level. See HOW TIDES ARE PRODUCED, p 670

Tien Shan (Tian Shan) *China/Kyrgyzstan* Mountain range on the border of north-western China and Kyrgyzstan, extending eastwards across the northern part of China's Xinjiang Uygur Autonomous Region. It has many peaks over 5000 m (16 500 ft); the highest is Pik POBEDY (Tomur Feng) at 7439 m (24 406 ft), on the Kirghiz-Chinese border.
Maps China – Ac; Kyrgyzstan – Ba

Tientsin *China* See TIANJIN

Tierra del Fuego *Argentina/Chile* Archipelago at the southern tip of South America, separated from the rest of the continent by the Strait of Magellan. The Portuguese explorer Ferdinand Magellan (1480-1521) arrived here in 1520, when he made passage through the strait and saw fires lit by the native Ona people – a people who are extinct – and indeed, the name 'Tierra del Fuego' means 'Land of Fire'. The land area is 73 700 km² (28 450 sq miles), two-thirds of which comprises the main island, also called Tierra del Fuego. This is divided between Chile and Argentina. Most of the other islands are Chilean. The main Chilean town is Porvenir; the main Argentinian town is Ushuaia – the continent's and the world's most southerly town. Tierra del Fuego's economy is based on oil, timber (though most natural forests have been cleared) and sheep farming.
Map Argentina – Cd

Tiflis *Russia* See T'BILISI

Tigray (Tigre) *Ethiopia* Province which covers 65 700 km² (25 360 sq miles) of the north-east, devastated by civil war and drought. The people of Tigray differ ethnically from the Ethiopians who are Amhara. In early 1986, the Tigray People's Liberation Front, resisting central government policies such as the imposition of Amharic, the Amhara language, was in control of most of the land. The capital is MEKELE.
Population 3 000 000
Map Ethiopia – Aa

Tigris *Middle East* Shared by Turkey, Syria and Iraq, the Tigris, which is 1840 km (1143 miles) long, is one of the major rivers of south-west Asia. It rises in a mountain lake in Turkish Kurdistan, and for a short distance forms the north-east border between Turkey and Syria. It then enters Iraq, passing through Mosul and BAGHDAD. At Al Qurnah it joins the EUPHRATES to form the SHATT AL ARAB. The Tigris has a number of large tributaries, and since ancient times it has been a valuable source of irrigation, especially in the area known as the FERTILE CRESCENT. It reaches its highest levels in April, following the spring thaw in Turkey, when it can rise by as much as 300 mm (12 in) an hour, and produce severe flooding. However, downstream the river's volume decreases rapidly due to the high rate of evaporation, and the extensive irrigation which taps its waters along its central course.
Map Middle East – Ba

Tihamah *Saudi Arabia/Yemen* Coastal plain running parallel to the Red Sea. It varies in width

HOW TIDES ARE PRODUCED

The endless cyclical movement in the great waters of the earth – the tides – is powered by the gravitational pull of the moon and, to a lesser extent, by that of the sun. Although the sun is 26 million times larger than the moon, it has about half as much gravitational pull on the earth because it is 390 times farther away from the planet.

In mid-ocean, the difference between high and low tide is only about 0.6 m (2 ft), but in shallow coastal seas and in estuaries the difference is much greater. The Bay of Fundy in eastern Canada, for example, has a tidal range of 15 m (49 ft).

At any time, there are two high tides on the earth – the direct tide on the side facing the moon, and the indirect tide on the opposite side. The direct tide is a bulge produced by the moon pulling the earth's water towards it more powerfully than it pulls the solid earth because the ocean – on the surface of the earth – is nearer to the moon. On the other hand, on the opposite side of the earth farthest from the moon, the solid earth is nearer the moon, and the

moon's pull on it is greater than on the water. So, as it were, the earth is drawn away from the water, and a second bulge, the indirect tide, is produced.

Since the moon orbits the earth once every 24 hours 50 minutes, there are two high tides and two low tides at any point in the world's oceans during that period. Local tidal timetables are, however, very much modified by coastal features such as bays, estuaries and promontories that speed or slow the water's movement.

The world's tides could be used to generate huge supplies of electricity, but only limited attempts have been made to harness this form of energy. One system in operation is the 24 megawatt generating station near St Malo in Brittany, France, opened in 1966. There a 750 m (2460 ft) tunnelled barrage blocks the River Rance estuary. When the tide comes in, the sea is fed through turbines in the dam, turning them to generate electricity, and filling the basin beyond. As the tide goes out, water stored in the basin is fed back through the turbines to generate more power.

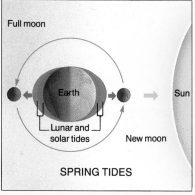

SPRING TIDES

The highest tidal ranges occur at new and full moon when the earth, sun and moon are in a straight line, the sun and moon reinforcing each other's gravitational pull.

NEAP TIDES

The lowest tidal ranges occur when the moon is in its first and last quarters. At these times, the pull of the sun and moon are at right angles to each other.

---

from 15 km (9 miles) to 80 km (50 miles). Agriculture dominates the Saudi Arabian area, with entire families, mostly descendants of African slaves, working in the fields. But temperatures are unbearably hot – 'Tihamah' means 'hot land' in Arabic – and rainfall is low. There is no oil, and the people are the poorest in the country. In Yemen, the plain is well irrigated and produces vegetables and cotton.
Map Saudi Arabia – Bc; Yemen – Aa

Tijuana *Mexico* Border city at the top of the Baja California Peninsula. Tijuana is busy throughout the year with American tourists who cross into Mexico for a few hours of cheap entertainment and shopping.

As a result, the city claims the highest per capita income in Mexico, and even the many surrounding shanty towns are gradually being transformed.
Population 743 000
Map Mexico – Aa

Tikal *Guatemala* Rediscovered in 1877, the largest, and probably the oldest, of the pre-Columbian Mayan cities in Central America. Tikal lies in the Petén rain forest of northern Guatemala, about 300 km (185 miles) north-east of the capital, GUATEMALA CITY. Forest and undergrowth cover most of what was one of the largest pre-industrial cities in the world, of which 16 km² (6 sq miles) have been cleared so far. Beside the great plaza in the city centre stand five temple pyramids, the tallest in the Americas, some almost 70 m (230 ft) high, with steep steps and carved chambers below the pinnacle sanctuaries.

The Mayas built at Tikal for more than 1100 years, creating in stone ever-taller sacrificial shrines, evolved from the basic design of a wooden hut. Traces of the oldest buildings have been dated to 600 BC. More than 3000 buildings and 200 stone monuments have been found at the site, but 10 000 earlier buildings are thought to lie beneath them. The largest temples are believed to date from about AD 600 to AD 800. Building

ended in about AD 900, when the Mayas seem to have abandoned the city, for reasons which are still unknown.
Map Guatemala – Ba

Tilburg *Netherlands* Spacious town, 90 km (55 miles) south-east of AMSTERDAM. It was an industrial centre until the 1960s when it went through a severe slump. It has since become a regional centre, focusing on service industries and educational institutions. Tilburg also has a university. Efteling Park, which is 10 km (6 miles) north of the town, has won an award as the best designed and most entertaining pleasure park in Europe.
Population (town) 162 400; (Greater Tilburg) 235 680
Map Netherlands – Bb

till See BOULDER CLAY

Timaru *New Zealand* Port and seaside resort on South Island, 150 km (95 miles) south-west of CHRISTCHURCH. It exports flour, wool and locally frozen meat.
Population 27 640
Map New Zealand – Cf

timberline See TREELINE

Timbuktu (Tombouctou) *Mali* Town on the edge of the Sahara. It was founded by the Tuareg people in the 11th century, and rapidly became a Saharan centre of the caravan trade because of its closeness to the Niger River, which laps at its walls when in flood. In the Middle Ages, its reputation as a centre for gold, ivory and salt had reached even Europe; an Italian, Beneditto Dei, visited the city in 1470. Tribal wars from the 16th century sent the town into decline. The French explorer René Caillé found it in 1828, and the French occupied the city in 1893.

Timbuktu has for centuries been an important Islamic centre, and had its own religious university and more than 100 schools in the 16th century. Its fortunes have been partly revived as a refuge for nomadic herdsmen who have been forced to abandon their lifestyle because of drought, and as a tourist centre. It is best reached by riverboat, four-wheel-drive vehicle or by air.
Population 20 000
Map Mali – Bb

time zones See WHAT TIME IS IT WHERE? p 671

Timgad *Algeria* Ruined Roman city in the northeast. Timgad was founded by Emperor Trajan in about AD 100, and destroyed by Berbers in the 7th century. Remains of the forum, capitol, theatre, several baths, an Arch of Trajan and well-preserved mosaics have been excavated.
Map Algeria – Ba

Timişoara (Temesvar; Temeschburg) *Romania* Major industrial city in the south-west of the country, 30 km (19 miles) east of the border with Yugoslavia. It dates from the 13th century and has an impressive 14th-century castle, a fine 18th-century town hall and several splendid religious buildings. Its products include textiles, processed food, metalwork, rolling stock, machine tools and chemicals.
Population 324 700
Map Romania – Aa

Timor *Indonesia* Island of some 30 775 km² (11 883 sq miles) in the Lesser Sunda Islands (NUSA TENGGARA) midway between Sulawesi and northern Australia. The island was divided until 1975 into a Portuguese east and an Indonesian west. The eastern half was taken over in 1975 from Portugal, but resistance still continues today from pro-independence guerrillas.
Population 3 085 000
Map Indonesia – Fd

Timor Sea *Indonesia/Australia* Situated between Timor, Indonesia and north-western Australia, the sea was explored by the Dutch navigator Abel Tasman in 1644 and later by the buccaneer William Dampier. The Timor Sea covers 450 000 km² (173 745 sq miles).
Map Pacific Ocean – Bc; Indonesia – Gd

Tindouf *Algeria* The most westerly town in the country, near the borders of Morocco and Mauritania. It is a tourist centre for desert travellers, and an airfield links the town to ALGIERS. The roads beyond Tindouf into neighbouring countries are very rough going.
Population 7000
Map Algeria – Ab

Tinian *Northern Marianas Island* Second largest island of the group, covering an area of 101 km² (39 sq miles). It is separated from SAIPAN, to the north, by a 6 km (4 mile) channel. After it was captured from the Japanese in June 1944, during the Second World War, the Americans constructed a huge air base on Tinian Island from which to bomb Japan. The aircraft that dropped the first atomic bombs, on Hiroshima and Nagasaki took off from here.
Population 2100
Map Pacific Ocean – Cb

Tínos *Greece* Hilly island of the CYCLADES group, south of Ándros. It was occupied by the Venetians for over 500 years (AD 1207-1712), longer than any other part of Greece, and an unusually large proportion of islanders are Roman Catholics. The pilgrimage church of Panayia Evangelistria is, however, Greek Orthodox. It was built in 1922 and has an icon said to work wonders. The island is known for its marble.
Population 7800
Map Greece – Dc

Tipperary (Tiobraid Árann) *Ireland* Market town, 160 km (100 miles) south-west of DUBLIN.

North of the town is the Golden Vale, a lush land with thriving dairy farms. The surrounding county of the same name covers some 4254 km² (1642 sq miles); Clonmel is the county town.
Population (town) 4770; (county) 132 770
Map Ireland – Cb

Tiranë (Tirana) *Albania* National capital founded by Turks in the early 17th century; it lies in what was originally woodland beneath a limestone ridge at the southern end of a fertile plain. By 1920, when it became the capital, Tiranë had a population of 10 000. Later, King Zog (1895-1961) – the first and only king of independent Albania – hired Italian architects to build a new city centre and government offices around Skanderbeg Square, named after a national hero who had fought unsuccessfully against the Turkish invasion in the 15th century.

By 1939 the population had risen to 35 000. Axis forces occupied the city during the Second World War; then, after the communists took over in 1946, a big expansion took place under Soviet influence. New factories were built, turning out a wide variety of products, from cigarettes to cement. New housing went up, a hydroelectric power station was constructed, and a modern

WHAT TIME IS IT WHERE?

Everyone regards midday as being the time when the sun is at its zenith, but because of the earth's rotation, this occurs at different times in different places. Using the sun's position as the arbiter of the clock, actual time varies with every kilometre travelled to east or west, and varies very considerably over great distances. The situation imposes timetable problems, as became apparent when the railways began. In 1884 the International Meridian Conference divided the 360 degrees of longitude into 15 degree time zones, one for each of the 24 hours. As the zero degree line ran through Greenwich, England, the basic time was established as Greenwich Mean Time (GMT).

Time to the east is ahead of GMT, that to the west behind it until, on the other side of the globe along the International Date Line – 180 degrees longitude – time is 12 hours different from GMT. Travellers crossing the line from east to west add 24 hours and omit a day. People travelling from west to east subtract 24 hours and repeat a day.

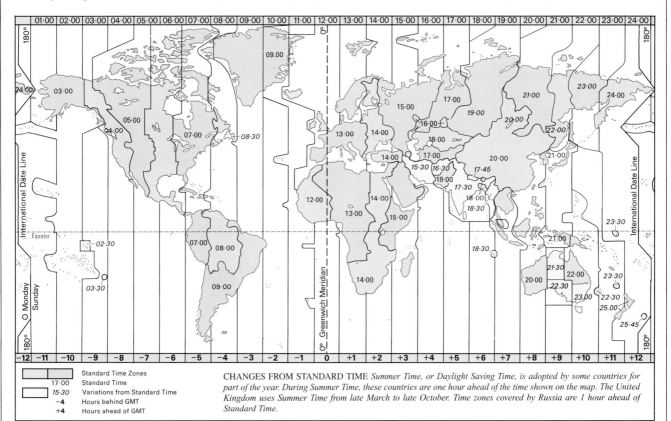

CHANGES FROM STANDARD TIME *Summer Time, or Daylight Saving Time, is adopted by some countries for part of the year. During Summer Time, these countries are one hour ahead of the time shown on the map. The United Kingdom uses Summer Time from late March to late October. Time zones covered by Russia are 1 hour ahead of Standard Time.*

	Standard Time Zones
17·00	Standard Time
15·30	Variations from Standard Time
−4	Hours behind GMT
+4	Hours ahead of GMT

water supply laid on. Rail links connect Tiranë to the country's main port of Durrës and the steel plant at Elbasan, where there is an airport.

Population 251 000

Map Albania – Bb

Tîrgovişte *Romania* City and former capital of the province of WALACHIA, 71 km (44 miles) north-west of Bucharest. It was a trading and market centre in the Middle Ages and although many of its medieval buildings were destroyed by the advancing Turks, some remain – including the 15th-century Chindia Tower which contains a museum devoted to Vlad Tepes, the 15th-century prince who is said to have been the model for the fictional character Count Dracula.

The nearby 16th-century Dealu Monastery contains the tombs of several Walachian rulers. In its crypt is the head of a Romanian national hero, Michael the Brave (1558-1601), who in 1600 briefly united the provinces of Walachia, TRANSYLVANIA and MOLDAVIA. The city manufactures petroleum products, rolling stock and steel.

Population 101 300

Map Romania – Bb

Tîrgu Mureş *Romania* Industrial and cultural centre in northern TRANSYLVANIA. The city stands on the Mureş River, some 80 km (50 miles) south-east of CLUJ-NAPOCA. Tîrgu Mureş has a large Hungarian-speaking population, and the city radiates from a large central square surrounded by a Hungarian-style inner town. Among its historic buildings are a 15th-century fortress, a 15th-century church and several Baroque houses. There is also a Jesuit monastery and two 18th-century Orthodox churches, one of them wooden. The city's products include metal goods, machinery, food, clothing and chemicals.

Population 166 000

Map Romania – Aa

Tiruchchirappalli (Trichinopoly) *India* City on the Kaveri River in the south-east, some 319 km (198 miles) south-west of the city of MADRAS. With the surrounding plains it is dominated by Rock Hill, 885 m (270 ft) high, whose fort (now ruined) was battled over by the British and the French in the 18th century.

Population 4 114 320

Map India – Ce

Tista *Bangladesh* River, 310 km (193 miles) long, that flows from Himalayan Sikkim onto the plains, and is liable to sudden floods. It has built a sandy alluvial fan in the Dinajpur and Rangpur districts, and shifted course several times during the past century. Any scheme to link the Indian sections of the BRAHMAPUTRA and GANGES rivers would have to cross the Tista in Bangladesh, presenting vast engineering and political problems.

Map Bangladesh – Bb

Tisza *Hungary/Ukraine/Yugoslavia (Serbia and Montenegro)* East European river, some 960 km (nearly 600 miles) long, which rises high in the Carpathians of Ukraine, and flows west, then south-west, across Hungary and Serbia to join the Danube 45 km (28 miles) north of the Serbian capital, BELGRADE. Between Chop on the Ukrainian border and SZEGED near the Serbian frontier, the river drops only 104 m (341 ft) in 440 km (273 miles), as it meanders sluggishly across the Great ALFÖLD.

Much of the river's course in Hungary is lined with levees (banks) built after devastating floods in 1879. There are towns along this part of the river only where natural terraces rise above the flood plain on both banks to provide easy foundations for bridges – at TOKAJ, SZOLNOK, Csongrád and Szeged.

Map Hungary – Bb

Titicaca, Lake *Bolivia/Peru* Second largest lake in South America, after Lake Maracaibo, and, at 3812 m (12 506 ft), the highest navigated lake in the world. It covers 8786 km^2 (3392 sq miles) and is shared by Bolivia and Peru. The average depth of the lake is 103 m (338 ft); its deepest known depth is 281 m (922 ft). The first steamer to ferry passengers across Lake Titicaca was built in England in 1862, dismantled on the coast of Peru, hauled up the Andes and reassembled on the lake to link the two countries.

Uru Amerindians live on the lake on floating 'islands' made out of the lake's totora reeds. Their homes on these huge rafts are made out of the same reeds, as are the boats from which they fish. Lake Titicaca was the holy lake of the Inca; ruins of the Inca and pre-Inca Tiahuanaco civilisations stand on the islands of the Moon and the Sun north of Copacabana Peninsula, and at the ruined city of TIAHUANACO just south of the lake.

Map Bolivia – Bb; Peru – Cb

Titograd *Yugoslavia* See PODGORICA

Tivoli *Italy* Hill town, 25 km (16 miles) east of ROME which was known as Tibur by the Romans. It has sumptuous Renaissance villas, such as the Villa d'Este and the Villa Gregoriana, set in magnificent gardens beside spectacular waterfalls. Emperor Hadrian (AD 76-138) built a summer resort here, and its remains, including some statues and mosaics, still surive.

Population 50 560

Map Italy – Dd

Tiwi *Philippines* See BICOL

Tizi Ouzou *Algeria* Tourist centre, some 88 km (55 miles) east of ALGIERS, lying in the Kabylie hill country. It is the administrative capital of the surrounding district. In recent years, the town, which is typical of the region, has expanded rapidly. Figs are processed locally, and olive oil is produced. The country's biggest textile factory is 10 km (6 miles) north-east of Tizi Ouzou, at Draa Ben Khedda.

Population 101 000

Map Algeria – Ba

Tlemcen (Tlemsen) *Algeria* Road and rail junction on the northern edge of the Tell Atlas, close to the Moroccan border. It is the capital of the district, exporting blankets, carpets, olive oil, leather and esparto grass through the port of Rashgun. Its manufactures include brassware, footwear, hosiery, carpets and furniture.

Originally founded by the Romans, Tlemcen later became the Moorish capital of the region, until it fell to the Turks in 1553, when it started to decline. In 1842, the French captured it. Architecturally, it retains its medieval Moorish character, with some of the best preserved buildings of the early Arab conquerors dating from the 12th century onwards. Among the numerous splendid mosques are the Grand Mosque and the

Mosque (and museum) of Bel Hassane. Other medieval remains include towers and minarets.

Population 146 000

Map Algeria – Aa

Toamasina *Madagascar* The republic's chief port, standing on the Indian Ocean, some 200 km (125 miles) north-east of the capital, ANTANA-NARIVO. Formerly called Tamatave, the city was devastated by a hurricane in 1927. It exports cloves, coffee, hardwood, meat and vanilla, and processes textiles, leather and tobacco. It had the country's only oil refinery, but this was destroyed by a cyclone in 1994.

Population 145 430

Map Madagascar – Aa

Toba *Indonesia* Volcanic crater lake situated in northern SUMATRA. The oval lake, formed after a massive eruption blew the top off a mountain 60 000 years ago, covers an area of 1786 km^2 (689 sq miles), which includes the 640 km^2 (247 sq mile) island, Samosir, in its centre. The island was actually a peninsula until the Dutch cut a canal through the narrow isthmus. In places the lake is 500 m (1640 ft) deep. Popular with tourists for its peaceful tropical setting, Toba Lake is the home of the Batak people, thought to be descended from 14th-century Chinese settlers.

Map Indonesia – Ab

Tobago *Trinidad and Tobago* Smaller and more northerly of the two Caribbean islands constituting the Republic of Trinidad and Tobago. It covers about 300 km^2 (116 sq miles) 32 km (20 miles) off the north-eastern tip of Trinidad, and was supposedly the home of Robinson Crusoe in Daniel Defoe's book. About half of it is covered with rain forest; the other half produces cocoa, coconuts and bananas, but agriculture has declined in recent years as many inhabitants have emigrated to Trinidad. Tourism is the island's main industry. SCARBOROUGH, on the south-east coast, is the capital.

Population 48 600

Map Trinidad and Tobago – Aa

Tobruk (Tubruq) *Libya* Town and port on the north-east Mediterranean coast, some 120 km (75 miles) from the Egyptian border. It has an airfield and an excellent harbour. Tobruk was occupied by the Italians in 1912 and was the scene of much fighting, including a siege of the British in the town, during the Second World War.

Population 59 000

Map Libya – Ca

Tocorpuri *Bolivia* Permanently snow-capped mountain of 5833 m (19 137 ft) in the Western Cordillera range on the border with Chile.

Map Bolivia – Bc

Togliatti *Russia* See TOL'YATTI

Togo See p 673

Togo, Lake *Togo* Former estuary of the Chio River, forming the country's largest lagoon. It covers 78 km^2 (30 sq miles) between the capital, LOMÉ, and Aného, and is a rich source of fish and phosphate. The phosphate, with reserves of 100 million tonnes, is Togo's chief export, making the country the third biggest producer in Africa.

Map Togo – Bb

Togo

ONCE GERMAN, TOGO NOW SAYS IT IN FRENCH – AND WAITS FOR ITS INVESTMENT IN TOURISM TO PAY OFF

The West African nation of Togo is unusual in being a French-speaking country that was once a German colony. Moreover, its geographical diversity is far greater than its tiny area and long, narrow shape would suggest.

Grassy plains in the north and south are separated by the Togo highlands, which run from south-west to north-east, here and there rising to more than 1000 m (about 3280 ft). High plateaus, particularly in the more southerly ranges, are heavily forested with teak, mahogany and bamboo.

The original colony of Togoland included part of present-day Ghana and was a German protectorate and, later, colony from 1884 until 1914. After the First World War the country came under British and French mandates, and in 1922 the League of Nations divided it up – the large, eastern region being administered by France. This region became the independent nation of Togo in 1960.

The main ethnic group, making up more than 60 per cent of the population, is the Ewe people, concentrated in the south, a traditionally better developed region. They regard themselves as the intelligentsia, and usually hold the important administrative posts. The Kabré, the dominant ethnic group of the north, are noted for their farming skills and have also dominated the country's army and government – the main source of political conflict between the two groups.

Traditionally, agriculture has been the mainstay of Togo's economy and it continues to involve about 78 per cent of the population. In the north, millet, sorghum and beans are subsistence crops and groundnuts are the main cash crop. In the highlands and the south, yams, cassava, maize, oil palms, vegetables and rice are grown for local consumption – while coffee, cocoa and cotton provide cash. Agriculture used to be the main export earner, but this gradually declined.

However, a 'Green Revolution' launched in 1977 has made the country self-sufficient in basic foods, except in times of drought. The drought has taken its toll of trees, but there has also been a failure to replant at regular intervals and to guard against disease – particularly in the cocoa plantations. Replanting schemes – as well as a number of other economic reform measures – are being financed by the World Bank to overcome problems of ageing, disease and decay in tree crops, but it will be some years before any benefits are reaped from this operation.

MINERAL WEALTH

Minerals, especially phosphates, are now the main export earners. In 1974, Togo launched several major developments, based on the income from phosphates, but by 1978 the bottom had fallen out of the market and Togo's sales were no longer sufficient to finance its development schemes.

The government had to borrow from abroad and is still doing so. Phosphate prices have since recovered considerably, and Togo is now the ninth biggest producer in the world. Phosphates bring in 40 per cent of foreign exchange earnings.

Most of the exports are handled through the port and capital, LOMÉ. Although the country made a heavy investment in tourism, with facilities being developed in all the major centres, the tourist industry is waiting for an upturn in the world economy, as well as a return of political stability.

After independence, General Gnassingbe Eyadéma staged a bloodless coup in 1967, which put an end to multi-party democracy. His military regime was stable until 1990, when the country was plunged into political crisis and violence as the people demanded a return to multi-party democracy. A national conference held in 1991 stripped the president of his executive powers and a transitional government was installed. A new constitution was approved in a referendum in September 1992. In August 1993 Eyadéma was re-elected in a presidential election boycotted by the main opposition parties.

In national assembly elections held in February 1994, the opposition won a majority of seats. The results were contested in court, but some parties refused to accept its finding. Eventually, after by-elections, Edem Kodjo, leader of one of the opposition parties and former secretary-general of the Organisation of African Unity, emerged as the new prime minister – and was faced with the challenge of making democracy work in a country ruled by a single party for almost 30 years.

TOGO AT A GLANCE	
Official name Republic of Togo	
Area 56 790 km² (21 922 sq miles)	
Population 4 300 000 **Per km²** 76 **(Per sq mile)** 196	
Capital Lomé	
Government Republic	
Currency 1 CFA franc = 100 centimes	
Languages French (official and commercial language), Ewe, Dagomba, and other West African languages	
Religions Indigenous (65%), Christian (25%), Muslim (10%)	
Climate Tropical and humid, with rainfalls declining farther north. Average annual temperature in Lomé is 28°C (82°F)	
Land use Grazing and other 50%, cultivation 26%, forest and woodlands 24%	
Main primary products Cassava, maize, millet, oil palms, groundnuts, cotton, cocoa, coffee, livestock; phosphates	
Major industries Mining, cement, agriculture, textiles, food processing	
Main exports Phosphates, cocoa, coffee	
Annual income per head (US$) 400	
Population growth (per thous/yr) 33	
Life expectancy (yrs) Male 50 Female 54	

Tokaj *Hungary* Town in the north-east of the country, 48 km (30 miles) east of the city of MISKOLC. It stands above the Tisza River on sunny volcanic slopes and has been Hungary's most famous wine-producing centre for more than 700 years. Tokay, a golden dessert wine, is sold in numerous wine cellars in the old town's narrow streets. The town has a wine museum which illustrates the history of wine production in the area.
Population 5170
Map Hungary – Ba

Tokelau *Pacific Ocean* New Zealand dependency consisting of three atolls – Atafu, Nukunonu and Fakaofu – some 480 km (nearly 300 miles) north of WESTERN SAMOA, and with a total land area of 12 km² (4.6 sq miles). The economy is based on fishing, various crops and livestock; some copra is exported. In 1965 an agreement was reached which stated that most of the population would be resettled in New Zealand – this plan has now been suspended at the request of the islanders.
Population 1580
Map Pacific Ocean – Ec

Tokushima *Japan* Virtually destroyed by Allied bombing in the Second World War, this city on the east coast of Shikoku Island is today a thriving commercial and industrial centre specialising in chemicals.
Population 263 000
Map Japan – Bd

Tokyo *Japan* National capital on HONSHU Island, and one of the world's largest cities – prosperous,

vast, overcrowded, vibrant, and a melting pot of Japanese and Western civilisation. Powerful ministries, corporation headquarters, the top universities, national theatres and museums, and a thousand shrines are concentrated here. The city has more telephones than the entire continent of Africa. More than a quarter of Japan's 124 million people live within 50 km (31 miles) of the walls and moats of the imperial palace. Land reclaimed from Tokyo Bay sustains an industrial juggernaut whose products range from books and electrical goods to heavy machinery.

Called Edo when it was founded in 1457, the city became the headquarters of Ieyasu, the first of the Tokugawa clan of shoguns, or warlords. Japan's ruling shogun in the early 17th century, Ieyasu kept an eye on fractious provincial warlords, and drained their purses, by requiring them to build and maintain elaborate residences in his capital. The city was renamed Tokyo (meaning 'Eastern capital') in 1868 after the last of the Tokugawa shoguns was deposed by Emperor Meiji. It twice rose from its own ruins. In 1923 it was devastated by an earthquake which killed about 143 000 people, and in 1945 it was severely damaged in the fierce firestorms caused by Allied bombing.

The modern capital has 26 satellite cities and is itself divided into distinct districts – including Marunouchi, Japan's financial and business heart, Kasumigaseki, the government quarter containing the marble Diet (parliament) building, and the neon-lit Chuo-Ku district with the Ginza shopping centre and the city's principal nightlife. Prices in hotels, restaurants and places of entertainment make Tokyo one of the most expensive cities in the world for the visitor, especially for business entertainment. Public transport is fast and frequent though crowded in rush hours. Farther from the metropolitan core, railway stations have become major shopping and entertainment complexes in their own right, at Shibuya, Ikebukuro and Shinjuku.

Among the city's main Buddhist temples is Sensoji, where a tiny golden image of Kannon, revered as the Buddhist goddess of compassion, is enshrined. The statue is said to have been found by fishermen in the Sumida River, on which the city stands, in AD 628. Tokyo has Japan's most spectacular festivals, including the Dezome Shiki, or Firemen's Parade, in which acrobats perform hair-raising stunts atop bamboo ladders.

Population (city) 8 278 000; (Greater Tokyo) 11 900 000

Map Japan – Cc

Tolbukhin (Dobrich) *Bulgaria* City recently renamed Dobrich, on the Varna-Constanţa railway, 40 km (25 miles) north of VARNA. Formerly called Dobrich, it was renamed in 1944 after the Soviet Marshal Tolbukhin who liberated it from the occupying Nazis. A market town for the fertile Dobrudja region, it processes foods and makes farm machinery.

Population 115 790

Map Bulgaria – Cb

Toledo *Spain* Historic city 65 km (40 miles) south-west of MADRID. Perched on a Tejo River-flanked granite hill, Toledo has a wealth of fine Moorish and medieval Christian architecture and Roman remains. It is the seat of the Primate of Spain and – appropriately for a town whose swords were the world's finest – manufactures firearms. The Cretan painter El Greco (1541-1614) spent most of his life at Toledo, where he painted his masterpieces. The surrounding province – also called Toledo – grows cereals and raises sheep and goats.

Population (city) 59 800; (province) 489 540

Map Spain – Cc

Toledo *USA* City in Ohio, standing on the south-western tip of Lake ERIE. It is one of the country's main inland coal and oil ports and also handles grain and timber. Its industries include machinery, food and metal processing, glass making, transport equipment, car parts and vehicle manufacture; it is the home of the Jeep.

Population (city) 332 900; (metropolitan area) 614 100

Map United States – Jb

Toluca *Mexico* The capital of the state of MEXICO, situated 56 km (35 miles) south-west of Mexico City. It is a thriving commercial and industrial city, manufacturing cars and industrial machinery. It is also the centre of an agricultural region which produces dairy products, sausages and dried meats, as well as pottery.

Population 487 630

Map Mexico – Cc

Tol'yatti (Togliatti) *Russia* Town on the Volga River, 65 km (40 miles) north-west of Samara. The former town of STAVROPOL' was drowned in the 1960s by the Samara Reservoir, created when the Volga was dammed nearby. A new town was built beside the new shoreline and named Tol'yatti after Palmiro Togliatti (1893-1964), leader of the Italian Communist Party, who died while on a visit to the USSR. The new town is an industrial centre with ship-repairing, engineering, car and food processing industries.

Population 655 000

Map Russia – Fc

tombolo Sand or shingle bar that links an island with the mainland, such as Chesil Beach in Dorset, England, which joins the Isle of Portland to the mainland west of the town of Weymouth.

Tombouctou *Mali* See TIMBUKTU

Tomsk *Russia* Siberian town on the Tom River, a tributary of the Ob', about 200 km (125 miles) north-east of NOVOSIBIRSK. Founded as a fort in 1604, it developed as a regional capital after the discovery of gold nearby in 1824. Today it is an industrial town producing electrical equipment, machinery, timber, paints, dyes and rubber goods. It is also an educational centre.

Population 506 000

Map Russia – Jc

Tongariro, Mount *New Zealand* Volcano rising to 1968 m (6457 ft) in North Island's Tongariro National Park, 230 km (140 miles) north-east of the capital, WELLINGTON. It has several volcanic lakes and hot springs. The national park, which covers 767 km² (296 sq miles), was founded in 1894 after the land around the volcano was presented to the nation by a Maori chief, Te Heuheu.

Map New Zealand – Ec

Tongatapu *Tonga* Main island of the kingdom and the name given to one of the three main island groups, lying in the south; the name means 'sacred Tonga' or 'sacred garden'. Tongatapu is a coral island with a land area of some 256 km² (99 sq miles); it is largely covered by coconut plantations. About two-thirds of Tonga's people live here. NUKU'ALOFA, the national capital, stands on the north coast.

Population 68 000

Map Pacific Ocean – Ed

Tonga Trench See KERMADEC-TONGA TRENCH

Tongking, Gulf of (Gulf of Tonkin) *China/Vietnam* Arm, 480 km (300 miles) long and 240 km (nearly 150 miles) wide, of the South CHINA SEA between north-eastern Vietnam, mainland China and the island of HAINAN. In August 1964 two United States destroyers reported an attack by North Vietnamese torpedo boats in the gulf. President Lyndon B. Johnson responded by

Tonga

THE PACIFIC'S FRIENDLY ISLANDS, IN NAME AND IN NATURE, ARE HEADED BY A SUITABLY FRIENDLY KING

Probably the first time that Tonga was represented on the world stage was when the coronation of Elizabeth II of England was televised in 1953 and one of the most engaging and popular personalities who emerged was Tonga's Queen Salote. She smiled broadly from an open carriage in the pouring rain, underlining the name Friendly Islands that Captain James Cook gave Tonga in the 1770s. Her equally friendly son, Taufa'ahau Tupou IV, succeeded her in 1965. Five years later Tonga, which had been a British protectorate since 1900, became an independent member of the Commonwealth.

Tonga has had hereditary monarchs for a thousand years, but recently a desire for greater democracy has surfaced. The royal palace stands in the capital, NUKU'ALOFA, on the largest island, TONGATAPU.

The Kingdom of Tonga forms a triangle with Fiji and Samoa, close to the international date line. The Pacific group of islands consists of about 170 islands and islets grouped in two chains running north to south. The two differ vastly in character: the western chain is volcanic, reaching peaks of over 1000 m (3280 ft), and emits sporadic streams of lava. Dense rain forests grow along the slopes. The eastern chain is a row of white coral islands overgrown with coconut palms. East of this group, the Tonga Trench plummets down to 10 000 m (32 800 ft) below sea level. Tonga often falls victim to tropical hurricanes.

The government owns all the land. Men over 16 are entitled to rent a town plot and a 3 hectare (8 acre) allotment for growing food, but population growth in the past – it has now begun to decline – has made land scarce.

TONGA AT A GLANCE		
Map Pacific Ocean – Ed		
Official name Kingdom of Tonga		
Area 748 km² (289 sq miles)		
Population 103 950 **Per km²** 139 **(Per sq mile** 360)		
Capital Nuku'alofa		
Government Constitutional monarchy		
Currency 1 pa'anga = 100 seniti		
Languages Tongan (official), English		
Religion Christian (mainly Protestant)		
Climate Tropical and humid; average annual temperature in Nuku'alofa is 26°C (79°F)		
Land use Cultivation 64%, forest 11%		
Main primary products Coconuts, sweet potatoes, cassava, taro, citrus fruits, vanilla, fish, cocoa, coffee, ginger, black pepper		
Major industry Coconut processing		
Main exports Copra, coconut oil, vanilla, watermelons and other fruits, taro		
Annual income per head (US$) 1350		
Population growth (per thous/yr) -5		
Life expectancy (yrs) Male 66 **Female** 70		

ordering air attacks on North Vietnam, and the US Congress's Tongking Gulf Resolution led to an increase in American involvement in the Vietnamese conflict.
Map Vietnam – Ba

Tonkin *Vietnam* Name of the northern MIEN BAC region in French colonial times.
Map (Mien Bac) Vietnam – Aa

Tonle Sap *Cambodia* Freshwater lake at the centre of the Cambodian lowland. Once a branch of the sea, the Tonle Sap ('Great Lake') has been formed by the silting up of the MEKONG River. It acts as a natural regulator of the Mekong's flow. For most of the year, its waters drain via the Tonle Sap River into the Mekong. At the height of the Mekong floods in October, however, the flow is reversed and the lake swells from 2600 km² (1000 sq miles) to 10 400 km² (4000 sq miles). This spreading of fertile silt makes the land by the lake Cambodia's richest rice-growing region.
Map Cambodia – Ab

Tønsberg *Norway* The country's oldest town, founded in AD 871 under King Harald Fairhair. Tønsberg was a whaling port until the curtailment of Norway's commercial whaling, and is still a timber port near the mouth of Oslofjord, 70 km (45 miles) south-west of OSLO. Industries include food processing and clothing manufacture.
Population 32 400
Map Norway – Cd

Toowoomba *Australia* City in Queensland, 110 km (nearly 70 miles) west of BRISBANE on the edge of the Great Dividing Range and the DARLING DOWNS. It is a market centre for the surrounding farming and grazing lands.
Population 75 990
Map Australia – Id

topaz Semi-precious stone that can be colourless, blue, yellow or pink, found in IGNEOUS ROCK and QUARTZ veins. It is composed of hydrous aluminium fluosilicate.

Topeka *USA* State capital of Kansas, on the Kansas River. Lying 90 km (55 miles) west of KANSAS CITY, it was founded in 1854. Today it is a railway centre and headquarters of the Atchison, Topeka and Santa Fe Railway. It has metal processing, printing and agriculture-based industries. The psychiatric hospital and research centre of the Menninger Foundation is based here.
Population (city) 119 900; (metropolitan area) 161 000
Map United States – Gc

tor Isolated mass of weathered rock, usually granite, on or near the summit of a large, rounded hill. Tor is a term local to south-west England, but similar features are present in many parts of the world.

Torbay *United Kingdom* Popular English tourist centre in Tor Bay, on the south coast of the county of Devon, 45 km (28 miles) east of PLYMOUTH. The area is famous for its mild climate, subtropical trees, water sports and fine beaches. The district's main resorts are Torquay, Paignton and Brixham, a fishing port. Torquay and Brixham harbours have marina complexes.
Population 122 600
Map United Kingdom – De

Torino *Italy* See TURIN

tornado (cyclone) Twisting column of fast moving air, usually accompanied by a funnel-shaped downward extension of a cumulonimbus cloud, whirling destructively at speeds of up to 480 km/h (300 mph). A tornado occurs at a cold front when humid, warm air suddenly releases a large amount of water as it cools. The energy liberated by the condensing water drives the tornado, producing a fast spinning column of cloudy air that leaves a trail of destruction as it moves, swirling up quite large objects – even cars and horses have been carried aloft. Over sea, a tornado produces a waterspout. Over sandy soils, it produces a whirlwind of dust. Tornadoes are short-lived, subsiding after an hour or two, and are usually no more than about 100 m (330 ft) in diameter. They commonly occur in the US, in the Great Plains and in the south-east of the country.

Torne (Torneälven; Tornionjoki) *Sweden/Finland* River marking the border between Sweden and Finland at the head of the Gulf of BOTHNIA. It is 510 km (317 miles) long and is noted for its salmon. Its Finnish name is Tornionjoki.
Map Sweden – Db

Toronto *Canada* Canada's largest city and capital of Ontario, standing on the north shore of Lake ONTARIO. It is the main industrial and commercial centre of Canada, and an important port, shipping grain, meat and livestock. Its industries include the manufacture of electrical equipment, farm machinery and aircraft, and the production of iron and steel. Toronto's central business area is close to the lake, and is dominated by many tall buildings, including the tallest freestanding structure in the world, the 553 m (1814 ft) CN tower, with a revolving restaurant at 347 m (1140 ft). The Canadian National Exhibition is held each year at Toronto's Exhibition Park, and the city's Royal Ontario Museum houses a famous collection of Chinese art. The University of Toronto (established in 1827) is Canada's largest university.

▲ **WHIRLING WATERS A waterspout cavorts off Key West, Florida. It formed when the downward extension of cloud in a tornado joined with a cone of spray thrown up by the stormy seas.**

Toronto is on the site of an old French trading post, Fort Rouille, founded in 1749. The British bought the site from native Americans in 1787 and chose Toronto as the capital of Upper Canada. The city has a rich and varied population mix, including people of Italian, Portuguese, Greek, Ukranian, Polish and Chinese descent.
Population (city) 2 275 770; (metropolitan area) 3 893 050
Map Canada – Hd

Torquay *United Kingdom* See TORBAY

Torremolinos *Spain* See MÁLAGA

Torrens, Lake *Australia* Lake covering some 5775 km² (2230 sq miles) west of the Flinders Range in South Australia, situated about 400 km (250 miles) north-west of ADELAIDE. It is normally a saline mud flat, but has been known to overflow after heavy rains.
Map Australia – Fe

Torreón *Mexico* City in the state of Coahuila, some 800 km (500 miles) north-west of MEXICO CITY. It is one of the main textile centres of the Laguna cotton-growing district. It also has breweries, flour mills and chemical factories.
Population 459 810
Map Mexico – Bb

Torres Strait *Australia/Papua New Guinea* The shallow strait which links the ARAFURA and CORAL seas. Its discovery in 1606 by the Spanish navigator Luis Vaez de Torres proved that New Guinea is an island and not part of a southern continent. It is little used by shipping today because of its dangerous shallows and reefs.
Map Australia – Ga; Papua New Guinea – Ba

Torres Strait Islands *Australia* Group of about 70 islands in the Torres Strait, off the far northern tip of QUEENSLAND, of which 30 or so are inhabited. They centre on Thursday Island, which lies 16 km (10 miles) off the mainland and covers 100 km² (38 sq miles).

Most of the population are Torres Strait Islanders, a Melanesian group more closely related to the people of Papua New Guinea, to the north, than to Australian Aborigines. The strait has beds of pearl-shell and *bêche-de-mer*, but world demand for these commodities is low and many of the islanders are now consequently unemployed or live in mainland Queensland.

Population 6300
Map Australia – Ga

Tórshavn *Denmark* Administrative centre and principal port of the FAEROE ISLANDS on the island of Streymoy (Stromo). It is the islands' main educational centre and a Faeroese Academy of Higher Learning was established here in 1965. There is an airport on the island of Vágar, some 30 km (20 miles) from Tórshavn.

Population 14 700
Map (Faeroe Islands) Europe – Cb

Tortola *British Virgin Islands* Main island of the Caribbean colony, covering an area of 54 km² (21 sq miles). It is mountainous, reaching 521 m (1710 ft) at Mount Sage, but fertile farmland grows sugar cane, sweet potatoes, avocados, bananas and other crops. The island's capital is Road Town; its British Virgin Islands Folk Museum has interesting artefacts from the Arawaks who were among the early settlers of the islands. Tourists are attracted to Tortola's white sandy beaches and clear turquoise water.

Population 13 570
Map Caribbean – Cb

Toruń (Thorn) *Poland* Industrial city 177 km (110 miles) north-west of the capital, WARSAW. Toruń, birthplace of Polish astronomer Nicolaus Copernicus (1473-1543), was a Hanseatic city during the Middle Ages and still retains much of its medieval walls. Among them is the Leaning Tower, 35 m (115 ft) high and 1.4 m (4.5 ft) out of true. Its industries include chemicals, artificial fibres, timber and printing.

Population 202 200
Map Poland – Cb

Toscana *Italy* See TUSCANY

Touggourt (Tuggurt) *Algeria* Oasis town in the north-east of the country at the end of the railway, 210 km (130 miles) south-west of ANABA. It lies in the Souf region of dunes and is an important centre of communications for the oil fields to the south. The Touggourt oasis produces dates.

Population 45 500
Map Algeria – Ba

Toulon *France* One of the nation's most important naval bases along with BREST, occupying a sheltered Mediterranean harbour some 50 km (30 miles) south-east of MARSEILLES. Much of the French fleet was scuttled here in 1942 to prevent it from falling into German hands. Shipbuilding and ship repairing are Toulon's major industries.

Population (town) 168 000; (metropolitan) 438 000
Map France – Fe

Toulouse *France* Capital of Haute Garonne Department, on the Garonne River, some 90 km (56 miles) from the Spanish border. Its characteristic red-brick houses have given it the nickname of *la ville rose* ('the pink town'). A centre of the French aviation industry, Toulouse was the birthplace of the French Concorde aircraft and the European airbus.

Population 560 000
Map France – De

Touraine *France* Area in the Loire Valley, centred on the city of TOURS and known as 'the Garden of France'. It is famous for rich crops of wheat and vegetables, and for its wines, produced at such villages as Vouvray, Touraine and Bourgueil.

Map France – Dc

Tourane *Vietnam* See DA NANG

Tourcoing *France* See LILLE-ROUBAIX-TOURCOING

Tournai (Tournay; Doornik) *Belgium* Medieval city 50 km (31 miles) south of GHENT. Its centre retains the circular outline of the medieval town, but the walls have been replaced by boulevards. It produces carpets and leather goods.

Population 70 700
Map Belgium – Aa

Tours *France* Capital of the Indre-et-Loire Department. It lies on a triangle of land between the Loire and Cher rivers, 210 km (130 miles) south-west of PARIS. Despite extensive damage during the Second World War, it possesses many historic buildings such as the fine 13th to 16th-century cathedral and two churches built during the same period. Engineering, electronics, wine-making and tourism are its main industries.

Population (town) 129 500; (metropolitan area) 272 000
Map France – Dc

Townsville *Australia* Port in Queensland, about 1125 km (nearly 700 miles) to the north-west of BRISBANE. It is the administrative, commercial and industrial capital of northern Queensland and home of James Cook University. It exports meat, wool and minerals from the interior, and sugar and timber from the fertile coastal region. Its industries include meat packaging, sawmilling and copper refining. Townsville is also a base for exploring the GREAT BARRIER REEF and nearby Magnetic Island.

Population 101 400
Map Australia – Hb

Toyama *Japan* Industrial city situated near the north coast of HONSHU Island, some 255 km (158 miles) north-west of the capital, Tokyo. Once famous for patent medicines and silk spinning, it is now noted for its expertise in robotics and pharmaceuticals.

Population 321 100
Map Japan – Cc

Trá Lí *Ireland* See TRALEE

Trâblous *Lebanon* See TRIPOLI

Trabzon (Trebizond) *Turkey* Port near the south-east corner of the Black Sea. It is the cap-ital of the province of the same name. The town's 13th-century Aya Sofia Church has some of Turkey's finest surviving Byzantine frescoes. The province, which covers an area of 4685 km² (1809 sq miles), produces fruits, beans, hazelnuts, and tobacco.

Population (town) 173 400; (province) 795 850
Map Turkey – Ba

trade winds Winds which blow from the subtropical high-pressure belts to the equatorial low-pressure belts. They blow from the north-east in the northern hemisphere and from the south-east in the southern hemisphere. See WEATHER AND WIND SYSTEMS

Trafalgar, Cape *Spain* The country's most south-westerly point, some 60 km (37 miles) west of GIBRALTAR. The waters off the cape were the scene of a decisive British naval victory over a French and Spanish fleet in 1805, during the Napoleonic Wars, in the course of which the British admiral, Lord Horatio Nelson, was killed.

Map (Gibraltar) Spain – Cd

Tralee (Trá Lí) *Ireland* County town of Kerry, 91 km (57 miles) north-west of the southern city of CORK. It is a tourist centre, and an industrial centre linked to the sea by canal. The song 'Rose of Tralee', written in the town by William Mulchinock (1820-64), has led to an August festival at which a Kerry girl is chosen as Rose of the Year. Sir Roger Casement (1864-1916), an Irish-born British consular official, was captured at McKenna's Fort, an earthwork on Tralee Bay, in 1916, when he landed here after attempting to gain German help for the Irish nationalists during the First World War. He was executed in London for treason.

Population 17 230
Map Ireland – Bb

tramontana Cold north wind sweeping down from the mountains in Italy and the Mediterranean region.

Tran Ninh Plateau *Laos* Upland plateau in the centre of the country, consisting of limestone and sandstone hills that were once covered with tropical monsoon rain forest. Shifting cultivation by the hill people has left the plateau covered by tropical grasslands, with scattered oaks and pines growing along the stream courses. The plateau has unworked deposits of alluvial gold, antimony, copper, lead, zinc and silver. The inner core of the plateau is known as the Plain of JARS.

Map Laos – Ab

Trans-Amazonian Highway *Brazil* Network of roads extending for 4800 km (nearly 3000 miles) across north-central Brazil. It runs from Cruzeiro do Sul near the Peruvian border to RECIFE on the Atlantic coast. Construction of the highway in 1970 opened up the Amazon River basin to new settlement and allowed the tapping of its vast and largely unexploited mineral wealth – including iron, copper, tin and petroleum – but also threatened its ecological and social balance.

The project, though costly, has not brought the desired results; instead of 100 000 poor families settling here, as envisaged, only 7000 came, and most of these have moved on – leaving extensive environmental damage in their wake because the soil could not sustain cultivation for long.

Transantarctic Mountains *Antarctica* Vast mountain range dividing the Antarctic into its eastern and western regions. The mountains extend more than 3200 km (2000 miles) from Victoria Land to the Weddell Sea, and reach a height of more than 4500 m (14 750 ft) in the Queen Maud Mountains. The range is cut by numerous glaciers, including the BEARDMORE.
Map Antarctica – d

Trans-Canada Highway *Canada* The world's longest national road. It runs for about 8000 km (nearly 5000 miles) from VICTORIA on the Pacific coast to North Sydney in Nova Scotia, and across the island of Newfoundland from Port Aux Basques to St John's on the Atlantic coast. It was completed in 1962.
Map (Victoria) Canada – Cd; (St John's) Canada– Jd

Transkei *South Africa* Former 'homeland', which covers 43 798 km² (16 910 sq miles), created for the Xhosa people in the Eastern Cape. It was declared an independent republic in 1976, but was recognised only by South Africa, on which it was economically dependent. In 1994 it was reintegrated into South Africa. Most of the population are subsistence farmers who produce maize, goats, wool and mohair, sheep and cattle. The northern coast, known as the Wild Coast, has attractive scenery and beaches and has developed a growing tourist trade. The former capital and principal town is UMTATA.
Population 2 539 000
Map South Africa – Cc

Trans-Siberian Railway *Russia* The world's longest railway line, stretching some 9300 km (5775 miles) from MOSCOW to the Pacific coast ports of VLADIVOSTOK and NAKHODKA. The main Siberian section was built between 1891 and 1904, with simultaneous construction east from the Urals city of Chelyabinsk and west from Vladivostok. This section of the line is 7380 km (4584 miles) long. It marked the turning point in Siberia's development, opening up vast areas to exploitation, settlement and industrialisation.

Through trains now travel four to seven times a week via Yekaterinburg, Novosibirsk, Irkutsk and Khabarovsk, where foreign travellers change for the Nakhodka train. The journey takes eight days, with 92 stops. There are also trains between Moscow and Beijing, China via Ulan Bator, Mongolia. A newer 3200 km (2000 mile) extension, the Baikal-Amur-Magistral line, completed in 1984, passes north of Lake Baikal to Komsomol'sk-na-Amure and the port of Sovetskaya Gavan', opening up more of Siberia.
Map (Moscow) Russia – Ec; (Vladivostok) Russia – Od

Transvaal *South Africa* See EASTERN TRANSVAAL; GAUTENG; NORTHERN TRANSVAAL

Transylvania (Ardeal; Siebenbürgen) *Romania* Region in the centre and north of the country, covering 55 166 km² (21 300 sq miles). It consists largely of a triangular plateau some 400-650 m (1310-2130 ft) high. It is enclosed on the north, south and east by the CARPATHIANS, and is cut by the Mureş River.

The region is largely agricultural, but it has a paper and furniture industry and large natural gas deposits, as well as coal and iron ore mining.

Transylvania was once part of the Roman province of Dacia. In the Middle Ages it was conquered first by the Hungarians, then by the Turks. It was incorporated into the Habsburg Empire in 1799 and became part of Hungary in 1867. In 1920 a plebiscite decided in favour of ceding Transylvania to Romania – although the Hungarian part of the population voted for union with Hungary.
Map Romania – Aa

Trápani *Italy* Port near SICILY's western end. Its closeness to Africa, 230 km (about 140 miles) away, made it an important port and naval settlement under Carthaginian, Roman, Arab and Spanish rule. Its products today include tuna fish, macaroni and wine.
Population 69 270
Map Italy – De

Trasimeno *Italy* Lake covering some 128 km² (49 sq miles) 16 km (10 miles) west of PERUGIA. Three small islands lie in the lake, which reaches a maximum depth of only 6 m (20 ft). The main towns, Castiglione and Passignano, are developing as popular bathing resorts. Trasimeno was the scene of the Carthaginian general Hannibal's crushing defeat of the Romans, under consul Gaius Flaminius, in 217 BC.
Map Italy – Dc

Travancore *India* Former state in the south-west which was ruled by Muslim princes. Following Indian independence in 1947, it merged with neighbouring Cochin to form a state which was renamed KERALA in 1956.
Map India – Be

Trebizond *Turkey* See TRABZON

Treblinka *Poland* Nazi concentration and extermination camp near the town of Malkinia, some 100 km (62 miles) north-east of the capital, WARSAW. The camp is preserved as a museum and memorial to the 750 000 people – mostly Jews and Polish partisans – who died here between July 1942 and November 1943.
Map Poland – Eb

tree line (timber line) Altitude above which trees do not grow. It is not a uniform line since it depends on local as well as general conditions of relief, climate and soil, but in general it rises with increasing temperatures, from sea level in the TUNDRA to 4000 m (13 000 ft) in East Africa.

Trelleborg *Sweden* Seaport on the south-west coast, 30 km (18 miles) south of MALMÖ. It was founded in the 12th century and is now the main ferry port link with Germany.
Population 34 100
Map Sweden – Be

tremor In geology, the term used for a minor earthquake with a low level of intensity.

trench See OCEAN TRENCH

Trengganu *Malaysia* See TERENGGANU

Trent *United Kingdom* River of central England, rising in the hills of STAFFORDSHIRE and flowing 270 km (167 miles) through the cities of Stoke-on-Trent and Nottingham into the HUMBER. The river is mainly used to provide cooling water for power stations on its banks.
Map United Kingdom – Ed

Trentino-Alto Adige *Italy* Mountainous region consisting of the Trentino and Alto Adige provinces covering 13 613 km² (5256 sq miles) in the Alps to the north of Verona. It was once Austrian territory and was incorporated into Italy in 1919. Many of the people, particularly in Alto Adige (formerly South Tyrol) in the north, still speak German. The region's highest peak is Ortles, at 3905 m (12 812 ft), near the Swiss border. At the heart of the region are the valleys of the ADIGE and Isarco rivers, which contain the provincial capitals of TRENTO and BOLZANO and the main railway and motorway from Verona to the BRENNER PASS into Austria.

Hydroelectricity is the chief natural resource of the region, which is also Italy's leading producer of timber and apples. In July 1985 a burst dam attached to a fluorite mine at Stava near the town of Tesero caused a flood of water and mud that killed more than 200 people.
Population 877 800
Map Italy – Ca

Trento (Trient) *Italy* Capital of the Trentino-Alto Adige region, about 120 km (75 miles) north-west of VENICE. It stands on the road to the BRENNER PASS into Austria. Indeed, Trento was governed as part of Tyrol from the 19th century until 1919, when it was transferred to Italy. Trento's oldest quarter is between the 12th to 13th-century cathedral and the imposing Castello del Buonconsiglio, which now houses a museum. The economy of the city is based on tourism and production of electrical products, chemicals, pottery and silk.
Population 100 000
Map Italy – Ca

Trenton *USA* State capital of New Jersey, standing at the limit of navigation on the Delaware River, located 40 km (25 miles) north-east of PHILADELPHIA. Trenton was first settled by Quakers in the late 17th century. Today, it has a wide range of industries, including steel cables, ceramics, electrical equipment and rubber.
Population (city) 88 700; (metropolitan area) 325 800
Map United States – Lb

Trèves *Germany* See TRIER

Treviso *Italy* Old fortified town, some 25 km (16 miles) north of VENICE. It has an 11th-century cathedral and a 13th-century palace. Its main industries centre on the production of textiles, ceramics and lamps.
Population 85 800
Map Italy – Db

Triassic First of the three periods in the Mesozoic era of the earth's timescale. See GEOLOGICAL TIMESCALE, p 256

Trichinopoly *India* See TIRUCHCHIRAPPALLI

Trient *Italy* See TRENTO

Trier (Trèves) *Germany* City on the Moselle River, some 10 km (6 miles) from the border with Luxembourg. It is the trading centre for the

Moselle-Saar-Rüwer wine-growing area and beneath its streets are kilometres of cellars, where more than 30 000 million litres (6600 million gallons) of wine are stored.

Trier is one of Germany's oldest towns. Julius Caesar conquered it and by about 16 BC it was a Roman city, Augusta Treverorum, later to become a summer residence of the emperors. Baths, a basilica and Emperor Trajan's amphitheatre, which seats 20 000 people, survive, as well as the Porta Nigra, the northernmost of the fortified gates. Karl Marx (1818-1883) was born in the city.
Population 97 200
Map Germany – Bd

Trieste *Italy* Seaport at the head of the Adriatic Sea, just west of the Slovenian border. It was founded as the Roman city of Tergeste. A theatre and a triumphal arch remain from those days. The cathedral church of San Giusto dates from the Middle Ages; the adjacent castle was built in the 15th to 18th century – the time when Trieste formed part of Austria. In 1919 the city was ceded to Italy. From 1946 to 1954, it was occupied by a joint British/United States military force. Most of the city's port trade is funnelled through to Austria and Eastern Europe. Its chief industries include steel, oil and tourism.
Population 231 000
Map Italy – Db

Triglav *Slovenia* See JULIAN ALPS

Trim (Baile Atha Troim) *Ireland* Town on the River Boyne, 40 km (24 miles) north-west of DUBLIN. It has Ireland's largest Anglo-Norman castle, a 13th-century fortress covering more than 1.2 hectares (3 acres), built on the site of an earlier castle visited by King John. Parliaments were held at the castle in the 15th century.
Population 1780
Map Ireland – Cb

Trincomalee (Trinkomali) *Sri Lanka* Port in the north-east, 230 km (140 miles) from COLOMBO. Its well-preserved fort, built on a peninsula by the Portuguese in the 17th century, guards Trinco-malee Bay. The British admiral Lord Horatio Nelson (1758-1805) described this as the finest harbour in the world. The sheltered, deep water bay covers an area of some 80 km² (31 sq miles). Trincomalee was a British naval base until 1957, when the Sri Lankan navy took it over. North of the town is one of the island's finest beaches, centred on the resort of Nilaveli.
Population 44 900
Map Sri Lanka – Ba

Trinidad and Tobago See p 679

Trinkomali *Sri Lanka* See TRINCOMALEE

Tripoli (Trâblous; Tarabulus el Sham) *Lebanon* Port and the country's second city, 72 km (45 miles) north-east of the capital, BEIRUT. Tripoli was founded by the Phoenicians, proba-bly in the 9th century BC, and was the capital of the Phoenician triple federation of city-states – the Greek word *tripolis* translates as 'with three cities'. The states forming the federation were Sidon, Tyre and Arvad. Despite its ancient pedigree, most of the city's monuments date no further back than the 14th and 15th centuries. Today, Tripoli is the terminus for the oil pipeline

from Iraq; much of the oil is refined locally. The city's chief manufactures are textiles, cement and soap. Its main exports are citrus fruit and tobacco. It also trades in silk; in the city centre, sweet shops abound.

Tripoli was damaged during the civil war, but not as badly as Beirut. Places of interest include a 12th-century Crusader castle as well as the 13th-century Great Mosque, Teinal Mosque and the Mosque of Emir Wartawi.
Population 175 000
Map Lebanon – Aa

Tripoli (Tarabulus) *Libya* Capital and principal city of Libya. It stands on the north-west coast, 200 km (124 miles) from the Tunisian border, and is a major port with an international airport. Industries include fishing, fish and tobacco processing, the manufacture of textiles, and the production of bricks, salt and leather goods.

The city was founded by the Phoenicians and developed by the Romans as one of the three main cities of the province (the other two being LEPTIS MAGNA and SABRAHTA), but there are few ancient remains except the Arch of Marcus Aurelius in the old city. The Spanish castle near the harbour dates from the 16th century and houses a museum which covers prehistory, archaeology, ethnology and natural history.

The old city, which is Moorish in character, has many narrow alleyways and souks (markets) for weavers, coppersmiths, goldsmiths and leatherworkers. There are several mosques containing rich art treasures and some marvellous 18th-century gardens.

The modern city owes much to the architec-ture of the Italian period and has a fine seafront and eucalyptus-lined avenues. It was heavily bombed during the Second World War but has expanded extensively since the 1969 revolution.
Population 989 000
Map Libya – Ba

Tripolitania *Libya* The smallest of Libya's three historical regions but the most productive and populous, containing 75 per cent of the country's population. Tripolitania's Mediterranean coastal plain, 10 km (6 miles) wide, is the country's most

fertile region. During the 2nd century AD the region supplied one-third of ancient Rome's grain imports. Its agriculture today is highly developed, with groundnuts, castor oil, tobacco, and fruit and vegetables as its main crops.
Map Libya – Ba

Tristan da Cunha *South Atlantic* Lonely island, only 104 km² (40 sq miles) in area, lying halfway between South Africa and South America. It is formed almost entirely from a 2060 m (6760 ft) high volcano, which unexpectedly erupted in 1961. The island was evacuated, but its people returned from Britain in 1963. Tristan da Cunha's main crop is potatoes; cattle, sheep and pigs are raised, and fishing is good. The island is British – part of the St Helena Dependencies – and is one of a group of four. The others are uninhabited.
Population 300
Map Atlantic Ocean – Ef

Trivandrum (Thiruvananthapuram) *India* Seaport and capital of KERALA State, standing on the south-west coast, 90 km (55 miles) north-west of the mainland's southernmost point. It has an 18th-century fort, museums and a university. The fine beach of Kovalam is 16 km (10 miles) from the city and there are hill resorts inland.
Population 520 100
Map India Be

Trnava *Slovakia* Market town and industrial centre, some 44 km (27 miles) north-east of BRATISLAVA. Trnava has many Gothic and Baroque churches and a university founded in 1635. It makes washing machines and hardware and processes food. The country's first nuclear power station opened at nearby Jaslovské Bohunice in 1972.
Population 67 600
Map Slovakia – Ab

▼ **ATLANTIC OUTPOST** Cottagers on lonely Tristan da Cunha, whose first settlers arrived from St Helena in the 19th century, speak a dialect based on Victorian English.

Trinidad and Tobago

COLOURFUL AND EXCITING, THESE ISLANDS CONSTITUTE A LAND OF CARNIVAL, CALYPSO AND CRICKET

Land of calypso, carnival and steel bands, tropical Trinidad and its tranquil neighbour TOBAGO lie off the ORINOCO River delta of north-eastern Venezuela. The islands, 31 km (19 miles) apart, have secluded palm-fringed beaches and bays sheltered by coral reefs. They form the third largest Commonwealth country in the West Indies, after the Bahamas and Jamaica.

The islands were sighted in 1498 by Christopher Columbus, who claimed them for Spain. Trinidad was settled by the Spanish from about 1532 and they retained the island until it was captured by the British in 1797. Tobago changed hands several times before it was finally ceded to Britain by France in 1814. The islands were linked to form a single administrative unit in 1889. They gained full independence from Britain in 1962. In 1976

Trinidad and Tobago was declared a republic. Since 1987 Tobago has, however, had full internal self-government, with a 15-member House of Assembly.

The people, largely English speaking and keen cricketers, are a complex racial and religious mixture; about 43 per cent are descended from African slaves, 40 per cent are of Asian origin, and the rest are of mixed or European descent. The Asians, both Hindu and Muslim, came to Trinidad from India as indentured labourers to work on the sugar estates after slavery was abolished in 1834.

The islands are the most southerly in the Lesser Antilles. Trinidad is crossed by three upland ranges: the mountainous Northern Range, a continuation of the Coastal Range of Venezuela, reaches 941 m (3088 ft) at Mount Aripo. Two undulating plains separate the much lower Central and Southern Ranges.

The climate is tropical. Temperatures vary little throughout the year, but a rainy season lasts from June to December. Trinidad has extensive forests, parts of which have been made into nature reserves and national parks.

The biggest attraction for tourists is the annual carnival between February and March, when the whole island throws a two-day party and the colourful troupes parading through the streets sometimes number thousands. The most lavish festivities are in PORT-OF-SPAIN.

Income from oil has provided Trinidad's people with a standard of living unmatched by other Commonwealth countries in the West Indies. It is one of the oldest oil-producing countries in the world. The first well was drilled in 1867, only eight years after the first American oil well, in Pennsylvania. Commercial production from fields in the south-east and south-west began in 1909. In the 1950s an offshore field near the south-west coast was put on stream, and in the 1970s, an oil and gas field south-east of the island began production.

Output is small by international standards, but oil and oil products provide 80 per cent of Trinidad's exports and most of its revenue. The fat 1970s are, however, over. Falling oil prices in the 1980s meant that the per capita income of the islanders almost halved and the government was forced to institute severe austerity measures, which resulted in soaring unemployment and, at times, widespread labour unrest. Rising world oil prices in the early 1990s brought some recovery. The government has meanwhile tried to reduce its dependence on oil exports, giving a higher priority to such industries as tourism.

Near La Brea in the south-west of Trinidad is the amazing PITCH LAKE, the world's main source of natural asphalt. Crude oil oozing through porous rocks from an underground reservoir formed the lake, which is some 87 m (285 ft) deep in the middle.

Agriculture has declined in importance in Trinidad, and food now accounts for about 10 per cent of the country's imports. Sugar is the main crop and is grown mostly on large estates in the west of the island. Cocoa is the second crop, cultivated chiefly by small farmers.

Tobago survives almost entirely on tourism. Visitors are attracted by its relaxed, friendly atmosphere, the hilly scenery and the fine sandy beaches. The capital is SCARBOROUGH, which consists mainly of hotels and guesthouses. Much of Tobago's agricultural land has been abandoned as farmers have sought a better living in tourism or in the oil industry in Trinidad. As a result, the main crops of cocoa and coconuts have declined considerably.

TRINIDAD AND TOBAGO AT A GLANCE

Official name Republic of Trinidad and Tobago

Area 5130 km² (1980 sq miles)

Population 1 313 800 **Per km²** 256 **(Per sq mile** 664)

Capital Port of Spain

Government Parliamentary republic

Currency 1 Trinidad and Tobago dollar = 100 cents

Languages English, Hindi, French, Spanish

Religions Christian (Roman Catholic 32%, Anglican and other 28%), Hindu (24%), Muslim (6%)

Climate Tropical and humid. Average temperature is 25°C (77°F) through the year

Land use Culvitavtion 31%, forest 44%, meadows and pastures 2%, other 23%

Main primary products Oil and natural gas; sugar, cocoa, coffee, rice, citrus fruits, bananas

Major industries Oil and gas production and refining, steel, fertilisers, agriculture, food processing, cement, paints, plastics, tourism

Main exports Crude oil, petroleum products, chemicals, sugar, cocoa, coffee, fruit

Annual income per head (US$) 3940

Population growth (per thous/yr) 11

Life expectancy (yrs) Male 68 **Female** 73

Trogir *Croatia* One of Dalmatia's loveliest towns, lying on a small island 20 km (12 miles) to the west of SPLIT. It is linked to the mainland by a bridge.

At one end of the island stands a Venetian *campanile* (bell tower) and a Romanesque-Gothic cathedral which has an elaborately carved door. Limestone houses, churches, monasteries and palaces lie between them and the Kastel-Kamerlengo, a ruined 15th-century fortress, at the island's other end. Between Trogir and Split are the remains of seven medieval castles, from which the 'Riviera of the Seven Castles' or 'Kastela Riviera' takes its name.
Population 6200
Map Croatia – Cc

Trois Rivières (Three Rivers) *Canada* Industrial town on the north bank of the St Lawrence River in southern QUEBEC. Abundant hydro-electricity has fostered industrial development, which includes the manufacture of paper, iron, textiles and electrical equipment.
Population (city) 49 430; (metropolitan area) 136 300
Map Canada – Hd

Trojmiasto *Poland* See GDAŃSK

Tromsø *Norway* Chief town of northern Norway, and administrative centre of Troms Province. Tromsø is on a small island 340 km (210 miles) north of the Arctic Circle. It is the country's

biggest fishing and fish processing centre, especially for herring, and is the supply port for Arctic expeditions and trading. It has a cathedral and the world's most northerly university, founded in 1968. Its museum has a Lapp research centre. The town also has an institute for the study of the AURORA borealis (northern lights).
Population 47 300
Map Norway – Ea

Trondheim *Norway* Third largest city of Norway, after Oslo and Bergen, and port located in central Norway on the south shore of Trondheimsfjord. The town, founded in AD 977 and first named Nidaros, was the country's first medieval capital. Its 12th-century cathedral is built over the tomb of King Olaf II, which is now a shrine. Olaf, who ruled Norway from 1015 to 1028, died fighting to regain his throne at the Battle of Stiklestad in 1030, and was canonised because of his campaign to convert his subjects to Christianity. He is now recognised as Norway's patron saint. The cathedral was a place of pilgrimage in the Middle Ages, attracting thousands and creating much wealth for the city; until 1906 Norway's kings were crowned here. The city's fortifications date from the 17th century.

Trondheim contains Stiftsgården, the largest wooden building in Norway, which dates from 1774 and covers 1150 m² (12 380 sq ft); it is now used as a royal residence. The city is a commercial and shipping centre, and manufactures wood products, electronic equipment and clothing.

Population 140 700
Map Norway – Cc

Troödos *Cyprus* Range running for 60 km (37 miles) across the south of the island. Its highest point – and that of the island itself – is Mount Olympus at 1951 m (6401 ft). Named after the Greek mountain which is the legendary home of the ancient Greek gods, Olympus is now the site of an early-warning radar scanner. The peak is also known as Mount Troödos.

The range contains copper and asbestos, many Byzantine monasteries and the rare mouflon (wild mountain sheep). Many of the tourist resorts in the hills are winter sports centres with excellent ski slopes.

Map Cyprus – Ab

tropical air An air mass that has originated in low latitudes, either over the ocean (in which case, it is known as 'tropical maritime') or over a continental interior (when it is referred to as 'tropical continental').

tropical climates Low latitude climates characterised by high temperatures throughout the year. There are two types: the 'tropical maritime' and 'tropical continental' climate. Tropical maritime climate has no dry season, but it has a summer maximum of rain associated with the arrival of the INTERTROPICAL CONVERGENCE ZONE. This climate covers east coast regions and islands in the tropics (see TROPICAL RAIN FOREST and EQUATORIAL CLIMATE). Tropical continental climate, on the other hand, has a distinct winter dry season. It is found on the west side of continents and in their interiors, corresponding with the SAVANNAH areas.

tropical cyclone Very low pressure area, 80 to 160 km (50-100 miles) in radius, which originates over the sea in tropical regions and is frequently marked by circulating winds of HURRICANE strength. At the centre of the circulating winds is a calm – the eye of the storm. Wind systems of this type are called 'hurricanes' in the West Indies and North America, 'cyclones' in Australia and India, and 'typhoons' in East Asia. They can cause great damage when they move inland, though they are generally short-lived over land.

tropical rain forest (selva) Dense, evergreen forest of prolific growth, also called 'selva', and dominated by tall, hardwood trees which are often festooned with trailing lianas. The tree crowns form a canopy beneath which are layers of smaller trees, shrubs and other plants. The large trees are also host to plants called epiphytes

which grow on them, but are not parasitic, such as orchids. The forest has no seasonal rhythm, hence its evergreen appearance, as each tree – or even branch – buds, flowers and sheds its leaves independently of the other vegetation. The tropical rain forest is typical of areas of hot, humid TROPICAL CLIMATES, such as the Amazon basin, Zaire basin, and the coastal lowlands of Southeast Asia, Central America and West Africa. See also TEMPERATE RAIN FOREST.

tropics Two lines of latitude, 23.5°N and 23.5°S of the Equator, which mark the limits of the zone within which the noon sun is overhead at some time of the year. The northern limit (Tropic of Cancer) is reached on 21 June and the southern limit (Tropic of Capricorn) on 21 December; the limits are referred to as 'solstices'.

tropopause The boundary between the troposphere and the stratosphere in the earth's atmosphere. It can be seen from high-flying jet aircraft, especially over oceans, appearing as a hazy layer which contains trails of CIRRUS CLOUD. See THE DIFFERENT LAYERS OF THE ATMOSPHERE, p 53

troposphere See THE DIFFERENT LAYERS OF THE ATMOSPHERE, p 53

trough See OCEAN TRENCH

trough of low pressure A narrow area of low pressure between two areas of higher pressure, often an extension of a DEPRESSION.

Troy (Truva; Ilium) *Turkey* Ruined city at the southern end of the Dardanelles Strait, 270 km (165 miles) south-west of ISTANBUL. Excavations in the late 19th century revealed nine levels of occupation dating back to the Bronze Age and extending for more than 4000 years. In 1300 BC the city was destroyed by an earthquake but it was

rebuilt and flourished again under the Romans and until the 13th century AD, when it fell to the Osmans and was finally destroyed. The largest remains are the House of Pillars and the city's fortifications and towers.

The story of the legendary Trojan War – which may have been based on a real war fought in the 13th century BC – is told in the *Iliad,* the epic poem by the Greek poet Homer.

Map Turkey – Ab

Troyes *France* City on the Seine River, 158 km (98 miles) south-east of PARIS. It gave its name to the troy system of measuring weights, probably first used at medieval trade fairs here and still used for weighing gems and precious metals. Textiles, knitwear and electrical goods are now the city's main industries.

Population 123 000
Map France – Fb

Trujillo *Peru* Industrial city and capital of La Libertad Department, about 500 km (310 miles) north-west of LIMA, close to the Pacific coast. It was founded in 1534 by Spanish colonists and refounded the next year by the Spanish adventurer Francisco Pizarro, who named it after his birthplace. It is a bustling modern city but retains much of its colonial heritage, including the large cathedral (twice rebuilt after earthquakes in the 17th and 18th centuries), the Monastery of San Carmen and the churches of La Compañía and San Francisco. The ruins of Chan-Chan, the former capital of the Chimú people, lie nearby.

Population 532 000
Map Peru – Ba

▼ **STEAMY HEAT** Swathes of mist rise to the tree-top canopy in Surinam's tropical rain forest. It covers three-quarters of the country.

Truk (Chuuk) *Federated States of Micronesia* Most populous state of the island federation, and home to more than half of the nation's total population. It consists of a number of atolls with nearly 300 small islands in all; the administrative centre is Moen Island in Truk Atoll. Truk, now also known as Chuuk, was a Japanese fortress and major naval base in the Second World War and was heavily bombed in 1944. In its lagoon are many sunken Japanese ships.
Population 38 000
Map Pacific Ocean – Cb

Truong Son (Annamite Chain) *Vietnam* See MIEN TRUNG

Trust Territory of the Pacific Islands *Pacific Ocean* Former United Nations trusteeship established in 1947, administered by the United States. The trusteeship, for all except PALAU, was terminated by the United Nations in December 1990, but the independent states have kept close economic and political ties with the United States. Palau became independent in 1994.
The trusteeship applied to four territories. Two of them – the MARSHALL ISLANDS and the Federated States of MICRONESIA – had in 1986 signed Compacts of Free Association, allowing the Americans to retain strategic defence rights; in September 1991 they obtained full membership of the United Nations. The path of the third territory, Palau, to independence was slowed by internal unrest and political instability, but Palau eventually obtained its independence in October 1994 when the United States ended its role as the island's administering power. The fourth, the NORTHERN MARIANA ISLANDS, created after the Second World War defeat of Japan, which from 1920 had administered them under a League of Nations mandate, decided to keep closer American links by becoming a Commonwealth of the United States in 1986, thus granting the Marianas population US citizenship.
Map Pacific Ocean – Cb

Tsaratanana Massif *Madagascar* Highlands in the north which include the island's highest peak, Maromokotro, at 2876 m (9436 ft).
Map Madagascar – Aa

Tsaritsyn *Russia* See VOLGOGRAD

Tsavo National Park *Kenya* Wildlife reserve covering 21 000 km² (8108 sq miles) of thorny scrubland on either side of the main road and railway between MOMBASA and NAIROBI. It was the setting for *The Man-eaters of Tsavo*, by Colonel John Patterson (1867-1947). Luxurious tourist lodges near waterholes such as Mzima Springs, some 40 km (25 miles) south-west off the road, offer close-up views of the reserve's game.
Map Kenya – Cb

Tselinograd *Kazakhstan* See AKMOLA

Tshikapa *Zaire* Town on the Kasai River, about 580 km (360 miles) south-east of the capital, Kinshasa. It is the centre of Zaire's gem diamond industry. Diamonds were first discovered in Tshikapa in 1907.
Population 100 000
Map Zaire – Bb

Tsien Tang Kiang *China* See FUCHUN JIANG

Tsuen Wan *Hong Kong* Most populous of the eight new towns established by Hong Kong in the NEW TERRITORIES to the north-west of KOWLOON. Founded in 1963, it is more self-sufficient than the other new towns and contains many factories which provide local employment opportunities. An industrial city of skyscraper apartment blocks and broad looping highways, it is connected to Kowloon and VICTORIA by the underground Mass Transit Railway. Tsuen Wan's container terminal, at Kwai Chung, is one of the largest in the world.
Population 700 000
Map Hong Kong – Ca

Tsukuba *Japan* A 'science city', some 60 km (37 miles) north-east of TOKYO, begun in the late 1960s and used as the site of the 1985 Expo industrial exhibition. It has been designed to house research and educational institutions which were moved from Tokyo; it has shopping centres and recreational facilities for the scientists and students. About one-third of Japan's national research institutes are situated here.
Population 150 100
Map Japan – Dc

tsunami (tidal wave) Very large ocean wave caused by an underwater EARTHQUAKE or volcanic eruption. TSUNAMI (Japanese for 'overflowing wave') increases in size as it approaches the coast and surges inland over coastal areas, causing much destruction. The largest recorded tsunami was one 85 m (278 ft) high off Ishigaki Island in the Ryukyu group of islands of south Japan in 1971. A tsunami some 37 m (120 ft) high which hit the coasts of Java and Sumatra after the Krakatoa eruption of 1883 drowned 36 000 people. It is also called a 'tidal wave'.

Tsuruoka *Japan* Old town near the west coast of northern HONSHU Island, 340 km (210 miles) north of Tokyo. The Chidokan, once a school for the children of samurai warriors, is now a national historic property.
Population 100 200
Map Japan – Cc

Tsushima *Japan* Two islands, Shimoshima and Kamishima, in Tsushima Strait, between KYUSHU Island and South Korea. The two, which cover an area of 703 km² (271 sq miles), are separated by a narrow channel of water. The Russian fleet was annihilated by the Japanese under Admiral Heihachiro Togo near the islands in 1905, a battle which established Japan as a world power.
Population 48 900
Map Japan – Ad

Tuamotu *French Polynesia* Archipelago of 76 islands, mostly low-lying coral atolls, stretching over more than 1500 km (932 miles) of the Pacific to the east and north-east of TAHITI. The islands produce pearls and pearl shell (especially at Hikueru), coconuts, breadfruit and pandanus (screw pines, whose leaves are used for thatch and baskets). Mururoa Atoll is used for French nuclear tests; Rangiroa is the largest atoll. *Kon-Tiki* – a raft built by the Norwegian explorer Thor Heyerdahl to support his theory that the Polynesian peoples could have migrated from South America – landed on Raroia Atoll at the end of its voyage from Peru in 1947.
Population 12 370
Map Pacific Ocean – Fc

Tübingen *Germany* Medieval university town on the Neckar River, 35 km (22 miles) south of STUTTGART. Among the students of the university, founded in 1477, have been the astronomer Johannes Kepler (1571-1630) and the philosopher Georg Hegel (1770-1831). This quaint old town attracts thousands of tourists each year. It has clothing and textile industries.
Population 75 000
Map Germany – Cd

Tubruq *Libya* See TOBRUK

Tubuai Islands (Austral Islands) *French Polynesia* A 1300 km (800 mile) chain of volcanic islands, atolls and reefs 600 km (375 miles) south of TAHITI. Rurutu and Tubuai are the main islands.
Population 6500
Map Pacific Ocean – Fd

Tucson *USA* City in southern Arizona, about 100 km (60 miles) north of the Mexican border. Lying in a desert valley surrounded by mountains, it was founded as a military outpost by the Spanish in 1776. The city (whose name is pronounced 'Too-zahn') attracts visitors and retired people with its dry, sunny climate. It is the seat of a state university. It has several high-technology industries and is a service centre for the surrounding copper-mining and agricultural area.
Population (city) 405 400; (metropolitan area) 666 900
Map United States – Dd

Tudmur (Palmyra; Tadmur) *Syria* Ruined oasis city situated just north of the Syrian Desert and 215 km (134 miles) north-east of DAMASCUS. It has been greatly restored and has a modern settlement outside its walls. The city is said to have been built by Solomon in the 1st century BC beside the oasis of Tudmur where date palms were an important crop. It became the main stopping-place along the caravan route from The GULF to the Mediterranean. The Romans colonised Tudmur in AD 217 and renamed it Palmyra, 'the City of the Palms'. As a desert fortress, it became a powerful buffer between Roman-occupied Syria and Persia (today, Iran). Later in the 3rd century Queen Zenobia of Palmyra led an armed revolt against Rome, and in AD 273 the city was pillaged by the Emperor Aurelian. In 1089 it was destroyed by an earthquake. It never fully recovered, although new fortifications were built and the city was occupied until the 17th century. Known today as 'the City of 1000 Columns', it is spread over some 6 km² (2.3 sq miles). Its most impressive remains are the Great Colonnade, which stretches for 1000 m (3280 ft) complete with buildings and monuments and the Great Temple of Bel (Baal).
Map Syria – Bb

tuff Rock formed from ash ejected from a volcano and fused together on the ground. The particles are mostly less than 2 mm (0.08 in) across.

Tugela *South Africa* River rising in the Drakensberg range near the Lesotho border, and flowing some 500 km (310 miles) east to reach the Indian Ocean about 75 km (45 miles) north-east of DURBAN. Formerly the traditional frontier between white-settled Natal and the kingdom of Zululand, the Tugela flows for much of its course through narrow gorges and over steep drops.

681

Tugela Falls, a series of five falls with a total drop of 948 m (3110 ft), include a single, uninterrupted fall of 411 m (1350 ft) – the highest in Africa.
Map South Africa – Db

Tula (Tollán) *Mexico* Archaeological site in the central state of Hidalgo. It is believed to be Tollán, the ancient capital of the Toltec civilisation which flourished in the 10th to 13th centuries AD. The site is dominated by the Quetzalcóatl with the Atlantes – giant carved columns of black basalt standing 5 m (15 ft) high on the central pyramid. The columns, representing Toltec warriors, once supported the roof of the priests' temple. Tula also contains a well-preserved frieze of jaguars, eagles and coyotes, the feather serpent, and a reclining statue of the god Chac-Mool.
Map Mexico – Cb

Tula *Russia* City situated 190 km (120 miles) south of MOSCOW. Founded in 1146, and the site of Tsar Peter the Great's arms factory in the 18th century, Tula is now an industrial town which produces weapons, sewing machines, musical instruments, machinery and samovars (Russian tea urns). Chemicals are produced from local deposits of coal, and iron and steel from local ore. Places of interest are the Museum of the History of Arms (built 1724), and opposite this, a kremlin (fortress) built in 1530.
Population 544 000
Map Russia – Ec

Tulcea *Romania* Main inland port of the Danube delta, 70 km (43 miles) west of the Black Sea. Tulcea is built on tree-clad slopes overlooking a graceful bend of the river. It is a halfway mark for sea-going cargo ships sailing to and from the upriver port of Brăila. Tulcea was noted as a port in the time of the Persian king Darius the Great (about 549-486 BC). Today it is also a manufacturing centre, producing food, timber and building materials. Its Museum of the Danube Delta houses many examples of the plants and wildlife found in the area's nature sanctuary.
Population 70 200
Map Romania – Ba

Tulsa *USA* City in north-east Oklahoma on the Arkansas River, about 160 km (100 miles) north-east of OKLAHOMA CITY. Standing in the heart of Midcontinent Oil Field, it is often called 'the Oil Capital of the World'. Besides being a major distribution and administrative centre, it is also important for aircraft production and maintenance, and engineering.
Population (city) 367 300; (metropolitan area) 709 000
Map United States – Gc

Tumbes *Peru* City, department and major oil-producing region on the border with Ecuador. The Spanish adventurer Francisco Pizarro landed on the coast in 1532 near the present city, which is surrounded by a fertile oasis growing tropical and subtropical crops such as rice, bananas and maize.
Population (city) 64 800 (department) 144 200
Map Peru – Aa

tundra Area between the perpetual snow and ice of the Arctic regions and the TREE LINE of northern latitudes. Tundra has a frozen subsoil and only low-growing vegetation, such as lichens, mosses, dwarf shrubs and stunted birch trees.

▲ RUINED METROPOLIS The capital of the Toltecs near Tula, Mexico, covered more than 8 km² (3 sq miles) and was home to more than 20 000 people.

Tungabhadra *India* River, measuring 640 km (400 miles), in southern India. It is formed from the Tunga and Varada Rivers which rise in the Western GHATS, and flows north-east to the Krishna River. The Tungabhadra Dam, 2441 m (8008 ft) long and 49 m (162 ft) high, is among the world's biggest masonry dams.
Map India – Bd

Tunguska *Russia* Two rivers, both tributaries of the Yenisey, in central Siberia. The Stony Tunguska (Podkamennaya) is 1550 km (960 miles) long; the Lower Tunguska (Nizhnyaya) is some 2989 km (1857 miles) long. The Tunguska plateau has a crater, 1.5 km (1 mile) wide and 190 m (625 ft) deep, created by a large meteorite that crashed to earth in June 1908.
Map Russia – Kb

Tunis *Tunisia* Capital of the country and the political, administrative, commercial and industrial centre. It is built on the west bank of the Lac de Tunis lagoon. An artificial harbour is linked by a deep-water channel to the Bay of Tunis where the new port at La Goulette serves the city. The ruins of CARTHAGE lie just north of La Goulette.

Founded in about 1000 BC, Tunis existed as a small town during the Carthaginian Empire. It remained unimportant until the 7th-century Muslim conquest and destruction of Carthage; in the 9th century, it became the administrative capital of Tunisia (replacing KAIROUAN). It consists of two adjacent districts, widely different in character – the Muslim old town (the medina) and the modern city, which encircles the medina.

The main entrance to the old, walled medina is through the Bab el-Bahar Gate. Inside is an impressive collection of medieval Islamic buildings – there are more than 700 ancient monuments including mosques, mausoleums, palaces and *madrassas* (Islamic places of learning) which have been turned into apartment houses. The medina is a place of winding alleys with *souks* (markets) selling their wares in vaulted and ceramic-tiled *zawias* (courtyards).

Various streets specialise in particular crafts – textiles, brass and copperwork, gold and silver jewellery, leatherware, perfumes and spices.

Other places of interest include the 9th-century Great Mosque of Zitouna and the Lapidary Museum of Sidi Bou Krissan, containing a collection of Arab tombstone inscriptions, as well as the 18th-century Palace of Dar Hussein, which houses the National Institute of Archaeology and Arts and the Museum of Islamic Art. About 5 km (3 miles) north of the city, in Le Bardo suburb, is the Bardo Museum, with one of the finest collections of Roman, Carthaginian and Islamic art in the world.
Population (Greater Tunis) 1 935 000
Map Tunisia – Ba

Tunisia See p 683

Tunja *Colombia* Capital of Boyacá department, lying at 2820 m (9250 ft) above sea level, in the eastern Andean Mountains 135 km (85 miles) north-east of BOGOTÁ. It was founded by the Spanish in 1539, on the site of a Chibcha capital. The decisive battle in the struggle for Colombia's independence was fought in 1819 by revolutionary troops under Simón Bolívar at Boyacá Bridge, 16 km (10 miles) south of the city.
Population 180 000
Map Colombia – Bb

tunnels See UNDER WATER AND UNDER GROUND BY TRAIN, p 684

Turin (Torino) *Italy* Second city of northern Italy, after MILAN and capital of the kingdom of Sardinia-Piedmont between 1861 and 1865 – capital now of the region and of Turin Province. It stands on the Po River, 125 km (78 miles) south-west of Milan. The 20th-century growth of Turin has been fuelled by the steady growth of the Fiat car empire, founded here by Giovanni Agnelli in 1899.

The heart of Turin, with its Roman grid street layout, is well ordered and picturesque, with fine squares and tree-shaded boulevards. Apart from a Roman gateway, some relics of the medieval citadel and its Romanesque cathedral, Turin's important buildings date almost entirely from the 17th and 18th centuries, *(continued on p 684)*

Tunisia

ONE OF THE MOST EUROPEANISED PARTS OF THE ARAB WORLD, TUNISIA IS A PRO-WESTERN AND PROGRESSIVE ISLAMIC SOCIETY

Phoenician traders first sailed to Tunisia's jasmine-scented coast in 1200 BC from Syria and Lebanon. In 814 the Phoenicians founded CARTHAGE, the seat of one of the greatest and richest ancient empires until the Romans razed it in 146 BC. Vandals, Byzantines, Arabs, Turks and the French have also invaded over the centuries. Now more than 3 million tourists invade every year, drawn by the beautiful beaches, the sunshine and the relics of great civilisations.

Tourism is one of the mainstays of the modern economy. Oil, produced in the SAHARA near the Algerian border, and phosphates are also great revenue earners and these three modern sources of wealth have underpinned the traditional agricultural economy. But since the fluctuations in international markets, depressing the oil and phosphate prices, agriculture has again picked up and now employs 40 per cent of the country's workforce.

Tunisia was dominated by the French from 1881 until independence on 20 March 1956. A year later the country overthrew its Bey (King) Sidi Lamine, to become a republic under President Habib Bourguiba, who had led the nationalist movement since the 1930s. He was deposed in a bloodless coup in 1987.

Northern Tunisia consists of plains, hills and valleys, and is divided south-west to north-east by the ranges which extend from the eastern end of the ATLAS MOUNTAINS. The hills are forested with evergreen oaks, including the cork tree whose outer bark is stripped for cork. The Cape Bon Peninsula east of Tunis is the country's market garden and is fragrant with orange and lemon groves, mimosas, magnolias, jasmine and roses. Farther south along the east coast are the orchards and olive groves around the resort of SOUSSE, and vast olive groves around the port of SFAX; Tunisia is the world's fifth biggest olive oil producer. Off the coast are small islands, including the popular resort island of JERBA in the hot south.

Inland, the *dorsale* (backbone) mountains separate the *tell* coastal zone from the central plains before the land drops to the *chott* country. *Chotts* are white salt pans, sometimes lower than sea level, which turn to quagmires after rain but are cracked and dry in summer, and glisten with a strange mirage effect. Around the *chotts* are fruitful oases – with luxury hotels. Tourist excursions run from the hotels into the Sahara in the southern half of Tunisia.

Most of the people of Tunisia are Arabic-speaking Muslims. Arabic is the country's official language but French is widely spoken, and two of the four daily Tunisian newspapers are in French. More than half the population live in towns, which are mainly along the east coast from BIZERTE to Sfax, and especially in the capital, TUNIS.

There are still some Europeans in the cities, particularly the descendants of French and Italian settlers. But most settlers left at the time of independence, which followed guerrilla warfare in the 1950s. Some Greeks, Spaniards and Sicilians still make a home here. Tunisia also has a Jewish community, accounting for 1 per cent of its population, around Tunis and Jerba, as well as a few Berbers who live by farming in the hills of the centre and south. Black descendants of slaves inhabit the oases of the desert.

Tunisia's varied resources have given its people a relatively good standard of living. There is still a huge gulf between rich and poor, but some of the government's projects have been to the general good. Health facilities have been improved in recent years and Tunisia now boasts one of the best educational systems in Africa, which is free all the way from primary level through to university.

INDUSTRIES NEW AND OLD

Industry is growing, making use of iron ore, lead and zinc deposits and substantial offshore reserves of natural gas, as well as oil and phosphates. But it is the ancient crafts that give Tunisia much of its colourful reputation. Women sitting on low cushioned seats weave beautiful carpets on hand looms, or embroider cloth with traditional patterns of flowers and leaves in brilliant colours. Ceramic tiles are patterned with elaborate interwoven floral shapes, in the familiar Islamic style which decorates the interiors of mosques.

It is in the medinas that most of the mosques are found. Although Muslim, Tunisia is not a fanatical country – its main fundamentalist party, Ennahda or Nahda ('The Awakening'), is illegal. Since independence Tunisia has liberated its women from the veil. However, the 9th-century Grand Mosque at KAIROUAN is one of the holiest places in the world for Muslims; to visit Kairouan seven times is an act of piety equal to a visit to MECCA. Tunisia is rich in Islamic architecture, with fine domed and minareted mosques in Tunis, Sousse and Sfax as well as Kairouan.

Tunisia also possesses exceptional Roman remains. Its countryside is dotted with forums, temples, baths, amphitheatres and colosseums – for example, at DOUGGA, EL JEM, UTICA and Maktar. The legends of Carthage draw the tourists here. Such was the hatred of the Romans for the civilisation they destroyed at Carthage, that they levelled the city, burnt the ground for seven days, and salted the earth to make it infertile. The ruins of the city the Romans built in its place are what remains.

MODERN UNCERTAINTIES

After many years of economic depression, Tunisia is once again experiencing economic growth. The economy had a growth rate of 8 per cent in 1992, at a time when the world economy was still going through a recession. Unemployment, as high as 20 per cent in the early 1990s, is down to less than 15 per cent and steadily decreasing. The government has been loosening its control on the economy and introducing a free market system. There is a slow but steady move towards political and economic liberalisation and the government wants to encourage further investment by proving itself the most stable country in the region. The European Union has become its main trading partner.

Presidential and legislative elections in 1994 brought victory for President Ben Ali and the ruling *Rassamblement Constitutionnel Democratique* (RCD), the political successor to Bourguiba's Socialist Constitution Party. Although six legal opposition parties took part in the elections not a single opposition candidate was elected.

TUNISIA AT A GLANCE	
Official name Republic of Tunisia	
Area 163 610 km² (63 170 sq miles)	
Population 8 900 000 **Per km²** 54	
(Per sq mile 141)	
Capital Tunis	
Government Republic	
Currency 1 Tunisian dinar = 1000 millimes	
Languages Arabic (official), French	
Religions Muslim (98%), Jewish and Christian minorities	
Climate Temperate on the coast, very dry inland. Average temperature in Tunis ranges from 10.2°C (50.4°F) in January to 26.2°C (79.2°F) in August	
Land use Cultivation 30%, grazing 19%, forest and woodlands 4%, other (mainly desert) 47%	
Main primary products Wheat, barley, olives, tomatoes, dates, timber, citrus fruits, livestock, fish, wine; crude oil and natural gas, phosphates, salt, iron, lead	
Major industries Mining, tourism, textiles, agriculture, oil refining, cement, steel, food and phosphate processing, beverages	
Main exports Crude oil and refined products, phosphates, textiles and clothing, leather goods, olive oil, wine, grains, fruit, fertilisers	
Annual income per head (US$) 1740	
Population growth (per thous/yr) 20	
Life expectancy (yrs) Male 70 **Female** 74	

UNDER WATER AND UNDER GROUND BY TRAIN

The world's first successful underwater transport tunnel – the Thames Tunnel – was built in London between 1825 and 1843 and runs for 1253 m (4111 ft) between Rotherhithe and Wapping. It is still used as a railway tunnel. By contrast, the world's longest underwater railway tunnel runs for 54 km (33.5 miles) and links the Japanese island of Hokkaido with the island of Honshu. It was opened in March 1988 after more than 24 years' work, and is used by the Shinkansen, or Bullet Train. The Channel Tunnel connecting France with Britain opened in 1994.

THE WORLD'S LONGEST TRANSPORT TUNNELS

Tunnel	Location	Length (km/miles)	Opened
Seikan (underwater rail)	Japan	54/33.5	1988
Channel Tunnel (underwater rail)	United Kingdom-France	50/31	1994
Moscow Metro	Russia	31/19 (Medvedkovo-Belyaevo)	1979
London Underground	United Kingdom	28/17 (East Finchley-Morden)	1939
Daishimizu (rail)	Japan	22/13.6	1982
Simplon II (rail)	Switzerland-Italy	20/12.4	1922
Shin-Kanmon (underwater rail)	Japan	19/11.8	1975
Great Appenine (rail)	Italy	18.5/11.5	1934
St Gotthard (road)	Switzerland	16/10	1980
Rokko (rail)	Japan	16/10	1972

when it was a ducal capital. Its link with the former Duchy of Savoy gives it an architectural tone more French than Italian.

Turin's most sacred possession, kept in the Santa Sindone Chapel of the cathedral, is a piece of cloth measuring 4.3 by 1.1 m (13.5 by 3.5 ft) that bears the faint image in negative of a man. Many Christians believe that this is the shroud in which Christ's body was wrapped after his Crucifixion. The shroud has been the subject of many investigations. In 1988, carbon-dating fixed its origin to between 1260 and 1390; the shroud was last put on display in 1978.

Modern Turin is one of Italy's leading centres of publishing, art, education and fashion. Its major products, apart from motor vehicles, include chemicals, textiles, metal goods, rubber, plastic, pharmaceuticals and television and radio sets, as well as vermouth and chocolate.
Population (city) 991 900; (province) 2 230 170
Map Italy – Ab

Turkana, Lake *Kenya/Ethiopia* Lake extending almost 300 km (190 miles) from north to south. Its maximum width is less than 55 km (35 miles) since it is hemmed in on the east and west by the walls of the GREAT RIFT VALLEY. The lake, most of which lies in Kenya, was formerly known as Lake Rudolf, and is often called 'the Jade Sea' because of its colour. Water flows into the lake from the Omo River in the north and the Turkwel and Kerio rivers in the south, but there is no outlet. The lake's level is maintained by evaporation in the fierce heat – a process that leaves the water salty. Fish caught commercially in the lake are sent some 450 km (280 miles) south to Nairobi, the Kenyan capital. On the lake's barren shores, tools and bones of hominids 2.5 million years old have been found. Turkana is the name of a large district west of the lake.
Map Kenya – Ca

Turkey See p 685

Turkmenistan See p 687

Turks and Caicos Islands *West Indies* British colony consisting of eight large islands and about 40 small islands south-east of the BAHAMAS. They cover a total area of 500 km² (193 sq miles). The islands were sighted by Christopher Columbus in 1492. For 300 years their main industry was the production of salt by evaporating sea water – an occupation that began after salt-rakers from BERMUDA settled here in 1678.

Fishing – particularly for lobster and conch – became an important export industry after demand for salt declined in the early 1960s, but since then tourism has taken over as the main employer. Indeed, direct flights have now been introduced from Miami to the international airport on the main tourist island of Providenciales. In addition, the islands' status as a tax haven has encouraged many offshore financial companies to open offices there.

Only six of the islands in the group are inhabited – among them Grand Turk, the site of the capital, COCKBURN TOWN. Most of the islanders today are descendants of African slaves brought from the United States in the 18th century and freed in the mid-19th century.
Population 11 700
Map Caribbean – Aa

Turku (Åbo) *Finland* The country's fourth largest city – after Helsinki, Espoo and Tampere – and a principal seaport, on the Gulf of Bothnia, 155 km (95 miles) north-west of HELSINKI. Turku is the administrative centre of the province of Turku and Pori. Founded in 1229, it was the capital of Finland until 1812. It was also the site of Finland's first university – founded in 1640 – which was transferred to Helsinki after most of Turku was destroyed by fire in 1827. Today there are two universities. There is a restored 13th-century castle, the oldest in Finland, and a late medieval cathedral built in the German Gothic style. The city celebrates its own history on Turku Day, held in mid-September each year. Turku's industries include shipyards and engineering.
Population 159 900
Map Finland – Bc

Turpan Hami *China* Basin of western China, covering 50 000 km² (19 305 sq miles), situated at the eastern end of the TIEN SHAN range. The centre of the basin lies 154 m (500 ft) below sea level, the lowest part of the country. The basin is renowned for its summer heat, when temperatures can rise above 47°C (117°F). It receives hardly any rain, but fruit, wheat and cotton are grown with the aid of irrigation.
Map China – Cc

Tuscany (Toscana) *Italy* Central region of 22 992 km² (8877 sq miles), whose capital is FLORENCE. It includes the Tuscan Islands, of which ELBA is the largest. Tuscany is the heartland of the nation's medieval history, with such historic places as LUCCA, PISA and SIENA; the Tuscan dialect is the root of modern Italian.

The Apennine Mountains form the eastern border of Tuscany. At its heart is the ARNO valley, where most of its main towns lie. The bare hills of Siena and the Maremma lie to the south. The terraced hillsides produce olives and vines for Chianti wine in this mainly farming area. Wheat and sugar beet are grown. The region has substantial deposits of iron, tin, mercury, copper and lead – and for this reason attracted the Etruscans, who made Tuscany their centre. Today there are some important producers of steel at Piombino, clothing in AREZZO Province, motor scooters at Pontedera, marble at CARRARA and wool at PRATO.
Population 3 562 500
Map Italy – Cc

Tutuila *American Samoa* Main island of the territory, covering 135 km² (52 sq miles). It is mountainous, rising to 653 m (2142 ft) at Mount Matafao. The beautiful natural harbour of PAGO PAGO almost divides the island in two.
Population 30 000
Map Pacific Ocean – Ec

Tuva Republic *Russia* Mineral-rich autonomous republic of the Russian Federation, covering 170 500 km² (65 800 sq miles), on the border with north-west Mongolia. Sixty-five per cent of the population are former nomadic herders who today live mostly in farming villages. The republic's capital is Kyzyl (population 153 000), which has leather, timber and food processing industries. Tuva's rich mineral deposits are unexploited because of the region's remoteness.
Population 309 000
Map Russia – Kc

Turkey

IN HALF A CENTURY, THE FORMER CENTRE OF THE OTTOMAN EMPIRE HAS BEEN TRANSFORMED INTO A MODERN WESTERN MUSLIM REPUBLIC

Turkey forms the bridge between Europe and Asia, with land on both continents. It guards the only sea passage between the Black Sea and the Mediterranean, which flows through the BOSPORUS, the Sea of MARMARA and the DARDANELLES, scene of First World War slaughter. For 1000 years Turkey was the hub of the Byzantine Empire; for nearly 500 years it was at the centre of the Ottoman Empire; recently it formed the south-eastern flank of the NATO alliance and aspires to be a member of the European Union (EU).

The Turks, who once controlled a quarter of Europe, have a reputation as a race of warriors, and violence lies near the surface in domestic politics. Three times since 1960 the military have taken over the government. The last time, in 1980, 20-30 people were being killed daily by extremists on the left and the right before the army moved in. In 1983, when three political parties were allowed to contest an election, the Motherland Party came to power with plans to reduce long-standing state controls.

THE FACE OF THE LAND

Modern Turkey covers an area of about 780 000 km² (301 000 sq miles) and has a population of 61 million. It is the most popu-lous country in the Middle East, consisting of THRACE on the European mainland, and the much larger area of ANATOLIA in Asia. European Turkey, at the tip of the Balkan Peninsula, is fertile agricultural land with a Mediterranean climate. Here, vast areas are covered with sunflowers, tobacco and wheat. Anatolia, which is shaped like a saucer, has a mountainous rim and a central plateau with an average height of 750 m (2460 ft) high. The Pontus Mountains in the north and the TAURUS MOUNTAINS in the south converge eastwards, rising to 5165 m (16 945 ft) at Mount ARARAT. This mountain is the legendary resting place of Noah's Ark, but expeditions to the site have found no concrete evidence of the Bible story.

The weather ranges from Mediterranean warmth on the coasts to the hot summers and bitterly cold winters of the central plains and the snows of the high mountains. There are occasional destructive earthquakes.

About a quarter of Turkey is covered by forests. These are mainly coniferous, but also include deciduous beech, poplar, walnut and oak. The north is the most densely wooded area and there are subtropical forests by the BLACK SEA. In the south and west, large areas are covered by thick Mediterranean scrub.

Animal life includes wild boars, bears, lynx and leopards. Wolves can be a menace in the mountains, and cause havoc among flocks of sheep when the snows drive them to lower ground. Each village has 'wolf dogs' to protect its sheep; they are smaller than sheepdogs but more powerful, and trained to attack wolves. Walkers can also be at risk in winter – hungry wolves may attack a lone traveller who is unwise enough to venture between remote villages. Migrant birds abound in the Bosporus in autumn, heading south in winter.

RICH MIXTURE OF PEOPLE

The first Turks came from the central steppes of Asia, but over the centuries the population has become a rich blend of peoples and cultures. They are of Mediterranean stock in the west, and Armenians from the CASPIAN area in the east; while in the south-east there are about 15 million Kurds, ethnically close to the Kurds in Iran and Iraq. A secessionist group among the Kurds has waged a war against government troops since 1984 and claimed responsibility for several terrorist bomb attacks. Ninety per cent of the people speak Turkish, a language of many dialects which was changed from Arabic script to the Roman alphabet in 1928. Most are Muslims.

The country has a substantial industrial sector, good agricultural land and mineral resources, including large deposits of chrome, coal, iron ore and copper. There are small oil reserves on the banks of the TIGRIS which meet some of the country's fuel needs. Industry is mainly around ISTANBUL and ANKARA, and in the south-east near ADANA and ISKENDERUN. Manufactures include iron, steel, textiles, motor vehicles, petrochemicals, processed food, tobacco, meerschaum pipes, pottery and Turkey's famous carpets. The two great rivers of the Middle East, the Tigris and EUPHRATES, both rise in Turkey, and hydroelectric plants supply much of the country's energy needs, as well as water for irrigation schemes.

Encouraged by the government, farmers have achieved the best rate of agricultural production in the Middle East, outside Israel. Turkey is largely self-sufficient in food production. Despite its dryness, the Anatolian plateau is the country's granary: the main crops are cereals, rice, cotton, fruit and tobacco; most of the country's sheep, goats and cattle are raised there. A variety of crops are grown along the coast and in the lowland regions.

Mechanisation and an increasing population have forced many peasants to seek jobs in the overcrowded cities. Others have emigrated to Western Europe. About 1.7 million Turks work in Germany, for instance, while many others are in the oil-rich Middle Eastern states. Their remittances home help the country's continual struggle to pay for its imports. The negative side of this migration is that the young generation of Turks who grew up in Western European countries are strangers in their own land, where traditions and values are firmly based on the Muslim religion.

HOW EMPIRES ROSE AND FELL

The first settlers came to Anatolia in about 7000 BC. The Hittites, warriors from Central Asia, founded an empire here about 1800 BC. In turn the Persians, Macedonians and Romans came. The Emperor Constantine moved the eastern Roman capital to Byzantium (now Istanbul) in AD 330, renaming it Constantinople. When the empire split, the eastern half continued as the Byzantine Empire. In 1453 the Ottomans, a Turkish tribe, took the city of Constantinople, and by the 16th century their Muslim empire stretched from the Danube (although they besieged Vienna they were never able to take it) to the GULF, from the Crimea to Morocco. Then a decline began.

During the First World War Turkey sided with Germany and defeated an Allied invasion force at GALLIPOLI on the Dardanelles, but its armies were beaten back in the Middle East. Its political alliance cost it the regions of Syria, Iraq and Palestine. The Ottoman Empire came to an end. Nationalists led by Mustafa Kemal, a successful general, rejected the proposed

▼ DOMES OF GLORY With the mosque of Bayezid II in the foreground, the mystery and beauty of Istanbul unfold. Within the city walls are byways with fascinating names like the Elephant's Path and the Street of the Chicken That Could Not Fly.

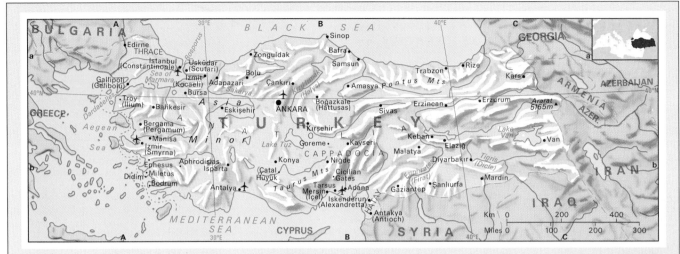

peace, which favoured Turkey's old rival, Greece. They set up a national assembly, deposed the Sultan and drove out the Greeks.

Kemal, who later took the surname Ataturk, became president of the republic in 1923 and ruled virtually as a dictator until 1938. He was the father of modern Turkey, bringing a backward Muslim state into the modern world. To Westernise his country he separated religion and the state, abolished polygamy, banned men from wearing the fez (the brimless felt cap of Muslims) and discouraged women from using the veil. He sent everyone under 40 to classes to learn the new Latin alphabet for the Turkish language and decreed that everyone should have a surname. Ankara became the new national capital.

During the Second World War Turkey remained neutral until February 1945, when it declared war on Germany. Turks also fought alongside Americans in the Korean War. It became a member of NATO in 1952. In 1974, after a Greek-inspired coup in CYPRUS, the Turks invaded the island, claiming to protect the Turkish minority. Other points of dispute still sour Turkish-Greek relations, including claims to minerals beneath the AEGEAN SEA.

After the 1980 coup, many leading figures were banned from politics and, according to Amnesty International, about 20 000 people were held as political prisoners, some of whom were tortured. The press was censored. The military government was increasingly criticised for its human rights violations and came under pressure to restore democracy. Since 1989 the president has been elected by the national assembly and in 1993 Tansu Çiller became Turkey's first woman prime minister.

LIFE OF THE PEOPLE

Urban life is centred on the three largest, most industrialised and Westernised cities: Istanbul, Ankara and IZMIR, where the professional and middle classes enjoy a comfortable lifestyle. But for most Turks life remains simple. The small café in town or village is the place where men meet to drink tea and talk, read the papers or play *tavla* (backgammon). The café reflects a male-dominated Islamic society. There have nevertheless been several women cabinet ministers, and in the cities

women take part in professional activities and are becoming more influential in public life. But generally far fewer jobs are open to them. Even a waitress is uncommon. In the country women remain subservient to men. They are tolerated in mosques, for example, but traditionally are expected to say their prayers at home. The custom of arranged marriages and payment of dowries still persists.

FOR THE VISITOR

Turkish food is distinctive – a happy blend of the culinary traditions of Central Asian pastoral people and the Mediterranean influence. The fact that the country produces all its own food is reflected in the daily freshness of its vegetables, fruit, meat and fish. The coffee is world famous, but has become increasingly expensive. The wine is often excellent.

For the adventurous tourist Turkey offers high mountains and rugged scenery to the north, south and east of the Anatolian Plateau, as well as the spectacular landscapes of the plateau itself. For the historically inclined the plateau has ancient Neolithic remains (ÇATAL HÜYÜK), Hittite (BOĞAZKALE), Greek (TROY) and Roman (ANTAKYA and KAYSERI), as well as fine examples of Byzantine art, and mosques and museums of the Ottoman Empire in many of the cities. For relaxation Turkey offers the beautiful Mediterranean and Aegean coasts. All this has made tourism another important source of foreign income.

The old city of Istanbul, destination of the famous Orient Express train from Paris, was built on seven hills around the Golden Horn, an inlet of the Bosporus. Together with its Asian section, it has a population exceeding 6 million. Beneath its glistening domes and minarets, there is continual urban bustle and movement: from shoppers and street sellers, to vehicles on its streets.

Topkapi was the great palace of the Ottoman sultans and is now a museum. The basilica of Saint Sophia, also a museum, was built by Constantine the Great and renovated by Emperor Justinian: its dome rises to 50 m (164 ft). The covered bazaar has a different street for each trade. A graceful suspension bridge, built in the 1970s, spans the Bosporus, linking Europe with Asia.

Ankara, Turkey's second largest city, had a population of only 75 000 when Kemal Ataturk made it his capital in the 1920s; now it has more than 2.5 million. It is very much his creation, containing many modern buildings, an abundance of trees, the Ataturk Museum and the Museum of Anatolian Civilisations with examples of Hittite art going back to 6000 BC. Izmir (Smyrna), probable birthplace of the Greek poet Homer, is an important port and a starting point for excursions to the ancient Greek towns of Pergamum (BERGAMA) and EPHESUS.

TARSUS, near the Mediterranean coast, was the birthplace of St Paul. The Aegean coast is rich in Greek remains, including the site of Troy near Canakkale. The Black Sea coast has such towns as TRABZON (the ancient Trebizond) with Greek and Byzantine remains.

TURKEY AT A GLANCE	
Official name Republic of Turkey	
Area 779 452 km² (300 946 sq miles)	
Population 60 897 900 **Per km²** 78	
(Per sq mile 202)	
Capital Ankara	
Government Parliamentary republic	
Currency 1 lira = 100 kurus (piastres)	
Languages Turkish (official), Arabic, Greek, Circassian, Armenian, Yiddish, Kurdish	
Religion Muslim (98%)	
Climate Mediterranean on the south and west coast, subtropical on the Black Sea coast and continental inland. Average temperature in Ankara ranges from 0°C (32°F) in January to 29°C (84°F) in August	
Land use Cultivated 34%, pastures 12%, forest and woodland 26%, other 28%	
Main primary products Wheat, barley, maize, cotton, sugar beet, pulses, cattle, sheep, goats, tobacco, fruit (olives, nuts, figs), vegetables, fish; coal, iron, crude oil, chrome, boron, copper	
Major industries Agriculture, steel, textiles, tobacco and food processing, oil refining, chemicals, paper, fishing, mining	
Main exports Textiles, cotton, nuts, fruit, tobacco, cereals, pulses	
Annual income per head (US$) 1950	
Population growth (per thous/yr) 21	
Life expectancy (yrs) Male 68 **Female** 73	

Turkmenistan

A NEW PRESIDENT HAS ESTABLISHED HIS OWN PERSONALITY CULT IN THIS DESERT STATE ON THE CASPIAN SEA

The area that makes up the desert state of Turkmenistan has been settled by nomadic people since ancient times and was on the caravan routes linking Asia with Europe.

It was under Persian rule in the 6th to 4th centuries BC, and was later invaded by Alexander the Great, by the Arabs, and by the Tatars and Mongols. It was ceded to Russia in the late 19th century and on 27 October 1924 was incorporated in the Soviet Union as the Turkmen Soviet Social Republic. But the country retained its own language and religion. Ethnic Turkmen today account for more than 73 per cent of the country's population. Russians make up less than 10 per cent.

Independence from the Soviet Union was gained in October 1991, 67 years after the Turkmen republic had been founded. Since independence the country has maintained good relations with Russia, which has several strategic military bases in the country and keeps troops here. A dual citizenship accord was signed with Russia in 1993, granting both Turkmen and Russian citizenship to the 400 000 Russians living in Turkmenistan.

Economically, Turkmenistan is one of the most independent former Soviet states. After independence it refused to attend important Commonwealth of Independent States (CIS) meetings, and it became a full member of the CIS only in December 1993. It preferred to go its own way economically, keeping closer relations with Iran and Saudi Arabia.

After independence, the Communist Party remained in power, though under the new name of Democratic Party of Turkmenistan. A new constitution was adopted in 1992, whereupon President Saparmurad Niyazov set about establishing a personality cult. Members of his parliament proposed life presidency for him, and so, in 1994, the presidential elections scheduled for 1997 were cancelled.

EXPANDING DESERT

Of all the former Soviet states, landlocked Turkmenistan is the fourth largest, covering an area of 488 100 km² (188 412 sq miles). In terms of population figures it ranks 11th, with a low population density of some 8 per km² (21 per sq mile). Most of the population is concentrated in irrigated oases.

More than 90 per cent of Turkmenistan's land area is taken up by desert and dry steppe. The Peski Karakumy (Kara Kum Desert) alone covers 340 000 km² (130 000 sq miles) – an area which is larger than Finland. Much of the desert receives an annual rainfall of about 100-150 mm (3.9-5.9 in), and different plant species grow there. Originally, most of it was covered with some sort of plant growth, protecting the soil from erosion by wind and rain. But during the past century the size of the herds grazed here gradually increased, until all the grass was eaten. The fertile topsoil was eroded away and shifting dunes established.

Only 3 per cent of the land area is used for cultivation. Nonetheless, agriculture employs about 42 per cent of the working population and the country ranked second among all the former Soviet republics in cotton production. Today Turkmenistan contributes about 15 per cent of the cotton produced by the CIS states. Other crops include grains, melons, grapes, and figs, as well as pomegranates from the Peski Karakumy oases. Karakul sheep and goats are reared on the fringes of the desert. Traditionally nomadic, today's herders transport their sheep to the mountains by truck.

The country is also an illicit producer of cannabis and opium, which are exported to CIS countries.

OIL AND SALT

Since the beginning of the 20th century, many canals have been built throughout the country to irrigate crops and provide water for the settlements. More than 90 per cent of the crops are under irrigation, mainly in the west, where the Karakumskiy Canal taps the Amurdar'ya River. The canal runs 1400 km (870 miles) from the Amudar'ya River, east of the desert, as far as the port of Turkmenbashy (Krasnovodsk) on the Caspian Sea. Here water is desalinated by atomic power.

The water level of the Caspian Sea has fallen considerably – the large streams which fed it in the past are all being tapped along their way. Consequently, the Kara-Bogaz-Gol, a vast bay linked to the Caspian Sea by a narrow canal, has been drained. In the past, an annual 5 billion m³ (176 billion cu ft) of water flowed into the bay and evaporated in the heat. Over the millennia, vast deposits of salt – Glauber's salt, potassium as well as iodine – were built up. These salts are processed and exported.

But the lowlands along the Caspian shores, which lie more than 50 m (164 ft) below sea level, contain other mineral resources besides salt. The country's vast petroleum, natural gas and sulphur resources from this region provide most of its foreign income.

Turkmenistan is the CIS's second most important exporter of natural gas, after Russia. While under Soviet rule, it produced 11 per cent of the total Soviet gas produced and 1 per cent of Soviet oil.

The country's mineral resources form the basis of its industry, and mining employs 10 per cent of its workforce. The capital, ASHGABAT, which lies in a fertile oasis, a mere 40 km (25 miles) from the Iranian border, is irrigated by the Karakumskiy Canal. Its industries include food processing, textiles, carpets and silks as well as leather, skins and hides.

Other industrial cities are Chardzhev (formerly Chardzhou), Dashkhovuz (formerly Tashauz), Turkmenbashy (formerly Krasnovodsk) and Mary.

The textile industry traditionally takes a leading place in the economy and cotton, wool and silk are important exports.

In 1992, the economy went into decline, due to shortages in spare parts and disputes – notably with the Ukraine – over the price of oil. Inflation rocketed to more than 50 per cent during the first quarter of 1993 and unemployment rose to 15-20 per cent.

The transition to a free market system has been extremely slow; however, a 10-year privatisation and land reform programme was announced in February 1993. Although most imports come from Europe and other non-CIS sources, Turkmenistan's exports go almost exclusively to CIS countries. But the country has now begun looking for partners in the West to exploit and help develop oil resources. In the mid-1990s, agreements were signed with Iran and Turkey for a pipeline to run across the territory to Europe.

TURKMENISTAN AT A GLANCE		
Official name Republic of Turkmenistan		
Area 488 100 km² (188 412 sq miles)		
Population 3 915 000 **Per km²** 8 **(Per sq mile** 21)		
Capital Ashgabat		
Government Republic		
Currency manat		
Languages Turkmen (official), Russian, Uzbek		
Religions Muslim (87%), Eastern Orthodox (11%)		
Climate Desert climate, with cool winters and hot summers, and little rainfall. Average temperature in Ashgabat ranges from 2.1°C (36°F) in January to 31.2°C (88°F) in July		
Land use Grazing 69%, cultivation 3% other 28%		
Main primary products Petroleum, natural gas, coal; cotton		
Major industries Mining, textiles, carpet making		
Main exports Natural gas, oil, textiles, carpets, cotton		
Annual income per head (US$) 1270		
Population growth (per thous/yr) 20		
Life expectancy (yrs) Male 61 **Female** 69		

Tuvalu

THE MAIN EXPORT FROM THIS GROUP OF TINY PACIFIC ISLANDS IS POSTAGE STAMPS

The nine coral atolls of Tuvalu have a total area of only 26 km² (10 sq miles); none of them rises above 4.5 m (15 ft) out of the South Pacific. The main island and capital, FUNAFUTI, covers a mere 2.4 km² (0.9 sq mile) and has a population of 2800. The other islands are scattered in a chain 600 km (375 miles) long, some 960 km (600 miles) north of Fiji.

In 1974 the Polynesian inhabitants of Tuvalu, then the Ellice Islands, voted to separate from the Gilbert Islands (now Kiribati, whose people are Micronesians), with which they had formed a joint British colony since 1916. In 1978 Tuvalu acquired independent status within the Commonwealth.

The atolls' natural vegetation is coconut palms, breadfruit trees and palm-like pandanus trees, with large buttress roots above ground and crowns of narrow leaves. The coconuts provide the main agricultural export of copra, but most of the revenue comes from licensing fees for fishing rights paid by Taiwan, Korea and the United States, from the sale of elaborate postage stamps to philatelists and from money sent back by Tuvaluans working abroad – mainly on Nauru.

There are about 9000 islanders, mostly living in thatched huts beside lagoons and producing their own food in gardens and by fishing. The population was once much higher, but between 1850 and 1875 the islands were raided by 'blackbirders' – Europeans who took most of the inhabitants, willingly or as slaves, to work the plantations of other Pacific islands. They reduced a population of 20 000 to a mere 3000. Another European invasion occurred during the Second World War, when the United States used the islands as air and naval bases.

TUVALU AT A GLANCE

Map Pacific Ocean – Dc	
Area 26 km² (10 sq miles)	
Population 9100 **Per km²** 350	
(Per sq mile 910)	
Capital Funafuti	
Government Parliamentary monarchy	
Currency 1 Tuvaluan and Australian dollar = 100 cents	
Languages Tuvaluan, English	
Religion Christian (mainly Protestant)	
Climate Tropical; average temperature is 25-32°C (77-90°F) all year	
Main primary products Coconuts, fruit, vegetables, fish	
Major industry Fishing	
Main exports Postage stamps, copra	
Annual income per head (US$) 560	
Population growth (per thous/yr) 17	
Life expectancy (yrs) Male 61 **Female** 63	

Tuxtla Gutiérrez *Mexico* Busy modern capital of the state of CHIAPAS, which replaced the less accessible San Cristóbal de las Casas as the seat of government in 1892. A distribution centre for the surrounding area's coffee and tobacco crops, Tuxtla also has markets selling varnished gourds, leatherwork and gold jewellery.

Population 295 620
Map Mexico – Cc

Tuzla (Dolnja Tuzla) *Bosnia-Herzegovina* Town in north-east Bosnia, 80 km (50 miles) north-east of SARAJEVO. Its salt mines have been exploited since before the Middle Ages. The town has developed into a major chemicals centre.

Population 121 700
Map Bosnia-Herzegovina – Ca

Tver' (Kalinin) *Russia* Industrial city and river port straddling the Volga, 160 km (100 miles) north-west of MOSCOW. It was founded in 1181, and was rebuilt after a fire in 1763. As the centre of the Tver Principality until 1485, it was originally called Tver', but was renamed Kalinin in 1932 in honour of its most renowned native son, Mikhail Kalinin (1875-1946), Soviet president from 1938 to 1946. After the collapse of the Soviet Union, the city was renamed Tver'. Its industries include textiles and rubber.

Population 455 000
Map Russia – Ec

Twante *Myanmar (Burma)* Canal linking the capital, Yangon, with the rice lands of the Ayeyarwady delta. It is the most important of a network of canals running through the delta.

Map Myanmar – Cc

Tychy *Poland* Industrial town, 16 km (10 miles) south of KATOWICE. It was once a small, sleepy town – indeed, its name means 'quiet' – but has grown since 1951 as a residential district for people working in Katowice, and since 1970 as a production centre for Polski Fiat cars.

Population 190 000
Map Poland – Cc

Tyne and Wear *United Kingdom* North-eastern county which is named after the two rivers that cross it. It covers 540 km² (208 sq miles) and was formed in 1974 from parts of the counties of Northumberland and Durham. It is centred on the city of NEWCASTLE UPON TYNE and includes the declining shipbuilding and industrial towns of Jarrow, Wallsend, Tynemouth, South Shields and SUNDERLAND. During the 1930s and again in the 1980s it was among the parts of the country worst hit by the recession.

Population 1 130 400
Map United Kingdom – Ec

typhoon Small, intense tropical CYCLONE in the China Sea and western Pacific, accompanied by strong winds of 160 km/h (100 mph) or more, torrential rain and thunderstorms.

Tyre (Sour) *Lebanon* Small seaport 65 km (40 miles) south of BEIRUT. It was fought over and badly damaged during the Israeli invasion of southern Lebanon in 1982. Founded in about 1500 BC, Tyre was the Phoenicians' main trading centre. Gradually, Tyre's trade spread throughout the Mediterranean world and reached the Atlantic coast of Morocco, where its merchants established colonies, and Tunisia, where they founded the city of Carthage. Tyre, like nearby SIDON, was renowned for its prosperity and for its glass and purple dye, called 'Tyrian purple'.

The Phoenician alphabet – adopted by the Greeks in 1000 BC – was attributed to Cadmus of Tyre, whose sister, the princess Europa, gave her name to the continent of Europe. The Babylonian king Nebuchadnezzar (died about 562 BC) besieged Tyre without success for 13 years, but in the 4th century BC Alexander the Great sacked the city. Tyre became important again in Roman times, and today most of the remains are Roman. They include a monumental archway, an aqueduct, baths, a theatre, a large hippodrome and a necropolis.

Population 14 000
Map Lebanon – Ab

Tyrol (Tirol) *Austria* Most mountainous of the country's states, covering some 12 647 km² (4882 sq miles), with the Bavarian Alps on the northern border and the Ötztaler Alps in the south and centre part. It is crossed by the Inn and Lech rivers. Its highest mountains are the Ötztaler Alps and the Zillertal Alps and HOHE TAUERN, which rise to between 3700 and 3800 m (12 140 and 12 467 ft). The capital, INNSBRUCK, is linked with Bolzano (Italy) by the BRENNER PASS. The state's main industry is tourism. South Tyrol was ceded to Italy in 1919 (see TRENTINO-ALTO ADIGE).

Population 630 360
Map Austria – Bb

Tyrone *United Kingdom* Largest of Northern Ireland's six counties, covering some 3136 km² (1211 sq miles). It is a mainly rural area, growing potatoes, turnips and oats. Tyrone lies west of Lough Neagh and rises to 683 m (2240 ft) in the Sperrin Mountains. The county was dominated by the O'Neills, Kings of Ulster, until the end of the 16th century when Hugh O'Neill was forced to flee to mainland Europe to escape the English. The county's chief towns are Omagh (population 17 280) and Strabane (population 11 670).

Population 157 570
Map United Kingdom – Bc

Tyrrhenian Sea (Mare Tirreno) Deepest basin, reaching nearly 3758 m (12 329 ft) at its deepest point, in the western MEDITERRANEAN SEA, bordered by the west coast of mainland Italy, Sicily, Sardinia and Corsica, and covering 155 400 km² (60 000 sq miles). It includes the volcanic EOLIAN ISLANDS in the south-east, resort islands, such as CAPRI and ISCHIA, and the mountainous island of ELBA in the north. Named after Tyrrhenus, legendary founder of the Etruscan Kingdom, it is generally placid in summer, though electric storms are common. The SIROCCO, a wind blowing from the Sahara in spring and early summer, can bring humid weather. In winter, depressions moving in from the Atlantic and a northerly wind, the GREGALE, cause choppy seas.

Map Italy – Cd

Tyumen' *Russia* Western Siberian town 300 km (185 miles) east of YEKATERINBURG. Founded in 1586, it was the first Russian town to be settled east of the Urals. It is the centre of a rich oil and natural gas region. Its industries include boatbuilding, engineering and petrochemicals.

Population 487 000
Map Russia – Hc

UAE See UNITED ARAB EMIRATES

Ubangi *Equatorial Africa* One of the largest tributaries of the Zaire River (also known as the Congo). It is formed by the meeting of the UELE and BOMU rivers and flows 1060 km (nearly 660 miles) west, then south into the Zaire River in the Republic of the Congo. It handles much of the Central African Republic's foreign trade.
Map Central African Republic – Bb

Ubud *Indonesia* Village on BALI, which is famous for its artistic traditions and contains studios and galleries of several European and Balinese painters who have depicted the way of life on Bali – on canvas.

Ubud's Puri Lukisan Palace of Paintings contains a fine collection of Balinese paintings and sculptures from the 1930s to the present day. There is also a school of traditional music and dancing where the visitor can see some of the finest performances on the island.
Map (Bali) Indonesia – Ed

Ucayali *Peru* River formed by the meeting of the Apurímac and Urubamba 360 km (225 miles) east of the capital, LIMA. It flows north through Pucallpa, the centre of thriving gas, oil and timber industries, to join the Marañón, forming the Amazon. The Ucayali is some 1928 km (1198 miles) long and is navigable by boats of up to 3000 tonnes for 1600 km (995 miles).

The department of the same name covers 129 658 km² (50 061 sq miles) of eastern Peru.
Population (department) 230 100
Map (River) Peru – Ba

Udagamandalam (Ootacamund) *India* Health resort and tourist centre. It lies at 2268 m (7440 ft) in the NILGIRI HILLS and was the colonial summer capital of the former state of Madras, when it was known as 'the Queen of Hill Stations'.
Population 63 000
Map India – Be

Udaipur *India* Walled city on the edge of the THAR Desert, some 800 km (nearly 500 miles) north-east of Bombay. Its many palaces and Hindu temples make it popular with tourists. The 16th-century maharajah's palace of glistening white stone and numerous other 17th and 18th-century marble palaces stand on islands in an artificial lake, called Pichola, at the foot of low hills.
Population 308 000
Map India – Bc

Udi Plateau (Donga Ridge) *Nigeria* South-eastern upland, on which the town of ENUGU is situated. Its north-eastern edge is an escarpment which, in parts, rises to more than 457 m (1500 ft) above the Cross River plains. The plateau's sandstone ridges run in a south-westerly direction from the escarpment, then east towards Cameroon.
Map Nigeria – Bb

Udine *Italy* City in the foothills of the Alps, about 60 km (37 miles) north-west of TRIESTE. Udine was ruled by Venice for centuries and its architecture, especially in the older parts of the city around the Piazza della Liberta, is a strong reminder of those days. The city has a 13th-century cathedral with an unfinished 15th-century

campanile (bell tower), a 15th-century town hall (restored after a fire in 1876) and a 16th-century castle. Inside the castle is a museum and picture gallery. Its economy is based on the production of chemicals, textiles, food, machines and leather, and the casting of bells.
Population 98 900
Map Italy – Da

Uele *Zaire* River, about 1200 km (750 miles) long, which rises north-west of Lake Albert and flows west to the Bomu River to become the UBANGI River. In 1870, the German botanist Georg August Schweinfurth became the first European to see the Uele, while exploring the Nile-Zaire divide.
Map Zaire – Ba

Ufa *Russia* Capital of Bashkiria, lying on the western fringe of the Ural Mountains, about 360 km (225 miles) south of PERM'. Founded in 1574 as a fort to guard the trade routes across the Urals, it grew as an industrial centre from the late 19th century and especially after the exploitation of the Volga-Urals oil fields in the 1950s. Its industries include machinery, chemicals, timber products, electrical equipment, telecommunications and office equipment.

The satellite town of Chernikovsk, some 32 km (20 miles) to the north-east, has several oil refineries and petrochemical works.
Population 1 097 000
Map Russia – Gc

Uganda See p 690

Uíbh Fhailí *Ireland* See OFFALY

Ujiji *Tanzania* See KIGOMA

Ujjain *India* Centre of early Hindu civilisation in the state of MADHYA PRADESH in northern India, 674 km (418 miles) north-east of Bombay. It is one of the seven holy Hindu cities where, every 12 years at a *Kumbha mela* (religious festival and fair), pilgrims bathe in the Sipra River.

Ujjain's history dates back to the 3rd century BC but no buildings remain of those days. The city has mosques dating from its time under Muslim rule (1235-1750), temples and palaces, and the ruins of an observatory built by the astronomer Jai Singh II (1699-1743), Maharajah of Jaipur.
Population 366 800
Map India – Bc

Ujszeged *Hungary* See SZEGED

Ujung Pandang (Makassar) *Indonesia* Port and largest city on SULAWESI. Formerly called Makassar, the city is in the extreme south-west of the island and is an important trading and cultural centre, with a university, an archbishopric and a 17th-century Dutch fortress which houses the island's museum. Ferries link it to the rest of Indonesia.
Population 709 100
Map Indonesia – Ed

Ujvidék *Yugoslavia* See NOVI SAD

UK See UNITED KINGDOM

Ukraine (Ukraina) See p 692

Ulan Bator (Ulaan Baatar; Ulaanbaatar) *Mongolia* Capital city, situated some 1300 m (4265 ft) up on the Mongolian plateau on the Tuul River, a tributary of the SELENGE.

The city dates back to the 17th century and the founding of the Temple of the Living Buddha. The settlement which grew around the temple became an important base for caravans crossing the Gobi Desert. By the 19th century, when it was known as Urga, it was one of the country's few substantial towns. Its name was changed to Ulan Bator – 'Red Hero' – in 1924 after the communist revolution in Mongolia.

Ulan Bator is a bleak city of wide streets and modern buildings, many of them reminiscent of Joseph Stalin's Russia. Many high-rise apartment blocks have been built here, especially since the early 1970s, and most buildings have steam central heating – essential in the bitterly cold winters. *Yurts* – the traditional portable felt homes of the nomadic herders – are still commonplace on the outskirts, but today they are supplied with electricity.

The city's State Central Museum has numerous prehistoric remains unearthed from the Gobi Desert, including two complete dinosaur skeletons, 20 cm (8 in) long dinosaur eggs, and the remains of a giant rhinoceros. Other attractions include the art museum, and the state circus, opera and ballet. There is also a university.

The main industries, along the Transmongolian railway line, are meat packaging and processing, flour-milling, printing and the manufacture of textiles, building materials and prefabricated houses.
Population 575 000
Map Mongolia – Db

Ulan-Ude *Russia* Capital of the Buryat Republic. It is a railway city 235 km (145 miles) east of IRKUTSK; until 1934 it was known as Verkhneudinsk. Ulan-Ude was founded as a fort in 1666 and came to prominence in 1900 as a junction on the Trans-Siberian Railway. A branch line to the Mongolian capital, Ulan Bator, 445 km (275 miles) to the south, was opened in 1949. Ulan-Ude's products include locomotives and rolling stock, glass, processed food and timber.
Population 359 000
Map Russia – Lc

Ulcinj *Yugoslavia* See BAR

Uleåborg *Finland* See OULU

Ullswater *United Kingdom* Second largest lake, after Windermere, in the LAKE DISTRICT of north-west England. It is nearly 13 km (8 miles) long and is a popular tourist attraction.
Map United Kingdom – Dc

Ulm *Germany* Industrial city, transport hub and university town on the Danube River, 70 km (43 miles) south-east of STUTTGART. Its products included vehicles, electrical goods, textiles, clothing and processed foods.

The beautiful old part of the city, founded before 800, has a 14th-century town hall and a Gothic cathedral with the highest spire in the world, at 161 m (528 ft). Ulm was the birthplace of the physicist Albert Einstein (1879-1955), originator of the theory of relativity.
Population 98 700
Map Germany – Cd

Uganda

A COUNTRY DEVASTATED BY AN INFAMOUS DICTATOR AND ETHNIC CONFLICT MAY ONCE AGAIN BECOME 'THE PEARL OF AFRICA'

Long ago, in colonial days, British states-man Winston Churchill (1874-1965) described Uganda as 'the pearl of Africa'. Then, the description was not unmerited. Although its only access to the sea was by road and rail through Kenya to Mombasa, Uganda was relatively prosperous. It was a British protectorate, not a colony, and was encouraged to manage its own affairs and develop its own resources.

For the most part, Uganda was and is a richly fertile land, well watered and with a kindly climate – though there is some soil erosion and deforestation. It also has fair deposits of copper and cobalt although these have not been exploited.

Much of the north and centre of the country is a high plateau, 1000-1400 m (3280-4590 ft) above sea level, lying between the east and west arms of the GREAT RIFT VALLEY system. It is a wide savannah belt, nurturing one of the last great wildlife communities of the African grasslands. To the west are the RUWENZORI Mountains, whose spiky peaks, reaching up to 5109 m (16 762 ft), are capped with snow and perpetually wreathed in drifting veils of mist. Below, in the Bushenyi region, plantations clothe the foothills.

The south-eastern border of Uganda runs through the basin of Lake VICTORIA. The lowlands around the lake were once forested, but have now mostly been cleared for cultivation. From the lake, the Victoria NILE runs northwest through central Uganda, into Lake KYOGA and onward to become the main course of the Nile itself.

Lakes and rivers between them, married to the elevation of the land, keep Uganda fairly cool despite its position athwart the equator. The chief exception is the KARAMOJA district in the north-east, where rainfall is uncertain and the terrain semi-desert. The inhabitants are nomadic herders, living only a step from disaster, as was apparent from the drought of 1979-81, when 30 000 people died.

Uganda gained independence on 9 October, 1962, an occasion greeted with celebrations and high hopes. Of all former colonial possessions in Africa, it had the highest percentage of Western-educated people, and was well supplied with hospitals, schools and centres of higher learning. Yet within a decade and a half, optimism crumbled into a tragic shambles – a ruined economy, internal strife and a frightened population.

The seeds of disaster were sown long before independence. Uganda was a portmanteau state of half a dozen kingdoms and tribal regions. The heart of the protectorate was the powerful and ancient kingdom of Buganda, with the weaker kingdoms of Bunyoro, Toro and Ankole lying to the west. To the north and east are tribal lands, including those of the Lango and Acholi, while far to the south-west is the Kigezi district. Culturally, linguistically and racially, the people living at one end of Uganda have about as much in common with people living at the other as the Irish have with the Chinese.

It would have been thought that the Bagandans of Buganda, as the largest, most prosperous and best-educated ethnic group – as well as being the possessors of the capital, KAMPALA – would have dominated the new nation in the same way as the Kikuyu dominated Kenya. As it turned out, the Bagandan leaders had little interest in the republic of Uganda and preferred, instead, to concentrate upon their own tribal affairs. Consequently, the reins of national power were taken by the leaders of the less powerful tribes. The first prime minister, Milton Obote, for example, was a Lango from the north.

THE YEARS OF TERROR

In 1966 Buganda attempted to secede from the union, which Obote used as an excuse to abolish the tribal monarchies, as well as the constitution, and set up a central government with himself as president. Without Bagandan support, however, his reign was uneasy, and during his absence at the Commonwealth Prime Ministers' Conference in 1971, he was deposed by the Minister of Defence, General Idi Amin. The general's move may have been accelerated by a court of inquiry which was investigating the loss of US$5 million from defence ministry funds.

So began the sordid reign of Idi Amin who proclaimed himself president for life. At first he was acclaimed as a strong man and his rise to power was also welcomed by the British government, who were becoming alarmed at Obote's nationalising of British assets. Ugandan cheers swiftly died away as Amin quashed all political activity and dispatched his military death squads to smell out opposition.

In 1972, chronically short of the cash necessary to maintain his repressive rule, he expelled all Asians with British passports – virtually the entire business community – and seized their property. Most other foreigners fled. The businesses were handed over to Ugandans who had no experience in running them, and the economy sagged still further. Amin then began to eliminate all those who offered the slightest challenge, real or imaginary, to his rule, and was prepared to condone, if not encourage, repeated atrocities by his soldiers.

Many of these monstrous acts were the settling of old scores, arising out of ancient tribal jealousies and hatreds. It is estimated that between 100 000 and 300 000 people were brutally slain. Amin's ruthlessness became notorious throughout the world. His downfall came at last in 1979, when he decided to annex a part of Tanzania. Tanzania's president, Julius Nyerere, instructed his forces not only to remove Amin's invaders but to remove Amin from Uganda, with the help of the Uganda National Liberation Army.

When Amin fled he left behind a devastated country in a desperate state. The economy was shattered, law and order had broken down, and there was a deep mistrust between the country's many factions. The two presidents who succeeded him in 1979 were speedily deposed. In 1980, a national election brought victory to the Ugandan People's Congress Party led by the returned Milton Obote.

But there were those who claimed that the election was rigged. One of the dissidents was Yoweri Museveni, who had been foreign minister in one of the short-lived post-Amin governments. He took to the bush with his National Resistance Army, formed out of units of the old Uganda National Liberation Army that had helped the Tanzanians to oust Amin. At the same time, trouble broke out in the Karamoja district of the north-east where heavily armed cattle rustlers clashed with government troops. All in all, probably well over 200 000 people died in the troubles after Amin's departure.

As always, inter-tribal rivalries were at the root of the problem. Amin was a Muslim of the northern Kakwa group, and so favoured Uganda's Muslims, a mere 10 per cent of the population. Obote promoted Lango soldiers of his own ethnic group – in order to give them an advantage in an army which, since the days of the colonial King's African Rifles, had been mostly recruited among the Acholi. In Obote's time, fighting frequently broke out around the capital between Acholi and Langi factions within the army. Then, in 1985, the army ousted

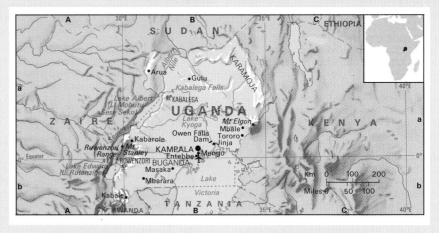

Obote in a coup supported by Ugandan exiles, and the army commander, Lieutenant-General Tito Okello, established a military caretaker government and became head of state.

Okello invited Museveni and the NRA to throw in their lot with the new regime. An accommodation between the two sides came to an end early in 1986 when NRA guerrillas drove the army out of Kampala and Museveni was sworn in as president. Since Museveni, the country has experienced political stability, and its economy a slow upward turn.

GRADUAL RECOVERY

Some parts of Uganda escaped the horrors of the 1970s and early 1980s. Kampala's oppressive governments could not directly influence all aspects of people's lives throughout the country. Only 10 per cent of the people live in towns and the country dwellers are widely scattered. In remote areas life has hardly changed. Most Ugandans still grow their own food, staving off widespread famine despite the struggling economy. Family ties are strong, and even urban dwellers are often able to obtain food from rural kin. Not that a great variety of foodstuffs is called for. In the south, the staple is a green banana called *matoke*, while in the north it is millet and sorghum.

Cotton and coffee were Uganda's main cash crops in colonial times. Agriculture is still important and accounts for nearly 60 per cent of gross domestic product (GDP), employing more than 80 per cent of the country's work-

▼ MURCHISON FALLS In a land of waterfalls, this is one of the most stupendous. The waters of the Victoria Nile seem to boil as they plunge 122 m (400 ft) in three thunderous cascades, of which Murchison is the first.

force. Most Ugandans are still subsistence farmers. Besides the cash crops of coffee, tea, cotton and tobacco, the people plant food crops such as potatoes, millet, maize, cassava and pulses. Coffee became, and still is, Uganda's most important export. But by the 1990s, the government began looking for ways to try to reduce the country's dependency on coffee.

During the 1970s, production seemed to decline sharply, but in fact large quantities of coffee were being smuggled into Kenya and Zaire. Coffee survives because it is a perennial plant that continues to produce whether it is tended or not. Cotton, on the other hand, must be newly planted each year, and shortages of seed, labour and pesticides resulting from the general breakdown considerably reduced the output. Many northern farmers, who once grew cotton, have reverted to subsistence farming, though the coffee growers of the south still remain relatively prosperous.

After Museveni came to power he reformed, modernised and stabilised Uganda's economy; he privatised, introduced currency reforms, and freed most prices. The inflation rate, which was running at more than 300 per cent in 1987, was down to about 23 per cent by 1994. Ugandan industries, such as sugar refining, brewing, tobacco, textiles and cement, were showing a growth rate of more than 7 per cent. Until the early 1990s, GDP and industrial production had stayed around the levels of the early 1970s.

Under Museveni, attempts have been made to restore enterprises which were under way at the time of independence in the 1960s. These have included the tea plantations of the far west, two large sugar plantations near Lake Victoria, a copper mine (also in the west) and new manufacturing industries based in Kampala and near the Owen Falls Dam and power station at JINJA. Asians have been

invited to return and their property has been restored to them. The Mombasa-Kampala railway is operating again, after many years of hostility between Kenya and Uganda. The tourism industry is also recovering.

On the political scene, the traditional monarchies were restored in 1993. A non-party assembly with powers to draft a new constitution was elected in March 1994. This development and the economic indicators point to a period of stability and growth ahead. Indeed, if the economy keeps improving, Uganda may again be considered one of the more fortunate African countries.

UGANDA AT A GLANCE	
Official name Republic of Uganda	
Area 236 040 km² (91 114 sq miles)	
Population 19 000 000 **Per km²** 80 **(Per sq mile** 208)	
Capital Kampala	
Government Republic	
Currency 1 Uganda shilling = 100 cents	
Languages English (official), Swahili, Luganda, other African languages	
Religions Christian (66%), indigenous beliefs (18%), Muslim (16%)	
Climate Equatorial, tropical; cooler in mountain areas. Temperature in Entebbe is 21°C (70°F) throughout the year	
Land use Cultivation 32%, forest 30%, grazing 25%, other 13%	
Main primary products Coffee, cotton, tea, sugar, livestock, millet, maize, bananas, sorghum, yams, timber; copper, cobalt	
Major industries Agriculture, textiles, food processing, cement, copper processing, metal processing, mining	
Main exports Coffee, cotton, tea	
Annual income per head (US$) 190	
Population growth (per thous/yr) 33	
Life expectancy (yrs) Male 38 **Female** 39	

Ukraine

A COUNTRY DOMINATED BY OTHER NATIONS FOR MORE THAN 900 YEARS OF ITS 1000-YEAR HISTORY

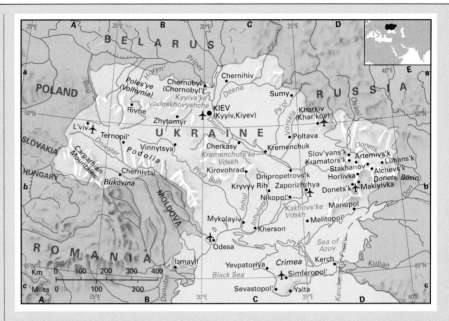

The Ukraine, the second largest European country, gained its independence on 24 August, 1991 – almost 338 years after being joined to Russia, its 'big brother'. Old ties, both economic and political, still link the Ukraine to its former coloniser. The general elections in 1994 showed that an overwhelming percentage of Ukrainians were still in favour of closer political and economic ties with Russia.

The elections also showed that the country was politically divided down the centre, with L'VIV in the west being the hub of pro-Western sentiment, while the Donets'k region in the east is the centre of pro-Russian sentiment. This split was most evident in the autonomous republic of CRIMEA, whose pro-Russian president Yuri Meshkov was locked throughout that year in a power struggle both with the central Ukrainian government and with the Crimean Supreme Council dominated by ethnic Ukrainians.

In its long history, the Ukraine was independent once for a short while during the 17th century. In 1653 it became part of the Russian Empire. The Crimea was the scene of war between Russia and the forces of France, England and Turkey in 1853-6, remembered particularly for the historic Charge of the Light Brigade (immortalised in verse by the Victorian poet Alfred, Lord Tennyson) and the work of the English nurse Florence Nightingale (1820-1910).

The Crimean Peninsula again hit the headlines in February 1945, when the famous Yalta Conference took place at the Black Sea coast resort of Yalta. Here Franklin Roosevelt, Winston Churchill and Joseph Stalin met and planned the final stages of the war and discussed the division of power over Europe after the war was ended.

Then, in April 1986, the world's worst civil nuclear disaster occurred at a power station at Chernobyl' near KIYEV (known at the time as Kiev), the country's capital. One of the plant's four nuclear reactors melted down, releasing tonnes of radioactive debris in a cloud that spread over large parts of the Ukraine and as far away as Britain and Canada.

IMMIGRANTS AND EMIGRANTS

Historically, the Ukraine consisted of a region to the south of Kiyev, on the left and right banks of the Dnieper River. But today the Ukrainians, descendants of East Slav tribes, have spread far beyond this central region. Almost 4 million Ukrainians live in Russia, and another 3 million live in the United States and Canada – descendants of Ukrainian immigrants to the New World.

The name 'Ukraina' means 'frontier land' – a most apt description because the land

between the Carpathian Mountains and the Crimea, and between the Pripyat' marshes and the Black Sea is a place where many cultures and natural regions meet and overlap. The Ukraine's history has been determined by its position as a frontier state between the Ottoman Empire in the south, Russia in the north and Poland in the west.

From the east, the region was repeatedly invaded by belligerent nomadic peoples, such as the Huns during the 4th and 5th centuries, the Avars in the 6th century, and the Tatars in the 13th century. They came from across the steppes of Asia to invade the fertile land that was sandwiched between the eastern European mixed forest zone and the belt of Eurasian steppe.

This wedge of land between the forest and the steppe can itself be divided into three natural regions, from west to east, each roughly taking up about one-third. In the north are the forests and marshes of the Ukrainian Palyessye (also known as Pripyat), a vast, marshy lowland to the left and right of the Pripyat' River which joins the Dnieper north of Kiyev.

Forest accounts for roughly 10 per cent of the central region, which was originally a tree steppe. The rest of the region is cultivated today, and deforestation is continuing. The southern region, extending to the Black Sea and the Sea of Azov, which was once expansive grass-and-wormwood steppes, is now largely cultivated.

Besides a tiny slice of Carpathian Mountains in the west, and the mountains on the Crimean Peninsula which rise to 1545 m (5069 ft) at Roman Kosh, the Ukraine is a lowland, with more than two-thirds at an altitude of 100-300 m (328-984 ft). The land is moulded into characteristic shields, or shallow plateaus, such as the Podolic shield, and the Dnieper and Azov plateaus. The Black Sea coast consists of a wide coastal plain, where sandy beaches and a sea of dunes are popular with tourists, particularly Russians coming to soak up the southern

sun. The rest of the country is shaped into gently undulating plateaus incised by a deep valley or interrupted by a vast lowland here and there. Large areas are covered with loess and debris deposited at the edge of glaciers as they pushed forward during the ice ages; and almost three-quarters of the Ukraine has fertile, wind-borne soil.

This deep, loose humus soil is the country's best natural resource. But thunderstorms and meltwaters wash away the fertile topsoil. Where the water collects to form large torrents, it erodes deep gullies into the ground. The higher-lying land especially is incised by a network of gulleys (*owragi*). Already 30 per cent of the cultivated land along the Dnieper has become unusable. Strong winds called *suchowej* on hot, dry summer days also carry away the topsoil.

In all, more than 10 per cent of arable Ukrainian land has been destroyed by soil erosion. In addition, there is substantial water pollution. A large area of farmland was also turned unusable by the Chernobyl' nuclear disaster. In 1994, the Ukrainian government appealed to the West for aid to shut down the Chernobyl' nuclear plant and help with the clean up of radiation-contaminated land.

Despite its environmental problems, the country is still the granary of Eastern Europe. Before the disintegration of the Soviet Union, 46 per cent of its agricultural products came from the Ukraine – although Ukrainians made up only 18 per cent of the total population and their country had only a 3 per cent share of the total area of the Soviet Union. More than 50 per cent of the Soviet Union's sugar beet and sunflowers were grown in the Ukraine which also produced 25 per cent of the Union's potatoes, vegetables and wheat.

Ukrainians have a reputation for being hardworking and proud people who preferred the hard but free life of the peasant to serfdom. Mismanagement by communist functionaries has in the past lowered agricultural production. Forced agricultural reform, including collectivisation, under Joseph Stalin resulted in famine which cost the lives of millions. But today the Ukraine still produces the bulk of the region's food needs. Many of its agricultural exports are handled through its Black Sea ports. The ports are also used as transshipment centres for cannabis and opium produced illegally in the country. (The government has done little to date to clamp down on illegal production.)

MINING AND INDUSTRY

The Ukraine was the second largest economic component of the Soviet Union after Russia. Its heavy industry was well developed, supplying other Soviet countries with raw materials. Industrial concerns, based on mineral wealth, had sprung up between the Donets River and the Dniester in the 19th century, at a time when Russia was still a largely agricultural nation. In the Second World War, Adolf Hitler's armies invaded the Ukraine, devastating the land, destroying industries – and killing hundreds of thousands of Jews in concentration camps. After 1945 it took many years of reconstruction to achieve pre-war levels of production.

Today the Ukraine has an even spread of industries and produces a wide variety of goods, unlike Russia whose industries are concentrated around a few isolated centres. Ukrainian industries produce sugar, tinned fish and other processed food, iron and steel, machinery, television sets and electric motors. The Donets basin is an industrial centre and one of the largest coal-mining areas of the region; industries are also concentrated along the Dnieper River near the mines of KRYVYY RIH. The country's coal and iron ore production has been declining since the collapse of the Soviet Union and the accompanying economic crisis, and since world market prices for the ores have been depressed. In its best years, the Donets basin produced more than 200 million tonnes of coal per year; in 1991-2, production was down to 132 million tonnes, and the annual output has declined since then.

After independence, the Ukrainian government started a gradual process of privatisation, especially of small and medium-size concerns. However, despite this, and despite some economic reform, the country's production slumped to pre-1990 levels, suggesting that the changeover to a free market economy would take many years. Between 1993 and 1994, industrial growth dropped by 50 per cent; by early 1994, the Ukraine had a monthly inflation rate of 80 per cent.

Little export was encouraged, with a resultant huge trade deficit. By late 1993, people had resorted to illegal trading and subsistence farming to survive.

RELATIONS WITH RUSSIA

The Ukraine has only small deposits of oil and natural gas and is heavily dependent on oil imports. Until 1991 Russia supplied it with an annual 57 million tonnes of oil; by 1994, only 40 million tonnes a year were being supplied by Russia and the country has had to look for oil imports elsewhere on the world market, at a higher price. Naturally,

this has produced some tension between the Ukraine and Russia.

One of the countries willing to supply oil cheaply to the Ukraine is Iran – but not without an ulterior motive. At independence the Ukraine still controlled part of the former Soviet Union's nuclear arsenal. A member of the Commonwealth of Independent States, the Ukraine initially intended to keep its military advantage – and so, came to loggerheads with Russia over the nuclear weapons and the Black Sea navy.

In 1992 Russia and the Ukraine reached an agreement on the Black Sea Fleet but in 1994, in the port of Odesa, the two countries quarrelled over the fleet when Ukrainian forces took over a Russian-controlled naval base. The base was later returned into Russian hands, but the Ukrainian-Russian relationship has been further strained over the issue of Crimea's independence.

Relations with the West, too, became lukewarm soon after independence, when the Ukrainians refused to sign any nuclear disarmament treaties – one reason being that Russia still had nuclear warheads, and maintained 2000 troops in the Ukraine to guard the remaining nuclear missiles. In 1994, however, the Ukraine, Russia and the USA finally agreed to a weapons treaty by which the Ukraine would transfer most of its nuclear warheads to Russia within the year.

Other significant Ukrainian developments have been the signing of a trade agreement with the European Union and the surprise victory in presidential elections of the progressive Leonid Kuchma, who soon afterwards initiated a radical economic reform package, including mass privatisation of land and property.

UKRAINE AT A GLANCE	
Area 603 700 km² (233 035 sq miles)	
Population 51 900 000 **Per km²** 86 **(Per sq mile** 223)	
Capital Kiyev (Kiev)	
Government Republic	
Currency 1 hryvnya = 100 kopecs	
Languages Ukrainian (65%), Russian (33%), Romanian, Polish	
Religions Christian (Ukrainian Orthodox, Ukrainian Catholic)	
Climate Temperate continental; subtropical, with little rain, in the south. Average temperature in Kiyev ranges from -5.3°C (22.5°F) in January to 19.3°C (66.7°F) in July	
Land use Cultivation 58%, meadows and pastures 12%, other 30%	
Mineral resources Iron ore, manganese, mercury, natural gas, oil, coal, salt, nickel, sulphur, titanium, uranium	
Main primary products Timber, grains, vegetables, sugar beet, meat, dairy products; iron ore, coal, oil, natural gas	
Major industries Machinery and transport equipment, coal, iron and steel, chemicals, electronics, food processing	
Main exports Iron and steel, metals, food, coal, electricals	
Annual income per head (US$) 2326	
Population growth (per thous/yr) -3	
Life expectancy (yrs) Male 65 **Female** 75	

Ulsan *South Korea* Industrial port on the southeast coast, 50 km (30 miles) from PUSAN. It is the heart of the country's special industrial area known as the Ulsan Industrial District. It has oil refineries and petrochemical plants, and manufactures motor vehicles and ships.

Population 683 000
Map Korea – De

Ulster (Cuige Ulaidh) *Ireland/United Kingdom* One of the ancient provinces of Ireland, Ulster was the northern kingdom, extending south to Louth and west to the Atlantic coast in Donegal. It was colonised by the British in the 16th and 17th centuries. In 1922, six of its nine counties (Antrim, Armagh, Fermanagh, Down, Londonderry and Tyrone) remained within the United Kingdom as the province of Northern Ireland; the other three – Donegal, Monaghan and Cavan – became part of what is now the Republic of Ireland. Northern Ireland is still called Ulster by some people.

Population (Northern Ireland) 1 589 400; (Republic) 232 200
Map United Kingdom – Bc

Ulundi *South Africa* Capital of the kingdom of Zululand during the reign of King Cetshwayo (1873-9), site of the final battle of the Anglo-Zulu War (1879) and capital of the self-governing territory of Kwa-Zulu (1977-1994).

Situated about 160 km (100 miles) north of Durban, Ulundi has developed into a modern administrative centre.

Population 11 100
Map South Africa – Db

Ul'yanovsk *Russia* City on the Volga River, 170 km (105 miles) south of KAZAN'. Set on the top of a hill, Ul'yanovsk is a major stage on the Volga riverboat route. It was formerly called Simbirsk, and was renamed in 1924 in honour of the revolutionary leader Vladimir Ilyich Lenin (1870-1924), who was born here as Vladimir Ilich Ulyanov. His birthplace is now a memorial museum. The city's industries include engineering, production of machinery, electronics, textiles, vodka and beer, and food processing.

Population 648 000
Map Russia – Fc

Umbria *Italy* Landlocked central region covering 8456 km² (3265 sq miles) in the Apennine Mountains to the north of Rome. There is a dense scatter of medieval hill towns, including ASSISI, GUBBIO, SPOLETO, ORVIETO, Todi and Città di Castello.

The region's green hills and carefully cultivated valleys have changed little over the centuries, producing grains, olives, grapes and fruit. Farming is the main occupation. Industry is centred on the two provincial capitals – steel and chemicals in TERNI, and textiles, clothing and confectionery in PERUGIA.

Population 822 800
Map Italy – Dc

Umeå *Sweden* Port and timber processing town on the Gulf of Bothnia, 676 km (420 miles) northeast of STOCKHOLM. It has Sweden's northernmost university, and is linked by ferry with the Finnish city of Vaasa.

Population 94 900
Map Sweden – Dc

Umm al Qaywayn *United Arab Emirates* One of the seven emirates, it stretches for 24 km (15 miles) along The GULF and covers 750 km² (290 sq miles). The sheikdom has large unexploited reserves of natural gas. In the chief town, also called Umm al Qaywayn, fine examples of old Arab architecture contrast with high-rise office and apartment blocks. An old fort serves as the police headquarters. The town has a small port which trades in bulk ore, transships containers, and can handle tankers.

Industries include cement manufacture (the mainstay of the emirate's economy), an asbestos factory and some small plants producing agricultural fertilisers and chemicals. Offshore are the Umm al Qaywayn Islands, with 18th-century watch towers and a wide variety of wildlife.

Population (emirate) 27 000; (town) 2900
Map United Arab Emirates – Ba

Umm Said (Musay'id) *Qatar* Important industrial town and port 40 km (25 miles) south of DOHA. It contains Qatar's main oil terminal, and has a number of heavy industries, including steel, fertilisers and petrochemicals.

Population 7000
Map Qatar – Ab

Umtali *Zimbabwe* See MUTARE

Umtata *South Africa* Capital of the former Republic of TRANSKEI (1976-94), and a marketing centre. Umtata is situated in the Eastern Cape, 190 km (120 miles) north-east of EAST LONDON. It was settled by whites in 1869. Today the city has a university and both Anglican and Roman Catholic cathedrals.

Population 45 000
Map South Africa – Cc

unconformity Break in a sequence of SEDIMENTARY ROCKS, representing an interval when no sediments were deposited. The older rocks may be tilted and eroded before deposition resumes. Another type of unconformity occurs when sediments are laid down on top of IGNEOUS ROCKS.

undertow An undercurrent moving down a beach, caused by a backflow of water deposited on the beach by a breaking wave.

United Arab Emirates (UAE) See p 695

United Kingdom (UK) See p 696

United Provinces (United Provinces of Agra and Oudh) *India* See UTTAR PRADESH

United States of America (USA) See p 702

Unzen-Amakusa *Japan* National park, covering 257 km² (99 sq miles), in western KYUSHU Island. The park includes the Shimabara Peninsula, where the Christian armies of Kyushu were finally defeated by the Tokugawa shogunate in 1638. After the battle a few Christians took refuge on the nearby Amakusa Islands, where their descendants kept their faith alive in secret for 250 years until the Tokugawa era ended.

Map Japan – Bd

Upemba National Park *Zaire* Reserve covering 11 730 km² (4529 sq miles) in central Shaba region, west of Lake Mweru. It contains swamp-land, with crocodiles and hippopotamuses, in the west; to the south and east is savannah country with animals such as antelopes, buffaloes, eland and zebras. The park was established in 1939.

Map Zaire – Bb

Upington *South Africa* Northern Cape Province market town on the Orange River, some 365 km (230 miles) east of KIMBERLEY. It is an important road and rail junction on the route to Namibia and a transit point on the way to the Augrabies Falls and Kalahari Gemsbok national parks. Upington, founded as a mission station in 1871, is set in an irrigated region that produces lucerne, fruits, vegetables, sheep, goats and cattle.

Population 36 320
Map South Africa – Bb

Upolu *Western Samoa* Second largest but most populous island of the Pacific nation, covering 1100 km² (425 sq miles). It is mountainous, rising to 1113 m (3652 ft) at Mount Fito; in parts it is covered by rain forests with high waterfalls. Coconuts are the main crop. The capital, APIA, is on the north coast.

Population 112 230
Map Pacific Ocean – Ec

Upper Austria (Oberösterreich) *Austria* Hilly, well-forested northern state noted for its agriculture. Tourism is an important source of income. Industry is concentrated at the capital LINZ, STEYR and Ranshofen.

Population 1 340 080
Map Austria – Ca

Upper Egypt *Egypt* See LOWER EGYPT

Uppland *Sweden* Province, the historical cradle of Sweden, covering 12 673 km² (4893 sq miles) in the east of the country, just north of Stockholm. It is rich in minerals. The iron-mining district of Dannemora is on its northern edge. UPPSALA is the provincial capital.

Population 1 129 000
Map Sweden – Cc

Uppsala *Sweden* Medieval city, 68 km (42 miles) north of STOCKHOLM. It has the country's oldest university (founded in 1477), which houses manuscripts going back to the 6th century. Sweden's kings were once crowned in its 13th to 15th-century cathedral. It also has a 16th-century castle, and to the north is Old Uppsala, with the prehistoric burial mounds of legendary kings.

Known as Östra Aros in the Middle Ages, Uppsala is regarded as the country's historical centre. It lost much of its importance during the 18th and 19th centuries, when the royal family moved to Stockholm. Today the city produces machinery, building materials, pharmaceuticals and processed foods.

Population 174 600
Map Sweden – Cd

up-valley wind See ANABATIC WIND

Ur *Iraq* Archaeological site of the ancient Sumerian town which was the home of the Old Testament patriarch Abraham. The site lies about 310 km (190 miles) south-east of the capital, BAGHDAD. Ur was discovered around 1853 by the British Consul in Basra. In its centre stands the temple of the moon god Nanna-Sin with a zig-gurat – a stepped tower which gives a bird's-eye view over the ruined city. The city, founded between 6000 and 5000 BC, fell into ruin in the 4th century BC after the course of the Euphrates changed and river trade by-passed the city. The city is mentioned in the Bible as Ur of the Chaldees (Chaldeans).

Map Iraq – Cc

Urals *Russia* North-south mountain range forming the geographical boundary between Europe and Asia. The Urals run for 2000 km (1240 miles) from the Kara Sea in north Russia to the Aral Sea and the Kazakh steppes in the south. Their average height is about 1000 m (3300 ft) and they are divided into three contrasting sections.

The Northern Urals are the narrowest and highest, culminating at 1894 m (6214 ft) in Mount Narodnaya. The mineral-rich Middle Urals consist mainly of a low, densely forested plateau forming the historic gateway to Siberia. They contain the Ural Industrial Region, an industrial and mining area possessing many minerals – including iron, bauxite, nickel, copper, asbestos, chrome, platinum and gold.

The region also has large oil fields. The Southern Urals, made up of rich pastureland, contain the source of the Ural River, which flows south-eastwards for 2535 km (1575 miles) into the Caspian Sea and is navigable from URAL'SK.

Map Russia – Gc

Ural'sk *Kazakhstan* Market town on the Ural River, 470 km (292 miles) north of the Caspian Sea. It trades mainly in cattle and grain; its chief products are processed food and leather goods.

Population 214 000
Map Kazakhstan – Ba

Urbino *Italy* Tourist town, 110 km (70 miles) east of FLORENCE. Its narrow streets, cheek-by-jowl buildings and lack of modern development seem to suggest that in Urbino, time has stood still since the Middle Ages. Raphael, one of the greatest of the Renaissance painters and one of the architects of St Peter's in Rome, was born here in 1483.

Urbino's pride is its vast, 15th-century ducal palace – which now houses the Marche National Gallery, containing the finest picture collection in the region, covering many Italian schools of painting. The town's university dates from 1506.

Population 15 900
Map Italy – Dc

Uri *Switzerland* German-speaking canton in central Switzerland said to have been the home of the legendary Swiss folk hero William Tell, who opposed Austrian rule. Uri was one of the three cantons of the infant Swiss state formed in 1291. Covering 1077 km² (416 sq miles), it consists largely of mountains, forests and glaciers. The canton's arms is a bull's head, and the name Uri probably comes from *Urochs*, meaning 'wild bull'.

The canton's main industries are construction and tourism, both largely consequences of Uri being a transit canton on the Gotthard route. The capital is the small town of Altdorf (population 8200), which has a statue of William Tell on the spot where he was supposed to have shot the apple from his son's head.

Population 35 500
Map Switzerland – Ba

United Arab Emirates

A FEDERATION OF SEVEN SHEIKDOMS THAT IS USING ITS FABULOUS WEALTH FROM OIL TO DIVERSIFY THE ECONOMY AND INVEST OVERSEAS

The coast of the United Arab Emirates (UAE), 1318 km (819 miles) long, was once known as the Pirate Coast. Armed with fearsome curved daggers (*kunjahs*) and scimitars (*qattaras*), Arab corsairs preyed on European ships trading in The GULF. Their ruined forts and watchtowers can still be seen along the shores.

At the beginning of the 19th century, RAS AL KHAYMAH had a population of 390 000. It had a fleet of 880 ships and 19 000 seamen to crew them. The pirates began foraging as far afield as the Indian Ocean, so that by the mid-19th century Britain was forced to interfere, to protect the merchant fleets of the West. Britain destroyed Ras al Khaymah, enforcing its reputation as the rulers of the seas. It imposed a truce in 1853 and for this reason the area became known as the Trucial Coast.

Now the sigh of the trade winds in the rigging and the fierce cries of battle have given way to the sound of the engines of oil tankers making their way along The Gulf.

SEVEN SHEIKDOMS

The UAE is a federation of seven sheikdoms formed in 1971 and 1972 when Britain left the region: ABU DHABI, DUBAI, SHARJAH, Ras al Khaymah, AL FUJAYRAH, UMM AL QAYWAYN and AJMAN. Since 1962 oil production has brought undreamed-of riches to the Emirates which once traded in spices, pearls and slaves. Abu Dhabi, the largest of the group, has the UAE's richest oil wealth.

The country, which covers 77 700 km² (29 993 sq miles), has a short coastline east of the MUSANDAM Peninsula, on the Gulf of OMAN, as well as that on The Gulf. The land is mainly flat, sandy desert but to the north on the peninsula, the HAJAR MOUNTAINS rise to 2081 m (6826 ft). Temperatures in the hot, humid summer can reach 49°C (120°F) in July and August, but from October to mid-May the weather is delightful, with warm, sunny days and pleasantly cool evenings. Scant rain is brought mainly by winter storms. The emirate of Ras al Khaymah, the coastal plain of Al Fujayrah and the oases are fertile, but most of the remaining land is desert.

As much as 80 per cent of the Emirates' estimated 1.9 million population consist of expatriates, mainly Iranians, Pakistanis, Indians and other East Asians. The development financed by oil wealth has drawn immigrants like a magnet with men outnumber women by two to one.

The seven states abolished their separate ministries in 1974 and the country is ruled by a council of ministers, which is appointed by the president, a member of the Supreme Council of Rulers, which in turn consists of the rulers from the seven states – each one an absolute monarch in his own state. At least five members of the Supreme Council must agree before a decision is passed.

There is also an appointed 40-member Federal National Council, an advisory body recently revived in response to criticism of the lack of democracy. The country is Muslim, strict but not fanatical, although fundamentalism has been gaining ground. Restrictions are not, however, imposed on Westerners.

INTO THE MODERN WORLD

Until the 1950s the coastal towns lived from trade in pearls and spices. Most of the population still live on the coast, but since the discovery of oil and gas, the UAE has used its wealth to diversify the economy at a staggering rate, to finance social improvements and to invest overseas. Abu Dhabi and Dubai are the main industrial centres. Among the ancient forts along the coast there now stand an oil refinery, gas liquefaction and fertiliser plants, a steel rolling mill and aluminium smelter, and factories making cement and other building materials. The bustling construction industry has turned tiny fishing ports into modern oil terminals and harbours for bulk carriers and other vessels.

New office blocks have replaced many of the traditional Arab buildings in the towns. Dubai has a 39-storey International Trade Centre, and its modern airports cater for international flights. Where once health and educational facilities were limited there are now hospitals and clinics, schools and universities. Modern roads link the sheikdoms to each other.

Near the coast, irrigation projects, based on giant desalination plants producing fresh water from sea water, have extended agriculture beyond the oases. Fields of wheat, tobacco and alfalfa lie among the sand dunes while thriving market gardens provide tomatoes, aubergines and melons for the home market.

In the past, fish were one of the few sources of protein and fishing is still an important industry – although farmers have been encouraged to raise cattle. The clear Gulf waters are rich in prawns, tuna, mackerel, anchovies and sardines, and for sport there are sailfish, sharks and marlin. The UAE can consequently meet 25 per cent of its own food needs.

Piracy has given way to the excitement of camel racing, falconry, wrestling and even ice skating on artificial rinks. Camel races usually take place soon after the cool desert dawn, either at established tracks or at impromptu meeting places in the desert. There is no betting, but cheering spectators follow the animals by truck until the winners emerge in a cloud of sand and dust. Falcons are trained for hunting – although the bustard, a common prey, is becoming scarce. As a complete contrast to this ancient sport, football has been imported; it is played and watched with passion – just another step towards the Western way of life for these ancient sheikdoms.

UNITED ARAB EMIRATES AT A GLANCE	
Official name United Arab Emirates	
Area 77 700 km² (29 993 sq miles)	
Population 1 909 000 **Per km²** 24.5 **(Per sq mile)** 63.5)	
Capital Abu Dhabi	
Government Non-elected federal council	
Currency 1 dirham = 100 fils	
Languages Arabic, English (commercial), Hindu, Urdu, Persian	
Religions Muslim (96%), Christian, Hindu and other (4%)	
Climate Desert with high humidity. Average temperature in Sharjah ranges from 18°C (64°F) in January to 34°C (93°F) in August	
Land use Pastures 2%, other (mainly desert) 98%	
Main primary products Goats, sheep, camels, cattle, fruit, vegetables, fish and shellfish; oil and natural gas	
Major industries Oil production, natural gas liquefaction, steel milling, cement, aluminium smelting, fishing, fish processing, fishmeal	
Main exports Crude oil, natural gas, shrimps, prawns, fishmeal	
Annual income per head (US$) 22 220	
Population growth (per thous/yr) 50	
Life expectancy (yrs) Male 70 **Female** 74	

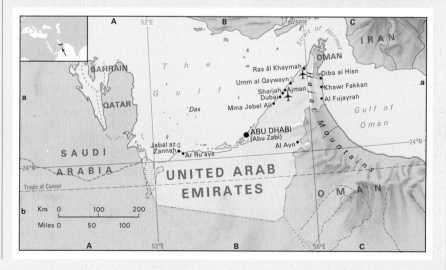

695

United Kingdom

THE CHANNEL TUNNEL HAS ENDED BRITAIN'S ISLAND STATUS – BUT ARGUMENT OVER POLITICAL AND ECONOMIC LINKS WITH CONTINENTAL EUROPE GOES ON

When, in the mid-19th century – that great golden age of engineering and bridge-building – people in the United Kingdom first began to envisage a tunnel link between themselves and Continental Europe, it was not without raising profound fears among the proud inhabitants of a long-independent island nation. What would stop plague-carrying rats from infecting the country in droves? Or enemy soldiers from making a silent surprise attack?

The real drawback was, however, economic. A large proportion of the UK's cross-Channel trade was in bulky grains that still would have to be carried by ship. Cheap fuels and super-lorries changed the economic argument. And since the end of the Second World War more and more people of all income groups have become accustomed to travelling and holidaying on the Continent. Members of today's younger generation certainly think of themselves as 'European' in a way that their grandparents could not have imagined.

Joint Anglo-French construction of the tunnel began in 1987, and in May 1994 the tunnel between Folkestone in Kent and Calais in Artois was officially opened. It was a historic moment – but one that somehow failed to capture the public's imagination. Although the United Kingdom joined the European Economic Community (now the European Union) in 1971 and confirmed its membership by a 2 to 1 majority in a referendum of 1975, the British people remain reluctant to cast their lot – politically and institutionally – with the rest of Europe.

The three main national political parties are united in their 'commitment to Europe', but the British are increasingly wary of ever closer European union and evermore powerful supra-national European laws and institutions. Nearly a thousand years have passed since the last successful invasion of Britain, and its people retain a fierce pride in their island independence.

BRITAIN'S GEOGRAPHY

The United Kingdom is, in reality, an archipelago of islands, stretching from Shetland and Orkney in the extreme north to the Isle of Wight and the Channel Islands in the south. England was cut off from the European landmass by the Strait of Dover about 7500 years ago, when the retreat of the last Ice Age allowed the North Sea to submerge the lowland forests and marshes that linked what was to become England to the Continent. Although it ranks by population as the 17th largest country in the world, in area it is about the size of New Zealand, half that of France. Britain rests on a continental shelf, and the shallow waters around its rugged, highly indented coastline have been a rich breeding-ground for fish.

England and Wales may be said to be divided into a highland and a lowland zone, the line of bisection running, roughly, from Exmouth in the south-west to the river Tees in the north-east. The highland zone contains four great uplands: the moors of the southwest peninsula (Cornwall and Devon); the mountains of Wales; the Pennine range, (commonly called 'the backbone of England') with its deep U-shaped valleys known as dales, that divides Lancashire from Yorkshire, and, wet and north of the Pennines, the beautiful Lake District of Cumbria.

The lowlands exhibit a remarkable diversity in a short space, including as they do the fens of East Anglia, the alluvial plain of the River Thames, and the rolling countryside of the North and South Downs. Across the scarped uplands and clay vales of southern England the imprint of more than 2000 years of cultivation can still be seen. Here and there, chalk downs bear the marks of terraces farmed in the early Iron Age, on Salisbury Plain, for example, and near the Uffington White Horse (itself perhaps 1000 years older) in Oxfordshire. Also in Oxfordshire, near North Leigh, the site of a Romano-British village is surrounded by obvious traces of the ridges and furrows of medieval communal cultivation.

Scotland, like England, has deeply contrasting highland and lowland regions, the smaller lowland zone in the south containing the vast majority of its population, concentrated in Glasgow and Edinburgh. The highlands have historically supported livestock farming – sheep on the hills and cattle in the fertile valleys – while the lowlands have come under the plough.

Farming is a year-round enterprise, thanks to an exceptionally mild and persistently moist climate. London lies nearly 9° of latitude north of Toronto, yet its winter temperature rarely dips below freezing. This mildness is the result of the warming influence of the Gulf Stream, over which blow the prevailing south-westerly winds.

Temperatures vary only slightly, from an annual average of 8°C (46°F) in the Hebrides off the west coast of Scotland to 11°C (52°F) in London and the south-east. Rainfall, fairly evenly distributed throughout the year, falls much more heavily in the mountainous north and west – about 1600 mm (more than 60 in) annually – than in the lowlands of the south and east, which receive about half that amount.

The Romans valued the island for its rich mineral resources, especially tin, iron and lead, which they were the first to mine. Together with coal, these resources fuelled the world's first industrial revolution, roughly from the mid-18th century, a revolution that was greatly assisted, in the days before the railways, by England's abundant rivers and by the close proximity sof its coastline to places of manufacture. Nowhere in the United Kingdom is more than 200 km (120 miles) from the sea.

Concern to protect the environment, greatly increased today by fears about the depletion of the world's rain forests, the polluting of air and water and the possibly disastrous consequences of alleged global warming, is not new. Legislation against the industrial pollution of the British environment goes back at least to the mid-19th century. More recently, the Clean Air Acts of 1956 and 1968 have been so effective in eliminating smog that 'pea-soupers' – choking, dense yellow fogs – have become a thing of the past.

Monitoring water pollution is the task of the National Rivers Authority, which increasingly is required to bring Britain's rivers and coastal waters up to the standards set by the European Commission. Between 1985 and 1992 the United Kingdom achieved a reduction of 69 per cent in mercury inputs into coastal waters.

LESS AIR POLLUTION

Another success has been the reduction of 40 per cent, since 1970, in emissions of sulphur dioxide into the atmosphere. In April 1995, the European Union's scheme of 'eco-management and audit' was due to become fully operational. The government has meanwhile set its own target for half of all recyclable household waste to be recycled by the year 2000.

Beyond the European sphere, Britain has pledged to contribute £89 million to the Global Environment Facility, a United Nations fund established to assist poor countries to participate. The protection of both natural and artificially created sites of lasting interest and beauty has long been a British priority. In the UK, there are scores of organisations whose purpose is to protect the countryside, wildlife and the country's architectural heritage.

The World Heritage Convention has listed 13 sites of outstanding value in Britain. These are: Canterbury Cathedral and its environs; Durham Cathedral and Castle; the Tower of London; the Palace of Westminster, Westminster Abbey and its environs; Blenheim Palace, Oxfordshire; Studley Royal Gardens and Fountains Abbey, Yorkshire; Ironbridge Gorge, Shropshire; Stonehenge and Avebury prehistoric stone circles, Wiltshire; City of Bath, Avon; Hadrian's Wall; Islands of St Kilda, Scotland; Edward I's castles and town walls, north Wales; Giant's Causeway and adjacent coast, Northern Ireland.

Though surrounded by water, Britain has, throughout its history, been subject to waves of immigration. Burial mounds still provide traces of Neolithic settlers who arrived about 6000 years ago. They settled at places like Avebury and Stonehenge, developed by later people of the Bronze Age, who built the great megalithic, or stone, circles that still stand. In the latter part of the Bronze Age (2200-650 BC) and the early Iron Age (650 BC-AD 43) came the Celts and other tribes. They built the large earthen encampments that still crown many uplands. Their languages survive in the form of Welsh and Gaelic, still used in what is sometimes called the 'Celtic fringe' beyond the borders of England.

The Romans were the next to make their

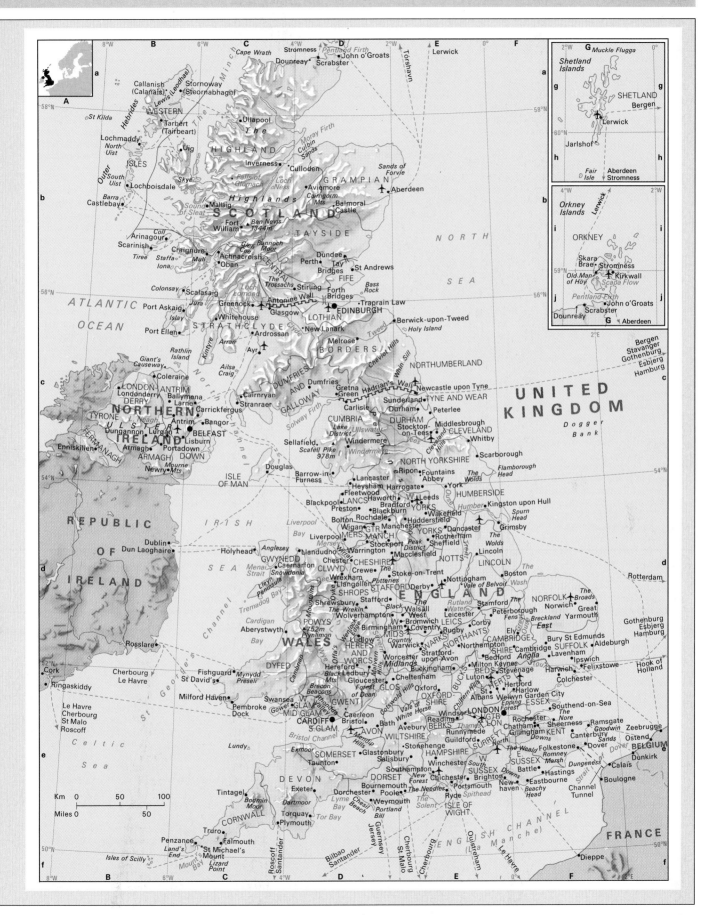

appearance. Julius Caesar led two expeditions to Britain's shores in 55 and 54 BC, but permanent Roman occupation began only in AD 43, when Emperor Claudius conquered Britain for the Roman Empire. That occupation lasted until 408, during which time the Romans established order, built a road network and laid out the first real towns – places such as Chester and Chichester.

'BARBARIAN' INVASION

The Roman withdrawal, completed by 409, was followed by the influx of Angles (from whom England derived its name), Saxons and Jutes – part of the great wandering of peoples in Europe known as the barbarian invasions. For the next five centuries the Anglo-Saxons kept the native Britons confined to the north and the west, while in the prosperous south they established the English language, laid the foundations of the English Christian church and slowly evolved the institutions of central kingship.

The Anglo-Saxons weathered the onslaughts of the Danish Vikings in the 9th and 10th centuries, though for a while they had to pay tribute money – or Danegeld – to the conquerors of the eastern part of the country. Then, in 1066, the last great invasion of England, undertaken by William, Duke of Normandy, overthrew the rule of King Harold and resulted in the Norman Conquest, sealed at the Battle of Hastings. By the time of Domesday Book in 1086 most of England had been pacified and brought under the rule of an imported Norman feudal aristocracy.

The Anglo-Norman conquest of Wales began in the 12th century; the last principality – Gwynedd – succumbed in 1282, though it was not until the Tudor Acts of Union of 1536 and 1542 that Wales was brought entirely under English law and governance. The kingdom of Scotland, formed in the 9th century by the union of Picts and Scots, resisted the Norman English more successfully, until James VI of Scotland, a great-great-grandson of Henry VII of England, succeeded to the English throne in 1603 and ruled both kingdoms. In 1707 the Scottish parliament voted to dissolve itself and merge with the parliament of England and Wales at Westminster.

Attempts to bring the Roman Catholic island of Ireland under Anglo-Norman control began in about 1170, intensified during the great 'plantation' of Ireland by English Protestant landowners in the 16th and 17th centuries, and came to their fruition in the Act of Union which created the United Kingdom of Great Britain and Ireland in 1801.

The people of the United Kingdom are thus the product of a rich mixture of stocks. Since 1945 the country has received one more wave of immigration, on a small scale, mostly from independent states of the former British Empire, especially India, Pakistan and the West Indies. The 1991 census, which for the first time asked questions about 'ethnic' status, revealed that 94.5 per cent of the population belonged to the so-called 'white' group, while the remainder, slightly more than 3 million people, described themselves as belonging to some other ethnic group. That 'non-white'

population is concentrated in industrial, urban areas, more than half of it in the south-east, especially Greater London, the rest chiefly in Leicester, the West Midlands and West Yorkshire. About half of that ethnic minority was born in the UK. There are now about 1.5-2 million Muslims in the United Kingdom, 320 000 Hindus and 300 000 Sikhs. There are also 300 000 Jews – the second largest Jewish grouping among European nation states.

The UK has two established Christian churches (that is, legally and financially underpinned by the state): the Anglican, episcopal Church of England and the Presbyterian Church of Scotland. The Irish Church was disestablished in 1869, the Welsh in 1920.

The number of nominal members of the Church of England far exceeds the declining number of regular churchgoers; the Church of Scotland had (in 1994) an adult membership of nearly 733 000. Among the Dissenting Christian churches the largest are the Roman Catholic (about 10 per cent of the population), the Methodist (410 000 adult members) and the Baptist (150 000 adult members).

THE CONSTITUTION

Britain is undoubtedly a more multi-cultural society than ever before in its history. It remains, even so, jealous of its distinctive political heritage, which ever since the kingship of Anglo-Saxon days and the emergence of parliamentary traditions in the High Middle Ages, has, as Lord Tennyson described, 'broadened down from precedent to precedent'.

The country has for centuries lived under the rule of law in a representative parliamentary monarchy, with sovereign authority vested in the Queen-in-parliament. It is often incorrectly stated that the UK has no written constitution; in fact, it does have a constitution, parts of which are written down in statute, parts of which derive from the common law and parts of which are mere conventions with no legal standing.

The spread of parliamentary institutions throughout the world, largely through the instrument of the British Empire, has, along with the spread of the English language itself,

been the United Kingdom's most evident contribution to the globe.

Of the three constituent parts of parliament – the monarch, the House of Lords and the House of Commons – only the Commons exercises effective power. The monarch, who is the Supreme Governor of the Church of England and by the Act of Settlement in 1700 must be a Protestant, retains certain prerogative powers – the choosing of a prime minister, the dissolving of parliament – but does not exercise them against the advice and will of ministers or the House of Commons.

The House of Lords, which in 1994 consisted of the two archbishops and 24 bishops, 773 hereditary peers and 399 life peers, has no power to amend bills sent to it from the lower house that deal exclusively with taxation or expenditure, and it is able to delay the passage of other legislation for a year only.

The House of Commons, which until the next redistribution of seats consists of 651 members, must be elected every five years and may be elected at much shorter intervals. All persons over the age of 18 – except for certain disqualified categories such as prisoners – are eligible to vote in national and local elections. Peers of the realm are, by a convention that evolved after the 1832 reform act, banned from taking any part in parliamentary elections.

Representation in the lower house is supposed to be by population; in fact, Scotland's 72 members, 11.5 per cent of the total, exceed its 8 per cent share of the population and so, to a lesser extent, do Wales's 38 members, 10 per cent of total seats for a 5 per cent share of the population. England has 524 seats and Northern Ireland 17. At the elections of 1992 the Conservative party won 336 seats, Labour

▼ NORTHERN LIGHTS Edinburgh – called 'The Athens of the North' – is the capital of Scotland, which has been ruled by the same monarch as England and Wales since 1603, when Scottish King James VI became James I of England.

271 and the Liberal Democrats 20. At the end of 1994 there were 60 women MPs, three Asian and three black MPs.

No statute requires it, but, with rare exceptions, the bulk of the cabinet, or executive, is drawn from the House of Commons. The day-to-day running of the country is in the hands of the prime minister and his cabinet colleagues, but constitutionally, at any rate, the system of responsible government that has pertained since the mid-19th century means that the cabinet's existence depends upon the continuing support of a majority of the House of Commons.

Since the Lord Chancellor, the senior law officer in the land, presides over the House of Lords, and the House of Lords itself is the supreme court of appeal in legal matters, it will be seen that the executive, legislative and judicial are intertwined – unlike the separation of those powers which is maintained in many nations, notably the United States.

In addition, the United Kingdom has 87 seats in the European parliament at Strasbourg, to which representatives are elected every five years. The European parliament is a weak institution; effective power rests with the European Commission, which initiates legislation, and with the Council of Ministers, which is the law-making body. The decisions of the Council have the force of statute in the UK and rulings of the European Court are binding on British courts.

NORTHERN IRELAND

The province of Northern Ireland, separated from the rest of the country by the Irish Sea, occupies a special constitutional position. When the southern part of Ireland won its independence from the UK in 1921, six counties of the north, popularly known by the ancient name of Ulster, remained part of Britain.

For the next 50 years Northern Ireland had its own regional parliament, but the minority Roman Catholic population felt permanently excluded from power and discriminated against in matters such as housing and employment. In 1921 Protestants outnumbered Roman Catholics by nearly 2 to 1; at the 1991 census some 51 per cent of the people called themselves Protestant, and 38 per cent said they were Roman Catholic.

The eruption of sectarian strife in the late 1960s – the beginning of the 'Time of Troubles' – led to the introduction of the British army to keep order in 1969, and the abolition of the Ulster assembly at Stormont in 1972. In the conflict that raged for the next 25 years between the Roman Catholic and nationalist Irish Republican Army (which seeks a united Ireland) and the several paramilitary terrorist groups supporting continued union with the United Kingdom, 3210 people were killed.

The IRA ceasefire of August 1994 followed the 1985 Anglo-Irish Agreement and the 1993 Downing Street Declaration. The Agreement created a permanent inter-governmental conference at which Irish and British ministers can discuss issues such as cross-border cooperation and security; the Declaration merely stated fundamental principles – notably the

right of the majority of Ulster people to decide the province's future – without advocating a solution to the Irish problem. In February 1995, however, the peace process was taken a further step forward when the British and Irish prime ministers jointly launched a 'framework document' setting out proposals for a Constitutional Settlement.

The United Kingdom was the world's first highly industrialised and densely urbanised nation. The 1851 census revealed that for the first time more people lived in towns than in the countryside. British prosperity in the high days of Queen Victoria's Empire was based on the heavy 'smokestack' industries – textiles, coal, shipbuilding, for example – and on the easy availability of a more or less captive imperial market overseas.

In the 20th century the British economy, like that of most countries of the western world, has been transformed. One statistic can tell the story: in 1955, there were 850 British coal mines in operation; by 1992 the number had dwindled to 50 and by 1995 to 31. Agriculture's place in the daily life of the people has continued to decline to the point at which, by 1994, only one per cent of the work force was employed in agriculture, forestry and fishing. Employment in the booming service industries, on the other hand, has doubled in the past half-century and stood in 1994 at 73 per cent of the work force.

ATTRACTING THE TOURIST

Jobs in the service industries have been spurred by the growth of tourism, which, since the mid-1980s, has been the second largest 'industry'. In 1994 more than 20 million tourists from overseas entered Great Britain. The income derived from tourism is now the sixth largest in the world. For tourists, the radical changes that have taken place in

the country are obscured by a history so long and so rich in incident that it has imposed itself upon almost every aspect of both town and country.

Great cathedrals, such as those at Canterbury, Durham, Ely and Lincoln, raised almost 1000 years ago, divert the visitor's eye from the present. The music sung in them is a basic ingredient of English culture, celebrated in the Three Choirs Festival held in turn at Gloucester, Worcester and Hereford cath-

▲ HUB OF REVOLUTION Arkwright's cotton mill at Cromford, Derbyshire is one of many 18th and 19th-century factories in which the technology of the Industrial Revolution generated the wealth that built a huge empire.

edrals. Other festivals of the arts include the great international event lasting three weeks at Edinburgh at the end of every summer and others at York, Harrogate, Bath, Cheltenham and Brighton.

The literary trail, too, has its powerful associations – the Lake District of Wordsworth, Coleridge and Southey; the romantic Exmoor of Lorna Doone; Burns's Ayrshire countryside, the Brontës' Yorkshire, and Hardy's Dorset. Above all, there is Stratford-on-Avon, the birthplace (and burial place) of William Shakespeare. The greatest tourist attraction of all remains the capital, London, with its many great art galleries, theatre tradition, shopping streets and its royal associations.

A great boost to the British economy (especially to its balance of trade) was the discovery in the 1960s of substantial offshore oil and gas reserves in the UK Continental shelf. By the end of 1993 more than 5000 wells had been, or were being, drilled in the shelf. The revenues from gas and oil contributed mightily to the boom years of the 1980s. During an eight-year period the economy grew without interruption, at its most rapid rate since the war; in a remarkable turn-around, the rate exceeded that of the rest of western Europe.

Prosperity was not evenly distributed. It was most noticeable in the financial sector and in the south. The North and the Midlands, more tied to the traditional heavy industries of the Industrial Revolution, were slow to feel the benefit.

The economic upsurge extended not only to places like the new city of Milton Keynes, home of the Open University instituted by Harold Wilson's government in the 1960s, and to other post-war new towns like Harlow, Stevenage and Crawley, but to older towns, like Swindon, which enjoyed a revival. From Hampshire to the valleys of the Chilterns, villages doubled in size, partly as the result of the rapid expansion of the electronics industry, which became a growth-leader in the new high-tech economy.

The same area was most heavily hit by the recession of the late 1980s and early 1990s, during which unemployment rose to more

▼ TUNNEL VISION An award-winning new international station at Waterloo is the London terminus of the rail link which takes passengers non-stop to Paris and Brussels through the Channel Tunnel, opened in 1994.

than 10 per cent of the work force. The slow-down was deeper and longer-lasting than anyone anticipated, but as Britain showed signs of emerging from slump at the end of 1992, it was expected that the economy would reap the rewards of improved management and labour relations, higher productivity and low inflation. The UK entered recovery with higher employment rates than most other countries in the European Union.

Manufacturing industry has for some years been under assault, partly because of low internal investment (public and private), partly because of increasingly fierce competition from overseas. Between 1971 and 1982 the sale of imported vehicles in the UK leapt from 15 per cent of the market to 45 per cent. For chemicals the rise was from 19 to 33 per cent, for electrical-engineering products from 18 to 49, and for textiles from 17 to 40.

The country has ceased to build ships; South Korea has replaced the great yards on the Tyne and the Clyde. And in 1994 the last great British car maker, the Rover Group, was sold to the German company, BMW – despite the fact that the number of cars on the roads in 1994 was about 20 million, ten times the number in 1952.

In the first half of this century the structure of British industry underwent a profound change, as ownership became concentrated in fewer and fewer hands. In 1909 half the country's manufacturers were produced by 2000

or more firms; by 1970 the same proportion of a much larger quantity of goods was being turned out by only 140 firms.

Recently there has been an effort to reverse that trend. By the end of 1991 there were in the region of 900 000 more small businesses than there had been in 1979.

YOUNG AND OLD

Concern about the long-term competitiveness of the British economy (and, on a larger scale, that of Europe itself) is accentuated by the implications for production and expenditure of recent population trends. The first is that the population is at nearly zero growth: forecasts are for it to rise from 58 million (mid-1993 estimate) to 59.9 million by 2001 and 61.3 million by 2011. The second is that, since people are living much longer, production has to support more and more people in their retirement.

The general death rate in 1992 was, at 11 per 1000 of the population, only slightly higher than the birth rate. (Analysts disagree about the consequences for society of the fact that illegitimate births continue to rise: 31 per cent of live births occurred outside marriage, in 1992.)

Circulatory disease now accounts for nearly 50 per cent of all deaths in the UK and death from heart disease itself remains high compared to the rate in most other developed countries. The figure remains high despite a marked

reduction in the smoking habit: 29 per cent of adult males and 28 per cent of adult females were smokers in 1992, compared with figures of 52 and 41 per cent 20 years earlier.

Poll after poll suggests that the greatly increased amount of leisure time afforded by longer lives, earlier retirements and shorter working weeks is chiefly filled by watching television and making home improvements. But Britain remains a country of outdoor pursuits. Angling is far and away the most popular participator sport in the country. But jogging, cross-country running, surfing, sailing, hang-gliding and climbing are increasingly popular.

Spectator sports, too, continue to thrive. Many of them were invented and developed in Britain. The world's first football association was formed in England in 1853; lawn tennis was first played there in 1873 and the first All-England championships were held at Wimbledon in 1877. Golf may not have been invented in Scotland, but it was there that the game took root and from there that it was carried all over the world. The Open Championship remains the most prestigious tournament in the professional golfer's year.

RECENT HISTORY

Two world wars and the great movements for national liberation that have swept over the globe have wrought drastic changes in the small island state that once held sway over the largest empire in the history of the world. The First World War left the United Kingdom's empire intact – one-quarter of the world's people owed allegiance to the British crown, although the white, settler colonies of Canada, Australia, New Zealand and South Africa had already gained independent, dominion status – but its economic ascendancy, the fruit of the Industrial Revolution of the 18th and 19th centuries, was already beginning to fade.

The Second World War left it seriously overstretched, militarily and financially, and the disintegration of the empire, though a traumatic experience, was a necessary one. By the mid-1960s most of the United Kingdom's overseas possessions had won their independence and most chose to remain within the looser association of the Commonwealth, from which the adjective 'British' had been dropped.

At the end of 1994 the United Kingdom retained political control of only a few very small territories: Anguilla; Bermuda; British Antarctic Territory; British Indian Ocean Territory; British Virgin Islands; Cayman Islands; Falkland Islands; Gibraltar; Hong Kong (to be returned to China in 1997); Montserrat; Pitcairn Islands; St Helena and its dependencies, Ascension and Tristan da Cunha; South Georgia and South Sandwich Islands; and Turks and Caicos Islands.

Britain entered the post-war years under the first majority Labour government in its history. Despite the persistent tendency of British politics to resolve itself essentially into a two-party fight, the Labour party, founded in 1893, has only twice won substantial electoral victories, in 1945 and 1966. But under Clement

Attlee, the Labour governments of 1945-51 not only began the retreat from empire by granting independence to India and Pakistan in 1947, but nationalised the steel, coal, gas, electric and railway industries and established the National Health Service to provide free medical treatment for all (on a system of compulsory contributions from wages and salaries).

For some time a consensus of sorts existed between the two main political parties who, despite their differences, broadly upheld the need to maintain the welfare state, the desirability of 'full employment' and the prudence of keeping the backbone industries and utilities in state hands. That consensus began to crumble in 1970, when the Conservative party increasingly proclaimed the virtues of the free market as the most important determinant of economic and social behaviour.

The so-called 'Thatcher revolution' of the 1980s – Margaret Thatcher, Conservative prime minister from 1979 to 1990, was the first woman to lead a political party in the United Kingdom and the first person to lead a party to three consecutive triumphs at general elections – was marked by the denationalisation, or privatisation, of gas, water and electricity, telephones, British Airways, coal and many other industries; and by a fall in the level of direct taxation (graduated by income) and an increase in indirect taxation (not graduated by income).

The chief priority of recent Conservative governments has not been to reduce unemployment, but to hold down inflation and reduce public borrowing (private debt has soared). Unemployment has remained very high, and the rate of inflation has been dramatically reduced. It was also a high priority to undermine the power of the trade unions. Legislation has outlawed secondary picketing and made secret strike ballots mandatory.

The joint effect of high unemployment and the anti-union climate fostered by the government has been to erode membership in trade unions: in 1979 it stood at 13.3 million, but by 1994 it had fallen to about 9 million. Over the same period the days lost to strikes or 'industrial action' fell from 29.5 million a year to fewer than 700 000 a year – one of the lowest rates in the developed world.

Questions which are filling the immediate political horizon in the United Kingdom are tending, one way and another, to be of a constitutional kind. Should Scotland have a parliament of its own? (The issue of devolution versus independence is the chief political issue in Scotland itself, which voted in favour of devolution in a referendum of 1979, but not by a majority large enough to meet the standards required by parliament.)

Should Wales and the English regions have their own parliaments? Should the House of Lords be abolished or, at least, reformed, with the hereditary element removed? What is the place of the monarchy, somewhat battered in recent years, in a modern democracy?

Does the UK require, as a curb on the sovereign power of parliament (and the so-called 'elected dictatorship' that is the executive), a new Bill of Rights to define citizenship

and protect citizens' rights? Above all, what is Britain's relationship with its neighbours on the Continent and what is now called the European Union?

FEARS OF A FEDERAL EUROPE

Specific difficulties have prevented the development of harmonious relations between the European Union and Britain. The continuing budget imbalance and corruption issuing from the Common Agricultural Policy has been one. Another is the failure of the Commission to extend the democratic powers of the European Parliament at Strasbourg and hence the democratic accountability of European edicts and legislation.

But something much more fundamental is at issue, or at any rate is believed to be at issue by those people called 'Euro-sceptics'. The prospect of a federal Europe, with a single central bank and a single currency, with all that such a development would imply for the traditional sovereignty of the British parliament (already much eroded by membership of the European Union) and for independent British parliamentary politics, has raised alarms in an island that is proud to have defended and nourished through the centuries of its history the ideals of freedom and justice under the law.

UNITED KINGDOM AT A GLANCE	
Area 241 752 km² (93 319 sq miles)	
Population 57 998 000 (mid-1992 estimate)	
Per sq km 239 (**Per sq mile** 619)	
Capital London	
Government Parliamentary monarchy	
Currency Pound sterling = 100 pence	
Languages English, Gaelic, Welsh	
Religions Christian (55% Protestant, 10% Roman Catholic), Muslim (2%), Jewish, Hindu and Sikh minorities	
Climate Temperate; average London temperature ranges from 2-6°C (36-43°F) in January to 13-22°C (55-72°F) in July	
Main primary products Wheat, barley, potatoes, sugar beet, fruit and vegetables, fish; oil and natural gas, coal	
Major industries Agriculture, electricity, gas and water supply, oil and gas extraction, mining and quarrying, construction, wholesale and retail distribution, tourism, transport, storage and communications, financial services, basic metals and metal products, including steel, mechanical engineering, electrical, electronic and instrument engineering, motor vehicles, aerospace, food processing and beverages, textiles and clothing, wood and paper, publishing and printing, construction	
Main exports Manufactures (82 per cent of total): machinery and transport equipment, chemicals, aerospace, electronics, crude oil and petroleum products, motor vehicles, food, beverages and tobacco products, iron and steel, metal manufactures, non-ferrous metals, scientific instruments and photographic apparatus, textiles, clothing and footwear	
Annual income per head (US$) 13 285 (England: 13 540)	
Population growth (per thous/yr) 3 ·	
Life expectancy (yrs) Male 73 **Female** 78	

United States of America

FOR TWO CENTURIES THE UNITED STATES, BORN WITH SOME OF HUMANITY'S NOBLEST CONCEPTS, HAS STRIVEN TO REMAIN TRUE TO THE IDEALS OF ITS FOUNDERS

This land is your land; This land is my land; From California to the New York island; From the redwood forest to the Gulf Stream waters; This land was made for you and me – so begins folk singer Woody Guthrie's impression of the American ideal of equality that has attracted millions since first it was formulated by the Declaration of Independence, drafted by Thomas Jefferson (1743-1826) in a room above a stable in PHILADELPHIA in the baking summer of 1776.

The ideals of that declaration have not always been fulfilled during the intervening two hundred and more years, but they have remained indestructible, embodying some of humanity's noblest concepts. They have also given birth to the 'American dream' offering all those who have flocked to settle there the inalienable rights to life, liberty and the pursuit of happiness free from governmental oppression. Those ideals may embody the essence of the nation formed by the 50 united states of America. But what city or region in that vast country can be said to typify it?

It is not NEW YORK with its Aladdin's Cave shops, stone and glass canyons that broil in summer and freeze in winter, and its mosaic of nationalities. Nor is it LOS ANGELES – the city that consists of several dozen 'suburbs in search of a city'; nor the spectacular coast of CALIFORNIA. SAN FRANCISCO has too distinct a character of its own; NEW ORLEANS, though it gave jazz to the modern world, is too European; while NEVADA, where LAS VEGAS and RENO blaze neon signs to the desert stars, is too brash. No one quite knows what to make of FLORIDA, with its emerald and sapphire seas, its dripping swamps, space shuttles, Disney World and resorts for rich and poor.

Neither is it CHICAGO, with its Hollywood movie image of gangsters, underworld violence and political chicanery – an image of the past, for today's city offers enough art, music and theatre to make it a cultural centre second only to New York. Nor is it TEXAS, that reminder of the Old South, a quarter bigger than France, with vast oil fields and cattle ranches. Not even NEW ENGLAND, the early centre of opposition to English rule, can be considered as typically American. The creeper-clad universities (hence Ivy League), the small ports, the simple brick or white weather-boarded churches and villages seem to hark back to the Old World.

Nor is it the image that so many city dwellers hold of the small town waiting for them some-

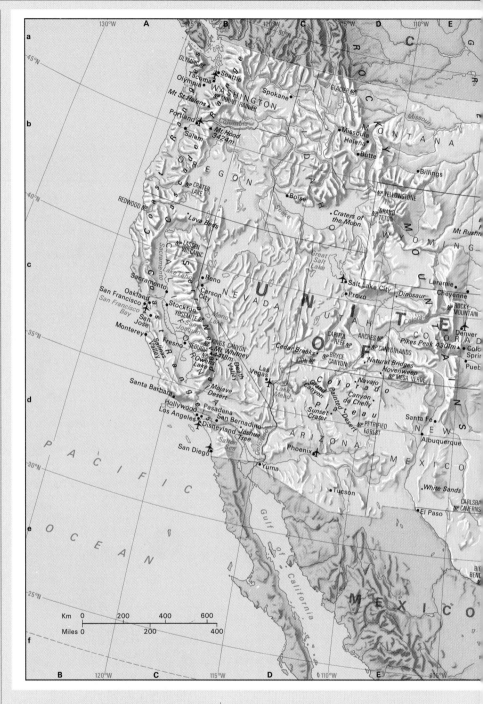

where, possibly in the MIDWEST, complete with main street, courthouse, church, war memorial, barber shop, drug store, pretty houses surrounded by lawns, maybe a Civil War cannon, and a general store.

None of these is truly representative of the United States of America. And together they can give only a slight idea of the immense diversity of a country whose sheer magnitude (it covers almost 9.4 million km², 3.63 million sq miles) and great distances impart a sense of freedom that takes the breath away.

A double ribbon of highway, the Interstate 80, runs for nearly 4700 km (2900 miles) from just outside New York City on the Atlantic to San Francisco on the Pacific coast with never a traffic light on the way. The older US Route 1 stretches almost half that distance from the Canadian border to Florida. The huge sweep of the interior plains covers almost 2 million km² (772 000 sq miles). The mighty MISSISSIPPI River drains half a continent on its 3779 km (2348 mile) journey to the Gulf of Mexico. The majestic ROCKY MOUNTAINS traverse the West for 4830 km (3000 miles).

So vast a land is the United States that dawn comes four hours later to its citizens who live on the westernmost islands of ALASKA than to

1	CONNECTICUT
2	MASSACHUSETTS
3	NEW HAMPSHIRE
4	RHODE ISLAND

those in eastern MAINE; even California is three hours behind the east coast, and the Pacific islands of Hawaii five to six hours, depending on the time of the year.

In between is a sub-continent embracing almost every kind of climate and terrain that the planet has to offer; polar tundra in northern Alaska to hot, sandy deserts in southern California; around the Gulf of Mexico it is humid and subtropical, while Arctic and sub-Arctic Alaska is intensely cold and arid.

The United States produces 70 per cent of the world's maize exports, 66 per cent of soya beans, 36 per cent of wheat and 31 per cent of cotton. It is also the foremost producer of industrial timber, poultry, beef and cheese. Despite this amazing output, agriculture today represents only 2 per cent of gross domestic product (GDP). A century ago, about 15 per cent of the population worked on the land. Now no more than about 2.5 per cent do so.

The high yields produced by such a small workforce are due to the 19th and 20th-century American genius for technology – in soil science, seed hybridisation, irrigation and pest control, for example – which also turned most of the rest of the population into urban workers. Apart from energy, which its cities and industries use at a colossal rate, the USA produces very much more than it can consume.

The country's abundance of natural resources has been the basis for its spectacular industrial growth. The USA leads the world in the production of gypsum, kaolin, molybdenum, magnesium, salt, phosphates and sulphur, and is close to being the premier producer of lead, iron, copper, coal and crude steel. It is also among the leading producers of aluminium.

Energy sources are among the country's greatest strengths. Although its oil reserves are only the seventh largest in the world it has the world's largest refining capacity, with more than 200 refineries. It also has the fifth largest reserves of natural gas and is a foremost producer of uranium, supplying 109 nuclear power stations, with more to come on line if environmental objections are resolved. Its rivers and dams produce hydroelectric power in abundance, and it is the world's leading producer of coal. Produced by whatever means, the USA generates and consumes more fuel and electricity than any other country in the world.

Much of that energy has gone into making America the world's leading industrial power, having the most diverse and technologically advanced economy, pre-eminent in, among others, aircraft and aerospace, motor vehicles,

American industry is lighter on its feet than that of many European countries, and it began to advance the development of the new, high-technology industries – such as those in California's Silicon Valley and other parts of the so-called Sun Belt.

Always a trendsetter in industry and lifestyles, the United States is now leading the way to a 'post-industrial' future in which manufacture is based on computers and automation, people live in smaller, more widespread communities and the world is being transformed by the communications revolution with developments such as satellites, computer information highways and virtual real-

claiming it for France. They founded trading posts along the St Lawrence River and along the upper Mississippi. Their first settlement was at today's New Orleans at the mouth of the Mississippi. But it was the British who were the most enthusiastic early colonisers of the North American territories. Many of the early British settlers fled political or religious persecution at home, others came in search of fortune or adventure. JAMESTOWN, the first permanent English settlement, was founded in what is today VIRGINIA in 1607. Thirteen years later the *Mayflower* landed the Pilgrims at PLYMOUTH in today's MASSACHUSETTS. With the Pilgrims came the Puritan work ethic that still underlies American industriousness and ambition today. The southern colonies, where the settlers were at first more Anglican, were found to be ideal for growing cotton, tobacco and other warm-climate cash crops, and from 1619 African slaves were brought in increasing numbers to work on the plantations.

By 1763, when Britain began to enforce a 'mercantile system' policy of economic exploitation, there were 13, mostly self-governing, colonies along the eastern seaboard. These colonists' steady determination to control their own economic and political destiny led, in 1773, to the celebrated Boston Tea Party, in which a band of patriots, dressed as native Americans, threw 342 chests of the British East India Company's tea into Boston harbour – a double protest against attempted British monopolies and British taxation.

This act was one of the opening moves of the War of Independence (1775-83) which played out on a larger scale what the Tea Party had implied: the former colonists were now and forever uniquely American. On 4 July, 1776 the 13 original colonies, through the Declaration of Independence, pronounced themselves the United States of America. After a long, hard conflict, the infant republic drove the British into capitulation at YORK-TOWN in 1781.

The 1777 Articles of Confederation, which tried but failed to bind the 13 states to one another, provided the stepping stone for the 1787 Constitution and, in 1791, for the first 10 amendments to it, known as the Bill of Rights. The Constitution established a federal republic with powers divided among the national government, the states and the people. The central (federal) government was, further set up with a system of checks and balances which ensured that no arm of government could hold sway over the others, and divided power between three branches of authority: the executive (the president), the legislative (the Congress), and the judiciary (the Supreme Court and federal courts). In 1789 George Washington became the first president.

With further amendments from time to time, the Constitution has proved remarkably far-sighted and durable. Elected every fourth year, the American president has become undoubtedly the most powerful man in the country, able to initiate and veto legislation and – perhaps more crucially – to guide the national mood by direct appeals to the people. He is also commander-in-chief of the armed forces. His power can never be absolute, however,

▲ GLITTERING DREAM Or neon nightmare? Downtown Las Vegas – self-proclaimed Entertainment Capital of the World – gleams with the wealth attracted, and generally lost, to its gambling tables and floorshows. It is one version of the American Dream.

telecommunications, chemicals, computers and electronics. It is also by far the richest, with a gross national product of about $6 trillion and per capita GDP of $23 400.

However, unemployment at a rate of around 6 per cent has affected the poorer Americans in particular – and social disparities have been steadily growing. Black Americans and Hispanics suffer most of all. The unemployment rate among black Americans (who make up about 12 per cent of the population) is more than twice as high as among their white fellow citizens; and their incomes are about 35 per cent lower on average. The situation is similar, although a little better, for Hispanics, who form 9 per cent of the population.

During the 1970-80s, heavy industries were affected particularly as a result of car import agreements advantageous to the Japanese – as were, consequently, the unskilled working populations of the northern inner cities. But

ity. The services sectors, already forming a major component of the economy, is expanding at an ever-faster rate. More than three-quarters of the work force are now engaged in communications, the wholesale and retail trades, health care, banking and finance, government, catering, and in leisure and other service industries.

CREATING A NATION

America, named after Florentine merchant, navigator and explorer Amerigo Vespucci, was first inhabited by people who crossed the BERING STRAIT from Asia over a land bridge about 25 000 years ago. They were the ancestors of the 'Indians' of South and North America and of the Inuit (the so-called 'Eskimos'). North America was sighted and briefly settled by Norsemen in the 10th and 11th centuries. Christopher Columbus never saw the northern mainland, but did open the route to the unknown continent, first for the Spanish then to a multiplicity of peoples.

In 1565 the Spanish founded St Augustine in what is now Florida, the first permanent European settlement in North America. A century later, the French, coming from New France (today's Canada), began exploring the Mississippi Valley, naming the entire drainage area LOUISIANA after King Louis XIV and

since it is restrained by the Constitution, the law and the other two branches of government: the Congress, consisting of the Senate and the House of Representatives, and the Supreme Court, whose nine justices are appointed for life and whose decisions can overturn any law that is found to be contrary to the Constitution.

It took a little more than 180 years for the original 13 states to become 50. The most important steps were the purchase, for $15 million, of the Mississippi basin from France in 1803 and the acquisition of Texas, which became a state in 1845, and of the south-west (including California) after the Mexican-American War (1846-8).

Throughout the 19th century, the USA expanded westwards as farmers and ranchers tamed new lands that were eventually admitted to the Union as fully fledged states. This expansion led to conflicts between homesteaders (who were granted free land under the 1862 Homestead Act), cattlemen (who claimed the right to graze their stock freely on the open range) and the Native Americans (who had been here long before either of the others, who depended on the free-ranging bison herds – and who died or were forced into reservations). Today there are about 2 million Native Americans left, about half of them living, either permanently or periodically, outside the reserves. These federal reservations and trust lands lie mostly in semi-arid or mountainous regions in the west and south-west.

The earliest westward pioneers crossed the country in wagon trains, but the completion of the first transcontinental railway route in 1869 caused a great surge of growth in the west. Between 1870 and 1900, helped by the flood of immigrants from Europe the country's population almost doubled to 76 million.

Founded on the principle that 'All men are created equal', the USA has prided itself in offering emigrés a new life and new opportunities. At the entrance to New York harbour stands France's tribute to America, the torch-bearing Statue of Liberty, on whose base are inscribed the words of Emma Lazarus, a wealthy young New Yorker who, horrified at Russian oppression of the Jews, became an ardent exponent of the American Dream: *Give me your tired, your poor/Your huddled masses yearning to breathe free... /I lift my lamp beside the golden door*. The words summarise the hope that brought millions of European immigrants to America in the decades around the 20th century.

With no money to travel farther, many of the newcomers stayed in the east, but by no means all. CLEVELAND in OHIO has more people of Hungarian descent than any city outside Budapest, and Chicago almost as many people with Polish names as Warsaw. Many immigrants, Irish and Italian especially, settled in Boston. There is a Chinatown in San Francisco, and a Little Italy in New York City. There are Greeks in Tarpon Springs, Florida; eastern European Jews in Kansas City, Missouri; Armenians in California's Central Valley. From 1820 to 1992 nearly 60 million immigrants entered the country; at the 1990 census 20 million citizens and legally resident foreigners had not been born there.

Despite its importance and size, New York did not become the capital. Washington, named after George Washington, was founded in 1790. The South won the location against northern claimants as part of a political compromise. The capital is referred to as Washington DC to distinguish it from the north-western state of Washington. The District of Columbia (DC) is not a state but a 'territory' originally carved out of neighbouring states.

South-west of New York lie the other middle-Atlantic states – New Jersey, PENNSYLVANIA, MARYLAND and DELAWARE. South and west of the capital lie the former slave-owning states that were the backbone of the breakaway Confederacy during the Civil War of 1861-5 – MISSISSIPPI, Louisiana, ALABAMA, Florida, GEORGIA, NORTH and SOUTH CAROLINA, and Virginia. The western counties of Virginia, a land of small farmers who had no interest in keeping slaves, voted against secession from

▲ GLEAM OF HOPE 'Miss Liberty' has welcomed arrivals to New York – millionaires on luxury liners and impoverished emigrés alike – since 1886; she was scrubbed and restored for her hundredth birthday. Behind rise the twin World Trade Center towers.

the Union in 1861 and became the new state of WEST VIRGINIA in 1863.

The secession and the Civil War that followed were not based solely upon the issue of slavery, which President Abraham Lincoln thought morally wrong and did not wish to see extended into the western territories and new states. They were also based upon the states' claim to sovereignty over issues not specifically denied them nor specifically allotted to the federal government by the Constitution. In this case the slave-owning states refused to accept any federal curb on the

Venezuela

UNCONTROLLED SPENDING AND A HUGE INTERNATIONAL DEBT HAVE LEFT VENEZUELA WITH IMMENSE ECONOMIC PROBLEMS

When Venezuela began to produce oil in 1917 it was transformed overnight from an economic backwater into the richest country in South America. Cities boomed, motorways were built, and Venezuela made a belated leap into the 20th century. But the poor were left behind. Successive governments failed to make sure that everyone shared in the prosperity, and even today, more than 70 years later, poverty is as much in evidence as wealth.

The country, which forms the northernmost crest of South America, is the size of France and Germany put together, and has more than 20 million inhabitants. The Caribbean coastline is the most densely populated and developed strip, but beyond this stretches a vast – and, in many places, little explored – interior.

To the north-west of the country, a spur of the snow-capped ANDES Mountains, the Cordillera de Mérida, with peaks reaching more than 5000 m (16 400 ft), runs south-west to north-east. It continues in the central highlands, covered by tropical forest, which lie behind the capital, CARACAS. In the centre of the country, north of the great ORINOCO River, lie undulating plains, the llanos, where herds of zebu cattle are reared. Farther south and east are the GUIANA HIGHLANDS containing some of the oldest mountains in South America. This largely unexplored area occupies nearly half the country. The region inspired Sir Arthur Conan Doyle's novel *The Lost World,* with its isolated plateaus and outcrops of solid sandstone rising suddenly from rolling savannah.

The extreme south, lying on the rim of the AMAZON basin, is covered with dense tropical jungle scattered with small frontier towns, Protestant missions and small communities of the original inhabitants, Amerindian tribes.

OIL WEALTH AND AMERINDIANS

The precious oil fields, which launched the country into wealth and which will, by 1998, be producing 3.5 million barrels a day, lie in the north-west, in and around Lake MARACAIBO – today a forest of oil derricks. Strictly speaking, the lake is a river estuary, forming a vast but slow-moving pool that flows past the narrow straits on which lies the metropolis of Maracaibo, the oil capital, to empty into the Caribbean. Since the 1920s, Maracaibo has grown into the economic centre of western Venezuela. It has a humid climate, and a population that is a brash mixture of oil executives, cattlemen, competitive merchants – and gentle nomadic Guajiro Amerindians, who come from the Guajira Peninsula to the west.

The Guajira Peninsula is semi-desert, with wild brushland and lagoons crowded with flamingos – an area which is largely unexplored and very possibly has mineral wealth, too. It is divided between Venezuela and Colombia; and border conflicts between the two are now as frequent as they are between Venezuela and its neighbour to the east, Guyana. The locals, the Guajiros, regard themselves as a nation and so do not take account of this frontier. They rear cattle and herd goats, and although they have a veneer of Spanish Catholicism, their way of life follows its own traditions. The men wear a headband of feathers to go visiting; the women are always dressed in long, flowing black or patterned cotton mantles.

South-west of Maracaibo lives a native South American tribe of very different character: the Motilones, a warlike people who violently resisted European colonisation, retaliating when their villages were burnt and firing arrows at geologists who came to explore for oil. Only recently have they been pacified, and they have now become part of the tourist scene – for instance, at the Tucuco Mission near Machiques.

North of Maracaibo City, on the Gulf of Venezuela, live the Paraujanos, in huts on stilts. It was these houses that gave Venezuela its name. When the Italian explorer and mapmaker Amerigo Vespucci first saw the country in 1499, the year after Columbus had claimed it for Spain, he thought the waterways that divided the homes looking out over water resembled his own Venice, and called the land 'Little Venice', or, in Spanish, 'Venezuela'.

The Cordillera de Mérida rises to 5007 m (16 427 ft) at Pico Bolivar and separates the oil fields around Maracaibo from the rest of Venezuela. The world's longest (12 km, 7.5 miles) and highest (4765 m, 15 633 ft) cable-car ride runs from the colonial, coffee-producing town of MÉRIDA to Pico Espejo.

East from Maracaibo stretch some 1700 km (1100 miles) of beaches, from coves with underwater wonderlands of coral reefs just offshore, past busy tourist centres near Caracas, to the island maze of the Orinoco delta where the hunting and gathering Guaraunos also live in houses on stilts. A favourite resort is MARGARITA Island, a four-hour ferry ride out into the Caribbean from CUMANA. It is named after and noted for its pearls (*margarita* is Spanish for 'pearl').

The capital, Caracas, stretching along a narrow, fertile strip between mountains and shore, is the focus of the country's culture, industry and commerce, except for oil, which is based in Maracaibo. A vast community of poor live in a maze of brick and corrugated iron boxes they built themselves on the steep hillsides around the capital. As many people live in these *barrios* as in the city itself, where most of the population lives in flats. Single family houses are affordable only for the very wealthy – or are built in makeshift fashion by the poor. Caracas's population has exploded over the past 50 years: it has the same population today as the whole of Venezuela had in 1920. Each year, thousands of people flock to the capital – it has 75 per cent of all Venezuelan industries, and wages are eight times higher than those paid for agricultural labour.

Appropriately enough for the birthplace of the great liberator Simón Bolívar, Venezuela's government is democratic, although from 1821 to 1958 military rule was virtually uninterrupted. When oil was discovered, the bloodthirsty Juan Gómez was dictator; he ruled intermittently from 1908 until his death in 1935. He is now referred to as *El Benemérito,* 'The Deserving', an ironic comment on his greed and ruthlessness.

The military stayed in power until 1945. After a brief democratic respite, they returned in 1948. Years of attempted coups, unrest and civil war followed. The last military dictator, Marcos Jiménez, was overthrown in 1958, and democratic government returned. It has continued to the present day. Indeed, voting is compulsory, resulting in a 90 per cent turnout in elections. But in spite of democracy there is still a wide gulf between rich and poor. The widespread poverty has made the situation more and more volatile, with persistent social discontent and periodic coup attempts.

Much of the problem revolves around Venezuela's oil wealth. For a long time the industry was controlled by multinational companies, most of which employed foreigners. Many of the profits were siphoned off to the

parent companies – a fact that has left Vene-zuelans suspicious of big foreign companies, particularly American ones. Determined to control its own resources and the profit from them, Venezuela nationalised the oil industry in 1976. An economic boom followed and many projects, such as agricultural land development, roads, railways and huge apartment blocks, were financed by the government.

However, administrative inefficiency and uncontrolled spending by the government produced a financial crisis. This was further compounded by an economic recession when world oil demand dropped. Dependence on oil has been considerable – 80 per cent of all government revenues come from it, although only 3 per cent of the population are employed in the industry – an important reason why national wealth has not been spread among the people. With the oil recession, Venezuela built up a huge national debt to international banks, amassed largely because imports far exceeded exports. Agricultural reforms were introduced in 1960 but the land distribution problem has not been resolved. More than half of Venezuela's peasants cultivate 2.5 per cent of the country's arable land. About 80 per cent of the land belongs to land barons and is not under cultivation. As a result, the country cannot produce enough food for its population, and has to import it at higher prices.

NEW RESOUCES OF MINERALS

Oil revenue was so easy to come by in the past that other sources of prosperity were not nurtured; for example, mineral resources such as iron ore, bauxite and coal have only recently been exploited. The country's vast timber resources, too, have barely been tapped. The financial crisis of the 1980s forced the government to revive its privatisation bill in 1992 and again encourage foreign investment in the oil industry. Foreign investment increased at first but, during the first half of 1993, it fell by 40 per cent. When economic growth slowed to 2.3 per cent in 1993, helped along by falling oil prices and high inflation, a value added tax was introduced, for the first time in Venezuela.

Its status as a democracy has benefited its diplomatic relations. With Colombia, Mexico and Panama, it is a member of the Contadora Group of Latin American countries, and it has an authoritative voice in the continent's affairs. (Only with neighbouring Guyana and Colombia are relations poor, due to a long-standing dispute over the border.) However, its internal stability suffered once again when President Carlos Andrés Pérez was suspended from office in May 1993 to face corruption charges. In presidential elections in December that year a former president, Rafael Caldera Rodriquez, was returned to power. Another severe blow came in January 1994 when mis-management, fraud and political corruption involving the country's second largest financial institution, Banco Latino, cost the government US$4 billion as well as the confidence of the electorate. A year of political disturbances and social unrest followed.

The region that is seen as the potential source of future prosperity is the great expanse of plain, the llanos, between the populated

▲ **TROPICAL HIGHLANDS The Cordillera Mérida, a spur of the Andes Mountains, runs diagonally across the north-west of Venezuela. Small Amerindian settlements lie in the foothills, and are surrounded by scrubland and fertile valleys.**

northern coastline and the rain forest in the south and east. This is the region that makes the country mostly self-sufficient in meat; it is Venezuela's Wild West, where cowboys, called *llaneros,* spend their lives herding cattle across the trackless plains. It is they who dance the national dance, the *joropo,* with its fancy footwork accompanied by a handheld harp.

The *llaneros* are of part Amerindian descent. Unlike some of the other countries of South America, Venezuela has no evidence of great pre-Columbian civilisations, but Amerindian influence is still felt in its culture, religion and folklore. A blend of Roman Catholicism and paganism is practised in the hills between Maracaibo and Caracas.

But the Venezuelans are not only a mixture of European and native South American. During colonial days, traders brought black slaves to the Caribbean coast and sold them for coffee and tobacco. Many Venezuelans have some African ancestry. The communities that have managed to stay out of this racial melting pot are the native South American tribes of the remote rain forest high up the Orinoco River, whose culture has been the same since the Stone Age. Some are threatened by such possible economic developments as exploiting the Orinoco still further to generate hydro-electric power. A hydroelectric power complex has already been built at the confluence of the Orinoco and Caroní rivers.

The east is fast developing into a third industrial region, after Caracas and Mara-caibo. Large deposits of oil have been discovered in the Orinoco region. But neither the rural population flocking to the cities in search of a better life, nor the geologists prospecting for oil are the first people seeking wealth in

Venezuela. European colonisers came in the 16th century, seeking gold.

With vast areas of uninhabited rain forest and open plains, Venezuela is the most sparsely populated country in South America. Its 20.3 million people are mostly concentrated in a few limited areas. But both the busy and the remote regions share a passion for festivities, which sweep the country on frequent religious holidays. The culturally rich mixture that is Venezuela's character becomes most apparent in the uninhibited revelry.

VENEZUELA AT A GLANCE	
Official name Republic of Venezuela	
Area 912 050 km² (352 143 sq miles)	
Population 20 300 000 **Per km²** 22 **(Per sq mile** 58)	
Capital Caracas	
Government Presidential republic	
Currency 1 bolívar = 100 céntimos	
Language Spanish	
Religion Christian (92% Roman Catholic)	
Climate Tropical; humid lowlands; cooling progressively with altitude. The rainy season is from May to November; the north-west is dry throughout. Temperature at Caracas ranges from 18.3°C (65°F) in January to 20.6°C (69°F) in July. Los Nevados (village in the Andes) is constantly cold: 14°C (57°F)	
Land use Forest 33%, grazing 19%, cultivation 4%, other 44%	
Mineral resources Oil, natural gas, iron ore, bauxite, coal, gold, diamonds, manganese	
Main primary products Livestock, coffee, cocoa, maize, rice, sugar, cotton, tobacco, fruit, vegetables; petroleum, iron ore, diamonds, manganese, bauxite, gold, coal	
Major industries Steel, transport equipment, ships, petroleum products, chemicals, food processing, textiles, cement, aluminium	
Main exports Oil and oil products, iron ore, aluminium, coffee, cocoa, sugar, some hardwoods	
Annual income per head (US$) 2900	
Population growth (per thous/yr) 26	
Life expectancy (yrs) Male 67 **Female** 73	

Verde, Cape (Cap Vert) *Senegal* Atlantic peninsula between the Senegal and Gambia rivers. It formed when a volcanic island became linked to the mainland by sediments. DAKAR, the national capital, is on the peninsula's south side; the western tip, known as Almadies Point, is mainland Africa's most westerly point. The peninsula's coastal road, known as the Route de la Corniche, follows the shore past fine sandy beaches, Dakar's colourful Soumbédioune market, the city's museum and university, the Mamelles (two extinct volcanoes) and Yoff international airport.
Map Senegal – Ab

Verdun *France* Small industrial town in eastern France, 65 km (40 miles) west of METZ. During the First World War the nearby fortifications were the scene of one of the longest and fiercest battles on the Western Front. The battle lasted from February to December 1916, and cost a total of 360 000 French and 335 000 German lives. Extensive war cemeteries mark the battlefield.
Population 20 800
Map France – Fb

Vereeniging *South Africa* Industrial town on the Vaal River 55 km (35 miles) south of Johannesburg. The Treaty of Vereeniging ending the Anglo-Boer War (1899-1902) was signed here. Its main industries are coal mining and steel.
Population 71 260
Map South Africa – Cb

Verkhoyansk *Russia* Tin and gold-mining centre in the Yakutsk Republic of north-east Siberia, on the Yana River, 120 km (75 miles) north of the Arctic Circle. The surrounding mountains, which rise to more than 2300 m (7550 ft), trap cold Arctic air in the river valley, causing winters of extreme severity; the lowest recorded temperature is –67.5°C (–89.5°F). Oymyakon, in the mountains some 630 km (390 miles) south-east, is even colder, with the Northern Hemisphere record of –72°C (–97.6°F).
Population 1600
Map Russia – Ob

Vermont *USA* North-eastern state in New England, called the 'Green Mountain' state after the forested range of the Appalachians, which bisects it from north to south and rises to 1339 m (4393 ft) in the north. The state covers 24 887 km² (9609 sq miles) between the Canadian border and Massachusetts and between New York State and New Hampshire. It was first settled in the 1720s, and became a state in 1791 – the first to do so after the original 13 states. Dairy cattle, maple syrup and apples are the main farm products.

Manufactures include electronics, computers, paper and other wood products. Although less important to the economy, marble, slate and granite are exploited in the state. Vermont is popular with tourists, particularly for skiing and its autumn colours. MONTPELIER is the state capital.
Population 562 800
Map United States – Lb

Verona *Italy* City about 100 km (60 miles) west of VENICE. It has a Roman grid street plan and contains some fine Roman remains, including a 22 000-seat 1st-century arena – the scene each July and August of Verona's famous opera festivals – a theatre carved into the hillside overlooking the Adige River and the Porta dei Borsari

▲ SUN FOR THE SUN KING In summer, palm and orange trees grace Versailles' Orangery, a courtyard sun-trap at the south end of the great palace leading down to the Swiss Lake (far left).

gateway. Many Romanesque churches were built in the 11th to 13th centuries, some of them unusually striped with red and white stone.

Verona reached its peak in the 14th century under the Della Scala dynasty; the family's palace, now the Prefecture, still survives. The fine Castelvecchio, now the art museum, overlooks the fortified Scala Bridge, which was rebuilt after its wartime destruction in 1945. The city was the birthplace of the artist Paolo Veronese (1528-88), some of whose paintings hang in the church of San Giorgio Maggiore.

The city's chief industries include textiles, leather goods, chemicals, paper, printing and wine. There is also a busy grain market.
Population 258 900
Map Italy – Cb

Versailles *France* Formerly a town, now a south-western suburb of Paris and the site of the magnificent 17th-century Palace of Versailles, built for Louis XIV, who became known as the *Roi Soleil* ('Sun King') for the brilliance of his court. Elegant gardens were laid out in 100 hectares (250 acres) of grounds around the palace, and a new town was built nearby for the palace's servants, soldiers and traders. The palace was ransacked in 1793 during the French Revolution, but was restored after the First World War. The Treaty of Versailles, which officially ended that war, was signed in the palace's Hall of Mirrors on 28 June, 1919. Each year almost 2 million people visit the palace, which is now a museum.
Population 87 800
Map France – Eb

Vert, Cap *Senegal* See VERDE, CAPE

vesicle Small cavity in volcanic rock formed by a bubble of gas trapped in the lava.

Vesterålen *Norway* Island group covering some 2370 km² (910 sq miles) in north Norway just north-east of the LOFOTEN ISLANDS. The group comprises four large islands (Hinnøya, Largøya, Andrøya and Hadseløya), and a number of smaller islands and islets. The main port and capital is Harstad.
Population 34 000
Map Norway – Da

Vestmannaeyjar *Iceland* See WESTMAN ISLANDS

Vesuvius (Vesuvio) *Italy* Volcano, about 15 km (9 miles) east of Naples. Its most violent eruption, in AD 79, destroyed the Roman towns of POMPEII and HERCULANEUM. There were major eruptions in 1631, 1794 and 1906, and Vesuvius erupted again in 1944. The road to the 1277 m (4190 ft) summit snakes through fertile vineyards and orchards and then across a black lava desert. A cableway goes up the final 500 m (1650 ft).
Map Italy – Ed

Veszprém *Hungary* Administrative centre of Veszprém County on the north side of Lake Balaton. The hilly city, 105 km (65 miles) south-west of the national capital, Budapest, grew on the southern flanks of the Bakony Mountains around a medieval castle perched on a limestone crag some 30 m (100 ft) high.

Veszprém has a largely rebuilt cathedral begun in the 11th century which contains fine collections of vestments and plate, a Baroque bishop's palace and a 13th-century chapel with superb wall paintings. New suburbs in the east are home to workers in the engineering and electronics industries. The Herend porcelain factory (and museum) is 10 km (6 miles) west of the city.
Population 64 730
Map Hungary – Ab

Viangchan *Laos* See VIENTIANE

Viareggio *Italy* Seaport and tourist resort about 80 km (50 miles) west of Florence. The English poet Percy Bysshe Shelley (1792-1822) drowned off the coast here. The resort, with its mild climate, attracts tourists throughout the year. An added attraction is its lively carnival with colourful processions.

Population 59 500
Map Italy – Cc

Viborg *Denmark* Town and capital of Viborg County, in north-central JUTLAND. With RIBE, Viborg is the oldest city of the kingdom. It was a medieval bishopric and has a heavily restored Romanesque cathedral. In the 17th century it was Jutland's largest town. It lies in the middle of the Jutland heaths and is the headquarters of the Danish Heath Society, responsible for afforestation, land reclamation and conservation.

Population 25 900
Map Denmark – Ba

Vicenza *Italy* City 60 km (40 miles) north-west of Venice. Vicenza's most famous resident was the architect Andrea Palladio (1508-80), originator of the Palladian style that became fashionable in Europe, and especially in Britain, in the mid-18th century. He gave the city some of his finest works, including the Basilica Palladiana, the Olympic Theatre, the Palazzo Chiericati, the Capitaniato Loggia and his own house.

More of Palladio's elegant villas are scattered around the countryside. The city's industries include iron and steel, machinery, food processing, textiles, furniture and glass.

Population 109 100
Map Italy – Cb

Vichy *France* Spa and health resort about 120 km (75 miles) north-west of LYONS. Its nine springs, used since Roman times, now serve a thriving mineral water industry. The factory where the water is bottled is open to the public.

In 1940, during the Second World War, the town was made capital of the unoccupied southern zone of France, and seat of the government under Marshal Henri Philippe Pétain (1856-1951). The government was disbanded in 1942 when Germany occupied the whole of France.

Population (town) 27 800; (metropolitan area) 62 400
Map France – Ec

Vicksburg *USA* Port city, at the junction of the Yazoo and Mississippi rivers in western Mississippi State. Its National Military Park has preserved the American Civil War fortifications of the 1862-3 naval and land campaign, culminating in a successful 47-day siege by the North, which then controlled the Mississippi River.

Population 20 900
Map United States – Hd

Victoria *Australia* State covering 227 600 km² (88 875 sq miles) in the extreme south-east. Formerly part of New South Wales, it became a separate colony in 1851. The state capital, MELBOURNE, lies on the coastal lowlands, backed by the Great Dividing Range. It was the seat of government until 1927. Beyond are the arable and grazing lands of the inland plains. Sheep, cattle and wheat are the state's agricultural mainstays.

Gold was discovered in 1851 at BALLARAT and BENDIGO, starting a rush, but today brown coal is the chief mineral, although gas and oil deposits are now being exploited. After New South Wales, Victoria is the country's second most important

industrial state. Nearly three-quarters of the state's population live in Melbourne.

Population 4 427 400
Map Australia – Gf

Victoria *Canada* Port city, university town and capital of BRITISH COLUMBIA, at the south-eastern end of VANCOUVER ISLAND. It is a business and tourist centre, the Pacific headquarters of the Canadian navy and the base of a deep-sea fishing fleet. Industries include timber processing, shipbuilding and paper making.

The city's many parks and gardens, its notable Victorian architecture and its mild climate have made it popular with tourists. It is linked by ferries to VANCOUVER on the mainland and to SEATTLE in the United States.

Population 299 550
Map Canada – Cd

Victoria *Hong Kong* Seaport city and capital of the British Crown Colony of Hong Kong, on the north-west coast of Hong Kong Island. One of the most prosperous cities in the Far East, Victoria began as a settlement on land reclaimed from the sea after Britain acquired Hong Kong Island in 1842. Its fortunes have been inextricably linked with Victoria Harbour, one of the greatest natural harbours in the world, which is used by more than 10 000 ocean-going ships every year as well as myriads of smaller craft engaged in local trade. Victoria is connected to KOWLOON, on the other side of the harbour, by ferries, a railway and a cross-harbour road tunnel.

Central Victoria contains the island's business section – the business and banking centre of the colony – while towards Kennedy Town in the west and Causeway Bay in the east, densely populated tenements are mixed with offices and factories. Several streets run parallel with the shoreline, each representing a successive stage of land reclamation. Behind the waterfront zone, narrow streets hemmed in by high tenements run up the steep slope of VICTORIA PEAK. These 'ladder streets', as they are sometimes called, are filled with dozens of curio stores, tiny pavement workshops and crowded open-air markets.

Higher up the slopes of the mountain are the colony's Botanic Gardens, Government House and Hong Kong University, all of which have superb views of the city and harbour lying below.

Population 1 100 000
Map Hong Kong – Cb

Victoria *Malta* Capital of the charming island of Gozo, and locally known as Rabat. The citadel hill, known as Gran Castello, was the centre of the Norman town and now houses the law courts, Gozo Museum, the 17th-century cathedral and the Old Bishop's Palace. The modern town is built around the main square, it-Tokk. The old town, a tangled knot of balconied streets, lies to the south and west. It has many goldsmiths' workshops.

Population 5410
Map Malta – Ca

Victoria *Seychelles* Capital, chief port and commercial and tourist centre of the Seychelles. It stands on the north-east coast of MAHE, against a backdrop of mountains. Most of the Seychelles' population lives in Victoria. The international airport of Pointe Larne is nearby.

Population 25 000
Map (Seychelles) Indian Ocean – Bc

Victoria, Lake *Kenya/Tanzania/Uganda* The world's third largest inland sheet of water after the Caspian Sea and Lake Superior. It covers 69 500 km² (26 800 sq miles) – about the size of the Republic of Ireland – but is nowhere deeper than 85 m (279 ft). It is one of the main sources of the Nile, the lake's only outlet, which flows out of the lake beside the Ugandan town of Jinja on the north shore.

The lake – reached in 1858 by the British explorer John Speke and named by him – has provided hydroelectric power since completion of the Owen Falls Dam just down the Nile in 1954. The dam covered the falls, which were 20 m (65 ft) high. The lake's main value to the people who live around its shores, however, is as a source of fish. Shipping services link the towns – such as KISUMU in Kenya, MWANZA in Tanzania and ENTEBBE in Uganda – and countries beside the lake.

Map Kenya – Bb; Tanzania – Ba; Uganda – Bb

Victoria Falls *Zimbabwe/Zambia* One of the world's great waterfalls, plunging 90-108 m (295-354 ft) on the Zambezi River near the north-west tip of Zimbabwe. The first European to see the falls, in 1855, was the Scottish missionary and explorer David Livingstone. The local name for them, *Mosi-oa-Toenja*, means 'the Smoke that Thunders', referring to the clouds of spray that soar up to 500 m (1600 ft) above the chasm into which the river tumbles.

Map Zimbabwe – Ba

Victoria Peak *Hong Kong* Mountain, 551 m (1808 ft) high in the north-west of HONG KONG ISLAND. It is the highest point on the island, and was the last place to be surrendered to the Japanese by the British in the fall of Hong Kong on Christmas Day 1941. From the summit, there are fine views over VICTORIA, KOWLOON and, on a clear day, the NEW TERRITORIES.

Map Hong Kong – Cb

Vidarba *India* See BIDAR

Vidin *Bulgaria* River port on the Danube near the Serbian border, 280 km (175 miles) west of RUSE, to which there is a connecting steamer service along the river. It trades in wine, grain and olives. A Shakespeare festival is held each July in the medieval citadel of Baba Vida, built on the site of the Roman town of Bononia.

Population 61 200
Map Bulgaria – Ab

Vienna (Wien) *Austria* Once the capital of the mighty Austro-Hungarian Empire (1867-1918), Vienna now rules over Austria alone and is a capital that is almost too large for its country. Set mostly on the right bank of the Danube in north-east Austria, Vienna is also a separate state – the country's smallest, with an area of 415 km² (160 sq miles), but its most populous. It was formerly capital of LOWER AUSTRIA. About one-fifth of Austria's population lives in Vienna. Divided into 23 districts, the city is grouped in two sweeping circles around the historic inner city.

The modern city grew after the removal, in the late 1850s, of the ramparts guarding the inner city and the building of the 4 km (2.5 mile) long Ringstrasse boulevard along their site. Most of the city's main and most imposing buildings line the Ringstrasse and include the handsome

opera house, the imperial Hofburg Palace (another palace stands in formerly suburban SCHÖNBRUNN), the neo-Gothic *Rathaus* (city hall), the Greek-style houses of parliament, the *Burgtheater* and the *Musikverein*, home of the Vienna Philharmonic Orchestra. In the centre of the city is the Cathedral of St Stephen, Austria's finest Gothic church, built in the 14th and 15th centuries. The cultural history museum, natural science museum and the university are also worth visiting.

For centuries, Vienna has been a meeting-place for East and West and it now houses the headquarters of several world agencies, including the Organisation of Petroleum Exporting Countries (OPEC) and the UN International Atomic Agency. Recently Vienna has expanded along the main route to the south. Large new shopping complexes rival the busy shopping street of *Mariahilferstrasse* and the traditional shops found in *Kärntnerstrasse* and *Am Graben*.

Although Vienna is noted for its industries, which produce clothing, electronics, paper, chemicals and beer, it is one of the world's great tourist centres. People flock there to savour the atmosphere of a city which began life as a Celtic settlement, was later a Roman camp, known as Vindobona, and from 1273 to 1918 was the home of the powerful Habsburg dynasty. It was during this era that Vienna grew to be a large city, designed to serve an empire much larger than today's Austria. It was then also that it attracted its mix of populations – Magyars, Germans, Czechs, Italians – which turned it into a cosmopolitan centre and a mosaic of cultures. At the end of the First World War it became the capital of Austria. In March 1938 the Nazis entered the city and Austria became part of Germany. Because of this, Vienna was intensively bombed by the Allies in the Second World War.

In 1945 it was split into four zones which were occupied until 1955 by Britain, the United States, the Soviet Union, and France. It was against this background that the novelist Graham Greene set his script for the thriller film *The Third Man*, with

▲ A CITY REBORN The fine dome of the Karlskirche towers over Vienna's magnificent Baroque centre, built after the devastating Turkish siege of 1683.

its catchy theme tune played on the zither. Indeed, Vienna is a city of music, the home of the waltz and the birthplace in 1825 of Johann Strauss the Younger, who wrote more than 400 waltzes, including 'The Blue Danube' and 'Tales from the Vienna Woods'. Musicians' Square, in the central cemetery, is the burial place of Beethoven, Brahms and Schubert, whose music is regularly heard in the city.

Population 1 531 200
Map Austria – Ea

Vientiane (Viangchan) *Laos* The country's capital city and northern province. The city was the royal capital before this was moved to LUANG PRABANG. The city is on the north bank of the Mekong River opposite Thailand. It is about the size of a French provincial town and, before the communist takeover in 1975, it had much of the flavour of France, with pavement cafés, vendors selling French patisseries, and even its own smaller Arc de Triomphe monument, modelled loosely on the arch in Paris. The city also contains about 30 pagodas, including the That Luang Pagoda which is said to contain relics of the Buddha, and several Buddhist temples and shrines. The province of Vientiane is mostly a low-lying plain which is partly irrigated by the Nam Ngum River and is the country's most densely populated area.

Population (province) 378 000; (city) 177 000
Map Laos – Ab

Vierwaldstätter See *Switzerland* See LUCERNE, LAKE

Vietnam See p 720

Vigo *Spain* Atlantic port and resort about 15 km (10 miles) north of the Portuguese border. Vigo

is the country's leading fishing port and has fish processing, shipbuilding, oil-refining and distilling industries.

Population 279 110
Map Spain – Aa

Viipuri *Russia* See VYBORG

Vijayanagar *India* Ruined capital of the great Hindu empire of the same name in southern India. It lies at Hampi, about 540 km (335 miles) southeast of BOMBAY. The empire, founded in 1336, fought off Muslims from the north for 200 years, but was finally destroyed by them in 1565. The ruins, a massive fort and temple complex, include palaces and the elephant stables of the royal retinue.

Map India – Bd

Vijayapura *India* See BIJAPUR

Vijayawada (Bezwada) *India* City in the southeast at the head of the Krishna River delta, 433 km (269 miles) north-east of MADRAS. It is the focus of an ancient system of irrigation canals which are still in use.

Population 543 000
Map India – Cd

Vijosë (Aóös Potamós) *Albania/Greece* Southern Albania's largest river, 269 km (167 miles) long. It rises in the Greek Píndhos Mountains and flows north-west into the Adriatic, north of VLORE. Its valley provides a comparatively easy route for a recently constructed international road from Yannina in Greece to Korçe, Vlorë and Durrës in Albania.

Map Albania – Cb

Vila *Vanuatu* See PORT VILA

Vila Pery *Mozambique* See CHIMOIO

Vila Viçosa *Portugal* Town situated 150 km (95 miles) east of the capital, LISBON. From the 15th century it was the seat of the dukes of Braganza. There is a 13th-century castle, modified in the 17th century. The ducal palace, dating from the 16th century, contains personal possessions of the royal family.

Population 4400
Map Portugal – Cc

Villa Cisneros *Western Sahara* See AD DAKHLA

Villahermosa *Mexico* Oil town on the Grijalva River and capital of the state of TABASCO. Until 1915 the town was known as San Juan Bautista, and its original charm as an agricultural centre was destroyed by the arrival of the Nationalised Mexican Petroleum Industry (PEMEX) in the late 1960s. Now ablaze with the burning chimneys of the oil processing plants, Villahermosa is noisy, crowded with migrant oil workers and, despite its wealth, shabby. Paradoxically, its name means 'beautiful town'. The open air museum has Olmec artefacts, among them carved basalt heads of up to 18 tonnes a piece.

Population 390 200
Map Mexico – Cc

Villány *Hungary* See PÉCS

Villmanstrand *Finland* See LAPPEENRANTA

Vilnius (Vilna; Wilno) *Lithuania* Capital city, about 150 km (95 miles) east of KAUNAS. Vilnius was founded in the 14th century and was the capital of independent Lithuania from 1323 to 1795. Its extensive old town surrounds the ruins of a 14th-century castle and there is also a mosque, a synagogue and almost 40 churches.

The city's university is the oldest in the Baltic region; it was founded in 1579, but was closed by the Russian government in 1832 because of the 'revolutionary activities' of the students. It was reopened by the Polish government, which ruled the city from 1919 until 1939, when it was occupied by Russian troops.

Vilnius, a road and railway hub, manufactures farm machinery, machine tools, electrical products, textiles, chemicals and processed food.

Population 593 000
Map Lithuania – Cb

Viña del Mar *Chile* Picturesque city with a casino and popular seaside resort, about 9 km (5.5 miles) north of VALPARAÍSO. The president's summer palace is among the splendid buildings on its palm-fringed avenues and plazas. An international song festival is held in the city's Quinta Vergara gardens every February.

Population 312 000
Map Chile – Ac

Vinnytsya *Ukraine* City in the Ukraine, situated among rich farmland 190 km (120 miles) southwest of KIYEV. The surrounding area is noted for its pottery, embroidery, woven goods and carpets. Vinnytsya has the Ukraine's largest sugar refineries and manufactures fertilisers, farm machinery, shoes and clothes.

Population 360 000
Map Ukraine – Bb

Vinson Massif *Antarctica* Highest peak in the Antarctic, rising to 4897 m (16 066 ft). It is part of the Ellsworth Mountains at the southern end of the Antarctic Peninsula.

Map Antarctica – Ec

Virgin Islands, British *West Indies* British Crown Colony in the Caribbean consisting of some 40 islands and cays in the LEEWARD ISLANDS, to the east of the US Virgin Islands. Their total area is 130 km² (50 sq miles).

Only 16 of the British Virgin Islands are inhabited, among them TORTOLA (the biggest), Virgin Gorda, Anegada and Jost Van Dyke. Christopher Columbus reached the islands in 1493. British settlers took control from the original Dutch colonists in 1666. There is some farming but tourism has become the chief employer since the mid-1960s, aided by the islands' position close to the United States and to the sheltered waters which provide some of the best sailing in the world. About two-thirds of the visitors stay on yachts.

The colony operates a strict conservation policy designed to preserve the beauty and diversity of the islands' coral reefs. The capital, ROAD TOWN, is on Tortola, which is also the main tourist island.

Population 16 750
Map Caribbean – Cb

Virgin Islands, US *West Indies* American dependency consisting of three main islands – ST THOMAS, ST CROIX and ST JOHN – and more than 50 mainly uninhabited cays and small islands in the Caribbean. They lie in the LEEWARD ISLANDS to the east of Puerto Rico; the main islands total 347 km² (134 sq miles). The capital is CHARLOTTE AMALIE. Christopher Columbus reached the isles in 1493 and they were partly settled by, in turn, the Dutch, British and French before they were acquired by Denmark.

In 1917 the United States bought them for US$25 million because of their strategic importance – they command the deep Anegada Passage from the Atlantic Ocean to the Caribbean Sea and the passage's approach to the Panama Canal.

The US Virgin Islanders were made American citizens in 1927, but constitutionally the islands are an 'unincorporated territory'. The inhabitants of both the US and British Virgin Islands are mostly descendants of African slaves, brought from the United States in the 18th century when the islands' economy depended on sugar.

Geologically, the islands are the peaks of a volcanic ridge running eastwards below sea level from Puerto Rico. Agriculture is limited by low and unreliable rainfall. There is some commercial fishing, but game fishing is more important. Watches, electronics, textiles, and pharmaceuticals are manufactured with imported materials, and rum is a valuable export. But tourism is by far the most important industry.

Cruise ships and aircraft bring large numbers of visitors, including day-trippers who take advantage of the duty-free shopping on St Thomas. Other islands are less busy and offer a quiet holiday with clear seas and good sailing, swimming and snorkelling.

The local population is far too small to cope with the large tourist influx, and as a result 55 per cent of residents are from the other Caribbean islands and also from Puerto Rico.

Population 101 800
Map Caribbean – Cb

Virginia *USA* Southern state covering some 105 611 km² (40 767 sq miles) mainly west of Chesapeake Bay on the Atlantic coast. It reaches from the Atlantic west into the Blue Ridge range of the Appalachians, which rise to 1746 m (5729 ft). Virginia is central to the history of the United States. Chartered by James I in 1606, it was the first English colony in America and became one the of the original 13 states.

The state is known as the 'Mother of Presidents' because eight American presidents – including the first, George Washington (1732-99), and seven of the first 12 – were born in Virginia. This impressive list includes Thomas Jefferson (1743-1826), chief author of the American Declaration of Independence and Bill of Rights.

Both the American War of Independence and the American Civil War ended in Virginia. In 1789 Virginia ceded territory – the District of Columbia (see WASHINGTON DC) for the new federal capital. During the Civil War (1861-5) the western part of Virginia remained loyal to the Union while the rest joined the Confederacy; as a result the separate state of WEST VIRGINIA was formed in 1863.

The state's attractive scenery and coastline and its many historic sites make it popular with tourists. RICHMOND is the state capital and NORFOLK the largest city.

Dairy and beef cattle, poultry, cured ham, peanuts and tobacco are the chief farm products. Coal is the main mineral exploited in the area.

Manufactures include cigarettes, food products, and chemicals.

Population 6 216 600
Map United States – Kc

Virunga National Park *Zaire* Reserve covering some 7600 km² (2934 sq miles) on the Ugandan and Rwandan borders. It includes the former Albert National Park, the RUWENZORI range, Lake EDWARD and the Parcs des Volcans, incorporating the volcanic Virunga Mountains, two of whose peaks (Nyiragongo and Nyamlagira) are active. Pygmies live in parts of the park, which shelters abundant wildlife, including elephants, giraffes, hippopotamuses, leopards, lions and okapis.

Map Zaire – Bb

Visayan Islands (Visayas) *Philippines* Island group forming the central part of the Philippine archipelago. Its main islands are BOHOL, CEBU, LEYTE, MASBATE, NEGROS, PANAY and SAMAR. Ferdinand Magellan's first landfall after crossing the Pacific Ocean in 1521 was on the small island of Homonhon, which lies in Leyte Gulf south of Samar.

Population 13 500
Map Philippines – Bc

Visby *Sweden* Main port and seaside resort on the Baltic island of GOTLAND. A Hanseatic town (see HAMBURG), Visby was one of the richest trading centres in northern Europe in the Middle Ages, controlling much of the Baltic, and even Russia and the Orient, and minting its own money. Its 3 km (2 miles) of medieval walls are studded with more than 40 massive towers. Within them the old city has many churches and a 12th to 13th-century cathedral. A pirates' stronghold in the 14th century, Visby is today a main tourist port.

Population 19 300
Map Sweden – Cd

Visegrád *Hungary* Historic village, with fewer than 3000 inhabitants. It overlooks the great Danube bend, where the river turns south, 43 km (27 miles) north of BUDAPEST. Once a thriving medieval town and royal residence, it is now popular with tourists for its fine views, ruined 14th-century citadel and, below it, the restored sections of a magnificent 350-room palace built by the Hungarian King Matthias Corvinus in the 15th century.

Map Hungary – Ab

Višegrad *Bosnia-Herzegovina* Town at the lower end of the Drina River gorge, 72 km (45 miles) east of SARAJEVO. Its graceful white bridge, built by the Turks in 1577, has 10 arches and is 180 m (591 ft) long. The town has timber and machinery factories, and is a tourist centre, especially for visitors to the lake behind the hydroelectric dam and for holidaymakers heading for the Tara Planina, a mountain resort area to the north-east.

Map Bosnia-Herzegovina – Cb

Viseu *Portugal* City about 75 km (45 miles) south-east of OPORTO, in the centre of the Dão wine region. The old, mainly medieval town, founded in Roman times, has attractive narrow streets and a cathedral built between the 16th and the 18th century.

Population 21 000
Map Portugal – Cb

Vietnam

ECONOMIC REVIVAL BEGINS IN A
COUNTRY UNIFIED UNDER
COMMUNISTS AFTER DECADES OF WAR

In this South-east Asian country, formerly the separate republics of North and South Vietnam, the United States for the first time lost a war. Its forces fought for eight years against the Viet Cong, a communist guerrilla movement supported by the North Vietnamese government.

The Americans dropped 7 million tonnes of bombs, sprayed defoliant chemicals on 16 200 km² (6250 sq miles) to strip the land of vegetation, and sacrificed 56 000 American lives. But the most modern machinery of death could not prevail against the determination of the North Vietnamese guerrillas.

About 2 800 000 Americans served in Vietnam during 1965-73. They went in to preserve pro-Western South Vietnam's independence from the communist north. The theory was that if South Vietnam fell, other South-east Asian states would follow, like toppled dominoes. In the slaughter that followed, well over 1 million Vietnamese – capitalist and communist – were killed, and as the weary years dragged on, America lost the will to win. After a temporary truce and two years after American forces withdrew, South Vietnam finally surrendered on 30 April, 1975. Vietnam was formally unified under communist control in 1976.

Whether for economic or political reasons, hundreds of thousands of Vietnamese fled the country in small boats after the communist takeover, mainly in 1978 and 1979. Many of these 'boat people' died in storms or at the hands of pirates, but the survivors posed serious refugee problems to neighbouring countries before they were resettled in places as far apart as Australia, Canada, France and the USA. Many tens of thousands still remain in UN refugee camps around Asia, after the world has absorbed more than 1 million Vietnamese refugees, and the UN High Commissioner for Refugees moved towards a policy of forced repatriation – backed by a stipend to start a new life back home. Remaining camps around Asia were to be closed by mid to late 1995.

ENDLESS FIGHTING

Vietnam has known war for almost half a century. In 1940, when it was a French colony, the Japanese invaded the country. They were defeated in 1945, and nationalists led by Ho Chi Minh took to arms to prevent the restoration of French rule. France had at first recognised Vietnam's independence but soon tried to colonise Indo-China anew. The Indo-Chinese War broke out in 1946 and the French only withdrew in 1954, after the Battle of DIEN BIEN PHU. However, international agreements left the country divided along the 17th parallel (17°N). In South Vietnam an anti-communist government came to power with substantial support from the USA. But the government failed to launch the economic reforms that the country desperately needed, and by 1960 internal resistance, with heavy Viet Cong support, had grown. The 20-year North-South Vietnam struggle erupted into a full-scale war in 1964.

From 1975 until 1989, Vietnamese forces occupied neighbouring CAMBODIA, where they dislodged the communist Khmer Rouge regime of Pol Pot and installed their own government. On Vietnam's northern border, China (which supported the Khmer Rouge) launched diversionary attacks in support of its Cambodian allies, at great loss to themselves.

This embattled history emphasises two important characteristics of modern Vietnam: first, the all-pervasive presence of the military in the country, for ever since 1975 Vietnam has maintained a large standing army which participates in all aspects of the country's life; second, the extraordinary patience, and resilience, of a whole generation of Vietnamese people who have never seen real peace, in the face of the country's desperate problems.

Vietnam is one of the world's poorest countries. For most of its 72 million people, life remains a struggle. The country's continued involvement in Cambodia drained its resources; in 1980, when the world's attention was drawn to the plight of the starving Cambodian people, Vietnam itself was suffering food shortages that brought reports of widespread malnutrition among the young.

Much of the country's industry – largely based in the north – was damaged by American bombing during the war and Chinese incursions in 1979. Add to all this the US embargo which was lifted only in 1994.

Only in the cities of former South Vietnam are consumer goods more widely available, particularly in the former capital Saigon, now renamed HO CHI MINH CITY after North Vietnam's great nationalist leader.

Vietnam today is very much a tale of two cities. In the south, Ho Chi Minh City still retains something of the atmosphere of pre-1975 times. Then, swollen with refugees from the countryside, Saigon was a brash and noisy place, full of the trappings of Western civilisation. The city made its living from the artificial economy created by the war.

Today the only foreigners are a few Western journalists and a handful of tourists. But the young people of Ho Chi Minh City can still be seen in the roadside cafés listening to popular Western music. The moped, introduced by the Americans, is still the typical means of transport. By contrast, HANOI, in the north, is drab and strait-laced, with little of the life of its southern counterpart. Few black market goods are available and bicycles provide transport.

Vietnam is also a tale of two rivers. The country, which lies on the eastern coast of Indo-China looking out over the south CHINA

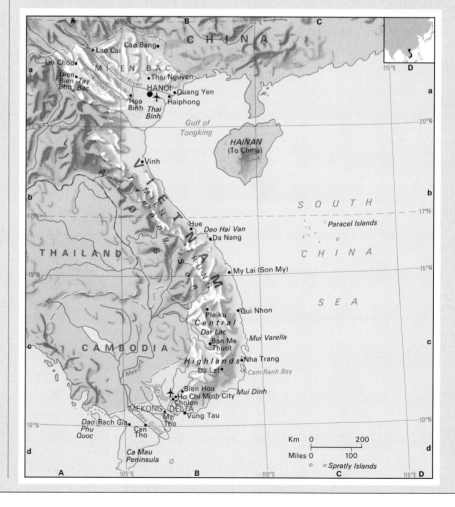

SEA, is shaped like a dumbbell, with a narrow central area linking the broader plains to the north and south. The plains are centred on the valleys of the RED and MEKONG rivers. The Red River (or Song Hong) is largely confined to Vietnam, but the Mekong starts in China and flows through Myanmar and Laos, along the border of Thailand and through Cambodia before reaching the sea south of Ho Chi Minh City. The two river plains are separated by a narrow, hilly central zone, which has always made communications between north and south a problem – a significant factor in Vietnamese history.

This narrow 'handle' of the dumbbell – once known as Annam and crossed by the 17th parallel – divides Vietnam not only geographically, but also culturally. The true Vietnamese are a Mongoloid people originating from GUANGDONG province of southern China. They settled in the Red River Valley (the area subsequently called TONKIN) more than 2000 years ago at a time when the south was dominated by the Cham and Khmer peoples.

It was only in the 17th century that the Vietnamese began to move southwards in any numbers, absorbing the Chams and pushing back the Khmers to the TONLE SAP basin to the west. Even then, the migration was limited before the arrival of the French in the middle of the 19th century. By 1867 the south had become the French colony of Cochin China and by 1884 Tonkin and Annam were French protectorates.

CONTRASTING RICELANDS

The valley of the Red River is the true heartland of Vietnam. Perhaps the resilience of the north Vietnamese over the past 50 years has something to do with their long experience of this wild and dangerous river which takes the name Red from the heavy load of silt it carries, far out of proportion to its length. Brought down from the hills of north-west Vietnam and China, this silt is redeposited as the river changes gradient in the northern plains of Vietnam. Over the years it has built up the bed of the Red River and its tributaries to 12 m (40 ft) above the level of the surrounding plain.

Throughout history, the Vietnamese have faced the danger of flooding. However, the waters of the river and its chief tributaries are easily diverted for use in agriculture, and in many parts of North Vietnam two crops of rice are grown each year. The higher areas are cultivated at the peak of the monsoon from May to November, while low-lying lands yield a winter crop after the floods have subsided.

Tonkin is heavily populated, with more than 1000 people per km² (2590 per sq mile) in the province of Thai Binh, and modern intensive rice-growing techniques imported from China are used. Intensive cultivation is also found in the coastal lowlands in central Vietnam. Here smaller streams which fall steeply from the hills of the ANNAMITE CHAIN provide irrigation water for the plains. In some places five crops of rice can be grown in two years.

Five years after North Vietnam became independent in 1954, agriculture in the area was organised into cooperatives. Groups of villages held and worked their land in common to fulfil production targets set by the country's central government. This system seemed to work in the war years, when food production in the north increased rapidly. But after the war farmers were reluctant to produce more food without incentives.

After 1979, farm prices were raised and the cooperatives turned over land to families on a contract basis. In 1992 farmers were also allowed to inherit and sell plots. Earlier, in 1989, the government had already begun to abandon all efforts at controlling agriculture.

The southern delta of the Mekong shares some features with the north, notably the importance of waterborne transport, but the landscape is very different. When the Mekong reaches southern Vietnam it is a wide and majestic river that has already begun to separate into different branches. Much of the land through which it passes is swampy, and sea water intrudes when the river's flow is at its lowest. The higher parts of the delta were drained by the French, and much of the new land was turned into large estates which produced rice for sale to the north and for export. The tradition of commercial farming has remained, and it proved difficult to impose cooperatives on this area; more than half are still privately farmed.

Since 1975 the government has tried to reduce overcrowding in the north by resettling people in the central hill areas. Despite opposition the plan has gone ahead. Such development needs outside help, and thus foreign exchange to pay for it. Since the war, rice production has increased significantly, and the country has been self-sufficient in this staple since 1989. Vietnam does, however, still import other grains, and agriculture accounts for about 50 per cent of gross national product (GNP), well below the 70 per cent during the 1980s. The north promises more favourable conditions for industrial development. There are rich deposits of coal and iron ore. Other industries today include textiles and timber.

Although the political system remains that of a single-party communist state, the Vietnamese government has been following a policy of economic liberalisation, gradually relinquishing centralised control and allowing market forces to rule. In 1994 the United States lifted its 19-year economic embargo and an

▲ RED REMEMBRANCE On Lenin's birthday in the Vietnamese capital Hanoi, workers pass portraits of the grim-faced former Russian leader. One sports a 1975 vintage Western suit – rarer in the strait-laced north than in freewheeling Saigon, now Ho Chi Minh City.

accelerating growth in GDP has drawn the interest of major French, American, Japanese and other foreign companies, which in 1994 pledged US$4 billion in new investment.

VIETNAM AT A GLANCE	
Official name Socialist Republic of Vietnam	
Area 329 560 km² (127 213 sq miles)	
Population 71 787 700 **Per km²** 217 **(Per sq mile)** 564	
Capital Hanoi	
Government Communist republic	
Currency 1 dong = 10 hao = 100 xu	
Languages Vietnamese (official), Chinese, French, English, Khmer, tribal languages	
Religions Buddhist (55%), Christian (7%), Muslim, indigenous beliefs	
Climate Tropical monsoon, with the rainy season from April to October; subtropical in the north. Average temperature in Hanoi ranges from 17°C (63°F) in January to 29°C (84°F) in June	
Land use Cultivated 24%, pastures 1%, forest and woodland 40%, other 35%	
Mineral resources Phosphates, coal, manganese, bauxite, chromate, offshore oil deposits, wolfram, zinc, lead	
Main primary products Rice, cassava, sweet potatoes, bananas, maize, pineapples, coffee, tea, rubber, tobacco, pigs, cattle, timber, fish, coal, anthracite, lignite, iron, chrome, titanium, manganese, gold, phosphates, limestone, salt	
Major industries Agriculture, mining, food processing, textiles, cement, cotton and silk manufacture, forestry, paper	
Main exports Coal, textiles, clothing, pharmaceuticals, seafood, oil, and other ores	
Annual income per head (US$) 230	
Population growth (per thous/yr) 19	
Life expectancy (yrs) Male 63 **Female** 67	

Vishakhapatnam (Vizagapatam) *India* Port on the east coast, 610 km (380 miles) north-east of MADRAS. It has India's most important shipyards, an oil refinery and fish processing plants and is a naval base.

Population 750 000
Map India – Cd

Visla *Poland* See VISTULA

Viso, Monte (Monviso) *Italy* Highest peak of the Cottian Alps, rising to 3841 m (12 602 ft) near the French-Italian border, about 65 km (40 miles) south-west of Turin. The River PO rises on its eastern slopes.

Map Italy – Ab

Vistula (Wisla; Visla; Weichsel) *Poland* The country's longest river, with a total length of 1047 km (651 miles). It rises near the Slovak border south-west of the city of CRACOW and flows roughly north and east. Near its mouth, a canal takes the river directly into the Gulf of Gdańsk, thus bypassing the Vistula Lagoon and Russian territory.

Map Poland – Cb-Dc

Vitebsk *Belarus* See VITSYEBSK

Viterbo *Italy* Well-preserved medieval town about 65 km (40 miles) north-west of Rome, about halfway between Lago di Bolsena and Lago di Vico. Among its splendours are the 12th-century Gothic cathedral, the 13th-century episcopal palace and town hall, and the 15th-century Farnese palace. There are a number of beautiful churches dating from the 9th to the 16th centuries and the town is known for its multitude of wells. Viterbo produces excellent cheeses and sausages, as well as *sambuca*, a herbal liqueur.

Population 57 800
Map Italy – Dc

Viti Levu *Fiji* Main island of Fiji, covering 10 429 km² (4027 sq miles). It is mountainous, rising to 1324 m (4344 ft) at Tomaniivi (Mount Victoria). On the eastern side, facing the prevailing trade winds, very heavy rain falls and there are dense rain forests. The western side is drier but still receives substantial rainfall; this is the main sugar-growing region, centred on LAUTOKA.

SUVA, Fiji's capital since 1882, is on the southeast coast. About one-fifth of all Fijians live in Greater Suva. There are many resorts along the Coral Coast west of the capital, on the low-lying islands off the west coast and in the vicinity of the international airport at Nadi.

Population 515 000
Map Pacific Ocean – Dc

Vitória *Brazil* Capital of Espírito Santo and port, 500 km (311 miles) north-east of Rio de Janeiro. The port, which is on a beautiful offshore island, exports coffee and cocoa, as well as iron ore and timber from the mountains of Minas Gerais.

The agricultural state of Espírito Santo, which covers 45 597 km² (17 605 sq miles), has a population of 2 524 000. Its main products are coffee and tropical hardwoods.

Population 287 000
Map Brazil – Dd

Vitoria *Spain* Basque city, capital of Álava Province, about 60 km (35 miles) south-east of

the port of Bilbao on the Bay of Biscay. It manufactures bicycles, furniture and agricultural machinery, and has sugar refining, tanning and distilling industries. Founded in 1181, the city has a 12th-century fortress-cathedral and a still unfinished New Cathedral. The British, under Sir Arthur Wellesley and helped by Spanish auxiliaries, defeated the French nearby during the Peninsular War in 1813.

Population 206 120
Map Spain – Da

Vitsyebsk (Vitebsk) *Belarus* City on the West Dvina River, about 200 km (125 miles) northeast of MINSK. It was a trading centre in the Middle Ages and was later occupied in turn by Lithuania, Poland and Russia. It became an important centre when the St Petersburg-Kiyev railway line, along which it lies, was opened. Vitsyebsk's industries include machine tools, radios, hosiery, furnishing and processed food.

Population 350 000
Map Belarus – Ca

Vlaanderen *Belgium* See FLANDERS

Vladimir *Russia* Administrative centre and industrial city about 160 km (100 miles) east of Moscow. Formerly a political, religious and cultural hub, visitors flock there to see its kremlin (citadel), its imposing Golden Gate – built in 1158-64 as a fortified entrance on the road from Moscow – and its two 12th-century cathedrals, St Dmitri and Uspenski Sobor, which were restored in the 19th century. The city today produces chemicals, engineering, electrical goods and textiles.

Population 353 000
Map Russia – Fc

Vladikavkaz (Ordzhonikidze) *Russia* Industrial city in the Caucasus Mountains, about

280 km (175 miles) south of GROZNYY. It was for a long time the main Russian military and political centre in the Caucasus. Renamed Ordzhonikidze in honour of a close associate of Stalin, it was renamed Vladikavkaz after 1991. Its products include zinc, lead, glass, chemicals, clothing, footwear and railway wagons.

Vladikavkaz is the capital city of the Autonomous North Ossetian Republic in the Russian Federation. Covering about 8000 km² (3090 sq miles), the republic has a total population of 634 000.

Population 303 000
Map Russia – Fd

Vladivostok *Russia* Largest port on the Russian Pacific coast. It is a naval base and shipyard only 50 km (30 miles) east of the Chinese border and 135 km (85 miles) from North Korea. Its products include machinery, timber, oil refining, beer, tea and flour. The city stretches around a sheltered harbour on the Sea of Japan.

Founded in 1860, the port became the base for the Russian Pacific Fleet in 1872; it also has fishing and whaling fleets. In 1903 it became the terminus of the Trans-Siberian Railway. It was a restricted military area at one time and foreigners had to use nearby NAKHODKA. Now a thriving business area, looking increasingly to its Asian neighbours for trade, Vladivostok is also the capital of the Russian Far East, also called Russia's 'Maritime Territory'. In 1994 inhabitants began pressing for the region's sovereignty.

Population 648 000
Map Russia – Od

▼ VOLCANIC PERFECTION Winter snows cover Mount Ngauruhoe, the strikingly symmetrical cone of an active volcano in Tongariro National Park on the Volcanic Plateau, New Zealand.

Vlissingen *Netherlands* See FLUSHING

vloer In South Africa, the term for a salty clay flat on an arid plateau.

Vlorë (Valona) *Albania* Port in the south which, like DURRËS, the country's chief port, has been occupied and influenced by a series of foreign rulers, including the Venetians and Turks. The main town developed amid olive groves, 3 km (2 miles) inland from a small fishing village.

After the communist takeover in the 1940s, the town was expanded to serve Albania's industrial development. Bitumen, oil, rice and cotton flowed out to be traded for imports of Soviet, and later Chinese, equipment.

Population 76 000
Map Albania – Bb

Vlotslavsk *Poland* See WLOCLAWEK

Vltava (Moldau) *Czech Republic* Bohemian river, 433 km (269 miles) long, which rises southwest of PRAGUE, near the German border. It forms the Austrian-Czech border for a while, then flows through Prague to join the Elbe. It is dammed for hydroelectric power at six places. The river inspired composer Frederic Smétana to write a famous symphonic poem.
Map Czech Republic – Bb

Vojvodina (Voyvodina) *Yugoslavia (Serbia and Montenegro)* Autonomous province of Serbia, covering 21 506 km² (8301 sq miles) and lying north of the Sava River. It is drained by the Danube and Tisza rivers. Apart from the hills of the Fruska Gora, the volcanic heights of Vrsac Planina and the sandy hills of Deliblatska Pescara between Vrsac and Belgrade, it is flat and extremely fertile. The province is Serbia's granary and sugar bowl, and a major producer of vegetables, sunflower seed, soya beans and meat. Apart from Serbs, there are many Croats, Slovaks and Romanians making up its diverse population.

The province's major towns include the capital NOVI SAD, Subotica, Senta, Zrenjanin and Vrsac, which is renowned for its September Grape Harvest and Folk Festival. The province is crossed by the 750 km (465 mile) long Veliki Backu-Vrsaki Canal.

Population 2 050 000
Map Yugoslavia – Bb-Cb

Volcanic Plateau *New Zealand* Area of volcanic mountains and hot springs around Lake Taupo in the centre of North Island. The most active volcano is Mount NGAURUHOE, 2291 m (7516 ft) high. In the 1920s and 1930s, 172 000 hectares (425 000 acres) of the plateau were planted with radiata pine trees and today the Kaingaroa Forest created by the planting is the basis of a flourishing timber industry. Pastoral farming also thrives in the area.
Map New Zealand – Ec

volcano A vent or fissure at a weak point in the earth's crust through which LAVA, ash and gases are erupted or expelled. The term, derived from Vulcan, the Roman god of fire, is applied both to the conical mountain and to the fissure type of eruptions. Volcanoes are classified as active, dormant or extinct. An active volcano is one which is either erupting or expected to erupt. A dormant volcano has not recently erupted, or is not expected to erupt, and an extinct volcano is one which has not erupted for a long time.

It is estimated that there are some 500 active volcanoes on earth; the other two categories are uncertain at best. A volcano may remain dormant for thousands of years – a short time in geological terms – and then, without warning, erupt violently. See HOW A VOLCANO IS CREATED, p 724

Volcano Islands *Japan* See IWO JIMA

Volga *Russia* Europe's longest river and a trade route between Europe and Asia since the Middle Ages. It rises in the VALDAI HILLS about 300 km (185 miles) north-west of Moscow and runs south-eastwards for 3688 km (2292 miles) – through Tver' Yaroslavl', Kazan, Ul'yanovsk, Saratov and Volgograd – into the Caspian Sea.

The Volga forms a delta at its mouth and drains an area of 1 360 000 km² (524 975 sq miles). It is navigable for most of its length, and was used by the Tsar Ivan the Terrible in 1552-6 to ship supplies and reinforcements to his troops as he expanded his empire along the river against fierce Tatar opposition.

Since 1930, dams have been built at several points on the Volga to generate power and irrigate nearby steppes, and to provide road and rail crossings. Canals built in the 1950s and 1960s link the Volga to the River Don and to the Baltic and White seas, enabling ocean-going ships to sail into Russia's industrial heartland in the Donets Basin.
Map Russia – Ec-Fc

Volgograd (Tsaritsyn) *Russia* Major port on the Volga River at the eastern end of the Volga-Don Canal, 440 km (275 miles) from the Caspian Sea. It was founded in 1589 as Tsaritsyn, a fort designed to protect trade along the river from marauding Cossacks. In 1925 it was renamed Stalingrad in honour of the Soviet leader Joseph Stalin (1879-1953), who had organised the defence of the city for the Bolsheviks against the White Army in the 1918-20 civil war. In 1961, five years after Stalin's rule was officially denounced, the city was renamed Volgograd.

During the Second World War, it was the scene of fierce fighting from August 1942 to February 1943 as the Russians gradually recaptured the city from the Germans, who had occupied parts of it after a siege lasting for more than two months. During the Battle of Stalingrad some 350 000 German troops were cut off and mostly killed and the city was virtually destroyed.

It was rebuilt after the war, and there is now a large park beside the Volga with monuments to the Russians who died in the battle. Volgograd's industries today include oil refining, petrochemicals, steel, shipbuilding, agricultural machinery and building materials.
Population 989 000
Map Russia – Fd

Volhynia *Ukraine* Fertile, wooded region in western Ukraine. Its rolling hills stretch west to Poland, which ruled the region in the 16th century. Previously it had been an independent principality, then, from 1340, it was ruled as part of Lithuania.

Volhynia became part of Russia in 1793-6 and from the early 1920s was shared by the Soviet Union and Poland. In 1939 the Russians seized the Polish sector and it formed part of the Soviet Union until 1991, when the Ukraine gained its independence. Volhynia's main town is L'VIV.
Map Ukraine – Ba

Vólos *Greece* City, port and industrial centre in eastern THESSALY. Founded in the 14th century, the city lies close to Mount PÍLION and is a centre for tourists exploring the ancient sites of Demetrias and Iolkos, the legendary port from which Jason and the Argonauts set out. Vólos has cement and steel works and produces tobacco and textiles.
Population 107 000
Map Greece – Cb

Volta *West Africa* River with three headstreams, which rise in Burkina and flow southwards into Ghana. There they merge in Lake Volta, which drains to the Indian Ocean some 110 km (70 miles) east of the capital, Accra. The Black Volta, which forms part of the Burkina-Ghana and Ghana-Ivory Coast borders, is the main headstream; it flows 1600 km (1000 miles) to the sea. The White Volta, which receives the Red Volta as a tributary, flows 885 km (550 miles) into the lake. Only a short section, between the lake and sea, is known by the name Volta. The Volta system is Ghana's main waterway; its basin covers nearly three-quarters of the country.
Map Ghana – Ab

Volta, Lake *Ghana* One of the world's largest artificial lakes. It lies behind the AKOSOMBO Dam on the Volta River, and covers 8480 km² (3250 sq miles) – or 3.5 per cent of Ghana's entire area. The dam, the lake, a modern aluminium smelter, and a bauxite (aluminium ore) quarry and processing plant make up West Africa's largest development scheme. Some 76 000 people had to be resettled when the lake formed after completion of the dam in 1966. The lake provides irrigation water for crops and livestock.

It has opened up transport in an area where communications were particularly poor and is a rich source of fish. However, there are now fears that bilharzia, river blindness, malaria and other waterborne diseases will increase.
Map Ghana – Bb

Volubilis *Morocco* Ruined capital of the ancient Roman province of Mauretania Tingitana, lying 30 km (19 miles) north of MEKNÈS. The 1800-year-old remains are 3 km (2 miles) from the holy city of MOULAY IDRISS. The most notable Roman remains in Morocco, they include a capitol, forum, basilica, baths, triumphal arch, several houses – many with intact mosaics – a bakery and 2nd-century walls. The remains of many oil mills are a reminder that the area was once densely planted with olive trees.
Map Morocco – Ba

Vorarlberg *Austria* The country's smallest and most westerly state, noted for its breathtaking Alpine scenery. It became part of Austria after the First World War. Covering 2601 km² (1004 sq miles), it is centred on the ARLBERG PASS. The state's prosperity comes from tourism (there is fine skiing), dairy farming and textiles. It is by far the most industrialised Austrian region apart from Vienna. Its capital is the resort of BREGENZ on the shore of Lake Constance.
Population 333 130
Map Austria – Ab

Voronezh *Russia* City on the Voronezh River, 450 km (280 miles) south of Moscow. It was founded as a fortress in the late 16th century. Today it is a largely industrial city manufacturing machinery, electrical products, chemicals, cigarettes and processed foods.

Population 900 000
Map Russia – Ec

Voroshilovgrad *Ukraine* See LUHANS'K

Vosges *France* Mountains along the Rhine Valley in eastern France. They are separated from the Jura Mountains to the south by the Belfort Gap. The Vosges' highest point is the Ballon de Guebwiller, at 1424 m (4671 ft).
Vosges is also the name of the department in eastern France on the western fringe of the Vosges Mountains. Agriculture is poor while traditional crafts, such as woodcarving and hand-made textiles, glass and paper, are still important.

Population 396 000
Map (mountains) France – Gb

Voss *Norway* Lakeside summer and winter resort on the Oslo-Bergen railway 70 km (45 miles) north-east of BERGEN. It has a medieval church.

Population 14 000
Map Norway – Fa

Vostok Base *Antarctica* Russian research station on the South Polar Plateau near the South Pole. It holds the record for the coldest temperature recorded on earth, –89.6°C (–129.3°F), in 1983.
Map Antarctica – Mc

Voyvodina *Yugoslavia* See VOJVODINA

Vridi Canal *Ivory Coast* Deepwater channel, 3 km (nearly 2 miles) long and 14 m (45 ft) deep, cut through a sandbar and linking Abidjan's PORT BOUET with the open sea. It was opened in 1950.
Map Ivory Coast – Bb

V-shaped valley Narrow, steep-sided VALLEY formed by stream erosion and mass movement, commonly indicating youthfulness, as opposed to the broader valleys with flood plains of more mature rivers, or to the U-shaped valleys formed by glacial erosion.

Vulcano *Italy* Volcanic holiday island – one of the EOLIAN ISLANDS to the north of Sicily – covering 21 km² (8 sq miles). It has three volcanoes: Vulcano Vecchio, rising to 500 m (1640 ft) and now extinct; Gran Cratere, which looms 391 m (1283 ft) over the harbour and last erupted in 1888; and the 123 m (404 ft) high Vulcanello to the north, extinct since 183 BC when it was formed during a submarine eruption. These volcanoes have created thermally heated sea water.

Population 430
Map Italy – Ee

Vyatka *Russia* See KIROV

Vyborg (Viipuri) *Russia* Fishing port 120 km (75 miles) north-west of ST PETERSBURG and only 30 km (18 miles) from the Finnish border. It is a gateway into Finland by road, rail and canal. Apart from fishing, its main industries are machinery, shipbuilding, timber and furniture.

Population 79 000
Map Russia – Db

HOW A VOLCANO IS CREATED

According to the theory of PLATE TECTONICS, the earth's crust or outer shell consists of a number of rigid plates which float upon the magma – the molten rock of the planet's interior. Where one plate collides with or slides beneath another, the crust may be weak enough to permit the magma – squeezed by immense pressures – to force its way up to the surface. Thus a volcano is created.

In many cases this is a symmetrical, conical mountain, such as Japan's Mount FUJI or Kenya's KILIMANJARO. These were formed by a single central vent pouring forth a mixture of lava – magma become runny through the release of pressure – and hot ash, dust and rock fragments called pyroclasts. As this material was forced up the vent, it rolled down the sides of the cone in all directions, building it ever higher but still symmetrical in shape.

If volcanic activity ceases for a time, the lava in the vent hardens into a plug. Renewed subterranean pressure may force this plug out, resulting in a cataclysmic explosion. In other cases the lava forces new exits or lateral intrusions through the strata of the mountainside. A violent eruption may create a huge empty chamber where once there was magma; the subsequent collapse of the volcano's peak can create a vast crater called a caldera – which often fills with water to form a lake or bay.

The behaviour of a volcano depends on such factors as the nature and structure of the crust in the vicinity and on how fluid the lava is. The latter depends in turn on its chemical composition; generally the higher the proportion of oxygen and silicon in a lava the less runny it is and the more violent an eruption is likely to be. Eruptions fall into several categories, named after typical examples.

The Plinian type, for example, is caused by an explosion of gases in the magma chamber below the volcano. This creates a chimney through which a column of red-hot ash shoots as through a gun barrel, rising high into the air before smothering the countryside downwind. Such was the fate of Pompeii and Herculaneum when VESUVIUS erupted in AD 79; among those killed was the Roman naturalist Pliny the Elder – whose name lives on.

Most violent of all is the Peléan, after Mont PELÉE on the island of Martinique. Here, in 1902, a previously slowly rising volcanic plug gave way in a series of violent explosions, sending a Plinian column of ash far up into the stratosphere, accompanied by fountains of lava and rocky debris from the disintegrated plug. At the same time, a cloud of superheated steam and cinders rolled down the mountainside, killing nearly all in nearby St Pierre.

The explosion that rent the island of KRAKATAU in 1883 was Peléan, too, but there the damage was due to nine huge waves (tsunami) caused by the subsequent collapse of the vocanic island into its own magma chamber. Such explosions can have long-term effects. The stratosphere dust produced by the eruptions of Mount ST HELENS in 1980 and Mount PINATUBO in 1991 is thought to have affected the global climate for several years.

Not all volcanic eruptions are violent. The

Strombolian, for example – named after Mount STROMBOLI in Sicily – is generally noisy and spectacular, with lava bombs shooting up a little way and falling back again, but does little damage. The HAWAIIAN type is usually a slow outpouring of very fluid basic lava from a wide, low crater. And volcanoes bring benefits as well as destruction – particularly to soil fertility.

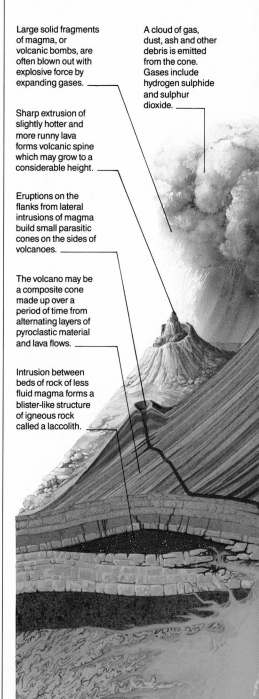

Large solid fragments of magma, or volcanic bombs, are often blown out with explosive force by expanding gases.

A cloud of gas, dust, ash and other debris is emitted from the cone. Gases include hydrogen sulphide and sulphur dioxide.

Sharp extrusion of slightly hotter and more runny lava forms volcanic spine which may grow to a considerable height.

Eruptions on the flanks from lateral intrusions of magma build small parasitic cones on the sides of volcanoes.

The volcano may be a composite cone made up over a period of time from alternating layers of pyroclastic material and lava flows.

Intrusion between beds of rock of less fluid magma forms a blister-like structure of igneous rock called a laccolith.

Plinian Violently explosive eruptions are named after the Roman writer Pliny, who died when Vesuvius erupted in AD 79.

Peléan A deadly avalanche of fiery gases and ash characterised the Mont Pelée eruption of 1902, which killed 30 000.

Strombolian Any explosion is of limited force and usually is accompanied by all-engulfing streams of white-hot lava.

Hawaiian Seldom explosive – rather a slow-spreading seepage of lava from a wide crater that creates a gently sloping cone.

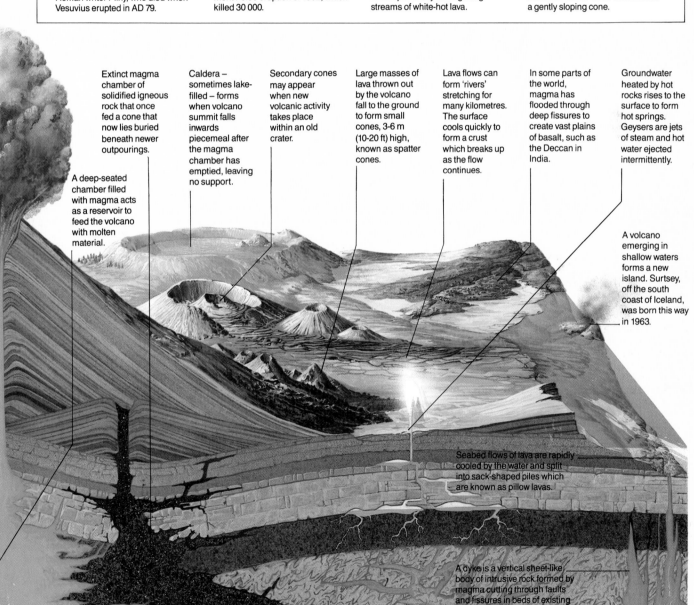

Extinct magma chamber of solidified igneous rock that once fed a cone that now lies buried beneath newer outpourings.

Caldera – sometimes lake-filled – forms when volcano summit falls inwards piecemeal after the magma chamber has emptied, leaving no support.

Secondary cones may appear when new volcanic activity takes place within an old crater.

Large masses of lava thrown out by the volcano fall to the ground to form small cones, 3-6 m (10-20 ft) high, known as spatter cones.

Lava flows can form 'rivers' stretching for many kilometres. The surface cools quickly to form a crust which breaks up as the flow continues.

In some parts of the world, magma has flooded through deep fissures to create vast plains of basalt, such as the Deccan in India.

Groundwater heated by hot rocks rises to the surface to form hot springs. Geysers are jets of steam and hot water ejected intermittently.

A deep-seated chamber filled with magma acts as a reservoir to feed the volcano with molten material.

A volcano emerging in shallow waters forms a new island. Surtsey, off the south coast of Iceland, was born this way in 1963.

Seabed flows of lava are rapidly cooled by the water and split into sack-shaped piles which are known as pillow lavas.

A dyke is a vertical sheet-like body of intrusive rock formed by magma cutting through faults and fissures in beds of existing rocks.

'W' National Park *West Africa* Huge wildlife reserve covering 10 230 km² (3950 sq miles) of savannah on the borders of Benin, Burkina and Niger. It gets its name from the huge W-shape the Niger River forms as it flows through the reserve. It contains a vast range of animals and birds, but is less developed than Benin's adjoining Pendjari National Park.

Tracks in the 'W' National Park are open only from January to June, when they have dried out after the end of the wet season.
Map Benin – Ba; Niger – Ab

Waal *Netherlands* See RHINE

Wad Medani *Sudan* Market town, some 160 km (100 miles) south-east of the capital, KHARTOUM, and the centre of the El Gezira cotton-growing region. The town has increased in importance since the Gezira irrigation scheme was established and is the headquarters of the research division of the Ministry of Agriculture.
Population 153 000
Map Sudan – Bb

Waddenzee *Netherlands* Shallow inlet of the NORTH SEA between the FRIESIAN ISLANDS and the Netherlands mainland, covering an area of some 10 000 km² (3860 sq miles).

It is used as a conservation area, a recreational area and a nursery for fish and shellfish. It was once part of the ZUIDERZEE, a former lake which became a salt-water gulf when it was flooded by the sea in the 13th century. A dyke completed in 1932 divided the Zuiderzee into two parts – the Waddenzee, which was left to the sea, and the IJSSELMEER, which has been largely reclaimed.
Map Netherlands – Ba

wadi In Arabic countries, a steep-sided VALLEY, GULLY or riverbed of desert terrain that usually remains dry, but sometimes contains a short-lived torrent after heavy rain.

Wadi el Natrun *Egypt* Salt-marsh valley, some 100 km (62 miles) south-west of ALEXANDRIA. It contains extensive deposits of sodium carbonate.
Map Egypt – Bb

Wagga Wagga *Australia* Prosperous agricultural and educational centre on the Murrumbidgee River in New South Wales, 386 km (240 miles) south-west of SYDNEY.
Population 54 220
Map Australia – Hf

Waikato *New Zealand* Longest river in the country, rising on the eastern slopes of Mount RUAPEHU in the centre of North Island and flowing some 425 km (264 miles) to the Tasman Sea, 40 km (25 miles) south of Auckland. It has been harnessed for hydroelectric power. The Huka Falls, 4 km (2.5 miles) from Lake Taupo, where the river is squeezed into a natural millrace and bursts over a 12 m (39 ft) drop like the jet from a giant hosepipe, is a popular tourist attraction.
Map New Zealand – Eb

Waipou Forest *New Zealand* See NORTHLAND

Wairarapa *New Zealand* Sheep and dairy-farming area covering 9800 km² (3784 sq miles) on North Island. The main market town is Masterton, about 80 km (50 miles) north-east of the capital, WELLINGTON. The fertile lowlands were among the first to be settled by Europeans, in the 1840s.
Population 38 900
Map New Zealand – Ed

Waitaki *New Zealand* South Island river rising in the Southern Alps and flowing some 153 km (95 miles) into the Pacific, 130 km (80 miles) north of the east coast port of DUNEDIN. Much of the water has been harnessed for hydroelectric power. The dam, lake and power station at Benmore, completed in 1965, is the largest of the many schemes along the river.
Map New Zealand – Cf

Waitangi *New Zealand* Historic North Island site on the Bay of Islands, 200 km (125 miles) north-west of AUCKLAND. A group of Maori chiefs signed the Treaty of Waitangi here in 1840, ceding sovereignty to the British, and in return were guaranteed the right to their land and other possessions. Today the site has a museum, a Maori meeting house and a Maori war canoe.
Map New Zealand – Ea

Waitemata Harbour *New Zealand* See MANUKAU HARBOUR

Waitomo Caves *New Zealand* North Island cave system discovered by Europeans in 1887, some 200 km (125 miles) south of AUCKLAND and now visited by more than 250 000 tourists a year. The most spectacular chamber is Glow-worm Grotto – so called because it is lit by the eerie blue light of millions of glow-worms. The Waitomo Caves were returned to Maori ownership in 1990.
Map New Zealand – Ec

Wakatipu *New Zealand* S-shaped lake in a glacier-carved valley on South Island, 160 km (100 miles) north-west of the port of DUNEDIN. It covers an area of 293 km² (113 sq miles) and is a major tourist attraction.
Map New Zealand – Bf

Wakayama *Japan* City on the south coast of HONSHU Island, 60 km (35 miles) south-west of OSAKA. Its 16th-century castle commands a view over the Kii Channel between Honshu and the island of Shikoku. Its industries include a steelworks.
Population 397 000
Map Japan – Cd

Wake Island *Pacific Ocean* Atoll of three islets between HAWAII and GUAM. American territory since 1898, it was a relay station on the transpacific telegraph, a submarine and air base (held by the Japanese in 1941-5) and a refuelling stop for military and commercial transpacific flights before the advent of long-range jets. Today it is used only for emergency air stops.
Map Pacific Ocean – Db

Wakefield *United Kingdom* Cathedral city in West Yorkshire, 12 km (7 miles) south of LEEDS. It was the area's chief town in the Middle Ages, but with the industrialisation of the wool trade at the end of the 18th century, factories in Leeds and Bradford quickly took over. Wakefield's cathedral dates from the 14th century.
Population 75 840
Map United Kingdom – Ed

Walachia *Romania* Region and former principality between the southern Carpathians to the north and the River Danube to the east which is crossed by the River Olt. Covering an area of 76 586 km² (29 570 sq miles), it consists mainly of plains, river valleys and fertile farmland. The national capital, BUCHAREST, is in the centre of the plain, and from it, roads lead to every corner of the region.

Walachia was united with Moldavia to form the new nation of Romania in 1861, but it still keeps its own folklore, and many of its rural homes retain a characteristic regional style, with deep verandahs and wooden-tiled roofs.

The region's fertile soil and warm climate make it one of Romania's best agricultural areas. Wheat, maize, sugar beet, sunflowers, tobacco, vines and fruit are grown, and there is some coal, oil, natural gas and salt production.
Map Romania – Ab

Walbrzych (Waldenburg) *Poland* Coal, iron and steel-producing town some 68 km (42 miles) south-west of the city of WROCLAW. Hochberg Castle, built near the town between the 14th and 16th centuries, was used as a retreat by the Nazi leader Adolf Hitler (1889-1945).
Population 141 000
Map Poland – Bc

Walcheren *Netherlands* Former island in the south-west, at the mouth of the Schelde River. It covers 212 km² (82 sq miles) and is joined to Zuidbeveland Island by POLDER. MIDDELBURG and the port of FLUSHING are the main towns. Most of Walcheren lies below sea level, and severe flooding occurred when its dykes were breached by military action in 1944 and by the sea in 1953. The Delta Plan for land reclamation, begun in 1956, is designed to minimise the danger.
Map Netherlands – Ab

Waldenburg *Poland* See WALBRZYCH

Wales *United Kingdom* Western constituent principality of the kingdom, covering 20 761 km² (8014 sq miles), and which was united with England in 1536-42. The sovereign's oldest son is generally Prince of Wales. CARDIFF is the capital, and Newport and SWANSEA the main industrial cities. Today its people have a strong feeling of nationalism; the Welsh language is known by about a quarter of them.

Wales was once known for its heavy industries which were centred on its coalfields; however, with the closure of the coal pits, its factories now specialise in high-technology products, pharmaceuticals and processed foods.
Population 2 891 500
Map United Kingdom – Dd

Wallace's Line See LOMBOK

Wallis and Futuna *Pacific Ocean* French overseas territory consisting of three main islands and many islets in the South Pacific to the north-east of Fiji. The Hoorn group, about 240 km (150 miles) from Fiji, consists of Futuna, which covers 64 km² (25 sq miles) and uninhabited Alofi, covering 51 km² (20 sq miles); both are mountainous, Futuna reaching 765 m (2510 ft) at Mount Puke. Wallis (Uvéa) Island, some 160 km (100 miles) farther north-east, covers 159 km² (61 sq miles) and is hilly, with a coral reef. Wallis

▲ RUGGED FASTNESS Two lovely lakes, Llyn Idwal and Llyn Ogwen beyond, lie below Devil's Kitchen, a craggy basin on the flanks of Glyder Fawr, rising to 999 m (3278 ft) in Snowdonia, Wales.

has 60 per cent of the population; the capital, Mata-Utu, is on its east coast. The people, who are Polynesians, grow coconuts, bananas and taro, and raise pigs. Many of its people have emigrated.

Population 13 700
Map Pacific Ocean – Ec

Wallsend *United Kingdom* See HADRIAN'S WALL; NEWCASTLE UPON TYNE

Walsall *United Kingdom* Industrial and market town in the English metropolitan county of West Midlands, situated 12 km (8 miles) north-west of BIRMINGHAM. It has numerous engineering and other industries and produces leatherware.

Population 178 850
Map United Kingdom – Ed

Walvis Bay *Namibia* The country's chief port, 400 km (250 miles) along the tarred main road south-west of the capital, WINDHOEK. It was given its name, meaning Whale Bay, by whalers who frequented it in the 19th century. Walvis Bay was annexed by Britain in 1878 and was transferred to the Cape Colony in 1884. It became part of South Africa in 1910 but was administered as part of Namibia from 1922. With negotiations for Namibian independence it was again administered as an integral part of South Africa until March 1994 when it was transferred to Namibia.

Population 25 000
Map Namibia – Ab

Wanganui *New Zealand* North Island river rising on Mount RUAPEHU and flowing about 288 km (180 miles) into the Tasman Sea at the port of Wanganui (population 41 210), 160 km (100 miles) north of the capital, WELLINGTON.

Map New Zealand – Ec

Wangaratta *Australia* See BEECHWORTH

Wankie *Zimbabwe* See HWANGE

Warasdin *Croatia* See VARAŽDIN

warm front Boundary plane between the warm and cold air in a DEPRESSION, at which the mass of warm air is rising above the cold air which it is overtaking. As the front moves, warm air replaces cold air on the ground, hence its name. Low nimbostratus clouds usually produce a belt of rain along the front.

warm sector Area of warmer air in the central part of an atmospheric DEPRESSION, bounded by a WARM FRONT and a COLD FRONT.

Warri *Nigeria* Port in the west of the Niger River delta, located about 270 km (170 miles) south-east of LAGOS. It dates back at least to the 15th century when it was visited by Portuguese missionaries. Later, it was a base for Portuguese and Dutch slave traders. Today the town has an oil refinery and it imports drilling equipment. It exports groundnuts and cotton products, which are brought down the Benue River from northern Nigeria, Cameroon and Chad, and cocoa, palm oil, rubber and timber from southern Nigeria.

Population 88 900
Map Nigeria – Bb

Warrington *United Kingdom* Beer-brewing and industrial town situated in Cheshire, in north-west England. It stands on the River Mersey and the Manchester Ship Canal, midway between MANCHESTER and LIVERPOOL.

Population 185 000
Map United Kingdom – Dd

Warsaw (Warszawa) *Poland* The country's capital and largest city. Its history reaches back to the Middle Ages, but it is also one of Poland's most modern cities, for when the Russians liberated it in January 1945, less than 1 in 10 of the buildings were fit for use, and 800 000 of its inhabitants – nearly two in three of the city's prewar population – were dead or deported.

In a massive rebuilding project the snugly walled, medieval and Renaissance Stare Miasto (Old Town) was painstakingly re-created. Its narrow streets are now for pedestrians only and teem with cafés and bars, echoing the lively atmosphere of the commercial and crafts centre

that existed before 1596, when Warsaw was made the national capital in place of CRACOW. Along the route towards Cracow spread the 17th and 18th-century expressions of Warsaw's new status – the royal palace (reopened in the 1980s), town palaces of the nobles, including the Radziwill and Potocki families, and centres of art and learning, which now form the city's university. Here also are churches and convents, recreational areas, including Lazienki Park and, 10 km (6 miles) from the centre, Wilanow, the royal summer residence, which was built for the Polish King Jan Sobieski in the 1680s.

The opening of the Vienna-Warsaw Railway in 1845 unleashed industrial development, both to the west of Muranow, where most Jewish immigrants settled, and to the east across the Vistula River in Praga. By 1939 there were some 1 290 000 Warsovians, 35 per cent of them Jewish, and nearly 30 per cent Yiddish-speaking. In 1940 the occupying Germans confined 450 000 Jews to a walled ghetto – on rations providing only 184 calories a day – to await transfer to death camps. Survivors were removed and the ghetto razed after an abortive Jewish rising in 1943. The following summer the Germans began a systematic destruction of the rest of the city after a rising led by the Polish Resistance failed, while the Russian Red Army was poised outside the city. By 1945 only 162 000 of the original 1 290 000 Warsovians remained in the city.

Rebuilding has created a spacious new city outside the old town. Arterial roads and a motorway have replaced the former maze of streets. The highways stretch out to modern high-rise suburbs, with plants producing steel, cement, cars, tractors, electronic goods and pharmaceuticals. Warsaw possesses numerous universities and colleges and is the cultural centre of Poland. Half of Poland's scientists and 90 per cent of the publishing houses are based in the city. Among Warsaw's many evocative monuments are the Warsaw *Nike*, which commemorates the city's war dead, and the memorial to the heroes of the ghetto at Muranow.

The Polish composer Frédéric Chopin (1810-49) was born in the village of Żelazowa Wola, 53 km (32 miles) to the west, on the edge of Kampinos Forest. There is a Chopin museum there at which recitals by world-famous pianists are given each summer.

Population 1 655 600
Map Poland – Db

Wartburg *Germany* Historic 11th-century castle in the extreme south-west. It is perched on a hill 172 m (564 ft) above the town of EISENACH and is accessible only on foot or by donkey. It has a magnificent Romanesque palace and some fine half-timbered houses. The religious reformer Martin Luther (1483-1546) was sheltered here in 1521-2 while he translated the Bible from Latin into German. The castle was used as a setting by the composer Richard Wagner (1813-83) in his opera *Tannhäuser*.

Map Germany – Dc

Warwick *United Kingdom* Town and county covering some 1981 km² (765 sq miles) in the Midlands of England. Farming and tourism are the county's main sources of income. There is an imposing medieval castle in Warwick town – 30 km (20 miles) south-west of Birmingham – and the remains of another castle are in nearby

Kenilworth. Leamington Spa, situated close by, was a fashionable 18th-century watering place. But the county's biggest attraction is STRATFORD-UPON-AVON, the birthplace and home of the poet and dramatist William Shakespeare (1564-1616).

Population (town) 22 000; (county) 489 200
Map United Kingdom – Ed

wash Fine material, particularly alluvium, found especially in the dry beds of streams that flow intermittently, and on desert surfaces where flowing water is rare. The term is also used for fine material moved down the slope of a mountain by rain, especially where there is little vegetation.

Wash, The *United Kingdom* Shallow inlet of the NORTH SEA between the counties of Lincolnshire and Norfolk on the east coast of England. The Great Ouse, Nene, Welland and Witham rivers empty into the bay through the surrounding Fenland. Much land has been reclaimed from The Wash and the FENS to the south and south-west by drainage and dyking projects carried out by surrounding landowners since Roman times. Today The Wash is 35 km (20 miles) long and about 25 km (15 miles) wide.

Map United Kingdom – Fd

Washington *USA* North-west state, covering an area of 176 521 km² (68 139 sq miles), bordering Canada in the north and the Pacific Ocean in the west. It is traversed from north to south by the Coast and Cascade ranges, the latter rising to 4392 m (14 410 ft) at Mount Rainier, and has spectacular scenery. Wheat, cattle and apples are the main farm produce and the region is rich in fish, especially salmon. More than half the state is forest so there is a large timber processing industry. Other industries include aerospace, aluminium and processed food. OLYMPIA is the state capital and SEATTLE the chief town and port.

Population 4 409 000
Map United States – Ba

Washington DC *USA* The national capital, lying in the District of Columbia – Federal territory now covering an area of 179 km² (69 sq miles) on the Potomac River between Maryland and Virginia, about 200 km (125 miles) from the Atlantic coast. The territory was ceded by Maryland and Virginia in 1788-9 and the Federal government moved here from Philadelphia in 1800.

The district, which today incorporates the once separate GEORGETOWN and spreads beyond its original boundaries into the adjoining states, has a spacious centre, with Capitol Hill (site of the Congressional legislative and administrative buildings, the Library of Congress and the Supreme Court) at one end of The Mall, and the Washington Monument, Lincoln Memorial and the Vietnam Veterans' Memorial at the other.

The Mall is flanked by public buildings, including the various museums of the Smithsonian Institution, the National Art Gallery, the Holocaust Memorial Museum and government buildings. To the north lies the White House, the official residence of the president. On the west bank of the Potomac, in Virginia, stands the Arlington National Cemetery and the Pentagon, the United States' military headquarters.

The Federal civil service is Washington's main employer, but there are also many American and foreign, professional, charitable and financial offices and headquarters. The city has more than 300 parks and smaller green areas, a cathedral, five major universities, many cultural establishments, including the John F. Kennedy Center for the Performing Arts, and two international airports. Washington is a popular tourist attraction, receiving some 20 million visitors a year.

Population (city) 606 900; (metropolitan area) 3 923 600
Map United States – Kc

water meadow Meadow on the flood plain of a river, which is irrigated and fertilised with silt by the periodic flooding of the river. In addition to meadows subject to natural flooding, from late medieval times artificial water meadows were created in Europe by farmers digging channels controlled by sluices. The running water did not often freeze, so grain continued to grow, providing early feed for livestock.

water table (saturation level) Boundary between the aeration zone – the region of soil and rock in which air and water mix – and the saturation zone, a region of porous rock in which all the pores are filled up with water (see GROUND WATER). The water table is uneven, depending on the nature of the pores or holes in the rock; it also varies according to the amount of rainfall, rising in wet weather and falling in dry weather. SPRINGS may appear on the surface at the junction of the porous rock with an underlying layer of impermeable rock. The rise and fall of the water table affects the flow from springs as well as the level of water in wells and ARTESIAN STRUCTURES.

waterfall Point where a river plunges more or less vertically over a steeply sloping rock face or an overhang. It may result from a layer or a DYKE of hard, wear-resistant rock lying across softer, more easily eroded strata; or from the arrival of a river at a coastal cliff, the edge of a plateau, a fault-line ESCARPMENT or the mouth of a HANGING VALLEY. See MIGHTY CATARACTS AND SPARKLING CASCADES (opposite).

Waterford (Port Láirge) *Ireland* City and port 135 km (84 miles) south-west of the capital, DUBLIN. Waterford was originally settled by Danish invaders but they were driven out by the Anglo-Normans in 1170. The quays of its busy port line the River Suir. Waterford manufactures food, metal and engineering products, pharmaceuticals and furniture, but it is best known for its light opera festival, in September, and for its crystal glass. The city has two cathedrals, two 13th-century monasteries and a tower (now a museum), built by the Danes in 1003. County Waterford covers 1837 km² (702 sq miles); its county town is Dungarvan (population 6920).

Population (city) 40 330 (county) 91 620
Map Ireland – Cb

Waterloo *Belgium* Town 16 km (10 miles) south of BRUSSELS. It lies near the scene of the great battle on 18 June, 1815, when the French Emperor Napoleon was defeated by the British under the Duke of Wellington, with the crucial help of the Prussians. The battlefield still draws tourists to the town, which is now a suburb of the capital. There is a museum in Wellington's former headquarters and there are several monuments in the village church.

Population 25 000
Map Belgium – Ba

watershed The divide lying between two different river systems. Watersheds may be marked by steep ridges but may also run across plateau surfaces and even lakes and bogs.

waterspout A whirling column of sea, spray and mist created by a TORNADO or similar whirlwind occurring over water.

Wau *Sudan* Provincial capital of Bahr el Ghazal, and major cattle market. It lies 1050 km (about 650 miles) south-west of the capital, KHARTOUM, to which it is linked by rail.

Population 58 100
Map Sudan – Ac

Wave Rock *Australia* Natural rock formation and tourist attraction in Western Australia, near the town of Hyden, 300 km (186 miles) south-east of the capital, PERTH. The granite rock, carved by wind and rainwater over tens of millions of years, is 15 m (50 ft) high and is part of a 2 km (1 mile) long chain of outcrops known collectively as Hyden Rock, which looks like a huge wave about to break over the dry, flat country surrounding it.

Map Australia – Be

Waza National Park *Cameroon* Nature reserve covering 170 000 hectares (420 000 acres) in the far north, on the Nigerian border. It is considered one of the best game parks in West Africa, and it is home to a wide range of wildlife.

Map Cameroon – Ba

Weald, The *United Kingdom* See DOWNS, THE

weather The day-to-day condition of the atmosphere at any place, with respect to the variety of elements, such as pressure and temperature, sunshine, wind, clouds, fog and precipitation. See WEATHER AND WIND SYSTEMS, p 730; for weather records, see BONE DRY AND SOAKING WET, p 729 and FROM BOILING HOT TO FREEZING COLD, p 731

weathering Processes of disintegration and decomposition by which rocks at or near the earth's surface are broken down into loose fragments. Weathering provides the materials with which the agents of EROSION work.

Weddell Sea *Atlantic Ocean* Part of the South ATLANTIC OCEAN between Coats Land and the ANTARCTIC PENINSULA. The British sealer and explorer James Weddell (1787-1834) discovered it in 1823. The sea is over 4000 m (13 125 ft) deep and extends across an area of 800 000 km² (308 880 sq miles), with huge masses of drift ice usually collecting in its centre. In 1915 Sir Ernest Shackleton's *Endurance* became trapped in this ice. After 10 months, Shackleton (1874-1922) abandoned the ship, recording: 'It was a sickening sensation to feel the decks breaking up under one's feet, the great beams bending and then snapping with a noise like gunfire.' Shackleton and his crew reached safety after an epic journey over the ice and across the ocean by small boats.

Map Antarctica – Cb

wedge of high pressure See RIDGE OF HIGH PRESSURE

Wei He *China* River rising in Shaanxi Province and flowing some 800 km (500 miles) into the Huang He (Yellow River). Its fertile valley in the

LOESSLANDS OF CHINA is the cradle of Chinese civilisation and contains many important archaeological sites. These include Lantien, site of the discovery of Lantien Man, a skull 500 000 years old; Banpo, site of a Neolithic (5000 BC) village of the Yangshao culture, Xi'an, capital of China under the Qin dynasty (221-206 BC), and the site of Chang'an, capital during the Han (206 BC-AD 220) and Tang dynasty (AD 618-906).
Map China – Fd

Weichsel *Poland* See VISTULA

Weimar *Germany* City, 80 km (50 miles) southwest of LEIPZIG. The poets Johann Wolfgang von Goethe (1749-1832) and Friedrich von Schiller (1744-1805) lived here and their fame made it the cultural centre of 18th-century Germany. The composer Johann Sebastian Bach was court organist and concert master to the Elector of the Duchy of Saxe-Weimar from 1708 to 1717.

Franz Liszt was the musical director in the city in the mid-19th century. During the second half of the 19th century Grand Duke Karl Alexander (1853-1901) supported the arts; an orchestra school was founded in the city in 1972, as was an arts college.

Today Weimar is an exquisitely restored city of half-timbered houses, painted stone buildings, castles and Baroque palaces. Politically, too, the city has played an important part in German history. In 1919 the German National Assembly met here to draw up a constitution and the town gave its name to the republic which survived until Adolf Hitler came to power in 1933.
Population 60 500
Map Germany – Dc

Weipa *Australia* Mining settlement situated in the extreme north of QUEENSLAND, on the west coast of Cape York Peninsula. It has one of the world's largest deposits of bauxite.
Population 2510
Map Australia – Ga

Welkom *South Africa* Established as recently as 1947, Welkom lies 150 km (93 miles) north-east of BLOEMFONTEIN and is the major town of the Free State goldfields; large quantities of uranium are also mined here.
Population (town) 68 110; (district) 248 190
Map South Africa – Cb

Welland Canal *Canada* Waterway linking Lake ERIE with Lake ONTARIO. The first canal, with 25 locks, was built between 1824 and 1829 to enable ships to avoid the 98 m (320 ft) drop of NIAGARA FALLS. The latest, with eight locks, built between 1913 and 1932, and then improved in 1973, today carries ships up to 221 m (725 ft) long with up to 7.9 m (26 ft) draught.
Map Canada – Hd

Wellington *New Zealand* National capital and port in the extreme south-west of North Island, built on a steep and often windswept slope around a deep, sheltered harbour. It exports various dairy produce, wool, meat and hides, and manufactures motor vehicles, footwear and chemicals.

The city was founded in 1840 by the New Zealand Company. It was named after the British military hero and prime minister, the Duke of Wellington (1769-1852), who gave the company his support. The city succeeded Auckland as the

capital in 1865. Many of Wellington's outstanding buildings date from the Victorian period; they include the Houses of Parliament (1877), Victoria University (1897), St Paul's Church, the cathedral (1866), and the Thistle Inn (1860).

The city's industries are concentrated in the nearby Hutt Valley and the town of Porirua.
Population 325 700
Map New Zealand – Ed

Welo *Ethiopia* Province covering 79 000 km² (30 500 sq miles), 160 km (100 miles) north of ADDIS ABABA. It was struck by famine in 1984-5.
Population 3 000 000
Map Ethiopia – Aa

MIGHTY CATARACTS AND SPARKLING CASCADES

Waterfalls are among nature's most awe-inspiring phenomena, whether tall and relatively slim or lower but broad like Niagara and Victoria Falls. With the flooding of the Guaíra Falls on the Alto Paraná of South America in 1982, the mightiest of all in terms of annual water flow are the Boyoma Falls of Zaire.

SOME OF THE WORLD'S HIGHEST WATERFALLS

Waterfall	Location	Height (m/ft)
Angel	Venezuela	979/3212
Yosemite (total drop)	United States	739/2425
Mardalsfoss	Norway	655/2150
Utigård	Norway	600/1970
Sutherland	New Zealand	580/1904
Gavarnie	France	422/1384
Tugela	South Africa	411/1350
Krimml	Austria	381/1250

BONE DRY AND SOAKING WET

The driest places on earth are the Dry Valleys in the cold desert landscapes near McMurdo Sound, opposite Ross Island in Antarctica. Scientists estimate that no rain has fallen here for 2 million years. Another of the world's driest places – Death Valley in California – is also one of the hottest, with temperatures reaching 56.6°C (134°F), and one of the lowest, 86 m (282 ft) below sea level.

The greatest recorded rainfall was in the Indian village of Cherrapunji in July 1891, when over 11 836 mm (466 in) of rain fell during the month. Mawsynram, also in India, holds the record for the highest rainfall over 24 hours – 989.6 mm (38.96 in) which fell on 10 July, 1952. But the wettest known place of all is Mount Waialeale in Hawaii, with an average rainfall of 12 344 mm (486 in).

SOME OF THE THE WORLD'S DRIEST PLACES

Place	Location	Rainfall (mm/in per year)
Dry Valleys	Antarctica	No rain for 2 million years
Death Valley	United States	3/0.1
Arica Desert	Chile	3/0.1
Gobi Desert	Central Asia	5/0.2
Sahara (parts)	North Africa	25/1
Lake Eyre Basin	Australia	101-152/4-6

AND SOME OF ITS WETTEST

Mount Waialeale	Hawaii	12 344/486
Mawsynram	India	11 873/467
Cherrapunji	India	11 314/445
Mount Cameroon	Cameroon	10 160/400
Sprinkling Tarn	Cumbria, England	6528/257
North-west Washington State	United States	2997/118

Weser *Germany* River formed by the confluence of the Fulda and Werra rivers, 25 km (15 miles) north-east of KASSEL. It flows 440 km (273 miles) northwards to the North Sea at Bremerhaven.
Map Germany – Cb

West Bank *Jordan* Area situated west of the River Jordan and the Dead Sea. Covering some 5858 km² (2262 sq miles), it contains some of Jordan's most fertile land. The area was seized by Israel during the Six-Day War of 1967. East Jerusalem was the only part that was formally annexed by Israel; the rest of the area was increasingly integrated into the Israeli economy. Under the Israelis, a number of (continued on p 731)

729

WEATHER AND WIND SYSTEMS

The fundamental cause of all weather is the temperature difference between the poles and the Equator. The Equator receives more heat from the sun than it loses into space; the poles lose more heat to space than they receive from the sun. The redistribution of heat from the Equator towards the poles is the action that produces all weather, modified in its operation by the spin of the earth. This spin tends to 'thicken' the atmosphere along the Equator and, other things being equal, would cause high pressure at the Equator. But other things are not equal.

The intense heat at the Equator causes hot, moisture-laden air to rise and creates a low pressure zone – the Intertropical Convergence Zone (ITCZ); the high pressure belt that should exist is divided into two and displaced to the tropics. The resultant pattern of pressure belts gives rise to wind systems.

Winds blow towards the Equator and towards the mid-latitudinal (23.5° – 66.5° Northern and Southern Hemispheres) low pressure systems from the high pressure belts at the tropics. Winds are also directed to the mid-latitudinal regions from the polar high pressure regions.

The spinning of the earth deflects the winds, for example, 'pulling' the trade winds of the tropics to the west. In the middle latitudes the spin gives rise to the steady westerlies that have such a marked effect upon the weather of Europe, North America and Australia.

This is the pattern at ground level, but high aloft some counter-currents operate. Circulation over the globe, which ultimately transports heat towards the poles, consists of a number of cells – Hadley Cells. If the earth were a smooth sphere, covered only by water or flat landscape, the movement of winds would be constant and unchanging, and all weather patterns predictable. However, air currents are diverted and distorted by mountains, valleys, ice caps and deserts – even the heat radiated by cities. A major influence is the contrast between land and sea, which affects maritime climates at all latitudes. Land warms and cools more quickly than water, so that it is hotter than the sea in summer and colder in winter.

This has a profound seasonal effect on wind and weather for the same reasons as for the daily shift that takes place on coasts. There, in the day, cool breezes off the sea move in to replace the hot air rising off the land; at night, as the sea retains its warmth and the land cools down, the whole system goes into reverse.

In middle latitudes, north and south of the tropics, the weather is governed by CYCLONES (depressions) and ANTICYCLONES. These are whirlpools of air set in motion by the rotation of the earth, occurring at low levels where warm air from mid-latitudes meets the polar air from the polar high. They are influenced by topographical and other features, and are therefore all the less predictable in their effect.

ANATOMY OF A HURRICANE

Hurricanes originate over tropical seas where the water surface has been heated to over 27°C-(80°F). An area of intense low pressure may form, and is thrown into a spin by the rotation of the earth. Winds bearing water vapour from the warm sea are dragged up into the spiral to create towering banks of cloud. Energy is released in torrential rains and winds of up to 300 km/h (190 mph).

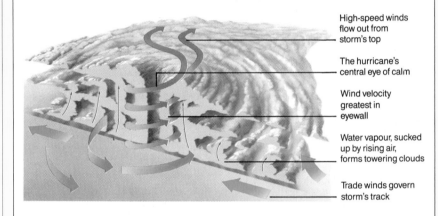

High-speed winds flow out from storm's top

The hurricane's central eye of calm

Wind velocity greatest in eyewall

Water vapour, sucked up by rising air, forms towering clouds

Trade winds govern storm's track

WHERE THE WIND BLOWS

Major circulation systems of rising and falling air, Hadley Cells, lie either side of the Equator. Winds within the cells are curved by the 'drag' of the earth's spin. The rotation rate in upper latitudes disrupts the cells, which otherwise would stretch as two great loops to the poles.

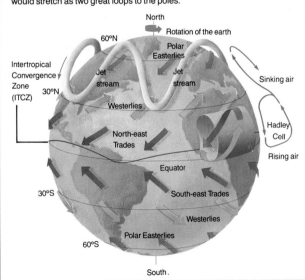

North
Rotation of the earth
60°N
Polar Easterlies
Intertropical Convergence Zone (ITCZ)
30°N
Jet stream
Jet stream
Sinking air
Westerlies
North-east Trades
Hadley Cell
Rising air
Equator
30°S
South-east Trades
Westerlies
Polar Easterlies
60°S
South.

DEVELOPMENT OF A DEPRESSION

Heavy, thundery rain from cumulonimbus clouds accompanies the cold front

Cold air descending

Prevailing wind

Cold front

Warm front

Warm air rising

Prolonged rain or drizzle precedes the warm front

Warm sector with thin cloud cover or clear skies

A depression is a low-pressure region into which whirlpools of air spiral. Cold polar air and warm tropical air converge in the depression and, as the depression moves, warm and cold air masses are forced against each other to form fronts. The warm front's passage is marked by drizzle and rain from dark, cloudy skies, the cold front's by heavy rain from towering cumulonimbus or thunder clouds.

FROM BOILING HOT TO FREEZING COLD

The highest shade temperature in the world – 58°C (136.4°F) – was recorded in the Libyan village of Al' Aziziyah, about 40 km (25 miles) south of Tripoli, one day in 1922. However, the world's consistently hottest place is Death Valley in California, where in 1917 maximum temperatures of more than 48°C (118°F) were recorded on 43 days in succession. The 'award' for the world's coldest recorded temperatures has been won three times running by the Russian Antarctic research station, Vostok Base. At recordings made between August 1958 and July 1983 the extremely low temperatures at the base dropped progressively from -87.4°C (-125°F) to -88.3°C (-126.9°F) to -89.6°C (-129.3°F).

controversial Jewish settlements were set up here, and 111 000 Jews moved to the West Bank. By 1992, 67 per cent of the land was controlled by the Israelis and 5 per cent by the Palestinians.

The Rabat summit of Arab heads of state in October 1974 declared the Palestinian Liberation Organisation (PLO) to be the sole legitimate representative of the Palestinian people (with the inference that the West Bank would eventually become an independent Palestinian Arab state); in July 1988 Jordan surrendered its claims to the West Bank in favour of the PLO.

On 15 November, 1988 the PLO proclaimed its own state, with East Jerusalem as its capital, a state which is recognised by many countries. In 1994 Israel granted the PLO limited autonomy in JERICHO and the GAZA STRIP, and it continues to negotiate suitable terms for handing the rest of the West Bank to the PLO.
Map Jordan – Aa

West Bengal *India* The western part of the former Indian state of Bengal, which was divided at Indian independence in 1947. (East Bengal became East Pakistan, now Bangladesh.) West Bengal covers 88 752 km² (34 258 sq miles) and has most of India's 75 million Bengali-speaking people. It produces rice, jute, wheat and coal. Its factories make steel, cars, aluminium and fertilisers, mainly around the capital CALCUTTA.
Population 67 982 700
Map India – Dc

West Bromwich *United Kingdom* Major industrial centre on the western outskirts of BIRMINGHAM, in the English county of West Midlands. Coal has been mined here since the late 18th century; industry – which today includes oil refining and chemicals – soon followed.
Population 155 000
Map United Kingdom – Ed

West Indies *Atlantic Ocean/Caribbean* Group of islands to the east of Central America. They separate the Caribbean and the Gulf of Mexico from the Atlantic Ocean. The West Indies consist of three main groups – the Bahamas in the northeast, the Greater Antilles, which includes CUBA, HISPANIOLA, JAMAICA and PUERTO RICO, in the middle, and the Lesser Antilles, which includes BARBADOS, TRINIDAD AND TOBAGO, and the LEEWARD, WINDWARD and VIRGIN ISLANDS in the south-east. Christopher Columbus reached the West Indies in October 1492; their name recalls the fact that he was searching for a western passage to India when he chanced upon these islands.
Map Atlantic Ocean – Ac, Bc-Bd

West Midlands *United Kingdom* Metropolitan county in central England, covering 899 km² (347 sq miles) and taking in the industrial area

known as 'the Black Country'. The large cities of BIRMINGHAM, COVENTRY and WOLVERHAMPTON are situated here.
Population 2 658 000
Map United Kingdom – Dd

West Palm Beach *USA* City on the east coast of Florida. It stands on Lake Worth, across from the resort town of PALM BEACH 110 km (68 miles) north of MIAMI. It is a commercial, fishing and business centre serving the Palm Beach tourist industry. Its Norton Art Gallery is noted for its collection of Chinese jade.
Population (city) 67 600; (metropolitan area) 863 500
Map United States – Je

West Point *USA* Site of the United States Military Academy, on the heights above the Hudson River, 80 km (50 miles) north of NEW YORK CITY. The academy, the US Army officers' training college, founded in 1802, was built on the site of American defence posts established in 1778 during the American War of Independence.
Map United States – L

▼ **FANTASTIC UNDERWORLD** Shapely stalactites and stalagmites lend an eerie beauty to a limestone cave at Yallingup, near Western Australia's south-western coast.

West Virginia *USA* Hilly southern state covering an area of some 62 772 km² (24 231 sq miles) between the Appalachian Mountains and the upper Ohio River.

It was originally part of Virginia, but its inhabitants opposed slavery and remained loyal to the Union during the American Civil War (1861-5). The rest of Virginia seceded in 1861; West Virginia was admitted as a state in 1863. Harpers Ferry National Historical Park at the junction of the Shenandoah and Potomac rivers in the north-east corner of the state commemorates John Brown's rising against slavery in 1859. CHARLESTON is the state capital.

Cattle, dairy products, timber, tobacco, chickens and apples are West Virginia's main agricultural products. The state is a leading producer of bituminous coal, and also produces oil and gas; these fuel its chemicals, rubber, metal processing, ceramics and glass industries.
Population 1 793 500
Map United States – Jc

westerlies Winds that blow from the subtropical high pressure systems to the temperate low pressure systems, from 35 to 65 degrees north and south of the equator. They blow from the southwest in the Northern Hemisphere and from the north-west in the Southern Hemisphere.

Western Australia *Australia* Largest state in the country, covering an area of some 2 525 500 km² (975 100 sq miles). It was the second British colony in Australia, founded in 1829. Over 70 per cent of its people live in the state capital, PERTH.

Much of the land is flat and arid; however, the PILBARA and the KIMBERLEY PLATEAU in the north-west are highland regions, and the former is extremely rich in minerals, notably iron. Farming relies on sheep and wheat, but the 'Mediterranean' south-west produces wheat and fruit, and in the far north attempts are being made to grow irrigated rice and cotton.
Population 1 665 900
Map Australia Bc

Western Cape *South Africa* One of nine provinces that were created in 1994. An area covering 129 386 km² (49 994 sq miles), it extends from the principal city, Cape Town, for some 420 km (670 miles) east along the south (Indian Ocean) coast and for 260 km (415 miles) north along the west (Atlantic) coast. Inland it includes fertile farming areas with a Mediterranean climate, as well as parts of the semi-arid Little and arid Great KAROO. The Cape's folded mountains are both a scenic feature and a profound influence on the distribution of rainfall. Agricultural products of the Western Cape include fruit, cereal and wine. Tourism is important to the economy.
Population 3 600 000
Map South Africa – Ac-Bc

Western Desert *Egypt* See SAHARA

Western Isles *United Kingdom* The Inner and Outer Hebrides off the west coast of Scotland, totalling 2898 km² (1119 sq miles). The Outer Hebrides stretch over 200 km (120 miles) from the Butt of Lewis in the north, through North and South Uist, to Berneray in the south. Lewis is the largest island, and has the only substantial town, Stornoway. The Inner Hebrides, nearer the Scottish mainland, include SKYE, MULL, Jura and Islay. Sheep farming, production of woollen goods, whisky distilling and fishing are the principal occupations. The islands are linked with each other and with the mainland ports by ferry.
Population 29 110
Map United Kingdom – Ba

Western Province (Barotseland) *Zambia* Region of 126 386 km² (48 798 sq miles) which borders Angola. Its provincial capital, MONGU, is 550 km (340 miles) west of the national capital, Lusaka. The state is the home of the Lozi people, who live mostly on the savannah-covered flood plain of the Zambezi River and fish and grow maize and pulses on small subsistence farms.
Population 607 000
Map Zambia – Bb

Western Sahara See p 733

Western Samoa See p 734

Westland *New Zealand* South Island district stretching along the coast west of the Southern Alps. It covers some 15 415 km² (5952 sq miles) and was a gold-mining area in the 19th century. Today its industries are farming, logging and coal mining. Westland's magnificent mountains, glaciers, lakes, forests and fine coastline make it a popular tourist area. Its main towns are Greymouth and HOKITIKA. Westland National Park and neighbouring Mount Cook National Park have been declared World Heritage areas.
Population 34 200
Map New Zealand – Ce

Westman Islands (Vestmannaeyjar) *Iceland* Group of 15 islands – covering in total 16 km² (6 sq miles) – 10 km (6 miles) off the south coast of Iceland. The group consists largely of lava and TUFF formations. The islands' main port, on the island of Heimaey, was partly destroyed by a volcanic eruption in 1973, but the town has since been rebuilt and the fishing harbour improved.
Population 5500
Map Iceland – Bb

Westmeath (An Iarmhí) *Ireland* Cattle-raising county covering 1763 km² (681 sq miles) of flatlands and lakes in central Ireland, about 50 km (30 miles) north-west of DUBLIN. The county town is Mullingar (population 8000).
Population 61 880
Map Ireland – Cb

Westminster *United Kingdom* See LONDON

Westmorland *United Kingdom* See CUMBRIA

Westphalia *Germany* See NORTH RHINE-WESTPHALIA

wet spell In the United Kingdom, the term for a period of at least 15 consecutive days during which the daily rainfall is 1 mm (0.04 in) or more. In other parts of the world, the criterion varies depending on climatic conditions.

Wewak *Papua New Guinea* Capital of East Sepik province on the north coast of New Guinea, west of the SEPIK River mouth. It is a port handling exports of coffee, cocoa and rubber; it is also the transit point for air travel to Irian Jaya, the Indonesian part of New Guinea. At Cape Wom, west of the town, Japanese forces formally surrendered on 13 September, 1945.
Population 23 800
Map Papua New Guinea – Ba

Wexford (Loch Garman) *Ireland* Market town, county town and port, 116 km (72 miles) south of DUBLIN. Wexford became a stronghold during the Anglo-Norman invasion of 1169. Its importance as a port has declined as the River Slaney has silted up, but industry has grown. The county of the same name, covering an area of 2351 km² (908 sq miles), is an arable plain, backed in the north-west by the Blackstairs Mountains.
Population (town) 9540; (county) 102 070
Map Ireland – Cb

Whangarei *New Zealand* Main city of NORTHLAND, situated on a fine natural harbour 135 km (85 miles) north of AUCKLAND. It is an industrial and regional market centre.
Population 44 180
Map (Northland) New Zealand – Da

whirlwind Column of rotating air centred on a small, local area of low ATMOSPHERIC PRESSURE. It is less violent than a TORNADO.

White Russia See BELARUS

White Sands National Monument *USA* Area of glistening white, gypsum dunes reaching heights up to 15 m (50 ft). It lies in southern New Mexico, 160 km (100 miles) north of EL PASO, Texas, and is 24 km (15 miles) west of Alamogordo, near the site where the first atomic bomb was tested on 16 July, 1945. The monument covers an area of 58 170 hectare (143 700 acres).
Map United States – Ed

White Sea (Bjeloje Morje) *Russia* Almost enclosed inlet of the BARENTS SEA off Russia, which largely freezes over in winter, although icebreakers keep its main ports open. It was used during the Second World War to get essential convoys with supplies into the Soviet Union. The sea, which covers 95 000 km² (36 670 sq miles),

is rich in fish. Timber is a major export from its chief port, ARCHANGEL (Arkhangel'sk). The northern DVINA and Onega rivers flow into the sea.
Map Russia – Eb

Whitehorse *Canada* Capital of YUKON TERRITORY, at the upper limit of navigation on the Yukon River, 84 km (52 miles) north of the border with British Columbia. It is the terminus of the White Pass and Yukon Railway to Skagway, 90 km (55 miles) south in the Alaskan 'panhandle', and lies just off the Alaska Highway. Whitehorse was an important staging point during the Klondike gold rush in the 1890s.
Population 17 930
Map Canada – Bb

whiteout Polar weather condition caused by a heavy cloud cover over the snow, in which the light coming from above is about equal to the light that is reflected from below. Whiteouts are characterised by the absence of shadow and the invisibility of the horizon.

Whitney, Mount *USA* Second highest peak in the country, after Mount McKinley in Alaska. It rises to 4418 m (14 494 ft) in the Sequoia National Park and Inyo National Forest in the Californian Sierra Nevada, 280 km (175 miles) north of Los Angeles. It is 130 km (80 miles) from DEATH VALLEY, the lowest point in the Americas.
Map United States – Cc

Whyalla *Australia* Major port and steel-making city in South Australia. It is situated on Spencer Gulf, 225 km (140 miles) north-west of ADELAIDE. It exports iron ore from open-cast mines inland.
Population 25 530
Map Australia – Fe

Whydah *Benin* See OUIDAH

Wichita *USA* City straddling the Arkansas River in southern Kansas. As a wide open 'cow town' in the 1870s and 1880s, it was a shipping port on the Chisholm Trail from Texas. The legendary Wyatt Earp was a local law officer in the 1870s and the city now has a Wichita Cow Town restoration project for visitors. It is a centre for aircraft production and a major processing and distribution centre for oil, grain and livestock.
Population (city) 304 000; (metropolitan area) 485 300
Map United States – Gc

Wicklow (Cill Mhantáin) *Ireland* Resort, market town and county town 45 km (28 miles) south of DUBLIN. The county of the same name covers 2025 km² (782 sq miles) and includes the Wicklow Mountains, rising to 926 m (3038 ft).
Population (town) 5850; (county) 97 270
Map Ireland – Cb

Wielkopolska *Poland* See POLAND, GREATER

Wiener Neustadt *Austria* Industrial city 40 km (25 miles) south of VIENNA. Most of it was destroyed in the Second World War. The Holy Roman Emperor Maximilian I (1459-1519) was born in the city and is buried here, although the memorial tomb to the emperor lies in INNSBRUCK. The city's products include aircraft and textiles.
Population 35 270
Map Austria – Eb

Western Sahara

THE EXISTENCE OF THIS MINERAL-RICH DESERT LAND AS A SEPARATE COUNTRY HAS LONG BEEN A MATTER OF DISPUTE

This territory of phosphates, desert and nomads has, for nearly two decades, been a matter of dispute and the scene of guerrilla warfare involving the Sahrawis (the indigenous people of the area) and the Moroccans, who have long claimed it.

The country, covering some 252 120 km² (97 321 sq miles) on the north-west coast of Africa, and bordering Morocco, Algeria and Mauritania, consists entirely of desert – mainly stony but with some sandy areas. There are a few scattered oases and a little poor pasture land. Annual rainfall is less than 50 mm (2 in). There is a fishing industry based at the ports of Boujdour and AD DAKHLA.

Most of the estimated 206 600 Sahrawis are a mixture of Arab and Berber, and almost all are Muslims. The area was colonised by Spain from 1884 and declared a Spanish province in 1960, but like the rest of Africa was affected by growing nationalism. In 1973 the Sahrawi liberation movement, the Polisario Front, demanded independence. After that, the major problems began.

PEACEFUL INVASION

Morocco has historical claims to the territory as part of 'Greater Morocco', and in 1975 King Hassan sent 350 000 unarmed Moroccans on a peaceful 'green march' into Western Sahara in an attempt to force Spain to hand it over. Apart from historical claims, Morocco's interest lay in Western Sahara's deposits of phosphate – the world's largest – at BOUKRA, just south of the chief town, LAÀYOUNE. With Morocco's own considerable output of phosphate (the raw material used for fertiliser), these deposits would give the Moroccans a dominant position in world markets.

The peaceful occupation was successful and in 1976 Spain withdrew from the colony, agreeing that it should be divided between Morocco and Mauritania, Morocco taking the northern 162 000 km² (about 62 500 sq miles), which contained the phosphates.

In the same year, however, the Polisario declared an independent Saharan, or Sahrawi, Arab Democratic Republic (SADR), which was recognised by 70 nations including some 30 in Africa. Mauritania eventually bowed out and reached a peace agreement with the Polisario Front in 1979.

Refusing to recognise an independent SADR, however, Morocco's King Hassan proceeded to incorporate the territory into Morocco in 1979, after Mauritania had completely withdrawn, and Western Sahara was subdivided into four Moroccan provinces: Laayoune (formerly El-Aaiún), including the town Laayoune; Boujdour, centred on the port by the same name; Es-Smara, including the town of Es-Smara; and Oued ed-Dahab, based on the southern port of Ad Dakhla (formerly Villa Cisneros).

The Polisario began their guerrilla war in earnest, but after eight years of fighting, Morocco was largely in control of most of the northern section, including Laayoune and the phosphate deposits. The Moroccans had built defensive fortifications which ran for 1600 km (1000 miles) from north to south across the territory to restrict the guerrillas' movements, but guerrilla incursions continued. Some 125 000 Sahrawi refugees were living in camps in south-western Algeria.

The Moroccans began investing heavily in the part of Western Sahara they controlled, improving transport, fisheries, water supplies and education. Phosphate exports could finance much more development, both social and economic. By 1991 a ceasefire had been agreed upon. In the years that followed, a United Nations-supervised referendum to settle the matter of independence or integration with Morocco was repeatedly postponed, first because of disagreement on who was eligible to vote and, more recently, because of the slow progress in identifying potential voters.

WESTERN SAHARA AT A GLANCE	
Area 252 120 km² (97 321 sq miles)	
Population 206 600 (estimate)	
Chief town Laàyoune	
Government Disputed territory administered by Morocco	
Currency 1 Moroccan dirham = 100 centimes	
Languages Arabic, Berber, Spanish	
Religion Muslim	
Climate Hot and arid. Average temperature on the coast ranges between 16°C (61°F) and 25°C (77°F); hotter inland	
Land use Grazing 19%, desert 81%	
Main primary products Phosphates, iron ore, fish	
Major industry Phosphate mining, fishing	
Main exports Phosphates	
Annual income per head (US$) 300	
Population growth (per thous/yr) 25	
Life expectancy Male 44 **Female** 46	

Wienerwald (Vienna Woods) *Austria* Forested hills, an extension of the eastern Austrian Alps, to the west and north-west of the capital. The villages of the region's eastern foothills are set among vineyards and have become commuter towns for VIENNA. The new *Heuriger* wines can be tasted in the courtyard bars of the converted farmhouses.
Map Austria – Da

Wiesbaden *Germany* Spa town and state capital of Hessen. It lies beside the Rhine, some 32 km (20 miles) west of FRANKFURT am Main. Rheumatics sufferers have visited its hot saline springs since Roman times. Wiesbaden is a publishing and film centre; other industries include sparkling wine, chemicals and building materials. Wiesbaden is an administrative town and houses Germany's institutes of Germanology and statistics. The state library of Hessen is housed here.
Population 260 300
Map Germany – Cc

Wigan *United Kingdom* English town, formerly in Lancashire, now in Greater Manchester, 25 km (16 miles) north-east of LIVERPOOL. There was a Roman fort on the site and it received its royal charter in 1246, but Wigan grew to industrial prominence with the 19th-century development of local coal mines – the last of which closed in 1967 – and the cotton textile industry (which is also in decline). Its many new industries include micro-electronics, paper products, printing, glass fibre, engineering, plastics and processed food.
Population 80 000
Map United Kingdom – Dd

Wight, Isle of *United Kingdom* Island county covering 381 km² (147 sq miles) off the south coast of England, and separated from the mainland by the Spithead and Solent channels. It is 36 km (22 miles) long, stretching from the town of Bembridge in the east to the NEEDLES chalk stacks in the west, and is a world-famous yachting centre, with a regatta at Cowes each August.

In the 19th century, Queen Victoria and the poet Alfred, Lord Tennyson (1809-92) had homes on the island; the queen died here in 1901. The island county has remained a popular tourist and retirement area. Newport is the county town.
Population 126 300
Map United Kingdom – Ee

Wigtown *United Kingdom* See DUMFRIES AND GALLOWAY

Wilhelm, Mount *Papua New Guinea* The country's highest peak, rising to 4509 m (14 793 ft) in the BISMARCK RANGE.
Map Papua New Guinea – Ba

Wilhelmshaven *Germany* North Sea port, 65 km (40 miles) north-west of BREMEN. It is the only German seaport with sufficient depth for loaded 250 000-tonne ships.

Wilhelmshaven was founded in the 19th century as a naval base and named after King Wilhelm I of Prussia. Since the Second World War it has become an oil-importing town, with a pipeline link to the industrial Ruhr. Today it makes petrochemicals and precision instruments.
Population 90 200
Map Germany – Cb

Willemstad *Netherlands Antilles and Aruba* Seaport of CURAÇAO and capital of the Netherlands Antilles, situated on the Schottegat on the south-west coast of Curaçao. It is one of the largest ports in the world in terms of the tonnage

Western Samoa

IN A TWICE-COLONISED NATION, TRADITIONAL LIFE THRIVES ONCE AGAIN UNDER A NATIVE KING

When the Pacific islands of Western Samoa gained independence in 1962, it was the second time that they had won freedom in 73 years. The mountainous tropical islands had been squabbled over by the Dutch, British, Germans and Americans since their charting by Europeans in the 18th century. In 1889 they were given independence under King Malietoa Laupepa, but after the king died in 1898, the islands came under German control, by agreement with the United States and Great Britain.

English influence had already been well established by the arrival in 1830 of missionaries, who quickly converted the people to Christianity. New Zealand occupied the German-ruled islands in 1914, received a post-war mandate in 1919 and proceeded to administer them until 1962. Elections for the 49-member legislative assembly were based on universal suffrage for the first time in 1991.

The islands lie in the Polynesian sector of the Pacific, about 720 km (450 miles) northeast of Fiji. The two main islands are SAVAI'I and UPOLU. Savai'i is largely covered with volcanic peaks and lava plateaus; Upolu has two-thirds of the population and the capital, APIA.

According to legend, Polynesians set out to colonise other parts of the Pacific from Samoa; and the traditional lifestyle still thrives. Most people today live in *fale* (traditional open-sided thatched houses) in coastal villages. Many tourists visit the grave of Robert Louis Stevenson and his home, Vailima, now the official home of King Malietoa Tanumafili II.

WESTERN SAMOA AT A GLANCE	
Official name Independent State of Western Samoa	
Map Pacific Ocean – Ec	
Area 2660 km² (1104 sq miles)	
Population 199 700 **Per km²** 75 **(Per sq mile** 181)	
Capital Apia	
Government Parliamentary monarchy	
Currency 1 tala = 100 sene	
Languages Samoan, English	
Religion Christian	
Climate Tropical; average temperature in Apia is between 29°C (82 and 84°F) all year	
Main primary products Bananas, coconuts, cocoa, taro, timber, fish	
Major industries Agriculture, food processing, forestry, fishing, tourism	
Main exports Copra, coconut oil, taro, cocoa, timber	
Annual income per head (US$) 650	
Population growth (per thous/yr) 23	
Life expectancy (yrs) Male 61 **Female** 65	

it can accommodate – nearly 100 million tonnes of shipping at a time – and has the largest dry dock in the Americas. It is also an important oil-refining centre, and handles crude oil from the Lake Maracaibo oil fields of Venezuela.

Population 80 000
Map Caribbean – Bc

Williamsburg *USA* Historic city on a peninsula between the James and York rivers in Virginia, about 180 km (110 miles) south of the national capital, Washington DC. It was the state capital from 1689 until 1780 and the centre of Virginia's colonial opposition to Britain before 1776. The College of William and Mary, founded in 1693, is the second oldest university in the United States, after Harvard.

Tourists flock to Colonial Williamsburg, a huge restoration project occupying the centre of Williamsburg, where guides wear period costume. The project has more than 500 buildings renovated and re-created in their original settings. It is connected to JAMESTOWN and YORKTOWN in the nearby Colonial National Historical Park by the Colonial Parkway.

Population 11 000
Map United States – Kc

Wilmington *USA* Industrial and port city on the west bank of the Delaware River, in north-east Delaware State, some 40 km (25 miles) south-west of PHILADELPHIA. It is one of the world's largest centres of the chemical industry and the home, since 1802, of the Du Pont multinational chemical firm.

Population (city) 71 500; (metropolitan area) 513 300
Map United States – Kc

Wiltshire *United Kingdom* County of central southern England, covering an area of 3481 km² (1344 sq miles). It contains two of Britain's oldest landmarks – STONEHENGE and AVEBURY stone circles – and the army training grounds of Salisbury Plain. Most of the county is agricultural, but Swindon – a former railway town whose locomotive workshops are now closed – has a number of computer and light industrial companies. The lovely cathedral of SALISBURY has the tallest spire in England, at 123 m (404 ft). The county town is Trowbridge.

Population 571 800
Map United Kingdom – De

Winchester *United Kingdom* Cathedral city in the English county of Hampshire, about 80 km (50 miles) south-west of LONDON. It was the capital of the Saxon kingdom of Wessex before the 7th century. There are remains of the Roman city of Venta Belgarum, and of a Norman castle, but Winchester's best known landmark today is its cathedral. It replaced a Saxon building, was begun in 1079 and was completed three centuries later, with further additions until 1525. Around the lawns of the cathedral close are the ruins of two bishop's palaces and several buildings dating from before 1500. They include some parts of Winchester College – one of the most famous of England's public schools, founded in 1382 by William of Wykeham, Bishop of Winchester.

A round table, said to have belonged to the legendary King Arthur, can be seen in the hall (the only remaining part) of the Norman castle; but the table dates only from about 1150, some six

centuries after Arthur died. It is another king – Alfred the Great (849-99), called 'the Founder of the English Nation' – who is chiefly commemorated in this historic city. Winchester was Alfred's capital and his main base in the struggle against the Danish invasion.

Population 35 660
Map United Kingdom – Ee

wind Current of air on the earth's surface induced by difference in ATMOSPHERIC PRESSURE, varying in strength from light air to a hurricane. Winds are identified by the direction from which they come, although many have local names. See PRESSURE GRADIENT and BEAUFORT SCALE.

wind deflection Difference in the direction of the wind from that suggested by the PRESSURE GRADIENT, caused by the rotation of the earth. The deflection is negligible at the equator and greatest at the poles. In the Northern Hemisphere winds are deflected to the right, causing them to circulate anticlockwise around low pressure systems and clockwise around highs. The opposites apply in the Southern Hemisphere. The deflection is apparent, rather than real, since it is the earth that moves and not the wind that changes direction.

Windermere *United Kingdom* Largest lake in England, and also a town in the LAKE DISTRICT of north-west England, some 110 km (70 miles) north of LIVERPOOL. The narrow lake is 16 km (10 miles) long, with many islands, and is a popular water sports and tourist attraction.
Map United Kingdom – Dc

Windhoek *Namibia* National capital and industrial city. It was founded as the military headquarters of German South West Africa in 1890 and still has many German buildings, including the *Alte Feste* (Old Fort) and the *Christuskirche*, the German Evangelical Lutheran church.

A German carnival is celebrated each year in the city, and local Herero women wear dresses modelled on those worn by the wives of 19th-century Rhenish missionaries. The local population mix is varied, and since independence in 1990, the city has attracted many other nationals.

Industries include food processing, brewing, manufacture of bone meal, engineering, diamond sorting and the processing of 'Persian lamb' pelts from karakul lamb pelts.

Population 165 000
Map Namibia – Ab

Windsor *Australia* With its twin town, Richmond (population 9310), Windsor is one of Australia's oldest settlements. Both were founded about 1810 on rich farming land by the Hawkesbury River, 50 km (30 miles) north-west of SYDNEY. They contain many fine colonial buildings.

Population 10 660
Map (Sydney) Australia – Ie

Windsor *Canada* Canada's most southerly city, on the Detroit River in south-west ONTARIO. It is linked by bridge and tunnel with the city of Detroit, across the United States border. It makes cars, machine tools and chemicals, has salt mines and distilleries, and lies at the centre of a rich farming region.

Population (city) 191 440; (metropolitan area) 260 700
Map Canada – Gd

▲ STORMY PAST The peaceful church of Fort Amsterdam, built by the Dutch at Willemstad from 1634, still has a cannonball embedded in its wall – a gift from English attackers in 1804.

Windsor *United Kingdom* Royal English town on the River Thames in Berkshire, some 35 km (22 miles) west of LONDON. There has been a castle at Windsor since the Norman conquest in 1066. The base of the great round tower dates from the 12th century, and St George's Chapel from the 16th. Most of the remainder of the huge castle, which covers over 5 hectares (12 acres), was built in the 19th century. The castle is one of the homes of the British royal family, and the burial place of several kings and queens. Its state apartments were severely damaged by fire in November 1992. The castle and the 1940 hectare (4795 acre) Great Park beside it enclose the town of Windsor on the east and south sides. Across the Thames is Eton, the famous public school founded in 1440 by Henry VI.
Population 31 000
Map United Kingdom – Ee

Windward Islands *Caribbean* See LEEWARD ISLANDS

Windward Islands *French Polynesia* See SOCIETY ISLANDS

Winnipeg *Canada* Capital of Manitoba, lying at the junction of the Assiniboine and Red rivers. First settled as a trading post in 1738, it became a city in 1873 with a population of fewer than 4000. It is now one of the largest cities of the prairie provinces and has one of the world's biggest wheat markets. It has two universities, the Royal Winnipeg Ballet and a symphony orchestra. In addition to grain elevators and flour mills, the city has meat packaging and food processing plants, clothing factories and railway workshops.
Winnipeg lies just south of Lake Winnipeg, which covers some 24 500 km² (9460 sq miles) in southern Manitoba. The lake drains via the Nelson and other rivers into HUDSON BAY. Like Lake Manitoba to the west, it formed part of an important transport route before the railways.
Population (city) 592 060; (metropolitan area) 652 360
Map Canada – Fc

Winterthur *Switzerland* Industrial town, 20 km (12 miles) north-east of the city of ZÜRICH and the centre of the Swiss machinery and metal industry. It has a world famous art collection and a 14th to 16th-century church in its well-preserved old town dating from the Middle Ages.
Population 85 600
Map Switzerland – Ba

Wirral *United Kingdom* Peninsula on the Irish Sea in north-west England, between the estuaries of the Mersey and Dee rivers. It is divided between the counties of Cheshire and Merseyside. The industrial and port conurbation of Birkenhead and Wallasey smothers the tip. The town of CHESTER is at the peninsula's southern end.
Map United Kingdom – Dd

Wisconsin *USA* Northern midwest state covering 145 439 km² (56 154 sq miles) between the Great Lakes and the upper Mississippi River. Dairy farming is the main agricultural activity; the state also produces cattle, maize, pigs and fruit. Machinery, cheese, meat packaging and beer – for which MILWAUKEE, the state's largest city, is famous – are the principal manufactures. MADISON is Wisconsin's capital.
Population 4 775 000
Map United States – Hb

Wisla *Poland* See VISTULA

Wittenberg *Germany* Cradle of the Reformation, this industrial town is midway between LEIPZIG and BERLIN. Plaques on the doors of the Schlosskirche (castle church) mark the place where in 1517 the priest and reformer Martin Luther (1483-1546) nailed his 95 theses against the Roman Catholic Church. Today, the town is a centre of Germany's chemical industry.
Population 50 100
Map Germany – Ec

Witwatersrand *South Africa* An 80 km (50 mile) ridge called 'the White Water Ridge' that runs roughly east to west and lies some 1750 m (5330 ft) above sea level. Gold was discovered here in 1886, launching South Africa into the era of deep-level gold mining and unprecedented industrialisation. JOHANNESBURG lies at the centre of the Witwatersrand (familiarly known as 'the Rand'). Towns to the west include Roodepoort

and Krugersdorp, with Boksburg, Benoni and Brakpan among those lying to the east. Relatively densely populated, it is the industrial and commercial heartland of South Africa.
Map South Africa – Cb

Wloclawek (Vlotslavsk) *Poland* Paper-making city, 140 km (86 miles) north-west of the capital, WARSAW. It was founded as a fortress in the 11th century, and has a 14th-century cathedral. The city also makes nitrate fertilisers.
Population 121 000
Map Poland – Cb

Wolds, The *United Kingdom* Series of rounded chalk hills in eastern England on both sides of the Humber estuary in Lincolnshire, Yorkshire and Humberside. The hills are not generally higher than 180 m (600 ft); they provide good grazing and arable land.
Map United Kingdom – Ed

Wolfe Creek *Australia* Site of a meteorite crater measuring some 840 m (2755 ft) across and 90 m (295 ft) deep. It lies near the town of Halls Creek in the KIMBERLEY PLATEAU area of Western Australia, 160 km (100 miles) from the state's border with the Northern Territory. The perfectly preserved circular crater – which is littered with glassy green balls of rock formed by the impact through partial melting – is thought to be about one million years old.
Map Australia Db

Wollomombi Falls *Australia* The highest waterfall in the country, at 455 m (1493 ft), near Armidale in the New England region of New South Wales, 370 km (230 miles) north of Sydney.
Map Australia – Ie

Wollongong *Australia* Port and major industrial city in New South Wales, 80 km (50 miles) south of SYDNEY. It is a coal-mining centre, and the urban area includes PORT KEMBLA, where Australia's biggest steel works is situated.
Population 235 000
Map Australia – Ie

Wolverhampton *United Kingdom* Town in the West Midlands of England, the so-called 'Black Country', situated 22 km (14 miles) north-west of BIRMINGHAM. It began as a settlement around a 10th-century monastery and grew, in the 19th century, into a coal-mining and manufacturing town, noted for its locks and safes.
Population 248 500
Map United Kingdom – Dd

Wonga Wongue National Park *Gabon* Nature reserve covering 3800 km² (1462 sq miles) north of the Ogooué River in western Gabon. It contains antelopes (including the rare situtonga), buffaloes, crocodiles, elephants, hippopotamuses and leopards. The unusual Bam Bam amphitheatres in the park are beautiful semi-circular basins eroded by torrential rain into the sides of hills that rise about 100 m (328 ft) above the surrounding plain. The scenery is reminiscent of the BADLANDS of the United States.
Map Gabon – Ab

Wonsan *North Korea* East coast industrial city and port, 140 km (87 miles) east of PYONGYANG. It was heavily bombed during the Korean War

▲ SEAT OF BISHOPS Medieval Marienburg Fortress, home of the ruling bishops of Würzburg for 500 years, overlooks the city and its vineyards – noted for exuberant, fine white wines.

(1950-3) but has since been rebuilt. Wonsan now has chemicals and oil-refining plants, as well as shipbuilding yards and factories making railway rolling stock.

Population 350 000

Map Korea – Cc

Woodlands-Kranji *Singapore* The largest new-town development on Singapore Island, begun in 1971 on the north side, near the southern end of the causeway that carries the road, railway and water pipeline links with the Malaysian mainland. It is planned to accommodate up to 300 000 people. The Japanese invasion force landed at the largely undefended site in December 1941, during the Second World War.

Population 68 600

Map Singapore – Aa

Woomera *Australia* Town in South Australia, situated in a semi-arid region about 160 km (100 miles) north-west of Port Augusta, from which water is piped to it. It was established in 1947 to house workers at the nearby missile testing site, which was operating in the 1950s and 1960s. The town derives its name from an Aboriginal word for a spear-thrower – a notched wooden stick used by hunters to extend the range of their weapons. The Joint Defence Space Communications Station, operated jointly by the United States and Australian Departments of defence, is 17 km (11 miles) south of Woomera.

Population 1600

Map Australia – Fe

Worcester *United Kingdom* City and former county of western England, the latter now amalgamated with its neighbour, Herefordshire. The city stands on the River Severn, 45 km (28 miles) south-west of BIRMINGHAM. Its cathedral was built between the 11th and the 14th century and shares the annual Three Choirs Festival with its counterparts in Hereford and Gloucester. The composer Edward Elgar was born in the nearby village of Lower Broadheath and lived in the area for much of his life. Royal Worcester porcelain and Worcestershire sauce are its famous products.

Population 76 000

Map United Kingdom – Dd

Worms *Germany* Cathedral city on the Rhine, 20 km (12 miles) north-west of MANNHEIM. It is one of the oldest German cities, founded by the Celts under the name of Borbetomagus and extended later by the Romans as a military station. In AD 420 it became the capital of Burgundy. It is known for its Dom, one of the most beautiful Romanesque cathedrals in Germany.

At Worms in 1521 Martin Luther (1483-1546), a priest already excommunicated by the Pope, appeared before the Imperial Diet, or assembly, and repeated his religious defiance to the Holy Roman Emperor Charles V, thus taking the Protestants another step towards their break with Rome. Worms lies in a grape-growing region which produces high-quality white wine.

Population 75 900

Map Germany – Cd

Wrath, Cape *United Kingdom* North-western tip of mainland Scotland, in the Highland region. Nearby are 260 m (850 ft) high cliffs, the tallest in Britain.

Map United Kingdom – Ca

Wrexham *United Kingdom* The main industrial and commercial centre of North Wales, lying in the county of Clwyd, 40 km (25 miles) south of LIVERPOOL. Its 19th-century prosperity was based on coal mining, and steel and brickmaking, but as these declined, new industries such as chemicals and textiles developed. Elihu Yale (1648-1721), whose donations helped to found Yale University in the American state of Connecticut, is buried in Wrexham churchyard.

Population 40 930

Map United Kingdom – Dd

Wroclaw (Breslau) *Poland* Industrial city 306 km (190 miles) south-west of the capital, WARSAW. Founded in the 10th century astride several tributaries of the Oder River, it prospered as a trading town. Under Austrian rule (1526-1741), craft industries developed, and the city's Renaissance and Baroque houses and churches date from that time. Wroclaw became an industrial centre under Prussian-German occupation (1741-1945). Although badly damaged in the Second World War, it is a major industrial centre today, producing electric railway locomotives, wagons, machinery, electrical appliances, chemicals, processed food and artificial silk.

Some 100 bridges span the 90 km (56 miles) of canals and rivers, giving the city a Venetian air. Near Wroclaw is Chelmsko Slaski, a small

Sudetenland town whose market square is surrounded by fine burghers' houses, a Baroque church and arcaded rows of weavers' wooden houses. All date from the 17th and 18th centuries, when the town was a major textile centre.

Population 631 300

Map Poland – Bc

Wuhan (Hankow) *China* Capital of HUBEI Province. It lies at the confluence of the Han Shui and Chang Jiang rivers and is made up of the cities of Wuchang, Hankow and Hanyang. In 1911, Wuhan was the scene of the Nationalist unrest which led to the overthrow of the Manchu emperors, the last of imperial China. This set off the long period of fighting between the nationalists and communists until the eventual triumph of the communists in 1949.

Today Wuhan has one of the most important concentrations of manufacturing in central China, with iron and steel, cement, papermaking, glass, rubber and cotton textile factories.

Population 4 200 000

Map China – He

Wuppertal *Germany* Textile-manufacturing city about 25 km (15 miles) east of DÜSSELDORF. An overhead tramway runs through the city and along the Wupper River.

Population 383 700

Map Germany – Bc

Würzburg *Germany* University city, 100 km (62 miles) south-east of FRANKFURT am Main. It was the seat of prince-bishops, who from 1650 created a Baroque city with a cathedral, bridges, a palace, vineyards and gardens. The city was badly damaged during the Second World War but has been restored.

Population 127 800

Map Germany – Cd

Wusul Jiang (Ussuri) *China/Russia* River rising in Russia and flowing 800 km (500 miles), into the Heilong Jiang (AMUR) River. It forms part of the boundary between China and Russia, where clashes between Chinese and Russian forces occurred in the mid-1960s.

Map China – Kb

Wuxi *China* City of JIANGSU Province on the north shore of Tai Hu Lake, 120 km (75 miles) west of SHANGHAI. Its major industries are cotton textiles and silk reeling.

Population 930 000

Map China – Ie

Wyoming *USA* Western state covering some 253 597 km² (97 914 sq miles). It is the ninth largest state in area but the smallest in population numbers. The Rocky Mountains cross the state diagonally from the north-west to south-central Wyoming, rising to some 4208 m (13 804 ft) at Gannett Peak. Plains more than 1000 m (3300 ft) above sea level stretch north-eastwards to the Black Hills on the South Dakota border.

The region's rainfall is scant. Cattle and sheep, wheat and sugar beet are the main farm products. The state has extensive deposits of uranium, coal, oil and natural gas. Its scenery – including that of YELLOWSTONE NATIONAL PARK – attracts millions of tourists annually. CHEYENNE is the capital.

Population 453 600

Map United States – Eb

Xai-Xai *Mozambique* Seaport, formerly called João Belo, on the mouth of the LIMPOPO River in southern Mozambique, 230 km (145 miles) from the South African border. Centre of a productive farming area – water for the 30 000 hectares (74 100 acres) under irrigation is tapped from the Limpopo Dam – the port exports cotton, maize, rice and sugar. Xai-Xai is the capital of Gaza Province.

Population (province) 1 030 500
Map Mozambique – Ac

Xauen *Morocco* See CHAOUÈN

xenolith Rock fragment which is different from the IGNEOUS ROCK mass in which it occurs.

Xi Jiang *China* Longest river in the south, rising in the mountains of YUNNAN Province then flowing eastwards for 2197 km (1370 miles) to the Zhu Jiang delta on the South China Sea. The Xi Jiang is an important transport artery and its water is used extensively for irrigation.

Map China – Gf

Xiamen (Amoy) *China* Port of FUJIAN Province, on the Taiwan Strait. It was the first Chinese port to be used by 16th-century Portuguese, Dutch and English traders, and it became a treaty port, where foreigners could live and trade, in 1842. It is one of China's 'special economic zones' – areas set up to attract overseas investment.

Population 588 000
Map China – Hf

Xi'an (Sian) *China* Capital of SHAANXI Province, 900 km (560 miles) south-west of the capital, Beijing. It is an ancient city in the Wei He Valley and, under the name of Chang'an ('Everlasting Peace'), was the capital of imperial China during the Tang dynasty (AD 618-906). Its town wall, 11 km (6.8 miles) long, is still intact, and it has a 13-storeyed pagoda built in the 8th century.

The famous SILK ROAD began in Xi'an. Among the many archaeological sites around Xi'an is the vast tumulus tomb, at Mount Li, of Qin Shi Huangdi (who ruled 221-210 BC), China's first emperor and founder of the Qin or Ch'in dynasty, from which China gets its name. The tumulus, discovered in 1974, contains an army of more than 7000 life-sized terracotta figures. Xi'an is a centre for textile and electrical manufactures.

Population 2 710 000
Map China – Gd

Xiang Jiang *China* River in south-central China, flowing 1150 km (719 miles) north through HUNAN Province to Dongting Lake and the CHANG JIANG (Yangtze) River.

Map China – Ge

Xigaze *China* Town in the XIZANG AUTONOMOUS REGION (Tibet), lying in the Yarlung Zangbo (Brahmaputra) Valley 230 km (140 miles) south-west of Lhasa. It is the site of the Tashi-Lhunpo Buddhist monastery. Before the Chinese moved in on Xizang in 1950 the monastery had 3000 monks. It still survives, although with fewer monks and without their spiritual leader, the Panchen Lama, whose authority was second only to that of the Dalai Lama.

Population 40 000
Map China – Ce

Xingang *China* Port in the north of the country, at the mouth of the Hsi He River, some 45 km (28 miles) east of TIANJIN. It has been developed since the 1950s as the main harbour for Tianjin, and is now one of the largest ports in China.

Map China – Hc

Xining *China* Capital of QINGHAI Province, about 200 km (125 miles) north-west of LANZHOU. It is a centre for the manufacture of iron and steel, farm machinery, fertilisers and textiles.

Population 927 000
Map China – Ed

Xinjiang Uygur (Sinkiang Uygur) *China* Autonomous region covering 1 600 000 km² (620 000 sq miles) of western China, historically known as Chinese Turkestan or Dzungaria. For centuries, nomadic herding was the mainstay of the region's economy, and Uygur, Kazakh, Kirghiz and Tadzjik herders still graze their animals on the region's vast pastures. The main town is Urümgi.

Since the mid-1950s, the great mineral wealth of the deserts has begun to be exploited. There is oil beneath the Junggar and Tarim basins, as well as minerals such as coal and metal ores. A railway connects the region with GANSU Province and northern China.

Population 15 810 000
Map China – Bc

Xishuangbanna *China* Region in YUNNAN Province, on the border with Myanmar and Laos. It is famous for its rubber, shellac varnish made from resin secreted by the lac insect, quinine derived from the cinchona bark, and rare plants.

Map China – Ef

Xizang Autonomous Region (Tibet; Bod Ranskyon-ljon) *China* One of the most remote parts of the world, an autonomous region of China since 1965. It lies north of the great Himalaya range and encompasses the vast Xizang Gaoyuan (Tibetan Plateau), the upper valley of the Yarlung Zangbo (BRAHMAPUTRA) River and part of the QAM'DO region. Much of the 1 222 000 km² (471 700 sq mile) plateau is uninhabited. The small population is concentrated mainly in the Yarlung Zangbo Valley, where barley, rye and wheat are grown, as well as fruit and vegetables. In the Qam'do area, tea is produced. Almost half of the population herds yaks, sheep and cattle.

Tibet was formerly a Buddhist kingdom, ruled from Lhasa by a priestly aristocracy headed by the Dalai Lama, the spiritual leader of Tibetan Buddhists. In earlier centuries China intermittently extended its rule over Tibet, which was absorbed into China during the Yüan dynasty (1279-1368) and again during the Qing dynasty. Chinese armies occupied Lhasa and Tibet was incorporated within the Chinese empire until the revolution of 1911.

In 1913 the Dalai Lama declared his country independent. China invaded in 1950. After an uprising in October 1959 quashed by the Chinese, the Dalai Lama and about 20 000 of his subjects fled to India and Nepal. The Dalai Lama was awarded the Nobel Peace Prize in 1989.

During the Cultural Revolution in China (1966-76) many of Tibet's monasteries and shrines were destroyed and practising Buddhists were persecuted. From 1976 China adopted a more moderate policy towards Tibet and after 1980 some monasteries and shrines were renovated and reopened. However, in 1987 strong tensions between the 120 000 ethnic Chinese and the 1 710 000 native Tibetans exploded into violent rioting in Lhasa. Unrest flared up again in March 1989.

Population 2 280 000
Map China – Ce

Xizang Gaoyuan (Tibetan Plateau) *China* An immense expanse of uninhabited highland, the biggest and highest plateau in the world, covering roughly 1 760 000 km² (679 500 sq miles) and lying 4000 m (15 000 ft) above sea level. Its climate is extremely harsh – even at the height of summer, temperatures rarely rise above freezing point. The plateau is rich in mineral resources but these have not been exploited because of its remoteness.

Map China – Cd

Xuzhou *China* Strategic city in north-west JIANGSU, 625 km (390 miles) south and slightly east of BEIJING. It is a transport hub and centre of a rich coal-mining area. Its products include machinery and textiles.

Population 910 000
Map (Jiangsu) China – Hd

Yakutiya (Sakha) Republic Autonomous republic of the Russian Federation in east-central Siberia. Covering an area of 3 103 200 km² (1 198 150 sq miles), it is almost as large as the whole of Western Europe. Its capital is Yakutsk.

The republic consists largely of mountains reaching between 2000 and 3000 m (6560 and 9840 ft), interspersed with barren, sub-Arctic plains and marshy pine forests. Farther north, pine forest merges with wooded tundra, which in the extreme north becomes treeless tundra.

The region's climate is appallingly severe. Its towns of Oymyakon and VERKHOYANSK are two of the coldest places on earth, with winter temperatures plummeting to below –70°C (–94°F). Only 1 per cent of the land is cultivated. The main occupations are fishing, hunting and herding, often of reindeer. Gold, tin and coal are mined although the severe temperatures make mining very difficult; apart from the operating difficulties, people refuse to live here. As a result most of the population are still Yakuts.

Population (republic) 1 109 000
Map Russia – Nb

Yakutsk *Russia* Capital of the Yakutiya Republic, lying 480 km (300 miles) south of the Arctic Circle. It was founded in 1632 as a fort beside the 10 km (6 mile) wide Lena River. The city is now a fur trading centre. Its main industries are leather tanning, brick making and timber milling.

Population 191 000
Map Russia – Nb

Yala *Sri Lanka* See RUHUNA

Yalta *Ukraine* Tourist and health resort in the Crimea, 50 km (30 miles) east of Sevastopol'. Stretched along a sheltered bay, Yalta is bedecked with flowering shrubs and plants. It has attracted holidaymakers and the ailing since the early 19th century and today has numerous hotels and sanatoriums. In February 1945 the Grand Palace in the small resort of Livadiya,

3 km (2 miles) to the south, was the setting for the historic Yalta Conference to plan the final stage of the Second World War and the division of power in postwar Europe.

The conference was attended by the British Prime Minister, Winston Churchill (1874-1965), the President of the United States, Franklin D. Roosevelt (1882-1945), and the Soviet leader, Joseph Stalin (1879-1953).

Population 85 000
Map Ukraine – Cc

Yalu *China/North Korea* River forming 795 km (500 miles) of the border between China and North Korea. It is an important source of hydro-electric power.
Map China – Ic; Korea – Cb

Yamagata *Japan* City on north HONSHU Island, about 320 km (200 miles) north-east of Tokyo. It is a commercial centre and specialises in processed foods and electrical goods.
Population 249 000
Map Japan – Dc

Yambol (Jambol) *Bulgaria* River port on the Tundzha River, 72 km (45 miles) east of STARA ZAGORA. It ships wool and wine. Its industries include metal goods, textiles and tanning.
Population 99 230
Map Bulgaria – Cb

Yamoussoukro *Ivory Coast* Birthplace of Felix Houphouët-Boigny, who was the country's president from independence in 1960 until his death in December 1993.

The town lies in the centre of the country, 220 km (137 miles) north-west of ABIDJAN. In 1983 the government announced plans to make Yamoussoukro the country's new capital in place of Abidjan. Work began on new administrative buildings here, and the transfer of government departments began the following year. Although the country is extremely poor, its new capital has luxurious palaces and administration buildings. In 1990 the biggest cathedral in Africa, the Basilica of Our Lady of Peace, modelled on St Peter's in Rome, was built here.
Population 130 000
Map Ivory Coast – Ab

Yamuna (Jumna) *India* Northern river, 1376 km (855 miles) long, rising in the Himalayas in north-west UTTAR PRADESH State. It flows generally south past Delhi, then south-east to enter the Ganges at ALLAHABAD.
Map India – Cb

Yan'an *China* Town in SHAANXI Province. The Chinese Communists set up their headquarters here in 1947 after their Long March through the mountains from the south. Yan'an's revolutionary museum commemorates the feat.
Population 268 000
Map China – Gd

Yangon (Rangoon) *Myanmar (Burma)* Port and national capital beside the Yangon River, just east of the Ayeyarwady (Irrawaddy) River delta; it is also the centre of a 10 171 km² (3927 sq mile) division of the same name. The city was developed as a port to replace Syriam by the 18th-century Burmese ruler King Alaungpaya, and became the administrative centre of Lower Burma in 1852

and the capital in 1886. Its name comes from the Burmese word *yangun*, meaning 'end of strife'. Some scholars believe that the name is linked with Dagon, the god whose enormous 100 m (328 ft) high Shwe Dagon pagoda dominates the city. The pagoda – a funnel-shaped solid brick spire – was originally built in the 6th century but was destroyed and rebuilt in 1564. It is covered with 8868 thin sheets of gold and is now the centre of a marble-floored Buddhist temple complex.

The city was redeveloped after Britain colonised southern Burma in the mid-19th century and became a spacious colonial capital, the seat of government and the main outlet for rice exports from the Ayeyarwady delta. Since independence in 1948, however, Myanmar's prosperity has dwindled and with it the money available to modernise the port and maintain the city's buildings. A handful of mosques and Hindu temples remain in the northern part of the city – legacies of the Indians who handled much of the administration under British rule – but the Indians themselves have mostly left since 1948.

The commercial hub of the city today is the Bogyoke Aung San market (named after Aung San, a prime minister under the British who was assassinated just before independence). Besides food and clothes, intricately painted lacquerware is on sale, as well as gold, silver, rubies and jade from Myanmar's mines. Near the Shwe Dagon Pagoda are the Royal Lakes, on whose shores are the villas of the country's elite.
Population (city) 2 513 000; (division) 4 300 000
Map Myanmar – Cc

Yangshuo *China* Town in Guangxi-Zhuang Autonomous Region, which is famous for its spectacular mountain scenery. It is the terminus for tourist boats down the Gui Jiang from Guilin.
Population 400 000
Map China – Gf

Yangtze Kiang *China* See CHANG JIANG

Yankari Game Reserve *Nigeria* Wildlife park on the Geji River, about 860 km (535 miles) north-east of LAGOS. It covers some 2224 km² (858 sq miles) of wooded savannah.
Map Nigeria – Cb

▲ CRIMEAN RIVIERA Yalta has as much sunshine as Nice in the south of France. The playwright Anton Chekhov (1860-1904) was a school teacher in the Crimean resort; his villa has been turned into a museum.

Yannina (Ioánnina) *Greece* Main town of EPIRUS region. It stands on a promontory in Lake Ioánnina, at an altitude of 520 m (1708 ft) and is overlooked by the Píndhos mountain range. The town was built by Normans in the 11th century. It fell first to the Serbs (1345), then to the Turks (1430-1913). Yannina has a university, a Norman fortress, a mosque and many monasteries. Local industries include fine filigree silverwork and jewellery crafts – reminders of Turkish rule – and the growing of tobacco and cereals. An annual literature and art festival is held in August.
Population 44 800
Map Greece – Bb

Yaoundé *Cameroon* Capital since 1916. It is set among beautiful hills on the edge of dense rain forest and is overlooked by Mont Febé. Western influences are strong in its luxury hotels, restaurants, nightclubs and sporting facilities, yet it is only minutes away from a traditional rural African environment. It has an international airport and a road and rail link to the port of Douala.
Population 768 000
Map Cameroon – Bb

Yap *Federated States of Micronesia* The most westerly of the federation's four states, consisting of Yap Island, its three adjoining islands and some 130 outliers. Yap retains much of its strong traditional culture, including stone money, used on special occasions, with 'coins' measuring up to 4 m (13 ft) across.
Population 12 000
Map Pacific Ocean – Bb

Yaracuy *Venezuela* North-western state, covering 7096 km² (2741 sq miles). Bauxite and copper are mined, but the economy is predominantly agricultural. Cattle, sugar and citrus fruits are the main products. The state's capital is San Felipe.
Population 384 540
Map Venezuela – Ba

yardang Sharp desert ridge up to 6 m (20 ft) high and separated from its parallel neighbour by a furrow. It results from the scouring effect of sand-laden winds on weakly consolidated sediments.

Yaren *Nauru* District in the south of the island containing the parliament and government buildings and also the airport.
Population 400
Map (Nauru) Pacific Ocean – Dc

Yarlung Zangbo *China* See BRAHMAPUTRA

Yarmuk *Jordan/Syria* River rising in the JABAL AD DURUZ region and flowing west to the Sea of Galilee. For part of its 80 km (50 mile) long course it forms the boundary between Jordan and Syria. It is an important source of irrigation water for both countries.
Map Jordan – Aa; Syria – Ab

Yaroslav *Poland* See JAROSLAW

Yaroslavl' *Russia* Industrial city and river port on the Volga River, 240 km (150 miles) north-east of Moscow. Founded in the early 11th century by Prince Yaroslav I, it became a market and crafts town, as well as a staging post on the North Russian fur trade routes. It has a fortified 12th-century monastery, a 16th-century cathedral and an 18th-century theatre – the oldest in Russia. What was then the country's largest textile mill was built in Yaroslavl' in 1772; it still produces linen. The city's main products today are petrochemicals, plastics, dyes, synthetic rubber, tyres, diesel engines, lorries, shoes and processed food.
Population 638 000
Map Russia – Ec

Yazd (Yezd) *Iran* Textile town about 520 km (325 miles) south-east of the capital, Tehran, on the edge of the Dasht-e-Kavir Desert. It produces woollen and silk fabrics and carpets and was known for its fine silks by the time of the 13th-century Venetian traveller Marco Polo. The Zoroastrian religion, founded in Persia in the 6th century BC, is still practised in the surrounding area: 'towers of silence', where the dead are put out for the vultures, stand on the desert's edge.
Population 275 000
Map Iran – Ba

Yekaterinburg (Sverdlovsk) *Russia* Industrial city on the eastern fringe of the URALS, about 1400 km (870 miles) east of Moscow. It was founded as an iron-making centre in 1721 by Tsar Peter the Great (1672-1725) and flourished after the building of the Great Siberian Highway (1783) and the Trans-Siberian Railway (1878).

The last of the Russian tsars, Nicholas II (1868-1918), and his family were imprisoned and shot by the Bolsheviks in Yekaterinburg (then known also as Ekaterinburg) in 1918, after the Russian Revolution. The city was renamed Sverdlovsk in honour of the Bolshevik leader Yakov Mikhaylovich Sverdlov (1885-1919) in 1924 but it regained its former name in 1991. Today it is the largest city of the Ural region. Its industries include heavy machinery, electrical equipment, copper, chemicals and ball bearings.
Population 1 375 000
Map Russia – Hc

Yellow River *China* See HUANG HE

Yellow Sea (Huang Hai; Hwang Hai) Arm, 640 km (400 miles) long and just as wide, of the PACIFIC OCEAN between China and Korea. It is named after the yellow loess or fine-grained silt deposited by the HUANG HE and other rivers. The mainland has shifted about 50 km (31 miles) into the sea during the past 4000-5000 years due to silting. At the present rate of deposition, the sea will silt up in about 24 000 years unless a rising sea level compensates. Its shifting sandbanks and fogs are navigational hazards.
Map China – Id

Yellowknife *Canada* Capital of the NORTHWEST TERRITORIES, on the Yellowknife River, at the north-west end of Great Slave Lake. It developed after gold was discovered in the area in 1934.
Population 15 180
Map Canada – Db

Yellowstone National Park *USA* The country's oldest national park, established in 1872, and its largest outside Alaska, covering some 8980 km² (3467 sq miles) of the Rocky Mountains. The park's coniferous forests are spread over parts of three states – north-western Wyoming, southern Montana and eastern Idaho. It has some 200 geysers, about 10 000 hot springs and bubbling mud pools, and spectacular alpine scenery.

There are canyons, lakes and two waterfalls, dropping 33 and 94 m (109 and 308 ft) on the Yellowstone River. The park has much wild life, including black and grizzly bears, elks and bison. Its Old Faithful geyser erupts at regular intervals of about 70 minutes, shooting a 46 m (150 ft) jet of water and steam into the air
Map United States – Db

Yemen See p 740

Yenangyaung *Myanmar (Burma)* Town whose name means 'smelly water creek', in central Myanmar, about 200 km (125 miles) south-west of MANDALAY. From 1886 to 1970 the town was the country's most important oil centre, but the oil fields became exhausted and Yenangyaung is now a market place for the surrounding agricultural region.
Population 128 000
Map Myanmar – Bb

Yenbo (Yanbu) *Saudi Arabia* Port on the Red Sea, which has been developed as a huge industrial centre. Lying 160 km (99 miles) west of MEDINA, it is the main entry place for pilgrims to that city. An east-west oil pipeline ending at Yenbo supplies its refineries and petrochemical plants. Until its modernisation the port exported mainly dates. Its occupations today include fishing, boatbuilding and salt panning.
Population 45 000
Map Saudi Arabia – Ab

Yendi *Ghana* Town about 440 km (275 miles) north of the capital, ACCRA. It developed from the 17th century as a trading centre for savannah products from the north and products from the forests of the south. However, Yendi is now overshadowed by Tamale, situated 85 km (53 miles) to the west.
Population 25 000
Map Ghana – Ab

Yengema *Sierra Leone* A diamond area, just east of the Nimini Hills and near the Guinea border. The gems – mostly of industrial quality and not suitable for jewellery– are found in river gravels.
Map Sierra Leone – Ab

Yenisey *Russia* River rising near the north-west border of Mongolia and flowing northwards through Siberia for 4092 km (2543 miles) into the Kara Sea. Its tributaries include the Angara and the Stony and Lower Tunguska. Although it is frozen over in parts for much of the year, it is navigable for most of its length. On the river stands the city of KRASNOYARSK, settled by Cossacks in the early 17th century.
Map Russia – Jb-Kc

▼ **QUIET CAPITAL** A ferryman crosses an inlet in Yangon, the former Rangoon, capital of Myanmar. An original fishing village was developed as a port on the Ayeyarwady delta, and became the busy capital of colonial Burma in 1886.

Yemen

AFTER CENTURIES OF DIVISION, THE REALM OF THE QUEEN OF SHEBA HAS BEEN REUNITED – BUT TENSIONS HAVE ALREADY ERUPTED IN CIVIL WAR

This country of rugged mountains, green hills and trackless desert at the heel of the Arabian Peninsula has a history that goes back to Biblical times. It is said that SAN'A, Yemen's capital, was founded by Noah's eldest son, Shem, and that Noah was buried in the mountains beyond.

East of San'a lies MARIB, the capital at the time of the Queen of Sheba. The Sabaeans ruled from here from the 7th to the late 2nd century BC; today it is a small oasis town.

In ancient times Yemen was a fertile and wealthy country, crossed by the legendary Incense Road. The country was part of the Himyarite kingdom from 100 BC to AD 525, when it was captured by the Ethiopians. In 630 it became an Islamic country. The Yemenis living in the northern highlands are largely Shiites; along the coastal belt along the RED SEA and in the south, they are Sunni. A long-term conflict between North Yemen and South Yemen was due partly to these religious differences; another factor was the ethnic differences between the light-skinned highlanders and the darker coastal people – many of whom came originally from Africa.

EARLY CONFLICTS

From 1517 until this century, the north formed part of the Ottoman Empire and fought alongside the Turks in the First World War; the south, on the other hand, was ruled by sheikhs and sultans who had signed treaties with the British who occupied the port of ADEN, situated strategically on the southern coast of the Red Sea, since 1839. This was the start of the North-South conflict. In 1937 Aden was made a British Crown Colony, and in 1918, with the disintegration of the Ottoman Empire, North Yemen became an independent kingdom.

Political developments in the 1960s exacerbated the differences. In 1962, civil war broke out in the north between royalists and republicans, and continued even after the monarchy had been abolished and the Arab Republic of Yemen was founded. Conflicts between the factions were partly resolved when, in 1970, the military took action and several royalist politicians were appointed government ministers.

In the south, uprisings forced the British to withdraw in 1967 and the leftist wing of the National Liberation Front came to power. South Yemen adopted a socialist constitution. The government introduced a centrally planned economy which stagnated – and it seemed that reuniting the two Yemens was the only solution to the country's problems.

By 1981 it seemed likely that the two countries would unite. During South Yemen's 22nd independence celebrations in December 1989 the two leaders came to a final agreement, and in January 1990 the North-South borders were opened; reunification was officially announced on 22 May, 1990, and the Republic of Yemen was founded. In May 1991 a national referendum was held to approve the country's new constitution, and finally in 1993 democratic elections were held to appoint 301 members to the Council of Representatives.

TWO-MONTH CIVIL WAR

Despite the political agreements, analysts expected it to be decades before the two Yemens would truly be one state, culturally and militarily. There were innumerable differences between the two – evident in a simple comparison between the two capital cities, San'a and Aden.

The ancient city of San'a still has an atmosphere redolent of *The Arabian Nights*. Despite modern development, within the old city wall visitors still find bazaars, mud-brick palaces, and white-domed mosques. In 1984 this part of town was declared a cultural monument by UNESCO.

Aden, the former capital of South Yemen, on the other hand, is a simple port, well situated but having no ancient buildings or monuments worth mentioning. Aden would be destined to become the commercial capital.

Then, in May 1994 vice-president Ali Salim al-Bid (previously president of South Yemen) broke away from the coalition government and set himself up – with loyal military brigades that had never been integrated – in Aden. The north attacked with war planes. The civil war that followed, during which Aden was under siege for a month, ended in early July with a crushing defeat for the south, whose leaders fled the country.

FERTILE MOUNTAIN LAND

The coastal region of this desert land forms part of the TIHAMAH, where rainfalls are low and temperatures can soar to 40°C (104°F) in summer in places such as HODEIDA, the second port. East of the coastal plain the mountain ranges rise steeply, peaking at some 3760 m (12 336 ft) at Hadur Shuaib. The arid lowlands are fertile around a few isolated oases, but vegetation increases with altitude.

Grains, citrus fruits, grapes and vegetables are grown at altitudes ranging between 1400 and 2300 m (4593 and 7546 ft), as well as cotton and coffee for export. The famous mocca coffee bean (which is synonymous with the word coffee in many parts of the world) was named after Mocha, a port situated on South Yemen's coast, which handled the region's coffee exports.

The south and east of reunited Yemen consists largely of desert. Whereas in the north, 14 per cent of the land is arable, in the south the corresponding figure is 1 per cent. Few industries are developed, and consequently few Yemenis are employed in any occupation except agriculture.

YEMEN AT A GLANCE	
Official name Republic of Yemen	
Area 527 968 km² (203 801 sq miles)	
Population 12 100 000 **Per km²** 23 **(Per sq mile** 60)	
Capital San'a	
Government Republic	
Currency 1 Yemen rial = 100 fils	
Languages Arabic (official), some English and Russian	
Religions Muslim (more than 90%), Christian and Judaic minorities	
Climate Desert, hot and humid along the coastal plain; extremely arid and hot in the interior; higher rainfalls with altitude. Average temperature in San'a ranges from 14°C (57) in January to 22°C (72) in July	
Land use Grazing 30%, forest and woodlands 7%, cultivation 6%, other 57%	
Main primary products Cotton, coffee, sorghum, millet, sesame, wheat, fish, fruit, crude oil	
Major industries Agriculture, oil refining, textiles, fishing, fish processing, leather goods, handicrafts	
Main exports Crude oil, coffee, qat (mildly narcotic shrub), cotton, dried and salted fish, hides and skins, vegetables	
Annual income per head (US$) 640	
Population growth (per thous/yr) 33	
Life expectancy (yrs) Male 50 **Female** 52	

Yerevan *Armenia* Capital of Armenia, situated only 12 km (7.5 miles) from the Turkish border amid orchards and vineyards. Standing on the ancient trade routes from Russia and Anatolia to Persia, it was built on the site of a fort dating from 783 BC. The city grew rapidly when thousands of Armenians returned to their home country at the beginning of the 20th century. Under Soviet rule, Yerevan acquired a university, several colleges, an academy of sciences, a theatre, orchestras and museums. Its main products today are chemicals, synthetic rubber, tyres, copper, aluminium, machinery, electrical appliances, clocks, food, wine and brandy.

The city has been shaken by earthquakes several times in its past. The last devastating earthquake was in 1988, when over 25 000 people were killed and many of the city's industries were destroyed.

Population 1 300 000
Map Armenia – Bd

Yerushalayim *Israel* See JERUSALEM

Yevpatoriya (Evpatoriya) *Ukraine* Seaport and resort on the west coast of the Crimea, 70 km (43 miles) north of Sevastopol'. Russia took Yevpatoriya from the Turks in 1783. In more recent times it has grown into a popular holiday place. Lying back from its beaches is the old quarter – a jumble of twisting alleyways, merchants' quarters, craft shops and a mosque.

In contrast, the new port has a gleaming array of fishing quays, sanatoriums, hotels, parks and gardens. Just to the east of the town is Lake Saki, whose medicinal mud is used to treat arthritis and nervous disorders.

Population 109 000
Map Ukraine – Cb

Yezd *Iran* See YAZD

Yin Shan *China* Range running some 1200 km (750 miles) from west to east through NEI MONGOL Autonomous Region (Inner Mongolia). The mountains, between 2000 and 2500 m high (6560 and 8200 ft), mark the southern edge of the Mongolian Plateau.
Map China – Gc

Yinchuan *China* Capital of the NINGXIA-HUI AUTONOMOUS REGION. Standing on the HUANG HE (Yellow River) close to the GREAT WALL of China, it is now a centre for processing hides, wool and tobacco.
Population 658 000
Map China – Fc

Yixing *China* City in Jiangsu Province, 55 km (34 miles) south-west of WUXI on the west side of Tai Lake. It is famous for its pottery and stoneware and for its impressive Shan Juan cave, 15 km (10 miles) south-west of the city. The cave extends for 5000 m² (53 820 sq ft). Its spectacular stalactite and stalagmite formations make it a popular tourist attraction.
Population 30 000
Map China – He

Yogyakarta (Jogjakarta) *Indonesia* City and former sultanate of central Java. It was the centre of the Indonesian revolution against Dutch rule in 1825-30 and of the fight for independence in 1947-9, when it was Indonesia's capital. It is the

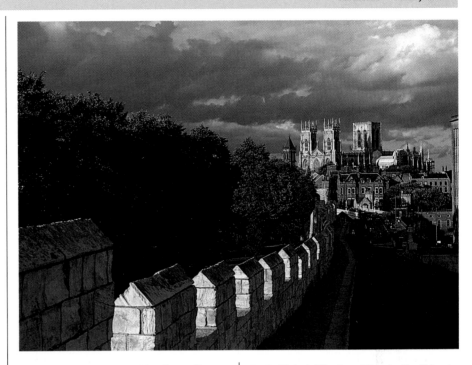

▲ **LIVING HISTORY York's city walls enclose the magnificent minster and fine buildings which chart the city's history – and that of England too.**

focus of Javanese culture, its *kraton* (sultan's palace) keeping alive court traditions and preserving valuable artefacts. The city's craftsmen produce batik, shadow puppets and silverware. It has two universities. There are many fine temples nearby, notably BOROBUDUR.
Population 400 000
Map Indonesia – Dd

Yoho *Canada* National park covering 1313 km² (507 sq miles) in eastern BRITISH COLUMBIA. It has glaciers, mountains and alpine meadows and a profusion of wildlife and flowers. The town of Field is the park's headquarters and tourist centre.
Map Canada – Dc

Yokkaichi *Japan* Industrial port in south HONSHU Island, situated on the Ise Bay, 105 km (65 miles) east of Osaka. In the 1960s its petrochemical plants were responsible for 'Yokkaichi disease', a form of asthma brought on by high levels of sulphur dioxide in the air.
Population 274 000
Map Japan – Cd

Yokohama *Japan* The country's leading port and second largest city after neighbouring Tokyo. A fishing port until the mid-19th century, Yokohama, on Tokyo Bay in south-east HONSHU Island, became the capital's deep-water port.

It was rebuilt after a disastrous earthquake in 1923 and again after Allied bombing during the Second World War. Its industries today include oil refining, shipbuilding and the manufacture of electrical machinery and automotive products. Many of its factories are built on land reclaimed from the bay. Yokahama's Chinatown, the most colourful in Japan, adds to the city's attractions.
Population 3 220 000
Map Japan – Cc

York *United Kingdom* Historic English city on the River Ouse in North Yorkshire. It was founded by the Romans as Eboracum in AD 71. Constantine the Great was proclaimed Emperor by his men in the city in 306. The Angles made it their capital and it became an archbishopric in 634. The Vikings burnt it and then made it their capital as Jorvik from 867.

But the city's chief claim to fame is its magnificent minster, or cathedral, which was completed in 1470 and is the largest medieval church in northern Europe. Its beautiful 15th-century stained-glass east window is one of the biggest in the world and another, dating from about 1150, is probably the oldest in Britain; its lantern tower is the tallest in England, at 60 m (198 ft). The roof was badly damaged by fire in 1984 and needed extensive repairs.

York's medieval city walls, built on Roman foundations, stretch for 5 km (3 miles). The city has a modern university and many museums, including the National Railway Museum (the only national museum outside London) and the Viking Museum. The Jorvik Viking Festival, with longship races as one of its features, is held annually in February. York's products include chocolate and other confectionery. The city's many historic buildings make tourism a significant industry.
Population 103 000
Map United Kingdom – Ed

York, Cape *Australia* Northernmost point on the Australian mainland, lying in Queensland across Torres Strait from Papua New Guinea. The peninsula on which it stands, Cape York Peninsula, extends north for about 750 km (470 miles) between the Coral Sea to the east and the Gulf of Carpentaria to the west. It is covered by tropical forest and grassland to the west. During the rainy season – called the 'wet' – in the summer, rivers become torrents, making overland travel impossible. On the west coast of the peninsula is the bauxite-mining and exporting centre of WEIPA.
Map Australia – Ga

Yorkshire *United Kingdom* Former county of northern England. The ancient county, centred on the city of YORK, was Britain's largest, covering 15 859 km² (6123 sq miles), and was divided into three ridings – named from an Old English word meaning 'a third'. These were reorganised in 1974 into several new areas whose boundaries correspond only here and there with those of the old ridings. Parts were cut off to form sections of the new counties of Cleveland and Humberside.

Today's North Yorkshire, covering 8317 km² (3211 sq miles) is the largest English county. It stretches from Britain's east coast to within 20 km (l2 miles) of the west. Much of it is moorland. The North York Moors (or Yorkshire Moors), with some of the country's loveliest and most open landscapes, lie in the east, and the Pennine Hills – cut by the Yorkshire Dales – in the west. Between them is the fertile Vale of York, an area of rich farmland. Tourism is concentrated on the seaside resort of SCARBOROUGH, the spa town of HARROGATE and the Pennine Hills. Northallerton is the county town.

South Yorkshire, extending over 1560 km² (602 sq miles), has traditionally been the centre of British coal mining and metalworking. Its principal city is SHEFFIELD, on the River Don, which flows past ROTHERHAM, a Saxon foundation overwhelmed by industry in the 19th century, and the railway town of DONCASTER. North of Sheffield lies BARNSLEY, a name connected perhaps more than that of any other British town with coal. Yet even in this heavily industrialised county there are wide areas of open moorland.

West Yorkshire, covering about 2040 km² (787 sq miles) is focused on the cities of LEEDS, BRADFORD and WAKEFIELD, and on a number of other large towns, such as HUDDERSFIELD and HALIFAX, which grew with the rise of the wool industry in the 19th century. These towns, with their huge mills and their vast town halls, reflected the civic pride of Yorkshire men and women. But many of the mills and the factories have disappeared, along with much of the employment.
Population 4 105 500
Map United Kingdom – Ec

Yorktown *USA* Historic town on the York River in south-east Virginia, 190 km (120 miles) southeast of the national capital, WASHINGTON DC. It was settled in 1631 and was an important tobacco port by 1750. During the American War of Independence it was blockaded by the French navy and taken from the British on 19 October, 1781 by General George Washington (1732-99), who later became the country's first president. This was the victory that finally won America its freedom. Yorktown lies, with JAMESTOWN, in the Colonial National Historical Park.
Map United States – Kc

Yorubaland *Nigeria* See SOUTH-WEST NIGERIA

Yosemite National Park *USA* Spectacular mountainous area of 3080 km² (1190 sq miles), about 320 km (200 miles) east of SAN FRANCISCO in the Sierra Nevada range of California.

Yosemite has deep gorges and ravines, towering granite cliffs and mountains – some rising to more than 3960 m (13 000 ft) high – many waterfalls and giant sequoia trees. The Yosemite Falls, the highest in the United States, tumble down a total of 739 m (2425 ft).
Map United States – Cc

Youghal (Eochaill) *Ireland* Port and resort town 42 km (26 miles) east of CORK. According to legend, Sir Walter Raleigh, the English explorer, who was mayor of Youghal in 1588-9, planted the first potatoes in Ireland – at Myrtle Grove in the town.
Population 5530
Map Ireland – Cc

Youth, Isle of (Isla de la Juventud) *Cuba* Island in the Caribbean Sea, 97 km (60 miles) off the south-west coast. Formerly called the Isle of Pines, it was renamed in 1978 in recognition of the Cuban young people's contribution to its agricultural development and educational courses. Once notorious for its large prison, it is now an important growing region for citrus fruits. The chief town is Nueva Gerona on the north coast.
Population 61 000
Map Cuba – Aa

Ypres (Ieper) *Belgium* Battlefield town in Flanders, 65 km (40 miles) west of the city of GHENT. In the 13th and 14th centuries it was one of Europe's largest cloth-making cities, with a population of 200 000. Later it fell into decline, only to find a different kind of fame after being flattened during three great battles in the First World War. Cemeteries now surround the rebuilt town – known to British troops, who struggled to pronounce its name, as 'Wipers'. The Menin Gate is a memorial to 58 000 Allied soldiers who died on the Ypres front and have no known grave.
Population 35 000
Map Belgium – Aa

▼ **DESOLATE ROAD** The Dempster Highway linking Dawson and Inuvik makes its way through the Richardson Mountains of the northern Yukon – a bleak, forbidding land in winter.

Yü Shan (Hsin-kao Shan) *Taiwan* Highest mountain peak in Taiwan and in the maritime fringes of East Asia, reaching 3997 m (13 113 ft). It stands almost on the Tropic of Cancer in the island's central mountains, 180 km (111 miles) south of T'AI-PEI. The vegetation on its flanks is extremely varied, ranging from subtropical camphor trees on the lower slopes to subalpine plants nearer the summit.
Map Taiwan – Bc

Yuan Jiang *China* See RED RIVER

Yucatán *Mexico* State on the Yucatán Peninsula at the south-east end of Mexico. It is a low-lying rain forest-covered region, and was the home of an advanced Mayan culture before the 10th century AD. It includes the archaeological sites of CHICHÉN ITZÁ, Kabah and Uxmal.

MÉRIDA, Yucatán's capital, is an important tourist centre, but the state's economy depends principally on the henequen plant, which yields fibres used in making ropes, nets and bedding.
Population 1 327 300
Map Mexico – Db

Yucatán Channel (Yucatán Strait) Channel, 217 km (135 miles) wide, between the Yucatán Peninsula of Mexico and Cuba, linking the CARIBBEAN SEA to the Gulf of MEXICO. Part of the westward-moving Equatorial Current flows through the channel from the Caribbean into the Gulf of Mexico and emerges from the gulf through the Straits of Florida as the GULF STREAM.
Map Mexico – Db

Yugoslavia (Serbia and Montenegro) See p 743

Yukon Territory *Canada* Vast area of mountain ranges, forests and plateaus in the extreme northwest of Canada, covering some 483 450 km² (186 660 sq miles). Named (continued on p 744)

Yugoslavia (Serbia and Montenegro)

BLAMED FOR EXCESSES IN A BITTER CIVIL WAR, SERBIANS FEEL UNDER THREAT WHILE THE WORLD POWERS REFUSE TO RECOGNISE THEIR UNION AS THE 'NEW YUGOSLAVIA'

The so-called 'land of the South Slavs' has disintegrated. Nevertheless, Serbia and Montenegro, two former members of the Socialist Federal Republic of Yugoslavia now claim to be the rightful successors to the title of Yugoslavia.

But for many countries, confusion still surrounds its status; most governments do not recognise Serbia and Montenegro. However, the United Kingdom has accepted the term 'Yugoslavia' as the correct term for the federation of the two states of Serbia and Montenegro, while the USA does not recognise the 'new' Yugoslavia at all – and neither country recognises this federation as the successor to the 'old' Yugoslavia.

When Marshal Tito died in 1980, he left behind his life's work and political legacy – the Socialist Federal Republic of Yugoslavia – a federation of six republics of 255 000 km² (93 493 sq miles) and 23 million people.

Today, Yugoslavia has shrunk to less than half its original land area and population, only seven decades after being founded as the kingdom of the Serbs, Croats and Slovenes. The country has splintered into five independent states, and since April 1992 has consisted of only Serbia and Montenegro. It is doubtful that the process of dissolution will end there.

The Muslim Albanian inhabitants of Kosovo, for instance, who form 90 per cent of the population, are demanding full independence. At present, Kosovo is ruled by Serbia. The Belgrade-based government tried to stop the break up of Yugoslavia by military action, first in Slovenia, then in Croatia, and finally in Bosnia-Herzegovina, but achieved only wars, international censure, and isolation for itself.

Ethnic tensions were evident when Tito was alive, but he played off ethnic groups against each other. Himself a Croat, he had no hesitation in cracking down on Croat nationalists. But the forces which reduced the country to the rump of Serbia and Montenegro were gathering pace during the 1980s.

The Serbs ended the autonomy of the province of Kosovo and the 90 per cent Albanian population were discriminated against in jobs and education. Dissidents were jailed. The Albanians set up a parallel unofficial system and observers asked themselves how long the Serbs could keep the lid on the region before it revolted.

But Serb politicians saw Kosovo as an opportunity rather than a threat. The 600th anniversary of Kosovo Folje in 1989 was celebrated as an exercise in triumphalism rather than a commemoration of the historic defeat which led to 500 years of subjugation under the Turks. Slobodan Milosevic, the star guest at the ceremonies, played the Serb card which ultimately led to the disintegration of the federation.

Even without this, Croatia may have sought complete independence. But the accession of Milosevic to the Serb leadership made such a move in Zagreb inevitable. To invoke Serb history was to revive the fears which Tito had contained. Suspicions were mutual. Memories of wartime atrocities carried out against them by pro-German fascist Croats helped bind the Serbs together in outrage when Bonn led the European Union into recognising the statehood of Croatia.

While his enemies portrayed Milosevic as a pan-Serb ogre, the prospects of his leading the way to a greater Serbia shrank. The Bosnian Serbs and those in Croatia operated with increasing independence from Belgrade, thus reinforcing the isolation of rump Yugoslavia.

Even in the much reduced state, wide discrepancies exist in the standards of living. In Kosovo unemployment is much higher and per capita income much lower than in Serbia proper. Kosovo has long been the poorest region of the former Yugoslavia. Before the break up of the federation, Kosovo lagged far behind Slovenia, the richest republic, on every economic indicator.

One reason for the economic inequalities within the former federation is the variation in natural resources. But social and political discrimination must also be blamed for the plight of the 2 million ethnic Albanians who live under Serb rule.

Before Yugoslavia broke up, it contained four large natural regions, which are still obvious in the great variations in landscape. Nearly all of Montenegro and the adjacent part of Serbia, including Kosovo, lie in the Dinaric Alps – a largely bare, arid mountain range with only the occasional *polje* of fertile soil.

By contrast, the low-lying region north of the Sava and Danube rivers has a continental climate and a deep, fertile loess soil and has for centuries yielded rich harvests, especially of wheat, maize and sugar beet. A fair portion of the Adriatic coast (from the Bay of Kotor going south) belongs to Montenegro. The landscape and its ports are typically Mediterranean.

Further inland, however – and especially in the depression starting at the Danube and Belgrade and continuing south to Macedonia – the summers are hot and the winters icy, as one would expect of the climate of the whole of the Balkan Peninsula.

The above depression is in an area in which regular earthquakes occur; these have proved useful, as they have lifted and exposed rich mineral deposits – for instance the copper ore deposit in the eastern Serbian Ore mountains, and lead and zinc west of the Morava river.

There are also rich deposits of coal, petroleum, natural gas, magnesite, antimony and bauxite scattered over the whole republic. Yet Yugoslavia depended on imports of raw materials, particularly of petroleum and petroleum

products, since it could only meet one-quarter of its demand from its own supplies.

This is the vulnerability which the UN Security Council and the EC chose to exploit when they imposed trade sanctions against Serbia and Montenegro in retaliation for the Serbs' war of ethnic cleansing against the Bosnian Muslims. The economic blockade was aimed at forcing rump Yugoslavia, which is given the bulk of the blame for the civil wars, to make concessions to the other states and negotiate a peace settlement in Bosnia.

The sanctions only served to persuade Serbs that the world had turned against them and that their nationhood was threatened. When NATO aircraft struck at Serbian artillery positions in Bosnia, Serb resolve hardened. Multiple meetings and negotiations in Geneva, along with countless broken truces brought no respite as the war continued.

▼ SERBIAN HEART Belgrade, capital of the Serbian-dominated rump of former Yugoslavia, was rebuilt after Second World War devastation. The city's name means 'white citadel'.

SERBIA AND MONTENEGRO AT A GLANCE	
Official name Serbia and Montenegro	
Area 102 350 km² (39 491 sq miles)	
Population 10 699 539 **Per km²** 105 (**Per sq mile** 271)	
Capital Belgrade	
Government Republic	
Currency 1 Yugoslav new dinar = 100 paras	
Languages Serbo-Croat, Albanian, Hungarian	
Climate Continental with hot and humid summers and cold winters. Average temperature in Belgrade ranges from –3 to 3°C (27-37°F) in January to 17-28°C (63-82°F) in July	
Land use Cultivation 35%, pastures 20%, forest 25%, other 20%	
Mineral resources Oil, natural gas, coal, lead, antimony, copper, zinc, nickel, gold, pyrite, chrome	
Main exports (before the war) machinery, transport equipment, manufactures, chemicals, food and livestock, fuels and lubricants, tobacco, electricity	
Annual income per head No statistics	
Population growth No statistics	
Life expectancy (yrs) No statistics	

after one of Canada's largest rivers, it contains the country's highest peak, Mount Logan, at 5951 m (19 524 ft). The Yukon is a land of great beauty with high mountains and sweeping valleys cut by great rivers. Summers are hot and winters cold, but the climate is more temperate than that of the adjoining NORTHWEST TERRITORIES.

The 1896 gold strike on a tributary of the territory's Klondike River led to the development of DAWSON. Today, gold, silver, lead, zinc and other minerals are mined. Hydroelectric power has been developed, there is a forestry industry and tourism is growing. Water and road links are used, but most people travel by air.
Population 27 800
Map Canada – Bb

Yumen *China* City in GANSU Province, once the eastern terminus of the SILK ROAD. It is now a centre for oil production. The Yumen oil field is one of the largest in China.
Map China – Ec

Yungas *Bolivia* Name of an area of hillside forests below about 2500 m (8200 ft). Tangled and mossy and often wreathed in cloud, the forests mark the transition from the treeless Andean highlands of Bolivia to the lush lowland Amazon rain forests to the east. The best developed of these forests are north-east of the capital, La Paz, at Coroico and around Cochabamba, where rich alluvial soil washed down from the Andes supports crops such as coffee, cacao, coca, sugar cane, maize, bananas, oranges and other citrus fruits.
Map Bolivia – Bb

Yungui Plateau *China* Plateau of roughly 160 000 km² (61 800 sq miles) covering western GUIZHOU and eastern YUNNAN provinces in southern China. It lies between 1000 and 2000 m (3300 and 6600 ft) above sea level. Much of the plateau has a cool, springlike climate all year round.
Map China – Fe

Yunnan *China* Province in the south-west of the country, covering 390 000 km² (150 000 sq miles) and bordering on Vietnam, Laos and Myanmar. It is a mountainous area, embracing the western half of the YUNGUI PLATEAU and the great ranges and canyons of the southern QAM'DO region. While the plateau surface has a mild, temperate climate, the south is subtropical and tropical. The mountains contain some bizarre scenery – including a 'forest' of tree-shaped 30 m (100 ft) high limestone crags, pitted with caves. Home to more than 15 000 plant species, it is also a botanist's paradise. The best agricultural land is around the capital city of KUNMING, where rice, wheat and maize are grown. Pu'er, in the south, is noted for its black tea. The outlying regions also produce some opium. Yunnan contains many minority ethnic groups related to the peoples of South-east Asia.
Population 38 320 000
Map China – Ef

Zabljak *Yugoslavia* See DURMITOR

Zabrze (Hindenburg) *Poland* Industrial city 24 km (15 miles) west of the city of KATOWICE. It produces coal, steel, coke, chemicals and machinery, and is linked to the Oder River by the Klodnica Canal.
Population 203 000
Map Poland – Cc

Zacatecas *Mexico* Mineral-rich state on the central plateau, crossed by the Sierra Madre Occidental Mountains. It is one of the best cattle-raising areas in the country. The capital of Zacatecas State, 600 km (375 miles) north-west of MEXICO CITY is also called Zacatecas, an Aztec word for 'Place of the Grasses'. Zacatecas made its fortune from the world's largest silver mine, which lies nearby at Real de Angeles. The city has several magnificent Spanish Baroque churches, including the Jesuit Church of Santo Domingo, and picturesque pink stone houses.

Population (state) 1 278 000; (city) 71 000
Map Mexico – Bb

Zadar (Zara) *Croatia* Ancient capital of Dalmatia, on a promontory above the Adriatic, 116 km (72 miles) north-west of the port of SPLIT. Despite the destruction of the Second World War and the 1990s war between former Yugoslavian states, several of the town's beautiful buildings remain, including the 13th-century Church of Sv Franje (St Francis), the 9th-century, circular, castle-like St Donat's Church, which was built in the remains of a Roman forum, and a Renaissance Bishop's palace. Zadar also has Dalmatia's largest church, the Romanesque Cathedral of Sv Stosija (St Anastasia). The city's Baroque Church of Sv Simun (St Simon) contains the saint's extravagant reliquary. Made from 250 kg (551 lb) of silver for the Hungarian monarch Queen Elizabeth in 1380, it is now supported by bronze angels cast from Turkish cannons.

Modern Zadar is a major port, shipbuilding centre and industrial city. Until civil war broke out in 1991 it was a popular tourist resort.

From the 9th century to the 14th century, Nin, 18 km (11 miles) to the north, was the first capital of the Croatian kings and bishops. Its Sveti Kriz (Holy Cross) Church is Croatia's oldest intact church, dating from the 7th century.

Population 134 700
Map Croatia – Bb

Zagazig *Egypt* Rail centre and canal junction on the Nile delta waterway system, about 65 km (40 miles) north-east of CAIRO. It trades in cotton and grain grown on delta farms.

Population 279 000
Map Egypt – Bb

Zagorje *Croatia* Beautiful region of rolling hills topped by Baroque churches, woods and castles, lying in north-west Croatia, between Zagreb and the Drava River. It has more than 100 castles, including that at Trakošćan.

Map Croatia – Ba

Zagorsk *Russia* See SERGIYEV POSAD

Zagreb (Zagrab; Agram) *Croatia* National capital and the country's largest city, about 120 km (75 miles) from the Adriatic coast. The city grew as twin towns on a hill overlooking the Sava River plain. Gradec (or Gornji Grad, meaning 'small town'), once a royal town, was fortified against Tatar attacks in the 13th century and some of its medieval walls and gates survive.

Zagreb has several fine Gothic and Baroque churches and palaces. Kaptol, a religious foundation lower down the hill, is dominated by the filigree spires of the cathedral and the fortified bishop's palace. Bitter rivalry between Gradec and Kaptol lasted through the centuries until

1850 when they united. The pastel-painted buildings of the two historic towns contrast with the grey stone of the 19th-century hub of the city, the Donji Grad, or 'lower town'. Since 1960 the focus has moved again, this time across the Sava, where new commercial and administrative offices, the university, shopping centres and high-rise flats and hotels have sprung up.

Zagreb has played its part in more recent history. The state of Yugoslavia was founded here in October 1918. During the civil war between former Yugoslavian states which broke out in 1991 the city became a vast refugee camp. Before the war it was a fine tourist centre, with the Zagorje hills, the Plitvice lakes, and ski resorts on Sljeme (1035 m; 3396 ft) within easy reach. It has several museums.

Population 1 195 000
Map Croatia – Bb

Zagros Mountains *Turkey/Iran* Range, nearly 1600 km (1000 miles) long, stretching south-east from eastern Turkey along The GULF to the Strait of Hormuz and the Gulf of Oman. The range consists of parallel chains of mountains cut by rivers and deep gorges, with large areas of forest and fertile oases on the highlands. The highest peak is Zard Kuh, at 4547 m (14 918 ft), west of the city of ESFAHAN.

Map Iran – Aa

Zaire See p 748

Zaire (Congo) *Zaire/Congo* Africa's second longest river, after the Nile, extending some 4670 km (2900 miles). Its drainage basin of 3 690 000 km² (1 424 000 sq miles) and its flow are second only to those of the Amazon. Its chief headstream is the LUALUBA River.

The Zaire's mouth was reached by a Portuguese mariner, Diego Cão, in 1482, but its course was not explored until the British adventurer Sir Henry Morton Stanley (1841-1904) sailed down the Lualaba River to the Boyoma Falls and on to Malebo Pool in 1876-7.

In the 16th century, the river was called the Zaire, a corruption of several words in local dialects meaning 'river'. It was renamed Rio Congo in the 17th century after the Kongo people along its lower course. The Zairean government reintroduced the original name in 1971.

The Zaire's upper course contains rapids, waterfalls and lakes. Its middle course starts at Kisangani and here the channel is broad and the volume of water is increased by such large tributaries as the Ubangi and Kasai. About 560 km (350 miles) from the sea, the river widens into Malebo Pool (formerly Stanley Pool), on which stand the cities of KINSHASA and BRAZZAVILLE. Below Malebo Pool, the river passes through the unnavigable Inga Rapids before reaching the head of its estuary at Matadi. More than 12 000 km (7500 miles) of the Zaire and its tributaries are navigable – and the rivers have an estimated one-sixth of the world's total hydro-electric potential.

Map Zaire – Ab

Zákinthos (Zante) *Greece* One of the IONIAN ISLANDS, off the west coast of the Peloponnese. It is a tourist resort of gardens and flowers with a rugged coast. Zákinthos was occupied by the Venetians from AD 1487 to 1797 and a Venetian castle stands on the site of the ancient acropolis

(citadel). The island's main business now is tourism, but it also produces melons, currants, white wine and olives. Its main town, also called Zákinthos, was rebuilt after an earthquake in 1953.

Population 30 000
Map Greece – Bc

Zakopane *Poland* The country's main tourist resort, which receives more than 2.5 million visitors a year. Lying at 800 m (2625 ft) in the Carpathian Mountains, 84 km (52 miles) south of the city of CRACOW, it is surrounded by magnificent scenery, with marked trails for walkers and climbers and good skiing facilities, including a cable car to Kasprowy Wierch, which rises to 1988 m (6522 ft) on the Slovakian border.

Zakopane's *gorale* (highlanders) wear regional costumes as a tourist attraction and keep alive folk crafts and customs. The town and the hillsides around it are dotted with traditional mountain chalets, with carved gables and steeply pitched roofs.

Population 29 700
Map Poland – Cd

Zamárdi *Hungary* See SIÓFOK

Zambezi *Southern Africa* Longest river of southern Africa, flowing 2700 km (1700 miles) south and east from north-western Zambia to the Indian Ocean. It crosses part of eastern Angola, forms a small part of the Zambian border with Namibia, touches the point where Zambia, Namibia, Botswana and Zimbabwe meet, continues east-north-east to form all of the Zambian-Zimbabwean border, and finally crosses central Mozambique. It is dammed for hydroelectricity, forming lakes at KARIBA (between Zambia and Zimbabwe) and CAHORA BASSA in Mozambique.

Map Zambia – Cb; Zimbabwe – Ba; Mozambique – Ab

Zambézia *Mozambique* See QUELIMANE

Zambia See p 750

Zamboanga *Philippines* Chief trading port of the southern Philippines, situated at the tip of Zamboanga Peninsula in south-western MINDANAO. Originally a fort built to protect Christian settlers against the local Muslim population, Zamboanga was rebuilt early this century. The local people speak a dialect consisting largely of Spanish words.

Zamboanga is known as the 'City of Flowers' and has a colourful market. It exports timber, copra, coconut oil, Manila hemp, rubber and other crops produced in the region, while *kumpits* (motorboats) carry barter goods to and from islands of the SULU ARCHIPELAGO and ports in BORNEO. Smaller craft called *vintas* are the homes of the Badjaos, or sea gypsies.

Population 442 000
Map Philippines – Bd

Zamora *Spain* Town 210 km (130 miles) north-west of MADRID whose many churches of the 12th and 13th centuries make it a living museum of Romanesque architecture. It trades in cereals and wine and has flour-milling, cement, textiles and soap industries. The surrounding province of the same name is a sheep-rearing area.

Population (town) 64 480; (province) 214 670
Map Spain – Cb

Zaire

A VAST LAND OF EQUATORIAL RAIN FORESTS AND SAVANNAHS IN THE HEART OF AFRICA IS DRAINED BY ONE OF THE WORLD'S GREAT RIVERS

Zaire lies in the heart of what was Darkest Africa to 19th-century Europeans, a country of nightmarish horror according to Joseph Conrad's 1899 novel, *Heart of Darkness*. Zaire is vast, nearly 77 times the size of Belgium, which ruled the country as the Belgian Congo from 1908 to 1960 and its 43.5 million people (little more than four times Belgium's population) have left large tracts of equatorial rain forest virtually uninhabited.

Europeans penetrated this forest in the 19th century by way of the great Congo River (now the ZAIRE). A narrow corridor, following the lower course of the river, reaches westwards from the capital KINSHASA to the Atlantic Ocean, where Zaire has a short coastline of 40 km (25 miles) and a port on the Zaire estuary at MATADI.

The Zaire and its tributaries, despite being interrupted in places by waterfalls and rapids, were essential lines of communication in the 19th century and are still important today. Although slow, river transport is cheap and rivers remain a vital part of the transport network, forming pathways through, in Conrad's words, the 'towering multitude of trees ... the immense matted jungle'.

For instance, one of the chief outlets for copper and other metals from the main mining province of SHABA (formerly Katanga) is a railway to ILEBO and then river transport down the KASAI and Zaire Rivers to Kinshasa. From there, goods are carried by rail to the coast, bypassing the INGA RAPIDS (formerly Livingstone Falls).

Central Zaire, which occupies a shallow depression in the central African plateau, has an equatorial climate, with abundant rain and high temperatures throughout the year. In the north and particularly in the high plateau country in the south, there are marked dry winter seasons. Here, luxuriant rain forest merges into wooded savannah. But nowhere is the contrast in vegetation as great as in the highlands, such as the RUWENZORI Mountains, that overlook the great lakes in the GREAT RIFT VALLEY on the eastern border.

Zaire's rain forests cover 55 per cent of the country and contain valuable hardwoods including mahogany and ebony. Ecologists hope that exploitation, now confined to under 2 per cent of the forest, will not be expanded.

Agriculture, forestry and fishing (mainly of inland waters) employ about 75 per cent of the workforce but most of the people live by subsistence farming. Yet cultivated land makes up only 3 per cent of the country and grazing land – limited to those savannah regions which tsetse fly does not afflict – 4 per cent. In the formal sector the largest employers are the mines and the government.

The main food eaten by the rural people and grown in most parts of Zaire is cassava, a starchy root. Other foods include plantains and various root crops in wet areas, rice in flooded valleys, and beans, groundnuts, guinea corn and maize in drier places. Many people practise shifting cultivation, clearing land for crops and moving on when the soil has become exhausted. Women do most of the farmwork.

Commercial agriculture produces export crops, especially coffee, palm oil and rubber, and also cocoa, cotton and tea.

Cash crops are still important to the economy, although it is mineral and extensive hydroelectric resources, and mining industries, that make Zaire potentially one of Africa's richest countries. Zaire leads the world as a cobalt producer and ranks second in diamonds (mainly industrial diamonds). Copper is the main export, accounting for more than half the country's total value of exports. Next in importance are diamonds (making up 11.9 per cent of exports), cobalt (7.6 per cent), crude oil (7.5 per cent) and coffee (5.7 per cent).

Apart from southern Kasai, where most of the diamonds are mined, the chief mining region is in Shaba Province. It includes KOLWEZI, Likasi, Kipushi and LUBUMBASHI. The area produces cadmium, cobalt, copper, germanium, silver, zinc and other metals, using power from hydroelectric stations on the Lufira and Lualaba rivers. But the largest hydroelectric project is at INGA DAM, not far from the Zaire estuary, downriver from fast growing Kinshasa.

FRENCH AS THE LINGUA FRANCA

About 100 000 pygmies live in the forests, descendants of Zaire's first inhabitants. They are divided into several groups, including the Mbuti of the north-eastern ITURI FOREST, who still live by hunting and gathering.

The Mbuti, who call themselves *bamiki ba 'ndura* (children of the forest), live in small bands, or 'families', of up to 30 households. Members of a band regard each other as kin, although they are linked by their struggle to make a living rather than by common blood.

The pygmies are one of the many ethnic and language groups in Zaire. More than 200 languages and dialects are spoken, so French remains the common (and official) language of the educated elite, with Lingala, Kikongo, Chiluba and Kiswahili the widely spoken vernaculars. About two-thirds of the people speak languages which belong to the Bantu family.

When Portuguese navigators first reached the Zaire River estuary in 1482, they made contact with the kingdom of Kongo, a powerful nation ruled from the royal capital Mbanza, 250 km (155 miles) inland. The Kongo people had elaborate legal and political institutions, financed by taxes. In the 16th century the Portuguese traded with the Kongo for slaves and drastically depleted the population.

Most rural people live in villages of 300 or so. Standards of living are low, diets are poor and limited, and malnutrition and disease take a heavy toll of life. There is an infant mortality rate of more than 11 per cent.

European influence in the interior did not begin until the late 19th century, after the explorations of Henry Morton Stanley in 1874-7. His later expeditions were financed by King Leopold II of Belgium who decided to claim the Congo Free State (as it was known) as his personal property. Leopold granted

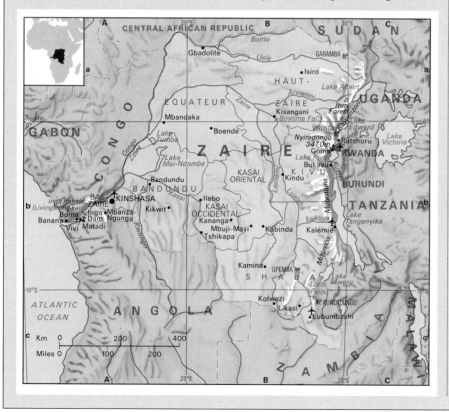

rights to exploit the land to Belgians of his choice, which in some cases led to severe exploitation of the native people of the region. This was exposed by the British consul, Roger Casement, and in 1908 Belgium took control of the Congo from Leopold. Casement was knighted in 1911 for his services, but five years later he was executed as a traitor for plotting an Irish uprising against Britain. Belgium's policies, although humanitarian, were similar to those of Leopold, in that private concessionaires were granted rights to collect ivory and forest products, and large companies were encouraged to establish plantations and mines.

This style of colonial development was successful in creating living standards that, by the 1950s, were among the highest in tropical Africa. But no great effort was made to prepare the country for independence – higher education, for instance, became available to black Africans only after 1954. Political parties, first permitted in the 1950s, were generally tribally or regionally based and many wanted independence from the rest.

The Belgians lost control of events and mounting unrest culminated in nationalist riots in 1959. Independence came on 30 June, 1960, but within a few days there was anarchy in the newly named Republic of the Congo. The army of black soldiers rebelled against their white officers; there were outbreaks of communal fighting and rebellions in the provinces of Kasai, Stanleyville and Katanga.

POWER STRUGGLES

In Katanga (now Shaba), the rebel government, backed by European business interests, seceded from the rest of the country and ran its own affairs until the breakaway was ended by United Nations troops in January 1963. But the Katangese never quite forgot their dream of independence. In December 1993 the people of this region again showed their determination to go it alone.

The two leading political figures at independence were Patrice Lumumba and Joseph Kasavubu, leaders of rival nationalist factions who became, respectively, prime minister and president. A power struggle ended with Lumumba's assassination in 1961.

In 1965 General Joseph Désiré Mobutu seized power – and held on to it for three decades, resisting repeated attempts to unseat him in successive conspiracies and invasions by exile armies into Shaba. After 1971, Zaire became a one-party nation. As leader of the *Mouvement Populaire de la Révolution* (MPR), Mobutu was automatically head of state.

In 1971 President Mobutu introduced his policy of 'authenticity', an attempt to return the country to its African origins, especially by changing many European names, both personal and geographical, to Zairean ones. The 'Congo' became 'Zaire', Léopoldville became Kinshasa, and citizens were required to drop their (European) Christian names. Mobutu changed his own name to Mobutu Sese Seko, dropping 'Joseph Désiré'.

Political reforms announced in 1990 led to a national conference which ended 18 months later after it had dissolved the National Assembly and elected a transitional High

▲ LIVING DEATH A dancer pretends that a giant knife has penetrated his head, as part of a traditional ceremony used for initiations and funerals. The dancers come from the Bandundu region in south-west Zaire.

Council of the Republic (HCR). By March 1993, the country had two rival governments – one with a prime minister appointed by the president, the other with the HCR's choice of prime minister, supported by opponents of the president. Violent ethnic conflict had meanwhile broken out in Shaba, Kasai and Kivu provinces, resulting in the death of more than 6000 people and the displacement of hundreds of thousands. The deadlock between the supporters of the president and the opposition was resolved by mid-1994 when a transitional parliament, re-constituted by Mobuto, elected a new prime minister, Kengo Wa Dondo, and towards the end of that year some semblance of political and economic stability appeared to have been restored.

In the first years after Mobuto's takeover, stability brought about by strong government (although without much respect for human rights) led to an economic revival, but one that was short-lived. A stark economic crisis, caused by ill-conceived nationalisation programmes in all branches of the economy, rising import costs and falling copper prices, corruption and excessive borrowing, hit Zaire in the mid-1970s.

Vast discrepancies between the haves and the have-nots, accompanied by vast enrichment of the president, contributed to the economic decline. An elite of about 5000 Zaireans earned about half of the population's total wages; the rest of the population lived – as they still do – in extreme poverty, with 80 per cent of them malnourished.

Today, the country's exports – agricultural products and minerals – fetch deflated prices on the world market. Added to this is a huge government deficit over the years and declining mineral production. Zaire by the early 1990s was among the poorest of nations. Monetary reform introduced in October 1993 proved disastrous; by early 1994 Zaire had an average annual inflation rate of 2000 per cent. People again resorted to barter, finding that Zairean bank notes were virtually worthless.

Tourism, too, has been negligible. However, for adventurous travellers, Zaire contains some of the world's most unspoilt places, with magnificent wildlife in its national parks and reserves. Its tourist attractions include scenic mountains and lakes in the east, thundering waterfalls, among them the majestic 340 m (1113 ft) Lofoi Falls, and tranquil rivers.

ZAIRE AT A GLANCE	
Official name Republic of Zaire	
Area 2 345 410 km² (905 354 sq miles)	
Population 43 500 000 **Per km²** 18 **(Per sq mile** 48)	
Capital Kinshasa	
Government Republic	
Currency 1 Zaire = 100 makuta	
Languages French (official), Lingala, Kikongo, Chiluba, Kiswahili	
Religions Christian 92% (Roman Catholic 47%, Protestant 28%), Muslim	
Climate Tropical; equatorial in centre. Average temperature in Kinshasa from 22°C (72°F) in July to 27°C (81°F) in February	
Land use Forest and woodlands 78%, grazing and other 19%, cultivation 3%	
Main primary products Cassava, plantains, maize, sugar cane, groundnuts, bananas, palm oil and kernels, coffee, rubber, cotton, cocoa, tea, timber; copper, oil and natural gas, cobalt, zinc, tin, diamonds, gold	
Major industries Agriculture, oil refining, food processing, textiles, clothing, mining, forestry	
Main exports Copper, cobalt, coffee, diamonds, oil, cocoa	
Annual income per head (US$) 210	
Population growth (per thous/yr) 30	
Life expectancy (yrs) Male 45 **Female** 49	

747

Zambia

*TO SURVIVE, ZAMBIA MUST DEVELOP
ITS VAST AGRICULTURAL POTENTIAL
NOW THAT RELIANCE ON COPPER
EXPORTS HAS LEFT IT EXPOSED ON A
FALLING MARKET*

A single product, copper, has governed the development of Zambia's economy, dictated where people live, and determined where the main roads and railways run. The mineral, a basic raw material used in industries worldwide, provides about 80 per cent of Zambia's export income and more than half of all government revenue. Indeed, Zambia is second only to the United States in reserves of the ore, and is one of the largest producers in the world.

Since independence from Britain in 1964 (before which the country was known as Northern Rhodesia), the mines have been nationalised, with the result that any changes in world copper prices directly affect Zambia's whole economy. In addition, two other factors influence the well-being of the country's copper industry: an efficient system for transporting the mineral out of the country (landlocked Zambia relies on railways to get its copper to the coast for export); and cheap, secure sources of power for mining and processing the ores.

THE CLOSED BORDER

The illegal Unilateral Declaration of Independence (UDI) by Southern Rhodesia in 1965 and the ensuing internal war, along with international sanctions against Southern Rhodesia, led to Zambia's closing of the mutual border. Lengthy liberation wars in Angola (1961-74) and Mozambique (1964-74), followed by continued fighting since their independence, also hit Zambia's outlets through these countries.

Large quantities of electrical power, the second factor affecting the copper industry, are needed by Zambia's mines. Before independence, when the country was Northern Rhodesia, it was a key partner in the 1953-63 Federation of Rhodesia and Nyasaland (now Malawi). The huge copper revenues helped to build the vital KARIBA Dam on the Zambezi, which provided power for industry in both Northern and Southern Rhodesia.

Unfortunately for Zambia, the hydroelectric power station was on the Southern Rhodesian side of the Zambezi, which forms the border between the two countries. During the UDI period, Southern Rhodesia was in a position to put pressure on Zambia against joining in sanctions, by threatening to cut off its power. Zambia built a power station on its side of the river, and another on the KAFUE.

Faced by these problems in the 1960s and 1970s, Zambia was forced to turn north for a solution. Here lay Tanzania, the only neighbouring state that was then in any way similar

– newly independent and black-ruled. In 1968, an oil pipeline was completed, running south from DAR ES SALAAM in Tanzania to Ndola on Zambia's Copperbelt in the north of the country. The links were strengthened with the completion of a railway in 1975, known as the Tazara (or Tanzam), which connected Zambia's rail system with Dar es Salaam, facilitating copper export. The Tazara was built with Chinese finance, under Chinese direction and partly with Chinese labour.

The Chinese involvement was less a reflection of Zambia's political leanings – Zambia is a non-aligned state – than the fact that the West rejected the project as an uneconomic investment. Indeed, port congestion at Dar es Salaam and operating problems on the railway reduced the potential of Tazara to release Zambia from its transport dependence on the south.

In 1978, after severe economic hardship, Zambia was forced to re-establish rail links through Southern Rhodesia (which became the independent Zimbabwe in 1980) and in 1982,

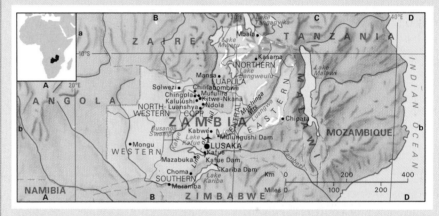

it again started using South African ports; in consequence, Zambia's transport situation has greatly improved. The accumulated debts of the 1960s, 1970s and 1980s, including that for the Tazara railway, were a huge drain on Zambia's foreign exchange and its ability to import. These problems, combined with a depressed market for copper – world prices have halved in real terms since 1964 – have led to enormous financial problems.

Throughout the 1980s the Zambian government tried, but was unable to keep to the strict austerity measures demanded by the International Monetary Fund (IMF). In 1990, when President Kenneth Kaunda finally did give in to IMF pressure and agreed to remove the government's subsidy of maize meal, thus doubling the price for the national staple, riots broke out nationwide.

In a bid to appease the Zambians, Kaunda promised elections and a multiparty constitution. Elections in October 1991 brought a landslide defeat to Kaunda and the United National Independence Party (UNIP), which had been the only legal political party since 1972.

Kaunda had been head of state from the time of independence. A much respected national figure, he was one of the key leaders of the so-called Front Line States, which played a major diplomatic role in bringing about inde-

pendence in Zimbabwe and opposed apartheid in South Africa.

He was replaced by President Frederick Chiluba, leader of the Movement for Multiparty Democracy (MMD) party. Chiluba's rule quickly ran into problems, too. In 1993 he declared a state of emergency after claiming to have uncovered a coup plot. Cabinet ministers resigned in 1993 and formed the new National Party. In early 1994 Chiluba announced a major cabinet reshuffle after some of his ministers were accused of corruption, including drug trafficking. The accusation had been made by donor agents, and Zambia, heavily dependent upon foreign aid, could not afford to estrange donor countries.

THE RURAL COMMUNITY

Most of Zambia is a high plateau, 900-1500 m (about 3000-5000 ft) above sea level. Bordering it to the south is the Zambezi River, and in the south-west the sands of the desolate KALAHARI Desert. The country has a number of other large rivers, including the Luangwa, and some large lakes – the largest is Lake BANGWEULU, but parts of lakes MWERU and TANGANYIKA are also in Zambia.

The country lies in the tropics but the height of the land reduces temperatures slightly. From May to August it is dry and cool, with temperatures averaging 9-23°C (48-73°F). September to November is hot and dry, with temperatures of 18-31°C (64-88°F). The hot, rainy season from December to April is only slightly cooler.

The natural vegetation is mainly savannah, ranging from dry grassland in the south to more luxuriant grassland with scattered trees and bushes, and some woodland, in the north. The country contains a great variety of wildlife; large game parks on the Luangwa and Kafue rivers have antelopes, Cape buffaloes, elephants, leopards, lions, giraffes, hippopotamuses, rhinoceroses and zebras. With nature parks, huge open spaces, and scenic attractions such as the VICTORIA FALLS on the Zambezi, the tourist industry is developing, much encouraged by the government.

Zambia's population of 9.5 million contains more than 72 different ethnic groups, each with its own language. About a third of the people are Bemba-speakers, with the Tonga, Nyanja and Lozi the next largest groups.

Although the population is growing fast – at 3 per cent a year – there is no great pressure on land, since the country covers 750 000 km² (about 290 000 sq miles) – more than three times the size of the United Kingdom.

Although 40 per cent of Zambians live in the Copperbelt in the north-east, 85 per cent of the Zambian workforce is employed in agriculture. About 70 per cent of the population are peasant farmers living in scattered villages. Large commercial farms also exist – particularly along the railway route between the Copperbelt and the Zimbabwe border – but only 7 per cent of the country's total area is systematically farmed. Many of the commercial farms are still owned by Europeans (who make up less than 1 per cent of the population) or large international companies.

However, the bulk of the country's agricultural output is grown by the peasants – maize is the staple crop – who also eat most of it. Their farming methods are primitive; hand hoes are used for cultivation, and bush-fallow techniques (by which much of the land is left unused and allowed to grow wild at any one time) are widespread. Women, as in most of Africa, do the bulk of the work.

Despite the number of people working on the land, and despite the extensive water resources and vast tracts of land, food shortages are constant, malnutrition is common, and both contribute to an infant mortality rate of 8.3 per cent of live births. The government has been forced to turn to expensive food imports to feed its people but, because of its huge debts, these are drastically limited.

In the 20 years following the decline in the price of copper, Zambia did little to use its huge agricultural potential as a substitute foreign exchange earner. Indeed, major irrigation schemes are still lacking, transport systems and marketing facilities for isolated areas are still inadequate, pricing policies unattractive, and produce marketing boards incompetent. As a result, in the past there was little incentive for the farmers to grow surplus for sale; and huge numbers of people left the countryside to seek new opportunities in the towns.

However, the government has now been taking steps to halt the slide. Farmers are encouraged to grow cash crops, and grow surpluses. Scattered communities have been brought together in larger village groups to make it more economical to provide services – health clinics, primary schools, cooperative marketing schemes, and better roads.

THE DRIFT TO THE TOWNS

The movement of people from the countryside to the towns has produced one of the most urbanised societies in Africa. About 56 per cent of Zambia's population now live in the cities. In particular, the string of Copperbelt towns (NDOLA, CHINGOLA, KITWE-NKANA, Chilila-Bombwe and Luanshya) form one of the largest urban concentrations in tropical Africa, though the capital, LUSAKA, to the south of the mining area, remains Zambia's largest city with nearly 1 million inhabitants. Many of the rural migrants have been unable to find paid jobs and have therefore set up in a variety of self-employed businesses.

Many townspeople retain strong links with their rural homelands. It is still common for some migrant workers who have lived in towns for many years to retire to their tribal villages, where the system of communal landholding allows them access to a plot of land to farm. Other townspeople – the children of migrants, born in the cities – regard themselves as permanent city dwellers and have lost their ethnic or old regional affiliations. For example, most of the original migrants to the Copperbelt were from Bemba-speaking regions in the north-west; now their descendants largely regard themselves as Copperbelters.

One significant effect of independence has been the steady takeover by Zambians of what

▲ RIVER OF SPLENDOUR The blue ribbon of the Luangwa winds through the browns and greens of the Zambian landscape. The Luangwa, 806 km (500 miles) long, joins the Zambezi on the Mozambique border. On its way it flows through two great national parks, each rich in wildlife and abounding in magnificent scenery.

were previously regarded as white jobs. During the colonial period, the mining companies employed single men as a ready source of unskilled, casual labour. But as mining techniques advanced, it was in the interest of the mines to take on a more permanent, skilled African labour force.

Thus the settled population of the copper towns began to grow. Certain jobs, however, were reserved for whites and fiercely guarded by the white trade unions, so that it was not until the 1950s that any concessions were made. After independence, the process of Africanisation of skilled jobs was speeded up and Zambians have taken the key positions. This has created a well-paid, well-educated Zambian elite.

Zambia, along with eight other African countries, is a member of the Southern Africa Development Community (SADC), which is

committed to rehabilitating and improving the railways and ports which have been the cause of so many of its economic problems. If this succeeds – and financial support from abroad is critical – the country's economic burden may be lightened. However, the poor market prospects for copper, a resource that will run out eventually, make it vital for Zambia to press on with developing its vast agricultural potential.

Droughts in 1991-2 again reduced agricultural output and inflation rocketed to 170 per cent in that year. The country was faced with yet another maize shortage in February 1994 – similar in scale to the one which had forced Kaunda to call multiparty elections in 1991 – elections which had cost him his presidency.

ZAMBIA AT A GLANCE

Official name Republic of Zambia

Area 752 610 km² (290 516 sq miles)

Population 9 500 000 **Per km²** 13 **(Per sq mile** 33)

Capital Lusaka

Government Republic

Currency 1 Zambian kwacha = 100 ngwee

Languages English (official), Bemba, Nyanja, Lozi, Tonga and other African languages

Religions Christian (75%), Muslim and Hindu, indigenous beliefs

Climate Tropical; cool in the highlands. Average temperature in Lusaka ranges from 16.1°C (61°F) in July to 24.4°C (76°F) in October

Land use Grazing and other 54%, forest and woodlands 39%, cultivation 7%

Main primary products Maize, cassava, sugar, tobacco, cotton, groundnuts, beef cattle; copper, cobalt, zinc, lead, coal, gemstones

Major industries Agriculture, mining, brewing, chemicals, cement

Main exports Copper, cobalt, zinc, lead, emeralds, amethysts, aquamarines, garnets, tourmaline, tobacco

Annual income per head (US$) 290

Population growth (per tho/yr) 30

Life expectancy (yrs) Male 45 **Female** 46

▲ MAGICAL CONSULTING ROOMS The motley paraphernalia of his trade adorn a witch doctor's premises on Zanzibar off the East African coast.

Zante *Greece* See ZÁKINTHOS

Zanzibar (Unguja) *Tanzania* Island situated in the Indian Ocean, about 50 km (30 miles) north of DAR ES SALAAM. In the 12th century Zanzibar was settled by Persians from Shiraz and by Arabs from Oman. It became a Portuguese trading place in the early 16th century. By 1700, Arabs had transformed the island into a thriving centre of the East African ivory and slave trades. It became a British protectorate in 1890 – as did the island of PEMBA 60 km (35 miles) to the north-east. The two islands became an independent country for four months in 1963-4. On 26 April, 1964 they joined mainland Tanganyika to form the United Republic of Tanzania.

Zanzibar covers 1660 km² (640 sq miles). Its chief products are cloves and copra. The Zanzibar Channel, a branch of the Indian Ocean, 31 km (19 miles) wide, separates Zanzibar Island from the mainland.

The port of Zanzibar on the west coast of the island is the main town. In the 19th century it was used as a base by European explorers and missionaries heading for the African interior. Today it has numerous small shops, bazaars, mosques, palaces, churches, former colonial mansions, courtyards and squares. The oldest part is the Stone Town, a labyrinth of narrow, winding streets lined with quaint whitewashed houses.

Population (island) 420 000; (town) 157 630
Map Tanzania – Ba

Zaozan *Japan* Mountainous area in northern HONSHU Island, about 260 km (160 miles) north of Tokyo. It rises to 1841 m (6040 ft) and is popular with visitors because of its volcanic scenery and, in winter, its skiing and curious *juhyo* – 'snow monsters' formed by ice and snow draped over the mountain pines.
Map Japan – Dc

Zaporizhzhya *Ukraine* City on the Dnieper River, 250 km (155 miles) south of KHAR'KIV. The nearby Dnieper hydroelectric dam powers an industrial complex which now makes iron and steel, aluminium, vehicles and machinery.
Population 891 000
Map Ukraine – Db

Zaragoza (Saragossa) *Spain* University city and former Spanish capital, on the Ebro River about 275 km (170 miles) north-east of Madrid. It has engineering, sugar-refining, flour-milling and wine-making industries.

Its province of the same name supports some agricultural products – sugar cane, cereals, grapes and olives – despite low rainfall and extreme temperatures. It was a centre of Moorish settlement and large parts of the lands were placed under irrigation by the Moors.
Population (city) 594 390; (province) 837 330
Map Spain – Eb

Zaria (Zegzeg) *Nigeria* Historic walled town, about 690 km (430 miles) north-east of the capital, LAGOS, in Kaduna State. It was founded in the 11th century as the capital of the Hausa Kingdom of Zazzan and was a centre of the slave trade. Its ancient monuments include the beautiful Friday Mosque and the emir's palace.

One of Nigeria's largest traditional emirates, it is a large-scale producer of cigarettes, palm oil, cotton textiles, printed materials, cosmetics and bicycles. Traditional weaving, dyeing and basket-making are also carried out.
Population 267 300
Map Nigeria – Ba

Zarqa (As Sarka) *Jordan* The country's second largest city and an important industrial centre, situated 20 km (12 miles) north-east of Amman. Zarqa developed from a 19th-century village and is now the administrative centre for the Zarqa region, which includes an extensive phosphate mining area. The town is on the Zarqa River, which flows into the Jordan and provides much of the water that irrigates the Jordan Valley.
Population 622 000
Map Jordan – Ba

Zealand (Sjaelland) *Denmark* The country's largest island, covering an area of 7014 km² (2708 sq miles), between the GREAT BELT channel and the Öresund, the main entrance to the Baltic Sea. Bridges span the waters between Zealand and its neighbouring islands, FALSTER and MØN. COPENHAGEN, the Danish capital, is on the island's eastern coast.

Zealand has a gently rolling landscape, which is humped with ridges left by retreating Ice Age ice sheets. Most of its coastline is serrated into bays and deep fjords. The island also embraces Denmark's largest lake, Arresø, extending over 41 km² (16 sq miles).

Zealand forms an important road and rail link between central and northern Europe. It is intensively cultivated, with 10 per cent of the land covered by well-managed woodlands. It is also the most urbanised part of Denmark. Although it has only about one-sixth of Denmark's total area, more than 30 per cent of the Danish people live here. There are numerous towns and settlements, from Kalundborg in the north-west, with its five-towered fortified church, to the old port of Køge in the east, with its richly timbered houses; from ELSINORE, with Hamlet's castle, in the north-east, to Korsør, with its naval base, in the west. Baronial homes such as Løvenborg, Borreby Castle and the royal hunting seat at Frederiksborg Castle, HILLERØD, recall Denmark's feudal past.

Many parts of Zealand have fine beaches, where there are large numbers of summer cottages. Zealand is also the site of 40 per cent of Denmark's industries, most of which are clustered near the capital.
Population 2 125 000
Map Denmark – Bb

Zeebrugge *Belgium* Ferry and cargo port linking BRUGES to the North Sea by a 13 km (8 mile) long canal. The waterway was opened in 1903 to provide deep-water access to the city.

Its harbour, used by German submarines in the First World War, was the scene of a daring naval raid by the British on 23 April 1918, which destroyed the mole and blocked the port.
Population 15 000
Map Belgium – Aa

Zeeland *Netherlands* South-western province consisting of the former southern islands of the Maas-Schelde delta and the mainland area south of the Schelde estuary. Most of its 3017 km² (1165 sq miles) are below sea level and are protected by dykes built under the Delta Plan for land reclamation, begun in 1956. Zeeland produces cereals, potatoes, flax and fruit. Its capital is Middelburg.
Population 359 200
Map Netherlands – Ab

Zegzeg *Nigeria* See ZARIA

Zelazowa Wola *Poland* See WARSAW

Zemlya Frantsa-Iosifa (Franz Josef Land) *Russia* Group of about 60 small Arctic islands north of NOVAYA ZEMLYA. The most northerly pieces of land in the Eastern Hemisphere, they cover an area of about 20 700 km² (8000 sq miles) – most of it ice-covered, with a few pockets of lichen. The islands were discovered by Austrian explorers in 1873 and claimed in the 1920s by the Russians, who established research stations here.
Map Arctic Ocean – a

Zenica *Bosnia-Herzegovina* Industrial town some 65 km (40 miles) north-west of SARAJEVO. In times of peace, it is a major coal, steel and paper-producing centre. To the west of Zenica lies the ancient town of Travnik.
Population 132 700
Map Bosnia-Herzegovina – Ba

zeolite Any of a group of silicate minerals which usually contain sodium, potassium, calcium or barium, or a mixture of these elements, used chiefly as filters or as water softeners.

Zermatt *Switzerland* Resort village in the southwest, 10 km (6 miles) north of the Italian border. It is ringed by mountains rising to some 4000 m (13 100 ft) or more – most notably the MATTERHORN – lies at an altitude of 1616 m (5302 ft) and is accessible only by rail. Zermatt is a centre for winter sports, mountaineering and summer walks.
Population 4900
Map Switzerland – Aa

zeuge Isolated flat mass of hard rock standing on a narrower pillar of softer underlying rock in a desert, or surviving as a relic of former climatic conditions. Zeugen are produced by sand-laden winds eroding the softer rock and undercutting the harder mass. They stand up to about 30 m (100 ft) high.

Zhambyl (Dzhambul) *Kazakhstan* Industrial town in the south, about 480 km (300 miles) west of ALMA-ATA. Its main products are metals, fertilisers, processed food, refined sugar and textiles.
Population 311 000
Map Kazakhstan – Db

Zhangjiakou *China* City in HEBEI Province. It was once a junction of caravan routes on the dry northern plateau of Hebei. Today it serves the nearby Xuanhua iron ore field.
Population 670 000
Map China – Hc

Zhanjiang *China* Industrial city in GUANGDONG Province, 400 km (250 miles) west of GUANGZHOU. It is a centre for the manufacture of chemical fertilisers and is also a base for offshore oil exploration in the South China Sea.
Map China – Gf

Zhdanov *Ukraine* See MARIUPOL'

Zhejiang *China* Province covering 100 000 km² (38 500 sq miles) south of the delta of the CHANG JIANG (Yangtze River). Hills and mountains make up more than 70 per cent of this subtropical province, and cultivated land is restricted to valleys and the narrow coastal plains. Rice, wheat, tea, maize, sugar cane and tangerines are grown. Zhejiang is also a leading producer of silk, for which mulberry trees are grown. Dense forests of pine, spruce and bamboo cover the mountainsides in the interior. They are exploited to serve a flourishing timber and paper industry. Despite its mountainous terrain, Zhejiang has a very high population density, even by Chinese standards.
Population 42 360 000
Map China – He

Zhengzhou *China* Capital of HENAN Province, lying south of the Huang He (Yellow River), on the North China Plain. It is the main railway junction of north China and an important centre for the manufacture of textiles and food products. During the course of its 4000-year history Zhengzhou has been destroyed several times by the floodwaters of the Huang He.
Population 1 660 000
Map China – Gd

Zhenjiang *China* City in JIANGSU Province, situated on the south bank of the Chang Jiang (Yangtze River). It manufactures machinery, chemicals and paper. The beautiful wooded hills in the southern suburbs are dotted with numerous Buddhist temples, some of which date back to the 6th century.
Population 345 600
Map China – Hd

Zhob *Pakistan* River valley, town and district in north-east BALUCHISTAN. Chromite (chromium ore) is mined here for export. The valley is rich in wild flowers and fruit orchards.
Population (town and district) 79 000
Map Pakistan – Cb

Zhu Jiang (Pearl River) *China* River formed by the triple confluence of the Xi Jiang, Bei Jiang and Dong Jiang rivers near GUANGZHOU. It flows 176 km (110 miles) into a broad delta, one of the richest lowlands in southern China and the most densely populated part of Guangdong Province. The British and Portuguese enclaves of HONG KONG and MACAU lie on either side of the main estuary.
Map China – Gf

Zhytomyr *Ukraine* City 130 km (80 miles) west of KIYEV. It is the centre of an important hopgrowing and beer-brewing region. It has large grain mills, timber mills and sugar refineries and manufactures machinery, furniture and clothing.
Population 296 000
Map Ukraine – Ba

Zibo (Tshifu; Yantai) *China* City of SHANDONG Province, lying on the Shandong Peninsula about 370 km (230 miles) south-east of BEIJING. Once a fishing village, it is today a large centre for the manufacture of glass, aluminium and petrochemicals.
Population 2 400 000
Map China – Hd

Zielona Góra (Grünberg) *Poland* Industrial city situated some 145 km (90 miles) north-west of WROCLAW.
It was a producer of fine cloth and wine in the Middle Ages and still holds an annual wine festival in early autumn after the grape harvest. The city also has metal, clothing, food processing, linen and printing industries.
Population 113 300
Map Poland – Ac

Ziguinchor *Senegal* River port and regional capital situated just north of the Guinea-Bissau border. It was founded by the Portuguese in the 16th century on the Casamance River, some 70 km (45 miles) from the Atlantic. However, it now has more of a French flavour, with elegant buildings and broad, palm-lined streets built during the French colonial period (1840-1960).
From the town, tourists can visit the beach of Cap Skirring, or go on trips into the surrounding tropical forests, or ride a boat upriver, past mangrove swamps and villages of the region's traditional conical huts.
Population 72 700
Map Senegal – Ab

Zihuatanejo *Mexico* Picturesque fishing port and popular beach resort on the Pacific coast of GUERRERO State. New developments, including an international airport, have not spoilt the town's natural beauty. Neighbouring Ixtapa, by contrast, is a purpose-built resort specialising in water sports.
Population 50 000
Map Mexico – Bc

Zikhron Ya'aqov *Israel* Village in the Sharon Valley, located about 25 km (15 miles) south of Haifa. The village was founded in 1882 by the Rothschilds, whose family tombs are here, and is the centre of a region producing grapes and wine. It is the site of a Carmelite nunnery.
Population 5000
Map Israel – Aa

Zilina *Slovakia* Industrial town situated 74 km (46 miles) south-east of OSTRAVA. It makes chemicals, timber products, textiles and processed foods. Its Old Town includes an arcaded Baroque market square, medieval buildings including the Romanesque Church of St Stephen and the parish church, and several theatres and museums. There are also a conservatory of music and a centre for the treatment of eye diseases.
Population 87 800
Map Slovakia – Bb

Zillertal Alps *Austria* Part of the Eastern Alps between the BRENNER PASS and the HOHE TAUERN range, on the border with Italy. Rising to 3510 m (11 516 ft) at the Hochfeiler, they are the site of several popular ski resorts, such as Mayrhofen, in the valley of the Zill River.
Map Austria – Bb

Zimbabwe See p 752

Zinder *Niger* Walled market town and former national capital, situated 750 km (465 miles) east of the present capital, NIAMEY, and at the southern end of a road that stretches northwards across the Sahara to the Algerian capital, Algiers.
Zinder ceased to be Niger's capital in 1926. Today its wares include groundnuts, skins, hides, leather goods and blankets. There is an air field located at Zinder.
The town is dominated by Muslims who have close family and trade links with the city of KANO in northern Nigeria. It has a dense and involved maze of alleyways, a former sultan's palace built in 1860, a mosque and a fort dating from the French occupation.
Population 60 000
Map Niger – Ab

Zimbabwe

A LAND WHICH HAS A TURBULENT HISTORY REMAINS RICH IN MINERALS AND TEEMING WILDLIFE

Everything about Zimbabwe is dramatic, from its spectacular landscapes and teeming wildlife to its turbulent history and recent past. In the north, the ZAMBEZI River flings itself over the mile-wide VICTORIA FALLS and flows through a lake nearly 300 km (186 miles) long; in the south, the 'great, grey-green, greasy' LIMPOPO River, as Kipling described it, marks the border with South Africa. Mountain ranges ridge the east, peaking in 2592 m (8504 ft) Inyangani; and to the west lies safari country, forests and savannah full of lions, elephants, buffaloes and giraffes.

Then there is the gold that attracted the English imperialist Cecil John Rhodes to the territory, which was named Rhodesia after him. He brought white colonial rule after a bloody war. Less than 100 years later, another war ravaged the country before blacks ruled again and called it Zimbabwe.

There never was gold in the quantities hoped for by Rhodes, already rich from his South African mining operations. But Zimbabwe, although it is a minor gold producer, is rich in other minerals and has a great wealth of farmland. It is also one of Africa's major manufacturing nations. The manufacturing sector – based on mining and agriculture – produces a quarter of gross domestic product (GDP).

Rhodes was one of the larger-than-life men whose names punctuate Zimbabwe's history, from the canny Matabele leader Mzilikazi and his son, Lobengula, to the most recent prime minister, the scholarly Robert Mugabe. They include Ian Smith, the last of the white rulers, who as prime minister decided to defy the British Government and made a Unilateral Declaration of Independence (UDI) in 1965. Not to mention the Scottish missionary-explorer David Livingstone, the first white man to reach the Zambezi, in 1851 – and four years later the Victoria Falls.

INDEPENDENCE UNDER BLACK RULE

Many of Zimbabwe's 225 000 whites left after the country became legally independent under black rule in 1980, but Mugabe, needing their skills and know-how, has tried hard to win their confidence and stop the exodus. About 100 000 remain, 4000 of them farmers. However, the 1992 Land Acquisition Act proved to be a new threat to this group of white farmers, forcing them to sell their land. Many blacks argue that in 14 years of independence, little has been done to redress the colonial imbalance of land ownership. But Mugabe has come under much criticism and international pressure over the apparent abuse of this Act. Initially, the land acquired for redistribution to peasant farmers, was meant to involve under-utilised land rather than commercial farms. It was soon discovered that the first farm to be

appropriated, a valuable 3000-acre estate, had been leased to a cabinet minister, instead of being apportioned to the needy. Eventually it emerged that all of the 98 farms compulsorily purchased had been leased to senior government officials and civil servants. Legal action by white farmers to challenge the Act as unconstitutional was dismissed by the courts.

Ninety-eight per cent of the 11.2 million population are black. The Shona people (who are known also as the Mashona), inhabitants of Zimbabwe since the 11th century or earlier, make up 71 per cent of the population. The rest are Ndebele (or Matabele), descendants of Mzilikazi's warrior tribe who broke from the Zulu nation in the 1830s, and smaller groups such as the Venda, Tonga and Shangaan.

Broadly speaking, Mugabe's Zimbabwe African National Union (ZANU) represented the Shona majority and Joshua Nkomo's Zimbabwe African People's Union (ZAPU) the Ndebele. The parties were born as guerrilla movements which fought and brought down the Smith regime.

The newly independent country took its name from GREAT ZIMBABWE, an historic site of massive stone ruins near Masvingo; the ruins are believed to date from a Shona civilisation that dominated the region from the 12th to the 15th centuries.

The Shona were mainly farmers. In the 1830s, the invading Ndebele, who lived largely on cattle, the spoils of raiding parties and tributes from those they had conquered, established a community in what is now BULAWAYO. Their descendants still live in the surrounding region, called MATABELELAND (or Ndebeleland). The Shona are in the north, around the modern capital, HARARE (formerly Salisbury), in the region known as MASHONALAND or Shonaland, and to the east in the Manicaland and Masvingo provinces.

Almost all this scenic country is more than 300 m (1000 ft) above sea level. A great central plateau, called the highveld, is some 1200-1500 m (4000-5000 ft) high and occupies 52 000 km^2 (about 20 080 sq miles). Huge granite outcrops called koppies dot the land-

scape, but this is a fertile savannah area, with temperatures averaging a pleasant 20°C (68°F) and an annual rainfall of 700-900 mm (28-35 in), most of it in summer storms.

The highveld runs roughly from north-east to south-west, and on either side the land falls away to the middleveld, again fine plateau country 600-1200 m (about 2000-4000 ft) high. Below 600 m is the lowveld, a narrow strip in the Zambezi valley and a broader band between the Limpopo and Sabi rivers.

Tobacco is the country's most important crop, accounting for about 20 per cent of export earnings. The biggest food crop is maize, grown mainly for internal consumption, followed by other grains, tea, coffee, sugar and citrus fruits. Cotton is another significant crop. In normal years, the country is self-sufficient in food production; however, in times of drought, as in 1991-2, yields are too small to feed the growing population.

A geological feature called the GREAT DYKE runs for 515 km (about 320 miles) across the highveld from north-west to south-east, and forms the low ridge on which Harare stands. In and around the Dyke are deposits of gold, chrome, nickel, platinum and other minerals.

The modern history of Zimbabwe began in the 1880s when emissaries from Cecil Rhodes persuaded King Lobengula to grant them mining rights. Rhodes decided that Lobengula's concession amounted to a licence to grab the entire country. He proceeded to do so, forming the British South Africa Company and sending a 'pioneer column' to invade in 1890.

The outraged Ndebele went to war against the company in 1893. A bloody conflict ended in the defeat and subsequent death of Lobengula. In 1896-7, the Ndebele rebelled again and were joined by the Shona. It was a final resistance to colonialism that ended with British regular troops being called in.

The territory so won became the colony of Southern Rhodesia, and in 1930 a Land Apportionment Act limited the areas in which Africans could own land, creating, in effect, native reserves. Whites ended up owning almost all the best land.

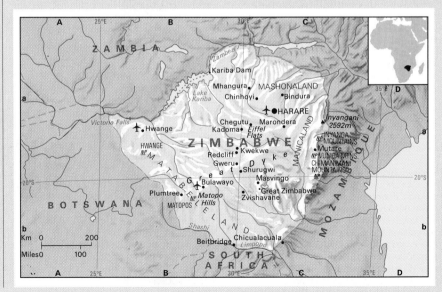

In 1953 Southern Rhodesia became part of a federation with Northern Rhodesia (now ZAMBIA) and Nyasaland (now MALAWI). But this broke up in 1963, primarily through black opposition to white rule. In 1964 Zambia and Malawi both achieved their independence with black governments, and since both had never had significant numbers of white settlers, independence was peaceful.

But in the 1960s about 300 000 whites lived in Southern Rhodesia and would not accept a black government. In 1965, led by Ian Smith and his Rhodesian Front Party, Southern Rhodesia rebelled against Britain and declared itself independent as Rhodesia. Britain cut off trade and financial links with the breakaway colony. The United Nations followed with international sanctions.

Neighbouring South Africa and Mozambique continued to allow vital supplies into Rhodesia, which already possessed a strong industrial base and was manufacturing a range of substitutes for blocked imports. In 1970 Smith declared a republic. Guerrilla action against him escalated rapidly into civil war.

Eventually the combined effects of the war, sanctions and diplomatic pressure persuaded Smith to make way for black majority rule. At a conference in London in 1979 a new constitution was thrashed out. Elections followed in 1980, Zimbabwe was born, and it rejoined the Commonwealth.

The years since have witnessed a post-independence boom gradually fade away, while a deepening rift grew between ZANU and ZAPU – especially during 1982-6. However, in 1987 the two parties merged under the name Zanu-Patriotic Front and won 117 of the 120 parliamentary seats in the 1990 general election.

Unemployment and tension over land ownership are mounting problems among a population which is growing by 1.3 per cent yearly and the controversy over resettlement continues to rage.

Mining employs 5 per cent of the workforce but accounts for 40 per cent of export earnings. Zimbabwe has a developed industry – helped along by wealth in raw materials and abundant hydroelectric energy.

Tourism is another major growth industry, which attracted more than half a million visitors in 1990. This beautiful country, with its pleasant climate, has much to offer. As well as the Victoria Falls and Great Zimbabwe, there are the glorious Eastern Highlands, laced with trout streams and home to leopards and eagles. Lake KARIBA, in the north, is 5000 km² (nearly 2000 sq miles) of blue, island-studded water offering fishing, sailing and game-rich shores. In the MATOPO HILLS, near Bulawayo, there are 2000 km² (about 800 sq miles) of grotesque, wind-sculptured granite hills, with caves that were painted by ancient peoples. Then there are national park and wildlife reserves which cover 44 700 km² (about 17 260 sq miles), or 11 per cent of the whole country. But in 1993 the total national budget allocation to nature conservation was drastically cut and environmentalists feared that mining companies would be granted prospecting rights in one of the country's biggest national parks.

Like several of its neighbours, Zimbabwe is landlocked; but it is nevertheless strategically situated on the transport routes between countries to the north and the major South African ports. Its nearest outlets to the sea are through Mozambique and the ports of Beira and Maputo, though much of its traffic is diverted on to the South African transport system.

ZIMBABWE AT A GLANCE

Official name Republic of Zimbabwe

Area 390 580 km² (150 768 sq miles)

Population 11 200 000 **Per km²** 29 **(Per sq mile)** 75)

▼ **TEMPLE OF MYSTERY This enclosure, known as the Temple, is part of the mysterious ruins of Great Zimbabwe, near Masvingo. It was built as a king's residence by the Shona Bantu-speaking people, who settled the area from about AD 300.**

Capital Harare

Government Parliamentary republic

Currency 1 Zimbabwe dollar = 100 cents

Languages English (official), Shona, Ndebele

Religions Christian (55%), indigenous (24%), Muslim and other

Climate Subtropical; temperatures governed by altitude. Average temperature in Harare ranges from 13.6°C (56°F) in July to 21.5°C (71°F) in October

Land use Forest and woodlands 52%, grazing 40%, cultivation 7%, other 1%

Main primary products Maize, tobacco, wheat, sugar, millet, cotton; gold, asbestos, coal, iron, chrome, nickel, copper, tin

Major industries Agriculture, cigarettes, fertilisers, cement, coke, iron and steel, textiles, mining, tourism

Main exports Tobacco, maize, cotton, sugar, coffee, ferrochrome, gold, asbestos

Annual income per head (US$) 570

Population growth (per thous/yr) 30

Life expectancy (yrs) Male 41 **Female** 44

Zion, Mount *Israel* See JERUSALEM

Zion National Park *USA* Area of some 593 km² (229 sq miles) in south-west Utah. The park contains multi-coloured rock formations and canyons, of which the Zion Canyon, some 16 km (10 miles) long and more than 610 m (2000 ft) deep, is the most spectacular.
Map United States – Dc

Zipaquirá *Colombia* Town situated about 50 km (30 miles) north of Bogotá. It has a huge rock salt mine, said to be able to supply the world's needs for 100 years. Parts of the mine are open to the public. One section was shaped in 1954 into a vast underground cathedral, with a salt block weighing about 18 tons as its altar.
Population 41 000
Map Colombia – Bb

Ziz *Morocco* See ERFOUD

Zlatibor *Yugoslavia (Serbia and Montenegro)* Thickly forested mountain massif rising to 1496 m (4908 ft) in west Serbia, some 140 km (about 85 miles) south-west of BELGRADE. The massif, which has an exceptionally clear atmosphere and 2000 hours of sunshine a year, is a popular summer and winter resort.
Map Yugoslavia – Bc

Zlatoust *Russia* City in the Ural Mountains, about 200 km (125 miles) south of YEKATERINBURG. It grew out of iron and copper works set up near local mines in 1754, and specialises in making cutlery, swords and engraved and stainless steels. It also processes timber and produces machine tools, clocks and watches.
Population 208 000
Map Russia – Gc

Zomba *Malawi* Capital of Nyasaland from 1885-1964 and of independent Malawi from 1964 to 1975, when it was displaced by LILONGWE, some 240 km (150 miles) to the north-west. It was originally the base from which the slave trade was destroyed in the 19th century. It is now the centre of a fertile farming region which produces coffee, cotton, tobacco and tung oil, used in making waterproof paints and varnishes.
Population 49 400
Map Malawi – Bc

Zouerate *Mauritania* Mining town situated about 650 km (405 miles) north-east of NOUAKCHOTT and connected by rail to the port of Nouadhibou. Its iron ore provides some 50 per cent of the country's total exports.
Population 50 000
Map Mauritania – Ba

Zrenjanin (Petrovgrad; Veliki Beckerek; Nagybecskerek) *Yugoslavia (Serbia and Montenegro)* One of the largest cities in the Serbian autonomous province of VOJVODINA. It lies on the Becej River, 48 km (30 miles) north-east of the province's capital, NOVI SAD. A market, transport and industrial centre, it produces farm machinery, beer, spirits, sugar and processed foods.
Population 139 300
Map Yugoslavia – Cb

Zug *Switzerland* Town and resort on the northeast shore of Lake Zug, 20 km (12 miles) south of ZÜRICH. It is capital of the canton of the same name. The town has a 16th-century town hall and a clock tower dating from 1480. It makes electrical goods and kirsch. The canton became a member of the Swiss confederation in 1352.
Population (town) 21 780; (canton) 87 100
Map Switzerland – Ba

Zugspitze *Germany* The country's highest mountain, rising to 2962 m (9718 ft). It is in the Bavarian Alps beside the Austrian border, 90 km (55 miles) south-west of MUNICH.
Map West Germany – De

Zuiderzee *Netherlands* See IJSSELMEER

Zulawy *Poland* Fertile farming region covering an area of 2515 km² (970 sq miles) in the delta of the Vistula River to the east of the Baltic port of Gdańsk. Wheat, sugar beet and oilseed rape are grown, Friesian cattle graze the rich fenland pastures, Dutch-type windmills survive and half-timbered churches and houses still abound. The Teutonic Knights began draining the area in the 14th century, and imported settlers from the Netherlands and German Friesland to farm it.
Map Poland – Ca

Zululand *South Africa* See KWAZULU-NATAL

Zürich *Switzerland* Capital of the canton of Zürich and the country's largest city, standing on a lake of the same name, about 20 km (12 miles) south of the German border. The site was a lakeside settlement even before the Romans occupied it in 15 BC.

Today Zürich is Switzerland's main financial and commercial centre, housing numerous insurance and international banking headquarters. The city, which is popular with tourists, has a 12th-century town hall, a Romanesque cathedral with two fine Gothic towers and a Romanesque-Gothic church dating from the 9th century which has windows created by Marc Chagall in 1970.

Swiss Protestantism was founded in Zürich by the religious reformer Ulrich Zwingli (1484-1531). The city embraced the new faith in 1523 after Zwingli – while a priest at the Grossmünster (now a Protestant church) – violently denounced Roman Catholicism. He was killed by the forces of the Catholic cantons in a battle near Zürich.
Population (city) 345 200; (canton) 1 158 100
Map Switzerland – Ba

Zutphen *Netherlands* Historic town on the IJssel River, situated about 95 km (60 miles) south-east of Amsterdam. The town was besieged by the Spanish during the 80 Years' War (1568-1648) and occupied by the Germans during the Second World War. It makes paper, leather, soap and glue.
Population 31 120
Map Netherlands – Ca

Zwickau *Germany* Industrial city situated about 65 km (40 miles) south of the city of LEIPZIG. It was the birthplace of the Romantic composer Robert Schumann (1810-56). Zwickau's products include porcelain, dyes, pharmaceuticals, textiles and cars.
Population 120 100
Map Germany – Ec

Zwolle *Netherlands* Capital of the province of Overijssel, some 75 km (45 miles) north-east of AMSTERDAM, on the Zwarte Water (river).

Within its canal-encircled medieval centre stands the market place, town hall and a 15th-century church. The Sassenpoort (gate) is almost all that remains of the medieval ramparts. Zwolle produces ships, iron and chemicals.
Population 98 320
Map Netherlands – Ca

▼ **SWISS BANKING CAPITAL** The Limmat glides into the lake near Zürich's lofty twin-towered cathedral. The city's Bahnhofstrasse, the 'Fifth Avenue of Switzerland', is a shopper's paradise for luxury goods.

Afghanistan

Albania

Algeria

Andorra

Angola

Antigua and Barbuda

Argentina

Armenia

Australia

Austria

Azerbaijan

Bahamas

Bahrain

Bangladesh

Barbados

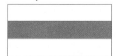
Belarus

Belgium

Belize

Benin

Bhutan

Bolivia

Bosnia-Herzegovina

Botswana

Brazil

Brunei

Bulgaria

Burkina

Burundi

Cambodia

Cameroon

Canada

Cape Verde

Central African Republic

Chad

Chile

China

Colombia

Comoros

Congo

Cook Islands

Costa Rica

Croatia

Cuba

Cyprus

Czech Republic

Denmark

Djibouti

Dominica

Dominican Republic

Ecuador

Egypt

El Salvador

Equatorial Guinea

Eritrea

Estonia

Ethiopia

Fiji

Finland

France

Gabon

The Gambia

Georgia

Germany

Ghana

Gibraltar

755

FLAGS OF THE WORLD

 Greece

 Greenland

 Grenada

 Guatemala

 Guinea

 Guinea-Bissau

 Guyana

 Haiti

 Honduras

 Hong Kong

 Hungary

 Iceland

 India

 Indonesia

 Iran

 Iraq

 Ireland

 Israel

 Italy

 Ivory Coast

 Jamaica

 Japan

 Jordan

 Kazakhstan

 Kenya

 Kiribati

 North Korea

 South Korea

 Kuwait

 Kyrgyzstan

 Laos

 Latvia

 Lebanon

 Lesotho

 Liberia

 Libya

 Liechtenstein

 Lithuania

 Luxembourg

 Macedonia

 Madagascar

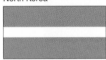 Malawi

 Malaysia

 Maldives

 Mali

 Malta

 Marshall Islands

 Mauritania

 Mauritius

 Mexico

 Federated States of Micronesia

 Moldova

 Monaco

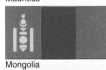 Mongolia

 Morocco

 Mozambique

 Myanmar

 Namibia

 Nauru

 Nepal

 Netherlands

 New Zealand

 Nicaragua

 Niger

 Nigeria

 Niue

Northern Marianas

Norway

Oman

Pakistan

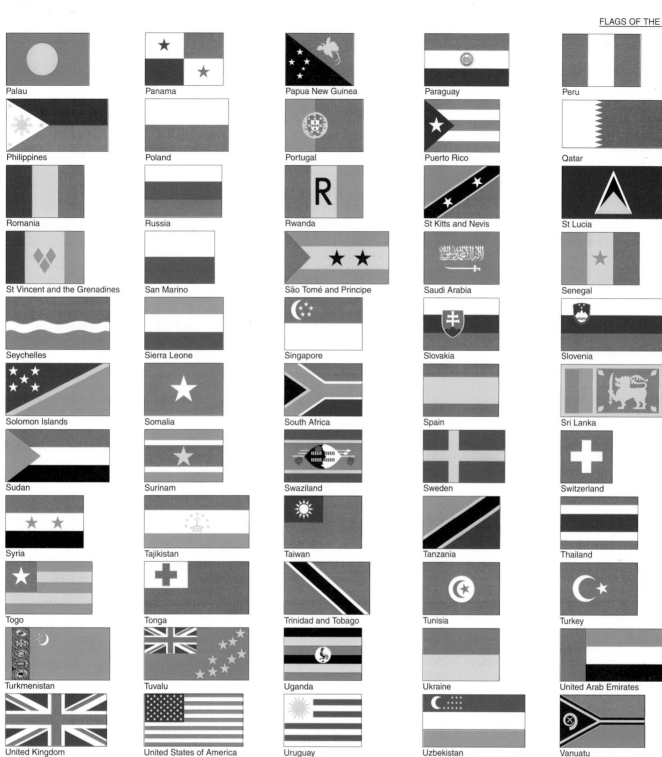

Palau

Panama

Papua New Guinea

Paraguay

Peru

Philippines

Poland

Portugal

Puerto Rico

Qatar

Romania

Russia

Rwanda

St Kitts and Nevis

St Lucia

St Vincent and the Grenadines

San Marino

São Tomé and Principe

Saudi Arabia

Senegal

Seychelles

Sierra Leone

Singapore

Slovakia

Slovenia

Solomon Islands

Somalia

South Africa

Spain

Sri Lanka

Sudan

Surinam

Swaziland

Sweden

Switzerland

Syria

Tajikistan

Taiwan

Tanzania

Thailand

Togo

Tonga

Trinidad and Tobago

Tunisia

Turkey

Turkmenistan

Tuvalu

Uganda

Ukraine

United Arab Emirates

United Kingdom

United States of America

Uruguay

Uzbekistan

Vanuatu

Vatican City State

Venezuela

Vietnam

Western Samoa

Yemen

Yugoslavia

Zaire

Zambia

Zimbabwe

International Organisations and Alliances

There are dozens of international alliances and agencies devoted to promoting international cooperation of one kind or another. Some are global, others regional; some are much more active and effective than others. Among the more important and vigorous are the following.

Association of South East Asian Nations (ASEAN)

Founded in 1967 by a group of non-communist nations of the region in order to aid economic, cultural and social development. Members are Brunei, Indonesia, Malaysia, the Philippines, Singapore and Thailand. Its headquarters are in Jakarta, Indonesia.

Caribbean Community (CARICOM)

Founded by four nations in 1973 and expanded the next year by taking in members of the Caribbean Common Market. Its aim is to coordinate foreign policy and social development. Antigua and Barbuda, Bahamas (not a member of the Caribbean Common Market), Barbados, Belize, Dominica, Grenada, Guyana, Jamaica, Montserrat, St Kitts and Nevis, St Lucia, St Vincent, and Trinidad and Tobago are members. British Virgin Islands, Turks and Caicos and US Virgin Islands are associate members. Its headquarters are in Georgetown, Guyana.

Colombo Plan

An economic development programme for South and South-east Asia and the Pacific region. It began in 1950 with seven members, and now has 24: Afghanistan, Australia*, Bangladesh, Bhutan, Burma, Cambodia, Fiji, India, Indonesia, Iran, Japan*, Laos, Malaysia, Maldives, Nepal, New Zealand*, Pakistan, Papua New Guinea, the Philippines, Singapore, South Korea, Sri Lanka, Thailand and the USA*. Canada and the United Kingdom withdrew in 1991-2. The four countries marked with an asterisk are donor members. Its headquarters are in Colombo, Sri Lanka.

The Commonwealth

One of the loosest but most far-reaching international organisations, with no formal charter but (in 1995) spanning 51 nations and a quarter of the world's population – all at one time ruled by Britain or a British protectorate. Originally defined in 1926 as a group of 'autonomous communities within the British Empire', what ties member states today is recognition of Queen Elizabeth II as head of the Commonwealth. More than 30 of the members are republics, five have their own monarchs and in only 16 is the Queen also head of state. Two nations which left the Commonwealth – South Africa (1961) and Pakistan (1972) – have rejoined – Pakistan in 1989 and South Africa in 1994. Several countries did not join on becoming independent.

Commonwealth heads of government meet every two years (in different capitals) to discuss common political and economic concerns, and there is a permanent secretariat in London. The Commonwealth Games are held every four years; the 1994 venue was in Canada, and the 1998 venue will be Kuala Lumpur, Malaysia.

Commonwealth of Independent States (CIS)

Founded in 1991 upon the dissolution of the USSR, it is a voluntary association of former USSR countries. Their aim is cooperation as they untangle their interrelationships and establish themselves as fully independent states while preserving a measure of economic cooperation. There are 12 members: Armenia, Azerbaijan, Belarus, Georgia, Kazakhstan, Kyrgystan, Moldova, Russia, Tajikistan, Turkmenistan, Ukraine and Uzbekistan.

European Union (EU)

Association of West European countries created in 1993 by the Maastricht Treaty of 1992 and replacing the earlier European Community (EC). The goal of the EU is to promote closer economic and political cooperation between its members, and to achieve monetary union by 1999. It has its origins in the 1950s, when France, West Germany, Italy, the Netherlands, Belgium and Luxembourg – 'the Six' – came together to form the European Coal and Steel Community (ECSC), the European Atomic Energy Community (Euratom, concerned with civilian nuclear power) and the European Economic Community (EEC). The EEC, founded in 1958 following the signing of the Treaty of Rome in 1957, sought to abolish tariff barriers and encourage trade between members in a 'Common Market'. It was fused with the ECSC and Euratom in the late 1960s to form the European Community. The EC developed common institutions and policies, including a Common Agricultural Policy (which regulates community farm prices) and European Monetary System (to stabilise exchange rates).

The original six EC members were joined in 1973 by Britain, Denmark and Ireland; Greece joined in 1981, and Spain and Portugal in 1986. Greenland, a Danish dependency, withdrew in 1985. Austria, Finland and Sweden became members in 1995. The EU's headquarters are in Brussels, where the European Commission – its executive – is based. Other EU bodies include the Council of Ministers (of member states); the European Parliament, which meets in Brussels and Strasbourg and is mainly consultative; and the European Court of Justice, in Luxembourg. The Treaty of Rome was amended by the Single European Act in 1986 which was designed to break down barriers to free movement of goods, services, people and finance. The Maastricht Treaty of 1992 extended community cooperation into the areas of foreign and security policy, and policing and home affairs.

European Free Trade Association (EFTA)

Formed as a tariff-free area by seven countries in 1960, following the failure of Britain's first attempt to join the Common Market, there are still seven countries, though their identity has changed. The present members are Austria, Norway, Sweden, Switzerland, Iceland, Finland and Liechtenstein. Following agreements between EFTA and the EU, there is now virtually free trade in industrial goods between members of the two bodies.

League of Arab States (Arab League)

Regional, primarily political, grouping founded in 1945 by seven Arab nations, and expanded by 1992 to 21 stretching from Mauritania and Morocco to Iraq and Oman. The League's secretariat is in Cairo, Egypt.

North Atlantic Treaty Organisation (NATO)

Main Western military alliance, founded in 1949 by Belgium, Canada, Denmark, France, Iceland, Italy, Luxembourg, the Netherlands, Norway, Portugal, the UK and USA. Subsequently, West Germany (now united Germany), Greece, Turkey and Spain also joined. NATO headquarters are in Brussels, and its military command – which in an emergency would assume control over members' armed forces except those of Iceland (which has none) and France (liaison status only) – is also in Belgium, near Mons. The disintegration of the USSR and old Eastern Bloc group of countries has caused NATO to undergo radical restructuring and orientation.

Organisation for Economic Cooperation and Development (OECD)

Formed in 1961 from members of the Organisation for European Economic Cooperation (OEEC) to promote world trade and economic development. Its members include all the main western European nations (including neutral Austria, Sweden and Switzerland) plus Australia, Canada, Japan, New Zealand and the USA. The organisation is based in Paris, France.

Organisation of African Unity (OAU)

Founded in 1963 by 32 nations at a conference in Addis Ababa, Ethiopia (where its headquarters remain), the OAU now has some 52 member states throughout the continent. Its aims are to promote African unity and development, and to eliminate colonialism. In 1991 the OAU declared its intention to create an African economic community by 2000.

Organisation of American States (OAS)

Founded in 1948 but tracing its origins back to the first International Conference of American States in 1890, this organisation now has more than 35 members in North, Central and South America and the West Indies. Its original emphasis was on regional security and independence, but it has increasingly promoted economic and social development. Its secretariat is in Washington DC.

758

Organisation of Petroleum Exporting Countries (OPEC)

Founded in 1960 to promote the interests of oil producers, OPEC was highly effective in raising crude oil prices in the 1970s and early 1980s. Its influence waned with the oil glut of the mid-1980s and the failure to control the output of major oil-producing non-members such as Mexico, Norway, the UK and USA. Its headquarters are in Vienna, Austria, but more than half its 12 members are in the Middle East.

South Pacific Forum

Regional political and economic grouping of 15 independent states: Australia, Cook Islands, Federated States of Micronesia, Fiji, Kiribati, Marshall Islands, Nauru, New Zealand, Niue, Papua New Guinea, Solomon Islands, Tonga, Tuvalu, Vanuatu and Western Samoa. The forum provides an opportunity to discuss common problems, for example, it has vocally opposed French nuclear testing in the Pacific. Its headquarters are in Suva, Fiji.

The United Nations (UN)

Founded in 1945 as the successor to the League of Nations to foster international peace, security and cooperation. Most countries are members – 184 in all in 1993. Exceptions are the two Koreas, Switzerland (on account of its neutrality) and Taiwan (replaced by mainland China in 1971). All members have a seat in the General Assembly, which meets annually, but only five (China, France, Russia, UK and USA) are permanent members of the Security Council, which functions continuously and is responsible for maintaining international peace and security; an additional ten members serve on the Security Council for two-year terms.

The main UN headquarters are in New York, but the International Court of Justice, another UN body, is based in The Hague. There are also numerous specialised agencies, but not all UN members participate in all of these. They include the Food and Agriculture Organisation (FAO), based in Rome, General Agreement on Tariffs and Trade (GATT; Geneva), International Atomic Energy Agency (IAEA; Vienna), International Labour Organisation (ILO; Geneva), International Monetary Fund (IMF; Washington DC), United Nations Educational, Scientific and Cultural Organisation (UNESCO; Paris) and World Health Organisation (WHO; Geneva).

Acknowledgments

Artwork on pages 506 (bottom), 542 (bottom) was commissioned for *The Reader's Digest Great World Atlas*. All other artwork was commissioned for this book.
The photographs in this book came from the sources listed below. Those commissioned by Reader's Digest appear in *italics*.

1 Michael Freeman. 2 B. Glinn/Magnum/John Hillelson Agency. 5 (TL) D. Turner/Colorific!; (TR) Preman Sotomayor; (B) V. Englebert/Susan Griggs Agency. 6-7 I. Morath/Magnum/John Hillelson Agency. 7 J. Yates/Susan Griggs Agency. 13 Mike Goldwater/Network. 16-17 R. & S. Michaud/John Hillelson Agency. 18 J. Gardey/Robert Harding Picture Library. 20-21 S. Kraserman/Bruce Coleman Ltd. 25 Robert Estall Photographs. 27 G. Tortoli/Zefa. 29 Michael Freeman. 33 Gruschwitz/Schapowalow. 37 Jack Piccone/Network. 45 P. Frey/The Image Bank. 47 Wolfgang Kunz/Bilderberg. 52 N. Westwater/Robert Harding Picture Library. 57 U. Seer/The Image Bank. 60 G. Ricatto/Zefa. 63 H. Munzig/Susan Griggs Agency. 64 Abbas/Magnum. 66 Dr. Jürgen Nittinger/Xeniel-Dia. 70 B. Barbey/Magnum/John Hillelson Agency. 75 R. Perry/Impact Photos. 76 Patrick Eagar. 78 V.C.L./Photo Access. 81 Urs F. Kluyver/FOCUS. 85 K. Benser/Zefa. 89 K. Benser/Zefa. 94 F. Breig/Zefa. 95 J. Cleare/Mountain Camera. 96 R. Smith/Tony Stone Worldwide. 99 T. Ives/Susan Griggs Agency. 100 B. Barbey/Magnum/John Hillelson Agency. 106 W. Rawlings/Robert Harding Picture Library. 109 M. Vautier/vdn Picture Library. 112 B. Barbey/Magnum/John Hillelson Agency. 116 P. Vauthey/Sygma/John Hillelson Agency. 117 Kummels/Zefa. 124 F. Damm/Zefa. 127 Larry Burrows/Aspect Picture Library. 132-133 A. Farquar/Daily Telegraph Colour Library. 134 T. Fincher/Colorific! 140 Tom Hanley. 146 A. Woolfitt/Susan Griggs Agency. 149 David Noble/FPG International/Photo Access. 151 B. Barbey/Magnum/John Hillelson Agency. 153 M. Koene/Colorific! 154-155 B. Barbey/Magnum/John Hillelson Agency. 163 A. Evrard/Susan Griggs Agency. 164-165 N. Tomalin/Bruce Coleman Ltd. 168 D. Beatty/Susan Griggs Agency. 172 N. Woldendorp/Susan Griggs Agency. 175 J. Gascoigne/Robert Harding Picture Library. 180-181 Jean-Pierre Amet/Sygma/Inpra. 182 W. Meier/Zefa. 183 B. Seed/John Hillelson Agency. 185 J. Davis/Robert Harding Picture Library. 188 *Malcolm Aird.* 189 E. Young/Robert Harding Picture Library. 195 A. Durand/Robert Harding Picture Library. 202 A. le Garsmeur/Impact Photos. 209 M. Beebe/The Image Bank. 210-211 K. Langley/Aspect Picture Library. 217 A. Deane/Bruce Coleman Ltd. 221 G. Gerster/John Hillelson Agency. 223 D. Barrault/Robert Harding Picture Library. 227 Les Stone/Sygma/Inpra. 229 D. Muscroft/Falklands Pictorial. 235 G. Davis/Colorific! 237 H. Wiesner/Zefa. 238-239 G. Mangold/The Image Bank. 241 Robert Estall Photographs. 242 H. Munzig/Susan Griggs Agency. 245

L. Arepi/The Image Bank. 246 V.C.L./Photo Access. 248 T. Okuda/Aspect Picture Library. 249 The Image Bank. 255 Stuart Dee/The Image Bank. 257 M. Fogden/Bruce Coleman Ltd. 261 Robert Estall Photographs. 262 Dario Mitidieri/Select. 263 I. Wandmacher/Fotex/Inpra. 267 I. Berry/Magnum/John Hillelson Agency. 270 P. Carmichael/Aspect Picture Library. 273 H. Hanser/The Image Bank. 275 J. Stage/The Image Bank. 277 A. Woolfitt/Susan Griggs Agency. 279 S. & D. Cavannagh/Robert Harding Picture Library. 280 W. Ferchland/Zefa. 284 Michael Freeman. 286 D. Reed/Impact Photos. 287 P. & C. Leimbach/Robert Harding Picture Library. 295 C. Joyce/Impact Photos. 299 Caroline Penn/Impact Photos. 303 G. Gerster/John Hillelson Agency. 307 UWF/Zefa. 310 K. de Francke/Bilderberg. 312 J. Poncar/Bruce Coleman Ltd. 314-315 R. & S. Michaud/John Hillelson Agency. 320 M. Vautier/vdn Picture Library. 325 V.C.L./Photo Access. 327 Les Stone/Sygma/Inpra. 329 P. Carmichael/Aspect Picture Library. 330-331 C. Molyneux/Bruce Coleman Ltd. 334 A. Havlicek/Zefa. 335 P. Carmichael/Aspect Picture Library. 338-339 Rainbird/Robert Harding Picture Library. 341 J. Lister/Hutchison Picture Library. 344 E. Bleicher/Zefa. 347 K. Kurita/Gamma/Frank Spooner Pictures. 351 A. Woolfitt/Susan Griggs Agency. 353 S. Sassoon/Robert Harding Picture Library. 355 C. Jopp/Robert Harding Picture Library. 358 J. Cleare/Mountain Camera. 363 B. O'Connor/Robert Harding Picture Library. 365 P. Baker/Photobank. 366-367 R. Gilmor/Bruce Coleman Ltd. 369 R. Harding/Robert Harding Picture Library. 375 Sally & Richard Greenhill. 379 Wilf James/Telegraph Colour Library/Photo Access. 380 August Sycholt/Silvestris. 387 C. Jones/Impact Photos. 392 D. Simpson/Hutchison Picture Library. 397 Gritscher/Bavaria. 400 M. Cator/Impact Photos. 405 Alain Nogues/Sygma/Inpra. 407 Serge Attal/Sygma/Inpra. 408 C. Jones/Impact Photos. 410 V.C.L./Photo Access. 415 J. Langley/Aspect Picture Library. 416 R. Smith/Tony Stone Worldwide. 419 G. Cubitt/Bruce Coleman Ltd. 421 R. Harding/Robert Harding Picture Library. 425 M. Vautier/vdn Picture Library. 427 I. Griffiths/Robert Harding Picture Library. 428 A. Woolfitt/Susan Griggs Agency. 429 D. Houston/Bruce Coleman Ltd. 433 M. Andrews/Susan Griggs Agency. 440 R-N. Guidicelli/Hutchison Picture Library. 441 V.C.L./Photo Access. 445 B. Glinn/Magnum/John Hillelson Agency. 447 Hoagland/Liaison/Frank Spooner Pictures. 449 B. Davis/Aspect Picture Library. 451 G. Lincoln/Aspect Picture Library. 455 Stephanie Colasanti. 457 J. Highet/Hutchison Picture Library. 458 Sylvain Grandadam/Tony Stone Worldwide/Color Library. 460 Harald Sund/The Image Bank. 461 D. Levenson/Colorific! 463 J. Jackson/Robert Harding Picture Library. 464-465 C. Bonnington/Bruce Coleman Ltd. 466 Heinz Koch/IFA-Bilderteam. 469 S. Meiselas/Magnum/John Hillelson Agency. 471 L. Woodhead/Hutchison Picture Library. 474 M. Reardon/Tony Stone

Worldwide. 479 D. Bayes/Aspect Picture Library. 480-481 M. Cator/Impact Photos. 485 A. Hutchison/Hutchison Picture Library. 486 K. Benser/Zefa. 489 G. Hunter/Bruce Coleman Ltd. 497 G. Gerster/John Hillelson Agency. 503 R. Lomas/Zefa. 504 Tony Stone Worldwide. 508 T. Nebbia/Aspect Picture Library. 510 R. Smith/Zefa. 513 R. Ellis/Robert Harding Picture Library. 515 J. Langley/Aspect Picture Library. 521 G. Gerster/John Hillelson Agency. 523 J-G. Jules/John Hillelson Agency. 528 F. Mayer/John Hillelson Agency. 531 Maroon/Zefa. 533 Cameraman/Zefa. 535 M. Koene/Colorific! 537 B. Coates/Bruce Coleman Ltd. 541 M. Berge/Bruce Coleman Ltd. 544 J. Cleare/Mountain Camera. 545 A. le Garsmeur/Impact Photos. 548 G. Gerster/John Hillelson Agency. 549 J. Margesson/Robert Harding Picture Library. 553 Stephanie Maze/Woodfin Camp, Inc. 558 A. le Garsmeur/Impact Photos. 562 M. Collier/Robert Harding Picture Library. 567 G. Mather/Robert Harding Picture Library. 572 N. Devore/Bruce Coleman Ltd. 574 Starfoto/Zefa. 578-579 V.C.L./Telegraph Colour Library. 585 Tony Stone Worldwide. 591 T. Okuda/Aspect Picture Library. 593 J. Belog/Black Star/Colorific! 598 G. Tortoli/Colorific! 599 P. Toler/Impact Photos. 600-601 A. Hander/Daily Telegraph Colour Library. 605 J. Silvestris/Frank Lane Agency. 606 L. Taylor/Hutchison Picture Library. 613 M. Shrimpton/Tony Stone Worldwide. 625 Peter Pickford. 629 L. Myers/Bruce Coleman Ltd. 632 A. Woolfitt/Susan Griggs Agency. 633 C. Weckler/The Image Bank. 635 R. Singh/John Hillelson Agency. 637 J. Glover/Tony Stone Worldwide. 639 J. Langley/Aspect Picture Library. 641 S. Errington/Hutchison Picture Library. 644 M. Yamashita/Colorific! 645 P.I. Productions/Photo Access. 647 W. Ferchland/Zefa. 651 M. St. Maur Sheil/Susan Griggs Agency. 653 M. Hackforth-Jones/Robert Harding Picture Library. 660 J. Rowan/Aspect Picture Library. 661 J-P. Ferrero/Ardea, London. 663 J. Fry/Bruce Coleman Ltd. 667 D. Hiser/The Image Bank. 675 R. Chesher/Planet Earth Pictures. 678 P. Steyn/Ardea, London. 680 F. Lanting/Bruce Coleman Ltd. 682 Michael Holford. 685 R. & S. Michaud/John Hillelson Agency. 691 I. Berry/Magnum/John Hillelson Agency. 698 Patrick Thurston. 699 Clive Coote. 700 Phil Conrad/Aspect. 704 Veronica Jones/Photo Access 705 J. Langley/Aspect Picture Library. 711 J. Korman/Bruce Coleman Ltd. 715 P. Dickerson/Susan Griggs Agency. 716 A. Burman/Aspect Picture Library. 718 J. Bulmer/Susan Griggs Agency. 721 Nakamura/Sygma/John Hillelson Agency. 722 W. Ruth/Bruce Coleman Ltd. 727 *Patrick Thurston.* 731 A. Williams/Planet Earth Pictures. 735 M. Freeman/Bruce Coleman Ltd. 736 Zefa. 738 I. Bradshaw/Colorific! 741 A. Woolfitt/Susan Griggs Agency. 742 S. Kraserman/Bruce Coleman Ltd. 744 Franco Zecchin/Magnum. 747 P. Maitre/Gamma/Frank Spooner Pictures. 749 R. Campbell/Bruce Coleman Ltd. 750 Bullaty/Lomeo/The Image Bank. 754 P. Justitz/Zefa. 755-757 Flag Images © 1995 to the Flag Institute. All rights reserved.

The publishers acknowledge their indebtedness to the following books, which were consulted for reference.

Africa: A Natural History by Leslie Brown (Hamish Hamilton); *Africa on a Shoestring* by Geoff Crowther (Lonely Planet, Australia); *Albania* (New Albania Magazine, Tiranuë); *All-Asia Guide* (Far Eastern Economic Review, Hong Kong); *The Alps* (National Geographic Society, USA); *Amnesty International Report* (annual); *Arab Gulf States on a Shoestring* by Gordon Robinson (Lonely Planet, Australia); *The Atlas of Mankind* (Mitchell Beazley); *The Atlas of the Universe* (Mitchell Beazley and George Philip); *Australia A Travel Survival Kit* by Tony Wheeler (Lonely Planet, Australia). *Baedeker's Germany*; *Bali and Lombok A Travel Survival Kit* by Mary Coverton and Tony Wheeler (Lonely Planet, Australia); *Berlitz Guides* to various parts of the world; *British Columbia* by G. E. Mortimore (Collins, Toronto). *Chambers's Biographical Dictionary* (Chambers); *Chambers's World Gazetteer* edited by T. C. Collocott and J. O. Thorne (Chambers); *Collins Atlas of the World*; *The Companion Guide to Jugoslavia* by J. A. Cuddon (Collins, London; Prentice-Hall, USA); *Continents Adrift: Readings from Scientific American* (W. H. Freeman). *A Dictionary of Geography* by F. J. Monkhouse (Edward Arnold). *Eastern Europe: A Geography of the Comecon Countries* by Roy E. H. Mellor (Macmillan Press); *Encyclopedia Americana*; *Encyclopedia Britannica* and its *Books of the Year*; *An Encyclopedia of World History* edited by William L. Langer (Harrap); *Explore Australia* (George Philip and O'Neil, Australia). *Facts About Germany* (Lexikothek Verlag, Gütersloh); *Fielding's Travel Guide to Europe* (Fielding, New York); *Fodor's Guides* to various parts of the world (Hodder and Stoughton, London). *Gazetteer of the British Isles* (Bartholomew); *The Geographical Digest* (George Philip; annual); *A Glossary of Geographical Terms* edited by Sir Dudley Stamp and Audrey N. Clark (Longman); *The Great Cities* series, including *Dublin* by Brendan Lehane (Time-Life Books); *The Great Geographical Atlas* (Mitchell Beazley); *Guide to East Africa* by Nina Casimati (Travelaid, London; Hippocrene Books, New York); *Guide to South African Game and Nature Reserves* by Chris and Tilde Stuart (Struik, Cape Town); *The Guinness Book of Records* (Guinness Superlatives; annual); *Hong Kong, Macau and Canton* by Carol Clewlow (Lonely Planet, Australia). *The Hutchinson Pocket Encyclopedia* (Helicon, Oxford) *Indonesia Handbook* by Bill Dalton (Moon Publications, California); *Indonesia, Malaysia and Singapore Handbook* (Trade and Travel Publications, annual); *Insight Guides* to various countries, notably *Malaysia* and *Philippines* (APA Productions, Hong Kong; Harrap, London; Lansdowne, Australia; Prentice-Hall, USA); *Iraq: A Tourist Guide* (State Organisation for Tourism, Baghdad); *Isles of the South Pacific* (National Geographic

Society, USA). *Japan Handbook* by J.D. Bisignani (Moon Publications, USA) *Korea Guide* by Edward B. Adams (Seoul International Publishing Co). *Library of Nations* series, notably *Arabian Peninsula, Australia, China, Germany, The Soviet Union* and *The United States* (Time-Life Books); *Life Nature Library* series, notably *South America* (Time-Life Books). *Man's Religions* by John B. Moss (Macmillan, New York); *Meteorological Glossary* (HMSO for the Meteorological Office); *Mexico and Central America Handbook* edited by Ben Box (Trade and Travel Publications, annual); *Michael's Guide to South America* by Michael Shichor (Inbal Travel Information, Tel Aviv); *Middle East on a Shoestring* (Lonely Planet, Australia); *Middle East Review* (Kogan Page and Walden, London); *The Milepost: All-the-North Travel Guide* (Alaska Northwest Publishing Co, Anchorage); *Monuments of Civilization* series, notably *India* by Maurizio Taddei (Cassell and Reader's Digest). *National Geographic Atlas of the World* (National Geographic Society, USA); *New African Yearbook* (IC Publications, London); *The New Columbia Encyclopedia* Columbia University Press); *The New Larousse Encyclopedia of the Earth* by Leon Bertin (Hamlyn); *North Sea Fields: Facts and Figures* (Shell UK). *Pacific Islands Yearbook* edited by Stuart Inder (Pacific Publications, Sydney); *The Penguin Dictionary of Geography* by W. G. Moore (Penguin Books); *The Penguin Dictionary of Geology* by D. G. A. Whitten with J. R. V. Brooks (Penguin Books); *Penguin Travel Guides* to various parts of the world; *Philips' Pocket Guide to the World* by Bernard Stonehouse (George Philip); *The Physical Earth* (Mitchell Beazley Joy of Knowledge Library); *Planet Earth* series, including *Continents in Collision* and *Volcanoes* (Time-Life Books); *Poland* by Marc Heine (B. T. Batsford); *Post Guides* to various parts of East and South-East Asia (South China Morning Post, Hong Kong). *Rand McNally Cosmopolitan World Atlas* (Rand McNally, USA); *Regional Surveys of the World* (Europa Publications, London); *The Rough Guide to Yugoslavia* by Martin Dunford and Jack Holland (Routledge and Kegan Paul). *Saudi Arabia: a MEED Practical Guide* (Middle East Economic Digest); *The Shell Guide to Europe* edited by Diane Petry (Michael Joseph); *South America and Central America: A Natural History* by Jean Dorst (Hamish Hamilton); *The South American Handbook* edited by John Brooks (Trade and Travel Publications; annual); *South Asian Handbook* by edited by Robert Bradnock (Trade and Travel Publications, annual); *South-East Asia on a Shoestring* (Lonely Planet, Australia); *South Pacific Handbook* by David Stanley (Moon Publications, California); *Spirit of Asia* by Michael Macintyre (BBC Publications); *The Statesman's Yearbook* edited by John Paxton (Macmillan Press); *Statistical Yearbook* and various other UN publications (United Nations); *Switzerland* by Francois Jeanneret, Walter Imber

and Franz auf der Maur (Kümmerly and Frey, Berne); *Syria Today* by Jean Hureau (Editions J. A., Paris). *This is your Country* (Ministry of Information, Saudi Arabia); *The Times Atlas of China* (Times Books); *The Times Atlas of the Oceans* (Times Books); *The Times Atlas of the World* (Times Books); *The Times World Index Gazetteer* (Times Publishing Co); *Travel Guide to the People's Republic of China* by Ruth Lor Malloy (William Morrow, New York); *Traveller's Guide to West Africa* (IC Magazines). *United Nations List of National Parks and Protected Areas* (IUCN Switzerland and UK). *Volcanoes* (HMSO for the Institute of Geological Sciences). *Webster's New Geographical Dictionary* (Merriam-Webster); *Whitaker's Almanack* (J. Whitaker; annual); *The World Factbook* (Central Intelligence Agency; annual); *The World in Figures* (The Economist); *World Reference Atlas* (Dorling Kindersley, London); *World Tables* (Johns Hopkins University Press for the World Bank); *The World's Great Religions* by the editorial staff of Life (Collins); *The World's Wild Places* series (Time-Life Books); and official guides, surveys and statistical information published by numerous national tourist and information offices and national and regional governments.

Various editions of the following newspapers, magazines, periodicals and journals were also consulted: *Africa Insight; The Economist; The Financial Times; Geo; The Geographical Journal* (Royal Geographical Society); *The Geographical Magazine; Keesing's Contemporary Archives* (Longman); *National Geographic; Newsweek International; The Observer* and its colour magazine; *South; The Sunday Times* (London) and its colour magazine; *Telegraph Sunday Magazine; Time* (Atlantic Edition); *The Times* (London).

Also of great value were a number of books published by regional Reader's Digest offices and their associated companies, including the following: *America the Beautiful* (USA); *Antarctica: Great Stories from the Frozen Continent* (Australia); *Atlas of Australia* (Australia); *Atlas of Canada* (Canada); *Atlas of Southern Africa* (South Africa); *Book of British Towns* (Drive Publications, UK); *Great World Atlas* (UK); *Illustrated Guide to Britain* (Drive Publications, UK); *Illustrated Guide to Britain's Coast* (Drive Publications, UK); *Library of Modern Knowledge* (UK); *Natural Wonders of the World* (France/USA); *Reader's Digest Almanac and Yearbook* (USA); *Reader's Digest Book of the Great Barrier Reef* (Australia); *Southern Africa: Land of Beauty and Splendour* (South Africa); *Trésors de France* (France); *Vie et Payasages des Montagnes de France* (France); *Wild Australia* (Australia); *Wild New Zealand* (Australia).

Words from the song 'This Land is Your Land' on p 702 used by permission. Words and music by Woody Guthrie. © Copyright 1956 (renewed 1984), 1958 (renewed 1986) and 1970 Ludlow Music, Inc., New York, N.Y.

Paper: C. Townsend Hook Paper Co. Ltd. Snodland
Printing and Binding: Fabrieken Brepols n.v., Turnhout, Belgium
Cartography: Map Creation Ltd., Maidenhead, Berkshire

40-042-4